2010 Twenty-Fifth Annual IEEE Applied Power Electronics Conference and Exposition

(APEC 2010)

Palm Springs, California, USA
21 – 25 February 2010

Pages 1542-2327

IEEE Catalog Number: CFP10APE-PRT
ISBN: 978-1-4244-4782-4

Copyright © 2010 by the Institute of Electrical and Electronic Engineers, Inc
All Rights Reserved

Copyright and Reprint Permissions: Abstracting is permitted with credit to the source. Libraries are permitted to photocopy beyond the limit of U.S. copyright law for private use of patrons those articles in this volume that carry a code at the bottom of the first page, provided the per-copy fee indicated in the code is paid through Copyright Clearance Center, 222 Rosewood Drive, Danvers, MA 01923.

For other copying, reprint or republication permission, write to IEEE Copyrights Manager, IEEE Service Center, 445 Hoes Lane, Piscataway, NJ 08854. All rights reserved.

This publication is a representation of what appears in the IEEE Digital Libraries. Some format issues inherent in the e-media version may also appear in this print version.

IEEE Catalog Number: CFP10APE-PRT
ISBN 13:   978-1-4244-4782-4
Library of Congress No.: 90-643607
ISSN: 1048-2334

Additional Copies of This Publication Are Available From:

Curran Associates, Inc
57 Morehouse Lane
Red Hook, NY 12571 USA
Phone: (845) 758-0400
Fax: (845) 758-2633
E-mail: curran@proceedings.com
Web: www.proceedings.com

2010 Twenty-Fifth Annual IEEE Applied Power Electronics Conference and Exposition (APEC 2010)

Palm Springs, California, USA
21-25 February 2010

IEEE Catalog Number: CFP10APE-POD
ISBN: 978-1-42444-782-4

TABLE OF CONTENTS

Session A1L-A: DC-DC Converter I
Tuesday, February 23, 8:30 - 10:10
Session Chairs: Van Niemela, *Fairchild Semiconductor*
Haidong Yu, *Phoenix International*

**Minimum Deviation Digital Controller IC for Single and Two Phase DC-DC
Switch-Mode Power Supplies** 1
Aleksandar Radić, *University of Toronto, Canada*
Zdravko Lukić, *University of Toronto, Canada*
Aleksandar Prodić, *University of Toronto, Canada*
Robert de Nie, *NXP Semiconductors, Netherlands*

Modeling and Design Considerations of Coupled Inductor Converters 7
Guangyong Zhu, *Auscom Engineering, Inc., United States*
Kunrong Wang, *Dell, Inc., United States*

**Design Procedure for High Frequency Operation of the Modified Series Resonant
APWM Converter with Improved Efficiency and Reduced Size** 14
Darryl J. Tschirhart, *Queen's University, Canada*
Praveen K. Jain, *Queen's University, Canada*

Expiremantal Results and Study of a Modified Adaptive Bus Voltage Controller 19
Jaber A. Abu Qahouq, *University of Alabama, United States*
Gautam Muralidhar, *University of Alabama, United States*

Session A1L-B: AC-DC Power Factor Correction Topologies I
Tuesday, February 23, 8:30 - 10:10
Session Chairs: Gerry Moschopoulos, *University of Western Ontario*
Omer Onar, *Illinois Institute of Technology*

Bridgeless Buck PFC Rectifier 23
Yungtaek Jang, *Delta Products Corporation, United States*
Milan M. Jovanović, *Delta Products Corporation, United States*

**An Active-Clamped Full-Wave Zero-Current-Switched Quasi-Resonant Boost
Converter in Power Factor Correction Application** 30
E. Firmansyah, *Kyushu University, Japan*
S. Abe, *Kyushu University, Japan*
M. Shoyama, *Kyushu University, Japan*
S. Tomioka, *TDK-Lambda Corporation, Japan*
T. Ninomiya, *Nagasaki University, Japan*

Novel Adaptive Master-Slave Method for Interleaved Boundary Conduction Mode (BCM) PFC Converters 36
Hangseok Choi, *Fairchild Semiconductor, United States*

A Novel Bridgeless Single-Stage Half-Bridge AC-DC Converter 42
Woo-Young Choi, *Virginia Polytechnic Institute and State University, United States*
Wen-Song Yu, *Virginia Polytechnic Institute and State University, United States*
Jih-Sheng Lai, *Virginia Polytechnic Institute and State University, United States*

Session A1L-C: Power Electronics for Utility Interface I
Tuesday, February 23, 8:30 - 10:10
Session Chairs: Zareh Soghomonian, *BMT Syntek Technologies*
Jin Wang, *Ohio State University*

Power Quality Improvement at Medium-Voltage Grids Using Hexagram Active Power Filter 47
Jun Wen, *University of California, Irvine, United States*
Liang Zhou, *University of California, Irvine, United States*
Keyue Smedley, *University of California, Irvine, United States*

A Generalized Capacitor Voltage Balancing Scheme for Flying Capacitor Multilevel Converters 58
Mostafa Khazraei, *Missouri University of Science and Technology, United States*
Hossein Sepahvand, *Missouri University of Science and Technology, United States*
Keith Corzine, *Missouri University of Science and Technology, United States*
Mehdi Ferdowsi, *Missouri University of Science and Technology, United States*

An Active Damping Technique for a Current Source Inverter Employing a Virtual Negative Inductance 63
Ahmed Salah Morsy, *Texas A&M University at Qatar, Qatar*
Shehab Ahmed, *Texas A&M University at Qatar, Qatar*
Prasad Enjeti, *Texas A&M University at Qatar, Qatar*
Ahmed Massoud, *Qatar University, Qatar*

Maximum Solar Power Transfer in Multi-Port Power Electronic Interface 68
Wei Jiang, *University of Texas at Arlington, United States*
Babak Fahimi, *University of Texas at Arlington, United States*

Session A1L-D: Passive Devices I
Tuesday, February 23, 8:30 - 10:10
Session Chairs: Laura Lyle, *Wright Patterson Air Force Base*
Mike Schutten, *General Electric*

SMD Inductors Based on Soft-Magnetic Powder Compacts 74
Etsuo Otsuki, *Toho Zinc Co., Ltd., Japan*
Kenichiro Ishii, *Toho Zinc Co., Ltd., Japan*
Shinya Nakano, *Toho Zinc Co., Ltd., Japan*

High Density Low Profile Coupled Inductor Design for Integrated Point-of-Load Converter 79
Qiang Li, *Virginia Polytechnic Institute and State University, United States*
Yan Dong, *Virginia Polytechnic Institute and State University, United States*
Fred C. Lee, *Virginia Polytechnic Institute and State University, United States*

**Relationship of Quality Factor and Hollow Winding Structure of
Coreless Printed Spiral Winding (CPSW) Inductor** 86
Y.P. Su, *Virginia Polytechnic Institute and State University, United States*
Xun Liu, *ConvenientPower HK Ltd., China*
C.K. Lee, *Hong Kong Polytechnic University, China*
S.Y.R. Hui, *City University of Hong Kong, China*

**Modeling of Adaptable-Diameter Burners Formed by Concentric Planar Windings for
Domestic Induction Heating Applications** 92
Jesus Acero, *University of Zaragoza, Spain*
Claudio Carretero, *University of Zaragoza, Spain*
Ignacio Millan, *University of Zaragoza, Spain*
Oscar Lucía, *University of Zaragoza, Spain*
Jose-Miguel Burdío, *University of Zaragoza, Spain*
Rafael Alonso, *University of Zaragoza, Spain*

Session A1L-E: Controls in Motor Drives I
Tuesday, February 23, 8:30 - 10:10
Session Chairs: Jonathan Kimball, *Missouri S&T*

**Flux Concentration and Pole Shaping in a Single Phase Hybrid Switched
Reluctance Motor Drive** 98
Uffe Jakobsen, *Aalborg University, Denmark*
Kaiyuan Lu, *Aalborg University, Denmark*

**Parameter Independent Maximum Torque Per Ampere (MTPA) Control of
IPM Machine Based on Signal Injection** 103
Sungmin Kim, *Seoul National University, Korea, South*
Young-Doo Yoon, *Seoul National University, Korea, South*
Seung-Ki Sul, *Seoul National University, Korea, South*
Kozo Ide, *Yaskawa Electric Corporation, Japan*
Koji Tomita, *Yaskawa Electric Corporation, Japan*

Performance Analysis of Three-Phase Capacitor Motor in Frequency Control System 109
Zheng-Feng Ming, *Xidian University, China*
Guang-Zheng Ni, *Zhejiang University, China*
Bing-Zhong Yang, *Chongqing University, China*

**Efficiency Improvement by Changeover of Phase Windings of Multiphase
Permanent Magnet Synchronous Motor with Outer-Rotor Type** 112
Young-Gook Kim, *Pusan National University, Korea, South*
Chae-Bong Bae, *Pusan National University, Korea, South*
Jang-Mok Kim, *Pusan National University, Korea, South*
Hyun-Cheol Kim, *Agency for Defense Development, Korea, South*

Session A1L-F: Digital Controls in DC-DC Converters I
Tuesday, February 23, 8:30 - 10:10
Session Chairs: Dragan Maksimović, *University of Colorado at Boulder*
Jason Neely, *Purdue University*

Digital Power Controller with Non-Linear Variable Switching Frequency 120
Jaber A. Abu Qahouq, *University of Alabama, United States*

**Digital Charge Balance Controller with an Auxiliary Circuit for Superior Unloading
Transient Performance of Buck Converters** 124
Eric Meyer, *Queen's University, Canada*
Dong Wang, *Queen's University, Canada*
Liang Jia, *Queen's University, Canada*
Yan-Fei Liu, *Queen's University, Canada*

One-Step Digital Dead-Time Correction for DC-DC Converters 132
April Zhao, *University of Toronto, Canada*
Armin Akhavan Fomani, *University of Toronto, Canada*
Wai Tung Ng, *University of Toronto, Canada*

**The Practical Aspects of Utilizing Digital Power Controller for
Monitoring of Power Supply Operation** 138
Oleg Volfson, *Intersil Corporation, United States*

Session A1L-G: Wind Power
Tuesday, February 23, 8:30 - 10:10
Session Chairs: Morgan Kiani, *University of Texas at Arlington*

**A Unity Power Factor, Maximum Power Point Tracking Battery Charger for
Low Power Wind Turbines** 143
Gustavo Gamboa, *University of Central Florida, United States*
John Elmes, *University of Central Florida, United States*
Christopher Hamilton, *University of Central Florida, United States*
Jonathan Baker, *University of Central Florida, United States*
Michael Pepper, *University of Central Florida, United States*
Issa Batarseh, *University of Central Florida, United States*

**Maximum Power Point Tracking of a Wind Energy Conversion System Using
Adaptive Nonlinear Approach** 149
Majid Pahlevaninezhad, *Queen's University, Canada*
Suzan Eren, *Queen's University, Canada*
Alireza Bakhshai, *Queen's University, Canada*
Praveen Jain, *Queen's University, Canada*

A Hybrid Wind-Solar Energy System: a New Rectifier Stage Topology 155
Joanne Hui, *Queen's University, Canada*
Alireza Bakhshai, *Queen's University, Canada*
Praveen Jain, *Queen's University, Canada*

Dynamic Operation and Control of a Hybrid Wind-Diesel Stand Alone Power Systems 162

A.M.O. Haruni, *University of Tasmania, Australia*
A. Gargoom, *University of Tasmania, Australia*
M.E. Haque, *University of Tasmania, Australia*
M. Negnevitsky, *University of Tasmania, Australia*

Session A2L-A: DC-DC Converter II
Tuesday, February 23, 10:40 - 11:55
Session Chairs: Van Niemela, *Fairchild Semiconductor*
Haidong Yu, *Phoenix International*

SystemC-AMS Modeling and Simulation of Digitally Controlled DC-DC Converters 170

Matteo Agostinelli, *University of Klagenfurt, Austria*
Robert Priewasser, *University of Klagenfurt, Austria*
Mario Huemer, *University of Klagenfurt, Austria*
Stefano Marsili, *Infineon Technologies Austria AG, Austria*
Dietmar Straeussnigg, *Infineon Technologies Austria AG, Austria*

Modeling of Digitally Controlled Voltage Regulator Modules 176

Yi Sun, *Linear Technology, United States*
Fred C. Lee, *Virginia Polytechnic Institute and State University, United States*
Jian Li, *Linear Technology, United States*

Design and Comparison of Digital Control Loops Analytical Models, Laboratory Measurements, and Simulation Results 183

Philip Cooke, *Infineon Technologies, United States*
Thomas G. Wilson, Jr., *SIMPLIS Technologies, United States*
Rohan Samsi, *Primarion, United States*

Session A2L-B: AC-DC Power Factor Correction Topologies II
Tuesday, February 23, 10:40 - 11:55
Session Chairs: Gerry Moschopoulos, *University of Western Ontario*
Omer Onar, *Illinois Institute of Technology*

Digital Control for Efficiency Improvements in Interleaved Boost PFC Rectifiers 188

Fu-Zen Chen, *University of Colorado at Boulder, United States*
Dragan Maksimović, *University of Colorado at Boulder, United States*

Reduction of the Output Capacitor in Power Factor Correctors by Distorting the Line Input Current 196

Diego G. Lamar, *Universidad de Oviedo, Spain*
Javier Sebastián, *Universidad de Oviedo, Spain*
Manuel Arias, *Universidad de Oviedo, Spain*
Arturo Fernández, *Universidad de Oviedo, Spain*

Universal-Input Single-Stage PFC Flyback with Variable Boost Inductance for High-Brightness LED Applications 203

Yuequan Hu, *Delta Products Corporation, United States*
Laszlo Huber, *Delta Products Corporation, United States*
Milan M. Jovanović, *Delta Products Corporation, United States*

Session A2L-C: Power Electronics for Utility Interface II
Tuesday, February 23, 10:40 - 11:55
Session Chairs: Zareh Soghomonian, *BMT Syntek Technologies*
Jin Wang, *Ohio State University*

High Frequency High Efficiency Bidirectional DC-DC Converter Module Design for 10 kVA Solid State Transformer 210
Haifeng Fan, *Florida State University, United States*
Hui Li, *Florida State University, United States*

Synchronization of Three-Phase Converters and Virtual Microgrid Implementation Utilizing the Power-Hardware-in-the-Loop Concept 216
O. Vodyakho, *Florida State University, United States*
C.S. Edrington, *Florida State University, United States*
M. Steurer, *Florida State University, United States*
S. Azongha, *Florida State University, United States*
F. Fleming, *Florida State University, United States*

A Single-Stage Grid-Connected Inverter with Wide Range Reactive Power Compensation Using Energy Storage System (Ess) 223
Liming Liu, *Florida State University, United States*
Zhichao Wu, *Florida State University, United States*
Hui Li, *Florida State University, United States*

Session A2L-D: Passive Devices II
Tuesday, February 23, 10:40 - 11:55
Session Chairs: Laura Lyle, *Wright Patterson Air Force Base*
Mike Schutten, *General Electric*

Polymer Bonded Soft Magnetics for EMI Filter Applications in Power Electronics 231
S. Egelkraut, *University of Erlangen-Nürnberg, Germany*
L. Frey, *University of Erlangen-Nürnberg, Germany*
M. Rauch, *Fraunhofer Institute for Integrated Systems and Device Technology, Germany*
A. Schletz, *Fraunhofer Institute for Integrated Systems and Device Technology, Germany*
M. März, *Fraunhofer Institute for Integrated Systems and Device Technology, Germany*

Lead-Acid Battery Modeling and State of Charge Monitoring 239
J.F. Araujo Leão, *Universidade Federal de Campina Grande, Brazil*
L.V. Hartmann, *Universidade Federal de Campina Grande, Brazil*
M.B.R. Corrêa, *Universidade Federal de Campina Grande, Brazil*
A.M.N. Lima, *Universidade Federal de Campina Grande, Brazil*

Voltage and Current Ripple Considerations for Improving Lifetime of Ultra-Capacitors Used for Energy Buffer Applications at Converter Inputs 244
Supratim Basu, *Bose Research Pvt. Ltd., India*
Tore M. Undeland, *Norwegian University of Science and Technology, Norway*

Session A2L-E: Controls in Motor Drives II
Tuesday, February 23, 10:40 - 11:55
Session Chairs: Jonathan Kimball, *Missouri S&T*

Implementation and Operational Investigations of Bipolar Gate Drivers 248
Jean-Christophe Crebier, *Grenoble Institute of Technology, France*
Manh Hung Tran, *Grenoble Institute of Technology, France*
Jean Barbaroux, *Grenoble Institute of Technology, France*
Pierre-Olivier Jeannin, *Grenoble Institute of Technology, France*

A Method for Impact Assessment of Faults on the Performance of Field-Oriented Control Drives: a First Step to Reliability Modeling 256
Ali M. Bazzi, *University of Illinois at Urbana-Champaign, United States*
Alejandro Dominguez-Garcia, *University of Illinois at Urbana-Champaign, United States*
Philip T. Krein, *University of Illinois at Urbana-Champaign, United States*

A Fault Tolerant Control System for Hexagram Inverter Motor Drive 264
Liang Zhou, *University of California, Irvine, United States*
Keyue Smedley, *University of California, Irvine, United States*

Session A2L-F: Digital Controls in DC-DC Converters II
Tuesday, February 23, 10:40 - 11:55
Session Chairs: Dragan Maksimović, *University of Colorado at Boulder*
Jason Neely, *Purdue University*

Power Analog to Digital Converter for Voltage Scaling Applications 271
M.C. Gonzalez, *Universidad Politécnica de Madrid, Spain*
M. Vasić, *Universidad Politécnica de Madrid, Spain*
P. Alou, *Universidad Politécnica de Madrid, Spain*
O. Garcia, *Universidad Politécnica de Madrid, Spain*
J.A. Oliver, *Universidad Politécnica de Madrid, Spain*
J.A. Cobos, *Universidad Politécnica de Madrid, Spain*
H. Visairo, *Intel Corporation, Mexico*

A Digital Pulse-Width Modulator for Phase-Shift Operation of Full-Bridge Isolated DC-DC Converters 277
L. Corradini, *University of Colorado at Boulder, United States*
D. Maksimović, *University of Colorado at Boulder, United States*

Digitally Controlled Integrated Pseudo-CCM SIMO Converter with Adaptive Freewheel Current Modulation 284
Yi Zhang, *University of Arizona, United States*
Dongsheng Ma, *University of Arizona, United States*

Session A2L-G: Fuel Cells
Tuesday, February 23, 10:40 - 11:55
Session Chairs: Morgan Kiani, *University of Texas at Arlington*

Analysis of Pulse-Link DC-AC Converter for Fuel Cells Applications Operated in Zero-Current-Slope Mode 289
Kentaro Fukushima, *Kyushu University, Japan*
Isami Norigoe, *I.N. Laboratory, Japan*
Masahito Shoyama, *Kyushu University, Japan*
Tamotsu Ninomiya, *Nagasaki University, Japan*
Yosuke Harada, *Ebara Densan Ltd., Japan*
Kenta Tsukakoshi, *Ebara Densan Ltd., Japan*

A Minimum Power-Processing Stage Fuel Cell Energy System Based on a Boost-Inverter with a Bi-Directional Back-Up Battery Storage 295
Minsoo Jang, *University of Sydney, Australia*
Vassilios G. Agelidis, *University of Sydney, Australia*

Power Conditioning System for Fuel Cell with 2-Stage DC-DC Converter 303
Byung M. Han, *Myongji University, Korea, South*
Jun-Young Lee, *Myongji University, Korea, South*
Yu-Seok Jeong, *Myongji University, Korea, South*

Session B1L-A: DC-DC Converter III
Wednesday, February 24, 8:30 - 10:10
Session Chairs: Alireza Khaligh, *Illinois Institute of Technology*
Sheldon Williamson, *Concordia University*

Real-Time FPGA-Based Hardware-in-the-Loop Development Test-Bench for Multiple Output Power Converters 309
O. Lucía, *University of Zaragoza, Spain*
O. Jiménez, *University of Zaragoza, Spain*
L.A. Barragán, *University of Zaragoza, Spain*
I. Urriza, *University of Zaragoza, Spain*
J.M. Burdío, *University of Zaragoza, Spain*
D. Navarro, *University of Zaragoza, Spain*

Oversampled Digital Controller IC Based on Successive Load-Change Estimation for DC-DC Converters 315
Zdravko Lukić, *University of Toronto, Canada*
Aleksandar Radić, *University of Toronto, Canada*
Aleksandar Prodić, *University of Toronto, Canada*
Simon Effler, *University of Limerick, Ireland*

Novel Nonlinear Control of Dual Active Bridge Using Simplified Converter Model 321
Diogenes D. Molina Cardozo, *University of Arkansas, United States*
Juan Carlos Balda, *University of Arkansas, United States*
Derik Trowler, *University of Arkansas, United States*
H. Alan Mantooth, *University of Arkansas, United States*

A Novel Digital Single-Wire Quasi-Democratic Stress Share Scheme for Paralleled Switching Converters 328
Karl Rinne, *Powervation Ltd., Ireland*
Anthony Kelly, *Powervation Ltd., Ireland*
Eamon O'Malley, *Powervation Ltd., Ireland*

Session B1L-B: AC-DC Conversion Control Strategies
Wednesday, February 24, 8:30 - 10:10
Session Chairs: Alireza Khaligh, *Illinois Institute of Technology*
Omer Onar, *Illinois Institute of Technology*

Minimum-Sensing Current Control of Three-Phase PFC Converters 336
Zhonghui Bing, *Rensselaer Polytechnic Institute, United States*
Jian Sun, *Rensselaer Polytechnic Institute, United States*

Direct Power Control of a Dual Converter Operating As Synchronous Rectifier 343
José Restrepo, *Universidad Simón Bolívar, Venezuela*
José M. Aller, *Universidad Simón Bolívar, Venezuela*
Alexander Bueno, *Universidad Simón Bolívar, Venezuela*
Julio C. Viola, *Universidad Simón Bolívar, Venezuela*
Alberto Berzoy, *Universidad Simón Bolívar, Venezuela*
Thomas Habetler, *Georgia Institute of Technology, United States*

A Low-Cost Adaptive Multi-Mode Digital Control Solution Maximizing AC-DC Power Supply Efficiency 349
Yong Li, *iWatt Inc., United States*
Jerry Zheng, *iWatt Inc., United States*

Average Modeling and Control for Three-Phase Three-Level Non-Regenerate Rectifier with Unbalanced DC Loads 355
Rixin Lai, *GE Global Research Center, United States*
Fred Wang, *University of Tennessee - Knoxville and Oak Ridge National Laboratory, United States*
Rolando Burgos, *ABB Inc., United States*
Dushan Boroyevich, *Virginia Polytechnic Institute and State University, United States*

Session B1L-C: Active Power Filter
Wednesday, February 24, 8:30 - 10:10
Session Chairs: Jingjun Liu, *Xi'an Jiaotong Univ.*
Jin Wang, *Ohio State University*

A Waveform Control Technique for High Power Shunt Active Power Filter Based on Repetitive Control Algorithm 361
Zhiqiang Wang, *Zhejiang University, China*
Chuan Xie, *Zhejiang University, China*
Chao He, *Zhejiang University, China*
Guozhu Chen, *Zhejiang University, China*

A Combined Series-Parallel Active Filter System Implementation Using Generalized Non-Active Power Theory 367

Mehmet Ucar, *Kocaeli University, Turkey*
Sule Ozdemir, *Kocaeli University, Turkey*
Engin Ozdemir, *Kocaeli University, Turkey*

A Novel Control Method for Unified Power Quality Conditioner (UPQC) Under Non-Ideal Mains Voltage and Unbalanced Load Conditions 374

Metin Kesler, *Kocaeli University, Turkey*
Engin Ozdemir, *Kocaeli University, Turkey*

Resonant Current Regulation for Transformerless Hybrid Active Filter to Suppress Harmonic Resonances in Industrial Power Systems 380

Tzung-Lin Lee, *National Sun Yat-sen University, Taiwan*
Yen-Ching Wang, *National Sun Yat-sen University, Taiwan*
Josep M. Guerrero, *Technical University of Catalonia, Spain*

Session B1L-D: Semiconductor Devices
Wednesday, February 24, 8:30 - 10:10
Session Chairs: Carl Blake, *Transphorm*
Chuck Mullett, *ON Semiconductor*

Performance Evaluation of High Voltage Super Junction MOSFETs for Zero-Voltage Soft-Switching Inverter Applications 387

Sung-Yeul Park, *University of Connecticut, United States*
Pengwei Sun, *Virginia Polytechnic Institute and State University, United States*
Wensong Yu, *Virginia Polytechnic Institute and State University, United States*
Jih-Sheng Lai, *Virginia Polytechnic Institute and State University, United States*

New 1.7kV IGBT Chip with Fine Pattern and Optimized Buffer Layer 392

John F. Donlon, *Powerex, Inc., United States*
Eric R. Motto, *Powerex, Inc., United States*
K. Satoh, *Mitsubishi Electric Corp, Japan*
K. Suzuki, *Mitsubishi Electric Corp, Japan*
Y. Yoshihiura, *Mitsubishi Electric Corp, Japan*
T. Takahashi, *Mitsubishi Electric Corp, Japan*

Novel Thermally Enhanced Power Package 398

Juan A. Herbsommer, *Texas Instruments, United States*
Jonathan Noquil, *Texas Instruments, Philippines*
Chris Bull, *Texas Instruments, United States*
Osvaldo Lopez, *Texas Instruments, United States*

Recent Advances in Silicon Carbide MOSFET Power Devices 401

Ljubisa D. Stevanovic, *GE Global Research, United States*
Kevin S. Matocha, *GE Global Research, United States*
Peter A. Losee, *GE Global Research, United States*
John S. Glaser, *GE Global Research, United States*
Jeffrey J. Nasadoski, *GE Global Research, United States*
Stephen D. Arthur, *GE Global Research, United States*

Session B1L-E: Sensorless Techniques in Motor Drives
Wednesday, February 24, 8:30 - 10:10
Session Chairs: Patrick Chapman, *University of Illinois*

Start-Up Transient Improvement for Sensorless Control Approach of PM Motor 408
Dong Jiang, *Virginia Polytechnic Institute and State University, United States*
Rixin Lai, *Virginia Polytechnic Institute and State University, United States*
Fred Wang, *University of Tennessee - Knoxville, United States*
Rolando Burgos, *Virginia Polytechnic Institute and State University, United States*
Dushan Boroyevich, *Virginia Polytechnic Institute and State University, United States*

Sensorless Position Control of Skewed Rotor Induction Machines Based on Multi Saliency Extraction 414
T.M. Wolbank, *Vienna University of Technology, Austria*
M.K. Metwally, *Vienna University of Technology, Austria*

Fuzzy Gain Scheduling PI Controller for a Sensorless Four Switch Three Phase BLDC Motor 420
Chung-Wen Hung, *National Yunlin University of Science and Technology, Taiwan*
Jen-Ta Su, *National Taiwan University, Taiwan*
Chih-Wen Liu, *National Taiwan University, Taiwan*
Cheng-Tsung Lin, *DynaPack Co., Ltd., Taiwan*
Jhih-Han Chen, *National Yunlin University of Science and Technology, Taiwan*

Equivalent EMF Based Position Observers for Sensorless Synchronous Machines 425
Jingbo Liu, *Rockwell Automation, United States*
Thomas Nondahl, *Rockwell Automation, United States*
Peter Schmidt, *Rockwell Automation, United States*
Semyon Royak, *Rockwell Automation, United States*
Mark Harbaugh, *Rockwell Automation, United States*

Session B1L-F: Modeling, Simulation & Control I
Wednesday, February 24, 8:30 - 10:10
Session Chairs: Mahesh Krishnamurthy, *Illinois Institue of Technology*

An Improved Winding Loss Analytical Model of Flyback Transformer 433
Wei Yuan, *Zhejiang University, China*
Xiucheng Huang, *Zhejiang University, China*
Peipei Meng, *Zhejiang University, China*
Guoxing Zhang, *Zhejiang University, China*
Junming Zhang, *Zhejiang University, China*

Identification of the Material Properties Used in Domestic Induction Heating Appliances for System-Level Simulation and Design Purposes 439
Jesus Acero, *University of Zaragoza, Spain*
Oscar Lucía, *University of Zaragoza, Spain*
Ignacio Millan, *University of Zaragoza, Spain*
Luis Angel Barragán, *University of Zaragoza, Spain*
Jose-Miguel Burdío, *University of Zaragoza, Spain*
Rafael Alonso, *University of Zaragoza, Spain*

A Retrofit 60 Hz Current Sensor for Non-Intrusive Power Monitoring at the Circuit Breaker 444

Zachary Clifford, *Massachusetts Institute of Technology, United States*
John J. Cooley, *Massachusetts Institute of Technology, United States*
Al-Thaddeus Avestruz, *Massachusetts Institute of Technology, United States*
Zack Remscrim, *Massachusetts Institute of Technology, United States*
Dan Vickery, *Massachusetts Institute of Technology, United States*
Steven B. Leeb, *Massachusetts Institute of Technology, United States*

Session B1L-G: Vehicle Electronics I
Wednesday, February 24, 8:30 - 10:10
Session Chairs: Ali Emadi, *Illinois Institute of Technology*

Feasibility of Capacitor Voltage Regulation and Output Voltage Harmonic Minimization in Cascaded H-Bridge Converters 452

Hossein Sepahvand, *Missouri University of Science and Technology, United States*
Mostafa Khazarei, *Missouri University of Science and Technology, United States*
Mehdi Ferdowsi, *Missouri University of Science and Technology, United States*
Keith Corzine, *Missouri University of Science and Technology, United States*

Examination of a PHEV Bidirectional Charger System for V2G Reactive Power Compensation 458

Mithat C. Kisacikoglu, *University of Tennessee, United States*
Burak Ozpineci, *Oak Ridge National Laboratory, United States*
Leon M. Tolbert, *University of Tennessee and Oak Ridge National Laboratory, United States*

Optimal Selection and Design of the Supercapacitor Module for Fuel Cell Vehicles 466

Sang-Hyun Kim, *Soongsil University, Korea, South*
Tae-Hoon Kim, *Soongsil University, Korea, South*
Wook Kim, *Soongsil University, Korea, South*
Jong-Hak Lee, *Soongsil University, Korea, South*
Woojin Choi, *Soongsil University, Korea, South*

Efficiency Evaluation of a 55kW Soft-Switching Module Based Inverter for High Temperature Hybrid Electric Vehicle Drives Application 474

Pengwei Sun, *Virginia Polytechnic Institute and State University, United States*
Jih-Sheng Lai, *Virginia Polytechnic Institute and State University, United States*
Hao Qian, *Virginia Polytechnic Institute and State University, United States*
Wensong Yu, *Virginia Polytechnic Institute and State University, United States*
Chris Smith, *Azure Dynamics Inc., United States*
John Bates, *Azure Dynamics Inc., United States*
Beat Arnet, *Azure Dynamics Inc., United States*
Alexander Litvinov, *Powerex Inc., United States*
Scott Leslie, *Powerex Inc., United States*

Session B2L-A: DC-DC Converter IV
Wednesday, February 24, 14:00 - 15:40
Session Chairs: Jin Wang, *Ohio State University*
Wayne Weaver, *Michigan Technological University*

Real-Time Hybrid Model Predictive Control of a Boost Converter with Constant Power Load 480
Jason Neely, *Purdue University, United States*
Steve Pekarek, *Purdue University, United States*
Ray DeCarlo, *Purdue University, United States*
Nir Vaks, *Purdue University, United States*

Predictive Control of Buck Converter Using Nonlinear Output Capacitor Current Programming 491
Victor Sui-pung Cheung, *City University of Hong Kong, China*
Henry Shu-hung Chung, *City University of Hong Kong, China*
Huai Wang, *City University of Hong Kong, China*

Analysis of a High Performance Voltage Regulator with Non-Linear Multi-Mode Control: Bandwidth and Large Transient Response 499
S. Pan, *Queen's University, Canada*
P.K. Jain, *Queen's University, Canada*

Multi-Output Synchronously-Rectified Forward Converter with Load Transient Considered ... 507
K.I. Hwu, *National Taipei University of Technology, Taiwan*
Y.T. Yau, *National Taipei University of Technology, Taiwan*

Session B2L-B: System Integration I
Wednesday, February 24, 14:00 - 15:40
Session Chairs: Shamala Chickamenahalli, *Intel*

Symmetric Current Balancing Circuit for Multiple DC Loads 512
Sungjin Choi, *Samsung Electronics Co., Ltd., Korea, South*
Pankaj Agarwal, *Samsung Electronics Co., Ltd., Korea, South*
Teahoon Kim, *Samsung Electronics Co., Ltd., Korea, South*
Joonhyun Yang, *Samsung Electronics Co., Ltd., Korea, South*
Baikhee Han, *Samsung Electronics Co., Ltd., Korea, South*

A Simple Method for Configuring Multi-PWM Channels for Multi-Level Converter Applications Based on PWM IP Core 519
Haibing Hu, *Nanjing University of Aeronautics and Astronautics, China*
Xiaodong Ding, *Nanjing Guojun Electric Co., Ltd., China*
Tao Xue, *Nanjing Sute Electric Co., Ltd., China*
Wenxi Yao, *Zhejiang University, China*
Zhengyu Lu, *Zhejiang University, China*

Technology Roadmapping for Power Supply in Package (PSiP) and Power Supply on Chip (PwrSoC) 525
Raymond Foley, *University College Cork, Ireland*
Finbarr Waldron, *Tyndall National Institute, Ireland*
John Slowey, *University College Cork, Ireland*
Arnold Alderman, *Anagenesis Inc., United States*
Brian Narveson, *Texas Instruments, United States*
Cian Ó'Mathúna, *Tyndall National Institute, Ireland*

Technology Road Map for High Frequency Integrated DC-DC Converter 533
Qiang Li, *Virginia Polytechnic Institute and State University, United States*
Michele Lim, *Virginia Polytechnic Institute and State University, United States*
Julu Sun, *Virginia Polytechnic Institute and State University, United States*
Arthur Ball, *Virginia Polytechnic Institute and State University, United States*
Yucheng Ying, *Virginia Polytechnic Institute and State University, United States*
Fred C. Lee, *Virginia Polytechnic Institute and State University, United States*
K.D.T. Ngo, *Virginia Polytechnic Institute and State University, United States*

Session B2L-C: Resonant DC-DC Converters I
Wednesday, February 24, 14:00 - 15:40
Session Chairs: Dustin Becker, *Emerson Network Power*
Russell Spyker, *USAF*

A New Valley-Detection Method for the Quasi-Resonance Switching 540
Gwan-Bon Koo, *Fairchild Semiconductor, Korea, South*
Sang-Cheol Moon, *Fairchild Semiconductor, Korea, South*
Jin-Tae Kim, *Fairchild Semiconductor, Korea, South*

Secondary-Side Control of a Constant Frequency Series Resonant Converter Using Dual-Edge PWM 544
Darryl J. Tschirhart, *Queen's University, Canada*
Praveen K. Jain, *Queen's University, Canada*

A Non-Insulated Resonant Boost Converter 550
Peng Shuai, *RWTH Aachen University, Germany*
Yales R. De Novaes, *ABB, Switzerland*
Francisco Canales, *ABB, Switzerland*
Ivo Barbi, *Federal University of Santa Catarina, Brazil*

Analysis and Design of a Low-Profile Resonant LCC Converter 557
A. Pawellek, *University of Erlangen-Nürnberg, Germany*
A. Bucher, *University of Erlangen-Nürnberg, Germany*
T. Duerbaum, *University of Erlangen-Nürnberg, Germany*

Session B2L-D: Miscellaneous Applications
Wednesday, February 24, 14:00 - 15:40
Session Chairs: Alejandro Dominguez-Garcia, *University of Illinois*

ZVS and ZCS DC-DC PWM Full-Bridge Fuel Cell Converters 564
Ahmad Mousavi, *University of Western Ontario, Canada*
Pritam Das, *University of Western Ontario, Canada*
Gerry Moschopoulos, *University of Western Ontario, Canada*

Effective Switching Mode Power Supplies Common Mode Noise Cancellation
Technique with Zero Equipotential Transformer Models 571
Yick Po Chan, *University of Hong Kong, China*
Man Hay Pong, *University of Hong Kong, China*
Ngai Kit Poon, *University of Hong Kong, China*
Chui Pong Liu, *University of Hong Kong, China*

50W Power Device (PD) Power in Power Over Ethernet (PoE) System with Input
Current Balance in Four-Pair Architecture with Two DC-DC Converters 575
Haimeng Wu, *Zhejiang University, China*
Zhengshi Wang, *Zhejiang University, China*
Jiande Wu, *Zhejiang University, China*
Xiangning He, *Zhejiang University, China*
Yan Deng, *Zhejiang University, China*

High-Resolution Physically-Windowed Sensors for Power Electronics Applications 580
Warit Wichakool, *Massachusetts Institute of Technology, United States*
James Paris, *Massachusetts Institute of Technology, United States*
Al-Thaddeus Avestruz, *Massachusetts Institute of Technology, United States*
Steven B. Leeb, *Massachusetts Institute of Technology, United States*

Session B2L-E: LED Lighting I
Wednesday, February 24, 14:00 - 15:40
Session Chairs: Regan Zane, *University of Colorado*

Edison Revisited: Impact of DC Distribution on the Cost of
LED Lighting and Distributed Generation 588
Brinda A. Thomas, *Carnegie Mellon University, United States*

A Novel Passive Off-Line Light-Emitting Diode (LED) Driver with Long Lifetime 594
S.Y.R. Hui, *City University of Hong Kong, China*
S.N. Li, *City University of Hong Kong, China*
X.H. Tao, *City University of Hong Kong, China*
W. Chen, *City University of Hong Kong, China*
W.M. Ng, *City University of Hong Kong, China*

Improving Current Regulation for Offline LED Driver 601
Jianwen Shao, *STMicroelectronics, United States*

LED Driver Circuit with Inherent PFC 605
D. Aguilar, *University of Minnesota, United States*
C.P. Henze, *Analog Power Design Inc., United States*

Session B2L-F: Power Electronics for Utility Interface III
Wednesday, February 24, 14:00 - 15:40
Session Chairs: Hui Li, *Florida State University*
Miaosen Shen, *United Technologies Research Center*

A New Circuit Design and Control to Reduce Input Harmonic Current for a Three-Phase AC Machine Drive System Having a Very Small DC-Link Capacitor 611
Hyunjae Yoo, *Samsung Heavy Industries Co., Ltd., Korea, South*
Seung-Ki Sul, *Seoul National University, Korea, South*

State-Space Modeling, Analysis, and Implementation of Parallel Inverters for Microgrid Applications 619
Chien Liang Chen, *Virginia Polytechnic Institute and State University, United States*
Jih-Sheng Lai, *Virginia Polytechnic Institute and State University, United States*
Daniel Martin, *Virginia Polytechnic Institute and State University, United States*
Yuang-Shung Lee, *Fu-Jen Catholic University, Taiwan*

Efficiency Improvement of Grid-Tied Inverters at Low Input Power Using Pulse Skipping Control Strategy 627
Haibing Hu, *University of Central Florida, United States*
Wisam Al-Hoor, *University of Central Florida, United States*
Nasser Kutkut, *University of Central Florida, United States*
Issa Batarseh, *University of Central Florida, United States*
John Shen, *University of Central Florida, United States*

Phase Locked Loop for Unbalanced Utility Conditions 634
Carlos D. Rodríguez-Valdez, *Rockwell Automation, United States*
Russ J. Kerkman, *Rockwell Automation, United States*

Session B2L-G: Isolated DC-DC Converters I
Wednesday, February 24, 14:00 - 15:40
Session Chairs: Alexis Kwasinski, *The University of Texas at Austin*
Sheldon Williamson, *Concordia University*

Analysis and Design Considerations for EMI and Losses of RCD Snubber in Flyback Converter 642
Peipei Meng, *Zhejiang University, China*
Xinke Wu, *Zhejiang University, China*
Jianyou Yang, *Zhejiang University, China*
Henglin Chen, *Zhejiang University, China*
Zhaoming Qian, *Zhejiang University, China*

A High Output Power Density 400/400V Isolated DC-DC Converter with Hybrid Pair of SJ-MOSFET and SiC-SBD for Power Supply of Data Center 648
Rejeki Simanjorang, *National Institute of Advanced Industrial Science and Technology, Japan*
Hiroshi Yamaguchi, *National Institute of Advanced Industrial Science and Technology, Japan*
Hiromichi Ohashi, *National Institute of Advanced Industrial Science and Technology, Japan*
Takashi Takeda, *NTT Facilities Inc., Japan*
Mikio Yamazaki, *NTT Facilities Inc., Japan*
H. Murai, *NTT Facilities Inc., Japan*

A 500 W Push-Pull DC-DC Power Converter with a 30 MHz Switching Frequency 654
John S. Glaser, *GE Global Research, United States*
Juan M. Rivas, *GE Global Research, United States*

Input-Series Connnected High Frequency DC-DC Converters with One Transformer 662
Deshang Sha, *Beijing Institute of Technology, China*
Zhiqiang Guo, *Beijing Institute of Technology, China*
XiaoZhong Liao, *Beijing Institute of Technology, China*

Session B3L-A: Renewable Energy
Wednesday, February 24, 16:10 - 17:25
Session Chairs: Chris Edrington, *Florida State University*
Alex Huang, *North Carolina State University*

Simple Photovoltaic Solar Cell Dynamic Sliding Mode Controlled Maximum Power Point Tracker for Battery Charging Applications 666
Emil A. Jimenez-Brea, *University of Puerto Rico-Mayaguez, Puerto Rico*
Eduardo I. Ortiz-Rivera, *University of Puerto Rico-Mayaguez, Puerto Rico*
Andres Salazar-Llinas, *University of Puerto Rico-Mayaguez, Puerto Rico*
Jesus Gonzalez-Llorente, *University of Puerto Rico-Mayaguez, Puerto Rico*

An Enhanced Circuit-Based Model for Single-Cell Battery 672
Jiucai Zhang, *University of Nebraska-Lincoln, United States*
Song Ci, *University of Nebraska-Lincoln, United States*
Hamid Sharif, *University of Nebraska-Lincoln, United States*
Mahmoud Alahmad, *University of Nebraska-Lincoln, United States*

A High Frequency Battery Model for Current Ripple Analysis 676
Jin Wang, *Ohio State University, United States*
Ke Zou, *Ohio State University, United States*
Chingchi Chen, *Ford Motor Company, United States*
Lihua Chen, *Ford Motor Company, United States*

Session B3L-B: System Integration II
Wednesday, February 24, 16:10 - 17:25
Session Chairs: Shamala Chickamenahalli, *Intel*

A Novel Power Line Communication Technique Based on Power Electronics Circuit Topology 681
Jiande Wu, *Zhejiang University, China*
Chushan Li, *Zhejiang University, China*
Xiangning He, *Zhejiang University, China*

Compact Temperature Compensation of Inductive Fly-back Clamps for Integrated Power Switches Using a High-Voltage Base-Current-Compensated V_{be} Multiplier 686
Timothy P. Duryea, *Texas Instruments, United States*
Hoi Lee, *University of Texas at Dallas, United States*

Optimal Design for the Damping Resistor in RCD-R Snubber to Suppress Common-Mode Noise 691
Peipei Meng, *Zhejiang University, China*
Henglin Chen, *Zhejiang University, China*
Sheng Zheng, *Zhejiang University, China*
Xinke Wu, *Zhejiang University, China*
Zhaoming Qian, *Zhejiang University, China*

Session B3L-C: Resonant DC-DC Converters II
Wednesday, February 24, 16:10 - 17:25
Session Chairs: Dustin Becker, *Emerson Network Power*
Russell Spyker, *USAF*

A High-Efficient LLCC Series-Parallel Resonant Converter 696
Christian P. Dick, *RWTH Aachen University, Germany*
Furkan Kaan Titiz, *RWTH Aachen University, Germany*
Rik De Doncker, *RWTH Aachen University, Germany*

Accurate Switching Loss Model and Optimal Design of a Current Source Driver Considering the Current Diversion Problem 702
Jizhen Fu, *Queen's University, Canada*
Zhiliang Zhang, *Nanjing University of Aeronautics and Astronautics, China*
Andrew Dickson, *Queen's University, Canada*
Yan-Fei Liu, *Queen's University, Canada*
P.C. Sen, *Queen's University, Canada*

Bidirectional Operation of Resonant Voltage Divider 710
K.I. Hwu, *National Taipei University of Technology, Taiwan*
Y.T. Yau, *National Taipei University of Technology, Taiwan*

Session B3L-D: RF Applications
Wednesday, February 24, 16:10 - 17:25
Session Chairs: Alejandro Dominguez-Garcia, *University of Illinois*

Multiple-Input Buck Converter Optimized for Accurate Envelope Tracking in RF Power Amplifiers 715
M. Rodríguez, *University of Oviedo, Spain*
P.F. Miaja, *University of Oviedo, Spain*
A. Rodríguez, *University of Oviedo, Spain*
J. Sebastián, *University of Oviedo, Spain*

Switching Capacities Based Envelope Amplifier for High Efficiency RF Amplifiers 723
M. Vasić, *Universidad Politécnica de Madrid, Spain*
O. García, *Universidad Politécnica de Madrid, Spain*
J.A. Oliver, *Universidad Politécnica de Madrid, Spain*
P. Alou, *Universidad Politécnica de Madrid, Spain*
D. Diaz, *Universidad Politécnica de Madrid, Spain*
J.A. Cobos, *Universidad Politécnica de Madrid, Spain*

High Efficiency Power Amplifier for High Frequency Radio Transmitters 729

M. Vasić, *Universidad Politécnica de Madrid, Spain*
O. García, *Universidad Politécnica de Madrid, Spain*
J.A. Oliver, *Universidad Politécnica de Madrid, Spain*
P. Alou, *Universidad Politécnica de Madrid, Spain*
D. Diaz, *Universidad Politécnica de Madrid, Spain*
J.A. Cobos, *Universidad Politécnica de Madrid, Spain*
A. Gimeno, *Universidad Politecnica de Madrid, Spain*
J.M. Pardo, *Universidad Politécnica de Madrid, Spain*
C. Benavente, *Universidad Politécnica de Madrid, Spain*
F.J. Ortega, *Universidad Politécnica de Madrid, Spain*

Session B3L-E: LED Lighting II
Wednesday, February 24, 16:10 - 17:25
Session Chairs: Regan Zane, *University of Colorado*

Applying One-Comparator Counter-Based Sampling to Current Sharing Control of Multi-Channel LED Strings 737

K.I. Hwu, *National Taipei University of Technology, Taiwan*
Y.T. Yau, *National Taipei University of Technology, Taiwan*

High Frequency PWM Dimming Technique for High Power Factor Converters in LED Lighting 743

D. Gacio, *University of Oviedo, Spain*
J.M. Alonso, *University of Oviedo, Spain*
J. Garcia, *University of Oviedo, Spain*
L. Campa, *University of Oviedo, Spain*
M. Crespo, *University of Oviedo, Spain*
M. Rico-Secades, *University of Oviedo, Spain*

A RGB-Driver for LED Display Panels 750

Jaber Hasan, *University of Arkansas, United States*
Do Hung Nguyen, *University of Arkansas, United States*
Simon S. Ang, *University of Arkansas, United States*

Session B3L-F: Power Electronics for Utility Interface IIII
Wednesday, February 24, 16:10 - 17:25
Session Chairs: Hui Li, *Florida State University*
Miaosen Shen, *United Technologies Research Center*

A Low Investment Single-Phase to Three-Phase Converter Operating with Reduced Losses 755

José A.A. Dias, *Instituto Federal de Educação, Ciência e Tecnologia da Paraíba, Brazil*
Euzeli C. dos Santos, *Universidade Federal de Campina Grande, Brazil*
Cursino B. Jacobina, *Universidade Federal de Campina Grande, Brazil*

Voltage and Power Balance Control for a Cascaded Multilevel Solid State Transformer 761
Tiefu Zhao, *North Carolina State University, United States*
Gangyao Wang, *North Carolina State University, United States*
Jie Zeng, *North Carolina State University, United States*
Sumit Dutta, *North Carolina State University, United States*
Subhashish Bhattacharya, *North Carolina State University, United States*
Alex Q. Huang, *North Carolina State University, United States*

Grid-Connected Voltage Source Inverter for Renewable Energy Conversion System with Sensorless Current Control 768
Suzan Eren, *Queen's University, Canada*
Majid Pahlevaninezhad, *Queen's University, Canada*
Alireza Bakhshai, *Queen's University, Canada*
Praveen Jain, *Queen's University, Canada*

Session B3L-G: Isolated DC-DC Converters II
Wednesday, February 24, 16:10 - 17:25
Session Chairs: Alexis Kwasinski, *The University of Texas at Austin*
Sheldon Williamson, *Concordia University*

Design of an 99%-Efficient, 5kW, Phase-Shift PWM DC-DC Converter for Telecom Applications 773
U. Badstuebner, *ETH Zurich, Switzerland*
J. Biela, *ETH Zurich, Switzerland*
J.W. Kolar, *ETH Zurich, Switzerland*

DC-DC Transformer Multiphase Converter with Transformer Coupling for Two-Stage Architecture 781
M.C. Gonzalez, *Universidad Politécnica de Madrid, Spain*
P. Alou, *Universidad Politécnica de Madrid, Spain*
O. Garcia, *Universidad Politécnica de Madrid, Spain*
J.A. Oliver, *Universidad Politécnica de Madrid, Spain*
J.A. Cobos, *Universidad Politécnica de Madrid, Spain*
H. Visairo, *Intel Corporation, Mexico*

A Comparison of Classical Two Phase (2L) and Transformer – Coupled (XL) Interleaved Boost Converters for Fuel Cell Applications 787
Kevin J. Hartnett, *University College Cork, Ireland*
Marek S. Rylko, *University College Cork, Ireland*
John G. Hayes, *University College Cork, Ireland*
Michael G. Egan, *University College Cork, Ireland*

Session C1L-A: Applications of DC-DC Converter I
Thursday, February 25, 8:30 - 10:10
Session Chairs: Chuck Mullett, *ON Semiconductor*
Kevin Parmenter, *Freescale*

Design Considerations for Narrow Vdc Based Power Delivery Architecture in Mobile Computing System 794
Xiaoguo Liang, *Intel Asia-Pacific Research & Development Ltd., China*
Gnanavel Jayakanthan, *Intel Asia-Pacific Research & Development Ltd., China*
Meng Wang, *Intel Asia-Pacific Research & Development Ltd., China*

Active Clamp Boost Converter with Switched Capacitor and Coupled Inductor 801
Yi Zhao, *Zhejiang University, China*
Wuhua Li, *Zhejiang University, China*
Bo Yang, *Zhejiang University, China*
Xiangning He, *Zhejiang University, China*

**Unified Modulation for Three-Phase Current-Fed Bidirectional DC-DC Converter
Under Varied Input Voltage** 807
Zhan Wang, *Florida State University, United States*
Hui Li, *Florida State University, United States*

**Integrated Switched-Capacitor Voltage Doubler with Clock Transition Periods
Boosting and Transfer Blocking Techniques** 813
Phong Ngo, *University of Arizona, United States*
Dongsheng Ma, *University of Arizona, United States*

Session C1L-B: AC-DC Conversion Misc. Topics I
Thursday, February 25, 8:30 - 10:10
Session Chairs: Frank Cirolia, *Emerson Network Power*
Alireza Khaligh, *Illinois Institute of Technology*

A Novel Class of Multipulse Converters Based on High-Frequency-Operated Transformers ... 818
Sheng Zheng, *Zhejiang University, China*
Dong Chen, *Zhejiang University, China*
Hai Lin, *Zhejiang University, China*
Yousheng Wang, *Zhejiang University, China*
Zhaoming Qian, *Zhejiang University, China*
Fang Z. Peng, *Michigan State University, United States*

A High Efficiency Flyback Converter with New Active Clamp Technique 823
Xiucheng Huang, *Zhejiang University, China*
Weijing Du, *Zhejiang University, China*
Wei Yuan, *Zhejiang University, China*
Junming Zhang, *Zhejiang University, China*
Zhaoming Qian, *Zhejiang University, China*

**Analysis and Design of a Novel Integrated Three-Phase Single-Stage
AC-DC PWM Full-Bridge Converter** 829
Dunisha Wijeratne, *University of Western Ontario, Canada*
Gerry Moschopoulos, *University of Western Ontario, Canada*

**Three-Phase Voltage Doubler Rectifier Based on Three-State Switching Cell for
Uninterruptible Power Supply Applications Using FPGA** 837
Raphael A. da Câmara, *Universidade Federal do Ceará, Brazil*
P.P. Praça, *Universidade Federal do Ceará, Brazil*
C.M.T. Cruz, *Universidade Federal do Ceará, Brazil*
R.P. Torrico-Bascopé, *Universidade Federal do Ceará, Brazil*
C.E.A. Silva, *Universidade Federal do Ceará, Brazil*
D.S. Oliveira, Jr., *Universidade Federal do Ceará, Brazil*
L.H.S.C. Barreto, *Universidade Federal do Ceará, Brazil*

Session C1L-C: Grid Interconnection I
Thursday, February 25, 8:30 - 10:10
Session Chairs: Ali Bazzi, *University of Illinois*
Patrick Chapman, *University of Illinois*

Multi-Loop Control Algorithms for Seamless Transition of Grid-Connected Inverter 844
Qin Lei, *Michigan State University, United States*
Shuitao Yang, *Michigan State University & Zhe Jiang University, United States*
Fang Z. Peng, *Michigan state University, United States*

Digital Controller Development for Grid-Tied Photovoltaic Inverter with Model Based Technique 849
Zhigang Liang, *North Carolina State University, United States*
Larry Alesi, *MegaWatt Solar Inc., United States*
Xiaohu Zhou, *North Carolina State University, United States*
Alex Q. Huang, *North Carolina State University, United States*

High-Performance and Cost-Effective Multiple Feedback Control Strategy for Standalone Operation of Grid-Connected Inverter 854
Qin Lei, *Michigan State University, United States*
Shuitao Yang, *Zhejiang University, United States*
Fang Z. Peng, *Michigan State University, United States*

Current Control Optimization for Grid-Tied Inverters with Grid Impedance Estimation 861
Guoqiao Shen, *Zhejiang University, China*
Jun Zhang, *Zhejiang University, China*
Xiao Li, *Zhejiang University, China*
Chengrui Du, *Zhejiang University, China*
Dehong Xu, *Zhejiang University, China*

Session C1L-D: Inverter I
Thursday, February 25, 8:30 - 10:10
Session Chairs: Russell Spyker, *USAF*
Haidong Yu, *Phoenix International*

A New Direct Peak DC-Link Voltage Control Strategy of Z-Source Inverters 867
Yu Tang, *Nanjing University of Aeronautics and Astronautics, China*
Jukui Wei, *Nanjing University of Aeronautics and Astronautics, China*
Shaojun Xie, *Nanjing University of Aeronautics and Astronautics, China*

High Performance Voltage Regulation of Current Source Inverters 873
S.A.S. Grogan, *Monash University, Australia*
D.G. Holmes, *Monash University, Australia*
B.P. McGrath, *Monash University, Australia*

Development of a New Voltage Source Inverter (VSI) Average Model Including Low Frequency Harmonics 881
S. Ahmed, *Virginia Polytechnic Institute and State University, United States*
D. Boroyevich, *Virginia Polytechnic Institute and State University, United States*
F. Wang, *University of Tennessee - Knoxville, United States*
R. Burgos, *ABB US Corporate Research Center, United States*

Realization and Improvement of Repetitive Control in Rotating Frame for Active Power Filter System 887

Baifeng Chen, *Wuhan University, China*
Xiaoming Zha, *Wuhan University, China*
Jinwu Gong, *Wuhan University, China*
Suxuan Guo, *Wuhan University, China*
Jianjun Sun, *Wuhan University, China*

Session C1L-E: PWM in Motor Drives I
Thursday, February 25, 8:30 - 10:10
Session Chairs: Dionysios Aliprantis, *Iowa State University*

Current Constraints of PWM Rectifier Under Unbalanced Voltage Supply 895

Miroslav Chomat, *Institute of Thermomechanics, Czech Rep.*
Ludek Schreier, *Institute of Thermomechanics, Czech Rep.*
Jiri Bendl, *Institute of Thermomechanics, Czech Rep.*

Space Vector PWM for a Direct Matrix Converter Based Open-End Winding AC Drives with Enhanced Capabilities 901

Ranjan K. Gupta, *University of Minnesota, United States*
Apurva Somani, *University of Minnesota, United States*
Krushna K. Mohapatra, *University of Minnesota, United States*
Ned Mohan, *University of Minnesota, United States*

Evaluation of the Hybrid Four-Level Converter Employing Half-Bridge Modules for Two Different Modulation Schemes 909

Alessandro L. Batschauer, *Santa Catarina State University, Brazil*
Arnaldo J. Perin, *Federal University of Santa Catarina, Brazil*
Samir A. Mussa, *Federal University of Santa Catarina, Brazil*
Marcelo L. Heldwein, *Federal University of Santa Catarina, Brazil*

A Comparative Study of Space Vector PWM Strategy for Dual Three-Phase Permanent-Magnet Synchronous Motor Drives 915

Yanhui He, *Xi'an Jiaotong University, China*
Yue Wang, *Xi'an Jiaotong University, China*
Jinlong Wu, *Xi'an Jiaotong University, China*
Yupeng Feng, *Xi'an Jiaotong University, China*
Jinjun Liu, *Xi'an Jiaotong University, China*

Session C1L-F: Magnetics in DC-DC Converters
Thursday, February 25, 8:30 - 10:10
Session Chairs: Arnold Alderman, *PSMA*

A Novel Coupled Inductor for Interleaved Converters 920

Qianhong Chen, *Nanjing University of Aeronautics and Astronautics, China*
Ligang Xu, *Nanjing university of aeronautics and astronautics, China*
Xiaoyong Ren, *Nanjing University of Aeronautics and Astronautics, China*
Lingling Cao, *Nanjing University of Aeronautics and Astronautics, China*
Xinbo Ruan, *Nanjing University of Aeronautics and Astronautics, China*

Transformer's Capacitance Effect on the Operation of Triangular-Current Shaped Soft-Switched Converters 928
Ilya Zeltser, *Ben-Gurion University of the Negev, Israel*
Sam Ben-Yaakov, *Ben-Gurion University of the Negev, Israel*

An Input and Output Ripple Free Converter with a Four-Winding Coupled Inductor 935
Zhuomin Feng, *Zhejiang University, China*
Zhe Zhang, *Zhejiang university, China*
Duo Li, *Zhejiang University, China*
Min Chen, *Zhejiang University, China*
Zhaoming Qian, *Zhejiang University, China*

Investigation on Transformer Design of High Frequency High Efficiency DC-DC Converters 940
Dianbo Fu, *Virginia Polytechnic Institute and State University, United States*
Fred C. Lee, *Virginia Polytechnic Institute and State University, United States*
Shuo Wang, *Virginia Polytechnic Institute and State University, United States*

Session C1L-G: Photovoltaics I
Thursday, February 25, 8:30 - 10:10
Session Chairs: Robert Balog, *Texas A&M University*

A DSP-Based Single-Stage Maximum Power Point Tracking PV Inverter 948
Wen Long Yu, *National Taiwan University of Science and Technology, Taiwan*
Ting-Peng Lee, *National Taiwan University of Science and Technology, Taiwan*
Guan-Hong Wu, *National Taiwan University of Science and Technology, Taiwan*
Qing Su Chen, *National Taiwan University of Science and Technology, Taiwan*
Huang-Jen Chiu, *National Taiwan University of Science and Technology, Taiwan*
Yu-Kang Lo, *National Taiwan University of Science and Technology, Taiwan*
Frank Shih, *Macroblock Inc., Taiwan*

A Simple Mixed-Signal MPPT Circuit for Photovoltaic Applications 953
P. Mattavelli, *University of Padova, Italy*
S. Saggini, *University of Udine, Italy*
E. Orietti, *University of Padova, Italy*
G. Spiazzi, *University of Padova, Italy*

Low-Power Maximum Power Point Tracker with Digital Control for Thermophotovoltaic Generators 961
Robert C.N. Pilawa-Podgurski, *Massachusetts Institute of Technology, United States*
Nathan A. Pallo, *Massachusetts Institute of Technology, United States*
Walker R. Chan, *Massachusetts Institute of Technology, United States*
David J. Perreault, *Massachusetts Institute of Technology, United States*
Ivan L. Celanovic, *Massachusetts Institute of Technology, United States*

11-Level Cascaded H-Bridge Grid-Tied Inverter Interface with Solar Panels 968
Faete Filho, *University of Tennessee, United States*
Yue Cao, *University of Tennessee, United States*
Leon M. Tolbert, *University of Tennessee, United States*

Session C2L-A: Applications of DC-DC Converter II
Thursday, February 25, 10:40 - 11:30
Session Chairs: Chuck Mullett, *ON Semiconductor*
Kevin Parmenter, *Freescale*

**A Life Prediction Scheme for Electrolytic Capacitors in
Power Converters without Current Sensor** 973
H.M. Pang, *University of Hong Kong, China*
M.H. Bryan Pong, *University of Hong Kong, China*

**Load-Interactive Steered-Inductor DC-DC Converter with
Minimized Output Filter Capacitance** 980
S.M. Ahsanuzzaman, *University of Toronto, Canada*
Amir Parayandeh, *University of Toronto, Canada*
Aleksandar Prodić, *University of Toronto, Canada*
Dragan Maksimović, *University of Colorado at Boulder, United States*

Session C2L-B: AC-DC Conversion Misc. Topics II
Thursday, February 25, 10:40 - 11:30
Session Chairs: Frank Cirolia, *Emerson Network Power*
Alireza Khaligh, *Illinois Institute of Technology*

**EMI Filter Design for High Switching Frequency Three-Phase/Level
PWM Rectifier Systems** 986
M. Hartmann, *ETH Zurich, Switzerland*
H. Ertl, *Vienna University of Technology, Austria*
J.W. Kolar, *ETH Zurich, Switzerland*

**Self-Driven AC-DC Synchronous Rectifier for Power Applications – A Direct
Energy-Efficient Replacement for Traditional Diode Rectifier** 994
W.X. Zhong, *City University of Hong Kong, China*
W.C. Ho, *ConvenientPower HK Ltd., China*
X. Liu, *ConvenientPower HK Ltd., China*
S.Y.R. Hui, *City University of Hong Kong, China*

Session C2L-C: Grid Interconnection II
Thursday, February 25, 10:40 - 11:30
Session Chairs: Ali Bazzi, *University of Illinois*

A Robust Control Scheme for Grid-Connected Voltage Source Inverters 1002
Shuitao Yang, *Zhejiang University and Michigan State University, China*
Qin Lei, *Michigan State University, United States*
Fang Z. Peng, *Michigan State University, United States*
Zhaoming Qian, *Zhejiang University, China*

**Application of Active NPC Converter on Generator Side for
MW Direct-Driven Wind Turbine** 1010
Jun Li, *North Carolina State University, United States*
Alex Q. Huang, *North Carolina State University, United States*
Subhashish Bhattacharya, *North Carolina State University, United States*
Wei Jing, *China University of Mining and Technology, United States*

Session C2L-D: Inverter II
Thursday, February 25, 10:40 - 11:30
Session Chairs: Russell Spyker, *USAF*
Haidong Yu, *Phoenix International*

Nonlinear Modeling of Switched Reluctance Motor Using Different Methods 1018
Jun Cai, *Nanjing University of Aeronautics and Astronautics, China*
Zhiquan Deng, *Nanjing University of Aeronautics and Astronautics, China*
Zeyuan Liu, *Nanjing University of Aeronautics and Astronautics, China*

Simplified Synchronous Reference Frame Control of the
Three Phase Grid Connected Inverter 1026
Abad Lorduy, *Carlos III University of Madrid, Spain*
Antonio Lázaro, *Carlos III University of Madrid, Spain*
Andrés Barrado, *Carlos III University of Madrid, Spain*
Cristina Fernández, *Carlos III University of Madrid, Spain*
Isabel Quesada, *Carlos III University of Madrid, Spain*
Carlos Lucena, *Carlos III University of Madrid, Spain*

Session C2L-E: PWM in Motor Drives II
Thursday, February 25, 10:40 - 11:30
Session Chairs: Dionysios Aliprantis, *Iowa State University*

A Novel Direct Digital SPWM Method for Multilevel Voltage Source Inverters 1034
Wanmin Fei, *Nanjing Normal University, China*
Yanli Zhang, *Nanjing Normal University, China*
Bin Wu, *Ryerson University, Canada*

Weight Oriented Optimal PWM in Low Modulation Indexes for
Multilevel Inverters with Unbalanced DC Sources 1038
Damoun Ahmadi, *Ohio State University, United States*
Ke Zou, *Ohio State University, United States*
Jin Wang, *Ohio State University, United States*

Session C2L-F: Measurement and Testing
Thursday, February 25, 10:40 - 11:30
Session Chairs: Patrick Chapman, *University of Illinois*

Oscillation-Test Technique for Buck Voltage Regulator 1043
Jing-Yi Huang, *National Cheng-Kung University, Taiwan*
Chun-Hsun Wu, *National Cheng-Kung University, Taiwan*
Le-Ren Chang-Chien, *National Cheng-Kung University, Taiwan*
Soon-Jyh Chang, *National Cheng-Kung University, Taiwan*

Core Loss Predictions for General PWM Waveforms from a
Simplified Set of Measured Data 1048
Charles R. Sullivan, *Thayer School of Engineering at Dartmouth, United States*
John H. Harris, *Thayer School of Engineering at Dartmouth, United States*
Edward Herbert, *FMTT, Inc., United States*

Session C2L-G: Photovoltaics II
Thursday, February 25, 10:40 - 11:30
Session Chairs: Robert Balog, *Texas A&M University*

High-Efficiency Inverter with H6-Type Configuration for Photovoltaic Non-Isolated AC Module Applications 1056

Wensong Yu, *Virginia Polytechnic Institute and State University, United States*
Jih-Sheng Lai, *Virginia Polytechnic Institute and State University, United States*
Hao Qian, *Virginia Polytechnic Institute and State University, United States*
Chris Hutchens, *Virginia Polytechnic Institute and State University, United States*
Jianhui Zhang, *National Semiconductor Corporation, United States*
Gianpaolo Lisi, *National Semiconductor Corporation, United States*
Ali Djabbari, *National Semiconductor Corporation, United States*
Greg Smith, *National Semiconductor Corporation, United States*
Tim Hegarty, *National Semiconductor Corporation, United States*

Analyzing the Optimal Matching of DC Motors to Photovoltaic Modules via DC-DC Converters 1062

Jesus Gonzalez-Llorente, *University of Puerto Rico-Mayaguez, Puerto Rico*
Eduardo I. Ortiz-Rivera, *University of Puerto Rico-Mayaguez, Puerto Rico*
Andres Salazar-Llinas, *University of Puerto Rico-Mayaguez, Puerto Rico*
Emil Jimenez-Brea, *University of Puerto Rico-Mayaguez, Puerto Rico*

Session C3L-A: Load Management Interface I
Thursday, February 25, 14:00 - 15:40
Session Chairs: Siamak Abedinpour, *Freescale*
Jonathan Kimball, *Missouri S&T*

Performance Analysis of an Interleaved High Step-Up Converter with Voltage Multiplier Cell 1069
Wuhua Li, *Zhejiang University, China*
Yi Zhao, *Zhejiang University, China*
Yan Deng, *Zhejiang University, China*
Xiangning He, *Zhejiang University, China*

FPGA-Based Multi-Phase Digital Pulse Width Modulator with Dual-Edge Modulation 1075
Martin Scharrer, *University of Limerick, Ireland*
Mark Halton, *University of Limerick, Ireland*
Tony Scanlan, *University of Limerick, Ireland*
Karl Rinne, *University of Limerick, Ireland*

Phase Doubler for High Power Voltage Regulators 1081
Chun Cheung, *Intersil Corporation, United States*
Weihong Qiu, *Intersil Corporation, United States*
Emil Chen, *Intersil Corporation, United States*
Greg Miller, *Intersil Corporation, United States*

Automatic Multi-Phase Digital Pulse Width Modulator 1087
Simon Effler, *University of Limerick, Ireland*
Mark Halton, *University of Limerick, Ireland*
Karl Rinne, *University of Limerick, Ireland*

Session C3L-B: Power Electronics in Motor Drives I
Thursday, February 25, 14:00 - 15:40
Session Chairs: Chris Edrington, *Florida State University*
Patrick Chapman, *University of Illinois*

A Simple Current Sharing Scheme for Dual Three-Phase Permanent-Magnet Synchronous Motor Drives 1093
Yanhui He, *Xi'an Jiaotong University, China*
Yue Wang, *Xi'an Jiaotong University, China*
Jinlong Wu, *Xi'an Jiaotong University, China*
Yupeng Feng, *Xi'an Jiaotong University, China*
Jinjun Liu, *Xi'an Jiaotong University, China*

Multilevel Current Source Inverter Topologies Based on the Duality Principle 1097
Jianyu Bao, *Ningbo Institute of Technology, Zhejiang University, China*
Weibing Bao, *Zhejiang University of Science and Technology, China*
Siran Wang, *Zhejiang University, China*
Zhongchao Zhang, *Zhejiang University, China*

3-Level Power Converter with High-Voltage SiC-PiN Diode and Hard-Gate-Driving of IEGT for Future High-Voltage Power Conversion Systems 1101
Kazuto Takao, *Toshiba Corporation, Japan*
Yasunori Tanaka, *National Institute of Advanced Industrial Science and Technology, Japan*
Kyungmin Sung, *Ibaraki National College of Technology, Japan*
Keiji Wada, *Tokyo Metoropolitan University, Japan*
Takashi Shinohe, *Toshiba Corporation, Japan*
Takeo Kanai, *Toshiba Mitsubishi-Electric Industrial Systems Corporation, Japan*
Hiromichi Ohashi, *National Institute of Advanced Industrial Science and Technology, Japan*

18 kW Three Phase Inverter System Using Hermetically Sealed SiC Phase-Leg Power Modules 1108
Hui Zhang, *Tuskegee University, United States*
Leon M. Tolbert, *University of Tennessee, United States*
Jung Hee Han, *Global Power Electronics, United States*
Madhu S. Chinthavali, *Oak Ridge National Laboratory, United States*
Fred Barlow, *University of Idaho, United States*

Session C3L-C: DC-DC Converter V
Thursday, February 25, 14:00 - 15:40
Session Chairs: Frank Ciriola, *Emerson*
Arnold Alderman, *PSMA*

Multiphase Optimal Response Mixed-Signal Current-Programmed Mode Controller 1113
Jurgen Alico, *University of Toronto, Canada*
Aleksandar Prodić, *University of Toronto, Canada*

Switching Loss Analysis of Closed-Loop Gate Drive 1119
Lihua Chen, *Michigan State University, United States*
Fang Z. Peng, *Michigan State University, United States*

Modeling and Analysis of Closed-Loop Gate Drive 1124
Lihua Chen, *Michigan State University, United States*
Baoming Ge, *Michigan State University, United States*
Fang Z. Peng, *Michigan State University, United States*

**Black-Box Modeling of DC-DC Converters Based on Transient Response
Analysis and Parametric Identification Methods** 1131
V. Valdivia, *Carlos III University of Madrid, Spain*
A. Barrado, *Carlos III University of Madrid, Spain*
A. Lázaro, *Carlos III University of Madrid, Spain*
C. Fernández, *Carlos III University of Madrid, Spain*
P. Zumel, *Carlos III University of Madrid, Spain*

Session C3L-D: Transportation
Thursday, February 25, 14:00 - 15:40
Session Chairs: Jaber Abu Qahouq, *University of Alabama*
Dionysios Aliprantis, *Iowa State University*

**Harmonic and Balance Compensation Using Instantaneous Active and
Reactive Power Control on Electric Railway Systems** 1139
A. Bueno, *Universidad Simón Bolívar, Venezuela*
J.M. Aller, *Universidad Simón Bolívar, Venezuela*
J. Restrepo, *Universidad Simón Bolívar, Venezuela*
T. Habetler, *Georgia Institute of Technology, United States*

**Review of Non-Isolated Bi-Directional DC-DC Converters for Plug-in Hybrid Electric
Vehicle Charge Station Application at Municipal Parking Decks** 1145
Yu Du, *North Carolina State University, United States*
Xiaohu Zhou, *North Carolina State University, United States*
Sanzhong Bai, *North Carolina State University, United States*
Srdjan Lukic, *North Carolina State University, United States*
Alex Huang, *North Carolina State University, United States*

**Control of Plug-in Hybrid Electric Vehicles for Mobile Power Generation and
Grid Support Applications** 1152
Gui-Jia Su, *Oak Ridge National Laboratory, United States*
Lixin Tang, *Oak Ridge National Laboratory, United States*

Interface Issues of Mining Haul Trucks Operating on Trolley Systems 1158
Joy Mazumdar, *Siemens Industry Inc., United States*
Walter Koellner, *Siemens Industry Inc., United States*
Rohit Moghe, *Georgia Institute of Technology, United States*

Session C3L-E: Power Converter Applications I
Thursday, February 25, 14:00 - 15:40
Session Chairs: Vajapeyam Sukumar, *Maxim Integrated Products*

Regenerative AC Electronic Load with One-Cycle Control 1166
In Wha Jeong, *University of California, Irvine, United States*
Mikhail Slepchenkov, *University of California, Irvine, United States*
Keyue Smedley, *University of California, Irvine, United States*
Franco Maddaleno, *University of California, Irvine, United States*

A High Efficiency Regulated Charge Pump Over Wide Input and Load Range 1172
Rong Guo, *North Carolina State University, United States*
Liyu Yang, *North Carolina State University, United States*
Alex Huang, *North Carolina State University, United States*
John Endredy, *RF Micro Devices, United States*

High Performance, High-Power Capacitor Charging:
Focus on Pulse-to-Pulse Repeatability 1177
A. Pokryvailo, *Spellman High Voltage Electronics Corporation, United States*
C. Carp, *Spellman High Voltage Electronics Corporation, United States*
C. Scapellati, *Spellman High Voltage Electronics Corporation, United States*

Generalized AC-DC Single-Phase Boost Rectifier 1183
C.B. Jacobina, *Universidade Federal de Campina Grande, Brazil*
Euzeli dos Santos, *Universidade Federal de Campina Grande, Brazil*
Nady Rocha, *Universidade Federal de Campina Grande, Brazil*

Session C3L-F: Utility Interface Applications
Thursday, February 25, 14:00 - 15:40
Session Chairs: Ali Davoudi, *University of Illinois*

Parallel Connection of Two Shunt Active Power Filters with Losses Optimization 1191
E.C. dos Santos, Jr., *Universidade Federal de Campina Grande, Brazil*
C.B. Jacobina, *Universidade Federal de Campina Grande, Brazil*
A.M. Maciel, *Universidade Federal de Campina Grande, Brazil*

Design and Implementation of an Improved Controller for Parallel-Connected
400 Hz Frequency Converters 1197
B. Tamyurek, *Eskisehir Osmangazi University, Turkey*
E. Birdane, *Kaynak Electronic Machine Industry and Trade Co. Ltd., Turkey*
Adil Ceyhan, *Kaynak Electronic Machine Industry and Trade Co. Ltd., Turkey*

Study on the Impact of the Complex Impedance on the Droop Control Method
for the Parallel Inverters 1204
Wei Yao, *Zhejiang university, China*
Mingzhi Gao, *Zhejiang University, China*
Zheng Ren, *Zhejiang university, China*
Min Chen, *Zhejiang University, China*
Zhaoming Qian, *Zhejiang University, China*

A Three-Phase Adaptive Approach to Extract Harmonic and Reactive Currents 1209
D. Yazdani, *Queen's University, Canada*
A. Bakhshai, *Queen's University, Canada*
P.K. Jain, *Queen's University, Canada*

Session C3L-G: Soft Switching Techniques I
Thursday, February 25, 14:00 - 15:40
Session Chairs: Jason Neely, *Purdue University*
Wayne Weaver, *Michigan Technological University*

Analysis, Optimized Design and Adaptive Control of a ZCS Full-Bridge Converter Without Voltage Over-Stress on the Switches 1214
Xin Zhang, *Nanjing University of Aeronautics and Astronautics, China*
Henry Shu-hung Chung, *City University of Hong Kong, China*
Xinbo Ruan, *Huazhong University of Science and Technology, China*
Adrian Ioinovici, *Holon Institute of Technology, Israel*

Analysis and Design of a Novel ZVS-PWM DC-DC Converter for Bidirectional Applications with Steep Conversion Ratio 1222
Pritam Das, *University of Western Ontario, Canada*
Ahmad Mousavi, *University of Western Ontario, Canada*
Gerry Moschopoulos, *University of Western Ontario, Canada*

Three-Level Phase-Shift ZVS-PWM DC-DC Converter with High Frequency Transformer for High Performance Arc Welding Machines 1230
Tomokazu Mishima, *Kure National College of Technology, Japan*
Hisayuki Sugimura, *Daihen Corporation, Japan*
Khairy Fathy Sayed, *Kyungnam University, Korea, South*
Soon Kurl Kwon, *Kyungnam University, Korea, South*
Mutsuo Nakaoka, *Kyungnam University and Yamaguchi University, Japan*

Fully Soft-Switched Bidirectional Resonant DC-DC Converter with a New CLLC Tank 1238
Wei Chen, *Zhejiang University, China*
Siran Wang, *Zhejiang University, China*
Xiaoyuan Hong, *Zhejiang University, China*
Zhengyu Lu, *Zhejiang University, China*
Shaoshi Ye, *Delta Electronics (Shanghai) Co., LTD., China*

Session C4L-A: Load Management Interface II
Thursday, February 25, 16:10 - 17:25
Session Chairs: Siamak Abedinpour, *Freescale*
Jonathan Kimball, *Missouri S&T*

Optimal Phase Changing Frequency Determination for Multiphase Voltage Regulator Modules 1243
Anand Ramamurthy, *North Carolina State University, United States*
Subhashish Bhattacharya, *North Carolina State University, United States*
Chris Thompson, *Intersil Corporation, United States*
Jon Day, *Intersil Corporation, United States*

A New Digital Adaptive Voltage Positioning Technique with Dynamically Varying Voltage and Current References 1248
S. Pan, *Queen's University, Canada*
P.K. Jain, *Queen's University, Canada*

A Three-Level Buck Converter and Digital Controller for Improving Load Transient Response 1256

Zhenyu Zhao, *Exar Corp, Canada*
Aleksandar Prodić, *University of Toronto, Canada*

Session C4L-B: Power Electronics in Motor Drives II
Thursday, February 25, 16:10 - 17:25
Session Chairs: Chris Edrington, *Florida State University*
Patrick Chapman, *University of Illinois*

Trends in MW-Rated VSI Technology and Reliability for Adjustable Speed Drives 1261

Hiromi Hosoda, *Toshiba Mitsubishi-Electric Industrial Systems Corporation, Japan*
Mostafa Al Mamun, *Toshiba Mitsubishi-Electric Industrial Systems Corporation, Japan*
Teruo Yoshino, *Toshiba Mitsubishi-Electric Industrial Systems Corporation, Japan*

Development of a Compact 750KVA Three-Phase NPC Three-Level Universal Inverter Module with Specifically Designed Busbar 1266

Jun Wang, *Zhejiang University, China*
Binjian Yang, *Zhejiang University, China*
Jing Zhao, *Zhejiang University, China*
Yan Deng, *Zhejiang University, China*
Xiangning He, *Zhejiang University, China*
Xu Zhixin, *Zhejiang University of Science and Technology, China*

Common Mode Voltage in DC-Fed Motor Drive System and its Impact on the EMI Filter 1272

Fang Luo, *Virginia Polytechnic Institute and State University &*
Huazhong University of Science and Technology, United States
Shuo Wang, *GE Aviation Systems, United States*
Fred Wang, *University of Tennessee - Knoxville and Oak Ridge National Laboratory, United States*
Dushan Boroyevich, *Virginia Polytechnic Institute and State University, United States*
Nicolas Gazel, *Virginia Polytechnic Institute and State University, United States*
Yong Kang, *Huazhong University of Science and Technology, China*

Session C4L-C: DC-DC Converter VI
Thursday, February 25, 16:10 - 17:25
Session Chairs: Arnold Alderman, *PSMA*
Frank Cirolia, *Emerson Network Power*

Black-Box Modeling of Three Phase Voltage Source Inverters Based on Transient Response Analysis 1279

V. Valdivia, *Carlos III University of Madrid, Spain*
A. Lázaro, *Carlos III University of Madrid, Spain*
A. Barrado, *Carlos III University of Madrid, Spain*
P. Zumel, *Carlos III University of Madrid, Spain*
C. Fernández, *Carlos III University of Madrid, Spain*
M. Sanz, *Carlos III University of Madrid, Spain*

Digital Autotuning of DC-DC Converters Based on Model Reference Impulse Response ... 1287

A. Costabeber, *University of Padova, Italy*
P. Mattavelli, *University of Padova, Italy*
S. Saggini, *University of Udine, Italy*
A. Bianco, *STMicroelectronics, Italy*

High-Fidelity and High-Speed Modeling and Simulation for Power Conversion Systems ... 1295
Chunchun Xu, *GE Global Research, United States*
Luis Garces, *GE Global Research, United States*
Paul Szczesny, *GE Global Research, United States*

Session C4L-D: Aerospace
Thursday, February 25, 16:10 - 17:25
Session Chairs: Jaber Abu Qahouq, *University of Alabama*
Dionysios Aliprantis, *Iowa State University*

Electrical Power Distribution System (HV270DC), for Application in More Electric Aircraft .. 1300
D. Izquierdo, *EADS, Spain*
R. Azcona, *EADS, Spain*
F.J. López del Cerro, *EADS, Spain*
Carlos Fernández, *EADS, Spain*
Bernardo Delicado, *EADS, Spain*

Supercapacitor-Based Energy Management for Future Aircraft Systems 1306
R. Todd, *University of Manchester, United Kingdom*
D. Wu, *University of Manchester, United Kingdom*
J.A. dos Santos Girio, *University of Manchester, United Kingdom*
M. Poucand, *University of Manchester, United Kingdom*
A.J. Forsyth, *University of Manchester, United Kingdom*

Buck Boost Regulator (B²R) for Spacecraft Solar Array Power Conversion 1313
Olivier Mourra, *European Space Agency, Netherlands*
Arturo Fernandez, *European Space Agency, Netherlands*
Ferdinando Tonicello, *European Space Agency, Netherlands*

Session C4L-E: Power Converter Applications II
Thursday, February 25, 16:10 - 17:25
Session Chairs: Vajapeyam Sukumar, *Maxim Integrated Products*

Quadratic Power Conversion for Industrial Applications 1320
Gerry Moschopoulos, *University of Western Ontario, Canada*

Multiple-Output Resonant Inverter Topology for Multi-Inductor Loads 1328
O. Lucía, *University of Zaragoza, Spain*
J.M. Burdío, *University of Zaragoza, Spain*
I. Millán, *University of Zaragoza, Spain*
J. Acero, *University of Zaragoza, Spain*

**Variable Frequency Pulse Density Modulation[1] for Efficient High Frequency
Operation of Series Resonant Converters Operating As Voltage Regulators** 1334
Darryl J. Tschirhart, *Queen's University, Canada*
Praveen K. Jain, *Queen's University, Canada*

Session C4L-F: General Lighting
Thursday, February 25, 16:10 - 17:25
Session Chairs: Ali Davoudi, *University of Illinois*

Flexible-Controlled High Power-Density Automotive HID Electronic Ballast Using Full-Digital Control Mode 1340
Xinyi Yang, *Zhejiang University, China*
Biwen Xu, *Zhejiang University, China*
Chongguang Ma, *Zhejiang University, China*
Min Chen, *Zhejiang University, China*
Zhaoming Qian, *Zhejiang University, China*

A "Class-A2" Ultra-Low-Loss Magnetic Ballast for T5 Fluorescent Lamps 1346
S.Y.R. Hui, *City University of Hong Kong, China*
D.Y. Lin, *City University of Hong Kong, China*
W.M. Ng, *City University of Hong Kong, China*
W. Yan, *City University of Hong Kong, China*

Simple Triac Dimmable Compact Fluorescent Lamp Ballast and Light Emitting Diode Driver 1352
Andre Tjokrorahardjo, *International Rectifier, United States*

Session C4L-G: Soft Switching Techniques II
Thursday, February 25, 16:10 - 17:25
Session Chairs: Jason Neely, *Purdue University*
Wayne Weaver, *Michigan Technological University*

High Efficiency Soft-Switched Step-Up DC-DC Converter with Hybrid Mode LLC+C Resonant Tank 1358
Wei Chen, *Zhejiang University and Delta Electronics (Shanghai) Co., LTD., China*
Xiaoyuan Hong, *Zhejiang University, China*
Siran Wang, *Zhejiang University, China*
Zhengyu Lu, *Zhejiang University, China*
Shaoshi Ye, *Delta Electronics (Shanghai) Co., LTD., China*

A Family of Zero Current Switching Switched-Capacitor DC-DC Converters 1365
Dong Cao, *Michigan State University, United States*
Fang Zheng Peng, *Michigan State University, United States*

Analysis and Design of the Half Bridge Magnetizing Inductor Resonant(L_mC) DC-DC Converter 1373
B.-C. Hyeon, *Seoul National University, Korea, South*
B.-H. Cho, *Seoul National University, Korea, South*

Session B4P-H: AC-DC Conversion
Thursday, February 25, 11:30 - 13:30

A High Power Density Single Phase PWM Rectifier with Active Ripple Energy Storage 1378
Ruxi Wang, *Virginia Polytechnic Institute and State University, United States*
Fred Wang, *University of Tennessee - Knoxville and Oak Ridge National Laboratory, United States*
Dushan Boroyevich, *Virginia Polytechnic Institute and State University, United States*
Puqi Ning, *Virginia Polytechnic Institute and State University, United States*

Design Considerations for High Efficiency Buck PFC with Half-Bridge Regulation Stage .. 1384
Bernard Keogh, *Texas Instruments (Cork) Ltd., Ireland*
George Young, *Texas Instruments (Cork) Ltd., Ireland*
Hagen Wegner, *Texas Instruments (Cork) Ltd., Ireland*
Colin Gillmor, *Texas Instruments (Cork) Ltd., Ireland*

A Novel Variable Frequency Soft Switching Method for Flyback Converter with Synchronous Rectifier 1392
Xiucheng Huang, *Zhejiang University, China*
Weijing Du, *Zhejiang University, China*
Wei Yuan, *Zhejiang University, China*
Junming Zhang, *Zhejiang University, China*
Zhaoming Qian, *Zhejiang University, China*

Optimal Design of a Compact 99.3% Efficient Single-Phase PFC Rectifier 1397
J. Biela, *ETH Zurich, Switzerland*
J.W. Kolar, *ETH Zurich, Switzerland*
G. Deboy, *Infineon Technologies Austria AG, Austria*

DCM Boost PFC Converter with High Input PF 1405
Kai Yao, *Nanjing university of aeronautics and astronautics, China*
Xinbo Ruan, *Nanjing University of Aeronautics and Astronautics, China*
Xiaojing Mao, *Nanjing University of Aeronautics and Astronautics, China*
Zhihong Ye, *Lite-on Technology Corp., China*

Interleaved Forward Converter with Ripple-Free Circuit for Humane Killer Poultry Applications 1413
S.-Y. Tseng, *Chang Gung University, Taiwan*
T.-Y. Chiang, *Chang Gung University, Taiwan*
K.-C. Wang, *Chang Gung University, Taiwan*
S.-A. Chuang, *Chang Gung University, Taiwan*

The Optimal Control Strategy for Rectifier Side of Low Switching Frequency Back-to-Back Converter 1419
Kai Tan, *Chinese Academy of Sciences, China*
Qiongxuan Ge, *Chinese Academy of Sciences, China*
Zhenggang Yin, *Chinese Academy of Sciences, China*
Congwei Liu, *Chinese Academy of Sciences, China*
Yaohua Li, *Chinese Academy of Sciences, China*

Transformer Structure and its Effects on Common Mode EMI Noise in Isolated Power Converters 1424
Pengju Kong, *Virginia Polytechnic Institute and State University, United States*
Fred C. Lee, *Virginia Polytechnic Institute and State University, United States*

A Single-Stage Single-Phase Bi-Directional Grid Interface Circuit with Digital Lookup Table Based Control 1430
Evan Reutzel, *University of California, Berkeley, United States*
Seth Sanders, *University of California, Berkeley, United States*

Session B4P-J: DC-DC Converter VII
Thursday, February 25, 11:30 - 13:30

A High Performance Dual Output DC-DC Converter Combined the Phase Shift Full Bridge and LLC Resonant Half Bridge with the Shared Lagging Leg 1435
Yu Chen, *Huazhong University of Science and Technology, China*
Xuejun Pei, *Huazhong University of Science and Technology, China*
Li Peng, *Huazhong University of Science and Technology, China*
Yong Kang, *Huazhong University of Science and Technology, China*

Dual Output DC-DC Converter with Shared ZCS Lagging Leg 1441
Yu Chen, *Huazhong University of Science and Technology, China*
Li Peng, *Huazhong University of Science and Technology, China*
Xuejun Pei, *Huazhong University of Science and Technology, China*
Yong Kang, *Huazhong University of Science and Technology, China*

A Novel ZVS Full-Bridge Converter with Auxiliary Circuit 1448
Zhong Chen, *Nanjing University of Aeronautics and Astronautics, China*
Biao Ji, *Nanjing University of Aeronautics and Astronautics, China*
Feng Ji, *Nanjing University of Aeronautics and Astronautics, China*
Lei Shi, *Nanjing University of Aeronautics and Astronautics, China*

An Active Clamp ZVT Converter with Input-Parallel and Output-Series Configuration 1454
Yi Zhao, *Zhejiang University, China*
Wuhua Li, *Zhejiang University, China*
Weichen Li, *Zhejiang University, China*
Xiangning He, *Zhejiang University, China*

A Parallel Front-End LCL Resonant Push-Pull Converter with a Coupled Inductor for Automotive Applications 1460
Yuan Yisheng, *East China Jiaotong University, China*
Chen Min, *Zhejiang University, China*
Qian Zhaoming, *Zhejiang University, China*

A Novel Full Bridge Dual Output DC-DC Converter with Complementary Pulse Widths and Frequency Modulation 1464
Yu Chen, *Huazhong University of Science and Technology, China*
Xuejun Pei, *Huazhong University of Science and Technology, China*
Li Peng, *Huazhong University of Science and Technology, China*
Yong Kang, *Huazhong University of Science and Technology, China*

Analysis and Design Considerations of an Improved ZVS Full-Bridge DC-DC Converter 1471
Zhong Chen, *Nanjing University of Aeronautics and Astronautics, China*
Biao Ji, *Nanjing University of Aeronautics and Astronautics, China*
Feng Ji, *Nanjing University of Aeronautics and Astronautics, China*
Lei Shi, *Nanjing University of Aeronautics and Astronautics, China*

A New Resonant Gate Driver for Switching Loss Reduction of High Side Switch in Buck Converter 1477
Xin Zhou, *North Carolina State University, United States*
Zhigang Liang, *North Carolina State University, United States*
Alex Huang, *North Carolina State University, United States*

Switching Loss Analysis Considering Parasitic Loop Inductance with Current Source Drivers for Buck Converters 1482
Zhiliang Zhang, *Nanjing University of Aeronautics and Astronautics, China*
Jizhen Fu, *Queen's University, Canada*
Yan-Fei Liu, *Queen's University, Canada*
P.C. Sen, *Queen's University, Canada*

Improved Asymmetric Space Vector Modulation for Voltage Source Converters with Low Carrier Ratio 1487
Di Zhang, *Virginia Polytechnic Institute and State University, United States*
Fred Wang, *University of Tennessee - Knoxville and Oak Ridge National Laboratory, United States*
Said El-Barbari, *GE Global Research, Germany*
Juan Sabate, *GE Global Research, United States*
Dushan Boroyevich, *Virginia Polytechnic Institute and State University, United States*

A Hybrid Switching Scheme for LLC Series-Resonant Half-Bridge DC-DC Converter in a Wide Load Range 1494
Woo-Young Choi, *Virginia Polytechnic Institute and State University, United States*
Bong-Hwan Kwon, *Pohang University of Science and Technology, Korea, South*
Jih-Sheng Lai, *Virginia Polytechnic Institute and State University, United States*

Session B4P-K: Motor Drives & Inverters I
Thursday, February 25, 11:30 - 13:30

Industrial Servo Applications of Linear Induction Motors Based on Dynamic Maximum Force Control 1498
Haidong Yu, *Phoenix International, United States*
Babak Fahimi, *University of Texas at Arlington, United States*

A Soft-Switching Interleaved Three-Level Inverter 1503
Yuan Yisheng, *East China Jiaotong University, China*
Chen Min, *Zhejiang University, China*
Qian Zhaoming, *Zhejiang University, China*

Reducing Common-Mode Voltage in Three-Phase Sine-Triangle PWM with Interleaved Carriers 1508
Jonathan W. Kimball, *Missouri University of Science and Technology, United States*
Maciej Zawodniok, *Missouri University of Science and Technology, United States*

Dynamic DC-Bus Voltage Control Strategies for a Three-Phase High Power Shunt Active Power Filter 1514
Zhiqiang Wang, *Zhejiang University, China*
Chuan Xie, *Zhejiang University, China*
Jing Zhang, *Zhejiang University, China*
Guozhu Chen, *Zhejiang University, China*

A Simplified Three Phase Three-Level Zero-Current-Transition Active Neutral-Point-Clamped Converter with Three Auxiliary Switches 1521
Jin Li, *Xi'an Jiaotong University and Virginia Polytechnic Institute and State University, China*
Jinjun Liu, *Xi'an Jiaotong University, China*
Dushan Boroyevich, *Virginia Polytechnic Institute and State University, China*

Comparison and Implementation of a 3-Level NPC Voltage Link Back-to-Back Converter with SiC and Si Diodes 1527
Mario Schweizer, *ETH Zurich, Switzerland*
Thomas Friedli, *ETH Zurich, Switzerland*
Johann W. Kolar, *ETH Zurich, Switzerland*

A Novel PWM Control Method to Eliminate the Effect of Dead Time on the Output Waveform for Hybrid Clamped Multilevel Inverters 1534
Jing Zhao, *Zhejiang University, China*
Xiangning He, *Zhejiang University, China*
Yunlong Han, *Zhejiang University, China*
Yan Chen, *Zhejiang University, China*
Rongxiang Zhao, *Zhejiang University, China*

Study on Wide Range Robust Speed Sensorless Control of Medium Voltage Induction Motor 1542
Siran Wang, *Zhejiang University, China*
Zhengyu Lu, *Zhejiang University, China*

Fault Detection and Diagnostics for Non-Intrusive Monitoring Using Motor Harmonics 1547
Uzoma A. Orji, *Massachusetts Institute of Technology, United States*
Zachary Remscrim, *Massachusetts Institute of Technology, United States*
Christopher Laughman, *Massachusetts Institute of Technology, United States*
Steven B. Leeb, *Massachusetts Institute of Technology, United States*
Warit Wichakool, *Massachusetts Institute of Technology, United States*
Christopher Schantz, *Massachusetts Institute of Technology, United States*
Robert Cox, *Massachusetts Institute of Technology, United States*
James Paris, *Massachusetts Institute of Technology, United States*
James L. Kirtley, Jr., *Massachusetts Institute of Technology, United States*
Les K. Norford, *Massachusetts Institute of Technology, United States*

Reliability Evaluation of Three-Level Inverters 1555
Yi Ding, *Nanyang Technological University, Singapore*
Poh Chiang Loh, *Nanyang Technological University, Singapore*
Kuan Khoon Tan, *Nanyang Technological University, Singapore*
Peng Wang, *Nanyang Technological University, Singapore*
Feng Gao, *Nanyang Technological University, Singapore*

Parallel Operation of PWM Inverters for High Speed Motor Drive System 1561
Un-Kwan Cho, *Seoul National University, Korea, South*
Jung-Sik Yim, *Seoul National University, Korea, South*
Seung-Ki Sul, *Seoul National University, Korea, South*

Session B4P-L: Active Components
Thursday, February 25, 11:30 - 13:30

Reverse Conduction of a 100 a SiC DMOSFET Module in High-Power Applications 1568
R.A. Wood, *US Army Research Lab, United States*
D.P. Urciuoli, *US Army Research Lab, United States*
T.E. Salem, *US Naval Academy, United States*
R. Green, *US Army Research Lab, United States*

Investigation of 1.2 kV SiC MOSFET for High Frequency High Power Applications 1572
Honggang Sheng, *Monolithic Power Systems, United States*
Zheng Chen, *Virginia Polytechnic Institute and State University, United States*
Fred Wang, *University of Tennessee - Knoxville, United States*
Alan Millner, *MKS Instruments, United States*

Comparative Analysis of Power Stage Losses for Synchronous Buck Converter in Diode Emulation Mode Vs. Continuous Conduction Mode at Light Load Condition 1578
Yang Chen, *International Rectifier Corp., United States*
Peyman Asadi, *International Rectifier Corp., United States*
Parviz Parto, *International Rectifier Corp., United States*

Controllable dv/dt Behaviour of the SiC MOSFET/JFET Cascode an Alternative Hard Commutated Switch for Telecom Applications 1584
Daniel Aggeler, *ETH Zurich, Switzerland*
Juergen Biela, *ETH Zurich, Switzerland*
Johann W. Kolar, *ETH Zurich, Switzerland*

Integral Micro-Channel Liquid Cooling for Power Electronics 1591
Ljubisa D. Stevanovic, *GE Global Research, United States*
Richard A. Beaupre, *GE Global Research, United States*
Arun V. Gowda, *GE Global Research, United States*
Adam G. Pautsch, *GE Global Research, United States*
Stephen A. Solovitz, *Washington State University Vancouver, United States*

3000V, 25A Pulse Power Asymmetrical Highly Interdigitated SiC Thyristors 1598
Ahmed Elasser, *GE Global Research, United States*
Peter Losee, *GE Global Research, United States*
Stephen Arthur, *GE Global Research, United States*
Zachary Stum, *GE Global Research, United States*
Jerome Garrett, *GE Global Research, United States*
Michael Schutten, *GE Global Research, United States*

Session B4P-M: System Integration III
Thursday, February 25, 11:30 - 13:30

Low Inductance Power Module with Blade Connector 1603
Ljubisa D. Stevanovic, *GE Global Research, United States*
Richard A. Beaupre, *GE Global Research, United States*
Eladio C. Delgado, *GE Global Research, United States*
Arun V. Gowda, *GE Global Research, United States*

Design of Multi-Turn LTCC Inductors for High Frequency DC-DC Converters 1610
Laili Wang, *Xi'an Jiaotong University, China*
Yunqing Pei, *Xi'an Jiaotong University, China*
Xu Yang, *Xi'an Jiaotong University, China*
Xizhi Cui, *Xi'an Jiaotong University, China*
Zhaoan Wang, *Xi'an Jiaotong University, China*
Guopeng Zhao, *Xi'an Jiaotong University, China*

Session B4P-N: Utility Interface
Thursday, February 25, 11:30 - 13:30

**Topological Research and Comparison of Low Harmonic Input Three-Phase
Rectifier with Passive Auxiliary Circuit** 1616
Zhong Chen, *Nanjing University of Aeronautics and Astronautics, China*
Yingpeng Luo, *Nanjing University of Aeronautics and Astronautics, China*
Yinyu Zhu, *Nanjing University of Aeronautics and Astronautics, China*

**The Reactive Power Compensation and Harmonic Filtering and the
Over-Voltage Analysis of the ITER Power Supply System** 1622
L. Xu, *Chinese Academy of Sciences, China*
Z. Sheng, *Chinese Academy of Sciences, China*
P. Fu, *Chinese Academy of Sciences, China*
G. Gao, *Chinese Academy of Sciences, China*
I. Benfatto, *Iter Organization, France*
A.D. Mankani, *Iter Organization, France*
J. Tao, *Iter Organization, France*

Optimal Design Method of Three-Phase Rectifier with Near-Sinusoidal Input Currents 1627
Zhong Chen, *Nanjing University of Aeronautics and Astronautics, China*
Yingpeng Luo, *Nanjing University of Aeronautics and Astronautics, China*
Yinyu Zhu, *Nanjing University of Aeronautics and Astronautics, China*
Shunqing Wang, *Nanjing University of Aeronautics and Astronautics, China*

**An Analysis on the Influence of Interface Inductor to STATCOM System with
Phase and Amplitude Control and Corresponding Design Considerations** 1633
Guopeng Zhao, *Xi'an Jiaotong University, China*
Jinjun Liu, *Xi'an Jiaotong University, China*

**Vector Oriented Control of Voltage Source PWM Inverter As a Dynamic VAR
Compensator for Wind Energy Conversion System Connected to Utility Grid** 1640
Mahmoud M.N. Amin, *Florida International University, United States*
O.A. Mohammed, *Florida International University, United States*

**Control System Design for Bi-Directional Power Transfer in Single-Phase
Back-to-Back Converter Based on the Linear Operating Region** 1651
Janeth Alcalá, *Universidad Autónoma de San Luis Potosi, Mexico*
Víctor Cárdenas, *Universidad Autónoma de San Luis Potosi, Mexico*
Emanuel Rosas, *Universidad Autónoma de San Luis Potosi, Mexico*
Ciro Núñez, *Universidad Autónoma de San Luis Potosi, Mexico*

Comparative Analysis of Low-Pass Output Filter for Single-Phase Grid-Connected Photovoltaic Inverter 1659
Hanju Cha, *Chungnam National University, Korea, South*
Trung-Kien Vu, *Chungnam National University, Korea, South*

Design and Development of Generation-I Silicon based Solid State Transformer 1666
Subhashish Bhattacharya, *North Carolina State University, United States*
Tiefu Zhao, *North Carolina State University, United States*
Gangyao Wang, *North Carolina State University, United States*
Sumit Dutta, *North Carolina State University, United Kingdom*
Seunghun Baek, *North Carolina State University, United States*
Yu Du, *North Carolina State University, United States*
Babak Parkhideh, *North Carolina State University, United States*
Xiaohu Zhou, *North Carolina State University, United States*
Alex Q. Huang, *North Carolina State University, United States*

Power Calculation Method Used in Wireless Parallel Inverters Under Nonlinear Load Conditions 1674
Zheng Ren, *Zhejiang University, China*
Mingzhi Gao, *Zhejiang University, China*
Qiong Mo, *Zhejiang University, China*
Kun Liu, *Zhejiang University, China*
Wei Yao, *Zhejiang University, China*
Min Chen, *Zhejiang University, China*
Zhaomin Qian, *Zhejiang University, China*

A Real-Time Fault Diagnosis System for UPS Based on FFT Frequency Analysis 1678
Won-Sul Shim, *Kangwon National University, Korea, South*
Gi-Taek Kim, *Kangwon National University, Korea, South*
Ha-Jin Jung, *Powertron Engineering Co. Ltd., Korea, South*
Deuk-Soo Kim, *Powertron Engineering Co. Ltd., Korea, South*

Control Strategy for a Buck-Boost Type Direct Interface Converter Using an Indirect Matrix Converter with an Active Snubber 1684
Koji Kato, *Nagaoka University of Technology, Japan*
Jun-Ichi Itoh, *Nagaoka University of Technology, Japan*

A PI Control Algorithm of Three-Level APF with Little Static Misadjustment for Tracking Harmonic Current 1692
Yingjie He, *Xi'an Jiaotong University, China*
Jinjun Liu, *Xi'an Jiaotong University, China*
Zhaoan Wang, *Xi'an Jiaotong University, China*
Yunping Zou, *Huazhong University of Science and Technology, China*

A Novel Topology of LLC Resonant Inverter with Two Resonant Tanks for Power Conditioning System 1698
Eun-Soo Kim, *Jeonju University, Korea, South*
Kwang-Ho Lee, *Jeonju University, Korea, South*
Bong-Gun Chung, *Jeonju University, Korea, South*
Joo-Hoon Kim, *Jeonju University, Korea, South*
Moon-Ho Kye, *Powerplaza, United States*

Analysis and Realization of a Fast Repetitive Controller in Active Power Filter System 1704
Jinwu Gong, *Wuhan University, China*
Xiaoming Zha, *Wuhan University, China*
Suxuan Guo, *Wuhan University, China*
Baifeng Chen, *Wuhan University, China*
Jianjun Sun, *Wuhan University, China*

Session B4P-P: Modeling, Simulation & Control II
Thursday, February 25, 11:30 - 13:30
Session Chairs: Jonathan Kimball, *Missouri S&T*
Omer Onar, *Illinois Institute of Technology*

**On Extended Kalman Filters with Augmented State Vectors for the
Stator Flux Estimation in SPMSMs** 1711
T.J. Vyncke, *Ghent University, Belgium*
R.K. Boel, *Ghent University, Belgium*
J.A.A. Melkebeek, *Ghent University, Belgium*

State Equations Based Resonant Converters Modeling Technique 1719
Yingqi Zhang, *GE Global Research, China*
P.C. Sen, *Queen's University, Canada*

**Design Considerations and Expiremantal Results of an Adaptive Frequency
Controller Under Variable Line and Load Conditions** 1723
Jaber A. Abu Qahouq, *University of Alabama, United States*
Wisam Al-Hoor, *University of Central Florida, United States*
Issa Batarseh, *University of Central Florida, United States*

**Modeling and Mitigation of Dynamic Load Beat-Frequency Oscillation in
Multiphase Voltage Regulators with High-Gain Peak Current Control Scheme** 1727
Chen-Hua Chiu, *National Taiwan University, Taiwan*
Dan Chen, *National Taiwan University, Taiwan*
Ching-Jan Chen, *National Taiwan University, Taiwan*
Wei-Hsu Chang, *RichTek Technology Corporation, Taiwan*

Half-Wave Symmetry SHE-PWM Method for Multilevel Voltage Inverters 1732
Wanmin Fei, *Nanjing Normal University, China*
Xiaoli Du, *Nanjing Normal University, China*
Bin Wu, *Ryerson University, Canada*

PI Type Dynamic Decoupling Control Scheme for PMSM High Speed Operation 1736
Hao Zhu, *Tsinghua University, China*
Xi Xiao, *Tsinghua University, China*
Yongdong Li, *Tsinghua University, China*

**High Performance Positive and Negative Sequence Filters in
Stationary Frame Based on Complex Transfer Function** 1740
Jingxin Mao, *Beijing Jiaotong University, China*
Fei Lin, *Beijing Jiaotong University, China*
Hong Li, *Beijing Jiaotong University, China*
Xiaojie You, *Beijing Jiaotong University, China*
Trillion Q. Zheng, *Beijing Jiaotong University, China*

Simulation Study of Parameter Influence on Dynamic Voltage Rise Control 1745
Ming Li, *Xi'an Jiaotong University, China*
Xiong Fang, *Xi'an Jiaotong University, China*
Yue Wang, *Xi'an Jiaotong University, China*
Leqiang Zhang, *Xi'an Jiaotong University, China*
Ke Wang, *Xi'an Jiaotong University, China*
Guopeng Zhao, *Xi'an Jiaotong University, China*

Shaping of the Noise Spectrum in Power Electronic Converters 1749
Cristian Lascu, *University of Nevada, Reno, United States*
Andrzej M. Trzynadlowski, *University of Nevada, Reno, United States*
R. Lynn Kirlin, *University of Victoria, Canada*

**Grid Interactions and Stability Analysis of Distribution Power Network with
High Penetration of Plug-in Hybrid Electric Vehicles** 1755
Omer C. Onar, *Illinois Institute of Technology, United States*
Alireza Khaligh, *Illinois Institute of Technology, United States*

**Rapid Simulation of Fourth-Order Multi-Resonant LLCC Converters with
Capacitive Output Filter** 1763
A. Bucher, *University of Erlangen-Nürnberg, Germany*
T. Duerbaum, *University of Erlangen-Nürnberg, Germany*

**FHA-Based Voltage Gain Function with Harmonic Compensation for
LLC Resonant Converter** 1770
Hong Huang, *Texas Instruments, United States*

Session B4P-Q: Aerospace & Transportation
Thursday, February 25, 11:30 - 13:30

**Analysis and Design of LCC Resonant Inverter for the
Tranportation Systems Applications** 1778
Mohamed Youssef, *Bombardier Transportation Inc., Canada*
Jaber A. Abu Qahouq, *University of Alabama, United States*
Mohamed Orabi, *South Valley University, Egypt*

**A Multi-Resolution Control Strategy for DSP Controlled 400Hz Shunt
Active Power Filter in an Aircraft Power System** 1785
Haibing Hu, *Nanjing University of Aeronautics and Astronautics, China*
Wei Shi, *Nanjing University of Aeronautics and Astronautics, China*
Jianren Xue, *Nanjing Sute Electric Co., Ltd., China*
Ying Lu, *Nanjing University of Aeronautics and Astronautics, China*
Yan Xing, *Nanjing University of Aeronautics and Astronautics, China*

Battery Discharge Regulator for Space Applications Based on the Boost Converter 1792
A. Fernandez, *European Space Agency, Netherlands*
F. Tonicello, *European Space Agency, Netherlands*
J. Aroca, *European Space Agency, Netherlands*
O. Mourra, *European Space Agency, Netherlands*

Electromagnetic Compatibility Results for an LCC Resonant Inverter for the Tranportation Systems 1800
Mohamed Youssef, *Bombardier Transportation Inc., Canada*
Jaber A. Abu Qahouq, *University of Alabama, United States*
Mohamed Orabi, *South Valley University, Egypt*

Torque Impulse for Experimental Modal Analysis in Transmitted Vibration Study of Engine-Generators 1804
Elias Ayana, *Cummins Power Generation & University of Minnesota, United States*
Steve Seidlitz, *Cummins Power Generation, United States*
Sze Kwan Cheah, *Cummins Power Generation, United States*
Ned Mohan, *University of Minnesota, United States*

Session B4P-R: Power Converters & Applications
Thursday, February 25, 11:30 - 13:30

Review and Analysis of the AC-DC Converter of ITER Coil Power Supply 1810
P. Fu, *Institute of Plasma Physics, China*
G. Gao, *Institute of Plasma Physics, China*
L.W. Xu, *Institute of Plasma Physics, China*
Z.Q. Song, *Institute of Plasma Physics, China*
Z.C. Sheng, *Institute of Plasma Physics, China*
I. Benfatto, *ITER Organization, France*
J. Tao, *ITER Organization, France*
A.D. Mankani, *ITER Organization, France*
J.S. Oh, *National Fusion Research Institute, Korea, South*
C. Neumeyer, *Princeton Plasma Physics Laboratory, United States*

Fault Tolerance on Interleaved Inverter with Magnetic Couplers 1817
K. Guépratte, *Grenoble Electrical Engineering Laboratory, France*
D. Frey, *Grenoble Electrical Engineering Laboratory, France*
P.-O. Jeannin, *Grenoble Electrical Engineering Laboratory, France*
H. Stephan, *Grenoble Electrical Engineering Laboratory, France*
J.-P. Ferrieux, *Grenoble Electrical Engineering Laboratory, France*

Latest Practical Developments of Triplex Series Load Resonant Frequency-Operated High Frequency Inverter for Induction-Heated Low Resistivity Metallic Appliances in Consumer Built-in Cooktops 1825
Hideki Sadakata, *Panasonic Corporation, Japan*
Atsushi Fujita, *Panasonic Corporation, Japan*
Shinichiro Sumiyoshi, *Panasonic Corporation, Japan*
Hideki Omori, *Panasonic Corporation, Japan*
Bishwajit Saha, *Kyungnam University / Yamaguchi University, Korea, South*
Tarek Ahmed, *Kyungnam University / Yamaguchi University, Korea, South*
Mutsuo Nakaoka, *Kyungnam University / Yamaguchi University, Korea, South*

A Study of Novel Flyback Converter with Very Low Power Consumption at the Standby Operating Mode 1833
Eun-Soo Kim, *Jeonju University, Korea, South*
Bong-Gun Chung, *Jeonju University, Korea, South*
Sang-Ho Jang, *Jeonju University, Korea, South*
Mun-Gi Choi, *LG Innotek, Korea, South*
Moon-Ho Kye, *Powerplaza, United States*

Improved Two-Stage DC-Coupled Gate Driver for Enhancement-Mode SiC JFET 1838
Robin Kelley, *SemiSouth Laboratories Inc., United States*
Andrew Ritenour, *SemiSouth Laboratories Inc., United States*
David Sheridan, *SemiSouth Laboratories Inc., United States*
Jeff Casady, *SemiSouth Laboratories Inc., United States*

Design and Implementation of Multi-Channel Land Fowls Stunner with Current Sharing Controller 1842
S.-Y. Fan, *Wufeng Institute of Technology, Taiwan*
S.-Y. Tseng, *Chang Gung University, Taiwan*
Y.-H. Su, *Chang Gung University, Taiwan*
W.-C. Wu, *Wufeng Institute of Technology, Taiwan*

High Voltage Generator Using Boost/Flyback Hybrid Converter for Stun Gun Applications .. 1849
S.-Y. Tseng, *Chang Gung University, Taiwan*
C.-M. Yang, *Chang Gung University, Taiwan*
K.-C. Wang, *Chang Gung University, Taiwan*
G.-W. Hsu, *Chang-Gung University, Taiwan*

Session C5P-H: DC-DC Converter VIII
Thursday, February 25, 11:30 - 13:30

A Method to Analysis and Design for Long Life Power Converter 1857
H.M. Pang, *University of Hong Kong, China*
M.H. Bryan Pong, *University of Hong Kong, China*

DC-DC Converter for Gate Power Supplies with an Optimal Air Transformer 1865
Christoph Marxgut, *ETH Zurich, Switzerland*
Jürgen Biela, *ETH Zurich, Switzerland*
Johann W. Kolar, *ETH Zurich, Switzerland*
Reto Steiner, *ABB, Switzerland*
Peter K. Steimer, *ABB, Switzerland*

A Digitally Controlled DC-DC Buck Converter Using Frequency Domain ADCs 1871
Hani Ahmad, *Arizona State University, United States*
Bertan Bakkaloglu, *Arizona State University, United States*

Low-Dropout (LDO) Regulator Output Impedance Analysis and Transient Performance Enhancement Circuit 1875
Sungkeun Lim, *North Carolina State University, United States*
Alex Q. Huang, *North Carolina State University, United States*

A Design for Small Time-Delay Control Circuit for DPWM- POL 1879
Yoichi Ishizuka, *Nagasaki University, Japan*
Yusuke Yamada, *Nagasaki University, Japan*
Fumitoshi Hirose, *Nagasaki University, Japan*
Mariko Nishi, *Nagasaki University, Japan*
Hirofumi Matsuo, *Nagasaki University, Japan*

Low Profile LLC Series Resonant Converter with Two Transformers 1885
Eun-Soo Kim, *Jeonju University, Korea, South*
Joo-Hoon Kim, *Jeonju University, Korea, South*
Sung-In Kang, *LG Innotek, Korea, South*
Jun-Ho Park, *LG Innotek, Korea, South*
Jae-Sam Lee, *LG Innotek, Korea, South*
Dong-Young Huh, *LG Innotek, Korea, South*
Yong-Chae Jung, *Namseoul University, Korea, South*

Adaptive Frequency Control for ZVS Synchronous Boost Converters Operated in Average Current Mode 1890
Ben York, *Virginia Polytechnic Institute and State University, United States*
Rae-Young Kim, *Virginia Polytechnic Institute and State University, United States*
Jih-Sheng Lai, *Virginia Polytechnic Institute and State University, United States*

Power Saving Control Strategies and Their Implementation in DC-DC Converter for Data and Telecommunication Power Supply 1897
Rais Miftakhutdinov, *Texas Instruments Inc., United States*

Session C5P-J: DC-DC Converter IX
Thursday, February 25, 11:30 - 13:30

Analysis and Optimized Design of an Efficient High-Voltage Converter with High Output Capacity 1904
Huai Wang, *City University of Hong Kong, China*
Henry Shu-hung Chung, *City University of Hong Kong, China*
Adrian Ioinovici, *Holon Institute of Technology, Israel*

A Novel Three-Phase Three-Level ZVS PWM DC-DC Converter 1911
Eloi Agostini Junior, *Federal University of Santa Catarina, Brazil*
Ivo Barbi, *Federal University of Santa Catarina, Brazil*

Optimize the Synchronous Rectifier for LCC Converters 1919
Feng Zheng, *Xidian University, China*
Zhengfeng Ming, *Xidian University, China*

Digital Control Scheme for Robust Clock Tuning and PWM Phase Synchronization in Digitally Controlled Multi-POL Applications 1922
Eamon O'Malley, *Powervation Ltd., Ireland*
Karl Rinne, *Powervation Ltd., Ireland*
Anthony Kelly, *Powervation Ltd., Ireland*
Basil Almukhtar, *Powervation Ltd., Ireland*
Paul Kelleher, *Powervation Ltd., Ireland*

Control Scheme and Transient Performance of Sigma VR 1927
Pengjie Lai, *Virginia Polytechnic Institute and State University, United States*
Julu Sun, *Virginia Polytechnic Institute and State University, United States*
Fred C. Lee, *Virginia Polytechnic Institute and State University, United States*

A Three-Phase Current-Fed Push-Pull DC-DC Converter with Active Clamp for Fuel Cell Applications 1934
Sangwon Lee, *Seoul National University of Technology, Korea, South*
Sewan Choi, *Seoul National University of Technology, Korea, South*

Resonant Voltage Divider with Startup Considered 1942
K.I. Hwu, *National Taipei University of Technology, Taiwan*
Y.T. Yau, *National Taipei University of Technology, Taiwan*

LLC Resonant Converter with Two Resonant Tanks 1949
Eun-Soo Kim, *Jeonju University, Korea, South*
Joo-Hoon Kim, *Jeonju University, Korea, South*
Kwang-Ho Lee, *Jeonju University, Korea, South*
Yong-Seog Jeon, *Jeonju University, Korea, South*
Jae-Sam Lee, *LG Innotek, Korea, South*
Dong-Young Huh, *LG Innotek, Korea, South*

Session C5P-K: Motor Drives & Inverters II
Thursday, February 25, 11:30 - 13:30

A Digital Control Strategy for Brushless DC Generators 1957
Nikola Milivojevic, *Illinois Institute of Technology, United States*
Igor Stamenkovic, *Illinois Institute of Technology, United States*
Mahesh Krishnamurthy, *Illinois Institute of Technology, United States*
Ali Emadi, *Illinois Institute of Technology, United States*

**Space Vector Based PWM Scheme Without Sector Identification for a 4-Level
Dual Inverter Fed Induction Motor Drive with Asymmetrical DC Link Voltages** 1963
G. Shiny, *College of Engineering Trivandrum, India*
M.R. Baiju, *College of Engineering Trivandrum, India*

Control Method for a Novel Converter Topology for Permanent Magnet Drives 1970
Philip Brockerhoff, *Universität der Bundeswehr München, Germany*
Martin Schulz, *Universität der Bundeswehr München, Germany*

**A Voltage Controlled Adjustable Speed PMBLDCM Drive Using a Single-Stage
PFC Half-Bridge Converter** 1976
Sanjeev Singh, *Indian Institute of Technology Delhi, India*
Bhim Singh, *Indian Institute of Technology Delhi, India*

**Comparison of HF Signal Injection Methods for Sensorless Control of
PM Synchronous Motors** 1984
Eisenhawer de M. Fernandes, *Universidade Federal de Campina Grande, Brazil*
Alexandre C. Oliveira, *Universidade Federal de Campina Grande, Brazil*
Cursino B. Jacobina, *Universidade Federal de Campina Grande, Brazil*
Antonio M.N. Lima, *Universidade Federal de Campina Grande, Brazil*

A Robust Sensorless Fault Diagnosis Algorithm for Low Cost Motor Drives 1990
Seung-deog Choi, *Texas A&M University, United States*
Bilal Akin, *Texas Instruments Inc., United States*
Mina M. Rahimian, *Texas A&M University, United States*
Hamid A. Toliyat, *Texas A&M University, United States*

High Dynamic Performance Constrained Optimal Control of Induction Motors 1995
Sébastien Mariéthoz, *ETH Zurich, Switzerland*
Alexander Domahidi, *ETH Zurich, Switzerland*
Manfred Morari, *ETH Zurich, Switzerland*

PMSM Control Based on Edge Field Measurements by Hall Sensors 2002
Sungyoon Jung, *Pohang University of Science and Technology, Korea, South*
Beomseok Lee, *Pohang University of Science and Technology, Korea, South*
Kwanghee Nam, *Pohang University of Science and Technology, Korea, South*

Bridged-T Speed Controller for High Performance Switched Reluctance Motor Drives 2007
Gregory Pasquesoone, *University of Akron, United States*
Iqbal Husain, *University of Akron, United States*
Robert J. Veillette, *University of Akron, United States*

**Reducing Losses in Multilevel Coupled Inductor Inverters Using
Interleaved Discontinuous SVPWM** 2013
Behzad Vafakhah, *University of Alberta, Canada*
Andy Knight, *University of Alberta, Canada*
John Salmon, *University of Alberta, Canada*

A Novel Elevator Load Torque Identification Method Based on Friction Mode 2021
Xiaoyuan Hong, *Zhejiang University, China*
Zhe Deng, *Zhejiang University, China*
Siran Wang, *Zhejiang University, China*
Lijun Hang, *Zhejiang University, China*
Wuhua Li, *Zhejiang University, China*
Zhengyu Lu, *Zhejiang University, China*

**A Novel Digital Current Control Strategy for Torque Ripple Reduction in
Permanent Magnet Synchronous Motor Drives** 2025
Haidong Yu, *Phoenix International, United States*

Session C5P-L: Passive Components
Thursday, February 25, 11:30 - 13:30

**Evaluation of a SiC Power Module Using Low-on-Resistance IEMOSFET and
JBS for High Power Density Power Converters** 2030
Kazuto Takao, *Toshiba Corporation, Japan*
Takashi Shinohe, *Toshiba Corporation, Japan*
Shinsuke Harada, *National Institute of Advanced Industrial Science and Technology, Japan*
Kenji Fukuda, *National Institute of Advanced Industrial Science and Technology, Japan*
Hiromichi Ohashi, *National Institute of Advanced Industrial Science and Technology, Japan*

**A Novel Integrated Power Inductor in Silicon Substrate for
Ultra-Compact Power Supplies** 2036
Mingliang Wang, *University of Florida, United States*
Jiping Li, *University of Florida, United States*
Khai D.T. Ngo, *Virginia Polytechnic Institute and State University, United States*
Huikai Xie, *University of Florida, United States*

A Class of Coupled Inductors Based on LTCC Technology 2042
Laili Wang, *Xi'an Jiaotong University, China*
Yunqing Pei, *Xi'an Jiaotong University, China*
Xu Yang, *Xi'an Jiaotong University, China*
Xizhi Cui, *Xi'an Jiaotong University, China*
Zhaoan Wang, *Xi'an Jiaotong University, China*
Guopeng Zhao, *Xi'an Jiaotong University, China*

Optimising the High Frequency Bandwidth and Immuntity to Interference of Rogowski Coils in Measurement Applications with Large local dV/dt 2050
Christopher R. Hewson, *Power Electronic Measurements Ltd., United Kingdom*
William F. Ray, *Power Electronic Measurements Ltd., United Kingdom*

PFC Inductor Selection Made Easy by "PL Product" 2057
Welly Chou, *Precision Incorporated, United States*

Evaluation of LTCC Capacitors and Inductors in DC-DC Converters 2060
Laili Wang, *Xi'an Jiaotong University, China*
Yunqing Pei, *Xi'an Jiaotong University, China*
Xu Yang, *Xi'an Jiaotong University, China*
Bo Song, *Xi'an Jiaotong University, China*
Zhaoan Wang, *Xi'an Jiaotong University, China*
Guopeng Zhao, *Xi'an Jiaotong University, China*

Session C5P-M: Vehicle Electronics II
Thursday, February 25, 11:30 - 13:30

Bi-Directional Charging Topologies for Plug-in Hybrid Electric Vehicles 2066
Dylan C. Erb, *Illinois Institute of Technology, United States*
Omer C. Onar, *Illinois Institute of Technology, United States*
Alireza Khaligh, *Illinois Institute of Technology, United States*

Session C5P-N: Renewable Energy Systems
Thursday, February 25, 11:30 - 13:30
Session Chairs: Robert Balog, *Texas A&M University*

Multi-Channel Three-Port DC-DC Converters As Maximum Power Tracker, Battery Charger and Bus Regulator 2073
Zhijun Qian, *University of Central Florida, United States*
Osama Abdel-Rahman, *ApECOR, United States*
Haibing Hu, *University of Central Florida, United States*
Issa Batarseh, *University of Central Florida, United States*

A Smart and Simple PV Charger for Portable Applications 2080
Weichen Li, *Zhejiang University, China*
Yuzhen Zheng, *Zhejiang University of Science and Technology, China*
Wuhua Li, *Zhejiang University, China*
Yi Zhao, *Zhejiang University, China*
Xiangning He, *Zhejiang University, China*

RTDS-Based Real Time Simulations of Grid-Connected Wind Turbine Generator Systems .. 2085
Gyeong-Hun Kim, *Changwon National University, Korea, South*
Young-Ju Kim, *Changwon National University, Korea, South*
Minwon Park, *Changwon National University, Korea, South*
In-Keun Yu, *Changwon National University, Korea, South*
Byeong-Mun Song, *Baylor University, United States*

Investigation of Fully Digital Controlled Li-Ion Battery Power Recovery System 2091
Siran Wang, *Zhejiang University, China*
Xia Zhou, *Zhejiang University, China*
Jifeng Chen, *Zhejiang University, China*
Wenxi Yao, *Zhejiang University, China*
Zhengyu Lu, *Zhejiang University, China*

A Novel Control System for Harmonic Compensation by Using Wind Energy Conversion Based on DFIG Technology 2096
Grazia Todeschini, *Worcester Polytechnic Institute, United States*
Alexander E. Emanuel, *Worcester Polytechnic Institute, United States*

A Transformerless Modular Permanent Magnet Wind Generator System with Minimum Generator Coils 2104
Xibo Yuan, *Tsinghua University, China*
Yongdong Li, *Tsinghua University, China*
Jianyun Chai, *Tsinghua University, China*

Small-Signal Modeling and Analysis of the Double-Input Buckboost Converter 2111
Deepak Somayajula, *Missouri University of Science and Technology, United States*
Mehdi Ferdowsi, *Missouri University of Science and Technology, United States*

A Novel Power Distribution Strategy for Parallel Inverters in Islanded Mode Microgrid 2116
Xuan Zhang, *Xi'an Jiaotong University, China*
Jinjun Liu, *Xi'an Jiaotong University, China*
Ting Liu, *Xi'an Jiaotong University, China*
Linyuan Zhou, *Xi'an Jiaotong University, China*

Direct Power Control of Doubly-Fed Generator Based Wind Turbine Converters to Improve Low Voltage Ride-Through During System Imbalance 2121
Murali M. Baggu, *Missouri University of Science and Technology, United States*
Luke D. Watson, *Missouri University of Science and Technology, United States*
Jonathan W. Kimball, *Missouri University of Science and Technology, United States*
Badrul H. Chowdhury, *Missouri University of Science and Technology, United States*

Active Damping for Torsional Vibrations in PMSG Based WECS 2126
Hua Geng, *Ryerson University, Canada*
Dewei Xu, *Ryerson University, Canada*
Bin Wu, *Ryerson University, Canada*
Geng Yang, *Tsinghua University, China*

Voltage and Frequency Stabilization Using PI-Like Fuzzy Controller for the Load Side Converters of the Stand Alone Wind Energy Systems 2132
Ameen Gargoom, *University of Tasmania, Australia*
Abu Mohammad Osman Haruni, *University of Tasmania, Australia*
Md. Enamul Haque, *University of Tasmania, Australia*
Michael Negnevitsky, *University of Tasmania, Australia*

Dual-Stage Converter to Improve Transfer Efficiency and Maximum Power Point Tracking Feasibility in Photovoltaic Energy-Conversion Systems 2138
Sairaj V. Dhople, *University of Illinois at Urbana-Champaign, United States*
Ali Davoudi, *University of Illinois at Urbana-Champaign, United States*
Patrick L. Chapman, *University of Illinois at Urbana-Champaign, United States*

A Novel Approach of Maximizing Energy Harvesting in Photovoltaic Systems Based on Bisection Search Theorem 2143
Peng Wang, *Nanyang Technological University, Singapore*
Haipeng Zhu, *Nanyang Technological University, Singapore*
Weixiang Shen, *Nanyang Technological University, Singapore*
Fook Hoong Choo, *Nanyang Technological University, Singapore*
Poh Chiang Loh, *Nanyang Technological University, Singapore*
Kuan Khoon Tan, *Nanyang Technological University, Singapore*

Simple Control Design for a Three-Port DC-DC Converter Based PV System with Energy Storage 2149
Sixifo Falcones, *Arizona State University, United States*
Raja Ayyanar, *Arizona State University, United States*

A Self-Powered Power Management Circuit for Energy Harvested by a Piezoelectric Cantilever 2154
Na Kong, *Virginia Polytechnic Institute and State University, United States*
Travis Cochran, *Virginia Polytechnic Institute and State University, United States*
Dong Sam Ha, *Virginia Polytechnic Institute and State University, United States*
Hung-Chih Lin, *National Tsing Hua University, Taiwan*
Daniel J. Inman, *Virginia Polytechnic Institute and State University, United States*

A Maximum Power Point Tracker Implementation for Photovoltaic Cells Using Dynamic Optimal Voltage Tracking 2161
Emil Jimenez-Brea, *University of Puerto Rico-Mayaguez, Puerto Rico*
Andres Salazar-Llinas, *University of Puerto Rico-Mayaguez, Puerto Rico*
Eduardo Ortiz-Rivera, *University of Puerto Rico-Mayaguez, Puerto Rico*
Jesus Gonzalez-Llorente, *University of Puerto Rico-Mayaguez, Puerto Rico*

Development of the Novel Control Algorithm for the Small Proton Exchange Membrane Fuel Cell Stack Without External Humidification 2166
Tae-Hoon Kim, *Soongsil University, Korea, South*
Sang-Hyun Kim, *Soongsil University, Korea, South*
Wook Kim, *Soongsil University, Korea, South*
Jong-Hak Lee, *Soongsil University, Korea, South*
Woojin Choi, *Soongsil University, Korea, South*

Session C5P-P: Modeling, Simulation & Control III
Thursday, February 25, 11:30 - 13:30
Session Chairs: Jonathan Kimball, *Missouri S&T*
Omer Onar, *Illinois Institute of Technology*

Stabilization of Constant-Power Loads by Passive Impedance Damping 2174
Mauricio Céspedes, *Rensselaer Polytechnic Institute, United States*
Troy Beechner, *Rensselaer Polytechnic Institute, United States*
Lei Xing, *Rensselaer Polytechnic Institute, United States*
Jian Sun, *Rensselaer Polytechnic Institute, United States*

An Adaptive External Ramp Control of the Peak Current Controlled Buck Converters for High Control Bandwidth and Wide Operation Range 2181
Liyu Yang, *North Carolina State University, United States*
Jinseok Park, *North Carolina State University, United States*
Alex Q. Huang, *North Carolina State University, United States*

Masterless Multirate Control of Parallel DC-DC Converters 2189
Anthony Kelly, *Powervation Ltd., Ireland*
Karl Rinne, *Powervation Ltd., Ireland*
Eamon O'Malley, *Powervation Ltd., Ireland*

FPGA-Based Spectral Envelope Preprocessor for Power Monitoring and Control 2194
Zachary Remscrim, *Massachusetts Institute of Technology, United States*
James Paris, *Massachusetts Institute of Technology, United States*
Steven B. Leeb, *Massachusetts Institute of Technology, United States*
Steven R. Shaw, *Montana State University, United States*
Sabrina Neuman, *Massachusetts Institute of Technology, United States*
Christopher Schantz, *Massachusetts Institute of Technology, United States*
Sean Muller, *Massachusetts Institute of Technology, United States*
Sarah Page, *Massachusetts Institute of Technology, United States*

Sigma-Delta Modulation of Multi-Phase High Frequency Converters 2202
Jonathan W. Kimball, *Missouri University of Science and Technology, United States*
Kyle Roger Eckler, *Missouri University of Science and Technology, United States*
Luke Watson, *Missouri University of Science and Technology, United States*

**Specialized Digital Signal Processor for Control of Multi-Rail/Multi-Phase
High Switching Frequency Power Converters** 2207
James Mooney, *University of Limerick, Ireland*
Mark Halton, *University of Limerick, Ireland*
Abdulhussain E. Mahdi, *University of Limerick, Ireland*

**Computer-Aided Design for Class-E Switching Circuits Taking into
Account Optimized Inductor Designs** 2212
Natsumi Sagawa, *Chiba University, Japan*
Hiroo Sekiya, *Chiba University and Wright State University, Japan*
Marian K. Kazimierczuk, *Wright State University, United States*

Characterization of IGBT Modules for System EMI Simulation 2220
Tao Qi, *Rensselaer Polytechnic Institute, United States*
Jeff Graham, *Fairchild Controls Corporation, United States*
Jian Sun, *Rensselaer Polytechnic Institute, United States*

**A Mathematical Model for Online Electrical Characterization of Thermoelectric
Generators Using the P-I Curves at Different Temperatures** 2226
Eduardo I. Ortiz-Rivera, *University of Puerto Rico-Mayaguez, Puerto Rico*
Andres Salazar-Llinas, *University of Puerto Rico-Mayaguez, Puerto Rico*
Jesus Gonzalez-Llorente, *University of Puerto Rico-Mayaguez, Puerto Rico*

**A Novel Method for Permanent Magnet Demagnetization Fault Detection and
Treatment in Permanent Magnet Synchronous Machines** 2231
Amir Khoobroo, *University of Texas at Arlington, United States*
Babak Fahimi, *University of Texas at Arlington, United States*

Session C5P-Q: Alternative Energy Applications
Thursday, February 25, 11:30 - 13:30

Series Connection of IGBT 2238
The-Van Nguyen, *Grenoble Institute of Technology, France*
Pierre-Olivier Jeannin, *Grenoble Institute of Technology, France*
Eric Vagnon, *Grenoble Institute of Technology, France*
David Frey, *Grenoble Institute of Technology, France*
Jean-Christophe Crebier, *Grenoble Institute of Technology, France*

**Three Phase Linear Permanent Magnet Energy Scavenger Based on
Foot Horizontal Motion** 2245
Igor Stamenkovic, *Illinois Institute of Technology, United States*
Nikola Milivojevic, *Illinois Institute of Technology, United States*
Cong Zheng, *Illinois Institute of Technology, United States*
Alireza Khaligh, *Illinois Institute of Technology, United States*

**Bidirectional Communication Techniques for Wireless Battery Charging
Systems and Portable Consumer Electronics** 2251
W.P. Choi, *ConvenientPower HK Ltd. and City University of Hong Kong, China*
W.C. Ho, *ConvenientPower HK Ltd., China*
X. Liu, *ConvenientPower HK Ltd., China*
S.Y.R. Hui, *City University of Hong Kong, China*

**Proposal of a DC-DC Converter with Wide Conversion Range Used in
Photovoltaic Systems and Utility Power Grid for the Universal Voltage Range** 2258
Jonas Reginaldo de Britto, *Universidade Federal de Uberlândia, Brazil*
Fábio Vincenzi Romualdo da Silva, *Universidade Federal de Uberlândia, Brazil*
Enane Antônio Alves Coelho, *Universidade Federal de Uberlândia, Brazil*
Luiz Carlos de Freitas, *Universidade Federal de Uberlândia, Brazil*
Valdeir José Farias, *Universidade Federal de Uberlândia, Brazil*
João Batista Vieira, Jr., *Universidade Federal de Uberlândia, Brazil*

Characterization of a 5 kW Solid Oxide Fuel Cell Stack Using Power Electronic Excitation .. 2264
John J. Cooley, *Massachusetts Institute of Technology, United States*
Eric Seger, *Montana State University, United States*
Steven Leeb, *Massachusetts Institute of Technology, United States*
Steven R. Shaw, *Montana State University, United States*

**Photovoltaic Parallel Resonant DC-Link Soft Switching Inverter Using
Hysteresis Current Control** 2275
Young-Ho Kim, *Sungkyunkwan University, Korea, South*
Jun-Gu Kim, *Sungkyunkwan University, Korea, South*
Young-Hyok Ji, *Sungkyunkwan University, Korea, South*
Chung-Yuen Won, *Sungkyunkwan University, Korea, South*
Yong-Chae Jung, *Namseoul University, Korea, South*

**Supercapacitor-Based Hybrid Storage Systems for Energy Harvesting in
Wireless Sensor Networks** 2281
S. Saggini, *University of Udine, Italy*
F. Ongaro, *University of Udine, Italy*
C. Galperti, *Politecnico di Milano, Italy*
P. Mattavelli, *University of Padova, Italy*

The Faulty Module Bypass for Thermoelectric Generation 2288
Wei Qian, *Michigan State University, United States*
Fang Z. Peng, *Michigan State University, United States*
Sangmin Han, *Michigan State University, United States*

Maximum Power Point Tracking Feasibility in Photovoltaic Energy-Conversion Systems . 2294
Sairaj V. Dhople, *University of Illinois at Urbana-Champaign, United States*
Ali Davoudi, *University of Illinois at Urbana-Champaign, United States*
Gerald Nilles, *University of Illinois at Urbana-Champaign, United States*
Patrick L. Chapman, *University of Illinois at Urbana-Champaign, United States*

Session C5P-R: Lighting Applications
Thursday, February 25, 11:30 - 13:30

Realization of a General LED Lighting System Based on a Novel Power Line
Communication Technology 2300
Chushan Li, *Zhejiang University, China*
Jiande Wu, *Zhejiang University, China*
Xiangning He, *Zhejiang University, China*

Solid-State Lamp with Integral Occupancy Sensor 2305
John J. Cooley, *Massachusetts Institute of Technology, United States*
Dan Vickery, *Massachusetts Institute of Technology, United States*
Al-Thaddeus Avestruz, *Massachusetts Institute of Technology, United States*
Amy Englehart, *Massachusetts Institute of Technology, United States*
James Paris, *Massachusetts Institute of Technology, United States*
Steven B. Leeb, *Massachusetts Institute of Technology, United States*

A 0.9 PF LED Driver with Small LED Current Ripple Based on
Series-Input Digitally-Controlled Converter Modules 2314
Qingcong Hu, *University of Colorado at Boulder, United States*
Regan Zane, *University of Colorado at Boulder, United States*

A Novel Dimmable Electronic Ballast for Compact Fluorescent Lamps Using
Phase-Cut Incandescent Lamp Dimmers with Wide Dimming Range and
Low Dimming Level Lamp Ignition Capability 2321
John Lam, *Queen's University, Canada*
Praveen K. Jain, *Queen's University, Canada*

Author Index

Study on Wide Range Robust Speed Sensorless Control of Medium Voltage Induction Motor

Siran Wang, Zhengyu Lu (Senior member, IEEE)
Department of Electrical Engineering
Zhejiang University
Hangzhou, Zhejiang, China
Siran.Wang09@gmail.com

Abstract—**This paper presents a robust speed sensorless vector control system of medium voltage induction motor. Rotor speed estimation based on Luenberger observer is analyzed as well as the influence of the motor parameter variation to the estimation accuracy. The characteristics of the motor parameter influence are generalized. According to the different influence of parameter variation at different operating speed, an optimizing strategy for parameter compensation is proposed to balance the accuracy of rotor speed estimation in the whole motor operating speed range. Speed sensorless vector control scheme based on Luenberger observer with parameter identification is applied to a medium voltage induction motor drive, and the system is simulated to verify the optimizing strategy for speed estimation with parameter compensation.**

I. INTRODUCTION

Induction motors (IM) are very attractive in many applications owing to their simple structure, low cost, and robust construction. Furthermore, in contrast to the commutation dc motor, it is suitable for aggressive environment and high power applications since there are no problems with spark and corrosion[1].

Recently the interest for sensorless drives of IM has been constantly rising. The advantages of speed sensorless IM drives are reduced hardware complexity and lower cost, reduced size of the drive machine, elimination of the sensor cable, better noise immunity, increased reliability, and less maintenance requirements. Operation in hostile environments mostly requires a motor without speed sensor[2]. However the main drawback of the sensorless techniques is the lower speed range, which limits its use to those industrial applications. A variety of speed sensorless solutions have been developed in the past few years, including open loop estimators, model reference adaptive systems (MRAS), Luenberger observers, Kalman filtering techniques, etc[3]. Among these methods, estimation based on Luenberger observer has been proved to be a good compromise between accuracy and complexity, and is able to work at wide speed range.

In this paper, the speed sensorless vector control based on Luenberger observer is presented. The influence of parameter variation on Luenberger observer is analyzed. Then

simultaneous estimation of rotor speed and resistances of IM is discussed, and an optimizing strategy for parameter compensation is proposed. At last, the performance of the speed sensorless vector control system of IM with parameter identification is evaluated in the whole operating speed range.

II. INDUCTION MOTOR MODEL

The model of IM[1] is given in stationary reference frame by the following expressions.

$$px(t) = Ax(t) + Bu(t)$$
$$y(t) = Cx(t) \tag{1}$$

$$A = \begin{bmatrix} -\dfrac{R_s}{\sigma L_s} - \dfrac{L_m^2}{\sigma L_s L_r T_r} & 0 & \dfrac{L_m}{\sigma L_s L_r T_r} & \dfrac{L_m \omega_r}{\sigma L_s L_r} \\ 0 & -\dfrac{R_s}{\sigma L_s} - \dfrac{L_m^2}{\sigma L_s L_r T_r} & -\dfrac{L_m \omega_r}{\sigma L_s L_r} & \dfrac{L_m}{\sigma L_s L_r T_r} \\ \dfrac{L_m}{T_r} & 0 & -\dfrac{1}{T_r} & -\omega_r \\ 0 & \dfrac{L_m}{T_r} & \omega_r & -\dfrac{1}{T_r} \end{bmatrix}$$

$$, B = \begin{bmatrix} \dfrac{1}{\sigma L_s} & 0 & 0 & 0 \\ 0 & \dfrac{1}{\sigma L_s} & 0 & 0 \end{bmatrix}^T , C = \begin{bmatrix} 1 & 0 & 0 & 0 \\ 0 & 1 & 0 & 0 \end{bmatrix}^T$$

where $x = \begin{bmatrix} i_{s\alpha} & i_{s\beta} & \psi_{r\alpha} & \psi_{r\beta} \end{bmatrix}^T$,

$u = \begin{bmatrix} v_{s\alpha} & v_{s\beta} \end{bmatrix}^T$, $y = \begin{bmatrix} i_{s\alpha} & i_{s\beta} \end{bmatrix}^T$, $\sigma = 1 - \dfrac{L_m^2}{L_s L_r}$,

$\eta = \dfrac{L_m}{\sigma L_s L_r}$, $T_r = \dfrac{L_r}{R_r}$, $v_{s\alpha}$, $v_{s\beta}$, $i_{s\alpha}$, $i_{s\beta}$ represent $\alpha\beta$-axis stator voltage and current component respectively. $\psi_{r\alpha}$, $\psi_{r\beta}$ are $\alpha\beta$-axis rotor flux component. R_s, R_r represent stator and rotor resistance. L_m, L_s, L_r represent mutual, stator, and rotor inductance. ω_r is rotor speed.

III. SPEED ESTIMATION WITH LUENBERGER OBSERVER

For convenience, the motor model (1) is rewritten in the following form.

$$px(t) = \begin{bmatrix} A_{11} & A_{12} \\ A_{21} & A_{22} \end{bmatrix} x(t) + Bu(t) \tag{2}$$

$$y(t) = Cx(t)$$

where $A_{11} = a_{r11}I$, $A_{12} = a_{r12}I + a_{i12}J$, $A_{21} = a_{r21}I$,

$$A_{22} = a_{r22}I + a_{i22}J, \quad I = \begin{bmatrix} 1 & 0 \\ 0 & 1 \end{bmatrix}, \quad J = \begin{bmatrix} 0 & -1 \\ 1 & 0 \end{bmatrix}.$$

The Luenberger observer which estimates the stator currents and the rotor flux is [3][4]

$$p\hat{x}(t) = A\hat{x}(t) + Bu(t) + L[C\hat{x}(t) - y(t)]$$
$$= [A + LC]\hat{x}(t) + Bu(t) - Ly(t) \tag{3}$$

Fig.1 shows the structure of Luenberger observer for speed and flux estimation of IM.

The Luenberger matrix gain L is chosen so that the poles of the characteristic matrix $A_L = A + LC$ to be stable. So all eigenvalues of A_L should have negative real parts. A traditional alternative way of doing this is based on the fact that the gain matrix can be analytically computed so that the observer poles are proportional to the motor poles. If the observer poles are located closer to the origin than the poles of the motor, it takes a long time to estimate the actual values of the state variables. So the observer poles should be located further from the origin than the poles of the motor. However, the control system could be unstable by the measurement noise if the observer poles are too far from the origin. For general rules, the observer poles are usually 5 to 6 more fast then the motor poles. Thus the matrix gain L can be calculated by the following equation.

$$L = \begin{bmatrix} g_1 & g_2 & g_3 & g_4 \\ -g_2 & g_1 & -g_4 & g_3 \end{bmatrix}^T \tag{4}$$

where $g_1 = (k-1)(a_{r11} + a_{r22})$, $g_2 = (k-1)a_{i22}$,

$$g_3 = (k^2 - 1)(\frac{1}{\eta}a_{r11} + a_{r21}) - \frac{1}{\eta}(k-1)(a_{r11} + a_{r22}),$$

$$g_4 = -\frac{1}{\eta}(k-1)a_{i22}.$$

From (3), the estimation error $e = x - \hat{x}$ of the state variable is described by the following equation.

$$pe = Ax - \hat{A}\hat{x} + L(Cx - C\hat{x}) = (A + LC)e - \Delta A\hat{x} \tag{5}$$

The Lyapunov function is defined as

$$V = e^T e + (\hat{\omega}_r - \omega_r)^2 / \lambda \tag{6}$$

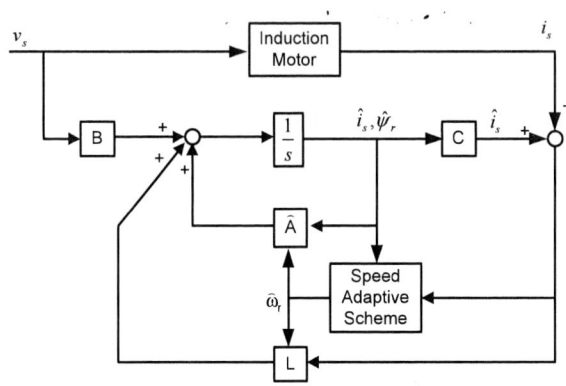

Figure 1. Block diagram of Luenberger observer.

where λ is a positive constant. Assuming that $e_{is\alpha} = i_{s\alpha} - \hat{i}_{s\alpha}$, $e_{is\beta} = i_{s\beta} - \hat{i}_{s\beta}$. The derivative of V becomes

$$pV = e^T \{(A + LC)^T + (A + LC)\}e$$
$$- 2\eta\Delta\omega_r(e_{is\alpha}\hat{\psi}_{r\beta} - e_{is\beta}\hat{\psi}_{r\alpha}) + 2\Delta\omega_r p\hat{\omega}_r / \lambda \tag{7}$$

The adaptation mechanism for the rotor speed estimation can be found by equalizing the last two terms of (7) to zero.

$$p\hat{\omega}_r = \lambda\eta(e_{is\alpha}\hat{\psi}_{r\beta} - e_{is\beta}\hat{\psi}_{r\alpha}) \tag{8}$$

Usually the following proportional and integral adaptation mechanism is used practically in order to improve the response of the speed estimation.

$$\hat{\omega}_r = K_p(e_{is\alpha}\hat{\psi}_{r\beta} - e_{is\beta}\hat{\psi}_{r\alpha})$$
$$+ K_i \int(e_{is\alpha}\hat{\psi}_{r\beta} - e_{is\beta}\hat{\psi}_{r\alpha})dt \tag{9}$$

IV. INFLUENCE AND COMPENSATION OF PARAMETER VARIATION

Since the parameters, especially the stator resistance R_s and rotor resistance R_r, vary with the operating state of IM. It is very hard to set the correct parameters in the control scheme. Fig.2 and Fig.3 illustrate some of the simulation results which show the estimation error caused by the parameter variation. Through plenty of simulation results, conclusions about the influence of parameter variation are derived.

1. Both estimated rotor speed and rotor flux of Luenberger observer have good robustness to the variation of R_s at nominal rotor speed. However, as the operating frequency goes down, the influence of the stator resistance variation becomes more and more significant. If the variation of R_s is large enough, Luenberger observer may even become unstable at zero speed.

2. The variation of R_r has no effect on the rotor flux estimation but only on the speed estimation. The error caused by the variation of R_r will not change with the operating frequency, which is different from the situation of R_s. In other words, a certain variation of R_r will cause a certain error in rotor speed estimation in spite of operating frequency.

978-1-4244-4782-4/10 $26.00 © 2010 IEEE

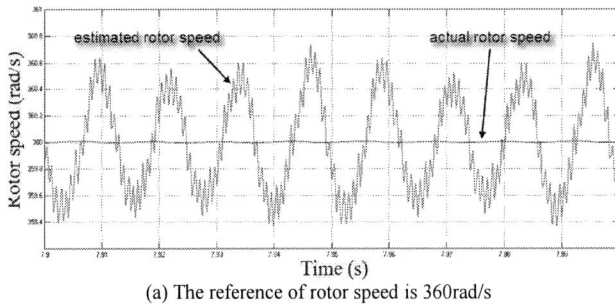

(a) The reference of rotor speed is 360rad/s

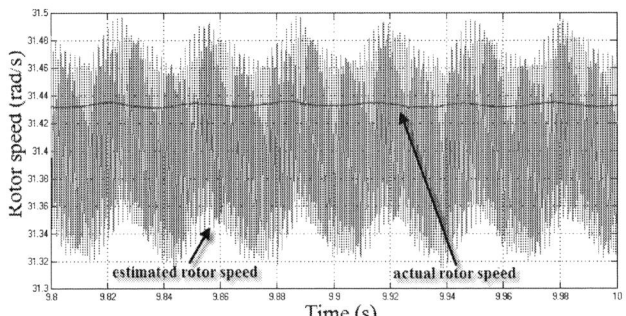

(b) The reference of rotor speed is 31.4rad/s

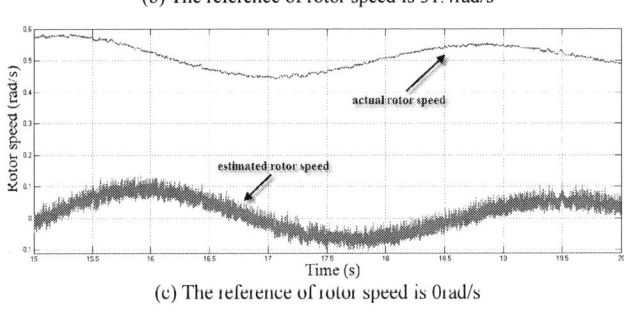

(c) The reference of rotor speed is 0rad/s

Figure 2.   Rotor speed(rad/s) when R_s used in Luenberger observer is 1.5 times as the actual value.

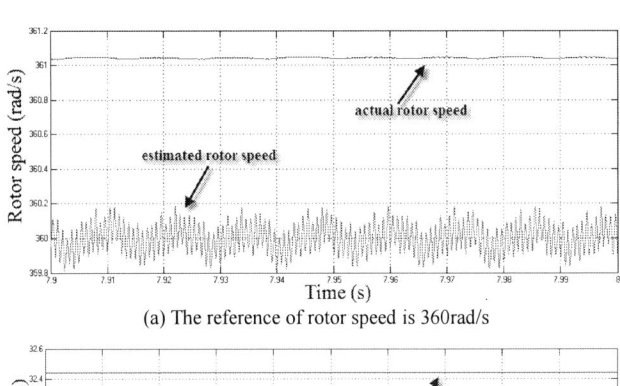

(a) The reference of rotor speed is 360rad/s

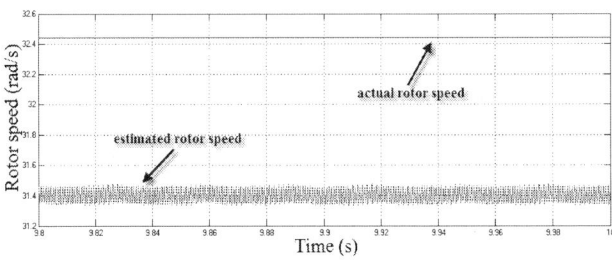

(b) The reference of rotor speed is 31.4rad/s

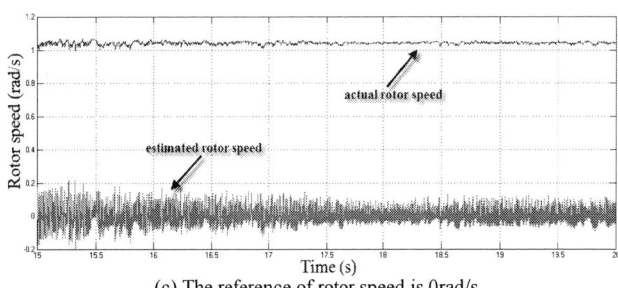

(c) The reference of rotor speed is 0rad/s

Figure 3.   Rotor speed(rad/s) when R_r used in Luenberger observer is 1.5 times as the actual value.

So variation of both stator and rotor resistances have significant influence on the estimation accuracy, especially at low rotor speed. For good performance of Luenberger observer at a wide operating speed range, parameter compensation is definitely needed.

The stator and rotor resistance are both included in the coefficient matrix A of IM model[5], which is the same as the rotor speed. So they can also be estimated with certain adaptive scheme. The adaptive mechanisms of stator and rotor resistance can be deduced by the same theory, and the results are directly shown in equations (10) and (11).

$$\hat{R}_s = -K_p(e_{is\alpha}\hat{i}_{s\alpha} + e_{is\beta}\hat{i}_{s\beta}) - K_i \int (e_{is\alpha}\hat{i}_{s\alpha} + e_{is\beta}\hat{i}_{s\beta})dt \quad (10)$$

$$\hat{R}_r = K_p \mathbf{e}_{is}\left(\frac{L_m}{\sigma L_s L_r^2}\hat{\boldsymbol{\psi}}_r - \frac{1}{\sigma L_r}\hat{\mathbf{i}}_s\right)$$
$$+ K_i \int \mathbf{e}_{is}\left(\frac{L_m}{\sigma L_s L_r^2}\hat{\boldsymbol{\psi}}_r - \frac{1}{\sigma L_r}\hat{\mathbf{i}}_s\right)dt \quad (11)$$

In this adaptive scheme, only two components of stator current are compared with the actual sampled values. As a result, only two of the three parameters (rotor speed, stator and rotor resistance) can be estimated simultaneously. Actually there is an alternative way. Because the variation of resistance is mainly caused by the temperature shift, the variation of the stator and rotor resistances should be almost the same with each other. According to the conclusion about parameter variation influence mention above, at higher operating speed the influence of rotor resistance is more significant, while at lower operating speed the influence of stator resistance is more significant. So it would be a good compromise to estimate rotor speed and rotor resistance simultaneously at higher speed and rotor speed and stator resistance simultaneously at lower speed. Then the variation of the other resistance can be deduced from the estimated one, since the thermal models of both resistances are the same.

V.   SIMULATION RESULTS

The overall system which is illustrated in Fig.4 is simulated in Matlab/Simulink. Model of a 1500HP, 8-pole, 4160V, 60Hz, 3 phase induction motor provided by EATON

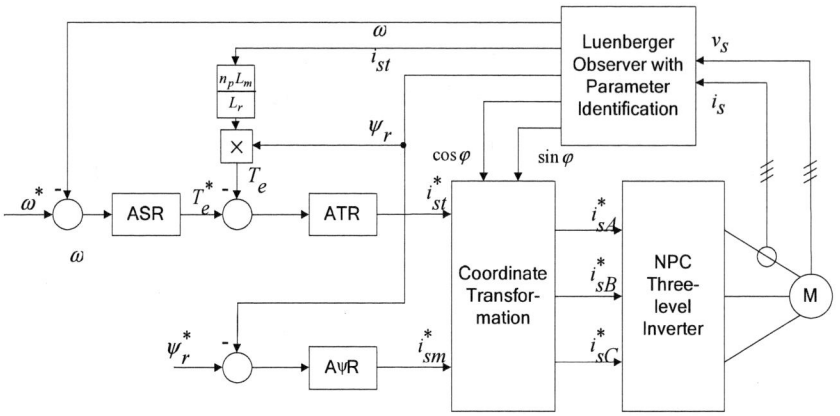

Figure 4. Block diagram of the speed sensorless vector control system.

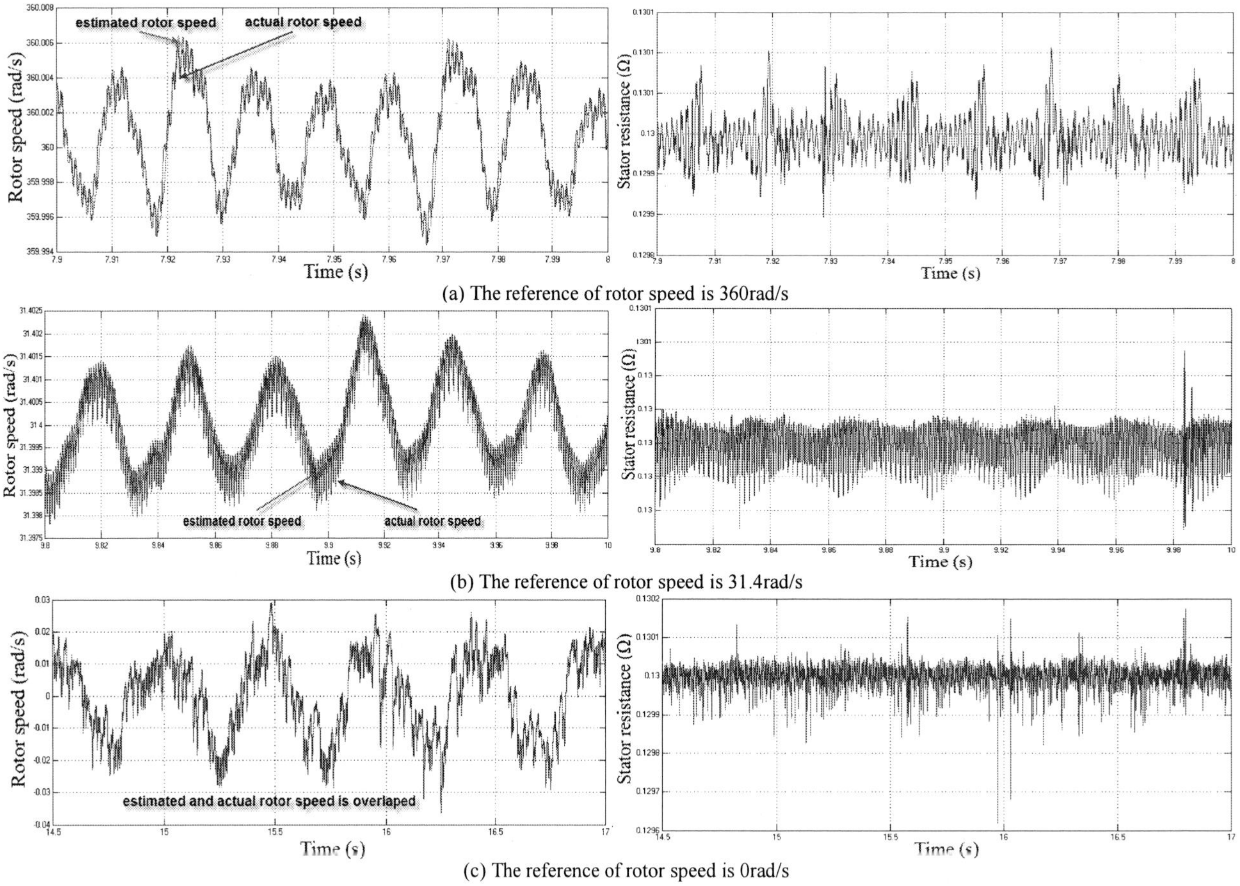

(a) The reference of rotor speed is 360rad/s

(b) The reference of rotor speed is 31.4rad/s

(c) The reference of rotor speed is 0rad/s

Figure 5. Simulation results of sensorless control system based on Luenberger observer with R_s identification (the actual value of R_s is 0.13Ω).

is used in the simulation. The motor is fed by a PWM NPC three-level inverter with the switching frequency of 600Hz, and the input is from a 12-pulse rectifier. Field oriented vector control is adopted, and Luenberger observer with stator/rotor resistance identification is used to calculate the rotor speed, the rotor flux, and the rotor flux angle for orientation. Simulation has been done in the whole operating speed range (from zero to nominal rotor speed). Some of the representative simulation results are illustrated in Fig.5 and Fig.6.

According to the simulation results illustrated above, Luenberger observers with different resistance identification give different performance. Luenberger observer with stator resistance identification has good performance on estimation accuracy in the whole operating speed range, while Luenberger observer with rotor resistance identification is not able to give accurate estimation at low operating speed. So the strategy about different resistance identification at different operating speed is confirmed by the simulation results.

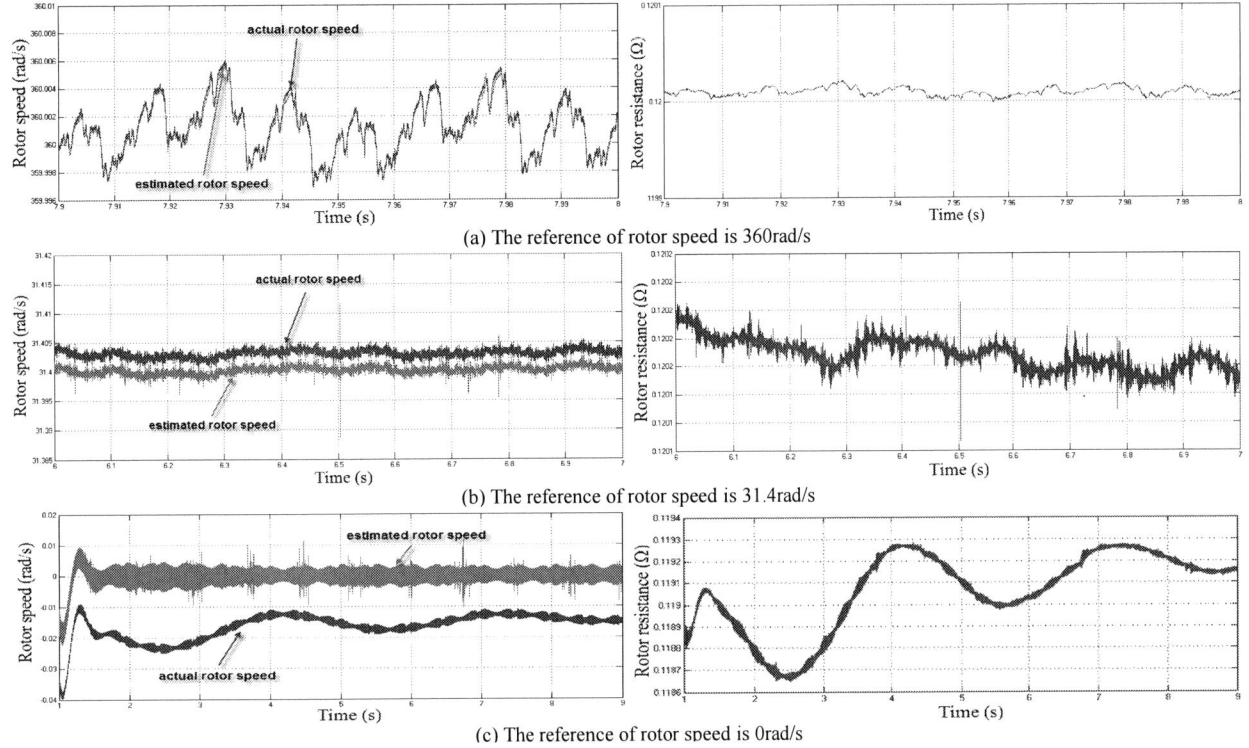

(a) The reference of rotor speed is 360rad/s

(b) The reference of rotor speed is 31.4rad/s

(c) The reference of rotor speed is 0rad/s

Figure 6. Simulation results of sensorless control system based on Luenberger observer with R_r identification (the actual value of R_r is 0.12Ω).

At higher operating speed, Luenberger observer with rotor resistance identification is adopted, because the Luenberger observer with rotor resistance identification also has good performance at high operating speed. Moreover, even though the calculation of R_s through thermal model is not accurate, the estimation accuracy of rotor speed will not be influenced.

At lower operating speed, Luenberger observer with stator resistance identification is adopted, because the Luenberger observer with stator resistance has better performance at low operating speed. And the error of the R_r calculation through thermal model will have limited influence on influence on the estimation accuracy of the rotor speed.

VI. CONCLUTIONS

This paper focuses on speed sensorless vector control of medium voltage IM drives. Rotor speed estimation method based on Luenberger observer is presented. Through plenty of researches, the influences of parameter variation on speed estimation are generalized. Based on these conclusions, a robust Luenberger observer with online parameter identification is presented. The described system is built in Matlab, and has been proved to have good performances both on steady state and transient response in a wide operating speed range. This system is to be implemented in the future work on TMS320F2812.

ACKNOWLEDGMENT

The support of EATON Innovation Center and Zhejiang University – SANTAK (an EATON brand) Joint Laboratory is gratefully acknowledged.

REFERENCES

[1] Bimal K.Bose, Modern Power Electronics and AC Drives, 1st ed. Beijing: CHINA MACHINE PRESS, 2003, pp. 266-321.

[2] Holtz.J, "Sensorless Control of Induction Machines—With or Without Signal Injection" IEEE Transactions on Industry Applications, Vol. 41, No. 2, pp. 591-598, March 2005.

[3] Shanshan Wu, Yongdong Li, Zedong Zheng, "Speed Sensorless Vector Control of Induction Motor Based on Full-Order Flux Observer" CES/IEEE 5th International Power Electronics and Motion Control Conference, Vol3, No. 14, pp. 1-4, August 2006.

[4] Zhiwu Huang, Weihua Gui, Xiaohong Nian, Xinhao Liu and Yongteng Shan, "A Novel Stator Resistance Identification for Speed Sensorless Induction Motor Drives Using Observer" IEEE international Symposium on Industrial Electronics, Vo3, pp. 2211-2216, July 2006.

[5] Han Li, Wen Xuhui, Chen Guilan, "General Adaptive Schemes for Resistance and Speed Estimation in Induction Motor Drives" IEEE Workshops on Computers in Power Electronics, No.16, pp. 173-178, July 2006.

[6] Lazhar Ben-Brahim, Susumu Tadakuma, Alper Akdag, "Speed control of induction motor without rotational transducers" IEEE Transactions on Industry Applications, Vol35, No. 4, pp. 591-598, July 1999.

[7] Ben Hamed Mouna and Sbita Lassaad, "Direct Stator Field Oriented Control of Speed Sensorless Induction Motor" IEEE International Conference on Industrial Technology, No.15-17, pp. 961-966, December 2006.

[8] Ben Hamed Mouna and Sbita Lassaad, "Speed Sensorless Indirect Stator Field Oriented Control of Induction Motor Based On Luenberger Observer" IEEE international Symposium on Industrial Electronics, Vol3, pp. 2473-2478, July 2006.

[9] Mohamed Rashed, Fraser Stronach and Peter Vas, "A new stable MRAS-based speed and stator resistance estimators for sensorless vector control induction motor drive at low speeds" Industry Application Conference, Vol2, pp.1181-1188, July 2006.

978-1-4244-4782-4/10 $26.00 © 2010 IEEE

Fault Detection and Diagnostics for Non-Intrusive Monitoring using Motor Harmonics

Uzoma A. Orji, Zachary Remscrim, Christopher Laughman, Steven B. Leeb,
Warit Wichakool, Christopher Schantz, Robert Cox,
James Paris, James L. Kirtley, Jr., Les K. Norford

Abstract—Harmonic analysis of motor current has been used to track the speed of motors for sensorless control. Algorithms exist that track the speed of a motor given a dedicated stator current measurement, for example [1–5]. Harmonic analysis has also been applied for diagnostic detection of electro-mechanical faults such as damaged bearings and rotor eccentricity [6–17]. This paper demonstrates the utility of harmonic analysis for fault detection and diagnostics in non-intrusive monitoring applications, where multiple loads are tracked by a sensor monitoring only the aggregate utility service. An optimization routine is implemented to maintain accuracy of speed estimation while using shorter lengths of data.

I. SMART MONITORING

At any point in the life of a system, mechanical and electrical equipment may be poorly operated. For example, as buildings age, both the electro-mechanical actuators and associated mechanical components wear, cease to function properly, and eventually fail, via myriad processes that are often undetected. Valves do not close fully, filters clog, air-conditioning system dampers stick, refrigerant leaks, heating and cooling coils – from the smallest refrigerator to the largest building air-conditioning system – become fouled with dirt and debris, and belts slip. Energy waste and excessive plant wear are often exacerbated by closed-loop control. Under active control, damaged but still functioning equipment will operate by extending run times or operating points to meet user commands, leaving few obvious signs of compromised operation.

For example, a number of surveys of airflow faults in buildings hint at the range and extent of these problems. One compendium of fault surveys [18], which examined 503 rooftop air-conditioning units in 181 buildings in five states in the Western U.S. from 2001-2004, found that the airflow was out of the specified range in approximately 42% of the units surveyed. A separate study [19] of $4,168$ commercial air-conditioners in California reported that 44% of the surveyed units had airflow that was out of specifications. Studies of 29 new homes in Washington State [20] found that average duct leakage rates to the exterior ranged from 687 to 140 cubic feet per minute (CFM). Extrapolating from such fault surveys, one estimate for the total energy consumed by duct leakage is $5 billion/year [21].

When "failure is not an option," the performance of important electro-mechanical loads on mission-critical systems like warships or power plants is often tracked by dedicated monitoring equipment [22]. An extensive sensing network can provide obvious advantages for fault detection, diagnosis, and prognosis. However, a large sensing network can be expensive and difficult to maintain.

Smart Grid and Smart Meter initiatives hope to allow energy providers and consumers to intelligently manage their energy needs through real-time monitoring, analysis, and control of electrical power usage. The U.S. Department of Energy has identified "sensing and measurement" as one of the "five fundamental technologies" essential for driving the creation of a "Smart Grid" [23]. The methods described in this paper could be used to provide detailed energy score keeping and diagnostic measurements for motors for new metering schemes. These methods can be used by a non-intrusive load monitor (NILM) that determines the operating schedule of major electrical loads from measurements taken from an aggregate power feed serving multiple loads [24].

II. NILM BACKGROUND

The non-intrusive load monitor has been demonstrated [25–28] as an effective tool for evaluating and monitoring electro-mechanical systems through analysis of electrical power data. The power distribution network can be pressed into "dual-use" service, providing not only power distribution but also a diagnostic monitoring capability based on observations of the way in which loads draw power from the distribution service. A key advantage of the non-intrusive approach is the ability to reduce sensor count by monitoring collections of loads.

Non-intrusive electrical monitoring has been described in [29, 30] among other publications. The systems that are described in these papers can be split into two broad categories: transient and steady-state approaches. The transient approach [30] finds loads by examining the full detail of their transient behavior. Reference [25] describes a platform for transient-based non-intrusive load monitoring appropriate for many applications.

The NILM detects the operation of individual loads in an aggregate power measurement by preprocessing measured current and voltage waveforms to compute spectral envelopes [31]. Spectral envelopes are short-time averages of the line-locked harmonic content of a signal. These spectral envelopes may be recognized as the coefficients of a time-varying Fourier series of the current waveform. For transient event detection on the ac utility, the time reference is locked to the line so that fundamental frequency spectral envelopes correspond to real and reactive power in steady state. Higher spectral envelopes

978-1-4244-4782-4/10 $26.00 © 2010 IEEE

correspond to line frequency harmonic content. A high performance transient event detection algorithm [30, 32] is available to disaggregate the fingerprints or spectral envelope signatures of individual loads in the aggregate measurement.

As will be shown below, subtle harmonics associated with the rotor of a motor are generally not found at integer multiples of the line frequency, although they may be spaced periodically in multiples of the line frequency. These rotor frequency harmonics, if detected, can be used to determine the operating speed of a motor. A NILM could provide this information by first detecting the activation of a motor of interest in the observed, line-locked spectral envelopes. Special attention can then be paid to the aggregate current frequency content just before and just after a turn-on transient to identify key harmonics that are indicative of rotor speed, and also pathological conditions in the motor.

III. OVERALL BLOCK DIAGRAM

This paper will introduce a new algorithm that can detect the speed-related slot harmonics in the NILM environment. The block diagram for this algorithm is shown in Fig. 1.

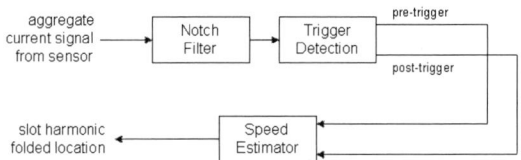

Fig. 1. Block diagram illustrating the processes of the algorithm. The important blocks include a notch filter, a trigger detection block and the final speed-estimation block.

A line-frequency notch filter improves the resolution of the relevant range of the input current. A pre-trigger and post-trigger detection block creates two streams of data from the output of the notch filter. Afterward, the outputs are used to estimate the speed for the corresponding motor.

IV. ROTOR SLOT HARMONICS

Rotor slot harmonics are widely used in speed detection algorithms. Other harmonics appear in the stator current of a motor which often make the task of searching for these rotor slot harmonics more difficult. Motors can experience various kinds of mechanical failures or faults that inject their own set of harmonics into the stator current. The utility line supplying current to the stator may have its own distortions that present troublesome harmonics, especially in the NILM environment.

Rotor slot harmonics present in the stator current of a motor arise from the interaction between the permeance of the machine and the magnetomotive force (MMF) of the current in the stator windings. As the motor turns, the rotor slots alter the effective length of the airgap sinusoidally thereby affecting the permeance of the machine. This sinusoidal behavior is seen in the flux, which is the product of the MMF and the permeance across the airgap. The odd harmonics present in the stator current introduce additional harmonics. Static and dynamic eccentricity harmonics also appear in the stator current as the rotor turns irregularly in relation to the stator.

The slot harmonics, including the principal slot harmonic (PSH), are located at frequencies

$$ f_{sh} = f\left[(kR + n_d)\frac{1-s}{p} + \nu \right] \quad (1) $$

where f is the supply frequency; $k = 0, 1, 2...$; R is number of rotor slots; $n_d = 0, \pm 1, ...$ is the order of rotor eccentricity; s is the per unit slip, p is the number of pole pairs and $\nu = \pm 1, \pm 3, ...$ is the stator MMF harmonic order [5, 33].

Previous research has been done to study the effects of mechanical faults on the stator current spectrum [6–17, 34]. These faults, including broken rotors bars, damaged bearings, rotor eccentricity, rotor asymmetry, bearings failures, and shaft speed oscillations, produce distinct harmonics in the current spectrum. One mechanical fault of interest is the shaft speed oscillation, which can be enhanced when an imperfectly balanced fan is attached to the motor shaft. The frequencies of interest [8] are predicted by

$$ f_{sso} = f\left[k\left(\frac{1-s}{p}\right) \pm \nu \right]. $$

These harmonics from the shaft speed oscillations are present in the stator current, complicating the detection of the principal slot harmonics.

The ac utility line is a potentially distorted sinusoid containing only the fundamental frequency f and harmonic multiples of this frequency. Depending on the time of day, the amount of loading on the utility line can vary substantially and can cause large distortions on the utility line. Also, line impedances create voltage distortions at frequencies determined by other loads in the system. These distortions, like the mechanical faults, introduce extra harmonics on the current spectrum.

A substantial amount of literature makes use of Eq. (1) for speed detection [1–3, 5, 35, 36]. The current spectrum of a motor with its slot harmonics Eq. (1) are shown in Fig. 2 for different values of ν. The motor used was a three-phase machine with $R = 48$ rotor slots and $p = 3$ pole pairs loaded by a dynomometer to run at $s = 0.0171$ or 1180 rpm. Typically, the slot harmonics with $k = 1$ and with $n_d = 0$ are the most pronounced in the current spectrum [1, 37]. For a given n_d, the slot harmonics differ exactly by $2f$ in Eq. (1).

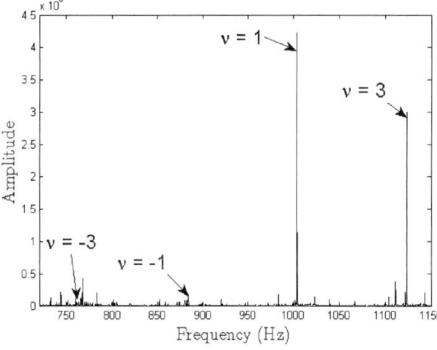

Fig. 2. Slot harmonics for a motor with the following parameters: f=60, k=1, R=48, n_d=0, $s = 1.71\%$ or 1180 rpm. The harmonics shown in the figure are labeled with the corresponding value of ν.

Analog techniques [35, 36] have been implemented to track these harmonics. The performance of these approaches can be limited in terms of accuracy, linearity, resolution, speed range, or speed of response [4]. Any analog filtering can require extremely complex circuitry and any output signal can be corrupted by noise. Digital techniques employing the Fast Fourier Transform (FFT) were developed [2] to overcome the flaws in the analog methods. These FFT methods were limited by the uncertainty principle, i.e. the trade-off between high frequency resolution and the response time to changes of speed that deteriorates with long data records. Parametric estimations [1, 3, 5] of the current spectrum were used to overcome the limitations of FFT but they require digital filters which make these methods less robust than the FFT. With a high stator frequency, the longer computations times can reduce any advantage these parametric methods may have over the FFT.

Some research has been done to combine the FFT and parametric estimation methods [1, 5]. In [1], the techniques do a successful job in tracking the slot harmonics in estimating speeds in a controlled environment. The authors make use of the periodicity of the slot harmonics by aliasing the spectrum such that these harmonics line up to increase detectability. We observe that the principal slot harmonic (PSH) is the most pronounced slot harmonic in the motors used in our experiments. The methods in this paper, therefore, will only search for the PSH, which simplifies the complexity of the algorithm.

There are certain trade-offs that must be made when deciding on the proper methods for detecting these speed-dependent rotor slot harmonics. The choices are often dictated by the practical setting.

V. DATA ACQUISITION

NILM experiments from previous research collect data from a current sensor with only analog filtering for anti-aliasing. There is a large 60 Hz line frequency component that dominates the current signal, making detection of the smaller slot harmonics more difficult. To improve the detectability of the harmonics, a 60 Hz notch filter is implemented to remove the large line frequency component before the data acquisition hardware in the NILM samples the current.

In Fig. 3, the stator current signal is sent through a 60 Hz notch filter. The output is amplified by a gain stage and later filtered by a passive antialiasing filter. The output buffer drives the input of the NILM data acquisition hardware. The notch filter stage allows for improved signal detectability of the smaller slot harmonics. By removing the large dominant line frequency, the smaller harmonic signals can then be amplified, increasing the overall signal-to-noise ratio by reducing the effect of quantization noise in the ADC of the NILM.

VI. PRACTICAL LIMITATIONS ON SLIP

For high efficiency induction motors, the slip s usually does not exceed 5% and possibly less. This assumption leads to interesting simplifications when searching for the principal slot harmonic (PSH). The PSH refers to the slot harmonic with $\nu =$

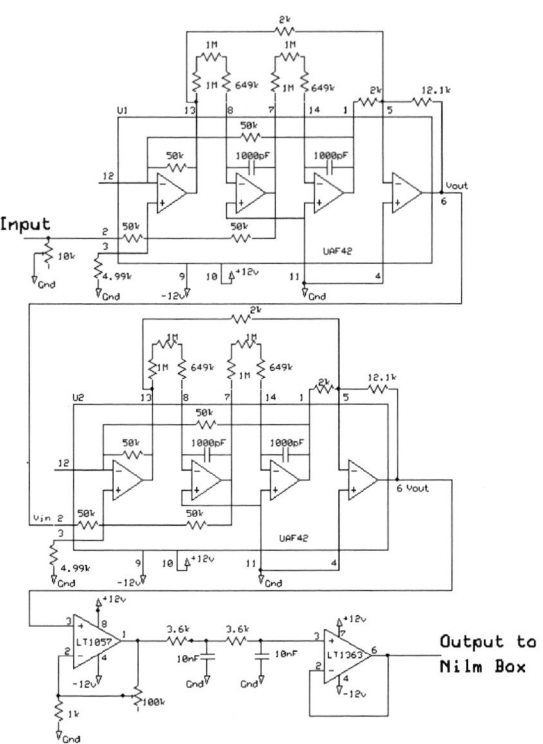

Fig. 3. Schematic of the 60 Hz notch filter circuit. The circuit notches the 60 Hz frequency, amplifies the signal and sends the signal through an antialiasing filter.

TABLE I
MAXIMUM SLIP FOR UNAMBIGUOUS SPEED ESTIMATION FOR SEVERAL VALUES OF R AND p

R	p			
	1	2	3	4
16	0.1247	0.2494	0.3742	0.4989
20	0.0997	0.2000	0.3000	0.4000
24	0.0833	0.1667	0.2492	0.3333
28	0.0714	0.1428	0.2142	0.2856
32	0.0622	0.1244	0.1875	0.2489
34	0.0586	0.1172	0.1758	0.2344
40	0.0500	0.1000	0.1500	0.1989
44	0.0453	0.0906	0.1358	0.1811
48	0.0417	0.0833	0.1250	0.1667
52	0.0383	0.0767	0.1150	0.1533
56	0.0356	0.0711	0.1067	0.1422
60	0.0333	0.0667	0.1000	0.1333

1, $k = 1$, and $n_d = 0$ in Eq. (1) and is used in speed detection algorithms since it is often the most pronounced [1, 37]. The slot harmonics for different values of ν differ by $2f = 120$ Hz. If the principal slot harmonic was confined to a 120 Hz window under these practical limitations of slip, there would be no ambiguity in determining the window of the PSH.

For example, Table I tabulates the maximum slip for different values of R and p for which the PSH would be confined to a 120 Hz window.

For a motor with $R = 60$ and $p = 1$, a maximum slip of $s = 0.033$ would need to be assumed in order to constrain the PSH to a 120 Hz wide window. This assumption would be unreasonable because it would be possible for the motor to be running with a slip of 0.04. In this situation, there would be some ambiguity in determining in which window the PSH lies. On the other hand, a motor with $R = 48$ and $p = 3$ can have a maximum slip of $s = 0.125$ to unambiguously determine the window of the PSH. This slip satisfies any reasonable low-slip assumptions.

VII. Speed Estimation via Slot Harmonics

Using a "low-slip" assumption in which the PSH is restricted to a single 120 Hz wide frequency window, this section will describe the application of the slot harmonics in determining speed of operation just after startup in a multi-load/multi-motor environment. To demonstrate the effectiveness of the algorithm, the speed estimation method was conducted with the following two motors. The first motor (Motor 1) was a three-phase motor from an HVAC evaporator in an air-handling unit. Motor 1 had $R = 48$ rotor slots and $p = 3$ pole pairs. The second motor (Motor 2) is a single-phase line-to-line machine from a fresh-air ventilation unit with $R = 34$ rotor slots and $p = 1$ pole pairs.

Consider an illustrative example in Fig. 4 in which the transient responses of induction motors are shown as they turn on and off. The region labeled A in 4(a) is when Motor 1 (the evaporator motor) turns on. Zooming in on Region A shows the transient response of Motor 1 as shown in Fig. 4(b). The region labeled B marks the region in which Motor 2 turns on so that both motors are running. The spectral envelope corresponding to "real" power as calculated by the NILM is shown in Fig. 5.

For the motors used in these experiments, Table I confirms that they satisfy the low-slip assumption. The PSH for Motor 1 will be between 900 Hz and 1020 Hz while the PSH for Motor 2 will be between 1980 Hz and 2100 Hz using Eq. (1).

As Motor 1 turns on, the NILM characterizes this new load from its transient turn-on response and spectral envelope. The aggregate current frequency contents before and after this transient event are recorded and are used to estimate the speed of Motor 1.

The window of interest to locate the PSH for Motor 1 is shown in Fig. 6(a). Using 5 seconds of current data, sampled at 7800 Hz, the location of the PSH can be estimated to be 993 Hz by finding the location of the maximum value within the window. Using Eq. (1), the speed can be estimated to be 1166.25 RPM.

By simply taking the maximum value within the PSH window, the resolution of the estimate is limited to the size of the FFT frequency bins, which is determined by the total duration of sampling. The estimate of the PSH (and the motor's speed) can be refined even further. As shown in Fig. 6(b), the energy of the PSH is actually spread over several frequency bins implying that there is not enough resolution to determine the PSH precisely [38]. One way to obtain finer resolution would be to sample data over a longer interval, which is unattractive as the speed of the motor may vary during the interval. A different approach, is to use the information in the frequency bins near the peak.

Consider the illustrative example in Figure 7 of the FFT of a pure sinusoid wave with a frequency of 994.34 Hz. In

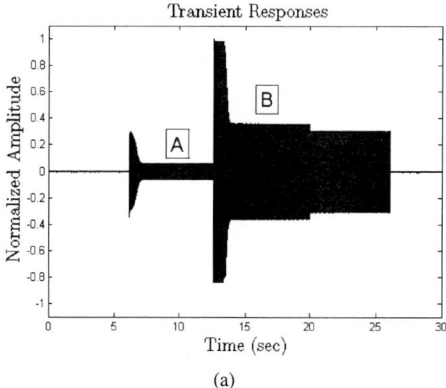

(a)

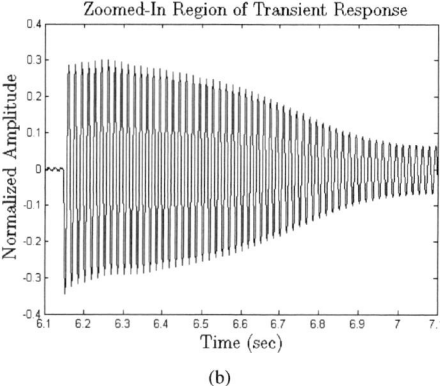

(b)

Fig. 4. In (a), the current data stream with transient responses. The turn-on transient when Motor 1 turns on (Region A) is shown in (b)

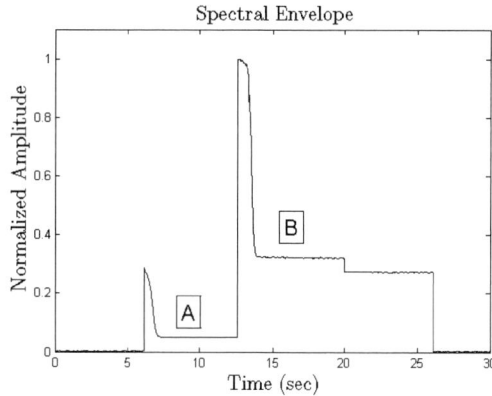

Fig. 5. Spectral envelope calculation.

Fig. 7(a), the FFT of 5 seconds of this pure sine wave is shown. Compare this result with that of Fig. 7(b) and Fig. 7(c) in which the data length is reduced to .5 seconds and .05 seconds respectively. The energy of the signal is spread over an increasingly wider frequency band as the data duration decreases.

When dealing with the short, .05 second long data, each bin of the resulting FFT is 20Hz wide. Estimating the frequency (994.34 Hz, in this example) from simply looking at the peak within a window could only hope to provide resolution of 20 Hz. The situation can be improved by producing .05 seconds

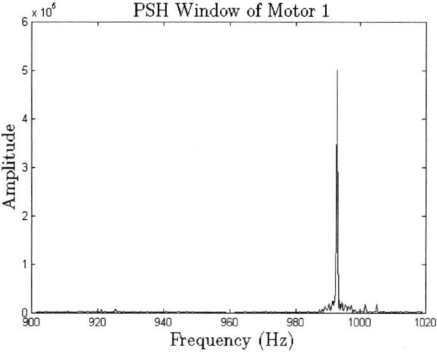

(a) Aggregate current spectrum w/ Motor 1 running.

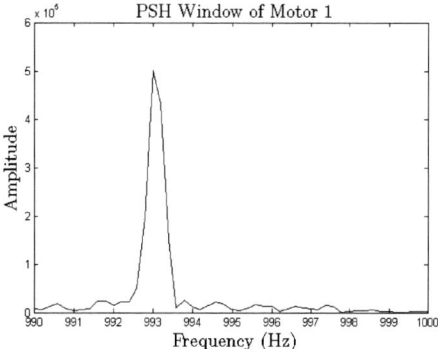

(b) Current spectrum with w/ Motor 1 running. The frequency content of the PSH is spread over multiple bins.

Fig. 6. In (a), the frequency content of the PSH is actually spread over several frequency bins as shown in Fig. 6(b).

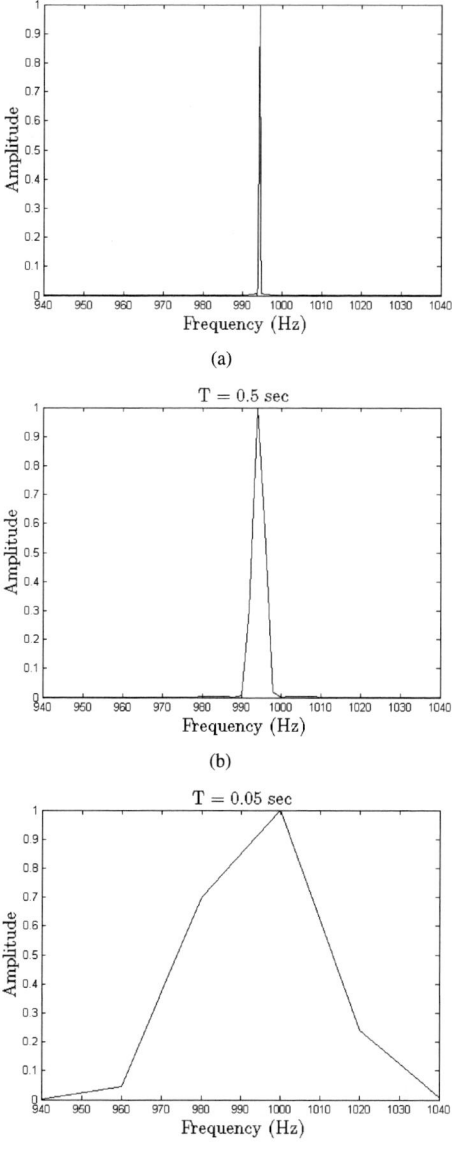

Fig. 7. In (a), the FFT is shown for the sinusoid with a sampling time of 5 seconds. The energy in the peak spreads out over several frequency bins as the sampling duration decreases as shown in (b) and (c).

of data from each of a family of sine waves with frequencies in the neighborhood of the observed peak. An optimization routine can pick the frequency of a sinusoid whose FFT is the best fit (in the minimum mean square sense) for the observed data over the entire frequency neighborhood around the peak. This routine makes the assumption that there is only a single, pure sinusoid in the window responsible for all observed frequency content. In this case, the optimization routine selects exactly 994.34 Hz as there is no noise to corrupt the data. Essentially, a finer estimate can be made by using the energy that has spread over several bins as opposed to only making use of the content in just one bin.

This same idea can be applied to the actual data taken from Motor 1. All of the following data is taken at a sampling frequency of 7800 Hz. Using the full 5 seconds length of data, the estimate of the PSH for Motor 1 is 993 Hz. A speed estimate of 1166.250 RPM is then calculated from this estimate of the PSH. The results of applying the optimization routine to shorter lengths of data are shown in Fig. 8.

The frequency content for 0.5 seconds of data of Motor 1 near the PSH is shown by the solid line in Fig. 8(a). The optimization routine finds the the frequency of a sinusoid whose FFT best matches the observed data, in this case 992.7 Hz. Again, the optimization routine is making the assumption that there is only a single, pure sinusoid responsible for all observed frequency content. The dotted line in Fig. 8(a) is the

FFT of a 0.5 second sinusoid with a frequency of 992.7 Hz. The FFT of this sinusoid closely matches that of the observed data (solid line) in the region of interest. A speed estimate of 1165.870 RPM is then calculated from the optimized estimate of the PSH. This is repeated for 0.05 seconds of data in Fig. 8(b). The optimized PSH is 992.08 Hz and the corresponding speed estimate is 1165.100 RPM.

This optimization routine maintains accurate speed estimates despite using a shorter duration of data. By utilizing more data from the nearby frequency bins, and the assumption that only a single sinusoid is responsible for the observed frequency content, the routine can predict reliable speed estimates on a smaller data set. Table II displays the results of

978-1-4244-4782-4/10 $26.00 © 2010 IEEE

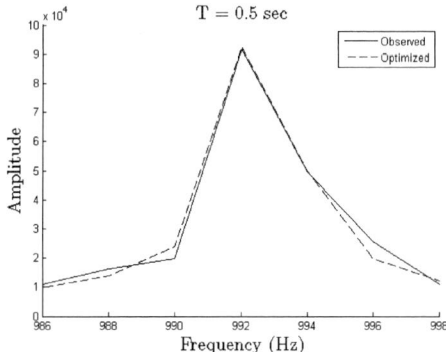

(a) The observed PSH is shown in the solid line. The dotted line is the FFT of the best-fit sinusoid. The sampling time is 5 seconds.

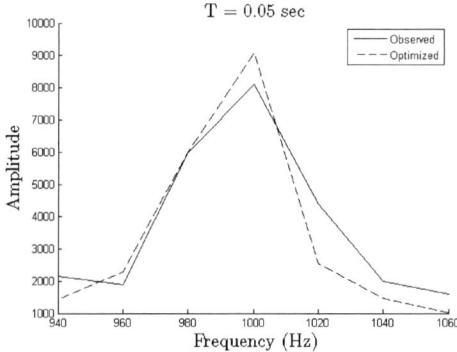

(b) The observed PSH is shown in the solid line. The dotted line is the FFT of the best-fit sinusoid. The sampling time is 0.5 seconds.

Fig. 8. The observed PSH and FFT of the best-fit sine waves for a sampling time of $T = 5$ seconds (a) and $T = 0.5$ (b).

TABLE II
OPTIMIZATION ROUTINE PSH AND SPEED ESTIMATES AT DIFFERENT
SAMPLING TIMES

T (sec)	Line Cycles	PSH (Hz)	Speed (RPM)
5	300	993.080	1166.350
3	180	992.929	1166.161
1	60	992.760	1165.950
0.5	30	992.700	1165.870
0.1	6	992.960	1166.200
0.05	3	992.080	1165.100
0.0333	2	997.200	1171.500
0.0166	1	990.140	1162.675

the optimizing routine for different sampling times.

The speed detection algorithm runs as new loads turn on. When Motor 2 turns on, there are scenarios that may complicate tracking speeds. If both motors are identical in parameters, the worst case scenario would have both motors running at the same speed. In such a case, the principal slot harmonic (PSH) of each motor would be at the exact same location. To prevent this, if possible, the motors can be selected so that such a scenario could not occur. In this experiment, the parameters of the motors were chosen such that the PSH of each motor would appear in separate windows.

Also complicating speed detection, the utility line has its own set of harmonics which are visible in the current spectrum.

The 17th harmonic at 1020 Hz could be troublesome in trying to identify the PSH of Motor 1. Fortunately, as these undesirable harmonics are located at integer multiples of 60 Hz, we can easily filter them out using the digital filter $y[n] = x[n] - x[n - N]$ where N is the number of sample points per 60 Hz line cycle (in this paper, all data is sampled at 7.8 KHz, which corresponds to $N = 130$). This filter will notch out all the integer multiples of the line frequency.

One problem, which shows up in this experiment as Motor 2 turns on, is the addition of eccentricity harmonics [6]. Motor 2 is loaded with a fan which exacerbates these eccentricities. Figure 9(a) shows the window of interest for the PSH of Motor 1 when both motors are running. In this paper, the motors examined had slot harmonics that are larger in amplitude than any of the eccentricity harmonics. Tracking these eccentricity harmonics may be possible by first estimating the speed of the eccentric motor. Once the slip is estimated, all possible eccentricity harmonics can be tabulated and tracked. Changes in the observed amplitudes of these eccentricity harmonics can be used to diagnose the health of the motor. Also, the above algorithm can use knowledge of eccentricity harmonics to make better speed predictions by properly including the effects of the eccentricity harmonics in the observed frequency content. The PSH for Motor 2 was determined to be between 1980 and 2100 Hz. The current spectrum at these frequencies is shown in Fig. 9(b). The optimized PSH for Motor 2 is estimated at 2046.28 Hz which corresponds to 3505.20 rpm.

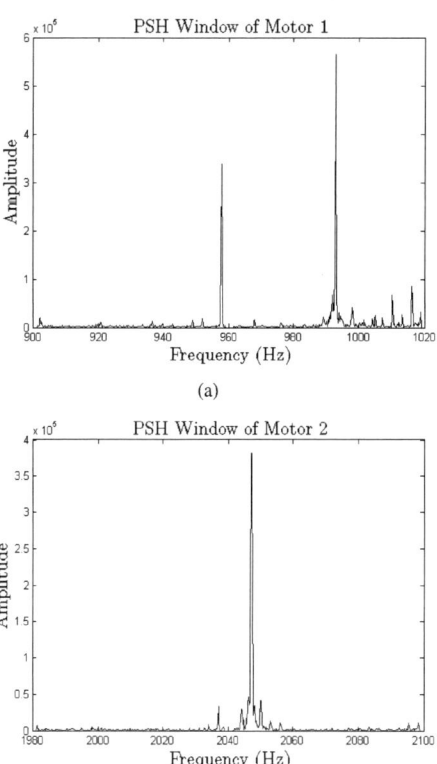

Fig. 9. Aggregate current spectrum with both motors running. In (a), the current spectrum in the PSH window of Motor 1 and in (b), the spectrum in the PSH window of Motor 2.

VIII. AIRFLOW DIAGNOSTICS APPLICATION

As an example of the utility of knowing motor speed in addition to power consumption for load diagnostics, consider ventilation systems in residential or commercial buildings. A system that is able to monitor the state of airflow and detect faults in ventilation systems would fulfill a significant need in contemporary buildings, due to the prevalence of airflow faults. One widely used ventilation system employs air-side distribution systems for air-conditioning units typically called air handlers, air handling units, or AHUs. A picture and a visual representation of the AHU used in [39] are shown in Fig. 10.

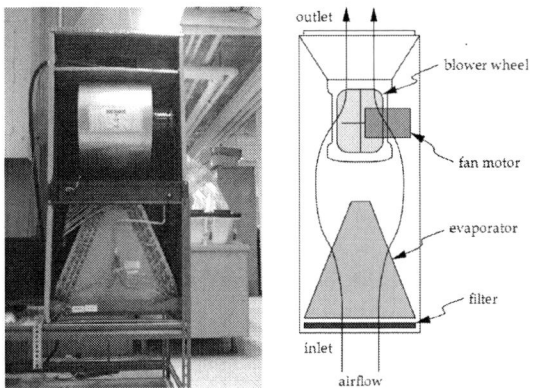

Fig. 10.   Schematic diagram and picture of air handler.

A common fault in these systems occurs when the filter to the air handler, or the evaporator itself, is clogged, causing the airflow through the fan to be reduced. While the most notable effect of such a fault will be on the reduction in the air delivered to the building occupants, this fault can also potentially chill the volume of air flowing through the AHU further than is intended. A dramatically reduced flow rate could also affect the system health of the overall air-conditioning system, and of the compressor more specifically; if little air is traveling through the evaporator, the cooling load on the evaporator could be substantially reduced, causing the amount of refrigerant evaporated in the evaporator to be much smaller than required by the design specifications. This could potentially result in liquid refrigerant entering the compressor through the suction line, causing the ingestion of liquid refrigerant by the compressor, permanently damaging it. Also, the accumulation of material on the filter or the evaporator can also effect the health of building occupants. The accumulation of bacteria or mold on these surfaces can affect people breathing the air.

The architecture of the airflow estimation method in [39] is dependent upon the estimation of three related quantities: the mechanical torque applied to the fan τ_f, the speed of the fan blades ω_f, and the fan curve at the operating point of the fan. Since the fan curve is measured empirically by the manufacturer, it is necessary to develop a method to determine ω_f and τ_f from the motor electrical variables V_m and I_m. The block diagram below shows the estimation scheme employed in [39].

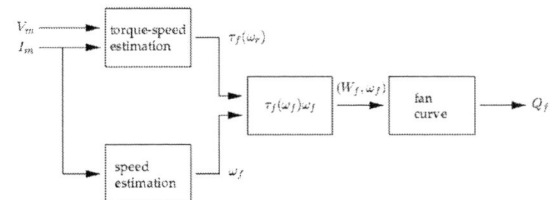

Fig. 11.   Structure of the airflow estimation method.

To identify the airflow Q_f, the speed ω_f and the torque-speed curve $\tau_f(\omega_r)$ are first estimated independently from the electrical variables. The methods introduced in this paper describe a way that the speed ω_f of the fan blades can be non-intrusively estimated. The torque $\tau_f(\omega_r)$ at the motor's present operating point can then be identified by monitoring the voltage and current supplied to the fan. With these estimates of ω_f and τ_f, the operating mechanical shaft power W_f can be identified. The mechanical power and the operating speed are then used to identify the point on the fan curve that describes the fan's current state, thereby generating an estimate of the volumetric airflow Q_f through the fan. An estimate of Q_f (cfm) can be recorded many times after many starts of the fan. These estimates can be collected in a histogram and tracked and trended over time as shown in Fig. 12 [39] for diagnostic purposes.

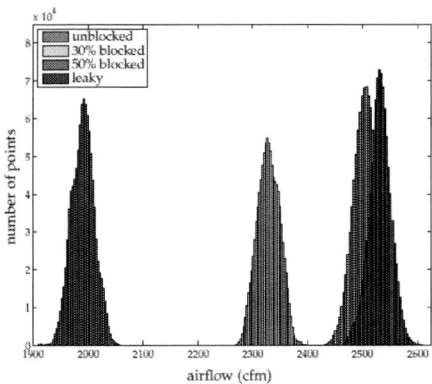

Fig. 12.   Illustration of airflow detectability using torque-speed curves that are generated from minimization against the motor current and the torque-speed curve, as collected for each blockage condition.

Fig. 12 shows experimentally derived histograms for the airflow of the AHU when the intake filter is unblocked, 30% blocked, 50% blocked and leaky. As expected, the airflow estimates are indicative of the mechanical condition of the AHU.

IX. CONCLUSIONS

It is often reasonable to assume that basic information about an electro-mechanical plant will be available for residential and industrial applications. For example, the torque-speed curve of an air handling motor or pump is often known, or could be obtained by or from a manufacturer. Similar information, e.g., a fan curve relating shaft power and speed to air flow, can be found for common fans and, similarly,

978-1-4244-4782-4/10 $26.00 © 2010 IEEE

for pump heads. The experimental results presented here demonstrate how a non-intrusive load monitor can determine electrical power and mechanical shaft speed from a reasonably chosen aggregate power measurement where other loads may also be operating. In situations where data like torque-speed curves are known *a priori*, a NILM could use this information along with its measurements of electrical power consumption and motor operating speed to perform fault detection and diagnostics on critical energy consumers like air conditioning and air handling systems.

This paper illustrates how the known behavior of motor harmonics could be exploited with reasonable assumptions about operating conditions to estimate speed in a non-intrusive setting. Spectral envelope computations are used to characterize the operating schedule of loads as then turn on. Once recognized as "on", the current before and after the transient can be analyzed to estimate the speed of the new motor joining the collection of operating loads on the monitored service. A fine estimate of speed can be calculated by employing an optimization routine to find the optimal frequency of a sinusoid that closely matches the spectral content of the observed data. This optimization routine allows for smaller sampling windows to obtain desirable frequency resolution. This technique can be extended to estimate the speeds of multiple motors from on aggregate current signal.

Of course, an unfortunate collection of loads could hinder the NILM's ability to detect operating speeds. In critical situations where non-intrusive monitor is desirable, motors might be selected during the design of a plant to enhance detection.

X. Acknowledgments

This work was supported in part by the MIT SeaGrant program, The Grainger Foundation, the National Science Foundation and the BP-MIT Research Alliance.

References

[1] K. D. Hurst and T. G. Habetler, "Sensorless Speed Measurement Using Current Harmonic Spectral Estimation in Induction Machine Drives," *IEEE Trans. Power Electron.*, vol. 11, no. 1, pp. 66–73, 1996.

[2] A. Ferrah, K. G. Bradley, and G. M. Asher, "Sensorless Speed Detection of Inverter Fed Induction Motors Using Rotor Slot Harmonics and Fast Fourier Transform," vol. 1. IEEE Power Electronics Specialist Conference, Jun. 29 - Jul. 3 1992, pp. 279–286.

[3] A. Ferrah, P. J. Hogben-Laing, K. J. Bradley, G. M. Asher, and M. S. Woolfson, "The Effect of Rotor Design on Sensorless Speed Estimation Using Rotor Slot Harmonics Identified by Adaptive Digital Filtering Using The Maximum Likelihood Approach," in *Industry Applications Conference, 1997. Thirty-Second IAS Annual Meeting, IAS'97., Conference Record of the 1997 IEEE*, vol. 1, Oct. 1997, pp. 128–135.

[4] A. Ferrah, K. J. Bradley, P. J. Hogben-Laing, M. S. Woolfson, G. M. Asher, M. Sumner, J. Cilia, and J. Shuli, "A Speed Identifier For Induction Motor Drives Using Real-Time Adaptive Digital Filtering," *IEEE Trans. Ind. Applicat.*, vol. 34, no. 1, pp. 156–162, 1998.

[5] K. D. Hurst and T. G. Habetler, "A Comparison of Spectrum Estimation Techniques For Sensorless Speed Detection in Induction Machines," *IEEE Trans. Ind. Applicat.*, vol. 33, no. 4, pp. 898–905, 1997.

[6] M. E. H. Benbouzid, "A Review of Induction Motors Signature Analysis as a Medium For Faults Detection," *IEEE Trans. Ind. Electron.*, vol. 47, no. 5, pp. 984–993, 2000.

[7] G. B. Kliman, R. A. Koegl, J. Stein, R. D. Endicott, and M. W. Madden, "Noninvasive Detection of Broken Rotor Bars in Operating Induction Motors," *IEEE Trans. Energy Conversion*, vol. 3, no. 4, pp. 873–879, 1988.

[8] G. B. Kliman and J. Stein, "Methods of Motor Current Signature Analysis," *Electric Machines and Power Systems*, vol. 20, no. 5, pp. 463–474, 1992.

[9] S. Nandi, H. A. Toliyat, and X. Li, "Condition Monitoring and Fault Diagnosis of Electrical Motors - A Review," *IEEE Trans. Energy Conversion*, vol. 20, no. 4, pp. 719–729, 2005.

[10] N. M. Elkasabgy, A. R. Eastham, and G. E. Dawson, "Detection of Broken Bars in the Cage Rotor on an Induction Machine," *IEEE Trans. Ind. Applicat.*, vol. 28, no. 1 Part 1, pp. 165–171, 1992.

[11] W. T. Thomson and M. Fenger, "Current Signature Analysis to Detect Induction Motor Faults," *IEEE Ind. Appl. Mag.*, vol. 7, no. 4, pp. 26–34, 2001.

[12] R. R. Schoen, T. G. Habetler, F. Kamran, and R. G. Bartfield, "Motor Bearing Damage Detection Using Stator Current Monitoring," *IEEE Trans. Ind. Applicat.*, vol. 31, no. 6, pp. 1274–1279, 1995.

[13] R. R. Schoen and T. G. Habetler, "Effects of Time-Varying Loads on Rotor Fault Detection in Induction Machines," *IEEE Trans. Ind. Applicat.*, vol. 31, no. 4, pp. 900–906, 1995.

[14] R. R. Schoen, B. K. Lin, T. G. Habetler, J. H. Schlag, and S. Farag, "An Unsupervised, On-line System For Induction Motor Fault Detection Using Stator Current Monitoring," in *Conference Record of the 1994 IEEE Industry Applications Society Annual Meeting, 1994.*, 1994, pp. 103–109.

[15] M. E. H. Benbouzid, M. Vieira, and C. Theys, "Induction Motors' Faults Detection And Localization Using Stator Current Advanced Signal Processing Techniques," *IEEE Trans. Power Electron.*, vol. 14, no. 1, pp. 14–22, 1999.

[16] J. R. Cameron, W. T. Thomson, and A. B. Dow, "Vibration and Current Monitoring for Detecting Airgap Eccentricity in Large Induction Motors," *IEE Proceedings B [see also IEE Proceedings-Electric Power Applications] Electric Power Applications*, vol. 133, no. 3, pp. 155–163, 1986.

[17] H. Guldemir, "Detection of Airgap Eccentricity Using Line Current Spectrum of Induction Motors," *Electric Power Systems Research*, vol. 64, no. 2, pp. 109–117, 2003.

[18] A. Cowan, "Review of Recent Commercial Rooftop Unit Field Studies in the Pacific Northwest and California," New Buildings Institute, PO Box 653, White Salmon,WA, 98672, Tech. Rep., Oct. 8 2004.

[19] R. J. Mowris, A. Blankenship, and E. Jones, "Field measurements of air conditioners with and without TXVs," 2004.

[20] D. Hales, A. Gordon, and M. Lubliner, "Duct Leakage in New Washington State Residences: Findings and Conclusions," *ASHRAE Transactions-American Society of Heating Refrigerating Airconditioning Engineers*, vol. 109, no. 2, pp. 393–402, 2003.

[21] K. Srinivasan, "Measurement of Air Leakage in Air-Handling Units and Air Conditioning Ducts," *Energy & Buildings*, vol. 37, no. 3, pp. 273–277, 2005.

[22] *Field Demonstration of a Real-Time Non-Intrusive Monitoring System for Condition-Based Maintenance.* National Harbor, Maryland: Electric Ship Design Symposium, Feb. 2009.

[23] U. S. Department of Energy, "The Smart Grid: An Introduction," World Wide Web electronic publication. [Online]. Available: http://www.oe.energy.gov/1165.htm

[24] G. W. Hart, "Nonintrusive Appliance Load Monitoring," *Proc. IEEE*, vol. 80, no. 12, pp. 1870–1891, Dec. 1992.

[25] S. R. Shaw, S. B. Leeb, L. K. Norford, and R. W. Cox, "Nonintrusive Load Monitoring and Diagnostics in Power Systems," *IEEE Trans. Instrum. Meas.*, vol. 57, no. 7, pp. 1445–1454, Jul. 2008.

[26] J. S. Ramsey *et al.*, "Shipboard Applications of Non-Intrusive Load Monitoring," in *ASNE Conference on Survivability and Reconfiguration*, Feb. 2005.

[27] T. DeNucci, R. Cox, S. B. Leeb, J. Paris, T. J. McCoy, C. Laughman, and W. C. Greene, "Diagnostic indicators for shipboard systems using non-intrusive load monitoring," *IEEE Electric Ship Technologies Symposium*, pp. 413–420, Jul. 2005.

[28] W. Greene, R. J. S., R. Cox, and T. DeNucci, "Non-intrusive monitoring for condition-based maintenance," *Proc. ASNE Reconfiguration and Survivability Symposium*, Feb. 16 2005.

[29] S. B. Leeb, "A Conjoint Pattern Recognition Approach to Nonintrusive Load Monitoring," Ph.D. dissertation, Massachusetts Institute of Technology, Cambridge, MA, Feb. 1993.

[30] S. B. Leeb, S. R. Shaw, and J. L. Kirtley Jr., "Transient Event Detection in Spectral Envelope Estimates For Nonintrusive Load Monitoring," *IEEE Trans. Power Delivery*, vol. 10, no. 3, pp. 1200–1210, Jul 1995.

[31] S. R. Shaw, "System identification techniques and modeling for nonintrusive load diagnostics," Ph.D. dissertation, Massachusetts Institute of Technology, Cambridge, MA, Feb. 2000.

[32] R. Cox, S. B. Leeb, S. R. Shaw, and L. K. Norford, "Transient Event Detection For Nonintrusive Load Monitoring and Demand Side Management Using Voltage Distortion," Mar. 2006.

[33] S. Nandi, S. Ahmed, and H. A. Toliyat, "Detection of Rotor Slot and Other Eccentricity Related Harmonics in a Three Phase Induction Motor With Different Rotor Cages," *IEEE Trans. Energy Conversion*, vol. 16, no. 3, pp. 253–260, 2001.

[34] C. Hargis, B. G. Gaydon, and K. Kamash, "The Detection of Rotor Defects in Induction Motors," in *Proc IEE EMDA Conf, London*, 1982, pp. 216–220.

[35] M. Ishida and K. Iwata, "A New Slip Frequncy Detector of an Induction Motor Utilizing Rotor Slot Harmonics," *IEEE Trans. Ind. Applicat.*, pp. 575–582, 1984.

[36] D. S. Zinger, F. Profumo, T. Lipo, and D. W. Novotny, "A Direct Field-Oriented Controller for Induction Motor Drives Using Tapped Stator Windings," *IEEE Trans. Power Electron.*, vol. 5, no. 4, pp. 446–453, 1990.

[37] S. Nandi, "Modeling of Induction Machines Including Stator and Rotor Slot Effects," in *Industry Applications Conference, 2003. Conference Record of the 38th IAS Annual Meeting.*, vol. 2, 2003.

[38] A. V. Oppenheim, R. W. Schafer, and J. R. Buck, *Discrete-time Signal Processing.* Prentice Hall Englewood Cliffs, NJ, 1989.

[39] C. Laughman, "Fault Detection Methods for Vapor-Compression Air Conditioners Using Electrical Measurements," Ph.D. dissertation, Massachusetts Institute of Technology, Cambridge, MA, Sep. 2008.

Reliability Evaluation of Three-Level Inverters

Yi Ding, Poh Chiang Loh, Kuan Khoon Tan, Peng Wang, and Feng Gao
School of Electrical and Electronic Engineering
Nanyang Technological University
50 Nanyang Avenue
Singapore 639798

Abstract—**To date, many inverter topologies have been proposed in the literature with each exhibiting certain advantages and disadvantages. These inverters are however mostly proposed with their reliability indexes left unexplored. The main reason for a lack of reliability assessment might be a lack of existing quantitative techniques that can be used for computing a mathematical index for associating with the inverter topology under consideration. Without the relevant mathematical index, comparing of inverter reliability is generally impossible or merely based on qualitative reasoning, which at times is subjective to individual preferences. Aiming to address that issue on reliability assessment, this paper proposes an appropriate method based on multi-state computation, uniquely fine-tuned for inverter (or general converter) evaluation. To demonstrate its applications, the formulated method is indifferently used for computing the reliability indexes / models of existing single-phase and three-phase three-level inverters to provide a common ground for their reliability evaluation. For inverters with possibilities of raising their indexes, topological and modulation modifications are proposed with their practicality and performance verified in simulation and experimentally.**

I. INTRODUCTION

Modern industrial and commercial systems usually use some forms of converters for processing energy, before channeling it to connected loads for consumption. Because of this proliferation of converters, different topological variants have since surfaced with each claimed to have its own advantages and disadvantages. Advantages would generally include factors like low cost, small size, lower harmonic distortion and electromagnetic interference (EMI), while disadvantages are mainly linked to factors like more complex implementation with higher computational burden, higher switching loss and bulkiness.

In spite of these well-documented characteristic features, an important performance indicator needed for judging converters is still at present not widely investigated, and that is to quantitatively assess the reliabilities of converter circuitries [1]. Without such a measure, effectiveness of fault ride-through schemes previously proposed in [2-5] with different compensating natures cannot be compared on a common numerical basis. Take for example, [2, 3] propose the addition of extra pre-installed switches for replacing failed switches in a timely manner, while [4, 5] suggest the selection of appropriate redundant states for riding through single

This project is financially sponsored by the Agency for Science, Technology and Research, Singapore, under the Intelligent Energy Distribution System Program.

semiconductor fault using no extra hardware. These efforts are discussed with no quantitative reliability indexes in view, meaning that comparison between them is at present not possible, other than knowing obviously that the former is qualitatively more expensive, and can tolerate more types of fault with lesser compromises.

To address the above concern, a method for quantitatively assessing converter reliabilities and deriving their relevant models is proposed in this digest. The method is developed from the multi-state system (MSS) concept discussed in [6], and is uniquely fine-tuned here for individual converter reliability evaluation. It is unlike existing techniques found in [7] and [8], where separate reliability analyses, treating the inverters as independent subsystems, are performed on photovoltaic inverters and paralleled inverters (two simple voltage-source inverters in parallel). Rather than treating the inverters as "black-box" entities, the method proposed here examines each individual converter topology, allowing different reliability indexes to be computed for representing different inverter circuitries, depending on the types of semiconductor failure that they can tolerate. This method is deemed as more appropriate, since it can help to identify truly reliable converter topologies that are suitable for critical applications such as in military ship propulsion systems and water purification systems, where system security and satisfactory performance need to be always maintained.

To demonstrate this usefulness, the proposed method is now applied to evaluate reliability indexes of existing three-level dc-ac inverters for comparison purposes, and where appropriate, topological modifications are also proposed with their reliability indexes recalculated for showing the percentage improvement achieved. The quantified improvement can then be weighed against factors like cost, efficiency and waveform quality, before deciding on whether the proposed modifications (whose practicalities are verified in simulation and experimentally) are suitable for the applications in view.

II. MULTI-STATE RELIABILITY MODELING TECHNIQUE

According to [6], a system with fault-tolerant capabilities can conceptually be modeled as a MSS. By that, it means that the system failure criterion cannot be formulated as an "all or nothing" arrangement when its performance or reliability index needs to be calculated. Reason for that limitation is simple, and is explained by the fact that failure of some system components will now not lead to complete system failure, but would rather have a wide-ranging effect on the

Fig. 1. Series RBD representation of a basic inverter with no fault-tolerant feature.

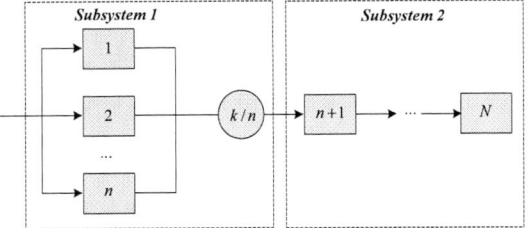

Fig. 2. RBD representation of an inverter with fault-tolerant features.

system overall performance. The system is therefore stated to have a finite number of performance levels, whose state probabilities are computed using the Markov model and some straightforward stochastic processing methods. However in real applications, the so-called straightforward stochastic processing methods are not that straightforward to implement after all because of the "dimension damnation" problem.

Therefore, instead of stochastic processing, an alternative method with reduced computational complexity is deemed as more appropriate for system reliability evaluation, where a possibility to adopt is the technique of reliability block diagram (RBD) [9, 10]. In concept, RBD is a convenient technique used for describing the functional relationships between a system and its individual components, and can generally be viewed as a technique for evaluating MSS reliability with reduced computational dimension. It encompasses basic series, parallel and series-parallel reliability structures that can collectively be used as building blocks for representing more complex, general network reliability structures. For illustration, a simple series RBD that can be used for representing an arbitrary inverter without fault ride-through option is shown in Fig. 1, where N system components (mainly IGBTs, diodes and capacitors) are noted to be in series.

By drawing them in series, it simply means that the system will work if and only if all the components are functioning well. The system reliability can then be computed as:

$$R(t) = \prod_{i=1}^{N} R^i(t) \tag{1}$$

where $R^i(t)$ is the reliability of component i (IGBT, diode or capacitor), which is usually treated as a binary state component since it either enters the successful or failure state. For this component, its reliability can be computed as [9]:

$$R^i(t) = e^{-\lambda^i \cdot t} \tag{2}$$

where λ^i is the failure rate of component i, which for an IGBT, it is written as (3) according to the stringent MIL-HDBK-217F military standard.

$$\lambda_{IGBT} = \lambda_{IGBT\_b} \cdot \pi_T \cdot \pi_Q \cdot \pi_E \tag{3}$$

where λ_{IGBT_b}, π_T, π_Q and π_E are the base failure rate of the IGBT, its temperature factor, quality factor and environment factor, respectively [11]. Referring to the same MIL-HDBK-217F standard, the diode failure rate can similarly be written as:

$$\lambda_{diode} = \lambda_{diode\_b} \cdot \pi_T \cdot \pi_S \cdot \pi_C \cdot \pi_Q \cdot \pi_E \tag{4}$$

where λ_{diode_b}, π_T, π_S, π_C, π_Q and π_E are the base failure rate of the diode, its temperature factor, voltage stress factor,

contact construction factor, quality factor and environment factor, respectively [11]. Besides semiconductor failures, failure rate of capacitors must also be determined, since they are usually the system bottlenecks with the weakest reliability indexes. To compute that, (5) can be used, where λ_{CP_b}, π_{Cv}, π_Q and π_E are the base failure rate of the capacitor, its capacitance factor, quality factor and environmental factor, respectively [11].

$$\lambda_{CP} = \lambda_{CP\_b} \cdot \pi_{Cv} \cdot \pi_Q \cdot \pi_E \tag{5}$$

Surely, the series RBD representation shown in Fig. 1 is an oversimplified example for modeling inverter reliability. To present a more involved representation, while at the same time clarifying the parallel structure, the RBD representation shown in Fig. 2 for modeling a fault-tolerant inverter is now explained. As seen, the overall RBD representation is modeled by two subsystems connected in series. The first subsystem consists of n components in parallel for representing a k-out-of-n: G subsystem. By that, it means that this subsystem will continue to operate well, if and only if at least k out of n components are operating. Indeed, this flexible setting of operating criteria has to date allowed the k-out-of-n parallel structure to find wide-ranging applications in both industrial and military systems for representing redundancy in fault-tolerant systems. In theory, it can similarly be used for representing inverters, especially multilevel inverters where multiple switches and redundant states are available, but has so far not been attempted. For a simple illustration of how the k-out-of-n representation can be applied, an inverter with arbitrary configuration and twelve switches (e.g. IGBTs) is considered. By further assuming that the inverter can still operate continuously to satisfy the load requirements even if any one of its switches fails short-circuit, albeit at a slight drop in voltage magnitude or waveform quality for example, it can then be written as a 11-out-of-12: G subsystem.

From [9], upon forming the k-out-of-n: G subsystem, its reliability index can then be written as:

$$R_{Sub1}(t) = \sum_{i=k}^{n} (-1)^{i-k} \binom{i-1}{k-1} \cdot \sum_{s_1 < s_2 < \ldots < s_i} \prod_{l=1}^{i} R^{s_l}(t) \tag{6}$$

where $R^{s_l}(t)$ represents the reliability index of any component block found in the parallel subsystem, and represented by symbol S_l ($S_1 < S_2 < \ldots < S_i < \ldots < S_n$). If S_i consists of a number of other subcomponents in series, then its reliability index can again be calculated using (1) for series-connected RBD. Suppose next that the inverter has additional

978-1-4244-4782-4/10 $26.00 © 2010 IEEE

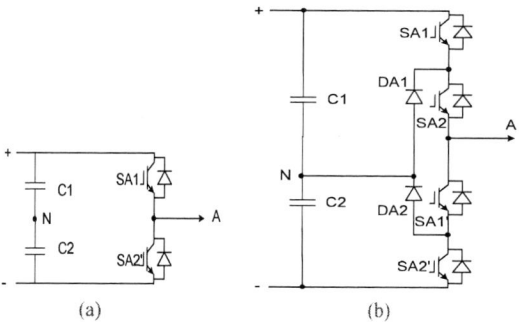

(a)   (b)

Fig. 3. Single-phase (a) two-level and (b) three-level inverters.

$N-n$ components (where N is the total number of components found in the overall system) like capacitors, whose failures will lead to complete system failure in the sense that load requirements can no longer be satisfied, a second series subsystem is needed, and is also drawn in Fig. 2. As indicated, the second subsystem has $N-n$ components connected in series, inferring that its reliability index can be calculated using (1), except with minor modifications made to the terms of multiplication to give the equation listed as follows:

$$R_{Sub2}(t) = \prod_{i=n+1}^{N} R^i(t) \qquad (7)$$

Using (6) and (7), the overall reliability index of the two subsystems in series can then be calculated as:

$$R(t) = R_{Sub1}(t) \cdot R_{Sub2}(t) \qquad (8)$$

Upon deriving the full set of equations detailed above, they can now be applied to specific inverter topologies for calculating their reliability indexes for comparison purposes. For this digest, computations are performed on all existing three-level dc-ac inverters, and where possible, fault-tolerant modifications are proposed with their reliability indexes calculated again using the derived equations for clearly illustrating the percentage gain in performance.

III. RBD REPRESENTATIONS OF EXISTING THREE-LEVEL INVERTERS

Following a systematic developmental procedure, the single-phase two-level inverter shown in Fig. 3(a) is first considered with two capacitors and two IGBT switches connected together. Being a simple circuit, failure of any component would cause the inverter to fail completely since no ac output can be produced. Because of that, the single-phase two-level inverter can be modeled using the series RBD diagram shown in Fig. 1 with four components connected in series. Moving next to the single-phase neutral-point-clamped (NPC) inverter shown in Fig. 3(b), where two capacitors, four IGBTs and two clamping diodes are connected together, failures of some components will now not cause the system to fail completely (only short-circuit failures are considered at present since they are after all the most common failure mode). For example, when SA1 or SA2' (or both) fails short-circuit, the inverter modulation can be reconfigured into two-level switching between the positive and negative dc-link terminals, by grouping {SA1, SA2} and {SA1', SA2'} as two complementary pairs.

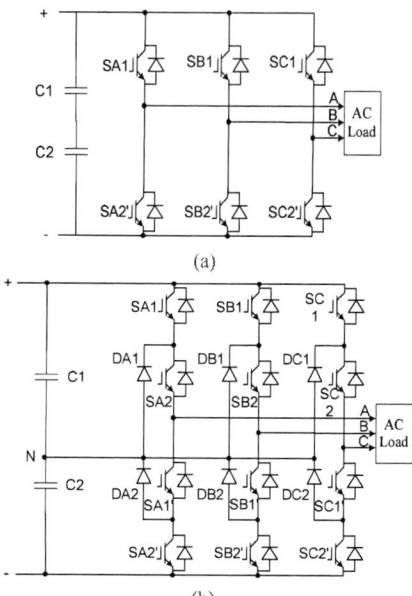

(a)

(b)

Fig. 4. Three-phase (a) two-level and (b) three-level inverters.

Doing so gives rise to the same output voltage amplitude, except for a slight degradation in waveform quality, and will therefore not cause the loads to trip. On the other hand, if SA2, SA1', DA1, DA2 or any of the capacitors fails short-circuit, the inverter would then not be able to produce the same output voltage amplitude, causing sensitive loads to trip off in turns. Based on this description, the single-phase three-level inverter should then be represented by the single RBD system shown in Fig. 1 with SA2, SA1', DA1, DA2 and the two dc-link capacitors connected in series. Since failure of SA1, SA2' or both will not affect the overall system reliability, they are not included in the RBD representation.

Extending the analyses to cover those three-phase cases shown in Fig. 4, the RBD representation for the two-level case can be drawn with eight components in series, comprising two dc-link capacitors (C1 and C2 can be combined into one if desired) and six IGBTs. The series representation is again used here because any failure of one component would result in the inverter not being able to produce the desired set of three-phase ac line voltage amplitudes.

For improved reliability, the three-phase three-level NPC inverter with two capacitors, twelve IGBTs and six clamping diodes shown in Fig. 4(b) is a better option, since [4] has previously shown that short-circuit failure of any single IGBT or clamping diode can smoothly be ridden through by using its designed fault-tolerant modulation scheme, although some dips in line voltage amplitudes cannot be avoided. If the de-rated conditions can indeed be tolerated by the connected ac loads, the inverter controlled by the suggested modulation scheme in [4] can be modeled as a 17-out-of-18: G subsystem for tolerating any single failure of IGBT or clamping diode (12 IGBTs plus 6 clamping diodes in total) connected to two series-connected capacitive blocks.

For more sensitive ac loads where slight dips in voltages cannot be tolerated, the 17-out-of-18: G subsystem

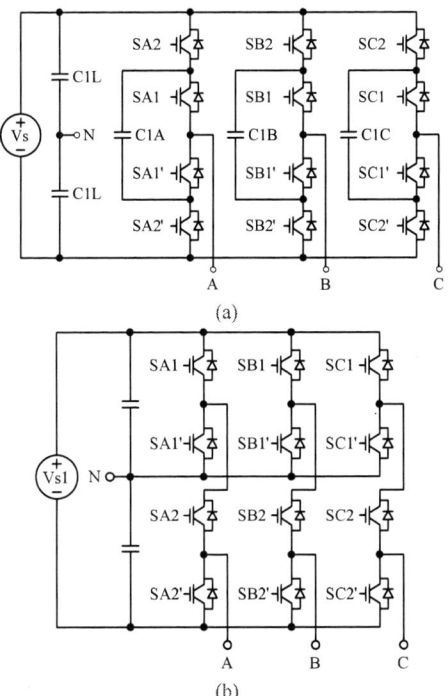

(a)

(b)

Fig. 5. Existing three-phase (a) flying-capacitor and (b) dc-link cascaded three-level inverters.

representation is no longer valid. Instead, the RBD representation should now include SX2, SX1', DX1, DX2 (X = A, B or C) and the two capacitors in series, since any failure in these components would lead to some dips in voltages that can trip off the sensitive loads. On the hand, short-circuit failure of any of the outermost switches (SX1, SX2' or both) can smoothly be ridden through without any dips in voltages, simply by grouping and controlling {SX1, SX2} and {SX1' and SX2'} as complementary two-level switching pairs. Because of their immunity, these switches are not considered when deriving the RBD representation of the three-phase three-level inverter. The RBD representation for the three-phase three-level case can be drawn with fourteen components in series, comprising two dc-link capacitors, six IGBTs and six clamping diodes.

The similar procedures can be used to analyze the RBD representations of other three-level inverters like the flying-capacitor, dc-link cascaded, cascaded, active NPC and conergy inverters (diagrams for the former two are shown in Fig. 5(a) and (b)). The corresponding reliability indexes are calculated in section V.

IV. MODIFIED FAULT-TOLERANT THREE-LEVEL INVERTERS AND THEIR RBD REPRESENTATIONS

A note brought forward from Section 3 is that any failure of an inner switch or clamping diode would cause the inverter to suffer dips in voltages. Take for example the case of SA2 failure, which would then make turning ON of SA1' and SA2' impossible since capacitor C2 would be shorted, inferring that the negative potential can never appear at the inverter output terminal.

To prevent this capacitive short-circuit from happening, an additional switch NP can be added to break the neutral connection, as shown in Fig. 6(a) for the NPC circuitry (since NP does not need to commutate at high frequency, a triac is sufficient for realizing it). Upon breaking the neutral connection and by again switching {SA1, SA2} and {SA1', SA2'} as two complementary pairs, the inverter can now smoothly ride through the fault with no voltage dips, except for a transition from three-level to two-level switching.

With NP added, any single failure of SA1, SA2 and DA1 (phase A for example) will not prevent obtaining the negative potential or positive potential at the inverter output terminal. Therefore SA1, SA2 and DA1 constitute a 2-out-of-3: G subsystem for tolerating any single failure of three components. Similarly SA1', SA2' and DA2 constitute another 2-out-of-3: G subsystem. These two 2-out-of-3: G subsystems are connected in series to give the overall RBD representation for one phase of the modified NPC inverter.

Expanding to three phases, if the NP is working properly the RBD representations for each phase are connected in series, before being connected to the two capacitive blocks in series. A point to note here is that since NP is shared among the three phases unlike the case of single-phase inverter, its reliability index must be adjusted before reliability index of the overall G subsystem is computed using (6).

The similar procedures can be used to evaluate the reliabilities of other modified three-level inverters (two of which are shown in Fig. 6(b) and (c)), which are presented in section V.

V. ILLUSTRATIVE EXAMPLES AND RESULTS

Using the RBD models developed earlier for traditional two-level, NPC and modified NPC inverters in both single and three-phase configurations, Fig. 7 and Fig. 8 show their respective reliability indexes calculated over a span of five years. From the presented results, it is clear that the two-level and NPC inverters have quite close reliability indexes (e.g.

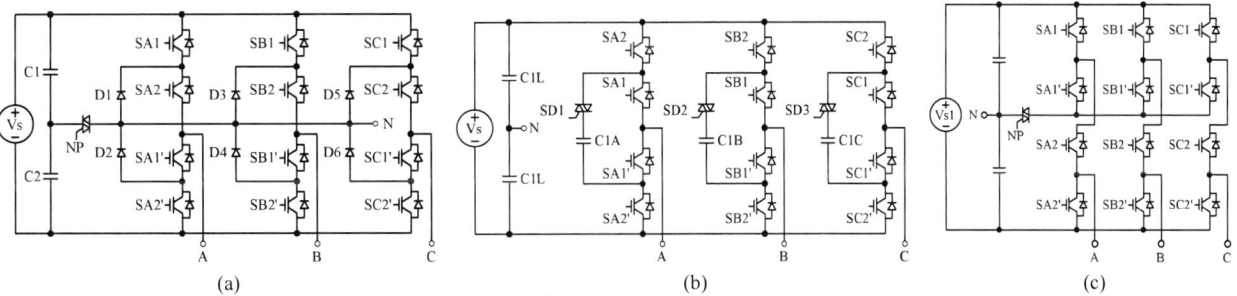

(a)　　　　　　　　　　　(b)　　　　　　　　　　　(c)

Fig. 6. Modified three-phase (a) NPC, (b) flying-capacitor and (c) dc-link cascaded three-level inverters.

978-1-4244-4782-4/10 $26.00 © 2010 IEEE

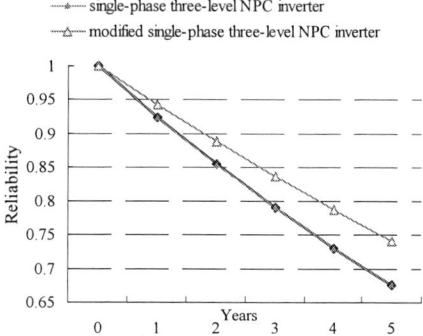

Fig. 7: Annual reliability indexes computed for different single-phase inverters over a span of 5 years.

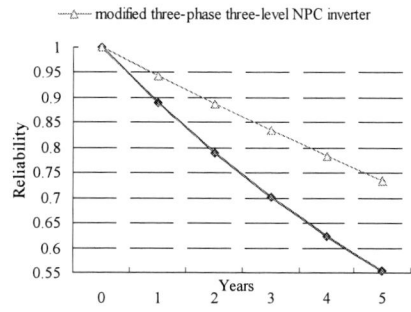

Fig. 8: Annual reliability indexes computed for different three-phase inverters over a span of 5 years.

0.6749 and 0.6747 in year 5 for single-phase inverters, respectively), inferring that using more switches in the NPC topology will not degrade the inverter overall reliability. That then means that whether a NPC circuitry should be used is determined by its overall cost, performance advantages gained, and not decided by its reliability consideration.

Comparing now with the modified NPC inverter, improvement in reliability is obvious with the modified inverter having a (0.73 – 0.55) / 0.55 = 32.7 % higher index in year 5, as compared to its traditional three-phase three-level counterpart.

For demonstrating that the modified NPC inverter indeed functions as demanded, Fig. 9 shows the simulated results with a short-circuit fault introduced to SA2 at 25 ms. Clearly, the line voltage waveform changes from five-level to three-level switching, while the phase voltage changes from three-level to two-level switching. In addition to that, the current waveform shows no significant dip in fundamental amplitude, inferring that sensitive loads can ride-through the fault smoothly. To further demonstrate that the modified NPC inverter performs as demanded practically, a laboratory prototype is currently being built (see Fig. 10), whose results, together with those of the other modified topologies, are at present not yet available, and will therefore be presented in another paper in the future.

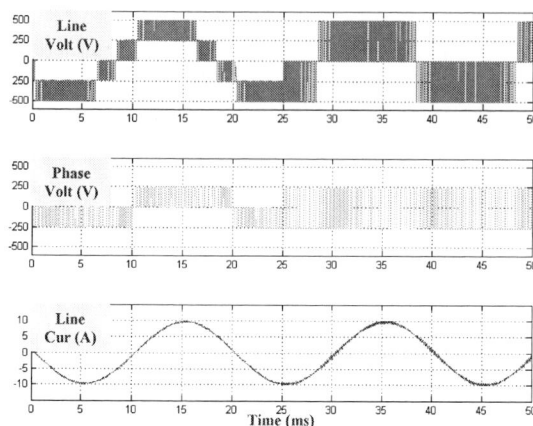

Fig. 9. Simulated waveforms illustrating the fault ride-through ability of the modified NPC inverter.

Fig. 10. Flexible experimental setup currently in construction for testing proposed modified three-level inverters.

Fig. 11 illustrates reliability indexes for traditional and modified three-phase flying-capacitor three-level inverters calculated over a span of five years. In year 5, the reliability index of the modified flying-capacitor inverter increased 26.0% comparing with the traditional counterpart.

The reliability indexes for traditional and modified three-phase dc-link cascaded three-level inverters are presented in Fig. 12. The modified dc-link cascaded inverter having higher index 14.4% in year 5, as compared to the traditional dc-link cascaded inverter.

Obviously the topological modifications of inverters can greatly increase their reliabilities. From Fig. 8, Figs. 10 and 11, we can observe that the modified three-phase NPC inverter can obtain the highest reliability comparing with other three-phase inverters.

VI. CONCLUSION

A method for assessing converter reliabilities is presented, and applied to different three-level dc-ac inverters. The computed reliability indexes then allow the inverters to be compared quantitatively, where it is specifically shown that the NPC inverter, using more switches, is not less reliable than the traditional two-level inverter. Where improvement in reliability is possible, suitable topological modifications are also proposed with their reliability indexes recalculated. From the calculations, it is clear that by adding an additional triac to

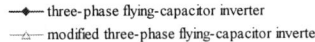

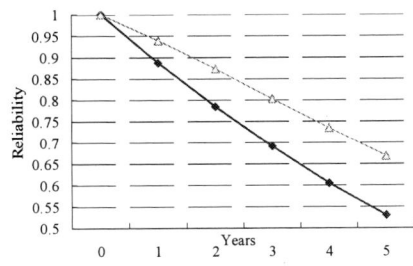

Fig. 11: Annual reliability indexes computed for three-phase flying-capacitor three-level inverters over a span of 5 years.

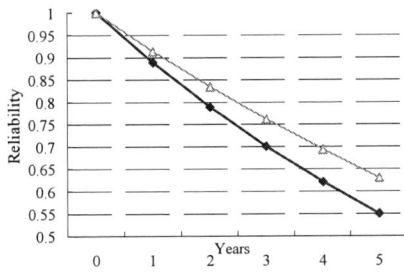

Fig. 12: Annual reliability indexes computed for three-phase dc-link cascaded three-level inverters over a span of 5 years.

certain three-level circuitries, their reliabilities can be raised greatly with that of NPC inverter raised by 32.7 %.

REFERENCES

[1] L. G. Franquelo, J. Rodríguez, J. I. Leon, S. Kouro, R. Portillo, and M. A. M. Prats, "The age of multilevel converters arrives", *IEEE Ind. Electron. Magazine*, pp. 28–39, Jun. 2008.

[2] R. L. A. Ribeiro, C. B. Jacobina, E. R. C. da Silva, and A. M. N. Lima, "A fault tolerant induction motor drive system by using a compensation strategy on the PWM–VSI topology", in *Proc. IEEE PESC'01*, 2001, pp. 1191–1196.

[3] M. B. R. Correa, C. B. Jacobina, E. R. C. Silva, and A. M. N. Lima, "An induction motor drive system with improved fault tolerance", *IEEE Trans. Ind. Appl.*, vol. 37, no. 3, pp. 873–879, May/Jun. 2001.

[4] S. Li and L. Xu, "Strategies of fault tolerant operation for three-level PWM inverters", *IEEE Trans. Power Electron.*, vol. 21, no. 4, pp. 933–940, Jul. 2006.

[5] F. Gao, P. C. Loh, F. Blaabjerg, and D. M. Vilathgamuwa, "Performance evaluation of three-level Z-source inverters under semiconductor-failure conditions", *IEEE Trans. Ind. Appl.*, vol. 45, no. 3, pp. 971–981, May/Jun. 2009.

[6] A. Lisnianski and G. Levitin, *Multi-State System Reliability Assessment, Optimization, Applications*, World Scientific, 2003.

[7] A. Ristow, M. Begovic, A. Pregelj, and A. Rohatgi, "Development of a methodology for improving photovoltaic inverter reliability", *IEEE Trans. Ind. Electron.*, vol. 55, no. 7, pp. 2581–2592, Jul. 2008.

[8] A. L. Julian and G. Oriti, "A comparison of redundant inverter topologies to improve voltage source inverter reliability", *IEEE Trans. Ind. Appl.*, vol. 43, no. 5, pp. 1371–1378, Sep/Oct. 2007.

[9] W. Kuo and M. J. Zuo, *Optimal Reliability Modeling Principles and Applications*, John Wiley & Sons, 2003.

[10] A. Lisnianski, "Extended block diagram method for a multi-state system reliability assessment", *Reliability Engineering and System Safety*, vol. 92, no. 12, pp.1601–1607, 2007.

[11] Military Handbook MIL-HDBK-217F, *Reliability Prediction of Electronic Equipment*, U.S. Dept. Defense, Washington, DC, Dec. 2, 1991.

Parallel operation of PWM inverters for high speed motor drive system

Un-Kwan Cho, Jung-Sik Yim, Seung-Ki Sul
School of Electrical Engineering and Computer Science
Seoul National University
Seoul, Korea
ukcho@eepel.snu.ac.kr, caesar@eepel.snu.ac.kr, sulsk@plaza.snu.ac.kr

Abstract— **This paper describes a topology with a parallel operation of PWM inverters for high speed motor drive systems. High speed motors have been widely used in industries to reduce system size and improve power conversion efficiency. However, the high speed motors sometimes suffer from core losses caused by PWM current ripples; noting that the phase inductance, L_s, of a high speed motor is smaller than that of ordinary motors, it is significant that the current ripple generated by a Pulse Width modulation(PWM) inverter becomes noticeable in the high speed motor. In the proposed topology, three PWM inverters are connected in parallel through nine coupled inductors. Compared to the PWM current ripple of the conventional single inverter system, that of the proposed scheme can be conspicuously reduced without the voltage drop at the inductors. In this paper, a theoretical analysis of the output voltage of the proposed topology is presented, and then the validity of the proposed method is verified by experimental results.**

I. INTRODUCTION

In some applications like turbo compressors and blowers, high speed motors have been used and it is directly coupled to the impeller. This structure has a number of merits compared to the conventional geared structure for high speed operation. It has smaller system size, better power conversion efficiency and lower maintenance cost due to the elimination of the gear box. Thanks to the advantages, the directly coupled structures with high speed motors become more popular in high power applications [1], [2] and their validities have been proved for years.

The high speed motors need higher fundamental frequency than ordinary motors. Since, however, the switching frequency of high power inverters is limited, parallel operations are needed in the high speed and high power applications. Besides, the high speed motors sometimes suffer from core losses caused by Pulse Width Modulation(PWM) current ripples; noting that the phase inductance, L_s, of high speed motors is smaller than that of ordinary motors, and the PWM current ripple of the high speed motor becomes larger [3].

Moreover, this problem becomes severe in high power applications. In order to reduce the current ripple, additional inductors can be connected between the inverter output and the motor terminal. In this configuration, however, voltage drop in the additional

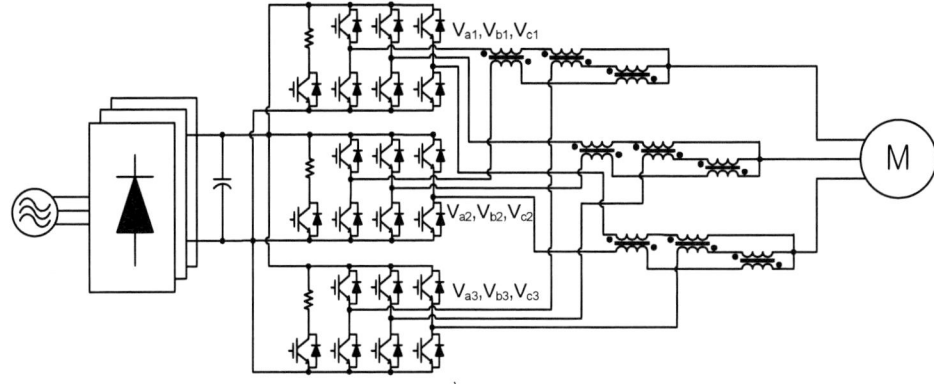

Fig.1 Proposed topology of parallel inverter with coupled inductor

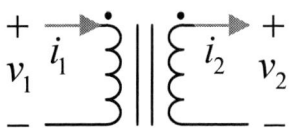

Fig. 2 Schematic diagram of coupled inductor

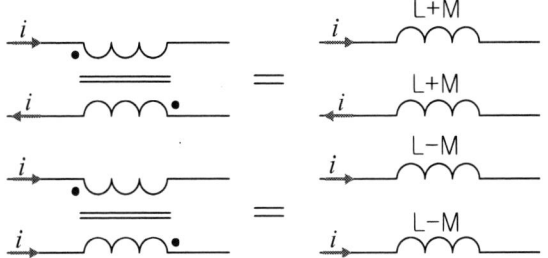

Fig. 3 Effective inductance of coupled inductor

inductors may be a problem.

In this paper, a topology for the parallel operation of PWM inverters is proposed to reduce the PWM current ripple. The proposed topology is shown in Fig. 1, where three PWM inverters are connected in parallel through nine coupled inductors. The diode rectifier in the figure may consists of many units to handle current capacity of the drive system In the proposed topology, the parallel inverters do the interleaving operation; the PWM carriers of the parallel inverters are shifted by 1/3 of the switching period away from each other. Thanks to the interleaving operation, the PWM current ripple of the motor can be conspicuously reduced. Moreover, because of the common mode operation of the coupled inductor, the fundamental currents of the machine are affected only by its leakage inductance, and the voltage drops of the coupled inductor are negligible while the coupled inductors reduce the circulating currents caused by the interleaving operation [4]. This is because the magnitude of the circulating currents depends on the differential mode inductance of the coupled inductor. A theoretical analysis of the output voltage of the proposed topology is also presented in this paper.

II. COUPLED INDUCTOR

Fig. 2 shows the schematic diagram of coupled inductor. v_1, i_1 and v_2, i_2 represent the voltage and current of first and secondary winding of coupled inductor and voltage equations can be described as (1).

$$v_1 = L\frac{di_1}{dt} - M\frac{di_2}{dt}, \quad v_2 = L\frac{di_2}{dt} - M\frac{di_i}{dt} \quad (1)$$

where L and M mean the self and mutual inductance of the coupled inductor, respectively. From this equation,

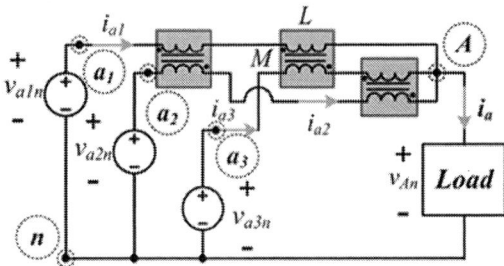

Fig. 4 Equivalent circuit

the effective inductance of the coupled inductor according to the direction of the flow of current can be derived.

As shown in Fig. 3, when currents flow in the opposite direction, the coupled inductor has the equivalent inductance as the sum of self and mutual inductance. On the other hand, the directions of current flow of the first and second winding are same, the coupled inductor equivalently act as leakage inductance which means the difference between the self inductance and mutual inductance. Using these characteristics of the coupled inductors, the proposed topology can minimize the effect of inductance for the fundamental current component, and maximize the effect of inductance for the ripple component of the circulation current due to interleaving operation.

III. ANALYSIS WITH EQUIVALENT CIRCUIT

A. Small signal analysis

Fig. 4 shows the per phase equivalent circuit of the proposed scheme. v_{a1n}, v_{a2n} and v_{a3n} mean pole voltages of the parallel inverters, and v_{An} means the equivalent pole voltage to the load. The relations between the pole voltages and output currents of inverters, i_{a1}, i_{a2}, and i_{a3} can be expressed as follows.

$$v_{a1n} - v_{An} = 2L\frac{di_{a1}}{dt} - M\frac{di_{a2}}{dt} - M\frac{di_{a3}}{dt},$$

$$v_{a2n} - v_{An} = 2L\frac{di_{a2}}{dt} - M\frac{di_{a1}}{dt} - M\frac{di_{a3}}{dt}, \quad (2)$$

$$v_{a3n} - v_{An} = 2L\frac{di_{a3}}{dt} - M\frac{di_{a1}}{dt} - M\frac{di_{a2}}{dt}.$$

Since the voltages and the currents have two frequency components, which are a fundamental frequency of the motor and a PWM switching frequency of each inverter, (2) can be rewritten as follows.

$$\hat{v}_{a1n} + \tilde{v}_{a1n} - \left(\hat{v}_{An} + \tilde{v}_{An}\right)$$
$$= 2L\frac{d}{dt}(\hat{i}_{a1} + \tilde{i}_{a1}) - M\frac{d}{dt}(\hat{i}_{a2} + \tilde{i}_{a2}) - M\frac{d}{dt}(\hat{i}_{a3} + \tilde{i}_{a3}),$$

$$\hat{v}_{a2n} + \tilde{v}_{a2n} - \left(\hat{v}_{An} + \tilde{v}_{An}\right)$$
$$= 2L\frac{d}{dt}(\hat{i}_{a2} + \tilde{i}_{a2}) - M\frac{d}{dt}(\hat{i}_{a1} + \tilde{i}_{a1}) - M\frac{d}{dt}(\hat{i}_{a3} + \tilde{i}_{a3}), \qquad (3)$$

$$\hat{v}_{a3n} + \tilde{v}_{a3n} - \left(\hat{v}_{An} + \tilde{v}_{An}\right)$$
$$= 2L\frac{d}{dt}(\hat{i}_{a3} + \tilde{i}_{a3}) - M\frac{d}{dt}(\hat{i}_{a1} + \tilde{i}_{a1}) - M\frac{d}{dt}(\hat{i}_{a2} + \tilde{i}_{a2})$$

where '^' means the components of the fundamental frequency and '~' means those of the PWM frequency. Considering two frequencies, the equations of the fundamental and PWM frequency component can be derived as (4) and (5), respectively.

$$\hat{v}_{a1n} - \hat{v}_{An} = 2L\frac{d\hat{i}_{a1}}{dt} - M\frac{d\hat{i}_{a2}}{dt} - M\frac{d\hat{i}_{a3}}{dt},$$

$$\hat{v}_{a2n} - \hat{v}_{An} = 2L\frac{d\hat{i}_{a2}}{dt} - M\frac{d\hat{i}_{a1}}{dt} - M\frac{d\hat{i}_{a3}}{dt}, \qquad (4)$$

$$\hat{v}_{a3n} - \hat{v}_{An} = 2L\frac{d\hat{i}_{a3}}{dt} - M\frac{d\hat{i}_{a1}}{dt} - M\frac{d\hat{i}_{a2}}{dt}.$$

$$\tilde{v}_{a1n} - \tilde{v}_{An} = 2L\frac{d\tilde{i}_{a1}}{dt} - M\frac{d\tilde{i}_{a2}}{dt} - M\frac{d\tilde{i}_{a3}}{dt},$$

$$\tilde{v}_{a2n} - \tilde{v}_{An} = 2L\frac{d\tilde{i}_{a2}}{dt} - M\frac{d\tilde{i}_{a1}}{dt} - M\frac{d\tilde{i}_{a3}}{dt}, \qquad (5)$$

$$\tilde{v}_{a3n} - \tilde{v}_{An} = 2L\frac{d\tilde{i}_{a3}}{dt} - M\frac{d\tilde{i}_{a1}}{dt} - M\frac{d\tilde{i}_{a2}}{dt}.$$

B. common mode operation of coulpled inductor

Since all inverters have same voltage references to regulate output current and the impedances of the coupled inductors are balanced, voltages and currents equations of the fundamental component can be described as (6) and (7). From them, (4) can be rewritten as (8).

$$\hat{v}_{a1n} = \hat{v}_{a2n} = \hat{v}_{a3n}. \qquad (6)$$

$$\hat{i}_{a1} = \hat{i}_{a2} = \hat{i}_{a3} = \frac{1}{3}\hat{i}_a. \qquad (7)$$

$$\hat{v}_{a1n} - \hat{v}_{An} = \frac{2}{3}(L - M)\frac{d\hat{i}_a}{dt}. \qquad (8)$$

Generally, the leakage inductance of a coupled inductor is much smaller than the mutual inductance. So,

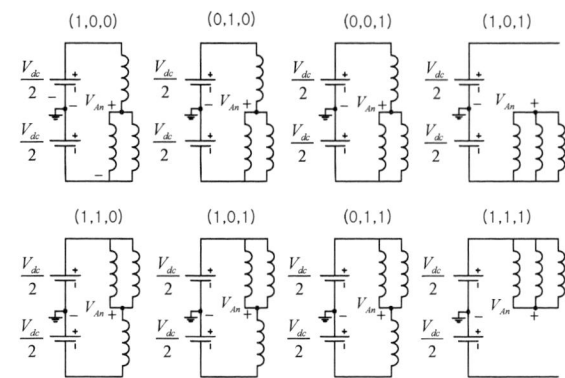

Fig. 5 Equivalent circuit according to switching state

(8) means that the voltage drops of the fundamental frequency currents on the coupled inductors would be negligible .And, it is possible to reduce the current ripples without fundamental component voltage drop by the coupled inductors.

C. Differential mode operation of coupled inductor

Instantaneous pole voltage differences due to the interleaving operation of the paralleled inverter result in circulating currents through the inverters. And, the common dc-link are return path of these circulating currents. The voltage differences can be expressed using (5) as follows.

$$\tilde{v}_{a1n} - \tilde{v}_{a2n} = \left(2L + M\right)\frac{d}{dt}\left(\tilde{i}_{a1} - \tilde{i}_{a2}\right). \qquad (9)$$

The circulating current can be suppressed by a coupled inductor whose equivalent inductance is (2L+M). And, the circulating current can be reduced conspicuously.

D. Output voltage with interleaving operation

In a single inverter system, only two kinds of pole voltages, $-V_{dc}/2$ and $V_{dc}/2$ are available. In the proposed topology with three inverters, however, the pole voltage has four voltage levels due to the parallel operation of inverters under the assumption that the impedances of the coupled inductors are balanced. Fig. 5 shows per phase equivalent circuits of the proposed topology according to switching states, and the output pole voltage of each state are listed as in Table I.

Four kinds of pole voltages, $-V_{dc}/2$, $-V_{dc}/6$, $V_{dc}/6$, $V_{dc}/2$ can be used to synthesize the output voltage using the proposed topology. Combining these pole voltages, output line-line voltage, V_{AB} has seven voltage levels, namely, $-V_{dc}$, $-2V_{dc}/3$, $-V_{dc}/3$, 0, $V_{dc}/3$, $2V_{dc}/3$, V_{dc}.

Table I Output pole voltage

State	S_{a1}, S_{a2}, S_{a3}	V_{An}
0	(0, 0, 0)	$-V_{dc}/2$
1	(0, 0, 1)	$-V_{dc}/6$
2	(0, 1, 0)	$-V_{dc}/6$
3	(1, 0, 0)	$-V_{dc}/6$
4	(0, 1, 1)	$V_{dc}/6$
5	(1, 0, 1)	$V_{dc}/6$
6	(1, 1, 0)	$V_{dc}/6$
7	(1, 1, 1)	$V_{dc}/2$

Table II Parameters of coupled inductor

Relative permeability of core	125
Number of turns	12
Self inductance	56.0 µH
Mutual inductance	55.5 µH

Therefore, the output voltage of the proposed topology could have much less harmonic contents compared to harmonics of the single inverter. In result, the proposed method can reduce current ripples of the output current furthermore.

IV. EXPERIMENTAL RESULTS

To verify the feasibility of the proposed method, some experiments were carried out. The experimental configuration is shown as Fig. 6. Fig. 6(a) shows a proto type coupled inductor with the high-flux core material and Litz wire. The parameters of the inductor measured by an RLC meter are listed as Table II. It can be noticed that the leakage inductance is less than 1 % of the self inductance of the coupled inductor.

Table III Parameters of induction machine

Rated power	22 kW
Rated voltage	220 V
Rated current	74.6 A
Rated speed	1765 r/min
Pole	4

Table IV Operating condition

Input voltage(line-to-line rms)	220 V
Operating speed	1,000 r/min
Switching frequency	10 kHz
Sampling frequency	20 kHz
Output power	3 kW

With these coupled inductors, a prototype inverter system was implemented as shown in Fig. 6(b). The system consisted with three inverters, and three diode rectifier units connected in parallel and nine coupled inductors according to the proposed circuit arrangement as shown in Fig. 1.

An induction motor was driven with the proposed inverter system and a DC machine was used as a load. Parameters of the induction machine are listed in Table III. Operating condition of the induction machine is shown as in Table IV.

Under this condition, experiments with two kinds of inverter systems were performed. One is a single inverter system and the other is the proposed one. The phase currents of the induction machine and their FFT (Fast Fourier Transform) results of two cases were demonstrated as in Fig. 7(a) and (b), respectively. When the induction machine was driven with the single PWM inverter, the phase current of the motor has some

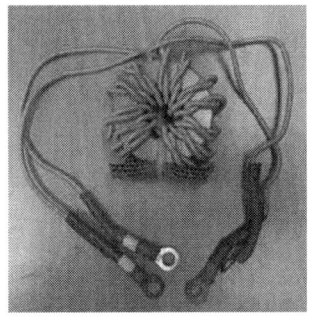

Coupled Inductor

dc-link Capacitor

Diode Rectifier

(a)   (b)

Fig. 6 Experimental setup;
(a) coupled inductor and (b) proto type inverter system

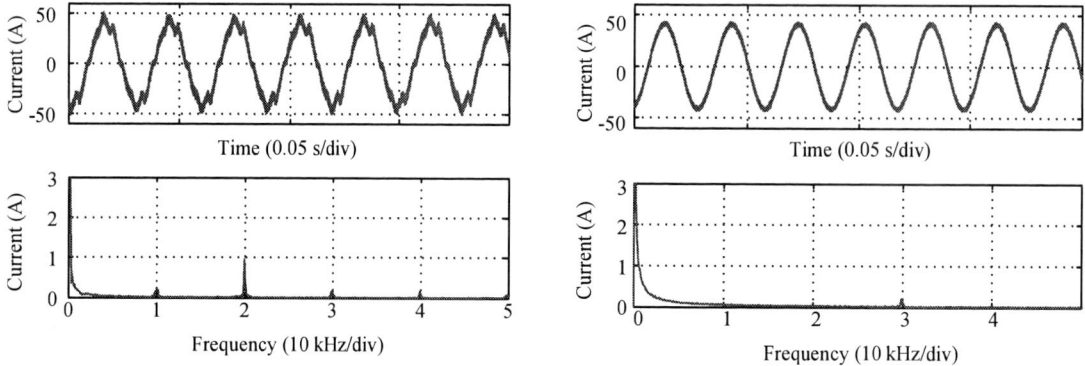

Fig. 7. Motor phase current and FFT waveforms
(a) wth single PWM inverter, (b) with proposed topology

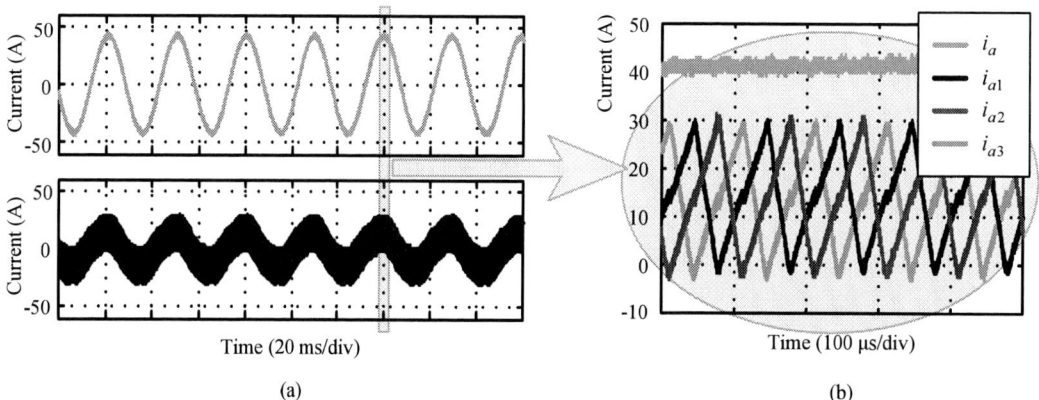

(a)   (b)

Fig. 8. Current waveforms of proposed topology

(a) waveforms of motor phase current, i_a, and inverter output current i_{a1}

(b) zoom-in waveforms of Fig. 8 (a) including i_{a2} and i_{a2}

distortions and the FFT waveform shows that ripple currents is quite large.

When the proposed topology is used to drive induction machine under the same condition as the previous one, the phase current of the motor are shown as shown Fig. 7(b). The current waveform has no distortion and its FFT result shows that PWM frequency component of the phase current is remarkably reduced. This experimental result clearly demonstrate that the proposed topology has advantage in reducing PWM current ripple compared over a single PWM inverter system.

The waveforms of the motor phase current, i_a, and the output currents of each parallel inverters, i_{a1}, i_{a2} and i_{a3} are displayed in Fig. 8(a). And, Fig. 8(b) shows the zoom-in waveform of the Fig. 8(a). The output currents of three inverters have the ripple of about 30 A and waveforms of currents are shifted 1/3 of the

switching frequency away from each other due to interleaving operation. Thanks to the interleaving operation, the resultant motor current, which is the sum of output currents of three inverters, has significantly reduced ripples.

Fig. 9 shows the waveforms of the circulating currents, which flow through the inverters. Fig. 9(a) and (b) show that the current when output powers are 0 kW and 3 kW, respectively. Both of them show similar waveforms regardless of output powers. This is because the currents come from the instantaneous voltage difference between inverters and it does not depend on the output power. The magnitude of the circulating currents can be estimated by (9) and expressed as follows.

$$i_{cir} = \frac{V_{dc}}{2\pi f_{sw}(2L + M)} \approx 31.1 \qquad (10)$$

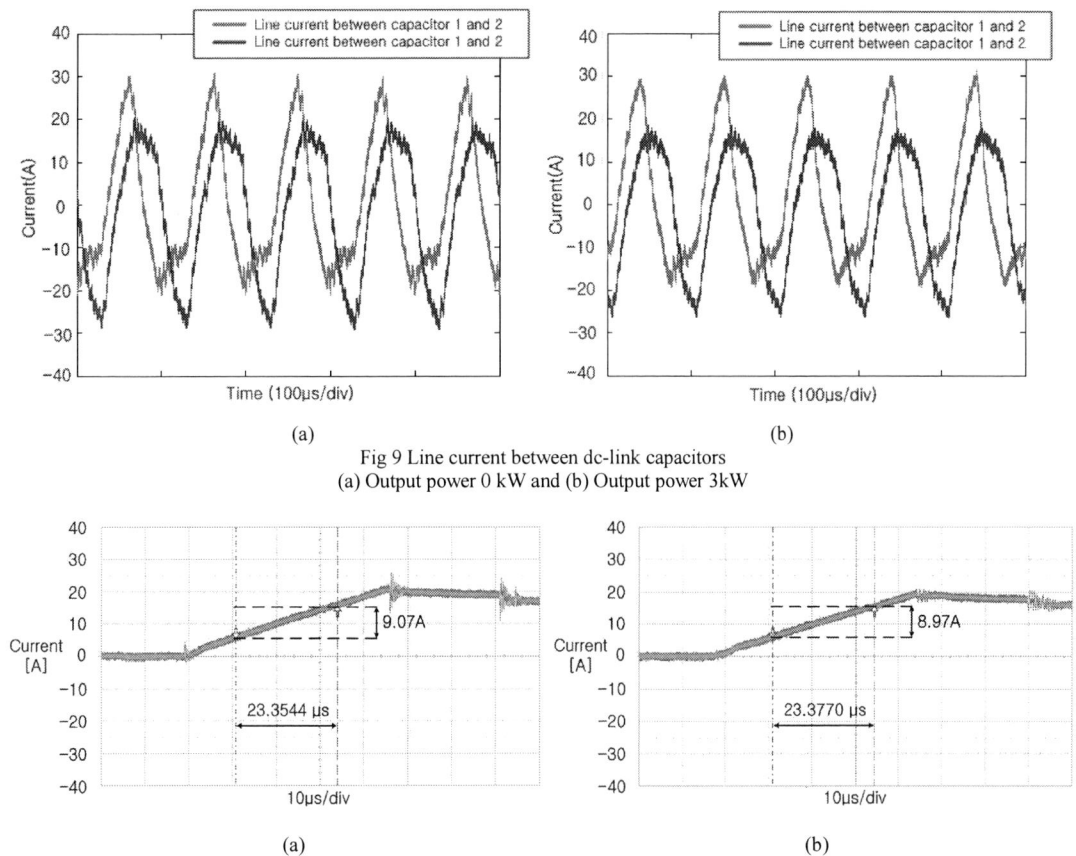

(a)   (b)

Fig 9 Line current between dc-link capacitors
(a) Output power 0 kW and (b) Output power 3kW

(a)   (b)

Fig. 10. Current waveform of the motor when apply'ing a step voltage;
(a) with single PWM inverter and (b) with proposed topology

Table V Measured inductance

	Single inverter	Proposed topology
Δt	23.4 μs	23.4μs
Δi	9.1 A	9.0 A
L_σ	514 μH	520 μH

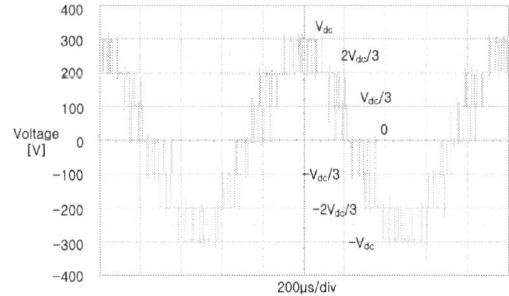

Fig. 11 Output line voltage of proposed topology

From Fig 9, it can be seen that the magnitude of the circulating current is almost same to the estimated value by (10). This means that the magnitude of the circulating currents is determined based on the voltage equation of (9) and the currents can be reduced by the value of the coupled inductors in (9).

In order to evaluate the voltage drop on the additional coupled inductors, the output inductance of the load including coupled inductors was measured with the single PWM inverter and the proposed scheme. The inductance was measured by applying a pulse voltage to the input terminal of motor including coupled inductors and measuring the output current as shown in Fig. 10.

The measured inductances with two systems are listed in Table V.

The difference of the measured inductance between two cases is about 6 μH. The difference may come from leakage inductances of the coupled inductors and parasitic inductances of wiring cables. The additional inductance due to the coupled inductor is negligible. Therefore, it can be concluded that the proposed method can reduce the ripple current with negligible voltage

978-1-4244-4782-4/10 $26.00 © 2010 IEEE

drop.

Fig. 11 shows the output line-line voltage, V_{AB} when the reference is a sine wave whose magnitude is 150 V and frequency is 1 kHz. As mentioned before, the proposed topology has seven levels of line-line output voltage, and the output current ripple can be reduced furthermore by these multilevel output voltages.

V. CONCLUSIONS

This paper describes a topology with a parallel operation of PWM inverters for high speed motor drive systems. In the proposed topology, three PWM inverters are connected in parallel through nine coupled inductors. Thanks to the interleaving operation of parallel inverters, the current ripples are remarkably reduced compared to that of single PWM inverter systems. On the other hand, the voltage drops on the coupled inductor are negligible due to the common mode operation of the coupled inductor. And, the coupled inductors also reduce the circulating currents caused by the interleaving operation. The output voltage of the proposed topology is analyzed and the feasibility is also verified experimentally.

REFERENCES

[1] W. L. Soong et al., "Novel high-speed induction motor for a commercial centrifugal compressor," *IEEE Transactions of Inductry Application*, May/Jun. 2000.

[2] M. Mekhiche, J. L. Kirtley, M. Tolikas, E. Ognibene, J. Kiley, E. Holmansky, and F. Nimblett, "High speed motor drive development for industrial applications," in *Conf. Rec. IEMD'99*, 1999

[3] Bon-Ho Bae et al., "Implemetation of Sensorless Vector Control for Super-high-Speed PMSM of Turbo-Compressor," *IEEE Transaction on Industry Application,* May/June 2003.

[4] Chen Liangliang, Xiao Lan, Hu Wenbin, Yan Yangguang, "Application of coupled inductors in parallel inverter system," Electrical Machines and Systems, 2003. ICEMS 2003. Sixth International Conference

[5] Itkonen, T. Luukko, J. Laakkonen, T. Silventoinen, P. and Pyrhonen, O, "Switching effects in directly paralleled three phase AC/DC/AC converters with separate DC links" Power Electroincs Specialists Conference, 2008. PESC 2008. IEEE June 2008

Reverse Conduction of a 100 A SiC DMOSFET Module in High-Power Applications

R. A. Wood
U. S. Army Research Lab
2800 Powder Mill Road
Adelphi, MD, USA

D. P. Urciuoli
U. S. Army Research Lab
2800 Powder Mill Road
Adelphi, MD, USA

T. E. Salem
U. S. Naval Academy
105 Maryland Ave
Annapolis, MD 21402 USA

R. Green
U. S. Army Research Lab
2800 Powder Mill Road
Adelphi, MD, USA

Abstract - **Numerous research efforts over the past few years have documented the enhanced capabilities that Silicon Carbide (SiC) offers over Silicon based power electronic devices. Additional research work has led to vast improvements in the manufacturing of SiC based components. As a result, SiC power electronic components, primarily diodes, are now readily available and this technology promises to have widespread market impact as more complex device structures are commercially realized. Recently, the development of a 1200 V 50 A SiC DMOSFET device and its use in a 100 A power module has been documented [1]. This paper extends that research work to report on the reverse conduction characteristics of the SiC DMOSFET and the system-level benefits for high-power applications that can be achieved by operating these devices in this manner. Experimental data is presented on the 100 A module consisting of two, 50 A SiC DMOSFETs and two, 50 A SiC JBS anti-parallel free-wheeling diodes used in a high-power bi-directional DC-DC converter during buck mode operation.**

I. INTRODUCTION

The design and development of electric and hybrid electric vehicles has been widely researched and documented, for example [2-4]. These vehicles afford unique challenges for power electronic systems such as reliable operation in a high-temperature environment while minimizing mass and volume [3]. In these vehicles, power electronics are used in inverter applications to directly drive the traction machines, and in DC-DC converters to efficiently manage power transfers between the energy storage system and the electric loads. This latter function of managing power transfer is typically realized using a bi-directional DC-DC converter [5-7]. Due to the power electronic devices used, most bi-

directional converter topologies do not use a complementary switching scheme, which is a system-level inefficiency. A thorough overview on the specific challenges for bi-directional converters has recently been presented in which the need for suitable switching devices was highlighted [8-9]. To address these issues, a 50 A SiC DMOSFET has been developed and its use in a 400 A all SiC phase leg module has been proposed [10]. This work reports on the system-level efficiency improvements achieved by using the reverse conduction capabilities of the SiC DMOSFET. Both experimental and simulation data is presented for a 100 A module consisting of two, 50 A SiC DMOSFETs and two, 50 A SiC JBS anti-parallel free-wheeling diodes used in a high-power bi-directional DC-DC converter.

II. SiC DMOSFET MODULE

The SiC DMOSFET, shown in Fig. 1, is a vertically structured device that can conduct very large currents and block very high voltages during high-temperature operation. During forward operation, electrons flow laterally from the source through the inverted p-well and into the JFET region and then vertically into the drift layer and out through the drain contact. When the drain voltage polarity is negative, an intrinsic parasitic body diode formed by the p-n junction of the p-well and the n⁻ drift layer is forward-biased and conducts current. Reverse current conduction through the SiC DMOSFET is possible when the gate of the SiC DMOSFET is turned-on while the drain voltage polarity is negative. During reverse conduction, electron flow is from drain-to-source via the same path as described for forward

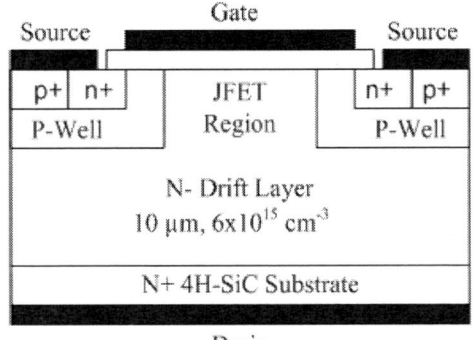

Figure 1: Simplified 1200 V SiC DMOSFET.

Figure 2: Picture of the 100 A SiC Module in the Bi-Directional Converter.

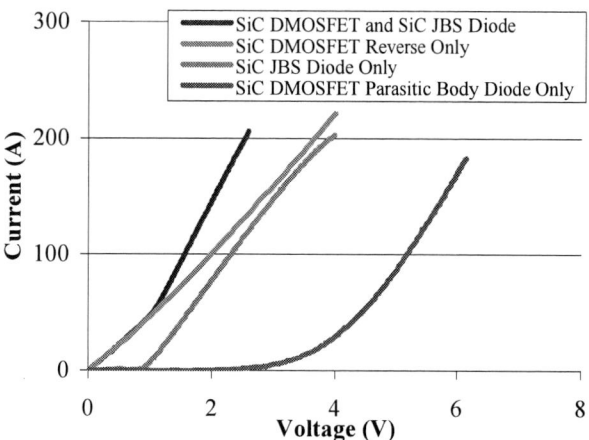

Figure 3: I-V Characteristics for the 100 A SiC Power Module at 25 °C.

conduction. Operating a MOSFET in reverse conduction while in a circuit is referred to as synchronous rectification and has been recently documented for a low-power SiC transistor [11-12].

The previously described 100 A SiC module shown in Fig. 2 had its current-voltage (I-V) characteristic curves measured by a Tektronix 371B high-power curve tracer. Fig. 3 shows the conduction characteristics of the SiC DMOSFET body diodes with a gate voltage of –10 V (DMOSFET turned off), the SiC JBS diodes only, and both the SiC JBS diodes and the SiC DMOSFETs turned on with a gate voltage of 15 V operating in reverse conduction. The gradual turn-on characteristic of the SiC DMOSFET body diode in comparison to the SiC JBS diode indicates that most of the current will flow through the SiC JBS diode path when the SiC DMOSFET is turned off. However, with the SiC DMOSFET turned on, the I-V characteristic indicates that at lower currents the SiC DMOSFET will have all of the flow until the SiC JBS diodes conduct at a reverse voltage of approximately 1.0 V. At each operating condition shown in Fig. 3, the product of voltage and current results in power loss in the device, and it is apparent that the minimum power loss always occurs when both the SiC MOSFET and the SiC JBS diode are available to conduct.

III. TEST PROCEDURE

Fig. 4 shows a circuit for a bi-directional converter that can operate either in boost or buck mode. Both modes produce similar waveforms and losses on the switching devices. This circuit was operated in buck mode with a commercial IGBT as switch S_1 and the 100 A SiC DMOSFET module as switch S_2. To demonstrate efficiency gains possible through synchronous rectification, S_2 was operated in two different switching schemes. First, S_2 was turned off and the SiC JBS diode conducted all of the S_2 current. Then a complimentary switching scheme was utilized which allowed reverse conduction in the SiC DMOSFET while the SiC JBS Diode was also conducting. In this second scheme, a dead time of 4.8 µs between the gating of S_1 and S_2 prevented shoot-through of the high side power supply (V_{HS}).

The experimental bi-directional circuit was operated at a switching frequency of 10 kHz and pulse widths for S_1 and S_2 were 45.2 µs. Gate signals for both switches were 14.7 V and –7.6 V for turn on and turn off, respectively. The magnitude of V_{HS} was varied to adjust the amount of current and power that flowed through the circuit. The SiC DMOSFET module (S_2) was mounted on a liquid-cooled heat exchanger and was cooled by propylene glycol and water.

As shown in Fig. 5, thermal images of the SiC DMOSFETs and the SiC JBS diodes under test were taken to determine the amount of power loss for each device. This was accomplished by first characterizing the temperature rise of the module components during DC operation for which voltage drop and current directly result in thermal power. Then, during operation of the module in the bi-directional converter, temperature rise measured by the infrared camera was correlated to the DC data to obtain the thermal power loss in each device.

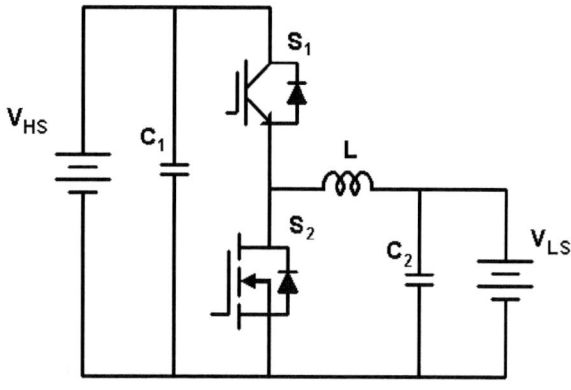

Figure 4: Bi-Directional Converter Circuit Used for Reverse Conduction of SiC DMOSFET.

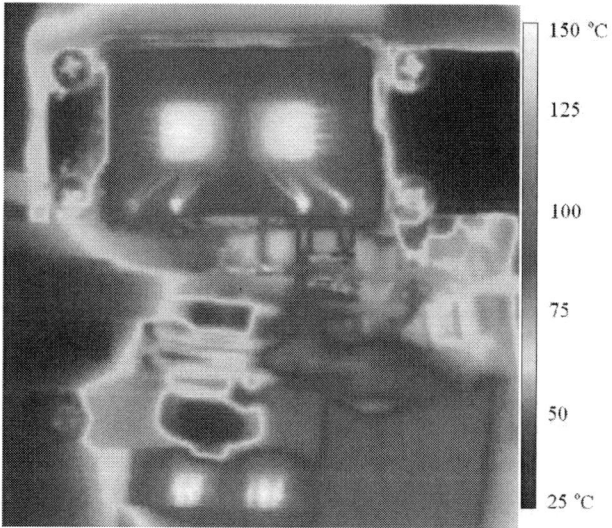

Figure 5: Thermal Image of Module at 23 kW Operation and 80 °C Coolant MOSFETs on Top, Diodes on Bottom

978-1-4244-4782-4/10 $26.00 © 2010 IEEE

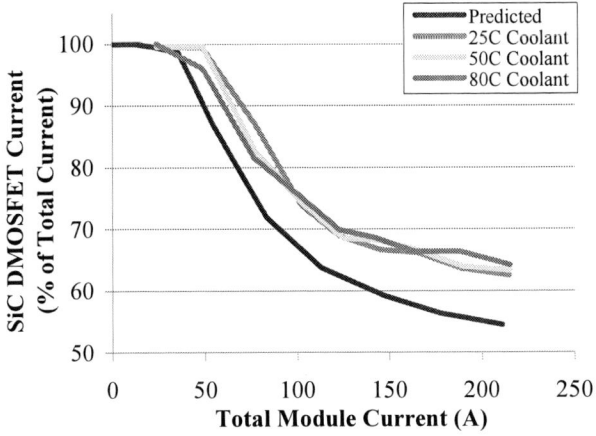

Figure 6: Current Sharing Between Devices During Reverse Conduction.

IV. SIMULATION RESULTS

To predict module performance during operation, current sharing among the devices needs to be known. Utilizing the I-V characteristics of the devices shown in Fig. 3, current sharing among the devices was predicted for different module currents. Shown in Fig. 6 is the predicted sharing curve between the MOSFETs and the diodes along with experimental data obtained from the bi-directional converter at three coolant temperatures. The I-V curve indicates that the MOSFET will conduct 100% of the module current below 40 A. At higher currents the MOSFET conducts 55% of the module current. The experimental MOSFET current was measured directly and showed that the MOSFET conducted a slightly higher current proportion (60%) than predicted. Furthermore, current sharing was found to be independent of the coolant temperature.

Next, the bi-directional converter circuit was simulated using the predicted current sharing to determine the losses in the MOSFET and the diode. There is minimal switching loss in the MOSFET since the device voltage is very small during MOSFET transitions. Therefore, the loss in each device was determined by estimating the conduction loss of the devices

during typical buck operation when operating as the free-wheeling path.

These simulation results demonstrated two distinct benefits for employing the MOSFET in reverse conduction. The first was an increase in efficiency as the loss experienced in the switch was estimated to decrease 40-50% when exercising the MOSFET in reverse. The second benefit was additional output power capability of the converter. When only using the JBS diode as free-wheeling path, the output power of the converter was limited to 7.4 kW due to high diode temperatures. When the MOSFET is conducting in reverse the output power could be safely increased three-fold to 22.5 kW.

V. EXPERIMENTAL RESULTS

Experiments using the bi-directional converter with the diode only and both the diode and MOSFET in reverse conduction were done using coolant loop temperatures of 25 °C, 50 °C, and 80 °C. The input voltage to the converter (V_{HS}) was varied from 30 V to 260 V which produced S_2 module currents from 15 A_{RMS} to 138 A_{RMS} and resulted in output powers up to 22.5 kW. When operating in diode only mode, the output power was limited to 7.4 kW because the diode temperature would have exceeded the maximum allowable junction temperature of 200 °C at the next higher operating condition.

Fig. 7 shows the current waveforms for the maximum operating point while using the diode only for free-wheeling. The power in this case was limited by the diode temperature to 7.4 kW. One item to note is that current flowed through the SiC DMOSFET even when the device was off due to the parasitic body diode. The body diode initially took a portion of the module current, and then settled to a lower value after the SiC JBS diode was sufficiently biased.

Fig. 8 shows the current waveforms for the maximum operating point with both the JBS diode and MOSFET conducting which produced a peak module current of approximately 250 A. The inductor current increased while S_1 conducted, after which the SiC JBS diode turned on and conducted most of the current during the switching dead time

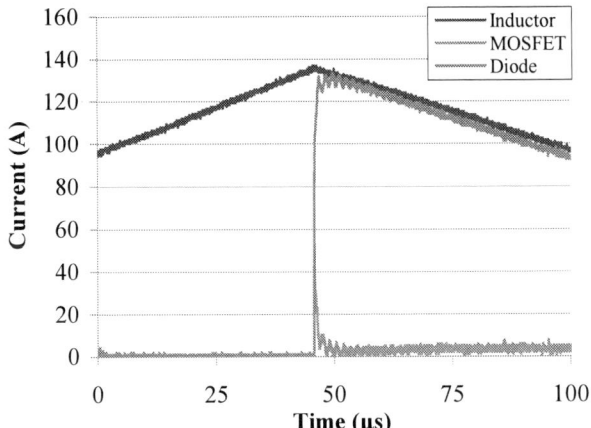

Figure 7: Current Waveforms at 7.4 kW Operation without Reverse Conduction and 80 °C Coolant.

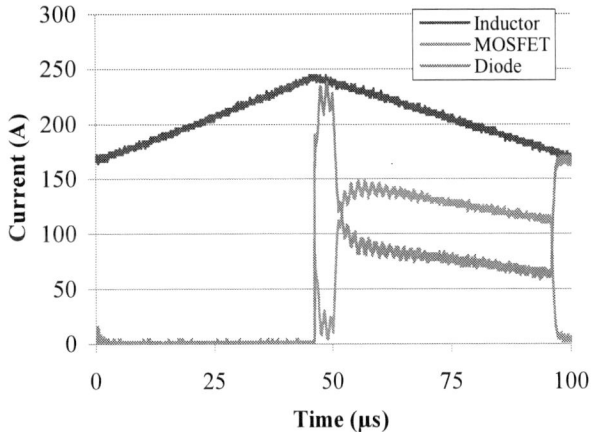

Figure 8: Current Waveforms at 22.5 kW Operation with Reverse Conduction and 80 °C Coolant.

TABLE I. EXPERIMENTAL RESULTS AT THREE COOLANT TEMPERATURES SHOWING DEVICE LOSSES WHILE UTILIZING THE MOSFET IN REVERSE AND USING THE DIODE ONLY

Output Power (kW)	Module Current (A_{RMS})	25°C Coolant			50°C Coolant			80°C Coolant		
		MOSFET & DIODE Loss (W)	Diode Only Loss (W)	Reduction in Loss (%)	MOSFET & Diode Loss (W)	Diode Only Loss (W)	Reduction in Loss (%)	MOSFET & Diode Loss (W)	Diode Only Loss (W)	Reduction in Loss (%)
1.1	30.9	26.9	33.8	20.4	25.0	31.9	21.7	27.6	35.2	21.7
2.9	49.5	46.0	61.8	25.6	45.1	62.6	28.0	48.2	70.6	31.7
5.1	65.3	66.8	101.6	34.2	65.3	106.2	38.5	71.8	122.2	41.2
7.4	78.8	87.5	154.8	43.5	87.0	165.7	47.5	96.9	195.4	50.4
10.1	92.6	113.0	Not attainable due to excessive diode temperatures		113.4	Not attainable due to excessive diode temperatures		126.4	Not attainable due to excessive diode temperatures	
13.2	106.0	144.3			145.9			163.9		
17.6	121.8	190.8			193.2			221.1		
22.5	137.5	248.7			256.1			293.6		

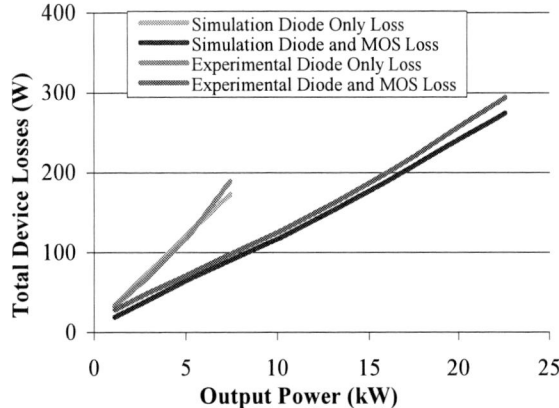

Figure 9: Device Loss vs. Output Power at 80 °C Coolant.

of 4.8 µs. The MOSFET current showed similar characteristics as before, taking some of the initial current spike and then settling to a low value as the body diode shared current with the JBS diode. After the dead time, S_2 turned on and current was shared between the SiC JBS diode and the SiC DMOSFET which was in reverse conduction.

Fig. 9 shows both the experimental and simulation data for the device losses at each operating point while operating at 80 °C coolant. As predicted by the simulation, losses in the module were reduced by 50% at higher power levels, although benefits at lower power levels are not as significant. Coolant temperature does have some effect on efficiency gain as observed in Table I, which summarizes the results from all of the experiments. Maximum benefits for reverse conduction occur at higher coolant temperatures.

VI. CONCLUSION

This paper documents a two-fold system-level benefit achieved by utilizing the reverse conduction mode of operation of a 100 A SiC DMOSFET module. The first benefit, decreased device losses, was shown through a bi-directional converter utilizing a complementary switching scheme. Experimental and simulation results demonstrated reduced power loss in the free-wheeling path by 30-50% achieved by a reverse conduction scheme. The second

benefit was an increase in output power capability achieved by reducing device temperatures. The output power capability of the experimental module was increased from 7.4 kW while in diode only operation, to 22.5 kW while utilizing the reverse conduction ability of the MOSFET.

REFERENCES

[1] T.E. Salem, D.P. Urciuoli, R. Green, and G.K. Ovrebo, "High-Temperature High-Power Operation of a 100 A SiC DMOSFET Module," *24th Annual IEEE Applied Power Electronics Conference and Exposition*, 2009, pp. 653-657.

[2] J. M. Miller, "Power Electronics in Hybrid Electric Vehicle Applications," *18th Annual IEEE Applied Power Electronics Conference and Exposition*, 2003, Vol. 1, pp. 23-29.

[3] A. Emadi, Y. J. Lee, and K. Rajashekara, "Power Electronics and Motor Drives in Electric, Hybrid Electric, and Plug-In Electric Vehicles," *IEEE Transactions on Idustrial Electronics*, vol. 55, no. 6, pp. 2237-2245, June 2008.

[4] Hybrid Electric, and Plug-In Hybrid Electric Vehicles," *IEEE Transactions on Industrial Electronics*, Vol. 55, No. 6, 2008, pp. 2237-2245.

[5] H. Zhang, L. M. Tolbert, and B. Ozpineci, "Impact of SiC Devices on Hybrid Electric and Plug-in Hybrid Electric Vehicles," *IEEE Industry Applications Society Annual Meeting*, 2008, pp. 1-5

[6] M. Shen and F. Z. Peng, "Converter Systems for Hybrid Electric Vehicles," *Proceeding of International Conference on Electrical Machines and Systems 2007*, 2007, pp. 2004-2010.

[7] A. Saplin, A. Meintz, and M. Ferdowsi, "Parametric Study of Alternative EV1 Powertrains," *IEEE Vehicle Power and Propulsion Conference*, 2008, pp. 1-5.

[8] K. Acharya, S. K. Mazumder, and P. Jedraszczak, "Efficient, High-Temperature Bidirectional Dc/Dc Converter for Plug-in-Hybrid Electric Vehicle (PHEV) using SiC Devices," *24th Annual IEEE Applied Power Electronics Conference and Exposition*, 2009, pp. 642-648.

[9] J. Lai and D.J. Nelson, "Energy Management Power Converters in Hybrid Electric and Fuel Cell Vehicles," *Proceedings of the IEEE*, Vol. 95, No. 4, April 2007, pp. 766-777.

[10] D. P. Urciuoli, R. A. Wood, T. E. Salem, and G. K. Ovrebo, "Design and Development of a 400A, All Silicon-Carbide Power Module," *NDIA-MI 2008 Ground-Automotive Power & Energy Workshop*, 2008.

[11] B. Baliga, *Fundamentals of Power Semiconductor Devices*, Springer, 2008.

[12] T. Funaki, M. Matsushita, M. Sasagawa, T. Kimoto, and T. Hikihara, "A Study on SiC Devices in Synchronous Rectification of DC-DC Converter," *22th Annual IEEE Applied Power Electronics Conference and Exposition*, 2007, pp. 339-344.

Investigation of 1.2 kV SiC MOSFET for High Frequency High Power Applications

Honggang Sheng
Monolithic Power Systems
San Jose, CA, USA 95120

Zheng Chen
CPES, Virginia Tech.
Blacksburg, VA, USA 24060

Fred Wang
The Univ. of Tennessee
Knoxville, TN, USA 37996

Alan Millner
MKS Instruments
Wilmington, MA, USA 01887

Abstract—**SiC is among the most promising materials for next generation power electronic devices due to its superior physical properties to Si and relative mature technology. SiC MOSFET is expected to offer performance improvement over Si counterpart. This paper presents the characterization of 1.2 kV SiC MOSFET, including its static and dynamic characteristics, and its high-frequency (1 MHz), high-power (1.2 kW) zero-voltage switching (ZVS) operation in a half-bridge parallel resonant converter. In comparison with SiC JFET and Si CoolMOS, the advantages and disadvantages of the SiC MOSFET are summarized.**

I. INTRODUCTION

Si-based power semiconductor devices have dominated the power electronics applications for decades. However, with increasing demand for high-frequency, high-voltage and high-temperature applications, Si power devices are now facing material limits. Wide Band Gap (WBG) materials, defined as semiconductors with bandgaps greater than 1.7 eV, promise superior performance in power devices. Among different WBG materials, in light of the development status, SiC is the most promising alternative to Si. SiC has an order of magnitude higher breakdown electric field than conventional materials, Si or GaAs, indicating that the power devices made of SiC can withstand much higher voltage stress with much lower on-resistance. Meanwhile, the large band gap energy of SiC also results in a much higher temperature capability and higher radiation hardness. Furthermore, SiC has a higher thermal conductivity, which leads to important benefits for power dissipation, and higher power handling capability [1]-[7].

Although SiC materials have superior characteristics, the system impact of SiC devices must be carefully studied, given their high cost, different size and package characteristics. Currently, only SiC Schottky diodes can be obtained commercially. Among the SiC active switches, SiC JFET, has been reported in many literatures, including modeling, characterization, and system applications [8]-[14]. Due to its normally-off feature and successful applications of Si power MOSFET, SiC MOSFETs are becoming the center of attention and expected to deliver superior performance over Si counterparts. In order to investigate SiC MOSFET for high-frequency, high-power applications, this paper fully characterized the static and dynamic characteristics of an 1.2 kV SiC MOSFET engineering sample, and compared it with a

state-of-the-art 600 V Si CoolMOS. A 1.2 kW 1 MHz parallel resonant converter is achieved with SiC MOSFET.

This paper starts with the static properties of SiC MOSFET characterization in Section II. The dynamic characteristics, including switching performance and body diode reverse recover performance, are presented in Section III. The operation and comparison between SiC MOSFET and Si CoolMOS resonant converters are provided in Section IV. The final comparison results are discussed and summarized in the final section

II. SiC MOSFET STATIC CHARACTERISTICS

The 1.2 kV SiC MOSFET prototype with 4.1 x 4.1 mm^2 shown in Fig. 1 is developed by Cree. Since very limited information of the prototype was provided with the device, the device is fully characterized. The key static characteristics, including on-resistance, device parasitic capacitance and gate threshold voltage, are presented in this section.

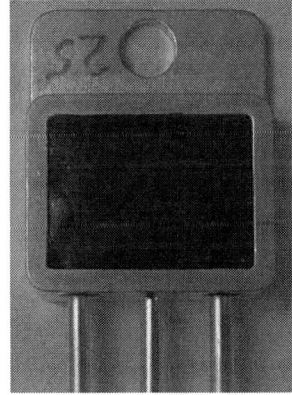

Figure 1. 1200V SiC MOSFET sample from Cree.

a) On-resistance

The on-resistance $R_{ds(on)}$ of the MOSFET measurement is based on the slope of the output characteristics in the linear region. Therefore, the measured resistance is the minimum possible resistance for a given V_{gs}. Unlike commercial high voltage Si MOSFET, the $R_{ds(on)}$ of SiC MOSFET is still significantly influenced by the V_{gs} when it is over 10V. Fig.2 shows the relationship of on-resistance vs. temperature at different gate voltages. High gate drive voltage is preferred for

This work was supported by MKS.

978-1-4244-4782-4/10 $26.00 © 2010 IEEE

low on-resistance, but it is limited by the gate breakdown voltage.

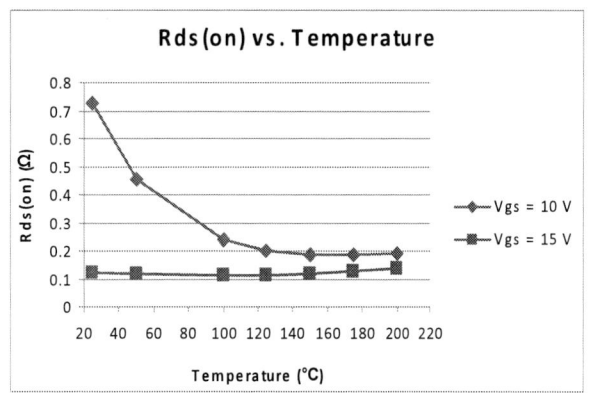

Figure 2. SiC MOSFET on-resistance vs. temperature at different gate voltages.

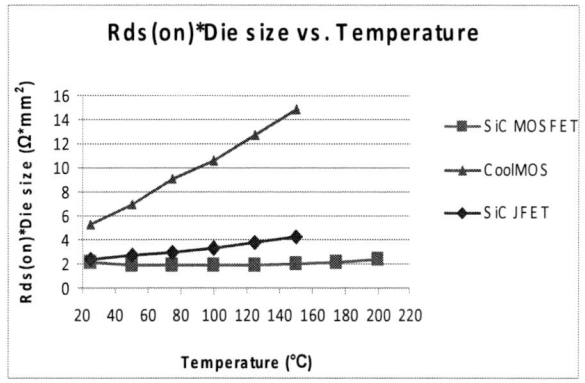

Figure 3. Comparison of specific on-resistance among 1.2 kV SiC MOSFET, 1.2 kV SiC JFET and 600 V CoolMOS.

Fig.2 also shows one of the interesting features of the SiC MOSFET, i.e. its high-temperature on-resistance. Fig. 3 shows the comparison of specific on-resistance at different junction temperatures among 1.2 kV SiC MOSFET, 1.2 kV SiC JFET(2.4 x 2.4 mm^2) [15] and 600 V Si CoolMOS [16]. Even compared with low voltage rating Si CoolMOS, SiC power devices still show their obvious advantage on on-resistance. The SiC MOSFET and SiC JFET of the same voltage rating have very close on-resistance at room temperature. However, the on-resistance of SiC JFET is doubled at 150 ^{o}C while the on-resistance of SiC MOSFET remains the same as at room temperature. For SiC MOSFET, there is an initial decrease in R$_{ds(on)}$ at low temperatures as the temperature increases, which is followed by an increase in R$_{ds(on)}$ at higher temperatures. This behavior in the R$_{ds(on)}$ of the 1.2 kV SiC MOSFET is due to the different behavior at different temperatures of the three primary resistances in the SiC MOSFET: 1) channel resistance; 2) JFET resistance; and 3) drift layer resistance. The channel resistance decreases with increasing temperature due to the reduction in threshold voltage and simultaneous increase in channel mobility. Both the JFET resistance and the drift layer resistance, on the other hand, increase with increasing temperature due to increased phonon (lattice vibration) scattering of the electrons. At lower temperatures, the channel resistance actually decreases faster with increasing temperature than the combined effect of increasing JFET

resistance and drift layer resistance, thereby causing in initial decrease in overall on-resistance for the SiC MOSFET. At higher temperatures, however, the increasing JFET resistance and drift layer resistance begins to dominate, resulting in an increase in resistance for the 1.2 kV SiC MOSFET.

b) Device capacitance (Ciss, Coss, Crss)

The input capacitance C_{iss}, output capacitance C_{oss}, and reverse transfer capacitance C_{rss} were measured using the impedance analyzer according to the methods introduced in [17]. The impedance analyzer applied a 50 mV, 1 MHz small signal for all measurements. Figure 4 shows the measured device capacitance at different drain-to-source voltages. The device capacitance is also measured at different temperatures. The variations of device capacitances are negligible at different temperatures.

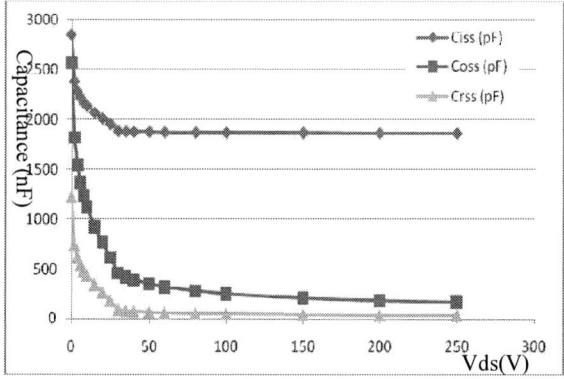

Figure 4. Non-linear capacitances, $C_{iss}/C_{oss}/C_{rss}$.

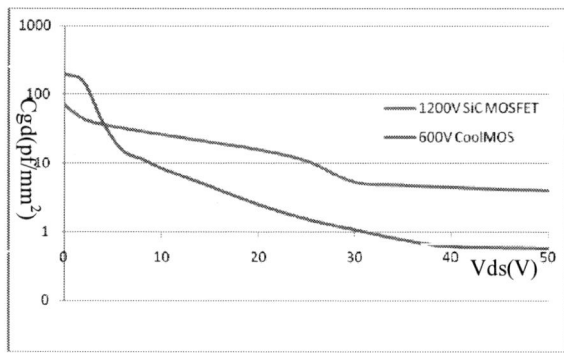

Figure 5 Comparison of Cgd capacitance densities.

The SiC MOSFET capacitances are compared with 600V CoolMOS after they are divided by die area. Figs 5-7 show each capacitance density (capacitance per mm^2) comparison of gate to drain capacitor (Cgd or Crss), gate to source capacitor (Cgs) and drain to source capacitor (Cds) between 1200 V SiC MOSFET and 600 V CoolMOS. Due to the large SiC breakdown electric field, the SiC power device can be made with much thinner drift region. Though the thin die help reduce the on-resistance, it increases the capacitances. Even the blocking voltage of SiC MOSFET is two times of Si CoolMOS in this comparsion, the SiC MOSFET capacitance densities of Cgd and Cds are much larger the Si CoolMOS. Since the blocking voltage has little effect on Cgs, the

978-1-4244-4782-4/10 $26.00 © 2010 IEEE   1573

capacitance densities of Cgs of the two devices are very close compared with Cgd and Cds. The relativly large Cgd will induce undesired large miller charge. With the mesured capacitance data, the curve can be described by a curve fitting equaiton. The charged energy is the integration of the capacitance curves over Vds. The ratio of miller charge(Qgd) to input charge (Qg) would be of use for gate drive design. Fig. 8 shows the comparsion of the ratios. Considering the high voltage applications, the SiC MOSFET has worse condtions for gate drive design.

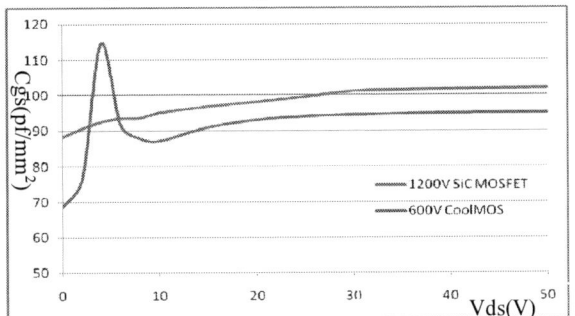

Figure 6. Comparison of Cgs capacitance densities.

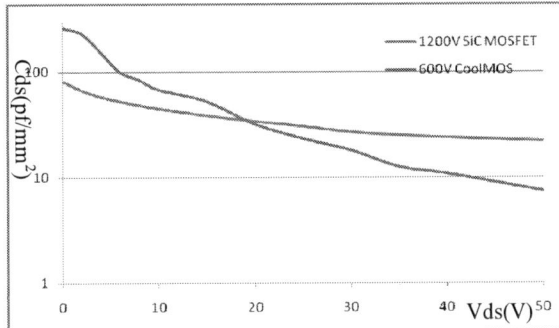

Figure 7. Comparison of Cds capacitance densities.

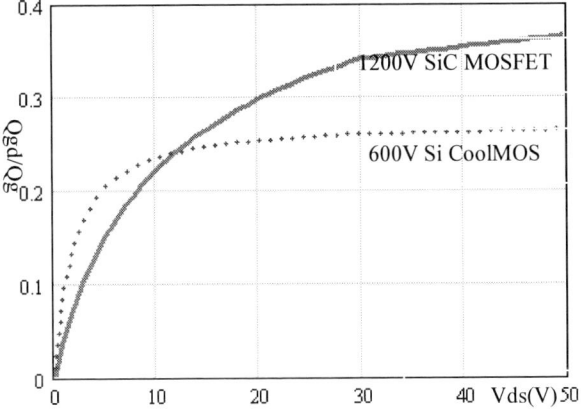

Figure 8. Comparison of the ratios of miller charge to input gate charge.

c) Threshold voltage

The unstable gate threshold voltage has been a concern for SiC MOSFET. The threshold voltage measurement has many criteria. A commonly used criterion for determining threshold voltage is the gate-source voltage level that produces 250 μA

of drain current when the drain and gate terminals are shorted. Using this criterion, the threshold voltage was determined as shown in Fig. 9. The lowest threshold voltage is around 1.5V at 150 °C. The low threshold voltage at high temperatures can explain why the on-resistance will reduce initially to some extent as the temperature increases. Though it is relatively low compared with commercial high voltage Si MOSFET (typically around 4V), it keeps stable at different temperatures. It should be noted that attention need to be paid on gate drive design to avoid false trigger when the gate threshold voltage is decreased and the ratio of Cgd/Cgs is increased.

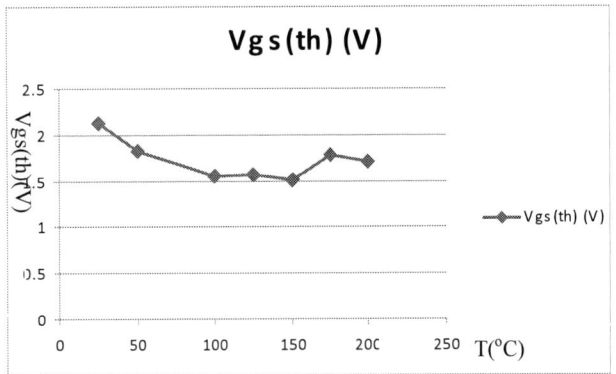

Figure 9. SiC MOSFET threshold voltages at varied temperatures.

III.   SiC MOSFET DYNAMIC CHARACTERISTICS

The static characteristics data is not sufficient for converter design and operation. The switching characterization is tested with double pulses under inductive load. During the first pulse the inductor current is charged to the desired value, and the falling edge of this pulse and the rising edge of the following pulse are corresponding to the turn-off and turn-on switching transients of the device under test (DUT), at any desired voltage and current levels The schematic of the double pulse test is illustrated in Fig. 10.

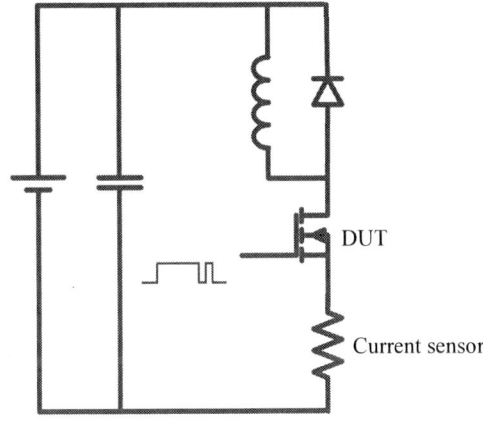

Figure 10. Schematic of double-pulse test.

When designing test circuit board, care must be taken to minimize parasitic inductance in the bus voltage path in order to avoid high switching spikes, and to minimize parasitic inductance in the gate drive path as well. High switching spikes can cause false trigger and also wrong measurements especially when gate resistance is small. Besides the parasitic inductance, the parasitic capacitances should also be minimized. The capacitances in parallel with the inductor, including the inductor stray capacitance, freewheeling diode parasitic capacitance and PCB board parasitic capacitance, influence the turn-on spike while the capacitance in parallel with the DUT influence the turn-off spike. In order to minimize the parasitic capacitance, low current rating freewheeling diode and a single layer inductor winding are desired.

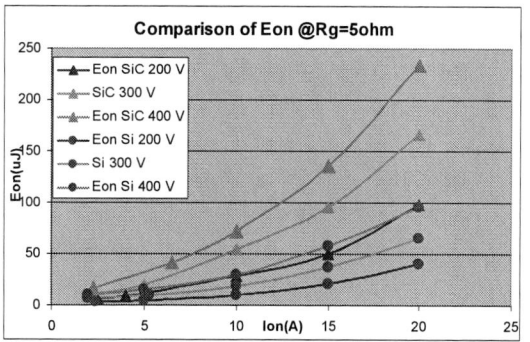

Figure 11. Turn-on loss energy comparison between SiC MOSFET and Si CoolMOS.

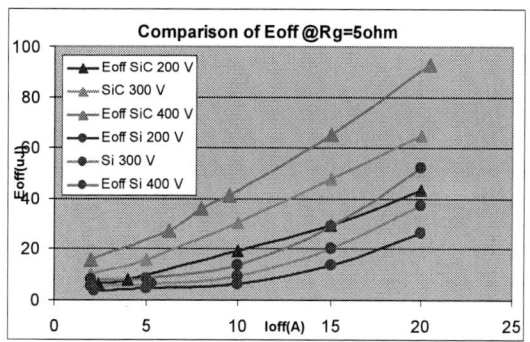

Figure 12. Turn-off loss energy comparison between SiC MOSFET and Si CoolMOS.

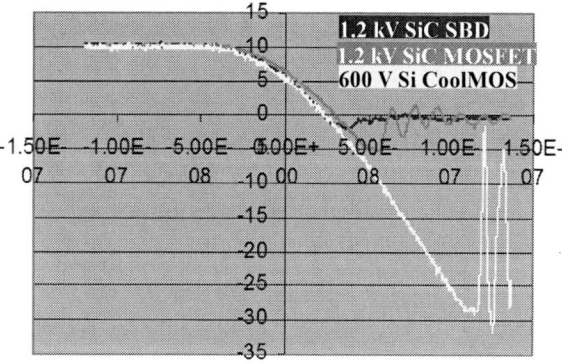

Figure 13. Body diode reverse recovery time (trr) comparison among 1.2 kV SiC SBD, 1.2 kV SiC MOSFET and 600 Si CoolMOS.

The switching characterization was done for voltage up to 800 V, current up to 20 A and gate resistance from 1 to 30 Ω. The switching power loss of SiC MOSFET is shown in Figs. 11 and 12, as well as that of Si CoolMOS for comparison. The switching power loss was calculated by the integration of the instantaneous power waveform (drain-source voltage waveform multiplied by drain current waveform). It is seen that the SiC MOSFET has a larger turn-on and turn-off switching loss than Si CoolMOS.

Another critical property is the SiC MOSFET body diode reverse recovery time (trr). The inferior performance of the Si MOSFET body diode is the main bottleneck for high-frequency application with ZVS operation [18]. At present, the commercial high-voltage Si MOSFET, even with a so-called fast reverse-recovery body diode, has a trr around hundreds of nanosecond at room temperature, which significantly restricts high-switching-frequency operation. The SiC MOSFET body diode reverse-recovery time measurements prove that it has a significant improvement on trr performance, with only about 40 ns reverse-recovery time at 25 °C which is even comparable with the performance of SiC Schottky diode (SBD), shown in Fig. 13. The low trr is mainly a result of the very small minority carrier lifetime as well as the thin drift layer. Furthermore, unlike the trr of Si device, the trr for SiC is relatively independent of temperature.

IV. HIGH FREQUENCY CONVERTER DEVELOPMENT AND OPERATION

Fig. 14 shows the topology used for power MOSFET system test. Since the purpose of the test is to evaluate the power switch's high-frequency operation, no rectifier bridge parallel resonant converter (PRC) is needed in the output. The converter with SiC MOSFET successfully operates at 1 MHz 1.2 kW with ZVS operation and 800V input voltage. The key waveforms of system test are shown in Fig. 15 and Fig. 16 for Si CoolMOS and SiC MOSFET respectively. Due to the low voltage rating of Si CoolMOS, the Si converter accordingly operates with 400 V input voltage while delivering the same power and voltage to load at 1 MHz switching frequency as the converter with SiC MOSFET. Figure 17 shows the ZVS turn-on waveform of SiC MOSFET and Figure 18 shows its turn-off waveform. With ZVS operation, the Vds and Vgs are very clean, but higher gate breakdown voltage is still desired for a reliable margin.

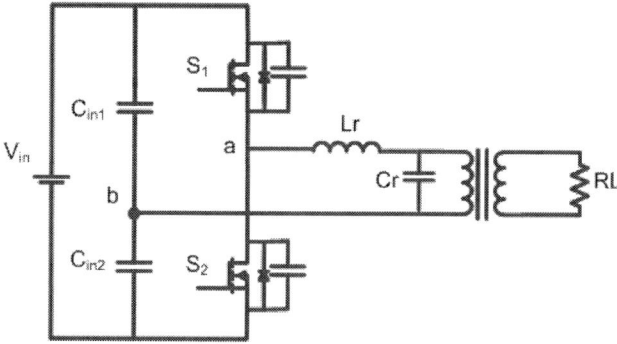

Figure 14. Half-bridge parallel resonant converter for SiC MOSFET high switching frequency operation.

Based on components temperature and efficiency measurements, Fig. 18 shows the power loss comparison between Si CooMOS and SiC MOSFET. Fig. 19 shows the junction temperature. The low junction temperature of SiC

MOSFET is mainly contributed by its metal case and SiC high thermal conductivity as well as slightly lower power loss.

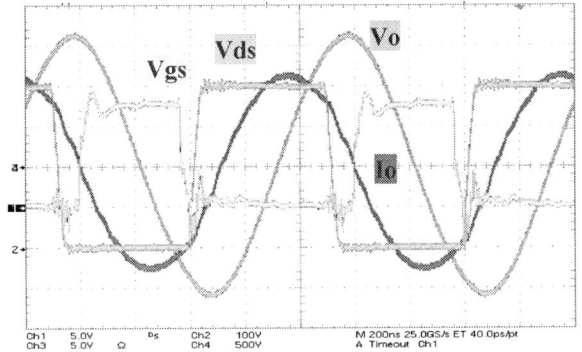

Figure 15. Main waveforms of converter with Si CoolMOS at 1.2 kW 1MHz.

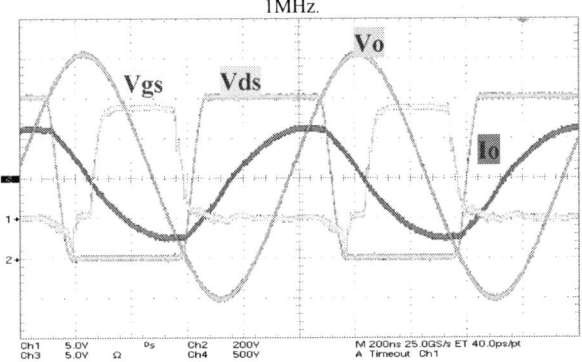

Figure 16. Main waveforms of converter with SiC MOSFET at 1.2 kW 1MHz.

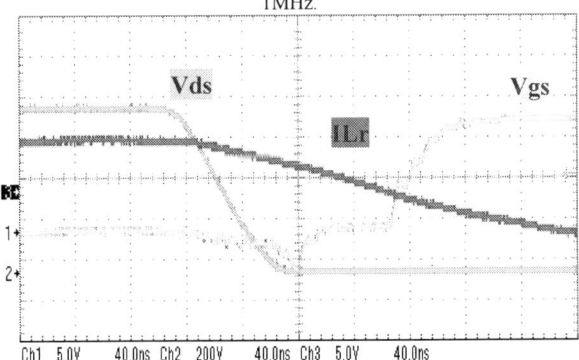

Figure 17. ZVS operation with SiC MOSFET at 1.2 kW 1MHz.

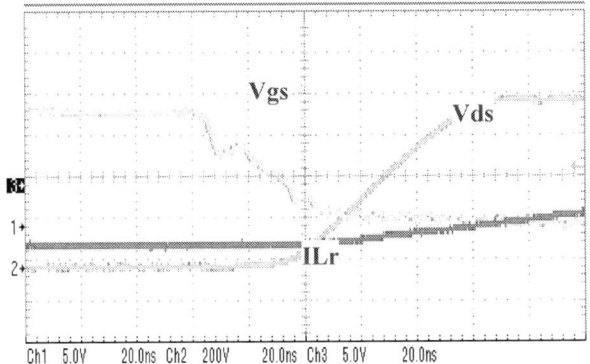

Figure 18. Turn-off waveform of converter with SiC MOSFET at 1.2 kW 1MHz.

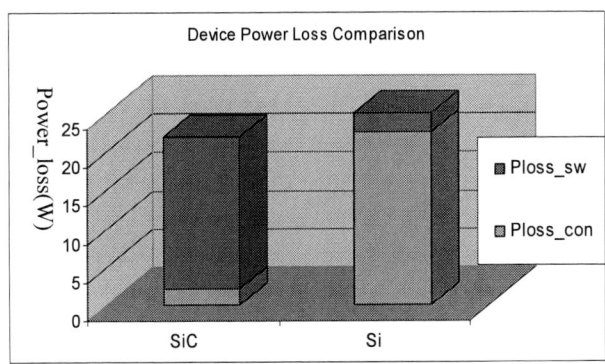

Figure 19. Power loss comparison between Si CoolMOS and SiC MOSFET.

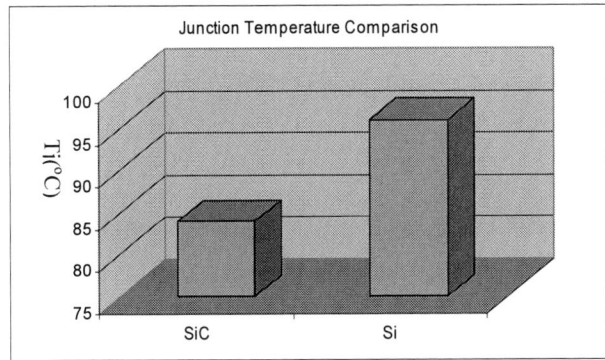

Figure 20. Junction temperature comparison between Si CoolMOS and SiC MOSFET.

Since the conduction loss can be calculated more accurately, the switching loss is obtained with the available total power loss and conduction loss. As the device characterization test shows, the SiC MOSFET has an obvious advantage on conduction loss due to low on-resistance and low conduction current due to the high input voltage when the converters are designed for the same power level and different input voltages, but it suffers high switching loss compared with Si CoolMOS. The system test results are consistent with the device testing results. Higher power or higher switching frequency can be expected when utilizing the high junction temperature margin of SiC MOSFET.

V. CONCLUSIONS

The 1.2 kV SiC MOSFET is fully characterized and applied successfully in a 1 MHz 1.2 kW converter with ZVS operation. For high switching frequency operation, thermal constraint is the practical issue. Although SiC MOSFET suffers relatively large switching loss partly due to larger capacitance density (capacitance per die size), it has a lower total power loss with its superior on-resistance. In addition, SiC MOSFET has a higher power loss handling capability due to its high thermal conductivity and high junction temperature which is limited by the current package technology. When ZVS is desired in high switching frequency applications, the fast reverse recovery time of SiC MOSFET body diode makes it an unbeatable alternative to Si counterpart for high switching frequency ZVS operation.

ACKNOWLEDGMENT

This work made use of ERC Shared Facilities supported by the National Science Foundation under Award Number EEC-9731677.

REFERENCES

[1] J.L. Hudgins, G.S. Simin, E. Santi, M.A. Khan, "An assessment of wide bandgap semiconductors for power devices," *IEEE Trans Power Electron.* , vol. 18, issue 3, May 2003, pp.907 – 914.

[2] Hajime Okumura, "Present Status and Future Prospect of Widegap Semiconductor High-Power Devices," Japanese journal of applied physics, vol.45, No.10A, Oct. 2006, pp. 7565-7586.

[3] B. Ozpineci, L. M. Tolbert, S. K. Islam, M. Chinthavali, "Comparison of Wide Bandgap Semiconductors for Power Applications," 10th European Conference on Power Electronics and Applications, Sept. 2003, Toulouse, France.

[4] A.M. Abou-Alfotouh, A.V. Radun, Hsueh-Rong Chang, C. Winterhalter, "A 1-MHz hard-switched silicon carbide DC-DC converter," *IEEE Trans Power Electron.* , vol. 21, issue 4, Jul. 2006, pp.880 – 889.

[5] Peter Friedrichs, "Silicon carbide power devices - status and upcoming challenges," European Conference on Power Electronics and Applications, Sept. 2007, pp.1 – 11.

[6] M. Ostling, H.-S. Lee, M. Domeij, C.-M. Zetterling, "Silicon carbide devices and processes - present status and future PERSPECTIVE," *in proc.* Mixed Design of Integrated Circuits and System (MIXDES), Jun. 2006, pp. 34 – 42.

[7] J. Millan, "Wide band-gap power semiconductor devices," IET Circuits, Devices & Systems, issue 5, Oct. 2007, pp. 372 – 379.

[8] Daniel Domes, Wilfried Hofmann, "SiC JFET in Contrast to High Speed Si IGBT in Matrix Converter Topology," *in proc. IEEE-PESC*, Jun. 2007, pp. 54 – 60.

[9] Tiefu Zhao, Jun Wang, Alex Q. Huang, Anant Agarwal, "Comparisons of SiC MOSFET and Si IGBT Based Motor Drive Systems," *in proc. IEEE-IAS*, Sept. 2007, pp.331 – 335.

[10] V. Bondarenko, M.S. Mazzola, R. Kelley, Cai Wang, Yi Liu, W. Johnson, "SiC devices for converter and motor drive applications at extreme temperatures," *in proc. IEEE Aerospace Conference*, Mar. 2006.

[11] S. Mounce, B. McPherson, R. Schupbach, A.B. Lostetter, "Ultra-lightweight, high efficiency SiC based power electronic converters for extreme environments," *in proc IEEE Aerospace Conference*, Mar. 2006.

[12] Callaway J. Cass, Yi Wang, Rolando Burgos, T.Paul Chow, Fred Wang, Dushan Boroyevich, "Evaluation of SiC JFETs for a Three-Phase Current-Source Rectifier with High Switching Frequency, " *in proc. IEEE-APEC*, Feb. 2007, pp.345 – 351.

[13] Y. Durrani, E. Aeloiza, L. Palma, P. Enjeti, "An Integrated Silicon Carbide (SiC) Based Single Phase Rectifier with Power Factor Correction,"*in proc IEEE PESC*, 2005, pp.2810 – 2816.

[14] S. Round, M. Heldwein, J. Kolar, I. Hofsajer, P. Friedrichs, "A SiC JFET driver for a 5 kW, 150 kHz three-phase PWM converter," *in proc. IEEE-IAS*, vol. 1, Oct. 2005, pp. 410 – 416.

[15] http://www.siced.com

[16] http://www.microsemi.com

[17] International Rectifier, "*Measuring HEXFET MOSFET Characteristics*", Application Note AN-957, International Rectifier, El Segundo, CA.

[18] A. Fiel, T. Wu, "MOSFET failure modes in the zero-voltage-switched full-bridge switching mode power supply applications," *in proc. IEEE-APEC*, vol. 2, 2001, pp. 1247 – 1252.

Comparative Analysis of Power Stage Losses for Synchronous Buck Converter in Diode Emulation Mode vs. Continuous Conduction Mode at Light Load Condition

Yang Chen, Peyman Asadi, Parviz Parto
EPBU POL, International Rectifier Corp.
8845 Irvine Center Drv., Suite 101,
Irvine, CA, 92618 U.S.A.
ychen1@irf.com

Abstract—**Light load efficiency attracts more attention nowadays and its requirement becomes stringent. To improve the light load efficiency in a synchronous Buck converter, several methods, such as pulse skipping, constant on-time, and Diode Emulation (DE), can be used. Compared to other methods, DE has the advantage of keeping its switching frequency unchanged and simple control circuits. Previous literatures present power loss estimation of DE in an ideal case, when Synchronous Field Effect Transistor (FET) is turned off at inductor current zero-crossing point. However, there is always deviation in practical circuit realization and this deviation may cause efficiency to drop by up to 2%. This paper first presents the comparative power stage loss analysis in DE vs. Continuous Conduction Mode (CCM) operation, then investigates the non-ideal DE situation, and follows with the extra power loss estimation. Simulation and experimental results verify the validity of the analysis.**

I. INTRODUCTION

Light load efficiency of Synchronous Buck converter becomes more important nowadays and its requirement becomes stringent in circuit design, especially for Point of Load (POL) applications. In light load condition, gate driver loss, body diode conduction and reverse recovery loss, switching loss and metal-oxide-semiconductor field-effect transistor (MOSFET) output capacitance loss are the major portions of total power loss, while MOSFET conduction loss is minor due to the small current in those low turn-on resistance MOSFETs which are optimized for heavy load efficiency. Several methods have been reported to reduce power losses at light load. Pulse-skipping and variable switching frequency methods are used in [1-3] to lower the switching frequency so as to reduce gate driver and switching losses. However, their switching frequency is variable and thus brings difficulty in the front-end EMI filter design. In [4], a combination of different MOSFETs and drivers in parallel are used and optimized for light, medium, and heavy load conditions respectively. Although this method improves the efficiency effectively, using redundant hardware is not cost-effective. In [5,6], gate drive voltage and deadtime are controlled to achieve better light load efficiency while keeping switching frequency fixed. Paper [1] and [7] present another simple control method, Diode Emulation (DE), to improve light load efficiency. For DE operation in a synchronous Buck converter as in Fig. 1, the synchronous (Sync) FET is turned off when the inductor current (I_L) drops to zero. A simple method to detect zero-crossing current, also known as $R_{ds(on)}$ sensing, is used to sense voltage across the Sync FET (or the switch node voltage V_{sw}). This method has fixed switching frequency and the current sensing method is lossless and cost-effective. Compared with Continuous Conduction Mode (CCM), where inductor current can go in reverse direction at light load, DE has less voltage and current ripples, and losses, such as FET and body diode conduction losses, switching loss, and output capacitance loss. Papers [1][7-9] give detailed power loss calculation, however, non-ideal situation, when Sync FET is not exactly turned off at zero-crossing point due to comparator offset and circuit delay, is not considered. This turn-off point deviation will cause extra power losses on top of the ideal case and may shift the total efficiency by up to 2% lower.

In this paper, the detailed comparative analysis of power stage losses in DE vs. CCM are presented in section II, then the non-ideal case of Sync FET turn-off is investigated and

978-1-4244-4782-4/10 $26.00 © 2010 IEEE

extra losses are estimated in section III. Section IV gives out simulation and experimental results to verify the validity of the above analysis. And finally the conclusions are summarized in section V.

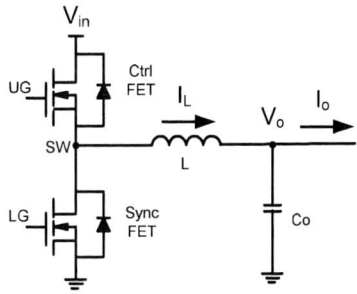

Figure 1. Synchronous Buck Converter

II. POWER STAGE LOEESE

Synchronous Buck converter is operating at the same switching frequency (F_s) for both CCM and DE, so they have almost the same gate driver loss, which is not covered in this paper. Their difference exists in power stage, where the reverse inductor current (I_L) incurs more losses in CCM. The following is the detailed comparison of power stage losses for DE vs. CCM with the assumption that Sync FET is turned off at I_L zero-crossing point in the load range of $0A < I_o < \Delta i_{(CCM)}$ (as in Fig.2). Non-ideal situation is analyzed in section III.

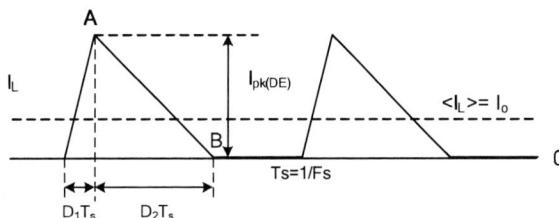

(a) DE Operation

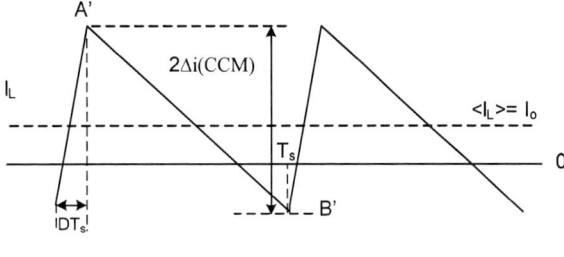

(b) CCM Operation

Figure 2. Inductor Current in DE vs. CCM

Power stage losses comparison between DE and CCM:

- MOSFET Conduction Loss: the inductor ripple current in DE is smaller than that in CCM under the same light load condition[10]. This results in a lower rms current for DE, which causes less conduction loss. Additionally, the Sync FET and inductor conduction loss can be further reduced by turning off LG when I_L

starts flowing in reverse direction. Conduction loss can be calculated as following: where $R_{ds(on)\text{-}ctrl}$ and $R_{ds(on)\text{-}sync}$ are the turn-on resistance of the control (Ctrl) FET and Sync FET, respectively.

DE :

$$P_{cond\_ctrl(DE)} = R_{ds(on)\_ctrl} \cdot I^2_{pk(DE)} \cdot \frac{D_1}{3}$$

$$P_{cond\_sync(DE)} = R_{ds(on)\_sync} \cdot I^2_{pk(DE)} \cdot \frac{D_2}{3} \quad (1)$$

$$\text{where, } D_1 = \sqrt{\frac{V_o^2 K}{V_{in}(V_{in}-V_o)}}, \; D_2 = \frac{K \cdot V_o}{D_1 \cdot V_{in}},$$

$$K = \frac{2L \cdot I_o \cdot F_s}{V_o}$$

CCM :

$$P_{cond\_ctrl(CCM)} = R_{ds(on)\_ctrl} \cdot (I_o^2 + \frac{1}{3}\Delta i^2_{(CCM)}) \cdot D$$

$$P_{cond\_sync(CCM)} = R_{ds(on)\_sync} \cdot (I_o^2 + \frac{1}{3}\Delta i^2_{(CCM)}) \cdot \quad (2)$$

$$(1 - D - 2t_{dead} \cdot F_s)$$

where, $D = V_o / V_{in}$, and t_{dead} is the deadband time.

- MOSFET Switching Loss: for DE, Ctrl FET turns off with peak current, which is the only point to incur switching loss; for CCM, Ctrl FET has turn-off loss as in DE, and Sync FET also has it because the reverse inductor current is cut off from Sync FET and flows into the bode diode of Ctrl FET. The Sync FET turn-off loss decreases with the increase of the load and finally drops to zero when $I_o = \Delta i_{(CCM)}$. Additionally, DE has a smaller peak current, which helps to reduce switching loss further. So the switching loss of DE is smaller. Their switching losses can be calculated as below:

DE :

$$P_{switch\_ctrl(DE)} = \frac{1}{2}V_{in} \cdot I_{pk(DE)} \cdot t_{f\_ctrl} \cdot F_s \quad (3)$$

CCM :

$$P_{switch\_ctrl(CCM)} = \frac{1}{2}V_{in} \cdot (I_o + \Delta i_{(CCM)}) \cdot t_{f\_ctrl} \cdot F_s \quad (4)$$

$$P_{switch\_sync(CCM)} = \frac{1}{2}V_{in} \cdot | I_o - \Delta i_{(CCM)} | \cdot t_{f\_sync} \cdot F_s$$

where, $t_{f\text{-}ctrl}$ and $t_{f\text{-}sync}$ are the fall time of Ctrl FET and Sync FET, respectively.

- MOSFET Body diode conduction loss: for DE, Sync FET body diode conduction loss exists; for CCM, both Ctrl FET and Sync FET have body diode conduction loss because of the bidirectional flow of the inductor current:

DE :

$$P_{bd\_sync(DE)} = V_{SD\_sync} \cdot I_{pk(DE)} \cdot t_{dead} \cdot F_s \quad (5)$$

CCM :

$$P_{bd\_ctrl(CCM)} = V_{SD\_ctrl} \cdot |I_o - \Delta i_{(CCM)}| \cdot t_{dead} \cdot F_s \qquad (6)$$

$$P_{bd\_sync(CCM)} = V_{SD\_sync} \cdot (I_o + \Delta i_{(CCM)}) \cdot t_{dead} \cdot F_s$$

where, $V_{SD\text{-}sync}$ and $V_{SD\text{-}ctrl}$ are the diode forward voltage drop of Sync FET and Ctrl FET, respectively.

- MOSFET Reverse Recovery Loss: neither DE nor CCM has body diode reverse recovery loss because of the zero-voltage turn-on of these FETs in this load current range.

- MOSFET Output Capacitance Loss: when Buck converter works at a high switching frequency, the losses incurred by charging and discharging the output capacitance (C_{oss}) become more important. For both DE and CCM, Ctrl FET dissipates such power:

$$P_{Coss\_ctrl} = 0.5 \cdot Q_{oss\_ctrl} \cdot V_{in} \cdot F_s \qquad (7)$$

where, Q_{oss_ctrl} is the output charge of Ctrl FET.

- Inductor Conduction and Core Losses:

DE :

$$P_{cond\_ind(DE)} = DCR \cdot I^2_{pk(DE)} \cdot \frac{D_1 + D_2}{3} \qquad (8)$$

where , DCR is the dc resistance of inductor

CCM :

$$P_{cond\_ind(CCM)} = DCR \cdot \left(I_o^2 + \frac{1}{3}\Delta i^2_{(CCM)}\right) \qquad (9)$$

Inductor core loss for both cases can be calculated as [11]:

$$P_{core\_ind(DE)} = 0.004203 \cdot (Fs)^{1.84} \cdot (0.05914 \cdot I_{pk(DE)})^{2.28} \text{ (mW)} \qquad (10)$$

$$P_{core\_ind(CCM)} = 0.004203 \cdot (Fs)^{1.84} \cdot (0.05914 \cdot 2 \cdot \Delta i_{(CCM)})^{2.28} \text{ (mW)}$$

III.   ANALYSIS OF NON-IDEAL DE OPERATION

Based on the above analysis, CCM operation has more power stage losses than DE in the range of $0 < I_o < \Delta i_{(CCM)}$, provided that Sync FET is turned off at zero-cross point. However, since the propagation delay and the offset voltage of the zero-crossing comparator always exist, Sync FET may be turned off before or after the true zero-crossing point. Even though this offset voltage remains in millivolt level, the deviation of I_L at Sync FET turn-off point would be possible in half ampere range because the current sensing is based on the small resistance $R_{ds(on)}$. So further investigation was conducted to find out the extra losses caused by this non-ideal DE operation.

Figure 3 shows I_L waveform in a switching cycle. Point D is the Sync FET turn-off point with a positive I_L value ($=I_{pk2}$) and F is the one with negative I_L value ($=I_{pk3}$), while E is the real zero-crossing. Power loss analysis is given below with respect to case D and F separately with the assumption that I_o and the peak inductor current remain almost the same.

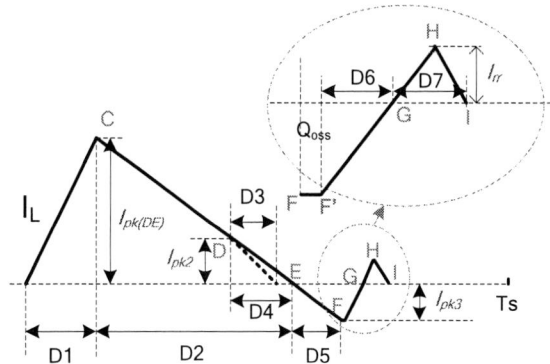

Figure 3.   Inductor Current in Non-ideal DE Operation

For case D, since Sync FET is turned off earlier, its conduction losses decreases while its body diode turns on and dissipates extra power. The extra losses caused in this case can be estimated as:

$$\Delta P_{cond\_sync} = R_{ds(on)\_sync} \cdot (I^2_{pk(DE)} + I_{pk(DE)} \cdot I_{pk2}$$

$$+ I^2_{pk2}) \cdot \frac{(D_2 - D_4)}{3} - R_{ds(on)\_sync} \cdot I^2_{pk(DE)} \cdot \frac{D_2}{3} \qquad (11)$$

$$\Delta P_{bd\_sync} = 0.5 \cdot V_{SD\_sync} \cdot I_{pk2} \cdot D_3$$

where,

$$D_3 = \frac{L \cdot I_{pk2} \cdot F_s}{(V_o + V_{SD\_sync})}, \quad D_4 = \frac{L \cdot I_{pk2} \cdot F_s}{V_o}$$

For case F, there will be several extra losses caused by the negative I_L value:

- Conduction losses in Sync FET and L from point E to F;

- Sync FET switching loss at point F;

- From point F to F′, Sync FET output capacitance C_{oss_sync} will be charged and Ctrl FET output capacitance C_{oss_ctrl} will be discharged, V_{sw} voltage rises. If $|I_{pk3}|$ is large enough to cause V_{sw} to go above V_{in}, Ctrl FET body diode turns on and I_L moves to zero from point F′ to G. The extra power losses in this range are: Ctrl FET and Sync FET output capacitance losses, Ctrl FET body diode conduction loss, and inductor conduction losses;

- After I_L reaches zero, the reverse recovery current of Ctrl FET body diode flows through L and increases to its peak I_{rr} (point H), after that it drops to zero (point I). This will cause reverse recovery loss of Ctrl FET body diode and the corresponding inductor conduction loss;

The above extra power losses can be estimated in (12):

$$\Delta P_{cond\_sync} = R_{ds(on)\_sync} \cdot I_{pk3}^2 \cdot \frac{D_5}{3}$$

$$\Delta P_{cond\_ind} = DCR \cdot \frac{I_{pk3}^2}{3} \cdot (D_5 + D_6) + DCR \cdot \frac{I_{rr}^2}{3} \cdot D_7$$

$$\Delta P_{switch\_sync} = 0.5 \cdot V_{in} \cdot |I_{pk3}| \cdot t_{f\_sync} \cdot F_s$$

$$\Delta P_{Coss\_ctrl} = 0.5 \cdot Q_{oss\_ctrl} \cdot V_{in} \cdot F_s \qquad (12)$$

$$\Delta P_{Coss\_sync} = 0.5 \cdot Q_{oss\_sync} \cdot V_{in} \cdot F_s$$

$$\Delta P_{bd\_ctrl} = 0.5 \cdot V_{SD\_ctrl} \cdot |I_{pk3}| \cdot D_6$$

$$\Delta P_{rr\_ctrl} = Q_{rr} \cdot V_{in} \cdot F_s$$

where,

$$D_5 = \frac{L \cdot |I_{pk3}| \cdot F_s}{V_o}, \ D_6 = \frac{L \cdot |I_{pk3}| \cdot F_s}{V_{in} + V_{SD\_ctrl} - V_o}$$

$$D_7 = t_{rr} \cdot F_s, \quad I_{rr} \approx \frac{2 \cdot Q_{rr}}{t_{rr}}$$

t_{rr} – Ctrl FET body diode reverse recovery time

Q_{rr} – Ctrl FET body diode reverse recovery charge

An exemption of the above calculation is that if $|I_{pk3}|$ is close to zero, V_{sw} won't go up to V_{in}. The output capacitance losses of both FETs are reduced. And there is no Ctrl FET body diode conduction loss or its reverse recovery loss since Ctrl FET body diode doesn't turn on.

Figure 4 shows calculation results of the power stage loss of case D vs. F. Obviously, when $|I_{pk3}|$ becomes large in case F, Ctrl FET body diode will conduct and extra reverse recovery loss will be incurred, which explains the power loss jump around -0.15A. On the right side of the picture, turning Sync FET off earlier than zero-cross still dissipates extra power, which rises with the increase of I_{pk2}. The optimal turn-off point E has no extra power loss. +0.2A/-0.05A peak current range is also acceptable with low losses in this example. Considering the propagation delay of logic and driver circuits, the beneficial region for DE operation is on the side of case D, namely turning off Sync FET earlier than point E. The desired switch node threshold to activate DE operation can be estimated as:

$$V_{sw\_th} = -R_{ds(on)\_sync} \cdot \frac{V_o}{L} \cdot t_{delay} \qquad (13)$$

where, t_{delay} is the overall propagation delay of logic and driver circuits.

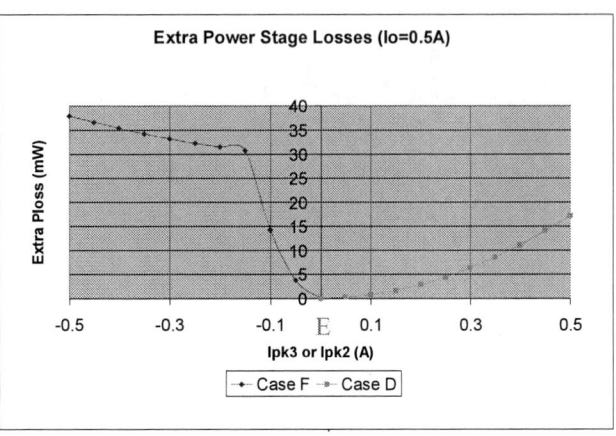

Figure 4. Extra Power Losses Caused by Non-ideal DE Operation

IV. EXPERIMENTAL VERIFICATION

Figure 5 shows the power stage losses calculation and experimental results of a Synchronous Buck converter in DE operation vs. CCM. Test conditions are: V_{in}=12V, V_{out}=1V, L=0.3uH, F_s=640kHz. The calculation is based on a negative I_{pk3}=-0.59A which is obtained from experiment. The critical point between DE and CCM is round Io=1.0A. It can be seen before the critical point, DE operation has less power stage losses than CCM as expected. Then DE merges into CCM operation, so they have the same losses. CCM has an almost flat power loss curve because in this light load region dominant factors are switching loss, body diode loss and L core loss, which are mainly determined by the peak inductor current instead of I_o. Calculation curves generally match experimental ones quite well, except for a few points at very light load, where the typical MOSFET parameters used in estimation may deviate from actual values at that condition. Figure 6(a) shows an example of each portion of power losses at I_o=0.5A. Calculation shows that the non-ideal turn-off point causes more than half of the total loss at I_o=0.5A as shown in Fig. 6(b).

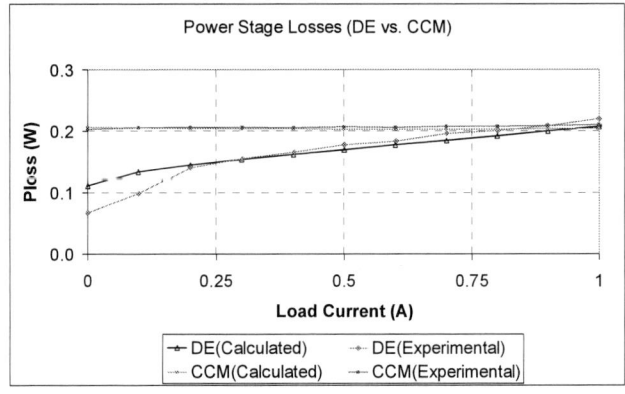

Figure 5. Power Stage Losses DE vs. CCM

978-1-4244-4782-4/10 $26.00 © 2010 IEEE

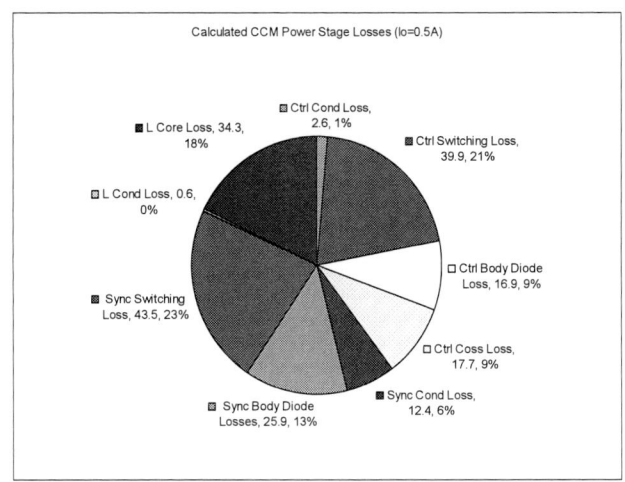

(a) Segments of CCM Power Stage Losses (in mW)

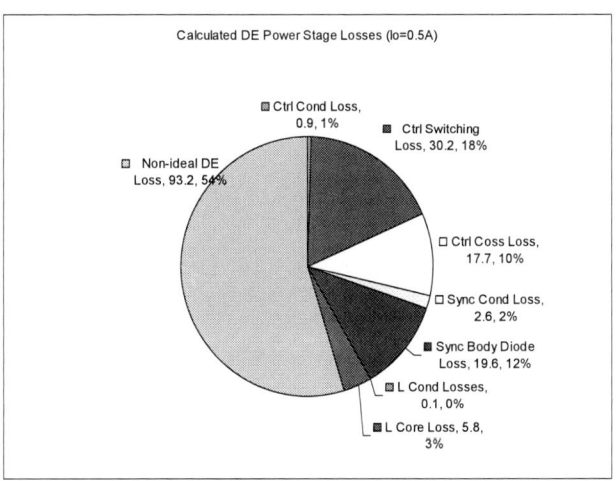

(b) Segments of DE Power Stage Losses (in mW)

Figure 6.   Power Stage Losses: DE vs. CCM at I_o=0.5A

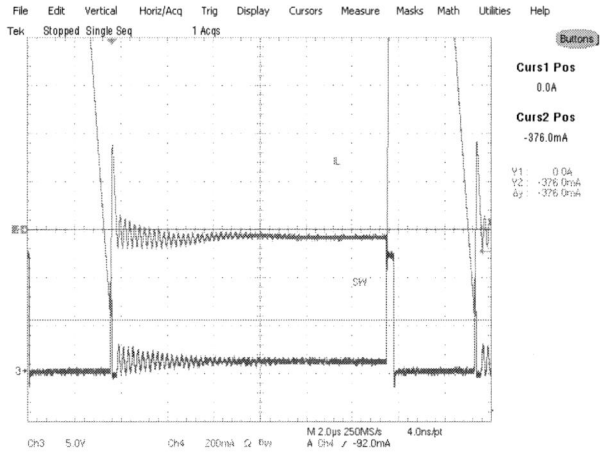

(a) Ipk3 = -376mA, I_L (upper) and SW(lower)

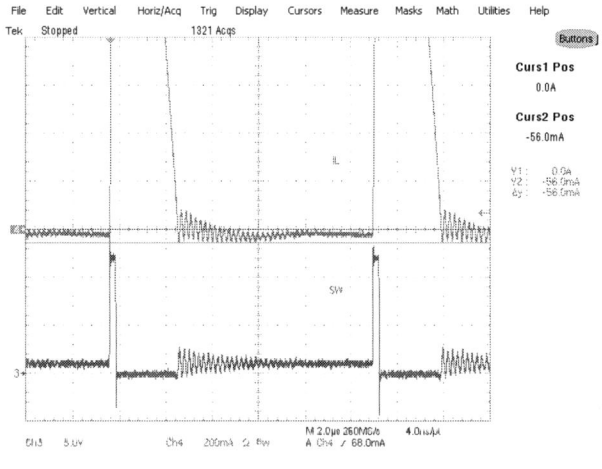

(b) Ipk3 = -56mA, I_L (upper) and SW(lower)

Figure 7. I_L and SW waveforms comparison in DE

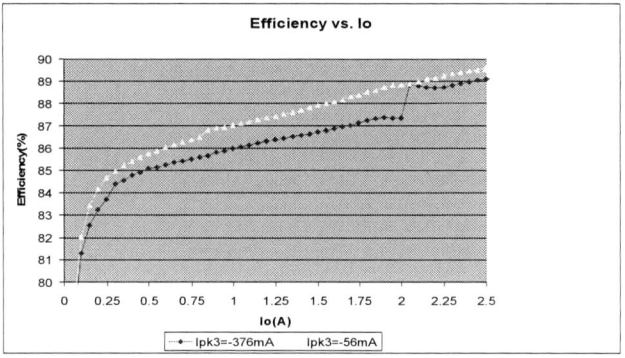

Figure 8.   System efficiency vs. I_o in DE

The above experiment shows the importance of moving Sync FET turn-off point close to zero in DE. In the following experiments, two different thresholds for zero-crossing current detecting are selected for comparison. Figure 7 (a) and (b) show the inductor current and the switching node voltage. The negative peak current in Fig.7(a) is -376mA, which is sufficient to turn on Ctrl FET body diode. Thus an extra voltage spike, close to the negative I_L spike, can be seen on V_{sw} waveform. In Fig.7(b), I_{pk3} is almost zero and Ctrl FET body diode doesn't turn on. Figure 8 shows the system efficiency comparison between these two cases. Test conditions are: V_{in}=12V, V_{out}=1.05V, L=0.68uH. With the reduced $|I_{pk3}|$, the efficiency can be boosted by 1~2%. Around I_o=2A, DE merges into CCM. The two efficiency curves approach each other.

V.   CONCLUSIONS

The operation of synchronous Buck converter with Diode Emulation vs. CCM was investigated and their power stage losses were compared at light load. Both analysis and

978-1-4244-4782-4/10 $26.00 © 2010 IEEE

experiments show DE has less power losses than CCM under the same light load condition. However, Sync FET turn-off point variation causes extra power losses in DE, thus makes efficiency lower. Estimation of the extra power losses is presented and verified by experimental results. The beneficial region is identified for better DE efficiency.

REFERENCES

[1] G X.W. Zhou, M. Donati, L. Amoroso, F.C. Lee, "Improved light-load efficiency for synchronous rectifier voltage regulator module", IEEE Trans. on Power Electronics, vol. 15, Issue 5, pp. 826 – 834, Sept. 2000

[2] A.V. Peterchev, S.R. Sanders, "Digital Multimode Buck Converter Control With Loss-Minimizing Synchronous Rectifier Adaptation", IEEE Trans. on Power Electronics, vol. 21, Issue 6, pp.1588 – 1599, Nov. 2006

[3] M. Gildersleeve, H.P. Forghani-zadeh, G.A. Rincon-Mora, "A comprehensive power analysis and a highly efficient, modehopping DC-DC converter", IEEE Asia-Pacific Conference on, pp.153 – 156, Aug. 2002

[4] O. Abdel-Rahman, J.A. Abu-Qahouq, L. Huang, I. Batarseh, "Analysis and Design of Voltage Regulator With Adaptive FET Modulation Scheme and Improved Efficiency", IEEE Trans. on Power Electronics, vol.23, no.2, pp.896-906, March 2008

[5] M.D. Mulligan, B. Broach, T.H.Lee, "A constant-frequency method for improving light-load efficiency in synchronous buck converters", IEEE Power Electronics Letters, vol, 3, Issue 1, pp.24 – 29, March 2005

[6] R. Miftakhutdinov, J. Zbib, "Synchronous Buck Converter with Increased Efficiency", IEEE Applied Power Electronics Conference, pp.714 – 718, Feb.-March 2007

[7] X.W. Zhou, T.G. Wang, F.C. Lee, "Optimizing design for low voltage DC-DC converters", IEEE Applied Power Electronics Conference, vol. 2, pp.612 – 616, Feb. 1997

[8] P. Markowski, "Estimating MOSFET switching losses means higher performance buck converters", http://www.planetanalog.com/showArticle.jhtml?articleID=12802296

[9] Y. Chen, P. Asadi, P. Parto, "Analysis of Diode Emulation in Light Load Condition of Synchronous Buck Converter", Power Electronics Intelligent Motion Power Quality, pp.602-607, May 2009

[10] Vitec Electronics Corporation: Datasheet of SMD High Frequency Power Inductor 59P9874N http://www.viteccorp.com/data/af4263.pdf

Controllable dv/dt Behaviour of the SiC MOSFET/JFET Cascode
An Alternative Hard Commutated Switch for Telecom Applications

Daniel Aggeler, Juergen Biela, Johann W. Kolar
Power Electronic Systems Laboratory (PES)
ETH Zurich, Switzerland
Email: aggeler@lem.ee.ethz.ch

Abstract—Switching devices based on SiC offer outstanding performance with respect to operating frequency, junction temperature and conduction losses and enable a significant improvement of the system performance. There, the cascode consisting of a MOSFET and a JFET additionally has the advantage of being a normally off device and offering a simple control via the gate of the MOSFET.

Without dv/dt control, however, the transients with hard commutation reach values of up to 45 kV/μs, which could lead to EMC problems and especially in drive systems to problems related to earth currents (bearing currents) due to parasitic capacitances. Therefore, new dv/dt control methods for the SiC MOSFET/JFET cascode as well as measurement results are presented in this paper. Based on this new concepts the outstanding performance of the SiC devices can be fully utilised without impairing EMC.

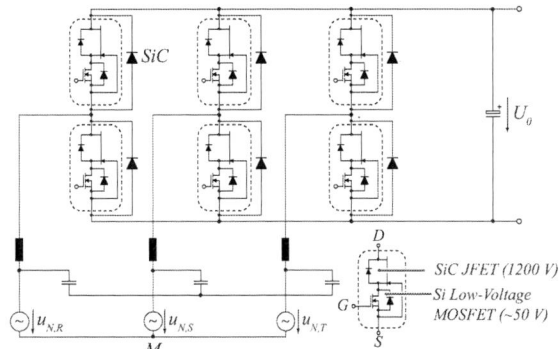

Fig. 1: Bidirectional three-phase boost topology with input and output filter and the SiC MOSFET/JFET cascode.

I. INTRODUCTION

The trend for the design of power electronic systems applied for example in telecom applications is towards higher power density and higher efficiency values. In order to reduce the system volume and achieve a higher power density, first the appropriate topology for the intended application must be chosen. Second, the design parameters must be chosen so that minimal volume and/or efficiency results. Due to large number of design parameters and coupling between these parameters this is advantageously done with an optimization procedure as presented in [1]. There, usually a high switching frequency is required for minimizing the volume of the passive components and the conduction as well as switching losses of the semiconductors must be low for achieving a high efficiency.

In the voltage range up to 600 V, high performance MOSFET switches capable of working at high switching frequency (e.g. COOLMOS) with low switching losses are applied in PFC converter systems (e.g. *VIENNA* rectifier [2],[3]). However, these devices offer a poor performance with respect to conduction losses in the 1200V range, so that usually IGBTs are used, which have significantly higher switching losses. This limits the reasonable operating frequency in this voltage range and therewith also the achievable power density. In order to overcome this limitation new devices based on SiC could be used.

The 1200 V SiC JFET [4] from SiCED offers very fast transients of up to 45 kV/μs (**Fig.** 2) with a blocking voltage of 1200V due to its vertical structure and the resulting low input/miller capacitance. However, the normally-on behaviour

of the SiC JFET prevented that the switch is fully accepted for industry applications although improved gate drive circuits have been developed [5].

A normally off behavior could be achieved by using a cascode configuration with a low-voltage Si MOSFET in series with the 1200 V SiC JFET, without losing the excellent characteristics of the SiC device. In this configuration only the low-voltage MOSFET is actively controlled whereas the SiC JFET is inherently controlled by the drain-source voltage of the MOSFET. The gate drive circuit for the SiC cascode is a standard IGBT/MOSFET driver and therefore the currently used switches could directly replaced by the SiC MOSFET/JFET cascode as shown for a three-phase boost topology in **Fig.** 1.

As a result of extremely fast voltage edges, which are

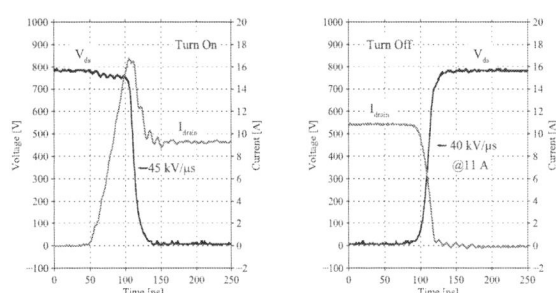

Fig. 2: Measurement result @ $V_{dc} = 800\,\text{V}/I_{drain} = 9\,\text{A}/11\,\text{A}$ (hard switching) with the SiC MOSFET (IRF2804)/JFET cascode.

978-1-4244-4782-4/10 $26.00 © 2010 IEEE

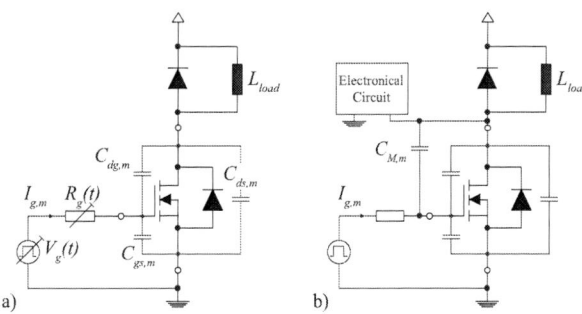

Fig. 3: Often used dv/dt limitation methods for MOSFET and IGBT switches; a) Varying gate resistors and 2- or 3-step controlled gate voltage. b) Additional drain source Capacitor $C_{M,m}$ causes an increased negative miller feedback.

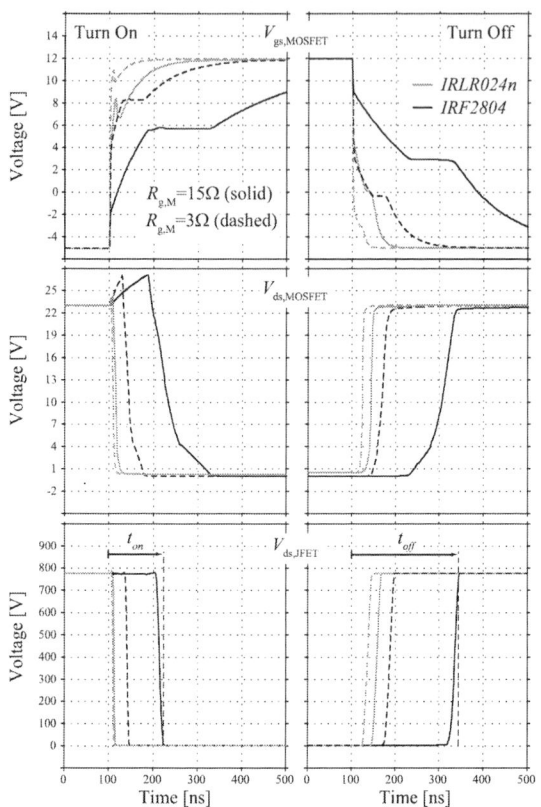

Fig. 4: Simulation results of applied conventional dv/dt limitation techniques with different gate resistors ($3\,\Omega$ dashed line,$15\,\Omega$ solid line) of the low-voltage MOSFET. Conventional techniques are not useful for the SiC cascode as illustrated by the equal dv/dt of the JFET drain source voltage.

achieved with the SiC cascode in hard commutated switching actions Fig. 2, the effort for the design of a low inductive layout avoiding switching related overvoltages is increasing. In [6] overvoltages are occurring due to parasitic and not avoidable module and layout inductances. Furthermore the desire for a controllable dv/dt of the cascode switching transients is high also due to EMI/EMC filtering [7].

In this Paper novel methods to control and adjust the dv/dt behaviour of the SiC MOSFET/JFET cascode are presented. Standard and well know techniques for MOSFET and IGBT semiconductors are shortly discussed in **Section** II. The application of the conventional techniques controlling dv/dt is evaluated related to the cascode configuration and the novel dv/dt controlling methods are detailed described in **Section** III. In **Section** IV experimental results of fast and controlled transients voltage edges of the cascode are shown. Finally, the switching energy of the SiC cascode is discussed and an exemplary calculation for a three-phase boost topology consisting of the SiC cascode is given.

II. CONVENTIONAL dv/dt LIMITATION TECHNIQUES

For currently often used semiconductors as Si MOSFET and IGBT devices several techniques [8] to reduce and control the dv/dt at fast switching edges are well known as shown in **Fig.** 3. The most simple and applied one is the control with the external gate resistor. The optimal gate resistance has to be selected to the corresponding switching device. In [9] the evaluation between varying gate resistors and an active gate voltage control (2-or 3-step gate voltage) is presented. There are different and more complex active gate control methods published as in [10] where the additional external ('artifical') miller capacitance is electronically adjusted to the effective gate to drain capacitance. An advanced methode is introduced in [11] where the current of the external miller capacitance is electronically controlled and at the same time the optimal point for minimal switching losses is calculated.

Most of these dv/dt limitation methods are based on the miller effect of an increased input capacitance at the switching events. For each turn on and turn off switching the gate

source and the drain source capacitance has to be charged and discharged.

Applying the conventional dv/dt limitation techniques to the SiC cascode results not in the desired behaviour of reducing the fast switching voltages edges. The reason is the serial connection of the low-voltage MOSFET and the SiC JFET. The conventional methods just influence the behaviour of the active controlled low-voltage MOSFET.

A. SiC Cascode

To investigate the influence of the conventional methods for the SiC cascode configuration a simulation setup (cf. schematic of experimental setup in **Fig.** 8) with $Simplorer^{TM}$ has been performed. There standard Spice models supplied by the manufacturers have been used for the low-voltage MOSFETs [12] and the freewheeling SiC Diode [13]. The applied SiC JFET model has been investigated and the Spice parameters have been extracted from experimental measurement in [14]. The simulation has been performed with different gate resistors ($R_{g,M} = 15\,\Omega$ and $R_{g,M} = 3\,\Omega$) of the various conventional techniques to control dv/dt behaviour. The same results could also be assumed with a 2 or 3-step

	IRLR024n	IRF2804
V_{DSS}	55 V	40 V
$I_D @ T_c = 25\,\mathrm{C}$	17 A	75 A
$R_{DS(on)} @ V_{GS} = 10\,\mathrm{V}$	65 mΩ@$I_D = 10$ A	2 mΩ@$I_D = 75$ A
$C_{iss, V_{ds}=0}$	680 pF	7800 pF
$C_{oss, V_{ds}=0}$	480 pF	5000 pF
$C_{rss, V_{ds}=0}$	230 pF	2100 pF

TABLE I: Main characteristics of the selected low-voltage MOSFETs from International Rectifier.

voltage or than an additional drain gate capacitor (artifical increasing of the miller capacitor) of the MOSFET.

The conventional techniques influence only the low-voltage MOSFET behaviour as shown in **Fig.** 4. Illustrated are the turn on switching behaviour of 4 A (hard switching) and the turn off switching at 7.5 A for two different low-voltage MOSFETs (cf. **Table** I; *IRLR024n* and *IRF2804*). Depending on the capacitance values, C_{iss},C_{oss} and C_{rss}, of the MOSFETs an increased value of gate resistance influence the charge and discharge behaviour of the gate source capacitance drastically. Therefore also the drain source voltage of the MOSFET is influenced. However there is almost no change in dv/dt behaviour.

The standard/conventional method to control the dv/dt value of the SiC cascode has no significant influence on the drain source voltage edge of the JFET as shown in the third simulation result. The influence by the different gate resistors is the time to start the switching action. In this case the delay time (t_{on},t_{off}) can be controlled by the conventional techniques. The slope of the drain source voltage keeps to be the same independent of the MOSFET type and also from the conventional techniques.

III. Novel dv/dt limitation methods - SiC Cascode

For the cascode topology novel methods to control the dv/dt has been investigated. Resulting are two concepts to slow down the very fast voltages edges at turn on as well as at turn off. In **Fig.** 5 the novel topologies for the cascode configuration to control the dv/dt behaviour are shown and in the following investigated.

A. Drain Gate Capacitor $C_{dg,M}$

The dv/dt controlling concept with the additional drain gate capacitor $C_{dg,M}$ is based on the conventional method of the MOSFET. The operating principle to control the dv/dt of the drain source voltage is explained in four time periods for the turn on characteristics as for the turn off characteristics as shown in **Fig.** 6. In the following the influence of the capacitor $C_{dg,M}$ in the cascode topology is investigated and detail discussed for the turn on behaviour.

1) Period T_1: During T_1 a positive voltage is applied to the gate source voltage of the MOSFET and the corresponding capacitance $C_{gs,M}$ is charged. This result in a marginal increase of the MOSFET drain source voltage. The cascode switch is still turned off and the behaviour is comparable with a single MOSFET device.

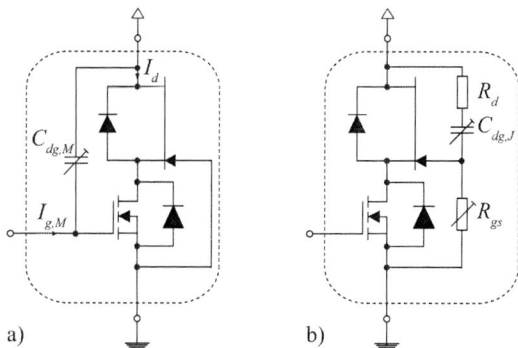

Fig. 5: The novel dv/dt controlling concepts for the SiC MOS-FET/JFET cascode; a) Additional drain gate capacitance resulting in an increased negative feedback to the MOSFET gate. b) RC-circuit between drain and gate of the JFET and a gate resistance for the JFET.

2) Period T_2: At the beginning of T_2 the gate source voltage achieves the miller level, where the drain current is equal to the load current and the current stops to conduct through the freewheeling diode. The $V_{ds,MOSFET}$ decreases fast to a level, which could be called cascode JFET miller level $V_{cascode,JFET,miller}$, and keeps the level almost constant until the cascode drain source voltage is decreased to the on voltage. The MOSFET voltage level of 16 V is depending on the drain current I_d of the JFET. The drain current can be approximatively calculated to

$$I_d = I_{g,M} + I_{load} \tag{1}$$

assuming that $\frac{V_{gs,M}}{dt} = \frac{V_{dg,M}}{dt} = 0$. The gate current $I_{g,M}$ is defined by the gate resistor R_g to

$$I_{g,m} = \frac{V_g - V_{gs,M}}{R_g} = C_{dg,M} \cdot \frac{V_{ds,J}}{dt}. \tag{2}$$

The cascode miller level is decreasing with a higher load current. Responsible for this load current cascode miller level is the JFET characteristics which has to open the channel to conduct the load current and consequently the gate source voltage of the JFET ($V_{ds,M} = V_{gs,J}$) has to decrease. At the end of period T_2 there is a fast small drop of I_d (cf. Fig. 6) where the capacitor $C_{dg,M}$ is completely discharged ($I_{g,M} = 0$).

3) Period T_3: Across the SiC JFET there is just the voltage drop caused by the $R_{ds(on)}$ of the channel applied at the beginning of period T_3. The gate source voltage is still in a stable level, in this case it's the well known miller level of a single MOSFET.

4) Period T_4: The cascode switch is completely in conduction mode and therefore the gate source voltage is increasing to the nominal gate voltage V_g applied from the gate driver. Furthermore the inductive load current increases in dependence of the load voltage and the inductance value.

The main part to limit the dv/dt is the time period T_2. There, only a gate driver limited current is flowing through the

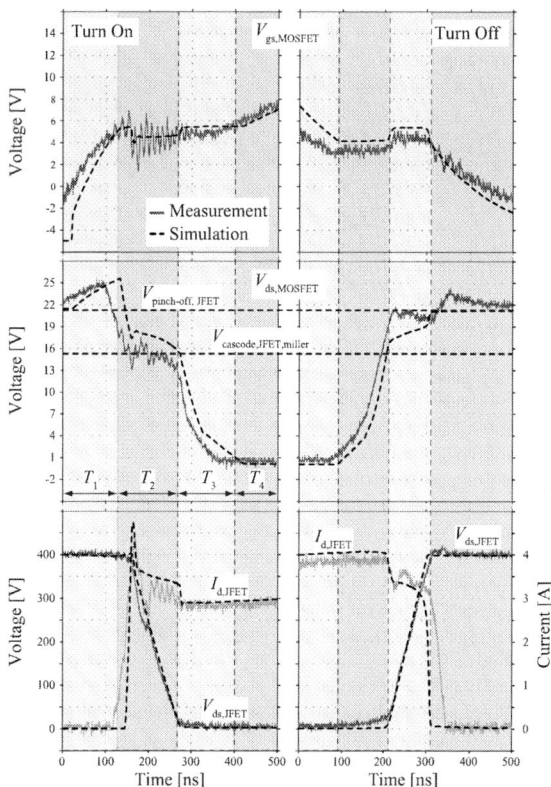

Fig. 6: Measurement and simulation results of the SiC cascode illustrate the influence of the $C_{dg,M}$ @400 V/3 A(turn on) and 4 A(turn off). Concept A. with $C_{dg,M} = 100$ pf and $R_g = 20\,\Omega$.

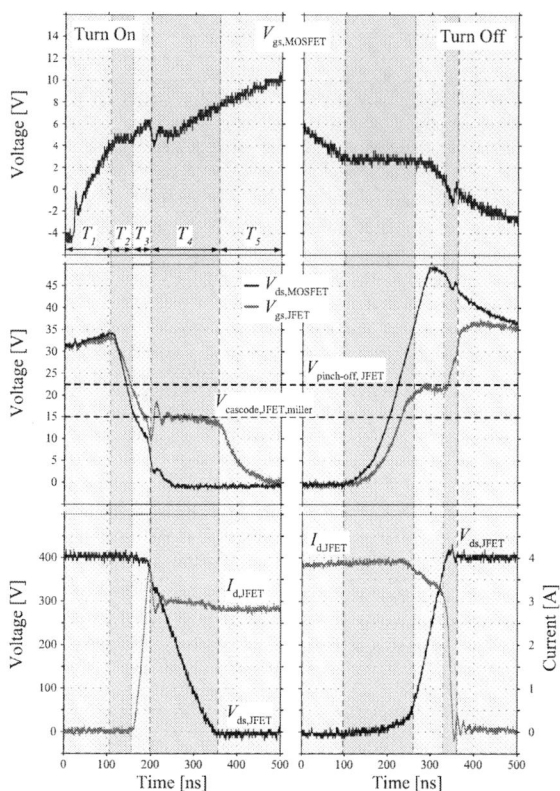

Fig. 7: Measurement results of the SiC cascode illustrate the influence of the RC-circuit and the JFET gate resistor @400 V/3 A(turn on) and 4 A(turn off). Concept B. with $R_d = 100\,\Omega$, $C_{dg,J} = 100$ pF, $R_{gs} = 47\,\Omega$ and $R_g = 20\,\Omega$.

capacitor $C_{dg,M}$ and therefore the length T_2 is controllable. At turn off the detail described turn on behaviour is analog and therefore the MOSFET blocks first to the called cascode miller level. Then, the main dv/dt rating event starts with charging the additional capacitor and finally the whole voltage is across the cascode switch.

B. RC-Circuit and JFET Gate Resistance

An alternative concept to control the dv/dt consists of a RC-circuit and an additional resistor R_{gs}. The detailed description of the dynamic behaviour is described based on **Fig. 7**.

1) Period T_1: Period T_1 is equal to the first control method described in (A.). The gate source voltage is applied and the MOSFET gate source capacitor is charged until the miller level is achieved. Remarkable is the level of the drain source voltage of the MOSFET because it's above the pinch-off voltage of the JFET and it's nearly the avalanche voltage of the MOSFET. This high level will be explained in detail in part four in particular at the turn off switching.

2) Period T_2: The gate source voltage of the MOSFET is equal to the miller level at the beginning of this time period. Hence, the drain source voltage of the MOSFET is decreasing rapidly and also the $V_{gs,JFET}$ is decreasing. At the same time the drain source voltage of the JFET keeps at the same level

because the gate source junction is still pinched-off as shown with the measurements.

3) Period T_3: The MOSFET drain source voltage is still decreasing while the gate source voltage of the MOSFET keeps the miller level. With decreasing $V_{ds,m}$ the gate source voltage of the JFET is also decreasing. At the beginning of T_3, $V_{gs,JFET}$ achieves the $V_{pinch-off,JFET}$ and continues to decrease. Therefore the SiC JFET channel opens and the drain current of the JFET increases fast with a small capacitive peak current. Achieving the value of the load current the freewheeling diode is turn off and the whole load current is flowing through the cascode.

4) Period T_4: At the beginning of T_4 the cascode JFET miller level is achieved and controlled by the load current. The JFET drain source voltage starts to decrease. Additional to the load current the limited current which discharges the capacitance $C_{dg,J}$ is flowing through the switch. At the end of the time period T_4 the gate source voltage is already increasing to the nominal gate voltage because the MOSFET is turned on earlier than the JFET.

5) Period T_5: The JFET gate source voltage is decreasing to zero volt and the cascode switch is turned on completely.

The dv/dt limitation in this concept of the cascode topol-

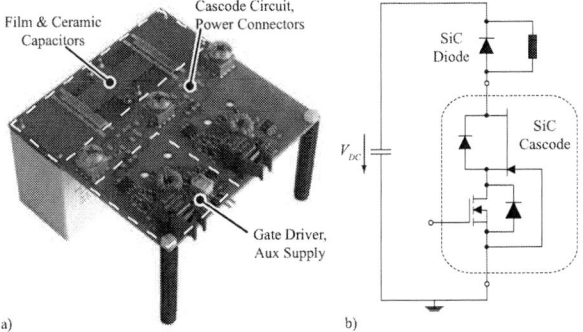

Fig. 8: a) Experimental setup to verify the concepts of the dv/dt controlling of the SiC MOSFET/JFET cascode. b) Schematic of the experimental topology (buck-topology).

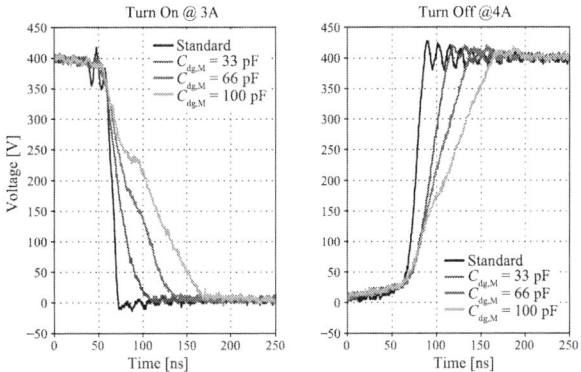

Fig. 9: Measurement results of the dv/dt concept A. @ 400 V with different values of the capacitance $C_{dg,M}$.

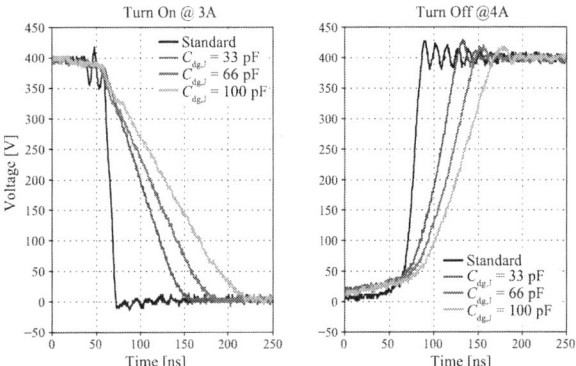

Fig. 10: Measurement results of the dv/dt concept B. @ 400 V with different values of the capacitance $C_{dg,J}$ and $R_{gs} = 47\,\Omega$.

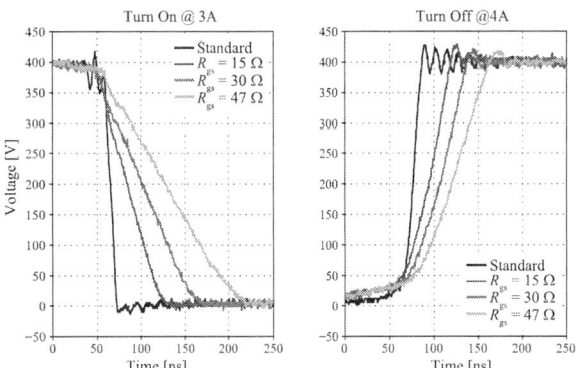

Fig. 11: Measurement results of the dv/dt concept B. @ 400 V with different values of the resistance R_{gs} and $C_{dg,J} = 100\,\mathrm{pF}$.

ogy takes place in the fourth time period where the discharge/charge of the capacitor is occurring. Resistance R_d is necessary due to damping gate drive oscillation and helps in the same way to limit the discharging/charging current. Therefore, the dv/dt limitation can be controlled by two or three parameters respectively. There is also the gate resistance R_{gs} of the JFET which builds an RC-circuit together with the gate source capacitance of the JFET. Resulting is a challenging behaviour at turn off as shown in the measurement (cf. Fig. 7). The MOSFET drain source voltage is increasing until the pinch-off voltage of the JFET is achieved. At this point the drain source voltage of the JFET starts to increase while the drain source voltage of the MOSFET is further increasing until the whole voltage is blocked by the cascode. During this dynamic behaviour the gate source diode of the JFET is pushed into avalanche. Depending on the MOSFET blocking voltage also the MOSFET is in avalanche mode a short time. After turn off there is a static balancing avoiding the avalanche. In case of continuous operation at high frequency the static balancing time will be too short and both junctions of the JFET and the MOSFET operating continuously in avalanche mode. Due to no available accurate avalanche Spice model of the avalanche behaviour of the MOSFET and the JFET gate source diode no simulation results are presented in Fig. 7.

Both concepts are working and reduces the dv/dt rating of the cascode. Advantageously of the first concept is the proper operation in the nominal and specified ranges of the devices. In the second concept there are more parameters to control the dv/dt but also the avalanche mode operation. Furthermore, both concepts have additional losses due to the additional capacitors and the decreased dv/dt of the cascode voltage edges. The resulting energy losses are shortly discussed in Section IV. Depending on the application of the cascode a combination of both concepts additional with conventional techniques lead to an optimized switching behaviour. In the following Section IV the controllable dv/dt is verified with measurements.

IV. EXPERIMENTAL RESULTS

An experimental setup shown in **Fig.** 8 has been built for the SiC MOSFET/JFET cascode to verify the concept of controlling the dv/dt behaviour at turn on as well as at turn off. The experimental testing has been performed with a buck topology.

Therefore, the setup consists of capacitors stabilizing the

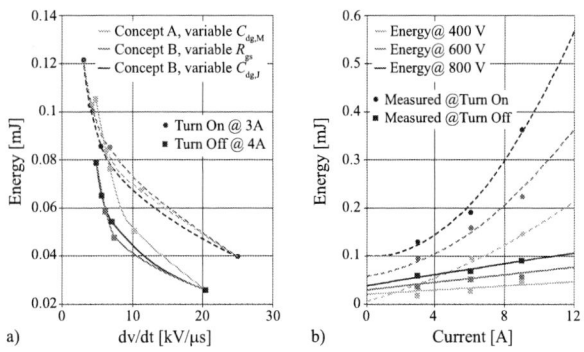

Fig. 12: Switching energy losses for a) the different dv/dt concepts @ 400 V and b) the SiC MOSFET/JFET cascode @ 400 V, 600 V and 800 V measured in a half-bridge configuration.

dc-link voltage, a halfbridge of SiC MOSFET/JFET cascodes and a standard gate drice circuits for MOSFET switches. Due to high dc-link voltage, the transfer of the gate signal is made with fiber optic transmitter/receiver. Furthermore, auxiliary power supplies are on the board feeding both gate drives and the fiber optic receivers.

In the following measurement results are presented to verify, with different values of the parameters ($C_{dg,M}$, $C_{dg,J}$ and R_{gs}), both concepts which are discussed in Section III . The measurement labeled as standard means the cascode topology with a low-voltage MOSFET and the SiC JFET without additional components. The gate resistance for this configuration has been selected to $4.7\,\Omega$. For all the other measurement verifying the both concepts a gate resistance of $20\,\Omega$ has been selected. **Fig.** 9 shows the measurement result of concept A. with three parameters $C_{dg,M} = 33\,\text{pF}, 66\,\text{pF}, 100\,\text{pF}$. In **Fig.** 10 and **Fig.** 11 measurements with concept B. are shown. The experimental measurement verify the controllable dv/dt with different parameters.

Analysis of the measurement results show the switching energy dependency on the dv/dt of the SiC cascode as pictured in **Fig.** 12 a). With a controlled dv/dt the energy losses are increasing. Thereby, almost no difference of the energy losses

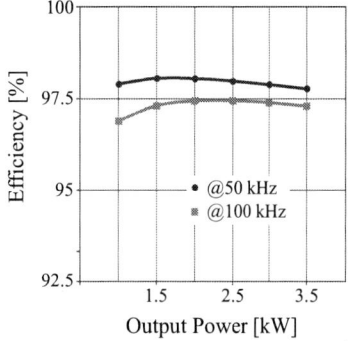

Fig. 13: Efficiency calculation of the three-phase boost converter (cf. Fig. 1) with a constant dc-link voltage of 800 V and a peak line voltage of 325 V.

is remarkable for turn on events. However, at turn off events concept A shows more energy losses than concept B. This additional amount of energy losses could be explained by the drain source voltage of the cascode (cf. Fig. 9). At turn off the drain source voltage is increasing fast up to 150 V compared to the turn off voltages edges achieved with concept B.

As an application for the SiC cascode is exemplary a three-phase boost topology (cf. Fig. 1) selected. Due to the high switching frequency possibility of the SiC material the input inductance of the three-phase boost converter is decreasing and the power density is increasing. Fig. 12 b) shows the energy characteristic of the standard SiC cascode for different voltages and currents. Therefore, the performance of the power stage (without input inductor, input and output filter) results in a maximal efficiency of 98 % with an output power of 1.5 kW as shown in **Fig.** 13 for 50 kHz and 100 kHz.

V. CONCLUSION

In this paper novel concepts/methods to control and adjust the dv/dt of the SiC MOSFET/JFET cascode has been presented. It has been shown with experimental measurements that the cascode switch could decrease the fast voltages edges. Therefore EMC problems as layout caused overvoltages could be handled and the SiC cascode could be applied as an alternative hard commutated switch.

ACKNOWLEDGMENT

The authors are very much indebted to the ABB Corporate Research Center, Baden-Daettwil, Switzerland, for supporting research on future SiC power semiconductor applications at the Power Electronic Systems Laboratory, ETH Zurich.

REFERENCES

[1] J. Biela, U. Badstuebner, and J. Kolar, "**Design of a 5kW, 1U, 10kW/ltr. resonant DC-DC converter for telecom applications,**" in *Proc. 29th International Telecommunications Energy Conference INTELEC 2007*, Sept. 30 2007–Oct. 4 2007, pp. 824–831.

[2] J. Kolar and H. Ertl, "**Status of the techniques of three-phase rectifier systems with low effects on the mains,**" in *Proc. 21st International Telecommunications Energy Conference INTELEC '99*, 6–9 June 1999, p. 16pp.

[3] S. Round, P. Karutz, M. Heldwein, and J. Kolar, "**Towards a 30 kW/liter, Three-Phase Unity Power Factor Rectifier,**" in *Proc. Power Conversion Conference - Nagoya PCC '07*, 2–5 April 2007, pp. 1251–1259.

[4] P. Friedrichs, "**Silicon carbide power devices - status and upcoming challenges,**" in *Proc. European Conference on Power Electronics and Applications*, 2–5 Sept. 2007, pp. 1–11.

[5] S. Round, M. Heldwein, J. Kolar, I. Hofsajer, and P. Friedrichs, "**A SiC JFET driver for a 5 kW, 150 kHz three-phase PWM converter,**" in *Fourtieth IAS Annual Meeting Industry Applications Conference Conference Record of the 2005*, vol. 1, 2–6 Oct. 2005, pp. 410–416.

[6] T. Nussbaumer, "**Netzrueckwirkungsarmes Dreiphasen-Pulsgleich-richtersystem mit weitem Eingangsspannungsbereich,**" Ph.D. dissertation, ETH Zurich, 2004.

[7] M. Moreau, N. Idir, P. Le Moigne, and J. Franchaud, "**Utilization of a behavioural model of motor drive systems to predict the conducted emissions,**" in *Proc. IEEE Power Electronics Specialists Conference PESC 2008*, 15–19 June 2008, pp. 4387–4391.

[8] P. Lefranc and D. Bergogne, "**State of the art of dv/dt and di/dt control of insulated gate power switches,**" in *Proceedings of the Conference Captech IAP1*, 2007.

[9] N. Idir, R. Bausiere, and J. Franchaud, "**Active gate voltage control of turn-on di/dt and turn-off dv/dt in insulated gate transistors**," vol. 21, no. 4, pp. 849–855, July 2006.

[10] S. Park and T. Jahns, "**Flexible dv/dt and di/dt control method for insulated gate power switches**," *IEEE J IA*, vol. 39, no. 3, pp. 657–664, May–June 2003.

[11] J. Kagerbauer and T. Jahns, "**Development of an Active dv/dt Control Algorithm for Reducing Inverter Conducted EMI with Minimal Impact on Switching Losses**," in *Proc. IEEE Power Electronics Specialists Conference PESC 2007*, 2007, pp. 894–900.

[12] "**International Rectifier**." [Online]. Available: http://www.irf.com

[13] "**CREE**." [Online]. Available: http://www.cree.com

[14] Y. Wang, C. J. Cass, T. Chow, F. Wang, and D. Boroyevich, "**SPICE Model of SiC JFETs for Circuit Simulations**," in *Proc. IEEE Workshops on Computers in Power Electronics COMPEL '06*, 16–19 July 2006, pp. 212–215.

Integral Micro-channel Liquid Cooling for Power Electronics

Ljubisa D. Stevanovic, Richard A. Beaupre, Arun V. Gowda, Adam G. Pautsch and Stephen A. Solovitz[1]

General Electric Global Research Center
Niskayuna, NY 12309
stevanov@ge.com

[1] Washington State University Vancouver,
School of Engineering and Computer Science
Vancouver, WA 98686

Abstract — **A novel integral micro-channel heat sink was developed, featuring an array of sub-millimeter channels fabricated directly in the back-metallization layer of the direct bond copper or active metal braze ceramic substrate, thus minimizing the material between the semiconductor junction and fluid and the overall junction-to-fluid thermal resistance. The ceramic substrate is bonded to a baseplate that includes a set of interleaved inlet and outlet manifolds for uniform fluid distribution across the actively cooled area of the heat sink. The interleaved manifolds greatly reduce the pressure drop and minimize temperature gradient across the heat sink surface. After performing detailed simulations and design optimization, a 200 A, 1200 V IGBT power module with the integral heat sink was fabricated and tested. The junction-to-fluid thermal resistivities for the IGBTs and diodes were 0.17°C*cm²/W and 0.14°C*cm²/W, respectively. The design is superior to all reported liquid cooled heat sinks with a comparable material system, including the micro-channel designs. It is also easily scaleable to larger heat sink surfaces without compromising the performance.**

I. INTRODUCTION

One of the key power electronics challenges is in the competing requirements of electrical and thermal design. The problem is most acute when designing a multi-chip power module for the packaging of semiconductor devices. A typical high-power module needs to provide electrical isolation between power devices and cooling fluid while maximizing thermal coupling to the fluid. This challenge is reconciled by attaching power devices to a direct bond copper (DBC) or active metal braze (AMB) ceramic substrate, which provides the requisite electrical insulation and thermal coupling. The substrate is soldered to a structural metal baseplate for mechanical mounting of the module to a heat sink. To avoid thermally inefficient dry surface contact between the baseplate and the heat sink, a thin layer of thermally conductive paste is applied at the interface. Even so, the paste is typically the biggest contributor to the overall junction-to-fluid thermal resistance and it diminishes the benefits of higher performance heat sinks. Therefore, significant improvement in the module thermal performance is not possible with inefficient thermal interface layers.

Micro-channel cooling employs sub-millimeter scale fluid passages with high surface-area-to-volume ratios, promising very high thermal performance. It has been actively researched for over three decades, with greater than 750 W/cm² performance demonstrated in previous experiments [1-6]. The outstanding thermal performance notwithstanding, the widespread commercial adoption of the micro-channel cooling has been impeded by several practical limitations: a) the outstanding convective heat transfer in micro-channels does not address the conductive thermal resistance of the material system between the semiconductor devices and the fluid, which could be an order of magnitude higher if it includes thermal interface materials, such as thermal paste; b) high pumping power requirement stemming from the high pressure drop of micro-channel designs, which is an exponential function of inverse hydraulic diameter of the channel; c) fluid filtering and maintenance to address the risk of micro-channel clogging; d) higher manufacturing cost due to sub-millimeter dimensions.

This paper describes micro-channel heat sink featuring an array of sub-millimeter channels fabricated directly in the back-metallization layer of the ceramic substrate (DBC or AMB). The integral heat sink, shown in Fig. 1, eliminates several high-resistance thermal layers, while maintaining electrical insulation between power devices and cooling fluid.

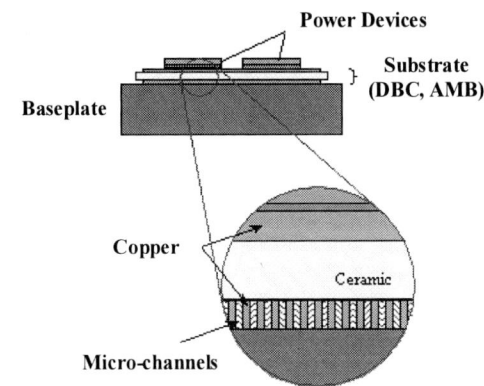

Figure 1. Schematic cross-section of the integral micro-channel heat sink.

978-1-4244-4782-4/10 $26.00 © 2010 IEEE

Combined with the outstanding convective heat transfer in micro-channels, the integral heat sink design is capable of junction-to-fluid thermal resistivities approaching 0.1°C*cm²/W. The substrate is bonded to a baseplate that includes a set of parallel inlet and outlet manifolds for uniform fluid distribution across the entire heat sink. By interleaving the manifolds with inlet-to-outlet spacing (the effective micro-channel length) of 2 mm, the pressure drop is minimized. The interleaved manifolds also help minimize the temperature gradient across the active area of the heat sink. The risk of clogging is addressed by selecting less-aggressive channel geometries, between 0.1 mm and 1 mm, and by having thousands of millimeter-long channels in parallel. Finally, the cost is partially addressed through a streamlined heat sink structure. Although the heat sink implementation reported here does not meet the cost target, alternate fabrication and material choices are expected to significantly reduce the cost. These refinements are a focus of ongoing R&D activities and will be reported in the near future.

The following contains descriptions of the integral micro-channel heat sink architecture, analysis, fabrication, testing, and conclusions.

II. HEAT SINK ARCHITECTURE

The new micro-channel heat sink design is based on an integral structure [7-8] that incorporates the micro-channels into the bottom copper layer of the ceramic substrate, as shown in Fig. 1. This effectively removes several high-resistance layers in a traditional power module, namely: thermal paste, module baseplate, and substrate-to-baseplate solder layer. The design consists of three types of fluid passages, denoted micro-channels, manifolds, and plenums, as shown in Fig. 2 and Fig. 3. Cold fluid enters through an inlet plenum and splits across a number of tapered inlet manifolds, which in turn feed a large number of micro-channels. The inlet manifolds are interleaved with tapered outlet manifolds, which drain heated fluid from the micro-channels. From the outlet manifolds, the fluid exits the heat sink via an outlet plenum.

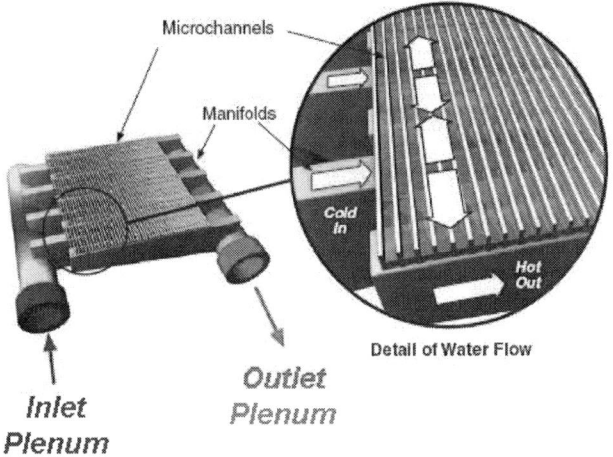

Figure 2. Schematic view of the heat sink design, including the three types of fluid passages: micro-channels, manifolds and plenums.

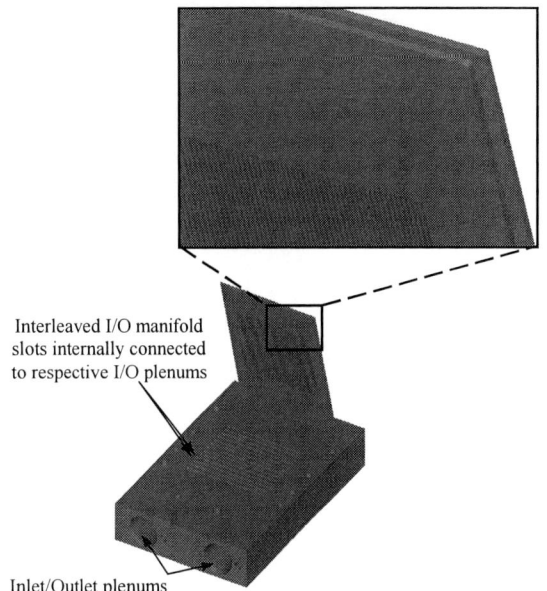

Figure 3. Exploded view of the integral heat sink. The baseplate shows interleaved manifolds across the actively cooled area, while the substrate features micro-channels in the backside metallization layer (detail).

Although the heat transfer occurs primarily at the smallest-scale micro-channels, the manifolds perform several important functions. First, they distribute the fluid evenly across the entire cooled surface, thus minimizing the temperature gradient across the heat sink. Second, the fluid is manifolded in and out of the micro-channels in multiple 2 mm-long flow paths connected in parallel, which decouples the length of the flow paths from the overall length of the channels. Consequently, the heat sink's pressure drop is significantly reduced compared to conventional designs with flow channels spanning the length of the cooled surface (from 2 cm to 20 cm long). Third, the manifolds alter the fluid direction from parallel to orthogonal with respect to the cooled surface, thus promoting flow mixing and improving the convective heat transfer.

In order to obtain the best performance for this design, each of the three fluid passages must be optimized for both heat transfer and flow performance.

III. DETAILED HEAT SINK DESIGN

Fig. 4 shows a layout of IGBT and diode devices of the 200 A, 1200 V power module with the integral micro-channel heat sink. The IGBTs and diodes are each rated at 100 A. The devices are located over the actively cooled area that is 26.9 mm wide and 25 mm long. The flow and thermal analyses were performed using Icepak™, a commercial CFD (computational fluid dynamics) simulation tool, using a mesh with a minimum of four elements across each fluid gap and two elements across each solid gap. The computational grid had approximately 1.2 million nodes, of which about 0.6 million were in the fluid passages. This grid was still somewhat coarse in some locations, but finer meshes only showed deviations of less than 10% in the pressure and less than 5% in the temperature.

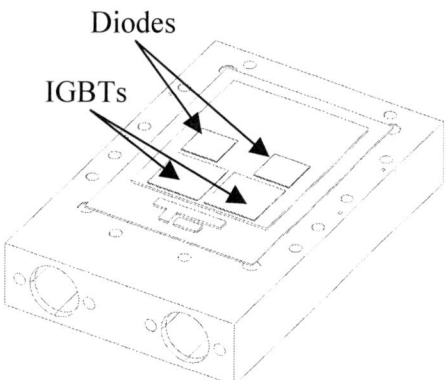

Figure 4. Design of 200 A, 1200 V power module with integral heat sink.

Following the design methodology described in [7], each of the three types of passages was optimized in turn, considering first the optimal micro-channel dimensions, then the manifolds, and finally the plenums. For each case, an individual DOE (design of experiments) was conducted, examining a range of channel dimensions to develop a transfer function for the pressure drop and the device temperature rise. These functions were optimized to determine the geometry with the minimum temperature rise within the following manufacturability and application constraints: minimum channel width of 0.1 mm, maximum channel depth of 0.3 mm, and maximum pressure drop of 345 kPa at water flow rate of 8 liters per minute (LPM).

The optimized heat sink design consists of a series of 135 parallel, 0.1 mm wide, 0.3 mm deep, and 25 mm long micro-channels cut into the bottom-side copper of the substrate (DBC or AMB). Each inlet and outlet manifold is 1 mm wide, 26.9 mm long, and has a tapered depth from 0.5 mm to 5 mm. There are total of six inlet and seven outlet manifolds. Finally, the plenums have a diameter of 4.8 mm.

Fig. 5 shows thermal simulations of the module with 1 LPM water flow rate, inlet temperature of 20°C, and input power to diodes of 500 W, or approximately 470 W/cm^2. The simulations show the maximum junction temperature of 98°C, resulting in the diode thermal resistivity of 0.166°C*cm^2/W.

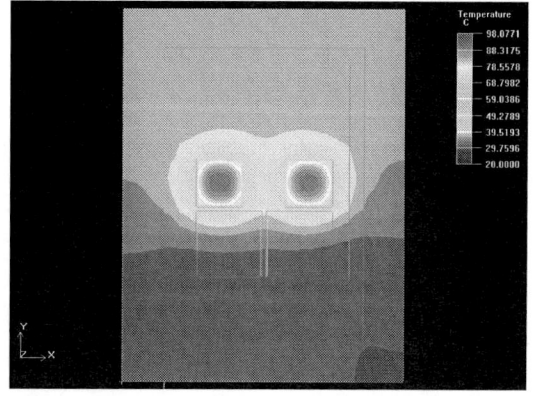

Figure 5. Thermal simulation results at 500 W input power to diodes, 1 LPM water flow rate and 20°C inlet temperature.

IV. MODULE FABRICATION AND ASSEMBLY

A. Micro-channel Fabrication

Based on the simulation results, the optimized micro-channels were fabricated into the bottom copper layer of a Si$_3$N$_4$ AMB substrate using laser ablation. Owing to the superior mechanical properties of silicon nitride, a thinner ceramic layer of 320 μm was used, compared to the 635 μm needed when using aluminum nitride substrates. Thicknesses of the top and bottom copper layers were 400 μm and 300 μm, respectively. An experiment was performed to determine the laser processing parameters to obtain the desired channel dimensions and to prevent the laser from encroaching the ceramic layer. Fig. 6 shows backside of the substrate with laser-ablated micro-channels. The substrate cross-section along the A-A' direction denoted in Fig. 6 is shown in Fig. 7.

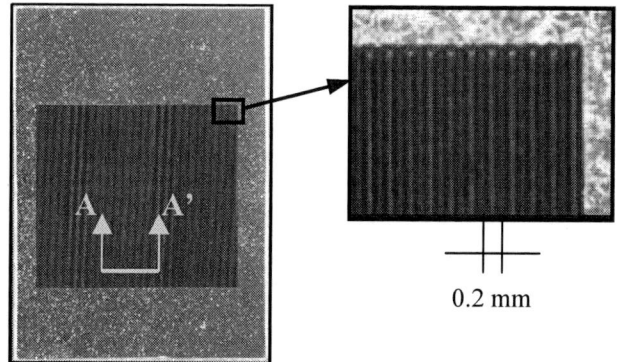

Figure 6. Backside of the substrate with laser-ablated micro-channels.

Figure 7. Cross-sectional view of laser-ablated micro-channels along the A-A' direction denoted in Fig. 6.

Clearly, the laser ablation yields triangular instead of rectangular channel cross-sections. The hydraulic diameter and the area contacting the fluid of the triangular cross-section is approximately 75% of the desired rectangular geometry. The discrepancy was deemed significant enough to justify experimental comparison of the two geometries; rectangular and triangular. To that end, rectangular micro-channels of the optimized geometry were machined in the back metallization layer of the substrate by micro-milling, as shown in Fig. 8.

978-1-4244-4782-4/10 $26.00 © 2010 IEEE

Figure 8. Cross-sectional view of micro-channels machined by micro-milling.

Heat sinks were fabricated from substrates with the two channel geometries and thermal and hydraulic performance was compared. The experimental results are shown in Fig. 9. The rectangular milled channels yield 14% lower thermal resistance and 45% lower pressure drop. Although beneficial to the heat sink performance, the micro milling fabrication process proved to be less practical than the laser ablation due the exceptionally long processing time required. Hence, subsequent substrate samples were processed using laser ablation. The low-cost implementation of the rectangular channels is subject of an ongoing R&D program and will be reported in the near future.

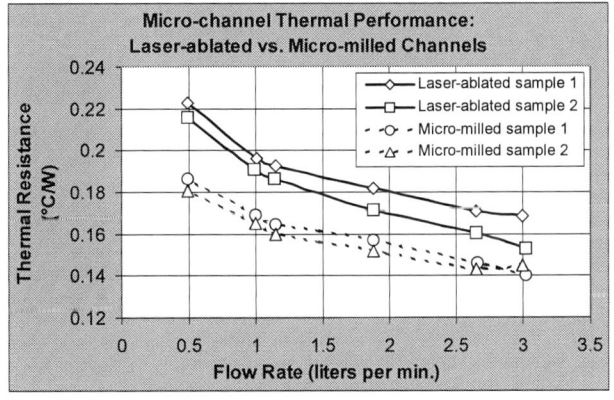

Figure 9. Heat sink thermal resistance with two types of micro-channels: the laser-ablated and the micro-milled.

B. Heat Sink Baseplate Fabrication

The baseplate was machined of an aluminum-graphite (Al-graphite) composite material with nickel-gold surface finish as a hermetic barrier to the fluid and as a solderable surface. The choice of the Al-graphite was made after initial temperature cycling tests with copper baseplates showed mechanical degradation of solder joint between the AMB substrate and the copper baseplate, resulting in fluid leaks. The failure of the solder joint was due to solder fatigue resulting from significant differences in the coefficient of thermal expansion (CTE) between the substrate (effective CTE value of 9 ppm/°C) and copper (18 ppm/°C). The composition of the Al-graphite material was designed to have a CTE of 9 ppm/°C. Since the heat transfer is dominated by convection in micro-channels, the lower thermal conductivity of the Al-graphite material (~200 W/m°C in-plane, 125 W/m°C out-of-plane), compared

to copper (390 W/m°C), was found to have a negligible effect (less than 3% based on simulations).

C. Power Module Assembly

Fig. 10 shows the power module assembly prior to soldering the substrate to the baseplate and power devices to the substrate. The devices are shown without (upper) and with (lower) power overlay topside interconnect [9]. Both soldering steps, the substrate to the baseplate and power overlay to the substrate, were performed simultaneously using a 63%Sn/37%Pb solder. Also attached during this one-step operation were two connectors, power and gate (shown in the lower photograph).

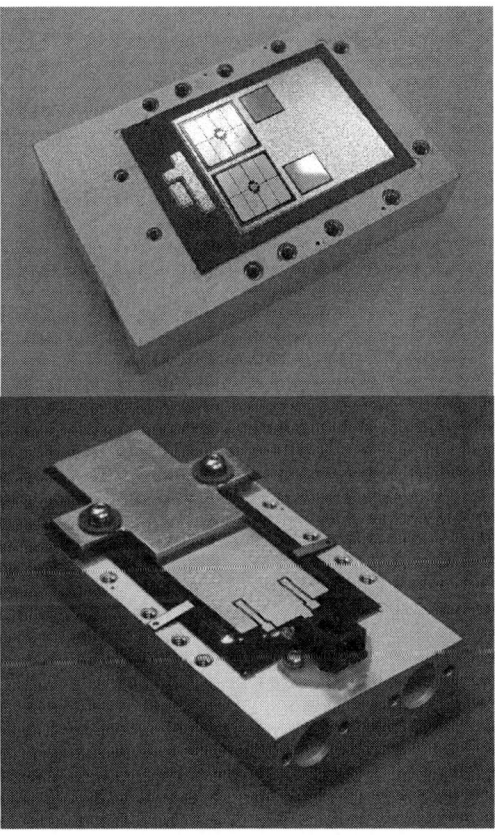

Figure 10. Power module assembly prior to soldering the substrate to the baseplate and power devices to the substrate. The devices are shown without (upper) and with (lower photograph) power overlay top-side interconnect.

V. EXPERIMENTAL RESULTS

A. Test Equipment

The heat sink evaluation test stand is shown schematically in Fig. 11. It consisted of a chiller (labeled as 1 in the schematic), various valves for safety and flow rate control, a differential pressure transducer (2) to measure the pressure drop across the heat sink, a high accuracy turbine flow transducer (3), and a needle valve for fine control of the flow rate. Another needle valve in a by-pass line was used for coarse fluid control. The rest of the setup comprised of two high accuracy RTDs for measuring fluid inlet and outlet temperatures, a carrier/demodulator (4), a data logger (5), an overhead IR camera (6), the heat sink device under test (7), a PC controller (8), and a DC power supply (9) for powering either diodes or IGBTs.

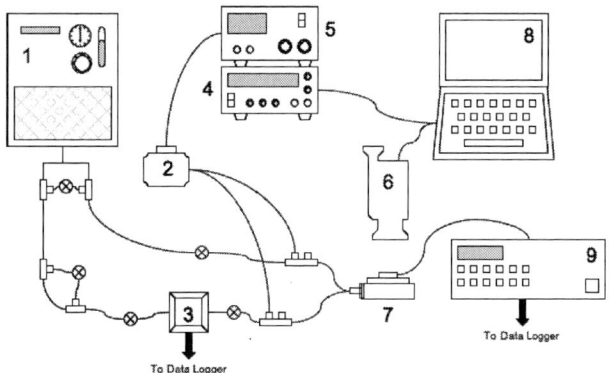

Figure 11. Schematic of the heat sink evaluation test.

A photograph of the experimental test stand is shown in Fig. 12. The chiller is seen in the left foreground. Under the bench are power supply and computer; above the bench are data acquisition system, digital delay generator, and oscilloscope. On the bench to the left is the flow measurement and control system, containing the valves, flow meter, pressure transducer, and temperature measurements. At the center of the image is the device under test with the overhead IR camera. A close-up image of the device under test is shown in Fig. 13.

Figure 12. Photograph of the heat sink evaluation test stand.

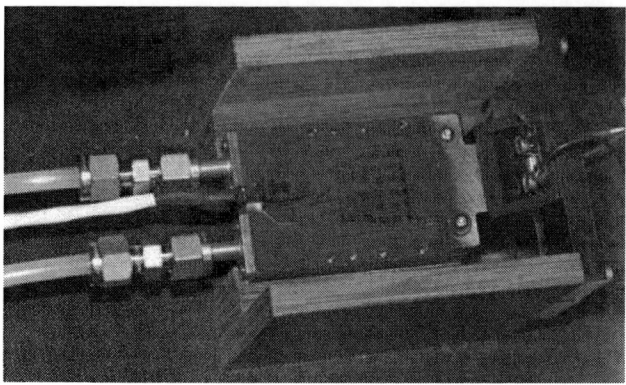

Figure 13. A close-up view of the test fixture with heat sink device under test. The fluid connections and gate driver cable are on the left, the electrical power and test terminals are on the right.

B. Test Results

A batch of ten GE power modules was built and experimentally characterized using infrared (IR) thermal measurements. IGBTs or diodes were used as power sources. The IGBTs were Infineon SIGC109T120R3L devices with a 10.47 mm × 10.44 mm chip size. The diodes were Infineon SIDC53D120H6 with a 7.3 mm × 7.3 mm chip size. Both chip types were 120 μm thick. Power levels were measured both as electrical input and the caloric rise of the coolant temperature, which were in good agreement. The junction-to-fluid thermal resistance:

$$R = \Delta T/P \qquad (1)$$

is conservatively calculated using the temperature difference ΔT between the maximum junction temperature T_{j_max} measured by the IR camera and the water inlet temperature T_{in} (maintained at 20°C by the chiller), divided by the applied power P. The heat sink's thermal resistivity is defined as:

$$R" = \Delta T/q" \qquad (2)$$

where $q"$ represents the heat flux applied to either diodes or IGBTs. The measurements were performed at flow rates from 0.5 – 3 LPM and several power levels, up to 350 W/cm^2. The plots of thermal resistivity and resistance versus flow rate for the IGBTs are shown in Fig. 14. Thermal performance of the diodes is shown in Fig. 15. Multiple data points for each flow rate represent measurements at different input power levels. As expected, the heat sink's performance is independent of the applied heat flux.

Using error propagation methods, the uncertainty of the thermal resistance was computed from the uncertainty of the measurements of temperature by the RTDs, temperature by the IR camera, and the power measurement made electrically. The total error in R was calculated to be 0.0004°C/W, or less than 1/2 percent relative error. Since the chip size tolerances are even smaller, the relative error of $R"$ is the same as for R.

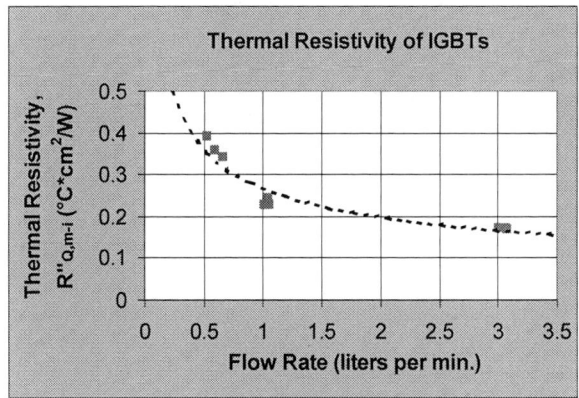

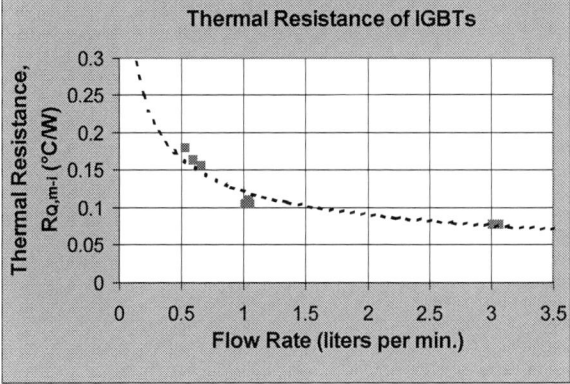

Figure 14. Thermal resistivity and resistance vs. flow rate for IGBTs, based on mesured maximum junction temperature rise above the 20°C inlet temperature. Measurements were performed at several values of input power to IGBTs, represented by multiple data points for each flow rate. The relative uncertainty in the thermral resistance is better than 0.5 %.

At the flow rate of 3 LPM, the junction-to-fluid thermal resistance for the IGBTs is 0.076°C/W and the thermal resistivity is 0.171°C*cm²/W. Similarly, the diodes thermal resistance is 0.136°C/W and the thermal resistivity is 0.143°C*cm²/W. Fig. 16 shows IR images of IGBTs and diodes at the flow rate of 1 LPM and input power of 500 W and 450 W, respectively[1]. According to the temperature scale alongside each IR image, the junction-to-inlet temperature rise for the IGBTs is 52°C and for the diodes 75°C.

Predicted and experimental values for the diode thermal resistivity vs. pressure loss are shown in Fig. 17. The experimental results were obtained at the flow rates of 1 LPM, 2 LPM and 3LPM and are in very good agreement with the simulations. In addition, the very low pressure loss values across the heat sink at the flow rates of interest (17 kPa @ 1

---

[1] One of the benefits of the power overlay wirebondless device interconnect can be observed in these thermal images. There is an absence of localized hotspots typically found in conventional wirebonded modules, which are caused by wirebond resistances. The power overlay significantly reduces the interconnect resistances and the associated power dissipation, thus eliminating the hotspots. Another benefit of the power overlay stems from planarity of the interconnect's top surface. The flat metallic surface atop the devices offers the possibility of adding a topside heat sink to the power module, resulting in double-sided cooling. The practical implementation and thermal benefits of the power overlay for double-sided cooling were explored under this program and the results were recently reported [10].

LPM, 66 kPa @ 2 LPM, 133 kPa @ 3 LPM), validate the benefits of interleaved manifolds. The thermal and the hydraulic performance were consistent across the population of ten GE power modules fabricated for this study.

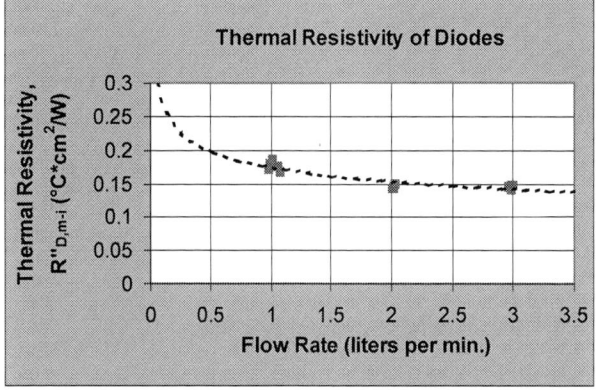

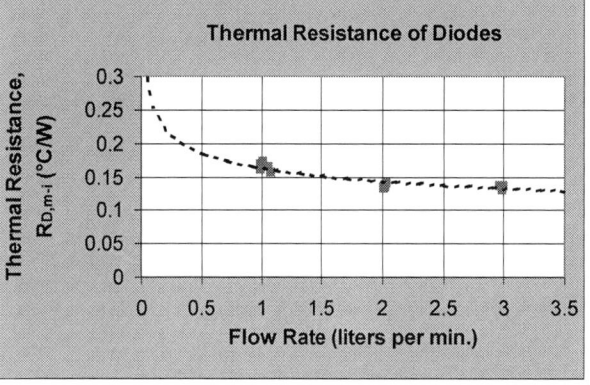

Figure 15. Thermal resistivity and resistance vs. flow rate for diodes, based on mesured maximum junction temperature rise above the 20°C inlet temperature. Measurements were performed at several values of input power to diodes, represented by multiple data points for each flow rate. The relative uncertainty in the thermral resistance is better than 0.5 %.

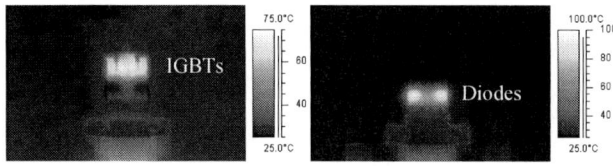

Figure 16. Thermal images taken with IGBTs and diodes powered at 500 W and 450 W, respectively. With the water inlet temperature of 20°C and the flow rate of 1 LPM, the junction-to-inlet temperature rise for the IGBTs is 52°C and for the diodes 75°C.

The GE integral micro-channel cooling offers the best-in-class thermal performance for a power electronics module, as shown in Fig. 18. The extremely low junction-to-fluid thermal resistance values for both IGBTs and diodes could enable significantly higher power densities, both at the module and converter level. For example, assuming the maximum junction-to-fluid temperature rise of 100°C, each 100 A IGBT could dissipate up 660 W on the GE heat sink. The maximum diode power dissipation under the same assumptions would be 370 W.

978-1-4244-4782-4/10 $26.00 © 2010 IEEE

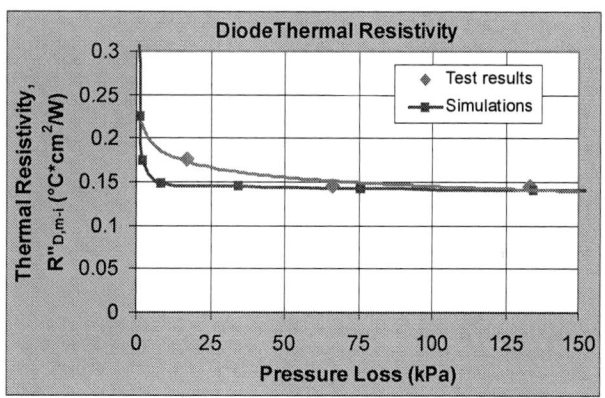

Figure 17. Diode thermal resistivity vs. pressure loss. The measured values are in very good agreement with simulations. The pressure loss vs. flow rate was also measured: 17 kPa @ 1 LPM, 66 kPa @ 2 LPM, 133 kPa @ 3 LPM.

VI. CONCLUSIONS

A 200 A, 1200 V IGBT power module with integrated micro-channel cooling has been designed, simulated, fabricated and experimentally validated. The design incorporates micro-channels fabricated on the backside copper of the ceramic substrate. The substrate is soldered to a baseplate that evenly distributes fluid to the micro-channels via interleaved inlet and outlet manifolds. Experiments were performed at power densities up to 350 W/cm^2 and flow rates up to 3 LPM. The junction-to-fluid thermal resistance for the IGBTs was 0.076°C/W and for the diodes 0.136°C/W. The thermal resistivity values for the IGBTs and diodes were 0.17°C*cm^2/W and 0.14°C*cm^2/W, respectively. The design is superior to all reported liquid cooled heat sinks with a comparable material system, including the micro-channel designs. It is also easily scaleable to larger heat sink surfaces without compromising the performance.

Ongoing R&D efforts are focused on development of micro-channel heat sinks that can achieve high performance at a cost comparable to that of conventional heat sinks. To accomplish this goal, three processes are being developed: 1) batch fabrication of micro-channels by a ceramic substrate manufacturer 2) molded, rather than machined, baseplate and 3) batch process for bonding the substrate to the baseplate.

ACKNOWLEDGMENT

Research was sponsored by the Army Research Laboratory and was accomplished under Cooperative Agreement Number W911NF-04-2-0045. The views and conclusions contained in this document are those of the authors and should not be interpreted as representing the official policies, either express or implied of the Army Research Laboratory or the U.S. Government. The U.S. Government is authorized to reproduce and distribute reprints for Government purposes not withstanding any copyright notation hereon.

Thanks are due to Paul McConnelee and Eladio Delgado of GE Global Research for their assistance during fabrication and testing. We also thank Prof. Albert Shih at the University of Michigan for fabrication of the micro-milled channels.

REFERENCES

[1] D.B. Tuckerman and R.F.W. Pease, "High-performance heatsinking for VLSI", IEEE Electronic Device Letters, EDL-2, pp. 126-129, 1981.

[2] R.W. Knight, D.J. Hall, J.S. Goodling, and R.C. Jaeger, "Heat sink optimization with application to microchannels," IEEE Transactions on Components, Hybrids, and Manufacturing Technology, vol. 15, no. 5, pp. 832-842, October 1992.

[3] S.V. Garimella, C.B. Sobhan, "Transport in microchannels – a critical review," Annual Reviews of Heat Transfer, vol. 13, pp. 1-50, 2003.

[4] B.C.Y. Cheong, P.T. Ireland, and A.W. Siebert, "High performance single-phase liquid coolers for power electronics," Proceedings of the 3rd International Conference on Microchannels and Minichannels, Paper # ICMM2005-75027, Toronto, ON, June 2005.

[5] J. Schulz-Harder, K. Exel, A. Meyer, T. Licht, and M. Loddenkötter "Micro channel water cooled power modules," Proceedings of the Power Conversion and Intelligent Motion Power Quality Conference, Nuremberg, Germany, June 2000.

[6] J. Valenzuela, T. Jasinski, and Z. Sheikh, "Liquid Cooling for High-Power Electronics," Power Electronics Technology Magazine, February 2005 issue.

[7] S.A. Solovitz, L.D. Stevanovic, and R.A. Beaupre, "Micro-channel thermal management of high power devices" IEEE 2006 Applied Power Electronics Conference, APEC 2006, pp. 885-891.

[8] S.A. Solovitz, L.D. Stevanovic, R. Beaupre, "Microchannels take heatsinks to the next level," Power Electronics Technology Magazine, November 2006 issue.

[9] R. Fillion, E. Delgado, R. Beaupre, P. McConnelee, "High power planar interconnect for high frequency converters," Proceedings of the 6th Electronics Packaging Technology Conference, EPTC 2004, pp. 18-24, Singapore, 2004.

[10] A.G. Pautsch, A. Gowda, L.D. Stevanovic and R.A. Beaupre, "Double-sided microchannel cooling of a power electronics module using power overlay," Proceedings of International Electronic Packaging Technical Conference and Exhibition, IPACK 2009, San Francisco, CA, July 19-23, 2009.

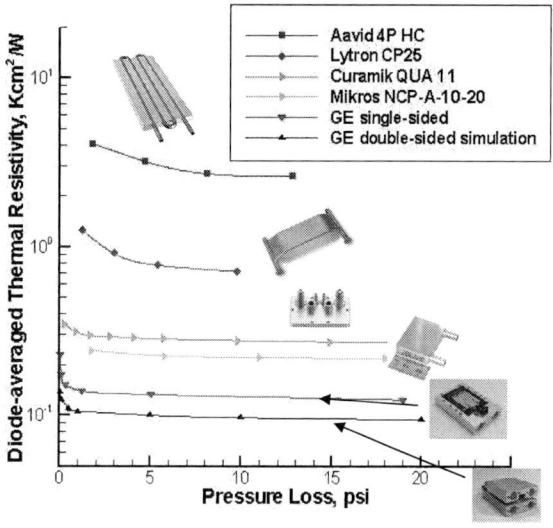

Figure 18. Thermal performance of the GE micro-channel heat sinks compares favorably to all other liquid cooled heat sink designs.

3000V, 25A Pulse Power Asymmetrical Highly Interdigitated SiC Thyristors

Ahmed Elasser, Peter Losee, Stephen Arthur, Zachary Stum, Jerome Garrett, and Michael Schutten

GE Global Research, Niskayuna, NY, 12309
Email: Ahmed.elasser@ge.com

Abstract- **A 3000V, 25A asymmetrical Silicon Carbide (SiC) Thyristor for pulse power applications is described here. It was fabricated on ultra low micropipe density 4H-SiC wafers. The device design, fabrication, wafer testing, packaging, static and dynamic characteristics are presented. The devices' chip area is 4mmx4mm, the yield after screening for blocking voltage, leakage current, forward drop, and latching is over 80%, a very high yield by the standards of SiC power devices. The chips are packaged in a 200°C capable, low profile, surface mount package with a low junction to case thermal resistance. The chips were tested both on-wafer and as packaged parts and exhibit very low leakage current (less than 1μA at 3000V and 25°C, less than 10μA at 3000V and 150°C), low forward drop at 25A (4.55V at 25A). Several devices were packaged and subsequently tested over several days under high current with no forward voltage degradation observed. While the devices were designed for 3kV blocking voltage, the actual packaged parts had a breakdown voltage >3.5kV. Pulse tests were performed to determine the devices' maximum pulse current, di/dt, and dv/dt performance.**

I. INTRODUCTION

Over the past two decades, SiC substrates and material have been evaluated for semiconductor power devices due to their inherent material characteristics that allow device designers to make very high voltage devices with a tenth of the blocking layer thickness required for similarly rated Silicon power devices [1]. Similarly, application engineers are attracted to this material for its ability to switch at ten times higher frequency than its silicon counterpart and to operate at very high junction temperatures (as high as 300°C), hence enabling a substantial reduction in the size of the passive elements such as capacitors, inductors, transformers, and heat sinks. With these valuable properties and application benefits, much work to develop SiC power devices has been done [2].

Substrate and material efforts took the front stage with the goal of making defect free substrates and large area wafers. Several defects plagued the material, chief among them were the micropipes, which proved to be killer defects, and a major limitation to making large area chips required for high current power devices [3]. Other defects such as screw

dislocations, and basal plane dislocations were also present in the material. The latter defects became the primary culprits in limiting the progress of bipolar devices, since they were found to cause the so-called forward voltage drift or as it is commonly referred to "Bipolar Degradation" [4]. Unipolar devices such as the SiC MOSFET are also facing many challenges [5].

Bipolar SiC devices such as the PiN diode, Thyristor, and GTO have been fabricated and explored since the early days of SiC [6]. Bipolar degradation, micropipes, and initial material cost have severely limited the progress of these devices and while many papers were published on these devices, yields were often very low, and performance is not on par with the material entitlement. SiC Thyristors were initially explored for pulse power applications, since their potentially high current densities, high blocking voltage, high switching frequency, and high junction temperature are ideal for applications such as the electromagnetic gun. This application requires currents of one million Amps. peak, turn-off times on the order of few microseconds, and very high junction temperatures [7-9].

Other bipolar devices that are derived from the basic four-layer thyristor structure such as the GTO have also been reported in the literature [10]. Both the Thyristors and GTOs use an N-type SiC Substrate to build the four-layer structure, therefore they require a negative gate current to turn-on. The devices reported in this paper have been under investigation over the past 5 years and several iterations were performed to improve the overall yield, to meet voltage, current, and other performance metrics such as low leakage current, minimal degradation, high di/dt and dv/dt, and low forward drop voltage under nominal and pulse power conditions.

II. DEVICE DESIGN, FABRICATION, AND CHIP TESTING

Unlike Silicon, where P-type substrates are readily available for high power devices, only N-type substrates are commercially available for SiC. This is partially due to low demand since P-type SiC is plagued by low hole mobility and low ionized carrier concentrations due to deep acceptor energy levels (>200meV). Therefore, to make a four-layer device, we start with an N-substrate and build an NPNP structure. The resulting structure is shown in Fig. 1 and it has an N-gate layer, hence the device requires a negative gate-anode current to turn it on. The structure is formed by first etching through the anode layer down to the N-base in closely spaced lines to create a highly interdigitated pattern of gate and anode connections. The highly interdigitated design enables the high di/dt and dv/dt needed for a high performance pulse power device. The lightly doped P-type blocking layer is 30μm thick and is doped at 10^{15}cm^{-3} to safely achieve a 3000V blocking voltage (Maximum parallel plane breakdown voltage, BV$_{PP}$ is 4.9kV).

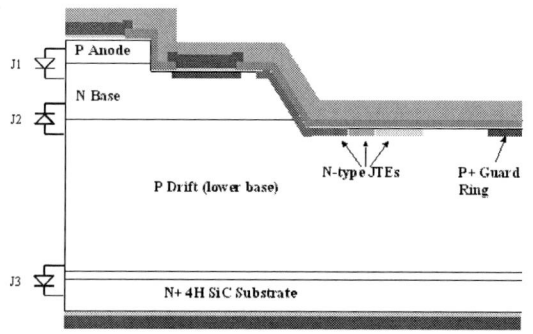

Figure 1: 3000V, 25A SiC Thyristor Device Structure Cross Section

A second, sloped etch is performed through the N-base at the edge of the device for the edge termination. Subsequent thick oxide layers are patterned and used as ion implant masks to create highly-doped regions for p-type and n-type contacts for the anode and gate, respectively, and to form the 3-zone junction termination extension (JTE). A p-type guard ring is formed at the same time as the anode implant. The three-zone, implanted N-type JTE is used to achieve high forward blocking voltage while being robust to process variations. The devices are asymmetrical, blocking high voltage only in the forward direction, and their reverse blocking voltage is on the order of 50V. For many pulse applications, only forward blocking is required.

Although the total chip area is 16mm^2, most of the area is consumed by the terminations and contact pads. The device's active area is about 1/3 of the total area, hence the device current density is quite high. Interconnect metal and interlayer dielectrics are used to connect the interdigitated fingers together while keeping the gate and anode isolated from each other. Thick pad metal is used to enable high pulse currents, and high-voltage passivation is used for device protection. The device P-contact anode resistance was optimized to reduce the forward drop during pulse conditions, where the anode-cathode current exceeds 1000A. A lower ohmic contact resistance is achieved.

Several hundred chips were tested at the wafer level on a probe tester using the Tektronix 370 and 371 curve tracers. The parts were first screened for latching, leakage current at 1kV, and forward drop at a few Amps. A second screen is performed where the parts are tested at a very high voltage (up to 2000V) on the 370 curve tracer and then up to 3200V on the 371 curve tracer while immersed in Fluorinert to prevent arcing. At the chip level, the blocking voltage yield was 90% and the composite yield when other factors such as leakage current, and forward drop are accounted for is above 80%. Better SiC substrates, epitaxial material as well as a number of process and design improvements have all led to an efficient, high-yield fabrication process [10].

The 3.2kV limit is imposed by the maximum voltage of the 371 curve tracer. When the devices were packaged in a high temperature package, they were tested using a standalone high voltage tester. The packaged devices were passivation limited to a maximum blocking voltage of 3.5kV. The chips were also stressed for several days under very high current densities (>100A/cm^2) to test for the bipolar degradation effect. No drift in the forward voltage was observed.

III. DEVICE PACKAGING, TESTING, AND STATIC CHARACTERISTICS

After the chips were screened for blocking voltage, leakage current, forward drop, and gate characteristics, the selected chips were packaged in a low profile, high temperature, surface mount package by Powerex. Figure 2 shows the device in its package; the chips are connected via a wirebond to the gate terminal and through two straps to the anode fingers. The cathode is connected to its terminal on the back of the chip.

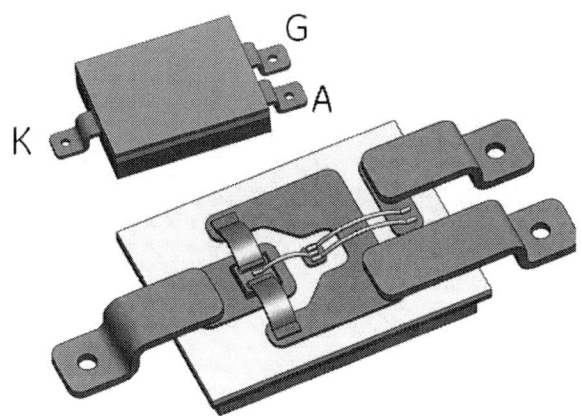

Figure 2: 3000V, 25A Surface Mount SiC Thyristor Package

This package configuration allows for ample separation of the gate/anode topside from the cathode backside for electrical isolation. A cavity is formed by the package sidewalls (not specifically illustrated), which allows for the chip and wires to be immersed in silicone gel. The packaged devices were subsequently tested under static conditions. Figure 3 shows a photograph of the packaged devices.

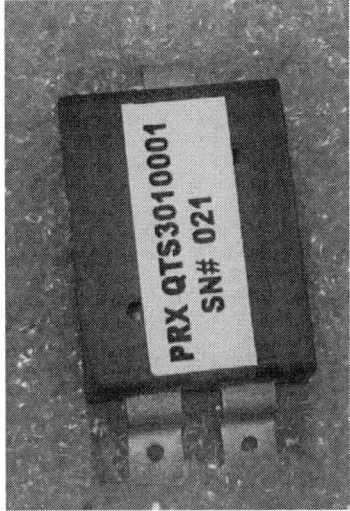

Figure 3: 3000V, 25A SiC Thyristor packaged in a high voltage, high temperature (250°C) surface mount package

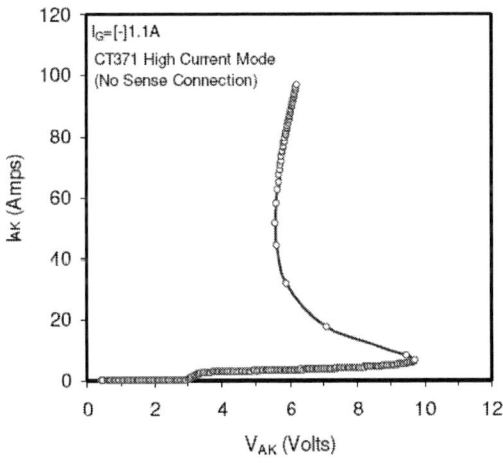

Figure 4: I-V Characteristics at Igate=-1.1A

Figure 4 shows the forward conducting I-V characteristic including the classic thyristor transition, snap-back, into the latched state. The on-state I-V characteristic is taken at a 100A pulse current using a 371-curve tracer. The forward voltage drop of Vak=6.3V@ Iak=100A (Jak=1.6kA/cm2) is recorded including lead and contact resistance. Meanwhile, the differential on-resistance of Ron,diff<1mΩ-cm2 illustrates significant conductivity modulation and good p-type anode contact resistance.

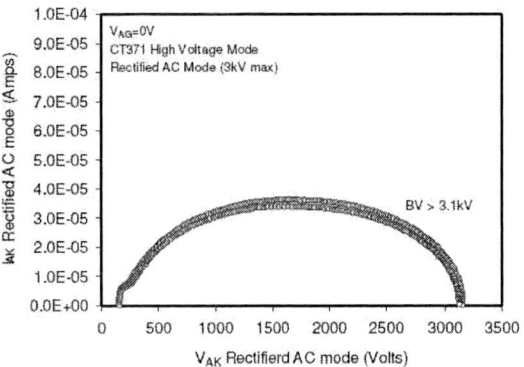

Figure 5: Forward Blocking Characteristics in the AC Loop Mode

Figure 5 shows the off-state I-V characteristics, the forward blocking curve, where only curve tracer capacitive looping is detectable in the voltage range measured. Figure 6 shows the DC leakage current at 3kV for junction temperatures from 25°C to 150°C. The devices exhibit a very low leakage current at room temperature and at high temperature as shown in Fig. 6. A maximum leakage current of 1μA is measured at RT on several packaged parts.

At 150°C, the maximum leakage current increases to 10μA. The high temperature (150°C) SiC leakage current is at least two to three orders of magnitude lower than the leakage current of an equivalent 3kV Silicon thyristor at 125°C junction temperature. At temperatures above 150°C, the SiC thyristors still exhibit very low leakage currents. A high

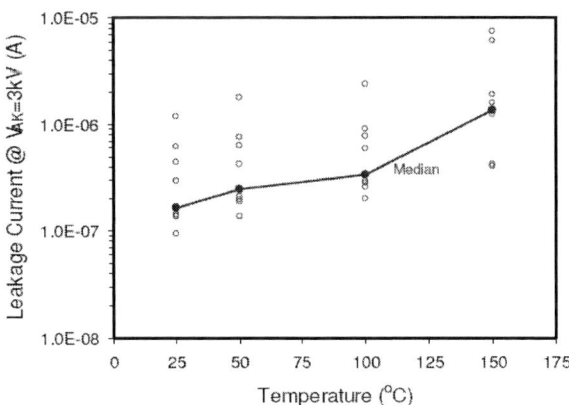

Figure 6: Leakage Current at 3000V and at junction temperatures from 25°C to 150°C

temperature package that is capable of 250°C and above is needed to take advantage of the SiC high temperature capabilities. In pulse applications, the switching device operates at low duty cycles and holds the blocking voltage most of the time, hence the need for very low leakage current.

IV. DYNAMIC CHARACTERISTICS UNDER PULSE CONDITIONS

Under pulse testing, the SiC parts have a much lower forward drop than a similarly rated Si part due to the thinner drift layer which is highly conductivity modulated, hence has a low differential on-state resistance. Figures 7 and 8 show a comparison between the SiC Thyristor and a 1500V Si device (Two of these Si devices need to be in series to meet the 3000V blocking voltage). In Figure 7, the SiC thyristor is pulsed to 390A for a 10us pulse width, its forward drop is only 22V at peak current. For the 1500V Si thyristor, under the same conditions, the forward drop is 25.6V at peak current as shown in Fig. 8. For an equivalent 3000V Si thyristor, the forward drop will be on the order of 51V, more than two times the SiC thyristor forward drop.

When the pulse width of the current waveform is reduced from 10µs to 3µs, the difference between the Si and SiC part increases. The SiC Thyristor still exhibits a low forward drop on the order of 22.8V at peak current as shown in Fig. 9. The Si thyristor exhibits a large forward drop on the order of 64V at peak current as shown in Fig. 10. The Si part did not reach conductivity modulation state with the short pulse width. For an equivalent 3000V Si thyristor, the forward drop is on the order of 128V at peak current. Note also that the Si Thyristor operating current is lower than its SiC counterpart.

The results shown so far were for our first generation of SiC parts. A second generation of SiC parts was designed and fabricated in 2008 and tested in 2009.

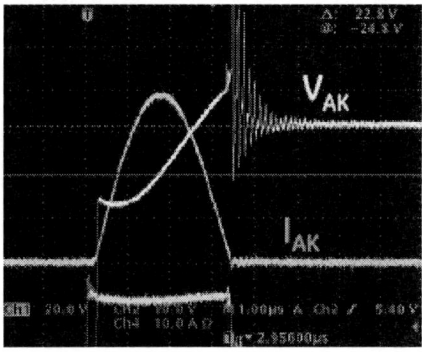

Figure 9: 3000V SiC Thyristor pulse testing, Iakp=390A (6240 A/cm2), Vak@Iakp=22.8V, pulse width=3µs – Green waveform: Current Iak (100A/div), Yellow waveform: Voltage Vak (20V/div), Timescale: 1µs/div

Figure 11 shows the results of a pulse test in which the SiC thyristor peak current is 1200A and the dc bus voltage is 1000V. The equivalent current density under pulse current is 19.2kA/cm^2. These high current densities are an order of magnitude higher than equivalent Si Thyristor current densities.

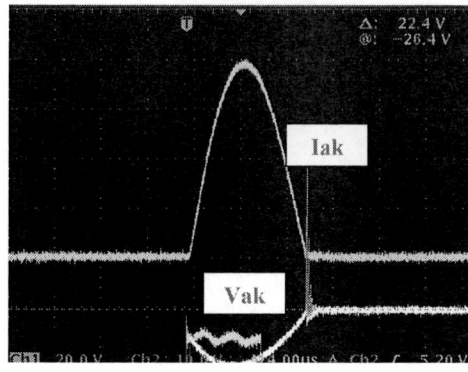

Figure 7: 3000V SiC Thyristor pulse testing, Iakp=390A (6240 A/cm2), Vak@Iakp=22V, pulse width=10µs – Green waveform: Current Iak (100A/div), Yellow waveform: Voltage Vak (20V/div), Timescale: 4µs/div

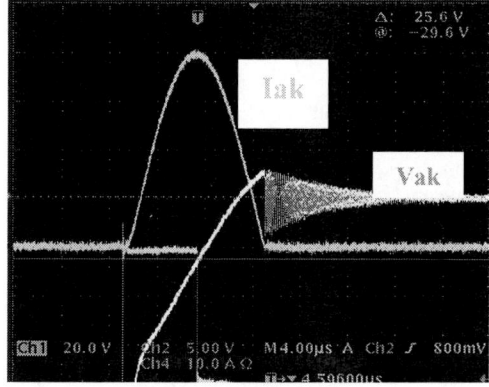

Figure 8: 1500V Si Thyristor pulse testing, Iakp=390A (2550 A/cm2), Vak@Iakp=25.6V, pulse width=10µs – Green waveform: Current Iak (100A/div), Yellow waveform: Voltage Vak (20V/div), Timescale: 4µs/div

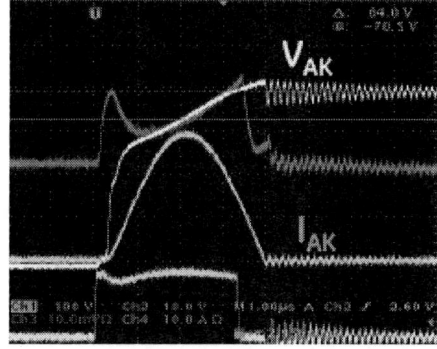

Figure 10: 1500V Si Thyristor pulse testing, Iakp=290A (1896 A/cm2), Vak@Iakp=64V, pulse width=3µs – Green waveform: Current Iak (100A/div), Yellow waveform: Voltage Vak (20V/div), Timescale: 1µs/div

978-1-4244-4782-4/10 $26.00 © 2010 IEEE

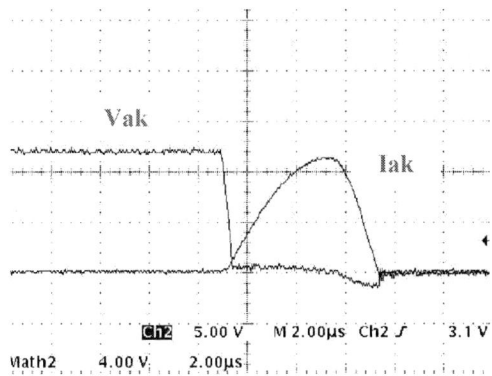

Figure 11: Gen. 2 SiC Thyristor devices' voltage and current under pulse conditions (Vak: 400V/div, Iak: 500A/div, 2μs/div)

The forward drop of the SiC thyristor at 1200A is about 20V. This improvement from Generation 1 devices to Generation 2 is due to the improvement in the design and fabrication processes. Although the di/dt is about 200A/μs, the devices are capable of much higher di/dts and dv/dts.

V. SUMMARY AND CONCLUSION

A relatively large area high voltage SiC Thyristor that exhibits excellent performance characteristics at the on and off-state, very low leakage current at high junction temperatures, and a high yield is presented. Pulse testing results show that the device is capable of very high current densities, and high peak currents. It also exhibits very low forward voltage at high pulse currents due to the thinner drift layer thickness. The devices were packaged in a high voltage, high temperature surface mount package.

For applications where high voltage, high pulse current, and high temperature are required, the SiC thyristor simplifies the circuit design, and enables reduced size and weight, hence compact power converters. The SiC parts significantly reduce conduction, leakage, and switching losses, thereby dramatically reducing cooling requirements and improving converter efficiency. The high current density under pulse conditions will lead to very high pulse current once large area SiC defect free substrate and material become a reality.

This progress is made possible by recent improvements in the quality of the SiC substrates (ultra low micropipe density), excellent epitaxial material (highly uniform with low defects and reduced internal stresses), great advances in device fabrication, packaging, and testing. With the continuous improvement in SiC substrates and material, and with the availability of SiC wafers from multiple suppliers, the cost of SiC devices will continue to decrease and will ultimately become competitive in the power device marketplace.

ACKNOWLEDGMENT

The authors acknowledge Dr. Scott Leslie, Duane Prussia and the Powerex packaging team for developing custom high voltage packages for the SiC thyristors discussed in this paper. The GE Global Research cleanroom team effort and dedication is highly appreciated. Special thanks to the thyristor fabrication and testing team: James Kretchmer, Peter Gipp, Patti Sicluna, Jason Galea, James Elson, Margaret Lazzeri, Timothy Vandenbriel, James Schermerhorn, Matt Edmonds, Michael Mero, Donald Roy, Derrick Brewer, Justin Welch, and Tammy Johnson.

REFERENCES

[1] B.J. Baliga, "Power Semiconductor Devices," PWS Publishing Company, Boston (1996)

[2] M. Bhatnagar, and B.J. Baliga, "Analysis of Silicon carbide Power Device Performance," Proceedings of the 3rd International Symposium on Power Devices and ICs, pp. 176-180, April 1991.

[3] P.G. Neudeck, and J.A. Powell, "Performance Limiting Micropipe Defects in Silicon Carbide Wafers," IEEE Electron Device Letters, Vol. 15, Issue 2, pp. 63-65, 1994.

[4] M. Zhang, P. Pirouz, and H. Lendenmann, "Transmission Electron Microscopy investigation of dislocation in forward-biased 4H-SiC p-I-n diodes," Applied Physics Letters, Vol. 83, Issue 16, pp. 3320-3322, 2003.

[5] K. Matocha, "Challenges in SiC Power MOSFET Design," International Symposium on Semiconductor Device Research, pp. 12-14, 2007.

[6] T. Burke, K. Xie, J.R. Flemish, H. Singh, and J. Carter, "Silicon Carbide Thyristors for Power Applications," 10th International Pulsed Power Conference, Vol. 1, pp. 327-335, 1995.

[7] T. Burke, K. Xie, H. Singh, T. Podlesak, and J.R. Flemish, "Silicon Carbide Thyristors for Electric Guns," IEEE Transactions on Magnetics, Vol. 33, Issue 1, pp. 432-437, 1997.

[8] H. O'Brien, W. Shaheen, and S.B. Bayne, "Evaluation of 4mmx4mm Silicon Carbide Thyristors," IEEE Transactions on Dielectrics and Electrical Insulation, Vol. 14, Issue 4, pp. 986-993, 2007.

[9] A.K. Agarwal, B. Damsky, J. Richmond, S. Krisnaswami, C. Capell, S.H. Ryu, and J.W. Plamour, "The First Demonstration of the 1cmx1cm SiC Thyristor Chip," Proceedings of the 17th International Symposium on Power Devices and ICs, pp. 195-198, May 2005

[10] J. Milan, "Wide Bandgap Power Semiconductor Devices," IET Circuits, Devices, & Systems, Vol. 1, Issue 5, pp. 372-379, 2007.

Low Inductance Power Module with Blade Connector

Ljubisa D. Stevanovic, Richard A. Beaupre, Eladio C. Delgado, and Arun V. Gowda

General Electric Global Research Center
Niskayuna, NY 12309
stevanov@ge.com

Abstract — A novel single-switch power module has been developed, featuring a laminated blade connector for low inductance interconnect to a busbar. The module was designed, optimized and experimentally validated as part of a high frequency three-phase converter, demonstrating parasitic inductances of less than one nano henry for the module and as low as five nano henries for the converter phase-leg commutation loop. The flexible plug-in hardware facilitated direct comparison of switching performance between three different chipsets, including a 150A and a 300A hybrid designs using the fastest 1200V silicon IGBTs with silicon carbide (SiC) Schottky diodes, as well as a 150A all-SiC module with emerging SiC MOSFETs. The results were also compared with switching performance of standard modules. First, the impact of parasitic inductance on switching performance was quantified by testing the same 300A hybrid chipset in an industry-standard module. Compared to the low inductance blade POL module, the standard module had 65% higher voltage overshoot and 30% higher total switching losses. Second, the switching performance of the 150A, 1200V fast IGBT, in either standard silicon or the hybrid blade module, was compared with the all-SiC blade module under the same test conditions. The IGBT switching losses of the standard silicon module were 3.5 times higher, while the hybrid blade module losses were 2.5 times higher than those of the all-SiC module. The new low inductance blade module is an excellent package for the new generation of fast silicon IGBTs and the emerging SiC power devices. The module will enable efficient power conversion at significantly higher switching frequencies and power densities.

I. INTRODUCTION

The workhorse of today's multi billion-dollar industrial power electronics sector is the silicon IGBT device. For medium and high power applications requiring devices with blocking voltages between 1.2kV and 6.5kV, the IGBT offers the best combination of performance, cost, ruggedness, and ease of control compared to all other competing devices. Since the introduction of IGBTs in the late 1980's, there have been as many as five generations of IGBT designs, offering varying degrees of improvement in performance, cost, and ruggedness. The one fundamental shortcoming of the IGBT is switching speed. Unlike a majority carrier device, such as MOSFET, the minority carrier IGBT is limited in switching speed. Emerging silicon-carbide (SiC) power MOSFETs can be switched at much faster rates than the IGBTs of the same ratings. When

compared to the Si IGBT, the SiC MOSFET offers significant reduction in conduction and switching losses, as well as operation at higher temperatures [1,2]. It is, therefore, expected that the SiC MOSFET will replace the Si IGBT in high-end applications, such as high-density converters.

As the next level of assembly for semiconductor power devices, the power module facilitates integration of devices with the rest of the converter. The power module technology has evolved at a slower pace than the power semiconductor devices. The trend has primarily been in the direction of standardized interfaces and footprints. This convergence has resulted in significant cost reduction due to direct competition and ever-increasing manufacturing volumes. In turn, the lower component costs have made power electronic system solutions more competitive when compared with alternatives, driving up the volumes and reinforcing the virtuous cycle. For example, lower-cost power electronics make the wind farms cost competitive with fossil power plants, increasing penetration of wind power in the power grid. One consequence of the convergence trend is that the standard modules have been optimized for slower switching IGBTs. In an effort to satisfy the competing electrical, thermal, and mechanical requirements of module designs, there was not enough emphasis on reducing parasitic inductances. The large parasitic inductance values, typically in the range from 20 to 100 nH, render the standard module unsuitable for fast switching Si and SiC devices.

A more radical departure from the standard module is presented in this paper. Fig. 1 shows a photograph of the advanced all-SiC power module with low inductance blade power connector.

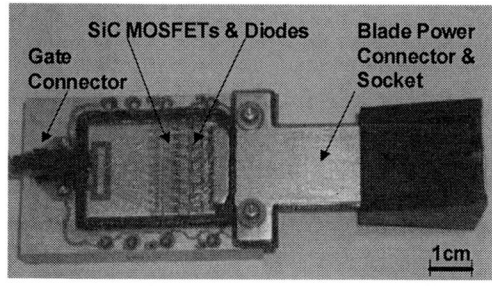

Figure 1. The all-SiC 150A, 1000V low inductance power module with blade connector. The module has 10 GE SiC MOSFETs and 15 SiC diodes.

978-1-4244-4782-4/10 $26.00 © 2010 IEEE

The module parasitic inductance is in the range of 1 nH. Section II of the paper describes the new module design and implementation of the ultra-low inductance phase-leg consisting of the two blade modules. Section III explores optimization of module design and layout, including two types of topside device interconnect: wirebonds and wirebond-less. Section IV of the paper describes the module fabrication. Section V describes experimental setup, including the use of high frequency inverter for switching tests. Experimental results are presented in Section VI. Finally, Section VII concludes the paper.

II. DESIGN APPROACH

Fig. 2 shows a cut-away view of the standard module with screw-type main power terminals. The wide spacing between terminals contributes majority of the module's inductance.

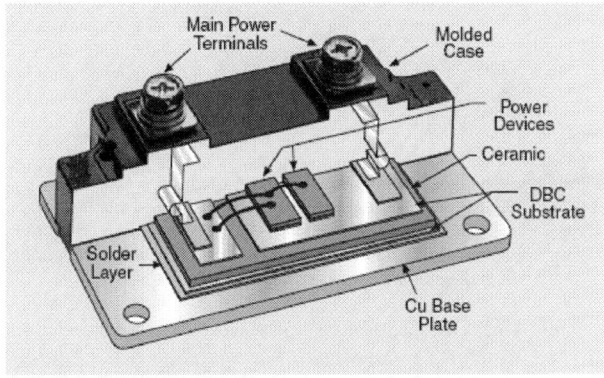

Figure 2. Cut-away view of a standard power module with screw-type power terminals. Wide spacing between the terminals is responsible for large parasitic inductance, typically in the range of 20 to 100 nH.

Fig. 3 shows a schematic of an IGBT phase-leg with parasitic inductances L_{Q1} and L_{Q2} associated with switches Q1 and Q2, parasitic inductances L_{D1} and L_{D2} associated with diodes D1 and D2, and busbar parasitic inductance L_{DC}.

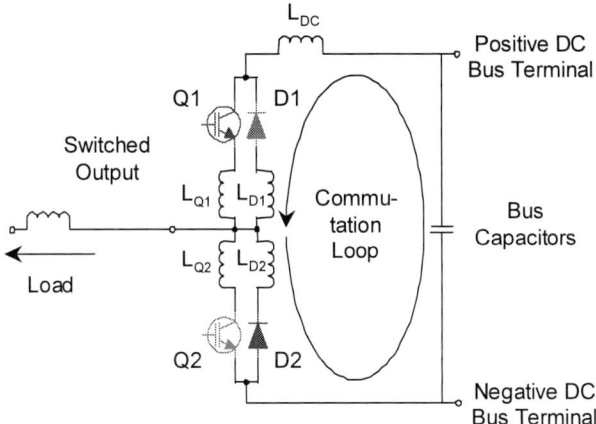

Figure 3. Schematic of the switching phase-leg shows parasitic inductances that contribute to the total commutation loop parasitic inductance. The current circulates in a commutation loop during switching between S1 and S2 (Q1 and D2, or Q2 and D1).

In a typical PWM inverter circuit, the load current commutates between Q1 and D2 when the current is positive, or between Q2 and D1 when the current is negative. Each commutation induces a current transient along the commutation loop, as indicated in Fig. 3. For commutations between Q1 and D2, their respective parasitic inductances L_{Q1} and L_{D2}, together with L_{DC} contribute to the total parasitic inductance of the commutation loop

$$L_{Q1D2} = L_{Q1} + L_{D2} + L_{DC.} \qquad (1)$$

Likewise, the total parasitic loop inductance during commutation between Q2 and D1 is

$$L_{Q2D1} = L_{Q2} + L_{D1} + L_{DC.} \qquad (2)$$

In most module layouts these two inductances are similar in value

$$L_{Q1D2} = L_{Q2D1} = L_{loop}. \qquad (3)$$

The voltage overshoot V_S across the switch during the turn-off transient is proportional to the loop inductance L_{loop} and the rate of current change $\Delta I/\Delta t_{OFF}$:

$$V_s = L_{loop} \cdot \frac{\Delta I}{\Delta t_{OFF}}. \qquad (4)$$

For a converter with the parasitic inductance $L_{loop} = 100$ nH and operating at 600V DC bus, a moderately fast turn-off transient $\Delta I/\Delta t_{OFF} = 5000A/\mu s$ would cause a 500V spike, leaving very little margin for the 1200V-rated devices. Additionally, large parasitic inductance increases switching losses and causes electromagnetic interference and compatibility (EMI/EMC) problems. Finally, mismatched inductances between paralleled devices could lead to uneven switching delays and poor current sharing during transients. This could lead to increased stresses, low reliability and device failures.

Previous attempts to reduce the module parasitic inductances [3-7] have been constrained by the desire to maintain compatibility with standard screw-type power terminals, focusing instead on internal layout of the module. By failing to address the inductance of power terminals, these designs have achieved limited improvements.

The faster SiC MOSFET devices require a paradigm shift away from the standard power module and the emphasis on reducing parasitic inductances throughout the switching circuit. The new low inductance power module is shown in Fig. 1. The module features a blade power connector with two laminated power terminals for plugging the module into a standard high current socket mounted to the busbar. The low-inductance blade connects the power devices inside the module with appropriate busbar layers. A thin (254 μm) layer of dielectric material separates the two laminated connector (copper) plates. The bottom plate of the connector is soldered to the top-metallization layer of the module's direct bond copper (DBC) substrate, interconnecting the bottom side of all power devices (diode cathode, IGBT collector, or MOSFET drain). The top side of the devices (anode, emitter, or source) is directly wire-bonded to the top of the blade connector. Alternatively, a wirebond-less device interconnect, such as GE Power OverLay (POL) [8], has been used for connecting the device top side to the connector.

The new design incorporates a symmetrical layout and interconnect of the semiconductor power devices within the switch module for excellent static and dynamic current sharing.

The blade module concept of parallel conductor plates separated by a thin insulating material readily extends throughout the entire converter power stage, as shown conceptually in Fig. 4. The laminated busbar consists of a switched output layer sandwiched between positive and negative DC bus layers. Several low inductance capacitors, such as multi-layer ceramic (MLC) capacitors, are connected across the back of DC busbar to minimize the parasitic inductance L_{DC}. On the face of the busbar are high current sockets to plug in the blade modules. This enables the low inductance phase-leg assembly consisting of two blade modules. A variety of converter topologies and V/I ratings can be implemented with appropriate number of modules and busbar layout.

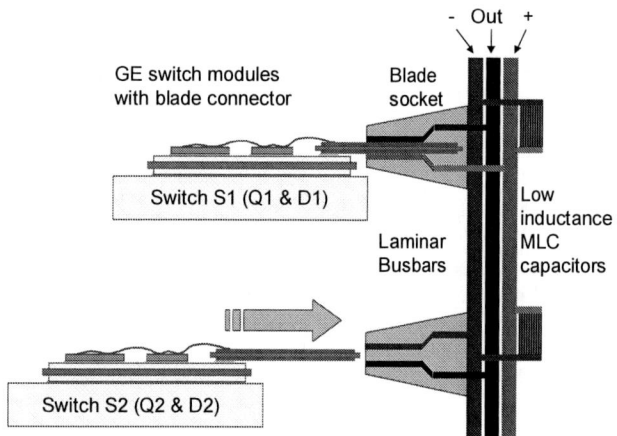

Figure 4.   Conceptual cross sectional diagram (not to scale) of the phase-leg assembly consisting of two blade modules with sockets mounted on laminated busbars.

III.   DETAILED DESIGN

Ansoft Q3D Extractor® electromagnetic finite element analysis (FEA) tool [9] was used to guide the module design and optimization efforts. In order to calculate the total parasitic inductance of the commutation loop

$$L_{loop} = L_Q + L_D + L_{DC} \qquad (5)$$

the diode and the IGBT parasitic inductances L_Q and L_D were each simulated. The module model included the following geometries: blade connector, topside interconnect

(wirebonded and wirebond-less POL), power devices (IGBTs or diodes), and top copper layer of the DBC substrate. Fig. 5 shows simulated geometries for the two versions of the topside interconnect.

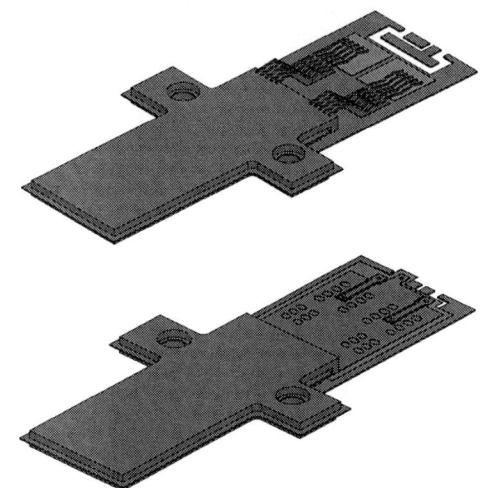

Figure 5.   The blade module design was simulated and optimized using Ansoft Q3D Extractor® electromagnetic FEA tool. The analysis considered two types of topside interconnect: wirebonds (top) and wirebond-less Power OverLay (bottom). The simulated geometries captured high level of detail, including the wirebond and copper via geometries.

Table 1 provides a summary of parasitic inductance values extracted from the simulations of the module and phase-leg geometries in Fig. 4 and 5. The predicted module inductance values were 2.5 nH for the wirebonded and 1 nH for the POL, while the phase-leg inductances were 8 nH with the wirebonded and 5 nH with the POL modules.

TABLE I.   CALCULATED AND EXPERIMENTAL VALUES OF PARASITIC INDUCTANCES

Parameter	Simulated POL (nH)	Simulated Wirebond (nH)	Experi-mental (nH)
Diode, including anode interconnect & blade connector	0.94	2.27	----*
IGBT, including emitter interconnect & blade connector	1.02	2.67	0.8*
Elcon connector (2 per phase-leg)	0.85	0.85	0.9
U-shaped busbar	2.09	2.09	2.5
Ceramic decoupling capacitors (6 in parallel)	0.23	0.23	0.1
Total Loop Inductance L_{loop}	**5.13**	**8.11**	**5**

* Measurements performed on a 200A wirebonded module with one shorted IGBT device.

For the busbar-mounted high current socket, several available products were evaluated based on the module current and voltage ratings. The Elcon dual CROWN CLIP™ socket [10] was selected.

Finally, detailed design of the three-phase converter was completed, including the laminated U-shaped busbar with high frequency ceramic and DC link film capacitors, gate drive and control circuits. Fig. 6 shows a high-frequency prototype converter with six blade modules (obscured by mounting brackets and covered by gate drive board on the left side of the image). The right side of the image shows the U-shaped busbar (white) with high frequency MLC and DC link film capacitors (black), with control interface board in the foreground. The converter continuous power output is over 100kVA.

Figure 6. A high-frequency converter with six blade modules (obscured by mounting brackets and covered by gate drive board on the left side of the image) has continuous output power rating over 100kVA. Right side of the image shows U-shaped busbars (white) with high frequency MLC and DC link film capacitors (black), with control interface board in foreground.

IV. MODULE FABRICATION

The following modules are compared in this paper:

- 300A, 1200V standard Infineon hybrid module with fast KS4 IGBTs and SiC Schottky anti-parallel diodes

- 300A, 1200V GE POL (wircbond-less) hybrid blade module with fast KS4 IGBTs and SiC Schottky anti-parallel diodes

- 300A, 1200V standard Powerex Si module with fast NFH IGBTs and Si P-i-N diodes

- 150A, 1200V GE hybrid blade module with fast NFH IGBT and SiC Schottky anti-parallel diodes

- 150A, 1000V GE all-SiC blade module with GE SiC MOSFETs and SiC Schottky anti-parallel diodes

Several samples of each blade module design were fabricated for this test. The standard Infineon hybrid module was procured and tested, while the data for the Powerex silicon module (CM300DU-24NFH) was scaled per 150A chip based on the module datasheet.

In order to directly quantify impact of parasitic inductance on switching waveforms and losses, the nearly-identical[1] 300A, 1200V hybrid chipsets (fast KS4 IGBTs + SiC diodes) were used in two different packages: the standard EconoDUAL™ module from Infineon and the GE POL blade

---

[1] The Infineon EconoDUAL™ 300A hybrid module uses three 100A KS4 IGBTs and twelve 15A Infineon SiC Schottky diodes per switch, while the GE 300A POL module uses two 150A KS4 IGBTs and twenty-four 10A Cree SiC Schottky diodes per switch.

module shown in Fig. 7.

POL replaces aluminum wire bonds with a planar copper film featuring electroplated via connections to the devices. The POL fabrication process is described in [8]. The module includes an integral heat sink that provides a significantly better thermal performance over cold plates used in conventional modules [11]. However, since the comparisons were made using double-pulse switching tests, there is no effect of the heatsink performance on the electrical results.

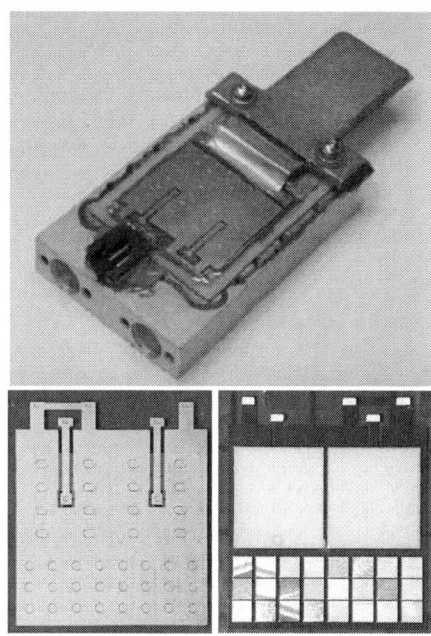

Figure 7. A 300A, 1200V hybrid blade module features two 150A fast IGBTs (Infineon KS4) and 24 of the 10A SiC diodes from Cree. The challenge of wirebonding many small devices is circumvented by using POL. Photographs of both sides of the POL sub-assembly are also shown.

The prototype 300A, 1200V GE POL modules were assembled at GE Global Research using a 63Sn/37Pb solder alloy in a single-step reflow process. The following solder joints were formed during the process: POL-to-AlN DBC substrate, the substrate-to-baseplate, the blade connector bottomside-to-substrate, gate connector-to-POL, and POL-to-blade connector topside.

The two wirebonded blade modules, the 150A hybrid and the 150A all-SiC, were fabricated using a slightly different fabrication process. An AlN DBC was first bonded to a baseplate/integrated heatsink using high temperature solder (92.5Pb/5Sn/2.5Ag). Devices were then soldered onto the top metallization using SAC (95.5Sn/3.8Ag/0.7Cu) solder. The gate connector and power blade connector were soldered using a lower temperature solder (63Sn/37Pb). The devices were then wirebonded using aluminum wire to the blade connector. Finally, the devices were encapsulated in a silicone gel.

V. EXPERIMENTAL SETUP

First, impedance measurements of the converter power stage were performed to determine parasitic inductance contributions from various components and geometries. The measurements were performed at 1 MHz using an Agilent 4294A impedance analyzer. The experimental inductance values, summarized in Table 1, were in agreement with simulation results. The blade module and the phase-leg loop inductance values were 0.8 nH and 5 nH, respectively.

Next, one phase-leg of the three-phase converter was used for module inductive switching loss measurements. The setup, shown in Fig. 8, had been used for direct comparison of all three types of blade modules reported in this paper. The phase-leg consisting of a busbar with two blade modules is located in the middle of the photo. The devices under test (DUTs) are wrapped in thermal blankets for testing at elevated temperatures. Resistive heater controllers, located in front and to the right of the DUTs, are used for adjusting the module temperature. A multi-meter behind the controllers is for monitoring the DC link voltage. Two adjustable gate drivers, located in front and to the left of the DUTs, were used for testing the modules. The drivers are capable of producing up to +20V and -10V output waveforms. Gate resistances were optimized for each type of module. Inductive switching was performed using a 149 µH air-core inductor, shown to the left of the DUTs. On the upper shelf are DUT temperature indicators (center) and gate pulse generator (right). Not shown in the image are power supplies and oscilloscope.

A five milliohm, low inductance precision shunt resistor from T&M Research Products, Inc. was installed on the busbar to measure the emitter/source current of the lower switch. The busbar was designed to allow testing with and without the shunt, while ensuring that there was no significant parasitic inductance added to the setup.

Figure 8. Photograph of the switching loss test setup for direct comparison of blade modules with three different chipsets reported in this paper.

VI. EXPERIMENTAL RESULTS

First, the same 300A, 1200V hybrid chipset was evaluated in two different packages: the standard EconoDUAL™ module from Infineon and the GE POL blade module. Both modules were tested under the same conditions: V_{CE} = 600V, I_C = 300A, T_j = 125°C. Turn-off and turn-on switching waveforms are shown in Fig. 9 and 10, respectively.

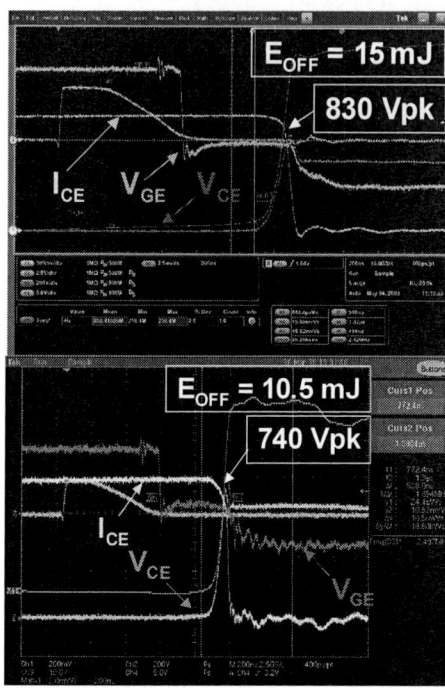

Figure 9. Impact of parasitic inductance on switching waveforms and losses is quantified by testing the same hybrid chipset in two different packages. Significantly higher parasitic inductance of the standard EconoDUAL™ module (top) results in 65% higher voltage overshoot and 43% higher turn-off losses compared to low inductance blade module with POL (bottom). The timescale was 200 ns/div in both cases. Test conditions: V_{CE} = 600V, I_C = 300A, T_j = 125°C.

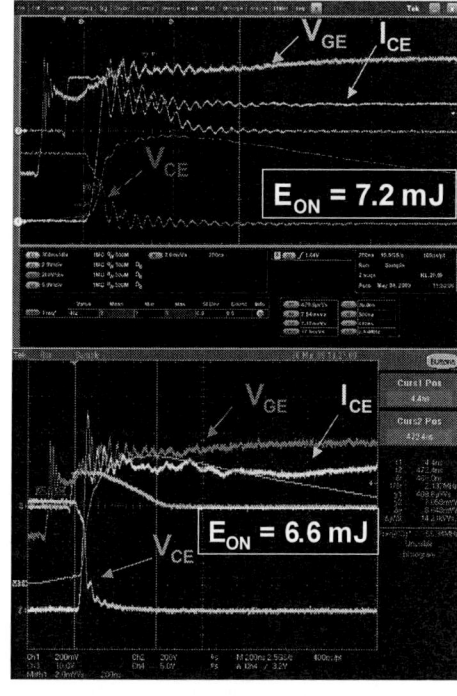

Figure 10. Comparison of turn-on losses between the two hybrid modules: commercially available (top) and the low inductance POL (bottom). The timescale was 200 ns/div in both cases. Test conditions: V_{CE} = 600V, I_C = 300A, T_j=125°C.

Significantly higher parasitic inductances of the standard module result in 65% higher voltage overshoot during turn-off and 30% higher total switching losses compared to the low inductance blade module. There's also significant ringing during the turn-on transient of the standard module that could result in EMI/EMC problems. This will either require slower switching or additional snubbers and filters, further increasing the losses and overall converter complexity.

Next, the published data for the standard silicon module using NFH-series IGBTs (Powerex CM30DU-24NFH, scaled to a 150A chip) was directly compared with the 150A GE hybrid blade module incorporating the same IGBT device. Fig. 11 shows inductive turn-off (upper) and turn-on (lower) waveforms for the 150A hybrid module. The timescale was 80 ns/div in both cases. The turn-off energy was 3.3 mJ and the turn-on energy was 4.9 mJ per pulse, measured under the following conditions: V_{CE} = 600V, I_C = 150A, T_j=125°C.

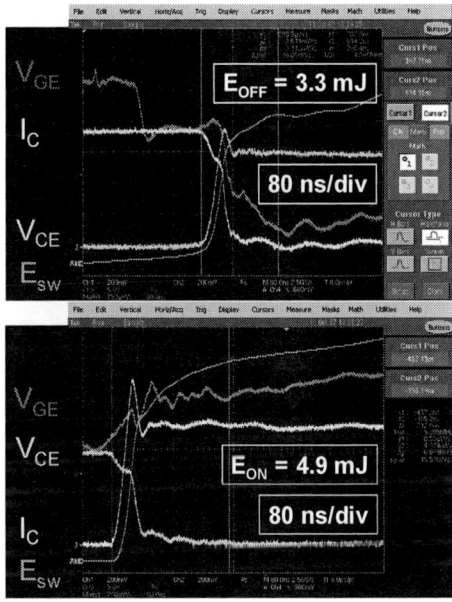

Figure 11. Inductive turn-off (upper) and turn-on (lower) waveforms for the blade module with hybrid chipset (one 150A NFH-series Si IGBT and three 50A SiC diodes). Testing was performed under the following conditions: V_{CE}=600V, I_C=150A, T_j=125°C.

Finally, the switching performance of the 150A, 1000V all-SiC blade module shown in Fig. 1 was measured. The turn-off and turn-on switching waveforms are captured in Fig. 12. The timescales are 200 ns/div and 80 ns/div, respectively. The turn-off energy was 1 mJ and the turn-on energy was 2.3 mJ per pulse measured under the following conditions: V_{DS} = 600V, I_D = 150A, T_j = 25°C. The tests were also repeated at temperatures of 125°C and 175°C. The variation of the losses with temperature was less than 10%.

The turn-off and turn-on losses for the three 150A modules are summarized in Figure 13. The total switching losses (E_{ON} + E_{OFF}) at 150A, 600V were: 11.7 mJ for the standard silicon module, 8.2 mJ for the hybrid blade module, and 3.3 mJ for the all-SiC module.

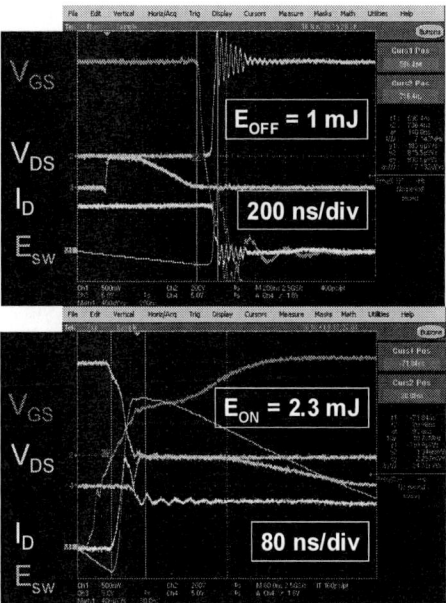

Figure 12. Inductive turn-off (upper) and turn-on (lower) waveforms for the 150A, 1000V all-SiC blade module shown in Fig. 1. Testing was performed under the following conditions: V_{DS} = 600V, I_D = 150A, T_j=25°C.

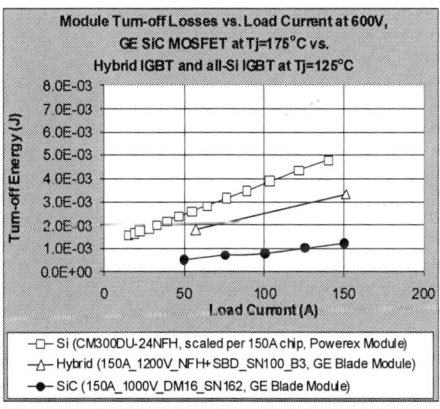

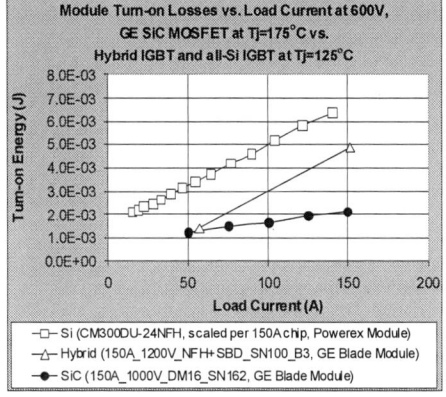

Figure 13. Turn-off (upper) and turn-on (lower) energies vs. load current under inductive switching. The graphs show comparison of three types of modules: silicon IGBT in a standard module (Powerex CM30DU-24NFH), hybrid chipset in GE blade module (the same IGBT chip as the Powerex module + SiC diode), and all-SiC chipset in GE blade module. Test conditions: V_{DC} = 600V, T_j=125°C for IGBTs, T_j=175°C for all-SiC.

978-1-4244-4782-4/10 $26.00 © 2010 IEEE   1608

VII. CONCLUSIONS

The low inductance blade module has been developed and experimentally validated as part of a high frequency three-phase converter, demonstrating parasitic inductances of less than one nano henry for the module and as low as five nano henries for the converter phase-leg. Switching performance was compared across three high frequency chipsets: the fast silicon IGBTs with either silicon P-i-N or SiC Schottky diodes, as well as the all-SiC chipset with emerging SiC MOSFETs. When the same 300A, 1200V hybrid chipset was tested in two different modules, the standard module showed 65% higher voltage overshoot and 30% higher switching losses than the blade POL module. When comparing switching losses across the three chipsets, the all-SiC blade module was the most efficient (E_{ON} + E_{OFF} = 3.3 mJ), the hybrid blade module was the distant second (8.2 mJ) and the standard silicon IGBT module was, not surprisingly, the worst[2] (11.7 mJ).

In summary, the main benefits of the low inductance blade module are: significantly lower switching losses, lower device stresses and less switching noise. The blade module is an excellent package for the new generation of fast silicon IGBTs and emerging SiC power devices. The module will enable efficient power conversion at significantly higher switching frequencies and power densities.

ACKNOWLEDGMENT

Thanks are due to Dr. Adam Pautsch of GE Global Research for his assistance during module testing, Jerry Sherbondy, Jr. of Powerex for fabrication of wirebonded hybrid blade modules and Bob Conte of Sensitron for fabrication of all-SiC blade modules.

---

[2]   Comparison of switching losses across all module types was based on the switch (IGBT or MOSFET) turn-off and turn-on losses, without including the diode switching losses. Because the SiC Schottky diodes have no reverse recovery losses, the comparison among hybrid and all-SiC modules is consistent. On the other hand, reverse recovery losses of the silicon P-i-N diode can be very significant. For example, the Powerex module has the reverse recovery losses of 15 mJ (scaled to 150A), making the standard silicon module even less competitive with the hybrid and all-SiC designs.

REFERENCES

[1] P. Losee, K. Matocha, S. Arthur, J. Nasadoski, Z. Stum, J. Garrett, M. Schutten, G. Dunne, L. Stevanovic, "DC and transient performance of 4H-SiC double-implant MOSFETs", *IEEE Transactions on Electron Devices*, v. 55, pp. 1824-1829, 2008.

[2] P. Losee, K. Matocha, S. Arthur, E. Delgado, R. Beaupre, A. Pautsch, R. Rao, J. Nasadoski, J. Garrett, Z. Stum, L. Stevanovic, R. Conte and K. Monaghan, "100 amp, 1000 volt class 4H- silicon carbide MOSFET modules," *Materials Science Forum Vols. 615-617 (2009) pp 899-902*.

[3] T. Tsunoda, T. Matsuda, Y. Nakadaira, H. Nakayama, Y. Sasada, "Low-inductance module constructed for high-speed, high current IGBT module suitable for electric vehicle application" *Proceedings of the 5th International Symposium on Power Semiconductor Devices and ICs*, ISPSD 1993, pp. 292 - 295, Tokyo, Japan, 1993.

[4] L. Schnur, G. Debled, S. Dewar, J. Marous, "Low inductance, explosion robust IGBT modules in high power inverter applications," *IEEE 1998 Industry Applications Society Conference*, IAS 1998, pp. 1056 - 1060.

[5] T. Ohi, T. Horiguchi, T. Okuda, T. Kikunaga, H. Matsumoto, "Analysis and measurement of chip current imbalances caused by the structure of bus bars in an IGBT module," *IEEE 1999 Industry Applications Society Conference*, IAS 1999, pp. 1775 – 1779.

[6] Osamu Usui, Hiroshi Nakatake and Takeshi Ohi, "Analysis of the dynamic characteristics of a power semiconductor module, considering the influence of electromagnetic coupling between wiring," *Proceedings of the 11th European Conference on Power Electronics and Applications*, EPE 2005, pp. P.1 – P.6.

[7] A. Wintrich, P. Beckedahl, T. Wurm, "Electrical and thermal optimization of an automotive power module family," *Proceedings of the 2nd International Conference on Automotive Power Electronics*, APE 2007, Paris, France, 26 - 27 September, 2007.

[8] R. Fillion, E. Delgado, R. Beaupre, P. McConnelee, "High power planar interconnect for high frequency converters," *Proceedings of the 6th Electronics Packaging Technology Conference*, EPTC 2004, pp. 18 – 24, Singapore, 2004.

[9] http://www.ansoft.com/products/si/q3d_extractor/

[10] http://www.tycoelectronics.com.cn/catalog/minf/en/560?BML=10576,17560,17685,17694

[11] L.D. Stevanovic, R.A. Beaupre, A.V. Gowda, A.G. Pautsch and S.A. Solovitz, "Integral Micro-channel Liquid Cooling for Power Electronics," *Proceedings of the 25th Applied Power Electronics Conference*, APEC 2010, Palm Springs, CA, Feb 21-25, 2010.

Design of Multi-turn LTCC Inductors for High Frequency DC/DC Converters

Laili Wang, Yunqing Pei, Xu Yang, Xizhi Cui, Zhaoan Wang, Guopeng Zhao
Electrical Engineering Department, Xi'an Jiaotong University
Xi'an, 710049, China
Email: l.l.wang@stu.xjtu.edu.cn

Abstract—This paper focuses on designing of multi-turn spiral inductors with multilayer parallel-connected conductors for high frequency DC/DC converters based on low temperature co-fired ceramic (LTCC) technology. Effects of conductors' arrangements on performance of LTCC inductors are analyzed. Based on observation and analysis, a model to approximate inductance of LTCC inductors is derived. Simulation results are provided to verify accuracy of the model. At last, a prototype is fabricated and tested in a 8MHz switching-mode buck converter.

I. INTRODUCTION

There is a growing interest of developing low profile, better performance switching-mode power supplies for Portable Electronic Products. Converters in these products are often required to be in very compact packages. With the development of semiconductor technology, power devices of a new generation such as GaN HEMTs, SiC JFET could operate from several Megahertz to hundreds of Megahertz [1-2]. DC/DC converters operating at higher frequency could be designed if these advanced devices are combined with resonant topologies [3-7]. Higher switching frequency will results in volume reduction of passives, which provides new chances for passive integration, especially magnetic integration.

LTCC technology which is originally used in fabrication of RF circuits becomes suitable for integration of DC/DC converters when passives are miniaturized at high frequency. In LTCC technology, ferrite and capacitor tapes could be co-fired together to fabricate integrated planar passive substrates. Moreover, LTCC materials have the same temperature expansion coefficient with silicon (Si), so hybrid integration could be realized through semiconductor wire bonding. Besides advantages stated above, high-thermal conductivity metal or ceramic substrates (e.g., AlN) could be co-fired together with LTCC tapes, helping to dissipate heat generated by power loss.

For advantages of applying LTCC technology to high frequency DC/DC converters described above, in recent years,

researchers have begun to focus on design and fabrication of planar inductors for low power DC/DC converters based on this technology [8-13]. However, no detailed analysis is executed on relationship of conductor arrangements and inductor performance. [14] presented a method for calculating inductance of spiral inductors, nevertheless, this method is too complicated to be used in practice. An effective model for calculating inductance should be derived with the purpose of designing LTCC inductors used in high frequency DC/DC converters.

This paper is focusing on design of power inductors for high frequency DC/DC converters based on LTCC technology. Section II of the paper presents arrangements of conductors in LTCC power inductors and explores their impacts on inductance and AC resistance. Factors constraining number of parallel conductors in LTCC inductors are also explored. Section III of the paper derives an effective analytical model to calculate inductance of single turn planar inductors. 2D FEA simulation results are provided to verify accuracy of the model. Under a reasonable assumption, we further extend this model to inductance calculation of multi-turn spiral inductors. Section IV of the paper demonstrates an inductor prototype fabricated with LTCC technology. We provide measurement results to verify the model, and apply the prototype to a high frequency DC/DC converter to evaluate its performance. Section V concludes the paper.

II. ARRANGEMENT OF CONDUCTORS IN LTCC POWER INDUCTORS

A LTCC spiral inductor is made of laminated ferrite tapes with screen printed conductor paste among them. Fig 1. shows

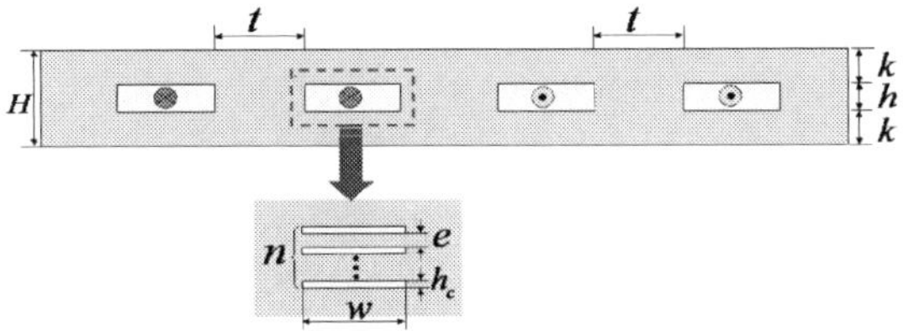

Figure 1. Structure of a two-turn n-layer spiral LTCC inductor

---

This work was supported in part by key project of Supporting Tech. From the Ministry of Science and Technology of China under Grant 2007BAA12B01.

the cross-section view of a two-turn n-layer spiral inductor and its major dimension parameters.

The function of LTCC inductors in high frequency DC/DC converters is very different from that in RF circuits. Inductors in DC/DC converters are employed to process power instead of signal, and equivalent series resistance of their windings inevitably causes conduction loss; therefore, cross-section area of windings should be large enough to insure that series resistance is as low as possible. The effective cross-section area A_e of one turn winding in a LTCC inductor can be expressed by (1)

$$A_e = n \cdot h_c \cdot w \qquad (1)$$

Where h_c is the thickness of single-layer conductor; w is width of the conductor; n is number of conductors connected in parallel. In (1), h_c can be seen as a constant and can not increase arbitrarily, because too much thickness of conductor paste will cause crack of LTCC substrates when co-fired together. As a result, two methods could be employed to increase the cross-section area of windings. First, increase width of conductor w; second, increase number of parallel layers n. Generally, two methods are combined together when designing windings of LTCC inductors. In practice, A_e is determined by the allowed maximum current together with distributed current density, therefore, A_e can be considered as a constant. We further analyze variation trend of inductance by increasing n and decreasing w. A 2D FEA simulation is set up to demonstrate the trend. The core size for simulation is 1mm thick and 15mm wide; relative permeability of ferrite tapes μ_r is 50; thickness of a single-layer conductor is 20μm. Simulation results are listed in Table I. As the results show, inductance of per unit length increases with number of parallel conductors, although the trend is slowing down. This attributes to length reduction of magnetic path as shown in Fig. 2. For LTCC inductors, especially those used to process power, it is a good way to reduce conductor width while increasing parallel-connected conductors under precondition that total cross-section area is unchanged. In this way, magnetic reluctance is reduced and inductance is increased. However, number of parallel conductors could not arbitrarily increase without any limitations. In fact, there are at least two limitations: thickness of inductors and AC resistance.

Ferrite tapes will shrink both in XY and Z direction when they are co-fired. Therefore, if too many layers of ferrite tapes are laminated to increase thickness of inductors, their differences of shrinkage probably cause crack of whole substrate when they are co-fired together. This limitation of LTCC inductors further constrains increase of parallel conductors, because flux will be forced to concentrate in the narrow area above and below conductors when number of parallel conductors is increased and thickness of inductors can not be increased any more. Then, width of the area can be expressed by (2)

$$k = \frac{H - n \cdot h_c - (n-1) \cdot e}{2} \qquad (2)$$

where e is layer thickness of ferrite tape. As described by (2), increasing n will cause reduction of k, which inevitably results in increase of magnetic reluctance and saturation of ferrite core.

Current flowing through power inductors in high frequency DC/DC converters is composed of DC and high frequency AC ripple. DC uniformly distributes in parallel-connected conductors. Suppose length of conductors is l, then, DC resistance is

$$R_{dc} = \frac{l}{n \cdot \gamma \cdot w \cdot h_c} \qquad (3)$$

where γ is electrical conductivity of conductors. However, distribution of AC is highly influenced by proximity effect and skin effect. For parallel planar conductors in a LTCC inductor, high frequency AC mainly flows on top and bottom conductors due to proximity effect. Take arrangement of 6 parallel conductors in Table I as an example, simulation result shows AC loss of top and bottom conductors accounts for 85% of total AC loss when 1 A sinusoidal current source is applied. Besides proximity effect, skin effect deteriorates the situation further more, due to skin effect, AC concentrates on surface of conductors. Consequently, AC resistance can be roughly approximated by

$$R_{ac} = \frac{l}{2\gamma \cdot w \cdot d} \qquad (4)$$

Where d is skin depth of the conductor. (4) indicates that with the increase of conductor layers (decrease of conductor width), effective area for AC current conduction in fact gets smaller and smaller. Therefore, AC resistance is another factor constraining number of conductor layers.

III. MODEL FOR INDUCTANCE CALCULATION

To design a LTCC spiral inductor, a model for approximately calculating the inductance should be derived by a very simple but effective means. Here, for the purpose of easy determination of inductance, the distance between turns t

TABLE I. INDUCTANCE VS NUMBER OF PARALLEL CONDUCTORS

Number of layers	1	2	3	4	5	6
Width (mm)	5.00	2.50	1.67	1.25	1.00	0.83
Inductance (μH/m)	2.84	4.78	6.08	6.92	7.39	7.66

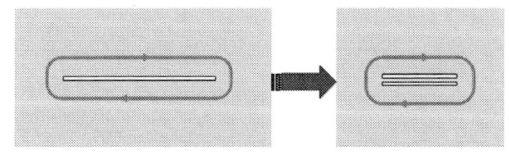

Figure 2. Reducing length of magnetic path by dividing one wide conductor into two parallel narrow ones.

is assumed to be larger than the ferrite thickness 2k, so that the flux generated by one turn winding does not couple with other turns. This assumption is actually realizable and reasonable when designing and fabricating LTCC inductors. Given this assumption, inductance approximation of a multi-turn spiral inductor can be simplified to that of a inductor. By calculating inductance of per unit length (1m) conductor buried in planar ferrite core, inductance of a single turn spiral inductor could be obtained.

The flux generated by current flow through a conductor can be expressed by (5)

$$\varphi = I \cdot P \tag{5}$$

Where I is the current, P is the total permeance of magnetic path. By observing flux lines in FEA simulation, we can approximate magnetic path around conductor by rounded corner rectangle shown in Fig. 3.

Therefore, total permeance can be seen as a summary of its rounded corner rectangle shaped differential elements dP which could be approximated by (6)

$$dP = \frac{\mu \cdot dx}{2(w + h_c + \pi \cdot x)} \tag{6}$$

By integrating infinite number of differential elements dP, whose magnetic paths are parallel to each other, we could obtain total permeance (7)

$$P = \int_0^k \frac{\mu \cdot dx}{2(w + h_c + \pi \cdot x)} \tag{7}$$

For a single turn structure, the inductance per unit length (8) can be expressed in the form of total flux divided by current which generated the flux

$$L_0 = \frac{\varphi}{I} \tag{8}$$

Substitute (5) and (7) into (8), yielding inductance per unit length of a single turn planar inductor (9)

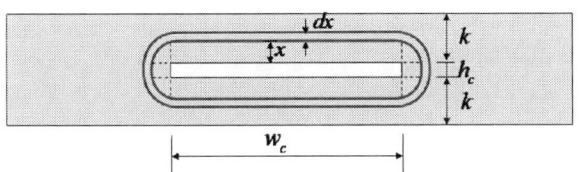

Figure 3. Differential magnetic path

$$L_0 = \frac{\mu}{2\pi} \cdot \ln\left(1 + \frac{k \cdot \pi}{w + h_c}\right) \tag{9}$$

A FEA simulation is set up to evaluate the accuracy of this formula. Table III listed parameters used in simulation and calculation. By increasing thickness of ferrite above and below the conductor k, a series of simulation and calculation inductance values are obtained, the results are shown in Fig. 4. It can be seen that calculated values are a little smaller than simulated values.

Although the derived model for approximating inductance is based on the structure of single layer conductor buried in planar ferrite core, it's also applicable to modeling of inductors with multilayer parallel-connected conductors when thickness of a LTCC inductor is much larger than distance between conductor layers [15]. In such a case, flux between conductors is partially canceled, and major magnetic path is still rounded corner rectangle as shown in Fig. 5. Thus, multilayer parallel conductors in the model can be substituted with a single-layer virtual conductor whose thickness is

$$h = n \cdot h_c + (n-1) \cdot e \tag{10}$$

And inductance model for multilayer parallel conductors buried in planar ferrite core can be

$$L_0 = \frac{\mu}{2\pi} \cdot \ln\left(1 + \frac{k \cdot \pi}{w + n \cdot h_c + (n-1) \cdot e}\right) \tag{11}$$

TABLE II.   PARAMETERS USED FOR SIMULATION AND CALCULATION

Parameter	k	w	h_c	μ_r
Value	0.1-2 mm	4 mm	20μm	50

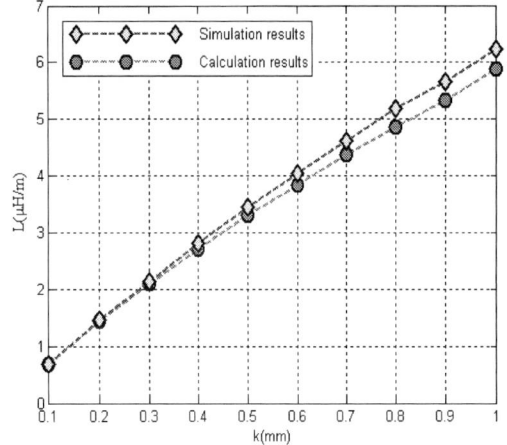

Figure 4. Comparison of inductance values obtained by simulation and calculation

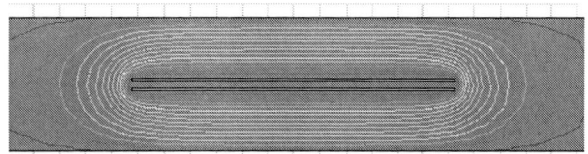

Figure 5. Flux lines generated by current flowing though two parallel conductors

Inductance values of both parallel conductors and single-layer virtual conductor buried in planar ferrite cores are simulated to verify the substitution. Inductance values calculated by (11) are also provided to validate the model. The core size for simulation is 2mm thick and 10mm wide; relative permeability of ferrite tapes μ_r is 50; thickness of single-layer conductor is 20μm. Results of both simulation and calculation are listed in Table III. It indicates that the proposed model is available and convenient for inductance calculation of LTCC inductors with multilayer parallel conductors.

Since the model is derived on the assumption that t >2k, magnetic flux does not couple between turns. In this way, inductance of a multi-turn inductor depends only on winding length rather than its arrangement.

IV. OPTIMIZATION FOR LIGHT LOAD EFFICIENCY

In a POL DC/DC converter, the inductor is used as a buffer for energy transmission. Value of inductance is trade off of two basic contradictory requirements of converters: current ripple and transient response. Small value inductor could improve transient response performance, however, it will also increase current ripple, which inevitably lead to high switching loss and low light load efficiency. Large value inductor could sufficiently suppress output current ripple, but it also constrains transient response time when heavily loaded. A good way to solve this problem is to use a nonlinear inductor, which shows high inductance value at light load, but small inductance value at heavy load. Conventional way of realizing this idea is to use two separate inductors which are connected in series. One is a saturated inductor; another one is a small value linear inductor. Based on LTCC technology, a totally integrated nonlinear inductor optimized for light load efficiency could be manufactured. Fig. 6 shows its cross-section view. More high permeability ferrite tapes are added between conductor and low permeability ferrite tapes to increase inductance value at light load. With the increase of output current, high permeability tapes gradually become saturated. Then, the inductance value is determined by low permeability ferrite tapes.

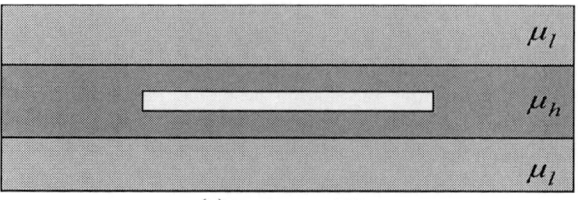

(a) Two-permeability

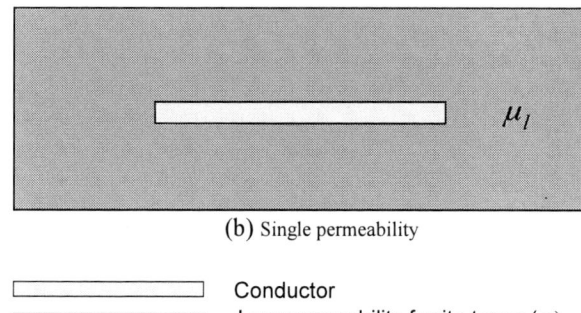

(b) Single permeability

▭	Conductor
▬	Low permeability ferrite tapes (μ_l)
▬	High permeability ferrite tapes (μ_h)

Figure 6. Cross-section views of LTCC inductors

V. EXPERIMENTAL RESULTS

A planar LTCC inductor is designed according the model developed in this paper. Its parameter values are presented in Table IV. Materials used to fabricate the LTCC inductor are NiZn ferrite tape 40010 whose permeability is 50 and silver/platinum (Ag/Pt) paste, both of them are from Electro-science Labs.

With the help of films which is used in PCB manufacturing process, via holes are punched on each layer, then silver paste is used to fill them for three dimensional connections and form spiral windings on middle layers. Finally, all layers are laminated and put into a furnace for sintering. Winding arrangement of the prototype is shown in Fig. 7. Measurement results and designing specifications are listed in Table V for comparison. Inductance value of the prototype is a little smaller than what is desired. Error between them is within 5%, which is acceptable in practice. AC resistance is larger than the specification; because screen printed conductors have sharp angles which could result in much more AC loss.

To further verify performance of the newly fabricated inductor, a 8MHz 4.2V input 3V output DC/DC converter is set up to test it. The schematic is shown in Fig. 8. Voltage

TABLE III
INDUCTANCE OF PARALLEL CONDUCTORS OBTAINED BY SIMULATION AND CALCULATION

Number of layers	2	3	4	5	6
Width of conductors (mm)	2.50	1.67	1.25	1.00	0.83
Simulated inductance values of parallel conductors (μH/m)	8.69	10.8	12.2	13.1	13.6
Simulated inductance values of single virtual conductor (μH/m)	8.63	10.7	12.1	12.8	13.2
Calculated inductance values by analytic model (μH/m)	7.69	9.48	10.6	11.2	11.4

TABLE IV
PARAMETERS OF EXPERIMENTAL INDUCTOR

Parameters	H	w	h_c	l	n
Value	0.30mm	4mm	10μm	20cm	1

TABLE V
COMPARISON OF DESIGN SPECIFICATIONS AND MEASUREMENT RESULTS

	$L(\mu H)$	$R_{ac}(\Omega)$ at 2 MHz	$R_{dc}(\Omega)$
Specifications	0.210	0.180	0.150
Measurement results	0.201	0.216	0.144

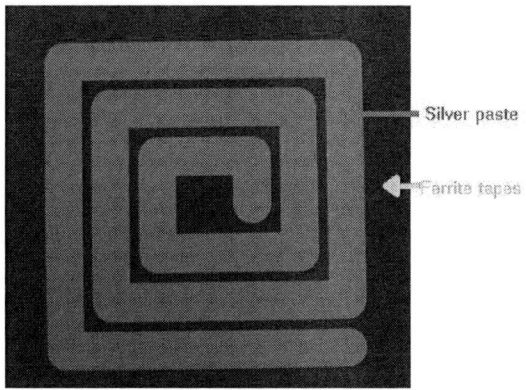

(a)   Winding of the inductor

(b) The LTCC inductor with shielding Pads

Figure 7. Pictures of the LTCC inductor prototype

and current waveforms of the inductor are shown in Fig. 9.

VI.   CONCLUSION

This paper explores designing of planar inductors for high frequency DC/DC converters based on LTCC technology. Impacts of multilayer parallel conductors' arrangements to inductance and AC resistance are analyzed first. Then, By observing distribution of flux lines, an effective model is derived to calculate the inductance of LTCC inductors. Finally, a LTCC inductor is fabricated and applied to a high

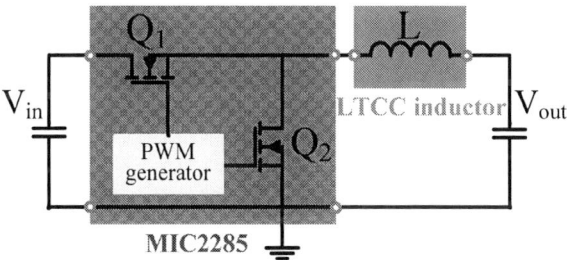

Figure 8. Schematic of a high frequency DC/DC converter with a LTCC inductor

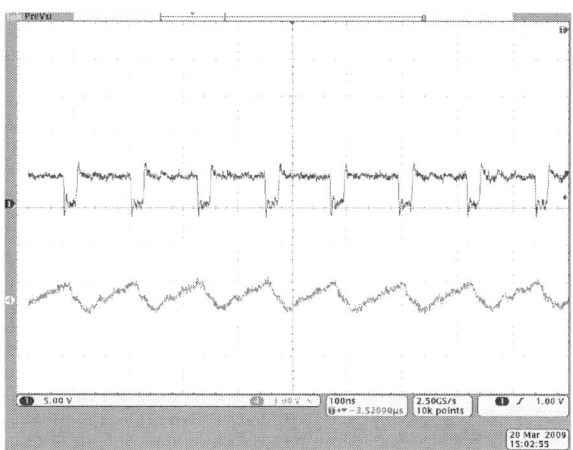

Figure 9.   Inductor voltage (top) and current (bottom) waveforms

frequency DC/DC converter to verify its performance; results of measurement and experiment are provided at the end of this paper.

REFERENCES

[1] Zhang Naiqian, V. Mehrotra, "Large area GaN HEMT power devices for power electronic applications: switching and temperature characteristics," in Proc. IEEE PESC, June 2003, vol.1, pp. 233-237.

[2] M.A. Khan, G. Simin, S.G. Pytel, A. Monti, E. Santi and J.L. Hudgins, " New developments in Gallium Nitride and the impact on power electronics," in Proc. IEEE PESC, June 2005, vol. 1, pp. 15-26.

[3] J. D. van Wyk, F. C. Lee, Z. Liang, R. Chen, S. Wang, and B. Lu, "Integrating active, passive and EMI-filter functions in power electronics systems: a case study of some technologies," IEEE Trans. Power Electron., vol. 20, pp. 523-536, May 2005.

[4] J. opovic, J A. Ferreira, "Design and evaluation of highly integrated dc-dc converters for automotive applications," in Proc. IEEE IAS, Oct. 2005, Vol. 2, pp. 1152-1159.

[5] D. Fu, F. C. Lee, Y. Qiu, F. Wang, "A Novel High-Power-Density Three-Level LCC Resonant Converter with Constant-Power-Factor-Control for Charging Applications", IEEE Transaction on Power Electronics, 2008.

[6] D. Fu, Y. Liu, F.C Lee, M. Xu, "A Novel Driving Scheme for Synchronous Rectifiers (SR) for LLC Resonant Converters", IEEE Transaction on Power Electronics, 2009.

[7] D. Huang, D. Fu, Fred C. Lee, "High Switching Frequency, High Efficiency CLL Resonant Converter with Synchronous Rectifier", in Proc. IEEE ECCE, 2009.

[8] A. Fathy, F. McGinty, "Low temperature co-fired ceramic on metal (LTCC-M) packaging technology," in Proc. IEEE EPEP, Oct. 2000, pp. 261-264.

[9] Chan-Young Kim; Hee-Jun Kim, Jong-Ryoul Kim, "An integrated LTCC inductor," IEEE Trans. Magn., vol. 41, pp3556-3558, Oct. 2005.

[10] Hee-Jun Kim, Young-Jin Kim, Jong-Ryoul Kim, "An Integrated LTCC Inductor Embedding NiZn Ferrite," IEEE Trans. Magn., vol. 42, pp. 2840-2842, Oct. 2006.

[11] M. H. Lim, J. Dong, J. D. van Wyk, "Shielded LTCC Inductor as Substrate for Power Converter," in Proc. IEEE PESC, June 2007, pp. 1605-1611.

[12] M. H. Lim and J. D. van Wyk, "A Class of Ceramic-Based Chip Inductors for Hybrid Integration in Power Supplies," IEEE Tran. Power Electro., vol. 23, pp. 1556-1564, May 2008.

[13] L. Wang, Y. Pei, X. Yang, X. Cui, and Z. Wang, "Three-dimensional integration of high frequency DC/DC converters based on LTCC technology," in Power Electronics and Motion Control Conference, 2009. IPEMC '09. IEEE 6th International, 2009, pp. 745-748.

[14] Roshen W.A, "Effect of finite thickness of magnetic substrate on planar inductors," IEEE Trans. Magn., vol. 26, pp. 270-275, Jan. 1990.

[15] M. J. Prieto, A. M. Pernia, J. M. Lopera, J. A. Martin, and F. Nuno, "Design and analysis of thick-film integrated inductors for power converters," IEEE Trans. on Industry Applications, vol. 38, pp. 543-552, 2002.

Topological Research and Comparison of Low Harmonic Input Three-Phase Rectifier with Passive Auxiliary Circuit

Zhong Chen, Yingpeng Luo and Yinyu Zhu
Aero-Power Sci-tech Center
Nanjing University of Aeronautics & Astronautics
Nanjing, P.R. China
Email: chenz@nuaa.edu.cn

Abstract—**A novel passive correction approach is proposed in this paper. By adding a passive auxiliary circuit to the ac side of the rectifier with dc-side C filter, the input currents and power factor can be improved. According to distinctive configurations of the auxiliary circuits, eight low harmonic input rectifier topologies are derived. Three rectifier topologies with better comprehensive performance are deeply analyzed and compared in operation principle, auxiliary circuit design and characteristic discussion. Experimental results from the prototypes are shown to confirm the validity of the analysis and the feasibility of the proposed approach.**

I. INTRODUCTION

Conventional three-phase rectifiers with dc-side C filter are widely used as the interface circuits between the grid and power electronic equipments for their simplicity and reliability. Because of the nonlinear characteristic of diodes, large mount of harmonics are drawn from the grid, leading to the severely distorted input currents and low power factor [1].

Reducing input current harmonics by modifying the rectifier itself mainly includes the following techniques: PWM rectifiers could improve the input currents effectively, but the shortcoming of large switching loss and unsatisfied efficiency limit its application especially in the high power application [2], [3]; Multiple pulse rectifiers could achieve acceptable input currents in relatively low cost and good EMI, but satisfied input current usually accompanies increase of pulse number, leading to heavier weight of the equipment [4], [5]; Rectifiers applying current injection contribute to the reduction of input current by injecting third order current harmonics into the input terminal of rectifier, but the method does not get widely used [6], [7]. Some novel rectifiers, such as RNSIC (rectifier with near-sinusoidal input currents) are proposed in recent years [8]-[9]. RNSIC rectifiers have caught increasing attention for their simple fabrication and

high reliability, but research on them is inadequate and unsystematic.

Based on the former research, a novel passive correction approach is proposed in the paper. By adding an inductor capacitor auxiliary circuit to the input terminal of diode bridge rectifier, a family of low harmonic input three-phase rectifier is derived. They are only composed of passive components, preferable to work in high power application with high EMI requirement. Some important conclusions are drawn from the comparative analytical results of three typical rectifiers in operation principle and characteristic discussion. Experimental results from seven prototypes built in the laboratory are shown to confirm the validity of the analysis and the feasibility of the proposed correction approach.

II. TOPOLOGY OF LOW HARMONIC INPUT THREE-PHASE RECTIFIER WITH AUXILIARY CIRCUIT

A. Proposal of Novel Passive Correction Approach

By adding a passive auxiliary circuit to the input terminal of conventional uncontrolled rectifier (as Fig. 1(a) shown), a novel low harmonic input three-phase rectifier (as Fig. 1(b) shown) is obtained.

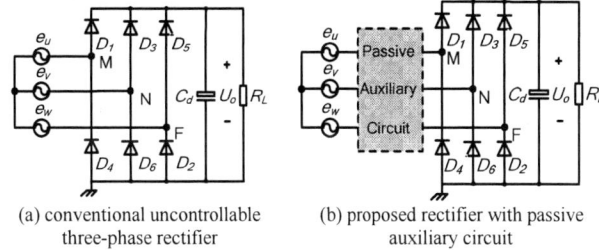

(a) conventional uncontrollable three-phase rectifier

(b) proposed rectifier with passive auxiliary circuit

Figure 1.   Three-phase rectifier with dc-side C filter.

There are four basic configurations of the inductor and capacitor auxiliary circuit (as Fig. 2 shown). Auxiliary circuit

---

This work was supported by Doctoral Fund of Ministry of Education of China under Award 200802871033 and Aeronautical Science Foundation of China under Award 2009ZC52030.

978-1-4244-4782-4/10 $26.00 © 2010 IEEE

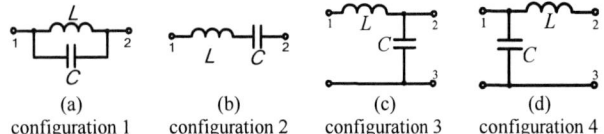

(a)   (b)   (c)   (d)

configuration 1   configuration 2   configuration 3   configuration 4

Figure 2. Four Basic configurations of *LC* auxiliary circuit.

1 [10-11] and 2 [12] are two-terminal network, while auxiliary circuit 3 and 4 are three-terminal network.

Fig. 3 illustrates eight topologies of low harmonic input three-phase rectifier with passive auxiliary circuit. By series connecting auxiliary circuit 1 and 2 to the ac-side of diode rectifier, topologies 1 and 2 will be arisen (as Fig. 3(a) and (b) shown); by connecting auxiliary circuit 3 and 4 to the ac-side of conventional rectifier, topologies 3 and 4 will be arisen (as Fig. 3(c) and (d) shown); by connecting terminal 1

and 2 of auxiliary circuit 3 and 4 to the ac-side of diode rectifier, terminal 3 to the dc bus of diode rectifier, topologies 5 [13] and 6 will be arisen (as Fig. 3(e) and (f) shown). Symmetrical versions of topologies 5 and 6 are named as topologies 7 and 8 (as Fig. 3(g) and (h) shown).

B. Comparisons Between the Eight Topologies

Auxiliary circuit in topology 1 which is actually a series passive filter can be considered as the dual form of parallel tuning filter [14]. The series passive filter consists of parallel-resonant LC circuit, each tuned at a harmonic frequency, thus acting as a harmonic current dam. On the other hand, if satisfied harmonics suppression performance is desired, several series passive filters with different tuning frequency must be set, thus increasing the number and weight of the passive components and limiting the usage of this topology.

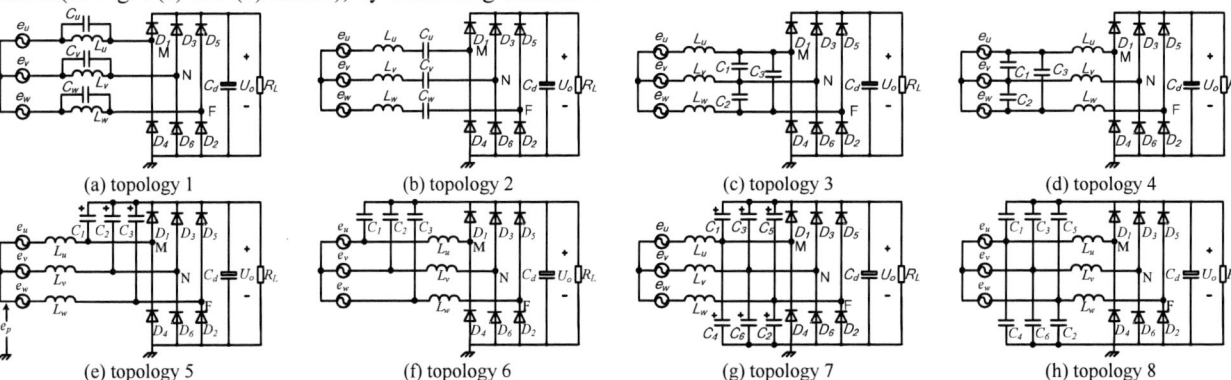

(a) topology 1   (b) topology 2   (c) topology 3   (d) topology 4

(e) topology 5   (f) topology 6   (g) topology 7   (h) topology 8

Figure 3. A family of low harmonic input three-phase rectifiers with passive auxiliary circuit.

Auxiliary circuit in topology 2 is another series passive filter, working as a band pass filter. This circuit is tuned at the fundamental frequency, exhibiting near zero impedance to the fundamental current but high impedance to the harmonic current. Following shortcomings limit its application: large value of passive components because of the small tuning frequency; unsatisfied harmonic suppression under load variation application because bandwidth of the filter is determined by the load. But it is a simple and effective unity power factor rectification in some special application [12].

In topology 4, 6 and 8, commutation capacitors of the auxiliary circuit which are parallel connected to the input terminal of rectifier contribute to high frequency harmonics filtering and reactive power compensation. But value of the capacitors is not easy to determine in practical application.

For topology 3, 5 and 7, former researches show that near sinusoidal input currents and high power factor will be achieved with optimal design of the auxiliary circuit [8-9, 13]. These three rectifier topologies catch more and more attention in the recent years.

In this paper, topology 3, 5 and 7 are picked up to be comparatively investigated and analyzed in operation principle, and characteristic discussion to reveal some general rules of this family of rectifiers. For convenient analysis, topology 3, 5 and 7 are named as RNSIC-2, RNSIC-3, and RNSIC-1.

III. OPERATION PRINCIPLE

As Fig. 3 shows, RNSIC converters are composed of three series inductors L_u, L_v, L_w of equal inductance values L and commutation capacitors of equal capacitance values C. Commutation capacitors are C_1-C_6 in RNSIC-1 but C_1-C_3 in RNSIC-2 and RNSIC-3.

For the sake of simplicity, following simplifications are perfumed: The mains currents are purely sinusoidal, no phase displacement between voltages and current.

$$e_u = U_M \sin \omega t, \quad i_u = I_M \sin \omega t \qquad (1)$$

$$e_v = U_M \sin(\omega t - \frac{2}{3}\pi), \quad i_v = I_M \sin(\omega t - \frac{2}{3}\pi) \qquad (2)$$

$$e_w = U_M \sin(\omega t + \frac{2}{3}\pi), \quad i_w = I_M \sin(\omega t + \frac{2}{3}\pi) \qquad (3)$$

Where U_m is the amplitude of phase voltage, I_m is the amplitude of input currents.

Zero time point ($t=0$) is defined as the moment when the current i_u crosses zero from negative to positive, and t_1 is defined as the moment when the capacitor C_1 finishes discharging. As t_1 decreases, operation mode of RNSIC converters changes from small current mode to medium current mode and large current mode. In RNSIC converters, commutation capacitors start charging or discharging when the currents through diodes cross zero. In RNSIC-1 and RNSIC-3, when diodes conduct, currents through them are the input

978-1-4244-4782-4/10 $26.00 © 2010 IEEE   1617

TABLE I. START TIME, END TIME AND DURATION OF THE CURRENT DISCHARGING C_1

Converter	Operation Mode	Start time	End time (ωt_1)	Duration
RNSIC-1 RNSIC-3	Large Current	0	$(0,\pi/3)$	$(0,\pi/3)$
	Medium Current		$(\pi/3,2\pi/3)$	$(\pi/3,2\pi/3)$
	Small Current		$(2\pi/3,\pi)$	$(2\pi/3,\pi)$
RNSIC-2	Large Current	0	$(0,\pi/3)$	$(0,\pi/3)$
	Medium Current	$(0, \pi/6)$	$(\pi/3,\pi/2)$	$\pi/3$
	Small Current	$\pi/6$	$(\pi/2,5\pi/6)$	$(\pi/3,2\pi/3)$

currents, so capacitors start charging or discharging when the input currents cross zero; while in RNSIC-2, when diodes conducts, currents through them comprise the sum of input currents and the currents through commutation capacitors. Start time, end time and duration of the currents discharging commutation capacitors vary with the load, as Table I shown.

A. Large Current Mode

When RNSIC converters work in the large current mode, t_1 varies in the interval $[0, \pi/(3\omega)]$. As Fig. 4 illustrates, in large current mode, diodes of RNSIC converters conduct in the same sequence. Two distinct operation stages in which two or three diodes conduct exist and conduction time of the diodes is $\pi/\omega - t_1$. As Fig. 5 illustrates, in RNSIC-1 and RNSIC-2, the dc side current i_o has the same waveform, a twelve-pulse waveform with different width; while in RNSIC-3, i_o is a six-pulse waveform with different width.

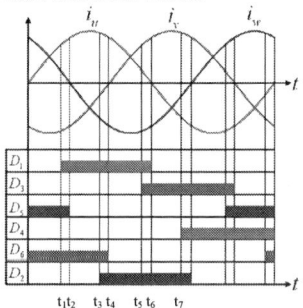

Figure 4. Key waveforms of RNSIC converters in large current mode.

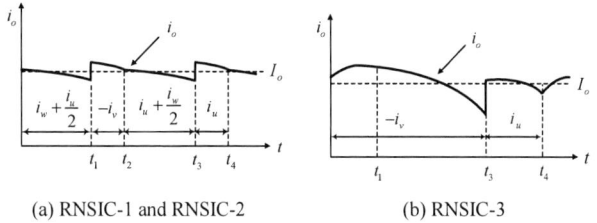

(a) RNSIC-1 and RNSIC-2   (b) RNSIC-3

Figure 5. I_o of RNSIC converters in large current mode.

I_o, the mean value of i_o in RNSIC-1 and RNSIC-2 could be expressed as follow:

$$I_o = \frac{3}{\pi}[\int_0^{\omega t_1}(I_m \sin(\omega t + \frac{2}{3}\pi) + \frac{1}{2}I_m \sin \omega t)d\omega t$$
$$+ \int_{\omega t_1}^{\frac{\pi}{3}} -\frac{1}{2}I_m \sin(\omega t - \frac{2}{3}\pi)d\omega t] \tag{4}$$

In RNSIC-3, I_o could be expressed as follow:

$$I_o = \frac{3}{2\pi}[\int_0^{\frac{\pi}{3}+\omega t_1} -I_m \sin(\omega t - \frac{2\pi}{3})d\omega t + \int_{\frac{\pi}{3}+\omega t_1}^{\frac{2\pi}{3}} I_m \sin \omega t d\omega t] \tag{5}$$

By simplifying (4) and (5), I_o in three RNSIC converters will be expressed as follow:

$$I_o = \frac{3}{2\pi}I_m(1+\cos \omega t_1) \tag{6}$$

B. Medium Current Mode

When RNSIC converters work in medium current mode, t_1 varies in the interval $[\pi/(3\omega), 2\pi/(3\omega)]$ for RNSIC-1 and RNSIC-3, but in the interval $[\pi/(3\omega), \pi/(2\omega)]$ for RNSIC-2.

As Fig. 6 illustrates, in RNSIC-1 and RNSIC-3, diodes conduct in the same sequence, two distinct operation stages in which one or two diodes conduct exist; while in RNSIC-2, two diodes conducts at any time, conduction time of the diodes and capacitors discharging duration fix to be $2\pi/(3\omega)$ and $\pi/(3\omega)$.

As Fig. 7 illustrates, i_o has distinctive waveform in RNSIC converters. In RNSIC-1, i_o is a twelve-pulse waveform with different width; in RNSIC-2, i_o is six-pulse waveform with same width; in RNSIC-3, i_o gets discontinuous. In RNSIC-3, when the only conducting diode is the upper diode, i_o becomes discontinuous for no path exists for the input currents flowing. Expressions of I_o in this mode are given in Table II.

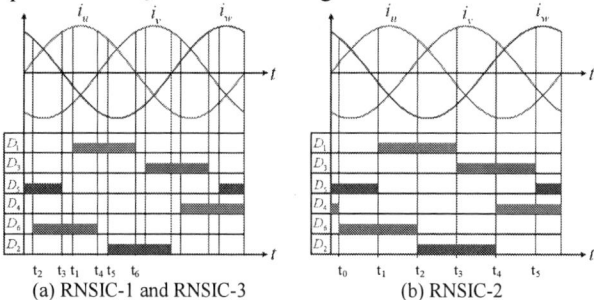

(a) RNSIC-1 and RNSIC-3   (b) RNSIC-2

Figure 6. Key waveforms of RNSIC converters in medium current mode.

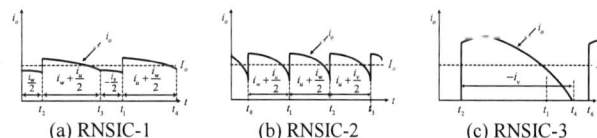

(a) RNSIC-1   (b) RNSIC-2   (c) RNSIC-3

Figure 7. I_o of RNSIC converters in medium current mode.

C. Small Current Mode

When RNSIC converters work in the small current mode, t_1 varies in the interval $[2\pi/(3\omega), \pi/\omega]$ for RNSIC-1 and RNSIC-3, but in the interval $[\pi/(2\omega), 5\pi/(6\omega)]$ for RNSIC-2. As Fig. 8 illustrates, in RNSIC-1 and RNSIC-3, diodes conduct in the same sequence, two distinct operation stages in

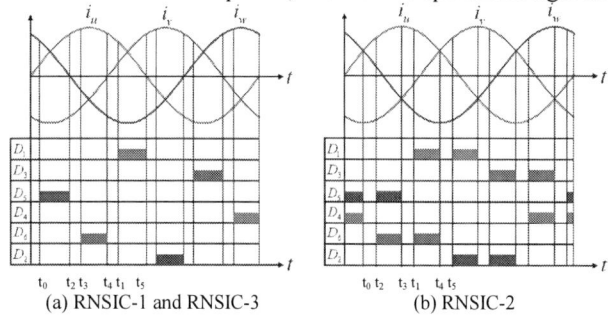

(a) RNSIC-1 and RNSIC-3   (b) RNSIC-2

Figure 8. Key waveforms of RNSIC converters in small current mode.

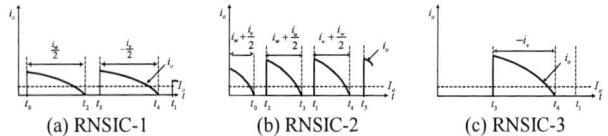

(a) RNSIC-1 (b) RNSIC-2 (c) RNSIC-3

Figure 9. I_o of RNSIC converters in small current mode.

TABLE II.   EXPRESSIONS OF I_o IN RNSIC CONVERTERS

Converter	Operation Mode	Expressions of I_o
RNSIC-1 RNSIC-3	Large Current	$I_o = \dfrac{3}{2\pi} I_m (1 + \cos \omega t_1)$
	Medium Current	
	Small Current	
RNSIC-2	Large Current	
	Medium Current	$I_o = \dfrac{3\sqrt{3}}{2\pi} I_m \cos(\omega t_1 - \dfrac{\pi}{6})$
	Small Current	$I_o = \dfrac{3\sqrt{3}}{2\pi} I_m (1 + \cos(\omega t_1 + \dfrac{\pi}{6}))$

which none or one diode conducts exist, and conduction time of the diodes is $\pi/\omega - t_1$. While in RNSIC-2, two distinct operation stages in which none or two diodes conducts exist, and conduction time of the diodes is $5\pi/(3\omega) - 2t_1$.

As Fig. 9 illustrates, i_o becomes discontinuous in RNSIC converters. In RNSIC-1 and RNSIC-2, i_o is a six-pulse waveform; while in RNSIC-3, i_o is three-pulse waveform. Expressions of I_o in this mode are given in Table II.

From above analysis, following conclusions are drawn:

1) For RNSIC converters, three operation modes exist. When working in large current mode, RNSIC-1 and RNSIC-2 have same equivalent circuit, input characteristic and output characteristic. In RNSIC-1 and RNSIC-3, diodes conduct in the same sequence, mean value of the dc side current i_o has the same expression, but different waveforms does i_o have.

2) In RNSIC converters, dc side current i_o would become discontinuous in at least one operation mode. But the reason why i_o become discontinuous is different in RNSIC-2 and RNSIC-3. In RNSIC-2, commutation capacitors are located in ac side of the rectifier, so the grid provides power to the load when at least two diodes conduct. In RNSIC-3, commutation capacitors are only parallel connected to the upper diode, so i_o become discontinuous when no lower diode conducts.

IV.   CHARACTERISTIC DISCUSSION

A. Optimal Value of Auxiliary Circuit

Fig.10 presents the variation of the optimal capacitance C as a function of the input voltage U_m in RNSIC converters.

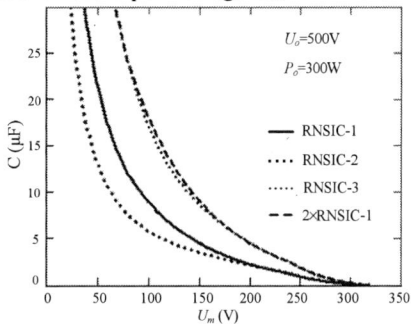

Figure 10. Variations of C as a function of U_m in RNSIC converters.

The values of P_o and U_o are adopted as 300 W and 500V. Optimal capacitances of auxiliary circuit have similar characteristic in three RNSIC converters. As the input voltage increases from low to high, optimal value of capacitance will decrease. Optimal capacitance is the smallest in RNSIC-2 but largest in RNSIC-3. Optimal capacitance of RNSIC-3 is nearly twice time as large as that of RNSIC-1. Moreover, as the output power increases, optimal capacitance will increase.

Fig.11 presents the variation of the optimal inductance L as a function of the input voltage U_m and the output power P_{out} in RNSIC converters. The values of U_o are adopted as 500 V, and the values of P_o are adopted as 300W and 1000 W, respectively. As U_m increases from low to high, optimal value of inductance will increase. But after RNSIC rectifiers enter into large current mode, the optimal inductance will decrease. As P_o is increased, the optimal inductance will decrease.

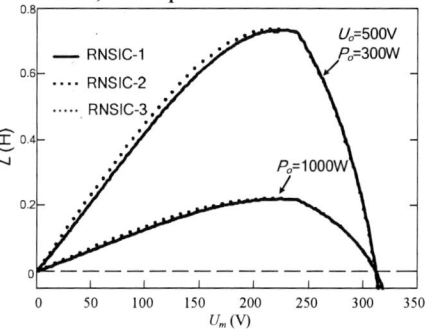

Figure 11. Variations of L as a function of U_m in RNSIC converters.

B. External Characteristic

Dc side current ripple is a key issue for the selection of electrolytic capacitor. Equation (7) and (8) give the expression of root mean square current ripper factor.

$$i_o = I_o + \sum_{n=1}^{\infty} I_n \sin(n\omega t + \varphi_n) \qquad (7)$$

$$\frac{I_{AC}}{I_o} = \frac{\sqrt{i_o^2 - I_o^2}}{I_o} \qquad (8)$$

Fig.12 shows the variation of dc side current ripple factor and normalized output voltage of RNSIC converters. As Fig. 12(a) illustrates, dc side current ripple of RNSIC converters has following properties: *1)* as ωt_1 increases, dc side current ripple will be increased; *2)* dc side current ripple is smallest in RNSIC-1 under same I_o. As Fig. 12(b) illustrates, output voltage of RNSIC converters have similar properties: As load resistor is increased, output voltage will increase.

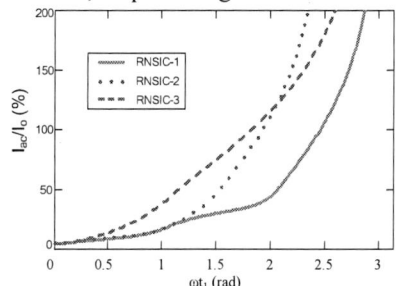

(a) Dc side current ripple factor I_{ac}/I_o as a function of ωt_1

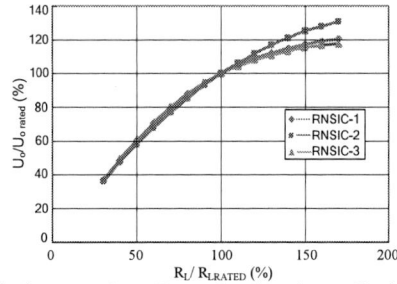

(b) Normalized output voltage U_o as a function of normalized load resistor

Figure 12. Output characteristics of RNSIC converters.

Fig.13 presents the variation of input current THD and power factor of RNSIC converters as a function of normalized load resistor. Input current THD is slightly influenced by the load resistor, all below 2%; input power factor is strongly influenced by load resistor. When the load resistor deviates from the rated value, power factor decreases drastically.

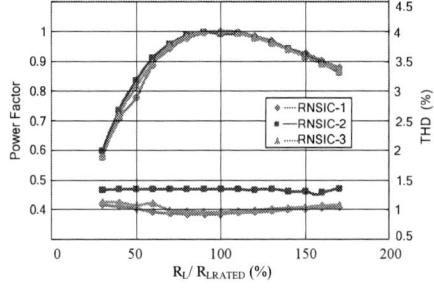

Figure 13. Input characteristics of RNSIC converters.

V. EXPERIMENTAL RESULTS

In order to verify the feasibility of the analysis derived above, seven RNSIC prototypes are built and tested in laboratory. Detail specifications are given in Table III. Topologies of prototype 1-5 are RNSIC-1, while topologies of prototype 6 and 7 are RNSIC-2 and RNSIC-3, respectively. Prototype 1, 4, 6 and 7 all work in large current mode, with the same design value of ωt_1 corresponding to 0.307π; Prototype 2 and 3 work in medium current mode and small current mode, with the design value of ωt_1 corresponding to 0.518π and 0.741π, respectively.

TABLE III. SPECIFICATION OF SEVEN RNSIC PROTOTYPES

Prototype	U_m /V	U_o / V	P_o / kW	L / mH	C / μF
1	250	500	7.18	27.7	22.46
2	150	500	6.55	27.7	98.7
3	55	500	2.52	27.7	159.2
4	250	500	4.79	41.5	14.92
5	250	515	7.95	27.7	27.7
6	250	500	7.18	27.7	22.46
7	250	500	7.18	27.7	44.88

Fig. 14 illustrates the experimental waveforms of i_u and e_u in seven prototypes. Input currents are nearly sinusoidal, and the displacement factors are all near unity. Input currents total harmonic distortion (THD) of prototype 1 is 4.656% (as Fig. 14.(h) shown), input currents THDs of other prototypes are all below 6% which are not given in Fig. 14. Additionally, power factors of seven prototypes are all higher than 0.99.

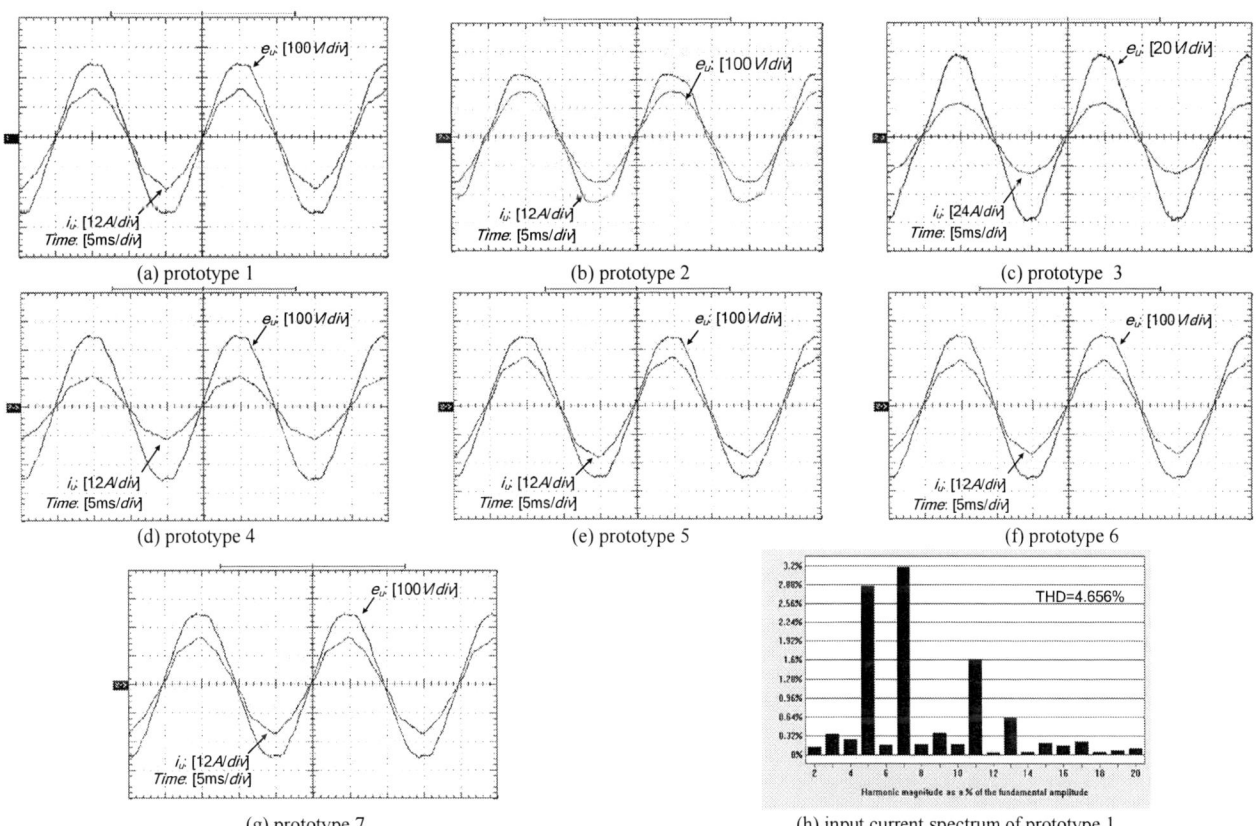

Figure 14. Experimental waveforms of phase voltage and input current.

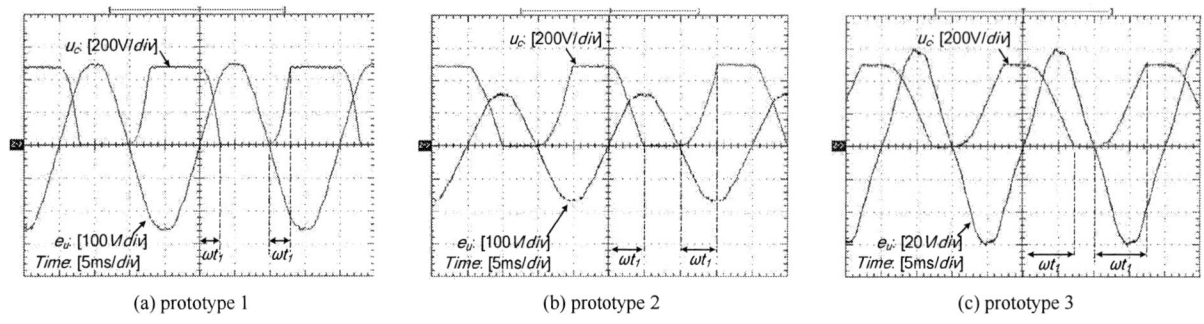

(a) prototype 1   (b) prototype 2   (c) prototype 3

Figure 15. Experimental waveforms of commutation capacitor voltage u_c.

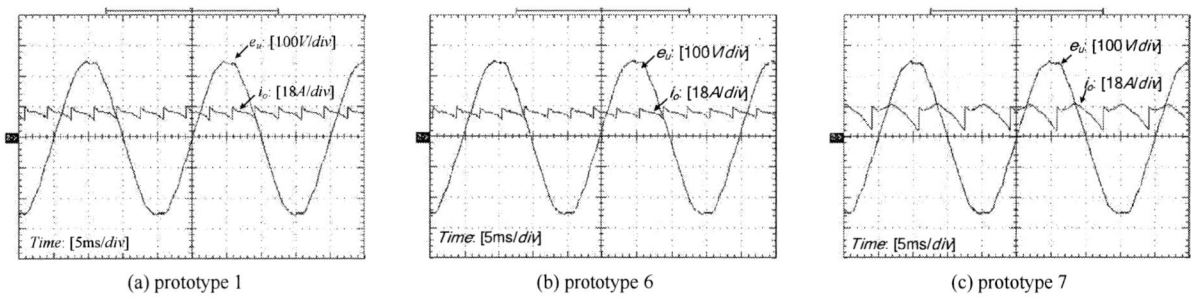

(a) prototype 1   (b) prototype 6   (c) prototype 7

Figure 16. Experimental waveforms of dc-side current i_o.

Fig. 15 illustrates the experimental waveform of u_c, commutation capacitors voltage in prototype 1, 2, and 3. Observed values of ωt_1 in the three prototypes are about 0.3π, 0.5π, and 0.7π, coinciding well with the design values.

Fig. 16 illustrates the experimental waveform of dc side current i_o in prototype 1, 6 and 7. When working under same application and in large current mode, i_o is the same 12-pulse waveform in RNSIC-1 and RNSIC-2, but 6-pulse waveform with large ripple in RNSIC-3, agreeing with above analysis.

VI. CONCLUSION

Based on four configurations of LC auxiliary circuit, a family of low harmonic input three-phase rectifiers with passive auxiliary circuit is proposed. Three topologies with better performance are selected to be deeply analyzed. Experimental results show that this family of rectifier with simple fabrication could effectively suppress the input current harmonics and improve power factor. Analytical results and character discussion in this paper provide reference for the analysis and application of other passive three-phase rectifier.

REFERENCES

[1] H. Akagi, E. H. Watanabe, and M. Aredes, *Instantaneous Power Theory and Applications to Power Conditioning*, New Jersey: John Willey & Sons, 2007, pp. 4-5.

[2] M. Malinowski, M.Jasinski, and M. P. Kazmierkowski "Simple Direct Power Control of Three-Phase PWM Rectifier Using Space-Vector Modulation (DPC-SVM)," *IEEE Trans. Ind. Electron.*, vol. 51, pp. 447–454, Apr. 2004.

[3] H. Mao, F.C. Lee, D. Borojevie, and S. Hiti, "Review of High-Performance Three-Phase Power-Factor Correction Circuits," *IEEE Trans. Ind. Electron.*, vol. 44, pp. 437–446, Aug. 1997.

[4] B. Singh, G. Bhuvaneswari, and V. Garg, "T-Connected Autotransformer -Based 24-Pulse AC-DC Converter for Variable Frequency Induction Motor Drives," *IEEE Trans. Energy Conv.*, vol. 21, pp. 663-672, Sep. 2006.

[5] F.J.M. de Seixas and I. Barbi, "A 12kW three-phase low THD rectifier with high frequency isolation and regulated dc output," *IEEE Trans. Power Electron.*, vol. 19, pp. 371-377, Mar. 2004.

[6] N. Vázquez, H. Rodriguez, C. Hernández, E. Rodriguez, and J. Arau, "Three-Phase Rectifier With Active Current Injection and High Efficiency," *IEEE Trans. Ind. Electron.*, vol. 56, pp. 110-119, Jan. 2009.

[7] P. Pejovic and Z. Janda, "An analysis of three-phase low harmonic rectifiers applying the third-harmonic current injection," *IEEE Trans. Power Electron.*, vol. 14, pp. 391-407, May 1999.

[8] D. Alexa, A. Sîrbu, and D. Dobrea, "Topologies of three- phase rectifiers with near sinusoidal input currents," *Proc. Inst. Electr. Eng.—Elect. Power Appl.*, vol. 151, no. 6, pp. 673–678, Nov. 2004.

[9] D. Alexa, A. Sîrbu, and A. Lazăr, "Three-Phase Rectifiers with Near Sinusoidal Input Current and Capacitors Connected on the AC side," *IEEE Trans. Ind. Electron.*, vol. 53, pp. 1612–1620, Oct. 2006.

[10] A. R. Prasad, P. D. Ziogas, and S. Manias, "A novel passive waveshaping method for single-phase diode rectifiers," *IEEE Trans. Ind. Electron.*, vol. 37, pp. 521–530, Dec. 1990.

[11] J. S. Lai, D. Hurst, and T. Key, "Switch-Mode Power Supply Power Factor Improvement Via Harmonic Elimination Methods," in *Proc. APEC'91*, 1991, pp. 415-422.

[12] V. Vorperian and R. B. Ridley, "A Simple Scheme for Unity Power-factor Rectification for High Frequency AC Buses," *IEEE Trans. Power Electron.*, vol. 5, pp. 70–87, Jan. 1990.

[13] Y. P. Luo, Z. Chen, and Y. Y. Zhu. "Three-phase rectifier with near-sinusoidal input currents and capacitors parallel connected with the upper diodes," in proc. *IEEE IPEMC 2009*, 2009, pp. 1697-1702.

[14] F. Z. Peng, "Harmonic sources and filtering approaches," *IEEE Ind. Appl. Mag.*, vol. 7, pp. 18–25, 2001.

The Reactive Power Compensation and Harmonic Filtering and the Over-voltage Analysis of the ITER Power Supply System

L. Xu, Z. Sheng, P. Fu, G. Gao
Institute of Plasma Physics, Chinese Academy of
Sciences Hefei 230031, China
xulw@ipp.ac.cn

Benfatto, A. D. Mankani, J. Tao
ITER International Organization
13067 Saint Paul Lez Durance, France

Abstract—The ITER Coil Power Supply System is characterised by a very large AC/DC conversion plant which produces a relatively high amount of reactive power and harmonic currents. Consequently, the ITER power supply system will include substantial reactive power compensation and harmonic filtering (RPC & HF) to reduce the reactive power and harmonic current down to the levels acceptable to the French/European 400 kV grid. This paper reports on the present design of the RPC & HF system taking account of the load power and grid requirements. The steady state and transient analysis has been performed using a detailed computer model which includes all main electrical parameters. The normal operating conditions and several different faults have been analysed and the main results are presented. The results show that the fault condition of fast load rejection may cause a relatively high overvoltage. Possible solutions to reduce the overvoltage are discussed.

I. INTRODUCTION

The ITER pulsed power distribution system (PPDS) will be connected to a powerful high-voltage (HV) grid capable of producing the large pulsed power needed to feed the superconducting coils and the H&CD systems. The HV line at 400kV supplies AC power to three Intermediate Voltage (IV) busbars at 66kV with step-down transformers each rated for 300MVA, continuous power, with 12% short circuit impedance. The three IV busbars supply the thyristor converters for all superconducting coils, the Neutral Beam (NB) power supply system and the Medium Voltage (MV) busbars at 22kV [1]. These high power ac/dc converter systems, using 12-pulsed thyristor bridges, will produce a relatively high amount of reactive power and harmonic

currents. The high reactive power will bring about a large drop and a remarkable fluctuation of the grid voltage. Also the harmonic will be harmful to the safe operation of the power system and the devices themselves [2]. Consequently, the ITER power supply system will include substantial reactive power compensation and harmonic filtering (RPC&HF) system to reduce the reactive power and harmonic currents. In 2001 design, the total RPC&HF system rated capacity of 540Mvar has been selected [3]. However, the reactive power provided by the 400KV grid is limited within 200Mvar, while it was 400Mvar in 2001. In addition, the maximum reactive power of the loads in initial phase increases to 930Mvar [4]. So the overall capacity of the RPC&HF system has to be increased. Moreover, in 2001 design, the main stepdown transformer second winding (66kV) supplies all pulsed loads, and third winding is connected to the APS (Active Power Shedding). The 50MVA, 66/22kV transformer connected on each 66kV busbar supplies loads less than 20MVA. However, the space is not enough for the 66/22kV transformer, and APS system is no more required, so the tertiary winding could be explored to supply 22 kV loads. The two options should be studied.

This paper reports on the present design of RPC&HF system taking account of the load power and grid requirements. A detailed computer model which includes all main electrical parameters has been built in PSCAD/EMTDC. The main simulation results are presented. The over-voltage caused by fast load rejection has been analyzed and possible solutions to reduce the over-voltage are discussed.

II. THE ITER RPC&HF SYSTEM

The ITER RPC & HF system consists of 3 identical units in accordance with the 3-busbar configuration. Each unit rated for 265Mvar (at 66kV) consists of six LC-filters and a three-phase TCR. The TCR is directly connected to the 66kV busbar. The main LC-filters are tuned to the 3rd (high pass), 5th, 7th, 11th, 13th and 22nd (high pass) harmonic of 50 Hz. Moreover, a relatively small 3rd harmonic high pass filter has been included to compensate the parallel resonance between the short circuit impedance of the main step-down transformer and the capacitors of the RPC unit [3]. The main parameters of the filters are given in Table 1. The one-line diagram of the ITER RPC&HF system on one 66kV busbar is shown in Fig.1.

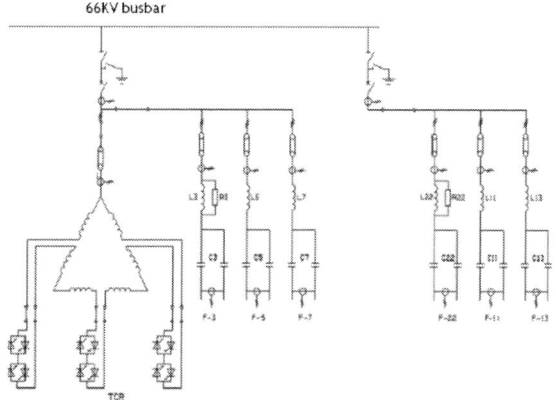

Fig. 1 One-line Diagram of the ITER RPC&HF System on one 66kV Busbar

Table 1 Main Parameters of the Filters

Parameter	Unit	Filter Circuit					
Order of harmonic		3	5	7	11	13	22
Reactive power (at 50 Hz)	Mvar	34	45	45	57	57	27
Inductance	mH	50.9	12.8	6.41	2.02	1.45	1.01
Capacitance	μF	22.08	31.57	32.21	41.31	41.40	19.7
Resistance	Ω	144					21.55
Quality factor		3	50	50	50	50	3

III. Harmonic Analysis

We suppose that all the loads generate the same harmonic current ratings. The individual harmonics are listed in the following table:

Two options are considered:

Option 1: 22kV loads are supplied by 66/22kV transformer.

Option2: 22kV loads are supplied by the tertiary winding of main step-down transformer.

Table 2 Harmonic Ratings

Harmonic Order	1	5	7	11	13	17	19	23	25
I_n/ I_1 (%)	100	3	1.2	7	6.5	0.5	0.45	3.5	3.2

The load distribution used in the simulation

66 kV: P = 350 MW, Q = 840 MVar

22 kV: P = 150 MW, Q = 90 MVar

The total harmonic distortions of voltage (THDU) with and without harmonic filters (HF) are shown in table 3 and 4. In the simulation, the short circuit power of 10GVA and an equal power sharing among the three 66kV busbars have been considered.

Table 3 THDU without HF

THDU on Bus	Option 1	Option 2
400kV	15.1%	15.1%
66kV	35.2%	35.4%

Table 4 THDU with HF

THDU on Bus	Option 1	Option 2
400kV	0.32%	0.3%
66kV	0.75%	0.73%

From the above tables, we can know that the THD on both busbars are very higher before. After the installation of HF, the THD are limited within a lower value.

IV. Transient Analysis of System

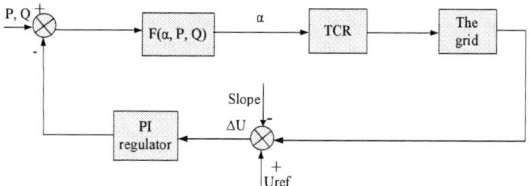

Fig. 2 The Control System of SVC

Fig. 2 shows the control system of SVC. The SVC control system consists of a PI AC voltage feedback controller and a direct compensation of reactive power and active power of the load. The direct compensation improves transient responses

[5]. To reduce the reactive power rating and prevent the SVC from reaching its reactive power limits too frequency [6], [7], a voltage droop of 5% has been considered.

In order to test the response of SVC control system, we have a simulation of step response, in the simulation, the load is a fixed three phase R-L load with the reactive power of 150Mvar and the active power of 90MW (rated at 66kV) on 66kV. It is controlled by an ideal breaker, which closes at 500ms, and breaks at 600ms. The simulation results are shown in fig.3 and fig. 4.

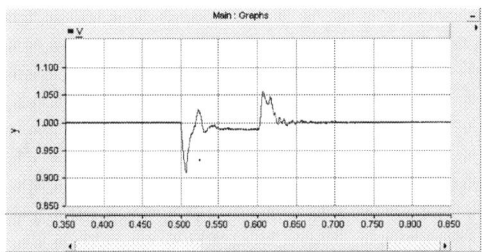

Fig.3 The Response Curve of 66kV Line Voltage

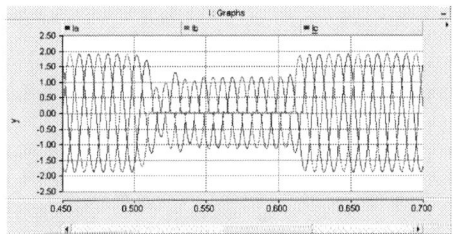

Fig.4 The Currents Wave of TCR

The simulation results show that the response time of the SVC is less than 10ms. The SVC response speed depends on the voltage regulator integral gain, proportional gain, system strength, droop and the measurement time constant.

V. The Over-voltage Analysis due to Fast Load Rejection

In case of a failure or a quench, the load power should be transferred immediately. It is often realized by fast discharge or converter bypass to avoid damage to the coils. This kind of fast load rejection often happens in a few milliseconds. Thus a relative high overvoltage will occur on the 66kV and 400kV busbar. In the analysis, we only consider the worst case. The pulsed power supply of 930Mvar and 500MW are rejected. The results are shown in table 5 and table 6.

Table 5
The Over-voltage of 66kv and 400kv (Within 7.5ms)

Overvoltage grid	Without TCR	TCR in normal operation
66kV busbar	47%	35.8%
400kV busbar	20.1%	14.8%

Table 6
The Over-voltage of 66kv and 400kv (Within 2ms)

Overvoltage	Without TCR	TCR in normal operation
66kV busbar	64.3%	54.9%
400kV busbar	27.5%	22.7%

From the tables, we can see that in the case of sudden lose of load, the overvoltage will occur in 5~10ms, the TCR do not compensate this fast voltage fluctuation. The over-voltage both on the 66kV busbar and 400kV grid is very high. The time constant of lost load is shorter, the over-voltage is higher.

In order to limit this kind of over-voltage, controlled load rejection is proposed. As shown in fig. 5, the loads shut down step by step. First the H&CD load shuts down in 7.5ms. After 50ms, the load on the first 66kV busbar rejects, then the load on the second 66kV busbar, at last the load on the third 66kV busbar. In this case, the overvoltage on 400kV busbar is 3.75%, and 14.7% on the 66kV busbar. The TCR can compensate this kind of overvoltage.

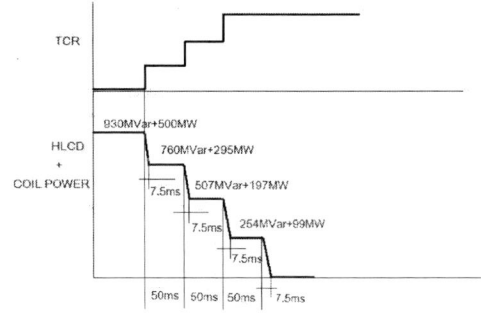

Fig. 5 The controlled sequence of load rejection

VI. Fault Analysis

A. The single line-to-ground fault analysis [8],[9]

The single line-to-ground fault is a very common fault type, when the single line-to-ground fault is happened, the non-faulted phases would bear the maximum line-to-line voltage, but the line-to-line voltage would remain the same as normal voltage. The system may still operate during a short time before the fault is removed. However, further study should be done to make sure that the system can operate safely. Fig.6 and fig.7 show the line-to-ground voltages and the line-to-line voltages during a single line-to-ground fault on 66kV busbar. The single line-to-ground fault on 22kV busbar is similar.

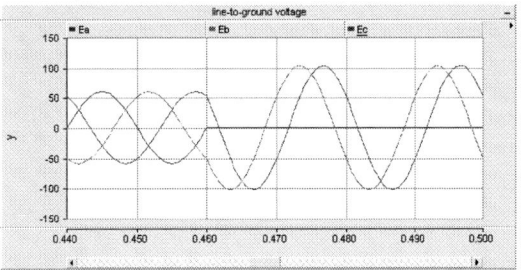

Fig.6 The line-to-ground voltage waves on 66kV busbar

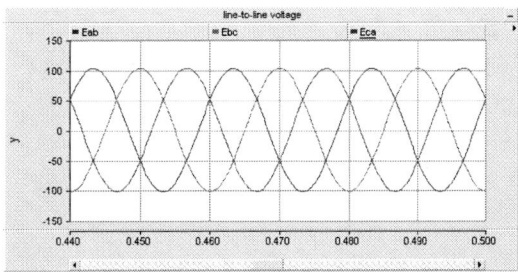

Fig.7 The line-to-line voltage waves on 66kV busbar

B. The three-phase-to-ground fault analysis

The three-phase-to-ground fault is a much less fault type; however, it is a severe fault to be interrupted by high-voltage circuit breakers. This kind of faults often cause very large short circuit currents, it should be carefully analyzed. The results will be used to perform the sizing of the cable.

In the simulation, we consider that loads do not contribute to short circuit currents. This assumption is justified by the fact that most of the loads are supplied by converters which do not contribute to short circuit current. The short circuit current values are therefore the same for every 66kV busbar and every 22kV busbar respectively, so we consider two options. In the simulation, the short circuit power of 11.78GVA and an overvoltage of 10% in the 400kV grid have been considered, the X/R ratio of short circuit impedance is 10. The results are listed in table 7.

From this table, we notice that the short circuit current at 22kV busbar is much higher in Option 2 than in Option 1. This result is because the line impedance is much higher than the

first configuration option, due to the both transformers connected in series.

Table 7 Peak Short Circuit Currents

Current Type	Option 1	Option 2
Ip at the 66kV busbar	54.1kA	54.1kA
Ip at the 22kV busbar	37.5kA	81.3kA

In case of line-to-line fault and double line-to-ground fault, there are also large short circuit currents and voltage drop, although they are much lower, these two kinds of fault should be interrupted immediately.

VII CONCLUSION

The strategy of ITER RPC&HF system and the control method are proposed. The passive filters and TCR can compensate the reactive power, and the harmonic distortion can also be reduced. The two power distribution options on 22kV busbar are analyzed. Several faults conditions have been analyzed, the results show that the short circuit currents in option 2 are much higher than in option 1. Overvoltage due to fast load rejection is also analyzed. In order to reduce the overvoltage, a controlled load rejection method is proposed. Future work will be done to verify the effectiveness of the harmonic filters and the grid. Further study should be done to investigate the feasibility of controlled load rejection.

REFERENCES

[1] **P. L. Mondino**, "The ITER pulsed power supply system", Fusion Engineering, 1997. 17th IEEE/NPSS Symposium, Volume 1, 6-10 Oct. 1997 pp. 491 – 496.

[2] J. Tao et al., "Reactive Power Compensation and Harmonic Suppression for Power Supply System of HT-7U Superconductive Tokamak," Plasma Science & Technology, Vol. 3(2001), No.1, pp. 629-634.

[3] " DDD4.1 Pulsed Power Supply " , ITER **Final Report.**

[4] J. Tao, "Voltage, Current and Power requirements in Plasma Scenarios".

[5] **K. Kahle, D. Jovcic, "Static VAR Compensator for CERN's Proton Synchrotron Accelerator" Proc. Securing Critical Infrastructure, Grenoble, 25-27 October, 2004.**

[6] **T.J.E Miller, "Reactive Power Control in Electric Systems", John Willey and Sons, New York, 1982.**

[7] Nang Sabai et al., "Voltage Control and Dynamic

Performance of Power Transmission System Using Static Var Compensator", Proceedings of World Academy of Science, Engineering and technology, Vol. 32(2008), pp.455-459.

[9] ANSI/IEEE C67.92-1987, "IEEE Guide for the Application of Neutral Grounding in Electrical Utility System".

[8] J.C. Das, "Power System Analysis: Short-Circuit Load Flow and Harmonics",Marcel Dekker, New York, 2002.

Optimal Design Method of Three-Phase Rectifier with Near-Sinusoidal Input Currents

Zhong Chen, Yingpeng Luo, YinyuZhu and Shunqing Wang
Aero-Power Sci-tech Center
Nanjing University of Aeronautics & Astronautics
Nanjing, 210016, P.R. China
Email: chenz@nuaa.edu.cn

Abstract—**In this paper, an optimal design method of three-phase rectifier with near-sinusoidal input currents (RNSIC) is proposed. By using the proposed method, high power factor could be achieved as well as low input current harmonics, making the converter more practical. Moreover, based on the operation principle analysis of RNSIC, characteristic of the rectifier is discussed. Experimental results form three prototypes with different specifications are shown to confirm the validity of the analysis and the feasibility of the proposed optimal design method.**

I. INTRODUCTION

Due to the rapid development of the power electronic technology, a number of power electronic equipments have been utilized. The conventional uncontrolled three-phase rectifiers with dc-side C filter are widely used as interface circuits between the utility grid and the power electronic equipments for their simplicity and reliability.

A large amount of harmonic components of current drawn from ac mains by the nonlinear load bring serious problems [1]. Active power filter (APF) technology has attracted more and more attention for its excellent harmonic filtering performance. However, the active filters are slightly inferior in cost and efficiency to the passive filters [2].

The passive power factor correction (PPFC) technology is more widely used for its low cost, high reliability and simple fabrication. Traditional method to improve three-phase power factor is using passive LC filters, which can be placed in ac side or dc side of the rectifiers. However, ac-side LC filter bring about dc voltage varying in a wide range and dc-side LC filter hardly obtain satisfactory performance [3], [4]. At present, some other PPFC technologies have been proposed and utilized. Multiple pulse rectifiers eliminate low-order harmonics through autotransformer, making the input current near-sinusoidal [5]-[7]. Rectifiers applying current injection contribute to the reduction of other harmonics present in input current, where the frequency of the injecting current is equal to the triple of the line frequency [8], [9]. Novel topology rectifiers, such as three-phase diode rectifiers with LC

resonance in commercial frequency [10], three-phase rectifier with near sinusoidal input currents (RNSIC) have caught increasing attention in recent years [11]-[14].

RNSIC was proposed by D. Alexa in [13] in 2004. Compared with the conventional three-phase full-bridge rectifier with passive filter, the RNSIC has following attractive characteristics: sinusoidal input current for large variations of the load, lower size volume and cost of the passive components. But the design method given in [13] is not precise theoretically or is presented without enough derivation details, e.g., the reason why the values of inductors L and capacitors C fulfill the relation $0.05 \le LC\omega^2 \le 0.1$ is not explained. As such, an interested reader will not feel comfortable to use them for design purposes. All above facts limit the application of RNSIC.

This paper analyzes the operation principle of RNSIC in detail. Based on the analysis, a criterion to determine the operation mode of RNSIC converter is given. Furthermore, a time-weighted-averaging method valid for the parameter design of the rectifier is proposed, achieving near sinusoidal input currents and near unity displacement factor. According to the proposed method, optimal value of the passive components will be obtained, making RNSIC more practical. Three prototypes working in different operation modes are built and tested to verify the validity of the analysis and the feasibility of the method.

II. OPERATION PRINCIPLE

Fig.1 shows the configuration of RNSIC, which is composed of three series inductors L_u, L_v, L_w of equal inductance values L and six commutation capacitors C_1-C_6 of equal capacitance values C, where C_d is the output filter capacitor.

The phase voltages and input currents are supposed as follows:

This work was supported by Doctoral Fund of Ministry of Education of China under Award 200802871033 and Aeronautical Science Foundation of China under Award 2009ZC52030.

$$e_u = U_m \sin \omega t \qquad\qquad i_u = I_m \sin \omega t$$

$$e_v = U_m \sin(\omega t - \frac{2}{3}\pi) \qquad i_v = I_m \sin(\omega t - \frac{2}{3}\pi) \qquad (1)$$

$$e_w = U_m \sin(\omega t + \frac{2}{3}\pi) \qquad i_w = I_m \sin(\omega t + \frac{2}{3}\pi)$$

Where U_m is the amplitude of phase voltage, I_m is the amplitude of input current.

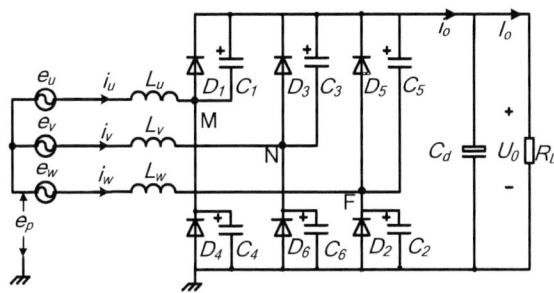

Figure 1. Topology of three-phase rectifier with near sinusoidal input currents.

In order to analyze the operation principle, the following assumptions are made:

- ideal diodes and passive components

- stead state already

Different durations of the current charging the commutation capacitor, which vary with the load, determine the distinct operation modes of the rectifier: large current mode, medium current mode and small current mode.

The duration of the current charging or discharging the capacitors is defined as t_1. Because capacitors begin to charge or discharge when the input currents cross zero, the conduction intervals of each diodes equal to $\pi/\omega - t_1$ due to the parallel connection of the capacitors and diodes.

In small current mode ($2\pi/3 < \omega t_1 < \pi$), none or one diode is conducting at any time. The conduction intervals of diode increase when the load current increases, as well as t_1 decreases. Rectifier changes into the medium current mode ($\pi/3 < \omega t_1 < 2\pi/3$), in which one or two diodes are conducting. With the load current increasing continuously, rectifier changes into the large current mode ($0 < \omega t_1 < \pi/3$), in which two or three diodes are conducting. The waveforms of the phase currents and conduction intervals of the diodes are shown in Fig. 2(a), 3(a) and 4(a), from which differences among the three modes can be deduced.

Fig. 2(b), 3(b) and 4(b) illustrate the dc-side current i_o of the rectifiers working in different modes. Current i_o is discontinuous when the rectifier is working in the small current mode, whereas i_o become a 12-pulse waveform when the rectifier is working in the medium current mode and large current mode.

In what follows, operation principle of RNSIC working in the medium current mode will be discussed in detail.

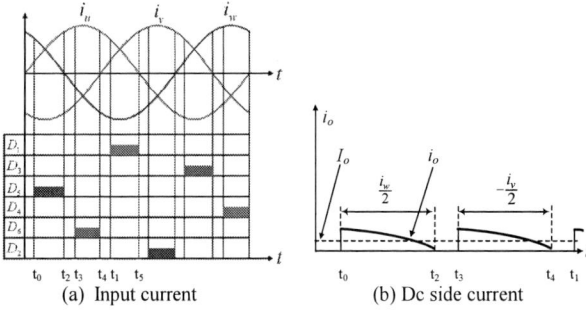

Figure 2. Key waveforms of RNSIC working in small current mode.

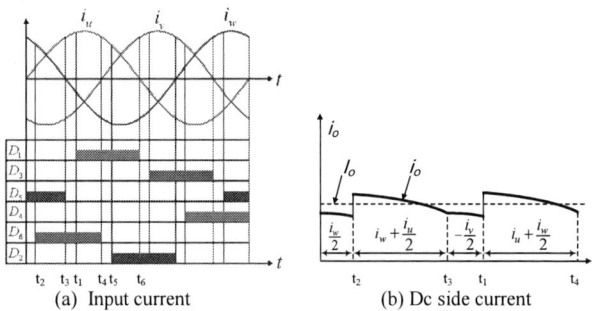

Figure 3. Key waveforms of RNSIC working in medium current mode.

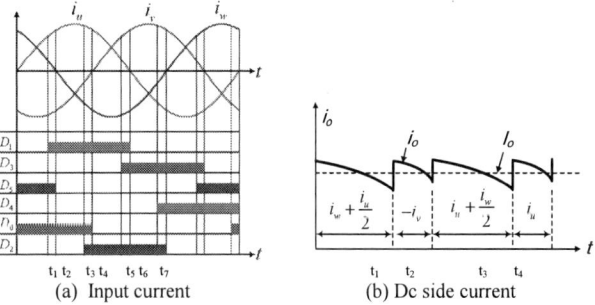

Figure 4. Key waveforms of RNSIC working in large current mode.

A. Medium Current Mode

As Fig. 5 shows, four different operation stages exist in the medium current mode.

Stage 1 [0, t_2]: Prior to 0, D_4 and D_5 conducts, $u_{c4}=0$, $u_{c1}=U_o$. The input current i_v which is negative, charges the capacitor C_3 and discharges the capacitor C_6, simultaneously.

In the stage, the input current i_u which becomes positive at time point 0 charges the capacitor C_4 and discharges the capacitor C_1, and diode D_5 keeps conducting, as shown in Fig.5 (a). In this stage, the dc side current of the rectifier i_o is given by

$$i_O = i_u/2 + i_v/2 + i_w = i_w/2 \qquad (2)$$

Stage 2 [t_2, t_3]: At time point t_2 when the voltage across C_6 reaches zero, diode D_6 starts conducting. In the stage, the input current i_u which is positive keeps charging the capacitor C_4 and discharging the capacitor C_1, simultaneously. In this stage, the dc side current of the rectifier i_o is given by

$$i_o = i_w + i_u / 2 \qquad (3)$$

Diode D_5 gets blocked at time point t_3 (corresponding to $\pi/3\omega$), when current i_w crosses zero, indicating the end of stage 2.

Stage 3 [t_3, t_1]: At time point t_3, the input current i_w whose direction is changed into negative, begin to charge the capacitor C_5 and discharges the capacitor C_2, simultaneously. In the stage, the input current i_u which is positive keeps charging the capacitor C_4 and discharging the capacitor C_1. Diode D_6 keeps conducting, as shown in Fig. 5(c). In this stage, the dc side current of the rectifier i_o is given by

$$i_o = i_u / 2 + i_w / 2 = -i_v / 2 \qquad (4)$$

Stage 4 [t_1, t_4]: At time point t_1, the voltage across C_1 reaches zero, diode D_1 starts conducting. In the stage, the input current i_w which is negative keeps charging the capacitor C_5 and discharging the capacitor C_2, simultaneously. In this stage, the dc side current of the rectifier i_o is given by

$$i_o = i_u + \frac{1}{2}i_w \qquad (5)$$

Diode D_6 gets blocked at time point t_4 (corresponding to $2\pi/3\omega$), when current i_v crosses zero, indicating the end of stage 4.

The dc side current i_o of RNSIC working in the medium current mode is given in Fig. 3(b). The output current I_o, the mean value of i_o, could be expressed as follow:

$$I_o = \frac{3}{2\pi}[\int_0^{\omega t_1 - \frac{\pi}{3}} \frac{1}{2}i_w d\omega t + \int_{\omega t_1 - \frac{\pi}{3}}^{\frac{\pi}{3}} (i_w + \frac{1}{2}i_u)d\omega t$$
$$+ \int_{\frac{\pi}{3}}^{\omega t_1} -\frac{1}{2}i_v d\omega t + \int_{\omega t_1}^{\frac{2\pi}{3}} (i_u + \frac{1}{2}i_w)d\omega t] \qquad (6)$$

The equation simplifies to

$$I_o = \frac{3}{2\pi}[I_m(1 + \cos \omega t_1)] \qquad (7)$$

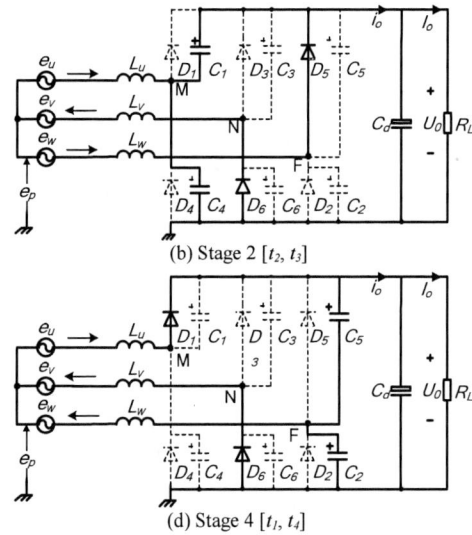

(a) Stage 1 [0, t_2]

(b) Stage 2 [t_2, t_3]

(c) Stage 3 [t_3, t_1]

(d) Stage 4 [t_1, t_4]

Figure 5. Equicalent operation stage circuits of RNSIC working in medium current mode.

B. Small and Large Current Mode

As Fig. 3 and 4 depict, none or one diode conducts in the small current mode and two or three diodes conducts in the large current mode. Operation principle of the rectifier in these two operation modes could be derived by using the same analysis.

The output current I_o in the small current mode could be expressed as follow:

$$I_o = \frac{3}{2\pi}(\int_{\omega t_1 - \frac{2}{3}\pi}^{\frac{\pi}{3}} \frac{1}{2}i_w d\omega t + \int_{\omega t_1 - \frac{\pi}{3}}^{\frac{2\pi}{3}} -\frac{1}{2}i_v d\omega t)$$
$$= \frac{3}{2\pi}[I_m(1 + \cos \omega t_1)] \qquad (8)$$

The output current I_o in the large current mode could be expressed as follow:

$$I_o = \frac{3}{2\pi}[\int_0^{\omega t_1} (i_w + \frac{1}{2}i_u)d\omega t + \int_{\omega t_1}^{\frac{\pi}{3}} -\frac{1}{2}i_v d\omega t$$
$$+ \int_{\omega t_1 + \frac{\pi}{3}}^{\frac{2\pi}{3}} (i_u + \frac{1}{2}i_w)d\omega t + \int_{\omega t_1 + \frac{\pi}{3}}^{\frac{2\pi}{3}} i_u d\omega t] \qquad (9)$$
$$= \frac{3}{2\pi}[I_m(1 + \cos \omega t_1)]$$

Very interesting, expressions of I_o of RNSIC are same in these three operation modes.

III. PARAMETER DESIGN

A. Operation Mode Selection

The first issue need to be solved in rectifier design is to determine its operation mode, according to the requirement of application field (the input voltage, output voltage and output power).

The input and output power is given as

$$P_{in} = 3\frac{U_m}{\sqrt{2}}\frac{I_m}{\sqrt{2}} = 3\frac{U_m I_m}{2} \qquad (10)$$

$$P_o = U_o I_o \qquad (11)$$

Where U_o is the output voltage value, neglecting the voltage ripple.

Neglecting the power losses of the converter gives:

$$P_{in} = P_o \qquad (12)$$

The relationship between U_o and U_m could be expressed as follow:

$$\frac{U_m}{U_o} = \frac{2I_o}{3I_m} \qquad (13)$$

Substituting (7) for I_o/I_m into (13) gives:

$$\frac{U_m}{U_o} = \frac{1 + \cos \omega t_1}{\pi} \qquad (14)$$

Solving (14) for ωt_1 gives the expression of ωt_1 as a function of U_m and U_o in RNSIC converter.

$$\omega t_1 = \arccos(\frac{\pi U_m}{U_o} - 1) \qquad (15)$$

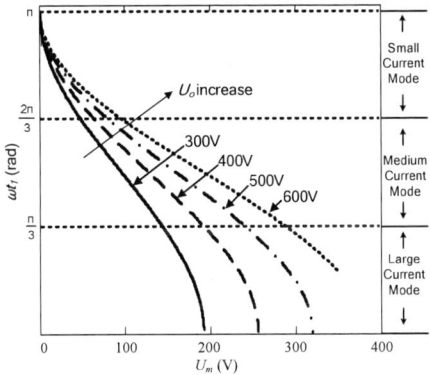

Figure 6. Variations of ωt_1 as a function of U_m and U_o.

Fig. 6 presents the variation of ωt_1 as a function of U_m and U_o in RNSIC converters. For the output voltage U_o, we have

adopted the value of 300 V, 400V, 500 V, 600 V. From the analysis given above, which operation mode RNSIC converters work in is determined by t_1. Actually, Fig. 6 gives us the criterion which operation mode RNSIC converters work in.

From this diagram, we could find that RNSIC converters have following characteristic: ① which operation mode RNSIC converters work in is only determined by the input voltage and output voltage, the output power does not affect their operation mode; ② in the application with low input voltage and high output voltage, RNSIC converter should be designed to work in the small current mode, whereas in the application with high input voltage and relatively low output voltage, RNSIC converters suit be designed to work in the large current mode.

B. Time Weighted Averaging Based Parameter Design Method

In this paper, passive components are designed by using time weighted average method. Final optimal parameter value is derived by weighted averaging multiple parameter values which are calculated in different operation stages.

Two distinct operation stages in which different diodes conduct exist in every 1/6 period for RNSIC converter. Moreover, operation stages of RNSIC converter repeat every 1/6 period. So the auxiliary circuit could be designed as follow:

--First, according to the operation mode of RNSIC converter, choose one 1/6 period, calculate the parameter value of auxiliary circuit in two operation stages, respectively.

--Second, obtain the conclusive optimal value by time weighted averaging the two values derived above in 1/6 period.

Take the design of RNSIC working in medium current mode for example.

The relationship between the ac side voltage and current of the rectifiers could be represented by employing KVL law as:

$$e_p + e_u = L\frac{di_u}{dt} + u_M \qquad (16)$$

$$e_p + e_v = L\frac{di_v}{dt} + u_N \qquad (17)$$

$$e_p + e_w = L\frac{di_w}{dt} + u_F \qquad (18)$$

Considering the symmetric of phase voltage, the voltage of neuter point e_p could be expressed as follow:

$$e_p = \frac{u_M + u_N + u_F}{3} \qquad (19)$$

1 Operation stage [t_2, t_3]

In the operation stage of [t_2, t_3], diode D_5 keeps conducting, input current i_u keeps charging C_4 and discharging C_1 from zero time point. So the voltage of point

M, N, F could be expressed as follows:

$$u_M = u_{c4} = \frac{1}{C}\int_0^t \frac{1}{2}I_m \sin\omega t\, dt = \frac{I_m}{2C\omega}(1-\cos\omega t) \quad (20)$$

$$u_N = 0 \quad (21)$$

$$u_F = U_o \quad (22)$$

Substituting (20) for u_M, (21) for u_N and (22) for u_F into (19) gives:

$$e_p = [\frac{I_m}{2C\omega}(1-\cos\omega t)+0+U_o]/3 \quad (23)$$

Subtracting (16) from (17) gives:

$$e_v - e_w = L(\frac{di_v}{dt}-\frac{di_w}{dt})+(u_N-u_F) \quad (24)$$

Substituting (1) for i_v and i_w, (21) for u_N, and (22) for u_F, into (24) gives:

$$e_v - e_w = L[d(I_M \sin(\omega t-\frac{2\pi}{3}))/dt$$
$$-d(I_M \sin(\omega t+\frac{2\pi}{3}))/dt]+(0-U_o) \quad (25)$$

Integrating both sides of (25) on the angel interval [ωt_2, ωt_3] and simplifying gives the value of optimal inductance as follow:

$$L_1 = \frac{\int_{\omega t_1-\frac{\pi}{3}}^{\frac{\pi}{3}}(e_w-e_v-U_o)d\omega t}{I_m\omega[-\sin(\omega t_1+\frac{\pi}{3})-\sin\omega t_1+\frac{\sqrt{3}}{2}]} \quad (26)$$

Here, ωt_2 corresponds to ωt_1-$\pi/3$, ωt_3 corresponds to $\pi/3$.

Substituting (21) for u_N, and (23) for e_p into (17) gives:

$$[\frac{I_m}{6C\omega}(1-\cos\omega t)+\frac{U_o}{3}]+e_v = L_1\frac{di_v}{dt}+0 \quad (27)$$

Integrating both sides of (27) on the angle internal [ωt_2, ωt_3] and simplifying gives the expression of C_1:

$$C_1 = \int_{\omega t_1-\frac{\pi}{3}}^{\frac{\pi}{3}}I_m(1-\cos\omega t)d\omega t/6\omega[L_1\omega I_m(\sin\omega t_1$$
$$-\frac{\sqrt{3}}{2})-\frac{U_o}{3}(-\omega t_1+\frac{2\pi}{3})-\int_{\omega t_1-\frac{\pi}{3}}^{\frac{\pi}{3}}e_v d\omega t] \quad (28)$$

2 Operation stage [t_3, t_1]

In the operation stage of [t_3, t_1], input current i_u keeps charging C_4 and discharging C_1, diode D_6 keeps conducting So the voltage of point M and N have the same expression with the one in the last stage and the voltage of point F could be expressed as follows:

$$u_F = u_{c2}\Big|_{t=t_3}+\frac{1}{C}\int_{t_3}^t \frac{1}{2}I_m(\sin\omega t+\frac{2}{3}\pi)dt$$
$$=U_o - \frac{I_m}{2C\omega}(1+\cos(\omega t+\frac{2\pi}{3})) \quad (29)$$

Equ. (23) should be modified as follow:

$$e_p = \frac{1}{3}[\frac{-I_m}{2C\omega}(\cos(\omega t+\frac{2\pi}{3})+\cos\omega t)+U_o] \quad (30)$$

After obtaining the voltage of point e_p and u_F, optimal inductance and capacitance in the stage of [t_3, t_1] could be derived by using the above similar method. Expression C_2 of and L_2 are derived as follow:

$$C_2 = \frac{\int_{\frac{\pi}{3}}^{\omega t_1}I_m[1+\cos(\omega t+\frac{2\pi}{3})]d\omega t}{3\omega[\int_{\frac{\pi}{3}}^{\omega t_1}(\frac{2U_o}{3}+e_v)d\omega t+I_m L_1\omega\sin(\omega t_1-\frac{2\pi}{3})]} \quad (31)$$

$$L_2 = \{\int_{\frac{\pi}{3}}^{\omega t_1}(e_v-e_w+U_o)d\omega t-\int_{\frac{\pi}{3}}^{\omega t_1}\frac{3[1+\cos(\omega t+\frac{2\pi}{3})]}{\sin\omega t_1+\sin(\omega t+\frac{2\pi}{3})-\frac{\sqrt{3}}{2}}\cdot$$
$$[U_m(\frac{1}{2}-\cos(\omega t_1-\frac{2\pi}{3}))+\frac{U_o}{3}(\omega t_1-\frac{\pi}{3})]d\omega t)\}/\{I_m\omega[\sin(\omega t_1-$$
$$\frac{2\pi}{3})-\sin(\omega t_1+\frac{2\pi}{3})+\frac{\sqrt{3}}{2}]-\int_{\frac{\pi}{3}}^{\omega t_1}\frac{3I_m\omega[1+\cos(\omega t+\frac{2\pi}{3})]}{\sin\omega t_1+\sin(\omega t+\frac{2\pi}{3})-\frac{\sqrt{3}}{2}}\cdot$$
$$[\sin(\omega t_1-\frac{2\pi}{3})+\frac{\sqrt{3}}{2}]d\omega t\} \quad (32)$$

Optimal inductance and capacitance are derived as follows:

$$L = \frac{L_1(\frac{2\pi}{3}-\omega t_1)+L_2(\omega t_1-\frac{\pi}{3})}{\frac{\pi}{3}} \quad (33)$$

$$C = \frac{C_1(\frac{2\pi}{3}-\omega t_1)+C_2(\omega t_1-\frac{\pi}{3})}{\frac{\pi}{3}} \quad (34)$$

IV. EXPERIMENTAL VERIFICATION

In order to verify the feasibility of the analysis and the proposed design method, three prototype systems working in different modes were built with the following specification (as Tab.1 shown).

Fig. 7 shows the experimental waveforms of the phase voltage and input current of the prototypes working in the large current mode, medium current mode and small current mode, respectively. It can be seen that the input currents are practically sinusoidal and the displacement factor is near unity.

TABLE I.   SPECIFICATION OF THREE RNSIC PROTOTYPES

Prototype	U_m / V	U_o / V	P_o / kW	L / mH	C / µF	Operation Mode
1	250	500	7.18	27.7	22.46	large current mode
2	150	500	6.55	27.7	98.7	Medium current mode
3	55	500	2.52	27.7	159.2	small current mode

Fig. 8 shows the spectrum of the input current of the three prototypes, where the THDs are 4.916%, 5.042%, 5.643%, respectively.

CONCLUSION

The RNSIC converter catches increasing attention for its simple configuration, high reliability as well as the reduced cost. This paper has proposed a practical design method of the RNSIC on the basis of operation principle analysis. Experimental results that the THD lower than 10% and power factor higher than 0.99, which are obtained from three prototypes work with different modes have verified both the validity of theoretical analysis and the feasibility of the design method.

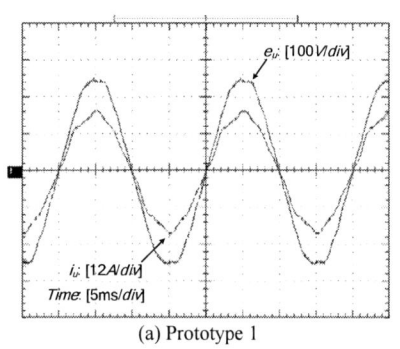 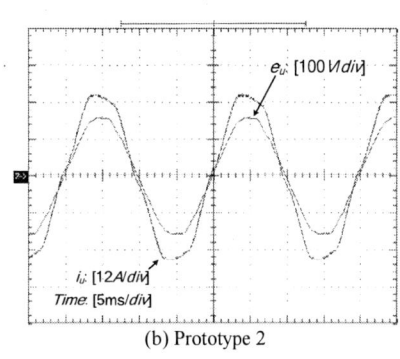 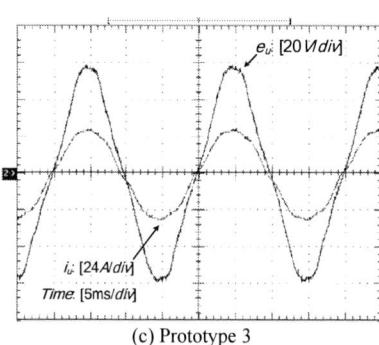

(a) Prototype 1   (b) Prototype 2   (c) Prototype 3

Figure 7.   Experimental waveforms of phase voltage $e_u(t)$ and current $i_u(t)$.

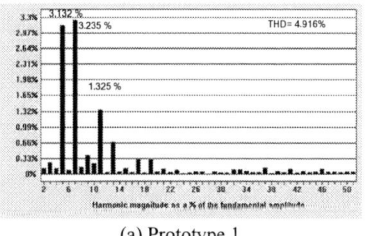 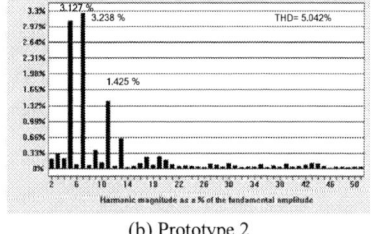 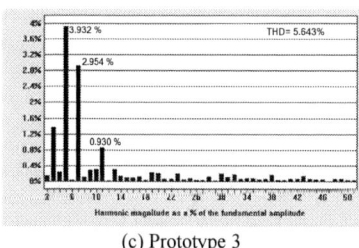

(a) Prototype 1   (b) Prototype 2   (c) Prototype 3

Figure 8.   Nomilized line current harmonics of experimental prototypes.

REFERENCES

[1] H. Akagi, E. H. Watanabe, and M. Aredes, Instantaneous Power Theory and Applications to Power Conditioning, New Jersey: John Willey & Sons, 2007, pp. 4-5.

[2] H. Akagi, "The state-of-the-art of active filters for power conditioning," EPE 2005-11th European Conference on Power Electronics and Applications, 2005, pp. 1-15.

[3] A. W. Kelley and W. F. Yadusky, "Rectifier Design for Minimum Line-Current Harmonics and Maximum Power Factor," IEEE Trans. Power Electronics, Vol. 7, No. 2, pp. 663-672, Apr. 1992.

[4] M. Sakui and H. Fujita, "An Analytical Method for Calculating Harmonic Current of a Three-phase Diode bridge Rectifier with dc Filter," IEEE Trans. Power Electronics, Vol. 9, No. 6, pp. 631-637, Nov. 1994.

[5] B. Singh, G. Bhuvaneswari, and V. Garg, "T-Connected Autotransformer - Based 24-Pulse AC-DC Converter for Variable Frequency Induction Motor Drives," IEEE Trans. Energy Conversion, Vol. 21, No. 3, pp. 663-672, Sep. 2006.

[6] R. P. Burgos, A. Uan-Zo-li, F. Lacaux, A. Roshan, F. Wang, and D. Boroyevich, "Analysis of New Step-Up and Step-Down 18 Pulse Direct Asymmetric Autotransformer Rectifiers," in Proc. IEEE Conf. IAS, 2005, Vol. 1, pp. 145–152.

[7] F.J.M. de Seixas and I. Barbi, "A 12kW three-phase low THD rectifier with high frequency isolation and regulated dc output," IEEE Trans. Power Electronics, Vol. 19, No. 2, pp. 371-377, Mar. 2004.

[8] P. Pejović, P. Božović, and D. Shmilovitz, "Low-Harmonic, Three-Phase Rectifier That Applies Current Injection and a Passive Resistance Emulator," IEEE Power Electronics Letters, Vol. 3, No. 3, pp. 96-100, Sep. 2005.

[9] P. Pejović and Z. Janda, "An Analysis of Three-Phase Low Harmonic Rectifiers Applying the Third-Harmonic Current Injection," IEEE Trans. Power Electronics , Vol. 14, No. 3, pp. 397-407, May 1999.

[10] I. Yamamoto, K. Ohtsuka, K. Matsui, and Y. Yao, "A Novel Three-Phase Diode Rectifier with LC Resonance in Commercial Frequency," in Proc. IEEE IECON'01 Conf., 2001, pp. 1350-1356.

[11] D. Alexa and A. Sîrbu, "Optimized Combined Harmonic Filtering System," IEEE Trans. Industrial Electronics, Vol. 48, No. 6, pp. 1210-1218, Dec. 2001.

[12] D. Alexa, A. Sîrbu, and D. Dobrea, "Topologies of Three-Phase Rectifiers with Near Sinusoidal Input Currents," in Proc. IEE Electr. Power Appl., Vol. 151, No. 6, pp. 673-678, Nov. 2004.

[13] D. Alexa, A. Sîrbu, and D. Dobrea, "An Analysis of Three-Phase Rectifiers with Near-Sinusoidal Input Current," IEEE Trans. Industrial Electronics, Vol. 51, No. 4, pp. 884-891, Aug. 2004.

[14] D. Alexa, A. Sîrbu, and A. Lazăr, "Three-Phase Rectifiers with Near Sinusoidal Input Current and Capacitors Connected on the AC side," IEEE Trans. Industrial Electronics, Vol. 53, No. 5, pp. 1612-1620, Oct. 2006

An Analysis on the Influence of Interface Inductor to STATCOM system with Phase and Amplitude Control and Corresponding Design Considerations

Guopeng ZHAO, Jinjun LIU

School of Electrical Engineering, Xi'an Jiaotong University
Xi'an, China 710049
guopengzhao@ieee.org

Abstract—Previous publications regarding the design and specification of interface inductor and DC side capacitor for STATCOM usually deal with the interface inductor and the DC side capacitor only, and seldom pay attention to the influences of interface inductor and DC side capacitor on STATCOM systems performance. Phase-shift Control and Phase and Amplitude Control (PAC) are two basic control methods. In this paper, the Phase and Amplitude Control is considered as the STATCOM control strategy. And, a detailed analysis on the influences of interface inductor and DC side capacitor on STATCOM systems and corresponding design consideration are presented. First, the models of the current loop and the voltage loop in STATCOM system by using two modeling methods are carried out. Second, the transfer functions of each control loop are presented. Through frequency domain methods, such as Bode plots, the influences of interface inductor and DC side capacitor on the STATCOM system stability are extensively investigated. Third, based on the analysis, the design considerations based on phase margin for interface inductor and DC side capacitor are discussed.

I. INTRODUCTION

Static Synchronous Compensator (STATCOM) is one of FACTS (flexible AC transmission system) devices. It is an advanced shunt compensator. The STATCOM can be utilized to regulate voltage, control power factor, stabilize power flow and improve dynamic performance of power systems [1]. The compensator has been gaining wide attentions in recent years [2-5]. There is a voltage source converter connected to the grid through an interface inductor in the STATCOM system. The interface inductor is a low pass filter.

The design of the interface inductor and the DC side capacitor is very important. Many papers discussed the choice methods of the STATCOM parameters. For example, in [6, 7], the interface inductor must satisfy the ripple current requirement, and, in [8], the voltage drop across the inductor must be considered to design an interface inductor. In practical application, the filtering performance is usually the main aspect considered. And then, by using the experience of practical application and many experimental results, the

parameters of the interface inductor are usually designed. About the DC side capacitor, many papers reduced the DC side capacitor. For example, in [8], the paper used high reliability film capacitors to minimize capacitor size and cost. This paper also mentioned that the design approaches of DC side capacitor often depended on three aspects: the maximum ripple current capability of the capacitor, the maximum allowable voltage ripple and the desired ride through capability during grid failure or voltage sags. In [9-11], the methods to reduce the DC side capacitor were presented, and the minimum capacitor was obtained. In [12-14], the papers presented the method of reducing the voltage ripple to design the DC side capacitor. Previous publications regarding the design and specification of interface inductor and DC side capacitor for STATCOM usually deal with the interface inductor and the DC side capacitor only, and seldom pay attention to the influences of interface inductor and the DC side capacitor on STATCOM systems performance. When the interface inductor and the DC side capacitor are designed, the influences of these parameters on the STATCOM system performance should be considered. How will the interface inductor and the DC side capacitor influence the stability of system? How to design the interface inductor and the DC side capacitor more reasonably and more optimized?

In this paper a detailed analysis on the interface inductor and the DC side capacitor which influence the stability of STATCOM system and corresponding design consideration are presented. In STATCOM system, especially in high voltage and high power situation, two basic control methods are used. One is phase-shift control [15-18], the other is Phase and Amplitude Control (PAC) [19-22]. In this paper, the Phase and Amplitude Control is considered as the STATCOM control strategy. First, the models of STATCOM is carried out, second, the transfer functions are presented, and the influences of the interface inductor and the DC side capacitor on the STATCOM system stability are extensively investigated. Third, based on the analysis, the design considerations for the interface inductor and the DC side capacitor are discussed, which leads to parameters different to traditional design.

Sponsored by grants from the Power Electronics Science and Education Development Program of Delta Environmental & Educational Foundation.

II. System Configuration

The main circuit of STATCOM is shown in Figure 1. $u_k(k=a,b,c)$ — source voltage, i_{Lk} — load current, i_{sk} — source current, i_{ck} — output current, U_{dc} — DC side voltage. The single-phase equivalent circuit is shown in Figure 2. The variables are shown as follow. u_s — source voltage, u_L — voltage of inductor, i_c — output current, u_1 — output voltage of converter, $Z=j\omega L$ — impedance of inductor. The losses of the converter are regarded as the active power in the resistor of the inductor [23-25].

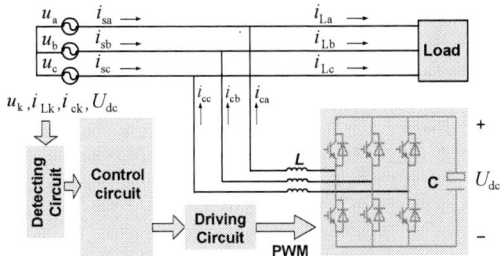

Figure 1. the main circuit of STATCOM

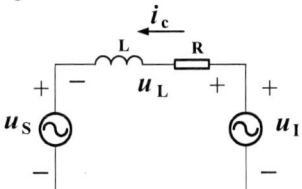

Figure 2. the single-phase equivalent circuit

III. System Model and Control Strategy

The mathematical instantaneous three-phase model of STATCOM is shown as equation (1) and (2). By using the transformation, equation (1) can be transformed to equation (3) in the synchronously rotating reference frame. With equation (4) and equation (5), the system model can be carried out as equation (6), where δ is the phase angle between the source voltage and output voltage of converter. The variable θ is determined by the amplitude of output voltage. The d axis in the synchronously rotating reference frame has the same direction with the phasor of source voltage, that is, the initial rotating angle is zero. The q axis is perpendicular to d axis. So, $u_{sq}=0$. The parameter k is decided by the DC side voltage and the peak value of output voltage.

$$L\frac{d}{dt}\begin{bmatrix} i_{ca} \\ i_{cb} \\ i_{cc} \end{bmatrix}+\begin{bmatrix} Ri_{ca} \\ Ri_{cb} \\ Ri_{cc} \end{bmatrix}=\begin{bmatrix} u_1 \\ u_2 \\ u_3 \end{bmatrix}-\begin{bmatrix} u_{sa} \\ u_{sb} \\ u_{sc} \end{bmatrix} \quad (1)$$

$$\frac{d}{dt}u_{dc}=-\frac{1}{C}i_{dc} \quad (2)$$

$$\frac{d}{dt}\begin{bmatrix} i_d \\ i_q \end{bmatrix}=\frac{1}{L}\begin{bmatrix} u_d-u_{sd} \\ u_q-u_{sq} \end{bmatrix}-\begin{bmatrix} \dfrac{R}{L}i_d-\omega i_q \\ \dfrac{R}{L}i_q+\omega i_d \end{bmatrix} \quad (3)$$

$$u_{dc}i_{dc}=\frac{3}{2}\left(u_d i_d+u_q i_q\right) \quad (4)$$

$$u_d=ku_{dc}\cos\delta \quad (5)$$

$$\frac{d}{dt}\begin{bmatrix} i_d \\ i_q \\ u_{dc} \end{bmatrix}=\begin{bmatrix} -\dfrac{R}{L} & \omega & \dfrac{k}{L}\cos\delta \\ -\omega & -\dfrac{R}{L} & \dfrac{k}{L}\sin\delta \\ -\dfrac{3}{2C}k\cos\delta & -\dfrac{3}{2C}k\sin\delta & 0 \end{bmatrix}\begin{bmatrix} i_d \\ i_q \\ u_{dc} \end{bmatrix}-\frac{1}{L}\begin{bmatrix} u_{sd} \\ 0 \end{bmatrix}$$

$$(6)$$

The Phase and Amplitude Control diagram of STATCOM is shown in Figure 3.

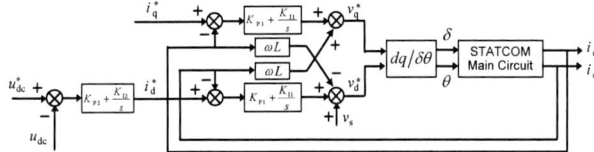

Figure 3. the Phase and Amplitude Control diagram of the STATCOM

The control principle is shown in Figure 4, and the technique of PWM can also be used instead of the quadrate waveform in order to eliminate the harmonics. With the Fourier analysis, the fundamental part is expressed as equation (7). The relationship between δ/θ to u_d/u_q is expressed in equation (8).

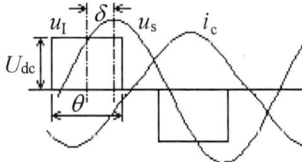

Figure 4. the control principle of the STATCOM

$$\begin{cases} u_d=\dfrac{4}{\pi}u_{dc}\sin\dfrac{\theta}{2}\cos\delta \\ u_q=\dfrac{4}{\pi}u_{dc}\sin\dfrac{\theta}{2}\sin\delta \end{cases} \quad (7)$$

$$\begin{cases} \theta=2\arcsin\left(\dfrac{\pi}{4u_{dc}}\sqrt{u_d^2+u_q^2}\right) \\ \delta=\arctan\left(\dfrac{u_q}{u_d}\right) \end{cases} \quad (8)$$

A. Model of the Main Circuit

By using the equation (6), the main circuit model is presented in equation (9). The small signal model is shown in equation (10). Where, the steady operation points are u_{dc0}, v_{d0}, v_{q0}, i_{d0} and i_{q0}. And, $k=[4\sin(\theta/2)]/\pi$.

$$\frac{d}{dt}\begin{bmatrix} i_d \\ i_q \\ u_{dc} \end{bmatrix} = \begin{bmatrix} -\dfrac{R}{L} & \omega & \dfrac{4}{\pi L}\sin\dfrac{\theta}{2}\cos\delta \\ -\omega & \dfrac{-R}{L} & \dfrac{4}{\pi L}\sin\dfrac{\theta}{2}\sin\delta \\ -\dfrac{6}{C\pi}\sin\dfrac{\theta}{2}\cos\delta & -\dfrac{6}{C\pi}\sin\dfrac{\theta}{2}\sin\delta & 0 \end{bmatrix}\begin{bmatrix} i_d \\ i_q \\ u_{dc} \end{bmatrix} - \frac{1}{L}\begin{bmatrix} u_{sd} \\ 0 \end{bmatrix} \tag{9}$$

$$\frac{d}{dt}\begin{bmatrix} \Delta i_d \\ \Delta i_q \\ \Delta u_{dc} \end{bmatrix} = \begin{bmatrix} -\dfrac{R}{L} & \omega & \dfrac{4}{\pi L}\sin\dfrac{\theta_0}{2}\cos\delta_0 \\ -\omega & -\dfrac{R}{L} & \dfrac{4}{\pi L}\sin\dfrac{\theta_0}{2}\sin\delta_0 \\ -\dfrac{6}{\pi C}\sin\dfrac{\theta_0}{2}\cos\delta_0 & -\dfrac{6}{\pi C}\sin\dfrac{\theta_0}{2}\sin\delta_0 & 0 \end{bmatrix}\begin{bmatrix} \Delta i_d \\ \Delta i_q \\ \Delta u_{dc} \end{bmatrix}$$
$$+ \begin{bmatrix} -\dfrac{4}{\pi L}u_{dc0}\sin\dfrac{\theta_0}{2}\sin\delta_0 & \dfrac{2}{\pi L}u_{dc0}\cos\dfrac{\theta_0}{2}\cos\delta_0 \\ \dfrac{4}{\pi L}u_{dc0}\sin\dfrac{\theta_0}{2}\cos\delta_0 & \dfrac{2}{\pi L}u_{dc0}\cos\dfrac{\theta_0}{2}\sin\delta_0 \\ \dfrac{3}{\pi C}\left(i_{d0}\sin\dfrac{\theta_0}{2}\sin\delta_0 - i_{q0}\sin\dfrac{\theta_0}{2}\cos\delta_0\right) & -\dfrac{3}{\pi C}\left(i_{d0}\cos\dfrac{\theta_0}{2}\cos\delta_0 + i_{q0}\cos\dfrac{\theta_0}{2}\sin\delta_0\right) \end{bmatrix}\begin{bmatrix} \Delta\delta \\ \Delta\theta \end{bmatrix} \tag{10}$$

B. Model of dq/δθ

The small signal model of $dq/\delta\theta$ is shown as equation (11) which is obtained by using equation (8).

$$\begin{cases} \Delta\delta = -\dfrac{\pi}{4u_{dc0}\sin\dfrac{\theta_0}{2}}\sin\delta_0\Delta u_d + \dfrac{\pi}{4u_{dc0}\sin\dfrac{\theta_0}{2}}\cos\delta_0\Delta u_q \\[4mm] \Delta\theta = 2\dfrac{\pi}{4u_{dc0}\cos\dfrac{\theta_0}{2}}\cos\delta_0\Delta u_d + 2\dfrac{\pi}{4u_{dc0}\cos\dfrac{\theta_0}{2}}\sin\delta_0\Delta u_q - \dfrac{2\Delta u_{dc}\sin\dfrac{\theta_0}{2}}{u_{dc0}\cos\dfrac{\theta_0}{2}} \end{cases} \tag{11}$$

C. The Model of System

From A and B, the small signal model from reference voltages to output currents is carried out as equation (12).

$$\frac{d}{dt}\begin{bmatrix} \Delta i_d \\ \Delta i_q \\ \Delta u_{dc} \end{bmatrix} = \begin{bmatrix} -\dfrac{R}{L} & \omega & 0 \\ -\omega & -\dfrac{R}{L} & 0 \\ -\dfrac{3u_{q0}}{2Cu_{dc0}} & -\dfrac{3u_{q0}}{2Cu_{dc0}} & \dfrac{3}{2Cu_{dc0}^2}\left(i_{d0}u_{d0}+i_{q0}u_{q0}\right) \end{bmatrix}\begin{bmatrix} \Delta i_d \\ \Delta i_q \\ \Delta u_{dc} \end{bmatrix} + \begin{bmatrix} \dfrac{1}{L} & 0 \\ 0 & \dfrac{1}{L} \\ -\dfrac{3i_{d0}}{2Cu_{dc0}} & -\dfrac{3i_{q0}}{2Cu_{dc0}} \end{bmatrix}\begin{bmatrix} \Delta u_d \\ \Delta u_q \end{bmatrix} \tag{12}$$

D. Another Modeling Method

Because the transfer $dq/\delta\theta$ has no time delay part and it is just a mathematical transfer, the model from reference voltages to output currents can be presented directly by using equation (4). The small signal analysis is shown in equation (13). Then, the same signal model can be obtained as equation (12). The two modeling methods have the same small signal model.

$$u_{dc0}\frac{d\left(\Delta u_{dc}\right)}{dt} = -\frac{3}{2C}\left(u_{d0}\Delta i_d + \Delta u_d i_{d0} + u_{q0}\Delta i_q + \Delta u_q i_{q0}\right) \tag{13}$$

IV. INFLUENCES OF PARAMETERS ON THE STATCOM SYSTEM CHARACTERISTIC AND DESIGN METHODS

There are two types of closed loops. One is current loop, and the other is DC voltage control loop. The designs of the two control loop and the influences of main power parameters on system stability are discussed. And then, the designs of parameters are also presented.

A. Current Control Loop

The current control loop is shown in Figure 5. The two current loops are decoupled in Figure 3. Because a period of source current is used to calculate the RMS of current, the calculate part can be expressed as a first-order inertia element and a proportion element model [26, 27]. The time delay is T_f /2(T_f is the current period 20ms).

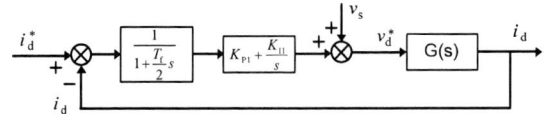

Figure 5.the current control loop of the STATCOM

When the P regulator is used, the transfer function of current control loop is shown in equation (14).

$$G_{open}(s) = \frac{k_{P1}}{\dfrac{LT_f}{2}s^2 + \left(\dfrac{RT_f}{2}+L\right)s + R} \tag{14}$$

A 50MVA/10kV system is considered to be analyzed. In traditional process, when the model of system and parameters are known, then the P regulator can be used to design a closed loop system. However, the parameters of the interface inductor and the DC side capacitor which are designed by using many methods aren't precise enough. In design process, the change of the interface inductor and the DC side capacitor isn't convenient. So, we should choose a proper parameter of P regulator, the influences of main circuit parameters on system should be analyzed, and then, it can provide a more optimized method to design the interface inductor and the DC side capacitor. The parameters are shown as follow. C=0.1F, L=3mH, ω=314, R=0.08Ω, k_p=0.2. The open loop Bode diagram and closed loop Bode diagram are shown in Figure 6 and Figure 7. The current loop is stable. The cut-off frequency is approximately 10Hz.

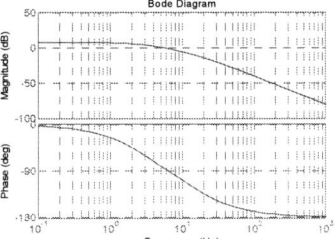

Figure 6.the Bode diagram of the open loop

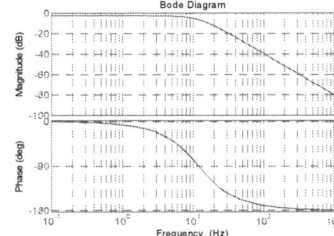

Figure 7. the Bode diagram of the closed loop

The influences of system parameters on the system stability are presented in Figure 8 and Figure 9. From Figure 8, parameter R has a big influence on the magnitude of low frequency, and hasn't an influence on the magnitude of medium frequency and high frequency. When the resistor R increases, the magnitude of low frequency decreases. From Figure 9, parameter L affects the magnitude of medium frequency and high frequency, and doesn't affect the magnitude of low frequency. When the inductor L increases, the cut-off frequency decreases and the effect on noise attenuation becomes better. The phase margin of the system increases. The relationship between the phase margin γ and the inductor L is illustrated in equation (15). Where, ω_0 is crossing frequency, and it can be obtained according to equation (16).

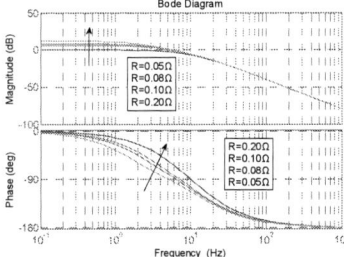

Figure 8. the influence of R on the open loop

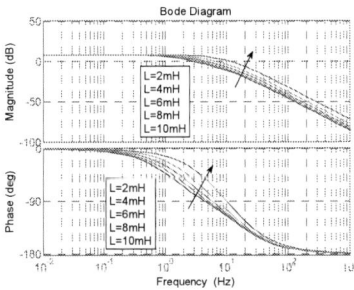

Figure 9. the influence of L on the open loop

$$\gamma = 180° - \arctan \frac{\omega_0 T_\mathrm{f}}{2} - \arctan \frac{\omega_0 L}{R} \quad (15)$$

$$\left(1 + \frac{T_\mathrm{f}}{2}\omega_0\right)\left(L\omega_0 + R\right) = k_{\mathrm{P1}} \quad (16)$$

The phase margin γ considered as one of input filter' parameters can be used to design the interface inductor. It is good to the stability and reliability of the whole STATCOM system. The interface inductor can be designed by using equation (17). The more the inductor increases, the bigger the phase margin is. An example is shown in Figure 10, where $k_{\mathrm{P1}}=0.2$, $L=5\mathrm{mH}$, $R=0.08\Omega$, $C=80000\mu\mathrm{F}$.

$$L = \frac{R}{\omega_0}\tan\left(180° - \arctan\frac{\omega_0 T_\mathrm{f}}{2} - \gamma\right) \quad (17)$$

When the regulator is PI regulator, the same conclusions are obtained, because the integration part increases the magnitude of the low frequency and affects the low frequency.

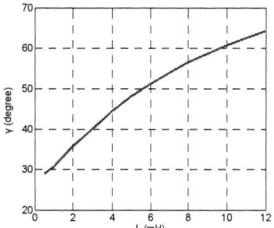

Figure 10. the relationship between phase margin γ and interface inductor L

B. DC Side Voltage Control Loop

The DC side voltage control loop is shown in Figure 3. From equation (12), the transfer functions can be presented as equation (18). The relationships between the variables in synchronously rotating reference frame and the variables in stationary reference frame (*abc* frame) are shown in equation (19) and equation (20), where E_m is the peak value of source voltage in *abc* frame, I_m is the peak value of active current in *abc* frame. Then, the small signal model is expressed as equation (21).

$$\begin{cases} \dfrac{\Delta u_\mathrm{dc}}{\Delta i_\mathrm{d}} = \dfrac{-\dfrac{3}{2Cu_\mathrm{dc0}}\left(Li_\mathrm{d0}s + u_\mathrm{sd0} + 2Ri_\mathrm{d0}\right)}{s} \\[4mm] \dfrac{\Delta u_\mathrm{dc}}{\Delta i_\mathrm{q}} = \dfrac{-\dfrac{3}{2Cu_\mathrm{dc0}}\left(Li_\mathrm{q0}s + u_\mathrm{sq0} + 2Ri_\mathrm{q0}\right)}{s} \end{cases} \quad (18)$$

$$u_\mathrm{d0} = \sqrt{3}\sqrt{\frac{2}{3}}U_\mathrm{a} = E_\mathrm{m} \quad (19)$$

$$i_\mathrm{d0} = \sqrt{3}\sqrt{\frac{2}{3}}I_\mathrm{ca} = I_\mathrm{m} \quad (20)$$

$$\frac{\Delta u_\mathrm{dc}}{\Delta i_\mathrm{d}^*} = -\frac{\dfrac{3k_\mathrm{P1}}{2Cu_\mathrm{dc0}}\left(E_\mathrm{m} + 2RI_\mathrm{m} + LI_\mathrm{m}s\right)}{s\left(\dfrac{LT_\mathrm{f}}{2}s^2 + \left(\dfrac{RT_\mathrm{f}}{2} + L\right)s + R + k_\mathrm{P1}\right)} \quad (21)$$

Because the cut-off frequency of DC side voltage control loop is much larger than that of the current control loop, the current control loop can be considered as one. The open loop transfer function of DC side voltage control loop is shown in equation (22).

$$G_\mathrm{V}(s) = \frac{\dfrac{3k_\mathrm{P1}k_\mathrm{P2}}{2Cu_\mathrm{dc0}}\left(E_\mathrm{m} + 2RI_\mathrm{m} + LI_\mathrm{m}s\right)}{s} \quad (22)$$

The open loop Bode diagram and the closed loop Bode diagram are shown in Figure 11 and Figure 12. The cut-off frequency is approximately 0.6Hz, where $k_\mathrm{P1}=0.2$, $k_\mathrm{P2}=1$, $L=5\mathrm{mH}$, $R=0.08\Omega$, $C=0.1\mathrm{F}$, $E_\mathrm{m}=8164\mathrm{V}$, $I_\mathrm{m}=-164\mathrm{A}$, $u_\mathrm{dc0}=2.5\mathrm{kV}$. From Figure 13, there is a Right-half Plane Zero. But, the Right-half Plane Zero is very large and it has a little influence on the system stability. The influence of system parameter C on the system stability is presented in Figure 14. From Figure 14, parameter C has a big influence on the magnitude characteristic, and doesn't affect the phase characteristic. When the capacitor C increases, the cut-off frequency decreases, and the effect on noise attenuation becomes better. The phase margin of the system increases.

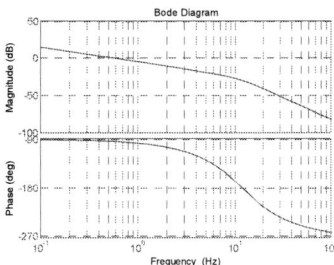

Figure 11. the Bode diagram (open loop)

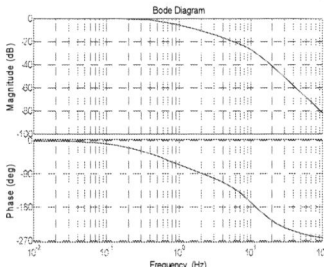

Figure 12. the Bode diagram (closed loop)

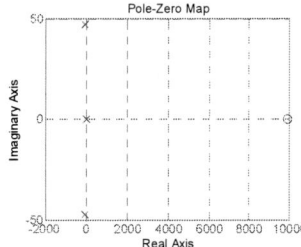

Figure 13. the pole-zero map of open loop

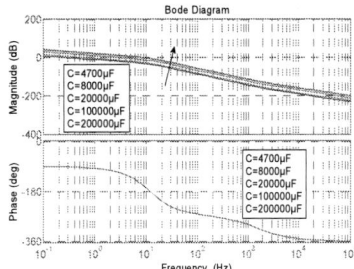

Figure 14. open loop Bode diagram(change C)

The relationship between the phase margin γ and the DC side capacitor C is illustrated in equation (23). Where, ω_0 is crossing frequency, and it can be obtained according to equation (24).

$$\gamma = 90^\circ + \arctan \frac{\omega_0 L I_m}{E_m + 2R I_m} \quad (23)$$

$$\omega_0 = \frac{3k_{P2}}{2Cu_{dc0}} E_m \Bigg/ \sqrt{1 - \left(\frac{3k_{P2}}{2Cu_{dc0}} L I_m\right)^2} \quad (24)$$

The phase margin γ considered as one of variables can be used to design the capacitor. It is good to the stability and reliability of the whole STATCOM system. The capacitor can be designed by using equation (25). The more the capacitor increases, the bigger the phase margin is. An example is shown in Figure 15, where k_{P1}=0.2, k_{P2}=0.8, L=5mH, R=0.08Ω, E_m=8164V, I_m=-164A.

$$C \approx \frac{3k_{P2}\sqrt{1 + \tan^2\left(\gamma - 90^\circ\right)}}{2u_{dc0}\tan\left(\gamma - 90^\circ\right)} \quad (25)$$

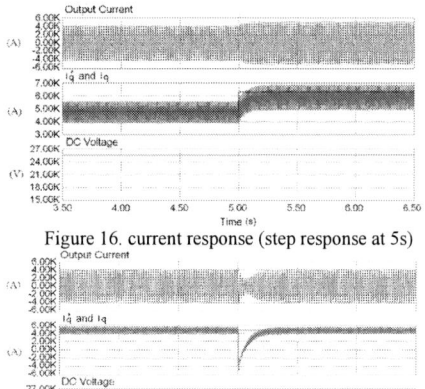

Figure 15. the relationship between phase margin γ and capacitor C

When the regulator is PI regulator, the same conclusions are obtained, because the integration part increases the magnitude of the low frequency and affects the low frequency.

V. SIMULATION AND EXPERIMENTAL RESULTS

In order to verify the analysis and specification, the simulation and hardware experimental investigation were carried out. The current response and DC side voltage response were shown in Figure 16 and Figure17. From the simulation results, the models of analysis were verified.

Figure 16. current response (step response at 5s)

Figure 17. DC side voltage response (step response at 5s)

The low voltage hardware experiment was presented in Figure 18 to Figure 26. About the current control loop, from Figure 18 and Figure 19, the output current can track the reference current. In the experiment, the Agilent 4395A Network/Spectrum/Impedance Analyzer was used to analyze the current control loop in Figure 20. Because the lowest

978-1-4244-4782-4/10 $26.00 © 2010 IEEE   1637

frequency of the equipment was 10Hz, we can just know that the cut-off frequency was approximately 10Hz. There was a time delay of 37 degrees, because of the calculation time. The experimental results verified the analysis on the current control loop.

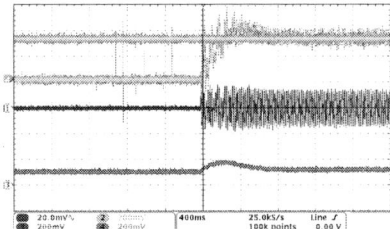

Figure 18. current response(dynamic process)
(Ch1:output current in *abc* frame; Ch2:reference current in synchronously rotating reference frame; Ch3 : DC side voltage; Ch4:output current in synchronously rotating reference frame)

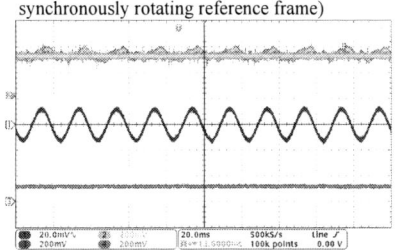

Figure 19. current response(steady-state)
(Ch1:output current in *abc* frame; Ch2:reference current in synchronously rotating reference frame; Ch3 : DC side voltage; Ch4:output current in synchronously rotating reference frame)

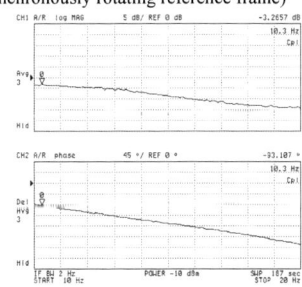

Figure 20. Bode diagram of current control loop

Figure 21 and Figure 22 verified the equation (15). When the interface inductor was 2.2mH, the system was unstable and the output current oscillated. When the interface inductor was increased to 5.5mH, the system was stable. It verified the conclusion in equation (15), that was, the more the inductor increased, the bigger the phase margin was.

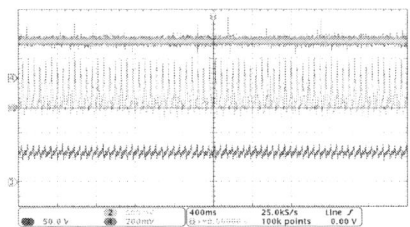

Figure 21. unstable current control loop(*L*=2.2mH)
(Ch2:reference current in synchronously rotating reference frame; Ch3 : DC side voltage; Ch4:output current in synchronously rotating reference frame)

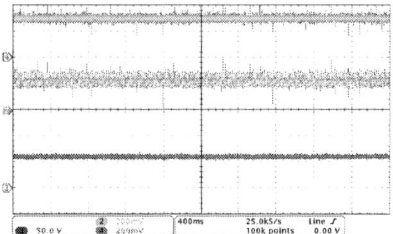

Figure 22. stable current control loop(*L*=5.5mH)
(Ch2:reference current in synchronously rotating reference frame; Ch3 : DC side voltage; Ch4:output current in synchronously rotating reference frame)

About the DC side voltage control loop, from Figure 23 and Figure 24, the DC side voltage can track the reference voltage. The experimental results verified the analysis on DC side voltage control loop.

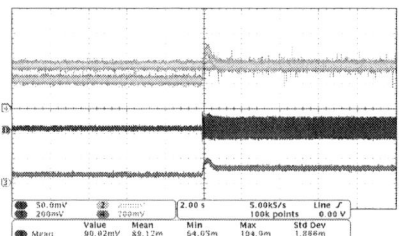

Figure 23. DC side voltage response(dynamic process)
(Ch1:output current in *abc* frame; Ch2:reference voltage in synchronously rotating reference frame; Ch3 : DC side voltage; Ch4:DC side voltage in synchronously rotating reference frame)

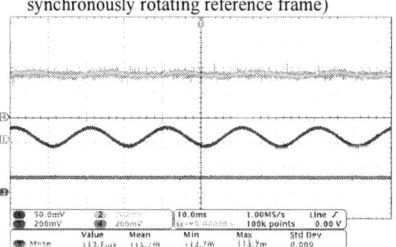

Figure 24. DC side voltage response(steady-state)
(Ch1:output current in *abc* frame; Ch2:reference voltage in synchronously rotating reference frame; Ch3 : DC side voltage; Ch4:DC side voltage in synchronously rotating reference frame)

Figure 25 and Figure 26 verified the equation (23). When the DC side capacitor was 3333μF, the system was unstable and the voltage oscillated. When the DC side capacitor was increased to 6666μF, the system was stable. It verified the conclusion in equation (23), that was, the more the DC side capacitor increased, the bigger the phase margin was.

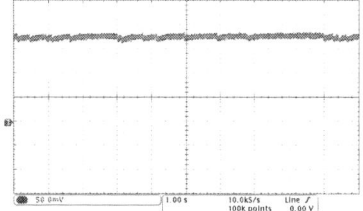

Figure 25. unstable voltage control loop(*C*=3333μF)

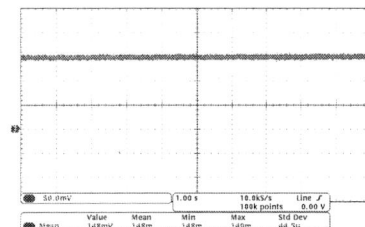

Figure 26. stable voltage control loop(C=6666μF)

VI. CONCLUSIONS

In this paper, the model of Phase and Amplitude Control is presented. A detailed analysis on the influences of interface inductor and DC side capacitor on the STATCOM systems and corresponding design consideration are presented. A method based on the phase margin to design the interface inductor and the DC side capacitor is presented in this paper.

REFERENCES

[1] N.G.Hingorani, L. Gyugyi, "Understanding FACTS-concepts and technology of flexible AC transmission systems," IEEE Press, New York, 1999.

[2] C. Schauder, M. Gernhardt, E. Stacey, T. Lemak, L. Gyugyi, T.W.Cease, A. Edris, "Development of a ±100 MVAr static condenser for voltage control of transmission systems," Power Delivery, IEEE Transactions on, Volume 10, Issue 3, July 1995 Page(s):1486 - 1496.

[3] Liu Wenhua, Liang Xu, Lin Feng, Luo Chenglian, Gao Hang, "Development of 20 MVA static synchronous compensator," Power Engineering Society Winter Meeting, 2000. IEEE Volume 4, 23-27 Jan. 2000 Page(s):2648 - 2653 vol.4.

[4] T. Nakajima, "Operating experiences of STATCOMs and a three-terminal HVDC system using voltage sourced converters in Japan," Transmission and Distribution Conference and Exhibition 2002: Asia Pacific. IEEE/PES Volume 2, 6-10 Oct. 2002 Page(s):1387 - 1392 vol.2.

[5] K. K. Sen, "STATCOM-STATic synchronous COMpensator: theory, modeling, and applications," Power Engineering Society 1999 Winter Meeting, IEEE. Volume 2, 31 Jan-4 Feb 1999 Page(s):1177 - 1183 vol.2.

[6] T.C.Y. Wang, Zhihong Ye, Gautam Sinha, Xiaoming Yuan, "Output filter design for a grid-interconnected three-phase inverter," Power Electronics Specialist Conference, 2003. PESC '03. 2003 IEEE 34th Annual. Volume 2, 15-19 June 2003 Page(s):779 - 784 vol.2.

[7] Yongqiang Lang, Dianguo Xu, S.R.Hadianamrei,Hongfei Ma, "A Novel Design Method of LCL Type Utility Interface for Three-Phase Voltage Source Rectifier," Power Electronics Specialists Conference, 2005. PESC '05. IEEE 36th 2005 Page(s):313 – 317.

[8] Liang Yiqiao, C. O. Nwankpa, "A new type of STATCOM based on cascading voltage-source inverters with phase-shifted unipolar SPWM," Industry Applications, IEEE Transactions on, 1999, 35 (5): 1118 1123.

[9] J. G. Hwang, P.W. Lehn, M. Winkelnkemper, "Control of grid connected AC-DC converters with minimized DC link capacitance under unbalanced grid voltage condition," Power Electronics and Applications, 2007 European Conference on. 2007: 1-10.

[10] J. Hobraiche, J.P. Vilain, C. Plasse, "Offline optimized pulse pattern with a view to reducing DC-link capacitor application to a starter generator," Power Electronics Specialists Conference, 2004. PESC 04. 2004 IEEE 35th Annual. 2004: 3336-3341 Vol.3335.

[11] B.G. Gu, K. Nam, "A Theoretical minimum DC-link capacitance in PWM converter-inverter systems," Electric Power Applications, IEE Proceedings -, 2005, 152 (1): 81-88.

[12] Gu Bon-Gwan, Nam Kwanghee, "A DC-link capacitor minimization method through direct capacitor current control," Industry Applications, IEEE Transactions on, 2006, 42 (2): 573-581.

[13] A. Kotsopoulos, J.L. Duarte, M.A.M. Hendrix, "A predictive control scheme for DC voltage and AC current in grid-connected photovoltaic inverters with minimum DC link capacitance," Industrial Electronics Society, 2001. IECON '01. The 27th Annual Conference of the IEEE. 2001: 1994-1999 vol.1993.

[14] Ching-Tsai Pan, Jenn-Jong Shieh. "New space-vector control strategies for three-phase step-up/down AC/DC converter," Industrial Electronics, IEEE Transactions on, 2000, 47 (1): 25-35.

[15] M. Liserre, F.Blaabjerg, S.Hansen, "Design and control of an LCL-filter-based three-phase active rectifier," Industry Applications, IEEE Transactions on, Volume 41, Issue 5, Sept.-Oct. 2005 Page(s):1281 – 1291.

[16] C. Schauder, "Vector analysis and control of advanced static VAr compensators," AC and DC Power Transmission, 1991., International Conference on 17-20 Sep 1991 Page(s):266 – 272.

[17] B.M.Han,G.G.Karady,J.K.Park,S.I.Moon, "Interaction analysis model for transmission static compensator with EMTP," Power Delivery, IEEE Transactions on Volume 13, Issue 4, Oct. 1998 Page(s):1297 – 1302.

[18] K.R.Padiyar, A.M.Kulkarni, "Analysis and design of voltage control of static condenser," Power Electronics, Drives and Energy Systems for Industrial Growth, 1996., Proceedings of the 1996 International Conference on Volume 1, 8-11 Jan. 1996 Page(s):393 - 398 vol.1.

[19] J.B.Choo, J.S.Yoon,B.H.Chang,B.Han,K.K.Koh, "Development of FACTS operation technology to the KEPCO power network-detailed EMTDC model of 80 MVA UPFC,"Transmission and Distribution Conference and Exhibition 2002: Asia Pacific. IEEE/PES Volume 1, 6-10 Oct. 2002 Page(s):354 - 358 vol.1.

[20] Z.Yang, C. Shen,L. Zhang,M.L.Crow, S.Atcitty, "Integration of a StatCom and battery energy storage," Power Systems, IEEE Transactions on Volume 16, Issue 2, May 2001 Page(s):254 – 260.

[21] L.Dong, M.L.Crow, Z.Yang, C.Shen, L.Zhang, S.Atcitty, "A reconfigurable FACTS system for university laboratories," Power Systems, IEEE Transactions on Volume 19, Issue 1, Feb. 2004 Page(s):120 – 128.

[22] J. W. Kolar, F. C. Zach, F. Casanellas, "Losses in PWM inverters using IGBTs," Electric Power Applications, IEE Proceedings-Volume 142, Issue 4, July 1995 Page(s):285 – 288.

[23] Boon Teck Ooi,G.Joos,Xiaogang Huang,"Operating principles of shunt STATCOM based on 3-level diode-clamped converters," Power Delivery, IEEE Transactions on Volume 14, Issue 4, Oct. 1999 Page(s):1504 – 1510.

[24] Dong Shen, Xu Liang,Yingduo Han, "A modified per-unit STATCOM model and analysis of open loop response time," Power Engineering Society Winter Meeting, 2000. IEEE Volume 4, 23-27 Jan. 2000 Page(s):2624 - 2629 vol.4.

[25] Y.Jiang,A.Ekstrom, "Applying PWM to control overcurrents at unbalanced faults of forced-commutated VSCs used as static VAr compensators," Power Delivery, IEEE Transactions on Volume 12, Issue 1, Jan. 1997 Page(s):273 – 278.

[26] V. Blasko, V. Kaura, "A new mathematical model and control of a three-phase AC-DC voltage source converter," Power Electronics, IEEE Transactions on, 1997, 12 (1): 116-123.

[27] V. Blasko, V. Kaura, "A novel control to actively damp resonance in input LC filter of a three-phase voltage source converter," Industry Applications, IEEE Transactions on, 1997, 33 (2): 542-550.

Vector Oriented Control of Voltage Source PWM Inverter as a Dynamic VAR Compensator for Wind Energy Conversion System Connected to Utility Grid

Mahmoud M. N. Amin, *Student Member, IEEE*, O. A. Mohammed, *Fellow, IEEE*

Florida International University, ECE Dept.

Energy Systems Research Laboratory

Miami, FL 33174 USA

Abstract— This paper presents analysis, design, and simulation for vector oriented control of three-phase voltage source pulse-width modulation (PWM) inverter which aims to optimize the utilization of wind power injected into the grid. To realize this goal, a digitally controlled converter-inverter system is proposed which provides economic utilization of the wind generator by insuring unity power factor operation under different possible conditions, and full control of active and reactive power injected into the grid using a digitally controlled voltage source inverter. The wind energy conversion system (WECS) is represented by three -phase self-excited induction generator (SEIG) driven by a variable-speed prime mover (VSPM) such as a wind turbine for the clean alternative renewable energy in rural areas. Mathematical models for both the converter and inverter are presented and simulated. A vector oriented control scheme is presented in order to control the energy to be injected into the grid. Closed-loop control of the converter/inverter utilizes a conventional proportional integral (PI) controller. In order to examine the dynamic performance of the system, its model is simulated and results are analyzed. Simulation results for different disturbance conditions show good performance of this proposed control algorithm. The simulation results are given using MATLAB7/SIMULINK program. Experimental results using DSpace-1104 confirm that the good performance of the proposed control system and agree with the simulation results to a great extend.

Index Terms— Self-excited induction generator (SEIG), vector oriented control (VOC), voltage source inverter (VSI), wind energy conversion system (WECS).

I. INTRODUCTION

RECENTLY, the wind generation system is attracting attention as a clean and safety renewable power source. The three-phase induction machine with a squirrel cage rotor or a wound rotor could work as a three-phase induction generator either connected to the utility ac power distribution line or operated in the self-excitation power generation mode with an additional stator terminal excitation capacitor bank. [1-6].

Variable speed operation of wind turbines has many advantages that are well documented in the literature [7-9].
Induction machines have many advantageous characteristics such as high robustness, reliability and low cost. The induction machines may be used as a motor or a generator. Self-excited

induction generators (SEIG) are good candidates for wind-power electricity generation especially in remote areas, because they do not need an external power supplies to produce the excitation magnetic fields. [10-12].

Three-phase voltage-source AC/DC/AC (PWM\) converters have been increasingly used for many applications such as uninterruptible power supply (UPS) systems, boost converters and wind energy conversion systems. The attractive features of them are constant dc-bus voltage, low harmonic distortion of the utility currents, bidirectional power flow, and controllable power factor [13]-[15].

Conventional thyristor phase-controlled converters with PWM technique have the inherent drawbacks that the power factor decreases as the firing angle increases and that harmonics of the line current are relatively high [16]. Also, the fast power semiconductors used [usually, MOSFET or insulated gate bipolar transistor (IGBT)], are free to switch at frequencies much higher than the mains frequency, enabling the voltage controller to provide an output voltage with fast dynamic response [17].

Nowadays, the voltage-oriented control (VOC), which guarantees high dynamics and static performance via internal current control loops, has become very popular and has constantly been developed and improved [18- 25]. However, this method depends on using conventional proportional and integral (PI) compensators in the rotating reference frame to produce its control input commands [26]. However, the conventional PI- controllers have the inherent drawbacks that its response is somewhat slower for very fast transients and its control range is limited because its fixed gains [27].

A conventional proportional and integral (PI) compensator can be applied to the variables in the rotating reference frame so as to achieve a zero steady-state error in response to step commands. Then, variables in the rotating reference frame must be restored in the stationary three-phase reference frame using the inverse– transformation.

In this paper, a vector oriented control strategy for a three-phase voltage-source PWM inverter is proposed in the synchronous d–q frame. The mathematical models of the converter in different frames including the synchronous frame are presented. The vector oriented current control technique which is similar to the field oriented control in induction machine is used to control the inverter operation. A Software Phase Locked Loop (SPLL) for phase angle detection of the generator voltage in synchronous reference frame is proposed.

978-1-4244-4782-4/10 $26.00 © 2010 IEEE

The proportional plus integral (PI) current controllers in d-q axes are designed and analyzed to meet the time domain specification: minimum overshot, minimum settling time and minimum steady-state error.

A vector current controlled grid connected voltage source inverter (VSI) is proposed as a dynamic VAR compensator system. An analytical model for the VSI connected to the grid with L and LC filter and the operating principle of the proposed vector controllers are introduced. Simulation results for VSI connected to the grid with L and LC filter are presented.

This paper is organized as follows. In section I, an introduction to the subject is presented. In Sections II and III a mathematical description and modeling of the system is introduced. In Section IV, simulation and experimental results are presented and analyzed. Finally, Section V presents some conclusions.

II. SYSTEM DESCRIPTION

Variable speed wind turbine (VSWT) systems are preferred for the higher output power generation. On the other hand, due to the variations of wind speed, direct interfacing of wind energy system to the utility grid gives rise to problems such as voltage fluctuations and flickering. Therefore, synchronization with the grid and stability concerns must be fulfilled. Power electronics systems playing an important position in the overall generation system. The classical scheme of AC/DC/AC conversions is used.

In general, the electrical components of VSWT system are generator, rectifier and inverter, and isolation transformer for grid connection. Fig. 1 shows a typical electrical system utilized for grid connected VSWT with the utility grid. The paper will focus on the power electronic conversion scheme that could be used to inject VSWT power into utility grid.

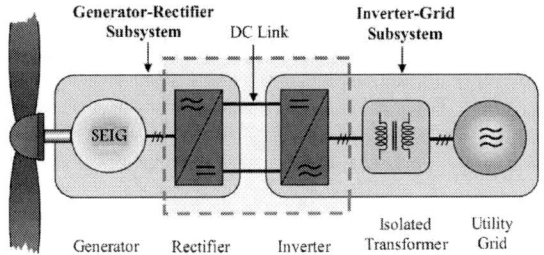

Figure 1. Electrical system for grid connected wind energy with the grid

III. SYSTEM MODELING

A. Modeling of Self-Excited Induction Generator

The model of an induction generator is helpful to analyze all its characteristics. The induction machine used as SEIG is a three-phase squirrel-cage machine.

In this paper the model, shown in Fig. 2, is used because it provides a complete solution (transient and steady state) of the self-excitation process.

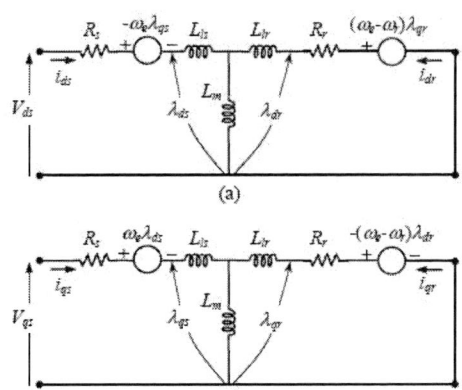

Figure 2. D-Q model of an induction machine in the synchronously rotating reference frame (a) d-axis (b) q-axis.

In the d-q model of the induction machine shown in Fig.2 v_{ds} and v_{qs} are the generated voltages along the d-axis and q-axis, respectively. The voltage of an induction machine in the synchronous reference frame, which can be obtained from Fig. 2 as follows:

$$v_{ds} + R_s i_{ds} - \omega_e \lambda_{qs} + L_{ls} p i_{ds} + L_m p i_{ds} + L_m p i_{dr} = 0 \quad (1)$$

$$K_d - (\omega_e - \omega_r)\lambda_{qr} + R_r i_{dr} + L_{lr} p i_{dr} + L_m p i_{ds} + L_m p i_{dr} = 0 \quad (2)$$

$$v_{qs} + R_s i_{qs} + \omega_e \lambda_{ds} + L_{ls} p i_{qs} + L_m p i_{qs} + L_m p i_{qr} = 0 \quad (3)$$

$$K_q + (\omega_e - \omega_r)\lambda_{dr} + R_r i_{qr} + L_{lr} p i_{qr} + L_m p i_{qs} + L_m p i_{qr} = 0 \quad (4)$$

Where

$$\begin{aligned} \lambda_{ds} &= L_s i_{ds} + L_m i_{dr} \\ \lambda_{dr} &= L_m i_{ds} + L_r i_{dr} \\ \lambda_{qs} &= L_s i_{qs} + L_m i_{qr} \\ \lambda_{qr} &= L_m i_{qs} + L_r i_{qr} \end{aligned} \quad (5)$$

The voltage equation of the d-q model is based on the stator and rotor currents are given as:

$$\begin{bmatrix} 0 \\ 0 \\ 0 \\ 0 \end{bmatrix} = \begin{bmatrix} (R_s + pL_s) & \omega_e L_s & pL_m & \omega_e L_m \\ -\omega_e L_s & (R_s + pL_s) & -\omega_e L_m & pL_m \\ pL_m & (\omega_e - \omega_r)L_m & (R_r + pL_r) & (\omega_e - \omega_r)L_r \\ -(\omega_e - \omega_r)L_m & pL_m & -(\omega_e - \omega_r)L_r & (R_r + pL_r) \end{bmatrix}$$

$$\begin{bmatrix} i_{qs} \\ i_{ds} \\ i_{qr} \\ i_{dr} \end{bmatrix} + \begin{bmatrix} V_{qs} \\ V_{ds} \\ K_q \\ K_d \end{bmatrix} \quad (6)$$

$$V_{qs} = \frac{1}{C} \int i_{qs} dt + V_{cq}\Big|_{t=0} \quad (7)$$

$$V_{ds} = \frac{1}{C} \int i_{ds} dt + V_{cd}\Big|_{t=0} \quad (8)$$

The electromagnetic torque is given by:

$$T_e = -\frac{3}{2}\frac{P}{2}\frac{L_m}{L_r}\left(\lambda_{dr} i_{qs} - \lambda_{qr} i_{ds}\right) \quad (9)$$

Where, v_{qs}, v_{ds}, i_{qs}, and i_{ds} are the stator voltages and currents, respectively. v_{qr}, and v_{dr} are the rotor voltages. λ_{qr}, and λ_{dr} are the rotor fluxes. R_s, L_s, R_r, and L_r are the resistance and the self inductance of the stator and the rotor, respectively. L_m is the mutual inductance. The mathematical equation that relates the wind turbine output torque with the electromagnetic torque of the induction generator is given by:

$$T_m = J \frac{d}{dt}\omega_m + \beta\omega_m + T_e \tag{10}$$

where ω_m, J, and β are the mechanical angular speed of wind turbine, the effective inertia of the wind turbine and the induction generator, and friction coefficient, respectively. From (6)-(10), the state equations of the SEIG and turbine can be accomplished as in (11) and (12).

$$\frac{d}{dt}\begin{bmatrix} i_{qs} \\ i_{ds} \\ i_{qr} \\ i_{dr} \end{bmatrix} = \frac{1}{L}\begin{bmatrix} -R_rL_r & (\omega_e-\omega_r)L_m^2-\omega L_sL_r & R_rL_m & -\omega L_mL_r \\ \omega L_sL_r-(\omega_e-\omega_r)L_m^2 & -R_sL_r & \omega L_mL_r & R_rL_m \\ R_sL_m & \omega L_sL_m & -R_sL_s & \omega L_m^2-(\omega_e-\omega_r)L_sL_r \\ -\omega L_sL_m & R_sL_m & (\omega_e-\omega_r)L_sL_r-\omega L_m^2 & -R_sL_s \end{bmatrix}$$

$$\begin{bmatrix} i_{qs} \\ i_{ds} \\ i_{qr} \\ i_{dr} \end{bmatrix} + \frac{1}{L}\begin{bmatrix} L_mK_q-L_rV_{qs} \\ L_mK_d-L_rV_{ds} \\ L_mV_{qs}-L_sK_q \\ L_mV_{ds}-L_sK_d \end{bmatrix} \tag{11}$$

$$\frac{d}{dt}\omega_m = \frac{1}{J}T_m - \frac{\beta}{J}\omega_m - \frac{1}{J}T_e \tag{12}$$

B. Modeling of Vector Oriented Controlled PWM Converter

A power circuit and of a per-phase equivalent circuit of a PWM voltage-source converters are shown in Fig. 3. It is assumed that a resistive load R_L is connected to the output terminal.

Where, (R, L) is the line inductor between generator and the converter terminal, (e_a) is the generator phase voltage, (v_a) is the bridge converter voltage controllable according to the demanded DC voltage level, (i_a) is the line current, (i_{dc}) is the converter DC output current, (v_{dc}) is the converter DC output controlled voltage, (i_c) is the DC-link capacitor current, (R_L) is the load resistance, and (i_L) is the load current.

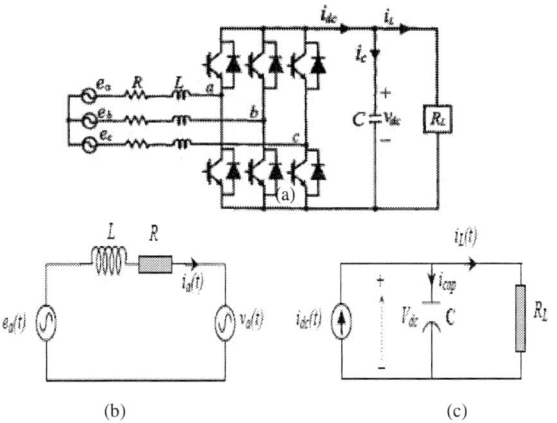

Figure 3. AC/DC PWM converter. (a) Power circuit. (b) Per-phase input equivalent circuit. (c) Per-phase output equivalent circuit.

From the equivalent circuit shown in Fig.4.b, assuming generator voltage is 3-phase balanced voltage source and its state equations are

$$\begin{bmatrix} e_a(t) \\ e_b(t) \\ e_c(t) \end{bmatrix} = \begin{bmatrix} E\sin\omega t \\ E\sin(\omega t - 2\pi/3) \\ E\sin(\omega t + 2\pi/3) \end{bmatrix} \tag{13}$$

where E is the maximum amplitude of the generator AC voltage.

From Fig.3.b. , the dynamic equation for the input side of the three phase converter system can be written as,

$$e_{abc}(t) - v_{abc}(t) = L\frac{di_{abc}(t)}{dt} + Ri_{abc}(t) \tag{14}$$

Referring to Fig.3.c. , the dynamic equation for the output side of the three phase converter system can be written as,

$$i_c(t) = i_{dc}(t) - i_L(t) = C\frac{dv_{dc}(t)}{dt} \tag{15}$$

The line voltage, the phase current, and the terminal voltage of the PWM converter can be transformed to a synchronous reference frame using the transformation matrix $H(\theta)$ as follows,

$$v_{abc}(t) = H(\theta).v_{dq}(t) \tag{16}$$

$$v_{dq}(t) = H^{-1}(\theta).v_{abc}(t)$$

$$H^{-1}(\theta) = \begin{bmatrix} \cos\theta & \frac{1}{\sqrt{3}}\sin\theta & \frac{-1}{\sqrt{3}}\sin\theta \\ \sin\theta & \frac{-1}{\sqrt{3}}\cos\theta & \frac{1}{\sqrt{3}}\cos\theta \end{bmatrix} \tag{17}$$

And θ is the rotating angle of transformation.

Now, the dynamic equation (14) can be transformed directly from the abc frame to the synchronous reference frame resulting in a mathematical model for the PWM converter in a synchronous rotating coordinates,

$$e_{dq}(t) - v_{dq}(t) = R.i_{dq}(t) + L\frac{di_{dq}(t)}{dt} - L.\omega.M.i_{dq}(t) \tag{18}$$

where

$$M = \begin{bmatrix} 0 & 1 & 0 \\ -1 & 0 & 0 \\ 0 & 0 & 0 \end{bmatrix}$$

For more simplification, we assume that the three-phase voltage source is balanced without zero sequence component. Therefore, (18) can be written as,

$$v_d(t) = e_d(t) - L\frac{di_d(t)}{dt} - R.i_d(t) + L.\omega.i_q(t)$$

$$v_q(t) = e_q(t) - L\frac{di_q(t)}{dt} - R.i_q(t) - L.\omega.i_d(t) \tag{19}$$

978-1-4244-4782-4/10 $26.00 © 2010 IEEE   1642

The converter system in S-domain can be obtained by applying Laplace transformation to the dynamic equations in the synchronous frame directly such as,

$$i_d(s) = \frac{1}{Ls+R} \cdot \left[e_d(s) - v_d(s) + \omega.L.i_q(s) \right]$$

$$i_q(s) = \frac{1}{Ls+R} \cdot \left[e_q(s) - v_q(s) - \omega.L.i_d(s) \right] \tag{20}$$

For the converter output side,

$$v_{dc} = \frac{R_L}{1+CR_L s} \cdot i_{dc} \tag{21}$$

The converter plant model in S domain representation is shown in Fig. 4.

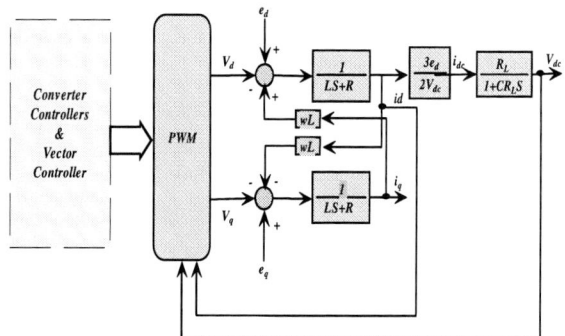

Figure.4. Converter model block diagram

According to the converter dynamic equations in synchronous frame,(19), there are coupling terms between these equations which degrade the dynamic performance (slow the controller transient and cause high overshoots) of the system. These terms are the coupling q current componenet (ωLi_q) and generator voltage on the d-axis equation, e_d, while coupling d current component (ωLi_d) and e_q on the q-axis. The vector controller will decouple these terms, giving the ability to control each current component separately without any effect from the other component. The block diagram of the vector controlled PWM converter is shown in Fig. 5.

IV. SYSTEM MODELING

A. Vector Oriented Controlled PWM Converter

The main circuit of the VSI connected to a three phase public grid is shown in Fig. 5. An inductance L works as line filter is mounted between the utility grid and the VSI having an internal resistance R. The phase potentials of the VSI denoted as $v_a(t)$, $v_b(t)$, and $v_c(t)$. The phase potentials of the utility grid denoted $u_a{}^g(t)$, $u_b{}^g(t)$, and $u_c{}^g(t)$. The currents flowing from the DC-link to the VSI denoted as $i_a(t)$, $i_b(t)$, and $i_c(t)$, while the DC-link current and voltage are denoted as $i_{dc}(t)$, and $V_{dc}(t)$ respectively.

In order to design a VSI control systems, mathematical models are important tools for predicting dynamic performance and stability limits of different control lows and system parameters. The system to be modeled is shown in Fig. 6. In the figure, the grid inductance (L_g) is assumed to be zero.

The assumption of the balanced state of the grid is presented, therefore, it can be represented by the state equation as,

$$\begin{bmatrix} u_a^g(t) \\ u_b^g(t) \\ u_c^g(t) \end{bmatrix} = \begin{bmatrix} U \cdot \sin \omega_g t \\ U \cdot \sin(\omega_g t - 2\pi/3) \\ U \cdot \sin(\omega_g t + 2\pi/3) \end{bmatrix} \tag{22}$$

where ωg is the angular frequency of the grid voltage vector.

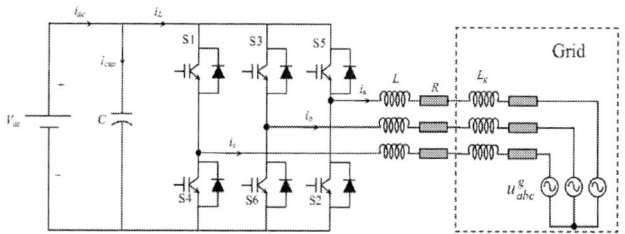

Fig. 5. The main electric circuit of the VSI connected to the utility grid

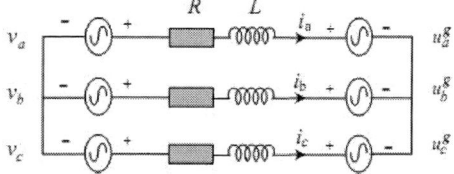

Fig. 6. Simplified circuit of a grid connected VSI, where the focus is on the L-filter.

The AC side of the inverter system is modeled by the differential equations for 3-phases such that,

$$L\frac{di_{abc}(t)}{dt} + Ri_{abc}(t) = v_{abc}(t) - u_{abc}^g(t) \tag{22}$$

By using vector notation, the last equation can be written in the αβ stationary frame such as,

$$L\frac{di_{\alpha\beta}(t)}{dt} + Ri_{\alpha\beta}(t) = v_{\alpha\beta}(t) - u_{\alpha\beta}^g(t) \tag{24}$$

Equation (24) can be written in the rotating reference frame synchronized with grid voltage as,

$$L\frac{di_{dq}(t)}{dt} + (R + j\omega_g L)i_{dq}(t) = v_{dq}(t) - u_{dq}^g(t) \tag{25}$$

The decoupled equation can be written in the state space form as,

$$\frac{dX}{dt} = A \cdot X - B \cdot Y \tag{26}$$

where the state vector and the input vector are defined by

$$X = \begin{bmatrix} i_d & i_q \end{bmatrix}^T, \tag{27}$$

$$Y = \begin{bmatrix} V_d & V_q & u_d^g & u_q^g \end{bmatrix}^T \tag{28}$$

respectively, the system matrix and the input matrix are given by

$$A = \begin{bmatrix} -R/L & \omega_g \\ -\omega_g & -R/L \end{bmatrix}, \qquad (29)$$

$$B = \begin{bmatrix} 1/L & 0 & -1/L & 0 \\ 0 & 1/L & 0 & -1/L \end{bmatrix} \qquad (30)$$

The DC side of the system is modeled by the equation,

$$C \frac{dV_{dc}(t)}{dt} = i_{dc}(t) - i_L(t) \qquad (31)$$

V. THE PROPOSED CONTROL SYSTEM

The schematic diagram of the VSI connected to the grid and its vector current controller is shown in Fig. 7.

A. Proposed VOC VSI as dynamic VAR compensator

Until recently, most wind power plant and utility have utilized capacitor banks to correct power factor to near unity. The capacitors are switched in and out by means of mechanical contactors. Unfortunately, because these contactors are relatively slow, they are unable to react to sudden momentary dips in voltage commonly seen in weak grid and can add greater stress to the utility grid. Vector oriented control VSI is proposed here as a dynamic VAR compensator system.

Dynamic VAR systems detect and instantaneously compensate for voltage disturbances by injecting leading or lagging reactive power at key points on power transmission grids. Through the VAR control system, reactive power is supplied to the grid in a fraction of second, regulating the system voltage and stabilizing weak grid. A controller measures utility line voltage compares it to the desired level, and compute the amount of reactive power needed to bring the line voltage back to the specified range.

The active and reactive power in the synchronous frame given by,

$$P(t) = \frac{3}{2}[u_d^j(t) \cdot i_d(t) + u_q^j(t) \cdot i_q(t)]$$

$$Q(t) = \frac{3}{2}[u_q^j(t) \cdot i_d(t) - u_d^j(t) \cdot i_q(t)] \qquad (32)$$

The power of the DC side is given by,

$$P_{dc} = V_{dc} \cdot i_{dc} \qquad (33)$$

The basic principle of the vector oreinted control method is to control the instantaneous active and reactive grid currents and, consequently, the active and reactive power, by separate controllers independently of each other. The grid voltages and currents are first sensed. By means of the software phase locked loop (SPLL), the grid phase angle and frequency can be detected in order to synchronize the VSI output with grid. The demanded amount of power is first estimated from the utility grid at the desired power factor, in consequence, the reference currents in a synchronous frame synchronized with grid

voltage are calculated. Consequently, the current controllers are trying to bring the actual currents to its references. So far is how the controller could be constructed.

The reference currents i_q^r, and i_d^r could be calculated from the power equation (32), such that,

$$i_d^r(t) = \frac{2}{3\prod}\left(u_d^j(t) \cdot P^r(t) + u_q^j(t) \cdot Q^r(t)\right),$$

$$i_q^r(t) = \frac{2}{3\prod}\left(u_q^j(t) \cdot P^r(t) - u_d^j(t) \cdot Q^r(t)\right) \qquad (34)$$

Where $\prod = u_d^{j^2}(t) + u_q^{j^2}(t)$, and $P^r(t)$, $Q^r(t)$ are the active and reactive power commands. With the assumption of zero reactive power command, the current command equations can be simplified such that

$$i_d^r(t) = \frac{2 \cdot P^r(t)}{3 \cdot u_d^j(t)} = \frac{2 \cdot V_{dc} \cdot i_{dc}^r(t)}{3 \cdot u_d^j(t)}, \qquad (35)$$

$$i_q^r(t) = 0$$

From the state equation (7), the current controllers can be constructed such that

$$v_d^r(t) = u_d^j(t) - \omega_g L i_q(t) + R \cdot i_d(t) + U_d^v,$$

$$v_q^r(t) = u_q^j(t) + \omega_g L i_d(t) + R \cdot i_q(t) + U_q^v \qquad (36)$$

where $v_d^r(t)$ and $v_q^r(t)$ are the d-axis and q-axis voltage commands respectively, U_d^v and U_q^v are the effective voltage commands. The coupling term between the d-axis and q-axis is $\omega_g L i_d(t), \omega_g L i_q(t)$ cancelled out by feed-forward controller. The effective voltage commands are obtained by using PI controller such that

$$U_d^v = k_p^i \cdot [i_d^r(t) - i_d(t)] + k_i^i \cdot \int [i_d^r(t) - i_d(t)]dt$$

$$U_q^v = k_p^i \cdot [i_q^r(t) - i_q(t)] + k_i^i \cdot \int [i_q^r(t) - i_q(t)]dt \qquad (37)$$

where k_p^i, and k_i^i are the proportional and integral gains of the current controllers respectively.

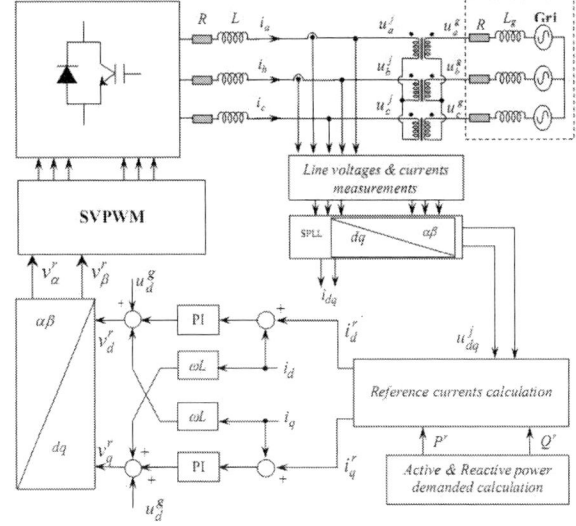

Fig. 7.   Schematic diagram of the proposed VOC VSI connected to the grid

B. Grid-connected VSI with LC filter

The per-phase equivalent circuit for the VSI connected to the grid with LC filter is shown in Fig. 8. The inverter dynamic should have additional voltage controller in front of the current controller to control the filter capacitor voltage.

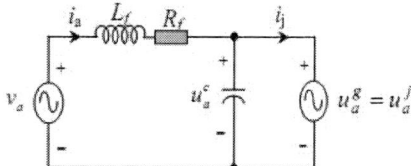

Fig. 8. VSI with LC filter equivalent circuit.

According to the equivalent circuit, the system dynamic equations can be written as,

$$L_f \frac{di_{abc}(t)}{dt} = v_{abc}(t) - R_f i_{abc}(t) - u_{abc}^c(t) \tag{38}$$

$$C_f \frac{du_{abc}^c(t)}{dt} = i_{abc}(t) - i_{abc}^j(t) \tag{39}$$

The three phase model of the system transformed to a synchronous reference frame, where the d axis is oriented with the grid voltage vector, can be described by the state equations,

$$L_f \frac{d}{dt} i_d(t) = v_d(t) - R_f i_d(t) - u_d^c(t) + \omega_g L_f i_q(t), \tag{40}$$
$$L_f \frac{d}{dt} i_q(t) = v_q(t) - R_f i_q(t) - u_q^c(t) - \omega_g L_f i_d(t),$$

$$C_f \frac{d}{dt} u_d^c(t) = i_d(t) - i_d^j(t) + \omega_g C_f u_q^c(t), \tag{41}$$
$$C_f \frac{d}{dt} u_q^c(t) = i_q(t) - i_q^j(t) - \omega_g C_f u_d^c(t),$$

The system utilizes an inner control loop to control the current through the filter inductor, and an outer control loop to control the filter capacitor voltage, which in turn will be applied to the primary terminals of the isolation transformer between the grid and the inverter.

The capacitor voltage could be controlled by controlling the filter inductor current since the injected current, i_{dq}^j, may be considered as a disturbance. Therefore, the controller could be constructed such that,

$$i_d^r(t) = i_d^j(t) - \omega_g C_f u_q^c(t) + U_d^c \tag{42}$$
$$i_q^r(t) = i_q^j(t) + \omega_g C_f u_d^c(t) + U_q^c$$

where $i_d^r(t)$ and $i_q^r(t)$ are the d-axis and q-axis current commands respectively, U_d^c and U_q^c are the effective current commands.

If *PI* controller is used, $u_{dq}^c(t)$ could be controlled such that,

$$U_d^c = k_p^c \cdot [u_d^{c\prime}(t) - u_d^c(t)] + k_i^c \cdot \int \left[u_d^{c\prime}(t) - u_d^c(t) \right] dt \tag{43}$$
$$U_q^c = k_p^c \cdot [u_q^{c\prime}(t) - u_q^c(t)] + k_i^c \cdot \int \left[u_q^{c\prime}(t) - u_q^c(t) \right] dt$$

where k_p^c, and k_i^c are the proportional and integral gains of the capacitor voltage controllers respectively.

The inner current controller has the same dynamics as the controller of L filter discussed in section A, (Eqns.34, and 35),

so it can be used here. The overall VSI controllers are shown in Fig. 9.

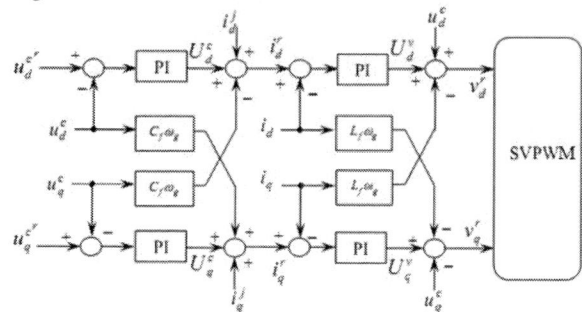

Fig. 9. VSI with LC filter controller's diagram.

VI. SIMULATION AND EXPERIMENTAL RESULTS

In order to investigate the VSI connected to the grid with the proposed VOC algorithm, a simulation program using Simulink™ was carried out using simulation parameters shown in Table I. The SVPWM presented in this paper has been used with a 5 KHz switching frequency. VOC strategy discussed in section III is utilized in the simulation. Two inner PI controllers are used as current controllers; one for the d-component and the other for the q-component. The outer PI is the voltage controller. A low pass filter with LC elements and 280 Hz cut-off frequency is connected before the isolation transformer in order to have clean power injection. Fig. 10 shows Schematic diagram of the overall proposed WECS control system connected to grid. The induction machine is listed in Table II in the appendix. Photograph for the experimental proposed WECS connected to the utility grid is shown in Fig. 11.

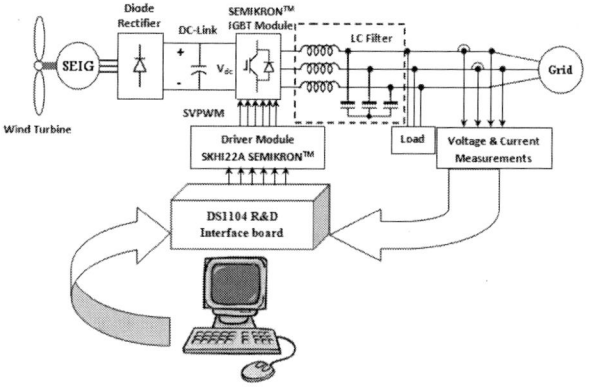

Fig. 10. Schematic diagram of the overall proposed WECS control system connected to grid.

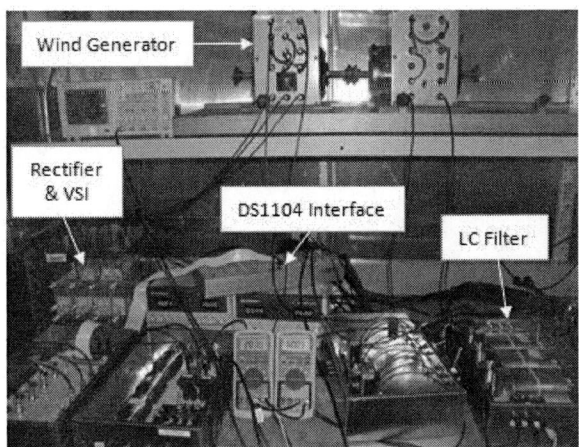

Fig. 11. Photograph for the experimental proposed WECS connected to the utility grid.

A. Wind speed variation

In order to evaluate the performance of proposed emulator in turbulent wind speed condition an experimental test has been carried out. The wind speed could be constant, or varying in a form of pulses, sinusoidal, or step change. Actually wind speed almost has a random variation according to the wind turbine location and its atmospheric conditions, but they can be set to operate within a given variation of speed. Fig. 12 shows the monthly average wind speed at La Venta station. After making fitting to these real data then we can get approximation for Wind speed versus time as shown in Fig. 13.

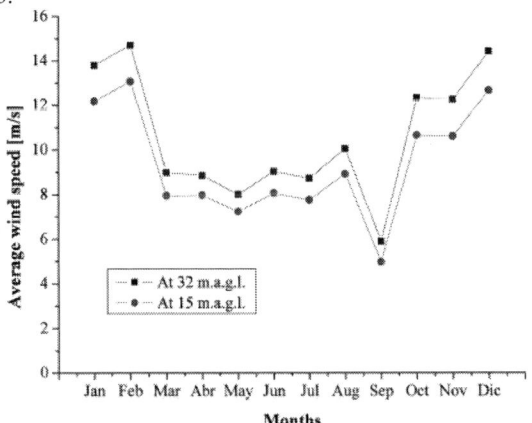

Fig. 12. Monthly average wind speed at La Venta station.

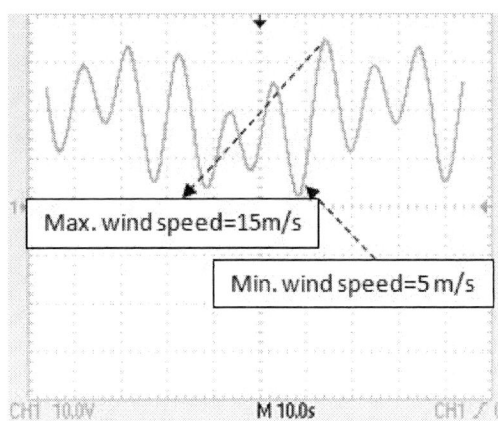

Fig. 13. Reference wind speed applied to emulator.

B. grid-connected VOC VSI with L filter

In order to estimate the power injection characteristics, two types of tests are carried out. The first is active power injection without any reactive power compensation. The second is injection of active power with reactive power compensation (lagging or leading).

The terminal phase and line voltage of the VSI in abc frame are shown in Fig. 14a,b, and Fig. 15a,b respectively. However the wind speed is variable according to the previous wind characteristics in Fig. 13, the output voltage has the rating of 208 v, 60 Hz which is synchronized with the utility grid rating.

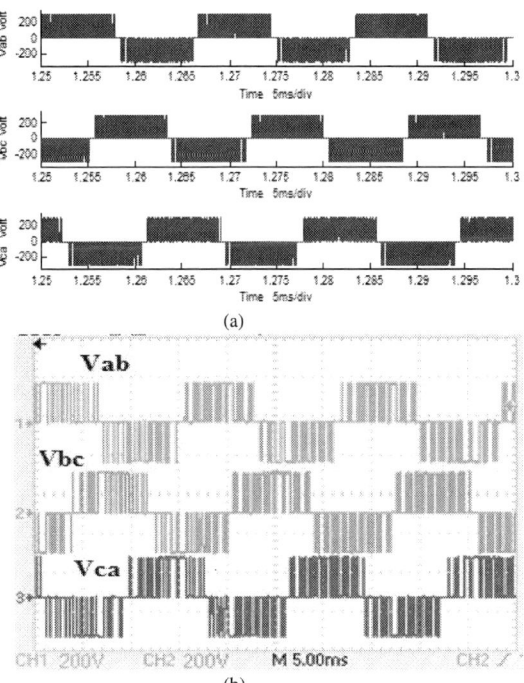

Fig. 14. Phase voltage of the VSI in *abc* frame (a) simulation. (b) experimental.

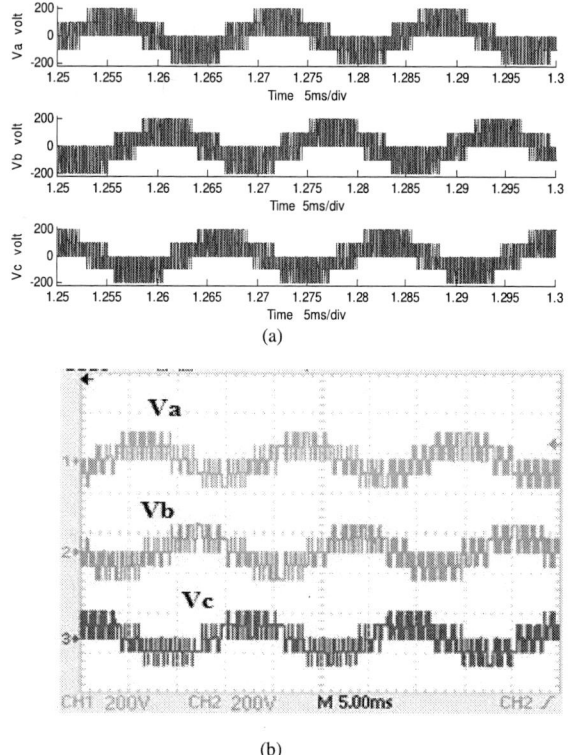

(a)

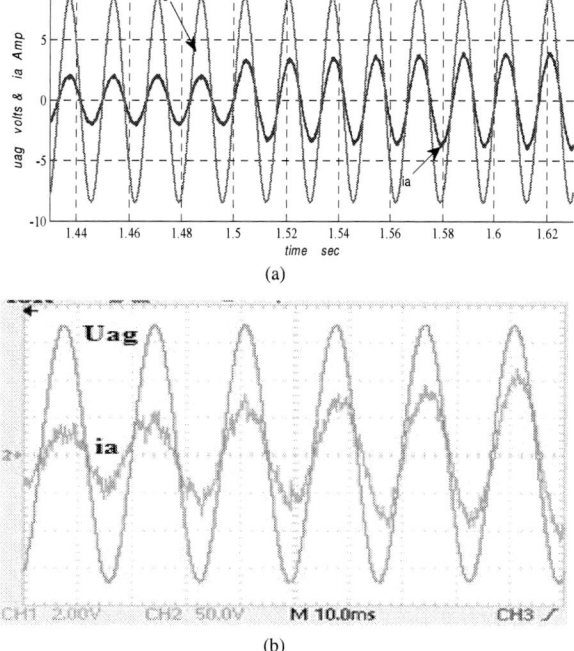

(b)

Fig. 15. Line voltage of the VSI in *abc* frame (a) simulation. (b) experimental.

Figure 16 shows the first test of the injection power. A step change to the demanded active power and the actual is shown in Fig. 16 keeping the reactive power equal zero.

(a)

(b)

Fig. 16. Grid phase voltage and injected current. (a) simulation. (b) experimantal

Because the injected power is only active, the current appears in phase with grid voltage as shown in Fig. 16.

The harmonic spectrum using fast Fourier transform (FFT) analysis for the injected current are shown in Fig. 17.

From Fig. 17, it is clearly appear that the current harmonics is small (THD=1.9%) due to using the L-filter.

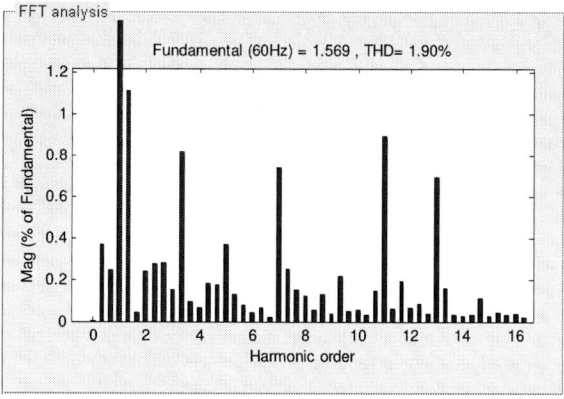

Fig. 17 The harmonic spectrum of the injected current.

The VSI phase voltage accompanied with the switching harmonics and its harmonic spectrum are shown in Fig. 18, 19.

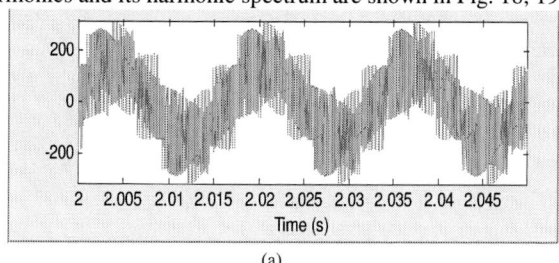

(a)

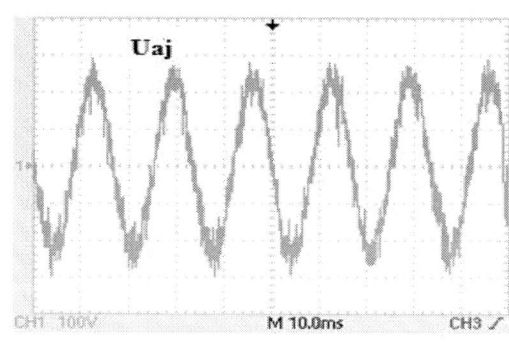

(b)

Figure 18. Inverter phase voltage to be connected to the grid with only L filter

From Fig. 19, it is noticed that the phase voltage has a large harmonic contents (THD=34.83%) due to the high switching frequency (5 KHz) of the VSI.

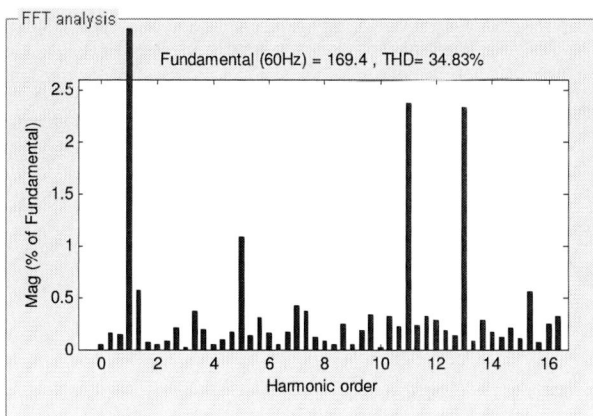

Fig. 19   The harmonic spectrum of the phase volatge.

In the second test to investigate the controller performance and the capability of the system assumption of demanded active and reactive power is considered such that,

$P^r = 500$ *watt*   over the test period.

$Q^r = 500$ VAR from the start till 2 sec, and reversed to a positive reference ($Q^r = -500$ VAR) after that.

The change between the modes of supplying and extracting reactive power is very clear with grid current and voltage waveforms as shown in Fig. 20. From the start until 2 sec, the VSI is considered as a capacitive load supplies reactive power to the grid. After that, the VSI absorbs reactive power from the grid causing the current to lag the voltage.

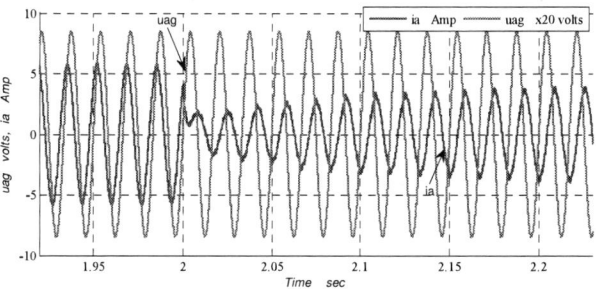

(a)

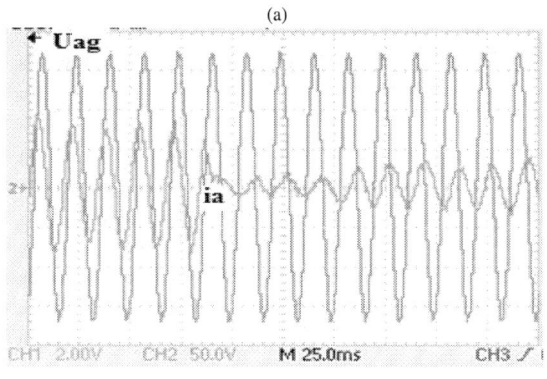

(b)

Fig. 20.   Step response in the reactive power to be injected to the grid. (a)simulation. (b) experimental.

C.   Grid-connected VOC VSI with LC filter

The capacitor voltage controller should give the value to be injected through the isolating transformer. For simplification, the transformer is assumed to have a ratio of 1:1. An LC filter with cut-off frequency of 205 Hz is used in the simulation. The simulation parameters are shown in Table II.

The in phase injected current with grid and capacitor voltages (i_a^j, u_a^g, u_a^c), the harmonic content of the injected current and filter capacitor voltage are shown in Figs. 21,22. Not like the L filter, the voltage waveforms are very clean.

Fig. 21 shows the synchronization between the grid voltage and the filter capacitor voltage where they are almost the same.

From Fig. 22, it is clearly shown that there is a significant change in the harmonic content for the output voltage of the WECS where the THD% decreases from 34.83% with L-filter to be almost negligible as 0.28% with LC filter.

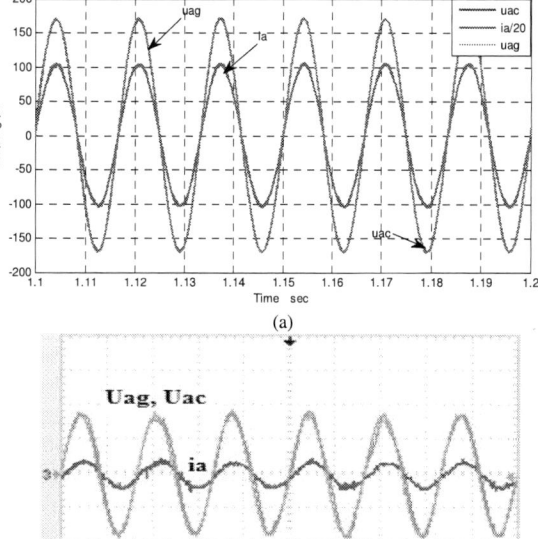

(a)

(b)

Fig. 21.   VSI response with LC filter for the grid and capacitor voltage with the injected line current to be connected to the grid . (a) Simulation. (b) experimental

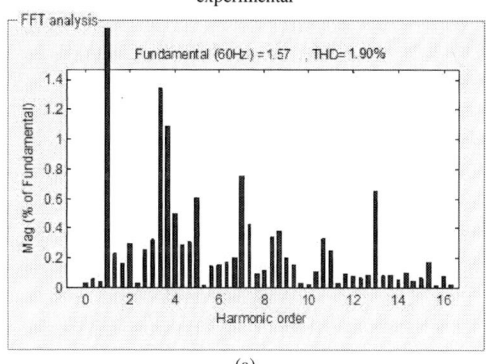

(a)

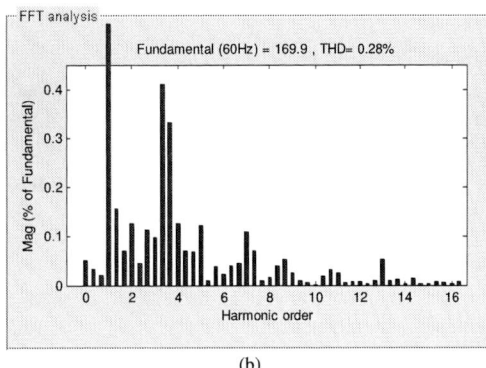

(b)

Fig. 22. The harmonic spectrum analysis with LC filter to be connected to the grid. (a) Injected current harmonic content. (h) Filter capacitor voltage harmonic content.

VII. CONCLUSION

In this paper; the dynamic mathematical model of the wind energy conversion system has been derived. The vector oriented control grid connected voltage source inverter has been investigated for high performance control operation. The simulation results showed how the control scheme succeeded in injecting the wind power as active or reactive power in order to compensate the grid power state. All results obtained confirm the effectiveness of the proposed control system for the SEIG feeding Three-phase bridge rectifiers and connected to grid through VOC VSI.

An experimental setup has been designed and implemented including a DS1104 R&D controller board as an interface element, and all the interfacing circuits to the analog power circuits. The complete wind energy conversion system has been tested and the experimental results showed good transient and steady state performance for each section as well as the overall system and agree with the simulation results to a great extend.

APPENDIX

TABLE I

SIMULATION PARAMETERS FOR VSI WITH L & LC-FILTER

Symbol	Quantity	Value
f_g	Grid frequency	60 Hz
V_{dc}	DC link voltage	350 v
L_f	Filter inductance	24 mH
R_f	Filter resistance	2 Ω
C_f	Filter capacitance	40 μt
f_{sw}	Switching frequency	5 KHz
$k_p{}^i$	Current controller gain	100
$k_i{}^i$	Current controller gain	1000
$k_p{}^c$	Voltage controller gain	0.07
$k_i{}^c$	Volatge controller gain	0.7

TABLE II

INDUCTION MOTOR IM PARAMETERS

Symbol	Quantity	Value
hp	Output power	1/2
V_n	Nominal voltage,	208 v L-L
rpm	Nominal speed	1800
I_n	Nominal curmt	3.5 A
V_f	Field voltage	125 v
R_r	Rotor resistance	3.592 Ω
R_s	Stator resistance	6.294 Ω
L_{ls}	Stator self-inducatnce	0.0168 H
L_{lr}	Rotor self-inducatnce	0.0168 H
L_m	Mutual inductance	0.464 H
J	Moment of inertia	0.03338 Kg.m^2

REFERENCES

[1] O. A. Mohammed, Fellow, IEEE, Z. Liu, and S. Liu, Senior Member, IEEE" A Novel Sensorless Control Strategy of Doubly Fed Induction Motor and Its Examination With the Physical Modeling of Machines", *IEEE Trans.on Magnetics,* Vol. 41, NO. 5, pp. 1852-1855, May, 2005.

[2] Lahcene Quazene and George McPherson " Analysis of The Isolated Induction Generator", *IEEE Trans. on Power Apparatus and Systems,* Vol.PAS-102, No.8, pp.2793-2798, August, 1983.

[3] N.H.Malik and S.E.Haque "Steady-state Analysis and Performance of An Isolated Self-Excited Induction Generator", *IEEE Trans. on Energy Conversion,* Vol.EC-1, No.3 pp.134-139, September, 1986.

[4] A.K.Aljabri and A.I.Alolah, "Capacitance Requirement for Isolated Self-Excited Induction Generator", *IEE Proceedings,* part B, Vol.137, No.3, pp.154-159, May, 1990.

[5] L.Shridhar, B.P.Singh and C.S.Jha," A Step Towards Improvements in The Characteristics of Self-Excited Induction Generator", *IEEE Trans. On Energy Conversion,* Vol.8, No.1, pp.40-46, March, 1993.

[6] S.P.Singh, M.P.Jain and Bhim Singh, "A New Technique for Analysis of Self-Excited Induction Generator", *Electric Machine and Power Systems,* Vol.23, pp.647-656, 1995.

[7] M. Steinbuch, "Optimal multivariable control of a wind turbine with variable speed," *in Proc. Eur. Wind Energy Conf.,* 1986, pp. 623–628.

[8] J. A. M. Bleij, A. W. K. Chung, and J. A. Rudell, "Power smoothing and performance improvement of wind turbines with variable speed," *in Proc. 17th British Wind Energy Assoc.,* 1995, pp. 353–358.

[9] S. M. B.Wilmshurst, "Control strategies for wind turbines," *Wind Eng.,* vol. 12, no. 4, pp. 236–249, 1988.

[10] D. Seyoum, C. Grantham and M. F. Rahman, "The dynamic charteristics of an isolated self-excited induction generator driven by wind turbine," *IEEE Trans. on Ind.App.,* Vol. 39, No. 4, pp. 936-944, July/Aug. 2003.

[11] R. Leidhold,G. Garcia, and M.I.Valla , "Field-oriented controlled induction generator with loss minimization ," *IEEE Trans. on Ind. Elec. ,* Vol. 49, No. 1, Feb. 2002.

[12] D. Seyoum, C. Grantham and M. F. Rahman, "A novel analysis and modeling of an isolated self-excited induction generator taking iron loss into account," *IEEE Trans. on Ind.App.,* Vol. 35, No 3, Apr. 2003,

[13] J. W. Dixon and B. T. Ooi, "Indirect current control of a unity power factor sinusoidal boost type 3-phase rectifier," *IEEE Trans. Ind. Electron.,* vol. 35, pp. 508-515, Nov. 1988.

[14] R.Wu, S. B. Dewan, and G. R. Slemon, "Analysis of an ac to dc voltage source converter using PWM with phase and amplitude control," *IEEE Trans. Ind. Applicat.,* vol. 27, pp. 355-364, Mar./Apr. 1991.

[15] A. Draou,Y. Sato, and T. Kataoka, "A new state feedback based transient control of PWM ac to ac voltage type converters," *IEEE Trans. Power Electron.,* vol. 10, pp. 716-724, Nov. 1995.

[16] H. K"om"urc"ugil, and 0. K"ukrer, "A Novel Current-Control Method for Three-Phase PWM ACIDC Voltage-Source Converters ," *IEEE Tran. On Ind. Electron. ,* Vol. 46, NO. 3, June 1999.

978-1-4244-4782-4/10 $26.00 © 2010 IEEE

[17] J. Fernando Silva, "Sliding-mode control of boost-type unity- power-factor PWM rectifiers," *IEEE Trans. On Ind. Electron..*, vol. 46, No. 3, Jun. 1999.

[18] M. P. Kazmierkowski, M. A. Dzieniakowski, and W. Sulkowski, "The three phase current controlled transistor DC link PWM converter for bi-directional power flow," *in Proc. PEMC Conf,* Budapest, Hungary, 1990, pp. 465-469.

[19] H. Kohlmeier, 0. Niermeyer, and D. Schroder, "High dynamic four quadrant AC-motor drive with improvedpower-factor and on-line optimized pulse pattern with PROMC," *in Proc. EPE Conf,* Brussels,Belgium, 1985, pp. 3.173-178.

[20] 0. Niermeyer and D. Schroder, "AC-motor drive with regenerative braking and reduced supply line distortion," *in Proc. EPE Conf,* Aachen, Germany, 1989, pp. 1021-1026.

[21] B. T. Ooi, J. C. Salmon, J. W. Dixon, and A. B. Kulkarni, "A 3- phase controlled current PWM converter with leading power factor," *in Conf Rec. IEEE-IAS* Annu. Meeting, 1985, pp. 1008-1014.

[22] B. T. Ooi, J.W. Dixon, A. B.Kulkarni, and M. Nishimoto, "An integrated AC drive system using a controlled currentPWMrecti erlinverter link," *in Proc. IEEE PESC'86,* 1986, pp. 494-501.

[23] F. F. M. El-Sousy and H. Godah," High-Performance Control of Self-Excited Induction Generator for Wind Energy Conversion System Using Rotor Flux Orientation ", *1st IFAC International Workshop on Advanced Control Circuits and Systems (ACCS05)*, Cairo, Egypt, March 6-10, 2005. CD-ROM.

[24] F. F. M. El-Sousy, M. Orabi and H. Godah," Maximum Power Point Tracking Control Scheme for Grid Connected Variable Speed Wind Driven Self-Excited Induction Generator ", *The Korean Institute of Power Elctronics (KIPE), Journal of Power Electronics (JPE)*, Vol. 6, No. 1, pp.52-66, January, 2006.

[25] Mahmoud M. Neam, Fayez F. M. El-Sousy, Mohamed A. Ghazy and Maged A. Abo-Adma'' DC-Bus Voltage Control of Three-Phase AC/DC PWM Converters for Renewable Energy Applications'',*IEMDC'09 Conf.,* Miami, USA, pp.1682-1691, 3-6 May, 2009.

[26] S. Fukuda, and T. Yoda, "A Novel Current-Tracking Method for Active Filters Based on a Sinusoidal Internal Model," *IEEE Tran. On nd. App.,* Vol., 37, No.3, pp.888-895, May /June 2001.

[27] V. Blasko and V. Kaura, "A new mathematical model and control of a three-phase ac-dc voltage source converter," *IEEE Trans. Power Electron.*, vol. 12, pp. 116-123, Jan. 1997.

[28] H. M. Godah, "A Digital control system for optimum utilization of renewable energy sources with the utility grid," Ph.D. dissertation, Faculty of Engineering, Al Azhar Univ., Cairo, 2003.

Control System Design for Bi-directional Power Transfer in Single-Phase Back-to-Back Converter Based on the Linear Operating Region

Janeth Alcalá, Víctor Cárdenas, Emanuel Rosas and Ciro Núñez
Centro de Investigación y Estudios de Posgrado
Universidad Autónoma de San Luis Potosí
San Luis Potosí, SLP
janethalcala@ucol.mx, vcardena@usalp.mx, emanuel_r_h@alumnos.uaslp.mx, calberto@uaslp.mx

Abstract—This paper presents two control strategies for a Single-Phase Back-to-Back (SPBTB) converter for control bi-directional power transfer based on the linear operating region of the topology. The control strategies presented allow an independent control of active/reactive power. The first scheme proposes to cancel the non-linear expressions associated to the linearized model. Therefore, it is obtained a suitable steady state and dynamic response of the system. The second control strategy uses a non-linear method that ensures a better response, since it does not linearize around a specific point, obtaining a global control law for the all operating region. The performance is verified with simulation and experimental results.

I. INTRODUCTION

It is well kwon that electrical grids restrain active and reactive power flow in the point of common coupling (PCC), which causes over-load and/or sub-utilization problems. Over-load implies higher energy consumption between both interconnected systems. More power transfer though the existing system implicates that some of the lines might operate closer to their capacity limits. Besides, there is a risk that the system operates around to its instability boundaries. Introducing additional interconnections to solve the problem will become the system more complex to operate. Significant benefits can be obtained achieving a load-balancing by introducing topologies that allow a better control of power transfer. In this aspect, voltage source converters (VSC) are a growing technological alternative [1]. Specifically through Back-to-Back (BTB) configuration is possible to accomplish these objectives. BTB converter is formed by two identical VSC connected by a common dc-link. This topology presents several advantages in terms of processing power and allows bidirectional power flow with almost sinusoidal currents at near-unity power factor. Therefore, the power flows can be changed and the thermal limits will not be exceeded. Moreover, losses can be minimized and stability margins are

not increased. BTB converter is widely used in the power system connected between the line and the load. The load can be active, passive or even another network, Fig. 1(a). The dc-link in the middle provides decoupling between both converters; as a result, they could be driven independently. Therefore, it is possible to have a fast and independent control of active and reactive power for both converters and improve the operation of the system. To simultaneously achieve these performances is important to explore control strategies which allow obtaining the desired control objectives. Previous works have presented control strategies were the current controllers are based on PI schemes [2-5]. However, these schemes exhibit coupling restrictions and do not provide an independently control of active/reactive power. In this paper two laws to the control of the BTB converter are discussed based on the *dq* mathematical model. The first technique presented cancels the coupled terms that appear in the mathematical model of the BTB converter and eliminate locally the nonlinearities [6-7]. The second solution, an input-output linearization technique allows to fully linearize the BTB converter. VSC_1 and VSC_2 are controlled separately; this can be accomplished because both converters are decoupled through the dc-link capacitor. Nevertheless is not the unique solution to the control of the BTB converter, but significantly simplifies the control of the two supply current components, which allows the SPBTB to transfer bi-directional active power and regulate the reactive power independently. Therefore, this configuration is used to the development of the control strategies. Moreover, the operating capacity of the topology is analyzed to establish the operating active/reactive region of the topology [8-9]. The knowing of the physical restrictions allow identifying what can be done by the control strategy and what cannot be obtained because of the inherent characteristic of the converter. Therefore, to avoid over-modulation conditions and to take advantage of the SPBTB response, an analytical procedure is developed to obtain the operating region.

978-1-4244-4782-4/10 $26.00 © 2010 IEEE

To the control of the BTB converter outer and inner loops are proposed. The designs of the outer loops are similar in both techniques unlike the inner loops which are different. The inner loops are used to control the current in VSC$_1$ and VSC$_2$. The outer loops in VSC$_1$ control the dc-link voltage and the reactive power and in VSC$_2$ they control active and reactive power. The variables tracking are utilized to prove the local and global tracking capability with suitable steady-state error in both techniques. The simulated and experimental results obtained with both techniques are presented. A SPBTB prototype using IGBT power switches is built and experiments are presented to validate the performance of the proposed controllers.

II. SYSTEM MODELING AND PRINCIPLES OF OPERATION

The configuration of the proposed system is showed in Fig. 1(a) and the state equivalent model circuit in Fig. 1(b).

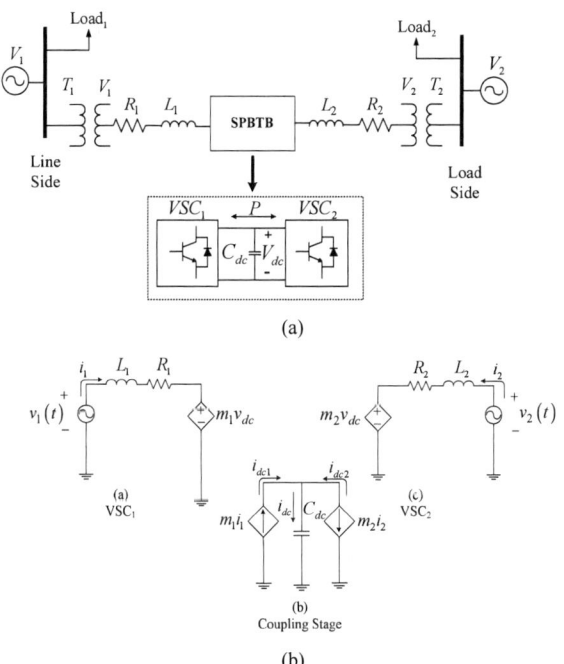

(a)

(b)
Coupling Stage

(b)

Fig. 1. SPBTB. (a) Power topology.

Fig. (b). Equivalent circuit.

The mathematical model of the SPBTB converter is a nonlinear-type because the state variables are multiplied by the control inputs. Theoretical results report that it is difficult to obtain global stability or a suitable tracking for sinusoidal waveforms. Therefore, the single-phase dq transformation [10] is used and the sinusoidal voltages and currents are expressed in rotating coordinates by two constant dq variables which provide phase and amplitude, respectively. As a result, through the dq model the control objectives are simplified and the sinusoidal tracking problem becomes a regulation problem. Hence, the control objectives can be achieved by regulating the dq variables to the desired

constant value. The dq model is given by (1) – (5), where m_1^{dq} and m_2^{dq} are the modulating signals, which must fulfill the constraints[1] $\left(m_{1,2}^d\right)^2 + \left(m_{1,2}^q\right)^2 < \left(u_T\right)^2$. ω is the angular frequency of ac mains; u_T is the peak amplitude of the triangular carrier.

$$\frac{d}{dt}i_1^d = \omega i_1^q + \frac{1}{L_1}v_1^d - \frac{1}{u_T L_1}v_{dc}m_1^d \qquad (1)$$

$$\frac{d}{dt}i_1^q = -\omega i_1^d - \frac{1}{u_T L_1}v_{dc}m_1^q \qquad (2)$$

$$\frac{d}{dt}i_2^d = \omega i_2^q - \frac{1}{L_2}v_2^d + \frac{1}{u_T L_2}v_{dc}m_2^d \qquad (3)$$

$$\frac{d}{dt}i_2^q = -\omega i_2^d - \frac{1}{u_T L_2}v_{dc}m_2^q \qquad (4)$$

$$C_{dc}\frac{d}{dt}v_{dc} = \frac{1}{2u_T}\left[\left(m_1^d i_1^d + m_1^q i_1^q\right) - \left(m_2^d i_2^d + m_2^q i_2^q\right)\right] \qquad (5)$$

A. Operation requirements

The SPBTB converter must keep the dc-link voltage level constant and the active power transmitted through the dc-link should be adjustable. As no reactive power is transmitted through the dc-link, the reactive power compensation is attributed to the VSC corresponding to each side control. One VSC must control the voltage level in the dc-link and the other the active power flow. Both converters can perform any of the two tasks; therefore the decision is subjective. In the SPBTB it was decided VSC$_1$ will be in charge of controlling the voltage level in the link and VSC$_2$ must control the active power flow. Based on the linear operating region of the SPBTB this task is guaranteed because the active/reactive power boundary of the converter will not be overcome.

B. Linear Operating Region

The operation region of the SPBTB converter is the graphic representation of all the equilibrium points, which are defined by the constraints imposed to the modulation indexes and by the topology characteristic [8]. The linear operating region guarantees power transfer between both converters in the SPBTB without over-modulation and a lower THD$_V$ (Total Harmonic Voltage Distortion). To define the linear operating region, an analytical procedure is studied. The system input restrictions and the constraint imposed by the voltage are:

$$\left(m_{1,2}^d\right)^2 + \left(m_{1,2}^q\right)^2 < u_T^2 \qquad V_{dc} > \sqrt{2}V_1 \qquad (6)$$

Solving (1)-(5) at the equilibrium point, yields to the following[2]:

$$I_1^d = -\frac{M_1^q V_{dc}}{\omega L_1} \qquad I_1^q = \frac{M_1^d V_{dc} - V_1^d}{\omega L_1} \qquad (7)$$

$$I_2^d = -\frac{M_2^q V_{dc} - V_2^q}{\omega L_2} \qquad I_2^q = \frac{M_2^d V_{dc} - V_2^d}{\omega L_2} \qquad (8)$$

---

[1] Sub-indices 1, 2 are referred to each one of the converters respectively.
[2] Capital letters are used to represent the steady state and peak values.

General expressions for active and reactive powers transmitted for both converters are defined by:

$$P_{1,2} = \tfrac{1}{2}\left(V_{1,2}^d I_{1,2}^d + V_{1,2}^q I_{1,2}^q\right) \qquad (9)$$

$$Q_{1,2} = \tfrac{1}{2}\left(V_{1,2}^q I_{1,2}^d - V_{1,2}^d I_{1,2}^q\right) \qquad (10)$$

$V_{1,2}^d$ are taken as references for VSC$_{1,2}$. Therefore, $V_{1,2}^q = 0$. The (11-14) expressions are obtained by evaluating (7) and (8) at the maximum and minimum values of the control inputs.

$$-\frac{V_1^d V_{dc}}{2\omega L_1} < P_1 < \frac{V_1^d V_{dc}}{2\omega L_1} \qquad (11)$$

$$\frac{(V_1^d - V_{dc})}{2\omega L_1} V_1^d < Q_1 < \frac{(V_1^d + V_{dc})}{2\omega L_1} V_1^d \qquad (12)$$

$$-\frac{V_2^d V_{dc}}{2\omega L_2} < P_2 < \frac{V_2^d V_{dc}}{2\omega L_2} \qquad (13)$$

$$\frac{(V_2^d - V_{dc})}{2\omega L_2} V_2^d < Q_2 < \frac{(V_2^d + V_{dc})}{2\omega L_2} V_2^d \qquad (14)$$

By mapping the operating region of the dq active and reactive current components into P and Q domain it is possible to determine the power transfer limits for a given dc-link voltage. As example, Fig. 2 shows the power transfer limits for VSC$_1$. The system parameters used to plot the linear operating region are shown in Table I. The dc-link capacitor was selected to transfer the maximum active power obtained from the graphics of the active operating regions.

The colors gradient gives the value of the transmitted power from V_1 to VSC$_1$ according to the dq coordinates for the modulating indexes $m_d \times m_q$. As example, from Fig. 2, VSC$_1$ and VSC$_2$ should have $M_1^d = 0.386$, $M_1^q = -0.132$ to transmit $200W$ from V_1 to V_2 at unitary power factor in both converters.

Fig. 2(c) shows the modulating indexes for VSC$_1$ depending on the power factor. The blue – green surface and the red – orange surface represents the lagged and leaded power factor for the VSC$_1$ respectively. It is observed that both converters are able to operate in a wider range in the inductive power factor region than in the capacitive one. The operation of the system closer to the periphery is equivalent to work with over-modulation conditions and all the problems that this involves. Therefore, to take advantage of a SPBTB response it must operate under its boundary region. For the system proposed the maximum active power that the converter is able to transfer from the mains to the load is around 1.5kW as can be note in Figs. 2(a) and 2(b).

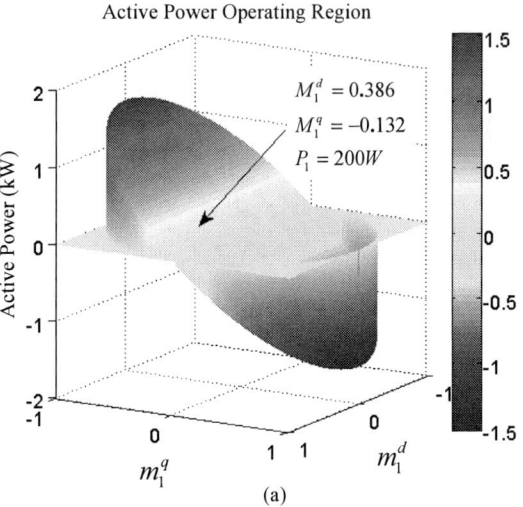

(a)

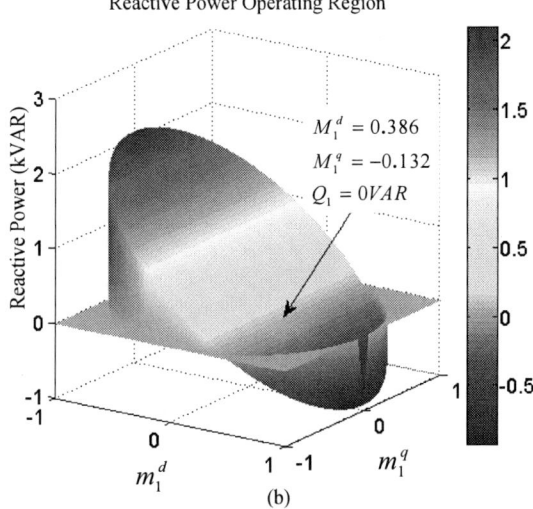

(b)

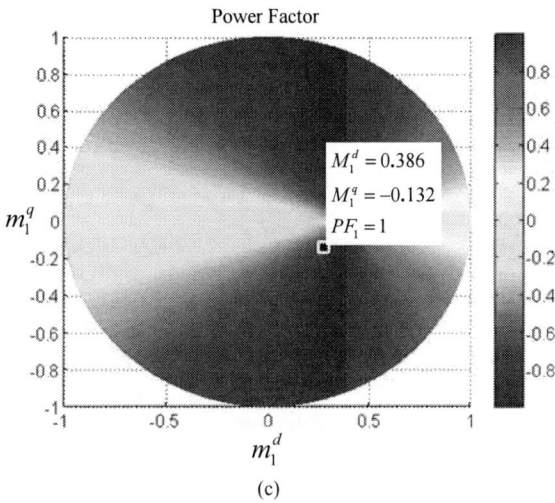

(c)

Fig. 2. VSC$_1$ Operating Region. (a) Active power transfer. (b) Reactive power transfer. (c) SPBTB Power Factor.

TABLE I
SYSTEM PARAMETERS

Parameter	Value
V_1^d, V_2^d	30 Vrms
ω	377r/s
L_1, L_2	4.1 mH
R_1, R_2	284mΩ
V_{dc}	110V

III. CONTROL STRATEGIES

The control objectives assigned to the SPBTB converter are: dc-link voltage v_{dc} regulated and active/reactive power flow control. The proposed control block is shown in Fig. 3; the outer loops connected in cascade with the inner loops provide the dq current references. Two control strategies are proposed for the inner current loops. The first one is based on a scheme that allows decoupling active/reactive components of the input current. The second one proposes use the input-output linearization method.

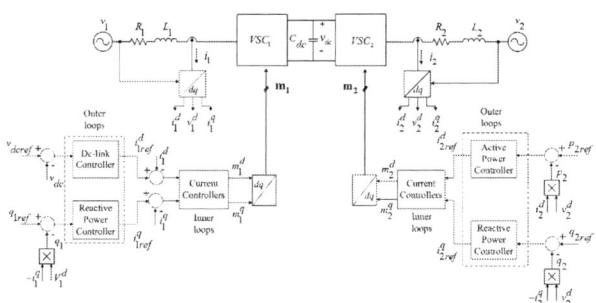

Fig. 3 Proposed Control Scheme for the SPBTB.

A. Active and reactive current decoupled control strategy

From equations (1)-(5) can be seen that the active/reactive line currents components are coupled and do not allow an independently control of the active and reactive power. The control block diagram for VSC$_1$ is shown in Fig. 4.

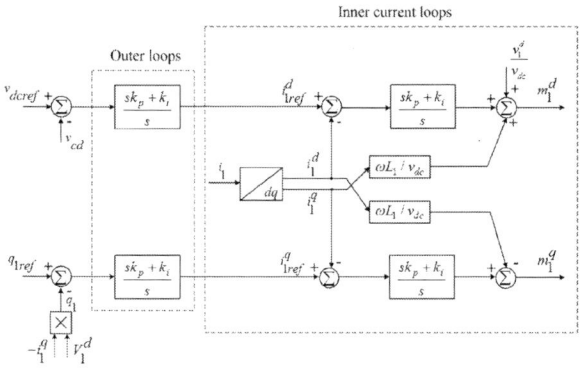

Fig. 4. Control Block Diagram for VSC$_1$.

Therefore, to cancel the non-linear expressions the following control law is proposed [11]:

$$m_1^d = k_p\left(i_{1ref}^d - i_1^d\right) + k_i \int\left(i_{1ref}^d - i_1^d\right)dt + \frac{u_T v_1^d}{v_{dc}} + \omega i_1^q \frac{u_T L_1}{v_{dc}} \quad (15)$$

$$m_1^q = k_p\left(i_{1ref}^q - i_1^q\right) + k_i \int\left(i_{1ref}^q - i_1^q\right)dt - \omega i_1^d \frac{u_T L_1}{v_{dc}} \quad (16)$$

From (15) and (16) is obtained the closed loop transfer function of the active/reactive current, this yields the system to a linear decoupled system around the operation point.

$$\frac{i_1^d(s)}{i_{1ref}^d(s)} = \frac{i_1^q(s)}{i_{1ref}^q(s)} = \frac{-k_p v_{dc}s - k_i v_{dc}}{u_T L_1 s^2 - k_p v_{dc}s - k_i v_{dc}} \quad (17)$$

Two outer loops based on PI Controllers are added for VSC$_1$. One controls the reactive power by manipulating the reactive current component and the second one controls the dc-link voltage. These control loops will be regulated to the desired level indirectly once the line current tracking is achieved. The transfer function for v_{dc} and q_1 is given by (18) and (19). Besides, another two outer loops to control active/reactive power in VSC$_2$ are used. The p_2 and q_2 transfer function are given by (20) and (21). To guarantee the decoupled between the inner current loop and outer loops the bandwidth of the outer loop must be a decade lower than the inner loop. The control system parameters are shown in Table II. The parameters are chosen in order to obtain a satisfactory dynamic response.

$$\frac{v_{dc}(s)}{v_{dcref}(s)} = \frac{k_p v_1^d s + k_i v_1^d}{2C_{dc} v_{dcref} s^2 + k_p v_1^d s + k_i v_1^d} \quad (18)$$

$$\frac{q_1(s)}{q_{1ref}(s)} = -\frac{v_1^d(k_p s + k_i)}{(2 - v_1^d k_p)s - v_1^d k_i} \quad (19)$$

$$\frac{p_2(s)}{p_{2ref}(s)} = \frac{v_2^d(k_p s + k_i)}{(2 + v_2^d k_p)s + v_2^d k_i} \quad (20)$$

$$\frac{q_2(s)}{q_{2ref}(s)} = -\frac{v_2^d(k_p s + k_i)}{(2 - v_2^d k_p)s - v_2^d k_i} \quad (21)$$

B. Non-linear control strategy

SPBTB model in steady - state can be represented as:

$$\begin{bmatrix} \dfrac{di_1^d}{dt} \\ \dfrac{di_1^q}{dt} \\ \dfrac{dv_{dc}}{dt} \end{bmatrix} = \begin{bmatrix} \omega i_1^q - R_1 i_1^d + \dfrac{v_1^d}{L_1} \\ -\omega i_1^d - R_1 i_1^q \\ \dfrac{-i_{dc2}}{C_{dc}} \end{bmatrix} + \begin{bmatrix} -\dfrac{v_{dc}}{u_T L_1} & 0 \\ 0 & -\dfrac{v_{dc}}{u_T L_1} \\ \dfrac{i_1^d}{2C_{dc}} & \dfrac{i_1^q}{2C_{dc}} \end{bmatrix}\begin{bmatrix} u_1 \\ u_2 \end{bmatrix}, \ x = \begin{bmatrix} i_1^d \\ i_1^q \\ v_{dc} \end{bmatrix}, u = \begin{bmatrix} m_1^d \\ m_1^q \end{bmatrix} (22)$$

According to the theory of input–output linearization [12-13], an input-output system is input-output linearize while exist a control law, such as:

$$u = -E^{-1}(x)\begin{bmatrix} L_f^{\rho-1}h_1(x) \\ ... \\ L_f^{\rho m}h_m(x) \end{bmatrix} + E^{-1}(x)\begin{bmatrix} v_1 \\ ... \\ v_m \end{bmatrix} \quad (23)$$

Where $E(x)$ is called decoupling matrix of the MIMO system, ρ is the relative degree of the system and v denote the auxiliary control variable. Therefore, calculating the Lie Derivates (24) is obtained. This control law exists for all $v_{dc} \neq 0$.

$$\begin{bmatrix} L_1\dot{x}_1 \\ L_1\dot{x}_2 \\ C_{dc}\dot{x}_3 \end{bmatrix} = \begin{bmatrix} \omega L_1 x_2 - R_1 x_1 + v_1^d \\ -\omega L_1 x_1 - R_1 x_2 \\ i_{dc2} \end{bmatrix} + \frac{1}{u_T}\begin{bmatrix} -x_3 & 0 \\ 0 & -x_3 \\ \frac{x_1}{2} & \frac{x_2}{2} \end{bmatrix}\begin{bmatrix} u_1 \\ u_2 \end{bmatrix} \quad (24)$$

$$y = \begin{bmatrix} x_1 & x_2 \end{bmatrix}^T$$

Where $x = \begin{bmatrix} i_1^d & i_1^q & v_{dc} \end{bmatrix}$, $u = \begin{bmatrix} u_1^d & u_1^q \end{bmatrix}$.

The control law is obtained using (23)

$$u = \frac{L_1 u_T}{x_3}\begin{bmatrix} \omega x_2 - \frac{R_1}{L_1}x_1 + \frac{v_1^d}{L_1} - v_1 \\ -\omega x_1 - \frac{R_1}{L_1}x_2 - v_2 \end{bmatrix} \quad (25)$$

The original nonlinear system (22) is transformed into the (input-output) linearized system such that:

$$\begin{bmatrix} \dot{y}_1 \\ \dot{y}_2 \end{bmatrix} = \begin{bmatrix} v_1 \\ v_2 \end{bmatrix} \quad (26)$$

The plant in the Laplace domain is seen through the control law as (27), which is a first – order plant. Therefore it is possible to achieve a zero steady-state error in the tracking toward its references using a simple proportionally controller. Fig. 5 shows the proposed control scheme.

$$\frac{Y_1(s)}{v_1(s)} = \frac{Y_2(s)}{v_2(s)} = \frac{1}{s} \quad (27)$$

It is then possible to choose suitable constants such that the error dynamics are stable and achieve the tracking of the constant reference by usual linear pole placement technique.

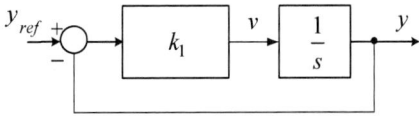

Fig. 5. Control scheme for the linearized plant.

The parameter of the controller $k_1 = -1535$ is chosen to have an acceptable dynamic response. The closed loop transfer function (28) is a first – order system.

$$\frac{i_1^d(s)}{v_1(s)} = \frac{i_1^q(s)}{v_2(s)} = \frac{-k_1}{s - k_1} \quad (28)$$

As a result, it is possible to control independently the active and reactive components of the line current and according to the linear relations between the power quantities:

$$p_1 = \frac{v_1^d i_1^d}{2}, \quad q_1 = \frac{v_1^d i_1^q}{2}, \quad q_2 = \frac{v_2^d i_2^q}{2} \quad (29)$$

Hence, the controls of the system powers correspond to the control of the corresponding currents, because v_1^d and v_2^d are constants. Moreover, the SPBTB is able to supply active/reactive power without modify the transfer of reactive/active power in both steady state and transient conditions. The outer control loops are designed as (18) – (21).

IV. SIMULATION AND EXPERIMENTAL RESULTS

The proposed control strategies have been simulated using Matlab/Simulink under the following conditions: $V_{1,2} = 30V_{rms}$, $f = 60Hz$, $L_{1,2} = 4.1mH$, $R_{1,2} = 284mH$, $U_T = 5V$, $V_{dc} = 110V$, $C_{dc} = 1050\mu F$, $P_{1,2} = 200W$, $PF_{1,2} = 1$ to verify their feasibility. Figs. 6 and 7 show the relevant waveforms for a transferred active power of *200W* with a 100% load reversed transient at *t=0.4s*. The results show that the *dq* current components present a good tracking for both control strategies. Nevertheless, there are differences in the transient response, which are mainly due to the second order dynamic that presents the decoupled active/reactive current control strategy. These differences can be observed in the oscillations of the currents. The input-output linearization strategy reduces the system to a first order type. Therefore, the control response converges exponentially and do not present the oscillation that appears in the first control strategy. Besides, the settling time is lower for the second proposed control strategy (*t=10ms*). Nevertheless, the difference is not significant (less than 5ms). The first control strategy have a similar performance the second one for the simulation parameters, but it is important to emphasize that this strategies decoupled locally around a set-point; this disadvantage is not presented by the non-linear control strategies, because the input-output linearization provides *global* stability.

It means that the system is decoupled for all the operating range. In order to validate the operating region presented in section II, simulations were done. From Fig. 2(a) is observed that the active power boundary for the SPBTB converter is around 1.5kW. Therefore, a power ramp was applied to the converter using the input – output linearization technique to guarantee global stability.

The objective of the simulation was to overcome the physical power constraints of the converter given by the active power operating region (Fig. 2(a)). Fig. 8 shows the dc-link voltage, the active power, the d current component and its reference, the dq components of the control signals and finally the control signal in the abc frame for VSC$_1$. It can be observed that close to the 1.35ms (when the converter reaches 1.5kW) the control signals are saturated and consequently the d current is unable to follow its reference. In turn, the system becomes unstable.

From the simulated results can be conclude that even if the control dynamic is very good it is not possible to achieved the control objectives because of the inherent characteristic of the SPBTB converter. Therefore, the operating region provides essential information that can be used to evaluate the performance of a controller and with the information discarded if the desired goals are not obtained because of the control technique used or if the problem is due to the own behavior of the topology.

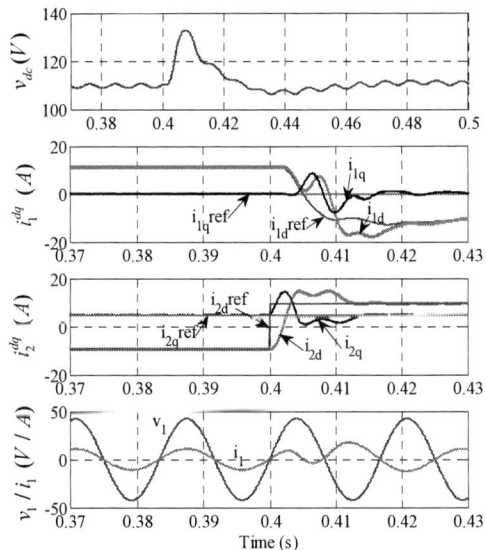

Fig. 6. Simulated results for a 100% load reversed. Active/reactive current decoupled control strategy. (a) v_{dc}. (b) i_1^{dq} currents tracking.

(c) i_2^{dq} currents tracking. (d) Supply voltage v_1 and current i_1.

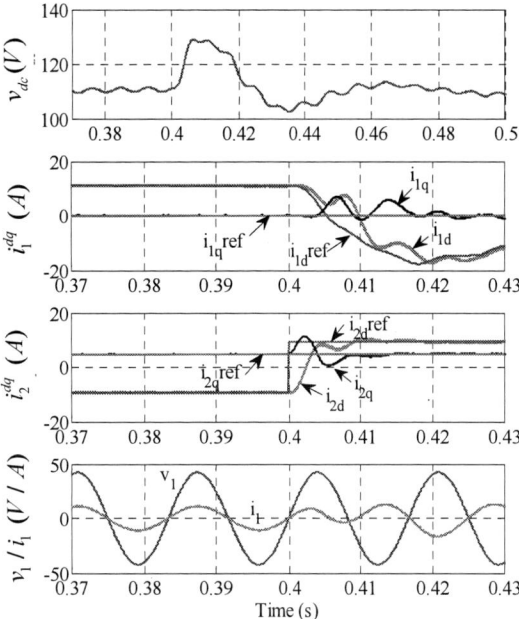

Fig. 7. Simulated results for a 100% load reversed. Input-output current decoupled control strategy. (a) v_{dc}. (b) i_1^{dq} currents tracking.

(c) i_2^{dq} currents tracking. (d) Supply voltage v_1 and current i_1.

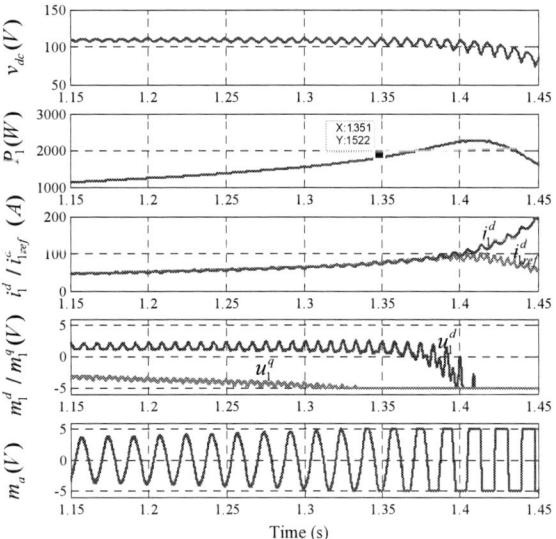

Fig. 8. VSC$_1$. (a) Dc-link voltage. (b) Active power ramp. (c) d current component and its reference. (d) dq control inputs components. (e) Control input component in the abc frame.

The control strategies were implemented and tested in an experimental SPBTB prototype under the same simulated conditions. In Figs. 9(a) and 10(a), CH1-CH4 show v_1 vs. i_1 (top) and v_2 vs. i_2 (bottom) when the SPBTB converter is transferring $200W$ from v_1 to v_2 at unitary power factor. It can be observed that the power transferred is sustained. A power step was applied to the converter, in order to evaluate the dynamic response. Figs. 9(b) and 10(b) show the dc-link voltage response to the power step when the dc-link. The power variation in the experiment was from $150W - 200W$. It ca n be notes that the dc-link voltage does not fluctuate significantly with both control schemes. In fact, the response of the decoupled active/reactive control strategy has a good performance considering that this control law only provides locally stability. The SPBTB converter was tested under regeneration, when the power is negative and flows into the dc-link, making dq current components negative. The line current change from $5.67A_{rms}$ to $-5.67A_{rms}$, experimental results are shown in Figs. 9(c) and 10(c); it is observed that the reversal response in the current is fast. Therefore, it can be conclude that both controllers present a good dynamical performance under reversal of load current, ensuring bi-directional power flow.

VI. CONCLUSIONS

This paper has presented and compared two control strategies of the SPBTB converter for providing an independent active/reactive power control. Moreover, in order to find the linear operating region of the SPBTB an analysis was derived, it was found that depending on dc-link voltage and the systems parameters (inductor link and line voltage), there is a limited range of active/reactive power that can be managed by the topology. These region boundaries allow identifying the physical restrictions of the system. Therefore, it is possible to assure from the control response if the desired results are not been reached because of the control strategy or by the restrictions imposed by the system. Based on the results two control strategies were proposed: a decoupled active/reactive current control strategy and an input-output linearization control strategy. The decoupling control has a good stability response. However, the PI parameters must be designed to a specific operating point. Nevertheless, experimental result show that the system recovers rapidly from power steps and reversed transient. The input – output linearizing current control algorithm, developed, linearizes the SPBTB converter to first – order ones and removes the cross – coupling between d and q current components. Therefore, it is possible to control independently active and reactive power, which improves the operation of the converter. Theoretical and experimental results show that both control strategies have a good dynamic of controlled variables. Moreover, by the experimental results it was shown that the control strategies allow bi-directional power flow and ensures power balancing control.

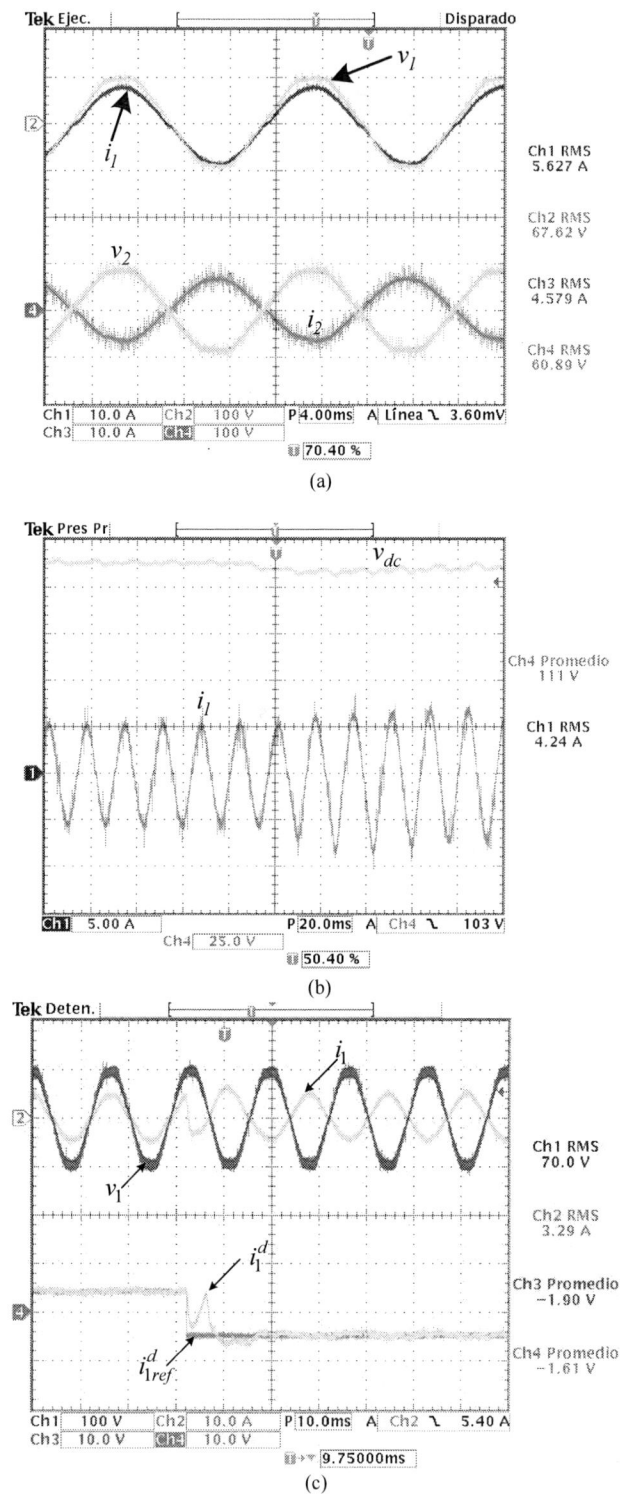

Fig. 9. Experimental results for the active and reactive decoupled control strategy: (a) Voltage and current supply in VSC $_1$ (top) and VSC$_2$ (bottom) when active power flows from V$_1$ to V$_2$. (b) Dc-link voltage and supply current response in VSC$_1$ for a power step (150W-200W). (c) Voltage and current supply in VSC$_1$ for a 100% (200W to -200W) reversed power transient (under regeneration) (top) and d component and its reference (bottom).

978-1-4244-4782-4/10 $26.00 © 2010 IEEE

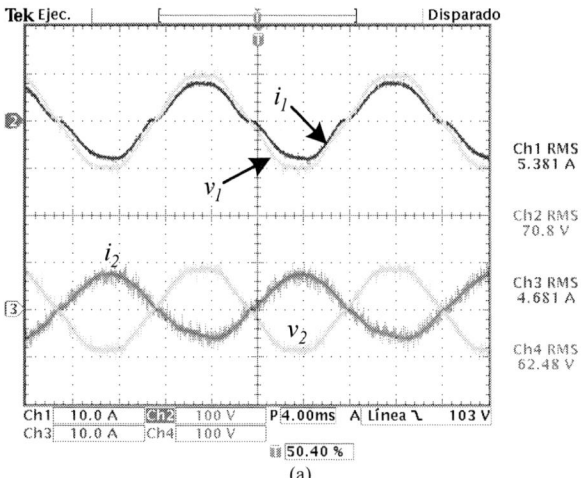

(a)

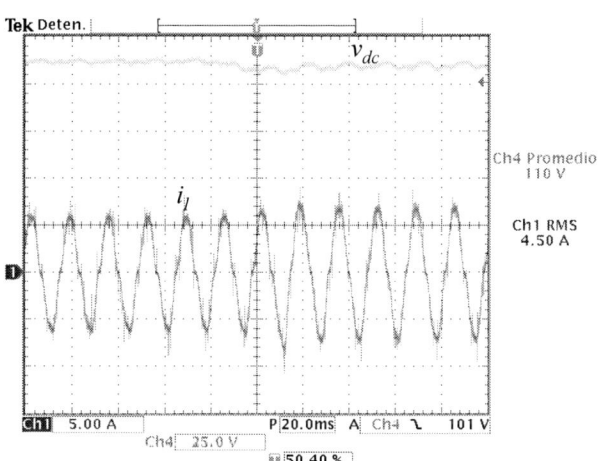

(b)

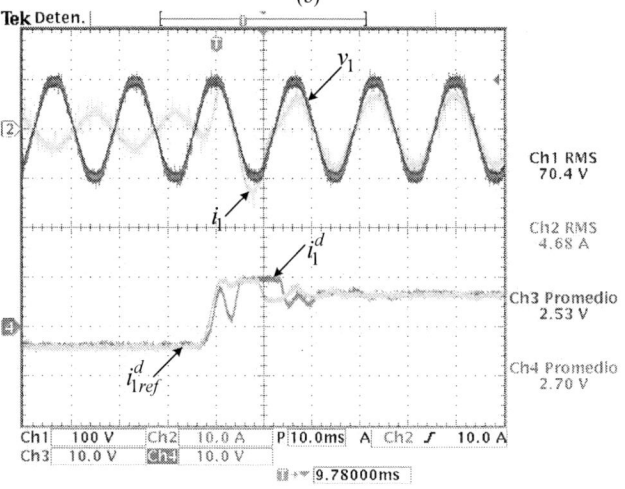

(c)

Fig. 10. Experimental results for the input-output linearization control strategy: (a) Voltage and current supply in VSC$_1$ (top) and VSC$_2$ (bottom) when active power flows from V$_1$ to V$_2$. (b) Dc-link voltage and supply current response in VSC$_1$ for a power step (150W-200W). (c) Voltage and current supply in VSC$_1$ for a 150% (200 W to -300W) reversed power transient (under regeneration) (top) and d component and its reference (bottom).

TABLE II. CONTROL PARAMETERS

PI	Gain			
Parameters	v_{dc}	q_1	p_2	q_2
k_p	1	-0.001	0.001	-0.001
k_i	100	-15.153	15.153	-15.153
τ	10(ms)	66(us)	66(us)	66(us)

REFERENCES

[1] Toledo P. F. and Söder L., "Power Flow Control in City Center Infeed", IEEE PowerTech Conference Proceedings. 2003. IEEE Bologna, vol. 3, pp. 1-8. 2003.

[2] Nasiri A. and Emadi A., "Different Topologies for Single-Phase Unified Power Quality Conditioners", Industry Applications Conference, 2003. 38th. IAS Annual Meeting. Conference Record, vol. 2, pp. 976-981. 2003.

[3] Li F., Zou Y. P., Wang C.Z., Chen W., Zhang Y.C. and Zhang J., "Research on AC Electronic Load Based on back to back Single-phase PWM Rectifiers", Applied Power Electronics Conference and Exposition 2008, APEC 2008, pp. 630-634. 2008.

[4] Wen-Song Ch. and Ying-Yu T., "Analysis and Desing on the Reduction of DC-Link Electrolytic Capacitor for AC/DC/AC Converter Applied to AC Motor Drives" IEEE Trans. on Industrial Electronics., vol. 49, no. 1, pp. 275–279, 1998.

[5] Jaya A. L., G. Tulasi Ram Das, K. Uma Rao, Sreekanthi K. and Rayudu K., "Different control strategies for Unified Power Qulity Conditioner at load side", Industrial Electronics and Applications, 2006, 1st IEEE Conference on Publication, pp. 1-7, 2006.

[6] Choi J. and Sul S., "Fast current controller in three-phase ac/dc boost converter using d-q axis crosscoupling", IEEE Trans. on Power Electron, vol., 13, pp. 17-185, 1998.

[7] Zhaoqing H., Chengxiong M. and Jiming L., "A novel Control Strategy for VSC based HVDC in Multi-Machine Power Systems", Journal of Electrical & Electronics Engineering". vol. 4, no. 2, pp. 1183-1190, 2004.

[8] Espinoza J., Joós G., Pérez M. and Morán T. L., "Operating Region in Active-Front-End Voltage/Current Source Rectifiers" IEEE Power Electronics Specialists Conference, PESC 2005, vol. 2, pp. 1726-1731. 2005.

[9] Alcalá J., Cárdenas V., Rosas E., Visario N. and Sierra R., "Linear Operating Region of a Single- Phase BTB Converter to Bidirectional Power Transfer", 2009 6th International Conference on Electrical Engineering, Computing Science and Automatic Control (CCE 2009), Toluca, Mexico. November 10-13, 2009, pp. 198-203. 2009.

[10] Gonzalez, M., Cárdenas V., and Pazos F., "DQ transformation development for single-phase systems to compensate harmonic distortion and reactive power". Power Electronics Congress, 2004. CIEP 2004. 9th IEEE International. pp. 177- 182, 2004.

[11] E. Rosas, Cárdenas V., Alcala J. and Núñez C., "Active and Reactive Current Decoupled Control Strategy Applied to a Single Phase BTB Converter ", 2009 6th International Conference on Electrical Engineering, Computing Science and Automatic Control (CCE 2009), Toluca, Mexico. November 10-13, 2009, pp. 204-209. 2009.

[12] Khalil H., Nonlinear control, New Jersey: Prentice Hall, Second Edition 1996, Chapter 12.

[13] Lee, D. C., Lee, G. M., and Lee, K. D., "DC Bus Voltage Control of Three-Phase AC/DC PWM Converters Using Feedback Linearization," IEEE Transactions on Industrial Applications, Vol. 36,No. 3, 2000, pp. 826-833.

Comparative Analysis of Low-pass Output Filter for Single-phase Grid-connected Photovoltaic Inverter

Hanju Cha and Trung-Kien Vu

Department of Electrical Engineering, Chungnam National University
Daejon, Korea
hjcha@cnu.ac.kr, vukien@gmail.com

Abstract - **Nowadays, the LCL-filter type becomes an attractive grid interfacing for grid-connected Voltage Source Inverter (VSI). LCL-filter can render the current harmonics attenuation around the switching frequency by using smaller inductance than L-filter. Moreover, system using LCL-filter does not depend on the grid impedance and has a better output response while comparing with LC-filter. Firstly, an analysis and design procedure of output LCL-filter for single-phase grid-connected Photovoltaic (PV) inverter system is presented in this paper. Due to the theoretical analysis, a comparison between the designed LCL-filter with L-filter and LC-filter based single-phase grid-connected PV inverter system is carried out. The comparison results are given to validate the theoretical analysis and effectiveness of filters.**

I. INTRODUCTION

To eliminate the current harmonics around the switching frequency and comply with the standards (i.e IEEE 1547), the grid-connected inverter for renewable energy source requires an output low-pass filter to interface with the grid. Ideally, the filter with low cut-off frequency and high attenuation at the high switching frequency is better to eliminate switching ripple effectively.

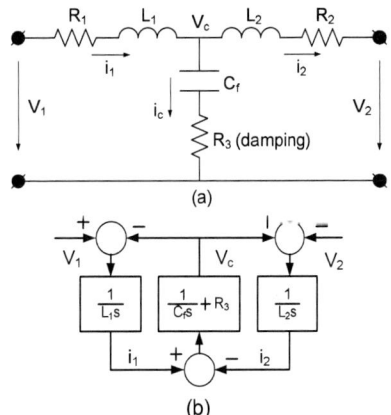

(a)

(b)

Fig. 1 (a) Equivalent circuit diagram and (b) model of LCL-filter

However, the filter design needs the trade-off when considering the switching loss/efficiency and fundamental voltage drop.

In the literature, L-filter, LC-filter and LCL-filter designs have been proposed and have the following properties.

Firstly, although a single inductor L-filter is popular and simple use, it has a low attenuation and high inductance value. The voltage drop across the inductor makes a poor system dynamics, hence causing a long-time response. By using L-filter, the inverter switching frequency must have a high value in order to sufficiently attenuate the harmonics [1].

Secondly, since the L-filter achieves low attenuation of the inverter switching components, a shunt element is needed to further attenuate the switching frequency components. A capacitor is selected to produce low reactance at the switching frequency and presents high magnitude impedance within the control frequency range.

The LC-filter is suited to configurations where the load impedance across the capacitor is relatively high at and above the switching frequency. To reduce the losses and cost, the capacitance should be high and then, the inductance can be reduced. But a very high capacitance is not recommended since the system may face with inrush current, high reactive current fed on capacitor at the fundamental frequency, possible resonance phenomenon at the grid side, etc. If a system is connected to the grid via LC-filter, the resonance frequency varies over time as the inductance value of the grid varies [2].

Thirdly, comparison with the previous filter topologies, LCL-filter produces better attenuation at the inverter switching frequency. LCL-filter can provide a better decoupling between the filter and the grid impedance. A lower ripple current distortion across the grid-side inductor since the current ripple is reduced by the capacitor. LCL-filter can provide a good attenuation ratio even with small L and C

978-1-4244-4782-4/10 $26.00 © 2010 IEEE   1659

values. However, the three-order LCL-filter design has to consider various constraints, such as the resonance phenomenon, the current ripple through inductors, the total impedance of the filter, the current harmonics attenuation at switching frequency and the reactive power absorbed by capacitor, etc.

In this paper, after the LCL-filter theoretical analysis and design, a comparison between the designed LCL-filter with L-filter and LC-filter based Photovoltaic (PV) inverter system is carried out. The simulation and experimental comparison results are given to validate the theoretical analysis and show the effectiveness of filter design methodology.

II. LCL-FILTER ANALYSIS AND DESIGN

The LCL-filter equivalent circuit diagram is shown in Fig.1 (a) and its equivalent model is shown in Fig.1 (b), where V_1 and V_2 are inverter and grid voltage, L_1, L_2, R_1, R_2 are the filter inverter-side and grid-side inductor and its equivalent resistors, respectively. A damping resistor R_3 is in series with capacitor C_f.

Based on the equivalent model of LCL-filter in Fig.1 (b), the transfer function of LCL filter using the inverter current as feedback can be derived by assuming that the value of R1 and R2 are small enough to be neglected:

$$G(s) = \frac{i_2(s)}{v_1(s)} = \frac{R_3 C_f s + 1}{L_1 L_2 C_f s^3 + (L_1 + L_2) R_3 C_f s^2 + (L_1 + L_2) s} \quad (1)$$

The resonant frequency is $\omega_{res} = \sqrt{\dfrac{L_1 + L_2}{L_1 L_2 C_f}}$ (2)

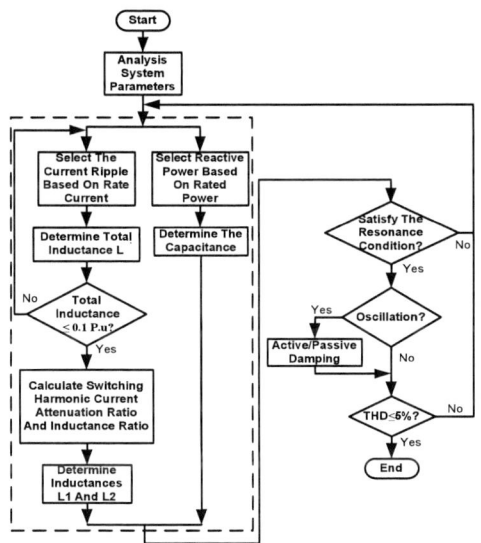

Fig. 2 LCL-filter design flow chart

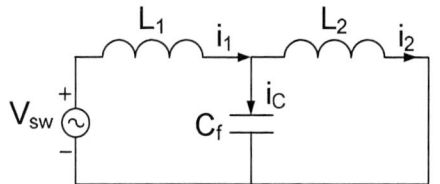

Fig. 3 LCL-filter equivalent circuit (considering only switching harmonic component)

The main LCL-filter design steps are summarized in Fig. 2. There are some limits on the parameter values such as [3]:

- The total inductance should be less than 0.1 (p.u). Because it results in the ac drop voltage during operation. Otherwise, a higher dc-link voltage will be required and this results in higher switching losses.

- The capacitance is limited by the reactive power factor (normally this factor is less than 5%).

- The resonant frequency should be in range: $10\omega_0 \leq \omega_{res} \leq \dfrac{\omega_{sw}}{2}$ to avoid resonance problems, where ω_0 is the utility frequency (rad/s), ω_{res} is the resonant frequency (rad/s) and ω_{sw} is the switching frequency (rad/s).

- The damping element losses cannot be high as to reduce efficiency.

A. Filter Inductance Calculation

Assuming the system has unity power factor and the unipolar Pulse Width Modulation (PWM) is adopted, the total filter inductance can be calculated by considering the maximum current ripple in the switching period T_s as following [4]:

$$L = \frac{V_{dc}}{4 I_{rated} \Delta_{ripple} f_s} (1 - m_a) m_a \quad (3)$$

where I_{rated} is the rated utility current; Δ_{ripple} is the maximum ripple magnitude percentage (5% - 25%); V_{dc} is the dc-link voltage; L is the total filter inductance; f_s is the switching frequency (Hz); m_a is the modulation index.

By adding a capacitor, the total filter inductance is divided into two parts: the inverter-side inductance L_1 and grid-side inductance L_2. These inductance values have the following relationship:

$$L_1 = a L_2 \quad (4)$$

where a ($a \geq 1$) is the inductance index can be calculated by using the switching harmonic current attenuation ratio [5].

As shown in Fig. 3, the switching harmonic current attenuation ratio can be defined as:

978-1-4244-4782-4/10 $26.00 © 2010 IEEE

$$\sigma = \frac{i_2}{i_1} = \frac{(1+a)r}{a(1-r)-r} = \frac{r}{\dfrac{r}{1+a}\dfrac{a}{r}(1-r)-\dfrac{r}{1+a}} \qquad (5)$$

where

$$r = \frac{1}{L_2 C_f \omega_{sw}^2} \text{ is constant} \qquad (6)$$

$$\frac{r}{1+a} = \frac{\Delta_{ripple} V_{rated} \omega_0}{2\pi^2 V_{dc} f_{sw} \alpha} \text{ is constant} \qquad (7)$$

Δ_{ripple} is the maximum ripple magnitude percentage (5% - 25%); V_{rated} is the rated utility voltage; α is the reactive power factor(<5%), ω_0 is utility frequency (rad/s).

B. Filter Capacitance Calculation

The filter capacitance C_f can be determined by considering the reactive power absorbed in filter capacitor as following:

$$C_f = \frac{Q_{re}}{\omega_0 V_{rated}^2} = \frac{\alpha P_{rated}}{\omega_0 V_{rated}^2} \qquad (8)$$

where Q_{re} is the reactive power absorbed by filter capacitor; P_{rated} is the rated power.

The more capacitance, the more reactive power flowing into capacitor, the more current demand from filter inductor and switches so that the overall filter inductor current will tend to increase. Hence the efficiency will be low. But the capacitance cannot be too small because a large inductance causes a high voltage drops across the filter inductor.

C. Damping Resistance Calculation

The LCL-filter can contain a damping resistor to avoid the resonance phenomenon. The damping resistance R_3 should be one third of the filter capacitor and resonant frequency as:

$$R_3 = \frac{1}{3\omega_{res} C_f} \qquad (9)$$

III. SINGLE-PHASE GRID-CONNECTED PV INVERTER SYSTEM

The block diagram of single-phase grid-connected PV inverter system is shown in Fig.4 (a) and equivalent current control block diagram is shown in Fig.4 (b), with rated power $P_{rated} = 3$ (kW), rated RMS line-to-line voltage is $V_{rated} = 220$ (V), dc-link voltage $V_{dc} = 400$ (V), switching frequency $f_{sw} = 10$ (kHz) and the rated current $I_{rated} = 13.64$ (A).

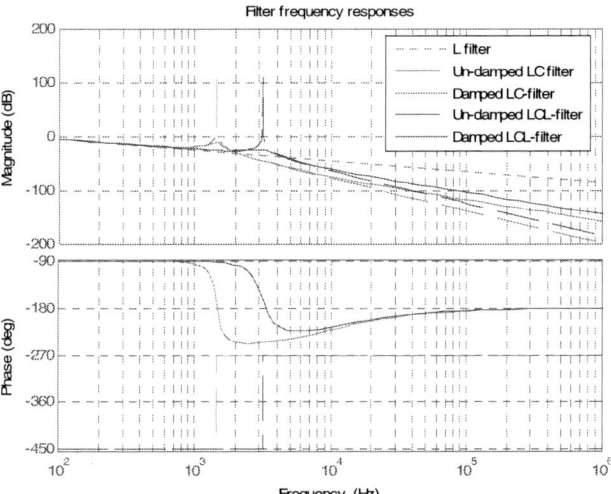

Fig. 5 Frequency responses of filters

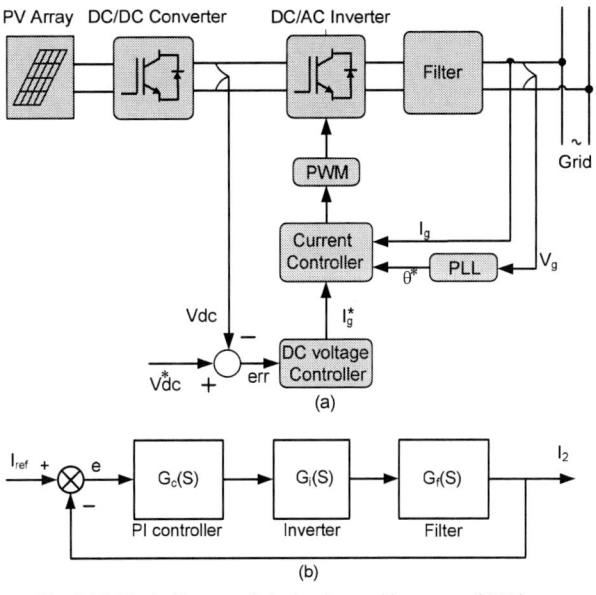

Fig. 4 (a) Block diagram of single-phase grid-connected PV inverter system and (b) equivalent current control block diagram

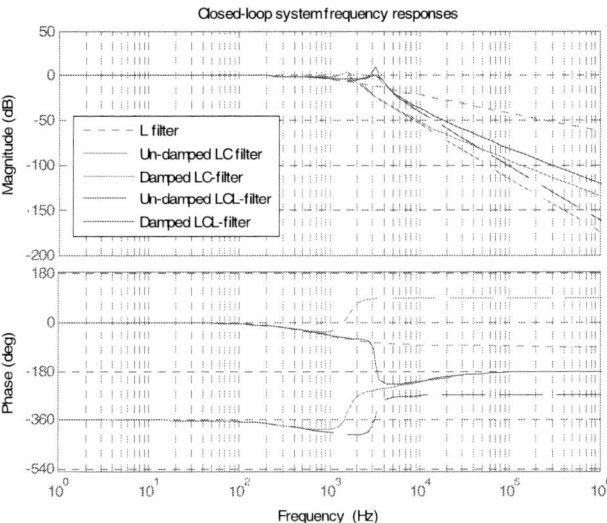

Fig. 6 Frequency responses of closed-loop system using different filters

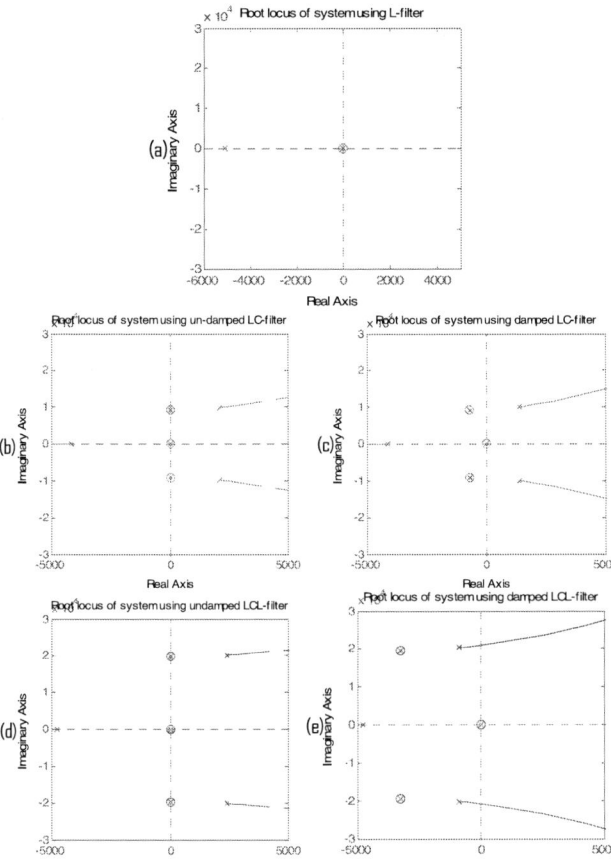

Fig. 7 Root locus diagram of closed-loop current control system by using (a) L-filter, (b) un-damped LC-filter, (c) damped LC-filter, (d) un-damped LCL-filter and (e) damped LCL-filter

Adopting I_{rm} = 4%.I_{rated} and Q_{re} = 2.5%.P_{rated}, we can calculate the filter inductances L_1 = 2.038(mH), L_2 = 0.896(mH), the filter capacitance C_f = 4.11(μF), the resonance frequency ω_{res} = 19772(rad/s) or f_{res} = 3.1468(kHz) and the damping resistance R_3 = 4.1(Ω).

For filter comparison, in this paper, we choose the inductance of L-filter is L = 2.934(mH); the inductance and capacitance of LC-filter are L = 2.934(mH) and C_f = 4.11(μF).

The single-phase grid connected PV inverter control system diagram is shown in Fig.4 (a) and its current control block diagram is shown in Fig.4 (b), where:

- $G_c(s) = K_p + \dfrac{K_i}{s}$ is the PI controller transfer function;

- $G_i(s) = K_{inv}$ is the transfer function of the inverter. Assuming the switching frequency is high enough to neglect the inverter dynamics, the PWM inverter can be represented by a gain for simplicity of analysis (due to relatively high switching frequency).

- $G_f(s)$ is the transfer function of filter as shown in

equation (1) in case of LCL-filter. $G_f(s) = \dfrac{1}{Ls}$ in case of

L-filter, $G_f(s) = \dfrac{1}{Ls} \dfrac{R_3 C_f s + 1}{LC_f s^2 + R_3 C_f s + 1}$ in case of LC-

filter.

The frequency responses of filters are shown in Fig.5. The bandwidths of filters are almost same below the resonant frequency ω_{res}. With damping resistor, the resonance phenomenon is almost eliminated.

Fig.6 shows the frequency response of the closed-loop system corresponding to each filter.

The peak of system resonance frequency is reduced by using a suitable damping resistance. Increase the damping resistance can depress the resonant peak but tends to reduce the attenuation.

Fig.7 shows the root locus diagram of closed-loop system using L-filter, un-damped and damped LC filter, un-damped and damped LCL filter, respectively.

Without damping resistor, the system using LC-filter and LCL-filter are on the boundary of stability. The system will be moved into the stable region by using a suitable damping resistance. Furthermore, the stability of closed-loop control is ensured with control algorithm.

IV.   SIMULATION RESULTS

By using the system parameters shown in the previous section, simulation model of 3kW single-phase PV inverter system with grid connection is shown in Fig. 8. The inverter current waveform is shown in Fig.9 (a) and the grid currents using L-filter, LC-filter and LCL-filter are shown in Fig.9 (b), (c), (d) respectively. Fig 9 (e) ~ (h) are zoomed in from Fig.9 (a) ~ (d). Fig. 9 shows that with small inductance value, the L-filter cannot eliminate the current ripples completely. By using a capacitor as a shunt component, LC-filter and LCL-filter have good ripple eliminations.

Fig.10 (a), (b), (c) and (d) show the frequency analysis and THD values of the inverter current and grid currents using L-filter, LC-filter and LCL-filter, respectively. L-filter cannot reduce the harmonics as well as LC-filter and LCL-filter do.

V.   EXPERIMENTAL RESULTS

The overall system of the 3kW single-phase grid-connected PV inverter is shown in Fig. 11 [6].

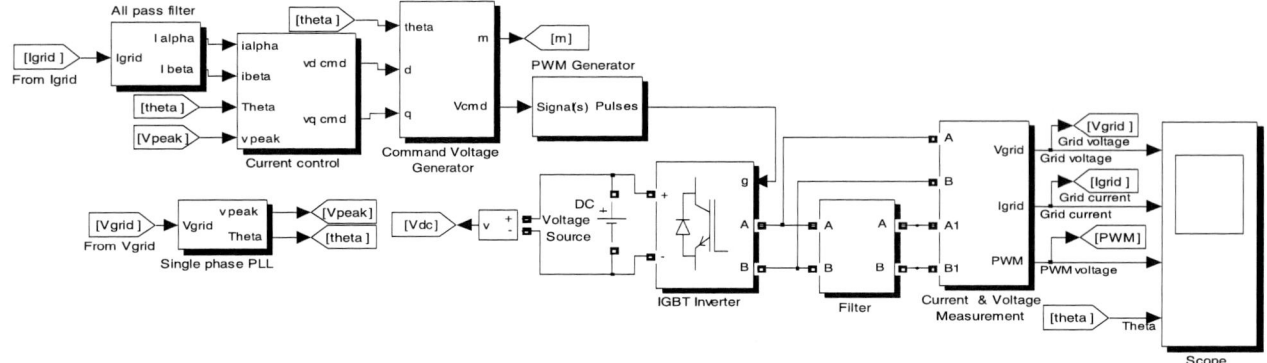

Fig. 8 Matlab/Simulink simulation model of 3kW PV inverter using low-pass filter

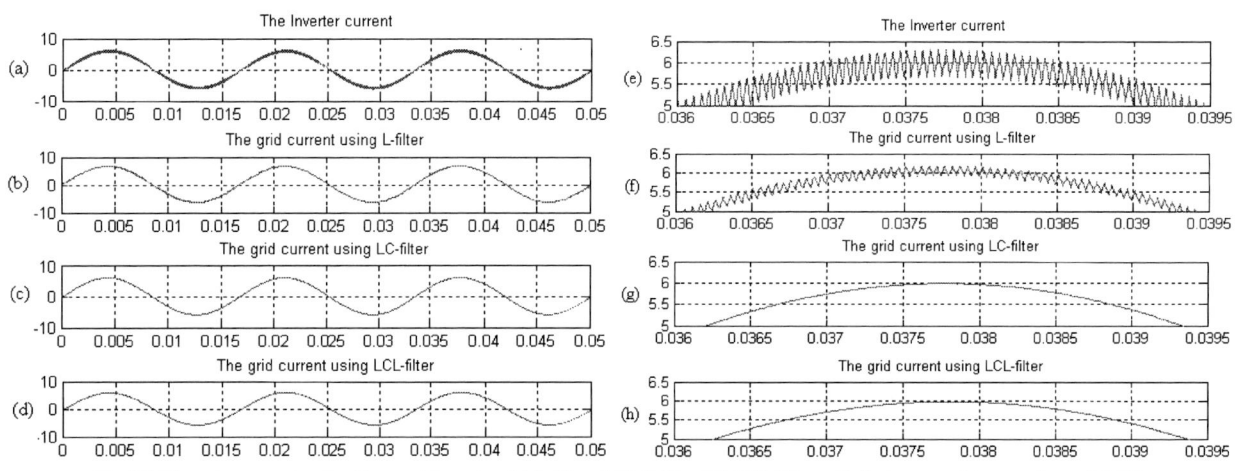

Fig. 9 (a) Inverter current and grid current by using (b) L-filter, (c) LC-filter (d) LCL-filter with their zoomed in waveforms (e)~(h)

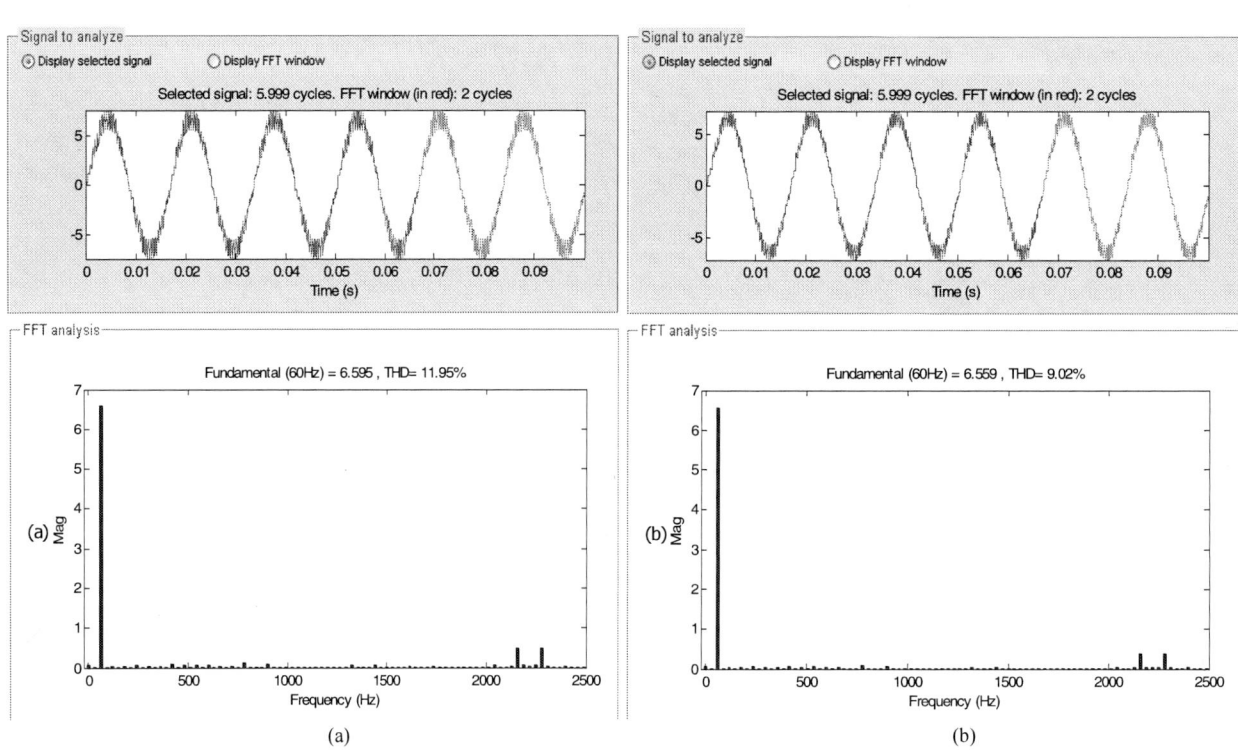

(a)   (b)

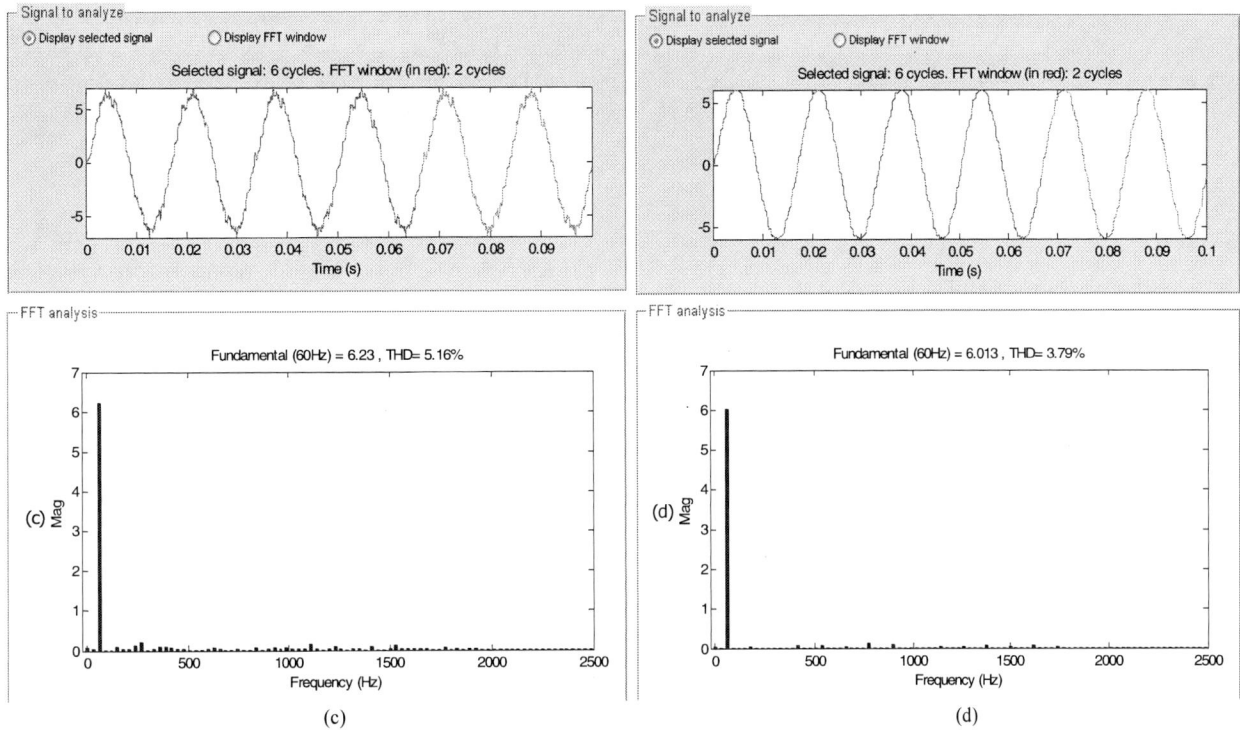

(c)   (d)

Fig. 10 Frequency analysis's and THDs of (a) inverter current, (b) grid current using L-filter, (c) grid current using LC-filter, (d) grid current using LCL-filter

Fig. 11 Experimental prototype of 3kW single-phase grid-connected PV inverter

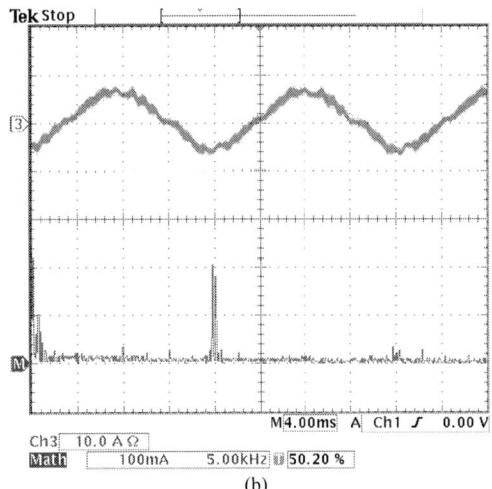

(b)

Fig. 12 (a) Inverter current and (b) grid current using L-filter

The current controller (using PI controller) is adopted fully in software with a 32-bit fixed-point DSP TMS320F2812 and the PWM pulses are generated through the internal pulse generator of the DSP. The experimental results are shown in Fig. 12 – Fig. 13.

In Fig. 12(a), the inverter current is shown in channel "2" and its frequency spectrum is shown in channel "M". The grid current using L-filter and its frequency spectrum are shown in Fig. 12(b) at channel "3" and channel "M", respectively. The grid current using L-filter almost has no ripple and harmonic mitigation.

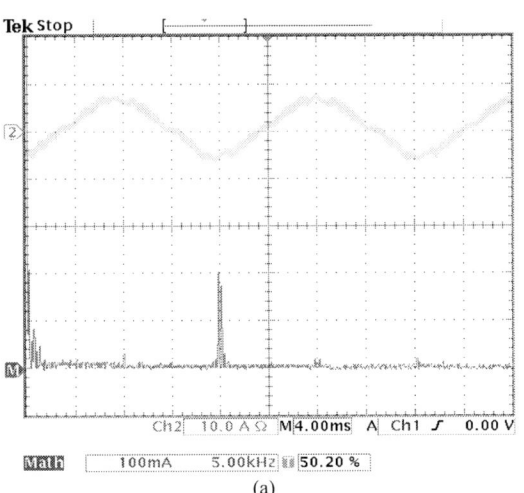

(a)

978-1-4244-4782-4/10 $26.00 © 2010 IEEE   1664

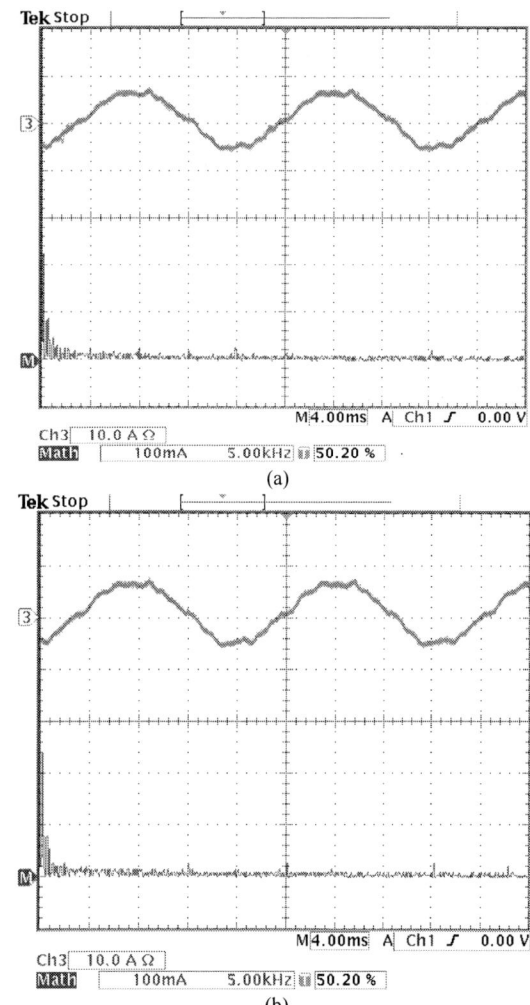

Fig. 13 Grid current using (a) LC-filter and (b) LCL-filter

others in case of the same inductance, capacitance and damping resistance values. Furthermore, the used methodology in this paper can be extended to three-phase and applied to such distributed generation application as micro-turbine, fuel cell, wind power, etc.

REFERENCES

[1] E-Habrouk M., Darwish M.K., Mehta P., "Active power filters: a review", IE Proceedings-Electric Power Applications, Vol. 147, Iss. 5, pp. 403-413, 2000.

[2] Akagi H., "Active harmonic filters", Proceedings of the IEEE, Vol. 93, Iss. 12, pp. 2128-2141, 2005.

[3] Marco Liserre, Frede Blaabjerg, Steffan Hansen, "Design and Control of an LCL-Filter-Based Three-Phase Active Rectifier", IEEE Transactions on Industry Application, Vol. 41, No. 5, pp. 1281-1291, 2005.

[4] Hyosung Kim, Kyoung-Hwan Kim, "Filter design for grid connected PV inverters", IEEE International Conference on Sustainable Energy Technologies (ICSET2008), pp. 1070-1075, 2008.

[5] Y.Lang, D.Xu, Hadianamrei S.R, H.Ma, "A novel design method of LCL type utility interface for three-phase voltage source rectifier", Proc. of PESC'05, pp. 313-317, 2005.

[6]Cha, Hanju and Lee, Sanghoey, "Design and Implementation of Photovoltaic Power Conditioning System Using a Current Based Maximum Power Point Tracking", Industry Application Society Annual Meeting (IAS2008), pp. 1-5, 2008.

The grid currents using LC-filter and LCL-filter are shown in Fig. 13(a) and Fig. 13(b), respectively. The grid currents, as shown in channel "3", has a good current ripple reduction and channel "M" shows the harmonic. By using the LCL-filter, the grid current ripple elimination and the harmonic mitigation are better than LC-filter. The damping resistor has no effect on the ripple reduction but the oscillation mitigation.

VI. CONCLUSIONS

In this paper, a LCL-filter design process for single-phase grid-connected PV inverter has been described. The filter design considers on the constraints of parameter determinations and the overall process can be done by step-by-step equations solving. The design procedure ensures that the parasitic of the filter components are kept as low as possible. And a feature comparison of LCL-filter with L-filter and LC-filter has been investigated and its result shows that the LCL-filter has a better effective performance than the

Design and Development of Generation-I Silicon based Solid State Transformer

Subhashish Bhattacharya, Tiefu Zhao, Gangyao Wang, Sumit Dutta, Seunghun Baek, Yu Du,
Babak Parkhideh, Xiaohu Zhou, Alex Q. Huang
FREEDM Systems Center, Department of Electrical and Computer Engineering
North Carolina State University, Raleigh, NC

Abstract—**The Solid State Transformer (SST) is one of the key elements proposed in the National Science Foundation (NSF) Generation-III Engineering Research Center (ERC) "Future Renewable Electric Energy Delivery and Management" (FREEDM) Systems Center. The SST is used to enable active management of distributed renewable energy resources, energy storage devices and loads. In this paper, the Generation-I SST single-phase 20kVA, based on 6.5kV Si-IGBT is proposed for interface with 12kV distribution system voltage. The SST system design parameters, overall system efficiency, high frequency transformer design, dual active bridge converter, auxiliary power supply and gate drives are investigated. Design considerations and experimental results of the prototype SST are reported.**

Index Terms--**Solid state transformer, FREEDM system**

I. INTRODUCTION

The Solid State Transformer (SST) is one of the key elements in the proposed Future Renewable Electric Energy Delivery and Management (FREEDM) Systems. In the electric configuration of the FREEDM system shown in Fig.1, low voltage (120V) residential class Distributed Renewable Energy Resource (DRER), Distributed Energy Storage Device (DESD), and loads are connected to the distribution bus (12kV) through a power electronics based Intelligent Energy Management (IEM) subsystem.

The solid state transformer is within the IEM and used to enable active management of DRER, DESD and loads, rather than a 60Hz conventional transformer. The SST has the features of instantaneous voltage regulation, voltage sag compensation, fault isolation, power factor correction, harmonic isolation and DC output [1-3]. The SST will have a 400V DC port that will facilitate more efficient connection of certain classes of DRERs and DESDs. Acting very much like an energy router, each SST will have bi-directional energy flow control capability allowing it to control active and reactive power flow and to manage the fault currents on both the low voltage and high voltage sides. Its large control bandwidth provides the plug-and-play feature for distributed resources to rapidly identify and respond to changes in the system.

In order to direct interface with the 12kV distribution voltage level, series devices or multilevel converter modules

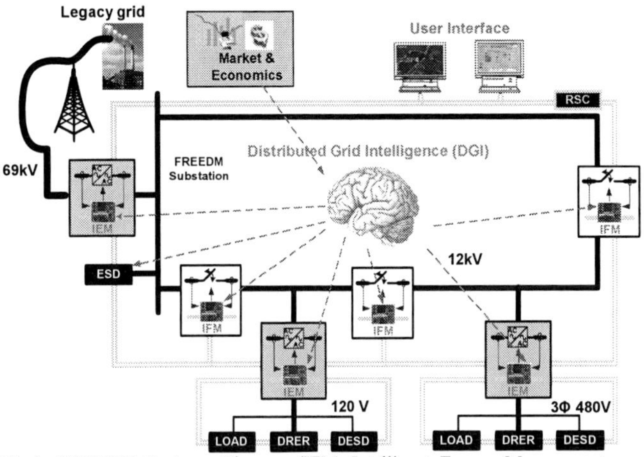

This work was supported by ERC Program of the National Science Foundation under Award Number EEC-08212121.

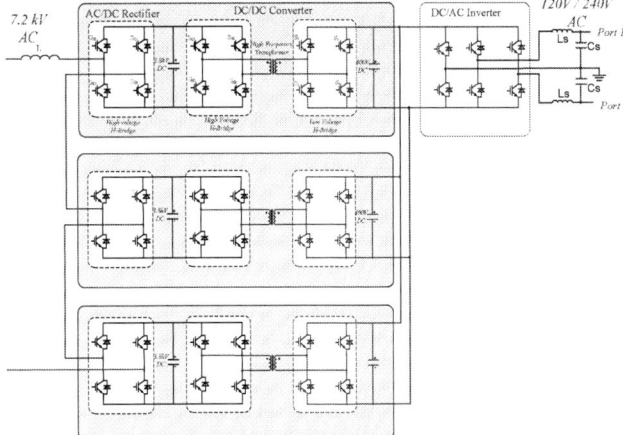

Fig.1. FREEDM Systems Diagram (IEM: Intelligent Energy Management, IFM: Intelligent Fault Management, DRER: Distributed Renewable Energy Resource, DESD: Distributed Energy Storage Device)

are still required due to today's semiconductor voltage level (6.5kV for silicon device and 10kV for SiC MOSFET).

The SST converts the voltage from AC to AC for step-up or step-down with the function same as the conventional transformer. However, the traditional 60Hz transformer is replaced by a high frequency transformer to provide isolation and step up/down function plus the power electronics converters, which are key to achieve size and weight reduction and the power quality improvement [1-7].

Fig. 2 Topology of proposed 20kVA Solid State Transformer

The basic configuration of the proposed 20kVA SST interfaced to 12kV distribution voltage, with center-tapped 120V single-phase output is shown in Fig.2. The SST is rated as single phase input voltage 7.2kV, 60Hz, output voltage

240/120V, 60Hz, single-phase/3 wires. The solid state transformer consists of three stages, an AC/DC rectifier, a Dual Active Bridge converter (DAB) with a high frequency transformer and a DC/AC inverter. The SST consists of a cascaded high voltage high frequency AC/DC rectifier that converts 60Hz, 7.2 kV AC to three 3.8 kV DC buses, three high voltage high frequency DC-DC bi-directional converters that convert 3.8 kV to 400V DC bus and a voltage source inverter (VSI) that inverts 400V DC to 60Hz, 240/120 V, single-phase/3 wires. The switching devices in high voltage H-bridge and low voltage H-bridges in Fig. 2 are 6.5kV silicon IGBT and 600V silicon IGBT respectively. The switching frequency of the high voltage silicon IGBT devices is 1 kHz, and the low voltage IGBT in the VSI switches is 15 kHz. The 20 kVA SST unit is envisioned as a building block of IEM and also for construction of a larger rated and three-phase 100kVA SST.

The switching device for high voltage side is a newly packaged 6.5kV, 25A H-bridge IGBT module as shown in Fig.3, while for the low voltage side is commercially available 600V IGBTs are used.

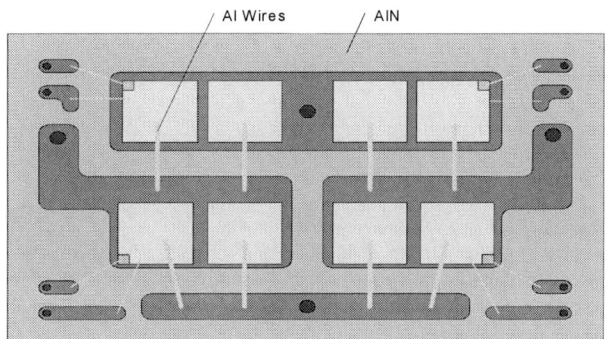

Fig. 3. 6.5kV 25A H-Bridge IGBT module layout

II. SST SYSTEM PARAMETERS

This section details the design of the SST components. The size of dc bus capacitor depends on the magnitude of ripple voltage and capacitor current. With charge and discharge balance, we have

$$I_{a-rms}\Delta t = C_d \Delta V_{dc}$$

$$\Delta t = \frac{1}{2} * \frac{1}{120} s, \Delta V_{dc} = 10\% V_{dc} = 380V, I_a = 2.8A$$

$$Cd = 30uF$$

When determining the AC line inductance, considerations are needed for AC current harmonics, maximum ripple and inrush current [8]. The current THD can be obtained with the spectral analysis results with a given switching frequency. It is given by

$$THD = \frac{\sqrt{\sum(\frac{u_i}{\omega_i L})^2}}{I_1}$$

Where, u_i is the amplitude of the harmonic voltage with order i, ω_i is the harmonic frequency, L is the harmonic inductor and I_i is the fundamental current.

Another consideration for the line inductance design is the instantaneous switching ripple. In this 20kVA SST design, the rated load current is only 2.8A rms and therefore, the ripple current design consideration becomes very significant. It should be suppressed to a reasonable level to guarantee the control feasibility and the proper operation of the switching device. The current ripple will vary as the mains input voltage varies over a fundamental cycle. The inductor design is based on the peak current ripple given as:

$$\frac{di_a}{dt} = \left|\frac{v_{in} - V_{dc}}{L_f}\right|$$

$$L_f = D_m \frac{|V_m - V_{dc}|}{2 f_{sw} \Delta i_a}$$

$$L = 0.89 * \frac{(3.8k*3 - 7.2k\sqrt{2})}{2*(3*1k)*(4*10\%)} = 0.23H$$, "3" accounts for the three cascaded H-bridges.

III. SST EFFICIENCY ESTIMATION

The SST system overall efficiency can be estimated by calculating the losses of each stage. Table 1 gives the losses of each component at 20kW rated power rating. As Fig. 2 shows, six 6.5kV/25A H-Bridge IGBT modules are used as high voltage switching devices. The following equations give the switching energy loss and diode recovery energy loss of the IGBT module [9]:

$$Eisw(Ic) = 4.25 \cdot 10^{-7} \cdot Ic^2 + 5.312 \cdot 10^{-4} \cdot Ic + 0.0282 \quad \dots (1)$$

$$Ediode\_rec(Ic) = -2.625 \cdot 10^{-7} \cdot Ic^2 + 4.475 \cdot 10^{-4} \cdot Ic + 0.03575 \cdots (2)$$

For the rectifier, when the switching frequency is 1080Hz, AC input is 60Hz, there are 9 switching cycles over half input voltage sine cycle, and the half cycle energy loss is the summation of these nine switching energy losses and diode recovery energy losses at different collect-emitter current plus the integration of the conduction energy loss. Then the total power loss equals half cycle energy loss multiplied by 2 times frequency, as given in Table 1.

TABLE 1 COMPONENT LOSS

PARTS		Loss Per Device(W)	No. of Device	Total Loss(W)
Rectifier Stage (1080Hz)	input inductor	75	2	578.4
	IGBT	15.3	12	
	Diode	20.4	12	
DAB Stage (3kHz)	high voltage side IGBT	14.1	12	566.7
	high voltage side Diode	0.2	12	
	Transformer	58.1	3	
	low voltage side IGBT	3.2	12	
	low voltage side Diode	15.2	12	
Inverter Stage (15kHz)	IGBT	86.2	4	535.0
	Diode	10.0	4	
	Output Inductor	75.1	2	

For the soft-switching DAB stage, the switching frequency

is 3 kHz, and the transformer leakage inductance is 61uH. As shown in Fig. 4, ZVS can be achieved for both primary and secondary IGBTs, so turn-on loss of the IGBTs will be almost zero, and the parallel diodes reverse recovery energy loss will also be almost zero. Therefore, the losses considered here are IGBT turn-off loss, IGBT conduction loss and diode conduction loss.

For the inverter stage, the outputs are two 120Vac voltages with 180 degree phase shift, if the load is balanced, there will no current through neutral line, so when calculating the losses, only four IGBTs are included. The losses for control circuit and cooling fans are estimated to be around 150W, and therefore, the total loss will be 1830W. Consequently, the SST overall efficiency is 91.62%.

IV. DUAL ACTIVE BRIDGE ANALYSIS

The Dual Active Bridge (DAB) is operated under soft switching to minimize the turn on losses [10]. The current and the voltage waveform through the transformer are shown in Fig. 4. In Fig. 4, the DAB power flow is from primary to secondary under ZVS. The transformer magnetizing inductance is neglected.

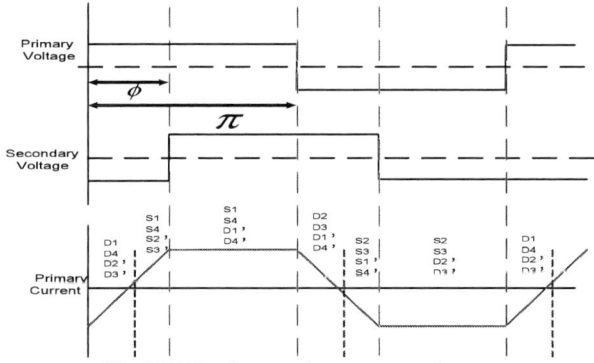

Fig 4 DAB voltage and current waveforms

The current waveform is given by the following equation:

$$i_p(\theta) = \frac{\left[v_p(\theta) - v_s'(\theta)\right]}{\omega L}(\theta - \theta_i) + i_p(\theta_i)$$

Using this equation, we enforce ZVS condition for the input and output bridges:

$$i_p(0) \le 0 \qquad (for\ input\ bridge)$$
$$i_p(\phi) \ge 0 \qquad (for\ output\ bridge)$$

And, we get the following constraints:

$$d \le \frac{1}{1 - \frac{2\phi}{\pi}} \qquad for\ leading\ input\ bridge$$

$$d \ge \frac{\pi - 2\phi}{\pi} \qquad for\ lagging\ output\ bridge$$

Where d is given by

$$d = \frac{V_o'}{V_p}$$

If we keep $d = 1$ by choosing the transformer turns ratio = 9.5 (input dc voltage of 3.8kV and output dc voltage of 400V), the

ZVS condition for the input and the output bridge is well satisfied for all values of ϕ.

A. Influence of magnetizing inductance:

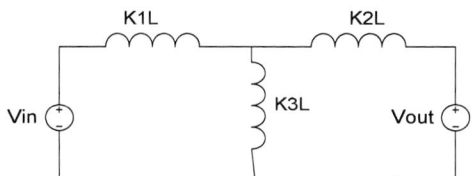

Fig. 5. DAB Transformer Model

We have considered the T model for accurately modeling the transformer. K_1, K_2, K_3 are the fractions of leakage inductance that appears on the primary, secondary and the magnetizing branch. From this circuit, we get the following equations:

$$\frac{di_p}{d\theta} = C_1 \frac{V_{in}}{\omega L} - C_2 \frac{V_s'}{\omega L} \quad C_1 = \frac{\frac{1}{K_2} + \frac{1}{K_3}}{1 + \frac{K_1}{K_2} + \frac{K_1}{K_3}} \quad C_2 = \frac{\frac{1}{K_2}}{1 + \frac{K_1}{K_2} + \frac{K_1}{K_3}}$$

$$\frac{di_s}{d\theta} = C_3 \frac{V_{in}}{\omega L} - C_4 \frac{V_s'}{\omega L} \quad C_3 = \frac{\frac{1}{K_1}}{1 + \frac{K_2}{K_1} + \frac{K_2}{K_3}} \quad C_4 = \frac{\frac{1}{K_1} + \frac{1}{K_3}}{1 + \frac{K_2}{K_1} + \frac{K_2}{K_3}}$$

Using this and enforcing the ZVS condition we get the following constraints:

1. $d \le \frac{C_1}{C_2}\left(\frac{\pi}{\pi - 2\phi}\right)$ (for leading input bridge)

2. $d \ge \frac{C_3}{C_4}\left(1 - \frac{2\phi}{\pi}\right)$ (for lagging output bridge)

TABLE II INFLUENCE OF MAGNETIZING INDUCTANCE

Leakage Inductance (mH)	Switching Frequency (Hz)	$\phi_{MIN\ for\ ZVS}$ (degrees)	Power transferred at ZVS (kW)
60	3000	5.143	0.978
50	3000	5.78	1.32
40	3000	6.428	1.83
30	3000	7.04	2.7

Hence we see that reducing the leakage inductance, or effectively increasing the magnetizing inductance to the leakage inductance ratio, forces the DAB out of ZVS. This is justified from the fact that if the magnetizing inductance is smaller, the primary side will see a more lagging load and hence ZVS will be easily achieved.

B. Influence of device snubber capacitance

From the device datasheet of the high voltage Si-IGBT device [9] we can calculate the emitter to collector capacitance of the device. This capacitance determines the minimum current required for ZVS on the primary side. From the datasheet, we have $C_{CE} = 36$ pF. During the dead band period t_m the inductor current is divided amongst the two capacitors.

$$i_p = 2C \frac{dv_c}{dt}$$

978-1-4244-4782-4/10 $26.00 © 2010 IEEE

These capacitors and the transformer leakage inductance forms a resonant tank that forces the current to zero and pushes the capacitor voltages to the positive and negative dc bus voltage rails. Assuming a lossless circuit, and performing energy balance, we have:

$$E_{inductor} = \frac{1}{2}LI_{min}^2$$

$$E_{delivered} = \int_0^{t_m} V_o' i_p dt = 2CV_o' \int_0^{t_m} dv_c = 2CV_o'V_i \quad (enforcing\ ZVS)$$

Equating the two expressions we get the minimum current for ZVS:

$$I_{min} = 2\sqrt{\frac{C}{L}\left(V_o'V_i\right)} = .187\,A$$

V. DAB TRANSFORMER DESIGN

The high frequency transformer in the DC-DC stage in solid state transformer plays an important role for the performance and overall efficiency of the SST system. In order to draw the best performance, in addition to the consideration about size, efficiency and cost as traditional transformer design, there are two significant issues to design the High-Voltage and High-Frequency (HVHF) transformer for SST. First of all, the leakage inductances of the transformer in a dual active bridge (DAB) converter play a crucial role as an element to determine the amount of power transfer between primary and secondary side. The leakage and magnetizing inductance can be adjusted by geometry and winding method so the optimized specification is chosen to reduce the size and show the best performance. Second, different insulation strategy is introduced to reduce the electric stress to avoid air breakdown between parts because this HVHF transformer is designed as oil-free type. There are countless possible cases of transformer design so the best selection for the material and specification is proposed and the summary of the optimized design and comparison is discussed on the basis of simulation and experimental results.

A. Core material selection

In general, there are several types of magnetic core material considered for the high frequency transformer, such as nanocrystalline and amorphous alloy. Silicon steel, one of the most common materials for use in soft magnetic applications, is not considered due to high power loss at high frequency, even though they have a high saturation flux density. In case of amorphous alloy, the cost is lower than nanocrystalline cores and the performance/cost factor is excellent in terms of the frequency range for the SST transformer. In addition, cut C-cores with large geometry and I-cores are commercially available, which provide design flexibilities and air gaps can be added to prevent core saturation due to DC bias current from dual active bridges. Based on these advantages and the designed switching frequency of 3 kHz for DAB possible with Silicon 6.5kV HV-IGBTs, amorphous alloy 2605SA1 is chosen for each of the three 7kVA HVHF transformer for 20kVA SST application shown in Fig. 2. For higher DAB switching frequencies, nanocrystalline and ferrite cores will be more suitable from the core loss considerations.

B. Bac optimization

Saturation flux density and core loss is one of the key trade-offs to design transformers. High saturation flux density can reduce size, weight and the number of turns, but it also leads to high eddy current loss in cores. The optimum Bac occurs when the derivative of the total loss is zero. The winding loss equation was simplified for convenience by ignoring skin and proximity effect because the skin and proximity effect is not significant at the operating switching frequency of 3kHz for Generation-I 20kVA SST. The resulting loss comparison as a function of Bac is shown in Fig. 6.

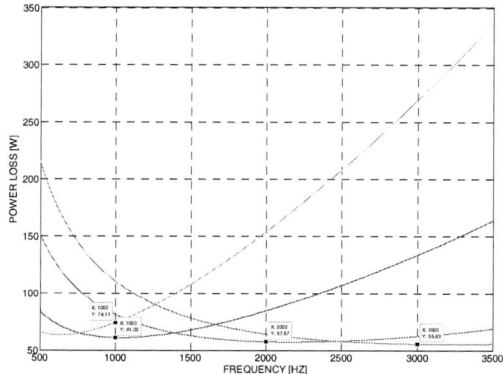

Fig. 6. Loss comparison at the optimal Bac: Bac 0.45 (blue), Bac 0.23 (green), Bac 0.16 (red), Bac 0.7 (light blue)

$$B_{ac}^{n+1} = \frac{Vin \cdot 10^4 \cdot 1000^m \cdot MLTp \cdot \rho_p \cdot I_p^2 + Vout \cdot 10^4 \cdot 1000^m \cdot MLTs \cdot \rho_s \cdot I_s^2}{4 \cdot A_c \cdot f^{m+1} \cdot k \cdot n \cdot total\ mass}$$

The transformer optimized design at the frequency of 1 kHz, 2 kHz and 3 kHz based on 2605SA1 C-core are shown in Table III. The temperature rise of the transformer is simulated with ePhysics module of Maxwell 3D FEA software based on the loss calculation results, as shown in Fig. 7.

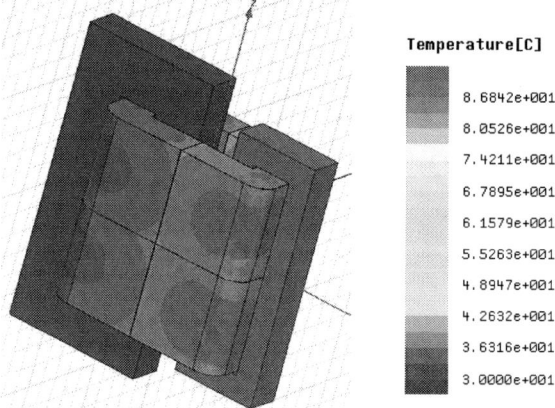

Fig. 7. FEA simulation results for temperature rise (with natural air cooling)

C. Leakage inductance estimation

The DAB DC-DC converter, in the solid state transformer requires a sufficient inductance to optimize the efficiency of dual active bridges and facilitate ZVS of the IGBT devices. The leakage inductance of the high voltage high frequency transformer is compared based on two types of winding layout. The coaxial winding structure in Fig. 8 (left) is the typical high voltage winding structure. The low voltage winding is in inner layer around the core and high voltage windings surround low voltage windings with insulation space. Based on Maxwell 3D FEA simulation, the leakage inductance with 0.1mm air gap is 10mH and the magnetizing inductance is 2.7H. This leakage inductance is not high enough, compared to the designed value. In case of the other design in Fig. 8 (right), the high and low voltage winding surround each leg. With this structure, the leakage inductance increases to 65mH. The relationship with magnetizing inductance and the thickness of air gap is shown in Fig. 9. It is important to mention that if air gap thickness is higher than 1mm, magnetizing current can be comparable to load current.

D. Insulation Consideration

One of the oil-free insulation strategies is proposed and shown in Fig 10. It adds a thin insulation layer, such as polypropylene film or plate, into the air gap between two C cores. The insulation layer mitigates the isolation requirement of primary winding to core. The air gap can also prevent DC-bias saturation of magnetic cores. Then the primary core is electrically connected to negative terminal of DC-link in each stage. The secondary core is electrically connected to output neutral point or the negative terminal of DC-link of its lower level cascaded stage. Thus, the common mode high voltage referred to ground is evenly distributed through primary winding, primary core, secondary core and secondary winding. Electric field distribution is simulated in Maxwell 3D to identify the high electric field position in transformer in Fig 11. High electric field occurs at the corner and edges area of air gap, which might potentially cause corona effect. These parts should be sealed with solid insulation potting material.

TABLE. III TRANSFORMER DESIGN TABLE

	1kHz	2kHz	3kHz
Bac	0.700	0.350	0.235
Cross-sectional Area, Ac	48.1 [cm^2]		
Window Area, Wa	42 [cm^2]		
Ap	2020[cm^4]		
Primary Wire (Rdc)	35/32, FEP insulation (15.8 [ohm/km])		
Secondary Wire (Rdc)	350/32 (1.58 [ohm/km])		
# of turns on HVS	280	280	280
# of turns LVS	30	30	30
Skin Depth [cm]	0.2386	0.1687	0.1378
Required leakage+external inductance (primary referred)[mH], phase shift	250, π/6	125, π/6	83, π/6
Winding Loss [W]	24.45	24.78	25.55
Core Loss [W]	49.68	42.36	39.07
Total loss [W]	74.13	67.15	64.62

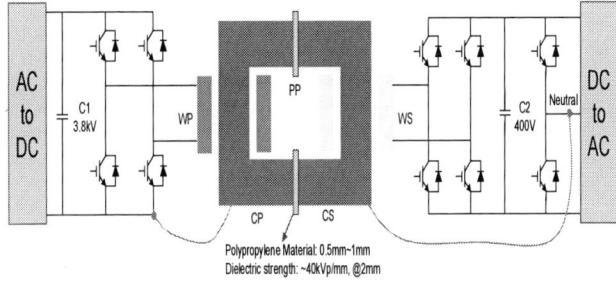

Fig. 10. Proposed oil-free insulation strategies

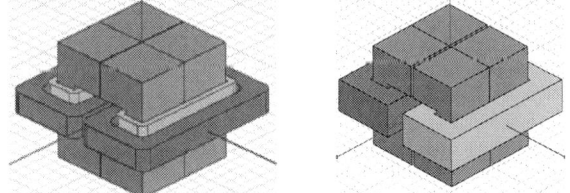

Fig. 8 Winding structures: Co-axial (left) and separate (right) windings

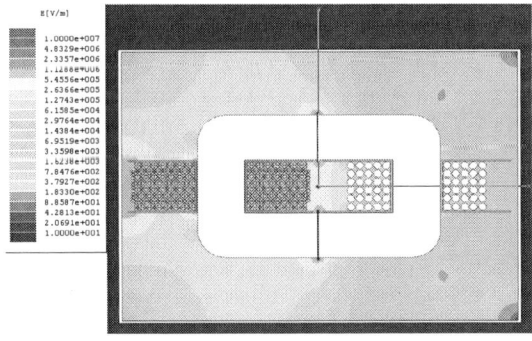

Fig.11. MAXWELL 3D electric field distribution for transformer

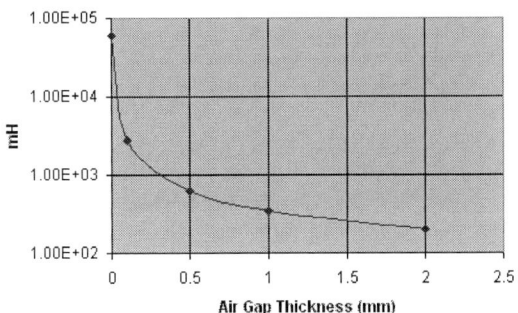

Fig. 9 The simulated magnetizing inductance (referred to high voltage side) with variation of air gap thickness

Fig. 12. MAXWELL 3D transient analysis with nonlinear B-H curve of amorphous alloy (2605SA1) - primary current (red), primary voltage (brown), secondary voltage (blue)

Fig. 12 shows MAXWELL 3D transient analysis of the designed transformer with non-linear B-H curve of amorphous alloy (2605SA1) core characteristics, operating under DAB circuit primary and secondary nominal voltage conditions.

VI. AUXILIARY POWER SUPPLY AND GATE DRIVES

The overview of the gate driver for high voltage side IGBTs is shown in Fig 12. The isolation between controls and power stage is obtained through optical fiber. This type of isolation, although more expensive than other alternatives, is necessary for 6.5kV voltage level. In this project, commercially available gate driver ICs are used, and to achieve higher power for the gate driver, two different discrete buffer circuits are designed.

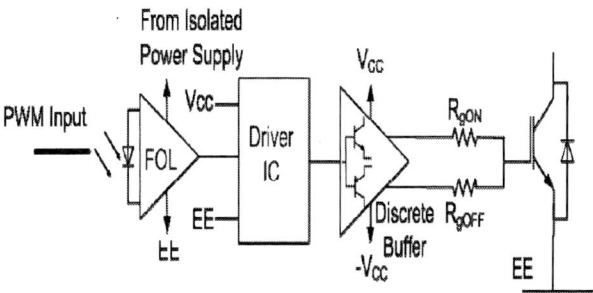

Fig.12. Overview of the gate driver circuit without protection

One of the main challenges is the power supply embodiment for the SST. This power unit should be designed to supply all the controls, gate drivers, and sensors, which are estimated to be around 100 W. The challenge is that the SST is to operate as a stand-alone unit. In other words, it does not have access to any external power supply. Based on the SST structure shown in 2, there are always two voltage supplies available at start-up of SST; one is the high voltage AC side, and the other one is the high voltage DC link.

Comparing the possible solutions of direct conversion, i.e. power electronics-based converter, linear regulator, supply frequency transformer, and passive divider, we have selected the power electronics-based converter option.

Power electronics-based converter option provides the required isolation in one or multiple stages. The major drawbacks for this option are the need for the HV switches and the fact that the current switches are optimized for low on-resistance. The latter point should be taken into account since the conventional power supplies are designed for much higher switching frequencies than that of the available HV IGBTs.

Fig. 13 depicts one of the proposed methods to supply and distribute the power among different SST components. Different isolation stages can be implemented with this method; however, as mentioned before, HV IGBTs are needed. The initial power for the converter is provided through linear regulator or passive divider.

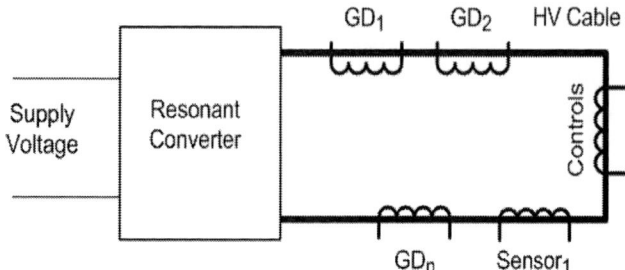

Fig. 13. A method to supply and distribute auxiliary power for SST

The other option is to provide power from the HV DC link for each stage. This design reduces the required isolation for the power supply for each stage. The considered converter topology is flyback for about 50 W/stage. The isolation can be implemented in multiple stages with other upfront converters; however, each level's isolation should be as high as > 12 kV. Dry type isolation transformers filled with potting compounds will be used for space concerns and reliability.

The general concerns about the selected option can be summarized as relatively low switching frequency operation of power supply, isolation implementation, spacing, initial start-up stage control like inrush current if the DC link voltages are used directly instead.

VII. INVERTER STAGE DESIGN

Among the voltage source inverter control methods, capacitor current feedback control method can achieve better voltage output with a low cost current transformer to sense the current from the output capacitor [11]. Thus capacitor current feedback control method is used as the inner current loop together with conventional PI controller as the outer voltage loop. The control block diagram is shown in Fig. 14, in three-leg topology there are two output filter capacitor feedback loops and each loop controls one bridge.

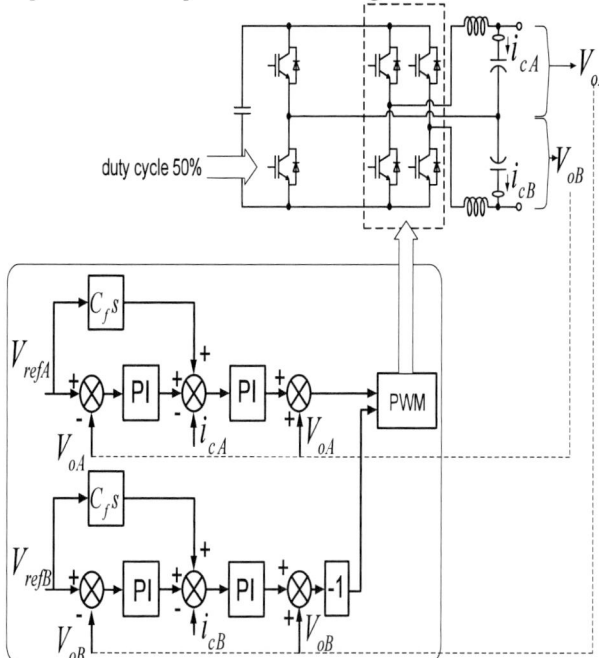

Fig.14. Output capacitor current control of the VSI

VIII. EXPERIMENTAL RESULT

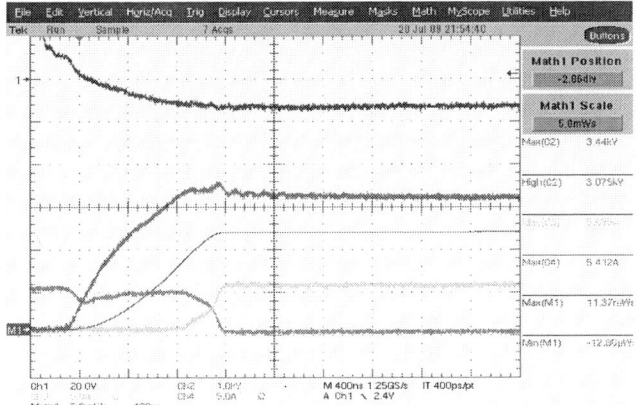

Fig. 15. 6.5kV Si-IGBT turn off at 5A with 25A packaged Si-IGBT die at 3kV dc voltage: Ch 1 (blue) - Vge (20V/div), Ch 2 (red) - Vce (1000V/div), Ch 4 (purple) - Ic (5A/div), Math 1 (blue) - Turn off loss (Eoff) = 11.37mJ

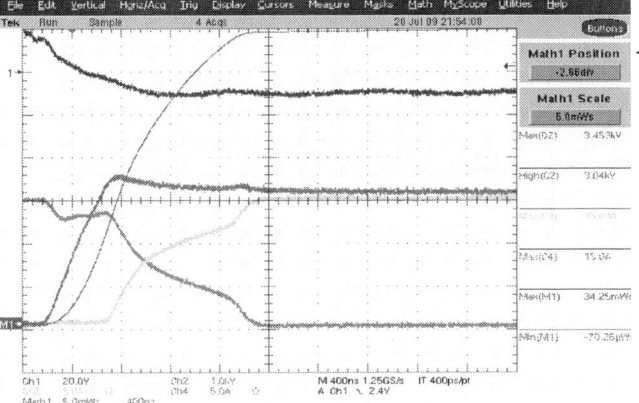

Fig. 16. 6.5kV Si-IGBT turn off at 15A with 25A packaged Si-IGBT die at 3kV dc voltage: Ch 1 (blue) - Vge (20V/div), Ch 2 (red) - Vce (1000V/div), Ch 4 (purple) - Ic (5A/div), Math 1 (blue) - Turn off loss (Eoff) = 34.25mJ

The 6.5kV, 25A packaged IGBT module (layout as shown in Fig.3) switching turn-off test results at 3kV dc voltage are shown in Fig. 15 and Fig. 16 for 5A and 15A respectively. These devices are used for the rectifier and HV-side of the DAB in the SST. The Eoff respective measurements of 11.37mJ and 34.25mJ are shown and provide a good estimate for the loss and efficiency estimation of the SST.

To verify the SST control strategy and performance, a scale-down SST 2kVA prototype is implemented as shown in Fig. 17 with the rectifier and DAB parameters as in Table IV.

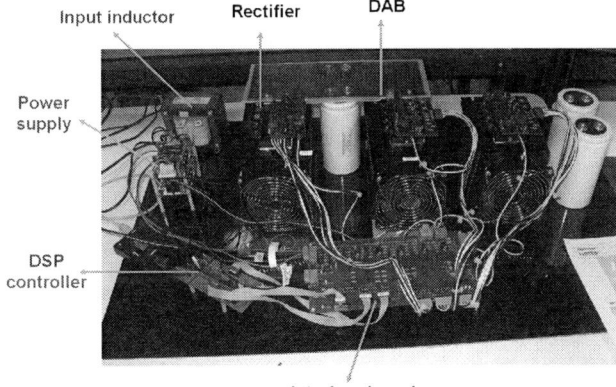

Fig. 18 Scale-down 2kVA solid state transformer prototype

Table IV. SST Prototype Parameters

Line inductance Ls	2.5mH
Primary DC Capacitor	900uF
Primary DC voltage reference	400V (designed)
Secondary DC Capacitor	900uF
Secondary DC voltage reference	400V (designed)
Transformer turns ratio	1:1
Transformer magnetizing inductance	27mH
Transformer leakage inductance	3mH
Rectifier Switching frequency	10kHz
DAB Switching frequency	4kHz
DC load	240W

TABLE V. SPECIFICATION OF SCALE DOWN 2kVA (1:1) TRANSFORMER

Core		Wire	
Core	AMCC250 (Metglass)	**Gauge**	AWG15
Ac	9.3 cm^2	**Diameter**	1.45 mm
Window Area	22.5cm^2	**Sectional area**	1.65 mm^2
Mass	209.5g	**# of turns**	78

Fig. 19. (Left): 7kVA High Voltage (3.8kV input and 400V output at 3kHz) and High Frequency (HVHF) Transformer; (Right-lower) scale-down SST prototype 2kVA transformer (400V input and 400V output upto 10kHz); (Right-upper) MAXWELL 3D of 2kVA transformer (400V:400V)

The SST module prototype is designed as single phase input voltage 60 Hz, 240V, and DC output 400V. It consists of an AC/DC rectifier, a DC/DC DAB converter that regulates the output dc voltage to 400V with 2kVA (1:1) transformer. The SST module 400V output will connect to an inverter with 120/240V AC output. The SST module prototype is implemented with 600V, 75A Intelligent Power Modules (IPM) and the controller is implemented in DSP TMS320F28335. The 2kVA (1:1) DAB transformer and MAXWELL 3D analysis is shown in Fig. 19. For the 20kVA SST with 7.2kV single-phase ac input (as shown in Fig. 2), each DAB stage 7kVA High Voltage (3.8kV input and 400V output at 3kHz) and High Frequency Transformer with Metglass (2605SA1) has been built and is being tested as shown in Fig 19. Table V gives parameters of 2kVA (1:1) high frequency transformer.

The single phase SST cascaded three stage rectifier d-q vector controller is implemented in DSP shown in Fig. 20. Fig. 21 shows experimental results of three stage cascaded rectifier with input ac voltage of 132V peak and DC voltage reference of 150V (50V per stage) at 240W load operating under unity displacement power factor. The dc bus voltage regulation of second H-bridge is shown with 120Hz voltage ripple present for single phase rectifier. The input ac voltage and current are in phase and the PWM rectifier seven-level voltage is shown and verify the single phase d-q vector control for the SST.

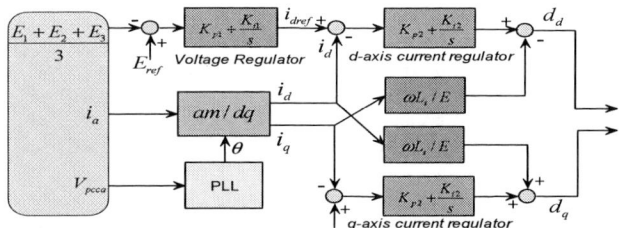

Fig. 20. Cascaded rectifier stage single phase d-q vector controller

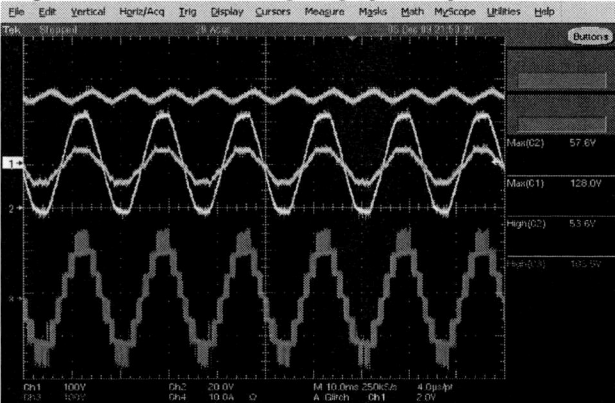

Fig. 21. SST rectifier stage results with input voltage peak 132V (93V rms): Ch 1 (yellow): Input ac voltage (100V/div), Ch 1 (blue): DC bus voltage of second H-bridge (20V/div), Ch 3 (purple): PWM rectifier voltage (100V/div), Ch 4 (green): Input ac current (10A/div)

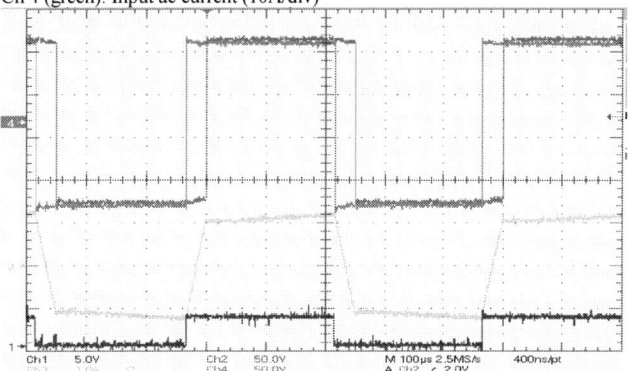

Fig. 21. DAB stage results with input and output dc voltage of 100V: Ch 1 (blue): Primary DAB gating signal (5V/div), Ch 2 (red): Primary DAB voltage (50V/div), Ch 3 (green): Primary DAB current (1A/div), Ch 4 (orange): Secondary DAB voltage (50V/div) 10A/div)

Fig. 21 shows the experimental results of the DAB stage with input primary side and output secondary side dc voltage of 100V and primary side current. The DAB is operated in open loop phase shift control with output dc voltage regulation and power flow from primary to secondary at switching frequency of 4kHz. Fig. 22 shows the inverter experimental results with 400V dc bus with regulated 240V ac output at 1.2kW with 15kHz switching frequency. The ac output voltage regulation is shown for both 240V and 120V output with the load current for 240V ac output. The two-level PWM voltage output is also shown with 400V dc bus voltage.

IX. CONCLUSION

The unique requirements for Generation-I SST single-phase 20kVA, based on 6.5kV Si-IGBT is proposed for interface with 12kV distribution system voltage and enable active power management of DRER, DESD and loads in the

Future Renewable Electric Energy Delivery and Management (FREEDM) System. The SST system design considerations and parameters, overall system efficiency, high frequency transformer design, dual active bridge converter, auxiliary power supply and gate drives are investigated and reported in this paper. Experimental results of the prototype SST are presented as verification for the SST sub-systems and to validate the SST controller performance.

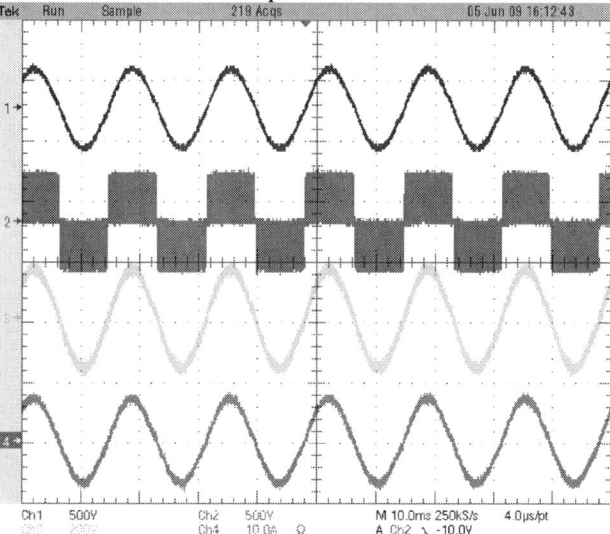

Fig. 22. Inverter operation with 240V ac output: Ch 1 (blue): Output ac voltage (500V/div), Ch 2 (red): Output PWM voltage (500V/div), Ch 3 (green): Output 120V ac voltage (200V/div), Ch 4 (purple): Load current for 240V output (10A/div)

X. REFERENCES

[1] J. L. Brooks, "Solid state transformer concept development," in *Naval Material Command. Port Hueneme*, CA: Civil Eng. Lab., Naval Construction Battalion Center, 1980.

[2] E. R. Ronan, S. D. Sudhoff, S. F. Glover and D. L. Galloway, "A Power Electronic-Based Distribution Transformer", *IEEE Transactions on Power Delivery*, vol. 17, pp. 537 - 543 , April 2002.

[3] Jih-Sheng Lai, A. Maitra, A. Mansoor and F. Goodman, "Multilevel Intelligent Universal Transformer for Medium Voltage Applications", *Conference Record of Industry Applications Conference, Fourtieth IAS Annual Meeting.* vol. 3, pp: 1893 – 1899, Oct, 2005.

[4] Tiefu Zhao, Liyu Yang, Jun Wang, Alex Q. Huang, "270 kVA Solid State Transformer Based on 10 kV SiC Power Devices," *IEEE Electric Ship Technologies Symposium*,21-23 May, 2007, pp.145 – 149

[5] D. Wang, C. Mao, J. Lu, S. Fan, and F. Peng, "Theory and application of distribution electronic power transformer, " *Electric Power System Research Journal*, vol.77, Issues 3-4, pp. 219 – 226, March 2007

[6] HQS Dang, Alan Watson, Jon Clare, etc al, "Advanced Integration of Multilevel Converters into Power System", *The 34th Annual Conference of IEEE Industrial Electronics, 2008*, pp.3188 – 3194.

[7] H. Iman-Eini, J. Schanen, S. Farhangi, J. Barbaroux, J. Keradec, "A power electronic based transformer for feeding sensitive loads," *IEEE Power Electronics Specialists Conference, 2008*, pp. 2549 – 2555

[8] Rixin Lai, F. Wang, Yunqing Pei, R. Burgos, B. Dushan, Minimizing Passive Components in High-frequency High-density AC Active Voltage Source Converters, IEEE Power Electronics Specialists Conference, 17-21 June 2007 pp. 672 - 677

[9] ABB HiPak IGBT 5SNA 0400J650100 module datasheet

[10] Mustansir H. Kheerlawala, Randal W. Gascoigne and Deepakraj M. Divan, "Performance Characterization of a High-Power Dual Active Bridge dc-to-dc Converter "

[11] M.J. Ryan, W.E. Brumsickle, and R.D. Lorenz, "Control topology options for single-phase UPS inverters," IEEE Trans. Ind. Appl., vol. 33, pp. 493–501, Mar./Apr. 1997.

978-1-4244-4782-4/10 $26.00 © 2010 IEEE

Power Calculation Method Used In Wireless Parallel Inverters Under Nonlinear Load Conditions

Zheng Ren[1], Mingzhi Gao[1], Qiong Mo[1], Kun Liu[2], Wei Yao[1], Min Chen[1], Zhaomin Qian[1]

[1]Collage of Electrical Engineering, Zhengjiang University, Hangzhou, Zhejiang, illgg@zju.edu.cn
[2]Collage of Science, Zhengjiang University, Hangzhou, Zhejiang, kunliu@zju.edu.cn

Abstract—**Droop control method is widely used in wireless parallel inverters control strategy. In this application, inverters output active power and reactive power need to be calculated accurately for control of the angular frequency and amplitude of output voltage. Traditional calculation methods have a slow and oscillating transient response and could be easily impacted by disturbance and variation of load. To overcome these limitations, this paper suggests an more suitable method of power calculation to increase calculating speed and improve the dynamic response especially under nonlinear load conditions.**

I. INTRODUCTION (HEADING 1)

Distributed power generation and Smart Grid are the future trend due to the ability to accommodate variety of renewable, alternative energy sources, the potential to improve the energy efficiency and power system capability, and the promise for power reliability and security [1]. Paralleling technology of inverters is developed to obtain N+I redundant power and to create a modular power distribution system [2]. The basic control objective in parallel inverters is to achieve accurate power sharing while maintaining close regulation of the output voltage magnitude and frequency [3]. Droop Control is widely used in wireless parallel inverters to adjust the frequency and amplitude of output voltage for each inverter [4-6].The control precision and the accuracy of power calculation are closely coupled. Traditional power calculation methods use DQ pattern or multiplies output voltage and current signals, then power value can be obtained through low-pass-filter (LPF) [7-9]. The LPF limits the transient response of calculating, especially in non-linear load conditions, because the output current is worsely distorted and the results of calculating have lots of harmonics. Based on theoretical analyses and experimental comparison of the three main methods of power calculation, this paper presents one fast and accurate calculating method which is very suitable for nonlinear load conditions. The whole calculating

process of it could be accomplished within the time of $1/2f$ s in all conditions, which is usually about 0.01s with 50/60Hz line input frequency. This method occupies much smaller DSP space because of the absence of sine tables and LPF, and it could cut down calculation greatly.

The "f" is the output voltage frequency. This method use simple multiplication and iterative operation to make out the result in high speed and the program can be simplified. For the result is the average power only, there is no need of LPF and the impact of it could be established.

II. COMPARING OF THREE POWER CALCULATION METHODS

A. Shortage of Two Traditional Power Calculation Methods

The output current has a large number of odd harmonics when the load is nonlinear, equation (1) shows the fourier transform of it.

$$i = I_{pk1}\sin(\omega t + \theta_1) + I_{pk3}\sin(3\omega t + \theta_3) + I_{pk5}\sin(5\omega t + \theta_5) + \cdots$$

$$(1)$$

Form Fig. 1, the instantaneous active power can be obtained by (3):

$$p = u \cdot i = u(i_1 + i_3 + i_5 + \cdots)$$
$$= U\sin\omega t \cdot \left[I_{pk}\sin(\omega t + \theta_1) + I_{pk3}\sin(3\omega t + \theta_3) + I_{pk5}\sin(5\omega t + \theta_5) + \cdots\right]$$
$$= U_{rms}\{I_{rms1}[\cos\theta_1 - \cos(2\omega t + \theta_1)] + I_{rms3}[\cos(2\omega t + \theta_3) - \cos(4\omega t + \theta_3)]$$
$$+ I_{rms5}[\cos(4\omega t + \theta_5) - \cos(6\omega t + \theta_5)] + \cdots\}$$

$$(2)$$

$$p = U_{rms}I_{rms1}\cos\theta_1 + P'_{pk2}\cos(2\omega t + \theta'_2) + P'_{pk4}\cos(2\omega t + \theta'_4) + \cdots$$

$$(3)$$

It can be seen that the calculating result has DC component and even harmonics components, and LPF is needed to cancel

---

This paper is supported by the National Natural Science Foundation of China under award number 50907061

the harmonic parts. LPF could slow down the system's dynamic response. Equation (4) shows the principle of PQ method:

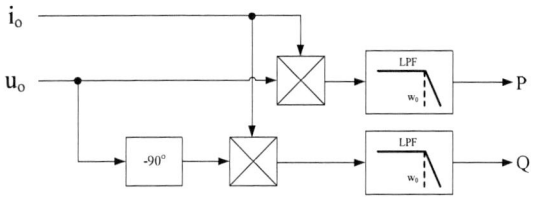

Figure 1 Common Calculation Method

$$\begin{bmatrix} p \\ q \end{bmatrix} = \frac{1}{2} \cdot \begin{bmatrix} v_\alpha\, v_\beta \\ v_\beta\, {-}v_\alpha \end{bmatrix} \begin{bmatrix} i_\alpha \\ i_\beta \end{bmatrix} = \begin{bmatrix} \bar{p} \\ \bar{q} \end{bmatrix} + \begin{bmatrix} \tilde{p} \\ \tilde{q} \end{bmatrix}$$

(4)

Where, $v\alpha$ and $i\alpha$ are output voltage and output current. $v\beta$ and $I\beta$ have a phase delay of $\pi/2$ from $v\alpha$ and $i\alpha$. Substituting (1) into (4), we have the expression of active power as

$$
\begin{aligned}
p &= \frac{1}{2}(v_\alpha i_{1\alpha} + v_\beta i_{1\beta}) + \frac{1}{2}(v_\alpha i_{3\alpha} + v_\beta i_{3\beta}) + \frac{1}{2}(v_\alpha i_{5\alpha} + v_\beta i_{5\beta}) + \cdots \\
&= \frac{1}{2}U_{pk}\left\{ \sin\omega t \cdot I_{pk1}\sin(\omega t + \theta_1) + \sin\omega(t - \frac{\pi}{2\omega})\cdot I_{pk1}\sin\left[\omega(t - \frac{\pi}{2\omega}) + \theta_1\right]\right\} \\
&\quad + \frac{1}{2}U_{pk}\left\{ \sin\omega t \cdot I_{pk3}\sin(3\omega t + \theta_3) + \sin\omega(t - \frac{\pi}{2\omega})\cdot I_{pk3}\sin\left[3\omega(t - \frac{\pi}{2\omega}) + \theta_3\right]\right\} \\
&\quad + \frac{1}{2}U_{pk}\left\{ \sin\omega t \cdot I_{pk5}\sin(5\omega t + \theta_5) + \sin\omega(t - \frac{\pi}{2\omega})\cdot I_{pk5}\sin\left[5\omega(t - \frac{\pi}{2\omega}) + \theta_5\right]\right\} + \cdots \\
&= \frac{1}{2}U_{pk}I_{pk1}\cos\theta_1 - \frac{1}{2}U_{pk}I_{pk3}\cos(4\omega t + \theta_3) + \frac{1}{2}U_{pk}I_{pk5}\cos(4\omega t + \theta_5) + \cdots
\end{aligned}
$$

(5)

Then p can be written as

$$p = U_{rms}I_{rms}\cos\theta_1 + P_{pk4}''\cos(4\omega t + \theta_4'') + P_{pk8}''\cos(8\omega t + \theta_8'')$$

(6)

Equ. (6) shows that the result of PQ method without LPF has lots of 4n times harmonics and it also needs an LPF to obtain the DC part. Therefore it also slows down dynamic response of the distributed power system with nonlinear load.

B. Principle of The Suitable Method of Power Calculation

According to the definition of active power, we have:

$$\bar{p} = \int_t^{t+T} u \cdot i \cdot dt$$

(7)

Where p is the average power value of inverter. Equ (8) is the discretization format of (7). Where Isn and Usn are the samp-ling current and voltage of DSP:

$$
\begin{aligned}
\bar{p}' &= f \cdot \left(I_{s1}U_{s1}T_s + I_{s2}U_{s2}T_s + I_{s3}U_{s3}T_s + \cdots + I_{sN}U_{sN}T_s \right) \\
&= fT_s\sum_{n=1}^{N} I_{sn}U_{sn} = \frac{f}{f_s}\sum_{n=1}^{N} I_{sn}U_{sn} \\
&= \frac{f}{f_s}\sum_{n=1}^{f_s/f} I_{sn}U_{sn}
\end{aligned}
$$

(8)

f is output voltage frequency and fs is sampling frequency of DSP. commonly, fs is much larger than f, as fs>>f, therefore the following equation is valid and output power can be written as

$$\bar{p} = \lim_{f_s \to \infty} \bar{p}'$$

(9)

This method could obtain the power value within 0.01s (about 1 2 f s), because only a half cycle signals need to be sampled. LPF is not needed at all as the computing result has DC component only. So the transient response of distribution system can be obviously improved. Fig.2 shows the flow cha-rt of the active power calculation program.

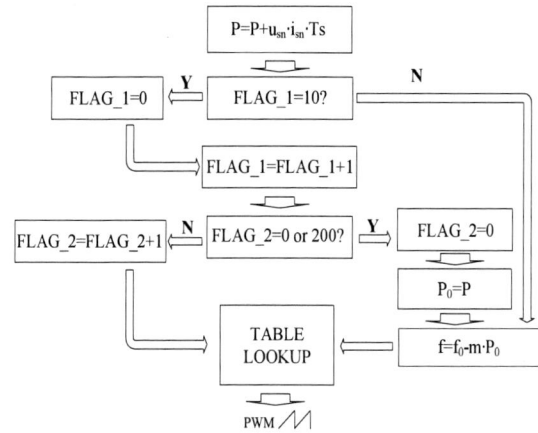

Figure 3 Flow Chart of Program

III. SIMULATION RESULTS

Fig.3 and Fig.4 show the FFT analysis of results of the first two methods. It can be seen that the simulation results are consistent with the theoretical analysis conclusions. So LPF is needed for getting the power value and it slows down the response of distribution system with nonlinear load. Fig.5 shows the simulation result of the suitable power calculation method. In this program, the sampling frequency is 200 kHz and the switching frequency is 20 kHz.The value of active power comes out after 0.01s. Fig.6 shows the FFT analysis of computing result of this method without LPF and it can be seen that the result has DC part only.

978-1-4244-4782-4/10 $26.00 © 2010 IEEE

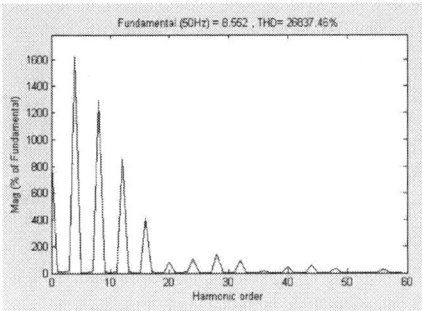

Figure 3 FFT Analysis of Computing
Result of Common Method Without LPF

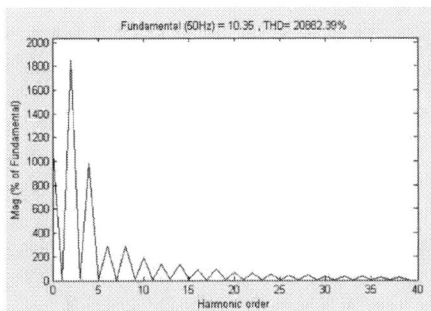

Figure 4 FFT Analysis of Computing
Result of PQ Method Without LPF

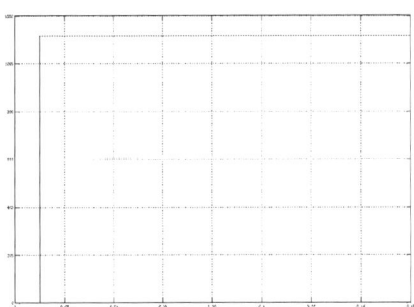

Figure 5 Computing Process of
The Improved Method

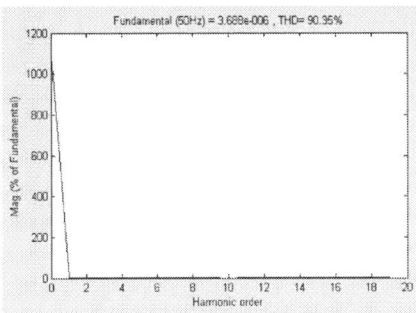

Figure 8 FFT Analysis of Computing
Results of Suitable Method Without LPF

IV. EXPERIMENTAL RESULTS

To compare computing speeds of the various methods of

power calculation, a DSP TMS320x2808 based single phase inverter platform with rectified load is used here and parameters are shown in Table 1. Fig.7 shows the experimental result with the DQ method which is illustrated as (2). It can be seen that the parallel system needs 5 to 6 periods to achieve load sharing. More detailed experimental results of the third method will be provided in the final paper.

Output Voltage (peak)	150V
Frequency	50Hz
Capacitance	470μ
Lod	13.3Ω

Table 1

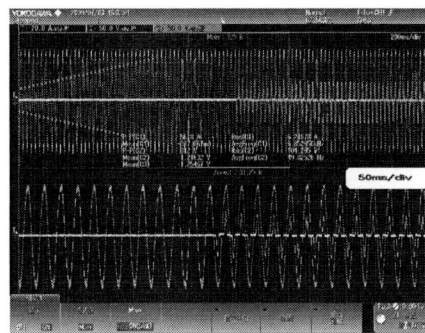

Figure 4 Experimental Results
of DQ Method

V. CONCLUTIONS

In this paper, three main power calculation methods have been analyzcd and compared to find a suitable one for droop control under nonlinear load conditions. By comparing experimental results of these three methods, it can be seen that the method this paper presented can effectively ameliorate the transient response with rectifier load. For this method does not need division calculation and LPF, it can give full play to the advantages of DSP operations to reduce the instructions to prepare. And its response time is very short even the load is nonlinear, so it could make droop control adjust frequency and output voltage fast on the basis of precise calculation which the other two methods can not do. There is another advantage of this method is that the computing time will be reduced by the increment of output frequency. It can be seen that this method is very suitable for droop control with nonlinear load.

REFERENCES

[1] A.F. Zobaa and C. Cecati' , "A comprehensive review on distributed power generation," *SPEEDAM 2006. International Symposium* on 23-26 May 2006 pp (s):514 - 518

[2] Chen Liangliang, Xiao Lan, Yan Yangguang, "A Novel Parallel Inverter System Based on Coupled Inductors," *Telecommunications Energy Conference, 2003. INTELEC '03. The 25th International,* 19-23 Oct. 2003, pp: 46 – 50.

[3] Yasser Abdel-Rady Ibrahim Mohamed and Ehab F. El-Saadany, "Adaptive Decentralized Droop Controller to Preserve Power Sharing Stability of Paralleled Inverters in Distributed Generation Microgrids," *IEEE Transactions on Power Electronics*, Vol. 23,

Issue 6, Nov. 2008, pp: 2806 – 2816.

[4] K. De Brabandere, B. Bolsens, J. Van den Keybus, A. Woyte, J. Driesen, R. Belmans and K.U. Leuven, "A voltage and frequency droop control method for parallel inverters," *Power Electronics Specialists Conference, 2004. PESC 04. 2004 IEEE 35th Annual,* Vol. 4, 2004 pp: 2501 - 2507

[5] Po-Tai Cheng, Chien-An Chen, Tzung-Lin Le and Shen-Yuan Kuo, "A Cooperative Unbalance Compensation Method for Distributed Generation Interface Converters," *in IEEE 2007 Industry Applications Conference,* 23-27 Sept. 2007 pp:1567 – 1573

[6] J.M. Guerrero, L.G. de Vicuna, J. Matas, M. Castilla and J. Miret, "A wireless controller to enhance dynamic performance of parallel inverters in distributed generation systems," *IEEE Transactions on Power Electronics,* Vol. 19, Issue 5, Sept. 2004 pp :1205 – 1213

[7] S.A. Oliveira da Silva, R. Novochadlo and R.A. Modesto, "Single-phase PLL structure using modified p-q theory for utility connected systems," *in IEEE 2008 Power Electronics Specialists Conference,* 15-19 June 2008 pp :4706 - 4711

[8] Hongliang Wang, Xiumei Yue, Xuejun Pei and Yong Kang, "A new method of power calculation based on parallel inverters," ***in IEEE 2009*** *Power Electronics and Motion Control Conference,* 17-20 May 2009 pp :1573 - 1576

[9] Wei Yu, Dehong Xu and Kuian Ma, "A Novel Accurate Active and Reactive Power Calculation Method for Paralleled UPS System," *in IEEE 2009 Applied Power Electronics Conference and Exposition,* 15-19 Feb. 2009 pp :1269 - 1275

A Real-Time Fault Diagnosis System for UPS Based on FFT Frequency Analysis

Won-Sul Shim,   Gi-Taek Kim
Dept. of Electronics & Communication Eng.
Kangwon National University
Chunchon, Korea
gikim@kangwon.ac.kr

Ha-Jin Jung,   Deuk-Soo Kim
Powertron Engineering Co. Ltd.
Seoul, Korea
dskim@powertron.co.kr

Abstract— **UPS provides emergency power when utility power is not available, so the reliability of UPS is more important than inverter drive systems. In this paper, a fault diagnosis system for UPS is proposed using FFT frequency analysis of output current of inverter side of UPS under linear and nonlinear load conditions. Software PLL for precise synchronization of one period sampling and double buffer memory for real time processing are proposed. Experimental results show the increase of even harmonics including dc offset in case of fault conditions such as increase of resistance and delay or misfiring of IGBT turn-on, and prove the possibility of UPS fault diagnosis system if the criteria for fault decision are well defined.**

I.   INTRODUCTION

An Uninterruptible Power Supply(UPS) provides emergency power and line regulation as well to connected equipment by supplying power using battery source when utility power is not available and, depending on the topology, line regulation as well to connected equipment by supplying power from a separate. UPS can be used to provide uninterrupted power to critical equipment until an auxiliary power supply can be turned on, utility power restored, or equipment safely shut down. While not limited to safeguarding any particular type of equipment, a UPS is typically used to protect computers, data centers, telecommunication equipment or other electrical equipment where an unexpected power disruption could cause injuries, fatalities, serious business disruption or data loss[1-3].

As business area of IT field increased, IDC center has more and more servers on a large scale and essential UPS still needs more bulky capacity to supply power for server. Therefore, the importance of reliability increases. In contrast, technology development speed of UPS is slow and there are problems of maintenance, administration, etc. General fault of average 12 - 20% happens in communication power supply equipment, and interrupts fatal damage happens. In order to improve the reliability, the fault diagnosis techniques for inverter drive systems have been studied recently, using ac current space vector method and dc current method[4-6], but

there are few reports for UPS because UPS has same inverter in output stage. UPS is used in case of power failure, so the reliability of UPS is more important than inverter systems. Generally UPS is used for linear loads as well as the nonlinear load for computer power supplies consisting of transformer, rectifier and capacitor filter loads. This causes the gross current waveforms dominated by switched mode power supply units (SMPS), which makes more difficult to detect faults.

Most of UPS use IGBT and MOSFET as the switching devices in inverter stage. IGBT is rugged, but it suffers degradation due to excess electrical and thermal stress [7-9]. The IGBT's are commonly used in the low and medium power static converters. In its discrete or module form, it has become a device of the highest importance in the high power converters. However, the phenomenon of ageing of the IGBT's has formed the subject of only few investigations essentially due to the complexity of their internal structure and the multiplicity of involved phenomena. Some recent and interesting studies have been published on this problem [7-8].

The most common faults in the inverters are those occurred in the IGBTs. IGBT and MOSFET faults can be broadly categorized as increase of resistance and delay, ringing, or misfiring of turn-on[10-11]. Among the techniques used to detect these faults, more of them are based on mathematical transform such as Fourier Fast Transform (FFT). The advantage of these techniques is that they can locate the device damaged with only analysis of a measured variable. In this paper, a fault diagnosis system for UPS is proposed based on the comparison of even harmonics using real time FFT analysis.

II.   DESCRIPTION OF SYSTEM

A typical UPS system is shown in Fig. 1, which consists of output stage of inverter and LC filter bank for supplying AC power to load, and the input stage of rectifier for charging battery storage bank. Normally UPS supplies the utility AC

This work was supported by the Korean Ministry of Small and Medium Business Administration.

power to load directly, and when the incoming utility voltage falls below a predetermined level the UPS turns on its internal inverter circuitry, which is powered from an internal storage battery. In some case another type of UPS can be employed, which is operating all the time without direct connection between input AC line and output load, in order to supply AC voltage free of over and under voltage and voltage sag or spikes of AC utility line, as shown in Fig. 2.

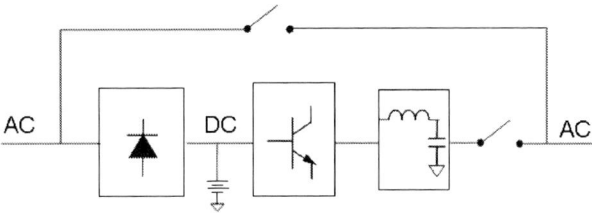

Figure 1   Typical UPS system

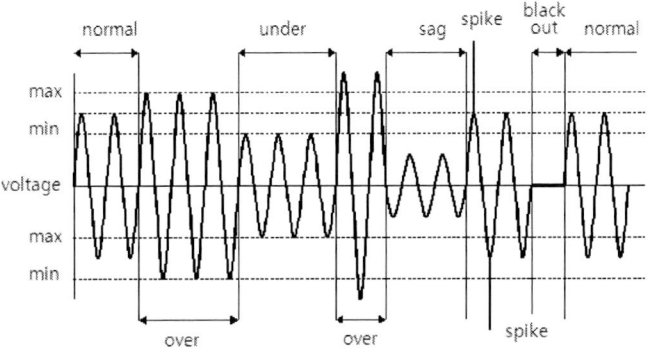

Figure 2   Under/over voltage, sag, and spike

UPS is used for linear loads as well as the nonlinear load for computer power supplies consisting of transformer, rectifier and capacitor filter loads, as shown in Fig. 3.

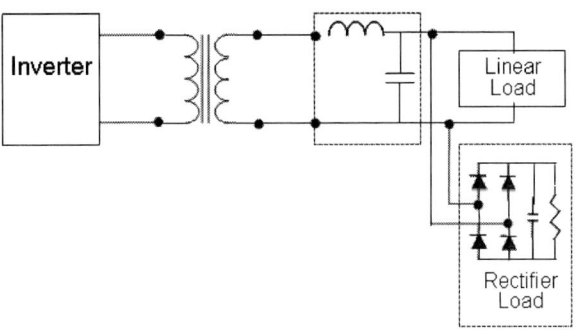

Figure 3   Linear and nonlinear load of UPS

The block-diagram and prototype of the proposed fault diagnosis system are shown in Fig. 4 and Fig. 5, respectively. In the output current after filter, the characteristics caused by IGBT fault may disappear, so output current of inverter before filter is inputted directly to the DSP processor (TI

TMS320F28335) using current sensor and A/D converter (Analog Device AD7328). The DSP is a 32bit floating point processor with throughputs of 150MIPS and 300MFlops on 150MHz clock, and has on chip 512KB Flash ROM, 68KB RAM, and I/O such as communication port and timers. A/D converter has 13bit resolution with 1M samples per second conversion speed.

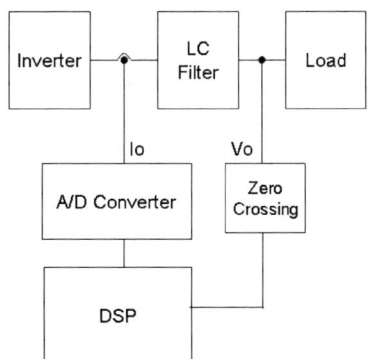

Figure 4   UPS fault diagnosis system

Figure 5   Prototype of the system

III.   PROPOSED SOFTWARE ALGORITHM

In order to obtain accurate sampling timing for one period, the zero crossing signal of output voltage is also inputted to DSP processor. The waveform of current has high frequency harmonics due to switching as well as the characteristics of fault. For the broad bandwidth of current waveform and the radix-2 FFT algorithm, the sampling rate should be high enough and the number of samples over one period be an integer power of 2 ($N=2n$). Considering the switching frequency(10kHz) and FFT processing time of DSP processor, $N=2,048(=2^{11})$ samples for one period(60Hz) is selected, so the sampling rate is $2,048 \times 60 = 122,880$Hz. In order to synchronize the sampling rate with the supply frequency, the software PLL(Phase Locked Loop) is used, as shown in Fig. 6. The phase difference between zero crossing of supply voltage and the output of frequency divider of sampling frequency by 2,048 is fed to PI controller to determine the timing of timer interrupt service routine (ISR).

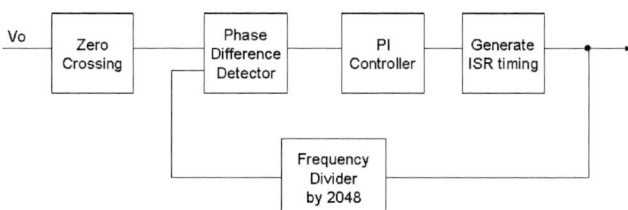

Figure 6   Software PLL for synchronization

Double buffer memory system for A/D conversion signals is employed in order to implement the real time system, as shown in Fig. 7. Firstly, A/D sampled signals of output current in the ISR routine, synchronized and initiated by software PLL, are stored to Buffer 1. After finishing one period of sampling, the flag is set and Buffer1 is switched to Buffer 2. Then, sampled signals are stored to Buffer 2 in ISR routine and FFT and diagnosis algorithm are executed in main program as a background job. In every period flag is set and reset, so buffers are switched and algorithm is executed without loss of one period.

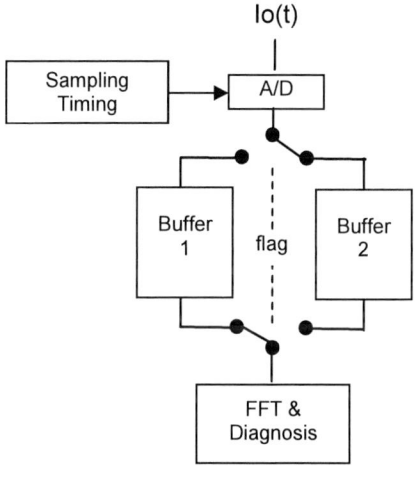

Figure 7   Double buffer system

The pseudo code of C language style of the interrupt routine is shown in Fig. 8. In every ISR routine the A/D sample is stored in buffer in a bit reversal order for Decimation In Time (DIT) FFT algorithm. After reaching 2,048 samples, timing information is stored for PLL and a flag is set or reset to indicate one period of samples is completed. In main program, shown in Fig. 9, run as a background job, if the flag is changed, PLL algorithm is executed with the timing of the zero crossing of supply voltage, and the FFT is then executed on the buffer of samples. The execution time of FFT algorithm is about 10msec, less than one period, and about one hundred times faster than DFT. UPS diagnosis is then performed based on FFT data.

```
x=ad_conversion();
i=i+1;
k=bit_reversal(i);
buffer[k]=x;
if(i>2048) {
    phase_sample=timer;
    flag ^= 1;
    i=0;
}
```

Figure 8   Interrupt service routine

```
if(flag != flag0) {
    phase_vo=zero_crossing();
    phase_dif=phase_vo-phase_sample;
    period=PI_picontroller(phase_dif);
    X=fft(buffer);
    UPS_diagnosis(X);
    Flag0 = flag;
}
```

Figure 9.   Main program

The FFT, which is a fast algorithm for calculation of the Discrete Fourier transform (DFT), transforms time domain data into frequency domain data. It makes possible to frequency domain analysis of power systems, such as the measurement and analysis of power system harmonics and control system. DFT formula is shown in eq. 1.

$$X[k] = \sum_{n=0}^{N-1} x[n]\, W_N^{kn} \qquad k = 0, 1, \ldots, N-1 \tag{1}$$

$$W_N^{kn} = e^{-j2\pi kn/N}$$

$$X[k] = \sum_{n=0}^{N-1} [x[n] Re(W_N^{kn}) + jx[n] Im(W_N^{kn})]$$

$$= \sum_{n=0}^{N-1} [x[n]\cos(\frac{2\pi kn}{N}) - jx[n]\cos(\frac{2\pi kn}{N})]$$

For DFT calculation N times of complex multiplication and (N-1) times of complex addition, which is NN times of real multiplication are needed. The radix-2 family of the FFT algorithm using the symmetry shown in eq. 2, is the most dominant form of all FFT application. They can reduce the NN calculation to $(N/2)\log_2 N$, and N has to be the power of 2, that is 2^M. So the sampling time must meet the requirement to be 2^M and synchronized to one cycle period. They are the most compact and efficient algorithms for calculating the DFT. Most of the other algorithms are not as compact and require more computation time and soon became dormant in the fast developing world of the FFT. The whole world of the FFT revolves around the radix-2 family. Cumbersome techniques are usually introduced to adapt the radix-2 FFT to all applications.The DIT algorithm for 8-point FFT is illustrated in Fig. 10, and program code is shown in Fig. 11.

978-1-4244-4782-4/10 $26.00 © 2010 IEEE   1680

$$W_N^{k[N-n]} = W_N^{-kn} = (W_N^{kn})^*$$

$$W_N^{kn} = W_N^{k(n+N)} = W_N^{(k+N)n} \qquad (2)$$

$$X[N-k] = X[k]^*$$

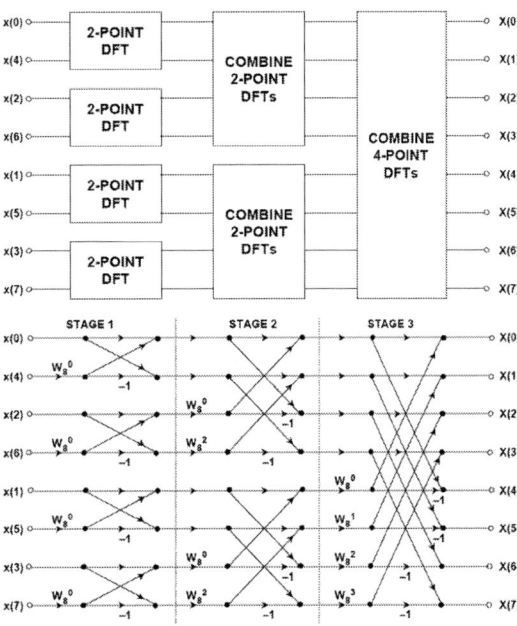

Figure 10 8-point DIT algorithm

```
radix2fft(x, w)
{
    n = length(x);
    t = log2(n);
    for(q=1;q<t;q++) {
        L = 2^q; r = n/L; L2 = L/2;
        for(k=0;k<r;k++) {
            for(j=0;j<L2;j++) {
                temp=w(L2-1+j+1)*x(k*L+j+L2+1);
                x(k*L+j+L2+1)=x(k*L+j+1)-temp;
                x(k*L+j+1)   =x(k*L+j+1)+temp;
            }
        }
    }
}
```

Figure 11   FFT routine using DIT algorithm

IV. EXPERIMENTAL RESULTS

Experiment of UPS diagnosis with current sensor are carried out, as shown in Fig. 12. Fig. 13 shows the performance of software PLL. In this case of 60Hz AC system the error of frequency is 0.3Hz, that is initially about 10usec period error is bounded in less than 2usec.

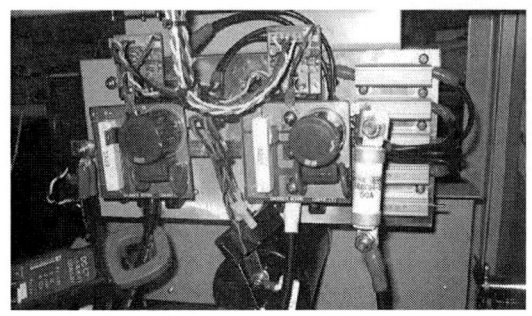

Figure 12.   Experiment with current sensor

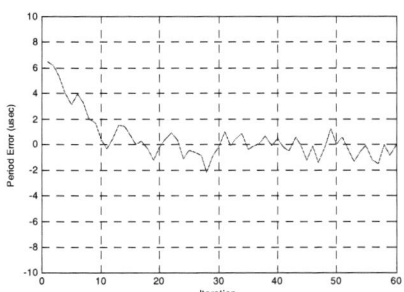

Figure 13   Period error of software PLL

In normal operation, even in the nonlinear rectifier load condition, the even harmonics including dc component are very small, but under fault conditions, the even harmonics become larger, so that fault condition can be detected.

Experiments are carried out in case one leg of single phase IGBT inverter of UPS is degraded. Summation of even harmonics is carried out up to 5kHz, half of switching frequency. Fig. 14 and Fig. 15 show waveforms and frequency spectrums normalized to fundamental component under linear and rectifier load conditions, respectively. One can see that there are very small even harmonics although there are odd harmonics in rectifier load. Fig. 16 and Fig. 17 show the waveforms and spectrums in case of increase of resistance and delay or misfiring of turn on, respectively. One can see the increase of dc offset and summation of even harmonics in these cases, as shown in Table 1.

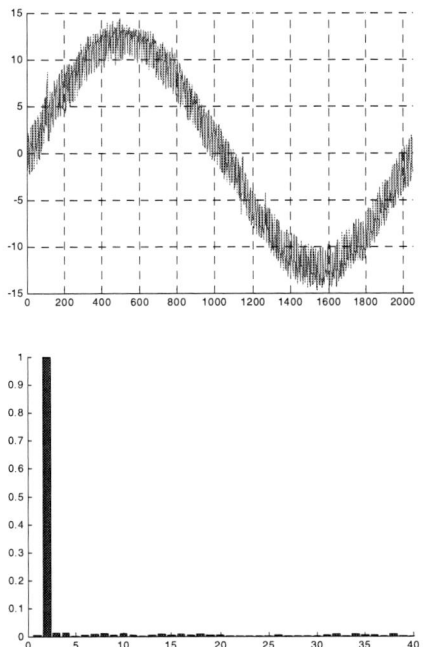

Figure 14 Waveform and FFT of normal(linear) load case

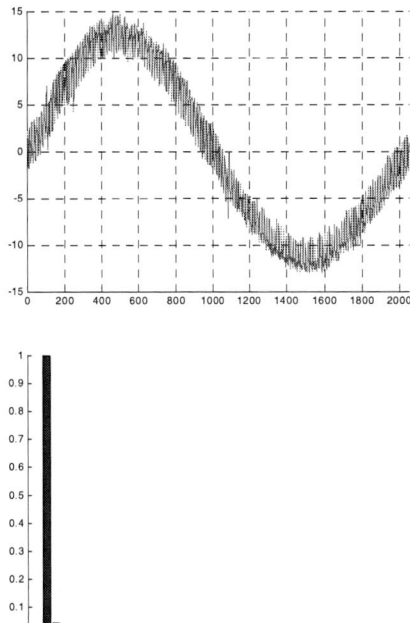

Figure 16 Waveform and FFT of turn-on resistance increase case

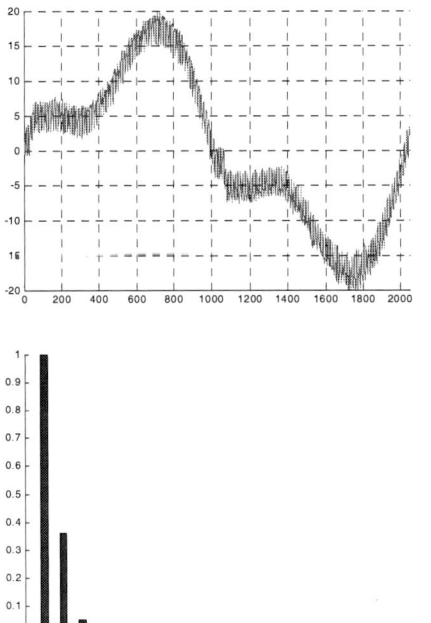

Figure 15   Waveform and FFT of rectifier(nonlinear) load case

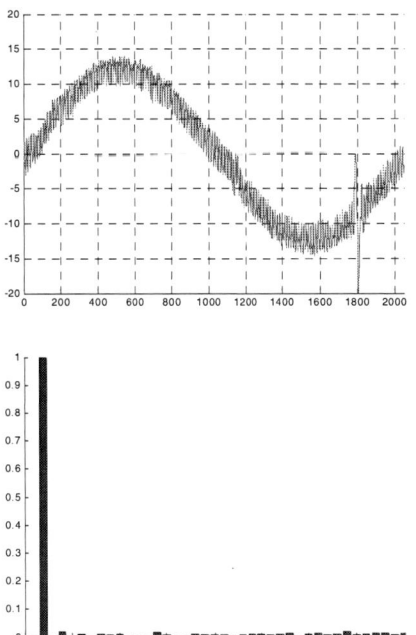

Figure 17. Wave form and FFT of turn-on delay and misfiring case

TABLE 1. Comparison of harmonics in cormal and fault condition

case	dc offset(%)	odd harmonics(%)	even harmonics(%)
Normal (linear load)	0.1	15.2	5.3
Normal (rectifier load)	0.1	56.6	5.6
turn-on resistance increase	2.2	15.9	18.8
turn-on dealy, misfiring	0.2	47.6	40.2

V. Conclusions

In this paper, the fault diagnosis system for UPS is proposed using FFT frequency analysis of output current of inverter side of UPS under the linear and nonlinear load conditions. Software PLL for precise synchronization of one period sampling and double buffer memory for real time processing are proposed. The system consists of DSP processor and A/D converter, and provides real time and very accurate frequency analysis using synchronized 2,048 samples to supply voltage with software PLL and radix-2 FFT algorithm executed within one period. Experimental results show the increase of even harmonics including dc offset in case of fault conditions, and prove the possibility of UPS fault diagnosis system if the criteria for fault decision are well defined.

References

[1] bowler, P., "UPS specifications & performance". Uninterruptible Power Supplies, IEE Colloquium on 1994, pp. 1/1 - 1/13.

[2] Heng Deng O. S., "High-performance control of a UPS inverter through iterative learning based on zero-phase filtering, "Industrial Electronics Society, 2004. IECON 2004. 30th Annual Conference of IEEE, pp. 1845-1850.

[3] Skok S.M., Vrkic N., "Electrical performance test procedure for uninterruptible power supplies," Industrial Technology, 2004, IEEE ICIT '04. 2004 IEEE International Conference, pp. 667-671.

[4] C. Kral and K. Kafka, "Power electronics monitoring for a controlled voltage source inverter drive with induction machines," IEEE Power Electron Specialists Conf., vol. 1, 2000, pp.213-217.

[5] M. A. Rodriguez, A. Claudio, D. Theilliol and L. G. Vela, "A new fault detection technique for IGBT based on gate voltage monitoring," IEEE Power Electronics Specialists Conf., 2007, pp.1001-1005.

[6] A. Ginart et. al., "Modeling aging effects of IGBTs in power drives by ringing characterization," International Conference on Prognostics and Health Management, 2008, pp.1-7.

[7] C. O. Maiga et. al., "Behaviour of punch-through and non-punch-through Insulated gate bipolar transistors under high temperature gate bias stress," IEEE International Symposium on Industrial Electronics, 2004, pp. 1035-1040.

[8] L. Dupont et. al, "Ageing Test Results of low voltage MOSFET Modules for electrical vehicles," European Conference on Power Electronics and Applications, 2007, pp. 1-10.

[9] A. Bernieri, G. Betta, C. De capua, A.pietrosanto, "Fault diagnosis and prediction on uninterruptible power systems," Instrumentation and Measurement Technology Conference, 1993. IMTC/93 Conference Record, IEEE, pp. 570 - 573.

[10] Bin Lu and Santosh Sharma, "A survey of IGBT fault diagnostic methods for three-phase power inverters," 2008 International Conference on Condition Monitoring and Diagnosis, 2008, pp. 756-763.

[11] Lu.I.D, Lee.P, "Use of mixed radix FFT in electric power system studies," Power Deliver, IEEE Transactions on 1994, pp.1276-1280.

Control Strategy for a Buck-boost Type Direct Interface Converter Using an Indirect Matrix Converter with an Active Snubber

Koji Kato
Nagaoka University of Technology
1603-1 Kamitomioka-cho Nagaoka City Niigata, Japan
katok@stn.nagaokaut.ac.jp

Jun-ichi Itoh
Nagaoka University of Technology
1603-1 Kamitomioka-cho Nagaoka City Niigata, Japan
itoh@vos.nagaokaut.ac.jp

Abstract— **This paper proposes a novel control method for an interface converter that is using an indirect matrix converter (IMC) with a DC power supply. The proposed converter is constructed based on an indirect matrix converter and a converter that combined the function of a boost type converter and a step-down type converter. The proposed converter connects an active snubber to the DC link part of IMC along with a boost type converter. The voltage transfer ratio from the AC power sources to the load can be improved by this active snubber. In addition, the proposed control method able to control the power distribution ratio and the output voltage at a same time. The basic operation of the proposed method has confirmed by simulation and experimental results.**

I. INTRODUCTION

Recently, renewable energies and hybrid electric vehicle (HEV) systems are receiving significant interest, with considerations of global warming and environmental problems. There are two types of power sources for renewable energies; AC power sources such as a wind turbine and DC power sources such as batteries, and fuel cells. In these systems, interface power converters have been intensely studied. A conventional power converter system, which consists of a pulse width modulation (PWM) rectifier, a DC/DC converter and an inverter, requires an electrolytic capacitor. However, this electrolytic capacitor in the conventional system is facing with several subjects, such as size, life time and costs. Particularly this large electrolytic capacitor has a problem in HEV applications because a power converter is installed at a high temperature area. Therefore, it is difficult to use electrolytic capacitors for HEV applications in terms of the lifetime that is affecting by the high temperature.

On the other hand, there is an AC/AC direct converter without a large energy buffer such as a matrix converter, which uses nine bidirectional switches, and an IMC, which consists of a current source rectifier and a voltage source inverter without a capacitor in the DC link part[1]-[8]. The utilization of these AC/AC direct converter in renewable energy systems can achieve the following advantages; downsizing, long-life cycle and low costs.

There are some studies about how to interface the matrix converter with a DC power supply as mentioned in [9]-[10]. A voltage source inverter is connected to the input or output side of the matrix converter. In this system, the matrix converter operates in parallel with the inverter. However, the drawback is size will be extended because of constituting many components. The inverter is used to interconnect with a DC power source because the matrix converter has no DC component. However, this turns out to be a high cost structure for the matrix converter.

On the other hands, there are two types of interface converters using an IMC have been proposed, which are classified as the boost type converter[11] and step-down type converter[12]. Since the IMC contains of a DC link part, it is easy to interconnect with a DC power supply from that connection. The boost type converter connects a DC/DC converter using a chopper to the DC link part of the IMC. The control strategy of the voltage source converter, which consists of the inverter and the added DC/DC converter, is implemented as a four-phase voltage source inverter. With regard to the voltage relationship between the DC power supply voltage and the DC link voltage, the proposed circuit provides a boost-up operation for the DC power supply. However, the proposed boost type converter does not improve the voltage transfer ratio of the IMC. That is, the voltage transfer ratio of IMCs, which defines the ratio between the output voltage and the input voltage, is well known as being constrained to 0.866.

The step-down type converter interfaces the DC power supply using an active snubber circuit in the IMC. The active snubber circuit is composed by a capacitor connected to a switching device in series. This snubber circuit pairs with a switching device to work as a step-down chopper for the DC power supply. In addition, the voltage transfer ratio can be improved by using the DC power supply because the DC power supply should be higher than the DC link voltage to avoid short circuit. However, when the battery is used for the DC power supply in this system, a lot of batteries have to be connected in series. In this case, the battery volumes and costs will increase.

This paper proposes a converter which is combined by a boost type converter and a step-down type converter. The proposed converter connects an active snubber to the DC link part in the boost type converter. The voltage transfer ratio from the AC power source to the load can be improved by the active snubber. Besides, the high voltage battery is not required because the DC voltage source is boosted by the boost type DC/DC converter. That is, the proposed system covers the disadvantage of the boost type converter and step-down type converter from one to each other. The PWM generator is implemented based on the space vector modulation (SVM). The proposed control method controls the DC/DC converters and the inverter independently. The fundamental operation of the proposed method has confirmed by experimental results and loss analysis results.

II. PROPOSED SYSTEM CONFIGURATION

A. Back to Back converter system

Figure 1 shows the AC and DC power supply interface systems using the back to back converter. A conventional interface system consists of a PWM rectifier, a DC/DC converter and an inverter, as shown in Fig. 1. This system requires a large electrolytic capacitor in the DC link part in order to smooth the DC link voltage. This system is very flexible in term of voltage condition among the input and output side because of using voltage type converters in both sides. However, the electrolytic capacitor in DC link part gives problem such as large volume, short lifetime in high temperature ambience and high cost. Particularly, the large electrolytic capacitor has a problem for the HEV application because the power converter is installed at a high temperature area. Therefore, the electrolytic capacitor is not suitable for the HEV applications.

B. Matrix converter system

Figure 2 shows a AC and DC power supply interface system using the matrix converter[9] [10]. In Ref.[9], the inverter is connected to the input side of the matrix converter. The characteristic of this system is that an input filter can be replaced by the inverter. In addition, the voltage transfer ratio of the matrix converter is improved by using the inverter. However, the efficiency between the battery and the load is not high because this system requires twice power conversion between a battery and load.

Figure 2(b) shows the interconnection system, which has been proposed in Ref. [10]. This system is a kind of delta configuration system which is connected by the rectifier, the inverter and the matrix converter. Furthermore, the power can be converted directly among the three sources; the generator, the battery and the motor. Therefore, the efficiency of this proposed system becomes higher than the BTB system. Moreover the proposed system does not require the interconnection transformers or reactors at the connection point between the matrix converter and the inverter because

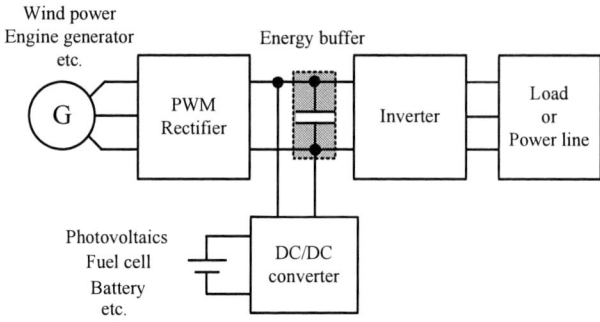

Fig. 1. Block diagram of the conventional AC and DC power supply interface system.

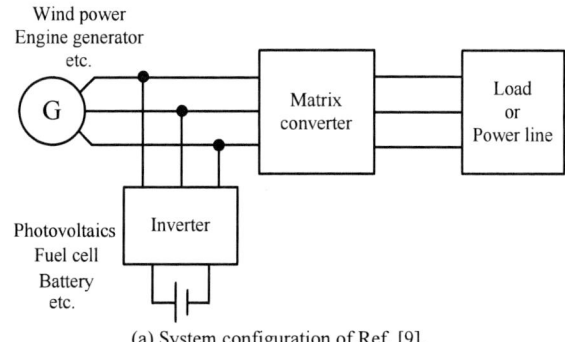

(a) System configuration of Ref. [9].

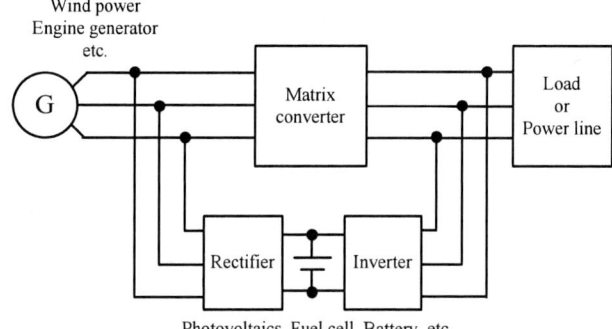

(b) System configuration of Ref. [10].

Fig. 2. Block diagram of the AC and DC power supply interface system using matrix converter.

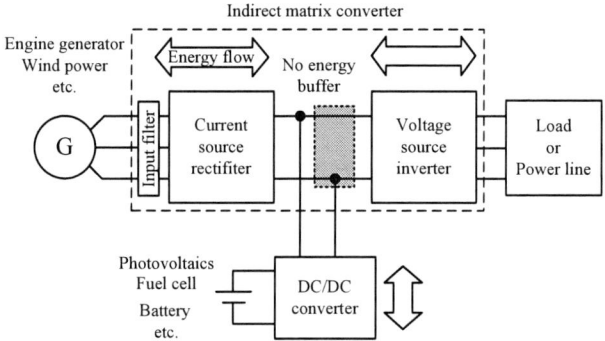

Fig. 3. Block diagram of the proposed system.

the operation of these two converters share the same time in one carrier cycle. However, it seems that the Ref. [9]-[10] system requires many components due to the insertion of the inverter and rectifier.

C. Proposed system

Figure 3 shows the block diagram of the proposed direct interface converters for the energy management system. The proposed interface converter is constructed based on an indirect matrix converter without the electrolytic capacitor as a large energy buffer. The DC/DC converter connects to the DC link of the indirect matrix converter. The proposed system can achieve higher efficiency than the conventional BTB system. In addition, the cost structure for the indirect control is lower than the matrix converter system because the number of components for the proposed system is less than the matrix converter system. Therefore, in this proposed circuit, since the size is smaller, the cost is lower as well and yet long life time.

III. CIRCUIT TOPOLOGY USING IMC

A. Boost type configureation

Figure 4 shows a boost type interface converter proposed in ref. [11]. A chopper is connected to the DC link part in the IMC. In this system, the DC link voltage becomes higher than the DC power supply. Therefore, it is referred to as a "boost type AC/DC/AC direct converter" in this paper. The chopper leg is controlled as a fourth leg of the voltage source inverter. Therefore, the control is implemented in a way that the voltage command of the chopper compares with the inverter carrier to obtain the desired PWM pattern.

However, the proposed converter has a problem which is the voltage transfer ratio between the power grid to the motor is constrained to 0.866 because the current source rectifier does not have a boost function. As a consequence, the output current of the IMC is higher than the BTB converter under the same output power rating. As a result, the motor loss and the converter loss are increased. Besides, the low voltage transfer ratio limits the applications of an IMC.

B. Step-down type AC/DC/AC direct converter.

Figure 5 shows a step-down type interface converter proposed in ref. [12]. A DC/DC converter is constructed with an active snubber circuit in the IMC. An IGBT is connected anti-parallel to the snubber circuit diode. This snubber circuit with the IGBT is used as a step-down chopper of the DC power supply. In this case, the DC power supply voltage of the snubber circuit should be higher than the peak of the AC input line voltage, because a rush current occurs between the AC and the DC input power supplies when the peak of the AC input line voltage is higher than the DC power supply voltage. Therefore, this converter is referred to as a "step-down type AC/DC/AC direct converter". The DC/DC converter is controlled as the fourth leg of the rectifier side converter. Time sharing is applied to the control of the snubber circuit.

Figure 6 shows the relationship between the DC link voltage and the active snubber duty ratio for the step-down type converter. For example, when the D_b becomes zero, the step-down type converter operates as a conventional IMC. The

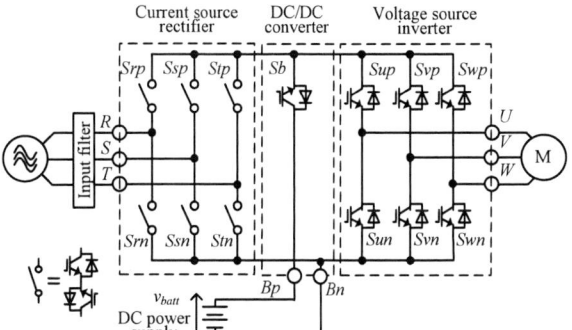

Fig. 4. Boost type AC/DC/AC direct converter.

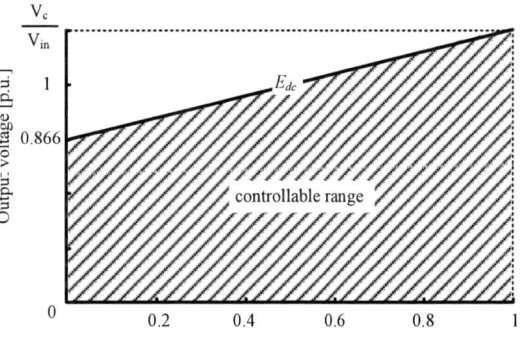

Fig. 5. Step down type AC/DC/AC direct converter.

Fig. 6. Control range for each converter.

output voltage is limited to $0.866v_{in}$. On the other hand the step-down type converter operates as a conventional inverter when the D_b reaches "1". That is, the proposed circuit operates as a conventional IMC or a conventional inverter alternately in a sequence to interface the AC and the DC power supply. Therefore the DC link voltage is decided by the average value of the output voltage between the rectifier stage and the DC/DC converter. The average value of the DC link voltage E_{dc} in the proposed circuit can be expressed as,

$$E_{dc} = D_{rec}^{*}v_{in} + D^{*}v_b \qquad (1),$$

where D_{rec}^{*} is the rectifier stage converter duty ratio, and D_b^{*} is the DC/DC converter duty ratio.

In the step-down type converter, DC power supply voltage becomes higher than the DC link voltage. Therefore, a lot of batteries need to be connected in series to provide high voltage to DC power supply. In this case, the volume of the battery will increase.

C. Buck-boost type AC/DC/AC direct converter.

Figure 7 shows the proposed circuit configuration. As previously discussed, the voltage transfer ratio of the boost type AC/DC/AC direct converter is constrained to 0.866, and for the step-down type AC/DC/AC direct converter, the DC power supply requires a lot of batteries to obtain high DC voltage.

In Fig. 7, the proposed converter combines both the boost type AC/DC/AC direct converter and the step-down type AC/DC/AC direct converter into one converter. In other words, the proposed buck-boost type AC/DC/AC direct converter is composed by a boost type AC/DC/AC direct converter with an active snubber circuit. The proposed converter can operate as the boost type AC/DC/AC direct converter and the step-down type AC/DC/AC direct converter at a same time. The voltage transfer ratio in this proposed circuit is improved by using the active snubber circuit.

Figure 8 shows the equivalent circuits of the proposed converter. The rectifier stage converter is similar to a four phase current source rectifier including the active snubber. Therefore, the switches in the three-phase PWM rectifier and active snubber must be separately turned on in order to avoid a short circuit between the AC power supplies and the capacitor. When the DC/DC converter switch S_c is turned off, the proposed converter operates as a boost type AC/DC/AC direct converter. In this case, the DC/DC converter becomes a conventional snubber circuit, as shown in Fig. 8(a). On the other hand, the proposed converter operates as a conventional inverter with a boost chopper when the active snubber switch S_c is turned on as shown in Fig. 8(b). Note that all the switches in the rectifier will be turned off in this mode.

That is, the proposed circuit alternately operates as a boost type AC/DC/AC direct converter or an inverter with the boost chopper under a simultaneous control. The proposed converter can achieve a wide control range by controlling the active snubber circuit.

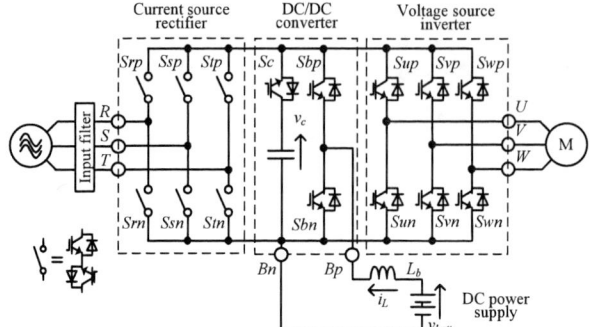

Fig. 7. Proposed circuit.

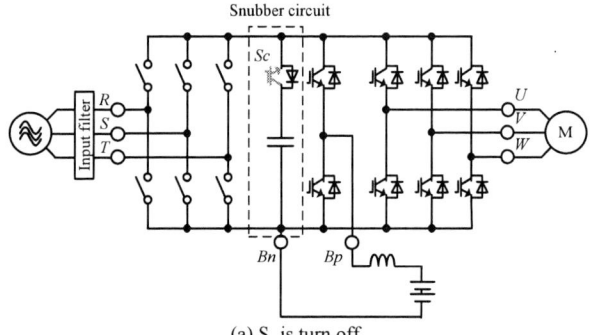

(a) S_c is turn off.

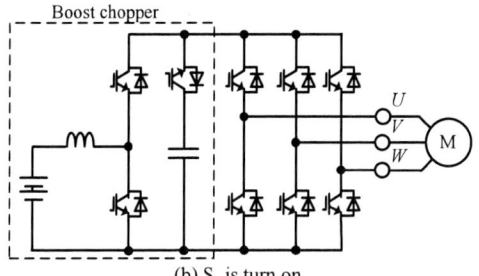

(b) S_c is turn on.

Fig. 8. Equivalent circuit of the proposed circuit.

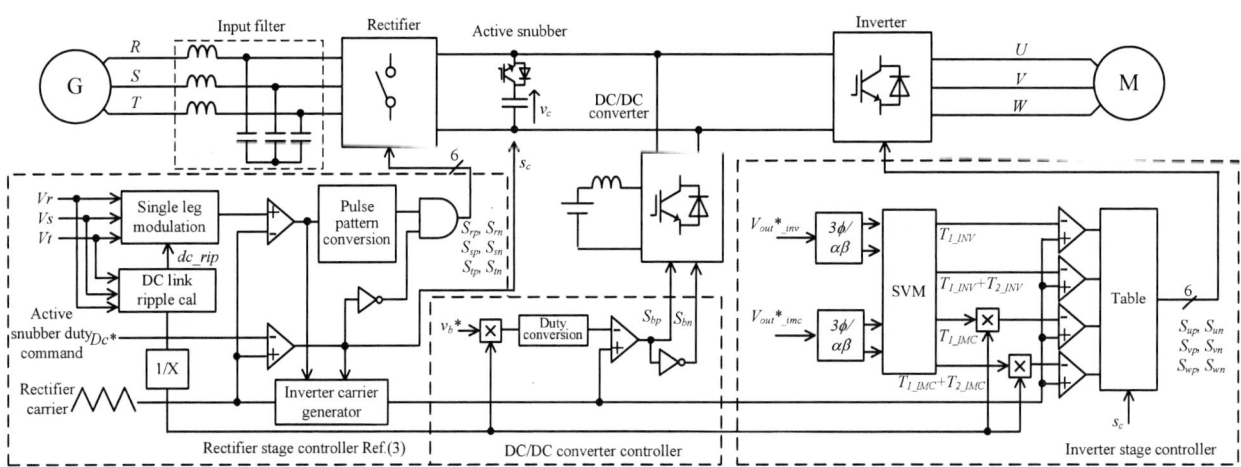

Fig. 9. Control block diagram of proposed converter.

978-1-4244-4782-4/10 $26.00 © 2010 IEEE   1687

IV. CONTROL STRATEGY

Figure 9 shows the control block diagram of the proposed converter. The proposed control method of the inverter stage is based on a space vector modulation. Therefore it is difficult to be added a DC/DC converter, because the space vector modulation is expressed by three dimensions[13].

Figure 10 following items in half control period; switching mode for each of the converter, and the carrier for inverter and DC/DC converter. The PWM pulse of the inverter and DC/DC converter carrier are obtained by comparison carriers and voltage command, which is calculated by the SVM. In the proposed method, the DC/DC converter applied the carrier comparison method using the carrier as shown in fig.10. The zero voltage vector of the inverter and the DC/DC converter lower arm switch s_{bn} have to be synchronized with the rectifier switching timing. Therefore, the inverter carrier is inverted when the active snubber switch is turned on. The ZCS of the rectifier stage is achieved at all time. That is, the proposed control method can control DC/DC converter independently.

Figure 11 shows the relationship between an output voltage command and a power distribution ratio for the proposed converter. The proposed converter has two types of operation modes. The operation mode is changed by the output voltage command v_{out}^* as follows.

A) $v_{out}^* \geq 0.866 v_{in}$: boost type operation mode

B) $v_{out}^* < 0.866 v_{in}$: buck-boost type operation mode

When the output voltage command is lower than 0.866 of the input voltage v_{in}, the proposed circuit operates as the boost type AC/DC/AC direct converter. In this case, the active snubber circuit is the same as a conventional snubber circuit. In other words, an active snubber duty command D_c^* is set to '0'. In addition, the DC/DC converter uses an ACR (Automated current regulator) to regulate battery current. The power distribution ratio is decided by the DC/DC converter current command.

On the other hand, the proposed converter operates as a buck-boost type AC/DC/AC direct converter when the output voltage command is higher than 0.866 of the input voltage v_{in}. In the proposed method, the power distribution ratio is decided by the active snubber duty ratio and the inverter stage output voltage v_{out}^*. In order to control the output voltage and the power distribution ratio at the same time, the output voltage command and the active snubber duty command is described as follows.

In the buck-boost type operation, the power distribution ratio is controlled by the operation time division of each operation mode. The output voltage is obtained by the average output voltage of inverter stage for the boost type mode and step-down type mode. Thus, when the active snubber duty is defined as D_c, the output voltage is obtained by (2).

$$v_{out} = \lambda_{sdt}^* D_c^* \frac{E_c}{2} + \lambda_{bt}^* (1 - D_c^*) \frac{E_{rec}}{2}$$
$$= D_b^* v_{sdt\_out} + (1 - D_c^*) v_{bt\_out} \tag{2}$$

where, v_{out} is the inverter stage output voltage of the proposed circuit, v_{bt_out} is the inverter stage output voltage of the boost

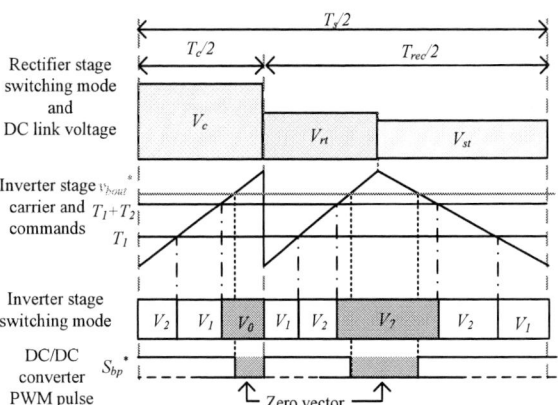

Fig. 10. Relation between inverter carrier, rectifier pulse and DC/DC converter pulse.

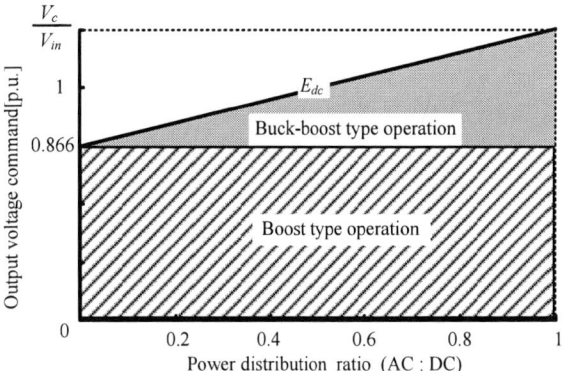

Fig. 11. Relation between output voltage command and power distribution ratio.

type operation, v_{sdt_out} is the inverter stage output voltage of step-down type operation, λ_{sdt} is the modulation index of the boost type operation, λ_{bt} is the modulation index of the step-down type operation, v_c is the capacitor voltage of active snubber, E_{rec} is the output voltage of rectifier stage and the subscript '*' represents the command.

The power distribution ratio is calculated by multiplying the output voltage of each operation and the active snubber duty ratio D_c^*. Then, the power distribution ratio is expressed by (3) when the instantaneous power is assumed to constant during one carrier cycle.

$$P_{sdt}^* : P_{bt}^* = D_c^* v_{sdt\_out}^* : (1 - D_c^*) v_{bt\_out}^* \tag{3}$$

where, P_{std} is the output power of the boost type operation and P_{bt} is the output power of the boost type operation.

The operation time ratio of the active snubber circuit D_c can be determined by (4).

$$D_c^* = \frac{v_{bt\_out}^* P_{sdt}^*}{v_{sdt\_out}^* P_{bt}^* + v_{bt\_out}^* P_{sdt}^*} \tag{4}$$

Thus, the output voltage of the proposed converter is constrained by the power distribution ratio. In order to control the output voltage and the power distribution ratio at the same time, the output voltage command $v_{out}{}^{*}$ and the active snubber duty command $D_c{}^{*}$ of the active snubber circuit is formed in (5) from (2) and (4).

$$v_{sdt\_out}{}^{*} = \frac{v_{out}{}^{*}P_{sdt}{}^{*}}{P_{sdt}{}^{*} + P_{bt}{}^{*}(1 - \frac{v_{out}{}^{*}}{v_{bt\_out}{}^{*}})} \qquad (5).$$

The modulation index of the inverter can be determined by (6) from (5).

$$\lambda_{sdt}{}^{*} = \frac{v_{out}{}^{*}P_{sdt}{}^{*}}{P_{sdt}{}^{*} + P_{bt}{}^{*}(1 - \frac{v_{out}{}^{*}}{v_{bt\_out}{}^{*}})}\frac{2}{v_c} \qquad (6).$$

In addition, an active snubber capacitor voltage v_c is controlled by AVR (Automated voltage regulator). It is noted that the modulation index λ_{bt} is fixed at '1' to output a maximum voltage. Therefore, the power distribution ratio and the output voltage can be controlled at the same time. The relationship between the output voltage and the active snubber duty ratio is shown in Fig. 11.

V. EXPERIMENTAL RESULTS

Table 1 provides the parameters and conditions for the simulation and the experimental. The operation of the proposed circuit is demonstrated by the simulation and experimental results.

First, the basic operation waveforms of the previous proposed converters are shown in Figure 12. Fig. 12(a) demonstrates the waveforms of the boost type AC/DC/AC direct converter; and Fig. 12(b) shows the waveforms for the step-down type AC/DC/AC direct converter. The input current and the output current show good sinusoidal waveforms for both the operation modes, and the DC current represents the battery current. As a result, in Fig. 12(a), it is confirmed that the total harmonic distortion (THD) of the input current, the output current and DC output current is 2.4%, 1.9% and 1.9% respectively. The input power factor is 99% and the efficiency is 96.3%. On the other side, the step down type converter shows the THD of the input current, the output current and the DC output current THD is 2.4%, 1.9% and 1.9%, the input power factor is 99% and the efficiency is 95.2% as shown in Fig. 12(b).

Figure 13 shows the waveform of the proposed control strategy. The dot line in the middle shows the change of the operation. At first, the proposed converter operates as the boost type AC/DC/AC direct converter. After that, the proposed converter operates as the buck-boost type AC/DC/AC direct converter. Sinusoidal waveforms without distortion are obtained for the input current and the output voltage. In Fig. (13), there is no sag occurs in the voltage and current when the operation mode is attempting a change. The output voltage of the boost type operation is 161 V. On the other hands the output voltage of the buck-boost type

Table 1. Experimental parameters.

Input voltage	200[V]	LC filter	2 [mH]
Input frequency	50[Hz]		6.6 [μF]
Carrier frequency	7.5[kHz]	Cut-off frequency	1[kHz]
Output frequency	40[Hz]	load	R-L
DC power supply	100[V]	Commutation time	2.5 [μs]

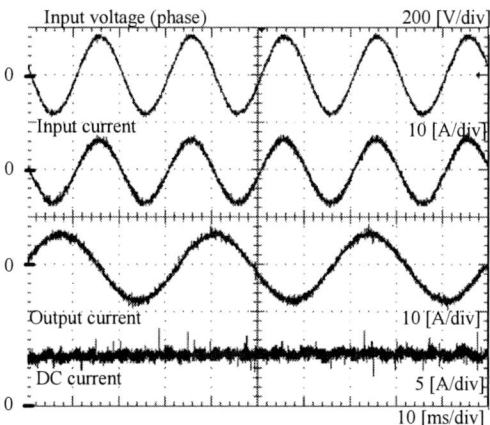

(a) Boost type AC/DC/AC direct converter.

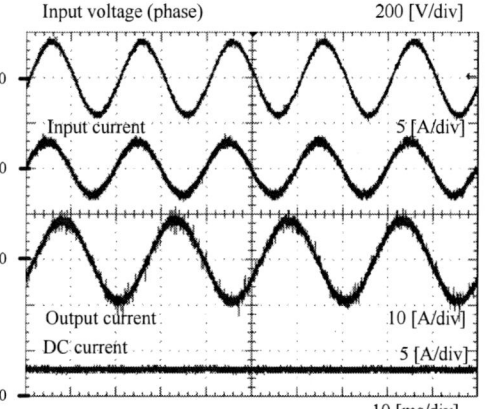

(b) Step down type AC/DC/AC direct converter.
Fig. 12. Experimental waveform

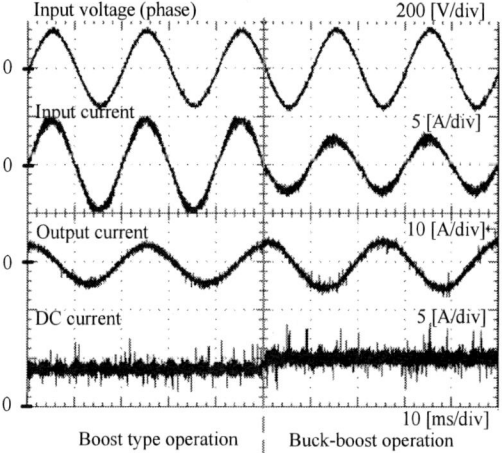

Fig. 13. Experimental waveform of proposed circuit.

operation is 180 V. The output voltage range is confirmed has been improved in the proposed circuit.

Figure 14 shows the THD of the input current, the output current and the DC input current which is 1.4%, 1.8% and 2.3%, respectively. In addition, the dead time error compensation is already applied in the FPGA control as reported in Ref. [11]. Therefore the proposed converter can obtain a low THD values for these current. It is noted that the DC current THD is defined by (7).

$$I_{dc\_THD} = \frac{I_{dc\_H}}{I_{dc}} \qquad (7),$$

where, I_{dc} is the RMS value in the DC current and I_{dc_H} is the RMS value in the DC current harmonics.

Figure 15 shows the efficiency and the input power factor of the proposed circuit under two circumstances. In Fig.15(a), all the switches in the chopper are turned off, battery voltage is neglected in this condition. That is, the converter consists of the indirect matrix converter and the active snubber only. The obtained data shows that the input power factor is 99% and efficiency is 95.4%. It is noted that the efficiency of a conventional multi-power supply interface converter with a large electrolytic capacitor is approximately 93%. Fig.15(b) shows the second circumstances where the inverter side is taken off from the converter by switching off all the IGBTs in the inverter. Then, the DC/DC converter is operating in boost type mode. The data shows the input power factor is over 99% and the maximum efficiency is 93.7%.

In this result, the efficiency drops due to the increment of the conduction loss in the DC/DC converter. The conduction loss becomes larger because of the output voltage of the DC/DC converter is lower than the inverter output voltage, which is set to 100 V. For that reason, the current in the DC/DC converter will force to increase. In addition, the efficiency of the indirect matrix converter can be improved by the applying the reverse blocking IGBTs into the rectifier stage.

Figure 16 shows the relationship between the efficiency and the power distribution ratio. In this result, the total output power is about 1.2kW, the DC/DC converter operates in charge mode and the inverter operates in motoring mode. The maximum efficiency is 95.4 % when the power distribution ratio is "0", that is the generator is 100% supplying the input power. On the other hand, when the power distribution ratio is "1", where the battery is providing 100% of the input power, the efficiency is 93.7%. In this experimental condition, the total efficiency of the proposed circuit becomes higher when the power distribution ratio is lower.

Figure 17 shows the loss analysis comparison between the proposed circuit and the BTB system that is simulated from a circuit simulator (PSIM, Powersim Technologis Inc) and DLL file (Dynamic Link Library)[12]. In this result, the power distribution ratio is "0.5". The switching losses do not occur in the rectifier stage of the proposed system because of applying the zero current switching. The proposed system decreases the switching loss in the rectifier stage by about 2/3 in compared with the rectifier stage of the BTB system. In the view of

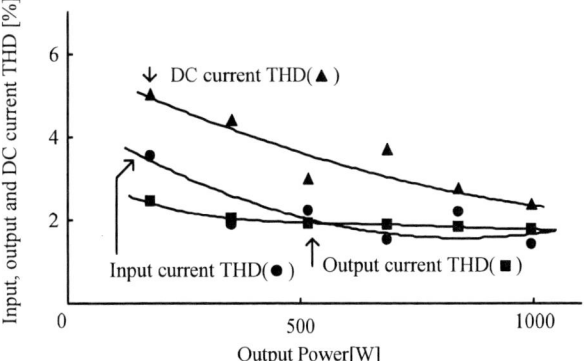

Fig.14. THD of input, output and DC current.

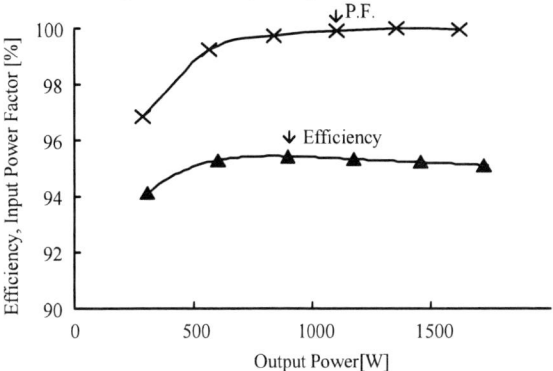

(a) DC/DC converter leg is stopped.

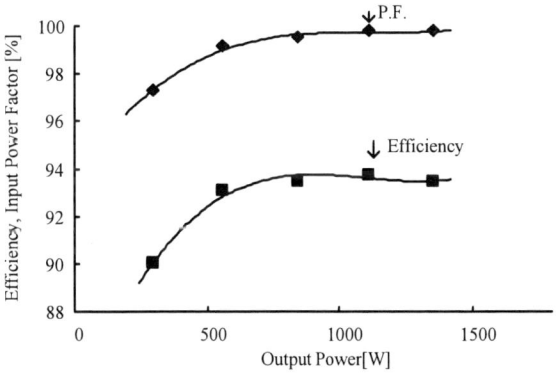

(b) Inverter leg is stopped.
Fig.15. Efficiency and input power factor.

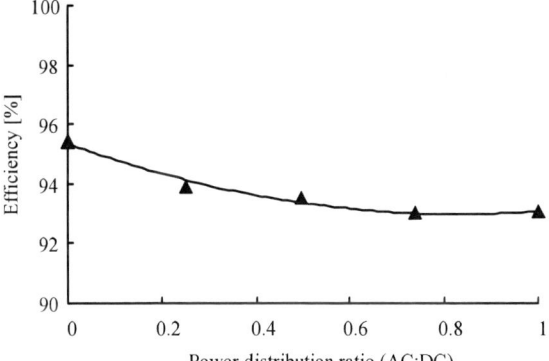

Fig.16. relation between efficiency and output power ratio.

inverter and DC/DC converter, the losses are almost the same at both systems since the output power of the inverter and the DC/DC converter is about the same. From the comparison, the proposed converter can obtain higher efficiency in compared to the BTB system.

VI. CONCLUSIONS

This paper proposes the novel control method and the converter which is combined by the boost type converter and step-down type converter. The proposed converter connects an active snubber to the DC link of the boost type converter. The voltage transfer ratio of the boost type converter can be improved by this active snubber. In addition, proposed control method is based on the space vector modulation (SVM). The proposed control method can control DC/DC converter and the inverter independently. The validity of the proposed strategy was confirmed by both the simulation and experimental results. As a result, it is confirmed that the AC input, output current, and the DC output current THD are 1.4%, 1.8%, 2.3%, respectively, and the input power factor is over 99% and the maximum efficiency is 95.4%.The basic operation of the proposed method has confirmed by experimental results and loss analysis results.

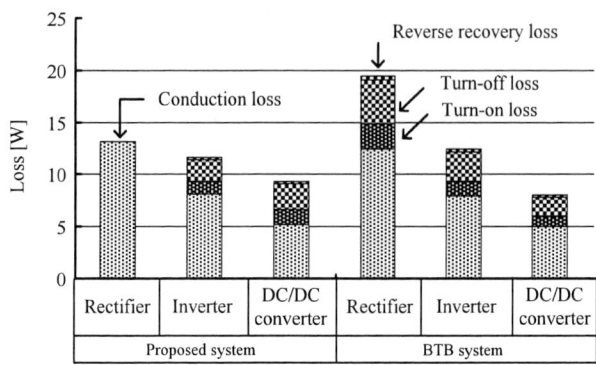

Fig.17. relation between efficiency and output power ratio.

REFERENCES

[1] P. W. Wheeler, J. Rodriguez, J. C. Clare, L. Empringham: "Matrix Converters: A Technology Review" IEEE Transactions on Industry Electronics Vol. 49, No. 2, pp274-288, 2002

[2] J.Itoh, I.Sato, H.Ohguchi, K,Sato, A.Odaka, N.Eguchi:"A Control Method for the Matrix Converter Based on Virtual AC/DC/AC Conversion Using Carrier Comparison Method"IEEJ Vol.124-D No.5,2004(in Japanese)

[3] J.Itoh, I.Sato, A.Odaka, H.Ohguchi, K.Kodachi:" A Novel Approach to Practical Matrix Converter Motor drive System with RB-IGBT" Power Electronics Specialists Conference 2004

[4] J. W. Kolara, T. Friedli, F. Krismer and S. D. Round:" The Essence of Three-Phase AC/AC Converter Systems" Power Electronics Motion Control Conference 2008 pp.27-42

[5] L.Wei, Y.Matsusita, T.A.Lipo: "Investigation of Dual-bridge Matrix Converter Operating under Unbalanced Source Voltage" IEEE Power Electronics Specialist Conference 2003, 1293(2003).

[6] J.W.Kolar, M.Baumann, F.Schafmeister, H.Ertl: "Novel Three Phase AC-DC-AC Sparse Matrix Converter", IEEE APEC 2002

[7] Y. Minari, K. Shinohara, R. Ueda: "PWM-rectifier/voltage-source inverter without DC link components for induction motor drive", IEE Proc. on Electric Power Applications, Vol. 140, No. 6, pp. 363 .368, Nov. 1993.

[8] C. Klumpner, T. Wijekoon, P. Wheeler: " Active Compensation of Unbalanced Supply Voltage for Two-Stage Direct Power Converters Using the Clamp Capacitor" PESC'05, pp.2376 - 2382, 2005

[9] S. Goto, S. Ogasawara, and H. Funato, "A New Power Converter Circuit Combining an Inverter with a Matrix Converter," IEEJ SPC-06-101, 2006.(in japanese)

[10] Jun-ichi Itoh, Hiroshi Tamura: "A Novel Control Strategy for a Combined System Using Both Matrix Converter and Inverter without Interconnection Reactors", IEEE PESC2008, pp.1741-1747, 2008

[11] Koji Kato, Jun-ichi Itoh, "A Novel Control Method of Direct Interface Converters for Several Power Supplies" 12th European Conference on Power Electronics and Applications 2007

[12] Koji Kato, Jun-ichi Itoh, " Control Method for a Three-Port Interface Converter Using an Indirect Matrix Converter with an Active Snubber Circuit", THE 13th IEEE Power Electronics and Motion Control Conference,448, 2008.

[13] D. Katsis, P.W. Wheeler, J.C. Clare, L. Emprongham, "A Utility Power Supply Based on a Four-Output Leg Matrix Converter" IEEE IAS 2005, pp.2355-2359

[14] J. Itoh, T. Iida, A. Odaka:" Realization of High Efficiency AC link Converter System based on AC/AC Direct Conversion Techniques with RB-IGBT" Industrial Electronics Conference, Paris, ,PF-012149,2006

A PI Control Algorithm of Three-Level APF With Little Static Misadjustment For Tracking Harmonic Current

Yingjie He[1], Jinjun Liu[1], Zhaoan Wang[1] and Yunping Zou[2]

[1] School of Electrical Engineering
Xi'an Jiaotong University
Xi'an, China

[2] School of Electrical and Electronic Engineering
Huazhong University of Science & Technology (HUST)
Wuhan, China

Project sponsored by the National Key Technology R&D Program of china (2007BAA12B03) ,by National Science Foundation of China (50907052) and by Delta Educational Development Foundation（DREG2009003）

Abstract—**Obviously, tracking harmonic current fast and precisely is one of the keys to design the APF. This paper researches the control method of three-level active power filters. The conventional current state feedback decoupling PI control is analyzed in this paper in detail. This paper pointes out the limitation of the conventional PI conditioner in the research field of APF and presents a novel PI control method. Canceling the delay of one sampling period and the misadjustment for tracking the harmonic current is key problem of this PI control. In this PI control, the predictive output current value is obtained by the state predictor. The delay of one sampling period is remedied in this digital control system by the state predictor. The predictive harmonic command current value is obtained by the repetitive predictor synchronously. The repetitive predictor can achieve better prediction of the harmonic current. By this means, the misadjustment of the conventional PI control for tracking the harmonic current is cancelled. The experiment results indicate that the steady-state accuracy and dynamic response are both satisfying when the proposed control scheme is implemented.**

Index Terms—**active power filter, three-level NPC inverter, PI control, repetitive predictor, state predictor**

I. INTRODUCTION

Harmonics cause serious problems at power conversion system. Passive filters that consist of *L* and *C* are generally implemented to attenuate harmonics generated in the power system. However the passive filter has disadvantages that the source impedance affects to filtering characteristics and amplification of the harmonic currents caused by the parallel resonance. Control of harmonic perturbations by active power filters(APF) has become a hot topic in the power engineering field[1]-[3].

But due to the limitation of voltage capability of power devices, it is very difficult to handle nonlinear loads for the traditional APF with two-level inverter in high voltage grid. In recent years, multilevel technology has become an effective and practical solution for high-voltage high-power application field. As described in many literatures, using multilevel technology, the voltage stress on switches will be reduced, the shape of output waveform will be improved and the rate of voltage and power can be increased too[4][5]. It is believed to have wider application for harmonic restriction and reactive power

compensation especially. Among all multilevel topologies, three-phase three-level neutral-point-clamped (NPC) PWM inverter is the most widely used and investigated topology at present. For this reason, the APF with three-level NPC inverter has received more and more attention in recent years[6]-[12].

Obviously, tracking the command current fast and precisely is one of the keys to design this APF. High quality system current waveform of power system evaluated with little total harmonic distortion (THD) is a basic requirement for APFs. In the past, the current state feedback decoupling PI control is an effective means of three-phase system in the current control of constant voltage constant frequency inverter and high frequency PWM reversible rectifier[13]-[15]. Because the fundamental signal is converted into DC quantity through synchronous rotating transformation, the conventional PI controller has little static misadjustment. However, the command currents of APFs include a large number of harmonic. The bandwidth of conventional PI controller is not far from enough for effective tracking the command harmonic current. The conventional PI controller has static misadjustment for tracking the command harmonic current and system current waveform filtered has large THD.

The conventional current state feedback decoupling PI control is analyzed in this paper in detail. This paper concludes that canceling the delay of one sampling period and the misadjustment for tracking the harmonic current is key problem of this PI control. This paper presents a novel PI control method. In this PI control, the predictive output current value is obtained by the state predictor. The delay of one sampling period is remedied in this digital control system by the state predictor. The predictive harmonic command current value is obtained by the repetitive predictor synchronously. The repetitive predictor can achieve better prediction of the harmonic current. By this means, the misadjustment of the conventional PI control for tracking the harmonic current is cancelled. This PI control method has fast dynamic convergence rate and little static misadjustment. The experiment results illustrate that the performance of the proposed approach is satisfactory.

II. MATHEMATIC MODEL AND CONTROL SYSTEM

The operating principle of the APF is shown in Fig.1. The

978-1-4244-4782-4/10 $26.00 © 2010 IEEE

periodic non-sinusoidal current which flows over nonlinear load (NL) can be resolved into the fundamental current component i_l and the summation of all harmonic current components i_h. The reference directions of the currents are defined as shown in Fig.1. u_{sa}, u_{sb} and u_{sc} are three-phase voltage, i_{ca}, i_{cb} and i_{cc} are the currents of APF, i_{la}, i_{lb} and i_{lc} are nonlinear load currents. After load currents flow through current transformers, desired currents which APF should compensate are first obtained by the detecting circuit of harmonic currents. Then desired currents are injected into the power system by the APF controlling the PWM inverter. It can be seen from Fig.1 that i_l is provided with both source current i_s and output current i_c of APF, i.e., $i_l = i_s + i_c$. Since $i_c = i_h$, then the power source only supplies i_l. Hence i_s after compensation becomes a sinusoidal current which has identical frequency with u_s. This is just the basic operating principle of APF.

The three-level NPC inverter applied to the APFs, as shown in Fig. 1, is built-up of twelve switches, each one with its freewheeling diode, and six power diodes that allow the connection of the phase outputs to the middle voltage. From Fig. 1 it shows that each phase of the NPC three-level inverter consists of four switches. The blocking voltage of each switch is 1/4 of the DC voltage. Each arm of the inverter can be clamped to the DC terminals P, N and Q, and produce three switching states. When the upper two switches T_1,T_2 are switched on ,the output of this phase is connected to the DC terminal P. When the middle two switches T_2,T_3 are switched on, the output of this phase is connected to middle point N. Similarly when the lower two switches T_3,T_4 are switched on, the output of this phase is connected to Q. The allowed logic configurations of NPC switches are able to provide the three different output voltage values furnished by each NPC inverter phase. The states of T_1,T_2,T_3 and T_4 are complementary. The two switches of each phase of the NPC inverter are closed, whilst the other two are opened at every time instant[16]-[17].

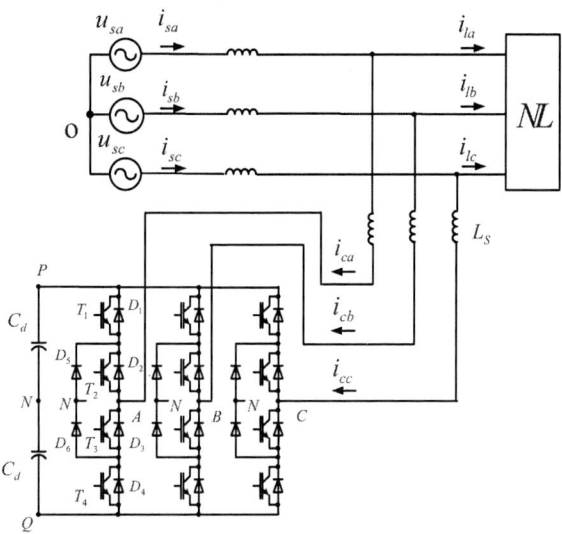

Fig.1 Schematic diagram of the active power filter with

three-level inverter

In this paper, the loss of the switching devices and snubber circuits is ignored. Based on the concept of switching function (SF), the equivalent switch-circuit of three-level NPC APF can

be obtained as Fig.2. The power loss on the resistor R is assumed to include all filter's losses.The three phase source voltage is assumed to be balance and symmetrical. The switching state of phase A, S_a of power devices T_{1a}-T_{4a}, is defined as [13]:

$$S_a = \begin{cases} 1 , & T_{1a}, T_{2a} \text{ON and } T_{3a}, T_{4a} \text{OFF} \\ 0 , & T_{2a}, T_{3a} \text{ON and } T_{1a}, T_{4a} \text{OFF} \\ -1 , & T_{3a}, T_{4a} \text{ON and } T_{1a}, T_{2a} \text{OFF} \end{cases} \quad (1)$$

The switch states are coded by symbols -1, 0 and 1 identifying the three voltage levels on each inverter phase .To get the mathematic model of the APF, S_a is decomposed as:

$$\begin{cases} S_{1a}=1, S_{2a}=0, S_{3a}=0 & S_a = 1 \\ S_{1a}=0, S_{2a}=0, S_{3a}=1 & \text{when, } S_a = 0 \\ S_{1a}=0, S_{2a}=1, S_{3a}=0 & S_a = -1 \end{cases} \quad (2)$$

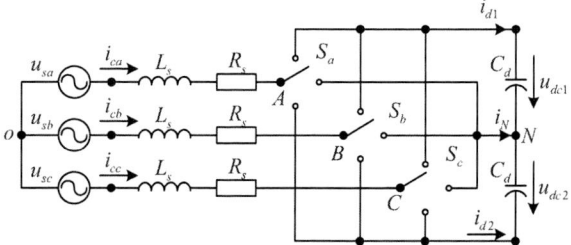

Fig.2 The equivalent circuit of the active power filter with

three-level inverter

Using the same method, S_b and S_c is decomposed too. Then, the comprehensive mathematic model of three-level inverter in stator coordinates is established as follow:

$$Z\dot{X} = A \cdot X + B \cdot e \quad (3)$$

where:

$$Z = diag[\begin{matrix} L_s & L_s & L_s & C_d & C_d \end{matrix}]$$

$$X = [\begin{matrix} i_{ca} & i_{cb} & i_{cc} & u_{dc1} & u_{dc2} \end{matrix}]^T$$

$$A = \begin{bmatrix} -R_s & 0 & 0 & -S_{1a}+\dfrac{S_{1a}+S_{1b}+S_{1c}}{3} & S_{2a}-\dfrac{S_{2a}+S_{2b}+S_{2c}}{3} \\ 0 & -R_s & 0 & -S_{1b}+\dfrac{S_{1a}+S_{1b}+S_{1c}}{3} & S_{2b}-\dfrac{S_{2a}+S_{2b}+S_{2c}}{3} \\ 0 & 0 & -R_s & -S_{1c}+\dfrac{S_{1a}+S_{1b}+S_{1c}}{3} & S_{2c}-\dfrac{S_{2a}+S_{2b}+S_{2c}}{3} \\ S_{1a} & S_{1b} & S_{1c} & 0 & 0 \\ -S_{2a} & -S_{2b} & -S_{2c} & 0 & 0 \end{bmatrix}$$

$$B = diag[\begin{matrix} 1 & 1 & 1 & 0 & 0 \end{matrix}]$$

$$e = [\begin{matrix} u_{sa} & u_{sb} & u_{sc} & 0 & 0 \end{matrix}]^T$$

To simplify the mathematic model, synchronous rotating coordinate transformation is used. A mathematical model of the tri-level system in the rotating dq frame is expressed as follows:

$$Z\dot{X} = A \cdot X + B \cdot e \quad (4)$$

where:

$$Z = diag\begin{bmatrix} L_s & L_s & C_d & C_d \end{bmatrix}$$

$$X = \begin{bmatrix} i_{cd} & i_{cq} & u_{dc1} & u_{dc2} \end{bmatrix}^T$$

$$A = \begin{bmatrix} -R_s & \omega L_s & -S_{d1} & S_{d2} \\ -\omega L_s & -R_s & -S_{q1} & S_{q2} \\ S_{d1} & S_{q1} & 0 & 0 \\ -S_{d2} & -S_{q2} & 0 & 0 \end{bmatrix}$$

$$B = \begin{bmatrix} 1 & 1 & 0 & 0 \end{bmatrix}$$

$$e = \begin{bmatrix} u_{sd} & u_{sq} & 0 & 0 \end{bmatrix}$$

By this means, three-phase alternating signal is changed into two-phase alternating signal. Thus the mathematic model of this APF is predigested. Base on the mathematic model, the control system of the APF is designed. The diagram of control system is shown in Fig.3. The control system includes the outer-loop voltage controller and the inner-loop current controller[14]. The outer-loop controller is used to produce the command active current to regulate DC voltage. The inner-loop current controller is divided into the detection module of calculating harmonic current and the control module of tracking the command current and restraining neutral-point voltage imbalance. The method for detecting harmonic currents based on the instantaneous reactive power theory is adopted in the detection module. The control module of tracking the command current deals with shaping the current waveforms to track the references. The command current of the control module is composed of DC voltage regulator component and harmonic current compensator component. In order to control the shape of output current waveform, a novel PI control technology is presented. Neutral-point voltage imbalance is restrained with selecting the small vectors that will move neutral-point voltage in the direction opposite from the direction of unbalance from the space vector PWM point of view in the control module[18].

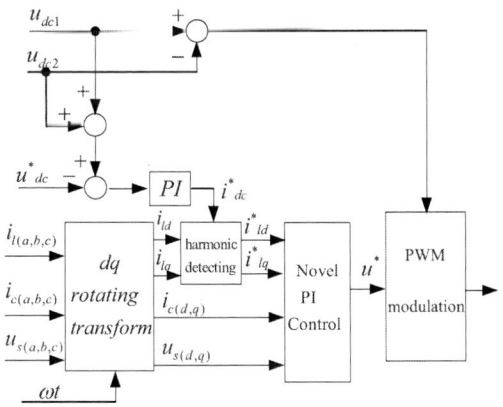

Fig.3 Block diagram of control system of the active power filter
with three-level inverter

III. NOVEL PI CONTROL METHOD

From (4), the control system model of tracking the command current in the inner-loop current controller is established as follow:

$$\begin{cases} L_s \dfrac{di_{cd}}{dt} = -R_s i_{cd} + \omega L_s i_{cq} - u_d + u_{sd} \\ L_s \dfrac{di_{cq}}{dt} = -R_s i_{cq} - \omega L_s i_{cd} - u_q + u_{sq} \end{cases} \quad (5)$$

Traditionally, the current state feedback decoupling PI control is an effective means of three-phase system in the current control of high frequency PWM reversible rectifier. From (5), the block diagram of the PI control system is proposed as shown in Fig.4. Through this PI control method, the equivalent decoupling control diagrams for i_d and i_q can be derived as shown in Fig.5. K_P and K_I is the proportional and integral parameter of PI controller respectively. The control plant is first-order. The current control closed-loop with PI controller is the typical second-order system[14].

$$C(s) = \frac{i_c(s)}{i_c^*(s)} = \frac{(K_P / L_S)s + K_I / L_S}{s^2 + [(K_P / L_S)s]/L_S + K_I / L_S} \quad (6)$$

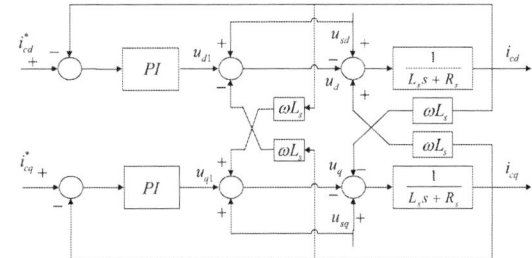

Fig.4 Block diagram of current state feedback
decoupling PI control

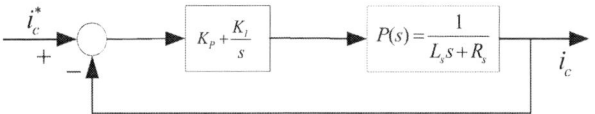

Fig.5 Block diagram of equivalent decoupling control

Obviously, the current control closed-loop is the typical second-order system which can be corrected by PI controller into a first-order element by means of Siemens method. Siemens method is used widely in the actual tuning of PI parameters. The core thought of this design is to let the zero of the controller offset the pole of the plant in the s-plane. Using this method, the PI parameter K_P and K_I is calculated as follow:

$$K_P = L_s / T_s, K_I = R_s / T_s \quad (7)$$

T_s is sampling\switch periods. Substituting equation (7) into (6) yields the following results,

$$C(s) = \frac{i_c(s)}{i_c^*(s)} = \frac{K_P}{K_P + L_S s} = \frac{1}{1 + (L_S / K_P)s}$$
$$= \frac{1}{1 + T_P s} = \frac{1}{1 + T_S s} \quad (8)$$

From the expressions (8), we find that the closed-loop current control system is corrected into a first-order element. The time constant T_P of the closed-loop current control is equal to the switch periods T_s.

The bode map of $C(s)$ is as Fig.6. From the bode map of $C(s)$, $C(s)$ has close-to-unit gain at low frequency band. Because the fundamental signal is converted into DC quantity through synchronous rotating transformation, the conventional PI controller which is designed with above-mentioned method has little static misadjustment in PWM reversible rectifier.

However, the command currents of APFs include a large number of harmonic. From the bode map of $C(s)$, the magnitude of $C(s)$ degrades in high and medium frequency band.

The bandwidth of this PI controller is not far from enough for effective tracking the command harmonic current. The PI controller has static misadjustment for tracking the command current and system current waveform filtered has large THD.

In order to make the output current track the compensation harmonic reference well, we may increase gain of PI controller, or increase sampling and switching frequency to make the value of the current loop response delay time T_P shorter. However, all these means cannot improve the performance of harmonic compensation effectively and may cause other problems such as oscillation and instability of the APF and much more switching loss. In addition, the value of T_r cannot be too small, due to the limit by the operating frequency of the devices.

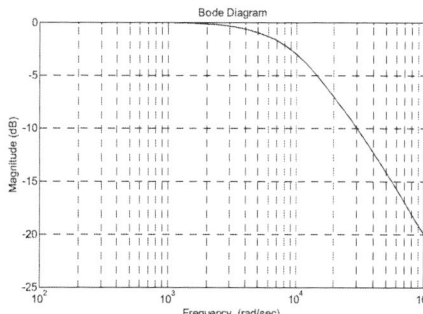

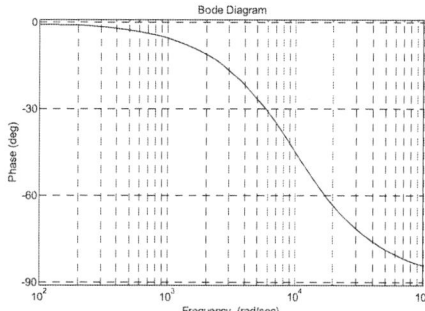

Fig.6 bode map of C(s)

From the expressions (8), the discrete equivalent equation of the closed-loop current control system can be represented as,

$$T_s \frac{i_c(k+1) - i_c(k)}{T_s} + i_c(k) = i_c^*(k) \qquad (9)$$

Simplifying the expressions (9), the answer can be represented as

$$i_c(k+1) = i_c^*(k) \qquad (10)$$

From the expressions (10), we find that the output compensation current *at (k+1)th* sampling instant is equal to the command harmonic current at *kth* instant in above-mentioned digital PI control system. Thus the controller has static misadjustment. If we predict the command harmonic currents $i_c^*(k+1)$ at *kth* instant and substitute $i_c^*(k+1)$ for $i_c^*(k)$ in the equation (10), we may obtain the following results,

$$i_c(k+1) = i_c^*(k+1) \qquad (11)$$

By this means, the misadjustment of the conventional PI control for tracking the harmonic current is cancelled. So predicting harmonic currents at the coming sampling instant pre-

cisely is key problem to improve this controller.

Moreover, the execution of the control program needs time, the calculation of the command voltage should be carried out a sampling period ahead, otherwise the maximum output pulse width will be significantly reduced. So remedying the delay of one sampling period is another key problem to improve this controller.

Therefore, the control process should be like this: at *(k-1)th* instant, sample the output current $i_c(k-1)$ of the APF, then estimate $i_c(k)$. Meanwhile sample the load current $i_l(k-1)$ and calculate the harmonic current $i_h(k-1)$ then predict $i_h(k+1)$. Finally the predicted value $i_h(k+1)$ is put into the PI controller of the closed-loop current control system. Using this approach mentioned above ,the control system can track the command current with no static misadjustment .

A good solution for the estimation of APF output currents $i_c(k)$ is a state observer[19]. A full order state observer for this APF is as follows:

$$\hat{X}(k+1) = G \cdot \hat{X}(k) + H \cdot U(k) + T(X(k) - \hat{X}(k)) \quad (12)$$

Where T is the observer gain. The state predictor is as fig.7. $\hat{X}(k)$ and $\hat{X}(k+1)$ denote the estimated values of output currents at *kth* and *(k+1)th* sampling instant. The observer gain is decided so that a stable pole placement is realized in the observer.

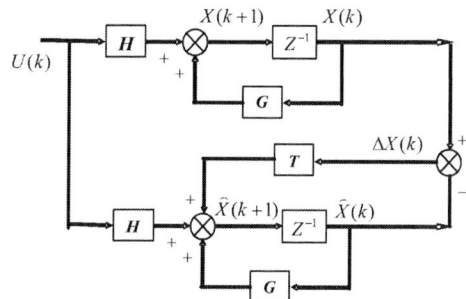

Fig.7 The state predictor

From (5) and (12), the error system is:

$$e_x(k+1) = (G-T) \cdot e_x(k) \qquad (13)$$

Where $e_x(k+1) = X(k) - \hat{X}(k)$ is the observer error. The poles of the observer can be arbitrarily placed by adjusting the feedback gain T. In practice it is easy to make the observer error well close to zero by setting the poles of the observer suitably.

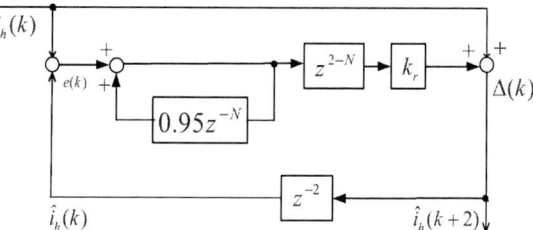

Fig.8 The repetitive predictor

It seems impossible to model the load current disturbance precisely when considering the diversity of load types. Since the harmonic is repetitive in most time, it makes sense to adopt a repetitive algorithm to predict the seemingly unpredictable load current[19]. The basic theorem of repetitive harmonic predictor is show as Fig.8.

The algorithm can be looked as a plug-in repetitive compensator for the basic predictor. The basic predictor is a unit delay: z^{-2}, which means it just takes the value of harmonic current at *kth* sampling instant (namely $i_{hd}(k)$, $i_{hq}(k)$ for harmonic current values at *(k+2)th* sampling instant since the *(k+2)th* sampling process has not yet begun. The input of the compensator is *e*, which is defined as the difference between the value of the actual current and the estimated value. It's assumed that harmonic current is sampled *N* times in a fundamental cycle. In order to repetitive compensate the error of basic prediction, *N* memory cells are prepared to store the correction values (*namely* △*(k)*, $k \in [1,N]$) in one fundamental cycle. The predicted value of harmonic current at *(k+2)th* sampling instant (namely $\hat{i}_h(k+2)$) can be obtained by adding △*(k) to* the value of harmonic current at *kth* sampling instant, as (14). The correction value △*(k)* for correcting the basic prediction value $i_h(k)$ to predict $\hat{i}_h(k+2)$ is modified periodically by adding the prediction error to the current correction value △*(k)*. The prediction error is obtained by subtracting the predicted value of harmonic current $\hat{i}_h(k+2)$ from the actual harmonic current value $i_h(k+2)$.

$$\hat{i}_h(k+2) = i_h(k) + \Delta(k) \qquad (14)$$

In face of unperfected modeling, a bandlimit filter is often included in a repetitive controller. Such a filter can improve stability of the filter system with eliminating the error *e* by halves. This filter equals 0.95 (the parameter was proved suitable in most of our simulations). In this case, the basic estimator is just a simple and fixed algorithm, and not a real physical system where measuring errors and parameter drift are inevitable. As for the gain k_r, when it is set to be unity, the error will be compensated nearly 100% in the second cycle under repetitive input. However in practice, the load current does not exactly repeat itself from cycle to cycle. For smooth convergence of the error, k_r should be chosen as less than 1. The simulation results show that the value k_r =0.98 appear to be a good choice for all the experiments.

IV. EXPERIMENT RESULT

Theoretic analysis results are necessary to further verify by the actual circuits. The diagrammatic sketch of experiment power circuits is as Fig.1. The parameters of the circuit and control system model for the experiment are listed in Table I. All control algorithms are implemented in digital-signal processors (DSP) TMS320F2407.The nonlinear load is the three-phase bridge rectifier with the inductance and resistance. The source voltage is u_s=110V/50H, and the impedance of voltage source is L_S=1mH. P-MOSFET 2SK1020 is selected as the switching device. Some important experimental oscillograms are shown respectively as follows:

TABLE I The parameters of the circuit and control system

Symbol	Experiment	Explanation
U_s	110V	the RMS phase voltage value
f	50Hz	Power frequency
L_s	2mH	the filter inductance
R_s	0.5 Ω	the resistor of APF output filter
u_{dc}	360V	DC voltage of the filter
C_d	4700 μ F	DC capacitor of the filter
f_s	9.6kHz	sampling \switch frequency
k_p	1.6	DC controller parameter k_p
k_I	64	DC controller parameter k_I
kr	0.98	repetitive predictor parameter

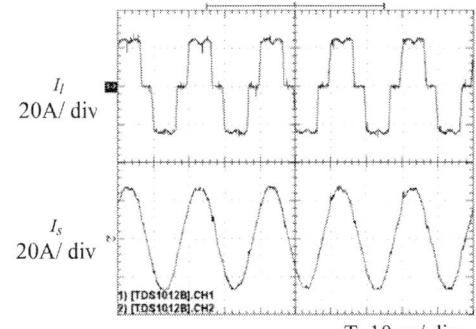

Fig.9 Experimental steady waveforms of load current and system supply current

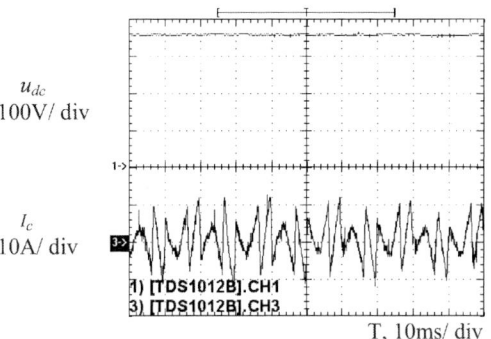

Fig.10 Experimental steady waveform of DC voltage and APF output current

Fig.9 shows experimental steady waveforms of load current and system supply current, where waveform 1 is load current i_l, waveform 2 is system supply current i_s through the APF filtering. The THD (total harmonic distortion) of load current i_l is calculated to be 22.54%. The THD of i_s is 3.30%. Fig.9 has proved that the proposed APF has good steady filtering capability. Fig.10 shows experimental steady waveforms of DC voltage and APF output current, where waveform 1 is DC voltage, waveform 2 is APF output current through the APF filtering. As shown in Fig.10, DC capacitor voltages are regulated to fluctuate in a small range about the reference voltage 360V during the APF steady operation. Fig.11 presents dynamic experimental waveforms when nonlinear load is added abruptly, where waveform 1 is load current i_l, waveform 2 is system supply current i_s through the APF filtering. As shown in Fig.11 we can see that the control method tracks the command current very quickly. Fig.12 shows dynamic experimental waveforms when nonlinear load is removed abruptly, where waveform 1 is load current i_l, waveform 2 is system supply current i_s. Fig.12 has proved that the proposed APF has good dynamic response under the load removing process. It is obvious that the proposed APF is controlled very well by the PI control. The experimental waveforms of u_{AN} and u_{AB} is illustrated in Fig.13, where waveform 1 is u_{AN}, waveform 2 is u_{AB}. It is obvious that the output voltage u_{AN} furnished by each NPC inverter phase is three-level and the line-line output voltage u_{AB} is five-level. The experiment result mentioned above shows that the proposed APF is fit for high-voltage high-power application field.

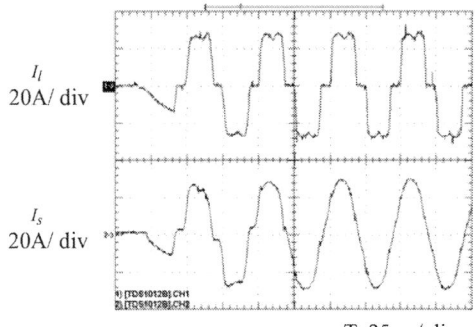

T, 25ms/ div

**Fig.11 Experimental dynamic waveforms of load current and
system supply current when nonlinear load is added abruptly**

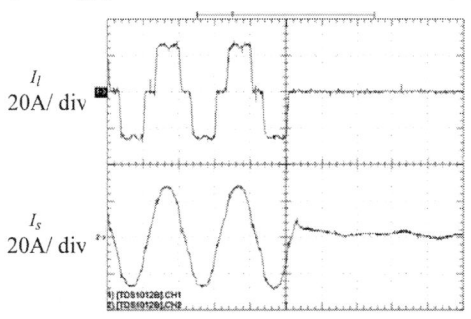

T, 10ms/ div

**Fig.12 Experimental dynamic waveforms of load current and
system supply current when nonlinear load is removed abruptly**

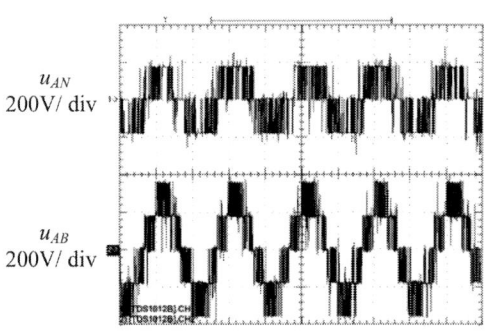

T, 10ms/ div

Fig13 Experimental waveforms of u_{AN} and u_{AB}

V. CONCLUSION

This paper researches control problem for active power filters with three-level NPC inverter and proposes a novel PI control algorithm for tracking harmonic command current. This novel PI control algorithm can suppress the periodic error in the whole system to achieve zero steady error tracking.In this scheme, the state variables are estimated with a state observer to cancel the delay of one sampling period in this digital control system. Harmonic current is predicted with a repetitive algorithm simultaneity, which makes use of the repetitive nature of load current. The controller is analyzed and designed in the paper, and the experiment results illustrate that this APF can be controlled in a satisfactory way.

REFERENCES

[1] Akagi H. New trends in active filters for power conditioning. IEEE Trans Industry Applications, 1996, 32(6):1312-1322.

[2] Dixon J W,Venegas G,Moran L A.A series active power filter based on a sinusoidal current-controlled voltage-source inverter.IEEE Trans Industrial Electronics,1997,44(5): 612-620.

[3] Kim S, Enjeti P N.A new hybrid active power filter topology. IEEE Trans Power Electronics, 2002, 17(1): 48-54.

[4] Akira N, Takahashi I, Akagi H. A new neutral-point-clamped PWM inverter. IEEE Trans Industry Applications, 1981, 17(3): 518-523.

[5] Rodriguez J, Lai J S, Peng F Z. Multilevel Inverters: A Survey of Topologies, Controls, and Applications. IEEE Trans Industry Applications, 2002, 49(4): 724-738.

[6] Rudnick H, Dixon J, Moran L. Delivering clean and pure power. IEEE Power and Energy Magazine.2003, 1(5):32-40.

[7] Aburto V, Schneder M, Moran L et al. A active power filter implemented with a three-level NPC voltage-source inverter. Proceedings of the 1997 IEEE PESC, 1997, 1121-1126.

[8] Tolbert L M., Peng F Z, Habetler T G. A multilevel converter-based universal power conditioner. IEEE Trans Industry Applications, 2000, 36(2): 596-603.

[9] Wong M C, Tang J, Han Y D. Cylindrical coordinate control of 3-D PWM technology in 3-phase 4-wire tri-level inverter. IEEE Trans Power Electronics, 2003, 18 (1): 208 -220.

[10] Lin B R, Huang C H, Yang T Y et al. Analysis and implementation of shunt active power filter with three-level PWM scheme. Proceedings of the 2003 IEEE PEDS , 2003 ,1580-1585.

[11] Jin T T, Wen J, Smedley K. Control and topologies for three-phase three-level active power filters. Proceedings of the 2005 IEEE APEC ,2005 , 655 - 664.

[12] Dai N Y, Wong M C , Han Y D. Application of a three-level NPC inverter as a three-phase four-wire power quality compensator by generalized 3DSVM.IEEE Trans Power Electronics, 2006,21(2): 51-58.

[13] Zhan C J, Han Y, Xie T et al. Mathematical model and dual-DSP control of tri-level PWM reversible rectifier. Proceedings of the 1999 IEEE PEDS, 1999, 174-179.

[14] Zhan C J, Wong M C, Han Y, et al. Universal custom power conditioner(UCPC) in distribution networks . Proceedings of the 1999 IEEE PEDS, 1999, 1025-1029.

[15] Wei X L, Dai K, Fang X, Geng P, Kang Y. Performance Analysis and Improvement of Output Current Controller for Three-Phase Shunt Active Power Filter. Proceedings of the 2006 IEEE IECON ,2006, 1757-1762.

[16] Bellini A, Bifaretti S, Costantini S. Implementation on a microcontroller of a space vector modulation technique for NPC inverters. Proceedings of the 2004 IEEE ICIT, 2004, 935-940.

[17] Ge Q X, Wang X X, Zhang S T et al. A high power NPC three-level inverter equipped with IGCTs. Proceedings of the 2004 IEEE IPEMC, 2004, 1097-1100.

[18] Celanovic N, Voroyevich D. A comprehensive study of neutral-point voltage balancing problem in three-level neutral-point-clamped voltage source PWM inverters. IEEE Trans Power Electronics, 2000, 15(2): 242-249.

[19] Zhang K, Kang Y, Xiong J, Chen J. Deadbeat control of PWM inverter with repetitive disturbance prediction. Proceedings of the 1999 IEEE APEC ,1999 , 1026-1031.

A Novel Topology of LLC Resonant Inverter with Two Resonant Tanks for Power Conditioning System

Eun-Soo Kim, Kwang-Ho Lee, Bong-Gun Chung,
Joo-Hoon Kim

Dept. of Electrical Engineering, Jeonju University
1200, Hyoja-Dong, Wansan-Gu, Jeonju
Jeonbuk, 560-759, South Korea
E-mail : eskim@jj.ac.kr

Moon-Ho Kye

Powerplaza USA
1705 E. Garry Ave #102, CA
92705, USA
E-mail: rkye@powerplazaus.com

Abstract— **Recently, Power Conditioning System (PCS) tends to require more compact, higher efficiency, higher power density with better performance. To meet these requirements, a novel topology consisted of LLC resonant converter with two resonant tanks for low power losses and Low Frequency (LF) cyclo-converter for sine wave filtering is proposed. The operating schemes are analyzed and described. A 400W proto product is built, tested and verified the performances by connecting the 110Vac 60Hz utility line. Test results are described with its high efficiency and good performance.**

I. INTRODUCTION

Recently, electric power is generated thru several resources such as wind, solar cell, and fuel cell. As those kinds of generations become more, PCS are needed more and more. PCS tends to require more compact, higher efficiency, higher power density with better performance and lower price through simpler circuits. To meet these requirements, PCS consisted of high frequency Zero Voltage Switching (ZVS) DC/DC converter and PWM inverter is widely studied. [1, 2, 3] A high frequency link grid connected PCS with ZVS DC/DC converter would make an isolation transformer smaller and lighter as showed in Fig. 1. However, PWM inverter connected to the utility line makes it difficult to reduce the size and volume in the utility line connecting circuits because the circuits have big input capacitor (C_d), 60Hz low frequency inductor (L_s) and big power loss caused by PWM hard switching in the circuits.

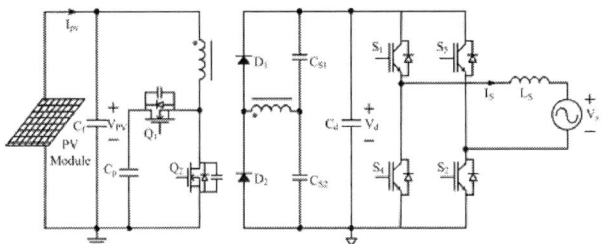

Figure 1. A high frequency link grid connected PCS [2]

In this paper a novel PCS topology consisted of LLC resonant inverter with two resonant tanks and LF (Low Frequency) cyclo-converter for sine wave filtering and connecting to the utility line is proposed. The two resonant tanks are consisted of two transformers and its secondary side leakage inductance is designed for filter inductance. The proposed topology is analyzed, designed and tested for 400W proto-type PCS unit. The unit is connected to the utility line of 110Vac 60Hz. A low cost structure and high compactness are good points, too.

II. CONFIGURATION OF PROPOSED LLC RESONANT CONVERTER AND LF CYCLO-CONVERTER

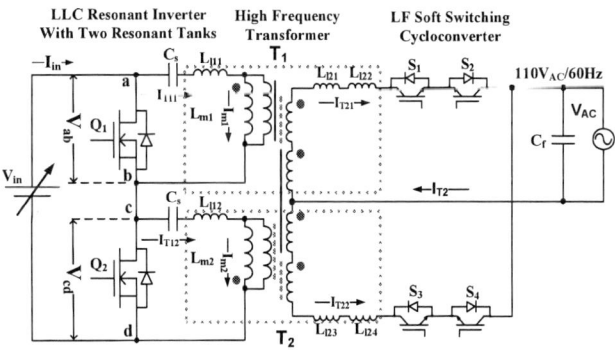

Figure 2. Configuration of proposed PCS topology consisted of LLC resonant inverter with two resonant tanks and LF Cycro-converter

Fig. 2 shows general configuration of the proposed PCS topology consisted of LLC resonant inverter with two resonant tanks and LF Cyclo-converter. LLC resonant inverter is made up of two separated high frequency transformers with their components. These can make compactness easier. A resonant tank 1 has a transformer T_1 and resonant capacitor C_s, and the other one does a transformer T_2 and resonant capacitor C_s independently.

Because the secondary windings are connected in series, the current unbalance issue can be resolved. So high voltage gain characteristics are achieved and high leakage inductance in the secondary side is able to use as LF (Low Frequency) filter inductor. As the circuit becomes simpler with smaller components numbers, the high compactness of PCS is able to achieve.

A. Equivalent circuit of the proposed LLC resonant inverter

Fig. 3(a) shows an equivalent circuit of the proposed topology. When a superposition theory is applied, Fig. 3(b) and Fig. 3(c) are two separate equivalent circuits with independent voltage sources (V_{ab}, V_{cd}) in each loop.

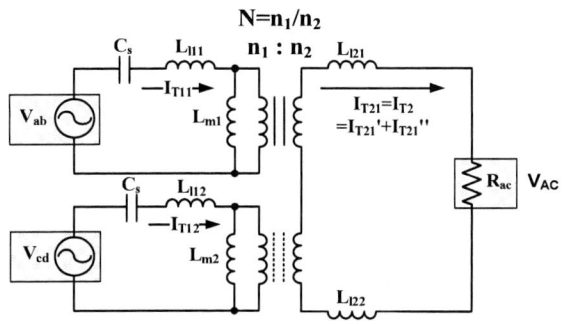

(a) Equivalent circuit of the proposed topology

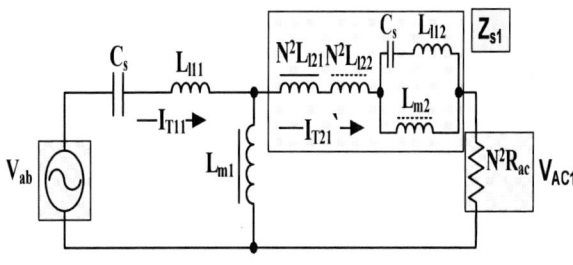

(b) Equivalent circuit for voltage source (V_{ab})

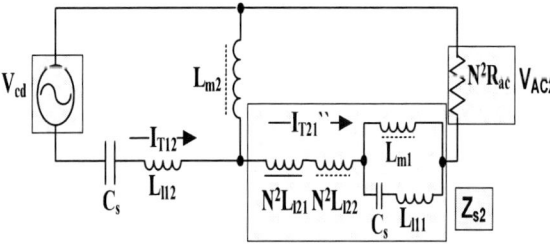

(c) Equivalent circuit for voltage source (V_{cd})

Figure 3. Equivalent circuit of the proposed circuit with superposition theory

Equation (2) & (3) show voltage equations (G_{V1}, G_{V2}) with voltage sources (V_{ab}, V_{cd}) and output voltages (V_{AC1}, V_{AC2}).

Equation (1) is to sum of two equations and explains the overall circuit of Fig. 3(a). The impedance related equations are shown in equation (4) & (5) and equation (6) & (7) with normalized values.

$$G_V = G_{V1} + G_{V2} \tag{1}$$

$$G_{V1} = \frac{V_{AC1}}{V_{ab}} \tag{2}$$

$$= \frac{1}{N} \left| \frac{1}{S_1 \left\{ \frac{L_{p1}}{L_{m1}} - \frac{1}{\omega^2 L_{m1} C_s} + j\left(\omega L_{l11} - \frac{1}{\omega C_s}\right)\left(\frac{1}{Z_{s1} + N^2 R_{ac}}\right)\right\}} \right|$$

$$G_{V2} = \frac{V_{AC2}}{V_{cd}} \tag{3}$$

$$= \frac{1}{N} \left| \frac{1}{S_2 \left\{ \frac{L_{p2}}{L_{m2}} - \frac{1}{\omega^2 L_{m2} C_s} + j\left(\omega L_{l12} - \frac{1}{\omega C_s}\right)\left(\frac{1}{Z_{s2} + N^2 R_{ac}}\right)\right\}} \right|$$

$$Z_{s1} = N^2 j\omega L_{ls1} + \frac{j\omega L_{m2}\left(j\omega L_{l12} - j\frac{1}{\omega C_s}\right)}{j\omega L_{m2} + j\omega L_{l12} - j\frac{1}{\omega C_s}} \tag{4}$$

$$Z_{s2} = N^2 j\omega L_{ls2} + \frac{j\omega L_{m1}\left(j\omega L_{l11} - j\frac{1}{\omega C_s}\right)}{j\omega L_{m1} + j\omega L_{l11} - j\frac{1}{\omega C_s}} \tag{5}$$

$$L_{ls1} = L_{l21} + L_{l22} \tag{6}$$

$$L_{p1} = L_{m1} + L_{l11}, \quad L_{p2} = L_{m2} + L_{l12} \tag{7}$$

Where

$$S_1 = 1 + \frac{Z_{s1}}{NR_{ac}}, \quad S_2 = 1 + \frac{Z_{s2}}{NR_{ac}} \tag{8}$$

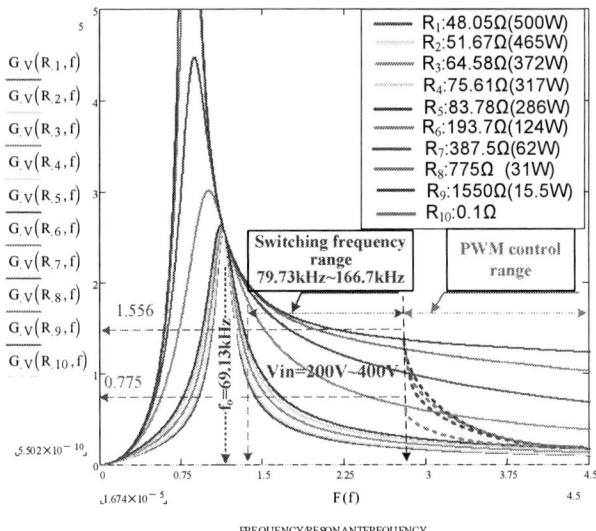

Figure 4. Voltage gain characteristics due to load variation (500W~ 15.5W)

B. Control scheme of the proposed LLC resonant inverter

Gain characteristics of LLC resonant inverter with two resonant tanks are shown in Fig. 4 and the method to connect the inverter to the utility line by using the gain characteristics is described. As shown in Fig. 5(a), the configuration of control circuit works as following steps.

1) Line voltage (V_{ac}) steps down through PT, and this is used as a reference signal.
2) Square waveform is made from window comparator for the line voltage
3) And the switches in the cyclo-converter turn on/off alternatively per the polarity of line voltage.
   (Positive Polarity: S_1, S_3 Turn-on, S_2,S_4 Turn-off,
   Negative polarity: S_2, S_4 Turn-on, S_1, S_3 Turn-off)
4) At the same time, the reference signal is rectified into |ksinωt|.
5) The error signal (I_o) between the CPU current reference (I_{ref}) and sensed input current signal (I_{IN-fed}) multiplies with the rectified reference signal(|ksinωt|) droved step (4) in the outer loop.
6) The multiplier output is used as a current commend signal (I_{com}).
7) In the inner-loop current compensator, the difference of output current (I_{T2_fed}) through feedback from the current commend signal (I_{com}) makes error signal (I_{err1}) to synchronize the output current to the line voltage
8) MC34067 IC changes the switching frequency as for this synchronized signal.
9) The output current (I_{T2}) is synchronized to the utility line and controlled as load varies.
10) Here, the maximum switching frequency control to reduce the gain (G_V) and output current (I_{T2}) is decided in according to the error signal (I_{err1}). However even

through the switching frequency is increased to the maximum, the gain shown in Fig. 4 has the constant value in the light load. Especially, waveform of the output current (I_{T2}) in the low line voltage (V_{ac}) and the light load can be distorted due to the constant dc gain characteristics. Therefore, to improve the waveform of the output current (I_{T2}) in the low line voltage (V_{ac}) and the light load, PWM control is needed in the maximum switching frequency as shown in Fig. 4 and Fig. 5.

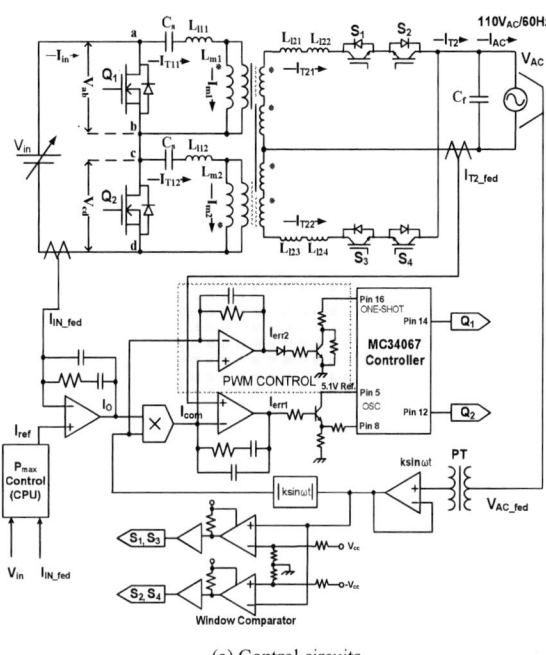

(a) Control circuits

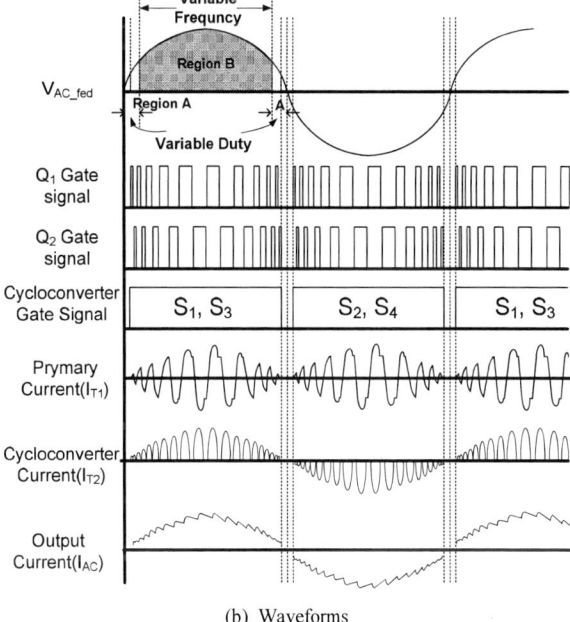

(b) Waveforms

Figure 5. Control circuits and expected waveforms of the proposed topology

978-1-4244-4782-4/10 $26.00 © 2010 IEEE

III. EXPERIMENTAL RESULTS

The specification of a proto product and its conditions are shown in table I & II. The proposed topology is designed to synchronize the unit with input voltage from 200Vdc to 400Vdc to the utility line of 110Vac 60Hz based on the analysis of equation (1) thru equation (8) and Fig. 4.

TABLE I.   PRINCIPAL SPECIFICATION OF THE PROPOSED LLC RESONANT CONVERTER

Input Voltage (V_{in})	200Vdc ~ 400Vdc
Resonant Frequency (f_r)	69.13kHz
Output Capacitor (C_f)	3uF
Output Voltage (V_{ac})	110Vac 60Hz
Switching frequency (f_s)	79.73kHz ~ 166.7kHz
Series Resonant Capacitor (C_s)	27nF
Switching Components (Q_1, Q_2)	SD20N60(600V, 20A)
LF Cyclo-converter IGBT	11N120CND(1200V 43A)

TABLE II.   PROPOSED LLC RESONANT INVERTER PARAMETERS

	LLC resonant inverter parameters	
	Transformer (T_1)	Transformer (T2)
Primary leakage inductance(L_{l11}, L_{l12})	89.68uH	88.28uH
Secondary leakage inductance(L_{l12}, L_{l22})	115.3uH	113.2uH
Magnetizing inductance(L_{m1}, L_{m2})	309.8uH	310.4uH
Equivalent leakage Inductance(L_{eq1}, L_{eq2})	197.776uH	195.133uH
$N(n_1/n_2)$	42/35	42/35
Cores	EFD5258: H:7mm L:59.6mm	

The values of series capacitors are calculated by the equivalent leakage inductances, and resonant frequency. The primary leakage inductance, secondary leakage inductance and magnetizing inductance are induced from the normalized equations. The parameters shown in table II and the voltage gain characteristics of equation (1) are applied to the load of 48.39.Ω(500W) ~ 2419Ω(15.5W) at the input voltage from 200V_{DC} to 400V_{DC}. When the series resonant frequency is 69.6kHz and the maximum load is 500W, the minimum switching frequency is limited by 69kHz and max frequency is by 300kHz in order to get all switching components soft switched.

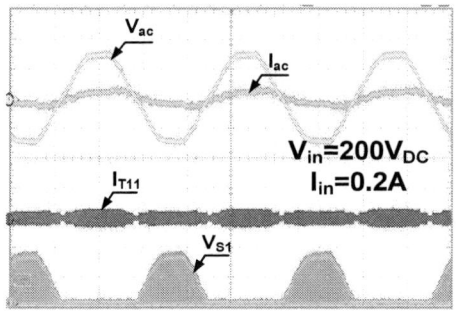

(a) I_{in} = 0.2A (100V/div., 2A/div., 5A/div., 200V/div., 5ms/div.)

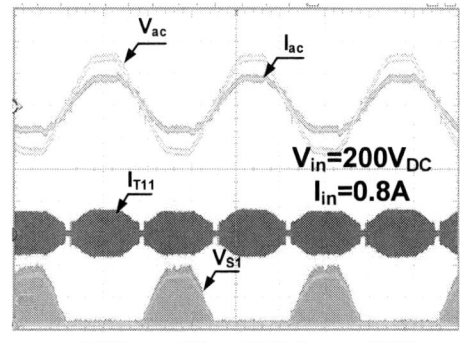

(b) I_{in} = 0.8A (100V/div., 2A/div., 5A/div., 200V/div., 5ms/div.)

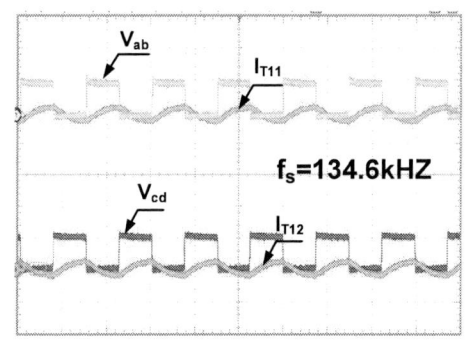

(c) I_{in} = 0.2A (200V/div., 5A/div., 200V/div., 5A/div., 5us/div.)

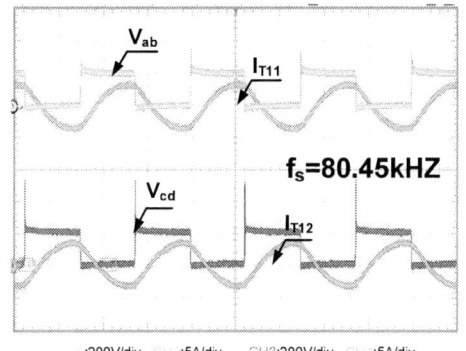

(d) I_{in} = 0.8A (200V/div., 5A/div., 200V/div., 5A/div., 5us/div.)

Figure 6.   Experimental waveforms when utility line (110Vac, 60Hz) is connected (V_{in} = 200Vdc)

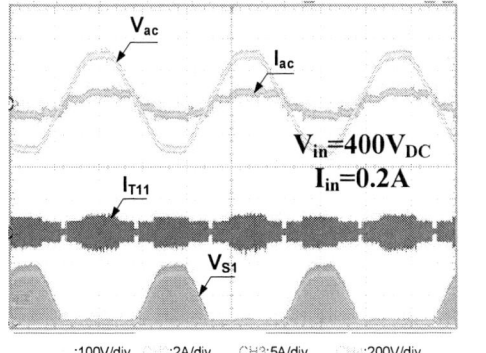

(a) I_{in} = 0.2A (100V/div., 2A/div., 5A/div., 200V/div., 5ms/div.)

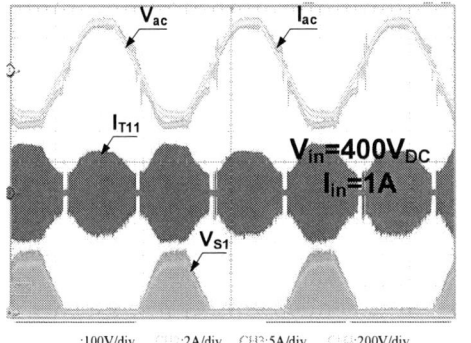

(b) I_{in} = 1A (100V/div., 2A/div., 200V/div., 5A/div., 5ms/div.)

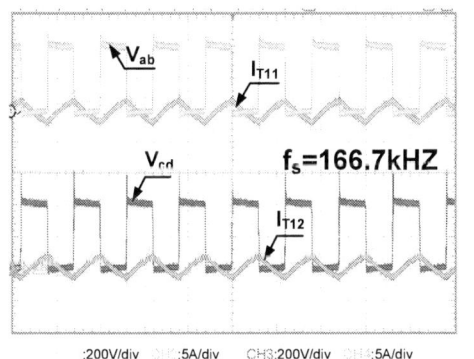

(c) I_{in} = 0.2A (200V/div., 5A/div., 200V/div., 5A/div., 5us/div.)

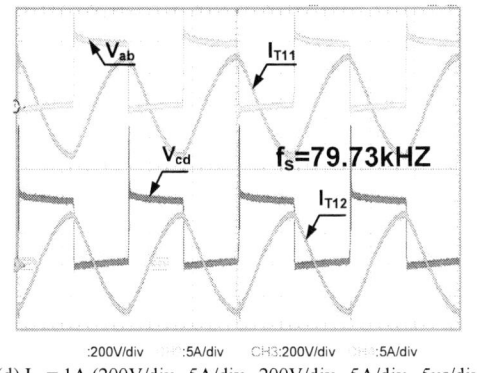

(d) I_{in} = 1A (200V/div., 5A/div., 200V/div., 5A/div., 5us/div.)

Figure 7. Experimental waveforms when utility line (110Vac, 60Hz) is connected (V_{in} = 400Vdc)

Fig. 6, Fig. 7 and Fig. 8 show the waveforms in the PCS circuit connecting to the line (110Vac 60Hz) thru constant current control per input=200Vdc and 400Vdc. The proto product is synchronized to the utility line as load is changed from 0.2A to 1A. The waveforms show that MOSFET's in the primary and IGBT's in the cyclo-converter got soft switching.

When the low output voltage (line voltage(Vac)) shown in Fig. 5(b) or light load is required, switching frequency is decided per the voltage gain characteristics shown in Fig. 4 and the output current (I_{ac}) becomes squared wave as shown in Fig. 8(a). To avoid this situation, the dead-time is adjusted for the controlled duty, output current (I_{ac}) is increased smoothly and becomes a sine wave.

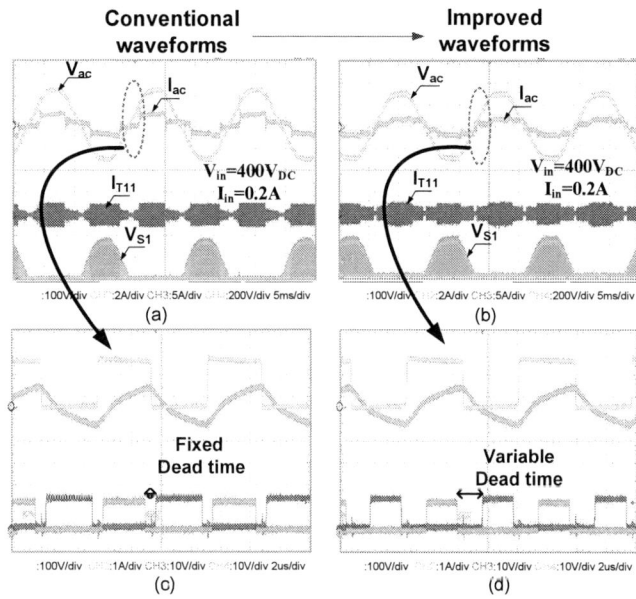

Figure 8. Experimental waveforms when utility line (110Vac, 60Hz) in the light load condition is connected (V_{in} = 400Vdc)

IV. CONCLUSION

This paper proposed a novel PCS topology consisted of LLC resonant converter with two resonant tanks and lower frequency cyclo-converter consisted of simple circuits for high power density, good performance, low cost. Technical operating analysis was explained and a 400W PCS proto product was designed, built, tested and verified. It was synchronized to 110Vac / 60Hz utility line. The primary and the secondary switches got soft-switchings to reduce switching losses and noise generation. The unit was very stable and worked well as the theoretical analysis was described. The test results showed that the proposed topology is good to use for improved Power Condition System.

ACKNOWLEDGMENT

This work was supported by the Korea Research Foundation Grant funded by the Korean Government (KRF-2008-313-D00369)

REFERENCES

[1] M. Cacciato, A. Consoli, etc., "A Digitally Controlled Double Stage Soft Switching Converter for Grid-connected Photovoltaic Applications"' IEEE APEC, pp 141~147, 2008.

[2] Hyun-Woo Seo, Jung-Min Kwon, Eung-Ho Kim, Bong-Hwan Kwon, "Modular Line-connected Photovoltaic PCS", The Transaction of The Korean Institute of Power Electronics, Vol. 13, No. 2, pp119~127, 2008. 4.

[3] Joy Mazumdar, Issa Batarseh, Nasser Kutkut, Osman Demirci, "High Frequency Low Cost DC-AC Inverter Design with Fuel Cell Source for Home Applications", IEEE IAS Volume 2, pp 789~794, 2002.

[4] Songquan Deng, Hong Mao, Joy Mazumdar, Issa Batarseh, Kazi Khairul Islam, "A New Control Scheme for High-frequency Link Inverter Design", IEEE APEC'03, pp 512~517, 2003.

[5] Jee-Hoon Jung, Jong-Moon Choi and Joong-Gi Kwon, "Design Methodology for Transformers Including Integrated and Center-tapped Structures for LLC Resonant Converters", Journal of Power Electronics, Vol. 9, No. 2, pp 215~223, March 2009

Analysis and Realization of a Fast Repetitive Controller in Active Power Filter System

Jinwu Gong, Xiaoming Zha, *Member, IEEE,* Suxuan Guo, Baifeng Chen, and Jianjun Sun

School of Electrical Engineering
Wuhan University
Wuhan, China
Gtmobile@foxmail.com, Xmzha@whu.edu.cn

Abstract— The repetitive control and resonant control are widely used in grid-connected voltage-source converters recently, such as uninterruptible power supplies (UPS), photovoltaic inverter (PV) and active power filter (APF). This paper introduces a new repetitive control strategy in APF system, whose repetitive period is only one half of the traditional and could eliminate all odd harmonics as a bank of resonant controllers. Compared to traditional repetitive control, proposed control strategy has faster tracking speed and smaller steady-state error; Compared to traditional PR control, it could be simply implemented in digital processor. Theory analysis with simulations and experiments demonstrate the validity and superiority of proposed fast repetitive control strategy.

Index Terms--Repetitive control; active power filter; fast repetitive controller.

I. INTRODUCTION

Active power filter (APF) has been demonstrated to be a very effective mechanism to alleviate the harmonic currents due to the daily growing nonlinear loads in the power system [1-2]. In the industry application of APF, Implementing a fast and accurate current controller is generally viewed as an important design work. Simple linear proportional–integral (PI) controller has the characteristics of fast tracking speed and easy for implementation, but it also has the drawbacks: steady-state error in the stationary frame and the need to decouple phase dependency in three phase systems [3].

To overcome the shortage of the PI controller, the repetitive control, which is originated from internal-model principle of control theory [4], is put forward and widely used in wind, hydro and solar energy system; uninterruptible power supplies (UPS), dynamic voltage restorer (DVR), and APF systems. Repetitive control's transfer function is usually described as

$$H_{\mathrm{RE}} = K_{\mathrm{r}} + \frac{K_{\mathrm{s}}}{\mathrm{e}^{\mathrm{s}T} - 1} \qquad (1)$$

In the (1), K_r and K_s represent the gain coefficients of direct feedback control and repetitive control respectively.

Repetitive control can incorporate feedback and feedforward into the controller of APF and it also has the characteristics of good robustness and self-stabilizing [5]. The block diagram of repetitive controller and theoretical Bode plots of the repetitive loop are shown in Fig.1 (a) and (b).

Fig. 1 (b) shows that repetitive control has infinite gain at frequency point of all harmonics, which means the repetitive control could eliminate all harmonic signals. But the repetitive control is based on a time delay T, which would influence the dynamic response.

For overcoming the computational burden and still achieving virtually similar frequency response characteristics as a synchronous frame PI controller, the proportional–resonant (PR) controller was developed to track reference in the stationary frame [6-7]. The PR controller's transfer function is usually described as

$$H_{PR} = K_{\mathrm{p}} + \frac{K_{\mathrm{i}}\mathrm{s}}{\mathrm{s}^2 + \omega_0^2} \qquad (2)$$

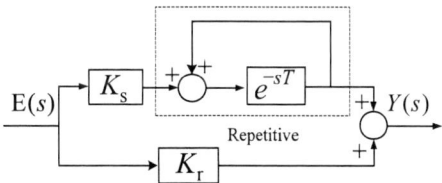

(a) Repetitive controller

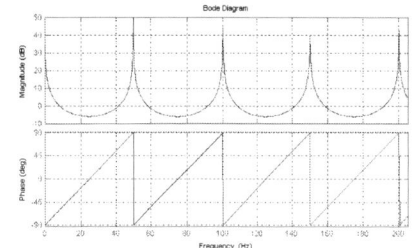

(b) Bode plots of repetitive loop

Fig. 1. Block diagram and Bode plots of repetitive controller

978-1-4244-4782-4/10 $26.00 © 2010 IEEE

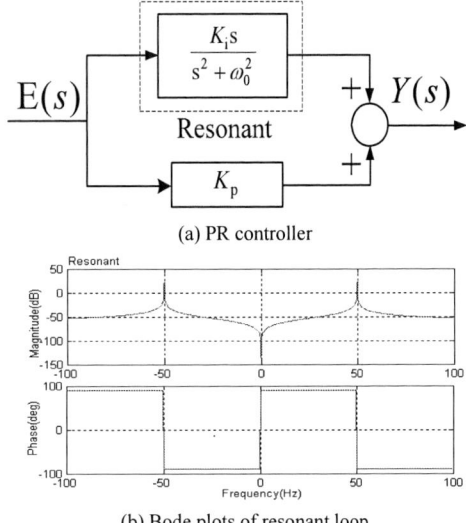

(a) PR controller

(b) Bode plots of resonant loop

Fig. 2. Block diagram and Bode plots of PR controller

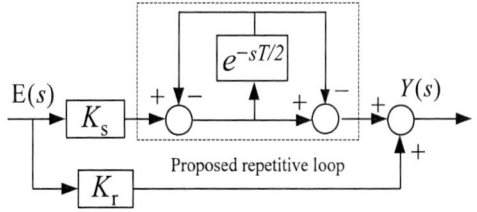

(a) Block diagram of proposed repetitive controller

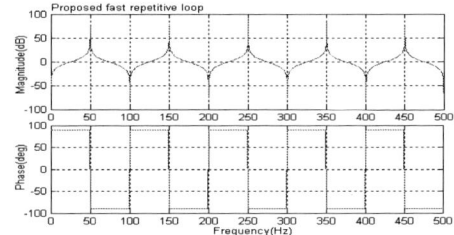

(b) Bode plots of the proposed repetitive loop

Fig. 3. Block diagram and Bode plots of poposed controller

In the (2), K_p and K_i represent the proportional and resonant gains respectively. The block diagram of PR control and theoretical Bode plots of resonant loop are shown in Fig.2 (a) and (b). As shown in Fig. 2(b), the PR controller has a theoretically infinite resonant peak at special frequency and could eliminate the steady-state error to zero at that frequency, which is therefore similar to an integrator who has an infinite gain for DC signals. The resonant loop of the PR controller can therefore be viewed as a generalized AC integrator, as proven in [8]. So, by using multiple PR controllers, the APF could eliminate selectively order harmonics [9-10]. This control strategy also has been used in UPS and single-phase photovoltaic inverters [11]. Although the PR control could realize zero steady-state error for AC signals, the inclusion of a series of resonant controllers in digital processor would undoubtedly complicate the overall implementation of current controller [12].

To avoid the disadvantages of the PI controller, conventional repetitive controller and PR controller, this paper introduces a new repetitive control strategy for eliminating all odd harmonics in three-phase APF system. This new control strategy not only blends the characteristic of the PR controller, but also has faster tracking speed and smaller steady-state error when compared with traditional repetitive controller.

II. PROPOSED CONTROL STRATEGY

A. Proposed Control Scheme

The transfer function of proposed repetitive control scheme is shown in (3) and the block diagram of controller is shown in Fig. 2(a).

$$\frac{Y(S)}{E(S)} = \frac{1-e^{-sT/2}}{1+e^{-sT/2}} = \frac{4}{T}\left\{\sum_{k=1}^{\infty}\frac{2s}{s^2+\left[(2k-1)\omega\right]^2}\right\} \quad (3)$$

Obviously, the (3) resembles a bank of resonant controllers for all odd harmonics. The Bode plots of the proposed control scheme is shown in Fig. 3(b), at $(2k\pm1)*50Hz, k=1,2,3...$, the proposed control strategy has infinite gain at the frequency of all odd harmonics, which could be used to eliminate harmonics in power system. The block diagram presented in Fig.3 only needs a time delay block, so the control strategy could be easily implemented in the DSP or FPGA [13-18].

B. Convergence Analysis

For analysis the dynamic characteristic and stability characteristic of this new control strategy, block diagram of a typical APF system is presented in the Fig. 4. After detecting harmonic current signals i_{sh} at the system side, the repetitive controller generates the reference signals, which let the inverter output an appropriate harmonic current i_{AF} to compensate the harmonic component current at load side. The block diagram of proposed repetitive controller based closed-loop control system is shown in Fig. 5.

In Fig. 5, $Id(s)$ and $e(s)$ represent the harmonic current at load side and system side respectively. $G_{AF}(s)$ stands for the transfer function of PWM inverter and the second-order lower pass filter for the harmonic detection, $G_{SP}(s)$ stands for the output filter. K_r and K_s represent the gain coefficient of the direct feedback control and the repetitive control respectively.

The repetitive controller could eliminate all periodic error caused by periodic disturbance $e(t)$, when periodic error occurs, the repetitive loop could track the real-time error and compensate it in the next period. So, for the repetitive loop, the dynamic response of repetitive loop is based on the repetitive period T/2. The analysis for convergence process is taken out in the discrete domain, which is base on the repetitive period T/2. T is the fundamental period, which is 0.02s here. And the k denotes the counting of the repetitive period, k=1, 2, 3... are equal to 0.5T, T, 1.5T respectively. And the convergence process repetitive control could be described in (4) to (6):

978-1-4244-4782-4/10 $26.00 © 2010 IEEE   1705

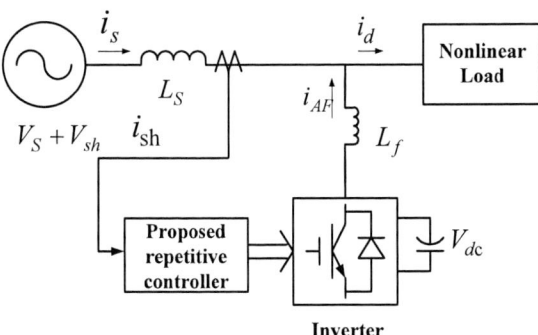

Fig. 4. Block diagram of APF configuration system

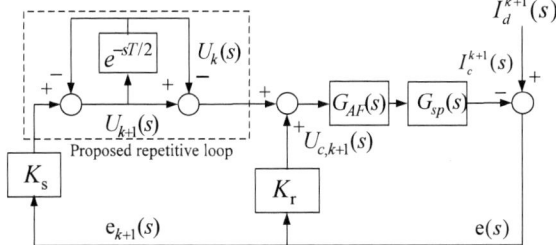

Fig. 5. Block diagram of the APF closed-loop control system

$$U_{k+1}(s) = K_s \cdot e_{k+1}(s) - U_k(s) \tag{4}$$

$$U_{c,k+1}(s) = K_r \cdot e_{k+1}(t) \tag{5}$$

$$e_{k+1}(s) = I_d^{k+1}(s) - (U_{k+1}(s) - U_k(s)$$

$$+ U_{c,k+1}(s)) \cdot G_{AF}(s) \cdot G_{sp}(s) \tag{6}$$

From (4) to (6), the relationship between $U_{k+1}(s)$ with $U_k(s)$ and $I_d^{k+1}(s)$ could be got as

$$U_{k+1}(s) = (1 - \frac{2(G_{AF}(s)G_{SP}(s)K_r + 1)}{1 + G_{AF}(s)G_{SP}(s)(K_s + K_r)})U_k(s)$$

$$+ \frac{K_s}{1 + G_{AF}(s)G_{SP}(s)(K_s + K_r)} I_d^{k+1}(s) \tag{7}$$

Similar with (7), the relationship between $U_k(s)$, $U_{k-1}(s)$ and $I_d^k(s)$ could also be presented as

$$U_k(s) = (1 - \frac{2(G_{AF}(s)G_{SP}(s)K_r + 1)}{1 + G_{AF}(s)G_{SP}(s)(K_s + K_r)})U_{k-1}(s)$$

$$+ \frac{K_s}{1 + G_{AF}(s)G_{SP}(s)(K_s + K_r)} I_d^k(s) \tag{8}$$

When the control system reaches the steady-state, the steady-state error would remain the same, which could be

expressed as $I_d^{k+1}(s) = -I_d^k(s)$. Together with (4) and (7) + (8), the convergence process could be deduced as

$$e_{k+1}(s) = (1 - \frac{2(G_{AF}(s)G_{SP}(s)K_r + 1)}{1 + G_{AF}(s)G_{SP}(s)(K_s + K_r)})e_k(s)$$

$$= \frac{G_{AF}(s) \cdot G_{SP}(s) \cdot (K_s - K_r) - 1}{1 + G_{AF}(s)G_{SP}(s)(K_s + K_r)} \cdot e_k(s) \tag{9}$$

At first, defining $G_{conF}(s)$ as convergence expression, which demonstrates the relationship of errors between two consecutive repetitive periods.

$$G_{conF}(s) = 1 - \frac{2(G_{AF}(s)G_{SP}(s)K_r + 1)}{1 + G_{AF}(s)G_{SP}(s)(K_s + K_r)} \tag{10}$$

In the dynamic process, the (9) also demonstrates the convergence characteristic of error. Only when $G_{conF}(s)$ serves

$$\|G_{conF}(s)\| < 1 \tag{11}$$

The control system would be stable. In other words, the error would convergence to zero when convergence expression is small than one. And the smaller the $G_{conF}(s)$ is, the faster convergence speed of error would be.

In the (7), the second part represents the error caused by $I_d(s)$, and the error transfer function of $I_d(s)$ in the closed-loop control system could be described as

$$\Phi_{Fe}(s) = \frac{e_{k+1}(s)}{I_d(s)} = \frac{1}{1 + G_{AF}(s)G_{SP}(s)(K_s + K_r)} \tag{12}$$

Equation (12) demonstrate that the steady-state error dependent on the value of $G_{AF}(s)$, $G_{SP}(s)$ and K_s+K_r.

III. TRADITIONAL REPETITIVE CONTROLLER

In order to compare traditional repetitive controller with the proposed fast repetitive controller, the block diagram of traditional repetitive control is shown in Fig. 6.

The meaning of each part in the Fig. 6 is same as the meaning in the Fig. 5. The analysis also is taken out in the discrete domain, which is base on the repetitive period T (T is the fundamental cycle, which is 0.02s here), and the k denotes the counting of the repetitive period, k=1, 2, 3…equal to $T, 2T, 3T$ respectively. The difference is that the counting of the repetitive period is equal to T, not 0.5T. Also the repetitive control process could be described in (13)-(15):

$$U_{k+1}(s) = K_s \cdot e_{k+1}(s) + U_k(s) \tag{13}$$

$$U_{c,k+1}(s) = K_r \cdot e_{k+1}(t) \tag{14}$$

$$e_{k+1}(s) = I_d^{k+1}(s) - (U_k(s) + U_{c,k+1}(s)) \cdot G_{AF}(s) \cdot G_{sp}(s) \tag{15}$$

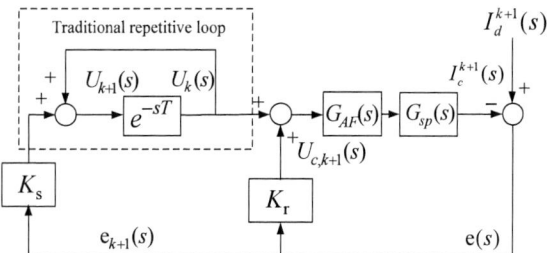

Fig. 6. Block diagram of traditional repetitive control system

From (13) to (15), the relationship between $U_{k+1}(s)$, $U_k(s)$ and $I_d^{k+1}(s)$ could also be expressed as

$$U_{k+1}(s) = (1 - \frac{K_s \cdot G_{AF}(s)G_{sp}(s)}{1 + K_r \cdot G_{AF}(s)G_{sp}(s)})U_k(s)$$

$$+ \frac{K_s}{1 + G_{AF}(s)G_{sp}(s) \cdot K_r} I_d^{k+1}(s) \qquad (16)$$

When the control system reaches the steady-state, the steady-state error would remain the same, which could be expressed as $I_d^{k+1}(s) = I_d^k(s)$. Together with (13) and (16), the convergence process could be deduced as

$$e_{k+1}(s) = (1 - \frac{K_s \cdot G_{AF}(s)G_{sp}(s)}{1 + K_r \cdot G_{AF}(s)G_{sp}(s)})e_k(s)$$

$$= \frac{1 - (K_s - K_r) \cdot G_{AF}(s) \cdot G_{sp}(s)}{1 + K_r \cdot G_{AF}(s) \cdot G_{sp}(s)} \cdot e_k(s) \qquad (17)$$

Besides, $G_{conT}(s)$ could also be defined as convergence expression for traditional repetitive controller. This expression also demonstrates the relationship of errors between two consecutive repetitive periods.

$$G_{conT}(s) = (1 - \frac{K_s \cdot G_{AF}(s)G_{sp}(s)}{1 + K_r \cdot G_{AF}(s)G_{sp}(s)}) \qquad (18)$$

In the (17), the second part represents the error caused by $I_d(s)$, and the error transfer function of $I_d(s)$ in the closed-loop control system is

$$\Phi_{Te}(s) = \frac{e_{k+1}(s)}{I_d(s)} = \frac{1}{1 + G_{AF}(s)G_{SP}(s) \cdot K_r} \qquad (19)$$

Equation (19) demonstrates that the steady-state error only dependent on the value of K_r. Comparing the error transfer function in (12) and (18), the proposed fast repetitive control is superior to the traditional repetitive control for smaller steady-state error.

The Convergence speed of repetitive control dependents on two factors, the first one is the repetitive period, and the second one is the convergence expression.

When considering the repetitive period, the fast repetitive control is superior to the traditional repetitive control, for the repetitive period is 0.5T, which is just one half of the traditional repetitive controller. So the proposed fast repetitive controller has fast dynamic response time.

When considering the convergence expression, a comparison between $G_{conF}(s)$ and $G_{conT}(s)$ is carried out as following

$$\left| \frac{G_{conF}(s)}{G_{conT}(s)} \right| = \left| \frac{1 + G_{AF}(s)G_{sp}(s)K_r}{1 + G_{AF}(s)G_{SP}(s)(K_s + K_r)} \right| < 1 \qquad (20)$$

Equation (20) means the fast repetitive controller not only has the small repetitive period but also has a smaller convergence expression. That is the proposed fast repetitive controller has better dynamic response.

IV. PRACTICAL APPLICATION ISSUES

The traditional repetitive loop is shown in Fig. 7(a), and transfer function is given out in (20):

$$\frac{Y(S)}{E(S)} = \frac{e^{-sT}}{1 - e^{-sT}} = \frac{1}{e^{sT} - 1} \qquad (21)$$

In Equation (21), $\omega_0 = 2\pi f_0 = 2\pi / T$, $f_0 = 50$Hz, T=0.02ms. The poles of repetitive loop can be found from $e^{sT} = 1$ ($\omega = \pm k\omega_0$, k=0, 1, 2...). The poles locations are shown in Fig. 7(c).

The proposed fast repetitive loop is shown in Fig. 7(b), and transfer function is

$$\frac{Y(S)}{E(S)} = \frac{1 - e^{-sT/2}}{1 + e^{-sT/2}} \qquad (22)$$

In the (22), the poles can be found by $e^{-sT/2} = -1$ ($\omega = \pm(2k+1)\omega_0$, k=0, 1, 2...). The zeros can be found by $e^{-sT/2} = 1$ ($\omega = \pm 2k\omega_0$, k=0, 1, 2...). The poles and zeros locations are shown in Fig. 7(d).

As commonly known, repetitive loop may bring unstable problems to control system, to make controllers more realizable in practice, this paper introduces a forgetting factor (FF) K_f immediately after the delay block for introducing some damping and enhancing robustness to the control path. The K_f ads damping to all the poles by slightly shifting them to the left of the imaginary axis by a distance of $\sigma > 0$. For the traditional repetitive loop, this shifting process is shown in Fig. 7(c) and could also be mathematically expressed as

$$\frac{Y(s+\sigma)}{E(s+\sigma)} = \frac{e^{-sT}}{1 - K_f e^{-sT}} = \frac{1}{K_f} \cdot \frac{e^{-(s+\sigma)T}}{1 - e^{-(s+\sigma)T}} \qquad (23)$$

$$0 \le K_f = e^{-\sigma \cdot T} \le 1, s = j\omega \qquad (24)$$

978-1-4244-4782-4/10 $26.00 © 2010 IEEE

$$0 \le K_f = e^{-\sigma \cdot T/2} \le 1 \qquad (26)$$

After the changing, the block diagram of traditional repetitive control and Bode plots of repetitive loop are shown in Fig. 7(a), (b). And the changing process of poles-zeros and Bode plots of proposed fast repetitive loop are shown in Fig. 7(d), (f).

The K_f would increase system stability by shifting the poles to the left of the imaginary axis. Compared with traditional repetitive control, the proposed fast repetitive control strategy has following advantages during this shifting process.

Firstly, regardless of the change of K_f, the proposed fast repetitive control loop has numbers of notches located in the middle of two consecutive peaks, thus, these notches improve the selective nature of the whole control system, which will in turn allow bigger gains and better performance. However, the traditional controller does not have notches and the minimum magnitude gain is no less than 1/2 (-6.0 dB). So the proposed control strategy has better selectivity and inhibitory.

Secondly, influenced by K_f, the Bode plots in Fig. 7 (e) and (f) clearly show that as K_f decreases from 1 to 0.9, the peak amplitude is reduced while the bandwidth of each peak increases. The traditional repetitive controller's peaks, originally of infinite magnitude, would have a maximum magnitude of $1/(1-K_f)$ and a minimum magnitude $1/(1+K_f)$. However, the proposed fast repetitive controller has a maximum magnitude of $(1+K_f)/(1-K_f)$ at original peaks, and a minimum magnitude of $(1-K_f)/(1+K_t)$ at original notches. So the proposed fast repetitive controller is superior to traditional controller when K_f is used to improve the robustness with respect to frequency variations.

Another modification recommended is to add a simple first order low pass filter (represented by $LPF = 1/(\tau s + 1)$) after the delay block. This alternation is introduced to remove switching noises from the control path and improve the overall system stability. But the low pass filter in current detection loop is enough for removing switching noises. So when equal to 0.95, the Kf would be sufficient for control system stability. The introduction of LPF and the mathematical analysis is not needed since high-order switching noised would be eliminated by the low-pass filter in harmonic detection loop [19].

V. SIMULATION AND EXPERIMENTS

A. Simulation

For verifying the validity of this proposed fast repetitive control strategy, the simulations for proposed repetitive control strategy and traditional repetitive control strategy are carried out on the MATLAB/simulink. The configuration of the APF system is shown in Fig. 8, the parameters of the APF system are shown in the TABLE 1. The simulation results are shown in the Fig. 9.

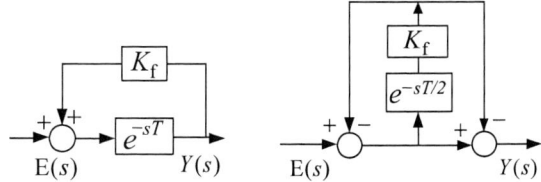

(a) Traditional repetitive loop   (b) Proposed repetitive loop

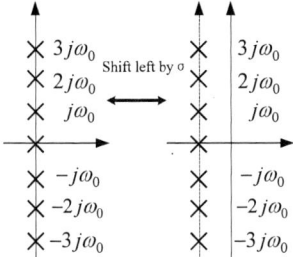

(c) Poles-zeros location of traditional repetitive control loop

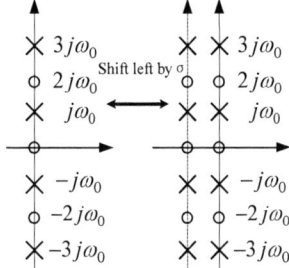

(d) Poles-zeros location of proposed repetitive control loop

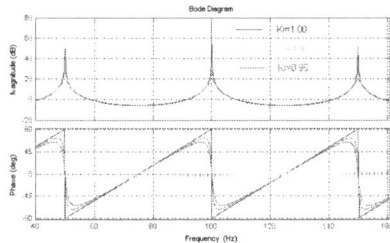

(e) Bode plots of traditional repetitive loop

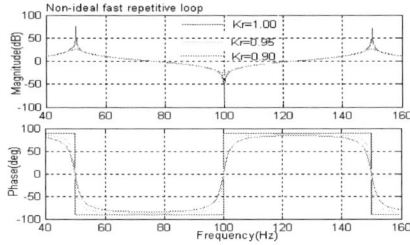

(f) Bode plots of proposed repetitive loop

Fig. 7. Block diagram, poles-zeros location and Bode plots

For the proposed fast repetitive loop, this shifting process is shown in Fig. 7(d) and it could also be mathematically expressed as

$$\frac{Y(s+\sigma)}{E(s+\sigma)} = \frac{1 - e^{-(s+\sigma)T/2}}{1 + e^{-(s+\sigma)T/2}} = \frac{1 - K_f e^{-sT/2}}{1 + K_f e^{-sT/2}} \qquad (25)$$

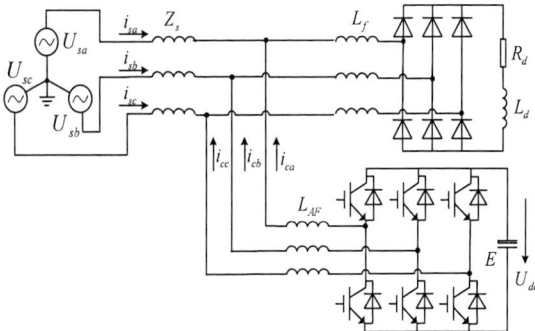

Fig. 8. Configuration the APF system

TABLE I
SIMULATION PARAMETERS

System side	Value	APF	Value	Load side	Value
system voltage	220V	U_{dc}	1000V	R_d	3 Ω
f	50Hz	L_{AF}	0.25 mH	L_d	1 mH

In the simulation of traditional repetitive controller, the APF starts to work at 0.06s. In the first compensation period, the repetitive loop is still out of work, and the repetitive loop would start work at 0.08s. In the first compensation period (0.06-0.08s), the system side current has four significant harmonic peaks, which could only be eliminated in the second compensation period (0.08-0.1s). When the control system reaches the steady-state, the Total Harmonic Distortion (THD) is reduced to 2.20%.

In the simulation of proposed fast repetitive controller, the APF starts to work at 0.06s. In the first half fundamental period (0.06-0.07s), the system side current has two significant harmonic peaks, which could be reduced in the second half fundamental period (0.07-0.08s). When the control system reaches the steady-state, the Total Harmonic Distortion (THD) is reduced to 1.96%.

B. Experiment

To test the validity of proposed fast repetitive controller and compare it with the traditional repetitive controller, these two controllers are used in the control of a 100 kVA shunt active filter, which is shown in Fig. 10. The control circuit is fully digitized and based on FPGA and DSP. FPGA mainly performs harmonic detection, repetitive control realization, and PWM generation and DC voltage controlling. DSP mainly performs man-machine interface function, data analysis and device protection.

The experiments results are shown in the Fig.11. The CH1 channel represents 80A/div (current); the CH2 channel represents 100V/div (voltage).

Without compensation, system has 35.5 A harmonic current and the THD is 34.5%, with the compensation of traditional repetitive controller, the THD is reduced to 5.24%. However, with the compensation of proposed fast repetitive controller, the THD is reduced to 5.16%.

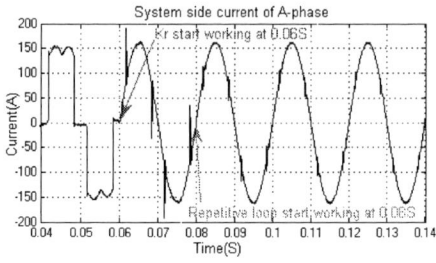

(a) Traditional repetitive controller

(b) FFT results

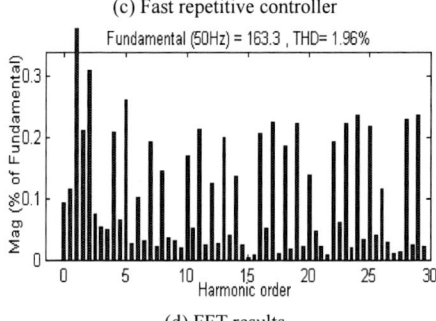

(c) Fast repetitive controller

(d) FFT results

Fig. 9. Simulation results

Fig. 10. Experiment Platform

978-1-4244-4782-4/10 $26.00 © 2010 IEEE

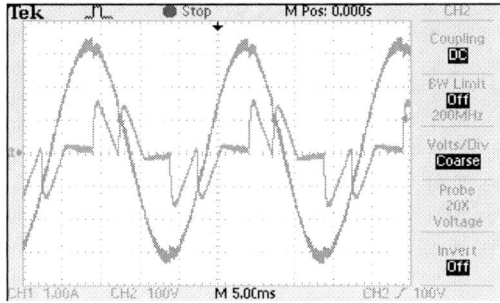

(a) Load side current of experiment

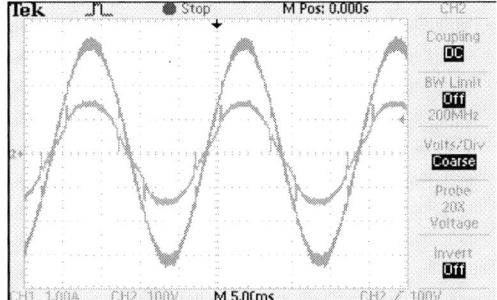

(b) Traditional repetitive controller

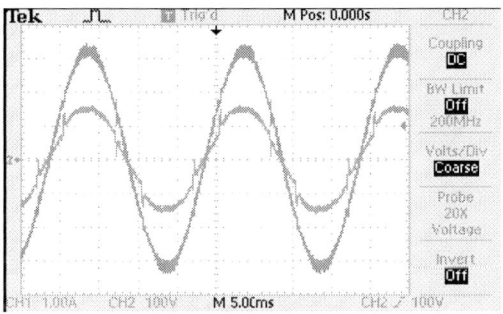

(c) Proposed fast repetitive controller

Fig. 11.   Experiment results

VI.   CONCLUSION

A fast repetitive controller is proposed in this paper, the detail analysis in closed-loop control system and frequency domains are also carried out. The theory analysis together with simulation and experiment results show that:

1) Similar with traditional repetitive controller, this controller could be easily implemented in DSP or FPGA.

2) Having the same function of a bank of resonant controllers, proposed controller could eliminate all odds harmonics.

3) Compared to traditional repetitive controller, this new controller has faster tracking speed and smaller steady-state error.

4) Compared to traditional repetitive controller, the proposed repetitive controller reflects better adaptability when control loop employs a K_f for practical application.

The proposed repetitive controller was successfully applied in the three-phase three-wire shunt APF system.

Through further analysis and design, the proposed fast repetitive control could also be applied into DVR, UPS, PV, STATCOM and other grid-connected converters.

REFERENCES

[1]   IEEE PES Working Group, FACTS Applications, IEEE Press, Publ. No. 96-TP-116, 1996.

[2]   J. Shlabbach, D. Blume and T. Stephanblome, "Voltage quality in electrical power systems," IEE power series No. 36, England, 2001.

[3]   M. Kazmierkowski., R. Krishnan, and F. Blaabjerg, "Control in power electronics. Selected problems" (Academic Press, 2002)

[4]   M Santolo, A Perfetto, "Comparison of different control techniques for active filter applications," Fourth IEEE International Caracas Conference on Devices, Circuits and Systems, Aruba, 2002.17-19.

[5]   X. M. Zha, Q. Tao, and J. J. Sun, "Development of iterative learning control strategy for active power filter," IEEE CCECE'02, Winnipeg, Manitoba, vol. 1, pp. 245-250, May 2002.

[6]   D. N. Zmood, D. G. Holmes and Bode, "Frequency domain analysis of three-phase linear current regulator", IEEE Trans. Ind. Appl., 2001, 37, pp. 601‒610

[7]   D. N. Zmood and D. G Holmes. "Stationary frame current regulation of PWM inverters with zero steady-state error", IEEE Trans. Power Electron., 2003, 18, pp. 814‒822

[8]   W. X. Yuan, and J. Allmeling. "Stationary frame generalized integrators for current control of active power filters with zero steady-state error for current harmonics of concern under unbalanced and distorted operating conditions", IEEE Trans. Ind. Appl., 2002, 38, pp. 523‒532

[9]   P. Mattavelli, "A closed-loop selective harmonic compensation for active filters", IEEE Trans. Ind. Appl., 2001, 37, pp. 81‒89

[10]   P. Mattavelli, "Synchronous-frame harmonic control for highperformance AC power supplies", IEEE Trans. Ind. Appl., 2001, 37, pp. 864‒872

[11]   R. Teodorescu, F. Blaabjerg, "A new control structure for grid-connected PV inverters with zero steady-state error and selective harmonic compensation".

[12]   J. Kauraniemi, T. I. Laakso and I. Hartimo, "Delta operator realizations of direct-form IIR filters," IEEE Trans. Circuits Syst., vol. 45, pp. 41‒51, Jan. 1998.

[13]   R Teodorescu, F. Blaabjerg and M. Liserre, "Proportional-resonant controllers and filters for grid-connected voltage-source converters," IEE Proc.-Electr. Power Appl., Vol. 153, No. 5, September 2006

[14]   G. Escobar, A.A. Valdez, J. Leyva-Ramos and P. Mattavelli, "A repetitive-based controller for UPS using a combined capacitor/load current sensing", Proc. IEEE-PESC, Recife, Brazil, 2005, pp. 955‒961

[15]   G. Escobar, P. R. Martinez, J. Leyva-Ramos and M. Hernandez-Gomez, "A repetitive-based controller for a powerfactor precompensator with harmonic compensation". Proc. IEEEPESC, Recife, Brazil, 2005, pp. 2363‒2369

[16]   G. Escobar, J. Leyva-Ramos, P. R. Martinez, and A. Valdez, "A repetitive-based controller for the boost converter to compensate the harmonic distortion of the output voltage", IEEE Trans. Control Syst. Tech., 2005, 13, pp. 500‒508

[17]   J. Leyva-Ramos, G. Escobar, P. R. Martinez and P .Mattavelli, "Analog circuits to implement repetitive controllers for tracking and disturbance rejection of periodic signals', IEEE Trans. Circuits Syst. II, 2005, 52, pp. 466‒470

[18]   G. Escobar, R. Ortega, and A. Astolfi, "A repetitive based controller for a shunt active filter to compensate for reactive power and harmonic distortion", Proc. 44th IEEE Conf. on Decision and Control and European Control Conf. (ECC), Seville, Spain, 2005, pp. 6480‒6485

[19]   G. Escobar, A.A. Valdez, J. Leyva-Ramos , P. Mattavelli, "A repetitive-based controller for UPS Inverter to compensate unbalance and harmonic distortion", IEEE Transactions on Industrial Electronics, Vol. 54, No. 1, February 2007, pp. 504-510..

On Extended Kalman Filters with Augmented State Vectors for the Stator Flux Estimation in SPMSMs

T.J. Vyncke, R.K. Boel and J.A.A. Melkebeek

Department of Electrical Energy, Systems and Automation (EESA)

Ghent University (UGent), Sint-Pietersnieuwstraat 41, B-9000 Gent, Belgium

phone: +32 (0)9 264 3442, fax: +32 (0)9 264 3582, e-mail: Thomas.Vyncke@UGent.be

Abstract—The demand for highly dynamic electrical drives, characterized by high quality torque control, in a wide variety of applications has grown tremendously during the past decades. Direct torque control (DTC) for permanent magnet synchronous motors (PMSM) can provide this accurate and fast torque control. When applying DTC the change of the stator flux linkage vector is controlled, based on torque and flux errors. As such the estimation of the stator flux linkage is essential. In the literature several possible solutions for the estimation of the stator flux linkage are proposed. In order to overcome problems associated with the integration of the back-emf, the use of state observers has been advocated in the literature. Several types of state observers have been conceived and implemented for PMSMs, especially the Extended Kalman Filter (EKF) has received much attention. In most reported applications however the EKF is only used to estimate the speed and rotor position of the PMSM in order to realize field oriented current control in a rotor reference frame. Far fewer publications mention the use of an EKF to estimate the stator flux linkage vector in order to apply DTC. Still the performance of the EKF in the estimation of the stator flux linkage vector has not yet been thoroughly investigated. In this paper the performance of the EKF for stator flux linkage is studied and simulated. The possibilities to improve the estimation by augmenting the state vector and the consequences of these alterations are explored. Important practical aspects for FPGA implementation are discussed.

I. INTRODUCTION

The use of highly dynamic electrical drives in a wide variety of applications has increased steadily in recent years. Within this market AC machines, and recently especially permanent magnet synchronous machines (PMSM's), have obtained dominance due to their characteristics of high efficiency, high power density and reliability. These highly dynamic electrical drives have to provide accurate and fast torque control together with the highest possible efficiency.

Rotor flux field oriented control has become an industry standard to control the torque and flux levels of AC machines. For induction motors (IM's) direct torque control (DTC) was proposed as an alternative control strategy in [1] and became very popular in the past two decades [2]. DTC for induction machines is inherently motion-state sensorless. In the past decade several authors [3]–[6] have proposed ways to adapt DTC to work with PMSM's.

To derive the principles of Direct Torque Control (DTC) the equation for electromagnetic torque T of a surface PMSM:

$$T = \frac{3N_p}{2L_s} \left| \underline{\Psi}_f \right| \left| \underline{\Psi}_s \right| \sin \delta \qquad (1)$$

is considered, where δ denotes the load angle between the stator flux linkage $\underline{\Psi}_s$ and permanent magnet flux linkage $\underline{\Psi}_f$ vectors in the stationary $\alpha\beta$ frame. The number of pole pairs is denoted by N_p and L_s is the stator inductance. From (1) can be seen that for constant stator flux linkage, the torque is changed by changing the load angle δ. The stator flux vector can be changed by applying from the inverter the voltage vector with the most appropriate radial and tangential components.

The switching decision is based on the estimated torque $T = \frac{3}{2}N_p(\Psi_{s,\alpha}I_\beta - \Psi_{s,\beta}I_\alpha)$, the stator flux linkage magnitude $\left| \underline{\Psi}_s \right|$ and stator flux linkage angle θ_{Ψ_s}; which are all determined by the estimation of the stator flux linkage components $\Psi_{s,\alpha}$ and $\Psi_{s,\beta}$. Thus the stator flux linkage estimation is crucial for a correct operation of the drive, as shown in figure 1.

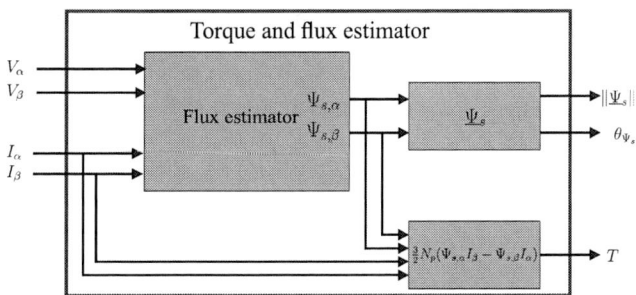

Figure 1: Torque and flux estimator schematic for DTC

Several estimation techniques have been reported in the literature [7]–[14]. Some include improvements on the back-emf integration such as low-pass-filtering [8] and stabilizing the integrator with a PI-corrector [12] or current offset [11]. Others use the current model of the PMSM, which often implies the use of a position sensor or needs an added, separate position estimation [14]; both are preferably avoided. State observers, such as the Extended Kalman Filter (EKF), that estimate the stator flux linkage vector by using its components as state variables or by calculating the flux components from other state components, and estimate rotor speed and position simultaneously, are another possibility. The EKF is often discussed for the sensorless control of PMSM's, however focused on the sensorless position estimation needed for field oriented control in the rotor flux reference frame. Few publications discuss the EKF for the estimation of the stator flux linkage vector [9], [10], [15]. Especially a thorough discussion addressing

978-1-4244-4782-4/10 $26.00 © 2010 IEEE

the effect of incorrect parameters (resistance, inductance, rotor flux magnitude) lacks.

The paper is organized as follows: in section II the EKF is discussed, using two different sets of state variables for an SPMSM, and it is shown that estimation errors occur with incorrect motor parameters. In section III both EKF versions are expanded to include parameter estimations. Simulated results with these EKF implementations are given in section IV. Important aspects of the practical implementation for FPGA are discussed in section V.

II. REDUCED-ORDER EKF FOR STATOR FLUX LINKAGE ESTIMATION

A. Stator Flux Linkage Estimation

In theory the integration of the back-emf can be used for this estimation when stator voltages and currents are measured:

$$\underline{\Psi}_s = \int_0^t (\underline{V}_s - R_s \underline{I}_s) dt + \underline{\Psi}_{s|t=0} \qquad (2)$$

The use of a pure open-loop integration however has many disadvantages, as DC-offsets in the measurements make the integration drift and resistance variations decenter the estimation. Still it is a simple method, relying on only one parameter R_s and independent of the rotor position. An overview and comparison of several improved methods based on this principle is given in [15], where also current model based methods are discussed and compared.

The current model is defined in the stationary $\alpha\beta$ reference frame for an SPMSM by:

$$\Psi_{s,\alpha} = L_s I_\alpha + \Psi_f \cos(\theta) \qquad (3)$$
$$\Psi_{s,\beta} = L_s I_\beta + \Psi_f \sin(\theta). \qquad (4)$$

As is clear from equations (3-4), these methods are dependent on the rotor position θ, stator inductance L_s and permanent magnet flux Ψ_f. The resulting need for a position sensor is, especially in DTC which is an inherently position sensorless method, considered as a major disadvantage. Also the increased parameter dependence on the inductances is, considering the saturation, a disadvantage. To reduce the parameter dependence and to perform the rotor position estimation needed in the current model a state observer can be used. Several observers have been proposed in literature, a short overview is given in [7], [8]. In this paper the extended Kalman filter is selected for elaboration.

B. Reduced-order EKF

The Kalman filter is a stochastic recursive optimum-state estimator. For nonlinear systems an extended Kalman filter (EKF) can be used to obtain unmeasurable states (e.g. speed and rotor position) by using a model for the dynamical system, measured states and statistics of the system and measurement noise. By means of the noise input it is possible to take account of both measuring errors and modelling errors. The EKF is a two-step method as shown in figure 2. With the measured

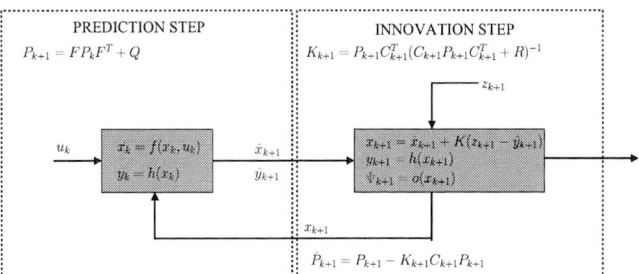

Figure 2: EKF scheme

inputs $\mathbf{u}_k$ and machine model ($\mathbf{f(x,u)}$ and $\mathbf{h(x)}$) the next state of the machine $\hat{\mathbf{x}}_{k+1}$ is predicted (prediction step). From this state the next output $\hat{\mathbf{y}}_{k+1}$ is calculated and compared to the measured value $\mathbf{z}_{k+1}$. The error on the output, together with the covariance values of measurement noise $\mathbf{R}$ and system $\mathbf{Q}$ are used to correct the state values in the next step. Often the covariance matrices are chosen to be diagonal. In this correction or innovation step the Kalman gain matrix $\mathbf{K}_{k+1}$ is calculated as well.

In this paper two implementations of the EKF are studied. The same nonlinear state-space model for the PMSM is used, the difference between the two methods is based on the selection of the state variables. In EKFC the current components in the stationary reference frame are selected as state variables, as in [7], [9]. In EKFF the stator flux linkage components in the stationary reference frame are selected as state variables, as in [10]. This means that for EKFC the state vector consist of four measurable quantities, however due to the preference for a motion-state sensorless drive only two state variables are assumed to be measurable. For EKFF the state vectors then consists of four unmeasurable quantities.

In both cases the voltage $\mathbf{u} = \begin{bmatrix} V_\alpha & V_\beta \end{bmatrix}^T$ and current components in the stationary reference frame $\mathbf{y} = \begin{bmatrix} I_\alpha & I_\beta \end{bmatrix}^T$ are selected as input and output respectively. The EKF is of reduced order as the inertia is assumed to be infinite so that the mechanical equation is omitted. This is very advantageous as the load torque and inertia in the mechanical equation typically are not known. Because the speed ω is in the state vector the EKF will correct this modelling error if a good value is chosen for the covariance. More details about the tuning of EKF's can be found in [16].

In this paper we define, besides the output function $\mathbf{y}$ that is used to correct the estimation, an additional output function $\mathbf{o(x)}$ which expresses the 'useful' output (the stator flux components) as a function of the state components.

1) EKFC: EKF with current components:
The state vector $\mathbf{x}$ is chosen with the current components in the stationary reference frame as state variables as in [7], [9]

$$\mathbf{x} = \begin{bmatrix} x_1 & x_2 & x_3 & x_4 \end{bmatrix}^T \qquad (5)$$
$$= \begin{bmatrix} I_\alpha & I_\beta & \omega & \theta \end{bmatrix}^T, \qquad (6)$$

where ω and θ denote rotor speed and position respectively. The system function $\mathbf{f(x,u)}$, output function $\mathbf{h(x)}$ and Jacobians $\mathbf{F} = \frac{\mathbf{f(x,u)}}{\partial \mathbf{x}}$ and $\mathbf{C} = \frac{\mathbf{h(x)}}{\partial \mathbf{x}}$ are:

$$\mathbf{f}(\mathbf{x}, \mathbf{u}) = \begin{bmatrix} -\frac{R_s}{L_s} x_1 + \frac{\Psi_f}{L_s} x_3 \cos x_4 + \frac{u_1}{L_s} \\ -\frac{R_s}{L_s} x_2 + \frac{\Psi_f}{L_s} x_3 \sin x_4 + \frac{u_2}{L_s} \\ 0 \\ x_3 \end{bmatrix} \quad (7)$$

$$\mathbf{h}(x) = \begin{bmatrix} x_1 \\ x_2 \end{bmatrix} \quad \mathbf{o}(\mathbf{x}) = \begin{bmatrix} L_s x_1 + \Psi_f \cos x_4 \\ L_s x_2 + \Psi_f \sin x_4 \end{bmatrix} \quad (8)$$

$$\mathbf{F} = \frac{\mathbf{f}(\mathbf{x}, \mathbf{u})}{\partial \mathbf{x}} = \begin{bmatrix} -\frac{R_s}{L_s} & 0 & \frac{\Psi_f}{L_s} \cos x_4 & -\frac{\Psi_f}{L_s} x_3 \sin x_4 \\ 0 & -\frac{R_s}{L_s} & \frac{\Psi_f}{L_s} \sin x_4 & \frac{\Psi_f}{L_s} x_3 \cos x_4 \\ 0 & 0 & 0 & 0 \\ 0 & 0 & 1 & 0 \end{bmatrix}$$
$$(9)$$

$$\mathbf{C} = \frac{\mathbf{h}(\mathbf{x})}{\partial \mathbf{x}} = \begin{bmatrix} 1 & 0 & 0 & 0 \\ 0 & 1 & 0 & 0 \end{bmatrix} \quad (10)$$

2) EKFF: EKF with flux components:

The state vector is chosen with the flux components in the stationary reference frame as state variables as in [10]

$$\mathbf{x} = [x_1 \; x_2 \; x_3 \; x_4]^T \quad (11)$$
$$= [\Psi_{s,\alpha} \; \Psi_{s,\beta} \; \omega \; \theta]^T, \quad (12)$$

System and output functions are:

$$\mathbf{f}(\mathbf{x}, \mathbf{u}) = \begin{bmatrix} -\frac{R_s}{L_s} x_1 + \frac{R_s}{L_s} \Psi_f \cos x_4 + u_1 \\ -\frac{R_s}{L_s} x_2 + \frac{R_s}{L_s} \Psi_f \sin x_4 + u_2 \\ 0 \\ x_3 \end{bmatrix} \quad (13)$$

$$\mathbf{h}(\mathbf{x}) = \begin{bmatrix} \frac{x_1 - \Psi_f \cos x_4}{L_s} \\ \frac{x_2 - \Psi_f \sin x_4}{L_s} \end{bmatrix} \quad \mathbf{o}(\mathbf{x}) = \begin{bmatrix} x_1 \\ x_2 \end{bmatrix} \quad (14)$$

$$\mathbf{F} = \frac{\mathbf{f}(\mathbf{x}, \mathbf{u})}{\partial \mathbf{x}} = \begin{bmatrix} -\frac{R_s}{L_s} & 0 & 0 & -\frac{R_s \Psi_f}{L_s} \sin x_4 \\ 0 & -\frac{R_s}{L_s} & 0 & \frac{R_s \Psi_f}{L_s} \cos x_4 \\ 0 & 0 & 0 & 0 \\ 0 & 0 & 1 & 0 \end{bmatrix} \quad (15)$$

$$\mathbf{C} = \frac{\mathbf{h}(\mathbf{x})}{\partial \mathbf{x}} = \begin{bmatrix} \frac{1}{L_s} & 0 & 0 & \frac{\Psi_f}{L_s} \sin x_4 \\ 0 & \frac{1}{L_s} & 0 & \frac{\Psi_f}{L_s} \cos x_4 \end{bmatrix} \quad (16)$$

When inspecting the equations for EKFC and EKFF it is clear that the choice of state vector components for EKFF would appear as the more natural one (this is most obvious in $\mathbf{o}(\mathbf{x})$). Furthermore it is important to notice that the speed ω is not needed in the equations of EKFF (its use to estimate θ is not necessary), this means that we could further reduce the order of EKFF and omit $x_3 = \omega$ as a state variable and estimate θ directly. It is however retained for two reasons. Firstly because using the same order for EKFC and EKFF simplifies some practical implementation aspects as we can reuse matrix manipulations. Secondly because in an DTC drive the knowledge of the speed is advantageous for different control purposes (switching over from one voltage vector selection algorithm to another, select flux reference value). When inspecting the Jacobians $\mathbf{F} = \frac{\mathbf{f}(\mathbf{x},\mathbf{u})}{\partial \mathbf{x}}$ and $\mathbf{C} = \frac{\mathbf{h}(\mathbf{x})}{\partial \mathbf{x}}$ it is obvious that the expression of $\mathbf{F}$ is more complicated for EKFC than EKFF while the reverse is true for $\mathbf{C}$.

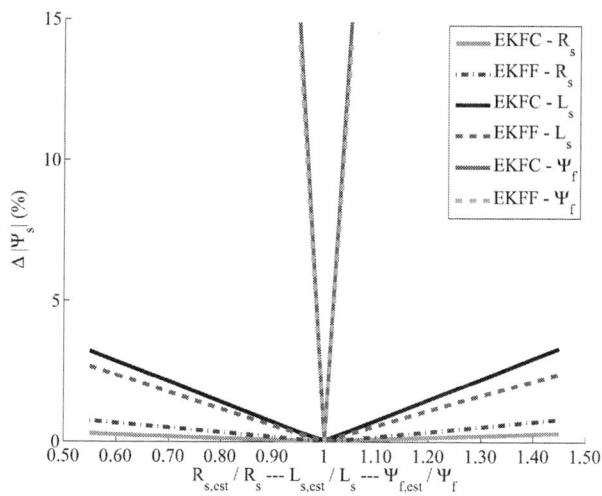

Figure 3: Sensitivity of EKFC and EKFF performance to parameter deviations

C. Influence of parameter variations

First the effect of an incorrect value for the stator resistance is evaluated. The simulations are done with the motor data found in appendix. The SPMSM is Direct Torque-Controlled and after a run-up it runs at half the rated speed. In figure 3 the RMS error is shown for the stator flux vector magnitude during steady state where $\frac{R_{s,est}}{R_s}$ is varied. Clearly the RMS error of the EKF estimators is very small, even for large deviations of R_s. In steady-state the RMS errors of the stator flux angle are also very small [15]. While only the steady state errors have been considered here, it has to be noted that wrong parameter values affect the (starting) transients even more, with errors that are even much larger than in steady state. The EKFs are methods based on the current model and thus also dependent on L_s and Ψ_f. In figure 3 the RMS-values of the errors are found. As shown in [15], the estimations with EKFC and EKFF do not yield better results than the open-loop current model with measured position. This means that the only remaining advantage is the sensorless fashion in which the estimation is executed. A thorough comparison between the performance of the EKFs (EKFC and EKFF) and several other stator flux estimators (most of them based on an integrator) is given in [15]. There the effects of parameter changes in R_s and L_s are discussed in more detail.

Clearly the EKF estimators can cope very well with errors in R_s, but variations in L_s and Ψ_f are more troublesome. The estimators remain stable but show a considerable steady-state deviation of both flux magnitude and angle, comparable to the case of on an open-loop current model (although one has to consider the fact that for an SPMSM the influence of L_s still is rather small). The correction in the estimation when R_s is varied is the result of the fact that the EKF estimators can correct for the modeling inaccuracies by the feedback loop. For variations in L_s and Ψ_f however this is not the case as the parameter L_s and Ψ_f, unlike R_s, are not only used

Table I: Equations for EKFC and EKFF with augmented state vector

Name of EKF	Added state	x	f(x,u)	h(x)	o(x)
EKFCA1	L_s	$\begin{bmatrix} x_1 \\ x_2 \\ x_3 \\ x_4 \\ x_5 \end{bmatrix} = \begin{bmatrix} I_\alpha \\ I_\beta \\ \omega \\ \theta \\ \frac{1}{L_s} \end{bmatrix}$	$\begin{bmatrix} -R_s x_1 x_5 + \Psi_f x_3 x_5 \cos x_4 + x_5 u_1 \\ -R_s x_2 x_5 + \Psi_f x_3 x_5 \sin x_4 + x_5 u_2 \\ 0 \\ x_3 \\ 0 \end{bmatrix}$	$\begin{bmatrix} x_1 \\ x_2 \end{bmatrix}$	$\begin{bmatrix} \frac{x_1}{x_5} + \Psi_f \cos x_4 \\ \frac{x_2}{x_5} + \Psi_f \sin x_4 \end{bmatrix}$
EKFCA2	L_s, R_s	$\begin{bmatrix} x_1 \\ x_2 \\ x_3 \\ x_4 \\ x_5 \\ x_6 \end{bmatrix} = \begin{bmatrix} I_\alpha \\ I_\beta \\ \omega \\ \theta \\ \frac{1}{L_s} \\ R_s \end{bmatrix}$	$\begin{bmatrix} -x_1 x_5 x_6 + \Psi_f x_3 x_5 \cos x_4 + x_5 u_1 \\ -x_2 x_5 x_6 + \Psi_f x_3 x_5 \sin x_4 + x_5 u_2 \\ 0 \\ x_3 \\ 0 \\ 0 \end{bmatrix}$	$\begin{bmatrix} x_1 \\ x_2 \end{bmatrix}$	$\begin{bmatrix} \frac{x_1}{x_5} + \Psi_f \cos x_4 \\ \frac{x_2}{x_5} + \Psi_f \sin x_4 \end{bmatrix}$
EKFCA3	L_s, R_s, Ψ_f	$\begin{bmatrix} x_1 \\ x_2 \\ x_3 \\ x_4 \\ x_5 \\ x_6 \\ x_7 \end{bmatrix} = \begin{bmatrix} I_\alpha \\ I_\beta \\ \omega \\ \theta \\ \frac{1}{L_s} \\ R_s \\ \Psi_f \end{bmatrix}$	$\begin{bmatrix} -x_1 x_5 x_6 + x_3 x_5 x_7 \cos x_4 + x_5 u_1 \\ -x_2 x_5 x_6 + x_3 x_5 x_7 \sin x_4 + x_5 u_2 \\ 0 \\ x_3 \\ 0 \\ 0 \\ 0 \end{bmatrix}$	$\begin{bmatrix} x_1 \\ x_2 \end{bmatrix}$	$\begin{bmatrix} \frac{x_1}{x_5} + x_7 \cos x_4 \\ \frac{x_2}{x_5} + x_7 \sin x_4 \end{bmatrix}$
EKFFA1	L_s	$\begin{bmatrix} x_1 \\ x_2 \\ x_3 \\ x_4 \\ x_5 \end{bmatrix} = \begin{bmatrix} \Psi_{s,\alpha} \\ \Psi_{s,\beta} \\ \omega \\ \theta \\ \frac{1}{L_s} \end{bmatrix}$	$\begin{bmatrix} -R_s x_1 x_5 + R_s \Psi_f x_5 \cos x_4 + u_1 \\ -R_s x_2 x_5 + R_s \Psi_f x_5 \sin x_4 + u_2 \\ 0 \\ x_3 \\ 0 \end{bmatrix}$	$\begin{bmatrix} x_5(x_1 - \Psi_f \cos x_4) \\ x_5(x_2 - \Psi_f \sin x_4) \end{bmatrix}$	$\begin{bmatrix} x_1 \\ x_2 \end{bmatrix}$
EKFFA2	L_s, R_s	$\begin{bmatrix} x_1 \\ x_2 \\ x_3 \\ x_4 \\ x_5 \\ x_6 \end{bmatrix} = \begin{bmatrix} \Psi_{s,\alpha} \\ \Psi_{s,\beta} \\ \omega \\ \theta \\ \frac{1}{L_s} \\ R_s \end{bmatrix}$	$\begin{bmatrix} -x_1 x_5 x_6 + \Psi_f x_5 x_6 \cos x_4 + u_1 \\ -x_2 x_5 x_6 + \Psi_f x_5 x_6 \sin x_4 + u_2 \\ 0 \\ x_3 \\ 0 \\ 0 \end{bmatrix}$	$\begin{bmatrix} x_5(x_1 - \Psi_f \cos x_4) \\ x_5(x_2 - \Psi_f \sin x_4) \end{bmatrix}$	$\begin{bmatrix} x_1 \\ x_2 \end{bmatrix}$
EKFFA3	L_s, R_s, Ψ_f	$\begin{bmatrix} x_1 \\ x_2 \\ x_3 \\ x_4 \\ x_5 \\ x_6 \\ x_7 \end{bmatrix} = \begin{bmatrix} \Psi_{s,\alpha} \\ \Psi_{s,\beta} \\ \omega \\ \theta \\ \frac{1}{L_s} \\ R_s \\ \Psi_f \end{bmatrix}$	$\begin{bmatrix} -x_1 x_5 x_6 + x_5 x_6 x_7 \cos x_4 + u_1 \\ -x_2 x_5 x_6 + x_5 x_6 x_7 \sin x_4 + u_2 \\ 0 \\ x_3 \\ 0 \\ 0 \\ 0 \end{bmatrix}$	$\begin{bmatrix} x_5(x_1 - x_7 \cos x_4) \\ x_5(x_2 - x_7 \sin x_4) \end{bmatrix}$	$\begin{bmatrix} x_1 \\ x_2 \end{bmatrix}$

Table I: Equations for EKFC and EKFF with augmented state vector

in $\mathbf{f}(\mathbf{x}, \mathbf{u})$. For EKFC the state vector $\mathbf{x}$ will converge to the correct values, but due to the use of L_s and Ψ_f in $\mathbf{o}(\mathbf{x})$ to determine Ψ_α and Ψ_β from $\mathbf{x}$ the output is incorrect. For EKFF L_s is used in $\mathbf{h}(\mathbf{x})$ and thus the state vector $\mathbf{x}$ will not converge to the right value.

III. ADDING PARAMETER ESTIMATIONS TO THE EKF

As demonstrated in the previous section, the EKFs fail to estimate the stator flux linkage vector correctly if certain motor parameters (those used in $\mathbf{h}(\mathbf{x})$ and $\mathbf{o}(\mathbf{x})$) deviate from the true values. One possibility to overcome this problem is to estimate the most important motor parameters in the EKF as well. This can be done by augmenting the state vector with the parameters to be estimated, where parameter variations are given no dynamics (i.e. the corresponding row of $\mathbf{f}(\mathbf{x},\mathbf{u})$ is 0).

In Table I the expressions are given for EKFC and EKFF with added parameter estimations. Three cases are considered. In the first case L_s is added to estimate as this is a parameter that, due to saturation, can vary strongly during operation of the drive. Furthermore it is present in either $\mathbf{h}$ or $\mathbf{o}$ and so errors in L_s propagate through the estimation. It has to be noted that in order to take the variation of L_s into account, actually $\frac{1}{L_s}$ is added to the state vector. This is advisable because in $\mathbf{f}(\mathbf{x}, \mathbf{u})$ (both for EKFC and EKFC) and $\mathbf{h}(\mathbf{x})$ (for EKFF) the stator inductance always is present as $\frac{1}{L_s}$. Choosing L_s as a state component would thus, due to the partial differentiation, result in mathematical expressions for the Jacobians $\mathbf{F}$ and $\mathbf{C}$ that are much more complex and could be prohibitive for

an actual implementation. In the second case both R_s and L_s are estimated as R_s can vary greatly with temperature. Finally, in the third case, the estimation of all three relevant parameters L_s, R_s, Ψ_f is performed. The estimated values of the parameters can be used outside the EKF as well, e.g. in the control algorithm.

For the sake of brevity, the expressions for the Jacobians $\mathbf{F}$ and $\mathbf{C}$ have been omitted in Table I. However it is clear that the increased complexity of the augmented state vectors is even more easily seen in the first $\mathbf{F}$ and $\mathbf{C}$ compared to $\mathbf{f}(\mathbf{x}, \mathbf{u})$ and $\mathbf{h}(\mathbf{x})$, due to the increased non-linearity of the model. To this end it is useful to compare the resulting increase in complexity for the families of EKFCA and EKFFA filters. For EKFCA1 and EKFFA1 $\mathbf{F}$ and $\mathbf{C}$ are given in Table II.

Obviously every addition of a parameter to estimate (where we assume no dynamics) first of all results in a row of zeros for $\mathbf{F}$ both in EKFCA and EKFFA. More important however is the fact that the first two rows of $\mathbf{F}$ now contain additional elements which augment the computational load considerably. Furthermore some elements of $\mathbf{F}$ are now no longer constants, but are dependent on the added state variable.

The changes in $\mathbf{C}$ are very different for EKFCA1 and EKFFA1: for EKFCA1 only a column of zeros is added, where the expression for EKFFA1 now contains the added state variable and a column with elements which again increase the computational load.

This of course also means that the computation of the estimation covariance matrix $\mathbf{P}$ and the Kalman correction

978-1-4244-4782-4/10 $26.00 © 2010 IEEE

EKFCA1

$$\mathbf{F} = \begin{bmatrix} -R_s x_5 & 0 & \Psi_f x_5 \cos x_4 & -\Psi_f x_5 x_3 \sin x_4 & -R_s x_1 + \Psi_f x_3 \cos x_4 + u_1 \\ 0 & -R_s x_5 & \Psi_f x_5 \sin x_4 & \Psi_f x_5 x_3 \cos x_4 & -R_s x_2 + \Psi_f x_3 \sin x_4 + u_2 \\ 0 & 0 & 0 & 0 & 0 \\ 0 & 0 & 1 & 0 & 0 \\ 0 & 0 & 0 & 0 & 0 \end{bmatrix}$$

$$\mathbf{C} = \begin{bmatrix} 1 & 0 & 0 & 0 & 0 \\ 0 & 1 & 0 & 0 & 0 \end{bmatrix}$$

EKFFA1

$$\mathbf{F} = \begin{bmatrix} -R_s x_5 & 0 & 0 & -R_s \Psi_f x_5 \sin x_4 & R_s(-x_1 + \Psi_f \cos x_4) \\ 0 & -R_s x_5 & 0 & R_s \Psi_f x_5 \cos x_4 & R_s(-x_2 + \Psi_f \sin x_4) \\ 0 & 0 & 0 & 0 & 0 \\ 0 & 0 & 1 & 0 & 0 \\ 0 & 0 & 0 & 0 & 0 \end{bmatrix}$$

$$\mathbf{C} = \begin{bmatrix} x_5 & 0 & 0 & \Psi_f x_5 \sin x_4 & x1 - \Psi_f \cos x_4 \\ 0 & x_5 & 0 & \Psi_f x_5 \cos x_4 & x2 - \Psi_f \sin x_4 \end{bmatrix}$$

Table II: Expressions of $\mathbf{F}$ and $\mathbf{C}$ for EKFCA1 and EKFFA1 respectively

matrix $\mathbf{K}$ become increasingly complex (see figure 2) as the matrix operations (especially multiplication) become more computationally demanding with higher matrix sizes. For n_s state variables the size of the matrices is given in Table III. One important remark however is the fact that the matrix to invert (needed to calculate the Kalman gain) stays a 2×2 matrix independently from n_s. Obviously the computational effort to calculate the matrix $\mathbf{CPC}^T + \mathbf{R}$ however will strongly depend on n_s.

Matrix or array	Size		
$\mathbf{f(x, u)}$	n_s	$\times$	1
$\mathbf{h(x)}$	2	$\times$	1
$\mathbf{Q}$	n_s	$\times$	n_s
$\mathbf{R}$	2	$\times$	2
$\mathbf{F}$	n_s	$\times$	n_s
$\mathbf{C}$	2	$\times$	n_s
$\mathbf{P}$	n_s	$\times$	n_s
$\mathbf{CPC}^T + \mathbf{R}$	2	$\times$	2
$\mathbf{K}$	n_s	$\times$	2

Table III: Matrix and array size for an EKF with n_s state vector components

IV. RESULTS IN SIMULATION

For the same motor and control scheme as before the behavior of the augmented EKFs is simulated. In figure 4a the flux amplitude error is shown for a low-dynamics drive cycle. The results shown are for EKFFA2 and EKFFA3, both with an initial 25% error on R_s and L_s and no error on Ψ_f. The angular error is shown in figure 4b and the evolution of the parameters in figure 4c and d. It is clear that EKFFA2 estimates both parameters correctly and results in good flux estimations. EKFFA3 should be able to cope with the errors in a similar way, but now also corrects Ψ_f during the transient and R_s only to a lesser extent. This leads to the observed drop in estimated flux magnitude in 4a. Similar observations

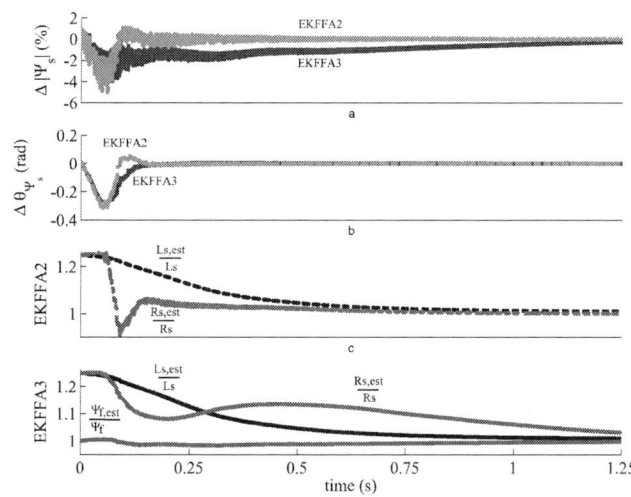

Figure 4: EKFFA2 and EKFFA3 with 25% error in R_s and L_s. a) Flux magnitude error b) flux angular error c) estimated parameters EKFFA2 d) estimated parameters EKFFA3

are obtained with the other formulations of the filter : adding parameters to the estimation can work, but great care should be taken as unwanted cross coupling effects of parameter variations can occur with a poorly chosen covariance matrix $\mathbf{Q}$. This further complicates the tuning of the EKF, which is often reported as one of the major drawbacks of this state observer (see also [16], where the problem for a 'standard' PMSM EKF is discussed). When implementing an EKF, one should carefully consider which parameters are the most likely to vary and at what rate. Clearly it is important to refrain from putting too little confidence (high values in $\mathbf{Q}$) in the model

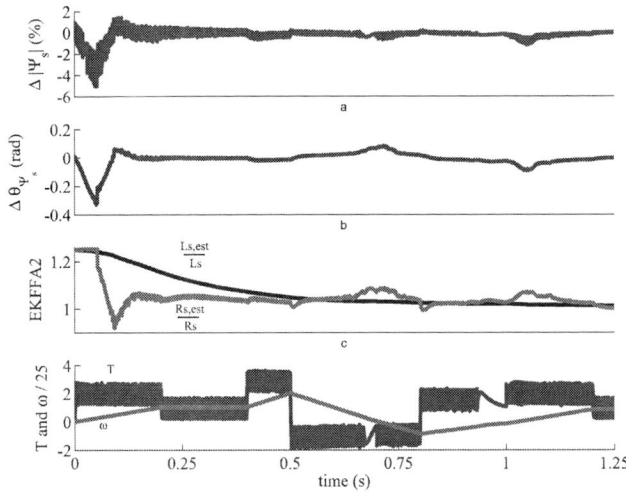

Figure 5: Performance of EKFFA2 with 25% error in R_s and L_s, dynamic drive cycle. a) Flux magnitude error b) flux angular error c) estimated parameters d) torque and speed (scaled by 25)

as this could induce overcompensation of the parameters and thus result in poor performance.

In figure 5 the performance of EKFFA2 with the same initial errors as before is shown in a highly dynamic drive cycle. It is clear that with a good choice of covariance matrix and no additional erroneous parameters EKFFA2 offers good estimations.

V. PRACTICAL ASPECTS FOR FPGA IMPLEMENTATION

A. Per unit formulation and covariance matrices

Normalizing the state vector and the equations not only allows an easier conversion to fixed-point format, it also allow an easier setting of the covariance matrices as discussed in [16]. A per unit system with base quantities $V_b, I_b, \omega_b, \theta_b$ is used here. The base system can be selected in such a way that the state components always are smaller than 1 so that purely fractional fixed-point arithmetic can be used for most operations. However it does not ensure that the intermediate results (especially those resulting from the matrix inversion) stay smaller than 1. The flexibility offered by the FPGA and the Xilinx tools to program it however allow to cope with this.

All of the EKFs are initialized with a zero matrix for **P**, while the covariance matrices are for EKFC and EKFF respectively:

$$\mathbf{Q} = diag(0.0012 \quad 0.0012 \quad 0.015 \quad 0.02) \quad \mathbf{R} = diag(0.1 \quad 0.1) \tag{17}$$

$$\mathbf{Q} = diag(0.0027 \quad 0.0027 \quad 0.05 \quad 0.1) \quad \mathbf{R} = diag(0.1 \quad 0.1) \tag{18}$$

These have been selected based on the method discussed in [16] and refined by the results from simulation. For the groups of EKFCA and EKFFA additional elements have to be selected to estimate the parameters. As said before, this can be tricky. Up till now they have been used as additional tuning

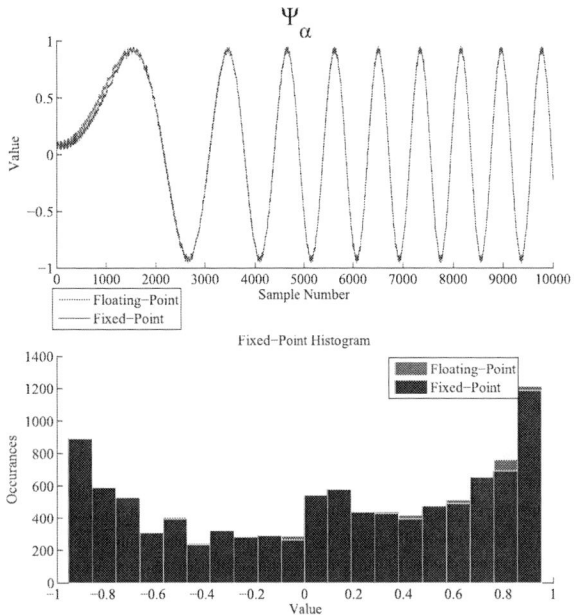

Figure 6: Estimation of the stator flux linkage component $\Psi_{s,\alpha}$ with the Matlab algorithm in floating-point and in fixed-point format by AccelDSP to implement it in the FPGA

parameters, but a method that expands the procedure of [16] would be desirable.

B. Implementation using AccelDSP

For the digital implementation of the stator flux linkage estimators two evaluation boards from Digilent Inc. are considered. As a rather low-cost option the Spartan 3E Starter Board, based on the Xilinx Spartan 3E FPGA (500K gates) clocked at 50MHz, is chosen. The Virtex II Pro board based on the XC2VP30 with a 32 or 100 MHz clock is considered here as a high-performance option. One goal is to optimize the implementation of EKFC and EKFF to the degree that it can easily be implemented on the Spartan 3E with a sampling frequency of 20kHz and enough resources left to implement the rest of the control (DTC in this case). Another goal is to implement the EKFs with augmented state vector. Due to the higher complexity in this case the specifications of a Spartan 3E board could be too limited (due to the increased degree of non-linearity it becomes increasingly harder to realize all the calculations with the 20 dedicated multipliers of the Spartan 3E). For the exploration of the possibilities to implement these EKFS the Virtex II Pro board is used.

For both options the configuration of the FPGA is programmed in Matlab/Simulink with the System Generator tool from Xilinx. Some specific functions are written in VHDL and interfaced through the Black Box block. The EKF algorithm is implemented with the AccelDSP tool from Xilinx. Here the implementation is briefly discussed. The AccelDSP tool takes a tested Matlab m-file with the algorithm to be implemented

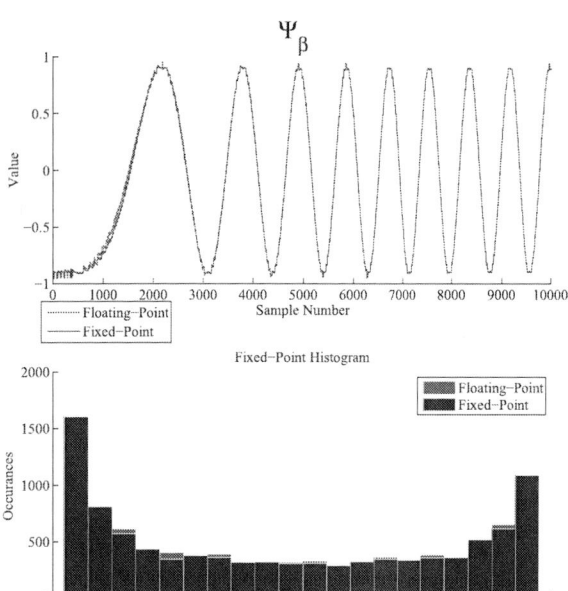

Matrix or array	no. cycles	slices	18x18 mult.
Prediction of state vector	24	3297	5 (4%)
Prediction of covariance matrix	139	1201	12 (9%)
Calculation of the argument of the matrix invert	99	1392	12 (9%)
Matrix invert	85	2069	18 (13%)
Kalman gain calculation and update of covariance matrix	90	1372	14 (10%)
Innovation of state vector	84	1283	18 (13%)
clock cycles	521		
total slices		8545 (62%)	

Table IV: Number of FPGA clock cycles and slices needed per EKF module for EKFF on Virtex II Pro, 'naive' implementation clocked at 32 MHz

Matrix or array	clock cycles	slices	18x18 mult.
Prediction step	47	2978	23 (17%)
Kalman gain calculation	138	2596	24 (18%)
Innovation of state vector	10	1958	18 (13%)
clock cycles	195		
total slices		7532 (55%)	

Table V: Number of FPGA clock cycles and slices needed for EKFF on Virtex II Pro, better implementation

Figure 7: Estimation of the stator flux linkage component $\Psi_{s,\beta}$ with the Matlab algorithm in floating-point and in fixed-point format by AccelDSP to implement it in the FPGA

and assists the programmer during the conversion to a fixed-point version as a first step, the realisation of an RTL version as a second step and finally the creation of an HDL module or System Generator block.

In figures 6 and 7 the results are shown for the AccelDSP fixed-point implementation of EKFF. The loss of precision during the transition from floating-point to fixed-point format can be noticed but is rather small. Most importantly the round-off errors during the matrix operations do not result in instability of the EKF. More background on the detrimental effects of round-off errors on the performance of Kalman filters can be found in [17]. In this research it was attempted to retain enough precision to avoid these effects, whilst keeping the data types small enough to be implemented. Here the strength of the AccelDSP tool comes into play, as we can adjust the data type as desired for every mathematical operation.

For a 'naive' implementation, where little attention is given to the optimization of the implementation, Table IV gives the needed clock cycles, FPGA slices and embedded 18x18 multipliers when realized by AccelDSP for the Virtex II Pro. Given the number of slices needed, this version can not be implemented on the Spartan 3E board. This EKF can run in under 15 μs which is sufficiently fast for the proposed sampling frequency of 20kHz. However faster results can be obtained as in [18], also on a Virtex II Pro board.

When more attention is given to the optimization (re-using of calculated values) both a smaller and faster implementation is obtained, the results are given in Table V. Still the number of slices and especially the number of embedded multipliers is too high to be implemented on Spartan 3E (maximum values

are 4658 and 20 for the Spartan 3E compared to 13696 and 136 for the Virtex II Pro). Clearly an improvement can be made by rolling some operations (in this implementation matrix multiplications are fully unrolled).

Besides the Spartan 3E implementation, the implementation of the augmented EKFs has to be explored. The FPGA implementation of the matrix invert can be re-used in all EKFCA and EKFFA versions, as the size of the matrix to invert is always 2×2. This means that the computational effort for this particular part of the algorithm will remain the same. However, the other computational effort needed for the other parts depends heavily on the value of n_s. Due to the increased non-linearity of the augmented versions the cycle time for calculations with $\mathbf{F}$ (and for the EKFF-family $\mathbf{C}$ as well) increases heavily. This is the result of the fact that not only the size increases but also more elements within the matrix $\mathbf{F}$ or $\mathbf{C}$ will be variable as constants are replaced by state components at several positions in the matrix. This implies that more dedicated multipliers will be used and less optimization by the AccelDSP tool (for example changing multiplications by constants to shift operations) will be performed.

Given the results in Table IV and V we can expect that the implementation of augmented extended kalman filters for Virtex II Pro should be possible, but the feasibility to do this on the Spartan 3E is not so certain.

C. Improvements for the FPGA implementation

In order to optimize the FPGA implementation (and to fit the implementation on a Spartan 3E) a further optimization of the fixed-point data format used in the calculations is required. Further improvements can be made by assuming that the Kalman gain $\mathbf{K}$ and the covariance matrix $\mathbf{P}$ are symmetrical, resulting in a significant reduction of the elements that need to be calculated. A further improvement is the use of RAM-blocks to store the different values during a calculation cycle instead of calculating them at several positions in separate matrices. Depending on the resources needed for rest of the control algorithm and the desired cycle period the EKF calculations could be further rolled or unrolled in AccelDSP to optimize either the number of FPGA cells or the number of clock cycles needed. Especially for the implementation on a Spartan 3E the number of multipliers used should be kept under control.

VI. Conclusions

In this paper it is shown that the Extended Kalman Filter, although stable in the case of parameter variations, will produce a stator flux linkage vector estimation that deviates strongly if incorrect motor parameters are used. Three cases can be considered. The first case is the one where the wrong value of a parameter of only R_s is not a large problem as there are no additional errors in calculating the output, so that the feedback correction loop is able to handle this. In the second case, which occurs in EKFF, a wrong value for L_s or Ψ_f would, even for a correct state estimate, lead to an incorrect output $\mathbf{y}$ and thus correction. The third case occurs in the EKFC: state vector $\mathbf{x}$ and output $\mathbf{y}$ can be correct, but an incorrect flux estimate is obtained through $\mathbf{o}(\mathbf{x})$.

To mitigate this problem several formulations of the EKF are given where the state vector is augmented with the parameters that need to be estimated. When implementing these, great care should be taken. It is shown that an uncareful selection of the parameters to be estimated and their covariance elements results in strongly divergent EKFs. If a good choice is made for the covariance, or if additional information about some of the parameters is available, a high quality estimation can be obtained.

Some aspects and caveats of the FPGA implementation are discussed. Specifically the process of translating the floating-point Matlab algorithm to an IIDL or System Generator module by using the AccelDSP tool from Xilinx is addressed.

Acknowledgment

T. Vyncke wishes to thank the Research Foundation-Flanders for his grant as Ph. D. fellowship of the Research Foundation - Flanders (FWO).

This work was supported by the GOA project BOF 07/GOA/006. The research was performed as part of the Interuniversity Attraction Poles programme IUAP P6/21 financed by the Belgian government.

Appendix

R_s	2.875 Ω	L_s	8.5 mH	J	0.008 kgm^2
Ψ_f	0.175 Wb	N_p	4	F	0.001 Nms

Parameters of SPMSM used in simulation

References

[1] I. Takahashi and T. Noguchi, "A new quick-response and high-efficiency control strategy of an induction motor," *IEEE Trans. Ind. Applicat.*, vol. 22, no. 5, pp. 820–827, Sept./Oct. 1986.

[2] G. S. Buja and M. P. Kazmierkowski, "Direct torque control of PWM inverter-fed AC motors – a survey," *IEEE Trans. Ind. Electron.*, vol. 51, no. 4, pp. 744–757, Aug. 2004.

[3] L. Zhong, M. F. Rahman, W. Y. Hu, and K. W. Lim, "Analysis of direct torque control in permanent magnet synchronous motor drives," *IEEE Trans. Power Electron.*, vol. 12, no. 3, pp. 528–536, May 1997.

[4] M. Niemelä, J. Luukko, and J. Pyrhönen, "Position sensorless PMSM DTC-drive for industrial applications," in *Conf. Proc. EPE*, Graz, 2001, p. 10.

[5] M. F. Rahman, L. Zhong, M. E. Haque, and M. Rahman, "A direct torque-controlled interior permanent-magnet synchronous motor drive without a speed sensor," *IEEE Trans. Energy Conversion*, vol. 18, no. 1, pp. 17–22, Mar. 2003.

[6] D. Swierczynski and M. P. Kazmierkowski, "Direct torque control of permanent magnet synchronous motor (PMSM) using space vector modulation (DTC-SVM) – simulation and experimental results," in *Conf. Proc. IEEE 28th Annual Conference of the Industrial Electronics Society (IECON'02)*, vol. 1, Nov. 5–8, 2002, pp. 751–755.

[7] P. Vas, *Sensorless Vector and Direct Torque Control*. New York: Oxford University Press, 1998, pp. 122–178.

[8] M. F. Rahman, M. E. Haque, L. Tang, and L. Zhong, "Problems associated with the direct torque control of an interior permanent-magnet synchronous motor drive and their remedies," *IEEE Trans. Ind. Electron.*, vol. 51, no. 4, pp. 799–809, Aug. 2004.

[9] A. Llor, J. Rétif, X. Lin-Shi, and S. Arnalte, "Direct stator flux linkage control technique for a permanent magnet synchronous machine," in *Conf. Rec. IEEE 34th Annual Power Electronics Specialists Conference (PESC'03)*, vol. 1, June 15–19, 2003, pp. 246–250.

[10] V. Comnac, M. Cernat, F. Moldoveanu, and I. Draghici, "Sensorless speed and direct torque control of surface permanent magnet synchronous machines using an extended kalman filter," in *Conf. Proc. 9th Mediterranean Conference on Control and Automation (MED'01)*, Dubrovnik, Croatia, June 27–29, 2001, p. 6.

[11] J. Luukko, M. Niemelä, and J. Pyrhönen, "Estimation of the flux linkage in a direct-torque-controlled drive," *IEEE Trans. Ind. Electron.*, vol. 50, no. 2, pp. 283–287, Apr. 2003.

[12] G. D. Andreescu and A. Popa, "Flux estimator based on integrator with DC-offset correction loop for sensorless direct torque and flux control," in *Proc. 15th Int. Conf. on Electrical Machines ICEM 2002*, Bruges, Belgium, Aug. 2002, p. 6.

[13] B. Lang, W. Liu, and G. Luo, "A new observer of stator flux linkage for permanent magnet synchronous motor based on kalman filter," in *Proc. 2nd IEEE Conf. on Industrial Electronics and Applications*, Harbin, China, May 2007, pp. 1813 – 1817.

[14] G. D. Andreescu, C. I. Pitic, F. Blaabjerg, and I. Boldea, "Combined flux observer with signal injection enhancement for wide speed range sensorless direct torque control of ipmsm drives," *IEEE Trans. Energy Conversion*, vol. 23, Issue 2, pp. 393 – 402, June 2008.

[15] T. J. Vyncke, R. K. Boel, and J. A. Melkebeek, "A comparison of stator flux linkage estimators for a direct torque controlled pmsm drive," in *IEEE Industrial Electronics Conference 2009*, Porto, Portugal, Nov.3-5 2009, pp. 967–974.

[16] S. Bolognani, L. Tubiana, and M. Zigliotto, "Extended Kalman filter tuning in sensorless PMSM drives," *IEEE Trans. Ind. Applicat.*, vol. 39, no. 6, pp. 1741–1747, Nov./Dec. 2003.

[17] M. S. Grewal and A. P. Andrews, *Kalman Filtering, Theory and Practice Using MATLAB*. New York: Wiley-Interscience, 2001, pp. 202–271.

[18] L. Idkhajine, E. Monmasson, and A. Maalouf, "Fully FPGA-based sensorless control for AC drive using an extended kalman filter," in *IEEE Industrial Electronics Conference 2009*, Porto, Portugal, Nov.3-5 2009, pp. 2939–2944.

State Equations Based Resonant Converters Modeling Technique

Yingqi Zhang
Precision Power Conversion Lab
GE R&D Center -- Shanghai
P.R.China
yingqi.zhang@ge.com

Dr. P.C.Sen, Life Fellow, IEEE
Department of Electrical and Computer Engineering
Queen's University
Kingston, Ontario, Canada
K7L 3N6
senp@queensu.ca

Abstract — State equations for resonant converters are developed by using orthogonal circuit synthesis. Low frequency state variables are selected for the resonant circuit, such as envelopes of resonant inductor current and resonant capacitor voltage, phase angle between inverter output voltage and output current. For a given resonant tank, its orthogonal counterpart is constructed. These two orthogonal tanks are combined into a complex resonant tank. By applying Kirchhoff's Voltage Law (KVL) and Kirchhoff's Current Law (KCL) to the complex circuit, complex differential equations are derived. By separating real and imaginary parts of the complex differential equations, state equations for the resonant tank are obtained. For resonant DC/DC converters, other state equations are derived from rectifier stages. These derived state equations for resonant converters contain only low frequency state variables and can predict large signal transitions. By perturbing these equations around the DC operating point, transfer functions such as input-to-output, control-to-output are derived. The proposed method is verified by SIMPLIS simulation results. This method can aid closed-loop control design for resonant converters.

I. INTRODUCTION

To accurately control a resonant converter, a dynamic model of the power converter is necessary. Resonant converters are more complicated than PWM converters. The resonant inductor current and the resonant capacitor voltage are high frequency variables. Since these sinusoidal variations are large in magnitude, State Space Averaging is not applicable in the analysis of resonant converters. Fourier Series Analysis [1] was used to analyze steady state operation of resonant converters. The principle of superposition was employed to obtain the response. State Plane analysis [2], [3] can predict the behavior of resonant converters, but this method is limited to the second order resonant converters. Sampled-data modeling [4], [5] can predict the large signal response of resonant converters but suffer from long computation time. Average modeling methods for the mixed

systems [6], [7], [8] were proposed to analyze the resonant converters. They were based on the slowly varying amplitude and phase transformations. State equations were derived with low frequency state variables. However, these methods are complicated. PSPICE macro models [9], [10], [11] for resonant converters were also derived. When dc, ac and transient analysis were performed, time and frequency domain responses were obtained. This method is also complicated. In [12], general modeling methods for resonant converter modeling have been proposed.

II. DERIVATION OF STATE EQUATIONS FOR RESONANT CONVERTERS

The resonant inductor current or resonant capacitor voltage can be expressed as a high frequency variable with its amplitude and angle modulated by low frequency functions. We can select the envelopes of resonant inductor current and resonant capacitor voltage as state variables, since they are low frequency variables with DC operating points. Also the phase angle between inverter output voltage and output current can be selected as one state variable.

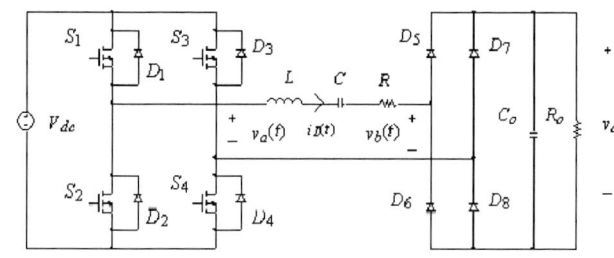

Figure 1.   Series resonant DC/DC converter

The resonant tank is simplified in Fig.2 a. It orthogonal counterpart is shown in Fig.2 b. The two resonant converters can be combined into a complex circuit, with its real part and imaginary part of its response being equal to the circuits in Fig.2b and Fig.2a respectively.

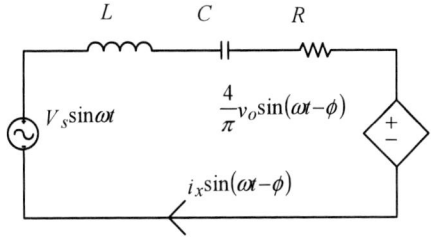

(a) original resonant circuit

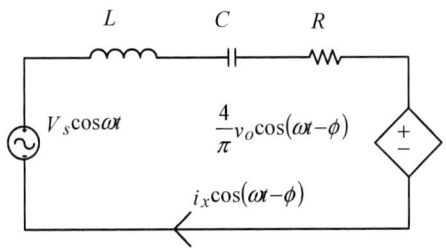

(b) orthogonal counterpart of the resonant circuit

Figure 2. two orthogonal resonant tanks

Two orthogonal circuits are combined into a complex resonant tank as shown in Figure 3. The rectifier stage is shown in Fig.3(b), with the equivalent current source as $\frac{2}{\pi}i_x$.

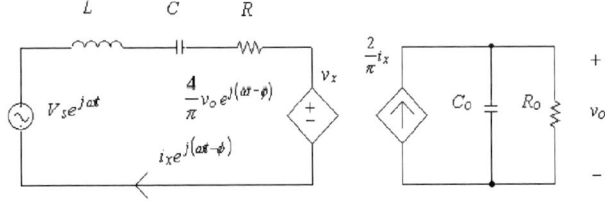

(a) complex resonant tank   (b) rectifier stage

Figure 3. Complex circuit for the series resonant DC/DC converter

In the complex circuit, the inductor current is:

$$i_L = i_x e^{j(\omega t - \phi)}.$$ (1)

The voltage across the resistor is:

$$i_R = R \cdot i_x e^{j(\omega t - \phi)}$$ (2)

From the definition, the resonant inductor voltage:

$$v_L = L\frac{d i_L}{dt}$$

Substitution of (1) into the above equation,

$$v_L = L\frac{d i_x}{dt}e^{j(\omega t - \phi)} + j(\omega - \frac{d\phi}{dt})L\,i_x e^{j(\omega t - \phi)}$$ (3)

The resonant capacitor voltage:

$$v_C = \frac{1}{C}\int i_x e^{j(\omega t - \phi)}\,dt$$

$$v_C = \left(\frac{1}{C\omega^2}\frac{d i_x}{dt} - j\frac{i_x}{\omega C} - j\frac{i_x}{C\omega^2}\frac{d\phi}{dt}\right)e^{j(\omega t - \phi)}$$ (4)

The input voltage of the rectifier is:

$$v_x = \frac{4}{\pi}v_o e^{j(\omega t - \phi)}$$ (5)

Apply Kirchhoff 's Voltage Law (KVL) to the complex resonant tank.

$$V_s e^{j\omega t} = v_L + v_C + v_R + v_x$$ (6)

Substitute equations (2-5) into (6), and take $e^{j\omega t}$ away from both sides of the equation.

$$V_s e^{j\phi} = \left(L\frac{d i_x}{dt} + j\omega L i_x - jL i_x\frac{d\phi}{dt}\right) +$$
$$+ \left(\frac{1}{C\omega^2}\frac{d i_x}{dt} - j\frac{i_x}{\omega C} - j\frac{i_x}{C\omega^2}\frac{d\phi}{dt}\right) + R i_x + \frac{4}{\pi}v_o$$ (7)

By separating the real and imaginary parts of (7), two state equations of resonant converters are obtained.

$$L_a\frac{d i_x}{dt} = V_s\cos\phi - R i_x - \frac{4}{\pi}v_o$$ (8)

$$L_a\frac{d\phi}{dt} = Z_\omega - \frac{V_s\sin\phi}{i_x}$$ (9)

From the rectifier stage,

$$\frac{d v_o}{dt} = \frac{1}{C_o}(\frac{2}{\pi}i_x - \frac{v_o}{R_o})$$ (10)

Where

$$L_a = L + \frac{1}{C\omega^2},$$

$$Z_\omega = \omega L - \frac{1}{\omega C}$$

The state equations (8), (9), (10) contain only low frequency state variables, resonant inductor current envelope i_x, phase angle θ, and output voltage v_o. These state equations can predict large signal transitions such as startup process or step load response.

By perturbing the state equations around the steady state operating point and taking Laplace transformation, transfer functions of input to output, control to output are derived. The steady state operating point is obtained by letting derivatives equal to zero in the state equations.

III. VALIDATION OF THE PROPOSED MODELING METHOD

One series resonant DC/DC converter has the following parameters: $L=130\mu H$, $C=22nF$, $R=0.5\Omega$, $C_o=47\mu F$, $R_o=8\Omega$, $V_{dc}=150V$, and switching frequency $f=100kHz$.

The state equations of this resonant converter are first derived. These state equations are valid for any operating point. They can be used to predict any large signal transient responses, such as startup process, step load transient, step input transients.

1) Startup process of series resonant DC/DC

The output voltage and the resonant inductor current envelope by the proposed method are shown in Fig.4a. The simulation results are shown in Fig.4b.

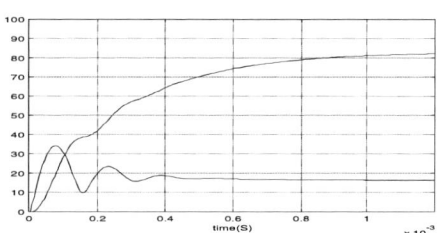

(a) start-up process by the proposed method

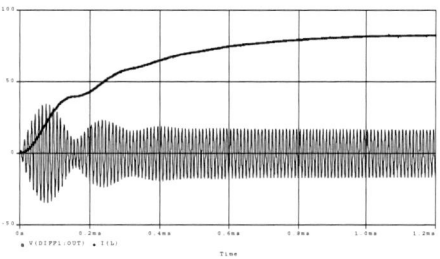

(b) start-up process by PSPICE simulation

Figure 4.   Start-up process of the resonant converter

(2) Transfer functions of series resonant DC/DC

By perturbing the state equation around the operating point, transfer functions are derived analytically. The developed transfer functions agree with those by SIMPLIS simulation very well.

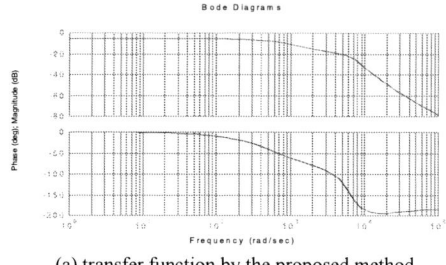

(a) transfer function by the proposed method

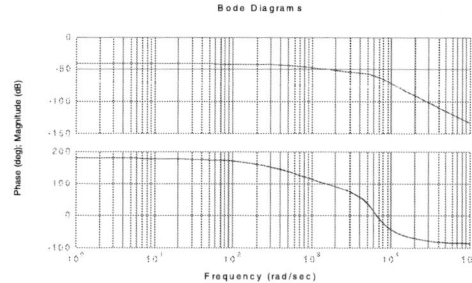

(b) transfer function by SIMPLIS

Figure 5.   Transfer function of $\dfrac{v_o(s)}{v_{dc}(s)}$ for series resonant DC/DC

Similarly, the transfer function of the switching frequency to output voltage can be derived from the state equations.

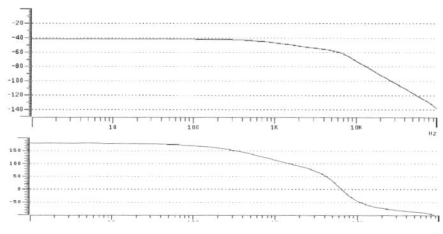

(a) transfer function by the proposed method

(b) transfer function by SIMPLIS simulation

Figure 6.   Transfer function of $\dfrac{v_o(s)}{f(s)}$ for series resonant DC/DC converter

IV. CONCLUSIONS

In this paper, state equations are developed for resonant converters. The derived state equations contain only low frequency state variables and can predict large signal transitions. By perturbing the state equations around the operating point, transfer functions are derived. The proposed method is verified by SIMPLIS simulation results. This modeling technique is applicable to other resonant converters. It can aid closed-loop design for resonant converters.

REFERENCES

[1] R.J.King, "A Fourier analysis for a fast simulation algorithm" IEEE Transactions on Power Electronics, Vol.4, No.4, October 1989, pp.434-441

[2] Ramesh Oruganti and Fred C. Lee, "Resonant power processors- state plane analysis" IEEE Transactions on Industry Applications, Vol.21, November/December 1985, pp. 1453-1471

[3] Nasser H. Kutkut, C.Q. Lee and Issa Batarseh, "A generalized program for extracting the control characteristics of resonant converters via the state-plane diagram" IEEE Transactions on Power Electronics, Vol.13, No.1, January 1998, pp. 58-66

[4] George c. Verghese, Malik E.Elbuluk and John G. Kassakian, " A General Approach to Sampled-Data Modeling for Power Electronic circuits" Conference Record of IEEE PESC 1984, pp.316-330

[5] M.E.Elbuluk, G.C.Verghese and John G. Kassakian, "Sampled-data modeling and digital control of resonant converters", IEEE Transactions on Power Electronics, Vol.3, No.3, July 1988, pp. 344-354.

[6] Jian Sun and Horst Grotstollen, "Averaged modeling and analysis of resonant converters" Conference Record of IEEE PESC 1993, pp.707-713.

[7] S.R.Sanders, J.M.Noworolski, X.Z.Liu and G.C.Verhese, "Generalized Averaging Method for Power Conversion Circuits", Conference Record of IEEE PESC 1990, pp.333 -340

[8] M. Castilla, L.G. de Vicuna, M. Lopez, and V. Barcons, "An averaged large-signal modeling method for resonant converters" Conference Record of IEEE IECON 1997, pp. 447-452

[9] Wong, S.C. and Brown, A.D. "Parallel resonant converter as a circuit simulation primitive" Conference Record of IEE Circuits, Devices and Systems, 1995, pp. 379 -386

[10] Siu-Chung Wong and A.D. Brown, "Analysis, modeling, and simulation of series- parallel resonant converter circuits" IEEE Transactions on Power Electronics, Vol.10, No.5, September, pp. 605 - 614

[11] Sam Ben-Yaakov, Stanislav Glozman and Raul Rabinovici, "Envelope Simulation by PSPICE Compatible Methods of Electric Circuits Driven by Modulated Signals" Conference Record of IEEE APEC 2000, pp. 811-817

[12] Yingqi Zhang, Ph.D dissertation "New Techniques in PWM and Resonant Converters", 2003, Queen's University, Canada

Design Considerations and Expiremantal Results of an Adaptive Frequency Controller Under Variable Line and Load Conditions

Jaber A. Abu Qahouq*, Wisam Al-Hoor** and Issa Batarseh**

**The University of Alabama
Department of Electrical and Computer Engineering
Tuscaloosa, Alabama 35487, USA*

***University of Central Florida
School of Electrical Engineering and Computer Science
Orlando, Florida 32816, USA*

Abstract —**This paper presents critical design guidelines and consideration for the adaptive step size function of the previously proposed ASSAT controller that tracks the optimum switching frequency with improved convergence speed and accuracy to maximize power converter efficiency. It is revealed how the adaptive step size function is affected by several factors and it is shown how the function should be designed to guarantee good convergence stability, speed, and accuracy. The design is verified by experimental results.**

I. INTRODUCTION

Power conversion efficiency of a power converter is a function of several design variables such as switching frequency, output inductance, switching devices characteristics, input and output voltages, and Synchronous Rectifiers (SR) dead-time [1-12] and is a function of several other surrounding factors such as temperature and component aging [8, 9, 10, 11]. Adaptive efficiency optimization schemes are able to auto-tune and auto-design power converter parameters in order to maintain the highest possible power converter efficiency for a given design. Examples of these methods are presented in [8, 9, 10, 11].

In [11], adaptive-step-size control method and algorithm, namely ASSAT controller, for a variable switching frequency controller is presented. The ASSAT method adaptive step size result significant convergence speed and accuracy to obtain the optimum switching frequency for highest efficiency when compared to the fixed step size controller presented in [10]. In [11], the variable adaptive step size is realized by a proposed equation (briefly reviewed in the next section) which has a design parameter called μ. This paper discusses how μ affects the convergence speed and accuracy of the ASSAT controller and provide the theoretical design guidelines and criteria to select μ value [12]. The paper also presents experimental results.

Next section reviews the ASSAT controller. Section III presents the design considerations of the ASSAT controller and its variable step size function. Section IV present experimental results and the conclusion is given in Section V.

II. REVIEW OF THE ADAPTIVE STEP SIZE FUNCTION

Fig. 1 shows the ASSAT controller with the adaptive step size function. The adaptive step size is determined by the following equation:

$$f_{sw\_step} = \xi \cdot \mu_{max} \cdot \nabla I_{in} \tag{1}$$

Where ∇I_{in} is the gradient function of the input current and the sign of ∇I_{in} (positive or negative) indicates the direction of the movement on the semi parabolic curve. The gradient function is given by:

$$\nabla I_{in} = \partial I_{in} / \partial f_{sw} \tag{2}$$

Where I_{in} is the input current and f_{sw} is the switching frequency.

μ_{max} is a parameter that controls the maximum step-size that should not be exceeded for a given power converter system design and ξ is a constant (less than one) which determines the tradeoff between the adaptive controller speed and convergence error. μ should satisfy the following condition in order for the controller to converge:

$$0 < \mu < \frac{1}{\lambda} \tag{3}$$

The design issue is the fact that the μ varies as a function of converter parameters especially load current and input voltage. This is because λ is a parameter that depends on the converter operation parameters. λ can be approximated by:

$$\lambda = \frac{\nabla I_{in}}{2 \cdot (f_{sw} - f_{SW-O})} \tag{4}$$

Where f_{sw} is the current switching frequency and f_{SW-O} is the optimum switching frequency that results in the minimum input current. One objective of this paper is to show the design guidelines for selecting the value of μ based on the above.

III. DESIGN CONSIDERATIONS AND DESIGN OF THE ADAPTIVE STEP SIZE FUNCTION

Consider the following design example: Single phase DC-DC buck converter with $V_{in} = 12V$, $V_o = 1V$, output Inductor: $L_o = 400nH$, output capacitance $C_o = 5mF$, upper FET: NTD40N03R, two in parallel., lower FET: NTD85N02R, two in parallel. Using a power loss model (programmed in MathCad® for example), the efficiency curve and the input current of the power converter can be approximated with good accuracy. The equations used in the power loss model can be found in [3, 6, 7]. It is found using the model that the optimum frequency exists near 180kHz. It is assumed, as a hardware design parameter, that the maximum frequency that the

978-1-4244-4782-4/10 $26.00 © 2010 IEEE

prototype can go to is 550kHz. The design guidelines and considerations that should be followed are as follows:

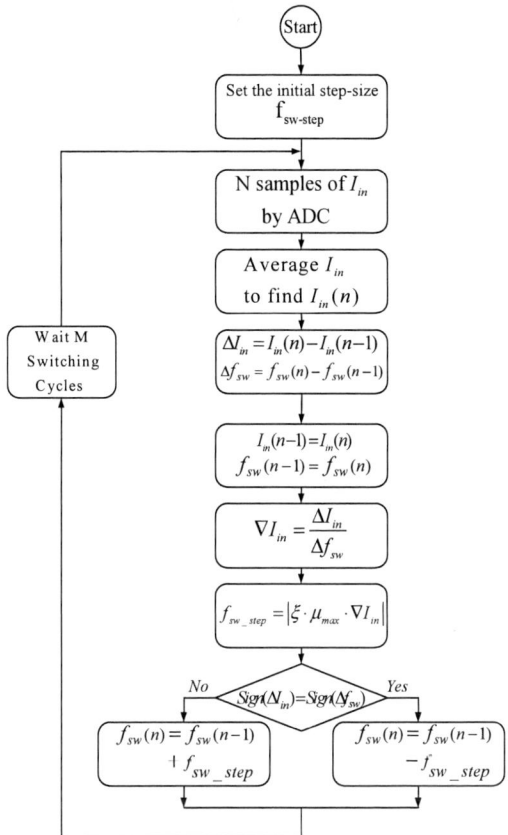

Fig. 1: ASSAT Controller Algorithm for Switching Frequency Auto-Tuning

(1) ξ, which is a constant less than one, is selected to result in $\mu = \xi \cdot \mu_{max} < \mu_{max}$ in Eq. (1). Since μ_{max} is an approximated parameter and can deviate based on the initial theoretical assumptions, the smaller ξ value is, the safer is the ASSAT controller from approaching the actual μ_{max} at a given set of operating conditions. While smaller ξ leads to improved stability and convergence safety margin, it results in slower convergence speed (but higher convergence accuracy), which makes this selection a designer choice based on the designed system conditions.

(2) μ_{max} depends on the gradient function value, which is approximated based on a line slope by assuming that the initial frequency is only far from the optimum frequency by a maximum given amount ($550 \times 10^3 - 180 \times 10^3$ in this design example). This is demonstrated by Fig. 2. The actual initial frequency is usually closer to the optimum point. As the optimum

switching frequency value is approached, the slope decreases, and therefore, the actual system μ_{max} increases. This means that that ASSAT controller becomes more immune to any possible instability or divergence as the optimum point is reached, which is an important characteristic of the ASSAT algorithm.

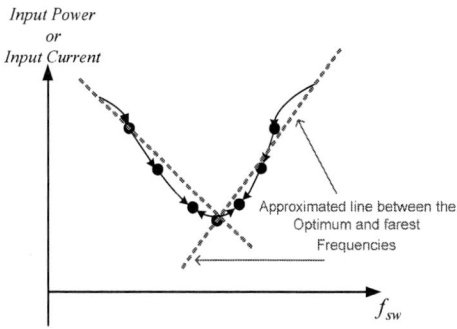

Fig. 2: Demonstration plot for the ASSAT design considerations

(3) The worst case design value for μ_{max} exists at the highest possible current and at the lowest possible input voltage that the power stage can work with without malfunctioning because of the components rating or/and regulation closed loop design range. This is because at higher load currents, it is expected that the same change (or step) in the switching frequency will result in higher change in the input current and input power, and vice versa, since the effect of the same percentage power loss has higher magnitude at higher output current or output power. The same can be concluded for the lowest input voltage, since for the same power, the input current magnitude is higher at lower input voltages. Table 1 shows additional verification for this behavior. It will be shown how the values of Table 1 were calculated later in this section.

(4) Therefore, given a range of load current, a range of input voltage, and a range of possible optimum switching frequencies (not exact values, just a rough range is used) that a power stage can operate with as maximum limits, a worst case value for μ_{max} can be selected (the smallest one). The selected value here is $\mu_{max} = 1.68 \times 10^{12}$. Adding to this using a $\xi < 1$ in Eq. (1), leads to large safety margin, while still obtaining excellent performance compared with the fixed step-size.

(5) To obtain the μ_{max} value, there is no need to have an accurate knowledge of the power stage efficiency. An approximated idea on the expected efficiency curve at the highest load current, lowest input voltage and the most far operating frequency value form the optimum one coupled with selecting a smaller μ_{max} value and using $\xi < 1$ provides sufficient safety margin.

Table 1: $\mu = 1/\lambda$ behavior as a function of load current and input voltages

Load Current	Parameter	Input Voltage	
		12V	6V
5A	$1/\lambda$	3.561×10^{12}	3.229×10^{12}
10A	$1/\lambda$	2.464×10^{12}	2.34×10^{12}
15A	$1/\lambda$	1.888×10^{12}	1.79×10^{12}

(6) The switching frequency range should be bounded or accounted for such that the output voltage ripple requirements are met at all switching frequency values and such that the operating ranges of components used in the design are not exceeded. While the possible switching frequency range is assumed to be wide (550kHz in this design example) in order to guaranteed convergence in the design calculations, in the actual operation, the optimum switching frequency will exist in a narrow range (130kHz to 190kHz in this design example).

The following is an example of how the values of Table 1 are calculated: By Assuming that the optimum switching frequency is somewhere near $f_{sw-o} \approx 180kHz$ and the maximum start frequency is $f_{sw-start-max} \approx 550kHz$, ∇I_{in} can be calculated at input voltages 6V and load currents 15A as:

$$\nabla I_{in}(n) = \frac{I_{in}(f_{sw-start-max}) - I_{in}(f_{sw-o})}{f_{sw-start-max} - f_{sw-o}}$$
$$= \frac{2.912 - 2.759}{550 \times 10^3 - 180 \times 10^3} = 4.135 \times 10^{-7}$$

This ∇I_{in} value represents a line slope approximation of Fig. 2 curve which is connected between the approximated optimum switching frequencies and the most far switching frequency (from the optimum frequency or from the range of optimum frequencies) that the controller is allowed to operate at.

λ can be then calculated using Eq. (4) as:

$$\lambda = 5.588 \times 10^{-13}$$

μ_{max} can be calculated using Eq. (3) as follows:

$$0 < \mu < \frac{1}{\lambda} \Rightarrow \mu < \frac{1}{5.588 \times 10^{-13}} \Rightarrow \mu < 1.79 \times 10^{12}$$

This value represents the maximum value for μ_{max} above which, the adaptive loop may not converge. The considerations of the final selection of this value that will be used in the hardware are presented next section.

If $\mu_{max} = 1.68 \times 10^{12}$ is selected (which is lower than the above calculated value for adding a safety margin) and it is desired to calculate the next iteration size-size at $\xi = 0.1$ at a given point, Eq. (1) can be used. For example, consider the case when the converter is operating at 10A load current and 12V input voltage, and the ASSAT controller initial frequency is $f_{sw_{(0)}} = 455kHz$, the initial step-size used is $f_{sw-step_{(0)}} = 35kHz$ and the optimum switching frequency is $f_{sw-o} \approx 180kHz$ (approximated from the power loss model). It is desired to calculate the next step-size. From the loss model it can be calculated that the average input current is $I_{in}(n) = 1.001A$ at $f_{sw}(n) = 420kHz$ and the previous iteration average input current is $I_{in}(n-1) = 1.011A$ at $f_{sw}(n-1) = 455kHz$. Then the gradient $\nabla I_{in}(n)$ can be approximated as:

$$\nabla I_{in}(n) = \frac{I_{in}(n) - I_{in}(n-1)}{f_{sw}(n) - f_{sw}(n-1)} = \frac{1.001 - 1.011}{420 \times 10^3 - 455 \times 10^3} = 2.8571 \times 10^{-7}$$

$$f_{sw\_step} = \left| \xi \cdot \mu_{max} \cdot \nabla I_{in} \right|$$
$$= \left| 0.1 \times 1.68 \times 10^{12} \times 2.8571 \times 10^{-7} \right| \approx 48 \text{ kHz}$$

The next iteration new approximated switching frequency value is $420 \times 10^3 - 48 \times 10^3 = 372 \times 10^3 Hz$. By repeating the same process, the next step-sizes and gradient values can be estimated and used as design guidance for the experimental adaptive loop design.

IV. EXPERIMANTAL REULTS

Experimental results have been obtained from a prototype with the specifications discussed in the previous section design example. Table 2 shows a summary of some of the results obtained for different μ values with the variable step size function. The table also shows a comparison with fixed step size case for different fixed step sizes. It can be observed how the μ value affects the convergence speed and accuracy of the controller as discussed earlier in this paper.

V. CONCLUSION

This paper presents critical design guidelines and consideration for the adaptive step size function of the ASSAT controller that tracks the optimum switching frequency with improved convergence speed and accuracy to maximize the efficiency. It is revealed how the adaptive step size function is affected by several factors and it is shown how the function should be designed to guarantee good convergence stability, speed, and accuracy. The design is verified by experimental results. The experimental results also compares the effect of step size and μ on the convergence speed and accuracy.

978-1-4244-4782-4/10 $26.00 © 2010 IEEE

Table 2: Summary and Comparison of sample Experimental Results

Load/Optimum Frequency	Criteria	Fixed-Step-Size			Adaptive-Step-Size		
		5 kHz	10 kHz	30 kHz	$\mu = 0.084 \times 10^{12}$	$\mu = 0.168 \times 10^{12}$	$\mu = 0.252 \times 10^{12}$
5A / 144kHz	Number of Iterations	80	42	15	33	19	12
	~% Error in f_{sw-o}	7.86% (155kHz)	2.77% (140kHz)	9.72% (130 kHz)	0.7% (143kHz)	2.1% (141kHz)	2.77% (148kHz)
10A / 165kHz	Number of Iterations	80	40	14	31	16	12
	~% Error in f_{sw-o}	6.06% (155kHz)	3.03% (160kHz)	3.03% (160kHz)	0.6% (164kHz)	1.2% (163kHz)	1.8% (168kHz)
15A / 173kHz	Number of Iterations	74	38	14	23	12	8
	~% Error in f_{sw-o}	6.93% (185kHz)	4.04% (180kHz)	7.15% (160kHz)	0.58% (172kHz)	1.73% (176kHz)	2.3% (169kHz)

REFERENCES

[1] X. Zhou, T. Wang, and F. Lee, " Optimizing design for low-voltage DC-DC Converters," Twelfth Annual Applied Power Electronics Conference and Exposition, APEC '97 Conference Proceedings, Volume 2, 23-27 Feb. 1997 Page(s):612 - 616 vol.2, 1997.

[2] B. Arbetter, R. Erickson and D. Maksimovid, "DC-DC Converter Design for Battery-Operated Systems" IEEE Power Electronics Specialists Conference, PESC'95, Vol.1,Pages:103-109, June 1995.

[3] J. Klein, "Synchronous buck MOSFET loss calculations with Excel model", Fairchild Semiconductor Application Notes AN-6005, 2006.

[4] B. Sahu and G. A. Rincón-Mora, "An Accurate, Low-Voltage, CMOS Switching Power Supply With Adaptive On-Time Pulse-Frequency Modulation (PFM) Control, " IEEE Transactions on Circuits and Systems-I: Regular Papers, Vol. 54, No. 2, pp. 312-321, Febreuary 2007.

[5] X. Zhou, M. Donati, L. Amoroso, and F. Lee, "Improved Light-Load Efficiency for Synchromous Rectifier Voltage Regulator Module," IEEE Transactions on Power Electronics, Volume 15, pp. 826 – 834, September 2000.

[6] M. Gildersleeve, H. Forghani-zadeh, and G. Rincon-Mora, "A Comprehensive Power Analysis and a highly efficient Mode-Hopping DC-DC Converter," IEEE Proceedings of Asia-Pacific Conference on ASIC, 2002, pp. 153–156.

[7] J. Abu-Qahouq, O. Abdel-Rahman, L. Huang and I. Batarseh, "On Load Adaptive Control of Voltage Regulators for Power Managed Loads: Control Schemes to Improve Converter Efficiency and Performance," IEEE Transactions on Power Electronics, Vol. 22, No. 5, September 2007.

[8] J. Abu-Qahouq, H. Mao, H. Al-Atrash, and I. Batarseh, "Maximum Efficiency Point Tracking (MEPT) Method and Dead Time Control," IEEE Transactions on Power Electronics, Vol. 21, Issue 5, pages: 1273-1281, Sept. 2006.

[9] V. Yousefzadeh and D. Maksimovic, "Sensorless optimization of dead times in DC-DC converters with synchronous rectifiers," IEEE Transactions on Power Electronics, Vol. 21, Issue 4, pages: 994-1002, July 2006.

[10] W. Al-Hoor, J. Abu-Qahouq, L. Huang, and I. Batarseh, "Adaptive Variable Switching Frequency Digital Controller Algorithm," IEEE International Symposium on Circuits and Systems, ISCAS'2007, May 2007.

[11] Jaber Abu-Qahouq, Wisam Al-Hoor, Wasfy B. Mikhael, Lilly Huang and Issa Batarseh, "Variable-Step-Size Auto-Tuning Algorithm for Digital Power Converter with Variable-Switching-Frequency," IEEE Power Electronics Specialists Conference, PESC'2007, Page(s): 105-111, June 2007.

[12] Jaber Abu-Qahouq, Wisam Al-Hoor, Wasfy Michael, Lilly Huang and Issa Batarseh, "Analysis and Design of an Adaptive-Step-Size Digital Controller For Switching Frequency Auto-Tuning," IEEE Transactions on Circuits and Systems I - Regular Papers, Vol. 56, No. 12, December 2009.

Modeling and Mitigation of Dynamic Load Beat-Frequency Oscillation in Multiphase Voltage Regulators with High-Gain Peak Current Control Scheme

Chen-Hua Chiu[1], Dan Chen[1], Ching-Jan Chen[1], and Wei-Hsu Chang[2]

[1] Electrical Engineering Dept.
National Taiwan University
Taipei, Taiwan

[2] RichTek Technology Corporation
Chupei City, Hsinchu, Taiwan

Abstract—**High-frequency dynamic load may cause phase current oscillation in multiphase interleaved voltage regulators. A model is presented in this paper to investigate this phenomenon in the recently reported high-gain peak current control (HGPCC) voltage regulators. Based on the analysis, it is concluded that HGPCC scheme suffers from beat-frequency oscillation problems, similarly to conventional peak current control scheme. A modified HGPCC scheme is then proposed and verified to mitigate this problem. A comparison of the three schemes will be made from the viewpoint of beat-frequency oscillation.**

I. INTRODUCTION

Multiphase interleaved buck converter configuration is often used in the voltage regulators (VR) for central processing unit (CPU) power applications. There have been numerous papers about this general subject in the past decade. Recently, a new issue about this subject has arisen, that is, interleaving may cause serious phase current oscillation under a high-frequency dynamic load such as a CPU [1-4]. The repetitive load changes ranges from several kilohertz to several megahertz. When the dynamic load frequency approaches the buck converter switching frequency, a significant beat (difference) frequency oscillation may occur, which hurts efficiency and may even cause the destruction of the main semiconductor switches.

Fig. 1 shows the circuit diagram of a two-phase buck regulator using conventional peak-current control (PCC) scheme and the resultant inductor current waveforms due to beat-frequency oscillation. In the circuit diagram, V_{ID} is the reference voltage specified by CPU, H_V is the voltage-loop compensator, and R_i is the current-sensing gain. As can be seen from the waveforms, the sum of the two inductor currents looks normal but each phase current oscillates with high amplitude.

A high-gain peak current control (HGPCC) scheme has been reported to achieve adaptive voltage positioning (AVP)

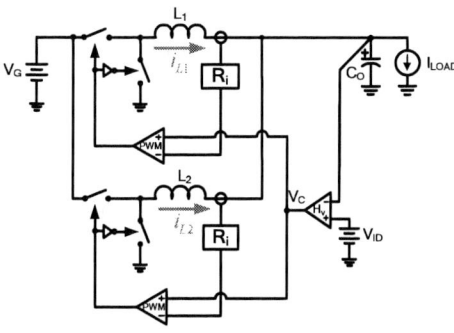

(a). Circuit diagram

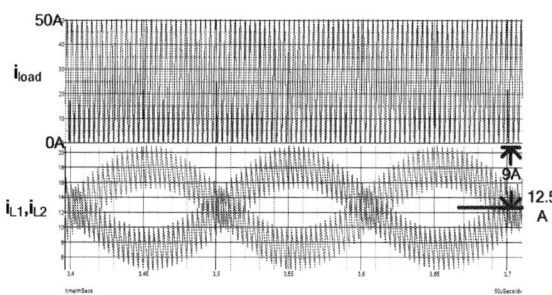

(b). Beat-frequency oscillation in inductor currents

Fig. 1: An example of beat-frequency oscillation. Simulated results for a 300 kHz two-phase PCC VR (I_{load} = 0~50A, f_{load} = 295 KHz, f_{beat}=5kHz)

in multiphase interleaved buck converters [5-8]. This scheme retains the advantages but overcomes the drawbacks of a conventional PCC scheme. That is, the performances of the converter DC output voltage offset and the low-frequency line regulation are much improved using a HGPCC scheme.[7-8]. Fig. 2 shows the circuit diagram of a two-phase voltage regulator using HGPCC scheme. Compared to a conventional PCC scheme, the key difference lies in the fact that there is a

978-1-4244-4782-4/10 $26.00 © 2010 IEEE

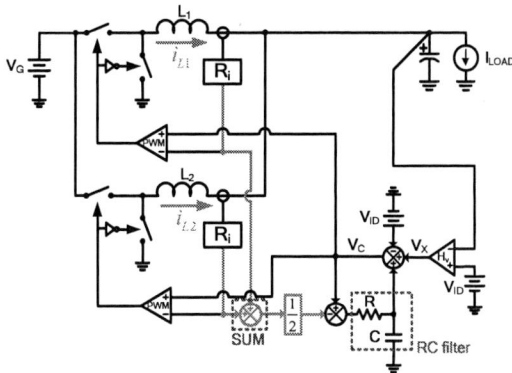

Fig. 2: Circuit diagram of a two-phase HGPCC VR

RC filter used to retrieve the DC information in the feedback loop in a HGPCC. The focus of the present paper is to analyze the HGPCC scheme from the perspective of beat-frequency oscillation. From the analytical results, a corrective measure will be proposed to mitigate the problem. A comparison of the three schemes will be made from the perspective of beat-frequency oscillation. Simulation results will be used to confirm the validity of the proposed scheme.

II. SMALL-SIGNAL MODEL FOR THE ANALYSIS OF BEAT-FREQUENCY OSCILLATION OF HGPCC SCHEME

The beat-frequency component is generated by the non-linear characteristic of a PWM modulator. In the two-phase interleaved PWM model shown in Fig. 3, the excitation frequency is the load frequency ω_{load}. Because of the nonlinear characteristics of the PWM modulator, two frequency components are generated [9]. One component is the load frequency ω_{load}, and the other is the component of beat(difference) frequency ω_{beat}. Equation (1) shows the relationships of the three frequencies.

$$\omega_{beat} = \omega_{sw} - \omega_{load} \qquad (1)$$

As shown in Fig. 3, there are four transfer functions correlating $\hat{v}_c$ to each of the two duty cycles $\hat{d}_1$ and $\hat{d}_2$. Since the focus in this paper is on the behavior of the beat-frequency component, only the transfer functions relating to ω_{beat} will be

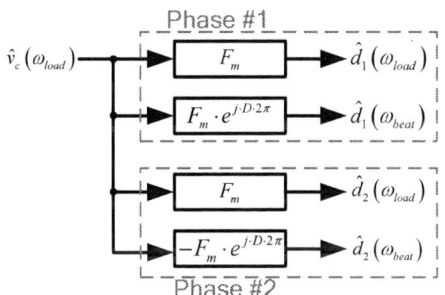

Fig. 3: Multy-frequency model of a two-phase interleaved PWM modulator

used in modeling the HGPCC scheme. If the dynamic load frequency approaches switching frequency, the beat frequency ω_{beat} is in low frequency range. Since every block in such a system has low-pass nature, the effects of ω_{beat} becomes prominent. It is also noticed that $\hat{d}_1(\omega_{beat})$ and $\hat{d}_2(\omega_{beat})$ are 180-degree out of phase for a 2-phase interleaved configuration.

A. Beat-frequency oscillation of HGPCC in its present form

A block diagram of the model for beat-frequency analysis of HGPCC schemes is shown in Fig. 4, where $Z_{OC}(\omega_{load})$ represents closed-loop output impedance of the VR [8], F_m represents PWM modulator gain, D represents the steady-state value of duty cycle, and $H_e(\omega_{beat})$ represents the current-loop sample-and-hold effect [10]. Transfer function $G_F(s)$ is related to the RC filter which will be illustrated later. $G'_{id}(\omega_{beat})$ is the beat-frequency duty-cycle-to-inductor-current transfer function of power stage.

A physical explanation of the model is given as follows. High-frequency load current variation, $\hat{i}_{load}(\omega_{load})$, causes the output voltage variation, $\hat{v}_o(\omega_{load})$, through transfer function $Z_{OC}(\omega_{load})$. This variation is then fed into control circuits, including $H_V(\omega_{load})$ and $G_F(\omega_{load})$, to generate $\hat{v}_c(\omega_{load})$. Through the PWM modulators, $\hat{v}_c(\omega_{load})$ then generate beat-frequency duty variation $\hat{d}(\omega_{beat})$. The beat frequency duty variation generate beat-frequency inductor currents. The beat-frequency inductor current information is fed back through the current loops. The current loop gain in beat frequency component, $T'_i(\omega_{beat})$, is defined as the product of four transfer functions, as shown by equation (2).

$$T'_i(\omega_{beat}) \equiv G'_{id}(\omega_{beat}) \cdot R_i \cdot H_e(\omega_{beat}) \cdot F_m \qquad (2)$$

Based on the model shown in Fig. 3, there is 180 degree phase difference between the beat frequency duty cycles

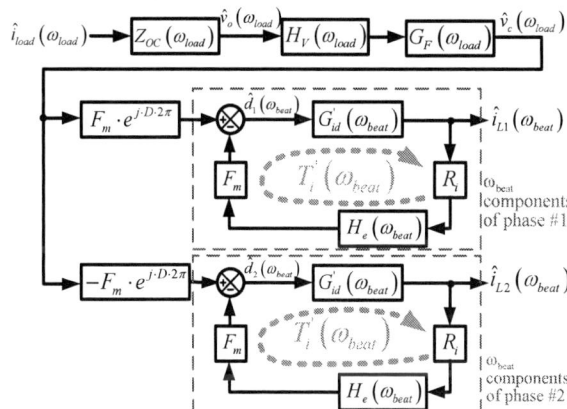

Fig. 4: Simplified model for the analysis of beat-frequency oscillation for HGPCC VR

978-1-4244-4782-4/10 $26.00 © 2010 IEEE

$\hat{d}_1(\omega_{beat})$ and $\hat{d}_2(\omega_{beat})$. Therefore, the summation of the two inductor currents $\hat{i}_{L1}(\omega_{beat})$ and $\hat{i}_{L2}(\omega_{beat})$ become zero and cause zero $\hat{v}_o(\omega_{beat})$. Thus, the voltage-loop is not presented for the analysis of beat-frequency oscillation [2]. Simultaneously, the transfer function of duty cycle to inductor current for beat-frequency component is expressed by equation (3), where $Z_{L1}(\omega_{beat})$ represents the impedance of each inductor at beat-frequency [2].

$$G'_{id}(\omega_{beat}) \triangleq \frac{\hat{i}_{L1}(\omega_{beat})}{\hat{d}_1(\omega_{beat})} = \frac{\hat{i}_{L2}(\omega_{beat})}{\hat{d}_2(\omega_{beat})} = \frac{V_g}{Z_{L1}(\omega_{beat})} \quad (3)$$

It is reported in [8] that the small-signal model of HGPCC scheme boosts the gain of voltage-loop and current-loop simultaneously with the $G_F(s)$, which is shown in equation (4), where ω_{LPF} represents the bandwidth of the RC filter. It can be seen that $G_F(s)$ has high gain for frequency below ω_{LPF}, but unity gain above ω_{LPF}. Thus, HGPCC shows high DC accuracy for output voltage and better audio susceptibility compared to PCC scheme.

$$G_F(s) = \frac{\omega_{LPF}}{s} \cdot (1 + \frac{s}{\omega_{LPF}}) \quad \text{where } \omega_{LPF} \equiv \frac{1}{RC} \quad (4)$$

From the model shown in Fig. 4, the relationship between the input $\hat{i}_{load}(\omega_{load})$ and the output $\hat{i}_{L1}(\omega_{beat})$ can be represented by a factor S which can be derived as (5), where $T'_i(\omega_{beat})$ is the current-loop gain at beat frequency. $G(\omega_{load}, \omega_{beat})$ is shown by (6).

$$S \equiv \frac{\hat{i}_{L1}(\omega_{beat})}{\hat{i}_{load}(\omega_{load})} = \frac{G(\omega_{load}, \omega_{beat})}{1 + T'_i(\omega_{beat})} \quad (5)$$

$$G(\omega_{load}, \omega_{beat}) \\ \equiv Z_{OC}(\omega_{load}) H_1 \cdot (\omega_{load}) G_F(\omega_{load}) \cdot F_m \cdot e^{j \cdot D \cdot 2\pi} \cdot G'_{id}(\omega_{beat}) \quad (6)$$

The factor S represents the circuit susceptibility to the beat-frequency oscillation due to load excitation. It can be seen that the beat frequency current-loop $T'_i(\omega_{beat})$ can

suppress beat-frequency oscillation.

From Fig. 2, the current signals of each phase are summed at SUM and fed through the RC filter to drive the two PWM modulators. Because the beat-frequency components of the two inductor current signals are 180-degree out of phase, the summation cancels out the beat-freq components. $G_F(\omega_{beat})$ therefore does not show in $T'_i(\omega_{beat})$ equation. That is to say, the addition of RC filter in HGPCC, although improves converter performances compared to conventional PCC converter, do not help the problem of beat-frequency oscillation.

B. A proposed HGPCC for mitigating beat-frequency oscillation

A modified version of HGPCC was then proposed to mitigate the problem. Fig. 6 shows the circuit diagram of the proposed modified HGPCC in which a RC filter is used for each phase. Each inductor current signal is fed through its own RC filter and drives its PWM modulator. There is no summation of inductor-current signals in this case. Thus, $G_F(\omega_{beat})$ appears in the $T'_i(\omega_{beat})$ as shown in Fig. 7. The low frequency gain of $T'_i(\omega_{beat})$ is therefore boosted, which mitigates beat-frequency oscillation.

Applying the same modeling approaches to the three control schemes mentioned above, those are PCC, HGPCC and the proposed HGPCC, equations can be obtained for comparison purposes. Tables I and II summarize the equations. Using equation (5) and the terms in Table I & II, a comparison of susceptibility from beat-frequency oscillation is shown in Fig. 8 for all the three schemes. As can be seen from the figure, there is no difference between the PCC and the HGPCC, but the proposed HGPCC exhibits a clear mitigation of oscillation at beat-frequencies below the corner frequency of the RC filter. This difference is due to the presence of $G_F(\omega_{beat})$, which boosts the current-loop gain $T'_i(\omega_{beat})$ in the proposed HGPCC.

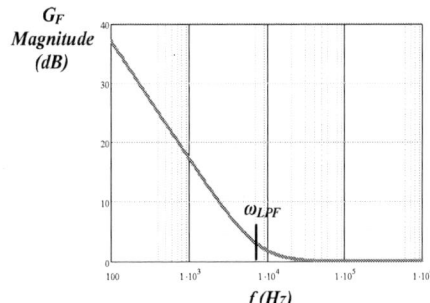

Fig. 5: Magnitude plot of $G_F(s)$

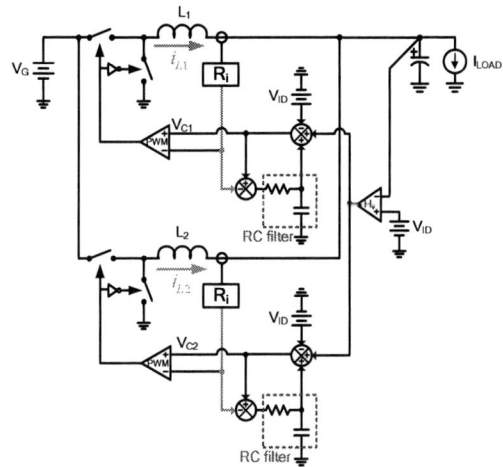

Fig. 6: Circuit diagram of the proposed modified two-phase HGPCC VR

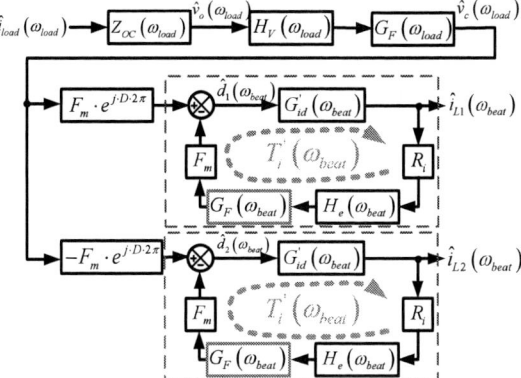

Fig. 7: Simplified model for the analysis of beat-frequency oscillation of the proposed modified HGPCC VR

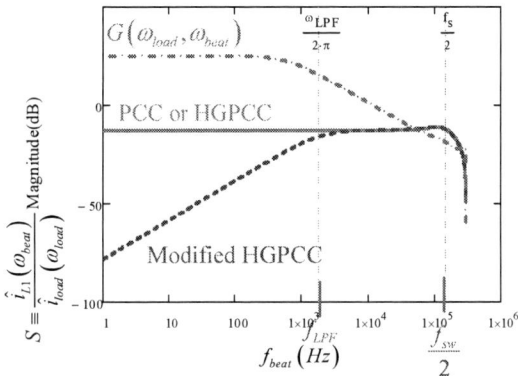

Fig. 8: Susceptibility from beat-frequency, S, oscillation for three schemes

III. SIMULATION RESULTS

As far as the IC designers are concerned, the simulated are as good as the experimental results. SIMPLIS simulations were performed to confirm the theoretical susceptibility S for the three schemes. The details of the circuit component value and working conditions are tabulated in Table III. Fig. 9 shows the time-domain waveforms of beat-frequency oscillation for the three schemes at two different beat frequencies. When the beat-frequency was at 10 KHz, which is above RC filter bandwidth, no difference was observed among the three schemes. However, at 100 Hz beat-frequency (299.9 kHz load frequency), the proposed HGPCC scheme shows a clear advantage of lower the susceptibility from beat-frequency oscillation for each phase. The simulation results agree with the theoretical results shown in Fig. 8.

A plot of susceptibility S for the modified HGPCC scheme for both the theoretical and the simulated results is shown in Fig. 10. Hardware for the modified HGPCC has not been built yet because IC implementation takes time, but the simulation results are normally well taken by IC community in general. The simulated results were obtained by point-by-point waveform simulations. It can be seen from the figure that the theoretical model predicts the experimental results.

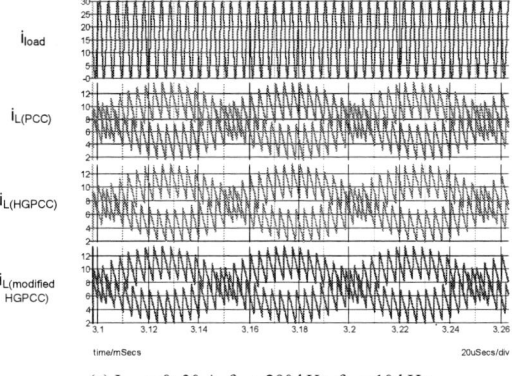

(a) I_{load} = 0~30 A, f_{load}=290 kHz, f_{beat}=10 kHz

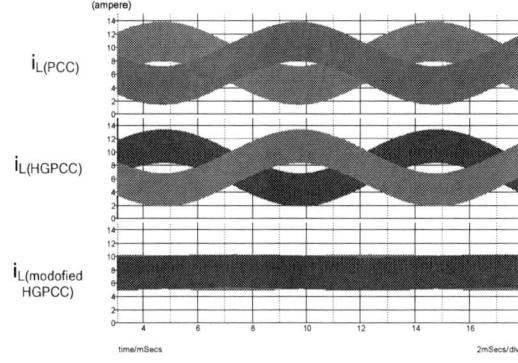

(b) I_{load} = 0~30 A, f_{load}=299.9 kHz, f_{beat}=100 Hz

Fig. 9: Waveforms of inductor currents under high-frequency dynamic load

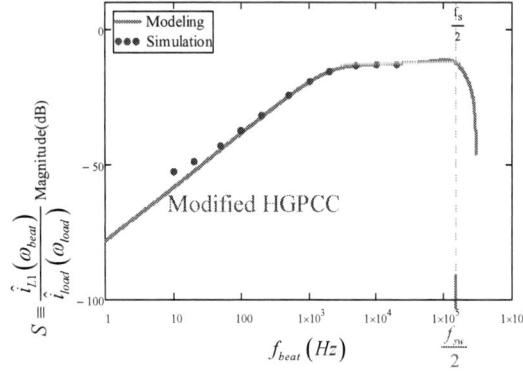

Fig. 10: Simulation of S for HGPCC scheme

IV. CONCLUSIONS

Models were presented to investigate the beat-frequency oscillation under high-frequency dynamic load for three types of current-mode AVP schemes. The results show that the conventional PCC scheme and the HGPCC scheme share the same degree of beat-frequency suppression for the same control bandwidth. However, with the proposed modified HGPCC scheme, the degree of suppression is improved when the beat-frequency is within the bandwidth of the RC filter. Therefore, the modified HGPCC offers the best combination of converter feedback performances and beat-frequency suppression.

TABLE I. THE TRANSFER FUNCTIONS OF $G\left(\omega_{load},\omega_{beat}\right)$

	$G\left(\omega_{load},\omega_{beat}\right)$
PCC	$Z_{OC}'\left(\omega_{load}\right)H_{V}\left(\omega_{load}\right)\cdot F_{m}\cdot e^{j\cdot D\cdot 2\pi}\cdot G_{id}'\left(\omega_{beat}\right)$
HGPCC	$Z_{OC}'\left(\omega_{load}\right)H_{V}\left(\omega_{load}\right)G_{F}\left(\omega_{load}\right)\cdot F_{m}\cdot e^{j\cdot D\cdot 2\pi}\cdot G_{id}'\left(\omega_{beat}\right)$
Proposed HGPCC	$Z_{OC}'\left(\omega_{load}\right)H_{V}\left(\omega_{load}\right)G_{F}\left(\omega_{load}\right)\cdot F_{m}\cdot e^{j\cdot D\cdot 2\pi}\cdot G_{id}'\left(\omega_{beat}\right)$

TABLE II. THE TRANSFER FUNCTIONS OF $T_{i}'\left(\omega_{beat}\right)$

	$T_{i}'\left(\omega_{beat}\right)$
PCC	$G_{id}'\left(\omega_{beat}\right)\cdot R_{i}\cdot H_{e}\left(\omega_{beat}\right)\cdot F_{m}$
HGPCC	$G_{id}'\left(\omega_{beat}\right)\cdot R_{i}\cdot H_{e}\left(\omega_{beat}\right)\cdot F_{m}$
Proposed HGPCC	$G_{id}'\left(\omega_{beat}\right)\cdot R_{i}\cdot H_{e}\left(\omega_{beat}\right)G_{F}\left(\omega_{beat}\right)\cdot F_{m}$

TABLE III. DETAILS OF COMPONENT VALUES USED IN SIMULATION

Two-phase interleaved buck converter	
$f_{S} = 300$ kHz	Switching frequency
$V_{G} = 12$ V	Input voltage
$V_{O} = 1.15$ V	Output voltage
$C_{O} = 1320$ nF	Output capacitance
$r_{Co} = 1.5$ mΩ	Series resistor of C_{O}
$L = 720$ nF	Inductance of each phase
$r_{L} = 2.8$ mΩ	Series resistor of L
$Ri = 28$ mΩ	Current-sensing gain
$f_{LPF} = 1.85$ kHz	Corner frequency of the RC filter

ACKNOWLEDGMENT

This work was supported by research grant #97~S-C48 to Taiwan University from Richtek Technology Corporation, Taiwan. The authors would also like to thank Transim Technology Corporation, U.S.A. for providing SIMPLIS simulation tool.

REFERENCES

[1] "Voltage Regulator-Down (VRD) 11.0, Processor Power Delivery Design Guidelines for Desktop LGA775 Socket," Nov. 2006, Intel.

[2] J. Sun, Q. Yang, M. Xu, F.C. Lee, "High-frequency dynamic current sharing analyses for multiphase buck VRs," *IEEE Trans. Power Electronics*, vol.22, no.6, pp.2424-2431, Nov. 2007.

[3] J. Sun, F.C Lee, M.. Xu, Q. Yang, "Modeling and analysis for beat-frequency current sharing issue in multiphase voltage regulators," in *Proc. IEEE Power Electronics Specialists Conference*, 2007, pp.1542-1548.

[4] W. Guo, P.K. Jain, "Analysis and modeling of voltage mode controlled phase current balancing technique for multiphase voltage regulator to power high frequency dynamic load," in *Proc. Applied Power Electronics Conference and Exposition*, 2009, pp. 1190-1196.

[5] K. Yao, M. Xu, Y. Meng, and F.C. Lee, "Design considerations for VRM transient response based on the output impedance," *IEEE Trans. Power Electronics*, vol. 18, pp. 1270-1277, Nov. 2003.

[6] M. Lee, D. Chen, K. Huang, E. Tseng ,and B. Tai, "Comparisons of three control schemes for adaptive voltage position (AVP) droop for VRMs applications," in *Proc. IEEE EPE-PEMC*, 2006, pp.206~211.

[7] J. R. Huang, S. C.-H. Wang, C. J. Lee, E. K.-L. Tseng, and D. Chen, "Native AVP control method for constant output impedance of DC power converters," in *Proc. IEEE Power Electronics Specialists Conf.*, 2007, pp. 2023-2028.

[8] C.-J. Chen, D. Chen, M. Lee, E. K.-L. Tseng, "Design and modeling of a novel high-gain peak current control scheme to achieve adaptive voltage positioning for DC power converters," in *Proc. IEEE Power Electronics Specialists Conference*, 2008, pp.3284-3290.

[9] Yang Qiu; Ming Xu; Kaiwei Yao; Sun, J.; Lee, F.C., "Multifrequency Small-Signal Model for Buck and Multiphase Buck Converters," *Power Electronics, IEEE Transactions on* , vol.21, no.5, pp.1185-1192, Sept. 2006.

[10] R. B. Ridley, "A new continuous-time model for current-mode control," *IEEE Trans. Power Electronics*, vol. 6, pp. 271~280, April 1991.

[11] R. B. Ridley, B.H. Cho, and F.C. Lee, "Analysis and interpretation of loop gains of multiloop-controlled switching regulators," *IEEE Trans. Power Electronics*, vol. 3, pp. 489-498, Oct. 1988.

Half-Wave Symmetry SHE-PWM Method for Multilevel Voltage Inverters

Wanmin Fei, Xiaoli Du
School of Electrical and Automation Engineering
Nanjing Normal University, Nanjing, China

Bin Wu
Department of Electrical and Computer Engineering
Ryerson University, Toronto, Ontario, Canada

Abstract—**Half-wave symmetry SHE-PWM method has a large number of valid solutions, which are beneficial to optimization design. This paper proposes a novel generalized formulation of half-wave symmetry SHE-PWM problems for multilevel inverters. The advantages of the proposed formulation include simplicity in format, flexibility in PWM waveforms and a broad solution space. Methods to obtain initial values for the SHE-PWM equations are discussed. Take a five-level inverter with M=0.75 as an example, six sets of solutions for two typical initial phase angles of 0 and 90 degrees are presented. Simulations and experiments are carried out. It is demonstrated that the experimental results agree well with simulated ones, which proves the validity and practicability of the new method proposed.**

I. INTRODUCTION

SHE-PWM techniques offer several advantages over other modulation methods including acceptable performance with low switching frequency to fundamental frequency ratios, direct control over output waveform harmonics, and the ability to leave triplen harmonics uncontrolled to take advantage of circuit topology in three phase systems, and therefore have drawn great attention in recent years. The main challenge of SHE-PWM techniques is to obtain the solutions. Much effort has been made in obtaining effective initial values and many algorithms have been proposed to deal with the problem of finding the desired solutions [1]-[3].

By now, most of the research on SHE-PWM method is based on an assumption that the PWM waveform is quarter-cycle symmetrical, which simplified the problem to a certain extent. However the quarter-wave symmetry constraint limits the solution space. On abolishing the symmetry requirements, more generalized techniques of SHE-PWM method can be formulated [4] [5], and which brings about infinite solutions. Some solutions may have merit relative to others with respect to system losses, ripple characteristics, distribution of un-eliminated harmonics, THD, or some other system aspect. However, four times the number of switching angles and harmonics elimination constraints compared with quarter-symmetry method result in asymmetry SHE-PWM methods

too complicated to be practical. With the same harmonics elimination and fundamental control tasks, if the quarter-wave symmetry constraint is relaxed to half-wave symmetry, half the number of switching angles and harmonics elimination constraints in contrast with asymmetry SHE-PWM method are needed. The complexity is reduced and the property of infinite solutions remains. The main purpose of this paper is to propose a generalized formulation of half-wave symmetry SHE-PWM problems for multilevel voltage inverters, and based on which to research the half-wave symmetry SHE-PWM technique.

II. A NOVEL FORMULATION OF HALF-WAVE SYMMETRY MULTILEVEL SHE-PWM PROBLEMS

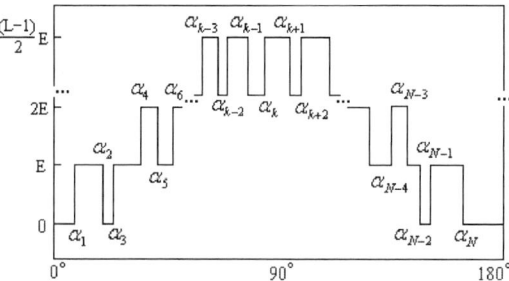

Fig. 1. PWM waveform in the first half-cycle for definition of switching angles

Switching angles in the first half-cycle of a typical half-wave symmetry SHE-PWM waveform of a multilevel inverter with L levels is shown in Fig. 1. If the PWM waveform in the second half-cycle is obtained with an assumption of $f(\pi + 2\pi f t) = -f(t)$, Then the problem under consideration is to find appropriate switching angles, namely $\alpha_1, \alpha_2, \alpha_3, \ldots, \alpha_N$ so that the $(N-1)/2$ nontriplen odd and even harmonics can be eliminated and control of the fundamental is also achieved. The Fourier series expansion of the SHE-PWM waveform is given by:

$$V(\omega t) = \sum_{n=1}^{\infty} (a_n \cos(n\omega t) + b_n \sin(n\omega t)) \qquad (1)$$

where ω is the radian frequency of the output voltage. Owing to the PWM waveform characteristics of half-wave symmetry, $a_n = b_n = 0$ when n is even. When n is odd, the generalized expressions of a_n and b_n for any number of switching angles and any number of voltage levels (even or odd, provided that the waveform is physically correct and can be implemented) are given by:

$$\begin{cases} a_n = -\dfrac{2E}{n\pi} \sum_{k=1}^{N} p_k \sin n\alpha_k \\ b_n = \dfrac{2E}{n\pi} \sum_{k=1}^{N} p_k \cos n\alpha_k \end{cases} \quad (2)$$

where E is the voltage of a level, and the coefficient of p_k is 1 for rising edges and -1 for falling edges.

For a multilevel inverter with L output levels, given a desired fundamental V_1, a corresponding modulation index M can be expressed as: $M = 2V_1/((L-1)E)$. The generalized equations of the half-wave symmetry SHE-PWM problems for multilevel voltage inverters can be shown as:

$$\begin{cases} \dfrac{2E}{\pi} \sum_{k=1}^{N} p_k \sin \alpha_k = -A_1 \\ \dfrac{2E}{\pi} \sum_{k=1}^{N} p_k \cos \alpha_k = B_1 \\ \dfrac{2E}{n\pi} \sum_{k=1}^{N} p_k \sin n\alpha_k = 0, n = 5,7,11,13\ldots \\ \dfrac{2E}{n\pi} \sum_{k=1}^{N} p_k \cos n\alpha_k = 0, n = 5,7,11,13\ldots \end{cases} \quad (3)$$

- N is the number of switching angles in a quarter cycle, and N is even for half-wave symmetry multilevel inverters;
- L is the number of levels of the output phase-voltage;
- α_k is the kth switching angles with a constraint:

$$0 \leq \alpha_1 \leq \alpha_2 \leq \alpha_3 \leq \cdots \leq \alpha_N \leq \pi \quad (4)$$

For a given fundamental control or a definite modulation index M, there are infinite corresponding arrays composed of constant A_1 and B_1 with a constraint expressed as follows: $\sqrt{A_1^2 + B_1^2} = \dfrac{L-1}{2}ME$. For a given goal of fundamental control and harmonic elimination, there are infinite sets of equations about the SHE-PWM problem. If one set of equations has several sets of solutions, there must be infinite sets of solutions to the SHE-PWM problems, which supply abundant opportunities for concrete optimization.

III. SOLUTIONS TO THE HALF-WAVE SYMMETRY SHE-PWM PROBLEMS FOR MULTILEVEL INVERTERS

A. Methods for obtaining initial values

For half-wave symmetry SHE-PWM problems, the phasing of fundamental is free to vary. The equations and initial values are dependent on the initial phase of the output fundamental. An effective method for obtaining initial values is using multi-carrier based SPWM, where the reference sinusoidal signal should have the same initial phase angle with that of the desired output fundamental voltage and the

amplitude should be close to that determined by the modulation index M.

Most of the initial values obtained with above methods are sufficiently close to the exact solution to guarantee convergence of the numerical iteration. Trial and error method can be adopted when above methods are invalid or more sets of solutions are wanted.

B. Resolving multilevel SHE-PWM equations

The procedure of resolving the non-linear equations is as follows: Firstly, the number of switching angles in a half-cycle, a modulation index M and the two constants A_1 and B_1 in Equation (3) are decided, and then a series of initial values composed of rising edges and falling edges can be obtained. Secondly, the multilevel SHE-PWM equations according to the rising edges and falling edges are formulated. Thirdly, the equations are solved with the fsolve () function provided by Matlab65. Finally, a judgment concerning whether the numerical result is a true solution to the multilevel SHE-PWM problem should be made. When a numerical result is in an ascending sequence strictly as the initial values, it satisfies the constraint shown in Eq. (4) and is a true solution undoubtedly. Otherwise, it still may be a true solution. The criterion is as follows: suppose there is an integer K within the range of $(-(L-1)/2, (L-1)/2)$, where L is the number of levels of the inverter,

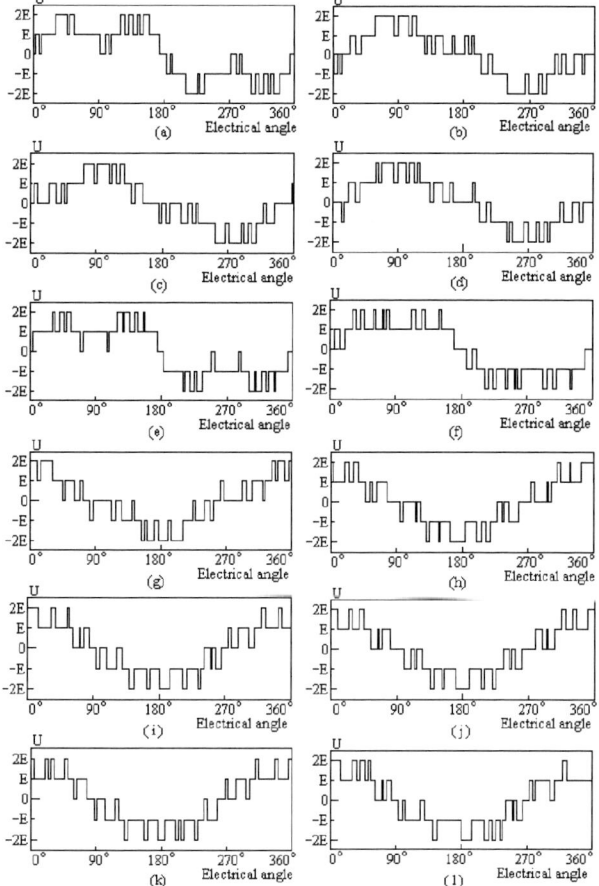

Fig. 2. PWM waveforms corresponding to the 12 sets of solutions listed in Table I

and the voltage is KE when t=0, with the increase of time, the PWM waveform increase a level E whenever a rising edge is got across and decrease a level E whenever a falling edge is got across. If all the PWM waveform in a half-cycle is within $(-(L-1)E/2, (L-1)E/2)$ for an L-level inverter, it is a true solution to the L-level SHE-PWM problems since it can be realized by an L-level inverter. Otherwise, if there isn't such an integer K, the numerical result is not a true solution to the SHE-PWM problems.

C. Solutions to SHE-PWM equations

Take a five-level inverter and 20 switching angles in half cycle as an example, let M=0.75 and the initial phase angle of the fundamental of output voltage be 0° and 90°, respectively, the SHE-PWM non-linear equations are formulated. Nine low order harmonics, including the 5th, 7th, 11th, 13th, 17th, 19th, 23rd, 25th, and 29th, are to be eliminated. Six solution sets are obtained and listed in Table I for initial phase angle of the fundamental of 0° and 90°, respectively. The switching angles with positive/negative signs in Table I are the rising/falling edges of the SHE-PWM wave. The SHE-PWM waveform corresponding to each set of solution is shown in Fig. 2.

The sets of solutions S_{a-f} are obtained with one group of equations in which (A1, B1) in Equ. 4 are (1.5, 0), and Sg-l are obtained when (A1, B1) in Equ. 4 are (0, 1.5). From the

numerical iteration process we found that for half-cycle symmetry SHE-PWM problems there are a lot of alterations in the sequences of the switching angles of the solutions, which do not satisfy the constraint expressed by Eq. (4). However, most of the solutions of the equations are true solutions to the SHE-PWM problems because they are applicable by judging with the criterion described above. This phenomenon results in a variety of SHE-PWM waveforms, part of which are shown in Fig. 2, obtained with one group of equations.

IV. SIMULATION

To verify the validity of the solutions obtained with the new formulation proposed in this paper, simulations based on Psim and a three-phase five-level cascade inverter are carried out. The parameters for simulation are as follows: voltage of a level $E=100$V, frequency of the output voltage $f=50$ Hz and resistance of the load $R=136$ ohm. Solutions S_a and S_g which are both the first group of solution we obtained for the initial phase angle of fundamental 0 and 90 degrees as $M=0.75$ are adopted in the simulations. The simulation waveforms of the line-to-neutral voltage, line-to-line voltage and their frequency spectrums are shown in Fig. 3, where, subplots (a), (b) and (c) show the case of solution S_a with a initial phase angle of fundamental of 0 degree, and (d), (e) and (f) show

TABLE I Twelve sets of solutions to half-wave symmetry SHE-PWM problems for five-level inverters at M=0.75

Solution Set	Switching Angles (in degrees)									
S_a	+3.2485	-8.4003	+11.6699	+29.4162	-45.7615	+48.4664	-55.5114	-92.2382	+99.1235	-104.908
	+107.245	+120.187	-126.898	+131.816	-137.406	+143.346	-146.736	+151.434	-159.901	-173.979
S_b	-1.0798	+5.4112	-10.2475	+12.6574	+23.5831	-32.2263	+38.6708	+58.9421	-83.7690	+89.6088
	-103.385	+108.481	-114.549	-125.2619	+129.701	-135.075	+140.080	-151.595	+162.833	-165.836
S_c	+4.5855	- 9.84980	+25.6050	-34.4928	+40.5663	-45.8163	+47.8760	+72.5358	-86.2786	+ 91.5694
	-107.574	+110.425	-116.882	+121.360	-127.779	-137.743	+143.184	- 152.881	-177.123	+178.934
S_d	-14.3034	+15.7220	+22.6346	-31.8407	+37.5625	+60.2993	-63.7096	+68.0273	-83.2316	+89.6901
	-101.684	+106.700	-112.396	+116.648	-119.8661	-133.774	+139.019	-151.798	+161.406	-165.061
S_e	+3.2304	+29.4345	-32.2707	+39.1778	-45.2232	+48.0714	-55.5813	-68.3771	+71.6572	-105.654
	+107.823	+120.315	-127.505	+129.089	-137.508	+143.739	-147.382	+155.471	-156.456	-174.133
S_f	+6.9551	-13.5299	+19.8079	+31.8811	-35.4178	+41.6939	-47.0931	+58.9422	-61.8143	+73.6191
	-74.5762	+77.0432	-81.4455	+111.7598	-116.8519	+125.1276	-130.6557	+148.1050	-150.8017	-168.4684
S_g	-9.7646	+14.9345	-29.8725	-44.6789	+47.9975	-59.3234	+68.5481	-72.9466	-82.1169	+86.5703
	-111.872	+118.794	-125.802	+140.162	-144.298	-153.503	+158.273	-161.378	+171.354	-176.682
S_h	-3.6337	+19.5525	-25.0890	+31.6489	-37.5817	-47.5529	+53.2027	-56.5219	+62.4425	-77.0617
	-94.366	+99.4725	-116.2673	+116.8441	-120.9853	-129.7346	+134.6243	-147.2681	+148.9293	-162.2676
S_i	-14.5899	+30.9052	-37.5279	+53.4864	-56.3022	-61.0191	+71.0029	-72.1781	+77.5244	-84.3627
	-93.8212	+99.3336	-107.6094	+122.7085	-129.4583	-139.8132	+146.3072	-162.4405	+167.8515	-179.779
S_j	-9.3099	+24.9151	-29.0225	+39.2111	-44.9304	-54.9592	+63.4982	-65.8335	+70.8127	-80.592
	-99.3993	+109.1665	-114.1489	+116.4818	-125.0277	-135.0557	+140.7745	-150.9688	+155.0723	-170.6745
S_k	-4.5575	+18.821	-23.7803	+27.2309	-32.2112	+45.4607	-49.9607	-58.3293	+63.7796	-76.7117
	-88.8059	+91.1163	-103.1969	+116.4057	-121.8191	-130.0847	+134.4828	-155.7791	+160.9827	-175.586
S_l	-13.926	+30.1670	-36.7687	+41.5689	-47.0078	+52.3672	-55.3433	-60.8433	+70.5706	-71.9277
	+76.8152	-83.9159	-93.3168	+98.4713	-101.7068	+122.1378	-128.9279	-139.1829	+145.5888	-179.3607

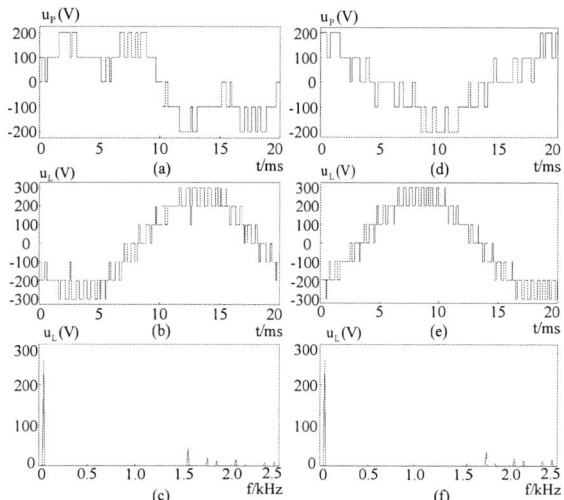

Fig. 4. Simulation results. (a) Phase-voltage with solution S_a. (b) Line voltage with solution S_a. (c) Frequency spectrum of the line voltage with solution S_a. (d) Phase voltage with solution S_g. (e) Line voltage with solution S_g. (f) Frequency spectrum of the line voltage with solution S_g.

that of solution S_g with a initial phase angle of fundamental of 90 degree.

From the simulation waveforms, it can be seen that all the harmonics lower than 30th of the line-voltages have been

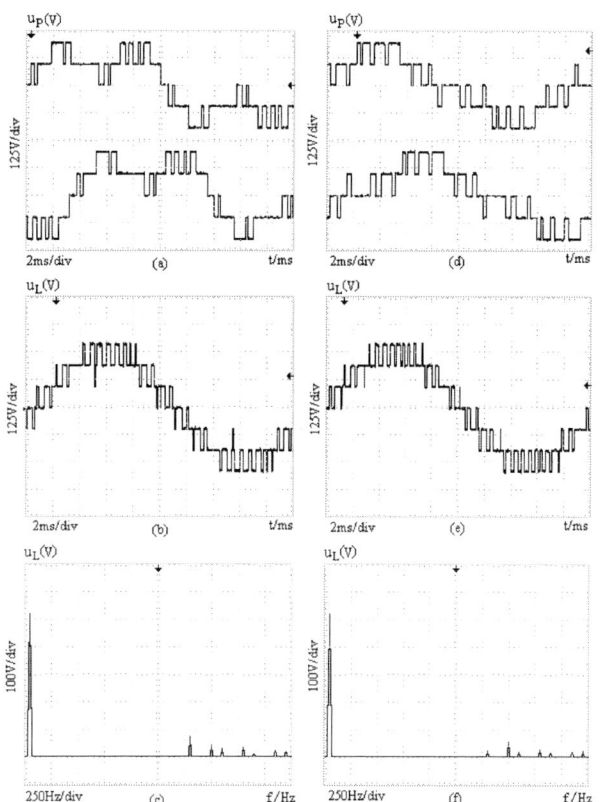

Fig. 4. Experiment results. (a) Phase-voltage of S_a. (b) Line-voltage of S_a. (c) Frequency spectrum of the line-voltage with solution S_a. (d) Phase-voltage of S_g. (e) Line-voltage with solution S_g. (f) Frequency spectrum of the line-voltage of S_g.

eliminated and the goals of fundamental control are realized in both case.

V. EXPERIMENT

To further verify the validity of the new method, a five-level three-phase cascade inverter prototype is constructed. The parameters concerned and the switching angles are the same as that employed in the simulation. The experimental waveforms of the line-to-neutral voltage, line-to-line voltage and their frequency spectrums are shown in Fig. 4, where, subplots (a), (b) and (c) show the case of solution S_a with a initial phase angle of fundamental of 0 degree, and (d), (e) and (f) show that of solution S_g with a initial phase angle of fundamental of 90 degree.

The experiment results agree well with the simulations and prove the validity of the new method proposed.

VI. CONCLUSIONS

A novel generalized formulation of half-wave symmetry SHE-PWM problems for multilevel voltage-source inverters is proposed in this paper. This method is very simple in formulation and brings great flexibility to the SHE-PWM waveform design, and thus substantially broadens the solution space. Half-wave symmetry SHE-PWM problems have a large number of unique solutions, which provide opportunities for optimization design. Many kinds of solutions can be obtained with one group of equations, and this made the process of solving the equations much simpler. It is demonstrated by simulation and experiments that nine low-order harmonics in the line voltage of the multilevel inverter are eliminated, which verifies the effectiveness of proposed SHE-PWM method.

REFERENCES

[1] V. G. Agelidis, A. Balouktsis et al, "A five-level symmetrically defined selective harmonic elimination PWM strategy: Analysis and experimental validation," *IEEE Trans. Power Electron.*, vol. 23, no. 1, pp. 19–26, Jan. 2008.

[2] J. N. Chiasson, L. M. Tolbert et al, "A complete solution to the harmonic elimination problem," *IEEE Trans. Power Electron.*, vol. 19, no. 2, pp. 491–499, March 2004.

[3] J. Sun, S. Beineke, and H. Grotstollen, "Optimal PWM based on real time solution of harmonic elimination equations," *IEEE Trans. Power Electron.*, vol. 11, no. 4, pp. 612–621, Jul. 1996

[4] J. R.Wells, B. M. Nee et al, "Selective harmonic control: A general problem formulation and selected solutions," *IEEE Trans. Power Electron.*, vol. 20, no. 6, pp. 1337–1345, Nov. 2005.

[5] M. S. A. Dahidah, V. G. Agelidis, and M. V. Rao, "On Abolishing Symmetry Requirements in the Formulation of a Five-Level Selective Harmonic Elimination Pulse-Width Modulation Technique", *IEEE Trans. Power Electron.*, vol. 21, no. 6, pp. 1833-1837, Nov. 2006

PI Type Dynamic Decoupling Control Scheme for PMSM High Speed Operation

Hao Zhu, Xi Xiao and Yongdong Li

Department of Electrical Engineering, Tsinghua University
Haidian District, Beijing, China 100084
h-zhu06@mails.tsinghua.edu.cn

Abstract—**Motor current decoupling control is of great importance for high dynamic performance of permanent magnet synchronous motor, especially when motor is at high speed range. Yet convention decoupling schemes do not perform well due to parameter sensitivity. In this paper, a PI type dynamic decoupling control scheme is proposed. Additional PI controllers are used to provide signals for motor d- and q- axes cross-coupling voltage compensation. Then the proposed decoupling control structure is revised into a simple structure in order to minimize the influence of d- and q- axes reference currents. Simulation and experimental results are presented to verify the proposed control scheme.**

I. INTRODUCTION

Permanent magnet synchronous motors (PMSM) are receiving increased attention for drive applications because of their high torque to inertia ratio, superior power density, and high efficiency. In most applications, it is desirable to control motor torque and flux separately so that maximum possible torque per ampere of stator current can be obtained under transient and steady-state conditions. This will permit a better utilization of the current capability of the power conditioning unit being used. As a mature technology, field oriented control technology is then developed for such a high performance operation of PMSM. Motor current is transformed into d- and q- axis currents in the rotor synchronous coordination frame[1], and these two currents are used to control motor stator flux and motor torque respectively. The voltages exerted to motor are generated using two PI regulators with the current signals, which is known as torque loop and magnetizing loop. And generally they are considered as having no influence with each other. Yet at high speed range, d- and q- axis currents will create remarkable coupling influence, which will degrade motor overall performance.

In order to overcome this problem, various decoupling methods have been proposed, such as feed-back decoupling, feed-forward decoupling, and Inverse System - Based decoupling control, etc[2]-[8]. Yet these methods are either sensitive to parameter inaccuracy, or complicated in calculation. In this paper, a PI type dynamic decoupling control scheme is proposed[9]. Two additional PI controllers

are used to compensate d/q axis current coupling terms. In further analysis a simplified decoupling structure is also discussed. Detailed simulation and experimental results are given in order to verify the proposed method.

II. PMSM DECOUPLING CONTROL

The machine model for the IPMSM on the synchronously rotating d/q reference frame with an electrical angular velocity we can be represented in the matrix form as follows[1]:

$$u_d = Ri_d + L_d \frac{di_d}{dt} - \omega_r L_q i_q \tag{1}$$

$$u_q = Ri_q + L_q \frac{di_q}{dt} + \omega_r (\psi_m + L_d i_d) \tag{2}$$

Where L_d, L_q and R are the inductances and the resistance of the stator, u_d, u_q, i_d, i_q are the d-and q- axis components for voltage and current, ω_r and ψ_m are motor electrical speed and rotor permanent magnet respectively. In (1) and (2), $\omega_r L_d i_d$ and $\omega_r L_q i_q$ are d-/q- axes current coupling terms. If motor speed is high, their influence can not be omitted. In order to eliminate the coupling part when control motor control, these two terms need to be subtracted in advance. This idea is proved to be an effective way, and Fig.1 gives the scheme diagram of decoupling control.

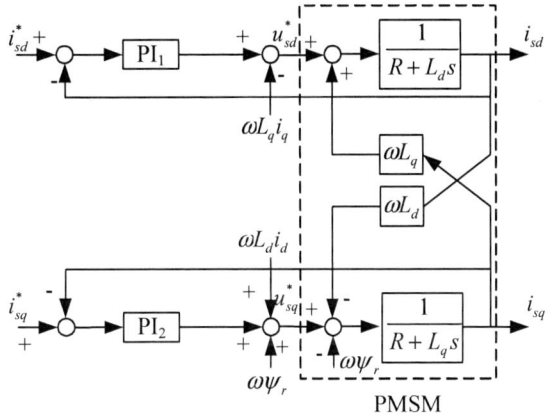

Figure.1 PMSM space vector control and current decoupling control

The primary objective of decoupling control is to provide fast dynamic response as well as good steady-state performance. But generally we do not know the exact values of L_d and L_q. If there are parameter deviation with motor inductances, and an assumed value of $\hat{L}_d$ and $\hat{L}_q$ is used instead, the performance of the decoupling control can not be properly assured. Based on motor mathematical equations when $\hat{L}_d$ and $\hat{L}_q$ are used for the decoupling control as is shown in Fig.1, the current transfer function in the rotor synchronous frame becomes

$$\begin{bmatrix} i_d \\ i_q \end{bmatrix} = \frac{1}{\Delta}\begin{bmatrix} g_{dd} & g_{dq} \\ g_{qd} & g_{qq} \end{bmatrix}\begin{bmatrix} i_d^* \\ i_q^* \end{bmatrix} \quad (3)$$

where g_{dq} and g_{qd} are the coupling gains,

$$g_{dq} = PI_2\left(\omega L_q - \omega \hat{L}_q\right)$$

$$g_{qd} = -PI_1\left(\omega L_d - \omega \hat{L}_d\right)$$

and

$$\Delta = \left(PI_1 + R + L_d s\right)\left(PI_2 + R + L_q s\right) + \omega^2\left(L_d - \hat{L}_d\right)\left(L_q - \hat{L}_q\right)$$

In (3), g_{dq} and g_{qd} indicate the gains of the coupling terms that come from motor parameter error. If there are no parameter error, g_{dq} and g_{qd} are equal to zero. But if there's parameter error, which is always the case, the coupling errors will remain. Especially, when motor speed is high their influence will be quite obvious, i.e. conventional decoupling control scheme is vulnerable to parameter uncertainty. In order to overcome this problem, in this paper a PI type decoupling controller is proposed. Because the current coupling terms comes from motor d- and q- axes currents, we use two PI controllers(PI$_3$ and PI$_4$ in Fig.2) to generate additional control information. And the output of PI$_3$ and PI$_4$ is added to d/q referenced voltages in order to perform a complete motor voltage control.

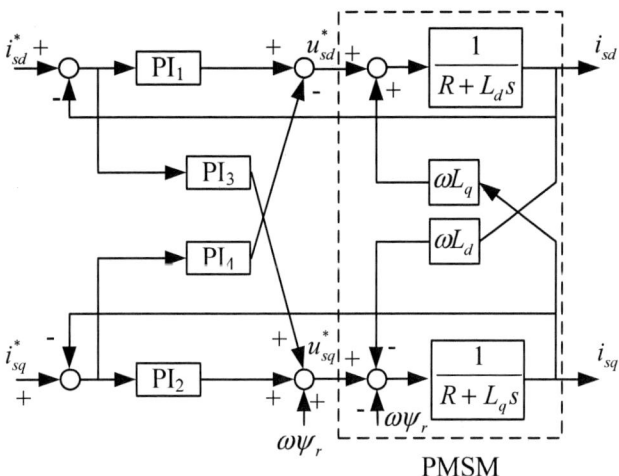

Figure.2 Block diagram of the PI type decoupling control

The transfer function of the newly designed decoupling control scheme becomes

$$\begin{bmatrix} i_d \\ i_q \end{bmatrix} = \frac{1}{\tilde{\Delta}}\begin{bmatrix} \tilde{g}_{dd} & \tilde{g}_{dq} \\ \tilde{g}_{qd} & \tilde{g}_{qq} \end{bmatrix}\begin{bmatrix} i_d^* \\ i_q^* \end{bmatrix} \quad (4)$$

Where $\tilde{g}_{dq}$ and $\tilde{g}_{qd}$ are the new coupling gains,

$$\tilde{g}_{dq} = PI_4\left(PI_2 + R + L_q s\right) - PI_2\left(PI_4 - \omega L_q\right)$$

$$\tilde{g}_{qd} = PI_3\left(PI_1 + R + L_d s\right) - PI_1\left(PI_3 + \omega L_d\right)$$

and

$$\tilde{\Delta} = \left(PI_1 + R + L_d s\right)\left(PI_2 + R + L_q s\right) - \left(PI_3 + \omega L_d\right)\left(PI_4 - \omega L_q\right)$$

In order to nullify their influence, let $\tilde{g}_{dq} = 0$ and $\tilde{g}_{qd} = 0$. Then we can get

$$PI_3 = -\frac{\omega \hat{L}_d}{R + \hat{L}_d s}PI_1 \quad \text{and} \quad PI_4 = -\frac{\omega \hat{L}_q}{R + \hat{L}_q s}PI_2$$

Fig.3 gives the simplified scheme diagram of the PI type decoupling controller. Only two PI regulators are needed, and they are more robust than the feedforward decoupling controller when there is parameter mismatch [9].

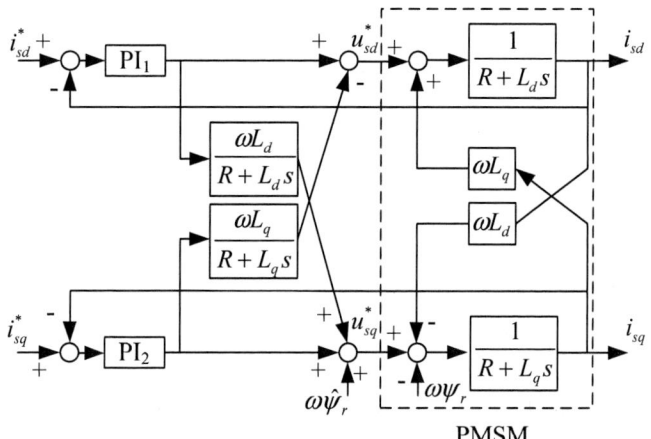

Figure.3 Equivalent block diagram of the proposed decoupling controller

III. SIMULATION AND EXPERIMENTAL REUSLTS

In order to evaluate the performance of the method, the decoupling control system is performed by Matlab simulation. In the simulation an interior permanent magnet synchronous motor (IPM) is used. Table.1 gives its detailed parameters. The coefficients of PI$_1$ and PI$_2$ are set as $K_p = 10$, $K_i = 100$, and motor is running with a 3Nm load. Fig.3 to Fig.7 show the simulation results.

Table.1 Motor parameters

Parameter	Value
Stator resistance R_s	2.259 Ω
Stator inductance L_d, L_q	20.74 mH
Rotor inertia J	0.0048 Kgm^2
Magnetic flux-linkage	0.411 Wb
Number of pole pairs	2

Fig.4 is the speed step response where motor speed rises from 0 to 188rad/s. Comparing to conventional space vector control scheme(Fig.4(a)), the result with PI type decoupling controller has a quicker starting time (Fig.4(b)).

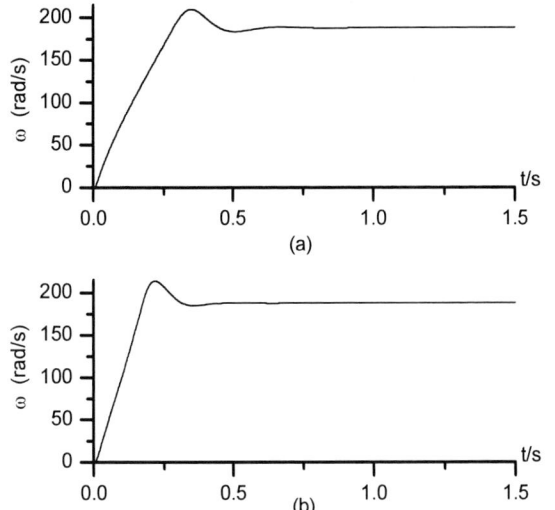

(a)

(b)

Figure.4 Motor speed response. (a) without decoupling control (b) with PI type decoupling control

The speed response of feed-back decoupling control has a similar result as PI type decoupling control which is shown in Fig.4(b). Then, Fig.5 and Fig.6 compares their current control performance. In the simulation a same parameter setting is used as is in Fig.4. Fig.5 shows the d- and q- axis current response using conventional feed-back decoupling control scheme. In the starting period, the d- axis current varies and its peak value is over 1.2A. The q- axis current varies greatly which means a great electric-magnet torque is need. In Fig.6, the proposed PI type decoupling controller is used. Comparing to Fig.5 we can see that the d- axis current is much more stable. During the transient period its peak value is no greater than 0.7A. Also, the q- axis current is no bigger than that is shown in Fig.5.

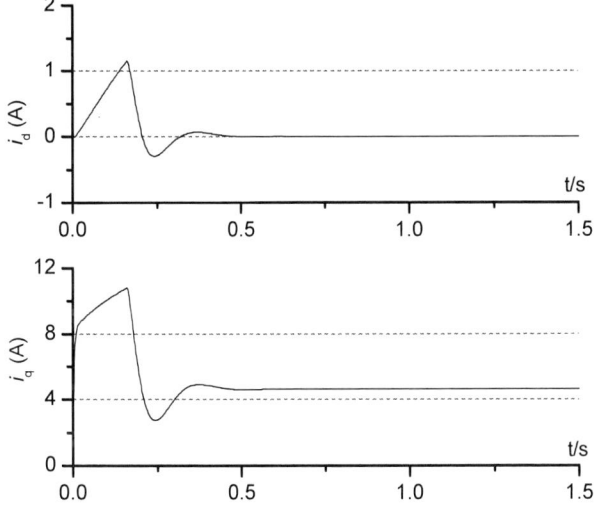

Figure.5 d- and q- axis current response (feed-back decoupling control)

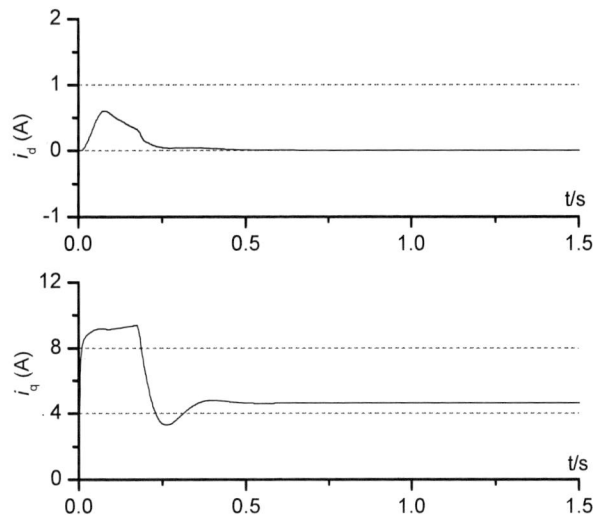

Figure.6 d- and q- axis current response (PI type decoupling control)

Finally, we need to point out that if the values of K_p and K_i are set properly, there's no big difference between conventional decoupling scheme and the proposed scheme. Then if K_p and K_i change, the result may changes radically. Fig.7 is the simulation result when K_p changes from 10 to 1, and K_i remains unchanged. From Fig.7(a) we can see that conventional decoupling scheme can not assure a stable starting. As a contrast, using the proposed PI type controller motor still starts correctly (Fig.7(b)).

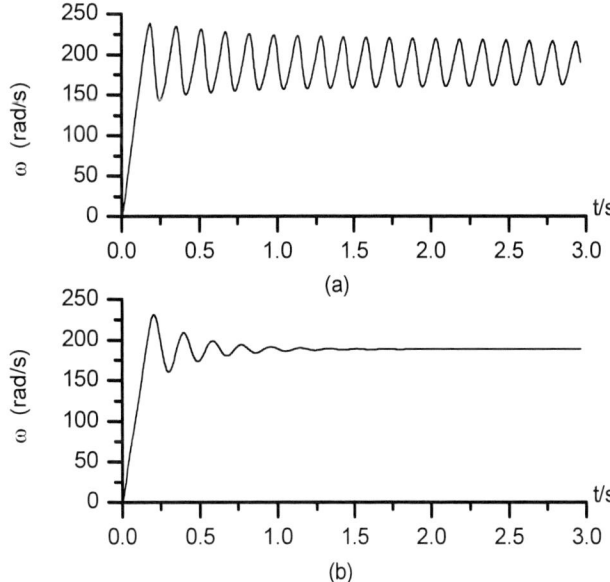

(a)

(b)

Figure.7 Motor speed response. (a) feed-back decouplingcontrol (b) PI type decoupling control

In experimental condition, a prototype of 2kW PMSM is used, cooperating with the DSP TMS320LF2812 to implement the proposed control technique. Fig.8 gives the no load test result of the q-axis current i_q. In the experiment, motor speed rises from 0.02 to 0.6 of the rated value. During the accelerating period the q-axis reference current i_q^* has a positive output and at the steady state it remains nearly zero. In order to protect the motor, i_q^* has an upper limiter which is 1.8 times of the rated current. Fig.8 shows that when using the PI type decoupling controller, the motor has a shorter dynamic period comparing to conventional feed-back decoupling control. Also, the maximum q-axis current (1.7 times of rated current in Fig.8(b)) is less than that using conventional method (2.0 times of rated current in Fig.8(a)).Then if the values of K_p and K_i changes radically, when using the proposed method a similar result can be observed like that in Fig.8(b), while conventional decoupling scheme leads to starting turbulence.

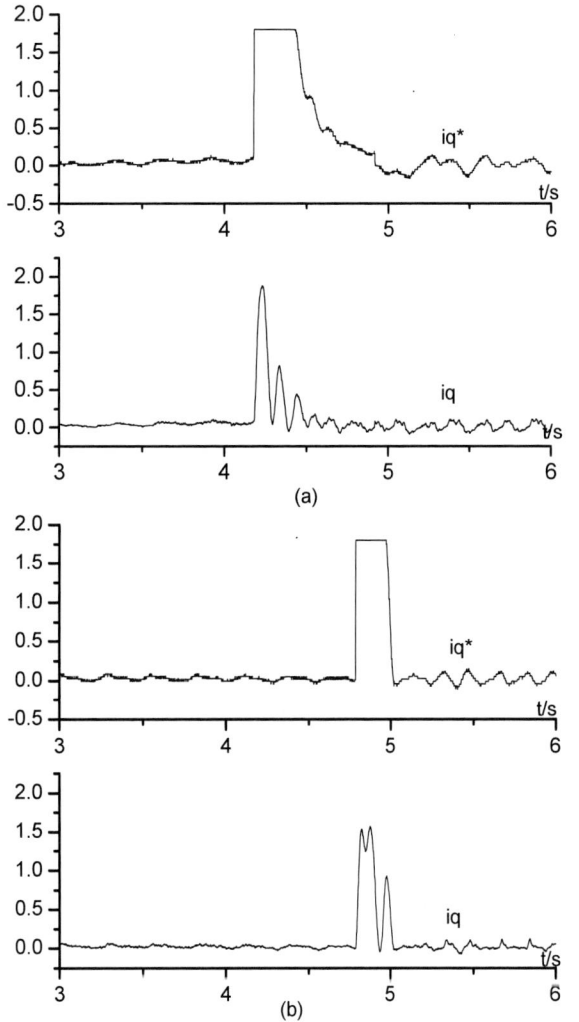

Figure.8 Referenced and real q-axis current(nominal value). (a) feed-back decoupling control (b) PI type decoupling control

IV. CONCLUSIONS

This paper proposed a PI type current decoupling control scheme where two additional PI regulators are used to compensate the d- and q- axis cross-coupling voltages. In the paper the transfer function of the controller is discussed in detail, and then a simplified structure which needs only two PI regulators is proposed. Simulation and experimental results has proved that the proposed controller is superior to conventional decoupling scheme, and is more robust to PI regulator parameter uncertainty. With a simple structure, this control scheme is easy for application.

ACKNOWLEDGMENT

This work was supported by National High Technology Research and Development Program of China (Project No. 2007AA04Z216).

REFERENCES

[1] T. M. Jahns, G. E. Kliman, and T. W. Neumann, "Interior permanentmagnet synchronous motors for adjustable-speed drive," *IEEE Trans. Ind. Applcation*, vol. IA-22, pp.738-747, July 1986.

[2] Edward Y. Y. Ho, Paresh C. Sen, "Decoupling Control of Induction Motor Drives," *IEEE Transactions on Industrial Electronics*, vol.35, no.2, pp.253-262, 1988.

[3] F. Briz, M. W. Degner, D. Lorenz, "Analysis and Design of Current Regulators Using Complex Vectors," *IEEE IAS Annual Meeting*, pp.1504-1511. 1997.

[4] Zhou Yuanshen, Jiang Jianguo, "Dynamic Decoupling Control for Asynchronous Motor," *S&M Electric Machines*, 2001, vol.02, pp.22-26.

[5] Zhou Yuanshen, "Survey on Decoupling Control Strategy for Induction Motor," *S&M Electric Machines*, 2005, vol.06, pp.56-64.

[6] Cao Jianrong, Yu Lie, "Inverse System - Based Decoupling Control of Induction Motor," *Transactions of China Electrotechnical Society*, 1990, vol.01, pp.7-11.

[7] Li Qing, Yang Liyong, "Stator Flux and Torque Decoupling Control of Induction Motor Using Inverse System Method," *Proceedings of the CSEE*, 2006, vol.06, pp.146-150

[8] Wang Wentao, Tang Xianyue, "Decoupling control of induction motor based on differential geometry theory," *Journal of Shenyang University of Technology*, 2006, vol.06, pp.623-627.

[9] Jinhwan Jung, Kwanghee Nam, "A Dynamic Decoupling Control Scheme for High Speed Operation of Induction Motors," *IEEE Transactions on Industrial Electronics*, 1999, vol.46, no.1, pp.100-110.

High Performance Positive and Negative Sequence Filters in Stationary Frame Based on Complex Transfer Function

Jingxin Mao, Fei Lin, Hong Li, Xiaojie You, Trillion Q Zheng
School of Electrical Engineering, Beijing Jiaotong Univ.
Beijing, China
E-mail: flin@bjtu.edu.cn

Abstract—**This paper proposed two high performance filters for extracting the positive- and negative-sequence components. The three-phase system of space vector is modeled by complex transfer function (CTF). Two filters are designed and analyzed according to frequency domain theory based on CTF. The performance of two filters is compared detailed by theory and simulation, under unbalanced, distorted, and frequency variable conditions. The effectiveness and validity of filters are verified through experiments.**

I. INTRODUCTION

For all kinds of voltage source converter under unbalanced and distorted grid conditions, in order to decrease the over current, current distortion and the ripple of the dc voltage, it is usually necessary to detect the fundamental positive and negative sequence components of the ac voltage and current in real time[1-4].

Most of the sequence components extraction methods are based on synchronized dq frame, using notch-filter, band-stop (BS) and low-pass filter (LP) [5], or phase-looked loop (PPL) [4]. Several methods for estimating positive and negative sequence have been propose: an improve approach for estimating using a complex Kalman filter was proposed in [7], and [6] proposed a observer-based source voltage unbalance control method in PWM Voltage-Source Converter that can compensate voltage unbalance in higher-power system with slower switching frequency and limited current controller-bandwidth. Then the dual controller is adopted in a double synchronous reference frame (one for each rotating sequence). Recently, some VSC control schemes based on stationary frame are proposed without rotational transformation. Therefore, the positive and negative sequence components detection is also in stationary frame for simplicity [8-10].

Using complex transfer function model [11-12], this paper proposes two filters for extracting the positive and negative sequence in stationary frame under unbalanced and harmonic conditions. First, the complex transfer function is introduced for three-phase systems. Following that, two filters are derived

and their positive and negative frequency domain characteristics are analyzed. Finally, simulations and experiments are carried out to verify the effectiveness of the proposed filters. This paper is supported by grants from the Power Electronics Science and Education Development Program of Delta Environmental & Educational Foundation.

II. THE COMPLEX TRANSFER FUCTION MODEL OF THREE PHASE SYSTEMS

A. The Concept of Space Vector for An Unbalanced and Distorted Three-phase Systems

The three-phase system variables under unbalanced and distorted conditions in abc frame are expressed as,

$$
\begin{aligned}
u_a &= u_a^+ + u_a^- + \sum_n \left(u_a^{n+} + u_a^{n-} \right) \\
u_b &= u_b^+ + u_b^- + \sum_n \left(u_b^{n+} + u_b^{n-} \right) \\
u_c &= u_c^+ + u_c^- + \sum_n \left(u_c^{n+} + u_c^{n-} \right), n \geq 2
\end{aligned}
\tag{1}
$$

Where, "+" and "-"denote positive and negative sequence respectively, n is the harmonic number. Positive-, negative-sequence and harmonic components is,

$$
\begin{aligned}
\{u_a^+, u_b^+, u_c^+\} &= \{\hat{U}_+ \cos\omega t, \hat{U}_+ \cos(\omega t - 2\pi/3), \hat{U}_+ \cos(\omega t - 4\pi/3)\} \\
\{u_a^-, u_b^-, u_c^-\} &= \{\hat{U}_- \cos\omega t, \hat{U}_- \cos(\omega t - 4\pi/3), \hat{U}_- \cos(\omega t - 2\pi/3)\} \\
\{u_a^{n+}, u_b^{n+}, u_c^{n+}\} &= \{\hat{U}_{n+} \cos n\omega t, \hat{U}_{n+} \cos(n\omega t - 2\pi/3), \hat{U}_{n+} \cos(n\omega t - 4\pi/3)\} \\
\{u_a^{n-}, u_b^{n-}, u_c^{n-}\} &= \{\hat{U}_{n-} \cos n\omega t, \hat{U}_{n-} \cos(n\omega t - 4\pi/3), \hat{U}_- \cos(n\omega t - 2\pi/3)\}
\end{aligned}
\tag{2}
$$

This paper is supported by grants from the Power Electronics Science and Education Development Program of Delta Environmental & Educational Foundation.

By the Clarke transformation, the $[u_a,\ u_b,\ u_c]$ can be modeled as a space vector $\boldsymbol{u} = u_\alpha + ju_\beta$.

B. The Complex Transfer Functions

It is a double-input-double-output system for the plant modeled in stationary $\alpha\beta$ frame. However, using the complex space vectors, it can be considered as a SISO system as

$$a_0 \frac{d^n}{dt^n} c(t) + a_1 \frac{d^{n-1}}{dt^{n-1}} c(t) + ... + a_{n-1} \frac{d}{dt} c(t) + a_n c(t)$$

$$= b_0 \frac{d^m}{dt^m} r(t) + b_1 \frac{d^{m-1}}{dt^{m-1}} r(t) + ... + b_{m-1} \frac{d}{dt} r(t) + b_0 r(t) \qquad (3)$$

By the Laplace transformation, it is expressed as

$$G(s) = \frac{C(s)}{R(s)} = \frac{b_0 s^n + b_1 s^{n-1} + ... + b_{n-1}s + b_n}{a_0 s^n + a_1 s^{n-1} + ... + a_{n-1}s + a_n} \qquad (4)$$

Where $G(s)$ is the complex transfer function because its coefficient a_i and b_i (i=0,1,...,n)are all complex numbers. The $G(s)$ can also be expressed as

$$G(s) = G_\alpha(s) + jG_\beta(s) \qquad (5)$$

For general complex transfer function, the system gain

$$\frac{|y|}{|r|} = \sqrt{\frac{r^H G(j\omega)^H G(j\omega) r}{r^H r}} \qquad (6)$$

is not equal to $|G(j\omega)|$. But for the symmetric three-phase system, the gain and phase shift are given as

$$\frac{|y|}{|r|} = |G(\pm j\omega)|, \varphi = \arg|G(\pm j\omega)| \qquad (7)$$

Where, "+" and "-"denote positive and negative sequence input variables respectively.

Proof: Suppose the closed-loop poles of (3) is $-p_1$, $-p_2$, ..., $-p_n$, $-p_j (j=1,2...,n)$, so $G(s)$ can be expressed as

$$G(s) = N(s) \Big/ \prod_{i=1}^{n} (s+p_i) \qquad (8)$$

for the symmetric three-phase system, the input signal is

$r(t) = Ae^{\pm j\omega t}$, its Laplace transform is

$$R(s) = A\frac{s}{s^2 + \omega_e^2} \pm jA\frac{\omega_e}{s^2 + \omega_e^2} = \frac{A}{s \mp j\omega_e} \qquad (9)$$

so the output signal is

$$Y(s) = G(s)R(s) = \frac{N(s)}{\prod_{i=1}^{n}(s+p_i)} \frac{A}{s \mp j\omega} = \sum_{i=1}^{n} \frac{k_i}{s+p_i} + \frac{k_c}{s \mp j\omega} \qquad (10)$$

After inverse Laplace transformation, the output is

$$y(t) = \sum_{i=1}^{n} k_i e^{-p_i t} + k_c e^{\pm j\omega t} \qquad (11)$$

when $t \to \infty$, $\sum_{i=1}^{n} k_i e^{-p_i t} \to 0$, $y_s(t)$ is the steady component of $y(t)$

$$y_s(t) = \lim_{t \to \infty} y(t) = k_c e^{\pm j\omega t} \qquad (12)$$

Where k_c is

$$k_c = G(s) \cdot \frac{A}{s \mp j\omega} \cdot (s \pm j\omega)\Big|_{s = \pm j\omega} = A \cdot G(\pm j\omega) \qquad (13)$$

$$y_s(t) = k_c e^{\pm j\omega t} = A \cdot G(\pm j\omega) e^{\pm j\omega t} = |G(\pm j\omega)| e^{j\varphi(\pm \omega)} A \cdot e^{j \pm \omega t} \quad (14)$$

so the frequency-domain-based analysis and synthesis methods are applicable for complex transfer function.

III. THE DESIGN OF PSPF AND NSPF

A. Filter I

The proposed positive-sequence-pass filter (PSPF) contains a band-pass and a band-reject filter.

The band-pass filter is designed to achieve a open-loop gain of

$$T = \frac{\omega_0}{s - j\omega_{50}} \qquad (15)$$

At the frequency of ω_{50}, $T = \frac{\omega_0}{s - j\omega_{50}}\Big|_{s=j\omega} \to \infty$, so the closed-loop transfer function get a unity gain at ω_{50}.

So this PSPF is designed to provide a unity gain for the 50Hz quantity of the positive-sequence component and entirely eliminates the 50Hz quantity of negative-sequence. The open-loop transfer function is

$$G_{ol-p} = \omega_0\left(s + j\omega_{50}\right)/2j\omega_{50}\left(s - j\omega_{50}\right) \quad (16)$$

and the PSPF is

$$G_{p1}(s) = 1\left/\left(1 + \left(\frac{2j\omega_{50}\left(s - j\omega_{50}\right)}{\omega_0\left(s + j\omega_{50}\right)}\right)\right)\right. \quad (17)$$

Where ω_0 determines the cut-off frequency. Therefore as shown in Fig.1, by decreasing the parameter ω_0, the attenuation rate for other frequency components is increased, however decreased response time. At the same time, it can be seen that the magnitude change and phase shift increased when the input signal vary around the central frequency.

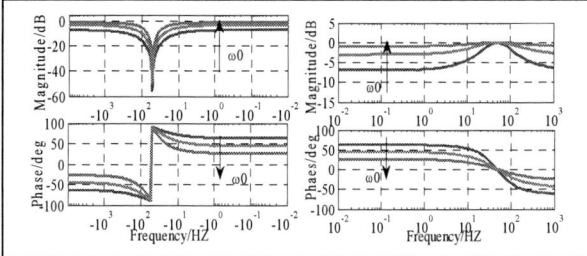

Figure 1.   Filter I plots in positive and negative-sequence frequency domain with different ω_0

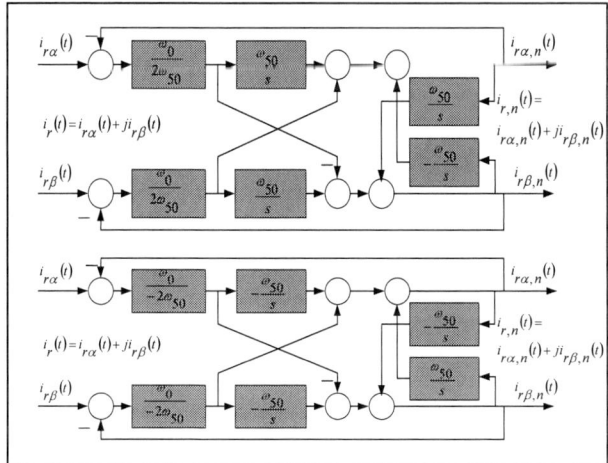

Figure 2.   Block diagram of PSPF and NSPF

The negative-sequence-pass filter (NSPF) can be derived by changing the ω_{50} to $-\omega_{50}$ in PSPF,

$$G_{n1}(s) = 1\left/\left(1 + \left(\frac{-2j\omega_{50}\left(s + j\omega_{50}\right)}{\omega_0\left(s - j\omega_{50}\right)}\right)\right)\right. \quad (18)$$

The PSPF and NSPF are shown as Fig.2.

B. Filter II

As shown by the Fig.1, the harmonic could decrease the performance of the filter I. So the filter II is proposed to extract fundamental positive and negative sequence components. The PSPF of filter II is described as

$$G_{p2}(s) = 1\left/\left(\left(\frac{s - j\omega_{50}}{\omega_n}\right)^2 + \sqrt{2}\frac{(s - j\omega_{50})}{\omega_n} + 1\right)\right. \quad (19)$$

where ω_n is the cutoff frequency.

It is similar to Butterworth filter to some extent. The Bode plot of this filter is shown as Fig.3. It is seen that only the fundamental positive-sequence components have a unity gain and zero phase-shift. All the other components are filtered.

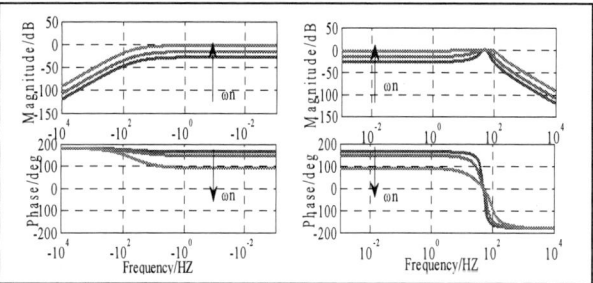

Figure 3.   Filter II plots in positive and negative-sequence frequency domain with different ω_n

It is worth noticing that the proposed filters are robust to small frequency variations. This can be explicitly analyzed from the frequency response characteristics, as shown in Fig.4.

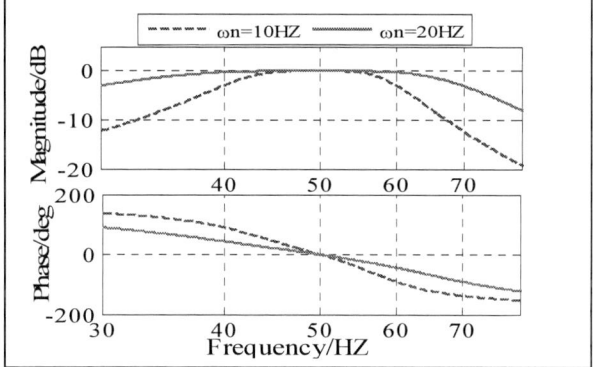

Figure 4.   Effects of small frequency variation

The block diagram of PSPF of filter II is shown as Fig.5. Changing the ω_{50} to $-\omega_{50}$ in PSPF, the NSPF is obtained.

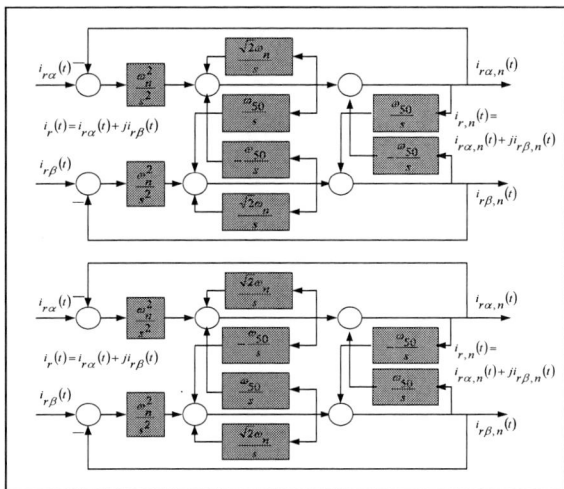

Figure 5. Block diagram of PSPF of filter II.

C. Digital Implementation

To implement the filters in DSP, different methods of discretization are compared. According to the references [13], forward difference transformation make the system instable and backward difference transformation make the frequency characteristic of the system distorted. Though zero-pole matching transformation is similar with bilinear transformation, it is suitable for the functions that take the zero-poles form. So finally the bilinear transformation is adopted:

$$G(s) \Leftrightarrow G(z) \; s = \frac{2}{T} \times \frac{1 - z^{-1}}{1 + z^{-1}} \qquad (20)$$

IV. SIMULATION AND EXPERIMENTAL RESULTS

A. Simulations

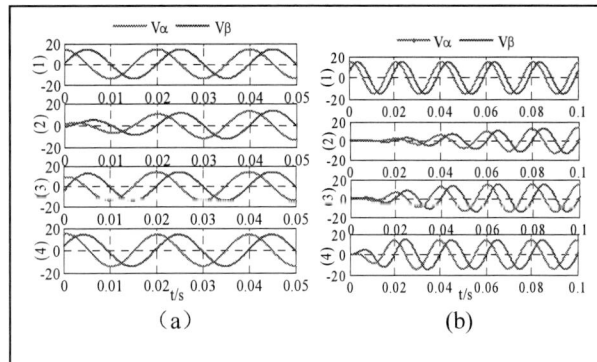

Figure 6. (a) two-phase input and output voltage of filter I(b) two-phase input and output voltage of filter II.

First, the simulations of two filters with unbalanced 50Hz three-phase input are carried out. As seen in fig 6, both the filter I and filter II can detect the positive and negative sequence components very well. Figures of (1) in (a) and (b) are the input unbalanced signal. The parameter ω_0 of filter I is 20π, 160π, 400π respectively of (2)(3)(4) in (a) and ω_n of filter II is 20π, 40π, 100π of (2)(3)(4) in (b), it can be seen that ω_0 and ω_n decide the response time.

Second, the simulation results with small frequency variation of 7% are given, the first filter's input positive sequence and output was shown in Fig.7 (a), it can be seen that the positive-sequence can be completely separated from other components. The second filter as shown in Fig.1 (b), where $\omega_n=40\pi$, 140π, 200π, provide a unity gain but a small phase shift for the positive-sequence. By increasing ω_n the phase shift is decreased, but at the same time the elimination capacity of the negative-sequence will be worse.

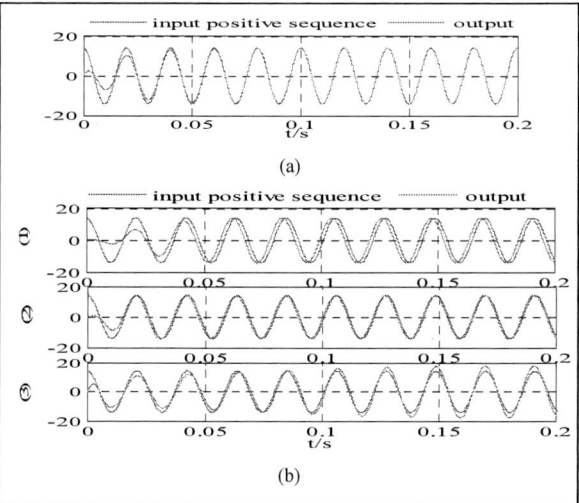

Figure 7. (a) input and output voltage of filter I(b) input and output voltage of filter II.

Then the 5th and 7th harmonics are added; only the filter II can extract the fundamental positive and negative sequence components. Suppose that the distorted input voltage consists of fundamental positive- and negative-sequence components and 20% of 5th, 14% of 7th harmonics. Fig.8 (a)(1) shows the unbalanced and distorted input voltage, and (2)(3) is output voltage of filter I with $\omega_0 = 40\pi$ and 2π, respectively. Fig.8 (b) shows the input and output voltage of filter II. The attenuation ability of the filter I is inferior to the filter II, only when $\omega_0 \leq 2\pi$ it can basically remove the harmonic. It also provide a longer response time than filter II.

978-1-4244-4782-4/10 $26.00 © 2010 IEEE   1743

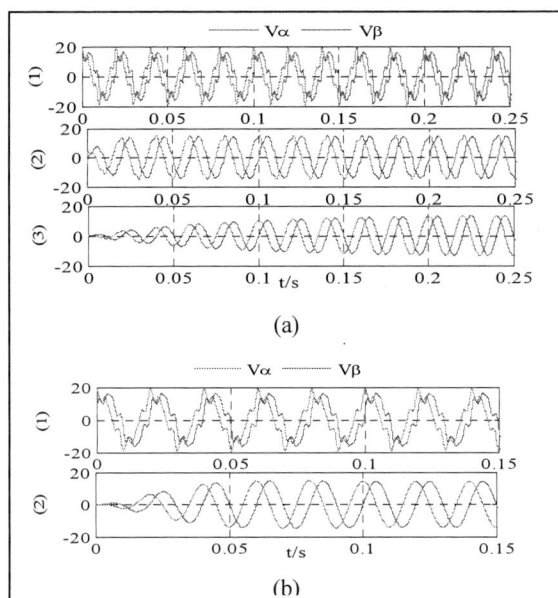

Figure 8.   (a) two-phase input and output voltage of filter I(b) two-phase input and output voltage of filter II.

B. Experiments

Experiments have been carried out for verification purposes. The filter is built in TI-DSP28335 and the sampling frequency is 5 kHz. Fig.9 (a) and (b) show the unbalanced and distorted voltage with 100% 50Hz positive, 40% 50Hz negative, 33% 150Hz positive，20% 250HZ positive and 14.2% 350Hz positive components in the abc frame, and αβ frame respectively. Fig.9 (c) and (d) show the fundamental positive and negative sequence components obtained by the filter II.

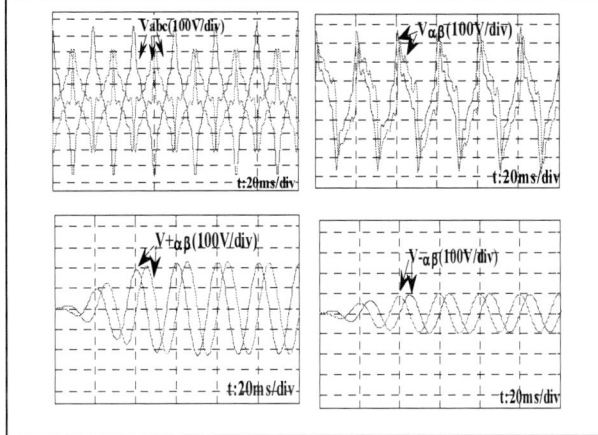

Figure 9.   Experimental results (a) three-phase voltage; (b) two-phase voltage; (c) positive sequence components; (d) negative sequence components

V.   CONCLUSIONS

This paper proposed two filters that can obtain the fundamental positive and negative sequence components. Using the complex transfer function concept, the filter is easily designed according classical frequency domain method in stationary frame. The first filter extracts the positive and negative sequence components fast, and the second filter could be used in distorted grid conditions. The influence of filter parameters and small frequency variations is also analyzed. Finally, the effectiveness of these two proposed filters is verified by simulations and experiments.

REFERENCES

[1]   Yongsug Suh, and Thomas A. Lipo, "Modeling and analysis of instantaneous active and reactive power for PWM ACDC converter under generalized unbalanced network," IEEE Transactions on Power Electronics, vol. 21, no. 3, pp. 1530-1540, July. 2006.

[2]   Yongsug Suh, and Thomas A. Lipo, "Control scheme in hybrid synchronous stationary frame for PWM ACDC converter under generalized unbalanced operating conditions," IEEE Transactions on Industry Applications, vol. 42, no. 3, pp. 825-837, May/June. 2006.

[3]   P. Rioual, H. Pouliquen, and J. P. Louis, "Regulation of a PWM rectifier in the unbalanced network state using a generalized model," IEEE Trans. Power Electron., vol. 11, no. 3, pp. 495–502, May 1996

[4]   A.Yazdani, and R.Iravani, "A unified dynamic model and control for the voltage-sourced converter under unbalanced grid conditions," IEEE Transactions on Power Delivery, vol. 21, no. 3, pp. 1620-1671, July. 2006.

[5]   Giuseppe Saccomando, Jan Svensson, "Transient operation of grid-connected voltage source converter under unbalanced voltage conditions," in IEEE 2001 IAS, 2001, pp. 2419–2424

[6]   Kevin Lee, T.M.Jahns, T.A.Lipo, and V Blasko, "New observer-based source voltage unbalance control methods in PWM voltage-source converters," in Proc. IEEE 2008 Power Electron.Spec. Conf., 2008, pp. 1509–1514

[7]   R.A.Flores, I.Y.H.Gu, T.A.Lipo, and M.J.H.Bollen, "Positive and negative sequence estimation for unbalanced voltage dips," in IEEE 2003 Power Engineering Society General Meeting, 2003, pp. 2498–2503

[8]   D.Roiu, R.Bojoi, L.R.Limongi, A.Tenconi,"New Stationary Frame Control Scheme for Three Phase PWM Rectifiers under Unbalanced Voltage Dips Conditions," in IEEE 2008 IAS, 2008, pp. 1-7

[9]   S.Karimi, P.Poure, S.Saadate,"High performances reference current generation for shunt active filter under distorted and unbalanced conditions," in Proc. IEEE 2008 Power Electron.Spec. Conf., 2008, pp. 195-201

[10] H.Chen,Z.Xu,,and F.Zhang,"Adaptive Detecting Method for Fundamental Positive Sequence, Negative Sequence Components and Harmonic Component Based on Space Vector," in IEEE 2006 Transmission and Distribution Conference and Exhibition, 2006, pp. 726-732

[11] S. Gataric and N. R. Garrigan, "Modeling and design of three-phase systems using complex transfer functions," in Proc. IEEE1999 Power Electron.Spec. Conf., Jun./Jul. 1999, vol. 2, pp. 691–697.

[12] L.Harnefors, "Modeling of Three-Phase Dynamic Systems Using Complex Transfer Functions and Transfer Matrices," IEEE Transactions on Industrial Electronics, vol. 54, no. 4, pp. 2239-2249, August. 2007.

[13] Yuxi Zhu, Ruchun Cui,Xiaolei Kuang, "Computer Control Technology," Pulishing house of electriconics industry,Beijing ,2005

Simulation Study of Parameter Influence on Dynamic Voltage Rise Control

Ming LI, Xiong FANG, Yue WANG, Leqiang ZHANG, Ke WANG and Guopeng ZHAO
School of Electrical Engineering
Xi'an jiaotong university
Xi'an, China
mean.lee@mail.xjtu.edu.cn

Abstract—**High power IGBTs are widely used in wind power generation systems. An advanced IGBT gate driver with Dynamic Voltage Rise Control (DVRC) is used to suppress the turn-off over-voltage of the IGBT. Mass simulations are done to evaluate the performance, showing influence of the DVRC parameters, in order to select the right parameters for applications.**

I. INTRODUCTION

While designing a high power IGBT converter, whose switching frequency is much higher than any other power semiconductor devices in such high current and voltage rating, gate driver circuit plays a very important role in the whole system operation and reliability. It is well-known that when turning off an IGBT, there is a peak appeared at the VCE waveform. This peak voltage, namely turn-off over voltage, which is caused by the energy stored in the power stage stray inductance, is obviously high when turning off a current higher than 200A, if the converter is suffered from an over load condition or short circuit fault condition, the peak voltage on IGBT may exceed the break down voltage[2,3]. In order to guarantee IGBTs' operating in safe operation area (SOA), a lot of turn-off over-voltage suppression methods are proposed. The simplest way is turn off IGBTs with a high value gate resistor, it generates a lot of additional turn-off time and loss. A lot of Active Gate Control (AGC) methods are proposed to gain better control and thermal performance. Dynamic Voltage Rise Control (DVRC) [13] uses a capacitor to sense the IGBT turn-off speed and to initiate a amplified recharging of the gate to avoid VCE over-voltage. Because of the fast amplifier recharging circuit, it is possible to achieve lower over-voltage without significant drawbacks. In this paper, DVRC is deeply analyzed with Saber simulator in order to provide parameters selection guide. Single switch hard switching experiment and short-circuit experiments are performed to verify the simulation results.

Sponsored by National Science Foundations of china, No.50707025

II. DYNAMIC VOLTAGR RISE CONTROL

Fig.1 shows a hard switching commutation circuit in a half-bridge circuit with inductive load, which is the most common commutation circuit in PWM converters. Looking into the v_{CE} waveform at turn off, we can find a turn-off over-voltage which can be expressed in (1).

$$\Delta V = L_q \times \frac{\mathrm{d}i_C}{\mathrm{d}t} \qquad (1)$$

Where, ΔV is the v_{CE} overshooting at turn off, L_q is the stray inductance in the commutation circuit, it could be come from 3 parts: DC-link capacitor bank, inter-media connections and IGBT itself. From (1), we could know that there are mainly 2 ways to suppress voltage overshooting at turn-off. First one, minimize the stray inductance; second one, slow the current falling.

In order to suppress turn-off over-voltage while reduce the expense of turn-off time and losses, driver with DVRC is investigated. DVRC senses and feedback the turn-off v_{CE} slope via a very small value high voltage capacitor to control a fast boost circuit recharging the IGBT's gate terminal, in this way, IGBT turn off transient is slowed down, hence, the turn-off over-voltage is suppressed. The circuit diagram and operation principle diagram of DVRC are shown in Fig.2 and Fig.3, respectively.

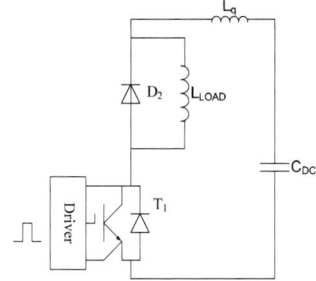

Figure 1. Commutation circuit

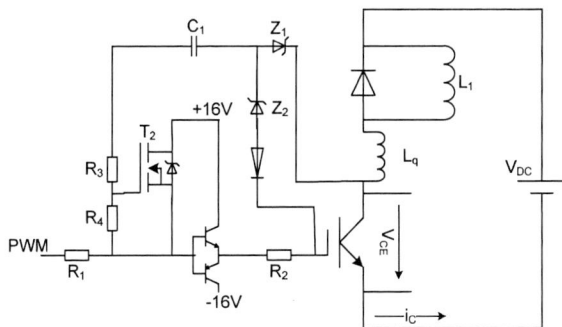

Figure 2. Circuit Diagram of DVRC

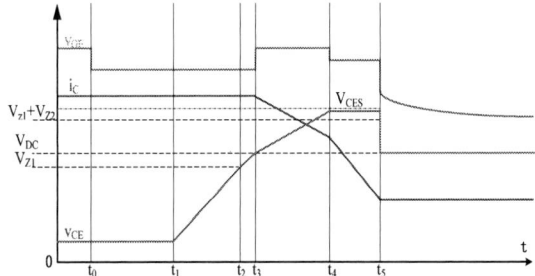

Figure 3. Operation principle of DVRC

Table I.   PARAMETERS INFLUENCE @ TURN OFF

Period	Reason	Result
t_0-t_1	R_G	Turn-off delay
t_1-t_2	R_G	dv_{CE}/dt
t_2-t_3	R_G, dv_{CE}/dt	dv_{CE}/dt
t_3-t_4	R_G, L_q, dv_{CE}/dt	dv_{CE}/dt, di_C/dt
t_4-t_5	v_{CE}	v_{CE}

Considered in the typical commutation circuit Fig.1, T_1 starts to turn off at t_0. After a turn off delay time determined by turn off gate resistor R_{G_off}, v_{CE} starts to rise at t_1, dv_{CE}/dt is defined by R_{G_OFF}. When v_{CE} reaches V_{Z1} at t_2, a current will flow through C_1, R_3 and R_4, and turn T_2 on. The current is given below.

$$i_{C1} = C_1 \times \frac{dv_{CE}}{dt} \qquad (2)$$

The current i_{C1} is boosted by T_2, causing a voltage drop on R_1, this voltage drop will be boosted by the bipolar push-pull circuit. This function will hold v_{GE} above the miller plateau, and force IGBT working in active region, then affect di_C/dt from T_1. A small i_{C1} could make good control performance of dv_{CE}/dt is an important feature of DVRC. If the limitation is not sufficient, such as overload conditions or short circuit fault conditions. At t_4, v_{CE} reaches $V_{Z1}+V_{Z2}$, then no current flow though C_1 any more, T_1 is off. At this time, the classical Active Voltage Clamping is active. This function makes the IGBT operated as a high power TVS, limits v_{CE} below the specified V_{CES}. This function will generate great heat both in clamping circuit and IGBT itself, while compared with driver

without DVRC, the thermal stress is much more reduced. The parameters affect the turn-off behavior is showed in Table I. It is noted that both V_{Z1} and V_{Z2} can be set independently, and DVRC function and AVC function will work independently, consequently. With the boost circuit, C_1 could be a very small value.

We can see from Fig.2, there are 3 parameters which decide the performance of DVRC, i.e. C_1, V_{Z1} and R_4. In order to do deeply analysis of DVRC and parameters selection, Saber simulator is used to evaluate DVRC performance with different parameters based on a guaranteed physical IGBT3 chip model. Each chip model is rated at 1700V, 150A. FZ2400R17KE3_B2 we used as power switch is represented by 16 chip models connected in parallel, this also accord with the real state inside IGBT module.

III.   SIMULATION RESULTS

Simulation circuit is built in Saber simulator, V_{DC}, L_q and R_{G_OFF} are set to constant value while the whole simulation, they are 900V, 200nH and 2Ω, respectively. Fig.5 shows the simulation result with different value of C_1 while set V_{Z1}=800V, R_4=36Ω. Fig.6 shows the simulation result with different value of V_{Z1} while set C_1=100pF, R_4=36Ω. Fig. 10 shows the simulation result with different value of R_4 while set V_{Z1}=800V, C_1=100pF. With the results above, it is shown that V_Z has less controllability on ΔV, only while C_1 is of extremely high value, V_Z has obvious influence on V_{CE-M}. Meanwhile, C_1 has almost the same influence as R_4, which is shown in Fig.11. In Fig.11, the product of $R_4 \times C_1$ is set constant, it is obvious that ΔV is almost unchanged with different value of C_1.

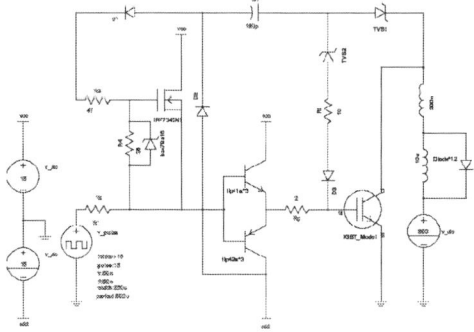

Figure 4. Simulation circuit diagram in Saber

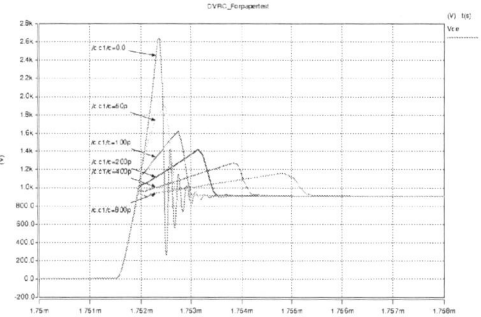

Figure 5. V_{CE} waveform with different C_1

978-1-4244-4782-4/10 $26.00 © 2010 IEEE   1746

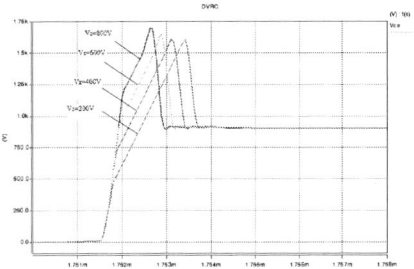

Figure 6. V_{CE} waveform with different V_{Z1}

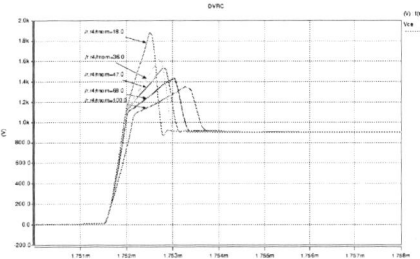

Figure 7. V_{CE} waveform with different R_4

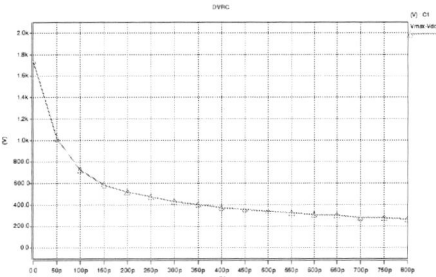

Figure 8. ΔV vs. C_1

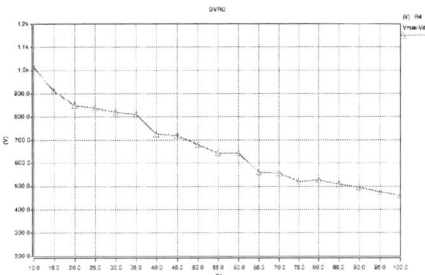

Figure 9. ΔV vs. R_4

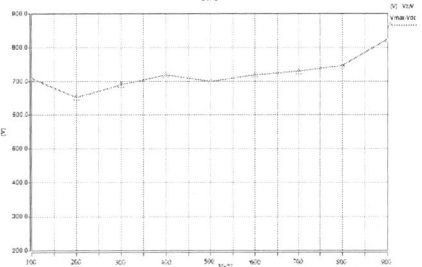

Figure 10. ΔV vs. V_Z

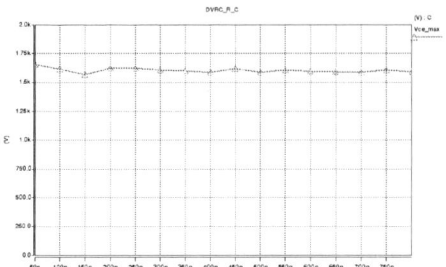

Figure 11. ΔV vs. different C_1 @ constant R4×C1

IV. FOM OF THE PARAMETERS

While evaluating several gate drive methods, it is hard to figure out which one is the best. Here, 3 parameters are selected to evaluate the methods, they are maximum value of V_{CE}, which corresponding to device safe operation area, turn-off time, which corresponding to dead time, and turn-off loss, which corresponding to thermal issue. A 100-mark evaluation criterion is proposed based on these 3 parameters to show the Figure Of Merit (FOM) of active gate drive methods. Results with nominal value gate resistor and extremely large value gate resistor are used to set the base value of the grading. After grading each parameter of different methods, 3 scores have to be integrated into one final score. In order to gain this, evaluation coefficients are needed. With consideration of the importance of safety of the device, the coefficients are set to 0.5 (V_{CE-M}), 0.1 (t_{doff}) and 0.4 (LOSS). The FOM is calculated using (3), the results are listed in Table II and III. Turn-off waveform of the best parameters is shown in Fig. 12.

$$\text{FOM} = \frac{A - A_0}{A_{100} - A_0} \times 100 \times a_1 + \frac{B - B_0}{B_{100} - B_0} \times 100 \times a_2 + \frac{C - C_0}{C_{100} - C_0} \times 100 \times a_3 \quad (3)$$

Table II. FOM of different parameters (R_4=36Ω)

C_1	100	200	310
t_{off}/μs	2.7943	3.128	3.461
V_{CE-M}/V	1447	1277.6	1205
LOSS/J	2.654	3.057	3.45
FOM	74.58	89.82	95.4399

Table III. FOM of different parameters (C_1=310pF)

R_4	10	25	36
t_{off}/μs	2.854	3.2611	3.461
V_{CE-M}/V	1425	1273	1205
LOSS/J	2.759	3.21	3.45
FOM	76.33	89.58	95.4399

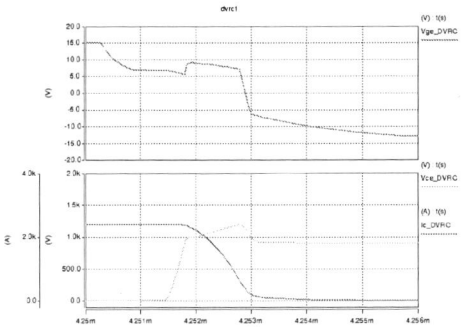

Figure 12. Turn-off waveform with the best parameter
(R_4=36Ω, C_1=310pF, V_Z=800V)

V. Conclusions

This paper reviewed some well-known active gate drivers for IGBTs, after that, a novel active driver DVRC is shown and studied with Saber simulations. Mass simulations were done to decide which parameter has the most influence on the turn-off over-voltage, and how they influence. Based on the proposed evaluation criterion, FOMs of different parameters are used to decide which one is the best. The results are comprehensive and could be a guide for industrial applications.

Acknowledgment

The Authors would thanks Dr. Leo Lorenz, Dr. P. Tuerkes, Mr. Andreas Volke, Ms. ZHENG Ziqing at Infineon for their support of Saber model, IGBT modules, Gate driver ICs and discussions. Special Thanks goes to Ms. LI at SYNOPSYS for her help in simulations.

References

[1] Deml, C.; Turkes, P. Physics-based models of power semiconductor devices for the circuit simulator SPICE, Power Electronics Specialists Conference, 1998. PESC 98 Record. 29th Annual IEEE, Volume 2, 17-22 May 1998 Page(s):1726 - 1731 vol.2

[2] Shen, Z.J. Briggs, D. Robb, S.P. An IGBT with Sustaining Voltage Determined By An Integrated Collector-Gate Clamp, This paper appears in: Electron Device Letters, IEEE, Mar 2000, Volume: 21, Issue: 3, On page(s): 119-122

[3] Phipps, J.P., Voltage Dependence of Self-Clamped Inductive Switching (SCIS) Energy Capability of IGBT's, Applied Power Electronics Conference and Exposition, 1990. APEC '90, Conference Proceedings, 11-16 March 1990 Page(s):476 - 480

[4] H. Rüedi, P. Köhli, HV-IGBT driver includes active-clamping function, PCIM Europe Magazine 3/2000, pp. 12-14.

[5] H. Rüedi, Dynamic Gate Controller (DGC) - A new IGBT Gate Unit for High Current / High Voltage IGBT Modules, PCIM 1995.

[6] M. Hornkamp, "Circuit Arrangement for Control of Semiconductor Circuit", WO Patent Application 002003085832A1, October, 2003.

[7] Palmer, P.R.; Rajamani, H.S., Active Voltage Control of IGBTs for High Power Applications, Power Electronics, IEEE Transactions on, Volume 19, Issue 4, July 2004 Page(s):894 – 901

[8] Bryant, A.T.; Yalan Wang; Finney, S.J.; Tee Chong Lim; Palmer, P.R., "Numerical Optimization of an active voltage controller for high-power IGBT converters", Power Electronics, IEEE Transactions on, Volume 22, Issue 2, March 2007 Page(s):374 – 383

[9] Lihua Chen; Peng, F.Z., "Closed-Loop Gate Drive for High Power IGBTs", Applied Power Electronics Conference and Exposition, 15-19 Feb. 2009 Page(s):1331 – 1337

[10] Norbert F. Will and Dr. Edmund Fischer, "New Electrolytic Capacitors with Low Inductance Simplify Inverter," In Proc. IAS, 2000, pp.3059-3062.

[11] INFINEON, Application Note, "Switching Behavior and Optimal Driving of IGBT3 modules", available at: http://www.infineon.com

[12] Chin-Chien Shen, Allen R. Hefner, Jr., Daving W. Berning, and Joseph B. Berstein, "Failure Dynamics of the IGBT During Turn-off for Unclamped Inductive Loading Conditions," IEEE Trans. Industrial Applications, Vol. 36, No.2, pp.614-624, March/April 2000.

[13] Semikron International, "Application Manual Power Modules," April 2000, Nurmberg, http://www.semikron.com.

[14] Ming Li, Yue Wang, Bin Wang, Leqiang Zhang, Gang Liu, "High power snubberless IGBT converter for direct-drive wind power generation system", Electrical Machines and Systems, 2008. ICEMS 2008. International Conference on, 17-20 Oct. 2008 Page(s):2512 – 2515

[15] Bin Wang, Yue Wang, Ming Li, Zhaoan Wang, Guochun Xiao, "Active Gate Control for High Power IGBTs in Wind Power Generation System", Applied Power Electronics Conference and Exposition, 15-19 Feb. 2009 page(s): 2068 – 2071

Shaping of the Noise Spectrum in Power Electronic Converters

Cristian Lascu
Andrzej M. Trzynadlowski
Department of Electrical and Biomedical Engineering
University of Nevada, Reno
Reno, NV 89557-0260, USA

R. Lynn Kirlin
Professor Emeritus
Department of Electrical and Computer Engineering
University of Victoria
Victoria, BC, V8P 5C2, Canada

Abstract—**Operation of switch-mode converters results in electromagnetic noise propagating throughout the power electronic system. PWM strategies with fixed switching frequency generate harmonics in the spectra of converter voltages and currents. Random pulse width modulation techniques allow elimination of the harmonics, resulting in a continuous spectrum of noise, that is, a spectrum retaining all frequency components. As a next step in advanced spectral shaping, a method presented in this paper produces spectral nulls at selected frequencies. This allows carving communication channels in the noise and removing spectral power at frequencies harmful to the system. Theoretical analysis, computer simulations, and experimental results are presented.**

I. Introduction

Electromagnetic noise generated in power electronic systems due to the switch-mode operation of constituent PWM converters is an insidious nuisance. The electromagnetic interference (EMI) disturbs operation of sensitive communication equipment, necessitating the inconvenient use of special filters [1].

Existing PWM techniques are implemented in digital modulators, in which the fixed sampling frequency coincides with the switching frequency of the controlled converter. Clusters of higher harmonics appear in the spectra of output voltage at multiples of those frequencies. To eliminate the harmonics and reduce the intensity of noise, random pulse width modulation (RPWM) methods have been developed, in which switching periods vary from one switching cycle to another [2]. In the so-called variable-delay RPWM, the average switching frequency is equal to the fixed sampling frequency of the modulator [3]. However, in the resultant continuous spectrum no frequency is exempt from the noise.

All PWM techniques, with both the fixed and variable switching frequency, share one trait: the length of the switching period T is independent of the duty ratio D of the converter. In the steady state, D is constant in the dc-dc converters, while in the ac-dc, ac-ac, and dc-ac PWM converters, D changes with the phase of the ac voltage.

Research described in this paper was sponsored by the US National Science Foundation, grant no. ECCS 0621603.

This paper shows that making T dependent on D opens new possibilities of spectrum shaping. Using this approach, a novel PWM method has been developed, such that in addition to harmonic elimination spectral nulls appear at specified frequencies. The resultant holes in the noise spectrum can be used for power line communications (PLC) utilizing power cables of the system. PLC is highly desirable in the currently developing smart national grid, which requires ubiquitous flow of information. Also, space-limited environments, such as oil rigs, electric vehicles, or residential energy systems could benefit from PLC. Another advantage is the removal of power at harmful resonant frequencies of the system.

The paper outlines the theoretical background of spectral nulls expanding the analyses published in [4]-[7]. It also presents results of computer simulations and experimental investigations of a chopper and a PWM inverter, both controlled using the spectral-null PWM strategy.

II. Theoretical Background

The purpose of RPWM is to eliminate all periodic components in the PWM switching functions, except for the fundamental. The frequency spectrum of RPWM pulse trains is both continuous and impulsive, with the spectral power distributed over a wide range of frequencies.

For stationary random processes, the power spectral density (PSD), $P(f)$ is defined as:

$$P(f) = \int_{-\infty}^{\infty} R(T) e^{-j2\pi f\tau} dT \qquad (1)$$

where $R(T)$ is the time autocorrelation function of the PWM signal $x(t)$, that is,

$$R(T) = \lim_{T_0 \to \infty} \frac{1}{2T_0} \int_{-T_0}^{T_0} E\left[x(t)x(t-T)\right] dt \qquad (2)$$

where $E(\dots)$ denotes statistical expectation, and $P(f)$ and $R(T)$ form a Fourier transform pair.

Generally, a pulse train can be represented by a length M block of random intervals corresponding to random switching frequencies f. The signal under consideration is created by concatenation of $2N+1$ such blocks. The start of the mth switching period in the nth block is labeled $t_{n,m}$. The pulse position within the period is represented by the delay $\Delta_{n,m}$ with respect to $t_{n,m}$. The pulse width is denoted by $\delta_{n,m}$. Using these definitions, the waveform $x(t)$ for the entire sequence may be described as

$$x_N(t) = \sum_{n=-N}^{N} \sum_{m=1}^{M} x\left[t - (t_{n1,m1} + \Delta_{n,m}); \delta_{n,m}\right] \quad (3)$$

where a pulse is zero outside the interval $0 \leq t\text{-}t' \leq \delta_{n,m}$:

$$x(t - t'; \delta_{n,m}) = \begin{cases} 1 & \text{for } 0 \leq t - t' \leq \delta_{n,m} \\ 0 & \text{otherwise} \end{cases} \quad (4)$$

From (2), the autocorrelation function is

$$R(\tau) = \lim_{N \to \infty} \frac{1}{(2N+1)MT_{av}} \int_{-\infty}^{\infty} \left\{ \sum_{n1=N}^{N} \sum_{m1=1}^{M} x\left[t - (t_{n1,m1} + \Delta_{n1,m1}); \delta_{n1,m1}\right] \times \sum_{n2=-N}^{N} \sum_{m2=1}^{M} x\left[t - (t_{n2,m2} + \Delta_{n2,m2} + T); \delta_{n2,m2}\right] \right\} dt \quad (5)$$

where $T_{av} = E(1/f)$ is the average switching period. Following some manipulations, the autocorrelation can be expressed as a double summation of partial autocorrelation functions:

$$R(\tau) = \sum_{m1=1}^{M} \sum_{m2=1}^{M} E[R_{m1,m2}(T)] \quad (6)$$

in which the partial autocorrelation functions are:

$$R_{m1,m2}(T) =$$
$$\lim_{N \to \infty} \frac{1}{(2N+1)MT_{av}} \int_{-\infty}^{\infty} \sum_{n1=-N}^{N} \sum_{n2=-N}^{N} \int_{-\infty}^{\infty} X(f_1; \delta_{n1,m1})$$
$$\times e^{-j\omega_1(t_{n1,m1}+\Delta_{n1,m1})} e^{j\omega_1 t} \times X(f_2; \delta_{n2,m2})$$
$$\times e^{-j\omega_2(t_{n2,m2}+\Delta_{n2,m2})} e^{j\omega_2 t} df_2 df_1 dt \quad (7)$$

where $\omega_{1,2} = 2\pi f_{1,2}$ are integration variables and $X(f;\delta)$ is the Fourier transform of $x(t; \delta)$.

Based on (7), the expectation of partial autocorrelation functions can be determined as:

$$E[R_{m1,m2}(T)] = \frac{1}{MT_{av}} \int_{-\infty}^{\infty} E\left\{ \sum_{k=-\infty}^{\infty} [X(f; \delta_{0,m1}) X^*(f; \delta_{k,m2}) \right.$$
$$\left. \times e^{-j\omega(t_{0,m1}-t_{k,m2})} e^{-j\omega(\Delta_{0,m1}-\Delta_{k,m2})}] e^{j\omega t} \right\} df \quad (8)$$

where $k = n_2 - n_1$ and X^* is the complex conjugate of X.

As the autocorrelation and PSD form a Fourier transform pair, the expression for the partial PSD function may be derived from (8) as:

$$P_{m1,m2}(f) =$$
$$\frac{1}{MT_{av}} E\left\{ \sum_{-\infty}^{\infty} \left[\begin{array}{c} X(f:\delta_{0,m1}) X^*(f; \delta_{k,m2}) \\ \times e^{-j\omega(t_{0,m1}-t_{k,m2})} e^{-j\omega(\Delta_{0,m1}-\Delta_{k,m2})} \end{array} \right] \right\} \quad (9)$$

and the PSD of the entire pulse train is given by

$$P(f) = \sum_{m1=1}^{M} \sum_{m2=1}^{M} S_{m1,m2}(f). \quad (10)$$

For a chopper that operates with a constant duty ratio and a random switching frequency (precisely, random switching periods), it is sufficient to assume $M = 1$. Then, (10) becomes

$$P(f) =$$
$$\frac{1}{T_{av}} E\left\{ \sum_{-\infty}^{\infty} [X(f; \delta_0) X^*(f; \delta_k) e^{j\omega(t_k-t_0)} e^{j\omega(\Delta_k-\Delta_0)}] \right\} \quad (11)$$

It is assumed that the pulses of switching variables are located at the centers of switching cycles. Individual switching periods T are determined from independent random actions, that is, there is no dependence between any two switching periods. Applying to (11) the so-called Bech's approximation proposed in [5] allows calculation of $S(f)$ from the formula

$$P(f) =$$
$$\frac{1}{T_{av}} \left\{ E[X^2(f; \delta)] + 2\left\{ \frac{E[X(f;\delta)e^{j\omega(1-\alpha)t}] E[X^*(f;\delta)e^{j\omega\alpha t}]}{1-E(e^{j\omega t})} \right\} \right\} \quad (12)$$

in which no infinite summation is needed. The variable α equals $(1 - D)/2$.

Taking into account the binary character of the switching variables, that is,

$$x_k(t) = \begin{cases} 1 & \text{for } 0 \leq t \leq \delta_k \\ 0 & \text{otherwise} \end{cases} \quad (13)$$

the PSD for the case of randomly varying switching periods can be expressed as

$$P(f) =$$
$$\frac{1}{T_{av}} \left(\frac{V_{dc}}{\pi f} \right)^2 \left\{ E[\sin^2(\pi f DT)] + 2\mathcal{R}e\left\{ \frac{E^2[\sin(\pi f DT)e^{j\pi ft}]}{1-E(e^{j2\pi ft})} \right\} \right\} \quad (14)$$

where V_{dc} denotes the dc supply voltage of the converter.

To extend the results to dc-ac converters (inverters), note that the fundamental frequencies are typically some two orders of magnitude lower than the average switching frequencies. Consequently, duty ratios can be considered practically constant from one switching cycle to another, and the PSD can be calculated from (14) with a sufficient degree of accuracy (see Appendix).

III. Spectral Nulls

Because there is no cyclic component of the switching interval sequence, no harmonics are generated in the voltage spectrum. Also, the PSD equals zero for a selected frequency $f = f_0$ and its multiples when

$$\sin(\pi f_0 DT) = 0 \quad (15)$$

which occurs when f_0DT is an integer, k. In other words, to create a spectral null, the switching period must be made dependent on the duty ratio, which is a novel and unique property in methods of pulse width modulation.

Based on (15), in a spectral-null PWM technique, individual switching periods are calculated as

$$T = \frac{k}{f_0 D} \qquad (16)$$

where consecutive values of k are randomly drawn from a given set K of integers. For a three-phase inverter, in which the spectral nulls are to be generated in the spectrum of line-to-line voltage v_{AB}, the duty ratio D is defined as

$$D = \frac{1}{2}|D_A - D_B| \qquad (17)$$

where D_A and D_B are duty ratios of phases A and B, respectively.

Practical limits on the allowable range of values of T define boundaries of K. Any PWM method employing the concept of switching periods can be modified to generate spectral nulls (note that the classic programmed PWM strategies do not belong in that class).

IV. COMPUTER SIMULATIONS

Selected results of computer simulations of a chopper are shown in Figs. 1 to 3. The noise spectrum of output voltage with a fixed switching frequency of 20 kHz is shown in Fig. 1. In all simulations, the modulation index was set to 0.9.

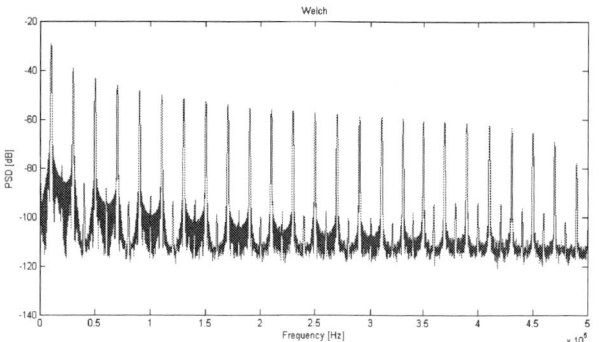

Fig. 1.   Noise spectrum of output voltage of a chopper with fixed switching frequency of 20 kHz.

When consecutive switching periods were made random, in the 50-470 µs range, the harmonics have disappeared from the spectrum, shown in Fig. 2. Results illustrated in those two figures are well known from numerous publications on random PWM. Here, they have been shown for reference only.

The set $K = 3, 5, \ldots, 31$ of prime numbers was used for generation of spectral nulls at $f_0 = 200$ kHz and its multiples. The resultant noise spectrum is shown in Fig. 3.

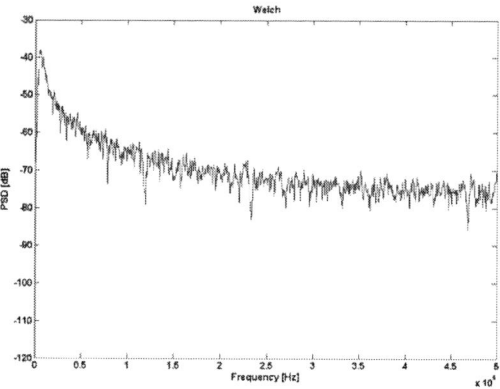

Fig. 2.   Noise spectrum of output voltage of a chopper with standard RPWM (no spectral nulls).

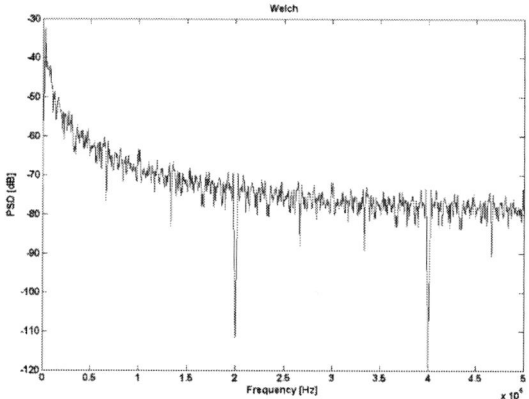

Fig. 3.   Noise spectrum of output voltage of a chopper with SN-RPWM: spectral nulls at multiples of 200 kHz.

Noise spectra of the line-to-line output voltage of a three-phase inverter are shown in Figs. 4 to 6. The classic space vector PWM method is employed. The case of fixed switching frequency of 10 kHz is illustrated in Fig. 4.

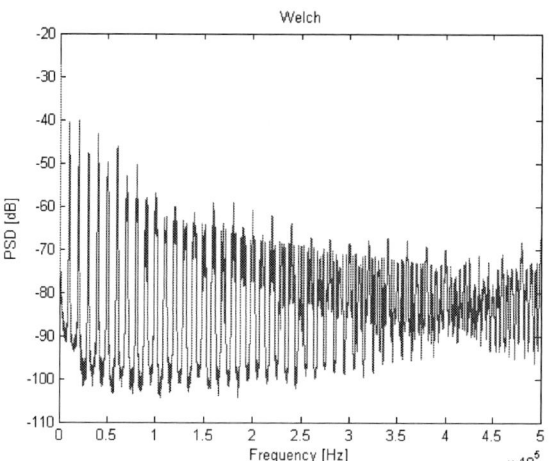

Fig. 4.   Noise spectrum of line-to-line output voltage of an inverter with fixed switching frequency of 10 kHz.

Fig. 5 shows the noise spectrum of the inverter voltage when the standard RPWM technique is used. As in the chopper simulations, the switching periods are randomized within the 50-470 µs range. Again, these well-known spectra have only been provided for reference. Finally, spectral nulls at multiples of 150 kHz are shown in Fig. 6. The same set K of prime numbers as in the case of the chopper was employed.

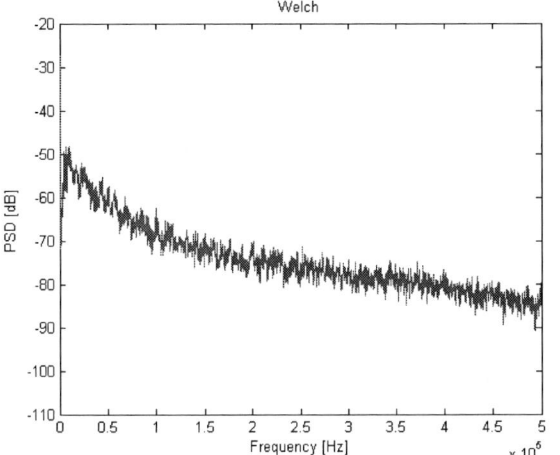

Fig. 5. Noise spectrum of output voltage of an inverter with standard RPWM (no spectral nulls).

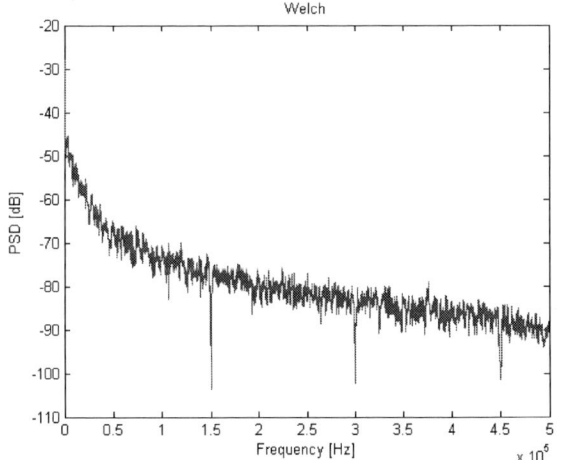

Fig. 6. Noise spectrum of output voltage of an inverter with SN-RPWM: spectral nulls at multiples of 150 kHz.

V. EXPERIMENTAL SETUP

To verify results of the theoretical considerations and computer simulations, an experimental setup was assembled from a 2.2 kW, 380 V, FC302 inverter from Danfoss and a 0.75 hp, 230 V induction motor. The original pulse width modulator of the inverter was bypassed with an eZdsp-F2812 digital signal processor (DSP) control board from Spectrum Digital. The board employs the TMS320F21812 DSP from Texas Instruments. The setup, whose block diagram is shown in Fig. 7, also included an analog to digital conversion (ADC) interface, a voltage-source inverter (VSI) interface, a spectrum analyzer, and a desktop computer.

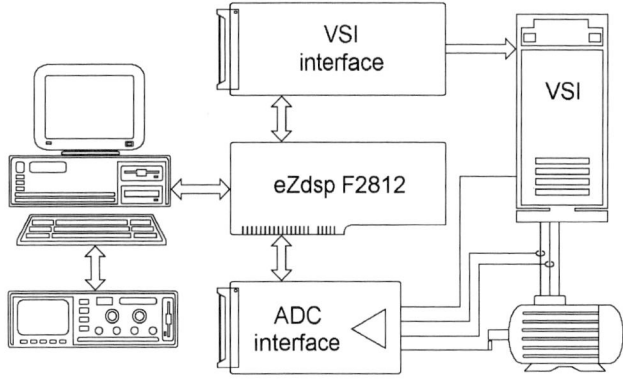

Fig. 7. Experimental setup.

VI. EXPERIMENTAL RESULTS

For experiments with a chopper, one leg of the inverter was disabled. In all cases presented in this section the modulation index, whose impact on the spectra is weak, was 0.9 Noise spectrum of the output voltage of the chopper, with the fixed switching frequency of 10 kHz is shown in Fig. 8.

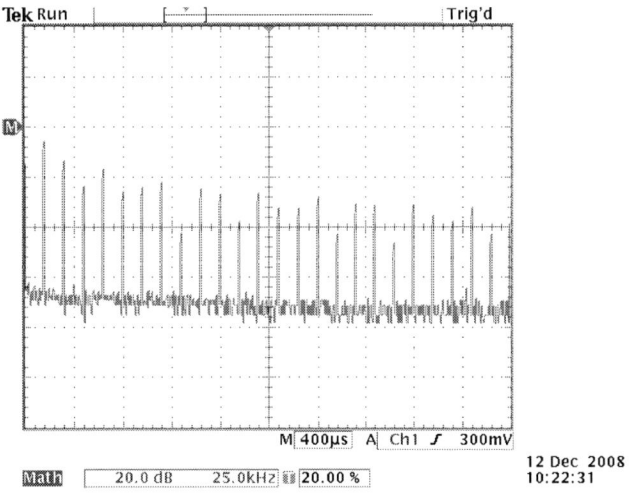

Fig. 8. Experimental spectrum of voltage noise of the chopper with fixed switching frequency of 10 kHz.

When switching periods were randomized within the 50-470 µs range, noise spectrum in Fig. 9 was obtained. Spectral nulls at multiples of 100 kHz are shown in Fig. 10.

Voltage noise spectra of output voltage of the three-phase inverter are shown in Figs. 11 to 13. The case of fixed switching frequency of 10 kHz is represented by Fig. 11, quite similar to Fig. 8. Randomization of switching periods in the 50-470 µs range, yielded spectrum in Fig. 12. Finally, generation of spectral nulls at multiples of 100 kHz, using the K set of prime numbers specified in Section IV, is illustrated by the oscillogram in Fig. 13. Similarity of the simulated and experimental spectra confirms validity of the proposed method. The dips in the noise spectra are deep and well defined.

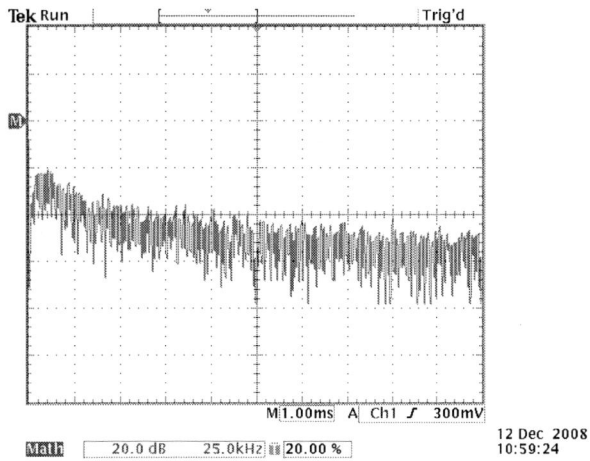

Fig. 9. Experimental spectrum of voltage noise of the chopper with variable switching frequency.

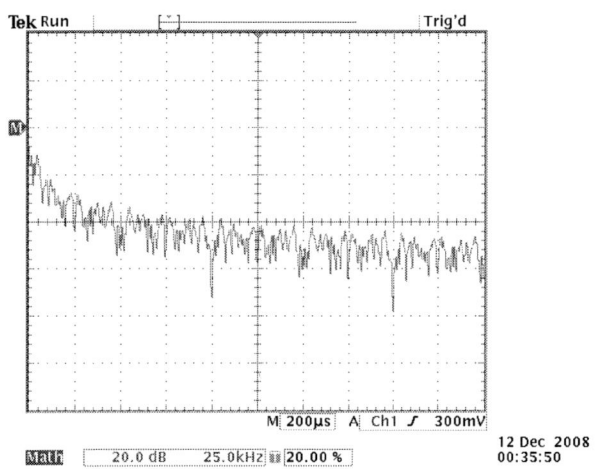

Fig. 10. Experimental spectrum of voltage noise of the chopper with spectral nulls at multiples of 100 kHz.

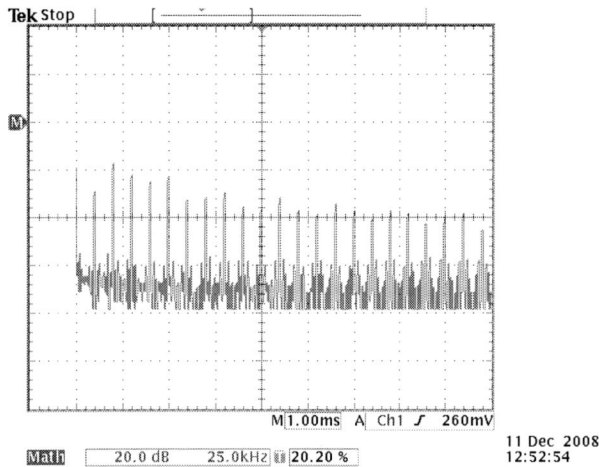

Fig. 11. Experimental spectrum of voltage noise of the inverter with fixed switching frequency of 10 kHz.

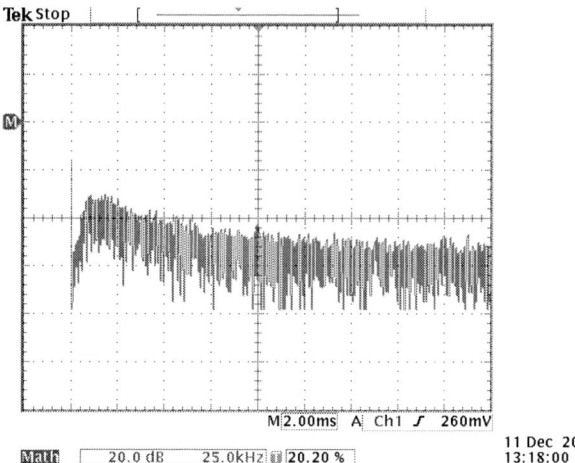

Fig. 12. Experimental spectrum of voltage noise of the inverter with variable switching frequency.

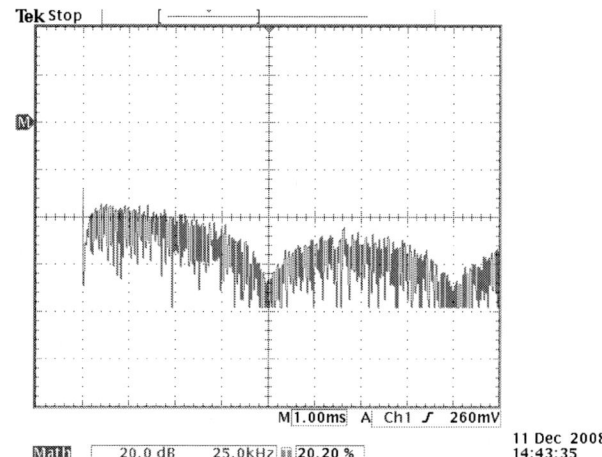

Fig. 13. Experimental spectrum of voltage noise of the inverter with spectral nulls at multiples of 100 kHz.

VII. CONCLUSION

The novel PWM method has been shown to effectively suppress specified frequency components in the spectra of voltage noise of dc-dc and dc-ac power electronic converters. It is accomplished by making individual switching periods dependent on the duty ratios of the converters. The method belongs in the class of random PWM techniques, retaining the known feature of harmonic elimination. The spectral-null strategy may find a number of applications, particularly the power line communications in the smart grid of tomorrow. The concept of spectral nulls can be extended on other types of PWM power electronic converters

Regarding the inverter, the spectral nulls were generated in one line-to-line voltage. Theoretically, all three line-to-line voltages could simultaneously be subjected to the spectral nulls procedure, but there are significant practical problems with such expansion. Either the switching periods would have to be excessively long, or digital implementation of the modulation strategy would require an excessive amount of real-time computations. Further technological progress in the areas of fast semiconductor power switches and digital processors may render those approaches feasible in the foreseeable future.

REFERENCES

[1] L. Tihanyi, "Electromagnetic Compatibility in Power Electronics," *IEEE Press*, New York, 1995.

[2] A. M. Trzynadlowski, F. Blaabjerg, J. K. Pedersen, R. L. Kirlin, and S. Legowski, "Random pulse width modulation techniques for converter-fed drive systems - A review," *IEEE Trans. Ind. Appl.*, vol. 30, no. 5, pp. 1166-1175, Sept./Oct. 1994.

[3] A. M. Trzynadlowski, K. Borisov, Y. Li, and L. Qin, "A novel random PWM technique with low computational overhead and constant sampling frequency for high-volume, low-cost applications," *IEEE Trans. Power Electron.*, vol. 20, no. 1, pp. 116-122, Jan. 2005.

[4] R. L. Kirlin, J. Wang, and R. M. Dizaji, "Study on spectral analysis and design for DC/DC conversion using random switching rate PWM", Proc. *2000 IEEE SSAP Worksh.*, Pocono, PA, pp. 378-382, Aug. 2000.

[5] M. M. Bech, "Random Pulse-Width Modulation Techniques for Power Electronic Converters," PhD dissertation, *Aalborg University*, Aalborg, Denmark, 2000.

[6] R. L. Kirlin, M. M. Bech, and A. M. Trzynadlowski, "Analysis of power and power spectral density in PWM inverters with randomized switching frequency," *IEEE Trans. Ind. Electron.*, vol. 49, no. 2, pp. 486-499, 2002.

[7] A. M. Stankovic and H. Lev-Ari, "Randomized modulation in power electronic converters," *Proc. IEEE*, vol. 90, no. 5, pp. 782-799, 2002.

APPENDIX

For reference, diagrams of the well-known power circuits of the chopper and inverter dealt with in the paper are shown in Figs. 14 and 15, respectively.

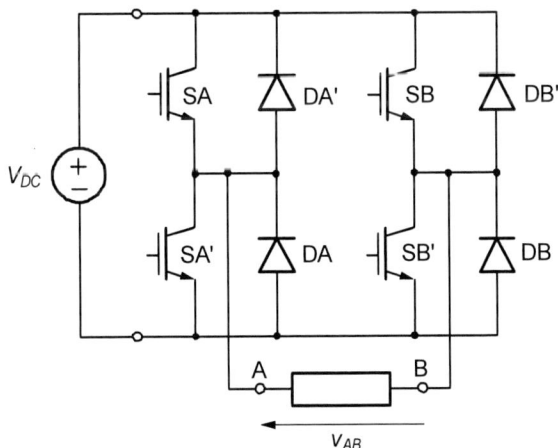

Fig. 14. Diagram of power circuit of the chopper.

Topological similarity of the chopper and inverter allows a similar approach to control of these two converters. The chopper is simply a three-phase inverter with one phase removed. Consequently, the line-to-line voltage v_{AB} of the inverter within a single switching cycle can be assumed identical with the output voltage v_{AB} of the chopper when both converters share the same duty ratio D. Of course, D in the inverter changes from one switching cycle to another, while the duty ratio of the chopper is maintained constant.

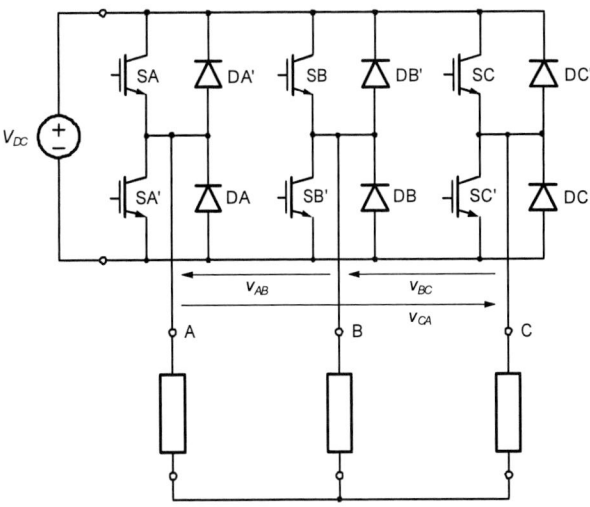

Fig. 15. Diagram of power circuit of the inverter.

Grid Interactions and Stability Analysis of Distribution Power Network with High Penetration of Plug-In Hybrid Electric Vehicles

Omer C. Onar, *Student Member, IEEE*, and Alireza Khaligh, *Senior Member, IEEE*,

Energy Harvesting and Renewable Energies Laboratory (EHREL),
Electric Power and Power Electronics Center, Electrical and Computer Engineering Department,
Illinois Institute of Technology, 3301 S. Dearborn St., Chicago, IL, USA
EML: oonar@iit.edu, khaligh@ece.iit.edu; URL: www.ece.iit.edu/~khaligh

Abstract—This study focuses on the effects of plug-in hybrid vehicles (PHEV) on the power system network. PHEVs are expected to be commercialized in near future. PHEVs will introduce a new type of load with extraordinary characteristics than the regular household loads. PHEVs may have significant effects on the existing power system networks, if they do not charged during off-peak load periods. A new peak demand period, distribution line capacities, electricity market prices, and power system level problems due to the harmonics and reactive power problems associated with PHEV chargers are the possible issues that may cause problems in the power systems. A new peak demand period, the need for additional distribution lines, additional harmonics and reactive power injection are among the issues that need be investigated due to integration of PHEVs. In this study, stability of a distribution power network integrated with PHEVs is analyzed. In this study it is been assumed that PHEVs do not necessarily need to be charge during off-peak load demand times..

I. INTRODUCTION

THE transportation market is going through a major restructuring to present new alternatives to the internal combustion engine (ICE) powered vehicles due to the increasing gasoline prices, depletion of fossil fuels, and environmental concerns. In PHEVs, the electric motor contribution to the overall propulsion power can be increased by employing a larger battery and a larger electric motor. This battery is used to capture and store the regenerative energy during braking periods. Moreover, by utilizing power electronic converters, the battery pack can be charged from the utility outlets. In addition, the power electronic converter can be bi-directionally operated and the energy storage system (ESS) of vehicle can be utilized as a distributed resource unit to sell power back to the distribution network.

ICE, electric propulsion system, power electronic interfaces, and ESS are the main components of a plug-in hybrid electric vehicle. The optimized operation of the vehicle can be achieved by a combination of these three systems allowing better fuel economy and resulting in less

energy consumption [1]. In PHEVs, batteries are either charged through regenerative braking, or through the grid. An AC/DC rectifier commonly followed by a DC/DC converter is used to attain DC power at an appropriate voltage level from the utility grid [2]. In first generation PHEVs, it is prospected that the energy flow will be unidirectional as power is taken from grid to charge the battery pack. Hence, a simple battery charger with a large battery pack will suffice for HEV to PHEV conversion [3]-[14]. Generally, the battery size is increased by adding more batteries in parallel in PHEVs in order to increase the energy storage capacity.

The PHEVs will present an additional battery charging load to the utility distribution system [15]. On the other hand, charging the PHEVs possibly will show a new peak period in the early off-peak period. The potential implication of PHEVs to the distribution networks can be (i) different peak demand, (ii) the need for additional distribution line capacity, (iii) additional harmonics and reactive power injection due to PHEV chargers, and (iv) energy cost variations.

Electric Power Research Institute (EPRI) has reported that PHEVs would be recharged during off-peak hours overnight since more that 40% of U.S. generating capacity operates at reduced load overnight [16]. According to the EPRI report, 8% increase in electricity generation is required if half of all vehicles on the roads are replaced with PHEVs by year 2050. Another study indicated that for 4 million PHEVs corresponding to 25% penetration for the Californian system, the existing generation capacity is already sufficient to charge the PHEV batteries if charging is done overnight [17]. PHEV charging during off-peak periods might be achieved by coordinated or programmed charging of multiple PHEVs in residential distribution grids [18] or demand management of grid connected PHEVs [19]. However, this proper or optimal use of the power grid is not simple since the PHEVs will introduce a significantly new load on the current primary and secondary distribution networks. Many of these networks do not have the monitoring and automation capability and even enough spare capacity [20].

Either an existing secondary distribution transformer in a residential neighborhood or a transformer/circuit connected to a distribution feeder will supply the additional charging load in a power system level. As the penetration level of PHEVs increases, this additional charging load will be added to distribution networks. Since many distribution systems in the U.S. were designed decades ago considering the load levels at that time, major changes might be required in the distribution system components due to the new load levels, patterns, and load characteristics, especially if they are not charged during off-peak load times [20]. Unbalanced conditions, reactive power requirements of the battery charger, and the current harmonics drawn by the battery chargers could cause power quality issues. Increased harmonics, line losses, transformer and line heating problems, and increased reactive power consumption may result in potential damage to the customer devices and power system equipments. These issues may cause overall power system and voltage stability problems, and even collapses.

Although a few studies were done on the load level impacts of PHEV penetration in the power systems [21]-[23], the effects of PHEV battery charging on the power systems due to their special load characteristics [24]-[26] has not been studied so far.

In this study, a network based on two distributed generation units with a load bus is modeled. The generators are used as swing busses and a load bus is used to supply power to a residential neighborhood with 20 houses, considering an IEEE 3-bus test system. Each house is considered to consist of one PHEV load that is an additional load to the residential neighborhood. The power system stability of the residential neighborhood is analyzed with and without PHEV loads under a three phase - ground fault and after fault clearance. Results of each case are compared in order to determine the effects of PHEV loads by means of rotor speed, rotor speed deviations, and power system eigenvalue trajectories. In order to make a fair comparison and to show effects of the characteristics associated with the PHEV loads, the active and reactive power demand of the PHEVs are added to the residential power demand as linear active and reactive loads in order to show that this is beyond a capacity problem. Furthermore, the load characteristics of PHEVs affecting the distribution system stability has been investigated and analyzed.

II. CHARACTERISTICS OF PHEVS AS AN ELECTRICAL LOAD

The plug-in vehicles require around 0.3 kWh of charged energy for one mile driving [20]. More than 50% of the cars in the U.S. are driven 25 miles a day or less [17]. Assuming 40 miles/day driving, the battery energy capacity for a PHEV should be 12 kWh. There is not any certain value for the PHEV battery voltage; however, it can vary from 96 V to 600 V [27]-[28]. In this study, a 400 V NiMH battery pack is selected with 30 Ah capacity in order to meet the overall energy storage of 12 kWh.

It is assumed that the residential neighborhood consists of 20 houses. Half of the PHEVs are being charged from the utility grid. For example, for 6 hours of charging of each PHEV, the charger can draw 2 kW of power. This is the typical charging power of PHEVs [29]. Assuming half of the vehicles are connected to the grid, the overall active power demand of all of the PHEVs is 20 kW. The charging circuit of the PHEV consists of a diode bridge followed by a boost converter. Although 10 single-phase individual chargers could be considered, a three-phase PHEV charger with all batteries, connected in parallel, is used for the massive PHEV load modeling. This PHEV load is connected to the secondary distribution transformer at 240 Vrms phase-to-phase. The massive model of the PHEV load is illustrated in Fig. 1.

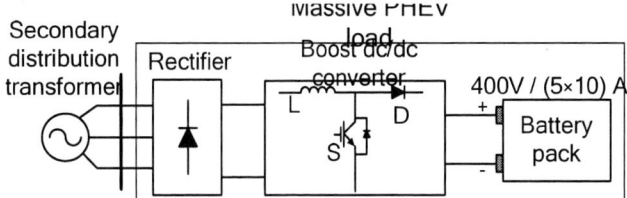

Fig. 1. PHEV load model with AC/DC rectifier, DC/DC converter, and battery pack.

Not only this configuration will add up 20 kW of active power to the distribution network, but also this powerful additional load exhibits a nonlinear load characteristic due to the harmonic currents drawn from the network. Moreover, it draws some amount of reactive power from the grid. Therefore, the PHEV load causes harmonic distortions in the system voltage and draws reactive power resulting in effects on the power system stability. A single phase current that is drawn by the massive PHEV loads and the distorted load bus voltage waveforms are shown in Fig. 2. This current is drawn by the massive PHEV load shown in the box in Fig. 1 and the voltage is measured from the secondary distribution transformer bus which gets distorted due to the harmonic currents.

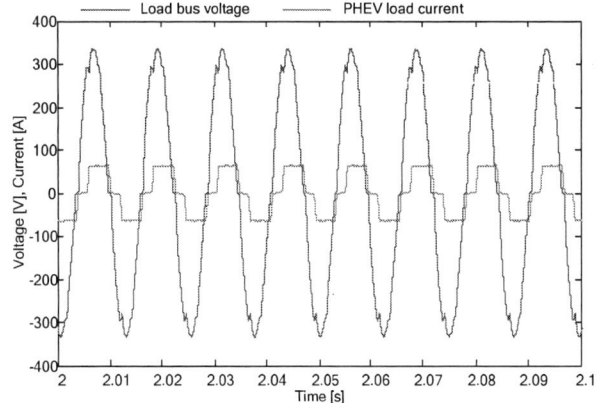

Fig. 2. PHEV load current and distorted load bus voltage due to harmonics.

The harmonic effects of battery chargers were presented in previous studies [24]-[26], [30], [31]. Upcoming PHEV charging technologies may contribute to high harmonic distortion in the distribution systems during simultaneous charging periods. According to the California Energy Commission's Public Interest Energy Report [30], a battery charger's input current total harmonic distortion (THD) may vary from 2.36% to 28% through a charging cycle. Additionally, the on-board battery charger of Ford's station wagon has 59.6% input current THD [31]. General effects of vehicular battery chargers are the harmonics and overloading problems while the affected distribution equipments are transformers, cables, circuit breakers, and fuses [26]. On the other hand, they also have impact on distribution network stability. Since the load characteristics have major influences on voltage stability and power system stability [32], [33], the effect of these characteristics should be carefully investigated. This is due to the fact that there is a trend to introduce PHEVs to the market in few years and their penetration in the existing distribution systems will increase day by day.

III. TEST SYSTEM DESCRIPTION

In this study, the 3-bus IEEE test system is considered [37]. This test system consists of electrical generators acting as swing busses and a load bus. The load bus is located in between these generator busses and connected through distribution lines. A residential neighborhood with 20 houses is considered as the load. Power consumption varies from house to house, but to simulate a realistic application, residential load is based on a 2500 ft^2 typical U.S. Midwest household including all electrical appliances as well as the air conditioning and heating [38], [39]. According to power profiles, demand varies from few hundreds of watts and may reach up to 7 kW. However, the average load demand is less than 2 kW. In this study, each house is assumed to draw 2.5 kW active power and 0.5kVAr reactive power from the utility grid during the simulation time. This constitutes the total power demand of the residential neighborhood to be 50 kW and 10 kVAr household load and 20 kW of PHEV load. The single line diagram of the 3-bus test system is shown in Fig. 3.

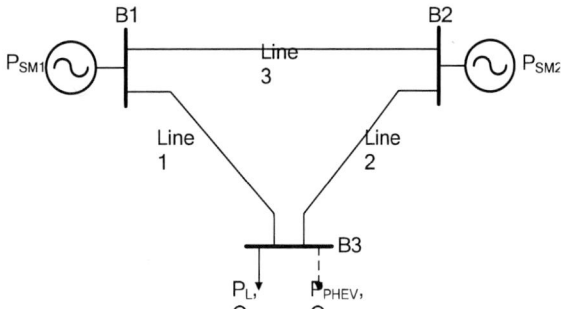

Fig. 3. IEEE 3-bus test system representing the power network integrated with residential and PHEV loads.

IV. POWER SYSTEM AND VOLTAGE STABILITY

Power system stability is defined as the ability of an electric power system to regain the state of operating at equilibrium points after being subjected to a disturbance [34]. Power system stability has shown its importance due to the major blackouts caused by this phenomenon [35], [36]. The power systems have recently experienced increased growth in interconnections, penetration of renewable energies, use of new technologies and controls. Addition of new loads such as

PHEVs will present other stability criteria to the electrical power systems.

A system defined by

$$x = f(t, x) \tag{1}$$

can be considered for stability analyses where x is the state vector, $\dot{x}$ is its derivative, and f is sufficiently differentiable function whose domain includes the origin. First, it is implied that the system is operating in a pre-disturbance equilibrium set X_n, which is an initially balanced condition. Then, a disturbance acts on the system which can be a short circuit on the distribution lines or any other physical disturbance. The system dynamics are analyzed after a disturbance such as a fault. These analyses are done with respect to the post-disturbance equilibrium set X_p which might be different from X_n. The system behavior is characterized with respect to X_p when the system initial condition belongs to a starting set x_p i.e., if the system trajectory will remain inside the technical viable set Ω_p. If the system dynamic response turns out to be stable, X_p and X_n are said to be stable. If the system motion crosses the boundary of the technically viable set $\partial\Omega_p$, an instability is detected.

Voltage stability is the electrical power system's capability to maintain acceptable operating conditions under normal and adverse conditions. Furthermore, voltage stability is also associated with providing adequate reactive power supply under both steady-state and transient conditions [40], [41]. Both active and reactive power demand affect system's voltage stability [33], and load characteristics play an important role in voltage stability [42].

In order to determine the critical parameters for voltage stability analysis, active and reactive power transfer equations can be used. For instance, power flow from bus B1 to bus B3 can be expressed as

$$P_{B13} = \frac{V_1 V_3}{X} \sin \delta , \tag{2}$$

$$Q_{B13} = \frac{V_1^2 - V_1 V_3 \cos \delta}{X} . \tag{3}$$

where P_{B13} and Q_{B13} are the active and reactive power flow from bus B1 to B3, V_1 and V_3 are the voltages of the buses B1 and B3, X is the equivalent impedance from bus B_1 to B_3, and δ is the phase angle between two bus voltages (V_1 and V_3).

By using Newton-Raphson power flow analysis, the Jacobian matrix can be built for considering the singularity to determine the critical values. Equations (2) and (3) are equalized to zero in order to obtain the Jacobian matrix by

$$f_1(V_{B1}, V_{B3}, \delta) = P_{B13} - \frac{V_1 V_3}{X} \sin \delta \tag{4}$$

$$f_2(V_{B1}, V_{B3}, \delta) = Q_{B13} - \frac{V_1^2 - V_1 V_3 \cos \delta}{X} . \tag{5}$$

The Jacobian matrix can be formed using Equations (4) and (5) by

$$\begin{bmatrix} \Delta P_{B13} \\ \Delta Q_{B13} \end{bmatrix} = \underbrace{\begin{bmatrix} \dfrac{\partial f_1}{\partial \delta} & \dfrac{\partial f_1}{\partial V_3} \\ \dfrac{\partial f_2}{\partial \delta} & \dfrac{\partial f_2}{\partial V_3} \end{bmatrix}}_{J} \begin{bmatrix} \Delta \delta \\ \Delta V \end{bmatrix} \qquad (6)$$

The stability limit, also known as critical point is designated to be the points where the power flow Jacobian matrix is singular [41], [42]. This singularity points can be calculated by equalizing the determinant of the Jacobian matrix to zero, i.e.,

$$\det[J] = 0 . \qquad (7)$$

The determinant is zero if

$$\frac{\partial f_1}{\partial \delta} \frac{\partial f_2}{\partial V_3} - \frac{\partial f_1}{\partial V_3} \frac{\partial f_2}{\partial \delta} = 0 . \qquad (8)$$

If Equations (4) and (5) are substituted in (8), then

$$V_1 = 2V_3 \cos \delta . \qquad (9)$$

Therefore, the relationship between the bus voltages at the critical point is obtained. The critical angle can be determined by using V_1 and δ is used instead of V_3.

The relationship between the active and reactive powers (P_L, Q_L) that are drawn from the load bus may be expressed as

$$Q_L = P_L \tan \varphi \qquad (10)$$

where φ is the load phase angle in radians. Combining equations (2), (3), and (10), the critical angle can be expressed in terms of the load angle that is

$$\tan(2\delta) = \frac{-1}{\tan \varphi} . \qquad (11)$$

Consequently, critical values for the angle and power can be obtained from equations (2), (9), (10), and (11) as

$$\delta_{critical} = \frac{1}{2} \arctan\left(\frac{-1}{\tan \varphi} \right), \qquad (12)$$

$$P_{L-critical} = \frac{V_1^2 \tan \delta_{critical}}{2X}, \qquad (13)$$

and

$$Q_{L-critical} = \frac{V_1^2 \tan \delta_{critical}}{2X} \tan \varphi . \qquad (14)$$

V. RESULTS AND DISCUSSIONS

For the stability analyses, a three-phase to ground short circuit fault on Line 1 as a large signal disturbance is applied to the power system shown in Fig. 3 with and without including PHEV loads. Three-phase to ground fault is applied at t=4s and lasted for 5 periods and cleared. The effect of this fault on the phase currents measured from B1 bus is shown in Fig. 4. The rotor speed, rotor speed deviation, and rotor angle deviation variations are shown in Figs. 5-7, respectively. Figures show the effects of disturbance with and without massive PHEV load connected to the load bus B3.

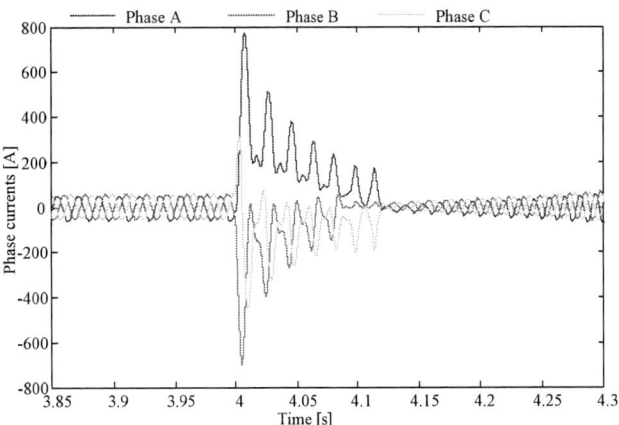

Fig. 4. Three-phase line current waveforms at bus B1 under short circuit effect.

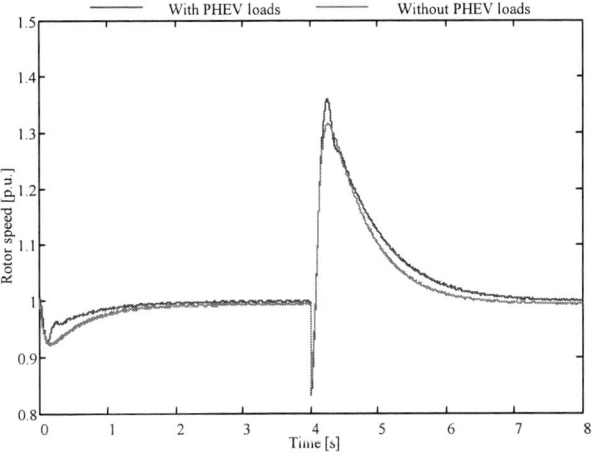

Fig. 5. Rotor speed of PSM1.

Fig. 6. Rotor speed deviation of PSM1.

According to these figures, it is observed that the distribution network without PHEV loads is more stable than the one with PHEV load. System with PHEV loads reacts with greater amplitude variations in rotor speed. Moreover, after the disturbance is cleared, system with PHEV loads need more time for settling back to the pre-disturbance conditions. Therefore, it can be stated that systems with PHEV loads have increased sensitivity on power system disturbances and have higher risks of instability and even collapses. Fig. 7 also shows that the rotor angle stability is in a better condition for the system without PHEV loads since it has less deviation for the post-disturbance state.

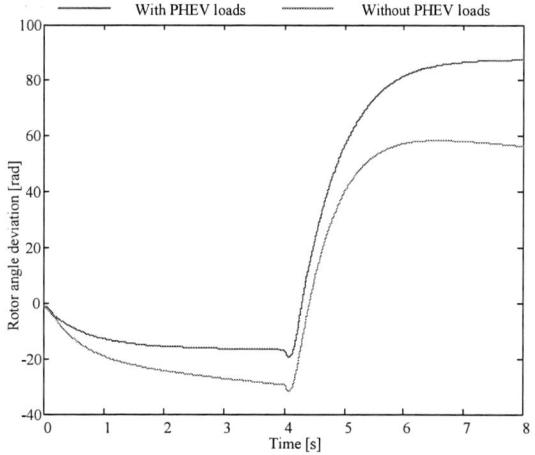

Fig. 7. Rotor angle deviation of PSM1.

The critical load for the systems with and without PHEV loads are calculated based on equations (12) and (13). The calculated critical load values for these cases are presented in Fig. 8. As shown in Fig. 8, the system without PHEV loads has higher critical load value. Therefore, the system with PHEV loads can reach its critical loading condition easier

than the system with PHEV loads; hence, it would be less stable. In the critical load value calculation, the settling time to the stead-state values are mainly due to the delays of the complex phase difference algorithm used for the load angle (φ) detection.

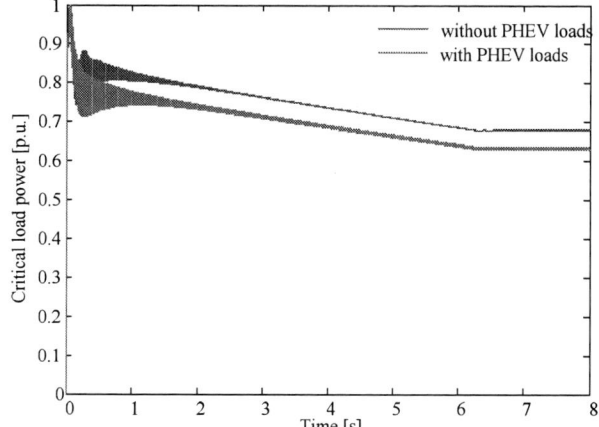

Fig. 8. Critical load power variation with and PHEV loads.

In order to theoretically prove and compare the stability conditions of the systems with and without PHEV loads, the equivalent state-space model of each power system is obtained by taking one of the short-circuit currents as input and the load bus voltage (B3) as the output. If the disturbance input to the system is a three-phase to ground fault current, definitely both systems will be unstable. However, systems can be compared in terms of being more stable and survivable to the disturbances. Therefore, Nyquist stability criterion is applied to the equivalent state-space model of both systems. Nyquist frequency responses of the systems are shown in Fig. 9.

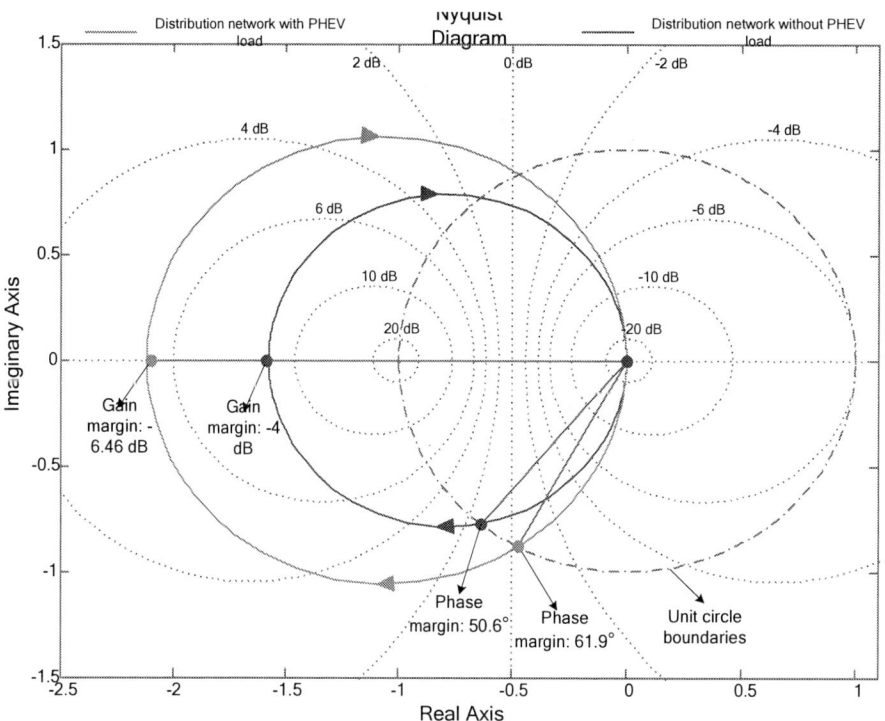

Fig. 10. Nyquist diagrams of systems with and without PHEV loads.

978-1-4244-4782-4/10 $26.00 © 2010 IEEE

The gain margins of the system are -6.46 dB with PHEV loads and -4 dB without PHEV loads. The phase margins for the systems are 61.9 degrees and 50.6 degrees for the systems with and without PHEV loads, respectively. Therefore, the system with PHEV loads are less stable . This is due to the fact that the Nyquist diagram of the system with PHEV loads is farther away from the unit circle boundaries. However, the Nyquist diagram of the system without PHEV loads is closer to the unit circle boundary, i.e., stability limits.

VI. CONCLUSIONS AND FUTURE WORK

The effects of PHEVs on distribution network stability have been investigated in this study. It is shown that the systems with PHEV connections are more sensitive to the power system disturbances and behave to be less stable in both the magnitude deviations and the time required to reach the pre-disturbance equilibrium conditions. Nyquist criterions also indicate that systems without PHEV loads have less phase and gain margins in their frequency responses; therefore, they are more resistible and survivable to the physical power system disturbances. This is mainly due to the special load characteristics of the PHEV chargers that draw current harmonics and require reactive power. In future work, a new bi-directional integrated power electronic converter will be designed that will eliminate the power quality problems associated with the overall power system stability, via effectively reducing the harmonic content and operating at unity power factor. Therefore the emerging PHEV loads that will get connected to the existing distribution systems problem will be reduced to a basic capacity problem.

ACKNOWLEDGMENT

This work has been supported in part by the U.S. National Science Foundation under Grant number 0801860, which is greatly acknowledged.

APPENDIX

The state-space model of the system with PHEV loads can be expressed as;

$$\dot{x}(t) = Ax + Bu \qquad (15)$$
$$y(t) = Cx + Du$$

where the state-space matrices are obtained as:

$$
C = \begin{bmatrix}
0 & 0 & 0 & 0 & 0 & 0 & 0 & 0 & 0 & 0 & 0 & 0 & 0 & 0 & 0 & 0 & 0 & 0 & 0 & 0 & 0 & 0 & 0 & 0 & 0 \\
0 & 0 & 0 & 0 & 0 & 0 & 0 & 0 & 0 & 0 & 0 & 0 & 0 & 0 & 0 & 0 & 0 & 0 & 0 & 0 & 0 & 0 & 0 & 0 & 0 \\
0 & 0 & 0 & 0 & 0 & 0 & 0 & 0 & 0 & 0 & 0 & 0 & 0 & 0 & 0 & 0 & 0 & 0 & 0 & 0 & 0 & 0 & 0 & 0 & 0 \\
0 & 0.0039 & 0 & 0 & 0 & 0 & 0 & 0 & 0 & 0 & 0 & 0 & 0 & 0 & 0 & 0 & 0 & 0 & 0 & 0 & 0 & 0 & 0 & 0 & 0 \\
0 & 0.0130 & 0 & 0 & 0 & 0 & 0 & 0 & 0 & 0 & 0 & 0 & 0 & 0 & 0 & 0 & 0 & 0 & 0 & 0 & 0 & 0 & 0 & 0 & 0 \\
0 & 0.0130 & 0 & 0 & 0 & 0 & 0 & 0 & 0 & 0 & 0 & 0 & 0 & 0 & 0 & 0 & 0 & 0 & 0 & 0 & 0 & 0 & 0 & 0 & 0 \\
0 & 0.0130 & 0 & 0 & 0 & 0 & 0 & 0 & 0 & 0 & 0 & 0 & 0 & 0 & 0 & 0 & 0 & 0 & 0 & 0 & 0 & 0 & 0 & 0 & 0 \\
0 & 0.0130 & 0 & 0 & 0 & 0 & 0 & 0 & 0 & 0 & 0 & 0 & 0 & 0 & 0 & 0 & 0 & 0 & 0 & 0 & 0 & 0 & 0 & 0 & 0 \\
0 & 0.0130 & 0 & 0 & 0 & 0 & 0 & 0 & 0 & 0 & 0 & 0 & 0 & 0 & 0 & 0 & 0 & 0 & 0 & 0 & 0 & 0 & 0 & 0 & 0 \\
0 & 0.0130 & 0 & 0 & 0 & 0 & 0 & 0 & 0 & 0 & 0 & 0 & 0 & 0 & 0 & 0 & 0 & 0 & 0 & 0 & 0 & 0 & 0 & 0 & 0 \\
0 & 0.0039 & 0 & 0 & 0 & 0 & 0 & 0 & 0 & 0 & 0 & 0 & 0 & 0 & 0 & 0 & 0 & 0 & 0 & 0 & 0 & 0 & 0 & 0 & 0 \\
\vdots & & & & & & & & & & & & & & & & & & & & & & & & \vdots \\
0 & 0 & 0 & 0 & 0 & 0 & 0 & 0 & 0 & 0 & 0 & 0 & 0 & 0 & 0 & 0 & 0 & 0 & 0 & 0 & 0 & 0 & 0 & 0 & 0
\end{bmatrix}
$$

$$
D = 10^3 \times \begin{bmatrix}
-0.0166 & -0.0166 & -0.0166 & 0 & 0 & 0 & 0 & 0 & 0 & 0 & 0 & 0 & 0 & 0 & 0 & 0 \\
-0.0166 & -0.0166 & -0.0166 & 0 & 0 & 0 & 0 & 0 & 0 & 0 & 0 & 0 & 0 & 0 & 0 & 0 \\
-0.0166 & -0.0166 & -0.0166 & 0 & 0 & 0 & 0 & 0 & 0 & 0 & 0 & 0 & 0 & 0 & 0 & 0 \\
0 & 0 & 0 & 0.8626 & 0.4294 & 0.4294 & 0.4294 & 0.4294 & 0.4294 & 0.4294 & 0.8626 & 0 & 0 & 0 & 0 & 0 \\
0 & 0 & 0 & 0.4294 & -1.8879 & 1.4455 & 1.4455 & 1.4455 & -1.8879 & 1.4455 & 0.4294 & 0 & 0 & 0 & 0 & 0 \\
0 & 0 & 0 & 0.4294 & 1.4455 & -1.8879 & -1.8879 & -1.8879 & 1.4455 & -1.8879 & 0.4294 & 0 & 0 & 0 & 0 & 0 \\
0 & 0 & 0 & 0.4294 & 1.4455 & -1.8879 & 1.4455 & -1.8879 & 1.4455 & 1.4455 & 0.4294 & 0 & 0 & 0 & 0 & 0 \\
0 & 0 & 0 & 0.4294 & 1.4455 & -1.8879 & -1.8879 & -1.8879 & 1.4455 & -1.8879 & 0.4294 & 0 & 0 & 0 & 0 & 0 \\
0 & 0 & 0 & 0.4294 & -1.8879 & 1.4455 & 1.4455 & 1.4455 & -1.8879 & 1.4455 & 0.4294 & 0 & 0 & 0 & 0 & 0 \\
0 & 0 & 0 & 0.4294 & 1.4455 & -1.8879 & -1.8879 & -1.8879 & 1.4455 & -1.8879 & 0.4294 & 0 & 0 & 0 & 0 & 0 \\
0 & 0 & 0 & 0.8626 & 0.4294 & 0.4294 & 0.4294 & 0.4294 & 0.4294 & 0.4294 & -0.8626 & 0 & 0 & 0 & 0 & 0 \\
0 & 0 & 0 & 0 & 0 & 0 & 0 & 0 & 0 & 0 & 0 & 0 & 0 & 0 & 0 & 0 \\
0 & 0 & 0 & 0 & 0 & 0 & 0 & 0 & 0 & 0 & 0 & 0 & 0 & 0 & 0 & 0 \\
0 & 0 & 0 & 0 & 0 & 0 & 0 & 0 & 0 & 0 & 0 & 0 & 0 & 0 & 0 & 0 \\
0 & 0 & 0 & 0 & 0 & 0 & 0 & 0 & 0 & 0 & 0 & 0 & 0 & 0 & 0 & 0 \\
0 & 0 & 0 & 0 & 0 & 0 & 0 & 0 & 0 & 0 & 0 & 0 & 0 & 0 & 0 & 0 \\
0 & 0 & 0 & -0.8587 & 0.4424 & 0.4424 & 0.4424 & 0.4424 & 0.4424 & -0.4424 & -0.8587 & 0 & 0 & 0 & 0 & 0 \\
-0.0166 & -0.0166 & -0.0166 & 0 & 0 & 0 & 0 & 0 & 0 & 0 & 0 & 0 & 0 & 0 & 0 & 0 \\
-0.0166 & -0.0166 & -0.0166 & 0 & 0 & 0 & 0 & 0 & 0 & 0 & 0 & 0 & 0 & 0 & 0 & 0 \\
-0.0166 & -0.0166 & -0.0166 & 0 & 0 & 0 & 0 & 0 & 0 & 0 & 0 & 0 & 0 & 0 & 0 & 0 \\
0 & 0 & 0 & 0 & 0 & 0 & 0 & 0 & 0 & 0 & 0 & 0 & 0 & 0 & 0 & 0 \\
0 & 0 & 0 & 0 & 0 & 0 & 0 & 0 & 0 & 0 & 0 & 0 & 0 & 0 & 0 & 0 \\
-0.0166 & -0.0166 & -0.0166 & 0 & 0 & 0 & 0 & 0 & 0 & 0 & 0 & 0 & 0 & 0 & 0 & 0 \\
-0.0166 & -0.0166 & -0.0166 & 0 & 0 & 0 & 0 & 0 & 0 & 0 & 0 & 0 & 0 & 0 & 0 & 0 \\
-0.0166 & -0.0166 & -0.0166 & 0 & 0 & 0 & 0 & 0 & 0 & 0 & 0 & 0 & 0 & 0 & 0 & 0 \\
-0.0166 & -0.0166 & -0.0166 & 0 & 0 & 0 & 0 & 0 & 0 & 0 & 0 & 0 & 0 & 0 & 0 & 0 \\
\vdots & & & & & & & & & & & & & & & \vdots \\
0 & 0 & 0 & 0 & 0 & 0 & 0 & 0 & 0 & 0 & 0 & 0 & 0 & 0 & 0 & 0
\end{bmatrix}
$$

REFERENCES

[1] A. Emadi, M. Ehsani, and J. M. Miller, Vehicular electric power systems: land, sea, air, and space vehicles, New York: Mercek Dekker, Dec. 2003.

[2] B. Kramer, S. Chakraborty, and B. Kroposki, "A review of plug-in vehicles and vehicle-to-grid capability," in *Proc., 34th Annual Conference of IEEE Industrial Electronics (IECON)*, Nov. 2008, pp. 2278-2283, Florida, USA.

[3] W. D. Jones, "Take this car and plug it [plug-in hybrid vehicles]," *IEEE Spectrum*, vol. 42, no. 7, pp. 10-13, July 2005

[4] S. S. Williamson, "Electric drive train efficiency analysis based on varied energy storage system usage for plug-in hybrid electric vehicle applications," in *Proc., IEEE Power Electronics Specialists Conference (PESC)*, pp. 1515-1520, June 2007.

[5] J. Wu, A. Emadi, M. J. Douba, and T. P. Bohn, "Plug-in hybrid electric vehicles: Testing, simulations, and analysis," in *Proc., IEEE Vehicle Power and Propulsion Conference (VPPC)*, pp. 469-476, Texas, USA, Sept. 2007.

[6] A. Emadi, Y. J. Lee, and K. Rajashekara, "Power electronics and motor drives in electric, hybrid electric, and plug-in hybrid electric vehicles," *IEEE Transactions on Industrial Electronics*, vol. 55, no. 6, pp. 2237-2245, June 2008.

[7] X. Li and S. S. Williamson, "Efficiency and suitability analyses of varied drive train architectures for plug-in hybrid electric vehicle

(PHEV) application," in *Proc., IEEE Vehicle Power and Propulsion Conference (VPPC)*, pp. 1-6, Harbin, China, Sept. 2008.

[8] Y. Gao and M. Ehsani, "Design and control methodology of plug-in hybrid electric vehicles," in *Proc., IEEE Vehicle Power and Propulsion Conference (VPPC)*, pp. 1-6, Harbin, China, Sept. 2008.

[9] A. F. Burke, "Batteries and Ultracapacitors for electric, hybrid, and fuel cell vehicles," *Proceedings of the IEEE*, vol. 95, no. 4, pp. 806-820, April 2007.

[10] S. J. Moura, D. S. Callaway, H. K. Fathy, and J. L. Stein, "Impact of battery sizing on stochastic optimal power management in plug-in hybrid electric vehicle," in *Proc., IEEE International Conference on Vehicular Electronics and Safety*, pp. 96-102, Ohio, USA, Sept. 2008.

[11] R. Ghorbani, E. Bibeau, P. Zanetel, and A. Karlis, "Modeling and simulation of a series hybrid electric vehicle using REVS," in *Proc., American Control Conference*, pp. 4413-4418, New York, USA, July 2007.

[12] J. Voelcker, "Plugging away in a Prius," *IEEE Spectrum*, vol. 45, no. 5, pp. 30-48, May 2008.

[13] P. Anumolu, G. Banhazl, T. Hilgeman, and R. Pirich, "Plug-in hybrid vehicles: An overview and performance analysis," in *Proc., IEEE Systems, Applications, and Technology Conference*, pp. 1-4, May 2008, Long Island, USA.

[14] X. Hui and D. Yunbo, "The study of plug-in hybrid electric vehicle power management strategy simulation," in *Proc., IEEE Vehicle Power and Propulsion Conference (VPPC)*, Harbin, China, pp. 1-6, Sept. 2008.

[15] G. B. Shrestha and S. G. Ang, "A study of electric vehicle battery charging demand in the context of Singapore," in *Proc., International Power Engineering Conference (IPEC)*, pp. 64-69, Dec. 2007.

[16] "Technology primer: the plug-in hybrid electric vehicle," Technical Report by Electric Power Research Institute (EPRI), pp. 1-2, 2007.

[17] L. Sanna, "Driving the solution, the plug-in hybrid vehicle," *EPRI Journal*, pp. 8-17, Fall 2005.

[18] K. Clement, E. Haesen, and J. Driesen, "Coordinated charging of multiple plug-in hybrid electric vehicles in residential distribution grids," in *Proc., IEEE Power Systems Conference and Exposition*, pp. 1-7, Washington, USA, March 2009.

[19] M. D. Galus and G. Andersson, "Demand management of grid connected plug-in hybrid electric vehicles (PHEV)," in *Proc., IEEE Energy 2030 Conference*, pp. 1-8, Georgia, USA, Nov. 2008.

[20] A. Ipakchi and F. Albuyeh, "Grid of the future," *IEEE Power and Energy Magazine*, vol. 7, no. 2, pp. 52-62, March-April 2009.

[21] C. Roe, F. Evangelos, J. Meisel, A. P. Meliopoulos, and T. Overbye, "Power system level impacts of PHEVs," in *Proc., 42nd Hawaii International Conference on System Sciences (HICSS)*, pp. 1-10, Jan. 2009, Hawaii, USA.

[22] C. Roe, F. Evangelos, J. Meisel, A. P. Meliopoulos, and T. Overbye, "Power system level impacts of plug-in hybrid electric vehicles using simulation data," in *Proc., IEEE Energy 2030 Conference*, pp. 1-6, Georgia, USA, Nov. 2008.

[23] Y. Lee, A. Khaligh, and A. Emadi, "Advanced Integrated Bi-Directional AC/DC and DC/DC Converter for Plug-in Hybrid Electric Vehicles," *IEEE Transactions on Vehicular Technology,* vol. 58, Oct. 2009, pp. 3970-3980.

[24] S.H. Berisha, G. G. Karady, R. Ahmad, R. Hobbs, and D. Karner, "Current harmonics generated by electric vehicle battery chargers," in *Proc., International Conference on Power Electronics, Drives, and Energy Systems for Industrial Growth*, vol. 1, pp. 584-589, Jan. 1996.

[25] M. Basu, K. Gaughan, and E. Coyle, "Harmonic distortion caused by EV battery chargers in the distribution systems network and its remedy," in *Proc., 39th International Universities Power Engineering Conference (UPEC)*, vol. 2, pp. 869-873, Sept. 2004.

[26] J. C. Gomez and M. M. Morcos, "Impact of EV battery chargers on the power quality of distribution systems," *IEEE Transactions on Power Delivery*, vol. 18, no. 3, pp. 975-981, July 2003.

[27] M. Bojruo, P. Karlsson, M. Alakula, and B. Simonsson, "A dual purpose battery charger for electric vehicles," in *Proc., Power Electronics Specialists Conference (PESC)*, vol. 1, pp. 565-570, May 1998, Fukuoka, Japan.

[28] M. S. Duvall, "Battery evaluation for plug-in hybrid electric vehicles," in *Proc., IEEE Vehicle Power and Propulsion Conference (VPPC)*, pp. 1-6, Sept. 2005.

[29] S. W. Hadley, "Evaluating the impact of plug-in hybrid electric vehicles on regional electricity supplies," in *Proc., iREP Symposium on Bulk Power System Dynamics and Control - VII. Revitalizing Operational Reliability*, pp. 1-12, August 2007, South Caroline, USA.

[30] F. Lambert, "Secondary distribution impacts of residential electric vehicle charging," *Public Interest Energy Res. (PIER), California Energy Commission*, Project no. 373, 1999.

[31] M. M. Morcos, C. R. Mersman, G. D. Sugavanam, and N. M. Diliman, "Battery chargers for electric vehicles," *IEEE Power Engineering Review*, vol. 20, no. 11, pp. 8-11, Nov. 2000.

[32] Y. V. Makarov, D.J. Hill, and J.V Milanovic, "Effect of load uncertainty on small disturbance stability margins in open-access power systems," in *Proc., Hawaii International Conference on System Sciences*, vol. 5, pp. 648-657, Jan 1997, Hawaii, USA.

[33] P.Kundur, Power System Stability and Control, New York: McGraw-Hill, 1994.

[34] P. Kundur, et. al., IEEE/CIGRE Joint Task Force on Stability Terms and Definitions, "Definition and classification of power system stability," *IEEE Transactions on Power Systems*, vol. 19, no. 2, pp. 1387-1401, May 2004.

[35] The US blackout timeline, *IEEE Power Engineer*, vol. 17, no. 5, pp. 11, Oct.-Nov. 2003.

[36] G. S. Vassell, "Northeast blackout of 1965," *IEEE Power Engineering Review*, vol. 11, no. 1, pp. 4, Jan. 1991.

[37] M. Reza, D. Sudarmadi, F. A. Viawan, W. L. Kling, and L van der Sluis, "Dynamic stability of power systems with power electronic interfaced DG," in Proc., IEEE Power Systems Conference and Exposition (PSCE), pp. 1423-1428, Oct-Nov. 2006.

[38] V. Marano and G. Rizzoni, "Energy and economic evaluation of PHEVs and their interaction with renewable energy sources and the power grid," in *Proc., IEEE International Conference on Vehicular Electronics and Safety*, pp. 84-89, Ohio, USA, Sept. 2008.

[39] M. Uzunoglu and M. S. Alam, "Dynamic modeling, design, and simulation of a combined PEM fuel cell and ultracapacitor system for stand-alone residential applications," IEEE Transactions on Energy Conversion, vol. 21, no. 3, pp. 767-775, Sept. 2006.

[40] C.S. Indulkar, B. Viswanathan, and S.S. Venkata, "Maximum power transfer limited by voltage stability in series and shunt compensated schemes for AC transmission systems," IEEE Transactions on Power Delivery, vol. 4, no. 2, pp. 1246-1252, April 1989.

[41] M. Uzunoglu, "Harmonics and voltage stability analysis in power systems including thyristor controlled reactor," Academic Proceedings in Engineering and Science - Sadhana, vol. 30, no. 1, pp. 57-67, Feb. 2005.

[42] M.K. Pal, "Voltage stability conditions considering load characteristics," IEEE Transactions on Power Systems, vol. 7, no. 1, pp. 243-249, Feb. 1992.

Rapid Simulation of Fourth-Order Multi-Resonant LLCC Converters with Capacitive Output Filter

A. Bucher   T. Duerbaum

Chair of Electromagnetic Fields
Friedrich-Alexander-University Erlangen-Nuremberg
Erlangen, Germany
a.bucher@emf.eei.uni-erlangen.de

Abstract— **In case of resonant converters, designers have to face an analysis which is in general more cumbersome than in case of PWM converters, as often a multitude of modes has to be regarded and resonant oscillations with durations impossible to calculate directly are encountered. Therefore, approximations are required in order to facilitate the design process. An extended approximation in the frequency domain for the steady-state solution of the multi-resonant LLCC converter with capacitive output filter is presented in this paper, with a new degree of simplification regarding the resulting closed-form solution. By means of this approximation, the design of the fourth-order resonant LLCC converter is significantly simplified, providing a tool for rapid simulation combined with a very high accuracy. The accuracy of the investigated approach was double-checked by means of exact calculations.**

I. INTRODUCTION

A. Existing approaches

Concerning the analysis of resonant converters, a large number of publications has been published following different approaches in order to derive their steady-state solution [1-6]. In case of the fourth-order multi-resonant LLCC converter shown in Fig. 1, several resonant frequencies can be identified, with a high number of possible switching modes, making an exact analysis very tedious [7-9]. Furthermore, numerical problems are encountered as most often transcendental equations have to be solved in order to obtain the steady-state solution. Time-discrete models on the other hand can deliver the steady-state solution, with the disadvantage of high requirements regarding computing time and thus making parameter variations cumbersome.

In order to overcome the problems associated with solutions in the time domain, several authors have investigated approximations in the frequency domain, with the **First Harmonic Approximation** (FHA) from [10] as the most known approach. With regard to resonant converters of higher order, this approximation is suffering from inaccurate results, whereas the

basic characteristics of resonant converters can be well understood by means of this approach [11-13]. The origin of these inaccuracies can be found in the equivalent model of the output rectifier bridge, with the regular FHA deriving a simple equivalent resistor representing the diode bridge together with the output filter and the load. In order to increase the accuracy of this approximation, extensions have been published, deriving equivalent impedances for the nonlinear load (rectifiers + output filter + load) together with the parallel reactances of the resonant tank [14-16]. One of the first approaches [17] dealing with this approach for the LLCC limited this approach to frequencies with the parallel capacitor dominating the impedance across the input of the rectifier bridge. A more general approach was published in [18], resulting in very good agreements with measurements but with high numerical requirements. Based on the latter approach a refined closed-form solution is derived in this paper, resulting in only one transcendental equation. These findings significantly reduce the numerical complexity of the corresponding implementation within a design program. For regular converter designs, the derived solution is valid for a very wide frequency range.

B. Assumptions

In order to analyze the converter, simplifications and assumptions are necessary. All semiconductors and passives are assumed to be ideal. The MOSFET bridge is driven with a duty-cycle of 50 %, neglecting dead times. All passive components are assumed to be loss-free. The actual transformer is modeled by a cantilever model [19] with its equivalent components L_s, L_h and n. The voltage of the parallel resonant tank across C_p is rectified by a full-wave rectifier,

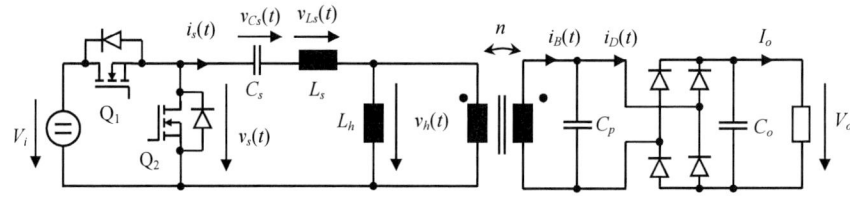

Figure 1. Schematic of the fourth-order multi-resonant LLCC converter with capacitive output filter, driven by a half-bridge MOSFET configuration.

978-1-4244-4782-4/10 $26.00 © 2010 IEEE

with the output capacitor assumed to be large enough for V_o to be ripple-free. For the operation of the converter, only modes with two subintervals are considered, which in fact covers the most part of the reasonable operating points of the converter.

C. Normalization

In order to simplify the obtained results, a normalization is carried out. Due to this normalized representation, the derived results can be easily applied to any converter design. Following the nomenclature of previous authors, normalized quantities are represented by replacing their letter while keeping the same subscript. The letter V denoting voltages is replaced by the letter M, I denoting currents is replaced by J. Table I shows the resulting normalized terms. The normalization voltage can be chosen freely, but if V_i is selected as normalization voltage, the dc conversion ratio $M = V_o/V_i$ is automatically included in the derived results.

II.   THE EXTENDED FIRST HARMONIC APPROXIMATION

A. Basic idea

Following the ideas of the regular FHA, it can be stated that for a large amount of operating points the series resonant tank current $i_s(t)$ is quasi-sinusoidal. An actual measurement of input waveforms with a switching frequency above the resonant frequency determined by L_s and C_s in case of the LLCC is shown in Fig. 2 (a), with the major part of the energy at the input indeed being transferred by the fundamental frequency. In contrast to that, the calculated waveforms at the converter's output are given in Fig. 2 (b), with a clipped voltage waveform typical for resonant converters with a parallel resonant capacitance across the diode bridge. As can clearly be seen, the output current $i_D(t)$ delivering energy to the load is by far not sinusoidal for capacitive output filters as illustrated in Fig. 2 (b). However, this is an assumption the regular FHA depends on. Furthermore, a significant phase angle between the fundamental components of the involved parallel voltage $v_p(t)$ and the output current $i_D(t)$ is observed, with the regular FHA completely neglecting this phase shift. Thus the proposed approach focuses on this phase shift in

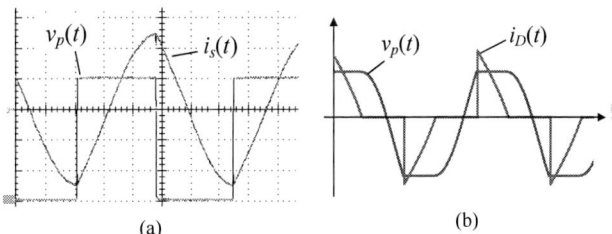

Figure 2.   LLCC waveforms; (a) measured input waveforms, (b) schematic output waveforms

order to model the interaction between the resonant tank and the nonlinear output stage more accurately as demonstrated in [18]. Thus, instead of an equivalent resistance, an equivalent impedance $\underline{Z}_{ac}$ is derived in order to model the nonlinear rectifier in combination with the load.

B. Analysed mode

With regard to the derivation of this equivalent impedance $\underline{Z}_{ac}$, a time domain analysis has to be carried out in the first place. The simplified equivalent converter model for this analysis is given in Fig. 3. As discussed above, the eFHA relies on the same assumption of a sinusoidal current $i_s(t)$ in the series resonant tank with

$$i_s(t) = \hat{I}_s \sin(\omega_s t - \nu) \tag{1}$$

as in case of the regular FHA, with the angle ν representing the phase shift of the input current $i_s(t)$ vs. the parallel voltage $v_p(t)$. Under this assumption, the switching action of the input bridge is directly modeled by the first harmonic of the input current. Thus only two switching states can occur within the eFHA which are determined by the output rectifiers. If energy is transferred to the output, two of the four rectifier diodes conduct, resulting in the equivalent circuit of Fig. 4. In between two switching states with conducting rectifiers, a resonant subinterval has to occur in order to recharge the parallel capacitance C_p from $+V_o$ to $-V_o$ or the other way round. During this subinterval all four diodes are reverse-biased as illustrated in Fig. 5. The waveform of the parallel voltage $v_p(t)$ during this subinterval is given by

$$v_p(t) = A\sin(\omega_{0p}t - \theta) + B\cos(\omega_s t - \nu) \qquad 0 \le t \le t_1 \tag{2}$$

with

$$B = n/C_p \cdot \hat{I}_s \omega_s / (\omega_{0p}^2 - \omega_s^2). \tag{3}$$

The current $i_p(t)$ flowing through the parallel capacitor is given

TABLE I   CIRCUIT AND NORMALIZED TERMS

Circuit variable	Symbol	Normalized variable
Characteristic impedance R_0	$R_0 = \sqrt{L_h/(C_p/n^2)}$	
Series resonant frequency f_{0s}	$\omega_{0s} = 1/\sqrt{L_s C_s}$	
Parallel resonant frequency f_{0p}	$\omega_{0p} = 1/\sqrt{L_h C_p/n^2}$	
Output voltage	V_o	$M_o = V_o/V_i$
Output current	I_o	$J_o = I_o Z_0/V_i$
Switching frequency	$f_s = \omega_s/(2\pi)$	$F = f_s/f_{0p}$.
Load resistor	R_L	$Q = R_L/Z_0$

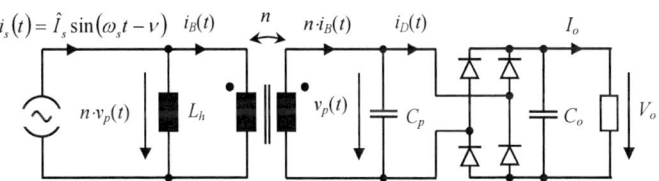

Figure 3.   Schematic of the fourth-order multi-resonant LLCC converter with capacitive output filter, driven by a half-bridge MOSFET configuration

978-1-4244-4782-4/10 $26.00 © 2010 IEEE

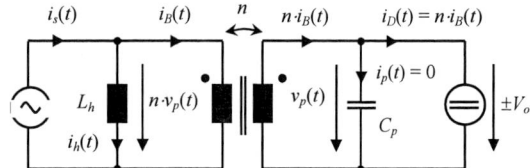

Figure 4. Equivalent circuit during subintervals with clamped parallel voltage $v_p(t) = \pm V_o$.

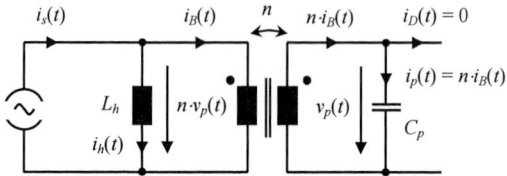

Figure 5. Equivalent circuit during subintervals with clamped parallel voltage $v_p(t) = \pm V_o$.

by the derivation of (2) with

$$i_p(t) = C_p\left[A\omega_{0p}\cos(\omega_{0p}t - \theta) - B\omega_s\sin(\omega_s t - \nu)\right] \quad 0 \leq t \leq t_1 \,.(4)$$

The magnetizing current $i_h(t)$ can be expressed in terms of the source current $i_s(t)$ and the primary transformer current with

$$i_h(t) = i_s(t) - i_B(t). \tag{5}$$

During subintervals with conducting diode bridge, the parallel voltage is clamped to the output voltage, resulting in

$$v_p(t) = -V_o \qquad t_1 \leq t \leq T_s/2 \tag{6}$$

and

$$i_p(t) = 0 \qquad t_1 \leq t \leq T_s/2. \tag{7}$$

The magnetizing current $i_h(t)$ during this subinterval is a linear function with

$$i_h(t) = i_h(t_1) - nV_o \cdot (t - t_1)/L_h \qquad t_1 < t \leq T_s/2. \tag{8}$$

The analysis discussed in this paper is limited to one mode consisting of four subintervals. This mode is most commonly encountered for a large range of configurations. Other possible modes, especially for switching frequencies below f_{0s} with a higher number of subintervals are neglected. For switching frequencies above the resonant frequency f_{0s} with the desirable feature of ZVS this mode is most likely to occur for regular converter designs. Therefore it is sufficient to limit the analysis to the aforementioned two subintervals, since the following two subintervals are anti-symmetric to the first two. Nevertheless, the described methodology can be applied to other operating points consisting of more than four subintervals accordingly.

C. Steady-State analysis

The complete eFHA waveforms concerning the analyzed operating points are shown in Fig. 6. In order to derive the equivalent impedance $\underline{Z}_{ac}$ by which the series branch is loaded with, the amplitude $\hat{I}_s$ of the current source driving the resonant tank is assumed to be known in the first step. First of all, the unknown parameters A, θ and ν have to be expressed in terms of the given variables. In this context it has to be noted that the time reference t_0 of the analysis is chosen as the moment in which the resonant capacitor C_p starts to be discharged from $v_p(0) = V_o$. Thus the first equation describing the closed-form solution can be identified based on (2) with

$$v_p(0^-) = v_p(0^+) \quad \Rightarrow \quad A\sin\theta = B\cos\nu - V_o. \tag{9}$$

The second subinterval ends at t_1 when $v_p(t)$ reaches the negative value of the output voltage, starting a clamped subinterval in the opposite direction with

$$v_p(t_1^-) = -V_o \Rightarrow A\sin(\omega_{0p}t_1 - \theta) + B\cos(\omega_s t_1 - \nu) = -V_o. \tag{10}$$

In order to simplify the resulting equations, the switching frequency f_s is normalized with respect to the resonant frequency f_{0p} of the parallel tank with the primary reflected parallel capacitance C_p/n^2 as shown in Table I. As illustrated in Fig. 6, the duration of the resonant subinterval is depicted as $\varepsilon = \omega_s t_1$ and (10) can be rewritten as

$$A\sin(\varepsilon/F - \theta) + B\cos(\varepsilon - \nu) = -V_o. \tag{11}$$

The second state variable describing the eFHA waveforms is the magnetizing current $i_h(t)$ defined in (5), which has to be steady at $t = t_1$, resulting in

$$i_{h1} = \hat{I}_s\sin(\varepsilon - \nu) - \\ - C_p/n \cdot \left[A\omega_{0p}\cos(\varepsilon/F - \theta) - B\omega_s\sin(\varepsilon - \nu)\right]. \tag{12}$$

The switching condition of the network is given by the rectifier current $i_D(t)$. As illustrated in Fig. 6, the output current delivering energy to the load vanishes at $t = 0$ or $t = T_s/2$, respectively. Based on this condition a further

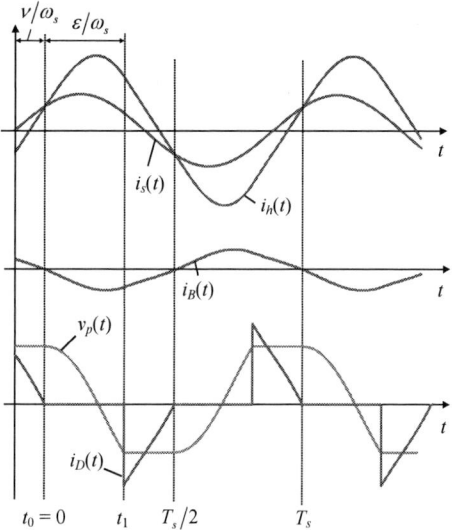

Figure 6. eFHA waveforms

978-1-4244-4782-4/10 $26.00 © 2010 IEEE

equation can be derived with

$$i_B\left(0^+\right) = 0 \quad \Rightarrow \quad A = -BF\sin v/\cos\theta. \tag{13}$$

Substitution of (13) in (9) yields

$$\theta = \arctan\left\{\left[V_o/(B\cos v)-1\right]/F\tan v\right\}. \tag{14}$$

Taking (9), (13) and (14) into account, (11) can be simplified

$$A\left[\sin(\varepsilon/F)\cos\theta - \cos(\varepsilon/F)\sin\theta\right] + B\cos(\varepsilon-v) = -V_o$$

$$\Rightarrow \quad \frac{V_o}{B} = \frac{F\sin v\sin(\varepsilon/F)+\cos(\varepsilon/F)\cos v - \cos(\varepsilon-v)}{\cos(\varepsilon/F)+1}. \tag{15}$$

Thus, for a given eFHA source current amplitude defining B in (3), the output voltage can be calculated by means of (15). The quotient V_o/B can now be substituted (14) and the angle θ is obtained as a function of the normalized switching frequency, the duration of the resonant subinterval ε as well as the phase angle v between $v_p(t)$ and $i_s(t)$ with

$$\theta = \arctan\left[\frac{-(1+\cos\varepsilon)/\tan v + F\sin(\varepsilon/F)-\sin\varepsilon}{F(\cos(\varepsilon/F)+1)}\right]. \tag{16}$$

In order to describe the closed-form solution of the investigated eFHA, the angles ε and v remain as the last unknown variables, if the amplitude $\hat{I}_s$ of the source current is assumed to be known. For the purpose of eliminating one of these unknown angles, a last equation is derived based on the switching condition at $t = T_s/2$ with

$$i_B\left(T_s/2^-\right) = 0 \quad \Rightarrow \quad \hat{I}_s\sin(\pi-v) = i_{h1} - \frac{nV_o}{\omega_s L_h}(\pi-\varepsilon). \tag{17}$$

Together with (3), (12), and (16) one obtains

$$\sin v\left[1 - F^2(1+\cos(\varepsilon/F))\right] - \sin(\varepsilon-v) + \\ + V_o/B\cdot\left[\pi-\varepsilon-F\sin(\varepsilon/F)\right]+F\sin(\varepsilon/F)\cos v = 0. \tag{18}$$

Substitution of (15) in (18) yields

$$v = \arctan(N/Z) \tag{19}$$

with $\quad K = \left[\pi-\varepsilon-F\sin(\varepsilon/F)\right]/\left[\cos(\varepsilon/F)+1\right]$ and

$N = -F\sin(\varepsilon/F)+\sin\varepsilon - K\left[\cos(\varepsilon/F)-\cos\varepsilon\right]$ and

$Z = 1-F^2\left(1+\cos\left(\dfrac{\varepsilon}{F}\right)\right)+\cos\varepsilon + K\left[F\sin\left(\dfrac{\varepsilon}{F}\right)-\sin\varepsilon\right].$

As shown in (19), the phase angle v is solely a function of the normalized switching frequency and the duration of the resonant subinterval ε. Thus, the steady-state of the converter can be described for a given combination of ε and $\hat{I}_s$. The derivation of these parameters for a given converter configuration and operating point is discussed in the next sections.

D. Equivalent impedance $\underline{Z}_{ac}$

As discussed in the previous sections, the steady-state waveforms under eFHA assumptions are determined by the angle ε and the amplitude $\hat{I}_s$ of the driving current source $i_s(t)$. Following the methodology described in [14-16,18], an equivalent impedance is derived in the next step, representing all parallel components (L_h, C_p/n^2) as well as the nonlinear rectifier bridge together with the load resistor. With a sinusoidal series current as the basic eFHA assumption, the active power delivered by the input is transferred by the first harmonics of $nv_p(t)$ and $i_s(t)$ only. Thus the impedance $\underline{Z}_{ac}$ can be expressed as

$$\underline{Z}_{ac} = n\cdot\hat{\underline{V}}_{p(1)}/\hat{\underline{I}}_{s(1)} \quad \text{with} \quad \hat{\underline{I}}_s = -j\hat{I}_s\mathrm{e}^{-jv}, \tag{20}$$

with the index 1 denoting the phasor of the first harmonic. The phasor of the first harmonic of the series current $i_s(t)$ is directly given by the definition of the corresponding waveform in (1). The phasor of the parallel voltage $v_p(t)$ has yet to be determined. A Fourier series expansion is carried out, neglecting all higher harmonics of the parallel voltage similar to the regular FHA. Nevertheless, the phase shift between the two phasors determining $\underline{Z}_{ac}$ in (20) is taken into account. Considering the anti-symmetric waveform of the converter's state variables, one obtains

$$\hat{\underline{V}}_{p(1)} = \frac{4}{T_s}\left\{\int_0^{t_1} A\sin(\omega_p t-\theta)\cdot\mathrm{e}^{-j\omega_s t}\mathrm{d}t \\ + \int_0^{t_1} B\cos(\omega_s t-v)\cdot\mathrm{e}^{-j\omega_s t}\mathrm{d}t + \int_{t_1}^{T_s/2}-V_o\cdot\mathrm{e}^{-j\omega_s t}\mathrm{d}t\right\} \tag{21}$$

These three integrals can be solved in terms of the unknown duration of the resonant subinterval ε. The final result can be obtained in terms of

$$\underline{Z}_{ac} - f(\varepsilon, v, F, R_0) \tag{22}$$

with R_0 representing the characteristic impedance of the parallel tank which is given by

$$R_0 = \sqrt{L_h/\left(C_p/n^2\right)} = 2\pi f_{0p}L_h = \left(\omega_{0p}C_p/n^2\right)^{-1}. \tag{23}$$

Keeping (19) in mind, the angle v can be expressed in terms of ε, thus it is found that the equivalent eFHA impedance $\underline{Z}_{ac}$ given in (22) is a function of R_0, the normalized switching frequency F as well as duration of the resonant subinterval represented by the angle ε alone. The influence of the series tank is not directly visible, as it is inherently included in the derivation of ε. In order to account for the impact of the series tank on the input impedance of the converter, a frequency domain analysis follows as described in the next section.

E. Frequency domain analysis

Based on (22) and the discussed eFHA assumptions, the equivalent circuit shown in Fig. 7 is obtained. This equivalent circuits represents the complete converter in the frequency domain. The amplitude $\hat{I}_s$ of the source current is excited by the first harmonic of the square-wave voltage $v_s(t)$ generated by the half-bridge shown in Fig. 1. The input impedance limiting this current consists of the series tank as well as the derived equivalent impedance $\underline{Z}_{ac}$ representing the parallel tank together with the nonlinear load. In order to analyze a given operating point, this amplitude has to be calculated. The amplitude of the phasor of the first harmonic of $v_s(t)$ is given by a Fourier series expansion of a square-wave signal with

$$\left|\underline{\hat{V}}_s\right| = 4V_g/\pi . \tag{24}$$

The amplitude V_g of the input square wave voltage is identical to the DC input voltage in case of a full-bridge configuration, in case of a half bridge switch configuration $V_g = V_i/2$ is obtained, with the DC component of $v_s(t)$ as additional voltage drop across the series capacitor C_s. The analysis of the equivalent circuit of Fig. 7 finally yields an expression for the current amplitude $\hat{I}_s$ as a function of the duration of the resonant subinterval ε with

$$\hat{I}_s = \left|\underline{\hat{I}}_s\right| = \left(4V_g/\pi\right)/\left|j\omega_s L_s + 1/\left(j\omega_s C_s\right) + \underline{Z}_{ac}\right| \tag{25}$$

Eq. (25) can be rewritten in normalized form as

$$\hat{J}_s = \dfrac{4/\pi}{\left|jF\lambda + \dfrac{1}{jF\zeta} + \dfrac{\underline{Z}_{ac}}{R_0}\right|} \quad \text{with} \quad \lambda = \dfrac{L_s}{L_h}, \quad \zeta = \dfrac{C_s}{C_p/n^2} . \tag{26}$$

Substituting (19) and (22) in (26), one obtains the missing link to the steady-state analysis of the eFHA within the time domain. As already mentioned, the converter's output voltage and the input current are linked by (15). This equation is thus used in order to derive the last unknown variable ε.

F. Output plane

The output plane of a resonant converter [6] is a graphical representation of the possible combination of given output voltage M_o vs. the corresponding output current. Keeping in mind that M_o is a given variable from this point of view, (17) can be rearranged as

$$\hat{J}_s = \dfrac{nM_o\left(1/F - F\right)\cdot\left[\cos\left(\varepsilon/F\right) + 1\right]}{F\sin\nu\sin\left(\varepsilon/F\right) + \cos\left(\varepsilon/F\right)\cos\nu - \cos\left(\varepsilon - \nu\right)} . \tag{27}$$

By substitution of (27) in (26), the steady-state solution describing the converter's output is found. Since M_o is known in terms of the output plane, a transcendental equation is found for the duration of the resonant subinterval ε. A solution for this equation has to be found in the range of $0 \le \varepsilon \le \pi$ by numerical means. Since the range of valid values for ε is limited, the solution of the obtained transcendental equation can be easily found.

In order to derive the output plane as diagram describing the converter's output characteristic it is necessary to calculate the DC output current I_o by integrating the output current $i_D(t)$. As there cannot be a DC current through the output filter capacitor C_o one obtains

$$J_o = \dfrac{n}{\pi}\Big\{\hat{J}_s\left[\left(\pi - \varepsilon\right)\sin\nu - \cos\nu - \cos\left(\varepsilon - \nu\right)\right] + nM_o\left(\pi - \varepsilon\right)^2/\left(2F\right)\Big\} . \tag{28}$$

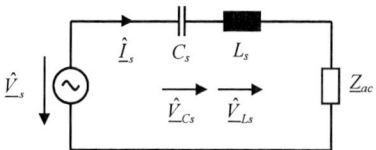

Figure 7. Eqivalent frequency domain circuit.

With the determined value for the angle ε, the amplitude of the source current is calculated by means of (26). With a known value for M_o, J_o can then be calculated using (28).

G. Control plane

In terms of controlling the output voltage within resonant converters, the switching frequency of the input bridge is used in order to adapt the converter's characteristic to input voltage or load variations. Thus the dc characteristics of resonant converters are often illustrated by means of the resulting output voltage M_o for a given load resistance vs. switching frequency F. Keeping the chosen normalization in mind, one obtains

$$J_o = M_o/Q \quad \text{with} \quad Q = R_L/R_0 . \tag{29}$$

Equation (29) together with (27) and (19) can be substituted in (28) resulting in a transcendental equation for the angle ε in terms of the control plane. For a given value of the load represented by Q, the corresponding output voltage for this load value can then be calculated by means of (27) in combination with (26) which has to be solved for M_o.

III. VERIFICATION

As the investigated approach in this paper is based on ideal assumptions, ideal analysis methods were used in order to evaluate its accuracy. Furthermore, measurements were carried out which, of course, show differences as losses occur in a prototype that are not accounted for within the described calculation.

A. Output plane

The exact solution with respect to the output plane for the ideal multiresonant LLCC converter shown in Fig. 1 was obtained by means of a time-domain analysis. For this purpose, several different modes have to be analyzed, with each mode having a set of transcendental equations that has to be solved in order to obtain the steady-state solution. It is evident that this approach is by far more time-consuming than the investigated eFHA. An exemplary comparison of output planes for $\xi = \lambda = 1.5$ is shown in Fig. 8 with the corresponding curves for the angle ε in Fig. 9.

For all displayed frequencies, the obtained eFHA results are very close to the exact solution. In case of the predicted no load voltage with $J_o = 0$, the eFHA calculates a lower value. The corresponding value for the duration ε under no-load conditions is π as shown in Fig. 9, as no power is transferred to the output and thus the resonant subintervals continue for the complete switching cycle. For short-circuit conditions with $M_o = 0$ on the other hand, the predicted output current is in

978-1-4244-4782-4/10 $26.00 © 2010 IEEE

very good agreement with the results obtained by the exact time domain model. The duration of the resonant subintervals is small for points of operation with high output currents as the parallel capacitance is recharged quickly. Altogether it can be stated that the eFHA is a suitable tool in order to calculate the converter's output characteristics.

B. Control plane

The verification of the results in terms of the control plane was done by means of SPICE. The voltage conversion ratio as a function of the normalized switching frequency is shown in Fig. 10. Additionally, FHA results are displayed in order to compare these results to the obtained eFHA predictions. Depending on the load of the converter represented by the quality factor Q, notable differences occur between the regular FHA results and the exact results obtained by SPICE. In contrast to that, the obtained eFHA results are visibly closer to the exact solution. The same behaviour is observed in terms of other important design quantities such as the rms series inductor current shown in Fig. 11.

C. Measurements

A prototype was built in order to verify the derived eFHA results. The measured output plane is shown in Fig 12. For a fixed frequency, the measured output characteristic differs from the predicted one as losses occur within the prototype. For resonant converters, these losses can be approximated within the calculation routine as demonstrated in [20]. The corresponding corrected output plane is shown in Fig. 12. By means of this correction, conduction and diode losses can be approximatively taken into account. However, hytersesis losses are neglected. First improvements to the original prototype indicate that this loss mechanism plays a significant role. An upgrade from ferrite material with medium hf properties to a more suitable material grade showed a significant decrease in the total converter losses. Due to the strong nonlinearity of hysteresis losses, special attention has to be paid to their calculation as discussed in [21]. It will be investigated if core losses can be integrated into the eFHA steady-state calculation by means of this described equivalent

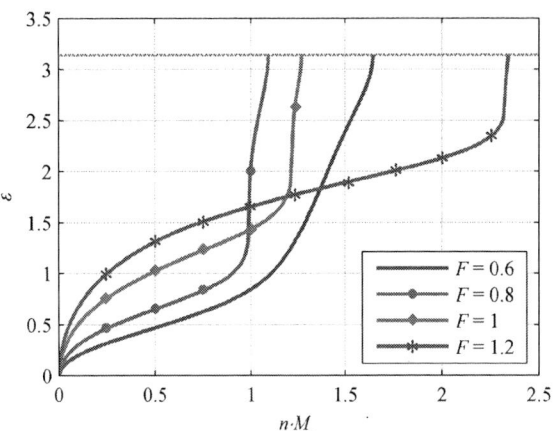

Figure 9. Duration ε of the resonant subintervals corresponding to the output plane shown in Figure 8 with $\xi = \lambda = 1.5$.

switching frequency approach. Furthermore, attempts will be taken in order to take winding losses into account as shown in [22,23].

CONCLUSIONS

An advanced method for the rapid calculation of multi-resonant converters is described in this paper. Offering a significantly reduced mathematical complexity compared to other approaches, this methodology is very well suited to be implemented within calculation routines for the design of resonant LLCC converters. Furthermore, numerical problems are minimized since the resulting closed-form solution can be simplified down to one transcendental equation. The accuracy of the results obtained by the investigated approach was double-checked in comparison to exact modeling approaches and a very good agreement was observed. Compared to measurements, differences are encountered, as losses occur within the prototype that are not yet taken into account. However, in order to increase the accuracy of the eFHA results with respect to practical designs, these losses should be included in the eFHA approach in order to achieve an even better correspondence with measurements.

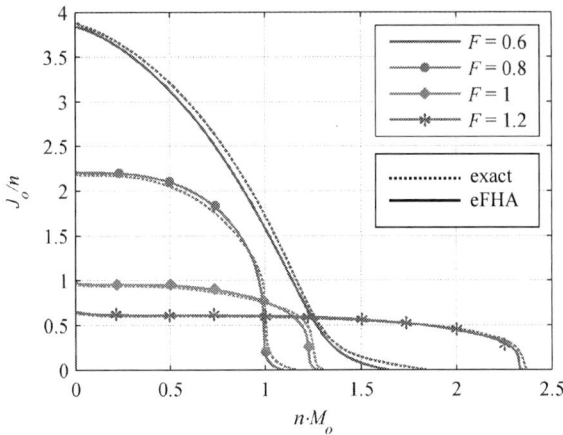

Figure 8. Comparison of eFHA vs. exact output plane with $\xi = \lambda = 1.5$.

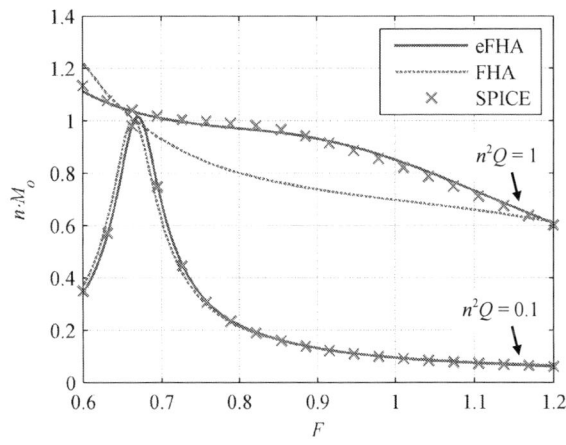

Figure 10. Comparison of eFHA vs. SPICE control plane with $\xi = \lambda = 1.5$.

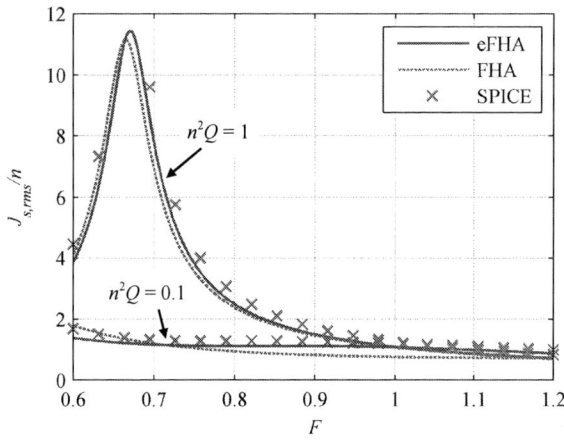

Figure 11. Comparison of eFHA vs. SPICE rms series inductor current $J_{s,rms}$ with $\xi = \lambda = 1.5$.

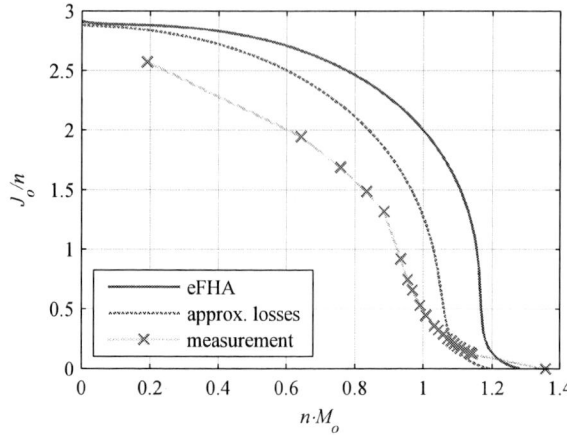

Figure 12. Measured output plane with $\lambda = 0.76$, $\xi = 2.02$ and $F = 1.01$.

REFERENCES

[1] A. Wittulski and R. Erickson, "Steady-state analysis of the series resonant converter," *IEEE Transactions on Aerospace and Electronic Systems*, vol. AES-21, no. 6, pp. 791–799, 1985.

[2] R. Oruganti, "State-plane analysis of resonant converters," Ph.D. dissertation, Virginia Polytechnic Institute and State University, 1987.

[3] M. Foster, H. Sewell, C. Bingham, D. Stone, D. Hente, and D. Howe, "High-speed analysis of resonant power converters," *IEE Proceedings - Electric Power Applications*, vol. 150, no. 1, pp. 62–70, 2003.

[4] I. E. Batarseh, "Analysis and design of high-order parallel resonant converters," Ph.D. dissertation, University of Illinois at Chicago, 1990.

[5] V. Vorperian, "Analysis of resonant converters," Ph.D. dissertation, California Institute of Technology, Pasadena, California, May 1984.

[6] S. Johnson and R. Erickson, "Steady-state analysis and design of the parallel resonant converter," *IEEE Transactions on Power Electronics*, vol. 3, no. 1, pp. 93–104, 1988.

[7] I. Batarseh and C. Lee, "Steady-state analysis of the parallel resonant converter with LLCC-type commutation network," *IEEE Transactions on Power Electronics*, vol. 6, no. 3, pp. 525–538, 1991.

[8] J. Hayes, M. Egan, J. Murphy, S. Schulz, and J. Hall, "Wide-load-range resonant converter supplying the SAE J-1773 electric vehicle inductive charging interface," *IEEE Transactions on Industry Applications*, vol. 35, no. 4, pp. 884–895, 1999.

[9] J.-H. Cheng and A. Witulski, "Analytic solutions for LLCC parallel resonant converter simplify use of two and three-element converters," *IEEE Transactions on Power Electronics*, vol. 13, no. 2, pp. 235–243, 1998.

[10] R. Steigerwald, "A comparison of half-bridge resonant converter topologies," *IEEE Transactions on Power Electronics*, vol. 3, no. 2, pp. 174–182, 1988.

[11] A. Bucher, T. Durbaum, D. Kubrich, and A. Stadler, "Comparison of different design methods for the parallel resonant converter," in *Proc. 12th International Power Electronics and Motion Control Conference EPE-PEMC 2006*, 2006, pp. 810–815.

[12] A. Bucher, T. Duerbaum, and D. Kuebrich, "Comparison of methods for the analysis of the parallel resonant converter with capacitive output filter," in *Proc. European Conference on Power Electronics and Applications*, 2007, pp. 1–10.

[13] A. Bucher, T. Duerbaum, D. Kuebrich, and S. Hoehne, "Multi-resonant LCC converter - comparison of different methods for the steady-state analysis," in *Proc. IEEE Power Electronics Specialists Conference PESC 2008*, 2008, pp. 1891–1897.

[14] G. Ivensky, A. Kats, and S. Ben-Yaakov, "An RC load model of parallel and series-parallel resonant dc-dc converters with capacitive output filter," *IEEE Transactions on Power Electronics*, vol. 14, no. 3, pp. 515–521, 1999.

[15] A. Forsyth, G. Ward, and S. Mollov, "Extended fundamental frequency analysis of the LCC resonant converter," *IEEE Transactions on Power Electronics*, vol. 18, no. 6, pp. 1286–1292, Nov. 2003.

[16] M. Foster, C. Gould, A. Gilbert, D. Stone, and C. Bingham, "Analysis of CLL voltage-output resonantconverters using describing functions," *IEEE Transactions on Power Electronics*, vol. 23, no. 4, pp. 1772–1781, 2008.

[17] J. Hayes and M. Egan, "Rectifier-compensated fundamental mode approximation analysis of the series parallel LCLC family of resonant converters with capacitive output filter and voltage-source load," in *Proc. 30th Annual IEEE Power Electronics Specialists Conference PESC 99*, vol. 2, 1999, pp. 1030–1036.

[18] Y. Ang, C. Bingham, M. Foster, D. Stone, and D. Howe, "Design oriented analysis of fourth-order LCLC converters with capacitive output filter," *IEE Proceedings -Electric Power Applications*, vol. 152, no. 2, pp. 310–322, 2005.

[19] R. Erickson and D. Maksimovic, "A multiple-winding magnetics model having directly measurable parameters," in *Proc. PESC 98 Record Power Electronics Specialists Conference 29th Annual IEEE*, vol. 2, 1998, pp. 1472–1478.

[20] A. Bucher, T. Duerbaum, D. Kuebrich, and M. Schmid, "Consideration of conduction losses for the series resonant converter by means of a simple extension to the SPA approach," in *Proc. 13th Power Electronics and Motion Control Conference EPE-PEMC 2008*, 2008, pp. 244–249.

[21] T. Duerbaum and M. Albach, "Core losses in transformers with an arbitrary shape of the magnetizing current," in *Proc. 6th European Power Electronics Conference EPE*, 1995, pp. 1171–1176.

[22] M. Albach, "Two-dimensional calculation of winding losses in transformers," in *Proc. IEEE 31st Annual Power Electronics Specialists Conference PESC 00*, vol. 3, 2000, pp. 1639–1644.

[23] M. Albach and H. Rossmanith, "The influence of air gap size and winding position on the proximity losses in high frequency transformers," in *Proc. PESC Power Electronics Specialists Conference 2001 IEEE 32nd Annual*, vol. 3, 2001, pp. 1485–1490.

FHA-Based Voltage Gain Function with Harmonic Compensation for LLC Resonant Converter

Hong Huang

Texas Instruments
50 Phillippe Cote Street
Manchester, NH 03101 USA

Abstract - **This paper discusses how to improve the accuracy of the gain expression obtained from the First Harmonic Approximation (FHA) for LLC resonant half bridge converter design. By reviewing the approximation, a gain compensation method is developed to improve the accuracy based on harmonic compensation. Theoretical analysis, computer simulation and bench test are presented to demonstrate the effectiveness of the proposed method.**

I. INTRODUCTION

Higher efficiency, higher power density and higher component density are now the trend and practice of power supply design and application. To adapt this trend, resonant power conversion, especially with the LLC resonant half bridge configuration, is receiving renewed interest due to its potential of achieving both higher switching frequency and lower switching losses. But designing such converters presents many challenges, among them the fact that the LLC resonant half bridge converter performs power conversion with frequency modulation, instead of pulse width modulation, requiring a different design approach.

Design method based on First Harmonic Approximation (FHA) [1 3] has successfully eased and greatly simplified the design process among these challenges. In fact, the voltage gain expression based on FHA plays an important role as a powerful tool in today's LLC resonant half bridge converter design. This expression shows the physical concept of each parameter in the formula and connects a frequency modulated switch-mode converter to the well developed sinusoidal ac circuit analysis and design. Recent published work [4-7] has further demonstrated that the FHA method is easy-to-handle and simple-to-follow when designing an LLC resonant half bridge converter, especially to initiate a new design. The FHA method makes design straight forward with physically meaningful steps.

However, recent work also suggests accuracy improvement for the FHA-based gain expression from circuit simulations and the bench measurements [6-7], although the accuracy issues are already indicated in publication [2] that

deviation exists from real measurement and can be more pronounced depending on input and output voltage and output power. Publication [7] has made efforts to enhance accuracy through transformer turns-ratio adjustment in order to adapt a transformer physical model. Publication [6] discloses improving the gain expression by considering transformer leakage inductance in order to accurately represent the converter circuit. This becomes a must for a design using magnetic integration. With these efforts, the result still presents observable differences compared to test result with either bench work or computer simulation [6]. The accuracy gets even worse in the so-called below-resonance region. To improve the accuracy, publication [6] makes further adjustment with attainable peak gain curves based on simulated result. While this is effective, the proposed methods become less easy-to-handle and simple-to-follow, and simulation work may become a necessity in a design process, because of the gain dependency of the working conditions [2]. In our recent evaluation, it is also found that gain varies with input voltage and output capacitance along with its ESR. This implies that gain correction cannot be made without assistance of simulation analysis. Although it is good to replace bench measurements with simulation, it is more desirable to get physical meaning from the expressions, to know where the approximations are from, and possibility to improve accuracy while keeping the simple expression format to reduce amount of simulation work, if an alternative gain expression or compensation method can be developed. This paper develops such a method by proposing a gain correction with harmonic compensation.

This paper is organized as following. Section I provides an introduction. Section II reviews currently used FHA-based gain expressions. Analysis is made using the FHA-based gain expressions and supported with simulation and bench test results to show the differences. Based on the theoretical analysis, simulation and bench test, possible improvements from gain compensation approaches are proposed in Section III. Following that, Section IV describes the compensated gain in explicit form. Section V presents the effectiveness of improving the gain accuracy by the proposed method. A 300W LLC half bridge converter is tested to show the gain

978-1-4244-4782-4/10 $26.00 © 2010 IEEE

accuracy improvement. Then the paper is concluded in Section VI with proposed future work.

II. REVIEW FHA-BASED GAIN FUNCTION EXPRESSION

Let's start with a review how the FHA-based gain function is obtained [1-7]. Figure 1 shows a simplified typical LLC resonant half bridge converter configuration. With the help of FHA method, the circuit shown in Figure 1 can be approximated into an equivalent sinusoidal ac circuit. Circuit in Figure 2 represents the LLC resonant half bridge converter in Figure 1 where

a) The resonant circuit composed of L_m, L_r, and C_r, receives a square voltage $v_{sq}(t)$;

b) The transformer, functioned as element for electrical insulation and the turns ratio, n, to deliver the required voltage level to the output, its primary winding receives a bipolar square wave voltage $v_{so}(t)$, and

c) The load R'_L is referred to the primary from the secondary including the load R_L in Figure 1 together with the losses from the transformer and output rectifiers.

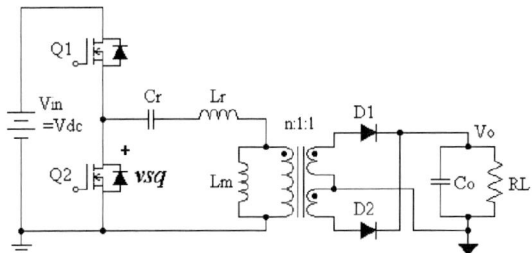

Figure 1. LLC resonant half bridge converter configuration.

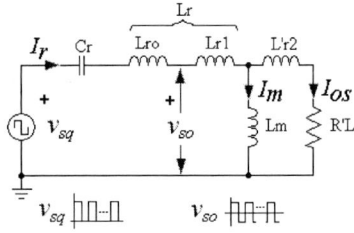

Figure 2. Switching circuit of the LLC resonant converter.

The both square voltage $v_{sq}(t)$ and $v_{so}(t)$ can be expanded in Fourier series. On the input side,

$$v_{sq}(t) = \frac{V_{dc}}{2} + \frac{2}{\pi} V_{dc} \sum_{k=1,3,5...} \frac{1}{k} \times \sin(k\omega t) \tag{1}$$

On the output side,

$$v_{so}(t) = \frac{4}{\pi} \times n \times V_o \sum_{k=1,3,5..} \frac{1}{k} \times \sin(k\omega t - \varphi_{vk}) \tag{2}$$

where φ_{vk} is the phase angle between $v_{sq}(t)$ and $v_{so}(t)$ of its k-th order of harmonic frequency component, and ω is switching frequency f_{sw} of its angular form,

$$\omega = \omega_{sw} = 2\pi f_{sw} \tag{3}$$

Depending on how to make approximation of the transformer leakage inductance, the resulted ac equivalent circuit may present two configurations, illustrated in Figures 3 and 4. Figure 3 shows the resulted equivalent ac circuit with consideration of the transformer leakage inductances while Figure 4 shows the ac equivalent circuit with the leakage inductance ignored. In Figures 3 and 4, the total ac components of $v_{sq}(t)$ is represented by $v_{ge}(t)$.

Based on FHA-method, only fundamental component is used for the ac equivalent circuit. Then the input voltage $v_{ge}(t)$ may be approximated by its fundamental component $v_{ge,1}(t)$ in Figures 3 and 4,

$$v_{ge,1}(t) = \frac{2}{\pi} \times V_{dc} \times \sin(\omega t) \tag{4}$$

On the output voltage $v_{so}(t)$, its fundamental component $v_{so,1}(t)$ is

$$v_{so,1}(t) = \frac{4}{\pi} \times n \times V_o \times \sin(\omega t - \varphi_{v1}) \tag{5}$$

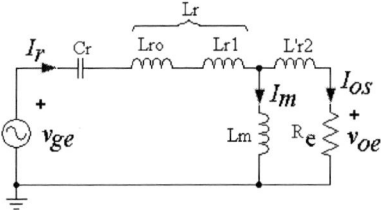

Figure 3. Equivalent ac circuit with the leakage inductance in consideration.

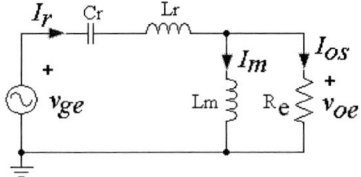

Figure 4. Equivalent ac circuit with the leakage inductance ignored.

For Figure 4 where the transformer winding leakage inductances are ignored, the load ac voltage $v_{oe}(t)$ fundamental component equals $v_{so,1}(t)$,

$$v_{oe,1}(t) = \frac{4}{\pi} \times n \times V_o \times \sin(\omega t - \varphi_{v1}) \tag{6}$$

For Figure 3 where the leakage inductances are under consideration, the corresponding load voltage $v_{oe,1}(t)$ is obtained with $v_{so,1}(t)$ through an impedance divider.

The rms voltage of $v_{ge}(t)$ and $v_{oe}(t)$ then is expressed in the below,

$$V_{ge,1} = \frac{\sqrt{2}}{\pi} V_{in} \tag{7}$$

$$V_{oe,1} = \frac{2\sqrt{2}}{\pi} \times n \times V_o \tag{8}$$

As $v_{so}(t)$ is a square voltage applied across resistive load, its corresponding current is reasonably taken as the same square wave shape. Then, the equivalent load resistance R_e is obtained as,

978-1-4244-4782-4/10 $26.00 © 2010 IEEE

$$R_e = \frac{V_{oe}}{I_{oe}} = \frac{8}{\pi^2} \times \frac{n \times V_o}{I_o/n} \tag{9}$$

Due to the leakage inductances, the series resonant frequency of Figure 3 and 4 is different because of the different equivalent series inductance, L_r, although the same formula is used to calculate this resonance,

$$f_0 = \frac{1}{2\pi\sqrt{L_r C_r}} \tag{10}$$

In the circuit of Figure 4, the series resonant inductance is from the series inductor value L_{r0}, *i.e.*,

$$L_r = L_{r0} \tag{11}$$

But in the circuit of Figure 3, as the transformer leakage inductances exist, its primary side winding leakage inductance L_{r1} takes effect on the series resonance and L_r is determined by (12) below,

$$L_r = L_{r0} + L_{r1} \tag{12}$$

In practice, it is often to use only L_{r1} for L_r, as called magnetic integraton. Appendix C provides how to obtain physical L_{r1} from usually used open/short circuit test, noticed that L_{r1} cannot be obtained directly from the measurement based on the open and short circuit tests.

A. Voltage Gain Function

The input to output voltage gain function is defined as,

$$M_g^{dc} = \frac{n \times V_o}{V_{in}/2} \tag{13}$$

Based on the FHA method, the gain function is approximated by their fundamental voltages,

$$M_{g,1}^{ac} = \frac{v_{oe,1}}{v_{ge,1}} \tag{14}$$

Noticed that (14) becomes (13) when $f_{sw} = f_0$ with $L'_{r2} = 0$ which makes $\varphi_{v1} = 0$ in (5). In fact,

$$\frac{v_{oe,1}}{v_{ge,1}} = \frac{(4/\pi) \times n \times V_o \sin(\omega t - \varphi_{v1})}{(2/\pi) \times V_{in} \sin(\omega t)}\bigg|_{\varphi_{v1}=0} = \frac{n \times V_o}{V_{in}/2} \tag{15}$$

Hence, Equation (14) is an approximation of (13). At the series resonance, (14) accuracy is good while away from the resonance, the accuracy becomes worse.

For the circuit of Figure 3, its FHA-based gain function can be expressed with circuit parameters as,

$$M_{gl}^{ac} = \left| \frac{(j\omega L_m)//(R_e + j\omega L'_{r2})}{(j\omega L_m)//(R_e + j\omega L'_{r2}) + j\omega L_r + 1/j\omega C_r} \times \frac{R_e}{(R_e + j\omega L'_{r2})} \right| \tag{16}$$

B. Gain Function with Leakage Inductance Ignored

If the transformer leakage inductances are ignored, the gain function is then modified by $L'_{r2} = 0$,

$$M_{g,1} = M_{g,1}^{ac} = \left| \frac{(j\omega L_m)//R_e}{(j\omega L_m)//R_e + j\omega L_r + (1/j\omega C_r)} \right| \tag{17}$$

Equation (17) can be expressed in a normalized format. To do this, we define a normalized switching frequency,

$$f_n = \frac{f_{sw}}{f_0} = \frac{2\pi f_{sw}}{2\pi f_0} = \frac{\omega_{sw}}{\omega_0} = \omega_n \tag{18}$$

and define an inductance ratio,

$$L_n = \frac{L_m}{L_r} \tag{19}$$

Also, noticed that quality factor of the series resonant circuit is defined as,

$$Q_e = \frac{\sqrt{L_r/C_r}}{R_e} \tag{20}$$

Then, the normalized gain function is obtained,

$$M_{g,1} = \left| \frac{L_n \times f_n^2}{[(L_n + 1) \times f_n^2 - 1] + j[(f_n^2 - 1) \times f_n \times Q_e \times L_n]} \right| \tag{21}$$

From the engineering viewpoint, (22) is required in order to ignore the leakage inductances, and details can be found in Appendix B,

$$L_{r2} \le \frac{1}{10} \times \frac{8}{\pi^2} \times \frac{R_L}{2\pi \times f_{sw,max}} \tag{22}$$

Leakage inductances introduce additional effect to the gain function as shown in Appendix A. But the effect is not related to the FHA, we will use (17) for the remaining discussion.

C. FHA Approximations

Examining the development of the input to output voltage gain function reveals that there exist five approximations in the voltage gain function development:

*   (Aprx-1) use frequency fundamental component to represent input dc voltage;

*   (Aprx-2) use frequency fundamental component to represent output dc voltage;

*   (Aprx-3) assume square waveform of secondary winding voltage;

*   (Aprx-4) use equivalent output resistance to convert non-linear rectifier circuit to a linear circuit;

*   (Aprx-5) no consideration is made to output capacitor and its ESR.

Logically speaking, the voltage gain accuracy can be improved by reducing the effect from these approximations.

III. GAIN ACCURACY IMPROVEMENT BY HARMONIC COMPENSATION

A. Circuit Linearization

FHA-based gain expression only uses the fundamental component on both input and output side from two approximations Aprx-1 and Aprx-2. Aprx-3 and Aprx-4 help to make the circuit linearization as shown in Figure 2. By analyzing the linear circuit model, it is found that both v_{ge} and v_{oe} can be approximated by more than just fundamental components. That is to say introducing high order frequency components should improve the gain function accuracy. The introduction of high order frequency components can be done based on the linear system superposition property.

B. Precisely Modeling

Based on the superposition theorem and the Fourier series, precise relationship between v_{oe} and v_{ge} can be established when C_r, L_r, L_m, and R_e are treated as linear and time-invariant parameters. The gain expression corresponding to k-th order harmonic component is then obtained,

$$M_{g,k} = \frac{v_{oe,k}}{v_{ge,k}} =$$

$$\left| \frac{L_n \times (k \times f_n)^2}{[(L_n+1) \times (k \times f_n)^2 - 1] + j[((k \times f_n)^2 - 1) \times (k \times f_n) \times Q_e \times L_n]} \right|$$

(23)

$$k = 1,3,5,...$$

Using the superposition theorem in frequency domain first, and back to time domain, $v_{oe}(t)$ can be obtained

$$v_{oe}(t) = \sum_{k=1,3,5,...} (M_{g,k} \times v_{ge,k}(t))$$

(24)

Since all $v_{oe,k}(t)$ in time domain are still in sinusoidal form with integer number multiplier of the fundament frequency, $v_{oe}(t)$ total rms value can then be obtained by orthogonal property of trigonometric functions for ωt over the interval $[0, 2\pi]$; refer to Appendix D for such property,

$$V_{oe}^2 = \sum_{k=1,3,5,...} (M_{g,k} \times V_{ge,k})^2$$

(25)

This obtained V_{oe}, in rms value, is an accurate response to its input excitation $v_{ge}(t)$.

C. Concerns to Precise Modeling

However, there are several concerns. First, (25) is no longer an FHA-based gain expression. As a matter of fact, the gain function is not expressed in an explicit form. In other words, the total gain is not expressed in a definition of V_{ge}-V_{oe}. Second, R_e is still a non-linear component although it is treated as a linear one. The third concern is the formula of (25) does not consider the output capacitance and associated ESR.

D. Explicit FHA Form of Precise Model

In the following, we will discuss how to approximate the V_{ge}-V_{oe} in (25) in an explicit form. On the R_e non-linear effect and output capacitor's effect, more future work will be needed such that this paper makes no further discussions in regarding to these two concerns.

IV. COMPENSATING FHA-BASED GAIN WITH HARMONICS

The following approximation can make the total gain expression in an explicit form. Noticed that $V_{ge,k}$ in Fourier series is expressed by

$$V_{ge,k} = (\frac{2}{\pi} V_{dc}) \times \frac{1}{k} \times \sin(k\omega_{sw}t)$$

$$= \frac{V_{ge,1}}{k} \times \sin(k\omega_{sw}t)$$

(26)

$$k = 1,3,5...$$

Magnitude-wise on each harmonic component, we can write this relationship based on (26),

$$V_{ge,k} = \frac{V_{ge,1}}{k}$$

(27)

As a result of (27), (25) can be modified as,

$$V_{oe}^2 = V_{ge,1}^2 \times \sum_{k=1,3,} (M_{g,k} / k)^2$$

(28)

If we make the same approximation from the original FHA-based gain expression, the gain can then be expressed as,

$$M_g = M_{g,1} \times \sqrt{1 + \sum_{k=3,5,7,...} \left(\frac{M_{g,k}/k}{M_{g,1}} \right)^2}$$

(29)

where $M_{g,k}$ is determined by (23).

Equation (29) may be named as FHA-based voltage gain function with harmonic compensation, or harmonic compensated voltage gain function.

In the following section, the harmonic-compensated gain is discussed based on simulation and test of a 300W LLC half bridge converter.

V. DISCUSSION ON FHA-BASED HARMONIC GAIN EXPRESSION

A 300W LLC resonant half bridge converter was built to examine the accuracy improvement from the compensated-gain of (29). The converter is designed based on the gain expression of (17) with $390V_{in}$ and $12V_o$. To well confine circuit parameters, the design was made without integrating the series inductance into the transformer.

The converter has the electrical specifications as follows:

- Input voltage: 375Vdc to 405Vdc
- Output power: 300W
- Output voltage: 12Vdc
- Output current: 25A
- Output voltage line regulation: ≤1%
- Output voltage load regulation: ≤1%
- Output voltage peak-to-peak ripple: ≤120mV

The converter circuit diagram is shown in Figure 5. On the LLC controller, UCC25600 was selected as a good choice. UCC25600 is an 8-pin device with built-in state of the art efficiency boosting features and high-level protection features to provide cost effective solutions for LLC resonant half-bridge converter applications. Note that for clarity, some auxiliary functions are not included in the schematic below. Detailed design can be found in [8].

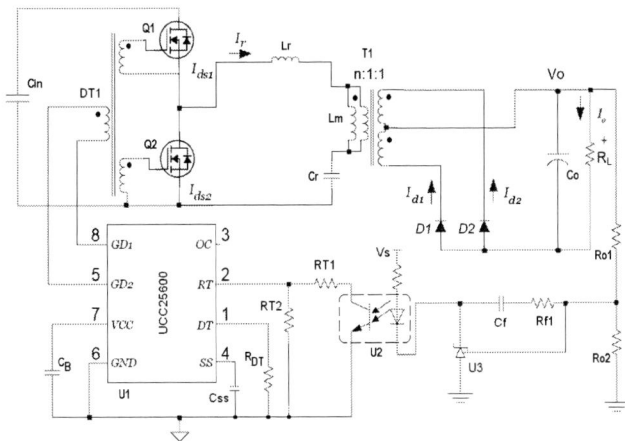

Figure 5 Circuit diagram used for test.

The resonant circuit parameters are designed as below:

- Transformer
  a. Turns ratio: n:1:1=17:1:1
  b. Primary magnetizing inductance, $L_m = 280$ μH
  c. Primary leakage inductance, $L_{r1} = 2.5$ μH
- Series resonant inductance, $L_{r0} = 60$ μH
- Series resonant capacitance, $C_r = 24$ nF

These are translated into normalized variables as

- $f_0 = 130$ kHz; $L_n = 4.48$; $Q_e = 0.453$;

- $R_e = 112.6$ Ω at full load

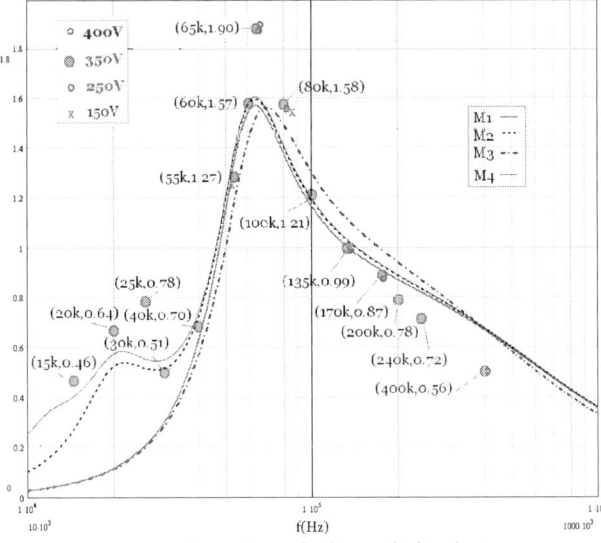

Figure 6 Comparison of voltage gain functions.

The improvement from compensated gain can be seen by comparing it with the un-compensated one. Bench tests together with simulation were made in the comparison work. Figure 6 depicts the comparison result. In Figure 6, M1 through M4 represent four results from different gain expressions. Figure 6 is in absolute values. Hence, the horizontal axis in Figure 6 is the frequency in hertz, and the vertical axis is magnitude of the gains.

- M1 is plotted based on (17), *i.e*, the gain with the transformer inductances ignored.

- M2 is plotted based on Equation (29) where 3rd– harmonic is added to the gain function as compensation.

$$M_g = \sqrt{M_{g,1}^2 + (M_{g,3}/3)^2}$$

- M3 is calculated based on (16) as reference for comparison with magnetic integration.

- M4 is based on (29) with 3rd and 5th harmonic are added as compensation.

$$M_g = \sqrt{M_{g,1}^2 + (M_{g,3}/3)^2 + (M_{g,5}/5)^2}$$

- Test supported simulation results are discretely marked.

Based on Figure 6, the compensated gain expression improves the accuracy overall in the whole frequency range with most compensated area being the capacitive region. This can be seen from curves M2 and M4 with frequency below 60k Hz. The least compensated area is around the peak gain area and the area of higher frequencies than the series resonance. Further work is required to understand why such significant deviations still exist in these areas.

VI. CONCLUSIONS

Harmonic compensation is proposed to improve the accuracy of FHA-based gain function after reviewing the FHA method in its theoretical development. Computer simulation and bench test are made to compare the proposed method with existing ones. Future work is needed to understand why significant deviations still exist in the vicinity of peak gain and in the area with frequencies much higher than the series resonance.

ACKNOWLEDGMENTS

The author of this paper expresses his sincere gratitude to the individuals below for their valuable discussions during the work of this paper preparation and writing: Richard Garvey and Bob Neidorff.

REFERENCES

[1] R.L. Steigerwald, "A Comparison of Half Bridge Resonant Converter Topologies," IEEE Trans. on Power Electronics, vol. xxx, April 1988, pp. 174–182.

[2] Thomas Duerbaum, "First Harmonic Approximation Including Design Constraints," Telecommunications Energy Conference, INTELEC 1998, pp. 321 – 328.

[3] James F. Lazar, Robert Martinelli, "Steady-state Analysis of The LLC Resonant Converter," Applied Power Electronics Conference and Exposition, 2001. APEC 2001. pp. 728 – 735.

[4] Bing Lu, Wenduo Liu, Yan Liang, Fred C. Lee, Jacobus D. van Wyk, "Optimal Design Methodology for LLC Resonant Converter," APEC 2006. pp.533-538

[5] S. De Simone, C. Adragna, C. Spini and G. Gattavari, "Design-oriented Steady State Analysis of LLC Resonant Converters Based on FHA," SPEEDAM 2006. pp.s41-16 –s41-23

[6] Hangseok Choi, "Analysis and Design of LLC Resonant Converter with Integrated Transformer," APEC2007. pp. 1630-1635

[7] De Simone, C. Adragna, C. Spini, "Design Guideline for Magnetic Integration in LLC Resonant Converters," SPEEDAM 2008. pp. 950-957

[8] Texas Instruments, *UCC25600EVM User Guide*, doc-#: *sluu361*, http://focus.ti.com/lit/ug/sluu361/sluu361.pdf, April 2009, 20p.

Appendix A. COMPARISON BETWEEN THE GAIN EXPRESSIONS

For the same design and operation conditions, the gain from $M_{g\ell}$ and from $M_{g,1}$ is different. A typical $M_{g,1}$ versus $M_{g\ell}$ is shown. The below observations can be made as below on the assumption $L_r = L_{r1}$, and $L_p = L_m + L_{r1}$ which definitions can be found in Appendix C.

a) Peak gain from $M_{g,1}$ is lower than $M_{g\ell}$;

b) The frequency corresponding to peak gain of $M_{g,1}$ is lower than of $M_{g\ell}$;

c) The unity gain is always achieved at $f_n = 1$ from $M_{g,1}$ while from $M_{g\ell}$ at $f_n = 1$, its gain is shifted up by ΔM that is estimated as

$$\Delta M = \sqrt{L_p/(L_p - L_r)} > 1 \tag{A-1}$$

d) However, the entire curve from $M_{g\ell}$ is not shifted by amount of ΔM. As such, we cannot simply multiply ΔM to $M_{g,1}$ to get $M_{g\ell}$.

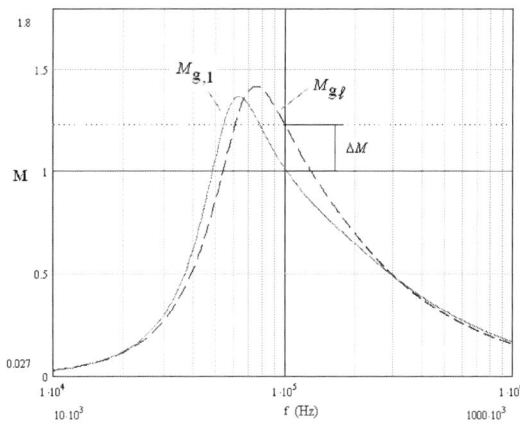

Figure A- 1 Comparison of different voltage gain functions

Appendix B. TRANSFORMER LEAKAGE INDUCTANCE IGNORABLE CONDITION

From the engineering viewpoint, the condition to ignore the secondary leakage inductance is

$$(2\pi f_{sw}) \times n^2 \times L_{r2} = (2\pi f_{sw}) \times L'_{r2} \ll R_e \tag{B-1}$$

As R_e shown in previous equation can be also expressed in load resistance R_L, then the above condition becomes,

$$(2\pi f_{sw}) \times L_{r2} \ll \frac{8}{\pi^2} \times R_L, \quad \forall (f_{sw}, R_L) \tag{B-2}$$

For a constant resistive load, R_L is a constant, then the above condition becomes,

$$(2\pi f_{sw}) \times L_{r2} \ll \frac{8}{\pi^2} \times R_L, \quad \forall f_{sw} \tag{B-3}$$

In other words, $M_{g\ell}$ becomes $M_{g,1}$ if the above condition is satisfied. From engineering practice, the above condition can be presented as,

$$L_{r2} \leq \frac{1}{10} \times \frac{8}{\pi^2} \times \frac{R_L}{2\pi \times f_{sw,max}} \tag{B-4}$$

Similarly, in order to ignore the primary leakage inductance, it requires,

$$L_{r1} \ll L_r \tag{B-5}$$

Appendix C. PHYSICAL TRANSFORMER MODELING AND VOLTAGE GAIN ACCURACY

A transformer physical model is shown below with all critical inductances and turns ratio. Typically, a transformer is designed close to an ideal one with leakage inductances minimized as much as possible both on the primary and the secondary. A bench model can then be easily obtained with two tests, the open circuit test and the short circuit test.

The open circuit test yields the result L_p while noticed that L_p is not the magnetizing inductance, instead, it is the sum of the magnetizing inductance L_m and the primary winding leakage inductance L_{r1},

$$L_{r1} + L_m = L_p \tag{C-1}$$

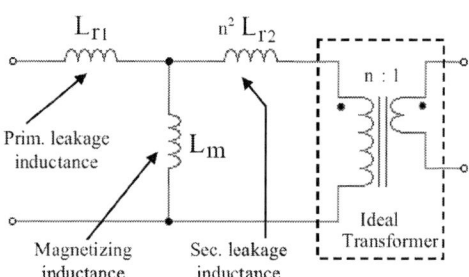

Figure C- 1 Transformer physical model with n, L_m, L_{r1} and L_{r2}.

The short circuit test produces test result, L_s, which is the combination of leakage inductances together with magnetizing inductance, noticed that the L_s is *neither* the primary winding leakage inductance L_{r1}, *nor* the secondary side winding L_{r2}. The L_s is a combined value can be described by (C-2),

$$L_{r1} + L_m // (n^2 L_{r2}) = L_s \tag{C-2}$$

A simplified transformer bench model is configured by these two measurements as shown in Figure C-2. This circuit model is valid since if we do a same open/short circuit test on this circuit model, we will obtain the same L_p and L_s.

This bench model has been used in many applications successfully. But concerns to this model are arising when it is used in LLC resonant converter, especially when magnetic integration is used in a design. The magnetic integration is to combine the series resonant inductance and the transformer together by enlarging the transformer leakage inductances. Also it is often the case in LLC resonant converter design that the magnetizing inductance is made comparable to the leakage inductances.

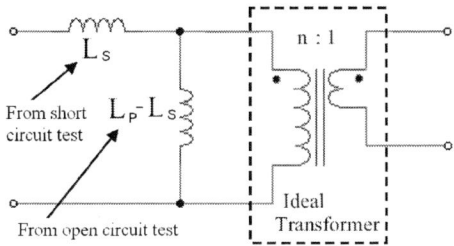

Figure C- 2 Transformer bench model.

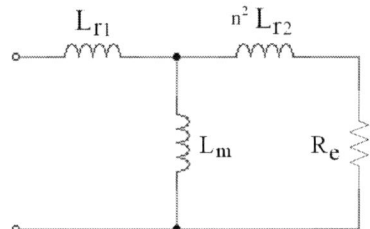

Figure C- 3 R_e connected in series with the leakage inductance.

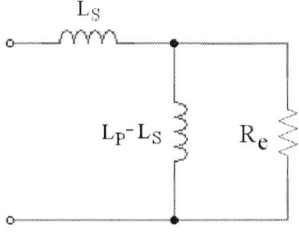

Figure C- 4 R_e connected in parallel with L_p

From these unique differences, the simple bench model introduces tolerance concerns. As in the transformer physical model of Figure C-1, the load equivalent resistance R_e is connected in series with the secondary winding leakage inductance as shown Figure C-3. But with a bench model, the R_e can only be connected in parallel with L_p as shown in Figure C-4. The resulted voltage function is different from the two different connections despite using the same physical transformer. It is obvious the connection of Figure C-4 is valid only if the transformer leakage inductance can be ignored while this is usually not the case for LLC resonant converter design with magnetic integration. As such, it is

needed to establish the model of Figure C-1 with the measurement results from often used open/short circuit test.

From (C-1) and (C-2), we may solve these two equations to get the parameters needed for Figure C-1 while we need add one more equation since there are three un-known variables while only two equations. To add the third equation, (C-3) shown below is a reasonable one from physics and engineering practice,

$$(n^2 L_{r2}) = L_{r1} \qquad (C-3)$$

After (C-3) is added, the solution for Figure C-1 circuit parameters can be obtained,

$$L_{r1} = n^2 L_{r2} = L_p - \sqrt{\left(L_p^2 - L_p L_s\right)} \qquad (C-4)$$

$$L_m = L_p - L_{r1} \qquad (C-5)$$

$$(n^2 L_{r2}) = L_{r1} \qquad (C-6)$$

Another often used transformer circuit model is shown in Figure C-5. This model is obtained by assumption of (C-3). This model is acceptable as long as the magnetizing inductance is much larger than the secondary winding leakage inductance as that L_m in (C-2) can be ignored,

$$L_{r1} + L_m // (n^2 L_{r2}) \approx L_{r1} + n^2 L_{r2} = 2 \times L_{r1} = L_s \qquad (C-7)$$

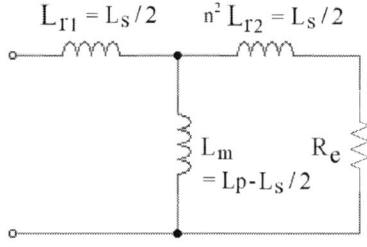

Figure C- 5 Transformer model when L_m much larger than the leakages.

From engineering viewpoint, (C-7) is acceptable when

$$L_m \geq 10 \times (n^2 L_{r2}) \qquad (C-8)$$

Otherwise, the model of Figure C-5 will cause tolerance concerns although it is better than the model of Figure C-4.

Appendix D. ORTHOGONALITY OF TRIGONOMETRIC FUNCTION SET

Set V formed by trigonometric functions is orthogonal over interval $[0, 2\pi]$,

$$V = \{1, \cos x, \sin x, \cos 2x, \sin 2x, ..., \cos nx, \sin nx, ...\} \qquad (D-1)$$

with its inner product defined as,

978-1-4244-4782-4/10 $26.00 © 2010 IEEE

$$\langle w, y \rangle := \int_0^{2\pi} w \cdot y \, dx, \quad \forall w, y \in V \tag{D-2}$$

Set V exists the below property,

$$\int_0^{2\pi} 1 \cdot \cos(nx) \, dx = 0 \tag{D-3}$$

$$\int_0^{2\pi} 1 \cdot \sin(nx) \, dx = 0 \tag{D-4}$$

$$\int_0^{2\pi} \sin(nx) \cdot \cos(nx) \, dx = 0 \tag{D-5}$$

$$\int_0^{2\pi} \sin(mx) \cdot \sin(nx) \, dx = 0, \ \forall m, n, m \neq n \tag{D-6}$$

$$\int_0^{2\pi} \cos(mx) \cdot \cos(nx) \, dx = 0, \ \forall m, n, m \neq n \tag{D-7}$$

$$\int_0^{2\pi} \sin(mx) \cdot \cos(nx) \, dx = 0, \ \forall m, n, m \neq n \tag{D-8}$$

where m and n are any positive integer numbers, *i.e.*,

$$m, n \in \{1, 2, 3 \dots\} \tag{D-9}$$

Example: calculate total rms value of $v_q(\omega t)$,

$$v_q(\omega t) = \sqrt{2} V_{qu} \sin(u\omega t - \varphi_{qu}) + \sqrt{2} V_{qv} \sin(v\omega t - \varphi_{qv})$$

Based on definition, $v_q(\omega t)$ total rms value is calculated as,

$$V_{q,rms} = \sqrt{\frac{1}{2\pi} \int_0^{2\pi} v_q^2(\omega t) \, d(\omega t)} = \sqrt{V_{qu}^2 + V_{qv}^2}$$

and

$$u, v \in \{1, 2, 3 \dots\}$$

Analysis and Design of LCC Resonant Inverter for the Tranportation Systems Applications

Mohamed Youssef*, Jaber A. Abu Qahouq** and Mohamed Orabi***

* Bombardier Transportation Inc. Research and Development Kingston, Canada Mohamed.Youssef@ieee.org	** The University of Alabama Department of Electrical and Computer Engineering Tuscaloosa, Alabama 35487, USA	*** South Valley University Department of Electrical Power and Machines Aswan, Egypt

Abstract — The analysis and design of an LCC-type power resonant inverter for transportation application in a train system is presented in this paper. The inverter powers the linear induction motor (LIM) used in urban transit applications. It has low components stresses compared with the conventional series and series-parallel resonant topologies. The voltage regulation is achieved by using a self-oscillating controller that results in high power efficiency for wide range of current motor operation. Moreover, by using an optimized transformer leakage inductance value, Zero-Current Zero-Voltage (ZCZVS) operation is achieved for the primary-side IGBTs, which allows for high switching frequency operation with lower poser losses, resulting in increased power density. The ZCZVS operation is achieved without using an auxiliary circuit by utilizing the switches capacitances and the transformer leakage inductance. The paper presents the theoretical analysis, which is used in the design of the power inverter. Experimental results based on 750Vdc/700Vac 400kW design are presented to verify and evaluate the LCC-type resonant inverter.

I. INTRODUCTION

The push for higher energy efficiency continues to increase in most applications in our life. This trend is motivated by different reasons that includes, but are not limited to, reducing energy cost, increasing battery life for battery powered applications, reducing system size, and improving systems friendliness with the environment. Transportation is one of the most important examples among these applications [1-2, 19].

Power inverters are very important part of an electric transportation system. The requirements for increased energy efficiency have raised the bar for higher inverter energy efficiencies over wide operating ranges and conditions used in rail transit industry [1-3]. Moreover, such inverters in transportation applications are required to have high power density while achieving higher efficiency, which imposes a challenge. Electromagnetic Compatibility (EMC) requirements and standards should be also met [3-5].

Recent advancements in the semiconductor industry allow the operation of high power IGBTs at higher switching frequencies [6]. Unlike the Pulse Width Modulated (PWM) converters, resonant converters have the advantage of operating at higher frequencies with reduced Electromagnetic Interference (EMI) levels [7].

Moreover, the higher frequency operation of resonant inverters allow a reduced real estate for the inverter in the under floor ceiling footprint and layout, which is a significant advantage [1]. This allows for additional space for other battery resources and other auxiliary powering devices to accommodate the need for more on-board luxury devices such as security cameras, which are now required by most transportation authorities, and advertisement panels, among others [1].

There are drawbacks of the conventional PWM inverters that include the degraded efficiency at high switching frequency and the possible need of external resonant circuitry during the switch transition period to realize zero voltage switching (ZVS). This led to the complexity and unreliability of such inverters. On the other hand, in order to obtain a reduced size inverter, the switching frequency must increase. Based on the fact that resonant converters and inverters can operate at higher frequencies, industry's attention is going back to such inverters and converters.

The series resonant converters can achieve ZVS due to the series resonance principle for the switches of the half bridge but the output voltage regulation is lost at light loading conditions [8]. The parallel resonant converters are suitable for high-voltage low-current applications because of the considerable conduction losses of the circulating current of the parallel capacitor [9]. The series-parallel resonant converters possess the advantages of the series and parallel resonant circuits and can regulate the output voltage for a wide loading range [10]. However, it needs a wide range of regulating frequency under the variable frequency control (VF). This will cause bulky input/output filters and hence a large size of the converter [2-5, 11-18].

The LLC resonant converters were introduced to offer a new topology that gathers the advantages of the series and parallel resonant converters with a high efficient operation with almost no switching losses [10]. These converters are suitable for integrated a magnetic solution, which offers a more compact

978-1-4244-4782-4/10 $26.00 © 2010 IEEE

size converter. The regulating frequency range is reduced for this topology, which enhances the reduction of the size of the input filter [4-5, 11].

In this paper, an LCC resonant inverter, as shown in Fig. 1, is discussed and analyzed. The output voltage is controlled by the variable frequency self-oscillating technique, provided that it is always under the resonant frequency, to guarantee the zero current switching operation (ZCS) of the IGBTs. The outer loop helps regulating the output voltage. Both the speed and the thrust are predicted in the control loop, which compares the thrust to the reference thrust signal and modulates the switching frequency of the inverter to adjust for the required output voltage to the LIM (Linear Induction Motor). The

presented inverter single phase circuit and overall schematics with protection circuitry is shown in Fig. 1 and Fig. 2 respectively.

This Paper presents part of the theoretical, simulation and experimental results for the LCC resonant inverter and its controller. Section II presents the steady-state characteristics of the LCC resonant inverter under self-sustained oscillation mode. Section III presents the design procedure for the inverter parameters and the primary side power factor angle optimization. Experimental development and results are presented in Section IV and the conclusion is given in Section V.

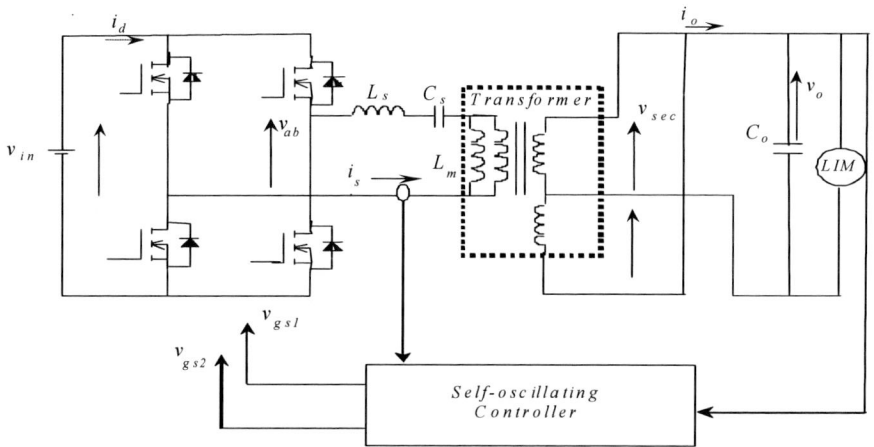

Fig. 1: Single-phase diagram of the proposed 3-phase resonant inverter

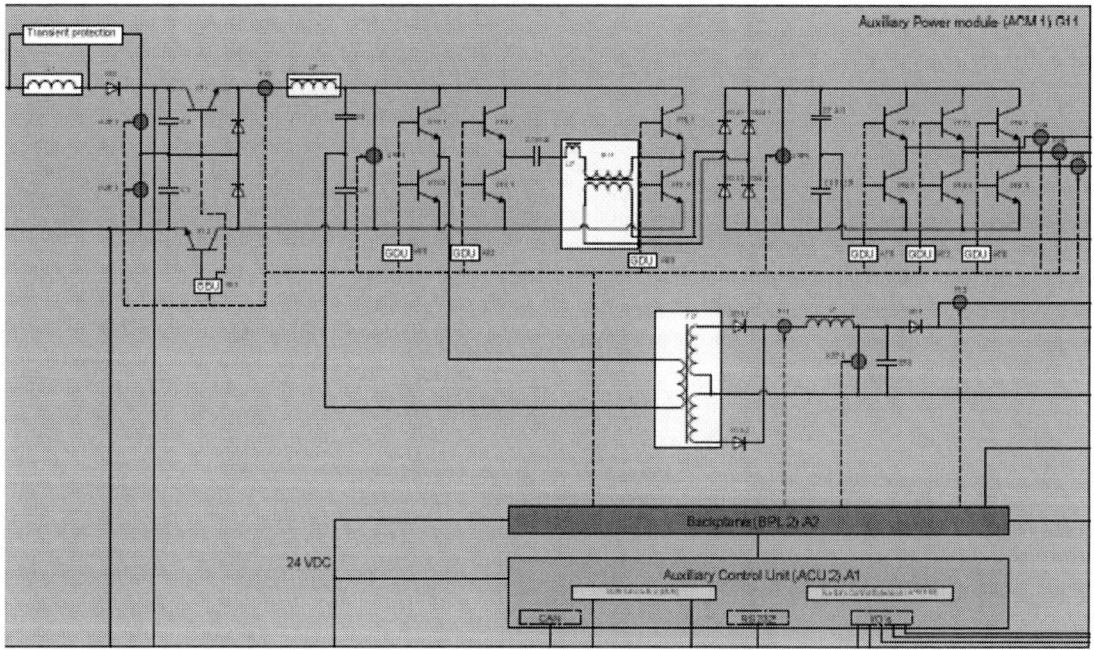

Fig, 2: Overall schematics of the powering system with two inverters powering two LIM

II. STEADY STATE CHARACTERISTICS OF THE LLC RESONANT INVERTER UNDER SELF-SUSTAINED OSCILLATION MODE

The operating principle of the SSOC (Self-Sustained Oscillation Controller) is that the operation of the inverter must satisfy the following relation [1]:

$$G(j\omega) = -\frac{1}{Ne(A,\gamma)} \tag{1}$$

Where $G(j\omega)$ is the open loop transfer function of the LLC resonant converter between the feedback variable and the inverter output voltage (Vab), (A) is the amplitude of the feedback variable, and (γ) is the control angle. The describing function of the converter plus the inverter is expressed by the following describing function:

$$Ne(A,\gamma) = \frac{4 \cdot V_g}{\pi \cdot A} \cdot [\cos(\gamma) - j \cdot \sin(\gamma)] \tag{2}$$

In the case of the inductor current (is) as a feedback variable, the open loop transfer function is given by equation (2).

$$G_{is\_vab}(s) = \frac{2}{\pi^2} \cdot \frac{s \cdot (s + k_L \cdot \omega_o \cdot Q_s)}{s^3 + s^2 \cdot (1 + k_L \cdot Q_s) + \frac{s}{\omega_o^2} + k_L \cdot \omega_o^3 \cdot Q_s} \tag{3}$$

As a first step, the equality (2) will be represented as two nonlinear equations. One is associated with the real part of (2) and the other with the imaginary part:

$$[2 - \alpha^2] \frac{\pi^2 \cdot \omega_o \cdot L_s \cdot \hat{i_s}}{V_g} = -8(1 + k_L) \cdot \left[-\frac{\alpha^2 \cdot \pi^2}{8 \cdot Q_s} \cdot \cos(\gamma) - \alpha \cdot \sin(\gamma) \right] \tag{4}$$

and:

$$[4 - \alpha^2] \cdot \frac{\pi^2 \cdot \omega_o \cdot L_s \cdot \hat{i_s}}{V_g} = -\frac{16 \cdot Q_s}{\pi^2} (1 + k_L) \cdot \left[\cos(\gamma) - \frac{\alpha \cdot \pi^2}{8 \cdot Q_s} \cdot \sin(\gamma) \right] \tag{5}$$

Where ω_o is the series resonant frequency:

$$\omega_o = \frac{1}{\sqrt{L_s \cdot C_s}}$$

k_L is the ratio between the resonant inductor (Ls) to the magnetizing inductance (Lm):

$$k_L = \frac{L_s}{L_m}$$

Q_s is the series quality factor:

$$Q_s = \frac{\omega_o \cdot L_s}{R_{eq}}$$

R_{eq} is the equivalent load resistance of the LIM phase:

$$R_{eq} = \frac{\pi^2}{8} \cdot n^2 \cdot R_L$$

R_L is the load resistance and n is the transformer turns ratio.

$$\alpha = \frac{\omega_o}{\omega_s}$$

$\omega_s = 2 \cdot \pi \cdot f_s$ is the radian switching frequency and f_s is the switching frequency in Hz.

The simultaneous solution of the nonlinear algebraic equations (4) and (5) result in possible operating points of the converter. It is worth noting that the solution must satisfy the constraint on the normalized frequency and this must be considered. The condition is ($\alpha < 1$) to guarantee ZCS.

Consider the design specifications shown in Table 1. The design parameters of Table 1 are obtained using the design procedure described next section. Fig. 3 shows a plot of frequency ratio (α) versus the control angle (γ) and Fig. 4 shows series resonant inductor current ratio versus the control angle. The inductor current ratio is defined as the actual inductor current to the inductor current at the resonance frequency.

It can be observed from Fig. 4 that the regulating switching frequency ratio needed for ZCS is < 3 around the no-load condition, which is still much better than the variable frequency technique (VF) [2-5, 11-18]. This fact will also be shown in experimental results in Section V.

Table 1: Design and Test Parameters

Parameter	Value
v_{in}	750V±20%
v_{out}	700Vac±5%
IGBT switches	FF1000R17IE4
I_o	750A (Full-load)
L_s	15.7 μH
L_m	320 μH
$n_1{:}n_2$	14:10
C_s	1.1 μF
Snubber Capacitors	6.8 nF
dv/dt capacitor filter for the motor	0.15μF
Switching frequency	25 kHz

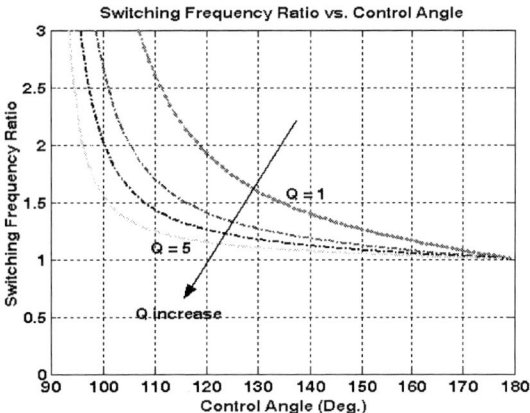

Fig. 3: Frequency ratio (α) vs. the control angle (γ)

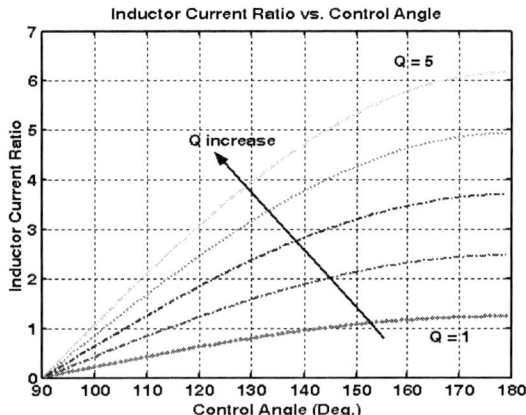

Fig. 4: Series resonant inductor current ratio vs. the control angle

III. INVERTER DESIGN PROCEDURE

Fig. 5 shows a plot of the resonant capacitor voltage ratio versus the control angle for different Q_s values and Fig. 6 shows a plot of the inductor current ratio versus the control angle for different Q_s values that can be used for the resonant inductor rating selection.

It can be observed from Fig. 5 that at heavy loading conditions (high Qs), the stress on the resonant capacitor becomes very high. Fig. 5 shows the stress on the resonant inductor when the control angle is 180° which corresponds to minimum input voltage and full-load current operation. It can be observed that a reasonable value for Qs at full-load is around 2.5 can be chosen for the resonant tank, otherwise higher Qs values may result in a narrower range of input voltage to control the stresses on the resonant tank elements. However, for a prototype of an input voltage of 750V±20% and an output voltage of 700V with a full-load output current of 560Arms, the values of the transformer ratio is n = 1.4, the

series resonant inductance (Ls) is 15.7µH, and the series resonant capacitors (Cs) is 1.1µF for a resonant frequency of around 30.5kHz.

The resonant inductor used is made of PQ2625 shape with a core material of PC44. It has 14 turns each of 2 strands made of 300X48 Litz wire to get a current carrying capacity of 4.16Arms. The series resonant capacitor is composed of 2 capacitors in parallel, one of them is 660nF. The transformer is made of 3F35 ferrite cores and they are selected for their low magnetic losses up to 2MHz and can withstand core temperatures of 120°C. The leakage inductance of the transformer was found to be 15.8µH and was used to replace the value of the series resonant inductance. The magnetizing inductance (Lm) was measured to be 300µH.

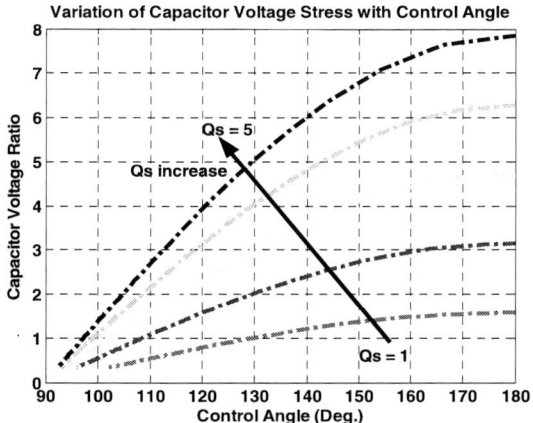

Fig. 5: Resonant capacitor voltage ratio vs. the control angle for different Q_s values

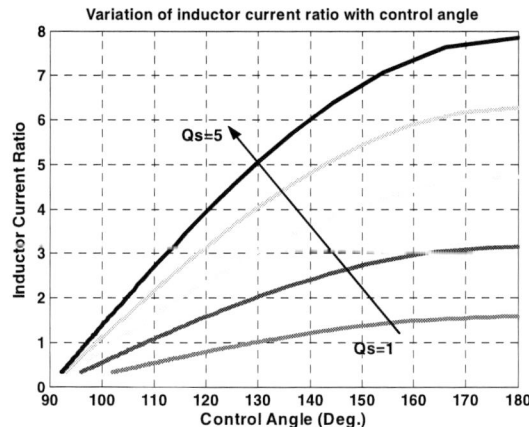

Fig. 6: Inductor current ratio vs. the control angle for different Q_s values

In order to optimize the converter working efficiency, the conduction power loss resonant tank should be minimized. This can be achieved by the minimization of the primary side power factor angle, under the constraint that it should be sufficient to achieve ZCS. In order to guarantee ZVS for the half bridge switches, the output capacitor of the switch should be completely discharged by the resonant current (is) during a defined dead time (td), while the other output capacitor of the other switch should be completely charged.

To attain ZCS, the following condition should be satisfied:

$$v_c = \frac{1}{C_{os}} \cdot \int_{-\frac{t_d}{2}}^{\frac{t_d}{2}} -\frac{1}{2} \cdot i_s(t) \cdot dt \le v_{in} \qquad (6)$$

Where v_c is the voltage across the switch output capacitor, C_{os} is the output capacitor of the switch, and v_{in} is the converter input voltage, and $i_s(t) = i_{s-\max} \cdot \sin(\omega_s \cdot t - \varphi)$ and φ is the power factor angle.

The following equation can be obtained from (6):

$$\sin(\varphi) \ge \frac{2 \cdot C_{os} \cdot v_{in}}{i_{s-\max} \cdot t_d} \qquad (7)$$

Using the first harmonic approximation, the input power of the resonant tank inverter can be given by:

$$P_{in} = \frac{v_{in}}{\pi} \cdot i_{s-\max} \cdot \cos(\varphi) \qquad (8)$$

By substituting Equation (8) into Equation (7), the following design constraint can be deduced:

$$\tan(\varphi) \ge \frac{2 \cdot C_{os} \cdot v_{in}^2}{\pi \cdot P_{in} \cdot t_d} \qquad (9)$$

It can be observed from Equation (9) that the smaller the switch output capacitor (C_{os}) the smaller the power factor angle. Consequently a reduced conduction loss can be achieved.

IV. EXPERIMANTAL RESULTS

Fig. 7 shows a sample picture for the developed experimental prototype. The design parameters discussed earlier in this paper and as shown in Table 1 are used.

Fig. 8 shows the waveforms of the resonant current and collector-emitter voltage of the IGBT at high loading conditions. The resonant current always lags the inverter output voltage and crosses zero inside of the non-zero voltage pulse at output of the inverter, resulting in the required current to reset the snubber capacitors and provide ZCS. The IGBT current was measured using an AC current clamp transducer with a factor of 1:30A, attached to the IGBT phase in the inverter box on the train.

Table 2 shows the overall working efficiency of the prototype as a function of the output power at nominal input voltage of 750Vdc. It can be observed that as the output power increases, the efficiency increases. This is because the circulating current losses in the resonant tank at light loading conditions results in reduced efficiency.

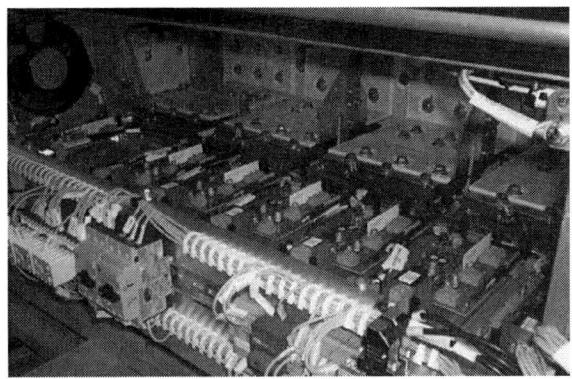

Fig. 7: A sample picture for the developed experimental prototype.

Table 2: Overall Inverter Efficiency at Different Loading Conditions

Load (%)	V_{in} (DC input voltage)	I_{in} (DC input current)	Efficiency ($\eta\%$)
5%	750V	100A	94.9
25%	750V	200A	96.5
50%	750V	300A	97.1
75%	750V	400A	98.2
100%	750V	500A	99

Moreover, the efficiency was measured at low input voltage of 600Vdc and full loading conditions, which is the worst operating condition, as the wayside voltage may drop down by 20%. The recorded efficiency was about 85.6%, which is still acceptable.

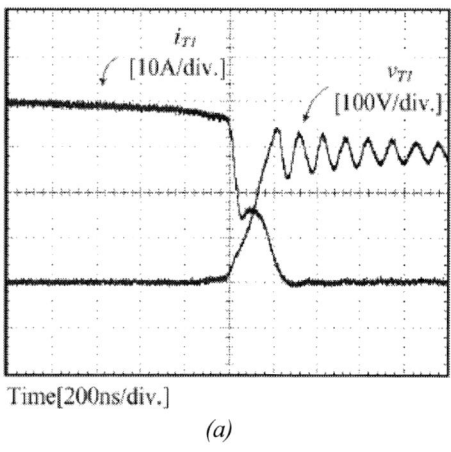

Time[200ns/div.]

(a)

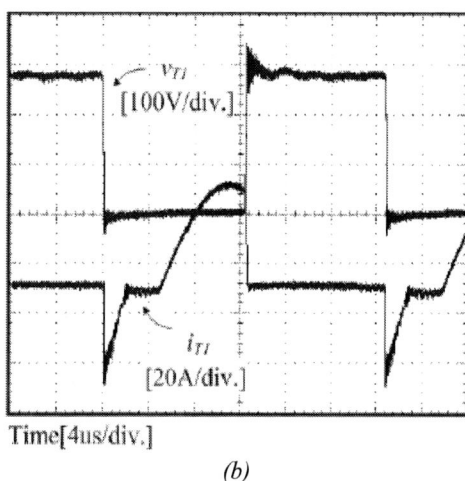

Time[4us/div.]

(b)

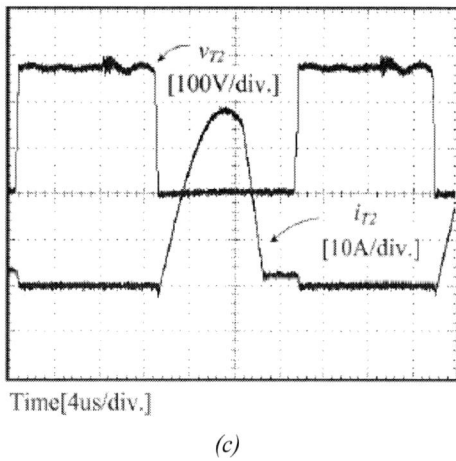

Time[4us/div.]

(c)

Fig. 8: Experimental results: (a) ZCS at 5% load (train at 1m/s speed), (b) ZCS at medium load level (train is at 5m/s) and (c) ZCS at full-load (train is at 20m/s)

V. CONCLUSION

Analysis and design of a high power LCC resonant inverter used to power a train LIM used in urban transit applications is presented in this paper. The proposed inverter topology combined with a self-oscillating controller to achieve voltage regulation and soft-switching results in high power efficiency, high power density, and good EMC performance. The inverter smaller under floor footprint and layout in the train allows additional space for other battery resources. The high efficiency offered by the presented LCC inverter, ranging from 99% at full load to 94.9% at 5% load results in lower energy consumption and therefore lower energy bill. Moreover, the paper presents experimental prototype design and results thayt validated the operation and advantages of the LCC inverter.

REFERENCES

[1] http://www.theclimateisrightfortrains.com/?lang=en

[2] http://www.windfair.net/transport.bombardier/welcome.html.

[3] H. Kuhn, T. Koneke, and A. Mertens," Consideration for a Digital Gate Unit in High Power Applications," Proceedings of the IEEE Power Electronics Specialist Conference, June 2008. pp. 2784-2790

[4] T. Kejellqvist, S. Ostlund, and S. Norrga, "Active Snubbber Circuit for Source Commutated Converters Utilizing the IGBT in the linear region" IEEE Transactions on Power Electronics, vol. 23, September 2008. pp. 2595-2601

[5] K. Fuji, P. Kollensperger, and R. W. Doncker, "Characterization and Comparison of High Blocking Voltage IGBTs and IEGTs under Hard and Soft Switched Conditions, "IEEE Transactions on Power Electronics, vol. 23, January 2008. pp. 172-179

[6] L. Lorenz, "Key power Semicondictor deices and Development Trends," Proceeding of the International Conference on Electrical Machines and Systems (ICEMS), October 2008, pp. 1137-1142.

[7] F. C. Lee, S. Wang, P. Kong, C. Wang and D. Fu, "Power Architecture design with Improved System Efficiency, EMI and power Density," Proceeding of the IEEE Power Electronics Specialists Conferece, June 2008, pp. 4131-4137.

[8] K. Shyu, C. Lai, K. Jwo, M. Pan and C. Ku, "Using Automatic Frequency Shifiting Technique for LLC-SCR Output Volatge Regulation," Proceeding of IEEE 5th International Power Electronics and Motion Control Conference, August 2006, pp. 1-5.

[9] A. Bucher, T. Durbaum, D. Kubrich, A. Stadler, "Comparison of Different Design Methods for the Parallel resonant Converter," Proceeding of 12th International power Electronics and Motion Control Conference, August 2006, pp. 810-815.

[10] M. Z. Youssef and P. K. Jain, "Modeling Techniques of resonsnat Converters: Merits and Demerits," Proceeding

of Candian Conference on Electrical and Computer Engineering, 2004, pp. 241-244.

[11] E. S. Kim and Y. H. Kim, "A ZVZCS PWM FB DC/DC Converter using a Modified Energy Recovery Snubber," IEEE Transactions on Industrial Electronics, vol. 48, July 2002. pp. 1120-1127

[12] J. Dudrik, P. Spanik, and N. D. Trip, "Zero Voltage and Zero Current Switching Full-bridge dc-dc Converter with Auxiliary Transformer," IEEE Transactions on Power Electronics, vol. 21, 2006. pp. 1328-1335

[13] M. Z. Youssef, H. Pinheiro, and Praveen K. Jain," Closed Loop Implementation and Design of a Novel Controller for 48V Voltage Regulator Module with Self-sustained Oscillation Controller", Proceedings of the Applied Power Electronics Conference, APEC, March 2006. pp. 1113-1120

[14] J. E. Slotine, and W. Li," Applied Nonlinear Control", 1st edition, Prentice Hall, 1991

[15] Ja. Z. Tsypkin," Relay Control Systems", 1st edition, Cambridge University Press, 1984

[16] M. Z. Youssef, and Praveen K. Jain,"Series-parallel Resonant Converters with the High Frequency Effects: Analysis, Modeling, and Design", IEEE Transactions on Industrial Electronics, vol. 54, issue 3, July 2007. pp. 1329-1341

[17] M. Z. Youssef, and M. Orabi,"Sampled-data Modeling of A New Ultra Fast 48V Voltage Regulator Module", Proceedings of the IEEE Applied Power Electronics Conference, APEC, February 2008. pp. 182-192.

[18] S. Zheng, and D. Czaarkowski," Modeling and Control of a Phase Controlled Series Parallel Resonant Converter", IEEE Transactions on Industrial Electronics, vol. 54, issue 3, July 2007. pp. 707-715

[19] W. J. Lee, C. E. Kim, G. W. Moon and S. K. Han," A New Phase-shifted Full-bridge Converter with Voltage-doubler Type Rectifier for High-efficiency PDP Sustaining Power Module", IEEE Transactions on Industrial Electronics, vol. 55, issue 3, July 2008. pp. 2450-2458

978-1-4244-4782-4/10 $26.00 © 2010 IEEE

A Multi-resolution Control Strategy for DSP Controlled 400Hz Shunt Active Power Filter in an Aircraft Power System

Haibing Hu, Wei Shi
Aero-Power-Sci-Center
Nanjing University of Aeronautics and
Astronautics, Nanjing, 210016, China
huhaibing@163.com

Jianren Xue
Nanjing SUTE ELECTRIC Co., LTD
Nanjing, 210016, China

Ying Lu, Yan Xing
Aero-Power-Sci-Center
Nanjing University of Aeronautics and
Astronautics, Nanjing, 210016, China

Abstract: **A multi-resolution control strategy is proposed for a DSP-controlled 400Hz Active Power Filter(APF) to reduce the real-time computational requirements. In this manner, the current loop bandwidth is extended for better harmonic compensating performance. By rearranging the computational elements into high and low computational groups, the proposed control strategy takes best advantages of the DSP computation resources to increase the control frequency for the high computational group. Based on the control bandwidth, analysis is presented in determining the sampling frequency for the low computational group. A 20kVA prototype is setup to verify the validity of the proposed strategy. Experimental results show that the proposed control strategy achieves good current compensating effects and dynamic performance.**

I. INTRODUCTION

In today aircraft industry, electrical aircrafts are increasingly employed to achieve better efficiency, cost reduction and better performance[1-5]. As a result, the number of electrical loads equipped on-board is increasing and the on-board power capacity is getting larger. However, most of the DC loads are typically supplied by the uncontrolled diode rectifier converters, causing the power quality problem in the aircraft power system. The high harmonic current distortion and poor power factor would be a major concern when the percentage of total system power processed by uncontrolled rectifiers is high. To address this problem, some standards such as DO-160D and ISO-1540 have been revised to keep strict limits on harmonic currents which user equipment can draw from the ac source[2-3]. To meet these requirements, some techniques such as active power filter (APF), power factor correction and multiphase (12- or 18-pulse) transformer-rectifier are employed [2][4-8]. Among these techniques, the shunt APF is the best solution due to its advantages in lower volume and lighter weight over other techniques, which are among most key issues in aerospace applications. Unlike other two techniques, another merit of shunt APF is that it does not alter the existing power stage and the failure of the APF does not threaten the system power supply and normal operation, which means installation of APF will not do harm to the reliability of the original power system.

Extensive research has been conducted on the APFs applied to the 50Hz or 60Hz commercial power systems[6]. Although the principle for APF applied to the commercial utility and to the aircraft power system is the same, there is a great challenge in designing high performance 400Hz APF due to the computational limits of DSPs suitable for power conversion. In this regard, there has been few publications on 400Hz APFs applied to the aircraft power system[7-8]. In this paper, the computation limits is addresses for achieving higher current control bandwidth using the mutli-resolution control strategy.

II. PRINCIPLE OF SHUNT APF

As seen from Fig.1, the control system consists of four main functional blocks: (1) Software phase-locked loop; (2) DC bus voltage control; (3) Low pass filter; (4) Current control loop. The measured load phase currents(i_{La}, i_{Lb}) are transformed into the synchronous reference frame to obtain i_{qL} and i_{dL}. The synchronous reference frame phase angle can be obtained by processing the measured system voltages with software phase-locked loop algorithm. Low-pass filters are applied to filter out the DC components in the synchronous reference frame, which represents the fundamental frequency components of the load currents. The harmonic components are easily obtained by a simple subtraction of the filtered components($\bar{i}_{qL}$, $\bar{i}_{dL}$)from the transformed components(i_{qL}, i_{dL}). Due to losses in power devices and load transient, the real power is consumed or transferred between the DC bus and the grid, which would result in DC voltage decrease or fluctuations. To keep the DC voltage constant, DC voltage regulator is used by adding DC regulating value Δi_d to the d-axis filter current. All these computations are implemented in a Digital Signal Processor(DSP).

Research was sponsored by the Doctoral Program of Higher Education of China (200802871040).

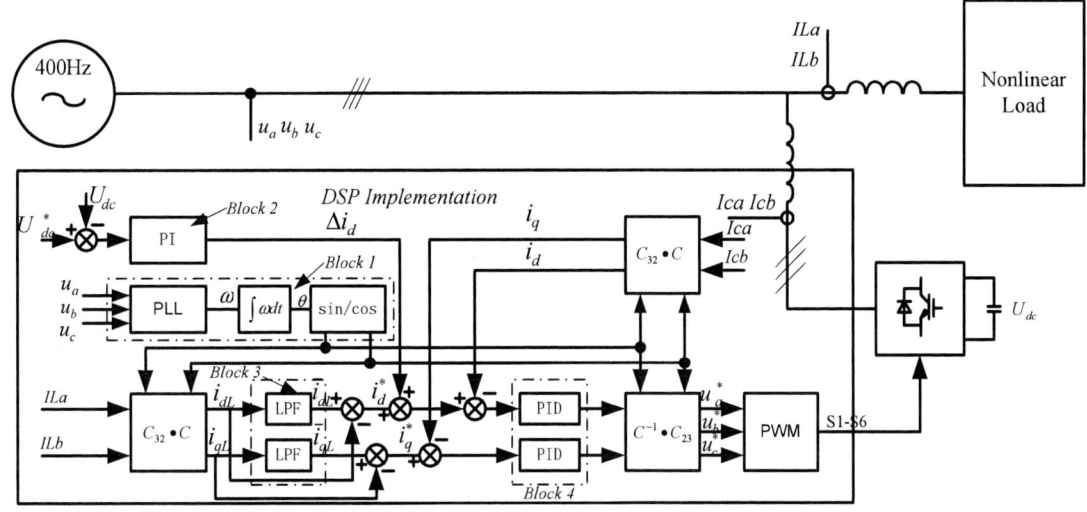

Figure 1: Structure of APF

III. EFFECT OF CONTROL DELAY ON CURRENT LOOP

In a balanced three-phase system, the converter model in the d-q rotating reference frame can be described as:

$$\begin{bmatrix} L\dfrac{di_d}{dt} \\ L\dfrac{di_q}{dt} \end{bmatrix} = \begin{bmatrix} -R & \omega L \\ -\omega L & -R \end{bmatrix}\begin{bmatrix} i_d \\ i_q \end{bmatrix} - \begin{bmatrix} u_d \\ u_q \end{bmatrix} + \begin{bmatrix} e_d \\ e_q \end{bmatrix} \tag{1a}$$

$$C\frac{du_{dc}}{dt} = \frac{3e_d}{u_{dc}}i_d \tag{1b}$$

$$\begin{bmatrix} e_d & e_q & e_o \end{bmatrix}^T = T\begin{bmatrix} e_a & e_b & e_c \end{bmatrix} \tag{1c}$$

$$\begin{bmatrix} i_d & i_q & i_o \end{bmatrix}^T = T\begin{bmatrix} i_a & i_b & i_c \end{bmatrix} \tag{1d}$$

$$\begin{bmatrix} u_d & u_q & u_o \end{bmatrix}^T = T\begin{bmatrix} d_a & d_b & d_c \end{bmatrix}u_{dc} \tag{1e}$$

Where T is the 3s/2r transformation matrix given by

$$T = \frac{2}{3}\begin{bmatrix} \sin\omega t & \sin(\omega t - \dfrac{2\pi}{3}) & \sin(\omega t - \dfrac{4\pi}{3}) \\ \cos\omega t & \cos(\omega t - \dfrac{2\pi}{3}) & \cos(\omega t - \dfrac{4\pi}{3}) \\ \dfrac{3}{2} & \dfrac{3}{2} & \dfrac{3}{2} \end{bmatrix}$$

In 1(e), vector $\begin{bmatrix} d_a & d_b & d_c \end{bmatrix}$ represents the control duty cycles of the inverter for the three legs respectively. The converter plant depicted in expression 1(a), which is valid for large signal operation, is a nonlinear and coupling system. With the feed-forward and dc-coupling terms, the transfer functions of the current loop being controlled can be linearized as:

$$G_{d\_c}(s) = \frac{1}{R + Ls} \tag{2a}$$

$$G_{q\_c}(s) = \frac{1}{R + Ls} \tag{2b}$$

Due to the computational delay in digital control system, a pure control delay is unavoidably introduced to the system, which has negative effect on the control bandwidth. The delay time T_d depends on the current loop control strategy, whose transfer function can be expressed as:

$$G_{delay}(s) = e^{-sT_d} \tag{3}$$

Due to its fast response, a P regulator is employed in the current loop control, whose diagram is illustrated in Fig.2. The parameters of the APF are listed in Tab. I.

Tab.I: Key parameters of the shunt APF converter

Parameters	Values
Input phase voltage	115
System fundamental frequency	400
L$_1$/mH	0.1
Lc/mH	0.1
DC capacitor Cdc/uF	1000
DC voltage udc/v	400
Switching frequency fc/kHz	50

The bandwidth of the current loop will decide the compensating effect of the APF. To achieve the better compensating effect, the current bandwidth should be designed as wide as possible under the condition of system stability. For simplification, the controller was designed based on continuous linear system without introducing the pure delay unit. In this application case, the cutoff frequency f_c of the current open loop is designed at 10kHz, which is one-fifth of the switching frequency 50kHz, and the phase margin is designed at 90^0 to make sure the current loop has enough stability margin when introducing a pure delay unit.

In a delay system, the reduction in phase margin can be accounted for by the pure time delay T_d, which translates to a phase lag as

$$Phase_{lag} = \omega T_d = 2\pi f_c T_d \qquad (4)$$

Where $f_c = 10kHz$ is the cutoff frequency at which the phase lag can be calculated.

Fig.3 shows the relationship between delay time and phase margin reduction, from which we can see that the current loop will fall in the unstable region when the delay time T_d is greater than 25us. As aforementioned, the delay time T_d depends on the DSP computation time and the duty cycle updating scheme. Many multisampling methods have been proposed to reduce delay time in the digital controller for PWM converters. However all these methods are unavoidable to sample the switching ripples, resulting in the degradation of the expected dynamic response, and they also require more hardware resources and computation efforts. Due to these reasons, multisampling schemes are not suitable for the high speed and computational high-demanding system-400Hz APF. In this paper, two typical "synchronous sampling methods", as illustrated in Fig.4, were chosen to make comparisons of their effects on the harmonic compensation. Two schemes have Ts and Ts/2 delay time respectively. With introducing the delay unit, the current close loop transfer function can be expressed as:

$$G_{i\_close} = \frac{k_p k_{pwm} e^{-T_d s}}{R + Ls + k_p k_{pwm} e^{-T_d s}} \qquad (5)$$

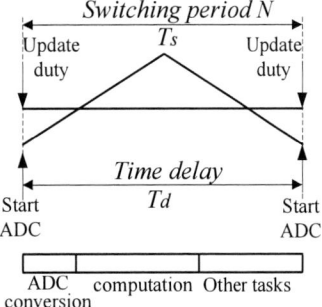

Figure 3: The relationship of phase margin and delay time

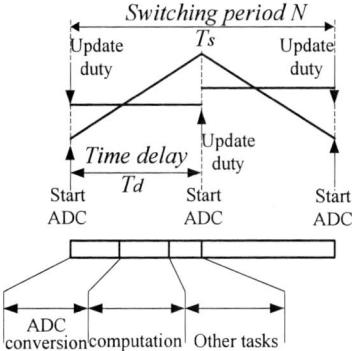

(b) Scheme 2

Figure 4: Two synchronous sampling schemes and their delay time

According to the close loop transfer function, we can calculate the phase lag and amplitude for each harmonic as illustrated in Fig.5. For an ideal close loop transfer function, the phase lag should be equal to 0 and the gain should be equal to unity. As seen from Fig.5(a), due to the current limited bandwidth(10kHz), the phase lag increases as harmonics order increases in all three conditions(no delay, 10us delay and 20us delay).Fig.5(b) shows the gain changes as harmonic order increases. It should be noted: (1) In the 10kHz bandwidth design, although the phase lag performance of no delay is better than that of 10us delay, the gain performance of 10us delay is better than that of no delay; (2) Due to limited current bandwidth, the harmonic orders more than 15th will be impossible to achieve the good compensating effect. In this design, the APF is not expected to compensate these high order harmonics, since these high order harmonics can be easily filtered out using small reactive filter installed at load side. As for sampling scheme, we choose synchronous sampling scheme 2 to implement the digital control. Generally, for APF a smaller time delay in current control loop is preferred to achieve the better current compensating performance. How to decrease the delay time in the current loop for a fixed computational DSP system? One feasible method is to reduce other control/computation frequencies in order to acquire the more computational resource for the current loop, which is referred to as the multi-resolution control strategy in this paper.

(a) Scheme 1

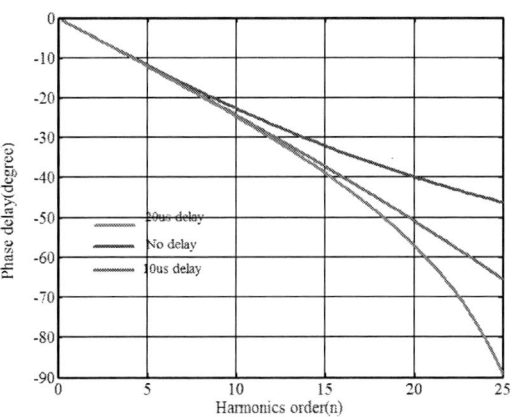

(a) Phase lag changes with harmonic order increase

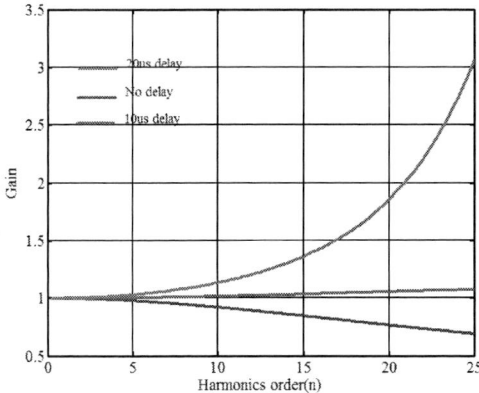

(b) Amplitude changes with harmonics order increase.

Figure 5: Phase and amplitude change as the harmonic order increases

IV. MULTI-RESOLUTION CONTROL STRATEGY

In this paper, all the computations are executed on the DSP, TMS320F2812, whose maximal operating frequency is 150MHz. The computational elements and their execution time are listed in Tab. II.

Table II: Summary of computational elements and their execution time

Computation task number	Computational elements	Execution time(us)
1	DC bus regulator	0.8
2	Software phase-locked loop	1.2
3	Computation of $\sin\theta$ and $\cos\theta$	1.2
4	3/2 Transformation of load currents	0.6
5	Low-pass filter	2.3
6	3/2 Transformation	0.6
7	Current loop regulator	1.2
8	2/3 transformation and PWM update	1.2
9	Protection	0.5
10	AD conversion for 7channels	2.1
	Total	11.7

As aforementioned in Section III, to achieve the better compensation effect, the current control frequency should reach up to 100kHz when employing the "synchronous sampling" scheme 2. However due to the limits of the DSP computation throughput, all these computational elements, whose total execution time is 11.7us as seen in Tab II, can not be completed within a control period of 10us.

As well known, the bandwidth of the different plants decides their control frequencies. In the APF system, current loop with wide bandwidth needs fast control to achieve good current tracking performance, while DC bus voltage with support of bulky capacitors can be well regulated even with slow control and software PLL, which usually takes several fundamental cycles(400Hz) to lock the phase, can be implemented in slow control as well.

To address the computation limitation issue in the 400Hz digitally controlled APF applications, the multi-resolution control strategy partitions the APF controller into high and low frequency computation groups based on their control bandwidth. The high frequency computation group, which executes twice in one switching period, has to complete its computations before underflow point and overflow point, respectively. Low frequency group, which executes once in n switching periods, is allocated to one switching period, during which the DSP completes all the computation for both groups. The diagram of timing sequence for the multi-resolution control strategy is illustrated in Fig.6. Computational elements, (3,4,6,7,8 and 9) in Tab.II, which are related to the current loop calculation, operate with 100kHz control frequency, and have total execution time of 6.56us(less than 10us). In this manner, the high frequency group can be completed in one high frequency control cycle, while other computational elements can be executed with much lower frequency.

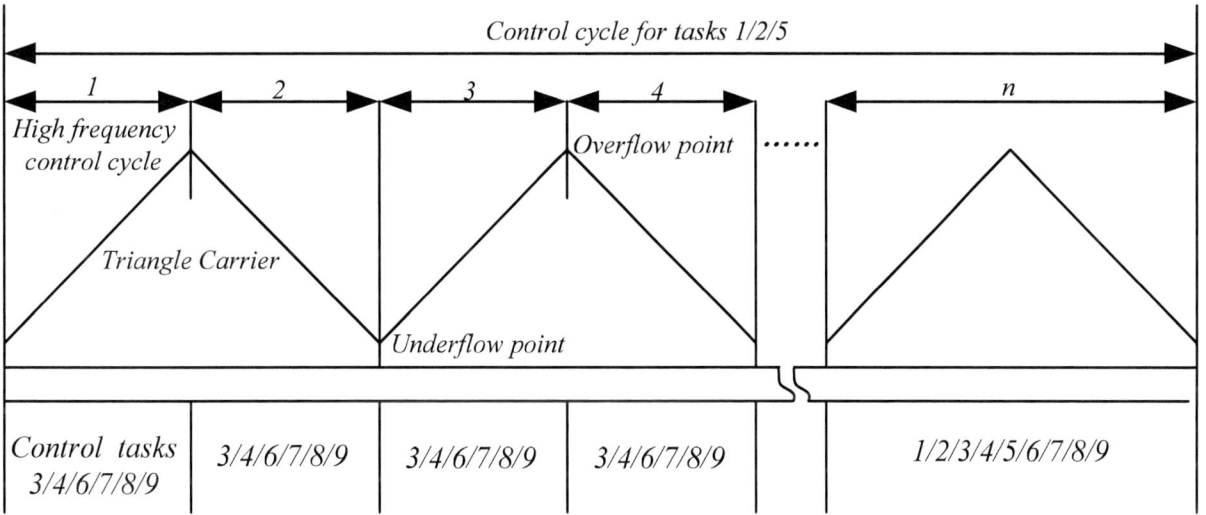

Figure 6: The diagram of control timing sequence for multi-resolution strategy

V. DETERMINATION OF CONTROL CYCLES FOR LOW FREQUENCY GROUP

A. The Control Frequency for DC Voltage Regulator

DC voltage regulator is designed to maintain the dc capacitor voltage at a constant value. During steady state operations, the DC capacitor voltage will not change much due to no real power transferred across the active power filter. The only pass to discharge the DC capacitor are the losses in switching devices and other components. In this case, only a small amount of real power is needed to charge the DC capacitor. The DC voltage regulator could be designed to be low cutoff frequency. The diagram of the DC voltage control loop is shown in Fig.7. In the proposed multi-resolution control strategy, one switching cycle delay($2T_d$) is supposed to be introduced in the voltage control loop. Since in this design we want to decide the control frequency for the voltage regulator, the open loop bode plot without PI regulator is plotted as shown in Fig.8. As seen from Fig.8, the cutoff frequency of the $G_v G_{i\_close}$ is around 55Hz, which will be fast enough to regulate the DC capacitor voltage when operating in steady states. It seems that the control frequency of DC voltage loop can be operated in 2kHz due to its great attenuation (-70db) at this frequency point.

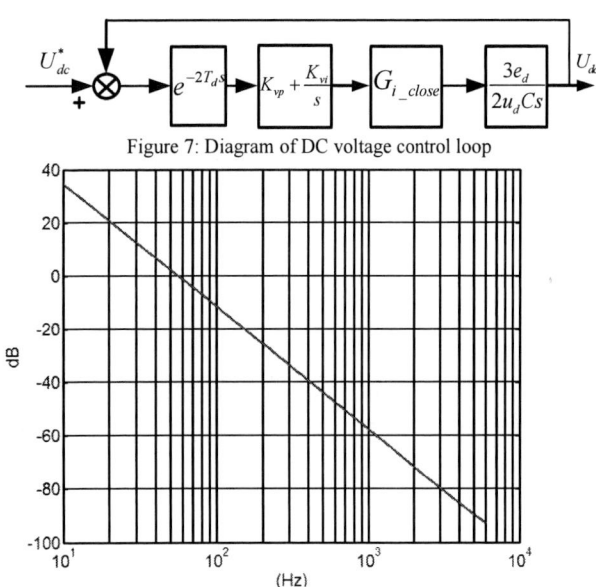

Figure 7: Diagram of DC voltage control loop

Figure 8: Bode diagram of open voltage loop

However, during APF transient response, a great amount of real power will be transferred back and forth across the active power filter due to the fact that LPFs "treat" the transient currents caused by sudden load changes as "harmonics", which are real power and will charge or discharge the DC capacitors, resulting in the great voltage fluctuations across DC capacitor.

Fig. 9 shows the diagram of real power transfer during load step-down transience. When the load steps down, the output current drops rapidly. The d-axis current in d-q rotating reference frame, which is fed to LPF to filter out the DC components, will drop correspondingly. Since the LPFs

are designated to extract the DC components from the load currents, the bandwidth of LPFs is design to be low(in next subsection, a detailed design for the bandwidth of LPFs is given). Therefore, the transient currents will be filtered out and be treated as "harmonics". Real power is needed to compensate these "harmonics" and the DC capacitor will be charged or discharged during the real power transfer. If the bandwidth of the DC voltage loop is designed too low, the disturbance from load changes can not be restrained to a low level, resulting in high voltage fluctuations across the DC capacitor and thus a higher voltage stress on switches and the DC capacitor. In order to lower the voltage fluctuations, a higher bandwidth is desired to better track the disturbance. During steady state operation, the harmonics compensation will cause the DC voltage ripples, whose frequency should be kept out of the bandwidth of the voltage loop to maintain the DC voltage stable. In this design, we made a trade-off decision on the bandwidth of the voltage loop as the wider bandwidth needs high sampling and control frequency. A 600Hz bandwidth is selected for the voltage open loop in this design as shown in Fig. 10. The attenuation at 5kHz is -50dB, which indicates that the 10kHz sampling and control frequency would be enough without additional anti-aliasing filters according to Shannon sampling theorem.

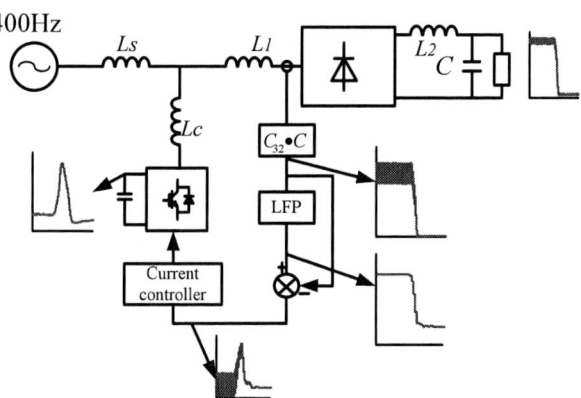

Figure 9: DC voltage fluctuations caused by load change

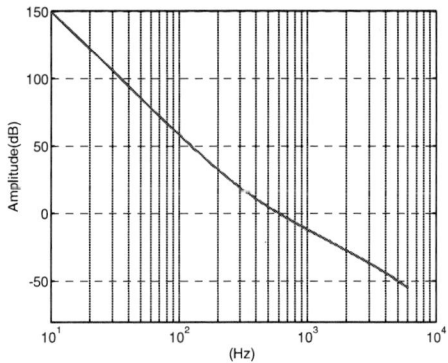

Figure 10: The Bode diagram of open loop with PI regulator

B. Low pass filter

A low pass filter with bandwidth 300Hz is designed to remove the harmonic components. Since in three phase system, the typical harmonic components are (6n±1) orders, after abc to dq transformation, the harmonic components will be transformed into [(6n±1)-1]. This means that the 13th order harmonic will be changed to 12th order after transformation. In the aircraft power system, the harmonic components are concentrated below 13th order as harmonic components above 13th order are easily filtered out by reactive components. In the application, the APF is designed to eliminate the harmonic components below 13th order. Therefore, based on Shannon sampling theorem, the sampling frequency should be greater than 9600kHz with high attenuation of high order harmonics by using the input anti-aliasing filters (twice the frequency of the highest harmonic order2x12x400Hz)

C. Phase-locked loop

Fig. 11 shows the block diagram of three-phase PLL system, whose linearized model is depicted as Fig.12. The phase tracking performance would be improved by designing a controller with a wider bandwidth. However, the use of higher bandwidth does not provide better results in practical applications, due to the great distorted voltages in the 400 Hz power system. The tracking performance deteriorates as the control bandwidth increases. Therefore, the selection of the control bandwidth is a compromise between various factors. In this application, a PLL controller with 400Hz bandwidth is designed, whose sampling frequency is selected to be below 9.6kHz.

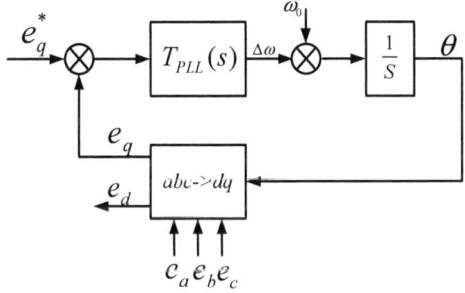

Figure 11: Block diagram of three-phase PLL system

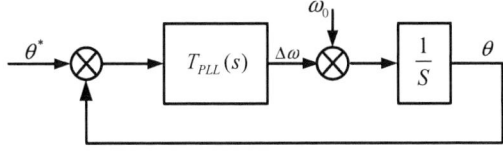

Figure 12: Linearized model of three-phase PLL

Based on above analysis and simple realization, a sampling frequency of 12.5kHz is selected for implementing the low frequency computation group.

VI. EXPERIMENT RESULTS

A 20kVA APF prototype was setup in the laboratory as shown in Fig. 13. Considering current stress on the power devices, two inverter modules were paralleled with common DC bus and same PWM signals. The key parameters are same as listed in table I. Uncontrolled diode rectifier with LC filter and R functions as a nonlinear load, whose parameters are 0.1mH, 330uF and 3.6 Ω, respectively.

Fig. 14 shows the phase A voltage and current waveform without operation of APF. The Total Harmonic Distortion(THD)of the current reached up to 23.5%. Fig.15 shows the phase A voltage and current waveform with operation of APF. The current THD is lowered down to 4.5%. With wide bandwidth of the voltage loop control, the DC voltage is well controlled during load change from 20kW to 3.5kW as illustrated in Fig.16, in whose case the DC voltage fluctuation is controlled within 30V. Fig.17 and 18 show the dynamic performances with load change from 3.5kW to 20kW, and vise versa. From Fig.17, it can be inferred that, during transient state from 3.5kW to 20kW the DC bus voltage is still maintained the same. This means, during the transient state the DC voltage regulator with its low control period has the capability to keep the DC bus stable. As seen from Fig. 18, APF current settling time is less than 4 fundamental periods, indicating good dynamic performance.

Figure 13: 20kVA prototype

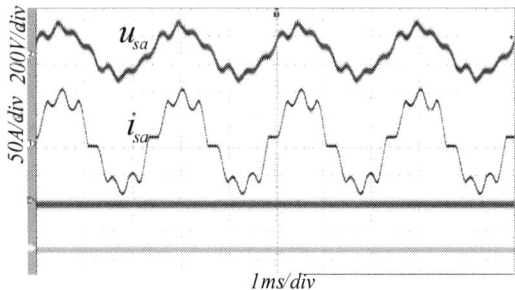

Figure 14: Phase voltage and current before APF operation

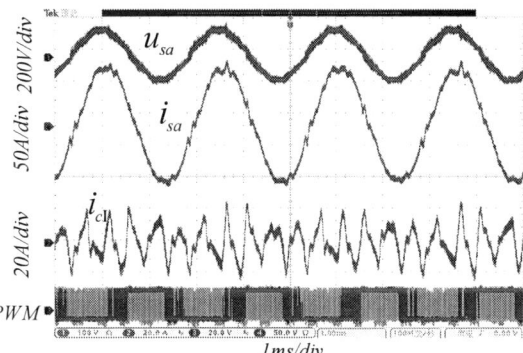

Figure 15: Phase voltage and current after APF operation

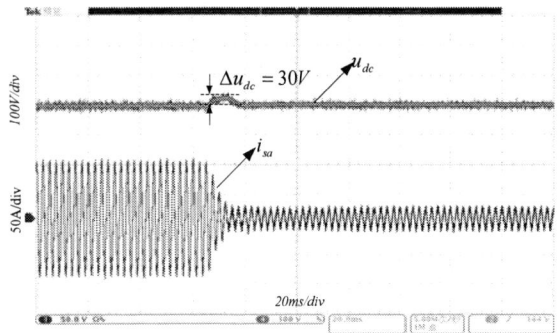

Figure 16: DC voltage transient response during load change from 20kW to 3.5kW

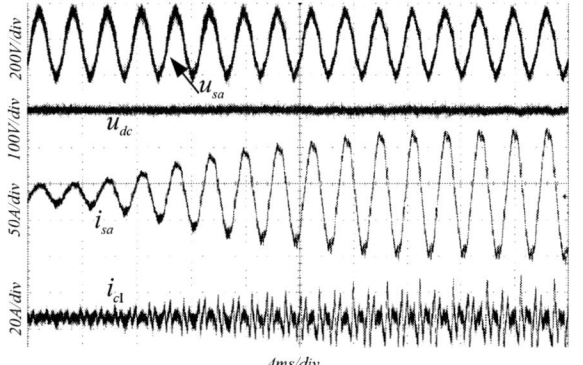

Figure 17: Phase voltage and current during load change from 3.5kW to 20kW

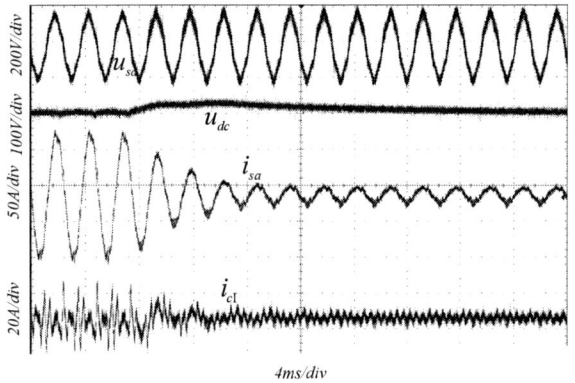

Figure 18: Phase voltage and current during load change from 20kW to 3.5kW

VII. CONCLUSION

The proposed multi-resolution control strategy divides the computational elements of APF into high and low frequency computational groups. The High frequency computational group is executed twice each switching period with new PWM value updating either at overflow point or underflow point. After high frequency computational elements are executed 8 times with updating PWM value only at underflow point, a switching period is allocated to all computations. The proposed method overcomes the limitation of the DSP computation resources in the 400Hz APF control system by rearranging the computation elements based on

their different requirements of control bandwidth. The experimental results verified the proposed control strategy, indicating that the control system has good harmonic current compensating effect, in addition to exhibiting a good dynamic response.

[1] Rosero,J.A. ;Ortega,J.A.; Aldabas, E.; Romeral,L. Moving towards a more electric aircraft. IEEE A&E Systems Magazine. Vol.22, no.3. March, 2007 pp.3-9.

[2] Jian Sun. Aircraft Power System Harmonics Involving Sigle-Phase PFC Converters. IEEE Trans. On Aerospace anf Electronic Systems, Vol.44, no.1, Jan. 2008, pp.217-226.

[3] Environmental conditions and test procedures for airborne equipment. DO-160D, Section 16-Power Input.

[4] Athalye.P, Maksimovic.D, High-performance front-end converter for avionics applications. IEEE Trans. on Aerospace and Electronic System, 2003, 39(2):462-470.

[5] G.Gong, M.L.Heldwein, U.Drofenik, etal. Comparative evaluation of three-phase high power factor AC-DC converter concepts for application in future more electric aircrafts. in proc. IEEE APEC 2004, pp.1152-1159.

[6] EI-Habrouk M. Darwish, M.K; Mehta,P. Active Power Filters: a review. IEE Proceedings Electric Power Applications. Vol.147, no.5,Setp.2000,pp.403-413.

[7] Donghua Chen, Tao Guo, Shaojun Xie, Bo Zhou. Shunt Active Filters Applied in the Aircraft Power Utility, Proc. IEEE power Electronics Specialists Conference, 2005,pp.59-63.

[8] Milijana Odavic, Pericle Zanchetta, Mark Sumner. A Low Switching Frequency High Bandwidth Current Control for Active Shunt Power Filter in Aircrafts Power Networks. The 33rd Annual Conference of the IEEE Industrial Electronics Society, 2007,pp.1863-1868.

Battery Discharge Regulator for Space Applications based on the Boost Converter

A. Fernandez, F. Tonicello, J. Aroca, O. Mourra
Energy & Power Conversion Division
European Space Agency
Keplerlaan 1, 2201 AZ Noordwijk, The Netherlands
Arturo.Fernandez@esa.int

Abstract— When a DC/DC converter is used to control the power on the main bus of a spacecraft it needs to comply with highly demanding requirements from the dynamic point of view. Hence, converters with very good dynamic performance, i.e. with no zero on the right-half plane are typically used. The conventional boost converter has not been used in European space applications mainly because of its poor dynamics. This paper addresses the way of using a boost converter in this application and demonstrates that this converter can cope with the very demanding requirements needed. A prototype has been built and tested to proof it.

I. INTRODUCTION

The power system is one of the key subsystems of a spacecraft (SC). Efficiency and robustness are the key requirements needed. Moreover, the SC power bus should be as ideal as possible because it will be loaded with very heterogeneous kinds of electric loads. In practice, this means that the designer should target as high control bandwidth as possible and also as low output impedance as possible. In the case of European SCs, there is a series of standards called European Cooperation for Space Standardisation (ECSS) in which the main requirements for the different subsystems are established. In the case of Electrical and Electronic circuits, the relevant regulation is the ECSS-E-ST-20C [1].

There are two key requirements that a converter used in a fully regulated bus system must comply with: the phase margin should be at least 60° and the output impedance must be kept below the template shown in Fig. 12. This template is derived from the requirement that the bus voltage transients shall not exceed 1% of its nominal value for load transients up to 50%. As can be seen, the requirements are very demanding and, as a consequence, the bus voltage quality will be good enough to cope with any kind of load that may be plugged to it. The down side of this is that such demanding requirements prevent "bad dynamics" converters from being used in the power system. If the battery voltage is below the bus voltage, the battery discharge regulator (BDR) needs a boost topology. Fig. 1 shows the scheme of a fully regulated bus power system. In this scheme, when the solar power is enough to

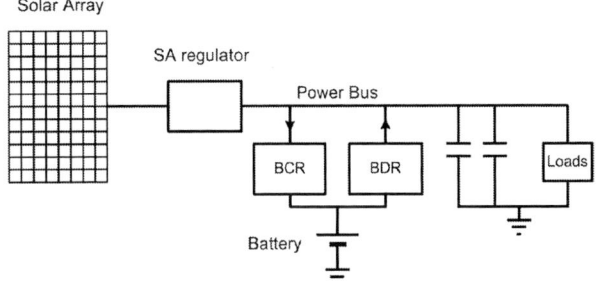

Fig. 1: Regulated bus power system scheme

supply all the loads the Solar Array Regulator (SAR) controls the bus voltage. If needed, the Battery Charge Regulator (BCR) charges the battery. When the solar array power is not enough (e.g. in eclipse), the BDR supplies the loads from the battery. Thus, the bus voltage is controlled by this BDR.

As far as the conventional boost converter (Fig. 2a) has a right half plane zero (RHPZ), it has been traditionally considered as not suitable for this application. As a consequence, other alternatives as the double inductor boost converter (Fig. 2b) and the Weinberg converter (Fig. 2c) have been used. These two topologies don't have the RHPZ and hence, their dynamic performance is quite good. However, both topologies are more complex than the boost and their size and weight is higher. Furthermore, since the stress on the semiconductors is higher than in the boost converter, their efficiency is lower. This paper addresses the way to enhance the boost dynamics as much as possible to cope with the ECSS requirements and to enable the use of this converter in the SC power system.

II. REVIEWING AVERAGE CURRENT MODE CONTROL

It is very well known that current mode control achieves higher bandwidths than voltage mode control. The key idea is that, when the current loop is closed, the converter behaves as an almost ideal current source. As the dynamics of the inductor are hidden within the current loop, the overall behaviour of the converter corresponds to a kind of first order system. Hence, the loop is much easier to close and the

978-1-4244-4782-4/10 $26.00 © 2010 IEEE

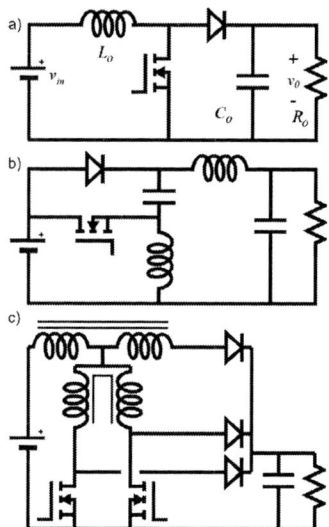

Fig. 2: a) Boost converter; b) double inductor boost converter; c) Weinberg converter

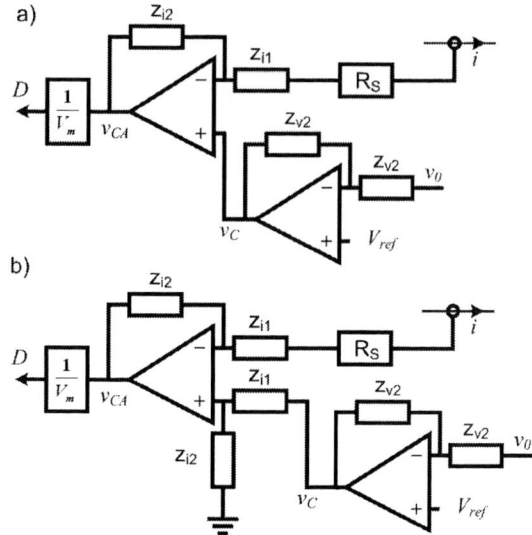

Fig. 3: a) conventional current controller; b) controller based on a differential amplifier

bandwidth that can be achieved is much higher than with a voltage mode control system [2-13]. Fig. 3a shows the typical current control loop. In this figure R_S represents the V/A ratio of the current sensor, V_m is the amplitude of the sawtooth waveform, v_{CA} is the output of the current amplifier and v_C is the control voltage coming from the voltage control loop. The expression of v_{CA} is:

$$v_{CA} = \left(1 + \frac{z_{i2}}{z_{i1}}\right) \cdot v_C - \frac{z_{i2}}{z_{i1}} \cdot R_S \cdot i \qquad (1)$$

This leads to the following closed loop current transfer function:

$$G_{ivC} = \frac{\hat{i}}{\hat{v}_C} = \frac{\left(1 + \frac{z_{i2}}{z_{i1}}\right) \cdot G_i \cdot \frac{1}{V_m}}{1 + \frac{z_{i2}}{z_{i1}} \cdot G_i \cdot \frac{R_S}{V_m}} \qquad (2)$$

In this expression, G_i is the transfer function between the duty cycle d and the current that is being controlled i. This current can be the inductor, the switch or the diode current. At low and medium frequencies the denominator of (2) is not able to cancel the numerator and hence, the transfer function is not exactly a first-order-like function as can be seen in Fig. 4. This may have an important impact on the design of the current loop, especially when an extreme design is targeted. When the controller is built around this transfer function, the crossover frequency will be very likely close to the resonance peak point of the function. As a consequence, the low frequency gain will be penalized because it is quite difficult for the controller to compensate such gain lack. Thus, the output voltage overshoot under load steps will be relatively high.

To improve this, the controller shown in Fig, 3b is proposed. As can be seen, if z_{i1} and z_{i2} are placed also on the non inverting input and a differential amplifier is built. The transfer function will actually be slightly simpler:

$$v_{CA} = -\frac{z_{i2}}{z_{i1}} \cdot \left(v_c - R_S \cdot i\right) \qquad (3)$$

and hence, the closed loop current transfer function will be:

$$G_{ivC} = \frac{\hat{i}}{\hat{v}_C} = \frac{\frac{z_{i2}}{z_{i1}} \cdot G_i \cdot \frac{1}{V_m}}{1 + \frac{z_{i2}}{z_{i1}} \cdot G_i \cdot \frac{R_S}{V_m}} \qquad (4)$$

This small modification has an interesting effect on the transfer function. As can be seen, when the loop gain is greater than 1, the numerator and denominator are almost cancelled and the expression simplifies to $G_{iLvC} = 1 / R_S$. Then, the gain will decrease with a -20dB/dC as in the case of a first order

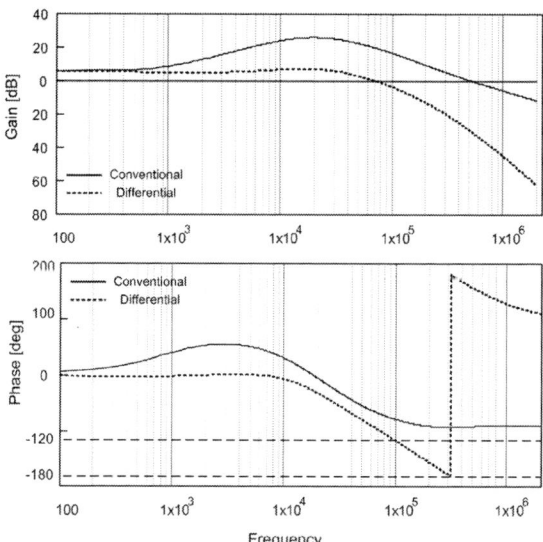

Fig. 4: Bode plot of the closed loop current loop (input current) with the conventional arrangement and with the differential arrangement

978-1-4244-4782-4/10 $26.00 © 2010 IEEE   1793

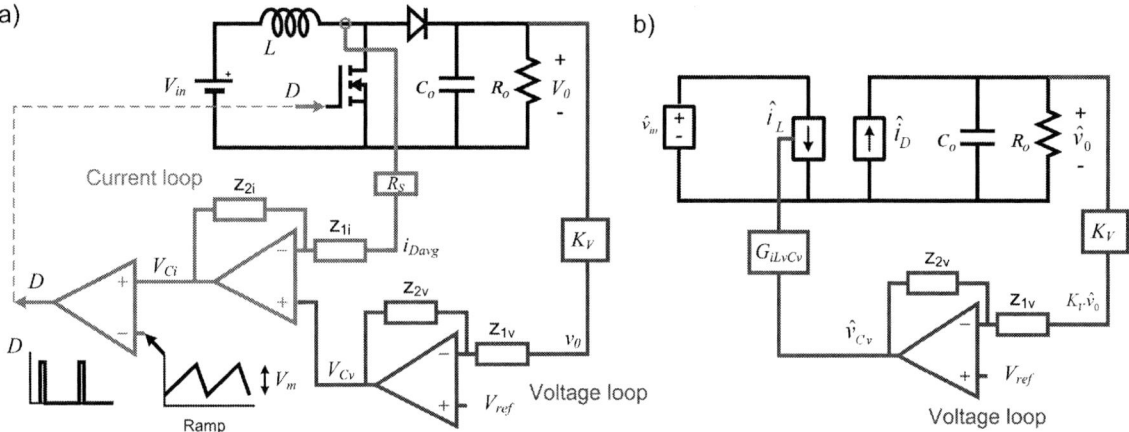

Fig. 5: a) Boost converter with input current mode control, b) equivalent small signal circuit

system as is shown in Fig. 4. This flat transfer function is easier to control than (2) and hence, the dynamic performance achieved will be typically better. The converter behaves basically as a resistor and when the loop is closed around it, the low frequency gain is not penalized so much. In the example shown in Fig. 4, the gain needed to achieve a 20 kHz crossover frequency with the conventional controller is around -20 dB while -5 dB are needed with the proposed one. This difference will be consequently reflected in the output impedance and on the output voltage performance under load steps.

It should be noted that the latter is the transfer function that is typically described when current mode control is explained, which is not actually achieved with the conventional controller shown in Fig, 3a.

III. CONTROLLING INPUT, OUTPUT OR SWITCH CURRENT

Average current control enables the possibility of controlling almost any current in the converter: inductor, switch or diode current. In fact, what is actually controlled is the averaged value. Obviously, depending on the current chosen to control, the transfer function will be different and the dynamic performance can be potentially different as well. The three currents will be used in this study so that the performance of all of them can be compared. Intuitively, the output current would seem to be the right choice to control the output voltage. However, this is difficult to say up front and the three options will be considered.

A. Input current control

The transfer function between the duty cycle and the inductor current is:

$$G_{iL} = \frac{\hat{i}_L}{\hat{d}} = \frac{2I_O}{\left(1-D\right)^2} \cdot \frac{1+\frac{1}{2}R_O C_O s}{1+\frac{L_e s}{R_O}+L_e C_O s^2} \quad (5)$$

I_0 is the output current, D is the duty cycle, R_0 is the output load and C_0 is the output capacitor. L_e corresponds to $L/(1-D)^2$ and is usually called effective inductance. As can be seen, there is no RHPZ in this function. However, the transfer

function between the output current and the duty cycle is also needed to close the voltage loop. Fig. 5a shows the scheme of the power plant and the control circuit. First, the closed loop current transfer function between the inductor current i_L and the voltage loop control voltage v_{Cv} will be calculated:

$$G_{iLvCv} = \frac{\hat{i}_L}{\hat{v}_{Cv}} = \frac{\left(1+\frac{z_{i2}}{z_{i1}}\right) \cdot G_{iL} \cdot \frac{1}{V_m}}{1+\frac{z_{i2}}{z_{i1}} \cdot G_{iL} \cdot \frac{R_S}{V_m}} \quad (6)$$

As can be seen in Fig. 5b, the diode current i_D is the one actually needed to calculate the output voltage. Thus, the transfer function between i_D and the control voltage is needed to close the voltage loop. This function has the following expression:

$$G_{i_D v_{Cv}} = \frac{\hat{i}_D}{\hat{v}_{Cv}} = G_{i_L v_{Cv}}\left(s\right) \cdot \left(\left(1-D\right)-\frac{I_L}{G_{iL}\left(s\right)}\right) \quad (7)$$

As can be seen, the term $(1-D)-I_L/G_{iL}$ does the conversion from i_L to i_D and eventually gives place to a RHPZ. Finally, the voltage loop gain will be:

$$T_{V_{i_L}} = K_V \cdot G_{CV}\left(s\right) \cdot G_{iLvCv}\left(s\right) \cdot \frac{R_O}{1+R_O \cdot C_O \cdot s} \quad (8)$$

K_V is the voltage divider ratio and $G_{CV}(s)$ is the voltage controller transfer function.

B. Output current control

The transfer function between the diode current and the duty cycle is:

$$G_{iD} = \frac{\hat{i}_D}{\hat{d}} = \frac{1}{R_O\left(1-D\right)^2} \cdot \frac{\left(V_{in}-I_L L_e s\right)\left(1+R_O C_O s\right)}{1+\frac{L_e s}{R_O}+L_e C_O s^2} \quad (9)$$

As can be seen, the RHPZ appears directly on this transfer function. This means that the current loop will be slightly more difficult to close with the output current than with the inductor current. On top of this, this current has a pulsed shape and as its average value should be obtained, a low pass filter will be needed in the current loop. This means that the bandwidth of this loop will be lower than in the case of using the inductor current. Note that the inductor current waveform will be relatively smooth and the filter needed is much smaller than the one needed to filter the diode or the MOSFET current. The closed loop transfer function will be:

$$G_{i_D v_{Cv}} = \frac{\hat{i}_D}{\hat{v}_{Cv}} = \frac{\left(1 + \dfrac{z_{i2}}{z_{i1}}\right) \cdot G_{iD} \cdot \dfrac{1}{V_m}}{1 + \dfrac{z_{i2}}{z_{i1}} \cdot G_{iD} \cdot \dfrac{R_S}{V_m}} \qquad (10)$$

And the voltage loop gain will be:

$$T_{V_{i_D}} = K_V \cdot G_{CV} \cdot G_{i_D v_{Cv}} \cdot \frac{R_O}{1 + R_O \cdot C_O \cdot s} \qquad (11)$$

C. MOSFET current control

This is the third possibility to control this converter and the procedure to follow to obtain the transfer functions is similar to the previous cases. The transfer function between the switch current and the duty cycle is:

$$\frac{\hat{i}_M}{\hat{d}} = G_{iM} = \frac{\dfrac{V_0 D}{Ls}}{1 + \dfrac{(1-D)^2}{Ls} \cdot \dfrac{R_0}{1 + R_0 C_0 s}} \qquad (12)$$

The closed loop current transfer function has then the following expression:

$$G_{i_M v_{Cv}} = \frac{\hat{i}_M}{\hat{v}_{Cv}} = \frac{\left(1 + \dfrac{z_{i2}}{z_{i1}}\right) \cdot G_{iM} \cdot \dfrac{1}{V_m}}{1 + \dfrac{z_{i2}}{z_{i1}} \cdot G_{iM} \cdot \dfrac{R_S}{V_m}} \qquad (13)$$

As in the case of the diode current, the MOSFET current has also a pulsed shape and it has to be filtered to be properly controlled. Again, the low pass filter needed for this purpose will compromise the controller bandwidth and the performance will be worse than in the case of the inductor current control. The relationship between the output current and the control voltage is needed to close the voltage loop. Hence, a transformation is needed to account for this:

$$G_{i_D v_{Cv}} = \frac{\hat{i}_D}{\hat{v}_{Cv}} = G_{i_M v_{Cv}}(s) \cdot \frac{\dfrac{V_0}{G_{iM}(s) Ls} - 1}{1 + \dfrac{1-D}{Ls} \dfrac{R_0}{1 + R_0 C_0 s}} \qquad (14)$$

Finally, the voltage loop gain will be:

$$T_{V_{i_M}} = K_V \cdot G_{CV} \cdot G_{i_M v_{Cv}} \cdot \frac{R_O}{1 + R_O \cdot C_O \cdot s} \qquad (15)$$

D. Output Impedance

The output impedance is another key parameter that will be needed to design the control loop and achieve a good

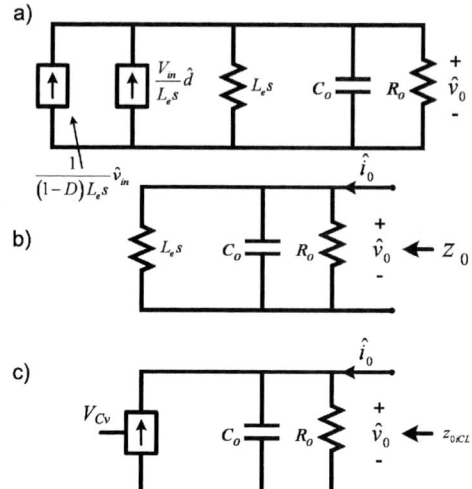

Fig. 6: a) small signal model of the output current and the output filter, b) small signal model of the open loop output impedance, c) small signal model of the closed loop output impedance

performance in transient response. The small signal model of the output current is:

$$\hat{i}_D = \left((1-D)\hat{v}_{in} - (1-D)^2 \, \hat{v}_0 + V_{in}\hat{d}\right)\frac{1}{L_0 s} \qquad (16)$$

In open loop, the circuit representing the output of the boost converter is then as shown in Fig. 6a. From the dynamic point of view, the effective inductance value L_e is higher that the physical one. Hence, the dynamic behaviour is slower than in the case of e.g a buck converter with the same inductor in which the dynamic and the physical value of the inductance are the same. Note also that the dynamic value depends on the operating point of the converter because it depends on the duty cycle value. To calculate the output impedance we assume that $\hat{v}_{in} = \hat{d} = 0$. Then, the equivalent circuit will be as shown in Fig. 6b. Then, the open loop output impedance will be:

$$Z_0(s) = \frac{\hat{v}_0}{\hat{i}_0} = \frac{L_e s}{1 + \dfrac{L_e}{R_0} s + L_e C_0 s^2} \qquad (17)$$

As can be seen, the inductance value is seen in the impedance expression when the converter is in open loop. When the current loop is closed, the circuit becomes slightly different as shown in Fig. 6c. The inductor is hidden in the current control loop and, when the open loop impedance is calculated now, we get:

$$Z_{0iCL}(s) = \frac{\hat{v}_0}{\hat{i}_0} = \frac{R_0}{1 + R_0 C_0 s} \qquad (18)$$

Finally, the closed loop value will be:

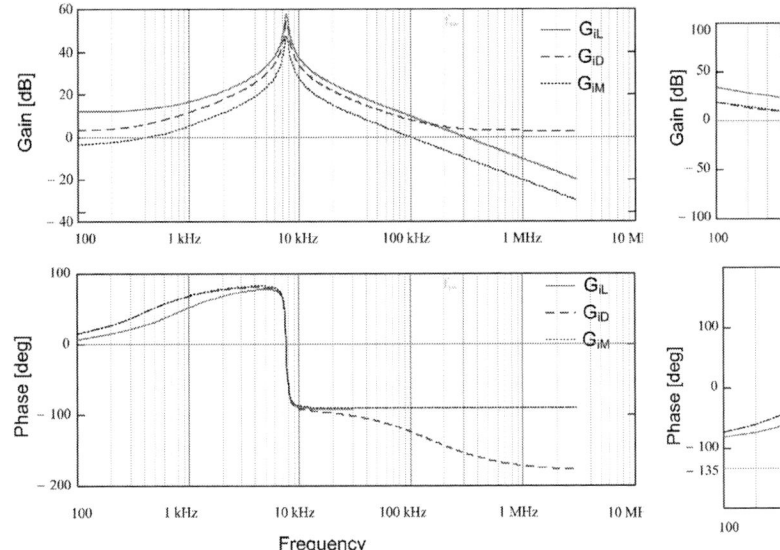

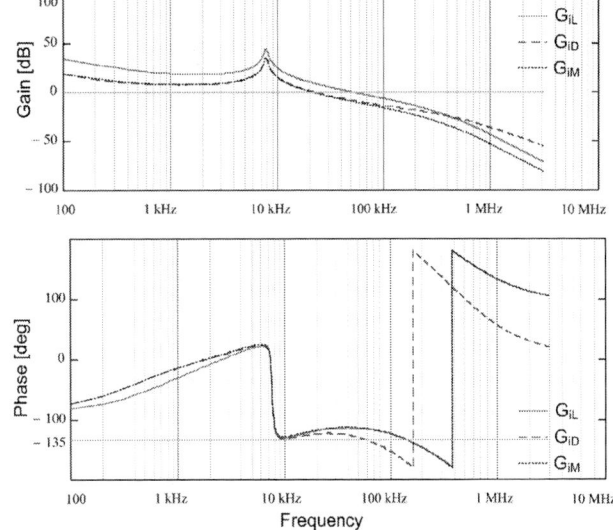

Fig. 7: Bode plots of the transfer functions between the three currents and

Fig. 8: Bode plots of the current loops of the three options

$$Z_{0CL}(s) = \frac{Z_{0iCL}(s)}{1+T_v(s)} \quad (19)$$

Where $T_v(s)$ is the open loop gain (8), (11) or (15) depending on the current that is controlled.

IV. PERFORMANCE COMPARISON

Fig. 7 shows the Bode plots of the transfer functions between the inductor current and the duty cycle (G_{iL}) (5), the diode current and the duty cycle (G_{iD}) (9) and the MOSFET current and the duty cycle (G_{iM}) (12). As can be seen G_{iL} and G_{iM} are very similar. However, G_{iL} has a higher gain than G_{iM}. On the other hand, G_{iD} shows an important difference: the RHPZ at around 100 kHz. As can be seen, the gain increases by an extra 20dB/dC and hence remains constant while the phase decreases by some extra 90 deg down to -180 deg. It should be noted that G_{iD} is already the current that flows into the output filter while G_{iL} and G_{iM} only account for the inductor and the MOSFET currents and still need to be transformed as was shown before. This issue is important for the current controller. These transfer functions are the one used to close the current loop. Hence, it will be slightly more difficult to close the loop in the case of the diode current than in the other two cases.

When closing the current loop, the crossover frequency has been tuned in such a way that there are no sub-harmonic oscillation problems. This occurs when the gain at the switching frequency is very high and the amplified sensed current "overpasses" the sawtooth waveform of the PWM generator. The conservative limit is to set:

$$f_{Ci\_max} = \frac{f_{sw}}{2\pi D} \quad (20)$$

Being f_{Ci_max} the maximum crossover frequency and f_{sw} the switching frequency. The worst case occurs at D = 1 and hence, the lowest frequency possible for the crossover will occur at f_{Ci_max} =1/6 approximately. Note that a symmetrical

sawtooth may double this limit. If the maximum duty cycle is lower, this frequency can be pushed further. On top of this, both the diode and the MOSFET current controllers have to filter the current waveform to work properly. Note this control system is supposed to work with the average value of the current. As a consequence, these two controllers will be designed in such a way that the gain at the switching frequency is at least -20 dB. Fig. 8 shows the Bose plots of the three current loop options. As can be seen, the best performance is achieved by the inductor current loop. Both the gain and the bandwidth are higher than in the other two cases.

It is also interesting to take a look to the closed current loop transfer functions to illustrate the difference between the conventional controller and the differential one. Fig. 9 shows

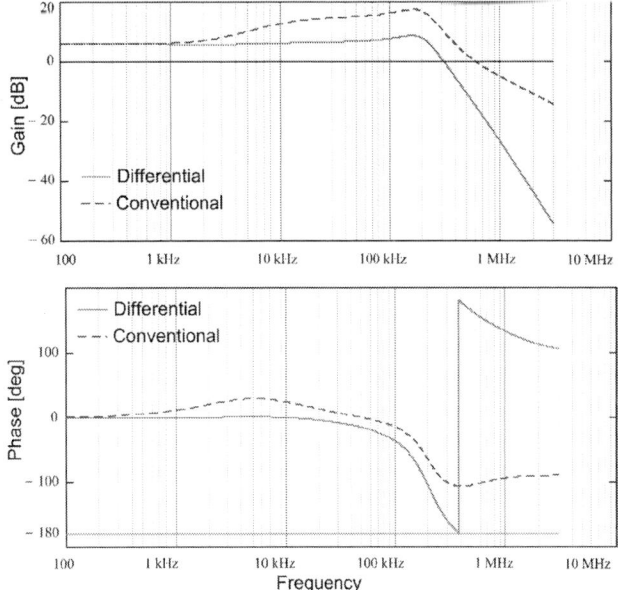

Fig. 9: Closed current loops when the inductor is controlled with a conventional and a differential controller

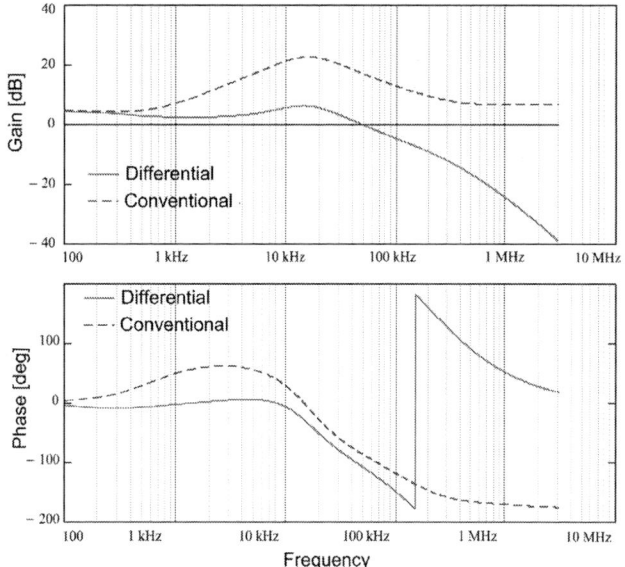

Fig. 10: Closed current loops when the diode current is controlled with a conventional and a differential controller

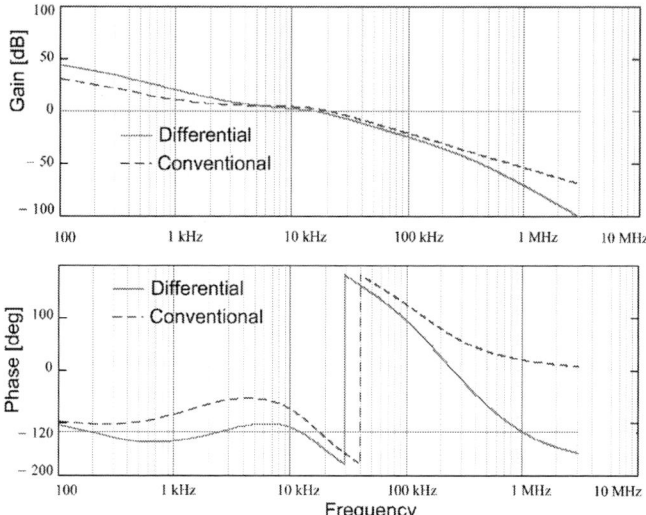

Fig. 11: total loop gain when the input current is controlled with a differential or a conventional controller

this in the case where the inductor current is the one used to control the converter. As can be seen, with the conventional controller the closed loop function is far from being a first order system. On the other hand, the differential controller gives place to a flat gain system until the crossover frequency. There is also a difference in the phase plot. In this case, the conventional scheme has a better performance as the phase is kept at -90 deg. It should be noted that the key figures for the voltage loop will be though the values at voltage loop crossover frequency, which will be placed at around $f_{sw}/10$. At this value, or lower, the figures are very similar.

The MOSFET transfer functions look quite similar to the inductor ones. However, the diode current transfer function has a significant difference. Fig. 10 shows its Bode plot. As can be seen, the results are quite different with both controllers. With the conventional one, the gain doesn't decay at high frequency and there is a kind of "resonance peak" around 10kHz. In the case of the differential controller, the Bode plot is similar to the one of the inductor current with a flat region until the crossover frequency. Again, the phase is better with the conventional controller. The effect of the RHPZ is clearly seen as the phase goes down to -180 deg instead of -90 deg as in the case of the inductor current. In the case of the differential controller the gain between 100 Hz and 10kHz is not completely flat and around 1 kHz reaches a minimum value. This will be important afterwards when calculating the output impedance.

Fig. 11 shows the final voltage loop gain (8) when the inductor current is controlled. Both conventional and differential cases are shown. As can be seen, both achieve a crossover frequency around 30kHz ($f_{sw}/10$) and their performance is quite similar. In both cases, the phase margin is 60 deg or higher. The differential controller achieves a higher gain at low frequencies and hence, the output impedance will be lower. The conventional controller has a better phase margin though. The low frequency gain is very

important to calculate the output impedance. In the range between 1 kHz and 10 kHz the loop gain plays a major role to keep the impedance within the ECSS mask shown in Fig. 12.

To summarise the results, Fig. 12a shows the output impedance of the boost converter with the three different currents used as control parameter. In this case, the figure shows the performance obtained with the conventional current controller. The option based on the MOSFET current is the worst one while the option based on the inductor current achieves the best performance. Even though the impedance plot is not completely inside the mask, the performance is very

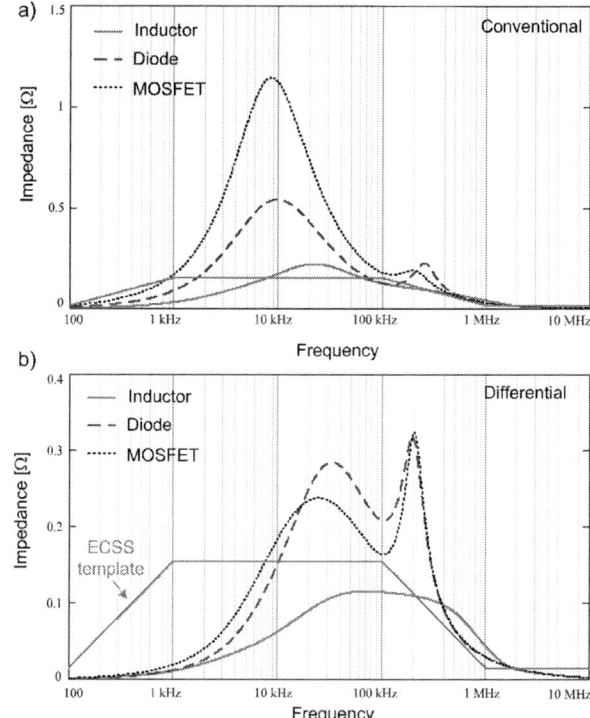

Fig.12: Output impedance when the three different currents are controlled with: a) a conventional controller, b) a differential controller

good. In general, the differential controller achieves a better performance with all the options as is shown in Fig. 12b. Among the three options, the one based on the inductor current is again the best one. Hence, we can conclude that the best option to control the boost converter is to use a differential amplifier and to control the input current. With this configuration, the boost converter could be used in European space applications as its performance is compliant with the ECSS requirements.

Fig. 13 shows some simulation results of the output voltage under 50% load steps with all the control systems. Fig. 13a shows the result when the input current is controlled with the conventional controller and Fig. 13b shows the result with the differential controller. In both cases the overshoot is less than 1%. However, the differential controller has a better result. Actually, the overshoot is around 30% smaller than in the former case. Fig. 13c and 13d show the case in which the output current is controlled and Fig. 13e and 13f show the result of controlling the MOSFET current. The latter is the worse one as the overshoot is 2.5 times the value obtained with the inductor current controller.

IV. EXPERIMENTAL RESULTS

A boost converter prototype has been built to verify the design proposed before. The input voltage range corresponds to a 6 cell Li-Ion battery used in a range from 50% to 100% of State of Charge (SoC). The minimum voltage is 19V while the maximum voltage is 25.2 V. The output voltage is 28 V and the maximum output power is 100W. The output capacitor is 80 μF. It should be noted that the bus capacitance is typically sized according to the bus impedance requirements set in the ECSS ($C = P_{max} / 400\pi V_{bus}^{2}$). The switching frequency (f_{sw}) of the prototype is 220 kHz.

The switching frequency plays an important role in this design. The inductor and the capacitor of the converter are basically designed to "filter" the switching frequency. Hence, the higher f_{sw} is the smaller the reactive components are. As a consequence, the RHPZ also moves to a higher frequency and there is more room to make a good controller design without being bothered by the extra phase and extra gain. The drawback, obviously, is that the efficiency is lower when the switching frequency increases. However, the semiconductor components available nowadays allow the use of relatively high switching frequencies without penalising too much the efficiency.

A set of tests has been done to measure the efficiency at different switching frequencies since the idea is to find the minimum frequency at which the converter dynamics can still meet the ECSS requirements and maximise the efficiency. The inductor value was changed accordingly in order to keep the same current ripple. Table I shows the results obtained. We can roughly say that 1% was lost every 100 kHz, which is very interesting. Between 200 kHz and 300 kHz the efficiency was kept within 94.3% to 95.9% which is a very good performance.

Another interesting issue is the inductor design. With low inductor values the converter enters into DCM and hence, the peak currents are higher and also the losses. On the other hand, very large inductors minimise the current ripple and the losses decrease. Fig. 14a shows a computation of the efficiency of a boost converter in which the following issues have been taken into account: switching frequency, inductor value, current ripple, losses on the MOSFET (switching, conduction and driver losses), losses on the diode (switching and conduction losses) and losses on the inductor (copper and magnetic losses). The actual operation in DCM or DCM for each particular case is also taken into account.

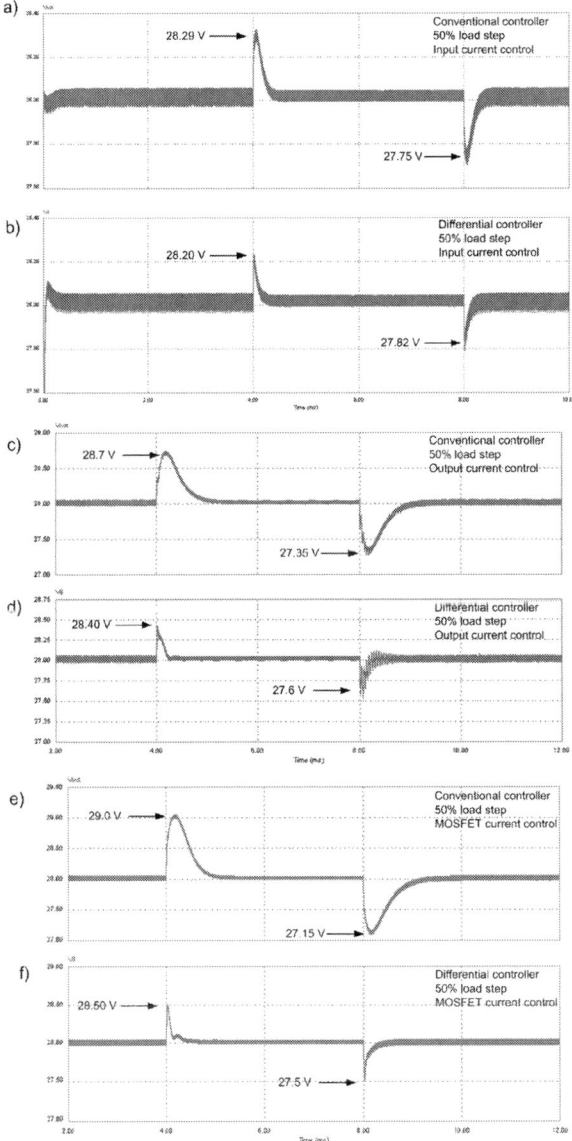

Fig. 13: output voltage response under a 50% load step. a) input current conventional controller, b) input current differential controller, c) output current conventional controller, d) output current differential controller, e) MOSFET current conventional controller, f) MOSFET current differential controller

TABLE I: EFFICIENCIES AT DIFFERENT SWITCHING FREQUENCIES

Frequency [kHz]	L_O [μH]	C_O [μF]	Efficiency [%]
150	7.5	100	95.8 – 97.0
200	5.6	78	94.8 – 95.9
300	3.8	60	94.3 – 95.0
400	2.83	42	93.8 – 94.7

Even though the actual efficiency of the converter would not be exactly the same, this computation shows the influence of all these parameters and the overall effect. As can be seen, the efficiency increases with higher inductor values. However, there is a point after which the increase ratio is very low. This means that, even though the ripple and the peak current values are decreasing, the influence on the efficiency is very small. Hence, it is not worth to have a bigger inductor after that point. Typically, this point is reached when the inductor is designed in such a way that the converter enters in DCM between P_{max} / 2 and P_{max}/1.5

In our case, the ECSS gives another hint in this issue as the load step overshoot has to be measured with a 50% load step. Hence, the design of the inductor will be done in such a way that the converter operates in CCM from P_{max} to P_{max} / 2. Thus, the inductor is minimised while keeping the efficiency very high. Moreover, the converter will operate in CCM during the full 50% load step and the control system will be able to keep the overshoot under control. As was already mentioned, the fact of optimising the inductor value keeps the RHPZ at relatively high and harmless values.

The efficiency of the converter was also measured keeping the frequency still (~220 kHz) and changing the inductor value. Fig. 14b shows the experimental results. The prototype was finally built with a 5.6 µH inductor in order to operate in CCM up to P_{max} / 2 and to achieve a good trade-off between efficiency and dynamic performance. As can be seen, the nominal efficiency is around 96%. Fig. 15 shows the output voltage under a 50% load step. As can be seen, the overshoot is kept below 1%. The performance is very good and the dynamic behaviour is in line with the ECSS requirements. These results fully enable the possibility of using this simple converter in a regulated bus power system.

V. CONCLUSIONS

The conventional boost converter has not been traditionally used in European SCs due to its poor dynamics. The RHPZ leads to lower bandwidths and hence, its dynamic

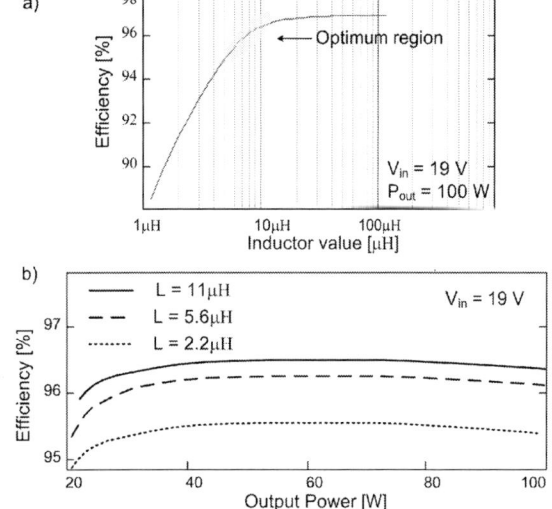

Fig. 14: a) theoretical efficiency at 200 kHz as a function of the inductor value, b) experimental measurements of the efficiency

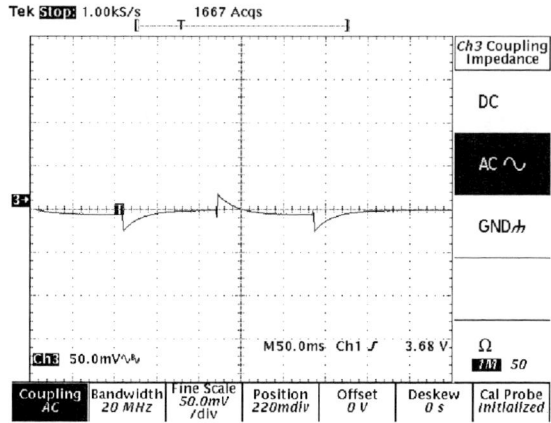

Fig. 15: Output voltage when the load goes from 100 W to 50 W

performance is worse than in the case of a buck derived topology. A simple modification on the current controller has been proposed. Hence, the converter can cope with the ECSS requirements and enables the possibility of using the boost converter in a fully regulated bus space power system. Apart from the loop itself, the switching frequency is another factor to include in the trade-off. A higher switching frequency enables a further decrease of the output impedance so that it can be kept below the limits imposed by the ECSS. However, the efficiency will be lower as the switching frequency increases. 200 kHz seems to be a good trade-off to cope with the ECSS dynamic requirements using a reasonable output capacitor and keeping the efficiency as high as 96%.

REFERENCES

[1] ECSS-E-ST-20C, 31 July 2008.

[2] R.D. Middlebrook, "Topics in Multiple Loop Regulators and Current-Mode Programming", IEEE PESC proceedings, June 1985

[3] D. O'Sullivan, H. Spruyt, A. Crausaz, "PWM Conductance Control", IEEE PESC 88, pp. 351 - 359

[4] W. Tang, F. C. Lee, and R. B. Ridley, "Small Signal Modeling of Average Current-Mode Control", IEEE Trans. on Power Electronics, Vol. 8, No. 2, April 1993, pp.12-19.

[5] W. Tang, "Average Current-Mode Control and Charge Control for PWM Converters", Ph.D. Dissertation, Virginia Polytechnic Institute and State University, October 1994.

[6] R.B. Ridley, "A New, Continuous-Time Model for Current-Mode Control", Proceedings of the Power Conversion and Intelligent Motion, October 16-19, 1989, pp. 455-464.

[7] R.B. Ridley, "A New Small Signal Model for Current Mode Control", Ph.D. Dissert., Virginia Poly. Inst. and State University, Nov. 1990.

[8] D. J. Perreault, G. C. Verghese, "Time-Varying Effects and Averaging Issues in Models for Current-Mode Control", IEEE Trans. on Power Electronics, Vol. 12, No. 3, May 1997, pp. 453-461.

[9] J. Sun and R. Bass, "Modeling and Practical Design Issues for Average Current Control", IEEE Applied Power Electronics Conference, March 1999, pp. 980-986.

[10] R.W. Erickson, "Fundamentals of Power Electronics", Chapman & Hall, 1997.

[11] F. C. Lee and B. H. Cho, "VPEC Power Electronics Professional Seminar Course 1 - Control Design", Virginia Polytechnic Institute and State University, August, 1991.

[12] L. H. Dixon, "Average Current Mode Control of Switching Power Supplies", Unitrode Power Supply Design Seminar, SEM-700, 1990, SEM- 800, U- 140.

[13] P.Cooke, "Modeling Average Current Mode Control", IEEE PESC 2000, pp. 256-262.

Electromagnetic Compatibility Results for an LCC Resonant Inverter for the Tranportation Systems

Mohamed Youssef*, Jaber A. Abu Qahouq** and Mohamed Orabi***

* Bombardier Transportation Inc. Research and Development Kingston, Canada	** The University of Alabama Department of Electrical and Computer Engineering Tuscaloosa, Alabama 35487, USA	*** South Valley University Department of Electrical Power and Machines Aswan, Egypt

Abstract — This paper presents the Electromagnetic Compatibility (EMC) testing setup and results for a DC-AC resonant inverter used to power a train linear induction motor (LIM) used in urban transit applications. While the inverter is based on the LLC-type resonant converter as it possesses much lower component stresses compared with the conventional series and series-parallel resonant topologies, EMC performance needs to be evaluated to meet industry standards. Electromagnetic compatibility (EMC) is a major challenge in the inverter design process in the transportation industry. EMC test results shows that the inverter is qualified against the EN50121-3 standards for revenue service purposes.

I. INTRODUCTION

There is an increasing trend to increase the energy efficiency of transportation systems in order to achieve greener transportation infrastructure, reduce energy cost, and increase reliability [1-14]. Therefore, improving the inverter efficiency used in transportation applications is important, however, Electromagnetic compatibility (EMC) is a major challenge in the Inverter design process in the transportation industry [1]. EMC performance needs to be evaluated to meet industry standards.

Recent advancements in the semiconductor industry allow the operation of high power IGBTs at higher switching frequencies up to 20 kHz. Resonant converters have the advantage of operating at higher frequencies without any major Electromagnetic Interference (EMI) levels challenge, unlike the pulse width modulated (PWM) converters. In addition, the higher frequency operation of resonant inverters allow a reduced real estate for the inverter in the under floor layout or the ceiling layout, which is a big advantage [1]. The reduced layout allocation in the car under floor or roof has allowed more space for other auxiliary powering devices to accommodate the need for more on-board luxury devices such as security cameras and advertisement panels, which is now required by most transportation authorities [1]. Due to the well-known drawbacks of the PWM converters like the degraded efficiency at high switching frequency and the need of external resonant circuitry during the switch transition period to realize zero voltage switching (ZVS). This led to the complexity and

unreliability of such inverters. In order to get a reduced size power supply, the switching frequency must increase. Based on the fact that resonant converters can operate at higher frequencies; industry's attention is going back to these converters. The series resonant converter can achieve ZVS due to the series resonance principle for the switches of the half bridge but the output voltage regulation is lost at light loading conditions. The parallel resonant converter is suitable for high voltage/low current applications because of the considerable conduction losses of the circulating current of the parallel capacitor. The series-parallel resonant capacitor possesses the advantages of the series and parallel resonant circuits and can regulate the output voltage for a wide loading range. However, it needs a wide range of regulating frequency under the variable frequency control (VF). This will cause bulky input/output filters and hence a large size converter [2-10]. The LLC resonant converter was introduced to offer a new topology that gathers the advantages of the series and parallel resonant converters with a high efficient operation with almost no switching losses. Moreover, these converters are suitable for integrated a magnetic solution, which offers a more compact size converter. The regulating frequency range is reduced for this topology, which enhances the reduction of the size of the input filter [4-6].

In [14], analysis and design of a high-power soft-switching LCC resonant inverter used to power a train LIM used in urban transit applications is presented in this paper. The proposed inverter topology combined with a self-oscillating controller to achieve voltage regulation and soft-switching results in high power efficiency, high power density, and good EMC performance. This paper presents EMC testing results of the inverter presented in [14]. The EMC test results show that the inverter is qualified against the EN50121-3 standards for revenue service purposes. Next Section reviews the inverter and Section III presents the EMC test setup and results. The Conclusion is given in Section IV.

II. LCC INVERRTER REVIEW

Fig. 1 shows the circuit diagram of the LCC resonant inverter which is considered for EMC testing in this paper. Fig. 2 shows a picture of the inverter prototype. The operation and design of the inverter is discussed in [14].

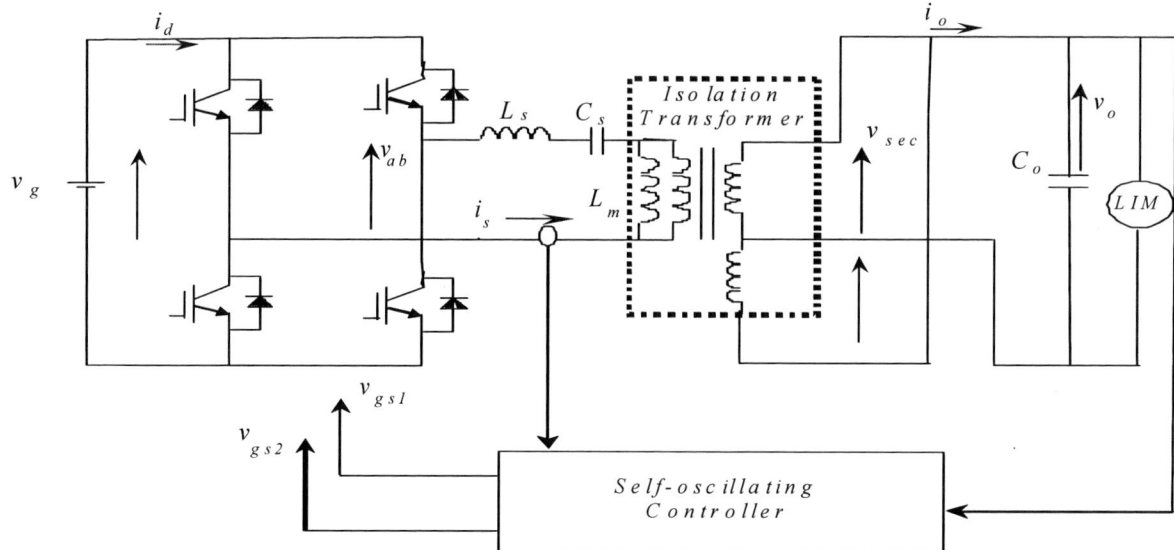

Fig. 1: Single-phase diagram of the proposed 3-phase resonant inverter

Fig. 2: Picture of the developed experimental prototype.

III. ELECTROMAGNETIC COMPATIBILITY TEST

Fig. 3 shows the EMC testing setup pictures in the wayside substation of a revenue service system. Fig. 4 shows the radiated EMC test results compared with the required limits.

The inductive test shows that the signal to noise ratio is above 40 dB, which is better than the 20 dB requirement. The measured conductive EMC levels were almost 1.5 mA, which is below the 20 mA threshold at the same Automatic Control System (ATC) frequency. Radiated EMC testing was performed from 9 kHz up to 3 GHz with full compliance to the EN50121-3-2 levels. EMC test results shows that the inverter is qualified against the EN50121-3 standards for revenue service purposes.

IV. CONCLUSIONS

The test EMC test results shows that the proposed LCC resonant inverter meets the EMC required standards and is qualified against the EN50121-3 standards for revenue service purposes. This is in addition to the many advantages the inverter has including high efficiency over wide load of operation and high power density.

(a) Test Setup

(b) Test setup at the wayside substation

(c) Antenna's position on the track

Fig. 3 Pictures of the EMC testing setup in the wayside substation of a revenue service system

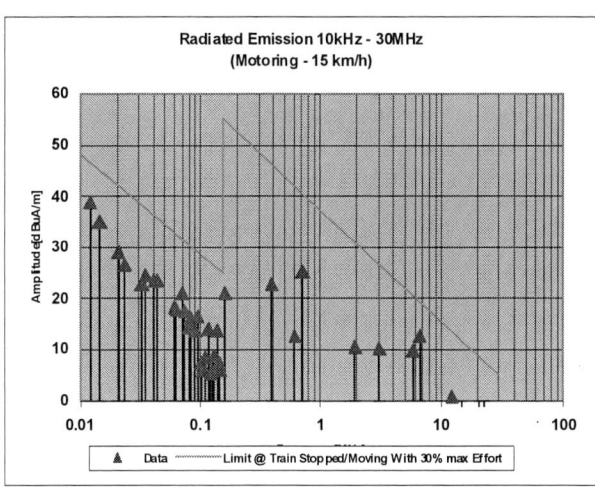

(x-axis: Frequency, y-axis: Amplitude in dBuA/m)

Fig. 4: Radiated EMC test results
(x-axis: Frequency, y-axis: Amplitud in dBuA/m)

REFERENCES

[1] http://www.theclimateisrightfortrains.com/?lang=en

[2] H. Kuhn, T. Koneke, and A. Mertens," Consideration for a Digital Gate Unit in High Power Applications," Proceedings of the IEEE Power Electronics Specialist Conference, June 2008. pp. 2784-2790

[3] T. Kejellqvist, S. Ostlund, and S. Norrga, "Active Snubbber Circuit for Source Commutated Converters Utilizing the IGBT in the linear region" IEEE Transactions on Power Electronics, vol. 23, September 2008. pp. 2595-2601

[4] K. Fuji, P. Kollensperger, and R. W. Doncker, "Characterization and Comparison of High Blocking Voltage IGBTs and IEGTs under Hard and Soft Switched Conditions, "IEEE Transactions on Power Electronics, vol. 23, January 2008. pp. 172-179

[5] E. S. Kim and Y. H. Kim, "A ZVZCS PWM FB DC/DC Converter using a Modified Energy Recovery Snubber," IEEE Transactions on Industrial Electronics, vol. 48, July 2002. pp. 1120-1127

[6] J. Dudrik, P. Spanik, and N. D. Trip, "Zero Voltage and Zero Current Switching Full-bridge dc-dc Converter with Auxiliary Transformer," IEEE Transactions on Power Electronics, vol. 21, 2006. pp. 1328-1335

[7] M. Z. Youssef, H. Pinheiro, and Praveen K. Jain," Closed Loop Implementation and Design of a Novel Controller for 48V Voltage Regulator Module with Self-sustained Oscillation Controller", Proceedings of the Applied Power Electronics Conference, APEC, March 2006. pp. 1113-1120

[8] J. E. Slotine, and W. Li," Applied Nonlinear Control", 1st edition, Prentice Hall, 1991

[9] Ja. Z. Tsypkin," Relay Control Systems", 1st edition, Cambridge University Press, 1984

[10] M. Z. Youssef, and Praveen K. Jain,"Series-parallel Resonant Converters with the High Frequency Effects: Analysis, Modeling, and Design", IEEE Transactions on Industrial Electronics, vol. 54, issue 3, July 2007. pp. 1329-1341

[11] M. Z. Youssef, and M. Orabi,"Sampled-data Modeling of A New Ultra Fast 48V Voltage Regulator Module", Proceedings of the IEEE Applied Power Electronics Conference, APEC, February 2008. pp. 182-192.

[12] S. Zheng, and D. Czaarkowski," Modeling and Control of a Phase Controlled Series Parallel Resonant Converter", IEEE Transactions on Industrial Electronics, vol. 54, issue 3, July 2007. pp. 707-715

[13] W. J. Lee, C. E. Kim, G. W. Moon and S. K. Han," A New Phase-shifted Full-bridge Converter with Voltage-doubler Type Rectifier for High-efficiency PDP Sustaining Power Module", IEEE Transactions on Industrial Electronics, vol. 55, issue 3, July 2008. pp. 2450-2458.

[14] M. Youssef, J. A. Abu Qahouq, and M. Orabi," Analysis and Design of LCC Resonant Inverter for the Tranportation Systems Applications," 2010 IEEE Applied Power Electronics Conference, APEC 2010, Feb. 2010.

Torque Impulse for Experimental Modal Analysis in Transmitted Vibration Study of Engine-generators

Elias Ayana
Cummins Power Generation
University of Minnesota
Minneapolis, Minnesota
elias@umn.edu

Steve Seidlitz, Sze Kwan Cheah
Cummins Power Generation
Minneapolis, Minnesota

Professor Ned Mohan
University of Minnesota
Minneapolis, Minnesota

Abstract—**Experimental modal analysis is used to measure the dynamic characteristics of a structure. Traditionally, excitation is provided with either a shaker or an impulse hammer. An alternative approach is to use power electronics for powering the generator; generating a torque impulse. This torque excitation replicates much of the engine's excitation, emphasizing the vibrational modes that are assorted with the generator set's transmitted vibration. Basic system setup for modal analysis using torque impulse, comparison with conventional impulse hammer test, and illustrations of some of the flexibilities inherent in torque impact testing are included.**

I. INTRODUCTION

Experimental Modal Analysis (EMA) is used to extract the frequency response of a structure. The underlying assumption is that the structure could be described by a lumped-parameter model and the system driven by the test input is in its linear range [1]. Through these assumptions, the goal is to obtain the frequency response of the structure by exciting the system with an input that has enough bandwidth in the frequency of interest. The general modal analysis setup includes an exciter, a structure under test, several transducers to measure response, signal conditioner and data acquisition module. The regularly used exciters are the impact hammer and the shaker (electromagnetic or electro hydraulic.) A load cell is used to measure the input force; be it a built-in sensor in the impact hammer tip or a separate load cell in use with a shaker. System responses are typically measured with numerous piezoelectric accelerometers.

In an engine driven generator (generator set) design, a mounting system is designed to reduce the vibration transmitted to the foundation. Softer mounting systems will have a very good attenuation at higher frequencies; however,

the mechanical integrity of the mounting system poses a limit as to how soft the mounting system can be made. If the transmitted vibration performance is not acceptable, one could consider using an engine with more cylinders and/or a secondary isolation system. These solutions will result in higher cost, greater volume, and more weight. In certain cases where the mounting system is not sufficient to attenuate specific vibration modes, it may be possible to avoid the excitation by running the engine at a higher speed. This usually compromises the noise performance. Yet another solution would be to attenuate the roll motion (around the crank shaft axis) of the engine by using Active Torque Cancellation (ATC) [2]. If the roll motion of the engine is the cause for exciting the offending vibration mode, ATC could be used to cancel the roll motion at the specific frequency using an electric machine that is coupled to the crankshaft. In certain applications such as generator set or hybrid vehicle, the machine and the power electronics are already part of the system; hence, making this approach practical. In order to understand which modes get excited in the system, EMA is an invaluable tool.

Torsional exciters have been proposed as early as 1945 [3] for the purpose of airplane propeller testing instead of using an aircraft engine. The exciter included a single-phase induction machine whose speed is adjusted to generate an alternating torque. Similarly, other torsional exciters have been proposed that impart continuous torque excitation to a rotating member [4-6]. In all of these studies the exciter is run continuously with a frequency sweep. In addition, the exciter is an additional part on the system under test; hence, mass/inertial loading of the structure is a point of concern as it may affect vibration modes. This paper focuses on a new method of torsionally exciting a generator set by using the generator as a

978-1-4244-4782-4/10 $26.00 © 2010 IEEE

motor. Said test results are used to understand the vibration modes that result from roll motion of the engine block, which is a consequence of the engine's firing pulses. We specifically show the frequency response results from a torque impulse using a permanent magnet synchronous machine (PMSM) and compare it to an impact hammer test. Section 2 describes roll torque and engine vibration. Section 3 outlines the torque impulse methodology and shows simulation results. Section 4 presents the prototype system setup. Results and discussion from the experimental modal analysis studies on the prototype using an impact hammer and a torque impulse are shown in Section 5. Finally, we present concluding remarks in Section 6.

II. ROLL TORQUE AND ENGINE VIBRATION

The two primary causes of vibration in engines are cylindrical pressure and inertial forces. In four stroke engines, the crankshaft rotates twice to complete the four strokes. During the power stroke, the cylinder pressure results in a huge torque on the crankshaft. The reverse occurs during the compression stroke. The reciprocating masses of the engine result in a cyclic torque on the crankshaft that is a function of the square of the speed. A typical crankshaft torque for an odd firing twin-cylinder engine operating at 1200rpm at no load is shown in Fig. 1.

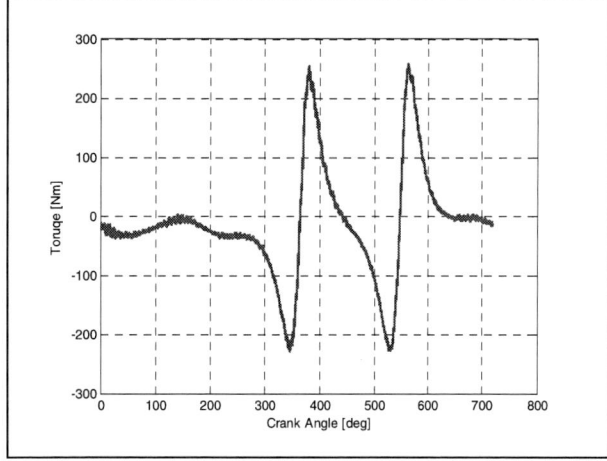

Figure 1. Engine Roll Torque

By taking the Fourier Transform of this torque wave shape, the frequency contents are shown in Fig. 2. The first engine order is the speed of the engine. For a four stroke engine, two revolutions are necessary to complete an engine cycle. Therefore the frequency contents, also called orders, are given in half multiples of the engine rpm. For example, when the engine is running at 1200rpm, the 0.5 order is at 0.5*1200rpm/ 60s = 10Hz.

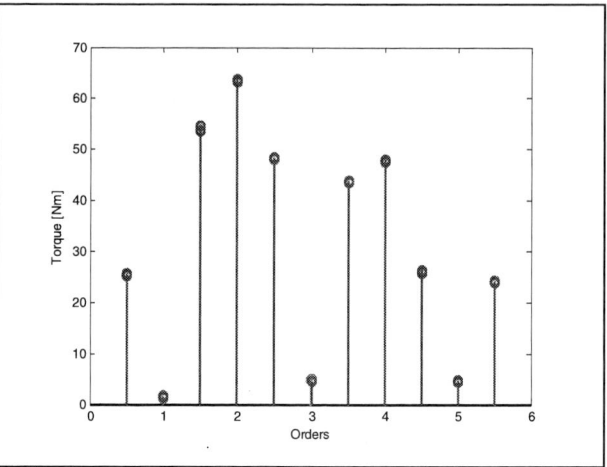

Figure 2. Engine Roll Torque Orders

Table 1 below shows the first three orders for engine speeds of 1100-2300rpm.

TABLE I. ENGINE ROLL TORQUE FREQUENCIES

Speed (rpm)	Frequency [Hz]		
	0.5 Order	1st Order	1.5 Order
1100	9.17	18.33	27.50
1300	10.83	21.67	32.50
1500	12.50	25.00	37.50
1700	14.17	28.33	42.50
1900	15.83	31.67	47.50
2100	17.50	35.00	52.50
2300	19.17	38.33	57.50

As shown in Fig. 2, for this two-cylinder engine, the 1st order has a very low roll torque magnitude. Also, it is expected that the 1.5 and higher orders have frequency contents that are high enough so that their vibrational effect is attenuated well by the mounting system. Thus, we are interested in learning from the EMA, which of these frequency contents will excite a mode that will result in a significant transmitted vibration. Here, the idea is to find modes that could be excited by the roll motion of the engine which will be eliminated by ATC. Note that the generator set will be moving in six degrees of freedom; however, ATC can only affect the roll motion.

III. TORQUE IMPULSE METHODOLOGY

Fig 3. outlines the system topology used to apply torque impulse.

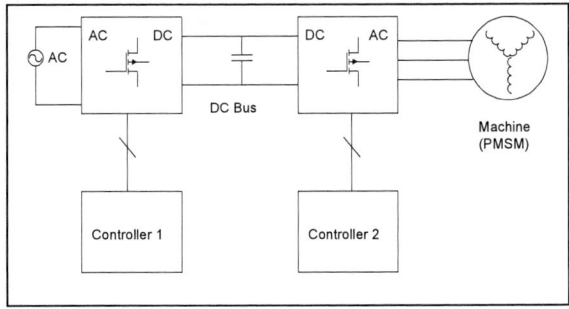

Figure 3. Power Electronics Topology for Torque Impulse

DC bus capacitors are charged from an AC source to have energy storage for delivering a short and tall torque pulse. Controller 2 uses a field oriented control technique [6] to impart a torque pulse using a PMSM. The rotor is rigidly bolted to the crankshaft of the engine and a Hall Effect sensor is used to determine the crank angle from the flywheel teeth. Two PI loops are used to control the q-axis and d-axis currents. The control strategy is shown in Fig. 4.

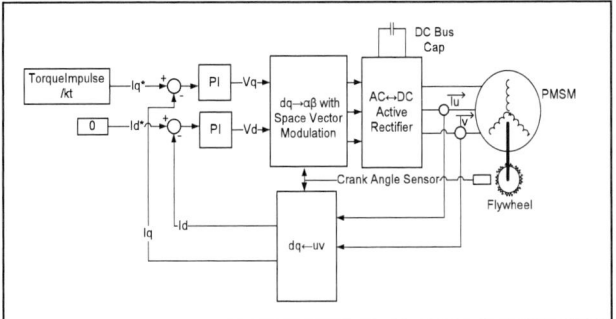

Figure 4. Field Oriented Control for Torque Impulse

A PSIM simulation is run along with a fixed point implementation of the control loop algorithm in a dll block. Tuning of the current lops is done using [8]. A 10ms, 150Nm torque is generated as shown in Fig. 5 below.

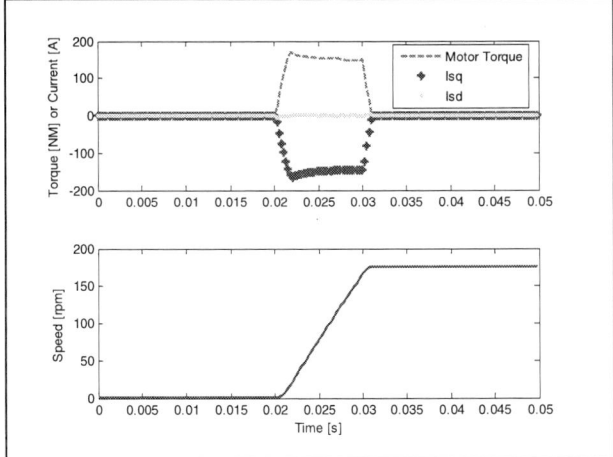

Figure 5. Torque Impulse Simulation

This torque shape will have a sinc frequency response with a first null close to 100Hz. As shown in Table 1, the 0.5 order frequencies are below 20Hz. Since there is no first order roll motion, the bandwidth in the 10ms pulse should more than suffice. Shorter pulses could be used should a broader bandwidth is required.

IV. PROTOTYPE SYSTEM AND SETUP

The generator set prototype, shown in Fig. 6, is built with a Kubota Z482 engine which is a two-cylinder odd firing engine with a 0.48lit total displacement. The alternator is a 7.5kW PMSM that is rigidly coupled to the crankshaft. The torque waveform on the crankshaft as shown in Fig. 1 and Fig. 2 has dominant 0.5, 1.5, 2 and 2.5 order components. The low pass action of the generator set mounting system reduces the amplitudes of higher orders so that they are effectively insignificant when we look at transmitted vibration. Three locations are used on the engine for accelerometer placement and hammer impact as shown in Fig. 6.

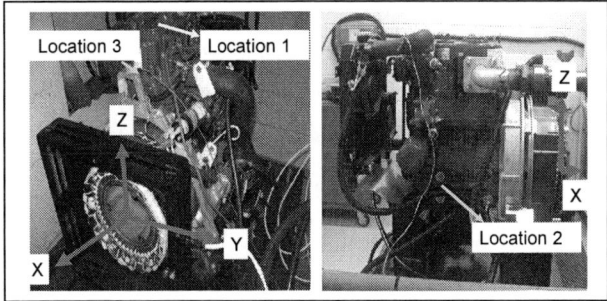

Figure 6. Prototype Generator Set

The control is implemented on a fixed-point Texas Instrument's F2810 DSP with a 20 kHz PWM switching. Attached to the rotor of the generator is angular accelerometer Endevco 7302B which measures the angular acceleration of the system. The system's angular acceleration times its rotational inertia yields the excitation torque. Piezoelectric based linear accelerometers are used to measure the response from the engine frame and from the foundation in the vicinity of the generator set. The impact hammer has a piezoelectric based force transducer with a soft tip to measure the impact force. The SCADAS III acquisition system is used to acquire the data at 4096 samples/sec.

V. RESULTS AND DISCUSSION

A. Impact Hammer

The hammer is hit at three locations in four directions, one in the x-direction, two in the y-direction and one in the z-direction. The input frequency response of the impact force is shown in Fig. 7 below. It is important to note that the impact has a broadband characteristic that is very attractive for modal analysis; however, impact hammer could be incapable of delivering enough energy to the structure in some scenarios. And sometimes the application or the installation may make it hard to find a good location for the hammer hit.

978-1-4244-4782-4/10 $26.00 © 2010 IEEE

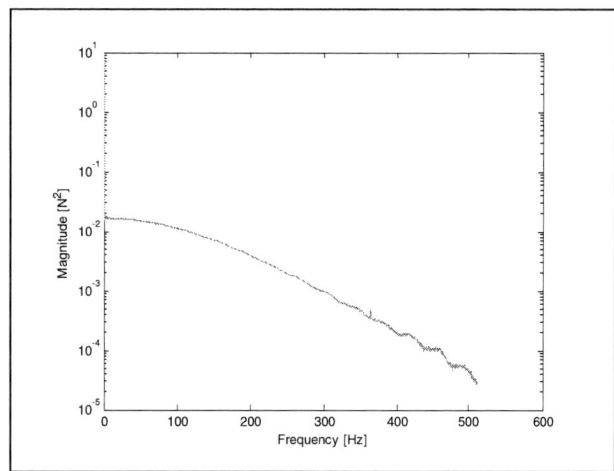

Figure 7. Impact Hammer Force Power Spectrum

The next plot, Fig. 8, shows the Frequency Response Function (FRF) at the impact location, also called driving points, in the direction of the impact. An FRF is the spectrum of the output (acceleration response) divided by the spectrum of the input (forcing function). Focusing on the FRF shown in Fig. 8, it is known that the modes below 5Hz are mainly due to the foundation structure; thus, we will only be concerned about responses above 5Hz in all modal analysis to follow.

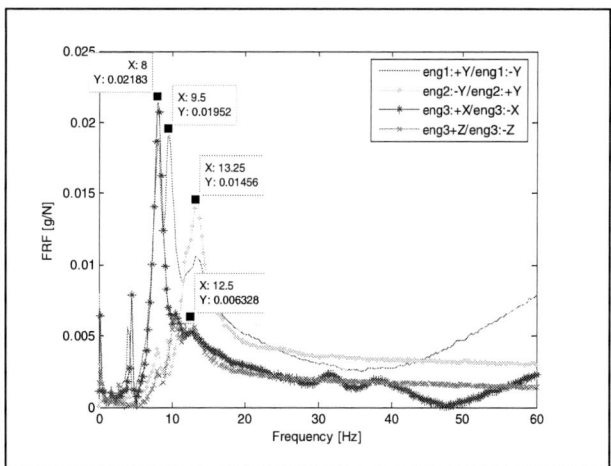

Figure 8. Frequency Response at Driving Points

At the driving points, the modes that are excited are at 8Hz, 9.5Hz, 13.25Hz and 12.5Hz. It is clear that the modes excited by a hit in one direction are not necessarily excited by another hit in a different direction. For instance, the 12.5Hz mode is apparent in the z-direction hit at location 3; however, this mode is coupled with the 13.25Hz mode on all hits from other directions.

Fig. 9 and Fig. 10 show the FRF at engine location 1 in the three directions for a –y-direction hammer impact at location 1 and a –x-direction hammer impact at location 3, respectively.

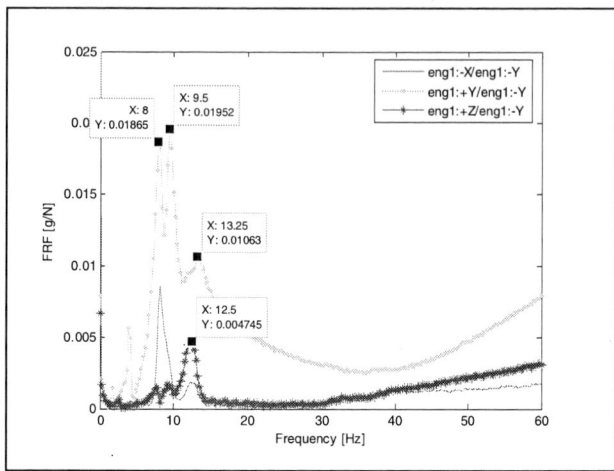

Figure 9. FRF's at Engine Location 1 for a Hammer Impact in –y-direction at Location 1

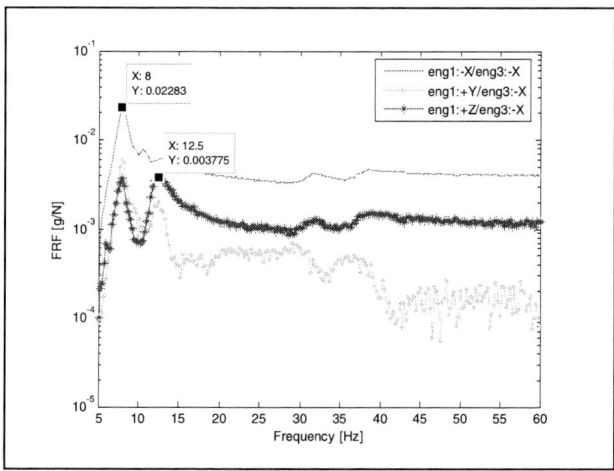

Figure 10. FRF's at Engine Location 1 for a Hammer Impact in –x-direction at Location 3

In the above two figures, the hit is delivered at two different locations, location 1 and location 3, respectively. The 12.5Hz mode is excited by both hits; however, the modes around 9.5Hz and 13.25Hz are not prevalent in the -x-direction hit at location 3.

Fig. 11 shows the response at the vicinity of the generator set on the foundation.

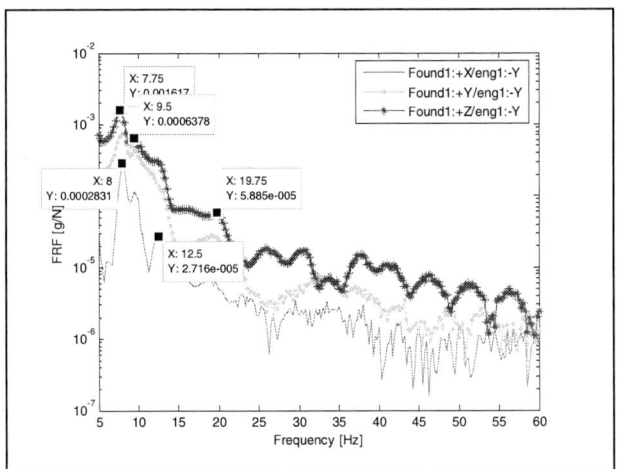

Figure 11. FRF's at the Foundation for a Hammer Impact in –y-direction at Location 1

The previous modes that were found on the engine block are also present on the foundation. Above 20Hz, the magnitudes of the FRF's decrease by two orders when compared to the highest peak of the 7.75Hz mode. This confirms the fact that for the purpose of ATC, the 0.5 order excitation is of highest concern.

B. Torque Impulse

The power spectrum of the 10ms, 150Nm torque impulse is shown in Fig. 12.

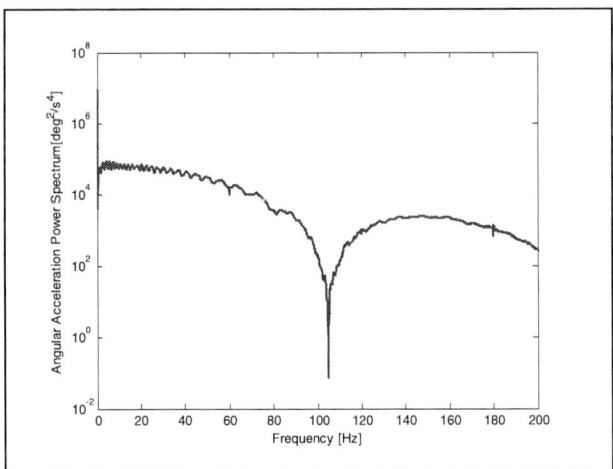

Figure 12. Angular Acceleration Power Spectrum for a Torque Impulse

The above plot shows that the torque impulse is a good exciter to about 90Hz where the magnitude drops off by one order of magnitude. The frequency could be pushed farther if higher bandwidth is needed by shortening the torque pulse below 10ms.

Fig. 13 shows the response from location 1 of the engine due to the torque impulse. This FRF is given by the response in g's divided by angular acceleration in deg/sec^2. As previously mentioned, angular acceleration should be directly proportional to the input torque. The rotor and the crankshaft

are not rotationally constrained to the rotor and the engine block. As the torque impulse is generated between the rotor and stator the system will react based on Newton's Third law. The torque impulse causes the rotor/crankshaft assembly to rotate in one direction and the reaction torque will rotate the engine frame in the opposite direction. Since the reaction torque is proportional to the angular acceleration of the rotor, we are able to normalize the response with respect to angular acceleration.

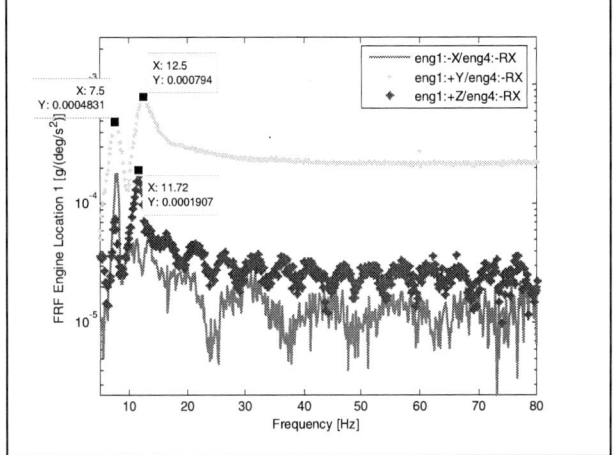

Figure 13. FRF's at Engine Location 1 for a Torque Impulse

Fig. 14 shows the response on the foundation from the torque impulse excitation.

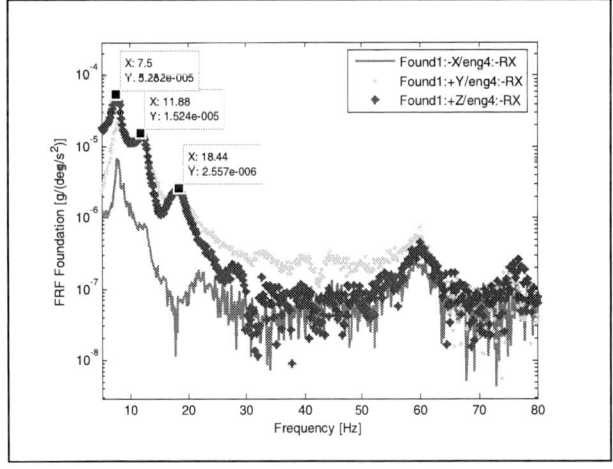

Figure 14. FRF's at the Foundation for a Torque Impulse

Similar to the result obtained using impact hammer test, high amplitude modes are contained below 20Hz. Also, the excited frequencies are very close to the hammer impact result except the 11.88Hz mode seen in Fig. 14 and the 9.5Hz mode seen in Fig. 11. We can conclude from these tests that the roll excitation of the engine results in reliable mode excitations. The torque impulse modal analysis shows us that by attenuating roll motion of the engine using torque control, it will be possible to reduce the motion of the modes and transmitted vibration. For instance, we have seen that there is a 12.5Hz mode in the generator set. That mode could be excited by a 0.5 order of the roll torque when the engine is running at 1500rpm. By applying ATC at the 12.5Hz, we could reduce the roll motion of the engine and hence reduce the transmitted vibration to the foundation. Fig. 15 shows the result of 0.5 order cancellation at no load.

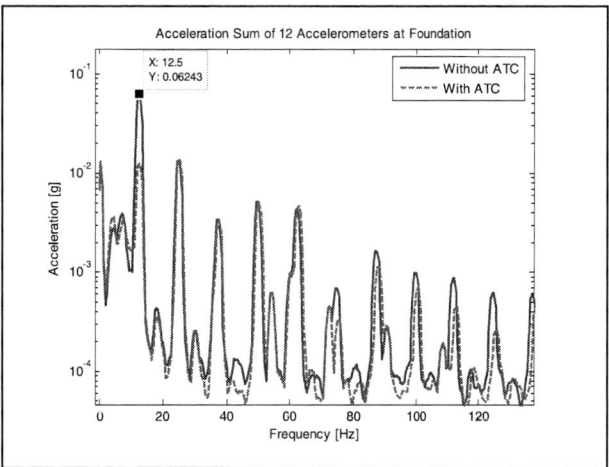

Figure 15. 0.5 Order Cancellation for 1500rpm using ATC

VI. CONCLUSION

It is well known in modal analysis that multiple force excitation points are required to excite all the modes of the system. For Active Torque Cancellation, the key degree of freedom that is being controlled is roll motion. Traditional modal analysis extracts mass, stiffness and damping to characterize the system but does not capture the inertial response of the system explicitly. By providing a torque impulse into the system, the contributing dynamics due to the roll are captured. Moreover, transfer functions could be generated from responses anywhere on the system to the torque input. This could be beneficial when the interest lies in finding transfer function from engine torque input to acceleration response at a location of interest.

ACKNOWLEDGMENT

The authors would like to thank Cummins Power Generation for supporting the research.

REFERENCES

[1] D. Inman, Vibration with Control. England: John Wiley & Sons Ltd.,2006.

[2] E. Ayana, P. Plahn, and Krzysztof Wejrzanowski, "Active Torque Cancellation for transmitted vibration reduction of low cylinder count Engine," IEEE VPPC, September 2009, pp. 325-329.

[3] A.M. Dudley, "Engine vibraiton generator," US Patent 2,384,987, Westinghouse Electric Corporation, 1945.

[4] S.J. Drew, and B.J. Stone, "Torsional (Rotational) vibration: Excitaion of small rotating machines," Jouranl of Sound and Vibraiton, 201(4), pp. 437-463, 1997.

[5] S.J. Drew, D.C. Hesteman, B.J. Stone, "The torsional excitation of variable inertia effects in a reciprocating engine," Mechanical Systems and Signal Processing, 13(1), pp. 125-144, 1999.

[6] C. Sihler, "A novel torsional exciter for modal vibration testing of large rotating machinery," Mechanical Systems and Signal Processing, 20, pp. 1725-1740, 2006.

[7] N. Mohan, Advanced Electric Drives. Minneapolis MN: MNPERE, 2001.

[8] V. Blasko, and V. Kaura, "A new mathematical model and control of a three-phase AC-DC voltage source converter," IEEE Trans. On Power Electronics, vol. 12, pp. 116-123, January 1997.

REVIEW AND ANALYSIS OF THE AC/DC CONVERTER OF ITER COIL POWER SUPPLY

P. Fu, G. Gao, L. W. Xu, Z. Q. Song, Z. C. Sheng,

ITER Chinese Party Team,
The Institute of Plasma Physics,
Hefei, China, 230031
fupeng@ipp.ac.cn

I. Benfatto, J. Tao, A. D. Mankani,

ITER International Organization,
13067 Saint Paul Lez Durance, France

J. S. Oh

ITER Korea Party Team,
National Fusion Research Institute,
Gwahango 113, Daejeon, Korea

C. Neumeyer

Princeton Plasma Physics Laboratory (PPPL),
PO Box 451, Princeton, New Jersey, USA

Abstract---**In this paper, firstly, ITER power supply and its AC/DC converter in 2001 design have been introduced. Some import criteria such as FSC and internal bypass are reviewed and researched. The system overvoltage and low frequency oscillation have also been studied. Then some proposals to improve the ITER design and baseline have been presented, and these proposals have been validated and supported by the scientists and experts from two Expert Groups organized by IO.**

Keywords--ITER coil power supply, ac/dc converter, bypass, FSC, overvoltage, low frequency oscillation

I. INTRODUCTION [1], [2]

ITER (International Thermonuclear Experimental Reactor) is an international research and engineering project aims to demonstrate the scienctific and technical fessibility of fusion power. ITER was formally agreed to fund 5 billon € and built in Cadarache, France, in Nov. 2006 by seven participants : China, Europe, India, Japan, Korea, Russia, and USA.

The power supply system shown in figure 1 comprises a 400 kV/900 MW HV substation, 66 kV/800 Mvar reactive power compensation system (RPC), 1.6 GVA AC/DC converters system, switching network unit to produce 20kV voltage to initiate plasma, and fast discharge units as quench protection.

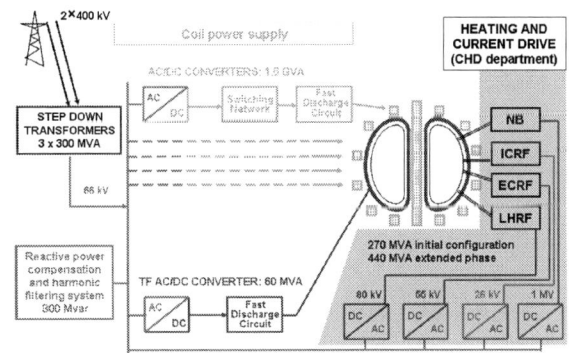

Figure 1.   Configuration of ITER power supply

In ITER power supply system, there are 39 AC/DC converters to feed several tens superconductive coils as shown in table 1. The connection of converter and its coil are presented as in figure 2.

TABLE I.   CONVERTER UNIT PARAMETERS

	No load voltage	Rated current (kA)	Unit numbers
MC1	2kV	±45	8
MC2	2kV	±55	4
TC	0.9kV	68	1
VC	4kV	±22.5	2
BC	2.8kV	±10	8
CC1	0.9kV	±10	3
CC 2	0.1kV	±10	6

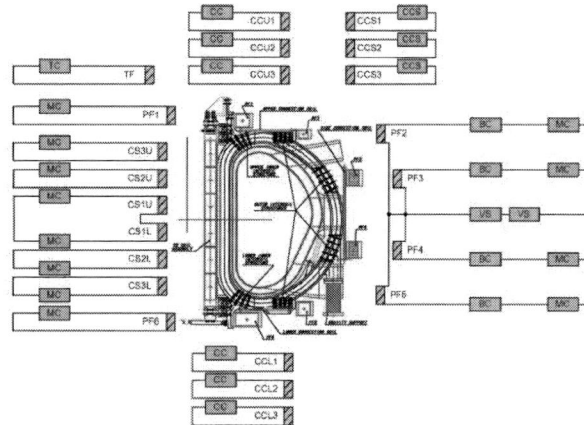

Figure 2.   ITER AC/DC converter circuit connected with coils

II.   THE AC/DC CONVETER AND ITS BASELINE IN FDR DESIGN [1], [2], [3]

The requirement on AC/DC converter in ITER includes: 20kV insulation to ground; four quadrant operation with circularting current; real-time current control; current path to superconductive coil at any case; fast voltage response and high reliability.

Before 2001, some scientists and experts from Europe, Japan and Russia including some industrial company spent several years and completed the AC/DC converter design and baseline (FDR 2001 design).

To minimize the cost and space, the baseline of converter design in FDR includs:

- Fault suppression capability (FSC). The FSC criteria allows fuse melting only in the case of thyristor failure. The fault will be cleared in the first AC cycle by the electronic protection and if it fails, the fuses and thyristors can not be damaged during 4 AC cycles (additional 3 cycles for AC breaker opening).

- Internal bypass: The current path in case of fault is formed by the thyristor arms connected to the same phase of converter transformer inside the bridge.

- High power density: Thyristor back to back connection for 4-quadrant operation and ±45kA/2kV converters in one assembly

The FSC and internal bypass has been used in HVDC, but there was no such experience on thyristors parallel connetion converters nor superconduction coil power supply.

To verify the reasonability of AC/DC converter configuration in FDR design, a converter prototype at half voltage of main converter was manufactured and tested under rated current and two-phase short-circuited conditions (powered by a two-phase transformer) by Europe home team in 1998. The tests proved that the current imbalance factor can achieve 1.4 under both conditions. It also proved the converter could withstand the first AC cycle of the fault without any component broken. However the test can not prove the two most important design criterion: internal bypass and. FSC under misfiring reverse thyristor fault.

III.   AC/DC CONVERTER REVIEW AND ANALYSIS

Since 2005, the design review by CN(China) and KO (Korea) teams under the request of IO focuses on FSC, internal bypass, overvoltage and low frequency oscillation caused by the large reactive power compensation system.

A. Fault Suppression Capability [2], [4],[5],[6],[7]

FSC will be achieved if the thyristor junction temperature is under the limit in the worst case: 1 cycle in DC short-circuit and 4 cycles in misfiring reverse thyristor fault for ITER. In misfiring, the faulted arm will carry the maximal current without AC voltage reapplied; the non-faulted arm will carry a reletivaly lower current but with AC voltage reapplied.

Based on the theory of calorifics and electrics, the model of the transient thermal impedance of a semiconductor device can be expressed by a RC circuit network as shown in figure 3.

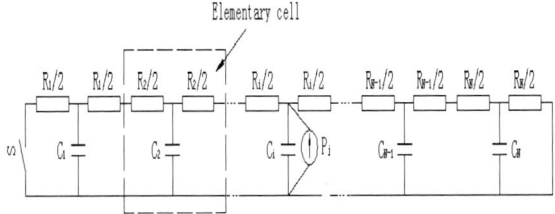

Figure 3. The thermal network of semiconductor device (Cauer network)

In the figure, R, C can be expressed respectively as:

$$C_i = C_{thi}\rho_i V_i \quad [J/K] \qquad (1)$$

$$R_i = \frac{1}{\sigma_i} \int_{x_i}^{x_i+\Delta x} \frac{dx}{S_{eff\_i}(x)} \quad [K/W] \qquad (2)$$

Here, C_{thi} :specific heat of material; ρ_i :the density of material; V_i :volume of cell; σ_i :thermal conductivity; Δx

Δx :thickness of cell; S_{eff-i} :effective region of heat flux.

Transient thermal impedance of thyristor junction to case can be also written as follows:

$$Z_{thjc}(t) = \sum_{i=1}^{n} R_{thi}\left(1-e^{-t/\tau_i}\right) \qquad (3)$$

According to thyristor required in FDR design, ABB thyristor: 5STP 52U5200 could be one option, its parameters are:

$V_{dsm,rsm}$=5200V, $V_{drm,rrm}$=4400V, I_{TAVM}=5200a, V_{T0}=1.03V r_T=0.110m Ω I_{tsm}=82.5kA T_{vjmax}=125 ℃ R_{thjhs}=4.8 k/kW (double-side-cooled) I^2t=34000 kA^{2}s

The transient thermal impedance (junction to case) is shown in figure 4. The values are shown in Table2.

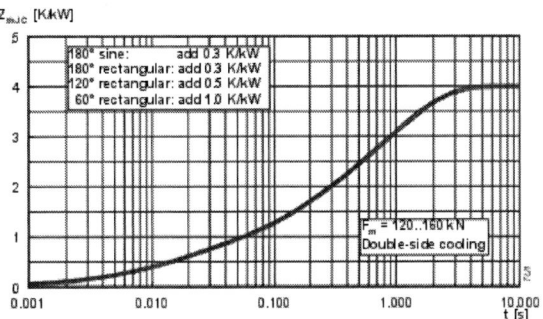

Figure 4. The thermal impedance (junction to case)

In generally, four former polynomial loops in equation (3) can express the transient thermal impedance of the thyristor accurately, thus the transient thermal network can be shown as in figure 5. Where,T_c;case temperature T_{vj} :virtual junction temperature;P_{th}:total power loss of thyristor.

In Fig.7, if thyristor is on-state, then

$$P_{th} \approx P_T = \frac{1}{T} \int_0^T i_T u_T dt \qquad (4)$$

TABLE II.   VALUES OF TRANSIENT THERMAL IMPEDANCE

i	1	2	3	4
$R_{thi}(K/kW)$	2.54	0.8	0.43	0.23
$\tau_i(s)$	0.9692	0.1571	0.0222	0.007

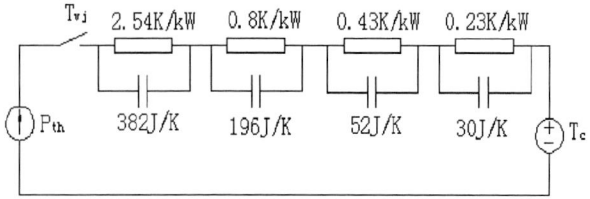

Figure 5.   Thermal network of 52U5200 thyristor junction to case

All possible faults are performed by the simulation software PSCAD/EMTDC. As a conclusion, the most severe fault is misfiring fault with fault current up to 352 kA; the second worst case is short circuit upstream the DC reactor with the maximal current at 262 kA.

At misfiring fault, the faulted arm thyristor will carry the fault current without the reapplied AC voltage, while the healthy arm will pass lower current, but it should withstand the reapplied AC voltage because the turn-off of that arm. Sometimes the healthy arm is easier to be damaged. Figure 6 shows the fault current, voltage and junction temperature of the non-fault arm thyristors at misfiring fault.   Table 3 shows the relation between thyristor junction temperature (T_{cj}) and parallel thyristor number of each arm at different fault cases.

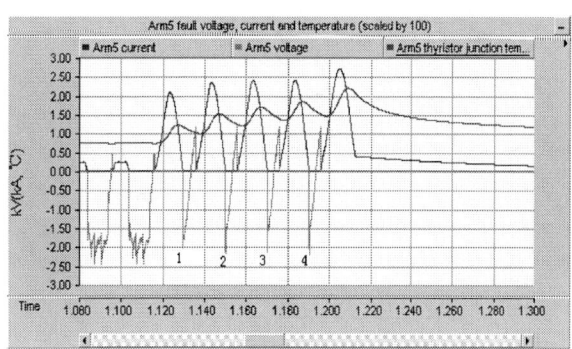

Figure 6.   Current, voltage and junction temperature of the most overloaded thyristor among non-misfired thyristors

TABLE III.   THYRISTOR JUNCTION TEMPERATURE (CURRENT IMBALANCE FACTOR K=1.4).

Parallel thyristor numbers		NP=7	NP=10
T_{cj} in short circuit of DC terminal		163 °C	115 °C
T_{cj} in misfiring fault	Thyristors With reapplied Voltage	184 °C	125 °C
	Thyristor without reapplied voltage	348 °C	215

It can be concluded:

- The misfiring fault is the worst case in ITER AC/DC converter of FDR design. When it occurs, the junction temperature of the faulted thyristor will reach 348°C; while to the healthy thyristor arm that will withstand the reapplied voltage, the temperature will reach 184°C.

- Circuit short in DC terminal is the second worst case, the thyristors in each arm will not withstand the applied voltage, and its maximal junction temperature is 163 °C.

The converter configuration in FDR design can not meet the ITER requirement on FSC. Solutions should be implemented:

- Back to back bridges instead of back to back, the parallel thyristor number in each arm can keep as in the original design.

- Increase the parallel thyristors in each arm from 7 to 11, however this solution will cause higher cost and space.

B. Internal Bypass [2], [4], [5], [7]

In ITER, the load, superconductive coils, stores a huge energe. The current path is essential to separate the load from AC source and ensure the load current is not interrupted in any case .

Internal bypass has been widely used in HVDC and suggested in FDR design for ITER operation.But there are many differences between ITER and HVDC. The main differences cover load (superconduction coil or not), coverter structre(parallel thyristors for each arm V.S. series connection), measurement and fault diagnose system(simple system VS. complicated system), and application (converter fault V.S. converters start, stop and fault signal has been detected.)

Even a right bypass pair could be set, it is not easy to ensure the DC current balance in the bypass, unless the converter arms are manufactured both for AC operation and DC operation, with cost incresing. In addition, because the ITER converter consists of two parallel 6-pulse bridges, the two internal bypasses in two bridges can not ensure the current balance.

Therefore an external bypass connected at converter terminal is suggested as in figure 7.

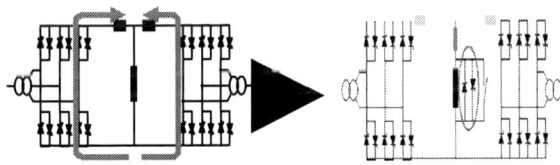

Figure 7.   bypass configuration from internal bypass (FDR design) to external bypass

C. Overvoltage and Low Frequency Oscillation of Power Supply [2], [3], [4], [8]

ITER power supply system consists of 400kV grid, 66kV grid, RPC and converters. In operation, the voltage and current of converter changes quickly in real time according to plasma controller. The Thyristor controlled reactor (TCR) and Fixed capacitors (FC) up to 800 Mvar will be adopted in ITER project.

One of worst case in this power supply system is overvoltage in case of sudden load lose due to the 800Mvar capacitve reactive power connected and lower response of TCR. Figure 8 shows the overvoltage in 66 kV grid in this case. The highest voltage will be around 142 kV, 52.2% high than rated voltage, and lasts 65ms.

In addition, there is a low frequency oscillation between the grid and the PRC that will amplify the 2nd harmonic current and result in grid system's instability. The impendance and phase of ITER power supply and the grid is shown in figure 9.

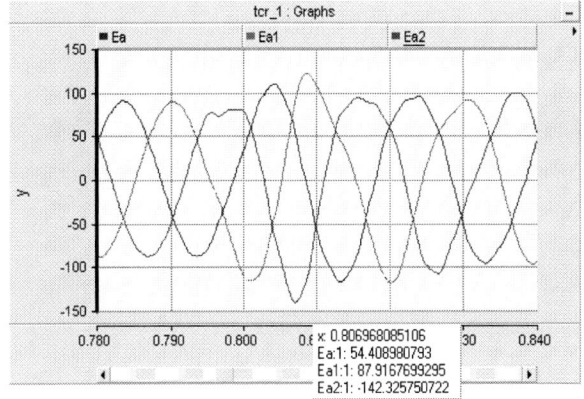

Figure 8.   Voltage on 66 kV line in case of sudden lose of load (142 kV max.)

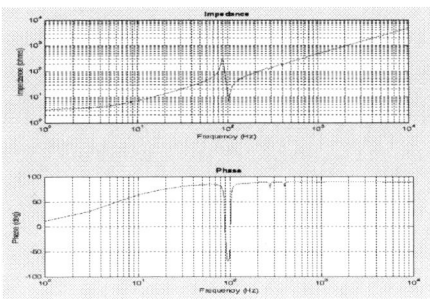

Figure 9.   Impedance and phase of ITER power supply in 400kV grid

To improve this overvoltage and low frequency oscillation, some proposals have been presented.

- Decrease the total reactive power at converter level such as split one 2kV converter to two series 1kV converters and implement sequence control. This improvement will reduce the system reactive power by several hundreds Mvar, so that the overvoltage will be decreased and the system oscillation point can be moved from about 100 Hz to near 150 Hz.

- Or STATCOM instead of TCR+FC to improve time response of PRC system.

D. *Prototype Test Review* [2], [3], [7]

To verify the reasonability of AC/DC converter configuration in FDR design, Europe home team manufactured a converter prototype and performed some test before 1998. The development was focused on: obtaining a current imbalance factor of less than 1.4, both in normal and fault conditions, in systems with several parallel thyristors per arm; proving the feasibility of the FSC in a converter with several parallel thyristors; demonstrating the converter capability of withstanding the electro-mechanical stresses in case of fault currents up to 305 kA.

Two kinds of test were performed. First a series tests at rated current and adjustments to improve the current sharing were carried out by factory; the second, after the optimization and test at rated current, the prototype has been subjected to short circuit tests that were performed at the short circuit test laboratory of Ferraz (Lyon, France). Figure 10 shows the test circuit. The current sharing, the FSC, the selectivity of the fuse intervention, and the capability to withstand the electrodynamic stresses have been tested.

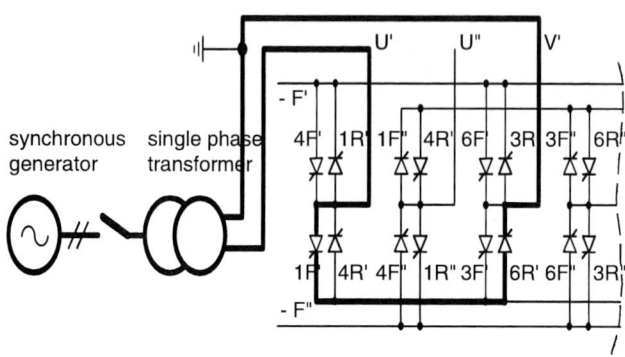

Figure 10. Electrical diagram of the short circuit test

The prototype test achieved some very useful fruits. It has proven that the main drivers of the current imbalance are the self and mutual inductances of the thyristor current paths, not only in case of normal operation, but also under fault conditions, the imbalance factor can reach 1.4.

However the comments from the review should be clarified: the test can not prove the most two important design criterion applied in FDR converter design: FSC and the internal bypass.

In FDR design, FSC concept requires the converter can not be damaged in the worst fault case, therefore the maximal test current should reach 350kA in short circuit test, and a repeatable AC voltage should be applied to the no-fault arms to judge thyristor's withstanding of temperature and voltage. However this test only applied two phase AC power supply, and the test current was only 305 kA; therefore FSC test can not be verified.

In addition, because the internal bypass test was not performed, it can not be proven.

IV. CONCLUSION [9], [10]

The experts from USA, Europe, India and Japan has been organized to review the baseline in 2008, and they agree with CN and KO teams on:

- The internal bypass cannot be formed under all fault conditions, and it can not be used as a general-purpose coil protection device;

- ITER active and reactive power demand is very large (unprecedented) and very dynamic such that grid operator's limits on voltage fluctuation may be violated during normal and/or fault operation;

- Parallel thyristor numbers in converter arm should be increased to 10+1.

- More extensive prototype testing, using the actual planned thyristor devices could be used as a means to justify the baseline, even though it is aggressive.

In March 2009, a team joined by IO, CN and KO is stabilished to improve the AC/DC converter configuration and baseline. The work is to be completed in Sep, 2009.

V. ACKNOWLEDGMENTS

The authors wish to thank two Expert Groups for the helpful discussions and suggestions during the assessment of the ITER ac/dc conversion system; also thank IO power supply division for providing a lot of beneficial and necessary information and discussion. And thank the ITER executive center of China for the finacial assistance.

REFERENCE

[1] I. Benfatto et al., "AC/DC Converters for the ITER poloidal system" in Proc of the 16th SOFE, 1995. Champaign, IL, USA, pp658-661.

[2] IO team, "Design description document (DDD 4.1), ITER pulsed power supply", ITER report, 2001

[3] P. L. Mondino, et al., "ITER R&D: Auxiliary Systems: Coil Power Supply components," Fusion Engineering and Design, 55 (2001), p.325-330

[4] "PSCAD/EMPDC", Version 4.2, 2007.

[5] P.Fu, et al, "Quench protection of poloidal field superconducting coil system for the EAST tokamak ", Nuclear Fusion, 46 (2006) S85-S89, February, 2006.

[6] P.Fu, et all, "Poloidal field power supply systems for the HT-7U steady-state superconducting tokamak", Fusion Science and Technology, vol.42, pp155-160, July.2002.

[7] P.Fu, et all, "Design and test results for the PF power supply system of EAST", Fusion Science and Technology, vol.54, pp1003-1009, Nov. 2008

[8] Liuwei Xu, et all, "The reactive power compensation and harmonic filtering and the overvoltage analysis of the ITER power supply stem". APEC, 2010

[9] Expert Group 1, "Final report, ITER Coil Power Supply & Distribution System", ITER assessment report, Dec. 12, 2008

[10] Expert Group 2, "Final report, ITER Coil Power Supply & Distribution System", ITER assessment report, Mar. 26, 2009

Fault Tolerance on Interleaved inverter with Magnetic couplers

K.Guépratte, D.Frey, P-O.Jeannin, H. Stephan, J-P.Ferrieux

Kevin.guepratte@g2elab.inpg.fr, pierre.olivier.jeannin@g2elab.inpg.fr

Grenoble Electrical Engineering (G2Elab), Thales Systemes Aeroportes

Abstract- The paper focuses on a new control strategy for improving the availability of power electronic converters based on interleaved structures. By using this strategy, the power electronic converters can continue to work (with reduced output power) in case of power component failure. The paper describes how to adapt the magnetic output filtering structure for this original control strategy. This structure is based on a monolithic coupler or a coupling transformer. Which are usually employed to minimize in a significant way the mass of the converters. They are normally sized to work with a fixed number of phases. Our control strategy induces new constraints on magnetic component, especially saturation problems. To reduce this problem some extra switch are added. Finally an experimental Power electronic converter driven by an FPGA is presented with the experimental results. It shows some experimental results with a 6 phases converter who work with 5 or 4 phases, to simulate one or two converter leg breakdown.

Index Terms-Power converters, Interleaved, Power efficiency

I. CONTEXTE, INTRODUCTION

The reliability is one of the challenges for power electronic converters. For high critical systems, the key point is to keep the electronic converter working even in case of breakdown of a power switch component. Due to its modular structure, an interleaved converter offers naturally the possibility of redundancy. This paper presents an original control strategy that uses this advantage for improving the operational availability and therefore the fault-tolerance. This fault tolerance implies the capacity to limit the impacts of the defaults.

If this original control strategy is used without any special adaptation of the power electronic structure, it will lead to saturation of the magnetic core of the interleaved converter. Part 2 will show the specific constraints on the magnetic device in these structures. It will show how power electronic structure has been adapted to take into account this phenomenon of magnetic saturation.

In order to validate the approach, an experimental setup has been built. Part 3 presents the demonstrator and experimental results on 4 and 6 cells single phase 115V-17A inverter. The control part of this power electronic converter is based on an FPGA.

II. POWER ELECTRONICS STRUCTURES

This part presents various power electronic structures which are used. In this paper, we are interested in interleaved structures, they can be carried out into simple or multilevel [2] [5]. Interlacing brings many structures of filtering making it possible to decrease the global volume of the converter. This structure of filtering will also have a significant impact on the phase currents of the converter.

A. Interleaved structure.

This structure is composed of different paralleled cells. The driving signal of each cell is phase shifted. This structure allows to reduce by a factor N (N is the number of cells) the output current ripple and increases the apparent switching frequency by the same factor ($N*F_{switching}$) [6]. For a defined output ripple frequency, the interleaved structure needs a smaller switching frequency than traditional power converters. Another advantage is that the output power is distributed between the different cells. This allows reducing the current rating of the different transistors. The switching and conduction losses are therefore smaller than for classical structures.

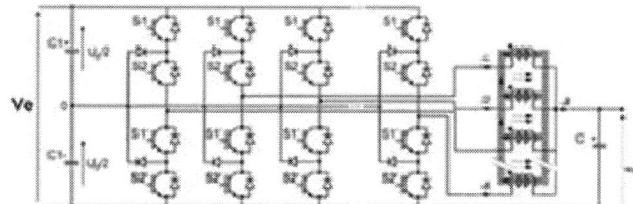

Figure 1: Interleaved structure

B. Magnetic filtered output voltage

In interleaved converters with filtered output voltage, the filters represent an important part of the converters weight. To solve this problem, it is necessary to find the best compromise between the weight of the input and output filters and the weight of power electronic elements.

The classical solution uses independent inductors as shown on Figure 4. This method imposes a very important weight and the magnetic device conception is therefore crucial. Multi-level and/or interleaved structures allow interesting opportunities.

978-1-4244-4782-4/10 $26.00 © 2010 IEEE

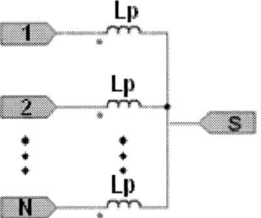

Figure 2: Filtered output voltage with independent inductors [10]

In this structure, the currents ripple in each inductance keep large amplitude and its ripple frequency remains unchanged, Figure 3. If we want to decrease the response time of the converter with a lower inductance, the current ripple is strongly increased inside each inductance. Therefore, the losses due to a larger current amplitude will increase the inductance losses.

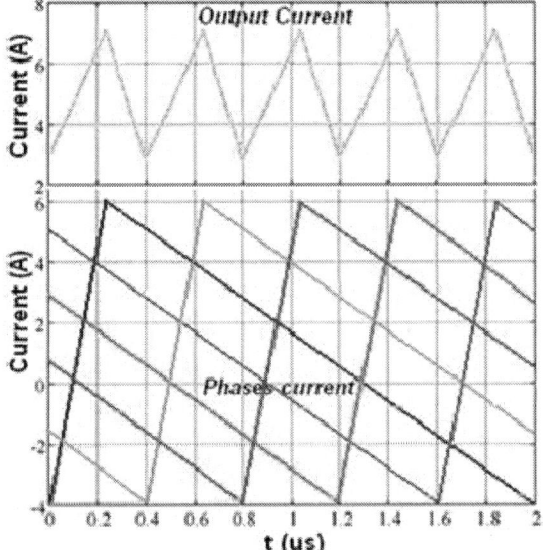

Figure 3: Current on independent inductors

These principal disadvantages can be overcome by using coupled inductances. There are several possibilities to realize **improved magnetic devices**. Two of the classical structures are presented below.

- Monolithic coupler [7] [8], which uses only one magnetic core Figure 5.
- Coupling transformers between the different output phases of the interleaved converter. On Figure 6, the cyclic cascade structure is presented [9]. Note that other coupling strategy using transformer can be used, like showed in Figure 16.

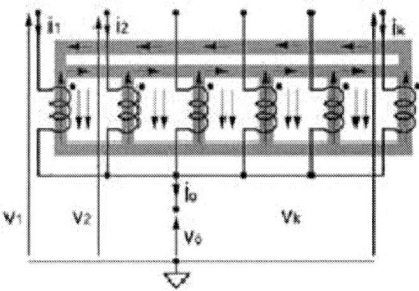

Figure 4: Interleaved structure with monolithic coupler [11]

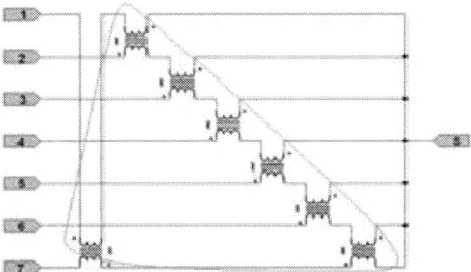

Figure 5: Interleaved structure with coupling transformers

Recent studies [4] showed that these structures generate a low output current ripple on the various converter phases, Figure 6. These currents have a ripple frequency N time the switching frequency. Thus, a faster response time can be reached without sacrificing the converter efficiency.

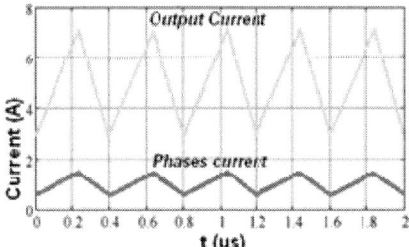

Figure 6: Current in coupled inductors

The choice of the output filtering structure must be made according two requirements: the magnetic losses and the capability of functioning with a variable number of cells. To minimize the magnetic structure and to reduce the losses, it is important to choose an output filtering structure that permit to keep a symmetrical distribution of magnetic flux in all the parts of the magnetic circuits.

III. DYSFUNCTIONS MODE

A. Dysfunctions of power electronic components

In case of dysfunction, the power electronic components can be in two different states: on or off. The on-state is generally (for MOSFETs or IGBTs) a result of the power component chip fail. The chip can't be controlled anymore and stays on. The off-state can be the result of different kind of troubles. If the current is too high, it can induce the destruction of the internal connections of the power module or a melting down of the power components. Another reason can be the destruction of the driver which isn't able anymore to turn on the transistor. This two types of dysfunction will induce the loss of an inverter leg which is the connected at the input of the magnetic structure.

B. Dysfunctions on coupling filters

The first studies of the "magnetic couplings" are based on transformer whose windings are connected like shown on Figure 5. The individual transformers have two windings and are produced with standard cores. This configuration makes it possible to transfer energy between each phase.

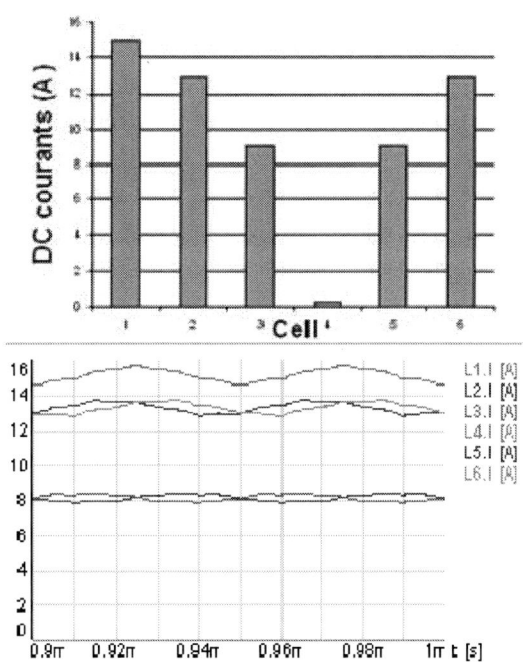

Figure 7: Simulated distribution of the DC part of the current in each winding, for a converter with 6 cells. In this structure the cell 4 is unused.

To operate, the connection of the transformers should be a closed loop. When a cell is not in use, the current can not circulate in the phase anymore. So the loop between all the transformers is open. This implies that the AC part of the current will be the same from one phase to another, but the DC part will be different [12], due the transformer (Figure 7). These current unbalances will have a significant effect on the magnetic field in the magnetic cores. Note that simulated

waveforms presented on Figure 7 do not take into account saturation phenomenons. These phenomenons can be observed on experimental waveforms on Figure 9.

In real case, the magnetic device will saturate and induce converter break down. Figure 8 presents time domain waveforms for the inverter with coupling transformers. The converter work with 6 stages and you can observe the current ripple bubble generate by the interleaved structure.

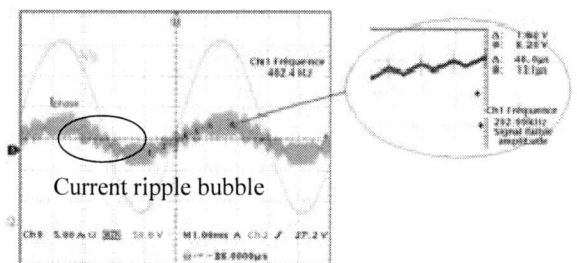

Figure 8: Experimental inverter waveforms with monolithic filter and 6 cells

If one phase breaks down, there will be only 5 phases left (on the six the total converter), therefore the magnetic core will saturate, Figure 9. This figure was realised with reduce voltage, at the beginning of core saturation. Belong this point, observations are very difficult. In this waveforms we can observed the current ripple bubble, it means that AC currents are transmit between the different phases in work.

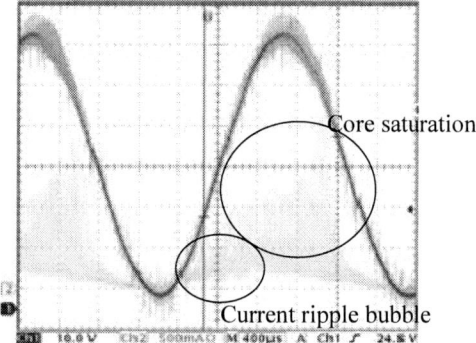

Figure 9: Experimental inverter waveforms with 5 cells in works and 1 cell not use.

C. Dysfunctions in monolithic structures

In monolithic structures, the different windings are on the same magnetic core. They have high filtering performances for a very low weight, whereas are more difficult to realise. The monolithic couplers can be produced many different ways. The most current topology is called in "scale". In this topology a problem of over sizing appear, because in some columns of the magnetic circuit there are strong concentrations of magnetic flow, Figure 10, [11] and [15].

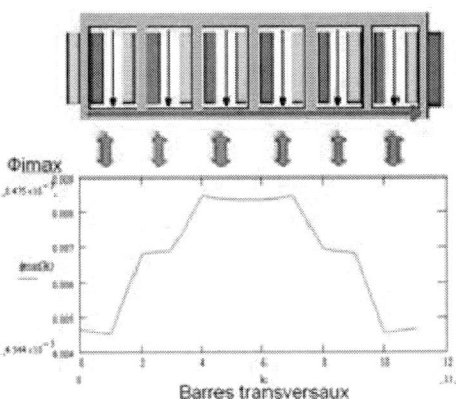

Figure 10: Distribution of the maximum induction value in a monolithic magnetic structure using a topology in "scale".

One drawback of this structure is in the case of electrical (and magnetic) unbalances. Indeed if a winding is open, the current in this winding will be null. So the magnetic reluctance will be very low (in this leg), and the magnetic flow will pass mainly by this leg. It will change the global flow distribution in the entire magnetic circuit, and it will cause local magnetic saturation. The magnetic circuit doesn't contain an air-gap, the flow generated by each phase is totally channelled by the unused leg and each phase can then be interpreted like a simple independent inductance Figure 11. The ripple of the resulting current in each phase will be at the switching frequency and an unequal concentration of flow occurs in the magnetic circuit. It is necessary to oversize the coupler according to the tolerated number of unused legs

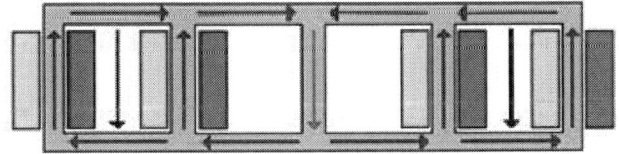

Figure 11: Flows channelled in monolithic structures with dysfunction on the central leg.

Figure 12 presents time domain waveforms for the inverter with monolithic coupling. The converter works with 6 stages and like the transformer coupling, the current ripple bubble appear.

If we use only 5 phases (instead of the six of the total converter), the magnetic filter is equivalent to five independent inductors with no air-gap. So the magnetic core will saturate, Figure 13. In these waveforms there are no current ripple bubbles, it means there is no coupling between the different phases.

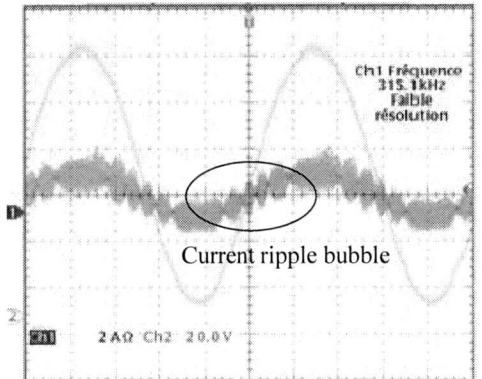

Figure 12: Experimental inverter waveforms with monolithic filter and 6 cells

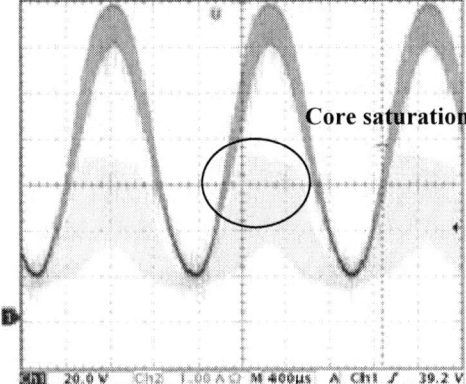

Figure 13: Experimental inverter waveforms with 5 working cells and 1 cell not in use.

IV. FAULT TOLERANCE AND CONTROL STRATEGY

In the paper [3], a method offers an interesting solution to increase the converter efficiency for output power much lower than its nominal value. The idea is to use only a part of the cells, so that each cell always works close to its nominal power as shown on figure 2 and 3. This solution is particularly interesting and can be adapted to ensure the converter operation with 1 or 2 legs out of order.

In the case of coupling filter, the iron section does not depend of the number of phases. The equation (2) always guaranteed a constant iron section even if the number of operating cells varies. So for magnetic flow saturation considerations, there is no need of increasing the switching frequency. However it can be interesting to modify the switching frequency to maintain the output switching frequency constant. This induces that the switching frequency of each cell changes, depending on the number of cells used. The **table 1** shows the switching frequency of each cell versus the number of cells used.

$$S_{fer} = \frac{E}{8 \times N_{spire} \times Be_{max} \times f_{min}} \quad (2) \ [18]$$

N of active cells	1	2	3	4	5	6
Switching frequency of each cell (kHz)	292	146	97	73	58	48

Table 1 : Switching frequency of each cell versus the number of cells used

The increase of the interleaved cell number with the output power induces a reduction of the switching frequency of each switch. Therefore the switching losses are reduced and the global efficiency will be higher.

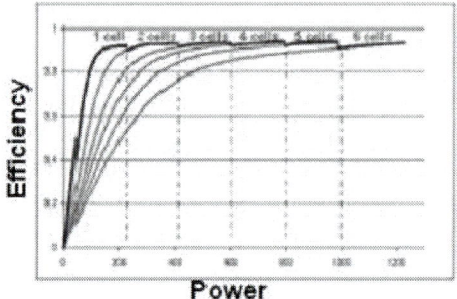

Figure 14: Efficiency optimization (blue), efficiency on variable cells (red).

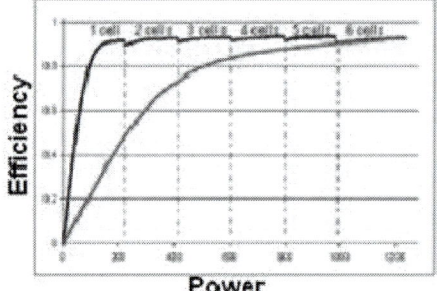

Figure 15: Efficiency optimization (blue), efficiency on classic inverter (red).

In case of control strategy for increase efficiency, all cells will not work simultaneously. To ensure the same lifetime and limit the constraint on the power transistors, the working time of each cell should be equal. A solution is to equalize the operating time of each output phase by making a circular bearing system of the cells. Of course, in case of dysfunctions the problem is different. The aim is to ensure the converter work utile its repair.

In the case of the interleaved structure with filter, it is necessary to ensure that the current can always pass through the different magnetic circuit. This should be done even if some converter cells are shut down. A solution consists in short-circuiting the phases of the magnetic coupler as presented on Figure 16.

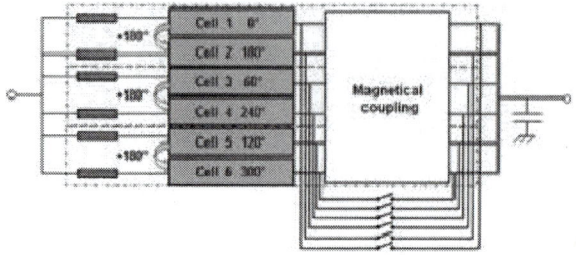

Figure 16: Additional switching to short-cut the unused phases of the magnetic output filter.

The main problem is the coupling between the phases. With this method, we can ensure to keep coupling even in case of dysfunctions. In practice, the coupling between phase n and output will generate additional losses. However, in case of reduced power this solution ensures a good compromise. The principal disadvantage of this system is the number of additional semiconductors required. The components must be controlled at turn on and turn off, be bidirectional in current and voltage. Several configurations of semiconductor devices can then be adopted (see Figure 17).

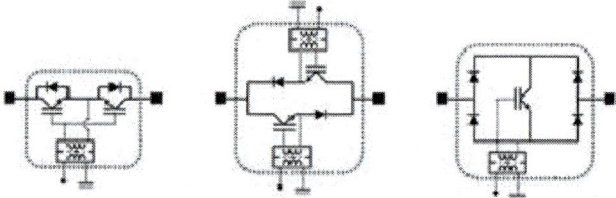

Figure 17: Switch semiconductors configuration with Transistors base.

In case of dysfunction, in order to insure a continuity of operation, the defective cell must be disconnected. To disconnect an invert arm, 3 switches are requested (red elements on Figure 18).

Many different technologies are available to insure this disconnection. Fuse technique [14] are generally found in literature, but power semiconductors can either be use. The advantages of semiconductors are that they can be electrically controlled, and they can be faster than fuse. With this flexibility and rapidity they will limit the disturbance due to the loss of one element.

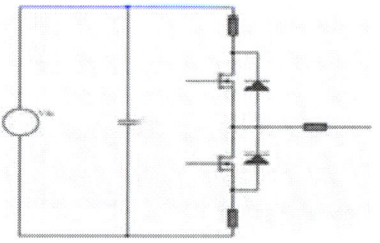

Figure 18 : Cell and switches (in red) in order to disconnect the cell

978-1-4244-4782-4/10 $26.00 © 2010 IEEE

This solution (Figure 16) works with both monolithic and coupling transformers structures. However physical phenomenons are different.

In case of coupling transformers, the short circuits on defective phase insure the continuity of current loop into the transformer (see Figure 5). The Figure 19 shows the example of the loss of phase 4, in this case the short circuit of phase 4 insures that current of phase 3 and 5 will be magnetically coupled.

The DC courant will be also equitably distributed into all the magnetic structure Figure 19 (b). This figure does not take into account the magnetic loss, in the real case this short circuit of phase 4 will introduce some extra losses.

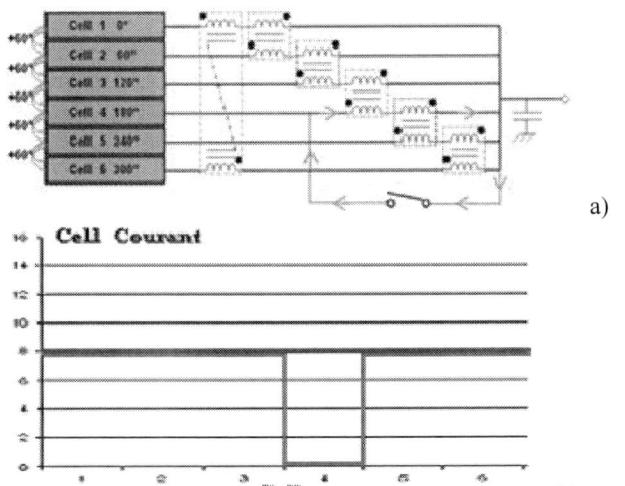

a)

b)

Figure 19: (a) Example of 6 cell structure with short-circuit component on cell 4. (b) In blue, courant into the different winding. In red, current going out each winding.

Figure 20 presents time domain waveforms of the inverter with transformer couplings. It shows the current ripple of the six-cell converter with dysfunction on one cell. On Figure 20, the converter works with 5 cells. One of the six cells is short-circuited and the FPGA (control circuit) has changed the phase shift. With one cell less, the courant ripple increase of a sixth. To keep it at the same level of a converter using 6 interleaved cells, we can increase the switching frequency like it is showed on table 1.

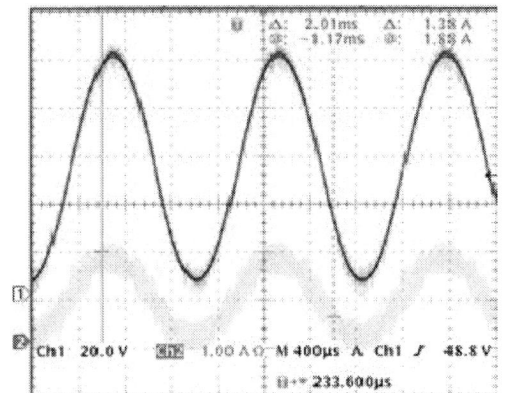

Figure 20: Experimental inverter waveforms with 5 cells

In the case of monolithic transformer, the explanation is based on flow analyze.

$$V = N \times \frac{d\Phi}{dt}$$

If V = 0 also $\quad N \times \frac{d\Phi}{dt} = 0 \quad \Rightarrow \quad \frac{d\Phi}{dt} = 0$

Based on Faraday's low, these equations show that flow variation is proportional to the voltage of the winding. If this voltage is null, there will be no flow variation. Therefore if a phase is broken, our system will close the switch on the corresponding winding; also the flow must be equilibrating between all other windings. Indeed, a current appear (Figure 22) that insure that the flow on the unused winding is null

In a real, case, due to unperfected coupling between windings, a stray inductance allows a way for the flow even on the short circuit wilding. It will cause some extra losses.

If we connect in short-circuit the unused or broken phases, a courant appear on the unused winding, Figure 22. The flows will take another way and will not use the arms in short-circuit.

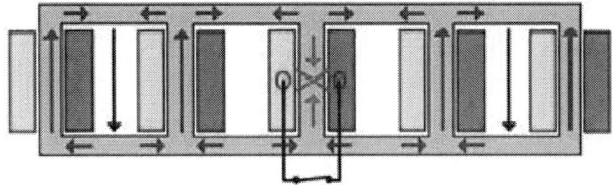

Figure 21: Circulation of flow in a coupler with 5 cells. 3 cells are used and short-circuit on the un-used phase.

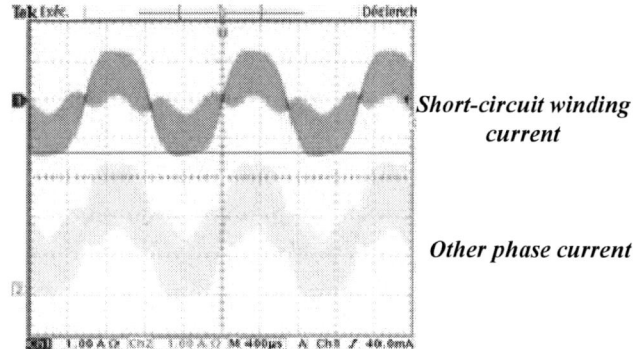

Short-circuit winding current

Other phase current

Figure 22: Experimental inverter waveforms with 2 cells and short-circuit.

For reduce iron section it can be use different phase shift between each cells, tables 2 & 3. Choose a phase shift as near as possible 180°betwen all adjacent winding (Table 3); allow reducing the flow on common iron section.

0	2π/6	4π/6	6π/6	8π/6	10π/6

Table 2 : Standard phase shift

0	4π/6	8π/6	2π/6	10π/6	6π/6

Table 3 : Permute phase shift

But phase shift between two phases vary versus the number of cell, and are more or less far from 180°. It result that an odd number of phase is better for volume reduction. The Figure 23 shows the iron losses for different cells number, the losses are directly proportionaly to volume.

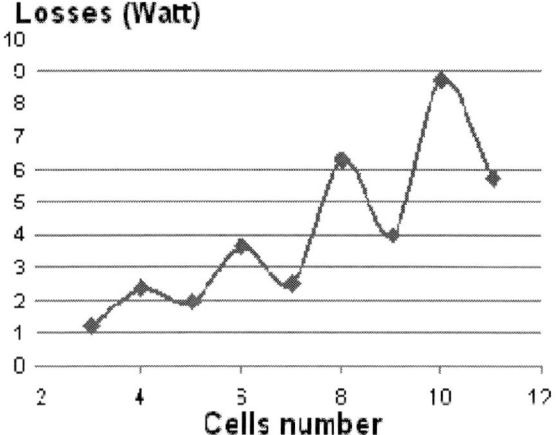

Figure 23: Iron losses for different cells number

In case of different number of phases and of various phase shifts, it can be necessary to increase switching frequency, in order to ensure a constant flow in magnetic core. In others case, some local flow saturation can appear.

V. EXPERIMENTS AND VALIDATION

Two tests benches have been developed; the first one is composed of 4 interleaved cells based on independent inductances and monolithic magnetic filters. Each cell is composed of a 3 level multi-level NPC converter. The second one is composed of 6 interleaved simple-level cells based on magnetic filters transformers. The whole system is driven by an FPGA. Figure 24 show one of these tests benches.

Each cell is able to provide a power of 400W. The global converter power is 1.6kW. The power electronic structure developed allows an auto-adaptive control. This control depends of the number of cells and the current and voltage measurements in the structure. The FPGA calculates the different control signals for each cell. The numerical structure into the FPGA is composed of a microprocessor NIOS II of Altera®.

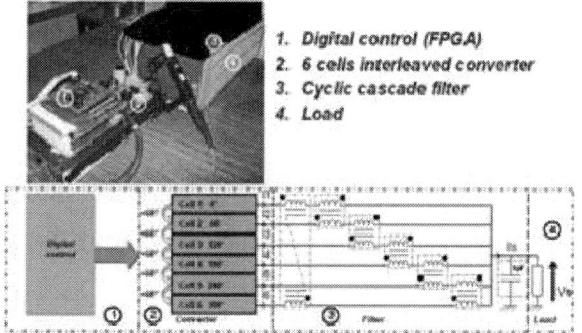

1. Digital control (FPGA)
2. 6 cells interleaved converter
3. Cyclic cascade filter
4. Load

Figure 24: 6 interleaved simple-level cells based on interphase transformers.

The processor analyzes the data coming from measurements (tension and current), it adapts PWMs in real time for functioning with variable number of phases versus the measured powers. His function is also to control the output voltage of the converter. The PWM module is a high DPWM resolution according to the model developed in [17]. A USB connection is also use with software of diagnostic and system requirements.

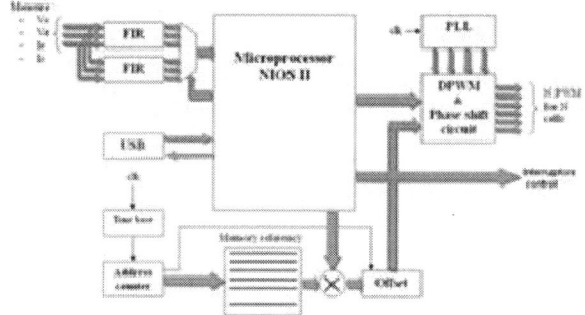

Figure 25: Numerical structure of the FPGA

VI. CONCLUSION

This paper shows an original control strategy that permits fault tolerance on an interleaved inverter with Magnetic couplers.

Part 2 presents how classical magnetic coupling Method (coupling transformers and monolithic coupler) can be used. Part 3 analyse the dysfunction mode and its effects on the coupling transformers or monolithic coupler.

Part 4 presents how power electronic structure must be adapted to take into account the fault tolerance control strategy.

Part 5 validate this approach by showing experimental setup.

Another advantage of using this control strategy is to permit the use of this king of converter with a reduced number of phases, and so to potentially increase the efficiency.

REFERENCES

[1] **J-A. ROOKS, A-K. WALLACE**, "Energy Efficiency of VSDs" IEEE Industry Applications Magazine May/June 2004

[2] **T. A. Meynard, H. Foch, P. Thomas, J. Courault, R. Jakob, M. Nahrstaedt**, "Multicell Converters: Basic Concepts and Industry Applications" , IEEE Transactions on Industrial Electronics 2002, vol.49, no.5, pp955-964

[3] **K.Guepratte, P.O.Jeannin, D.Frey, H.Stephan**, "High Efficiency Interleaved Power Electronics Converter for wide operating power range", IEEE Applied Power Electronics Conference, APEC, 2009

[4] **E. Labouré, A. Cunière, T. A. Meynard, F. Forest, E. Sarraute**, "A Theoretical Approach to InterCell Transformers, Application to Interleaved Converters", IEEE TRANSACTIONS ON POWER ELECTRONICS, VOL. 23, NO. 1, JANUARY 2008

[5] **S. Dieckerhoff, S. Bernet, D. Krug**, "Power Loss-Oriented Evaluation of High Voltage IGBTs and Multilevel Converters in Transformerless Traction Applications", IEEE transaction on power electronics, vol 20, no 6 novembre 2005

[6] **M.Stadler, J.Pforr**, "Multi-phase Converter for Wide Range of Input Voltages with Integrated filter Inductor", Power Electronics and Motion Control Conference, 2006. EPE-PEMC 2006. 12th International

[7] **Pit-Leong Wong,Q. Wu, Peng Xu, Bo Yang, F. C. Lee**, "Investigating Coupling Inductors in Interleaving QSW VRM", IEEE Applied Power Electronics Conference, APEC, 2000, pp. 973-978

[8] **Pit-Leong Wong,Q. Wu, Peng Xu, P. Yang, F. C. Lee**, "Performance Improvements of Interleaving VRMs with coupling inductors", IEEE Transactions on Power Electronics, 2001,pp.499-507

[9] **P. Zumel, O. Garcia, J. A. Cobos, J. Uceda**, "Magnetic Integration for Interleaved Converters", IEEE Applied Power Electronics Conference and Exposition, APEC, 2003, pp.1143-1149

[10] **In Gyu Park, Seon Ik Kim** , "Modelling and analysis of multi-interphase transformers for connecting power converters in parallel", IEEE 1997, pp1164-1170

[11] **F. Forest, T.A. Meynard, E. Labouré, V. Costan, E. Sarraute, A. Cunière, T. Martiré**, "Optimization of the supply voltage system in interleaved converters using intercell transformers", IEEE Transactions on Power Electronics, vol. 22, no.2, pp 934-942, 2007

[12] **P. Zumel, O. Garcia, J. A. Cobos, J. Uceda**, "Magnetic Integration for Interleaved Converters", IEEE Applied Power Electronics Conference and Exposition, APEC, 2003, pp.1143-1149

[13] **Valentin Costan, Thierry Meynard, François Forest, Eric Labouré**, "Core losses measurements in intercell transformers for interleaved converters", EPE 2007

[14] **B.A. Welchko, T.A. Lipo, T.M. Jahns, S.E. Schulz**, "Fault tolerant three- phase AC motor drive topologies; a comparison of features, costs, and limitations", IEEE Trans. on Power Electronics, vol. 19, pp. 1108-1116, July 2004.

[15] **T. A. Meynard, F. Forest, E. Labouré, V. Costan, A. Cunière, E. Sarraute**, "Monolithic Magnetic Couplers for Interleaved Converters with a High Number of Cells", CIPS, 2006

[16] **V. Costan, T.A. Meynard**, "Topologies Circulaires de Transformateurs Inter-cellules pour des Convertisseurs Parallèles Entrelacés", 2007, EF, Toulouse

[17] **Santa C. Huerta, A. de Castro, 0. Garcia, J.A. Cobos**, "FPGA based Digital Pulse Width Modulator with Time Resolution under 2 ns", IEEE 2008, pp4244-1668.

[18] **V. Costan**, "Etudes des pertes fer dans les transformateurs inter-cellules", thèse de doctorat en génie électrique, Laboratoire Plasma et Conversion d'Énergie, 2008.

Latest Practical Developments of Triplex Series Load Resonant Frequency-Operated High Frequency Inverter for Induction-Heated Low Resistivity Metallic Appliances in Consumer Built-In Cooktops

Hideki Sadakata, Atsushi Fujita,
Shinichiro Sumiyoshi, Hideki Omori
Home Appliances Company, Panasonic Co.
Kusatsu, Japan
e-mail: { sadakata.hideki, fujita.a}@jp.panasonic.com

Bishwajit Saha, Tarek Ahmed, Mutsuo Nakaoka
Kyungnam Univ. / Yamaguchi Univ.
Masan, Korea / Yamaguchi, Japan

Abstract— **This paper deals with the new generation prototype of one-stage boost-half bridge (B-HB) series load resonant soft-switching high-frequency inverter with a lossless snubbing edge resonant capacitor and two selective resonant capacitors in series with inductive load for consumer induction heating (IH) appliances. The B-HB high frequency inverter treated here is based upon a dual series load resonant frequency selection strategy, changed automatically in accordance with resistivity of metallic materials as IH loads. In the first place, the triplex series load resonant high frequency (three times of switching frequency) operated B-HB inverter is demonstrated for IH non-magnetic low resistivity metallic pans/utensils fabricated by aluminum, copper and multi-layer of aluminum and stainless steel. In the second place, the fundamental series load resonant frequency (switching frequency) operated B-HB high frequency inverter is also described for IH magnetic high resistivity metallic pans/utensils fabricated by iron, iron cast and stainless steel. Finally, the principle of operation and inherent unique features of the B-HB high frequency inverter employing dual series load resonant frequency selection scheme in accordance with either low resistivity or high resistivity IH loads is presented from an experimental point of view, along with its operating steady state performance. This new generation B-HB frequency inverter type built-in IH cooktop with two/three ranges is put into practice for home energy utilizations in all electricity residential systems.**

I. INTRODUCTION

In accordance with great advances of MOS gate power switching semiconductor devices, sensor interfaced devices and microprocessor-based digital controlling devices, as well as passive power circuit components, consumer power electronics relating to power conversion circuits and digital control implementations have become one major field technology in home energy utilization and management systems architecture. The electromagnetic-induction eddy current-based electric heating or induction heating processing systems technology employing high frequency inverters,

matrix converters and cycloconverters have attracted special interest for consumer induction heating applications.

The high frequency resonant inverter type built-in IH cooktops composed of PFC rectifier, high frequency inverter, planar type working coil, crystal spacer and induction heated pans/utensils. Direct heating of pans/utensils by the high frequency resonant inverter type built-in IH cooktops have some inherent remarkable advantageous points of high efficiency, easiness to control, cleaning, compactness, rapid cooking, comfortable environment, safety, high reliability and easy power management.

In recent years, the high frequency inverter type IH appliances; IH rice cooker and warmer, IH cooktop, IH hot water producer and IH super-heated steamer have been developed so far for home power applications from an energy saving point of view. The IH cooktop family with two/three ranges which incorporates voltage source high frequency resonant soft switching inverters as active-clamp, single-ended push pull, half bridge and full bridge circuit topologies has rapidly become more popular in Japan and Europe home use and business use.

However, most of high frequency inverter type built-in IH cooktop equipments could not heat effectively the metallic pans/utensils fabricated by non-magnetic and low resistivity metallic materials such as aluminum, copper and aluminum - stainless steel multi-layer, on the basis of induced eddy current based IH principle.

The authors have developed a novel prototype of cost-effective built-in IH cooktop family incorporating boost-half bridge series load resonant high frequency inverter with optimum resonant-frequency changing scheme, which is practically applicable for all metallic pans/utensils fabricated by low resistivity metallic materials and high resistivity ones and has just put into practice as commercial market.

This paper presents the triplex series load resonant frequency-operated B-HB inverter using the latest IGBT with anti-parallel FRD is demonstrated and designed, which can operate under the condition of a small damped oscillating current mode with three times of switching frequency under zero voltage soft switching for the IH aluminum, copper metal pans/utensils. The operating principle, control implementation and operating performances of B-HB high frequency inverter are described and discussed from an experimental point of view.

II. PRINCIPLE OF INDUCTION HEATING AND REQUIREMENTS FOR IH ALL METALLIC APPLIANCES

In induction heating process, a high frequency magnetic flux can be generated from planar type working coil by passing a high frequency current from an inverter operating at frequency of several tens of kHz into a planar working coil positioned below the pans/utensils, which act as the IH loads. The eddy current is induced in the bottom of the pans/utensils in proportional to the high-frequency magnetic fluxes caused on the basis of electromagnetic induction principle due to Farady's law. As a result, eddy current-based Joule heat corresponding to the product of the metallic pans/utensils electrical resistance and square of the eddy current. The pans/utensils themselves then heats up rapidly principle as shown in Fig. 1.

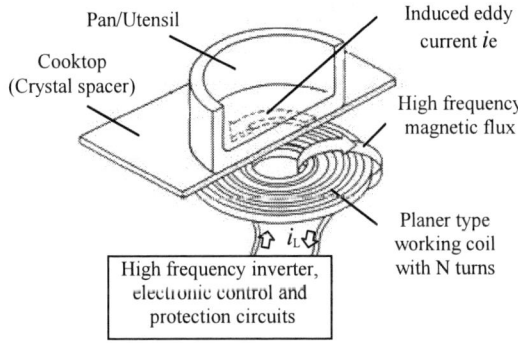

Fig. 1. A schematic architecture and Operating principle of induction heating cooktops.

In addition, in induction heating process, heating power P can be approximately expressed by,

$$P = R_s \cdot i_e^2 \propto \sqrt{\rho \cdot \mu \cdot f} \cdot (N \cdot i_L)^2 \qquad (1)$$

where Rs : skin effect resistance, ie : eddy current through the working coil with the metallic pan/utensil , ρ : resistivity of the pan/utensil, μ : permeability of the metallic pan/utensil, f : frequency of the working coil current, N : number of litz wire windings of the working coil and i_L : working coil current from high frequency inverter. The heating power P is proportional to Rs and the square of ie. The skin effect resistance Rs is proportional to $\sqrt{\rho \cdot \mu \cdot f}$ and ie is proportional to the intensity of magnetic field H, i.e., N and i_L.

Especially, because the aluminum material is non-magnetic substance, (μ_S=1, μ_0=4πx10⁻⁷) of aluminum is much

smaller than those of magnetic substances such as iron or stainless steel. Besides, ρ of aluminum is much smaller than that of iron or non-magnetic stainless steel. For example, as indicated in Table I, the product of ρ and μ_S in case of aluminum is 1/25 that of non-magnetic stainless steel. If the same induction heating power of aluminum pan/utensil as that of non-magnetic stainless steel pans/utensils is delivered to the load, then $\sqrt{f} \cdot (N \cdot i_L)^2$ must be kept to be 5.

In this IH appliance development, f of high frequency current to IH loads with planer working coil is designed approximately as three times of that of the previous development, N is increased up to approximately 1.7 times that conventionally used, if i_L is almost designed so as to be the same value. However the switching power loss can be increased as f is much higher. In order to reduce the increased switching power losses of power semiconductor devices, the triplex series load resonant frequency operated high frequency inverter operating under the condition of one third switching frequency is developed.

TABLE I.
PHYSICAL PARAMETERS OF ALL METAL IH COOKTOP

Physical value	Metallic Material	High Resistivity		Low Resistivity	
		Iron	Non-magnetic stainless steel	Aluminum	Copper
Resistivity ρ (μ Ωm)		0.17	0.7	0.027	0.017
Relative permeability μ_s		200	1	1	1
Previous developed IH cooktop		○	○	×	×
Newly developed IH cooktop		○	○	○	○

Remarks; ○ applicable × Not applicable

III. TRIPLEX RESONANT FREQUENCY OPERATED HIGH FREQUENCY INVERTER

A. Circuit Description

The inverter circuit configuration operating under the small damped resonant oscillation with triplex series load resonant frequency is shown in Fig.2. This triplex resonant frequency operated high frequency inverter is developed for IH cooktops applicable for all metallic pans/utensils.

The main power conversion circuit mainly consists of B-HB resonant inverter, high boost smoothing capacitor DC link, a passive PFC rectifier.

A B-HB resonant inverter with dual series load resonant frequency selection scheme consists of working coil L1, resonant capacitor C1 and Cr, the power semiconductor switching devices Q1(S1/D1) and Q2(S2/D2). The relay RL1 is used to change the series resonant capacitor C1 into

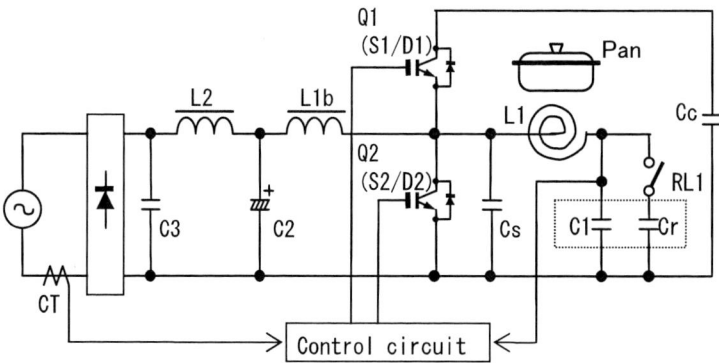

Fig. 2. Total circuit of the newly developed high frequency resonant inverter for IH cooktops applied for all metallic pans/utensils

C1+Cr according to the physical characteristics of the pans/utensils.

Boost converter in one stage B-HB inverter consists of boost choke coil L1b, S2, D1 and high-voltage DC smoothing capacitor Cc. Inverter and boost converter have common switch S2 of Q2. For soft-switching means of lossless snubbing edge-resonant capacitor Cs changeable by an undescribed switching device is connected in parallel with Q2 in parallel.

Moreover, especially when non-magnetic materials objects such as aluminum, copper pans/utensils are heated directly, a Lorenz force acts on heated pans/utensils by the high frequency magnetic field and the eddy current induced in the pans/utensils. Since variations of high frequency inverter supply voltage lead to the change of a Lorenz force and mechanical oscillation of pan-vibration sound. Thus the smoothing capacitor C2 of output side in passive PFC rectifier, which consists of electrolytic capacitor bank, and boost smoothing film capacitor Cc is designed so as to perform sufficient voltage-smoothing.

B. Operation Principle for Low-resistivity Non-magnetic Induction Heated Load

Let's consider the circuit operation in case that the IH loads as aluminum and copper of pans/utensil are low-resistivity and non-magnetic. The relay RL1 keeps open in this state. In this case, the C1 becomes a series resonant capacitor. The circuit parameters and design specifications of B-HB inverter become as shown in Table II. Figures 3 and 4 give the explanation of the circuit operation principle of the newly-developed B-HB high frequency resonant inverter using IGBTs (Q1, Q2). This circuit includes four operating modes during one switching period as indicated in Fig.4. The steady-state mode transition of triple resonant frequency operated series resonant high frequency inverter is described below by observing its relevant voltage and current waveforms.

■ *Mode 1*; (Q1:Off, Q2:On)

In mode 1, two loops in the circuit are composed of loop (1) and loop (2).In loop (1): C2 – L1b – Q2 – C2; the magnetic energy is stored into the boost inductor L1b.

And then in loop (2): C1 – L1 – Q2 – C1; The energy

stored in the resonant capacitor C1 is delivered to the pan with working coil. C1 is discharged resonantly.

■ *Mode 2*; (Q1:Off, Q2:On)

In mode 2, two loops in the circuit are composed of loop (1) described above and loop (3).In loop (3): C1 – D2 – L1 – C1; the energy stored in C1 is delivered to the pan. C1 is charged resonantly.

■ *Mode 3*; (Q1:On, Q2:Off)

In mode 3, two loops in the circuit are composed of loop (4) and loop (5).In loop (4): C1 – L1 – D1 – Cc – C1; the energy stored into Cc is delivered to the pan. C1 is discharged resonantly.In loop (5): C2 – L1b – D1 – Cc – C2; the magnetic energy stored into the boost inductor L1b is released by charging Cc via D1 of Q1.

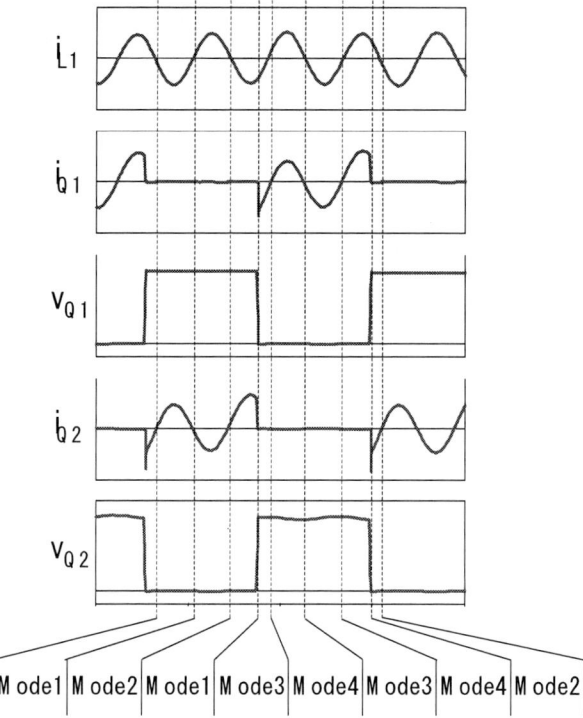

Figure. 3. Relevant voltage and current waveforms of the newly developed high frequency resonant inverter when low-resistivity non-magnetic materials pans; aluminum/copper are heated.

978-1-4244-4782-4/10 $26.00 © 2010 IEEE   1827

■ *Mode 4*; (Q1:On, Q2:Off)

In mode 4, two loops in the circuit are composed of loop (5) described above and loop (6).

In loop (6): C1 – Q1 – Cc – L1 – C1; the energy is delivered to the pan. C1 is charged resonantly through Q1.

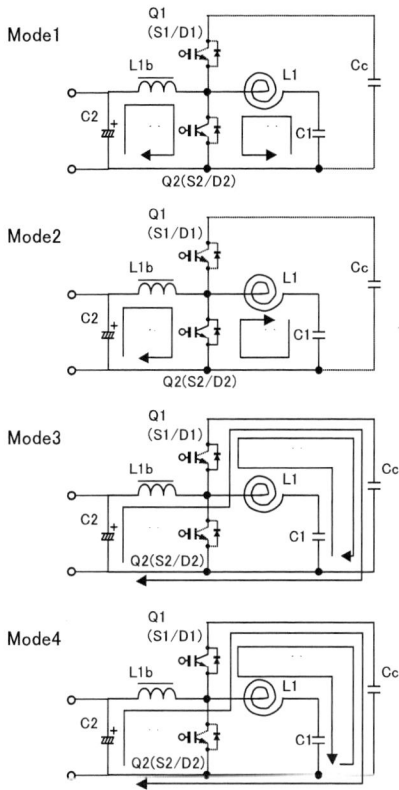

Fig. 4. Operating mode transitions and switching mode equivalent circuit of newly developed HF resonant inverter for IH low-resistivity non-magnetic materials pans.

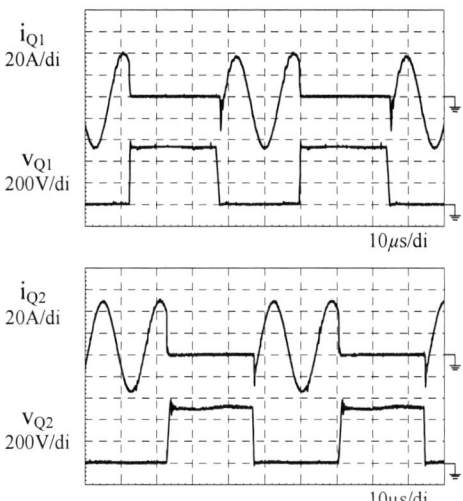

Fig. 5. Measured voltage and current operating waveforms for Q1(S1/D1) and Q2(S2/D2) when an aluminum pan/utensil in 2kW is heated by IH principle.

This circuit repeats the operating switching mode in order of mode 1, mode 2, mode 1, mode 3, mode 4, mode 3, mode 4, mode 2, and mode 1 as described in Fig. 3. The resonant frequency equal to natural frequency of L1 and C1 is designed so as to be approximate three times of the switching frequency (inverter output voltage frequency) of Q1 and Q2. Since the pan is non-magnetic and low-resistivity material, the resonant current with a small damping factor does not attenuate in the respective period of each Q1 and Q2 turning on. Figure 5 shows measured operating voltage and current waveforms of Q1 and Q2. As mentioned above, the switching frequency of Q1 and Q2 is set to be approximately 20kHz, so the triple resonant frequency of the oscillating current passing through L1 and C1 is estimated as approximately 60kHz.

So, this circuit manages output high frequency magnetic field corresponding to high frequency oscillating current which is more suitable for IH of aluminum pans/utensils and reduction of the switching losses of Q1 and Q2 due to lowering the switching frequency of Q1 and Q2. Heating effective power dissipated in IH load is controlled by varying switching frequency and minute change of duty ratio around 0.5 of Q1 and Q2.

TABLE II
CIRCUIT PARAMETERS OF THE NEWLY DEVELOPED INVERTER

Item	Symbol	Value		Unit
Working Coil with metallic pan	L1	with Aluminum Pan (at 60kHz)	227	μ H
			1.8	Ω
		with Iron Pan (at 23kHz)	312	μ H
			9.2	Ω
Boost Inductor	L1b	280		μ H
DC Filter	L2	2 (x 2pcs)		mH
Series Resonant Capacitor	C1	with Aluminum	0.03	μ F
	Cr	with Iron Pan	0.24	μ F
DC Filter	C2	470 (x 3pcs)		μ F
Boost Capacitor	Cc	10		μ F
Lossless Snubbing Edge Resonant	Cs	0.05 / 0.01 (changeable)		μ F

Remarks: fr:Resonant Frequency 60kHz
fo:Switching Frequency (Output Frequency) 20kHz

C. Operation Principle for IH Pans of High-Resistivity Magnetic Materials

Next, let's describe the circuit operation in case that the material of IH pan is high-resistivity and magnetic materials such as iron, iron cast and stainless- steel. In this case, the relay RL1 turns on to close (Cr is connected to C1 in parallel) and the circuit parameters become as indicated in Table II. Figures 6 and 7 provide the explanation of the circuit operation principle in steady state. Each switching operation mode for heating the iron pans/utensils is almost the same as the switching equivalent circuits for heating the aluminum, copper pans/utensils but mode transition order.

In this case, this B-HB inverter performs repeating its operating mode in order of mode 1, mode 2, mode 3, mode 4, and mode 1 as shown in Figs. 6 and 7. The fundamental resonant frequency of L1 and C1+Cr is designed so as to be approximately equal to the frequency of Q1 and Q2.

To heat aluminum or copper pans/utensils with lower coil current, the number of litz wire windings of L1 is designed so as to be 42, approximately 1.7 times as compared with that conventionally used. So, it is required for induction heating of pans with high-resistivity materials, such as iron and stainless steel, to apply high voltage to the B-HB inverter to achieve enough coil current. In this B-HB high frequency inverter circuit, that is easy because voltage across Cc is boosted by operation of L1b, S2, and D1. Figure 8 shows measured operating voltage and current waveforms of Q1 and Q2. In general, the conduction time of Q2 is longer than that of Q1. The voltage of boost operation depends on conduction time of Q2, so that, as the conduction time is longer, voltage across Cc tends to be higher. Figure 8 indicates that voltage across Cc is charged up to about 650V. Heating power is controlled by varying duty ratio of Q1 and Q2 with a constant switching frequency of 23kHz.

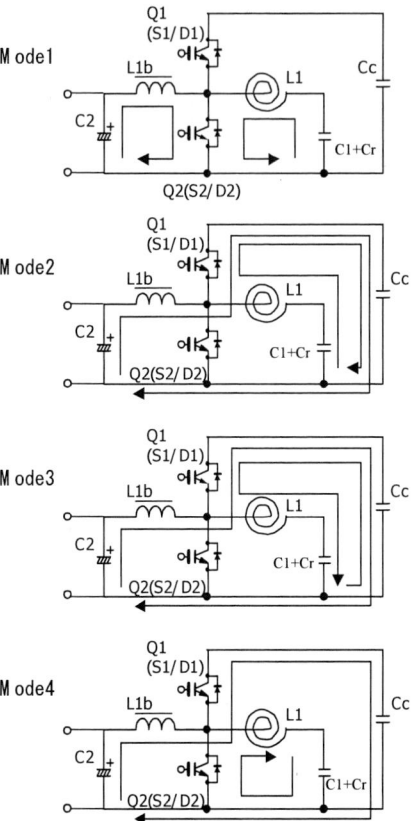

Fig. 7. Operating mode transitions and switching mode equivalent circuit of newly developed HF resonant inverter when the pan is heated for high-resistivity materials; the iron and stainless steel

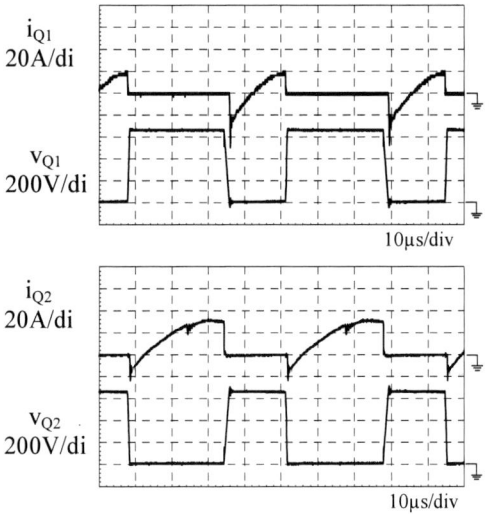

Fig. 8. Measured relevant voltage and current operating waveforms for Q1(S1/D1) and Q2(S2/D2) when an iron pan/stainless steel pan is heated in 2kW output.

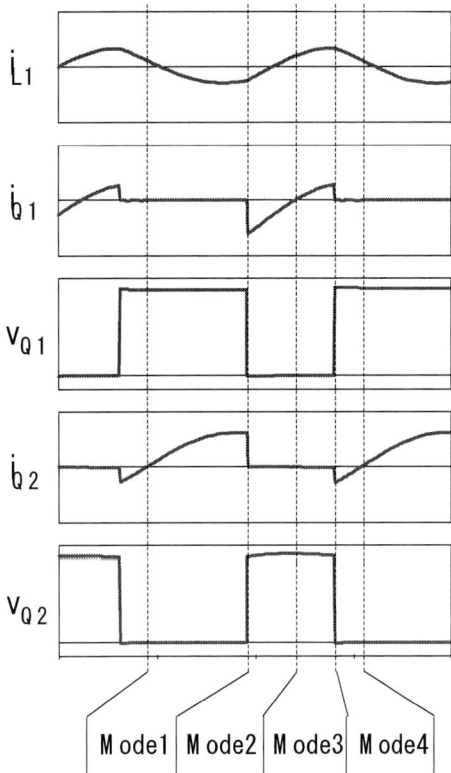

Fig. 6. Operating waveforms of newly developed inverter when the pan is heated for high-resistivity materials; the iron and stainless steel.

IV. OPERATING PERFORMANCES

A. High Frequency Inverter Implementation

Figure 9 shows the interior appearance of PC boards on the newly-developed B-HB series resonant high frequency inverter type IH cooktops designed for all metallic pans/utensils. This circuit mainly consists of two PC boards. Switching devices of B-HB inverter and control circuit are assembled on the same first PC board. Large size passive power components such as resonant capacitors, smoothing capacitors and relay control interface circuit are constructed on the second PC board. Moreover, heavy weight inductor with magnetic cores such as a boost choke inductor with powder core and a passive filter inductor with laminated Si-steel core are fixed on the holder supported by springs. The dimension of PC board size including heavy weight inductor holder is about 23cm in width x 23cm in depth x 4.5cm in height x 2pcs.

The 4th generation trench gate IGBTs (60A/950V) produced by Toshiba Semiconductor Co., Ltd as Q1 and Q2 in high frequency resonant inverter are incorporated.

Figure 10 shows the appearance of the specially-designed working coil for induction heated all metallic objects. Figure 11 shows the enlarged photograph of the litz wire for the working coil. To reduce the high frequency resistance of the working coil, the litz wire assembly consists of 1,620 entwined insulated copper wires in 50μm diameter.

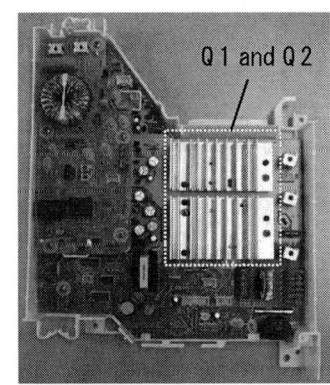

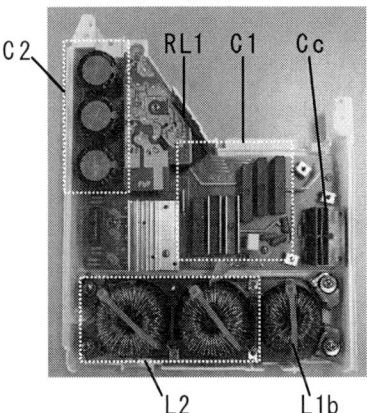

Fig. 9. Photographs of the PC boards of the newly developed high frequency resonant inverter type cooktops for IH all metallic pans/utensils.

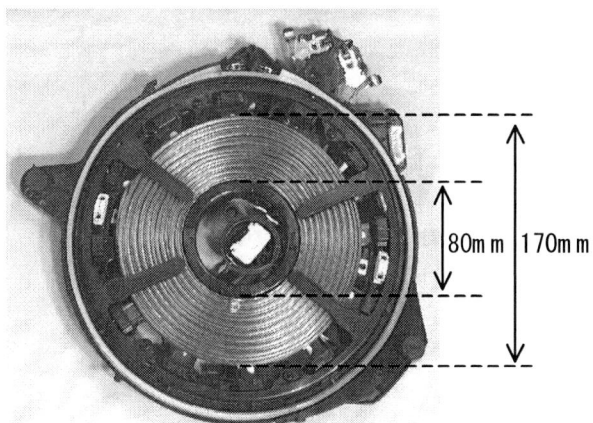

Fig. 10. Appearance of the working coil incorporated into the newly developed high frequency resonant inverter type cooktops for IH all metallic type pan/utensil.

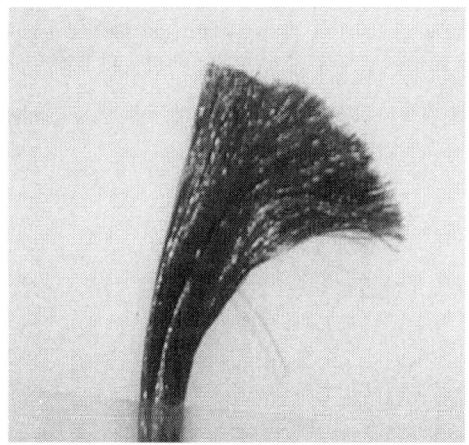

Fig. 11. Enlarged appearance of the litz wire for the working coil incorporated into the newly developed HF resonant inverter type cooktops for IH all metallic type pans/utensils.

B. Operating Characteristics

a) Power Regulation

Figure 12 illustrates the input power vs. the switching frequency of Q1 and Q2 characteristics when an aluminum pans/utensils are heated by IH principle with the pulse frequency modulation methods of the triple resonant frequency designed for three times as switching frequency operated mode. The small change of the switching frequency can realizes wide power regulation.

Figure 13 illustrates the input power vs. the duty ratio of Q2 characteristics when an iron pans/utensils are heated with the pulse width modulation methods of the fundamental resonant frequency equal to switching frequency operated mode. The more the duty ratio of Q2 becomes larger, the higher the charge-up capacitor Cc voltage across the boost capacitor Cc is boosted up, so the input power goes up.

b) Zero Voltage Soft Switching Operation

Figure 14 illustrates the observed switching voltage and current waveforms of Q1 and Q2 when aluminum pan is heated. Fig. 14, the enlarged switching voltage and current waveforms are illustrated when Q1 and Q2 turn off, respectively.

As can be seen in Fig. 14, the switch voltage waveform is going up with a slow slope by the charging effect of the lossless snubbing edge-resonant capacitor Cs. Moreover, this B-HB inverter circuit works with the switching frequency of about 20kHz in different from the triple resonant frequency of about 60kHz in this case. As a result, turn-on and off transition switching power dissipation in Q1 and Q2 are dramatically decreased.

The turn-off transition power dissipations in newly-developed B-HB inverter is estimated as 12W for Q1 and 19W for Q2 respectively, which are calculated from the actual voltage and current waveforms of Q1 and Q2.

Proposed B-HB inverter for IH cooktops applied for all metallic pans/utensils has passive PFC rectifier. In this way, input current waveform has harmonic distortion shown in Fig.15. From an energy saving point of view, it is necessary to improve a power factor by changing passive PFC rectifier to active PFC rectifier.

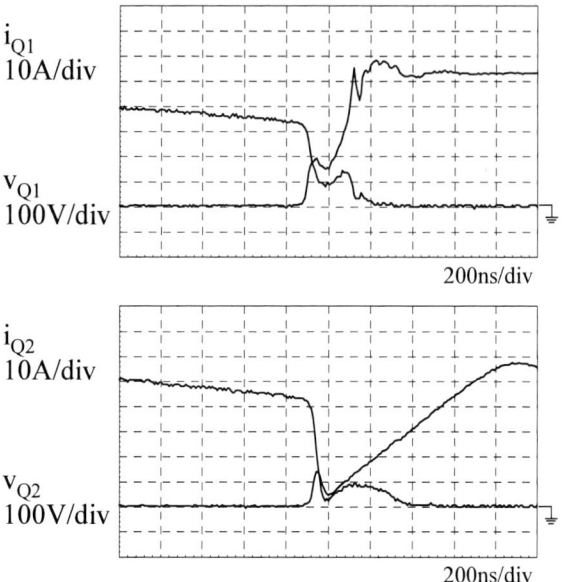

Fig. 14. Measured switching voltage and current waveforms for Q1(S1/D1) and Q2(S2/D2) when an aluminum pans/utensils in 2kW is heated by IH principle.

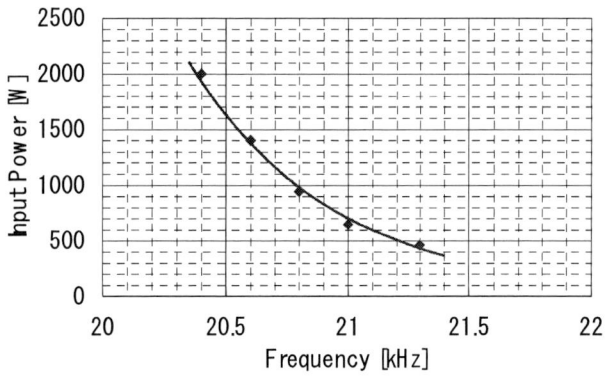

Fig. 12. Input characteristics of the newly developed high frequency resonant inverter when aluminum pans/utensils are heated.

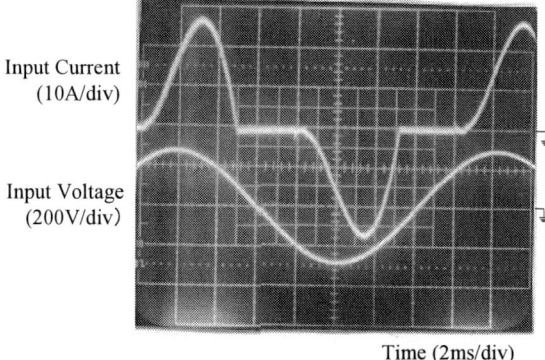

Fig. 15. Measured voltage and current waveforms for AC source when high resistivity metallic pans/utensils fabricated by iron cast in 2kW is heated by IH principle

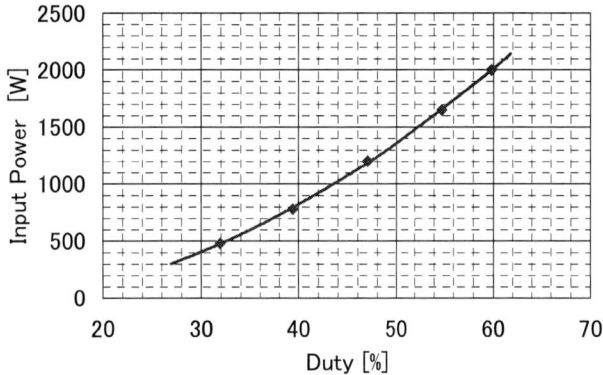

Fig. 13. Input characteristics of the newly developed high frequency resonant inverter when iron pans/utensils are heated.

Fig. 16. Whole appearance of the newly developed high frequency resonant inverter type cooktop for IH all metallic pans/utensils.

978-1-4244-4782-4/10 $26.00 © 2010 IEEE

TABLE III
GENERAL DESIGN SPECIFICATIONS OF NEWLY DEVELOPED HIGH
FREQUENCY RESONANT INVERTER TYPE IH COOKTOP

Power Supply	Single phase 200V$_{rms}$ AC (50Hz/60Hz)	
Rating Power	4.8kW	
	Left IH Heater (Conventional)	3kW (Except for low-resistivity non-magnetic pans/vessels)
	Right IH Heater (Newly developed)	2kW (For all-metallic pans/vessels)
Dimension	60cm (W) x 56cm (D) x 23cm (H)	
Weight	23kg	

V. CONCLUSIONS

In this paper, conceptual series load resonant high frequency inverter with optimum dual mode resonant capacitor selecting scheme in accordance with low resistivity or high resistivity metal pans/utensils, has been newly developed for consumer IH appliances, which includes one stage packed boost-half bridge soft switching pulse modulation inversion circuit with the intermediate high boost DC link, and simple front-end PFC rectifier. The operating principle, specific control and pulse modulation methods of the triple resonant frequency designed for three times as switching frequency operated high frequency resonant inverter and the fundamental resonant frequency equal to switching frequency operated one were described and confirmed respectively from an experimental point of view, together with its steady state operating performances.

The new generation cost-effective products of aforementioned series resonant high frequency inverter type built-in all metal IH cooktops family with two/three ranges has been demonstrated firstly in the world, which is applicable for not only low resistively metallic materials such as aluminum and copper but also high resistivity metallic materials such as iron and stainless-steel.

In the future, the advanced generation all metal IH appliances which incorporate into the high frequency resonant inverter and its related active PFC converter should be proceeded with the aid of (i) further the improvement of power semiconductor devices; Super-Junction MOSFETs, SiC/GaN MOSFETs, and IGBT with SiC-SBD, (ii) cost-effective circuit topologies and control schemes; magnetic integration, multi-phase/interleaving, time-sharing principle, (iii) optimum design of working coil fabricated by litz wire assembly, toward further improved requirements on high efficiency, high power density, and high performances.

REFERENCES

[1] T. Miyauchi, I. Hirota, H. Omori, and M. Nakaoka, "Induction heating cooking devices for aluminum pots", National Convention Record of IEEJ, vol. 5, p. 126, 1994. (in Japanese)

[2] K. Yasui, I. Hirota, T. Iwai, H. Omori, T. Ahmed, N. A. Ahmed, H. W. Lee, and M. Nakaoka, "Advanced Soft-Switching PWM High Frequency Power Supply with Charge-Up Function for Consumer IH Cooking Heater", the 2005 International Power Electronics Conference; IPEC NIIGATA 2005, S31-3, 2005.

[3] I. Hirota, H. Yamashita, H. Omori, and M. Nakaoka, "Historical Review of Electric Household Appliances using Induction Heating and Future Challenging Trends", IEEJ Transactions on Fundamentals and Materials, Vol. 124-A Number 8, pp. 713-719, 2004. (in Japanese)

[4] H. Sadakata, T. Kitaizumi, K. Yasui, T. Okude, and H. Omori, "Advanced Soft-Switching High Frequency Power Supply with Charge-up Function for IH Cooking Heater", JIPE-32-9, 2006. (in Japanese)

[5] J. Acero, J.M. Burdio, L.A. Barragan, D. Navarro, R. Alonso, J.R. Garcia, F. Monterde, P. Hernandez, S. Llorente, I. Garde, "The domestic induction heating appliance: an overview of recent research", IEEE Applied Power Electronics Conference and Exposition (APEC) 2008, 18-7, 2008

[6] Atsushi Fujita, Hideki Sadakata, Izuo Hirota, Hideki Omori and Mutsuo Nakaoka, "Latest Developments of High-Frequency Series Load Resonant Inverter Type Built-In Cooktops for Induction Heated All Metallic Appliances", pp2537-2544, 2009 IEEE 6th International Power Electronics and Motion ControlConference (IPEMC2009) in Wuhan, China, 17-20.May, 2009

A Study of Novel Flyback Converter with Very Low Power Consumption At the Standby Operating Mode

Eun-Soo Kim, Bong-Gun Chung, Sang-Ho Jang
Electrical Engineering, Jeonju University
1200, Hyoja-Dong, Wansan-Gu, Jeonju
Jeonbuk, 560-759, South Korea
E-mail : eskim@jj.ac.kr

Mun-Gi Choi
SMPS gr, LG innotek
379, Kasoo-dong, Osan,
Gyeonggi, 447-705, South Korea
E- mail : mgchoi@lginnotek.com

Moon-Ho Kye
Powerplaza USA
1705 E. Garry Ave #102, CA
92705, USA
E-mail : rkye@powerplazaus.com

Abstract— Recently, although the power consumption of green mode flyback converters at the light load and standby operating mode was minimized by the burst mode operation of PWM IC, the flyback converter has still about 0.3W~0.5W power consumption caused by the high magnetizing current flowing through magnetizing inductance of high frequency transformer in the primary side. PDP TV is required to reduce the standby power consumption by 0.2W. This paper proposes a novel topology of green mode flyback converter to decrease the standby power consumption by reducing the magnetizing current at the standby mode operation. A 70W prototype flyback converter for an auxiliary power module installed in 50" PDP TV is built, tested and verified. Its experimental results are described.

I. INTRODUCTION

Recently more researches to reduce standby power consumption have been performed for improving environmental issues caused by higher energy cost, and energy consumption. In Korea a report says that standby power consumption contains about 10% of the electric energy consumption [1]. This causes more energy generation and worse environmental issues. California Energy Commission (CEC) requires a mandatory regulation of standby power consumption of power supply lower than 250W should have less than 0.5W. These kinds of requirements become mandatory and severer in more countries including EU, USA & some Asian countries to use electric energy efficiently [2]. For these requirements new design and development of power supplies are more active [3], [4],[5]. Flyback converter is normally adopted for 60-80W AC-DC power supply for wide range input voltage, low cost and circuit simplicity. Recently green mode Flyback PWM IC is on the market to increase the efficiency from 75% to 84% and to reduce standby power consumption by using burst mode operation at the low load condition. Several semiconductor manufacturers show green mode IC's such as STR-W6252 (Sanken), ICE3BR0665J (Infineon), TEA1533 (NXP) and FAN6300 (Fairchild). These IC's work on burst mode operating at standby state and get 300mW to 500mW for standby power consumption. In burst

mode operation the transformer has high peak magnetizing current by low magnetizing inductance.

In this paper an improved topology from conventional flyback converter using green mode IC is presented and compared to get lower standby power consumption for power supply unit of 50" PDP TV. A 70W prototype green mode flyback converter is built, tested and verified that the topology is good to apply for an auxiliary power module in supply unit of 50" PDP TV. Reduction of standby power consumption is being focused on this paper.

II. PROPOSED TOPOLOGY AND ITS OPERATIONS

Fig 1 shows general configuration and block diagram of power supply unit (called PSU) for PDP TV. PSU for PDP TV is consisted of two DC-DC converters supplied by power factor correction (called PFC) output. A 600W LLC converter supplies major power for PDP TV and a 70W flyback converter does for auxiliary power. When remote controller turns on TV and PFC, flyback converter and LLC converter are working, this operation is defined by normal operating mode.

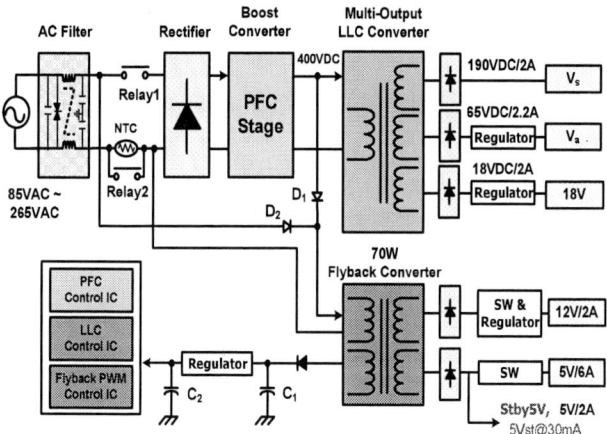

Figure 1.   Main configuration of PSU for PDP TV

But when remote controller turns off TV and PFC and LLC converter are not working anymore and this operation is called by standby operating mode which is await for a signal from remote turning on. So two operating modes named normal operating mode and standby operating mode are studied and described.

Fig 2 shows the conventional flyback converter to supply the multi-output auxiliary power output and standby power output and Fig 3 shows a proposed topology to reduce standby power consumption by separating standby mode from normal mode. The proposed topology has another winding (N_{1A}) with same turns with N_1 in the primary side and this winding is able to work at only standby operating mode. This extra winding makes higher magnetizing inductance and lower power consumption at burst operating mode. Obviously flyback converter has low primary inductance (L_m) to get discontinuous operation in wide input voltage range and the rated load. This low inductance makes magnetizing current bigger and higher peak current. Some power loss is induced at even standby mode.

In this paper, a standby mode is separated from a normal mode to evade this undesirable operation and both of the modes are described how to work and reduce magnetizing current with this higher magnetizing inductance at the standby operating mode of the flyback converter.

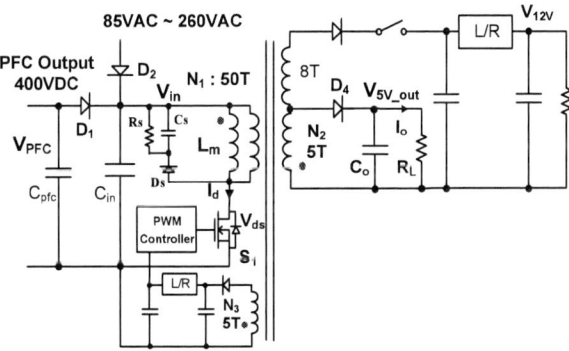

Figure 2.   Conventional flyback converter for the multi-output

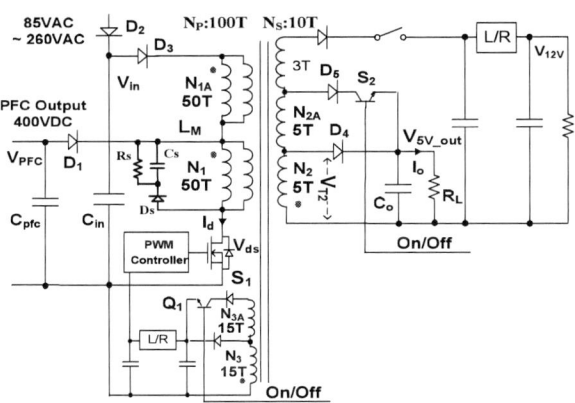

Figure 3.   Proposed topology to reduce the standby power consumption

II.1   OPERATION OF NORMAL MODE

Fig. 4 (a) and Fig. 4 (b) show normal operating mode per MOSFET switch on/off. As AC input (85Vac to 264Vac) supplies energy to a PFC and a flyback converter gets supplied thru D_1 by PFC output (400Vdc). The operation is same with the conventional flyback converter with turn ratio (N_1/N_2: 50T/5T) of the transformer. Detailed operation is described below when the switch S_1 is on and off each.

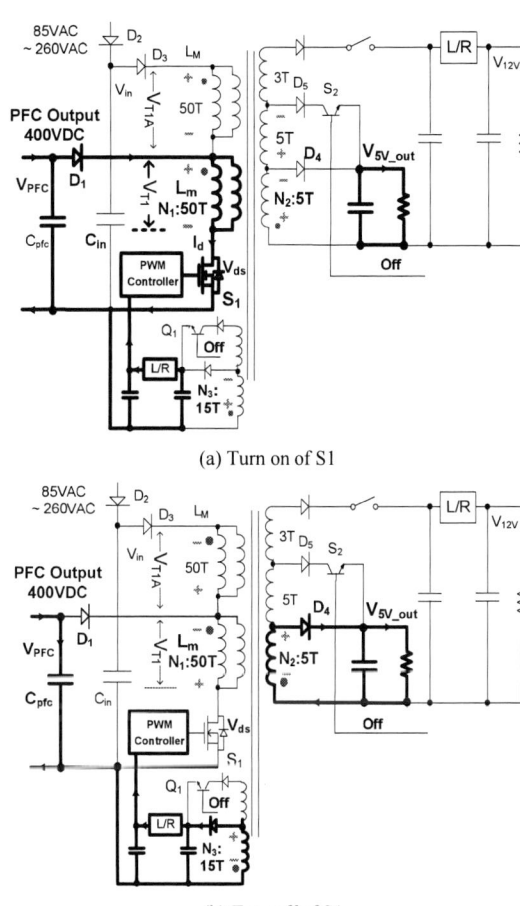

(a) Turn on of S1

(b) Turn off of S1

Figure 4.   Operations of Normal Mode

As shown Fig. 4(a), when the S_1 is on, the primary current flows from PFC output thru D_1 to the transformer (L_m). At this time the secondary is blocked by the diode (D_4) and the energy is stored in the primary inductance. In Fig. 4(b) as S_1 is off, the polarity of transformer is changing and the stored energy is transferred to the secondary side and the load. At this time S_1 gets the voltage stress of the PFC output voltage (V_{PFC}) plus reflected voltage ($V_{T1}=(N_1/N_2)V_{T2}$) from the secondary side. In normal mode, flyback converter has the low primary inductance (L_m) to get discontinuous operation in the minimum input voltage (V_{PFC_min}) and the rated load (R_{L_max}) as described in equation (1).

$$L_m < \frac{T \, D_{\max}^2 \, V_{PFC\_\min}}{2 \, P_{RL\_\max}} \qquad (1)$$

This low inductance (L_m) makes magnetizing current bigger and higher peak current. Some power loss is induced at even standby mode. To solve this issue, the standby mode is operated thru the separated circuits from the normal mode and has an increased inductance.

II.2 OPERATION OF STANDBY MODE

Standby operating mode is shown in Fig. 5 (a) and Fig. 5 (b). When relay 1 and relay 2 as shown in Fig. 1 are turned-off after the remote controller turn-off TV, PFC is not working and AC input (85Vac ~ 264Vac) supplies the energy to the flyback converter through D2 and D3. Burst mode operation is starting. As the windings ($N_p=N_{1A}+N_1$) of N_1 and N_{1A} are applied, the total magnetizing inductance (L_M) becomes four times of the winding N_1 inductance(L_m) and its power loss is decreased because of the increased turns and reduced magnetizing current.

In this mode turn ratio becomes the same with normal operating mode because the secondary side have another winding N_{2A}, too. So as the primary turns are changed from 50 turns to 100turns at standby mode, the secondary turns are changed from 5turns to 10 turns thru N3, too. With the increased magnetizing inductances (L_M) the turn on period becomes longer to keep $V_{5V\text{-out}}$ as constant 5V in discontinuous flyback mode. Equation (2) and (3) explain turn-on and Equation (4) explains the operation of turn-off period.

$$V_{5V\_out} = V_{in} \cdot T_{on} \cdot \sqrt{\frac{R_L}{2 \cdot L_M \cdot T}} \qquad (2)$$

$$T_{on} = \sqrt{\frac{2 \cdot L_M \cdot T}{R_L}} \cdot \frac{D}{D_A - D} \cdot \frac{N_s}{N_p} \qquad (3)$$

$$D_A - D = \sqrt{\frac{2 \cdot I_o \cdot L_M}{V_{5V\_out} \cdot T}} \cdot \frac{N_s}{N_p} \qquad (4)$$

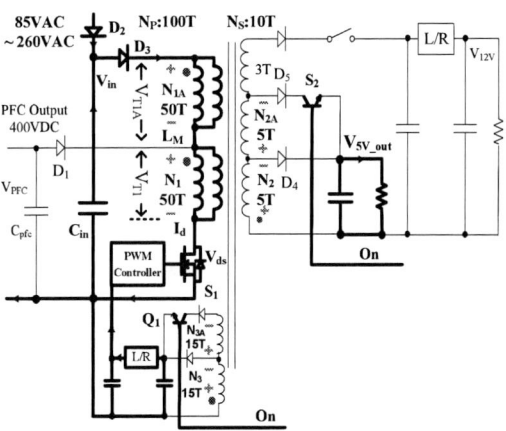

(a) Turn on of S1

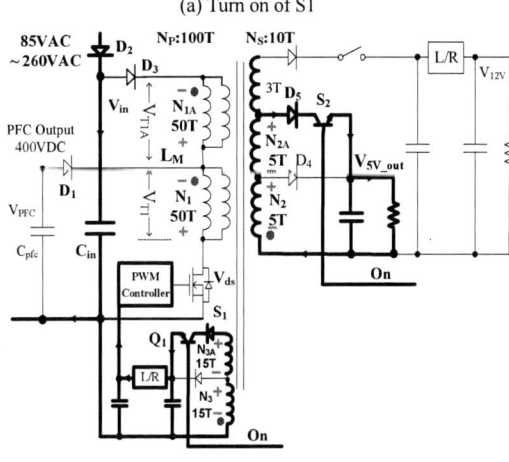

(b) Turn off of S1

Figure 5.　Operation modes for reducing standby power in the proposed converter

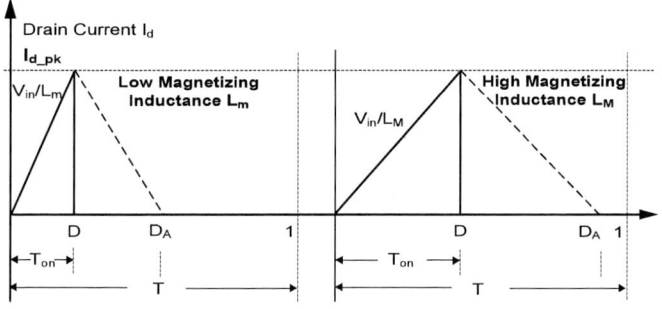

Figure 6.　Operation waveforms of the limited drain peak current (Id_pk) in the burst mode

Also as shown in Fig. 6 at the burst mode operation of Power IC (STR-W6252)[6], the peak current of MOSFET switch (I_{d_pk} : 400mA) is limited within 15% of normal mode (I_{d_pk} : 2.5A). This makes numbers of switching lower and amount of the stored energy more because the turn on period becomes longer to keep $V_{5V\text{-out}}$ as constant 5V per the increased magnetizing inductance as explained equation (2) and (4). Therefore the proposed flyback converter is able to reduce switching losses and current stress in switching devices.

III. EXPERIMENTAL RESULTS

The proposed flyback converter is tested and verified its performance for 70W output power with the parameters in table I and measured inductances of transformer in table II which are used in the testing. The waveforms are showing the comparison of the proposed topology and the conventional at standby operating mode.

TABLE I. DEVICES USED IN FLYBACK CONVERTER AND AC INPUT TERMINAL OF PDP TV PSU

Components		Part Number
Flyback Converter	Main Switch S_1	STR-W6252, 650V/2.7A
	Secondary Diode D_4	FCH30A10 100V/30A
Snubber	Snubber Diode Ds	1N4007 1000V/1A
	Snubber Capacitor Cs	2200pF
	Snubber Resistor Rs	75kΩ
PFC	PFC Diode D_2	GPP20M 1000V/2A
	AC input Diode D_1, D_3	
AC Input	Varistor	14D621K
	Thermistor	DSC-8D 8Ω 5.5A
	Rectifier Diode	GPP20M x 2 1000V/2A
	Filter Inductance	50uH
	Filter Capacitor	1000pF x 2

TABLE II. TRANSFORMER PARAMETERS DUE TO THE WINDING METHODS

Windings	N_1 : 50T / N_2 : 5T	N_p : 100T / N_s : 10T
Primary Inductance L_1	424.3uH	1.728mH
Secondary Inductance L_2	4.42uH	17.63uH
Primary Leakage Inductance	0.2uH	12.0uH
Equivalent Leakage Inductance L_{eq}	17.4uH	57.6uH

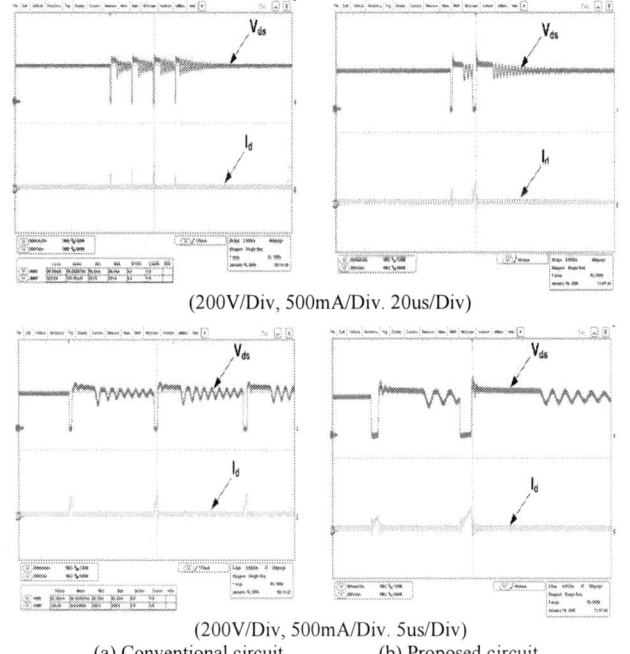

(200V/Div, 500mA/Div. 20us/Div)

(200V/Div, 500mA/Div. 5us/Div)
(a) Conventional circuit   (b) Proposed circuit

Figure 7. Experimental waveforms of voltage (V_{ds}) and current (I_d) of the main switch (S_1) at the standby mode operation of the conventional circuit and the proposed circuit

The waveforms in Fig 7 show the voltage (V_{ds}) and current waveforms (I_d) across the switch S_1 of the burst operation at the very light load condition which is in standby operation of

the conventional burst mode and the proposed one with 230Vac, 5Vdc, 30mA max.

Fig 7(b) shows the current slope goes up more slowly with the increased magnetizing inductance in the proposed topology. The higher inductance makes turn on period longer per equation (3). These make the switching frequency in burst operating fewer and MOSFET driving & switching losses lower.

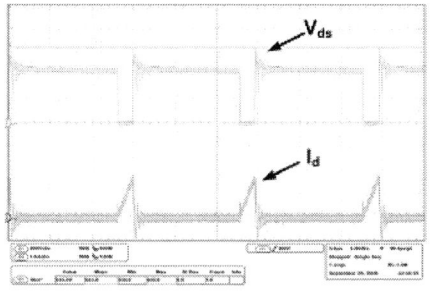

(200V/Div, 1A/Div. 5us/Div)

Figure 8. Experimental waveforms of main switch S_1 at normal mode operation of the proposed circuit (Input: 400VDC, Output: $5V_{DC}$/7A)

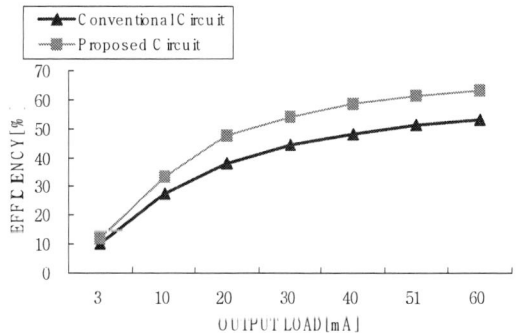

Figure 9. Efficiency characteristics under load conditions(5V, 3mA ~ 60mA with 230Vac input voltage)

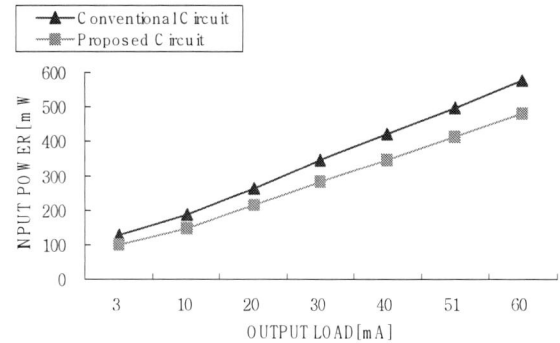

Figure 10. Measured the standby input powers(5V, 3mA ~ 60mA with 230Vac input voltage)

Fig. 9 and Fig. 10 show the comparison graphs of power consumption at light load variation (5V, 10mA ~ 60mA) at

978-1-4244-4782-4/10 $26.00 © 2010 IEEE

230Vac. In proposed topology the power consumption was 148mW comparing with 187mW at 5V 10mA in conventional burst mode operation and got an improved efficiency by 6.3% ~ 10% in this mode.

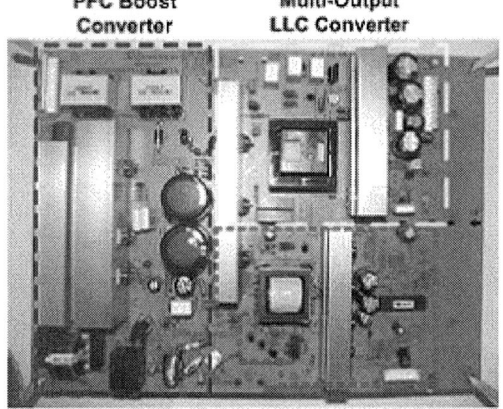

Figure 11. PSU main board of the 50 inch PDP TV

Figure 12. Standby power measurement photograph

IV. CONCLUSION

A novel flyback topology to reduce standby power consumption was proposed, analyzed, tested and compared with the conventional flyback converter in burst mode operation for an auxiliary power supply of 70W output power in PSU of PDP TV. For verification a green mode PWM IC of STR-W6252 (Sanken) was adopted. The power supply is working very stable in normal operating mode and got 84% average efficiency. In standby operating mode reduced power consumption were measured and verified. The power consumption is improved by 6.3% to 10% comparing with the conventional one.

ACKNOWLEDGMENT

This work is the outcome of a Manpower Development Program(2007-E-AP-HM-P-18-0000) for Energy & Resources supported by the Ministry of Knowledge and Economy (MKE)

REFERENCES

[1] Nam-Gyun Kim, " The Present and Future on the Standby Power of Home Appliance in Korea", Journal of The Korean Institute of Power Electronics, Vol. 11, No. 4, pp.22~25, 2006. 8

[2] Laurence McGarry, "The Standby Power Challenge", 2004 International IEEE Conference on Asian Green Electronics(AGEC), pp.56~62, 2004.

[3] Jin-ho Choi, Jung-won Kim, Dong-young Huh, "The new technique for the lowest power consumption in the stand-by of power supply", IEEE PESC2004, pp.741~746, 2004. 6

[4] Jin-ho Choi, Dong-young Hug, Young-seok Kim, "The improved burst mode in the stand-by operation of power supply", IEEE APEC'04. Nineteenth Annual IEEE Volume 1, pp.426 ~ 432 2004

[5] Bo-Teng Huang, Ko-Yen Lee, Yen-Shin Lai, "Design of a Two-Stage AC/DC Converter with Standby Power Losses Less than 1 W", Power Conversion Conference-Nagoya, 2007. PCC'07, 2007

[6] Sanken Electric Co., Ltd, "STR-W6200 Series Application Note", 2005. 11.

[7] EDN "Tightened power-efficiency regulations force power supplies to keep up", 2008. 2.21

Improved Two-Stage DC-Coupled Gate Driver for Enhancement-Mode SiC JFET

Robin Kelley, Andrew Ritenour, David Sheridan, and Jeff Casady

SemiSouth Laboratories, Inc, 201 Research Blvd., Starkville, MS 39759

Abstract — Normally-OFF SiC VJFETs have been proved to be advantageous as a "drop-in" replacement of MOSFETs and IGBTs in a variety of applications. As this device's acceptance continues to grow, developers are investigating optimized driver methods that will yield the best possible switching performance leading to higher system efficiencies. This paper presents new results for an alternative and more optimized gate driver to the capacitive coupled driver used in past literature. Additionally switching energy measurements are documented for the 50mOhm enhancement-mode SiC VJFET in the newly optimized two-stage, DC-coupled gate driver and compared against past results obtained using the initial driver design. Specific design guidelines are included for achieving the best possible results using the two stage gate driver design presented here.

I. INTRODUCTION

The first commercially available pure enhancement-mode silicon carbide power VJFET (EM SiC VJFET), introduced one year ago, has been gaining significant interest among power electronic designers [1]. This new three-terminal, enhancement-mode device provides designers with a normally-off solution that maintains all the functionality and benefits of its normally-on counterpart. This new EM SiC VJFET has a record low specific on-resistance of $2.8m\Omega\text{-cm}^2$ resulting in a die size of **9** mm^2 for a 50 mOhm device rated at 1.2 kV. This design enables very low parasitic charges of approximately 70 nC gate charge. It is also able to be offered very cost-effectively in higher volumes because of the physical device architecture. The EM SiC VJFET is in single die form, without any cascode and does not contain body diode or gate oxide. A cross-sectional drawing of the device is shown in Figure 1a, and a top view of a commercial TO-

247 packaged 1.2 kV, 30 A, 50 mOhm part is shown in Figure 1b. The device has a back-side drain contact, top side source, and recessed gate contacts. Unlike other JFET structures, the channel is completely vertical between the recessed trench gate regions.

Prior work has demonstrated the use of the EM SiC VJFET as a drop-in replacement for existing Si-MOSFETs and Si-IGBTs in high-side and low-side switch-mode applications [2, 3]. With only a direct replacement of the semiconductor power switches in both a commercial solar inverter and commercial PFC demo-board and addition of the RC gate driver with no other circuit optimization performed, incremental efficiency improvements were realized. These socket replacement demonstrated a first step toward device acceptance; however, further system efficiency improvements could be realized from power circuits and gate drivers specifically optimized for the SiC EM VJFET. The previously implemented RC driver (Figure 2) provides an inexpensive and easily implemented driver solution for new users but experience some application limitations. For best results with the capacitive coupled (RC) driver the bypass capacitor should be allowed to discharge fully prior to the

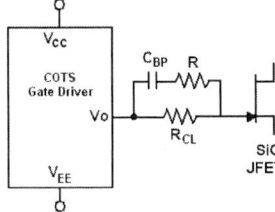

Figure 2. AC-coupled gate driver for SiC JFET.

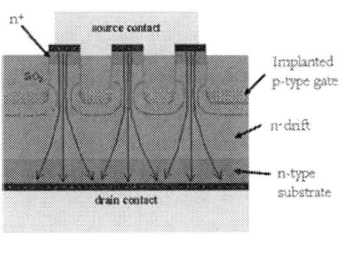

(a)   (b)

Figure 1. SiC VJFET (a) cross-section and (b) TO-247 package.

next switching event. As this discharge rate is reliant upon an RC time constant the maximum switching frequency is thus limited. Therefore it was necessary to develop a more optimal gate driver compatible with all applications that eliminates the limitations of the RC driver. Past publication introduced the idea of the two-stage DC-coupled driver that was not limited by duty cycle or ranges of switching frequencies [4]. This paper provides new results yielded by further optimization of the two-stage gate driver.

II. GATE DRIVER

Transition speed, turn-on and turn-off times, of the SiC JFET are ultimately limited by the device; however, the performance of the gate driver can impact this speed considerably. Specific to the SiC VJFET are two main requirements must be satisfied by the gate driver; delivery/removal of dynamic gate charge and sustainability of DC gate voltage and resulting gate-source current during the conduction period. The ability of the gate driver to quickly deliver/remove the necessary gate charge required by the internal gate-source and Miller capacitance of the device is the main factor that affects the time it takes for the device to transition between states. The gate drive should also be designed to efficiently maintain the steady state DC gate voltage and gate current required to maintain minimum $R_{DS(ON)}$ during conduction. A new two-stage, isolated, DC-coupled gate driver which is suitable for all applications and operating frequencies has been developed and further optimized to provide both of these requirements and results in record switching speeds. This two stage driver includes a first driver stage that delivers a high peak current pulse to the gate for quickly delivering the dynamic charge requirement required by the device input capacitances. The duration of operation for this stage last until the drain voltage of the device collapses. The second driver stage then reduces the drive voltage/current to satisfy the steady state gate requirements presented in the device datasheet [5]. While no negative bias is required by the EM SiC JFET to block voltage up to the full rating, a negative voltage is recommended when using the device in bridge configurations to aid in noise immunity against the "Miller effect".

The newly optimized isolated gate driver design, Figure 3, requires three user inputs; an isolated +/- 15V voltage supply and a user input PWM control pulse. An optocoupler is included at the input such that the necessary isolation is achieved if using this circuit in the high side position. The output of this optocoupler is then passed to a logic/timing circuit that generates the timing signals for the first and second driver stages. A short duration ON pulse (typically ~100ns) is input to first driver stage. An IXYS IXDD509 driver IC is used to deliver the high peak current (+/- 9A) necessary for quickly charging the device input capacitance thus quickly turning on the device. The output of the

IXDD509 is connected to the gate by a series low-ohm gate resistor (typically ~1-5 ohm). A second pulse that matches the full duration of the user input control pulse is inverted and applied to the base of a small PNP transistor used to provide the required on-state gate current of ≤ 200mA for maintaining a low $R_{DS(ON)}$ during the conduction period. The gate current is set by the value of a second gate resistor based on Equation 1. An optional +15V to +6V DC/DC converter is included to reduce the power loss in the second gate resistor during the conduction period. If the optional step down converter is not used then the second gate resistor will be required to drop more voltage at the same value of limited gate current thus increasing the power dissipated in the resistor. This will require a physically larger resistor with a higher power rating and could make the PCB layout more difficult. Figure 4 shows the measured gate current for the SJEP120R050 as driven by the optimized two-stage gate driver. As shown, a 5.5A peak current pulse is delivered to the gate of the device by the first driver stage for approximately 100ns. After which the second driver stage takes control and maintains the necessary DC gate current of ~200mA for the duration of the conduction period.

$$R_G = \frac{V_O - V_G}{I_{G(@VG)}} \tag{1}$$

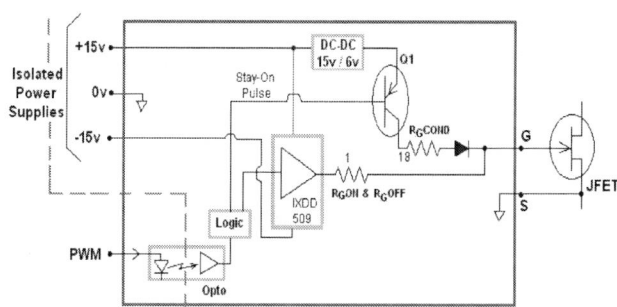

Figure 3. Isolated, Two-Stage Gate Driver for SJEP120R050

III. RESULTS

The switching losses for the SJEP120R050 SiC VJFET have been measured using the optimized two-stage DC-coupled gate driver solution presented here, Figure 3. A standard double pulse, clamped inductive load circuit with was used as the test circuit for evaluating the resulting device switching energy losses, Figure 4. The resulting gate current supplied by the two-stage gate driver was also measured to illustrate the functionality of the two driver stages. The test conditions were: $V_{DD} = 600$V, $I_{D\_PK} = 25$A. As mentioned in the gate driver description, a gate resistor for each driving stage was necessary. For the first driver stage small 1-5 ohm

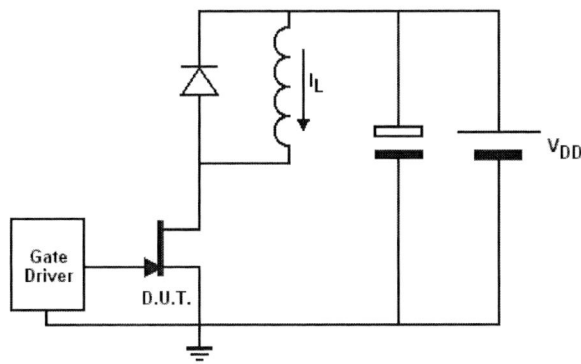

Figure 4. Standard double pulse, clamped inductive load circuit with switches arranged in a bridge configuration.

resistor will be required to damp oscillations during the high peak current charging stage. The value required will depend upon the amount of parasitic inductance in the actual prototyped circuit. The user will need to adjust the value as seen fit. For the prototype described here a 1 ohm resistor was satisfactory. The second driver stage requires a current limiting gate resistor sized according to equation 1 from the previous section. Since the option +15V to +6V DC/DC converter was included the resistor value was calculated as follows:

$$R_{G2} = \frac{6V - 3V}{200mA} = 15\Omega$$

Figure 5 shows the measured gate current supplied to the SJEP120R050 by the two-stage gate driver. While the first driver stage was active a peak current of 5.5A was supplied for ~100ns in order to quickly charge the device input capacitance. The second driver stage then takes control to maintain the steady state gate current of ~200mA during the conduction state of the JFET. Typical switching waveforms are shown in Figure 6 with measured results listed in Table 1. As shown the SJEP120R050 demonstrated a clean turn-on transistor with a fall time of only 25ns and turn-on energy loss of 137uJ. An RC snubber of 4.7nF and 22 ohms was connected across the DC bus to reduce ringing in the power

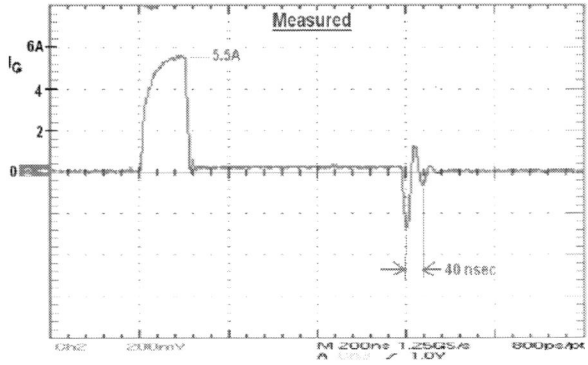

Figure 5. Measured Gate Current

circuit such that unnecessary oscillations were not feedback to the gate driver circuit.

Table1. Switching Losses for EM SiC JFET in Half Bridge Configuration using Two-Stage Gate Driver

Device	E_{ON} (uJ)	E_{OFF} (uJ)	$E_{ON} + E_{OFF} = E_{TS}$ (uJ)
SJEP120R050	137	120	257

IV. APPLICATION RECOMMENDATIONS

Since the EM SiC VJFET is a very fast switching device with very low intrinsic capacitances careful attention should be used when generated PCB layouts using these devices. Like MOSFET, high frequency oscillations between the device's internal capacitances and other circuit parasitics may be observed. A list of applications recommendations is provided

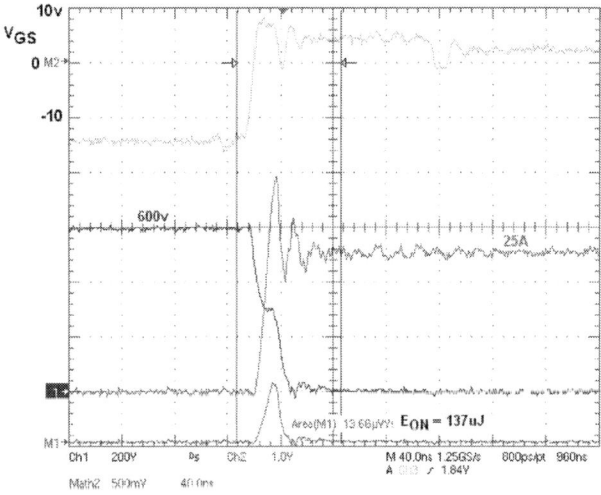

(a)

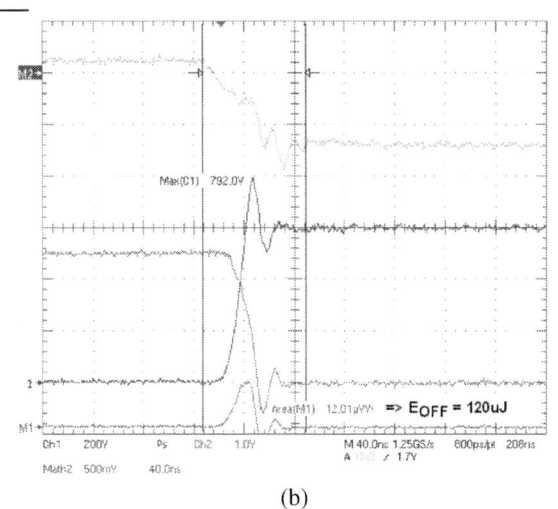

(b)

Figure 6. Typical Switching Waveforms for SJEP120R050 in the test circuit shown in Figure 5.

to assist users in fabricating the best possible circuit layout and test set-up for achieving maximum performance.

- The externally applied PWM input must be sufficiently isolated from the output of the optocoupler and the isolated power supplies of the gate driver.

- The gate current required by the JFET is set by the gate resistor at the output of the second driver stage (connected at emitter of the PNP transistor). The value will need to be adjusted based on the JFET in test as well as any differences in positive rail voltage of the second driver stage.

- If EMI concerns require the dV/dt of the device to be decreased, adjustment of the gate resistor at the output of the first driver stage will control the resulting dV/dt of the switched device.

- Excessive ringing on the gate voltage waveform is either caused by parasitic inductance in the main power circuit and/or the gate circuit. The loop area of the power circuit and the physical distance between the gate driver output and the gate terminal should be minimized as much as possible. A ferrite bead can be used but will result in slower turn-on losses.

- An RC snubber connected across the DC bus as close as possible to the device in test (or phase leg if a bridge configuration is used) is recommended for reducing ringing caused by inductance in the power circuit.

- A small capacitive clamp connected tightly across the gate-source terminals of each switch in a phase leg can help offset undesirable false trigging of the opposed switch due to the "Miller effect" when a bridge configuration is used.

- The duration of the high peak current pulse generated by the first driver stage should be adjusted such that it is turned off once the device is fully enhanced. For any duration after which the drain voltage has fully collapsed the high output current will be passed thru the gate-source diode resulting in unnecessary power loss in the gate of the device. Once the device is fully enhanced the second driver stage should immediately take control reducing the available gate current to the device.

- Decoupling caps should be place as close as possible to the gate driver circuit as high peak currents are involved. Also, a 4 layer PCB is recommended where one layer is dedicated to the 0V ground plane.

V. CONCLUSION

The newly optimized two-stage, DC coupled gate driver demonstrated record low switching losses for the SJEP120R050 EM SiC VJFET. Extremely fast turn-on speeds were observed with no resulting tail current at turn-off

like that of an IGBT. Also the absence of a body diode, like that of a MOSFET, means there are no problems operating the EM SiC JFETs in a bridge configuration. The resulting optimized two-stage gate driver circuit presented here has overcome the limitations of the AC coupled RC driver and further optimization led to a list of application recommendations that aid in achieving the best possible switching performance. The resulting circuit has also been simplified from past designs using common, low-cost COTS parts to provide users with a cost effective and easily implemented circuit capable of providing the lowest possible switching losses.

VI. REFERENCES

[1] D. C. Sheridan, A. Ritenour, V. Bondarenko, P. Burks, and J. B. Casady, "Record 2.8mΩ-cm^2 1.9kV enhancement-mode SiC VJFETs," *21st International Symposium on Power Semiconductor Devices & IC's*, (ISPSD '09), pp. 335-338, 2009.

[2] R. Kelley, M. Mazzola, S. Morrison, W. Draper, I. Sankin, D. Sheridan, and J. Casady, "Power Factor Correction using an Enhancement-mode SiC JFET," *Proc. 39th IEEE Power Elect. Specialist Conf.* (PESC '08), pp. 4766 – 4769, 2008.

[3] M. Mazzola, R. Kelley, W. Draper, and J. Casady, "Application of a Normally OFF Silicon Carbide Power JFET in a Photovoltaic Inverter," *Proc. 24th IEEE Applied Power Electronics Conf.* (APEC '09), pp. 649-652, 2009.

[4] R. Kelley, F. Rees, and D. Schwob, "Optimized Gate Driver for Enhancement-Mode SiC JFET," Conf. Proc. of Power Conversion Intelligent Motion (PCIM) 2009, 12-14 May 2009, Nuremberg, Germany.

[5] Datasheet, SJEP120R050, SemiSouth Labs, Available at <http://www.semisouth.com>

Design and Implementation of Multi-channel Land Fowls Stunner with Current Sharing Controller

*S.-Y. Fan, S.-Y. Tseng, Y.-H. Su, *W.-C. Wu

Department of Electrical Engineering
Wufeng Institute of Technology
Ming-Hsiung, Chia-Yi, Taiwan
E-mail: syfan@mail.wfc.edu.tw
TEL: 886-5-2267125

GreenPower Evolution Applied Research Lab (G-PEARL)
Department of Electrical Engineering
Chang Gung University
Kwei-Shan, Tao-Yuan, Taiwan, R.O.C
E-mail: sytseng@mail.cgu.edu.tw
Tel: 886-3-2118800; Fax: 886-3-2118026

Abstract—This paper presents an intelligent humane stunner for a multi-channel poultry stunning system with current sharing features. This proposed system consists of an interleaved buck-boost converter for driving the stunning system, a full bridge inverter for generating the stunning voltage, a half bridge inverter for generating dither voltages to expedite the breakdown of skin impedance, and a current sharing controller for providing uniform current to each channel of the stunning system. Hence, the slaughtering quantities of poultry can be increased, and also can reduce poultry stress and improve carcass quality during stunning interval. In this paper, a multi-channel poultry stunning system with ±160V stunning voltage and 320W output power is implemented for verifying its performance. Experimental results shown that the proposed stunning system can raise 10% more coma effectiveness over the one without current sharing algorithm, thus improve carcass quality. Hence, with this approach, fewer components are needed to achieve the better performance, resulting in a multi-channel poultry stunning system with lighter weight and smaller size.

Keyword: poultry stunner, single-capacitor turn-off snubber, half-bridge inverter, full-bridge inverter.

I. INTRODUCTION

From an animal welfare point of view, all slaughtered animals should be rendered unconscious and insensible, and remain in this state until a complete loss of brain responsiveness due to exsanguinations [1]. There are two methods general used for humane slaughter: Carbon dioxide (CO_2) and manual electrical stunning. [2] Since CO_2 method is subjected to many limitations and requires higher cost, the electrical method appear to be the most economic and effective method for stunning the land fowls. [3]

To render the poultry insensible to pain, stunning with electrical method induces unconsciousness by generating an epileptiform seizure characterized with by a tonic phase and clonic phase. Degrees of an epileptiform seizure produced are depended upon the amount and the area of the brain that current passing through.[2] In general, range of stunning current required for stunning a goose is between 300mA and 350mA,

and sustains at least 6s [4]. Conventionally, to generate the specified electrical quantities, rectified line voltage or battery is first chopped into square waveforms with power switches and then boost up with a low-frequency step-up transformer. In addition, as shown in Fig. 1(a), the conventional stunner uses only two electrodes which is easy to cause bone fractures and ecchymosis, resulting in low meat quality. For solving the mentioned problems, dc/dc converters with PWM control and a dither voltage generator for driving multi-electrode stunning systems are provided. [5,6] This converter can properly limit current level, regulate output voltage, and reduce poultry stress during stunning interval. In [4], a six-electrode stunning system is proposed and redrawn in Fig. 1(b). Due to the complicated controls and circuit structure, a poultry stunning system with four-electrode configuration shown in Fig. 1(c) has been proposed in this paper.

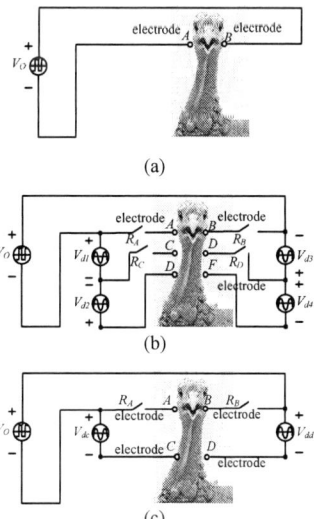

Figure 1. Illustration of stunning system (a) with conventional two-electrode configuration, (b) with six-electrode configuration, and (c) with four-electrode configuration.

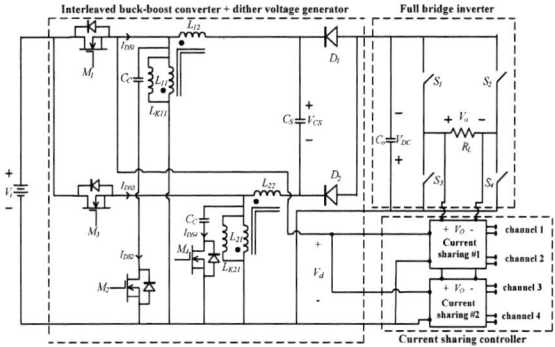

Figure 2. Schematic diagram of the proposed intelligent multi-channel humane stunning system.

To speed up slaughtering quantities of poultry, a multi-channel poultry stunning system is required. Since the amount of current passed through the brain of poultry determines the coma effectiveness, a current sharing controller for the multi-channel poultry stunning system to provide uniform stunning current to each channel can be an important role for promoting carcass quality and improving animal welfare. To achieve this goal, an intelligent multi-channel humane stunning system with current sharing controller shown in Fig. 2 is presented in this paper. This system consists of an interleaved buck-boost converter for driving the stunning system, a full bridge inverter for generating the stunning voltage, a half bridge inverter for generating dither voltages, and a current sharing controller for providing uniform current to each channel of the stunning system. Compared with the stunning system with six-electrode configuration, the proposed four-electrode stunning system has the advantages of small size, light weight, and low cost, as well as high conversion efficiency.

II. FUNDAMENTAL OF POULTRY STUNNING

Parameters of electrical stunning have highly effect on the coma effectiveness. To decide the desired electrical parameters, impedance properties of poultry and mechanism of poultry stunning are briefly described in the following.

A. Impedance properties of poultry

In poultry stunning, degrees of an epileptiform seizure are highly dependent on the amount of current passing through the brain of poultry. Voltage applied to the poultry must be enough to breakdown the poultry skin for providing the stunning current to the poultry. Comparing the tissue of poultry and human being, their electrical properties and equivalent circuit are similar except that their values are different. Thus, the properties of human skin are used for qualitative analysis the mechanism of poultry stunning. Electrical properties of human skin have been studied over a century and often characterized with impedance spectra shown in Fig. 3[7]. An equivalent circuit composed of resistor R_S series with the parallel combination of resistor R_{SC} and capacitor C_{SC} have been used by many investigators to represent the electrical properties of skin, as shown in Fig. 4. Both skin capacitance and resistance have been shown to be proportional to contact area [8]. The skin impedance versus voltage is shown in Fig. 5,

illustrating that the skin impedance is inversely proportional to the applied voltage. [9].

B. Mechanism of poultry stunning

In general, there are three main parts in neurons: soma, axon, and dendrite. A stimulation signal is sensed by sensory receptors which exist in dermis or subcutaneous layer. When sensory receptors receive a stimulation signal, it will transfer the sensory stimulation into nerve impulse. Its propagation direction is, in turn, through one neuron to the other neurons, in which they are connected in series by synapses, until the nerve impulse propagation is transmitted to a receiver of brain. That is, synapses can play a transducer role which is to transfer electrical stimulation signals to chemical signals, and vice versa, as illustrated in Fig. 6.

To explain the relationship between electrical stunning and a suppression of nerve impulse propagation, an equivalent circuit for describing the impulse signal propagation between neurons is shown in Fig. 7. In Fig. 7, since synapses play a role of transducer, it can be considered as a switch Q_1. When a stunning voltage E_i is applied to poultry, its skin impedance Z_0 enters a breakdown state, and then a current I_1 will pass the body of the poultry. In this stunning duration, if current I_1 is large enough, it will induce a high potential V_1 in the postsynaptic membrane to turn on switch Q_1. When switch Q_1 is turned on, nerve impulse signals will be bypassed. Thus, nervous system loses capability to propagate nerve impulse. On the other hand, nerve impulse can reach the sensory receivers B_1 of the brain through propagation impedance R_t of the neurons. As a result, poultry can sense a stimulation signal. As described previously, it can be observed that electrical stunning can inhibit nerve impulse propagation.

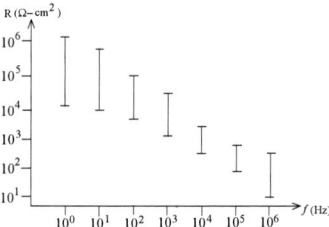

Figure 3. Illustration of the ranges of skin impedance versus frequency.

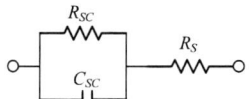

Figure 4. An equivalent impedane of skin.

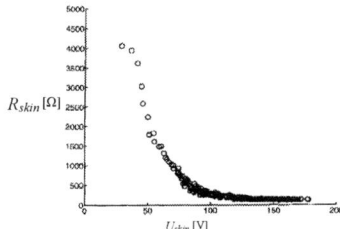

Figure 5. Plot of the skin impedance of poultry versus the applied voltage.

978-1-4244-4782-4/10 $26.00 © 2010 IEEE

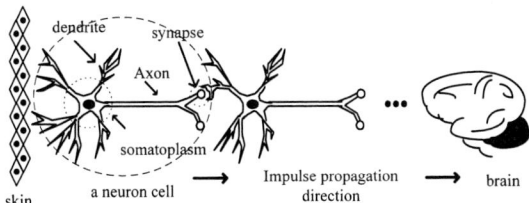

Figure 6. Illustration of impedance propagatrion between neurons.

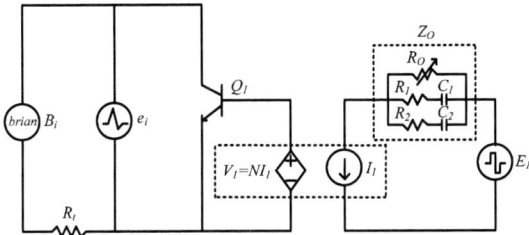

Figure 7. An equivalent circuit for describing the impulse signal propagation between neurons.

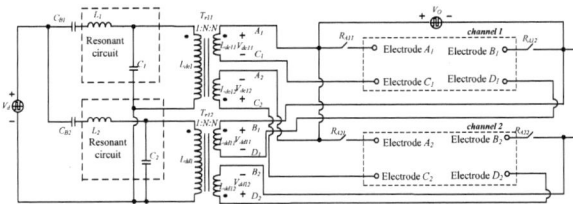

Figure 8. Schematic diagram of the proposed current sharing controller.

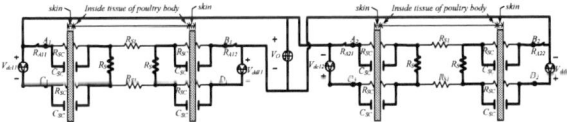

Figure 9. An equivalent circuit of the four-electrode configuration.

III. OPERATIONAL PRINCIPLE OF THE PROPOSED STUNNER SYSTEM

The proposed multi-channel stunner system shown in Fig. 2 consists of an interleaved buck-boost converter, a dither voltage generator, a full bridge inverter and a current sharing controller. Here the interleaved buck-boost converter includes single-capacitor turn-off snubbers and coupled inductors and is used for increasing conversion efficiency [5]. The dither voltage generator realized by a half-bridge inverter is adopted for generating dither voltage V_d to breakdown the skin impedance of poultry. To integrate the switches of the interleaved buck-boost converter and the interleaved buck-boost converter and the half bridge inverter, the synchronous switch technique presented in [10] is used. The full bridge inverter converters the DC output voltage of the converter V_{DC} into stunning voltage V_o.

Due to the different impedances between poultry in each channel, a current sharing controller is constructed for solving the current imbalance problems in the multi-channel poultry stunning system. The detail schematic diagram of current sharing controller with two channels is shown in Fig. 8. The current sharing controlled is constructed by resonant circuits and transformers for generating high frequency resonant dither voltages V_{dc11}, V_{dc12}, V_{dd11}, and V_{dd12} from the high frequency dither voltage V_d due to the coupling among these high frequency resonant voltages. To explain the operation principle of the proposed current sharing controller with four-electrode configuration, an equivalent circuit shown in Fig. 9 is illustrated. Based on the equivalent circuit modes shown in Fig.10 and the key waveforms shown in Fig.11, the operational principles of the proposed stunner system over a stunning interval are described briefly.

Mode 1 [Fig. 10(a); $t_0<t<t_1$]

Before t_0, the stunning voltage V_o is equal to 0, and the dither voltages V_{dc11}, V_{dc12}, V_{dd11}, and V_{dd12} are also equal to 0 due to the turn off of the switch R_{A11}, R_{A12}, R_{A21}, and R_{A22}. At $t= t_0$, stunning voltage V_o is applied across the electrodes A_1B_1 and A_2B_2. Since stunning voltage V_o is less than breakdown voltage of poultry skin, it does not induce an enough stunning current to achieve coma effectiveness. At this moment, switches R_{A11}, R_{A12}, R_{A21}, and R_{A22} are conducted, the dither voltages V_{dc11} and V_{dd11} are applied respectively to electrodes A_1C_1, and B_1D_1, and the dither voltages V_{dc12} and V_{dd12} are applied respectively to electrodes A_2C_2, and B_2D_2. During this interval, the amplitude of the dither voltages V_{dc11}, V_{dc12}, V_{dd11}, and V_{dd12} raise from 0 to a maximum value owing to the resonant effect. At this time, the dither current I_{dc11}, I_{dc12}, I_{dd11}, and I_{dd12} also increased from 0 to a certain value and flow through the current loop formed with the high impedance of skin C_{SC} and the low impedance of poultry tissue R_S.

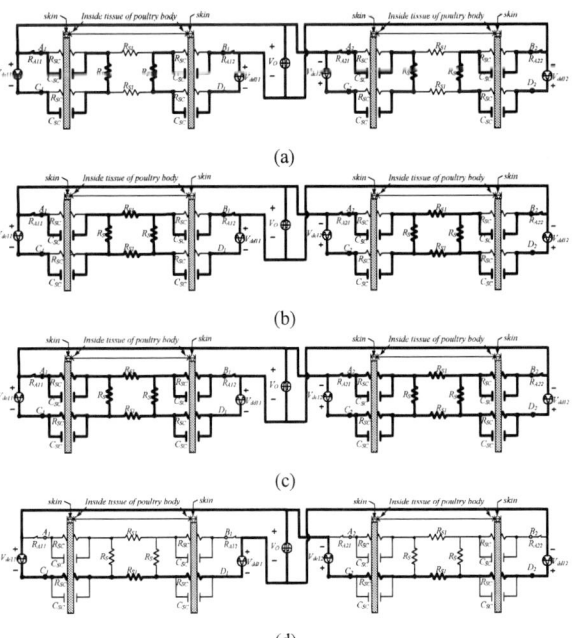

Figure 10. Equivalent circuit modes of the proposed stunner system over a stunning interval.

Mode 2 [Fig. 10(b); $t_1 < t < t_2$]

At $t = t_1$, switches R_{A11}, R_{A12}, R_{A21}, and R_{A22} are kept to be conducted. At this interval, impedance of poultry skin is breakdown gradually, causing the dither voltages V_{dc11}, V_{dc12}, V_{dd11}, and V_{dd12} to drop to its minimum and the dither current I_{dc11}, I_{dc12}, I_{dd11}, and I_{dd12} to rise to a stable value. The breakdown *of* poultry skin impedance lowered the value of impedances C_{SC} and R_{SC}, causing stunning current I_o to flow through the tissue inside poultry body and to be increased gradually to an enough value for stunning poultry.

Mode 3 [Fig. 10(c); $t_2 < t < t_3$]

At $t = t_2$, the skin impedance is breakdown completely. At this moment, the skin impedances C_{SC} and R_{SC} are close to 0, the voltages across the tissue inside poultry body are almost equal to the dither voltages V_{dc11}, V_{dc12}, V_{dd11}, and V_{dd12}. Under this condition, the dither voltages and dither *current* can be kept on stable values, and the stunning current can be stable on an enough value for stunning poultry.

Mode 4 [Fig. 10(d); $t_3 < t < t_4$]

At $t = t_3$, switches R_{A11}, R_{A12}, R_{A21}, and R_{A22} are turned off, resulting the voltages across electrodes A_1C_1 and B_1D_1, and electrodes A_2C_2, and B_2D_2 respectively to be 0 and the current currents flowed thereof being 0. At this moment, the dither voltages and stunning voltage can be considered to be connected in serial. On the other hand, due to the turn off of the switches R_{A11}, R_{A12}, R_{A21}, and R_{A22}, the dither voltages V_{dc11}, V_{dc12}, V_{dd11}, and V_{dd12} are increased slightly. Under this condition, if the impedances of the poultry in the two channels are equal, the voltages across electrodes C_1D_1 and C_2D_2 respectively are equal to stunning voltage V_o, thus the current flowed thereof are also the same. If the impedances of the poultry in the two channels are different, the dither voltages V_{dc11}, V_{dc12}, V_{dd11}, and V_{dd1} are varied accordingly. This variation caused the voltages across electrodes C_1D_1 and C_2D_2 respectively to be equal to $V_o + V_{dc11} - V_{dd11}$ and $V_o + V_{dc12} - V_{dd12}$ respectively, thus achieving the current sharing feature. At the end of time t_4, the stunning voltage V_o and the dither voltages V_{dc11}, V_{dc12}, V_{dd11}, and V_{dd1} are turn off for ending a poultry stunning interval.

IV. DESIGN OF THE PROPOSED CURRENT SHARING CONTROLLER

The proposed two channel current sharing circuit is shown in Fig. 8, illustrating that the proposed one is constructed by series resonant parallel load networks and transformers T_{r11} and T_{r12}. The quality factor Q of the resonant network is designed to be high enough to obtain high input to output transfer gain before skin impedance breakdown, which has been described in [6]. In the following, design of the proposed current sharing controller will be derived.

To analysis the proposed two channels current sharing controller, an equivalent circuit shown in Fig. 12 is provided. From Fig. 12, it can be seen that the stunning voltages V_{RS11} and V_{RS12} in channel 1 and channel 2 can be respectively expressed by

$$V_{RS11} = V_0 - V_{dc11} + V_{dd11}, \tag{1}$$

and

$$V_{RS12} = V_0 - V_{dc12} + V_{dd12} \tag{2}$$

where R_{S11} and R_{S12} are respectively skin breakdown impedances of poultry in channel 1 and channel 2, V_{dc11}, V_{dc12}, V_{dd11}, and V_{dd12} are the dither voltages.

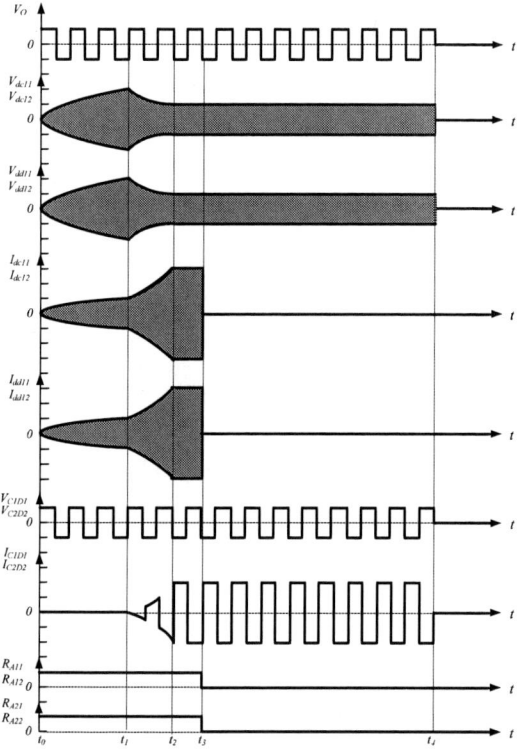

Figure 11. Key waveforms of the proposed stunner system over a stunning interval.

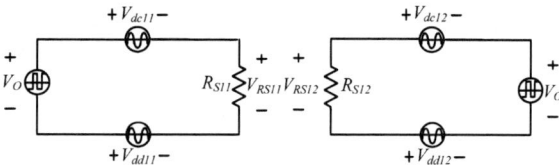

Figure 12. Equivalent circuit of the proposed two channels current sharing controller.

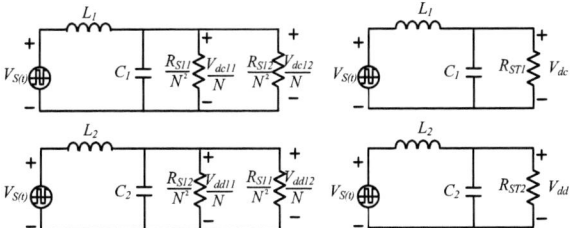

Figure 13. Equivalent circuit of the dither voltages for the current sharing controller with two channels.

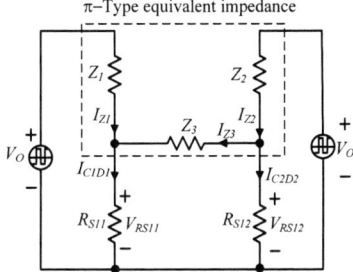

Figure 14. A π-Type equivalent impedance circuit of the two channel stunning system.

To derive the amplitudes of dither voltages V_{dc11}, V_{dc12}, V_{dd11}, and V_{dd12}, an equivalent circuit of the dither voltages for the current sharing controller with two channels is drawn in Fig. 13 based on Fig. 2. Here, R_{ST1} is the parallel connection of the load impedance R_{S11}, and R_{S12} in the secondary of channel 1 and channel 2 transferred to the primary and can be expressed as

$$R_{ST1} = R_{ST2} = \left(\frac{R_{S11}}{N^2} // \frac{R_{S12}}{N^2} \right) = \frac{R_{S11}R_{S12}}{N^2(R_{S11} + R_{S12})} \quad (3)$$

where N is the turn ratio of transformers T_{r11}.

If the resonant inductors (L_1 and L_2) and resonant capacitors (C_1 and C_2), as well as the load impedance (R_{ST1} and R_{ST2}) are the same for the series resonant parallel load networks shown in Fig. 8, the output to input voltage transfer function can be derived based on Fig. 13 and expressed as

$$\left| \frac{V_{dc}(t)}{V_s(t)} \right| = \left| \frac{V_{dd}(t)}{V_s(t)} \right| = \frac{1}{\sqrt{\left[1 - \left(\frac{w_s}{w_o} \right)^2 \right]^2 + \left(\frac{w_s}{w_o Q_p} \right)^2}} \quad (4)$$

where $V_{dc}(t) = V_{dc11}/N$, $V_{dd}(t) = V_{dd11}/N$, $Q_P = R_{ST1}/\sqrt{L_1 C_1}$ or $R_{ST2}/\sqrt{L_2 C_2}$, $\omega_O = 1/\sqrt{L_1 C_1}$ or $1/\sqrt{L_1 C_1}$, $\omega_S (= 2\pi f_S)$ is the angular velocity of the switching frequency. Voltage $V_s(t)$ represents the amplitude of basic sine wave transferred from a

square wave with amplitude of $\pm V_d/2$. From (4), it can been seen that the dither voltages $V_{dc}(t)$ and $V_{dd}(t)$ are only effected by resonant inductors and resonant capacitors. Hence, equal dither voltages can be achieved by making the resonant inductors and resonant capacitors are equal to each other in every channel.

From Fig. 12, it also can be revealed that dither voltage in the current sharing circuit will not affect the stunning voltage when the skin impedances between channels 1 and channel 2 are balance. If skin impedances between channels 1 and channel 2 are imbalance, dither voltages can be used to achieve the current sharing feature. For further deriving the current sharing controller, transformers T_{r11} and T_{r12} in Fig. 8 can be modeled with a π-type circuit and illustrated in Fig. 14. In Fig. 14, currents I_{Z1} and I_{Z2} can be separately expressed as

$$I_{z1} = I_{C1D1} - I_{Z3}, \quad (5)$$

and

$$I_{z2} = I_{C2D2} + I_{Z3}. \quad (6)$$

On the other hand, currents I_{Z1} and I_{Z2} can also be separately expressed as

$$I_{z1} = \frac{V_o - V_{RS11}}{Z_1}, \quad (7)$$

and

$$I_{z1} = \frac{V_o - V_{RS12}}{Z_2}. \quad (8)$$

Substituting (5) and (6) into (7) and (8) yield

$$V_o = I_{C1D1}(Z_1 + R_{s11}) - Z_1 I_{Z3}, \quad (9)$$

and

$$V_o = I_{C2D2}(Z_2 + R_{s12}) + Z_2 I_{Z3}, \quad (10)$$

where current I_{Z2} can be given as

$$I_{z3} = \frac{I_{C2D2}R_{S12} - I_{C1D1}R_{S11}}{Z_3}. \quad (11)$$

When the proposed two channel current sharing system is operated in current sharing condition, $I_{C1D1} = I_{C2D2}$ and $Z_1 = Z_2$. Hence, the following relation can be obtained by substituting (11) into (9) and (10):

$$Z_3 = -2Z_1. \quad (12)$$

From (12), it can be seen that when impedance Z3 of mutual inductance is equal to -2Z1, the proposed stunning system can achieve current sharing features.

V. MEASURED RESULTS

To verify the performance of the proposed intelligent stunner system, a prototype with the following specifications was implemented.

A. Stunner

- Input voltage V_i: 24 V_{ac},

- stunning frequency f_{S1}: 400 Hz,

- output voltage V_{AB}: 160 V_{ac},

- maximum output current $I_{o(max)}$: 2 A, and

- maximum output power $P_{o(max)}$: 320 W.

B. Dither voltage generator and current sharing controller

- Input voltage V_i: 12 V_{dc},

- switching frequency f_{S2}: 50 kHz,

- dither voltage V_{dc11} and V_{dc12} : 20~200 V_{ac},

- dither voltage V_{dd11} and V_{dd12} : 20~200 V_{ac}, and

- maximum output power $P_{o(max)}$: 20 W.

According to the properties of poultry skin impedance, it is very large before skin impedance breakdown and very small after skin impedance breakdown. In practice, skin impedance is about hundreds of kΩ before skin impedance breakdown and is about 300Ω to 500Ω after skin impedance breakdown. For verifying the current sharing features of the proposed stunner system, the skin impedance R_{SC} before breakdown is chosen to be 50 kΩ, and the skin impedance R_{S11} and R_{S12} after breakdown are chosen to be 300Ω and 500Ω. After applying ±160V/400Hz square-wave stunning voltage V_o to the two channel stunning system, measured waveforms of output voltages V_{RS11} and V_{RS12}, and output currents I_{C1D1} and I_{C2D2} are shown in Fig. 15. Here, Fig. 15(a) shows those waveforms without current sharing controller while Fig. 15(b) shows those waveforms with the proposed one. From Fig. 15, it revealed that the output currents I_{C1D1} and I_{C2D2} for these two channels can be uniform with the proposed current sharing controller.

From practical experimental results for stunning a turkey, the coma situation can sustain about 30s after applying 160VAC/400Hz stunning voltage and 280mA stunning current for 4 sec, which is long enough for bleeding. Fig. 16 shows measured waveforms of output voltage V_{RS11} and output current I_{C1D1} for channel 1, and output voltage V_{RS12} and output current I_{C2D2} for channel 2 during a turkey stunning interval. As shown from Fig. 16, the output currents for both channels are kept uniform after the skin impedances are breakdown.

From Fig. 15 and 16, it can be seen that the dither generator can generate a dither voltage to supply the current sharing controller for achieving uniform current in each stunning

channel. Therefore, the proposed stunner can reduce weight, size and volume, and improve carcass quality significantly.

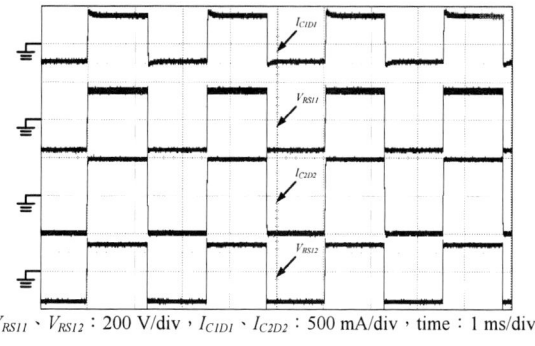

(V_{RS11}、V_{RS12}：200 V/div，I_{C1D1}、I_{C2D2}：500 mA/div，time：1 ms/div)

(a)

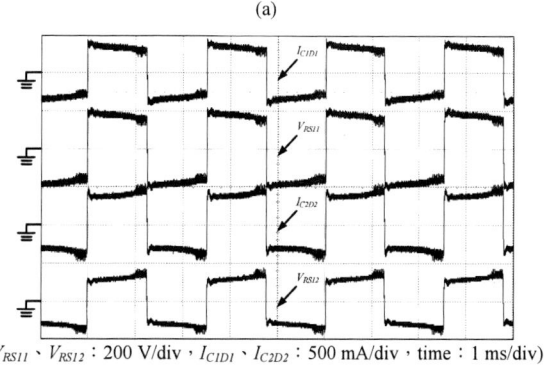

(V_{RS11}、V_{RS12}：200 V/div，I_{C1D1}、I_{C2D2}：500 mA/div，time：1 ms/div)

(b)

Figure 15. Measured waveforms of output voltages VRS11 and VRS12, output currents IC1D1 and IC2D2 for two channels stunning system, (a) without current sharing controller, and (b) with current sharing controller.

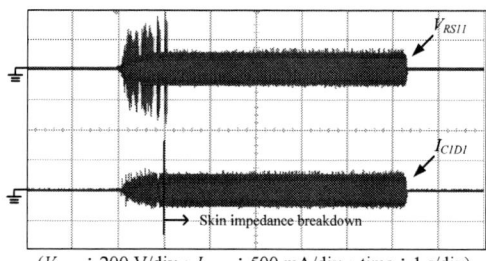

(V_{RS11}：200 V/div，I_{C1D1}：500 mA/div，time：1 s/div)

(a)

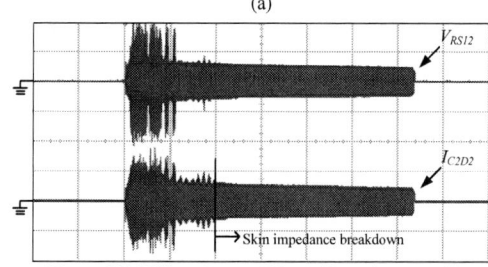

(V_{RS12}：200 V/div，I_{C2D2}：500 mA/div，time：1 s/div)

(b)

Figure 16. Measured waveforms of (a) output voltage VRS11 and output current IC1D1 for channel 1, and (b) output voltage VRS12 and output current IC2D2 for channel 2 during a turkey stunning interval.

VI. CONCLUSION

Coma mechanism of poultry with electrical stunning has been reviewed firstly in this paper. Then, operational principles and design of the proposed two-channel current sharing controller have been described. A prototype of the proposed poultry stunning system with two-channel current sharing controller was implemented and used for verifying current sharing feature. The measured results have shown that the stunning currents for these two channels can be kept uniformly during the stunning interval after the skin impedance breakdown. Finally, two turkeys have used for testing the proposed poultry stunning system, and the test results have been the same with the experimental ones. Hence, the proposed poultry stunning has been verified to have the current sharing features for multi-channel land fowls stunner.

ACKNOWLEDGEMENT

This work was supported by the National Science Council, Taiwan, R.O.C., under Project No. NSC-98-2221-E-182-062.

REFERENCES

[1] A. Velarde, *et al.*, "Effect of Electrical Stunning on Meat and Carcass Quality in Lambs," *Trans. on Meat Science*, 2003, pp. 35—38.

[2] H. A. Channon, A. M. Payne and R. D. Warner, "Comparison of CO_2 Stunninng with Manual Electrical Stunning (50Hz) of Pig on Carcass and Meat Quality," *Trans. on Meat Science*, 2002, pp.63—68.

[3] A. Velarde, *et al.*, "Effect of Electrical Stunning on Meat and Carcass Quality in Lambs," *Trans. on Meat Science*, 2003, pp. 35—38.

[4] S.-Y. Tseng, C.-T. Hsieh, and Y.-H. Su, "Active clamp flyback converter with current sharing scheme for multi-channel poultry stunning applications," *Proceedings of IEEE Applied Power Electronics Conference*, 2008, pp. 1989 – 1903.

[5] S-Y. Tseng, Y.-H. Su, J.-Z. Shiang, C.-M. Yang, and S.-Y. Fan, "Interleaved buck-boost converter with single-capacitor turn-off snubber using coupled inductor for stunning poultry applications," Proceedings of IEEE Power Electronics Specialists Conference, 2008, pp. 1964 – 1970.

[6] S.-Y. Tseng, C.-T. Hsieh, and H.-C. Lin, "Active clamp interleaved forward converter with single-capacitor turn-off snubber for stunning poultry applications," *Proceedings of IEEE Power Electronics Specialists Conference*, 2008, pp. 450 – 456.

[7] M. R. Prausnitz, "The Effects of Electric Current Applied to Skin: A Review for Transdermal Drug Delivery," Proceedings of the Advanced Drug Delivery Reviews, Vol. 18, 1996, pp. 395-425.

[8] U. Pliquett, R. Langer and J. C. Weaver, "Changes in the Passive Electrical Properties of Human Stratum Corneum Due to Electroporation,"Trans. On Biophysica Acta, 1995, pp.111-121.

[9] M. R. Prausnitz, "A Practical Assessment of Transdermal Drug Delivery by Skin Electroporation," Proceedings of the Advanced Drug Delivery Reviews, Vol. 35, 1999, pp. 61-7.

[10] T.-F. Wu, et al., "Unified Approach to Developing Single-Stage Power Converters," *IEEE Transactions on Aerospace and Electronic Systems*, Vol. 34, No. 1, pp. 211 – 223.

High Voltage Generator Using Boost/Flyback Hybrid Converter for Stun Gun Applications

S. -Y. Tseng, C. -M. Yang, K. -C. Wang and G. -W. Hsu

GreenPower Evolution Application Research Lab
(GPEARL)
Department of Electrical Engineering
Chang-Gung University
Kwei-Shan Tao-Yuan, Taiwan, R.O.C
E-mail: sytseng@mail.cgu.edu.tw
TEL: +886-3-2118800
FAX: +886-3-2118026

Abstract—**This paper presents high voltage generator for stun gun applications. The proposed high voltage generator consists of coupled-inductor boost and flyback converters, which adopts active clamp circuit to recover energy stored in leakage inductances of coupled inductor and transformer simultaneously. With this approach, coupled-inductor boost converter can boost voltage of battery up to high output voltage for driving flyback converter, while flyback converter is adopted to step up output voltage up to the desired voltage of capacitated paralysis for convicts conferring. Furthermore, the proposed generator can use single active clamp circuit to recover energies trapped in leakage inductors of two converters. As a result, the proposed generator can reduce cost, weight and size and increase conversion efficiency significantly. Compared with conventional stun gun, the proposed one can achieve 210 ten-second firings which is more than 1.4 times operation of the conventional one. Finally, a prototype under maximum output voltage of 30 kV and maximum output current of 5 mA has been implemented to verify feasibility of the proposed high voltage generator.**

Keywords: High voltage generator, Couple-inductor boost converter, active clamp flyback converter, stun gun.

I. INTRODUCTION

In the past few decades, remarkable advances in material science have been made. In particular, semiconductor technology is growing fast to quickly expand the consumer product market. Since performances of the consumer products require lighter, thinner, shorter and smaller, the portable consumer product is developed. Due to stable technology of the portable one, the stun gun, which is regarded as the electrical incapacitation weapon and uses battery as its voltage source, is growing in popularity among law-enforcement agencies.

The stun gun works with firing two tethered bards at a target and sending pulses or dc of high-voltage electricity through the tether wires to induce involuntary muscle contraction and intense pain. Since human uses electrical impulses to propagate sense signal, the stun gun technology uses similar electrical impulses to cause stimulation of the sensory and motor nerves. It can induce neuron muscular incapacitation (NMI) when a device is able to cause involuntary stimulation of both the sensory nerves and the motor nerves. Therefore, in order to effectively achieve safety of NMI, output current of the stun gun system must be limited within a safety range.

When electrical impulse is applied to human body, it will induce a NMI. However, when a higher electrical impulse is applied to human body, it will generate a ventricular fibrillation, resulting in a hurt of human body. According to standard of IEC479-1[1], the relationship between electric stunning duration versus electric stunning current is indicated the limit the stunning current for achieving an safety stunning electrical impulse of human. The plot of electric stunning duration versus electric stunning current is illustrated in Fig. 1 .where A_1~A_6 is defined in Table 1. From Table 1, it can be seen that operating area in area A_1 or A_2 is a safety area and can avoid to generate ventricular fibrillation. When ac voltage under frequency ranges from 15 to 100 Hz, in which its circuit structure is shown in Fig. 2, is applied to human body and its current value is below 5 mA, it does not generate ventricular fibrillation. When dc voltage is adopted to stun human, as shown in Fig. 3, and its value is below 20 mA, it does not induce the ventricular fibrillation. Therefore, the stun gun system using dc voltage to induce NMI is more safety than ac voltage. Additionally, since the stun gun uses two tethered bards and tether wires, it will induce extra losses in tether wires under ac output voltage. As mentioned above, the proposed stun gun system adopts dc output voltage to achieve NMI.

978-1-4244-4782-4/10 $26.00 © 2010 IEEE

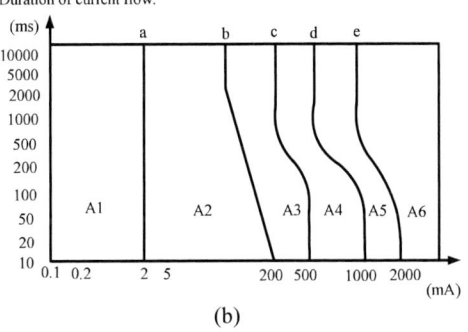

Duration of current flow.

(a)

Duration of current flow.

(b)

Fig. 1.Relationship among ventricular fibrillation, electric stunning duration and electric stunning current in the human body.

Table I. BEHAVIOR OF THE HUMAN PHYSIOLOGY IN DIFFERENT OPERATIONAL AREA

Operatio nal Area	Behavior of the human physiology
A_1	No behavior
A_2	No hurt behavior of the human physiology
A_3	Abnormal behavior of a ventricle
A_4	5% probability of ventricular fibrillation
A_5	50% probability of ventricular fibrillation
A_6	Ventricular fibrillation with probability more than 50%

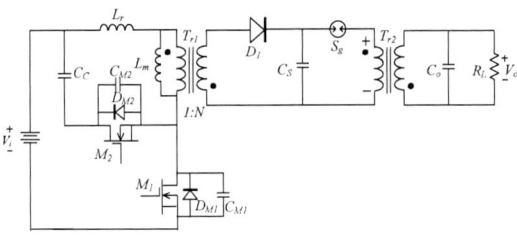

Fig. 2. Schematic diagram of the conventional high voltage generator with pulsating output voltage for stun gun applications.

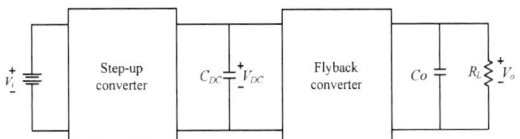

Fig. 3. Block diagram of high voltage generator with DC output voltage for stun applications.

To induce NMI, output voltage of the stun gun must penetrate through the skin and clothing barrier. In general, output electrical charge of the stun gun can penetrate up to 2 cumulative of clothing. Its requirements of parameters are under output voltage of 50 kV~100 kV and output frequency of 15~100 Hz. Since dc to ac equivalent factor, in which a ventricular fibrillation occurs at the same proportionality, is equal to 3.75, output voltage of the stun gun with dc voltage only needs 14 kV~30 kV to achieve the same functions with ac one. Moreover, maximum stunning current will be limited within 5 mA. To generate the specified electrical parameters, a coupled-inductor boost converter combined with a flyback converter is proposed [6]-[15], as show in Fig. 4. Since leakage inductances of coupled inductor and transformer will induce spike voltage and extra losses, two active clamp circuits are introduced into boost and flyback converters to recover energy trapped in leakage inductances. To further simplify circuit structure, synchronous switch technique [16] is used to integrate switches, as shown in Fig. 5. From Fig. 5, it can be seen that the proposed stun gun system uses less component counts to achieve the same functions and to increase conversion efficiency, significantly.

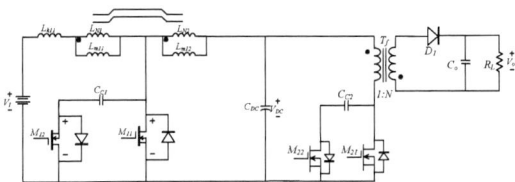

Fig. 4. Schematic diagram of high voltage generator with dc output voltage for stun gun applications.

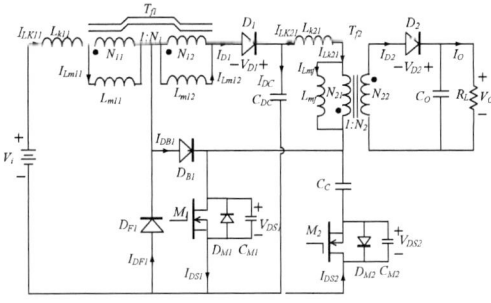

Fig.5. Schematic diagram of the proposed stun gun system.

II .MECHANISM OF HUMAN STUNNING

Human stunning quality (or NMI) is highly dependent on the parameters of electrical stunning. For determining the desired electrical parameters, mechanism of human stunning is briefly described. Human stunning must avoid an epileptic seizure in human body to achieve a safety and effective NMI. Since degrees of an epileptic seizure are dependent upon the amount of current passing the brain, an enough voltage source applied to human skin must overcome clothes and its skin impedance to attain an enough and limited stunning current for achieve an effective NMI of human. In the research of electrical properties for human skin, an equivalent circuit composed of resistor R_s series with the parallel combination of resistor R_{sc} and capacitor C_{sc} have been used by many investigators, as shown in Fig. 6. Both skin capacitance and resistance have been shown to be proportional to contact area [17]. The skin impedance versus voltage is shown in Fig. 7, illustrating that the skin impedance is inversely proportional to the applied voltage [18].

In general, there are three main parts to form neurons: soma, axon and dendrite. A muscle contraction impulse signal is generated by brain of human. Its propagation direction is, in turn, through upper motor neurons, lower motor neurons and node of ranvier to motor end plate for controlling muscle cells, as shown in Fig. 8. Since synapses in brain and neurons play a communication role and is to transfer electrical stimulation signals to chemical signals by unidirectional propagation, it can regarded as a passive switch. Moreover, propagation of muscle contraction impulse signal between nodes of ranviers adopts salutatory conduction which is controlled by potential amplitude. That is, nodes of ranviers are regarded as an active switch.

To explain the relationship between electrical stunning for human and an effective NMI in human body, an equivalent circuit for describing the muscle contraction impulse propagation between brain and muscle cells is shown in Fig. 9. In Fig. 9, since synapses in brain and each motor neuron are regarded as a passive switch, it can be considered as diode D_1. Moreover, the node of ranviers can be regarded as switch Q. When stunning voltage V_{HDC} is applied to human body, its skin impedance Z_S enters a breakdown state, and then current I_l will pass the body of human. In the contact zone of the stunning electrodes, the motor neurons will form an equivalent diode D_2. Within this stunning duration, if current I_l is large enough, it will induce a high potential V_l to turn on switch Q through diode D_2 and propagation impedance R_t. As a result, active impulse signal E_T for muscle contraction will be imposed on muscle cells to generate muscle contraction. As described previously, it can be observed that electrical stunning can generate muscle contraction for achieving NMI.

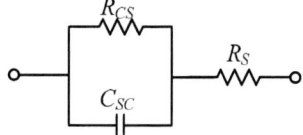

Fig.6. An equivalent impedance of skin

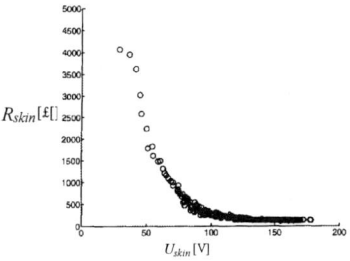

Fig.7. Plot of the skin impedance of poultry versus the applied voltage.

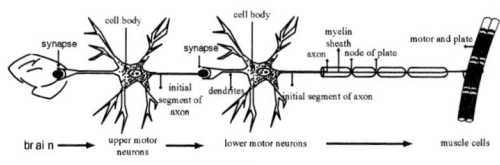

Fig. 8. An illustration of procedure for muscle contraction impulse propagation between brain and muscle cells.

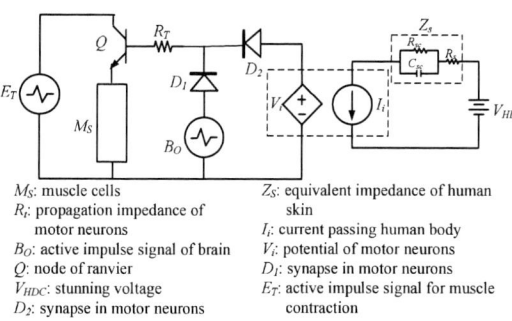

M_S: muscle cells
R_i: propagation impedance of motor neurons
B_O: active impulse signal of brain
Q: node of ranvier
V_{HDC}: stunning voltage
D_2: synapse in motor neurons

Z_S: equivalent impedance of human skin
I_i: current passing human body
V_i: potential of motor neurons
D_1: synapse in motor neurons
E_T: active impulse signal for muscle contraction

Fig. 9. An equivalent circuit for describing the muscle contraction impulse propagation between brain and muscle cells.

III. Derivation of the proposed stun Gun system

In the proposed stun gun system, a coupled-inductor boost converter with active clamp circuit and a flyback converter with active clamp one are integrated to generate a high dc voltage. The couple-inductor boost converter can boost battery voltage up to a desired output voltage for supplying flyback converter and use the active clamp circuit to recover energy trapped in leakage inductor. Furthermore, the flyback converter can generate a high dc voltage to induce NMI in human body and also adopt the active clamp one to recover energy trapped in transfer. In order to design the proposed stun gun system, derivation of the proposed one are briefly described as follows.

The proposed stun gun system is composed of an active clamp coupled-inductor boost converter and an active clamp flyback converter, as shown in Fig. 10(a). From Fig. 5(a), it can be seen that when switch pairs (M_1, M_3) and (M_2, M_4) are respectively operated in synchronous and switches in each switch pair share a common node. They can adopt synchronous switch technique [16] to integrate each switch pair. First, switch pair (M_1, M_3) adopts switch integration to simplify circuit structure. Its circuit structure is shown in Fig. 10(b). Since voltage across switch M_3 in the off state is higher than that across switch M_1. According to the simplified rules of synchronous switch technique, diode D_{B132} can be shorted, as shown in Fig.10(c). Moreover, body diode D_{M13} of switch M_{13} is connected in parallel with diode D_{F132}. Therefore, diode D_{F132} and switch M_{13} can be regarded as switch M_{13} and body diode D_{M13}, as shown in Fig .10(d).

Similarly, switch pair (M_2, M_4) can use switch integration to integrate circuit structure and its degenerated circuit is shown in Fig. 10(e). From Fig. 10(d), it can be seen that voltage across switch M_4 in the off state is higher than that across switch M_2. Therefore, diode D_{B241}

can be shorted. Its circuit structure is shown in Fig. 10(f). Since diode D_{F241} and body diode D_{M24} connect in parallel, they can be replaced by body diode D_{M24}, as shown in Fig. 10(g). In Fig. 10(g), transformer T_{f2} is required by a high isolation between primary and secondary windings, and an extreme high turns ratio, resulting in a very low coupling coefficient. As a result, the coupling coefficient of transformer T_{f2} is much larger than that of T_{f2}. That is, the energy trapped in leakage inductance L_{k21} is much greater than that trapped in L_{k11}. Capacitor C_{C1} can be neglected and removed, as shown in Fig. 10(h).

Although capacitor C_{C1} is removed, the energy stored in leakage inductor L_{k11} can be discharged through diode D_{B131} and diode D_{F242} to capacitor C_{C2} and be recovered by capacitor C_{C2}. Moreover, voltage across diode D_{F131} can be clamped to that across capacitor C_{C2}. From Fig. 10(h), it can be seen that current of capacitor C_2 is bidirectional through diode D_{F242}, and switch M_{24} in series with diode D_{B242}. Since switch M_{24} and its body diode D_{M24} can be regarded as a bidirectional switch, it can substitute for diode D_{F242}, and diode D_{B242} in series with switch M_{24}, as shown in Fig. 10(i). From Fig. 10(i), it can be found that the proposed stun gun system can adopt less component counts to achieve a high output voltage.

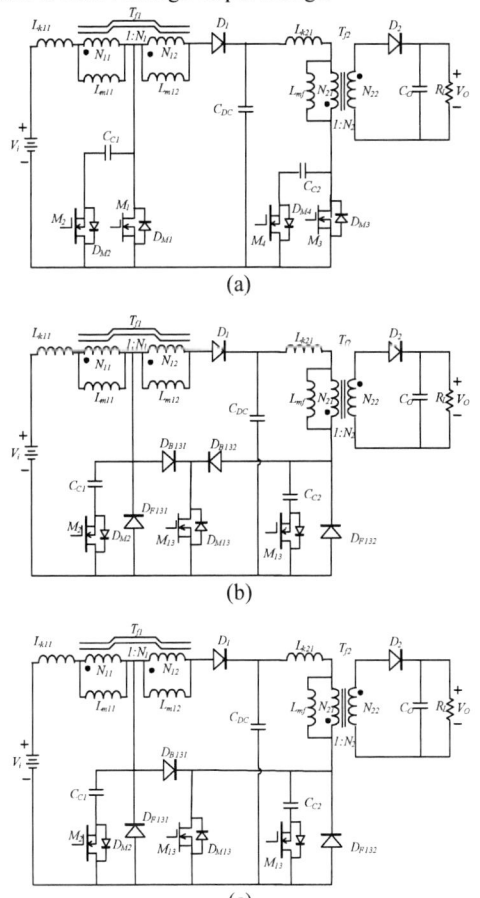

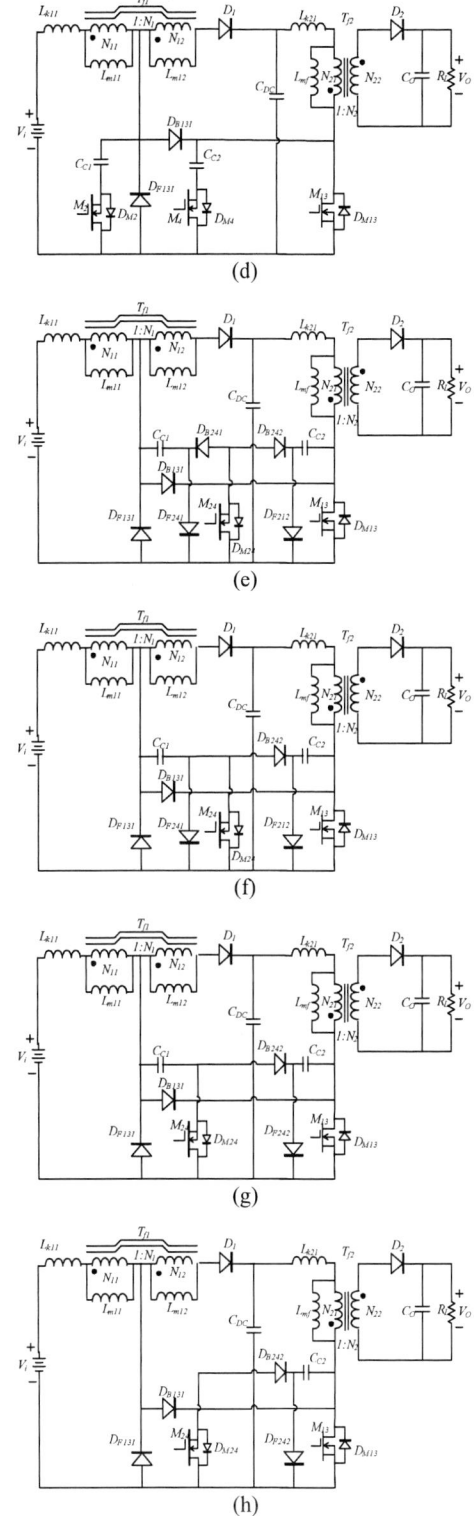

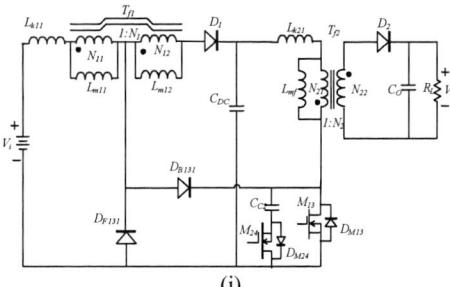

(i)

Fig.10. Derivation of the proposed stun gun system

IV. Design of the Proposed stun Gun System

To realize the proposed stun gun system systematically, design of duty ratios D_1 and D_2, couple inductor T_{f1}, transformer T_{f2} and the clamp capacitor C_C are illustrated as follows.

A. Duty Ratios D_1 and D_2

The proposed stun gun system consists of a coupled-inductor boost converter and an active flyback converter and their switches are integrated by switch integration technique. Although circuit structures of the couple-inductor boost converter and the active flyback converter are integrated with switch integration technique, their transfer functions M_1 and M_2 do not affect and their values are the same as the original converters. Therefore, transfer function M of the proposed stun gun system is equal to M_1M_2. In order to attain transfer ratio M, transfer ratios M_1 and M_2 are first determined. In the following, transfer ratios M_1 and M_2 are derived.

Since the couple-inductor boost converter is designed in continuous conduction mode (CCM), the relationship between input voltage V_i and voltage V_{DC} can be indicated with volt-second balance principle of coupled inductor T_{f1} as follows:

$$V_iD_1T_S+[-(V_{DC}-V_i)(1-D_1)/(N_1+1)]T_S = 0, \quad (1)$$

where N_1 is turns ratio of couple inductor T_{f1} and V_{DC} is dc bus voltage across capacitor C_{DC}. From (1), it can be observed that transfer function M_1 between input voltage V_i and voltage V_{DC} can be expressed as

$$M_1 = \frac{V_{DC}}{V_i} = \frac{N_1D_1+1}{1-D_1} \quad (2)$$

Furthermore, active clamp flyback converter is operated in discontinuous conduction mode (DCM) to generate high output voltage. Similarly, according to volt-second balance principle of transfer T_{f2}, transfer ratio M_2 can be determined by

$$V_{DC}D_1T_S+[-(V_O/N_2)D_2T_S] = 0 \quad (3)$$

where N_2 is turns ratio of transformer T_{f2} and D_2T_S is the completely discharge time interval of the energy stored in magnetizing inductor L_{mf}. Therefore, transfer ratio M_2 can be derived as

$$M_2 = \frac{V_O}{V_{DC}} = \frac{N_2D_1}{D_2} \quad (4)$$

As mentioned above, the total transfer ratio M can be determined as follows:

$$M = \frac{V_O}{V_i} = \frac{N_2D_1(N_1D_1+1)}{D_2} \quad (5)$$

According to (5), duty ratio D_1 can be derived by

$$D_1 = \frac{-(N_2+D_2)+\sqrt{N_2^2+2D_1N_2M+D_2^2M_2^2}}{2N_1N_2} \quad (6)$$

Based on the operational limitation of the converter operated in DCM, the value of ratio D_2 must be limited and it is less than $(1-D_1)$. According to (2) and (4), a large duty ratio D_1 corresponds to a smaller coupled inductor T_{f1} or transformer T_{f2} turn ratios N_1 and N_2, which attains a lower voltage stresses on freewheeling diodes D_1 and D_2 and a higher coupling coefficient between T_{f1} and T_{f2} for reducing switching losses in switches M_1 and M_2. However, in order to accommodate variations of load, line voltage and component value, it had better select an operating range between $D_1 = 0.7\sim0.8$.

B. Coupled Inductor T_{f1}

Once the duty ratio D_1 and voltage V_{DC} are selected, the turns ratio N_1 of coupled inductor T_{f1} can be determined from (2), which yields

$$N_1 = \frac{M_1-M_1D_1-1}{D_1} \quad (7)$$

By applying the Faraday's law, the number N_{11} of the turns at the primary winding can be determined as

$$N_{11} = \frac{D_1V_iT_S}{A_{C1}\Delta B_1} \quad (8)$$

where A_{C1} is the effective cross-section area of transformer T_{f1} core and ΔB_1 is the working flux density. According to (3) and (4), N_{12} can be therefore determined.

For the couple inductor T_{f1}, magnetizing inductor L_{m11} is determined by taking into account the current up slope, while corresponds to the on-time of switch M_1, and inductance L_{m11} must be large enough to maintain CCM operation. Fig.13 shows conceptual waveforms of currents of boost converter with coupled inductor operated in the boundary of CCM and DCM. From Fig.13, it can be seen that inductor current $I_{LK11(1)}$ can be expressed by

$$I_{LK11(1)} = \frac{2V_iD_1T_S}{L_{m11B}} \quad (9)$$

where L_{m11B} is the inductance of inductor L_{m11} operated in the boundary of CCM and DCM. Therefore, the inductance of L_{m11} must satisfy the following inequality:

$$I_{m11} \geq I_{m11B} = \frac{2V_iD_1T_S}{I_{LK11(1)}} \quad (10)$$

Since average current $I_{D1(av)}$ is equal to current I_{DC} of dc linkage, and current $I_{LK11}=(N+1)I_{LK11(2)}$, current $I_{LK11(1)}$ can be rewritten as

$$I_{LK11(1)} = \frac{2(N_1+1)I_{DC}}{(1-D_1)} \quad (11)$$

Substituting (9) and (11) into (10) yields

$$L_{m11} \geq L_{m11B} = \frac{V_iD_1(1-D_1)T_S}{(N_1+1)I_{DC}}. \quad (12)$$

When the maximum current $I_{DC(max)}$ is specified, the minimum inductance $L_{m11(min)}$ can be determined.

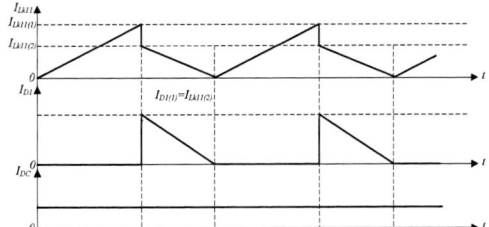

Fig. 13. Conceptual waveforms of currents of boost converter with coupled inductor operated in the boundary of CCM and DCM.

C. Transformer T_{f2}

According to (4), once the duty ratio D_1 is selected, the turns ratio N_2 of transformer T_{f2} can be determined which yields

$$N_2 = \frac{D_2 V_O}{D_1 V_{DC}} . \tag{13}$$

Since the maximum $D_{2(max)}$ is equal to $(1-D_1)$, N_2 must satisfy the following inequality:

$$N_2 \geq \frac{(1-D_1)V_O}{D_1 V_{DC}} . \tag{14}$$

By applying the Faraday's law, the number N_{21} of the turns at the primary winding can be determined as

$$N_{21} = \frac{D_1 V_{DC} T_S}{A_{C2} \Delta B_2} , \tag{15}$$

where A_{c2} is the effective cross-section area of the transformer T_{f2} core and ΔB_2 is the working flux density. According to (14) and (15), N_{22} can be therefore determined.

For the flyback converter, magnetizing inductor L_{mf} is determined by taking into account the current down slope, which corresponds to the off-time of switch M_1, and the inductor must be limited to maintain DCM operation. The inductance of L_{mf} can be expressed by

$$L_{mf} = \frac{V_{DC} D_2 T_S}{N_2^2 \Delta I_{D2(max)}} , \tag{16}$$

where $\Delta I_{D2(max)}$ is the maximum ripple of the secondary winding current of transformer T_{f2}. When ratio D_2 and the maximum current ripple is specified, the minimum magnetizing inductance L_{mf} can be determined.

D. Active clamp capacitor C_c

The active clamp capacitor C_c is used to achieve soft-switching feature. Since the energy trapped in leakage inductor L_{k21} is much greater than that trapped in L_{k11}, leakage inductor L_{k11} can be neglected when the active clamp capacitor C_c is designed. Moreover, the active clamp flyback converter can be operated in DCM or CCM, depending on output load. Therefore, the active clamp capacitor can help switch to achieve soft-switching feature and to recover the energy trapped in leakage inductor L_{k21} when the converter is operated in CCM. When the one is operated in DCM, the active clamp capacitor C_c only recovers the energy trapped in L_{k21}. According to transfer ratio M_2 of flyback converter, the boundary output voltage V_{OB} can be determined by

$$V_{OB} = \frac{N_2 D_1 V_{DC}}{(1-D_1)} . \tag{17}$$

To achieve a ZVS feature, the energy stored in inductor L_{k21} must satisfy the following inequality:

$$\frac{L_{K21}\left[I_{LK21(tv10)} - I_{LK21(tv8)}\right]^2}{2} \geq \frac{(C_{M1} + C_{M2})V_{DS1(max)}^2}{2} , \tag{18}$$

where $I_{LK21(tv8)}$ is the leakage inductor current at time t_8, $I_{LK21(tv10)}$ is that at time t_{10}, C_{M1} and C_{M2} are respectively the junction capacitors of switches M_1 and M_2, and $V_{DS1(max)}$ is the voltage across switch M_1 and its value is equal to $(V_{DC}+V_{OB}/N_2)$. Once C_{M1}, C_{M2}, $I_{LK21(tv8)}$ and $I_{LK21(tv10)}$ are specified, leakage inductor L_{K21} can be determined as

$$L_{K21} \geq \frac{(C_{M1} + C_{M2})V_{DS1(max)}^2}{(I_{LK21(tv10)} - I_{LK(tv8)})^2} . \tag{19}$$

To achieve ZVS feature using active clamp circuit, one half of the resonant period formed by L_{K21} and C_c should be equal to or greater than the maximum off time of M_1. Thus, capacitor C_c must satisfy the following inequality:

$$\pi\sqrt{L_{K21}C_C} \geq t_{off(max)} = (1-D_1)T_S . \tag{20}$$

Form (20), when L_{k21} is specified, the capacitance range of the clamp capacitor C_c can determined as

$$C_C \geq \frac{(1-D_1)^2 T_S^2}{\pi^2 L_{K21}} . \tag{21}$$

V. EXPERIMENTAL RESULTS

To verify the performances of the proposed stun gun system, as shown in Fig. 5, a prototype with the following specifications was implemented.

· input voltage V_i: 7.4V/1.65Ah (Lithium battery)
· switching frequency f_s: 25kHz,
· output maximum voltage Vo(max): 30kV
· output maximum current Io(max): 5mA, and
· output maximum power Po(max): 9W.

According to the previous specifications, components of the stun gun system are determined as follows:

· turns ratio of coupled inductor T_{f1}: 2,
· turns ratio of transformer T_{f2}: 100,
· magnetizing inductor L_{m11}: 24 uH,
· leakage inductor L_{k11}: 1.4 uH,
· magnetizing inductor L_{mf}: 106 uH,
· leakage inductor L_{k21}: 56 uH,
· Active clamp capacitor C_c: 0.3 uF,
· output capacitor C_o: 3 nF/40 kV,
· switch M_1 ; M_2: IRF59N60C,
· diode D_1 :STTH12R06D,
· diode D_2 :2CL2FM,
· diode D_B :DSEC60-03A, and
· diode D_{F1}: STTH05R06D

Measured voltage V_{DS} and current I_{DS} waveforms of switches M_1 and M_2 is shown in Figs. 14 and 15 under simulated load resistances of 100 MΩ and 500Ω, respectively. Fig. 14(a) and (b) shows those waveforms of switches M_1 and M_2 under load resistance of 100 MΩ, while Fig. 15(a) and (b) shows those waveforms of ones under load resistance of 500Ω. From Figs. 14 and 15, it can be seen that the proposed stun gun converter can be operated with ZVS at turn-on transition. Fig.16 shows measured waveforms of output voltage V_O and current I_O, illustrating that output voltage V_O can reach 30 kV. In order to observe rising time of output voltage V_O, Fig. 17 illustrates output voltage waveform system from power on to steady-state conditions. From Fig.17, it can be found that rising time of output voltage V_O is about 230ms. Moreover, when the skin and clothing barrier impedances occur in breakdown condition, the body resistance of human is about 600~1,750Ω which is posed in IEC 479-1. To simulate the body resistance of human, Fig.18 illustrates measured waveforms of output voltage V_O and output current I_O under step-load changes in which the simulated resistance varies from 100 MΩ to 500 Ω and 2 kΩ, respectively. Fig. 18(a) shows those waveforms under change of the simulated resistance from 100 MΩ to 500Ω, while Fig. 18(b) illustrates those waveforms under variation of the one from 100 MΩ to 2 kΩ. From Fig.18, it can be seen that when skin and clothing barrier impedances occur in breakdown condition, output current I_O can be kept in the desired constant current of 5 mA. From experimental results, we can find that the proposed stun gun system using Li-ion battery (7.4V/1.65 Ah) can achieve 210 ten-second firings. It is 1.4 times of the conventional stun gun which is made taser.

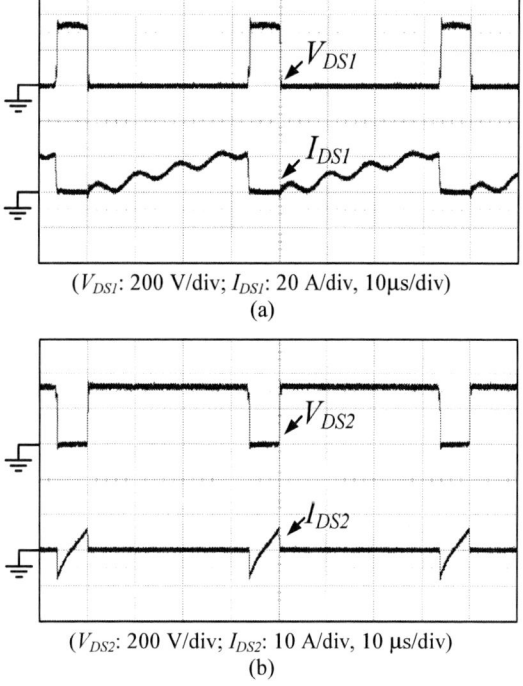

(V_{DS1}: 200 V/div; I_{DS1}: 20 A/div, 10μs/div)

(a)

(V_{DS2}: 200 V/div; I_{DS2}: 10 A/div, 10 μs/div)

(b)

Fig. 14. Measured voltage V_{DS} and I_{DS} current waveforms of (a) switch M_1 and (b) switch M_2 under the simulated load resistance of 100 MΩ.

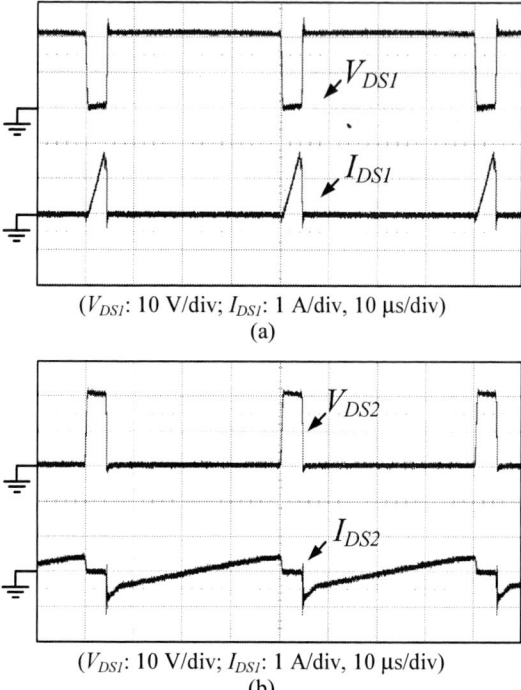

(V_{DS1}: 10 V/div; I_{DS1}: 1 A/div, 10 μs/div)

(a)

(V_{DS1}: 10 V/div; I_{DS1}: 1 A/div, 10 μs/div)

(b)

Fig. 15. Measured voltage V_{DS} and I_{DS} current waveforms of (a) switch M_1 and (b) switch M_2 under the simulated load resistance of 500 Ω.

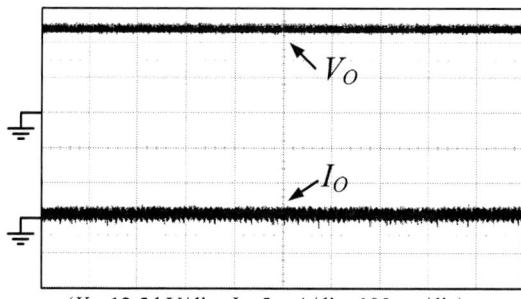

(V_O: 12.5 kV/div; I_O: 5 mA/div, 100 ms/div)

Fig. 16. Measure waveforms of output voltage V_O and current I_O under the simulated load resistance of 100 MΩ.

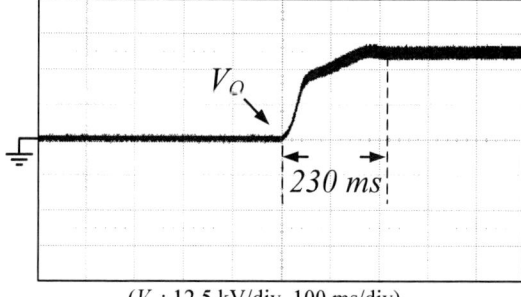

(V_O: 12.5 kV/div, 100 ms/div)

Fig. 17. Measured waveforms of output voltage V_O from power on to steady-state conditions of the proposed stun gun.

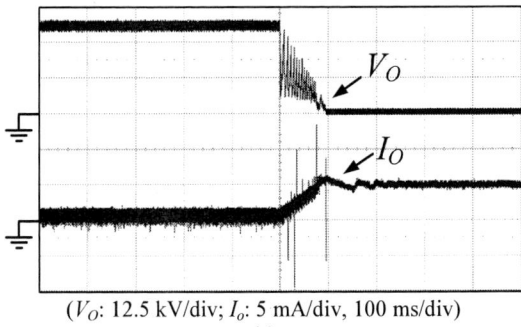

(V_O: 12.5 kV/div; I_o: 5 mA/div, 100 ms/div)

(a)

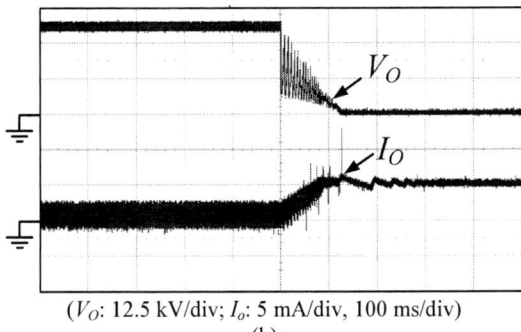

(V_O: 12.5 kV/div; I_o: 5 mA/div, 100 ms/div)

(b)

Fig. 18. Measured waveforms of output voltage V_O and current I_O under step-load change (a) from the simulated load resistance 100 MΩ to 500Ω and (b) from 100 MΩ to 2 kΩ.

VI. CONCLUSION

In this paper, NMI mechanism of human with electric shock has been briefly reviewed. Operational principle, steady-state analysis and design of the proposed system have been implemented to generate electric shock parameters, in which its maximum current is 5 mA and its maximum voltage is 30 kV. Moreover, the proposed coupled-inductor boost converter and active clamp flyback converter can be operated with ZVS at turn on. From experimental results, it can be also found that the proposed system has achieved 210 ten-second firings and it is 1.4 times of the conventional stun gun system.

ACKNOWLEDGEMENT

This work was supported by the Netional Science Council, Taiwan, R.O.C., under Project No. NSC-98-2221-E-182-062.

REFERENCES

[1] International Electrotechnical Commission IEC Report, 'Effects of Current Passing Through the Human Body, Part 1: General Aspects', IEC 479-1, 1984.

[2] Y. -K. Lo and J.-Y. Lin, "Active-Clamping ZVS Flyback Converter Employing Two Transformers", *IEEE Trans. on Power Electronics,* Vol. 22, Issue 6, 2007, pp. 2416 – 2423.

[3] Bor-Ren Lin and Fang-Yu Hsieh, "Soft-Switching Zeta–Flyback Converter With a Buck–Boost Type of Active Clamp", *IEEE Trans. on Industrial Electronics*, 2007, Vol. 54, Issue 5, pp. 2813 – 2822.

[4] Soo-Seok Kim, Dae-Kyu Choi, Su-Jin Jang, Tae-Won Lee and Chung-Yuen Won, "The Active Clamp Sepic-Flyback Converter" *IEEE Trans. on Power Electronics* , 2005, pp. 1209-1212.

[5] Jinno, M., Po-Yuan Chen and Kun-Chih Lin, "An efficient active LC snubber for multi-output converters with flyback synchronous rectifier" *IEEE Trans. on Power Electronics* , 2003, pp. 622-627.

[6] Bor-Ren Lin and Fang-Yu Hsieh, "Soft-Switching Zeta–Flyback Converter With a Buck–Boost Type of Active Clamp", *IEEE Trans. on Industrial Electronics*, 2007, Vol. 54, Issue 5, pp. 2813-2822.

[7] Jang, Y., Dillman, D.L. and Jovanovic, M.M, "A new soft-switched PFC boost rectifier with integrated flyback converter for stand-by power" *IEEE Trans. on Power Electronics* , 2004, pp. 413-419.

[8] Q. Zhao, F. Tao, F.-C. Lee, P. Xu, and J. Wei, "A simple and effective method to alleviate the rectifier reverse-recovery problem in continuous-current-mode boost converters", *IEEE Trans. on Power Electronics*, 2001, Vol. 16, Issue 5, pp. 649–658.

[9] A. S. Ba-Thunya, S. K. Pillai and D. Prasad, "Certain novel synchronized quasi-resonant topologies for achieving soft-switching DC/DC boost and flyback converters with minimum voltage stress across the switches" *IEEE Trans. on Power Electronics*, Vol. 1,1998, pp. 682-688.

[10] V. S. Murali and C. K. Tse, "Comparison of small-signal dynamics of BIFRED and single-stage cascaded boost-and-flyback PFC converters" *Proc. of Power Electronics*, Vol. 6,1998, pp. 1111-1117.

[11] Yousefzadeh, Vahid, Shirazi, Mariko, Maksimovic and Dragan, "Minimum Phase Response in Digitally Controlled Boost and Flyback Converters" *IEEE Trans. on Power Electronics*, 2007, pp. 685-670.

[12] D. M. Sable, B. H. Cho and R. B. Ridley, "Use of leading-edge modulation to transform boost and flyback converters into minimum-phase-zero systems", *IEEE Trans. on Power Electronics*, Vol. 6, 1991, pp. 704-711.

[13] M. Bierhoff, "Capacitor design issues for buck-boost and flyback converters" *Proc. of Power Electronics*, 2009, pp. 1-8.

[14] M. Rajeev, "An input current shaper with boost and flyback converter using integrated magnetics" *IEEE Trans. on Power Electronics*, Vol. 1 2003, pp. 327-331.

[15] M. A. E. Andersen, "Fast prediction of differential mode noise input filter requirements for flyback and boost unity power factor converters", *IEEE Trans. on Power Electronics*, Vol. 1,1997, pp. 230-234.

[16] Tsai-Fu. Wu, et al., "Unified Approach to Developing Single Stage Power Converters" *IEEE Trans. on Aerospace and Electronic Systems*, 1998, pp. 211-223.

[17] B.-Y. Shmuel, "Electrical Evaluation of the Taser M-26 Stun Weapon Final Report," *The Israel National Police, R&D Division*, 2006.

[18] E. Henneman, C. Olson, "Relations between structure and function in the design of skeletal muscle," *Trans on Neurophysiology*, 1965, pp. 581-589.

A method to analysis and design for long life power converter

Pang H. M.
Department of Electrical and Electronic Engineering
The University of Hong Kong
Hong Kong SAR, China

Pong M. H. Bryan, Senior Member IEEE
Department of Electrical and Electronic Engineering
The University of Hong Kong
Hong Kong SAR, China

Abstract—**Engineers are always looking for more reliable or longer life power converters. Electrolytic capacitor is the critical component to be considered. Using the powerful calculation tools like Mathcad and Matlab and allowable life model, this paper aims at proposing a method to optimize the circuit design to lower the current ripple through the electrolytic capacitor in order to reach for longer capacitor life. Comparison between two modes of operation and the two converter topologies, Forward and Flyback, are made as well. This work provides a guideline to design power converters for long life application like LED driver.**

I. Introduction

There is an increasing demand for long life power converters. Medical equipment, telecommunication system and LED lighting drivers need long life power converters at least with the life time of the load. Electrolytic capacitors are critical elements in Power Electronics circuits. They serve purposes like energy storage or filtering of rectified AC voltage ripple [1-4, 7]. An electrolytic capacitor is also one of the most expensive components in a power electronics circuit. Moreover, these capacitors must endure relatively high ripple current which lead to self heating. Most likely, the electrolytic capacitor is prone to have shorter life than other electronic devices in the circuit except mechanical fans [1-9]. For these reasons special attention is paid to factors affecting life of this component.

Several factors can cause failure of the electrolytic capacitor, such as extremely cold temperature, heat, high voltages, transients, extreme frequencies or reverse bias. The most influential factor is heating up of the electrolyte, which determines how fast the non-solid electrolytic solution is evaporated, causing the degradation in the electrical parameters. The dissipation heat generated by ripple current is one of the most important components [1, 5-6, 9]. The dissipation heat generated by the ripple current is an important component affecting the core temperature. Therefore, apart from ambient temperature, attention will also be given to factors like thermal resistance from core to ambient and the power loss by ripple current on the ESR.

Currently, few papers have worked on optimizing the electrolytic capacitor life from circuit design aspect. There is common view that capacitor life can be increased by operating below maximum ratings [1-5, 7, 9]. A. Riz et al. stated the useful life can be prolonged by lower operating voltage, current or ambient temperature and by cooling measures [4]. M. L. Gasperi showed the impact of ripple current on capacitor life, with experimental proof [5]. M. Huber et al. proposed ripple current reduction method of DC Link electrolytic capacitor by switching pattern optimization to prolong its life [9]. But there is yet optimization of switching power supply electrolytic capacitor, which this paper aims at prolonging the capacitor life through optimized circuit design.

II. Modeling

Analyzing and understanding the impact of each circuit parameter to electrolytic capacitor life is crucial and is achieved by an appropriate model. Electrolytic capacitor life model is the starting point, followed by converter model from which the switch current shape is calculated out. The section is completed by calculation of capacitor ripple current.

A. Electrolytic capacitor life model

Several life models for electrolytic capacitor have been published. Two common ones are ESR determination from capacitor inner pressure and life prediction from estimated core temperature [1-8]. However, only the latter one is a predictive approach that allows circuit optimization in design stage is proposed by M. L. Gasperi [5, 11]. The Arrhenius' theory is a common tool in capacitor industry for life prediction purpose.

$$\frac{L_p}{L_r} \approx 2^{\frac{T_r - T_p}{10}} \qquad (1)$$

where Lp: predicted life; Lr: rated operating life; Tr: rated core temperature; Tp: actual core temperature.

The core temperatures Tp in (1) is the sum of the self-generated heat gradient and ambient temperature in Kevin scale, as shown by (2).

$$T_{core} = T_A + \alpha \Delta T \qquad (2)$$

where Tcore: core temperature; TA: surface temperature; ΔT: surface heat rise; α: temperature factor.

Temperature rise is a function of the electrical resistance and the thermal resistance is calculated from (3) to (5). Thermal resistance is difficult to estimate but obtainable from manufacturers. It also depends on the internal dimensions [1, 3, 5-6, 11].

$$\Delta T = \frac{I^2 \times ESR}{\beta \times S} \qquad (3)$$

$$\beta = 2.3 \times 10^{-3} \times S^{-0.2} \qquad (4)$$

$$S = 2\pi(r^2 + rh) \qquad (5)$$

where I: ripple current (Arms); ESR: equivalent series resistance of capacitor; β: thermal resistance (W/°C·cm²); S: surface area of capacitor (cm²); h: capacitor height (cm); r: capacitor radius (cm).

ESR increases over long time operation but decreases with temperature and frequency rises [1-3, 5-6]. As the ESR increases, heat generated raises the temperature and deteriorates the capacitor. This factor has already been accounted for in the life parameters provided by manufacturers of quality capacitors [10]. Thermal and frequency dependence of the ESR are modeled by (6).

$$R = R_{ox}(f) + R_{sp}(T) \qquad (6)$$

ESR at specific condition (generally 20°C, 120Hz) can be calculated from the dissipation factor (DF) as (7), as worst case [3, 14].

$$R_{120} = \frac{DF}{2\pi f C} \qquad (7)$$

Manufacturer provides the frequency multiples for current ripple (Kf). The ESR at different frequencies can be calculated from these multiples [6, 10, 14].

$$R_{100K} = \frac{R_{120}}{Kf_{100K}^2} \qquad (8)$$

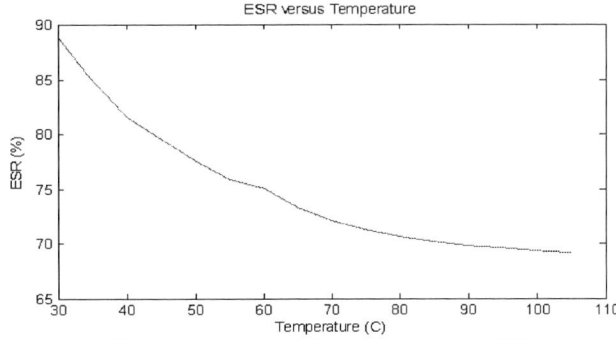
Figure 2. Experimental ESR vs Temperature (100kHz)

Temperature dependence of ESR is modeled by (9) [3, 5].

$$R(T) = R_{base} e^{\frac{T_{base} - T_{core}}{E}} \qquad (8)$$

where Rbase: R at base temperature; Tbase: base measurement temperature; E: temperature sensitivity factor.

The sensitivity factor (E) here is not easy to obtain. ESR against temperature plot shows that ESR begins to level off above 70°C. Taking average over this reasonable temperature range, the ESR can be estimated. Another approach is to find the approximate equation by curve fitting used in later section.

B. *Switch current waveform model (Flyback and Forward)*

Predicting the capacitor current waveform is one of the major steps to achieve optimization as the capacitor current greatly affect the amount of self-heat generated. Once the converter parameters are designed and input and output voltages are specified, the capacitor currents should be found. Switching current waveform is a reachable parameter in power converters. Deriving capacitor current waveform from the switching waveform is no more than processing simple calculations. A simple Flyback and Forward converter switch model can tell the switching current waveform in full load conditions.

Equations (9) and (10) are the steady state equation for CCM Flyback at switch on and switch off period respectively. Solving (9) and (10), the waveform parameters can be obtained.

a) CCM Flyback switch on period:

$$L \times fs \times \frac{Id}{D} = Vin - \frac{Iout}{n \times (1-D)} \times Rpri \qquad (9)$$

CCM Flyback switch off period:

$$L \times fs \times \frac{Id}{1-D} = Vout + Vf + \frac{Iout}{n \times (1-D)} \times Rsec \qquad (10)$$

where Id: the switch delta current; D: switch duty; fs: switch frequency; L: transformer inductance; n: transformer turn ratio; Vf: output diode voltage drop.

978-1-4244-4782-4/10 $26.00 © 2010 IEEE

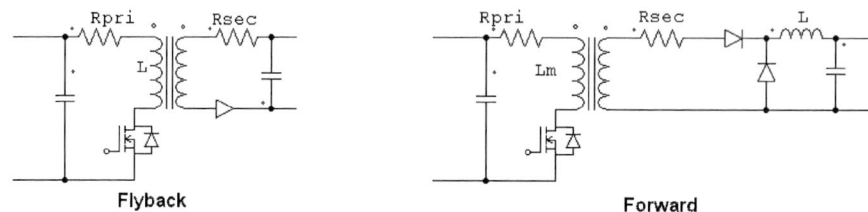

Figure 3. Generic circuit diagram for Flyback and Forward

Similarly, DCM Flyback steady state current can be worked out by (11) to (13). Equations (14) to (19) are those for a Forward converter. Fig.4 shows typical switch current waveforms and the parameters to be calculated.

b) DCM Flyback:

$$L \times fs \times \frac{2Ip}{D} = Vin - Ip \times Rpri \qquad (11)$$

$$L \times fs \times \frac{2Ip}{n \times D'} = Vout + Vf + n \times Ip \times Rsec \qquad (12)$$

$$n \times Ip \times D' = Iout \qquad (13)$$

c) CCM Forward:

$$L \times fs \times n \times \frac{Id}{D} = \frac{Vin}{n} - Vout - Vf - \frac{Iout}{n} \times Rpri - Iout \times Rsec \quad (14)$$

$$L * fs * n * \frac{Id}{1-D} = Vout + Vf + Iout * Rsec \qquad (15)$$

$$i_m = \frac{Vin \times D}{L_m \times fs} \qquad (16)$$

where L: output filter inductance; Lm: transformer magnetizing inductance; im: magnetizing current.

d) DCM Forward:

$$L \times fs \times \frac{2n \times Ip}{D} = \frac{Vin}{n} - Vout - n \times Rsec \times Ip - Rpri \times Ip \quad (17)$$

$$L \times fs \times \frac{2Ip}{n \times D'} = Vout + Rsec \times Ip \qquad (18)$$

$$D = -D' + \frac{Iout}{n \times Ip} \qquad (19)$$

$$i_m = \frac{Vin \times D}{L_m \times fs} \qquad (16)$$

C. Calculating capacitors ripple current

Switch current parameters, including lower peak (iL), higher peak (iH), switch duty (D) and switch off duty (D') in DCM, are the variables obtained from the converter models. With these known variables one can figure out the waveform shape [14]. Simple derivation can then help us to reach for input and output capacitor currents. Assumption made here is the current input and current output of the converter is constant, which can be easily justified. Input side of the converter always equips with common mode and differential mode filters than smoothen the input current. And the output constant load is a usual practice in steady state. Figures below show the switch current shape and how is it related to the input capacitor current waveform in CCM and DCM separately. Mathematical equations to work out the capacitors' current are shown as well. The ripple current RMS² can be calculated accordingly. Fig. 5 shows the relationship between the switch current and capacitors' current shape for Buck and Buck-boost type, the bases of Forward and Flyback converters. It can be easily seen that the RMS values of the input and output capacitors can be expressed in terms of the duty cycle D and current peaks. These values highly depend on the power converter design parameters and they can be designed to reduce the capacitor RMS values for long operation life.

a) CCM Flyback converter:
Input capacitor current RMS²:

$$\begin{aligned}I_{d\,rms}{}^2 = \frac{1}{12}[&(i_L{}^2 + i_H{}^2)(-3D^2 + 4D) \\ &+ i_L i_H (-6D^2 + 4D)]\end{aligned} \qquad (20)$$

Output capacitor current RMS²:

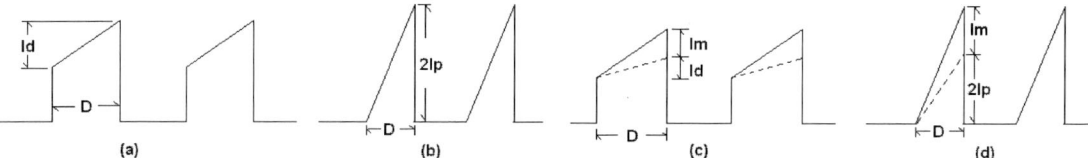

(a)   (b)   (c)   (d)

Figure 4. Switch current for a)CCM Flyback b)DCM Flyback c)CCM Forward d)DCM Forward

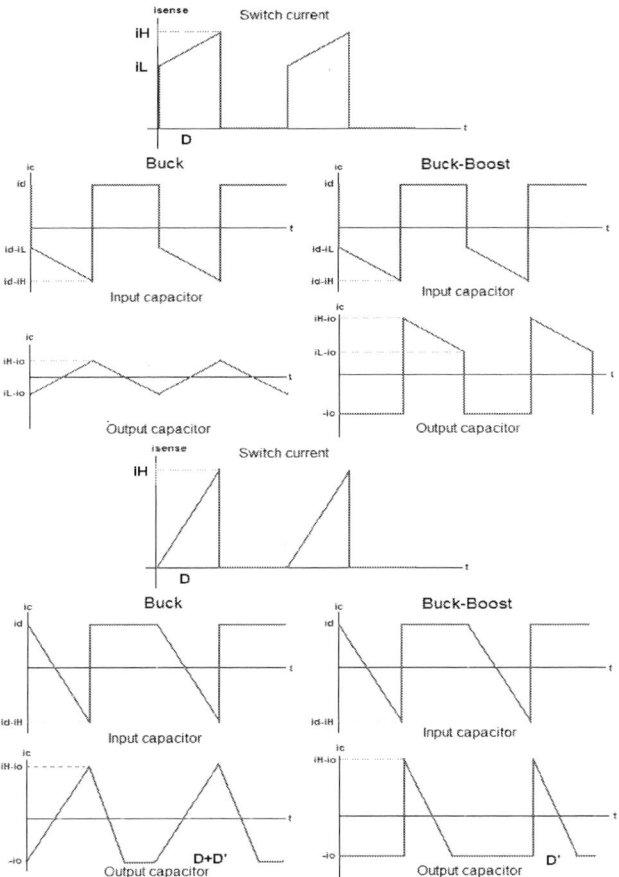

Figure 5. Switch current and capacitors current for CCM and DCM cases

$$I_{o\,rms}^2 = \frac{1}{12}[(i_{L\sec}^2 + i_{H\sec}^2)(-3D^2 + 2D + 1) + i_{L\sec}i_{H\sec}(-6D^2 + 8D - 2)] \quad (21)$$

where iLsec=n*iL; iHsec=n*iH.

b) CCM Forward converter:

$$I_{d\,rms}^2 = \frac{1}{12}[(i_L^2 + i_H^2)(-3D^2 + 4D) + i_L i_H(-6D^2 + 4D) + i_m^2 D_r] \quad (22)$$

$$I_{o\,rms}^2 = \frac{1}{12}(n(i_H - i_m) - ni_L)^2 \quad (23)$$

where Dr: reset duty.

It is assumed that the magnetizing current is much smaller than that of switch load current that equation (22) is accurate.

c) DCM Flyback converter:

$$I_{d\,rms}^2 = \frac{1}{12}i_H^2(-3D^2 + 4D) \quad (24)$$

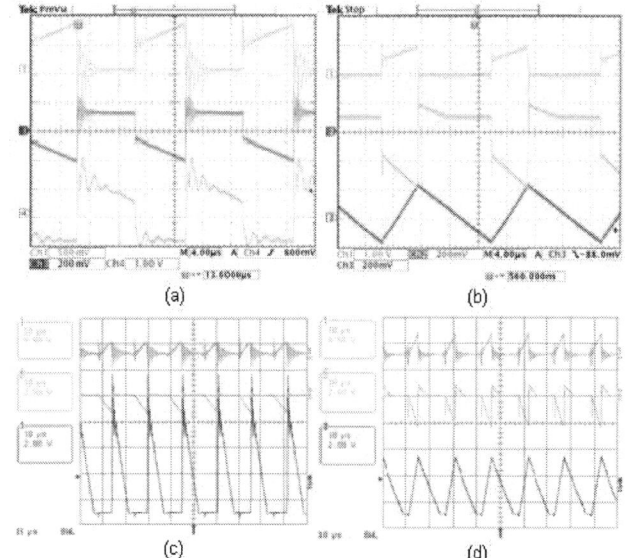

Figure 6. Experimental waveforms for a)CCM Flyback, b)CCM Forward, c)DCM Flyback, d)DCM Forward upper: switch current; middle: input capacitor current; lower: output capacitor current

$$I_{o\,rms}^2 = \frac{1}{12}i_{H\sec}^2(-3D'^2 + 4D') \quad (25)$$

d) DCM Forward converter:

$$I_{d\,rms}^2 = \frac{1}{12}i_H^2(-3D^2 + 4D) \quad (26)$$

$$I_{o\,rms}^2 = \frac{1}{12}n(i_H - i_m)^2[-3(D+D')^2 + 4(D+D')] \quad (27)$$

Fig. 6 shows the experiment captured waveforms which are comparable to the expected ones.

III. ANALYSIS OF CAPACITOR LIFE IN FLYBACK AND FORWARD CONVERTER

Putting (1) to (27) into program, the 3-D life plots as follow can be obtained. The variation of working conditions to life change of the converter can then be studied in a mathematical manner. Working temperature is the trivial factor to life and is assumed constant since it has not interest to circuit design. Line voltage and load current variations to a fixed design are the interesting factors to study. Comparison between Flyback and Forward converters lives is made. Afterward, circuit parameters variation to a fixed specification (defined input and output) is also studied in next section. All these issues can help to choose the more reliable configuration for suitable applications. Testing capacitors are as follow:

TABLE I.   TESTING CAPACITORS FOR COMPARISON

Flyack		
Input capacitor	Rubycon MXG series 450V 180uF	Measured ESR(50℃ 100kHz):

978-1-4244-4782-4/10 $26.00 © 2010 IEEE   1860

		104mΩ
Output capacitor	Rubycon ZL series 35V 1800uF	Measured ESR(50℃ 100kHz): 16mΩ
Forward		
Input capacitor	Rubycon MXG series 450V 180uF	Measured ESR(50℃ 100kHz): 104mΩ
Output capacitor	Rubycon ZL series 35V 680uF	Measured ESR(50℃ 100kHz): 21mΩ

Capacitance is selected according to the voltage ripple

Fig. 7 shows the line voltage and load variations to input and output capacitors' lives in general-designed CCM (left) and DCM (right) Flyback converters at an ambient temperature of 50℃, which is the appropriate temperature in a closed system. The converter is designed for 90V to 240V input and 12V 10A output. Several interesting points are noticed from these plots:

1) Input capacitor generally handles less current stress than output capacitor in the Flyback converter for both CCM and DCM. In other words, output capacitor tends to dominate the life of a Flyback converter. Note that this is not a straight rule and highly depend on the transformer design and the electrolytic capacitors selected. It can vary but is always true for general design.

2) The CCM Flyback design tends to last longer than the DCM one under the same loading conditions. For this example is around 100 times better.

3) Line voltage and load variations do not have much impact on input capacitor life in CCM and DCM Flyback;

4) In CCM Flyback, increased load current and reduced line voltage shorten the output capacitor life.

5) In DCM Flyback, output capacitor life is not affected by line voltage but reduced by increasing the load.

Transformer design: L=500uH, n=60/8, fs=65kHz (CCM); L=100uH, n=22/3, fs=65kHz (DCM).

The figures are alarming but explainable. The major factor is under full load condition Flyback secondary side current ripple is much higher than expected and far beyond the rated current ripple provided by the capacitor manufacturers. The enlarged capacitor self-generated heat greatly accelerates the electrolyte evaporation and sharply reduces its life. This also explains why there is essential to parallel many output capacitors with higher-than-required capacitance to share the current ripple in industrial products.

Same analysis is applied to the Forward converters. The converter is designed for 90V to 240V input and 12V 10A output, same as pervious Flyback. Working temperature is 50℃. Points to be noted are:

1) In CCM mode, output capacitor generally handles less current stress than input capacitor in the Forward converter. Input capacitor tends to dominate the life of a CCM Forward converter.

2) In DCM mode, however, output capacitor dominates the life of Forward converter.

3) The CCM Forward design lasts much longer than the DCM one especially under full load condition. Under full load condition it can be as much as a thousand times better (depending on the transformer design and the electrolytic capacitors selected).

4) Line voltage and load variations do not have much impact on both capacitors lives in CCM Forward.

5) In DCM Forward, increased load current and line voltage sharply shorten the output capacitor life.

Magnetic design: Lchoke=33uH, n=20/6, fs=65kHz (CCM); Lchoke=4uH, n=20/4, fs=65kHz (DCM).

The sharp reduction of capacitor life when forward operates in DCM can be explained by the sharply increased current ripple. In CCM, only small AC portion charges into and out of the capacitor as the filter choke is large enough and the DC bias level does no heating effect on the capacitor. In DCM the filter choke is however too small to maintain the DC level. Both Forward and Flyback converter has reduced life in DCM mode. Therefore, it is recommended to use CCM operation for life sensitive designs.

Comparison between Flyback and Forward converters lives is valuable. Flyback converter is always considered to be favorable isolated topology for low power output applications for its low cost, simplicity and multi-output availability. One of the familiar applications for Flyback is LED driver. However, in the analysis of the capacitor life between a Flyback and a Forward in CCM mode, it is found that the Forward tends to have longer life than the Flyback converter, especially when the load power is high. This gives out the idea

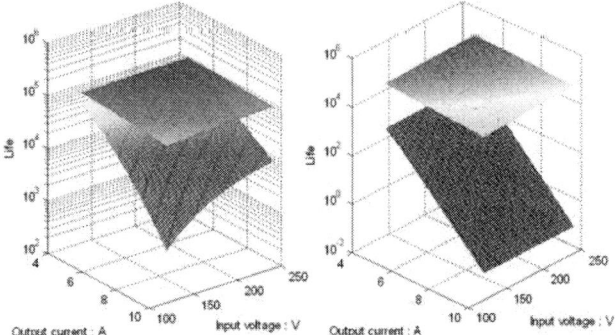

Figure 7. Life versus input and load variation for CCM (left) and DCM (right) Flyback upper: input capacitor; lower: output capacitor

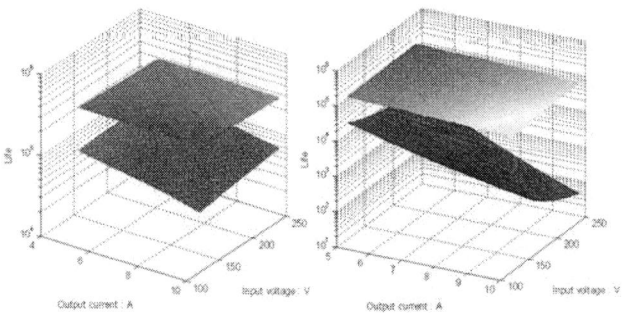

Figure 8. Life versus input and load variation for CCM (left) and DCM (right) Forward (Left) upper: output capacitor; lower: input capacitor (Right) upper: input capacitor; lower: output capacitor

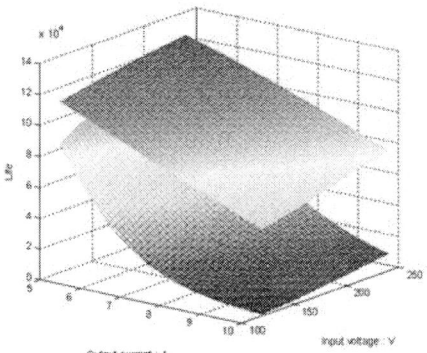

Figure 9. Life versus input and load variation for CCM
Forward (upper) and Flyback (lower)

why buck type topology (Forward) is more suitable for LED drivers. Forward is also more favorable for higher power applications (>100W) considering their reliability.

IV. OPTIMIZING THE CONVERTER DESIGN FOR LONGER LIFE

From previous deviations and analysis, it is well accepted that capacitor current ripple does have huge impact on capacitor life through self-heating. If the converter circuit is designed such that the current ripple stress on the electrolytic capacitors are reduced, it is possible to optimize the converter with longer working life. But what are the parameters affecting the current ripple value, which are controllable? Looking into the previous simplified circuit models, it can be easily told that those are inductance L, transformer turn ratio n and switch frequency fs. These parameters determine the peak magnitude and current shape, which in turn determine the RMS current value of the capacitors. Other parameters are either incontrollable or little impact on capacitor currents. The plots below can vary when circuit parameters, capacitor types or loading conditions change, but the trend and the design approach are the same. The test condition in this section for both converter topologies are operating in CCM to output 12V 5A with input voltage 120V.

Fig. 10 shows the frequency, inductance and transformer turn ratio variation to capacitor life changes in a Flyback (left) and a Forward (right) converter individually, assuming the other two parameters are constant. Increasing the switching frequency can increase the capacitor life. And the output capacitors of both topologies are more sensitive to the frequency change than their counterpart input capacitors. Therefore the life improvement to frequency increment is more significant to Flyback.

Comparing with frequency, transformer (choke) inductance and turn ratio shows greater impact on capacitor life. Moreover, the switching frequency is highly restricted by the product size, efficiency and the control. More effort would be paid to transform and choke design. Both greatly determine the current shapes. For both Flyback and Forward converters, there is a sharp drop in capacitors' lives in small inductance values. The converters run deeply into DCM mode in these values causing extremely high current ripple RMS. Cross point between the curves for the input and output capacitor life is the point where both input and output capacitors are under the same current stress. This is not the point with the longest

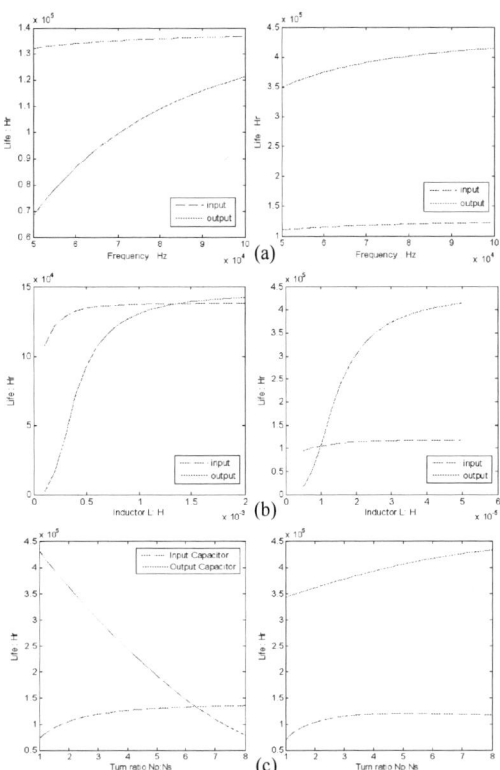

Figure 10. Life versus a)switching frequency, b)inductance c)turn ratio for CCM Flyback (left) and Forward (right)

converter life but above this point the life increment is almost level off. On the other hand, there are restriction to design high inductances, like the losses and core saturation. Therefore, it is recommended to design the inductance just above this point.

Transformer turn ratio affects the current peak magnitude as well as switch duty. Consider the CCM Flyback, increasing turn ratio increases the primary side current ripple while reducing secondary side current ripple. The cross point between two curves is the optimal point where converter life is longest. For the Forward, once it operates in CCM which high output choke, the current ripple is incomparable to rated ripple of output capacitor. Increasing turn ratio will further increase the life. Note again that figures above are only one of the possible design cases. The situation varies with design specifications but is true for generic design. Optimization in parameter choosing can still be done in a similar way.

Parameter variations affecting capacitor life are presented individually. In design process, however, these parameter values affect each other as well. Consider the switching frequency is determined beforehand by converter size, efficiency or else, the inductance and transformer turn ratio are organized into one three-dimension plot to help obtaining the best values of each, as shown by fig. 11. For Flyback converter, alone the boundary line is the combinations of inductance and turn ratio that give longer converter life than their neighbor points. For Forward converter, the boundary line is where the life starts to drop sharply due to DCM operation. For every inductance there is an optimal turn ratio. The higher the inductance gives the longer converter life but

978-1-4244-4782-4/10 $26.00 © 2010 IEEE   1862

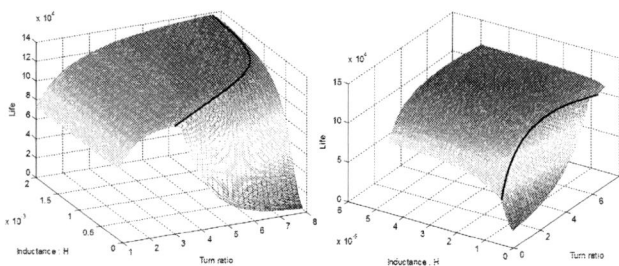

Figure 11. Life versus inductance and turn ratio for CCM Flyback (left) and Forward (right)

the difference is insignificant while adding cost and size. It is recommended to design the inductance just above this boundary.

V. EXPERIMENTAL RESULTS

A. Accelerated test on life model

Though there were numerous papers worked on verification of the life model, an accelerated test was conducted for completeness [5]. The testing media was a DCM flyback converter output capacitor, which was considered to be under high current stress. A current probe measures the capacitor current. The working conditions are listed in the table. The calculated life time is 63 hours. The result is acceptably accurate since manufacturer claims the equation is within 40% error range. Incomplete temperature control might also accelerate the capacitor death.

TABLE II.   ACCELERATED TEST CONDITIONS

Input voltage	127Vdc
Output voltage	12Vdc
Output current	7A
Capacitor current RMS	12.76Arms
Capacitor type	Rubycon ZL series 35V 1800uF
Measured ESR	16.8mΩ(65kHz, 25°C)
Ambient temperature	25°C

B. Design a long life converter for LED lamp

The verified model was demonstrated to design a 150W 2FET-Forward converter for LED street lamp. Input is 400Vdc and output is 24V 6.25A. The target converter life is around 50000 operation hours, which is comparable to LED lamp itself. Table shows the selected capacitor. ESR characteristics of the input capacitors obtained by experiment are shown in fig. 13. Curve fitting tool (Matlab) was used to found the mathematical expression and was put into the program to improve the model accuracy.

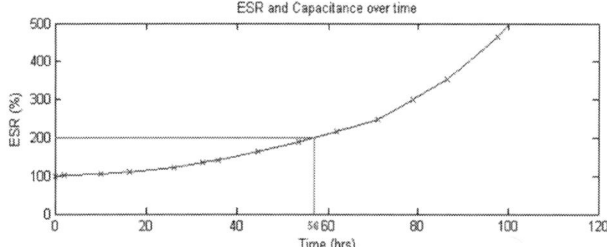

Figure 12. Experiment captured ESR versus time

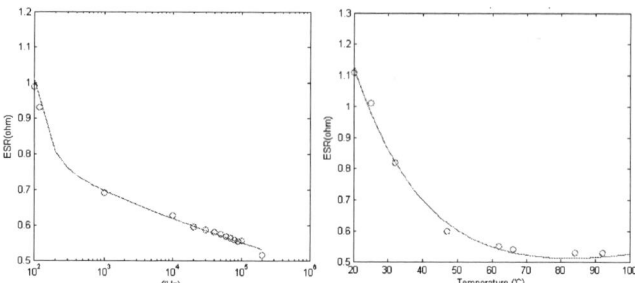

Figure 13. ESR versus frequency (left) and temperature (right) (AXW 450V 68uF)

TABLE III.   CAPACITORS FOR LED LAMP DRIVER

Input capacitor	Rubycon AXW series 450V 68uF
Output capacitor	Rubycon ZL series 35V 1800uF

Consider the switching frequency is fixed to be 100kHz, the magnetic component variation is shown by fig. 14. Duty plot was considered to ensure the duty is less than 0.5. Convert the life plot into 2-D plot, the boundary line is where the design is over 50000 working hours. Considering the efficiency and cost, the choke inductance and transformer turn ratio is selected to be 60uH and 5 respectively. The calculated life is around 66000 hours. The experiment captured waveforms are shown in fig. 15. The input capacitor ripple current is measured to be 0.723Arms. Substitute this value into the capacitor model, the received life is 63000hours. The discrepancy is due to circuit impedance.

VI. CONCLUSION

The operating life time of a power converter is often determined by the life of the electrolytic capacitor. The rms value of the capacitor currents should be reduced in order to maximize their life time. Here in this paper a new power converter design method is presented which aims to maximize

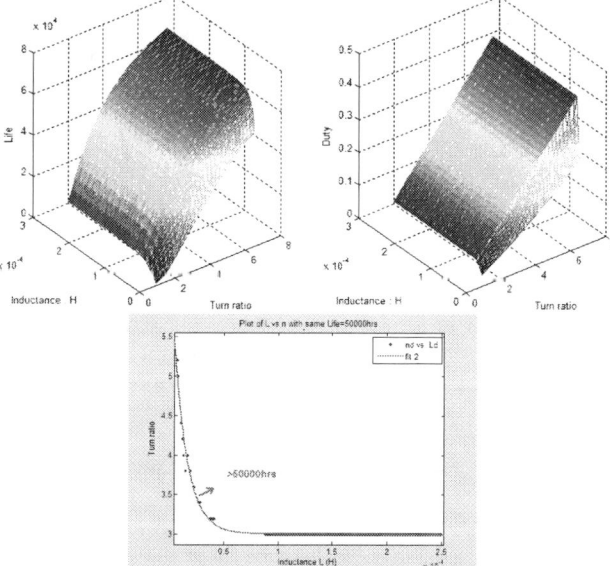

Figure 14. Life plot (left) and Duty plot (right) and 2-D life plot (bottom) for LED Driver

power converter life. Analysis and comparison of the two

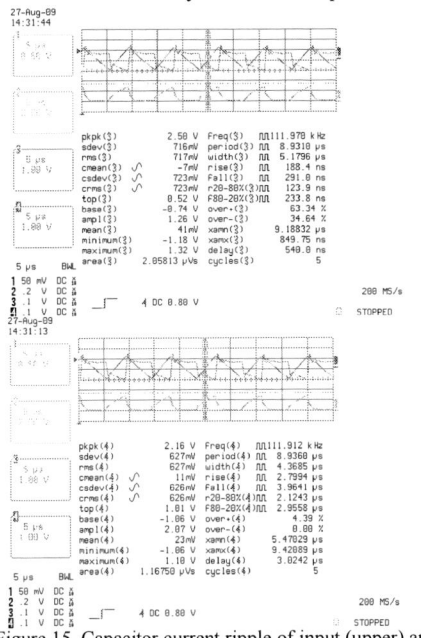

Figure 15. Capacitor current ripple of input (upper) and output (lower)

frequently used topologies (Flyback and Forward) in life aspect are made. Several parameters are optimized for long life design. They are the converter transformer turns ratio, the inductance and the duty cycle. A complete design example is also demonstrated to verify the method accuracy.

REFERENCES

[1] V. A. Sankaran, F. L. Rees and C. S. Avant, "Electrolytic capacitor life testing and prediction", Industry Applications Conference, 1997. Thirty-Second IAS Annual Meeting, IAS '97., Volume 2, 5-9 Oct.

1997, pp.1058-1065

[2] E. C. Aeloiza, J. H. Kim, P. Ruminot and P. N. Enjeti, "A Real Time Method to Estimate Electrolytic Capacitor Condition in PWM Adjustable Speed Drives and Uninterruptible Power Supplies", Power Electronics Specialists Conference, 2005. PESC '05. IEEE 36th, 16 June 2005, pp.2867-2872

[3] Hao Ma and Linguo Wang, "Fault diagnosis and failure prediction of aluminum electrolytic capacitors in power electronic converters", IEEE Industrial Electronics Society 31st Annual Conference, IECON 2005, 6-6 Nov. 2005, pp. 6 pp.-

[4] A. Riz, D. Fodor, O. Klug and Z. Karaffy, "Inner gas pressure measurement based life-span estimation of electrolytic capacitors", Power Electronics and Motion Control Conference, 2008. EPE-PEMC 2008. 13th, 1-3 Sept. 2008, pp. 2096-2101

[5] M. L. Gasperi, "Life prediction model for aluminum electrolytic capacitors", Industry Applications Conference, 1996. Thirty-First IAS Annual Meeting, IAS '96., Volume 3, 6-10 Oct 1996, pp.1347-1351

[6] M. L. Gasperi, "Life prediction modeling of bus capacitors in AC variable-frequency drives", IEEE Transactions on Industry Applications, Volume 41, Issue 6, Nov.-Dec. 2005, pp.1430-1435

[7] S. K. Maddula and J. C. Balda, "Lifetime of Electrolytic Capacitors in Regenerative Induction Motor Drives", Power Electronics Specialists Conference, 2005. PESC '05. IEEE 36th, 16-16 June 2005 pp.153-159

[8] Yaow-Ming Chen, Hsu-Chin Wu, Ming-Wei Chou and Kung-Yen Lee, "Online Failure Prediction of the Electrolytic Capacitor for LC Filter of Switching-Mode Power Converters", IEEE Transactions on Industrial Electronics, Volume 55, Issue 1, Jan. 2008, pp.400-406

[9] M. Huber, W. Amrhein, S. Silber, M. Reisinger, G. Knecht, G. Kastinger, "Ripple Current Reduction of DC Link Electrolytic Capacitors by Switching Pattern Optimisation", Power Electronics Specialists Conference, 2005. PESC '05. IEEE 36th, 16-16 June 2005, pp.1875-1880

[10] Maniktala, Sanjaya., "Switching power supply design & optimization", McGraw-Hill Professional, 2005, pp. 5-7, 324, 361-369

[11] Rubycon Corporation, "Technical Notes for Electrolytic Capacitor" www.rubycon.com

[12] CDE Cornell Dubilier, "Application Guide Aluminum Electrolytic Capacitors" www.cornell-dubilier.com

[13] Reliability Prediction of Electronic Equipment, Military Handbook 217 F, 1995.

[14] Maniktala, Sanjaya., "Switching power supply design & optimization", McGraw-Hill Professional 2005, pp5-7, 361-369

978-1-4244-4782-4/10 $26.00 © 2010 IEEE

DC-DC Converter for Gate Power Supplies with an Optimal Air Transformer

Christoph Marxgut*, Jürgen Biela*, Johann W. Kolar*, Reto Steiner[†] and Peter K. Steimer[†]

*Power Electronic Systems Laboratory, ETH Zurich
[†]ABB Switzerland Ltd., Power Electronics and MV Drives, Turgi, Switzerland
Email: marxgut@lem.ee.ethz.ch, www.pes.ee.ethz.ch

Abstract—Voltages in medium voltage (MV) systems as for example in MV drives, wind generation and smart grids are in the range of several kV. Hence, switches of medium power systems need gate drive power supplies which consider a galvanic isolation for safety reasons. In this paper a gate supply with a capacitive compensated air transformer for medium voltage systems is presented. This approach not only has the advantage of being capable in isolating almost arbitrarily high voltages but also is a compact, lightweight and cheap solution. The air transformer windings are realized as tracks on a circuit board (PCB). Furthermore, the air transformer has been optimized with respect to efficiency of the gate supply, which results in an optimal value of 85%. The optimization is accomplished by field simulation of the transformer and a circuit calculation to obtain the total losses. The simulation and optimization results are confirmed by a laboratory setup which is designed for an output voltage of 25 V at 100 W.

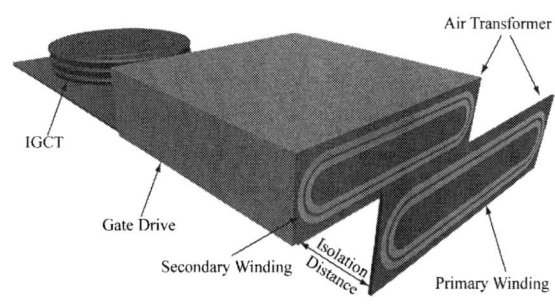

Fig. 1. IGCT setup with air transformer converter. The secondary compensation circuit and the diode rectifier can be integrated in the gate drive unit.

TABLE I
CONVERTER SPECIFICATIONS

V_{in}	48 V_{DC}
V_{out}	25 V_{DC}
P_{out}	100 W
Isolation Distance	20 mm
Max. Dimensions	150 mm × 45 mm

I. INTRODUCTION

System voltages of medium high voltage systems are in the range of 1 kV to 52 kV. Hence, switches of medium high voltage converters are usually realized by IGBTs, IGCTs or thyristors because of their high voltage capabilities. Due to the high voltages the gates drive power supplies of these switches need to consider a proper galvanic isolation for safety reasons.

Today, most supply units use dry-type cast coil transformers as they are able to withstand high voltages. These transformers, however, are bulky, heavy, expensive and even small fabrication defects can result in a severe malfunction. Since the operating voltages in medium power systems tend to increase these problems are getting worse. Typical dimensions of dry-type cast coil transformers which are capable to withstand the stress of $20\,kV_{rms}$ for 10 s without partial discharge are $20\,cm \times 20\,cm \times 20\,cm$ at a weight of about 5.5 kg [1], [2].

To overcome these drawbacks a topology with a capacitive compensated air transformer can be applied [3]. Fig. 1 shows a press pack setup with gate drive unit which is supplied via an air transformer. It can be seen that this approach is compact, lightweight, stackable and cost efficient. The distance between primary and secondary winding determines the electric isolation capability of the gate power supply. However, the leakage inductance is increasing with an increase of the distance and hence capacitors on both the primary and the secondary winding are required to compensate the leakage [4]–[6]. In this paper air transformers with a distance of 20 mm between primary and secondary winding are considered. A

setup with this distance is able to withstand $20\,kV_{rms}$ without having any partial discharges [7] which was tested in [3]. Further improvement in high voltage isolation can be obtained if the windings are coated with a laminating layer. In [8] the interrelation between the air transformer's cross section area and the distance is given. It is pointed out that the diameter of a winding has to be at least as large as the distance. This limits the distance between primary and secondary winding for a given cross section area which is often determined by the gate drive unit's dimensions (cf. Fig. 1).

Since the ohmic losses in the air transformer are the major contribution to the total losses of the converter an efficiency optimization of the air transformer setup has been done. The optimization process will be discussed in **Section II**.
Another important issue regarding air transformers is the EMI compatibility. To improve the EMI behavior magnetic sheets are placed on top and bottom of the air transformer as will be discussed in **Section III**.

To confirm the proposed optimization of the converter

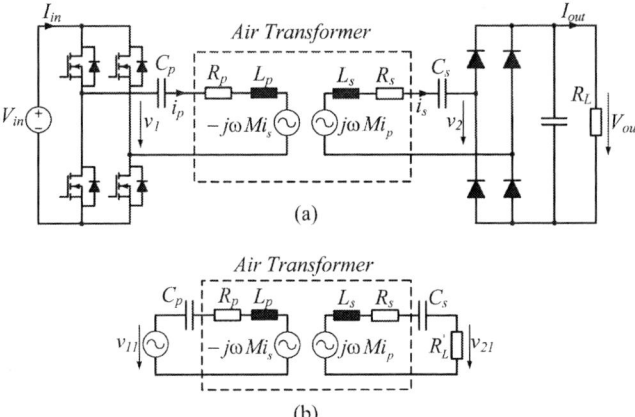

Fig. 2. (a) Circuit topology of the converter and (b) equivalent circuit for the fundamental component of the switching frequency.

system a laboratory setup has been build up which is shown in **Section IV**. The resonance circuit consisting of the resistive and inductive components of the air transformer as well as the compensation capacitors is fed by a full bridge inverter with a DC voltage V_{in} of 48 V and a switching frequency f_s of 400 kHz. The secondary voltage v_2 is rectified by a diode full bridge and an output capacitor. The circuit topology is shown in Fig. 2 (a). Typical values of gate drive power supplies are an output voltage of 25 V and an output power of 100 W. The converter specifications are given in Table I.

II. EFFICIENCY OPTIMIZATION

It is obvious that the air transformer represents the crucial element in this topology since the large leakage inductance and the winding resistances have major impact on the system behavior. Furthermore, a large primary current i_p, a large frequency ω and a large mutual inductance M would lead to a large secondary voltage as can be seen in the equivalent cirucit of the air transformer in Fig. 2 (a). The mutual inductance M is given by the transformer setup while the frequency is a degree of freedom. The primary current i_p can be maximized by decreasing the load of the input H-bridge which can be achieved by compensate the large primary inductance L_p by a capacitor C_p. To avoid a large voltage drop over the secondary leakage inductance a compensation capacitor C_s can be considered as well. This leads to a many parameters which have impact on the total efficiency of the DC-DC converter.

For that reason the electrical parameters of several air transformer designs have been evaluated. Based on these values the capacitors C_p and C_s are determined and a circuit calculation can be performed to obtain the system losses. This leads to an optimization loop of the gate supply.

The optimization process can be seen in Fig. 3. Starting point are the electrical specifications e.g. input and output voltages and the definition of the optimization space e.g. the maximum number of turns. Then an optimization loop starts which will be discussed in the following.

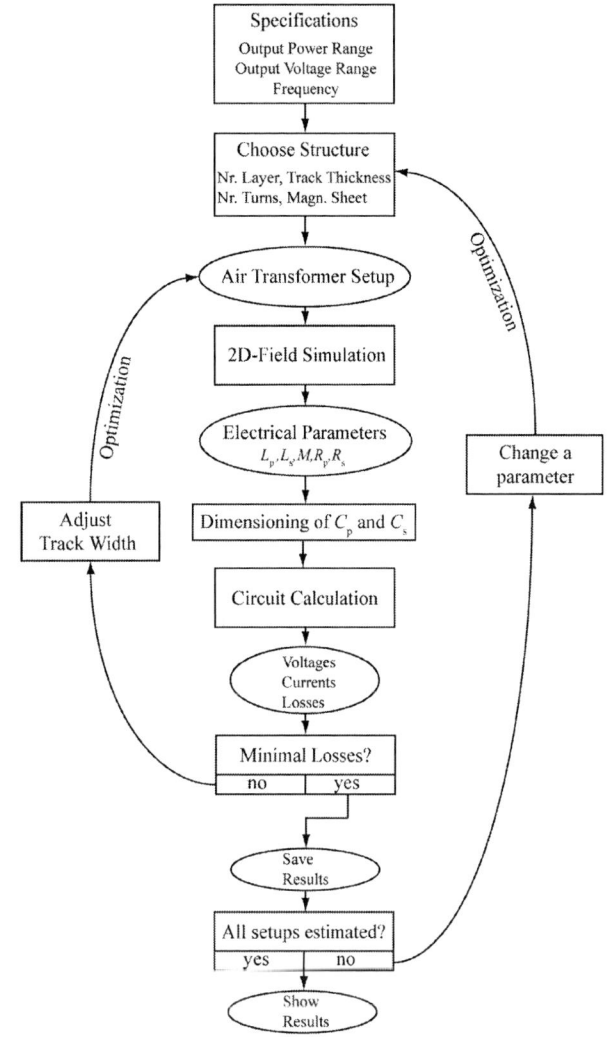

Fig. 3. Flow chart of the optimization process which starts with the electrical specifications and the limitations of the air transformer design parameters. Then for each setup an optimal track width with respect to efficiency is estimated. In the end the results are plottet to compare the setups.

A. Switching Frequency

The transfered power of an air transformer is proportional to the square of the switching frequency [4].

$$P = \Re\left\{\frac{\omega^2 M^2}{Z_s}\right\} \cdot i_p^2 \tag{1}$$

$$\text{whereas} \quad Z_s = R_s + j\omega(L_s - M) + \frac{1}{j\omega C_s} \tag{2}$$

Hence a high switching frequency leads to a good power transmission. On the other hand, a high frequency causes skin and proximity losses in the transformer windings. Furthermore, the gate drive losses increase. Therefore, in the following optimization a frequency of 400 kHz is chosen with is a trade-off between HF-losses and transmitted power. Switching losses can be neglected since soft switching is applied as will be discussed.

978-1-4244-4782-4/10 $26.00 © 2010 IEEE

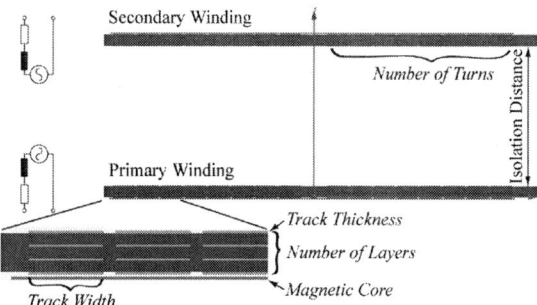

Fig. 4. Cross section of the air transformer with the design parameter: track width, track thickness, number of turns, number of layers and magnetic core material.

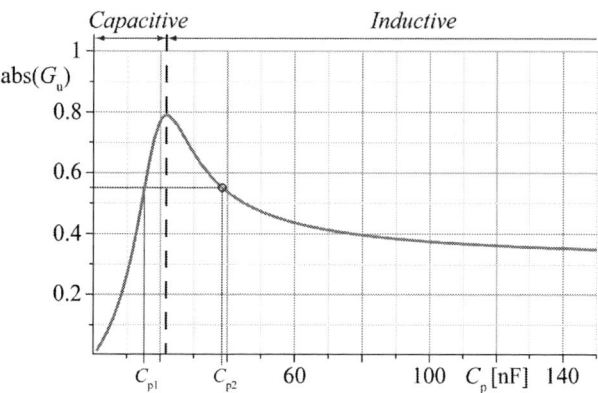

Fig. 5. Voltage transfer function G_u vs. capacitor C_p. It can be seen that two possible values for C_p would lead to the required voltage ratio. However, only the capacitor C_{p2} results in an inductive load to input H-bridge which enables ZVS.

B. Design Parameters of the Air Transformer

Due to the high switching frequency the proximity effect and the skin effect in the windings are considerable. The number and position of tracks affect the AC resistance while the track width as well as the track thickness determine the DC resistance of the windings. To improve the magnetic properties of the transformer ferrite sheets can be placed on top of each winding.

So the number of layers, the number of turns, the track thickness, the track width as well as magnetic sheets are the optimization parameters as is indicated in Fig. 4. These parameters are varied in 2D field simulations which are performed with Maxwell by Ansoft in order to obtain the mutual inductance M, the primary and secondary inductance L_p and L_s as well as the winding resistances R_p and R_s for each setup. The winding resistances already consider skin and proximity effects.

To limit the optimization space the track thickness is considered as either 35 µm or 105 µm which are conventional copper thickness values. Furthermore, the air transformer is supposed to be symmetrical which means that primary and secondary windings are identically. The number of layers is limited to six for reasons of economy and the geometric size of the air transformer is limited to 150 mm × 45 mm with respect to conventional thyristor and IGCT gate drive dimensions (cf. Fig. 1).

C. Dimensioning of the Compensation Capacitors

The schematic in Fig. 2 can be classified in the input full bridge, the resonant circuit including the air transformer, and the output rectifier bridge.

The input H-bridge produces a square wave voltage v_1 with a maximum duty cycle D_{max}. A spectral decomposition of this voltage leads to its fundamental component which can be considered as an equivalent voltage source v_{11}.

$$v_{11} = \frac{4}{\pi} \cdot D_{max} \cdot V_{in} \tag{3}$$

If a large output capacitor is considered the output voltage V_{out} is constant and v_2 has nearly a square wave form. Considering only the fundamental component of voltage v_2

an equivalent resistance R'_L of the diode rectifier, the output capacitor and the load R_L can be evaluated [9].

$$v_{21} = \frac{4}{\pi} \cdot V_{out} \tag{4}$$

$$R'_L = \frac{8}{\pi^2} \cdot R_L \tag{5}$$

Hence the circuit in Fig. 2 (a) is simplified significantly as can be seen in Fig. 2 (b). Capacitor C_s on the secondary is chosen to be in resonance with the secondary leakage inductance L_s-M of the air transformer at the switching frequency f_s.

$$C_s = \frac{1}{(2\pi f_s)^2 \cdot (L_s - M)} \tag{6}$$

Consequently, the voltage transfer function depends only on the primary capacitor C_p for a given transformer setup and hence C_p can be calculated for the required output specifications.

$$G_u = \frac{v_{21}}{v_{11}} = \frac{\frac{s \cdot M \cdot R'_L}{R'_L + R_s + s \cdot M}}{R_p + s \cdot L_p + \frac{1}{s \cdot C_p} - \frac{s^2 \cdot M^2}{R_s + R'_L + s \cdot M}} \tag{7}$$

In Fig. 5 the magnitude of the voltage transfer function G_u is shown as function of the primary capacitor C_p for a given parameter set. It can be seen that two solutions lead to the required voltage transfer ratio. If current i_p lags the voltage v_1 in Fig. 2 zero voltage switching (ZVS) of the H-bridge could be achieved. Consequently, it is beneficial to choose the capacitor C_p so that the H-bridge's load behavior is inductive. In Fig. 5 the border between inductive and capacitive behavior is indicated. C_{p2} has to be considered in order to achieve soft switching.

978-1-4244-4782-4/10 $26.00 © 2010 IEEE

TABLE II
OPTIMIZATION RESULTS

				Min.Losses	Trackwidth
Air Core	2 Layer	35 μm	$N=4$	88.25 W	3.2 mm
	2 Layer	105 μm	$N=4$	28.60 W	3.2 mm
	4 Layer	35 μm	$N=1$	46.10 W	6.95 mm
	6 Layer	35 μm	$N=1$	25.33 W	13.45 mm
	6 Layer	105 μm	$N=1$	17.53 W	12.85 mm
Epcos C351	2 Layer	35 μm	$N=4$	56.10 W	3.65 mm
	2 Layer	105 μm	$N=4$	28.66 W	3.35 mm
$\mu_r = 9$	4 Layer	35 μm	$N=1$	41.90 W	11.8 mm
$\sigma = 4\,\text{mS/m}$	6 Layer	35 μm	$N=1$	23.80 W	14.35 mm
	6 Layer	105 μm	$N=1$	16.66 W	13.6 mm
TDK IRJ08	4 Layer	35 μm	$N=1$	31.70 W	13.3 mm
$\mu_r = 100$	6 Layer	35 μm	$N=1$	22.20 W	16.05 mm
$\sigma = 4\,\text{mS/m}$	6 Layer	105 μm	$N=1$	15.50 W	16.05 mm

D. Loss Calculation

Since all circuit elements are determined a circuit calculation can be performed to obtain the circuit voltages, currents and losses. Due to the band-pass filter behavior of the compensated air transformer sinusoidal currents i_p and i_s are considered. This approximation is confirmed by measurements. The loss calculation includes the conduction losses in the air transformer P_{at}, the conduction losses in the input H-bridge P_{hb} as well as in the output rectifier bridge P_{rb}.

$$P_{at} = R_p \cdot i_{p,rms}^2 + R_s \cdot i_{s,rms}^2 \qquad (8)$$

$$P_{hb} = 4 \cdot (R_{DS(on)} \cdot i_{sw,rms}^2)$$
$$= 4 \cdot (R_{DS(on)} \cdot \frac{i_{p,rms}^2}{2}) \qquad (9)$$

$$P_{rb} = 4 \cdot (V_{fw} \cdot i_{diode,av}) = 4 \cdot (V_{fw} \cdot \frac{I_{out}}{2}) \qquad (10)$$

Here V_{fw} represents the forward voltage drop of the rectifier diodes.

Since soft switching is achieved no switching losses have to be considered. After the efficiency is calculated the track width is adapted until an optimal value is achieved. If an optimal track width is found either the number of turns or the number of layers is varied. The number of turns is limited by the maximum geometric dimensions of the transformer and the track width. The spacing between the tracks is kept constant at 0.5 mm. This procedure is done for an air transformer setup with no core sheet, Epcos C351 sheets ($\mu_r = 9$) [10] and TDK IRJ08 sheets ($\mu_r = 100$) [11]. A script file is written to automate the 2D simulation and the circuit calculation.

E. Optimization Results

The simulation results are presented in Fig. 6 and in Table II. Several air transformer designs lead to high resistances R_p and R_s so that the required voltage ratio can not be achieved. Therefore, these unfeasible setups are not shown in Fig. 6

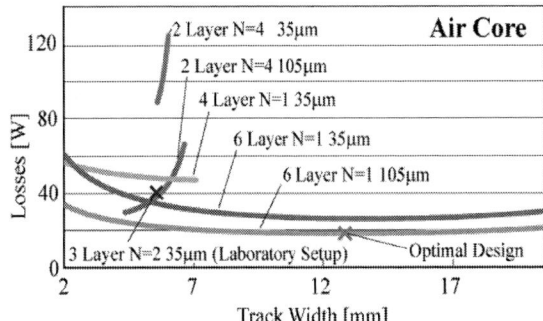

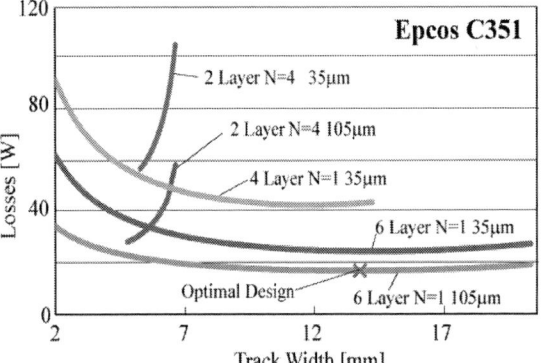

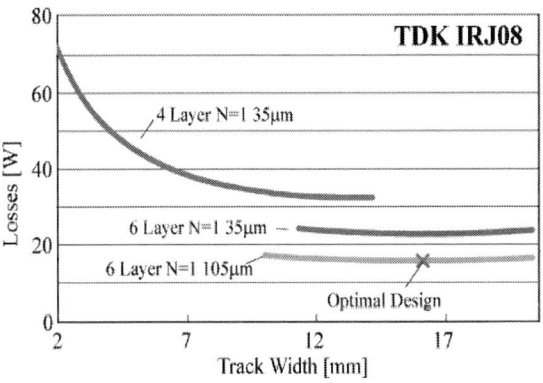

Fig. 6. Optimization Results.

and Table II but only those which achieve the voltage ratio. A further point of interest is that the optimal track width is equal for 35 μm and 105 μm. This can be explained by the fact that for six layers the proximity effect is irrelevant in the vertical direction (cf. Fig. 4). The skin depth of copper at a frequency $f_s = 400\,\text{kHz}$ is 104.5 μm. Hence, the resistance does not increase for this increase of the track thickness. Consequently, the simulation procedure has to be performed only for one track thickness.

It can be seen that the setup with 6 layers and 1 turn per layer is optimal and yields to minimal losses of 15.5 W. However, due to high manufacturing costs a suboptimal setup could be beneficial.

III. MAGNETIC PROPERTIES OF THE AIR TRANSFORMER

The core materials made by Epcos (C351) [10] and TDK (IRJ08) [11] differ in their relative permeability μ_r by a

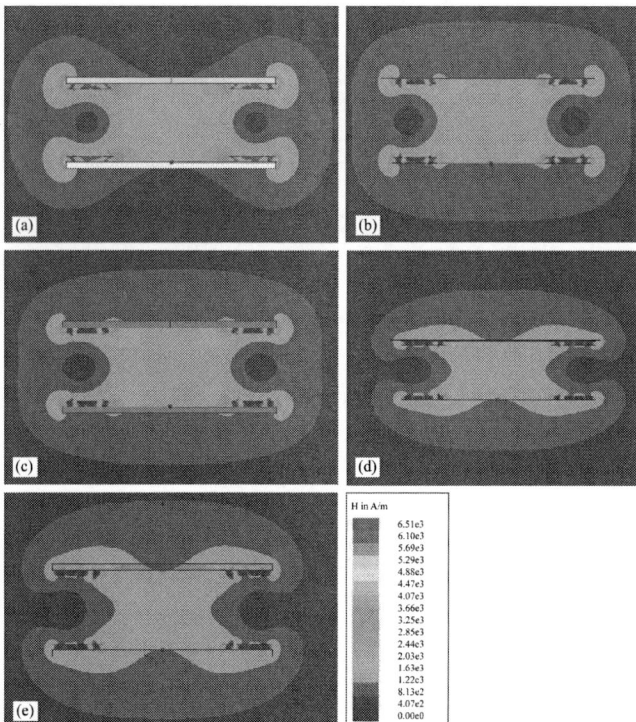

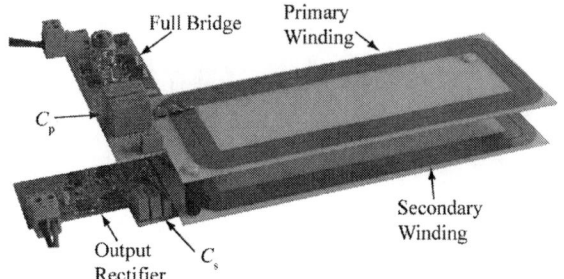

Fig. 8. Laboratory Setup whose parameters are given in Table III.

TABLE III
AIR TRANSFORMER PARAMETERS

Geometric Parameters	Electrical Parameters
Number of Layers: 6	$L_p = 6.75\,\mu H$
Number of Turns: 2	$L_s = 6.75\,\mu H$
Track Width: 6 mm	$M = 2.21\,\mu H$
Track Thickness: 35 μm	$R_p = 0.65\,\Omega$
No Core Material	$R_s = 0.65\,\Omega$
Distance: 20 mm	

TABLE IV
LABORATORY SETUP PARAMETERS

$L_p = 6.75\,\mu H$	$L_s = 6.75\,\mu H$
$M = 2.21\,\mu H$	$f_s = 400\,kHz$
$R_p = 0.65\,\Omega$	$R_s = 0.65\,\Omega$
$C_p = 33.3\,nF$	$C_s = 35\,nF$
MOSFETs - IR FR207Z	$R_{DS(on)} = 0.04\,\Omega$
Diodes - IRP705G	$V_{fw} = 0.65\,V$

Fig. 7. Magnetic field plots for: (a) 2 mm TDK IRJ08, (b) 0.25 mm TDK IRJ08, (c) 2 mm Epcos C351, (d) 0.25 mm Epcos C351 on top and bottom of the PCB, (d) without core sheets.

factor of ten. It is obvious that materials which have a higher permeability improve the coupling factor of the transformer setup. Hence, concerning magnetic coupling the TDK sheet is better than the Epcos sheet. However, as can be seen in Table II for some setups the Epcos material achieves better results since a higher permeability also implies a higher flux density and hence higher eddy current losses.

Because of the large air gap the increase of magnetic coupling due to the magnetic sheets is marginal. However, the emitted field is reduced considerably with magnetic sheets. This can be seen in Fig. 7 where magnetic field plots are shown.

In Fig. 7 (a) and (b) the TDK sheet is placed on top and bottom of the air transformer with a thickness of 2 mm and 0.25 mm, respecively, whereas in (c) and (d) magnetic field plots for Epcos sheets with the same thicknesses are shown. A comparison to the air transformer setup without any core material which is shown in Fig. 7 (e) points out that the shielding effect of the TDK sheet with 2 mm thickness is superior to the other cases. In particular, the setup with 0.25 mm Epcos sheets shows almost no improvement compared to the pure air transformer setup. In general, thicker sheets result in better EMI shielding. The required foil thickness can be obtained by stacking several sheets.

IV. LABORATORY SETUP

To confirm the optimization results a laboratory setup was built. The air transformer setup consists of three layers with each two turns. The track width is 6 mm and the track thickness is 35 μm. No core material is considered. This setup is not meant to be optimal as can be seen in Fig. 6 but is chosen due to its simple manufacturing. The simulation results can be seen in Table III while the setup is shown in Fig. 8. The measurement results for the given converter at an output power of 100 W can be seen in Fig. 9 (b) and a comparison with the simulated waveforms in Fig. 9 (a) shows good consistence. The input current i_p lags the input voltage v_1 so that soft switching is achieved. Furthermore, the sinusoidal approximation on which the loss calculation bases upon is justified. Electrical parameters of the setup are given in Table IV.

A breakdown of losses shows that the suboptimal air transformer causes 83 % of the total losses which can be seen in Fig. 10. The full bridge inverter and the diode rectifier generate 5 % and 12,% of the losses, respectively. Consequently, the optimal air transformer design is crucial for a good overall efficiency. The difference between calculation and measurment can be explaind by the fact that gate drive losses and losses in the auxiliary supply as well as high frequency current losses are not considered in the loss calculation. In Fig. 10 the breakdown of losses is shown for the optimal transformer setup. As can be seen the distribution of losses is more balanced. The losses in the diode rectifier bridge are constant

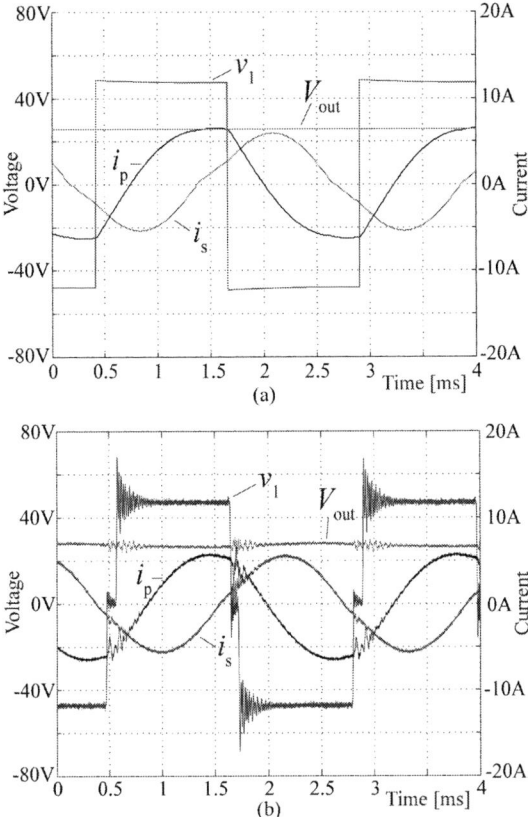

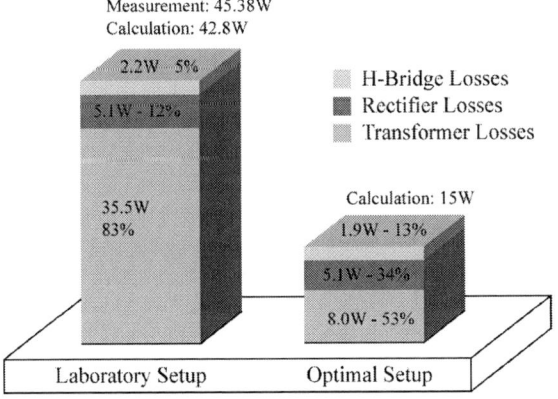

Fig. 9. (a) Circuit Simulation and (b) Measurement Results.

Fig. 10. Breakdown of losses for the laboratory setup and the optimal setup for an output voltage V_{out} of 25 V and an output power P_{out} of 100 W.

paper a distance of 20 mm between primary and secondary is considered which would withstand system voltages of $20 \, kV_{RMS}$ which is shown in [3]. Since the air transformer is responsible for the main part of the losses an optimization is performed to improve the converter efficiency. The optimization is accomplished by field simulations of several air transformer setups followed by a ciruuit calculation in which the compensation capacitors are obtained. After that the efficiency is calculated and the parameters are changed which is the initiation of a new calculation cycle. The optimization results shows that an overall efficiency of over 85 % at rated load can be obtained if a proper air transformer setup is chosen. The optimization method is confirmed by a laboratory setup.

This topology would further allow a gate signal transmission over the air transformer which would combine the isolation of the power supply and signal path. Hence a compact and cheap solution is obtained where the good galvanic isolation would also be applied for the gate signal.

REFERENCES

[1] "Siebel & scholl GmbH." http://www.siebel-scholl.de/index.html.
[2] A. van den Bossche and V. C. Valchev, *Inductors and Transformers for Power Electronics.* CRC, 1 ed., Mar. 2005.
[3] R. Steiner, P. K. Steimer, F. Krismer, and J. W. Kolar, "Contactless energy transmission for an isolated 100 W gate driver supply of a medium voltage converter," *Industrial Electronics, 2009. IECON 2009. 35th Annual Conference of IEEE,* no. 4, 2009.
[4] C. Wang, O. Stielau, and G. Covic, "Design considerations for a contactless electric vehicle battery charger," *Industrial Electronics, IEEE Transactions on,* vol. 52, no. 5, pp. 1308–1314, 2005.
[5] O. Stielau and G. Covic, "Design of loosely coupled inductive power transfer systems," in *Power System Technology, 2000. Proceedings. PowerCon 2000. International Conference on,* vol. 1, pp. 85–90 vol.1, 2000.
[6] X. Liu and S. Hui, "Optimal design of a hybrid winding structure for planar contactless battery charging platform," in *Industry Applications Conference, 2006. 41st IAS Annual Meeting. Conference Record of the 2006 IEEE,* vol. 5, pp. 2568–2575, 2006.
[7] W. Zaengl, S. Yimvuthikul, and G. Friedrich, "The temperature dependence of homogeneous field breakdown in synthetic air," *Electrical Insulation, IEEE Transactions on,* vol. 26, no. 3, pp. 380–390, 1991.
[8] E. Waffenschmidt and T. Staring, "Limitation of inductive power transfer for consumer applications," in *Power Electronics and Applications, 2009. EPE '09. 13th European Conference on,* pp. 1–10, 2009.
[9] R. Erickson and D. Maksimovic, *Fundamentals of Power Electronics.* Springer, 2nd ed., 2001.
[10] "Epcos AG." http://www.epcos.com.
[11] "TDK corporation." http://www.tdk.com.

since the output voltage and the output power is constant for both setups. The losses in the H-bridge decrease slightly since the primary current i_p which is required to transfer the power to the load decreases.

V. CONCLUSION

In this paper an isolated gate drive power supply for medium high voltage switches is presented. The electric isolation is realized by an air transformer whose distance between primary and secondary winding withstand the high voltages. In this

A Digitally Controlled DC-DC Buck Converter Using Frequency Domain ADCs

Hani Ahmad and Bertan Bakkaloglu
Ira A Fulton School of Engineering
Arizona State University
Tempe, AZ 85287-08406, USA
Hani.ahmad@asu.edu, Bertan.bakkaloglu@asu.edu

Abstract- **The design of a 0.18-μm CMOS digital control architecture for a buck converter is presented. Several features are implemented. These include: 1) Frequency-domain digitization technique based on first-order non-feedback Sigma-Delta frequency Discriminators (NF-SDFD); 2) a robust arrangement for the feedback ADCs to guard against false output voltage variation due to temperature and process variation; 3) A new improved hybrid Digital Pulse Width Modulator (DPWM) architecture. The proposed system has additional attractive futures such simplicity, scalability, low power, close to all digital implementation in addition to its capability of satisfying tight regulation requirements for wide range of applications. An 8-bit ADC resolution is achieved with less than 110 μA current consumption. A 9-bit DPWM consumes around 370 μA . A 2% output voltage regulation accuracy is achieved with less than 10 mVpp ripple.**

I. INTRODUCTION

A typical digital PWM DC-DC controller is shown in Fig. 1. The main building blocks of such controller are the ADC, compensator and digital PWM generator (DPWM) [1-3]. The ADC and DPWM blocks are typically the most challenging to design from the standpoint of power consumption, complexity and area. In this work, we present new frequency-domain digitization technique based on NF-SDFD [4-6]. Dual ADCs in the feedback loop is implemented to guard against false regulated voltage variation due to temperature, process or external effects. This digitization architecture is simple, scalable and can be implemented in standard digital CMOS

process. A high-frequency, high-resolution DPWM circuit is one of the critical blocks for successful practical realization of digital control for switching power converters A New hybrid DPWM architecture; DLL followed by a counter is presented. A charge-pump based DLL driving a counter is described to generate the required duty cycle. This solution combines the traditional advantages of the hybrid DPWM architecture [7] with guaranteed linearity and monotonicity and lower power consumption by using current-starved delay elements in the DLL.

II. PROPOSED ARCHITECTURE

The block diagram for the proposed architecture is shown in Fig. 2. The digitized scaled output voltage is compared to the digitized reference and the difference between the two (error signal) is then decimated and supplied to the compensator (PID). The PID calculates the required duty cycle to set the output voltage at a desired value. Finally, the DPWM converts this duty cycle value into a driver signal to drive the PFET and NFET via a gate driver. At regulation, the error signal should be within the zero bin error of the ADC. In this ADC architecture, when output voltage and reference voltage are equal, the VCOs and the frequency discriminators generate similar output and hence the difference is equal to zero.

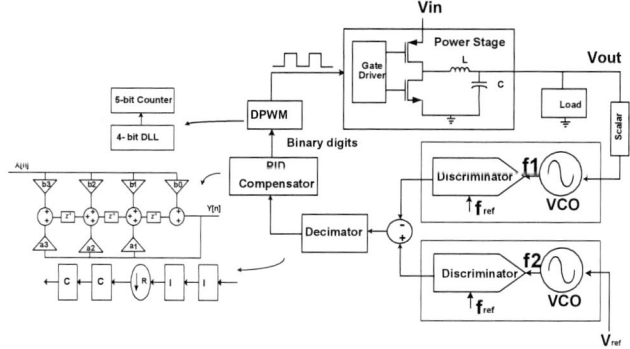

Figure 2 Proposed digitally controlled DCDC converter architecture

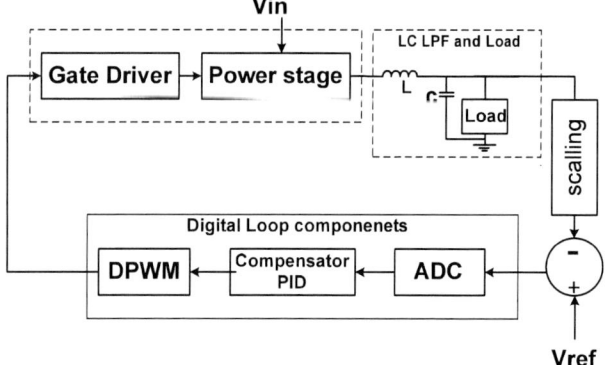

Figure 1 A typical digital PWM DC-DC controller

978-1-4244-4782-4/10 $26.00 © 2010 IEEE

The first-order NF-SDFD is shown in Fig. 3. It digitizes

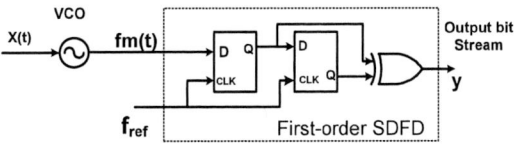

Figure 3 A first-order non-feedbacks SDFD

instantaneous frequency of a modulated carrier similar to an ADC that digitizes the amplitude of input signals [4-6]. This non-feedback SDFD is equivalent to the traditional $\Sigma-\Delta$ modulator in the sense that it performs the same three main functions on a signal similar to the traditional modulator. These functions are integration, quantization and differentiation. It accomplishes the integration via the FM modulator; the quantization via the detection of the FM phase zero-crossings position utilizing D-type flip flops (DFF) and the differentiation via the digital differentiator gate (XOR).

Fig. 4 depicts the block diagram of the proposed DPWM. At the beginning of every switching frame, the DPWM output is set. The selected delay of the DLL is used as input to the clock signal of the counter. Once the MSB bits match the count, the DPWM outputs get reset. This new architecture produces a duty cycle with high linearity and guaranteed monotonicity. With the DLL (Fig. 5) using current-starved inverter based delay elements, it aids lower power implementation. The DLL is designed to generate 16 taps with 3.9ns (1/16Mhz/16) phase delay between consecutive taps.

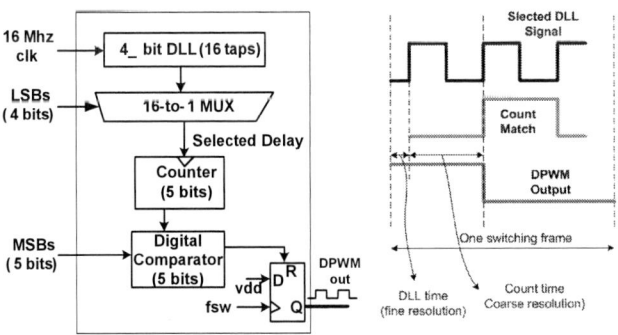

Figure 4 proposed DPWM architecture

The Gate drive shown in Fig. 6 is based on [8]. It has built in dead time to avoid shoot through current. The transistors in the gate drivers are designed to have enough strength to drive the power FETs and generate appropriate dead time to prevent shoot through current, and at the same time, minimize power loss during switching. The PFET and NFET are sized up based on 100m ohms on-resistance as a compromise for efficiency and silicon area.

The PID is designed to achieve 60 degrees of phase margin and maintain stability with wide input voltage range and load

variation. A typical design procedure is followed in constructing the PID transfer function and deriving the control law. For the decimator, A two-stage CIC structure is used with 16 MHz sampling frequency.

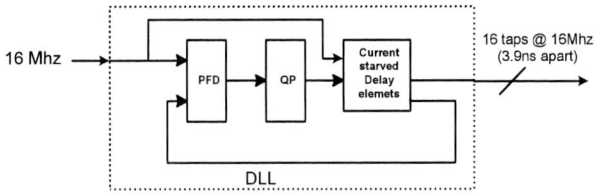

Figure 5 DLL architecture

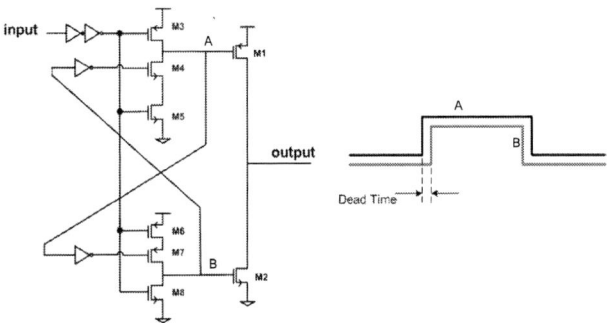

Figure 6 Gate driver with built-in dead time

III. SCHEMATICS AND SIMULATION RESULTS

The design parameters used in this implementation are listed in table 1 below.

Table I DESIGN PARAMETERS

Parameter	Value	Unit
Switching frequency	500	KHz
Cross-over frequency	50	KHz
f_{ref} (Sampling Frequency)	16	MHz
DPWM frequency	16	MHz
L	18.8	µH
C	22	µF
R in series with C	0.1	Ω
R in series with L	0.1	Ω
V_{out}	$1.8 \pm 2\%$	V
V_{in}	3.3	V
Imax	1	A
Load Transient	0.2-1	A

The schematic for the ADC is shown in Fig. 7. It is composed of a VCO followed by first-order NF-SDFD. The input control voltage x(t) at the input of the VCO represents the converter output or reference voltage. The VCO is designed to generate 5 MHz voltage at the set voltage level of 1.8 V. Its transfer function has a slope of 2 MHz/V. The transfer function of the VCO is shown in Fig. 8.

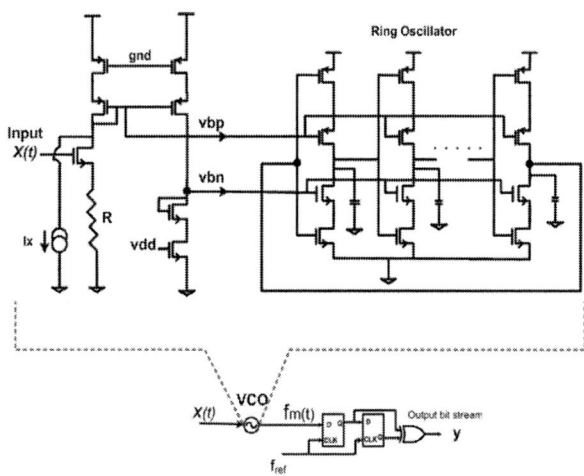

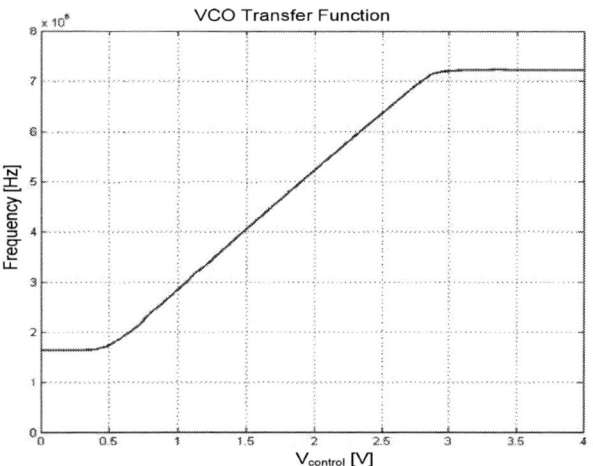

Figure 7 ADC schematic

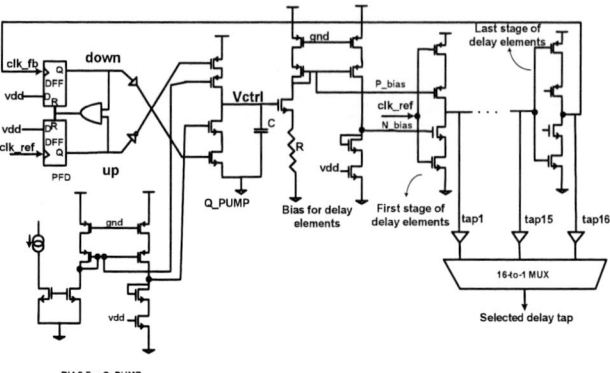

Figure 9 DLL schematic

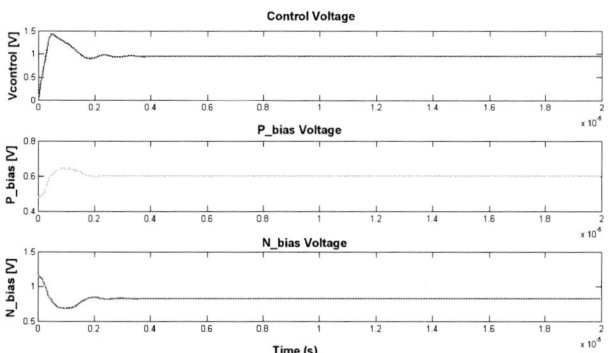

Figure 10 DLL in a lock state

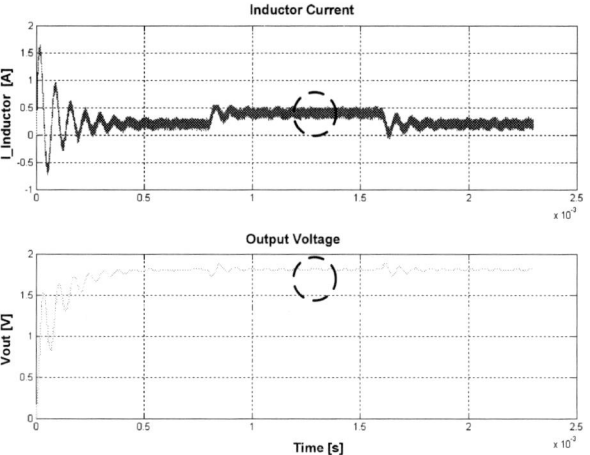

Figure 11 Inductor current and output voltage with 200 mA load transient

Figure 8 VCO transfer function

The DPWM is composed of 4-bit DLL followed by 5-bit counter. The DLL is used for the fine resolution and the counter is used for the coarse resolution. The schematic diagram for the DLL is shown in Fig. 9.

The DLL is considered in lock state when the reference clock and the feedback clock from tap 16 have close to zero phase shift, at which, the control voltage and the voltages to the delay elements remain constant as seen in Fig. 10.

The inductor current and the bias output voltage are shown in Fig. 11. An expanded view of the inductor current and the output voltage is shown in Fig. 12.

978-1-4244-4782-4/10 $26.00 © 2010 IEEE   1873

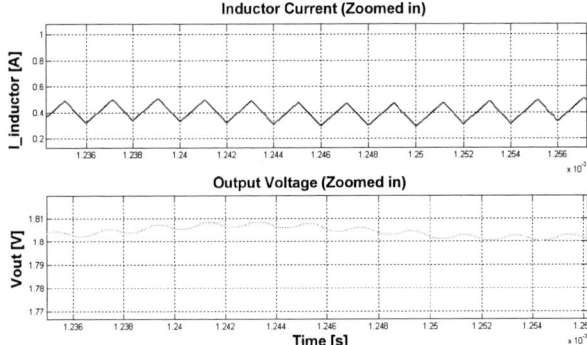

Figure 12 Expanded views of Inductor current and output voltage

IV. CONCLUSIONS

Frequency-domain DC-DC digital control architecture with new digitization technique and new hybrid DPWM architecture is presented. The proposed digitization technique is simple, scalable and can be implemented in standard digital CMOS process. The new hybrid DPWM guarantees high linearity and monotonicity of the duty cycle in addition to its low power implementation. An output voltage regulation with 2% accuracy has been achieved with less than 10 mVpp ripple. The 8-bit ADC resolution is achieved with less than 110 μA current consumption. The 9-bit DPWM consumes around 370 μA .

REFERENCES

[1] Prodic, D. Maksimovic and W. Erickson, "Design and Implementation of a Digital PWM Controller for a High-Frequency Switching DC-DC Power Converter", IECON'01: The 27th Annual Conference of the *IEEE Industrial Electronics Society*, vol. 2, pp. 893-898, Nov 2001.

[2] R .W. Erickson and D. Maksimovic, Fundamentals of Power Electronics, Second Edition, Kluwer Academic Publishers, 2000.

[3] Syed, E. Ahmad and D. Maksimovic, "Digital Pulse Modulator Architectures", 35th Annual *IEEE PESC*, Aashen, Germany, vol. 6, pp. 4689-4695, June 2004.

[4] M. Hovin, A. Olsen, T.S Lande and C. Toumazou, "Delta-Sigma Modulators using Frequency-Modulated Intermediate Values" *IEEE Journal of Solid-Sate Circuits*, vol. 32, no. 1, pp. 13-22, Jan 1997.

[5] M. Hovin, T. Saether, "A Narrow-band Delta-Sigma Frequency-to-Digital Converter" *IEEE proc. ISCAS*, vol. 1, pp.77-80, Jan 1997.

[6] D.T. Wisland, M E. Hovin, T.S. Lande, , "A Novel Multi–bit Parallel ΔΣ FM–to–digital Converter with 24–bit resolution", Proceedings of the 28th *European Solid-State Circuits Conference*, pp. 687-690, 2002

[7] Prodic, D. Maksimovic and W. Erickson, "Design of a Digital PID Regulator Based on Look-Up Tables for Control of High-Frequency DC-DC Converters ", *IEEE COMPEL*, pp. 18 – 22, June 2002.

[8] Changsik, "A CMOS Buffer Without Short-Circuit", IEEE TCAS—II, VOL. 47, NO. 9, September 2000.

978-1-4244-4782-4/10 $26.00 © 2010 IEEE

Low-Dropout (LDO) Regulator Output Impedance Analysis and Transient Performance Enhancement Circuit

Sungkeun Lim, and Alex Q. Huang
Semiconductor Power Electronics Center (SPEC)
North Carolina State University
Raleigh, NC 27606 USA

Abstract— **This paper presents a low-dropout regulator with a transient performance enhancement circuit. The transient performance enhancement circuit improves the transient response time by sinking a remaining current in a power delivery path. Due to the limited control bandwidth, traditional LDO could not respond rapidly to the load transients. As a result, a large output voltage spike can be occurred and the output voltage settling time is long during the load transients. In this paper, the stability conditions and the output impedance of LDO are discussed, and the output voltage spike and current distributions in the power delivery path are analyzed. The theoretical analysis will be confirmed by the cadence simulation results.**

I. INTRODUCTION

Low-dropout regulator is widely used to generate a stable and accurate supply voltage in portable device. The stability of the LDO regulator is determined by the output capacitance and ESR value, because there is no compensation network [1] [2]. A large value of the output capacitance and equivalent series resistance (ESR) is required to make a stable LDO regulator system. However the large capacitance and ESR generates one pole at low frequency side and limits the bandwidth of the LDO system. The output voltage spike is analyzed based on the output impedance of the LDO regulator in frequency domain [3] [4]. If the output impedance of the LDO regulator is constant from DC to high frequency, there is no output voltage spike during the load current transient. However, the output impedance is not constant in whole frequency range, because of the large capacitance and ESR value. As a result, the response time of the LDO regulator is long and a large output voltage spike occurs during the load current transients. To improve the load transient performance, the transient performance enhancement circuit is proposed. This circuit sinks a remaining current in the power delivery path to reduce the settling time during the load current transients. The theoretical analysis will be confirmed by the cadence simulation results.

II. STABILITY AND OUTPUT VOLTAGE SPIKE ANALYSIS BASED ON THE OUTPUT IMPEDANCE

A feedback loop is used in the LDO regulators to maintain a constant output voltage. To have a stable feedback loop, the phase margin of the feedback loop should be larger than $0°$ at the unity gain frequency. Figure 1 shows the simple LDO structure and the open loop gain [1].

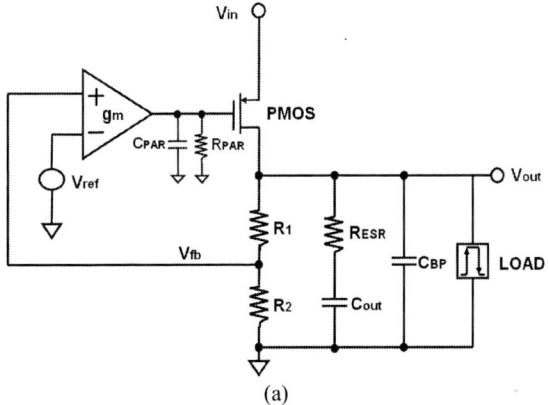

(a)

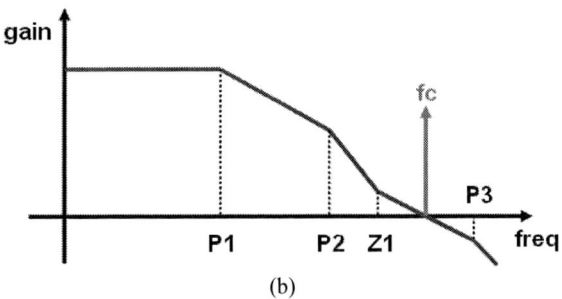

(b)

Fig. 1 LDO structure (a) and open loop gain (b)

$$G_{OL} = G_{EA} * G_{FB} * G_{PMOS} * \frac{1 + \dfrac{s}{2\pi f_{Z1}}}{(1 + \dfrac{s}{2\pi f\, P1}) * (1 + \dfrac{s}{2\pi f\, P2}) * (1 + \dfrac{s}{2\pi f\, P3})} \quad (1)$$

$$P1 = \frac{1}{2\pi(R_{O\_PMOS} + R_{ESR})C_{OUT}} \quad (2)$$

$$P2 = \frac{1}{2\pi R_{ESR} C_{BP}} \quad (3)$$

$$P3 = \frac{1}{2\pi R_{PAR} C_{PAR}} \quad (4)$$

$$Z1 = \frac{1}{2\pi R_{ESR} C_{OUT}} \quad (5)$$

978-1-4244-4782-4/10 $26.00 © 2010 IEEE

In the feedback loop, there are three poles and one zero, and the position of these poles and zero is determined by output capacitance (Cout), ESR (RESR), bypass capacitance (CBP), parasitic capacitance (CPAR), parasitic resistance (RPAR), and turn-on resistor of PMOS (RO_PMOS). Equation (1) is the open loop gain of LDO, and the equations (2~5) express how the position of the poles and zero are determined. Two poles (P1 and P2) and one zero (Z1) should be located within the unity gain frequency to stabilize the LDO system. Especially, the dominant pole (P1) and Z1, which are determined by the Cout and RESR, should be low frequency side. As a result, a large value of output capacitance and ESR are required to meet the stability condition. The values of the Cout and RESR should be larger than specific values to make a stable LDO regulator. When the Cout value or the RESR value is too small, the LDO regulator will be unstable because Z1 is beyond the unity gain frequency and two poles are within the unity gain frequency. Figure 2 shows the unstable cases.

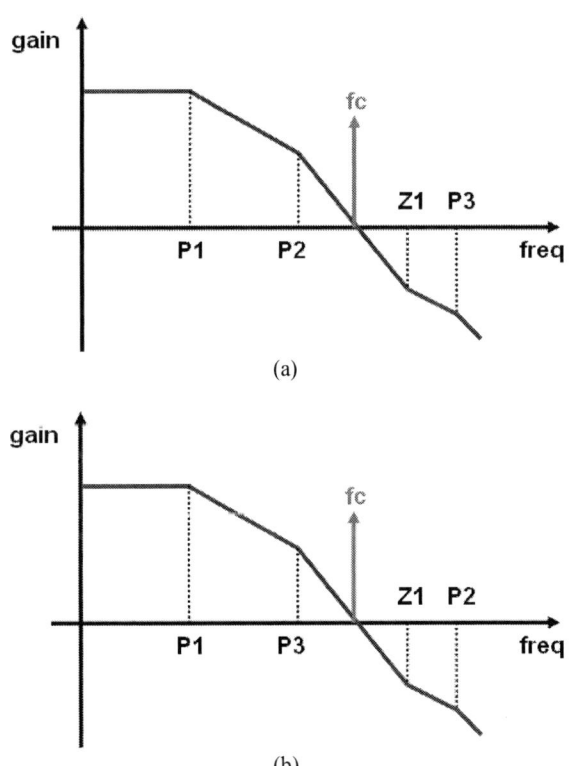

(a)

(b)

Fig. 2 Unstable LDO regulator with (a) small Cout and with (b) small RESR

To improve the accuracy of the output voltage regulation, a large pass element is used in LDO system which has a large parasitic capacitance at the gate node. However, the large Cout, CPAR, and RESR limit the loop gain bandwidth, and the large value of RESR causes a large output voltage spike and a slow output voltage settling time during the load current transient.

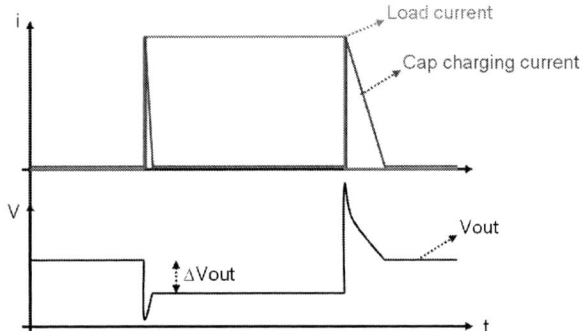

Fig. 3 current distributions in the power delivery path and output voltage spike during load transients

Figure 3 shows the current distributions in the power delivery path and the output voltage spike. The settling time of the output voltage spike during load current step down transient is longer than load current step up transient, because the PMOS turning off time is longer than the turning on time.

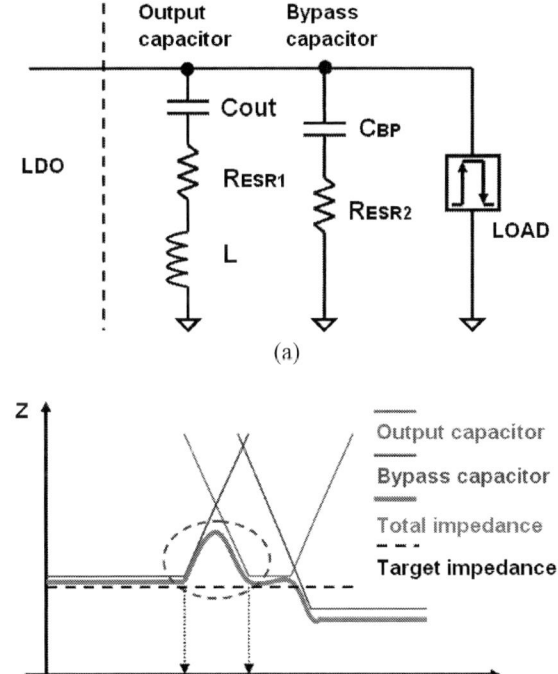

(a)

Fig. 4 LDO regulator (a) output node and output impedance in frequency domain

The impedance of the LDO regulator and the output capacitors can be analyzed in the frequency domain and figure 4 shows the total impedance of the LDO regulator system. If the impedance at the output node is constant from DC to high frequency, there is no output voltage spike during load current transient [3] [5]. The impedance from DC to the unity gain bandwidth (fc) of LDO is really low, and the impedance from the ESR frequency (fesr) to high frequency is determined by the Cout, CBP, L, RESR1, and RESR2 value. If the fc and fesr can be matched at same frequency point, the constant output

978-1-4244-4782-4/10 $26.00 © 2010 IEEE   1876

impedance in whole frequency can be achieved. To match the fc and the fesr, a large value of Cout and RESR1 is required. However, the Cout will increase the cost of the LDO regulator and the target impedance can not be achieved with the larger value of RESR1. As a result, the impedance between fc and fesr is much larger than the target impedance and a large output voltage spike is occurred during the load current transients.

III. TRANSIENT PERFORMANCE ENHANCEMENT CIRCUIT

To improve the transient response, several techniques are implemented [6~8]. In this paper, a transient performance enhancement circuit is proposed to improve the transient response. Additional NMOS is added at the output node to sink remaining current which charges the output capacitor. Figure 5 shows the proposed transient performance enhancement circuit. During the load current step down transient, the proposed circuit sinks the remaining current and reduces the output capacitor changing current. Finally, the output voltage settling time is reduced. The response time of the transient performance enhancement circuit is much faster than the response time of the LDO regulator, because the NOMS size is much smaller than the pass element PSOS size.

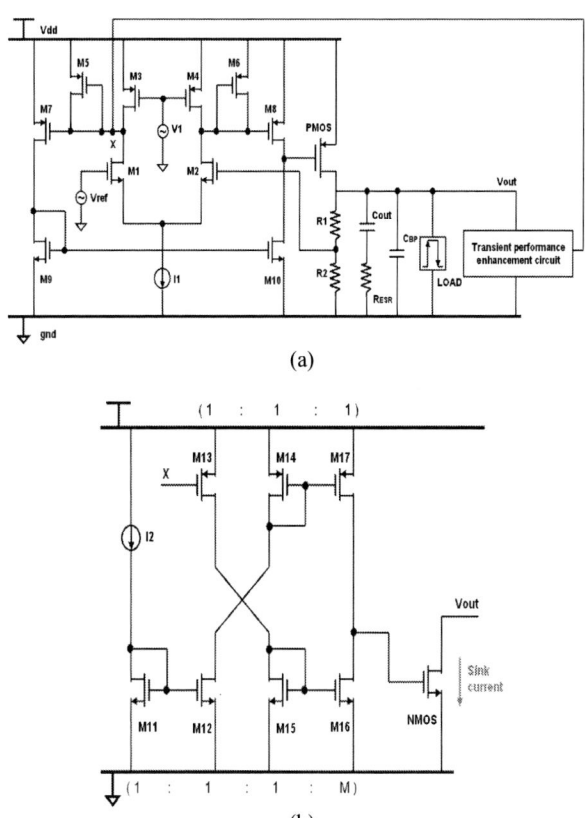

(a)

(b)

Fig. 5 LDO regulator (a) with proposed transient performance enhancement circuit (b)

IV. CADENCE SIMULATION RESULTS

A 5V input and 1.5V output LDO regulator with a 100mA load current transient is used in this cadence simulation. A 4.7uF output capacitor with 300mΩ ESR and a 0.1uF bypass capacitor are added at the output node. To improve the accuracy of the output voltage regulation, adaptive voltage position (AVP) is applied [9]. The AVP reduced 2.9mV DC different voltage for different load conditions and the output voltage comparison is shown in figure 6 (a).

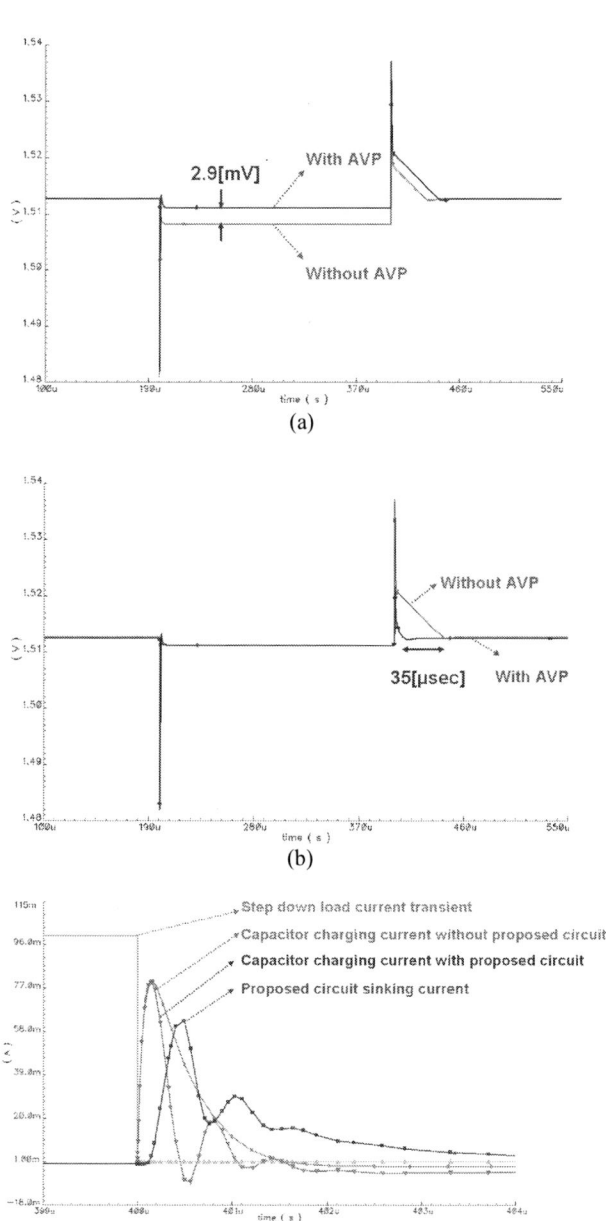

(a)

(b)

(c)

Fig. 6 Output voltage comparison with/without AVP (a), output voltage settling time comparison (b), and current distributions in the power delivery path during load current step down transient (c)

Figure 6 (b) shows the settling time comparison with/without the transient performance enhancement circuit and figure 6 (c) shows the current distributions in the power delivery path. During the load current step down transient, the settling time of the LDO regulator is reduced from 43μsec to 8μsec after applying the transient performance enhancement circuit.

V. CONCLUTIONS

This paper proposed a transient performance enhancement circuit to improve the output voltage settling time during the load current transients. The stability condition of the LDO regulator is discussed. The output impedance is analyzed in the frequency domain, and the output voltage spike and the current distributions in the power delivery path are analyzed. The results from cadence simulations for the transient performance enhancement circuit showed a good performance.

VI. ACKNOWLEDGEMENT

This work is supported by SPEC Power Management Consortium.

REFERENCES

[1] Texas Instruments, "Technical Review of Low-Dropout Linear Regulator Operation and Performance," application repost, 1999.

[2] G. A. Rincon-Mora, "Current Efficient, Low Voltage, Low Drop-Out Regulators," Ph.D. Dissertation, Dept. Elect. Eng., Georgia Institute of Technology., Atlanta, GA, 1996.

[3] Sungkeun Lim, "Design of a Transient Voltage Clamp (TVC) Based on Output Impedance Analysis," in *Proc. IEEE APEC,* 2007, pp.1636-1638.

[4] K. Yao, "High-Frequency and High-Performance VRM Design for the Next Generations of Processors," Ph.D. Dissertation, Dept. Elect. Eng., Virginia Polytechnic Institute and State Univ., Blacksburg, VA, 2004

[5] Intel's VRM 11.0 specification.

[6] H. Lee, P. K. T. Mok, and K. N. Leung "Design of low power analog drivers based on slew-rate enhancement circuit," *IEEE Trans. Circuits Syst. II, Exp. Briefs,* vol. 52, no. 9, pp. 563-567, Sep. 2005.

[7] G. A. Rincon-Mora and P. E. Allen, "A low-voltage, low quiescent current, low drop-out regulator," *IEEE J. Solid-State Circuits,* vol. 33, no. 1, pp. 36-44, Jan. 1998.

[8] R. K. Dokania and G. A. Rincon-Mora, "Cancellation of load regulation in Low drop-out regulators," *Inst. Electr. Eng. Electron. Lett.,* vol. 38, no. 22, pp. 1300-1302, Oct. 2002.

[9] K. Yao, K. Lee, M. Mu, and F. C. Lee, "Optimal design of the active droop control method for the transient response," in *Proc. IEEE APEC,* 2003, pp.718-723.

A Design for Small Time-Delay Control Circuit for DPWM- POL

Yoichi Ishizuka, Yusuke Yamada, Fumitoshi Hirose, Mariko Nishi and Hirofumi Matsuo
Faculty of Engineering
Nagasaki University
Nagasaki, Japan
isy2@nagasaki-u.ac.jp

Abstract— **Digital-controlled switching mode power supplies (SMPS) are popular in the fields of relatively large power supply system. The digital controller is consisted of DSPs or microprocessors performing with software. By contrast, such as on-board SMPSs, adapting digital controller isn't still popular because of its cost or response characteristics even if the controllers are constructed by FPGA or custom LSI which is hardware-logic base. This paper will discuss about the proposed hardware logic type digital controller for on-board SMPS, which is very simple and unique design. And, accuracy improved output voltage detection method for the proposed digital controller is proposed. The some experiments are done. And the results show the validity of the proposed circuits.**

I. INTRODUCTION

Digital electronic products have been spreading quickly by the advancement of the integrations technologies. ICs, DSPs and FPGAs requires a high performance and a high speed due to the trend. Along with the situations, their power consumption are increasing. To suppress the power consumption, the power supply voltage is getting lower toward to sub 1V. Because of the severe voltage margin by the lower power supply voltages, special SMPS, point of load (POL) is disposed very near to the load. The requirements of POL are relative low output voltage with large output current, high load change response, high-efficiency, low cost and etc... Also, the control circuit of POL is required to be high accurate, high speed, adaptive and low cost. We had proposed hardware-logic based digital PWM control circuit is effective to such requirements [1-3]. In this paper, the proposed DPWM control method's principles of operation and circuit configurations are described firstly. At the next, a piecewise linear output voltage detection method is proposed and discussed which can improve the output voltage detection accuracy. Finally, the proposed technique is confirmed with some experiments.

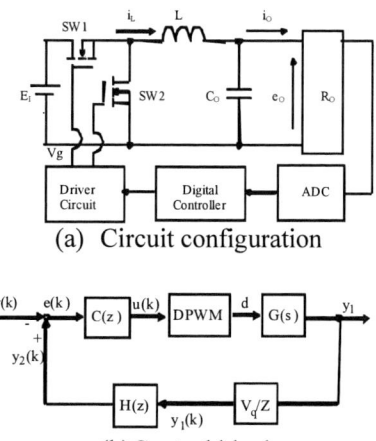

(a) Circuit configuration

(b)Control block
Fig. 1 General DPWM-POL

II. DPWM CONTROL METHOD FOR POL (DPWM-POL)

The circuit configuration of general DPWM controller for POL

The circuit configuration of general DPWM controller for POL is shown in Fig. 1[4-11]. This topology has two major time-delay problems. First, time-delay occurs at A/D converter with the conversion-delay. And, the calculation time of digital controller is another problem. Both of the time-delay directly effects on the response speed of the control circuit and influences stability of the control.

A. The circuit configuration of the proposed DPWM-POL

Figure 2 shows the main and the control circuit configuration of the proposed DPWM-POL in this research. Main POL circuit is a quite ordinary non-isolated buck converter. The control circuit is composed with D/A converter, analog comparator, digital controller and drive circuit.

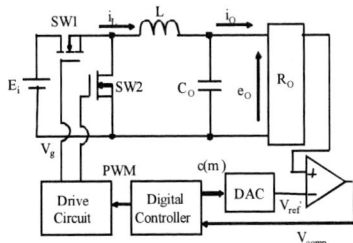

Fig. 2 Circuit configuration of the proposed DPWM-POL

B. The control circuit configuration of the proposed DPWM-POL

Figure 3 shows the control circuit configuration of the proposed DPWM-POL. All blocks are synchronized with the system clock f_{CLK}. Memory1 can store waveform values not only triangle or saw tooth but any waveforms. The output voltage e_o of POL is compared with the converted voltage of memory1's output in the ATC block, successively. The comparator's output is read out to the latch signal to the latch register at the timing that of e_o was sensed. And this latch signal is transferred to DPWM output block. And this latch signal is transfer to DPWM output block.

Also, the look-up table method is used for the duty ratio calculation with memory2, 3 and 4. Especially, the duty ratio information which is pre-calculated is stored corresponding to e_o, respectively, in memory2. The duty ratio information is read out from memory2 according to f_{CLK}. At the timing of e_o sensed, one of the duty ratio information is chosen in DFF_4. And the information becomes $u(k)$, where k is the switching term. At the digital comparator, $u(k)$ is converted to real-time analog PWM waveforms. These precise operation technique is described in [2,3]. Fig. 4 shows the waveform of V'_{ref} and V_{comp}.

As a result, comparing with a general digital control method, this proposed control technique doesn't need to calculate the duty ratio information in every switching term, and also the conversion time-delay at A/D converter doesn't exist. Therefore, time-delay drastically decreased to just the sum of calling and the loading time of the-memory2.

C. Piecewise Linear Output Voltage Detection Method (PLOVDM)

This controller can add the improvement in each block respectively. In this paper, the improvement of the output voltage detection accuracy in the ATC block without any cost-up is aimed. D/A converter used in the ATC block assumes the ladder type, and the output voltage V'_{ref} is provided between two potential of $V^{+}_{ref}=V_{ref}+\alpha$ and V_{ref}.

$$V'_{ref} = \frac{c(m)}{2^n}(V^{+}_{ref} - V^{-}_{ref}) + V^{-}_{ref} \qquad (2.1)$$

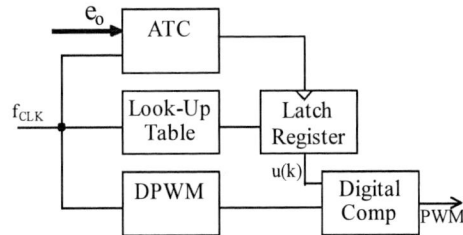

(a) Proposed control circuit blocks

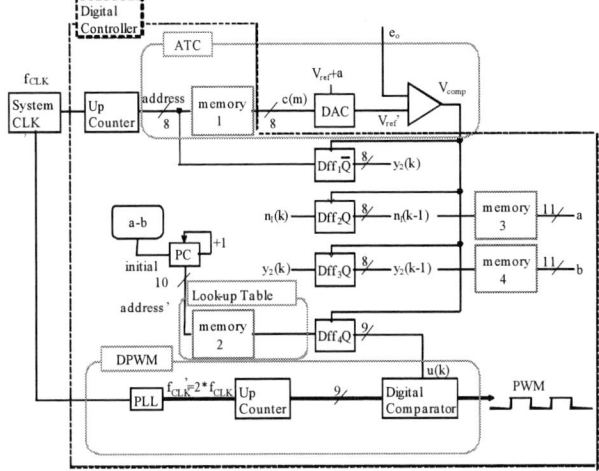

(b) Design descriptions of the control circuit blocks
Fig. 3 The system configuration of the part of the proposed control circuit

where $c(m)$ is waveform digital data prestored in memory1, m is clock timing and n is a number of bits. In this time, almost $n+1$ bits accuracy can be achieved by setting as $V_{ref}=V_{ref}/2$ in the n bits system. The detection time will be delay and it will be the trade-off between the accuracy and the time-delay. Therefore, another compensation technique is proposed to solve the trade-off problem. Thanks to the compensation technique, the trade off problem was solved. Precise descriptions will be revealed in proceeding.

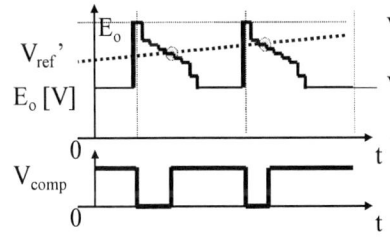

Fig. 4 The waveform of V'_{ref} and V_{comp}

III. EXPERIMENTS

We have done some experiments with the experiment parameters shown in Table 1. The control parameters can be easily modified by rewriting information in memory2. Fig. 5 shows the experimental waveforms. Fig. 6 shows I_O-E_O characteristics. It shows the deviation becomes within 5% in the range of the load change when the proportion gain K_P was larger than 3. Fig. 7 shows E_I-E_O characteristics, where K_P is over 5. The deviation becomes within 10%. Max power efficiency is about 91% shown in Fig. 8. Fig. 9 and Fig. 10 shows dynamic characteristic, where the load change between 0.5A and 5A with 10A/μs slew rate condition, respectively. The mixed-signal oscilloscope Textronix MSO4034 is used to measure analog and digital signal, simultaneously. Blue and Red line shows the output voltage and the output current, respectively. The 9 bits pulse waveforms shown at the bottom of Fig.9 are calculated

DPWM of FPGA. Fig. 9 shows the sudden load current increasing result. From this result, after the 1μs voltage drop, the output voltage immediately recovers to the reference voltage. Fig. 10 shows the sudden load current decreasing result. From this result, after the 1μs voltage rising, the output voltage immediately recovers to the reference voltage.

IV. SUMMARY

In this paper, the piecewise linear output voltage detection method for the proposed DPWM-POL is proposed. From the experiment results, it was confirmed that a comparatively steady control was possible by improving the detection accuracy not to accompany the time-delay.

Table. 1 Experiment parameters

Input voltage E_I	3-8V
Output reference V_{ref}	1.5V
Output current I_O	0-6A
Switching frequency f_S	130kHz
Choke inductor L	10μH
Output capacitor C_O	470μF
Proportion gain K_P	1,3,5,7,9
$V_{ref}+\alpha$	1.7V
V_{ref}^-	0.75V
system CLK f_{CLK}	33.3MHz

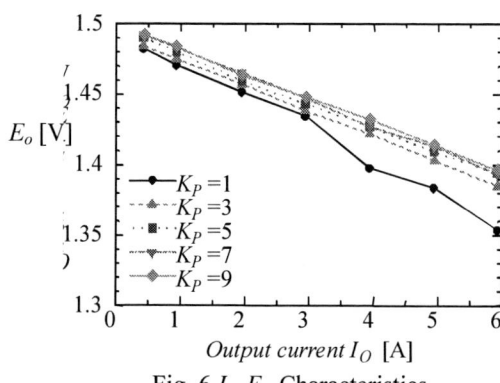

Fig. 6 I_O-E_O Characteristics

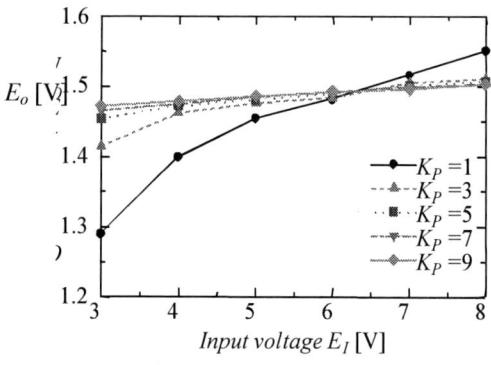

Fig. 7 E_I-E_O Characteristics

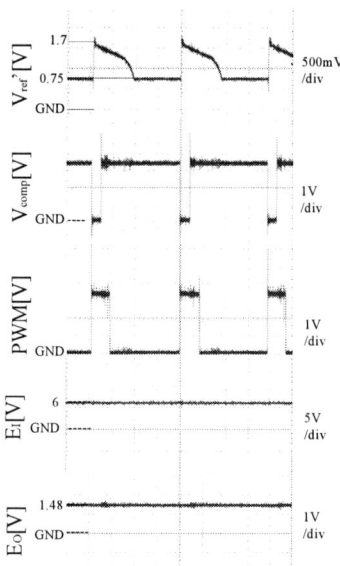

Fig. 5 Experimental waveforms

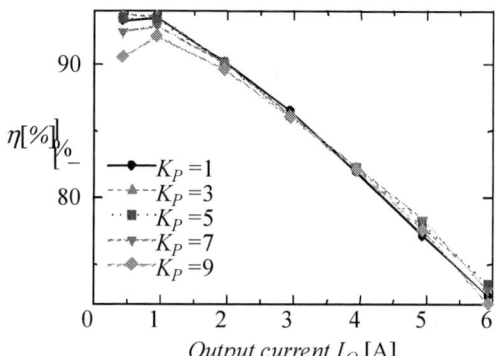

Fig. 8 I_O-η Characteristics

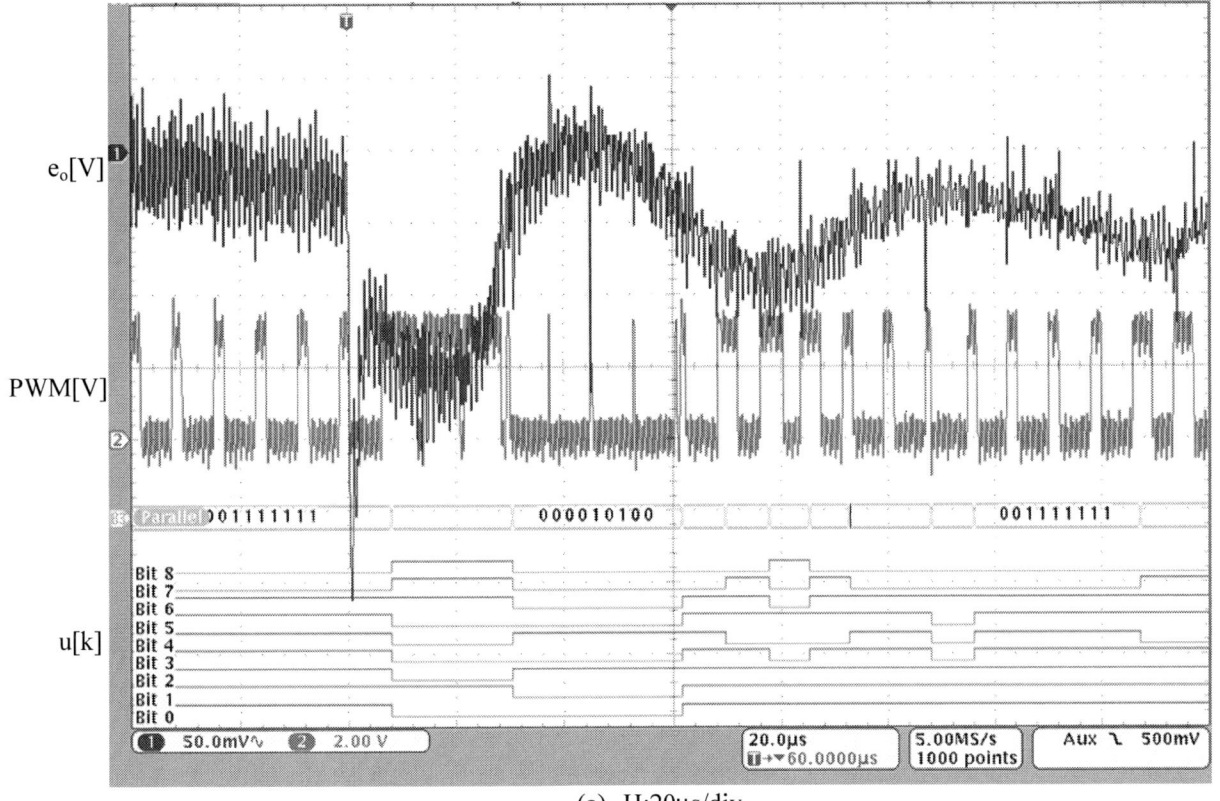

(a) H:20μs/div.

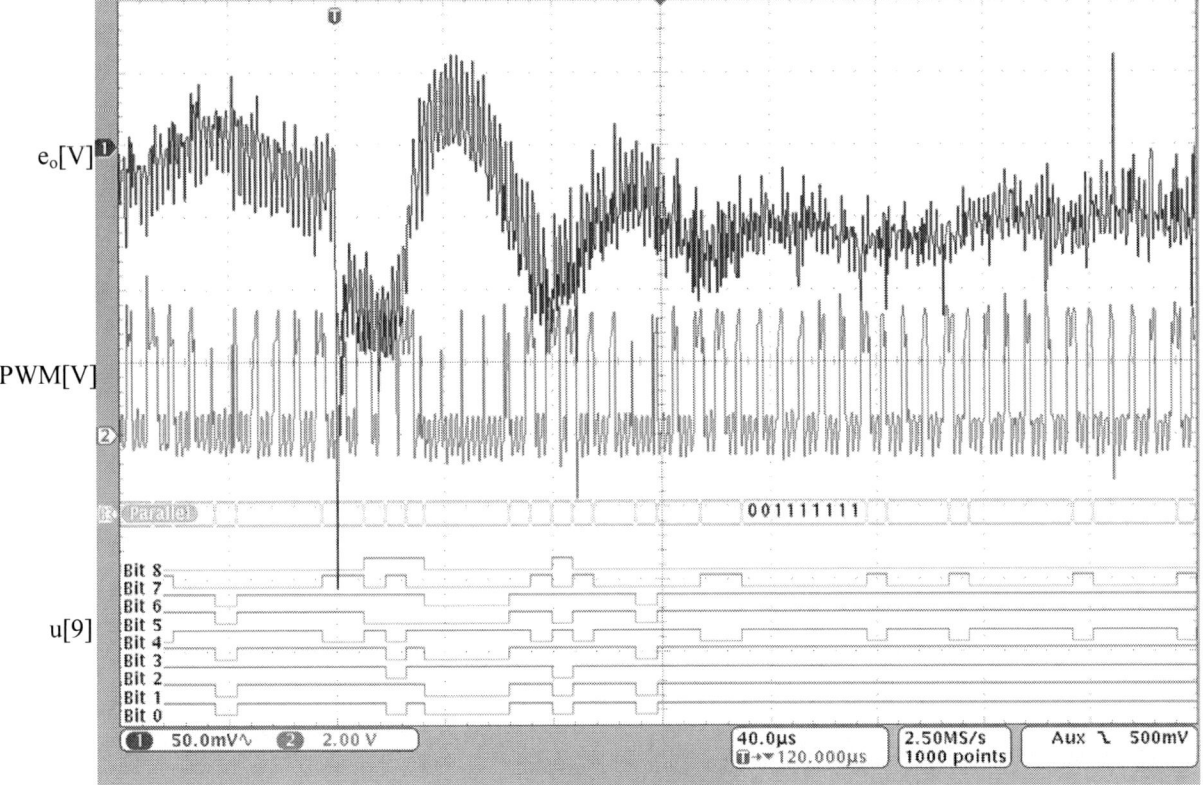

(b) H:40μs/div.

Fig. 9 Dynamic characteristics (from 0.5A to 5A)

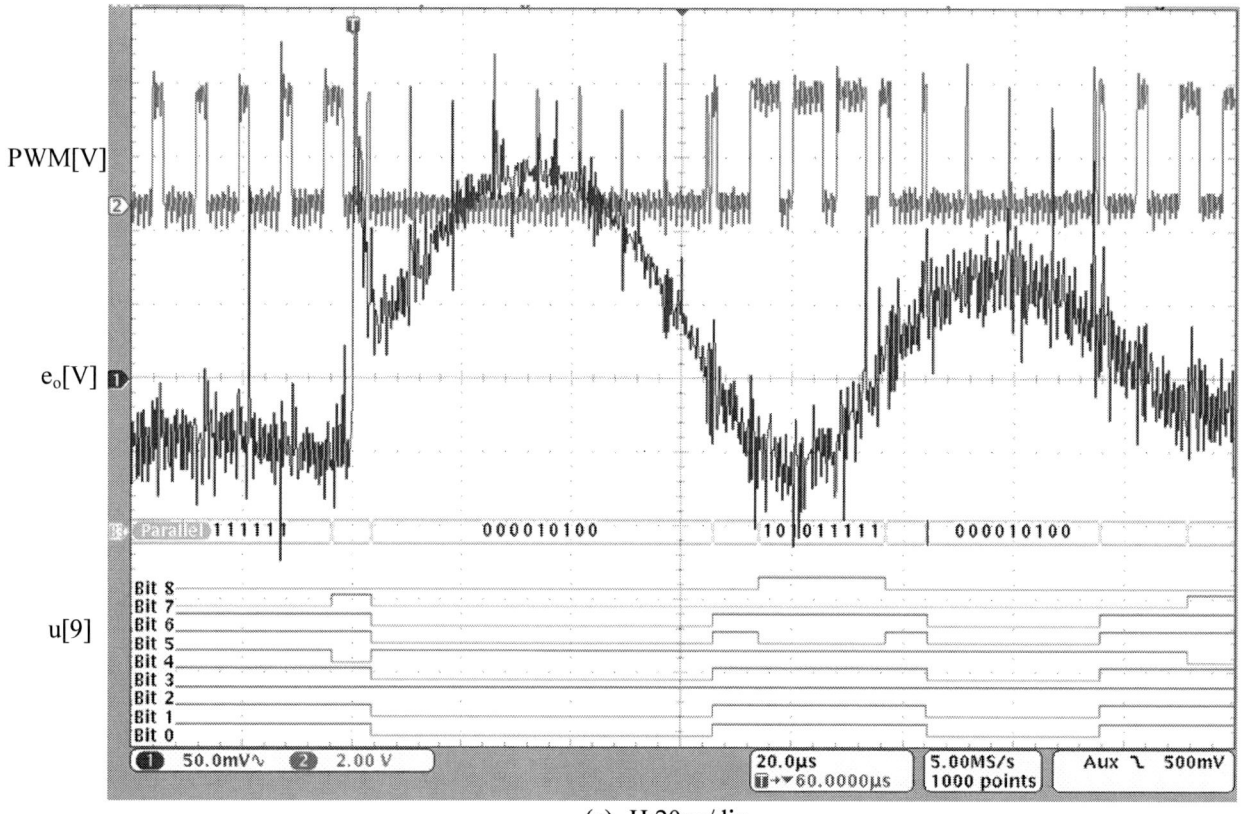

(a) H:20μs/div.

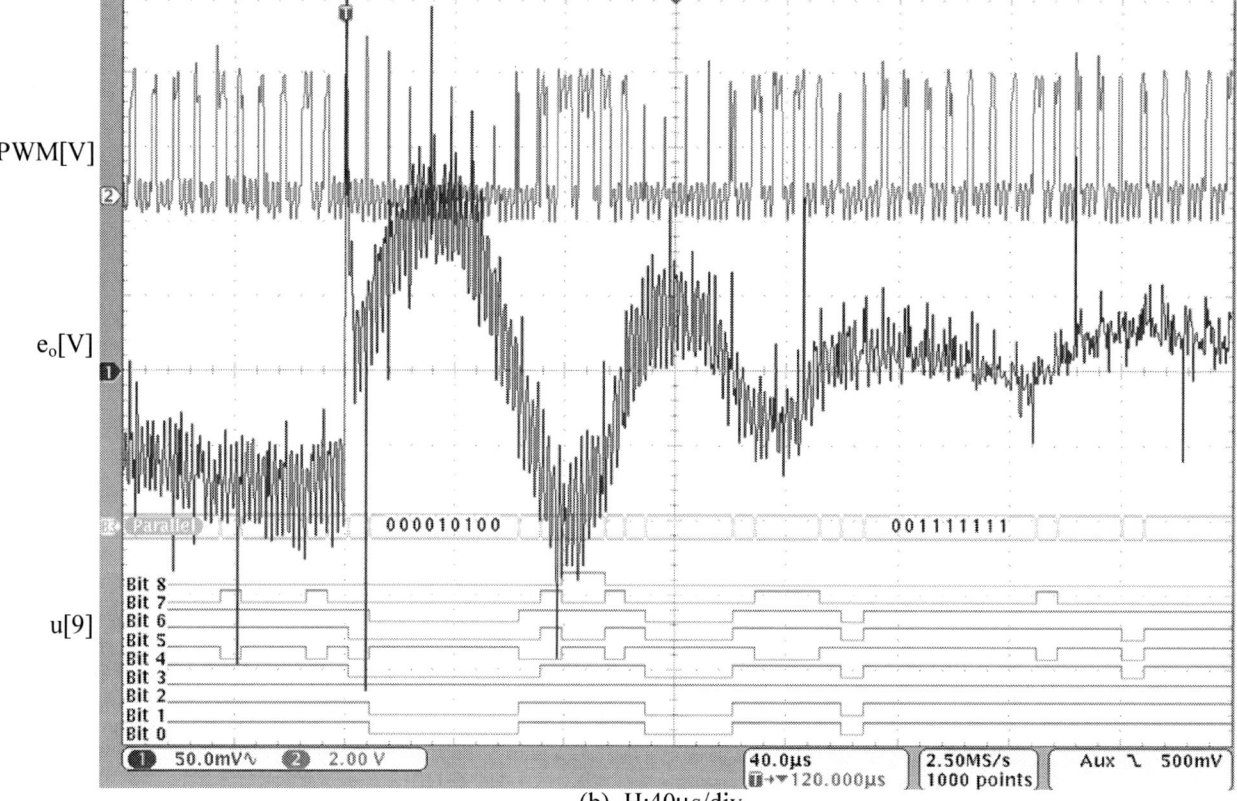

(b) H:40μs/div.

Fig.10 Dynamic characteristics (from 5A to 0.5A)

V. REFERENCES

[1] Akira Ichinose, Yoichi Ishizuka, Hirofumi Matsuo, "A Fast Response DC-DC Converter with Digital Pulse Width", IEICE vol. 105, no. 538, EE2005-58, pp. 67-71, January 2006.

[2] Yoichi Ishizuka, Masao Ueno, Ichiro Nishikawa, Akira Ichinose and Hirofumi Matsuo, "A Low-Delay Digital PWM Control Circuit for DC-DC Converters, "22nd Annual IEEE Applied Power Electronics Conference and Exposition, Anaheim, CA, USA, 15.7, pp.579-586 , 2007. 2

[3] Mariko Nishi, Yosuke Asako, Yoichi Ishizuka, Hirofumi Matsuo, "A control circuit composition and several characteristics of the proposed DPWM controlled POL", IEICE, vol. 107, no. 430, EE2007-46, pp. 13-18, January 2008.

[4] Edward Lam, Robert Bell and Donald Ashley, "Revolutionary Advances in Distributed Power Systems, " in Proc. IEEE APEC '03, 1.5, 2003.

[5] Angel V. Peterchev and Seth R. Sanders, "Quantization Resolution and Limit Cycling in Digitally Controlled PWM, IEEE Trans. Power Electronics, Vol. 18 No. 1, pp.301-308, January 2003.

[6] V. Peterchev, S. R. Sanders, "Quantization resolution and limit cycling in digitally controlled PWM converters", IEEE Transactions on Power Electronics, Vol. 18, No. 1, January 2003, pp. 301 - 308.

[7] B.J. Patella, A. Prodic, A. Zirger, D. Maksimovic, "High-frequency digital PWM controller IC for DC-DC converters", IEEE Transactions on Power Electronics, Vol. 18, January 2003.

[8] D. Maksimovic, R. Zane, R. Erickson, "Impact of Digital Control in Power Electronics", IEEE International Symposium on Power Semiconductor Devices & ICs, Kitakyushu, Japan, May 2004, pp. 13-22.

[9] Kaiwei Yao "High-Frequency and High-Performance VRM Design for the Next Generation of Processors", Doctor thesis of Virginia Polytechnic Institute and State University, April 14, 2004.

[10] S. Saggini, D. Trevisan, P. Mattavelli, "Hysteresis-Based Mixed-Signal Voltage-Mode Control for dc-dc Converters", IEEE Power Electronics Conference (PESC'07), Orlando, Florida, June 2007.

[11] S. Saggini, E. Orietti, P. Mattavelli, A. Pizzutelli, A. Bianco, "Fully-Digital Hysteretic Voltage-Mode Control for dc-dc Converters based on Asynchronous Sampling", IEEE Applied Power Electronics Conference (APEC'08), Austin, Texas, February 2008.

[12] Gene F. Franklin, J. David Powell, and Michael L. Workman, Digital Control of Dynamic Systems, Addison Wesley Longman Press, Menlo Park. CA, 1997.

[13] Z. Zhao, S. Ahsanuzzaman, A. Prodic "ESR Zero Estimation and Auto-Compensation in Digitally Controlled Buck Converters", IEEE Applied Power Electronics Conference (APEC'09), pp.247-251.

[14] Yu-Cheng Lin, Dan Chen, Yen-Tang Wang, Wei-Hsu Chang, "A Novel Loop Gain Correction Method for DigitallyControlled DC-DC Power Converters", 2009 IEEE Energy Conversion Congress & Expo pp3530-3535.

Low Profile LLC Series Resonant Converter With Two Transformers

Eun-Soo Kim, Joo-Hoon Kim

Dept. of Electrical Engineering,
Jeonju University
1200, Hyoja-Dong, Wansan-Gu, Jeonju
Jeonbuk, 560-759, South Korea
E-mail : eskim@jj.ac.kr

Sung-In Kang, Jun-Ho Park,
Jae-Sam Lee, Dong-Young Huh

SMPS gr, LG Innotek
379, Kasoo-dong, Osan
Gyeonggi, 447-705, South Korea
E- mail : sikang@lginnotek.com

Yong-Chae Jung

Dept. of Electronic Engineering
Namseoul University
Cheonan, Korea
E-mail : ychjung@nsu.ac.kr

Abstract — **In this paper, low profile LLC series resonant converter with two transformers is proposed for offering a slim PDP TV set. Design procedures and voltage gain characteristics on the proposed converter are described. Two transformers applied to LLC resonant converter are connected in series with the primary and in parallel with the secondary to reduce the unbalance of the primary current. Based on the theoretical analysis and simulation results, considering the characteristics of the voltage gain and load variation, a prototype 400W LLC resonant converter for 50 inch PDP TV power module is designed and built. And then, the design parameters and operational principles of the proposed converter are verified through the experimental results.**

I. Introduction

The LLC resonant converter is widely used and studied because of several advantages such as high efficiency, output regulations over a wide load range, no reverse recovery problem of output rectifier, cost effective and so on [1-6]. Nowadays, the big trend in flat panel TV market is slimming. In order to design the slim size power system for flat panel TV, the heights of heat sink, bulk capacitor, filter, transformer and so forth should be diminished. Among them, it is very difficult to reduce the height of the transformer.

In this paper, to solve the abovementioned problem, we propose LLC resonant converter with two transformers. Analysis and design procedures of the proposed circuit are described in detail. The usefulness of the proposed circuit is verified by simulations and experimental results with a prototype 400W LLC resonant converter for 50 inch PDP TV power module.

II. Low Profile LLC Resonant Converter

Fig. 1 shows the proposed low profile LLC series resonant converter with two transformers. The circuit is comprised of a half bridge inverter on the primary side and a

This work was supported by the cooperative research fund of LG Innotek

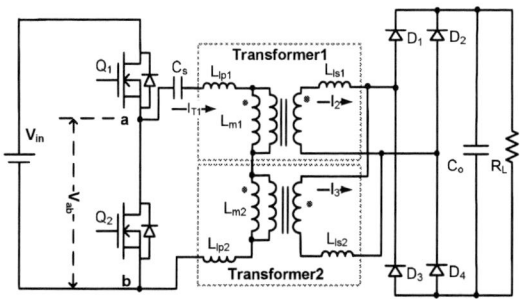

Fig. 1 LLC series resonant converter with two transformers

bridge rectifier on the secondary side. Two LLC transformers are connected in series with the primary and in parallel with the secondary to reduce the unbalance of the primary current. As the voltage applied to the primary of each transformer is reduced in half, the number of the primary turns is also reduced in half. Moreover, as the secondary windings of two transformers are in parallel, the wire thickness of the secondary windings can be reduced in half. Therefore, the advantage to cut down the core size can be obtained by lessening the transformer winding area.

In this paper, the equivalent circuit of LLC series resonant converter with two transformers is analyzed. Because the input-output voltage gain characteristics can make a difference according to minute differences of leakage inductances and magnetizing inductances of each transformer, the detailed analysis and optimal design method for the above-mentioned problem are presented.

III. Voltage Gain Analysis Using An Equivalent Circuit

Using the first harmonic approximation (FHA) [1], the equivalent circuit of the proposed LLC series resonant converter shown in Fig. 2 is derived to obtain input-output

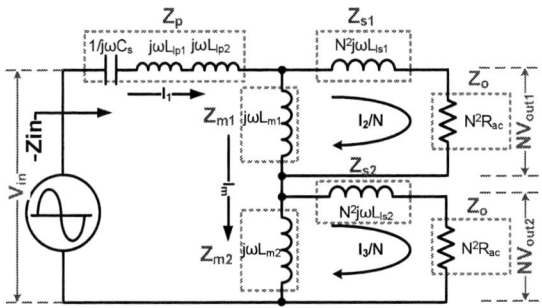

Fig. 2 Equivalent circuit of LLC series resonant converter with two transformers

response in frequency domain. C_s, L_{lp1}, L_{lp2}, N^2L_{ls1}, N^2L_{ls2}, L_{m1} and L_{m2} are series resonant capacitor, two primary leakage inductances, two secondary leakage inductances and two magnetizing inductances, respectively. N^2R_{ac} is the equivalent load resistance of the secondary side circuit shown in the primary side, where $R_{ac}=8R_L/\pi^2$ [6].

The voltage gain (G_V) can be expressed by the ratio of output voltage (NV_{out1}) of Transformer 1 to input voltage (V_{in}) as shown in (1). Each variable in (1) is given by (2) ~ (10).

$$G_v = \frac{NV_{out1}}{V_{in}} = \frac{Z_o I_2 / N}{Z_{in} I_1} = \left| \frac{1}{D_{re} + D_{im}} \right| \quad (1)$$

$$D_{re} = \left\{ 1 + \frac{L_{m2}}{L_{m1}}\delta + A_1 + \frac{L_{lp2}}{L_{m1}} + \left(\frac{\omega_r}{\omega}\right)^2 \alpha + \frac{L_{m2}}{L_{m1}}\beta \right\} \quad (2)$$

$$D_{im} = j\left(\frac{1+B_1}{N^2 R_{ac}}\right)\left\{\left(\omega L_{eq} - \omega L_{m2}\beta\right)\left(1 + \frac{L_{m2}^2}{L_{m1}}\gamma\delta\right) - \frac{1}{\omega C_s}\right\} \quad (3)$$

$$\alpha = A_1 + \frac{B_1}{1+B_1} \ , \quad \beta = A_2 + \frac{B_2}{1+B_2} \quad (4)$$

$$\gamma = \frac{A_2 + A_2 B_2 + B_2}{A_1 + A_1 B_1 + B_1} \ , \quad \delta = \frac{j\omega L_{m1} + N^2 j\omega L_{ls1} + N^2 R_{ac}}{j\omega L_{m2} + N^2 j\omega L_{ls2} + N^2 R_{ac}} \quad (5)$$

$$A_1 = \frac{L_{lp1}}{L_{m1}} \ , \quad B_1 = \frac{N^2 L_{ls1}}{L_{m1}} \quad (6)$$

$$A_2 = \frac{L_{lp2}}{L_{m2}} \ , \quad B_2 = \frac{N^2 L_{ls2}}{L_{m2}} \quad (7)$$

$$Z_{in} = \frac{1}{j\omega C_s} + j\omega L_{lp1} + \frac{j\omega L_{m1}(N^2 j\omega L_{ls1} + N^2 R_{ac})}{j\omega L_{m1} + N^2 j\omega L_{ls1} + N^2 R_{ac}}$$
$$+ j\omega L_{lp2} + \frac{j\omega L_{m2}(N^2 j\omega L_{ls2} + N^2 R_{ac})}{j\omega L_{m2} + N^2 j\omega L_{ls2} + N^2 R_{ac}} \quad (8)$$

$$L_{eq} = L_{lp1} + \frac{N^2 L_{ls1} L_{m1}}{N^2 L_{ls1} + L_{m1}} = L_{lp2} + \frac{N^2 L_{ls2} L_{m2}}{N^2 L_{ls2} + L_{m2}}$$
$$= L_{m1}\left(A_1 + \frac{B_1}{1+B_1}\right) = L_{m2}\left(A_2 + \frac{B_2}{1+B_2}\right) \quad (9)$$

Table 1. Specifications of LLC resonant converter

Input voltage(V_{in})	$320V_{DC}\sim400V_{DC}$
Output voltage (V_{s_out})	$195V_{DC}(185V_{DC}\sim200V_{DC})$
Output current (I_{s_out})	$0.1A\sim2A$
A (L_{lp}/L_m)	0.15
B (N^2L_{ls}/L_m)	0.1
Q ($\omega_r L_{eq}/N^2R_{ac}$)	1.2
Flux density (ΔB)	0.27T
Resonant frequency (f_r)	150kHz

$$\omega_r = 1/\sqrt{2L_{eq}C_s} \quad (10)$$

From above equations, assuming that each inductance of the two transformers is the same, i.e., $L_{lp}=L_{lp1}=L_{lp2}$, $N^2L_{ls}=N^2L_{ls1}=N^2L_{ls2}$, $L_m=L_{m1}=L_{m2}$, $A=A_1=A_2$ and $B=B_1=B_2$, the voltage gain and the equivalent inductance can be obtained as follows:

$$G_v = \left[2\left(1 + \frac{L_{lp}}{L_m} - \frac{1}{\omega^2 L_m C_s}\right) + j2Q\left(1 + \frac{N^2 L_{ls}}{L_m}\right)\left(\frac{\omega}{\omega_r} - \frac{\omega_r}{\omega}\right)\right]^{-1} \quad (11)$$

$$L_{eq} = L_m\left(A + \frac{B}{1+B}\right) \quad (12)$$

where the quality factor Q can be determined from

$$Q = \frac{\omega_r L_{eq}}{N^2 R_{ac}} \quad (13)$$

If $A=L_{lp}/L_m$, $B=N^2L_{ls}/L_m$ and $\omega_n=\omega/\omega_r$, the voltage gain of (11) can be represented by the following (14).

$$G_v = \left[2(1+A) - \frac{A+B+AB}{0.5\omega_n^2(1+B)} + j2Q(1+B)\left(\omega_n - \frac{1}{\omega_n}\right)\right]^{-1} \quad (14)$$

IV. DESIGN PROCEDURES

Based on the above derived equations, the design procedures of transformer and resonant capacitor are presented in this section. Table 1 shows the specifications of the proposed LLC resonant converter.

At first, assuming that two transformers have the same parameters, the values of A, B and Q can be determined from (6), (7) and (13). We use two EE4445 ferrite cores for two transformers designed by setting A=0.15 and B=0.1. In the case, as the values of A and B can be changed in accordance with bobbin shape or winding method, these two values are thoroughly investigated in the design to obtain the proper resonant characteristics.

Fig. 3 (a) and (b) show the voltage gain characteristics according to B in N=1. As can be seen in this figure, B raises the voltage gain totally and the DC gain as much as B/2 at resonant frequency (ω_n). Thus, turn ratio (N) can be obtained for maximizing the efficiency by operating at resonant frequency under heavy load conditions (static max ($I_{os}=2A_{ave}$),

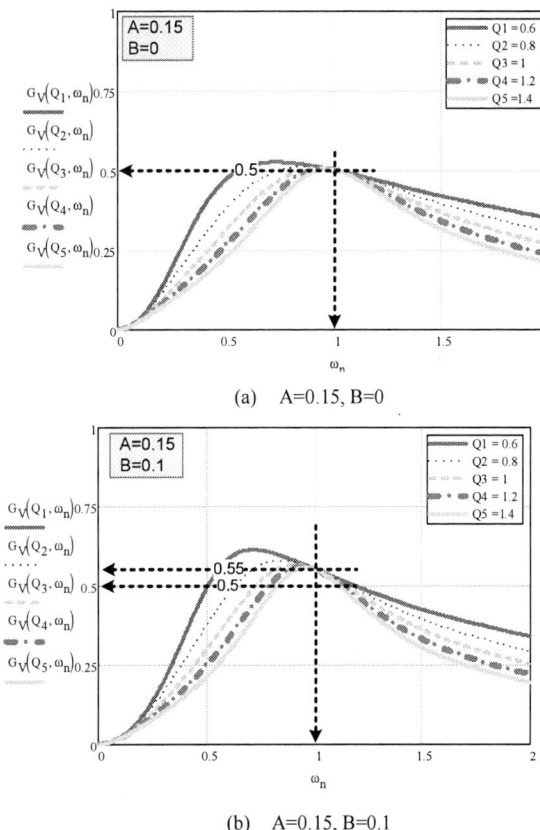

(a) A=0.15, B=0

(b) A=0.15, B=0.1

Fig. 3 Voltage gain characteristics according to B in N=1

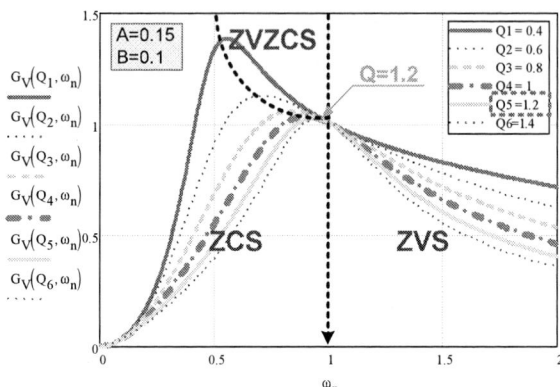

Fig 4. Voltage gain characteristics according to the variation of Q

Table 2. Calculated parameters

Primary self inductance (L_p)	58.5uH
Secondary self inductance (L_s)	185.1uH
Equivalent leakage inductance (L_{eq})	12.26uH
Magnetizing inductance (L_m)	50.89uH
$L_{lp}(AL_m)$	7.63uH
$N^2L_{ls}(BL_m)$	5.09uH
$N(n_1/n_2)$	0.545 (12/22)
Air-gap	0.36mm
Series capacitor (C_s)	46nF

Table 3. Measured parameters of two transformers

	TF 1	TF 2
Primary self inductance (Lp)	54.86uH	55.68uH
Secondary self inductance (Ls)	173.52uH	191.36uH
Equivalent leakage inductance (Leq)	12.49uH	12.53uH
A(Llp/Lm)	0.173	0.123
B(N2Lls/Lm)	0.104	0.149
N(n1/n2)	0.545 (12/22)	0.545 (12/22)

dynamic max (I_{os}=11A$_{peak}$)). Because DC gain goes up B/2 (0.05) in the influence of B at resonant frequency, turn ratio is determined by considering the raised gain at discontinuous mode operation like the following equation.

$$N = \frac{V_{in}}{V_{out}} \frac{(1+B)}{2} \qquad (15)$$

After the turn ratio is determined, the value of Q is selected with the condition the switch devices to realize ZVS at full load condition (1.5kW) like shown in Fig. 4. This figure shows the voltage gain characteristics according to the variation of Q. When Q varies from 0.4 to 1.4 with fixing A and B (A=0.15, B=0.1), the voltage gains by (14) are illustrated in this figure. To vary the value of Q means the variation of the equivalent leakage inductance (L_{eq}) at rated load. If the selected Q is too large, the voltage gain margin at rated load is lack and thus the required output can not be obtained. On the other hand, if the selected Q is too small, the circuit is designed enough to have the voltage gain margin. However, the circuit efficiency is reduced due to the circulating current increase. Therefore, the selection of Q value with the proper gain margin at the required output capacity is very important.

Determining the value of Q (Q=1.2), the equivalent leakage inductance (L_{eq}) can be also selected. Using the equivalent leakage inductance, the magnetizing inductance (L_m) and two leakage inductances (L_{lp} and N^2L_{ls}) are calculated by the following two equations.

$$L_m = \frac{L_{eq}}{A + B/(1+B)} \qquad (16)$$

$$L_{lp} = AL_m , \qquad N^2 L_{ls} = BL_m \qquad (17)$$

The equivalent resonant frequency of the proposed circuit is $\omega_r = 1/\sqrt{2L_{eq}C_s}$, i.e., the resonant frequency between the

Table 4. Specifications of LLC series resonant converter

Switching frequency range (fs)	143.6kHz ~156.9kHz
Resonant frequency (fr)	146.8kHz
Resonant capacitor (Cs)	47nF
Control IC	ST社 L6599D
Main switching devices	STW20NK50Z (500V, 20A)
Output rectifying diode	FCF10A40 (400V, 10A)

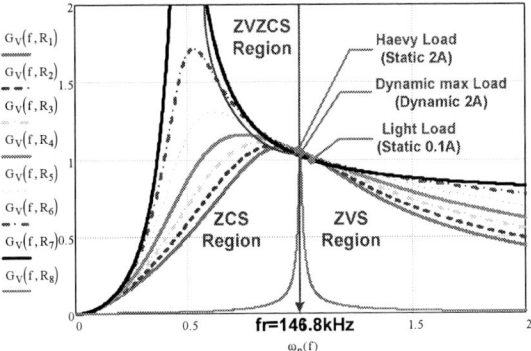

Fig 5. Operating frequency range according to the load condition .

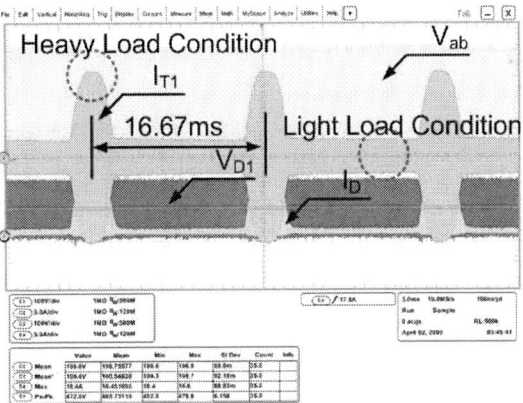

(a) Dynamic max: 2A

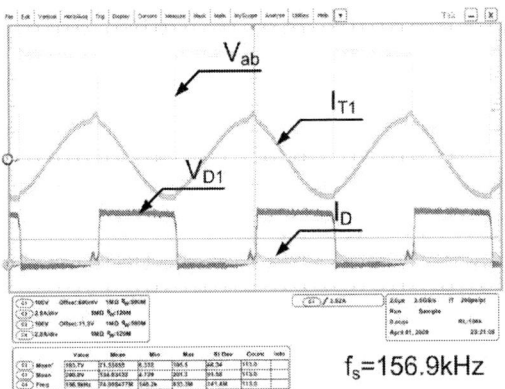

(a) Light load condition (I_o=0.1A_{ave})

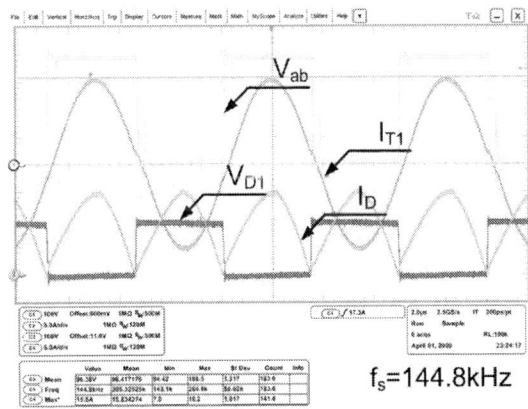

(b) Dynamic max (Extended)

Fig. 7 Experimental waveforms of primary terminal voltage (V_{ab}) and current (I_{T1}), the secondary diode voltage (V_{D1}) and the rectified current (I_{D1}) at dynamic load condition (CH1/CH3: 100V/div, CH2/CH4: 5A/div, 2us/div)

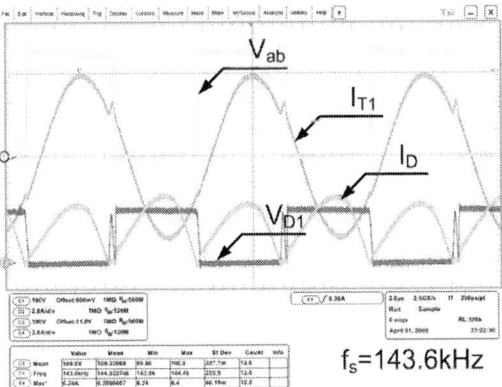

(b) Heavy load condition (I_o=2A_{ave})

Fig. 6 Experimental waveforms of the primary terminal voltage (V_{ab}) and current (I_{T1}), the secondary diode voltage (V_{D1}) and the rectified current (I_D) at static load condition (CH1/CH3: 100V/div, CH2/CH4: 2A/div, 2us/div)

equivalent leakage inductance (L_{eq}) and series capacitance (C_s). Thus, the series capacitance can be obtained by (18).

$$C_s = \frac{(1/\omega_r)^2}{2L_{eq}} \qquad (18)$$

Table 2 and Table 3 are calculated and measured parameters of the proposed converter, respectively.

V. EXPERIMENTAL RESULTS

To verify the theoretical analysis, a prototype LLC resonant converter for 50 inch PDP TV power module is designed and built by two transformers with the conditions of Table 3. Electronic loads are used as testing loads. When the output of Vs power part is even in static and dynamic load conditions, the switching devices in the primary should be

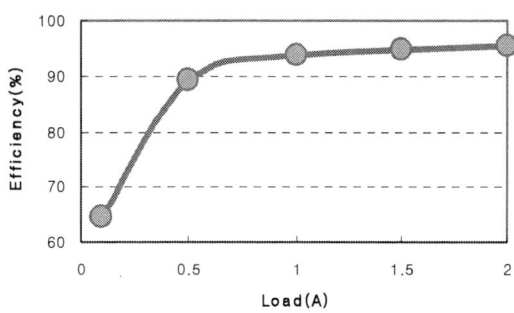

Fig. 8 Efficiency characteristic

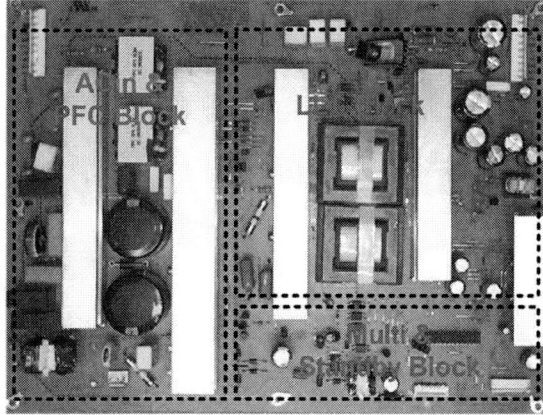

Fig. 9 Prototype of 50 HD PSU with two transformers

satisfied with ZVS operation as shown in Fig. 5. In this case, to provide the proposed LLC resonant converter with a similar load condition to the load condition of PDP TV module, the dynamic load is the load condition generated by 60Hz pulse type sinusoidal wave in electronic load. Table 4 shows specifications of LLC series resonant converter.

Fig. 6 and Fig. 7 show the experimental waveforms in the static load conditions and the dynamic load conditions, respectively. At each case, the switching frequency is displayed in bottom of the figure. These values coincide in the abovementioned analysis. The efficiency curve for different load conditions is shown in Fig. 8. The efficiency is more than 95.4% at full load condition. Fig. 9 shows a photograph of a 400W LLC resonant converter used in the experiments.

VI. CONCLUSIONS

To obtain the slim size power module for flat panel TV, LLC series resonant converter with two transformers is proposed. Two LLC transformers are connected in series with the primary and in parallel with the secondary to reduce the unbalance of the primary current. In this paper, analysis and design procedures of the proposed circuit are presented in detail. Through the experimental results, the usefulness of the proposed circuit can be verified.

REFERENCES

[1] A. K. S. Ghat, "Analysis and design of LCL-type series resonant converter," *IEEE Transactions on Industrial Electronics*, vol. 41, Issue 1, pp. 118-124, Feb. 1994.

[2] Bor-Ren Lin and Chao-Hsien Tseng, "Analysis of parallel-connected asymmetrical soft-switching converter," *IEEE Transactions on Industrial Electronics*, vol. 54, Issue 3, pp. 1642-1653, June 2007.

[3] B. Yang, F. C. Lee, A. J. Zhang and G. Huang, "LLC resonant converter for front end DC/DC conversion," *IEEE APEC'02*, Vol. 2, pp. 1108-1112, 2002.

[4] B. Lu, W. Liu, Y. Liang, F. C. Lee and J. D. van Wyk, "Optimal design methodology for LLC resonant converter," *IEEE APEC'06*, pp. 533-538, 2006.

[5] Y. Gu, L. Hang, U. Chen, Z. Lu, Z. Qian and J. Li, "A simple structure of LLC resonant DC-DC converter for multi-output applications," *IEEE APEC'05*, Vol. 3, pp. 1485-1490, 2005.

[6] Hang-Seok Choi, "Half-bridge LLC resonant converter design using FSFR-series Fairchild Power Switch," *Fairchild Semiconductor*, Application Note AN-4151, pp. 1-17, 2007.

Adaptive Frequency Control for ZVS Synchronous Boost Converters Operated in Average Current Mode

Ben York, *Student Member IEEE*, Rae-Young Kim, and Jih-Sheng Lai, *Fellow IEEE*
Virginia Polytechnic Institute and State University
Future Energy Electronics Center
Blacksburg, Virginia USA

Abstract— **For a zero-voltage switching (ZVS) synchronous boost converter, the power stage efficiency undergoes significant limitation when the load is outside the ideal range. In order to achieve optimum efficiency at a range of load conditions, the magnitude of the resonant current must be controlled. The most effective method to control resonant current is to adjust the converter's switching frequency. Though methods have been proposed as a solution to this issue, they are not compatible with average current mode control. In this paper, a two-loop solution is proposed, in which both the converter duty cycle and switching frequency are controlled in order to maintain the desired average inductor current under the ZVS condition. In support of this new adaptive frequency control, theoretical analysis, design criteria, as well as experimental validation are provided.**

I. INTRODUCTION

The zero-voltage switching (ZVS) synchronous boost converter is a quasi-square wave design, capable of high efficiency operation with a lower component count than many other resonant topologies. Under ZVS, active switches are turned on only after their drain-source voltage is zero. This virtually eliminates turn-on loss, and generally results in higher efficiency and lower noise operation than is possible with non-ZVS converters.

In ZVS synchronous converters, the standard topology's rectifier diode is replaced by an active switch, allowing current to flow through the inductor in both the forward and reverse directions during a single switching cycle. When the current flows in the reverse direction (toward the input source), it discharges the drain-source capacitance (C_{DS}) of the lower switch, which is initially off. This allows for a near loss-free turn-on when voltage is applied to the gate of that switch.

Though ZVS design practices can improve efficiency, there are definite drawbacks to these methods. One of the most significant difficulties is the management of the reverse, or resonant, current magnitude. If the magnitude of the resonant current is not large enough to charge the drain-source capacitance completely during a predetermined dead time, the transistor will not reach the ZVS condition, resulting in MOSFET turn-on loss, diode reverse recovery loss, and an

increase in switching noise. If the resonant current is too large, the excess current will circulate, causing an increase in conduction loss [1]. Potential ZVS converter designs generally have a very high efficiency under specific optimum conditions. Due to the difficulty in controlling the resonant current, however, when a ZVS converter is subjected to input conditions outside of that optimum range, its efficiency deteriorates significantly.

As a solution to this problem, a number of authors have proposed methods to adjust the switching frequency in order to control the resonant current, therefore improving the range of inputs over which optimum efficiency may be achieved. In [2], a resonant pole inverter with variable switching frequency showed the possibility of controlling resonant current [2]. In literature [3], the authors used a constant off-time approach to achieve a zero-voltage-switching. A voltage-controlled oscillator generated the switch timing based on the input voltage. This approach led to a constant current ripple, but did not directly control the resonant current. Therefore, a converter experienced an excessive amount of resonant current under light load. In [4], a bi-directional dc-dc converter was implemented with peak-current-mode control and a variable switching frequency. Active switches were triggered when a sensed inductor current reached the positive and negative limits of a window comparator. This concept was expanded to digital control techniques in [5]. The major issue found in [4] and [5] is the difficulty to control the average current precisely, which is an inherent problem in peak-current-mode control. Off-line variable switching frequency has also proposed to avoid noise issues [6].

However, little has been mentioned with regards to variable frequency within the context of average current mode control. While peak current mode control is a popular option for ZVS converters, the ability to utilize average current control methods allows for better compatibility with certain system strategies, such as maximum power point tracking (MPPT) in photovoltaic applications. Because the relationship between the peak and average current is variable, average current mode control allows more precise and effective operation under these system strategies.

978-1-4244-4782-4/10 $26.00 © 2010 IEEE

This paper presents a method in which the switching frequency and average current are designed to be controlled nearly independently. Reference values are determined both the average and resonant currents, and the resultant error is used to determine the converter duty cycle and switching frequency, respectively. Because a multi-loop system is generated, caution must be exercised to verify that the two loops are independent. In addition to analyzing loop interdependence, this paper will also address modeling and compensator design issues, as well as present experimental verification for these methods.

II. SYSTEM ARCHITECTURE AND OPERATION

A. Architecture

In the methods proposed in this paper, a ZVS synchronous boost converter is controlled via an external digital signal processor (DSP). The DSP includes both analog-digital (A/D) converters and dedicated pulse-width modulation (PWM) hardware. Currently, the converter only includes one phase, but may be expanded to multiple phases in the future. The system architecture is given in Figure 1:

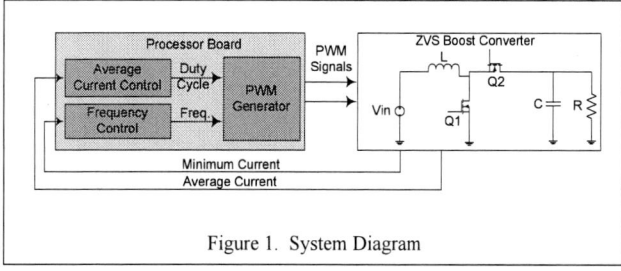

Figure 1. System Diagram

Notice that the DSP block includes both an average current control loop, as well as a frequency control loop. The average current control loop modulates the converter duty cycle (D), while the frequency control loop modulates the switching period (T_s). The resultant D and T_s are then combined in the PWM logic to generate the gating signals for the two switches. As a result, the "ON" switching intervals are given by:

$$T_{on,lower} = DT_s - d_t \qquad (1)$$

And

$$T_{on,upper} = (1-D)T_s - d_t \qquad (2)$$

Where d_t is the amount of dead time between the gate signals. Provided that the dead time is not too large, relative to T_s, it may be largely ignored in the proceeding derivations. The A/D sampling is synchronized to the same clock as the PWM signals, implying that each measured variable is sampled once per switching cycle.

B. Operation

As seen in Figure 2, the converter waveforms appear similar to a standard boost converter. However, the input inductor (L) is small enough that, at a given switching frequency, the current waveform (i_L) is allowed to drop

below zero. Though the inductor current is not continuously flowing in the positive direction (that is toward the output), the converter operation can be considered analogous to that of a continuous conduction mode (CCM) boost converter.

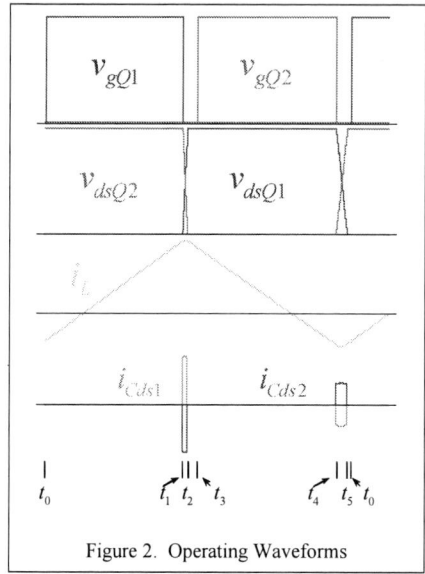

Figure 2. Operating Waveforms

In order to ensure proper ZVS operation on both switches, the drain source capacitances of both the upper and lower switches (C_{ds1} and C_{ds2}, respectively) must be allow to charge to the full output voltage during their respective charging periods ($t_1 \rightarrow t_2$ and $t_4 \rightarrow t_5$). Once the charging is completed, the parallel diode of the opposite switch begins conducting prior to the gate signal being applied. The length of the resonant charging/discharging intervals depends largely on the output voltage (V_{out}), the output current (i_{load}), and the magnitude of i_L. For the purposes of this paper, V_{out} and i_{load} will be considered fixed quantities.

Though a full model shall not be derived at this juncture, suffice to say that the length of time required to complete the resonant charging cycle is inversely proportional to the magnitude of i_L. The length of the interval between lower-switch turn-off and upper diode conduction, just prior to upper switch turn-on, ($t_1 \rightarrow t_2$) is thus inversely proportional to the peak value of i_L. Conversely, the length of the interval between upper-switch turn-off and lower diode conduction, just prior to lower-switch turn-on ($t_4 \rightarrow t_5$) is inversely proportional to the magnitude of the minimum of i_L. If positive net power is transferred, as it is in this case, from input to output, it stands to reason that the peak magnitude of i_L would be greater than its minimum magnitude. Therefore, if the magnitude of the minimum, or resonant, current can be controlled accurately, we can guarantee that the resonant charging cycle may be completed within a given dead time.

Control of the resonant current may be accomplished by modulating the switching period of the converter, which alters the peak-peak magnitude of the inductor current. At light loading conditions, a faster switching frequency may be utilized such that the resonant current is not overly large.

Likewise, under heavy load a slower switching frequency is preferable in order to ensure that the resonant charging period can complete in the amount of time allotted by the converter dead time.

III. Modeling Strategy

With two control loops implemented in the controller hardware, it becomes necessary to separate the resultant converter behavior into functional components which may be analyzed. As expected, the plant model may be split into four parts.

The duty-to-average current transfer function ($G_{ia,d}$) can be expressed as a continuous time transfer function in the s-domain, buffeted by a zero-order-hold at the input and a sampling effect at the output. This function may be derived in the s-domain, using state space averaging or another comparable technique, then discretized for controller design.

The frequency-to-resonant current transfer function ($G_{ir,f}$) represents the effect of the chosen switching period on the resonant current magnitude, and can be treated as an inherently discrete, or z-domain, transfer function.

Though it would be desirable that the two controllers belonged to independent control loops, there will inevitably be unintended coupling between the two loops. This may be represented by the frequency-to-average current function ($G_{ia,f}$) and the duty-to-resonant current function ($G_{ir,d}$).

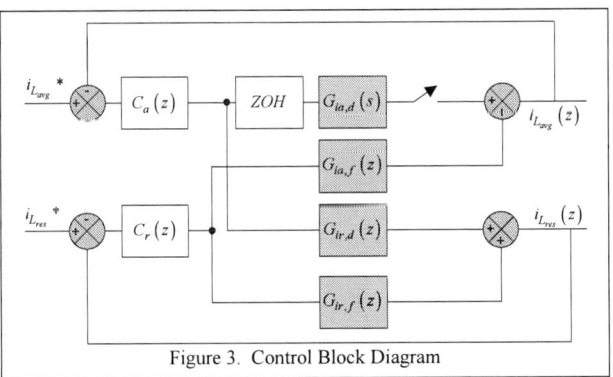

Figure 3. Control Block Diagram

The overall controller-converter block diagram is given in Figure 3. The transfer function $G_{ia,d}$ has been derived in several works such as [10], and therefore will not be derived here. The next three sections are devoted to the derivation and analysis of the remained three functions.

IV. Transfer Function Derivations

A. Frequency To Average Current Model Derivation

In an analog control solution, the switching frequency may be continuously modified throughout the switching cycle. A discrete controller, however, requires that the switching frequency not only be sampled, but also update at discrete points in the switching cycle. The sampling point is given by the definition of the resonant current, and must

therefore be sampled at or near the minimum point in the inductor current waveform, occurring just prior to the lower (main) switch turning on. The point in the switching cycle at which the switching frequency is updated, however, is at this point undetermined. In order to simplify controller design and improve transient operation it may be determined that an optimum location for the update point would maximize the gain of $G_{ir,f}$ while minimizing the gain of $G_{ia,f}$.

Natural locations for this update point would be generally at either the peak current, minimum current, or at the average current value. However, updating the switching frequency at various points in the switching cycle impacts the average output current differently. For instance, if the update occurs at the peak in the inductor current waveform (just before upper switch turn-on) the average current of the inductor may be given by the following relationship:

$$i_{Lavg}[n] - i_{Lavg}[n-1] = $$
$$\left(\frac{DV_{in}}{2L}\right)\left(T_s[n] + T_s[n-1]\right) - \left(D'T_s[n]\right)\left(\frac{V_{out} - V_{in}}{L}\right) \quad (3)$$

Where $D' = 1 - D$

Assuming the converter is allowed to reach steady state with no change to the duty cycle allows for the following assumption:

$$\frac{V_{out}}{V_{in}} = \frac{1}{D'} \quad (4)$$

Which simplifies (3) to:

$$i_{Lavg}[n] - i_{Lavg}[n-1] = \left(\frac{DV_{in}}{2L}\right)\left(T_s[n-1] - T_s[n]\right) \quad (5)$$

This indicates that if the switching frequency update coincides with the peak current, then there will be some appreciable difference in the steady-state operating point of the converter, not to mention in the transient case.

However, if the update point is moved so that it coincides with the average current (ideally located at the midpoint of the "ON" period for either switch), the average current equation may be written as:

$$i_{Lavg}[n] = i_{Lavg}[n-1] + \frac{(V_{in} - D'V_{out})\left(T_s[n] + T_s[n-1]\right)}{2L} \quad (6)$$

A graphical illustration of this location is given in Fig. 4. Once again, assuming the converter returns to steady-state operation:

$$i_{Lavg}[n] - i_{Lavg}[n-1] = 0 \quad (7)$$

Thus, there is no steady-state change in the average inductor current as a result of modifying the switching frequency at a point coinciding with the average inductor current. For the transient response, if we take the previous equation and convert it to the z-domain:

$$I_{Lavg}\left(1-z^{-1}\right) = \frac{\left(V_{in}-D'V_{out}\right)\left(T_s\left(1+z^{-1}\right)\right)}{2L} \tag{8}$$

$$G_{ia,f} = \frac{\hat{i}_{Lavg}}{\hat{T}_s} = \frac{\left(V_{in}-D'V_{out}\right)}{2L}\left(\frac{1+z^{-1}}{1-z^{-1}}\right) \tag{9}$$

From which we may gather that the further away the system deviates from its steady state operating point, the greater the impact that adjustments to the switching frequency have upon the average current.

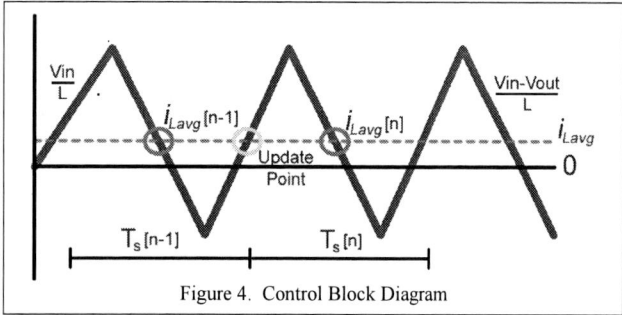

Figure 4. Control Block Diagram

B. Frequency to Minimum Current Model Derivation

If we consider the switching period update point to occur simultaneously with the average point of the inductor current, as indicated earlier, the relative expression of the minimum, or resonant current, may be expressed as:

$$i_{L\min}[n]-i_{L\min}[n-1] =$$
$$\frac{V_{in}\left(\left(1+D'\right)T_s[n]+DT_s[n-1]\right)-2\left(V_{out}\right)D'T_s[n]}{2L} \tag{10}$$

A graphical illustration of this procedure is given in Fig. 5. Once again, if we assume a small change in T_s:

$$i_{L\min}[n]-i_{L\min}[n-1] = \frac{V_{in}D\left(T_s[n-1]-T_s[n]\right)}{2L} \tag{11}$$

Representing the equation in the z-domain:

$$\frac{\hat{i}_{L\min}}{T_s} = \frac{V_{in}D\left(z^{-1}-1\right)}{2L\left(1-z^{-1}\right)} \tag{12}$$

Allowing pole-zero cancellation:

$$\frac{\hat{i}_{L\min}}{T_s} = \frac{-V_{in}D}{2L} \tag{13}$$

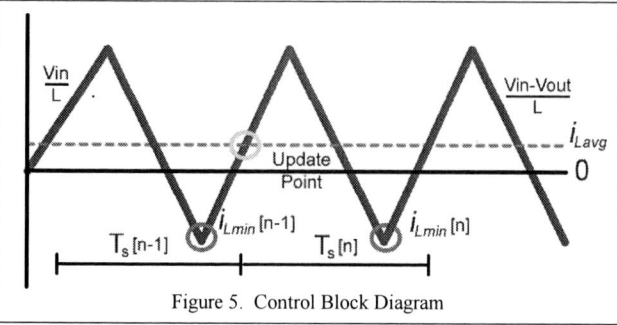

Figure 5. Control Block Diagram

This indicates, so long as the perturbation of the switching period remains small, that the transient response of the resonant current is time-independent. However, if the small-perturbation assumption is not valid, the resultant transfer function is given as follows:

$$G_{ir,f} = \frac{\hat{i}_{L\min}}{T_s} = \frac{1}{2L}\left(\frac{V_{in}Dz^{-1}+V_{in}\left(1+D'\right)-2V_{out}\left(D'\right)}{1-z^{-1}}\right) \tag{14}$$

C. Duty Cycle to Minimum Current Derivation

In a discrete control solution, as with the switching period adjustment, it is also required to select an update point for the duty cycle. In order to reduce impact on the minimum current, the optimum location would be to update the duty ratio just prior to the main switch turn-on, which would present a scenario described in the following equation:

$$i_{L\min}[n+1]-i_{L\min}[n-1] =$$
$$\frac{T_s}{L}\left(2V_{in}-\left(1-d[n-1]\right)V_{out}[n-1]-\left(1-d[n]\right)V_{out}[n]\right) \tag{15}$$

If we assume, once again, a small deviation in the duty cycle, we may make the following correlation:

$$V_{out}[n]\left(1-d[n]\right) = V_{in} \tag{16}$$

Thus,

$$i_{L\min}[n+1] = i_{L\min}[n-1]+\frac{T_s}{L}\left(2V_{in}-2V_{in}\right) = i_{L\min}[n-1] \tag{17}$$

This indicates that if the update point for the duty cycle is placed at the turn-on point for the main switch, there is negligible steady-state impact on the resonant current magnitude. However, in an actual system implementation, this is either a poor or completely unrealizable update point. A much more practical scheme updates the duty cycle and the switching period at the same point, simultaneous with the average current. This results in a transfer equation for the resonant current as follows:

$$i_{L\min}[n]-i_{L\min}[n-1] =$$
$$\frac{T_s}{L}\left(\frac{V_{in}}{2}\left(d[n-1]-d[n]\right)+V_{in}-\left(V_{out}\right)\left(1-d[n]\right)\right) \tag{18}$$

Assuming steady-state:

$$i_{L\min}[n] - i_{L\min}[n-1] = \frac{V_{in}T_s}{2L}\left(d[n-1] - d[n]\right) \qquad (19)$$

Applying the z-transform:

$$i_{L\min}\left(1 - z^{-1}\right) = \frac{V_{in}T_s}{2L}\left(d\left(z^{-1} - 1\right)\right) \qquad (20)$$

$$\frac{i_{L\min}}{d} = \frac{V_{in}T_s}{2L}\frac{z^{-1} - 1}{1 - z^{-1}} \qquad (21)$$

Allowing for pole-zero cancellation

$$G_{ir,d} = \frac{i_{L\min}}{d} = \frac{-V_{in}T_s}{2L}. \qquad (22)$$

Therefore, updating the duty ratio at a non-ideal point results in an undesirable disturbance in the minimum current. Proper design of the two control loops will be required to minimize the impact of this issue.

V. CONTROLLER DESIGN

A. Frequency Controller

For the frequency controller, the design goal is to compensate the frequency-to-resonant current transfer function $G_{ir,f}$ while limiting the disturbance caused by $G_{ir,d}$. For our purposes, $G_{ir,f}$ may be represented by the dc gain given in (14). Therefore, linear stability can be easily achieved. Because exact sample timing is difficult to achieve, some filtering will be necessary to negate the effect of numerical error. Ultimately, limiting the impact of $G_{ir,d}$ requires that the bandwidth of this loop be higher than that of the average current control loop.

For the design presented in this paper, a simple Type-I integrator system may be used for the compensator C_r.

B. Average Current Controller

With the frequency controller determined, the average current controller (C_a) may be designed to compensate $G_{ia,d}$. Once again, $G_{ia,d}$ may be defined in the s-domain according to the procedure similar to that of [10]. From there, the controller C_a may be defined either in the w-plane (according to the bilinear or Tustin transformation) and converted to the z-domain or defined directly in the z-domain.

Regardless of design procedure, it is important to note that the sampling frequency varies with the switching frequency. Because of this, the compensator C_a should be designed specifically to ensure appropriate gain and phase margin at the minimum switching frequency (or largest T_s) to ensure that stability criterion are met at this "worst-case" condition. As determined earlier, in order to lessen the effect of the disturbance $G_{ir,d}$, the bandwidth of the average current control loop should be less than that of the resonant control loop.

For the example presented in this paper, a type-II compensator provides sufficient gain and phase margin (in the w-plane) to ensure stability.

VI. EXPERIMENTAL RESULTS

For this experiment, the authors utilized a single phase of a four-phase ZVS boost converter prototype. The prototype was divided into two boards, one as a power stage shown in Fig. 6 and the other a DSP board (not shown). The DSP board employs a Texas Instruments TMS320F2808 fixed-point processor, on which the average current and resonant current control loops are implemented as well as the PWM logic.

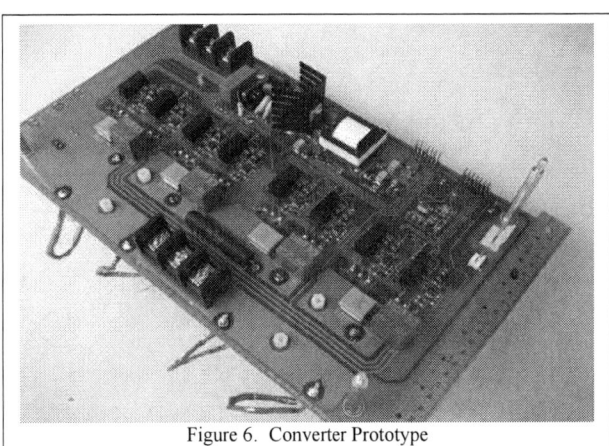

Figure 6. Converter Prototype

The boost converter prototype utilizes hall-effect type, isolated sensing for the inductor current. In order to effectively implement this control strategy, it is imperative that the selected current sensor and associated signal conditioning by of sufficient bandwidth and limited phase differential, so that the sample timing will remain accurate. In future work, it may be possible to implement another form of current sensing with a higher bandwidth than that of hall effect, therefore increasing the accuracy and performance of the designed converter.

Additionally, the inductor design, though the design is not as critical as in a fixed-frequency design, care must be taken to match the required input and load ranges with the desired switching frequency range. Generally, a smaller inductor requires a higher switching frequency to produce the same resonant current at a given input and load condition. The converter specifications are summarized in Table 1.

Finally, the minimum current reference i_{Lmin}^* must be selected, so that soft-switching can be maintained over a range of input conditions. The required minimum current varies very slightly with loading condition, and varies moderately with input.

Parameter	Value
Switching Freq.	20-67kHz
Dead Time	1.0μs
Input Capacitance	400μF
Output Capacitance	2000μF
Inductance	67μH
MOSFET	FCH20N60
Current Sensor	LEM LTS-25
Input Voltage	60-140V
Output Power	40-400W (per phase)

Table 1. Converter Specifications

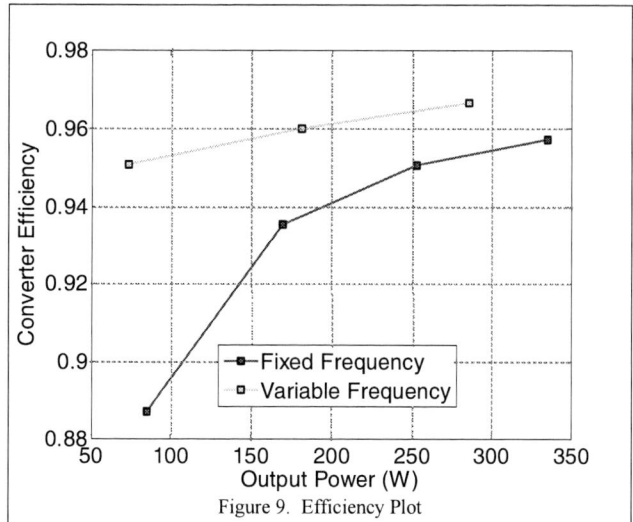

Figure 9. Efficiency Plot

From experimental results, it can be shown that the converter maintains the ZVS condition quite well over a range of operating conditions. Figure 7 displays a sample of the switching waveforms of the converter under nominal operating conditions. Figure 8 demonstrates the response of the converter to a change in the average input reference i_{Lavg}^{*}. Note that the minimum current remains nearly constant throughout the transient period.

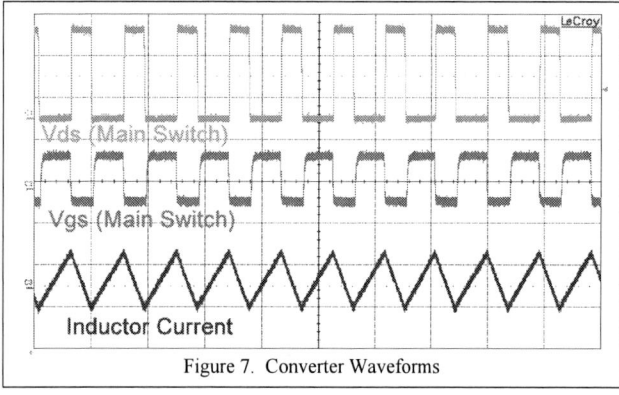

Figure 7. Converter Waveforms

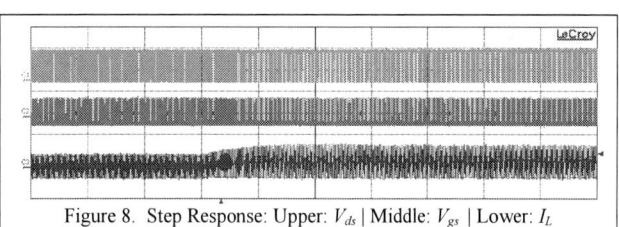

Figure 8. Step Response: Upper: V_{ds} | Middle: V_{gs} | Lower: I_L

VII. CONCLUSION

Ultimately, the goal of frequency variation in the synchronous boost converter is to improve operating efficiency. It may be observed from the efficiency plots in Fig. 9, that this control method accomplishes just that. Where a fixed frequency must be selected such that ZVS is achievable over most of the load range, especially at high power, the variable frequency method can improve efficiency at several load conditions. This is accomplished under light load by increasing the switching frequency in order to minimize the amount of circulating current through the converter. When the load is heavier, the converter decreases the switching frequency so that the ZVS condition is maintained. Control of the switching frequency is established through the monitoring of the magnitude of the minimum, or resonant, current. If very simple guidelines are observed, the resonant current control loop is nearly independent from the average current control loop, with only a need for the resonant current controller to have higher bandwidth in order to reduce the impact of a specific disturbance.

This paper is intended to indicate the feasibility of this type of variable-frequency control for ZVS boost converters. Its potential is derived from increased efficiency, more flexible design, as well as wide load and input ranges. However, several challenges must be faced in the future. As pointed out earlier, accurate and high-frequency current sensing is a must for this method. Also, the amount of current ripple at both the source and load puts a premium on operating the ZVS boost converter in a multi-phase arrangement. Therefore, it is imperative that this method be extended to multi-phase systems in the future.

REFERENCES

[1] G. Hua and F.C. Lee. "Soft-switching techniques in PWM converters" *IEEE Trans. on Industrial Electronics,* vol. 42, no. 6, pp. 595-603, 1995

[2] B. Acharya, D.M. Divan, and R.W. Gascoigne, "Active power filters using resonant pole inverters," *IEEE Trans. on Industry Applications,* vol. 15, no. 6, pp. 1269-1276, 1992

[3] W. Lau and S. R. Sanders, "An integrated controller for a high frequency buck converter," *in Proc. IEEE Power Electron. Spec. Conf.,* 1997, pp. 246-254

[4] S.-G. Yoon, J.-M. Lee, J.-H. park, I.-K. Lee, and B.H. Cho, "A frequency controlled bi-directional synchronous rectifier converter for HEV using super-capacitor," *in Proc. IEEE Power Electron. Spec. Conf.,* 2004, pp. 135-140.

[5] P. Andreassen, G. Guidi, and T. M. Undeland, "Digital variable frequency control of zero voltage switching and interleaving of synchronous buck converters," *in Proc. Power Electron. and Mot. Cont. Conf.,* 2006, pp. 184-188

[6] Wensong Yu and Jih-Sheng Lai. "Ultra high efficiency bidirectional dc-dc converter with multi-frequency pulse width modulation," *in Proc. Applied Power Electronics Conference,* 2008. pp.1079-1084

[7] Junhong Zhang, Jih-Sheng Lai, and Wensong Yu. Bidirectional DC-DC converter modeling and unified controller with digital implementation," in *Proc. Applied Power Electronics Conference,* 2008. pp. 1747-1753

[8] P. Andreassen and T.M. Undeland, "Digital control techniques for current mode control of interleaved quasi square wave converter," *in Proc. Power Electron. Spec. Conf,* 2005. pp. 910-914

[9] D. M. Sable, F.C. Lee, and B.H. Cho. "A zero-voltage-switching bidirectional battery charger/discharger for the NASA EOS satellite," *in Proc. Applied Power Electronics Conference,* 1992. pp. 614-621

[10] Bryant, B.; Kazimierczuk, M.K. "Small-signal duty cycle to inductor current transfer function for boost PWM DC-DC converter in continuous conduction mode" *Proceedings of the 2004 International Symposium on Circuits and Systems, 2004.* 2004. pp. 856-859

Power Saving Control Strategies and Their Implementation in DC/DC Converter for Data and Telecommunication Power Supply

Rais Miftakhutdinov
Power Supply Control Group
Texas Instruments Inc.
Cary, NC 27518, USA
r-miftakhutdinov1@ti.com, rais@ieee.org

Abstract— **The paper focuses on promising topologies and control strategies to meet strict power saving and efficiency requirements over wide output power range demanded by new regulations and trends in data and telecommunication industry. The optimal control algorithm for phase shifted ZVS DC/DC converter is described implemented in new IC controller. Some practical design recommendations are provided allowing zero voltage switching at all operating conditions. The advantages of suggested control strategy are verified by test results based on 95.3% efficient, 12-V, 660-W output DC/DC converter for server power supply.**

I. INTRODUCTION

Today's approach to data and telecommunication power supply design focuses on power saving and efficiency increase over wide output power range demanded by new regulations and industry trends [1], [2]. Section II of the paper provides brief review of new power saving regulations for server and telecom power supplies and sets the efficiency design goals for post-PFC isolated DC/DC converter. To meet these challenging efficiency and power saving requirements, few promising power stage topologies are available including asymmetrical half-bridge, LLC resonant and phase-shifted full-bridge [3]-[5]. Pros and cons of these topologies along with the review of possible control strategies are discussed in Section III. Section IV describes the new optimal control algorithm for the phase shifted ZVS converter providing power saving and high efficiency over entire output power range and the related IC controller, where this algorithm is implemented. Section V focuses on aspects of providing zero-voltage-switching (ZVS) over wide operating conditions for the phase shifted converter using new CoolMOS® or similar technology for primary-side high-voltage power FETs. Advantages of suggested control algorithm and validity of ZVS design recommendations are verified in Section VI by the test results of 95.3% efficient, 12-V, 660-W output DC/DC converter prototype designed for the server power supply.

II. STRIVE FOR EFFICIENCY AND POWER SAVING

High efficiency was always important for data and telecommunication power supply to achieve high power density and improve thermal performance. So far, only high efficiency at maximum power was required because it determines reliability, size and cost of equipment and cooling. Recently the focus is shifted on energy saving and high efficiency over the entire output power range. This approach reduces the overall power consumption, because data and telecommunication systems operate significant amount of time at mid and light loads. Worldwide movement for "Green power" and energy saving has resulted in new standards for increased efficiency over wide output power range. One example is ENERGY STAR®, which is a joint program of the U.S. Environmental Protection Agency and the U.S. Department of Energy. The version 1 of ENERGY STAR® Program Requirements for Computer Servers, which is effective since May 15, 2009, sets the efficiency and power consumption requirements for server power supplies [1]. Table I shows efficiency requirements at 10%, 20%, 50% and 100% output power of single-output AC/DC and DC/DC converters set force in this Program. The typical front-end server power supply includes PFC followed by an isolated DC/DC converter. Many server power supplies also include 5-V standby power supply. Thus, the PFC and standby power supply efficiency and losses have to be taken into account while setting the design goal for the post-PFC converter.

TABLE I. EFFICIENCY REQIREMENTS FOR SINGLE OUTPUT AC/DC OR DC/DC SERVER POWER SUPPLIES

Rated Output Power	10% Load	20% Load	50% Load	100% Load
≤ 500 W	70%	82%	89%	85%
501 – 1000 W	75%	85%	89%	85%
> 1000 W	80%	88%	92%	88%

978-1-4244-4782-4/10 $26.00 © 2010 IEEE

TABLE II. EFFICIENCY DESIGN GOALS FOR PFC, STANDBY POWER SUPPLY AND 660-W DC/DC CONVERTER IN SERVER POWER SUPPLY

Rated Output Power	10% Load	20% Load	50% Load	100% Load
Efficiency from Table 1	75%	85%	89%	85%
AC/DC Efficiency	80.8%	88.8%	90.5%	85.1%
Overall Power Consumption	89 W	157 W	376 W	788 W
PFC Efficiency	95.3%	96.4%	97.6%	97.7%
PFC Output Power	85 W	151 W	367 W	770 W
Standby Power	6 W	7 W	10 W	10 W
Standby Power Efficiency	80%	82%	85%	85%
Standby Power Consumption	7.5 W	8.5 W	12 W	12 W
DC/DC Input Power	77.5 W	142.4 W	355 W	758 W
DC/DC Output Power	66 W	132 W	330 W	660 W
DC/DC Efficiency Goal	85.2%	92.7%	93%	87.1%

Table II shows power distribution and efficiency goals for PFC, 660-W isolated DC/DC converter and 5-V, 2-A standby power supply as parts of the 670-W server power supply. The Table II is helpful to compare and optimize efficiency design goals of each part of the power supply and ensure the design meets overall power supply requirements. From Table II, the efficiency design goals for the post-PFC major DC/DC converter are 87.1% at 100% load, 93% at 50% load, 92.7% at 20% load and 85.2% at 10% load. Next step for the DC/DC converter design is the selection of proper power stage topology and control algorithm to meet the efficiency goals over the entire output power range.

III. DC/DC CONVERTER TOPOLOGIES AND CONTROL STRATEGIES

Topology selection of the post-PFC isolated DC/DC converter is critical for the overall power supply efficiency. The selection criteria must take into account the trade offs and challengers of maintaining required efficiency at mid and low output power range. It is desirable that the selected topology allows different control modes, optimal for the highest efficiency at specific output power conditions. Usually, the ZVS allowing topologies are preferable in such applications because of the relatively high input voltage range, from 350 V to 420 V. Potential candidates include phase shifted full-bridge, asymmetrical half-bridge, LLC resonant converter and variations of these topologies [3]-[5]. The optimal topology selection depends on the control strategy. Because of that consider first possible control algorithms allowing power saving and high efficiency over wide output power range. These control algorithms and approaches are listed below.

- Interleaving of few phases for better current and temperature distribution at maximum output power with gradual phase shedding when the load decreases;

- Synchronous rectification using MOSFETs with the diode emulation technique at light load to avoid current circulation. It could be beneficiary to shut off the drive circuit of rectifier MOSFETs at very light load where the drive losses exceed the conduction losses. Performance of synchronous rectifier significantly depends on accurate timing between the primary and secondary side switches;

- Proper use of ZVS and zero current switching (ZCS) technique to reduce switching losses in power MOSFETs. This requires optimal adaptive or predictable set of delays between switching events that depend on operating conditions;

- Optimal adjustment of intermediate bus voltage, drive voltage and other system parameters to maintain highest efficiency at different operating conditions;

- Smooth transition from one mode to another depending on operating conditions, for example from continuous mode to discontinuous, from fixed frequency to frequency foldback etc.;

- Proper use of pulse skipping or burst mode at light load and at no load to reduce the entire power consumption.

For the interleaved approach, the asymmetrical half-bridge topology is suitable best because of its relative simplicity per each phase [4]. One possible example of interleaving using 4 phases is shown in Fig. 1. The potential drawback of the asymmetrical half-bridge converter is its limited optimal input voltage range. This can be improved by selecting the unequal turn ratios of the power transformers [4].

In case of interleaving, the general challengers include current sharing between the phases and smooth phase shedding or adding depending on load change, especially at fast load current transients. Also, when the number of phases changes, it is desirable to adjust the phase shift angle between the control signals to maintain highest ripple cancellation effect. Based on these considerations, digital control with its programmability and flexibility is most suitable for such relatively complicated control algorithm. Although, the multi-phase interleaved approach has significant merits in achieving the best efficiency over wide output power range, the simpler attractive topologies for the 660-W converter are considered next.

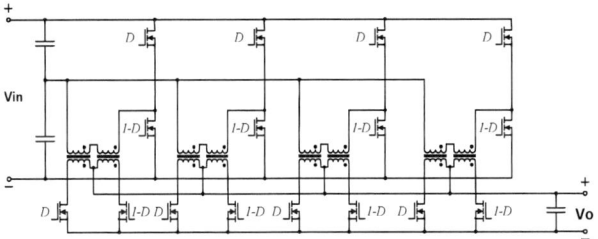

Figure 1. Four phases interleaved asymmetrical half-bridge

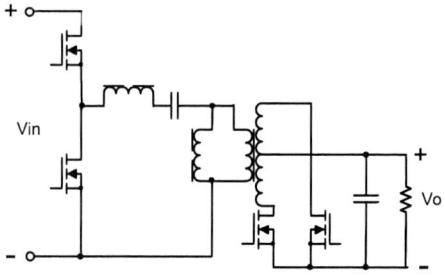

Figure 2. LLC resonant converter

Another strong candidate is the LLC resonant topology recently gaining popularity as post-PFC isolated DC/DC converter (Fig. 2). The main advantage of LLC resonant converter is ZVS for the primary side switches and ZCS for the secondary side synchronous rectifier MOSFETs [5]. This topology has the efficiency peak at the maximum input voltage, which is important advantage for the post-PFC converter, normally operating at the upper end of its input voltage range. Another advantage is the low-voltage rated secondary side synchronous rectifier FETs having lower Rdson versus the higher voltage rated counterparts. However, the variable switching frequency control, special attention to the light load operation and current sharing difficulties in case of interleaving are significant challenges to deal with this topology. Another drawback is the shorted output operation for some specified amount of time that sometimes required in the power system. Such operation requires significant operating frequency increase to limit the input current because there is no output inductor in such topology. Considering these trade offs and challengers, the use of such topology as part of the server power supply could have some limitations even if the high efficiency can be achieved.

The standard phase-shifted full-bridge topology, shown in Fig.3, has the long history of usage as the post-PFC converter [3]. The major advantages of this topology are ZVS of the primary power MOSFETs, fixed frequency PWM operation, reliable handling of the shorted output condition with the cycle-by-cycle current limiting and relatively wide input voltage operating range. The major drawback is the circulating current through the primary switches during PWM off time interval and need for the snubber or clamping circuit to prevent the ringing at the secondary side rectifiers.

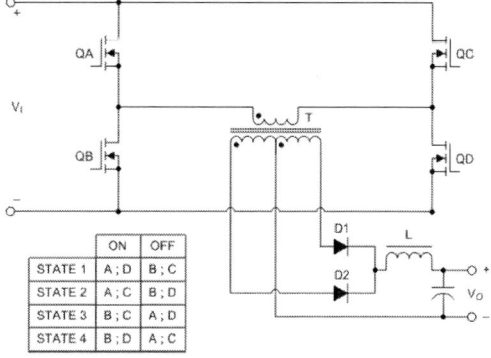

Figure 3. Phase-shifted full-bridge DC/DC converter

Nevertheless, to maintain high efficiency and reduce power consumption of the phase shifted converter over the wide output power range, an additional optimization and improvement of the control algorithm are needed. The following section discusses such optimization.

IV. PHASE SHIFTED DC/DC CONVERTER WITH ADVANCED CONTROL ALGORITHM AND RELATED IC CONTROLLER

The efficiency improvement of phase-shifted full-bridge DC/DC converter is achieved by using the synchronous rectification technique, control algorithm providing ZVS condition over the entire load current range, accurate adaptive timing of the control signals between primary and secondary FETs and special operating modes at light load for the highest efficiency and power saving [6]. The simplified electrical diagram of this converter is shown in Fig. 4. The controller IC is located on the secondary side of converter. Such location becomes popular trend allowing easier system level communication and better handling of some transient conditions that require fast direct control of the synchronous rectifier MOSFETs. The power stage includes primary side MOSFETs, QA, QB, QC, QD and secondary side synchronous rectifier MOSFETs, QE and QF. For the 12-V output converters in server power supplies use of the center-tapped rectifier scheme with L-C output filter is a popular choice.

To maintain high efficiency at different output power conditions, the converter operates in nominal synchronous rectification mode at mid and high output power levels, with transitioning to the diode rectifier mode at light load and further followed by the burst mode, as the output power becomes even lower. All these transitions are based on the current sensing on the primary side using the current sense transformer in this specific case.

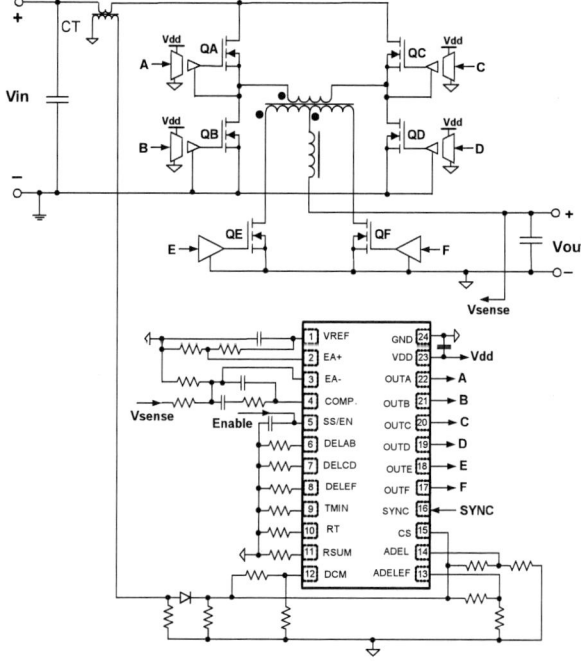

Figure 4. Phase shifted full-bridge converter with advanced controller

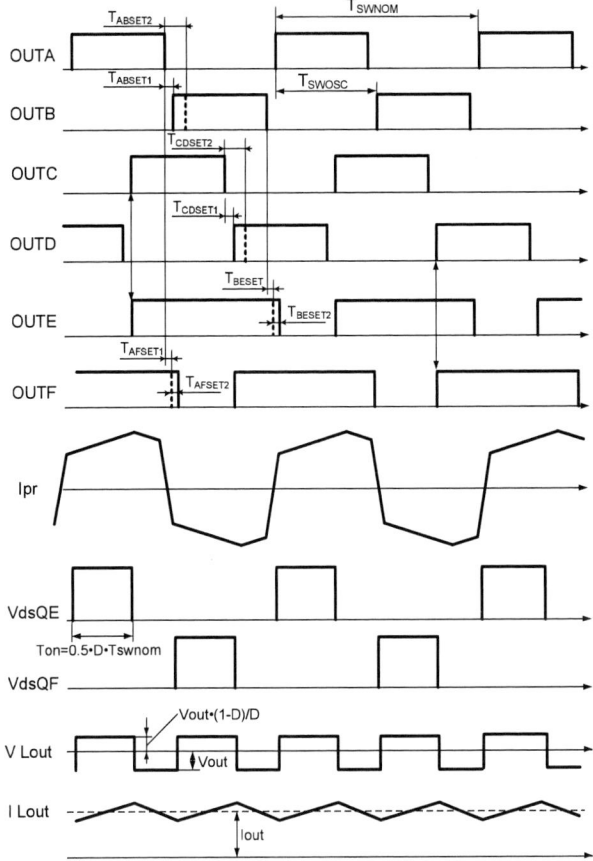

Figure 5. Major waveforms of phase-shifted converter at nominal mode

Major waveforms of the phase-shifted converter during nominal operation mode are shown in Fig. 5. Upper six waveforms in the Fig. 5 show the output drive signals of the controller. At nominal mode, the outputs OUTE and OUTF overlap during the part of the switching cycle when the both rectifier MOSFETs are conducting and the windings of power transformer are shorted. Current, Ipr, is the current flowing through the primary winding of power transformer. The bottom four waveforms show the drain-source voltages of rectifier MOSFETs, VdsQE and VdsQF, the voltage at the output inductor, V Lout, and the current through the output inductor, I Lout. Proper timing between the primary switches and synchronous rectifier MOSFETs is critical to achieve highest efficiency and reliable operation in this mode. The controller IC adjusts the turn OFF timing of rectifier MOSFETs as function of load current to ensure the minimum conduction time and reverse recovery losses of their internal body diodes.

ZVS is an important feature of relatively high input voltage post-PFC DC/DC converters to reduce switching losses associated with the internal parasitic capacitances of power switches and transformers. The controller ensures ZVS conditions over the entire load current range by adjusting the delay time between the primary MOSFETs switching in the same leg in accordance to the load variation. Controller also limits the minimum ON-time pulse applied to the power

transformer at light load, allowing the storage of sufficient energy in the inductive components of power stage for the ZVS transition.

As soon as the load current keeps reducing from the mid load current down to no-load condition, the controller selects the most efficient power saving mode by moving the converter from the nominal operation mode to the transition mode, then to the discontinuous-current diode-rectification mode and, eventually, at very light load and at no-load condition, to the burst mode. These modes and related output signals, OUTE, OUTF, driving the rectifier MOSFETs, are shown in Fig. 6.

It is necessary to prevent the reverse current flow through the synchronous rectifier MOSFETs and output inductor at the light load, during parallel operation and at some transient conditions. Such reverse current results in circulating of some extra energy between the input voltage source and the load and, therefore, causes increased losses and reduces efficiency. Another negative effect of such reverse current is the loss of ZVS condition. The suggested control algorithm prevents reverse current flow, still maintaining most of the benefits of synchronous rectification by modulating the drive signals of rectifier MOSFETs in a predetermined way. At some pre-determined load current threshold, the controller reduces the synchronous rectifier drive signals from the overlapping to 50%, and then gradually reduces their duration, until disabling at the second predetermined threshold. This operation is called transition mode.

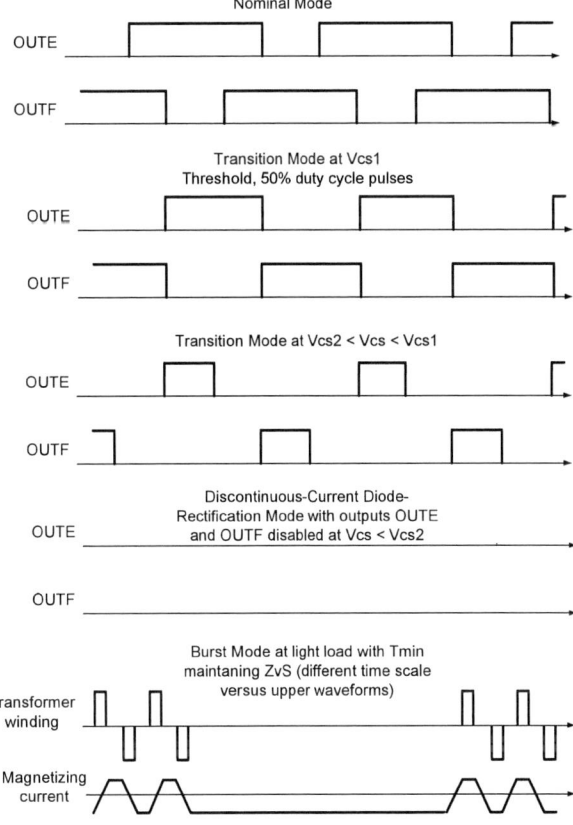

Figure 6. Timing diagrams and transitions between power saving modes

Synchronous rectification using MOSFETs requires some electrical energy to drive the MOSFETs. There is a condition below some light load threshold when the MOSFET drive related losses exceed the saving provided by the synchronous rectification. At such light load, it is beneficiary to disable the drive circuit and use the internal body diodes of rectifier MOSFETs, or external diodes in parallel with the MOSFETs, for more efficient rectification. This mode of operation is called discontinuous-current diode-rectification mode.

At the very light load and no load conditions, the duty cycle, demanded by the closed-feedback-loop control circuit for output voltage regulation, can be very low. This could lead to the loss of ZVS condition and increased switching losses. To avoid the loss of ZVS, the control circuit limits the minimum ON-time pulse applied to the power transformer. Therefore, the only way to maintain regulation at very light load and at no-load condition is to skip some pulses. The controller skips pulses in a controllable manner to avoid saturation of the power transformer. Such operation is called burst mode.

One example of the controller IC closely following the described control algorithm is the UCC28950 from Texas Instruments. The block diagram of this controller IC is shown if Fig. 7. This controller provides adaptive timing between primary and secondary MOSFETs switching and provides up to 10:1 delay time adaptive ratio for the primary switches in the same leg as function of the CS-pin signal. The light load efficiency management block provides optimal transition between different operating modes shown in Fig. 6 as the function of the CS-pin signal. The controller provides all major functions and features usually found in such ICs, like the accurate resistive switching frequency setting, user selectable voltage or current mode operation, closed loop soft start with enable, cycle-by-cycle current limit, PWM ramp slope compensation, under voltage shutdown, accurate reference regulator, synchronization pin for the interleaved operation of converters and others.

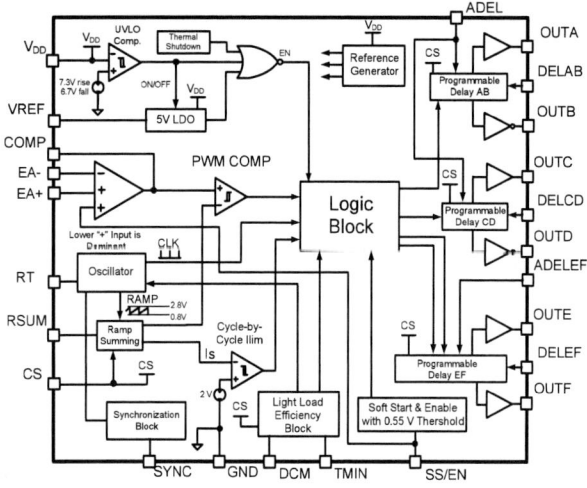

Figure 7. Block diagram of the advanced phase-shifted contoller IC UCC28950

V. ZVS PRACTICAL RECOMMENDATIONS

One of the most important aspects of post-PFC converter design is maintaining ZVS condition over entire load current range. For the accurate ZVS prediction, the switching energy of power FETs as function of the applied drain-source voltage Vds, needs to be known. The new super junction technology for high voltage MOSFETs significantly reduces Rdson, Cds and Cgs capacitances [7] providing lower conduction and switching losses. Because of significant non-linear behavior of drain-source capacitance, the super junction MOSFETs, like CoolMOS^{TM}, require new analytical model to estimate switching losses, junction capacitance energy and set proper ZVS conditions. New equation is suggested for the super junction FETs that is simple and provides accurate results:

$$Ecds = \frac{Coss}{Kc} \cdot (Vdsoss)^2 \cdot \ln(\frac{Vds+5V}{V}) + \frac{Cinit \cdot (Vds)^2}{2} \quad (1)$$

Here, *Ecds* is the energy stored, *Coss* is output capacitance from datasheet at *Vdsoss* = 25V, *Vds* is the voltage, where the switching energy has to be found. Parameters *Kc* and *Cinit* are constants for each specific MOSFET. For SPA11N60FCD type FET from Infineon, *Kc* = 2.2 and *Cinit* = 40 pF.

The plots in Fig. 8 compare calculated energy using (1) with the experimental plot provided in the datasheet.

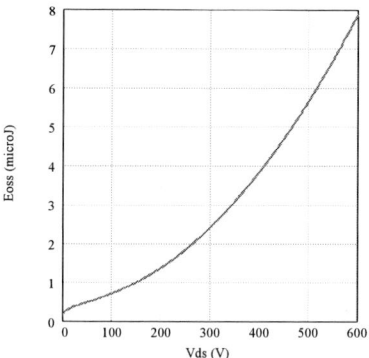

a) Analytically derived plot

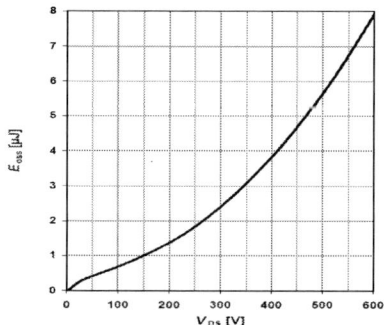

b) Experimental plot from datasheet

Figure 8. Comparison of analytical energy model Ecds over Vds for SPA11N60FCD type MOSFET

978-1-4244-4782-4/10 $26.00 © 2010 IEEE   1901

One can see that both plots correlate very accurately. The analytical equation is very helpful for accurate ZVS prediction.

The next step for the ZVS based design is to determine the leakage and magnetizing inductances of power transformer that guarantee ZVS switching over the entire load current range. The energy stored in the leakage and magnetizing inductances is compared with the energy required to discharge MOSFET capacitances in accordance to (1). At the mid and high load current conditions the ZVS is maintained by the energy stored in the leakage inductance. In some cases, it is beneficiary to use the special inductor in series with the primary winding of power transformer along with the clamped diodes as described in [8] to ensure ZVS. At light load the energy is provided by the magnetizing inductance of power transformer. Because the transition resonance frequency associated with the leakage inductance is significantly different from the resonance frequency associated with the magnetizing inductance, the delay time between primary FETs must be adjusted as function of the load current. Such adaptive delay time is provided by the controller in Fig. 7.

VI. EXPERIMENTAL RESULTS

The 660-W output power prototype of phase shifted full-bridge DC/DC converter has been designed and tested to verify the suggested control algorithm, meeting of the efficiency goals and the accuracy of theoretical efficiency and power losses prediction over the entire load current range. The input voltage is 350 V to 420 V and the output is 12 V, 55 A. The primary MOSFETs are SPA11N60CFD and the synchronous rectifier MOSFETs are FDP047AN08A0, two in parallel. The measured efficiency of the prototype (solid curve) and analytical prediction (dashed curve) along with the design goals (circles) over the load current are shown in Fig. 9. There is good correlation between the analytically predicted and measured efficiency in the load current range below 12% and above 50% of maximum current. The efficiency difference from 12% to 50% load is attributed to the partial ZVS loss in this range that adds additional switching losses.

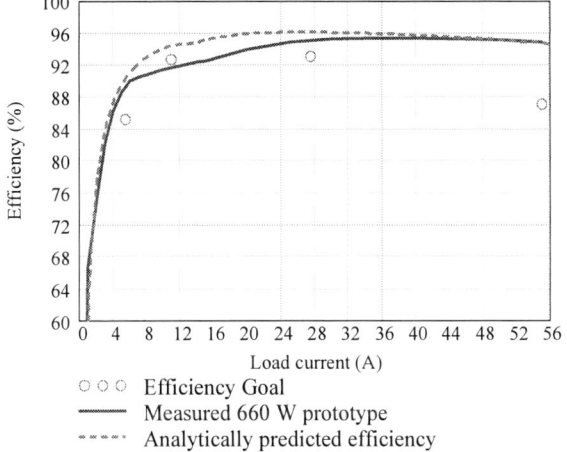

Figure 9. Measured (solid curve), analytically predicted (dashed curve) and efficiency design goals (circles) of the 660 W prototype

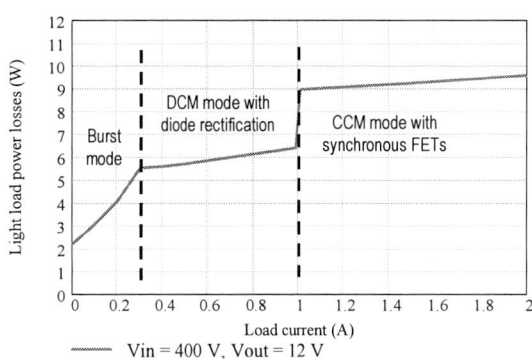

Figure 10. Power losses of 660 W DC/DC converter prototype at light load

One can see that there is good margin for the achieved efficiency at 10%, 50% and 100% load versus the design goals. However, the design goal at 20% load is not achieved and additional, at least 1% efficiency increase, is needed. There are few options to meet the design goal at 20% load. One option is to choose the higher Rdson FETs with the smaller die size and lower parasitic capacitances. The efficiency margin at 50% and 100% load allows sacrifice of some conduction losses for the sake of the reduced switching losses. However, this approach may not be the best one, because in some cases the industry standards exceed the regulations set by the government shown in Table 1. The better way is to reduce the magnetizing inductance and slightly increase the leakage inductance of the power transformer to achieve ZVS in the shown area. This example demonstrates that to meet such challenging requirements, additional tuning and design iteration might be needed.

Because of the power saving trend even at very light and no-load conditions, careful optimization of operation at light load condition of the prototype has been fulfilled to set the proper boundaries between different operation modes. The result of this optimization is shown in Fig. 10. This plot demonstrates the power savings while moving from the synchronous rectification mode above 1-A load current, into the discontinues current mode (DCM) with the diode rectification between 0.3-A and 1-A load current, and eventually into the burst mode operation at load current below 0.3 A.

VII. CONCLUSION

The new industry requirements and regulations for increased efficiency and power saving over wide output power range for the data and telecommunication power supply are outlined. Efficiency design goals for the post-PFC isolated DC/DC converter are derived as part of the 670-W server power supply that includes the PFC, the standby power supply with 5-V, 10-A output and the main DC/DC converter with 12-V, 55-A output. Few topologies including interleaved asymmetrical half-bridge, LLC resonant and phase-shifted full-bridge have been compared as candidates to meet the efficiency goals. The control strategies providing power saving and maintaining high efficiency control over the entire output power range are listed and discussed. Based on this, the phase-shifted full-bridge converter has been selected for further consideration and its advanced control algorithm presented along with the related IC controller. Some practical design rules to achieve the full ZVS and the new accurate analytical equations for the switching energy of the CoolMOS® technology based FETs are provided. Advantages of suggested control algorithm and the accuracy of analytical prediction to meet the efficiency goals are verified by the 660-W DC/DC converter prototype for the server power supply with the efficiency up to 95.3% achieved.

REFERENCES

[1]   "ENERGY STAR® program requirements for computer servers, version 1.0," Available:
http://www.energystar.gov/index.cfm?c=archives.enterprise_servers

[2]   R. Mammano, "Improving power supply efficiency - The global perspective," Texas Instruments Power Supply Design Seminar, Topic 1, SEM-1700, 2006.

[3]   J. Zhang, X. Xie, X. Wu and Z. Qian, "Comparison study of phase-shifted full bridge ZVS converters," in *PESC 2004 Conference Proceedings,* 2004, pp. 533-539.

[4]   R. Miftakhutdinov, A. Nemchinov, V. Meleshin and S. Fraidlin, "Modified asymmetrical ZVS half-bridge DC-DC converter," in *APEC 1999 Conference Proceedings,*1999, pp. 567-574.

[5]   D. Fu, B. Lu and F. Lee, "1MHz high efficiency LLC resonant converters with synchronous rectifier," in *PESC 2007 Conference Proceedings,* 2007, pp. 2404-2410.

[6]   R. Miftakhutdinov, "Power Saving Solutions in DC/DC Converter for Data and Telecommunication Power System," in *PEDS 2009 Conference Proceedings,* available on CD.

[7]   F. Bjoerk, J. Hancock and G. Devoy, "CoolMOS[TM] CP- How to make most beneficial use of the latest generation of super junction technology devices," *Infineon Appl. Note,* AN-CoolMOS-CP-01, February 2007.

[8]   R. Redl, N.O. Socal and L. Balogh, "A Novel Soft-Switching Full-Bridge DC/DC Converter: Analysis, Design Considerations, and Experimental Results at 1.5 kW, 100 kHz," in *IEEE Trans. on PE.,* vol. 6, No.3, pp. 408-418, July 1991

Analysis and Optimized Design of an Efficient High-Voltage Converter with High Output Capacity

Huai Wang[1], Henry Shu-hung Chung[2]
Centre for Power Electronics
City University of Hong Kong
Kowloon, Hong Kong
[1]huwang@student.cityu.edu.hk
[2]eeshc@cityu.edu.hk

Adrian Ioinovici
Department of Electrical and Electronics Engineering
Holon Institute of Technology
Holon, 58102, Israel
adrian@hit.ac.il

Abstract: **This paper presents the analysis and design of a new class of multiphase converter for high-to-low voltage applications. The converter features a V_i / n voltage stress on the primary-side switches, which are commutated under zero voltage switching (ZVS), and I_o / n current stress on the output inductors and rectifier diodes. A DC analysis allows for deriving the duty-cycle loss and the ZVS load range. A trade-off design is performed for optimizing these two conditions. The proposed converter is favorably compared with available solutions from the point of view of number of switches and the voltage stress to which they are submitted. Experimental results based on a 1500V/48V, 2 kW prototype verify the theoretical predictions.**

I. INTRODUCTION

Two of the challenges of high voltage dc-dc conversion are the unavailability of high frequency switching devices, like MOSFET for high voltages, and the high conduction loss induced by the on-resistances of the available switching transistors. If the switches are submitted to a high voltage, their on-resistance, which is proportional to V_{DS}^{k}, where $k \subset [1.6, 2.6]$, will take a large value. To overcome the above difficulties, solutions requiring a reduced voltage rating of switching devices were proposed, such as three-level (TL) converters [1-2] and input-series connected modular converters [3-4]. When the input voltage V_i is higher than 1000V, the voltage stress in TL converters, $V_i / 2$ is still high. And the modular converters need an increased number of switches and additional control strategies. A converter with $V_i / 3$ voltage stress on the primary-side switches, achieving zero-voltage-switching (ZVS) was proposed in [5].

A class of high voltage dc-dc converter has been proposed recently for high-to-low voltage applications [6]. It provides a single-step conversion solution for a DC-fed railway system to transfer power from high voltage dc transmission lines (600 V, 750 V, 1500 V and 3000 V) to low-voltage equipment (24 V, 32 V, 48 V and 64 V [7]). This converter has the following properties: a) a reduced number of switches, b) V_i / n voltage stress on the primary-side switches, c) switches commutation

under ZVS, (d) high output current capacity, (e) self-balanced input capacitor voltage and (f) reduced size of magnetic components.

The purpose of this paper is to provide a detail dc analysis and optimized design of the proposed converter. A 1500V/48V, 2 kW prototype has been built and evaluated to verify the theoretical predictions.

II. PROPOSED CONVERTER STRUCTURE AND OPERATION PRINCIPLE

A. Circuit structure

The proposed converter shown in Fig.1 is formed by n switch pairs in the primary-side, an n-phase isolation transformer, whose primary windings are connected with dc blocking capacitors C_{W1}, C_{W2}, ..., C_{Wn} and an n-phase rectifier. There are n dc-link capacitors C_1, C_2, ..., C_n of the same capacitance to split equally the input voltage V_i. The structure of the $(k-1)$-th and k-th switch pairs and corresponding rectifier branches is shown in Fig. 2. Each switch pair contains two switching devices, S_{kU} and S_{kD} (with the built-in diode-capacitor pairs D_{kU}-C_{kU} and D_{kD}-C_{kD}) operated in anti-phase. The switching patterns applied to the two switch pairs have a phase difference of $360^\circ / n$. The output current is shared by n identical parallel diode-inductor branches: D_1-L_{f1}, D_2-L_{f2},..., D_n-L_{fn} to reduce the current stress of the inductors to one-nth of the load current.

B. Operation Principle

The operation of the k-th switch pair in the k-th interval T_s / n of a switching cycle can be seen in Fig.3.

Mode k_1 [t_{k0}, t_{k1}]: At t_{k0}, S_{kD} is switched off with ZVS in the presence of C_{kD}. C_{kU} and C_{kD} are discharged from V_i / n to 0, and charged from 0 to V_i / n, respectively.

Mode k_2 [t_{k1}, t_{k2}]: The primary current increases linearly while the converter is still in freewheeling stage. This mode ends with the current $i_{pk}(t)$ reaching the reflected secondary current of $i_o / (mn)$. The duration of this mode t_{k12} was given by

---

The work described in this paper was fully supported by a grant from the Research Grants Council of the Hong Kong Special Administrative Region, China (Project No.: CityU 112406).

978-1-4244-4782-4/10 $26.00 © 2010 IEEE

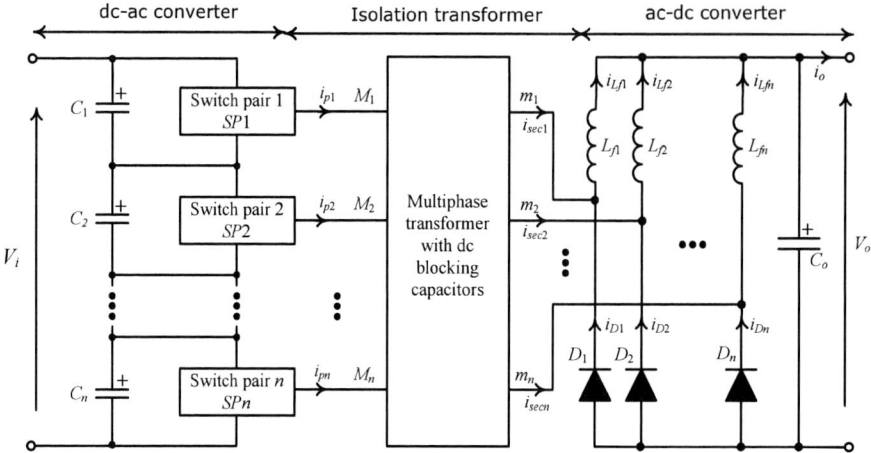

Figure 1. Circuit structure of the proposed class of high-voltage dc-dc converters.

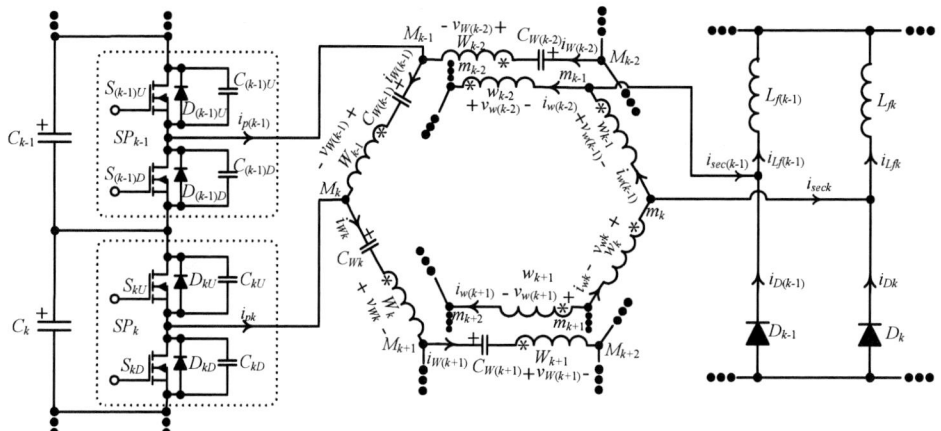

Figure 2. Circuit structure of the $(k-1)$-th and k-th switch pairs and rectifier branches.

$$t_{k12} = \frac{i_o L_{lk}}{mV_i} \tag{1}$$

Mode k_3 $[t_{k2}, t_{k3}]$: The converter enters the energy-transfer stage.

Mode k_4 $[t_{k3}, t_{k4}]$: S_{kU} is turned-off with ZVS at t_{k3}. The converter is transited from the energy-transfer stage to the freewheeling stage. C_{kU} and C_{kD} are charged and discharged by $i_{pk}(t)/2$. The mode ends when C_{kD} is fully discharged.

Mode k_5 $[t_{k4}, t_{k5}]$: S_{kD} turns on with ZVS. The converter operates in a new freewheeling stage.

III. DC ANALYSIS

A. Conversion ratio and duty cycle loss

From an input-to-output energy balance written for a T_s/n interval and by neglecting the very short ZVS commutation time t_{k01} and t_{k34}, one gets

$$\frac{V_i}{n} \frac{i_o}{mn} \left(\frac{DT_s}{n} - t_{k12} \right) = V_o i_o \frac{T_s}{n} \tag{2}$$

Therefore, according to (1) and (2), the conversion ratio M and duty cycle loss D_{loss} were given by

$$M = \frac{V_o}{V_i} = \frac{1}{mn^2} \left(D - \frac{ni_o L_{lk}}{mV_i T_s} \right) \tag{3}$$

$$D_{loss} = \frac{ni_o L_{lk}}{mV_i T_s} \tag{4}$$

B. ZVS voltage range

The ZVS conditions are guaranteed by that the energy stored in the leakage inductance L_{lk} involved in the transition period from freewheeling stage to energy-transfer stage is higher than the energy in the equivalent capacitance of the capacitors C_s. That is

$$\frac{1}{2} \left(\frac{i_o}{mn} \right)^2 \frac{L_{lk}}{2} > \frac{1}{2} \left(\frac{V_i}{n} \right)^2 2C_s \tag{5}$$

Beside this, the intervals of the left and right dead time

between the PWM signals of the anti-phase driven switches should meet the requirements:

$$t_{dl} > t_{k01} = \frac{1}{\omega} \sin^{-1} \left[2mV_i \Big/ \left(i_o \sqrt{L_{lk}/C_s} \right) \right] \qquad (6)$$

$$t_{dr} > t_{k34} = 2mV_i C_s / i_o \qquad (7)$$

where t_{dl} and t_{dr} are the left and right dead time respectively. $\omega = \sqrt{1/(L_{lk}C_s)}$ and C_s is the output parasitic capacitance of the MOSFETs.

C. Self-balanced voltages of the input capacitors

In the freewheeling stage of the k-th switch pair, Mode k_5, shown in Fig. 4, the voltage across $C_{W(k-1)}$ equals to V_{ik},

$$V_{CW(k-1)} = V_{ik} \qquad (8)$$

On the other hand, as the blocking capacitors have very large values, the voltage on each one can be considered constant within a switching cycle. So, the voltage across $C_{W(k-1)}$ can also be expressed as

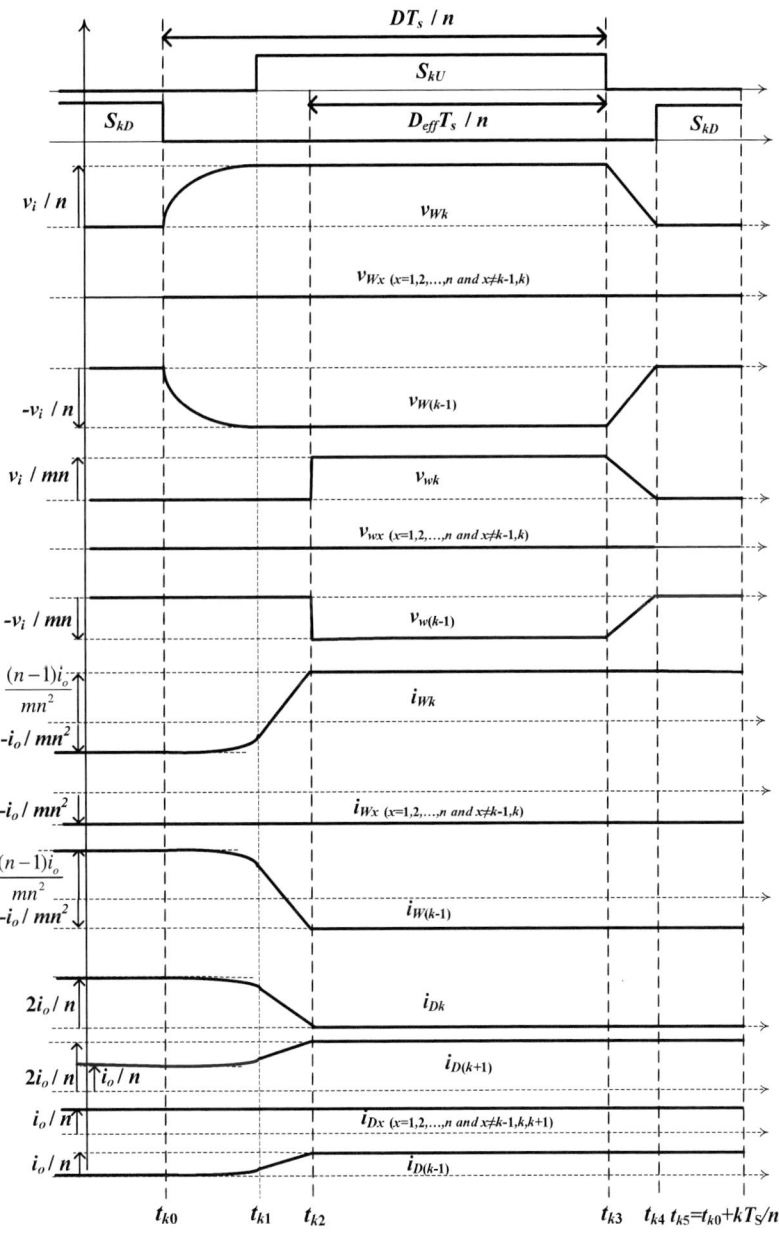

Figure 3. Timing diagram of *k-th* one-*n*th of a switching cycle.

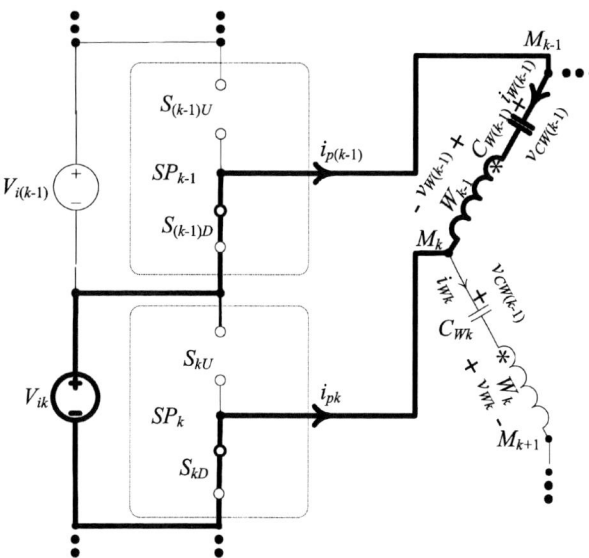

Figure 4. Freewheeling stage of k-th switch pair.

$$V_{CW(k-1)} = \frac{\left(T_s - D_k \dfrac{T_s}{n}\right)V_{ik} + D_{k-1}\dfrac{T_s}{n}V_{i(k-1)}}{T_s} \qquad (k \neq n)$$

$$= V_{ik} + \frac{1}{n}\left[D_{k-1}V_{i(k-1)} - D_k V_{ik}\right]$$

(9)

Therefore

$$\frac{V_{ik}}{V_{i(k-1)}} = \frac{D_{(k-1)}}{D_k} \qquad (10)$$

A simple voltage loop control scheme was applied to the proposed dc-dc converter. The output voltage was sampled and compared with the reference to generate a pair of PWM signals with corresponding duty cycle. This pair of PWM signals was then shifted by times of $360°/n$ to drive the other switching pairs accordingly. Therefore, by neglecting small differences that can appear in the shifting process, the duty cycle of the generated PWM driving signals is the same. As the typical total delay time of the MOSFETs and drivers is tens of ns, in the worst case, the difference in delay times between different drivers can be of the order of tens of ns, implying negligible differences between duty-cycles D_k, D_{k-1}. This makes an eventual voltage difference among different serial input capacitors within 10V. As a result, regardless of the capacitance values, the voltage distribution on each input capacitor will always reach parity after a few switching cycles from start-up or other dynamic perturbations.

IV. DESIGN CONSIDERATIONS

A. Design Specifications

Design considerations are presented here based on the prototype specified in Table I. The actual input voltage can vary in the range ± 10% of the rated value. Four switch pairs are used, i.e. $n=4$.

TABLE I.   DESIGN SPECIFICATIONS OF VIN/4 DC-DC CONVERTER

Items	Values	Units	Remarks
$V_{in,min}$	1350	V	minimum input voltage
$V_{in,rated}$	1500	V	rated input voltage
$V_{in,max}$	1650	V	maximum input voltage
V_o	48	V	output voltage
P	2000	W	rated output power
f_s	50	kHz	switching frequency
n	4		number of modules
ΔI_o	10	%	current ripple
ΔV_o	1	%	voltage ripple
T_h	20	ms	hold-up time from 1500V to 1200V
$D_{eff,designed}$	0.5		designed effective duty cycle

B. Turn Ratio of the Isolation Transformer (m)

According to (3),

$$m = V_i D_{eff} / \left(n^2 V_o\right) \qquad (11)$$

where $D_{eff} = D - n i_o L_{lk} / \left(m V_i T_s\right)$. By considering that $n=4$, $V_i=1500$ V, $V_o=48$ V, $D_{eff,min}=0.5$, it results in $m=0.977$, thus the value of m is chosen as 1.

C. Design of Leakage Inductance of the Isolation Transformer (L_{lk})

The leakage inductance is a very crucial parameter to determine the duty cycle loss and soft-switching range of the designed prototype. It can be observed from Fig. 5 and Fig. 6 that a lower inductance value will be beneficial for reducing the duty cycle while narrowing the soft-switching range. Therefore, a compromised design was made and L_{lk} was selected with the value of 20μH to achieve soft-switching from a 60% load and above. The duty cycle loss at nominal load will be 0.11.

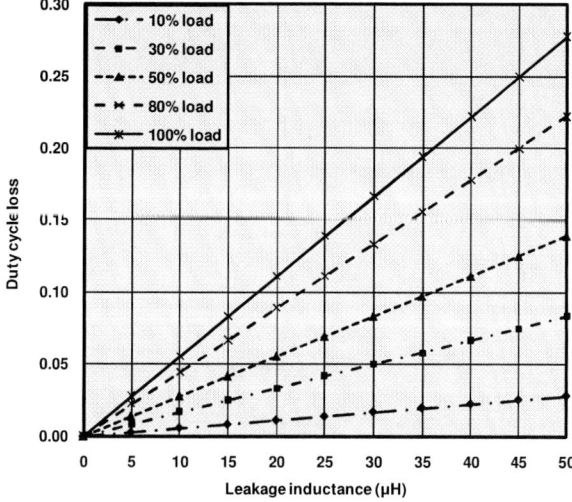

Figure 5. Duty cycle loss versus leakage inductance for different load conditions. ($V_{in}=1500$V, $I_{o,rated}=41.7$A, $m=1$, $n=4$, $f_s=50$kHz)

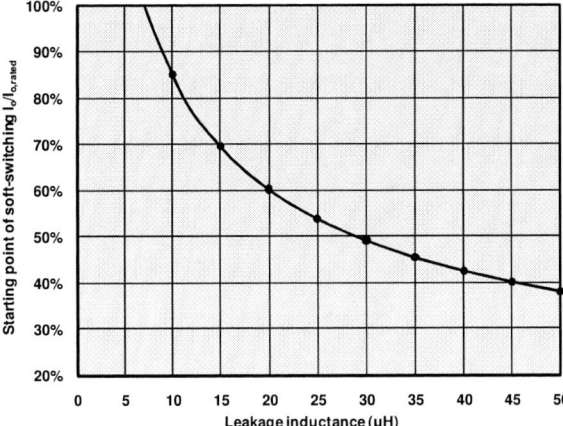

Figure 6. Starting point of soft-switching versus leakage inductance.
(V_{in}=1500V, $I_{o,rated}$=41.7A, m=1, n=4, f_s=50kHz)

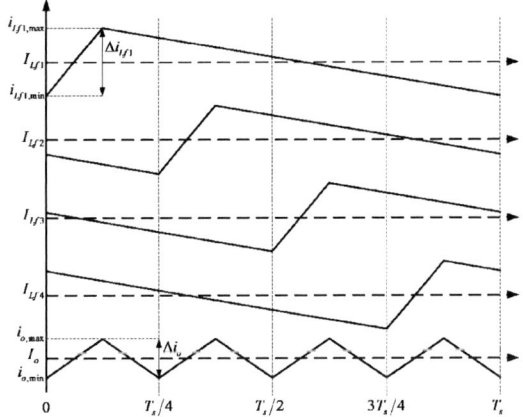

Figure 7. Phase inductor currents and output current. (n=4)

D. Design of Output Inductors (L_f)

According to Fig.7, the phase inductor current ripple (assuming that $L_{f1} = L_{f2} = \cdots = L_{fn} = L_f$) was given by

$$\Delta i_{Lfk} = \left(1 - \frac{D}{n}\right)\frac{V_o T_s}{L_f}, \qquad (n = 4) \qquad (12)$$

The output current ripple is given by

$$\Delta i_o = (1 - D)\frac{V_o T_s}{L_f} \qquad (13)$$

Therefore, the output current ripple is reduced to $n(1-D)/(n-D)$ times of the phase current ripple. For n=4, D=0.5, the output current ripple will be 4/7 of the phase current ripple. The output inductor was designed according to the specified requirement for Δi_o in Table I as

$$L_{f\min} \geq (1 - D_{\min})\frac{V_o T_s}{\Delta i_o} \qquad (14)$$

For each phase, an inductor with value of 124µH was chosen.

E. Design of Output Capacitor (C_o)

The output current ripple and RMS current of the capacitor are given by

$$\Delta v_o = \frac{\Delta i_{Lo}(T_s / n)}{8C_o} + \frac{\Delta i_{Lo} R_{ESR\_C_o}}{2} \qquad (15)$$

$$I_{Co,rms} = \frac{\sqrt{3}}{6}\Delta i_{Lo} \qquad (16)$$

The capacitance C_o, equivalent series resistance (ESR) R_{ESR_Co}, and ripple current rating I_{AC} of the capacitor are selected to guarantee that the output voltage ripple are less than 1% percent, and I_{AC} is no less than $I_{Co,rms}$. Accordingly, two capacitors with C_o=220µF, R_{ESR} = 212mΩ and I_{AC} =1.49A were paralleled as the output capacitors.

The power dissipation of the output capacitor is given by the following equation when leakage current effect is neglected.

$$P_{Co} = \frac{1}{12}\Delta i_{Lo}^2 R_{ESR} \qquad (17)$$

F. Design of Input Capacitors ($C_1, C_2,..., C_n$)

Two factors given in the following equations are considered here to select the proper input capacitors.

a) Input voltage ripple limitation:

$$C_{in} \geq \frac{\Delta i_{in}\Delta t}{\Delta V_{in}} + \frac{\Delta i_{in} R_{ESR\_C_{in}}}{2} \qquad (18)$$

b) Hold-up time limitation (20ms hold up time from 1500V to 1200V):

$$C_{in} \geq \frac{2P_o T_h}{\eta\left(V_{in\_norm}^2 - V_{in\_drop}^2\right)} \qquad (19)$$

where T_h is the designed hold-up time, V_{in_norm} is the normal operation input voltage, V_{in_drop} is the minimum dropped voltage within the hold-up time period, and η is the designed efficiency. Usually, the value of the C_{in} mainly depends on the hold-up time. The minimum calculated value of the input capacitor is 109.7µF. Practically, four 2*2 250V/680µF capacitor tanks were installed as the input capacitors.

G. Selection of MOSFET and Diode

The voltages stresses of the MOSFET are V_i/n and the current stresses are the reflected current of the secondary-side inductors. The voltage stresses of the secondary-side diodes are V_i/n while their average currents are I_o/n. For the specified prototype, 600V/35A MOSFET (SPW35N60CFD) and 600V/15A Diode (FFH15S60S) were utilized.

V. EXPERIMENTAL VERIFICATIONS

A 2kW 1500V/ 48V prototype with four switch pairs has been built up according to the specifications given in Table I. Experimental results are presented in Fig.8 to Fig. 11.

Fig. 8(a) and (b) show the switching waveforms of the upper and lower switch in the same switch pair respectively. It can be observed that both of the two switching transistors are turned on/off with zero-voltage, implying a negligible switching loss.

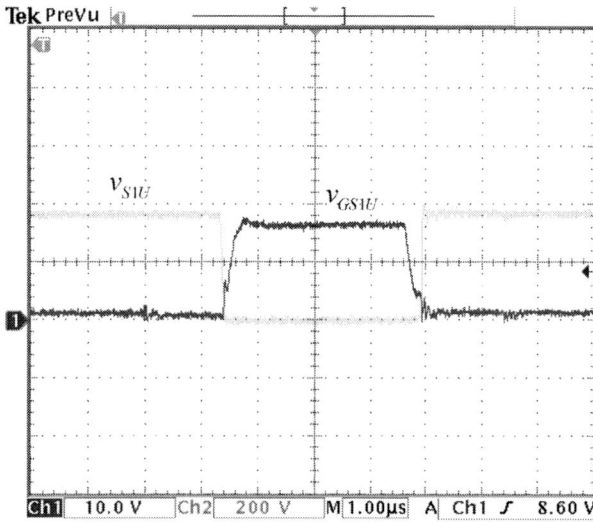

(a)   Switching waveforms of upper switch.

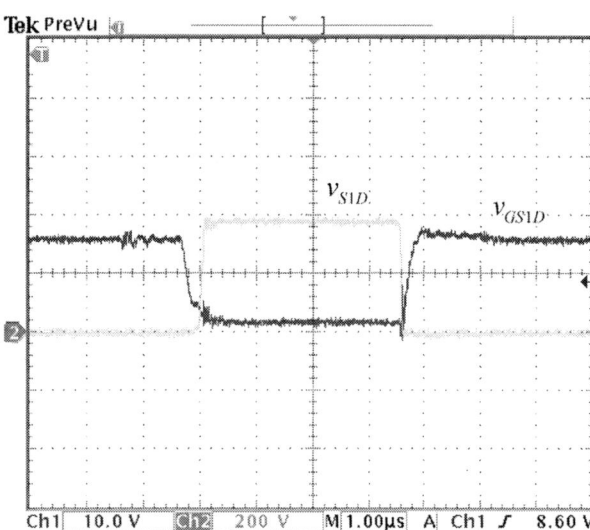

(b)   Switching waveforms of lower switch.

Figure 8.   Soft-switching of one switch pair (v_{GS1U} and v_{GS1D}: 10V/div, v_{S1U} and v_{S1D}: 200V/div, Timebase: 1.0μs).

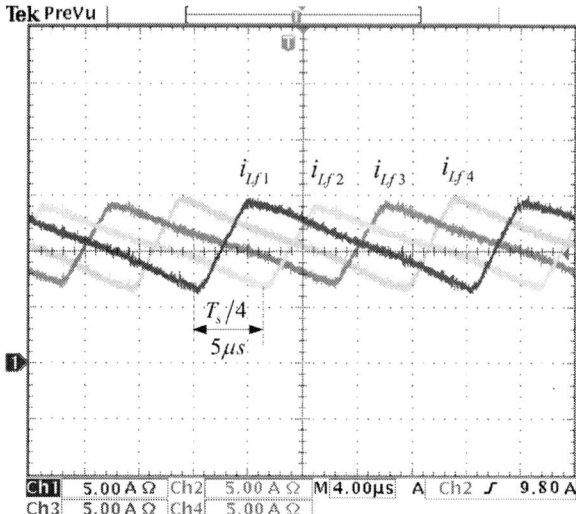

Figure 9.   Waveforms of output inductor currents
(i_{Lf1}, i_{Lf2}, i_{Lf3} and i_{Lf4}: 5A/div, Timebase: 4.0μs).

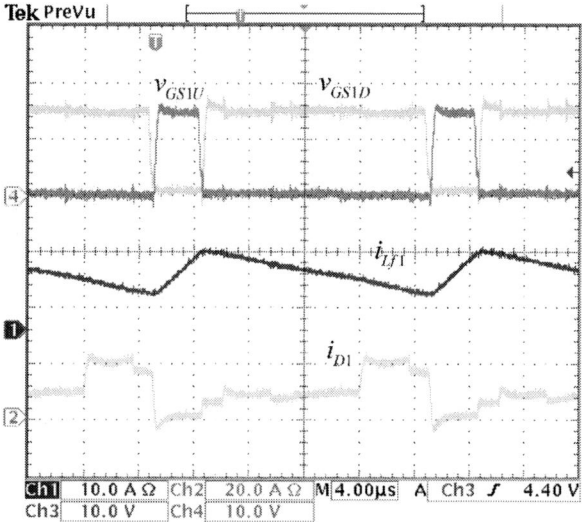

Figure 10.   Waveforms of driving signals and currents in a rectifier phase.
(v_{GS1U} and v_{GS1D}: 10V/div, i_{Lf1}: 10A/div, i_{D1}: 20A/div, Timebase: 4.0μs)

Fig. 9 presents the waveforms of the four output inductor currents. Due to a self-balance property of the input capacitor voltages, and practically almost identical isolation transformer windings and output inductors, the inductor currents in different rectification branches are well balanced with a phase-shift of $T_s/4$. Fig. 10 gives the inductor and diode currents in a rectifier branch; it can be seen that both the current stress of the inductor and average current of the diode are $1/n$ of the output current. The waveforms are in a good agreement with the theoretical analysis.

Fig.11 plots the efficiency of the built-up prototype. Efficiency at both light load, where ZVS was lost, and in the ZVS load range was measured. The efficiency at rated power is 88.23%.

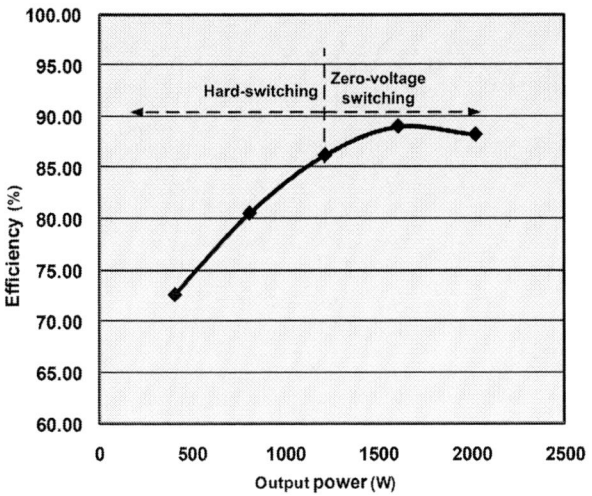

Figure 11. Measured efficiency under various load conditions.

TABLE II   COMPARISON OF NO. OF SWITCHES AND THEIR VOLTAGE
STRESSES OF FOUR KIND OF TOPOLOGIES

Topology	No. of switch	Voltage stress	Topology	No. of switch	Voltage stress
FB converter	4	V_i	ISOP converter	$4n$	V_i/n
TL converter	4	$V_i/2$	Proposed converter	$2n$	V_i/n

CONCLUSIONS

The analysis and design of a new dc-dc converter suitable for very high input voltages has been proposed. Both the voltage stress on the primary-side switches and current stress on the secondary-side inductors were reduced by n times ($n=4$ in the built prototype). The voltages across the input capacitors and the currents through the output inductors are well-balanced. Experimentally, an efficiency of 88.23% was obtained. Table II gives a comparison among four converter structures with reference to the number of switches and the voltage stress they have to withstand.

Comparing the proposed converter with a topology of input series output parallel connected converters (ISOP), one can see that in both of them the switches have to withstand the same voltage stress; however the number of switches is halved in the proposed solution.

REFERENCES

[1] Pinheiro J. and Barbi I., 1993, "The three-level ZVS-PWM DC-to-DC converter," *IEEE Trans. Power Electron.,* vol. 8, Oct, pp. 486-492.

[2] I. Barbi, R. Gules, R. Redl, and N. Sokal, "DC-DC converter: four switches $V_{pk} = V_{in} / 2$, capacitive turn-off snubbing, ZV turn-on," IEEE Trans. Power Electron., vol. 19, no. 4, pp. 918-927, Jul. 2004.

[3] R. Giri, V. Choudhary, R. Ayyanar, and N. Mohan, "Common-duty-ratio control of input-series connected modular DC-DC converters with active input voltage and load-current sharing," *IEEE Trans. on Ind. Appl.,* vol. 42, pp. 1101-1111, 2006.

[4] J. W. Kimball, J. T. Mossoba, and P. T. Krein, "A Stabilizing, High-Performance Controller for Input Series-Output Parallel Converters," *IEEE Trans. Power Electron.,* vol. 23, pp. 1416-1427, 2008.

[5] T. T. Song, H. S. H. Chung, and A. Ioinovici, "A high-voltage DC-DC converter with Vin/3 voltage stress on the primary switches," *IEEE Trans. Power Electron.,* vol. 22, no. 6, pp. 2124-2137, Nov. 2007.

[6] Huai Wang, Henry Chung, Saad Tapuchi and Adrian Ioinovici, "A Class of Single-Step High-Voltage DC-DC Converters with Low Voltage Stress and High Output Current Capacity," *IEEE Energy Conversion Congress and Exposition,* 2009, San Jose. CA.

[7] *IEEE Std. 1476-2000, IEEE Standard for Passenger Train Auxiliary Power Systems Interfaces,* New York: The Institute of Electrical and Electronics Engineers, 2000.

A Novel Three-Phase Three-Level ZVS PWM DC-DC Converter

Eloi Agostini Junior

Instituto de Eletrônica de Potência - INEP
Universidade Federal de Santa Catarina - UFSC
Florianópolis - Brazil
eloi@inep.ufsc.br

Ivo Barbi

Instituto de Eletrônica de Potência - INEP
Universidade Federal de Santa Catarina - UFSC
Florianópolis - Brazil
ivobarbi@inep.ufsc.br

Abstract— **A novel three-phase dc-dc converter based on the three-phase neutral point clamped (NPC) commutation cell is proposed. A static analysis is made for a particular mode of operation, allowing the development of a design procedure for the power stage. The small-signal analysis based on the phasor transformation is also proposed, providing fundamental knowledge for a satisfactory compensation in closed-loop operation. From the theoretical analysis carried out, a design procedure is elaborated, providing the values of all power stage components. Experimental results of a 10kW prototype are also presented to validate the theoretical considerations made in the paper.**

I. INTRODUCTION

Recently, it has been observed an increased interest in three-phase dc-dc conversion for high power applications [1], [2]. The major reason is that three-phase dc-dc converters can achieve lower power component current stresses and also drastically reduce input and output filter requirements, when compared to single-phase topologies. Moreover, high frequency three-phase power transformer can handle higher power levels than single-phase one, when both have the same size.

Another concern regarding high power dc-dc conversion is related to equipment size and weight. It is well known that frequency is a major parameter on determining the amount of material necessary for magnetics construction. In the case that frequency values are within the range of some decades of kilohertz, an increase in this parameter would certainly represent a reduction in magnetics size and weight. For hard-switched topologies it would not be convenient since the reduction achieved in magnetics should not compensate the need for more heat sink material to dissipate higher levels of switching losses. To overcome this issue an effort has been carried out on researching soft-switched topologies that would allow frequency increase without compromising thermal management and efficiency [3], [4], [5], [6].

The proposed three-phase three-level ZVS PWM (TPTL-ZVS-PWM) converter (Fig. 1) is very suitable for applications that require a high power density level and high efficiency,

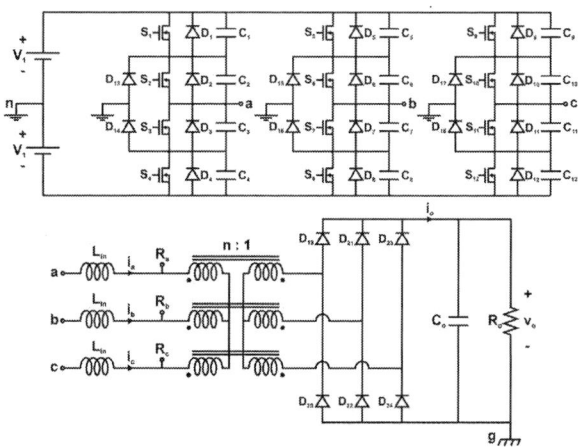

Figure 1. Proposed TPTL-ZVS-PWM converter.

such as aerospace, hybrid electric vehicles and renewable energy processing. It is also intended to be used at higher switching frequency than traditional topologies, since the reduced switch voltage stress allows the use of faster semiconductor devices.

The main converter features observed include:

- Switch voltage stresses are half input voltage value.

- Possibility of achieving ZVS in all switches.

- Output stage has voltage source characteristics.

- PWM employs symmetrical duty cycle.

II. PRINCIPLE OF OPERATION

The proposed TPTL-ZVS-PWM converter has thirteen operation modes: six in continuous conduction mode (CCM) and seven in discontinuous conduction mode (DCM). A given operation mode is classified as discontinuous if there is no current flowing through the inductor L_{in} during at least one operation stage. Otherwise, the mode is said continuous. Each operation mode is composed by eighteen operation stages, if

978-1-4244-4782-4/10 $26.00 © 2010 IEEE

commutation stages are not considered. In this work, only one continuous operation mode will be described since the paper size is limited. The other twelve modes can be described in a similar manner.

Fig. 2 shows nine operation stages corresponding to the mode in which the converter detailed in section VI operates under rated conditions and the commutation stages are not considered. The other nine stages are not depicted but they are symmetrically equivalent, simply permute inner and outer switches and diodes states in each converter leg, from the first nine stages.

III. STATIC ANALYSIS

The TPTL-ZVS-PWM converter static analysis will be made to provide the fundamental equations that are going to be used as a basis in a design procedure elaboration. Thus, all components in the power stage can be correctly chosen in order to meet all design specifications.

The voltages v_{an}, v_{bn}, and v_{cn} are imposed by the three-phase NPC commutation cell, and are graphically represented in Fig. 3.

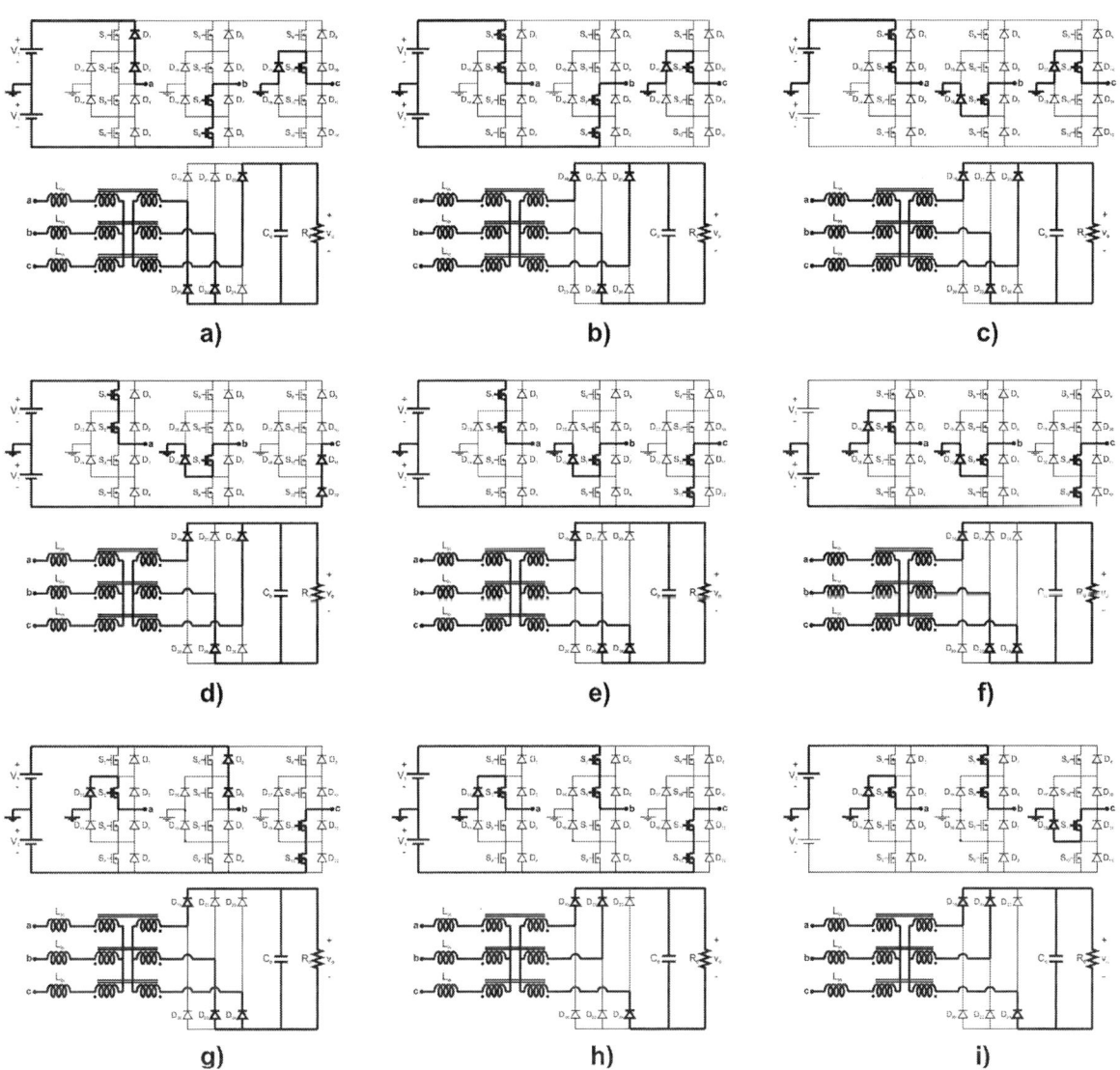

Figure 2. Operation stages for one continuous conduction mode.

Since voltages v_{an}, v_{bn}, and v_{cn} have the same waveforms, except for the fact that they are phase-shifted of 120°, the analysis can be performed for one of the three phases (here was chosen phase "a"), and the corresponding results of the other two phases are the same, if considered the mentioned phase-shift. It can be shown that the fundamental component of voltage v_{an}, here defined by v_{an1}, is given by (1). Its angular frequency is ω and its amplitude is controlled by duty cycle D. The angle ϕ represents the phase difference between the voltages v_{kn} and v_{Rkn}, where k = a, b, c. By convention, the voltage v_{Ran} fundamental component phase angle is set to 0°.

$$v_{an1} = \frac{4V_1}{\pi} sen\left(\frac{D\pi}{2}\right) sen(\omega t + \phi) \tag{1}$$

The definition of duty cycle D is given by (2).

$$D = \frac{\Delta \omega t}{\pi} \tag{2}$$

Equation (3) presents the phasor representation of voltage v_{an1}.

$$\dot{V}_{an1} = \frac{4V_1}{\pi} sen\left(\frac{D\pi}{2}\right)\bigg|\underline{\phi} \tag{3}$$

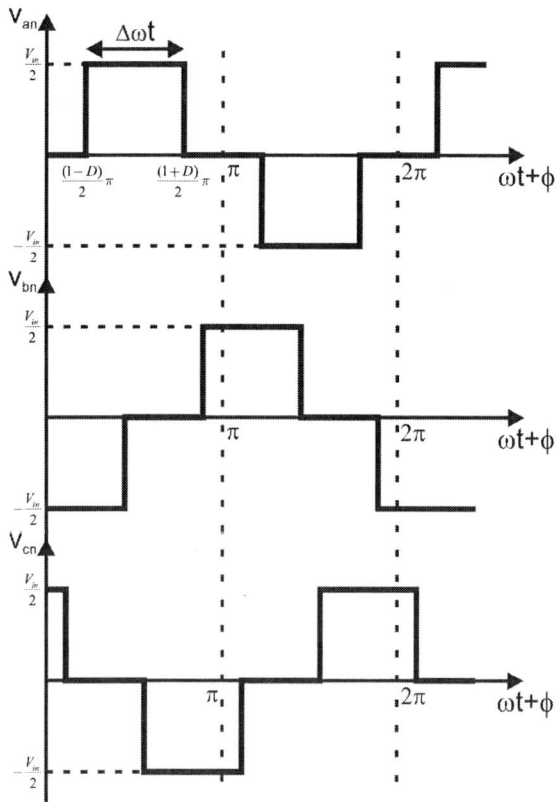

Figure 3. Waveform of voltages v_{an}, v_{bn} and v_{cn}.

The voltage between the node R_a and the common point of the wye connected primary windings of the transformer v_{Ran} is shown in Fig. 4, and its fundamental component is given by (4).

$$v_{Ran1} = \frac{2nV_o}{\pi} sen(\omega t) \tag{4}$$

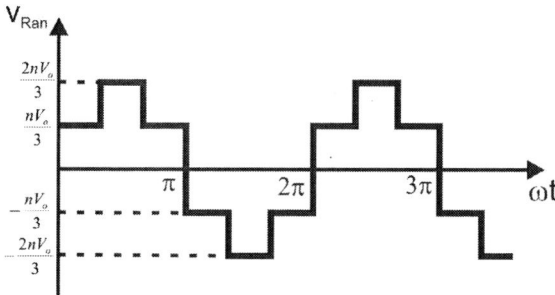

Figure 4. Waveform of voltage v_{Ran}.

The phasor representation of voltage v_{Ran1} is given by (5).

$$\dot{V}_{Ran1} = \frac{2nV_o}{\pi}\bigg|\underline{0^o} \tag{5}$$

Where:

- n: turns ratio between transformer's primary and secondary.

As proposed in [7], the three-phase bridge rectifier can be represented as a resistive load, when considering only the fundamental components of voltages and currents. Equation (6) presents the value of the three-phase bridge rectifier equivalent resistance.

$$R_{eq} = \frac{\omega L_{in} \left|\dot{V}_{Ran1}\right|}{\sqrt{\left|\dot{V}_{an1}\right|^2 - \left|\dot{V}_{Ran1}\right|^2}} \tag{6}$$

In case that only the fundamental components are considered, the TPTL-ZVS-PWM converter can be represented by the equivalent circuit presented in Fig. 5, for static analysis purpose.

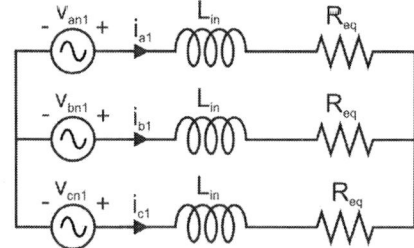

Figure 5. Equivalent circuit for TPTL-ZVS-PWM converter analysis.

It can be derived from Fig. 5 circuit analysis that the fundamental component phasor representation of current i_a is given by (7).

$$\dot{I}_{a1} = \frac{\left|\dot{V}_{an1}\right|}{\sqrt{R_{eq}^2 + (\omega L_{in})^2}}\underline{|0^o} = \frac{\sqrt{\left|\dot{V}_{an1}\right|^2 - \left|\dot{V}_{Ran1}\right|^2}}{\omega L_{in}}\underline{|0^o} \quad (7)$$

The output current mean value is given by (8), in the case that the currents flowing through inductors L_{in} have sinusoidal waveform.

$$I_o = \frac{3n}{\pi}\left|\dot{I}_{a1}\right| \quad (8)$$

Let the static gain be defined by (9) and the output parameterized current by (10).

$$q = \frac{nV_o}{2V_1} \quad (9)$$

$$\bar{I}_o = \frac{\pi^2 \omega L_{in}}{6nV_1} I_o \quad (10)$$

Using the several equations previously determined, it can be shown that equation (11) gives the relation between static gain and parameterized current, for a given duty cycle value.

$$q = \sqrt{sen^2\left(\frac{D\pi}{2}\right) - \frac{1}{4}\bar{I}_o^{\,2}} \quad (11)$$

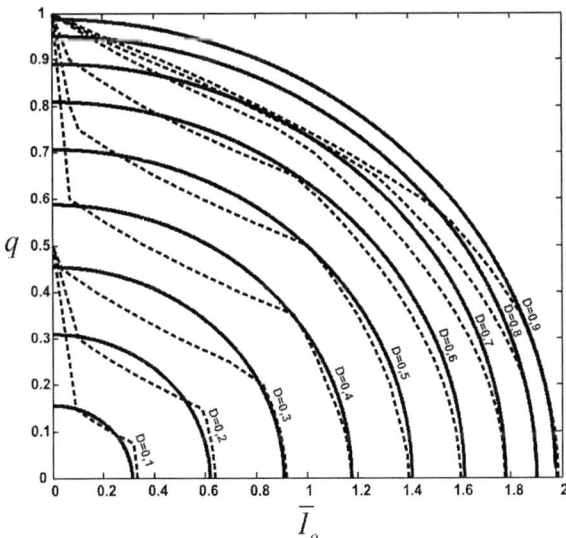

The TPTL-ZVS-PWM converter's output characteristic graph can be derived from (11), and it is presented in Fig. 6. It is important to notice that the theoretical analysis considering only the fundamental components of voltages and currents provides good results only when the converter operates in continuous conduction mode. Otherwise, harmonics present in voltages and currents will play an important role in converter operation, and since their influences are not considered in the proposed analysis, the results become compromised. Observing Fig. 6 one can notice that continuous and dotted lines are almost coincident only in a determined region of TPTL-ZVS-PWM converter output characteristic, which precisely corresponds to CCM operation.

IV. DYNAMIC ANALYSIS

Many applications require a converter capable of maintaining output voltage within certain limits. Looking at converter's output characteristic, it can be observed that output voltage value is dependent on load conditions, and therefore there is a need for output voltage closed-loop compensation. Thus, a controller has to be proposed in order to meet several design requirements, as load transients, closed-loop stability, response time, and so on. Basically, the main task in the dynamic analysis is to determine the transfer function that relates output voltage to duty cycle, obtained from a linearization performed on a given converter operating point.

Classical averaging approaches would certainly fail when applied to this kind of converter, since there is a large ripple in many important variables, as for example, in the current through inductors L_{in}. The methodology used to determine the control-to-output transfer function is based on the phasor transformation technique [8].

The differential equation (12) can be derived from Fig. 5 circuit analysis. The index "k" can assume the value "a", "b" or "c", depending on which system phase such equation is intended to refer to.

$$-\bar{v}_{kn1} + j\omega L_{in}\bar{i}_{k1} + R_{eq}\bar{i}_{k1} + L_{in}\frac{d\bar{i}_{k1}}{dt} = 0 \quad (12)$$

Applying the Laplace transformation to (12) and with some algebra, equation (13) can be determined.

$$\frac{\bar{i}_{k1}}{\bar{v}_{k1}} = \frac{\dfrac{1}{L_{in}}\left(s + \dfrac{R_{eq}}{L_{in}} - j\omega\right)}{s^2 + \dfrac{2R_{eq}}{L_{in}}s + \omega^2 + \dfrac{R_{eq}^2}{L_{in}^2}} \quad (13)$$

The instantaneous current phasors can be generically represented by (14).

$$\bar{i}_{k1} = I_{kR} + jI_{kI} \quad (14)$$

Where:

- I_{kR}: real component of instantaneous phasor $\bar{i}_{k1}$.

- I_{kI}: imaginary component of instantaneous phasor $\overline{i}_{k1}$.

Taking current i_{a1} as reference (0°), the components of $\overline{i}_{k1}$ are:

$$I_{kR} = I_s \quad and \quad I_{kI} = 0 \qquad (15)$$

Considering the reference adopted, it can be shown that the voltages imposed by the three-phase NPC commutation cell are given by (16).

$$\overline{v}_{kn1} = \left|\overline{v}_{kn1}\right|\left|\underline{\phi}\right. = v_s\left|\underline{\phi}\right. \qquad (16)$$

The angle ϕ can be evaluated by (17).

$$\phi = tg^{-1}\left(\frac{\omega L_{in}}{R_{eq}}\right) \qquad (17)$$

Equation (16) can be rewritten as:

$$\overline{v}_{k1} = \frac{v_s}{\sqrt{R_{eq}^2 + \left(\omega L_{in}\right)^2}}\left(R_{eq} + j\omega L_{in}\right) \qquad (18)$$

Substituting (18) into (13) follows that:

$$\overline{i}_{k1} = \frac{\dfrac{1}{L_{in}}\left(s + \dfrac{R_{eq}}{L_{in}} - j\omega\right)}{\left(s^2 + \dfrac{2R_{eq}}{L_{in}}s + \omega^2 + \dfrac{R_{eq}^2}{L_{in}^2}\right)}\frac{\left(R_{eq} + j\omega L_{in}\right)}{\sqrt{R_{eq}^2 + \left(\omega L_{in}\right)^2}}v_s \qquad (19)$$

It can be demonstrated that:

$$\left|\hat{\overline{i}}_{k1}\right| = \frac{I_{kR}\hat{i}_{k1R} + I_{kI}\hat{i}_{k1I}}{\left|I_k\right|} \qquad (20)$$

Where:

$$I_{kR} = \frac{V_s}{\sqrt{R_{eq}^2 + \left(\omega L_{in}\right)^2}} \quad and \quad I_{kI} = 0 \qquad (21)$$

$$\left|I_k\right| = \sqrt{I_{kR}^2 + I_{kI}^2} = \frac{V_s}{\sqrt{R_{eq}^2 + \left(\omega L_{in}\right)^2}} \qquad (22)$$

$$\hat{i}_{kR} = \frac{\dfrac{1}{L_{in}^2}\left[R_{eq}L_{in}s + R_{eq}^2 + \left(\omega L_{in}\right)^2\right]}{\sqrt{R_{eq}^2 + \left(\omega L_{in}\right)^2}\left(s^2 + \dfrac{2R_{eq}}{L_{in}} + \omega^2 + \dfrac{R_{eq}^2}{L_{in}^2}\right)}\hat{v}_s \qquad (23)$$

$$\hat{i}_{kI} = \frac{s\omega}{\sqrt{R_{eq}^2 + \left(\omega L_{in}\right)^2}\left(s^2 + \dfrac{2R_{eq}}{L_{in}} + \omega^2 + \dfrac{R_{eq}^2}{L_{in}^2}\right)}\hat{v}_s \qquad (24)$$

Substituting (21), (22), (23) and (24) into (20) yields the relation between $\left|\hat{\overline{i}}_{k1}\right|$ and $\hat{v}_s$, as presented in (25).

$$\left|\hat{\overline{i}}_k\right| = \frac{\dfrac{1}{L_{in}^2}\left[R_{eq}L_{in}s + R_{eq}^2 + \left(\omega L_{in}\right)^2\right]}{\sqrt{R_{eq}^2 + \left(\omega L_{in}\right)^2}\left(s^2 + \dfrac{2R_{eq}}{L_{in}} + \omega^2 + \dfrac{R_{eq}^2}{L_{in}^2}\right)}\hat{v}_s \qquad (25)$$

It is also possible to demonstrate the validity of (26).

$$\hat{i}_o = \frac{3}{\pi}\left|\hat{\overline{i}}_k\right| \qquad (26)$$

Thus:

$$\hat{i}_o = \frac{\dfrac{3}{\pi L_{in}^2}\left[R_{eq}L_{in}s + R_{eq}^2 + \left(\omega L_{in}\right)^2\right]}{\sqrt{R_{eq}^2 + \left(\omega L_{in}\right)^2}\left(s^2 + \dfrac{2R_{eq}}{L_{in}} + \omega^2 + \dfrac{R_{eq}^2}{L_{in}^2}\right)}\hat{v}_s \qquad (27)$$

Since the desired control variable is duty cycle "d", there is the need to investigate how variations in this parameter will affect the fundamental component of the voltages generated by the three-phase NPC commutation cell. Such relation is presented in (28).

$$\hat{v}_s = \frac{\partial\left|\dot{V}_{s1}\right|}{\partial D}\hat{d} = 2V_1 cos\left(D\frac{\pi}{2}\right)\hat{d} \qquad (28)$$

Using equations (27) and (28) the transfer function (29) can be determined.

$$\frac{\hat{i}_o}{\hat{d}} = \frac{\dfrac{6V_1 cos\left(D\dfrac{\pi}{2}\right)}{\pi L_{in}^2}\left[R_{eq}L_{in}s + R_{eq}^2 + \left(\omega L_{in}\right)^2\right]}{\sqrt{R_{eq}^2 + \left(\omega L_{in}\right)^2}\left(s^2 + \dfrac{2R_{eq}}{L_{in}} + \omega^2 + \dfrac{R_{eq}^2}{L_{in}^2}\right)} \qquad (29)$$

The relation between output voltage $\hat{v}_o$ and current $\hat{i}_o$ is still remaining to be obtained. It can be done by analyzing the converter's output filter, which yields (30).

$$\hat{v}_o = \left(\frac{R_o}{1 + sR_oC_o}\right)\hat{i}_o \qquad (30)$$

Finally, the transfer function between output voltage and duty cycle is given by (31).

$$\frac{\hat{v}_o}{\hat{d}} = \frac{\dfrac{6R_oV_1\cos\left(D\dfrac{\pi}{2}\right)}{\pi L_{in}^{2}\sqrt{R_{eq}^{2}+\left(\omega L_{in}\right)^{2}}}\left[R_{eq}L_{in}s+R_{eq}^{2}+\left(\omega L_{in}\right)^{2}\right]}{\left(1+sR_oC_o\right)\left(s^{2}+\dfrac{2R_{eq}}{L_{in}}+\omega^{2}+\dfrac{R_{eq}^{2}}{L_{in}^{2}}\right)} \quad (31)$$

A. Model Validation

In order to verify the validity of transfer functions (29) and (31), the TPTL-ZVS-PWM converter was simulated and the results were compared to models response, as shown in Fig. 7 and in Fig. 8.

It can be concluded observing Fig. 7 and Fig. 8 that the transfer functions derived from the proposed theoretical analysis provide a satisfactory representation of TPTL-ZVS-PWM dynamic behavior for CCM operation.

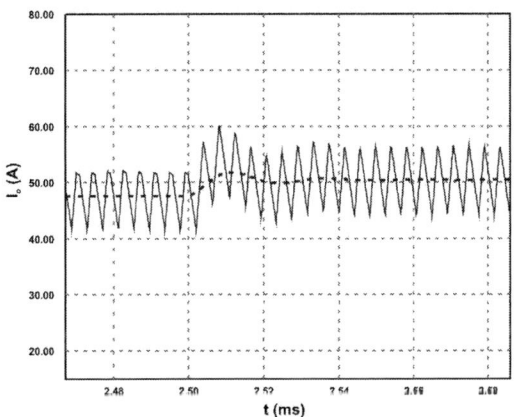

Figure 8. Output current dynamic responses: simulation (continuous line) and model (dashed line).

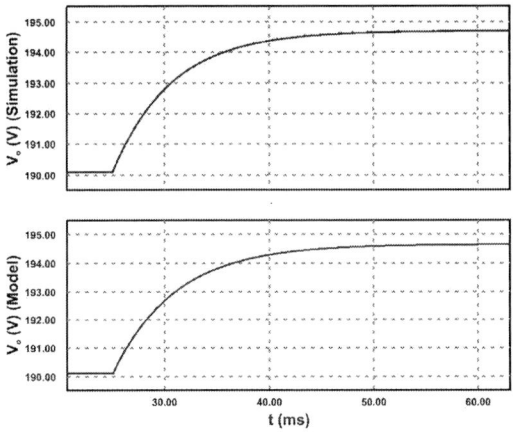

Figure 7. Output voltage dynamic responses: simulation and model.

V. THREE-PHASE HIGH FREQUENCY TRANSFORMER DESIGN CONSIDERATIONS

As previously mentioned, three-phase transformer can handle higher power levels than single phase one. Thus, for the same rated power it is expected that the choice for three-phase transformer provides a reduction in converter's volume and weight. In this section, the design of three-phase high frequency transformer will be detailed, where the following assumptions are made:

- Balanced load.

- Steady state operation analysis.

- The three-phase transformer has spatial symmetry.

Fig. 9 shows the spatial representation of the three-phase transformer under analysis.

Primary and secondary voltages are given by (32) and (33), respectively.

$$v_{pi}=N_p\frac{d\phi_i}{dt} \quad , i=1,2,3. \quad (32)$$

$$v_{si}=N_s\frac{d\phi_i}{dt} \quad , i=1,2,3. \quad (33)$$

Considering that the primary windings are identical, and so are the secondary, one can conclude that only half window area will be available to accommodate one primary and one secondary winding. Mathematically, this can be expressed by (34).

$$\frac{k_wA_w}{2}=\frac{N_pI_{pief}+N_sI_{sief}}{J} \quad , i=1,2,3. \quad (34)$$

Where:

- N_p, N_s: primary and secondary windings number of turns, respectively.

- I_{pief}, I_{sief}: RMS value of primary "i" and secondary "i" current, respectively.

- J: current density.

- A_w: window area.

- k_w: window area utilization factor.

The magnetic flux density in the transformer's leg "i" can be determined by **Erro! Fonte de referência não encontrada.**. Parameter A_e refers to the cross section area of a transformer leg.

$$B_i=\frac{\phi_i}{A_e} \quad , i=1,2,3. \quad (35)$$

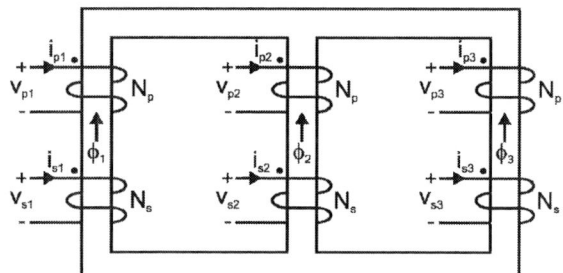

Figure 9. Spatial representation of three-phase transformer.

From Faraday's law and equation **Erro! Fonte de referência não encontrada.** follows that:

$$N_p = \frac{1}{B_{max}A_e}\int v_p(t)\,dt + constant \quad (36)$$

$$N_s = \frac{1}{B_{max}A_e}\int v_s(t)\,dt + constant \quad (37)$$

Solving integrals (36) and (37) for the waveform presented in Fig. 4, it is possible to obtain the values of N_p and N_s.

$$N_p = \frac{V_{pef}}{3\sqrt{2}\,B_{max}A_e f} \quad (38)$$

$$N_s = \frac{V_{sef}}{3\sqrt{2}\,B_{max}A_e f} \quad (39)$$

Where:

- V_{pef}, V_{sef}: RMS value of primary and secondary voltages, respectively.

- B_{max}: maximum magnetic flux density.

- f: frequency.

From the equations previously presented one can evaluate the product between A_e and A_w, which provides a good estimation for transformer dimensions.

$$A_e A_w = \frac{4 S_{3\phi}}{9\sqrt{2}k_w J B_{max} f} \quad (40)$$

Where:

- $S_{3\phi}$: total three-phase apparent power.

Using the several equations previously determined it is possible to design the high frequency three-phase transformer of TPTL-PWM converter. Comparing the estimation given by (40) to the dimensions expected from the design of three single phase transformers able to process the same amount of power, it can be concluded that a size reduction of approximately 33% is expected when using three-phase transformer.

VI. EXPERIMENTAL RESULTS

A TPTL-ZVS-PWM converter prototype was built in order to verify the analyses carried out. Table I contains the main converter parameters.

TABLE I.   CONVERTER PARAMETERS

Parameter	Value
Output Power (P_o)	10kW
Input Voltage (V_{in})	800V
Output Voltage (V_o)	200V
Switching Frequency (f_s)	40kHz

Six Semikron SKM100GB063D IGBT modules were used to realize the switches S_1-S_{12} and six Semikron SKK75F12 diode modules for the diodes D_{13}-D_{24}. Each inductor L_{in} was constructed using one Thornton IP12R EE-76/50/76 core and its inductance was adjusted to 65μH. The three-phase transformer uses three Thornton IP12R NC100/57/25 cores, with 2.2 of primary to secondary turns ratio.

The main experimental waveforms are depicted in Fig. 10. One can observe in Fig. 10 (d) that the converter injects in the output capacitive filter a relative low ripple current, thus output filter requirements are going to be low.

CONCLUSIONS

A novel three-phase three-level soft-switched PWM dc-dc converter was proposed. An approximated static analysis was performed, which results were satisfactory for CCM operation.

The main converter advantages observed include: minimum output filter requirements, zero voltage switching, possibility of operation at high frequency values, use of three-phase transformer, voltage-source output, switch voltage stresses are half input voltage value, resulting in a high power density equipment.

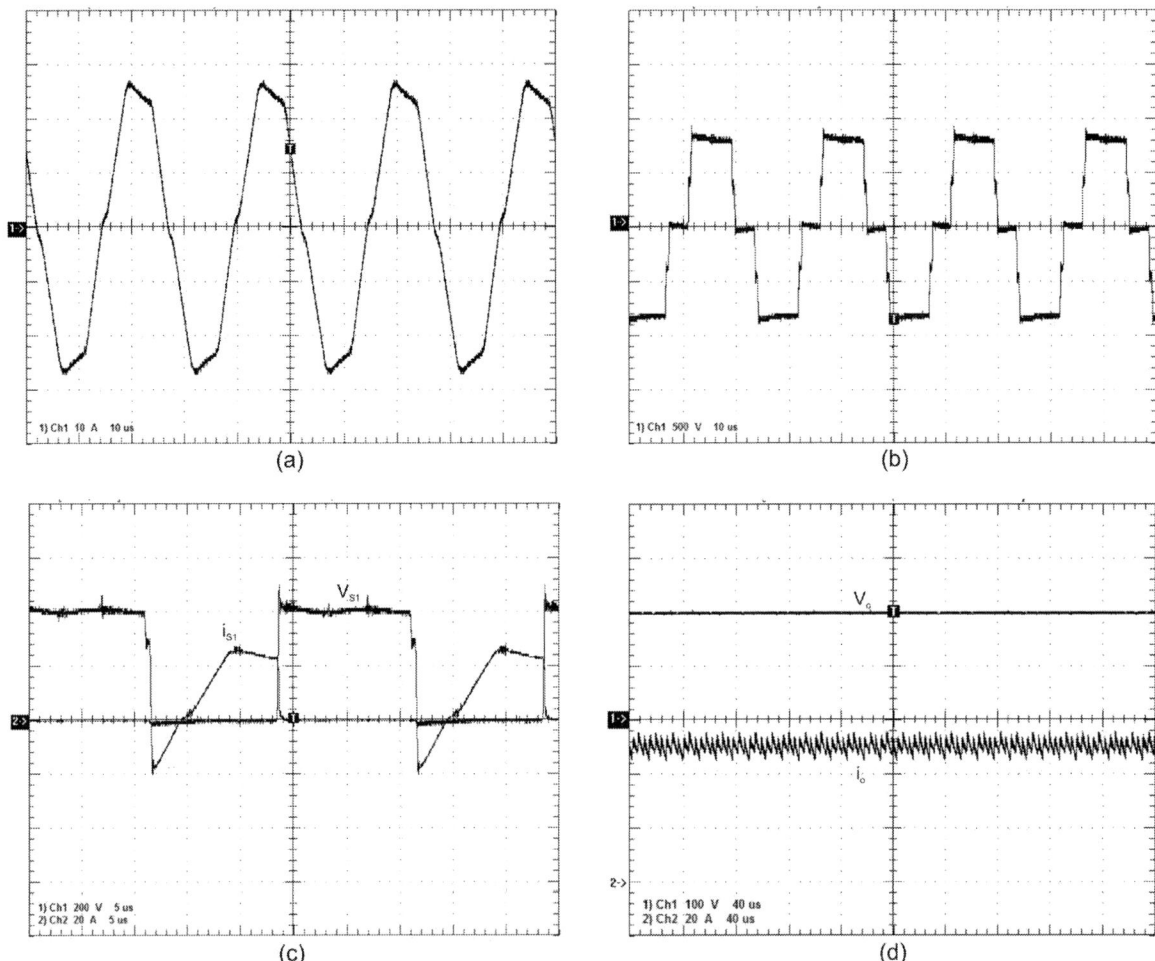

Figure 10. Experimental waveforms for: (a) current throught on inductor L_{in} I_a (10A / div); (b) voltage generated by the three-phase NPC commutation cell V_{ab} (500V / div); (c) voltage across S_1 V_{S1} (200V / div) and current through S_1 i_{S1} (20A / div); (d) output voltage V_o (100V / div) and output current i_o (20A / div).

REFERENCES

[1] P. D. Ziogas, A. R. Prasad, and S. Manias, "A Three-Phase Resonant dc/dc Converter", in Proc. *IEEE 22nd Annu. Power Electronics Specialists Conf. (PESC '91)*, Cambridge, MA, Jun. 24-27, pp. 463-473, 1991.

[2] R. W. A. A. De Doncker, D. M. Divan, and M. H. Kheraluwala, "A Three-Phase Soft-Switched High-Power-Density dc/dc Converter for High-Power Applications", in *IEEE Transactions on Industry Applications*, vol. 27, no. 1, pp. 63-73, Jan./Feb. 1991.

[3] D. S. Oliveira and I. Barbi, "A Three-Phase ZVS PWM DC/DC Converter With Asymmetrical Duty Cycle for High Power Applications", in *IEEE Transactions on Power Electronics*, vol. 20, no. 2, pp. 370-377, March 2005.

[4] H. Cha and P. Enjeti, "A Novel Three-Phase High Power Current-Fed DC/DC Converter with Active Clamp for Fuel Cells", in *Proc. IEEE Power Electronics Specialists Conference, PESC 2007*, pp. 2485-2489, June 2007.

[5] T. Song, H. S. H. Chung, S. Tapuhi, and A. Ioinovici, "A High Input Voltage Three-Phase ZVZCS DC-DC Converter with Vin/3 Voltage Stress on Primary Switches", in *Proc. IEEE Power Electronics Specialists Conference, PESC 2007*, pp. 350-356, June 2007.

[6] D. V. Ghodke, K. Chatterjee, and B. G. Fernandes, "Three-Phase Three Level, Soft Switched, Phase Shifted PWM DC-DC Converter for High Power Applications", in *IEEE Transactions on Power Electronics*, vol. 23, no. 3, pp. 1214-1227, May 2008.

[7] V. Caliskan, D. J. Perreault, T. M. Jahns, and J. G. Kassakian, "Analysis of Three-Phase Rectifiers With Constant-Voltage Loads", in *IEEE Transactions on Circuits and Systems – I: Fundamental Theory and Applications*, vol. 50, no. 9, pp. 1220-1226, September 2003.

[8] C. T. Rim and G. H. Cho, "Phasor Transformation and its Application to the DC/AC Analyses of Frequency Phase-Controlled Series Resonant Converters (SRC)", in *IEEE Transactions on Power Electronics*, vol. 5, no. 2, pp. 201-211, April 1990.

Optimize the Synchronous Rectifier for *LCC* Converters

Feng Zheng, *Member, IEEE*, Zhengfeng Ming, *Member, IEEE*
Electrical Engineering Department
Xidian University
Xi'an, China
e-mail: f.zheng@mail.xidian.edu.cn

Abstract— **This paper proposed a synchronous rectifier (SR) circuits for *LCC* converters. Using this circuits, the SR will working in both self and lossless driven mode. At the same time, parasitic parameters of SR MOSFET partly play the role of resonant elements and snubbers, and then the voltage ringing is mitigated. The problems of body-diode conduction losses are also addressed. A control scheme is thus presented to reduce the body-diode losses. Finally, a prototype is conducted to validate our methods.**

I. INTRODUCTION

Synchronous rectifier implementation will encounter two difficulties, namely, the right time to fire synchronous rectifier and the severe voltage ring caused by the body-diode recovery [1]-[2]. In order to solve the first difficulty, control strategies such as pre-firing, time compensation, precise control signal providing and current driving are implemented. Although using these methods reduces the body-diode conduction time, the control circuits become much complex. As to the second difficulty, the general method is paralleling snubber circuit across synchronous MOSFET's. However, snubbers will lead to power losses and slight changes in circuits' operation mode, let alone its design complexity and room occupied.

In this paper, we proposed a *LCC* topology. In this topology, the parallel resonant capacitors (PRC) play the role

of resonant elements, lossless snubbers and self driven signal providers. Due to its merits, we thoroughly analyzed its working principle using state plane method. Furthermore, in order to reduce the body diode conduction time, a control scheme is presented. Finally, a prototype of 50W output power was constructed to demonstrate the analysis.

II. PRINCIPLE OF THE PROPOSED CONVERTER

Fig. 1 shows the schematic of *LCC* converters proposed. The difference between the proposed and a common one is that PRC are placed at the secondary of isolate transformer.

We can benefit much from this slight change. First of all, as shown in Fig. 1, C_{prx}, where subscript x=1 or 2, parallels with the input capacitor of one synchronous rectifier MOSFET (SR) and the output capacitor of the other. Hence, the parasitic parameters of SR's can be used as resonant elements. Second, because the SR's are paralleled with PRC's, PRC's then play the role of snubbers. At the same time, the energy stored in the snubbers will be applied to the load, and this leads to the snubbers to be lossless. Above all, voltages across PRC's are in phase with SR's driven signals and both self-driven and lossless driven methods can be easily used.

The waveform of voltage across SR_1 is presented in Fig. 2. We can read that the body-diode conduction time takes a great

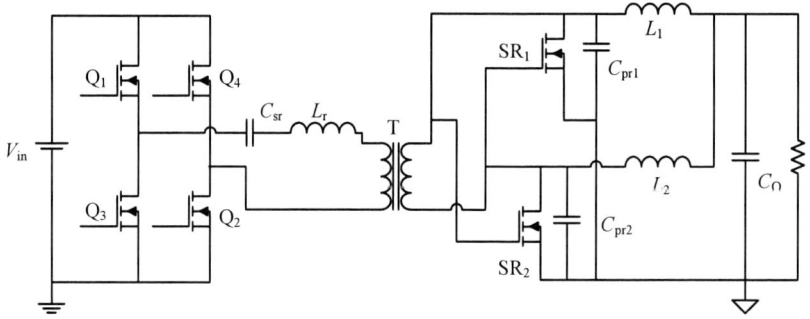

Fig. 1 Proposed *LCC* Converter with Synchronous Rectifier

This work is financially aided by the key project [Cx609014(4)] of Science and Technology Foundation of Xi'an, China.

part of SR_1 on time, and this will lower the efficiency of converters. In addition, when the body-diode is on, the body-diode of the other SR will also conduct. Resonant element C_{pr} will then be shorted, and the converter is changed form a LCC converter into a series resonant one.

Due to the current ripple through two output-filter inductor being small, we substitute the inductors with current sources.

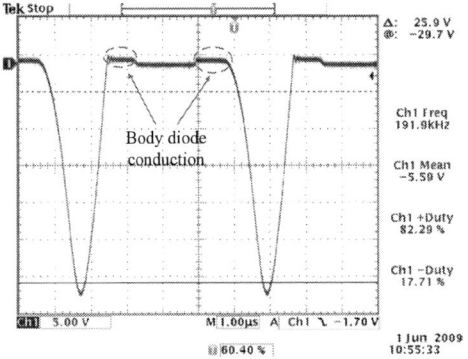

Fig. 2 Waveform of voltage across SR1

Then, Fig. 3 sketches all possible topologies of the converter during different cases, where C'_{prx} (x=1 or 2) and I_O are PRC's and I_O reflected to transformer primary, respectively.

Using the state-plane method and selecting current through L_r (i_L), voltages across C_{sr} (u_{CS})and C_{pr} (u_{CP}) as state variables, we obtain two sets of equations [3]-[5]. When the possible topologies are a), c), d), and f) the state equation is given by

$$
\begin{bmatrix} \dot{i}_L \\ \dot{u}_{CS} \\ \dot{u}_{CP} \end{bmatrix} = \begin{bmatrix} 0 & -\dfrac{1}{L} & -\dfrac{1}{L} \\ \dfrac{1}{C_S} & 0 & 0 \\ \dfrac{1}{C_P} & 0 & 0 \end{bmatrix} \begin{bmatrix} i_L \\ u_{CS} \\ u_{CP} \end{bmatrix} + \begin{bmatrix} \dfrac{1}{L} & 0 \\ 0 & 0 \\ 0 & -\dfrac{1}{C_2} \end{bmatrix} \begin{bmatrix} V_E \\ I_E \end{bmatrix} \quad (1)
$$

where V_E and I_E denote the voltage and current sources in circuits, respectively. Accordingly, cases b) and e) yield

$$
\begin{bmatrix} \dot{i}_L \\ \dot{u}_{CS} \end{bmatrix} = \begin{bmatrix} 0 & -\dfrac{1}{L} \\ \dfrac{1}{C_S} & 0 \end{bmatrix} \begin{bmatrix} i_L \\ u_{CS} \end{bmatrix} + \begin{bmatrix} \dfrac{1}{L} \\ 0 \end{bmatrix} V_E \quad (2)
$$

The typical waveforms of voltage expressed on resonant elements, currents through L_r, voltages across C_{sr} and C_{pr1} are sketched in Fig. 4. In order to simplify optimization, we take the positive crossover point of u_{CP} as start of a switching cycle and mark it t_0.

Following the analysis procedure presented in [6], we define a new state variable u_S as $u_{CS}+u_{CP}$ and normalize the voltages and currents in (1) and (2) by V_{in} and V_{in}/Z_o, respectively.

where

$$
Z_o = \sqrt{L_r/C_T}; \qquad C_T = \frac{C_{sr}C_{pr}}{C_{sr}+C_{pr}}
$$

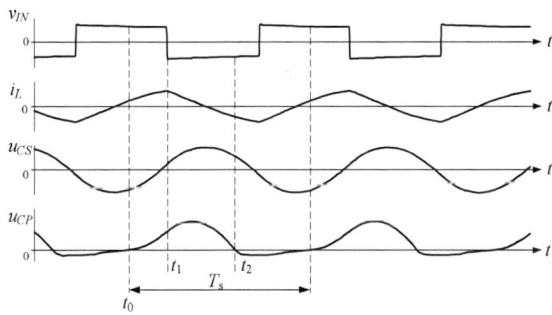

Fig. 4 Typical waveforms. Top to bottom: voltage expressed on resonant elements, current through L_r, voltages across C_{sr} and C_{pr1}

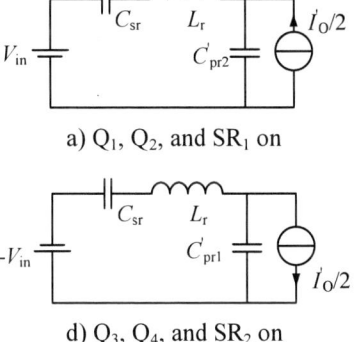

a) Q_1, Q_2, and SR_1 on

b) Q_1, Q_2, SR_1 and SR_2 on

c) Q_1, Q_2, and SR_2 on

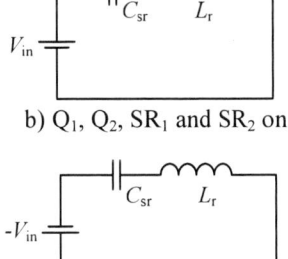

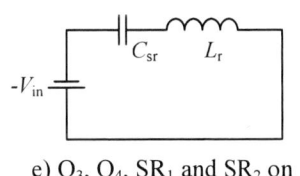

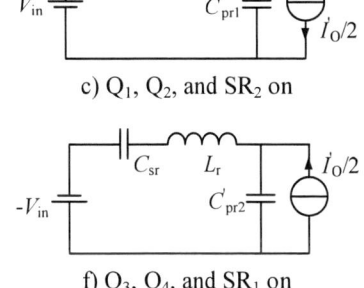

d) Q_3, Q_4, and SR_2 on

e) Q_3, Q_4, SR_1 and SR_2 on

f) Q_3, Q_4, and SR_1 on

Fig. 3 Possible topologies

We then obtain the steady state trajectory for all possible topologies as shown in Fig. 5. It must be noted that the trajectory from t_2 to $T_s/2$ is a part of ellipse, and two SR body diode conduct during this interval. The smaller γ is, the lower the body diode conduction losses are.

Since an *LCC* converter can be well controlled by β_1 and the radii of arc from t_0 to t_1, we then is able to control u_{CP}, e.g. the driven signal of SR's. Thus, the SR's will conduct in phase of current very well without complex control circuits.

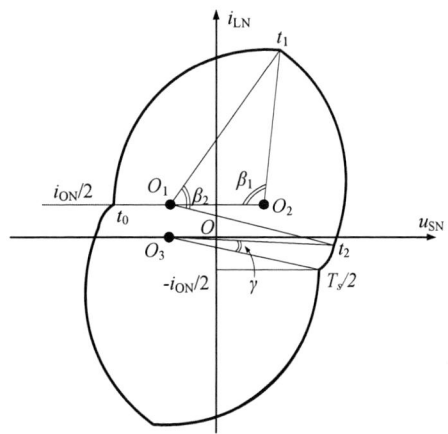

Fig. 5 Typical steady state trajectory for proposed LCC converter

The control scheme is sketched in Fig. 6. There are two control loops. The frequency loop is the outer loop, which control the output voltage. While, the inner loop is a current loop, which operation mode is similar to a peak current mode control, i.e., the transistors switch at the moments when resonant current is equal to the trigger value.

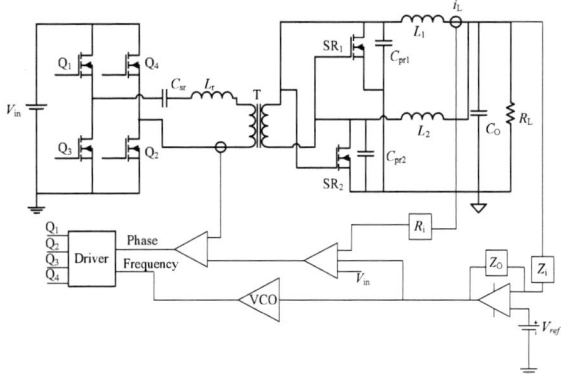

Fig. 6 Control scheme.

III. EXPERIMENTAL RESULTS

We validate the topology using a 50W (output voltage is 5V) prototype, which parameters are L_r=1μH; C_{sr}=1μF;

C_{pr}=0.6μF, and the turn ratio of transformer is 2. The waveforms of voltage across Cpr1 and Cpr2 are demonstrated in Fig. 7. It is clear the ringing on SR's are very small. Meanwhile, body diode conduction time is very small.

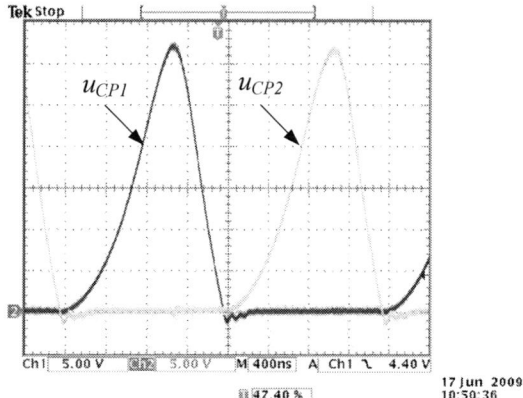

Fig. 7 Voltage waveforms

IV. CONCLUSION

This paper presents a novel *LCC* topology with SR's. The key points of this topology are: 1) its PRC is composed by SR's parasitic parameters and play the role of lossless snubbers.2) the SR's are worked in a self and lossless driven manner. Furthermore, working principles of the converter are analyzed using the state variable trajectory method. Then, the method on body-diode-conduction-loss reduction is clearly revealed. Finally, the topology is validated by a 50W prototype.

REFERENCES

[1] D. Fu, Y. Liu, F. C. Lee, and M. Xu, "A Novel Driving Scheme for Synchronous Rectifiers in *LLC* Resonant Converters," *Power Electronics, IEEE Transactions on*, vol. 24, pp. 1321-1329, 2009.

[2] J. Biela, U. Badstuebner, and J. W. Kolar, "Design of a 5-kW, 1-U, 10-kW/dm^3 Resonant DC-DC Converter for Telecom Applications," *Power Electronics, IEEE Transactions on*, vol. 24, pp. 1701-1710, 2009.

[3] R. Oruganti and F. C. Lee, "Resonant Power Processors, Part I - State Plane Analysis," *Industry Applications, IEEE Transactions on*, vol. IA-21, pp. 1453-1460, 1985.

[4] I. Batarseh, R. Liu, C. Q. Lee, and A. K. Upadhyay, "Theoretical and experimental studies of the *LCC* type parallel resonant converter," *Power Electronics, IEEE Transactions on*, vol. 5, pp. 140-150, 1990.

[5] C. Hao, E. K. K. Sng, and T. King-Jet, "Optimum Trajectory Switching Control for Series-Parallel Resonant Converter," *Industrial Electronics, IEEE Transactions on*, vol. 53, pp. 1555-1563, 2006.

[6] M. C. Tsai, "Analysis and implementation of a full-bridge constant-frequency LCC-type parallel resonant converter," *Electric Power Applications, IEE Proceedings -*, vol. 141, pp. 121-128, 1994.

Digital Control Scheme for Robust Clock Tuning and PWM Phase Synchronization in Digitally Controlled Multi-POL Applications

Eamon O' Malley, Karl Rinne, Anthony Kelly, Basil Almukhtar, Paul Kelleher

Powervation Ltd, Limerick, Ireland

Abstract—**This paper describes a novel clock tuning and subsequent PWM phase synchronization scheme for digitally controlled switching power converters. Its architecture and circuit blocks are presented and explained in detail. The scheme has been implemented in a commercially available digital controller integrated circuit (IC) using a standard CMOS process. Experimental results from a multi point-of-load (POL) application are presented.**

I. INTRODUCTION

Today's DC-DC POL conversion applications can have various types of load current demands such as network processors, field programmable gate arrays (FPGAs) and memory IC's, requiring both standalone and multi-phase interleaved pulse width modulated (PWM) configurations as shown in Fig. 1. When combining several PWM controller ICs on a system application printed circuit board (PCB) in either standalone or multi-phase configurations it is desirable to accurately phase shift their respective PWM signals to reap the benefits of reduced input filter design, lower output current ripple and hence benefit from higher efficiency while avoiding the phenomenon of beat frequencies [1]. In addition the ability to control the start/stop ramp rates of the various POL output voltages with respect to each other in a given time based order is often required for system sequencing [8-9, 11].

Digital PWM controllers attempt to generate precise control of PWM signals through having accurate on-chip clock generators (e.g. oscillators, phase locked-loops (PLLs)). However due to process, voltage and temperature (PVT) variations on-chip clock generators in separate digital PWM controllers can vary by as much as +/- 10 %. Hence multi-phase digital control applications remain reliant on having an external clock source (acting as an accurate time reference master device) to the digital PWM controller ICs (slave devices) in order to implement PWM phase synchronization [8-9] as shown in Fig 2 (top left).

The external master device issues a periodic pulse (SYNC signal in Fig 2 bottom) either at the PWM switching frequency (f_{sw}) or some multiple of it to which the slave PWM controllers synchronize their respective PWM signals to with/without phase offset delay (chosen by the power system designer for optimum system performance).

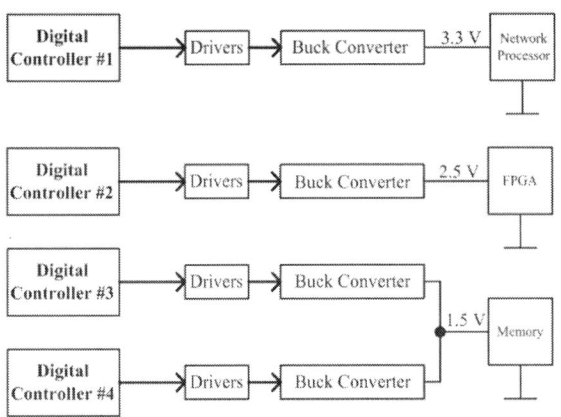

Figure 1 Multi-POL System Configuration Example.

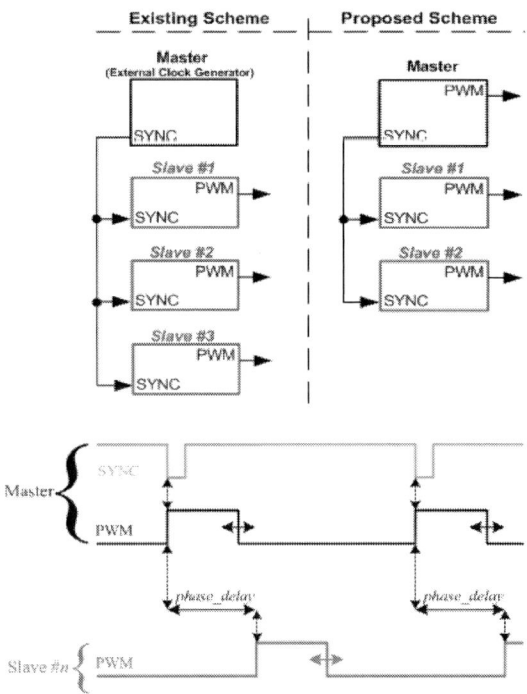

Figure 2 Existing and Proposed PWM Synchronization Schemes.

This paper addresses the mismatch in operating clock frequencies and absence of any inter-chip clock frequency information sharing between PWM controller ICs by way of a novel master-slave clock tuning scheme (see Fig. 2 top right). The scheme eliminates the requirement for additional external clock generator components while providing a robust PWM phase synchronization and accurate time base for system sequencing.

Section II of this paper gives a brief overview of PWM generation in digital controllers highlighting the difficulties in PWM phase synchronization while each controller's on-chip clock generator varies over PVT. Section III introduces the proposed master-slave clock tuning and PWM synchronization architecture and operation. Section IV discusses experimental results while Section V concludes the paper.

II. TYPICAL DIGITAL PWM CONTROLLER IC ARCHITECTURE AND ON-CHIP CLOCK GENERATION

Digital PWM controller ICs typically take the structure shown in Fig. 3 combining several key on-chip blocks such as an analog-to-digital converter (ADC), non-volatile memory (NVM), central processor unit (CPU), serial communications unit, a discrete-time compensator (DTC) in the form of look-up tables, state machines or a dedicated digital signal processing (DSP) unit [2-4, 8-9] and a digital PWM (DPWM) [5-8]. Several if not all of these key building blocks and most notably the DPWM require an accurate on-chip clock reference (typically provided by an on-chip oscillator (OSC) and PLL to generate precise PWM switching frequencies (grey shaded blocks in Fig. 3).

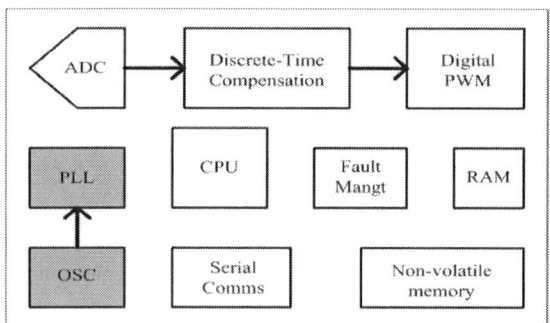

Figure 3 Digital PWM Controller IC Architecture.

As shown in Fig. 4 the DPWM typically comprises a clocked digital counter and comparator (comparing the counter value to a duty cycle command (*duty_cycle_cmd* in Fig.4) from the DTC block (see Fig. 3) to generate the PWM signal. The clock signal (*clk*) is generated from a combination of an on-chip OSC which provides a reference clock to a PLL which in turn multiplies the reference clock (by factor N in Fig 4.) to a frequency yielding a suitable DPWM resolution [5-6] given by,

$$clk = 2^M * f_{sw} \qquad (1)$$

where M is the number of bits of DPWM resolution. A tapped delay line structure can be added to this circuit to achieve higher resolution if required [2, 5-7].

As discussed in Section I variations over PVT and no sharing of clock frequency information between controller ICs in multi-POL applications leads to a chaotic situation of free-running, non-synchronized digital counters operating at different clock frequencies making PWM phase synchronization very difficult to implement.

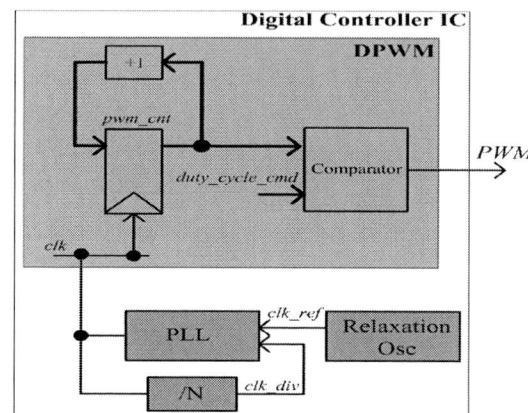

Figure 4 On-Chip Clock Generation and DPWM Architecture.

Fig. 5 shows the case for an application requiring two separate digital PWM controllers. The PWM signals from the respective DPWM sub-modules are misaligned (for this simple case controller #1 PWM leads controller #2 PWM by 4 (*Δpwm_cnt* in Fig. 5) PWM counter clock cycles.

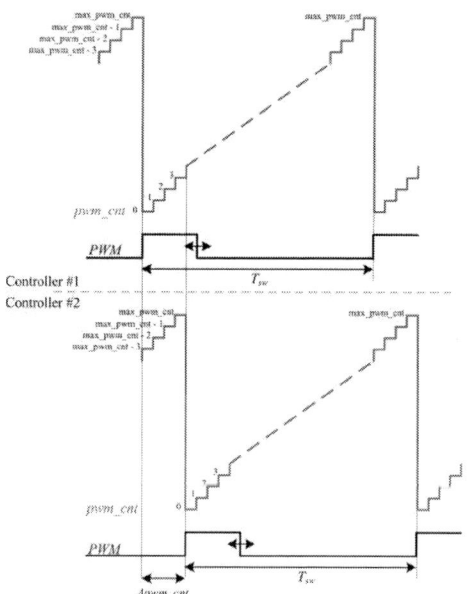

Figure 5 PWM Counter Phase Misalignment.

When a SYNC signal is received by a slave PWM device (see Fig. 2) it re-aligns its PWM signal to it by resetting its PWM counter to zero and re-starting the PWM switching cycle. Further and more catastrophic problems arise if the master SYNC generator operates at a lower f_{sw} than one or more slave controllers and their PWM duty cycles (on-times) are low (e.g. 5-15 %). This leads to the situation where double

PWM on-time pulses are generated per switching period (T_{sw}) conflicting with the DTC control law and hence unstable closed loop voltage regulation occurs [9].

The following section outlines a novel solution to the problems discussed above by utilizing a single-wire master-slave synchronization scheme (Fig. 2 top right). The scheme firstly tunes the on-chip clock generators in each digital PWM controller IC and subsequently uses the same signal information to implement accurate PWM phase synchronization between controllers.

III.   DIGITAL CLOCK TUNING AND PWM SYNCHRONIZATION ARCHITECTURE AND OPERATION

The proposed clock tuning and PWM synchronization technique allows the timing of the PWM signals between a nominated master and slave controller(s) operating at the same f_{sw} to be phase aligned through the use of a single-wire synchronization signal (SYNC in Fig. 2 top right) generated by the master and delivered to one or more slaves. The master device generates a high to low SYNC transition at the start of its primary PWM switching cycle. The slave devices listen for valid SYNC pulses and once detected they re-align their primary PWM phases with/without a specific phase delay relative to the time the SYNC pulse was received (see Fig. 6 bottom). The scheme operates by stepping through a 3-stage process as shown in Fig. 6 and explained below. The results of all three stages are reported continuously in various status registers in each slave device which can be polled by an external host. This allows appropriate action to be taken at the system level if required. It should be noted that all slave devices are operating (regulating) autonomously during stages 1-2 of the proposed scheme.

Stage 1 - During stage 1 of the proposed scheme all slave controllers check that the master and slave OSC clock frequencies are within a pre-defined frequency error range e.g. 2 PWM switching cycles. Stage 1 also determines that the master is generating valid SYNC pulses and that they are being received correctly by the slave devices.

As shown in Fig. 6 (see stage 1 portion) the valid SYNC signal being received correctly and OSC frequency error range check is implemented shortly after power-up through the use of a digital error counter in each slave device. Once valid SYNC pulses are received (by falling edge detection logic – signal *sync_pulse_fedge* in Fig. 6) by the slave from the master an internal counter *sync_ counter* is incremented. At the same time a separate PWM cycle counter *pwm_cycles* is incremented once per PWM switching cycle. At the end of *J* PWM switching cycles the *sync_counter* value *K* is compared to *J*. If,

- *K* +/- Frequency Error Range < *J* => The master clock is operating much slower than the slaves.

- *K* +/- Frequency Error Range = *J* => The master and slave clock frequencies are closely matched.

- *K* +/- Frequency Error Range > *J* => The master clock is operating much faster than the slaves.

If the slaves determine that the master is operating within the pre-defined frequency error range the scheme moves to stage 2 by activating the control signal *sync_receiving_ok* in Fig. 6. If not the slaves continue to operate autonomously in the knowledge that the master is either operating at too fast or slow a clock frequency for it to tune to.

Stage 2 - Tuning the OSC frequencies in each slave device to match the master device OSC frequency occurs during stage 2. This is primarily achieved through the use of a digital-to-analog converter (DAC) controlling current mirror circuits in a relaxation OSC. The control signal to the DAC is determined by a digital phase error detection circuit as shown in Fig. 6 (see Stage 2 portion). *Note:-* each device will have a default oscillator frequency configured during factory calibration - signal *osc_trim_default* in Fig. 6). As explained in Section II this default setting can vary over PVT.

On the detection of valid SYNC pulses each slave determines the difference between the present (*n*) and previous (*n-1*) switching cycle PWM counter values (*pwm_cnt[n]* and *pwm_cnt[n-1]* respectively) relative to when the SYNC pulse was received. The oscillator DAC is updated with a digital value (*osc_trim* – now sourced from signal *osc_trim_mod* not the factory calibrated default *osc_trim_default*) corresponding to the calculated phase difference. The phase difference results in either an increase (*inc*) or decrease (*dec*) of the OSC output frequency. As the OSC provides the clock reference (*clk_ref*) to the PLL (itself a second order closed loop control system) the control signal to the DAC is extended by 4 least significant bits (LSBs) so that increments/decrements can only occur every 16th switching cycle. This reduces the risk of a fast changing clock reference disturbing the PLL from a stable phase lock status.

This process continues until the frequency error range (FR in Fig. 6) between the master SYNC pulse (start of master PWM switching cycle) and the slave start of PWM switching cycle is within a pre-defined error range. The master and slave OSC clock frequencies are now tuned (locked). Once the slave device remains frequency locked for a pre-defined period of time the scheme enables stage 3 to begin PWM phase synchronization.

The logic at stage 1 continues to check in the background for OSC frequency errors between the master and slaves. If errors arise (e.g. the master gets disconnected or begins to transmit numerous SYNC pulses in a single PWM cycle) then all slaves return to autonomous mode and the process resets back to stage 1.

Stage 3 – PWM phase synchronization can now begin with the slave controllers leaving autonomous operating mode and adhering to the received master SYNC pulses. On detecting the falling edge of the received SYNC pulses the slaves align their PWM signals with/without a phase offset delay (*phase_delay* in Fig. 6, Stage 3).

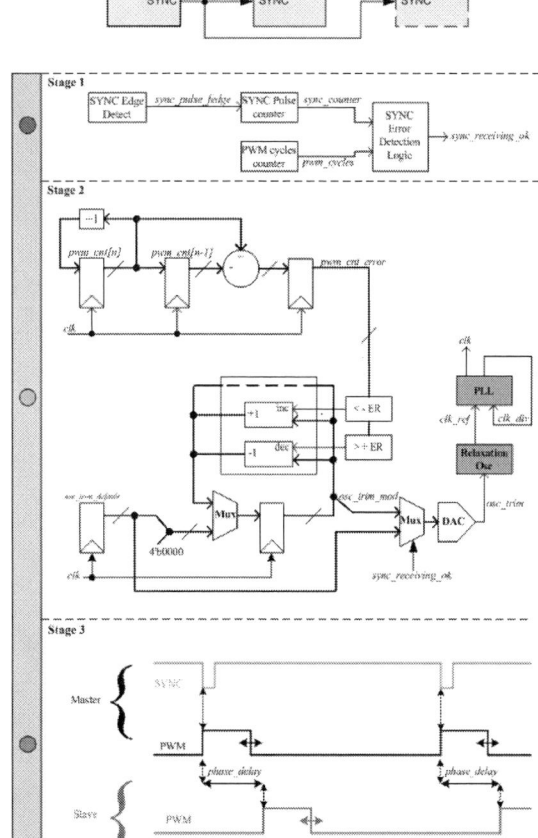

Figure 6 Three Stage Clock Tuning and PWM Phase Synchronization Architecture.

IV. EXPERIMENTAL RESULTS

This scheme was designed into a DPWM module in a dual phase digital PWM controller IC. The controller IC was designed using a standard CMOS process with the default OSC frequency trimmed to 8 MHz (scope plots below show a divide by 4 version of the OSC clock).

A 4-phase interleaved power stage was designed operating at 250 kHz switching frequency with the input voltage set to 12 V and the output voltage regulated to 1.0 V.

Fig. 7, Fig. 8 and Fig. 9 show how the scheme works for three use cases of when the master's OSC is running too fast, too slow and finally within the frequency error lock range of the slave controller.

For these scope plots: Ch4 = master OSC clock, Ch3 = slave OSC clock, Ch2 = slave primary PWM signal, Ch1 = master SYNC signal.

For stage 1 the SYNC pulse error check range value was selected between 10-14 while, J was set to 12. This allowed an OSC tuning range of $(T_{pwm\_slave} * (12/14))_{max} = 291.67$ kHz and $(T_{pwm\_slave} * (12/14))_{min} = 208.33$ kHz giving a +/- 16.67 % tuning range on the 8 MHz OSC frequency at 250 kHz PWM switching frequency.

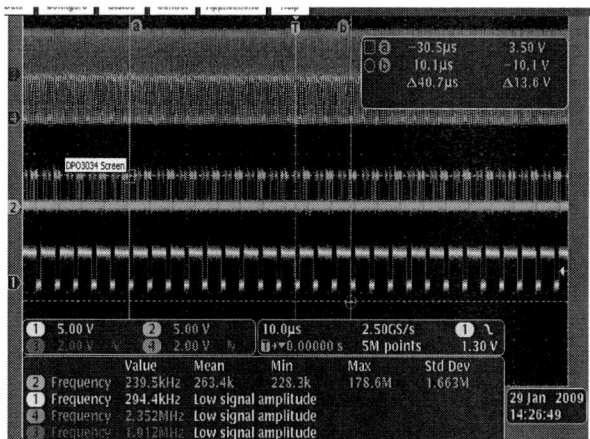

Figure 7 Master OSC Frequency Faster than Slave and outside the Tuning Error Range – Slave OSC Tuning and PWM Phase Synchronization does not occur.

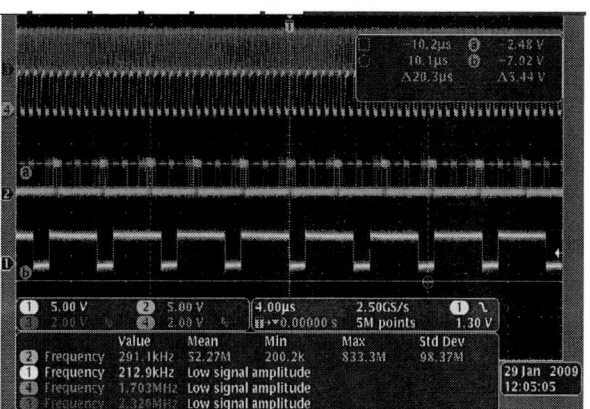

Figure 8 Master OSC Frequency Slower than Slave and outside the Tuning Error Range – Slave OSC Tuning and PWM Phase Synchronization does not occur.

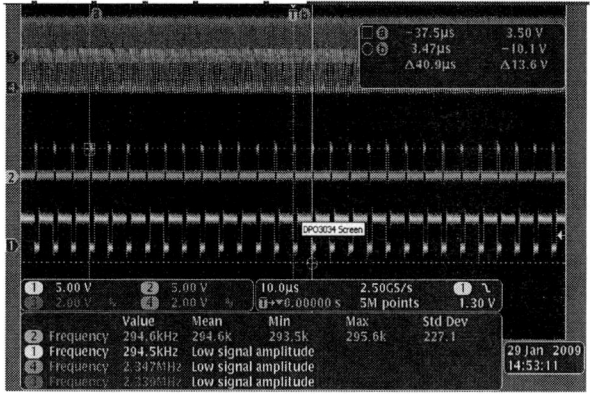

Figure 9 Master and Slave OSC Frequencies within the Tuning Error Range – Slave OSC tunes to the Masters OSC Frequency of ~2.34 MHz and PWM Phase Synchronization Occurs.

Fig. 10 shows the four PWM phases after the slave controller has tuned it's OSC to the masters OSC frequency (Fig. 9) and PWM phase synchronization/alignment takes place.

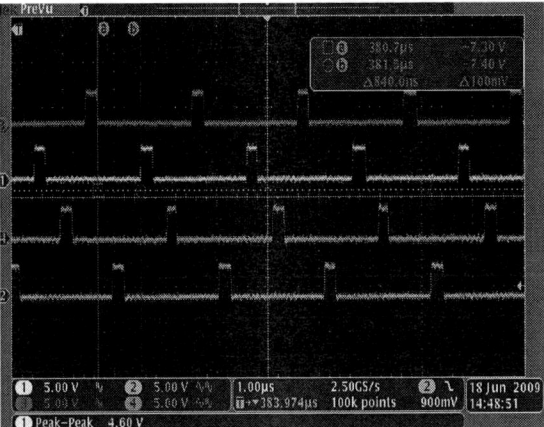

Figure 10 PWM Phase Synchronization between Master (top two traces) and Slave Controllers (bottom two traces).

V. CONCLUSIONS

A new digitally controlled master-slave clock tuning and PWM phase synchronization scheme has been introduced. Experimental results from an IC implementation prove the three stage scheme to operate correctly under different use cases of when on-chip clock frequencies vary between PWM controller ICs.

The scheme implements several key checks at all three stages to ensure a robust and reliable solution. Status information for all three stages is available if required by a system host.

The scheme eliminates the requirement for using external clock generators to facilitate PWM phase synchronization. In addition by tuning all controller OSC frequencies to be very closely matched accurate control of turn on/off delay times and ramp rates for system sequencing is automatically provided avoiding the need for an external discrete power sequencer IC [11].

REFERENCES

[1] Erickson, R.W. and D. Maksimovic, 2001. *Fundamentals of Power Electronics*. 2nd ed. Kluwer Academic Publishers (KAP), ISBN 0-7923-7270-0.

[2] Xiao J., A.V. Peterchev, and S.R. Sanders, 2001. "Architecture and IC Implementation of a Digital VRM Controller." In: *IEEE Power Electronics Specialists Conference*, Vancouver, Canada, June 17-22, 2001.

[3] Prodic, A., D. Maksimovic and R.W. Erickson, 2002. "Design of a Digital PID Regulator Based on Look-Up Tables for Control of High-Frequency DC-DC Converters." In: IEEE Workshop on Computers in Power Electronics, June 3-4, 2002.

[4] Xiao J., A.V. Peterchev, J. Zhang and S.R. Sanders, 2004. "An Ultra-Low-Power Digitally-Controlled Buck Converter IC for Cellular Phone Applications." In: *IEEE Applied Power Electronics Conference and Exposition*, Los Angeles, California, February 22-26, 2004.

[5] E. O'Malley, K. Rinne, " A Programmable Digital Pulse Width Modulator Providing Versatile Pulse Patterns and Supporting Switching Frequencies Beyond 15 MHz," in Proc. IEEE APEC'04 Conf., 2004, pp. 53 59.

[6] Patella, B.J., A. Prodic, A. Zirger, and D. Maksimovic, 2002. "High-Frequency Digital Controller IC for DC/DC converters." *IEEE Transactions on Power Electronics*, Vol. 18, No. 1, pp. 438-446, Jan 2003.

[7] Syed, A., E. Ahmed, D. Maksimovic and E. Alarcon, 2004. "Digital Pulse Width Modulator Architectures." In: *IEEE Power Electronics Specialists Conference*, 2004.

[8] ZL2005 Data Sheet, "Digital-DC^{TM} Integrated Power Management and Conversion IC Datasheet", 2006. Available from, www.zilkerlabs.com.

[9] Texas Instruments "Digital PWM System Controller Datasheet, July 2008. Available from, http://focus.ti.com/lit/ds/symlink/ucd9240.pdf.

[10] Maxim "Max1960/1961/1962 PWM Step-Down Controller Datasheets", 2003. Available from, www.maxim-ic.com.

[11] National Semiconductor "LM3881 Power Sequencer Datasheet", 2008. Available from, www.national.com.

Control Scheme and Transient Performance of Sigma VR

Pengjie Lai, Julu Sun, and Fred C. Lee

Center for Power Electronics Systems
The Bradley Department of Electrical and Computer Engineering
Virginia Polytechnic Institute and State University
Blacksburg, VA 24061 USA

Abstract— Sigma Voltage Regulator (VR) was proposed for achieving high efficiency. It takes advantage of the high efficiency unregulated converter to improve the overall efficiency and combines a low power Buck converter for voltage regulation. This paper mainly discusses the control scheme and transient performance of Sigma VR. The small-signal-model is analyzed first. Then the control strategy and the detailed control design of Sigma VR to achieve constant output impedance for AVP and transient are given. At the end, the simulation and experimental result demonstrates that Sigma VR can achieve very good transient performance with less output bulk capacitors compared with that of conventional Buck VR.

I. INTRODUCTION

According to Intel's roadmap, the future generation microprocessors are expected to be operated at lower voltage (0.8~1V) and will draw higher current (more than 120A) with higher current slew rate (20A/ns) [1]. Accordingly, the voltage regulator which provides the power of the microprocessors will have to meet the increasing challenges such as tighter voltage tolerance and faster transient requirement.

High efficiency is another challenging issue for future VR. Take server VR for example, 90% plus efficiency is expected for saving energy and electricity bills since there are so many microprocessors in data-server system.

The limitation of state of the art single-stage multi-phase Buck VR for high efficiency is the switching related losses. To reduce it and improve efficiency, two-stage architecture was proposed [1] and demonstrated in laptop VR application. But both the two stages have to handle the full power, the overall efficiency of the two-stage VR is the multiple of the first stage and the second stage which decreases the converter efficiency [2].

Besides the two-stage in series structure, another two-stage structure, the Sigma structure for higher efficiency performance is proposed in [2] which is shown in Fig 1. It is constructed by paralleling a high efficiency unregulated DCX and a Buck converter. And let the high efficiency DCX to handle most of the power while using the low power Buck

converter to achieve voltage regulation. Because the power is delivered in parallel and most power delivers through the high efficiency DCX, so that the overall efficiency can be improved compared with series two-stage structure. It was demonstrated that about 90% efficiency can be achieved by Sigma structure which is 4%~5% improved compared with current multi-phase Buck VR operating at 600 kHz [2].

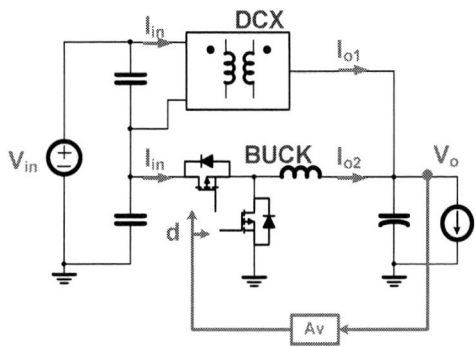

Fig 1. Sigma VR architecture

Except for the efficiency expectation in VR application, current and future VRs also are facing much more challenging dynamic transient requirement. In order to deal with the faster dynamic requirement, the speed of VR should be higher. It is well known that, using linear control method, the speed of the VR is in part determined by the control bandwidth which is limited by the switching frequency. Beyond the control bandwidth, external output bulk caps are required to handle the transient voltage spike. For example, for 600 kHz multi-phase Buck VR, assume that the control bandwidth can be pushed to 1/3 of switching frequency which is 200 kHz, six 330uF/6mohm SP bulk caps are required for transient dynamic requirement.

For Sigma structure, the power stage itself includes a fast dynamic DCX path. When load transient happens, if the DCX path can help to provide faster current response, then the requirement for the output bulk caps can be alleviated. From this basic concept, this paper mainly focuses on the transient analysis of Sigma VR. In section II, the power stage characteristics of the Sigma converter and the transfer functions are given. Based on the small-signal-model of Sigma structure, the control strategy for utilizing the faster dynamic DCX path is proposed in section III. In section IV, the Sigma VR simulation and experimental result is shown for validation. At last, a summary is given for the conclusion.

This work was supported primarily by Analog Devices, C&D Technologies, CRANE, Delta Electronics, HIPRO Electronics, Infineon, International Rectifier, Intersil, FSP-Group, Linear Technology, LiteOn Tech, Primarion, NXP, Renesas, National Semiconductor, Richtek, Texas Instruments. Also, this work make use of the ERC shared facilities supported by the National Science Foundation under Award Number EEC-9731677.

II. SMALL SIGNAL MODEL OF SIGMA VR

Figure 2 shows the power stage small-signal-model of Sigma VR where R_{out} and L_{out} represents the output impedance of DCX [7]. Through some approximation, the transfer functions of Sigma VR can be derived as shown in equation (1) to (6).

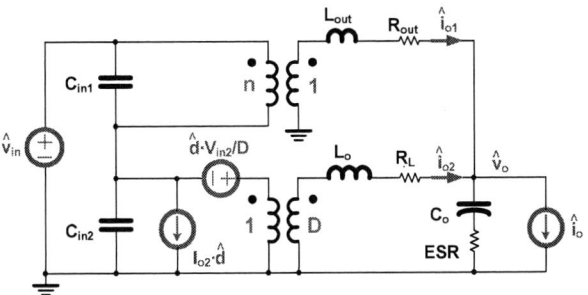

Fig 2. Small Signal Model of Sigma VR

In Fig 2, V_{in2} is the steady state input voltage of Buck converter and I_{o2} is the steady state Buck inductor current. In the following equations, G_{vd} is the transfer function of output voltage v_o to duty cycle d. G_{id2} is the transfer function of buck inductor current i_{o2} to duty cycle d. G_{id} is the duty cycle to total current $i_{o1}+i_{o2}$ transfer function. G_{ii2} is the transfer function of the buck inductor current i_{o2} to load current i_o. G_{ii} is the transfer function of total current $i_{o1}+i_{o2}$ to load current i_o. Z_o is the open loop output impedance of Sigma VR.

$$G_{vd}(s) = \frac{V_{in}}{(1+n \cdot D)^2} \cdot \frac{(1+\frac{s}{Q_{z1}\cdot\omega_{z1}}+\frac{s^2}{\omega_{z1}^2})(1+\frac{s}{\omega_{esr}})}{\Delta_{p1}\cdot\Delta_{p2}} \tag{1}$$

$$G_{id2}(s) = \frac{V_{in}\cdot(n^2C_{in}+C_o)}{(1+n\cdot D)^3} \cdot \frac{s\cdot(1+\frac{s}{Q_{z2}\cdot\omega_{z2}}+\frac{s^2}{\omega_{z2}^2})}{\Delta_{p1}\cdot\Delta_{p2}} \tag{2}$$

$$G_{id}(s) = \frac{V_{in}\cdot C_o}{(1+n\cdot D)^2} \cdot \frac{s\cdot(1+\frac{s}{Q_{z1}\cdot\omega_{z1}}+\frac{s^2}{\omega_{z1}^2})}{\Delta_{p1}\cdot\Delta_{p2}} \tag{3}$$

$$G_{ii2}(s) = \frac{1}{1+nD} \cdot \frac{(1+\frac{s}{Q_{z3}\cdot\omega_{z3}}+\frac{s^2}{\omega_{z3}^2})\cdot(1+\frac{s}{\omega_{esr}})}{\Delta_{p1}\cdot\Delta_{p2}} \tag{4}$$

$$G_{ii}(s) = \frac{(1+\frac{s}{Q_{z4}\cdot\omega_{z4}}+\frac{s^2}{\omega_{z4}^2})\cdot(1+\frac{s}{\omega_{esr}})}{\Delta_{p1}\cdot\Delta_{p2}} \tag{5}$$

$$Z_o(s) = \frac{n^2D^2R_{out}}{(1+nD)^2} \frac{(1+\frac{s}{Q_{z5}\cdot\omega_{z5}}+\frac{s^2}{\omega_{z5}^2})\cdot(1+\frac{s}{\omega_{z6}})\cdot(1+\frac{s}{\omega_{esr}})}{\Delta_{p1}\cdot\Delta_{p2}} \tag{6}$$

where

$$\Delta_{p1}\cdot\Delta_{p2} = (1+\frac{s}{Q_{p1}\cdot\omega_{p1}}+\frac{s^2}{\omega_{p1}^2})(1+\frac{s}{Q_{p2}\cdot\omega_{p2}}+\frac{s^2}{\omega_{p2}^2}) \tag{7}$$

$$\omega_{p1} = \frac{1}{\sqrt{(n^2C_{in}+C_o)(L_o/(1+n\cdot D)^2)}} \tag{8}$$

$$Q_{p1} \approx \frac{1}{(R_L+ESR)/(1+n\cdot D)^2} \sqrt{\frac{L_o/(1+n\cdot D)^2}{(n^2C_{in}+C_o)}} \tag{9}$$

$$\omega_{p2} = \frac{1}{\sqrt{(\frac{n^2C_{in}\cdot C_o}{n^2C_{in}+C_o})\cdot L_{out}}} \tag{10}$$

$$Q_{p2} \approx \frac{1}{R_{out}+ESR} \sqrt{\frac{L_{out}}{(\frac{n^2C_{in}\cdot C_o}{n^2C_{in}+C_o})}} \tag{11}$$

The other parameters are given in the appendix.

From the transfer functions described above, we can see the Sigma power stage is a fourth-order system. In normal condition, the L_{out} is much smaller than L_o. With this assumption, two separate pairs of double-pole can be observed in the denominator. The low frequency one is mainly determined by buck inductor and the equivalent cap which is $C_o+n^2(C_{in1}+C_{in2})$. This double-pole will move with different duty cycle, so that Sigma structure is not suitable for laptop case whose input voltage range is larger. The other one is located at high frequency range which is mainly determined by small L_{out} and the output cap C_o. In the later, you will see the influence from the L_{out}.

Another observation is that the DC gains of these transfer functions are smaller than buck converter case. The reason is that the power stage itself has close-loop feedback through the DC/DC transformer. And this feedback characteristic attenuates the DC gain of the transfer functions in contrast with the buck converter.

III. CONTROL STRATEGY AND DESIGN OF SIGMA VR

A. Active Droop Control of Buck VR

It is well known that CPU VR must achieve adaptive voltage positioning (AVP) to meet Intel load line specification. The active droop control scheme which is shown in Fig 3 is widely adopted by current power management IC companies to achieve AVP for Buck VR. The basic concept is to inject the inductor current information into feedback voltage information, so that AVP can be realized with an infinite DC gain compensator design for A_v. In addition, the constant output impedance concept is proposed for perfect transient response for VR. Based on active droop control scheme, the detailed control design guideline in order to achieve constant close-loop output impedance of Buck VR is described in [3].

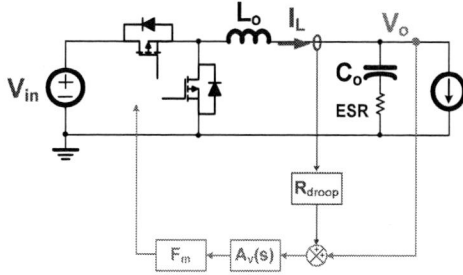

Fig 3. Active Droop Control Scheme for Buck VR

Normally, the current sensing coefficient is chosen to be specified R_{droop} which is the DC output impedance. Following the design guideline, first, proper A_v is designed to achieve high bandwidth current loop T_i. With high current loop bandwidth, the inductor will approximately behave like a controlled current source, so that the original second-order system can be reduced to the first-order system. The impedance Z_{oi} with closed high bandwidth current loop is just determined by the output cap which is shown in Fig 4(a).

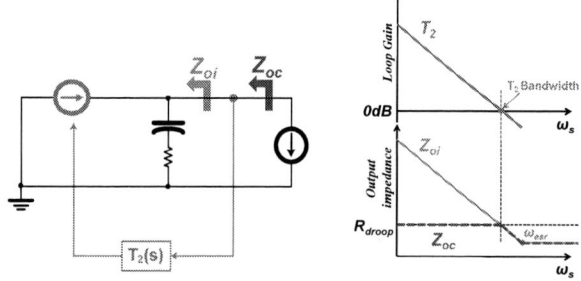

(a) Impedance Approximation   (b) Constant Output Impedance Design

Fig 4. Active Droop Control Design of Buck VR

Under high bandwidth current loop condition, the outer loop T_2 is closed to attenuate Z_{oi} to the target impedance R_{droop} (1mΩ) below the T_2 bandwidth. From Fig 4(b), we can see that the required T_2 bandwidth is determined by the intersect frequency point between Z_{oi} and R_{droop}. Here we assume that the ESR of the output cap is smaller than R_{droop} since in order to meet Intel impedance requirement, proper output bulk caps with that kind of ESR characteristics should be chosen.

B. Active Droop Control of Sigma VR

For Sigma VR, in order to achieve AVP, based on the concept of active droop control scheme, the total current information including both the DCX output current and Buck inductor current is needed to be injected into the feedback voltage information. With the specified R_{droop} current sensing coefficient, the AVP can be achieved. The structure is shown in Fig 5.

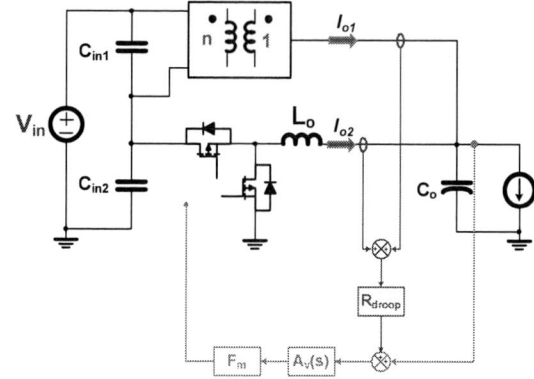

Fig 5. Basic Active Droop Control Scheme of Sigma VR

Actually, with this control strategy, we can also optimize the transient response by shaping close-loop output impedance as Buck VR does. However, there are two drawbacks by simply using this control strategy.

The first drawback is that it is relatively hard to sense the DCX output current compare with the integrated method to sensing Buck inductor current. The other one is that with high bandwidth current loop design, it cannot utilize the input cap to work with the output cap for reduced the required T_2 bandwidth which is the benefit of this Sigma Structure. The high bandwidth current loop will limit the behavior of fast dynamic DCX path which is shown in Fig (6).

From Fig 6(b), we can see that the required T2 bandwidth is also just determined by the output caps. For example, with 600 kHz switching frequency, we also need 6*330uF/6mohm output bulk caps which is the same as the Buck case for 200 kHz control bandwidth. The reason is that the faster dynamic DCX path is limited by current loop, so that input caps cannot be helpful. In order to free the DCX path and make the input cap help with the output cap, the new active droop control strategy of Sigma VR is needed.

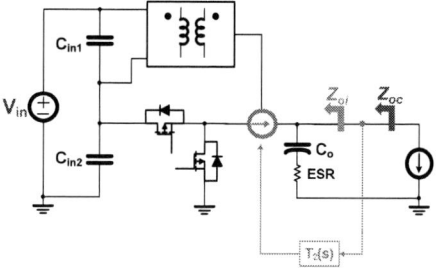

(a)   The Sum of Two Path is Approximated as a Current Source

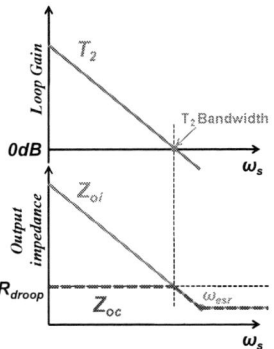

(b)   Constant Output Impedance Design

Fig 6. Basic Active Droop Control and Design of Sigma VR

C. Proposed Active Droop Control Scheme of Sigma VR

The proposed active droop control scheme of Sigma VR is shown is Fig 7. In order to utilize the input cap for reducing the required T2 bandwidth, we have to free the DCX path which is much lower impedance compare to Buck path. The active droop control strategy is shown in Fig7.

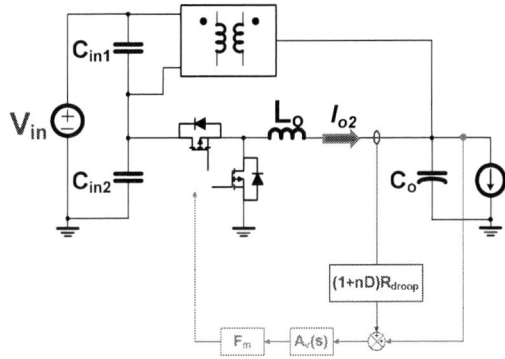

Fig 7. Proposed Active Droop Control Scheme of Sigma VR

From Fig 7, just the buck inductor current is sensed and injected into feedback voltage information. The current loop does not control the DCX path anymore but just for the Buck path. Because the current ratio between DCX and Buck is equal to the input voltage ratio of them, we set the current sensing coefficient to be $(1+nD) \cdot R_{droop}$ for DC close-loop output impedance.

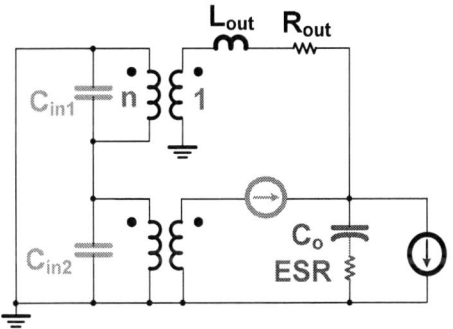

Fig 8. Buck Inductor Serves as a Controller Current Source

After designing high current loop bandwidth, the Buck inductor behaviors like a control current source, the power stage can be approximately becomes Fig 8. Reflect the input side cap into output side, the impedance Z_{oi} with high bandwidth current loop is shown in Fig 9(a). We can see that, the input equivalent input caps with DCX impedance are paralleled with output caps. But the capacitance is multiplied n^2 which effectively enlarges the capacitance.

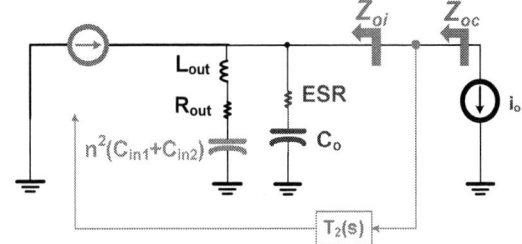

(a)   Equivalent Circuit with High BW Current Loop

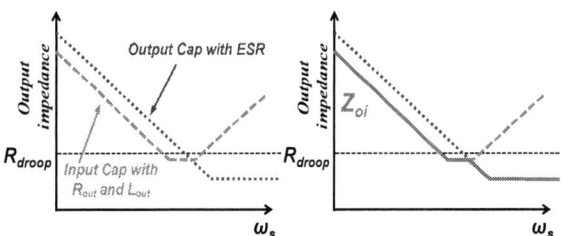

(b) Z_{oi} Impedance with High BW Current Loop Closed

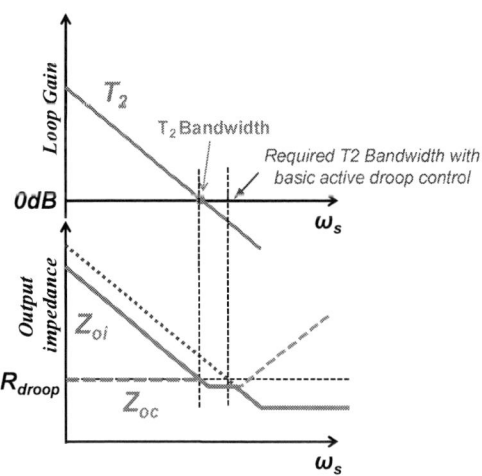

(a)   Required T2 BW reduced

Fig 9. Propose Active Droop Control and Design of Sigma VR

From Fig 9(b), we can see that Z_{oi} impedance is reduced during low frequency range because of the input capacitance effect. When the output impedance of the DCX R_{out} and L_{out} is small, the required T_2 bandwidth will be also reduced which can be clearly shown in Figure 9(c) compared with the basic active droop control scheme or Buck VR case. From another view angle, if we set the control bandwidth of Sigma VR and Buck VR to be the same, then the requirement for the

978-1-4244-4782-4/10 $26.00 © 2010 IEEE   1930

output bulk cap of Sigma VR can be alleviated. The input cap work with the output cap to handle the voltage spike when transient happens, thus not so many output bulk caps are necessary for Sigma VR.

D. Control Design Example of Proposed Control Scheme for Sigma VR

The design example is based on the following circuit parameters:

$V_{in}=12V$ $V_o=1.3V$
$C_{in1}=C_{in2}=44uF$ $C_o=4*300uF+18*22uF$
$DCX: n=6$ $R_{out}=1.6mohm$ $L_{out}=3nH$
$Buck: L_o=75nH$

Fig 11 shows the proposed control scheme block diagram.

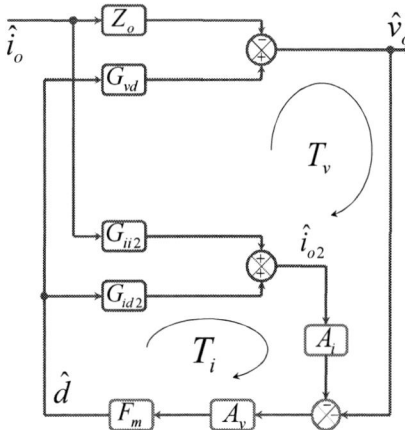

Fig 11. Control Blocks of Propose Control Scheme

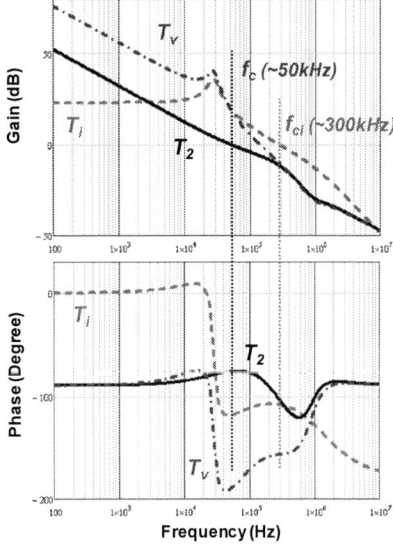

Fig 12. Control Loop Gain of Proposed Control Scheme of Sigma VR

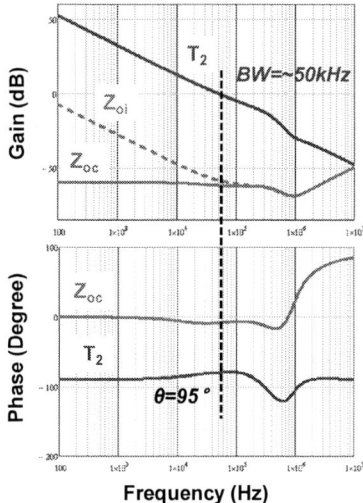

Fig 13. Control Bandwidth and Output Impedance of Sigma VR

From Fig 12, through design appropriate A_v to achieve high current bandwidth T_i which is around 300 kHz in this example, the closed high current loop output impedance Z_{oi} can be approximately as the input cap with output impedance of DCX paralleled with the output bulk cap which is shown in Fig 13.

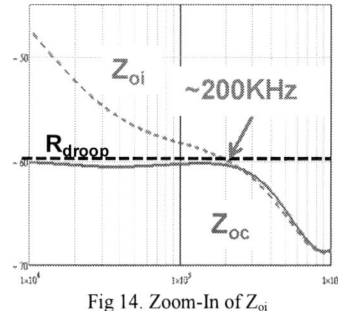

Fig 14. Zoom-In of Z_{oi}

Fig 14 is zoom-in diagram of Z_{oi} of this design. We can see that the required bandwidth should be around 200 kHz. But from Fig 13, just around 50 kHz bandwidth can also attenuate Z_{oi} into target impedance range. The reason is because T_2 phase margin is large which also gives some attenuation around bandwidth frequency. In addition, the Z_{oi} gain is also small at that frequency range.

IV. SIMULATION AND EXPERIMENT VERIFICATION

A. Simulation Result

The simulation result with the same circuit condition shown in Section III. And the simulation waveforms are shown in Fig 15. From the simulation result, we can see that with 4 330uF/6mohm SP output bulk cap, and around 50 kHz T_2 bandwidth, the transient response of sigma VR can meet the requirement.

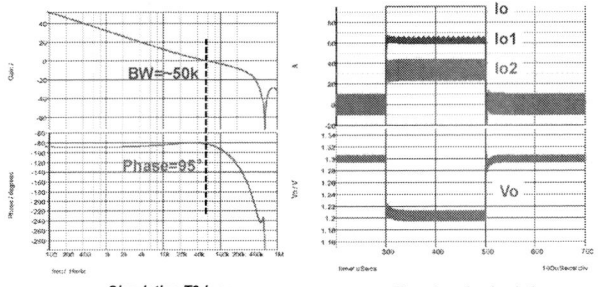

Simulation T2 loop   **Time domain simulation**

Fig 15. Simulation Result of Transient Performance of Sigma VR

B. Experimental Result

A Sigma VR board which is constructed by two-phase interleaved buck and two-phase interleaved DCX for a 12V to 1.3V/80A VR is design to verify theoretical analysis which is shown is Fig 16. The controller ISL6307 from Intersil is used. The switching frequency for Buck is 600 kHz and the inductor for each phase is set to be 150nH. Fig 17 shows the step-up and step-down transient response, the zoom-in of it is shown is Fig 18.

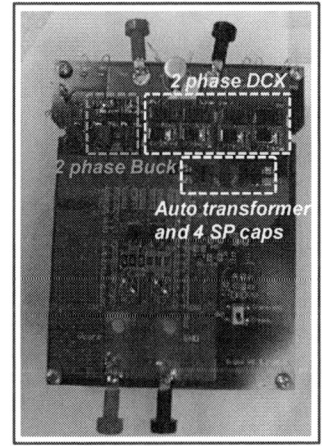

Fig 16. Sigma VR board

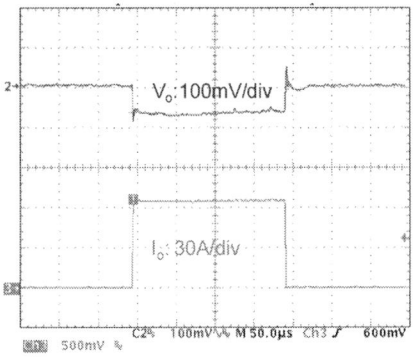

Fig 17. The Transient Response of Sigma VR with 4 SP Output Bulk Caps

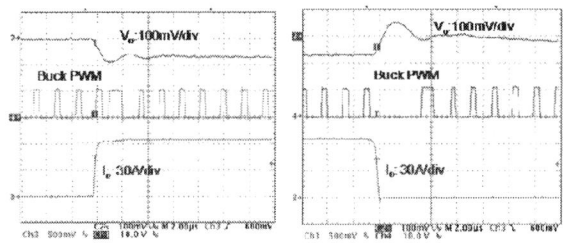

Fig 18. Enlarged Transient Response

With four 330uF/6mohm SP output bulk cap, Sigma VR can achieve good transient response. Compared with Buck VR case which required six that kind of output caps with the same control bandwidth, Sigma VR can achieve 2 SP caps reduction and keep higher efficiency performance. Actually, the cap reduction is not obvious. The reason is that in our prototype, all the devices and magnetic components are discrete, which result in large Lout and Rout. In theoretical analysis, if we can reduce the output impedance of DCX, the Sigma structure benefit for output cap reduction will be more obvious.

V. CONCLUSION

In this paper, the control strategy of Sigma VR which can utilize the input cap to help the transient response is analyzed. The experimental result suggests that Sigma VR can achieve good transient response with less output bulk cap compared with Buck VR. If we can further reduce the output impedance of DCX, the Sigma structure will save more output bulk caps.

REFERENCES

[1] Julu Sun, et al , " High Power Density, High Efficiency System Two-stage Power Architecture for Laptop Computers," *PESC 2006*
[2] Julu Sun, et al, " High Efficiency Quasi-Parallel Voltage Regulator," *2008 APEC*
[3] Kaiwei Yao, et al , " Adaptive Voltage Position Design for Voltage Regulators," *2004 APEC*
[4] Julu Sun, et al, " A Novel Input-Side Current Sensing Method to Achieve AVP for Future VRs," *2005 APEC*
[5] "Enterprise Platform Power Efficiency for Future Server System," IBM Sympotium technology 2004
[6] Kaiwei. Yao, et al, "Optimal Design of the Active Droop Control Method for the Transient Response," 2003 APEC
[7] Eric X. Yang, Fred C. Lee and Milan M. Jovanovic, "Small signal model of series and parallel resonant converters," PESC' 92

APPENDIX

The parameters of the transfer function of Sigma VR are given in this appendix:

$$\omega_{z1} = \frac{1}{\sqrt{(n^2 C_{in})\left(L_{out}/(1+n\cdot D)\right)}}$$

$$Q_{z1} = \frac{1}{R_{out}/(1+n\cdot D)}\sqrt{\frac{L_{out}/(1+n\cdot D)}{n^2 C_{in}}}$$

$$\omega_{z2} = \frac{1}{\sqrt{(n^2 C_{in} // C_o)L_{out}}}$$

$$Q_{z2} = \frac{1}{R_{out} + ESR}\sqrt{\frac{L_{out}}{(n^2 C_{in} // C_o)}}$$

$$\omega_{z3} = \frac{1}{\sqrt{(n^2 C_{in})\frac{L_{out}}{(1+nD)}}}$$

$$Q_{z3} = \frac{(1+nD)}{DCR}\sqrt{\frac{\frac{L_{out}}{(1+nD)}}{n^2 C_{in}}}$$

$$\omega_{z4} = \frac{1}{\sqrt{(n^2 C_{in})\frac{L_{out} + L_o}{(1+nD)^2}}}$$

$$Q_{z4} = \frac{(1+nD)^2}{DCR + R_{out}}\sqrt{\frac{\frac{L_{out}+L_o}{(1+nD)}}{n^2 C_{in}}}$$

$$\omega_{z5} = \frac{1}{\sqrt{\frac{C_{in}}{D^2}\cdot\frac{L_o n^2 D^2 L_{out}}{(L_o + n^2 D^2 L_{out})}}}$$

$$Q_{z5} = \frac{1}{\frac{R_L n^2 D^2 R_{out}}{(R_L + n^2 D^2 R_{out})}}\cdot\sqrt{\frac{\frac{L_o n^2 D^2 L_{out}}{(L_o + n^2 D^2 L_{out})}}{\frac{C_{in}}{D^2}}}$$

$$\omega_{z6} = \frac{R_L + n^2 D^2 R_{out}}{L_o + n^2 D^2 L_{out}}$$

A Three-Phase Current-Fed Push-Pull DC-DC Converter with Active Clamp for Fuel Cell Applications

Sangwon Lee and Sewan Choi, *IEEE Senior Member*

Seoul National University of Technology

Dept. of Control and Instrumentation Eng.

172 Kongneung-Dong, Nowon-Ku, Seoul, Korea

E-mail : schoi@snut.ac.kr

Abstract - In this paper a new active-clamped three-phase current-fed push-pull DC-DC converter is proposed for high power applications where low voltage high current input sources such as fuel cells are used. The proposed converter has the following features; active clamping of the transient surge voltage caused by transformer leakage inductances, natural ZVS turn-on of main switches using energy stored in transformer leakage inductor, small current rating and ZVZCS of clamp switches, no additional start-up circuitry for soft starting due to the operating duty cycle range between 0 and 1, ZCS turn-off of rectifier diodes leading to negligible voltage surge associated with the diode reverse recovery. Experimental results on a 5kW laboratory prototype are provided to validate the proposed concept.

I. INTRODUCTION

The step-up DC-DC converter with high frequency transformer has increasingly been used in high power applications such as fuel cell systems, photovoltaic systems, hybrid electric vehicles, and UPS where voltage step-up and galvanic isolation are required. The step-up DC-DC converter with high frequency transformer could be either voltage-fed or current-fed type. The advantages and disadvantages of the two types are detailed in [1]. Compared to the voltage-fed topology, the current-fed topology exhibits smaller input current ripple, lower diode voltage rating, lower transformer turns ratio and negligible reverse recovery problem at the rectifier side, in general [2]. Especially, lower transformer turns ratio leads to smaller duty cycle loss and transformer copper losses, which are important for efficient operation at high power levels. Direct and precise control of the input current is also possible with the current-fed topology. Therefore, the current-fed DC-DC converter is better suited to low voltage high current input application such as fuel cells [1].

Three-phase DC-DC converters have presented good performance in the high power applications where high device stresses are faced when implemented with the single-phase DC-DC converter[3-8]. Generally, the three-phase DC-DC converter has several advantages over its single-phase counterpart; easy MOSFETs selection due to reduced current rating, reduction of the input and output filters' volume due to increased effective switching frequency by a factor of three, reduction in transformer size due to better transformer

utilization. The isolated current-fed dc-dc converters can be classified by primary side configuration into three basic topologies; full-bridge[3, 4], L-type half bridge[5, 6, 7], and push-pull [8]. Among them, the current-fed push-pull converter [8] has the simplest structure in its gate drive circuits and power circuits. However, their active switches are hard switched, and the dissipative losses associated with passive clamp circuit for suppressing the voltage spikes caused by leakage inductance of the high frequency transformer are considerable. Therefore, it is not easy to achieve high efficiency and high power density at high power level.

In this paper, an active clamped three-phase current-fed push-pull DC-DC converter is proposed for high power applications where low voltage high current input sources such as fuel cells are used. In addition to the above mentioned advantages of the three-phase and current-fed topologies, the proposed converter has the following features;

- Active clamping of the transient surge voltage caused by transformer leakage inductances

- Natural ZVS turn-on of main switches using energy stored in transformer leakage inductor

- Small current rating and ZVZCS of clamp switches

- No additional start-up circuitry for soft starting due to the operating duty cycle range between 0 and 1

- ZCS turn-off of rectifier diodes leading to negligible voltage surge associated with the diode reverse recovery

II. OPERATING PRINCIPLES

The circuit topology of the proposed converter is basically a three-phase, current-fed, push-pull converter with an active clamp circuit, as shown in Fig. 1. The proposed converter includes an input filter inductor, three main switches S_{M1}, S_{M2}, and S_{M3}, and a clamp circuit consisting of three clamp switches S_{C1}, S_{C2}, and S_{C3} and a clamp capacitor C_C at the low voltage primary side, and a three-phase diode bridge at the high voltage secondary side. Note that a three-leg core must be used for proper operation of the proposed converter. The three-phase windings are configured in Y-Y connection. The neutral point of the three-phase primary winding is connected to input

978-1-4244-4782-4/10 $26.00 © 2010 IEEE

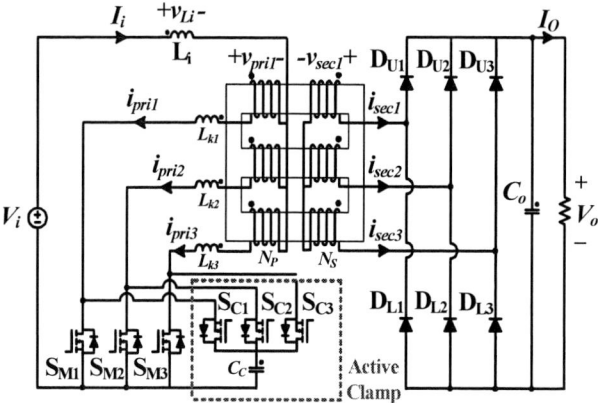

Fig. 1 Proposed three-phase push-pull converter with active clamp

source through the input inductor. The proposed active clamping method not only limits the transient surge voltage caused by transformer leakage inductances, but helps improve the efficiency by enabling soft switching of the main switches.

Output voltage control is achieved by applying the asymmetrical PWM switching to each main and clamp switch pair. The three switch pairs are interleaved with 120° phase shift, which leads to increased effective switching frequency resulting in smaller input current ripple. The duty cycle of each main switch is in the whole range between 0 and 1. The ideal voltage ratio of the proposed converter can be expressed as,

$$\frac{V_o}{V_i} = \frac{n}{(1-D)} \quad (0 < D < 1) \tag{1}$$

where $n = N_S/N_P$. Also, the voltage across the clamp capacitor C_C can be obtained by,

$$V_C = \frac{1}{(1-D)} \cdot V_i \tag{2}$$

In order to simplify the analysis of the steady-state operation several assumptions are made as follows,

• Input inductor L_i is sufficiently large so that it can be considered as a constant current source.
• Output capacitor C_o is sufficiently large so that it can be considered as a constant voltage source.
• Dead time between main and clamp switch pair is ignored.
• Magnetizing inductance is assumed to be infinity.
• All leakage inductance values are equal ($L_{k1}=L_{k2}=L_{k3}=L_k$).

A. Principle of operation

The proposed converter operates under three different regions based on the duty cycle; $D > 0.66$, $0.33 < D < 0.66$, and $D < 0.33$. The number of switches that simultaneously turn on is shown in Table I. The operating modes of the proposed converter are analyzed based on the three regions. In any case total number of switches that simultaneously turn on is three.

TABLE I. OPERATION MODES BASED ON DUTY CYCLE

Duty cycle	No. of main switches simultaneously on	No. of clamp switches simultaneously on
$D < 0.33$	up to 1	up to 3
$0.33 < D < 0.66$	up to 2	up to 2
$D > 0.66$	up to 3	up to 1

1) Operation in $D > 0.66$

Fig. 2 shows key waveforms of the proposed converter in the case of $D > 0.66$. The converter has five operating modes within each operating cycle per phase, and the equivalent circuits of the five operating modes are shown in Fig. 3.

Mode I [t_0, t_1]: At time t_0, all the diode currents become zero, and therefore the winding voltages become zero. Each of the primary winding current becomes identical.

$$i_{pri1} = i_{pri2} = i_{pri3} = \frac{1}{3}I_i \tag{3}$$

The output capacitor supplies the load during this mode.

Mode II [t_1, t_2]: At time t_1, main switch S_{M2} is turned off and current i_{pri2} is commutated to body diode of clamp switch S_{C2}. This causes current i_{pri2} to decrease and currents i_{pri1} and i_{pri3} to increase, leading to conducting of upper diode D_{U2} and lower diodes D_{L1} and D_{L3}, respectively, at the secondary. The voltages of the windings become,

$$v_{pri2} = \frac{1}{n} \cdot v_{sec2} = \frac{1}{n} \cdot \frac{2}{3}V_o \tag{4}$$

Then, the voltages across L_{k2} can then be obtained by,

$$V_{Lk,n} = \frac{2}{3} \cdot \left(V_C - \frac{V_o}{n} \right) \tag{5}$$

The current i_{pri2} is decreasing with the slope determined by $V_{Lk,n}/L_k$. It is seen that clamp switch S_{C2} is turned on with ZVS when the gate signal for S_{C2} is applied during this mode.

Mode III [t_2, t_3]: The current i_{pri2} reverses its direction of flow at t_2 and increases its magnitude while current i_{pri1} and i_{pri3} keep increasing linearly. Since the average current through each clamp switch is zero, $I_{pri2(t3)}$ which is the magnitude of the current i_{pri2} at the end of Mode III can be obtained by,

$$I_{pri2(t3)} = -\frac{1}{3}I_i \tag{6}$$

Then, current magnitudes $I_{pri2(t3)}$ and $I_{pri3(t3)}$ at t_3 can be obtained by,

$$I_{pri1(t3)} = I_{pri3(t3)} = \frac{2}{3}I_i \tag{7}$$

Mode IV [t_3, t_4]: At time t_3 clamp switch S_{C2} is turned off, and current i_{pri2} is commutated to body diode of main switch S_{M2}. This causes all primary currents i_{pri1}, i_{pri2}, and i_{pri3} to decrease. The voltages across L_{k2} can then be obtained by,

$$V_{Lk,p} = \frac{2V_o}{3n} \qquad (8)$$

The decreasing rate of current i_{pri2} is determined by $V_{Lk,p}/L_k$. It should be noted that main switch S_{M2} can be turned on with ZVS at this mode.

Mode V [t_4, t_5]: The current i_{pri2} reverses its direction of flow at time t_4 and increases its magnitude while currents i_{pri1} and i_{pri3} keep decreasing linearly until each of the primary winding current becomes identical, as shown in equation (1). It is also noted that the rectifier diodes D_{U2}, D_{L1}, and D_{L3} are turned off with ZCS. This is the end of one third of the cycle. The other parts of the cycle are repeated in the same fashion.

2) Operation in *0.33 < D < 0.66*

Fig. 4 shows key waveforms of the proposed converter in the case of *0.33 < D < 0.66*. The converter has four operating modes within each operating cycle per phase, and their equivalent circuits are shown in Fig. 5. At Modes I and IV two main switches and one clamp switch are conducting together while at Modes II and III one main switch and two clamp switches are conducting together.

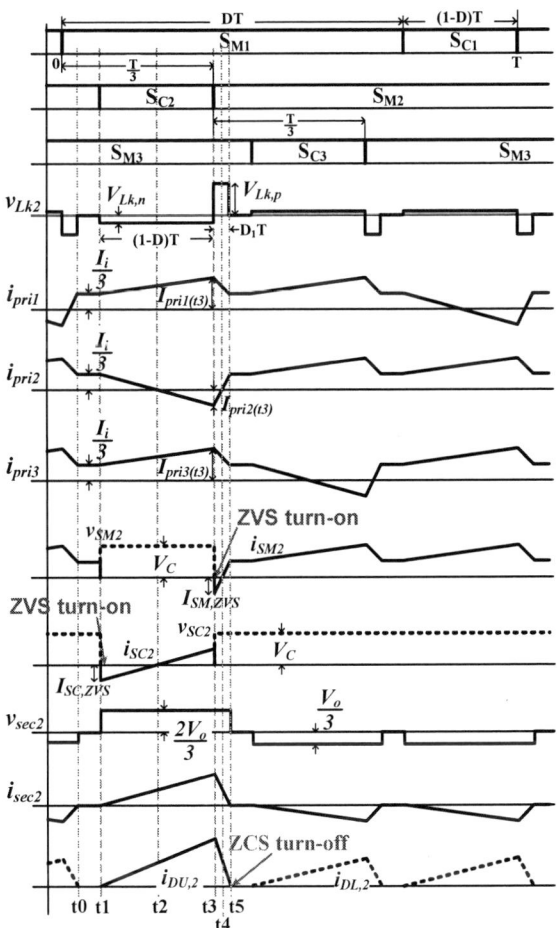

Fig. 2 Key waveforms of the proposed converter (*D>0.66*)

3) Operation in *D < 0.33*

Fig. 6 shows key waveforms of the proposed converter in the case of *D < 0.33*. The converter has five operating modes within each operating cycle per phase, and their equivalent circuits are shown in Fig. 7.

At Mode I all the main switches are being turned off, and all the clamp switches are conducting. At time t_1 clamp switch S_{C2} is turned off, and the current that was flowing through main channel of S_{C2} is commutated to its body diode as we can see in Mode II. When the gating signal is applied to main switch S_{M2} the current that was flowing through body diode of S_{C2} is commutated to main channel of main switch S_{M2}, resulting in

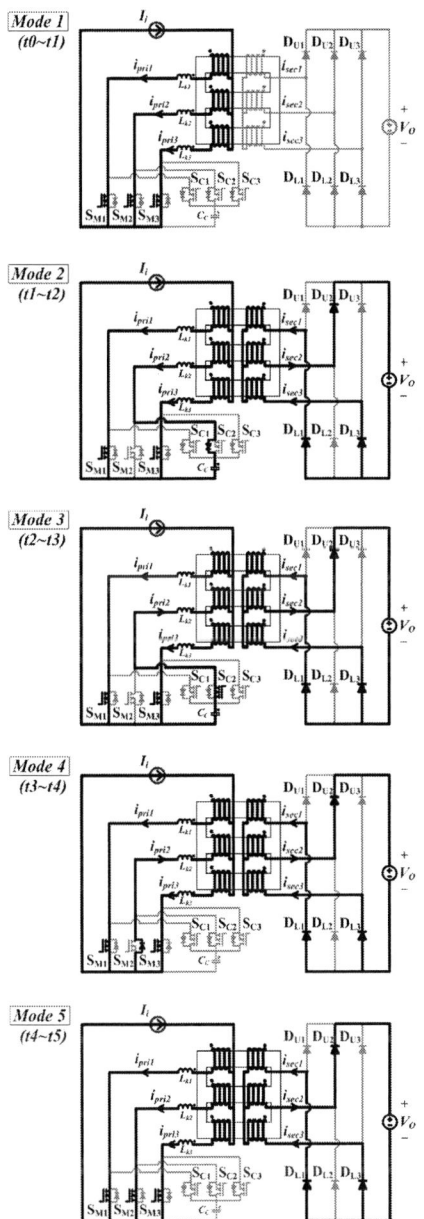

Fig. 3 Operation states of the proposed converter (*D>0.66*)

978-1-4244-4782-4/10 $26.00 © 2010 IEEE

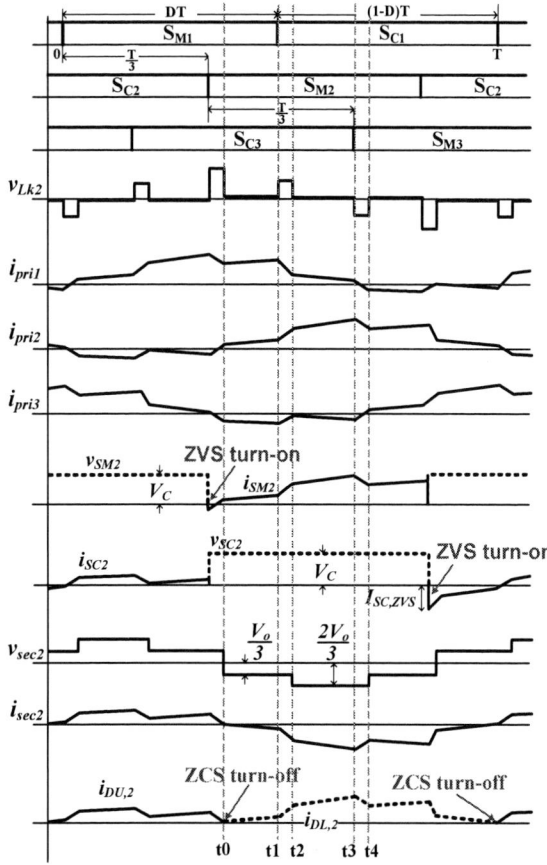

Fig. 4 Key waveforms of the proposed converter (*0.33<D<0.66*)

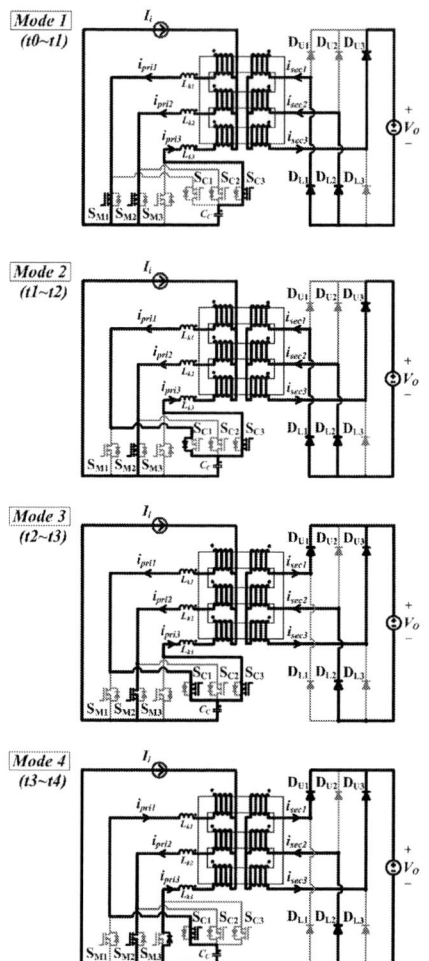

Fig. 5 Operation states of the proposed converter (*0.33<D<0.66*)

hard switching of main switch S_{M2}. Instead, clamp switch S_{C2} is turned off with ZCS.

Owing to the operation of clamp switches, the proposed converter can be operated with duty cycle less than 0.33, and therefore, no additional start-up circuit is required. In addition, this could improve dynamic characteristics of the closed-loop control system. Table II summarizes the soft switching condition for the proposed converter.

B. Voltage conversion ratio

The actual voltage conversion ratio of the proposed converter is derived in $D > 0.66$ case, considering the effect of voltage drop across the leakage inductor of the transformer.

Applying the voltage-second balance principle to leakage inductor L_{k2} from Mode I to Mode V, the following equation can be obtained (See waveforms of v_{Lk2} and i_{pri2} in Fig. 2),

$$V_{Lk,n} \cdot (1-D) \cdot T = V_{Lk,p} \cdot D_1 \cdot T \qquad (9)$$

From the waveforms of v_{Lk2} and i_{pri2}, it can be seen,

$$\frac{V_{Lk,p}}{L_k} = \frac{1}{D_1 T}\left(\frac{1}{3}I_i - I_{pri2(t3)}\right) \qquad (10)$$

Using equations (2), (5), (6), (8), (9) and (10), the voltage conversion ratio in D > 0.66 case can be obtained by,

$$V_o = n\left(\frac{V_i}{(1-D)} - \frac{P_o \cdot L_k \cdot f_s}{V_i \cdot (1-D) \cdot \eta}\right) \qquad (11)$$

where η is the converter efficiency, and $\eta \cdot I_i \cdot V_i = P_o$. In the similar way, the voltage conversion ratios in $0.33 < D < 0.66$ and $D < 0.33$ cases can be obtained by, respectively,

Table II. SUMMARY OF SOFT SWITCHING CONDITION

Duty cycle	Switching condition	Main Switches	Clamp Switches
D > 0.66	Turn On	ZVS	ZVS
0.33 < D < 0.66	Turn Off	None	None
D < 0.33	Turn On	None	ZVS
	Turn Off	None	ZCS

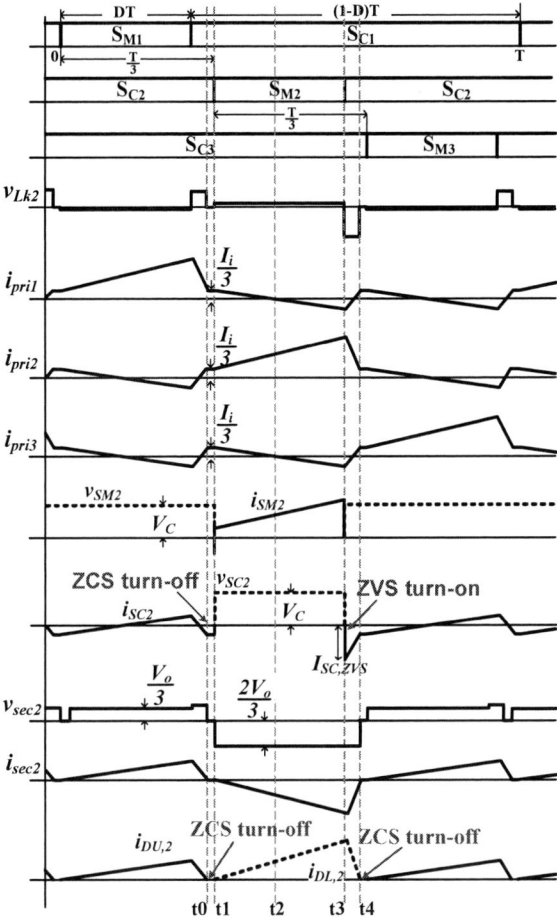

Fig. 6 Key waveforms of the proposed converter (D<0.33)

$$V_o = n \left(\frac{V_i}{(1-D)} - \frac{2 \cdot D \cdot P_o \cdot L_k \cdot f_s}{V_i \cdot (D - \frac{2}{9}) \cdot \eta} \right) \quad (12)$$

$$V_o = n \left(\frac{V_i}{(1-D)} - \frac{(1-D) \cdot P_o \cdot L_k \cdot f_s}{V_i \cdot D^2 \cdot \eta} \right) \quad (13)$$

From equations (11) to (13), the actual voltage conversion ratio of the proposed converter is plotted in Fig. 8 as a function of duty ratio D with different leakage inductances.

It is shown in Fig. 8 that at low duty cycle range the duty loss caused by leakage inductance of the transformer is significant, but at the duty cycle greater than 0.5, which is usually chosen to be the main operating range, the duty loss is negligible.

C. ZVS current and range

As shown in Fig. 2, the ZVS current of the main switch $I_{SM,ZVS}$ is the clamp switch current at turning off that is

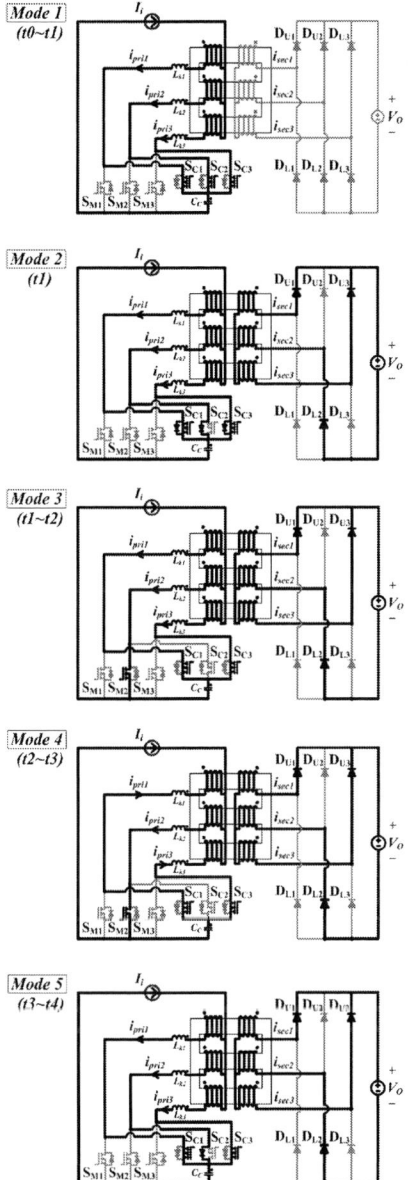

Fig. 7 Operation states of the proposed converter (D<0.33)

commutated to the main switch and used to discharge the output capacitance of the main switch and is determined by,

$$I_{SM,ZVS} = I_{pri2(t3)} = \frac{1}{3} I_i \quad (14)$$

To ensure the ZVS turn on of main switch the following condition should be satisfied,

$$\frac{1}{2} \cdot L_k \cdot I_{SM,ZVS}^2 > \frac{1}{2} \left(C_{os,M} + C_{os,C} \right) \cdot V_C^2 \quad (15)$$

where $C_{os,M}$ and $C_{os,C}$ are the output capacitances of main switch and clamp switch, respectively.

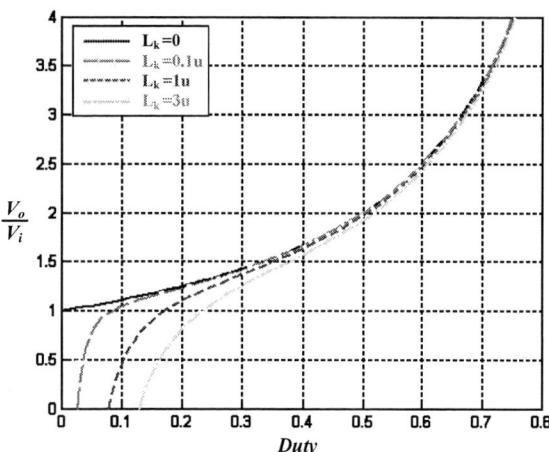

Fig. 8 Voltage conversion ratio as a function of duty cycle with different inductances ($P_o = 2kW$, $f_s = 50kHz$, $N_S/N_P = 1$, $\eta = 0.95$)

The ZVS current of the clamp switch $I_{SC,ZVS}$ is the main switch current at turning off that is commutated to the clamp switch and used to discharge the output capacitance of the clamp switch and is determined by,

$$I_{SC,ZVS} = \frac{1}{3} I_i \qquad (16)$$

To ensure the ZVS turn on of clamp switch S_{C2} the following condition should be satisfied,

$$\frac{1}{2} \cdot L_k \cdot I_{SM,ZVS}^2 > \frac{1}{2} \left(C_{os,M} + C_{os,C} \right) \cdot V_C^2 \qquad (17)$$

Using equations (14)-(17), the ZVS currents and ZVS ranges of main and clamp switches as the function of duty cycle and output power are plotted, respectively, as shown in Fig. 9. As shown in Fig. 9(a), the ZVS current of the main switch tends to

TABLE III. A COMPARISON OF CHARACTERISTICS OF THE PROPOSED AND CONVENTIONAL CONVERTERS

	Conventional converter[8]	Proposed converter
Number of switches	3	6
Clamping Method	Passive	Active
Switching Method	Hard switching	Soft switching
Operating Duty	0.33 ~ 1	0 ~ 1
Startup Circuit	Required	None
Diode Reverse Recovery Loss	Large	Negligible due to ZCS Turn Off

increase as the output power increases and/or the input voltage decreases(as the duty cycle increases). This means that the ZVS turn-on of the main switch can be more easily achieved under the condition of higher output power and lower input voltage. It is noted that the ZVS range of the main switch becomes broader for smaller total output capacitance $C_{os,tot} = C_{os,M} + C_{os,C}$ of MOSFETs. For example, if MOSFETs with total output capacitance $C_{os,tot}$ of 10nF are selected in this example, the ZVS turn-on of the main switch can be achieved with output power which is greater than 1000W at duty cycle of 0.5 (See Fig. 9(a)).

The ZVS current of the clamp switch tends to increase as the output power and input voltage decrease. It should be noted from Fig. 9(b) that the ZVS turn-on of the clamp switch can be achieved in the overall input voltage and output power ranges.

D. Summary of the features

Main characteristics of the proposed converter are compared to the conventional converter and are summarized in Table III. Compared to the conventional converter, number of switches and circuit complexity of the proposed converter are increased.

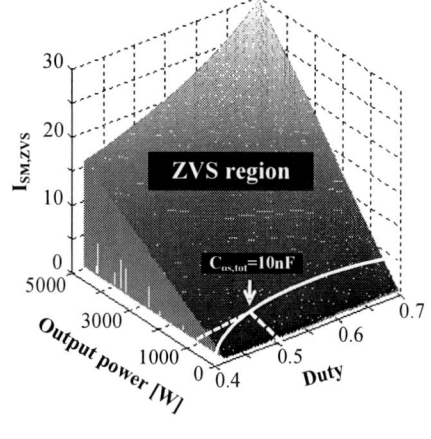

(a) Main switch

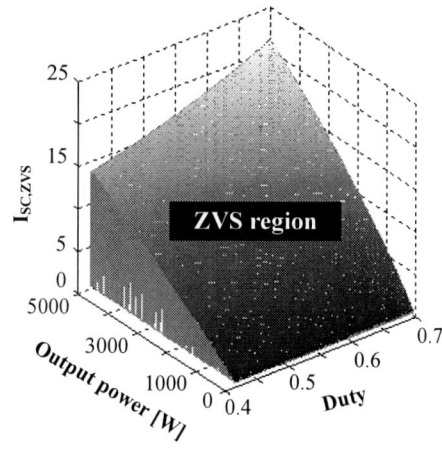

(b) Clamp switch

Fig. 9 ZVS current and ZVS region of switches

Nevertheless, lossless clamping of the surge caused by transformer leakage inductor, soft switching of both main and clamp switches and elimination of diode losses associated with reverse recovery characteristic by the active clamping technique result in an improvement of overall power conversion efficiency of the proposed converter. Further, the extended operating duty cycle not only improves the dynamic characteristics of the closed-loop system, but does not necessitate additional start-up circuit.

III. EXPERIMENTAL RESULTS

A 5kW laboratory prototype has been constructed, and the experimental results are presented to verify the operating principles of the proposed converter. The system specifications used in the experiment are given in the following,

- $P_o = 5\text{kW}$ • $V_i = 60\sim110\text{V}$ • $V_o = 380\text{V}$
- $\Delta I_i = 10\%$ • $\Delta V_o = 3\%$ • $f_s = 50\text{ kHz}$

The operating duty cycle for output voltage regulation is shown to be in the range $0.42<D<0.7$. The component ratings for the given specification and the respective selected devices from the manufactures are provided in Table IV. The voltage ratings of both the main and clamp switches and the clamp capacitor are 200V. A film capacitor was used for the clamp capacitor since the rms ripple current was 18A.

An off-the-shelf EI core of ferrite material was used for the three-phase transformer. Since the cross sectional area of the center leg of the off-the-shelf E core is twice them of the both side legs, the center leg is cut out so as to have equal width, as shown in Fig. 10. Even though the widths of all three legs are equal the magnetizing inductance of the center leg is still larger than them of the both side legs since the magnetic path length

TABLE IV. COMPONENT RATINGS AND SELECTED DEVICES

Design item		Rating	Selected devices
Main Switches	V_{pk}	200 V	IXTQ69N30
	I_{rms}	33 A	(300V, 69A)
Clamp Switches	V_{pk}	200 V	IXTQ69N30
	I_{rms}	9.5 A	(300V, 69A)
Diodes	V_{pk}	380 V	CSD20060D
	I_{rms}	9.5 A	(600V, 20A)
Clamp Capacitor	Capacitance	5 uF	Film 400V
	V_{dc}	200 V	10uF
Transformer	Number of Turns	10 : 20	Ferrite core EI-118
	Primary V_{rms}, I_{rms}	70 V, 37 A	
	Secondary V_{rms}, I_{rms}	178 V, 10 A	
	kVA	6540VA	
Input Inductor	Inductance	15 uH	Powder core 15uH
	I_{rms}	82A	
Output Capacitor	Capacitance	3 uF	Film 600V 15uF
	V_{dc}	380V	

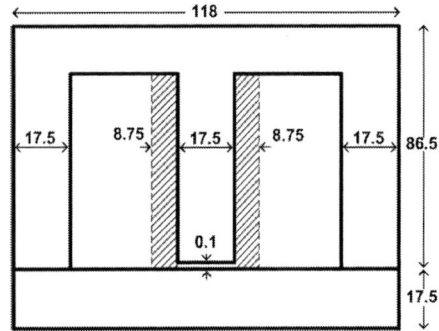

Fig. 10 Dimension of the three-phase core (mm)

Fig. 11 Photograph of the proposed converter

of the center leg is longer than them of the both side legs. This may cause more than 20% of unbalance in magnetizing inductances of the three legs[4]. In order to further decrease the unbalance in magnetizing inductances a small air gap was added in the center leg of the transformer. This led to less than 5% of unbalance in magnetizing inductances of the three legs.

Fig. 11 shows the photograph of the prototype. Experimental waveforms at three different duty cycles are shown in Fig. 12 to Fig. 14, respectively. It can be seen from Fig. 12 that the main switch is hard switched while the clamp switch is being turned on with ZVS and off with ZCS. Figs. 13 and 14 show that both the main and clamp switches are being turned on with ZVS.

IV. CONCLUSIONS

A new three-phase current-fed push-pull DC-DC converter featuring active clamping of the transient surge voltage, natural ZVS turn-on of main switches and clamp switches, unrestricted operating duty cycle range between 0 and 1, ZCS turn-off of rectifier diodes is proposed. The proposed three-phase converter could be a viable solution for high current high power application such as fuel cells. Experimental results on a 5kW laboratory prototype are provided to validate the proposed concept.

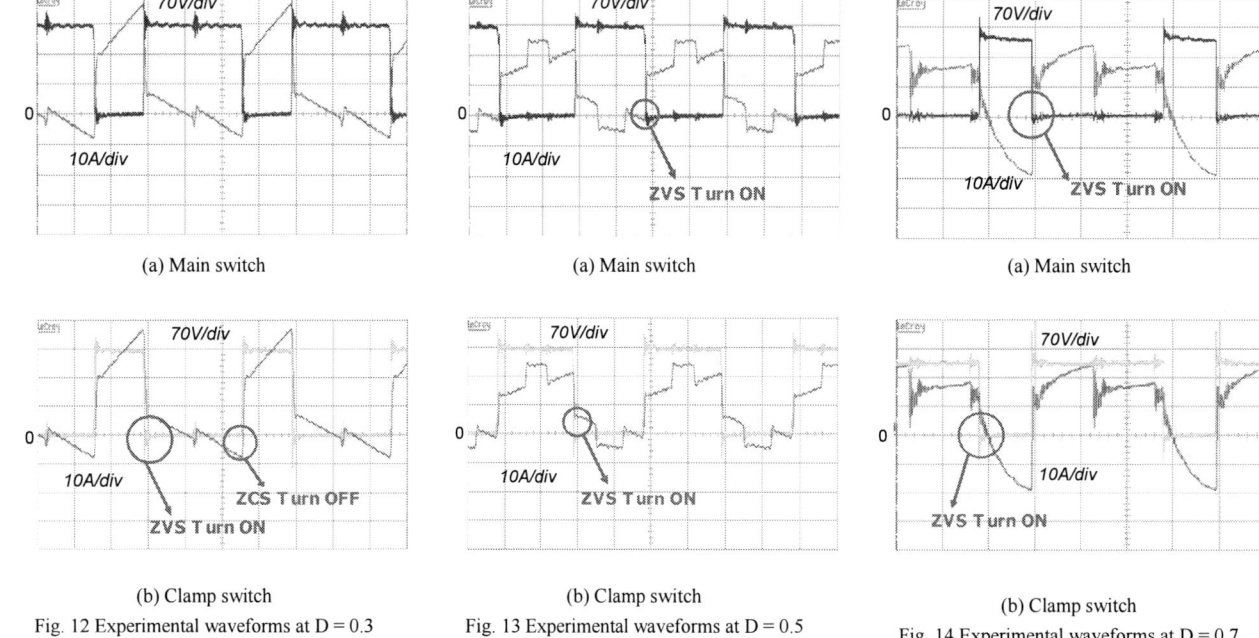

(a) Main switch	(a) Main switch	(a) Main switch
(b) Clamp switch	(b) Clamp switch	(b) Clamp switch
Fig. 12 Experimental waveforms at D = 0.3	Fig. 13 Experimental waveforms at D = 0.5	Fig. 14 Experimental waveforms at D = 0.7

REFERENCES

[1] X. Kong, L. T. Choi, and A. M. Khambadkone, "Analysis and Control of Isolated Current-fed Full Bridge Converter in Fuel Cell System," in *Proc. IEEE IECON*, Vol. 3, pp.2825-2830, Nov. 2004.

[2] V. Yakushev, V. Meleshin, and S. Fraidlin, "Full-Bridge Isolated Current Fed Converter with Active Clamp," in *Proc. IEEE APEC'99*, Vol. 1, pp.560-566, March 1999.

[3] S.V.G Oliveira, I. Barbi, "A three-phase step-up DC-DC converter with a three-phase high frequency Transformer", in *Proc. IEEE ISIE*, vol. 2, pp. 571-576, June 20-23, 2005.

[4] H. Cha; J. Choi, "Three-phase high frequency transformer design for a three-phase current-fed dc/dc converter with active clamp", ICEMS, pp. 204-208, Oct. 2007.

[5] S.V.G. Oliveira, C.E. Marcussi, I. Barbi, "An Average Current-mode Controlled Three-phase Step-up DC-DC Converter with a Three-phase High Frequency Transformer", in *Proc. IEEE PESC*, pp.2623-2629, June 2005.

[6] H. Cha, J. Choi, B. Han, "A new three-phase interleaved isolated boost converter with active clamp for fuel cells", in *Proc. IEEE PESC*, pp.1271-1276, June 2008.

[7] C. Yoon, S. Choi, "A Multi-Phase DC-DC converters using a boost half bridge cell for high voltage and high power applications", in *Proc. IEEE IPEMC*, pp.780-786, May 2009.

[8] R.L. Andersen, I. Barbi, "A Three-Phase Current-Fed Push–Pull DC–DC Converter", *IEEE Trans. on Power Electronics*, vol. 24, issue 2, pp. 358-368, Feb. 2009.

[9] D. Wang, X. He, R. Zhao, "ZVT Interleaved Boost Converters with Built-In Voltage Doubler and Current Auto-Balance Characteristic", *IEEE Trans. on Power Electronics*, vol. 23, issue 6, pp. 2847-2854, Nov. 2008.

[10] J. W. Kolar , G. R. Kamath , N. Mohan and F. C. Zach, "Self-adjusting input current ripple cancellation of coupled parallel connected hysteresis-controlled boost power factor correctors", in *Proc. IEEE PESC*, pp.164-173, 1995.

Resonant Voltage Divider with Startup Considered

K. I. Hwu, *Member, IEEE*, and Y. T. Yau, *Student Member, IEEE*

Department of Electrical Engineering, National Taipei University of Technology, Taiwan

Abstract–In this paper, the resonant voltage divider (RVD) is presented with the fixed switching frequency over all the load range, which is to be used as the first stage of the bus converter. The proposed resonant voltage divider without any control loop embedded can reach very high efficiency from the light load to the rated load, especially for the light load. Above all, it has a serious drawback that a large surge current at startup causes components to be saturated or destroyed. Consequently, how to reduce such a surge current based on the proposed variable switching frequency at startup is also taken into account. The basic operating principles of the RVD and how to suppress the startup surge current are illustrated in detail along with some simulated and experimental results to verify its effectiveness.

I. INTRODUCTION

Conventionally, for the converter with the high-voltage input to the low-voltage high-current output, the single-stage isolated converter has been widely used in the industry, such as the brick VRM (Voltage-Regulated Module) converter with the input voltage of 48V or 24V transferred to the output voltage between 1.2V and 3.3V. There are many literatures presented to discuss various types of structures for this converter, such as current double [1], synchronous rectification [2], resonance [3], etc. However, by doing so, the corresponding cost is expensive, because in general the overall system requires several or many different voltages. Especially, if such voltages come from individual isolated converters with the same input voltage, then the cost is too expensive and the size is too large. Consequently, the bus converter concept [4-7] is presented and hence two-stage converter is adopted. The first-stage is the bus converter whereas the second-stage is the VRM converter. The bus converter with high efficiency is presented, which is the isolated converter without the output voltage regulated, that is to say, the output voltage is varied with the input voltage. The purpose of the bus converter is used to perform isolation and voltage reduction and the purpose of the VRM converter is used to execute voltage regulation. Generally, the bus converter generally transfers 48V or 24V to 12V, then the VRM converter, generally implemented by the traditional buck converter, transfers 12V to 3.3V or lower.

However, most of today's VRM topologies are the multi-phase buck. Due to the very low output voltage, the buck converter has the extremely small duty cycle, which dramatically impacts the performance. Both the transient response and the efficiency suffer a lot [8-10]. Consequently, there are some literatures [11-15] presented to discuss how to increase duty cycles based on tapped inductors. After this, the two-stage structure for the VRM converter [16-25], based on the bus converter without isolation, is presented, which transfers the input voltage to between 6V and 8V in the first stage without any control and then to the desired output voltage

in the second stage with closed-loop control. It is found that the performance of the single-stage structure is worse than that of the second-stage structure in efficiency and size [1][4-7][11-15]. In [21][25], the voltage divider, used in the first-stage, is implemented by the charge pump. In this case, the efficiency of such a topology can reach up to 96% at full load and 97.5% at light load. Besides, the variable switching frequency, changed from 350kHz to 100kHz, is used at light load to further improve the efficiency up to 98%. In this paper, a resonant voltage divider is presented, where all the switches can operate in zero current switching (ZCS) under the fixed switching frequency. This makes the filter designed easily. By doing so, the performance in efficiency is improved, especially for light load up to 99%. However, it has a serious drawback that a large surge current at startup tends to cause components to be saturated or destroyed. Consequently, how to reduce such a surge current based on the proposed variable switching frequency at startup is also taken into consideration.

II. PROPOSED CIRCUIT CONFIGURATION

Fig. 1 shows the proposed resonant voltage divider (RVD), which is constructed by four switches Q_1, Q_2, Q_3, Q_4, one resonant inductor L_r and one resonant capacitor C_r. As this circuit operates at resonance, the output voltage is half of the input voltage. Fig. 2 shows the waveforms for the circuit operating under the ideal condition, the voltage on C_r swings between zero and the input voltage, which limits the maximum output transfer. In order to realize ZCS operation at resonance, the zero-current detector for I_{Lr} is used, thereby making the switching frequency entirely determined by the resonant frequency. It is noted that in theory the load current is large enough to let I_{Lr} force a negative value on V_{Cr}. If V_{Cr} is negative, then V_{Cr} will be over V_i next cycle. By doing so, the divergent current will occur after several cycles. And hence one diode D_c is added to remove the negative voltage on C_r, but D_c does not work most of time if the RVD operates in the normal condition.

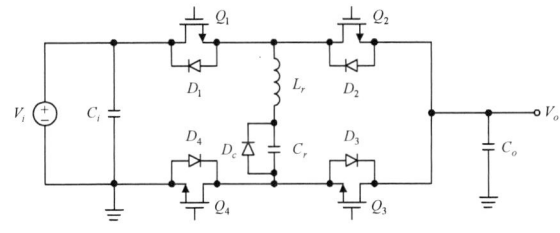

Fig. 1. Proposed RVD.

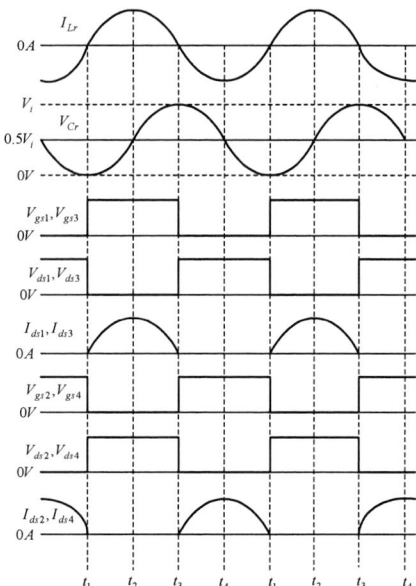

Fig. 2. Ideal waveforms of the proposed RVD operating.

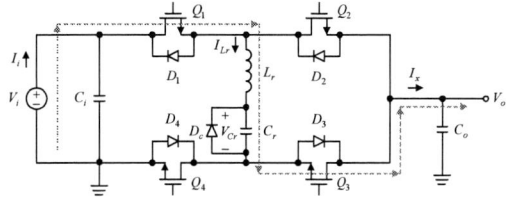

Fig. 3. Current flow of the proposed RVD without delay considered for modes 1 and 2.

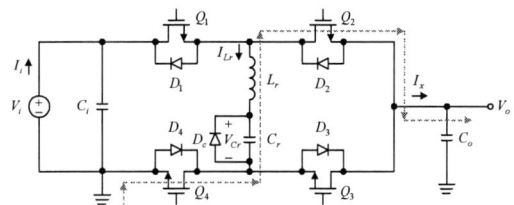

Fig. 4. Current flow of the proposed RVD without delay considered for modes 3 and 4.

III. BASIC OPERATING PRINCIPLES

Fig. 2 shows the operating waveforms of the ideal circuits of the proposed resonant voltage divider. According to Fig. 2, there are four operating modes to be described as follows.

A. Mode 1: $(t_1 \sim t_2)$

As displayed in Fig. 3, Q_1 and Q_3 are switched on, and Q_2 and Q_4 are switched off. In this mode, V_{Cr} is increasing from zero, whereas I_{Lr} is also increasing from zero. The current flow is from the input via Q_1 through L_r and then to C_r, Q_3 and the output. As soon as V_{Cr} reaches half of V_i or I_{Lr} reaches the maximum value, the operating mode proceeds to mode 2.

B. Mode 2: $(t_2 \sim t_3)$

As illustrated in Fig. 3, Q_1 and Q_3 are still in the on-state, and Q_2 and Q_4 are still in the off-state. In this mode, V_{Cr} is still increasing from half of V_i, whereas I_{Lr} is decreasing from the maximum value. The current flow is the same as that in mode 1. As soon as V_{Cr} reaches V_i or I_{Lr} is reduced to zero, the operating mode goes to mode 3. At this moment, Q_1 and Q_3 are switched off with ZCS.

C. Mode 3: $(t_3 \sim t_4)$

As depicted in Fig. 4, Q_1 and Q_3 are switched off, and Q_2 and Q_4 are switched on. In this mode, V_{Cr} is decreasing from V_i, whereas I_{Lr} is increasing in the opposite direction. The current flow is from the ground via Q_4 through C_r and then to L_r, Q_2 and the output. As soon as V_{Cr} is reduced to half of V_i or I_{Lr} reaches the minimum value, the operating mode proceed to mode 4.

D. Mode 4: $(t_4 \sim t_1)$

As described in Fig. 4, Q_1 and Q_3 are still in the off-state, and Q_2 and Q_4 are still in the on-state. In this mode, V_{Cr} is decreasing from half of V_i, whereas I_{Lr} is decreasing in the opposite direction. The current flow is the same as that in mode 3. As soon as V_{Cr} is reduced to zero or I_{Lr} reaches zero, the operating mode goes back to mode 1 and the next cycle is repeated. At this instant, Q_2 and Q_3 are switched off with ZCS.

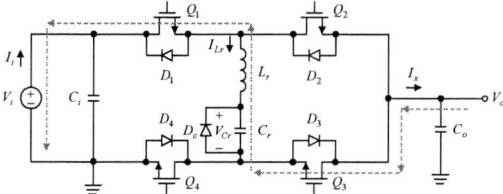

Fig. 5. Current flow of the proposed RVD with delay considered, inserted between mode 2 and mode 3.

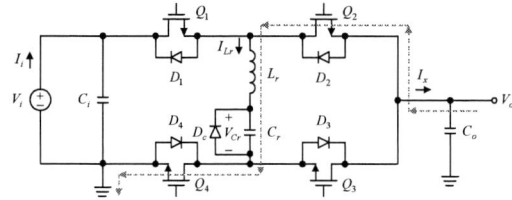

Fig. 6. Current flow of the proposed RVD with delay considered, inserted between mode 4 and mode 1.

On the other hand, in practice the delay time is indispensable in the system. This is because detecting the zero crossing of I_{Lr} will be delayed by the gate driver, FPGA calculation time, etc., thereby causing the switch not to be turned off at the zero crossing of I_{Lr} and hence the current flowing through L_r in the opposite direction to occur. Therefore, there are two required modes added to four ideal modes mentioned above. One mode shown in Fig. 5 is inserted between mode 2 and mode 3, and the other mode shown in Fig. 6 is inserted between mode 4 and mode 1.

978-1-4244-4782-4/10 $26.00 © 2010 IEEE

IV. CONTROL METHOD APPLIED

Fig. 7 shows the proposed overall system block diagram for the resonant voltage divider (RVD). The gate driving signals M_1, M_2, M_3 and M_4, used to drive the four main switches Q_1, Q_2, Q_3 and Q_4 respectively, are created from the field-programmable gate array (FPGA). And the current flowing through L_r is sensed by the current transformer (CT) and sent to one comparator COMP to get information on zero crossing of I_{Lr}, and the output result of COMP, ZCD, is sent to FPGA to get the desired gate driving signals. Besides, there are two half-bridge gate drivers used herein. One is for Q_1 and Q_2, and the other is for Q_3 and Q_4.

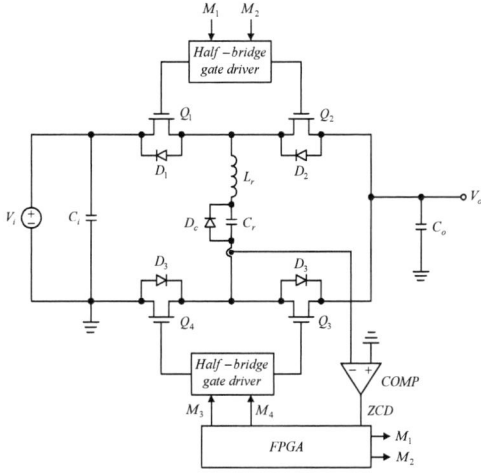

Fig. 7. Proposed overall system block diagram.

V. DESIGN OF KEY PARAMETERS

Prior to going into this topic, there are some specifications required as follows: (i) rated DC input voltage V_i is 24V; (ii) rated DC output voltage V_o is 12V; (iii) rated DC output current $I_{o\text{-}rated}$ is 10A; (iv) maximum DC output current $I_{o\text{-}max}$ is 11A; (v) switching frequency f_s is initially set to 120kHz; (vi) both values of the input and output capacitors are identical, with two 470μF electrolytic capacitors connected in parallel with two MLCC capacitors for each; (vii) product name of the half-bridge gate drivers is IR2011; (viii) product name of the switches is IRL3705ZS; (ix) product name of FPGA is EP1C3T100; (x) product name of COMP is LT1719; and (xi) product name of D_c is 1N5819.

Since the resonant capacitor C_r and the resonant inductor L_r are the key parameters of the proposed resonant voltage divider, how to design C_r and L_r is to be described as follows. Based on the basic operating principles of the resonant voltage divider, the following equation can be obtained

$$\frac{1}{2}V_o \cdot I_{o\text{-}max} = \frac{1}{2}C_r \cdot V_i^2 \cdot f_s \tag{1}$$

$$f_s = \frac{1}{2\pi\sqrt{L_r \cdot C_r}} \tag{2}$$

From (1) and (2), the values of C_r and L_r can be expressed by

$$C_r = \frac{V_o \cdot I_{o\text{-}max}}{V_i^2 \cdot f_s} \tag{3}$$

$$L_r = \frac{C_r \cdot V_i^2}{4\pi V_o^2 \cdot I_{o\text{-}max}^2} \tag{4}$$

Based on the given specifications, and (3) and (4), the value of C_r can be figured out to be 1.91μF and the value of L_r can be worked out to be 0.96μH. And eventually, the values of C_r and L_r are chosen to be 2μF and 1μH, respectively, and hence the corresponding switching frequency is 112kHz.

V. EXPERIMENTAL RESULTS

Figs. 8 to 15 show the experimental waveforms. Fig. 8 displays the resonant waveforms I_{Lr} and V_{Cr}. It is obvious that the maximum value of V_{Cr} approaches to 24V and its minimum value approaches to zero. This implies that this RVD operates close to the maximum power transfer. Fig. 9 depicts the waveforms of I_{Lr}, ZCD, V_{gs1} and V_{gs2}. Since Q_1 and Q_3 operate simultaneously whereas Q_2 and Q_4 work simultaneously, V_{gs3} and V_{gs4} are not shown again. During the positive cycle of I_{Lr}, ZCD is high whereas during the negative cycle of I_{Lr}, ZCD is low. And hence, based on the rising or falling edge of ZCD, FPGA can determine whether the switches operate or not. That is to say, Q_1 and Q_3 are turned off and Q_2 and Q_4 are turned on as I_{Lr} is negative, whereas Q_1 and Q_3 are turned on and Q_2 and Q_4 are turned off as I_{Lr} is positive. Figs. 10 and 11 display the associated waveforms I_{Lr}, ZCD, V_{gs1} and V_{gs2} in the neighborhood of the rising and falling edges of the zero crossing of I_{Lr}, respectively. It is obvious that the instants that switches are turned off are delayed by about 150ns to 200ns from the zero crossing of I_{Lr}. Under rated load, Figs. 12 to 15 show the resonant currents, the voltages on the switches and the PWM gate driving signals for Q_1, Q_2, Q_3 and Q_4. It is evident that each switch has ZCS operating characteristics.

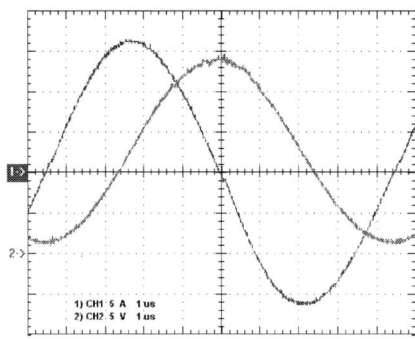

Fig. 8. Resonant waveforms at rated load: (1) I_{Lr}; (2) V_{Cr}.

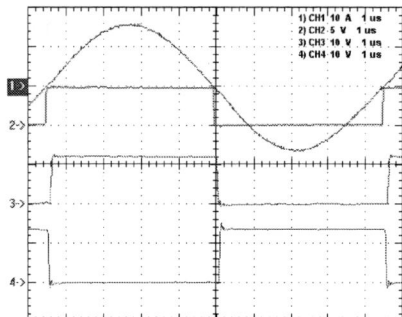

Fig. 9. Related ZCS waveforms at rated load: (1) I_{Lr}; (2) ZCD; (3) V_{gs1}; (4) V_{gs2}.

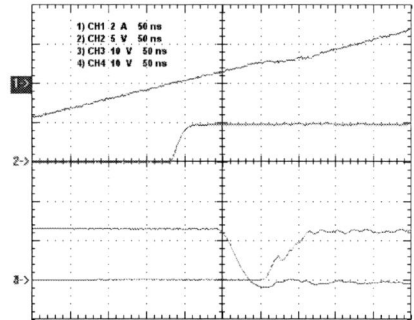

Fig. 10. Zoomed ZCS waveforms of Fig. 9 in the neighborhood of the falling edge of ZCD.

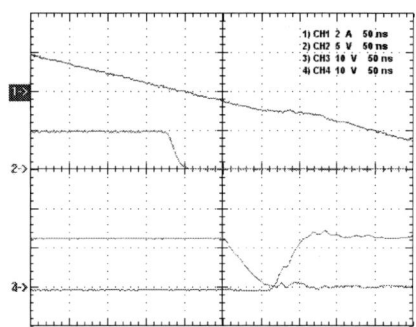

Fig. 11. Zoomed ZCS waveforms of Fig. 9 in the neighborhood of the rising edge of ZCD.

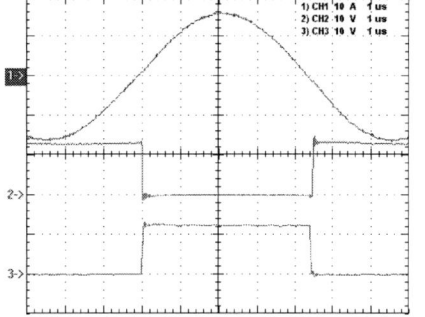

Fig. 12. Related ZCS waveforms at rated load for Q_1: (1) I_{Lr}; (2) V_{ds1}; (3) V_{gs1}.

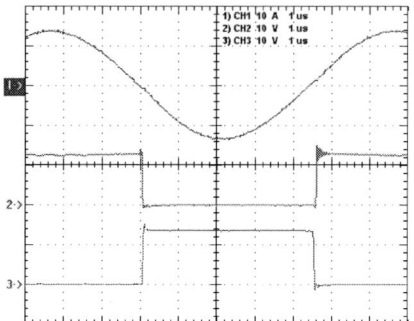

Fig. 13. Related ZCS waveforms at rated load for Q_2: (1) I_{Lr}; (2) V_{ds2}; (3) V_{gs2}.

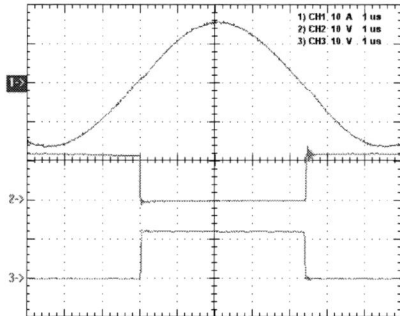

Fig. 14. Related ZCS waveforms at rated load for Q_3: (1) I_{Lr}; (2) V_{ds3}; (3) V_{gs3}.

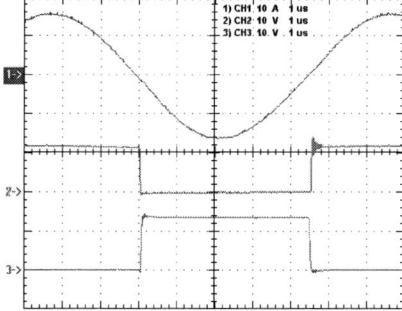

Fig. 15. Related ZCS waveforms at rated load for Q_4: (1) I_{Lr}; (2) V_{ds4}; (3) V_{gs4}.

Figs. 16 to 18 show the currents flowing through the resonant inductor L_r at no load, light load and rated load, respectively. It is obvious that the maximum peak current is about 55A for any load, and this is dangerous to the switches, the resonant inductor L_r and the resonant capacitor C_r. On the other hand, Figs. 19 to 21 show the input currents at startup. For each figure, the waveform shown at the top is the input voltage V_i, whereas the waveform shown at the bottom is the input current I_i with the first current spike created from the input capacitor. Since this circuit has functions of under-voltage lockout (UVLO) set to 20V and startup delay, the circuit does not work until V_i is stable. Hence, the input current appearing in the right side at the bottom still has the second current spike of about 9A, independent of the load current, and this occurs before the RVD operates and hence such a current spike is called working current spike. In order to overcome the

978-1-4244-4782-4/10 $26.00 © 2010 IEEE   1945

aforementioned problems, a variable switching frequency at startup is taken herein to realize soft start of the RVD, and achieved by a counter inside FPGA. Such a variable switching frequency is changed from high to low. As the variable switching frequency approaches the neighborhood of the resonant frequency f_r, the RVD is switched to the resonance mode. Since the variable switching frequency is reduced from the maximum frequency f_{max} to the minimum frequency f_{min} in the step of f_x, the associated frequency parameters are defined to be as follows under the system clock of 20ns:

$$f_{max} = \frac{1}{128 \times 20n} = 390.63\text{kHz} \tag{5}$$

$$f_{min} = \frac{1}{512 \times 20n} = 90.66\text{kHz} \tag{6}$$

$$f_x = \frac{1}{(128+x) \times 20n}, \ x = 0\text{~}384 \tag{7}$$

It is noted that x is increased by one per time span of Δt, where $\Delta t = 2^{14} \times 20\text{ns} = 327.68\mu\text{s}$. Therefore, the elapsed time for the soft start of the RVD at startup is 125.83ms. Since f_r is set to 110kHz, which is larger than f_{min}, another operating mode is that Q_1, Q_2, Q_3 and Q_4 are turned off with ZVS but without ZCS whereas Q_1, Q_2, Q_3 and Q_4 are turned on without ZVS and ZCS. Thus, the reason why f_{min} is chosen to be smaller than f_r, to be described as follows. The first reason is that there is a negligible influence on stable operation of the RVD because the time required for the soft start is short, and the second reason is that there exist tolerances in components. Hence, the variable switching frequency range at startup is enlarged.

At startup, Figs. 22 to 24 show the currents flowing through L_r under the soft start control based on the proposed variable switching frequency at no load, half load and rated load, respectively. Although the corresponding recovery times are longer than those in Figs. 16 to 18, the resulting maximum peak currents in L_r are up to about 38A at rated load, which is smaller than those in Figs. 16 to 18. Furthermore, the more the load current is, the less the maximum peak current, and this behavior is entirely different from that in Figs. 16 to 18 where the maximum peak currents are entirely independent of the load current. Namely, the proposed soft start control of the RVD can effectively reduce the maximum peak current at startup for any load. On the other hand, Figs. 25 to 27 show the input currents at startup. For each figure, the waveform shown at the top is the input voltage V_i, the waveform shown at the bottom is the input current I_i with the first current spike created from the input capacitor. Since the proposed soft start control is used herein, this circuit operates under UVLO without startup delay, and hence in these cases, the recovery times are reduced as compared to those in Figs. 19 to 21. Aside from this, the working current spikes are suppressed significantly for all load currents. By the way, as for the current spikes created from the input capacitor, they can be reduced via soft start of the output voltage of the front-end power supply which is used to feed the RVD. Besides, Fig. 28 shows that the highest efficiency, about 99%, occurs at light load and the efficiency can reach 94.4% at rated load.

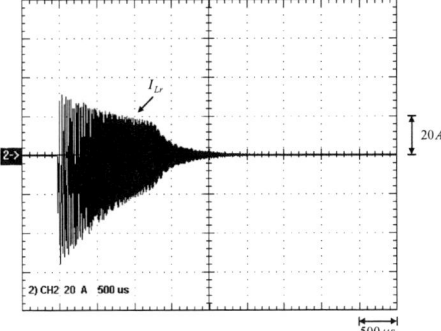

Fig. 16. I_{Lr} without soft start at no load.

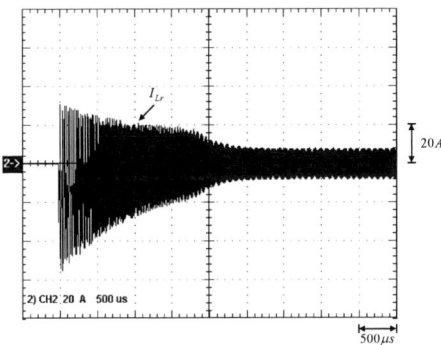

Fig. 17. I_{Lr} without soft start at half load.

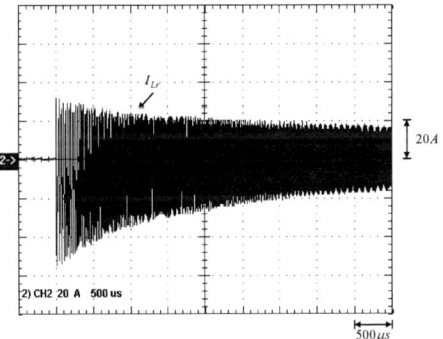

Fig. 18. I_{Lr} without soft start at rated load.

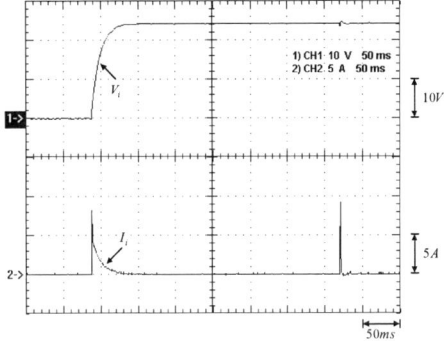

Fig. 19. I_i without soft start at no load.

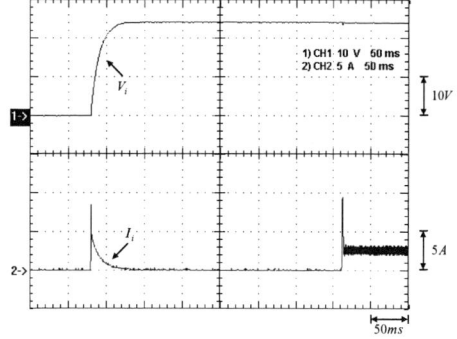

Fig. 20. I_i without soft start at half load.

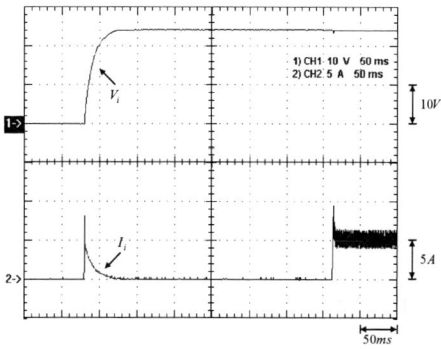

Fig. 21. I_i without soft start at rated load.

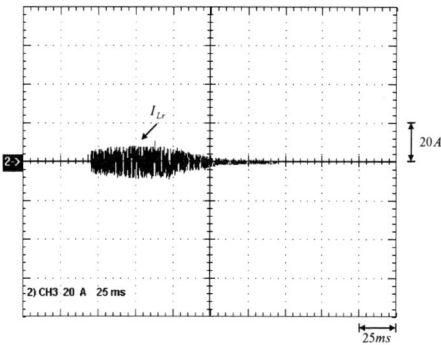

Fig. 22. I_{Lr} with soft start at no load.

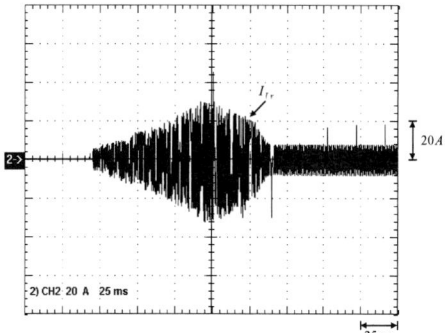

Fig. 23. I_{Lr} with soft start at half load.

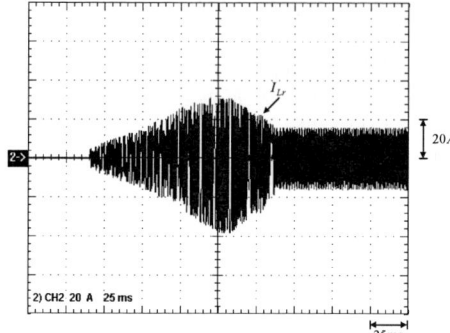

Fig. 24. I_{Lr} with soft start at rated load.

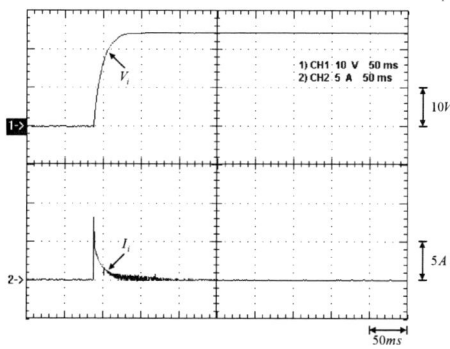

Fig. 25. I_i with soft start at no load.

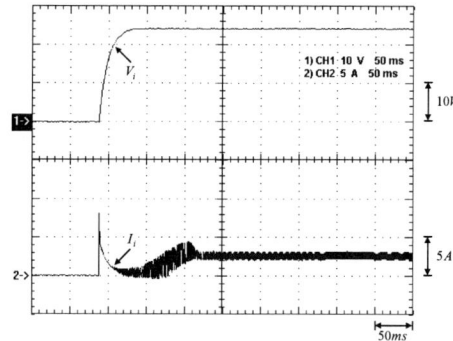

Fig. 26. I_i with soft start at half load.

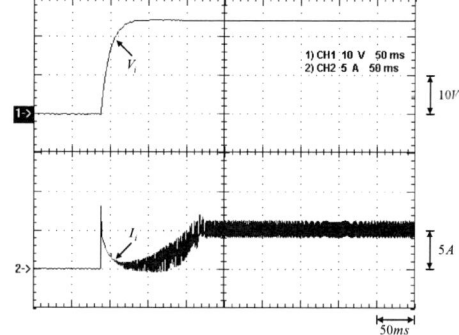

Fig. 27. I_i with soft start at rated load.

978-1-4244-4782-4/10 $26.00 © 2010 IEEE

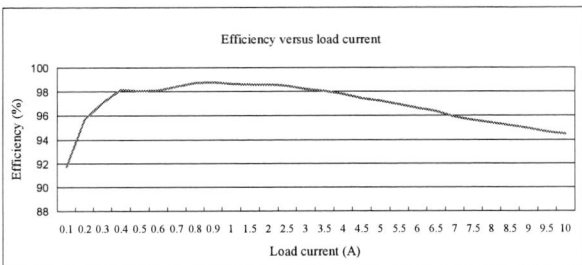

Fig. 28. Efficiency versus load current.

VI. CONCLUSION

In this paper, a simple RVD is presented, which possesses high efficiency and is very suitably used as a bus converter. From the experimental results, the efficiency at light load is up to about 99%. Besides, the switching frequency all over the load range is fixed, thereby causing the filter design to be easy. Above all, the soft start of the RVD based on the proposed variable switching frequency can effectively suppress the maximum peak currents in the resonant inductor and the working current spikes at the input.

REFERENCES

[1] Ming Xu, Yuancheng Ren, Jinghai Zhou and F. C. Lee, "1-MHz self-driven ZVS full-bridge converter for 48-V power pod and DC/DC brick," *IEEE Trans. Power Electron.*, vol. 20, no. 5, pp. 997-1006, 2005.

[2] Dianbo Fu, Bing Lu and F. C. Lee, "1MHz high efficiency LLC resonant converters with synchronous rectifier," *IEEE PESC'07*, pp. 2404-2410, 2007.

[3] M. M. Jovanovic, M. T. Zhang and F. C. Lee, "Evaluation of synchronous-rectification efficiency improvement limits in forward converters," *IEEE Trans. Ind. Electron.* vol. 42, no. 4, pp. 387-395, 1995.

[4] L. Balogh, "A new cascaded topology optimized for efficient DC/DC conversion with large step-down ratios," *Proc. Intel Symp.*, 2000.

[5] P. Alou, J. Oliver, J. A. Cobos, O. Garcia and J. Uceda, "Buck+half bridge (d=50%) topology applied to very low voltage power converters," *IEEE APEC'01*, vol. 2, pp. 715-721, 2001.

[6] Y. Ren, M. Xu, C.-S. Leu and F. C. Lee, "A family of high power density bus converters," *IEEE PESC'04*, vol. 1, pp. 527-532, 2004.

[7] Yuancheng Ren, Ming Xu, Julu Sun and F. C. Lee, "A family of high power density unregulated bus converters," *IEEE Trans. Power Electron.* vol. 20, no. 5, pp. 1045-1054, 2005.

[8] Xunwei Zhou, Pit-Leong Wong, Peng Xu, F. C. Lee and A. Q. Huang, "Investigation of candidate VRM topologies for future microprocessors," *IEEE Trans. Power Electron.*, vol. 15, no. 6, pp. 1172-1182, 2000.

[9] Y. Panvo and M. M. Jovanovic, "Design consideration for 12-V/1.5-V, 50-A voltage regulator modules," *IEEE Trans. Power Electron.* vol. 16, no. 6, pp. 776-783, 2001.

[10] R. Miftakhutdinov, "Optimal design of interleaved synchronous buck converter at high slew-rate load current transients," *IEEE PESC'01*, vol. 3, pp. 1714-1718, vol. 3, 2001.

[11] Kaiwei Yao, Yang Qiu, Ming Xu and F. C. Lee, "A novel winding-coupled buck converter for high-frequency, high-step-down DC-DC conversion," *IEEE Trans. Power Electron.* vol. 20, no. 5, pp. 1017-1024, 2005.

[12] Jinghai Zhou, Ming Xu, Julu Sun and F. C. Lee, "A self-driven soft-switching voltage regulator for future microprocessors," *IEEE Trans. Power Electron.*, vol. 20, no. 4, pp. 806-814, 2005.

[13] Yan Dong, Julu Sun, Ming Xu, Fred. C. Lee and Milan M. Jovanovic, "The light load issue of coupled inductor laptop voltage regulators and its solutions," *IEEE APEC'07*, pp. 1581-1587, 2007.

[14] Mao Ye, Ming Xu and F. C. Lee, "Tapped-inductor buck converter for high-step-down DC-DC conversion," *IEEE Trans. Power Electron.* vol. 20, no. 4, pp. 775-780, 2005.

[15] Yungtaek Jang, Milan M. Jovanovic, and Yuri Panov, "Multi-phase buck converters with extended duty cycle," *IEEE APEC'06*, pp. 38-44, 2006.

[16] Yuancheng Ren, Ming Xu, Kaiwei Yao, Yu Meng, F. C. Lee, Jinghong Guo and Y. Ren, "Two-stage approach for 12 V VR," *IEEE APEC'04*, vol. 2, pp. 1306-1312, 2004.

[17] Yuancheng Ren, Ming Xu, Kaiwei Yao and F. C. Lee, "Two-stage 48 V power pod exploration for 64-bit microprocessor," *IEEE APEC'03*, vol. 1, pp. 426-431, 2003.

[18] Yuancheng Ren, Ming Xu, Yu Meng and F. C. Lee, "12V VR efficiency improvement based on two-stage approach and a novel gate driver," *IEEE PESC'05*, pp. 2635-2641, 2005.

[19] Kisun Lee, Jia Wei, Ming Xu and F. C. Lee, "Adaptive bus voltage positioning system for two stage laptop voltage regulators," *IEEE PESC'07*, pp. 2-8, 2007.

[20] Yuancheng Ren, Kaiwei Yao, Ming Xu and F. C. Lee, "Analysis of the power delivery path from the 12-V VR to the microprocessor," *IEEE Trans. Power Electron.*, vol. 19, no. 6, pp. 1507-1514, 2004.

[21] Julu Sun, Ming Xu, F. C. Lee and Yucheng Ying, "High power density voltage divider and its application in two-stage server VR," *IEEE PESC'07*, pp. 1872-1877, 2007.

[22] Julu Sun, Ming Xu, Yucheng Ying and F. C. Lee, "High power density, high efficiency system two-stage power architecture for laptop computers," *IEEE PESC'06*, pp. 1-7, 2006.

[23] Julu Sun, Yuancheng Ren, Ming Xu and Fred C. Lee, "Light load efficiency improvement for laptop VRs," *IEEE APEC'07*, pp. 1120-126, 2007.

[24] Yuancheng Ren, Ming Xu, Kaiwei Yao, Yu Meng and F. C. Lee, "Two-stage approach for 12-V VR," *IEEE Trans. Power Electron.* vol. 19, no. 6, pp. 1498-1506, 2004.

[25] Ming Xu, J. Sun, J. and F. C. Lee, "Voltage divider and its application in the two-stage power architecture," *IEEE APEC'06*, pp. 499-505, 2006.

LLC Resonant Converter with Two Resonant Tanks

Eun-Soo Kim, Joo-Hoon Kim, Kwang-Ho Lee,
Yong-Seog Jeon

Dept. of Electrical Engineering, Jeonju University
1200, Hyoja-Dong, Wansan-Gu, Jeonju
Jeonbuk, 560-759, South Korea
E-mail : eskim@jj.ac.kr

Jae-Sam Lee, Dong-Young Huh

IPB gr, LG innotek
1271-0, Sal-Dong, Sangnok-Gu, Ansan
Gyeonggi, 426-171, South Korea
E-mail : jsleer@lginnotek.com

Abstract— **To cope with the high power density and low cost in switching power supply, LLC resonant converters with the two resonant tank circuits composed of resonance capacitors and two transformers are proposed in this paper. Each transformers used for the proposed resonant circuits are parallel connected in the primary and series connected in the secondary to reduce the current unbalance. The proposed LLC resonant converters are described and verified on 300W experimental prototype.**

I. INTRODUCTION

Recently, Power Supply Unit (PSU) for slim LCD TV tends to require more compact, higher efficiency, power density with better performance and lower price through simple circuits. To meet these requirements, LLC resonant converter with a single transformer as shown in Fig. 1(a) is widely used and studied because of several advantages such as high efficiency, output regulations over a wide load range, no reverse recovery problem of output rectifier, cost effective and so on [1, 2, 3]. However, it generates heat, especially when applied only by a single slim transformer because of significant restrictions in the transformer's window area and windings. This made the converter have restrictions in many areas such as reduction in size, application, etc.

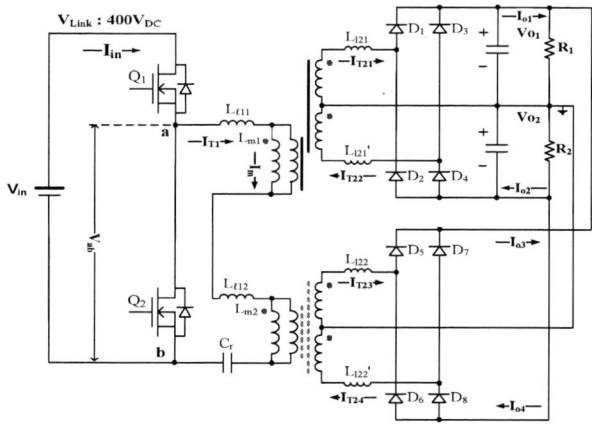

(b) Dual-output LLC resonant converter [4]

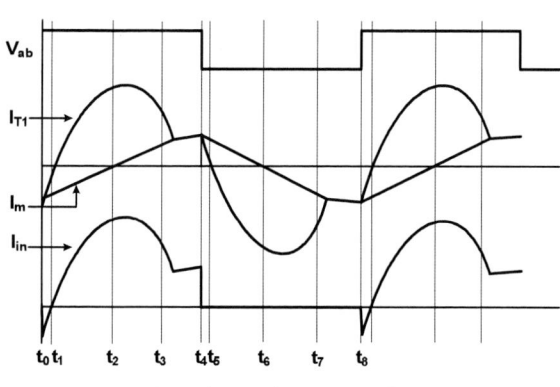

(c) Voltage and current waveforms

Figure 1. Main circuit and operating waveforms of the conventional LLC resonant converter

For the slimmer power supply and applied transformer's size reduction, a dual output LLC resonant converter, with two transformers that are connected in series and the secondary rectifier that are connected in parallel such as in Fig. 1(b), is being presented [4, 5]. However, the current stress of the input capacitor based on the input peak current (I_{in}) is large such as in Fig. 1(c).

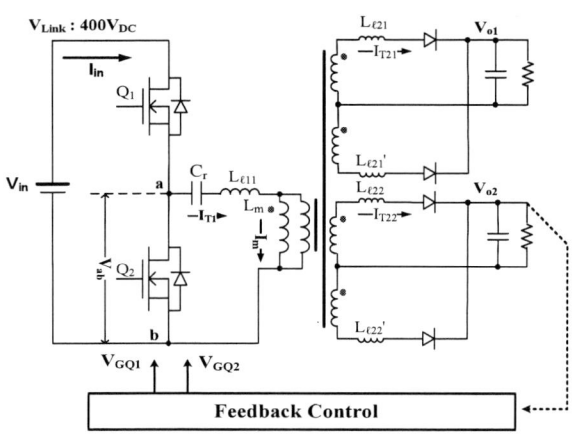

(a) Conventional LLC resonant converter

This work was supported by the cooperative research fund of LG Innotek

978-1-4244-4782-4/10 $26.00 © 2010 IEEE   1949

To cope with the recent trend of slimmer electronic display devices, this paper proposes a circuit for power supply with high power density. The proposed topology is LLC resonant converter with the two resonant tank circuits composed of resonance capacitors and two transformers. Each transformer used for the proposed resonant circuit is parallel connected in the primary and series connected in the secondary to reduce the current unbalance. Also, since the load variation has been coupled each other, the voltage of 12V output and 24V output could be controlled easily. The proposed LLC resonant converter is described and verified on 300W experimental prototype

II. PROPOSED LLC RESSONANT CONVERTER WITH TWO RESONANT TANKS

To solve the problem of the conventional LLC resonant converter, a new LLC resonant converter that has independent resonant tanks as shown in Fig. 2(a) and Fig 2(b) has been proposed. The proposed converters have been divided into two transformers (T_1, T_2) to make it slimmer. By making independent resonant tank circuits, the proposed LLC resonant converter's input current (I_{in}) flows from the input voltage source during all the cycles such as in Fig. 2(c), so the input maximum current (I_{in}) reduces in ½ comparing with the conventional LLC resonant converter's input current being delivered during half of the cycle such as in Fig. 1(c). Since the input current ripple is reduced to half, the input capacitor capacity can also be reduced. Also, since the connection method of primary resonant tank circuit is in parallel, small amount of magnetized current and resonant current will be flowing. The current stress can be reduced to half. Since the heat generated by the transformer coil and core can be minimized, it is convenient for integration. The secondary coil is connected in series between two transformers, and this has an advantage of having the ability to reduce the current unbalance problem due to the parameter difference. The proposed converter circuit 1 in Fig. 2(a) and the proposed converter circuit 2 in Fig. 2(b) are made up of two independent resonant tanks and have the same the operation characteristics. Therefore, this paper presents the circuit analysis and experimental results for the proposed converter circuit 1 from Fig. 2(a).

A. Operation modes of the proposed LLC resonant converter

The proposed LLC resonant converter of Fig 2(a) has two independent resonant tank circuit comprised of two transformers (T_1, T_2) and resonant capacitors (C_r).

Depending on the status (on/off) of the two switches (Q_1, Q_2), the two transformers (T_1, T_2) create an independent resonant tank circuit; transformer (T_1) forms a resonant circuit with the resonant capacitor(C_r) and transformer (T_2) forms another resonant circuit with the resonant capacitor (C_r).

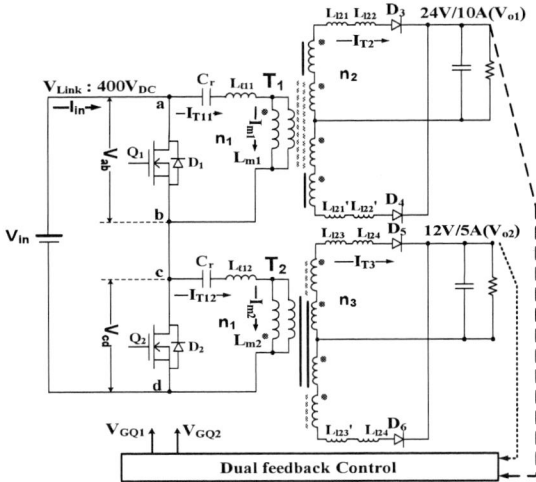

(a) Proposed converter

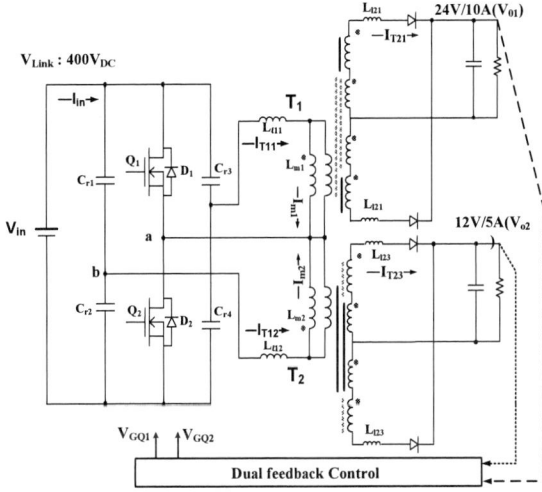

(b) Voltage and current waveforms of the proposed converter

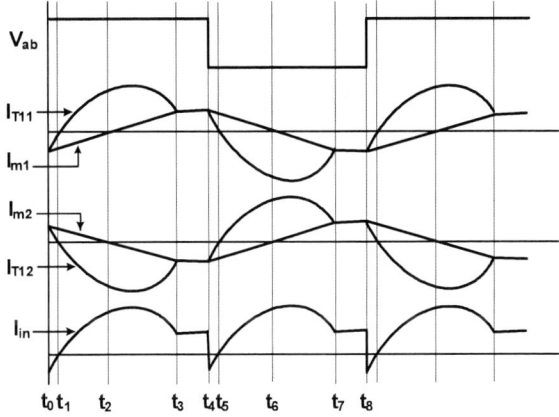

(c) Voltage and current waveforms of the proposed converter

Figure 2. Proposed circuits and voltage/current waveforms

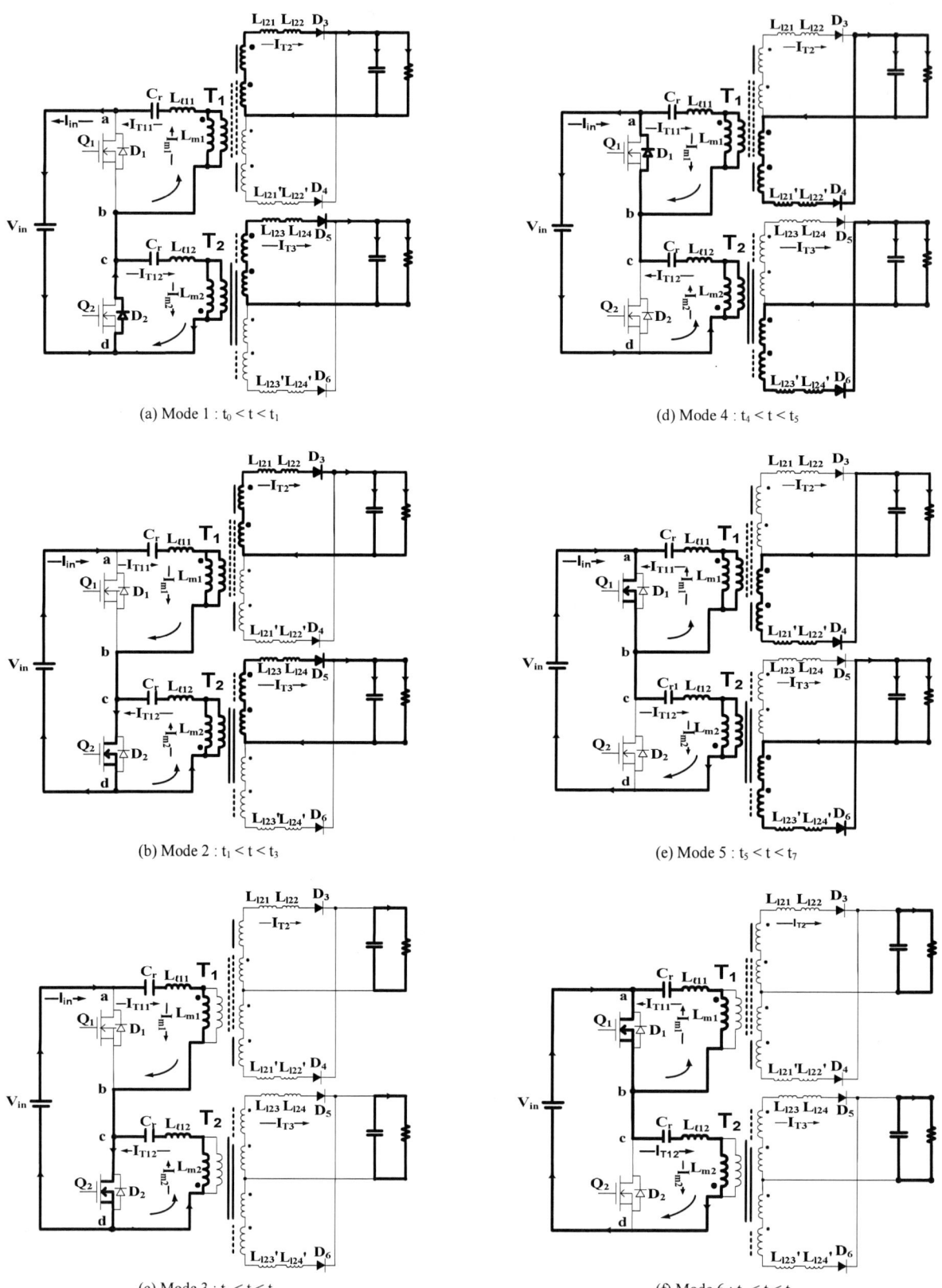

(a) Mode 1 : $t_0 < t < t_1$

(b) Mode 2 : $t_1 < t < t_3$

(c) Mode 3 : $t_3 < t < t_4$

(d) Mode 4 : $t_4 < t < t_5$

(e) Mode 5 : $t_5 < t < t_7$

(f) Mode 6 : $t_7 < t < t_8$

Figure 3. Operation modes of the proposed converter

The proposed LLC resonant converter, which operates in a discontinuous mode where switching frequency is below the resonant frequency (f_r), has 6 operation modes as seen in Fig. 3.

1) Mode 1 : $t_0 < t < t_1$

When Q_1 is turned off at t_0, according to the magnetizing current (I_{m1}, I_{m2}) of the transformers (T_1, T_2), output capacitance of Q_1 and Q_2 are charged and discharged to input voltage(Vin) and zero voltage, respectively, for a short dead time. At this time, the resonant current starts to flow because of the change in voltage polarity of the transformers (T_1, T_2). Thereafter, if Q_2 is turned on when the current is flowing in anti-parallel diode (D_2), Q_2 achieves Zero Voltage Switching (ZVS).

2) Mode 2 : $t_1 < t < t_3$

Mode 2 is the resonant mode interval flowing two independent resonant currents (I_{T11}, I_{T12}) through the two resonant tanks. When Q2 is turned on, the resonant current (I_{T11}) flows through the components of the leakage inductance (L_{l11}) and magnetizing inductance (L_{m1}) of transformer (T_1) and the resonance capacitor (C_r). Also, the resonant current (I_{T12}) flows through another independent resonant tank circuit comprised of the leakage inductance (L_{l12}) and magnetizing inductance (L_{m2}) of transformer (T_2) and the resonance capacitor (C_r).

3) Mode 3 : $t_3 < t < t_4$

At time t_3, Q_2 is kept turn-on and although this operation mode may look similar to operation mode 2, it is the interval where the resonant terminates and only the magnetizing currents(I_{m1}, I_{m2}) flow based on the magnetizing inductance (L_{m1}, L_{m2}) of the transformers (T_1, T_2). At this time, resonant takes place based on the magnetizing inductance (L_{m1}, L_{m2}) and resonant capacitor (C_r) of the transformers (T_1, T_2).

4) Mode 4 : $t_4 < t < t_5$

Operation of Mode 4 where the next half negative period begins to be operated has the similar operation mode such like Mode 1. When Q_2 is turned off at time t_4, according to the magnetizing current (I_{m1}, I_{m2}) of the transformers (T_1, T_2), output capacitance of Q_2 and Q_1 are charged and discharged to input voltage(V_{in}) and zero voltage during the short dead time respectively. At this time, the resonant current starts to flow because of the change in voltage polarity of the transformers (T_1, T_2). Thereafter, if Q_1 is turned on when the current is flowing in anti-parallel diode (D_1), Q_1 achieves Zero Voltage Switching (ZVS).

During the half negative period operation, Mode 5 ($t_5 < t < t_7$) and Mode 6 ($t_7 < t < t_8$) show in the same operation characteristics such like Mode 2 ($t_1 < t < t_3$) and Mode 3 ($t_3 < t < t_{\backslash 4}$).

In the case of a continuous operating mode, where it is operating at a frequency above the resonant frequency (f_r) of the proposed LLC resonant converter, it has remaining 4 operational modes excluding mode 3 and mode 6.

B. Equivalent circuit of the proposed LLC resonant converter

The proposed converter's main circuit in Fig. 2(a) is comprised of resonant circuit 1 made of transformer T_1 and resonant capacitor C_r, and resonant circuit 2 made of transformer T_2 and resonant capacitor C_r. Also, the proposed main circuit has a multi-output structure with 24V and 12V outputs.

Therefore, the input-output voltage gain characteristics (G_v) and equivalent resonant frequency (f_r) of the 24V output vary depending on the load condition of the 12V output, and because there is a change in resonant characteristics, one must consider the 24V output with great importance in circuit interpretation. Also, since secondary side connection method of transformer T_1 from resonant circuit1 and transformer T_2 from resonant circuit 2 is in series, one must consider the influence on the gain characteristics according to the coupling of T_1 and T_2.

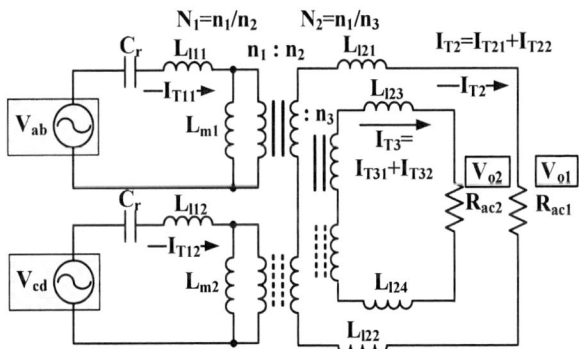

(a) Equivalent circuit of the proposed topology

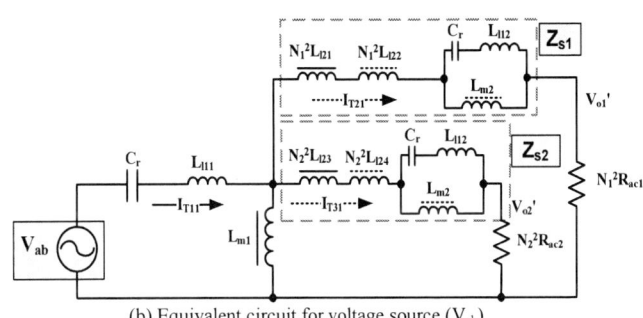

(b) Equivalent circuit for voltage source (V_{ab})

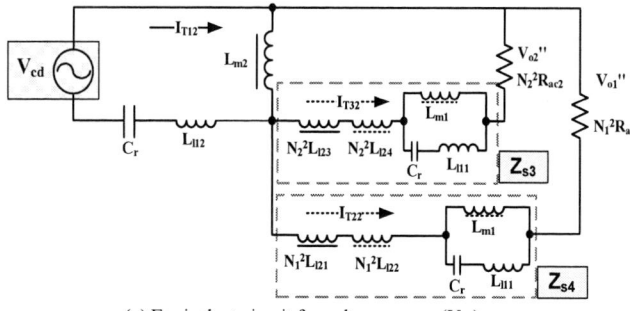

(c) Equivalent circuit for voltage source (V_{cd})

Figure 4. Equivalent circuit of the proposed circuit with superposition theory

First, the equivalent circuit of the proposed topology, as shown in Fig. 4(a), is a circuit including two separate voltage sources (V_{ab}, V_{cd}), a 12V output, and a 24V output. Therefore, in order to change the circuit from Fig. 4(a) to a T-shaped equivalent circuit, the Superposition Theory was applied. The principle of superposition could be used to transform the proposed circuit, that has two separate voltage sources (V_{ab}, V_{cd}), into a T-shaped equivalent circuit. Fig. 4(b) shows the equivalent circuit regarding to voltage source 1(V_{ab}). The voltage gain characteristics (G_{V1}) of the voltage source1 (V_{ab}) and a 24V load voltage (V_{o1}') can be shown as equation (1).

Also, since the secondary side connection method of T_1 and T_2 is in series, the impedance relation regarding to coupling relation that has a significant influence on the gain characteristic can be represented as equation (2) and equation (3), and the normalized value can be represented by below equations (4).

$$G_{V1} = V_{o1}'/V_{ab} =$$

$$\frac{1}{N}\left| \frac{1}{S_1\left\{ \frac{L_{p1}}{L_{m1}} + \omega^2 L_{m1}C_r + j\left(\omega L_{l11} - \frac{1}{\omega C_r} \right)\left(\frac{1}{Z_{s1} + N_1^2 R_{ac1}} + Z_{r1} \right) \right\}} \right| \tag{1}$$

$$Z_{s1} = N_1^2 j\omega L_{ls1} + \frac{j\omega L_{m2}\left(j\omega L_{l12} - j\frac{1}{\omega C_r} \right)}{j\omega L_{m2} + j\omega L_{lm2} - j\frac{1}{\omega C_r}} \tag{2}$$

$$Z_{s2} = N_1^2 j\omega L_{ls2} + \frac{j\omega L_{m2}\left(j\omega L_{l12} - j\frac{1}{\omega C_r} \right)}{j\omega L_{m2} + j\omega L_{lm2} - j\frac{1}{\omega C_r}} \tag{3}$$

Where $S_1 = 1 + \dfrac{Z_{s1}}{N_1 R_{ac1}}$, $Z_{r1} = \dfrac{1}{Z_{s2} + N_2^2 R_{ac2}}$, $L_{ls1} = L_{l21} + L_{l22}$

$$L_{ls2} = L_{l23} + L_{l24}, \quad L_{p1} = L_{m1} + L_{l11} \tag{4}$$

The equivalent circuit regarding voltage source 2(V_{cd}) with the principle of superposition applied, can be shown by Fig. 4(c) and the voltage gain characteristics (G_{V2}) of voltage source 2 (V_{cd}) and a 24V load voltage (V_{o1}'') can be found with equation (5). Also, since the secondary side connection method of T_1 and T_2 is in series, the impedance relation regarding to coupling relation that has a significant influence on the gain characteristic can be represented as equation (6) and equation (7), and the normalized value can be represented by below equations (8).

Therefore, the final voltage gain (G_V) of the proposed converter is the sum of equation (1) and equation (5) as shown in equation (9).

$$G_{V2} = V_{o1}''/V_{cd} =$$

$$\frac{1}{N_1}\left| \frac{1}{S_2\left\{ \frac{L_{p2}}{L_{m2}} + \frac{1}{\omega^2 L_{m2}C_r} + j\left(\omega L_{l12} - \frac{1}{\omega C_r} \right)\left(\frac{1}{Z_{s4} + N_1^2 R_{ac1}} + Z_{r2} \right) \right\}} \right| \tag{5}$$

$$Z_{s3} = N_1^2 j\omega L_{ls2} + \frac{j\omega L_{m1}\left(j\omega L_{l11} - j\frac{1}{\omega C_r} \right)}{j\omega L_{m1} + j\omega L_{m11} - j\frac{1}{\omega C_r}} \tag{6}$$

$$Z_{s4} = N_1^2 j\omega L_{ls1} + \frac{j\omega L_{m1}\left(j\omega L_{l11} - j\frac{1}{\omega C_r} \right)}{j\omega L_{m1} + j\omega L_{m11} - j\frac{1}{\omega C_r}} \tag{7}$$

Where $S_2 = 1 + \dfrac{Z_{s4}}{N_1 R_{ac1}}$, $Z_{r2} = \dfrac{1}{Z_{s3} + N_1^2 R_{ac2}}$

$$L_{p2} = L_{m2} + L_{l12} \tag{8}$$

$$G_V = G_{V1} + G_{V2} \tag{9}$$

Fig. 5 shows the characteristics of voltage gain over a large load range (from no load to short circuit) of the 24V output, where the load condition for 12V output was a medium range (12V/5A), and transformer parameters shown in Table (II) were used.

In the voltage gain characteristics (G_V) shown in Fig. 5, the corner frequency (f_o) represents the frequency ($f_o = \dfrac{1}{2\pi\sqrt{C_r L_{p1}}}$) when the 12V and the 24V output are open ($N_1^2 R_{ac1} = N_2^2 R_{ac2} = \infty$). The f_r is the resonant frequency when both 12V and 24V output are shorted ($N_1^2 R_{ac1} = N_2^2 R_{ac2} = 0$).

As shown in Fig. 5, since the operation is in discontinuous mode from light load (12V: 0.05A, 24V: 0.8A) to full load (12V: 5A, 24V: 10A) range, the switching components can achieve ZVS (Zero Voltage Switching) and ZCS (Zero Current Switching).

The peaking at corner frequency (f_o) is due to the difference values in inductance by coiling method of transformer T_1 and T_2. This peak will disappear when each transformer has the same inductance.

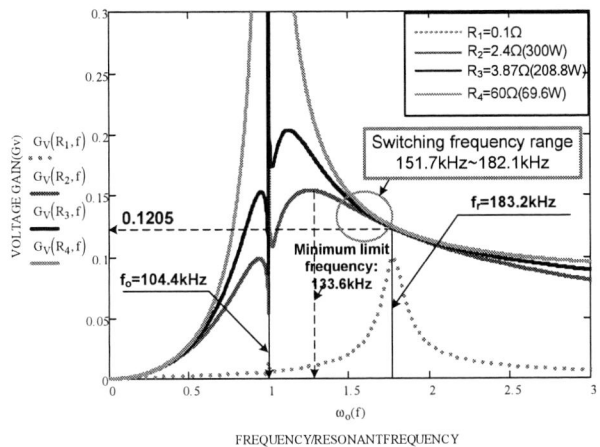

Figure 5. Voltage gain characteristics (G_V) due to the load variations

III. EXPERIMENTAL RESULTS

In this paper, experimental results based on a 300W laboratory prototype are presented to verify the proposed LLC resonant converter with 12V/5A, 24V/10A output under input voltage range of 340V~400V. Table I gives the experiment condition and specifications for the applied resonant converter from Fig 2(a). Fig. 6, Fig. 7 and Fig. 8 give the experimental waveforms of the resonant converter from Fig 2(a) and show the waveforms of each terminal voltage (V_{ab}, V_{cd}) and current (I_{T11}, I_{T12}) when the 12V output and the 24V output on the input voltage 340~400V are loaded with light load (12V/0.05A, 24V/0.8A), normal load (12V/3A, 24V/6.2A) and maximum load (12V/5A, 24V/10A), respectively.

We confirmed that the compared results of the experiment and the simulation of voltage gain characteristics (Gv) under the load variations from Fig. 5 were consistent.

By cross connecting the transformer's secondary coil of T_1 and T_2 in series, the current unbalance problems are reduced. Since the terminal current (I_{T11}, I_{T12}) based on terminal voltage (V_{ab}, V_{cd}) always has a lagging current flowing under zero to full load conditions, the switches always achieve ZVS (Zero Voltage Switching) even with maximum load condition (12V/5A, 24V/10A).

Fig. 9(a) shows the output voltage variation range based on the load conditions shown in Fig. 9(c), when the 24V output and 12V output were controlled by dual-feedback in the condition of input voltage 340V and 400V. Fig. 9(b) shows

the efficiency characteristics based on the same load variations when the input voltages were 340V and 400V.

Measured results indicate that the proposed converter with two tank circuits is within the output voltage variation range (±5%) in all of load variations.

Fig. 9(c) shows about 94% efficiency in the maximum load. By optimizing the proposed main circuit and transformer design, the efficiency can be improved and output voltage variation range can be minimized.

TABLE I.   Principal specification of the proposed LLC resonant converter

Input Voltage(V_{in})	340Vdc ~ 400Vdc	Output Voltage (V_o) And Current (I_o)	12V/5A, 24V/10A
Output capacity	300W	Switching frequency (f_s)	151.7kHz ~182.1kHz
Input/Output capacitor (C_{in} /C_f)	68uF×2(450V) 1000uF(35V)	Series Resonant Capacitor (C_r)	4.4nF
Switching Devices (Q_1, Q_2)		P9NK50ZFP 500V, 7.2A	
Output rectifier ($D_3 \sim D_6$)		FCQ20B06 60V, 20A	

TABLE II.   Proposed LLC resonant converter parameters

Transformer parameters				
Parameters	Transformer (T_1)		Transformer (T_2)	
Primary leakage inductance	L_{l11}	30.55uH	L_{l12}	33uH
Secondary leakage inductance	L_{l12}	0.373uH	L_{l22}	0.372uH
Magnetizing inductance	L_{m1}	497.3uH	L_{m2}	494.2uH
Equivalent leakage Inductance	L_{eq1}	166.3uH	L_{eq2}	168.3uH
N(n_1/n_2)		43/2		43/2
Cores	EFD3244H : H:5.1mm L:44mm			

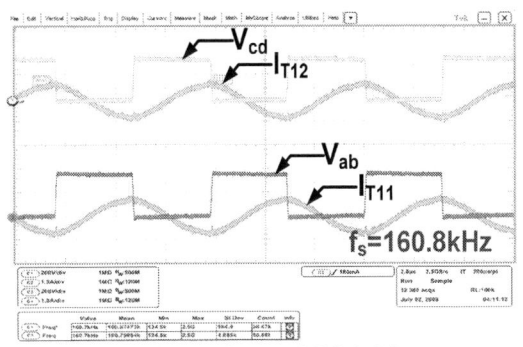

(a) V_{in}=340V

978-1-4244-4782-4/10 $26.00 © 2010 IEEE

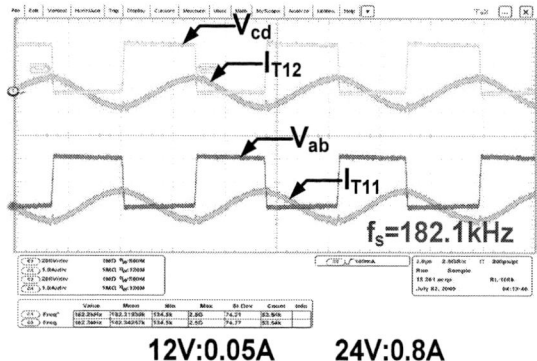

12V:0.05A   24V:0.8A

(b) V_in=400V

Figure 6.   Experimental waveforms of the terminal voltage (v_{ab},V_{cd}) and current(I_{T11},I_{T12}) on the input voltage(340~400V) variation and load conditions (12V/0.05A and 24V/0.8A). (200V/Div, 1A/Div, 2us/Div)

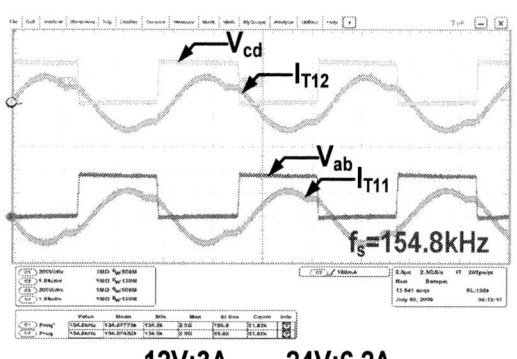

12V:3A   24V:6.2A

(a) V_in=340V

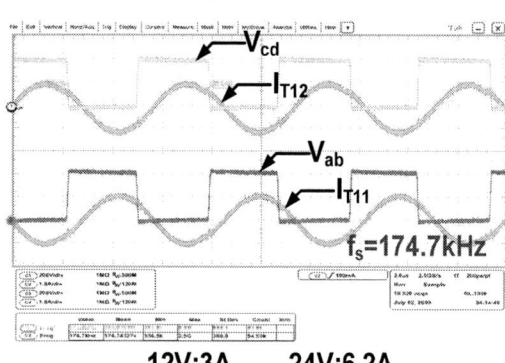

12V:3A   24V:6.2A

(b) V_in=400V

Figure 7.   Experimental waveforms of the terminal voltage (v_{ab},V_{cd}) and current(I_{T11},I_{T12}) on input voltage (340~400V) variation and load conditions (12V/3A and 24V/6.2A). (200V/Div, 1A/Div, 2us/Div)

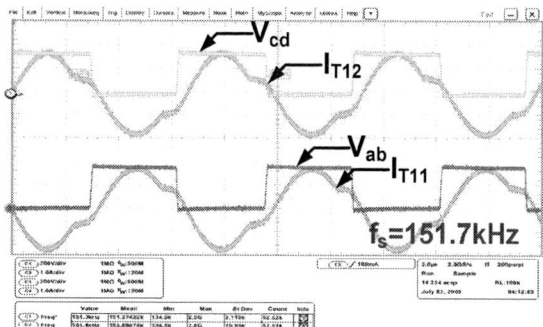

12V:5A   24V:10A

(a) V_in=340V

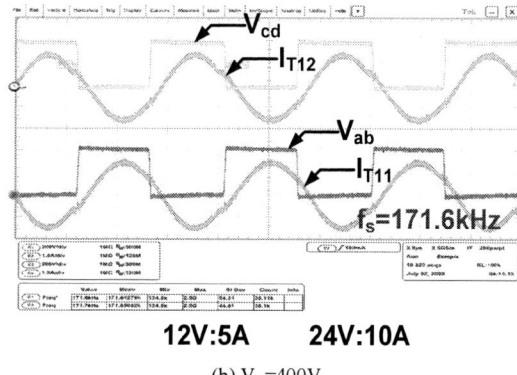

12V:5A   24V:10A

(b) V_in=400V

Figure 8.   Experimental waveforms of the terminal voltage (v_{ab},V_{cd}) and current(I_{T11},I_{T12}) on input voltage (340~400V) variation and load conditions (12V/5A and 24V/10A). (200V/Div, 1A/Div, 2us/Div)

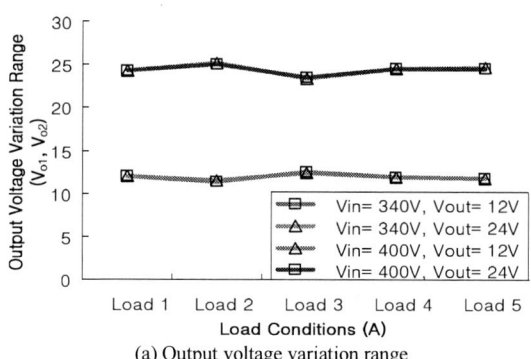

(a) Output voltage variation range

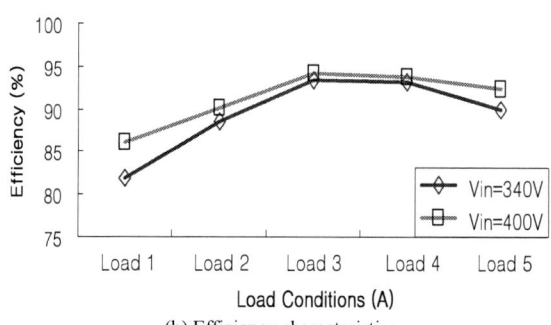

(b) Efficiency characteristics

load conditions	12V Output	24V Output
Load 1	0.05A	0.8A
Load 2	3A	0.8A
Load 3	0.05A	6.2A
Load 4	3A	6.2A
Load 5	5A	10A

(c) Load conditions

Figure 9.   Output voltage variation range and efficiency characteristics on the load conditions when input voltages are V_{in}=340V and V_{in}=400V

IV.   CONCLUSION

This paper proposes a new LLC resonant converter for the power supply's high power density, and describes the advantages of the proposed LLC resonant converter. Experimental results for a 300W laboratory prototype show that the output (12V, 24V), from dual-feedback controller under the input voltage (340~400V) condition is controlled. The experimental results are consistent with the simulation results of equation for voltage gain characteristics. This paper also confirms that the ripple of the input current is reduced and the current unbalance problem between the two transformers does not exist. Therefore, the proposed converter can be applied to the power supply with high power density.

REFERENCES

[1]   B. Yang, F. C. Lee, A. J. Zhang and G. Huang, "LLC resonant converter for front end DC/DC conversion," *IEEE APEC'02*, Vol. 2, pp. 1108-1112, 2002.

[2]   B. Lu, W. Liu, Y. Liang, F. C. Lee and J. D. van Wyk, "Optimal design methodology for LLC resonant converter," *IEEE APEC'06*, pp. 533-538, 2006.

[3]   Y. Gu, L. Hang, U. Chen, Z. Lu, Z. Qian and J. Li, "A simple structure of LLC resonant DC-DC converter for multi-output applications," *IEEE APEC'05*, Vol. 3, pp. 1485-1490, 2005.

[4]   Lin, B.-R., Chen, J.-J., Yang, C.-L, "Analysis and Implementation of Dual-Output LLC Resonant Converter", IEEE ICIT2008,

[5]   Bor-Ren Lin, Chao-Hsien Tseng, "Analysis of Parallel-Connected Asymmetrical Soft-Switching Converter." IEEE Transactions on Industrial Electronics, Vol. 54, No. 3, June 2007.

[6]   Marian K. Kazimierczuk, Dariusz Czarkowski "Resonant Power Converters" John Wiley & Sons, INC.

[7]   Yilei Gu, Lijun Hang, Chen, Zhengyu Lu, Zhaoming Qian, Jun Li ."A Simple Structure of LLC Resonant DC-DC Converter for Multi-output Applications" APEC 2005. Twentieth Annual IEEE, Vol. 3, pp. 1485-1490, 2005.

[8]   Li-jun HANG, Yi-lei GU, Zheng-yu LU, Zhao-ming QIAN, "Multi-Output LLC Resonant Converters With Symmetrical Auxiliary Output Structures", Telecommunications Energy Conference, 2006. INTELEC '06, 28th Annual international, pp.1-5, 2006, 10-14 Sept.

A Digital Control Strategy for Brushless DC Generators

Nikola Milivojevic, Student member, IEEE, Igor Stamenkovic, Student member, IEEE,
Mahesh Krishnamurthy, Member, IEEE, Ali Emadi, Senior member, IEEE

Electric Power and Power Electronics Center
Department of Electrical and Computer Engineering
Illinois Institute of Technology
Chicago, IL 60616, USA
kmahesh@ece.iit.edu

Abstract—**Control of BLDC drives in four-quadrant operation has been an area of active interest. One of the main challenges in generating mode applications is to allow a wide operating range for shaft speeds. This paper proposes a digital control strategy for BLDC generators that offers low maintenance costs, requires no additional hardware and is simple enough to allow implementation on an FPGA device instead of signal processors. This control strategy can either be part of existing control scheme/code, or as an embedded control module. Simulation results are validated and backed up by experimental results.**

Key terms- Digital Control, BLDC, Generator, FPGA

I. INTRODUCTION

Robustness and high-speed capabilities are major advantages of BLDC drives, when compared to widely used frequency controlled AC Drives. In addition, compared to standard vector controlled permanent magnet sine-wave motor drive, BLDC has higher torque for the same power density, and simpler machine design and lower manufacturing cost at the same time [1].

Conventional control algorithms that are used to control BLDC in generating mode have complex control routines and are not fault tolerant. Actually, in most of the applications, BLDC generators are controlled using DC/DC converter with diode bridge topology. Main reason for exploiting such solution is due to its simplicity and modular design, meaning that by adding Diode Bridge to any BLDC generator, different DC/DC converters may be employed with different sort of regulation (voltage, current, power). Common application for such systems is at small wind or hydro turbines (so called distributed energy systems).

However, it is possible to have vector controlled power converter running the BLDC generator. This makes control hardware of variable speed generator complex and expensive, but results in outstanding performance for small shaft speeds of the drive. This complex system can be found in regenerative braking applications in electric/hybrid vehicles, where dynamics of the system is of a big importance.

The paper proposes a simple digital control for speed and voltage regulation of BLDC drives in generating mode of operation. The technique uses standard Hall Effect sensors, which eliminates the need of a position encoder. Therefore such control strategy is different than standard Vector Control, which requires position sensor while using a complex algorithm for calculations when converting from one reference frame to another [2].

The paper has been presented in several parts- basic control of brushless DC machine working in generating mode is explained in the first section. Operating condition for exploit of active rectifier for control of PM generator is elaborated. The second section introduces basic principles of the digital control technique presenting two different options for implementing original control technique. Simulation results for various operating conditions are presented in the same section. Experimental set-up used for tests is described and experimental results are presented and commented in the following fourth section. These results are used to validate the simulation model and highlight the main benefits.

II. CONTROL OF BRUSHLESS DC GENERATORS

Recent developments in distributed generating (hydro and wind) as well as propulsion systems for hybrid and electric vehicles have significantly increased the popularity of PM generators. These machines have very high power density and field weakening capability compared to other machines topologies. In addition, PM drives can have very extended speed range that can go up to 5 times the rated speed. Such capability of the drive is possible to achieve by custom design brushless DC machine controlled by active rectifier. For applications where high torque is requirement, axial flux PM generator is commonly used, while for high speed applications, radial flux machine is employed [3].

978-1-4244-4782-4/10 $26.00 © 2010 IEEE

There are two major topologies used for controlling the BLDC machine in generating mode: Diode Bridge and active rectifier (inverter). Although the first solution is easier and simple to implement, employing an active rectifier gives better performance in terms of higher efficiency and wider speed range, mainly due to power factor correction capability. However, such solution includes complex hardware, which usually adds cost to the system. If a three phase diode rectifier is attached to its terminals, the phase back emf value depends on speed of the drive. Under this condition, the DC link voltage is defined is by equation 1:

$$U_d = \frac{3}{\pi}\sqrt{2}E - \frac{3}{\pi}I_d \cdot X_a \qquad (1)$$

where U_d, E and X_a, I_d, denote DC link voltage, phase back emf, phase armature reactance, and current in DC link. Therefore, it can be concluded that there is a preset condition for Diode Bridge to conduct the current, which depends on DC link voltage value [4-5].

In order to be able to control utilization factor as well as expand operating range of BLDC drives for speeds above rated, an active rectifier topology is typically utilized. However, the necessary condition for active rectifier to operate in generating mode is that back EMF has to be of lower value than the DC link voltage ($E \leq \frac{\pi}{3\sqrt{2}}U_d$).

Otherwise, diodes conduct all the time and switches cannot be controlled [6].

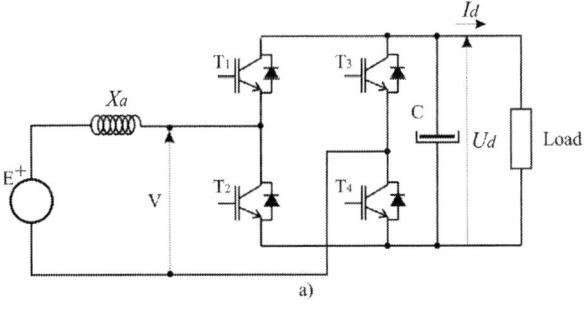

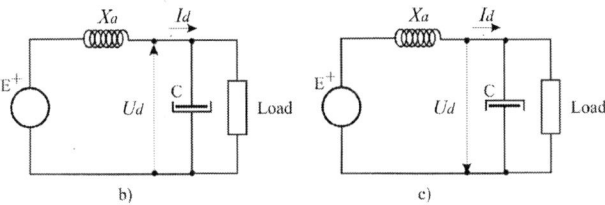

Figure 1: Operating principle of single phase active rectifier topology with BLDC drive in generating mode of operation: a) active rectifier topology, b)equivalent circuit when diodes are conducting, c)equivalent circuit when power switches are conducting

Control of power switches for BLDC drive in generating mode is shown in the fig 1. During positive current period (fig. 2b), diodes D1 and D4 conduct, supplying power to the DC link, while during negative period of phase current (fig.

2c), switches T2 and T3 are turned on, building–up the current [7 - 8].

Therefore, for higher speeds of the PM machine, the active rectifier converter is actually working as typical diode bridge rectifier (diodes are active only).

III. PRINCIPLES OF DIGITAL CONTROL

There are several conventional techniques for BLDC drive control that are commonly used in industry: hysteresis control, high-gain current regulated PWM (CRPWM), CRPWM with back emf compensation, etc. Such controlling techniques are very well adopted as industry standards and used for BLDC drives in various applications [10]-[11].

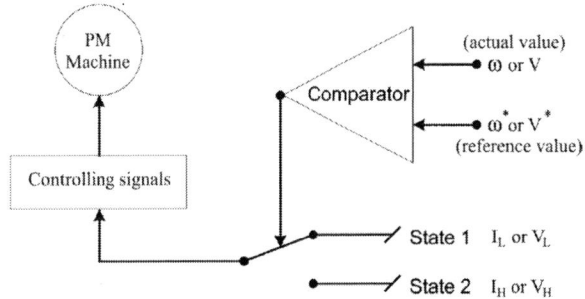

Figure 2: Fundamental principles of Digital control

The paper presents a non-complicated control technique for BLDC drives, which is easier to implement and does not require high computational resources. Such a control strategy opens up possibilities for cheap and trouble-free implementation. Digital control strategy regulates the speed of the machine by using time share of two predefined states of the system, so that one portion of time BLDC generator operates at one predefined state, while rest of the duty cycle time it operates at another predefined state. Predefined state actually corresponds to the reference signal that controls the system (shown on fig. 2) [12]-[13].

There are two different implementation of digital control: hysteresis and PWM digital control. Using current hysteresis control technique, generator's current is controlled in a way that it is switching between two predefined current reference values. Both reference values has very narrow hysteresis band, so instead of having only one reference for a current (as with conventional hysteresis technique), there are two predefined reference values in which the BLDC drive can operate. In this case current references for those two states are I_L and I_H (fig. 3).

Potential advantage of using two predefined reference values is in time share between them, which will result in much faster dynamic response than other industry known regulators. Time share between two references means that two references values are constant while duty cycle (operating time) for each of them is different and depends on regulated variables in the circuit. This concept offers considerably diverse way of looking onto variable speed drives and their speed/torque control for diverse industrial applications.

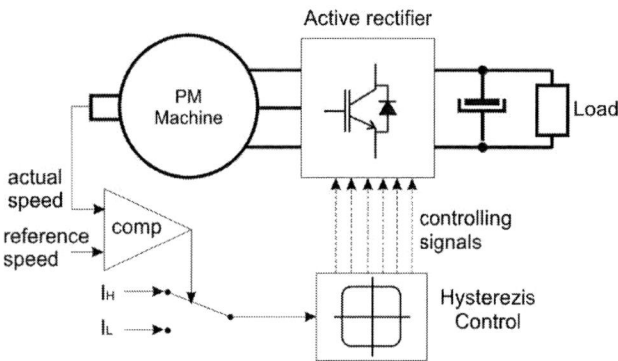

Figure 3: Current hysteresis Digital control used for speed regulation

According to the control strategy described in figure 3, a simple speed loop chooses one of two predefined states for consequent controlling period. For example, if the controller operates in state 1 (fig. 2), it results in reference signal I_L, while operating in state 2 results in reference signal I_H. It is not possible for reference signal of the generator to obtain any other value but I_L or I_H. This forms the basis of the Digital Control technique [14].

The dynamic equation for the brushless DC drive is given by equation (2):

$$T_{em} = k_t \cdot I = T_L + B \cdot \omega + J \cdot \frac{d\omega}{dt} \qquad (2)$$

where T_{em}, I, and T_L denote the electromagnetic torque at the shaft of the generator, generator phase current, torque from the prime mover respectively. When a high current reference is applied using digital control, is applied (I_H) equation 2 is modified to:

$$k_t \cdot I_H - T_L - B \cdot \omega = J \cdot \frac{d\omega}{dt} \qquad (3)$$

On another hand, when a lower current reference value (I_L) is applied, smaller electromagnetic torque will be developed on generator's shaft, which will cause speed dynamics defined by equation (4):

$$k_t \cdot I_L - T_L - B \cdot \omega = J \cdot \frac{d\omega}{dt} \qquad (4)$$

The following section describes the digital current hysteresis control of the brushless DC machine simulated for different constant values of speed. These tests used different values of prime mover torque. Since only hall effects are used for rotor position, there are only 6 points in one electrical revolution (very low band-width of speed regulator) when current reference can be changed. Fig. 4 shows phase currents and digital control current reference signals for three different shaft speeds (low, medium and high speed).

From the simulations, it can be seen that in order to regulate speed on the generator shaft at high value, the generator's phase current has small reference value – I_L, which means that developed electromagnetic torque is not

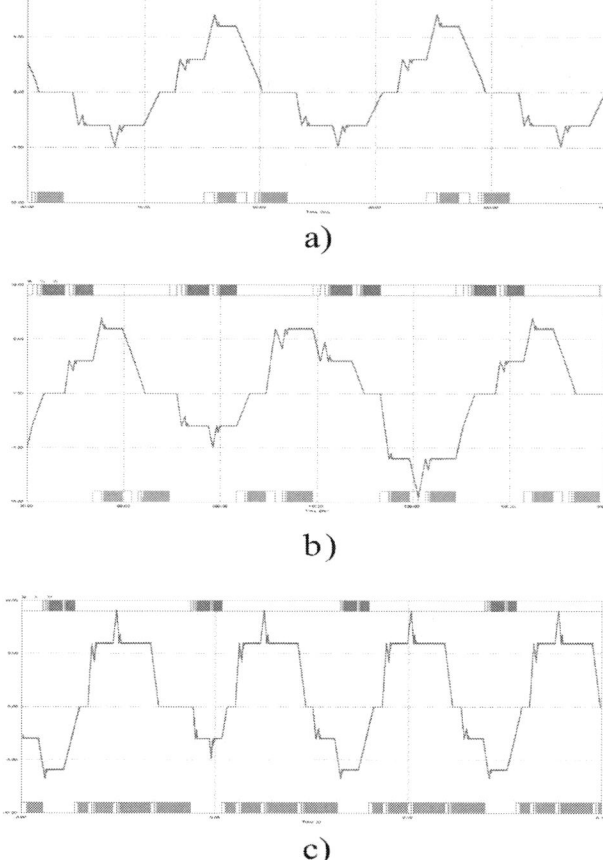

a)

b)

c)

Figure 4: Simulation results -waveform of phase current, and digital control reference current signals for I_H and I_L for different speeds: a)2700rpm, b) 1500rpm and c)600rpm

opposing the prime mover torque at full potential. On another hand, fig. 4c shows phase current that is almost always at reference value I_H, which maximizes the torque value; hence the speed is much smaller, around 20% of rated speed.

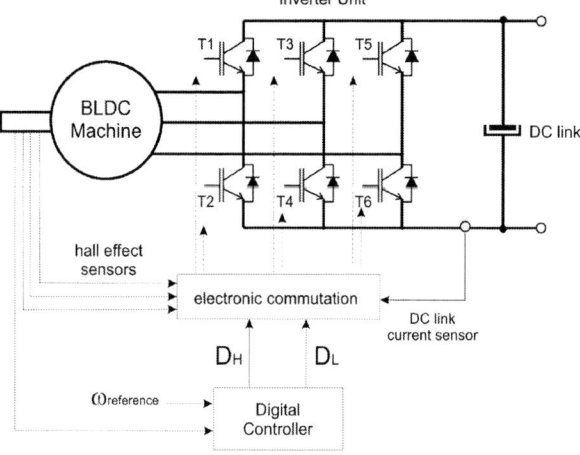

Figure 5: PWM Digital control used for speed regulation

In the same manner, the BLDC system can be controlled using digital PWM signals. There are two predefined states, PWM duty cycles, that are created using reference voltage values: V_H and V_L. When the generator speed needs to be reduced, reference signal V_H is used (Duty cycle corresponding V_H cycle is applied). When the generator speed needs to increase, reference signals V_L is applied (V_L PWM duty cycle signal controls switches of the active rectifier) [15].

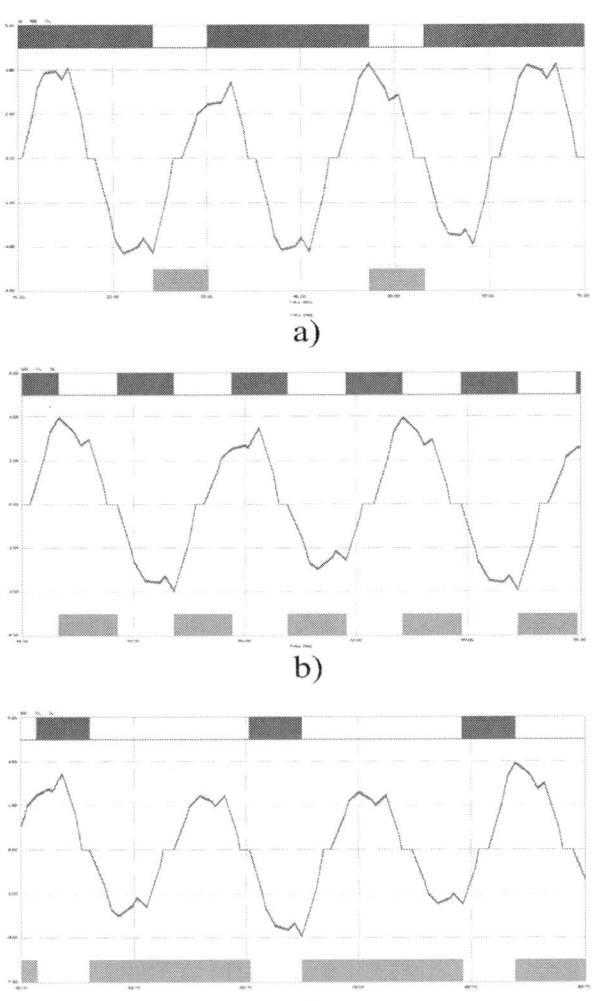

Figure 6: Simulation results -waveform of phase current, and digital control reference PWM duty-cycle signals DH and DL. for different speeds: a)2700rpm, b) 1500rpm and c)600rpm

Fig. 6 shows generator's phase current waveform, as well as reference values of DH and duty-cycles. For higher speeds (fig. 6a) , duty-cycle DL has much frequent occurrence, which results in smaller electromagnetic torque (opposing torque) on the machine's shaft.

On another hand, in order to keep the speed at lower value for the same prime mover's torque, duty-cycle DH is performed more frequently than DL. Choosing the correct values for duty cycles is related with operating range of the

brushless DC drive. In other words, these predefined values of duty cycles will be determined based on the application. Higher the duty cycle DL and lower the value of DH will results in small operating range of the drive, but at the same time will minimize torque ripple on the drive's shaft.

However, if wide speed range is a requirement, then value of DL has to be as small as possible, while DH should be at its maximum. Such conditions will lead toward high torque ripple, just because of significant difference in reference signal for the control drive.

On another hand, unlike current hysteresis digital control technique, high switching frequency can be implemented and actually improve the performance of the drive [16].

IV. EPERIMENTAL VERIFICATION

Figure 7 shows the setup used for experimental verification. The PMDC machine is controlled as a prime mover for testing 24V, 23Arms, and 451W BLDC generator.

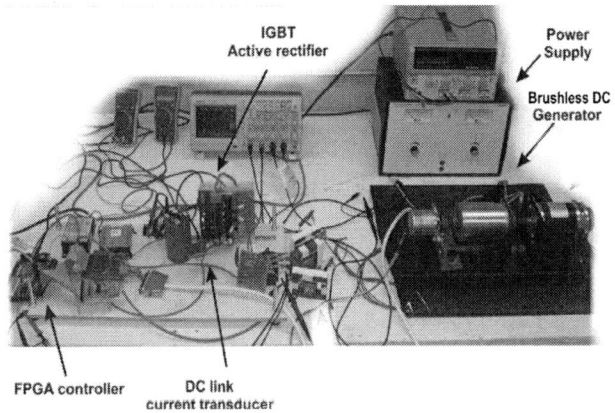

Figure 7: Experimental set-up consisting of Brushless DC generator and power converter controlled by digital control implemented in FPGA

Only one current sensor in DC link is used for digital hysteresis current control, while speed regulation is performed by using Hall Effect sensors only. Speed is calculated as a function of time between two hall-effect signals, which are utilized for switching pattern (commutation function) for power switches of active rectifier. Fig. 8 and 9 shows experimental results for both digital control techniques. Prime mover's load was kept constant, at 1.2Nm, in both scenarios. Results of experiment are shown for three different speeds, at which the brushless DC machine were generating power.

Fig. 8 presents the phase current waveform of the generator when controlled using current digital control algorithm. Simulation results show that for the higher speeds, phase current uses the lower value of the reference current I_L, majority of the time, therefore developing lower electromagnetic torque on the shaft, and extracting less power from the prime mover.

For the values of reference currents: I_L=2A, and I_H=10A, brushless DC generator was applied prime mover's torque of around 1Nm, and was controlled for three different speeds. It is interesting to mention that different combination of

978-1-4244-4782-4/10 $26.00 © 2010 IEEE   1960

reference currents is giving different phase current ripple, which is then causing variations in torque ripple as well.

At the moment when actual speed was exactly the same value as reference, the reference current was set-up to zero, meaning that no current is applied, in order to maintain the speed of the drive.

For higher speeds, both predefined reference values for current can have small current value because no high electromagnetic torque is needed. In order to control the brushless DC generator drive in low speed mode, for the same prime mover's torque, higher current reference I_H has to have high value just because it sets up high torque value on the shaft.

Fig. 9 presents experimental results for digital PWM control techniques. Again, waveforms of phase current and duty cycles of D_H and D_L are presented, for different rotational speeds of generator drive. The phase current in latter case is much smoother, resulting in significantly less acoustic noise while operating.

Duty cycles used for the experiment are: D_L=0.15 and D_H=0.8. One of the reasons for smoother control of PWM digital control is the fact that the difference in generated electromagnetic torque between predefined states for PWM control is significantly smaller and softer, than for current hysteresis control. Therefore, the latter kind of digital control has less impact over machines stress and mechanical losses.

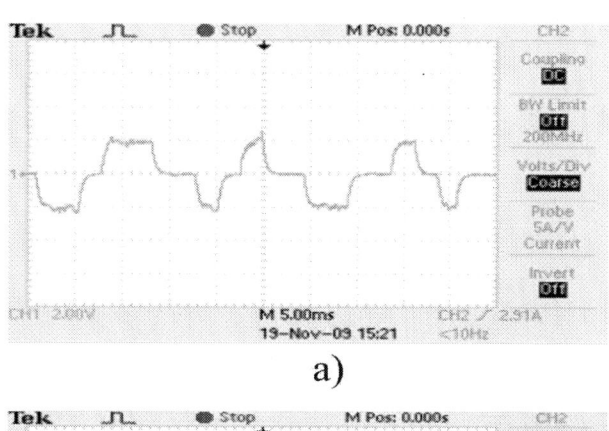

a)

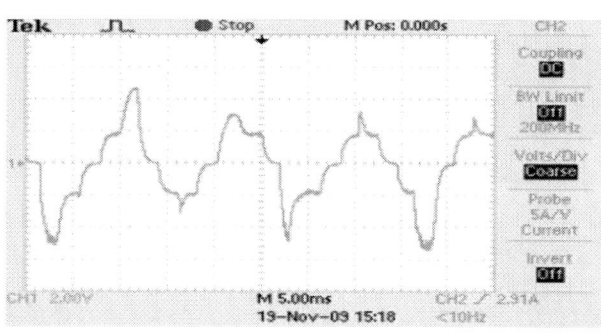

b)

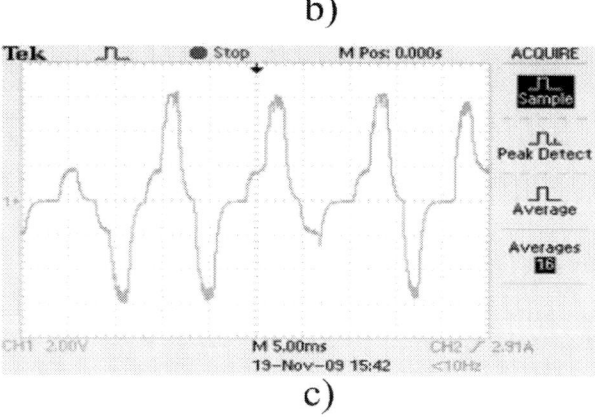

c)

Figure 8: Experimental results for digital current hysteresis control. Waveform of phase current for different rotational speeds for the same prime mover's torque value

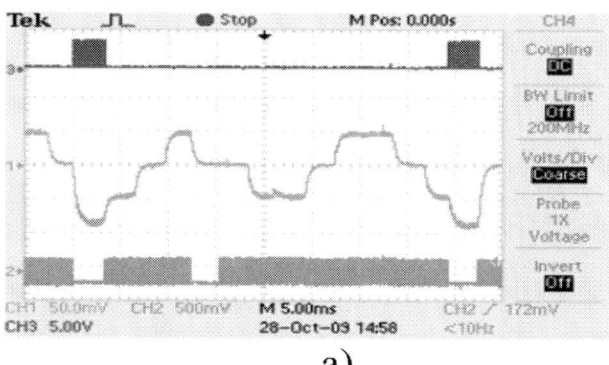

a)

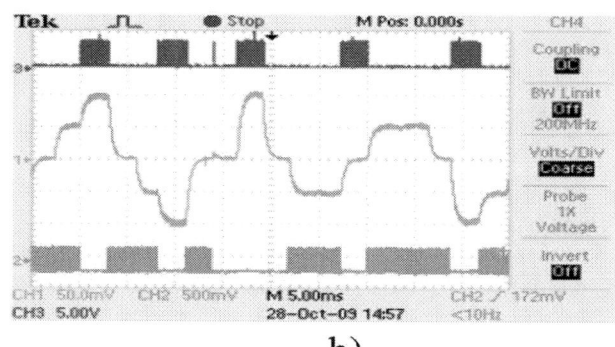

b)

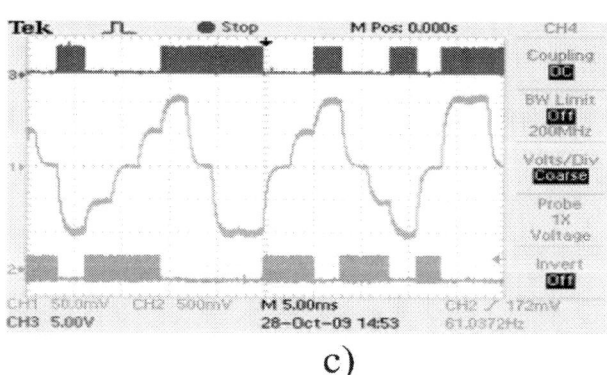

c)

Figure 9: Experimental results for digital PWM control. Waveforms of phase current and signals of D_H and D_L duty-cycles for different rotational speeds for the same prime mover's torque value

V. CONCLUSIONS

This paper presents a low-cost, easy to implement digital control strategy for BLDC generators that requires no additional hardware. It is implemented on an FPGA instead of digital signal processors. Simulation and experimental results are provided to validate the technique. The main advantage and merit of such control strategy is its simplicity and application to existing generated systems.

Several different controls can be achieved by using digital control concept. The paper presents the application of this strategy towards speed regulation, but it can also be used to regulate output voltage of the generator drive, or generated electromagnetic torque if needed. Such flexibility makes presented control strategy very suitable for variable speed drives.

REFERENCES

[1] P. Pillay, R. Krishnan, "Modeling, Simulation, and Analysis of Permanent-Magnet Motor Drives, Part I: The Permanent-Magnet Synchronous Motor Drive", *IEEE Transactions on Industry Applications*, vol. 25, no. 2, pp: 265-273, March/April 1989

[2] P. Pillay, R. Krishnan, "Modeling, Simulation, and Analysis of Permanent-Magnet Motor Drives , Part II: The Permanent-Magnet Synchronous Motor Drive", *IEEE Transactions on Industry Applications*, vol. 25, no. 2, pp: 274-279, March/April 1989

[3] L. McGrow, C. Pollock, "Low cost brushless generators", *Industry Applications Conference, 1999*. Thirty-Fourth IAS Annual Meeting. Conference Record of the 1999 IEEE, Volume 2, vol.2, pp: 1229 – 1236, 1999

[4] M. J. Khan, M. T Iqbal, "Simplified Modeling of rectified-coupled Brushless DC generators", *4th International Conference on Electrical and Computer Engineering ICECE 2006*, 19-21 December 2006, Dhaka, Bangladesh

[5] C. Z. Liaw, W. L. Soong, B. A. Welchko, N. Ertugrul, "Uncontrolled Generation in Interior Permanent-Magnet Machines", *IEEE Transactions on Industry Applications*, Vol. 41, no. 4, July/August 2005

[6] P. Pillay, R. Krishnan, "Application Characteristics of Permanent Magnet Synchronous and Brushless dc Motors for Servo Drives", *IEEE Transactions on Industry Applications*, Vol. 21, No. 5, September/October 1991.

[7] J. R. Rodríguez, J. W. Dixon, J. R. Espinoza, J. Pontt, P. Lezana, "PWM Regenerative Rectifiers: State of the Art", *IEEE Transactions on Industrial Electronics*, Vol. 52, No. 1, February 2005

[8] B. Meersman, S. Thielemans, K. De Gussem´e A. Van den Bossche, "Soft-Switch DC-DC Converter with a High Conversion Ratio for an Electrical Bicycle", *Vehicular Power and Propulsion Conference 2007.*

[9] P. Pillay, R. Krishnan, "Application Characteristics of Permanent Magnet Synchronous and Brushless dc Motors for Servo Drives", *IEEE Transactions on Industry* Applications, Vol. **21**, no. **5**, September/October 1991

[10] C. BI, C. S. Soh, "Influence of Transient Current to PM-AC Motor Driven By BLDC Mode", International Conference on Electrical Machines and Systems, ICEMS 2008, pp: 2761-2766, 17-20 Oct. 2008

[11] Y. –J. Lee, A. Emadi, "Integrated Bi-Directional AC/DC and DC/DC Converter for Plug-in Hybrid Electric Vehicle Conversion", *Vehicular Power and Propulsion Conference 2007*

[12] W. Hong, W. Lee ,B. –K. Lee, "Dynamic Simulation of Brushless DC Motor Drives Considering Phase Commutation for Automotive Applications", *IEEE International Electric Machines & Drives Conference*, Vol. 2, pp. 1377-1383, IEMDC '07.

[13] F. Rodríguez, A. Emadi, "A Novel Digital Control Technique for Brushless DC Motor Drives", *IEEE Transactions on Industrial Electronics*, Vol. 54, No. 5, October 2007

[14] P. C. Desai, A. Emadi, "A novel digital control technique for brushless DC motor drives: current control", *IEEE Electric Machines and Drives Conference*, pp. 326 – 331, 2005

[15] F. Rodriguez, P. Desai, and A. Emadi, "A novel digital control technique for trapezoidal Brushless DC motor drives", *Proceedings of Power Electronics Technology Conference Chicago* IL (Nov. 2004)

[16] Sathyan, A.; Milivojevic, N.; Young-Joo Lee; Krishnamurthy, M.; Emadi, A. "An FPGA-Based Novel Digital PWM Control Scheme for BLDC Motor Drives", IEEE Transactions on Industrial Electronics, , Volume 56, Issue 8, Aug. 2009 Page(s):3040 - 3049

Space Vector based PWM Scheme without Sector Identification for a 4-Level Dual Inverter fed Induction Motor Drive with Asymmetrical DC Link Voltages

Shiny G., M.R.Baiju
College of Engineering Trivandrum
Kerala, India
mrbaiju@ece.cet.ac.in

Abstract— **A Space Vector based PWM scheme for 4-level inverter is proposed. 4-level inversion is achieved by connecting two 2-level inverters in open-end winding configuration. The individual inverters are fed with asymmetric DC link voltages. The proposed scheme does not use sector identification method to find the sub hexagon, which encloses the tip of the reference space vector. During 3-level operation, the sub hexagon centers are directly identified from the instantaneous magnitude of 3-Phase reference sinusoid. The sub hexagon center for 4-level operation are determined from the generated sub hexagon centers for 3-level operation. The switching vectors for the two inverters are generated without using look-up tables. The scheme is implemented and tested with a 2-HP open-end winding induction motor drive. Experimental results including operation in over modulation region are presented.**

I. INTRODUCTION

Multilevel inverters based on Space Vector PWM (SVPWM) modulation is extensively used in high power industrial drive applications due to their ability to synthesize waveforms with higher voltage levels. Various multilevel inverter topologies like Neutral Point Clamped multilevel inverter, diode clamped multilevel inverter, flying capacitor multilevel inverter and multilevel inverter with separate DC sources have been proposed [1-4]. 3-level inverter, realized with dual 2-level inverters feeding an open-end winding induction motor is suggested in [5]. Two 2-level inverters connected in open-end winding configuration with asymmetric DC link voltage can realize a 4-level inverter configuration [6]. In this paper a simple algorithm for generation of SVPWM for 4-level inverter is presented. In Space Vector PWM, the reference space vector is approximated by switching amongst the three nearest voltage space vectors [6-14]. The SVPWM method presented in this paper requires no sector identification method to find the sub hexagon, which encloses the tip of the reference space vector. The scheme is experimentally

verified by using the dual inverter fed open-end winding 4-level inverter configuration.

II. DUAL INVERTER FED INDUCTION MOTOR DRIVE WITH OPEN-END WINDING

Fig. 1 shows a 4-level inverter configuration realized by connecting two 2-level inverters in open-end winding configuration. The inverters are supplied with asymmetric DC link voltages. The DC link voltage of Inverter-1 is $2V_{DC}/3$. Inverter-2 is fed with a DC link voltage of $V_{DC}/3$. The pole voltages of Inverter-1 are designated as V_{AO}, V_{BO} and V_{CO}. Similarly, the pole voltages of Inverter-2 are designated as $V_{A'O'}$, $V_{B'O'}$ and $V_{C'O'}$. Depending on the individual pole voltages of Inverter-1 and Inverter-2, the phase voltage of the 3-Phase induction motor can achieve four levels viz. $-V_{DC}/3$, 0, $V_{DC}/3$ and $2V_{DC}/3$. These four voltage levels are represented by vectors 0, 1, 2 and 3 respectively.

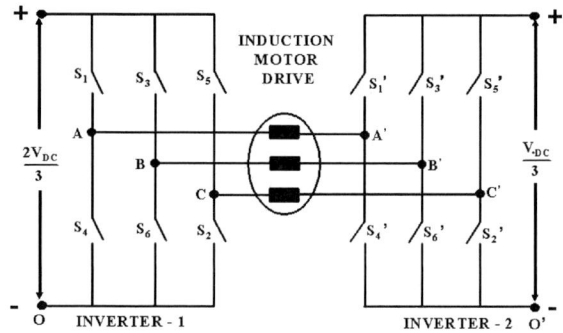

Figure 1. 4-Level Inverter with Asymmetric DC Link Voltage

Fig. 2 shows the space vector diagram of a 4- level inverter. For simplicity, all switching vectors are not shown in Fig. 2. The space vector locations of Inverter -1 are shown in Fig. 3 and Fig. 4 show the space vector locations of Inverter-2. Since each inverter can assume eight states independantly, altogether there are 64 switching vector combinations for a 4-level dual inverter configuration. The space vector combinations of the 4-level inverter scheme are shown in Fig. 5.

The space vector diagram of 4-level inverter can be viewed as a hexagonal structure. As can be seen from Fig. 5, the space vector diagram consists of one *inner* sub hexagon with a *sub hexagon center(SHC)* designated as O. There are six *middle* sub hexagons with *sub hexagon centers* A to F, which are centered on the apexes of the inner sub hexagon. Centered on the apexes of the middle sub hexagons, there are twelve *outer* sub hexagons with sub *hexagon centers* designated from G to S (except O). Each sub hexagon in the hexagonal structure is divided into small triangular region called sector. There are 54 such sectors in the space vector diagram of a 4-level inverter. As shown in Fig. 5, there are 37 switching vector locations for the dual inverter fed 4-level inverter.

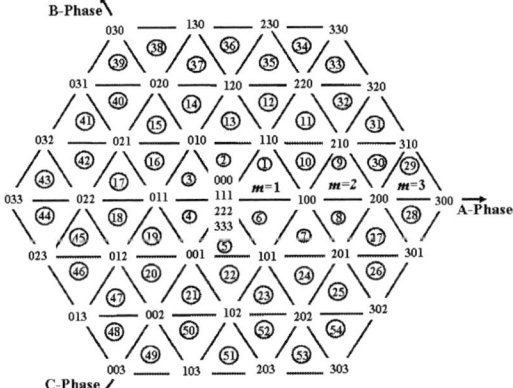

Figure 2.   Space Vector diagram of 4-level inverter

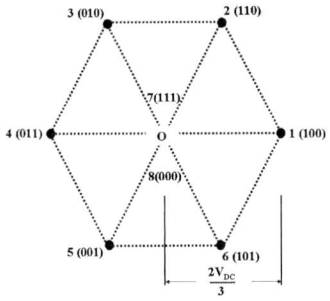

Figure 3.   Space Vector locations of Inverter-1

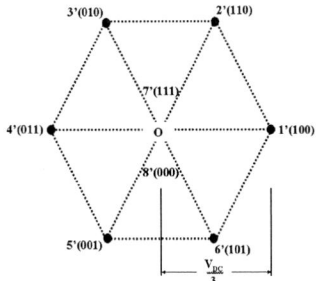

Figure 4.   Space Vector Combinations of Inverter-2

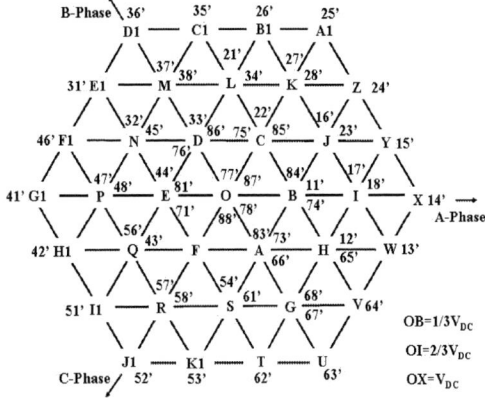

Figure 5.   Space Vector Combinations of 4-Level Inverter

III.   PRINCIPLE OF THE PROPOSED WORK

A.   Generation of Switching Vectors

For explaining the proposed scheme, consider a reference space vector OP shown in Fig. 6.

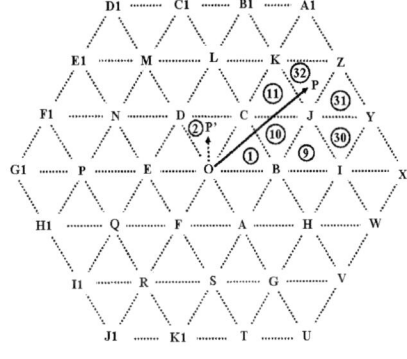

Figure 6.   Space Vector diagram showing reference vector OP and mapped vector OP'

The tip of the reference vector OP can be in any one of the 54 sectors. Depending on the position of the reference vector, there are four levels of operations namely 2-level operation, 3-level operation, 4-level operation and over modulation region of operation. As can be seen from Fig. 2, the space vector diagram of 4-level inverter consists of 3 layers of operation. The layers are designated as $m=1$, $m=2$ and $m=3$. The layer of operation, m can be calculated using the equation [15]

$$ m = 1 + \operatorname{int}\left\{ V_{j\max} \middle/ \left(\frac{\sqrt{3}}{2} \frac{V_{DC}}{n-1} \right) \right\} \qquad (1) $$

As shown in Fig. 2, when $m=1$, the inverter is operating as a 2-level inverter. In 2-level operation mode, the tip of the reference vector is confined to the inner sub hexagon (sector 1 to sector 6). In 3-level operation mode ($m=2$), the tip of the reference vector can lie in any one of the sectors between 7 to 24. If the tip of the reference vector lies in the outermost layer (sector 25 to 54), it is 4-level operation ($m=3$). If the reference vector lies outside all the hexagons, the inverter will be operating in over modulation region. In Fig. 6 shown, the tip of the reference vector OP lies in sector 32, which implies 4-level operation.

To realize the reference vector OP, we have to switch the vectors located at vertices J, K and Z. Once the switching vectors are determined, the switching time periods of the vectors are to be calculated. This can be achieved by the principle of mapping. In mapping, the sub hexagon center (for the reference vector OP, the sub hexagon center identified is the vector located at vertex J) is shifted to coincide with the center O of the inner hexagon using proper coordinate transformation [15]. In the proposed work, the sub hexagon center is generated without using sector identification method. The mapped reference vector is designated as OP' and it lies in sector 2 of the inner sub hexagon. Now the duration of the switching vectors can be determined by using the conventional equations for 2-level inverter [13].

B. Generation of Sub hexagon center(SHC) for 3-Level Operation

If the tip of the reference vector lies in the middle sub hexagon, $m=2$. In layer-2 operation, the sub hexagon center is directly identified from instantaneous magnitude of the 3 -Phase reference sinusoid [16].

Fig. 7 illustrates the scheme for finding the sub hexagon center. Let Va, Vb and Vc represent the instantaneous magnitude of the 3-Phase reference sinusoid. The sub hexagon centers of the *middle* sub hexagon are determined from the instantaneous magnitudes of Va, Vb and Vc. If the magnitude of the sine wave is positive, it is represented as "1" and if the magnitude is negative, it is represented as "0". For the 3-Phase reference sinusoid, during the time interval from $\omega t=0$ to $\omega t=60$, Va is positive, Vb is negative and Vc is positive. This corresponds to a sub hexagon center of 101 (+ − +). Similarly, during $\omega t=60$ to $\omega t=120$, Va is positive, Vb and Vc are negative which implies a sub hexagon center of 100 (+ − −). Similar procedure can be used to find the remaining sub hexagon centers.

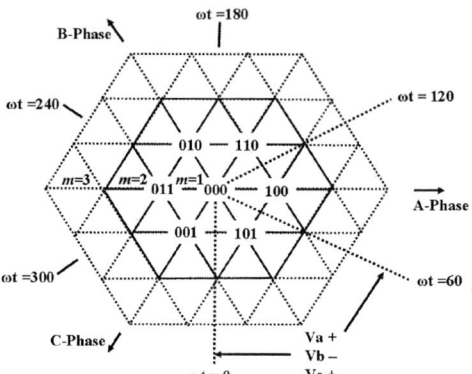

Figure 7.   Generation of Sub hexagon center for 3-Level operation

With this scheme, no sector identification method is required to find the sector which contains the tip of the reference vector in 3-level operation. In the proposed method, if the tip of the reference vector is confined to the *inner* hexagon, $m=1$. The sub hexagon center in this case is taken as 000.

C. Generation of Sub hexagon center (SHC) for 4-level Operation

From the generated sub hexagon centers for 3-level operation, we can obtain the sub hexagon centers for 4-level operation. This is achieved by adding the sub hexagon center obtained in 3-level operation with its preceding and succeeding vector. As shown in Fig. 8, for $m=3$ and $\omega t=0$ to $\omega t=60$, there can be three possible sub hexagon centers viz. 102, 202 and 201, which are generated using a simple arithmetic. As shown in Table-1, the SHC for $m=2$ and $\omega t=0$ to $\omega t=60$ is 101. For obtaining SHC vector 202, SHC 101 is added to itself (101+101=202). To get vector 102, SHC 101 is added with its preceding vector 001 (101+001=102). And to obtain 201, SHC 101 is added with its succeeding vector 100 (101+100=201). Out of these three new vectors, the vector which is closest to the tip of the reference vector is taken as the SHC for 4-level operation. Fig. 8 shows the schematic for SHC generation for 4-level operation.

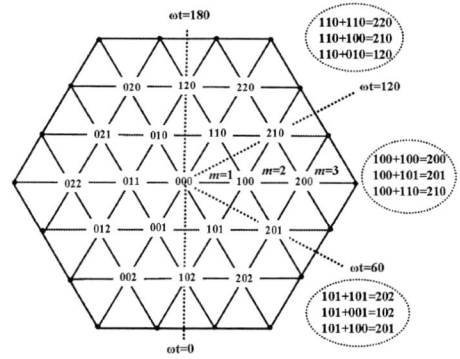

Figure 8.   Generation of Sub hexagon center for 4-Level operation

978-1-4244-4782-4/10 $26.00 © 2010 IEEE

TABLE I.   GENERATION OF SUB HEXAGON CENTER (SHC) FOR 4-LEVEL OPERATION

Angle ωt	0 - 60	60 - 120	120 - 180	180 - 240	240 - 300	300 – 360
SHC for 3-level operation	101	100	110	010	011	001
Preceding vector of SHC	001	101	100	110	010	011
Succeeding vector of SHC	100	110	010	011	001	101

D. Generation of Gate Signals for individual Inverters

Once the phase voltage timings T_{ga}, T_{gb} and T_{gc} corresponding to the three phases are computed [13], the vectors located at vertices J, K and Z are to be switched for realizing the reference vector. In the proposed method, this is done by adding the sub hexagon center with the mapped 2-level PWM signal as shown in Fig. 9. The PWM scheme is implemented using the dSPACE DS 1104 RTI platform and FPGA Xilinx Virtex. Table-2 shows the status of top switches of the individual inverters during different voltage levels (for A -Phase leg).

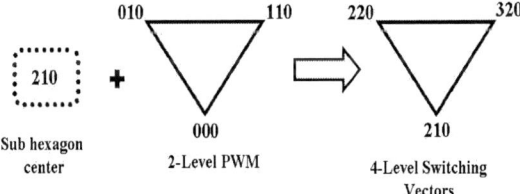

Figure 9.   Addition of SHC with 2-Level PWM signal generates switching vectors for 4-Level operation

TABLE II.   THE STATUS OF TOP SWITCHES OF THE INDIVIDUAL INVERTERS DURING DIFFERENT VOLTAGE LEVELS (FOR A – PHASE LEG)

Top Switch Status of INV-1 (S1)	Top Switch Status of INV-2 (S1')	Voltage Level in A-phase winding	Realized switching level
OFF	ON	$-V_{DC}/3$	0
OFF	OFF	0	1
ON	ON	$V_{DC}/3$	2
ON	OFF	$2V_{DC}/3$	3

When the tip of the reference vector lies outside the hexagons, the inverter will be operating in the over modulation region. During over modulation, the sub hexagon center vector will not be switching and the other two vectors will switch for the entire sample period. The scheme also works in over modulation region.

IV.   EXPERIMENTAL RESULTS

The proposed PWM scheme is experimentally verified by implementing the scheme on a 2 HP, 3 - Phase induction motor drive in open loop with v/f control for different modulation indices. The gating pulses for the two inverters are generated using the dSPACE DS 1104 RTI platform and FPGA Xilinx Virtex.

Experimental results for a modulation index of 0.45, corresponding to 3-level operation are shown in Fig. 10. Plot of T_{ga} is shown in Fig. 10(a). The gating signal used to drive the inverters are shown in Fig. 10(b). The pole voltage waveform of the inverters are shown in Fig. 10(c). Phase voltage and motor current for A-Phase are shown in Fig. 10(d) and Fig. 10(e). Experimental results for a modulation index of 0.75, corresponding to 4-level operation are shown in Fig. 11. Plot of T_{ga} is shown in Fig. 11(a). Gating signals shown in Fig. 11(b). The pole voltage waveform of the inverters are shown in Fig. 11(c). Phase voltage and motor current for A-Phase are shown in Fig. 11(d) and Fig. 11(e). The experimental results for over modulation region (modulation index=1.1) are shown in Fig. 12. Fig. 12(a) shows plot of T_{ga}. Gating signals for inverters are shown in Fig. 12(b). Pole voltages are shown in Fig. 12(c). Phase voltage and motor current are shown in Fig. 12(d) and Fig. 12(e).

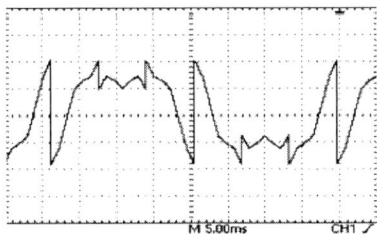

Figure 10(a).   Plot of T_{ga} for 3-Level operation
(Modulation index 0.45)
Scale: X-axis: 5ms/div
Scale: Y-axis: 2V/div

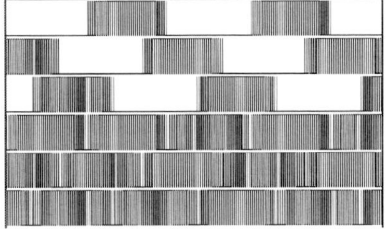

Figure 10(b).   Gating Signal for Inverters for 3- level Operation.
Upper Three Traces for Inverter-1
Lower Three Traces for Inverter-2
(Modulation index 0.45)

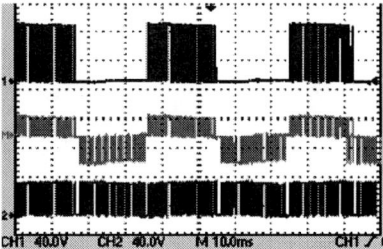

Figure 10(c). Experimental waveforms of
Pole Voltage for Modulation index 0.45
Upper Trace Pole Voltage of Inverter-1 (V_{AO})
Lower Trace Pole Voltage of Inverter-2 ($V_{A'O'}$)
Middle Trace Effective Pole Voltage ($V_{AO} - V_{A'O'}$)
Scale: X-axis:10ms/div; Y-axis: 40V/div

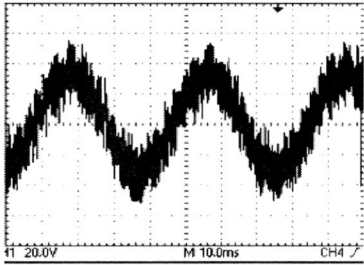

Figure 10(d). Experimental waveform of
Phase Voltage for Modulation index 0.45
Scale: X-axis:10ms/div; Y-axis: 20V/div

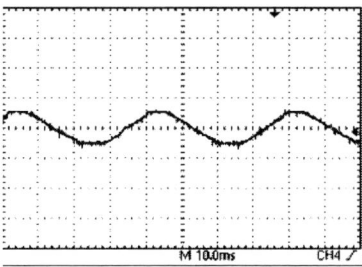

Figure 10(e). Experimental waveform of
Motor Current for Modulation index 0.45
Scale: X-axis:10ms/div

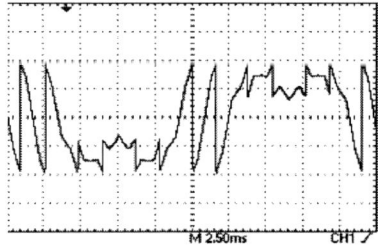

Figure 11(a). Plot of T_{ga} for 4-Level operation
(Modulation index 0.75)
Scale: X-axis: 2.5ms/div
Scale: Y-axis: 2V/div

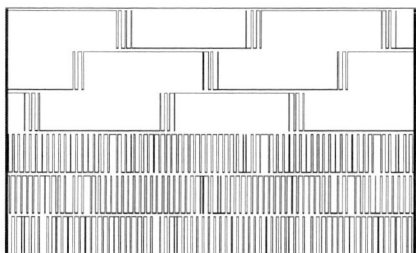

Figure 11(b). Gating Signal for Inverters for
4- level Operation.
Upper Three Traces for Inverter-1
Lower Three Traces for Inverter-2
(Modulation index 0.75)

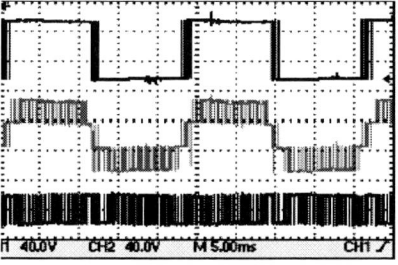

Figure 11(c). Experimental waveforms of
Pole Voltage for Modulation index 0.75
Upper Trace Pole Voltage of Inverter-1 (V_{AO})
Lower Trace Pole Voltage of Inverter-2 ($V_{A'O'}$)
Middle Trace Effective Pole Voltage ($V_{AO} - V_{A'O'}$)
Scale: X-axis: 5ms/div; Y-axis: 40V/div

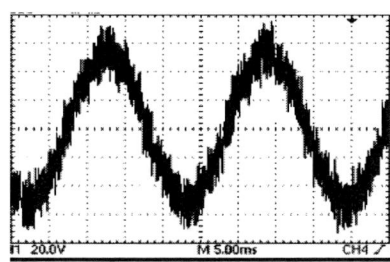

Figure 11(d). Experimental waveform of
Phase Voltage for Modulation index 0.75
Scale: X-axis: 5ms/div; Y-axis: 20V/div

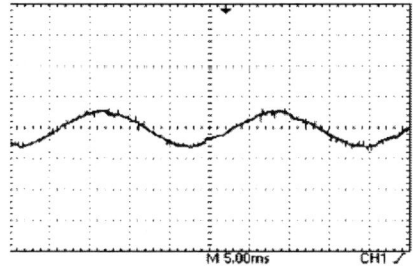

Figure 11(e). Experimental waveform of
Motor Current for Modulation index 0.75
Scale: X-axis: 5ms/div

978-1-4244-4782-4/10 $26.00 © 2010 IEEE

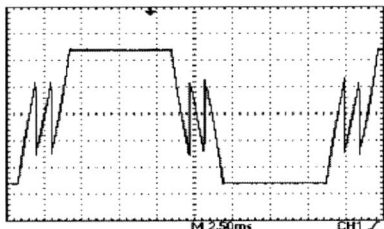

Figure 12(a). Plot of Tga for over modulation operation (Modulation index 1.1)
Scale: X-axis: 2.5ms/div
Scale: Y-axis: 2V/div

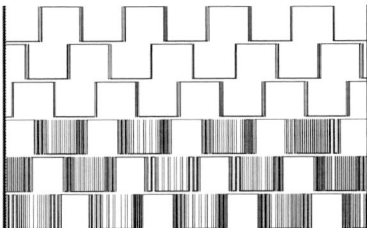

Figure 12(b). Gating Signal for Inverters for Over modulation Operation.
Upper Three Traces for Inverter-1
Lower Three Traces for Inverter-2
(Modulation index 1.1)

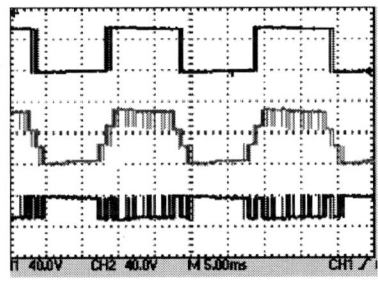

Figure 12(c). Experimental waveforms of Pole Voltage for Modulation index 1.1
Upper Trace Pole Voltage of Inverter-1(V_{AO})
Lower Trace Pole Voltage of Inverter-2 ($V_{A'O'}$)
Middle Trace Effective Pole Voltage (V_{AO} - $V_{A'O'}$)
Scale: X-axis: 5ms/div; Y-axis: 40V/div

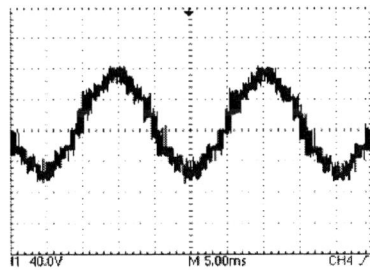

Figure 12(d). Experimental waveforms of Phase Voltage for Modulation index 1.1
Scale: X-axis: 5ms/div; Y-axis: 40V/div

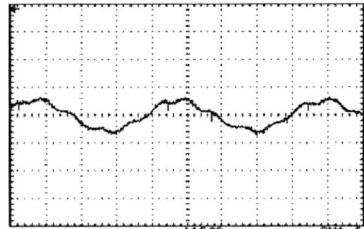

Figure 12(e). Experimental waveform of Motor Current for Modulation index 1.1
Scale: X-axis: 5ms/div

V. CONCLUSION

A simple SVPWM scheme for a 4-level inverter is proposed. 4-level inverter is realized by connecting two 2-level inverters in open-end winding configuration. The inverters use asymmetrical DC link voltages to realize 4-level operation. The proposed scheme does not require sector identification to find the sub hexagon, which encloses the tip of the reference vector. The sub hexagon center for 3-level operation is determined directly from the instantaneous magnitude of 3-Phase reference sinusoid. The proposed scheme is implemented and tested with 4-level inverter in open-end winding configuration and the experimental results are presented.

REFERENCES

[1] Nabae, I.Takahashi and H. Akagi, "A New Neutral Point Clamped PWM Inverter", IEEE Transactions on Industry Applications, vol.1A-17, No.5, September/October 1981, pp 518-523.

[2] Jih - Sheng Lai and Fang Zheng Peng, "Multilevel Converters - A New Breed of Power Converters", Proc. IAS'95 Conf., pp 2348-2356.

[3] Gautam Sinha and Thomas A Lipo, "A Four Level Rectifier – Inverter System for Drive Applications", ", IEEE Transactions on Industry Applications, vol.30, , July/August 1994, pp 938-944.

[4] N.Celanovic and Dushan Boroyevich, " A Fast Space-Vector Modulation Algorithm for Multilevel Three-Phase Converters",IEEE Transactions on Industry Applications, vol.37,No.2, March/April 2001,pp 637-641.

[5] Shivkumar E.G, Gopakumar. K, Sinha S.K, Andre Pittet, and Ranganathan V.T," Space Vector PWM control of dual inverter fed open- end winding induction motor drive", Proc. Applied Power Electronics Conf (APEC), 2001, pp 399-405.

[6] Shivkumar E.G, Somasekhar V.T, Krushna K, Mohapatra K, Gopakumar. K, Umanand L, and S.K. Sinha," A multi level space phasor based PWM strategy for an open- end winding induction motor drive using two inverters with different DC link voltages", Proc. IEEE PEDS 2001, pp 169-175.

[7] P.F Seixas, M.A. Severo Mendes, P.Donoso Garcia, A.M.N. Lima, " A Space Vector PWM Method for Three-Level Voltage Source Inverters", Proc. APEC, vol.1, 2000, pp 549-555.

[8] Heinz Willi Van Der Broeck, Hans-Christoph Skudelny, and Georg Viktor Stanke, " Analysis and Realization of a Pulsewidth Modulator Based on Voltage Space Vectors", IEEE Transactions on Industry Applications, vol.24, No.1, January/February 1988, pp 142-150.

[9] A.K.Gupta and A.M Khambadkone " A two-level inverter based SVPWM algorithm for a multilevel inverter", 30th Annual Conference of the IEEE Industrial Electronics Society, November 2-6, 2004,pp 1823-1828.

[10] A.K.Gupta and A.M Khambadkone " A Space Vector PWM Scheme for Multilevel Inverters Based on Two-Level Space Vector PWM", IEEE Transactions on Industrial Electronics, vol.53,No.5, October 2006,pp 1631-1639.

[11] Jae Hyeong Seo,Chang Ho Choi and Dong Seok Hyun, " A New Simplified Space Vector PWM Method for Three-Level Inverters", IEEE Transactions on Power Electronics, vol.16,No.4, July 2001,pp 545-550.

[12] Yo Han Lee, Bum-Seok Suh and Dong- Seok Hyun, " A Novel PWM Scheme for a Three-Level Voltage a Source Inverter with GTO Thyristors", IEEE Transactions on Industry Applications, vol.32,No.2, March /April1996,pp 260-268.

[13] Joohn-Sheok Kim, Seung-Ki Sul " A Novel Voltage Modulation Technique of the Space Vector PWM", IPEC Yokohama-95,pp 742-747.

[14] Anish Gopinath, Aneesh Mohamed A.S and M.R.Baiju, " Fractal Based Space Vector PWM for Multilevel Inverters – a Novel Approach", IEEE Transactions on Industrial Electronics, vol.52,No. 4, April 2009, pp 1230-1238.

[15] Aneesh Mohamed A.S, Anish Gopinath and M.R.Baiju, " A Simple Space Vector PWM Generation for any General N-Level Inverter", IEEE Transactions on Industrial Electronics, vol.56, No.5, May 2009, pp 1649-1656.

[16] S.Srinivas and V.T.Somasekhar, "Space vector based PWM switching strategies for a three - level dual-inverter fed open end winding induction motor drive and their comparative evaluation" IET- Electr.Power Appl., Vol 2, No. 1, January 2008, pp 19-31.

Control method for a novel converter topology for permanent magnet drives

Philip Brockerhoff, Martin Schulz
Institute for Power Electronics and Control
Universität der Bundeswehr München
Munich, Germany
Philip.Brockerhoff@unibw.de, Martin.Schulz@unibw.de

Abstract— **The paper describes a new converter concept for permanent magnet drives and its control. It is called multilevel integrated step-up (MIS) converter, because the new topology enables output voltages higher than the DC link voltage and a multilevel-like output voltage waveform, too. DC/DC converters to stabilize and increase the DC link voltage are not necessary, since DC-link voltage variations can be compensated by control of the MIS converter. The converter comprises internal switched "boost-capacitors". The machine inductance is used to generate a step-up conversion for the internal capacitor voltage. These points lead to additional degrees of freedom for controlling the system and optimization of important parameters e.g. current ripple or switching frequency. In comparison to standard flying capacitor topologies[1] no upper limit exists for the internal capacitor voltage. This is a unique feature of the MIS converter. The extended operating range is explained with respect to different machine characteristics with a low and high short circuit current ratio. The control scheme shows different ways to control the internal capacitor voltage in relation to machine's power and speed. Additionally, design considerations for the size of the internal boost capacitor are presented.**

I. INTRODUCTION

For mobile applications, like automotive and aerospace, there is a growing demand for new electric drive systems with very high power density [2, 3]. These drives need a high switching frequency and generally benefit from a multilevel output voltage. The permanent magnet (PM) synchronous machine is well suited for high power density and torque density[4, 5, 6]. To achieve the higher power densities, there is a need to increase the pole pair number of the machine. This increases the fundamental frequency of the machine and imposes high switching frequencies, when using standard

converters. Because the switching losses of power electronics in IGBT technology is become dominant above 20kHz, a new converter topology is introduced and studied.

In addition, the common DC/DC chopper for the energy storage to control the DC link voltage adds extra weight to the system.

The MIS converter generates a multilevel output voltage with possible amplitudes higher than the DC-link voltage. This enables an extended operating range, makes the DC/DC converter obsolete and reduces losses and torque ripple of the machine. A basic dimensioning procedure for a drive system design with the new converter was given in [7]. This paper will now focus on the extended operating range of the system and the control concept. Measurement examples are given for driving a passive load and a PM machine.

II. VOLTAGE GENERATION

A. General description

The new converter topology is shown in fig. (1). In the gray part the new converter leg with additional internal capacitor replaces a half bridge. It can be seen from table (1) that in certain switching states of S_2 and S_3, the internal

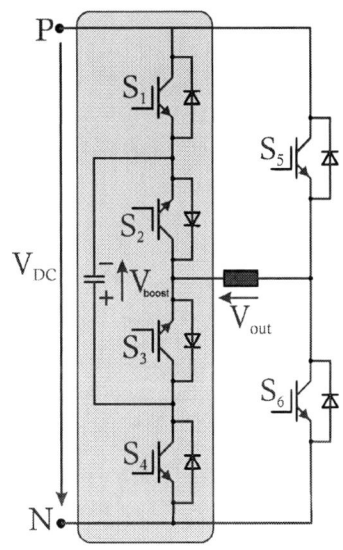

Figure 1. New converter concept for one phase

TABLE I. SWITCHING STATES

State No.	Switches						Output Voltage
	6	5	4	3	2	1	V_{out}
0	0	0	0	0	0	0	-
1	0	1	1	0	1	0	$V_{DC} + V_{boost}$
2	0	1	1	1	0	0	V_{DC}
3	0	1	0	1	0	1	$-V_{boost}$
4	0	1	0	0	1	1	0
5	1	0	1	0	1	1	V_{boost}
6	1	0	1	1	0	0	0
7	1	0	0	1	0	1	$-V_{DC} - V_{boost}$
8	1	0	0	0	1	1	$-V_{DC}$

978-1-4244-4782-4/10 $26.00 © 2010 IEEE

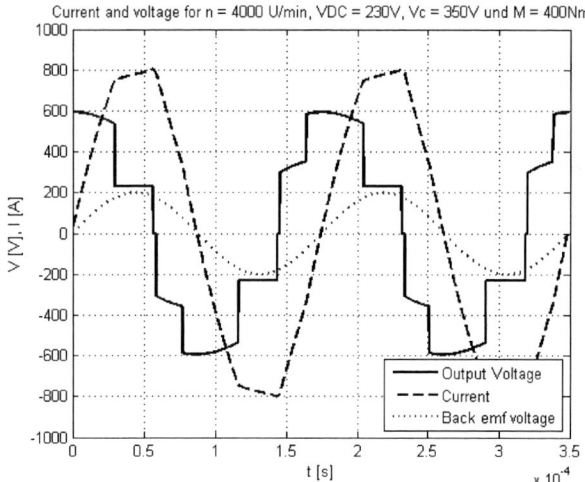

Figure 2. Output current waveform for high speed and high load with MIS converter.

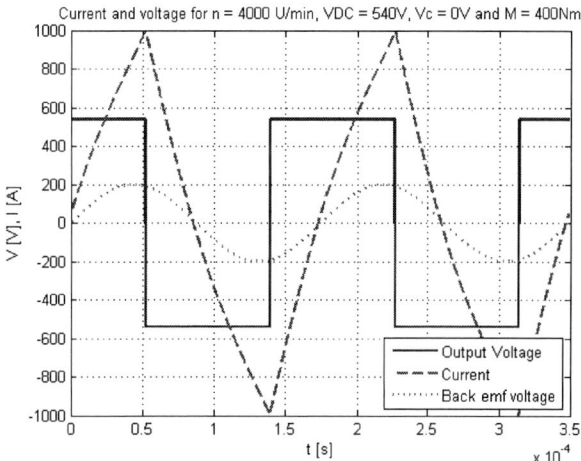

Figure 3. Output current waveform for high speed and high load with state of the art converter.

capacitor voltage (V_{boost}) is connected in series with the DC link voltage (V_{DC}). A seven level voltage form is applied to the load.

This even improves more, when it is taken into account, that the amplitude of the internal capacitor voltage is adjusted for every operating point regarding speed and torque. This leads to new degrees of freedom for the control concept.

B. Characteristic of the internal capacitor voltage

It has to be taken into account, that the internal capacitor cannot supply real power to the load and should be small in size. Therefore, the internal capacitor has to be switched symmetrically to the current zero crossing of the motor current in order to stabilize its voltage in steady state operation.

The amplitude of the internal capacitor voltage is controlled regarding speed and torque demand of the machine as shown later. Its voltage can be adjusted to any desired

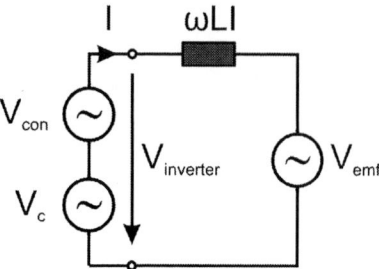

Figure 4. New converter concept for one phase

value, dynamically by using the machine's inductance for step up conversion. Thus the internal capacitor voltage can be even higher than the DC link voltage.

The internal capacitor voltage acts as a series voltage compensation of the machines reactive power. From tab. 1 there is no output state of $|V_{DC}|-|V_{boost}|$. V_{DC} has to deliver at high load/speed some of the reactive power, because at high duty cycle the voltage blocks would overlap.

C. No undesired feedback of energy to the dc link

The internal capacitor prevents the machine to feed undesired energy into the DC link, because the capacitor is always connected in series with the DC link. This is especially important for an idle running PM machine. The internal capacitors are loaded to:

$$V_{boost} \geq \frac{n}{n_0}\hat{V}_{emf} - V_{DC} \qquad (1)$$

n is the speed of the machine in rpm and V_{emf} the back emf rms voltage.

D. Output voltage example at high speed

Fig. (2) shows an output voltage example calculated in matlab for a machine with high pole pair number (large X, seen later) at high speed and high load. It can be seen that the current waveform is still acceptable even though the switching frequency is low compared to the fundamental frequency (f_1). No additional field weakening current is necessary and the motor current can be kept in phase with the back emf voltage [8].

To achieve the same power without the additional capacitor and without field weakening current, the DC link voltage has to be increased to 540V and the waveform becomes triangular (fig. (3)).

The switching frequency for this pulse pattern is f_1 for S1 to S_4 and $3 \cdot f_1$ for S_5 and S_6.

E. Operation at low speed

At start up and low speed the MIS converter is operated like a common full bridge converter. The pulse frequency is high compared to f_1 and the necessity for reactive power is comparatively low. V_{boost} is controlled to 0V and is kept discharged by appropriate switching patterns: S_2 is switched

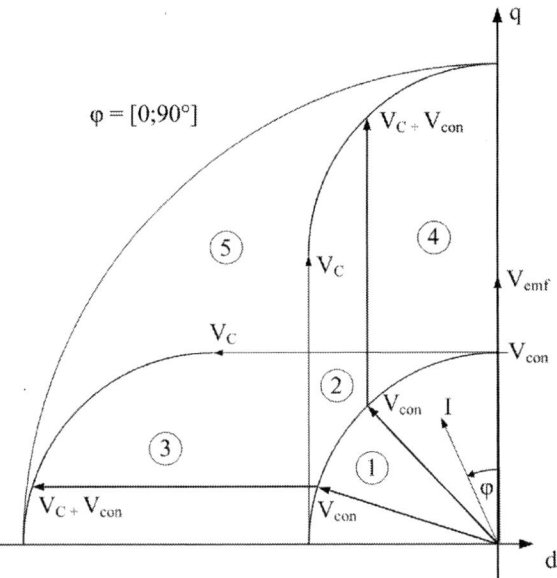

Figure 5. Vector diagram (dq) / operating area for new converter in motor mode.

on for positive current I or S_3 is switched on for negative I. This controlling mode does not add additional switching losses to the system, since the switches S_2 and S_3 are in zero current switching mode.

The converter benefits from the multilevel output waveform at low speed too, since a state of the art converter needs a higher DC link voltage to run the machine at high speed. This leads to higher ripple current at low speed or a higher switching frequency and larger losses.

III.   OPERATING AREA

The MIS converter generates an output voltage consisting of two components (fig. (4)): V_{con} is the rms component generated from V_{DC}. It can be active and reactive power. V_c is the rms component generated from V_{boost}. It can supply reactive power, only. Both voltage vectors add to the voltage $V_{inverter}$, which is the rms output voltage of the new MIS converter.

$$\vec{V}_{inverter} = \vec{V}_{con} + \vec{V}_c = \vec{V}_{emf} + \vec{V}_l \qquad (2)$$

With V_l the rms component of the inductance voltage $\omega L I$.

A.   Design considerations

As depicted in fig. (5) the operating range for a given DC link voltage is increased by the internal capacitor voltage V_c component. The machine is in motor operation. Therefore, the current is in the second quadrant, and back emf voltage V_{emf} is in the q-axis. For $\varphi = 0°$, V_c adds horizontal to V_{con} and extends the operating range to the left to area (3). If φ is about 90°, V_c adds vertical to V_{con} and extends the operating range to the top (area (4)).

Area (3) is considered as the high power region, since V_l is large and proportional to the machine's power and area (4) is the high speed region, because V_{emf} is large and proportional to speed.

There is no useful motor operation in area (5), because the motor current would be in quadrant 1, which would increase the motor voltage with a positive d current or in quadrant 3 and 4, which means generator operation. This is not topic of this paper, but generator operation is possible with the new converter topology, too.

Defining a parameter X:

$$X = \frac{L_{ind} \cdot I}{\Psi_{PM}} = \frac{I}{I_{short}} \qquad (3)$$

with L_{ind} the inductance of the machine, I the maximum current, Ψ_{PM} machine flux and I_{short} the short circuit current. This factor is used to distinguish between different machine classes to find their operating area in fig. (5).

For machines with small X the typical design will be an operating point in area (4) in fig. (5). For a machine with a high X, the DC link voltage will be small to achieve the desired power. Therefore, the typical operating point will be in area (3). For an optimal system design, the winding number of the machine can be changed to optimally utilize the converter for a given system voltage.

$$X = \frac{L_{ind} \cdot \left(\dfrac{w}{w_0}\right)^2 \cdot I \cdot \dfrac{w_0}{w}}{\Psi_{PM} \cdot \dfrac{w}{w_0}} \qquad (4)$$

By raising the winding number w in a machine the V_{emf} increases linearly and L_{ind} of the machine increases quadratic. The current I decreases linearly with w for constant power.

Therefore, X is independent of w. For machines with $X > 1$, the short circuit current is small compared to the nominal current.

In order to design a system for the new converter, the winding number will be designed to run the system with minimum current for the required machine power and DC link voltage. Therefore, the winding number will be chosen to have a back emf voltage in the range of the DC link at maximum speed ($V_{emf} = V_{DC}$) for a machine with large X or for the real power ($P = V_{DC} \cdot I$) for a machine with small X. The exact design rule for small X is derived from the vector diagram to

$$V_{emf}^{2} = V_{con}^{2} - \left(\left|V_c\right| + \left|V\right|_l\right)^2 \qquad (5)$$

978-1-4244-4782-4/10 $26.00 © 2010 IEEE   1972

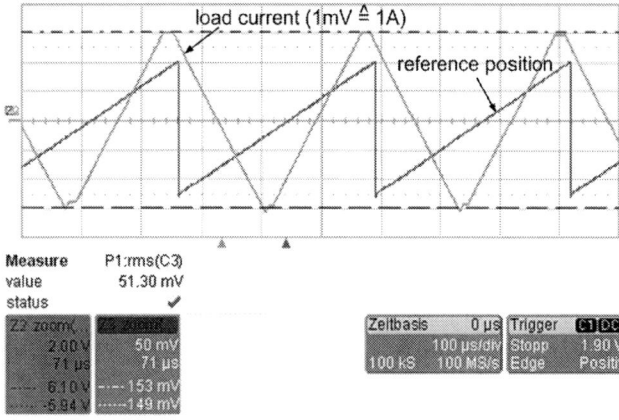

Figure 6. Measurement for passive load without internal capacitor.

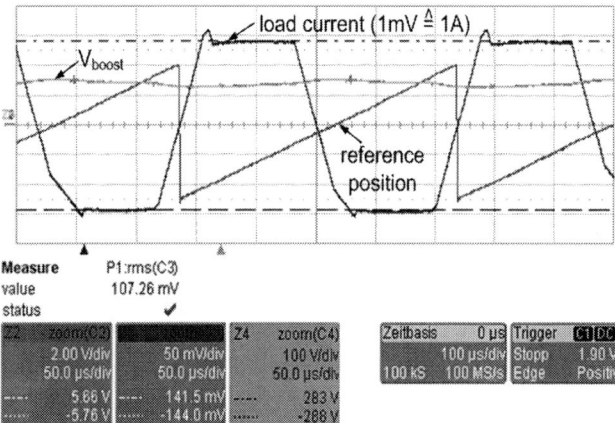

Figure 7. Measurement for passive load with new topology and internal capacitor.

B. Output example

Fig. (6) shows a measurement for an inductive load driven only by the DC link voltage for a block shaped target input.

Fig. (7) depicts how the additional internal capacitor voltage increases the current. In this example, V_{boost} is 120V and V_{DC} is 80V. The achievable rms current value is more than doubled without installing an additional DC/DC step up converter to raise the DC link voltage. The maximum di/dt is increased by 2.5 times. The slight slope on V_{boost} is generated by the charge and discharge cycle during every commutation.

IV. CONTROL

Control of V_{boost} should be chosen independent of torque, because load variations can vary quickly and it is not possible to change the capacitor voltage V_{boost} within one period substantially. Its amplitude is controlled proportional to speed, which does not vary within a few electric periods. Its value is calculated to be able to deliver always maximum power. This minimum case gives a linear map starting from a speed, where the DC link voltage is not sufficient any more to rise linearly to maximum speed.

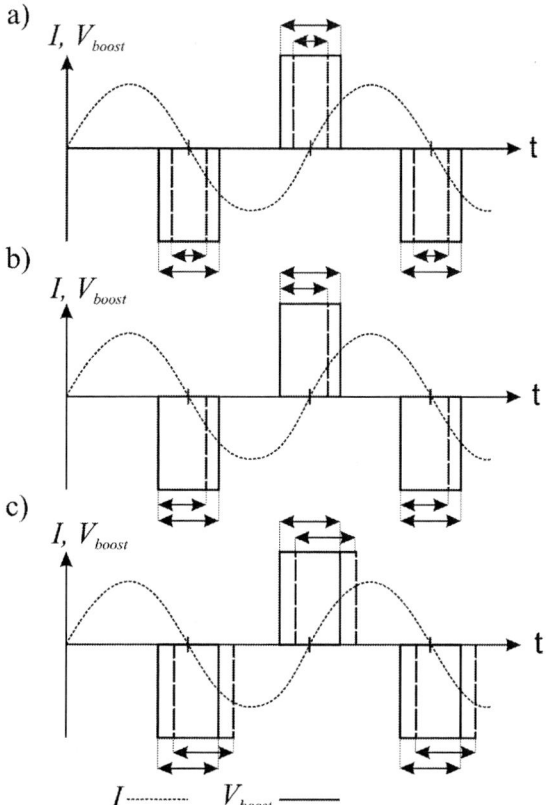

Figure 8. Different methods to control V_{boost}: a) Adjust the length of V_c by the duty cycle of V_{boost} b) Control V_{boost} by switching the block not symmetrical c) Control of V_{boost} by shifting the block of V_{boost}.

The length of V_c (vector diagram) can be adjusted dynamically by the duty cycle of the capacitor bridge for each electrical period as seen in fig. (8a). This is similar to a pulse width modulation strategy.

The amplitude of V_{boost} can be controlled in different ways. As shown in fig. (8b) it is possible to switch off the internal capacitor, if a certain voltage level is reached. A second method is to switch the V_{boost} block earlier or later as shown in fig. (8c). For the machine control, a V_{boost} block is calculated by aid of the machine model.

For control of the amplitude of the internal capacitor voltage, the rising edge of this block will be switched earlier for loading and the falling edge will be switched later for discharging.

The adaptation to speed adds a new degree of freedom to the seven levels of the new topology: Pulse amplitude modulation (PAM), which further minimizes total harmonic distortion (THD) of the output current.

A good practice to control V_{boost} is to have a fixed starting angle of the capacitor calculated by the machine model. The capacitor voltage amplitude control is implemented as a superimposed control for the switch off angle of the V_{boost}

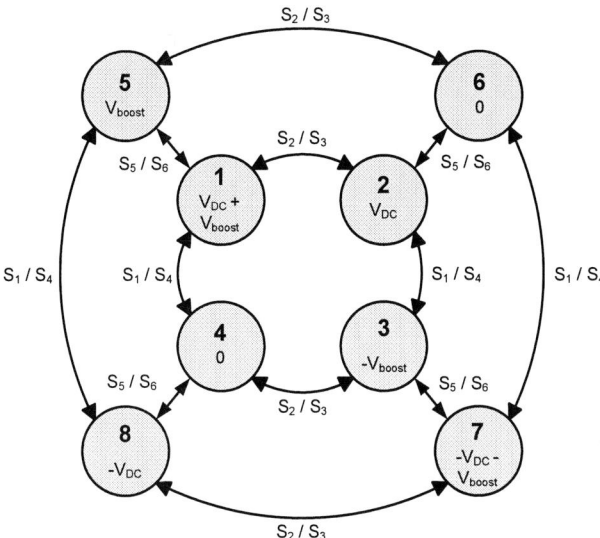

Figure 9. Output State machine for new converter topology.

block. The actuating element is limited to 15° and an anti reset wind-up element is implemented.

The DC link component is controlled with a hysteresis controller with feed forward control. It can react on small current disturbances caused by the internal capacitor amplitude control very quickly.

This control topology guarantees a superior dynamic behavior on machine current and DC link voltage disturbances. They are corrected online without the common dead time of one pulse width for pulse width modulated control algorithms. By the hysteresis bandwidth the current ripple and switching frequency can be controlled for stationary operation. If transients occur, the control can react immediately without deadtime and increasing the switching frequency for a short time until the transients decay.

A. State machine of the new converter

It has to be taken into account from the state machine in fig. (9), that some switching operations are used, in order to keep the capacitors discharged, when needed. E.g. going from state 4 (zero output voltage) to state 2 (V_d) the capacitor will be connected to the system during the dead time of the IGBTs for some microseconds and is charged over the free-wheeling diodes depending on the current direction.

Small IGBT dead times reduce this effect, but it has to be taken into account for the control strategy. This imposes an online control system that can switch the IGBT at a certain capacitor amplitude. An FPGA control, as implemented in the test bench, is recommended.

Fig. 10 shows a measurement with the MIS converter and a PM-machine with the electrical characteristics shown in tab. II at 2000rpm. Notice the short peak of V_{boost}, when the converter changes from state 2 (V_{DC}) to state 2 (0). For clarity only one of the two phases is shown.

TABLE II. CHARACTERISTICS OF PM MOTOR

Maximum Power	P_0	10kW
Maximum Torque between 0 and 2300 rpm	M_0	600Nm
Nominal Inductance (4 windings)	$L_0 = L_q = L_d$	24µH
Peak EMF at 4000rpm (4 windings)	V_{emf0}	170V
Maximum speed	n_0	4000rpm
Number of phases		2

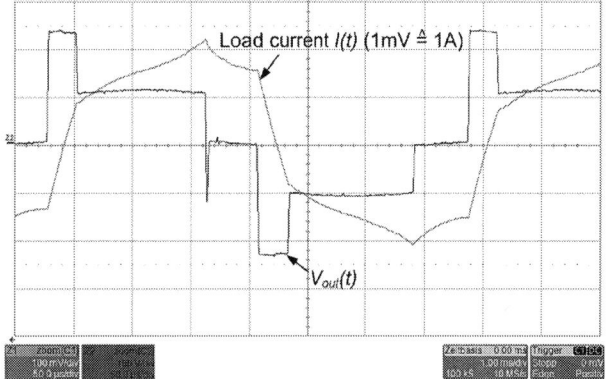

Figure 10. MIS converter with PM machine, $\hat{I} = 210$A, $V_{out} = 230$V.

The V_{boost} adds as a second voltage component during commutation to V_{DC} and increases the di/dt.

B. Size of the internal capacitor

C is sized related to I and the maximum voltage ripple in the capacitor V_{ripple}. For a given 600V IGBT, $V_{boostmax}$ is about 400V and a ripple of 20%. For a machine current of 1000A C is calculated to:

$$C = \frac{I \dfrac{\alpha}{360 \cdot f_p}}{V_{ripple}} = \frac{1000A \dfrac{180°}{360° \cdot 5kHz}}{80V} = 1.25mF \quad (6)$$

The energy stored in C is 100J at 400V. Compared to the DC-link for the desired machine power of 150kW this is negligible. The amplitude of the capacitor can be changed quickly.

Also the DC link capacitor is sized smaller with the new topology, since it has only to deliver the real power and not the apparent power. Additionally the capacitor voltage is smaller compared to state of the art converters, which allows the use of capacitors with lower voltage. Thus the additional capacitance for the whole system is smaller than the internal capacitor C.

The internal capacitor has to carry a large current ripple in the explained dimensioning. PCC foil capacitors or other technologies with low equivalent series resistance (ESR) are recommended for the MIS converter.

V. CONCLUSION

The new multilevel integrated step-up (MIS) converter topology offers new degrees of freedom in machine design and control.

It renders an additional DC/DC converter to increase the DC link voltage [9] unnecessary. The feature to control the amplitude of the internal capacitor voltage improves the seven level output voltage and adds a pulse amplitude modulation (PAM) characteristic to the converter control.

Special care has to be taken for the time critical control of the internal capacitor voltage, which is best achieved with an control implementation in a FPGA to guarantee fast online control with short deadtimes.

An additional field weakening current, which increases losses in the machine [6] can be avoided in the high speed operation (area (4)). The possible power of the machine is increased for operation in area (3)[8].

These benefits make the new MIS converter suitable for permanent magnet synchronous machines in mobile applications, for instance aerospace or automotive.

REFERENCES

[1] Hochgraf, C.; Lasseter, R.; Divan, D.; Lipo, T.A., "Comparison of multilevel inverters for static VAr compensation," Industry Applications Society Annual Meeting, 1994., Conference Record of the 1994 IEEE , vol., no., pp.921-928 vol.2, 2-6 Oct 1994

[2] Holl, E.; Neumann, G.; Piepenbreier, B.; Tolle, H.-J., "Water-cooled inverter for synchronous and asynchronous electric vehicle drives," Power Electronics and Applications, 1993., Fifth European Conference on , vol., no., pp.289-293 vol.5, 13-16 Sep 1993

[3] Nayeem Hasan S.M.; Husain Iqbal, "Power electronic interface with ultracapacitors and motor control for a fuel cell electric vehicle," Vehicle Power and Propulsion, 2005 IEEE Conference , vol., no., pp.815-822, 2005

[4] Hackmann, W.; „Systemvergleich unterschiedlicher Radnabenantriebe für den Schienennahverkehr: Asynchronmaschine, permanenterregte Synchronmaschine, Transversalflussmaschine" in german, dissertation Technische Universität Darmstadt, 2003

[5] Schneider, T.; Koch, T.; Binder, A., "Comparative analysis of limited field weakening capability of surface mounted permanent magnet machines," Electric Power Applications, IEE Proceedings - , vol.151, no.1, pp. 76-82, 9 Jan. 2004

[6] Hofmann, W.; Paul, M.; Tenberge, P., "Automatic gearbox continuously controlled by electromagnetic and electronic power converter," Power Electronics Specialists Conference, 2000. PESC 00. 2000 IEEE 31st Annual , vol.1, no., pp.521-526 vol.1, 2000

[7] Brockerhoff, P.; Ebert, M.; Marquardt, R., "A novel converter topology for permanent magnet drive systems," Power Electronics Specialists Conference, 2008. PESC 2008. IEEE , vol., no., pp.4657-4661, 15-19 June 2008

[8] Chen, C.H.; Cheng, M.Y., "Design and implementation of a high-performance bidirectional DC/AC converter for advanced EVs/HEVs," Electric Power Applications, IEE Proceedings - , vol.153, no.1, pp. 140-148, 1 Jan. 2006

[9] Nam, K, "Power Factor Selection for EV Traction Motors", Korea-Germany Joint Symposium on Power Electronics, Oct 2007

A Voltage Controlled Adjustable Speed PMBLDCM Drive using A Single-Stage PFC Half-Bridge Converter

Sanjeev Singh, Student *Member, IEEE*
e-mail: sschauhan.sdl@gmail.com

Bhim Singh, *Senior Member, IEEE*
e-mail: bhimsinghr@gmail.com

Electrical Engineering Department, Indian Institute of Technology Delhi, New Delhi -110016, India

Abstract— In this paper, a buck half-bridge DC-DC converter is used as a single-stage power factor correction (PFC) converter for feeding a voltage source inverter (VSI) based permanent magnet brushless DC motor (PMBLDCM) drive. The front end of this PFC converter is a diode bridge rectifier (DBR) fed from single-phase AC mains. The PMBLDCM is used to drive a compressor load of an air conditioner through a three-phase VSI fed from a controlled DC link voltage. The speed of the compressor is controlled to achieve energy conservation using a concept of the voltage control at DC link proportional to the desired speed of the PMBLDCM. Therefore the VSI is operated only as an electronic commutator of the PMBLDCM. The stator current of the PMBLDCM during step change of reference speed is controlled by a rate limiter for the reference voltage at DC link. The proposed PMBLDCM drive with voltage control based PFC converter is designed, modeled and its performance is simulated in Matlab-Simulink environment for an air conditioner compressor driven through a 1.5 kW, 1500 rpm PMBLDC motor. The evaluation results of the proposed speed control scheme are presented to demonstrate an improved efficiency of the proposed drive system with PFC feature in wide range of the speed and an input AC voltage.

Index Terms— PFC, PMBLDCM, Air conditioner, Buck Half-bridge converter, Voltage control, VSI.

I. INTRODUCTION

PERMANENT magnet brushless DC motors (PMBLDCMs) are preferred motors for a compressor of an air-conditioning (Air-Con) system due to its features like high efficiency, wide speed range and low maintenance requirements [1-4]. The operation of the compressor with the speed control results in an improved efficiency of the system while maintaining the temperature in the air-conditioned zone at the set reference consistently. Whereas, the existing air conditioners mostly have a single-phase induction motor to drive the compressor in 'on/off' control mode. This results in increased losses due to frequent 'on/off' operation with increased mechanical and electrical stresses on the motor, thereby poor efficiency and reduced life of the motor. Moreover, the temperature of the air conditioned zone is regulated in a hysteresis band. Therefore, improved efficiency of the Air-Con system will certainly reduce the cost of living

and energy demand to cope-up with ever-increasing power crisis.

A PMBLDCM which is a kind of three-phase synchronous motor with permanent magnets (PMs) on the rotor and trapezoidal back EMF waveform, operates on electronic commutation accomplished by solid state switches. It is powered through a three-phase voltage source inverter (VSI) which is fed from single-phase AC supply using a diode bridge rectifier (DBR) followed by smoothening DC link capacitor. The compressor exerts constant torque (i.e. rated torque) on the PMBLDCM and is operated in speed control mode to improve the efficiency of the Air-Con system.

Since, the back-emf of the PMBLDCM is proportional to the motor speed and the developed torque is proportional to its phase current [1-4], therefore, a constant torque is maintained by a constant current in the stator winding of the PMBLDCM whereas the speed can be controlled by varying the terminal voltage of the motor. Based on this logic, a speed control scheme is proposed in this paper which uses a reference voltage at DC link proportional to the desired speed of the PMBLDC motor. However, the control of VSI is only for electronic commutation which is based on the rotor position signals of the PMBLDC motor.

The PMBLDCM drive, fed from a single-phase AC mains through a diode bridge rectifier (DBR) followed by a DC link capacitor, suffers from power quality (PQ) disturbances such as poor power factor (PF), increased total harmonic distortion (THD) of current at input AC mains and its high crest factor (CF). It is mainly due to uncontrolled charging of the DC link capacitor which results in a pulsed current waveform having a peak value higher than the amplitude of the fundamental input current at AC mains. Moreover, the PQ standards for low power equipments such as IEC 61000-3-2 [5], emphasize on low harmonic contents and near unity power factor current to be drawn from AC mains by these motors. Therefore, use of a power factor correction (PFC) topology amongst various available topologies [6-14] is almost inevitable for a PMBLDCM drive.

Most of the existing systems use a boost converter for PFC as the front-end converter and an isolated DC-DC converter to produce desired output voltage constituting a two-stage PFC

978-1-4244-4782-4/10 $26.00 © 2010 IEEE

drive [7-8]. The DC-DC converter used in the second stage is usually a flyback or forward converter for low power applications and a full-bridge converter for higher power applications. However, these two stage PFC converters have high cost and complexity in implementing two separate switch-mode converters, therefore a single stage converter combining the PFC and voltage regulation at DC link is more in demand. The single-stage PFC converters operate with only one controller to regulate the DC link voltage along with the power factor correction. The absence of a second controller has a greater impact on the performance of single-stage PFC converters and requires a design to operate over a much wider range of operating conditions.

For the proposed voltage controlled drive, a half-bridge buck DC-DC converter is selected because of its high power handling capacity as compared to the single switch converters. Moreover, it has switching losses comparable to the single switch converters as only one switch is in operation at any instant of time. It can be operated as a single-stage power factor corrected (PFC) converter when connected between the VSI and the DBR fed from single-phase AC mains, besides controlling the voltage at DC link for the desired speed of the Air-Con compressor. A detailed modeling, design and performance evaluation of the proposed drive are presented for an air conditioner compressor driven by a PMBLDC motor of 1.5 kW, 1500 rpm rating.

II. PROPOSED SPEED CONTROL SCHEME OF PMBLDC MOTOR FOR AIR CONDITIONER

The proposed speed control scheme (as shown in Fig. 1) controls reference voltage at DC link as an equivalent reference speed, thereby replaces the conventional control of the motor speed and a stator current involving various sensors for voltage and current signals. Moreover, the rotor position signals are used to generate the switching sequence for the VSI as an electronic commutator of the PMBLDC motor. Therefore, rotor-position information is required only at the commutation points, e.g., every 60°electrical in the three-phase [1-4]. The rotor position of PMBLDCM is sensed using Hall effect position sensors and used to generate switching sequence for the VSI as shown in Table-I.

The DC link voltage is controlled by a half-bridge buck DC-DC converter based on the duty ratio (D) of the converter. For a fast and effective control with reduced size of magnetics and filters, a high switching frequency is used; however, the switching frequency (f_s) is limited by the switching device used, operating power level and switching losses of the device. Metal oxide field effect transistors (MOSFETs) are used as the switching device for high switching frequency in the proposed PFC converter. However, insulated gate bipolar transistors (IGBTs) are used in VSI bridge feeding PMBLDCM, to reduce the switching stress, as it operates at lower frequency compared to PFC switches.

The PFC control scheme uses a current control loop inside the speed control loop with current multiplier approach which operates in continuous conduction mode (CCM) with average current control. The control loop begins with the comparison of sensed DC link voltage with a voltage equivalent to reference speed. The resultant voltage error is passed through

a proportional-integral (PI) controller to give the modulating current signal. This signal is multiplied with a unit template of input AC voltage and compared with DC current sensed after the DBR. The resultant current error is amplified and compared with saw-tooth carrier wave of fixed frequency (f_s) in unipolar scheme (as shown in Fig.2) to generate the PWM pulses for the half-bridge converter. For the current control of the PMBLDCM during step change of the reference voltage due to the change in the reference speed, a voltage gradient less than 800 V/s is introduced for the change of DC link voltage, which ensures the stator current of the PMBLDCM within the specified limits (i.e. double the rated current).

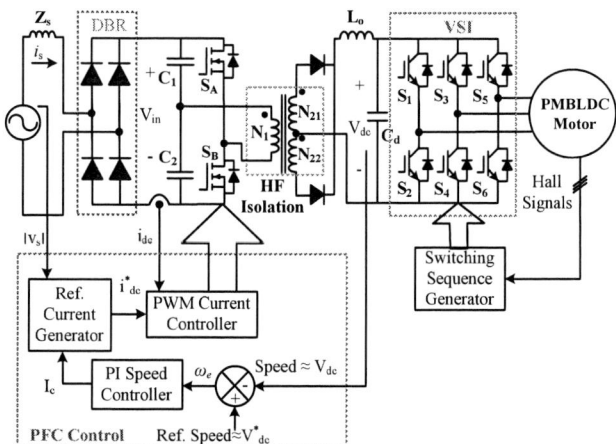

Figure 1. Control schematic of Proposed Bridge-buck PFC converter fed PMBLDCM drive

III. DESIGN OF PFC BUCK HALF-BRIDGE CONVERTER BASED PMBLDCM DRIVE

The proposed PFC buck half-bridge converter is designed for a PMBLDCM drive with main considerations on PQ constraints at AC mains and allowable ripple in DC link voltage. The DC link voltage of the PFC converter is given as,

$$V_{dc} = 2 (N_2/N_1) V_{in} D \text{ and } N_2 = N_{21} = N_{22} \qquad (1)$$

where N_1, N_{21}, N_{22} are number of turns in primary, secondary upper and lower windings of the high frequency (HF) isolation transformer, respectively.

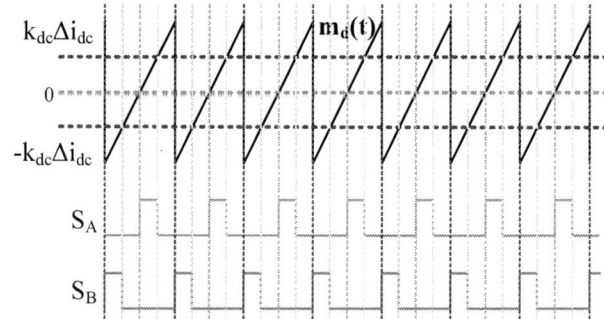

Figure 2. PWM control of the buck half-bridge converter

V_{in} is the average output of the DBR for a given AC input voltage (V_s) related as,

$$V_{in} = 2\sqrt{2}V_s/\pi \tag{2}$$

A ripple filter is designed to reduce the ripples introduced in the output voltage due to high switching frequency for constant of the buck half-bridge converter. The inductance (L_o) of the ripple filter restricts the inductor peak to peak ripple current (ΔI_{Lo}) within specified value for the given switching frequency (f_s), whereas, the capacitance (C_d) is calculated for a specified ripple in the output voltage (ΔV_{Cd}) [7-8]. The output filter inductor and capacitor are given as,

$$L_o = (0.5\text{-}D)V_{dc}/\{f_s(\Delta I_{Lo})\} \tag{3}$$

$$C_d = I_o/(2\omega\Delta V_{Cd}) \tag{4}$$

The PFC converter is designed for a base DC link voltage of $V_{dc} = 400$ V at $V_{in} = 198$ V from $V_s = 220$ Vrms. The turns ratio of the high frequency transformer (N_2/N_1) is taken as 6:1 to maintain the desired DC link voltage at low input AC voltages typically at 170V. Other design data are $f_s = 40$ kHz, $I_o = 4$ A, $\Delta V_{Cd} = 4$ V (1% of V_{dc}), $\Delta I_{Lo} = 0.8$ A (20% of I_o). The design parameters are calculated as $L_o = 2.0$ mH, $C_d = 1600$ μF.

TABLE I. VSI SWITCHING SEQUENCE BASED ON THE HALL EFFECT SENSOR SIGNALS

H_a	H_b	H_c	E_a	E_b	E_c	S_1	S_2	S_3	S_4	S_5	S_6
0	0	0	0	0	0	0	0	0	0	0	0
0	0	1	0	-1	+1	0	0	0	1	1	0
0	1	0	-1	+1	0	0	1	1	0	0	0
0	1	1	-1	0	+1	0	1	0	0	1	0
1	0	0	+1	0	-1	1	0	0	0	0	1
1	0	1	+1	-1	0	1	0	0	1	0	0
1	1	0	0	+1	-1	0	0	1	0	0	1
1	1	1	0	0	0	0	0	0	0	0	0

IV. MODELING OF THE PROPOSED PMBLDCM DRIVE

The main components of the proposed PMBLDCM drive are the PFC converter and PMBLDCM drive, which are modeled by mathematical equations and the complete drive is represented as a combination of these models.

A. PFC Converter

The modeling of the PFC converter consists of the modeling of a speed controller, a reference current generator and a PWM controller as given below.

1) Speed Controller: The speed controller, the prime component of this control scheme, is a proportional-integral (PI) controller which closely tracks the reference speed as an equivalent reference voltage. If at k^{th} instant of time, $V^*_{dc}(k)$ is reference DC link voltage, $V_{dc}(k)$ is sensed DC link voltage then the voltage error $V_e(k)$ is calculated as,

$$V_e(k) = V^*_{dc}(k) - V_{dc}(k) \tag{5}$$

The PI controller gives desired control signal after processing this voltage error. The output of the controller $I_c(k)$ at k^{th} instant is given as,

$$I_c(k) = I_c(k\text{-}1) + K_p\{V_e(k) - V_e(k\text{-}1)\} + K_i V_e(k) \tag{6}$$
where K_p and K_i are the proportional and integral gains of the PI controller.

2) Reference Current Generator: The reference input current of the PFC converter is denoted by $i_{dc}*$ and given as,

$$i^*_{dc} = I_c(k)\, u_{Vs} \tag{7}$$
where u_{Vs} is the unit template of the voltage at input AC mains, calculated as,

$$u_{Vs} = v_d/V_{sm}; \; v_d = |v_s|; \; v_s = V_{sm}\sin\omega t \tag{8}$$
where V_{sm} is the amplitude of the voltage and ω is frequency in rad/sec at AC mains.

3) PWM Controller: The reference input current of the buck half-bridge converter ($i_{dc}*$) is compared with its sensed current (i_{dc}) to generate the current error $\Delta i_{dc} = (i_{dc}* - i_{dc})$. This current error is amplified by gain k_{dc} and compared with fixed frequency (f_s) saw-tooth carrier waveform $m_d(t)$ (as shown in Fig.2) in unipolar switching mode [7] to get the switching signals for the MOSFETs of the PFC buck half-bridge converter as,

$$\text{If} \quad k_{dc}\,\Delta i_{dc} > m_d(t) \quad \text{then } S_A = 1 \text{ else } S_A = 0 \tag{9}$$

$$\text{If} \quad -k_{dc}\,\Delta i_{dc} > m_d(t) \quad \text{then } S_B = 1 \text{ else } S_B = 0 \tag{10}$$
where S_A, S_B are upper and lower switches of the half-bridge converter as shown in Fig. 1 and their values '1' and '0' represent 'on' and 'off' position of the respective MOSFET of the PFC converter.

B. PMBLDCM Drive

The PMBLDCM drive consists of an electronic commutator, a VSI and a PMBLDC motor.

1) Electronic Commutator: The electronic commutator uses signals from Hall effect position sensors to generate the switching sequence for the voltage source inverter based on the logic given in Table I.

2) Voltage Source Inverter: Fig. 3 shows an equivalent circuit of a VSI fed PMBLDCM. The output of VSI to be fed to phase 'a' of the PMBLDC motor is given as,

$$v_{ao} = (V_{dc}/2) \qquad \text{for } S_1 = 1 \tag{11}$$

$$v_{ao} = (-V_{dc}/2) \qquad \text{for } S_2 = 1 \tag{12}$$

$$v_{ao} = 0 \qquad \text{for } S_1 = 0, \text{ and } S_2 = 0 \tag{13}$$

$$v_{an} = v_{ao} - v_{no} \tag{14}$$
where v_{ao}, v_{bo}, v_{co}, and v_{no} are voltages of the three-phases and neutral point (n) with respect to virtual mid-point of the DC link voltage shown as 'o' in Fig. 3. The voltages v_{an}, v_{bn}, v_{cn} are voltages of three-phases with respect to neutral point (n) and V_{dc} is the DC link voltage. S= 1 and 0 represent 'on' and 'off' position of respective IGBTs of the VSI and considered in a similar way for other IGBTs of the VSI i.e. S_3- S_6.

Using similar logic v_{bo}, v_{co}, v_{bn}, v_{cn} are generated for other two phases of the VSI feeding PMBLDC motor.

3) PMBLDC Motor: The PMBLDCM is represented in the form of a set of differential equations [3] given as,

$$v_{an} = Ri_a + p\lambda_a + e_{an} \tag{15}$$

$$v_{bn} = Ri_b + p\lambda_b + e_{bn} \tag{16}$$

$$v_{cn} = Ri_c + p\lambda_c + e_{cn} \tag{17}$$
where p is a differential operator (d/dt), i_a, i_b, i_c are three-phase

currents, λ_a, λ_b, λ_c are flux linkages and e_{an}, e_{bn}, e_{cn} are phase to neutral back emfs of PMBLDCM, in respective phases, R is resistance of motor windings/phase.

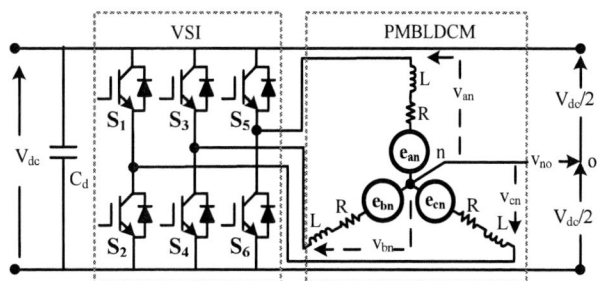

Figure 3. Equivalent Circuit of a VSI fed PMBLDCM Drive

The flux linkages are represented as,

$$\lambda_a = Li_a - M(i_b + i_c) \qquad (18)$$

$$\lambda_b = Li_b - M(i_a + i_c) \qquad (19)$$

$$\lambda_c = Li_c - M(i_b + i_a) \qquad (20)$$

where L is self-inductance/phase, M is mutual inductance of motor winding/phase. Since the PMBLDCM has no neutral connection, therefore,

$$i_a + i_b + i_c = 0 \qquad (21)$$

From Eqs. (14-21) the voltage between neutral terminal (n) and mid-point of the DC link (o) is given as,

$$v_{no} = \{v_{ao} + v_{bo} + v_{co} - (e_{an} + e_{bn} + e_{cn})\}/3 \qquad (22)$$

From Eqs. (18-21), the flux linkages are given as,

$$\lambda_a = (L+M) i_a, \quad \lambda_b = (L+M) i_b, \quad \lambda_c = (L+M) i_c, \qquad (23)$$

From Eqs. (15-17 and 23), the current derivatives in generalized state space form is given as,

$$pi_x = (v_{xn} - i_x R - e_{xn})/(L+M) \qquad (24)$$

where x represents phase a, b or c.

The developed electromagnetic torque T_e in the PMBLDCM is given as,

$$T_e = (e_{an} i_a + e_{bn} i_b + e_{cn} i_c)/\omega \qquad (25)$$

where ω is motor speed in rad/sec,

The back emfs may be expressed as a function of rotor position (θ) as,

$$e_{xn} = K_b f_x(\theta) \omega \qquad (26)$$

where x can be phase a, b or c and accordingly $f_x(\theta)$ represents function of rotor position with a maximum value ±1, identical to trapezoidal induced emf given as,

$$f_a(\theta) = 1 \qquad \text{for } 0 < \theta < 2\pi/3 \qquad (27)$$

$$f_a(\theta) = \{(6/\pi)(\pi-\theta)\}-1 \quad \text{for } 2\pi/3 < \theta < \pi \qquad (28)$$

$$f_a(\theta) = -1 \qquad \text{for } \pi < \theta < 5\pi/3 \qquad (29)$$

$$f_a(\theta) = \{(6/\pi)(\theta-2\pi)\}+1 \quad \text{for } 5\pi/3 < \theta < 2\pi \qquad (30)$$

The functions $f_b(\theta)$ and $f_c(\theta)$ are similar to $f_a(\theta)$ with a phase difference of 120° and 240° respectively.

Therefore, the electromagnetic torque is expressed as,

$$T_e = K_b\{f_a(\theta) i_a + f_b(\theta) i_b + f_c(\theta) i_c\} \qquad (31)$$

The mechanical equation of motion in speed derivative form is given as,

$$p\omega = (P/2)(T_e - T_L - B\omega)/(J) \qquad (32)$$

The derivative of the rotor position angle is given as,

$$p\theta = \omega \qquad (33)$$

where P is no. poles, T_L is load torque in Nm, J is moment of inertia in kg-m^2 and B is friction coefficient in Nms/Rad.

These equations (15-33) represent the dynamic model of the PMBLDC motor.

V. PERFORMANCE EVALUATION OF PROPOSED PFC DRIVE

The proposed PMBLDCM drive is modeled in Matlab-Simulink environment and evaluated for an air conditioning compressor load. The compressor load is considered as a constant torque load equal to rated torque with the speed control required by air conditioning system. A 1.5 kW rating PMBLDCM is used to drive the air conditioner compressor, speed of which is controlled effectively by controlling the DC link voltage. The detailed data of the motor and simulation parameters are given in Appendix. The performance of the proposed PFC drive is evaluated on the basis of various parameters such as total harmonic distortion (THD$_i$) and the crest factor (CF) of the current at input AC mains, displacement power factor (DPF), power factor (PF) and efficiency of the drive system (η_{drive}) at different speeds of the motor. Moreover, these parameters are also evaluated for variable input AC voltage at DC link voltage of 416 V which is equivalent to the rated speed (1500 rpm) of the PMBLDCM. The results are shown in Figs. 4-9 and Tables II-III to demonstrate the effectiveness of the proposed PMBLDCM drive in a wide range of speed and input AC voltage.

A. Performance during Starting

The performance of the proposed PMBLDCM drive fed from 220 V AC mains during starting at rated torque and 900 rpm speed is shown in Fig. 4a. A rate limiter of 800 V/s is introduced in the reference voltage to limit the starting current of the motor as well as the charging current of the DC link capacitor. The PI controller closely tracks the reference speed so that the motor attains reference speed smoothly within 0.35 sec while keeping the stator current within the desired limits i.e. double the rated value. The current (i_s) waveform at input AC mains is in phase with the supply voltage (v_s) demonstrating nearly unity power factor during the starting.

B. Performance under Speed Control

Figs. 4-6 show the performance of the proposed PMBLDCM drive under the speed control at constant rated torque (9.55 Nm) and 220 V AC mains supply voltage. These results are categorized as performance during transient and steady state conditions.

1) Transient Condition: Figs. 4b-c show the performance of the drive during the speed control of the compressor. The

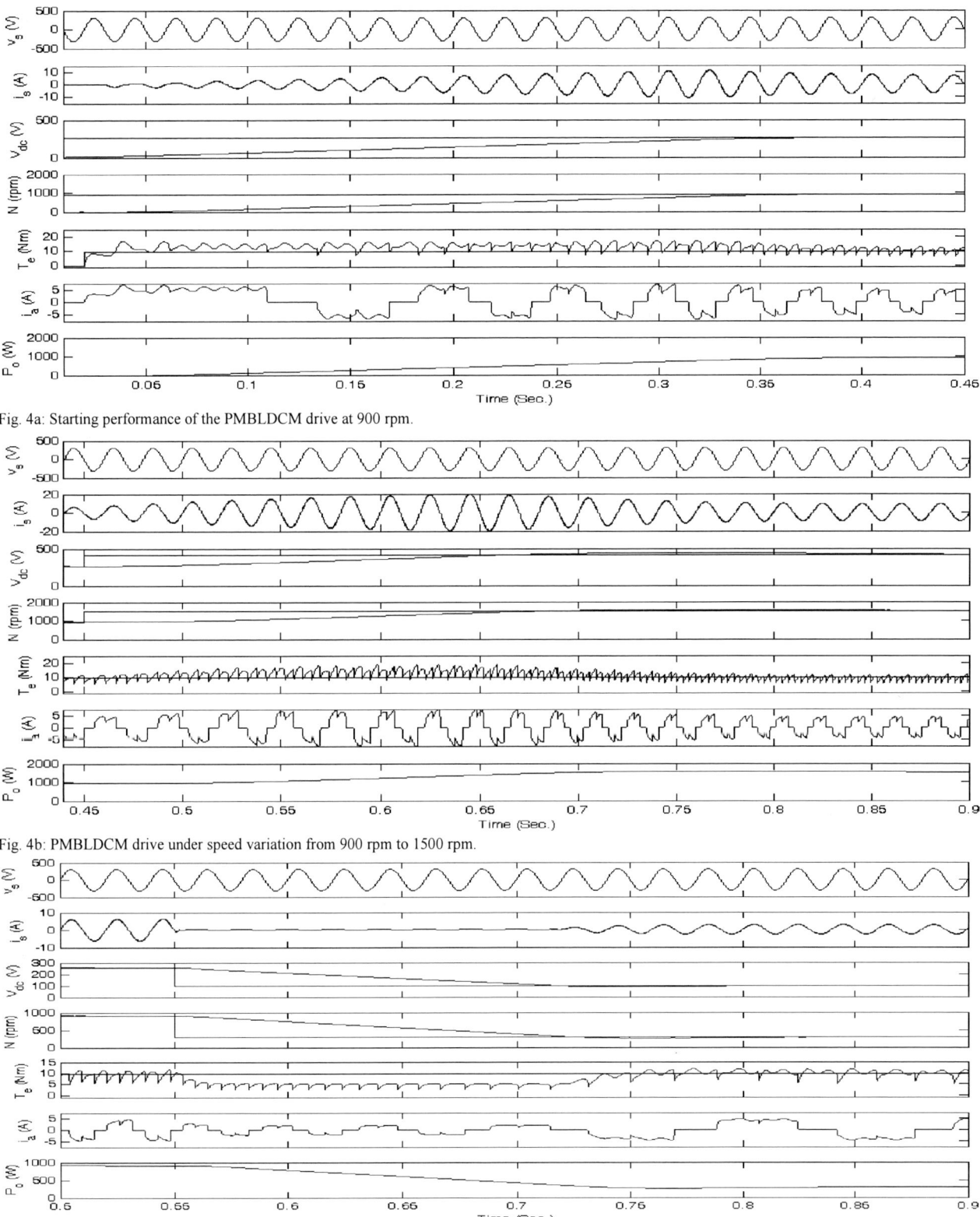

Fig. 4a: Starting performance of the PMBLDCM drive at 900 rpm.

Fig. 4b: PMBLDCM drive under speed variation from 900 rpm to 1500 rpm.

Fig. 4c: PMBLDCM drive under speed variation from 900 rpm to 300 rpm.

Figure 4. Performance of the Proposed PMBLDCM drive under speed variation at 220 VAC input.

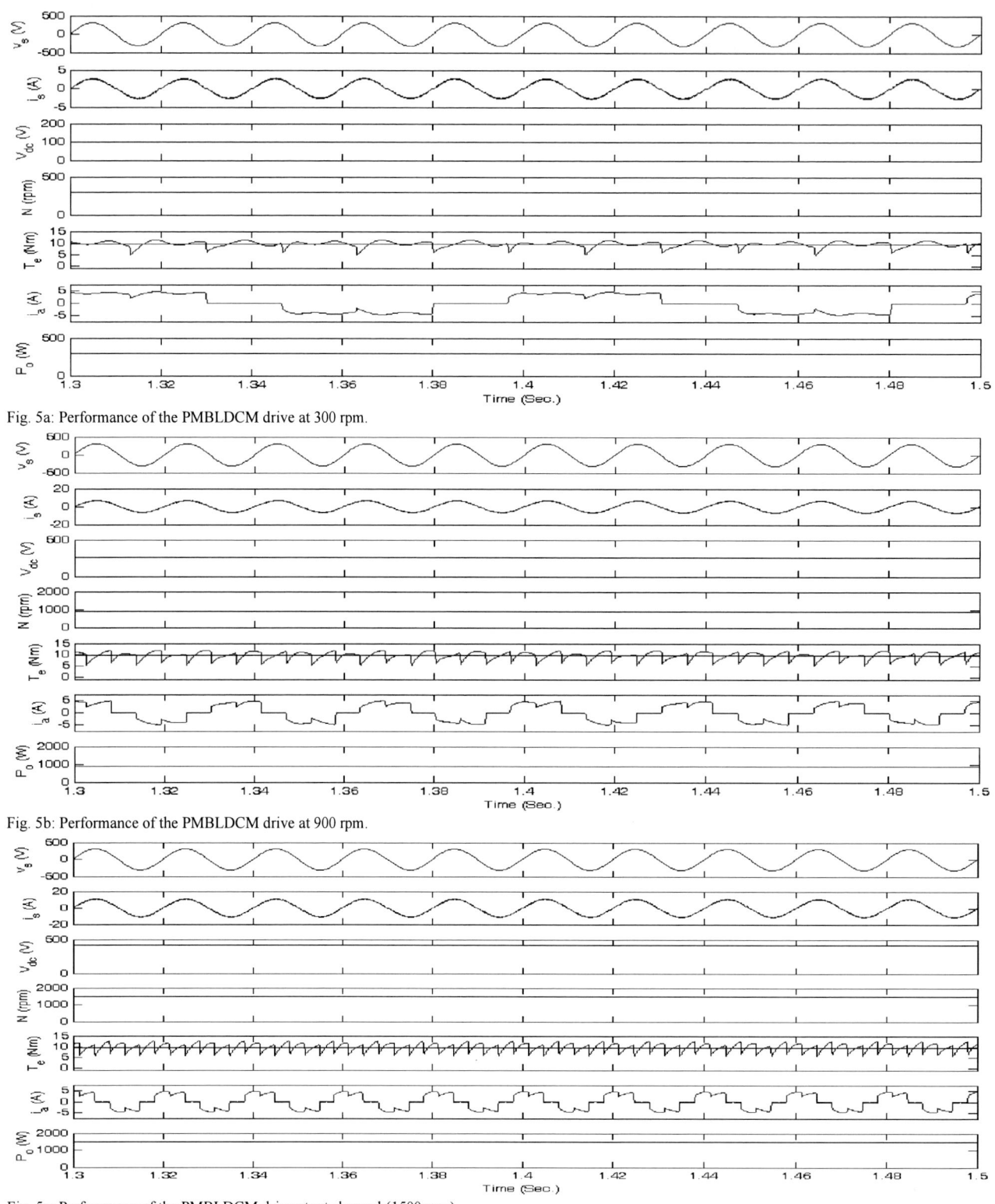

Fig. 5a: Performance of the PMBLDCM drive at 300 rpm.

Fig. 5b: Performance of the PMBLDCM drive at 900 rpm.

Fig. 5c: Performance of the PMBLDCM drive at rated speed (1500 rpm).

Figure 5.   Performance of the PMBLDCM drive under steady state condition at 220 VAC input.

978-1-4244-4782-4/10 $26.00 © 2010 IEEE

reference speed is changed from 900 rpm to 1500 rpm for the rated load performance of the compressor; from 900 rpm to 300 rpm for performance of the compressor at light load. It is observed that the speed control is fast and smooth in either direction i.e. acceleration or retardation with power factor maintained at nearly unity value. Moreover, the stator current of PMBLDCM is within the allowed limit (twice the rated current) due to the introduction of a rate limiter in the reference voltage.

2) Steady State Condition: The speed control of the PMBLDCM driven compressor under steady state condition is carried out for different speeds and the results are shown in Figs. 5-6 and Table-II to demonstrate the effectiveness of the proposed drive in wide speed range. Figs.5a-c show voltage (v_s) and current (i_s) waveforms at AC mains, DC link voltage (V_{dc}), speed of the motor (N), developed electromagnetic torque of the motor (T_e), the stator current of the PMBLDC motor for phase 'a' (I_a), and shaft power output (P_o) at 300 rpm, 900 rpm and 1500 rpm speeds. Fig. 6a shows linear relation between motor speed and DC link voltage. Since the reference speed is decided by the reference voltage at DC link, it is observed that the control of the reference DC link voltage controls the speed of the motor instantaneously. Fig. 6b shows the improved efficiency of the drive system (η_{drive}) in wide range of the motor speed.

C. Power Quality Performance

The performance of the proposed PMBLDCM drive in terms of various PQ parameters such as THD_i, CF, DPF, PF is summarized in Table-II and shown in Figs. 7-8. Nearly unity power factor (PF) and reduced THD of AC mains current are observed in wide speed range of the PMBLDCM as shown in Figs. 7a-b. The THD of AC mains current remains less than 5% along with nearly unity PF in wide range of speed as well as load as shown in Table-II and Figs. 8a-c.

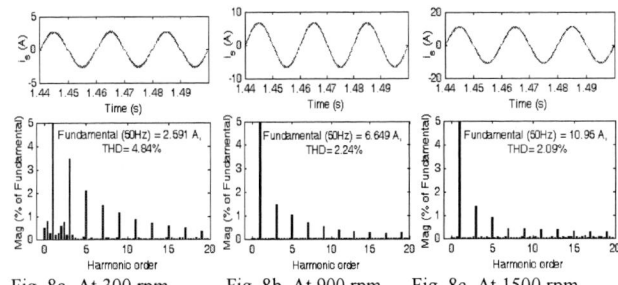

Fig. 8a. At 300 rpm   Fig. 8b. At 900 rpm   Fig. 8c. At 1500 rpm

Figure 8.   Current waveform at AC mains and its harmonic spectra of the PMBLDCM drive under steady state condition at rated torque and 220 V_{AC}

TABLE II.   PERFORMANCE OF DRIVE UNDER SPEED CONTROL AT 220 V AC INPUT

Speed (rpm)	V_{DC} (V)	THD_i (%)	DPF	PF	η_{drive} (%)	Load (%)
300	100	4.84	0.9999	0.9987	74.2	20.0
400	126	3.94	0.9999	0.9991	79.1	26.7
500	153	3.33	0.9999	0.9993	81.8	33.3
600	179	2.92	0.9999	0.9995	83.8	40.0
700	205	2.63	0.9999	0.9996	85.3	46.6
800	232	2.40	0.9999	0.9996	86.1	53.3
900	258	2.24	0.9999	0.9996	87.0	60.0
1000	284	2.16	0.9999	0.9997	87.6	66.6
1100	310	2.09	0.9999	0.9997	88.1	73.3
1200	337	2.03	0.9999	0.9997	88.1	80.0
1300	363	2.05	0.9999	0.9997	88.2	86.6
1400	390	2.07	0.9999	0.9997	88.1	93.3
1500	416	2.09	0.9999	0.9997	88.1	100.0

D. Performance under Variable Input AC Voltage

Performance evaluation of the proposed PMBLDCM drive is carried out under varying input AC voltage at rated load (i.e. rated torque and rated speed) to demonstrate the operation of proposed PMBLDCM drive for air conditioning system in various practical situations as summarized in Table-III.

Figs. 9a-b show variation of input current and its THD at AC mains, DPF and PF with AC input voltage. The THD of current at AC mains is within specified limits of international norms [5] along with nearly unity power factor in wide range of AC input voltage.

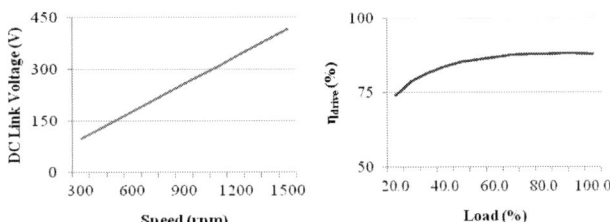

Fig. 6a. DC link voltage with speed   Fig. 6b. Efficiency with load

Figure 6.   Performance of the proposed PFC drive under speed control at rated torque and 220 V_{AC}

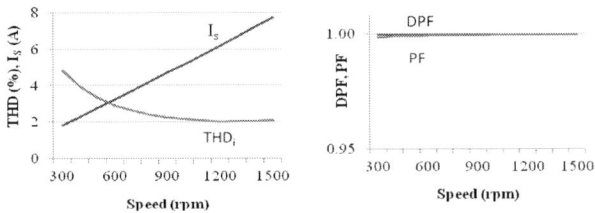

Fig. 7a. THD of current at AC mains   Fig. 7b. DPF and PF

Figure 7.   PQ parameters of PMBLDCM drive under speed control at rated torque and 220 V_{AC} input

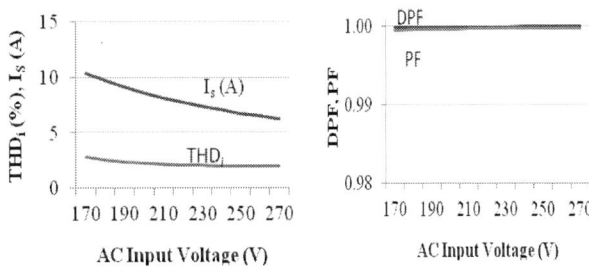

Fig. 9a. Current at AC mains and its THD   Fig. 9b. DPF and PF

Figure 9.   PQ parameters with input AC voltage at 416 V_{DC} (1500 rpm)

TABLE III. VARIATION OF PQ PARAMETERS WITH INPUT AC VOLTAGE (VS) AT 1500 RPM (416 VDC)

V_{AC} (V)	THD$_i$ (%)	DPF	PF	CF	I$_s$ (A)	η_{drive} (%)
170	2.88	0.9999	0.9995	1.41	10.4	84.9
180	2.59	0.9999	0.9996	1.41	9.7	85.8
190	2.40	0.9999	0.9996	1.41	9.2	86.3
200	2.26	0.9999	0.9996	1.41	8.6	87.2
210	2.14	0.9999	0.9997	1.41	8.2	87.6
220	2.09	0.9999	0.9997	1.41	7.7	88.1
230	2.07	0.9999	0.9997	1.41	7.4	88.2
240	2.02	1.0000	0.9998	1.41	7.1	88.4
250	1.99	1.0000	0.9998	1.41	6.8	88.7
260	2.01	1.0000	0.9998	1.41	6.5	88.7
270	2.01	1.0000	0.9998	1.41	6.2	89.0

VI. CONCLUSION

A new speed control strategy of a PMBLDCM drive is validated for a compressor load of an air conditioner which uses the reference speed as an equivalent reference voltage at DC link. The speed control is directly proportional to the voltage control at DC link. The rate limiter introduced in the reference voltage at DC link effectively limits the motor current within the desired value during the transient condition (starting and speed control). The additional PFC feature to the proposed drive ensures nearly unity PF in wide range of speed and input AC voltage. Moreover, power quality parameters of the proposed PMBLDCM drive are in conformity to an International standard IEC 61000-3-2 [5]. The proposed drive has demonstrated good speed control with energy efficient operation of the drive system in the wide range of speed and input AC voltage. The proposed drive has been found as a promising candidate for a PMBLDCM driving Air-Con load in 1-2 kW power range.

APPENDIX

Rated Power: 1.5 kW, rated speed: 1500 rpm, rated current: 4.0 A, rated torque: 9.55 Nm, number of poles: 4, stator resistance (R): 2.8 Ω/ph., inductance (L+M): 5.21 mH/ph., back EMF constant (K$_b$): 0.615 Vsec/rad, inertia (J): 0.013 Kg-m^2. Source impedance (Z$_s$): 0.03 pu, switching frequency of PFC switch (f$_s$) = 40 kHz, capacitors (C$_1$= C$_2$): 15nF, PI speed controller gains (K$_p$): 0.145, (K$_i$): 1.45.

REFERENCES

[1] T. Kenjo and S. Nagamori, *Permanent Magnet Brushless DC Motors*, Clarendon Press, oxford, 1985.

[2] T. J. Sokira and W. Jaffe, Brushless DC Motors: Electronic Commutation and Control, Tab Books USA, 1989.

[3] J. R. Hendershort and T. J. E. Miller, *Design of Brushless Permanent-Magnet Motors*, Clarendon Press, Oxford, 1994.

[4] J. F. Gieras and M. Wing, *Permanent Magnet Motor Technology – Design and Application*, Marcel Dekker Inc., New York, 2002.

[5] Limits for Harmonic Current Emissions (Equipment input current ≤16 A per phase), International Standard IEC 61000-3-2, 2000.

[6] B. Singh, B. N. Singh, A. Chandra, K. Al-Haddad, A. Pandey and D. P. Kothari, "A review of single-phase improved power quality AC-DC converters," *IEEE Trans. Industrial Electron.*, vol. 50, no. 5, pp. 962 – 981, oct. 2003.

[7] N. Mohan, T. M. Undeland and W. P. Robbins, "*Power Electronics: Converters, Applications and Design*," John Wiley, USA, 1995.

[8] A. I. Pressman, *Switching Power Supply Design*, McGraw Hill, New York, 1998.

[9] P.J. Wolfs, "A current-sourced DC-DC converter derived via the duality principle from the half-bridge converter," *IEEE Trans. Ind. Electron.*, vol. 40, no. 1, pp. 139 – 144, Feb. 1993.

[10] J.Y. Lee, G.W. Moon and M.J. Youn, "Design of a power-factor-correction converter based on half-bridge topology," *IEEE Trans. Ind. Electron.*, vol. 46, no. 4, pp.710 – 723, Aug 1999.

[11] J. Sebastian, A. Fernandez, P.J. Villegas, M.M. Hernando and J.M. Lopera, "Improved active input current shapers for converters with symmetrically driven transformer," *IEEE Trans. Ind. Appl.*, vol. 37, no. 2, pp. 592 – 600, March-April 2001.

[12] A. Fernandez, J. Sebastian, M.M. Hernando and P. Villegas, "Small signal modelling of a half bridge converter with an active input current shaper," in *Proc. IEEE PESC*, 2002, vol.1, pp.159 – 164.

[13] S.K. Han, H.K. Yoon, G.W. Moon, M.J. Youn, Y.H. Kim and K.H. Lee, "A new active clamping zero-voltage switching PWM current-fed half-bridge converter," *IEEE Trans. Power Electron.*, vol. 20, no. 6, pp. 1271 – 1279, Nov. 2005.

[14] R.T.Bascope, L.D.Bezerra, G.V.T.Bascope, D.S. Oliveira, C.G.C. Branco, and L.H.C. Barreto, "High frequency isolation on-line UPS system for low power applications," in *Proc. IEEE APEC'08*, 2008, pp.1296 – 1302.

Comparison of HF Signal Injection Methods for Sensorless Control of PM Synchronous Motors

Eisenhawer de M. Fernandes
Mechanical Engineering Department
Federal University of Campina Grande
Campina Grande-PB, Brazil
Email: eisenhawer@ee.ufcg.edu.br

Alexandre C. Oliveira and Cursino B. Jacobina and Antonio M. N. Lima
Electrical Engineering Department
Federal University of Campina Grande
Campina Grande-PB, Brazil
Email: jacobina@ee.ufcg.edu.br

Abstract— This paper presents a comparative study of high-frequency (HF) voltage signal injections used for rotor position estimation in sensorless control of AC motors. This work reviews these methods and evaluates the performance under sensorless speed control applied to a surface permanent-magnet synchronous motor (PMSM). Theoretical analysis is verified by simulations and confirms the effectiveness of these methods. The experimental results of speed control using an industrial PMSM drive system are presented for each of the injection methods.

I. INTRODUCTION

The vector control of permanent-magnet synchronous machine (PMSM) is based on the information of rotor position. Encoders and rotational transducers are commonly used to provide these signals, mounted on the shaft of the machine. However, the use of position sensors represent some drawbacks related to cost, size, complexity and to reduction of the reliability of the drive system [1]-[2]. Therefore, efforts have been focused on replacing these sensors by solutions based on obtaining the rotor position information indirectly, i.e., extracting the position information from the machine itself.

Sensorless control schemes can be classified into two types. The first class is based on fundamental-frequency models aimed to estimate the back-emf induced in the motor windings. These schemes present good performance for solutions in medium and high speed range because the magnitude is proportional to the rotor speed [1],[3],[4], [5]. However, this kind of methods exhibits degradation of the performance as the speed decreases leading to instability of the control system [2],[6], [7]. Besides this, back-emf estimators also present dependency of parameter variations due to temperature and magnetic saturation.

In the low speed region, the detection of rotor position can be carried out by using the injection of extra signals (current [8] or voltage [9]-[10]) superimposed to the fundamental excitation. The applied signals interact with the rotor saliency or magnetic anisotropy of the motor, and the resulting response (voltage or current) is processed to extract rotor position information. The injection of HF current presents some disadvantages like the estimation bandwidth is limited by the current controller bandwidth and additional difficulties of implementation in an industrial drive system [8]. On the other hand, in the voltage injection, the estimation bandwidth is

determined by the observer bandwidth [8].

The application of voltage signal can be [11],[12]: 1) persistent HF voltage, 2) discrete voltage pulses or 3) modified pulse-width modulation (PWM) pulses. Persistent high frequency voltage injection can be classified as rotating vector [13]-[14] or pulsating vector [2],[15]. In the first approach, a rotating voltage vector is superimposed to the fundamental excitation in the stationary reference frame. The pulsating injection is carried out by the application of a voltage vector along the estimated synchronous reference frame. It can be applied in the $\hat{d}$-axis [2], $\hat{q}$-axis [10] or a combination of both [9], [10], [11], [16]).

There are some papers that addressed a comparative analysis of the two voltage injection sensorless methods. Different features of the sensorless control at low or zero speed has been addressed like: influence of inverter dead-time [10],[17], effect of multiple saliencies [17], cross-saturation [17], ZCC effect [10] and sensorless torque control performance [14],[16]. In [14] is presented a comparison of HF signal injection applied to an IPMSM (Integrated Starter Generator - ISG). With experimental measurement for speed and torque control under load transients, the saturation-induced saliencies have been identified as a function of fundamental current. In [16] is investigated the performance and the constraints of the sensorless methods for direct torque control applied to an IM motor at low speed. In [17] is presented an analysis of the sensitivity to physical attributes (dead-time, multiple saliencies and cross-saturation) of the two injection voltage for an IPMSM, however, in relation to sensorless control performance, it was reported the rotor estimation error under constant load.

This paper presents a comparative study of the high frequency voltage injection for sensorless speed control of surface-mounted PM synchronous motor. The work presents a review of the principles of HF voltage injection techniques proposed in literature. Simulation and experimental results shown the performance under two circumstances: speed reference variation and application of load torque disturbance.

978-1-4244-4782-4/10 $26.00 © 2010 IEEE

II. HIGH-FREQUENCY VOLTAGE SIGNAL INJECTION METHODS

The model of a PMSM for rotor position estimation is derived from the fundamental voltage model under the following assumptions [2],[17]: (1) the frequency of the injected signal (ω_h) is high to the rotor frequency (ω_r), (2) at low speeds the back-emf and resistive voltage drop is negligible and (3) the presence of single magnetic saliency. Thus, in steady-state, the high frequency model in the rotor reference frame is

$$\begin{bmatrix} v_{sdh}^r \\ v_{sqh}^r \end{bmatrix} = \begin{bmatrix} j\omega_h l_d & 0 \\ 0 & j\omega_h l_q \end{bmatrix} \begin{bmatrix} i_{sdh}^r \\ i_{sqh}^r \end{bmatrix} \tag{1}$$

where v_{sdh}^r, v_{sqh}^r and i_{sdh}^r, i_{sqh}^r are the high frequency voltages and currents in the rotor reference frame, respectively. $j\omega_h l_d$, $j\omega_h l_q$ are the high frequency inductances in the rotor reference frame.

A. Rotating voltage injection

The approach superimposes a balanced polyphase voltage vector rotating at a high frequency ω_h and magnitude V_h. The voltage vector can be defined as (2). The diagram of the injection is illustrated in Fig.1(a). The machine saliency modulates the HF current amplitude, which can be written in the stationary frame as (3).

$$\mathbf{v}_{sdqh}^s = V_h e^{j\omega_h t} \tag{2}$$

$$\mathbf{i}_{sdqh}^s = I_{h1} e^{j\omega_h t} + I_{h2} e^{j(-\omega_h t + 2\theta_r)} \tag{3}$$

where: $I_{h1} = \frac{V_h \Sigma L_s}{\omega_h (\Sigma L_s^2 - \Delta L_s^2)}$, $I_{h2} = \frac{V_h}{\omega_h (\Sigma L_s^2 - \Delta L_s^2)}$

In expression (3), ΣL_s is the average inductance ($\frac{l_q + l_d}{2}$), ΔL_s is the differential inductance ($\frac{l_q - l_d}{2}$), thus, the component I_{h2} contains the rotor position information ($2\theta_r$). The carrier currents are extracted by bandpass filters. There are different schemes of demodulation to obtain the position information proposed in the literature, the most common choices is the use of synchronous frame filtering [16] or a heterodyning process followed by a closed-loop observer [9][see Fig.1(b)].

B. Pulsating voltage injection

The approach applies a voltage vector at a high frequency ω_h on one of the axes of the estimated synchronous reference frame ($\widehat{d}$-$\widehat{q}$). The voltage vector injected in the $\widehat{d}$-axis frame can be defined as (4). The diagram of the injection is illustrated in Fig.2(a).

$$\mathbf{v}_{sdqh}^{\widehat{r}} = \begin{bmatrix} v_{sdh}^{\widehat{r}} \\ v_{sqh}^{\widehat{r}} \end{bmatrix} = \begin{bmatrix} V_h \cos(\omega_h t) \\ 0 \end{bmatrix} \tag{4}$$

On the other hand, the injection in the $\widehat{q}$-axis frame can be defined as

$$\mathbf{v}_{sdqh}^{\widehat{r}} = \begin{bmatrix} v_{sdh}^{\widehat{r}} \\ v_{sqh}^{\widehat{r}} \end{bmatrix} = \begin{bmatrix} 0 \\ V_h \cos(\omega_h t) \end{bmatrix} \tag{5}$$

The resulting HF current can be written in the $\widehat{d}$-$\widehat{q}$ frame as

$$\mathbf{i}_{sdqh}^{\widehat{r}} = \frac{1}{2} \left(I_{h1} + I_{h2} e^{j2(\theta_r - \widehat{\theta}_r)} \right) \sin(\omega_c t) \tag{6}$$

where I_{h1} and I_{h2} were defined previously. In this type of injection, the useful position information is provided by the $\widehat{q}$-axis a HF current ($i_{sqh}^{\widehat{r}}$) which is proportional to the rotor position estimation error ($\theta_r - \widehat{\theta}_r$). The injection of pulsating vector in $\widehat{d}$-axis or $\widehat{q}$-axis are illustrated in Fig. 2(a). The HF currents are extracted by bandpass filters.

There are different schemes of demodulation to obtain the position error in the literature, the most common form is illustrated in Fig.2(b), a BPF filter, followed by a product with a orthogonal function and a LPF filter. Usually, PLL structures [15],[2] or a closed-loop observer [10] are used to force the estimation error to zero, reducing the angular displacement between the real and estimated synchronous frames [see Fig.2(b)]. The injection in $\widehat{q}$-axis and demodulation process are analogous to $\widehat{d}$-axis voltage injection. The main difference between the two injection shemes is the injection on the $\widehat{q}$-axis produces torque ripples and additional losses from the HF currents [2]. However, the injection of pulsating voltage in q-axis can be attractive if the application has non-linearity of VSI such as dead-time and ZCC effect, because the resulting HF currents presents robustness against these phenomena [10].

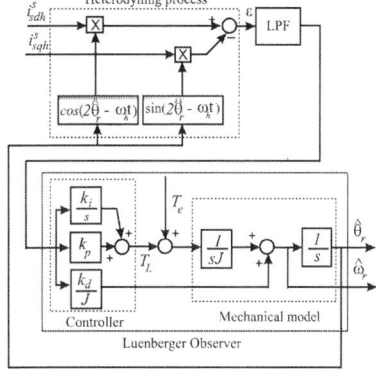

Fig. 1. Demodulation scheme based on heterodyning process and Luenberger observer.

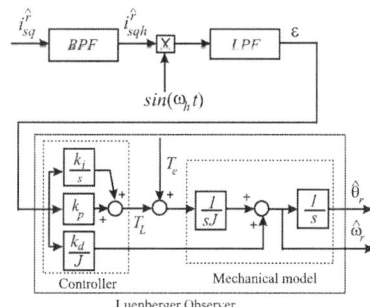

Fig. 2. Demodulation scheme and estimator structure - Luenberger style observer.

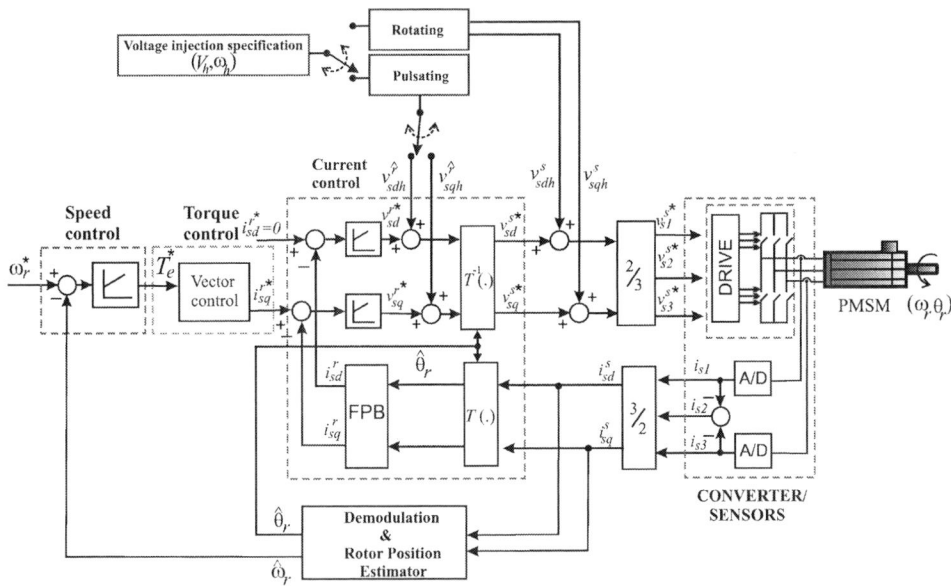

Fig. 3.   Control block diagram.

III. EXPERIMENTAL RESULTS

A. Configuration of Test Setup

In order to verify the simulation results, the HF voltage injection techniques have been implemented in a laboratory test setup (Fig. 4). Experiments on sensorless operation of the SMPM motor have been performed with the control strategy of Fig. 3. The test setup is composed by a microcomputer and two industrial servodrives systems (Converters 1 and 2). The microcomputer is equipped with a Renesas microprocessor programming board (SH7047F) used to control the tested motor (Motor 1). The servodrives consist of three-phase VSI converters with 10 kHz switching frequency and $1.0\mu s$ deadtime. The test motor is a surface mounted PMSM. The PMSMs data are listed in the Table I.

The sampling frequency for current control, speed control and position estimation has been set at 10kHz. The speed and current controllers bandwidth is 80Hz and 250Hz, respectively. The DC-link voltage is 200V. The Converter 2 and Motor 2 are used to provide different load conditions. The Motor 2 is operated in a torque-control mode with a torque constant of 0.70. The rotor position estimation performance is evaluated for two conditions: 1) Step variation in the reference speed and 2) load disturbance. A resolver attached to the motor shaft has been used to monitor the actual rotor position and speed.

B. Command speed variation

The performance of sensorless speed control at low speed is presented in Figs. 5, 6 and 7. Waveform data indicate, from the top, mechanical rotor speed (N), estimated rotor position ($\widehat{\theta}_r$), rotor position estimation error which is the input of the rotor position observer (ε), and the motor torque developed by the load machine (T_L). In the test, the speed reference is step changed from 20 rpm to 100 rpm, and back to 20 rpm under

Fig. 4.   Experimental test setup.

TABLE I

PARAMETERS OF THE PMSM MACHINES.

Parameter	Motor 1	Motor 2
Rated power (kW)	0.4	1.2
Rated voltage (Vrms)	220	220
Rated torque (N.m)	1.6	2.5
Resistance	6.63	0.7
l_d	24	1.93
l_q	33	2.2
Pole pairs	4	4
Rated current (Arms)	2.0	7.2
Maximum speed(rpm)	3,000	6,000
Back-emf constant (V/krpm)	56.16	26.5

constant load. During the speed variation, the load torque has been set to 20% of the rated torque of test motor.

Fig. 5 shows results of sensorless speed control operation based on $\widehat{d}$-axis pulsating injection. The specification of the HF voltage injection has been selected as 30V-1.0kHz. The rotor position observer bandwidth is set to 300Hz. It is observed that rotor position estimation error increases during speed transients but its maximum instant value is inferior to 0.5

radians.

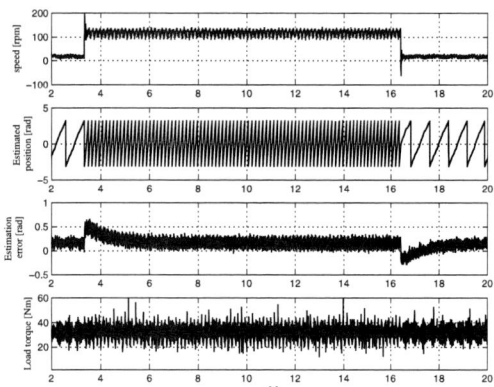

Fig. 5. Performance of rotor position estimation algorithm based on $\widehat{d}$-axis pulsating voltage injection during speed transient (ω_r) from 20 rpm to 100 rpm (from top to bottom: measured rotor speed (ω_r), estimated rotor position ($\widehat{\theta}_r$), rotor position estimation error ($\theta_r - \widehat{\theta}_r$), and load torque ($T_L$).

In Fig. 6 are presented the results for operation based on rotor position estimation based on $\widehat{q}$-axis pulsating injection. Again, the injection condition has been selected as 30V-1, 0 kHz. The observer bandwidth is set to 300 Hz. The rotor postion estimation error increases during speed transients but inferior to 0.5 radians. It can be noted that the performance of rotor position estimation based on pulsating injection on $\widehat{q}$-axis presented similar responses than that based on $\widehat{d}$-axis pulsating injection.

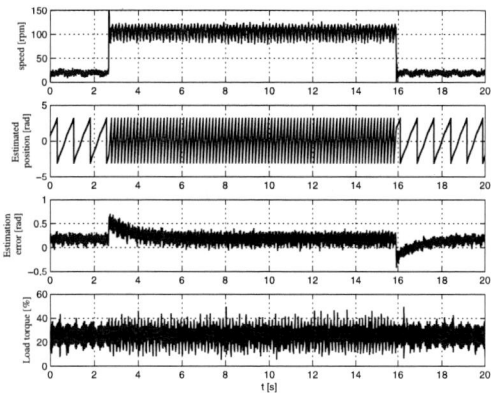

Fig. 6. Performance of rotor position estimation algorithm based on $\widehat{q}$-axis pulsating voltage injection during speed transient (ω_r) from 20 rpm to 100 rpm (from top to bottom: measured rotor speed (ω_r), estimated rotor position ($\widehat{\theta}_r$), rotor position estimation error ($\theta_r - \widehat{\theta}_r$), and load torque ($T_L$).

The results for operation based on rotor position estimation based on rotating voltage injection in the stationary reference frame are shown in Fig. 7. In this case, the injection condition has been selected as 50V-500 Hz. The observer bandwidth is set to 500 Hz, approximately. It can be noted that the rotor position estimation error increases during speed transients but the maximum instantaneous value is lower than 0.2 radians. One important feature of this result is that the observer tuning

provides a lower convergence time than the pulsating injection cases.

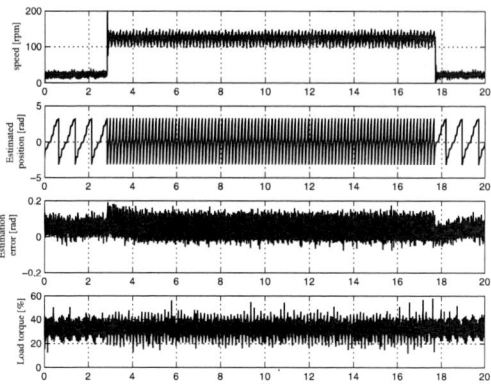

Fig. 7. Performance of rotor position estimation algorithm based on rotating voltage injection during speed transient (ω_r) from 20 rpm to 100 rpm (from top to bottom: measured rotor speed (ω_r), estimated rotor position ($\widehat{\theta}_r$), rotor position estimation error ($\theta_r - \widehat{\theta}_r$), and load torque ($T_L$).

C. Load torque disturbance

One important characteristic in sensorless speed control for PMSMs at low speeds is the performance when load torque is applied. In order to verify this property, rotor position estimation under load disturbance were carried out. The results are shown in Figs. 8, 9 and 10. Waveform data mean, from the top, mechanical rotor speed (N), estimated rotor position ($\widehat{\theta}_r$), rotor position estimation error which is the input of the rotor position observer (ε), and the motor torque developed by the load machine (T_L).

In the tests, the mechanical speed reference is set to 50 rpm (or 3, 33 Hz). Besides, the voltage injection condition and the observer bandwidth for the all injection methods are the same of the speed reference variation test. Under constant speed, the load torque (T_L) is varied from 0% to 60% of the rated torque of motor 1, and after $10s \sim 12s$ is removed.

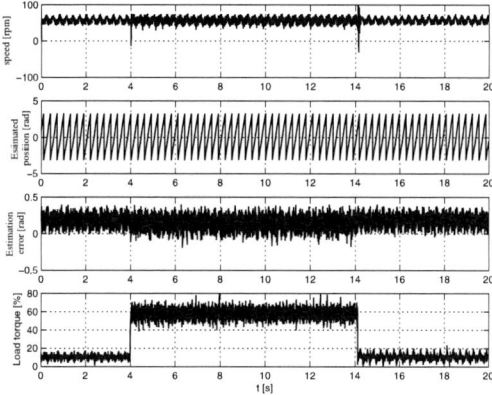

Fig. 8. Performance of rotor position estimation algorithm based on $\widehat{d}$-axis pulsating voltage injection under load toque variation (T_L) from 0 to 60% of rated torque (from top to bottom: measured rotor speed (ω_r), estimated rotor position ($\widehat{\theta}_r$), rotor position estimation error ($\theta_r - \widehat{\theta}_r$), and load torque ($T_L$).

Fig. 8 presents the results for operation based on rotor position estimation based on $\widehat{d}$-axis pulsating injection. It can be seen that during load transients rotor postion estimation error increases but its average value is lower than 0.25 radians. On the other hand, in Fig. 9 show the results for rotor position estimation based on $\widehat{q}$-axis pulsating injection. The rotor position estimation error increases during speed transients, but its maximum value is 0.5 radians, approximately.

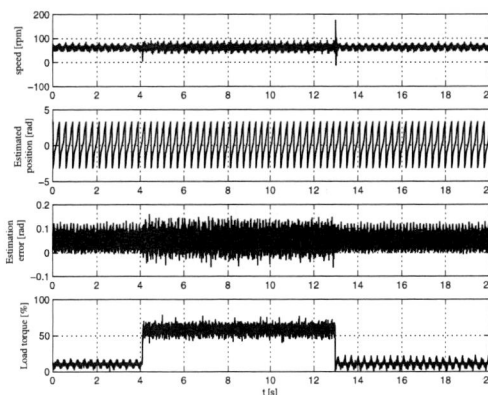

Fig. 10. Performance of rotor position estimation algorithm based on rotating voltage injection under load toque variation (T_L) from 0 to 60% of rated torque (from top to bottom: measured rotor speed (ω_r), estimated rotor position ($\widehat{\theta}_r$), rotor position estimation error ($\theta_r - \widehat{\theta}_r$), and load torque ($T_L$).

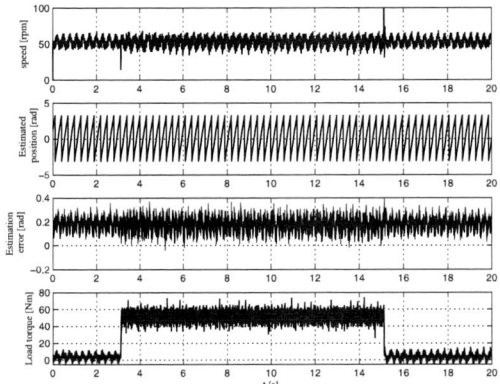

Fig. 9. Performance of rotor position estimation algorithm based on $\widehat{q}$-axis pulsating voltage injection under load toque variation (T_L) from 0 to 60% of rated torque (from top to bottom: measured rotor speed (ω_r), estimated rotor position ($\widehat{\theta}_r$), rotor position estimation error ($\theta_r - \widehat{\theta}_r$), and load torque ($T_L$).

The results for operation based on injection of rotating voltage in the stationary reference frame are shown in Fig. 10. Under the same load circumstances, the estimation error is lower than 0.2 rad and is confined nearly zero in the transient state as well as steady state. As it can be seen, the rotor position estimator based on rotating voltage injection provides the lower estimation error among the others approaches. This feature means that it presents a higher load disturbance rejection compared to that based on pulsating injection. As observed in the results, sensorless speed control strategies based on the saliencies tracking methods by means of HF voltage injection does not lose rotor estimation and allow for good control capability in the low speed region.

IV. CONCLUSIONS

The paper presents a comparative study of the HF voltage injection commonly used for rotor position estimation in sensorless control of AC motors. It was reviewed the theoretical fundamentals about each method and verified by simulations the performance for sensorless speed control of a surface mounted PMSM machine.

Experimental tests have been performed for sensorless speed control under reference speed variation and load torque disturbance. The experimental results show that the introduction of voltage vector on $\widehat{d}$-axis or $\widehat{q}$-axis present similar performance. However, according to simulation results, the HF pulsating injection on $\widehat{q}$-axis ripple torque are expected. The rotating voltage injection on stationary frame presents the best response under step load variation. Nevertheless, this kind of injection

requires a higher magnitude in order to produce a reasonable estimation.

It can be concluded that this category of rotor position estimation methods are suitable for operation at low speed region as an alternative for back-emf estimation strategies.

ACKNOWLEDGMENT

The authors would like to thank the support provided by Conselho Nacional de Desenvolvimento Cientfico e Tecnolgico (CNPq) and by the Coordenao de Aperfeioamento de Pessoal de Nvel Superior (CAPES).

REFERENCES

[1] R. Wu and G. R. Slemon, "A permanent magnet motor drive without a shaft sensor," *Conf. Rec. IEEE-IAS Annu. Meeting*, pp. 553–558, 1990.

[2] J.-H. Jang, J.-I. Ha, M. Ohto, K. Ide, and S.-K. Sul, "Analysis of permanent-magnet machine for sensorless control based on high-frequency signal injection," *IEEE Trans. Ind. Applicat.*, vol. 39, pp. 1595–2004, May/June 2003.

[3] S. Ichikawa, C. Zhiqian, M. Tomita, S. Doki, and S. Okuma, "Sensorless control of an interior permanent magnet synchronous motor on the rotating coordinate using an extended electromotive force," *Proc. IECON'01*, vol. 3, no. 29, pp. 1667–1672, Dec. 2001.

[4] S. Marimoto, K. Kawamoto, M. Sanada, and Y. Takeda, "Sensorless control strategy for salient-pole pmsm based on extended emf in rotating reference frame," *IEEE Trans. Ind. Applicat.*, vol. 38, pp. 1054–1061, July/Aug. 2002.

[5] H. Kim, M. C. Harke, and R. Lorenz, "Sensorless control of interior permanent-magnet machine drive with zero-phase lag position estimation," *IEEE Trans. Ind. Applicat.*, vol. 39, no. 6, pp. 784–789, Nov./Dec. 2003.

[6] L. Ribeiro, M.C.Harke, and R. Lorenz, "Dynamic properties of back-emf based sensorless drives," *Proc. of the IAS Annual Meeting*, vol. 1, pp. 2026–2033, Oct. 2006.

[7] R. W. Hejny and R. D. Lorenz, "Evaluating the practical low speed limits for back-emf tracking-based sensorless speed control using drive stiffness as a key metric," *Conf. Rec. ECCE, 2009*, vol. 4, pp. 2481–2488, Sept. 2009.

[8] L. Ribeiro, M. Degner, F. Briz, and R. Lorenz, "Comparison of carrier signal voltage and current injection for estimation of flux angle and rotor position," *Proc. of the IAS Annual Meeting*, vol. 1, pp. 452–459, Oct. 1998.

[9] M. J. Corley and R. D. Lorenz, "Rotor position and velocity estimation for a salient-pole permanent magnet synchronous machine at standstill and high speed," *IEEE Trans. Ind. Applicat.*, vol. 34, no. 4, pp. 784–789, July/Aug. 1998.

[10] C. Choi and J.-K. Seok, "Pulsating signal injection-based axis switching sensorless control of surface-mounted permanent-magnet motors for minimal zero-current clamping effects," *IEEE Trans. Industry Applications*, vol. 44, pp. 1741–1748, Nov./Dec. 2008.

[11] S. Shinnaka, "A new speed-varying ellipse voltage injection method for sensorless drive of permanent-magnet synchronous motors with pole saliency - new pll method using high-frequency current component multiplied signal," *IEEE Trans on Ind. Appl.*, vol. 44, pp. 776–788, May/June 2008.

[12] S. Ovrebo, "New self-sensing method based on inform, heterodyning and luenberger observer," *Conf. Rec. IEMDC, 2003*, vol. 3, p. June, Sept. 2003.

[13] A. Consoli, G. Scarcella, and A. Testa, "Industy application of zero-speed sensorless control techniques for pm synchronous motors," *IEEE Trans. Ind. Applicat.*, vol. 37, no. 2, pp. 806–812, March/April 2001.

[14] H. Kim and R. D. Lorenz, "Carrier signal injection based sensorless control methods for ipm synchronous machine drives," *Conf. Rec. IEEE-IAS Annu. Meeting*, vol. 2, pp. 977–984, Oct. 2003.

[15] M. Linke, R. Kennel, and J. Holtz, "Sensorless speed an position control of permanent magnet synchronous machines," *Proc. of the IECON'02*, vol. 34, no. 4, pp. 784–789, Nov. 2002.

[16] C. Caruana, G. M. Asher, and M. Sumner, "Performance of hf signal injection techniques for zero-low-frequency vector control of induction machines under sensorless conditions," *IEEE Trans. Ind. Electron.*, vol. 53, no. 2, pp. 195–206, Feb. 2006.

[17] D. Raca, P. Garcia, D. Reigosa, F. Briz, and R. Lorenz, "A comparative analysis of pulsanting vs. rotating vector carrier signal injection-based sensorless control," *Proc. of APEC 2008*, vol. 1, pp. 874–885, Feb. 2008.

A Robust Sensorless Fault Diagnosis Algorithm for Low Cost Motor Drives

Seung-deog Choi[1]
IEEE, Student Member

Bilal Akin[2]
IEEE, Member

Mina M. Rahimian[1]
IEEE, Student Member

Hamid A. Toliyat[1]
IEEE, Fellow

[1]Advanced Electric Machines & Power Electronics Laboratory
Department of Electrical & Computer Engineering
Texas A&M University,
College Station, TX 77843-3128
E-mail: Toliyat@ece.tamu.edu

[2]Texas Instruments Inc.
C2000 System & Applications
12203 Southwest Freeway,
Stafford, Texas 77477
E-mail: b-akin@ti.com

Abstract – **This paper presents a sensorless fault diagnosis technique for low cost AC drives. Currently, in order to achieve a reliable fault diagnosis, a high resolution speed sensor is employed to measure the frequencies of fault signature which depends on motor shaft speed. There is an increased tendency toward sensorless control of AC motor drives because of mounting problems, associated cost, etc. Therefore, the speed sensors become less common feedback tools in high performance motor control applications. The fault diagnosis becomes quite a challenging task in the absence of information from a speed sensor as highly precise motor speed estimation is needed. In this paper, a simple and efficient algorithm for fault diagnosis is proposed utilizing frequency tracking method which does not require speed sensor feedback. It is explicitly verified that the performance of the proposed algorithm is almost comparable to the cases where an accurate speed sensor is used. The algorithm is derived mathematically and its efficacy is proved experimentally using a 3-hp motor generator setup.**

I. INTRODUCTION

Innovative research has been increasingly pursued in fault diagnosis of induction machines. One of the most promising and well known techniques for diagnosis is the motor current signature analysis (MCSA) method [1-5]. Because MCSA techniques usually utilize the parameters and signals obtained in motor control such as the stator current signal, excitation frequency, rotor speed slip, etc., it is relatively easy to implement both motor control and fault diagnosis within a single microcontroller unit. In recent research, the development of hybrid diagnosis system based on motor drive theory and advanced diagnostic signal processing theory yields further reliability and accurate results for industrial systems applications [3-4].

In MCSA-based diagnosis approaches, it is assumed that the electrical or mechanical motor faults are reflected in the stator current due to electro-magnetic coupling between motor current and fault condition-dependent motor operation.

Therefore, the stator current spectrum exhibits abnormal harmonic amplitude modulation which the observed frequencies are generalized as so-called fault characteristic frequencies in many literatures [1-5]. By monitoring and tracking those harmonics irregularity at the expected frequencies, it becomes possible to perform more efficient diagnosis of a motor fault conditions [1-5]. Inspecting specific frequencies at which the fault signatures existed are expected to present reduced computational complexity of fault diagnosis techniques. Whereas, in whole spectrum monitoring technique such as fast Fourier transform (FFT) approach, the complex diagnosis technique is quite involved. In addition to reduced complexity, an effective noise immunization can also be achieved when the diagnosis computation focuses on a specific frequency point in the current spectrum [3]. The reduced complexity and improved noise immunization turns the proposed algorithms into practical tools which can be directly combined with motor control subroutine without violating the CPU utilization of the microprocessors [3-4].

As a tradeoff for low complexity and flexible implementation, a frequency offset caused by a tolerable speed feedback error for motor control might result in a serious error in the measurement of expected fault frequency. Those resolution errors might degrade the overall performance of the fault diagnosis. The reason is that the low complexity signature-based diagnosis algorithms are commonly optimized for precise signal detection at a specific frequency.

Performance degradation and diagnosis system failure becomes more common when the motor is driven without a position sensor. Although there are effective sensorless speed measurement techniques shown in the literature [8], their reliability is usually lower than the physical speed sensor feedback which results in deteriorated diagnosis performance. The computational burden of speed dependent fault signature

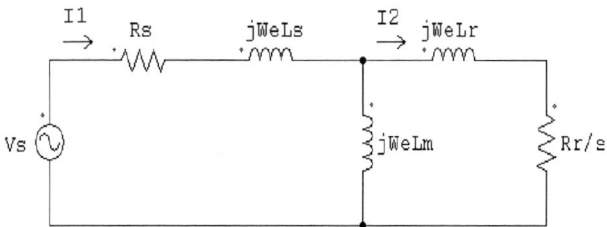

Fig. 1. single phase equivalent circuit of induction motor.

estimation based on speed observers significantly loads the processors' CPU. This also is a reason that speed dependent current signatures are rarely detected in sensorless motor control even though their existence has been widely known [6-7].

In this paper, a two-step procedure for diagnosis is proposed which consists of coarse slip estimation and fine adjustment. The coarse estimation with low complexity algorithm significantly reduces errors within a limited range, the performance of which usually determines the overall efficiency of a system. Fine adjustment with high performance Maximum Likelyhood (ML) algorithm is effectively applied to the coarsely estimated limited error region. The two-step procedure is expected to minimize the overall complexity and increase the accuracy of fault diagnosis algorithm due to effective tracking. It will be verified that the diagnosis using these tracking algorithms not only guarantee low complexity but show almost comparable performance with speed sensor-assisted systems.

II. COARSE SLIP ESTIMATION

It has been reported in many literatures that the fault frequency is dependent on the motor speed and, hence, the slip frequency of a motor. The expected fault frequency can be determined by utilizing motor speed feedback or advanced and sophisticated sensorless speed estimation methods [8]. In this paper, a simple and rough sensorless method for coarse slip estimation is employed. It will be verified the robust performance of the proposed diagnosis method even under continuous error condition which is assumed condition in low cost motor drive applications.

The coarse slip estimation is mainly performed based on the torque-slip relation where the torque is estimated from the line current amplitude. The per unit electromagnetic torque of the induction motor is derived in (1) by referring to the single-phase equivalent circuit of induction motor as shown in Fig.1.

$$T_e^{pu} = \frac{\left(I_2^{pu}\right)^2 r_r}{s\omega_e^{pu}} = \frac{\left(I_2^{pu}\right)^2 r_r}{\omega_{sl}^{pu}} \quad (1)$$

It is assumed that ω_{eL_m} is large enough and R_s is negligible. By simply assuming that $I_2^{pu} \approx T_e^{pu}$, the per unit slip frequency can approximately be derived as follows:

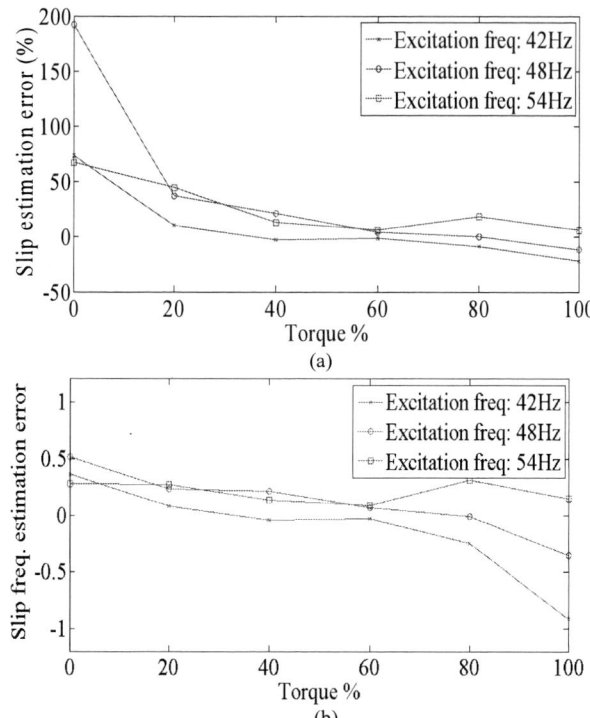

Fig. 2. (a) Slip frequency estimation error percentage (100*slip_esti./slip), and (b) frequency error for motor controlled with V/Hz.

$$1 \approx \frac{I_2^{pu} r_r}{\omega_{sl}^{pu}} \Rightarrow \omega_{sl}^{pu} \approx I_2^{pu} r_r \quad (2)$$

If the error of initial coarse slip estimation derived from (2) is localized with an acceptable range, then sophisticated fine adjustment can be efficiently applied focusing on the limited expected error range. In this paper, the frequency error is located around zero frequency, where coarse slip estimation in (2) is adjusted as follow:

$$\omega_{sl}^{pu} \approx I_2^{pu} r_r \Rightarrow \omega_{sl}^{pu} \approx I_2^{pu} r_r' \quad (3)$$

where $r_r' = kr_r$, k is the parameter which can be obtained through experiments of a test motor.

The slip frequency at the maximum torque, S_{MT} is assumed to be the maximum slip estimation error which is derived approximately as follows:

$$\omega L_s + \omega L_r = \frac{r_r}{S_{MT}} \Rightarrow S_{MT} = \frac{r_r}{\omega L_s + \omega L_r} \quad (4)$$

The performance of coarse slip estimation can be identified if the slip estimation error is sufficiently smaller than S_{MT} and it reduces the error range enough to be tracked later during the fine adjustment step.

In Fig. 2., the slip estimation error is shown assuming that the speed sensor feedback is precise. Fig. (2- a) shows that

978-1-4244-4782-4/10 $26.00 © 2010 IEEE

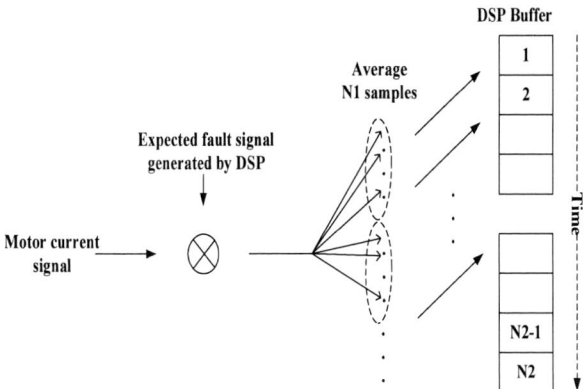

Fig. 3. Averaging scheme of line current signal

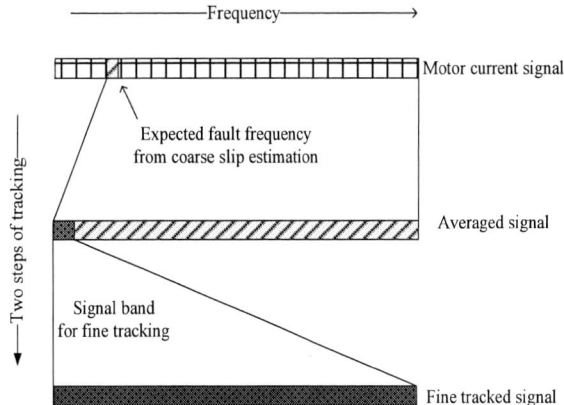

Fig.4. Two steps (coarse plus fine) tracking algorithm.

there are almost continuous slip estimation error with the coarse estimation method in (3). Fig. (2-b) shows an almost an uniform distribution of slip frequency estimation error ranging from -0.5 to 0.5 Hz. It is assumed that, at least, the slip estimation errors are localized around zero frequency through proper design of k in (3) which is shown in Table. 1.

The maximum slip error based on (4) with 42Hz, 48Hz, and 54Hz excitations are computed with a test motor as 1.12Hz, 0.98Hz, and 0.87Hz, respectively. In Fig. (2-b), it is observed that the maximum slip estimation error is sufficiently less than the assumed maximum slip error, which proves that the initial coarse estimation has been satisfied. The frequency offset from coarse estimation is usually less than 1Hz through experiments as shown in Fig. 2. (b).

III. FINE ERROR ADJUSTMENT

A. Signal Averaging

Precision of signal analysis is commonly limited by the memory space to store the signals in the physical DSP buffer. Due to the memory capacity limit, the signal is averaged and stored within a relatively small buffer size which is depicted in Fig. 3. Effective averaging with minimal loss of original information becomes possible if the target signal is at zero frequency (DC). It is because the averaging operation is effective the low pass filtering without interfering DC signal.

At first, the expected fault frequency is sent to zero frequency. This operation is performed by multiplying motor signal by the fault signal generated with fault frequency obtained through coarse slip estimation. It is the similar operation and the reason why the motor signal is converted to zero frequency signal (DC signal) in Park transform based motor control, which is much simplified in this paper. Through this operation, the fault signature is localized at zero frequency and the deviation of which depends on the slip error depicted in Fig. 2. The frequency offset between the coarsely expected fault frequency and the existing fault frequency due to slip estimation error is as follows:

$$f_{offset} = \mid f_{fault} - f_{coarse} \mid \qquad (5)$$

The offset is assumed to be less than 1Hz, which is within an acceptable range according to several coarse slip estimation experiments made under various conditions shown in Fig. 2.

Based on the assumption that fault signature is at zero frequency or at frequency within 1Hz error, the signal is effectively averaged with N1 consecutive sample time as shown in Fig. 3. The averaging becomes possible without interfering target signal information by assuming the fault signal period is much larger than N1 sample time. Therefore, only the irrelevant high frequency signals that the period is much smaller than N1 sample period is removed and noise is minimized through averaging.

B. The ML-Based Fine Slip Error Compensation

The Maximum Likelihood (ML) algorithm [3][9] is used to estimate the frequency offset remaining in the averaged signal which is the maximum of the periodogram and is given by,

$$\hat{f}_{ML} = \arg\max \left| \frac{1}{N} \sum_{n=1}^{N} x_n e^{j2\pi fn} \right| \qquad (6)$$

The maximum of the periodogram given in (6) yields the value of the compensated slip errors which are shown in Fig. 2. However, the complexity of the ML algorithm can be overwhelming for the low cost DSP applications and violates the CPU utilization as it requires relatively high computation complexity. In this paper, the complexity is reduced by limiting the complexity dependent parameters N and f in (6). N is limited by averaging shown in the previous section and f is limited by 1 Hz maximum error of coarse slip estimation assumption.

The overall two-step algorithm consists of coarse and fine tracking is shown in the frequency domain as depicted in Fig. 4. The expected fault frequency obtained utilizing coarse slip frequency estimation is sent to around zero frequency. The error due to the coarse estimation localized around zero frequency in averaged signal is analyzed in the fine tracking step focusing on the limited error region.

IV. EXPERIMENT RESULT

Table 1. System Environment.

Sampling Hz	25 kHz
Data acquisition board	NI-DAQmx
Motor/# of pole	3 hp IM / 4
$Rr/Ls(=Lr)$	0.0727 Ohm/0.079 H
K	0.48
Full load speed	1760
Input voltage	230V
$N1$	125
$N2$	200
Tracking resolution	0.0192 Hz

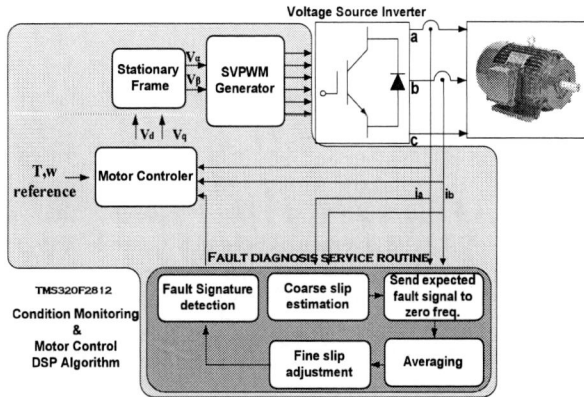

Fig.5. Overall algorithm of DSP based motor control and diagnosis.

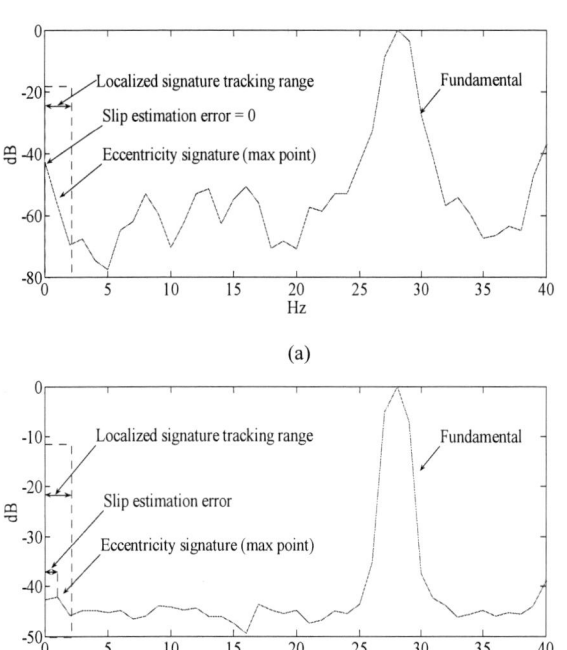

Fig. 6. FFT spectrum of averaged signal (excitation of 48Hz, torque 40%): (a) with speed feedback, and (b) with coarse slip estimation.

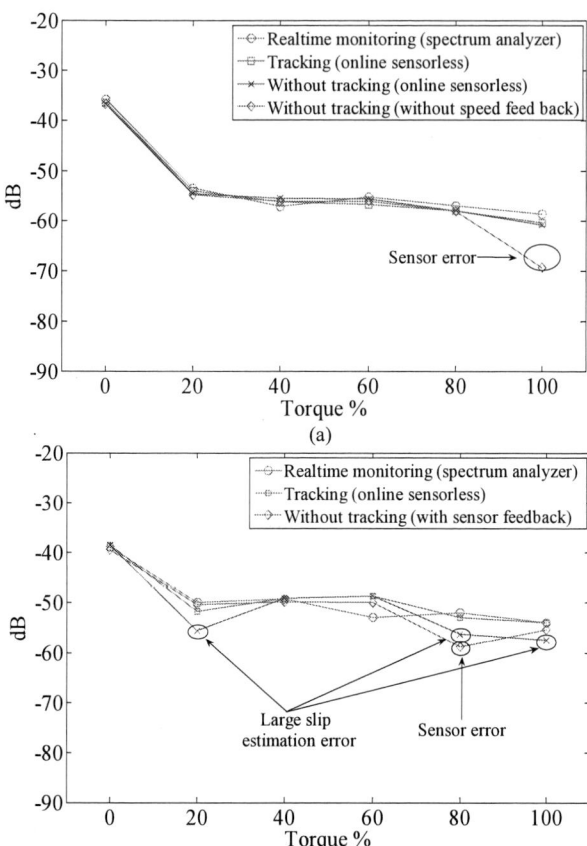

Fig. 7. Signature estimation (a) with excitation of 42Hz, and (b) excitation of 52Hz.

V/f motor control and online fault diagnosis are implemented within a digital signal processor (DSP) with 32-bit fixed-point, 150 MHz, and 12-bit embedded ADC. The proposed fault diagnosis service routine is added in the free memory space of DSP where the motor control is located without degrading control performance. Fig. 5. shows the overall algorithm of DSP-based motor control and proposed diagnosis service routine.

Eccentricity fault has been considered in this paper which is expected to clearly show the capability of the proposed algorithm under wide amplitude variation of fault signature versus load. Proposed algorithm is can be applied to various fault models such as the broken bar, and bearing faults in future studies.

The expected characteristic frequency of the eccentricity fault signatures in the motor stator currents are given by,

$$f_{eccentricity} = \left[1 \pm m\left((1-s)/(p/2)\right)\right]f = f \pm mf_r, \quad m=1,2,3,\cdots (7)$$

where s is the per unit slip, f is the fundamental frequency of the stator current, p is the number of poles, and f_r is the mechanical rotational frequency.

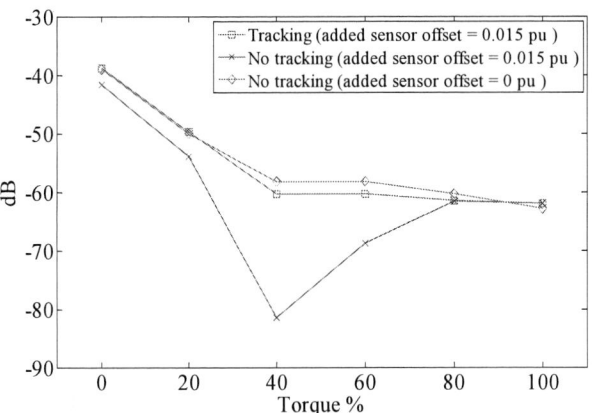

Fig. 8. Signature estimation with excitation of 54Hz..

Figs. (6-a) and (6-b) show that the signal in the averaging block buffer which fault signature is shown in the base frequency (0 Hz) band. The signature is located exactly at zero frequency with having the correct speed feedback shown in Fig. (6-a). The signature is slightly deviated from zero frequency with the coarse slip estimation in Fig. (6-b). Due to the effective localization of the prospective signature at low frequency and its small error region, the fine tracking can be achieved with low complexity.

Figs. (7-a) and (7-b) show the signature amplitude measured using Spectrum Analyzer as the reference, with fine tracking without position sensor, without fine tracking without position sensor, and without fine tracking with position sensor feedback. During the experimental measurements shown in Fig. (7-a), the coarse slip estimation error is usually recorded as less than 0.002 pu, which is quite low and does not affect detection performance. At 100% of the torque in Fig. (7-a), the detection with speed feedback shows an unreliably detected value which is assumed to be an instantaneous sensor error. It shows the need for frequency tracking detection even in a system with a speed sensor for high performance application. In the whole speed range, the sensorless diagnosis assisted by frequency tracking algorithm shows reliable result. In Fig. (7-b), large coarse slip estimation errors are monitored at 20%, 80%, 100% torques, where the tracking algorithm compensates these losses which can be confirmed with results obtained from the Spectrum Analyzer.

In Fig. 8, a speed sensor error of 0.015 pu is intentionally added to coarse slip estimation through the DSP software in the experiment. In this case, a significant deviation is observed versus entire torque range when compared to zero offset addition. Fig.8 also shows how an effective tracking algorithm compensates for distorted slip estimation and yields almost the same performance as with the zero added error case.

V. CONCLUSION

A low cost, high performance position sensorless diagnosis algorithm is presented in this paper. The proposed algorithm provides a solution for one of the fundamental problems in fault diagnosis when a speed sensor error arises unexpectedly. The paper also shows its effectiveness in position sensorless diagnosis with a performance almost comparable to the system with an accurate speed sensor measurement. The proposed algorithm can further extend MCSA-based diagnosis to low cost applications in industry.

ACKNOWLEDGMENT

This publication was made possible by a NPRP grant from the Qatar National Research Fund (a member of The Qatar Foundation). The statements made herein are solely the responsibility of the authors.

REFERENCES

[1] S. Nandi, H.A. Toliyat, and X. Li, "Condition Monitoring and Fault Diagnosis of Electrical Machines-A Review," *IEEE Trans. on Energy Conversion,* vol. 20, pp. 719-729, no. 4, Dec. 2005.

[2] M. El Hachemi Benbouzid, "A review of induction motors signature analysis as a medium for faults detection," *IEEE Trans. on Industrial Electronics ,* vol. 47, pp. 984 – 993, Oct. 2000.

[3] S.D. Choi, B. Akin, M.M. Rahimian, H.A. Toliyat, "Fault Diagnosis Implementation of Induction Machine based on Advanced Digital Signal Processing Techniques," *Proceeding of APEC '09,* Washington, DC, USA.

[4] B. Akin, H.A. Toliyat, U. Orguner, and M. Rayner, "Phase Sensitive Detection of Motor Fault Signatures in the Presence of Noise," *IEEE Trans. on Industrial Electronic,* Vol. 55, pp. 2539-2550, June, 2008.

[5] G.B. Kliman, W.J. Premerlani, B. Yazici, R.A. Koegl, J. Mazereeuw, "Sensorless, online motor diagnostics," *IEEE Computer Applications in Power,* pp. 39 – 43, Vol. 10, Issue 2, Apr. 1997.

[6] S. Nandi, S. Ahmed, H.A. Toliyat, R.M. Bharadwaj, "Selection criteria of induction machines for speed-sensorless drive applications," *IEEE Trans. Industry Applications,* pp. 704 – 712, Vol. 39, Issue 3, May-June 2003.

[7] G. Joksimovic, M. Djurovic, J. Penman, "Cage rotor MMF winding function approach," *IEEE Power Engineering Review,* Vol. 21, pp. 64 – 66, Apr. 2001.

[8] Guzinski, J.; Abu-Rub, H.; Toliyat, H.A.;, "An advanced low-cost sensorless induction motor drive," *IEEE Trans. Industry Applications,* Vol. 39, pp. 1757 – 1764, Nov.-Dec. 2003.

[9] A.J. Viterbi, *Principles of Coherent Communication,* McGraw-Hill New York, 1966.

High dynamic performance constrained optimal control of induction motors

Sébastien Mariéthoz, Alexander Domahidi and Manfred Morari
Automatic Control Laboratory, ETH Zürich
Physikstrasse 3, CH - 8092 Zürich, Switzerland
Email: mariethoz@control.ee.ethz.ch

Abstract—This paper deals with field oriented explicit model predictive control of induction motors. Constraints are inherently handled by model predictive controllers. Taking into account the inverter limitations and maximum admissible stator current magnitude, the torque and flux can be optimally controlled, yielding optimal dynamic performance and disturbance rejection. Casting the problem to solve as a constrained quadratic finite time optimal control problem, an explicit controller can be computed off-line, the resulting piecewise affine control law requiring to run a binary search-tree in order to select the optimal linear state feedback. An observer is used to estimate motor electrical state and disturbance, which allows to effectively takes into account the induced voltage and cross-coupling between the d and q axis. The resulting controller requires little tuning and can automatically be generated for different motor parameters.

I. INTRODUCTION

The ability to deal with constraints and switched systems renders model predictive control (MPC) very attractive for the control of power electronics systems. Two main categories of model predictive control approaches may be considered.

A. Finite switching state enumerative MPC

The converter switching states can be changed only at discrete time instants, which are generally the sampling instants. At each sampling period all the possible future switching states of the power converter are enumerated for future discrete time instants over the prediction horizon. This approach is very simple; it however suffers from several critical drawbacks. The time interval between the different discrete time instants directly limits the accuracy of the generated switching pattern. The number of switching state combinations to explore grows exponentially with the prediction horizon. As a consequence, the horizon length is practically very limited (1 to 3), which results in a poor control quality at steady state. Finally, it is practically impossible to control or limit the switching frequency and the only viable way seems to use a sampling frequency similar to the switching frequency [1].

B. (Continuous set) MPC

The converter switching state can be changed at any time instant, usually by selecting the optimal duty cycles of a PWM used to synthesize the (voltage/current) pattern to be generated by the power converter. As a consequence, the harmonic properties of the steady state solution are very good. The optimal duty cycles may however be difficult to obtain as they require solving a constrained (finite time) optimal control problem, which is usually referred as CFTOC problem in MPC literature. The feasibility and effectiveness of this MPC approach have been demonstrated experimentally for various power electronics applications [2]–[4] including drive control [5], [6] and high frequency applications [7].

C. Previous applications of MPC to electrical drives

In [8], a reduced order model is used to derive a controller where only the stator currents are considered as state variables. As rotor flux and speed are not featured in the model, this results in poor performance. In [9], the inverter switches are directly controlled using a high sampling rate (typically 5-20 μs). As consequences, a high computational power is required to run the MPC algorithm and the switching frequency is not directly controlled as it depends on the operating point. Extensive tests are required to ensure that there are no operating points for which excessive switching frequency and thus losses occur. In [10] an *unconstrained* optimal control also referred as GPC is used to directly control the speed and the flux. The performance is poor as the torque limit is not considered. [11] also follows the speed control approach but using explicit model predictive control, which allows incorporating constraints in the optimal controller. The main drawbacks of the direct speed control approach are that the knowledge of the mechanical dynamics is required and that it is difficult to embed a complex trajectory controller in such frameworks. It is therefore preferable to employ a cascade control approach and to provide an optimal torque controller that can be incorporated into a higher level speed or trajectory controller.

D. Objectives and contributions

This paper investigates the application of explicit model predictive control (EMPC) to the torque control of induction motors. The main objectives of this work are:

1) to obtain optimal performance and a controller effective for all operating points,
2) to take into account all constraints and disturbances in the controller synthesis in order to eliminate external control functions such as anti-windup and feed-forward,
3) to reduce the tuning time by deriving a controller, which depends only the machine parameters and on a minimum of additional tuning parameters,

4) to increase the algorithm reliability by having an automatic design procedure where only the data structure needs to be changed for new parameters.

The proposed control scheme manipulates the stator voltages through a PWM inverter. The EMPC determines the optimal duty cycles using a full order electrical model of the machine. The general concepts proposed in [5] for the control of a PM motor are applied here to an induction motor. The current and flux space vectors are projected on the rotor flux space vector and the model non-linear or time varying terms are approximated using an additive disturbance which augments the controller state.

II. PREDICTION MODEL DERIVATION

A. Induction motor dynamic equations

The first ingredient in model predictive control is the model used to predict the evolution of the system state variables, which are necessary to optimize the control objective over the control horizon. The dynamic equations of an electrical drive may be very complex. In particular the mechanical dynamic equations highly depend on the application both in structure and parameters. These dynamics are however slower than the electrical system dynamics and they can be neglected for most applications to predict the evolution of the electrical variables over a few switching periods. We therefore consider only the dynamic equations of the electrical part, which can be written in matrix form as:

$$\dot{x} = A\,x^f + A_s\,\omega_s^f\,x^f + A_r\,\omega_r^f\,x^f + B\,u_s^f \quad (1)$$

where the terms that depend on the flux, rotor and reference frame angular velocities have been separated from the time invariant dynamics. f denote the employed coordinate reference frame. Selecting the stator currents and rotor fluxes to form the state:

$$x = \begin{bmatrix} i_s^f \\ \psi_r^f \end{bmatrix} \quad (2)$$

The system matrices are:

$$A = \begin{bmatrix} -\frac{1}{\sigma\tau_s} - \frac{1-\sigma}{\sigma\tau_r} & 0 & \frac{1-\sigma}{\sigma\tau_r L_m} & 0 \\ 0 & -\frac{1}{\sigma\tau_s} - \frac{1-\sigma}{\sigma\tau_r} & 0 & \frac{1-\sigma}{\sigma\tau_r L_m} \\ \frac{L_m}{\tau_r} & 0 & -\frac{1}{\tau_r} & 0 \\ 0 & \frac{L_m}{\tau_r} & 0 & -\frac{1}{\tau_r} \end{bmatrix} \quad (3a)$$

$$A_s = \begin{bmatrix} 0 & 1 & 0 & -\frac{1-\sigma}{\sigma L_m} \\ -1 & 0 & \frac{1-\sigma}{\sigma L_m} & 0 \\ 0 & 0 & 0 & 0 \\ 0 & 0 & 0 & 0 \end{bmatrix} \quad (3b)$$

$$A_r = \begin{bmatrix} 0 & 0 & 0 & \frac{1-\sigma}{\sigma L_m} \\ 0 & 0 & -\frac{1-\sigma}{\sigma L_m} & 0 \\ 0 & 0 & 0 & -1 \\ 0 & 0 & 1 & 0 \end{bmatrix} \quad (3c)$$

$$B = \begin{bmatrix} \frac{1}{\sigma L_s} & 0 \\ 0 & \frac{1}{\sigma L_s} \\ 0 & 0 \\ 0 & 0 \end{bmatrix} \quad (3d)$$

B. Discrete-time control model

For the MPC, a discrete-time model equivalent to (1) is computed numerically in the rotor flux reference frame:

$$x_{k+1}^r = A_d\,x_k^r + A_{s,d}\,\omega_s^r\,x_k^r + A_{r,d}\,\omega_r^r\,x_k^r + B_d u_{s,k}^r \quad (4)$$

To make notations simpler, we will omit the reference frame marker r and assume rotor flux reference frame for the rest of the paper. Starting at each sampling period from the initial state x_k, the *open loop* evolution of the state over the horizon N is predicted. The sequence of states $x_{k+1}, x_{k+2}, \ldots x_{k+N}$ is thus expressed as a function of the initial state x_k and the sequence of control actions $u_{s,k}, u_{s,k+1}, \ldots u_{s,k+N-1}$ using model (4).

C. MRAS flux and speed observer

The only measured quantities are stator currents and DC-link voltage. The rotor flux is required in order to obtain the initial state x_k and to obtain the position of the reference frame. One of the simplest way to obtain relatively good estimates of the rotor flux and speed is to use a model reference adaptive system (MRAS) [12].

D. State and disturbance observer

In order to predict the evolution of the state over the control horizon, the knowledge of ω_s and ω_r is required. (4) can be rewritten:

$$x_{k+1} = A_d\,x_k + B_d u_{s,k} + w_k \quad (5a)$$
$$w_k = A_{s,d}\,\hat{\omega}_{s,k}\,x_k + A_{r,d}\,\hat{\omega}_{r,k}\,x_k \quad (5b)$$
$$x_k = \begin{bmatrix} i_{s,k} \\ \hat{\psi}_{r,k} \end{bmatrix} \quad (5c)$$

The rotor speed and flux estimates $\hat{\omega}_r$ and $\hat{\psi}_{r,k}$ are readily available from the MRAS observer. The flux velocity ω_s can be estimated or it can more simply be assumed that $\hat{\omega}_r \approx \hat{\omega}_s$. These estimated quantities are however corrupted by inverter non-linearities, modeling errors and measurement noise. As a consequence of these inaccuracies, the control offset at steady state cannot be eliminated without integral action. A very attractive alternative way to an integrator and to the "feed-forward" computation of w_k, which allows eliminating control offset, consists in estimating the disturbance w_k using a model-based observer and correcting it using the error between the measured and predicted currents. Theoretical foundation, existence of the solution, stability and convergence properties are discussed in [13]. We briefly describe here the main concept. At startup, $\hat{x}_0$ and $\hat{w}_0$ are initialized to zero. At each sampling instant state $\hat{x}_{k+1|k}$ is predicted using the model (5). At the next sampling instant, the error between $\hat{x}_{k|k-1}$ and x_k is used both to correct both $\hat{x}_{k|k-1}$ and $w_{k|k-1}$, resulting in the following observer algorithm:

$$\epsilon_k = x_k - \hat{x}_{k|k-1} \quad (6a)$$
$$\hat{w}_{k|k} = \hat{w}_{k|k-1} + L_w\,\epsilon_k \quad (6b)$$
$$\hat{x}_{k|k} = \hat{x}_{k|k-1} + L_x\,\epsilon_k \quad (6c)$$
$$\hat{x}_{k+1|k} = A_d\,\hat{x}_{k|k} + B_d u_{s,k} + \hat{w}_{k|k} \quad (6d)$$

The design of the optimal correction feedback $\boldsymbol{L}_w$ and $\boldsymbol{L}_x$ using Kalman filter framework is discussed in [13]. The rotor flux obtained from the MRAS observer is used as measured flux. With this approach, we need the MRAS observer only to build the rotor flux and the reference frame. The flux and rotor speeds are not anymore explicitly required for the torque control and (6) provides all the information required to derive and feed our MPC control scheme.

III. PROPOSED CONTROL SCHEME

A. MPC control approaches

MPC gives more flexibility than classical control to select the variables to control. Restricting the selection to field oriented control approaches, we can control:

- the d and q components of the stator current in order to indirectly control the rotor flux and the torque,
- directly the flux and the torque,
- directly the flux and the q component of the stator current.

The first approach is similar to the classical field oriented indirect approach adapted to the MPC framework. In the second approach, one expects to obtain better performance by controlling directly the objective variables. As the torque is a bilinear function (vector product) of the flux and stator currents, this considerably increases the complexity of the resulting optimal control problem, making this approach unpractical. Assuming that the rotor flux is well controlled, a nearly perfect torque control is obtained by controlling the q-component of the stator current. In the third approach, the flux is therefore directly controlled, while the torque is indirectly controlled through the q component of the stator current. By directly controlling the flux, one expects to better maintain it in face of the disturbances that occur during very fast transients. On the other hand the difficulty in this approach is that the d component of the stator current is not anymore directly controlled. The stator current could therefore exceed its maximum admissible value. It is thus necessary to add a constraint to the optimization problem in order to limit the stator current and ensure a safe operation of the motor and inverter.

In the following we will consider and discuss only the first and the third approaches that will be denoted EMPC-d-q and EMPC-flux-q.

B. Target reference tracking

The role of the controller is to bring the system state to an equilibrium point, which ideally brings the measured output to its reference value despite disturbances. The target reference is the vector formed by the state and stator voltages, which form an equilibrium point for the targeted output reference and given disturbance. It is the solution of the system:

$$
\begin{bmatrix} \boldsymbol{I} - \boldsymbol{A}_d & -\boldsymbol{B}_d \\ \boldsymbol{C} & 0 \end{bmatrix} \begin{bmatrix} \boldsymbol{x}_{\text{ref},k} \\ \boldsymbol{u}_{s,\text{ref},k} \end{bmatrix} = \begin{bmatrix} \boldsymbol{I} & 0 \\ 0 & \boldsymbol{I} \end{bmatrix} \begin{bmatrix} \boldsymbol{w}_k \\ \boldsymbol{y}_{\text{ref}} \end{bmatrix} \quad (7)
$$

which yields:

$$
\begin{bmatrix} \boldsymbol{x}_{\text{ref},k} \\ \boldsymbol{u}_{s,\text{ref},k} \end{bmatrix} = \begin{bmatrix} \boldsymbol{I} - \boldsymbol{A}_d & -\boldsymbol{B}_d \\ \boldsymbol{C} & 0 \end{bmatrix}^{-1} \begin{bmatrix} \boldsymbol{w}_k \\ \boldsymbol{y}_{\text{ref}} \end{bmatrix} \quad (8)
$$

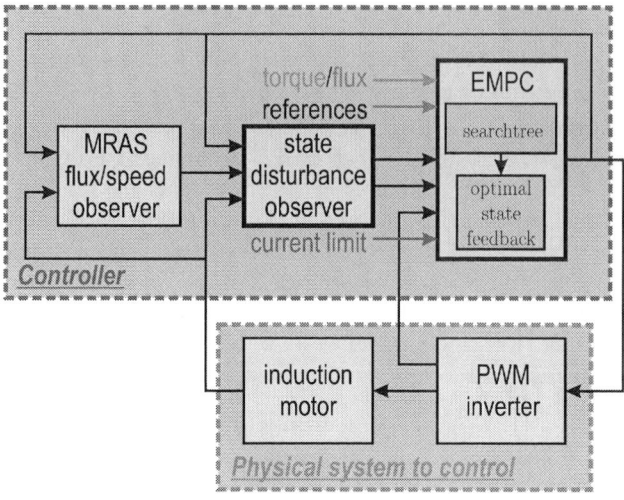

Fig. 1. Overview of the control scheme including the observers (coordinate transformations are not shown for simplicity)

Using a target reference allows determining the reference for the stator currents, rotor fluxes and stator voltages, as a function of the reference and applied disturbance. It improves tracking performance by predicting the equilibrium point of the system to reach. Using the same model for the target reference and for the disturbance estimation (6), *no integral action* is required in the controller even in presence of model mismatch as the observer identifies the disturbance, which brings the real system to the equilibrium.

C. Constrained finite time optimal control problem

The control objective is to minimize the tracking error. The main ingredient to achieve this is the cost function that measures the error between the predicted state sequence and the target reference:

$$
\begin{aligned}
J = \sum_{n=1}^{N} & \left(\boldsymbol{x}_{k+n} - \boldsymbol{x}_{\text{ref},k}\right)^T \boldsymbol{Q} \left(\boldsymbol{x}_{k+n} - \boldsymbol{x}_{\text{ref},k}\right) \\
& + \left(\boldsymbol{u}_{s,k+n-1}^r - \boldsymbol{u}_{s,\text{ref},k}\right)^T \boldsymbol{R} \left(\boldsymbol{u}_{s,k+n-1}^r - \boldsymbol{u}_{s,\text{ref},k}\right)
\end{aligned} \quad (9a)
$$

One of the main difference between MPC and other control approaches is the ability to incorporate constraints in the control objective. For our problem, the constraints are the maximum admissible current and the maximum available inverter voltage:

$$
\hat{\imath}_{s,d,k+n} < i_{\text{max},k} \quad \text{maximum admissible current} \quad (9b)
$$
$$
\boldsymbol{H}_u \boldsymbol{u}_{s,k+n-1}^r \leq \boldsymbol{K}_u \quad \text{i.e. } \boldsymbol{u}_{s,k+n-1}^r \text{ is within a polygon} \quad (9c)
$$

The cost function (9a) is employed for both MPC-d-q and MPC-flux-q controllers. The d-current value that brings the flux to its reference value is selected as reference. The reference value for the q-component of the stator current is obtained by dividing the torque reference by flux reference. (8) is used to compute the target state and stator voltage from this two reference values and from the identified disturbance. The cost function is selected to tailor the closed-loop system behaviour by capturing the control trade-offs through the matrix Q and R. In the EMPC-d-q scheme, there is a high penalty on the stator current tracking errors, while there is a high penalty on the rotor flux in the EMPC-flux-q scheme.

D. Explicit model predictive controller

The optimal stator voltage (or equivalently duty cycles) is found by solving the CFTOC problem defined by the cost function and the constraints formulated in (9) at each sampling instant. This procedure is time consuming and the run-time moreover depends on the system state. For (hybrid) linear constrained systems this problem can however be solved off-line. The resulting control law is piecewise affine [14] and the linear state feedback control law to be evaluated for a given state is found by running a binary search tree where a linear function is evaluated at each node [15]. The algorithm maximum run-time is given by the binary search-tree maximum depth.

The state is extended to feature all parameters required to solve the problem:

$$ x_k = \begin{bmatrix} \hat{x}_k \\ \hat{w}_k \\ i_{\text{ref},k} \\ i_{\max,k} \\ u_{DC,k} \end{bmatrix} \quad (10) $$

Resulting in a 11 dimension extended state, where the motor state $\hat{x}_{k+n}$ and disturbance $\hat{w}_k$ are directly obtained from (6).

The CFTOC problem is formulated and solved using high level software tools under matlab [16], [17], directly yielding the data structure that is used to automatically generate the controller. Changing motor parameters only requires rerunning a script that automatically generates the data structure used by the control algorithm.

IV. EXPERIMENTAL RESULTS

A. Set-up description

The proposed EMPC scheme is applied to the torque control of an induction motor rated 750 W (1 hp), 115 V, 60 Hz, 4 poles (Baldor EM3546). The simplified block diagram of the experimental rig is shown in Fig 2. The induction motor is supplied by a 2-level 3-phase inverter. A coupled synchronous motor of similar ratings serves as load. All control functions are implemented in C on a 32-bit 32 MHz floating point DSP (Sharc ADSP 21062). For the presented results, the inverter switching frequency is 2.5 kHz, the nominal DC-link voltage is 170 V.

B. MPC scheme evaluation

The control scheme is evaluated using the torque profile illustrated in Fig 3. Fig 4 zooms on a transient between triangular torque reference and pulsed torque reference. We verify on the duty cycle plot that the inverter full capability is used during very abrupt transients. Fig 5 zooms more around the

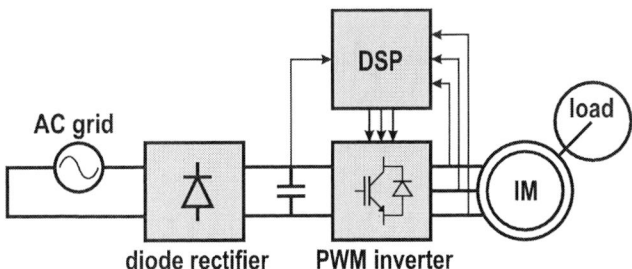

Fig. 2. Simplified block diagram of the experimental rig.

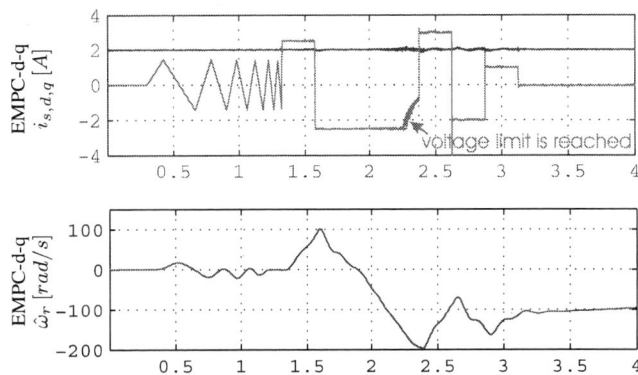

Fig. 3. Stator currents' profile. Results shown for the EMPC-d-q scheme.

abrupt zero current crossing. It can be seen on the duty cycle plot that the inverter available voltage is fully utilized during the transient. The flux and disturbance estimate are corrupted by the uncompensated inverter distortion around zero. This effect can be seen on the MRAS observer rotor flux and speed estimate (see corresponding plots). This eventually results in a significantly wrong state prediction and an overshoot. When comparing the speed signal to the stator current signal, it can moreover be seen that the MRAS speed estimate features a significant lag. Fig 4 zooms on the inverter limitation that is caused by overspeed.

C. Control schemes comparison

Several field oriented control strategies are compared using the same torque profile illustrated in Fig 3:

1) PI field oriented controller with anti-windup and feed-forward speed and cross coupling compensation (PIFF),
2) proposed EMPC control scheme with d-q stator current component control (EMPC-d-q),
3) proposed EMPC control scheme with direct flux and q stator current component control (EMPC-flux-q).

For all control schemes, the rotor flux and speed are estimated using the same MRAS observer. Fig 3 shows the overall result for the EMPC-d-q scheme.

1) Startup and triangular q-current profile: Fig 7 shows a startup where the flux is first built by applying open-loop a reference for the d stator current component. The PIFF and EMPC-d-q schemes impose the d-current component, while the EMPC-flux-q scheme penalizes more the flux error than

978-1-4244-4782-4/10 $26.00 © 2010 IEEE

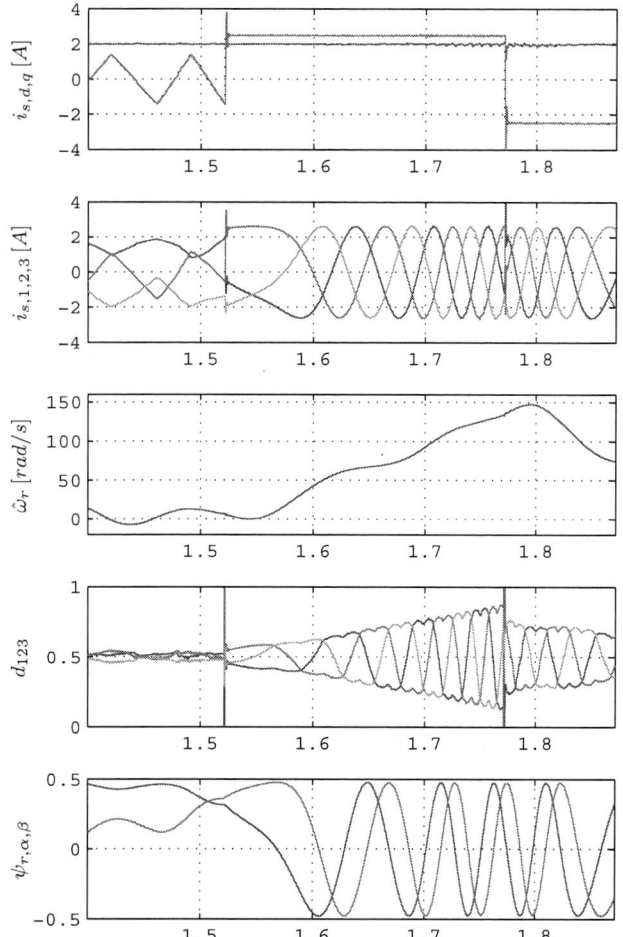

Fig. 4. Zoom on stator current transient between triangular reference and pulsed reference for the EMPC-d-q scheme.

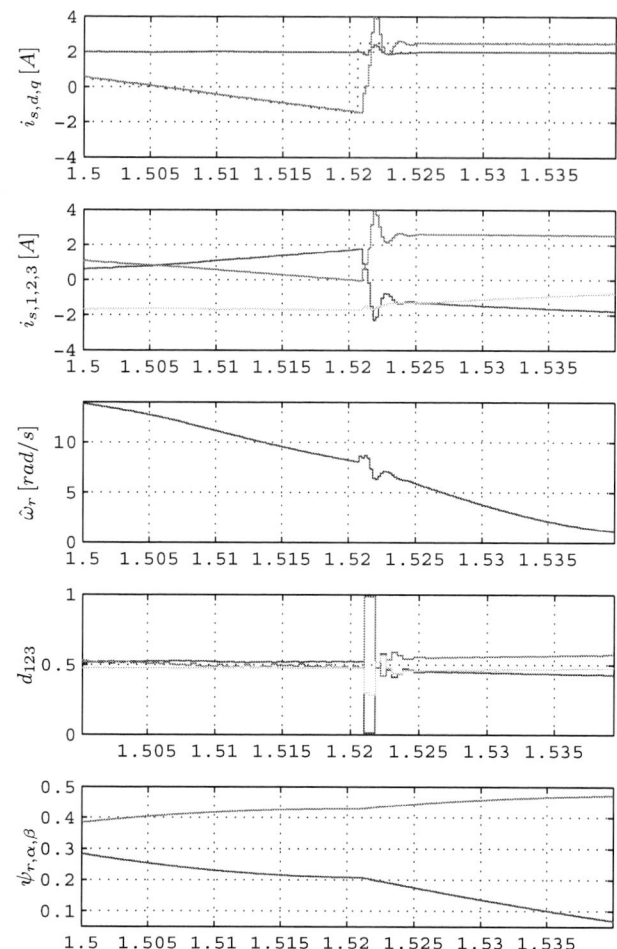

Fig. 5. Zoom on stator current transient between triangular reference and pulsed reference for the EMPC-d-q scheme. See flux and speed distortions around the current zero crossing.

the d-current error in the cost function. As a consequence, the d-current deviate from its nominal value and this controller is inherently the faster to build the rotor flux. The same functionality can of course be added to the other controllers at the cost of adding an additional cascade controller stage on their top. Once the flux is built, a triangular torque reference of increasing frequency is applied. For all methods, the stator currents can barely be distinguished from their reference, which indicates a good tracking.

2) Increasing frequency triangular q-current profile and pulsed torque profile: Fig 8 illustrates fast torque reference tracking. An increasing frequency triangular torque reference followed by a pulse torque pattern is tracked by the controllers. All controllers perform well on this test, however a small deviation can be observed on the d-current of the PIFF controller. The two variants of the proposed control scheme better maintain the d-current axis during these fast transients.

3) Pulsed torque profile and operation at high speed: In Fig 9, the controllers' behaviour is compared at high speed when the inverter voltage limitation becomes active. The demagnetization is not activated for these tests. The PIFF

controller features a much higher ripple than the two proposed EMPC controllers when the voltage limitation becomes active. The reason is that the EMPC controllers better handle the inverter saturation than the anti-windup system. This shows one of the big advantages of replacing the controller integral action by a disturbance observer.

D. MRAS observer limitations

These experimental results show that the MRAS observer is inappropriate in its simplest PI form to applications with very high dynamic performance. The observer is difficult to tune and unable to track very fast accelerations and fast varying accelerations. As a consequence, the dynamic performance is considerably affected by the observer performance. Moreover, the observer and the controller are strongly coupled as each affects the performance of the other.

V. CONCLUSION

A high dynamic performance torque control scheme has been proposed. The explicit optimal control law is a piecewise affine function of the state and control parameters. It is

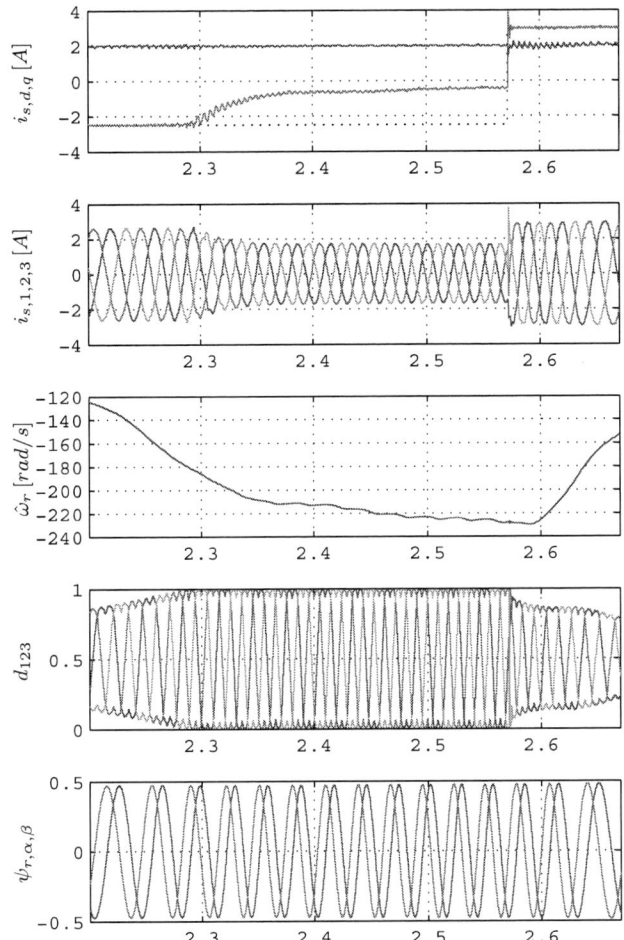

Fig. 6. Zoom on overspeed area for the EMPC-d-q scheme.

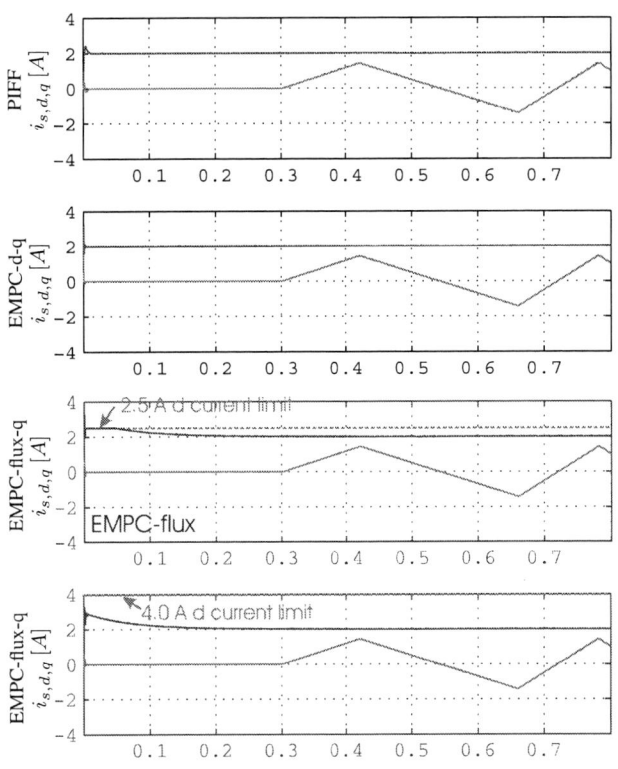

Fig. 7. Comparison of control schemes at start-up

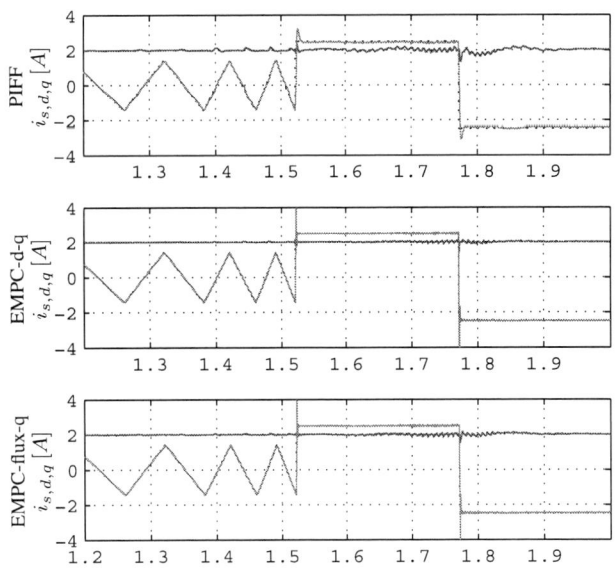

Fig. 8. Comparison of control schemes during fast torque and speed transients

obtained by solving off-line a constrained finite time optimal control problem. It is implemented efficiently online through a binary search-tree that selects the optimal linear control law. A model based observer is used to compensate the computation time delay by predicting the state evolution. The same observer is used to estimate the disturbance that is mostly due to the terms in the state update equation that depend on the flux and rotor speed. The proposed control scheme cancels tracking error by tracking the target reference which is an equilibrium point of the system for the torque and flux reference and identified disturbance. The flux control loop is inherently embedded into the controller by penalizing the flux tracking error. The effectiveness of the proposed control concepts is demonstrated experimentally. Experiments show that the dynamic performance limitations mostly stem from the flux observer performance. Dynamic performance could therefore be significantly improved by better compensating inverter distortion and selecting a more effective flux observer scheme. It is experimentally demonstrated that the proposed control scheme outperforms the state-of-the-art PI with feed-forward field oriented controller. The two proposed EMPC controller variants perform similarly well. The direct control

of the flux allows a slightly better control of the flux at startup and during fast transients.

REFERENCES

[1] M.W. Naouar, A.A. Naassani, E. Monmasson, and I. Slama-Belkhodja. FPGA-based predictive current controller for synchronous machine speed drive. *IEEE Trans. on Power. El.*, 23(4):2115–2126, 2008.

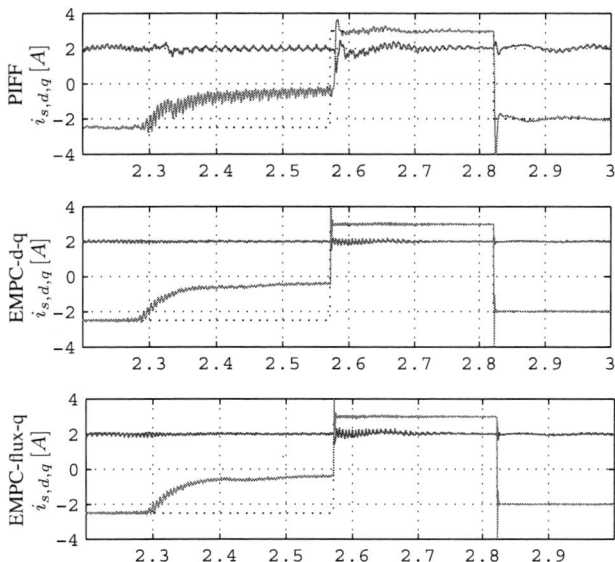

Fig. 9. Comparison of control schemes during inverter saturations

[2] M. Veenstra and A. Rufer. Control of a hybrid asymmetric multilevel inverter for competitive medium-voltage industrial drives. *IEEE Trans. on Ind. Applicat.*, 41(2):655–664, 2005.

[3] S. Mariethoz, M. Herceg, and M. Kvasnica. Model Predictive Control of buck DC-DC converter with nonlinear inductor. In *IEEE COMPEL Workshop on Control and Modeling for Power Electronics*, Zurich, Switzerland, August 2008.

[4] S. Mariethoz and M. Morari. Explicit Model-Predictive Control of a PWM Inverter With an LCL Filter. *IEEE Trans. on Ind. Electron.*, 56(2):389 – 399, February 2009.

[5] S. Mariethoz, A. Domahidi, and M. Morari. Sensorless Explicit Model Predictive Control of Permanent Magnet Synchronous Motors. In *IEEE IEMDC, Int. Electric Machine and Drive Conf.*, pages 1492 – 1499, Miami, FL, USA, May 2009.

[6] S. Mariéthoz, A. Domahidi, and M. Morari. A model predictive control scheme with torque ripple mitigation for permanent magnet motors. In *IEEE IECON, Industrial Electronics Conf.*, Porto, Portugal, November 2009.

[7] S. Mariéthoz, U. Mäder, and M. Morari. High-speed FPGA implementation of observers and explicit model predictive controllers. In *IEEE IECON, Industrial Electronics Conf.*, Porto, Portugal, November 2009.

[8] A. Linder and R. Kennel. Model predictive control for electric drives. In *Proc. Power El. Spec. Conf.*, pages 1793–1799, 2005.

[9] T. Geyer, G. Papafotiou, and M. Morari. Model Predictive Direct Torque Control — Part I: Concept, Algorithm and Analysis. *IEEE Trans. on Industrial Electronics*, 56(6):1894–1905, June 2009.

[10] E.S. Santana, E. Bim, and W.C. Amaral. A predictive algorithm for controlling speed and rotor flux of induction motor. *IEEE Trans. on Ind. Electron.*, 55(12):4398–4407, 2008.

[11] S. Bolognani, S. Bolognani, L. Peretti, and M. Zigliotto. Design and implementation of model predictive control for electrical motor drives. *IEEE Trans. on Ind. El.*, 56(6), 2009.

[12] S.M. Gadoue, D. Giaouris, and J.W. Finch. Performance evaluation of a sensorless induction motor drive at very low and zero speed using a MRAS speed observer. In *Proc. Conf. on Industrial and Information Systems*, Dec. 2008.

[13] Urban Maeder, Francesco Borrelli, and Manfred Morari. Linear offset-free model predictive control. *Automatica*, 45(10):2214 – 2222, 2009.

[14] A. Bemporad, F. Borrelli, and M. Morari. Model Predictive Control Based on Linear Programming - The Explicit Solution. *IEEE Trans. on Aut. Control*, 47(12):1974–1985, Dec. 2002.

[15] P. Tondel, T.A. Johansen, and A. Bemporad. Computation and approximation of piecewise affine control laws via binary search trees. In *In Proc. of the 41st IEEE Conf. on Decision and Control*, volume 3, pages 3144 – 3149, Dec 2002.

[16] J. Löfberg. YALMIP : A Toolbox for Modeling and Optimization in MATLAB. In *Proc. of the CACSD Conf.*, Taipei, Taiwan, 2004.

[17] M. Kvasnica, P. Grieder, and M. Baotić. Multi-Parametric Toolbox (MPT), 2004.

PMSM Control based on Edge Field Measurements by Hall Sensors

Sungyoon Jung, Beomseok Lee and Kwanghee Nam
Department of Electrical Engineering, POSTECH,
Hyoja San-31, Pohang, 790-784 Republic of Korea.
Tel:(82)54-279-2218,Fax:(82)54-279-5629,
E-mail:mokona79@postech.ac.kr, bestlbs@postech.ac.kr and kwnam@postech.ac.kr

Abstract— **Great merits of utilizing Hall sensors as a position detector lie in the cost effectiveness and relatively easiness in installation. In this work, the axial component of the rotor PM edge field is detected by linear Hall sensors. The measured values by the two Hall sensors look like $\sin\theta$ and $\cos\theta$, but contain a significant amount of third harmonics which causes angle estimation error. In this work, a phase locked loop (PLL) is utilized with the adaptive notch filter (ANF) for the purpose of filtering out the third order harmonics from the PLL input signals. The PLL yields both speed and angle estimates, and is used for speed feedback and coordinate change into the synchronous reference frame, respectively. Experiments showed satisfactory results in current and speed regulations. This kind of edge field detection method is advantageous, since the edge field is not strong enough to saturate the commercial Hall sensors and the edge field measuring is not affected by the stator current.**

I. INTRODUCTION

Generally, a resolver and an encoder are used for detecting a rotor position. But in some special applications, their attachment to the motor shaft is very difficult. Lots of efforts were invested to utilize Hall sensors for position sensing due to the low cost and the installation convenience of not requiring any mechanical connection to the shaft [1]- [3]. Brushless DC motors (BLDCMs) are the typical examples that utilize discrete hall sensors for rotor position detection. The Hall sensors monitor the rotor flux, and produce three phase square waves which are 120 degrees apart. These signals are used for the square wave current commutation. Unfortunately, discrete-type hall sensors yield very rough position information, (60° resolution), so that they cannot be used for sinusoidal excitation of the current. Significant research effort has been focussed on solving this problem. Morimoto et al. [1] interpolated the position between 60° interval using an average speed of a previous step. He said that the position estimation error increases as the speed decreases and the acceleration increases. Ozturk et al. [2] utilized an acceleration estimate for the rotor angle interpolation, and applied to the direct torque control of a washing machine. Capponi et al. [3] estimated the intermediate position using a vector-tracking phase-locked loop. A torque feedforward to the mechanical model improved the speed regulation in a low speed region.
Discrete hall sensors are also applied to Permanent magnet synchronous motors (PMSMs) for field-oriented vector control [4], [5]. Kim et al. [4] proposed a hybrid position

and speed estimator working over the entire speed range. He used the two discrete hall sensors for low speed acceleration and back-EMF for high speed operation. In order to achieve bumpless transitions, a different weight function was applied to the position and speed estimators. In [5], a dual observer(a position and a speed) was applied to a washing machine. Position sensor offset was compensated by injecting an additional negative d-axis current. Leading or lagging status compared to a real position was determined by q-axis current variation. In this work, Hall sensors are utilized for the rotor position detection. However, the main difference from the previous work is that linear-type Hall sensors are employed to detect the permanent magnet(PM) edge field. Specifically, two Hall sensors are positioned at one end side of the rotor with 90° (in electrical angle) apart. Thus, an interpolating algorithm is not needed between 90° interval. However the monitored signals contain high-order harmonics. To remove these harmonics, the ANF and a PLL are used in this work.

II. PM EDGE FIELD DETECTION BY HALL SENSORS

Recently, PMSMs are utilized for vacuum pumps for their higher power efficiency. In a vacuum pump, a can-type housing is preferred for vacuum sealing. In these environments, the motor shaft is supported by a single bearing and installation of encoder or resolver is very difficult.

Linear Hall-Effect sensors, (A1322: Allegro Microsystems Inc.) are utilized to measure the rotor position. Fig. 1 (a) shows cross sectional view of the PMSM used in this experiment, and Fig. 1 (b) shows longitudinal view of the PMSM along with an edge field detection device. Hall sensors and signal conditioning circuit are mounted at a torus shaped PCB. The two Hall sensors are separated 90 degree in an electrical angle, so that the sensors produces sine and cosine waveforms when the rotor turns around. Fig. 2 (a) shows monitored Hall sensor signals when the rotor rotates at a constant speed. Note that the two signals are separated about 90 degree. Also rotor angle estimates obtained by taking atan on the two Hall sensor signals are depicted. Note that the angle curve is not a straight line. It means the estimates contained errors originated from the harmonics in the Hall sensor signals. Fig. 2 (b) shows a FFT result which indicates that third harmonic components are significant. The harmonic components induces rotor position errors, leading to degradation of current and speed controls.

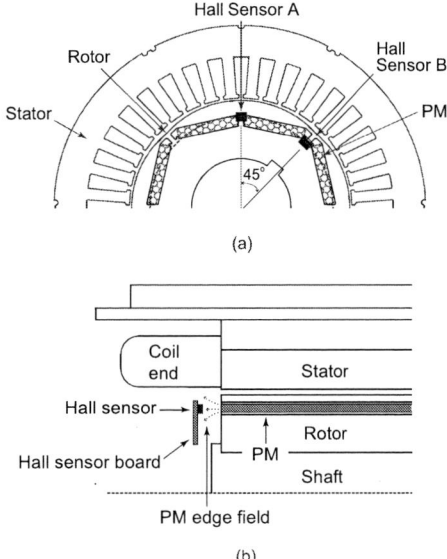

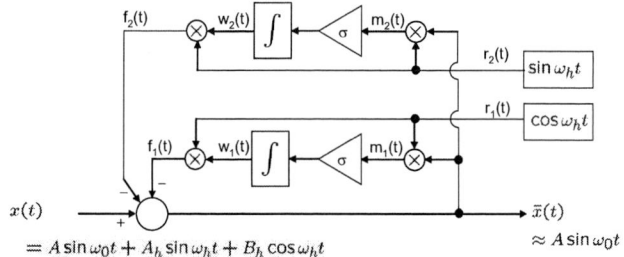

Fig. 3. Functional block diagram of the adaptive notch filter.

Fig. 1. Cross sectional (a) and longitudinal (b) views of the PMSM used for experiments.

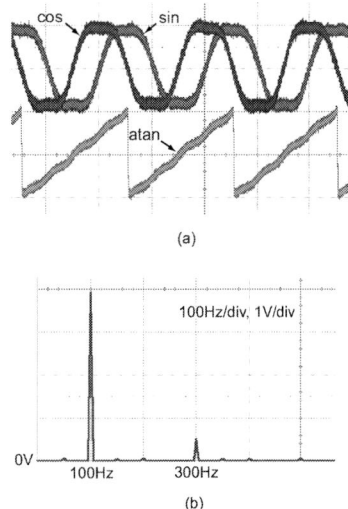

Fig. 2. (a) Distorted sine and cosine Hall sensor signals and their arctangent, (b) FFT of the Hall sensor signals when the rotor rotates at 100Hz, electrical frequency.

III. REVIEW OF ADAPTIVE NOTCH FILTER

The PLL is normally based on a fixed frequency. However, in the motor control applications, the base frequency changes as the motor speed changes. As was shown in Fig. 2, the main distortion comes from the third order harmonics. A purpose of utilizing a PLL filter is to extract the third order component. Fig. 3 shows that direct and quadrature components of a high order harmonic, ω_h are synthesized using the PLL output, and that the synthesized signals are used for notching out the components of $\sin\omega_h t$ and $\cos\omega_h t$ from the original signal. This notching filter is called "adaptive notch filter" [7]- [10].

The adaptive notch filter(ANF) can be constructed with a closed signal loop as shown in Fig. 3. It is assumed here that a desired signal, $A\cos\omega_0 t$ is mixed with a high frequency component, $A_h\cos\omega_h t$. Further, it is assumed that $\omega_h \gg \omega_0$ is known a priori. The objective is to filter out the specific harmonic component of ω_h without altering the magnitude of the desired signal. Fig. 3 shows two feedback loops for the two orthogonal harmonic components, $\sin\omega_h t$ and $\cos\omega_h t$. Hereforth, only the $\cos\omega_h t$ loop is discussed, since the same is true for $\sin\omega_h t$.

Note also that

$$
\begin{aligned}
W_1(s) &= \mathcal{L}[\bar{x}(t) \cdot \frac{\sigma}{2}(e^{j\omega_h} + e^{-j\omega_h})] \\
&= \frac{\sigma}{2s} \cdot \{\overline{X}(s+j\omega_h) + \overline{X}(s-j\omega_h)\}.
\end{aligned}
$$

Therefore, Laplace transform of $f_1(t)$ is

$$
\begin{aligned}
F_1(s) &= \frac{1}{2} \cdot \{W_1(s+j\omega_h) + W_1(s-j\omega_h)\} \\
&= \frac{\sigma}{4(s+j\omega_h)}\{\overline{X}(s) + \overline{X}(s+2j\omega_h)\} \\
&\quad + \frac{\sigma}{4(s-j\omega_h)}\{\overline{X}(s) + \overline{X}(s-2j\omega_h)\}. \quad (1)
\end{aligned}
$$

Similarly, Laplace transform of $f_2(t)$ is

$$
\begin{aligned}
F_2(s) &= \frac{\sigma}{4(s+j\omega_h)}\{\overline{X}(s) - \overline{X}(s+2j\omega_h)\} \\
&\quad + \frac{\sigma}{4(s-j\omega_h)}\{\overline{X}(s) - \overline{X}(s-2j\omega_h)\}. \quad (2)
\end{aligned}
$$

Note that

$$
\overline{X}(s) = X(s) - F_1(s) - F_2(s) = X(s) - \frac{\sigma s}{s^2 + \omega_h^2} \cdot \overline{X}(s).
$$

Therefore, the closed-loop transfer function of the ANF, is

$$
H(s) \equiv \frac{\overline{X}(s)}{X(s)} = \frac{s^2 + \omega_h^2}{s^2 + \sigma s + \omega_h^2} \quad (3)
$$

Equation (3) is exactly the same as the conventional second order notch filter. It is emphasized that σ is inversely proportional to Q-factor which determines the sharpness of a notch filter.

978-1-4244-4782-4/10 $26.00 © 2010 IEEE

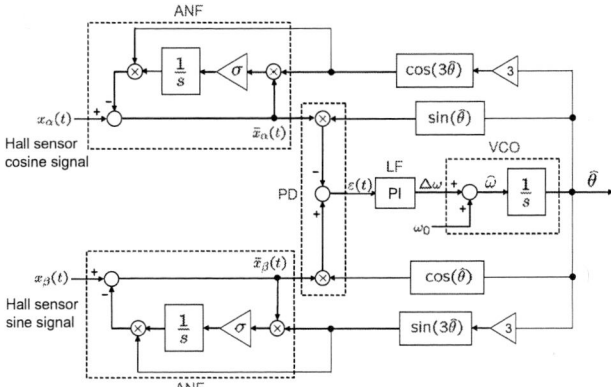

Fig. 4. The proposed PLL incorporating the adaptive notch filter.

IV. PLL INCORPORATING ADAPTIVE NOTCH FILTER

Fig. 4 is the proposed signal processing block diagram designed for extracting out the third order components from the Hall sensor signals. The PLL is used to synchronize the output signal with input signal in frequency as well as in phase [6]. Note that the input signals are the Hall sensor signals, x_α and x_β, and that the output signal are estimates of rotor angle, $\hat{\theta}$ and speed, $\hat{\omega}$. The PLL employs two ANFs for filtering out the third order harmonic components. Normally, FIR(finite impulse response) notch filter would be a good choice, if the fundamental frequency is fixed. However, in the motor application, the frequency changes as the rotor speed does. The notch frequency of the ANF is determined as $3\hat{\theta}$ in Fig. 4.

V. VECTOR CONTROL BASED ON HALL POSITION SENSORS AND EXPERIMENTAL RESULTS

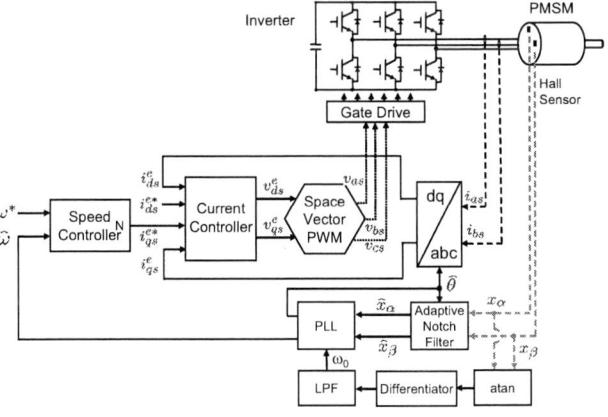

Fig. 5. The overall control block for PMSM utilizing the Hall position sensors.

Fig. 5 is the speed control block diagram of the PMSM that utilizes Hall position sensors. The rotor angle estimate, $\hat{\theta}$ and speed estimate, $\hat{\omega}$ are obtained by passing Hall sensor signals through the ANF and the PLL. The angle is used for

changing the stationary currents, i_{as} and i_{bs} into ones, i_{ds}^e and i_{qs}^e in the synchronous reference frame. The estimated speed, $\hat{\omega}$ is used for speed regulation. The speed feedforward to the PLL, ω_0 is also obtained from Hall sensor signals. Because the Hall sensor signal harmonics are amplified by differentiator, low pass filter bandwidth is restricted to reduce the estimated speed ripples. Slow dynamics of the speed control loop is compensated by the PLL tuning parameters(K_p, K_i).

A PMSM used in this experiments has the following parameters: number of poles : 4, rated speed : 6000 RPM, rated power : 10kW, rated phase voltage : 122V, rated phase current : 42.4A, rated torque : 16Nm.

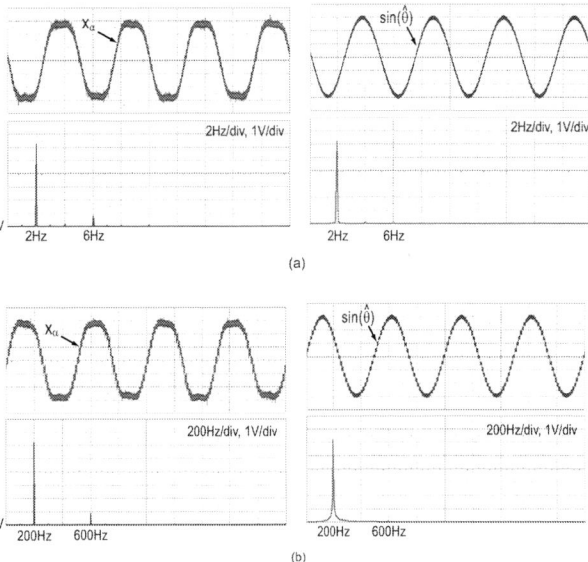

Fig. 6. Hall sensor signals and their FFT before and after the proposed signal processing: (a) 60RPM, (b) 6000RPM

When the PMSM was under the speed control, Hall sensor signals before and after processing were shown in Fig. 6. Note from Fig. 6(a) that the third harmonic(6Hz) component is greatly reduced, thus the filtered signal looks like a pure sinusoidal wave at 60 RPM. Fig. 6(b) is another result at 6000 RPM. In Fig. 7-11, the reference rotor position, θ_{enc} was provided by an encoder and the raw rotor position, θ_{raw} was provided by an arctangent of Hall sensor signals. As show in Fig. 7, the rotor position error, $\theta_{enc} - \hat{\theta}$ was less than 3° in both cases. The speed regulation performance for step change of load torque are shown in Fig. 8-9. In Fig. 8, a 3.2Nm load was applied to the PMSM at 60 RPM. Speed ripples shown in Fig. 8 are originated from the fluctuating speed feedback. This is because the ANF is hard to cancel out the third order harmonics completely in low speed region. These ripples are mitigated as the rotor speed is increased as show in Figs. 9. In Fig. 9, a 8.0Nm load was applied at 3000 RPM. Fig. 10 is a speed control performance when a rotating direction is changed. The PMSM is accelerated from 1000 RPM to -1000 RPM and vice versa with no load. However the position error is increased near zero speed, the whole

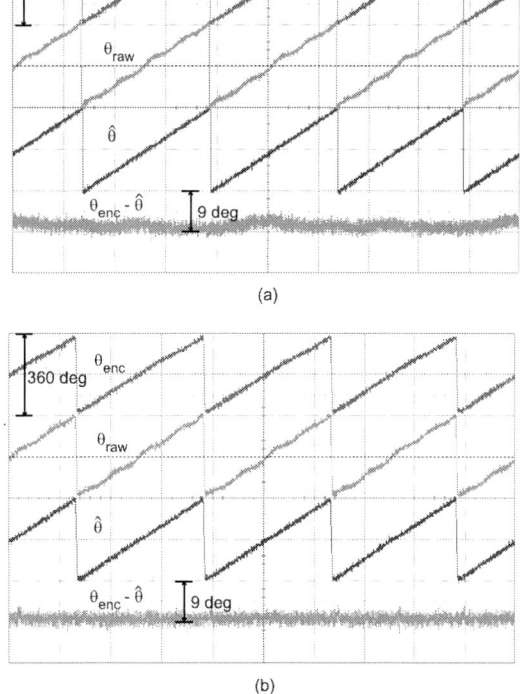

(a)

(b)

Fig. 7. Rotor position errors: (a) at 60 RPM(0.01pu) , (b) at 6000 RPM(1pu).

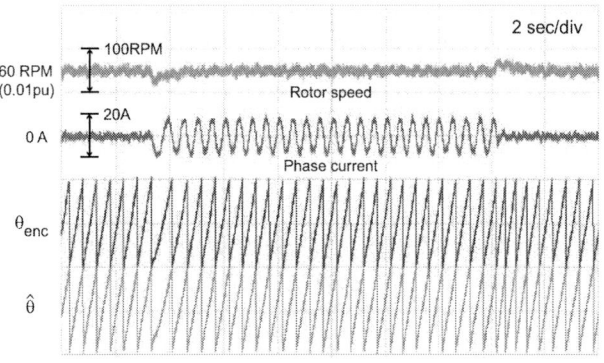

Fig. 8. Speed response for step change of load torque(0%→ 20%→ 0%) at 60 RPM(0.01pu).

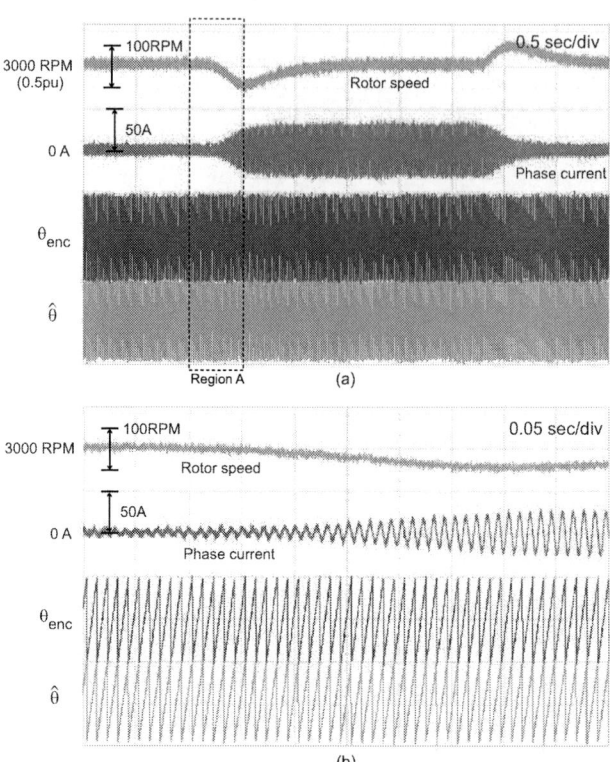

(a)

(b)

Fig. 9. (a) Speed response for step change of load torque(0%→ 50%→ 0%) at 3000 RPM(0.5pu) and (b) a magnified plot of region A.

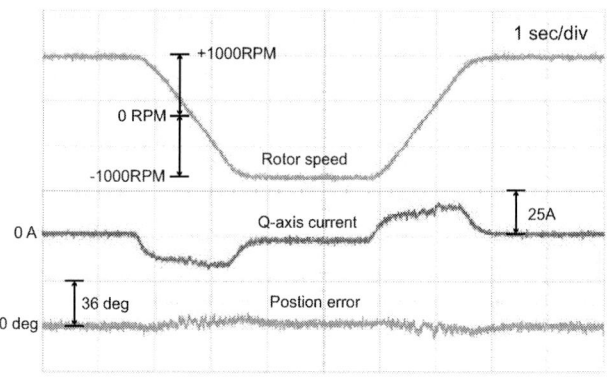

Fig. 10. Speed control performance with no load(1000 RPM→-1000 RPM→1000 RPM).

speed response is same as the case using an encoder or a resolver. Fig. 11 is a standstill starting performance with no load. Since the estimated rotor speed had some ripples in the low speed region, the PMSM was accelerated by torque control at standstill starting. In Figs. 11, 1.6Nm starting torque was applied until the rotor speed was closed to 1000 RPM. The torque control was shifted to the speed control gradually with decreasing the starting torque.

From these results, it can be said that the proposed Hall sensor signal process provides accurate rotor angle information although the Hall sensor signals have third harmonics. And it

shows a good speed control performance under the various load condition.

VI. CONCLUSION

In this work, we proposed the low-cost position sensor for PMSM using linear Hall sensors. These sensors monitor the edge field of PMs. The monitored signals look like sine and cosine waveform, but have third order harmonics. To remove these harmonics, we proposed the signal process using the ANF and the PLL. The ANF canceled out the third harmonics

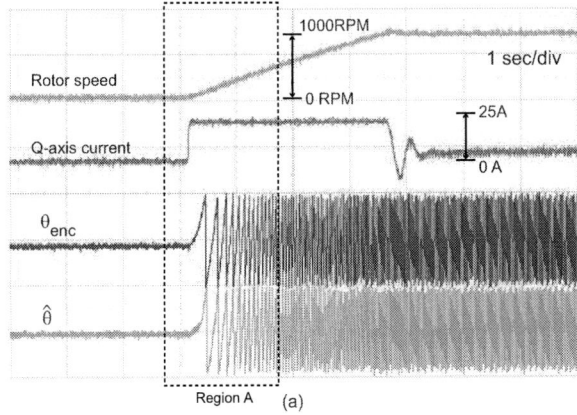

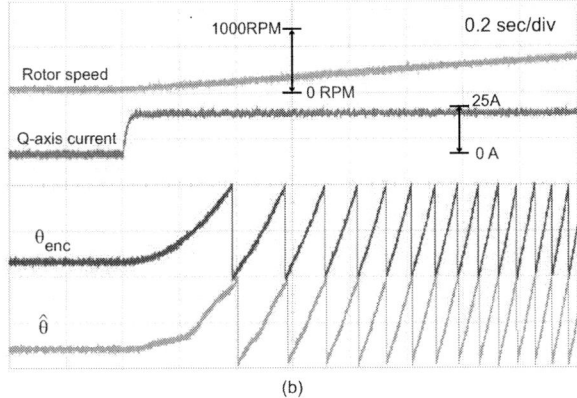

Fig. 11. (a) Standstill starting performance with no load(0 RPM→1000 RPM) and (b) a magnified plot of region a.

and then the PLL adjusted a frequency and a phase of the filtered Hall sensor signals to minimize position error. The proposed signal process was verified by experiments. Although, the experimental results showed some limits compared with an encoder or a resolver, the proposed position sensor is used for cost effective PMSM driving systems.

REFERENCES

[1] S. Morimoto, M. Sana da and Y. Takeda, "Sinusoidal current drive system of permanent magnet synchronous motor with low resolution position sensor," *in Proc. IEEE IAS Annu. Meeting,* Oct. 1996. pp. 9-13.

[2] S. B. Ozturk, B. Akin, H. A. Toliyat and F. Ashrafzadeh, "Low-cost direct torque control of permanent magnet synchronous motor using hall-effect sensors," *in Proc. IEEE Appl. Power Electron.Conf.(APEC),* pp.19-23, March, 2006.

[3] F. G. Capponi,G. De Donato, L. D. Ferraro, O.Honorati and M.C Harke and R.D. Lorenz, "AC brushless drive with low-resolution hall-effect sensors for surface-mounted PM machines," *IEEE Trans. Ind. Appl.,* vol. 42, no. 2, Mar./Apr. 2006, pp. 526-535.

[4] H. Kim, S. Yi, N. Kim and R.D. Lorenz, "Using low resolution position sensors in bumpless position/speed estimation methods for low cost PMSM drives," *in Proc. IEEE IAS Annu. Meeting,* Oct. 2005, vol. 4, pp. 2518-2525.

[5] A. Yoo, S. Sul, D. Lee and C. Jun, "Novel speed and rotor position estimation strategy using a dual observer for Low Resolution Position Sensors," *in Proc. IEEE PESC'08,* June 2008, pp. 674 - 653.

[6] Roland E. Best, *Phase-Locked Loops: Design, Simulation, and Applications,* McGraw-Hill, 2003.

[7] B. Widrow and S. D. Stearns, *Adaptive signal processing,* Prentice-Hall, 1985.

[8] B. Widrow, et al., "Adaptive noise cancelling: Priciples and apllications," Proc. IEEE, vol. 63, pp. 1692-1716, Dec. 1975.

[9] S. Luo, et al., "An adaptive detecting method for harmonic and reactive currents," IEEE Trans. on Ind. Elec., vol. 42, pp. 680-684, Feb. 1995.

[10] M. Karimi-Ghartemani, et al., "Measurement of harmonics/inter-harmonics of time-varying frequencies," IEEE Trans. on Power Delivery, vol. 20, pp. 22-31, Jan. 2005.

Bridged-T Speed Controller for High Performance Switched Reluctance Motor Drives

Gregory Pasquesoone, Iqbal Husain, IEEE Fellow, Robert J. Veillette, IEEE Member

Department of Electrical and Computer Engineering,
The University of Akron,
Akron, OH-44325-3904, United States of America
E-mail: gregory_pasquesoone@hotmail.com, ihusain@uakron.edu, veillette@uakron.edu

Abstract—A so-called bridged-T controller, which combines both open-loop and closed-loop control components is proposed for speed control of a switched reluctance motor. The open-loop controller provides a fast dynamic response and the closed-loop controller assures zero steady-state speed error by feedback of the error between the ideal open-loop response and the actual machine response. The bridged-T controller is easy to design, to implement, and to tune. A three-phase switched reluctance motor drive with this bridged-T controller has been simulated using Matlab-Simulink and implemented in real time using a digital signal processor. The simulated and actual responses match closely, and demonstrate a dynamic response superior to that which may be obtained using the standard closed-loop control alone.

I. INTRODUCTION

The switched reluctance motor (SRM) is a robust, low-cost, fault-tolerant machine suitable for several niche applications. Depending upon the application, the SRM may be controlled in a torque mode or in a speed mode. A given application may require both the speed and torque modes of operation. In the SRM drive, the torque is controlled by controlling the phase currents while accounting for the nonlinear current-torque relationship. For a standard closed-loop speed control, the speed is controlled in an outer loop, while the torque or current is controlled in an inner loop. The speed control loop, which may be based on a simple first-order transfer function that describes the speed response of the machine, can be implemented using a proportional-integral (PI) feedback [1,2]. For the closed-loop controllers reported in the existing literature, there is a tradeoff between the slow transient response and the overshoot of the speed response [3,4].

By use of an open-loop controller, a fast system response can be obtained without the problem of overshoot; however, the steady-state speed error cannot be eliminated because of unavoidable errors and unmodelled dynamics in the system model [5,6].

A bridged-T controller is the combination of an open-loop controller and a closed-loop controller. It has the advantage of being easy to design and to implement, as the open-loop and closed-loop control components can be designed and tuned independently. Combining the two control components

gives a system that has the fast response typical of an open-loop control system along with a null steady-state error typical of a standard closed-loop control system.

This paper presents the methods for the characterization of the system dynamics and the development of the open-loop and closed-loop controllers for the switched reluctance motor drive. Experimental results on a 2.7kW SRM and drive are presented in this paper.

II. SRM DRIVE STRUCTURE

The SRM drive used in this research uses a classical bridge power converter with the phase winding between the two switches of the same phase. The motor phase currents are synchronized with the rotor position to produce a positive torque for motoring operation. A current controller regulates the phase current in the machine windings. With a classical bridge converter three modes are possible: (1) Magnetization ($+V_{DC}$ applied to the phase winding), (2) Free-wheeling (0V applied across the phase winding) and (3) Demagnetization ($-V_{DC}$ applied to the phase winding as long as there is current in the winding). Figure 1 shows the classical bridge converter and the different equivalent circuits corresponding to the three modes applied to the SRM phase A only. The solid lines in the figure show the current conduction paths in the three modes. IGBT power devices have been used in this research for implementation of the power converter. Mode (2) or free-wheeling can be realized in two different ways using either the top or bottom IGBT for current conduction as shown. Alternating between the two free-wheeling modes reduces the switching frequency for each IGBT which is beneficial for high-current-rated devices.

Magnetization (M) and Free-wheeling (FW1, and FW2) are used to maintain the phase current close to the reference current by applying respectively $+V_{DC}$ and 0V to the machine phase winding. Demagnetization (D) is used to bring the phase current quickly to zero applying $-V_{DC}$ to the machine phase winding as long as the phase current is greater than zero then 0V is applied to the machine phase winding so that the next phase starts to conduct.

978-1-4244-4782-4/10 $26.00 © 2010 IEEE

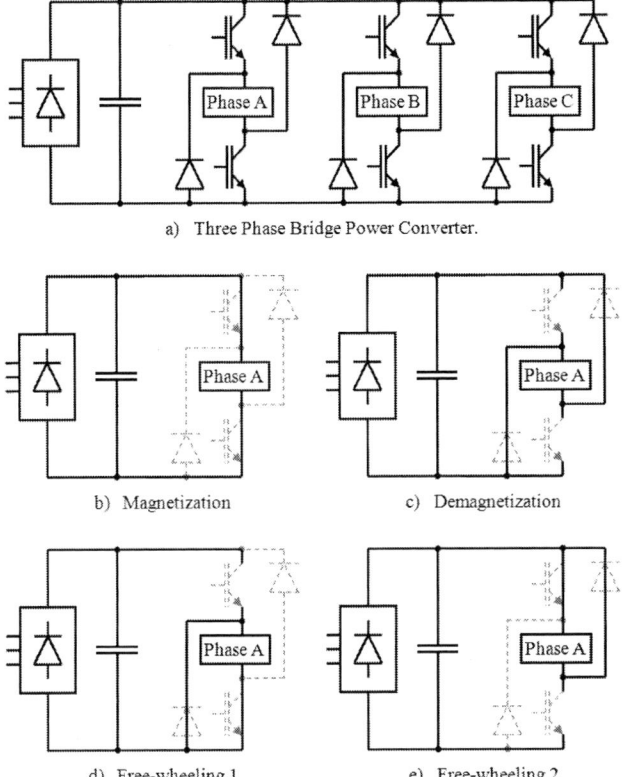

a) Three Phase Bridge Power Converter.

b) Magnetization   c) Demagnetization

d) Free-wheeling 1   e) Free-wheeling 2

Figure 1: (a) Classical SRM power converter; (b) magnetization mode; (c) demagnetization mode; (d) freewheel mode 1; and (e) freewheeling mode 2.

Figure 2 shows a typical phase current (I_a) and the two IGBT command signals (IGBT$_H$, and IGBT$_L$). The advantage of using the two free-wheeling (FW1, and FW2) paths is obvious from Figure 2. The commutation is shared by the top and bottom IGBTs. The frequency of commutation for either of the two IGBTs is reduced by a factor of two compared to the phase current switching frequency. This is advantageous in high power applications, where the IGBTs switching frequency limit can be as low as few kHz.

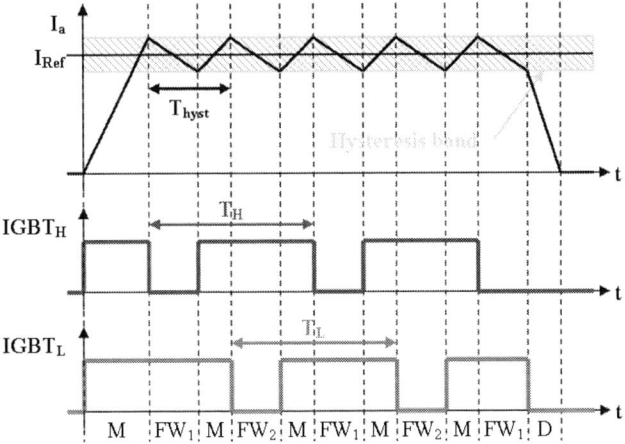

Figure 2: Phase A current & IGBTs commutation waves form.

The current controller implemented in this research is a hysteresis controller that uses the three converter modes (or four if we consider the free-wheeling 1 and 2 to be two different modes). The hysteresis controller is used to regulate the phase current in order to produce a constant torque by means of speed-dependent phase advancing and phase overlapping. The phase overlapping implies that two phases can be used for powering the SRM simultaneously. Since the phase current has to be controlled in two phases simultaneously, each phase must have its own hysteresis controller.

In terms of control, two controllers have been realized; the hysteresis controller that regulates the phase currents in the inner torque loop, and the speed controller that generates the torque loop command in the outer loop. The speed controller uses a combination of an open-loop (feed-forward) controller based on the speed command only, and a closed-loop (feedback) controller based on the error between the speed command and the actual speed information. Figure 3 gives an overview of the complete system with the power converters, electrical machine, sensors, and controllers (torque and speed).

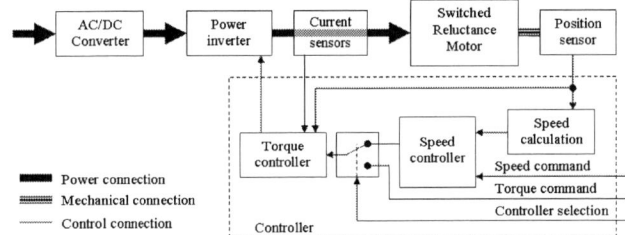

Figure 3: SRM drive structure.

III. SPEED CONTROLLER

A. System Identification

In order to realize a speed controller, the system transfer function needs to be identified. For an electrical machine, the machine dynamics is based mainly on its moment of inertia and damping coefficient, the mechanical response is much slower than the electrical response; hence, a first-order transfer function is sufficient to describe the plant dynamics. The transfer function can be found experimentally by applying a step torque command (corresponding to a current command) and measuring the resulting speed response of the system. From the test results, the system time constant and the gain can be obtained as

$$\tau : \text{rising time at 63\% of the steady-state speed,} \quad (1)$$

$$k = \frac{\text{Steady state speed}}{\text{Current command}}. \quad (2)$$

For a first-order system, these coefficients define the system transfer function (the plant) as

$$H_{SRM} = \frac{k}{\tau \times s + 1}. \quad (4)$$

From the transfer function obtained, the open-loop and closed-loop control components can be separately designed and tuned based on some required specifications, such as percentage of overshoot, saturation level of the actuator, and the maximum transient time.

B. Bridged-T controller

The bridged-T controller using the open-loop and closed-loop control components is shown in Figure 4. This bridged-T structure has been described in [7]. It is composed of three transfer functions: D1 the open-loop controller, D2 the closed-loop controller, and D3 the ideal open-loop model, which is a combination of the open-loop controller and plant transfer functions. The output of D3 is the ideal response of the system. This ideal response is compared with the actual machine response, and the error between the two is used as input to the closed-loop controller D2. This controller minimizes the error by a standard feedback design. Since D2 is designed to produce a stable closed-loop system, and the transfer functions D1 and D3 are themselves stable, the bridged-T controller is guaranteed to produce a stable system.

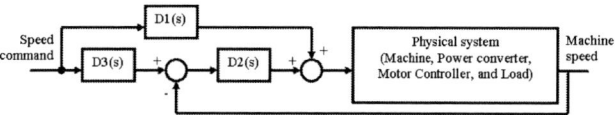

Figure 4: Bridged-T controller block diagram.

The open-loop controller D1 is designed to shape the control input to the plant in such a way as to obtain a fast and non-oscillatory response. This is done by using the zeros of D1 to cancel the plant poles that correspond to slow or oscillatory modes of response, and placing the poles of D1 in more favorable locations. As a result, the output of D1 commonly displays overshoot in response to step commands. Unfortunately, for large step changes in the command, the overshoot at the output of D1 may cause saturation in the plant actuator, resulting in a degraded system response.

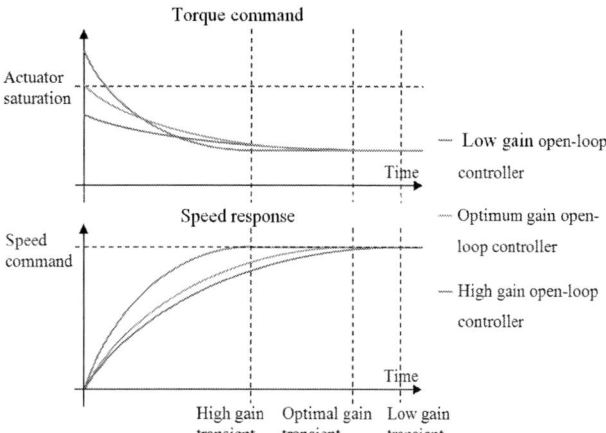

Figure 5: Low gain, high gain and optimal gain open-loop controller.

Figure 5 represents three different design of the open-loop controller. The design using the optimal gain has the best response that can be achieved with this type of controller. If the gain is lower, the torque command is lower and the speed response is longer. If the gain is higher, the speed response should be faster, but the torque command exceeds the actuator saturation level so this response is not possible to obtain.

C. Modified Bridged-T Controller

The principle of the modified bridged-T controller is to improve the performance of the bridged-T by replacing the linear open-loop controller transfer function (D1 in Figure 4) with a novel controller: the saturation controlled open-loop controller. The modified open-loop D1 controller has two modes of operation based on the speed feedback information which is otherwise used in the closed-loop controller (D2). The first mode of operation occurs whenever there is a speed command increase. In this mode, the open-loop controller applies the maximum torque command to the system, causing the speed error to be quickly reduced. The controller switches to the second mode of operation as soon as the system speed reaches the speed command value. In this mode, the controller applies the steady-state open-loop torque command proportional to the speed command to maintain the speed at the desired value. The open-loop controller is executed based on the rule

$$I_{CMD} = \begin{cases} 0 & \text{if} \quad \omega_{CMD} < \omega_{CMD\ MEMORY} \\ I_{MAX} & \text{if} \quad \omega_{CMD} > \omega_{CMD\ MEMORY} \\ SS_{CMD} & \text{if} \quad \omega_{CMD} = \omega_{CMD\ MEMORY} \end{cases} \tag{5}$$

A flowchart of the saturation controlled open-loop controller logic is given in Figure 6, where ω_{CMD} is the speed command, $\omega_{CMD\ MEMORY}$ is a previously saved speed command, I_{CMD} is the open-loop controller output, ω_{SENSED} is the speed feedback information, and SS_{CMD} is the current command corresponding to the steady-state speed ω_{CMD}.

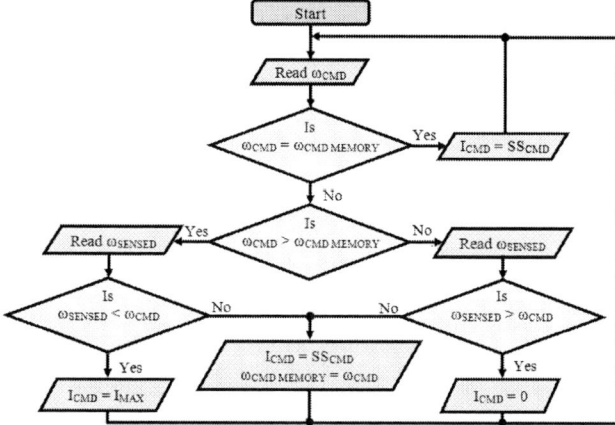

Figure 6: Saturation controlled open-loop controller flowchart.

This new controller is simple to design and easy to implement, since it is based on the same equation as the first-

978-1-4244-4782-4/10 $26.00 © 2010 IEEE   2009

order open-loop controller. Although it uses some feedback in its first mode of operation, it works basically as an open-loop controller. Figure 7 compares the speed response of the Optimum open-loop controller applied to a system with the modified (saturation controlled) open-loop controller applied to the same system. The actuator saturation limits the speed of the response. However, the modified open-loop controller in its first mode of operation produces the fastest response possible.

The new controller D1 cannot cause unstable operation because it applies the feedback only conditionally, in the initial acceleration phase after a change in the speed command. Further, because the plant considered for speed control is first-order, D1 theoretically provides deadbeat response with zero overshoot.

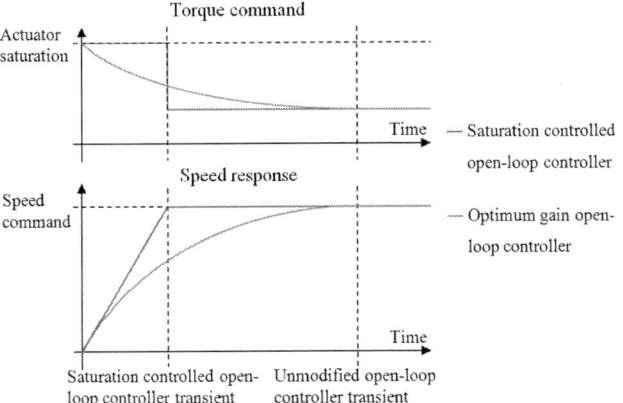

Figure 7: Saturation controlled open-loop controller vs. Optimum open-loop controller.

D. Example of Bridged-T Controller

The example in this section describes the design of a bridged-T controller for the SRM system considered here, which has the parameters listed in Table 1. The step response is shown in Figure 8. In the next section, simulation and experimental results are reported.

Table 1. SRM parameters

Power	2.7 kW
Base speed	75 rpm
May speed	120 rpm
Current	27 A
Voltage	500 V
Phase resistance	2.5156 ohm
Torque	346.1765 Nm
L aligned	381.8089 mH
L unaligned	30.6906 mH
Stator poles	18
Rotor poles	12

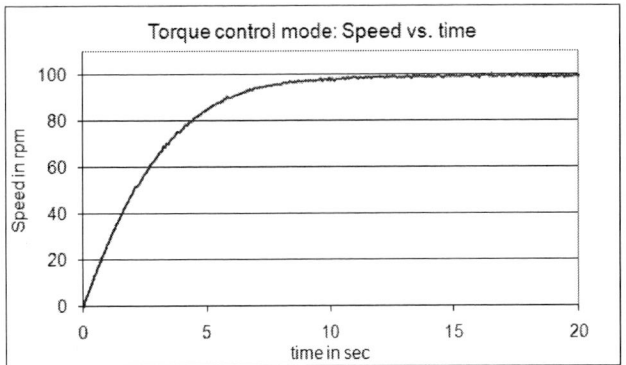

Figure 8: Speed response for a current step input of 1518 counts.

The system transfer function identification method gives the plant transfer function as

$$H_{SRM}(s) = \frac{1}{42.78 \times s + 15.17}. \tag{6}$$

The maximum machine current command is limited to 26A, which is equivalent to 4095 counts in the digital controller, and the top speed of the machine is 125rpm. The design of the open-loop transfer function D1 is based on pole-zero cancellation with the zero of D1 located at the pole of the plant and a new pole placed to set the new system transient. This open-loop transfer function can be written as

$$D1(s) = \frac{42.78 \times s + 15.17}{\gamma \times s + 1}. \tag{7}$$

The parameter γ is determined based on the system limitations. The maximum output of the open-loop controller occurs as the initial overshoot in response to an increase in the speed command, which can be calculated using the initial value theorem of the Laplace transform. The worst-case overshoot occurs when the speed command goes from no speed (0rpm) to the highest speed (120rpm), and has to be limited to 26A or 4095counts. Applying the initial value theorem, the condition is written as

$$\frac{42.78}{\gamma} \times 120 \leq 4095. \tag{8}$$

The above inequality sets the parameter γ to be at least 1.25363. For the open-loop controller, the parameter γ has been set to 1.28 in order to simplify the software implementation.

The transfer function D3 (9) is simply the product of the D1 (7) and the plant transfer function (6), which is found as

$$D3(s) = \frac{1}{1.28 \times s + 1}. \tag{9}$$

The closed-loop transfer function used in this bridged-T controller is a simple proportional integral (PI) controller which can be written as

$$D2(s) = \frac{K_P + K_I \times s}{s}, \tag{10}$$

where K_P and K_I are the proportional and the integral gains, respectively. The gain K_P has been set to 32 and K_I has been set to 7.8125.

For the modified bridged-T, the open-loop controller is replaced with the saturation controlled open-loop controller described in (5). The I_{MAX} corresponding to the maximum command in the example is 4095counts, and SS_{CMD} corresponding to the steady state speed command is

$$SS_{CMD} = \eta \times \omega_{CMD}, \tag{11}$$

where ω_{CMD} is the actual machine speed command (or machine reference speed) and $\eta = 15.17$ is the reciprocal of the DC gain of the plant.

IV. SIMULATION AND EXPERIMENTAL RESULTS

The simulation and experimental results are given for a three-phase, 2.7kW high-torque, low-speed SRM drive. The system, the open-loop, the closed-loop and the ideal response transfer functions have already been presented in the previous sections.

The simulations of the bridged-T controller and the modified bridged-T controller have been realized using Matlab-Simulink. In both simulations, a proportional integral controller has been used as a closed-loop controller. The simulation circuit of the bridged-T and the modified bridged-T are shown respectively in Figure 9 and Figure 10. In both cases, the results have been obtained for the ideal case where the plant and system transfer function are identical.

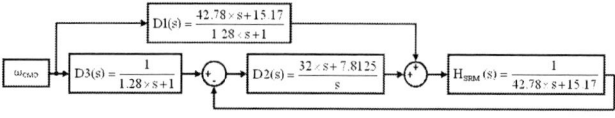

Figure 9: Bridged-T controller.

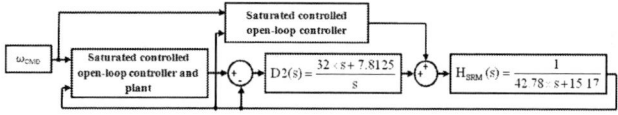

Figure 10: Modified bridged-T controller.

The modified bridged-T controller has also been implemented and tested in an experimental platform, as shown in Figure 11, using a 2.7kW SRM. The controller has been implemented using a digital signal processor (DSP) [8].

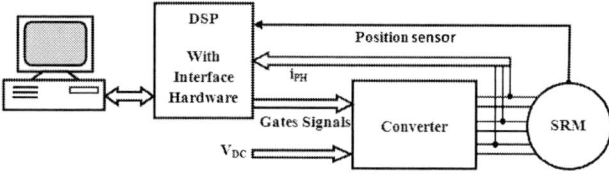

Figure 11: Experimental setup.

The simulation and the experimental results have been obtained by applying a speed command of 0 to 90 RPM as a step input. Figure 12 shows the simulation results for the unmodified and the modified (saturation controlled) bridged-T controllers, as well as the experimental result for the modified controller. Figure 13 is an oscilloscope capture of the three

phase currents for the experiment with the modified controller. The current scale is 10mV = 5A. On this graph, the two modes of operation can be clearly identified. In the first mode of operation, when the speed is lower than the speed command, the phase current is set at the maximum phase current of 26A. In the second mode of operation, when the speed has reached the command value, the phase current is set to 12A in order to keep a steady speed. This plot also shows the phase current commutation that increases with speed and the high frequency switching of the hysteresis controller in order to maintain a fixed phase current.

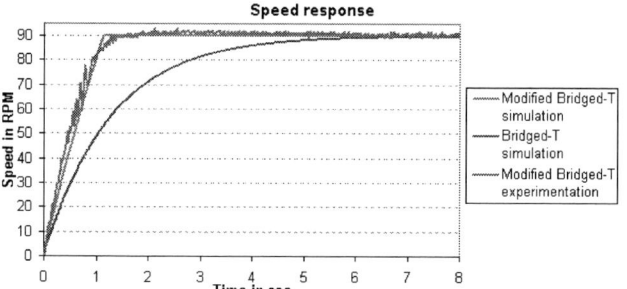

Figure 12: Bridged-T controller and Modified bridged-T controller simulation and Modified bridged-T controller experimental step response.

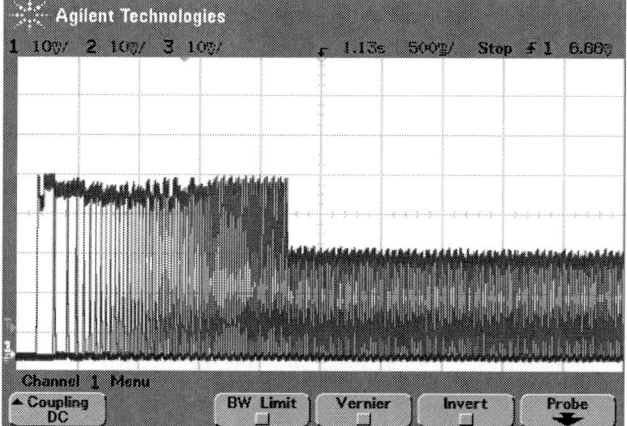

Figure 13: Phase currents vs. time.

The simulation and experimental results of the speed response for a command speed of 90 rpm are shown in Figure 12. The system has a faster response with the modified bridged-T controller than with the unmodified bridged-T controller. The simulation and experimental results obtained with the modified bridged-T controller are very close. Figure 13 shows the currents in the phase windings. One can clearly differentiate the two modes of operation of the saturation controlled open-loop controller defined on Figure 7. We can also deduce that the model transfer function is a good approximation of the plant since the steady state current command remains nearly constant with no correction needed from the closed-loop controller.

The plant dynamics may deviate from the modeled transfer function due to load changes or other secondary effects. The performance of the modified bridged-T controller

978-1-4244-4782-4/10 $26.00 © 2010 IEEE

in handling such deviations has been evaluated through simulation. Two perturbations in the plant transfer function are considered as

$$H_{SRM1}(s) = \frac{1}{50 \times s + 25}, \qquad (12)$$

$$H_{SRM2}(s) = \frac{1}{30 \times s + 5}. \qquad (13)$$

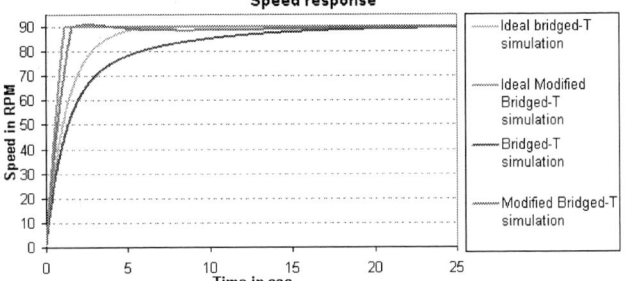

Figure 14: Bridged-T controller and Modified bridged-T controller simulation step response for the plant transfer function (12) and for the ideal plant.

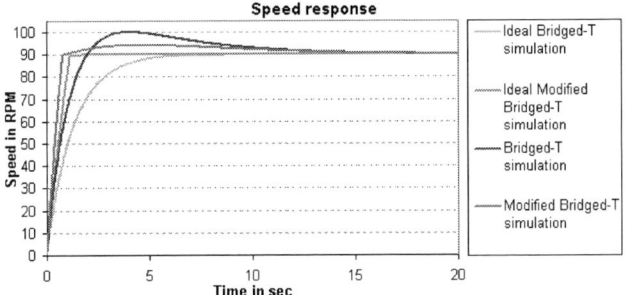

Figure 15: Bridged-T controller and Modified bridged-T controller simulation step response for the plant transfer function (13) and for the ideal plant.

Table 2: Bridged-T data

Inertia factor model error in %	0	-29.874	-41.561	16.877	28.565
Damping factor model error in %	0	-67.040	-34.080	64.799	97.759
Rising Time in sec	2.809	1.406	1.623	6.196	8.534
Speed Overshoot in %	0	11.222	1.444	0	0

Table 3: Modified Bridged-T data

Inertia factor model error in %	0	-29.874	-41.561	16.877	28.565
Damping factor model error in %	0	-67.040	-34.080	64.799	97.759
Rising Time in sec	0.912	0.631	0.589	1.251	1.553
Speed Overshoot in %	0	4.722	0.122	0.900	4.900

The speed response results with the bridged-T and modified bridged-T controllers for the above two perturbed plant transfer functions are shown in Figures 14 and 15. The responses for the ideal bridged-T and modified bridged-T

controllers that know the perturbed plant transfer functions are also plotted in the figures for comparison. The ideal results are based on the assumption that the controllers correctly account for the plant transfer function parameters.

The results, summarized in Tables 2 and 3, demonstrate that the modified bridged-T controller also has excellent response characteristics even when the plant parameters are not exactly known. Results in Figures 14 and 15 also show that the system has a faster response with the modified bridged-T controller than with the unmodified bridged-T controller. Also, for the case where the inertia of the model is smaller than the plant inertia, the modified bridged-T gives a speed response with a very small overshoot but the response is considerably faster than with the unmodified bridged-T controller. For the case where the inertia of the model is bigger than the plant inertia, there is less overshoot in the speed response (Figure 15). The closed-loop controller takes charge in this case to correct the steady-state error. The settling time and overshoot or undershoot depend on the differences in actual and assumed plant dynamics, and closed-loop PI-controller parameters.

V.   CONCLUSION

This paper introduces new speed controllers for electric machines called the bridged-T and the modified bridged-T controllers. The controllers have been simulated and implemented for speed control of an SRM drive. The experimental results demonstrate impressive dynamic speed response characteristics of electric motor drives. The simplicity of the method makes it very easy to implement. This new controller type can be implemented for any electric machine. It will be fairly simple to upgrade the existing closed-loop controller software to improve the dynamic performance. The modified bridged-T controller gives the optimum system response with almost no additional computation compared to the bridged-T controller.

REFERENCES

[1]   P. Pillay and R.Krishnam, "Control characteristic and speed controller design for a high performance permanent magnet synchronous motor drive," *IEEE Trans. on Power Electronics,*Vol. 5, No. 2, April 1990, pp.151-159,

[2]   K. Astom and T. Hagglund, *PID Controllers: Theory, Design and Tuning*, 2nd ed., Instrument Society of America: Research Triangle Park, 1995.

[3]   S. Paramasivam, R. Arumugam, and S. Balamurugam, "Implementation of digital controller for 6/4 pole switched reluctance motor drive," *2004 IEEE Region 10 Conference*, TENCOM 2004.

[4]   G. Yuan, *Speed Control of Switched Reluctance Motors*, PhD thesis, Hong Kong University of Science and technology, Hong Kong, 2000.

[5]   B. C. Kuo, and F. Golnaraghi, *Automatic Control System*, 3rd ed., John Wiley & Sons Inc., NY 2002.

[6]   S. Sarat and S. Hostetter, *Design of Feedback Control System*, 3rd ed., Saunders College, 1994.

[7]   T. T. Hartley, R. J. Veillette, and G. Cook, "Techniques in deadbeat and one-step-ahead control." In Control and Dynamic Systems – Advances in Theory and Application, vol. 79: *Digital Control System Implementation and Computation Techniques*, C. T. Leondes, ed., Academic Press, 1996, pp.117-157.

[8]   TMS320F2812 Datasheet, Texas Instruments, Houston, TX, 2002.

Reducing Losses in Multilevel Coupled Inductor Inverters Using Interleaved Discontinuous SVPWM

Behzad Vafakhah, Andy Knight, and John Salmon

Department of Electrical and Computer Engineering University of Alberta
Edmonton, AB Canada T6G 2V4
vafakhah@ece.ualberta.ca, knight@ece.ualberta.ca, salmon@ece.ualberta.ca

Abstract—This paper presents generalized interleaved Discontinuous Space Vector PWM (DSVPWM) schemes for 3-level 3-phase 6-switch split-wound Coupled Inductor Inverters (CII). A simplified approach is applied to the generation of interleaved DSVPWM schemes with the need to minimize high frequency current ripple and associated losses in the coupled inductor and the inverter. The performance of the CII is investigated and compared for DSVPWM schemes. The current ripple is analyzed based on the position of the discontinuous period for several DSVPWM schemes with different loads and power factors. Regardless of the loads, 60° discontinuous SVPWM around the positive and negative peaks of the fundamental reference voltage, DSVPWM1, lowers losses in the coupled inductor. The simulation and experimental results validate the drive performance.

I. INTRODUCTION

Numerous multilevel converter topologies have been introduced during the last few years [1-5]. If the application and power level justify the choice of multilevel topologies, one recent alternative multi-level topology uses coupled inductors at the output stage of a 6-switch inverter [6, 7], Figure 1. The Coupled Inductor Inverter (CII) produces 3-level output voltages from a single DC voltage supply at each inverter output terminal using a 3-phase split wound coupled inductor. This topology reduces the size of the output ac filter inductors and the output high frequency harmonic currents. Compared to other 3-level converters (i.e., 12-switch 3-level NPC), the CII topology uses half the number of switches to generate the same number of voltage levels. In addition, the significant advantage of the CII topology is that the requirement for dead-time to avoid shoot-through current is eliminated. Thus, traditional adverse effects of dead-time are avoided.

Pulse width modulation (PWM) strategies, including carrier-based PWM and Space Vector PWM (SVPWM) schemes used for a standard inverter, can be modified and applied to the CII topology. However, when using any modulation scheme, care must be taken (i) to minimize the winding current ripple and associated inductor losses and (ii) to balance the common mode winding current (voltage) over each switching period for the continuous mode of operation [7-10]. The coupled inductor current ripple in one inverter leg

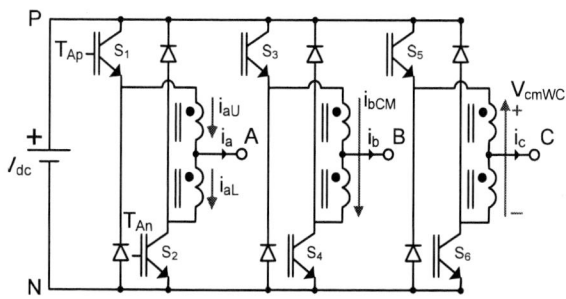

Figure 1. 3-level 3-phase coupled inductor inverter

can be affected by both the switching states in the same inverter leg and in the other inverter legs due to the magnetic coupling between the windings in a three-limb core. Since the coupled inductor winding configuration at each switching state can be changed with the condition of being on and off for each inverter switch, the current ripple is also dependent on the value of the effective inductance between the upper and lower switches in each inverter leg [7-10].

By using the interleaved PWM [11], the carrier-based modulations—Sinusoidal PWM (SPWM) and Discontinuous PWM (DWPM1)— intrinsically balance the common mode winding voltage during each switching cycle. Compared to the SPWM method, discontinuous PWM can also remove the lowest effective inductance switching states and, as a result, lowers the winding current ripple. DPWM eliminates the effects of the non-switching leg on the current ripple around the peak voltage and also reduces the effects of the coupling between the inductor windings in a three-limb core with only two inverter legs actively PWM at a given time. However, even with high quality output signals, the overall performance of the CII when using DPWM schemes is poor due to coupled inductor losses. The presence of switching states with a low effective winding inductance produces a large high frequency current ripple in the inductor windings with associated high winding and core losses, and, as a result, lowers the performance of the CII topology, especially noticeable at low modulation depths [9, 10].

Compared to carrier-based multilevel PWM schemes, the space vector techniques provide a wider variety of choices of the available switching states and sequences [12-14]. In

This research is funded by the Natural Sciences and Engineering Research Council of Canada.

978-1-4244-4782-4/10 $26.00 © 2010 IEEE

general, the fundamental output voltage produced by carrier-based PWM techniques can be identical to that produced by SVPWM. However, the optimal sequence of pulses within the sampling interval leads to the superior performance of the space-vector modulation technique. In [9], the interleaved DSVPWM strategy reduces the inductor losses and significantly improves the performance of the inverter drive by eliminating the low effective inductance switching states. The DSVPWM algorithm is the combination of DSVPWM1 for high modulation indices ($m_a > 0.5$) and DSVPWM3 for low modulation depths ($m_a < 0.5$). However, as will be shown in this paper, DSVPWM3 produces higher coupled inductor losses compared to other DSVPWM schemes and is not an optimal modulation scheme for the CII inverter.

In this paper, a very simple approach is used to investigate the effects of the position of the discontinuous period on the inductor winding current ripple. The proposed approach considers the relationship between the time-varying voltage across the inductor with the effective inductance and the time-varying current passing through; the duration of applying the PWM switching voltage (the average voltage) across each inductor winding in each switching state can also impact the current ripple magnitude.

Generalized interleaved discontinuous SVPWM strategies, DSVPWM0, DSVPWM1, DSVPWM2 and DSVPWM3, are presented to improve the CII performance by lowering coupled inductor and inverter losses. The performance of the CII inverter with the DSVPWM schemes is investigated for modulation depths lower than 0.5 (as the effect of the current ripple is more significant in this region). The same effective switching states are chosen for DSVPWM schemes to eliminate the variation of the current ripple due to the different value of the effective inductance switching states. The impacts of the load current and power factor on the winding current ripple are explored. The effectiveness of approach is verified by simulation and experimental tests. Additional experimental loss measurements are obtained to validate the comparison between DSVPWM methods.

II. INTERLEAVED DISCONTINUOUS SVPWM

A generalized DSVPWM method is designed and developed based on the interleaved switching sequence in [9]. The concept of the interleaved PWM is described for DSVPWM methods. The interleaved switching states with a high effective inductance are chosen to minimize the high frequency common mode current deviations.

The space vector block diagram of the CII is depicted in Figure 2. Each sector is divided into two 30 degree sections and each 30 degree section has two combinations of switching states. The combination of different switching states produces the phase displacement between the reference voltage and current for each DSVPWM strategy. For example, when the phase of the reference voltage is between 90° and 150°, the two sets of switching states for voltage vectors of V_0, V_1 and V_2 in triangle Δ_1 are shown in Figure 3 and Figure 4, respectively. Figure 3 demonstrates the switching states type-A. It can be seen that the phase A voltage is clamped to the positive DC bus voltage, and phase B and C have the interleaved PWM switching. For the same triangle, Figure 4 shows an alternative switching states type-B where the phase C voltage is clamped to the negative DC bus voltage, and

phase A and B have the interleaved PWM switching sequence. To demonstrate how these switching states can be used to generate different DSVPWM methods, the pattern of switching sequences for phase A is shown in Table I. The corresponding switching states is defined in Table II for two periods of the reference voltage.

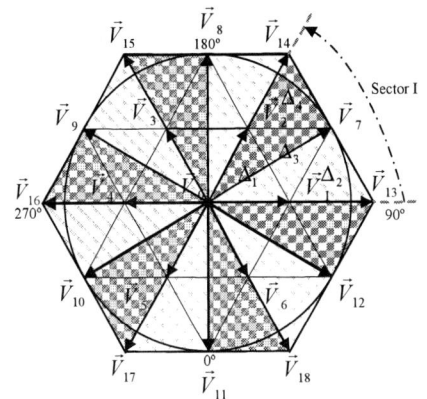

Figure 2. CII space vector block diagram

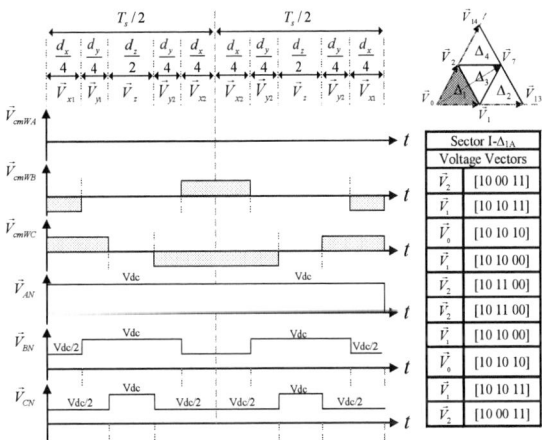

Figure 3. Switching states I and corresponding inductor winding voltage and phase voltage in sector I-Δ_1 in each leg, (Common mode voltages are $\pm V_{dc}$)

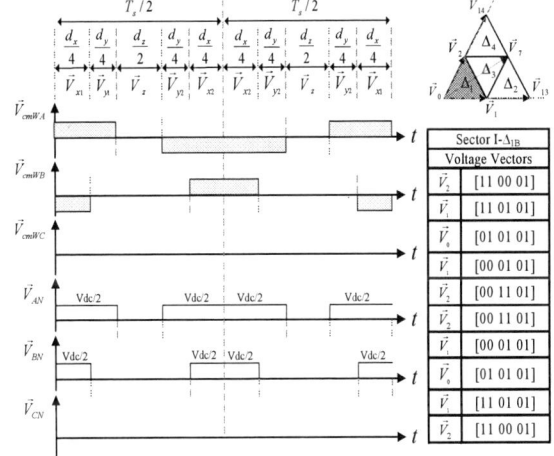

Figure 4. Alternative switching states II and corresponding inductor winding voltage and phase voltage in sector I-Δ_1 in each leg, (Common mode voltages are $\pm V_{dc}$)

Table I. DSVPWM switching sequence for Phase A

Reference Phase	Phase A			
	DSVPWM0	DSVPWM1	DSVPWM2	DSVPWM3
0-30	PWM	PWM	PWM	PWM
30-60	PWM	PWM	HIGH	HIGH
60-90	PWM	HIGH	HIGH	PWM
90-120	HIGH	HIGH	PWM	PWM
120-150	HIGH	PWM	PWM	HIGH
150-180	PWM	PWM	PWM	PWM
180-210	PWM	PWM	PWM	PWM
210-240	PWM	PWM	LOW	LOW
240-270	PWM	LOW	LOW	PWM
270-300	LOW	LOW	PWM	PWM
300-330	LOW	PWM	PWM	LOW
330-360	PWM	PWM	PWM	PWM

Table II. DSVPWM switching states for Phase A

Reference Phase	Phase A			
	DSVPWM0	DSVPWM1	DSVPWM2	DSVPWM3
90-120	Type-A	Type-A	Type-B	Type-B
120-150	Type-A	Type-B	Type-B	Type-A

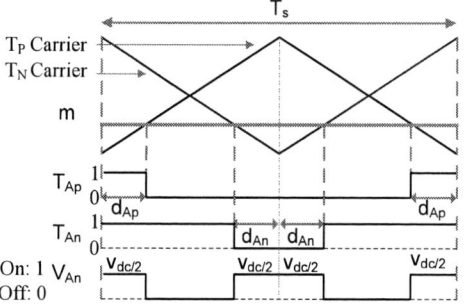

Figure 5. Interleaved switching scheme I (Active High)

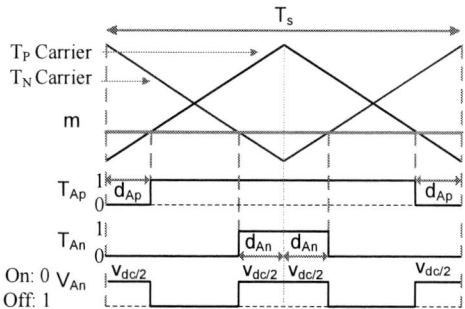

Figure 6. Interleaved switching scheme II (Active Low)

A. SVPWM Interleaved Switching

The interleaved PWM switching can be achieved by using either two modulating waves of the same magnitude and frequency but 180° out of phase compared with a common triangular carrier wave or two triangular carrier waves 180° out of phase compared with a common modulating wave. To easily explain and investigate the interleaved switching in DSVPWM, the interleaved approach based on two carrier waveforms is adopted here. Figure 5 demonstrates the interleaved switching scheme I when two triangular carrier waveforms, T_P and T_N, are 180° out of phase compared with m_a, the common modulation waveform value, at the switching cycle of T_s. T_{Ap} and T_{An} are the gating signals for the upper and lower switches, S_1 and S_2 in Figure 1, respectively.

The switching instances in which both the upper and lower switches overlap each other are interleaved at the beginning, middle, and end of the switching cycle. The interleaved property is obtained because of the unique characteristic of the CII topology where both the upper and lower switches can be either on "type-P" or off "type-N" at the same instant. In Figure 5, during the time interval of d_{Ap}, both switches are on, and during the time interval of d_{An}, both switches are off. Since the value of the modulation signal is the same for both carrier signals, d_{AP} is equal to d_{An}.

In Figure 5, the type-P and type-N switching states generate the same output voltage ($V_{dc}/2$). Thus, their position in a switching cycle can be interchanged. If this change happens, the new interleaved switching scheme (scheme II) can be obtained as shown in Figure 6. Both interleaved switching schemes I and II produce the same output voltage. Therefore, in designing a switching sequence, these interleaved switching states can be replaced in order to minimize the winding current ripple in DSVPWM schemes.

The type-P and type-N switching states are only switching states that impact on the winding common mode current (i_{bCM} in Figure 1). The type-P and type-N switching state energize and de-energize the winding coil, respectively. The balanced distribution of these two types of the switching states in each switching cycle can balance the winding voltage and, as a result, the winding common mode current [8-10].

The advantage of using the interleaved PWM in the CII topology is the possibility of obtaining the double effective switching frequency at the inverter outputs while balancing the common mode voltage (current) which is a significant issue for the successful operation of the CII topology. Figure 5 and Figure 6 reveal that while the actual switching frequency is f_c, the effective output voltage switching frequency is twice ($2f_c$). Since the gating signals are symmetrical to the midpoint line ($T_s/2$), the output voltage in the first half cycle of the switching cycle is identical to the second half of the switching cycle. While just one transition occurs in the gating signals from high to low or vice versa, two transitions occur in the output voltage. To obtain the double output frequency, it is not necessary to have d_{Ap} equal to d_{An}. The main reason why these time intervals should be equal is to balance the winding voltage and, as a result, to minimize the common mode dc current deviations.

The switching states type-A and type-B in Figure 3 and Figure 4 use the combination of both interleaved switching schemes. In type-A switching states, phase B and C employ the interleaved PWM switching scheme II and I, respectively and in type-B, phase A and B employ the interleaved PWM switching scheme I and II, respectively. The use of both interleaved switching schemes eliminates the low effective switching states and minimizes the winding current ripple [9].

978-1-4244-4782-4/10 $26.00 © 2010 IEEE

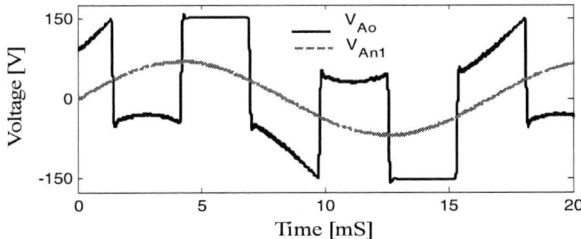

Figure 7. Simulated DSVPWM0: CII filtered terminal voltage (V_{Ao}) and load fundamental phase voltage (V_{An1}) (m_a =0.4)

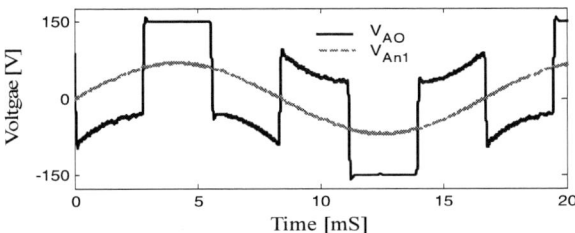

Figure 8. Simulated DSVPWM1: CII filtered terminal voltage (V_{Ao}) and load fundamental phase voltage (V_{An1}) (m_a =0.4)

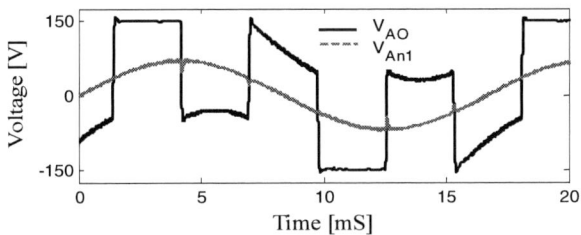

Figure 9. Simulated DSVPWM2: CII filtered terminal voltage (V_{Ao}) and load fundamental phase voltage (V_{An1}) (m_a =0.4)

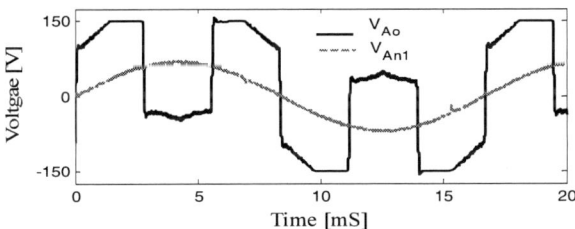

Figure 10. Simulated DSVPWM3: CII filtered terminal voltage (V_{Ao}) and load fundamental phase voltage (V_{An1}) (m_a =0.4)

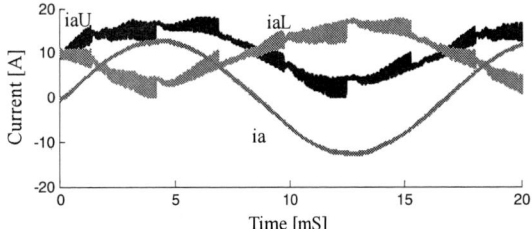

Figure 11. Simulated DSVPWM0: winding currents and phase A current (m_a =0.4, f_c= 15 kHz)

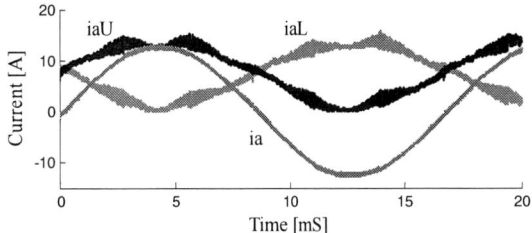

Figure 12. Simulated DSVPWM1: winding currents and phase A current (m_a =0.4, f_c= 15 kHz)

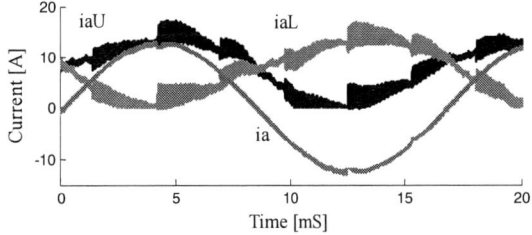

Figure 13. Simulated DSVPWM2: winding currents and phase A current (m_a =0.4, f_c= 15 kHz)

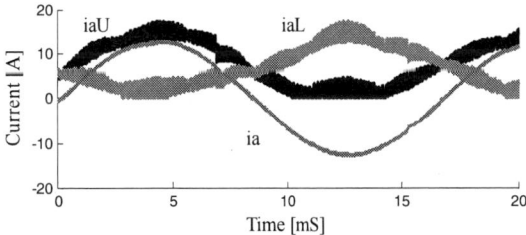

Figure 14. Simulated DSVPWM3: winding currents and phase A current (m_a =0.4, f_c= 15 kHz)

III. INTERLEAVED DSVPWM SIMULATION RESULTS

The operation of the CII topology with the DSVPWM, strategies was simulated in Simulink. The tests were carried by using an inverter with 300V DC link voltage and 15 kHz switching frequency, driving a 5.2 Ω;1 mH 3-phase load with the fundamental frequency of 60Hz.

By using the interleaved DSVPWM strategy, a 60° discontinuous period can be distributed in each positive and negative cycle of the fundamental reference voltage over a range of modulation indices. Generalized 60° discontinuous SVPWM strategies can be categorized as interleaved DSVPWM0, DSVPWM1, DSVPWM2 and DSVPWM3. The simulation results in Figure 7 through Figure 14 compare DSVPWM0, DSVPWM1, DSVPWM2, and DSVPWM3 methods when m_a is 0.4. The load fundamental phase voltage (V_{An1}) to the load neutral point (n) and the inverter filtered terminal voltage (V_{Ao}) to the DC bus midpoint (o) are shown

at m_a = 0.4 in Figure 7 to Figure 10 for the DSVPWM0, DSVPWM1 DSVPWM2, and DSVPWM3 schemes, respectively. 60° discontinuous regions (where the terminal voltage is clamped to either $\pm V_{dc}/2$ at the positive and negative cycles) are dispread for these schemes.

For a leading power factor load, the discontinuous period can be advanced by up to 30°, DSVPWM0 scheme shown in Figure 7. Similarly for a lagging power factor load, the DSVPWM2 scheme is obtained as shown in Figure 9. In Figure 8, 60° discontinuous SVPWM is located around the peak positive and negative cycles. The resulting waveform is called the DSVPWM1 scheme. However, in Figure 10, the discontinuous regions are divided into two 30° discontinuous regions around the peak, indicating the proposed DSVPWM is DSVPWM3.

The corresponding upper and lower winding currents and phase A currents are plotted in Figure 11 through Figure 14, respectively.

With the same optimal use of the high effective inductance switching states, the 60° discontinuous PWM at the positive and negative voltage peaks in DVSPWM1 produces a lower high-frequency winding current ripple compared with that of DSVPWM0, DSVPWM2 and DSVPWM3. The current ripple in DSVPWM1 is lower than that in DSVPWM0, DSVPWM2 and DSVPWM3 for the following reasons.

First, the current ripple in one leg due to its own winding effect is directly minimized when the discontinuous period is located at the positive and negative voltage peak cycles. For the 60° discontinuous periods, the winding is shorted, so the current ripple is not impacted by its own winding. The 60° PWM switching periods around the voltage zero crossing, provide the minimum average winding voltage compared to that if the PWM switching period is located in other parts of the voltage waveform. Thus, the minimum average voltage allocates the most dwell time in each switching cycle for the '00' and '11' switching states (instead of the '01' and '10' switching states) in each switching sequence, which minimize the current ripple.

Second, the current ripple due to the coupling of the other windings is also indirectly minimized. For the 60° discontinuous periods, the other (active) phases with PWM switching are not at around their peak voltages. In addition, since high-inductance states are selected (i.e., the flux produced from one winding is observed by the other winding) the effect of the coupling is also minimized. Thus, the minimum average voltage is induced in the shorted winding, which produces the minimum current ripple. For the 60° PWM switching periods, when switching states of '00' and '11' occur in the switching cycle, the windings in the other legs have a minimum impact on the current ripple since high-inductance states are selected with the minimum average voltage. When switching states of '01' and '10' occur in the switching cycle, the windings in the other legs have no impact on the current ripple.

The above conclusion is independent of the load magnitude and power factor since the coupled inductor topology operates based on volt per second and the load does not affect the operation of the coupled inductor [15].

In addition, compared to the DSVPWM0, DSVPWM2 and DSVPWM3 schemes, the DSVPWM1 scheme also can reduce inverter switching losses for a unity power factor load by eliminating PWM switching at the peak voltage when the load current is almost at the maximum value. If the discontinuous period coincides with the maximum load current, the switching losses are minimized [12, 16].

IV. INTERLEAVED DSVPWM EXPERIMENTAL RESULTS

In order to validate the proposed approach, the investigated PWM modulation algorithms are implemented using a TI TMS320F2812 DSP, and experimental tests were carried out using the small-power converter system illustrated in Figure 15, with the following parameters: $V_{DC} = 300$ V, $I_{pk} = 20$ A, $f_c = 15$kHz switching frequency, the split-wound coupled inductor magnetizing inductance (L_t) = 4.7 mH. Most of the computation for multilevel DSVPWM strategies is performed offline and tabulated as PWM-switching look-up tables, so that the on-line computation is minimized.

A. CII topology operation performance with DSVPWMs

The upper and lower winding currents and phase A currents are plotted in Figure 16 through Figure 19 for DSVPWM0, 1, 2, and 3 at $m_a = 0.4$, respectively.

Figure 15. Experimental set-up of the CII inverter, combined three limb core, interface boards and a TI DSP controller

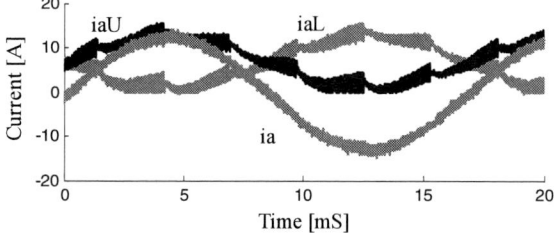

Figure 16. Experimental DSVPWM0: winding currents and phase A current (Unity power factor load, R= 5.2 Ω, m_a =0.4, f_c= 15 kHz)

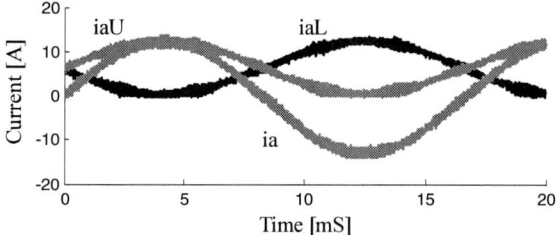

Figure 17. Experimental DSVPWM1: winding currents and phase A current (Unity power factor load, R= 5.2 Ω, ma =0.4, f_c= 15 kHz)

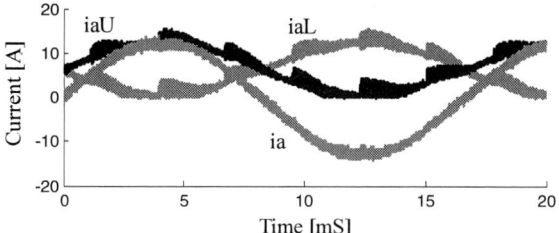

Figure 18. Experimental DSVPWM2: winding currents and phase A current (Unity power factor load, R= 5.2 Ω, m_a =0.4, f_c= 15 kHz)

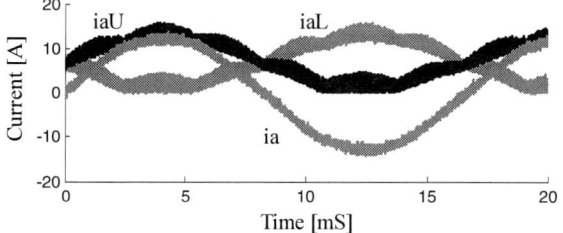

Figure 19. Experimental DSVPWM3: winding currents and phase A current (Unity power factor load, R= 5.2 Ω, m_a =0.4, f_c= 15 kHz)

The magnitude of the ripple in the output currents are the same. However, the inductor winding current ripple is different. The DSVPWM3 winding current has a large ripple distributed over the fundamental cycle where as the DSVPWM1 winding current has a small current ripple. The optimal use of the position of the discontinuous period in DVSPWM1 produces a much lower maximum high-frequency winding current ripple compared with that of DSVPWM0, DSVPWM2, and DSVPWM3: lower by a factor of 1.5 at $m_a =$ 0.4.

The DSVPWM0 upper and lower winding currents and phase A currents are plotted in Figure 20 and Figure 21 for a unity and o.86 (30° phase lag) power factor loads at $m_a = 0.4$, respectively. Similarly, the DSVPWM2 upper and lower winding currents and phase A currents are plotted in Figure 22 and Figure 23 for a unity power factor and o.86 (30° phase lag) loads at $m_a = 0.4$, respectively. The loads are chosen in order to have the same magnitude of the load current for both conditions. For a unity power factor load, the winding current waveforms and phase A current are in-phase. However, for the inductive load, the phase A current has a phase lag of 30 degree. The result indicates that the winding current ripple for a unity power factor and an inductive load are the same.

The high frequency winding ripple contributes to the winding losses and the high frequency common mode current component of the winding current produces core losses [7, 9, 15]. The common mode currents for DSVPWM methods are plotted for two different loads of 5.2 Ω and 10.4 Ω in Figure 24 and Figure 25, respectively. Under DSVPWM1, the magnitude of the high frequency common mode current is significantly reduced in both experiments. This result can lower coupled inductor core losses as shown in the following section. Regardless of the load value, the peak magnitude of the current ripple is approximately the same and the shape of the common mode current waveforms is followed by each other. One can predict that this result can lead to produce the same core losses.

The common mode currents for DSVPWM0 and 2 methods are plotted for a 0.86 power factor load in Figure 26. Compared to the DSVPWM0 and 2 common mode current waveforms in Figure 25, the common mode current waveforms in Figure 26 are shifted by 30 degrees. This phase displacement is occurred due to the power factor difference between these conditions. The peak ripples are the same and the common mode current waveforms are followed by each other. These results indicate that the common mode currents are in accordance and, as a result, similar core losses for different loads can be estimated.

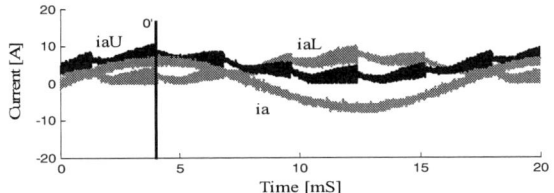

Figure 20. Experimental DSVPWM0: winding currents and phase A current (Unity power factor load, R=10.4Ω, m_a =0.4, f_c= 15 kHz)

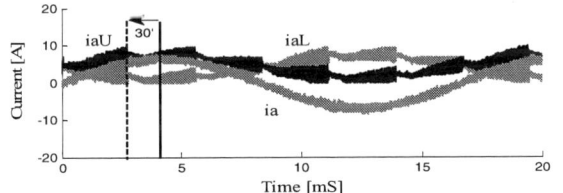

Figure 21. Experimental DSVPWM0: winding currents and phase A current (Load power factor = 0.868, L=10mH, m_a =0.4, f_c= 15 kHz)

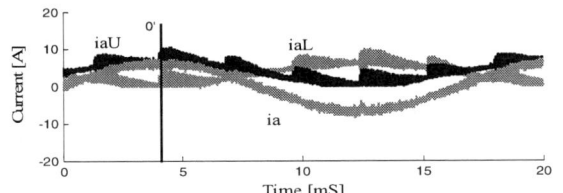

Figure 22. Experimental DSVPWM2: winding currents and phase A current (Unity power factor load, R=10.4Ω, m_a =0.4, f_c= 15 kHz)

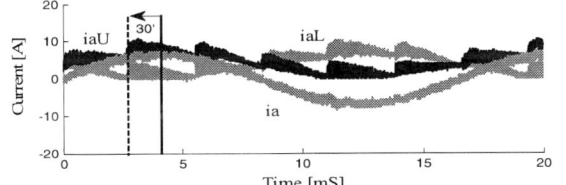

Figure 23. Experimental DSVPWM2: winding currents and phase A current (Load power factor = 0.866, L=10mH, m_a =0.4, f_c= 15 kHz)

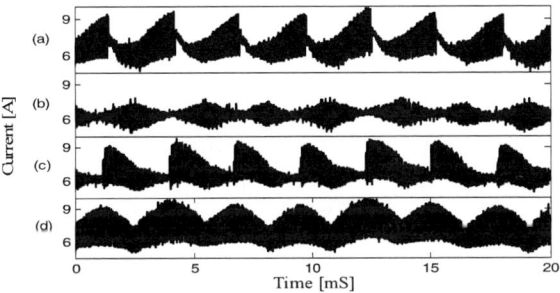

Figure 24. Experimental common mode current with (a) DSVPWM0 (b) DSVPWM1 (c) DSVPWM2 (d) DSVPWM3 (Unity power factor load, R= 5.2 Ω, m_a=.9, f_c= 15 kHz)

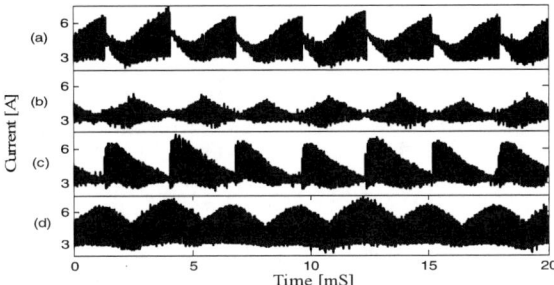

Figure 25. Experimental common mode current with (a) DSVPWM0 (b) DSVPWM1 (c) DSVPWM2 (d) DSVPWM3 (Unity power factor load, R= 10.4 Ω, m_a=.9, f_c= 15 kHz)

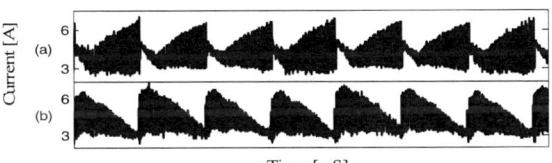

Figure 26. Experimental common mode current with (a) DSVPWM0 (b) SVPWM2 (Load power factor = 0.86, m_a=.9, f_c= 15 kHz)

B. Coupled indutor losses comparison for DSVPWMs

The effects of the winding current ripple reduction on coupled inductor losses are investigated with different DSVPWM methods. To magnify the difference between methods, the power loss measurements are carried out for m_a lower than 0.5. A lower winding current ripple produces lower rms, resulting in a lower inductor winding losses, and the lower common mode current ripple reduces the core losses, which are related to the peak of the high-frequency component of the common mode current. Thus, lower inductor losses and, as a result, lower inverter losses can be predicted for a DSVPWM scheme with a minimum winding current ripple.

Experimentally measured power losses are presented for the coupled inductor. Loss comparisons curves are compared when using different discontinuous SVPWM schemes. The tests were carried out with a 3-phase RL load under different conditions for the CII with a nominal power of 5kW. The coupled inductor power losses were obtained by measuring the voltage across each coil and the corresponding winding current. The coupled inductors power losses for a unity and 0.86 power factor loads are demonstrated as a function of m_a in Figure 27. The losses are increased as m_a increases due to two effects: core loss dependence on m_a and the load current value increasing with a lower m_a. For the same load power factor at each m_a, the coupled inductor losses generated by the DSVPWM1 are the lowest. In contrast, the DSVPWM3 produces very large power losses. The power losses for DSVPWM0 and DSVPWM2 are coincided for the unity power factor and the inductive load. This result indicates that the power losses are independent of the load power factor.

The coupled inductors' power losses for the CII topology are demonstrated at various rms load currents in Figure 28 when m_a is 0.50. The differences between the power losses are almost constant for each m_a and are mostly due to core flux losses. The results also indicate that the power losses for DSVPWM0 and DSVPWM2 are the same; the power losses in the DSVPWM0 and 2 schemes are located between the power losses in the DSVPWM1 and DSVPWM3 schemes, with the DSVPWM1 giving the lowest power losses.

V. CONCLUSION

The interleaved DSVPWM, categorized as DSVPWM0, DSVPWM1, DSVPWM2 and DSVPWM3 is presented. The difference between these schemes is the placement of the discontinuous period within each fundamental voltage cycle. The comprehensive analysis illustrates that the winding currents ripple in a CII is not only dependent to the selection of switching states with a high effective inductance but also the position of discontinuous PWM period in the inverter output voltage. The coupled inductor losses are minimized when the discontinuous period coincides with the maximum inverter terminal voltage. Regardless of loads, the DSVPWM1 scheme lowers the high-frequency current ripple compared to other DSVPWM schemes, minimizing the coupled inductor losses. The overall performance of the coupled inductor inverter, as validated by experimental loss measurements, is improved significantly. This benefit can lead to lowering the required core size for the split-wound inductor structure and to reducing the inductor cost.

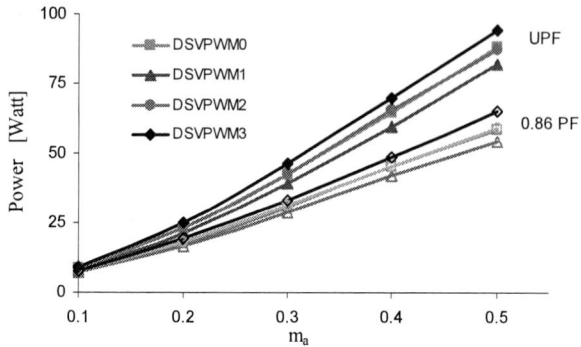

Figure 27. Experimental coupled inductor losses for DSVPWM0, 1, 2 and 3 schemes with a unity and 0.86 power factor load

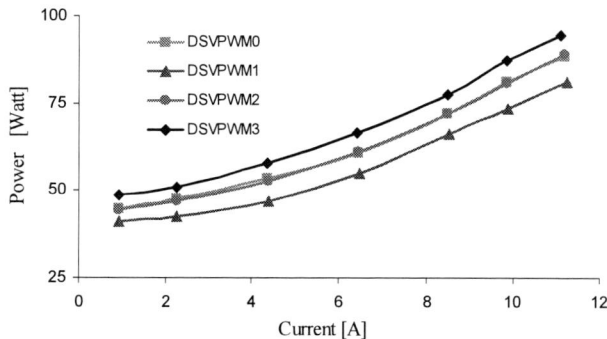

Figure 28. Experimental coupled inductor losses for various load currents with DSVPWM0, 1, 2 and 3 (Unity power factor, $m_a = 0.5$)

REFERENCES

[1] J. Rodriguez, L. Jih-Sheng, and P. Fang Zheng, "Multilevel inverters: a survey of topologies, controls, and applications," Industrial Electronics, IEEE Transactions on, vol. 49, pp. 724-738, 2002.

[2] L. G. Franquelo, J. Rodriguez, J. I. Leon, S. Kouro, R. Portillo, and M. A. M. Prats, "The age of multilevel converters arrives," Industrial Electronics Magazine, IEEE, vol. 2, pp. 28-39, 2008.

[3] M. H. Rahsid, Power Electronics Handbook: Academic Press.

[4] L. Jih-Sheng and P. Fang Zheng, "Multilevel converters-a new breed of power converters," Industry Applications, IEEE Transactions on, vol. 32, pp. 509-517, 1996.

[5] W. Bin, High-power converters and ac drives. New Jersey: John Wiley & Sons, Inc., 2006.

[6] J. Ewanchuk, J. Salmon, and A. Knight, "Performance of a High Speed Motor Drive System using a Novel Multi-level Inverter Topology," Industry Applications, IEEE Transactions on, vol. PP, pp. 1-1, 2009.

[7] J. Salmon, J. Ewanchuk, and A. Knight, "PWM Inverters Using Split-Wound Coupled Inductors," in Industry Applications Society Annual Meeting, 2008. IAS '08. IEEE, 2008, pp. 1-8.

[8] B. Vafakhah, M. Masiala, J. Salmon, and A. M. Knight, "Space-vector PWM for inverters with split-wound coupled inductors," in Electric Machines and Drives Conference, 2009. IEMDC '09. IEEE International, 2009, pp. 724-731.

[9] B. Vafakhah, J. Salmon, and A. M. Knight, "Interleaved Discontinuous Space-Vector PWM for A Multi-Level PWM VSI using a 3-phase Split-Wound Coupled Inductor," 2009.

[10] B. Vafakhah, J. Salmon, and A. M. Knight, "A New Space-Vector PWM with Optimal Switching Selection for Multi-Level Coupled Inductor Inverters," Industrial Electronics, IEEE Transactions on, 2009, in press.

[11] X. Kun, F. C. Lee, D. Borojevic, Y. Zhihong, and S. Mazumder, "Interleaved PWM with discontinuous space-vector modulation," Power Electronics, IEEE Transactions on, vol. 14, pp. 906-917, 1999.

[12] D. G. Holmes and T. A. Lipo, Pulse width modulation for power converters. New Jersey, 2003.

[13] B. P. McGrath, D. G. Holmes, and T. Lipo, "Optimized space vector switching sequences for multilevel inverters," Power Electronics, IEEE Transactions on, vol. 18, pp. 1293-1301, 2003.

[14] N. Celanovic and D. Boroyevich, "A fast space-vector modulation algorithm for multilevel three-phase converters," Industry Applications, IEEE Transactions on, vol. 37, pp. 637-641, 2001.

[15] A. M. Knight, J. Ewanchuk, and J. C. Salmon, "Coupled Three-Phase Inductors for Interleaved Inverter Switching," Magnetics, IEEE Transactions on, vol. 44, pp. 4119-4122, 2008.

[16] Y. Wu, C. Y. Leong, and R. A. McMahon, "A Study of Inverter Loss Reduction Using Discontinuous Pulse Width Modulation Techniques," in Power Electronics, Machines and Drives, 2006. The 3rd IET International Conference on, 2006, pp. 596-600.

A Novel Elevator Load Torque Identification Method Based on Friction Mode

Xiaoyuan Hong, Zhe Deng, Siran Wang, Lijun Hang, Wuhua Li, Zhengyu Lu

College of Electrical Engineering
Zhejiang University
Hangzhou, Zhejiang, China
Email: hongxiaoyuan@gmail.com

Abstract—**This paper presents a method to identify the elevator load torque when the brake is released. With the friction model, the identification process is converted to a searching question. The principle of the method is illustrated in detail, and the factors that influence the identification process and sliding distance are also analysed. Finally, the proposed method is implemented and verified by the results on a permanent magnet synchronous motor drived gearless elevator with an absolute encoder ECN413 from Heidenhain.**

I. Introduction

Nowadays more and more elevators are installed in the buildings all over the world[1-2]. Safety and stability are the basic and essential demands of the elevator system, but not enough. People pay more attention to riding comfort. Sliding when the brake is released is one of the subjects that greatly influences passengers' feel. In order to prevent sliding, a weighing sensor is mounted in the car of the elevator. With the weighing signal, a electromagnetic torque is produced to balance the load torque. It works well but costs a lot. Many papers[3-5] focused on the load identification when the motor was running to improve the dynamic performance. Observers were used based on the motor model in these papers, and they were parameters dependent. Moreover, acurate speed was requested to calculate accurate load torque which was not applicable to the occasion that the motor with an absolute encoder run in ultra low speed.

This paper presents a novel elevator load torque identification method based on the friction model. The identification process is prompt enough for the motor to produce a balanced torque that assures a short enough sliding distance. The whole process can be easily implemented in the DSP, and it is independent of the motor parameters.

II. Proposed Identification Method

A simplified model of an elevator is shown as Fig.1 where CW represents counterweight, Car reprensents both the car and the passengers, T_f represents the friction torque and T_e represents the electromagnetic torque respectively. So the total load torque can be expressed as:

This paper and its related research are supported by grants from the Power Electronics Science and Education Development Program of Delta Environmental & Educational Fundation, and National Natural Science Foundation of CHINA 110301-N10930.

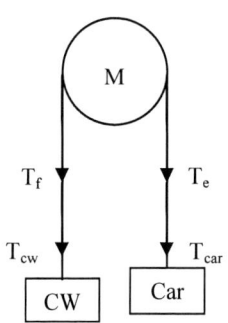

Fig.1: Simplified model of an elevator.

$$T_m = T_{cw} - T_{car}. \quad (1)$$

The mechanical function of the system can be expressed as:

$$T_e - T_m - T_f = J\frac{d\omega}{dt}, \quad (2)$$

where J is the total inertia of the elevator system and ω is the angle velocity of the motor. To keep the motor still, an

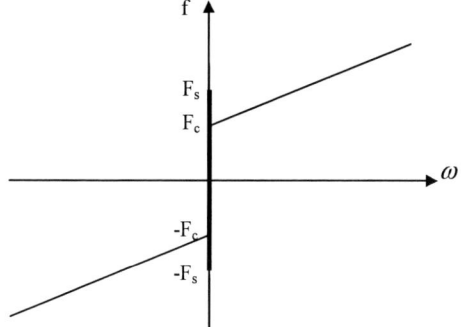

Fig.2 Friction model.

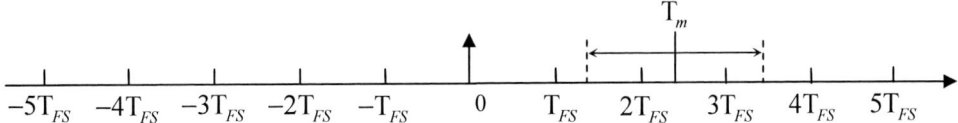

Fig.3 Quantization of the load torque range.

electromagnetic torque is produced as $T_e = T_m + T_f$.

As we know, the friction torque is not constant, especially at a standstill. The friction model[6] can be seen in Fig.2, where F_s is the breakout friction and F_c is the coulomb friction. The coulomb friction is smaller than the breakout friction. In order to start motor from standstill, T_e must conform to the following inequality:

$$|T_e - T_m| > T_{FS}, \qquad (3)$$

where T_{FS} represents the breakout friction torque. In other words, the produced torque that meets the demand to keep motor still is a range rather than a value. The range is twice as big as T_{FS}. It's easy to obtain the breakout friction by applying a gradually increasing torque to the motor. Thus the range of load torque can be devided into some torque units like Fig.3. Then the identification of the load torque can be converted to a searching question. According to linear search method, an electromagnetic torque which increases from $-T_n$ to T_n by step length of $2T_{FS}$ is implemented on the motor and there would be a step that keep motor still. (T_n represents the nominal load torque.) Although the resultant step is not exactly equal to the load torque, the goal to keep motor still is achieved and that's the purpose to identify the load torque.

III. REDUCING SLIDING DISTANCE AND ACCELERATING THE IDENTIFICATION PROCESS

According to (1), the electromagnetic torque is propotional to the acceleration. So the most straightforward method to verify which step is the resultant one is to caculate the acceleration. Assumed that the electrical time constant is neglectable and the encoder signal can be scaled infinitely, the torque is produced instantaneously without delay and the acceleration can be caculated soon after producing corresponding torque. Then the searching process can be finished with little rotation. But in fact, there are many nonideal factors that greatly deteriorate the performance. Among the nonideal factors, the finite resolution of the encoder must be the most important one. As shown in Fig.4, the pulses output from the encoder represent the rotor position. Deviding by the elapsed time between two pulses gets the average speed. And deviding by the elapsed time between two speeds gets the average acceleration. So, at least two scale units are needed to get the acceleration. If the searching process is implemented by increasing the electromagnetic torque from $-T_n$ to T_n step by step, the acceleration difference between two adjacent ones may not be obvious.

Taking into consideration of the response time of the torque current, the finite word length in DSP and the finite speed of the CPU, more scale units may be neccesary. It is hard to accurately evaluate the influence of the nonideal factors to the actual acceleration. In order to distinguish correctly the acceleration difference, five or more scale units are used. This is unacceptable due to the long sliding distance.

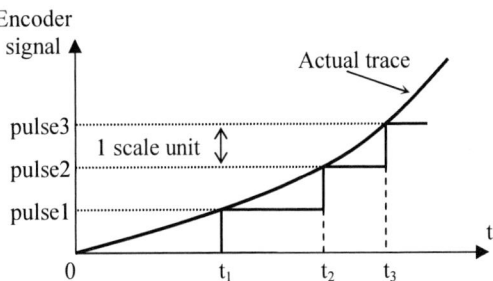

Fig.4 Relation between actual trace and encoder output

In order to reduce sliding distance, the sequence to choose T_e is changed into: $0 \rightarrow -T_n \rightarrow T_n \rightarrow -T_n + 2T_{FS} \rightarrow T_n - 2T_{FS} \rightarrow -T_n + 4T_{FS} \rightarrow T_n - 4T_{FS} \rightarrow \cdots$. With this sequence, the motor rotates bidirectionally to reduce the total sliding distance. And the acceleration caculation can be eliminated by just judging the rotating direction. For example, when the motor runs in the positive direction, the encoder pulse count increases. Then an opposite electromagnetic torque will be implemented on the motor. The speed goes down and across zero at last. The motor begins to run in negative direction. Only one scale unit is needed to identify the direction and this must be correct.

Since the first rotating direction determines the polarity of the load torque, the sequence to choose T_e can be changed into: $0 \rightarrow T_n \rightarrow 2T_{FS} \rightarrow T_n - 2T_{FS} \rightarrow 4T_{FS} \rightarrow T_n - 4T_{FS} \rightarrow \cdots$ or $0 \rightarrow -T_n \rightarrow -2T_{FS} \rightarrow -T_n + 2T_{FS} \rightarrow -4T_{FS} \rightarrow -T_n + 4T_{FS} \rightarrow \cdots$. This can accelerate the process of identification and reduce the sliding distance. Another factor that influences the identification process is the step length. Usually the breakout friction varies along the hoistway and it's unrealistic to measure the breakout friction of every single point in the hoistway. So some points are chosen and the least one is considered as T_{FS}. Then a step length smaller than $2T_{FS}$ is chosen to guarantee the success of the identification. However, smaller step length means deccelerating of the process and increasing the sliding distance. So there is a tradeoff.

978-1-4244-4782-4/10 $26.00 © 2010 IEEE

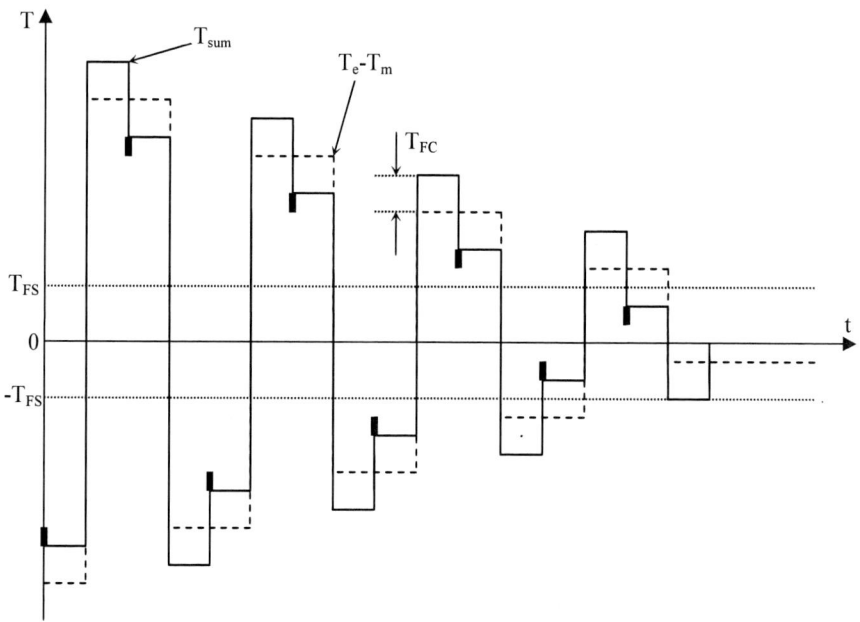

Fig.5 Realization process of the proposed method.

Fig.5 shows the identification process when $T_m > 0$. The real line represents synthetic torque to the motor which can be expressed as:

$$T_{sum} = T_e - T_m - T_f. \tag{4}$$

And the dotted line represents $T_e - T_m$. As shown in Fig.5, T_{sum} swings around $T_e - T_m$ due to the variation of the friction with the rotating direction. The swing instant corresponds to the bold line at which the speed equals zero. Each step in the searching sequence is composed of positive speed section, zero speed section and negative speed section. The process starts at $t = 0$ when the brake is released. Then T_e is produced according to the rotating direction and converges step by step with an presetting length. At last, T_e converges to a constant value that makes T_{sum} equal 0. And the motor comes into a standstill.

IV. SIMULATION AND EXPERIMENTAL RESULTS

The proposed method is implemented on a permanent magnet synchronous motor drived gearless elevator. The main parameters of the motor are listed below: 20 poles, 8.7kw nomial power, 450Nm nomial torque, 20kgm^2 inertia. An absolute encoder ECN413 from Heidenhain is mounted on the motor shaft. The encoder outputs 8192 pulses every revolution. The simulation results are shown in Fig.6 and Fig.7. $T_m = 100\text{Nm}$, $T_{FS} = 18\text{Nm}$, $T_{FC} = 15\text{Nm}$, step length is 30Nm in Fig.6 and 20Nm in Fig.7. The identification process in Fig.6 is faster than that in Fig.7. The experimental results are shown in Fig.8 and Fig.9. The step length is 11Nm in both Fig.8 and Fig.9. With light load (T_m=-40Nm) in Fig.8, the identification process is shorter than that with heavier load(T_m=-150Nm) in Fig.9. That is due to the fewer steps taken in lighter load. But the sliding distance equals in both loads. The motor slides only a pulse distance which means only 0.16mm distance in line. Passengers can not even feel it.

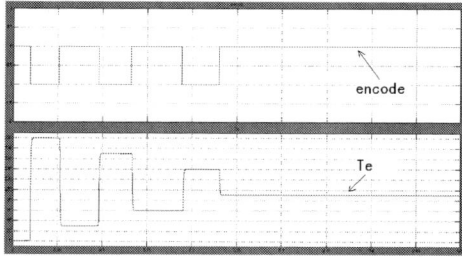

Fig.6 Simulation result1(encode: 0.5pulse/grid,Te: 20Nm/grid).

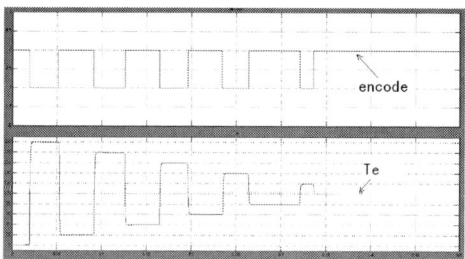

Fig.7 Simulation result2(encode: 0.5pulse/grid,Te: 20Nm/grid).

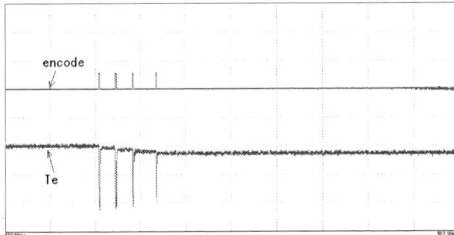

Fig.8 Tm=-40Nm(encode: 2 pulse/grid, Te: 150Nm/grid).

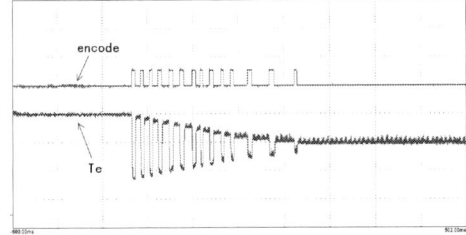

Fig.9 Tm=-150Nm(encode: 2 pulse/grid, Te: 150Nm/grid).

V. CONCLUSIONS

An elevator load torque identification method is proposed in this paper which can be easily implemented in the DSP. The sliding distance when the brake is released is controlled in a comfortable range with this method. Simulation and experimental results verify the feasibility of the method.

REFERENCES

[1] Hiroaki Mizuguchi, Toshiaki Nakagawa, Yoshiaki Fujita, "Breaking the 1,000 mpm barrier," *Elevator World*, pp.71-76, September 2005.

[2] Chung DW, Ryu HM, Lee YM, et al. "Drive systems for high speed gearless elevators," *IEEE INDUSTRY APPLICATIONS MAGAZINE*, vol. 7, no. 5, pp. 52-56, SEP-OCT 2001.

[3] Xu DG, Wang H, Shi JZ, "PMSM servo system with speed and torque observer," in *IEEE 2003 Power Electronics Specialists Conference*, 2003, pp.241-245.

[4] Rao FQ, Gu GB, "Load torque observer for minimizing torque ripple in PMSM," in *IEEE 2003 International Conference on Electrical Machines and Systems*, 2003, pp. 473-476.

[5] Nam-Joon Kim, Hee-Sung Moon, Dong-Seok Hyun, "Inertia identification for the speed observer of the low speed control of induction machines," *IEEE Transactions on Industry Applications*, vol. 32, no. 6, pp. 1371-1379, November/December 1996.

[6] Suraneni S, Kar IN, Bhatt RKP, "Adaptive stick-slip friction compensation using dynamic fuzzy logic system," in *IEEE 2003 Conference on Convergent Technologies for the ASIA-PACIFIC Region*, 2003, pp. 1470-1474.

A Novel Digital Current Control Strategy for Torque Ripple Reduction in Permanent Magnet Synchronous Motor Drives

Haidong Yu

Phoenix International – A John Deere Company
Fargo, ND, USA
yuhaidong@ieee.org

Abstract—**Permanent magnet synchronous motors (PMSM) are attractive candidates in various applications because of their high efficiency, high power density, and high position resolution. In some special applications such as military and aerospace, the torque ripple percentage from PMSM has to be maintained in an acceptable level. This objective drives motivations of either optimum design of electrical machines or optimum control of motor drives. On optimum control of motor drives, injecting higher order harmonics has been a dominant technique. This paper presents a novel digital current control strategy to regulate currents with feature of fast dynamic response and easy implementation. Compared with conventional current control loop in vector control of PMSM, a bank of resonant controllers is proposed in parallel connection of PI type controllers. As a result, both fundamental and harmonic components of current can be regulated using high bandwidth digital controllers. Simulation results have verified the control effect of the proposed strategy.**

I. Introduction

Permanent magnet synchronous motors (PMSM) are attractive candidates in various applications because of their high efficiency, high power density, and high position resolution. In some special applications such as military and aerospace, the torque ripple percentage from PMSM has to be maintained in an acceptable level. This objective drives motivations of either optimum design of electrical machines or optimum control of motor drives. In recent years, several papers have been published describing methods of optimum design of PMSM [1]-[2]. This results in high production cost of PMSM and difficult modelling of electrical machines. On optimum control of motor drives, injecting higher order harmonics has been a dominant technique. This technique is comprised of current reference generation and current control. The generation of optimum *d*-axis and *q*-axis current references is out of scope in this paper. [3]-[4] have proposed a multi-reference-frame method. This method expands the conventional single synchronous *dq* frame into multiple synchronous *dq* frames targeted for all harmonics of interest. However, the low pass filter in each synchronous frame limits

the bandwidth of the current controllers. In addition, the computation effort in microcontroller or DSP is increased significantly due to multiple frames. Combination of linear and non-linear controllers has been introduced in [5]-[6]. The control effect was satisfactory. However, the complexity of control design is a little high, which limits its application domain.

This paper presents a novel digital current control strategy to regulate currents with feature of fast dynamic response and easy implementation. Compared with conventional current control loop in vector control of PMSM, a bank of resonant controllers is proposed in parallel connection of PI type controllers. As a result, both fundamental and harmonic components of current can be regulated using high bandwidth digital controllers.

II. Digital Resonant Controller

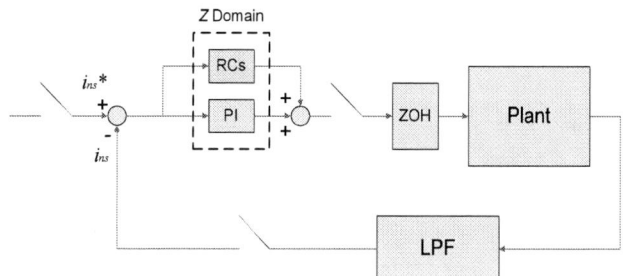

Figure 1.   Generalized current control loop with digital controllers

The concept of resonant controller [7]-[8] has been utilized in active power filters (APF) in recent years. The motivation of this type of applications is because PI type controllers can only provide satisfactory transient and steady state performance for dc frequency component. For medium and high frequency components, the control effect will be degraded. This results in phase shift and amplitude deviation, especially when local power system experiences big transient events, such as sudden load demand change. The existence of

---

978-1-4244-4782-4/10 $26.00 © 2010 IEEE

resonant controller can accommodate these issues. Due to similarity of applications, the concept of resonant controller can also be utilized in motor drives to reduce torque ripples from PMSM.

Typically, the most significant harmonic orders in a healthy three-phase motor drive system are 5th, 7th, 11th, and 13th. After Park transformation, these harmonics will be transformed into $\pm 6th$ and $\pm 12th$. Therefore one can add two resonant controllers in each current control loop to control these harmonics. The generalized control block with digital PI and resonant controllers is illustrated in figure 1. The subscript 'n' represents d or q in the synchronous frame. 'ZOH' is the zero order hold caused by the analog to digital conversion. The digital resonant controller is in the format as follows:

$$G_{RC}(z) = \frac{bz^2 - cz + d}{z^2 - az + 1} \qquad (1)$$

where a is corresponding to the harmonic frequency of interest. It directly determines the transient performance and stability of the resonant controller. The digital PI controller is in the following format:

$$G_{PI}(z) = \frac{k(z - e)}{z - 1} \qquad (2)$$

As a result, the closed loop transfer function in digital domain can be derived as follows:

$$G_{CLTF}(z) = \frac{(G_{RC}(z) + G_{PI}(z))Z\{G_{ZOH}(s)G(s)\}}{1 + (G_{RC}(z) + G_{PI}(z))Z\{G_{ZOH}(s)G(s)G_{LPF}(s)\}} \qquad (3)$$

where $G_{ZOH}(s)$, $G(s)$, and $G_{LPF}(s)$ are transfer functions of zero order hold, plant, and low pass filter respectively.

In adjustable speed motor drives, fundamental frequency varies in a wide band. Therefore, all harmonic frequencies are time-varying quantities. In order to achieve superior transient performance, the characteristics between all parameters in (1) with motor speed will be stored in look-up tables in the real implementation. Incorporating the digital resonant controllers, the complete functionality of vector control for PMSM is represented in figure 2. As can be seen, the implementation of resonant controllers does not change the structure of conventional vector control or increase any production cost.

III. SIMULATION RESULTS

For the PMSM drive system under investigation, the stator inductance ($Lq=Ld$) is 40 mH, stator resistance is 20 Ω, PWM frequency is 16 kHz, and bandwidth of low pass filter is 2 kHz. The fundamental frequency corresponding to base speed is about 50 Hz. Therefore, the frequency spectrum of 5th and 7th harmonics in synchronous frame will be around 300 Hz. As a result, the bandwidth of resonant controller for these harmonics is designed at 500 Hz, with damping factor about 0.9. For sake of simplicity, resonant controller of 11th and 13th harmonics was not conducted in the simulation. Figure 3 depicts the synchronous frame current response from the

simulation. During the simulation, at instant of 0.1 sec, a step change on the q-axis current was applied. Then, at instant of 0.2 sec, $\pm 6th$ harmonics were added into the current reference. At 0.3 sec, a step change on the reference was applied again. Finally, both a step change and $\pm 6th$ change happened at 0.4 sec. Figures 4-7 represent the zoomed response at those instants. From these figures, it is notable that the conventional PI controller caused significant phase shift and amplitude deviation on the $\pm 6th$ current response. The combination of PI and resonant controller reduces the phase shift and amplitude deviation significantly such that the current response follows the reference accurately and quickly. It is observed from figure 4 that the proposed configuration causes some oscillation in the response to a step command. It indicates that the design of the resonant controller has to be very careful in order to maintain the stability of the whole system.

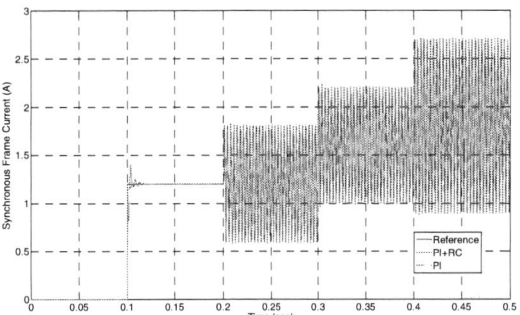

Figure 3. Current response

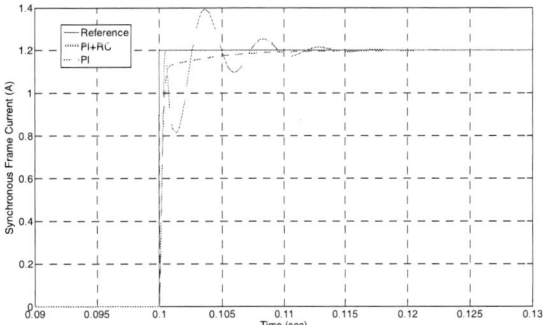

Figure 4. Zoom of instant 1

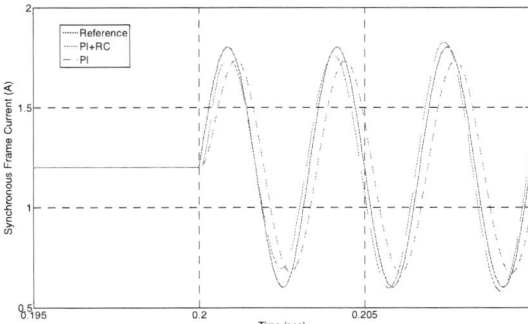

Figure 5. Zoom of instant 2

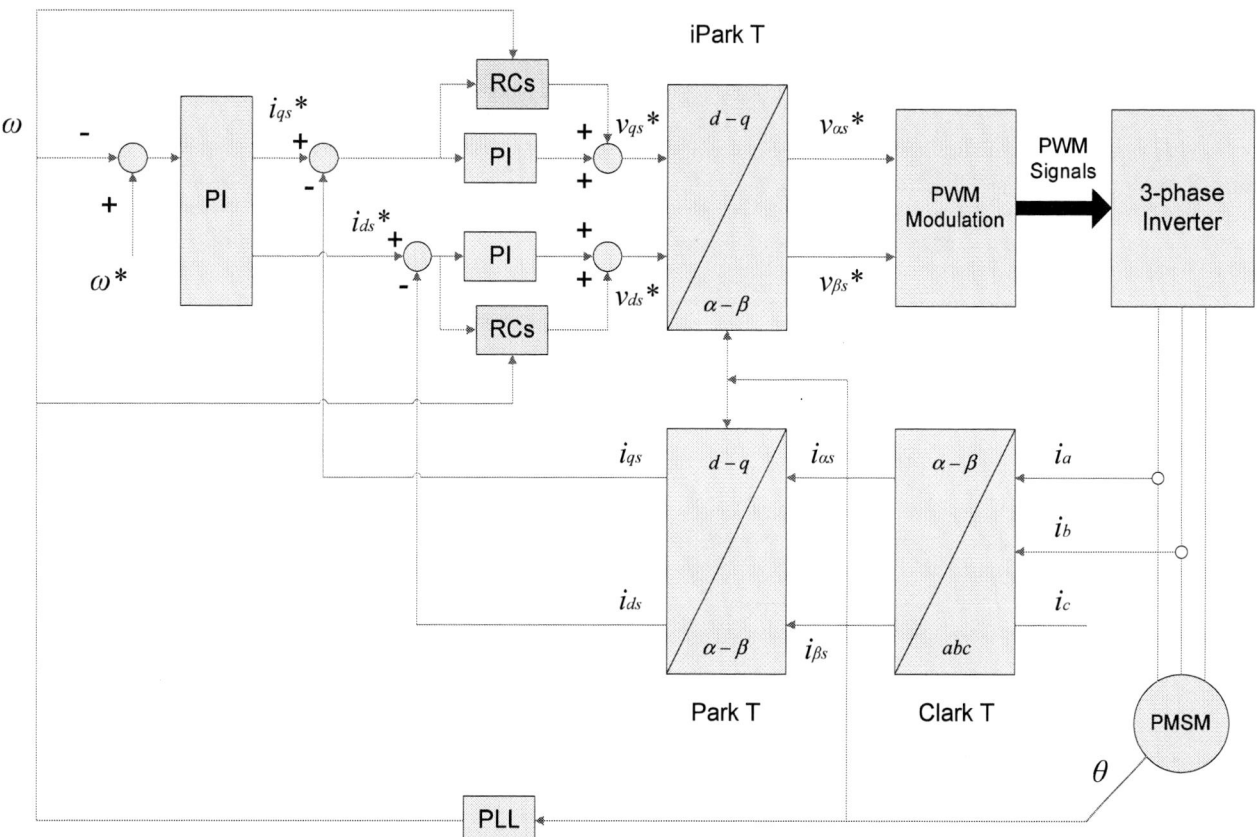

Figure 2. Complete proposed control scheme for PMSM

In the real implementation, subject to continuous operation duty and ambient temperature, the stator resistance may vary in a wide range. This results in the change of dominant pole in the open loop transfer function. In order to investigate the stability of the proposed control scheme, the stator resistance value was changed from 20 Ω to 30 Ω. Figures 8-12 depict the current response for the condition of stator resistance change. From these figures, it is notable that even though the stator resistance changed 50% both transient and steady state performance of the proposed control scheme are satisfactory. In addition, the control effect of conventional PI controller was further degraded due to the motor parameter change.

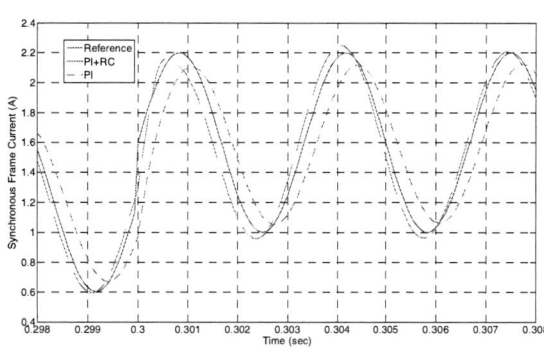

Figure 6. Zoom of instant 3

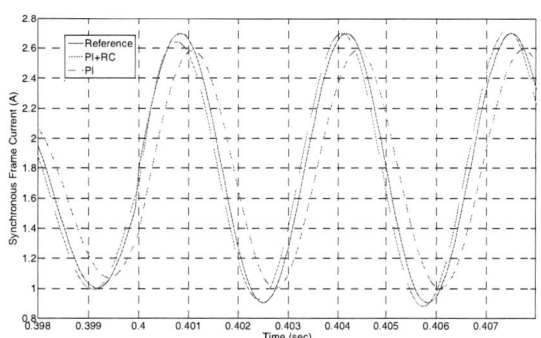

Figure 7. Zoom of instant 4

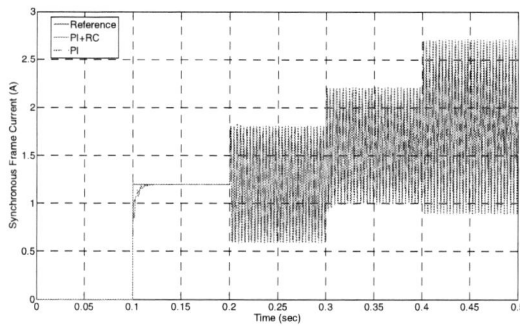

Figure 8. Current response with varying stator resistance

978-1-4244-4782-4/10 $26.00 © 2010 IEEE

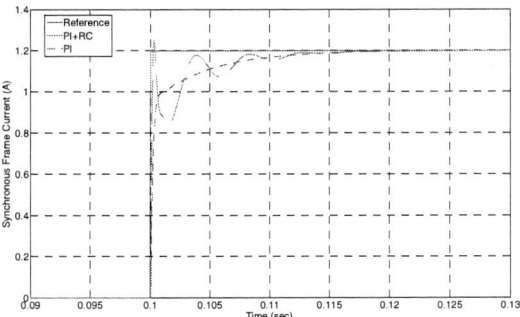

Figure 9. Zoom of instant 1

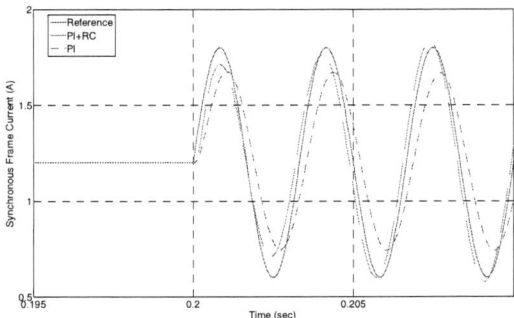

Figure 10. Zoom of instant 2

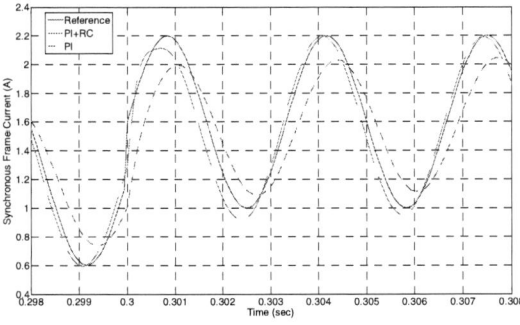

Figure 11. Zoom of instant 3

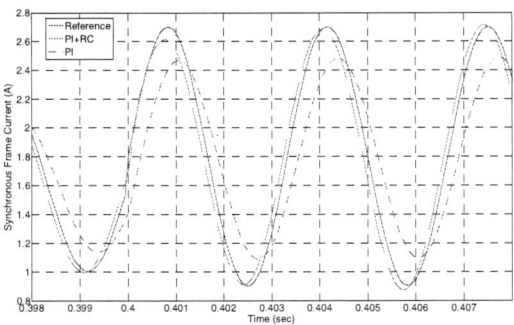

Figure 12. Zoom of instant 4

Finally, the disturbance rejection test has been done to verify the control stiffness of the resonant controller. In the simulation, at instant of 0.15 sec a disturbance signal of 400

Hz was injected into the system. Figures 13-17 illustrate the corresponding current response.

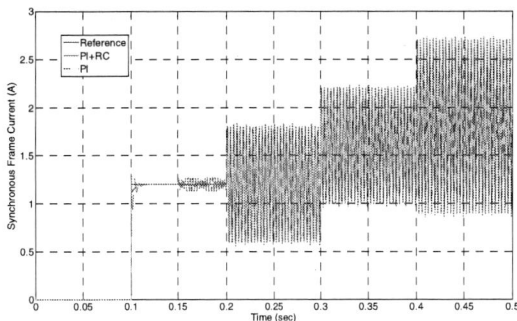

Figure 13. Current response with disturbance

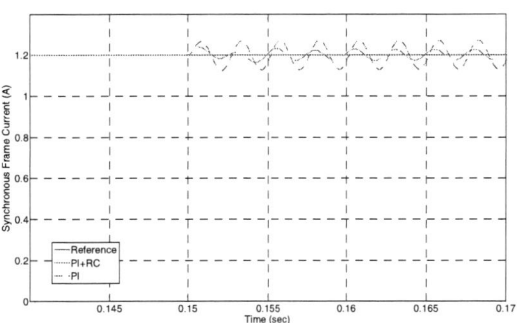

Figure 14. Zoom of 0.15 sec

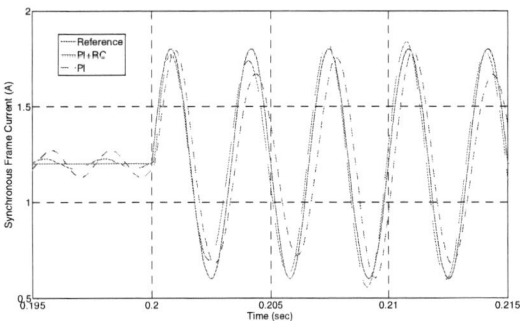

Figure 15. Zoom of instant 2

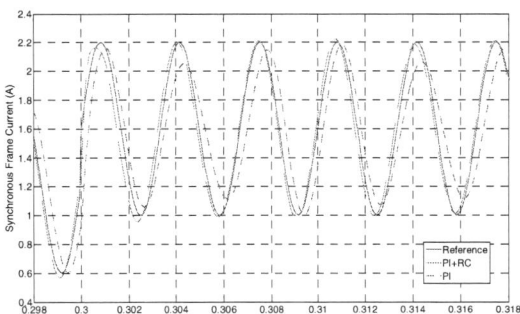

Figure 16. Zoom of instant 3

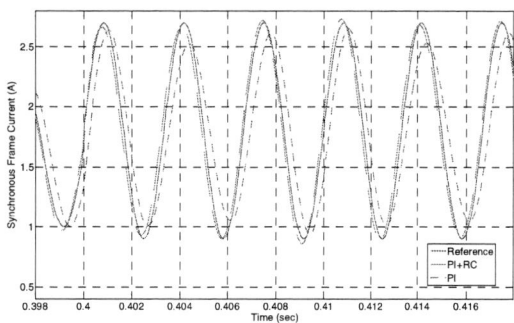

Figure 17. Zoom of instant 4

As can be seen, the control effect from conventional PI controller has been significantly degraded by the disturbance signal especially with amplitude fluctuation in the output signal. As a result, this may propagate to cause higher torque ripples in the PMSM. By contrast, the proposed method demonstrates excellent disturbance rejection performance, which guarantees the successful reduction of torque ripples in the motor drive systems.

IV. CONCLUSIONS

A novel digital current control strategy using resonant controllers to reduce torque ripples in PMSM has been proposed in this paper. The control structure and digital resonant controller design have been illustrated in details. The superior transient and steady state performance of the proposed control scheme has been verified by simulation studies. In addition, the proposed control method also demonstrated excellent results on immunity to motor parameter variance and disturbance rejection.

REFERENCES

[1] Borghi, C.A.; Casadei, D.; Cristofolini, A.; Fabbri, M.; Serra, G.; "Minimizing torque ripple in permanent magnet synchronous motors with polymer-bonded magnets," IEEE Transactions on Magnetics, Volume 38, Issue 2, Part 2, March 2002 Page(s):1371 – 1377.

[2] Joon-Ho Lee; Dong-Hun Kim; Il-Han Park; "Minimization of higher back-EMF harmonics in permanent magnet motor using shape design sensitivity with B-spline parameterization," IEEE Transactions on Magnetics, Volume 39, Issue 3, Part 1, May 2003 Page(s):1269 – 1272.

[3] Chapman, P.L.; Sudhoff, S.D.; Whitcomb, C.A.; "Multiple reference frame analysis of non-sinusoidal brushless DC drives," IEEE Transaction on Energy Conversion, Volume 14, Issue 3, Sept. 1999 Page(s):440 – 446.

[4] Chapman, P.L.; Sudhoff, S.D.; "A multiple reference frame synchronous estimator/regulator," IEEE Transaction on Energy Conversion, Volume 15, Issue 2, June 2000 Page(s):197 – 202.

[5] Weizhe Qian; Panda, S.K.; Jian-Xin Xu; "Torque ripple minimization in PM synchronous motors using iterative learning control," IEEE Transactions on Power Electronics, Volume 19, Issue 2, March 2004 Page(s):272 – 279.

[6] Mattavelli, P.; Tubiana, L.; Zigliotto, M.; "Torque-ripple reduction in PM synchronous motor drives using repetitive current control," IEEE Transactions on Power Electronics, Volume 20, Issue 6, Nov. 2005 Page(s):1423 – 1431.

[7] R. Magureanu, S. Ambrosii, D. Creanga, L. Bratosin, A. Draghici, "Active power filters advanced control," ATEE 2002.

[8] Haidong Yu, "DC Link Voltage Regulation and Current Control of Shunt Active Filters," M.Sc. thesis, University of Nottingham, 2005.

Evaluation of a SiC Power Module using Low-On-Resistance IEMOSFET and JBS for high power density power converters

Kazuto Takao, Takashi Shinohe
Corporative Research & Development Center,
Toshiba Corporation
1, Komukai-Toshiba-cho, Saiwaiku, Kawasaki, Japan

Shinsuke Harada, Kenji Fukuda, Hiromishi Ohashi
National Institute of Advanced Industrial Science and
Technology
1-1-1, Umezono, Tsukuba, Ibaraki, Japan

Abstract—**In high-power density power converter designs, power losses of power devices are essential design parameters. Silicon-carbide (SiC) power devices are expected as next generation power devices due to their superior performances compared to conventional silicon (Si) power devices. The power loss performances of a SiC power module using SiC Implantation and Epitaxial MOSFET (SiC-IEMOSFET) and Junction barrier controlled Schottky Diode (SiC-JBS) has been evaluated based on parameters of the junction temperature, current density, and switching frequency. The advantage of the SiC power module compared to a latest Si-IGBT and SiC-diode hybrid-pair module are discussed from the view point of the power loss reduction.**

I. INTRODUCTION

High power density power converters are key components for electric vehicles and power supply systems for information and telecommunications networks [1], [2]. Increase of the power density has been underpinned by the progress of power device performances using silicon (Si) technologies [3]. In view of the desire for higher power density, demand for improved power device characteristics is becoming more urgent.

Silicon-carbide (SiC) power devices have been actively developed as next-generation power devices owing to their superior performances compared to those of Si power devices. The driving force for the use of SiC power devices in power converters is the benefit to realize low-loss and very fast unipolar power devices with blocking voltage from 600 V up to 4000 V [3]. Since 2001 SiC Schottky diodes (SiC-SBDs) in the voltage classes of 300 V-1200 V have been commercially available [4], [5]. Power loss reductions of power converters have been demonstrated by using the SiC-SBDs with Si switching devices, which pair is called hybrid-pair [6]-[11].

SiC-MOSFETs are attractive SiC switching devices owing to their normally-off characteristics. Because SiC-MOSFETs have no tail currents, their switching losses are much smaller than those of Si-IGBT. Therefore, low-loss and higher-switching-frequency operation could be available with the SiC-MOSFETs and SiC-SBDs. Recently, prototype SiC power modules using SiC-MOSFETs and SiC-SBDs have been reported and demonstrated significant power loss reductions compared to conventional Si-IGBT modules [12]-[15].

It should be noted that the power loss reductions attributable to the use of SiC power modules are influenced by operation conditions (i.e. switching frequency, junction temperature, and current density). In the case of Si-IGBTs with a voltage rating of 600 V-1200 V, the future current density and junction temperature will approach 300 A/cm^2 and 200 °C, respectively [16], [17]. This means SiC-MOSFETs are also required to have the same current densities and junction temperatures as Si-IGBTs or higher ones. One should also consider the advantage of the SiC power modules compared with Si-IGBT and SiC-SBD hybrid pairs [18], [19] because of Si-IGBTs' potential for performance improvements based on a nanometer-scale trench gate structure [20].

This paper evaluates power loss performances of a 1200 V class SiC power module based on parameters of the junction temperature, current density, and switching frequency. The evaluated SiC power module comprises SiC-IEMOSFETs [21] and low-on-voltage SiC-JBSs. The SiC-IEMOSFET has low-on-resistance characteristics compared to those of SiC-DIMOSFETs having conventional structure. The temperature dependence of the switching characteristics of the SiC-IEMOSFET has been analyzed by an analytical model. Power losses of SiC power modules are estimated by using a power

---

A part of this work was supported by New Energy and Industrial Technology Development Organization (NEDO) project, Development of Future Power Electronics Technology, Japan.

loss model and compared to those of Si-IGBT and SiC-diode hybrid pairs in a three-phase PWM inverter.

II. CHARACTERISTICS OF A SiC-POWER MODULE

Power MOSFETs on 4H-SiC are attractive power switching devices owing to their high-speed and low-on-resistance characteristics. Reflecting the improvement in characteristics in recent years, a large-current-capacity device of 100 A has been demonstrated [22]. However, the specific on-resistance ($R_{on}A$) reported so far is still higher than the theoretical limit of SiC. This is due to the low channel mobility whose origin is the poor SiO_2/SiC interfacial quality [23], [24]. In order to overcome the above-mentioned issues, Implantation and Epitaxial MOSFET (IEMOSFET) has been developed [21].

A cross-sectional view of the SiC-IEMOSFET is shown in Fig. 1. The SiC-IEMOSFET has three special features. The first is utilization of the carbon face. The channel mobility of the lateral MOSFET fabricated on the carbon face is reported to be above 100 cm^2/Vs, which is much higher than that on silicon face [25]. The second feature concerns the device structure. The upper half of the p-well is formed by epitaxial growth without activation annealing to obtain the smooth surface. Using this technique, the degradation of the channel

mobility in conventional DIMOSFET can be prevented. The third feature is utilization of the buried channel structure.

Fig. 2 shows the SiC power module using SiC-IEMOSFETs and SiC-JBSs. The active areas of SiC-IEMOSFETs and SiC-JBS are 1.7 x 1.7 mm^2 and 2.0 x 2.0 mm^2, respectively. The temperature dependence of the $R_{on}A$ of the SiC-IEMOSFET at V_{gs} = 20 V is illustrated in Fig. 3. The $R_{on}A$ at 25 °C is 3.5 mΩcm^2. The $R_{on}A$ at 250 °C is 9.3 mΩcm^2 and 2.66 times larger than the value at 25 °C.

Fig. 4 shows the temperature dependence of the threshold voltage (V_{th}). V_{th} has a negative temperature coefficient. Since V_{th} is 3.76 V at 250 °C, the SiC-IEMOSFET keeps the normally-off characteristic in the high-temperature condition.

The switching characteristics of the SiC power module are measured under an inductive load condition. Fig. 5 shows the switching waveforms of the SiC-IEMOSFET at 25 °C and 125 °C. An external gate resistance is not utilized in the switching experiment because the tested SiC-IEMOSFET has the internal gate resistance of 11.36 Ω. As seen in Fig. 5, the temperature has little effect on the turn-off waveforms, whereas the turn-on fall time of the drain-source voltage (v_{ds})

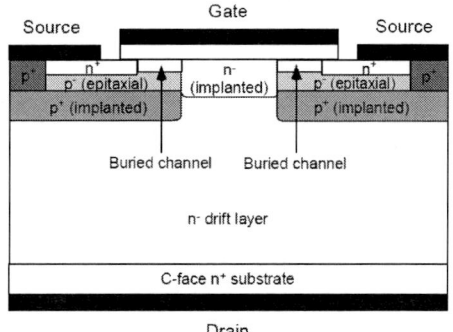

Fig. 1. Cross-sectional view of the SiC-IEMOSFET [21]

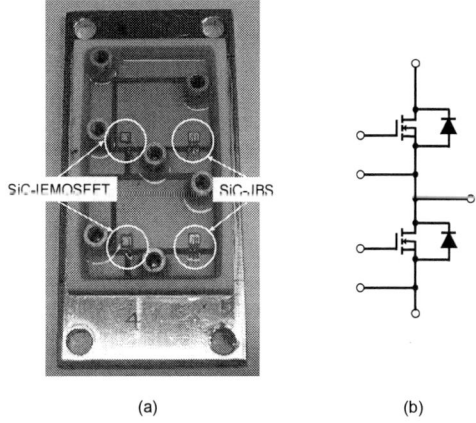

Fig. 2. SiC power module (a) Overview, (b) Equivalent circuit of the module.

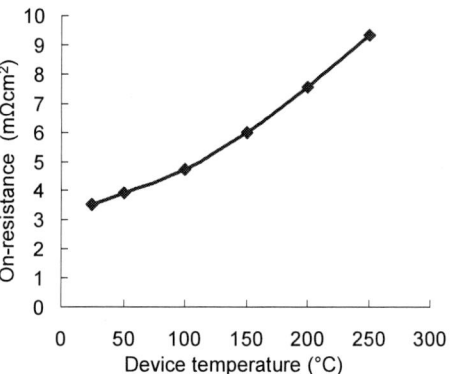

Fig. 3. Temperature dependence of the $R_{on}A$ of the SiC-IEMOSFET at V_{gs} = 20 V

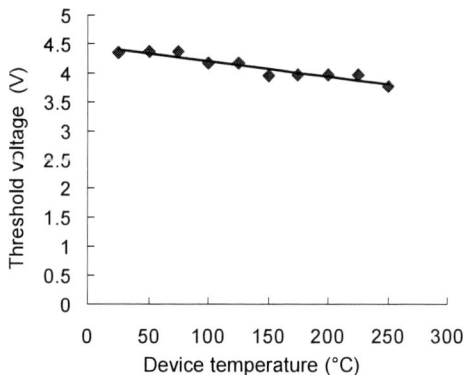

Fig. 4. Temperature dependence of the threshold voltage of the SiC-IEMOSFET

decreases with increasing temperature. Here, V_{ds} at turn-on is described as follow [26]:

$$v_{ds} = V_{cc} - \left\{ \frac{V_{GH} - \left(V_{th} + I_L\!\big/ g_m \right)}{R_g \cdot C_{gd}} \right\} \cdot t \qquad (1)$$

where, V_{cc} is the input DC voltage, V_{GH} is applied gate voltage at the on-state, I_L is the load current of the chopper, g_m is the transconductance, R_g is the gate resistance, and C_{gd} is the gate-source capacitance (Miller capacitance) of the SiC-

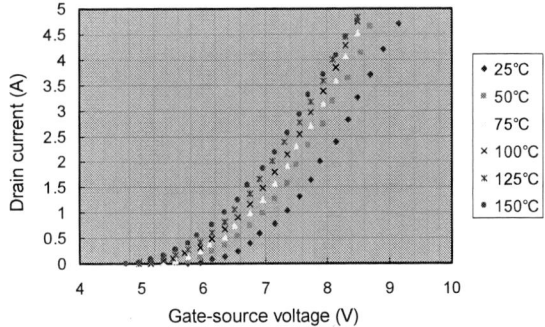

Fig. 6. $I_d - V_{ds}$ characteristic of the SiC-IEMOSFET

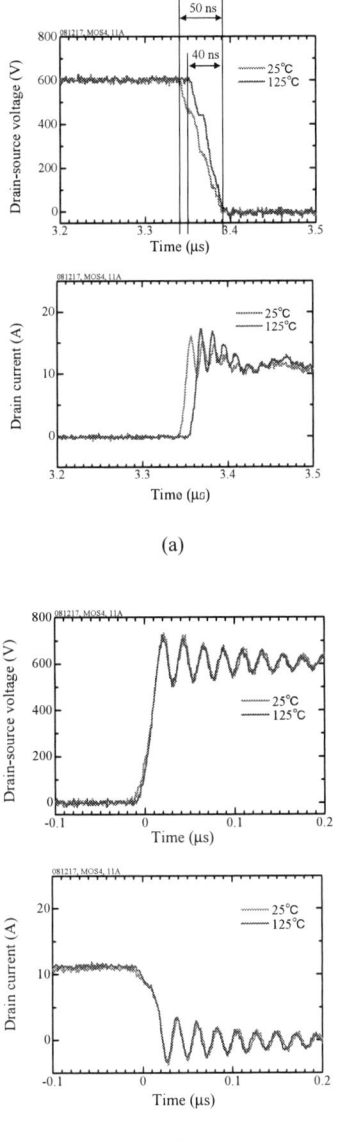

(a)

(b)

Fig. 5 Switching waveforms of the SiC-IEMOSFET at 25 °C and 125 °C. (a) Turn-on, (b) Turn-off

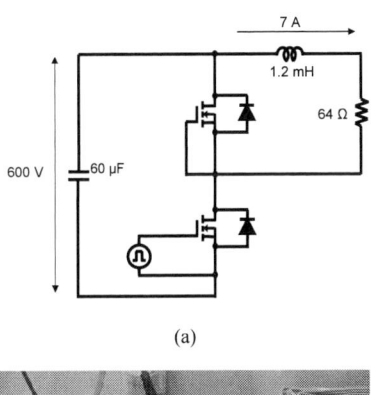

(a)

(b)

Fig. 7. Equivalent circuit and experimental setup for the Continuous operation tests of the SiC power module (a) equivalent circuit, (b) experimental setup

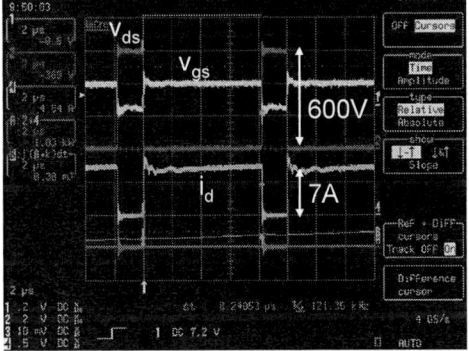

Fig. 8. Operation waveforms of the low side SiC-IEMOSFET at 100 kHz

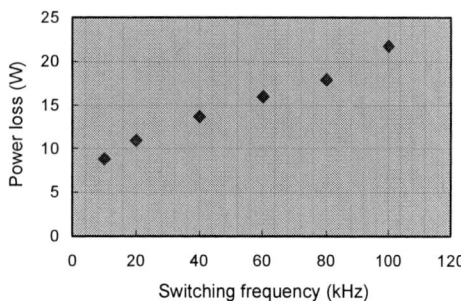

Fig. 9. Dependence of the power loss of the SiC power module on the switching frequency

IEMOSFET. In (1), V_{th} and g_m are temperature-dependent parameters, and the temperature dependence of the V_{th} is shown in Fig.4. The value in the brace of (1) is coincident with the slope of v_{ds} (dv_{ds} / dt) at the turn-on.

In order to investigate the temperature dependence of the g_m, the I_d-V_{gs} characteristic of the SiC-IEMOSFET is measured and shown in Fig. 6. As seen in Fig.6, estimated transconductances ($g_m = \Delta I_D$ / ΔV_{GS}) at $T_j = 25$ °C and 125 °C are in close agreement. The linearly approximated value of g_m is 2.5. In the case of the experimental conditions of $V_{GH} = 15$ V, $I_L = 11$ A, values of numerator in the brace in (1) at $T_j = 25$ °C and 125 °C are 4 and 5, respectively. Therefore, the dv_{ds} / dt is increased 1.25 times by increasing temperature from 25 °C to 125 °C. As seen in Fig. 5, the dv_{ds} / dt at $T_j = 125$ °C (600 V / 0.04 μs = 15 kV/μs) is 1.25 times higher than the dv_{ds} / dt at $T_j = 25$ °C (600 V / 0.05 μs = 12 kV/μs). The result is coincident with the calculated value by using (1). From the above-mentioned analysis, the temperature dependence of turn-on v_{ds} waveform is attributed to the temperature dependence of V_{th}.

Continuous operation tests of the SiC power module were implemented in a chopper configuration to measure the power loss. Fig. 7 shows the equivalent circuit and experimental setup. Because the duty factor of the low-side SiC-IEMOSFET is set to 0.8, the power loss of the high-side SiC-JBS is much smaller than that of the low-side SiC-IEMOSFET and the total power loss of the SiC power module is considered to be the power loss of the low-side SiC-IEMOSFET. In the experimental chopper, the load current is 7 A and output power is 3150 W. The switching frequency of the chopper is varied from 10 kHz to 100 kHz.

Operation waveforms of the low-side SiC-IEMOSFET at 100 kHz are shown in Fig. 8. The successful operation of the SiC power module has been confirmed. The dependence of the power loss of the module on the switching frequency is shown in Fig. 9. The power loss increases with increasing switching frequency.

III. POWER LOSSES OF SIC POWER MODULE

In order to clarify the advantage of the SiC power modules compared to Si-IGBT and SiC-diode hybrid-pair modules,

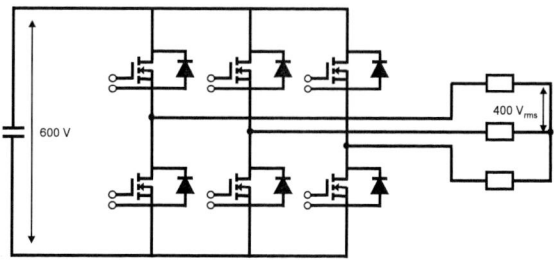

Fig. 10. Three-phase PWM inverter

power loss performance of the SiC power modules has been investigated based on parameters of the junction temperature and current density. Power losses of the SiC-IEMOSFET and SiC-JBS are calculated by using an analytical power loss model [27], [28]. The SiC device parameters, which are utilized for the power loss calculations, are extracted from the SiC-IEMOSFET and SiC-JBS in section II. In the power loss calculations, the application converter assumed is the three-phase PWM inverter shown in Fig. 10. In the hybrid-pair modules, 1200V-15A TrenchStop IGBTs produced by Infineon (chip area: 12.8 mm^2) and the SiC-JBSs (chip area: 4

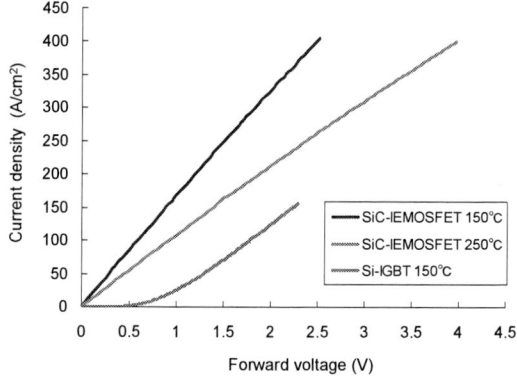

Fig. 11 Comparison of forward IV characteristics of the SiC-IEMOSFET and Si-IGBT

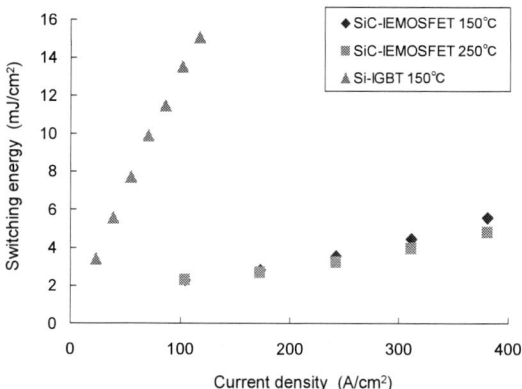

Fig. 12 Current density versus switching energies of the SiC-IEMOSFET and Si-IGBT

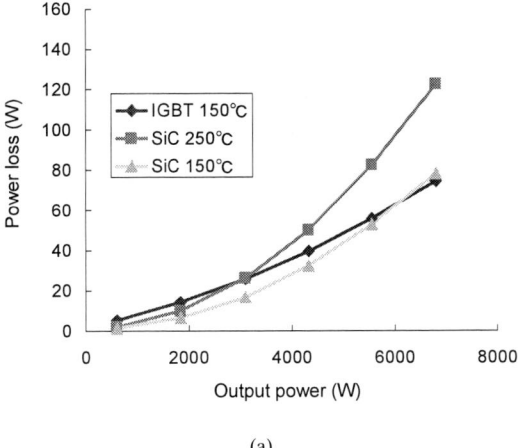

(a)

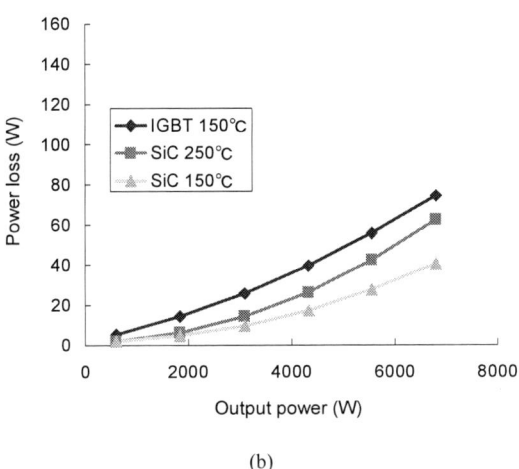

(b)

Fig. 13 Output power of the inverter versus total power losses of power devices at f_{sw} = 5 kHz. Chip areas of the SiC-IEMOSFET are (a)2.89 mm^2 and (b) 5.78 mm^2

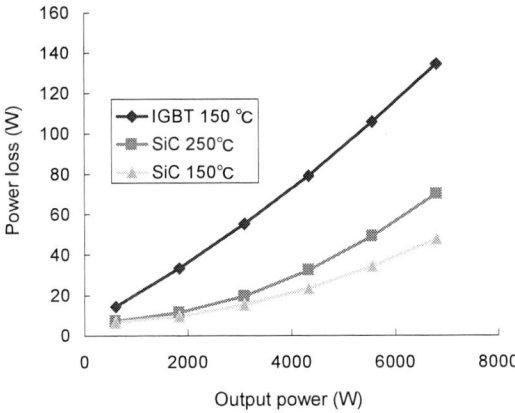

Fig. 14 Output power of the inverter versus total power losses of power devices at f_{sw} = 20 kHz. Chip areas of the SiC-IEMOSFET is 5.78 mm^2

mm^2) are utilized. The power losses of the hybrid pair are estimated by using measured static and switching characteristics.

Fig. 11 shows the comparison of forward *IV* characteristics of the SiC-IEMOSFET and Si-IGBT. In the same current density conditions, the on-voltage of the SiC-IEMOSFET is smaller than that of the Si-IGBT even if T_j = 250 °C. Fig. 12 shows the current density versus switching energies of the SiC-IEMOSFET and Si-IGBT. As seen in Fig. 12, the switching energy of the SiC-IEMOSFET is much smaller than that of the Si-IGBT at the same current densities, because the SiC-IEMOSFET has no the tail current.

In the assumed three-phase PWM inverter, the input DC voltage, output line-to-line AC voltage and power factor of the inverter are 600 V, 400 V and 0.85, respectively. Fig. 13 shows the output power of the inverter versus total power losses of power devices in the case that the switching frequency (f_{sw}) is 5 kHz. When the chip area of the SiC-IEMOSFET is 2.89 mm^2 (Fig. 13 (a)), the difference of power losses is small between the SiC module at T_j = 150 °C and the hybrid-pair module. In addition, the power loss of the SiC module at T_j = 250 °C is larger than that of the hybrid pair module in the range of the output power > 3000 W. The power losses of the SiC module with a chip area of 5.78 mm^2, which is about 1/2 that of the IGBT, are shown in Fig. 13 (b). By increasing the chip area the power loss of the SiC module at T_j = 250 °C is smaller than that of the hybrid-pair module at T_j = 150 °C. In the case of 5 kHz and 6800 W output power, the power loss of the SiC module at T_j = 150 °C is 54 % of that of the hybrid pair module (46% loss reduction). In the case of T_j = 250 °C, the power loss of the SiC module is 84 % of that of the hybrid-pairs (16% loss reduction).

It can be predict that the power loss reduction effect by using the SiC power module increases with increasing switching frequency because the switching energy of the SiC module is smaller than that of the hybrid-pair module. Fig. 14 shows the output power of the inverter versus total power losses of power devices at f_{sw} = 20 kHz. As seen in Fig. 14, the power loss of the SiC module at T_j = 250 °C is 52 % of that of the hybrid pair module at T_j = 150 °C in the case that the output power (P_{out}) is 6800 W. This result indicates that the SiC module is attractive for application in power converters that required high-switching frequency operations.

IV. CONCLUSION

In this paper, power loss performances of a 1200 V class SiC power module have been evaluated based on parameters of the junction temperature, current density, and switching frequency to fully utilize their advantage compared with a Si-IGBT and SiC-JBS hybrid-pair module. The evaluated SiC power module comprises SiC-IEMOSFETs and low-on-voltage SiC-JBSs. The SiC-IEMOSFET keeps the normally-off characteristic up to T_j = 250 °C. The switching characteristics of the SiC-IEMOSFET depend on the temperature dependence of the threshold voltage.

In a three-phase PWM inverter, significant power loss reduction with the SiC power module can be realized in the case that the chip area of the SiC-IEMOSFET is more than 1/2 that of the Si-IGBT. In the case of $f_{sw} = 5$ kHz and $P_{out} = 6800$ W, the power loss of the SiC-IEMOSFETs is 54 % compared to that of the hybrid-pair module. In the case of $f_{sw} = 20$ kHz, the power loss of the SiC module at $T_j = 250$ °C is 52 % of that of the hybrid pair module at $T_j = 150$ °C when the $P_{out} = 6800$ W. This result indicates that the SiC module is advantageous in high-switching-frequency operations compared with the hybrid pair module.

REFERENCES

[1] H. Ohashi, "Recent Power Devices Trend," in IEEJ, Vol. 12, No.3, 2002, pp. 168-171

[2] J. W. Kolar, U. Drofenik, J. Biela, M. L. Heldwein, H. Ertl, T. Friedli, and S. D. Round, "PWM Converter Power Density Barriers," in *Proc. 4th Power Conversion Conf. (PCC-Nagoya)*, 2007, pp.9-29

[3] L. Lorenz, "Power Semiconductor Devices - Development Trends and System Interactions," in *Proc. 4th Power Conversion Conf. (PCC-Nagoya)*, 2007, pp.348-354

[4] Infineon web site, http://www.infineon.com/cms/en/product/index.html

[5] Cree web site, http://www.cree.com/

[6] M. Tsukuda, I. Omura, T. Domon, W. Saito, and T. Ogura, "Demonstration of High Output Power Density (30 W/cc) Converter using 600 V SiC-SBD and Low Impedance Gate Driver," in Proc. *Int. Power Electronics Conf. (IPEC-Niigata)*, 2005, pp.1184-1189

[7] Z. Liang, B. Lu, J. D. van Wyk and F. C. Lee, "Integrated CoolMOS FET/SiC-Diode Module for High Performance Power Switching," *IEEE Trans. Power Electron.*, vol. 20, no. 3, pp. 679-686, May 2005.

[8] B. Eckardt, A. Hofmann, S. Zeltner, M. Maerz, "Automotive Powertrain DC/DC Converter with 25kW/dm³ by using SiC Diodes," *in Proc. 4th Int. Conf. on Integrated Power Systems (CIPS 2006)*, 2006, pp. 227-232.

[9] B. Ozpineci, M. S. Chinthavali, L. M. Tolbert, A. S. Kashyap, and H. A. Mantooth, "A 55-kW Three-Phase Inverter With Si IGBTs and SiC Schottky Diodes," *IEEE Trans. Industry Applications.*, vol. 45, no. 1, pp. 278–285, Jan. / Feb. 2009.

[10] J. S. Lai, W. Yu, H. Qian, P. Sun, P. Ralston, and K. Meehan, "High Temperature Device Characterization for Hybrid Electric Vehicle," in *Proc. 24th Appl. Power Electronics Conf. and Expo (APEC'09)*, 2009, pp.665-670

[11] B. Weis, D. Peters and M. Wölz, "A new 690 VAC Drive with SiC Schottky Freewheeling Diodes," *Power Electronics Europe*, issue 8, pp.33-35, Dec. 2006

[12] N. Miura, K. Fujihira, Y. Nakao, T. Watanabe, Y. Tarui, S. Kinouchi, M. Imaizumi, and T. Oomori, "Successful Development of 1.2kV 4H-SiC MOSFETs with the Very Low On-Resistance of 5 m□cm²," in *Proc. CD-ROM, IEEE 2006 International Symposium on Power Semiconductor Devices & IC's (ISPSD2006)*, 2006.

[13] S. Harada, Y. Hayashi, K. Takao, A. Kinoshita, M. Kato, M. Okamoto, T. Kato, S. Nishizawa, T. Yatsuo, K. Fukuda, H. Ohashi and K. Arai "Demonstration of motor drive with SiC normally-off IEMOSFET/SBD power converter," in *Proc. CD-ROM, IEEE 2007*

[14] O. Stalter, B. Burger, and S. Lehrmann "Silicon Carbide (SiC) D-MOS for Grid-Feeding Solar-Inverters," in *Proc. CD-ROM, 12th European Conference on Power Electronics and Applications (EPE 2007)*, 2007.

International Symposium on Power Semiconductor Devices & IC's(ISPSD2007), 2007.

[15] P. Zacharias, "Perspective of SiC Power Devices in Highly Efficient Renewable Energy Conversion Systems," in *Proc. CD-ROM, 7th European Conference on Silicon Carbide and Related Materials*, 2008.

[16] Z. J. Shen and I. Omura, "Power Semiconductor Devices for Hybrid, Electric, and Fuel Cell Vehicles," *Proceedings of the IEEE.*, vol. 95, no. 4, pp. 778–,789 April. 2007.

[17] U. Schlapbach, M. Rahimo, C. von Arx, A. Mukhitdinov, and S. Linder, "1200V IGBTs operating at 200 °C ? An investigation on the potentials and the design constraints," in *Proc. CD-ROM, IEEE 2007 International Symposium on Power Semiconductor Devices & IC's(ISPSD2007)*, 2007.

[18] K. Takao, C. Ota, J. Nishio, T. Shinohe, and H. Ohashi, "Design Consideration of High Power Density Inverter with Low-on-voltage SiC-JBS and High-speed Gate Driving of Si-IGBT," in *IEEE 2009 Applied Power Electronics Conference*, 2009, pp. 397-400.

[19] J. S. Lai, W. Yu, H. Qian, P. Sun, P. Ralston, and K. Meehan, "High Temperature Device Characterization for Hybrid Electric Vehicle," in *IEEE 2009 Applied Power Electronics Conference*, 2009, pp.665-670

[20] A. Nakagawa, "Theoretical investigation of silicon limit characteristics of IGBT," in *IEEE 2006 International Symposium on Power Semiconductor Devices & IC's(ISPSD2006)*, 2006, pp. 5-8.

[21] S. Harada, M. Kato, K. Suzuki, M. Okamoto, T. Yatsuo, K. Fukuda, and K. Arai, "1.8 mΩcm², 10A Power MOSFET in 4H-SiC," in *Proc. CD-ROM, IEEE International Electron Device Meeting*, 2006.

[22] T. E. Salem, D. P. Urciuoli, R. Green, and G. K. Ovrebo, "High-Temperature High-Power Operation of a 100A SiC DMOSFET Mudule," in *IEEE 2009 Applied Power Electronics Conference*, 2009, pp. 653-657.

[23] R. Schorner, P. Friedrichs, D. Peters and D. Stephani, "Significantly Improved Performance of MOSFET's on Silicon Carbide Using the 15R-SiC Polytype," *IEEE Electron Device Lett.*, vol.20, pp.241-244, May 1999.

[24] N. S. Saks, S. S. Mani and A. K. Agarwal, "Interface Trap Profile near the Band Edges at the 4H-SiC/SiO2 Interface," *Appl. Phys. Lett.*, vol. 76, pp. 2250-2252, April 2000.

[25] K. Fukuda, M. Kato, J. Senzaki, K. Kojima and T. Suzuki, "4H-SiC MOSFETs on (000-1) Face with Inversion Channel Mobility of 127 cm2/Vs," *Mater. Sci. Forum*, vols.457-460, pp. 1417-1420, 2004.

[26] B. J. Baliga, "Pwer Semiconductor Devices," PWS Publishing Company, 1996, pp. 387-394

[27] K. Takao, H. Irokawa, Y. Hayashi, and H. Ohashi, "Novel exact power loss design method for high output ower density converter," in *IEEE 2006 Power Electronics Specialists Conference*, 2006, pp. 2651-2655.

[28] K. Takao, Y. Hayashi, and H. Ohashi, "Study on advanced power device performance under real circuit conditions with an exact power loss simulator," in *Proc. CD-ROM, 12th European Conference on Power Electronics and Applications (EPE 2007)*, 2007.

978-1-4244-4782-4/10 $26.00 © 2010 IEEE

A Novel Integrated Power Inductor in Silicon Substrate for Ultra-Compact Power Supplies

Mingliang Wang[1*], Jiping Li[1], Khai D.T. Ngo[2] and Huikai Xie[1]

[1]Department of Electrical and Computer Engineering, University of Florida
[2]Department of Electrical and Computer Engineering, Virginia Tech

Abstract— A novel silicon-based inductor, power inductor in silicon, or *PIiS*, has been proposed and experimentally demonstrated. The *PIiS* is fabricated at wafer level using a silicon molding micromachining technique, in which 200 μm thick copper windings are embedded into a silicon substrate and both sides of the substrate are capped with a polymer-magnetic power composite. Through-silicon vias (*TSVs*) and copper routings are also added so that a PIiS can be directly used as a surface mountable packaging substrate. A 3×3×0.6 mm³ *PIiS* with a measured inductance of 390 nH has been fabricated. The Q factor of this *PIiS* is 10 at 6 MHz. An ultra compact buck converter has been made by surface mounting off-shelf power ICs and capacitors on a *PIiS*. The buck converter is 3×3×1.2 mm³, which has successfully delivered 500 mA at 1.8V with an 80% efficiency at 6 MHz.

I. INTRODUCTION

Portable electronics always demand smaller and cheaper power systems. The sizes of power ICs have been decreasing rapidly in recent years, leaving power passives the bottleneck for further reducing the size of power systems. Currently, Power System In Packaging (*PSiP*) [1] and Power System on Chip (*PSoC*) [2] are two most commonly-used approaches to integrate power passives to minimize the overall power system size. In PSiP, ICs are stacked with power inductors inside the package. Such power inductors include SMT inductors [3, 4], LTCC inductors [5, 6], inductors in PCB [7], and inductors in magnetic substrate [8]. These inductors generally have large power handling capability up to tens of amperes and high Q of over 10 even at MHz, but the size reduction is limited. PSoC, on the other hand, employs CMOS technology and/or microfabrication technology to fabricate power inductors directly on IC chips. Several PSoC power inductors have been reported, such as spiral air core inductors [9, 10], sputtered magnetic inductors [11, 12], electroplated magnetic inductors [2, 13-17] and ferrite magnetic inductors [18, 19]. These inductors are typically fabricated on the top of CMOS substrates. This approach may lead to the minimal system size by taking advantages of microfabrication and may further reduce the cost due to batch fabrication. However, this monolithic integration approach still needs to overcome the small inductance and/or low Q of the thin-film inductors to be practically useful.

In this paper, we report a microfabricated power inductor, called Power Inductor in Silicon, or *PIiS*. In the *PIiS*, thick copper windings are embedded into the silicon substrate for low DC resistance, and ferrite powders are used for the magnetic core to extend low core loss up to 10 MHz. This novel inductor has large inductance and high Q. Also with embedded copper routing and through-silicon vias, the *PIiS* can be used as a packaging substrate for surface mounted capacitors and ICs. A compact buck converter based on a *PIiS* has been demonstrated to verify the feasibility of this approach.

II. CONCEPT OF *PIiS* AND *PIiS*-ENABLED COMPACT DC-DC CONVERTER

A Power Inductor in Silicon (*PIiS*) is an integrated power inductor that is embedded into silicon substrate. Fig. 1 shows the concept of a *PIiS*, where a pot-core inductor is taken as an example (Fig. 1a). The copper winding and magnetic vias are

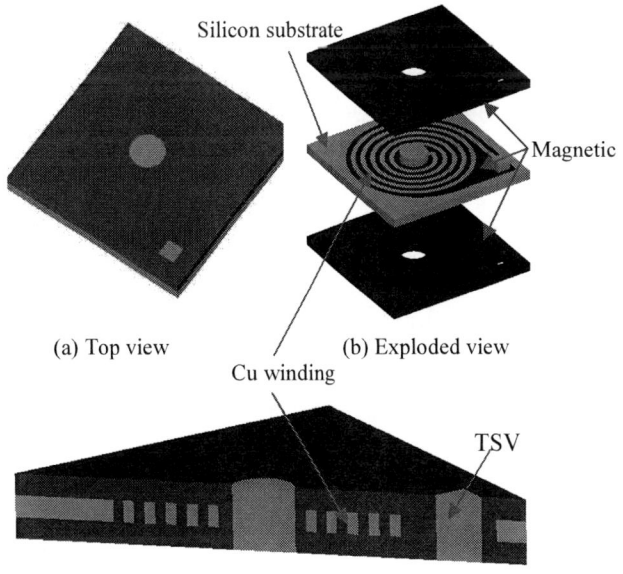

(a) Top view   (b) Exploded view

(c) Cross-section view
Figure 1: 3D models of a pot-core *PIiS*.

This project is partially supported by the National Science Foundation under award #0601294.

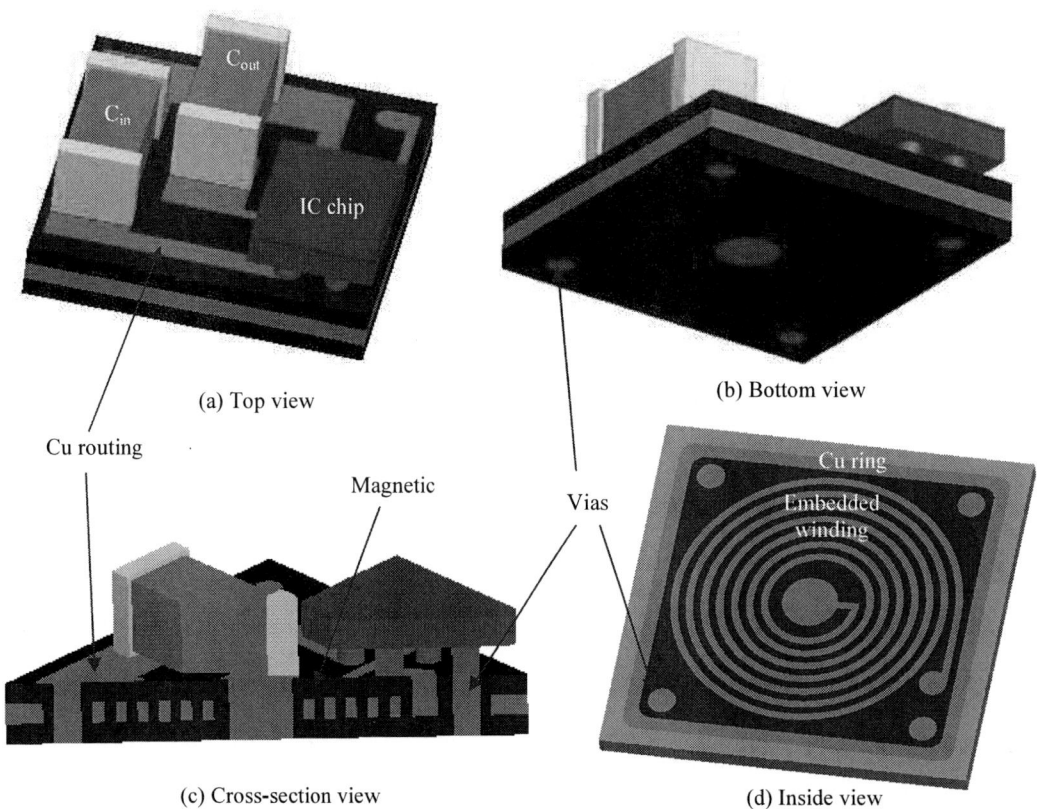

(a) Top view

(b) Bottom view

(c) Cross-section view

(d) Inside view

Figure 2: Schematic and 3D models of *PIiS* based converter.

embedded into the silicon substrate (Fig. 1c) with both sides capped by magnetic powders (Fig. 1b). The copper winding is formed inside the silicon substrate by electroplating. The electroplating molds are created by deep reactive ion etch (DRIE) of silicon [16]. Silicon DRIE offers high aspect ratios, typically about 20:1. Such high aspect ratio can minimize the width of the copper winding to its skin depth while still maintain low AC and DC resistance. This *PIiS* design can be fabricated using a silicon substrate molding technique [20]. The device fabrication is detailed in Section 4.

Fig. 2 shows the schematic of a compact DC-DC converter enabled by the *PIiS*, in which SMT capacitors and a power IC chip are mounted directly on the top of a *PIiS* by flip-chip bonding. No wire bonding is needed since through-silicon vias and copper routings are built in during the *PIiS* fabrication. A copper ring is also added to serve as mechanical frame and enhance the mechanical robustness. The through-silicon vias (TSVs) and the copper ring are formed along with the copper winding without any additional process steps. In this work, a *PIiS*-based buck converter has been fabricated to demonstrate this concept.

III. *PIiS* DESIGN

The IC chip TPS62601 (from TI, $1.3 \times 0.9 \times 0.6$ mm^3) and two SMT capacitors (4.7 µF, 0402 and 2.2 µF, 0402) are selected to be mounted on the top of a *PIiS* to demonstrate a compact DC-DC converter. TPS 62601 outputs 500 mA,

1.8 V power at 6 MHz. Based on these specifications, the targeted inductance, DC resistance and size of the *PIiS* are 0.3~0.5 µH, 120 mΩ and $3 \times 3 \times 0.6$ mm^3, respectively.

Toroidal-type inductors and spiral-type inductors such as pot-core inductors are the two most common micro-inductors [15]. In this work, the pot-core type is selected to verify the *PIiS* concept. A cross-sectional view of the pot-core *PIiS* inductor design is shown in Fig. 3. The thickness of the copper winding is 200 µm, which is the thickness of the silicon wafer used in the fabrication. The width of the copper winding is 60 µm since the skin depth of copper at 6 MHz is around 30 µm. The aspect ratio of the spacing is 5 based on available process equipment, which results in 40 µm wide spacings. The rest of the parameters are optimized by simulation after the magnetic material is characterized.

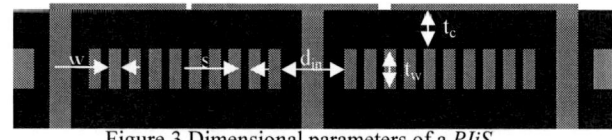

Figure 3 Dimensional parameters of a *PIiS*.

A composite of a fully sintered NiZn ferrite powder (FP350 from Powder Processing Technology, LLC) with 89wt% and Sylgard 184 PDMS (from Dow Corning) with 11wt% is used as the magnetic material. Using a vibrating sample magnetometer (VSM), the permeability (μ_r), coercive (H_c), and saturation flux density (B_{sat}) were characterized, which are 8, 15 *Oe* and 0.2 *T*, respectively, as shown in Fig. 4.

978-1-4244-4782-4/10 $26.00 © 2010 IEEE

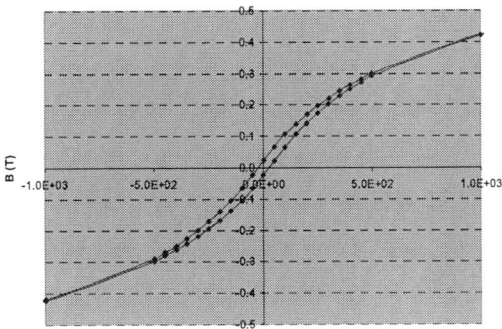

Figure 4 B-H curve of the composite of ferrite powder 89wt% and PDMS 11wt%

Table1: Parameters of *PIiS*

Parameters	Values
Number of turns (n)	10
Height of winding (t_w)	200 µm
Width of winding (w)	60 µm
Spacing of winding (s)	40 µm
Thickness of core (t_c)	200 µm
Inner diameter (d_{in})	300 µm
Inductance (L)	350 nH (μ_r=5)
	392 nH (μ_r=6)
Saturation Current (I_{sat})	> 7 A
DC resistance (R_{dc})	117 mΩ

By testing a hand-wound toroidal inductor with this magnetic composite, the measured permeability at 1~10 MHz was calibrated to be ~6.

Using the magnetic composite and silicon for the winding spacing is simulated and compared, as shown in Fig. 5. Based on the Maxwell simulation results, the inductance of a *PIiS* with magnetic spacings is 12% greater than that of a *PIiS* with silicon spacings. From the flux distribution in Fig. 5(b), it is seen that, with magnetic spacings, the flux leakages through spacings 3-6 can be ignored, while the flux through spacings 1-2 increases the effective core area. Therefore, the magnetic composite is also filled into the spacings. With the pre-set parameters and the characterized material properties, the rest parameters are optimized by simulation, which is listed in Table 1.

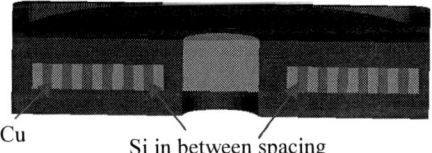

(a) With silicon in between the spacing

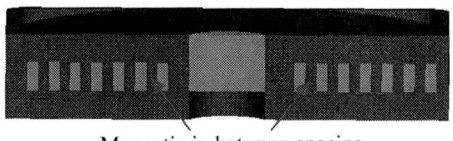

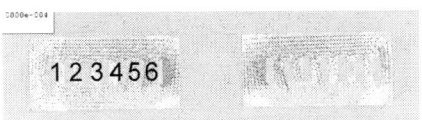

(b) With magnetic in between the spacing

Figure 5: Simulated flux distribution in *PIiS*.

IV. DEVICE FABRICATION

The fabrication process is shown in Fig. 6. First, a 10 µm copper layer is deposited on the backside of the substrate, which serves as the seed layer for the following electroplating and as a mechanical support for the silicon walls. The silicon substrate is etched through by deep reactive ion etching (DRIE) (Fig. 6(1)), and copper is electroplated with the over-plated copper being polished away (Fig. 6(2)). Then the substrate is etched through by DRIE again. A microscope picture of the device at this step is shown in Fig. 7(a). A magnetic composite is filled in the silicon trenches with the overfilled part being polished away (Fig. 6(3)). Next, solder balls are dispensed on the top of the substrate with silver epoxy as the bonder (Fig. 6(4)), which is shown in the microscope picture in Fig. 7(b). The magnetic composite is pressed on and polished until the solder balls are exposed (Fig. 6(5)) (see picture in Fig. 7(c)), followed by electroplating a 20 µm copper routing layer on top of the magnetic composite (Fig. 6(6)). Finally, the 10 µm copper seed layer on the backside is etched away and the process steps for solder balls and magnetic composite are repeated on the backside (Fig. 6(7)).

Fig. 8(a) shows a picture of a fabricated *PIiS* die, whose size is $3 \times 3 \times 0.6$ mm^3 and Fig. 8(b) shows a photo of the cross-section view of the *PIiS*. The fabricated *PIiS* is mounted on the testing PCB with solder balls and silver epoxy; and then the IC chip and capacitors listed in table 1 are mounted on the top of *PIiS*, both by flip-chip bonding. Fig. 9 shows photos of the integrated compact buck converter based on *PIiS* and its testing board.

V. TESTING RESULTS

By using a 4-probe station, the measured DC resistance R_{DC} of the *PIiS* was 140 mΩ, which was higher than the calculated 117 mΩ due to (1) the copper winding thickness reduction resulting from the multi polishing steps and the over etching when removing the backside copper seed layer and (2) the resistivity of the electroplated copper is higher than the ideal copper. The frequency dependence of the inductance (L), Q and AC resistance (R_{AC}) were measured using an HP Precision Impedance Analyzer, and the results are plotted in Fig. 10. At 6 MHz, the measured L is 390 nH,

978-1-4244-4782-4/10 $26.00 © 2010 IEEE

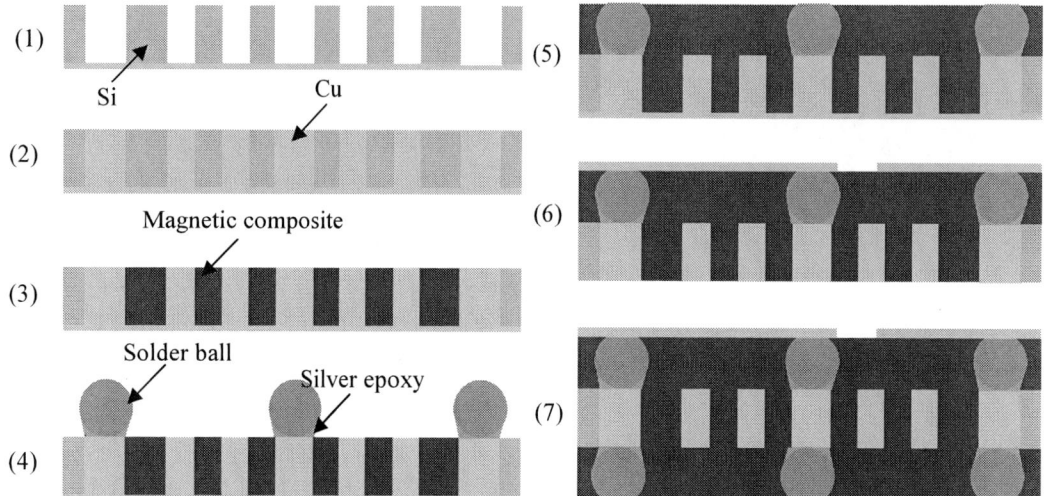

Figure 6 Cross-section view of process flow: (1) Deposit 10 µm Cu on backside, etch through 200 µm Si substrate; (2) Electroplate and polish Cu; (3) Etch through substrate, as in Fig. 9 (a), fill in magnetic composite and polish; (4) Bond 300 µm solder balls with silver epoxy, as in Fig. 9 (b); (5) Press magnetic composite and polish until solder ball is exposed, as in Fig 9 (c); (6) Electroplate 20 µm Cu routing on the top of magnetic composite, etch away the Cu layer on the backside with Cu routing on the frontside protected by photoresist; (7) repeat (4) and (5) on the backside.

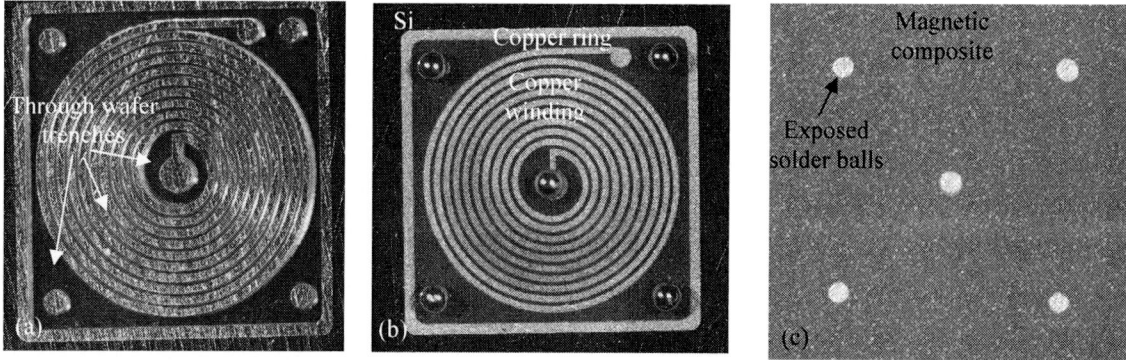

Figure 7 photo pictures of the device (a) after copper winding, copper ring and vias being embedded into silicon substrate and the silicon between copper winding being etched away; (b) after magnetic composite being filled into (a) and solder balls being bonded on the top; and (c) after magnetic composite being compressed on the top and being polished until solder sold being exposed.

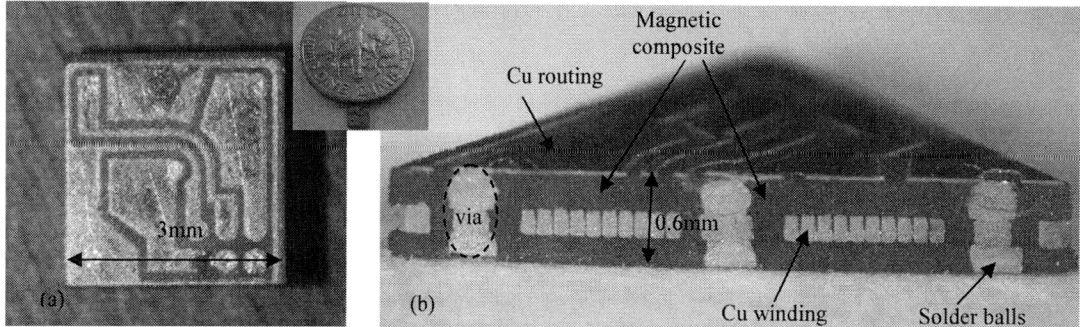

Figure 8: Photos of a fabricated *PIiS*. (a) A single *PIiS* die. (b) Cross-section view of a *PIiS*.

the Q is 10, and the R_{AC} is 1.13 Ω. The measured peak efficiency of this converter is about 80% as shown in Fig.11, and its efficiency only drops ~2% over a 100°C temperature increase, as shown in Fig.12.

978-1-4244-4782-4/10 $26.00 © 2010 IEEE   2039

Figure 9 photos of the integrated compact buck converter: (a) Close-up side view of the integrated compact buck converter bonded on testing board, (b) Testing board

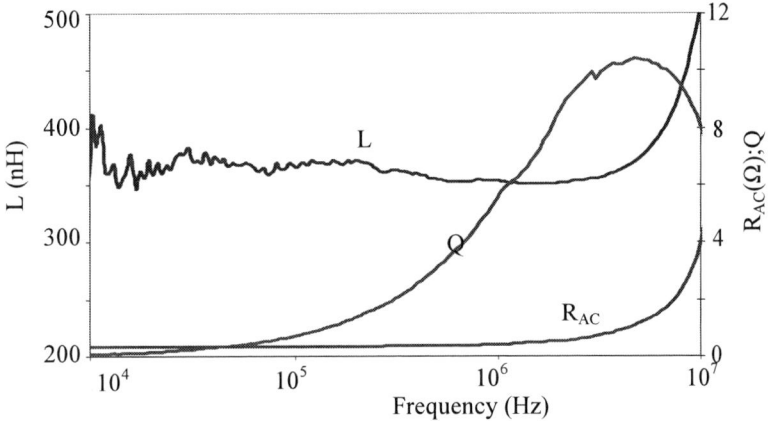

Figure 10: Measured frequency dependence of the Q, AC resistance and inductance of the *PIiS*. At 6 MHz, Q = 10, L = 390 nH, and R_{AC}= 1.15 Ω.

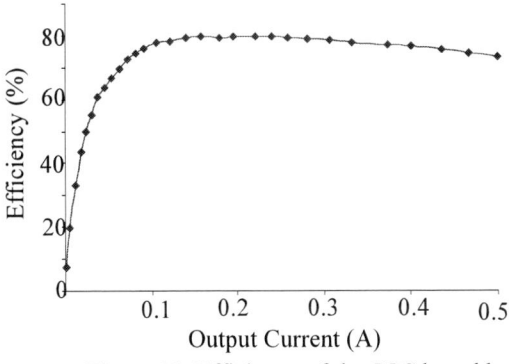

Figure 11: Efficiency of the *PIiS*-based buck converter with (Vin =3.6V; Vout =1.8V).

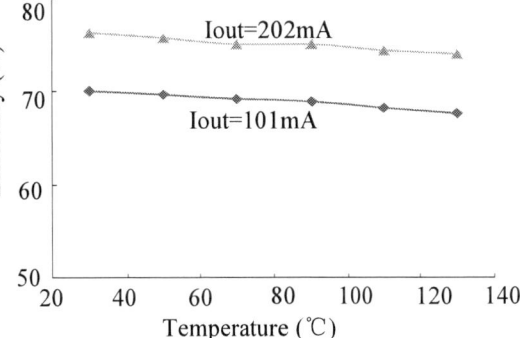

Figure 12: Temperature characteristics of the *PIiS*-based buck converter.

CONCLUSIONS

The concept of power inductors in silicon (*PIiS*) has been experimentally verified. The fabricated *PIiS* achieves Q as high as 10, resistance of ~1 Ω and high bandwidth up to 10 MHz. The *PIiS* is surface mounting ready and can be fabricated at wafer level, which can lead to extremely low cost. A compact buck converter based on a *PIiS* has also been successfully demonstrated, in which the built-in through-wafer vias (TSVs) serve as signal paths and thermal plugs. The results show that *PIiS* is a very promising approach for power system integration.

ACKNOWLEDGMENT

This project is partially supported by the National Science Foundation under award #0601294.

REFERENCES

[1] F. C. Lee, J. D. van Wyk, D. Boroyevich, L. Guo-Quan, L. Zhenxian, and P. Barbosa, "Technology trends toward a system-in-a-module in power electronics," *Circuits and Systems Magazine, IEEE*, vol. 2, pp. 4-22, 2002.

[2] S. C. O. Mathuna, T. O'Donnell, W. Ningning, and K. Rinne, "Magnetics on silicon: an enabling technology for power supply on chip," *Power Electronics, IEEE Transactions on*, vol. 20, pp. 585-592, 2005.

[3] Enpiron, "EN5322Q datasheet," 2007.

[4] "LM2832 datasheet," National Semiconductor.

[5] M. H. Lim, J. D. van Wyk, F. C. Lee, and K. D. T. Ngo, "A Class of Ceramic-Based Chip Inductors for Hybrid Integration in Power Supplies," *Power Electronics, IEEE Transactions on*, vol. 23, pp. 1556-1564, 2008.

[6] T. Mikura, K. Nakahara, K. Ikeda, K. Furukuwa, and K. Onitsuka, "New substrate for micro DC-DC converter," in *Electronic Components and Technology Conference. 2006. Proceedings. 56th*, 2006, p. 5 pp.

[7] H. J. Bergveld, R. Karadi, and K. Nowak, "An inductive down converter system-in-package for integrated power management in battery-powered applications," in *Power Electronics Specialists Conference, 2008. PESC 2008. IEEE*, 2008, pp. 3335-3341.

[8] Z. Hayashi, Y. Katayama, M. Edo, and H. Nishio, "High-efficiency dc-dc converter chip size module with integrated soft ferrite," *Magnetics, IEEE Transactions on*, vol. 39, pp. 3068-3072, 2003.

[9] P. Hazucha, G. Schrom, J. Jaehong, B. A. Bloechel, P. Hack, G. E. Dermer, S. Narendra, D. Gardner, T. Karnik, V. De, and S. Borkar, "A 233-MHz 80%-87% efficient four-phase DC-DC converter utilizing air-core inductors on package," *Solid-State Circuits, IEEE Journal of*, vol. 40, pp. 838-845, 2005.

[10] G. Schrom, P. Hazucha, F. Paillet, D. J. Rennie, S. T. Moon, D. S. Gardner, T. Kamik, P. Sun, T. T. Nguyen, M. J. Hill, K. Radhakrishnan, and T. Memioglu, "A 100MHz Eight-Phase Buck Converter Delivering 12A in 25mm2 Using Air-Core Inductors," in *Applied Power Electronics Conference, APEC 2007 - Twenty Second Annual IEEE*, 2007, pp. 727-730.

[11] Y. Katayama, S. Sugahara, H. Nakazawa, and M. Edo, "High-power-density MHz-switching monolithic DC-DC converter with thin-film

inductor," in *Power Electronics Specialists Conference, 2000. PESC 00. 2000 IEEE 31st Annual*, 2000, pp. 1485-1490 vol.3.

[12] L. Weidong, S. Yuqin, and C. R. Sullivan, "High-frequency resistivity of soft magnetic granular films," *Magnetics, IEEE Transactions on*, vol. 41, pp. 3283-3285, 2005.

[13] M. Wang, K. D. T. Ngo, and H. Xie, "SU-8 enhanced high power density MEMS inductors," in *Industrial Electronics, 2008. IECON 2008. 34th Annual Conference of IEEE*, Orlando, FL, 2008, pp. 2672-2676.

[14] J. Y. Park and M. G. Allen, "Integrated electroplated micromachined magnetic devices using low temperature fabrication processes," *Electronics Packaging Manufacturing, IEEE Transactions on [see also Components, Packaging and Manufacturing Technology, Part C: Manufacturing, IEEE Transactions on]*, vol. 23, pp. 48-55, 2000.

[15] C. H. Ahn and M. G. Allen, "Micromachined planar inductors on silicon wafers for MEMS applications," *Industrial Electronics, IEEE Transactions on*, vol. 45, pp. 866-876, 1998.

[16] M. Wang, I. Batarseh, K. D. T. Ngo, and H. Xie, "Design and Fabrication of Integrated Power Inductor Based on Silicon Molding Technology," in *Power Electronics Specialists Conference, 2007. PESC 2007. IEEE*, 2007, pp. 1612-1618.

[17] T. O'Donnell, N. Wang, R. Meere, F. Rhen, S. Roy, D. O'Sullivan, and C. O'Mathuna, "Microfabricated inductors for 20 MHz Dc-Dc converters," in *Applied Power Electronics Conference and Exposition, 2008. APEC 2008. Twenty-Third Annual IEEE*, 2008, pp. 689-693.

[18] K. W. Moon, S. H. Hong, H. J. Kim, and J. Kim, "A fabrication of DC-DC converter using LTCC NiZnCu ferrite thick films," in *Magnetics Conference, 2005. INTERMAG Asia 2005. Digests of the IEEE International*, 2005, pp. 1109-1110.

[19] I. Kowase, T. Sato, K. Yamasawa, and Y. Miura, "A planar inductor using Mn-Zn ferrite/polyimide composite thick film for low-Voltage and large-current DC-DC converter," *Magnetics, IEEE Transactions on*, vol. 41, pp. 3991-3993, 2005.

[20] M. Wang, K. D. T. Ngo, I. Batarseh, and H. Xie, "Integrated Power Inductor with Silicon Substrate Molding," in *Asia-Pacific Conference on Transducers and Micro-nano Technology APCOT* Singapore, 2006.

A Class of Coupled Inductors Based on LTCC Technology

Laili Wang, Yunqing Pei, Xu Yang, Xizhi Cui, Zhaoan Wang, Guopeng Zhao
Electrical Engineering Department, Xi'an Jiaotong University
Xi'an, 710049, China
Email: l.l.wang@stu.xjtu.edu.cn.

Abstract—**Using coupled inductors in multi-phase point of load converters is an effective way to improve system steady state and dynamic performance. Thus, package of coupled inductors for system three-dimensional integration is attracting great interest of research. This paper explores design, fabrication and analysis of planar coupled inductors based on low temperature co-fired ceramic technology. Four structures classified by winding forms and their distribution in the core are introduced and compared firstly. Then, an analytical model for approximating inductance value is derived. Two LTCC coupled inductors are fabricated and their measurement results are provided. Finally, a 12V input, 1.8V/40A output two-phase interleaved buck converter is construed for testing a LTCC coupled inductor.**

I. INTRODUCTION

High power density and fast dynamic response are two key factors for DC/DC converters used in portable electronic equipments. In today's electronic industry, portable electronic equipments are always demanded to provide more features, but to keep their volume as small as possible. Thus, on one hand, fast dynamic response power supplies are often used in these equipments, powering CPUs and application specific integrated circuits to execute high speed calculation and other complex functions. Buck or other non-isolated converters are often chosen to convert battery energy to power these equipments. On the other hand, to make sure that the whole equipment is compact, the space left for DC/DC converters is limited and sometimes converters are required to be in very planar form because they are used in very thin electronic equipments. In one word, development of portable electronic equipments makes higher and higher demands of DC/DC converters for power density and dynamic performance. One typical example is laptop computer. In laptop computer, usually, a two phase buck converter is used to power CPU, One effective way to satisfy the demand is to increase switching frequency. By increasing switching frequency, magnetic components which dominate the volume of converters could be largely reduced; the cross frequency of output filters could also be extended. Thus, both power density and dynamic performance of converters could be improved. However, with the increase of switching frequency,

switching loss and driving loss which is proportional to frequency also increase quickly. This will inevitably degrade efficiency of DC/DC converters. One solution to solve this contradiction is to improve dynamic performance by multiphase interleaving and to increase power density by package integration. When multiphase interleaving technique is adapted and the converter is designed properly, frequency of current ripple in the output filter is determined by its equivalent switching frequency which equals to frequency of single phase multiplied by number of phases. The inductor of each phase could be designed smaller because current ripple of each phase will be partly cancelled at the output capacitors. Thus, multiphase interleaving could improve transient response indirectly. Further research also exposed that by coupling inductors of multiphase converters, the transient response of DC/DC converters could be further improved when the coupling coefficient is set properly by controlling the air gap of magnetic cores [1-6]. However, prototypes made in these papers are constructed with modified commercial magnetic cores and PCB. This has at least two disadvantages. First, with the development of semiconductor and package technology, active devices become thinner and smaller, nevertheless, commercial magnetic cores are still bulky and much higher than other components in converters, which will reduce power density of DC/DC converters. Second, commercial magnetic cores can not be buried into passive substrate for mismatch of their temperature expansion coefficients. The two disadvantages of commercial magnetic cores make them not very convenient for system integration to increase power density. This paper explores designing and fabricating of planar coupled inductors using LTCC technology. LTCC technology is very suitable for passive integration of high frequency DC/DC converters [7-12]. There are several reasons. Firstly, it is very suitable for three-dimensional connection. This advantage makes it a good choice for three-dimensional integration. Secondly, by using ferrite tapes and paste, capacitor tapes and paste, planar embedded passives could be easily fabricated and integrated into passive substrates. Thirdly, all the passives and conductors are manufactured with co-firing or post-firing process. They show the same thermal characteristic. This is much different from those manufactured based on traditional

---

This work was supported in part by key project of Supporting Tech. From the Ministry of Science and Technology of China under Grant 2007BAA12B01.

PCB technology. Fourthly, temperature expansion coefficient of LTCC materials is the same with silicon material, so, hybrid integration could be easily realized. Fig.1 shows structures of isolated DC/DC converters based on different package technology. For the advantages of LTCC technology described above, in this paper, LTCC technology is adopted to fabricate coupled inductors in VRs. For simplicity, we just focus on design of two-phase coupled inductors. Section II of the paper describes and compares several structures of the two-phase coupled inductor. Section III develops an analysis model for calculating and designing of coupled inductors. Simulation results are also provided to verify accuracy of the model. Section IV set up a two-phase interleaved buck converter to test performance of coupled inductor prototypes. Final part concludes this paper.

II. STRUCTRUES

The basic materials in fabrication of LTCC components are tapes and conductor paste. Each layer of tapes is about 70 micrometers thick before co-firing and 50 micrometers thick after co-firing. Fabricating process based on LTCC technology can be described as follows. First, holes and slots are punched on ferrite tapes and they are filled with silver paste to form via holes and conductors. Second, different layers are laminated

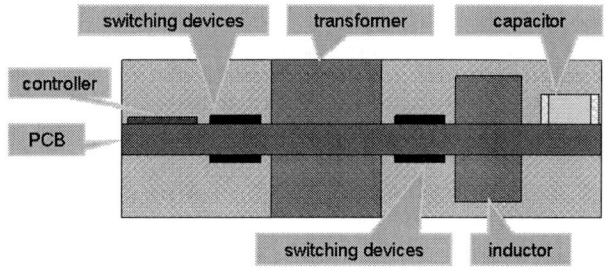

(a) A DC/DC converter constructed with commercial magnetic cores and PCB

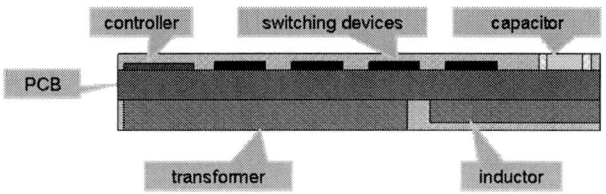

(b) A DC/DC converter constructed with planar LTCC magnetic components and PC

(c) A DC/DC converter constructed with LTCC passive substrate

Figure 1. Structure of DC/DC converters based on different package technology

together to make a three-dimensional connected LTCC substrate. The first two steps sometimes need repeat several times until all layers are laminated together. After lamination, the planar multilayer substrate is put into a furnace for co-firing. Following the fabricating process of LTCC substrate, various structures of two-phase coupled inductors can be designed. They could be classified according to winding forms and winding distribution in the substrate. Generally, three winding forms are used in planar inductors. They are spiral winding, solenoid winding and toroidal winding. For the planar LTCC coupled inductors, isolation of two windings and symmetry of coupling coefficient become difficult to be realized if toroidal winding form is adapted. Therefore, only spiral and solenoid winding forms are considered in this paper. Relative distribution of two windings in the substrate is another factor which highly determines the structure and characteristics of the coupled inductor. The two windings can distribute in vertical or horizontal direction to the substrate plane. Based on the description and classification above, four possible structures of a two-phase LTCC coupled inductor are developed as shown in Fig. 2. For convenience of analysis, they are called SV (spiral winding form, vertical distribution), SH (spiral winding form, horizontal distribution), NV (Solenoid winding form, vertical distribution), NH (Solenoid winding form, Horizontal distribution). Inductance and direct current resistance, coupling coefficient of the four structures will be compared according to magnetic circuit theory. For inductors used in high current POL DC/DC converters such as VRs, one turn winding is preferred to reduce winding loss. Thus, this paper focuses on one turn winding coupled inductor. Based on this assumption, coupling between different turns of the same winding no longer need to be considered and inductance value of a closed loop winding can be calculated by multiplying inductance of per unit length and total winding length. Inductance value is determined by winding length rather than its shape. To compare their direct current resistance, self-inductance and coupling coefficients of the four structures fairly, more assumptions should be set as follows:

There's no coupling between different parts of one winding. This is actually realizable as well as reasonable. For a straight conductor buried in ferrite core, there is no coupling between different parts. A spiral winding can be seen that a straight conductor is folded into a closed loop. But, if distance between different folded parts of the conductor is still larger than thickness of ferrite core, coupling between them can be omitted.

Permeability of magnetic core is much higher than that of air so that magnetic flux generated by flowing current is constrained in the core.

Volume of the four magnetic cores is the same, although their height, basal area are different.

Distance of two windings s is the same.

Thickness, length and width of windings which are buried in magnetic cores are the same.

Windings are in planar form; their width is much larger than height.

Magnetic cores are also in planar form.

A. Self-Inductance and direct current resistance

Self-inductance can be calculated by flowing current and flux linkage it generates. Fig. 3 illustrates flux path around a conductor which is buried in LTCC magnetic substrate. Generally, for planar LTCC inductors used in high current DC/DC converters, both width and length of a conductor are much larger than its height. That means a and c is much larger

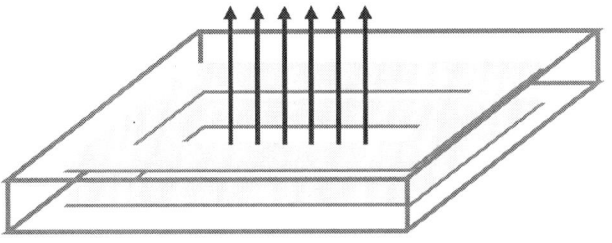

(a)　Spiral winding with vertical arrangement

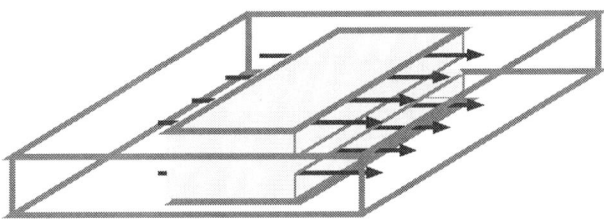

(b)　Solenoid winding with vertical arrangement

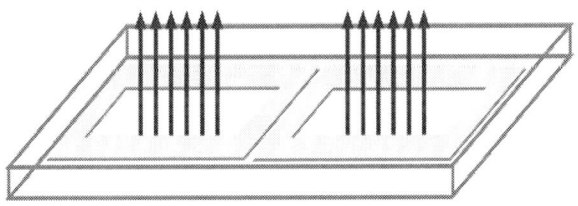

(c)　Spiral winding with horizontal arrangement

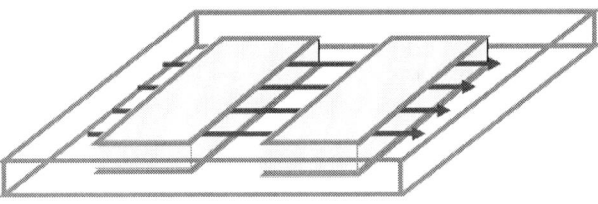

(d)　Spiral winding with horizontal arrangement

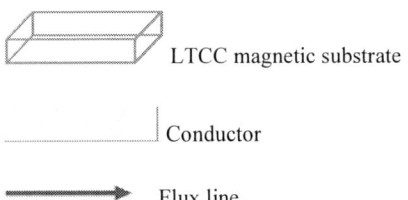

LTCC magnetic substrate

Conductor

Flux line

Figure 2.　Structures of planar LTCC coupled inductors

than h. Therefore, magnetic reluctance of region B is much larger than that of region A and C. Thus, magnetic reluctance of region A and C can be omitted when estimating magnetic reluctance around the conductor. Then, self-inductance is mainly determined by magnetic reluctance of Region B. Fig. 4. shows cross-section views of the four structures. SV and NV structures are two times as high as SH and NH structures. Therefore, their self-inductance of per unit length should be larger than those of SH and NH structures. For the same reason, flux distribution in SV and NV structures are also the same. Then, they all have the same self-inductance. But, as to direct current resistance, NV is two times as large as SV and NH because it has another conductor on the surface to form a closed loop with the buried one. NH structure has the lowest self-inductance value although it has the same direct current resistance with NH structure, because it has the shortest buried conductor compared with other three structures. Based on the analysis above, self-inductance and direct current resistance of the four structures should be sorted as (1) and (2).

$$L_{SV} = L_{NV} > L_{SH} > L_{NH} \qquad (1)$$

$$R_{NV} = R_{NH} > R_{SV} = R_{SH} \qquad (2)$$

B. coupling coefficient

Coupling coefficient is a structure related parameter; and is mainly determined by relative distribution of two windings in magnetic core. It can be expressed by (3)

$$k = \frac{M}{\sqrt{L_1 L_2}} \qquad (3)$$

Where M is the mutual inductance, L_1 is self-inductance of winding 1; L_2 is self-inductance of winding 2. Since the four structures described in this paper are all symmetrical, self-inductance of winding 1 and winding 2 equal to each other as in (4)

$$L_1 = L_2 = L \qquad (4)$$

Total mutual inductance could also be calculated by multiplying per unit length mutual inductance and length of windings which has coupling with each other. Coupling coefficient can then be expressed by (5)

$$k_i = \frac{M_i}{L_i} \qquad (5)$$

Where k_i (i=SV, NV, SH, NH) are coupling coefficient of the specific structures; M_i, L_i are mutual inductance and self-inductance.

For SV and HV structures, windings buried in the magnetic cores are in parallel with each other; and they have constant coupling. Per unit length mutual inductance of the two structures equals to each other for their flux distribution in magnetic cores are the same. Therefore, coupling coefficients of the two structures are the same as shown in (6).

$$k_{SV} = k_{NV} = \frac{M_{SV}}{L_{SV}} = \frac{M_{SVP} \times l_s}{L_{SVP} \times l_s} \qquad (6)$$

Where M_{iP}, L_{iP} are mutual inductance and self-inductance of per unit length; l_s is length of windings which determines self-inductance. Fig. 4 illustrates reluctance distribution along coupling and leakage flux path in the four structures. For SV structures, reluctance along coupling flux path is a little larger than that along leakage flux path. Or, they can be considered the same for approximation. Therefore, the two structures are subject to high coupling coefficient. SH and NH structures have low reluctance along leakage flux path, but high reluctance along coupling flux path. Relationship of per unit length mutual inductance and self-inductance could be expressed by (7)

$$\frac{M_{SVP}}{L_{SVP}} = \frac{M_{NVP}}{L_{NVP}} > \frac{M_{NHP}}{L_{NHP}} = \frac{M_{SHP}}{L_{SHP}} \qquad (7)$$

Compared with SV structure, one fourth of total windings has coupling in SH structure. For NH structure, the effective winding length for calculating self-inductance and mutual inductance is very short. It equals to thickness of magnetic substrate. Fig. 5 shows both leakage flux path and coupling flux path in SH and NH structures.

$$k_{SH} = \frac{M_{SH}}{L_{SH}} = \frac{M_{SHP} \times l_s / 4}{L_{SVP} \times l_s} \qquad (8)$$

$$k_{NH} = \frac{M_{NHP} \times h / 2}{L_{NHP} \times h / 2} \qquad (9)$$

For the purpose of fair comparison, distance between coupling windings in these four structures are set to be the same, which is half of magnetic substrate thickness. Then, according to analysis and approximation above, relationship of coupling coefficients of the four structures could be described by (10)

$$k_{SV} = k_{NV} > k_{NH} > k_{SH} \qquad (10)$$

A 3D FEA simulation is set up to verify analysis of the four structures. Five models are constructed for comparison, their dimensional parameters are shown in Table I. Simulation results are shown in Table II. Simulation results agree with

analysis results very well. Coupling coefficient of SH structure is very small and NH structure has the lowest self-inductance. SV has larger coupling coefficient than NV does; and its direct current resistance is only half of NV's. One strange phenomenon is that SV's self-inductance is smaller than NV's. That's because its effective winding length is less than NV's. NV structure should has smaller coupling coefficient than SV

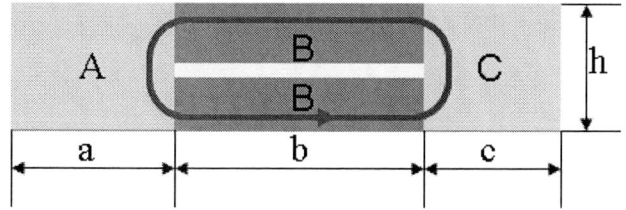

Figure 3.   Magnetic path of single conductor in planar core

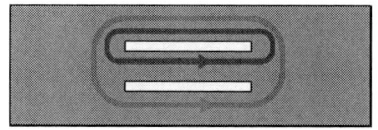

(a)   SV structure

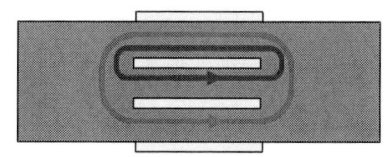

(b)   NV structure

(c)   SH structure

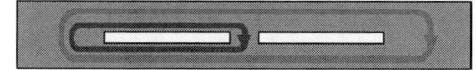

(d)   NH structure

Figure 4.   Cross-section view of conductors and flux distribution in four structures

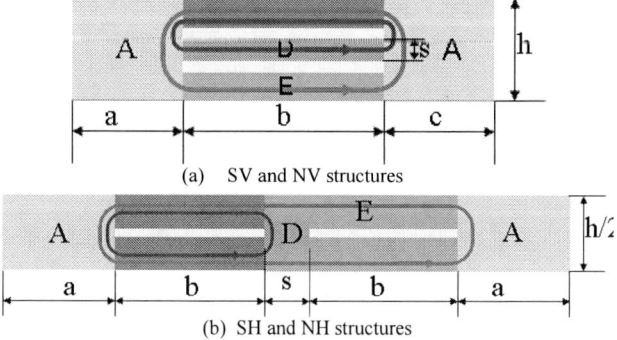

(a)   SV and NV structures

(b)   SH and NH structures

Figure 5.   Reluctance distribution of four structures

978-1-4244-4782-4/10 $26.00 © 2010 IEEE   2045

TABLE I. PARAMETERS OF COUPLED INDUCTORS FOR SIMULATION

Total volume (V)	2450mm^3
Width of winding (b)	6mm
Distance between coupling windings (s)	0.4mm
Thickness of SV and NV structures	2mm
Thickness of SH and NH structures	1mm
Outer edge length of windings in SV and SH structures	90mm
Length of straight windings in NV and NH structures	90mm

TABLE II. SIMULATION RESULTS

Structures	Direct current resistance(mΩ)	Self-inductance(nH)	Coupling coefficient
SV	17	251.6	0.598
NV	39.7	302.9	0.51
SH	17	142.3	0.005
NH	39.5	32.0	0.38

does. This is different from the approximation result. Because in SV structure, other conductors to form closed loop with windings can be seen to distribute in the air (usually, they are on PCB). But, in NV structure, the windings include co-fired conductors on the surface of magnetic core are used to form closed loops. Leakage flux in NV should be more than that in SV structure. Based on analysis and simulation results, a SV structure is selected to be modeled and taken into experiments for its better performance.

III. MODELING AND SIMULATION

Many experimental models for calculating inductance of spiral inductors have been published; however, most of them are derived for designing air core chip inductors. And they are not suitable for LTCC inductor fabricated with ferrite tapes. In this paper, we will derive a simple but effective model for LTCC inductor calculation according to magnetic circuit theory.

Theoretically, inductance value of a LTCC inductor could be obtained by complex three-dimensional magnetic field calculation. And the calculation is usually executed by computer finite element analysis (FEA) software. Nevertheless, it is a long-time-consuming process. Based on two assumptions in this paper, Inductance can be obtained by multiplying per unit length inductance and winding length. Then, the complex three-dimensional field calculation is simplified to two-dimensional calculation. Magnetic path theory then becomes suitable for inductance calculation. For specialty of LTCC technology itself and requirement of flowing high current through the conductor, conductor width w is usually much larger than thickness of the core H; thus, the major magnetic reluctance is determined by the ferrite region above and below the conductor. To calculate total magnetic reluctance of the core, its reciprocal-magnetic permeance should be firstly obtained. Total permeance can be obtained by

integrating finite permeance differential cells. Fig. 6. shows the differential path of permeance of a LTCC planar inductor. According to principle of flux continuity, flux flowing through ferrite tapes above the conductor should also flows through ferrite tapes below the conductor although thickness of the two groups of laminated ferrite tapes is not the same. Suppose flux distributes evenly in each group of ferrite tapes; then, differential step length in each group is proportional to thickness of ferrite tapes as expressed in (11)

$$n = \frac{h_3}{h_2} \tag{11}$$

Flux generated by the flowing current can be obtained by magnetic circuit model as expressed by (12)

$$\phi_0 = \frac{NI}{R_0} \tag{12}$$

Where N is number of turns, in this paper, $N=1$; R_0 is total magnetic reluctance around the conductor, it is the reciprocal of magnetic permeance as (13)

$$R_0 = \frac{1}{P_0} \tag{13}$$

Differential elements of P_0 can be expressed by (14)

$$dP_0 = \frac{\mu dx}{(w+2mx) \bullet (1+\frac{1}{n})} \tag{14}$$

Total permeance could be obtained by integrating all the elements as (15)

$$P_0 = \int_0^{h_2} \frac{\mu dx}{(w+2mx) \bullet (1+\frac{1}{n})} \tag{15}$$

For one turn structure, per unit length inductance can be expressed by (16)

$$L_0 = \frac{\phi_0}{I} \tag{16}$$

Substitute (12), (13), (15) into (16), yielding expression of per unit inductance value (17)

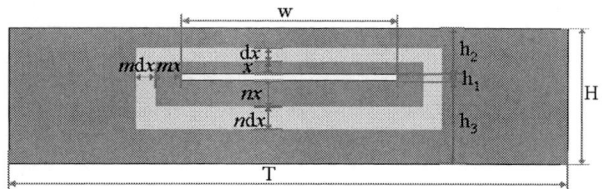

Figure 6. Differential magnetic path of a LTCC inductor

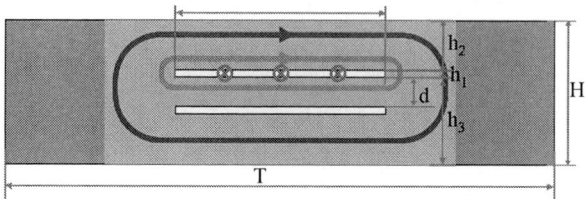

Figure 7. Differential magnetic path of a SV coupled LTCC inductor

$$L_0 = \frac{\mu n}{2(n+1)m}\left(1+\frac{2mh_2}{w}\right) \qquad (17)$$

For a two-phase coupled inductor of SV structure, flux generated by one current is divided into two groups as shown in Fig. 7. One group is self-coupling flux and the other one is mutual coupling flux. The two conductors distribute in the core symmetrically, therefore, leakage-inductance of each winding is the same, and it is determined by magnetic reluctance in the inner blue region. n should be re-defined by (8) because anther conductor is added in the magnetic core.

$$n = \frac{h_3 - h_1}{h_2} \qquad (18)$$

Similarly, Leakage inductance is determined by flux flowing through the intermediate region. Permeance along linkage flux path is (19)

$$P_{l0} = \int_0^{\frac{d}{n}} \frac{\mu dx}{(w+2mx)\cdot\left(1+\frac{1}{n}\right)} \qquad (19)$$

Then per unit length linkage inductance can be expressed by (20)

$$L_{l0} = \frac{\mu n}{2(n+1)m}\cdot\ln\left(1+\frac{2md}{nw}\right) \qquad (20)$$

The coupling coefficient thus can be obtained as (21)

$$k = \frac{\ln\left(\dfrac{nw+2mnh_2}{nw+2md}\right)}{\ln\left(1+\dfrac{2mh_2}{w}\right)} \qquad (21)$$

A 3D FEA simulation is set up to identify accuracy of the model for inductance and coupling coefficient approximation. Parameters used in the simulation are shown in Table III. To compare calculation and simulation results, keep distance of

two windings and conductor width constant; then, increase thickness of the whole inductor. A series of elf-inductance, leakage inductance as well as coupling coefficient values are obtained by increasing h_2. They are described in Fig. 8 by curve fitting. The range of h_2 is from 0.2mm to 1mm. This is enough for reasonable design of thin planar coupled inductors because distance of two windings in simulation is only 0.4mm.

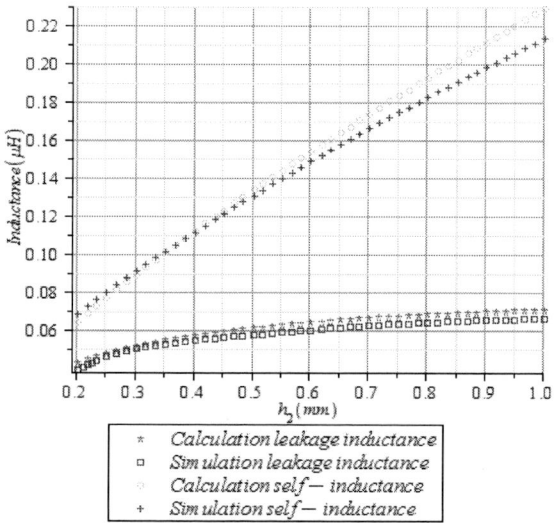

(a) Self-inductance and leakage inductance

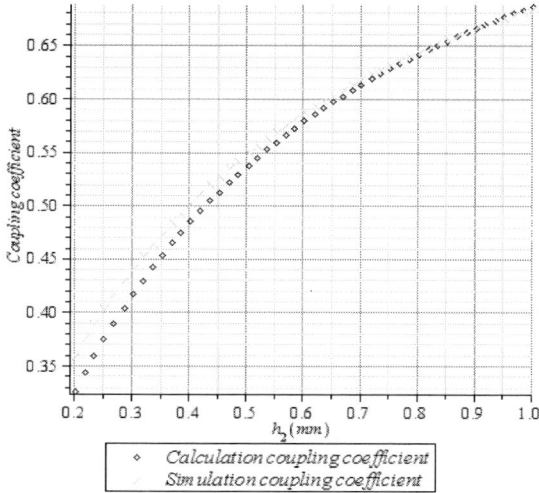

(b) Coupling coefficient

Figure 8. Comparison of simulation and calculation results

978-1-4244-4782-4/10 $26.00 © 2010 IEEE

TABLE III.   PARAMETERS OF COUPLED INDUCTORS FOR SIMULATION

Width of winding (b)	7mm
Distance between coupling windings (s)	0.4mm
length of windings	50mm

Error between simulation results and calculation results are within 10%. Actually, other simulations expose that when distance of two windings is smaller than 1mm (considering limitation of thickness in LTCC technology, this value is large enough for coupled inductor design), error between simulation and calculation is keeping in 20%.

IV. EXPERIMENT

Two LTCC coupled inductors are firstly fabricated One is SV structure; the other one is NV structure. Pictures of prototypes are shown in Fig. 9. The SV structure coupled inductor's volume is 35mm×35mm×1.3mm; The NV structure coupled inductor's volume is 15mm×45mm×2.2mm. Two windings of coupled inductor distribute symmetrically in the core. And their pads for soldering are separately laid out on two sides of inductors. Both calculation and measurement results of inductance values and coupling coefficients are provided in Table IV for comparison. Self-inductance values obtained by measurements are a little larger than those obtained by calculation for the existence of extended windings.

In the model, it is supposed that self-inductance and mutual inductance are contributed by parallel planar windings buried in magnetic cores. However, in practice, windings are extended to soldering pads. These extended windings contribute to self-inductance and thus influence coupling coefficients indirectly.

To evaluate performance of LTCC coupled inductors, a 12V input, 1.8V/40A output two-phase interleaved buck converter as shown in Fig. 10. The converter is constructed with HIP6302 from intersil and R2J20602NP from Renesas. HIP6302 is a two-phase interleaved synchronous buck controller. R2J20602NP is a Drmos which integrates two N-channel Mosfets and their driving circuit in a 8mm×8mm×0.95mm QFN package. Each phase of the converter operates at 1MHz frequency. Waveforms of two-phase AC current are shown in Fig. 11.

(a) Top view

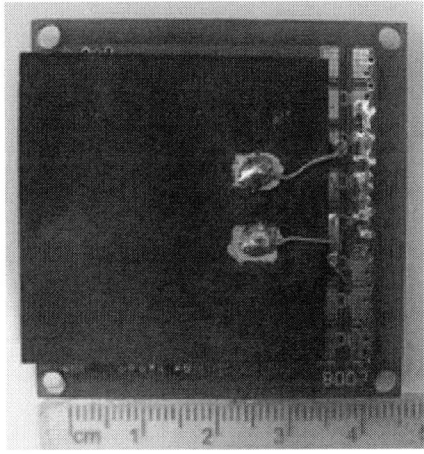

(b) Bottom view

Figure 10. Picture of a two-phase interleaved DC/DC converter with

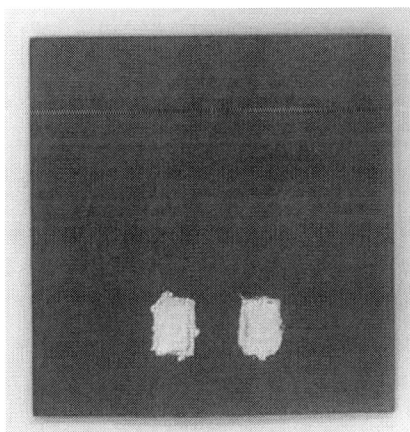

（a）SV structure

(b) NV structure

Figure 9.   Coupled inductor prototypes

TABLE IV.   MEASUREMENT RESULTS OF LTCC COUPLED INDUCTORS

Type of coupled inductors	Spiral coupled inductor			Solenoid coupled inductor		
Inductance	Self-inductance of winding 1	Mutual-indutance	Self-indutance of winding 2	Self-indutance of winding 1	Mutal-inductance	Self-indutance of winding 2
Measurement value	106.7nH	41.1nH	106.1nH	195.1nH	53.0nH	197.0nH
Calculation value	90nH	45nH	90nH	165nH	63nH	165nH

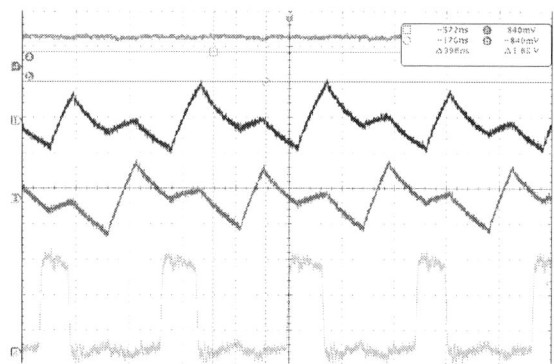

Figure 11.   AC current waveforms of a two-phase LTCC coupled inductor

V.   CONCLUSION

This paper explores designing of planar coupled inductors based on LTCC technology. Four structures of two-phase coupled inductors are proposed and compared. Prototypes of SV and NV coupled inductors are designed for their high coupling coefficients. They are applied in a 12V input, 1.8V/40A output two-phase interleaved buck converter to show their performance in a DC/DC converter. Further work on system integration of a DC/DC converter will be explored to show advantages of planar LTCC coupled inductors for three-dimensional integration.

REFERENCES

[1]   P.-L. Wong, P. Xu, P. Yang, and F. C. Lee, "Performance improvements of interleaving VRMs with coupling inductors," Power Electronics, IEEE Transactions on, vol. 16, pp. 499-507, 2001.

[2]   P. L. Wong, F. C. Lee, X. Jia, and D. Van Wyk, "A novel modeling concept for multi-coupling core structures," in Applied Power

Electronics Conference and Exposition, 2001. APEC 2001. Sixteenth Annual IEEE, 2001, pp. 102-108 vol.101.

[3]   S. Chandrasekaran, V. Mehrotra, and S. Han, "A new matrix integrated magnetics (MIM) structure for low voltage, high current DC-DC converters," in Power Electronics Specialists Conference, 2002. pesc 02. 2002 IEEE 33rd Annual, 2002, pp. 1230-1235 vol.1233.

[4]   S. Jian, K. F. Webb, and V. Mehrotra, "An improved current-doubler rectifier with integrated magnetics," in Applied Power Electronics Conference and Exposition, 2002. APEC 2002. Seventeenth Annual IEEE, 2002, pp. 831-837 vol.832.

[5]   J. Li, C. R. Sullivan, and A. Schultz, "Coupled-inductor design optimization for fast-response low-voltage DC-DC converters," in Applied Power Electronics Conference and Exposition, 2002. APEC 2002. Seventeenth Annual IEEE, 2002, pp. 817-823 vol.812.

[6]   J. Li, A. Stratakos, A. Schultz, and C. R. Sullivan, "Using coupled inductors to enhance transient performance of multi-phase buck converters," in Applied Power Electronics Conference and Exposition, 2004. APEC '04. Nineteenth Annual IEEE, 2004, pp. 1289-1293 vol.1282.

[7]   L. Wang, Y. Pei, X. Yang, X. Cui, and Z. Wang, "Three-dimensional integration of high frequency DC/DC converters based on LTCC technology," in Power Electronics and Motion Control Conference, 2009. IPEMC '09. IEEE 6th International, 2009, pp. 745-748.

[8]   M. H. F. Lim, L. Zhenxian, and J. D. van Wyk, "Low Profile Integratable Inductor Fabricated Based on LTCC Technology for Microprocessor Power Delivery Applications," Components and Packaging Technologies, IEEE Transactions on, vol. 30, pp. 170-177, 2007.

[9]   M. H. Lim, J. D. van Wyk, F. C. Lee, and K. D. T. Ngo, "A Class of Ceramic-Based Chip Inductors for Hybrid Integration in Power Supplies," Power Electronics, IEEE Transactions on, vol. 23, pp. 1556-1564, 2008.

[10]   Y. Li, Y. Liu, H. Zhang, L. Han, and Z. Yang, "A multilayer low pass filter fabricated by ferrite and ceramic cofiring system based on LTCC technology," in Electronic Packaging Technology & High Density Packaging, 2009. ICEPT-HDP '09. International Conference on, 2009, pp. 961-964.

[11]   H.-J. Kim, Y.-J. Kim, and J.-R. Kim, "An Integrated LTCC Inductor Embedding NiZn Ferrite," Magnetics, IEEE Transactions on, vol. 42, pp. 2840-2842, 2006.

[12]   C.-Y. Kim, H.-J. Kim, and J.-R. Kim, "An integrated LTCC inductor," Magnetics, IEEE Transactions on, vol. 41, pp. 3556-3558, 2005.

Optimising the high frequency bandwidth and immuntity to interference of Rogowski coils in measurement applications with large local dV/dt.

Dr Christopher R Hewson
Power Electronic Measurements Ltd
164 Lower Regent St, Beeston
Nottingham, UK
Email chris.hewson@pemuk.com

Mr William F Ray
Power Electronic Measurements Ltd
164 Lower Regent St, Beeston
Nottingham, UK
Email bill.ray@pemuk.com

Abstract—**Electrostatic interference through capacitive coupling onto a Rogowski coil causes unwanted pick-up noise on the output of the integrator. An electrostatic screen fitted to a Rogowski coil reduces this interference. This paper discusses how to maximise the high frequency (hf) bandwidth of a Rogowski transducer with a screened coil.**

I. INTRODUCTION

Rogowski current transducers, comprising a Rogowski coil and electronic integrator, are often used for measuring pulsed and transient currents in power electronic circuits [1][2]. They can be thin, flexible, clip-around and thus easy to use. The Rogowski coil has a very low insertion impedance and the overall transducer can have a high frequency bandwidth of greater than 10MHz.

There are an increasing number of power electronic applications that require a high bandwidth current measurement with excellent immunity to large local dV/dt transients. Such applications still need a thin, flexible, Rogowski coil. For example in power converters faster IGBT switching speeds and higher blocking voltages in increasingly compact designs place the Rogowski coil near to 'noisy' environments. Another example is the increasing power capability of solid state rf transmitters and amplifiers where currents of 100A, voltages of up to 1kV, at frequencies of several MHz are common.

The problem of electrostatic interference through capacitive coupling on to the Rogowski coil is analysed in [3][4][5]. Fig 1. shows the coupling mechanism of a voltage source V_x via stray capacitance C_x on to a Rogowski coil having self-inductance L, and outer winding to inner return conductor capacitance C (where from [1] $C'=(2/\pi)^2 C$). The worst case occurs if all the coupling capacitance is concentrated at the free end of the Rogowski coil. The

injected current will go through the entire coil impedance generating a voltage, which once integrated, appears as pick-up noise on V_{out}. Typically the capacitance is distributed uniformly along the coil, (e.g. where the conductor is a cable or cable bunch with an outer diameter approaching the inner coil diameter). This is approximated by assuming the coupling capacitance is concentrated at the mid-point of the Rogowski winding.

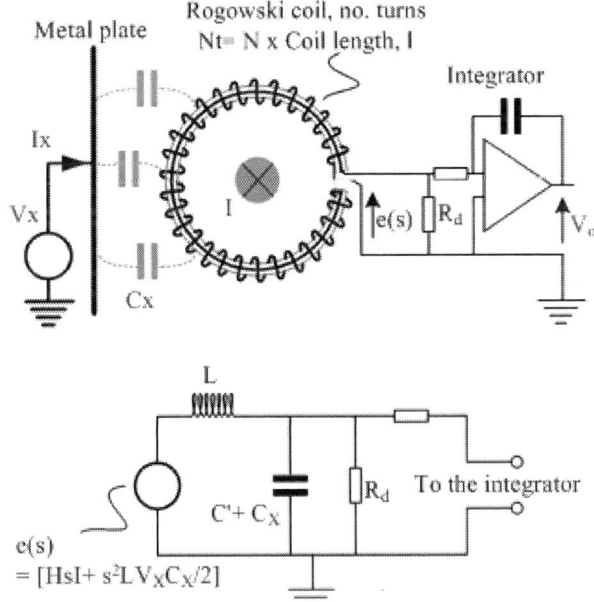

Fig. 1. Schematic and equivalent circuit showing capacitive coupling onto the Rogowski coil

From the schematic of Fig 1. a lumped parameter model is obtained. The voltage HsI is generated by the primary current I, where H is the coil sensitivity in (Vs/A). The coil winding is assumed to be close packed where $H=L/N_t$. In addition to the voltage HdI/dt induced by the primary current there is a voltage $(LC_x/2)(d^2V_x/dt^2)$ induced by the displacement current I_x due to the electrostatic field as shown in Fig. 1. Hence the equivalent error current is given by:

$$I_e = \frac{C_x N_t}{2}\frac{dV_x}{dt} \qquad (1)$$

To reduce the interference from the electrostatic field it is necessary to either bypass I_x by fitting an electrostatic screen or to minimise I_x by inserting a common mode impedance. Fitting the coil with an electrostatic screen will reduce the high frequency performance of the Rogowski transducer [5]. This paper examines how to maximise the high frequency bandwidth of a Rogowski coil, reject interference from high dV/dt fields, and maintain a thin flexible coil.

II. FITTING AN ELECTROSTATIC SCREEN

Screening the coil provides a low impedance path to ground to enable I_x to by-pass the coil winding. The screen introduces capacitance C_s between the coil winding and ground, which is in parallel with C' and which replaces C_x in Fig 1. Thus including the dynamics of the electronic integrator from Fig 1. the high frequency response of the Rogowski transducer is [2]:

$$\frac{V_{out}}{R_{SH}I} = \frac{1}{1+2\xi T'_c\, s + T'^2_c\, s^2}\frac{1}{1+T_b s} \qquad (2)$$

The integrator delay $T_b=1/2\pi GBW$, where GBW is the gain bandwidth product of the op-amp. The coil delay T'_c is increased by the additional screen capacitance i.e. $T'_c=\sqrt{L(C+C_s)}$' and $\xi=(\sqrt{L/(C+C_s)}')/2R_d$. To reduce any 'peaking' in the hf response of the Rogowski transducer due to C_s the terminating resistor R_d can be adjusted, but inevitably the addition of a screen for a given coil cross section and winding density results in a reduction of the hf bandwidth of the transducer.

A Optimising the bandwidth of a screened coil

From (2) it is clear that to maximise Rogowski coil bandwidth it is necessary to reduce coil inductance and the total coil and screen capacitance, but how is this optimised for a given high frequency bandwidth and coil thickness and what are the resulting limitations on overall Rogowski transducer performance?

If we wish to maintain a thin coil with a specific voltage insulation capability the outer screen diameter d_s is fixed. Similarly the Rogowski coil winding wire diameter, $d_{wire,}$ is fixed, being as thin as can practicably survive repeated opening and closing of a Rogowski coil. With reference to

Fig 2. both the diameter of the plastic former onto which the Rogowski coil is wound, d_f, and the pitch of the winding, x, can be manipulated to alter the self inductance of the coil L and the capacitance $C+C_s$.

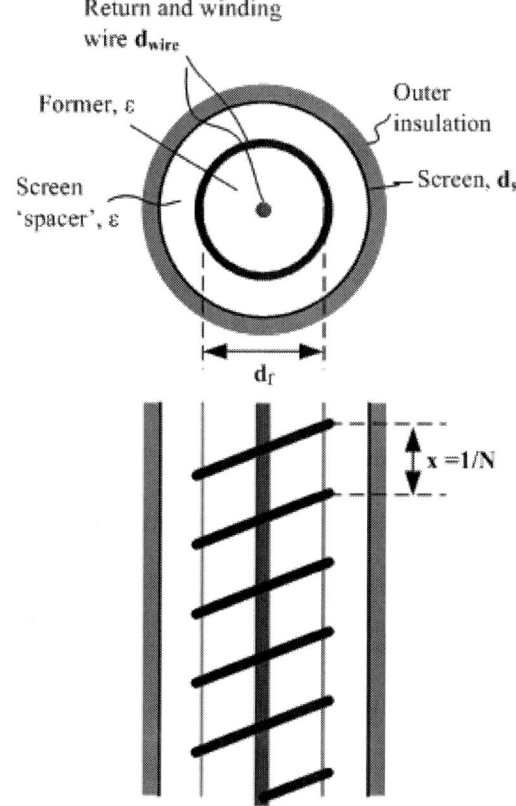

Fig 2. Construction of a Rogowski coil with screen

Assuming that the flux in each turn of the Rogowski coil is constant the coil sensitivity is

$$H = \mu_o NA \qquad (3)$$

where A is the cross sectional turn area $A=(\pi/4)(d_f+d_{wire})^2$ and $N=N_t$/coil length. The flux per turn is constant provided the Rogowski coil is formed into a closed loop around the conductor, the pitch is uniform, and the turn area A is << than the conductor radius, or if the conductor is very thin, it is spaced a distance d away from the edge of the coil where $d^2 > (d_f+d_{wire})^2$. With Rogowski coil and conductor insulation taken into account we need to ensure this is the case for typical currents of >100A, conductor diameters >10mm. There are other practical considerations in the choice of A and x. For power electronic applications the thinner the Rogowski coil the better as converter designs become more compact, therefore an overall coil cross section including screen and insulation of < 7mm is chosen, this limits d_f to typically 5mm. Assuming a wire radius of 0.2mm this sets the minimum former diameter to 1.5mm to enable sufficient wire bending radius. It is difficult to wind an accurate pitch for a loose spaced winding. An accurate pitch is essential for rejecting interference from currents external to the Rogowski coil. Given a d_f of between 1.5 and 5mm, experience of winding coils limits the pitch to < 5mm.

Reducing the coil winding area A and the winding density N to reduce the time constant T'_c will also reduce H. A low frequency and broadband noise, V_{noise}, is generated in the electronic integrator of a Rogowski transducer where $f_l V_{noise}$ α 1/H and f_l is the low frequency limit of the integrator [2]. Thus the effect of reducing T'_c is to either increase the transducer noise or limit the low frequency performance.

B. *Optimising T_c for a given coil sensitivity H*

The coil sensitivity H can be fixed to achieve a given low frequency Rogowski transducer performance. Additionally if the screen diameter d_s and winding wire diameter d_{wire} are fixed then, if the former diameter, d_f, is chosen, from (3) the turns density N must be chosen to achieve the specified H. The time delay of the Rogowski coil is given by

$$T_c = \sqrt{L(C+C_s)} \qquad (4)$$

As d_f and N are adjusted for a given H, so the inductance and capacitance of the Rogowski coil will vary. This section introduces approximate expressions for L and $C+C_s$ in terms of d_f and N=1/x. Ultimately these expressions are used to calculate the minimum T_c that can be achieved for a given H and thus optimise the bandwidth of the overall Rogowski transducer.

The inductance per m (H/m) of a long close packed Rogowski coil winding (effectively a solenoid) is well known and is given by

$$L = \mu_o N^2 A \qquad (5)$$

For a close packed winding, which we will define as a pitch of $x < 2.5 d_{wire}$, all the flux within the coil cross section area A of the winding links all the turns, as the pitch of the winding is increased there is additional flux that links just the wire itself and not all the turns thus the measured inductance is larger than predicted by (5). As the pitch, x, is increased the inductance can be adequately predicted by

$$L = \mu_o N^2 A + 2 \times 10^{-7} l_{wire} \ln [0.3667((2x/d_{wire})-2)] \qquad (6)$$

where $l_{wire}=(\pi h N+1)$ and $h=\sqrt{((d_f+d_{wire})^2+(1/\pi N)^2}$ for $x \geq 2.5 d_{wire}$.

Table 1a. and 1b. show a comparison of measured inductance values against those predicted by (5) and (6). The measurements are taken using an LCR bridge, model LCR400, the frequency of the measurement is 10kHz. At 10kHz the skin depth of copper wire is δ=0.65mm, thus for d_{wire}=0.22mm there is a distribution of flux inside the wire

adding an additional inductance to the measurement of $(\mu_o/8\pi)*l_{wire}$ which is added to (6) to give the total inductance 'Theory' in Table 1.

TABLE I. COMPARISON OF THEORETICAL AND MEASURED ROGOWSKI COIL SELF-INDUCTANCE

Pitch x=1/N (mm)	L, Inductance (µH/m) – for d_f = 3.74mm, d_{wire} = 0.22mm		
	Measured	Theory	Error (%)
0.22	315.20	322.65	-2.39
0.44	82.70	81.41	1.57
0.657	39.84	38.40	3.63
0.88	23.76	23.14	2.63
1.667	8.66	8.68	-0.22
4.545	3.22	3.03	6.3
5.857	2.46	2.58	-3.49
9.100	1.97	2.15	-9.17

1a.

Pitch x=1/N (mm)	L, Inductance (µH/m) – for d_f = 1.86mm, d_{wire} = 0.22mm		
	Measured	Theory	Error (%)
0.22	86.96	89.76	-3.22
0.44	23.19	22.85	1.47
0.88	7.11	7.28	-2.34
1.176	5.18	4.96	4.35
3.846	1.99	1.92	3.45
6.33	1.56	1 70	-8.65

1b.

For a pitch < 5mm the theoretical value is within 5% of the measured value and (5) and (6) provide a good estimation of inductance. At a pitch of > 5mm the winding is tending toward a configuration more akin to two parallel wire conductors and (6) is no longer valid, however as previously discussed x < 5mm also represents a reasonable practical limit in the construction of accurately wound Rogowski coils.

The capacitance of a screened close packed Rogowski coil winding is the parallel capacitance between the coil outer winding and the screen ,and the coil outer winding and the return conductor. For a coil with a pitch of d_{wire} this approximates to the co-axial case where

$$C_{total} = C_{s\_nom} + C = \frac{2\pi\varepsilon}{\ln\left(d_s/\left(2d_{wire}+d_f\right)\right)} + \frac{2\pi\varepsilon}{\ln\left(d_f/d_{wire}\right)} \qquad (7)$$

Table 2. shows the measurement of capacitance C_s for two different coil former diameters with a close packed winding $x=d_{wire}$ and a fixed d_{wire}=0.22mm and d_s=6.76mm where the screen is fitted on a PTFE former of inner diameter 4.7mm. The capacitance measurements were made on an LCR bridge, model LCR400, at 10kHz. The measured results and those predicted by (7) are in excellent agreement.

TABLE II. CAPACIATNCE C_s FOR CLOSE PACKED CASE WHERE X=D$_{WIRE}$

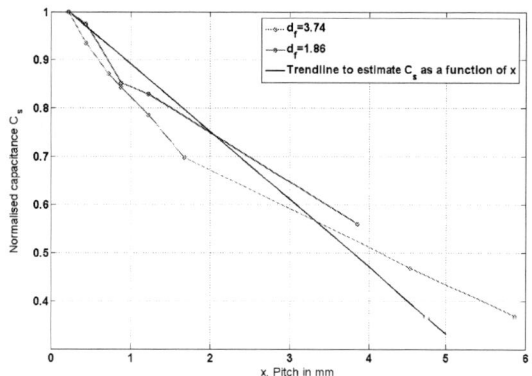	df (mm)	C_s Theory (pF/m)	C_s Measured (pF/m)
	3.74	191.6	192.7
	1.86	62.6	65.9

However for a given former diameter as the pitch is increased significantly it is no longer reasonable to assume an evenly distributed charge on the surface of the outer coil winding and (7) with its implication of two discrete capacitances is an approximation. In the limit the capacitance between the outer winding and the screen coil can be modelled as a single wire above an infinite plane to represent the screen, multiplied by the total length of the winding wire. Measurements verify this as a reasonably accurate estimation but only where the influence of screen to inner conductor capacitance is negligible, this makes the approximation of little practical use for our range of d_f=1 to 5mm and x<5mm.

In the absence of a reasonable theoretical model to determine the variation of C_{total} as x varies, measurements of the capacitance C_s for d_f=1.86mm and d_f=3.74mm as x varies from d_{wire} to 5mm were obtained. The capacitance measurements have been normalised to the close packed co-axial case where x=d_{wire}. The results are plotted in Fig. 3. The variation of C_s with x can be roughly estimated from the linear trend-line shown in Fig 3.

Fig 3. Variation in C_s with a variable pitch x and d_f

The empirical relationship (8) is used to obtain C_s for variations in x where $C_{s\ nom}$ is obtained for a given former diameter d_f from (7). The linear fit is quite poor but does provide a tolerable indication of the capacitance variation.

$$C_s = (1.031-0.14x)C_{s\_nom}$$
$$C_{total}= C_s+C \qquad (8)$$

There is a similar but smaller variation for C. However this value is typically very much smaller than C_s. Using (7) to calculate C is a reasonable approximation for the range of d_f and x investigated in this paper.

It should be noted that the measurement of L and C, and the approximate formulas derived for calculating L and C_{total} are based on measurements at 10kHz. In [6] these low frequency estimates of L and C_{total} are shown to be a reasonable indication of T_c up to f=1/4√(LC$_{total}$).

In (5),(6),(7) and (8) we have expressions for L and C_{total} in terms of d_f and x. From (3) it is possible to generate values of d_f and x for a given H and subsequently find $T_{c\ min}$. For example assume H=20nVs/A, with d_{wire}=0.22mm, d_s=6.76mm (the screen is fitted on a PTFE spacer of inner diameter 4.7mm and ε_r=2.1), d_f=1.5 to 5mm, x=0.22 to 5mm. We can plot the resultant (d_f, x) for H=20nVs/A and see how T_c varies. This is shown in Figure 4.

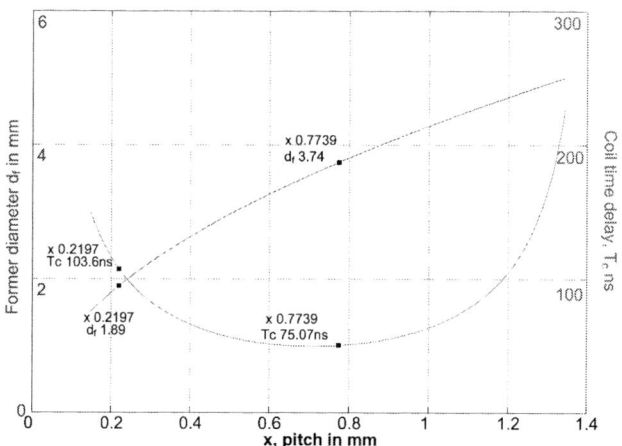

Fig 4. Variation in winding pitch x, former d_f, and coil delay T_c for H=20.0nVs/A, d_{wire}=0.22mm, d_f=6.76mm

C. Comparison of overall transducer performance for a specified H with a different T_c

From Fig. 4 it is informative to pick two points on the curve and compare the high and low frequency performance of two complete transducers with Rogowski coils of identical H but different T_c.

Both coils use the same integrator with T_b=10ns and a gain T_i = 2µs. This yields an overall transducer sensitivity of 10mV/A. The coil length is 445mm and the length of the cable between coil and integrator is 1m.

There is a small difference from the theoretical T_c and H predicted by Fig. 4 but not significant given our approximate formulas, the measured values for both coils are given in Table 3. where Z_o=√L/(C+C_s):

978-1-4244-4782-4/10 $26.00 © 2010 IEEE 2053

TABLE III.   MEASURED COIL VALUES

	Coil 1. df=3.74mm, x=0.78mm	Coil 2. df=1.86mm, x=0.22mm	
H	19.96	20.23	nVs/A
C_s	154	82.5	pF/m
C	30	39	pF/m
L	27	87	μH/m
R_d	388	836	Ω
T_c	70.5	102.8	ns/m
Z_o	383	846.2	Ω

In both cases $T_c > T_b$ so from (2) the high frequency performance is principally determined by the coil dynamics and not the integrator. Figure. 5 shows the response of the two transducers to a 10 to 90% step of approx 220ns with I_{peak} = 90A. The reference measurement is an 800MHz SDN-10 co-axial shunt with a nominal sensitivity of 100mV/A, this is in series with a 15T coil through which the Rogowski coil is threaded.

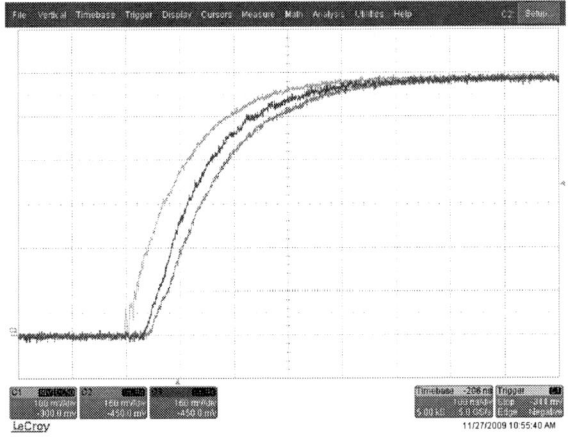

Figure 5.
C1 Yellow – SDN-100 800MHz co-ax shunt nominal 100mV/A, 100mV/div
C2 – Blue Transducer with coil 1 (15 x10mV/A), 150mV/div
C3 Red Transducer with coil 2 (15 x10mV/A), 150mV/div
Timebase: 100ns/div

The expected delay between the two traces is 14.5ns, which is evident at the start of the pulse, however the difference between the two traces increases as the current rises because T_c=46ns for the transducer with Coil 2. This is not sufficiently fast compared to the current rise time, thus some distortion of the measured waveform is evident due to the bandwidth limitation .

The low frequency limit of the two Rogowski transducers is set to be identical f_1=4.2Hz which yields a (3dB) bandwidth of 3.5Hz. Thus it is expected that the both the amplitude and phase shift should be identical measuring a 20Hz waveform, we would also expect the low frequency noise (centred around the low frequency bandwidth) to be identical. This is evidenced in Figures 6 and 7.

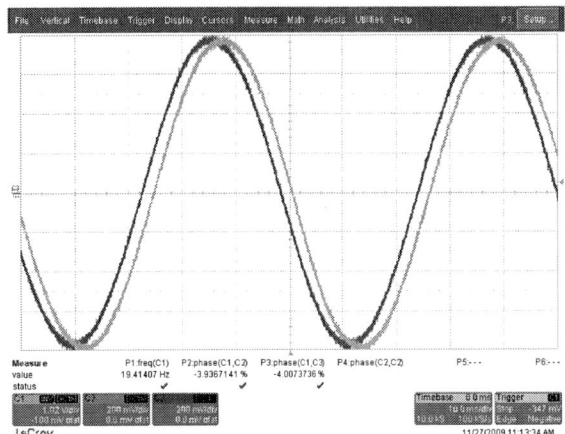

Figure 6. Measurement phase shift at 20Hz
C1 Yellow – SDN-10 400MHz co-ax shunt nominal 1.00V/A
C2 – Blue Transducer with coil 1 (through 20T x 10mV/A), 200mV/div
C3 Red Transducer with coil 2 (through 20T x 10mV/A), 200mV/div
Timebase: 10ms/div

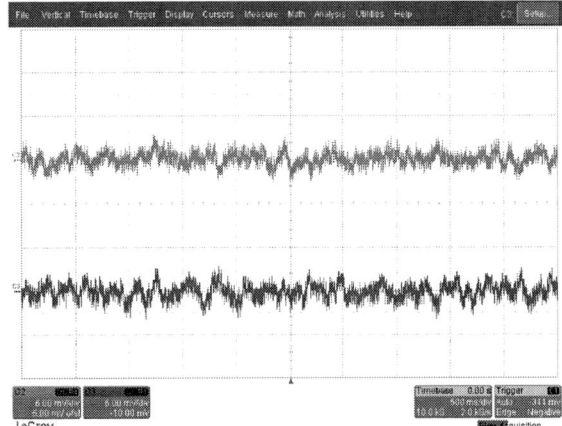

Figure 7. Low frequency noise measaurement
C2 – Blue Transducer with Coil 1 (10mV/A) 5.0mV/div
C3 Red Transducer with Coil 2 (10mnV/A) 5.0mV/div
Timebase: 500ms/div

D. A 30MHz transducer with a screened Rogowski coil

There is a degree of optimization of T_c for a given H that can be achieved, but ultimately for a specified H the additional capacitance of the screen will reduce the high frequency bandwidth. The following example shows how much the low frequency has to be 'sacrificed' to achieve a 30MHz (-3dB) bandwidth with a 445mm Rogowski coil. To quantify the effectiveness of the screen there are also results to show how the high frequency performance differs with the and without the screen.

A 30MHz (-3dB) can be achieved with T_c=6.6ns where the coil is terminated with a damping resistor R_d=152.5Ω and using an active integrator utilising an op-amp with T_b=0.8ns. Using the same coil length as the in sub-section C., and with a former diameter of d_f =3.74, and using the same d_{wire}=0.22mm and d_s=6.76mm, T_c=6.6ns can be achieved with x=5.88mm. The values for L, C and H are provided in Table 4 below.

TABLE IV   MEASURED COIL VALUES

	Coil 1. df=3.74mm, x=5.875mm	Coil 2. df=3.74mm, x=5.875mm	
H	2.58	*as coil 1.*	nVs/A
Cs	72.4	No screen	pF/m
C	18.9	*as coil 1.*	pF/m
L	2.42	*as coil 1.*	µH/m
R_d	152.5	359	Ω
T_c	14.9	6.8	ns/m
Z_o	162.8	357.8	Ω

The same coil and integrator is used for both the screened and unscreened versions where an integrator gain of T_i = 0.26µs yields an overall transducer sensitivity of 10mV/A as per the previous sub-section. The length of the cable between coil and integrator is 1m.

The test circuit in this case is the modified high di/dt pulse rig described in [7] used to verify the hf performance of Rogowski coils. In the original design great care was taken to ensure the excitation coil through which the Rogowski coil is threaded was largely at gnd potential and away from any dV/dt interference. In this case the circuit has been modified as shown in the schematic and photograph of Figure 8a and 8b so that the excitation coil is right next to the high dV/dt and furthermore the excitation turn area has been increased to ensure a large degree of coupling to the coil. Additionally each leg of the pulse rig has been reduced to a single excitation turn (c.f. 3T on the original) so that the transducer has to resolve a smaller current in the presence of a much higher dV/dt.

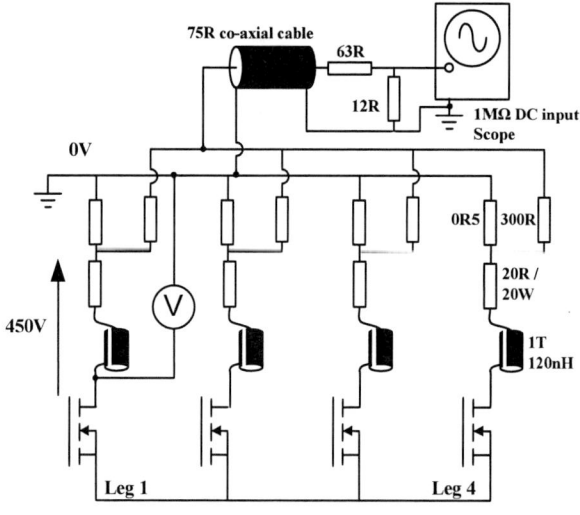

Figure 8a. Schematic of test circuit showing SMT resistor measurement and voltage measurement

Figure 8b. Photograph of the test set-up

The results are shown in Fig 9. (Screened coil 1.) and Fig 10. (Unscreened coil 2.)

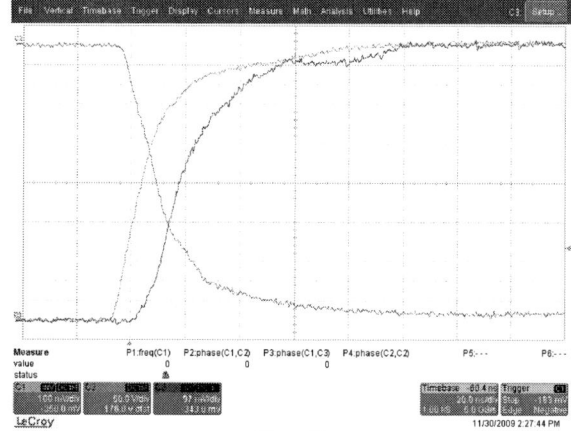

Figure 9.
C1 – Yellow – SMT resistor measurement, sensitivity 0.1V/A , 100mV/div
C2 Red –Voltage measurement 50V/div
C3- Blue – Rogowski measurement (4 x10mV/A), 97mV/div
Timebase: 20ns/div

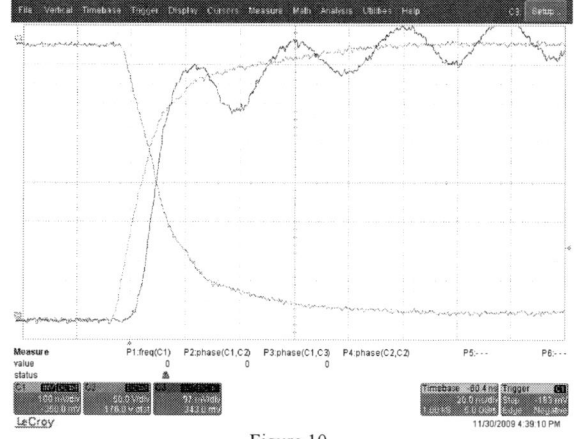

Figure 10.
C1 – Yellow – SMT resistor measurement, sensitivity 0.1V/A , 100mV/div
C2 Red –Voltage measurement 50V/div
C3- Blue – Rogowski measurement (4 x10mV/A), 97mV/div
Timebase: 20ns/div

978-1-4244-4782-4/10 $26.00 © 2010 IEEE

The 10 to 90% rise time of the measurement in Fig 9. is 40ns and the measured delay of the Rogowski transducer is 13ns, in excellent agreement with the calculated delay. The waveform also shows very little distortion due to the close coupled dV/dt, which is a high as 25kV/μs at the steepest point. There is a slight ripple in the Rogowski measurement at the top of the waveform, this is larger than that exhibited by the shunt measurement. This is conceivably due to the mis-match in damping resistor R_d and Z_o but more likely to be voltage pick up.

Figure 10. shows the measurement without a screen. The measurement should have less delay than the screened version and the initial output from the Rogowski transducer indicates that this is the case. However clearly the additional capacitance added to the Rogowski coil by the excitation coils and the coupled voltage causes unwanted ringing on the output of the measurement.

The aim of a 30MHz bandwidth transducer with a screened Rogowski coil has been achieved but in order to attain such a measurement the coil sensitivity has been reduced to 2.6nVs/A. This transducer has the same overall sensitivity (10.0mV/A) and the same coil and cable length as those discussed in sub-section C. so it is interesting to compare the low frequency performance of the 30MHz transducer with the transducers described in sub-section C. With the same low frequency noise as in Fig .7, the low frequency (-3dB) bandwidth of the Rogowski transducer with H=2.6nVs/A is 60.5Hz (cf 3.5Hz). This is due to the inherently poorer noise performance of the integrator op-amp with T_b=0.8ns (it is true to say that improved GBW products always comes at the expense of poorer lf and broadband noise figures) and the reduction in coil sensitivity from 20nVs/A to 2.6nVs/A.

CONCLUSIONS

Fitting a screen to a Rogowski coil reduces its susceptibility to measurement interference through capacitive coupling of dV/dt transients onto the coil winding. The additional capacitance between screen and coil winding reduces the high frequency performance.

Setting the coil sensitivity H largely determines the low frequency performance of a Rogowski transducer. Different winding configurations will achieve the same H but produce different values for the distributed inductance, L, and capacitance, C, of the coil. The LC product determines the coil hf performance of the Rogowski coil and this can be optimized for a given H. Practical results showing this optimization process have been presented. It is however fair to say that only modest gains in high frequency performance can be achieved.

Of more interest are the results of the Rogowski transducer with a screened coil length of 445mm and overall (-3dB) transducer bandwidth of 30MHz. This demonstrates two important results:

- The delay of the measurement transient is very close to the predicted delay (and this is when measuring a current step with a 40ns rise time). The delay is not affected by the close coupled voltage source.
- The pick-up interference though not completely attenuated is very small, this is in the context of measuring 60Apk in close proximity to a 25kV/μs transient.

The compromise in achieving a 30MHz hf bandwidth is limiting the low frequency performance, where to achieve a noise of typically 8.0mVp-p with a sensitivity 10mV/A, the lf (-3dB) bandwidth has to be increased to around 60Hz.

Finally, it was the authors intention to present some results on a method of using a high frequency common mode choke to minimize I_x. Initial results show this idea works and merits the submission of a full paper at a future date.

REFERENCES

[1] W.F. Ray and R.M. Davis, "High frequency improvements in wide-bandwidth Rogowski current transducers," European Power Electronics Conference, Conference Proceedings, Lausanne 1999

[2] W. F. Ray and C. R. Hewson, "High performance Rogowski current transducers'," IAS IEEE Industrial Applications Society, Conference Proceedings, Rome 2000

[3] J. Cooper, "On the high frequency response of a Rogowski coil" Journal of Nuclear Energy Part C, Vol. 5, pp 285 to 289

[4] V. Nassisi and A. Luches, "Rogowski coils theory and experimental results," Rev. Sci. Instrum. 50(7), July 1979.

[5] C.R. Hewson and W. F. Ray, "The effect of electrostatic screening of Rogowski coils designed for wide-bandwidth current measurement in power electronic applications" IEEE-PESC conference, Conference Proceedings, Aachen 2004

[6] W. F. Ray, C. R. Hewson and J M Metcalfe, "High frequency effects in current measurement using Rogowski coils", EPE 2005 conference proceedings, Dresden

[7] C.R. Hewson, W. F. Ray and R. M. Davis, "Verification of Rogowski current transducer's ability to measure fast switching transients", APEC IEEE conference Feb 2006, Dallas

PFC Inductor Selection Made Easy by "PL Product"

Welly Chou

Precision Incorporated
Minneapolis, MN 55430
Email: welly.chou@precision-inc.com

Abstract— **In recent years, power factor correction (PFC) has been gaining momentum. While the U.S. does not have any existing regulatory requirements, in Europe, the employment of PFC in power supplies is necessary to meet the European EN61000-3-2 regulatory standard. As the U.S. government continues to emphasize energy conservation and efficiency, speculation among power supply designers is such that the domestic requirement for PFC is not far from the horizon.**

Inductors play a critical role in the boost pre-regulator for PFC. Faced with today's challenge of balancing power requirements, switching frequency, and efficiencies, circuit designers find it necessary to understand a PFC inductor's operating characteristics and its effect on the PFC circuit as design criteria changes. This has been difficult due to the lack of an effective way for magnetic manufactures to relate the performance of PFC inductors.

To address this inadequacy, a figure of merit known as the PL product has been developed to better characterize PFC inductors over a range of output power. Based on desired circuit operating parameters, designers can calculate the PL product required for their application and correlate it to the PL curves of various PFC inductors. This will not only simplify the design process by making inductor selection easier, but also allow circuit and PFC inductor characteristics to be realized under various operating conditions.

In order to demonstrate and validate the applicability of PL product, a 400 Watt PFC inductor is tested in a continuous conduction mode (CCM) PFC circuit with output power ranging from 325 to 475 Watt. Measured data of temperature rise and ripple percentage are compared and verified against those calculated based on the PL product at 25 Watt increments.

I. INTRODUCTION

Traditionally, PFC inductors are usually developed based on a set of circuit parameters, i.e., minimum line voltage, switching frequency, ripple current, DC output voltage, etc. While such an application specific approach optimizes the inductor design, it makes it difficult to understand the relationship and the trade-offs among the circuit parameters and the inductor.

PL product simplifies this by separating the output power (P) and the Inductance (L) needed for a desired ripple current from the PFC circuit parameters, leaving designers with a

simple computation to derive the minimum PL required without delving into the intricacies of the inductor. Once the minimum PL product is determined, designers can navigate through families of PL curves of PFC inductors graphically and determine the suitable inductor by referencing to its PL curve. Performance trade-off information is also provided to designers to understand the impact on the circuit characteristics.

II. DERIVATION OF THE PL PRODUCT

PL product is defined as the product of power (Watt) and inductance (Henry). Power is the output power of the PFC circuit while L is the inductance of the PFC inductor. P and L are closely related. As output power varies, the inductance of a given PFC inductor changes based on core material characteristics. The required inductance of a PFC inductor in a continuous mode conduction PFC topology is calculated based on the following equation [1]:

$$L = \frac{\sqrt{2} \times V_{IN\min} \times D}{\Delta I \times f_s} \quad (1)$$

where V_{INmin} is the minimum low line RMS voltage (V), D is the maximum duty cycle, ΔI is the ripple current (A); L is inductance (H), and f_s is the switching frequency (Hz) of the PFC controller. Subsequently, D is calculated as follows

$$D = \frac{V_o - \sqrt{2} \times V_{IN\min}}{V_o} \quad (2)$$

ΔI is calculated as follows:

$$I_{LINEpk} = \frac{\sqrt{2} \times P}{\eta \times V_{IN\min}} \quad (3)$$

$$\Delta I = Ripple\% \times I_{LINEpk} \quad (4)$$

where η is the PFC circuit efficiencies (typically 0.92), P is the output power, I_{LINEpk} is the peak line input current (A), V_o is the DC output voltage (V), and Ripple % is the percent ripple of the I_{Linepk} and is ≤ 1. By substituting (3) and (4) into (1), the following PL product relationship is formed.

$$PL = \frac{V_{in\min}^2 \times \eta \times D}{f_s \times Ripple\%} \quad (5)$$

978-1-4244-4782-4/10 $26.00 © 2010 IEEE

The PL product derived from (5) is the *minimum* PL product required to achieve the ripple current % at the specified switching frequency, low line input voltage, and DC output voltage. This minimum PL product can then be correlated to families of PL curves of PFC inductors to select the suitable inductor.

III. PFC INDUCTOR PL CURVES

PFC inductors' PL curves are developed by applying the DC current handling capability of specific magnetic materials as shown in Fig. 1 to an inductor over a range of output power [2]. An example of a resulting PL curve is shown in Fig. 2.

MAGNETIC CHARACTERISTICS

Figure 1. Initial Permeability vs. DC Magnetizing Force

In Fig. 2, the blue line shows the minimum PL product required (PL = 0.15) from (5) with the following circuit parameters:

V_{INmin} = 85Vrms
η = 0.92
V_o = 375V
Ripple % = 30%
f_s = 100 kHz

The black line is the PL product curve of a 400W PFC inductor over the power range. The intersection of the PL curve and the minimum PL product required represents the optimum output power level (400W) to achieve 30% ripple current. If the inductor is used at a lower output power level, hence, to the left of the intersection, the ripple current percentage increases. Operating further downwards along the PL curve may require special considerations as the inductor will eventually enter discontinuous mode operation.

On the other hand, operating at a higher output level, hence, to the right of the intersection, will decrease the ripple current percentage. While it may seem desirable operating in this region, the trade off between higher output power and inductor temperature rise should be considered and, if necessary, reconciled. It is important to note that, within the typical switching frequency range of today's PFC controllers, the PL curve of an inductor is frequency independent. As

switching frequency changes, only the minimum PL required changes per (5).

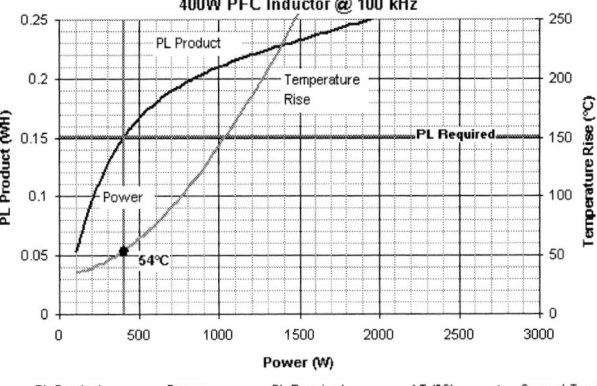

Figure 2. PL Product curve of the 400Watt PFC Inductor

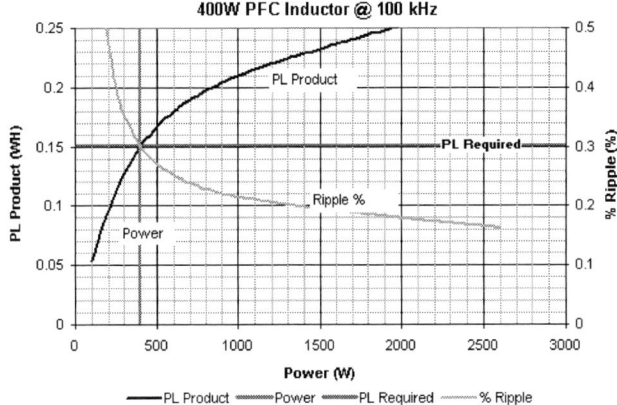

Figure 3. Ripple percentage and PL Product relationship

IV. REAL WORLD DATA VALIDATION

A 400W PFC inductor that was designed for 30% ripple at 85Vrms, 100kHz switching frequency and a DC output voltage of 375V is tested at low line condition (85Vrms) for output power ranging from 325W to 475W, in a 25W increments. Inductor temperature rise and the ripple current percentage are calculated based on the PL product of the inductor. Comparison of the calculated vs. measured data of the inductor temperature rise and ripple current percentage is summarized in Fig. 4 and 5.

A. Inductor temperature rise comparison

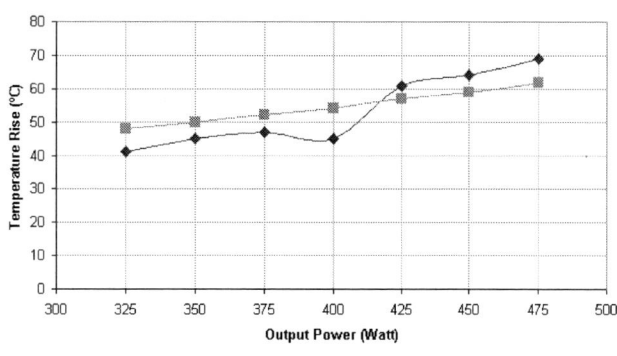

Figure 4. 400W PFC inductor temperature rise characteristics

Temperature rise of an inductor is a result of power loss caused by both core loss and winding losses [3]. Unfortunately, it is quite difficult to accurately estimate the temperature rise of an inductor. Even core loss data published by core manufacturers typically fluctuate ±15% [2]. As shown in Fig. 4, the calculated inductor temperature rise is within close approximation (±10°C) of the measured data. Such a result reveals that, by referencing the PL product of a given PFC inductor, circuit designers can closely estimate the temperature rise of the inductor and ensure proper functioning of the circuit as load condition varies.

B. Ripple current percentage comparison

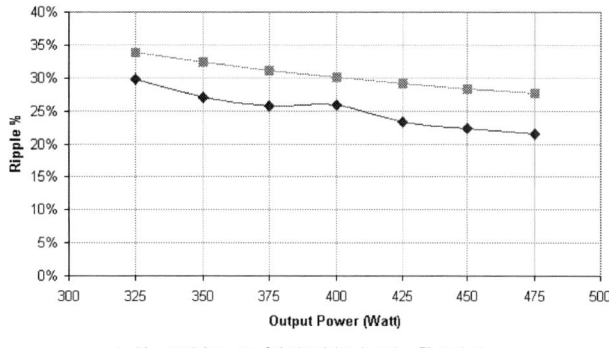

Figure. 5. 400W PFC inductor ripple percentage characteristics

Fig. 5 shows that the calculated ripple percentage runs parallel above the actual data, indicating that the actual ripple percentage is less than calculated. This could be explained by the tolerance of the AL value of the core material (±10%) and the typical tolerance of +20%, -10% of the curve shown in Fig. 1 [2]. A PFC inductor with higher AL value results in higher inductance which in turn lowers the ripple current percentage. The importance of this graph is that the calculated ripple current percentage tracks the measured value. Once again, by referencing the PL product curve, a circuit designer

can approximate the ripple current percentage within reasonable certainty as load condition varies.

V. CONCLUSIONS and FUTURE WORK

PL product empowers circuit designers to be able to select the right PFC inductor for their application. Furthermore, it gives them the insight into the relationship among output power, ripple current, switching frequency and the PFC inductor and allow educated trade-offs to be made. The real world results of the 400W PFC inductor in a CCM PFC application proves that the PL Product is applicable and can be beneficial to the practicing power electronic professional.

In the future, I would like to apply the concept of the PL Product to PFC inductors designed for other PFC topologies such as discontinuous conduction mode (DCM), critical mode conduction (CrM), and interleaved configuration.

ACKNOWLEDGMENT

I would like to thank the following individuals for their support and guidance on this project:

David Anderson, President of Precision Inc.

Thomas Emmons, TRE Engineering

Bruce Fishbeck, Senior Design Engineer at Precision Inc.

Bruce Runyan, Senior Design Engineer at Precision Inc.

Michael Shields, DMSI Engineering

REFERENCES

[1] P. C. Todd, "Boost power factor corrector design with the UC3853," Unitrode application note U-159.

[2] Micrometals, Inc. (2007, February). 200C series high temperature power cores for power applications issue C .[ONLINE].Available: http://www.micrometals.com/200cparts/200C_C.pdf.

[3] G. G. Orenchak, "Estimating temperature rise of transformers," Power Electronics, pp. 11-22, July 2004.

Evaluation of LTCC Capacitors and Inductors in DC/DC Converters

Laili Wang, Yunqing Pei, Xu Yang, Bo Song, Zhaoan Wang, Guopeng Zhao
Electrical Engineering Department, Xi'an Jiaotong University
Xi'an, 710049, China
Email: l.l.wang@stu.xjtu.edu.cn

Abstract—Low temperature co-fired ceramic (LTCC) technology has been proved to be an effective way to realize system integration. By using this technology, passive components in a DC/DC converter could be integrated into a passive substrate. This paper presents fabrication and evaluation of LTCC capacitors and inductors and summarizes challenges of applying LTCC technology to DC/DC converters. Four capacitors are made; two of them are made with single layer high permittivity thick film paste, and maximum 1.1µF capacitance is obtained; the other two are made with multilayer low permittivity LTCC tapes, maximum 30nF capacitance is obtained. A 70nH LTCC inductor is also fabricated with ferrite tapes. It is tested in a 1.2MHz buck converter and a 30MHz class Φ_2 DC/DC converter. Finally, challenges of conductor resistance, capacitor tape permittivity, and shrinkage matching are discussed and summarized.

I. INTRODUCTION

Passive integration is a powerful solution to increase power density and improve performance of DC/DC converters [1-8]. With the application of new generation power semiconductors for high frequency switching, it is feasible to integrate more passives into a LTCC substrate [9-11]. Design, construction and manufacturing of a conventional DC/DC converter will be revolutionarily simplified. This paper demonstrates fabrication of capacitors inductors based on LTCC technology. And they are evaluated in DC/DC converters for the purpose of system integration.

LTCC technology has many advantages for system integration. First, ferrite and capacitor tapes could be co-fired together with ceramic tapes. This makes it possible to embed passive components into planar passive substrates. Moreover, LTCC materials have the same temperature expansion coefficient (TEC) with silicon (Si); so, hybrid integration could be realized through semiconductor wire bonding. Finally, high-thermal conductivity metal or ceramic substrates (e.g., AlN) could be post-fired together with LTCC substrate, helping to dissipate heat generated by power loss.

Fabricating process based on LTCC technology is shown in Fig. 1. The process could be summarized in four steps as follows:

1. First vias are punched on ferrite tapes. These vias are later filled with silver paste for multilayer connection vertically. Silver paste is also screen printed on tapes for connection on each layer

2. Different layers are laminated together with hydraulic machine. The laminated block is then put into a furnace to co-fire, according temperature profile set before co-firing process.

3. Finally, active devices are mounted on surface of the block.

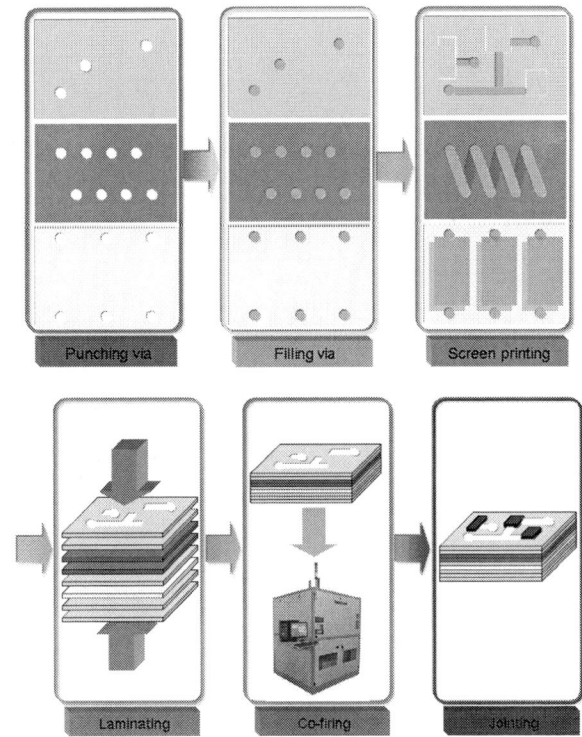

Figure 1. Fabricating process of integrated passive substrate based on LTCC technology

---

This work was supported in part by key project of Supporting Tech. From the Ministry of Science and Technology of China under Grant 2007BAA12B01.

II. CAPACITORS

Thick film and LTCC capacitors are firstly made to test their performance. The thick film capacitor is made with dielectric paste 4210-C from ESL. Permeability of the paste is 10000. High permeability makes it possible to design large capacitance in DC/DC applications.. The fabricating process of thick film capacitor is described as follows.

First, a layer of 10μm-thick conductor paste is screen printed on the surface a ceramic substrate. The conductor paste is then fired at 850℃ peak temperature to form first electrode. Capacitor paste is then screen printed on the first electrode; and they are together put into a furnace for firing at the same peak temperature. Finally, second electrode is made with the same method as the first one. In this experiment, only one layer capacitor is made. If a multi-layer capacitor is need to be made, repeat process described above. Two prototypes are made. One is made with silver paste as electrodes; the other is made with gold paste as electrodes. Their pictures are shown in Fig. 2. Effective dielectric area is 6cm^2.

The two capacitors are measured with RLC electric bridge and impedance analyzer. Capacitance and equivalent resistance are shown in Table I. The thick film capacitors made in this paper actually is the same with the one made in [12] except for larger capacitance value. During the fabricating process of thick-film capacitors, every time a new layer of paste is screen printed, the unfinished capacitor should be put into a furnace for firing. Thus, it must be a heavy workload when a multilayer capacitor is to be made. In contrast, as the process described in the first part, only once co-firing is enough when making a multilayer LTCC capacitor. Two LTCC capacitors are fabricated to test their electric characteristics. They have the same size and electrodes arrangement. What is different is their dielectric materials. One kind is 41260 from ESL. Its permeability is 250. The other one is 7240D from Ferro. Its permeability is 250. Pictures of the fired capacitors are shown in Fig. 3. They are also measured by RLC electric bridge. Measurement results are shown in Table II.

TABLE I.   MEASUREMENT RESULT OF THICK FILM CAPACITORS

Type of electrodes	Capacitance (μF)	ESR (mΩ)
Silver	0.67	67
Gold	1.1	60

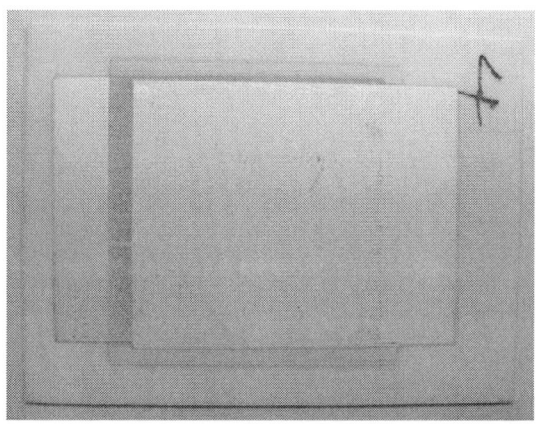

（a）Silver electrodes

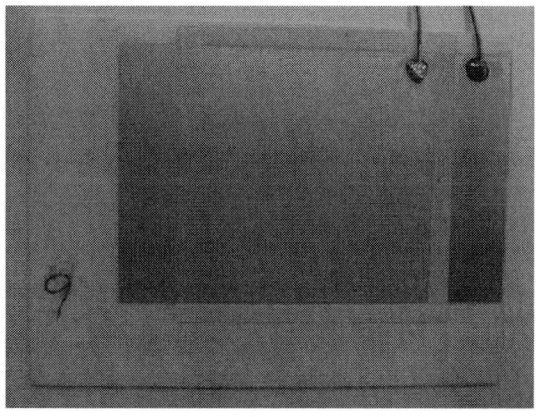

(b) Gold electrodes

Figure 2.   Picture of thick film capacitors

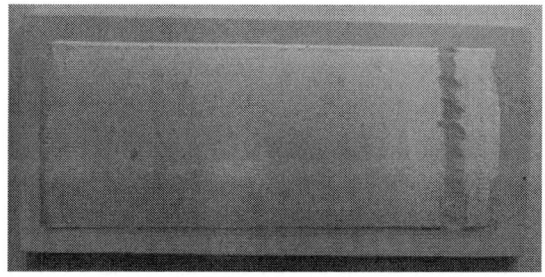

(a)   Multilayer LTCC capacitor made of 7240D

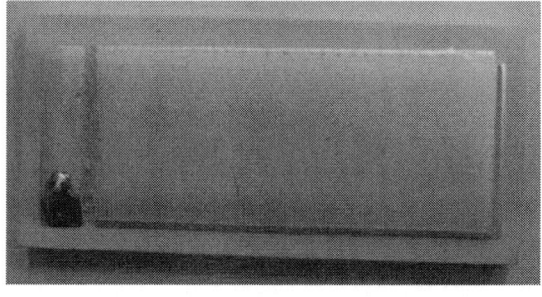

(b)   Multilayer LTCC capacitor made of 41260

Figure 3.   Picture of multilayer LTCC capacitors

TABLE II.   MEASUREMENT RESULT OF LTCC CAPACITORS

Capacitor tape	Capacitance (nF)	ESR (mΩ)
7240D	3.23	68
41260	37	97

Impedance curve of the 37nF capacitor is provided in Fig. 4. The resonant point is at as high as 38MHz. This means capacitors made based on LTCC technology is enough for high frequency DC/DC converters. Unfortunately, we only have two kinds of low permeability LTCC capacitor tapes. They are originally designed to be used in RF applications. Their permeability is too low to make a high value filter capacitor used in high frequency converters. If the capacitor tapes are optimized for high permittivity; then, filter capacitor could also be integrated into passive substrate. Both 0.67μF thick film capacitor and 37nF LTCC capacitor are tested in a 1.1MHz bootstrap circuit. They are used to charge a power MOSFET BSC020N03MS from Infineon. The input capacitor C_{iss} is 7200pF. Voltage waveforms across the fabricated capacitors and C_{gs} of MOSFET are shown in Fig. 5. The 37nF bootstrap capacitor is only five times as high as MOSFETF input capacitor; so, it has a voltage drop when charging the capacitor. The tested capacitors are connected to PCB board with copper wire. And parasitic inductance of the wire cause voltage ring when charging and discharging MOSFET input capacitor.

III. INDUCTOR

An LTCC inductor is fabricated according to the analytical model derived in [13]. The model can be expressed by (1)

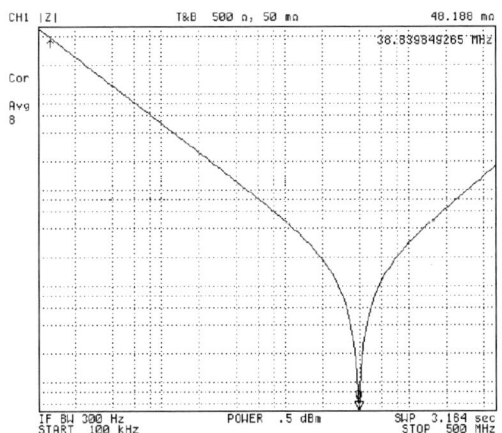

Figure 4.   Impedance curve of 41260 LTCC capacitor

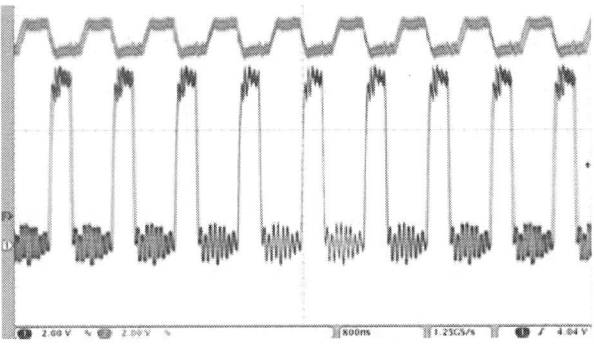

(a)   Voltage across 37nF bootstrap capacitor (top) and MOFET C_{gs} capacitor (bottom)

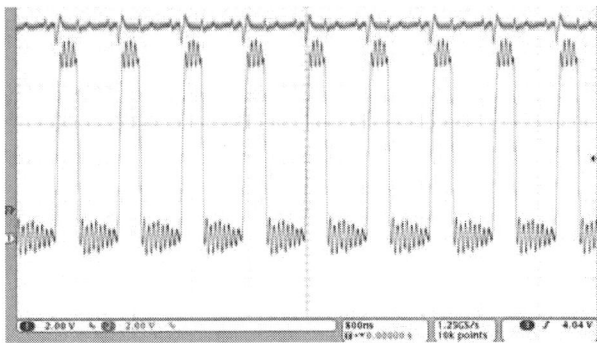

（b）Voltage across 0.67μF bootstrap capacitor (top) and MOSFET C_{gs} capacitor (bottom)

Figure 5.   Waveforms of LTCC capacitors in bootstrap test circuit

$$L = \frac{\mu l}{2\pi} \cdot \ln\left(1 + \frac{k \cdot \pi}{w_c + h_c}\right) \qquad (1)$$

Where l is length of winding. μ is permeability of ferrite tapes after co-firing. Meanings of other parameters are illustrated in Fig. 6. Ferrite tape used for fabrication is 40010 from ESL. Permeability of the tape is 50. The inductor is fabricated according the process [14]. As shown in Fig. 7, it

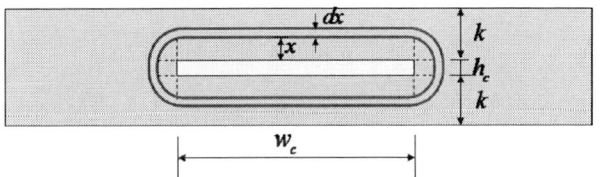

Figure 6.   Differential magnetic path

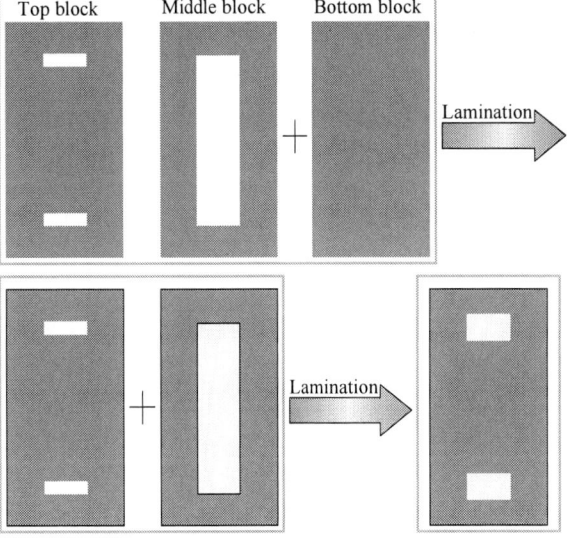

Figure 7.   Fabricating process

978-1-4244-4782-4/10 $26.00 © 2010 IEEE

can be described as follows: First, ferrite tapes are cut into small pieces, size of each piece is 30mm×15mm. They are firstly laminated into three blocks: top, middle and bottom blocks. The middle block consists of 8 layers; top and bottom blocks both consist of 16 layers. A 20mm×5mm slot is cut on the middle block; then, the middle block is laminated onto the bottom block. Conductor paste is then filled into the block to form inductor winding. Two 5mm×1mm slots are cut on the top block, distance of the two holes should equal to length of winding on middle layer. The top block is also laminated onto the middle one. And the holes on the top block should be filled with conductor paste to lead the windings to external circuit. Finally, they are laminated together for co-firing. Picture of the inductor is shown in Fig. 8. Its volume is 28×14×2mm after co-firing. The inductor is also measured by electric bridge and impedance analyzer. Inductance is 69nH and direct current resistance (DCR) is 0.4mΩ. Measured impedance curve is shown in Fig. 9.

A 1.1MHz 10V input 5V output buck converter is set up to evaluate performance of the fabricated inductor. Fig. 10 shows AC current waveform flowing through the inductor when the load current is 20A. As the current (load current plus positive AC peak current) goes up to 30A, the inductor is still unsaturated. This means small LTCC inductors could be used in high frequency, large output current DC/DC applications.

As the impedance curve shows, the LTCC inductor exhibits inductive characteristic up to 500MHz, this is much different from a multi-turn air core inductor. Impedance characteristic of an air core is greatly influenced by parasitic

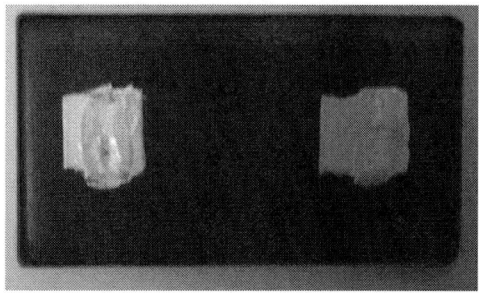

Figure 8.   Picture of LTCC inductor

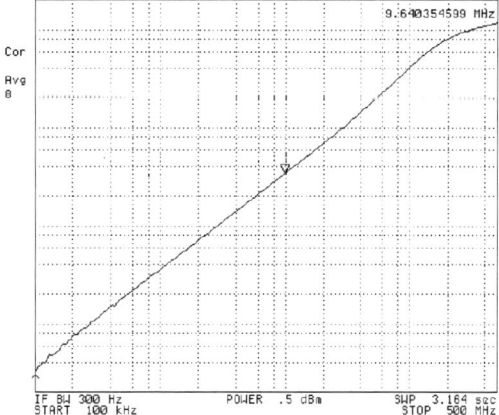

Figure 9.   Impedance curve of the LTCC inductor

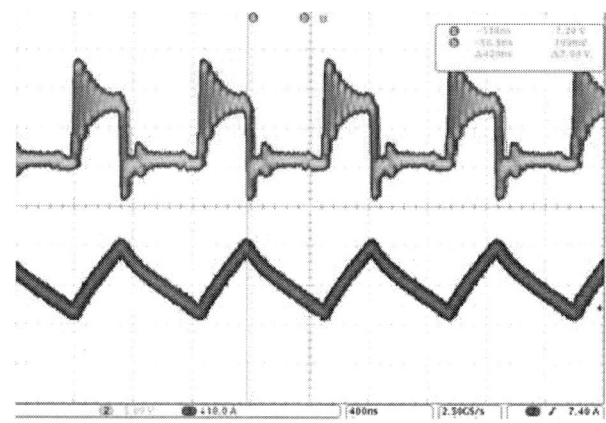

Figure 10.   Voltage waveform across switch (top) and AC current waveform (bottom) flowing through the inductor

capacitance between turns. Benefit from low permeability NiZn ferrite magnetic core, winding of a LTCC inductor used in very high frequency DC/DC converter is very short and do not need to be a multi-turn winding. Thus, it shows good impedance characteristic and low winding loss. Of course, magnetic material will cause additional core loss [15]. Trade off should be found when designing. To evaluate performance of the LTCC inductor at very high frequency, it is applied in a 30MHz class Φ_2 DC/DC converter. Current waveform is shown in Fig. 11.

IV.   CHALLENGES OF APPLYING LTCC TECHNOLOGY IN DC/DC CONVERTERS

A.   Conductor Resistance

Under most circumstances, LTCC technology only needs to deal with signals since it is originally used in RF circuit, Conductor paste used in these applications has higher resistivity than copper used in PCB technology. This may result in more power loss when high current flows through it. Fortunately, some company has begun to provide high conductivity paste to reduce conduction loss. Limit of conductor paste thickness is also a problem. Following conventional fabricating process, the conductor is often screen printed on the LTCC tapes. Thickness of paste is 15μm for

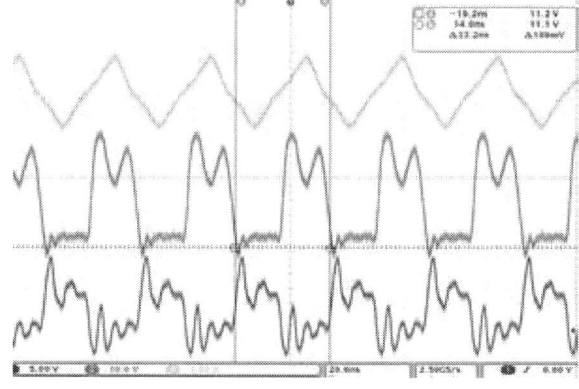

Figure 11.   Current waveform of a LTCC inductor （top） in 30MHz classΦ_2 DC/DC converter

each times printing. More times printing is not recommended for it will cause crack of LTCC substrate when co-firing. Therefore, if thickness of conductor needs to be increased further more. Slots should be cut with laser machine to increase thickness of conductor paste. This increases the complexity of fabrication.

B. Embeded planar passives

High permeability thick film paste could be used to realize large capacitance; however, it could only be post-fired on the surface of LTCC substrate. Capacitors laminated with LTCC tapes could be embedded into passive substrate; but, only small capacitance could be obtained with capacitor tapes in the market for their low permittivity. Compared with commercial ceramic capacitors, LTCC capacitors made with these tapes are even much larger in volume. This to some degree weakens the effect of integration for increasing power density. To solve this problem, new high permittivity capacitor tapes for making filter capacitors in high frequency DC/DC converters should be devised by LTCC material provider.

An inductor made with high permeability ferrite tapes inevitably has non-linearity. This property is useful when such an inductor is used as filter inductor. The reason is that at light load, a LTCC inductor has no saturation and exhibits high inductance value. Then, switching loss of converters could be largely reduced for very low current ripple. As the output current increases, magnetic material around conductors become saturated; and the inductance value decreases to its normal value [16]. Oppositely, the non-linearity of LTCC inductors make them not suitable to operate as a resonant inductor since it will cause the resonant point changing when increasing flowing current. This is not good for resonant converter design when output voltage gain is adjusted by varying switching frequency. For resonant converters, low permeability tapes should be chosen to prevent saturation of magnetic core. Resonant capacitors in these converters are often in the order of pF, they could be easily realized by existing capacitor tapes.

C. Shrinkage matching

A good scheme of power converter's integration based on LTCC technology is to integrate all the passives in the substrate and they are co-fired together. Dies of active devices are mounted on the surface and are connected with passives through wire bindings. This goal could be realized if certain obstacles cleared. Besides those stated above, there is still another challenge—shrinkage matching. Usually, LTCC tapes will shrinkage in XY and Z direction during co-firing. Take the ferrite tape 40010 from ESL as an example. It has 17% shrinkage in XY direction and 14% in Z direction. Unfortunately, different tapes often have different shrinkage even they come from the same company. When they are laminated and co-fired together, crack problem may happen. ESL has published papers to show co-firing of ferrite tapes, capacitor tapes, and ceramic tapes together although these tapes have different shrinkage in XY and Z direction [17].

V. CONCLUSION

This paper evaluates LTCC capacitors and inductors for the purpose of system integration. Impedance curve of them are provided to show frequency band they are effective. Both capacitors and inductors are tested to show their performance in DC/DC applications. Based on measurements and experiments, challenges of applying LTCC technology to DC/DC converters are summarized. It can seen that the LTCC technology is very promising for integration of high frequency DC/DC converters if these challenges are well solved.

REFERENCES

[1] J. D. Van Wyk, "Power electronic converters for motion control," Proceedings of the IEEE, vol. 82, pp. 1164-1193, 1994.

[2] J. A. Ferreira and J. D. Van Wyk, "Electromagnetic energy propagation in power electronic converters: toward future electromagnetic integration," Proceedings of the IEEE, vol. 89, pp. 876-889, 2001.

[3] F. C. Lee, J. D. van Wyk, D. Boroyevich, L. Guo-Quan, L. Zhenxian, and P. Barbosa, "Technology trends toward a system-in-a-module in power electronics," Circuits and Systems Magazine, IEEE, vol. 2, pp. 4-22, 2002.

[4] P. J. Wolmarans, J. D. van Wyk, J. D. van Wyk, Jr., and C. K. Campbell, "Technology for integrated RF-EMI transmission line filters for integrated power electronic modules," in Industry Applications Conference, 2002. 37th IAS Annual Meeting. Conference Record of the, 2002, pp. 1774-1780 vol.1773.

[5] C. Rengang, J. T. Strydom, and J. D. van Wyk, "Design of planar integrated passive module for zero-voltage-switched asymmetrical half-bridge PWM converter," Industry Applications, IEEE Transactions on, vol. 39, pp. 1648-1655, 2003.

[6] J. T. Strydom and J. D. van Wyk, "Volumetric limits of planar integrated resonant transformers: a 1 MHz case study," Power Electronics, IEEE Transactions on, vol. 18, pp. 236-247, 2003.

[7] S. C. O. Mathuna, P. Byrne, G. Duffy, W. Chen, M. Ludwig, T. O'Donnell, P. McCloskey, and M. Duffy, "Packaging and integration technologies for future high-frequency power supplies," Industrial Electronics, IEEE Transactions on, vol. 51, pp. 1305-1312, 2004.

[8] F. Blaabjerg, A. Consoli, J. A. Ferreira, and J. D. van Wyk, "The future of electronic power Processing and conversion," Power Electronics, IEEE Transactions on, vol. 20, pp. 715-720, 2005.

[9] M. Hui Fern Lim, J. D. van Wyk, and L. Zhenxian, "Internal Geometry Variation of LTCC Inductors to Improve Light-Load Efficiency of DC-DC Converters," Components and Packaging Technologies, IEEE Transactions on, vol. 32, pp. 3-11, 2009.

[10] M. H. F. Lim, L. Zhenxian, and J. D. van Wyk, "Low Profile Integratable Inductor Fabricated Based on LTCC Technology for Microprocessor Power Delivery Applications," Components and Packaging Technologies, IEEE Transactions on, vol. 30, pp. 170-177, 2007.

[11] L. L. Wang, Y. Pei, X. Yang, X. Cui, and Z. Wang, "Three-dimensional integration of high frequency DC/DC converters based on LTCC technology," in Power Electronics and Motion Control Conference, 2009. IPEMC '09. IEEE 6th International, 2009, pp. 745-748.

[12] F. Luo, M. H. F. Lim, R. Robutel, S. Wang, F. Wang, and D. Boroyevich, "Research on LTCC Capacitors and its Potential for High Power Converters," in Applied Power Electronics Conference and Exposition, 2009. APEC 2009. Twenty-Fourth Annual IEEE, 2009, pp. 2034-2038.

[13] L. Wang, "Design of LTCC inductors for high frequency DC/DC converters," in unpublished.

[14] M. H. Lim, J. D. van Wyk, F. C. Lee, and K. D. T. Ngo, "A Class of Ceramic-Based Chip Inductors for Hybrid Integration in Power Supplies," Power Electronics, IEEE Transactions on, vol. 23, pp. 1556-1564, 2008.

[15] H. Yehui, G. Cheung, L. An, C. R. Sullivan, and D. J. Perreault, "Evaluation of magnetic materials for very high frequency power

applications," in Power Electronics Specialists Conference, 2008. PESC 2008. IEEE, 2008, pp. 4270-4276.

[16] M. H. Lim, J. D. van Wyk, and L. Zhenxian, "Effect of geometry variation of LTCC distributed air-gap filter inductor on light load efficiency of DC-DC converters," in Industry Applications Conference, 2006. 41st IAS Annual Meeting. Conference Record of the 2006 IEEE, 2006, pp. 1884-1890.

[17] A. H. FEINGOLD, M. HEINZ, and R. L. WAHLERS, "Compliant dielectric and magnetic materials for buried components," in international symposium on microelectronics, 2002

Bi-Directional Charging Topologies for Plug-in Hybrid Electric Vehicles

Dylan C. Erb, Omer C. Onar and Alireza Khaligh

Energy Harvesting and Renewable Energies Laboratory (EHREL),
Electric Power and Power Electronics Center, Electrical and Computer Engineering Department,
Illinois Institute of Technology, 3301 S. Dearborn St., Chicago, IL, USA
E-mails: derb2@illinois.edu, oonar@iit.edu, khaligh@ece.iit.edu; URL: www.ece.iit.edu/~khaligh

Abstract—The automotive industry is going through a major restructuring, and automakers are looking for new generations of hybrid vehicles called plug-in hybrid electric vehicles (PHEVs). In the event that PHEVs become more available and the number of PHEVs on the road increases, certain issues will need to be addressed. One vital issue is the method by which these vehicles will be charged and if today's grid can sustain the increased demand due to more PHEVs. Although these vehicles appear to pose a large liability to the grid, if executed properly, they can actually become an even larger asset. The grid can benefit greatly from having reserves that can store or release energy at the appropriate times. Enabling PHEVs to fulfill this niche will require a bi-directional interface between the grid and each vehicle. This bi-directional charger must have the capability to charge a PHEV's battery pack while producing minimal current harmonics and also have the ability to return energy back to the grid in accordance with regulations. This paper will first review some of the power electronic topologies of bi-directional AC-DC and DC-DC converters that fulfill these requirements and then discuss the best choice for combining two topologies to form a bi-directional charger.

Index Terms—Bi-Directional Converters, Plug-in Hybrid Electric Vehicle, and Power Electronic Converters

I. INTRODUCTION

TODAY'S electrical grid has many inefficiencies that are both costly and wasteful. Some of these issues are simply a result of the fluctuations in demand that occur each day in addition to the need for voltage and frequency regulations [1]. When the demand placed on the grid exceeds the capacity of the base-load power plants, peak power plants, and sometimes spinning reserves, must be turned on [2]. During periods of low demand, the electricity usage drops below the output of the base-load power plants, and all the unused energy is wasted. Increased market penetration of PHEVs along with a vehicle-to-grid (V2G) network that enables coordinated charging could significantly reduce problems with electrical demand [3]. In fact, a PHEV market penetration of only 10% could take the place of 25% of the electrical generation capacity in most regions of the United States [4].

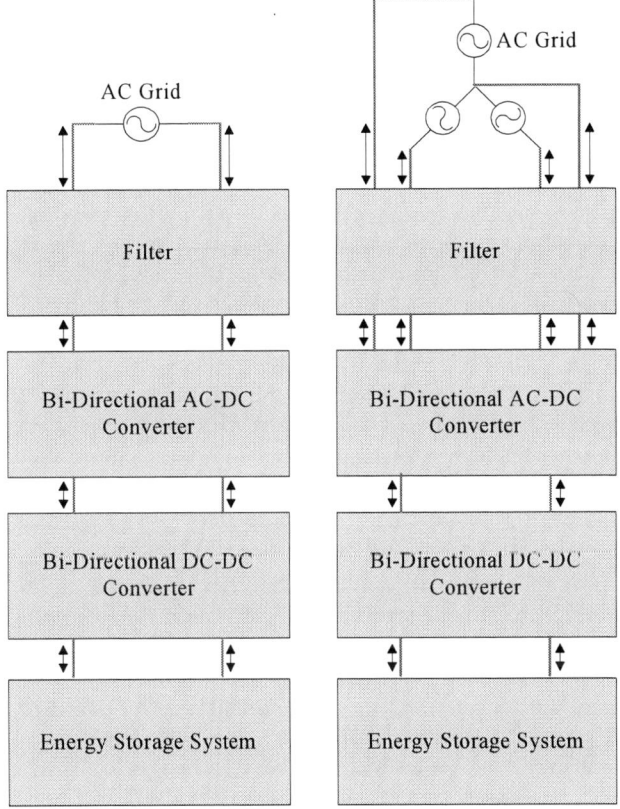

Fig. 1. General bi-directional charger topology for single-phase (left) and three-phase systems (right).

This huge potential for improving the efficiency of the grid by utilizing idle storage capacity in PHEVs can be unlocked through a bi-directional interface. The bi-directional charger will need to function smoothly in both directions. When operating in battery charge mode, it should draw a clean sinusoidal current in phase with the grid to avoid harmful harmonic currents and poor power factor. In battery discharge mode, the charger should return current in a similar sinusoidal form that complies with regulations [5]. Many different electronic circuits can complete this task, but they tend to follow the same general circuit topology shown in Figure 1.

978-1-4244-4782-4/10 $26.00 © 2010 IEEE

When working in the battery charging mode, the AC current first goes through a filter which helps remove unwanted frequency components. Next, the AC current is rectified into DC current as it passes through the bi-directional AC-DC converter. Since this AC-DC converter output voltage might not match the voltage of the DC energy storage, a bi-directional DC-DC converter ensures that the proper charging voltage is supplied to the energy storage unit. From the general viewpoint, the single-phase and three-phase topologies follow the same steps.

If the charger is in battery depletion mode, the process is reversed. The electric current leaves the energy storage unit and is changed back to the proper DC voltage with the bi-directional DC-DC converter. This DC current is then inverted into AC by the bi-directional AC-DC converter. Then, it passes through the filter, which smoothes out the AC current so it is suitable for injection back into the grid.

Many different circuits can accomplish the tasks outlined by the general topology, and there is no clear-cut combination of circuits that outperforms the others. To try and gain insight into creating the best combination of circuits for this task, different circuit topologies for bi-directional AC-DC and bi-directional DC-DC converters are reviewed in this paper. This paper will examine the half-bridge PWM, full-bridge PWM, and three-level PWM bi-directional AC-DC converters. Then, the dual active bridge (DAB) and two quadrant buck-boost bi-directional DC-DC converters will be discussed along with a special case of the bi-directional buck-boost converter that has an integrated high-voltage bus. Finally, the question of which individual topologies fit together best will be addressed.

II. BI-DIRECTIONAL AC-DC CONVERTERS

The bi-directional AC-DC converter must have the capability to turn alternating current into direct current during battery charge mode and convert direct current into alternating current in the battery discharge mode. This section will review three different topologies that accomplish this task with varying complexity and quality. The half-bridge PWM is the simplest of the group and will be reviewed first. With a slightly increased complexity, the full-bridge PWM will come next. Finally the most complex of the group, the three-level PWM, will come last. For each topology, both the single-phase and the three-phase system will be examined.

A. Half-Bridge PWM AC-DC Converter

The single-phase version of the half-bridge PWM converter (Fig. 2) consists of one inductor, one resistor, two capacitors, and two switches that have internal diodes [6]. During battery charge mode, the switches can be left in the open position and act as simple diodes. When the AC source is delivering a positive voltage, the internal diode of T1 turns on and the internal diode of T2 is off. This effectively connects C1 across the AC source when it is positive. When the AC source outputs a negative voltage, the opposite occurs and C2 is effectively connected across the AC source. The resultant DC voltage across the DC-DC converter is more or less equal to

the AC

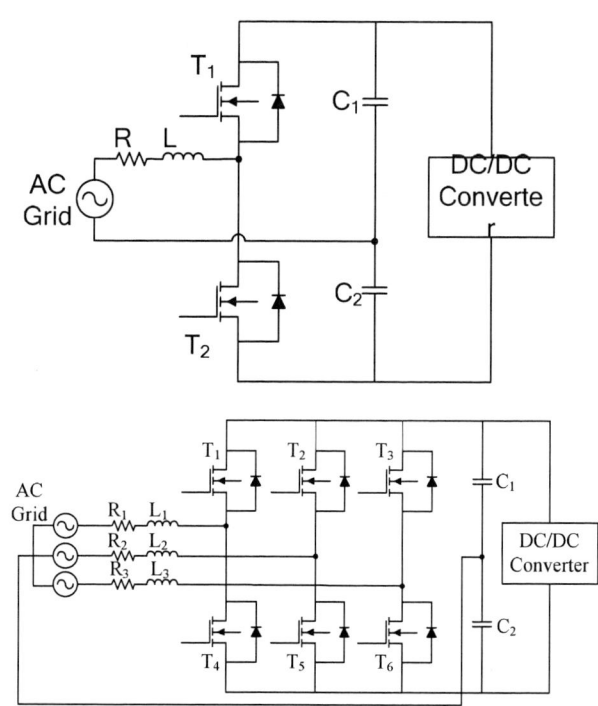

Fig. 2. Half-bridge PWM bi-direction AC-DC converter depicted in single-phase (top) and three-phase (bottom) configurations.

peak-to-peak voltage of the AC source [6]. For active rectification, the transistors can be switched on when the diodes would be forward biased.

On the other hand, when the single phase charger is running in battery discharge mode, the transistors are now switched on and off by PWM controllers. To avoid fault currents, T1 and T2 cannot be switched on at the same time. Assuming that C1 and C2 are identical, T1 and T2 can be switched on and off to deliver either $V_{DC-DC}/2$ or $- V_{DC-DC}/2$ across R1, L1 and the AC source [6]. A PWM controller varies the times that these two voltages are applied so that after being filtered by R1 and L1, the voltage delivered to the AC source is of an acceptable sinusoidal form.

The three-phase, half-bridge converter (Fig. 2) has three inductors, three resistors, two capacitors, and six switches with internal diodes [7]. The principle of operation is essentially the same for the three-phase version as it is for the single-phase. During battery charge mode, each AC source charges C1 when it outputs a positive voltage and it charges C2 when it outputs a negative voltage. The resultant voltage across the DC-DC converter is roughly the same as the single-phase version (peak-to-peak voltage of one source), except that the charging capacitors receive a larger charging current, so faster battery charging is possible. With three times the input current, three times the output current is possible, resulting in three times the maximum theoretical battery charging rate.

978-1-4244-4782-4/10 $26.00 © 2010 IEEE

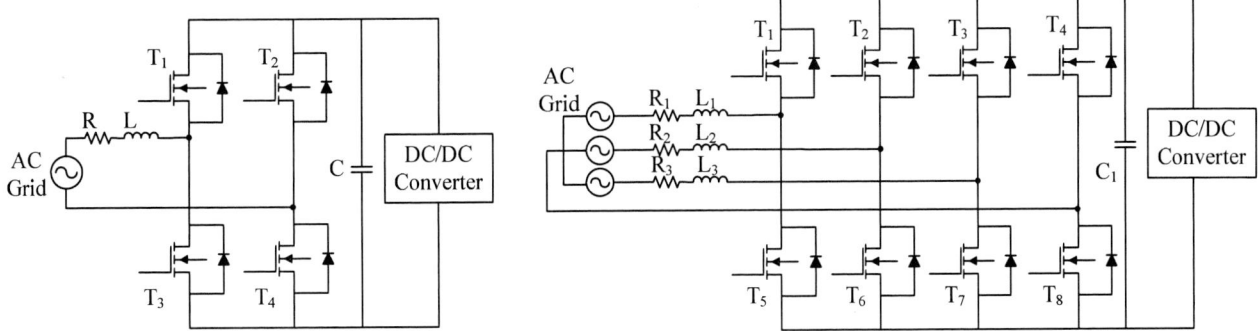

Fig. 3. Full-bridge PWM bi-directional AC-DC converter depicted in single-phase (left) and three-phase (right).

When the three-phase charger is operating in battery discharge mode, the bi-directional AC-DC converter acts as three independent single-phase circuits working simultaneously. T1 and T4, T2 and T5, and T3 and T6, each work together to deliver either VDC-DC/2 or -VDC-DC/2 across each respective set of one inductor, one resistor, and one AC source [6]. Just like the single-phase version, a PWM controls each transistor so the right combination of the two voltages is applied across the inductor, the resistor, and the AC source to result in a clean sinusoidal output voltage across the AC source.

One large benefit of the half-bridge topology is the simplicity of the design. This translates to fewer components, which in turn means lower cost of manufacturing. However, this topology exhibits high component stresses, which could require transistors to be connected in parallel and/or series to handle the high current and/or voltage [6].

Unless properly filtered, the half-bridge converter produces many harmonic currents which could be detrimental to the grid. A conventional passive filter suited to remove the harmonic currents would require bulky components. Although the low number of components makes the cost look low on paper, implementing this circuit at high power levels could require additional costs [6].

B. Full-Bridge PWM AC-DC Converter

The single-phase version of the full-bridge PWM AC-DC converter (Fig. 3) consists of one inductor, one resistor, one capacitor, and four switches with internal diodes [6]. Just like the half-bridge circuit, the switches remain open in the battery charging mode and the internal diodes rectify the current passively. When the AC source outputs a positive voltage, C1 is effectively connected in parallel with the AC source through the internal diodes of T1 and T4. During the times when the AC source outputs a negative voltage, the internal diodes of T2 and T3 conduct to allow C1 to once again be hooked in parallel with the AC source, but in a reversed configuration. In this way, the DC-DC converter acquires a DC input voltage more or less equal to the amplitudes of the AC source. For active rectification, the transistors can be switched on when the diodes would be forward biased.

When the single-phase, full-bridge circuit acts in battery discharge mode, T1 and T4 or T2 and T3 can be turned on to apply VDC-DC or - VDC-DC respectively across the AC source. Of course, switching T1 and T2 or T3 and T4 on at the same time would produce 0 volts. Through PWM control of each transistor, the proper switching pattern can be implemented. When this combination of VDC-DC, -VDC-DC, and 0 volt pulses is applied, the inductor and resistor smooth out the signal to resemble the desired AC sine wave [6].

The three-phase, full-bridge converter circuit (Fig. 3) has three inductors, three-resistors, one capacitor, and eight switches that each have internal diodes [7]. Once again, the three phase version of the circuit functions almost exactly like three separate single phase circuits that share the same capacitor and DC-DC converter. Having three-phases instead of one results in no change in the supplied voltage to the DC-DC converter under no load, but the three- phase circuit can support a much larger load. In other words, the maximum theoretical rate of charging is increased.

When operating in battery discharge mode, the three-phase version of the converter behaves like three independent, single-phase circuits that share the DC source. T1 and T5, T2 and T6, and T3 and T7 each work in combination with T4 and T8 to supply VDC-DC, - VDC-DC, or 0 volts across each respective inductor, resistor, and AC source. Through controlling each switch with the appropriate PWM signal, the right combination of pulses can be applied to each inductor, resistor, and AC source to deliver the desired sinusoidal voltage output to the AC source [6].

As compared to the half-bridge PWM converter, the full-bridge has one less capacitor and two more transistors with internal diodes. This suggests that the full-bridge would have a higher cost, since it has more components; however, it is important to keep in mind that the component stresses are lower in the full-bridge circuit. This means that despite having more switches, the cost of each switch is reduced. The cost of the power electronics could be somewhat similar, but the full-bridge will require twice as many PWM inputs. This will add to the complexity and cost of control circuitry.

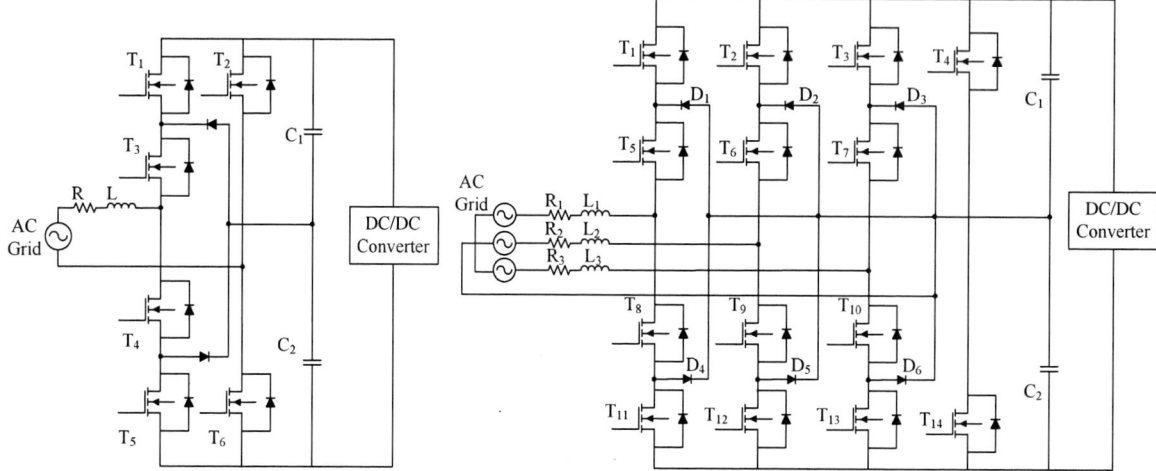

Fig. 4. Three-level PWM bi-directional AC-DC converter depicted in single-phase (left) and three-phase (right).

Similar to the half-bridge converter, the full-bridge rectifier can produce harmonic currents if not properly filtered. Also, power factor issues may need to be addressed [8]. It is interesting to note that the half-bridge circuit outputs a voltage of $\pm$ $V_{DC-DC}/2$, while the full-bridge can output a voltage of V_{DC-DC}. This could play a significant role in the choice of DC-DC converter, since the V_{DC-DC} must be twice as high with the half-bridge. In other words, for a given input voltage into the DC-DC converter, the output voltage must be twice as high when using the half-bridge AC-DC converter.

C. Three-Level PWM AC-DC Converter

The final and most complicated of the three bi-directional AC-DC converters presented here is the three-level PWM converter. It essentially combines the functionality of the half-bridge and the full-bridge PWM at the cost of added complexity. The single-phase three-level PWM (Fig. 4) circuit operates just like the full-bridge rectifier in battery charge mode [6]. When the AC source outputs a positive voltage, the internal diodes of T1, T3, and T6 turn on. When the AC source outputs a negative voltage, the internal diodes of T2, T4 and T5 conduct. During active rectification, the switches can be turned on in such a pattern to apply multiple voltage levels (as detailed in the next paragraph). Both passive and active methods charge the capacitors of C1 and C2 each up to half the amplitude of the AC source, resulting in the application of the amplitude voltage of the AC source across the DC-DC converter.

When T1, T3 and T6 or T2, T4, and T5 are switched on, a voltage of V_{DC-DC} or $-V_{DC-DC}$ can be applied to the load respectively. When T3 and T6 or T2 and T4 are activated, a voltage of $V_{DC-DC}/2$ or $-V_{DC-DC}/2$ can be applied to the load respectively. Like the full-bridge, 0 volts can also be delivered to the load [6].

Like the other two types of converters, the three-phase, three-level converter (Fig. 4) behaves in a similar fashion to its single-phase counterpart [7]. Each phase applies a voltage

equivalent to the amplitude voltage of the AC source across the DC-DC converter through the rectifier. Again, the three-phases allow for support of a much greater load.

The three-phase topology of the three-level PWM converter requires that each AC source receives an independent control signal. But, each of the three signals is delivered through the same means as the single-phase circuit. Each column of four transistors in the three-phase version acts like the column of four transistors in the single-phase circuit. The column of two transistors and two capacitors in the three-phase version works such as the right half of the single-phase circuit for all the phases in order to cater to the particular AC source that it is connected to. Each set of one resistor, one inductor, and one AC source receives a combination of $\pm V_{DC-DC}/2$, $\pm V_{DC-DC}$, and 0 volts as dictated by the PWMs controlling the transistors [6]. The inductors and resistors filter out the unwanted noise and sinusoidal voltage waveforms are delivered to the AC sources.

One clear concern with the three-level topology is the added complexity and additional number of components. These two issues add to the cost and magnitude of the required control circuitry. There are, however, many benefits gained with this design. The voltage waveform is much improved over the other designs, which allows for a much smaller and less expensive filter. The component stresses are lower, so smaller and cheaper components can be used for a given power level. Additionally, the electromagnetic interference and acoustic noise is reduced [7].

III. BI-DIRECTIONAL DC-DC CONVERTERS

The job of the bi-directional DC-DC converter is to couple the AC-DC converter to the energy storage unit. It must have the capability to convert the DC output voltage of the AC-DC converter into a suitable voltage to charge the batteries and vice-versa. This section will discuss three different topologies that accomplish this task. The dual active bridge, two quadrant (buck-boost), and a variation on the buck-boost converter will

978-1-4244-4782-4/10 $26.00 © 2010 IEEE

all be discussed. This last converter is not bi-directional, so the changes required to enable this feature will be presented in a modified version. After the presentation of each topology, the possible benefits and drawbacks of using each will be reviewed.

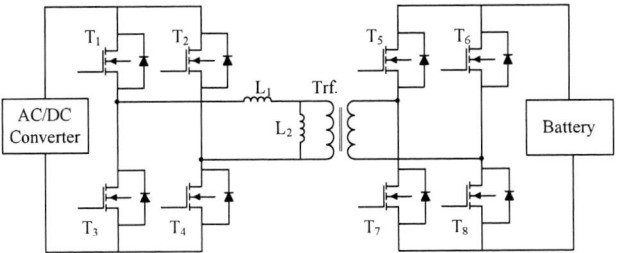

Fig. 5. Dual Active Bridge DC-DC Converter.

A. Dual Active Bridge DC-DC Converter

The dual active bridge DC-DC converter (Fig. 5) consists of two active bridges linked by a transformer. The version shown in the figure uses two full-bridges and requires two inductors, one transformer, and eight switches with internal diodes. When delivering energy from the AC-DC converter to the battery, the active bridge on the left side of Figure 5 acts as an inverter while the internal diodes of the switches on the right act as a full-bridge to rectify the AC back to DC. This topology also allows the rectifier to be switched actively. The transformer is fixed to some ratio that is suitable for the application.

When the converter runs in battery discharge mode, the bridge on the right inverts the DC current of the battery which induces an AC voltage in the left active bridge through the transformer. The internal diodes of the switches on the left hand side of Figure 5 act as a full-bridge to rectify the current back to DC that is usable by the vi-directional AC-DC converter. Once again, the rectifier switches can be actively switched if needed. In other words, this circuit converts DC to AC, and then rectifies AC back to DC. If the DC is inverted with a very high frequency, a fairly small transformer can be used.

It is important to note that because of the high number of switches contained in this circuit, ZVS and ZCS techniques might be implemented to reduce switching losses. The large number of components can also add to cost. While this circuit provides a high power density and fast control, device stresses become high when the voltage range exceeds 2:1 [9].

B. Two Quadrant DC-DC (Buck-Boost) Converter

As shown in Figure 6, the two quadrant DC-DC converter has two inductors, one capacitor, and two switches with internal diodes [10]. This converter works as a boost converter in one direction and a buck converter in the other. When the charger operates in battery charge mode, the converter must function as a buck converter. To activate this mode, T2 is left open as T1 is switched on and off with a PWM signal. The reduction ratio of the voltage depends on the duty cycle of the signal applied to T1. If the duty cycle is 100%, then the full

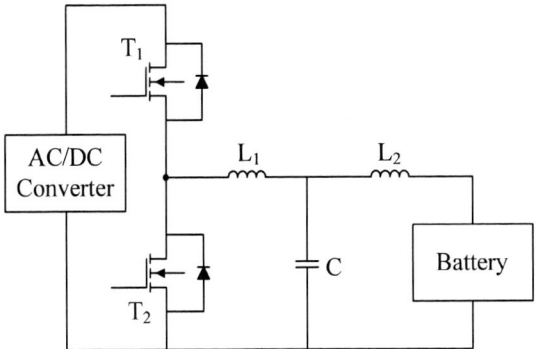

Fig. 6. Two Quadrant (Buck-Boost) DC-DC Converter

DC voltage from the AC-DC converter is applied to B1. If the duty cycle is 50%, then half the voltage is supplied, and so forth.

When the circuit is operating in battery discharge mode, T1 remains off while T2 is switched on and off with a PWM signal. In this mode, the circuit functions as a boost converter. The battery voltage is boosted by a factor of 1/(1-D), where D is the duty cycle of the signal applied to T2. A duty cycle of 50% doubles the voltage, and so on.

Compared to the dual active bridge, this circuit has 5 fewer components. In addition, only two switches must be controlled, which greatly simplifies the control circuitry. There are, however, two high current inductors that tend to be bulky and expensive. Also, this topology only has the capability to operate as buck in one direction and boost in the other, which would work for most applications, but situations might arise where the functionality of a buck and boost converter is required in both directions.

C. Integrated Buck-Boost DC-DC Converter

The final bi-directional DC-DC converter discussed here is the integrated buck-boost converter (Fig. 7) [11]. This original topology has a number of added capabilities integrated into the design that traditional DC-DC converters leave out, however, it is unable to deliver energy from the battery back to the AC-DC converter. Because this functionality is necessary for a bi-directional charger, a modified circuit is presented in Figure 8.

The modified circuit allows for plug-in charging of B1 through T1 and T5, boost DC-DC from B1 to the high voltage bus through T2, T4, and T5, regenerative charging from the high voltage bus to B1 through T3 and T6, and boost DC-DC from B1 to the AC-DC converter through T2, T5, and T7 [11]. Each of these integrated converters use the same inductor and act much like the two quadrant DC-DC (Buck-Boost) converter. This cuts down on the total number of high current inductors required for the circuit.

The added functionality from integrating the high voltage bus into the charger circuit does add some complexity to the circuit, but it eliminates the need for another DC-DC converter. The total number of high current inductors is

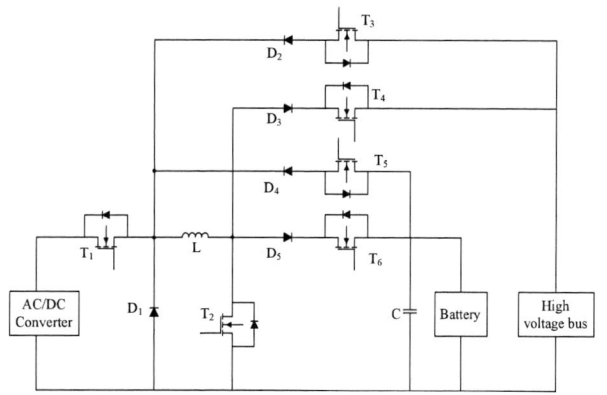

Fig. 7. Integrated Buck-Boost DC-DC Converter

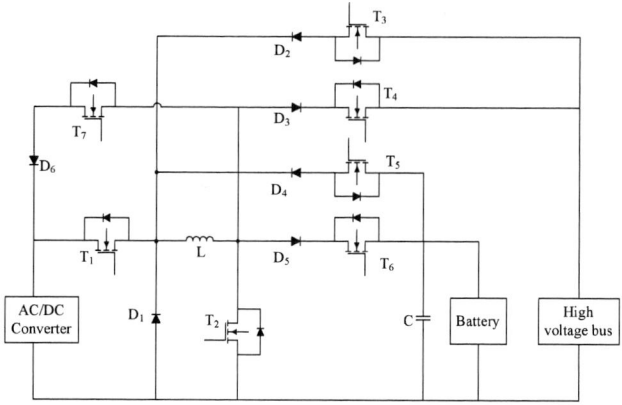

Fig. 8. Modified Integrated Buck-Boost DC-DC Converter

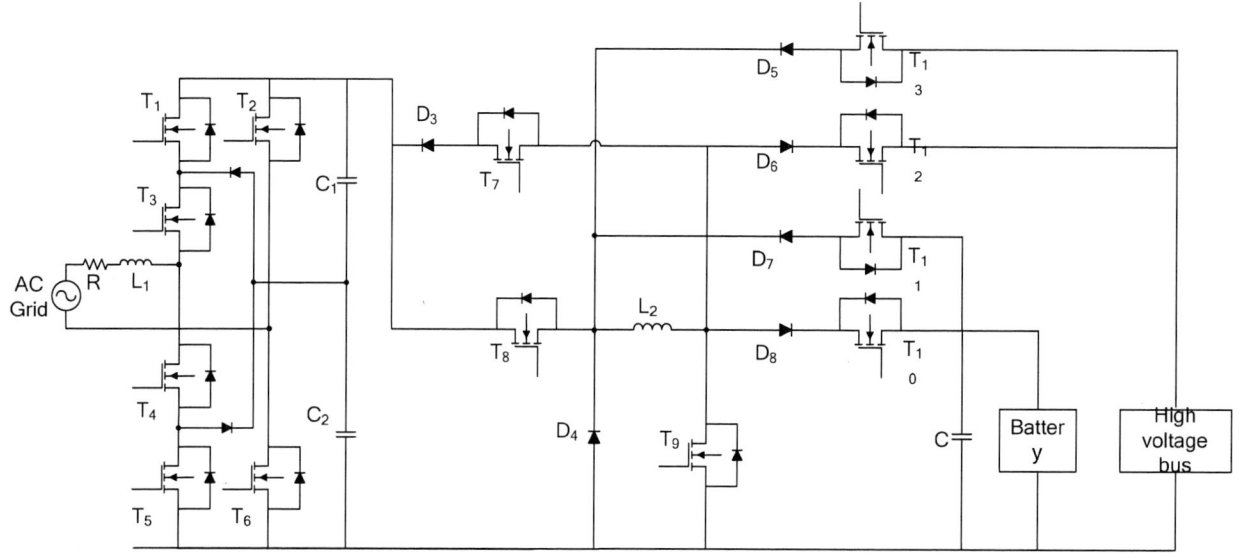

Fig. 9. Combined Three-Level AC-DC Converter and Modified Bi-Directional Buck-Boost DC-DC Converter

reduced at the cost of additional switches. These extra switches result in higher conduction losses, but these can be reduced through removing the diodes paired to the switches [11].

IV. BI-DIRECTIONAL COMBINED CHARGER

In order to implement the bi-directional charger, a bi-directional AC-DC topology must be combined with a bi-directional DC-DC topology. Ideally this bi-directional charger will have the capability to charge a PHEV's battery pack while producing a minimum of current harmonics, and also have the ability to return energy back to the grid in accordance with regulations.

Since the three-level AC-DC converter has an unfiltered voltage waveform that is much improved over the other designs and allows for a smaller and less expensive filter, it has been chosen to fill the AC-DC portion of the charger. In this topology, the component stresses are lower, so smaller and cheaper components can be used for a given power level [6]. It is more complicated than the other AC-DC topologies,

but the benefits are well worth the complexity for the high power application of PHEVs.

For the bi-directional DC-DC converter, the modified buck-boost design has been chosen because of the reduced number of high-current inductors and the integrated high voltage bus. Due to the built-in high voltage bus, this DC-DC topology is well suited for PHEV conversions, which could be a major market for bi-directional chargers. Combining these two topologies forms a fully functional bi-directional charger that has the capability to allow for charging of a plug-in vehicle's battery pack as well as providing an interface for V2G interactions. The combined charger is shown in Figure 9.

V. CONCLUSIONS

If more manufacturers begin to produce PHEVs and consumers purchase these PHEVs, the benefits of enabling V2G interactions and the side-effects of proceeding with uncoordinated charging will both continue to grow. With the implementation of V2G technology, the added demand from PHEVs can become a large resource instead of a costly burden for the grid.

If the bi-directional interfaces do not function well, the grid could suffer from the consequences of harmonic currents and poor power factor. In order to make V2G technology feasible, an efficient and well designed bi-directional charger that minimizes current harmonics will be required. If done properly, large scale coordinated charging could vastly improve grid efficiency and cut costs for utilities and ultimately for consumers.

This paper started with an explanation of the general topology of a bi-directional charger and then discussed the parts individually. Three bi-directional AC-DC and three bi-directional DC-DC topologies were discussed and compared. Finally, a complete charger topology was proposed that combines two of the discussed topologies. The proposed charger combines a three-level bi-directional AC-DC rectifier with a modified bi-directional DC-DC converter. In theory, this charger should exhibit many of the desired traits for a bi-directional PHEV charger.

ACKNOWLEDGMENT

This work has been supported in part by the U.S. National Science Foundation under Grant number 0852013, which is greatly acknowledged.

REFERENCES

[1] K. Clement, E. Haesen, and J. Driesen, "Coordinated charging of multiple plug-in hybrid electric vehicles in residential distribution grids," in *Proc. IEEE Power Systems Conference and Exposition,* pp.1-7, Mar. 2009.

[2] P. Kadurek, C. Ioakimidis, and P. Ferrao., "Electric Vehicles and their impact to the electric grid in isolated systems," in *Proc. Power Engineering International Conference on Energy and Electrical Drives,* pp.49-54, Mar. 2009.

[3] C. Guille and G. Gross, "Design of a Conceptual Framework for the V2G Implementation," in *Proc. Energy 2030 Conference,* pp.1-3, Nov. 2008.

[4] S. D. Jenkins, J. R. Rossmaier, and M. Ferdowsi, "Utilization and effect of plug-in hybrid electric vehicles in the United States power grid" in *Proc. IEEE Vehicle Power and Propulsion Conference,* pp. 1-5. Sept. 2008.

[5] B. Kramer, S. Chakraborty, and B. Kroposki, "A review of plug-in vehicles and vehicle-to-grid capability," in *Proc. 34th Annual Conference of Industrial Electronics,* pp.2278-2283, Nov. 2008.

[6] Bor-Ren Lin, Der-Jan Chen, Hui-Ru Tsay, "Bi-directional AC/DC converter based on neutral point clamped," in *Proc. IEEE International Symposium on Industrial Electronics,* vol.1, pp.619-624, 2001.

[7] Shijie Li and Yaohua Li, "Study and design considerations of three-phase bi-directional AC/DC converter," in *Proc. 2004 IEEE International Conference on Industrial Technology,* vol.1, pp. 400-407 Dec. 2004.

[8] S.Y.R. Hui, H. Chung, and S.C. Yip, "A bi-directional AC-DC power converter with power factor correction," in *Proc. 29th Annual IEEE Power Electronics Specialists Conference,* vol.2, pp.1323-1329, May 1998.

[9] Sangtaek Han and D. Divan, "Bi-directional DC/DC converters for plug-in hybrid electric vehicle (PHEV) applications," in *Proc. Twenty-Third Annual IEEE Applied Power Electronics Conference and Exposition,* pp.784-789, Feb. 2008.

[10] M. Bojrup, P. Karlsson, M. Alakula, and B. Simonsson, "A dual purpose battery charger for electric vehicles," in *Proc. 29th Annual IEEE Power Electronics Specialists Conference,* vol.1, pp.565-570, May 1998.

[11] Y. Lee, A. Khaligh, and A. Emadi, "Advanced Integrated Bi-Directional AC/DC and DC/DC Converter for Plug-in Hybrid Electric Vehicles," *IEEE Transactions on Vehicular Technology,* vol. 58, Oct. 2009, pp. 3970-3980.

Multi-channel Three-port DC/DC Converters as Maximum Power Tracker, Battery Charger and Bus Regulator

Zhijun Qian
Univ. of Central Florida
Orlando, FL, USA
zqian@knights.ucf.edu

Osama Abdel-Rahman
ApECOR
Orlando, FL, USA
sabdel@apecor.com

Haibing Hu
Univ. of Central Florida
Orlando, FL, USA
huhaibing@nuaa.edu.cn

Issa Batarseh
Univ. of Central Florida
Orlando, FL, USA
batarseh@mail.ucf.edu

Abstract - **This paper presents the design and analysis of a satellite platform power system, which utilizes several three-port converters to interface various independent solar panels, one battery to a regulated bus. The three-port converter features low component count and compact structure. The paralleled three-port converters can not only boost the power level, but also provide redundancy, which is important to such critical applications. The proposed output port "hybrid" current sharing (CS) method shows good transients while requiring no inter-channel level CS bus, meanwhile, the proposed passive CS method is well suited to both the MPPT algorithm for the input port and the battery charge control for the battery port. The analysis and design of the system level control strategy, the CS methods for three different ports, the MPPT and battery charging algorithms are verified by simulation and experiments.**

I. INTRODUCTION

Multi-port converters [1]-[11] can interface and control multiple energy sources and storage devices with one integrated topology. Therefore, it has the potential to replace the power system having loose structure of multiple two-port converters to save space and reduce weight. But compared with the efforts spent on the design and optimization of traditional two-port converters, less has been put on multi-port converters. Besides, most reports about multi-ports focused on topology and power stage design aspects, but lack of investigation on the challenging control design aspects which include so many interacting feedback loops due to the power stage integration issue. Therefore, the proper decoupling has to be made to prevent control loop interaction and furthermore to facilitate separate controller design.

This paper is a continuing research based on [1], [2] and [3], which clearly presented the power topology design considerations, control loop design for single three-port converter and current sharing design for two paralleled converters. This paper will be focused on the system level control strategy to achieve the power management control of multiple three-port converter channels, in addition to the current sharing.

The integrated three-port topology interfaces one solar panel input port, one bi-directional battery port and an isolated output port, which is shown in Fig 1. Two converters are connected at each port to form one channel. These independent channels have different solar sources, while the

battery ports and output ports are all connected together to interface with one battery pack and one distribution bus which delivers power to the satellite user power system. The distributed photovoltaic (PV) panel structure allows maximum solar power harvesting for each PV panel, resulting in maximal solar power input for the whole system.

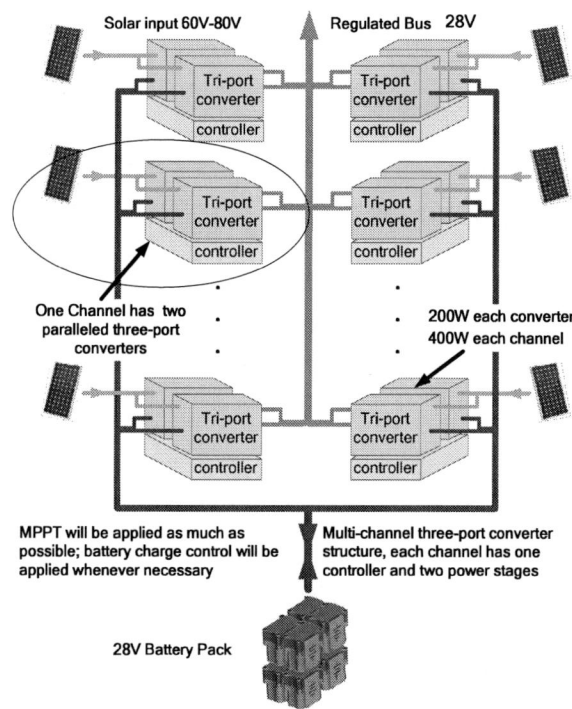

Figure 1. *Multi-channel converter structure*

Current Sharing control (CS) is necessary to equally distribute current or power at both intra-channel and inter-channel level. However, traditional one port CS control is not enough for paralleled three-port converters because unequal current distribution in the other two ports may occur, and sometimes the large current stress may break the circuit. Meanwhile, the added CS control should preserve the attractive features like MPPT and battery charge control while still maintaining system stability. There are basically two categories of CS method, active method [12]-[16] and passive droop method [17], [18]. Active CS has better transients but it requires at least one or two shared bus, meaning that the extra

978-1-4244-4782-4/10 $26.00 © 2010 IEEE

wiring connection is required to realize the real time communication among different channels, which is difficult to implement and susceptible to noise. On the other hand, the droop CS method requires no CS bus structure, and the droop rate can be programmed conveniently within the DSP instead of inserting real resistors which will dissipate power. The droop method can also achieve good CS performance given that the voltage tolerance is not tight. Fig 2 gives the different results with different CS approaches. In order to take advantage of both the active and passive CS method, a hybrid CS method is proposed for output port CS control strategy. "Hybrid" means active CS at intra-channel level and droop CS at inter-channel level. Therefore, compared with the droop method, the hybrid method has better transients and will have lower current limit setting to allow better circuit protections due to the inherent active CS structure (Fig 2(b)). For the input port and the battery port, the proposed CS functions are well suited to the existing MPPT or battery charge control.

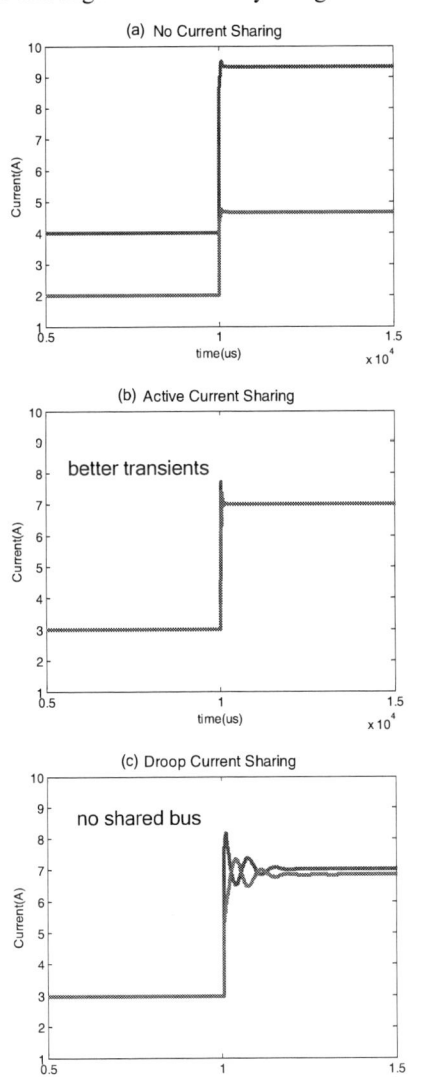

Figure 2. *Current sharing results with different approaches*

II. SINGLE CONVERTER CONTROL STRUCTURE, MODE TRANSITION AND DECOUPLING METHOD

Fig 3 shows a modified PWM half bridge topology which includes three basic circuit stages within a constant-frequency switching cycle to provide two independent control variables, namely duty-cycles *d1* and *d2*, respectively. This allows tight control over two ports, while the third port provides the power balance for the system. The detailed circuit operation was reported in [1]. The multi-objective control architecture includes various control loops, such as input voltage regulator (IVR), output voltage regulator (OVR), battery voltage regulator (BVR), etc. The OVR loop is simply a voltage loop, and duty cycle *d1* is used as its control input. As a result, *d2* is made to control either input port or battery port. The IVR loop is to regulate the solar panel voltage to the optimal operating voltage provided by an MPPT controller, this operational mode is defined as the battery-balanced mode (Mode 1). Otherwise, *d2* can be decided by the battery control loop which is defined as the battery-regulation mode (Mode 2), which has two paralleled controllers, BVR and BCR. Therefore, whether *d2* is commanded by IVR, BVR or BCR depends on which mode it is in and thus requires the mode transition control.

The autonomous mode transition is proposed in a competitive manner to allow smooth mode transitions. It saves precious DSP computation time which otherwise needs to detect and calculate complicated mathematics to determine the optimum operating mode for the converter. With this proposed mode transition control, BVR, BCR and IVR run in parallel to compete for the minimum value to win control over *d2*. It should be noted that if IVR loop loses control, MPPT function needs to be disabled accordingly because of MPPT algorithm's inherent noise issue. Furthermore, the proposed autonomous mode transition is valid not only for single channel but also for multi-channel operation when the same mode is enforced within one channel.

Decoupled transfer functions for different ports need to be obtained before different power ports' control loops can be analyzed separately. Since the detailed converter model was given in [2], the decoupling network is then introduced to obtain:

$$\begin{cases} v_o(s)/d_1(s) = g_{11} - g_{12} \cdot g_{21}/g_{22} \\ v_{in}(s)/d_2(s) = g_{22} - g_{12} \cdot g_{21}/g_{11} \end{cases} \tag{1}$$

Where these g_{xy} values denoting converter models has been derived. With these two transfer functions, output and input port control loops can be designed separately. Also, since input PV dynamics are relatively less stringent than output port which connects with fast transient load, the input port control loop is designed to have one tenth bandwidth of that of output loop to ensure minimum loop interactions. The decoupling of these control loops potentially provide great convenience to add different CS controllers to each port separately, while allowing equal current distribution for multiple paralleled converters.

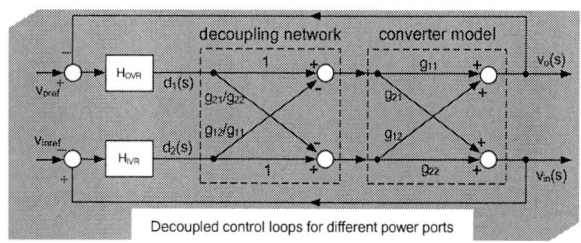

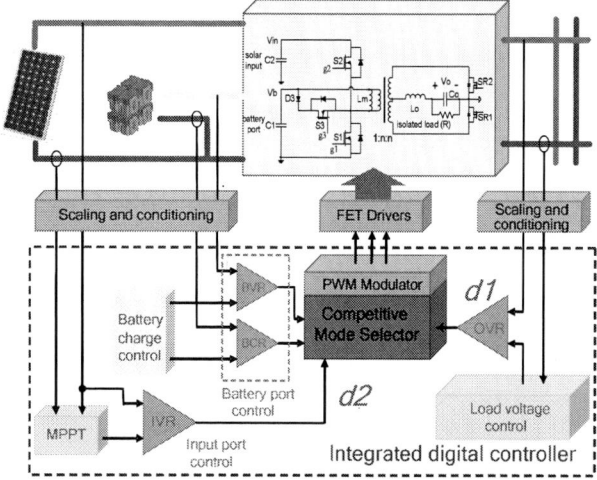

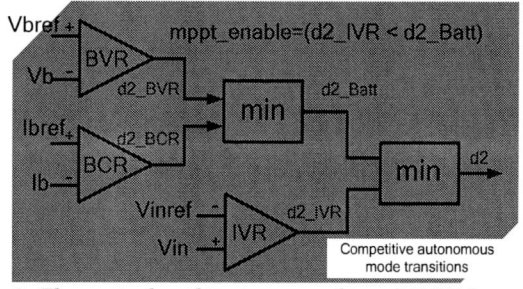

Figure 3. The control architecture to achieve MPPT for solar port, battery charge control for battery port and meanwhile always maintaining bus regulation for output port. The topology, autonomous mode transitions and decoupled transfer functions for input and output port are highlighted in green, purple and red, respectively.

III. INTRA-CHANNEL AND INTER-CHANNEL CURRENT SHARING FOR DIFFERENT PORTS

A. Output Port Hybrid Current Sharing

Unlike the conventional inner CS loop outer voltage loop or the inner voltage loop outer CS loop control structure, voltage loop and CS loop are constructed in parallel as shown in Fig 4. The major benefit is that the voltage loop and the CS loop are not heavily coupled together, which has been proved in [3]. The decoupling keeps the three-port control loops design simple. Otherwise, with inner and outer CS structure, those control loops are cross-coupled with each other. Moreover, outer voltage/CS loop's bandwidth is always limited by that of inner CS/voltage loop. Therefore one other benefit is that CS loop can be designed with relatively high bandwidth to gain

good transients with this parallel structure. Besides, the CS loop reference is democratically determined to have fault-tolerance feature. Then certain droop voltage is subtracted from the voltage reference to enable CS among different channels. With this hybrid CS structure, CS bus is not needed among different controllers, while the bus is required within one channel but it can be conveniently implemented within one DSP unit.

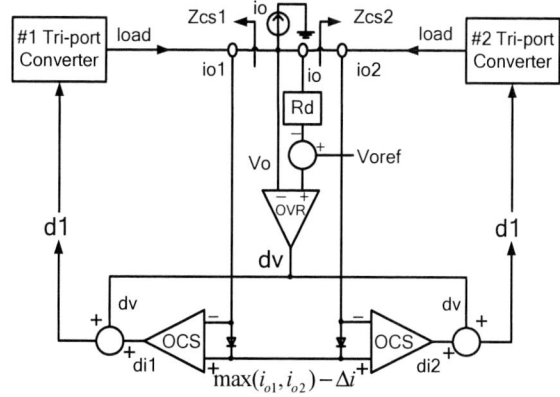

Figure 4. Output port hybrid CS structure

An effective method to judge CS performance is through the thevenin equivalent impedance, thus output impedance is to judge output CS performance. Its DC value determines steady state CS error, and the dynamic CS performance during transients is determined by its impedance value over the interested frequency range. To illustrate the impedance analysis approach, the power stage is represented by a thevenin equivalent circuit in series with an impedance as shown in Fig 5. We can understand Z_{OL}, Z_{OC}, Z_{CS} as follows: Z_{OL} represents the open loop impedance without considering OVR and CS loop; Z_{OC} represents the close loop impedance with OVR closed but CS loop left open; Z_{CS} represents the modified impedance with both OVR and CS closed. In other words, we can simply treat the converter as a black box only represented by this small signal impedance Z_{CS}. And Z_{CS} for each module can be derived as follows:

$$\begin{cases} Z_{CS1} = -\dfrac{v_o}{i_{o1}} = \dfrac{Z_{OL1}}{1 + H_{OVR} \cdot G_{vd1}} & (2) \\ Z_{CS2} = -\dfrac{v_o}{i_{o2}} = \dfrac{Z_{OL1} \cdot Z_{OL2} + Z_{OL1} \cdot H_{CS} \cdot G_{vd2}}{Z_{OL1} + (1 + H_{OVR} \cdot G_{vd1}) \cdot H_{CS} \cdot G_{vd2} + H_{OVR} \cdot G_{vd2} \cdot Z_{OL1}} \end{cases}$$

The expression for the admittance Y_{cs2} which is the reciprocal of $Zcs2$ can be further simplified as follows:

$$Y_{CS2} = (Y_{OC2} + Y_{OC1} \cdot T_{cs}) / (1 + T_{cs}) \qquad (3)$$

Where $T_{CS} = H_{CS} \cdot G_{vd2} / Z_{OL2}$ is defined as the CS loop gain.

Examination of this equation indicates that in the case of identical modules *(Yoc1=Yoc2)*, the term *(1+Tcs)* will cancel out by itself, meaning that CS will be achieved naturally for such identical modules. But unfortunately modules are not identical in reality, due to their variations of component parameters. So the objective is applying the CS control to modify terminal impedances to make them equal to each other in the desired frequency range. For instance, if the crossover frequency of CS loop gain Tcs is designed to be high, the impedance of different converter terminals will be altered to match each other up to that crossover frequency point as shown in Fig 6. As a result, CS transients of the hybrid CS will be improved compared with the conventional droop method as shown in Fig 7.

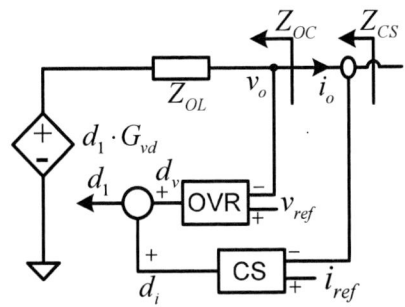

Figure 5. Thevenin equivalent circuit

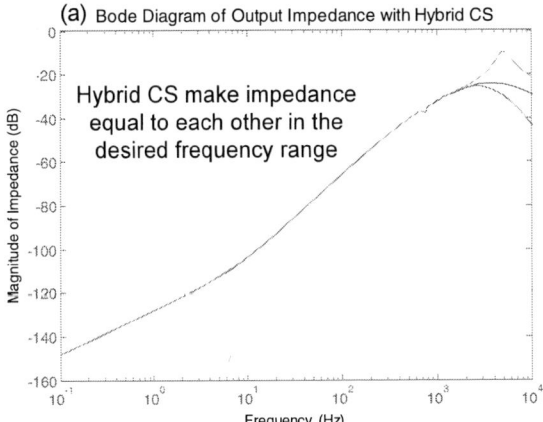

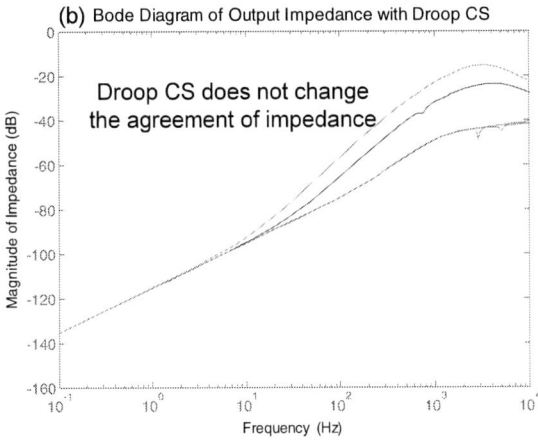

Figure 6. Output impedance with (a) the hybrid CS method and (b) the droop CS method.

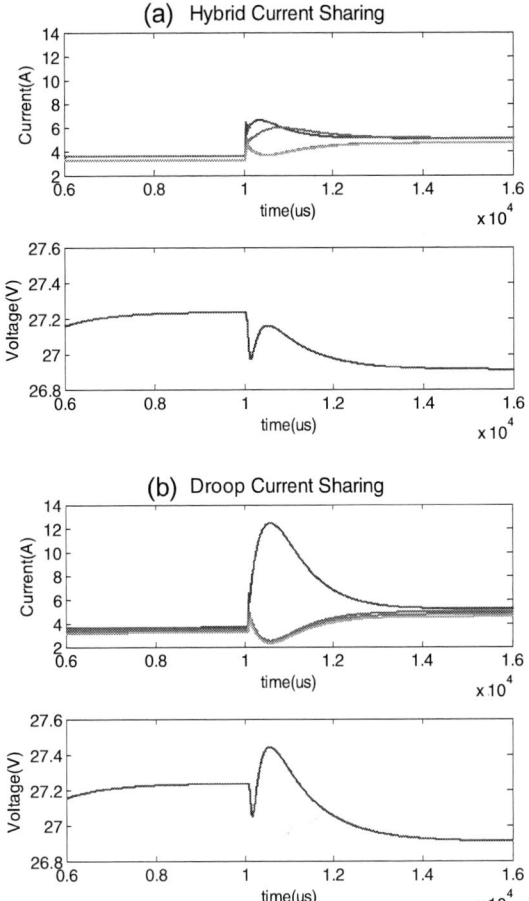

Figure 7. CS simulation results, (a) hybrid CS method; (b) droop CS method.

B. Input Port CS incorporated into MPPT algorithm

With available input voltage and current information of the two converters within one channel, it is possible to incorporate CS function into MPPT algorithm. It should be noted that only intra-channel level CS is required for input port since different channels have different PV sources. The Perturb and Observe MPPT method is used as shown in Fig 8. After input voltage reference $V_{ref(k+1)}$ is obtained, it is added to the product of coefficient K and the individual current measurement to derive different voltage references for each converter. For instance, if input current I_1 is larger than I_2, V_{ref1} will be greater than V_{ref2}. Since a higher input voltage will bring down its input current, simple current sharing without closed loop control is achieved. Most importantly, this CS is compatible with MPPT algorithm. But it should also be noted that the speed of CS control is dependent on MPPT controller speed. Due to the slow characteristics of PV, in most conditions, its current sharing control and MPPT control do not need to be very fast, which justifies the feasibility of this method for most applications.

978-1-4244-4782-4/10 $26.00 © 2010 IEEE   2076

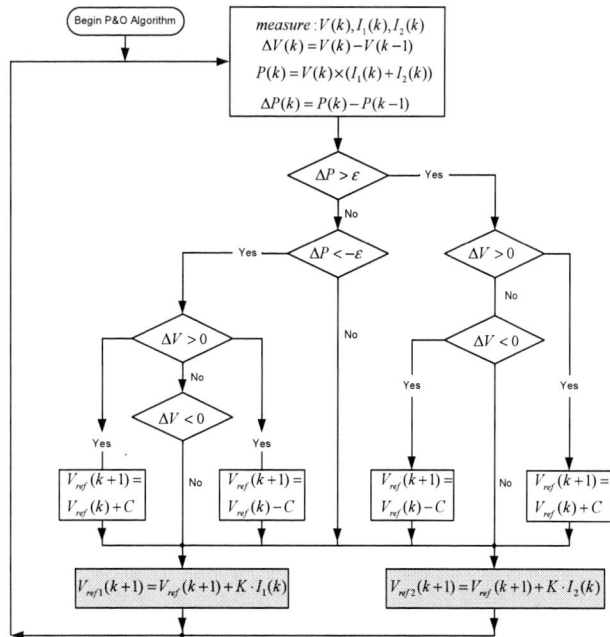

Figure 8. Perturb & Observe MPPT algorithm with CS

C. Battery Port CS incorporated into battery charging algorithm

As illustrated in Fig 9, the battery port CS function is incorporated into the battery charging algorithm. There are two battery charging steps. In the first step, the battery is charged by maximum available solar power which is the deficit of input and output power, so it can be taken as a constant power charging period, in which battery voltage rises gradually. When the battery upper voltage setting V_{bmax} is met, the converter will switch to regulate its voltage to prevent over-charging, which is the second charging step. With the proposed CS function incorporated, the converter will switch to battery voltage regulation at the point of V_{bmax} minus $I_b \cdot R_{droop}$, someplace earlier than V_{bmax}. So the converter achieves CS at the cost of being charged a little bit slower than the regular method. But CS function is critical to such three-port converters; tradeoff has to be made between the CS performance and the battery charging speed.

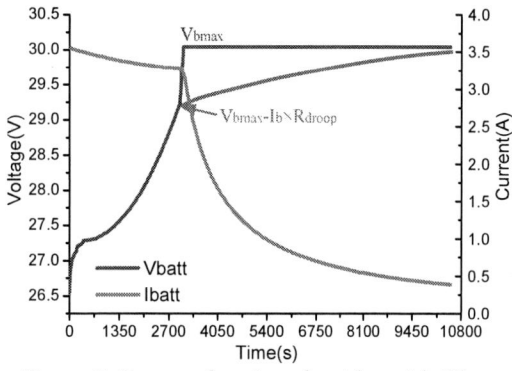

Figure 9. Battery charging algorithm with CS

D. Synchronization among different channels

Since each channel has its own DSP controller and each DSP has its own timing circuitry. When multiple channels are paralleled, the timing circuitry drifting can be observed. Furthermore, the switching frequency at the level of 100 kHz is drifting. Due to small impedance among different channels' ports, even small voltage ripple can cause large current ripple, which is shown in Fig 11(a). Therefore synchronization is necessary and the "wireless" solution would be preferred due to the noise issue of wiring and the freedom to place the converter channels at different locations closer to users. The block diagram implementation method is illustrated in Fig 10. The output voltage has the ripple actually including the switching frequency and exact switching point information. By processing this signal, the DSP synchronization could be achieved. The original output voltage signal V0 will be first filtered to obtain an average value of V1, deducted by V0 and then amplified by coefficient K to be V2, finally comparing with some preset value to generate a square waveform which feeds back into DSP to trigger the PWM counter. As a result, every DSP can be synchronized by the same signal, which is output voltage. In our design, the falling edge is used to trigger PWM counters.

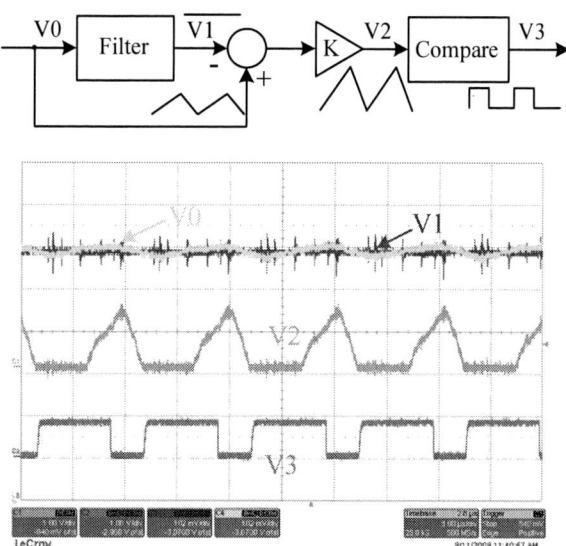

Figure 10. Implementation of synchronization with no wires

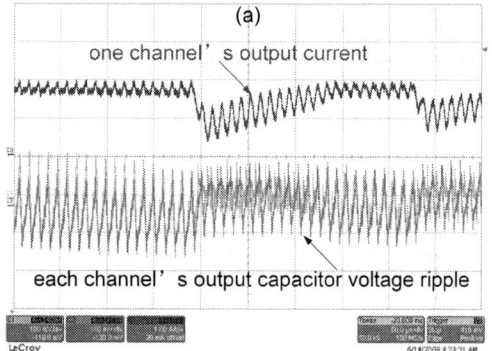

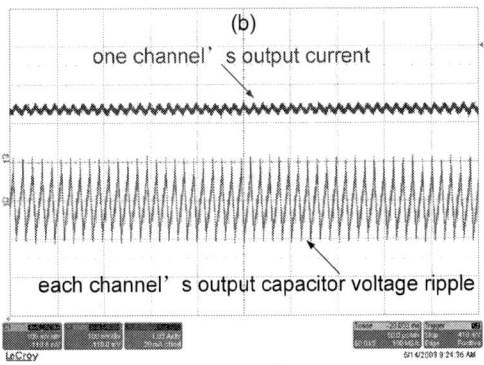

Figure 11. Signal synchronization, (a) without synchronization; (b) with synchronization

IV. EXPERIMENTAL RESULTS

The three-port converter system is verified through a two-channel four-converter prototype rated at 800W as shown in Fig 12. Fig 13 gives the output port CS performance. The hybrid CS (b) has better load transients than the conventional droop method (a). It should be noted that the proposed hybrid CS does not affect steady state CS performance as the droop rate is the same for both methods. And output voltage has no spike when one channel fails (c), which implies the fault-tolerant feature of the multi-channel converters. Fig 14 shows that input CS compatible with MPPT algorithm, and even during solar irradiance level changes, input currents still agree with each other. Fig 15(a) shows that both channels are working under MPPT to maximize solar power, while the battery port provides the power balance for the system when the load power changes, and two PV panel have very different maximum power points. Fig 15(b) shows that one channel goes to regulate the battery port first because its upper voltage limit has been reached, and then followed by the other channel when the other voltage limit is met. The reason is that although V_{bmax} and R_{droop} are the same, two channels have different I_b, as a result, their voltage limit settings are different. Therefore, according to Fig 15, the proposed autonomous mode transition allows smooth transition for independent channels under different conditions.

Figure 12. Prototype photo of two converter channels

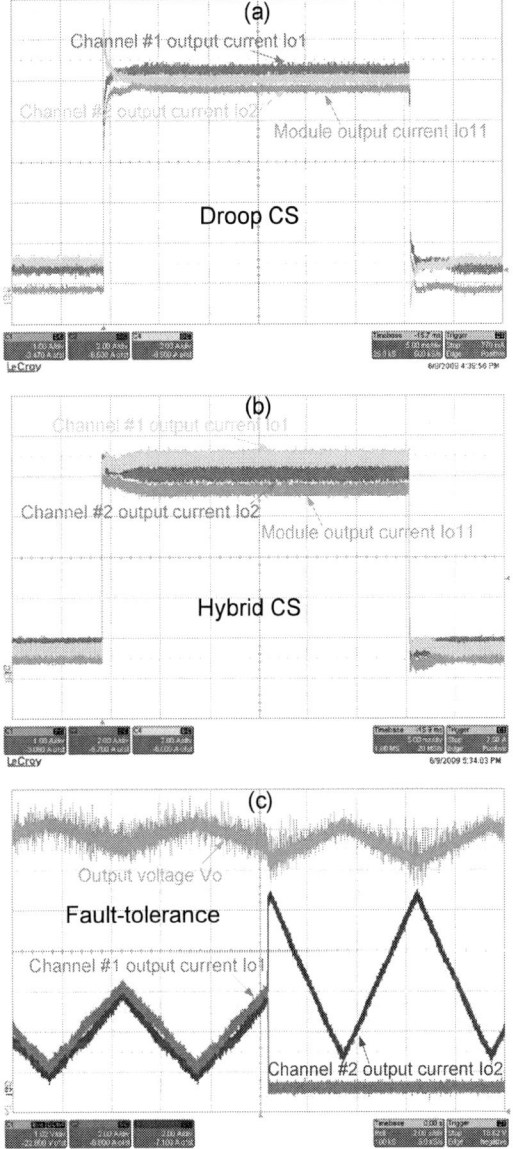

Figure 13. Output port CS performance, (a) droop CS; (b) hybrid CS with better transients; (c) one channel fails

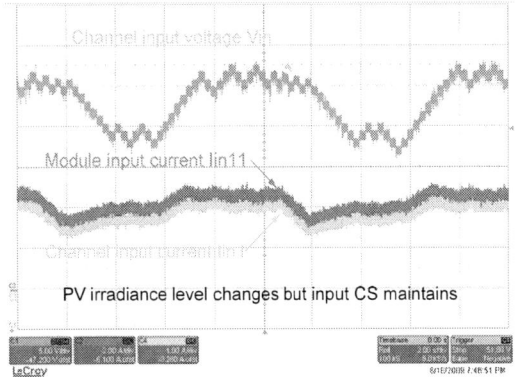

Figure 14. Input Port CS with MPPT

978-1-4244-4782-4/10 $26.00 © 2010 IEEE   2078

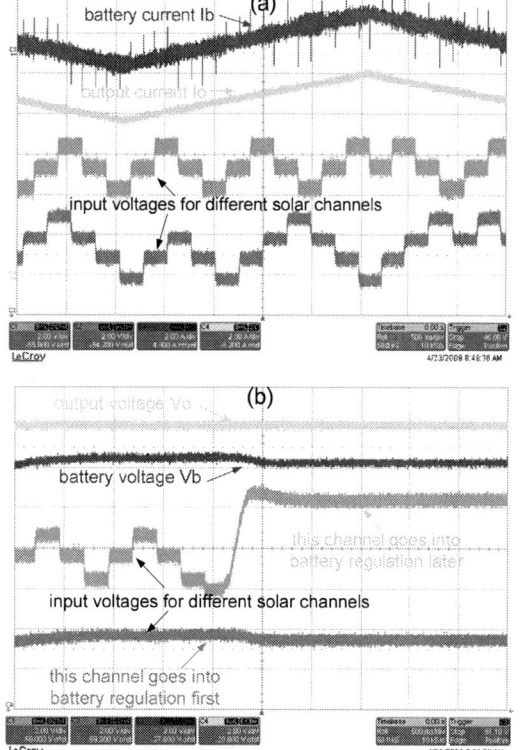

Figure 15. Autonomous mode transitions, (a) both with MPPT; (b) transit from one without MPPT to both without MPPT.

V. CONCLUSION

This paper presents a fault-tolerant two-channel three-port converter system with different channels being able to work independently under different modes of operation to maximize the solar power harvesting. Each channel is a separate autonomous sub-system with several converters connected together at all three power ports. When multiple converters are paralleled, CS control is always preferred, especially for three-port converters necessitating CS for at least two ports. The output port hybrid CS is implemented to achieve better transients than the droop CS method at both intra & inter channel levels. And the passive CS control is well fitted into the battery charging algorithm and the MPPT algorithm. To sum up, the proposed system control strategy is working well for the multi-channel multi-port paralleled converters with all the functionalities such as autonomous mode transitions, current sharing, MPPT, battery charging, etc. Finally, the system control strategy is verified by both simulation and experimental results.

REFERENCES

[1] H. Al-Atrash, F. Tian, I. Batarseh, "Tri-Modal Half-Bridge Converter Topology for Three-Port Interface," IEEE Transactions on Power Electronics, vol. 22, pp. 341 - 345, January 2007.

[2] Z. Qian, O. Abdel-Rahman, J. Reese, H. Al-Atrash, I. Batarseh, "Dynamic Analysis of Three-Port DC/DC Converter for Space Applications," in IEEE 2009 Applied Power Electronics Conference, 2009, pp.28 - 34.

[3] Z. Qian, O. Abdel-Rahman, M. Pepper, I. Batarseh, "Analysis and Design for Paralleled Three-port DC/DC Converters with Democratic Current Sharing Control," in IEEE 2009 Energy Conversion Congress and Exposition, 2009, pp. 1375-1382

[4] B.G. Dobbs, P.L. Chapman, "A Multiple-Input DC-DC Converter Topology," in IEEE Power Electronics Letters, vol. 1, pp. 6-9, March 2003.

[5] G.J. Su, and F.Z. Peng, "A Low Cost, Triple-Voltage Bus DC-DC Converter for Automotive Applications," in IEEE 2005 Applied Power Electronics Conference, 2005, pp.1015-1021.

[6] F.Z. Peng, H. Li, G.J. Su, J.S. Lawler, "A New ZVS Bidirectional DC-DC Converter for Fuel Cell and Battery Application," IEEE Trans. Power Electronics, vol. 19, pp. 54-65, January 2004.

[7] H. Tao, A. Kotsopoulos, J.L. Duarte, M.A.M. Hendrix, "Family of Multiport Bidirectional DC-DC converters," in Proc. IEEE Power Electronics Specialists Conf., 2008, pp.796 – 801.

[8] Y. M. Chen, Y. C. Liu, and F. Y. Wu, "Multi-input DC/DC Converter Based On the Multiwinding Transformer for Renewable Energy Applications," IEEE Trans. Industrial Applications, vol. 38, pp. 1096 - 1104, August 2002.

[9] A. Kwasinski, "Identification of Feasible Topologies for Multiple-Input DC-DC Converters," IEEE Trans. Power Electronics, vol. 24, pp. 856 - 861, March 2009.

[10] W. Jiang, B. Fahimi, "Multi-port Power Electric Interface for Renewable Energy Sources," in Proc. IEEE Applied Power Electronics Conference, 2009, pp. 347 - 352.

[11] C. Zhao, S.D. Round and J.W. Kolar, "An Isolated Three-Port Bidirectional DC-DC Converter with Decoupled Power Flow Management," IEEE Trans. Power Electronics, vol. 21, pp. 2443 - 2453, September 2008.

[12] Y. Panov, J. Rajagopalan and F.C. Lee, "Analysis and Design of N Parallel DC-DC Converters with Master-Slave Current-Sharing Control," in IEEE 1997 Applied Power Electronics Conference, 1997, pp.436-442.

[13] J. Sun, Y. Qiu, B. Lu, M. Xu, F.C. Lee, W.C. Tipton, "Dynamic Performance Analysis of Outer-loop Current Sharing Control for Paralleled DC-DC Converters," in IEEE 2005 Applied Power Electronics Conference, 2005, pp.1346 - 1352.

[14] Y. Zhang, R. Zane and D. Maksimovic, "System Modeling and Digital Control in Modular Masterless Multi-phase DC-DC Converters," in IEEE 2006 Power Electronics Specialists Conference, 2006, pp.1-7.

[15] Y. Zhang, R. Zane, D. Maksimovic, "Current Sharing in Digitally Controlled Masterless Multi-phase DC-DC Converters," in IEEE 2005 Power Electronics Specialists Conference, 2005, pp. 2722-2728.

[16] S. Luo, Z. Ye, R. Lin and F.C Lee, "A Classification and Evaluation of Paralleling Methods for Power Supply Modules," in IEEE 1999 Power Electronics Specialists Conference, 1999, pp.901-908.

[17] B.T. Irving and M.M. Jovanovic, "Analysis, Design, and Performance Evaluation of Droop Current-Sharing Method," in IEEE 2000 Applied Power Electronics Conference, 2000, pp.235-241.

[18] I. Batarseh, K. Siri, H. Lee, "Investigation of the Output Droop Characteristics of Parallel-connected DC-DC Converters," in IEEE 1994 Power Electronics Specialists Conference, 1994, pp. 1342-1351.

[19] W. Wu, N. Pongratananukul, W. Qiu, K. Rustom, T. Kasparis, I. Batarseh, "DSP-based multiple peak power tracking for expandable power system," in IEEE 2003 Applied Power Electronics Conference, 2003, pp. 525-530.

978-1-4244-4782-4/10 $26.00 © 2010 IEEE

A Smart and Simple PV Charger for Portable Applications

Weichen Li*, Yuzhen Zheng**, Wuhua Li*,Yi zhao*, Xiangning He*

* College of Electrical Engineering, Zhejiang University
Hangzhou, 310027, P.R. China
Email: woohualee@zju.edu.cn

** Department of Electrical Engineering
Zhejiang University of Science and Technology
Hangzhou, 310023, P.R. China

Abstract—**A smart and simple PV charger circuit is presented in this paper for the portable applications, which is only composed of several analog chips to simplify the system structure. A three-mode charging solution is employed to meet the demands of the PV array and the lithium battery, which includes the maximum power point tracking (MPPT) mode, the constant-voltage mode and the current-limited mode. Suitable charging mode can be achieved automatically by the smart switch to improve the utilization of the PV array and to protect the battery. The fractional open-circuit voltage method is employed to realize the MPPT performance due to its simple implementation performance. Furthermore, the proposed charger can be integrated into an IC chip to reduce the size and make it more attractive in the portable applications. At last, a 60W prototype is built and tested to verify the effectiveness of the proposed solution.**

I. INTRODUCTION

With the increase of the energy demand and the concern of environmental pollution around the world, photovoltaic (PV) power system is becoming more and more popular. The off-grid PV power generation system is widely used in the portable applications to provide clear and long energy with a high power density. As a part of the off-grid PV power generation system, the PV charger is used to charge the battery from the solar energy. Nowadays, the portable equipments relying on the solar energy as a power supply are widely used in the daily life. It is more suitable to make the PV charger cheaper, smaller and lighter in the portable applications. As a result, the PV charger should be simple and smart.

The traditional charger is the component, which transfers the power from the grid to the battery. In normal condition, the grid can supply continuing and stable energy to the battery. Therefore, the battery can be charged according to its self-condition by the traditional charger. Normally, the charging process includes two stages: in the first stage, the battery is charged in the constant-current mode. When the voltage of the battery is up to the floating voltage, the

charging mode is changed to the constant-voltage (CV) mode as the second stage [1]. During the whole charging process, only the charging current and the voltage of the battery need to be controlled.

Compared with the utility grid, the output power of the PV array is not stable. It is affected by the irradiance and the temperature in the surrounding environment. The insolation performance of a PV array is shown in Fig.1. The temperature performance of a PV array is given in Fig.2. It is shown that the output power of the PV array is different at different irradiance and temperature levels. At the particular irradiance and temperature level, the P-V curve of the PV array is nonlinear, and a maximum power point (MPP) exists. At this point, the output power of the PV array is the maximum. The P-V curve changes as the variation of the irradiance and the temperature. In order to make full use of the PV array, it is necessary to make the PV array operate at the MPP to obtain the most PV energy. Unfortunately, the position of the MPP is variable and the MPPT should be employed.

There are many methods to realize MPPT. In these methods, Constant Voltage Tracking (CVT), Fractional Open-Circuit Voltage (FOCV) and Hill Climbing (HC) are commonly used. In Fig.1 and Fig.2, It can be seen that the voltage of the MPP only changes a little although the position of the MPP varies. Therefore, the PV array can be controlled to operate near the MPP by the way of stabilizing its operating voltage at a constant one. CVT is realized easily and economically due to less sensor and simple control circuit. Consequently, CVT is suitable for the portable applications. Clearly, the voltage of the MPP of the PV array is about 80% of the open-circuit voltage. Fortunately, the ratio coefficient keeps constant as the temperature and the irradiance variation. Consequently, if the open-circuit voltage of the PV array is obtained, MPPT can be achieved by controlling the operating voltage of the PV array at 80% of its open-circuit voltage. Compared with CVT, the open-circuit voltage changes as the variation of the P-V curve, so the operating voltage of the PV array follows to alter other than the constant voltage in CVT. For this reason, applying FOCV, MPPT can be realized more accurately. The open-circuit voltage can be sensed by the resistors easily.

This work is sponsored by the National Nature Science Foundation of China (50907058), the Power Electronics S&E Development Program of Delta Environmental & Education Foundation (DREM2009001) and the China Postdoctoral Science Foundation (200902625).

This method can be achieved by only several analog chips. Due to simple and economical, FOCV is also suitable for the portable applications.

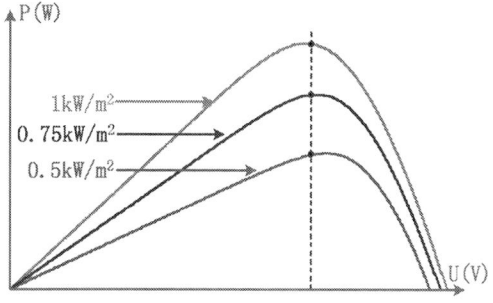

Figure 1.   Insolation performance of a PV array

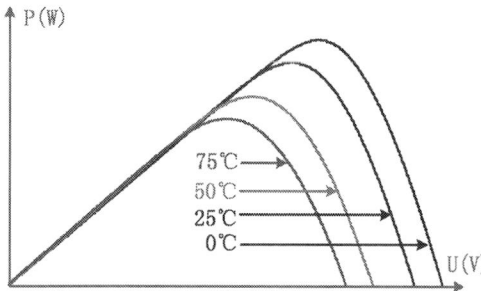

Figure 2.   Temperature performance of a PV array

HC involves a perturbation in the operating voltage of the PV array. Normally, the PV array is connected to the power converter. Consequently, the perturbation of the operating voltage can be achieved by perturbing the duty ratio of the converter. HC can track the MPP more accurately than CVT and FOCV. However, it is necessary to calculate the power by sensing the voltage and the current. Hence, on the one hand, Digital control is more suitable for HC. On the other hand, the current sensor and the voltage sensor are required. Although in PV charger, the voltage of the battery changes very slowly, so the power can be calculated through only sensing the charging current, the current sensor is necessary. Both the aspects above will lead to increasing the cost and complicating the circuit. Therefore, HC is not very suitable solution considering that the power level is low in the portable applications. Besides the methods mentioned above, there are some other methods to realize MPPT such as fuzzy logic control, neural network, ripple correlation control (RCC) and so on [2]. These methods are complex to be realized. Some of them need more sensors and result in high cost. They are not suitable for the portable applications.

The PV charger is different from the traditional charger. It transfers the energy from the PV array to the battery. Applying the solar energy to charge the battery, on the one hand, it's the best for the PV array to make it output the maximum power all the time. But in the later stage of the charging, the large charge current is harmful to the battery, which reduces the battery life. On the other hand, if charging the battery as the normal mode, it will waste the solar energy.

So the battery charging mode for PV array is different from the normal charging solution, because the utilization of the PV array should be considered in the PV charger. Meanwhile, the weight and the volume of charger are also very important in the portable applications.

In this paper, a smart and simple PV charger for the portable applications is presented, which is suitable to charge the battery using the PV array. The FOCV method is adopted to realize the MPPT. The charging process includes three modes. The first mode is the current-limited mode, which limits the charging current below the maximum charging current to protect the battery. The second mode is the MPPT charging mode. In this mode, the charger will track the maximum power point of PV array in order to make full use of the PV array. The third mode is the CV charging mode. In this mode, the charger only charges the battery in the floating way to improve the performance of the battery. The charging current in this mode is very low. The selection of the three modes is switched automatically by a smart circuit.

II.   PROPOSED SMART AND SIMPLE PV CHARGER

Fig.3 is the block diagram of the proposed smart and simple PV charger. The voltage of the PV array is selected from 16.5V to 21V to satisfy the compact requirements in the portable applications, which is higher than the voltage of battery. The lithium battery is used here to reduce the system size. The step-down Buck converter is employed as the charging circuit due to its simple structure and low cost.

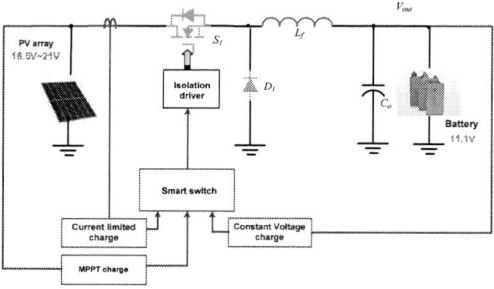

Figure 3.   Block diagram of the proposed smart and simple PV charger

The control circuit shown in Fig.3 is composed of the MPPT control loop, the current limited control loop, the CV control loop and a smart switch. The switch can choose one of the three control loops automatically according to the output power of the PV array and the battery voltage.

When the output power of the PV array is too large, the smart switch chooses the current limited control loop to make the charge current lower than the maximum charge current of the battery. When the output power of PV array becomes low and the battery voltage is still lower than the floating voltage, MPPT control loop is selected by the smart switch to make full use of the PV array. When the battery voltage reaches the floating voltage, CV control loop is employed by the smart switch to charge the battery with small current.

The FOCV method is employed here because only the PV array open-circuit voltage is sensed in this method, which is simple. And the MPPT circuit is easy to be integrated into an IC chip.

III. ADVANTAGES OF THE PROPOSED SOLUTION

The advantages of the proposed charger are as following:

1. The charger circuit is simple because the control circuit is composed of only several analog chips and the converter is only a single phase Buck converter. So the cost, the weight and the volume are low to meet the compact requirements in the portable applications.

2. The presented charger can realize MPPT to track the maximum power point of the PV array as the temperature and the insolation intensity change.

3. The proposed charger solution can be integrated into an IC chip to reduce the system size and to further improve the power density.

IV. DESIGN CONSIDERATIONS

A. Control circuit

Fig.4 is the detailed block diagram of the control circuit. The smart switch is achieved by two small diodes. At the beginning of the charging, the battery voltage is lower than the floating voltage, so the output of CV control loop is high. The diode D_2 is reversed to block the CV control loop. And the Buck converter is controlled by the MPPT control loop. When the battery voltage reaches the floating voltage, the CV control loop takes effect to keep the output voltage steady instead of the MPPT control loop. The current limited control loop is realized by a control IC with current peak mode (CPM). The slope compensation is added to provide the circuit stability. So whichever control loop is chosen, the charge current will not be higher than the maximum charge current of the battery.

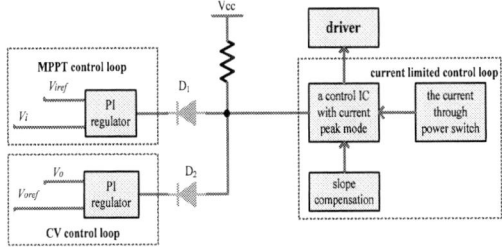

Figure 4. Detailed block diagram of control circuit

B. Sample & hold circuit

The FOCV method is selected in this paper to realize MPPT. As mentioned before, the voltage of the MPP of PV array is about 80% of the open-circuit voltage. Fortunately, the ratio coefficient keeps constant as the temperature and the insolation intensity variation. So 80% of the open-circuit voltage is used to in the MPPT control loop in order to realize MPPT [3].

How to obtain the PV array open-circuit voltage is a key consideration in the FOCV method. The block diagram of the sample & hold (S/H) circuit is shown in Fig.5. The open-circuit voltage is acquired by shutting down the power

switch in Buck converter periodically. The 555 timer is used to generate the sampling pulse signal periodically. When its output is a high level, on one hand, it affects the control IC with CPM to shut down the power switch. On the other hand, the high level affects the S/H chip LF398 to make its output equal to the input voltage of Buck. At this time, as the power switch is shut down, the input voltage of Buck converter is just right the PV array open-circuit voltage. Thus the output voltage of LF398 is the sampling open-circuit voltage of the PV array. After the short sampling time, the output of 555 timer is a low level, the converter operates as normal Buck converter. Furthermore LF398 is in hold state to keep the output voltage invariable no matter how the input voltage of Buck changes. The voltage held by LF398 is still the sampling open-circuit voltage of the PV array. Then it is easily to obtain 80% of the sampling open-circuit voltage by applying divided resistors. This voltage is used as the reference voltage in MPPT control loop. If the MPPT control loop takes effect, the input voltage of Buck converter, also the operating voltage of the PV array is controlled at 80% of the open-circuit voltage of the PV array to realize MPPT.

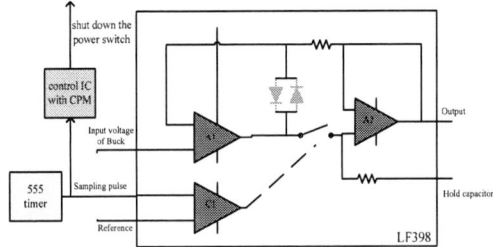

Figure 5. Block diagram of sample & hold circuit

The sampling period should be designed carefully. A short sampling period will cause large solar energy losses. On the contrary, a long sampling period will lead to an inaccurate sample data because the sampling voltage reduces. In this paper, the sampling period is about 10 seconds. Meanwhile, the sampling time is selected as 100 milliseconds due to the similar considerations.

The design considerations of the parameters for the constant voltage loop and the current-limited loop are similar to those of the conventional battery charging circuit.

C. Driver circuit

The MOSFET can't be driven as a normal one because its source is floated in the Buck converter. The gate driver methods can be divided as the isolation way and non-isolation way. In the non-isolation driver methods, the bootstrap is a good candidate. However, the load is the battery for PV charger. In this circumstance, before the converter starts, the voltage level of the source of the MOSFET is kept the same as the voltage of the battery. Therefore it is difficult to start the converter.

In the isolation driver methods, the optical coupler and the transformer are widely used. The switch speed will be restrained by applying the optical coupler. So the transformer driver circuit is the best choice.

The transformer driver circuit is shown in Fig.6. The transistors Q_1 and Q_2 are used to enhance the drive capability. The capacitors C_1 and C_2 are applied to strain away the dc component. D_1 is the freewheeling diode. The transistor Q_3 and the diode D_2 are helpful to accelerate to turn-off speed of the MOSFET. The duty ratio can be changed in a wide scope in this transformer driver circuit. The simulation waveforms and the experiment results are shown in Fig.7 and Fig.8 respectively.

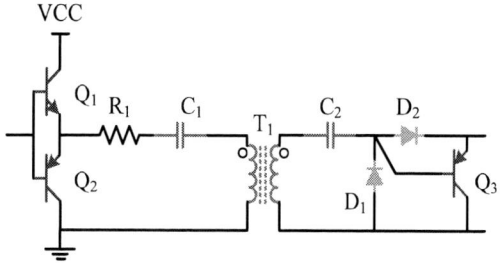

Figure 6.   Transformer driver circuit

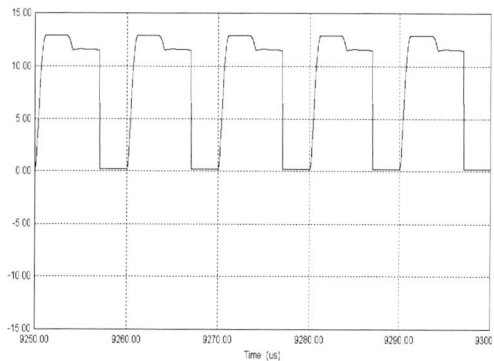

Figure 7.   Simulation waveforms of the driver circuit

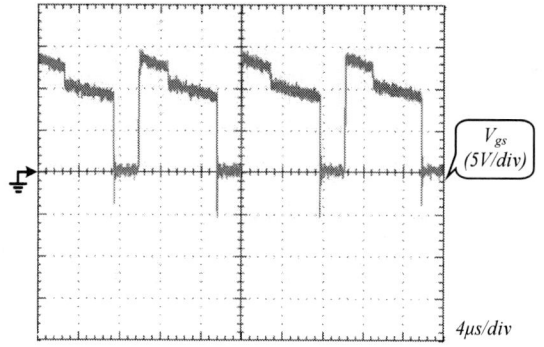

Figure 8.   Experimental waveforms of the driver circuit

V.   EXPERIMENTAL VERIFICATION

The parameters of the charger circuit are given in Tab.I. A 60W PV charger prototype is built to verify the effectiveness of the proposed solution. The open-circuit voltage of the PV array is from 16.5V to 21V and the output voltage of the lithium battery is from 9V to 12V with 10 Ah.

TABLE I.   PARAMETERS OF CHARGER CIRCUIT

PV Voltage	16.5V-21V	Battery Voltage	9V-12V
Operation Frequency	100kHz	Maximum output power of PV array	60W
Lithium Battery capacity	10Ah	MOSFET	IRFZ34N
Diode	MBR1045T	Output Capacitor	100µF
Filter Inductor	46µH	Control IC with CPM	UC3843

The experimental sampling waves are shown in Fig.9 and Fig.10. In Fig.9, the sampling voltage V_{sample} can track the open-circuit voltage of the PV array V_{oc} when it changes. In Fig.10, during the sampling time (the sampling pulse V_{pulse} is given), the MOSFET is shut down, and the input voltage of Buck converter V_{in} is equal to the PV array open-circuit voltage V_{oc}. During the holding time (V_{pulse} is low), the charger circuit works and V_{in} is equal to the operating voltage of PV array V_p. If the charger works in MPPT mode, V_p is equal to about 80% of V_{oc}.

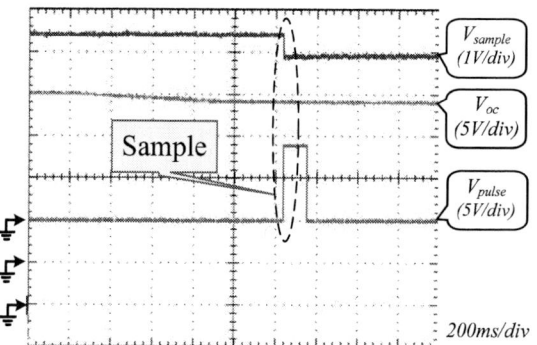

Figure 9.   Waves of sampling open-circuit voltage

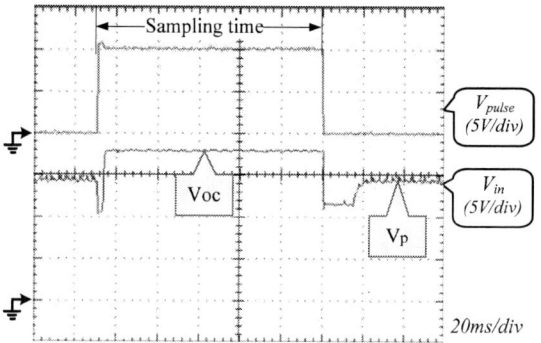

Figure 10.   Sampling waves when the charger works

The drain-source voltage V_{ds} and the gate-source voltage V_{gs} of the MOSFET are shown in Fig.11. The input voltage V_{in} and output voltage V_{out} of Buck converter are shown in Fig.12. Both of them demonstrate the proposed charger circuit works well.

The experimental results of the battery voltage and the charging current during the charging stage are shown in Fig.13. The battery voltage is charged from 11.55V to 11.82V. The charging current falls down. At last, as the battery voltage is a constant, which shows it goes into the CV charging mode.

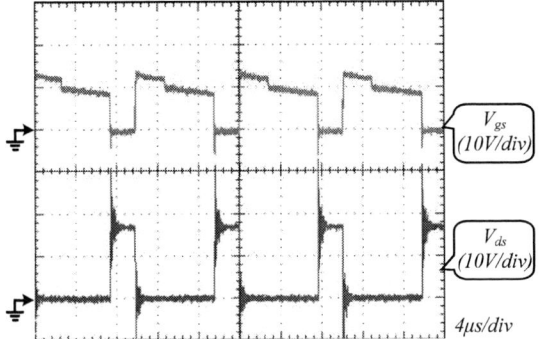

Figure 11. V_{gs} and V_{ds} of power MOSFET

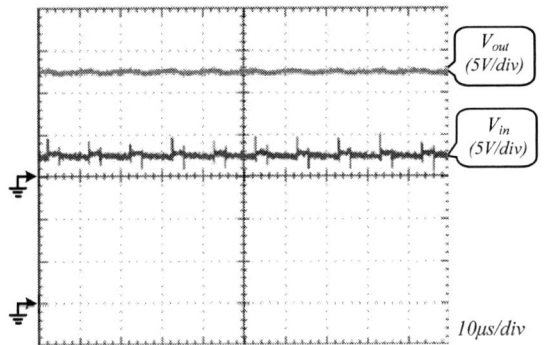

Figure 12.   Input and output voltages of Buck converter

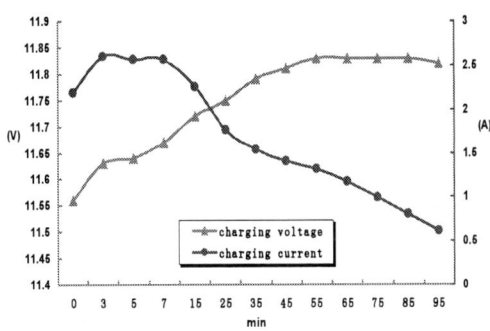

Figure 13.   Experimental results of charging

The experimental results of the ratio of the operation voltage V_{op} and the PV array open-circuit voltage V_{oc} V_{op}/V_{oc} is shown in Fig.14. At the beginning stage of the operation, the ratio is about 82%, which shows that the PV array works at the MPP. When the system changes to operate at CV charging mode, the ratio is bigger than 0.82, which shows that the PV does not work at the MPP any more to protect the battery.

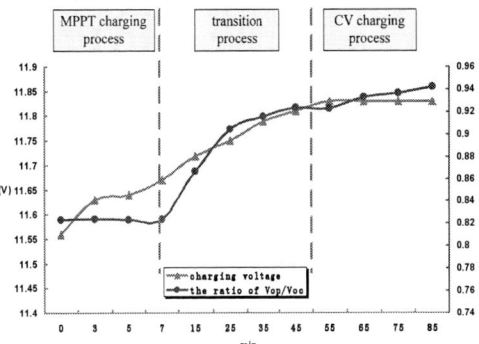

Figure 14.   Charging voltage and the ratio of V_{op}/V_{oc}

VI.   SUMMARY

In this paper, a smart and simple PV charger for portable applications is presented. The advantages of the proposed solution are: a) The proposed charger is simple, the cost, the weight and the volume are low; b) The proposed charger can realize three charging modes including the MPPT mode, the CV mode and the current limited mode. c) The proposed charger can be integrated into an IC chip to reduce the size due to the simple structure. At last, a 60W prototype is built and tested to confirm the validity and applicability of the proposed solution.

REFERENCES

[1]   Min Chen, Gabriel A. Rincon-Mora, "Accurate, Compact, and Power-Efficient Li-Ion Battery Charger Circuit," IEEE Transactions on Circuits and Systems, Vol.53, NO.11, November 2006, pp.1180-1184.

[2]   T. Esram, P. L. Chapman, "Comparison of Photovoltaic Array Maximum Power Point Tracking Techniques," IEEE Transactions on Energy Conversion, vol. 22, NO.2, June 2007, pp. 439-449.

[3]   Abu Tariq, M.S. Jamil Asghar, "Development of An Analog Maximum Power Point Tracker for Photovoltaic Panel," IEEE PEDS 2005, pp. 251-255.

RTDS-based Real Time Simulations of Grid-Connected Wind Turbine Generator Systems

Gyeong-Hun Kim, Young-Ju Kim, Minwon Park, and
In-Keun Yu
Department of Electrical Engineering
Changwon National University
Changwon, Korea
yuik@changwon.ac.kr

Byeong-Mun Song
Department of Electrical and Computer Engineering
Baylor University
Waco, Texas, USA
ben_song@baylor.edu

Abstract— This paper presents the modeling and analysis of grid-connected wind turbine systems using real time digital simulator (RTDS). The different modeled systems of wind turbine generator systems (WTGS) such as permanent magnet synchronous generator (PMSG), induction generator (IG), synchronous generator (SG) and doubly fed induction generator (DFIG) are simulated. Three control schemes for 3 MW WTGSs of squirrel cage induction generator (SCIG), DFIG with back-to-back converter and PMSG with back-to-back converter are designed for the modeled systems. The simulation results show the performances of each type of WTGS under the same wind speed pattern. It has been verified that the RTDS based real-time simulations of grid-connected wind turbine generator systems to be effectively employed to evaluate the control schemes and output performances of different types of WTGSs.

I. INTRODUCTION

Among various renewable energy resources such as solar, wind, hydraulic, etc., wind energy is the most desirable energy source from the view point of efficiency and affordability. Recently, wind power generator systems have been attracting interests all over the world [1]. Different wind turbine generator systems (WTGSs) have been developed and installed for wind energy systems. The squirrel cage induction generator (SCIG) is robust and has a lower cost but is a fixed speed WTGS. In spite, the use of the SCIGs has decreased from almost 70% in 1995 to about 24% in 2004 [2]. The doubly fed induction generator (DFIG) system is less robust and more expensive than the SCIG [3]. The power converter operates with a wind turbine at variable speeds. The permanent magnet synchronous generator (PMSG) system is also a variable speed WTGS. The PMSG system can directly drive for wind turbines, which are spared from losses, maintenance, and costs associated with a gearbox [4].

WTGSs are typically modeled utilizing an acronym for electromagnetic transients program (EMTP) simulator like PSCAD/EMTDC, or ATP/EMTP; however, the EMTP-based simulation software is not capable of testing real-time simulations. The RTDS is a fully digital electromagnetic

transient power system simulator that operates in real-time [5]. Furthermore, the RTDS provides fast, reliable, accurate, and cost-effective results of power systems with complex high-voltage alternating currents as well as high-voltage, direct current networks.

This paper presents the modeling and analysis of various grid-connected wind turbine systems using real time digital simulator (RTDS). Three control schemes for SCIG, DFIG with a back-to-back converter, and PMSG with a back-to-back converter are simulated using RTDS. The modeled systems using RSCAD/RTDS are characterized along with the subsystem model components. The effectiveness of the RTDS-based real time simulations is demonstrated by the performance results of the different type of 3 MW WTGS under the same wind speed conditions. All completed systems are simulated, and their simulation models are validated for the system operation. The detailed simulation results will be discussed in this paper.

II. WIND TURBINE GENERATOR SYSTEMS MODELING AND CONTROL

A. Basic Wind Turbine Model

The MOD-2 model [6] is adopted for the wind turbine modeling in the RSCAD, and the tip speed ratio is calculated by (1).

$$\lambda = R\omega / V_m \tag{1}$$

where, R is the blade radius (m), ω indicates the mechanical angular velocity of blade (rad/s), and V_m is the wind velocity (m/s). The MOD-2 model is characterized by the following equations [7].

$$C_P = \frac{1}{2}(\lambda - 0.022\beta^2 - 5.6)e^{-0.17\lambda} \tag{2}$$

This work has been supported by grant No. RTI04-01-03 from the Regional Technology Innovation Program of the Ministry of Knowledge Economy (MKE).

978-1-4244-4782-4/10 $26.00 © 2010 IEEE

$$T_W = \frac{1}{2}\rho\pi R^3 v_m{}^2 C_p(\lambda, \beta)/\omega \qquad (3)$$

where, C_p is the power conversion efficiency, and calculated by functions of λ and β, β is the blade pitch angle. The turbine toque is calculated by (3), where T_W is the turbine toque, and ρ is the air density (kg/m^3). Table I represents the MOD-2 wind turbine parameters.

1) Pitch Angle Control

The pitch angle controller with the wind turbine is used to limit the output power at the terminal of the induction generator when the wind speed is over the rated speed [8]. In the case of the variable speed WTGS, the output power is limited by the converter, resulting in the pitch angle controller limiting the rotating speed of the wind turbine. Fig. 1 shows the block diagram of the pitch angle controller. Generation power (G_p) is limited to 1.0 p.u in the SCIG; the limit of the rotation speed (Ω) for the DFIG is 1.3 p.u and the rotation speed limit is 1.5 p.u (in the case of the PMSG). The pitch angle (β) is limited to a value of 10°/s for all type systems in this study.

2) Maximum Power Point Tracking (MPPT)

In the case of the variable speed generator, MPPT control can be adopted [9]. The MPPT control is applied to the models of the DFIG and PMSG type in this paper. In the wind turbine, an optimum value of tip speed ratio leads to a maximum power coefficient. From (1), (2), and (3), the maximum power of the wind turbine is given by (4) [10]:

$$P_{\max} = \frac{1}{2}\frac{\rho\pi R^5 C_{p\_\max}(\lambda, \beta)}{\lambda_{opt}^3}\omega_{opt}^3 \qquad (4)$$

From (4), the target power is expressed by (5).

$$P_{t\,\arg et} = k_{opt}\omega_{opt}^3 \qquad (5)$$

where, $k_{opt} = \dfrac{1}{2}\dfrac{\rho\pi R^5 C_{p\_\max}(\lambda, \beta)}{\lambda_{opt}^3}$

Fig. 2 shows the maximum power extraction curve of the wind turbine under different wind speeds [11, 12]. When the target power (P_{target}) is expressed by a function of rotation speed as in (5), and becomes a reference power of inverter control, the inverter tracks the maximum power point according to the rotation speed variance caused by the torque mismatch between the turbine and the generator load.

TABLE I.   MOD-2 WIND TURBINE PARAMETERS

Items	Value
Propeller type	Horizontal propeller
Rated power	3.0 MW
Cut-in speed	5.8 m/s
Rated speed	11. 8 m/s
Cut-out speed	16.0 m/s
Rotor diameter	91.43 m

B. Squirrel Cage Induction Generaotr (SCIG)

Fig. 3 shows the overall configuration of a WTGS with SCIG model, consisting of a 3-MVA fixed-speed induction generator, a capacitor bank, and a pitch angle controller. The squirrel induction generator is connected in a 22.9 kV infinite bus. A capacitor bank is connected to the terminal of the wind generator to compensate reactive power demand for the generator. The induction generator parameters used in this work are shown in Table II.

C. Doubly Fed Induction Generator (DFIG) with Back-to-Back PWM Converter

Fig. 4 shows the DFIG model with back-to-back power converter. The use of two, 6-switch, hard-switched converters, with a common coupled capacitor, Fig. 5, has been explored. The rotor side converter (RSC) controls the electrical torque of generator and reactive power that flows into rotor winding from utility. The reactive power of the grid and the voltage of the common coupled capacitor are controlled by the grid side

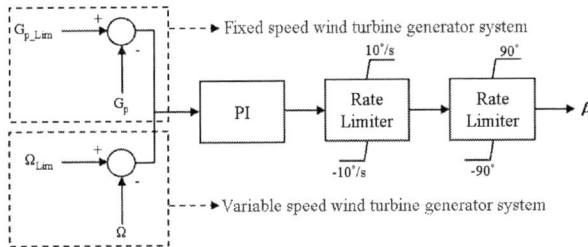

Figure 1. Block diagram of the pitch angle control.

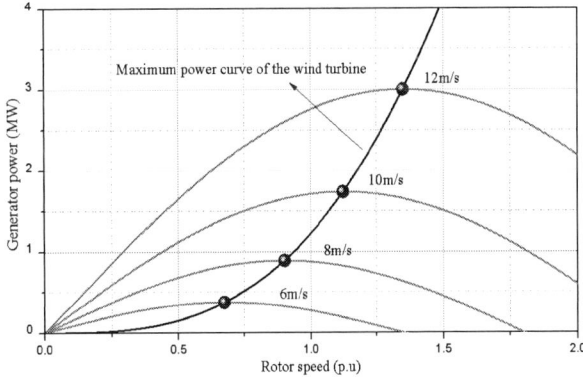

Figure 2. Maximum power extraction curve of the wind turbine under diiferent wind speeds.

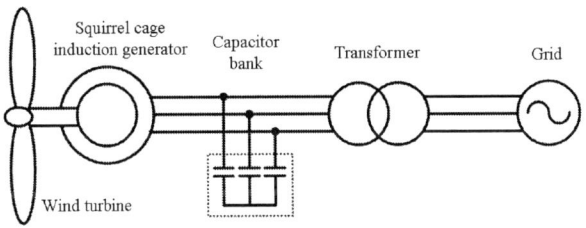

Figure 3. The wind turbine generator system model with a squirrel cage induction generator.

978-1-4244-4782-4/10 $26.00 © 2010 IEEE

converter (GSC). Parameters for the DFIG are shown in Table III.

1) Rotor side converter controller modeling

The RSC extracts the maximum power from the wind turbine through controlling the stator active power and controls reactive power consumption to zero. The stator active and reactive powers can be described as (6) and (7) [3], [13]:

$$P_s = V_s I_{qs} = -V_s \frac{M}{L_s} I_{qr} \tag{6}$$

$$Q_s = V_s I_{ds} = \frac{V_s \psi_s}{L_s} - \frac{V_s M}{L_s} I_{dr} \tag{7}$$

Using (5) and (6), the active power of the generator can be controlled by the q-axis rotor current component (i_{qr}). And the d-axis rotor current component (i_{dr}) is linked to the reactive power of the stator. Reference currents (i_{dr_ref} and i_{qr_ref}) are controlled by the proportional-integral (PI) controller, which generates the voltage reference. A pulse-width modulation (PWM) signal is generated from the voltage reference for the RSC [14]. Fig. 6 depicts the control block diagram of the RSC.

TABLE II. SQUIRREL CAGE INDUCTION GENERATOR PARAMETERS

Items	Value
Capacity	3.0 MW
r_a	0.01 p.u
x_a	0.1 p.u
r_{fd}	0.352 p.u
x_{fd}	0.07
x_{md0}	2.0 p.u
Inertia	3.0 Sec
Rated voltage (Lint to ground)	0.398 kV
Rated current	2.51 kA

TABLE III. DOUBLY FED INDUCTION GENERATOR PARAMETERS

Items	Value
Capacity	3.0 MW
Stator resistance	0.0054 p.u
Wound rotor resistance	0.00607 p.u
Magnetizing inductance	4.362 p.u
Stator leakage inductance	0.102 p.u
Wound rotor leakage inductance	0.11 p.u
Inertia	3.0 Sec
Rated voltage (Lint to ground)	0.398 kV
Rated current	2.51 kA

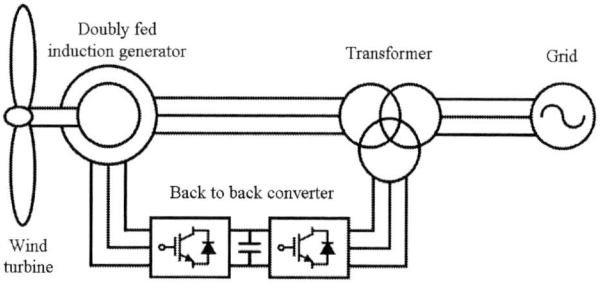

Figure 4. The wind turbine generaotr system model with a doutbly fed induction generator.

2) Grid side converter controller modeling

The GSC keeps the dc-link voltage constant. The reactive power between the GSC and the grid is set to zero [13]. Using the Park's transformation and instantaneous power theory, the active and reactive powers of the grid are expressed by (8) and (9):

$$P_g = V_{qg} I_{qg} \tag{8}$$

$$Q_g = V_{qg} I_{dg} \tag{9}$$

From (8) and (9), the dc-link voltage is controlled by the q-axis current component (i_{qg}), and the reactive power of the grid can be controlled by the d-axis current component (i_{dg}). The PI controller and the inverse Park's transformation generate the PWM signal for the GSC converter. The control block diagram of the GSC is shown in Fig. 7.

D. Permanent Magnet Synchronous Genrator (PMSG) with Back-to- Back PWM Converter

Fig. 8 shows the PMSG model with back-to-back PWM converter. The back-to-back converter has the same structure as in Fig. 5. In this case, the machine side converter (MSC) controls the torque of the generator. The GSC controls the voltage of the common coupled capacitor and the reactive power of the grid. The parameters of the modeled PMSG are presented in Table IV.

TABLE IV. PERMANENT MAGNET GENERATOR PARAMETERS

Items	Value
Capacity	3.0 MW
Stator resistance	0.005 p.u
Stator leakage reactance	0.154 p.u
Rated voltage (Lint to ground)	0.398 kV
Rated current	2.51 kA

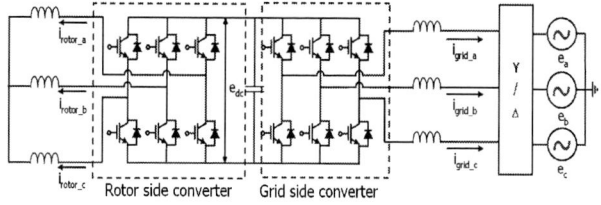

Figure 5. Back-to-back converter for doubly fed induction generator.

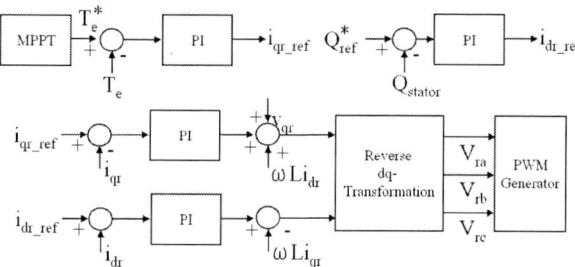

Figure 6. Control block diagram of the rotor side converter.

1) Machine side converter controller modeling

The electromagnetic torque is given by (10):

$$T_e = \frac{3}{2} p \Psi i_{Sq} \qquad (10)$$

where, p is the pole pair number and Ψ is the magnet flux. Using (10), the generator power can be controlled by the quadrature current component [10]. The linear rotating frame current controller is used for current (PI) control, and the power reference is calculated by (5). The control block diagram of the MSC is the same as in Fig. 6.

2) Grid side converter controller modeling

The objective of the GSC for the PMSG type is almost the same as the GSC for the DFIG type. The GSC for the PMSG type also controls the dc-link capacitor voltage and the reactive power between the GSC and the grid based on the same vector-control algorithm of the DFIG type.

III. REAL TIME SIMULATION SETUP

To validate the proposed system model, various simulations have been performed in the WTGSs. The RTDS hardware has different types of processor cards such as the Triple Processor Cards (3PCs), the RISC Processor Card (RPC), and the Giga Processor Card (GPC). The 3PC analyzes power network model and control systems with a typical time step of 50 μs. The GPC can solve the small time step (<2 μs) simulations of voltage source converters (VSC) with high switching frequency [14]. The VSC model is simulated in the small time step module. The VSC interface transformer is connected between the VSC model (small time step simulation) and the main power network (typical time step simulation).

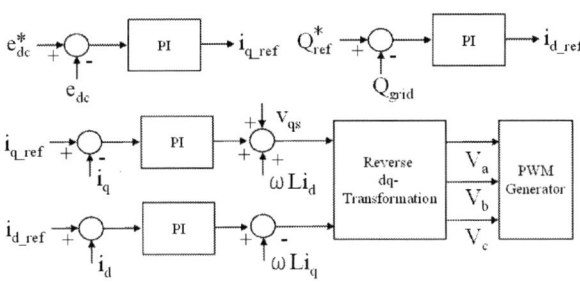

Figure 7. Control block diagram of the grid side converter.

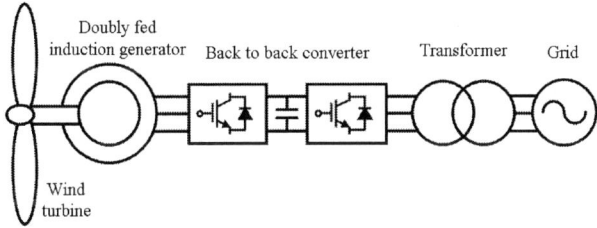

Figure 8. The wind turbine generaotr system model with a permanent magent synchornous generator.

In this simulation, three different types of WTGSs are simulated using the RTDS. The SCIG type WTGS does not include the VSC model. It is analyzed on the only 3PC with a typical time step. Fig. 9 shows the RTDS modules and the processor assignments for the SCIG type WTGS. In the case of DFIG type WTGS has the back-to-back power converter model and the DFIG model which is solved at small time step. This simulation needs the VSC interface transformer for being connected between the DFIG with the converter and the main power network. The RTDS modules and the processor assignments for the DFIG type WTGS is shown in Fig. 10.

If the temperature characteristic of permanent magnet is neglected, and assume the exciting current of the SG is constant, the SG can be the PMSG. The SG is simulated using the typical time step. The PMSG type WTGS needs two interface transformers because the back-to-back converter is

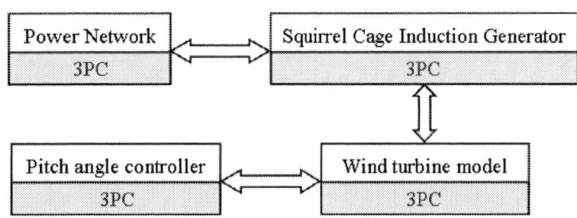

Figure 9. RTDS modules and the processor assignments for the SCIG type WTGS.

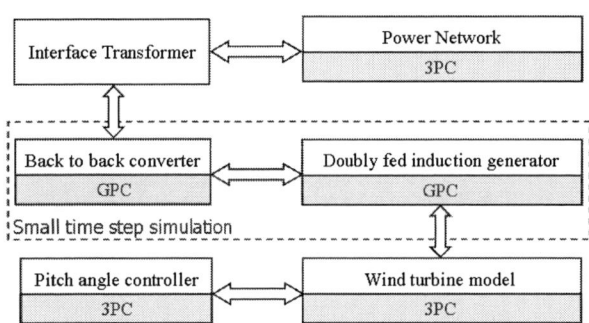

Figure 10. RTDS modules and the processor assignments for the DFIG type WTGS.

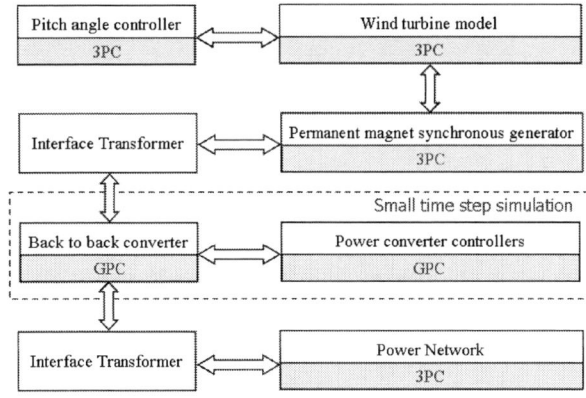

Figure 11. RTDS modules and the processor assignments for the PMSG type WTGS.

connected between main power network and the PMSG as shown in Fig. 8. The simulation structure of the PMSG system is shown in Fig. 11.

IV. SIMULATION RESULTS AND DISCUSSION

The SCIG, PMSG and DFIG models are implemented in the RSCAD and simulated using the RTDS. The output performances of three different types of WTGS are compared under the same wind speed pattern as shown in Fig. 12. The SCIG is fixed speed WTGS, on the other hand, the PMSG and the DFIG are variable speed WTGS. The MPPT control for wind turbines is applied to the PMSG and the DFIG. Fig. 13 shows the generation power of each type of the WTGSs. The variable speed systems generate more active power than the

fixed speed WTGS due to the MPPT control. The reactive power consumptions between the WTGSs system and the grid are shown in Fig. 14. The reactive power of the PMSG and DFIG types is controlled to zero by the power converter. The SCIG type consumes the reactive power compensated by the capacitor bank; moreover, it needs variable type of reactive power compensator such as STATCOM for the reactive power consumptions.

Fig. 15 depicts the power coefficient variations of the WPGSs in which the C_p of DFIG & PMSG is kept to maximum value by the MPPT control. The rotation speed of each generator is shown in Fig. 16. The rotational speeds of DFIG and PMSG change like MPPT curves; however, the speed of SCIG is fixed. The limit value of the DFIG's rotational speed is 1.3p.u. If wind speed increases, the

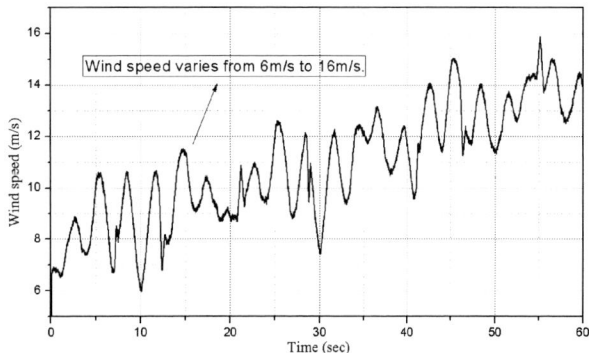

Figure 12. Wind speed pattern.

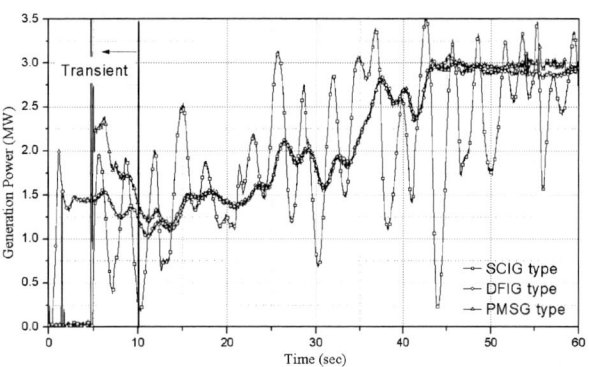

Figure 13. Generation power of the WTGSs.

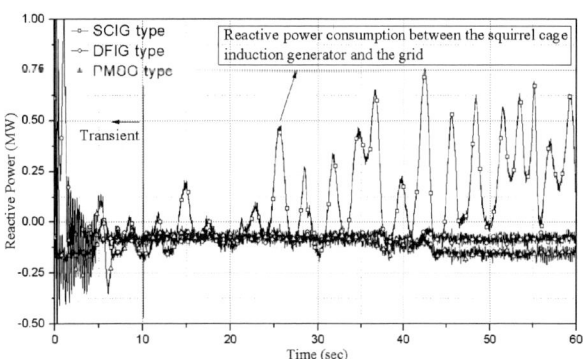

Figure 14. Reactive power consumptions between the each generator and the grid.

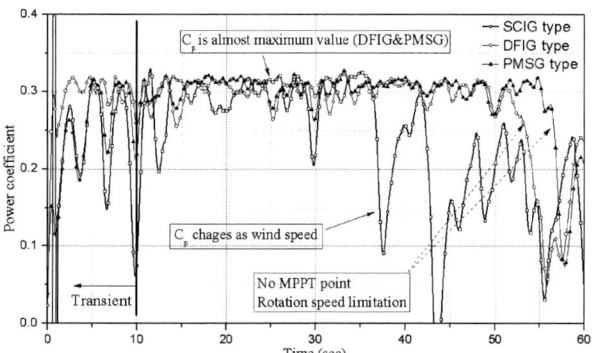

Figure 15. Power coefficient variations of the WPGSs.

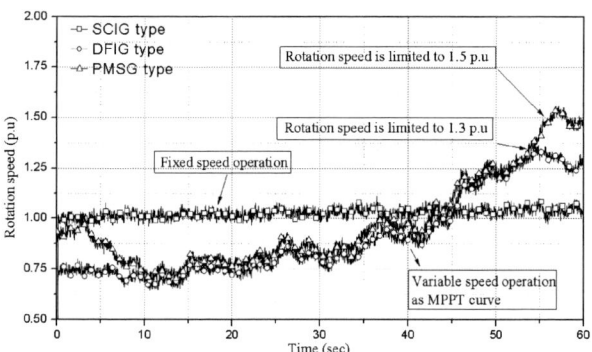

Figure 16. Rotational speed patterns of the WTGSs.

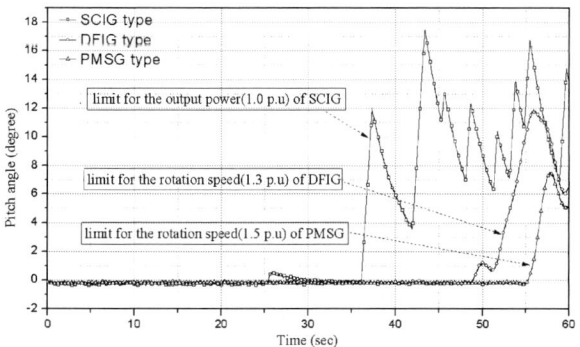

Figure 17. Pitch angle variations of the WTGSs.

978-1-4244-4782-4/10 $26.00 © 2010 IEEE

generation power is limited by the inverter controller. The rotation speed is increased by this power limit and the pitch angle control limits the rotational speed to 1.3p.u. The rotational speed of PMSG is also restricted by the same algorithm, but the limit value is 1.5p.u. This algorithm is represented Fig. 17 which shows the pitch angle variations of the WTGSs. The pitch angle controller of the SCIG type limits only the generation power.

V. CONCLUSION

Three different types of grid-connected WTGSs such as SCIG, DFIG and PMSG, are fully modeled and simulated in real time using the RSCAD/RTDS. The simulation results that the output performances of the WTGSs are compared under the same wind speed pattern. The SCIG type generates the lowest output power due to the fixed speed characteristic. In the case of variable speeds of WTGSs such as DFIG and PMSG, the output power depends on the wind velocity so that DFIG and PMSG generate more power than the SCIG. The proposed simulation method can effectively be applied to evaluate control schemes and output performances according to the types of WTGSs.

REFERENCES

[1] D. J. Park, Y. J. Kim, M. H. Ali, M. Park, and I. K. Yu, "A Novel Real Time Simulation Method for Grid-connected Wind Generator System by Using RTDS," *Conf. Rec. of ICEMS 2007*, pp. 1936-1941, Oct. 2007.

[2] H. Li, Z. Chen, "Overview of different wind generator systems and their comparisons," *IET Renewable Power Generation*, vol. 2, no. 2, pp. 123-138, 2008.

[3] F. Poitiers, T. Bouaouiche, M. Machmoum, "Advanced control of a doubly-fed induction generator for wind energy conversion," *Elsevier Science Ltd. On Electric Power Systems Research*, vol. 79, pp. 1085-1096, 2009.

[4] S. Eriksson, A.Solum, M. Leijon, and H.Bernhoff, "Simulations and experiments on a 12kW direct driven PM synchronous generator for wind power," *Eleservier Science Ltd. On Renewable Energy*, Vol. 33, pp. 674-681, 2008.

[5] P. Forsyth, T. Maguire, and R. Kuffel, "Real time digital simulation for control and protection system testing," in *Proc. 35th Annu. IEEE Power Electron. Spec. Conf.*, Jun. 20-25, pp. 329-335, 2004.

[6] P. M. Anderson and A. Bose, "Stability Simulation of Wind Turbine Systems," *IEEE Trans. Power Apparatus and Systems*, vol. PAS-102, no. 12, pp. 3791-3795, Dec. 1983.

[7] I. Schiemenz and M. Stiebler, "Control of a Permanent Magnet Synchronous Generator used in a Variable Speed Wind Energy System," in *Proceedings of IEEE IEMDC'01*, pp. 872-877, 2001.

[8] T. Senjyu, R. Sakamoto, N. Urakaki, H. Higa, K. Uezato, and T. Funabashi, "Output Power Control of Wind Turbine Generator by Pitch Angle Control Using Minimum Variance Control," *Electrical Engineering in Japan*, 154(2), pp. 10-17, 2006.

[9] Y. Y. Hong, S. D. Lu, and C. S. Chiou, "MPPT for PM wind generator using gradient approximation," *Elsevier Energy Conversion and Management*, vol. 50, pp. 82-89, 2009.

[10] M. Chinchillar, S. Arnaltes, and J. C.Burgos, "Control of Permanent-Magnet Generators Applied to Variable-Speed Wind-Energy Systems Connected to the Grid," *IEEE Transaction on Energy Conversion*, vol. 21, no. 1, pp. 130-135, March 2006.

[11] M. E. Mokadem, V. Courtecuisse, C. Saudemont, B. Robyns, and J. Deuse, "Fuzzy Logic Supervisor-Based Primary Frequency Control Experiments of a Variable-Speed Wind Generator," *IEEE Transaction on Power Systems*, vol. 24, no. 1, pp. 407-417, February 2009.

[12] E. Koutroulis and K. Kalaizakis, "Design of a Maximum Power Tracking System for Wind-Energy-Conversion Application," *IEEE Transaction on Industrial Electronics*, vol. 53, no. 2, pp. 486-494, Aplil 2006.

[13] Z. Lubosny, *Wind Turbine Operation in Electric Power Systems*, 1st ed., Springer – Verlag Berlin Heidelberg, pp. 132-162, 2003.

[14] W. Qiao, G. K. Venayagamoorthy, and R. G. Harley, "Real-Time Implementation of a STATCOM on a Wind Farm Equipped With Doubly Fed Induction Generators," *IEEE Transaction on Industry Application*, vol. 45, no. 1, pp. 1073-1080, January/February, 2006.

Investigation of Fully Digital Controlled Li-Ion Battery Power Recovery System

Siran Wang, Xia Zhou, Jifeng Chen, Wenxi Yao, Zhengyu Lu (Senior member, IEEE)

Department of Electrical Engineering
Zhejiang University
Hangzhou, Zhejiang, China
Siran.Wang09@gmail.com

Abstract—**This paper proposes a design scheme for grid-connected Li-ion battery power recovery system which is utilized for battery formation during the production process of Li-ion battery. The proposed scheme based on a compound dc-dc converter with cascade structure and a three-phase grid-connected inverter with LCL filter is able to interface between low dc voltage of batteries and ac grid. The system is digital controlled, and the design of the digital control scheme is discussed in detail. Voltage-oriented PI controller and active damping method are adopted for regulating the grid currents. The restrictions of the digital compensator design are concluded. The proposed system scheme and the design considerations are both proved by the implementation of a 5kVA prototype.**

Keywords—**Grid-connected system, Full bridge converter, LCL filter, Active damping, Li-ion battery**

I. INTRODUCTION

Distributed generation such as photovoltaic or fuel cell are recently more and more popular, as the energy crisis and environmental problems is extraordinarily serious now [1-6]. Li-ion battery power recovery system is also an implementation of distributed generation and power conservation. According to the production technology of Li-ion battery, batteries need to be charged and discharged for several times during the formation process. This power recovery system is able to transmit the charged power back to ac grid system, so that the power will not be consumed.

Since Li-ion battery is low voltage dc source, the power recovery system needs both dc-dc converter for voltage boost and three-phase dc-ac inverter for interfacing with ac grid. The dc-dc converter should convert the voltage of Li-ion battery to regulated 600VDC. The full-bridge topology is an appropriate option in which phase shift pulse width modulation (PWM) technique can be adopted to realize zero voltage switching (ZVS) [7]. Thus the switching loss of the dc-dc converter can be reduced. The control of dc-dc part is quite simple, because most function of the system is achieved by the dc-ac part. For grid-connected inverter, LCL filter gives much better performance on rejecting switching frequency harmonics than L filter does [8]. So inverter with LCL filter is included in the

Li-ion power recovery system for interfacing with ac grid. The control strategy of grid-connected inverter with LCL filter requires carefully design, since the resonant peak of the LCL filter brings stable problems [9]. Passive damping (PD) [10], active damping (AD) [11-12], or other compensation methods [13] should be adopted in addition to voltage-oriented PI current control.

In this work, a grid-connected system for Li-ion battery power recovery during the formation process is designed. Fully digital control is implemented in the system using a DSP TMS320LF2407A and a slave CPLD EPM7128. The control scheme of the grid-connected system which adopts voltage-oriented PI control strategy and AD method is discussed in detail. A 5kVA prototype which is able to operate stably with low harmonic distortion in the grid currents is made to verify the design.

II. DESIGN OF MAIN CIRCUIT

There are two stages in the proposed main circuit of Li-ion battery power recovery system. The first stage is dc-dc converter which boosts the voltage of batteries to the level of ac grid. The second stage is three-phase inverter for interfacing with ac grid.

A. DC-DC converter topology

Li-ion batteries are 24Vdc voltage source. In order to interface with 380Vac grid, the input voltage of the grid-connected inverter should be 600Vdc at least. However, it is difficult for a single stage converter with the step-up ratio more than 25 to guarantee high conversion efficiency. Furthermore, electrical isolation should be implemented within this stage. So a cascade structure is adopted here, as shown in Fig.1. The inputs of the modules are connected in parallel, while the outputs of the modules are connected in series. With this structure the step-up ratio for each module is degraded to 25/n. A single module in the cascade structure is constructed by full-bridge topology. With appropriate design of circuit parameters [14-17], ZVS can be realized in each module and the turn-on loss of all the switches in first stage can be limited.

Project Supported by Education Department of Zhejiang Province
(Y200804049)

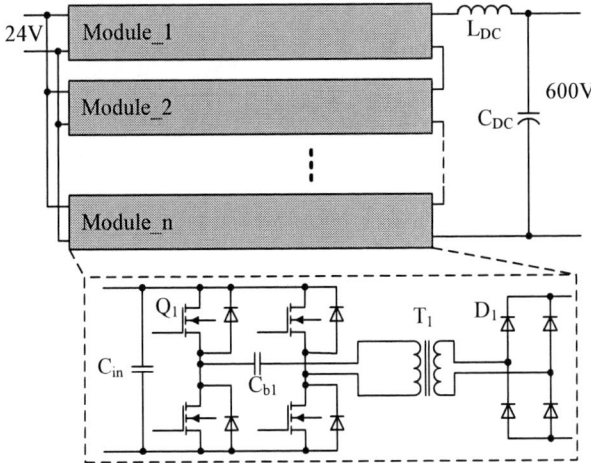

Figure 1. The first stage of the power recovery system.

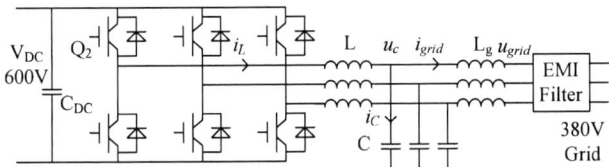

Figure 2. The second stage of the power recovery system.

B. Grid-connected inverter topology

As illustrated in Fig.2, the second stage of the system is three-phase grid-connected inverter with LCL filter. An EMI filter is also set between inverter and the ac grid for suppression of common mode noise. The design of the LCL filter is depending on the following considerations [8]:

1. Overall filter size, cost and losses.

2. Ripple current in filter components.

3. Resonant frequency and dynamic performance of the overall system.

III. DESIGN OF CONTROL STRATEGY

A. Open loop control of dc-dc converter

As the second stage of the system does not claim any accuracy in the input voltage of the inverter, simple open loop control is appropriate for the first stage. Moreover open loop control of the first stage will not bring any stable issues for the second stage, which is quite important for the system operation. The switching frequency is selected to be 100kHz, as it should be higher than then switching frequency of the second stage.

There are totally n modules in the first stage. The slave CPLD following the instructions of the main DSP will provide n groups of complementary PWM signals for driving the modules. For each module, ZVS is guaranteed by a phase shift between the two complementary PWM signals. The phase shift angle is calculated according to the circuit parameters. There is also a *180°/n* delay between the PWM signals of adjacent modules, which is able to reduce the switching ripple

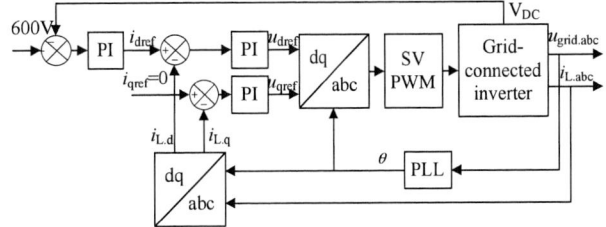

Figure 3. Voltage-oriented PI control scheme of the second stage.

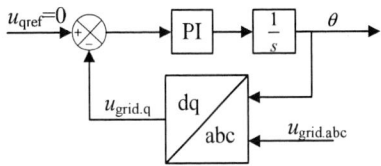

Figure 4. Digital PLL for three-phase grid voltage.

in the output voltage.

B. Voltage-oriented PI control of grid-connected inverter

Voltage-oriented PI control is adopted in the second stage, as shown in Fig.3. Grid voltage orientation is realized by a digital phase lock loop (PLL) illustrated in Fig.4.

There is a dc voltage control loop for regulating the input voltage of the inverter at desired level. The output currents of the inverter are regulated by PI controller in the dq coordinates rotating synchronously with the gird voltage. The output of the voltage control loop is used as the reference of the active current component. And the reference of the reactive current component is set to be 0, so that the output current of the inverter is able to exactly track the phase of the grid voltage.

C. Cancellation of the LCL resonant peak

Actually voltage-oriented PI control is well known in grid-connected inverters with L filter. The implementation of voltage-oriented PI control in grid-connected inverter with LCL filter calls for further considerations yet. The most important issue is the resonant peak of the LCL high order model.

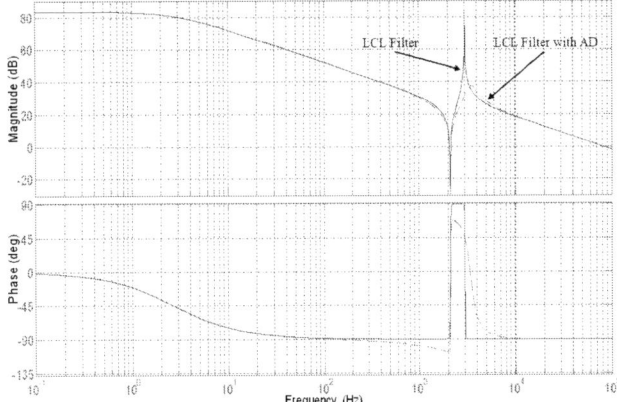

Figure 5. Bode diagram of $G_{plant}(s)$.

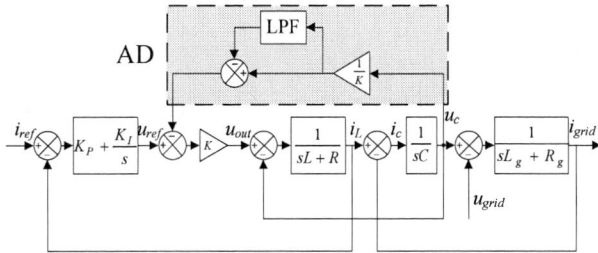

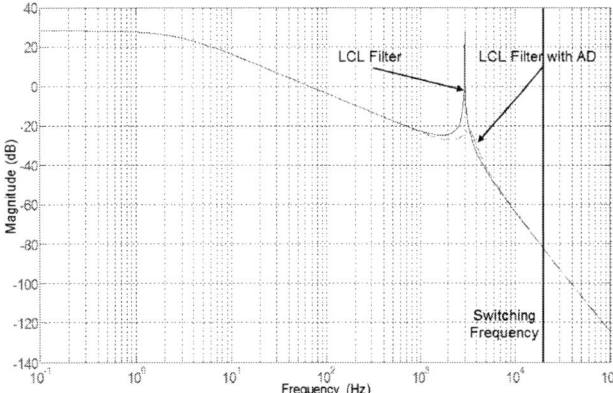

Figure 6. Scheme of voltage-oriented PI control with AD.

Figure 7. Bode diagram of $G_{out}(s)$.

Fig.5 illustrates bode diagram of $G_{plant}(s)$ which is the transfer function from input PWM signal of the inverter to the output current of the inverter. As the damping ratio is very low, there is a quite high resonant peak at the resonant frequency which is related to the LCL filter parameters. The resonant frequency is around 2kHz on condition that the switching frequency of the inverter is 20kHz. So the resonant peak will bring remarkable limitation for the design of the PI controller.

For cancellation of the LCL resonant peak, one of AD methods is adopted in the voltage-oriented PI control scheme. As shown in Fig.6, the measured capacitor voltages are fed back to the output of the PI controller through high pass filter (HPF) which is realized by the subtraction of all-pass and low pass filter (LPF). Comparing to the characteristic of the original LCL filter, bode diagram of $G_{plant}(s)$ with AD is also illustrated in Fig.5. It is obvious that the resonant peak is effectively attenuated by adoption of AD. With this control scheme, it is much easier to guarantee system stability.

Besides, AD have no influence on the attenuation ability of LCL filter at switching frequency, which is a significant advantage comparing to PD methods. Fig.7 illustrates bode diagram of $G_{out}(s)$ which is the transfer function from the output voltage of the inverter to the grid current. The diagram shows that the amplitude frequency response characteristic of LCL filter with AD is generally the same as that of original LCL filter at high frequency.

TABLE I. MAIN PARAMETERS OF GRID-CONNECTED INVERTER

Symbols	Values
Input dc voltage V_{DC}	600V
Grid line voltage u_{grid}	380V$_{rms}$
Filter inductance L, L_g	1.2mH
Filter capacitor C	4.7uF
Filter internal resistance R, R_g	0.02Ω
Switching frequency f_s	20kHz
Sampling period T_c	25us

D. Design of digital PI controller

The main parameters of the grid-connected inverter are listed in Table I.

According to the parameters, the plant model with AD of PI controller is

$$G_{plant}(s) = \frac{500000(s+18440)(s^2+16.67s+1.773e8)}{(s+16.67)(s+14300)(s^2+4554s+4.671e8)} \quad (1)$$

The discrete plant model $Gd_{plant}(z)$ can be obtained by Z transformation of $G_{plant}(s)$ and zero-order hold, with the sampling time of T_c.

The discrete PI controller can be expressed as

$$Gd_{PI}(z) = K_P \frac{(1+\omega_{PI}T_c)z-1}{z-1}, \quad \omega_{PI} = \frac{K_I}{K_P} \quad (2)$$

where K_P and K_I are the proportional and integral gains of the PI compensator, ω_{PI} is defined as the corner frequency of the PI compensator.

Taking digital system time delay T_{delay} which is one sampling period at least into consideration, the open loop pulse transfer function of the grid-connected system is

$$Gd_{open}(z) = z^{-T_{delay}/T_c} Gd_{PI}(z) Gd_{plant}(z) \quad (3)$$

There are totally 4 poles of the plant model, which is different from that of common voltage source inverters. The corner frequencies respectively related to the poles are

$$\omega_{p1} = 16.67; \omega_{p2} = 14300; \omega_{p3,p4} = \sqrt{4.671e8} = 21612 \quad (4)$$

At low frequency stage there is a single pole $p1$, so that the phase angle of $Gd_{plant}(z)$ is approximately -90° at frequency of $10\omega_{p1}$. The phase angle of $Gd_{PI}(z)$ is -90° at frequencies lower than $\omega_{PI}/10$. To make sure that the phase angle of $Gd_{open}(z)$ will not exceed -180° at low frequency stage, the value of ω_{PI} must meet the following equation.

$$\omega_{PI} < 100\omega_{p1} \quad (5)$$

At high frequency stage the two conjugate poles $p3$ and $p4$ are dominant. The amplitude of $Gd_{open}(z)$ at high frequency stage should be attenuated, so that the resonant peak at frequency $\omega_{p3,p4}$ is below 0. The following equation should be satisfied.

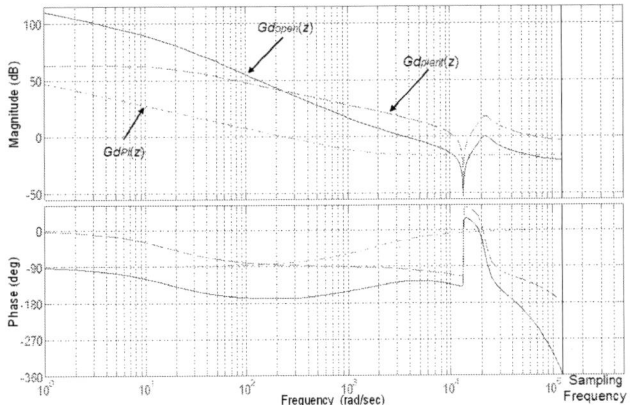

Figure 8. Bode diagram of $Gd_{plant}(z)$, $Gd_{PI}(z)$ and $Gd_{open}(z)$.

Figure 9. Prototype of Li-ion battery power recovery system.

$$\left| Gd_{open}\left(e^{j\omega_{p3,p4}T_c} \right) \right| < 1 \qquad (6)$$

According to equations (5) and (6), the parameters of PI compensator K_P and K_I can be calculated. The bode diagrams of $Gd_{plant}(z)$, $Gd_{PI}(z)$ and $Gd_{open}(z)$ are illustrated in Fig.8.

IV. EXPERIMENTAL PROTOTYPE

According to the design scheme proposed in this paper, a 5kVA prototype of Li-ion battery power recovery system has been built, as shown in Fig.9. There are 8 modules of full-bridge dc-dc converter in the first stage of the prototype, and the PWM signals for these dc-dc modules are generated by CPLD EPM7128, with the switching frequency of 100kHz. The grid-connected inverter of the second stage is designed with respect to the parameters in Table I, and is based on the Intelligent Power Module (IPM) PS22056 of MITSUBISHI SEMICONDUCTOR. The control algorithm of the second stage is implemented in DSP TMS320LF2407A which communicates with CPLD by I/O ports.

The prototype is able to operate stably. The experimental results under various load conditions show that the efficiency of the whole system is in the range of 88%-90%. To simulate the actual operating conditions of Li-ion battery power

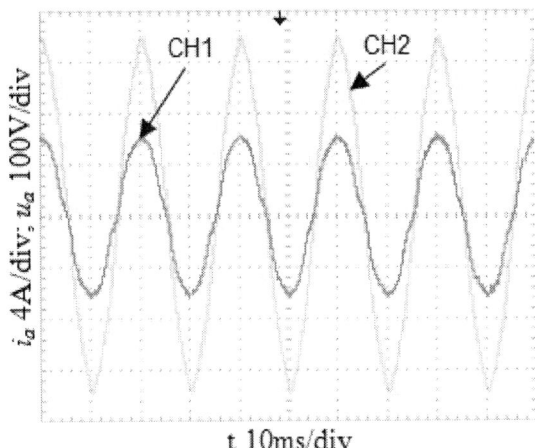

Figure 10. Waveform of grid current under half load condition. CH1 – Grid current i_a; CH2 – Grid voltage u_a.

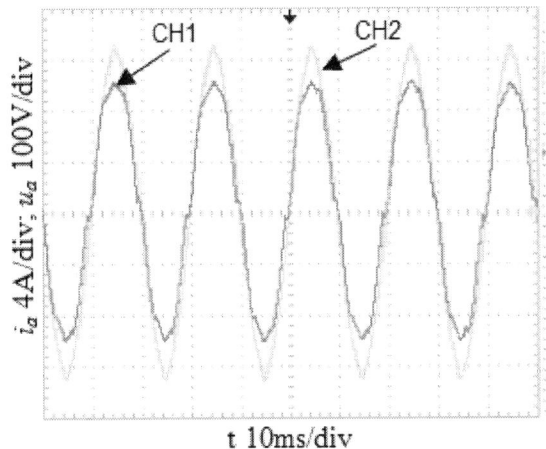

Figure 11. Waveform of grid current under full load condition. CH1 – Grid current i_a; CH2 – Grid voltage u_a.

recovery system, the outputs of the prototype are connected to the ac grid with distorted voltage in the experiments. According the previous literatures, the distortion in the grid voltage has severe impact on the output current of grid connected inverter [18-20].The waveforms of the grid current under half load and full load conditions are illustrated in Fig.10 and Fig.11 respectively. Due to the implementation of AD methods, the harmonics in the grid currents caused by the grid voltage distortion can be limited.

V. CONCLUSIONS

This paper focuses on the design and implementation of Li-ion battery power recovery system. A two stage structure employing a cascade dc-dc converter as the first stage and a grid-connected inverter with LCL filter as the second stage is proposed. And the control scheme based on voltage-oriented PI control and AD method is design for digital controller. The proposed design scheme is verified by a 5kVA prototype which has given acceptable performance both on the system efficiency and grid current quality. In the future work, the control strategy of grid-connected system will still be studied to improve the grid current quality under distorted grid voltage.

REFERENCES

[1] Alepuz. S, Busquets-Monge. S, Bordonau. J, Gago. J, Gonzalez. D, and Balcells. J. "Interfacing Renewable Energy Sources to the Utility Grid Using a Three-Level Inverter," IEEE Transactions on Industrial Electronics, vol. 53, no. 5, pp. 1504-1511 Oct. 2006.

[2] Khan. M.J, Iqbal. M.T, and Quaicoe. J.E. "Utility interactive fuel cell inverter for distributed generation: design considerations and experimental results," Electrical and Computer Engineering, May. 2005, pp. 583–586.

[3] Sangmin. Jung, Youngsang. Bae, Sewan. Choi, and Hyosung. Kim. "A Low Cost Utility Interactive Inverter for Residential Fuel Cell Generation," IEEE Transactions on Power Electronics, vol. 22, no. 6, pp. 2293–2298, Nov. 2007.

[4] Walker. G.R, and Sernia. P.C. "Cascaded DC-DC converter connection of photovoltaic modules," IEEE Transactions on Power Electronics, vol. 19, no. 4, pp.1130-1139, Jul. 2004.

[5] Shen. Z.B, and El-Saadany. E.F. "Novel Interfacing for Fuel Cell Based Distributed Generation," IEEE Power Engineering Society General Meeting, Jun. 2007, pp. 1-6.

[6] Godoy. R.B, Maia. H.Z, Jacques. F, Filho. T, Galotto. L, Onofre. J, Pinto. P, and Tatibana. G.S. "Design and Implementation of a Utility Interactive Converter for Small Distributed Generation," IEEE Industry Applications Conference, vol. 2, Oct. 2006 pp. 1032-1038.

[7] Zhang. J.M, Xie. X.G, Wu. X.K, and Zhaoming Qian. "Comparison study of phase-shifted full bridge ZVS converters," IEEE Power Electronics Specialists Conference vol. 1, Jun 2004 pp. 533-539.

[8] Karshenas. H.R, and Saghafi. H. "Basic Criteria in Designing LCL Filters for Grid Connected Converters," IEEE International Symposium on Industrial Electronics, vol. 3, Jul. 2006 pp. 1996-2000.

[9] Dannehl. J, Wessels. C, and Fuchs. F.W. "Limitations of Voltage-Oriented PI Current Control of Grid-Connected PWM Rectifiers With LCL Filters," IEEE Transactions on Industrial Electronics, vol. 56, no. 2, pp. 380-388. Feb. 2009.

[10] Liserre. M, Blaabjerg. F, and Hansen. S. "Design and control of an LCL-filter-based three-phase active rectifier," IEEE Transactions on Industry Applications, vol. 41, no. 5, pp. 1281-1291 Sep.-Oct.. 2005.

[11] Malinowski. M, and Bernet. S. "A Simple Voltage Sensorless Active Damping Scheme for Three-Phase PWM Converters With an LCL Filter," IEEE Transactions on Industrial Electronics, vol. 55, no. 4, pp. 1876-1880 Apr. 2008.

[12] Eric Wu, and Lehn. P.W. "Digital Current Control of a Voltage Source Converter with Active Damping of LCL Resonance," IEEE Transactions on Power Electronics, vol. 21, no. 5, pp. 1364-1373. Sep. 2006.

[13] Shen. G. Xu. D, Cao. L, and Zhu. X. "An Improved Control Strategy for Grid-Connected Voltage Source Inverters with an LCL Filter," IEEE Transactions on Power Electronics, vol. 23, no. 4, pp. 1899-1906. Jul. 2008.

[14] Jain. P.K, Wen Kang, Soin. H, and Youhao Xi. "Analysis and design considerations of a load and line independent zero voltage switching full bridge DC/DC converter topology," IEEE Transactions on Power Electronics, vol. 17, no. 5, pp. 649-657. Sep. 2002.

[15] Gwan-Bon Koo, Gun-Woo Moon, and Myung-Joong Youn. "Analysis and design of phase shift full bridge converter with series-connected two transformers," IEEE Transactions on Power Electronics, vol. 19, no. 2, pp. 411-419. Mar. 2004.

[16] Xinke. Wu, Zhang. J.M, and Zhaoming Qian. "Optimum design considerations for a high efficiency ZVS full bridge DC-DC converter," Telecommunications Energy Conference, Sep. 2004. pp. 338-344.

[17] Xinke Wu, Junming Zhang, Xiaogao Xie, and Zhaoming Qian. "Analysis and Optimal Design Considerations for an Improved Full Bridge ZVS DC-DC Converter with High Efficiency," IEEE

Transactions on Power Electronics, vol. 21, no. 5, pp. 1225-1234. Sep. 2006.

[18] Abeyasekera. T, Johnson. C.M, Atkinson. D.J, and Armstrong. M. "Suppression of line voltage related distortion in current controlled grid connected inverters," IEEE Transactions on Power Electronics, vol. 20, no. 6, pp. 1393-1401. Nov. 2005.

[19] Hyung Soo Mok, Gyu Ha Choe, Sang Hoon Kim, Jeong Min Lee, and In Young Suh. "Current THD reduction and anti-islanding detection in distributed generation with grid voltage distortion," IEEE International Conference on Sustainable Energy Technologies, Nov. 2008. pp. 989-994.

[20] Erika. Twining, and Holmes. D.G. 'Grid current regulation of a three-phase voltage source inverter with an LCL input filter," IEEE Transactions on Power Electronics, vol. 18, no. 3, pp. 888-895. May. 2003.

A Novel Control System for Harmonic Compensation by Using Wind Energy Conversion Based on DFIG Technology

Grazia Todeschini, *Student Member, IEEE*, and Alexander E. Emanuel, *Life Fellow, IEEE*

Abstract—This paper describes a novel control method for a WECS (Wind Energy Conversion System) meant to partially operate as an AF (Active Filter). The control is defined in an equivalent reference frame obtained by applying the Park Transformation to the three-phase quantities. The zero-sequence component is includd in the control system; this feature enables the compensation of zero-sequence harmonics. The effectiveness of the proposed methodology is quantified in terms of current spectra and reduction of the voltage and current THD at the Point of Common Coupling (PCC).

The proposed application results in higher power loss in the power converters that connects the generator rotor to the PCC with respect to sinusoidal operation. Nevertheless, it is proved that the WECS derating enables the reduction of the power converters' power loss. Voltage distortion at the windings terminals and power converters' terminals is a second consequence of the proposed application, and the peak stator and rotor voltage is an important design parameter in the choice of the solid state devices.

Index Terms—Doubly Fed Induction Generator, Control, $d-q$ theory, Machine Derating, Harmonic Compensation, Wind generation.

I. INTRODUCTION

A typical Wind Energy Conversion System (WECS) is shown in Fig. 1.a. The Doubly Fed Induction Generator (DFIG) stator terminals are connected to the point of common coupling (PCC) through a transformer and a feeder, represented by the equivalent resistance and inductance R_c and L_c. The feeder that connects the Non-Linear Load (NLL) to the PCC is represented by the equivalent resistance and inductance R_h and L_h. The DFIG rotor is supplied by two back-to-back connected converters: the rotor side converter (RSC) and the line side converter (LSC). The feeder that connects the LSC to the PCC has the equivalent resistance and inductance R_L and L_L. The power converters' transistors are driven by means of pulse width modulation (PWM) [1]. The dc-link voltage v_{dc} is the sum of a constant value V_{dc} and a ripple voltage $\tilde{v}_{dc}$:

$$v_{dc} = V_{dc} + \tilde{v}_{dc} \tag{1}$$

One can identify two main subsystems in the WECS (Fig. 1); each subsystems is named after the power converter responsible for its control:

1) RSC subsystem: includes the DFIG, the RSC and the transformer-line equivalent resistance and inductance R_c and L_c.
2) LSC subsystem: includes the dc-link, the LSC and the transformer-line equivalent resistance and inductance R_L and L_L.

Since the transfer functions of the two subsystems are independent of each other, the vectors and matrices that describe the mathematical models of the two subsystems are simpler than the ones that model the whole system. The control for both subsystem is implemented in an equivalent $d-q$ domain.

According to the topology shown in Fig. 1.a, the grid currents $i_{g,abc}$ contain harmonic components injected by the NLL. The WECS can be used as an active filter (AF) to sink the harmonics injected by the NLL. Two basic concepts are described in the literature. The first [2] uses the RSC to modulate the DFIG stator voltage in such a manner that the stator currents i_{sa}, i_{sb} and i_{sc} contain harmonic phasors equal and 180° out of phase with the harmonic phasors of the currents i_{ha}, i_{hb} and i_{hc}. The second approach [3] uses the LSC to modulate the currents i_{La}, i_{Lb} and i_{Lc} in such a way that they contain harmonic phasors equal and 180° out of phase with the harmonic phasors of the currents i_{ha}, i_{hb} and i_{hc}.

In the present work, the second approach is addressed. The new contributions of this study are:

1) A novel control system is designed for the simultaneous operation of the WECS as a power generator and a harmonic compensator. The original feature of the proposed control system is the inclusion of the zero-sequence current component, that corresponds to the triplen harmonic currents [4].
2) Derating of the DFIG is proposed as a method to reduce the excessive power loss in the LSC and RSC caused by harmonic currents injection;
3) Voltage distortion due to harmonic compensation is analyzed.

II. THE CONTROL SYSTEM

A. The $d-q$ transformation

The Park transformation from the abc system to the $dq0$ system is defined as follows:

$$[x_d \ x_q \ x_0]^T = [T(\vartheta)] \ [x_a \ x_b \ x_c]^T \tag{2}$$

with the transformation matrix:

$$[T(\vartheta)] = \sqrt{\frac{2}{3}} \begin{bmatrix} \cos\vartheta & \cos(\vartheta - \frac{2\pi}{3}) & \cos(\vartheta - \frac{4\pi}{3}) \\ \sin\vartheta & \sin(\vartheta - \frac{2\pi}{3}) & \sin(\vartheta - \frac{4\pi}{3}) \\ 1/\sqrt{2} & 1/\sqrt{2} & 1/\sqrt{2} \end{bmatrix}$$

978-1-4244-4782-4/10 $26.00 © 2010 IEEE

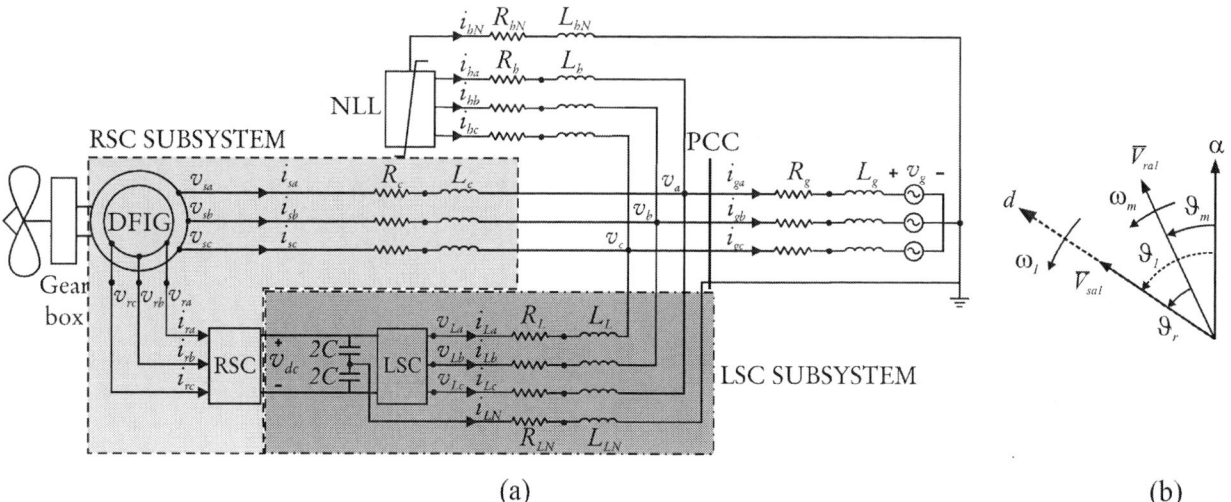

(a)   (b)

Fig. 1.   The studied system and the reference frame used for the Park transformation: (a) WECS (Wind Energy Conversion System) configuration; (b) Stator voltage reference frame: $\overline{V}_{as1}$ and $\overline{V}_{ar1}$ are the phasors corresponding to phase a stator and rotor fundamental voltages. ϑ_m is the angle between $\overline{V}_{as1}$ and $\overline{V}_{ar1}$; $\omega_m = d/dt\vartheta_m$; ϑ_1 is the angle between the rotating reference axis-d and the stationary reference axis-α; $\omega_1 = d/dt\vartheta_1$.

where ϑ is the transformation angle. In the present work, the stator fundamental voltage has been chosen as a reference frame, meaning that the d-axis is aligned with the stator voltage phasor $\overline{V}_{sa1}$ [5] and is rotating with synchronous speed ω_1. According to the notation used in Fig. 1.b, the transformation angle is $\vartheta = \vartheta_1$ for stator quantities and $\vartheta = \vartheta_r$ for rotor quantities. The Park transformation (2) includes the zero-sequence component x_0 that quantifies the current flow in the neutral.

B. RSC subsystem

The RSC subsystem is intended to control the power flow from the wind turbine to the grid and to regulate the power factor (PF) at the DFIG terminals. The stator instantaneous active power p_s and stator instantaneous imaginary power q_s [6] are the input for the control system and they are function of the PF. If γ is the angle between the stator fundamental voltage and the stator fundamental current phasors, the active and imaginary instantaneous powers at the stator terminals are obtained as follows [7]:

$$\begin{cases} p_s = \dfrac{3}{2}\hat{V}_{s1}\hat{I}_{s1}\cos(\gamma) \\ q_s = \dfrac{3}{2}\hat{V}_{s1}\hat{I}_{s1}\sin(\gamma) \end{cases} \qquad (3)$$

where $\hat{V}_{s1}$ and $\hat{I}_{s1}$ are the peak amplitude the stator voltage and current phasors.

1) Rotor reference currents:

The rotor reference current i_{dr1}, i_{qr1} are derived from the values of stator instantaneous active p_s and imaginary power q_s as follows [7]:

- the stator instantaneous active power p_s defines the fundamental rotor current d-component i_{dr1}:

$$i_{dr1} = -\frac{L_s}{L_m}\frac{p_s}{\overline{v}_s} \qquad (4)$$

where L_s is the stator winding self-inductance, L_m is the mutual inductance between stator and rotor windings, $\overline{v}_s = v_d$ is the space vector corresponding to the stator voltage.

- the stator instantaneous imaginary power q_s defines the fundamental rotor current q-component i_{qr1}:

$$i_{qr1} = -\frac{1}{L_m}\left(\lambda_s - L_s\frac{q_s}{v_s}\right) \qquad (5)$$

where λ_s is the flux linkage computed as $\lambda_s = L_s\, i_{qs1} + L_m\, i_{qr1}$.

2) RSC control block diagram:

The block diagram describing the RSC control is shown in Fig. 2. The upper branch describes the process for the d-axis variables and the lower branch deals with the q-axis variables.

On the top left corner of Fig. 2, the fundamental component i_{dr1} is obtained by applying (4). The error $\Delta i_{dr} = i_{dr1} - i_{dr,meas}$ is the input signal of the Proportional Integral controller PI_1 [8]. The output of the PI_1 controller is the auxiliary voltage v'_{dr}, that should not be confused with the reference voltage $v_{dr,ref}$. The relation between the auxiliary voltage v'_{dr} and the reference voltage $v_{dr,ref}$, in the stator flux reference frame [7], [9] is as follows:

$$v_{dr,ref} = v'_{dr} + \omega_1 L_m\, i_{ds} - (\omega_1 - \omega_r)\,(L_r\, i_{qr} + L_m\, i_{qs}) \quad (6)$$

where $v'_{dr} = v_{dr} - \dfrac{d}{dt}\, M\, i_{ds} + (\omega_1 - \omega_r)\,(L_r\, i_{qr} + L_m\, i_{qs})$.

The q-axis control process is similar to the d-axis. Since the system geometrical and electrical symmetry results in the same parameters for the equivalent d and q equivalent circuits, two identical controllers PI_1 are used for the upper and lower branches. The actual reference voltage $v_{qr,ref}$ is obtained from the auxiliary voltage as follows [7], [9]:

$$v_{qr,ref} = v'_{qr} + \omega_1 L_m\, i_{qs} + (\omega_1 - \omega_r)\,(L_r\, i_{qr} + L_m\, i_{qs}) \quad (7)$$

where $v'_{qr} = v_{qr} - s\, M\, i_{ds} + (\omega_1 - \omega_r)\,(L_r\, i_{qr} + L_m\, i_{qs})$.

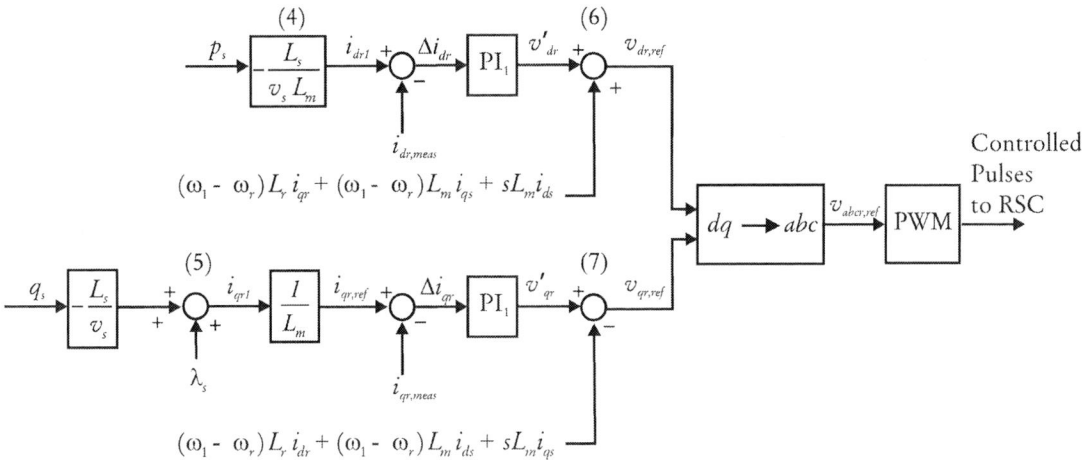

Fig. 2. The RSC control: block diagram. The inputs are the stator instantaneous active power p_s and stator instantaneous imaginary power q_s. The output are the three-phase rotor reference voltages $v_{rabc,ref}$.

By applying the inverse Park transformation to the reference voltages v_{qr} and v_{dr}, one obtains the three-phase reference voltages $v_{abc,ref}$ that are the input for the PWM process of the RSC.

The RSC control does not include the zero-sequence component from the Park transformation (2). Since the DFIG windings are never connected to the neutral [10], triplen harmonics corresponding to zero-sequence current cannot flow in the RSC subsystem.

C. LSC subsystem

The LSC subsystem is meant to keep a constant dc-link voltage, allowing the power flow in and out of the rotor [11] and to inject harmonic currents equal in magnitude and $180°$ out of phase with respect to the NLL harmonic currents. The LSC reference currents $i_{dL,ref}$ and $i_{qL,ref}$ are the sum of two components:

$$i_{dL,ref} = i_{dL1} + i_{dLh} \qquad (8)$$

$$i_{qL,ref} = i_{qL1} + i_{qLh} \qquad (9)$$

where i_{dL1} and i_{qL1} are the components at fundamental frequency ($h = 1$), while the sets i_{dLh}, i_{qLh} include the harmonic, interharmonic and subharmonic components ($h \neq 1$). The LSC reference currents $i_{0L,ref}$ is responsible for the compensation of zero-sequence triplen harmonics ($h = 3k$).

1) Fundamental LSC currents:

The reference current i_{qL1} at fundamental frequency is obtained from the imaginary power at the LSC terminals q_L [9]:

$$i_{qL1} = \frac{q_L}{v_d} \qquad (10)$$

where $q_L = v_d\, i_{qL1}$ in the stator voltage reference frame and i_{qL1} is obtained by applying the Park transformation (2) to the currents $i_{L,abc}$ (Fig. 1.a).

The fundamental reference current i_{dL1} is related to the variation of the dc-link voltage [9]:

$$i_{dL1} = \frac{2\sqrt{2}}{3}\, \frac{1}{m_a}\, \frac{dv_{dc}}{dt} \qquad (11)$$

where $0 < m_a < 1$ is the PWM amplitude modulation factor [1].

2) Harmonic LSC currents:

In the stator voltage reference frame, the operation of the LSC as an AF is is expressed as follows:

$$i_{dLh} = -i_{dh} \quad , \quad i_{qLh} = -i_{qh} \quad \text{and} \quad i_{0Lh} = -i_{0h} \qquad (12)$$

3) LSC control block diagram:

The LSC control block diagram is presented in Fig. 3; the upper branch of the diagram refers to the d-axis variables, the middle branch refers to the q-axis variables, the lower branch refers to the zero-sequence variables.

The d-axis variables are considered at first: in the top left corner of Fig. 3, the difference Δv_{dc} between the reference voltage and the measured voltage is the input for the dc-link PI$_2$ controller [9]. The PI$_2$ controller output is the LSC fundamental reference current $i_{dL,ref}$. The sum of the LSC fundamental reference current $i_{dL1,ref}$ and the negative of the load harmonic current $i_{dLh,ref}$ is the LSC reference current, according to (8) and to the second relation of (12).

The error $\Delta i_{dL} = i_{dL,ref} - i_{dL,meas}$ is the input for the PI$_3$ controller of the LSC. The output of the PI$_3$ controller is the auxiliary reference voltage v'_{dL}. The actual reference voltages $v_{dL,ref}$ is obtained from the auxiliary voltage v'_{dL}. According to [7], [9]:

$$v_{dL,ref} = v'_{dL} - \omega_1\, L_L\, i_{qL} + v_d \qquad (13)$$

where $v'_{dL} = v_s + \dfrac{d}{dt}\, L_L\, i_{qL} - v_{dL}$.

A similar procedure is applied to the q-axis variables except for reference current i_{qL} definition derived from the value of

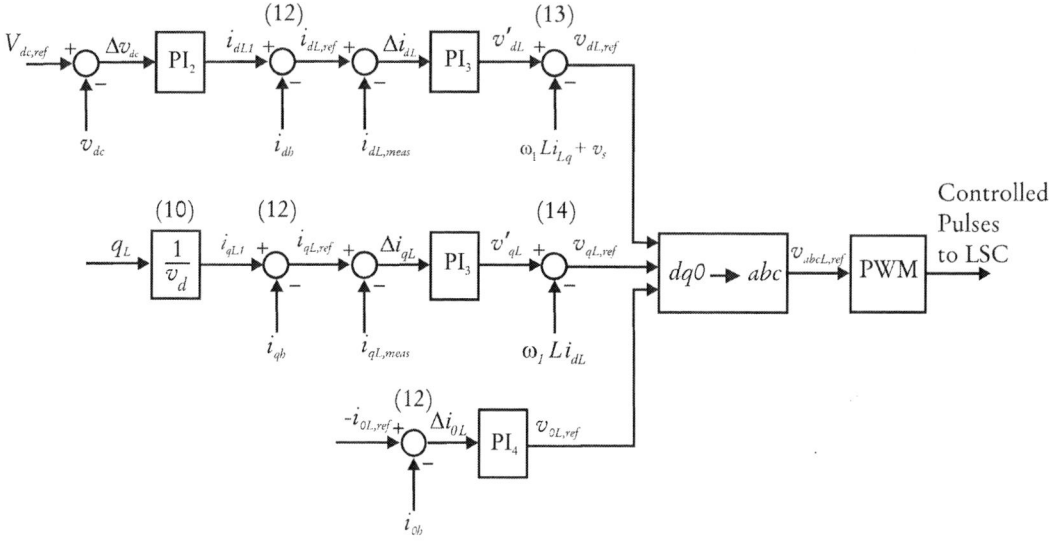

Fig. 3. The LSC control: block diagram. The inputs are dc-link voltage v_{dc}, the imaginary power q_L and the NLL equivalent currents i_{dh}, i_{qh} and i_{0h}. The output are the three-phase LSC reference voltages $v_{Labc,ref}$

q_L (10). The actual reference voltages $v_{qL,ref}$ is obtained from the auxiliary voltage v'_{qL} [7], [9]::

$$v_{qL,ref} = v'_{qL} + \omega_1 \, L_L \, i_{dL} \qquad (14)$$

where $v'_{qL} = -\dfrac{d}{dt} L_L \, i_{dL} + v_{qL}$.

The zero-sequence control system is shown in the lower branch. The zero-sequence control system does not include any auxiliary voltage because there is no coupling between the zero-sequence variables and the d and q variables. For the considered WECS (Fig. 1.a), the zero-sequence transfer function (TF) differs from the d and q transfer functions, because the zero-sequence current flow includes the neutral. Since the PI controller parameters are determined from the TF [8], it results that the zero-sequence controller PI$_4$ differs from the d and q axis controllers PI$_3$. The PI controller parameters are listed in the Appendix.

III. VOLTAGE HARMONICS

The flow of harmonic currents through the line that connects the LSC to the PCC implies voltage distortion at the LSC terminals. Simultaneously, the PWM control of the RSC implies voltage distortion at the rotor terminals; the rotor and the stator of the DFIG are mutually coupled, therefore voltage distortion at the rotor's terminals implies voltage distortion at the stator terminals. Voltage distortion, and mainly the peak voltage, deserves special attention, and should be within the limits that can be withstood by the machine windings' insulation and the solid state switches.

The voltage distortion a the stator terminals, at the PCC and at the output of the power converters will be analyzed qualitatively by considering a case study.

IV. POWER CONVERTERS LOSS AND DFIG DERATING

Since the same active power flows through the rotor, the RSC and the LSC, the power converters are rated for the rotor rated power (or slip power). For DFIGs, the rotor rated power is a ratio of the stator rated power [12] according to the expression:

$$P_{nr} = |s_{max}| P_{ns} \qquad (15)$$

where s_{max} is the maximum slip. In typical applications, the maximum slip is found in the range of 0.2 to 0.5. For the present WECS, the maximum slip is $s_{max} = +0.5$; therefore, the RSC and LSC rated power loss ΔP_n is determined at $s = +0.5$, when it reaches its maximum value.

In the present work, the RSC and LSC power loss ΔP_n is obtained as the difference between the input and the output active powers on the RSC and on the LSC [13].

The harmonic currents cause additional power loss in the power converters and the WECS derating is needed to maintain the power converters' loss equal or below the rated value. The WECS derating is tantamount to the DFIG derating that consists in a reduction of the fundamental power flow through the stator. In [7], it has been proved that DFIG stator derating results in an automatic rotor derating: in other words, the reduction of fundamental power flow through the stator results in a reduction of fundamental power flow through the rotor and the power converters. As a consequence, the power loss through the power converters is lowered.

V. SIMULATIONS

The studied WECS parameters are summarized in the Appendix. The NLL is simulated by means of an array of equivalent current harmonic sources. The Simulink® software [14] is used to model the system and a fixed-step discrete time ($T_s = 5\mu$s) simulation is chosen.

The harmonics and interharmonics currents injected by the NLL are listed in Table I. The corresponding NLL current spectrum is depicted in Fig. 4.a. Unity power factor (PF=1) is assumed for operation of the LSC, resulting $q_L = 0$ and (10) respectively.

TABLE I
HARMONIC INJECTED BY THE NLL FOR THE CASE STUDY.

h	0.05	3	5	7	7.5
I_h/I_n	0.10	0.20	0.20	0.14	0.10
Assumed Sequence	+	0	+	-	+

A. Current spectra and THD at the PCC

The current spectra shown in Fig. 4 prove the validity of the proposed control system.

The LSC current spectrum is depicted in Fig. 4.b. The current component at fundamental frequency is responsible for the active power flow through the rotor. The harmonic components of the LSC current are responsible for the cancellation of the harmonic currents injected by the NLL (Fig. 4.a). Other minor harmonic components appear due to the fact that the solid state switches are modeled as real devices, therefore the voltage drop across the switches affects the LSC and RSC current and voltage oscillograms. The stator current spectrum is depicted in Fig. 4.c: the first harmonic current is responsible for the fundamental power flow and the other minor components are due to the coupling with the RSC through the rotor.

The grid current spectrum with compensation is displayed in Fig. 4.d: it contains a dominant fundamental component at 60 Hz while the harmonic currents of the same order as the harmonics injected by the NLL are significantly reduced; other harmonic components are due to power converters' operation and to numerical truncation.

The reduction of harmonic pollution on the grid can be evaluated by comparing the THD (total harmonic distortion) of the grid current and grid voltage [15], [16] before and after the compensation by means of LSC modulation. The results are listed in column 2 and 3 of Table II and the percent reduction is shown in column 4. All the values are calculated for the DFIG operating at maximum power and maximum rotor speed; the reduction of current THD from 25% to 7.5 % is significant and proves the validity of the proposed application.

TABLE II
VOLTAGE AND CURRENT THD AT THE PCC FOR THE CASE STUDY.

		THD	No Compensation	LSC Modulation	% Reduction
(a)	Current		0.25	0.075	-333 %
(b)	Voltage		0.11	0.055	-50 %

B. RSC and LSC power loss

Three curves that corresponds to the normalized RSC and LSC power loss under different operating conditions are

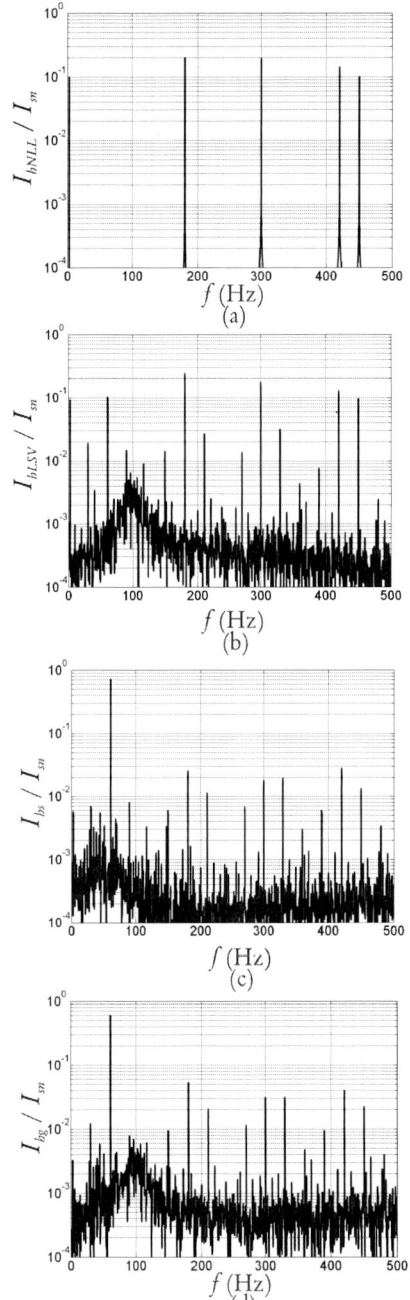

Fig. 4. Amplitude spectra: (a) NLL current, (b) LSC current, (c) stator current and (d) grid current.

presented as function of the normalized rotor speed ω_r/ω_1 in Fig. 5:

1) $\Delta P_n/P_n$: power loss ratio at rated conditions (sinusoidal operation). Maximum power loss $\Delta P_{n,max}/P_n \approx 11.8$ is obtained at maximum normalized rotor speed $\omega_r/\omega_1 = 1.5$.

2) $\Delta P_H/P_n$: power loss ratio when the DFIG is used as

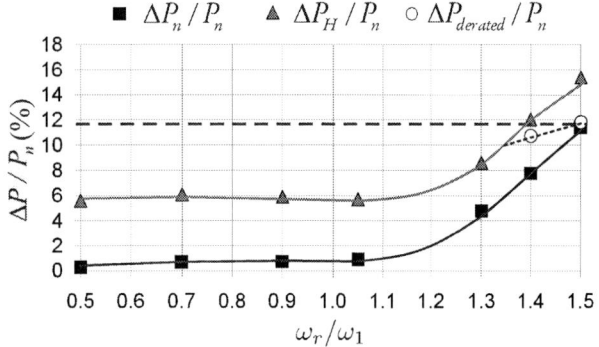

Fig. 5. RSC and LSC normalized power loss vs. normalized rotor speed. Base values are rated stator power and synchronous speed. $\Delta P_n/P_n$: power loss ratio at normal (sinusoidal) operation; $\Delta P_H/P_n$: power loss ratio when the DFIG is used as AF for the currents listed in Table I; $\Delta P_{derated}/P_n$: power loss ratio when derating is implemented according to Fig. 6.

AF for the currents listed in Table II. The RSC and the LSC power loss is higher when the LSC operates as an AF than for rated operation. This is due to the fact that additional harmonic currents flow in the power converters.

3) $\Delta P_{derated}/P_n$: power loss ratio when derating is implemented according to the curve illustrated in Fig. 6. If DFIG derating is applied, the power loss in the region $1.0 \leq \omega_r/\omega_1 \leq 1.5$ are reduced below the rated value, because the power flow through the converters is lowered.

C. Stator Fundamental Current Curve

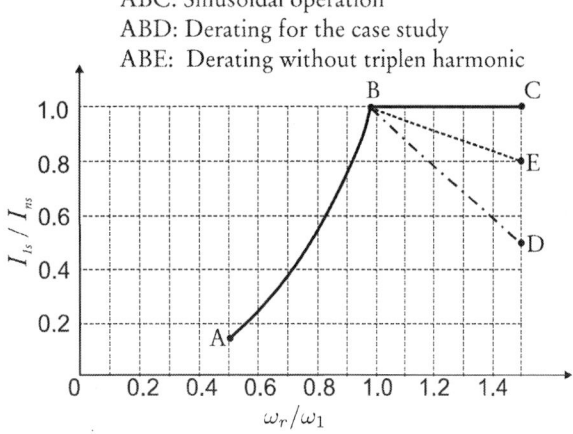

Fig. 6. Normalized stator fundamental current vs. normalized rotor speed, assuming PF=1 at the DFIG terminals. Base values are rated stator current and synchronous speed. Continuous line: fundamental stator current curve corresponding to the turbine tracking characteristic [17], [18]; Dot-dashed line: fundamental stator current curve when derating is applied; Dashed line: fundamental stator current for the same NLL current spectrum excluded the triplen harmonic component.

In Fig. 6, the normalized stator fundamental current I_{1s} as function of the normalized rotor speed is presented. Base

values are the rated stator current and the synchronous speed. The stator fundamental current is considered only because, as proved in [7], derating of the stator results in an automatic derating of the rotor. Since PF=1 at the stator terminals, the stator fundamental current is obtained as the ratio between the fundamental active power and the rated voltage.

For subsynchronous speeds ($0.5 \leq \omega_r \leq 1.0$), the fundamental stator current curve corresponding to the tracking characteristic and to derating overlap. For supersychronous speeds ($1.0 \leq \omega_r \leq 1.5$), the derating curves implies a decreasing value of the fundamental current from 1 to 0.5 in the region $1.0 \leq \omega_r \leq 1.5$, under the condition of constant power converters loss $\Delta P_{n,max}/P_n$.

D. Voltage oscillogram, spectra and THD

The voltage oscillograms and spectra are shown in Figs. 7-14. The voltage THD at the PCC are listed in row (b) of Table II.

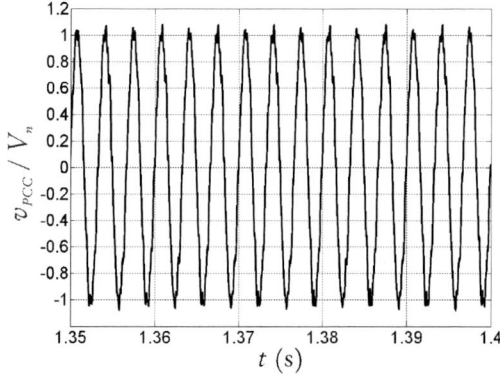

Fig. 7. Phase-to-phase PCC voltage oscillogram for the case study. Compensation by means of LSC modulation is applied.

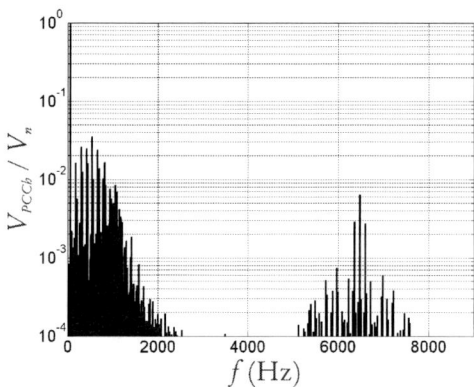

Fig. 8. Phase-to-phase PCC voltage amplitude spectrum for the case study. Compensation by means of LSC modulation is applied.

The voltage at the PCC is a slightly distorted sinusoid (Fig. 7). The distortion can be explained by analyzing the PCC voltage amplitude spectrum illustrated in Fig. 8. The first harmonic dominates; there are some low frequency harmonic

978-1-4244-4782-4/10 $26.00 © 2010 IEEE

components due to the harmonic voltage drop on the line impedance; other high frequency harmonic components are due to the PWM of the LSC (the carrier frequency is f_c=6450 Hz).

The stator voltage oscillogram (Fig. 9) and voltage amplitude spectrum (Fig. 10) are very similar to the PCC voltage oscillogram and voltage amplitude spectrum respectively; the stator voltage is more distorted due to the harmonic voltage drop on the feeder impedance that connects the stator terminals to the PCC.

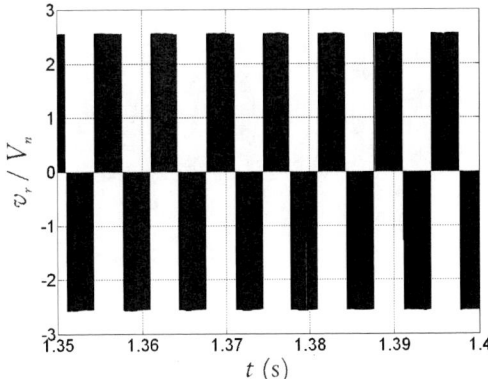

Fig. 11. Phase-to-phase rotor voltage oscillogram for the case study. Compensation by means of LSC modulation is applied.

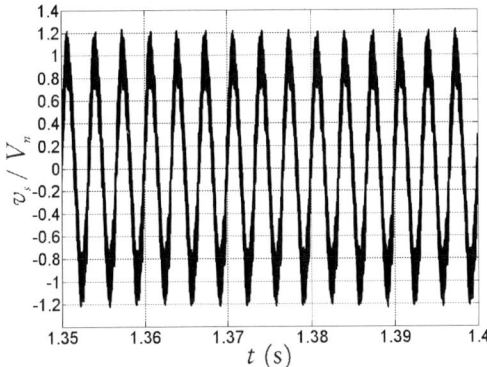

Fig. 9. Phase-to-phase stator voltage oscillogram for the case study. Compensation by means of LSC modulation is applied.

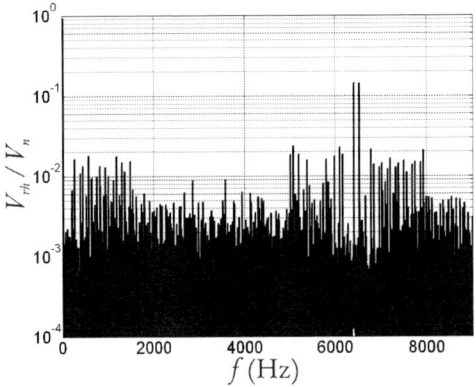

Fig. 12. Phase-to-phase rotor voltage amplitude spectrum for the case study. Compensation by means of LSC modulation is applied and s=0.5.

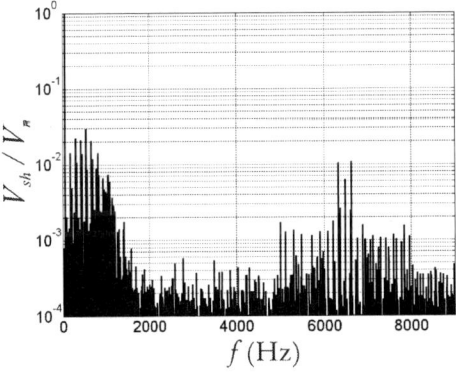

Fig. 10. Phase-to-phase stator voltage amplitude spectrum for the case study. Compensation by means of LSC modulation is applied.

The rotor voltage oscillogram (Fig. 11) is a typical output of a power converter controlled by means of PWM. The rotor voltage amplitude spectrum (Fig. 12) significant high frequency harmonics due to PWM. The fundamental voltage harmonic is a function of the rotor speed: in any induction machine the rotor current and voltage frequency are function of the slip s [19], [20]:

$$s = (\omega_1 - \omega_m)/\omega_1 \qquad (16)$$

where ω_m is the rotor speed.

Similarly to Fig. 12, the phase-to-phase LSC voltage oscillogram (Fig. 13) is a typical output of a power converter controlled by means of PWM. The fundamental harmonic is

at 60 Hz because the LSC three-phase terminals are connected to the mains through a feeder. The harmonic components cased PWM are located in the same frequency region as in the rotor voltage spectrum because the same carrier frequency is used. The main difference with respect to the rotor voltage spectrum (Fig. 12) is the presence of low frequency harmonics, due to the use of the LSC as an AF.

VI. CONCLUSIONS

A control strategy for the use of the LSC as an AF has been presented and the effects in terms of voltage distortion and power loss have been identified and evaluated by means of current and voltage spectra and by means of THD variation. Two side design issues are addressed:

1) Derating is needed to maintain the power loss in the DFIG windings and in the power converters equal or below the rated values.

2) Current harmonic sinking causes voltage distortion at the DFIG terminals and increased peak voltages impressed across the rotor and stator windings.

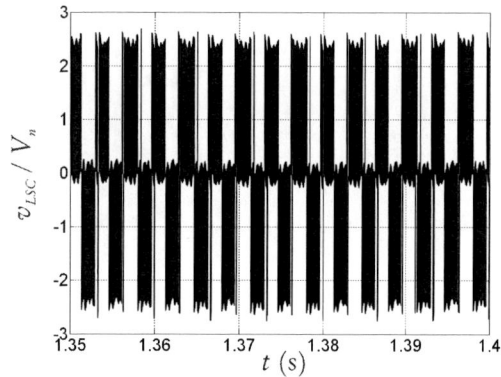

Fig. 13. Phase-to-phase LSC voltage oscillogram for the case study. Compensation by means of LSC modulation is applied.

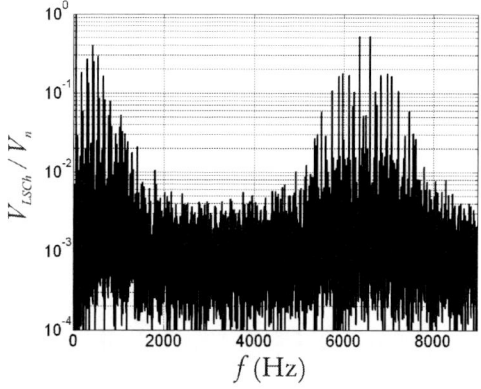

Fig. 14. Phase-to-phase LSC voltage spectrum for the case study. Compensation by means of LSC modulation is applied.

REFERENCES

[1] N. Mohan, T. M. Undeland, and W. P. Robbins, *Power Electronics: converters, applications and design, 3rd edition*, J. Wiley & Sons, 111 River Street, Hoboken, NJ 07030-5774, US, 2003.

[2] M. T. Abolhassani, H. A. Toliyat, and P. Enjeti, "An electromechanical active harmonic filter," in *Proc. Electric Machines and Drives Conference IEMDC 2001, IEEE Intern.*, Cambridge, MA, US, 17-20 June 2001, pp. 349 – 355.

[3] E. Tremblay, A. Chandra, and P.J. Lagac, "Grid-side converter control of DFIG wind turbines to enhance power quality of distribution network," in *Proc. Power Engineering Society General Meeting*, Montreal, Canada, 18-22 June 2006.

[4] W. D. Stevenson, *Elements of Power System Analysis*, McGraw-Hill, 6th Avenue, New York City, NY, 10001-6309, US, 1992.

[5] A. K. Jain, R. Tirumala, N. Mohan, T. Gjengedal, and R. M. Halet, "Harmonics and flicker control in wind farms," in *Wind Power and the Impacts on Power Systems Workshop*, Oslo, Norway, 17-18 June 2002, pp. 1–5.

[6] H. Akagi, E. H. Watanabe, and M. Aredes, *Instantaneous power theory and applications to power conditioning*, IEEE Press, 111 River Street, Hoboken 07030-5774, NJ, US, 2007.

[7] G. Todeschini and A. E. Emanuel, "Wind energy conversion system as an active filter: Design and comparison of three control systems (part I)," Companion paper.

[8] C. L. Phillips and R. D. Harbor, *Feedback Control Systems (Fourth Edition)*, Prentice Hall, Upper Saddle River, NJ 07548, US, 2000.

[9] R. Pena, J.C. Clare, and G.M. Asher, "Doubly Fed Induction Generator Using Back-to-Back PWM Converters and its Application to Variable-

Speed Wind-Energy Generation," *IEE Proc.-B, Electric Power Applicat.*, vol. 143, no. 3, pp. 231–241, May 1996.

[10] P. C. Krause, O. Wasynczuk, and S. D. Sudhoff, *Analysis of Electric Machinery and Drive Systems*, Wiley-IEEE Press, 111 River Street, Hoboken 07030-5774, NJ, US, 2002.

[11] F. Blaabjerg, Z. Chen, R. Teodorescu, and F. Iov, "Power electronics in wind turbines systems," in *Proc. CES/IEEE-PELS Int. Power Electronics and Motion Control Conference (IPEMC)*, Shangai, China, August 13-16 2006, pp. 1–11.

[12] I. Boldea, *Variable Speed Generators*, CRC Press, 2000 NW Corporate Blvd Boca Raton, FL 33431, US, 2006.

[13] E. F. Fuchs and M. A. S. Masoum, *Power Quality in Power Systems and Electrical Machines*, Academic Press, 6th Avenue, New York City, NY, 10001-6309, US, 2008.

[14] Simulink User Guide, "http://www.mathworks.com," 2008.

[15] IEEE Power Engineering Society Standard 519, *Recommended Practices and Requirements for Harmonic Control in Electrical Power Systems*, 1992.

[16] IEEE Power Engineering Society Standar 519, *Correction to IEEE Recommended Practices and Requirements for Harmonic Control in Electrical Power Systems*, 1992.

[17] T. Ackermann, *Wind Power in Power Systems*, John Wiley and Sons, Hoboken, NJ 07030-5774, US, 2005.

[18] T. Burton, D. Sharpe, N. David, and E. Bossanyi, *Wind Energy Handbook*, John Wiley & Sons Inc., Hoboken, NJ 07030-5774, US, 2001.

[19] A. S. Langsdorf, *Theory of Alternating Current Machinery*, McGraw-Hill, 6th Avenue, New York City, NY, 10001-6309, US, 1937.

[20] A. E. Fitzgerald, C. Kingsley, and S. D. Umans, *Electric machinery, 5th Edition*, McGraw-Hill, 6th Avenue, New York City, NY, 10001-6309, US, 1990.

APPENDIX
SYSTEM PARAMETERS

DFIG: $P_n = 1.50$ MW, $cos\varphi_n = 0.90$, $V_n = 575$ V, number of pole pairs = 3, $R_s = 0.00706$ p.u., $L_{ls} = 0.171$ p.u. , $R_r = 0.005$ p.u., $L_{lr} = 0.156$ p.u., $L_m = 2.9$ p.u., H (inertia constant)= 5.04 s, F (friction factor) = 0.01 p.u.

Supply line: $V_n = 600$ V, $f_n = 60$ Hz, $R_g = 1$ m Ω, $L_g = 50$ mH.

RSC,LSC and dc-link: $C = 0.6$ F, $V_{dc} = 1200$ V, $R_L = 3$ mΩ, $L_L = 300$ mH, $\Delta P_{n,RSC}/P_{nr} = \Delta P_{n,LSC}/P_{nr} \approx 0.06$.

PI controllers: RSC current controller (PI$_1$): k_{p1}=0.1, k_{i1}=4; dc-link voltage controller (PI$_2$): k_{p2}=0.025, k_{i2}=0.001; LSC current controller (PI$_3$): k_{p3}=4, k_{i3}=8, LSC zero-sequence controller (PI$_4$): k_{p4}=0.002, k_{i4}=400.

The turbine tracking characteristic is shown in Fig. 15.

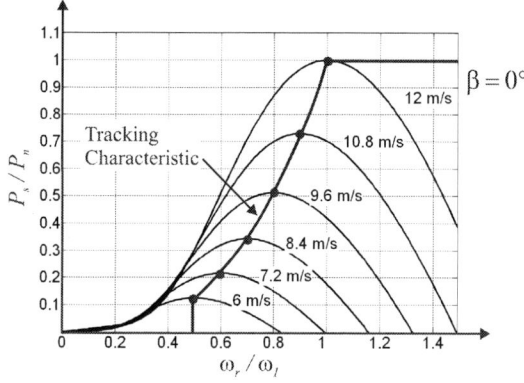

Fig. 15. Wind turbine tracking characteristic: the target normalized stator power for different wind speeds vs. normalized rotor speed.

A Transformerless Modular Permanent Magnet Wind Generator System with Minimum Generator Coils

Xibo Yuan, Yongdong Li and Jianyun Chai
Dept. of Electrical Engineering, Tsinghua University
Beijing, China
Email: yuanxb05@mails.tsinghua.edu.cn

Abstract—This paper has presented a modular converter for multi-coil direct-drive permanent magnet wind generator system. The converter modules are cascaded to achieve medium voltage output (6kV~33kV), thus eliminating the grid-side step-up transformer, which is desirable for both on-shore and off-shore wind turbines. Each converter module is composed of a rectifier, dc-link and an inverter. The generator coils with 90 degree phase shift are rectified through the power factor correction (PFC) circuit and connected in series to get unity power factor, stable dc-link power and higher dc-link voltage. The generator armature inductance is used as the AC-side PFC boost inductor, thus reducing the system size and cost. The inverter adopts a neutral point clamped (NPC) five-level converter to match the dc-link voltage level and are cascaded to achieve multilevel medium voltage output. The vector control scheme is used to regulate the converter active and reactive power transferred to the grid. Simulation results with a 1.5MW wind generator and converter system and experimental results with a scaled 3kW system validate the proposed topology and control method.

I. INTRODUCTION

The direct-drive permanent magnet wind generator interfaced to the grid through a full power converter is being increasingly adopted due to its higher power density, better controllability and reliability, especially so during the grid-faults[1]. The voltage level of wind generator and its power converter is usually in the range of 380V~690V due to the generator voltage rating and voltage limitation of the power electronics devices. And the power converter is usually connected to the grid via a step-up transformer to match the grid voltage level (6kV~35kV). For onshore wind turbines, generator and converter are usually put in the nacelle on top of the tower, while the grid-side step-up transformer is placed at the bottom. Electric power is transmitted down through flexible cables of high current rating which are expensive and can suffer from I^2R loss. An offshore wind turbine usually has to include the step-up transformer in the nacelle, which adds significantly to the mechanical loading of the tower [2]. In either case, a transformer-less, medium voltage power converter system would be an attractive technology for large wind turbines.

To that end, this paper developed a modular direct-drive permanent magnet wind generator system, aiming to eliminate the step-up transformer by cascading converter modules. Each module is composed of a rectifier fed from isolated generator coils, a dc-link and an inverter as shown in Fig.1. Previous papers have studied the design of the multi-pole modular direct-drive permanent magnet generator (PMG), which has a number of isolated coils [3-5]. Similar converter topologies are proposed in [6,7], without further considering the control and implementation issues. Whereas in [2,8], the authors mainly focus on the control of the cascaded H-bridge inverters as shown in Fig.1(b). Since the generator usually has limited isolated coils, it is preferable to reduce the number of required isolated coils in the generator. This can be achieved by increasing the dc-link voltage of each converter module, such that the high voltage (6kV~35kV) can be reached with minimum number of stages in Fig.1 (a). Therefore, this paper proposes to connect the two power factor correction (PFC) circuit in series to increase the dc-link voltage rather than in parallel as shown in Fig.1 (c). Accordingly, the inverter side adopts a neutral point clamped (NPC) type inverter to match the device rating and harmonics requirement. The control scheme of both the rectifier and inverter are then described in the paper, which is verified by simulation and experimental results.

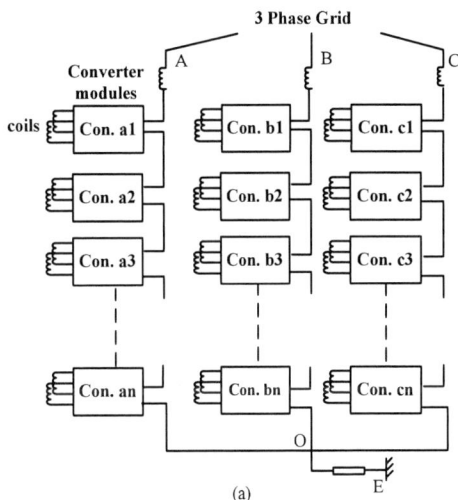

(a)

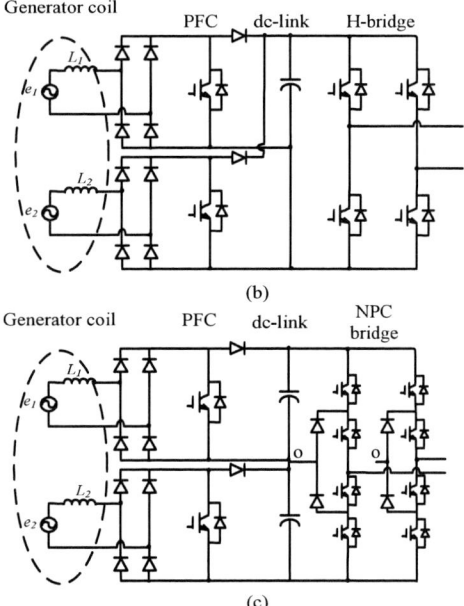

(b)

(c)

Fig.1. Configuration of the proposed system. (a) Electrical configuration of the wind generator and multilevel converter system; (b) Topology of the converter module with two PFC circuits in parallel and H-bridge inverter. (c) Topology of the converter module with two PFC circuits in series and NPC inverter.

II. PROPOSED TOPOLOGY AND CONTROL METHOD

A. Rectifier Topology and Control Scheme

In Fig.1 (c), each isolated generator coil is rectified through a PFC circuit to get the unit power factor and stable dc-link voltage. The boost-type PFC also improves the dc-link voltage to a certain level under different wind speed, thus matching the grid voltage with limited converter stages. It should be pointed that the output power of the single phase PFC circuit has an ac component which is twice of the generator stator frequency. This may cause the dc-link voltage ripple and affect the inverter output voltage. For the direct-drive PMG, since the stator frequency is relatively low (usually below 15Hz), it needs a large dc-link capacitor to reduce the voltage ripple. In this paper, the output of the two generator coils with 90 degree phase shift are rectified and connected in series to counteract the twice of the stator voltage frequency component, as indicated by (1)

$$p_{dc-link}(t) = \sqrt{2}V \sin(\omega t)\sqrt{2}I \sin(\omega t)$$
$$+ \sqrt{2}V \sin(\omega t + \frac{\pi}{2})\sqrt{2}I \sin(\omega t + \frac{\pi}{2}) = 2VI \quad (1).$$

Where, V and I are the voltage and current of the generator coil, respectively; $p_{dc-link}$ is the power fed into the dc-link, which is constant by connecting the two rectifiers in series; ω is the PMG stator frequency. Here, the generator armature inductance is used as the ac-side boost inductor, without extra inductance, thus reducing the system size and cost. The required range of the inductor value is mainly determined by the PMG stator current ripple constraint and the current zero-crossing distortion. Larger inductance will reduce the current ripple, however, causing larger the zero-crossing distortion [9-

11]. The generator inductance value may need to be designed to meet the system performance specifications.

The PFC circuit can also adopt other topologies, such as half-bridge, full-bridge or bridges-less PFC. Considering the unidirectional power flow, unity power factor operation and the number of active switch (the corresponding cost), the single-switch PFC is used. Note that, although the whole dc-link voltage does not have the low frequency power ripple from the generator side, the neutral point O still has the single phase pulsating power. In Fig.1(c), the neutral point of the NPC bridge inverter is connected to the same midpoint of the dc-link as the rectifier neutral point. In this way, the neutral point potential of the NPC bridge is actively clamped and balanced by the rectifier. The disadvantage here is the dc-link neutral point low frequency ripple may appear on the NPC bridge output and affect the output voltage waveform. This ripple can be filtered out passively by adding more dc-link capacitors. Another option is to let the NPC bridge have their own dc-link capacitors, whose neutral point is not connect to the rectifier neutral point. In this case, proper balancing method must be used for controlling the capacitor neutral point voltage. In order to simplify the PWM generation (using phase shifted PWM) and achieve the active clamping, this paper adopts the topology in Fig.1 (c).

The PFC control scheme enables the coil current to track the generator back-EMF, thus achieving unit power factor and sinusoidal current waveform. Fig.2 shows the basic control diagram of the PFC circuit, where the I_{ref} is the phase current reference, which is the output of the outer dc-link voltage loop. Note that the generator back-EMF, which is not accessible for measurement in the proposed topology, is reconstructed by the PMG rotor position obtained from the shaft encoder.

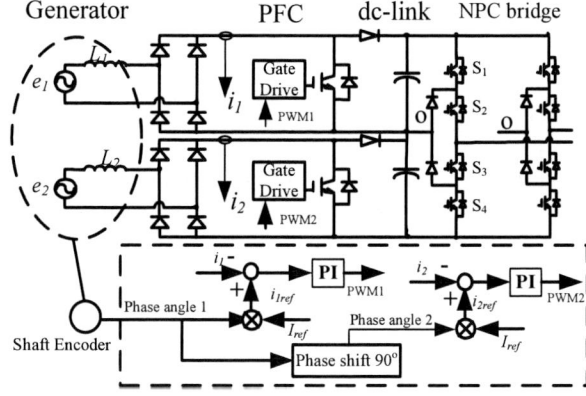

Fig.2 Control diagram of the PFC circuit (two rectifier in series)

As mentioned above, the rectifier (generator) side power is achieved constant by connecting two generator coils with 90 degree phase shift and associated converter in series. However, the grid side converter, the NPC bridge, is still single phase, which will also cause the dc-link voltage ripple with the frequency of the 100Hz or 120Hz for 50Hz or 60Hz grid, respectively. If the voltage loop in Fig.2 has relatively large bandwidth, the dc-link voltage loop will introduce the current ripple to the reference of the current loop. Subsequently, the generator stator current will have the distortion, which will affect the power factor and may increase the power losses. The affect of the dc-link voltage ripple can be filtered by adding a notch filter at the dc-link voltage feedback loop, the center frequency of the notch filter should be selected as twice of the grid frequency [12].

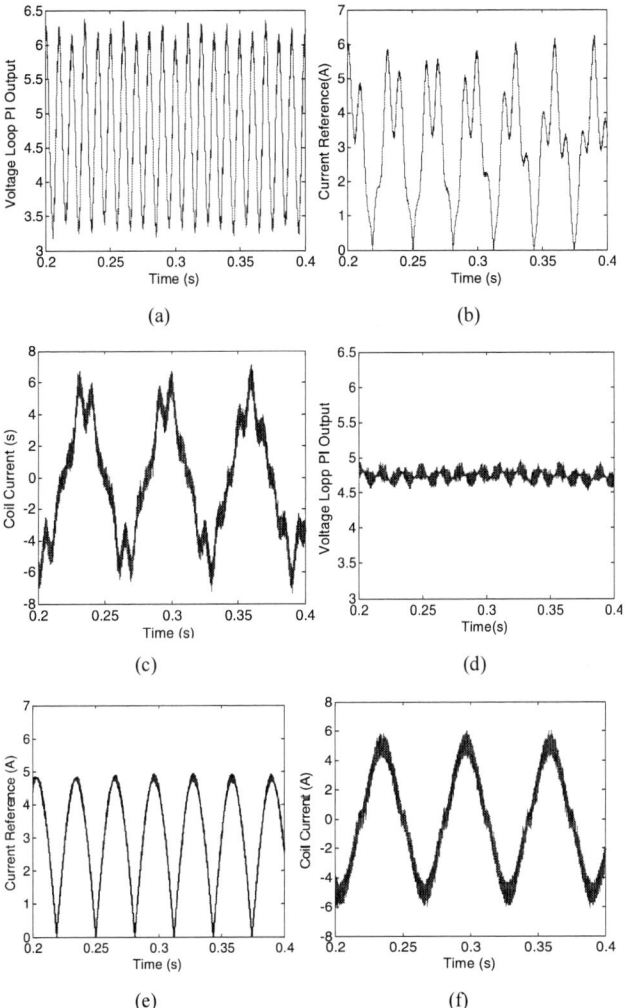

Fig.3 Simulation results: Without notch filter: a) voltage loop output, b) current reference, c) real generator current; With notch filter in the voltage feedback loop: d) voltage loop output, e) current reference, f) real generator current

Fig.3 shows the simulation results for the voltage feedback loop with and without the notch filter, for a 3kW PMG-converter system. Fig.3(a) shows the voltage loop controller output containing an 100Hz ripple for the 50Hz grid. Fig.3(b) and (c) shows the distorted current reference and the real generator current, where the real generator current is distorted due to the voltage ripple. In contrast, Fig.3 (d)(e)(f) shows the corresponding waveform with the notch filter added. As seen, the generator current remains sinusoidal and the dc-link voltage ripple is not reflected on the current loop.

B. NPC-bridge Inverter Control Scheme

The grid-side cascaded NPC-bridge inverter can be modeled on d-q frame which rotates synchronously with the grid voltage vector. If the d-axis of the rotating frame is oriented along the grid voltage, the active power P and reactive power Q can be formulated by (2).

$$P = u_d i_d + u_q i_q = u_d i_d$$
$$Q = u_d i_q - u_q i_d = u_d i_q \tag{2}$$

From (2), the grid-side active power and reactive power can be controlled independently by controlling the d-axis and q-axis current. The vector control diagram for the grid-side cascaded NPC-bridge inverter is shown in Fig.4. PI controllers are used here to compensate the current control loop. The so-called phase-shifted carrier PWM is the standard method for modulating the cascaded H-bridge inverter. For NPC bridge modulation, there are several PWM methods concerning about the neutral point balance [13-15]. Since the NPC bridge neutral point is already actively clamped by the rectifier in this paper, the phase-shifted carrier PWM is extended for cascaded NPC type inverter as in Fig.1 (c). The PWM strategy is described as follows: 1) The vector control scheme gives the voltage reference $U^*_{a,b,c}$ in three-phase stationary coordinate as shown in Fig.4; 2) The voltage reference $U^*_{a,b,c}$ is normalized by half of the dc-link voltage as in Fig. 2, thus the normalized reference voltage are within the range of 0~2 as shown in Fig.5. The modulation signal (voltage reference) is compared with two carrier signals (phase disposition) to get the gate signal for the NPC bridge of the converter first stage in Fig.1(a). Note that, for the switches in Fig.2, S1 and S3, S2 and S4 will switch complementary, and the fraction part of normalized voltage reference will determine the duty cycle. The NPC bridge right phase leg modulation signal has a 180^0 phase delay of the left phase leg modulation signal, thus enabling each NPC cell to output 5-level and better harmonic spectrum. 3) The PWM signal of the first stage is shifted by π/N to get the rest stages' PWM signal, where N is the number of H-bridge stages.

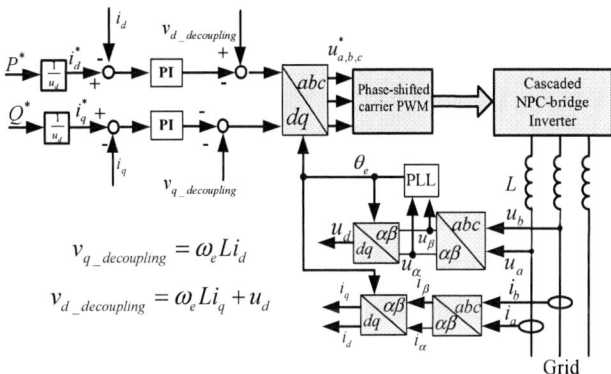

Fig.4 Vector control diagram of the grid-side cascaded NPC-bridge inverter

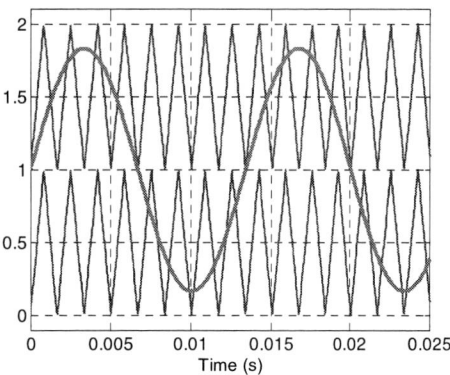

Fig.5 Modulation scheme for NPC bridge

Considering the harmonics analysis of the modulation scheme in Fig.5, [16] gives the spectrum analysis for one phase leg with natural sampled and carrier phase disposition modulation as indicated in (3)

$$
\begin{aligned}
V_{PD\_AN}(t) &= MV_{DC}\cos(\omega_0 t) \\
&+ \frac{4V_{DC}}{\pi^2}\sum_{m=0}^{\infty}\left\{\frac{1}{(2m+1)}\sum_{k=0}^{\infty}\left\{\frac{J_{2k+1}(2\pi(2m+1)M)}{2k+1}[1-2\cos(k\pi)\sin((2k+1)\phi)]\right\}\cos((2m+1)\omega_c t)\right\} \\
&+ \frac{V_{DC}}{\pi}\sum_{m=1}^{\infty}\sum_{\substack{n=-\infty\\n\neq 0}}^{\infty}\left\{\frac{1}{2m}J_{2n-1}(4m\pi M)\cos((n-1)\pi)\cos(2m\omega_c t+(2n-1)\omega_0 t)\right\} \\
&+ \frac{4V_{DC}}{\pi^2}\sum_{m=0}^{\infty}\sum_{\substack{n=-\infty\\n\neq 0}}^{\infty}\left\{\frac{1}{(2m+1)}\sum_{k=0}^{\infty}\left\{\frac{J_{2k+1}(2(2m+1)\pi M)}{(2k+1)^2-4n^2}\begin{bmatrix}(2k+1)(\cos(n\pi)+2\cos((k+1)\pi)\cos(2n\phi)\sin((2k+1)\phi))\\+4n(\cos(k\pi)\sin(2n\phi)\cos((2k+1)\phi))\end{bmatrix}\right\}\times\right. \\
&\left. \quad\cos((2m+1)\omega_c t+2n\omega_0 t)\right\}
\end{aligned}
$$

(3)

where $\phi = \cos^{-1}(\frac{1}{2M})$ $0\le\phi\le\frac{\pi}{4}$, the definitions of the variables in (3) can be found in [16].

For each NPC bridge, since the right phase leg has a 180 degree phase shift of the left phase leg, the right phase leg modulation harmonics spectrum can be derived by substituting $\omega_0 t$ with $\omega_0 t+\pi$ in (3). And the line voltage of the NPC cell will be the voltage difference between the left phase leg and the right phase leg as shown in (4)

$$
\begin{aligned}
V_{PD\_l}(t) &= 2MV_{DC}\cos(\omega_0 t) \\
&+ \frac{2V_{DC}}{\pi}\sum_{m=1}^{\infty}\sum_{\substack{n=-\infty\\n\neq 0}}^{\infty}\left\{\frac{1}{2m}J_{2n-1}(4m\pi M)\cos((n-1)\pi)\cos(2m\omega_c t+(2n-1)\omega_0 t)\right\}
\end{aligned}
$$

(4)

For N series NPC converters, the harmonics spectrum can be derived as in (5) (with phase shifted π/N between different stages)

$$
\begin{aligned}
V_{PD\_l}(t) &= 2MNV_{DC}\cos(\omega_0 t) \\
&+ \frac{2NV_{DC}}{\pi}\sum_{m=1}^{\infty}\sum_{\substack{n=-\infty\\n\neq 0}}^{\infty}\left\{\frac{1}{2m}J_{2n-1}(4Nm\pi M)\cos((n-1)\pi)\cos(2Nm\omega_c t+(2n-1)\omega_0 t)\right\}
\end{aligned}
$$

(5)

In this way, the harmonics due to the switching can be pushed beyond $2N\omega_c$, where ω_c is the converter switching frequency. As seen, the harmonics spectrum is similar to the cascaded H bridge case, and the side-band harmonics is mitigated.

III. SIMULATION AND EXPEIMENTAL VERIFICATION

The simulation model with 1.5MW wind generator and converter system for the 6kV grid is built up in

MATLAB/Simulink to verify the proposed topology and control method. The generator has 12 coils, which can form 6 pairs of coils with $90°$ phase shift. The dc-link voltage of each converter module is set at 3300V and the grid-side three-phase cascaded NPC-bridge inverter has 2 stages and can output 9 voltage levels. The active power reference is set as 830kW, which is transferred to the grid by the vector control. The PMG stator frequency is 16Hz and the dc-link capacitor is 1mF. Fig.6 (a) (b) (c) shows the grid-side converter output phase voltage, the grid voltage and grid current, indicating the converter is transferring power to the gird (the positive direction of the current is going into the grid). Fig. 6 (d) (e) (f) shows the capacitor voltages in Fig.1(c), where, (d) (e) shows the upper and lower capacitor voltages, which has both the low frequency ripple form the generator side and 100Hz ripple from the grid side. (f) shows the whole dc-link voltage of the NPC cell, which only contains the 100Hz voltage ripple.

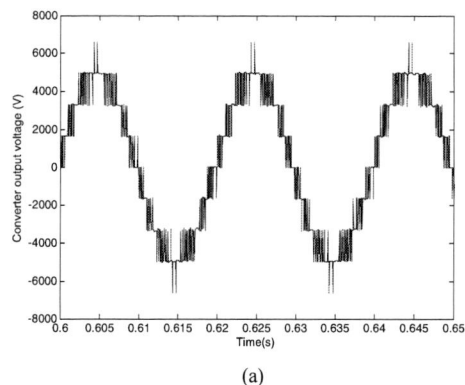

(a)

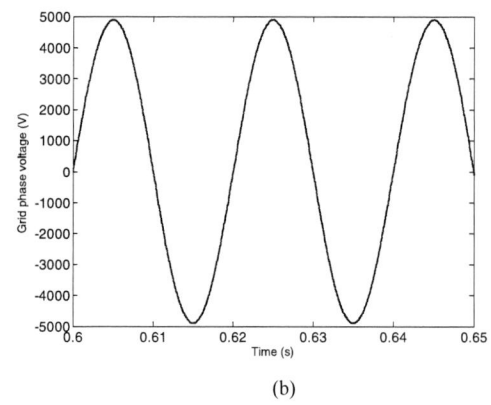

(b)

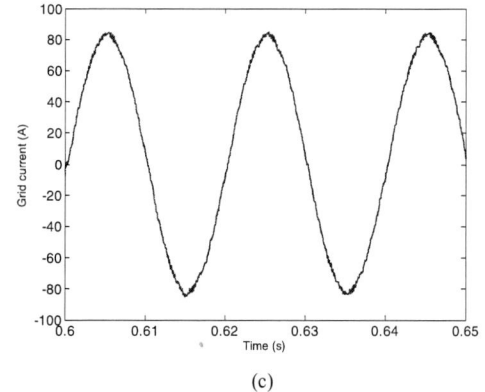

(c)

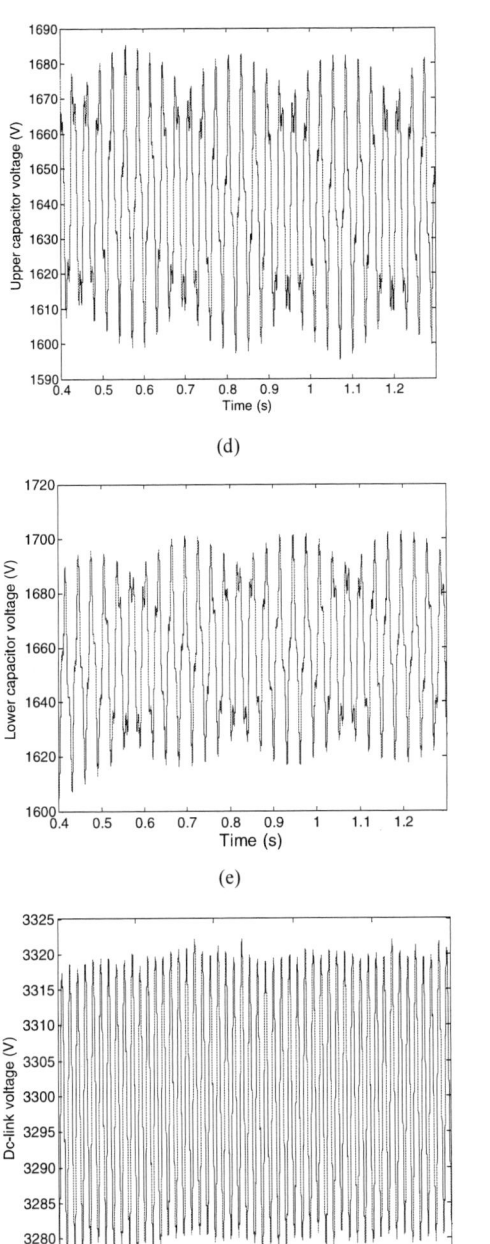

(d)

(e)

(f)

Fig.6 Simulation Results with a 1.5MW system: (a) converter output voltage (b) grid current (c) grid phase voltage (d) upper capacitor voltage (e) lower capacitor voltage (f) dc-link voltage

A 3kW experimental setup with a PMG driven by an induction motor has been setup in the lab and the converter system is shown in Fig.7 (a). The TMS320F2812 based DSP board is for implementing the algorithm and the CPLD MAXII 1270 is for generating the phase shifted PWM. The PMG is a two-pole six phase machine and the coils configuration is shown in Fig.7(b), which has three pairs of coils with 90 degree phase shift. The system under test uses

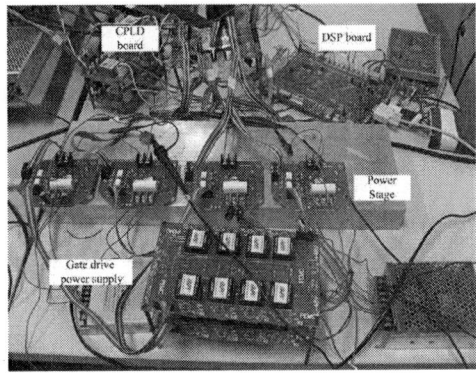

(a) Experimental setup

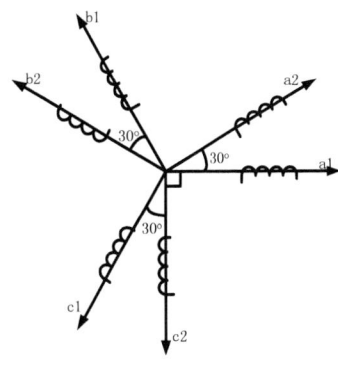

(b) Coil configuration

Fig. 7 Experimental setup and coil configuration

two pairs of the coils in Fig.7(b), thus forming a single phase system with two stages. The PMG armature inductance is 14.6mH and dc-link capacitor is 2200μF.

The experimental results are shown in Fig.8. Fig.8 (a) shows the converter dc-link voltage and the current waveform of the generator coil pairs with resistive load connected to the dc-link, which validates the PFC control scheme. The dc-link voltage is constant by connecting two coils with 90 degree phase shift and the associated PFC circuits in series. Fig.8(b) shows the output voltage of one NPC bridge cell and two cells in series, which has 5-level and 9-level output, respectively, thus validating the modulation scheme. The dc-link voltage of each cell is around 75V. Fig.8 (c) shows single phase experimental results with R-L load. The load resistor and inductor are 70ohm and 10mH, respectively. The PMG speed is around 100r/min. As seen, the single phase converter outputs 9-level voltages and the load current is sinusoidal. Fig.8(d) shows the whole dc-link voltage and the upper capacitor voltage of the converter module. As expected, the whole dc-link voltage has a 100Hz ripple, while the upper capacitor voltage has both the low frequency ripple due to the generator side and the 100Hz ripple from the grid side. Fig.8 (e) shows the harmonics spectrum of the output voltage with two NPC stages in series (nine levels). And the most lowest significant harmonics due to switching is around 32kHz and the switching frequency is 8kHz. The results agree with the theoretical analysis in the paper.

978-1-4244-4782-4/10 $26.00 © 2010 IEEE

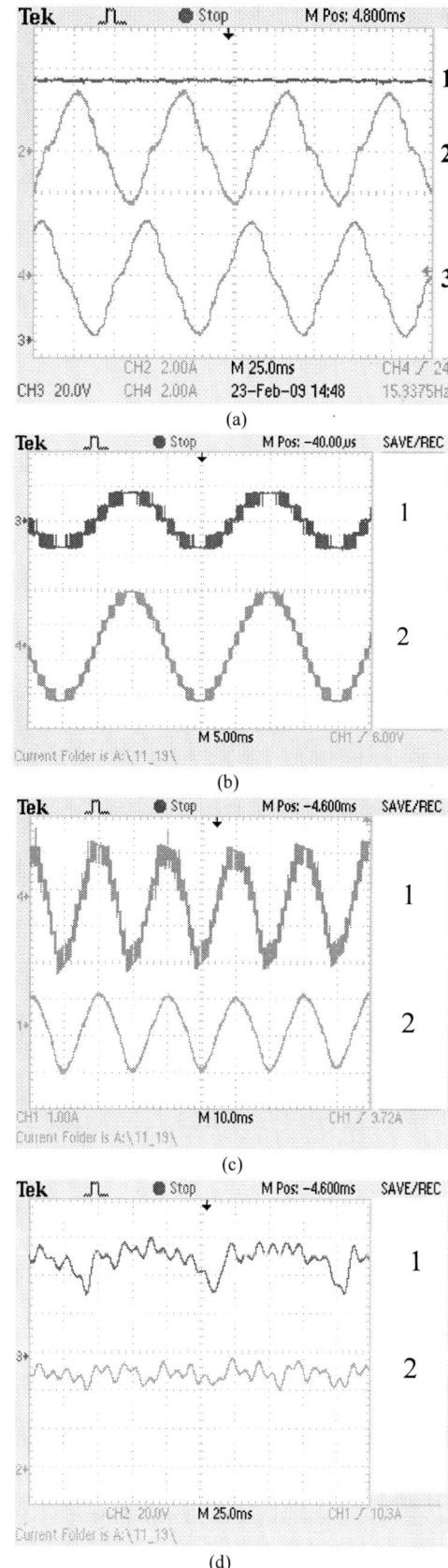

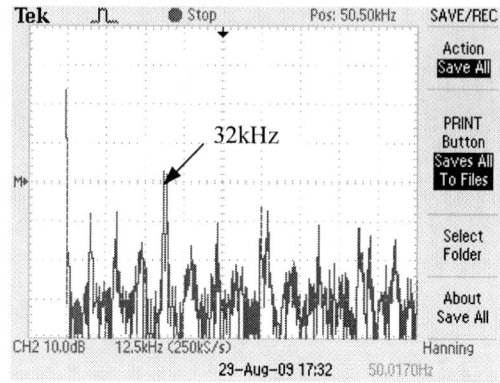

(e)

Fig.8 (a) 1) Dc-link voltage (20V/div), 2) Coil current (2A/div), 3) Coil current with 90⁰ phase shift (2A/div); (b) Converter output voltages: 1)One NPC bridge (50V/div), 2) Two NPC bridges in series (50V/div); (c) 1) Converter output voltage(50V/div), 2) Load current (1A/div); (d) 1) dc-link upper capacitor voltage (20V/div), 2) whole dc-link voltage (50V/div); (e) Harmonics spectrum analysis of the 9-level converter (12.5kHz/div)

IV. CONCLUSIONS

In order to reduce the required number of generator coils, this paper has discussed the way to improve the dc-link voltage of each converter cell, thus reducing the converter stages to match the grid voltage level. Two PFC circuits are connected in series and a NPC inverter is adopted accordingly. The phase shifted PWM method can be extended for modulating the NPC type cascaded converter and the harmonics analysis is provided in the paper. Simulation and experimental results shows that the converter can successfully transfer power from generator to the grid, which validate the proposed converter topology and control strategies. Other issues including the dc-link voltage ripple elimination and topology alternatives are also discussed in the paper.

REFERENCES

[1] F. Blaabjerg and Z. Chen, Power Electronics for Modern Wind Turbines. US: Morgan & Claypool Publishers, 2006, Chap. 4.

[2] C. Ng, M. Parker, P. Tavner "A multilevel modular converter for a large light weight wind turbine generator" IEEE Trans. Power Electronics, vol. 23, no. 3, pp.1062-1074, May 2008.

[3] Z. Chen, E. Spooner, "A modular, permanent-magnet generator for variable speed wind turbines" in Proc. IEEE Int. Elect. Mach and Drives Conf. vol.1, pp.453-457, Sept. 1995.

[4] E. Spooner, P. Gordon and C.D.French, "Lightweight, ironless-stator, PM generators for direct-drive wind turbines" in Proc. Int. Power Electronics, Machines and Drives Conf., vol.1, pp. 29-33, Mar. 2004.

[5] D.Vizireanu, X. Kestelyn and S. Brisset. "Polyphased Modular direct-drive wind turbine generator" in Proc. EPE'05 Conf., vol.1, pp.1-9, Sept. 2005.

[6] J. M. Carrasco, L.G. Franquelo, J.T. Bialasiewicz, "Power-electronic systems for the grid integration of renewable energy sources: A survey" IEEE Trans. Industrial Electronics. vol.53, no.4, pp.1002-1016, Aug. 2006.

[7] E. Cengelci, P. Enjeti, "Modular PM generator/converter topologies, suitable for utility interface of wind/Micro turbine and flywheel type electronmechanical energy conversion system" in Proc. IEEE IAS'00 Conf., vol.4, pp. 2269-2276,Oct. 2000.

978-1-4244-4782-4/10 $26.00 © 2010 IEEE

[8] M.A. Parker, C.H. Ng, L. Ran, "Power control of direct drive wind turbine with simplified conversion stage & transformerless grid interface", in Proc. 41st Int. Univ. Power Eng. Conf., vol.2, pp. 65-68, Sept. 2006.

[9] Y. Lo, S. Ou and H. Chiu, "On evaluating the current distortion of the single-phase switch-mode rectifiers with current slop maps", IEEE Trans. Industrial Electronics, vol.49, no.5, pp.1128-1137, Oct. 2002.

[10] J. Rodriguez, S. Bernet, B. Wu, et al. "Multilevel voltage-source-converter topologies for industrial medium-voltage drives", IEEE Trans. Industrial Electronics, vol.54, no.6, pp.2930-2945, Dec. 2007.

[11] L. Liu, Y. Ma, J. Chai, X. Sun, "Cascaded single phase PFC rectifier pair for permanent magnet wind generation system ", in Proc. IEEE IPEMC'08 Conf., vol.1, pp.2516-2520, Oct. 2008.

[12] Williams, J.B, "Design of Feedback Loop in Unity Power Factor AC to DC Converter" in Proc. IEEE PESC'89 Conf., vol.2, pp.959-967, Jun. 1989.

[13] A. Joseph, J. Wang, Z. Pan. "A 24-pulse rectifier cascaded multilevel inverter with minimum number of transformer windings", in Proc. IEEE IAS'05, vol.1, pp.115-120, Oct. 2005.

[14] US patent: Five level high power motor drive converter and control system. Patent Number, 6058031 2000.

[15] B. Ge, F. Peng, "An Effective SPWM Control Technique for 1MVA 6000V Cascaded Neutral Point Clamped Inverter", in Proc. IEEE IAS'08 Conf., vol.1, pp.1-6, Oct. 2008.

[16] McGrath, B.P.; Holmes, D.G, "A Comparison of Multicarrier PWM Strategies for Cascaded and Neutral Point Clamped Multilevel Inverters" in Proc. IEEE PESC'00 Conf., vol.2, pp.18-23, Jun. 2000.

[17] Holmes, D.G.; McGrath, B.P , "Opportunities for Harmonic Cancellation with Carrier Based PWM for Two-Level and Multi-Level Cascaded Inverters", in Proc IAS'99 Conf., vol.2, pp.781-788, Oct. 1999.

Small-Signal Modeling and Analysis of the Double-Input Buckboost Converter

Deepak Somayajula and Mehdi Ferdowsi
Missouri University of Science and Technology
Rolla, MO 65409 USA
http://power.mst.edu

Abstract— Multi-input dc-dc power electronic converters have been gaining popularity in applications such as renewable energy sources and electric-drive vehicles due to their reduced part count and flexibility in integration. In this paper, the small-signal model of the double-input buckboost converter is developed. The model is then used for a multiple loop feedback design. Considering a photovoltaic-battery hybrid power system, the control objectives are threefold. These include output voltage regulation, constant power demand from the PV panel, and load accommodation by the battery pack. Two feedback compensation networks are designed based on the developed small-signal model. It is also demonstrated that the two inputs of the double-input buckboost topology can be controlled independently. This offers greater flexibility for the compensator design. The results of the time domain analysis are consistent with those of the theoretical model.

Index Terms—Small-signal modeling, double-input converter, compensator design

I. INTRODUCTION

Wind and solar energy generation is on the rise along with other green energy sources. The intermittent nature of these energy sources is the main drawback which has prevented their complete integration into the mainstream energy generation. Therefore, combining various energy sources with each other to form a hybrid energy system is proposed in the literature [1]. In general, a dc-dc converter is required to integrate each energy related module into the system. Integrating each energy source with a dc-dc converter is expensive, bulky, less efficient, and hard to control. Instead, using a single dc-dc isolated or non-isolated multi-input converter is proposed [1-10]. Utilizing a single dc-dc multi-input converter to integrate all of the energy sources provides several advantages including reduced component count, potential reduction in weight, control simplicity, and flexibility in the integration of sources [7].

Several non-isolated dc-dc double-input power electronic converters (DIPEC) including double-input (DI) buck, buckboost, and buck-buckboost converters have been introduced, analyzed, and compared in the literature [7-12]. A comparative study between various dc-dc multi-input topologies is made in [13] where the authors compare the topologies based on the reliability, flexibility, cost, and modularity potential. Control aspects for specific DIPEC topologies are discussed in [14-17] as it forms an integral part of the development and widespread use of such topologies. In [14], the control objective is to keep one of the source currents constant along with the output voltage regulation in a DI buck-buckboost topology. In [15], one of the inputs of a DI buckboost converter is controlled using a maximum power point tracking algorithm and the other input is controlled through output voltage regulation while both of them together meet the load demand. In [16], the control of a DI buckboost converter is discussed when the control objective is optimal power sharing such that during load variations one of the sources, e.g., a battery, supplies constant power and the second source, e.g., an ultra-capacitor, has to meet the excess load demand. In these papers, the control objectives were successfully achieved; however, what is missing is a systematic controller design approach based on small-signal models. Development of such models is crucial for analyzing system stability and designing optimal compensators.

In this paper, a small signal model for the DI buckboost converter is developed in Section II. In Section III, the transfer functions which are important to meet the control objectives are developed. In Section IV, using the derived transfer functions, it is analytically proved that the two control loops which control the two inputs of a DI buckboost converter can be independently controlled with two different control objectives. This feature of the loops being independently controllable makes the compensator design much simpler as the two compensators can now be independently designed. Section V includes the simulation results and analysis for the DI buckboost converter with the compensators included in the system. The concluding remarks are presented in Section VI.

II. SMALL SIGNAL MODEL DEVELOPMENT FOR THE DOUBLE-INPUT BUCKBOOST CONVERTER

Power electronic converters are non-linear systems and they have to be linearized by carrying out small signal analysis. The feedback compensators can then be designed based on the developed linear time invariant (LTI) models

978-1-4244-4782-4/10 $26.00 © 2010 IEEE

inorder to meet various control objectives. Small signal analysis for single input dc-dc converters is very well established in the literature [18, 19]. However, small signal modeling for DIPEC topologies has not been reported yet. Although, the control of DIPEC topologies has been reported in the literture [14, 15], a systematic design procedure of compensators based on LTI models has not been reported. Thus, the development of small signal models for DIPEC topologies is necessary in order to optimize the compensator design and to provide a stable system which meets all the control objectives. This being the intention, the small signal modeling of the DI buckboost converter (shown in Fig. 1) is carried out in this section. Various transfer functions that are important in the control of the converter are obtained and feedback loop compensator design is carried out. Similar linearization techniques can be carried out for any DIPEC topology to aid in the compensation design. In the DI buckboost converter, the steady-state average equations governing the dynamic response of the system are:

$$L\frac{d\langle i_L(t)\rangle_{T_s}}{dt} = D_1 V_1 + D_2 V_2 - (1 - D_1 - D_2)\langle v_o(t)\rangle_{T_s} \quad (1)$$

$$C\frac{d\langle v_o(t)\rangle_{T_s}}{dt} + \frac{\langle v_o(t)\rangle_{T_s}}{R} = (1 - D_1 - D_2)\langle i_L(t)\rangle_{T_s} \quad (2)$$

where D_1 and D_2 are the duty ratios of switches S_1 and S_2, respectively. Equation (1) is obtained by averaging the voltage across the inductor during one switching period. Equation (2) is obtained by averaging the capacitor current waveform. Perturbing and linearizing (1) and (2) around a given operating point, neglecting the product of small signal perturbed ac terms, and converting the obtained equations into frequency domain using the Laplace Transformation would give [19]:

$$sL\hat{i}_L(s) = D_1\hat{v}_1(s) + D_2\hat{v}_2(s) + (V_1 + V_{out})\hat{d}_1(s)$$
$$+ (V_2 + V_{out})\hat{d}_2(s) - (1 - D_1 - D_2)\hat{v}_o(s) \quad (3)$$

$$sC\hat{v}_o(s) + \frac{\hat{v}_o(s)}{R} = -(\hat{d}_1(s) + \hat{d}_2(s))I_L$$
$$+ (1 - D_1 - D_2)\hat{i}_L(s) \quad (4)$$

The process of obtaining (3) and (4) is very similar to the process ascertained for single-input dc-dc converters and the only difference for DI topologies is that in their case there are two control inputs $\hat{d}_1, \hat{d}_2$ and also two disturbance inputs $\hat{v}_1, \hat{v}_2$. The small signal model shown in Fig. 2 is obtained by replacing the terms in (3) and (4) with current sources, voltage sources, current dependent sources, and voltage dependent sources. The input side of the small signal model which has the switch current perturbations can be obtained by perturbing the steady state switch current equations. The steady state average current equations for switch currents I_{s1} and I_{s2} are approximated by

$$I_{s1} = D_1 I_L \text{ and} \quad (5)$$

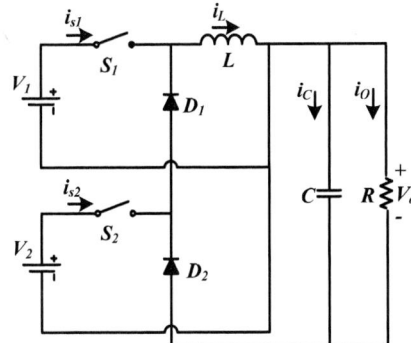

Fig. 1. Block diagram of the DI buckboost converter

$$I_{s2} = D_2 I_L \quad (6)$$

Equations (5) and (6) in perturbed form give

$$\hat{i}_{s1}(s) = D_1\hat{i}_L(s) + I_L\hat{d}_1(s) \quad (7)$$

$$\hat{i}_{s2}(s) = D_2\hat{i}_L(s) + I_L\hat{d}_2(s) \quad (8)$$

Therefore, the small signal model shown in Fig. 2 is obtained by considering (3), (4), (7), and (8).

III. TRANSFER FUNCTIONS OF THE DOUBLE-INPUT BUCKBOOST CONVERTER

It can be observed from Fig. 2 that the model has two control inputs $\hat{d}_1, \hat{d}_2$ and two disturbance inputs $\hat{v}_1, \hat{v}_2$ and all the other perturbations are dependent on these four inputs. The two control inputs $\hat{d}_1, \hat{d}_2$ can be controlled based on various control objectives like maximum power point tracking or optimal power sharing between the inputs. The transfer functions required to meet the various control objectives can be derived from the small signal model shown in Fig. 2. For instance, to study the effects of the perturbations in D_1 on the output voltage, one should find transfer function $G_{vd1}(s)$ which is control-1 ($\hat{d}_1$) to output transfer function. This transfer function can be obtained from (7) and (8) by assuming the disturbance inputs $\hat{v}_1 = \hat{v}_2 = 0$ and also the control-2 $\hat{d}_2 = 0$ resulting in

$$G_{vd1}(s) = \frac{\hat{v}_o(s)}{\hat{d}_1(s)}\bigg|_{\hat{d}_2=\hat{v}_1=\hat{v}_2=0} = \frac{(V_1 + V_{out})(1 - D_1 - D_2) - sLI_L}{s^2 LC + s\dfrac{L}{R} + (1 - D_1 - D_2)^2} \quad (9)$$

This transfer function is a second order system with a resonant pole pair and a right half plane (RHP) zero just like a single-input buckboost converter where the RHP zero limits the bandwidth of the system. If D_2 is zero, then the transfer function will be similar to the control-output transfer function of the single input buckboost converter [19]. Controlling the average inductor current for equal current sharing between inputs of a parallel connected dc-dc converter is discussed in [20]. However, if the objective is to maintain one of the average switch currents constant, the control-2 to switch current 2 transfer function is important in this context which

978-1-4244-4782-4/10 $26.00 © 2010 IEEE

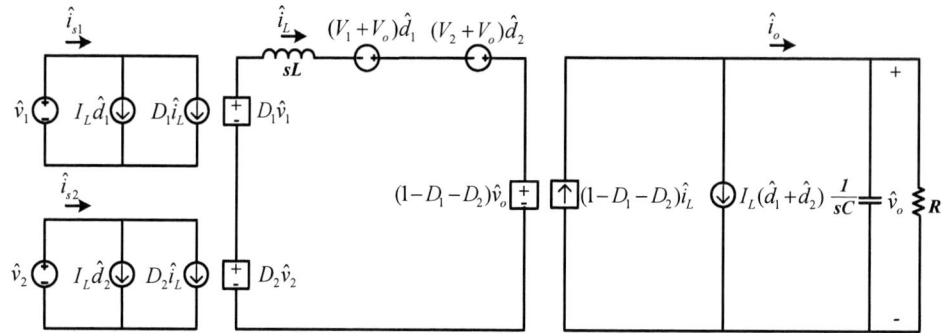

Fig. 2. Small signal model of a double-input buckboost converter without offset time control

has been developed for single-input topologies in [21] and similar analysis can be carried out for DIPEC topologies using Fig. 2. Control-2 to switch current-2 transfer function $G_{is2d2}(s)$ for a DI buckboost converter is shown here in which the control-2 to inductor current transfer function $G_{id2}(s)$ can be found from (3) and (4) by eliminating $\hat{v}_o(s)$ and settingthe control-1 $\hat{d}_1 = 0$ resulting in final value shown in (10).

From (9) and (10) it can be observed that the transfer function $G_{vd1}(s)$ and $G_{is2d2}(s)$ are important to maintain output voltage regulation and to maintain one of the switch currents constant and thereby supplying constant power from one of the sources irrespective of the load demand.

$$G_{is2d2}(s) = \frac{\hat{i}_{s2}(s)}{\hat{d}_2(s)}\bigg|_{\hat{d}_1 = \hat{v}_1 = \hat{v}_2 = 0} = I_L + D_2 \frac{\hat{i}_L(s)}{\hat{d}_2(s)} = I_L + D_2 G_{id2}(s)$$

$$G_{is2d2}(s) = \frac{\hat{i}_{s2}(s)}{\hat{d}_2(s)}\bigg|_{\hat{d}_1 = \hat{v}_1 = \hat{v}_2 = 0} = I_L + D_2 \frac{(V_2 + V_{out})(\frac{1}{R} + sC) + (1 - D_1 - D_2)I_L}{s^2 LC + s\frac{L}{R} + (1 - D_1 - D_2)^2} \quad (10)$$

IV. INDEPENDENT CONTROL OF THE TWO LOOPS

In this section, it is shown that two control inputs $\hat{d}_1$ and $\hat{d}_2$ can be independently controlled with each loop having a different control objective. Inner current control loop $T_i(s)$ is shown in the Fig. 3 and it can be observed that average switch current I_{s2} is being held constant through average switch current control. Transfer function $G_{is2d2}(s)$ responsible for this has been developed in (10). However, perturbations in $\hat{i}_{s2}(s)$ are also dependent on $\hat{d}_1$ and this dependency is given by transfer function $G_{is2d1}(s)$ as shown in the Fig. 3 and in (11). In order to reduce this dependency the perturbations in $\hat{d}_1$ are considered as disturbance signals for the inner loop once inner current control loop $T_i(s)$ is closed and this leads to (12). Equation (13) shows the dependency of $\hat{v}_o(s)$ on the two control inputs and in (14) $\hat{d}_2$ is replaced with corresponding perturbations in $\hat{i}_{s2}(s)$ from Fig. 3.

$$\hat{i}_{s2}(s) = G_{is2d1}(s)\hat{d}_1(s) + G_{is2d2}(s)\hat{d}_2(s) \quad (11)$$

$$\hat{i}_{s2}(s) = \frac{G_{is2d1}(s)}{1 + T_i(s)}\hat{d}_1(s) + \frac{T_i(s)}{1 + T_i(s)}\hat{i}_{ref2}(s) \quad (12)$$

$$\hat{v}_o(s) = G_{vd1}(s)\hat{d}_1(s) + G_{vd2}(s)\hat{d}_2(s) \quad (13)$$

$$\hat{v}_o(s) = G_{vd1}(s)\hat{d}_1(s) + G_{vd2}(s)\frac{G_{c2}(s)}{V_M}\hat{i}_{s2}(s) \quad (14)$$

From (12) $\hat{i}_{s2}(s)$ can be substituted in (14) to obtain:

$$\begin{aligned}\hat{v}_o(s) = G_{vd1}(s)\hat{d}_1(s) + G_{vd2}(s)\frac{G_{c2}(s)}{V_M} \\ \left(\frac{G_{is2d1}(s)}{1 + T_i(s)}\hat{d}_1(s) + \frac{T_i(s)}{1 + T_i(s)}\hat{i}_{ref2}(s)\right)\end{aligned} \quad (15)$$

When the average switch current I_{s2} is held constant then $\hat{i}_{ref2}(s) \approx 0$ and therefore (15) becomes

$$\hat{v}_o(s) = G_{vd1}(s)\hat{d}_1(s) + G_{vd2}(s)\frac{G_{c2}(s)}{V_M}\left(\frac{G_{is2d1}(s)}{1 + T_i(s)}\hat{d}_1(s)\right) \quad (16)$$

$$G_{new}(s) = \frac{\hat{v}_o(s)}{\hat{d}_1(s)} = G_{vd1}(s) + G_{vd2}(s)\frac{G_{c2}(s)}{V_M}\left(\frac{G_{is2d1}(s)}{1 + T_i(s)}\right) \quad (17)$$

From (9) and (17) it can be seen that transfer functions $G_{vd1}(s)$ and $G_{new}(s)$ are not exactly the same; however, it can also be observed from Figs. 5 and 6 that the gain and phase of the two transfer functions are very close over a wide frequency range for a given operating point. Therefore, it can be concluded that the two loops can be independently controlled as shown in Fig. 4 and compensator design for $G_{c1}(s)$ and $G_{c2}(s)$ for the two loops can also be carried out independently. The operating point around which the DI buckboost converter is linearized and the magnitude and the phase plot are drawn is described in the section V.

V. TIME DOMAIN SIMULATION RESULTS FOR THE CLOSED-LOOP RESPONSE

Closed-loop response of the system can be obtained when both the inner current control loop $T_i(s)$ and outer voltage

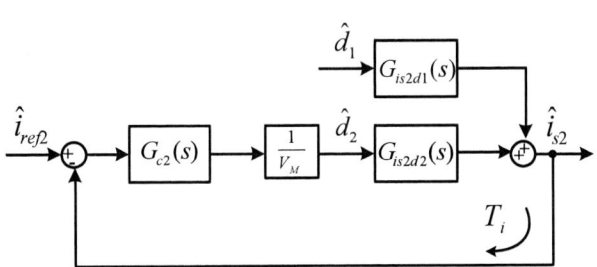

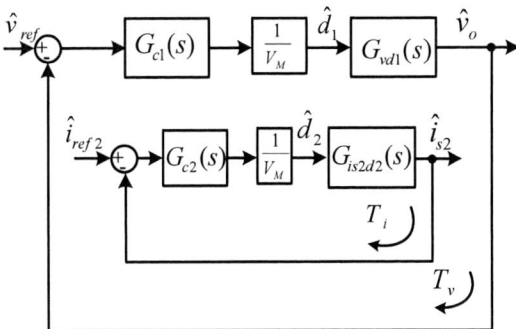

Fig. 3. Block diagram of the converter system with the inner current loop closed

Fig. 4. Small-signal control loop of the DI buckboost converter where I_{s2} and V_{out} are constant and the loops are independently controlled

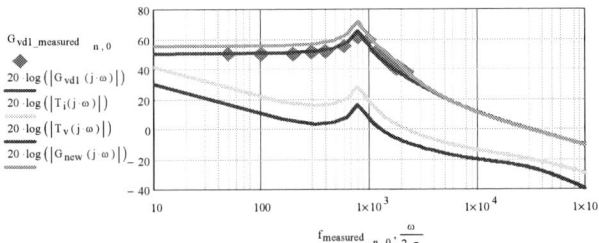

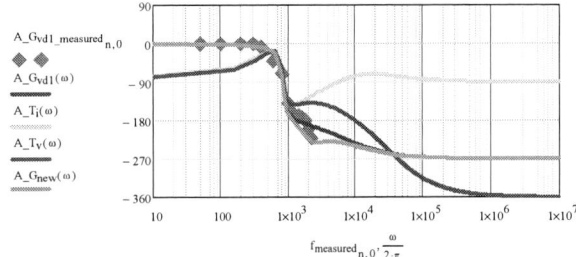

Fig. 5. Magnitude plot for the functions $G_{vd1}(s)$, $G_{new}(s)$, $T_i(s)$ and $T_v(s)$ and measured $G_{vd1}(s)$ indicated by dotted line

Fig. 6. Phase plot for the functions $G_{vd1}(s)$, $G_{new}(s)$, $T_i(s)$ and $T_v(s)$ and phase measured $G_{vd1}(s)$ indicated by dotted line

$$G_{c1}(s) = \frac{30}{s}\left(\frac{1+\dfrac{s}{2\pi * 575.311}}{1+\dfrac{s}{10\pi * 7356}}\right)\left(\frac{1+\dfrac{s}{2\pi * 575.311}}{1+\dfrac{s}{10\pi * 7356}}\right)$$

$$G_{C2}(s) = \frac{400}{s}\left(\frac{1+\dfrac{s}{6\pi * 575.311}}{1+\dfrac{s}{6\pi * 7356}}\right)$$

control loop $T_v(s)$ are closed. A stable current compensator $G_{c2}(s)$ is designed initially and then the voltage compensator $G_{c1}(s)$ is designed based on design methodology proposed in [22, 23]. The values for the compensators $G_{c1}(s)$ and $G_{c2}(s)$ along with the magnitude plots and the phase plots for the various transfer functions are shown in Figs. 5 and 6, respectively. $G_{vd1}(s)$ and $G_{new}(s)$ plots are obtained by linearizing the DI buckboost converter around the following operating point, V_1=40 V, V_2=70 V, D_1=0.2, D_2=0.4, V_o=90 V, I_{ref2}=9 A, and R=10 Ω. The plots for $T_i(s)$ and $T_v(s)$ are also obtained around the same operating point by using the compensators shown in Figs. 5 and 6, respectively. The magnitude and phase of $G_{vd1}(s)$ are measured and the plot is also shown in Figs. 5 and 6 with a dotted line and it can be seen that the measured function closely follows the actual function $G_{vd1}(s)$. Using the compensators, a time domain simulation of the system is carried out around the same operating point and the load is varied from 10 Ω to 5 Ω in order to test the stability and effectiveness of the system in meeting its control objectives. The results of the time domain simulation are shown in Figs. 7 and 8. It can be seen that during the step change in the load, output voltage V_{out}

remains relatively constant at 90 V, average switch current I_{s2} also remains relatively constant at 9 A, and the additional power requirements are met by Source 1 through changes in I_{s1}. This indicates that the required control objectives are effectively met through the independent control of the two control loops.

VI. CONCLUSION

A small-signal model for the DI buckboost converter was developed and the compensator design is carried out for the system. Two compensators are designed to meet the control objective of supplying constant power from one source (PV) and meeting the additional load demand through the other source (battery) during load variations. It is shown that the control objective can be achieved by the ndependent control of the two loops controlling the two switches. This independent control of the two loops simplifies the compensator design procedure. Therefore, the two loops are designed independently one to maintain output voltage regulation and another to maintain switch current from source 2 constant. The closed loop system is tested for load regulation using the designed compensators. The system is stable and has a good dynamic response. The developed models and the analysis can be used for the integration of renewable energy sources using DIPEC topologies.

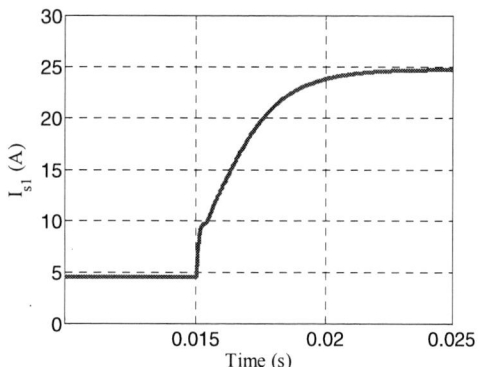

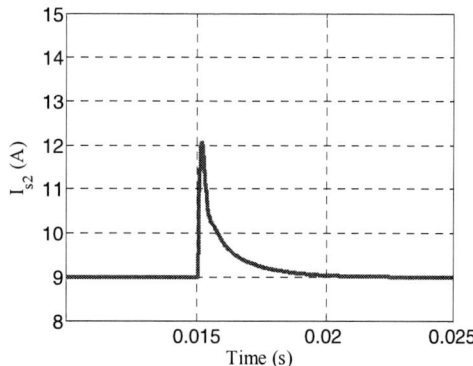

Fig. 7. Average current waveforms for both sources for a step change in the load from 10 to 5 Ω at t=0.015 s with both control loops closed.

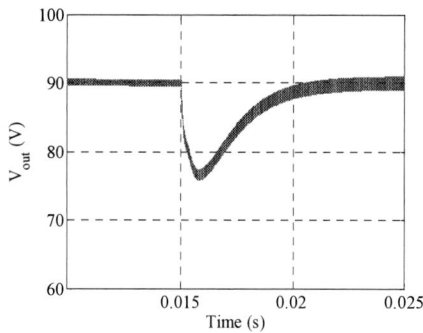

Fig. 8. Output Voltage waveform for a step change in the load from 10 to 5 Ω at t=0.015 s with the current and voltage loops closed.

REFERENCES

[1] Y. M. Chen, Y. C. Liu, and F. Y. Wu, "Multi-input converter with power factor correction, maximum power point tracking, and ripple-free input currents," *IEEE Trans. Power Electronics*, vol. 19, pp. 631-639, May 2004..

[2] A. D. Napoli, F. Crescimbini, S. Rodo, and L. Solero, "Multiple input dc-dc power converter for fuel-cell powered hybrid vehicles", in *Proc. IEEE Power Electronics Specialists Conf. (PESC)*, Jun. 2002, vol. 4, pp. 1685–1690.

[3] H. Li, Z. Du, K. Wang, L. M. Tolbert, and D. Liu, "A hybrid energy system using cascaded H-bridge converter," in *Proc. IEEE Industrial*

Applications Conf., vol. 1, Oct. 2006, pp.198-203.

[4] R. M. Schupbach, J. C. Balda, M. Zolot, and B. Kramer, "Design methodology of a combined battery-ultracapacitor energy storage unit for vehicle power management," in *Proc. IEEE Power Electronics Conf.*, 15-19 Jun. 2003, vol. 1, pp. 88-93.

[5] H. Tao, A. Kotsopoulos, J. L. Duarte, and M. A. M. Hendrix, "Family of multiport bidirectional dc-dc converters," in *Proc. IEEE Electric Power Applications*, vol. 153, May 2006, pp. 451 - 458.

[6] H. Tao, A. Kotsopoulos, J. L. Duarte, and M. A. M. Hendrix, "Triple-half-bridge bidirectional converter controlled by phase shift and PWM", in *Proc. IEEE Applied Power Electronics Conf. and Exposition (APEC), Dallas, TX*, USA, Mar. 2006,.pp. 1256-1262.

[7] K. P. Yalamanchili, M. Ferdowsi, and K. Corzine, "New double input dc-dc converter for automotive applications", in *Proc. IEEE Vehicle Power and Propulsion Conf.*, 6-8 Sept. 2006, pp 1-6.

[8] K. P. Yalamanchili and M. Ferdowsi, "Review of multiple input DC-DC converters for electric and hybrid vehicles," in *Proc. IEEE Vehicle Power and Propulsion Conf.*, 7-9 Sept. 2005, pp. 160-163.

[9] K. Gummi and M. Ferdowsi, "Synthesis of double-input dc-dc converters using single pole triple throw switch as a building block," in *Proc. IEEE Power Electronics Specialists Conf. (PESC), Greece*, Jun. 2008.

[10] K. Gummi and M. Ferdowsi, "Derivation of new double-input converters using H-Bridge cells as Building Blocks," *to appear in Proc. IEEE Industrial Electronics Conf., Orlando, FL, USA*, November 2008.

[11] A. Khaligh, J. Ciao and Y. J. Lee, "A Multiple-input dc-dc converter topology," *IEEE Power Electronic Letters*, vol. 24, pp. 862-868, Mar 2009.

[12] B. G. Dobbs and P. L. Chapman, "A multiple-input dc-dc converter topology," *IEEE Power Electronics Letters*, vol. 1, pp. 6-9, Mar. 2003.

[13] S. H. Choung and A. Kwasinki, "Multi-input dc-dc converter topologies comparison," in *Proc. IEEE Industrial Electronics Conf. (IECON)*, Nov. 2008, pp. 2359-2364.

[14] Y. M. Chen, Y. C. Liu, and S. H. Lin, "Double-input PWM dc/dc converter for high/low-voltage sources," *IEEE Trans. Industrial Electronics*, vol. 53, pp. 1538-1545, Oct. 2006.

[15] N. D. Benavides and P. L. Chapman, "Power budgeting of a multiple-input buck-boost converter," *IEEE Trans. Power Electronics*, vol. 20, pp. 1303-1309, Nov. 2005.

[16] D. Somayajula and M. Ferdowsi, "Power sharing in a double-input buckboost converter using offset time control," in *Proc. IEEE Applied Power Electronics Conf. (APEC)*, Feb. 2009, pp. 1091-1096.

[17] Y. M. Chen, Y. C. Liu, S. C. Hung, and C. S. Cheng, "Multi-input inverter for grid-connected hybrid pv/wind power system," in *Proc. IEEE Trans. Power Electronics*, vol. 22, pp. 1070-1077, May 2007.

[18] J. G. Kassakian, M. F. Schlecht, and G. C. Verghese, *Principles of Power Electronics*, 2nd ed, Addison Wesley, 1991.

[19] R. W. Erickson and D. Maksimovic, *Fundamentals of Power Electronics*, 2nd ed. Norwell, M. A. Kluwer, 2001.

[20] R. Ayyanarr, R. Giri, and N. Mohan, "Active input-voltage andload-current sharing on input-series and output-parallel connected modular dc-dc converters using dynamic input-voltage reference scheme," *IEEE Trans. Power Electronics*, vol. 19, pp.1462-1473, Nov 2004.

[21] W. Tang, F. C. Lee, R. B. Ridley and I. Cohen "Charge Control: Modeling, Analysis, and Design," *IEEE Transaction on Power Electronics*, vol. 8, Oct. 1993, pp.396-403.

[22] S. Angkititrakul, H. Hu and Z. Liang, "Active inductor current balancing for interleaving multi-phase buck-boost converter," in *Proc. IEEE Applied Power Electronics Conf. (APEC)*, Feb. 2009, pp. 527-532.

[23]http://dataweek.co.za/news.aspx?pklNewsId=31635&pklCategoryID=46 accessed on July 11, 2009.

A Novel Power Distribution Strategy for Parallel Inverters in Islanded Mode Microgrid

Xuan ZHANG, Jinjun LIU, Ting LIU, Linyuan ZHOU
School of Electrical Engineering, Xi'an Jiaotong University
Xi'an, China 710049
Xuanzhang@ieee.org

Abstract—This paper presents a novel wireless load-sharing control strategy, active droop positioning, for islanding parallel inverters in an ac-distributed system. Normally, the microgrid is intended to operate in islanded mode when the disconnection from the upstream MV network occurs. In this condition, the inverter-interfaced microsources acts as a voltage source, with the magnitude and frequency of the output voltage controlled through droops. However, under overload or underload circumstance, this droop characteristic may lead to large frequency deviation. The mild droop characteristic can avoid this large deviation, but it's hard to realize a good power sharing among inverters. Compared with the conventional droop control method, by automatically changing the droop position of some large capacity microsources, the proposed control strategy can bind system frequency deviations in an endurable range while ensure good current sharing performance among most microsources. Under this control strategy, almost all the micro sources in the microgrid can work around rated power, and when the load reduces dramatically, the distributed energy resources (like wind power and solar power) can also be taken full advantage. The analysis of the proposed method and design procedure are provided. Simulation and experimental results validate the proposed control strategy.

I. INTRODUCTION

Nowadays, penetration of distributed generation (DG) poses significant new challenges and benefits to the existing electricity market structure. Distributed energy resources (DER) management and the successful technology advancement have paved the way for the creation of the micro grid (MG) distribution network [1]. Microgrids are systems that have at least one distributed energy resource and associated loads and can form intentional islands in the electrical distribution systems [2]. The MG is intended to operate in and smoothly transit between the following two different operating conditions [3,4].

Normal interconnected mode: the MG is connected to a main MV network, either being supplied by it or injecting some amount of power into the main system.

Emergency mode: the MG operates autonomously, in a similar way to physical islands, when the disconnection from the upstream MV network occurs.

Most DER can not be connected into an MG directly due to the energy characteristics. Therefore, power electronic interfaces (dc/ac or ac/dc/ac) are required and inverter control is thus the main concern in MG operation. When an MG is working in islanded mode, the droop control concept is applied to control the inverters [3,5]. In [6, 7], the experimental results has shown how the MG disconnected from and reconnected to the strong grid, and how the operation mode of inverters in MG changed into droop control scheme automatically. But MG operates in an LV network in which line impedance is more like an R-L filter rather than a pure L filter. So the droop functions need some modifications [8-10].

By the modifications above, the conventional P-f and Q-U droops are changed into P-U and Q-f droops when the transmission line is pure resistive or into P-P'-f and Q-Q'-U droops when there is arbitrarily R/L rate in transmission line, thus the problem of accurate sharing among inverters in MG has solved. But when there is a large load change while islanding mode of MG, the problem that the frequency drifts in a large margin still exists. In [11], this kind of problem has been received attention, but the resolution is setting the working point of the key inverter by the superior controller in MG through signal cables rather than controlling the inverters wirelessly.

In this paper, a novel wireless droop control strategy, which adaptively controls the droop position of some large capacity inverters, is proposed for the inverter parallel operation in an MG while it's in islanding mode.

This greatly improves the stability of the whole MG system. The analysis of the proposed method and the design procedure are provided. The simulation based on PSCAD/EMTDC verifies the good performance of the proposed control strategy.

II. THEORETICAL BACKGROUND

A. The Conventional Droop Control

The active and reactive power flowing into the line at point A, as represented in Figure 1, are as follows:

$$P = \frac{U_1 \cdot U_2}{\omega L} \cdot \sin \delta \tag{1}$$

$$Q = \frac{U_1 \cdot (U_1 - U_2 \cos \delta)}{\omega L} \tag{2}$$

When the line is in usual length, the phase separation between two terminals of the line δ is quite small. So $\sin\delta \approx \delta$, $\cos\delta \approx 1$ and equations (1) and (2) become:

$$P = \frac{U_1 \cdot U_2}{\omega L} \cdot \delta \qquad (3)$$

$$Q = \frac{U_1 \cdot (U_1 - U_2)}{\omega L} \qquad (4)$$

Equation (3) and (4) show that the active power is dominantly depended by the power angle δ and the reactive power is dominantly depended by the voltage difference. The power angle δ can be controlled dynamically by controlling the frequency. Thus, if the active power and reactive power the load consuming are known, frequency and amplitude of the grid are determined. By these conclusions, the conventional frequency and voltage droop regulation through respectively active and reactive power can be obtained.

$$f - f_0 = -k_p (P - P_0) \qquad (5)$$

$$U - U_0 = -k_q (Q - Q_0) \qquad (6)$$

f_0 and U_0 are rated frequency and grid voltage respectively, and P_0 and Q_0 are the active and reactive power rating of the inverter. The frequency and voltage droop control characteristics are shown in Figure.2.

B. The Novel Active Power Management

Figure 3 shows the P-f droop control for two inverters. In Figure 3(a), the frequencies of output voltages at A_1, A_2 are controlled by the block diagram shown in Figure 3(c). In Figure 3(b), when the load changes from "load0" to "load1", the system frequency changes from f_0 to f_1. This control scheme

can make a good power sharing performance among inverters. But if f_1 is far from f_0, the load may not work very well and the system may become unstable.

The solution presented shown in Figure 4(a) is to change the P-f droop position of one inverter when the frequency is up or down to a border region, like $50.0 \pm 0.2\text{Hz}$. In the original state, the inverter and load P-f droop characteristics are represented by the curves a and L_0, when the load increased suddenly, the load P-f droop characteristics is changed into L_1, so the operating point of the inverter will move from A to C. but when the operating point moves to B, the system frequency can't be lower anymore due to the consideration of the system stability. So the P-f droop curve moves from a to a', a'', a''' ... , during this time, the operating point moves from B to D. when the curve moves from a to b and at the same time the operating point moves from B to D, we get the final P-f droop, and the

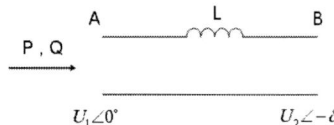

Figure 1. Active and reactive power through transmission line

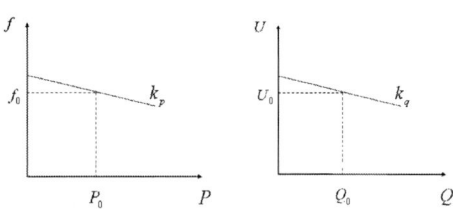

Figure 2. P-f and Q-U droop characteristic

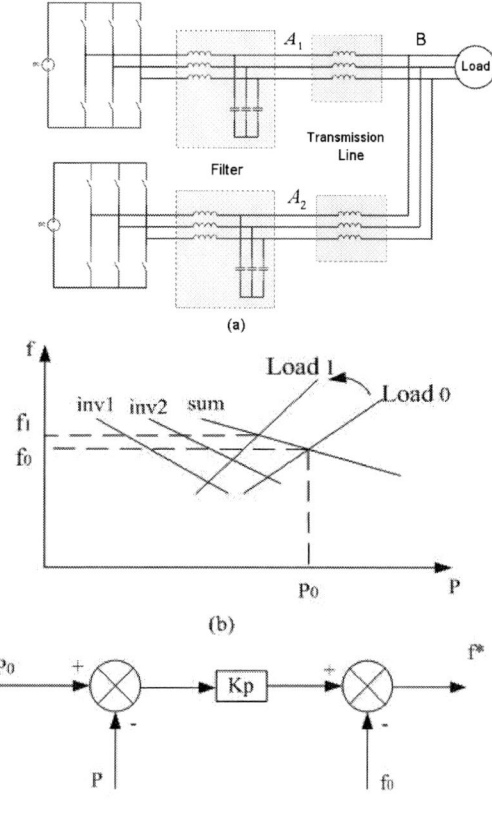

(a) Circuit diagram of two parallel inverters
(b) P-f droop characteristic of inverters and load
(c) Block diagram of droop control
Figure 3. P-f droop characteristic for parallel inverters

978-1-4244-4782-4/10 $26.00 © 2010 IEEE

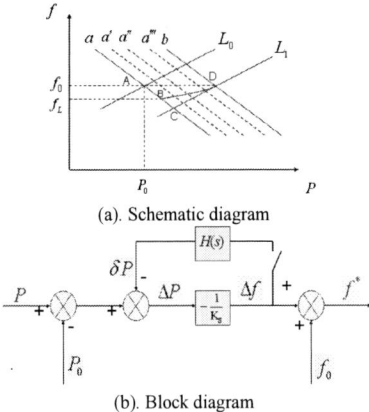

(a). Schematic diagram

(b). Block diagram

Figure4. Active Droop Positioning

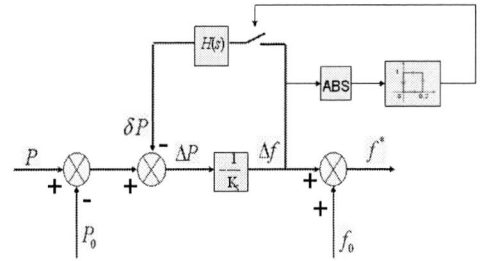

Figure 5. Control of switch S

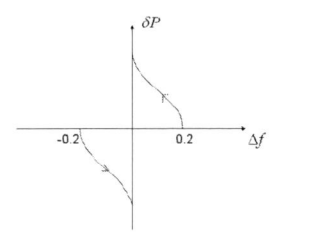

Figure6. The effect of adjustor H(s)

system frequency becomes 50.0Hz even. After that, when the load changes, the inverters can still share the power by the conventional way unless the frequency becomes too high or too low again.

The block diagram of the novel active power management is shown in Figure 4(b). When the $|\Delta f|$ is up to the threshold, the switch S closes. Then δP can be obtained by adjusting Δf, so the position of the droop changes, the final droop characteristic is given by following equation.

$$f - f_0 = -k_p[P - (P_0 - \Delta P)] \qquad (7)$$

III. IMPLEMENTATION

When a microgrid is in islanding mode, what kind of inverters in the microgrid can use this active management? How to control the time to turn on or turn off the switch S in Figure 4(b)? How to design the adjustor H(s) in Figure 4(b) to make sure the system frequency can go back to 50.0Hz even? Those problems can be solved in this section.

A. Choosing the Inverters

Microgrids consist of several basic technologies for operation. These include distributed generation (DG), distributed storage (DS), interconnection switches, and control systems[2]. Among all those, DG and DS are connected into the MG by inverters. In DS technology, the storage capacity can be categorized in terms of energy density requirements (for medium- and long-term needs) or in terms of power density requirements (for short- and very short-term needs). Due to the characteristics of this novel active power management, the inverter needs to change its output power in a wide range and have a big energy density. So this control scheme can be applied to long-terms-needs DS in MG.

B. Controlling the Switch

If the system frequency deviation is higher than 0.2Hz due to large-scale change of the load, the switch is turned on, and the P-f droop starts to change its position. When the output power meets the load at 50.0Hz, the switch is turned off. The block diagram to control the switch is in Figure 5.

When the P-f droop starts to move, an adjustor H(s) is used to make sure the deviation is zero. The adjustor can be expressed as follow:

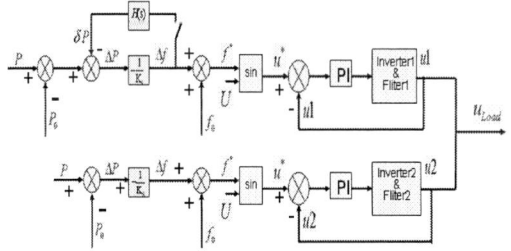

Figure7. Block diagram for two inverters

TABLE I. THE CONTROL PARAMETERS OF TWO INVERTERS

	P_o (W)	f_0 (Hz)	k_s (W/0.1Hz)
Inverter 1	300	50	200
Inverter 2	200	50	200

$$H(s) = \frac{k_1 + k_2 s}{s} \quad (8)$$

The effect of the adjustor can be illustrated in Figure 6.

IV. SIMULATION AND EXPERIMENTAL VALIDATION

The circuit topology in Figure 3(a) is used for simulation, the control method is shown in Figure 7, and PSCAD/EMTDC is used as simulation software. The main parameters are shown in TABLE I. Firstly, we simulate the conventional droop condition. Secondly, the active droop positioning is applied for comparison.

(b)Figure 8 shows the frequency deviations with load variations under conventional droop control scheme. The system becomes stable after 0.1s. At 0.3s, the load increases, these two inverters share the power proportionally, and the system frequency changes into 49.92Hz. At 0.7s, the load increase again, and this time the system frequency changes into 49.8Hz.

When one inverter is controlled by the active droop positioning, the result is shown in Figure 9. At 0.7s, the load increases, and the frequency starts to decrease. When the frequency decreases to 49.8Hz, which is assumed as the lower limit, the frequency starts to rise again. And finally the frequency of the whole system comes back to 50.0Hz even at 1.2s. At 1.5s, the load changes again, the inverters can also share the load power proportionally.

One inverter is used to verify the control strategy. The inverter has a power rating of about 5kW. In the experimental set-up, the LC PWM filter values are: L=2mH, C=25uF. The equivalent value of the transmission line is 0.5mH.

The rating output power at 50.00Hz is set as 900W, and the droop parameter is 200W/0.1Hz. The output data are shown by power quality analyzer (HIOKI 3197) in Figure 10. Figure 10(a) shows that when the output is 900W, the frequency is 50.00Hz. When the output power reduces to 450W, the frequency is up to 50.23Hz (Figure 10(b)), which creates a large deviation. Then the novel control strategy is used, and the

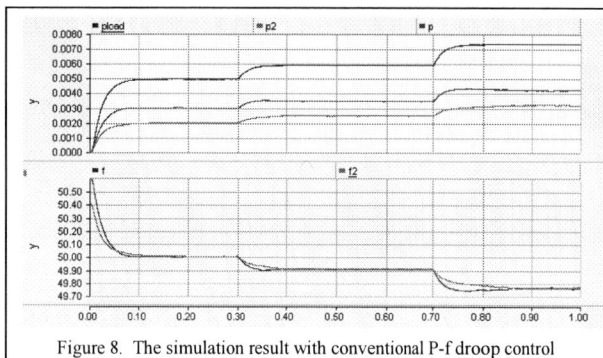

Figure 8. The simulation result with conventional P-f droop control

frequency rises to 50Hz (Figure 10(c)). So the whole system

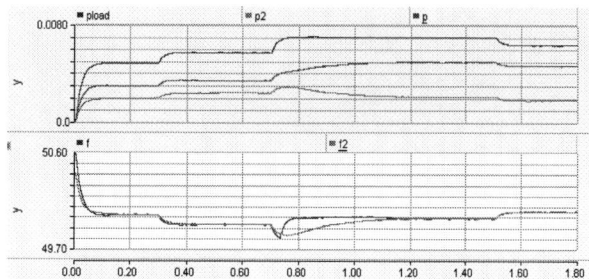

Figure 9. Simulation result with active droop positioning

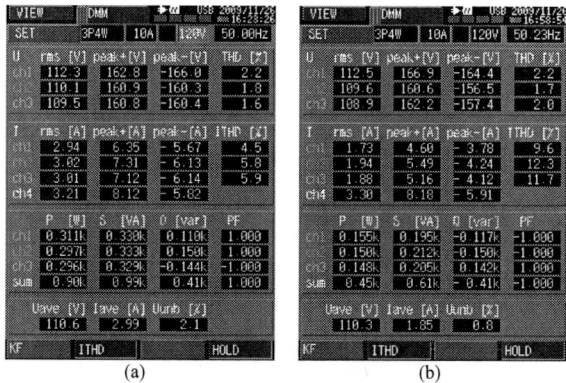

(a)   (b)

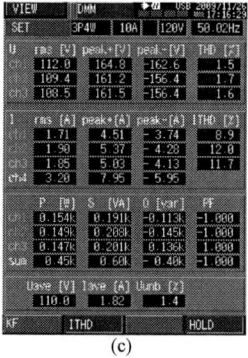

(c)

Figure 10. Output data in power quality analyzer

would be more stable if this control strategy is used in microgrid.

V. CONCLUSION

This paper presents a novel control strategy, active droop positioning, for parallel inverters in microgrid while it's under islanded mode. By automatically changing the droop position of some large capacity microsources, the proposed control strategy can bind system frequency deviations in a endurable range while ensure good current sharing performance among most microsources. Under this control strategy, almost all the micro sources in the microgrid can work around rated power, and when the load reduces dramatically, the new energy resources can also be taken full advantage. The choice of the inverter type, the control strategy, and the design are discussed. The validity of this strategy is tested by simulation.

REFERENCES

[1] Khaled. A. Nigim, Wei-Jen Lee, "Micro Grid Integration Opportunities and Challenges, " in IEEE 2007 Power Engineering Society General Meeting, June 2007, pp. 1-6.

[2] Kroposki. B., Lasseter. R., Ise. T., Morozumi. S., Papatlianassiou. S., Hatziargyriou. N., "Making microgrids work, " Power and Energy Magazine, vol. 6, pp. 40-53, May-June 2008.

[3] Farid. Katiraei, Reza. Iravani, Nikos. Hatziargyriou, Aris. Dimeas, "Microgrids Management, " Power and Energy Magazine, vol. 6, pp. 54-65, May-June 2008.

[4] J. A. Pecas Lopes, C. L. Moreira, A. G. Madureira, "Defining control strategies for analyzing microgrids islanded operation, " Power Tech, pp. 1-7, June 2005.

[5] Mukul.C. Chandorkar, Deepakraj. M. Divan, Rambabu. Adapa, "Control of parallel connected inverters in standalone AC supply systems," IEEE Transactions on Industry Applications, vol. 29, pp. 136-143, Jan.-Feb. 1993.

[6] Josep M. Guerrero, Nestor. Berbel, Jose. Matas, Jorge L. Sosa, Luis Garcia de Vicuna, "Droop Control Method with Virtual Output Impedance for Parallel Operation of Uninterruptible Power Supply Systems in a Microgrid, " in IEEE 2007 Applied Power Electronics Conference, 2007, pp. 1126-1132.

[7] T. Loix, K. De Brabandere, J. Driesen, R. Belmans, "A Three-Phase Voltage and Frequency Droop Control Scheme for Parallel Inverters" Industrial Electronics Society, 2007. IECON 2007. 33rd Annual Conference of the IEEE 5-8, Nov. 2007, pp.1662-1667.

[8] Hannu. Laaksonen, Pekka. Saari, Risto. Komulainen, "Voltage and frequency control of inverter based weak LV network microgrid, " in Future Power Systems 2005 International Conference, 2005, pp. 1-6.

[9] Josep. M. Guerrero, Jose. Matas, Luis Garcia de Vicuna, Miguel. Castilla, Jaume. Miret, "Decentralized Control for Parallel Operation of Distributed Generation Inverters Using Resistive Output Impedance, " IEEE Transactions on Industrial Electronics, vol. 54, pp. 994-1004, Apr. 2007.

[10] De Brabandere. Karel, Bolsens. Bruno, Van den Keybus, Jeroen, Woyte, Achim, Driesen, Johan, Belmans, Ronnie, "A Voltage and Frequency Droop Control Method for Parallel Inverters, " IEEE Transactions on Power Electronics, vol. 22, pp.1107-1115, July 2007 .

[11] Nuno. Jose. Gil, J.A.Pecas. Lopes, "Hierarchical Frequency Control Scheme for Islanded Multi-Microgrids Operation, " Power Tech, 2007 IEEE Lausanne 1-5, pp.473 - 478, July 2007.

978-1-4244-4782-4/10 $26.00 © 2010 IEEE

Direct Power Control of Doubly-Fed Generator Based Wind Turbine Converters to Improve Low Voltage Ride-Through during System Imbalance

Murali M. Baggu
Sr. R&D Engineer
Broadstar Wind Systems
muralimohan@ieee.org

Luke D. Watson
ECE Department
Missouri S&T
ldwxf5@mst.edu

Jonathan W. Kimball
ECE Department
Missouri S&T
kimballjw@mst.edu

Badrul H. Chowdhury
ECE Department
Missouri S&T
bchow@mst.edu

Abstract— **A novel control technique using direct active and reactive power control called Direct Power Control (DPC) is discussed for Low Voltage Ride Through (LVRT) of DFIG based wind turbine converters. This controller eliminates the conventional current loops and uses delta modulation comparators, which yields a faster response. The switching of the converter is done using a simple optimum switching table. This control achieves real and reactive power stability with simple active and reactive power control variables. A modified DPC algorithm is proposed to eliminate the current harmonics created by DPC during system disturbances. The practical verification of DPC is carried out by a scaled converter. The control is coded in C and implemented on a TMS320F2812 DSP. The converter using DPC is tested for system unbalance conditions created by an Industrial Power Corruptor (IPC) in the laboratory.**

I. Introduction

The new grid code requirements for wind power integration state that doubly fed induction generator (DFIG) controllers should be capable of overcoming temporary voltage disturbances [1]. They should remain online instead of tripping due to low voltage. This work addresses the design of controllers that would keep the wind turbine in stable operation during an external fault that causes the voltage to drop up to 60% for two seconds. The focus is on the control of the grid-side converter. If the grid-side converter is capable of controlling power flow and maintaining dc bus voltage, then the DFIG will stay online during the disturbance.

A fast acting controller that eliminates the conventional current control and acts directly upon the real and reactive power of the system is proposed by [2]. As this control acts directly according to the error in the real and reactive power flows, this control is called Direct Power Control (DPC). In DPC, the power required for the converter is commanded using the instantaneous voltages and currents whether they are balanced or not. A modified DPC algorithm is derived below

to reduce current harmonics that occur in the grid-side converter during disturbances.

II. DPC Operation And Simulation

The single line diagram of a PWM converter is shown in Fig. 1. v_{abc} represents the source or the line voltage and v_{a1b1c1} represents the voltage of the converter that can be controlled by the DC link. The voltage of the converter depends upon the switching sequence of the converter and the magnitude of the DC voltage. The series inductance of the line and the dc bus capacitance provides boost characteristics for the converter from the AC to the DC side. The flow of current from the source to the load is governed by the difference between the source voltage and the converter voltage. The line inductance provides stiff current characteristics to the source whereas the bus capacitance provides stiff voltage characteristics to the dc link. Typically the source voltage is assumed to be constant, although this is not true in case of system imbalance or during a transient. Hence the flow of the current in the circuit is governed by the magnitude and angle of the converter voltage.

A. DPC Switching States and Delta Modulation

The switching states of a voltage source inverter are based on the space vector modulation approach presented in [3]. The magnitude and angle of the voltage vector can either be increased or decreased by applying appropriate vectors in the α-β plane. In DPC, an optimal switching table based on the analysis in [2],[4] is used for the instantaneous control of real and reactive power. When applying the zero vector, the choice between U_0 (000) or U_7 (111) depends on the converter legs switching during change of states. The three phase instantaneous real and reactive power is estimated by the scalar and vector product of the instantaneous voltages and

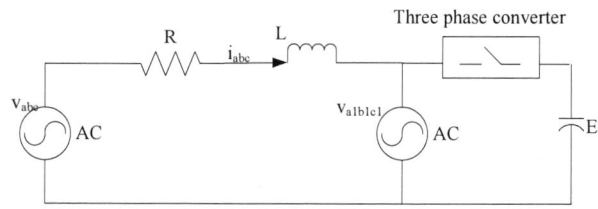

Figure 1.   Single line diagram of the PWM converter

This work was supported in part by the U.S. National Science Foundation under Grant ECS-0523897. Mr. Watson is being supported on a Fellowship from the US Dept. of Education.

978-1-4244-4782-4/10 $26.00 © 2010 IEEE

currents as in (1) and (2), respectively.

$$p = \left(v_a i_a + v_b i_b + v_c i_c\right) \qquad (1)$$

$$q = \frac{1}{\sqrt{3}}\left\{i_a\left(v_b - v_c\right) + i_b\left(v_c - v_a\right) + i_c\left(v_a - v_b\right)\right\} \qquad (2)$$

The commanded real and reactive powers are compared with the estimated power from (1-2). The error is digitized using a delta modulation as follows:

For reactive power: $d_q = 1$ for $q < q_{ref}$ $d_q = 0$ for $q > q_{ref}$
For active power: $d_p = 1$ for $p < p_{ref}$ $d_p = 0$ for $p > p_{ref}$

The vector position is estimated using the angle of the source voltage in the stationary reference frame, as indicated in (3). The digitized outputs along with the vector position as in (3) are fed to the optimum switching table. The vector position (θ) is calculated in the range of $-180°$ to $180°$ and the situation of $V_\alpha = 0$ is eliminated by using the abc to $\alpha\beta$ reference frame block in PLECS® block set.

$$\theta = \tan^{-1}\left(\frac{V_\beta}{V_\alpha}\right) \qquad (3)$$

A simulation was performed in Matlab/Simulink® using PLECS. The block diagram is shown in Fig. 2.

Several simulations have been performed and are shown in Fig. 3. In all cases, a 60% voltage dip occurs in phase A from 0.3 s to 0.8 s. The response of a conventional controller is shown in Fig. 3(a). Figs. 3(b) and 3(c) show the response with DPC. DPC has significantly less power ripple and dc voltage ripple, but also has some distortion in the phase currents.

III. MODIFIED DIRECT POWER CONTROL

In order to decrease the distortion in the currents of the DPC, a hybrid approach similar to that presented in [5] is proposed. In this approach the real and reactive power references for the DPC are appended with powers due to the imbalance. The negative sequence voltages contribute significantly to power imbalance. Neglecting the negative sequence current components, the real and reactive power equations due to the imbalance are modified as in (4) and (5).

$$P_{imb} = v_{dne}i_{dpe} + v_{qne}i_{qpe} \qquad (4)$$

$$Q_{imb} = v_{dne}i_{qpe} - v_{qne}i_{dpe} \qquad (5)$$

The partial simulation diagram showing the addition of unbalanced power commands is shown in Fig. 4. Notch filters are used for positive and negative sequence separation, and the negative sequence voltages are used to correct for imbalance.

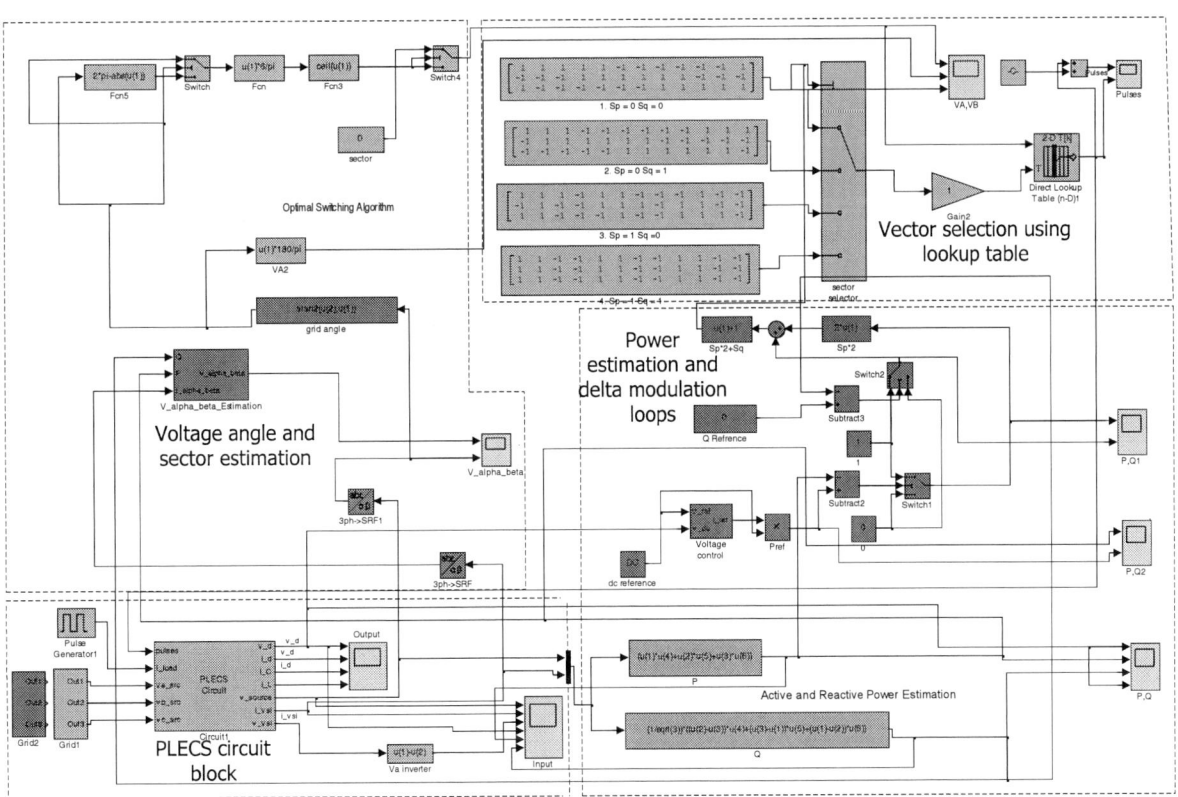

Figure 2. DPC structure for voltage source converter

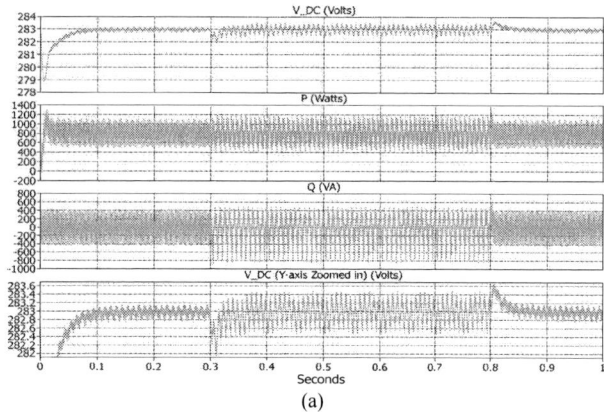

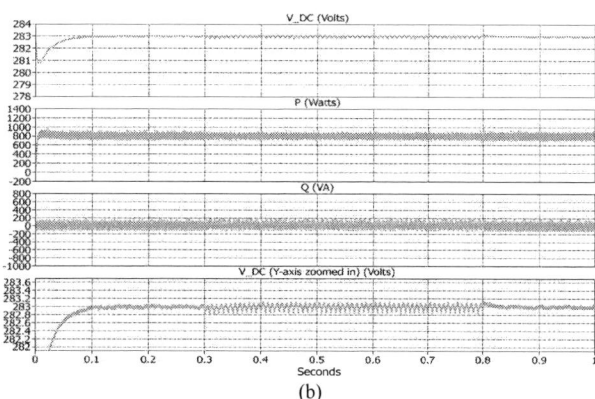

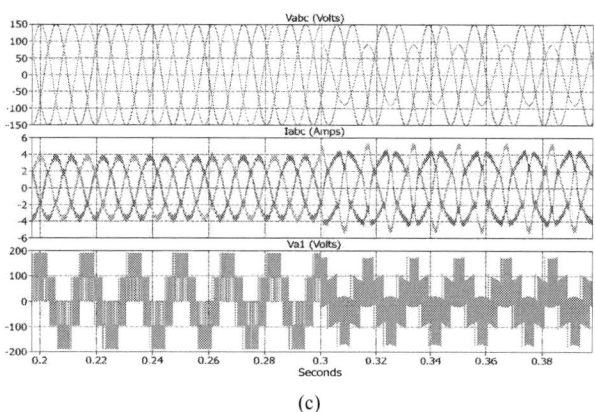

Figure 3. Grid side converter simulations for a voltage dip of 60%. (a) Conventional current control; (b) and (c) DPC. In (a) and (b), traces are DC bus voltage, real power, reactive power, and zoomed bus voltage. In (c), plots are three-phase source voltage, three-phase line current, and phase A line-to-neutral voltage.

The modified DPC was simulated for the same 60% voltage dip in the input as for the normal DPC. The simulation results are shown in Figure 5. Table 1 shows the comparison of peak-to-peak ripple DC link voltage, real power and reactive powers of the 60% voltage dip. The source currents are less distorted than the normal DPC due to added unbalance in real and reactive power compensation. The trade-off is that the DC link voltage, real power and reactive power have slightly higher oscillations than the normal DPC but have lesser oscillations than the conventional current control

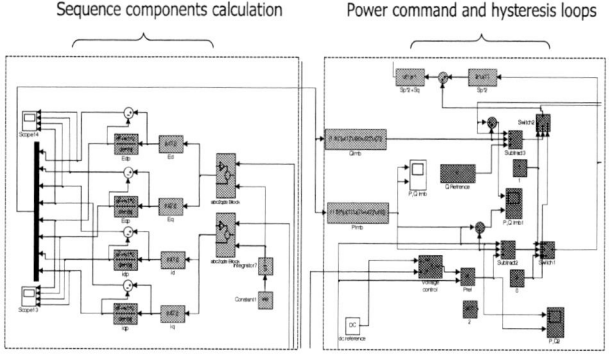

Figure 4. Partial block diagram of MPDC

TABLE I. COMPARISON OF RIPPLE WHEN USING DIFFERENT CONTROL METHODS.

Control Min/Max	DC Link Voltage (283V)	Real Power (800 Watts)	Reactive Power (0VA)
Conventional	282.4/283.4	500/1100	-800/400
DPC	282.8/283.1	730/870	-100/100
MDPC	282.4/283.2	600/1000	-200/200

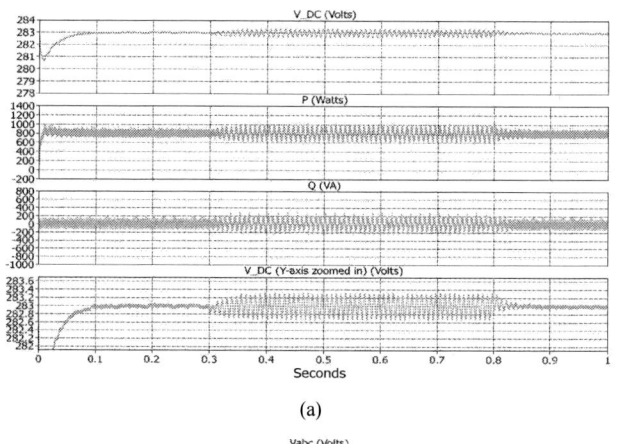

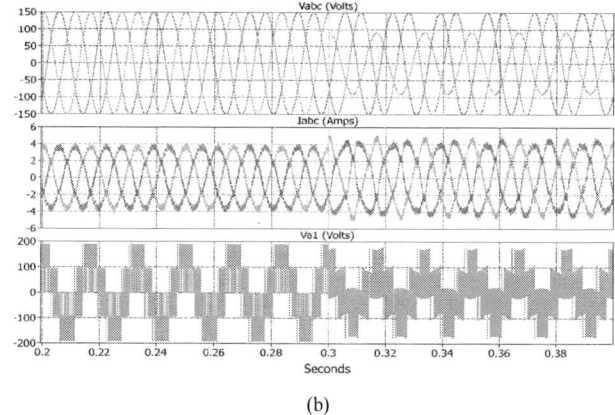

Figure 5. Grid side converter simulation results using MPDC. From top: (a) DC bus voltage, real power, reactive power, and zoomed bus voltage; (b) three-phase source voltage, three-phase line currents, and phase A switched output voltage.

technique. Imbalance power compensation retains the same settling time and overshoot as that of the normal DPC as there is no addition of new controllers. Four new notch filters are added to the circuit which adds computational complexity. A compromise between current distortion and power oscillations needs to be considered when choosing a normal DPC or a modified DPC.

IV. PRACTICAL IMPLEMENTATION

A diagram of the practical implementation is shown in Figure 6. Due to limitations associated with the protection circuitry, the 230 V 3-phase source is connected to a 3-phase variable transformer to step the voltage down. The output of the variable transformer is connected to a delta-wye transformer, which connects to the Industrial Power Corrupter (IPC). The IPC is capable of creating voltage sags to imitate actual conditions on the grid during a disturbance. The delta-wye transformer ensures that the uncorrupted voltage source is balanced.

Va, Vb, Vc, Ia, Ib, Ic, and VDC are retrieved by the Sensor Board and converted to 0-3V by the Analog Board. A TMS320LF2812 DSP samples the signals and calculates the appropriate vector to apply at a rate of 20 kHz. A sixpack IGBT module, International Rectifier CPV364MU, is used for switching.

The control was implemented using the TMS320LF2812 DSP. The experimental results in Figure 7 demonstrate that the coded algorithm is capable of maintaining the dc link voltage. The currents are more distorted in the experimental results and the powers have greater oscillations, but the real and reactive powers are controlled by the DPC. Fig. 7 shows the experimental results under no fault.

A voltage sag of 30% on Va is applied to the system for three cycles as shown in Figure 8. In this case the DC link voltage does not vary during the fault and stays flat. Hence the DPC is capable of riding through these post fault recovery sags and operates normally for voltages greater than 30%.

V. CONCLUSIONS

The controllers based on DPC and MDPC are fast acting; hence they are suitable for sudden grid and wind disturbances. Direct power control is capable of low voltage ride-through (LVRT) for wind turbines and can replace conventional current control. MDPC retains the advantages of DPC and improves current waveform fidelity.

ACKNOWLEDGMENT

This work was funded by the U.S National Science Foundation (NSF) under grant ECS-0523897 and the Electrical and Computer Engineering Department at Missouri University of Science and Technology, Rolla. Mr. Watson is being supported on a fellowship from the US Department of Education.

REFERENCES

[1] R. Zavadil, N. Miller, A. Ellis, E. Muljadi, E. Camm, and B. Kirby, "Queuing Up: Interconnecting Wind Generation into the Power System," *IEEE Power and Energy Magazine*, vol. 5 pp. 47-58, November/December 2005.

[2] T. Noguchi, H. Tomiki, S. Kondo, I. Takahashi, "Direct power control of PWM converter without power-source voltage sensors," *IEEE Trans. Ind. Appl.*, vol. 34, pp. 473-479, May/June 1998.

[3] H. W. Van der Broeck, H-C Skudelny and G V Stanke, "Analysis and realization of a pulsewidth modulator based on voltage space vectors," *IEEE Trans. Ind. Appl.*, vol 24, pp. 142-150, January/February 1988.

[4] M. Malinowski, M. P. Kazmierkowski, S. Hansen et, al "Virtual-Flux-based Direct power Control of Three-Phase PWM Rectifiers," *IEEE Trans. Ind. Appl*, . vol. 37, July/August 2001.

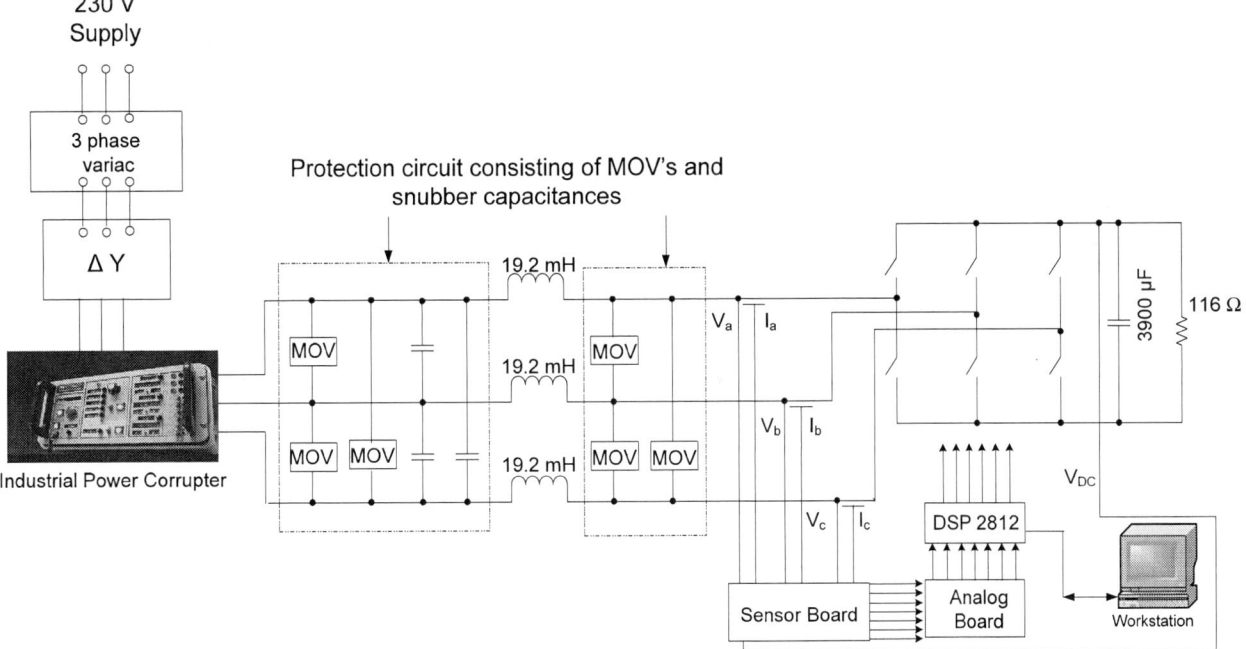

Figure 6. Experimental setup.

[5] J. Eloy-Garcia, S. Arnaltes, J. L. RodriguezAmenedo, "Extended direct power control of a three-level neutral point clamped voltage source inverter with unbalanced voltages," in *Rec. IEEE Power Electronics Specialists Conference* , pp.3396-3400, 2008.

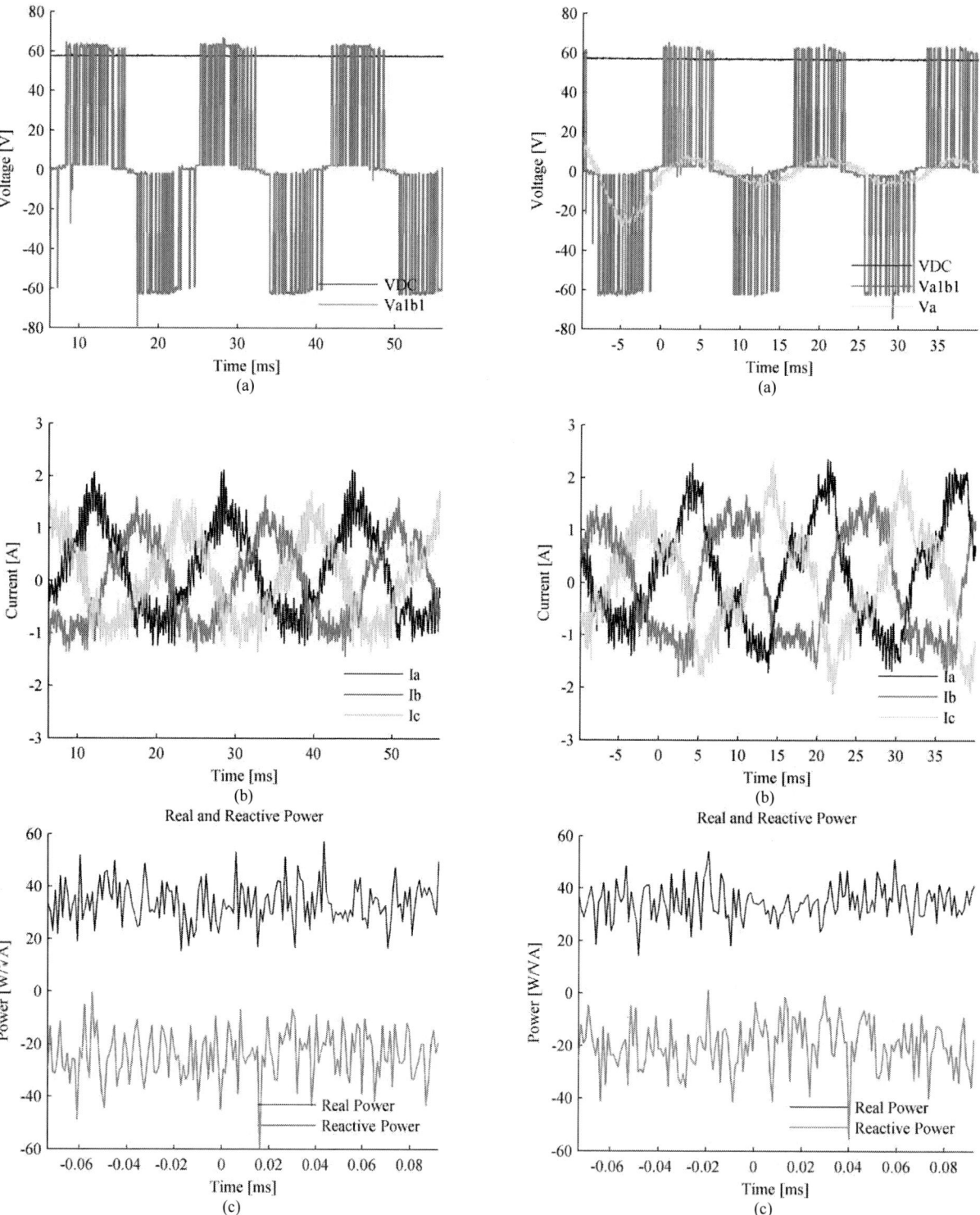

Figure 7. DPC during normal operation. (a) DC bus voltage and switched output voltage, (b) three-phase line currents, (c) real and reactive power.

Figure 8. DPC during a 30% sag on Va (which occurs at time 0 and lasts for 50 ms). (a) DC bus voltage and switched output voltage, (b) three-phase line currents, (c) real and reactive power.

Active Damping for Torsional Vibrations in PMSG based WECS

Hua Geng, Dewei Xu and Bin Wu
Department of Electrical and Computer Engineering
Ryerson University
Toronto, M5B 2K3, Canada
hgeng@ee.ryerson.ca

Geng Yang
Automation Department
Tsinghua Universiy
Beijing, 100084, P.R.China
yanggeng@tsinghua.edu.cn

Abstract — **Multi-pole permanent magnetic synchronous generator (PMSG) applied in the direct driven wind energy conversion system (WECS) usually has a relative "soft" drive train. Because of no internal damping winding, torsional vibrations are easily excited in the PMSG based WECS. For the maximum power extraction requirement, the vibrations can be automatically damped with optimal torque (OT) or power signal feedback (PSF) scheme implemented in the generator side. However, if the WECS is required to output a smoothed or constant power for the power regulation, the vibrations will not be damped with the control strategy. Thus additional active damping control is necessary. Active damping can be achieved with power control in both grid and generator sides or just dc link voltage control. With additional compensator in the power or dc voltage control loop, the damping torque is introduced to suppress the vibrations in the generator torque. Different active damping strategies for different output power requirements are designed and verified with simulation results in the paper.**

I. INTRODUCTION

By eliminating the gearbox, direct driven permanent magnetic synchronous generator (PMSG) based wind energy conversion system (WECS) significantly improve the attractiveness [1] due to efficiency and low maintenance. In order to provide high torque and low speed operation, direct driven configuration requires the generator to have a large number of poles. According to [2], the effective stiffness of the generator shaft is inversely proportional to the pole pairs, which means that the PMSG with larger capacity has a "softer" driven train. In the case, the torsional twist of the shaft for a multi-pole PMSG affects the operation of WECS significantly [2,3]. Because of the torsional characteristic of the drive train, the generator speed is prone to oscillations whenever the system gets excited by mechanical or electrical load changes [3]. The oscillations can result in the fluctuation in the output power and the increase of mechanical stress of the drive train. The typical frequency of the oscillation is as low as 0.1-10 Hz which tends to coincide with the frequencies associated with power system low frequency

sub-harmonic oscillations (0.1-2.5 Hz) [4]. As a result, stability issues are being concerned for the power system with high wind power penetrations if no damping technique is adopted for the PMSG based WECS.

In the traditional power plant, the wound-field synchronous generators are directly connected to the grid. The damper winding in the rotor can provide additional damping torque for the speed oscillations. However, it is not the case for the direct driven variable speed WECS. Usually, a full-scale converter is applied to connect the generator with the grid. Damper winding of the wound-field synchronous generators has no influence on counteracting the speed oscillations because there is no relative movement between the rotor field and the stator field resulting in no induced damping current [5]. The case is even worse for a PMSG based WECS as this configuration does not have a damper winding.

Active damping control provides an effective and flexible way to suppress the torque and power oscillations, which leads to a effective solution rather than passive methods [3,5,6,7]. This article designs of different active damping strategies for the torsional vibrations of PMSG based WECS with different output power requirements. Firstly, the small signal models of WECS with different control structures are presented and analyzed in the frequency domain. It is noticeable that the vibrations can be damped with the grid side inverter or generator side converter. Then, based on the power requirements of the power systems, different active damping strategies are designed and verified with simulations in the time domain.

II. SYSTEM MODELING AND ANALYSIS

A typical configuration of the direct driven PMSG based WECS is shown in Fig.1, which consists of the wind turbine, drive train, PMSG and a back-to-back converter.

The eigen frequency of the drive-train is rather low and within the bandwidth that is normally taken into account in power system dynamics simulations. A two-mass model representation of the drive-train is therefore essential in order to illustrate the dynamic impact of wind turbines on the grid properly. Because of no inherent damper in the PMSG, the damping factor of the shaft is negligible. The mechanical system of WECS can be expressed as follows [2,3,8].

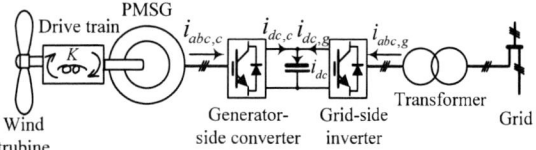

Fig.1. Configuration of PMSG-based WECS

$$\dot{\omega}_h = \frac{1}{J_h}\left(T_{wt} - K\theta\right)$$

$$\dot{\omega}_g = \frac{1}{J_g}\left(K\theta - T_g\right) \quad (1)$$

$$\dot{\theta} = \omega_h - \omega_g$$

where, ω_h, ω_g are the rotational speed of the wind turbine and the generator; J_h, J_g are the turbine inertia and generator inertia respectively; θ is the electrical angle of the shaft; K is the stiffness of the shaft and T_{wt} is the mechanical torque of the wind turbine which is a function of wind speed, turbine rotational speed and pitch angle.

$$T_{wt} = K_w C_q v^2 \quad (2)$$

where $K_w = \pi R^3 \rho / 2$, R, ρ are the turbine radius and air density, v is the wind speed and C_q is the turbine torque coefficient which is a nonlinear function of turbine rotational speed, wind speed and pitch angle.

The electrical system of WECS has much faster dynamics compared with that of the mechanical system because of the very large turbine and generator inertias. As a result, the electrical system can be simply modeled as follows.

$$\dot{T}_g = \frac{1}{T_s}\left(T_g^* - T_g\right)$$

$$\dot{V}_{dc} = \frac{1}{C_{dc} V_{dc}}\left(T_g \omega_g - P_{out}\right) \quad (3)$$

$$\dot{P}_{out} = \frac{1}{T_s'}\left(P_{out}^* - P_{out}\right)$$

where T_g, T_g^* are the generator electrical torque and its command; C_{dc}, V_{dc} are the dc link capacitance and dc voltage; P_{out}, P_{out}^* are the output power and its command of the grid side inverter; T_s, T_s' is the equivalent time constants of the generator torque and grid side inverter.

Usually, there are two control structures can be applied in this system: (i) utilize the generator side converter for the power control while the grid side inverter controls the dc link voltage to balance the input and output power and (ii) swap the functions of the grid and generator side converters above [6]. The control diagrams of the two structures are shown in Fig.2.

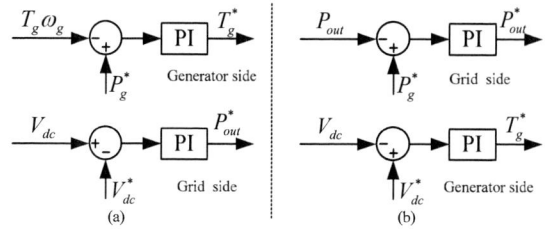

Fig.2. Control diagram of the two control structures (a) DC link is controlled by grid side converter, (b) DC link is controlled by the generator side converter

Considering the PI parameters of the power loop and dc link voltage loop as k_p, k_i and k_p', k_i' respectively, the small signal models of WECS with structure (i) and (ii) can be derived from (1)-(3).

$$\Delta \dot{x} = A\Delta x + B\Delta u \quad (4)$$

$$\Delta \dot{x} = A'\Delta x + B'\Delta u \quad (5)$$

where $\Delta x = \begin{bmatrix} \Delta\omega_h & \Delta\omega_g & \Delta\theta & \Delta V_{dc} & \Delta T_g & \Delta T_g^* & \Delta P_{out} & \Delta P_{out}^* \end{bmatrix}^T$, $\Delta u = \begin{bmatrix} \Delta T_{wt} & \Delta P_g^* & \Delta V_{dc}^* \end{bmatrix}^T$ and the matrixes A, B, A', B' in Eq.(4), (5) are illustrated in the Appendix.

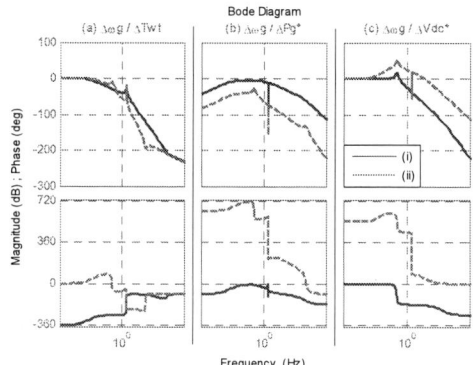

Fig.3. Bode diagram of the two control structures (i),(ii)

The bode diagrams of the small signal models are shown in Fig.3. One can see that the torsional vibrations (1.93 Hz) can be excited if the turbine torque changes. The vibrations are severer with structure (ii). With power control, the torsional vibrations can be damped in both structures. With structure (ii), the vibrations can be damped with the dc link voltage control as well, which has been fully studied in literature [3]. It should be noticed that phase criterion shown in Fig.3 needs to be satisfied in order to design an active damping strategy. Taking the structure (i) as an example, the vibration is suppressed if the small signal ΔP_{g}^{*} is proportional to $\Delta \omega_g$ and the phase different between ΔP_{g}^{*} and $\Delta \omega_g$ is around $0°$.

III. ACTIVE DAMPING STRATEGIES FOR TORSIONAL VIBRATIONS

Active damping scheme functions by adding a damping term to the generator torque so that the torsional vibrations are suppressed. However, the damping term is not necessary all the time. It depends on the power requirements of the grid and the power control strategies of the WECS. For the maximal power point tracking (MPPT) occasion, the optimal torque (OT) [9] or power signal feedback scheme (PSF) [10,11] with structure (i) has inherent vibration damping abilities because the command of generator power P_{g}^{*} or torque T_{g}^{*} is calculated based on the generator speed feedback. PSF scheme with structure (ii) and the tip speed ratio (TSR) scheme [12] cannot damp the vibration automatically as the phase criterion in Fig.3 is not satisfied. For the large wind power penetration situation, the grid usually prefer a smoothed power or even constant power injection to avoid the instability or power quality issues [13,14]. Power smoothing is achieved with a first order filter before the PSF controller [8]. In this case, the generator response slows down because of the filter effect, which results in the increase of the equivalent generator inertia. As the vibration frequency is determined by the shaft stiffness, turbine inertia and generator inertia, it will deviate from the natural resonant frequency of WECS and the vibrations are hard to damp. For the constant power production, no internal damping is provided by the control strategy and external active damping controller is necessary.

A. Active damping with power control

As shown in Fig.3, the vibrations can be damped with both structure (i) and (ii) if a compensator ΔP_{g}^{*} is included in the power command P_{g}^{*}. It is noticed that $\Delta P_{g}^{*}/\Delta \omega_g$ should satisfy the phase criterion shown in Fig.3 for the active damping. In order to obtain the small signal value $\Delta \omega_g$, a band pass filter with transfer function (5) is designed.

$$G_{filter}\left(s\right) = \frac{2\xi_n \omega_n s}{s^2 + 2\xi_n \omega_n s + \omega_n^2} \qquad (6)$$

where ξ_n, ω_n are the damping factor and the band pass frequency respectively.

The bode diagram of the band pass filter is shown in Fig.4. The band pass frequency is designed to be the vibration frequency which is 12.127 rad/s in this paper. The control diagram of the scheme is denoted in Fig.5(a). The phase compensator is introduced so that $\Delta P_{g}^{*}/\Delta \omega_g$ satisfies

the phase criterion. It can be seen from Fig.3 that the phase compensation is not necessary for structure (i) but is compulsory for structure (ii). The phase compensation can be easily achieved with a first-order filter.

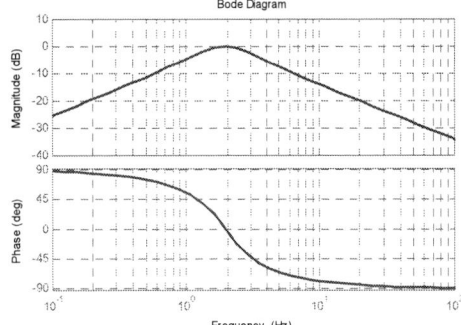

Fig.4. Bode diagram of the band pass filter

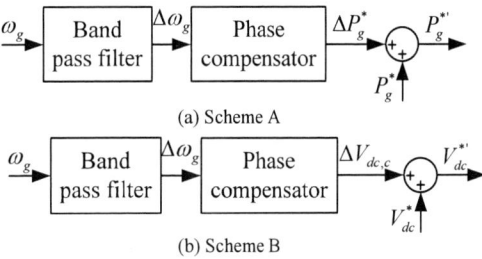

(a) Scheme A

(b) Scheme B

Fig.5. Control diagram of the active damping scheme

B. Active damping with dc link voltage control

The vibrations can be damped with the control of dc link voltage in structure (ii). Similarly with scheme A, a band pass filter is necessary for the oscillation detection. The control diagram of scheme B is shown in Fig.5(b). A 90° phase should be compensated in this scheme as denoted in Fig.3 [3].

IV. SIMULATIONS RESULTS

The simulations are carried out in MATLAB/SIMULINK. The PMSG is modeled as a three-order system in the dq reference frame. The back to back converter and its controller are modeled as a digital system with a 1 kHz switching frequency and a 10 kHz sampling frequency. Both generator side converter and grid side inverter are controlled with vector control techniques. The parameters of WECS are listed in the Appendix. Two kinds of wind profile shown in Fig.6, a step down and a constant wind profile, are adopted in the simulations.

A. Conventional power control without active damping

Fig.7 shows the output power, generator torque, turbine speed and generator speed of WECS under (a) conventional

PSF-MPPT control with structure (i), (b) PSF-MPPT control with structure (ii) and (c) power smoothing control. The wind profile for MPPT control is the step down wind shown in Fig.6(a). One can see that PSF-MPPT has inherent vibration damping ability only if implemented in the generator side. With power smoothing control, the vibration frequency is changed (from 1.93 Hz to 2.48 Hz) because of the deviation of equivalent generator inertia. Fig.8 is the system responses in constant power control with structure (i) and (ii) and the constant wind profile is adopted in the simulation. Unfortunately, the vibration can not be automatically damped any more. Vibrations are severer for structure (ii) because the grid side inverter is directly controlled to output the constant power and cannot provide additional damping.

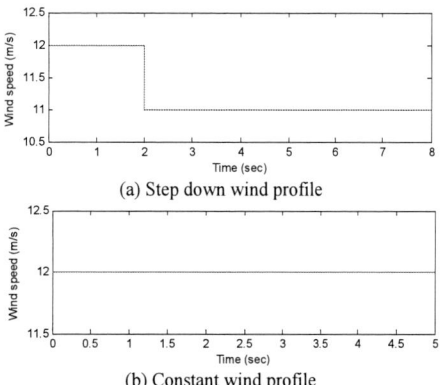

(a) Step down wind profile

(b) Constant wind profile

Fig.6. Wind speed profiles in the simulations

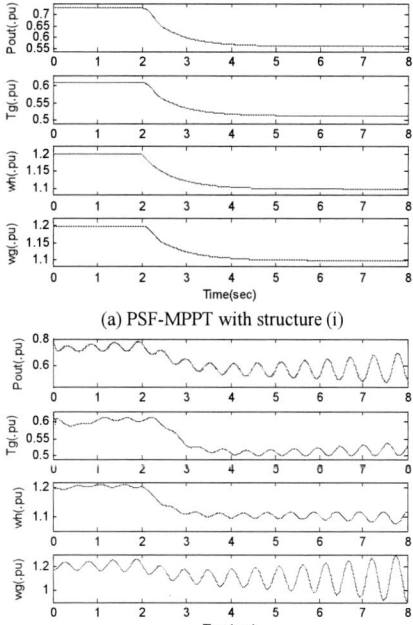

(a) PSF-MPPT with structure (i)

(b) PSF-MPPT with structure (ii)

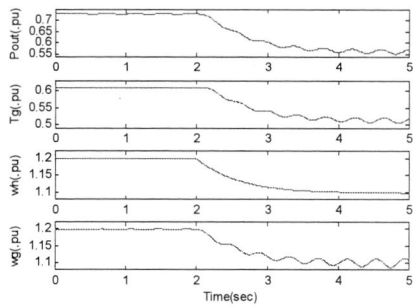

(c) Power smoothing with structure (i)

Fig.7. Responses of WECS without active damping

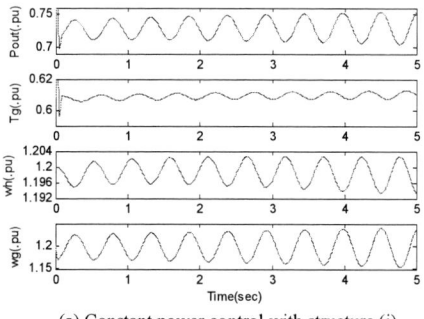

(a) Constant power control with structure (i)

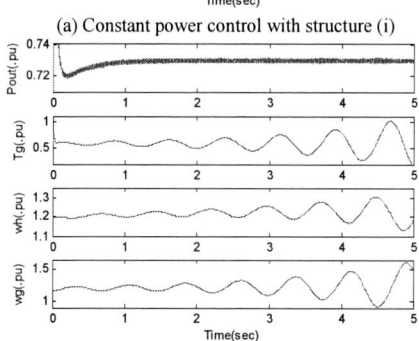

(b) Constant power control with structure (ii)

Fig.8. Responses of WECS without active damping

B. Active damping schemes

Active damping schemes are applied to the power control of WECS. Only the simulation results of constant power control are presented here because the control strategy cannot provide any inherent damping, which is the worst situation for the WECS operation. The system responses are shown in Fig.9. Obviously, the vibrations are dramatically suppressed. The damping is easier from the grid side because the grid has an "infinite inertia" compared with the generator. However, the output power of WECS has more oscillations with this structure. So structure (ii) with active damping scheme B has practical applications when large wind power is injected to the power systems.

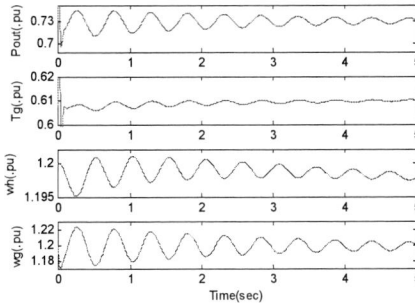

(a) Active damping scheme A with structure (i)

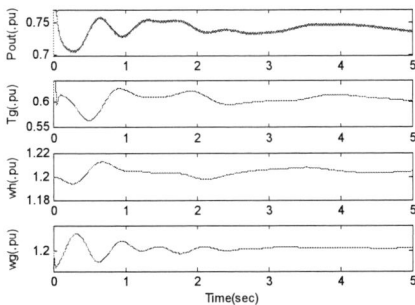

(b) Active damping scheme A with structure (ii)

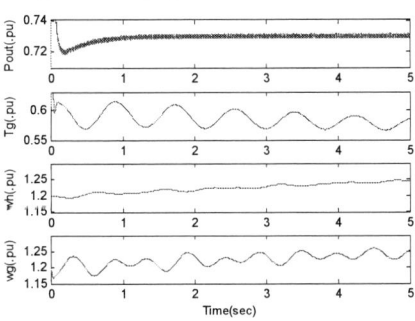

(c) Active damping scheme B with structure (ii)

Fig.9. Responses of WECS

V. CONCLUSIONS

Direct driven configuration with PMSG for WECS has many advantages. However, because no damping winding is included in the generator, torsional vibration and speed oscillations appear. This article designs of different strategies for active damping of torsional vibrations in PMSG based WECS. The active damping schemes can be chosen based on the output power requirements and the power controller structures. For the smoothed or constant power output occasion, active damping is necessary and can be applied in generator side or grid side. With additional compensator in the power or dc voltage control loop, a damping torque is produced to suppress the oscillations in the generator speed and the output power. As a result, the stability of the WECS and power systems can be improved.

REFERENCES

[1] Polinder, H., van der Pijl, F.F.A., de Vilder, G.-J., Tavner, P.J., "Comparison of direct-drive and geared generator concepts for wind turbines", IEEE Transaction on Energy Conversion, 21(3), 2006, pp.725-733.

[2] Akhmatov V., "Analysis of dynamic behavior of electric power systems with large amount of wind power", PhD Thesis, Ørsted DTU, 2003.

[3] Anca D. Hansen, Gabriele Michalke, "Modelling and control of variable-speed multi-pole permanent magnet synchronous generator wind turbine", Wind Energy, 2008, 11(5), pp. 537-554.

[4] S. Brownlees, B. Fox, D. Flynn, T. Littler, "Wind Farm Induced Oscillations", Proceedings of the 41st International Universities Power Engineering Conference, 2006, 1:118 – 122.

[5] Jauch C., "Stability and control of wind farms in power systems", PhD Thesis, Risø-PhD-24, Risø National Laboratory, 2006.

[6] Jingya Dai, Dewei Xu, Bin Wu, Zargari, N.R., Yongqiang Lang, "Dynamic performance analysis and improvements of a current source converter based PMSM wind energy system", 2008 Power Electronics Specialists Conference, June 2008, pp. 99-105.

[7] Rigby, B.S., Chonco, N.S., Harley, R.G., "Analysis of a power oscillation damping scheme using a voltage-source inverter", IEEE Transactions on Industry Applications, 38(4), 2002, pp.1105-1113.

[8] Rawn, B.G., Lehn, P.W., Maggiore, M., "Control Methodology to Mitigate the Grid Impact of Wind Turbines", IEEE Transaction on Energy Conversion, 2007, 22(2): 431-438.

[9] R. Cardenas, R. Pena, P. Wheeler, J. Clare, and G. Asher, "Control of the reactive power supplied by a wecs based on an induction generator fed by a matrix converter," Industrial Electronics, IEEE Transactions on, vol. 56, no. 2, pp. 429–438, Feb. 2009.

[10] S. Bhowmik, R. Spee, and J. Enslin, "Performance optimization for doubly fed wind power generation systems," Industry Applications, IEEE Transactions on, vol. 35, no. 4, pp. 949–958, Jul/Aug 1999.

[11] Z. Chen, J. M. Guerrero, and F. Blaabjerg, "A review of the state of the art of power electronics for wind turbines," Power Electronics, IEEE Transactions on, vol. 24, no. 8, pp. 1859–1875, Aug. 2009.

[12] K. Tan and S. Islam, "Optimum control strategies in energy conversion of pmsg wind turbine system without mechanical sensors," Energy Conversion, IEEE Transactions on, vol. 19, no. 2, pp. 392–399, June 2004.

[13] E.ON Netz GmbH "Grid Code High and Extra High Voltage", E.ON Netz. GmbH Bayreuth, August 2003, http://www.eon-netz.com.

[14] F. Iov, R. Teodorescu, F. Blaabjerg, B. Andersen, J. Birk, and J. Miranda, "Grid code compliance of grid-side converter in wind turbine systems," the 37th IEEE Power Electronics Specialists Conference, 2006.

APPENDIX

1. Parameters of the WECS in the simulations.

Wind turbine moment of inertia [kg.m2]	6.2506×10^6
Turbine rated speed [rad/s]	1.75
Generator moment of inertia [kg.m2]	6.5110×10^5
Shaft stiffness [Nm.s/rad)]	8.6727×10^7

Generator rated electromagnetic torque [N.m] 1.1411×10^6
Generator rated power [Mw]   2
Generator rated line voltage [kV]   3.3
Generator rated frequency [Hz]   10.6
Pole pairs   38
Generator inductance in dq reference frame [pu]   0.7
Generator stator resistance [pu]   0.01
DC link capacitance [pu]   1
Line inductance [pu]   0.1

2. The matrixes in Eq.(4) and (5).

$$
A = \begin{bmatrix}
0 & 0 & \dfrac{-K}{J_h} & 0 & 0 & 0 & 0 & 0 \\[2mm]
0 & 0 & \dfrac{K}{J_g} & 0 & -\dfrac{1}{J_g} & 0 & 0 & 0 \\[2mm]
1 & -1 & 0 & 0 & 0 & 0 & 0 & 0 \\[2mm]
0 & \dfrac{\overline{T}_g}{C_{dc}\overline{V}_{dc}} & 0 & 0 & \dfrac{\overline{\omega}_g}{C_{dc}\overline{V}_{dc}} & 0 & \dfrac{-1}{C_{dc}\overline{V}_{dc}} & 0 \\[2mm]
0 & 0 & 0 & 0 & -\dfrac{1}{T_s} & \dfrac{1}{T_s} & 0 & 0 \\[2mm]
0 & -k_i\overline{T}_g & -\dfrac{k_p\overline{T}_g K}{J_g} & 0 & -k_i\overline{\omega}_g + \dfrac{k_p\overline{\omega}_g}{T_s} + \dfrac{k_p\overline{T}_g}{J_g} & -\dfrac{k_p\overline{\omega}_g}{T_s} & 0 & 0 \\[2mm]
0 & 0 & 0 & 0 & 0 & 0 & -\dfrac{1}{T_s'} & \dfrac{1}{T_s'} \\[2mm]
0 & \dfrac{k_p'\overline{T}_g}{C_{dc}\overline{V}_{dc}} & 0 & k_i' & \dfrac{k_p'\overline{\omega}_g}{C_{dc}\overline{V}_{dc}} & 0 & -\dfrac{k_p'}{C_{dc}\overline{V}_{dc}} & 0
\end{bmatrix}
$$

$$
A' = \begin{bmatrix}
0 & 0 & \dfrac{-K}{J_h} & 0 & 0 & 0 & 0 & 0 \\[2mm]
0 & 0 & \dfrac{K}{J_g} & 0 & -\dfrac{1}{J_g} & 0 & 0 & 0 \\[2mm]
1 & -1 & 0 & 0 & 0 & 0 & 0 & 0 \\[2mm]
0 & \dfrac{\overline{T}_g}{C_{dc}\overline{V}_{dc}} & 0 & 0 & \dfrac{\overline{\omega}_g}{C_{dc}\overline{V}_{dc}} & 0 & \dfrac{-1}{C_{dc}\overline{V}_{dc}} & 0 \\[2mm]
0 & 0 & 0 & 0 & -\dfrac{1}{T_s} & \dfrac{1}{T_s} & 0 & 0 \\[2mm]
0 & -\dfrac{k_p'\overline{T}_g}{C_{dc}\overline{V}_{dc}} & 0 & -k_i' & -\dfrac{k_p'\overline{\omega}_g}{C_{dc}\overline{V}_{dc}} & 0 & \dfrac{k_p'}{C_{dc}\overline{V}_{dc}} & 0 \\[2mm]
0 & 0 & 0 & 0 & 0 & 0 & -\dfrac{1}{T_s'} & \dfrac{1}{T_s'} \\[2mm]
0 & 0 & 0 & 0 & 0 & 0 & -k_i' + \dfrac{k_p}{T_s'} & -\dfrac{k_p}{T_s'}
\end{bmatrix}
$$

$$
B = \begin{bmatrix}
\dfrac{1}{J_h} & 0 & 0 \\[2mm]
0 & 0 & 0 \\[2mm]
0 & 0 & 0 \\[2mm]
0 & 0 & 0 \\[2mm]
0 & 0 & 0 \\[2mm]
0 & k_i & 0 \\[2mm]
0 & 0 & 0 \\[2mm]
0 & 0 & -k_i'
\end{bmatrix},\
B' = \begin{bmatrix}
\dfrac{1}{J_h} & 0 & 0 \\[2mm]
0 & 0 & 0 \\[2mm]
0 & 0 & 0 \\[2mm]
0 & 0 & 0 \\[2mm]
0 & 0 & k_i' \\[2mm]
0 & 0 & 0 \\[2mm]
0 & 0 & 0 \\[2mm]
0 & k_i & 0
\end{bmatrix}
$$

Voltage and Frequency Stabilization using PI-Like Fuzzy Controller for the Load Side Converters of the Stand Alone Wind Energy Systems

Ameen Gargoom*, Abu Mohammad Osman Haruni, Md. Enamul Haque, and Michael Negnevitsky
Centre for Renewable Energy and Power System
School of Engineering - University of Tasmania
Hobart, TAS 7001, Australia
*Email: ameen.gargoom@ieee.org

Abstract— **In this paper, a robust control system is proposed for controlling the input DC voltage and the output AC voltage and frequency of the load-side converters in stand-alone wind energy systems. The proposed controller consists of voltage regulators based on a PI-like fuzzy technique and a current-mode controlled boost converter. The proposed controller is tested on several operation and fault conditions. The performance of the proposed controller is compared with a conventional PI-based controller in terms of the Total Harmonic Distortion of the output voltages.**

I. Introduction

Stand-alone Remote Area Power Systems (RAPS) are defined as autonomous systems that supply electricity without being connected to the electric grid. These systems are usually the main energy sources for remote area communities. Diesel generators are still predominantly used in these areas. However, with the recent concerns over the environmental and economical issues, combined with the recent advances in power electronics and control technologies, renewable energy, generally, and Wind Energy Systems (WES) in particular, have become vital alternatives for the fossil fuel-based systems. However, the main challenge of stand alone wind energy systems is their weakness and susceptibility to any sudden changes due to their low inertia [1].

Unlike grid connected WES, where the generated voltage and frequency are supported by the stiffness of the electric grid, stand-alone WES is considered the only source meant to produce constant voltage and frequency output to the customers. Therefore, voltage and frequency stabilization is crucial for the deployment of WES in RAPS. The stabilization of the voltage and frequency can be achieved by controlling power electronic interfaces which converts the variable energy from the wind into a controllable energy to the customers through two conversion stages: AC to DC and DC to AC.

Different topologies for power electronic interfaces for WES have been investigated in the literature, [2]-[4]. Among the studied topologies, back-to-back two-level voltage source

converters are the most used for the grid connected WES [5]. However, in the stand-alone mode, since the power is transferred in one direction from the generating units (or energy storage units) to the load, the DC link voltage which controls the transferred power, and the load voltage and frequency are very susceptible to any change in the wind speed and/or loading condition of the customers. Therefore, in this study, two-stage power electronic converter is proposed for compensating the effect of the wind and load variations.

Two control systems are usually associated with the WES: the machine-side converter controller and load-side converter controller. The former is responsible for controlling the torque and speed of the generating units in the system whereas the later is responsible for controlling the AC voltage and its frequency and the power, which is associated with the DC-link voltage control. The scope of this study is only limited to the load-side converter controller.

A Proportional Integral (PI) controller has been widely used with many proposed controllers. However, a major issue associated with PI controllers is the tuning of the controller gains especially in case of dynamic and nonlinear systems such as the wind energy systems. On the other hand, Fuzzy logic-based controllers have shown a promising performance in dealing with nonlinear systems. The fuzzy logic controller structure incorporates attractive features such as simplicity, good performance, and automation, while using a low-cost hardware and software implementation [6]. Therefore, in this paper a PI-like fuzzy based controller is proposed as an efficient control system for stand-alone WES. A comparison between the proposed controller and a conventional PI-based controller is performed to show the advantages of the proposed controller.

The configuration of the WES used for this study and its control system are shown in Fig. 1. As it is shown, the proposed control system consists of two controllers; the AC voltage and frequency controller and the DC voltage controller. Details of these controllers are described below.

This work is supported by the Australian research Council (ARC) and Hydro Tasmania Linkage Grant, K0015166.

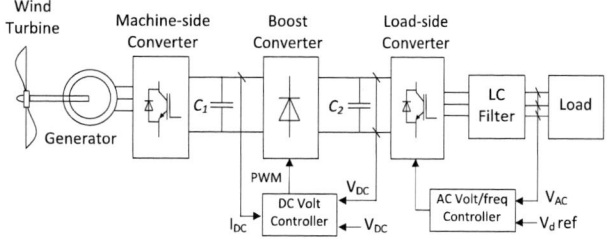

Fig. 1. A Block diagram of the controlled stand-alone wind energy system.

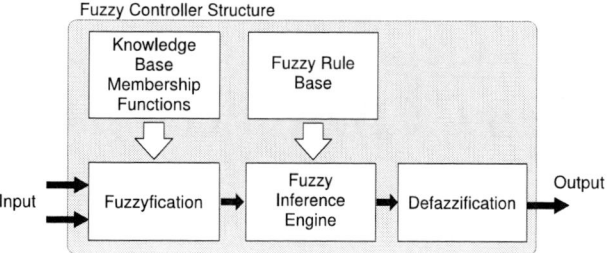

Fig. 3. Fuzzy control process.

II. AC VOLTAGE AND FREQUENCY PI-LIKE FUZZY CONTROLLER

The developed PI-Like Fuzzy controller is used to regulate the output voltage and frequency of a DC/AC two level converter to maintain a constant voltage and frequency to the output loads. The principle of the PI-like Fuzzy controller is based on the conventional PI technique which is represented as

$$u(t) = K_p \cdot e(t) + K_i \cdot \int e(t) \cdot dt \qquad (1)$$

where K_p and K_i are the proportional and integral gains of the PI controller respectively, $e(t)$ is the error function, and $u(t)$ is the control signal.

As (1) reveals, two inputs for the controller are identified, (i.e. the error function and its integral). However, since the integration might spans over a wide range of the output space (the universe of discourse), for the rule based fuzzy algorithms, it is not trivial to set rules based on an integral error. Therefore, to avoid the integration as input to the fuzzy controller, (1) has been reformed to [7]

$$\frac{du(t)}{dt} = K_p \frac{de(t)}{dt} + K_i \cdot e(t) \qquad (2)$$

Thus, the integration of the error in (1) is replaced with the error rate change in (2). The output of (2), however, has to be integrated to obtain the control signal $u(t)$ based on the PI approach. The schematic representation of (2) is shown in Fig. 2.

In case of the proposed AC voltage controller, the input signals are the error and the error change rate in the direct (and quadrature) axis component of the measured AC load voltage, V_d (and V_q), with respect to a reference set value(s). These input signals are represented in the discrete form as

$$e_{dq}(k) = V_{dqref}(k) - V_{dq}(k) \qquad (3)$$

$$ce_{dq}(k) = e_{dq}(k) - e_{dq}(k-1) \qquad (4)$$

where e_{dq} and ce_{dq} are the error and the error change rate in the d-axis (or q-axis) component of the load voltage respectively, and k is their sampling index.

The Fuzzy controller block shown in Fig. 2 involves three main processes: Fuzzification, Fuzzy Reasoning, and Defuzzification as illustrated in the block diagram in Fig. 3. These processes are explained below:

A. Fuzzification

The fuzzification is a process in which the actual input signals measured from the system under control are mapped into Fuzzy input variables (linguistic variables). A set of functions called *Membership Functions* (MF), are used for this purpose. With a proper design of Fuzzy rules, a simple set of MF can be used. Different MF can be used for each input to the controller.

In the proposed controller, the MF used for mapping the input variables e_{dq} and ce_{dq} are demonstrated in Fig. 4, top, (the universe of discourse of ce_{dq} is in the range $[-10^{-4}, +10^{-4}]$ instead of $[-1, +1]$). As shown in the figure, three linguistic Fuzzy inputs are identified from the MF are: Negative (N), Zero (Z), and Positive (P). For the output variable $(du(t)/dt)$, two additional linguistic variables have been added to ensure a smooth response of the controller as shown in Fig. 4, bottom. The variables in the figure are: Negative Big (NB), Negative Small (NS), Zero (Z), Positive Small (PS), and Positive Big (PB).

B. Fuzzy Reasoning

The Fuzzy reasoning process is a knowledge-based process in which a set of rules are used to control the output of the Fuzzy controller. Based on several experimental simulations, a set of rules have been constructed to generate the optimum controlling performance. These rules are given in a tabular form as shown in Table I.

TABLE I

IMPLEMENTED RULES IN THE PROPOSED PI-LIKE FUZZY CONTROLLER

e	ce		
	N	Z	P
N	NB	NB	Z
Z	NS	Z	PS
P	Z	PB	PB

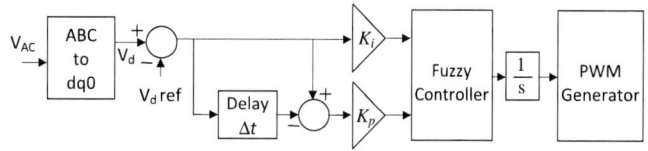

Fig. 2. A Block diagram of PI-Like Fuzzy-based voltage controller.

978-1-4244-4782-4/10 $26.00 © 2010 IEEE

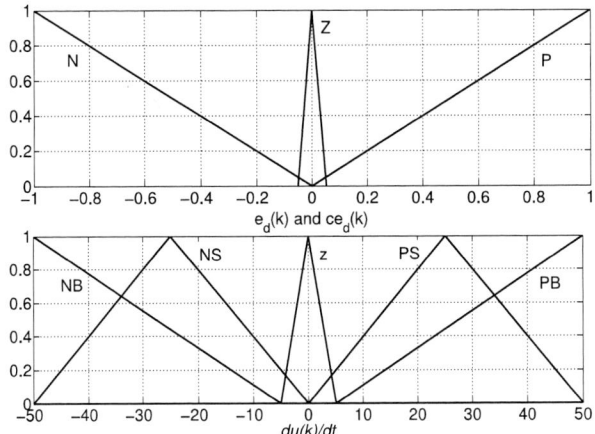

Fig. 4. Membership function of the error in the d-axis component of the voltage.

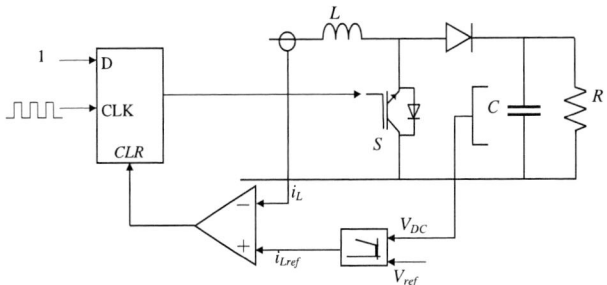

Fig. 5. Current mode controller for the proposed boost converter.

C. Defuzzification

The defuzzification is a process for mapping the linguistic outputs from the reasoning process into numerical values. The center of average method [8] is used to obtain the fuzzy controller's output which is represented as follows:

$$\frac{du(t)}{dt} = \frac{\sum_{i=1}^{N}(w_i \cdot c_i)}{\sum_{i=1}^{N} w_i} \qquad (5)$$

where, w_i is the membership function output of the variable c_i, and N is number of active rules.

In order to stabilize the load frequency, the controlled direct and quadrant components of the AC voltage are transformed into a reference three phase system with a predefined constant frequency using the inverse $d-q$ transformation. Based on the Pulse Width Modulation (PWM) technique, the fundamental component of the output voltage from the load-side converter will always have the frequency of the reference three-phase signals. Therefore, with a proper filter design, the load frequency will always follow the frequency of the reference three-phase signals.

III. DC Voltage Controller

The DC voltage on the DC link between the machine and load converters is controlled by the aid of a boost converter. Although there are several techniques to control boost converters, in this study a current-mode control technique is implemented due to its simplicity and robustness [9]. The block diagram of the controller is demonstrated in Fig. 5.

In the figure, the IGBT Switch, S, is controlled by a feedback control loop, which consists of the comparator and the D-type flip-flop with clear. The switch S is closed at every positive edge of the clock signal. The converter inductor current increases practically linearly until it approaches the reference value i_{Lref}. This clears the output of the flip-flop and opens the switch S (the off-state). During the off-state the current i_L decreases until the next clock pulse sets the flip-flop again. The reference current, i_{Lref}, is specified based on the required reference DC voltage using a PI comparator. Thus,

any variations in the DC voltage due to the wind speed are compensated by controlling the input DC current of the boost converter.

IV. Application and Results

A. Fuzzy-based Controller Performance

The proposed control system is tested on a 35kW 450V stand-alone wind energy system (WES) shown in Fig. 1. The controller and the WES are simulated in SIMULINK environment using a SimPowerSystems toolbox [10]. The parameters of the simulated system are given in Table II. Several operation and fault scenarios have been simulated to illustrate the time response of the proposed control system. Some of the test results are shown below.

1) Effect of Load and Wind Speed Variations: Figs. 6 to 10 demonstrates the system behavior in case of a sudden wind and load changes. In these figures, the WES was allowed to start at a wind speed of 15 m/s with a small load (3.5kW). Until $t = 0.4s$, the gating signals of the DC/DC boost and DC/AC load-side converters were blocked and the DC link capacitor (C_1 in Fig. 1) was allowed to be charged to about 600V. This blocking period is very important to avoid the unexpected behaviors of the controllers during the starting period. At $t = 0.4s$, the gating signals are released and and the control system takes

TABLE II
PARAMETERS OF THE SIMULATED WIND ENERGY SYSTEM

Permanent Magnet Synchronous Generator	
Number of pole pairs	4
Rated speed (rpm)	3000
Rated power (kW)	35
Stator resistance (Ω)	0.05
Direct Inductance L_d (mH)	0.635
Quadrature Inductance L_q (mH)	0.635
Torque constant (N $\cdot$ m/A_{peak})	1.152
Inertia ($J.kg.m^2$)	0.011
Wind Turbine	
Power (kW)	35
Base wind speed (m/s)	15
LC Filter	
Series Inductance (mH)	13
Branch Capacitance (μF)	16

978-1-4244-4782-4/10 $26.00 © 2010 IEEE

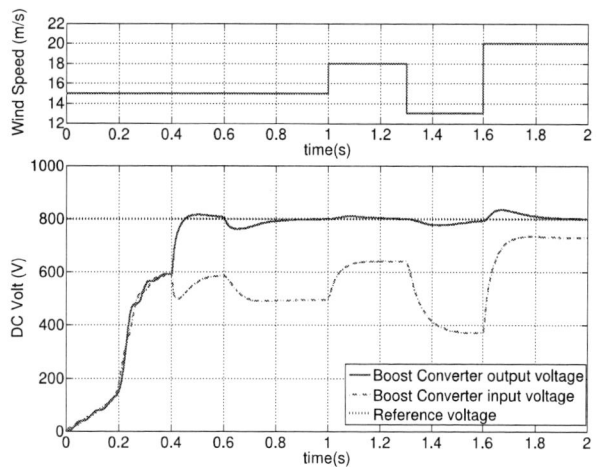

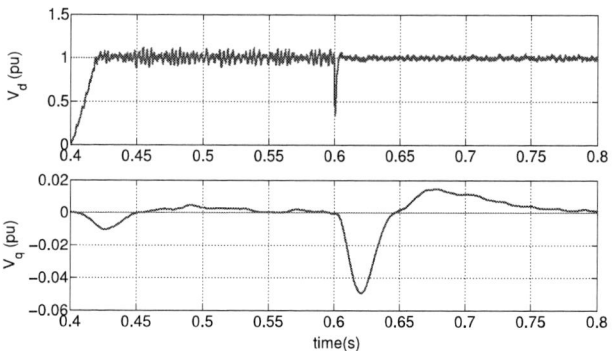

Fig. 8. Direct axis V_d (top), and quadrature axis V_q (bottom) components of the AC load voltage.

Fig. 6. A hypothetical wind speed pattern with the time response of the DC link voltage to this pattern.

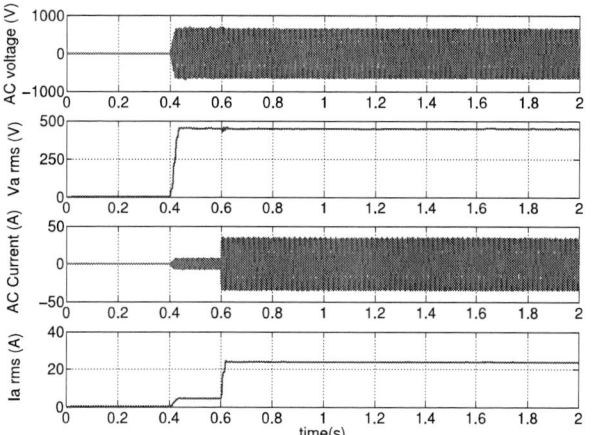

Fig. 7. The instantaneous and rms values of the load voltage and current.

over. At $t = 0.6s$, a 15kW load is connected to the system. Following this, step wind increase and decrease are introduced at times $1.0s$, $1.3s$, and $1.6s$ as shown in Fig. 6 (top). Although in practice, the change in wind speed is gradual, step changes have been introduced artificially to subject the system to the most severe conditions.

The time response of the boost converter input and output DC voltages are shown in Fig. 6 (bottom). The figure depicts the effects of the load and wind speed changes on the DC link voltage (the dotted plot in the figure). The figure shows also the performance of the DC/DC converter controller to maintain a constant 800 V DC voltage when the gate signal to the controller is released at $t = 0.4s$. Adding loads and wind speed changes were adequately regulated by the controller. One can see from the figure, in case of the system under test, a less than 400 V drop in the input voltage was efficiently regulated in around $0.2s$.

On the AC load side of the system, the recorded voltage and current signals are shown in Fig. 7. From the rms mea-

surements in the figure, it is depicted that voltage magnitude is regulated rapidly to its rated values by the PI-like Fuzzy controller. This has been achieved by controlling V_d to 1 pu and V_q to 0 to achieve a unity power factor. Closeup views of the responses of these values are shown in Fig. 8. The figure shows the fast response of the Fuzzy controller to achieve the reference values. The sudden change in the load from 10% to 40% of full load which occurred at $t = 0.6s$ was regulated in around $0.01s$ for the V_d signal and in around $0.2s$ for the V_q signal.

With this efficient control of V_d and V_q, low harmonic distorted voltage and current signals were obtained with the aid of the LC filter described in Table II. Snapshots of the instantaneous voltage and current signals together with their harmonic spectrums are shown in Fig. 9. The average Total Harmonic Distortion (THD) of the three phases of the voltage and current signals are measured at 57% loading condition as 2.47% and 2.52% respectively. Up to the 1998th harmonic order was considered for calculating the THD.

The instantaneous frequency response to the load and wind speed changes is also presented in Fig. 10. In the demonstrated test, the largest impact on the system frequency was due to the load change at $t = 0.6s$. The frequency was regulated to 50 Hz in around $0.2s$. It is noticeable that wind speed changes had insignificant effect on the load frequency since the load is decoupled from the wind turbine by means of the used converters.

2) Effect of Faults: The performance of the proposed PI-like Fuzzy controller is also tested in case of single- and three-phase faults. In Fig. 11, the WES was supplying a 15kW load when the load subjected to a single line to ground faults occurred on phase a at $t = 1.0s$. The fault is self cleared after $0.3s$. As it is seen in the figure, one of the healthy phases suffered an overvoltage as the Fuzzy controller was maintaining the V_d and V_q components to their reference values as shown in Fig. 12. However, a 100 Hz ripple component was experienced by the V_d component. This is a result of the AC voltage unbalance due to the fault. Filtering this component is possible by using a notch filter. However, this will affect the performance of the controller in terms of

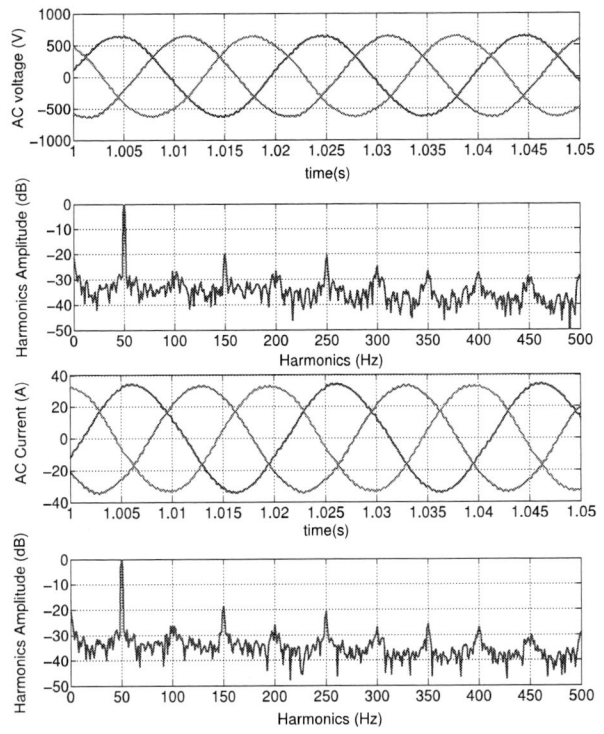

Fig. 9. The instantaneous three phase AC load voltage and current, and their frequency spectrum of one phase.

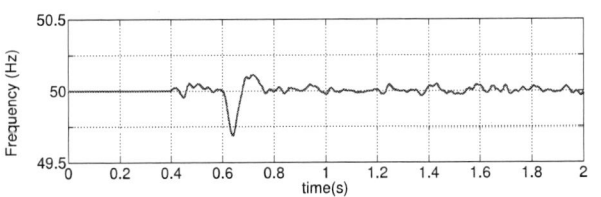

Fig. 10. Instantaneous system frequency response to load and wind variations.

the total harmonic distortion in normal operation condition. After the fault was cleared, the controller was able to restore the V_d and V_q components in around $0.1s$ and 0.2 respectively.

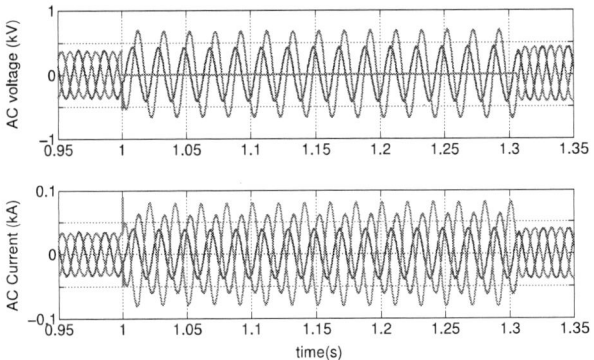

Fig. 11. The instantaneous three-phase voltages and currents during single-line to ground fault.

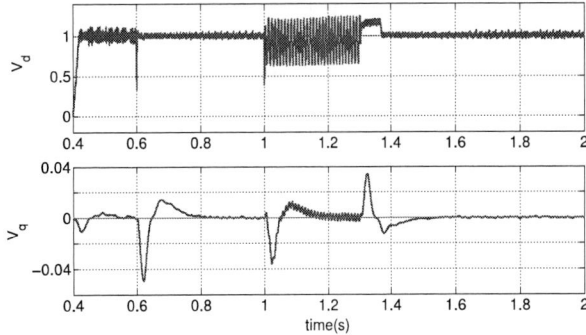

Fig. 12. The response of the V_d and V_q components to a single-line to ground fault.

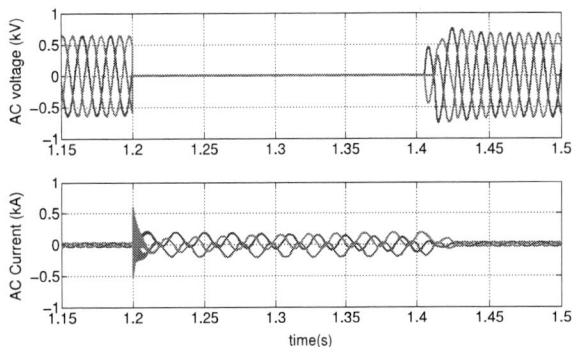

Fig. 13. The instantaneous three-phase voltages and currents during three phase to ground fault.

The control system is also tested in case of three-phase fault and the results are shown in Fig. 13. In this case, a three-phase to ground fault occurred at $t = 1.2s$, and automatically cleared after $0.2s$. Fig. 14 shows the Fuzzy controller response to regulate the V_d and V_q components. During the fault, these component were forced to zero due to the fault. However, when the fault was cleared, the controller responded rapidly to restore the reference values of V_d and V_q within $0.35s$ and $0.2s$ respectively. These results show a well recovering response of the controller following single and three phase faults.

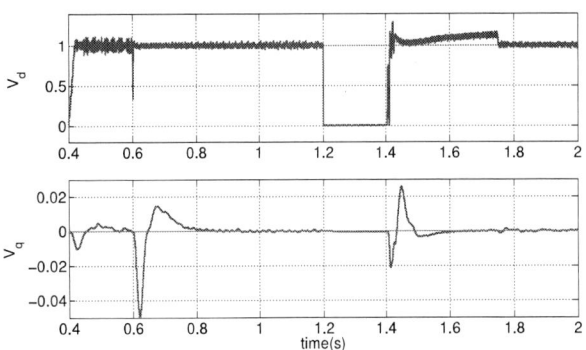

Fig. 14. The response of the V_d and V_q components to a three phase to ground fault.

978-1-4244-4782-4/10 $26.00 © 2010 IEEE

The frequency responses of the one of phase during and after the single and three phase faults are shown in Fig. 15, which illustrates two transient response of the frequency at the starting and clearing times of the the fault.

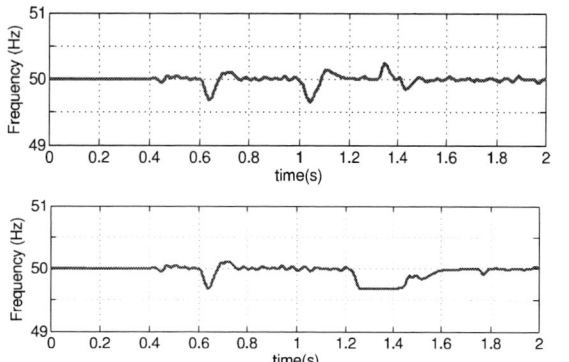

Fig. 15.   The instantaneous system frequency during single phase to ground fault (top), and three-phase to ground fault (bottom).

B. Comparison with PI-based Controller

For the sake of comparison, the PI-Like fuzzy controller is replaced by a conventional PI-based controller and tested on different loading conditions. Similar to the Fuzzy based controller, the PI controller consists of two PI loops for regulating the V_d and V_q components of the measured load voltage. The PI controllers were tuned using Ziegler-Nichols tuning method [11]. The comparison is performed in terms of the THD of the load output voltages in case of different loading conditions.

In the comparison, the WES was operated at constant wind speed (15 m/s) with loading condition varying from 10% (3.5 kW) to 100% (35.0 kW) of the WES rating. The THD of load voltage versus the loading condition for the Fuzzy based and PI based controller is shown in Fig. 16. The graph reveals that, based on the performance of the PI-based controller, satisfactory results can be obtained from the system in case of loading conditions more than 20% (less than 5% THD can be obtained). Whereas, in case of using the developed PI-Like Fuzzy controller, a higher voltage quality in terms of its THD can be generated at low loading as well as high loading conditions.

V. CONCLUSIONS

The paper offers a control system for stabilizing the DC voltage, the AC voltage and the frequency of the stand-alone wind energy systems. The proposed AC voltage controller is based on a PI-like fuzzy algorithm. Since the fuzzy controller is designated to control the direct and quadrant DC components of the AC voltage in case of the AC voltage regulator, only a small number of rules were sufficient for an efficient performance. The performance of the proposed controller is illustrated on a simulated stand-alone wind energy system with different operation and fault conditions. The performance of the proposed controller is also compared with a conventional

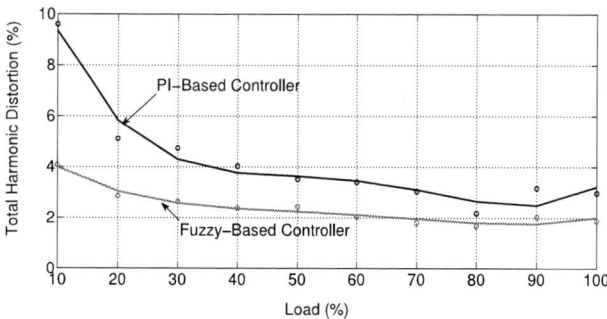

Fig. 16.   THD of load voltage at different loading condition when using PI- and Fuzzy-based controllers.

PI-based controller. It was shown from the results that, by using same analogue filter parameters, a higher voltage quality can be delivered from the system when using the developed Fuzzy controller as compared with PI-based controller.

REFERENCES

[1] Sharma, H.; Islam, S.; Nayar, C.V.; and Pryor, T.; 'Dynamic Response of a Remote Area Power System to Fluctuating Wind Speed,' IEEE Power Engineering Society Winter Meeting, Vol. 1, 23-27 Jan. 2000, pp.499 - 504.

[2] Bueno, E. J.; Cbreces, S.; Rodrguez, F. J.; Hernndez, A.; and Espinosa, F.; 'Design of a Back-to-Back NPC Converter Interface for Wind Turbines With Squirrel-Cage Induction Generator,' IEEE Transaction on Energy Conversion Vol. 23, Issue 3, Sept. 2008, pp. 932 - 945.

[3] Nikkhajoei, H.; and Lasseter, R.H.; 'Power Quality Enhancement of a Wind-Turbine Generator under Variable Wind Speeds using Matrix Converter,' IEEE Power Electronics Specialists Conference, PESC, 15-19 June 2008, pp. 1755 - 1761.

[4] Eun-Soo Kim; Hwan-Kook Song; Joo-Hun Kim; Hyun-Kwan Lee; and Yoon-Ho Kim; 'Efficiency Characteristics of a Half-Bridge Series Resonant Converter for the Contact-Less Power Supply,' Twenty-Third Annual IEEE Applied Power Electronics Conference and Exposition, APEC, 24-28 Feb. 2008 Papp.1555 - 1561.

[5] Ackermann, T.; Ed. 'Wind Power in Power System,' John Wiley & Sons Ltd., New York, 2005.

[6] Xinping Ding; Zhaoming Qian; Shuitao Yang; Bin Cui; and Fangzheng Peng; 'A Direct DC-link Boost Voltage PID-like Fuzzy Control Strategy in Z-source inverter,' IEEE Power Electronics Specialists Conference, PESC, 15-19 June 2008, pp. 405 - 411.

[7] Leon Reznik "Fuzzy Controllers Handbook: How to Design Them, How They Work", Newness, 1997.

[8] Liping Guo; Hung, J.Y.; and Nelms, R.M.; 'Experimental Evaluation of a Fuzzy Controller using a Parallel Integrator Structure for Dc-Dc Converters,' Proceedings of the IEEE International Symposium on Industrial Electronics ISIE, Vol. 2, 20-23 June 2005 pp.707 - 713.

[9] Chattopadhyay, S.; and Das, S.; "A Digital Current-Mode Control Technique for DC-DC Converters," IEEE Transactions on Power Electronics Vol. 21, Issue 6, Nov. 2006, pp. 1718 - 1726.

[10] SimPowerSystems User's Guide, Version 5.1, Hydro-Qubec, Math-Works, www.mathworks.com, 2009.

[11] Åström, K. J.; and Hägglund, T.; 'Revisiting the Ziegler-Nichols step response method for PID control,' Journal of Process Control, 15 (2005), pp. 371-382.

Dual-Stage Converter to Improve Transfer Efficiency and Maximum Power Point Tracking Feasibility in Photovoltaic Energy-Conversion Systems

Sairaj V. Dhople, Ali Davoudi, and Patrick L. Chapman
Grainger Center for Electric Machinery and Electromechanics
Department of Electrical and Computer Engineering
University of Illinois at Urbana-Champaign
Urbana, Illinois 61801, USA
sdhople2@illinois.edu

Abstract— **This work addresses transfer efficiency and tracking ineffectiveness in switch-mode, dc-dc converters for photovoltaic energy-conversion systems. Tracking ineffectiveness constrains maximum power point tracking algorithms due to dc-dc converter operational constraints. Transfer efficiency quantifies the effect of switching-frequency ripple on the usable power extracted from the photovoltaic source. A dual-stage boost buck-boost converter is proposed as a solution to alleviate tracking ineffectiveness and enhance transfer efficiency. Simulations and experimental results indicate that both concerns are addressed by the proposed topology.**

I. INTRODUCTION

Photovoltaic (PV) energy is typically harvested by switch-mode, dc-dc converters. In this work, boost front-end converters that implement maximum power point tracking (MPPT) are examined in the context of the energy-conversion chain depicted in Figure 1. Typically, the front-end converter is followed by other dc-dc or dc-ac converters based on the load type.

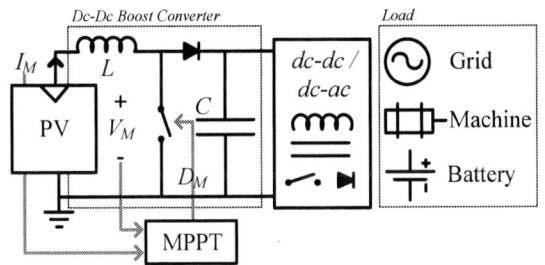

Figure 1. Boost front-end converter employed in a PV-energy system

A wide range of MPPT algorithms can be utilized to control the boost front-end converter. These algorithms vary based on ease of implementation, platform (analog/digital), complexity and convergence speed. Perturb and observe [1] and incremental-conductance methods [2] have been widely adopted, and they are generally executed on a digital platform. Ripple correlation control offers simple analog implementation [3]. Fractional open-circuit voltage [4] and short-circuit current [5] methods are easier to realize, but they do not guarantee operation at the true maximum power point. References [6]-[7] provide exhaustive comparisons of different MPPT techniques.

Boost-derived front-end converters offer numerous advantages in PV energy conversion. An in-depth comparison of boost and buck topologies in MPPT applications for a representative set of design specifications is provided in [8]. The buck topology requires a large capacitor to filter discontinuous current forced on the PV source. Additionally, the requirement of high-side gate drivers and blocking diodes for protection typically drives up cost. In contrast, the boost topology enjoys superior dynamic performance in terms of wide bandwidth and negligible resonance effects, due to the smaller input capacitance. Similar comparisons in [9]-[10] demonstrate that the boost converter enables higher energy harvest and is more efficient than a buck converter for typical MPPT applications.

Conventional design concerns for power-electronic circuits include input-output efficiency, dynamic response and steady-state performance. For PV applications, the metrics of tracking ineffectiveness and transfer efficiency also deserve scrutiny. Tracking ineffectiveness constrains MPPT feasibility due to converter operational traits or type of load [11]. In the context of a general PV energy-conversion chain (e. g., Figure 1), the power converter that realizes MPPT matches impedances for maximum power transfer. The PV impedance that yields maximum power varies with ambient conditions. Irrespective of MPPT algorithm, there are likely to be scenarios in which converter operational constraints hinder impedance matching.

Such constraints could be imposed because the set of feasible duty ratios is the closed set, [0, 1], which prevents impedance matching for certain loads, or due to operation in discontinuous current conduction mode (DCM) for certain ambient conditions or loads.

Transfer efficiency quantifies the power loss at the PV modules due to switching-frequency ripple introduced by power-electronic converters. Ripple in the sourced power causes the average harvested power to deviate from the true maximum [12]-[13]. By definition, transfer efficiency, $\eta_{TRANSFER}$, is the fraction of the true maximum power processed by the MPPT converter:

$$\eta_{TRANSFER} = \frac{P_{AVG}}{P_M(S,T)} = \frac{P_{AVG}}{I_M(S,T)V_M(S,T)} \tag{1}$$

The product of the MPP voltage, V_M, and MPP current, I_M, (functions of insolation and temperature), is denoted as the true-maximum power, P_M. The usable dc power processed by the MPPT converter is denoted as P_{AVG}.

A dual-stage converter is proposed that successfully addresses tracking ineffectiveness and degradation in transfer efficiency noticed in conventional boost converters. Section II introduces the concept of tracking ineffectiveness and highlights the reasons for low transfer efficiency in a boost converter. Section III formulates the dual-stage converter to solve the two concerns. Simulations and experimental results are presented in Section IV.

II. Design Concerns Addressed by Dual-Stage Converter

In this section, the concerns of tracking ineffectiveness and transfer efficiency are introduced and described with reference to the ideal boost converter depicted in Figure 2. The boost converter delivers power from a PV source to a resistive load, denoted as R. The analysis presented herein justifies a second conversion stage that follows the boost front-end converter.

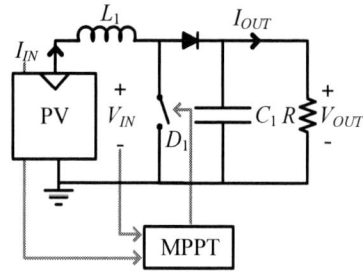

Figure 2.   Boost converter used to implement MPPT for a resistive load

A.   Tracking Ineffectiveness

The optimal PV-source impedance, R_M that yields maximum power is defined as the ratio of the maximum power point voltage, V_M to current, I_M:

$$R_M = \frac{V_M}{I_M} \tag{2}$$

MPPT algorithms match the front-end converter's input impedance to the optimal PV impedance. This extracts maximum power from the PV source. Assume that the boost converter depicted in Figure 2 operates in CCM, with duty cycle D_1. The average output voltage, V_{OUT}, is related to the input voltage, V_{IN}, through:

$$V_{OUT} = \frac{V_{IN}}{1-D_1} \tag{3}$$

Conservation of power dictates that the average output current, I_{OUT}, is the following function of the input current, I_{IN}:

$$I_{OUT} = I_{IN}(1-D_1) \tag{4}$$

Dividing (3) by (4) provides the following expression that relates the average input impedance, R_{IN}, to the output load, R:

$$R_{IN} = \frac{V_{IN}}{I_{IN}} - R(1-D_1)^2 \tag{5}$$

In this system, the MPPT controller seeks an optimal duty cycle, D_M, that ensures the input impedance of the boost converter is matched to the optimal PV impedance.

$$R_M = \frac{V_M}{I_M} = R(1-D_M)^2 \tag{6}$$

The optimal duty cycle for MPPT is solved from (6):

$$D_M = 1 - \sqrt{\frac{R_M}{R}} \tag{7}$$

Since the duty cycle represents the fraction of the switching period when the active switch is turned on, it is constrained to the closed set [0, 1]. Enforcing this limit on (7) provides the following bound on the output load:

$$R \geq R_M \tag{8}$$

Equation (8) suggests that realizing MPPT with boost converters is infeasible for output loads less than the optimal PV impedance.

B.   Transfer Efficiency

With MPPT implemented, the inductor current ripple, ΔI_{IN}, is approximated using a straight-line fit for the PV current during the period when the active switch conducts:

$$\Delta I_{IN} = \frac{V_M D_M}{f_1 L_1} \tag{9}$$

In the above expression, f_1 is the switching frequency, V_M is the optimal PV voltage and L_1 is the boost inductance. The formulation for the optimal duty-cycle in (7) indicates that the current ripple is expected to be high for large output loads. This contributes to degraded transfer efficiency.

978-1-4244-4782-4/10 $26.00 © 2010 IEEE

III. DUAL-STAGE CONVERTER

Since the boost converter can only step up input voltages, maximum power cannot be delivered to an arbitrary load for all ambient conditions. In addition, the predetermination of the duty ratio based on ambient conditions and load can degrade transfer efficiency. These problems can be addressed by a dual-stage converter that reinforces the boost front end, as is shown in Figure 3.

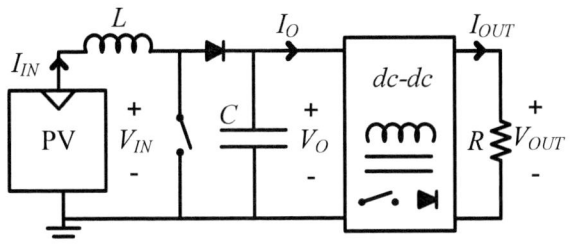

Figure 3.   Cascaded dc-dc converter to reinforce boost front-end converter

If operated in CCM, an ideal buck-boost converter is unaffected by tracking ineffectiveness. To prove this, consider the setup in Figure 4, where the buck-boost converter is expected to track the optimal PV impedance, R_M, and deliver power to the load resistance, R.

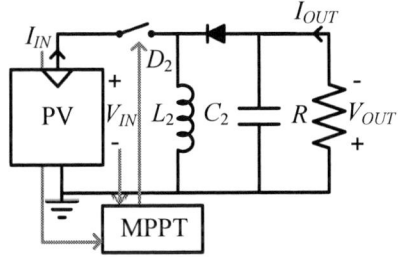

Figure 4.   Buck-boost converter implements MPPT for a resistive load

Assume that the buck-boost converter operates in CCM, with duty cycle D_2. The average output voltage, V_{OUT}, is related to the input voltage V_{IN} through

$$V_{OUT} = \frac{D_2}{1-D_2} V_{IN} \qquad (10)$$

Conservation of power dictates that the average output current, I_{OUT}, is the following function of input current, I_{IN}:

$$I_{OUT} = \frac{1-D_2}{D_2} I_{IN} \qquad (11)$$

Dividing (10) by (11) provides the following expression that relates the average input impedance, R_{IN}, of the buck-boost converter to load, R:

$$R_{IN} = \frac{V_{IN}}{I_{IN}} = R\left(\frac{1-D_2}{D_2}\right)^2 \qquad (12)$$

As with the boost converter, the MPPT controller seeks the optimal duty cycle, D_M, to ensure the converter input impedance is R_M:

$$R_M = \frac{V_M}{I_M} = R\left(\frac{1-D_M}{D_M}\right)^2 \qquad (13)$$

The optimal duty cycle for MPPT is solved from (13):

$$D_M = \left(1 + \sqrt{\frac{R_M}{R}}\right)^{-1} \qquad (14)$$

In the limit, for very low or very high loads,

$$\begin{cases} \lim_{R\to\infty} D_M = \lim_{R\to\infty}\left(1+\sqrt{\frac{R_M}{R}}\right)^{-1} = 1 \\[2mm] \lim_{R\to0} D_M = \lim_{R\to0}\left(1+\sqrt{\frac{R_M}{R}}\right)^{-1} = 0 \end{cases} \qquad (15)$$

Equation (15) indicates that when operated in CCM, the buck-boost converter can match any output load to the optimal PV impedance and does not have constraining limits such as (8). Hence, the buck-boost converter is considered a candidate topology to realize the second power stage in the dual-stage converter. The schematic of the proposed dual-stage converter is illustrated in Figure 5.

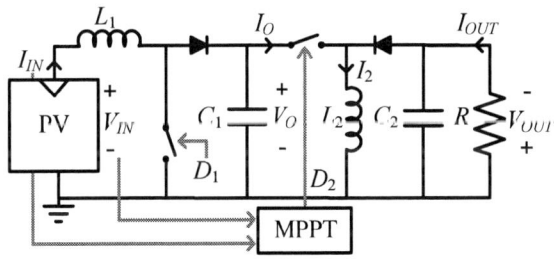

Figure 5.   Proposed dual-stage boost buck-boost converter

The two power stages independently address concerns of tracking ineffectiveness and transfer efficiency. If operated in CCM, the buck-boost converter ensures that maximum power can be delivered to any output load. Congruently, the boost converter can minimize PV current ripple. Alternately, the boost duty cycle can be set to control the voltage on the dc bus (voltage, V_O, across capacitor, C_1).

IV. SIMULATIONS AND EXPERIMENTAL RESULTS

Simulations and experimental studies are highlighted here to demonstrate that the dual-stage converter addresses tracking ineffectiveness and improves transfer efficiency. The PV module parameters and the specifications of the dual-stage converter are listed in Tables I and II, respectively.

TABLE I
PARAMETERS OF PV MODULE

Symbol	Quantity	Value
V_{OC}	Rated open-circuit voltage	44.8 V
I_{SC}	Rated short-circuit current	5.5 A
I_M	Rated current	5.2 A
V_M	Rated voltage	36.6 V
α	Temperature coefficient for current	0.065 % / °C
β	Temperature coefficient for voltage	160 mV / °C

TABLE II
CONVERTER SPECIFICATIONS - SIMULATIONS

Symbol	Quantity	Value
L_1	Boost converter inductance	4 mH
f_1	Boost switching frequency	10 kHz
L_2	Buck-boost converter inductance	1 mH
$C_{1,2}$	Filter capacitance	100 µF
R	Output load	2 Ω, 50 Ω

The parameters for the PV module in Table I indicate that the optimal maximum power point impedance, R_M, at 1000 Wm⁻² and 25 °C is approximately 7 Ω (2). The discussion in Section II suggests that MPPT is infeasible for output loads less than 7 Ω with a single boost converter. This is verified in Figure 6 which depicts the MPPT duty cycle in a single boost converter and the dual-stage boost buck-boost converter for a 2 Ω output load. The dual-stage converter is able to track the maximum power point, as it is unconstrained by tracking ineffectiveness.

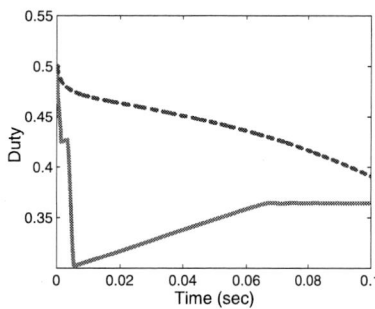

Figure 6.   Duty cycles for boost (▬ ▬) and dual-stage (▬) converters for a 2 Ω load to demonstrate tracking ineffectiveness

Figure 7 depicts PV currents with MPPT implemented for a 50 Ω load in a single boost converter and the proposed dual-stage converter. In each case, the parameters of the boost converters are the same. A single boost converter can only serve the purpose of MPPT, and the current ripple is predetermined based on load and ambient conditions. However, the duty-cycle of the dual-stage converter can be set independently to control PV-current ripple. As a result, the dual-stage converter harvests 182.85 W from the PV source while the single boost converter harvests 164.89 W. This translates to a 9.7% improvement in transfer efficiency.

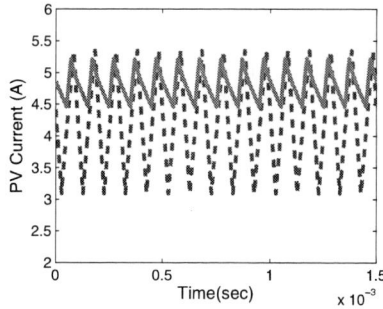

Figure 7.   PV currents in boost (▬ ▬) and dual-stage (▬) converters for a 50 Ω load to demonstrate improvement in transfer efficiency

The advantages of the proposed dual-stage converter are also verified by an experimental laboratory prototype. The goal of the experimental studies is to demonstrate the ability of the proposed dual-stage topology to minimize PV current ripple. Loads varying by a factor of 10 (50 Ω, 500 Ω) are employed, and low PV-current ripple is desired in addition to MPPT. From the analysis presented thus far, note that a single boost converter can not simultaneously satisfy both requirements for all ambient conditions. The relevant converter specifications for the experimental studies are shown in Table III. A limit is enforced on the maximum allowed boost-converter duty ratio to prevent the dc-bus voltage from exceeding the voltage rating of the capacitor C_1.

TABLE III
CONVERTER SPECIFICATIONS - EXPERIMENTS

Symbol	Quantity	Value
L_1	Boost converter inductance	4.45 mH
f_1	Boost switching frequency	10 kHz
D_{MAX}	Boost converter max. allowed duty	80 %
L_2	Buck-boost converter inductance	1 mH
$C_{1,2}$	Filter capacitance	100 µF
R	Output load	50 Ω/500 Ω

Figure 8 depicts the PV voltage and current for the 500 Ω load with the boost converter operated at D_{MAX}. Plotted in dashed lines on the same figure are the MPP voltage and current values. Compare the waveforms in Figure 8 to Figure 9, where the PV-current ripple is reduced while MPPT is simultaneously executed.

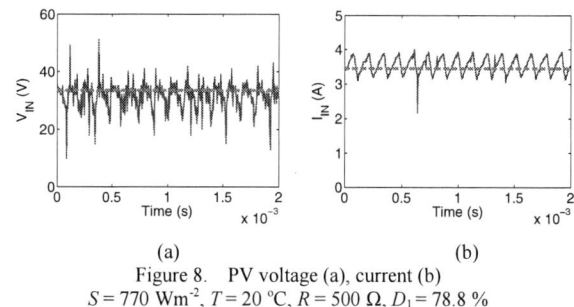

(a)   (b)

Figure 8.   PV voltage (a), current (b)
$S = 770$ Wm⁻², $T = 20$ °C, $R = 500$ Ω, $D_1 = 78.8$ %

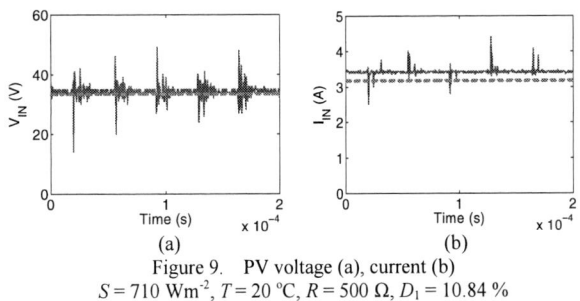

Figure 9. PV voltage (a), current (b)
$S = 710$ Wm^{-2}, $T = 20$ °C, $R = 500$ Ω, $D_1 = 10.84$ %

Figure 10 depicts the PV voltage and current for a 50 Ω load with low peak-to-peak PV-current ripple. In similar ambient conditions, if a single boost converter was used, the duty ratio required for MPPT would be around 50%. The ensuing current ripple would be much greater than that noticed in Figure 10. This is proven by the waveforms in Figure 11 that correspond to MPPT implemented with the front-end boost converter operating at a duty ratio of 51.61%.

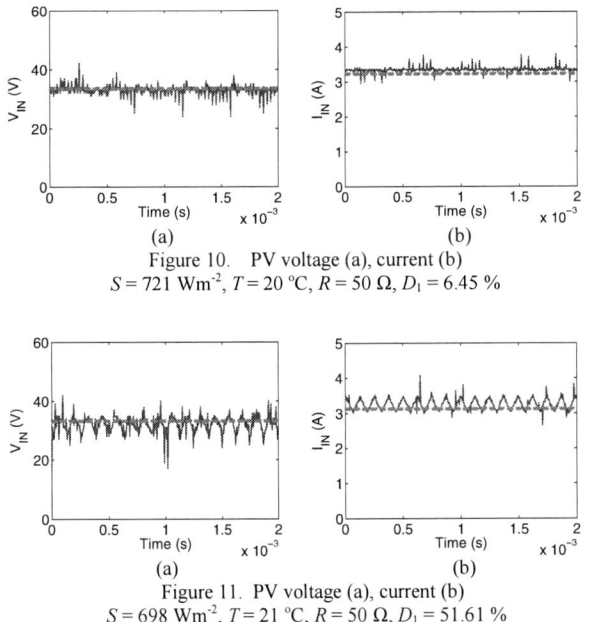

Figure 10. PV voltage (a), current (b)
$S = 721$ Wm^{-2}, $T = 20$ °C, $R = 50$ Ω, $D_1 = 6.45$ %

Figure 11. PV voltage (a), current (b)
$S = 698$ Wm^{-2}, $T = 21$ °C, $R = 50$ Ω, $D_1 = 51.61$ %

V. CONCLUSIONS

Tracking ineffectiveness and transfer efficiency are design concerns that need to be addressed in boost front-end converters for PV applications. Cascading the boost topology with a buck-boost converter is noted to improve MPPT feasibility and transfer efficiency. Experimental and simulation results demonstrate reduction in PV current ripple with the simultaneous execution of MPPT.

REFERENCES

[1] N. Femia, G. Petrone, G. Spagnuolo, and M. Vitelli, "Optimization of perturb and observe maximum power point tracking method," *IEEE Trans. Power Electronics*, vol. 20, pp. 963-973, July 2005.

[2] K. H. Hussein and I. Muta, "Maximum photovoltaic power tracking: an algorithm for rapidly changing atmospheric conditions," *IEE Proc. Generation, Transmission, and Distribution*, 1995, pp. 59-64.

[3] T. Esram, J. W. Kimball, P. T. Krein, P. L. Chapman, and P. Midya, "Dynamic maximum power point tracking of photovoltaic arrays using ripple correlation control," *IEEE Trans. Power Electronics*, vol. 21, pp. 1282–1291, Sep. 2006.

[4] J. J. Schoeman and J. D. van Wyk, "A simplified maximal power controller for terrestrial photovoltaic panel arrays," in *Proc. IEEE Power Electronics Specialists Conf.*, 1982, pp. 361-367.

[5] S. Yuvarajan and S. Xu, "Photo-voltaic power converter with a simple maximum-power-point-tracker," in *Proc. International Symp. On Circuits and Systems*, 2003, pp. 399-402.

[6] T. Esram and P. L. Chapman, "Comparison of photovoltaic array maximum power point tracking techniques," *IEEE Trans. Energy Conversion*, vol. 22, pp. 439-449, June 2007.

[7] D. P. Hohm and M. E. Ropp, "Comparative study of maximum power point tracking algorithms," *Progress in Photovoltaics: Research and Applications*, vol. 11, pp. 47-62, Jan. 2003.

[8] W. Xiao, N. Ozog, and W. G. Dunford, "Topology study of photovoltaic interface for maximum power point tracking," *IEEE Trans. Industrial Electronics*, vol. 54, pp. 1696-1704, June 2007.

[9] G. Walker, "Evaluating MPPT converter topologies using a Matlab PV model," *Australian Journal of Electrical and Electronics Engineering*, vol. 21, pp. 49-55, 2001.

[10] I. Glasner and J. Appelbaum, "Advantage of boost versus buck topology for maximum power point tracker in photovoltaic systems," in *Proc. IEEE Conv. Electr. Electron. Eng. Israel*, 1996, pp. 355-358.

[11] J.M. Enrique, E. Durán, M. Sidrach-de-Cardona and J.M. Andújar, "Theoretical assessment of maximum power point tracking efficiency of photovoltaic facilities with different converter topologies" *Solar Energy*, vol. 81, pp. 31-38, Jan. 2007.

[12] N. D. Benavides and P. L. Chapman, "Modeling the effect of voltage ripple on the power output of photovoltaic modules," *IEEE Trans. Industrial Electronics*, vol. 55, pp. 2638-2643, July 2008.

[13] S. B. Kjaer, J. K. Pederson, and F. Blaabjerg, "A review of single-phase grid-connected inverters for photovoltaic modules," *IEEE Trans. Power Electronics*, vol. 41, pp. 1292-1306, Oct. 2005.

A novel approach of maximizing energy harvesting in photovoltaic systems based on bisection search theorem

Peng Wang, Haipeng Zhu, Weixiang Shen, Fook Hoong Choo and Poh Chiang Loh and Kuan Khoon Tan
School of Electrical and Electronic Engineering
Nanyang Technological University, Singapore
Email: {epwang, HPZHU, wxshen, efhchoo, epcloh, TA0008ON}@ntu.edu.sg

Abstract— **This paper presents a new approach of maximizing energy harvesting in photovoltaic (PV) systems using bisection search theorem (BST). The fundamental of the BST and its application into maximum power point tracker (MPPT) in PV systems are described. A microcontroller is used to control a DC/DC boost converter to realize the MPPT function. Experimental results from solar array simulator show that the proposed technique can track maximum power point very fast within a few steps. The feasibility of the proposed MPPT is also verified in natural environment condition with two solar modules in parallel. Since the proposed technique is simple in computation, cheap in implementation and fast in tracking, it is expected to be widely used to replace conventional MPPT techniques in PV systems.**

I. INTRODUCTION

With ever-increasing concerns on environment and energy conservation, the research and development of photovoltaic (PV) system technology has been accelerated recently due to its free of pollution, silent operation, long life time and low maintenance [1]. Although the improvement of solar cell technologies and the increasing demand for PV systems have led to a reduction of the price of PV module [2], the costs of PV systems are still too high. Therefore, it is an important to design the PV systems which can maximize energy harvesting from Sun through solar modules. The output power of solar module varies as a function of solar radiance, temperature and operating point because of its nonlinear current-voltage (I-V) relationship [3]. Therefore, the maximum power point tracker (MPPT) is widely used to maximize the power output of the solar module. As such, many MPPT techniques have been developed and implemented. The techniques vary in sensors required, tracking speed, complexity and cost of hardware implementation, such as Hill Climbing/P&O and its variants [4-7], Incremental Conductance and its variants [8-9], Factional Open-Circuit Voltage [10], [11] and Fractional Short-Circuit Current [12]. For those techniques, the qualitative comparison and the quantitative comparison in terms of simulation and experiment have been conducted [13-15], respectively. The results show that Hill Climbing/P&O and Incremental Conductance are in general

the most efficient techniques. For these two techniques, the derivatives of voltage and power measured from the solar module are still required. In this paper, a novel MPPT technique based on bisection search theorem (BST) is proposed without the necessity of derivative computation. Thus, the new technique is even simpler in computation, cheaper in implementation and faster in tracking. The experimental results from solar array simulator in the laboratory show that the proposed technique can track maximum power point very fast within a few steps. The feasibility of the proposed technique is also verified under real solar modules at the presence of natural environmental conditions. Thus, it is expected to be widely used to replace conventional MPPT techniques in PV systems.

II. BISECTION SEARCH THEOREM AND MPPT APPLICATION

A. Prinple of bisection search theorem

The bisection search theorem is one of the bracketing methods for finding roots of equations [16], [17]. Assume that function $y = f(x)$ and an interval $[a, b]$ which contains a root x^* of $f(x)$ that lies somewhere in the interval as shown in Fig. 1 such that $f(c) = 0$.

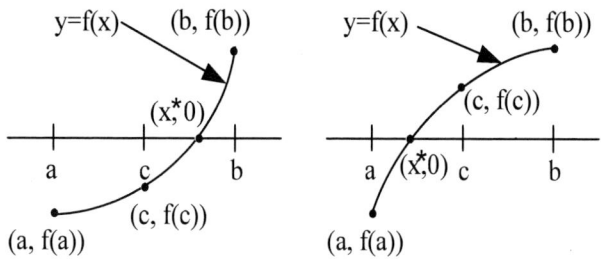

(a) if $f(c)$ and $f(b)$ have opposite sign, then squeeze from left

(b) if $f(a)$ and $f(c)$ have opposite sign, then squeeze from right

Fig. 1 Decision process for bisection search

This work is supported by The National Research Foundation (NRF) of Singapore through the research project NRF-G-CRP-2007-02.

The BST systematically moves the endpoints of the interval closer and closer together in the pace of halving interval for each step until an interval of arbitrarily small width that brackets the zero is obtained. The decision step for this process is first to choose the midpoint $c = (a+b)/2$ and then to analyze the three possibilities that might rise: 1). If $f(a)$ and $f(b)$ have opposite signs, a zero lies in $[a,c]$. 2) If $f(a)$ and $f(b)$ have opposite signs, a zero lies in $[c,b]$. 3) if $f(c)=0$, then the zero is c. If either case (1) or (2) occurs, an interval half as wide as the original interval that contains the root is found as shown in Fig.1. If the process continues, $c_1, c_2, c_3, ... c_n$ represents the sequence of midpoints which converges to the root x^* within a certain degree of accuracy (DOA), namely

$$\left| c_n - x^* \right| \le (b-a)/2^n \text{ for } n = 1, 2, ..., \tag{1}$$

From (1), it can be seen that increasing n can make c_n more closer to x^* at the DOA in terms of the value of $(b-a)/2^n$.

B. MPPT technique based on BST

In order to apply the BST into the MPPT technique in PV systems, the function of $y = f(x)$ and the variable x should be chosen carefully. Fig. 2 shows a real current-voltage $(I-V)$ curve of a solar module and its corresponding power-voltage $(P-V)$ curve tested in a sunny day in Singapore.

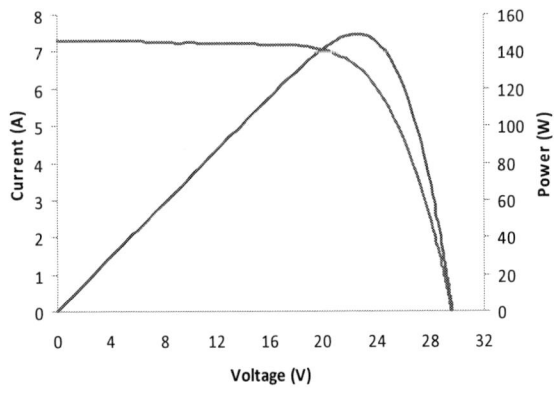

Fig. 2 Typical current-voltage and power-voltage curves

From the $P-V$ curve, it can be observed that the change in power ΔP with respect to voltage approaches zero at the maximum power point as illustrated in Fig. 3. Obviously, the powers at short circuit voltage (0 V) and open circuit voltage (V_{oc}) are zero, so maximum power should not happen in these two particular points even though the changes in power at these two points are also zero, which is caused by the small powers around these two points. Thus, tracking the maximum

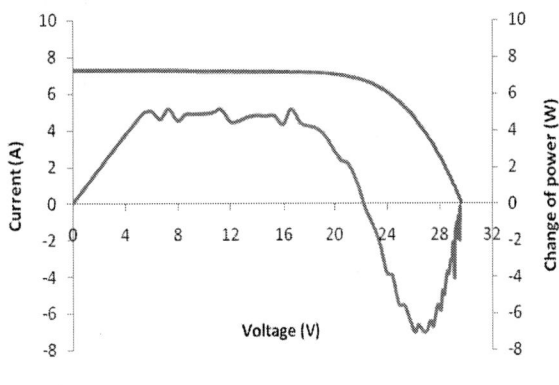

Fig. 3 Typical change in power versus voltage curve

power point is essential to find the root in the function ΔP by regulating the voltage of solar module or solar array. As a result, the function $y = f(x)$ can be regarded as the change in power ΔP, where the variable x is the voltage of solar module or solar array and can be written as:

$$y = \Delta P(V_m) \tag{2}$$

where V_m is the voltage across the solar module. To regulate the voltage V_m, converters are conventionally required as interface between solar modules and loads.

A DC/DC booster converter is adopted to implement the proposed MPPT technique in this paper. Fig. 4 shows the schematic diagram of the system. When the converter works at continuous mode, the relationship between the duty cycle and the voltage of solar modules at steady-state is written as:

$$V_m = (1-D)V_{out} \tag{3}$$

where V_{out} is the output voltage of the converter. Thus, the duty cycle D ($0 \le D \le 1$) can be used to regulate the voltage of solar module or solar array [18].

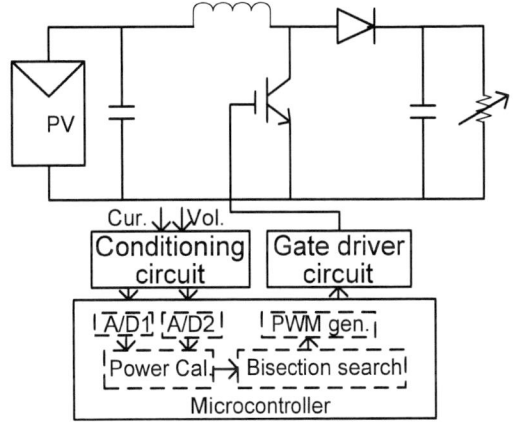

Fig. 4 DC-DC boost converter for MPPT

The working principle of the system can be described as follows. The entire system is controlled by a microcontroller. The voltage and current of solar modules are continuously sampled and the duty cycle of the converter is calculated by the microcontroller based on the proposed MPPT technique using the BST. The flowchart of the program embedded in the microcontroller is shown in Fig. 5.

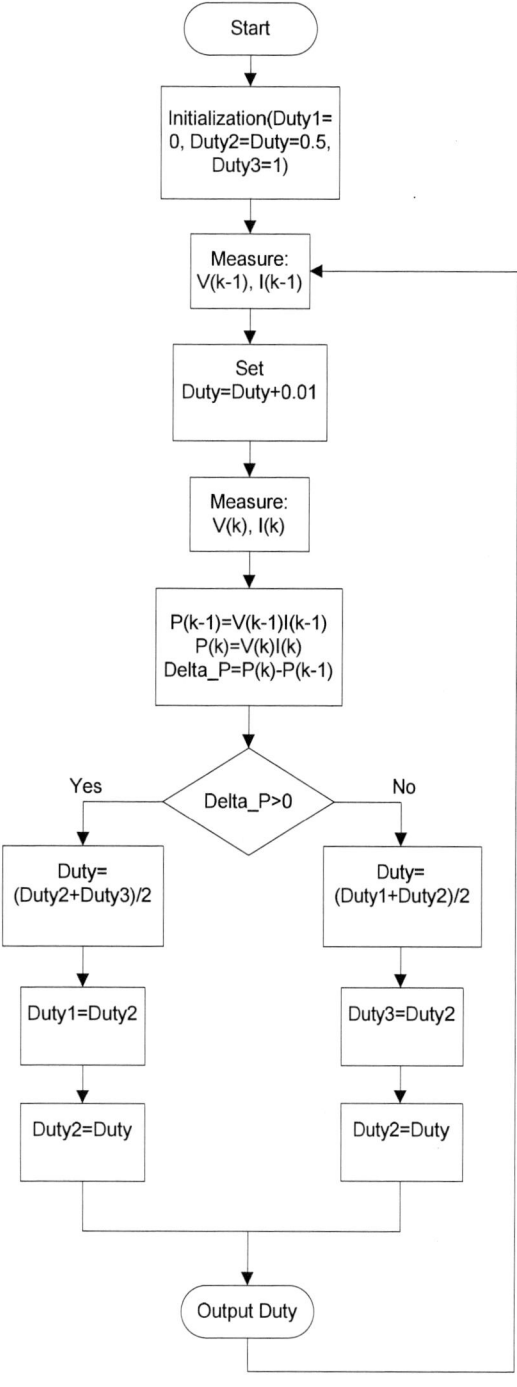

Fig. 5 Flowchart of program for MPPT implementation

III. EXPERIMENTAL SETUP

To verify the feasibility and effectiveness of the proposed MPPT technique, the hardware of the system has been set up. Two sources were used to test the proposed MPPT. One is a real solar module working under natural environmental conditions while the other is a solar array simulator which simulates different sizes of solar modules under different conditions.

A. Solar modules

The solar module is a power source of PV systems. In this experimental setup, solar array consists of two solar modules connected in parallel. The specifications of the solar module are shown in Table 1.

TABLE I. SPECIFICATIONS OF SOLAR MODULE

Item	Value
Nominal Maximum Output (Pin)	45W
Nominal Open Circuit Voltage (Voc)	18V
Nominal Short Circuit Current (Isc)	3.45A
Nominal Maximum Output Voltage (Vmpp)	14.5V
Nominal Maximum Output Current (Impp)	3.11A
Nominal Weight	5.3kg

These two solar modules have almost the same I-V characteristics, then the output voltage of the solar array is equal to the output voltage of one solar module, and the output current is twice as much as the output current of one solar module. Thus, for this solar array at a sunny day, the voltage at the maximum power should be around 14.5 V while the current at the maximum power should be around 6.2 A, as shown in Fig. 6.

Fig. 6 Two solar modules for testing MPPT

B. Solar array simulator

The solar array simulator (SAS) is the important tools to investigate the PV systems. It can create the I-V curve of different sizes of solar array under various environmental conditions. It can also test the response time of the MPPT when solar radiation changes from one into another. Fig. 7

shows the Agilent SAS model of E4360. It can output power up to 1200W with maximum open circuit voltage of 130V and maximum short circuit current of 10 A. Therefore, any size of solar array within the above-mentioned ranges can be simulated by using this SAS.

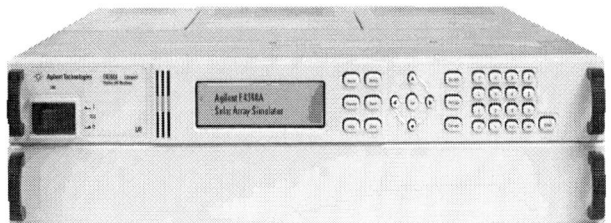

Fig. 7 Solar array simulator for testing MPPT

C. Hardware implementation of MPPT

The proposed MPPT based on the BST is realized by a Freescale MC9S08AW60 Microcontroller. The inputs of the microcontroller are the voltage and current sensed from solar array and the output of the microcontroller is the pulse width modulation (PWM) pulse which is used to control the duty-cycle of the IGBT in the DC-DC boost converter through a gate driver circuit as shown in Fig. 4, where the parameters of major components are illustrated in Table II. The C-language program is employed in this controller. The experimental setup of the proposed MPPT is shown in Fig. 8.

TABLE II.   PARAMETERS OF MAJOR COMPONENTS

Item	Value
Booster inductor, L (mH)	0.2mH
Smoothing capacitor, C (uF)	470uF
Switching frequency, fs (kHz)	40khz
IGBT (type)	IRG4PC50UD

Fig. 8 Experimental setup of MPPT based on BS theorem

IV.   EXPERIMENTAL RESULTS AND DISCUSSIONS

A.   Experimental results

Three types of tests have been conducted on the experimental setup of the MPPT technique based on BST.

The first type of test is to prove its feasibility and show the steps of how the proposed MPPT tracks the maximum power point under constant solar radiation, where the SAS repeat generating one I-V curve. The results of tracking process which shows variation of voltage, current and power are illustrated in Fig. 9. It can be seen that seven steps are approximately required for the proposed MPPT to track the maximum power which is equivalent to about 1.5 seconds.

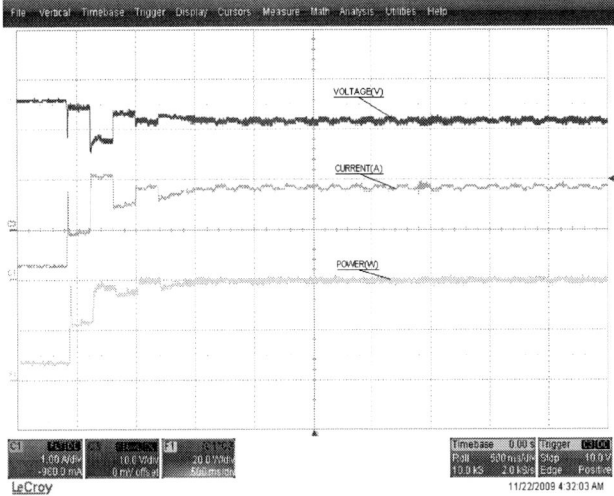

Fig. 9 Verification of feasibility and steps to track maximum power for the proposed MPPT

The second type of test is to investigate its tracking capability under slow variation of solar radiation, where the SAS generates two I-V curves repetitively with a defined interval in between. The results of tracking process which shows the variation of voltages, currents and powers under the conditions of the changing solar radiations are illustrated in Fig. 10. It indicates that the proposed MPPT can still track the maximum power point despite the changing solar radiation.

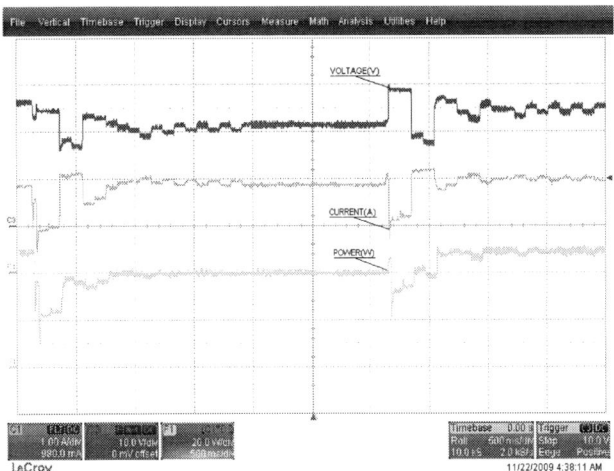

Fig. 10 Investigation of tracking capability of the proposed MPPT under variation of solar radiation

The third type of test is to operate the proposed MPPT under natural environmental conditions, where the MPPT controller is connected to two solar modules in parallel as shown in Fig. 6. The experimental results are illustrated in Table III. Under different times (or solar radiations), the duty cycle of power converter can stabilize at 40.6% and 32.7% (see Figs. 11 and 12), respectively, which are very close to the values of duty cycles: 41% and 33%. These two values are corresponding to the maximum powers for the solar modules at the times indicated in Table III.

TABLE III EXPERIMENTAL RESULTS FOR PROPOSED MPPT TO OPERATE UNDER TWO SOLAR MODULES IN PARALLEL AT DIFFERENT TIMES (DIFFERENT SOLAR RADIATIONS)

15:05, 8 May, 2009		
Duty cycle	Output Voltage	Output Power
80%	12.05V	4.84W
70%	13.16V	5.77W
60%	14.17V	6.69W
50%	15.29V	7.79W
41%	16.03V	8.57W
40%	15.83V	8.35W
30%	14.66V	7.16W
20%	11.43V	4.35W

16:34, 8 May, 2009		
Duty cycle	Output Voltage	Output Power
80%	13.74V	6.29W
70%	15.31V	7.81W
60%	17.13V	9.78W
50%	19.27V	12.38W
40%	21.13V	14.88W
33%	21.56V	15.49W
30%	21.30V	15.12W
20%	16.92V	9.54W

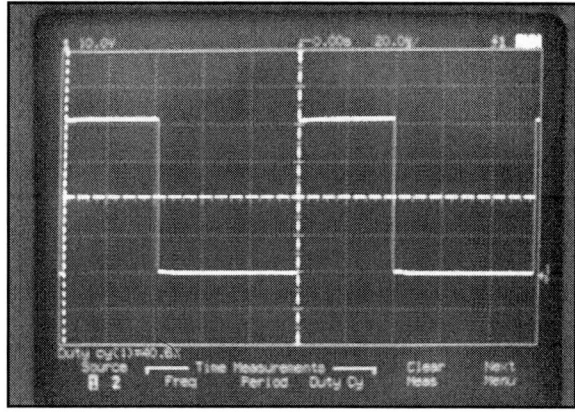

Fig. 11 Power converter operating at duty cycle of 40.6%

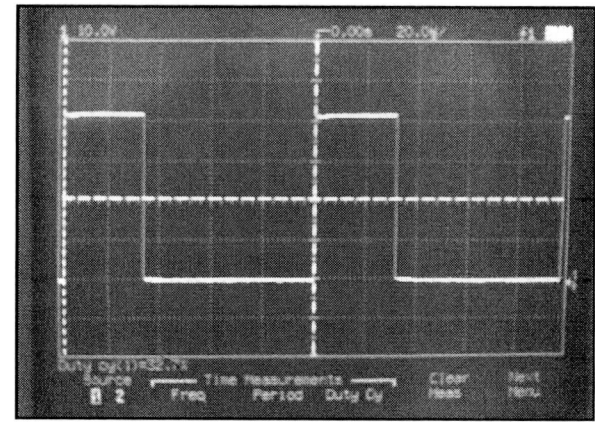

Fig. 12 Power converter at duty cycle of 32.7%

B. Discussions

According to the description of the proposed MPPT technique and its experimental results, some observations can be made and discussed here.

Firstly, there is no any requirement of derivatives of voltage and power measured from solar array, which reduces the complexity in computation and hence implementation. Therefore, the proposed MPPT technique is very suitable for the use of PV systems.

Secondly, the control signal of IGBT can be easily generated due to the fact that every time the new duty cycle is simply taken by halving the sum of its previous value and current value. Thus, tracking maximum power point can be very fast within a few steps.

Finally, although some ripples of voltage, current and power can be seen from the experimental data which is not evitable for the DC/DC booster converter, the proposed MPPT technique can still work well, which show a certain degree of robust and reliability of the proposed MPPT technique.

V. CONCLUSIONS AND FUTURE WORK

In this paper, a novel MPPT technique based on bisection search theorem has been presented. A microcontroller together with a DC-DC boost converter is applied to implement the proposed MPPT control system. Experimental results from solar array simulator show that the proposed MPPT technique can track the maximum power very fast under slow variation of solar radiation within a few steps. The feasibility of the proposed MPPT control system has also been verified in natural environmental conditions.

Further research can be conducted on how the proposed MPPT can track the maximum power at fast change of solar radiation and how it can identify the global maximum and local maxima under the partially-shaded solar modules.

REFERENCES

[1] Bialasiewicz, J.T., "Renewable Energy Systems with Photovoltaic Power Generators: Operation and Modeling," IEEE Trans. on Industrial Electronics, vol. 55, no. 7, pp. 2752-2758, July 2008.

[2] D. Poponi, "Analysis of diffusion paths for photovoltaic technology based on experience curves," Solar Energy, vol. 1, no. 74, pp.331-349, 2003.

[3] Tomas Markvart, Solar electricity, John Wiley & Sons Inc. 2000.

[4] E. Koutroulis, K. Kalaitzakis, and N. C. Voulgaris, "Development of a microcontroller-based, photovoltaic maximum power point tracking controlsystem," IEEE Trans. Power Electron., vol. 16, no. 21, pp. 46–54, Jan. 2001.

[5] O. Wasynczuk, "Dynamic behavior of a class of photovoltaic power systems," IEEE Trans. Power App. Syst., vol. 102, no. 9, pp. 3031–3037, Sep. 1983.

[6] S. Jain andV.Agarwal, "A newalgorithm for rapid tracking of approximatemaximum power point in photovoltaic systems," IEEE Power Electron. Lett., vol. 2, no. 1, pp. 16–19, Mar. 2004.

[7] N. Femia, G. Petrone, G. Spagnuolo, and M. Vitelli, "Optimization of perturb and observe maximum power point trackingmethod," IEEE Trans. Power Electron., vol. 20, no. 4, pp. 963–973, Jul. 2005.

[8] K. H. Hussein and I. Mota, "Maximum photovoltaic power tracking: An algorithm for rapidly changing atmospheric conditions," in IEE Proc. Generation Transmiss. Distrib., 1995, pp. 59–64.

[9] Y.-C. Kuo, T.-J. Liang, and J.-F. Chen, "Novel maximum-power-point tracking controller for photovoltaic energy conversion system," IEEE Trans. Ind. Electron., vol. 48, no. 3, pp. 594–601, Jun. 2001.

[10] G. W. Hart, H. M. Branz, and C. H. Cox, "Experimental tests of open loop maximum-power-point tracking techniques," Solar Cells, vol. 13, pp. 185–195, 1984.

[11] M. A. S. Masoum, H. Dehbonei, and E. F. Fuchs, "Theoretical and experimental analyses of photovoltaic systems with voltage and current-based maximum power-point tracking," IEEE Trans. Energy Convers., vol. 17, no. 4, pp. 514–522, Dec. 2002.

[12] S. Yuvarajan and S. Xu, "Photo-voltaic power converter with a simple maximum-power-point-tracker," in Proc. 2003 Int. Symp. Circuits Syst., 2003, pp. III-399–III-402. R. M. Hilloowala and A. M.

[13] T. Esram, P.L. Chapman, "Comparison of photovoltaic array maximum power point tracking techniques," IEEE Trans. on Energy Conversion, vol. 22, pp. 439-449, June 2007.

[14] R. Faranda, S. Leva, " Energy comparison of MPPT techniques for PV systems," WSEAS Transaction on Power systems, vol. 3, no. 6, pp. 446-455, June 2008

[15] M. Berrera, A. Dolara, R. Faranda, S. Leva, " Experimental test of seven widely-adopted MPPT algorithms," in Proc. 2009 Int. Conf. Power Tech, 2009, June 28th - July 2nd, Bucharest, Romania.

[16] Horst A. Eiselt, Carl-Louis Sandblom, Linear programming and its applications, Springer, 2007.

[17] John H. Mathews and Kurtis K. Fink, Numerical method using Matlab, Prentice-Hall Inc., Upper saddle river, New Jersey, USA, 2004.

[18] N. Mohan, T. M. Undeland, and W. P. Robbins, Power Electronics-Converters, Applications, and Design, John Wiley & Sons, 2002.

Simple Control Design for a Three-Port DC-DC Converter Based PV System with Energy Storage

Sixifo Falcones and Raja Ayyanar

School of Electrical, Computer and Energy Engineering
Arizona State University
Tempe, AZ 85281 USA
sfalcon@asu.edu, rayyanar@asu.edu

Abstract—**A simple control technique for a three-port DC-DC converter based PV system with storage is presented herein. The controller takes advantage of the fast response and cross coupling characteristics of the selected three port topology itself to obtain transient ride through support from the storage while using traditional control design techniques, e.g. SISO control design and Feedforward compensation. A gyrator average model for the selected three port topology is also presented. The performance of the controls is verified through simulations.**

I. INTRODUCTION

In commercial PV systems, the storage is usually connected either directly to the DC link, which makes it difficult to independently control the charging/discharging current, or through a separate and dedicated DC-DC converter, which increases the number of components. On the other hand, for a three-port DC-DC converter based topology, the storage and the source can act in parallel with independently controlled currents, thus using the storage to provide support during transients, while keeping the number of components low. [1]-[3] present different three-port DC-DC topologies.

From the family of three-port DC-DC converters, the triple active bridge (TAB) topology has been selected for the development of the PV system presented herein since it simplifies the control design [3]-[4]. It is based on a three-winding high frequency (HF) transformer and three bidirectional full-bridges connected at the HF transformer terminals. A PV system based on the TAB is shown in Fig. 1. The average power that flows from DC port i into DC port j of the TAB is calculated from

$$P_{ij} = \frac{V_i{}' V_j{}'}{2\pi f_S L_{ij}{}'} \varphi_{ij} \left(1 - \frac{|\varphi_{ij}|}{\pi} \right) \tag{1}$$

where the DC port voltages V_i, V_j and the equivalent link inductance L_{ij}, for convenience, are reflected onto the port 1 side by appropriately using the number of turns of the transformer N_1, N_2 and N_3, and

$$\varphi_{ij} = \varphi_i - \varphi_j \quad i, j = 1, 2, 3 \tag{2}$$

is the difference in phase between the two square voltages generated by the corresponding full-bridges [3]. Based on the above equations, a change in one phase angle will produce a change in the TAB power flow. Consequently, for the TAB based PV system, when the PV module increases its output power, the storage and the inverter can both sink the extra power simultaneously by adjusting a single phase angle, thus

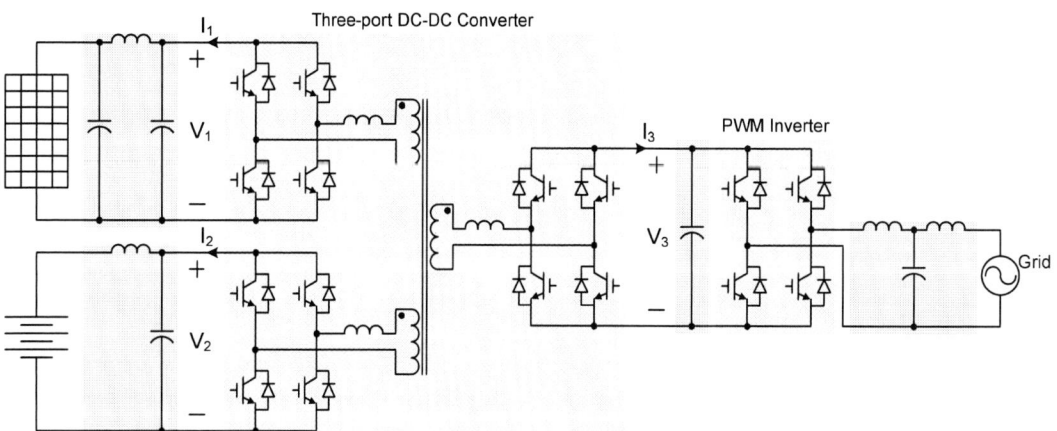

Figure 1. TAB based PV System.

978-1-4244-4782-4/10 $26.00 © 2010 IEEE

allowing for a better performance on the regulation of the inverter side DC link voltage.

II. TAB GYRATOR AVERAGE MODEL

After manipulating the set of power flow equations obtained from (1), it turns out that the cycle-by-cycle average (CCA) current at one DC port linearly depends on the voltages of the other two DC ports. Due to this, a gyrator average model for the TAB can be determined as seen in Fig. 2 [5]. The involved gyration gains g_{ij} can be read from

$$I_i = \sum_{j \neq i}^{n} g_{ij} V_j \quad (3)$$

$$g_{ij} = \frac{I_{ij}}{V_j} = \frac{N_1^2}{N_i N_j} \frac{1}{2\pi f_s L_{ij}} \varphi_{ij} \left(1 - \frac{|\varphi_{ij}|}{\pi}\right) \quad (4)$$

If φ_1 is set to zero, then the remaining two phase angles represent phase lags with respect to port 1 and can be used as control variables. The above gyrator model shows that the current at each DC port has a maximum and that a change in a single phase angle forces all these currents to change simultaneously. The latter is the basis to obtain ride through support from the storage during disturbances almost instantly.

III. PV SYSTEM AVERAGE MODEL

Since the current at each DC port of the TAB comes from the rectification of its inductor current, it has a large ripple. In order to prevent ripple currents from reaching the PV and storage, CLC and LC filters are connected in between, respectively, as seen in the switching model of the PV system illustrated in Fig. 1. Since the third port has two back-to-back full-bridges, only the DC link capacitor is needed.

If the stray capacitances and resistances within the TAB are neglected, the modes associated with the external filters are the predominant dynamics of the system; thus the control design is simplified. The average model of the PV system is

shown in Fig. 3. The average models of the stages with no associated dynamics are represented by Ψ functions.

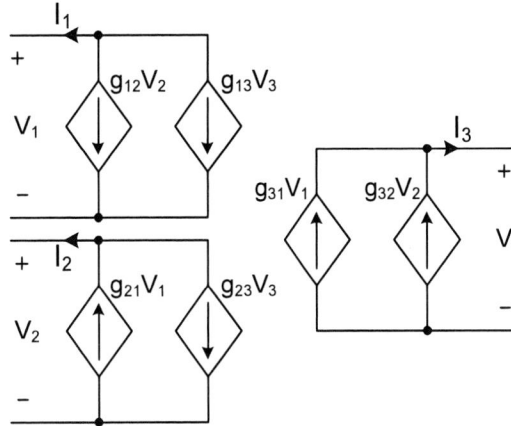

Figure 2. TAB gyrator average model.

IV. CONTROL STRATEGY

For the TAB, the CCA current at DC port 1, I_1, can be represented as function of the two phase angles φ_3 and φ_2 as seen in Fig. 4. The middle dotted line, where points A and B are located, represents the zero crossing of the battery current. If the PV voltage and the battery current are controlled by modulating φ_3 and φ_2, respectively, then, multiple-input multiple-output (MIMO) control techniques may be needed to deal with the TAB model cross-coupling terms [3]-[4]. A simple approach, presented in this paper, is to take advantage of the cross-coupled nature of the TAB itself while using the simpler and more intuitive single-input single-output (SISO) techniques for both controllers. Since following a reference signal from a maximum power point tracking (MPPT) algorithm requires more bandwidth than the one from a battery management algorithm, the PV voltage controller is designed to be faster than the battery current controller. Additionally, to ensure that the operating point (φ_3, φ_2) moves in the appropriate direction when the PV voltage controller acts, feedforward of φ_3 must be used to generate φ_2. The

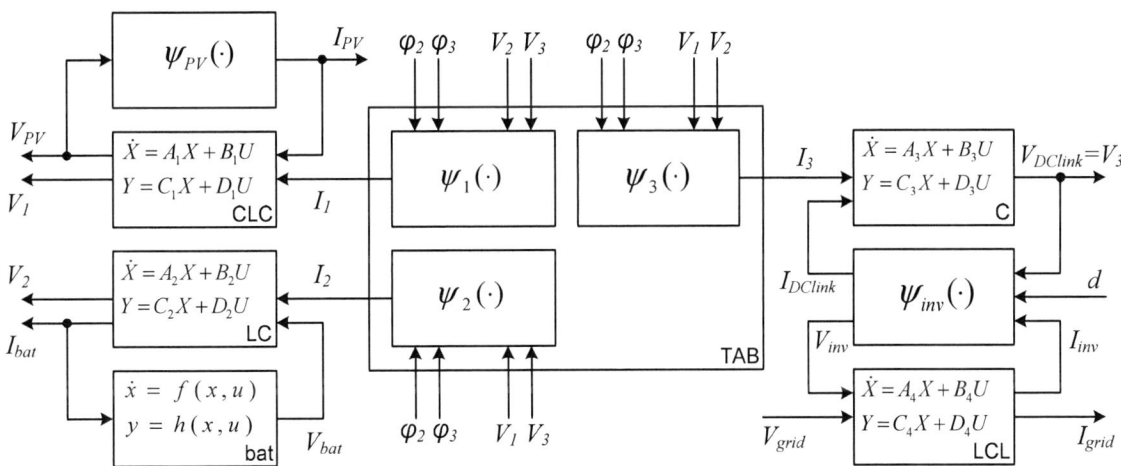

Figure 3. PV system average model.

feedforward gain can be calculated at each initial steady state operating point A from

$$k = \frac{\Delta\varphi_2}{\Delta\varphi_3} = \frac{\left.\dfrac{\partial I_1}{\partial \varphi_2}\right|_A}{\left.\dfrac{\partial I_1}{\partial \varphi_3}\right|_A} \qquad (5)$$

This forces the operating point (φ_3, φ_2) to move very fast towards the direction that maximizes I_1. As a result, the performance of the DC link voltage regulation on the inverter side is improved during transients on the PV side. Fig. 4 shows the hypothetical scenario where the PV voltage loop has infinite bandwidth. A-C-B and A-D-B represent the trajectories of the operating point with and without the feedforward of φ_3, respectively. The reference for the battery current is assumed to be 0. Once I_1 reaches its steady state value, after bringing the PV voltage back to its reference value, the battery current controller will slowly bring the operating point towards I_2 steady state value along the contour level designated by I_1.

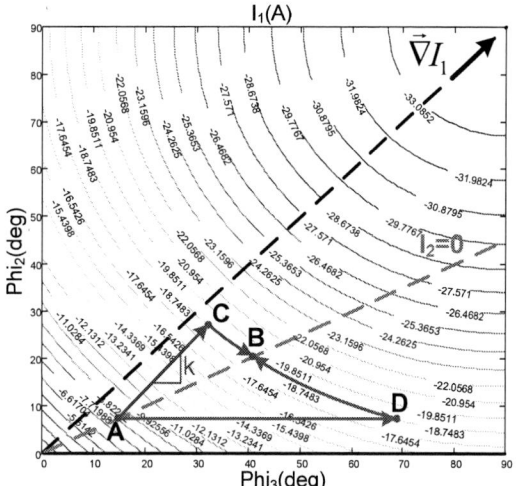

Figure 4. Contour plot of DC port 1 current as a function of the phase angles

The block diagram of the PV system controls is shown in Fig. 5. For the inverter, the duty cycle d is used to control the injected current into the grid. The PV voltage controller follows a reference command from an MPPT algorithm [6]. The battery current follows a reference command from a management algorithm that uses pricing information [7]. The inverter uses instantaneous reactive power (IRP) and orthogonal generation approaches to calculate the reference for the injected grid current, thus enabling the DC link voltage regulation and on-demand reactive power support to the grid [8].

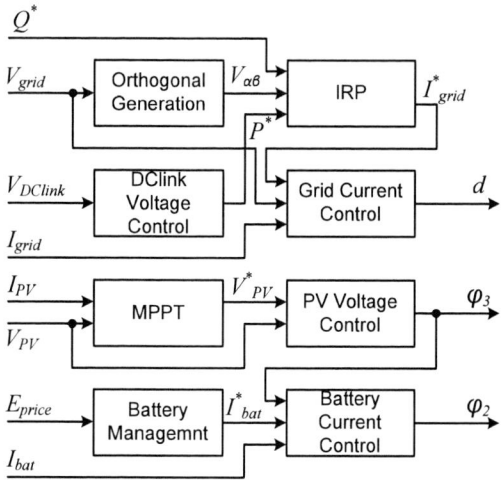

Figure 5. PV system controller.

V. SIMULATION RESULTS

The switch model of the PV system was developed in PLECS® and the control scheme in Simulink®. Fig. 6 shows the time response of the system during an insolation step on the PV panel for the approach without feedforward of φ_3. The battery current worsens the DC link voltage regulation due to the wrong trajectory of the phase angles.

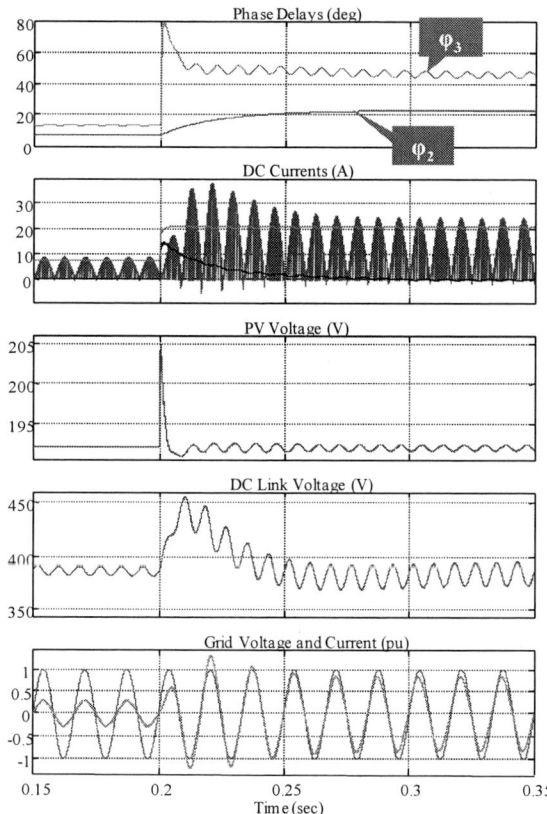

Figure 6. Time response to an insolation step on the PV panel without φ_3 feedforward

The time response of the system during an insolation step on the PV panel for the approach with feedforward of φ_3 can be observed in Fig. 7. It can be seen that, after the step event, the two phase angles experience the same increment for $k=1$ forcing the battery to store some of the extra energy. As a result, the performance of the DC link voltage regulation is improved. Additionally, thanks to the filters, the battery and PV currents are smooth.

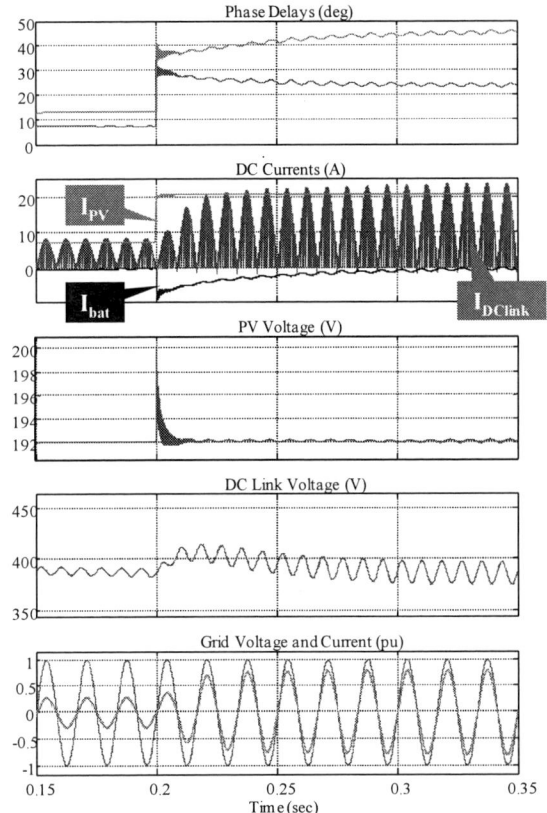

Figure 7. Time response to an insolation step on the PV panel with φ_3 feedforward

Fig. 8 shows the TAB switching waveforms corresponding to the square wave voltages generated by each full-bridge and the current through each leakage inductor.

VI. HARDWARE RESULTS

A 1kW prototype of the PV system has been developed. Fig. 9 shows the measured phase shifted square wave voltages generated by each full-bridge of the TAB as well as the inductor current at port 3. This agrees with the simulation results from Fig. 8. At the time of this publication, the system is being tested in closed loop. Its result will be included in a future publication.

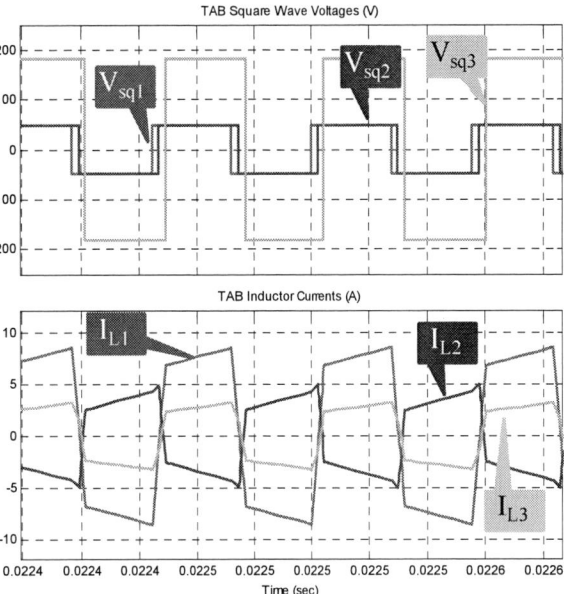

Figure 8. Time response to an insolation step on the PV panel with φ_3 feedforward

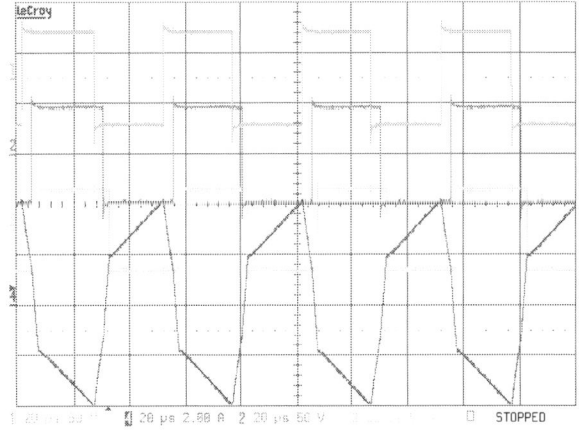

Fig. 9. Measured phase shifted square wave voltages and port 3 inductor current of the TAB

VII. CONCLUSIONS

This paper proposes a simple yet effective control strategy for the control of a PV system based on a TAB converter. The controller takes advantage of the cross coupling terms in the TAB model to implement fast transient ride through support from the battery. A gyrator average model is also developed. The performance of the designed controller is verified through simulations.

REFERENCES

[1] Tao, H.; Kotsopoulos, A.; Duarte, J.L.; Hendrix, M.A.M., "Family of multiport bidirectional DC-DC converters," *Electric Power Applications, IEE Proceedings*, vol.153, no.3, pp. 451-458, 1 May 2006

[2] Liu, Danwei; Li, Hui; Marlino, Laura D., "Design Of A 6 kW Multiple-Input Bi-directional DC-DC Converter With Decoupled Current Sharing Control For Hybrid Energy Storage Elements," *Applied Power Electronics Conference, APEC 2007 - Twenty Second Annual IEEE*, vol., no., pp.509-513, Feb. 25 2007-March 1 2007

[3] Zhao, C.; Kolar, J., "A Novel Three-phase Three-Port UPS Employing a Single High-Frequency Isolation Transformer," *Power Electronics Specialists Conference, 2004*, vol. 6, June 2004, pp. 4135 - 4141

[4] Duarte, J.L.; Hendrix, M.; Simoes, M.G., "Three-Port Bidirectional Converter for Hybrid Fuel Cell Systems", *Power Electronics, IEEE Transactions on*, vol. 22, Issue 2, March 2007, pp. 480-487

[5] M. Ehsani, I. Husain, M. O.Bilgic, "Power converters as natural gyrators", *IEEE Transactions on Circuits and Systems I: Fundamental Theory and Applications*, v 40, n 12, p 946-949, Dec, 1993.

[6] X. Mao, and R. Ayyanar, "Average and Phasor Models of Single Phase PV Generators for Analysis and Simulation of Large Power

Distribution Systems," in *Proc. IEEE APEC'09*, Washington, DC, Feb. 2009, pp. 1964-1970.

[7] Vosen SR, Keller JO. Hybrid energy storage systems for stand-alone electric power systems: optimization of system performance and cost through control strategies. Int J Hydrogen Energy 1999;24:1139–56.

[8] X. Mao, S. Falcones, and R. Ayyanar, "Energy-Based Control Design for a Solid State Transformer," in *Proc. FREEDM Annual Conference 2009*, Raleigh, NC, May. 2009, pp. 217-220.

[9] Qiang, M; Wei-yang, W; Zhen-lin, X, "A Multi-Directional Power Converter for a Hybrid Renewable Energy Distributed Generation System with Battery Storage," *Power Electronics and Motion Control Conference, 2006. CES/IEEE 5th International*, vol. 3, August 2006, pp. 1-5.

[10] Manwell JF, McGowan JG., "Lead acid battery storage model for hybrid energy systems," *Sol Energy*, vol. 50, May 1993, pp. 399-405.

978-1-4244-4782-4/10 $26.00 © 2010 IEEE

A Self-powered Power Management Circuit for Energy Harvested by a Piezoelectric Cantilever

Na Kong[1], Travis Cochran[1], Dong Sam Ha[1], Hung-Chih Lin[2], Daniel J. Inman[1]

[1]Virginia Polytechnic Institute and State University
Blacksburg, VA 24061 USA
{kongna, cochrant, ha, dinman}@vt.edu

[2]National Tsing Hua University,
Hsin-Chu, 30013 TAIWAN
hclin@larc.ee.nthu.edu.tw

Abstract-This paper presents development of a self-powered power management circuit for energy harvested by a piezoelectric cantilever. A full-wave rectifier followed by a buck-boost converter running in the discontinuous conduction mode rectifies the AC output, matches the source impedance, and generates a regulated DC output provided the input power is sufficient to charge up the load. A low power microcontroller unit is used for the maximum power point tracking and the output voltage regulation. Experimental results show that the circuit can harvest up to 3.5 mW with a 50x31.8 mm² piezoelectric cantilever under 0.5g (rms) base acceleration. Detailed loss analysis is presented for efficiency enhancement in the future.

Index Terms- Power management circuit, Energy harvesting, Piezoelectric cantilever

I. INTRODUCTION

Replacement or recharge of batteries is a major bottleneck for wide deployment of wireless sensor nodes (WSNs), and energy harvested from ambient sources offers a promising solution to the problem [1]-[2]. Energy harvesting for mechanical vibrations is suitable for structural health monitoring, which is the target application of this paper. Vibration-based power generators convert the mechanical energy of vibrating surfaces into electrical energy using a suitable mechanical-to-electrical energy converter (or generator) such as electromagnetic, electrostatic, or piezoelectric transduction devices. We use a piezoelectric cantilever, which offers relatively high power density [3].

Raw electrical power harvested from ambient sources should be conditioned and regulated to a desired voltage level to power up electronic devices. The key design issue for power conditioning circuits is impedance matching. It is complicated as the source impedance depends on the operating conditions. Ottman et al. adopted an adaptive buck converter for impedance matching, but the approach is limited to a harvester whose rectified AC voltage is higher than the output voltage of the system [4]-[5]. Lefeuvre et al. proposed use of a discontinuous conduction mode (DCM) buck-boost converter [6]. Their approach eliminates the previous limitation, but the impedance matching is not guaranteed due to adoption of the open-loop operation (which is intended to reduce the circuit complexity and hence the power consumption). A resonant converter, whose network matches its impedance to the parasitics of the piezoelectric disk, can achieve high efficiency for power conversion [7]-[9]. However, the operating frequency of a piezoelectric transformer is several MHz, which is much higher than a typical vibration frequency of structures (below 200 Hz). To match the impedance of a piezoelectric cantilever, it requires a prohibitively large inductance. Several researchers incorporated switches and inductors to directly shape the output voltage waveform to be in phase with outgoing current [10]-[13]. The non-linear treatment of direct shaping output voltage achieves the maximum power point (MPP) for a certain DC output voltage, which is usually too high for WSNs. So the approach often requires one or two additional voltage regulation stages for WSNs.

This paper presents a self-powered power management circuit, which is capable of handling a wide input voltage range and adjusting the input impedance dynamically through a closed-loop control. The output voltage of the proposed circuit is regulated to power up a WSN provided the input power is sufficient. Loss analysis for the proposed circuit is presented in detail, which can be useful to further enhance the efficiency.

II. THE PROPOSED POWER MANAGEMENT CIRCUIT

A. Source impedance of a cantilevered piezoelectric generator

A piezoelectric energy harvester is typically a cantilevered beam with one or two piezoceramic layers (a unimorph or a bimorph). Figure 1 shows a typical bimorph cantilever configuration [14], where S is strain, V voltage, M mass, and z vertical displacement. A mass is placed on the free end to tune the resonant frequency of the system. The two piezoceramic layers are poled oppositely along the 3 axis and electrodes are placed on the surfaces perpendicular to the 3 axis. The beam undergoes bending vibrations due to the motion of its base, and the driving vibrations are assumed to exist only along the 3 axis. Under the assumptions, the piezoelectric material experiences a one-dimensional state of stress along the 1 axis. Then, the dynamic strain induced in the piezoceramic layers generates an AC voltage output across the electrodes.

The equivalent circuit for the first vibrational mode of a bimorph piezoelectric cantilever is shown in Figure 2 [15].

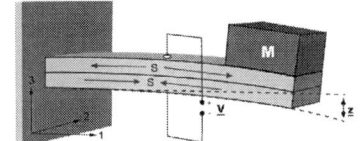

Figure 1. A bimorph cantilever configuration [14].

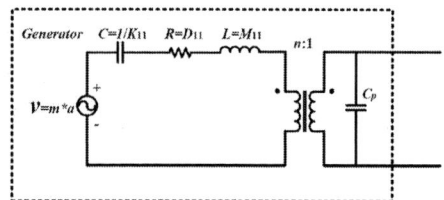

Figure 2. The equivalent circuit for the first mode piezoelectric generator.

The voltage generator $v = m^*a$ represents the force induced by the base vibration and is the only source in the electrical model, where m^* is the effective modal forcing term and a is the base acceleration amplitude. The equivalent inductor $L = M_{11}$ represents the modal mass of the first mode. The resistor $R = D_{11}$ and the capacitor $C = 1/K_{11}$ represent mechanical damping and compliance (reciprocal of stiffness), respectively. The electromechanical coupling is modeled as a transformer with the turns ratio of n. C_P is the equivalent inherent capacitance of the piezoceramic layers.

The equivalent source impedance is shown in Figure 3. The resonance frequency of the generator is usually tuned to match the vibration frequency of the base structure. In this case, the resistive load which matches the source impedance extracts maximal power.

B. Operation of the impedance matching circuit

Figure 4 shows the circuit diagram. A full-wave rectifier rectifies the AC output. A buck-boost converter running in the DCM is chosen for the second stage to
(i) accommodate a wide range of input voltage; and
(ii) behave as a lossless resistor to match the source impedance for the maximum power point tracking (MPPT) [16].

An ultra-low power microcontroller unit (MCU) MSP430 from Texas Instruments is configured to provide the MPPT and the output voltage regulation. A supercapacitor is chosen as a storage device for our system due to a less strict requirement for charging and virtually unlimited charging and discharging cycles. A backup battery is added for the initial start of the circuit (specifically for the MCU) as well as backup power in case of insufficient input power.

Waveforms for a half cycle of a harmonic base vibration are shown in Figure 5. The waveform of the rectified voltage v_{rect} is approximately sinusoidal with the peak value of $V_{in,peak}$. Since the base vibration frequency F_v is much lower (about 44 Hz for our case) than the switching frequency $F_S = 1/T_S$ (about 3 kHz), the input voltage to the buck-boost converter is

assumed as DC voltage for each switching period T_S. The effective input resistance of the buck-boost converter is obtained as (1).

$$R_{in} = \frac{v_{rect}}{\frac{1}{T_S}\int_0^{D_1 T_S} i_L dt} = \frac{v_{rect}}{\frac{1}{T_S}\int_0^{D_1 T_S} \frac{v_{rect}}{L} t dt} = \frac{2L}{D_1^2 T_s} \tag{1}$$

In order to achieve the resistive impedance matching, the effective input resistance R_{in} should be equal to the optimal resistive load $R_{in,opt}$ given in (1). Hence, the optimal duty cycle can be expressed as

$$D_{1,opt} = \sqrt{\frac{2L}{R_{in,opt}T_s}} . \tag{2}$$

The maximum inductor current $I_{L,max}$ and the first off-time D_2 during each switching cycle can also be assumed sinusoidal, i.e.

$$I_{L,\max}(n) = I_{L,\max,peak}\left|\sin\left(2\pi F_v(n-1+D_1)T_s\right)\right|, \tag{3}$$

where $n = 1, 2, ...$; $I_{L,\max,peak} = V_{in,peak}D_1 T_s / L$;

$$D_2(n) = D_{2,peak}\left|\sin\left(2\pi F_v(n-1+D_1)T_s\right)\right|, \tag{4}$$

where $n = 1, 2, ...$; $D_{2,peak} = I_{L,\max,peak}L / V_o T_s$.

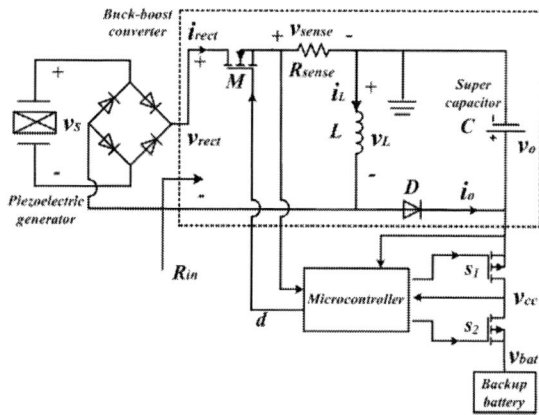

Figure 4. Circuit diagram of the self-powered energy harvesting system.

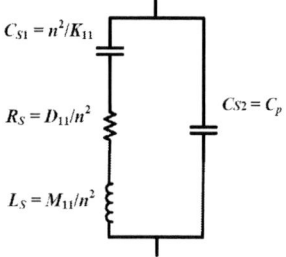

Figure 3. Equivalent source impedance of a piezoelectric generator.

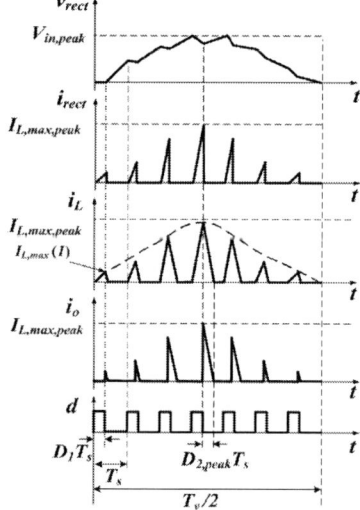

Figure 5. Waveforms during half cycle of a harmonic base vibration.

978-1-4244-4782-4/10 $26.00 © 2010 IEEE

III. MPPT AND VOLTAGE REGULATION

Instead of adopting multiple power conversion stages, we propose to use a single power stage with two different operation modes – MPPT and output voltage regulation – depending on the output voltage of the supercapacitor or the charge level of the supercapacitor. When the supercapacitor is charged initially, the capacitor voltage is lower than a predetermined high threshold voltage V_{HT}. The control loop adjusts the duty cycle of the switches to achieve the MPPT. When the output voltage reaches above V_{HT}, the MPPT control loop turns off, and another control loop turns on to regulate the output voltage of the supercapacitor. The status implies that it has harvested sufficient energy, and so it adjusts the duty cycle to regulate the output voltage rather than the impedance matching. When the input power harvested is insufficient, the output voltage decreases and eventually becomes lower than another a predetermined low threshold voltage V_{LT} (which is lower than V_{HT}). Then, the MPPT control loop turns on to maximize power harvesting through the MPPT.

The optimal duty cycle $D_{1,opt}$ corresponding to the maximum power point is obtained in (2). Based on the circuit parameters used for our experiments (to be given in Section V), the optimal duty cycle is around 3.5% or 10 µs switch on-time for the switching frequency of 3.5 kHz. 1% change of the optimal duty cycle ($\Delta D = 0.035\%$) for the traditional constant switching frequency modulation scheme requires $\Delta T_{on} = 100$ ns, which in turn requires 10 MHz of the clock frequency. In contrast, the - 1% change of duty cycle ($\Delta D = -0.035\%$) for the constant on-time modulation requires $\Delta T_{off} = -2.6$ µs), which can be achieved with 385 kHz of the clock frequency. The constant on-time modulation reduces the clock frequency by a factor around 26 compared with the constant switching frequency modulation, and is adopted for our circuit. The optimal switching frequency for a fixed switch on-time T_{on} can be derived from (2) and is given as

$$F_{S,opt} = \frac{2L}{T_{on}^2 R_{in,opt}} \tag{5}$$

For each switching cycle, the input power charges the inductor during switch on-time and releases the inductor energy fully to the load due to the DCM operation. Therefore, the average power delivered to the load for each switching cycle can be expressed as in (6).

$$P_{avg} = \frac{1}{2} L I_{L,\max}^2 F_S \tag{6}$$

Sensing only the inductor current is sufficient to obtain the average power harvested by the buck-boost converter. The constant on-time modulation makes it simple to sense the inductor current at the middle point of the switch on-time, which avoids sensing the noisy peak current.

Control-loop adjusts the switching period T_S to achieve the MPPT using hill-climbing method [17]-[18]. The detailed operation of the MPPT algorithm is shown in Figure 6. The switch on-time is fixed T_{on} and the initial switching period is perturbed with a small decrease. Since the maximum inductor current $I_{L,\max}$ is sinusoidal, the controller looks for the peak of

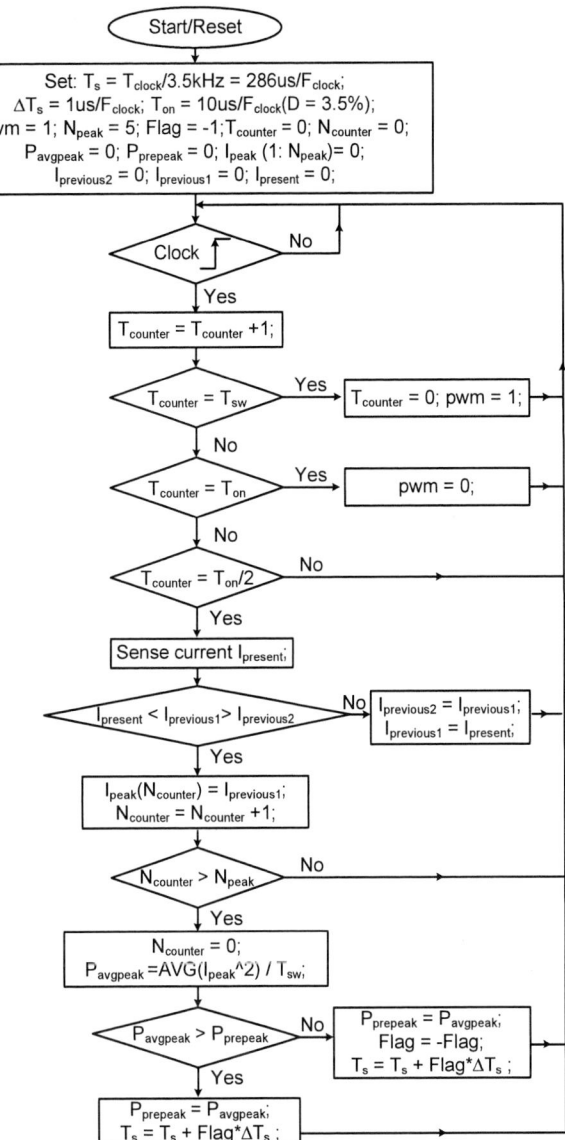

Figure 6. Flow chart of the MPPT algorithm.

the sinusoidal waveform and use (6) to calculate the average input power. To overcome the errors caused by noises, the calculation of average input power takes an average of five consecutive values. If the average input power increased, the switching period keeps decreasing; otherwise the switching period is increased. The hill-climbing process continues. Finally, the switching period will stable around the optimal operation point. Once the circuit reaches at the MPP, the controller runs at a low clock frequency to save power and updates its switching period less frequently.

IV. ANALYSIS OF POWER LOSSES

In order to figure out the sources of power dissipation, the losses in the proposed system are analyzed in this section. The target application for the proposed power management circuit is structural health monitoring, in which a sensor node is activated at a low duty cycle such as once or a few times a day. So, the load for the proposed power management circuit is assumed idle for our loss analysis. Further, the circuit is assumed in the steady state, in which the resistive impedance matching has achieved already by adjusting the switching period T_S.

Under the assumptions given above, the waveform of the rectified voltage V_{rect} is approximately $V_{in,peak}|\sin(2\pi F_v t)|$, where $V_{in,peak}$ is the peak voltage of the piezoelectric generator and F_v is the base vibration frequency. Also, the envelops of inductor current i_L and the super capacitor charge current i_o are approximately sinusoidal. The major power losses are due to power dissipation of six components – the full-bridge rectifier, the MOSFET M, the diode D, the inductor L, the sensing resistor R_{sense} and the MCU. Refer to Figure 4 and Figure 5 for notations used in expressions given below.

A. Rectifier

The first-order forward voltage drop of a diode is expressed as $v_F = ki_L + b$, where i_L is the forward current and k and b are two constants, and the power loss of a diode is $v_F \times i_L$. During the switching cycle corresponding to the inductor current $I_{L,max}$ achieving the peak value $I_{L,max,peak}$, the average conduction loss of the rectifier in this switching cycle can be calculated as

$$
\begin{aligned}
&P_{cond,rectifier,peak} \\
&= \frac{2}{T_s}\int_0^{D_1 T_s} V_F i_L dt \\
&= \frac{2}{T_s}\int_0^{D_1 T_s}\left(k\left(\frac{I_{L,max,peak}}{D_1 T_s}t\right)^2 + b\frac{I_{L,max,peak}}{D_1 T_s}t\right)dt \\
&= 2D_1\left(\frac{k}{3}\left(I_{L,max,peak}\right)^2 + \frac{b}{2}I_{L,max,peak}\right)
\end{aligned}
\tag{7}
$$

So the average conduction loss of the rectifier in the vibration cycle can be calculated as

$$
\begin{aligned}
&P_{cond,recfitier} \\
&= \frac{2}{T_v}\int_0^{T_v}2D_1\left(\frac{k}{3}\left(I_{L,max,peak}\sin(2\pi F_v t)\right)^2 + \frac{b}{2}\left(I_{L,max,peak}\sin(2\pi F_v t)\right)\right)dt \\
&= 2D_1\left(\frac{k}{6}\left(I_{L,max,peak}\right)^2 + \frac{b}{\pi}I_{L,max,peak}\right)
\end{aligned}
\tag{8}
$$

Smaller D_1 and/or $I_{L,max,peak}$ reduces the loss, but it degrades the system performance. Smaller k and b can be achieved by using higher performance diodes.

B. MOSFET M

Similarly, the conduction loss can be obtained as:

$$
\begin{aligned}
P_{cond,mosfet,peak} &= \frac{1}{T_s}\int_0^{D_1 T_s} i_L^2 R_{ds,on} dt \\
&= \frac{1}{T_s}\int_0^{D_1 T_s}\left(\frac{I_{L,max,peak}}{D_1 T_s}t\right)^2 R_{ds,on} dt \\
&= \frac{D_1}{3}\left(I_{L,max,peak}\right)^2 R_{ds,on}
\end{aligned}
\tag{9}
$$

$$
\begin{aligned}
&P_{cond,mosfet} \\
&= \frac{2}{T_v}\int_0^{T_v} P_{cond,mosfet,peak}\sin^2(2\pi F_v t)dt \\
&= \frac{1}{2}P_{cond,mosfet,peak}
\end{aligned}
\tag{10}
$$

where $R_{ds,on}$ is the drain-source on-resistance of the MOSFET. Many factors such as the gate-source voltage V_{GS} and the drain current I_D affect $R_{ds,on}$.

The switching loss is due to voltage-current overlap during the turning-off transition and the loss on output capacitance during the turning-on transition.

$$
\begin{aligned}
&P_{switching,mosfet} \\
&= \frac{1}{2}P_{switching,mosfet,peak} \\
&= \frac{1}{2}\left(\frac{1}{2}V_{in,peak}I_{L,max,peak}t_f + \frac{1}{2}C_{oss}\left(V_{in,peak}\right)^2\right)F_s
\end{aligned}
\tag{11}
$$

where F_s is the switching frequency, t_f is the falling time of the gate input signal, and C_{oss} denotes the output capacitance of the MOSFET. Reduction of t_f and C_{oss} decrease the switching loss without direct affect on the performance, but other parameters do affect.

C. Diode D

The current going through the diode is the off-time inductor current. Similar to the rectifier:

$$
\begin{aligned}
&P_{cond,diode,peak} \\
&= \frac{1}{T_s}\int_0^{D_{2,peak}T_s} V_F i_L dt \\
&= \frac{1}{T_s}\int_0^{D_{2,peak}T_s}\left(k\left(\frac{I_{L,max,peak}}{D_{2,peak}T_s}t\right) + b\right)\frac{I_{L,max,peak}}{D_{2,peak}T_s}t dt \\
&= D_{2,peak}\left(\frac{k}{3}\left(I_{L,max,peak}\right)^2 + \frac{b}{2}I_{L,max,peak}\right)
\end{aligned}
\tag{12}
$$

$$
\begin{aligned}
&P_{cond,diode} \\
&= \frac{2D_{2,peak}}{T_v}\int_0^{T_v}\left(\frac{k}{3}\left(I_{L,max,peak}\right)^2\left(\sin(2\pi F_v t)\right)^3 + \frac{b}{2}I_{L,max,peak}\left(\sin(2\pi F_v t)\right)^2\right)dt \\
&= D_{2,peak}\left(\frac{4k}{9\pi}\left(I_{L,max,peak}\right)^2 + \frac{b}{4}I_{L,max,peak}\right)
\end{aligned}
\tag{13}
$$

The switching loss of the diode is only due to the turning-on activity.

$$
P_{switching,diode,peak} = \frac{1}{2}C_j\left(V_o + V_{in,peak}\right)^2 F_s
\tag{14}
$$

$$
P_{switching,diode} = \frac{1}{2}C_j\left(V_o^2 + \frac{4}{\pi}V_o V_{in,peak} + \frac{1}{2}\left(V_{in,peak}\right)^2\right)F_s
\tag{15}
$$

where C_j is diode capacitance.

D. Inductor L

The loss is due to the parasitic resistance R_{dcr}, and occurs both during on- and off-times.

$$P_{cond,ind,peak}$$

$$= \frac{1}{T_s} \int_0^{T_s} i_L^2 R_{dcr} dt \qquad (16)$$

$$= \frac{1}{3} \left(I_{L,\max,peak} \right)^2 R_{dcr} \left(D_1 + D_{2,peak} \right)$$

$$P_{cond,ind}$$

$$= \frac{1}{3} \left(I_{L,\max,peak} \right)^2 R_{dcr} \left(D_1 \left(\frac{2}{T_v} \int_0^{\frac{T_v}{2}} \sin^2(2\pi F_v t) dt \right) + D_{2,peak} \left(\frac{2}{T_v} \int_0^{\frac{T_v}{2}} \sin^3(2\pi F_v t) dt \right) \right)$$

$$= \frac{1}{3} \left(I_{L,\max,peak} \right)^2 R_{dcr} \left(\frac{1}{2} D_1 + \frac{4}{3\pi} D_{2,peak} \right)$$

$$(17)$$

E. Sensing resistor

Conduction loss is similar to R_{dcr} of the inductor, but it occurs only during on-time.

$$P_{cond,Rsense} = \frac{1}{6} \left(I_{L,Max,peak} \right)^2 R_{sense} D_1 \qquad (18)$$

F. Controller

It is mainly attributed to the power dissipated by the MCU. The loss of the controller is hard to analyze, so we rely on actual measurements.

V. EXPERIMENTAL RESULTS

To verify the feasibility of the proposed circuit, experiments were performed using a cantilevered bimorph generator with a tip mass. The bimorph (manufactured by Piezo Systems, Inc. with model number T226-A4-503X) consists of two oppositely poled PZT-5A piezoelectric elements bracketing a brass substructure layer, and the two piezoelectric elements are connected in series. The base acceleration applied to the piezoelectric cantilever is $0.5g$ (rms) for the experiment. The optimal resistive load of the piezoelectric cantilever for a given frequency is the one which maximizes the average power output and was identified by tuning the load resistor. The experimental result is shown in Figure 7. The optimal resistor is 60 kΩ at the resonant frequency of 44 Hz in the figure.

The experiment setup to measure the performance of the power management circuit is shown in Figure 8. The electrical components used for the circuit are listed in Table I. The load of the circuit is a wireless sensor board (which is a TI MSP430 MCU evaluation board configured for structural health monitoring) with supply voltage of 3.5 V. Figure 9 shows the measured current going into the MCU (top), the charging profile of the supercapacitor and the supply voltage of the MCU (bottom) under $0.5g$ (rms) base acceleration . The MCU is in the active mode for 0.4 second and returns to sleep mode for 13 seconds. It consumes 3.5 mA in the active mode and 175 µA during the sleep mode. The top graph in Figure 9 shows sharp differences in current during the two different modes. The bottom graphs indicate that the circuit is in the MPPT mode for the first 210 seconds. The output voltages of the supercapacitor and the MCU rise continuously during the mode.

When the two voltages reach around 3.5 V, the circuit switches to the voltage regulation mode and tries to maintain the voltage.

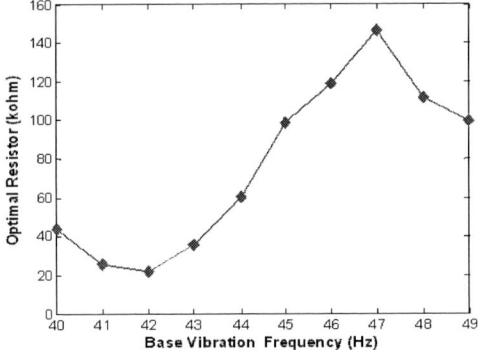

Figure 7. Optimal resistance versus excitation frequency.

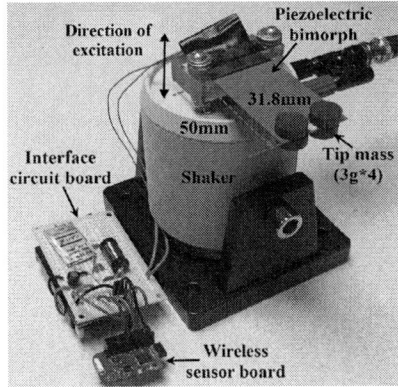

Figure 8. The experiment setup

Table I. Converter parameters

Component	Part Number	Notes
Rectifier	BAS3007	V_F =0.32V@10mA, 0.4V@ 100 mA.
MOSFET	2N7002	R_{dson} =5.3 Ω@V_{GS} = 4.5V;C_{oss}=40 pF.
Schottky Diode	PMEG4005	V_F=0.27mV@10mA,0.35V@100mA.
Inductor L	102K1R3	L = 10 mH; DCR = 9.1 Ω.
Super capacitor C	GW209F	C = 0.12 F; ESR = 70 mΩ.
Rsense	--	10 Ω

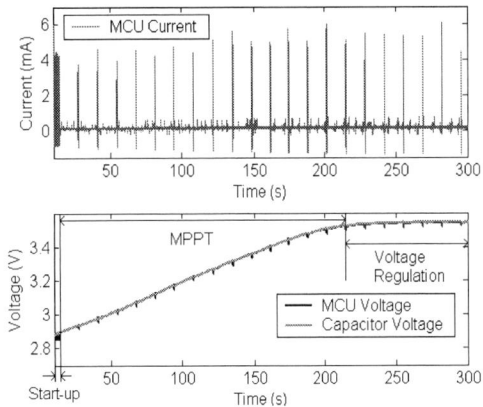

Figure 9. The measured MCU current and the output voltages

Figure 10 compares the maximal available power from the piezoelectric cantilever generator and the power extracted by our power management circuit. The maximum available power is measured as the power delivered to the optimal resistive load (of a given vibration frequency) connected directly to the piezoelectric generator, and the extracted power of our power management circuit is the power delivered to the supercapacitor while charging. The maximum available power of the piezoelectric generator under the acceleration of $0.5g$ is about 6.6 mW at the resonant frequency of 44 Hz. Our circuit harvests 3.5 mW, whose efficiency is about 53%. Some competing circuits report over 60% of efficiency in [6] under the acceleration of $0.5g$ (rms), and below 40% in [5] for the same amount of input power. However, direct comparison of those designs with ours should be judicious due to different environments. For example, [6] does not have a feedback controller, which attributes a substantial power consumption as described below. The circuit in [5] can manage input power as high as 50 mW, which results in a relatively low efficiency for low input power.

A breakdown of power losses at the resonant frequency of 44 Hz is shown in Figure 11. The power dissipation of the controller (specifically the MCU) and the diode are two major sources for the power loss and account for 64% of the total power loss. The next two sources of power losses are attributed to the inductor and the rectifier, especially their conduction losses.

VI. CONCLUSION

A self-powered power management circuit based on a DCM buck-boost converter is presented. An ultra-low power MCU is adopted for both the MPPT and the output voltage regulation. Experimental results indicate that the proposed system can harvest up to 3.5 mW power under $0.5g$ (rms) base acceleration for a piezoelectric cantilever and achieves 53% of the efficiency at the frequency. The sources of power losses are analyzed, and a breakdown of measured power losses is presented. Future works include further improvement of power efficiency and development of the efficient power management system in a monolithic IC.

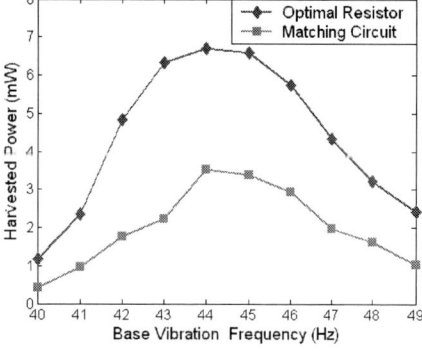

Figure 10. Average power harvested by optimal resistors and the proposed circuit.

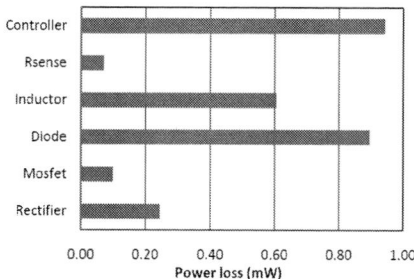

Figure 11. Power loss breakdown.

ACKNOWLEDGMENT

The authors gratefully acknowledge the support of the U.S. Department of Commerce, National Institute of Standards and Technology, Technology Innovation Program, Cooperative Agreement Number 70NANB9H9007.

REFERENCES

[1] N. S. Hudak and G. G. Amatucci, "Small-scale energy harvesting through thermoelectric, vibration, and radiofrequency power conversion," Journal of Applied Physics, vol. 103, pp. 101301-1, 2008.

[2] C. O. Mathuna, T. O'Donnell, R. V. Martinez-Catala, J. Rohan, and B. O'Flynn, "Energy scavenging for long-term deployable wireless sensor networks," Talanta, vol. 75, pp. 613-623, 2008.

[3] S. Roundy, E. S. Leland, J. Baker, E. Carleton, E. Reilly, E. Lai, B. Otis, J. M. Rabaey, P. K. Wright, and V. Sundararajan, "Improving power output for vibration-based energy scavengers," IEEE Pervasive Computing, vol. 4, pp. 28-36, 2005.

[4] G. K. Ottman, H. F. Hofmann, A. C. Bhatt, and G. A. Lesieutre, "Adaptive piezoelectric energy harvesting circuit for wireless remote power supply," IEEE Transactions on Power Electronics, vol. 17, pp. 669-676, 2002.

[5] G. K. Ottman, H. F. Hofmann, and G. A. Lesieutre, "Optimized piezoelectric energy harvesting circuit using step-down converter in discontinuous conduction mode," *IEEE Transactions on Power Electronics,* vol. 18, pp. 696-703, 2003.

[6] E. Lefeuvre, D. Audigier, C. Richard, and D. Guyomar, "Buck-boost converter for sensorless power optimization of piezoelectric energy harvester," IEEE Transactions on Power Electronics, vol. 22, pp. 2018-25, 2007.

[7] T. Zaitsu, O. Ohnishi, T. Inoue, M. Shoyama, T. Ninomiya, F. C. Lee, and G. C. Hua, "Piezoelectric transformer operating in thickness extensional vibration and its application to switching converter," in Power Electronics Specialists Conference, PESC '94 Record., 25th Annual IEEE, 1994, pp. 585-589 vol.1.

[8] P. Joung-hu, C. Sungjin, L. Sangmin, and B. H. Cho, "Gain-adjustment Technique for Resonant Power Converters with Piezoelectric Transformer," in Power Electronics Specialists Conference, 2007. PESC 2007. IEEE, 2007, pp. 2549-2553.

[9] F. Dianbo, L. Ya, F. C. Lee, and X. Ming, "A Novel Driving Scheme for Synchronous Rectifiers in LLC Resonant Converters," Power Electronics, IEEE Transactions on, vol. 24, pp. 1321-1329, 2009.

[10] C. Richard, D. Guyomar, D. Audigier, and G. Ching, "Semi-passive damping using continuous switching of a piezoelectric device," in Proceedings of SPIE - The International Society for Optical Engineering, 1999, pp. 104-111.

[11] D. Guyomar, A. Badel, E. Lefeuvre, and C. Richard, "Toward energy harvesting using active materials and conversion improvement by nonlinear processing," IEEE Transactions on Ultrasonics, Ferroelectrics and Frequency Control , vol. 52, pp. 584-595, 2005.

[12] S. Xu, K. D. T. Ngo, T. Nishida, C. Gyo-Bum, and A. Sharma, "Converter and controller for micro-power energy harvesting," in IEEE Applied Power Electronics Conference and Exposition, 2005, pp. 226-230 Vol. 1.

[13] A. Badel, A. Benayad, E. Lefeuvre, L. Lebrun, C. Richard, and D. Guyomar, "Single crystals and nonlinear process for outstanding vibration-powered electrical generators," Ultrasonics, Ferroelectrics and Frequency Control, IEEE Transactions on, vol. 53, pp. 673-684, 2006.

[14] S. Roundy and P. K. Wright, "A piezoelectric vibration based generator for wireless electronics," Smart Materials and Structures, vol. 13, pp. 1131-1142, 2004.

[15] N. G. Elvin and A. A. Elvin, "A general equivalent circuit model for piezoelectric generators," Journal of Intelligent Material Systems and Structures, vol. 20, pp. 3-9, 2009.

[16] W. Erickson and D. Maksimovic, "Fundamentals of Power Electronics," Norwell, MA: Kluwer, 2001.

[17] O. Wasynczuk, "Dynamic behavior of a class of photovoltaic power systems," IEEE transactions on power apparatus and systems, vol. PAS-102, pp. 3031-3037, 1983.

[18] N. Femia, G. Petrone, G. Spagnuolo, and M. Vitelli, "Optimization of perturb and observe maximum power point tracking method," IEEE Transactions on Power Electronics, vol. 20, pp. 963-973, 2005.

A Maximum Power Point Tracker Implementation for Photovoltaic Cells Using Dynamic Optimal Voltage Tracking

Emil Jimenez-Brea[*], Andres Salazar-Llinas[†], Eduardo Ortiz-Rivera[‡] and Jesus Gonzalez-Llorente[§]

Department of Electrical and Computer Engineering
University of Puerto Rico, Mayaguez, Puerto Rico 00681-9000
Email: emil.jimenez@ece.uprm.edu[*],andres.salazar@ece.uprm.edu[†],eduardo.ortiz@ece.uprm.edu[‡], jesus.gonzalez@ece.uprm.edu[§]

Abstract— **A maximum power point tracker (MPPT) for photovoltaic (PV) cells, PV modules (PVM) and PV arrays is presented using a dynamic optimal voltage estimator to estimate the voltage at which a PV cell generates its maximum power, and, using a DC-DC converter, to force the PV cell to reach and operate at voltage in a finite time and to stay there for all future time. The optimal voltage estimator reads the temperature at the surface of the PV array and the solar irradiance that reaches it surface to estimate the maximum power voltage point. A sliding mode controller, implemented in a low cost microcontroller, uses the estimated optimal voltage to generate a control signal which forces the PV cell to track and operate at this estimated optimal voltage for all future time. The procedures for the design, simulation, implementation and results are presented in this paper.**

I. INTRODUCTION

The use of renewable energy systems as an alternative way to produce electricity has been increasing during the past years [1]. The need of a cleaner, more efficient and cheaper method for generating electric power is helping this growth. Among all the renewable energy systems, the use of solar cells is one of the most common system. Photovoltaic is the technology that uses solar cells or an array of them to convert solar light directly into electricity. The power produced by the PV array depends directly of factors that are not controlled by the human being as the cell's temperature and solar irradiance.

PV arrays have only one operating point where the product of the voltage and the current results in a maximum power point(MPP). PV arrays, if connected directly to the load, will only operate at that maximum power point when the load is equal to the division of the values of voltage and current that results in a maximum power. Fig. 1 shows that for an specific PV array power curve, operating at standard conditions, the maximum power point will be reached only when a load equal to 3Ω is connected to the PV array ; otherwise, due to load mismatching, the PV array will operate at a suboptimal power point. The maximum power operation point depends on variable factors such as the cell temperature and solar irradiance which change the value of the load at which the PV array generates its maximum power (Fig. 2). Due to this, a device capable of tracking the maximum power operating point and force the PV array to operate at this point is required. A

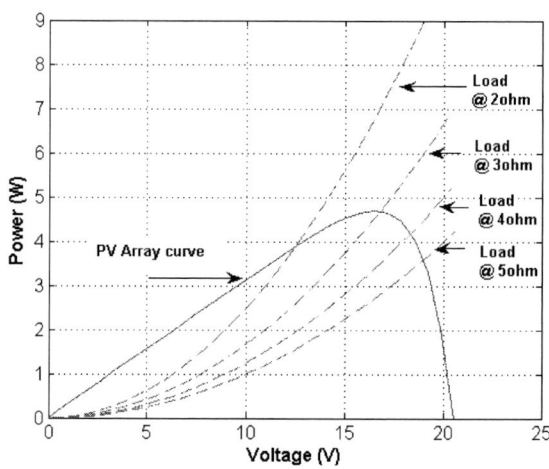

Fig. 1. PV Array Power Curve and different resistive load power curves in a direct matching.

maximum power point tracker is a device capable of detecting the maximum power point and forcing the system to reach and operate at this point.

In this paper a maximum power point tracker, using a sliding mode control algorithm, is presented. With this algorithm the system is always forced to track and reach the optimal voltage point, in a finite time, and stay there for all future time in order to optimize the power generation from any PV array. As previous results, simulations of this method can be found in [2], while the validation, for the PV equations used in this paper, is studied in [3].

II. EXISTING METHODS

A very common method used is the Perturb and Observe algorithm [4] [5]. Perturb and Observe algorithm measure the converter's output power in order to modify the input voltage by modifying the converter's duty cycle. Other common method is the hill-climbing method [6] [7]. This method is based on a trial and error algorithm where the voltage is increased until the voltage where the PV array exhibits a maximum power is reached. Other MPPT algorithms sample

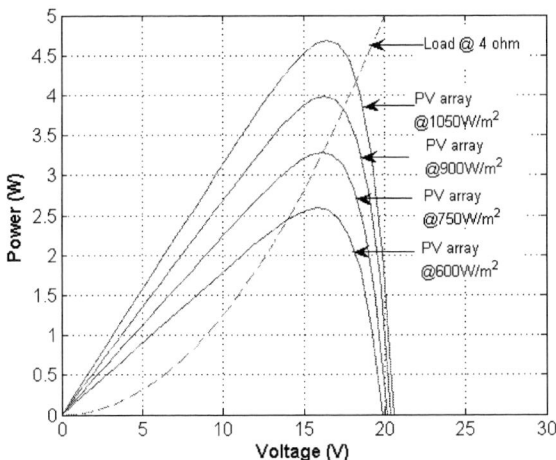

Fig. 2. PV Array Power Curve for different Irradiance Levels.

the open circuit voltage and operate the PV array at a fixed percent of this voltage. The incremental conductance algorithm is another method to track the MPP [8] [9] [10]. Other methods that have been used to obtain the MPP are parameters estimations [11], neural networks [12] and Linear Reoriented Coordinates Method (LRCM) [13], an implementation of this method can be found in "in press" [14].

III. PROPOSED METHOD

Fig. 3 shows the proposed scheme for the MPPT. This system uses a PV array composed of s in series and p in parallel PV modules. It is connected to a converter in order to decrease the desired voltage. After that, is connected directly to the load. Measurement of the PV array voltage, Irradiance and Temperature on the PV array surface are taken in order to estimate the optimal voltage for the maximum power, and then a non linear MPPT algorithm takes this value to produce the signal for driving the switching element of the DC/DC converter.

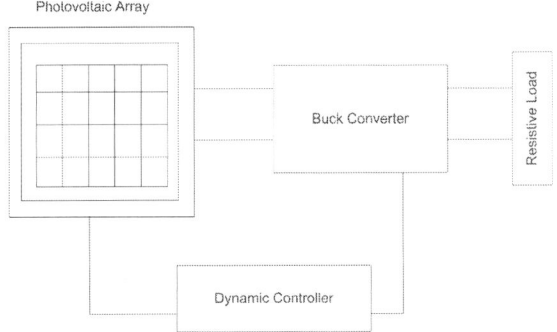

Fig. 3. General Scheme for the proposed method

A. Photovoltaic Cell Equations

Equations (1)-(4) presented in this work are based in [13]. These equations describe the behavior of the curve for any PV array under different values of temperature and solar irradiance and using values that can be obtained directly from any manufacturer's datasheet. I_x and V_x represent the short circuit current and open circuit voltage at a given temperature and solar irradiance. V is the PV array output voltage, T is the PV array temperature, T_N is the standard conditions temperature, E_i is the effective solar irradiance at the PV array, E_{in} is the standard condition solar irradiance, TCV is the open circuit voltage temperature coefficient and T_{Ci} is the short circuit current temperature coefficient. V_{max} is the open-circuit voltage at 25^oC and more than $1200W/m^2$. V_{min} is the open-circuit voltage at 25^oC and less than $1000W/m^2$. b is the characteristic constant and is unique to each PVM, s and p are numbers of in series and in parallels modules with the same electrical characteristics.

$$I(V) = \frac{pI_x}{1 - exp(\frac{-1}{b})} \left[1 - exp \left(\frac{V}{bsV_x} - \frac{1}{b} \right) \right] \quad (1)$$

$$I_x = p\frac{E_i}{E_{i_N}} \left[I_{sc} + TC_i(T - T_N) \right] \quad (2)$$

$$V_x = s\frac{E_i}{E_{i_N}} TCV\,(T - T_N) + sV_{max} -$$

$$s\,(V_{max} - Vmin)\,exp \left(\frac{E_i}{E_{i_N}} \ln \left(\frac{V_{max} - V_{op}}{V_{max} - Vmin} \right) \right) \quad (3)$$

$$while\ |b_{n+1} - b_n| > \epsilon$$

$$b_{n+1} = \frac{V_{op} - V_{oc}}{V_{oc} \ln \left[1 - \frac{I_{op}}{I_{sc}} \left(1 - exp \left(\frac{1}{b_n} \right) \right) \right]} \quad (4)$$

B. Optimal Voltage Equation

It can be seen in (1) that there is only one value of current for each value of voltage. As said before, the maximum power point correspond to a single voltage, V_{op}, or, since for each value of voltage there is a value of current, a single current, I_{op}. In this case we just have to find one of them. By multiplying (1) times V we obtain a power equation, presented in (5).

$$P(V) = V \cdot I(V) = \frac{V \cdot I_x}{1 - exp \left(\frac{-1}{b} \right)} \left[1 - exp \left(\frac{V}{bV_x} - \frac{1}{b} \right) \right] \quad (5)$$

By differentiating (5) with respect to V, equaling it to zero and solving it for V, (6) is obtained. Using this equation the optimal voltage for any solar cell or panel can be estimated.

$$Vop = b \cdot V_x \left(lambertw \left(2.7138exp^{\frac{1}{b}} \right) - 1 \right) \quad (6)$$

C. Sliding Mode Control

A sliding mode controller is a variable structure control where the dynamics of a non linear system is altered via the application of a high frequency switching control and the trajectories of the system are forced to reach a sliding manifold or surface, where it exhibit desirable features, in finite time and to stay on the manifold for all future time. Equation (7) presents a sliding surface that will accomplish the objective as a maximum power point tracker.

$$\sigma = V - V_{op} \qquad (7)$$

The sliding mode will be controlling the duty cycle of a switching device. So the switching device will have two operation state:

$$On, \ 1 \quad V - V_{op} > 0$$
$$Off, \ 0 \quad V - V_{op} < 0$$

A control law that guarantees us that our controller will behave in that way is given by (8).

$$u = \frac{1}{2} + \frac{1}{2} sign(V - V_{op}) \qquad (8)$$

IV. IMPLEMENTATION

The MPPT method presented in this article was easily implemented using a low cost microcontroller from Microchip family PIC 16F687X running at a frequency of 20 MHz. Lectures from the PVM surface temperature(T), PVM solar irradiance(E_i) and the PVM voltage(V) were done by ADC ports of the microcontroller. The Implementation of the mathematical functions such as division multiplication and exponential functions was made using a C language compiler, PIC C from $HI - TECH\ C$.

The schematics of the implemented circuit is shown in Fig. 4. A flow chart of the programming sequence for the microcontroller is presented in Fig. 5.

V. RESULTS

Two tests were done to verify the effectiveness of the MPPT implementation. Both of them were developed under controlled irradiance levels; showing the dynamical behavior of the system. Both Experimentations were developed using a BP Solar PVM, model: SX305M.

The first experiment consider load variations from 10Ω to 100Ω with a fixed irradiance level of $409W/m^2$ and an average temperature of 48^oC. Fig. 7 and Fig. 8 show the dynamic behavior in the PVM voltage and PVM Power, while Fig. 9 shows the power and voltage delivered to the load

In the second experiment a step change in the irradiance level, from $409W/m^2$ to $102W/m^2$, was introduced to the system when a resistive load of 100Ω is connected to it. Fig. 11 to Fig. 13 show the dynamical behavior in the PVM power, PVM voltage, PVM surface temperature, and Fig. 12 shows the power delivered to the Load.

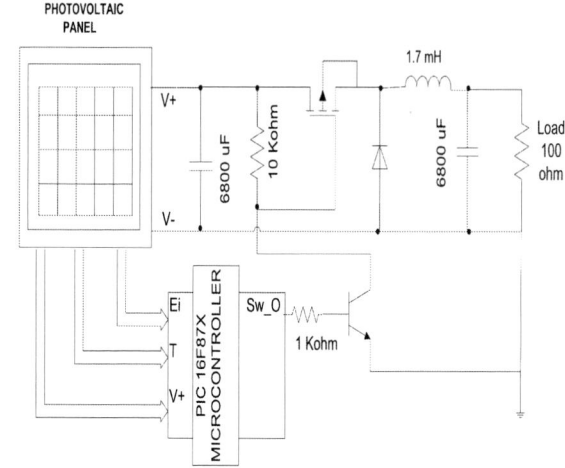

Fig. 4.   Circuit Schematic

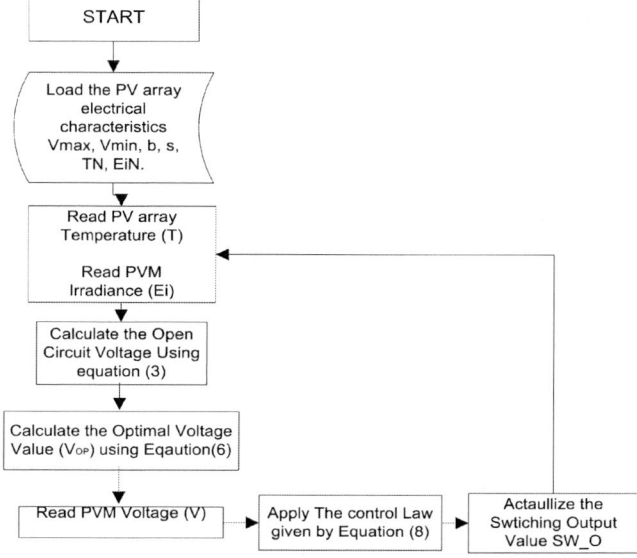

Fig. 5.   Flow Chart For the Microconntroller Application

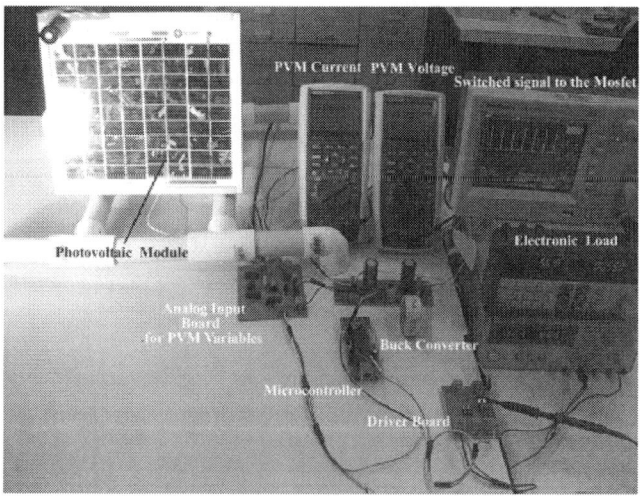

Fig. 6.   Experimental Setup

978-1-4244-4782-4/10 $26.00 © 2010 IEEE

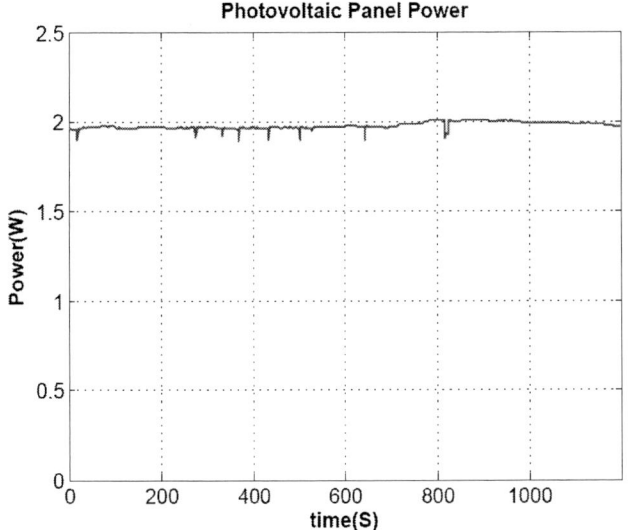

Fig. 7. PVM Power for different load changes

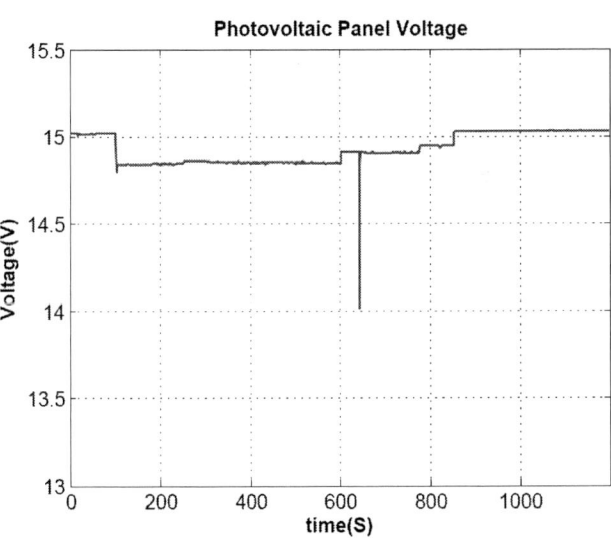

Fig. 8. PVM Voltage for diferent load changes

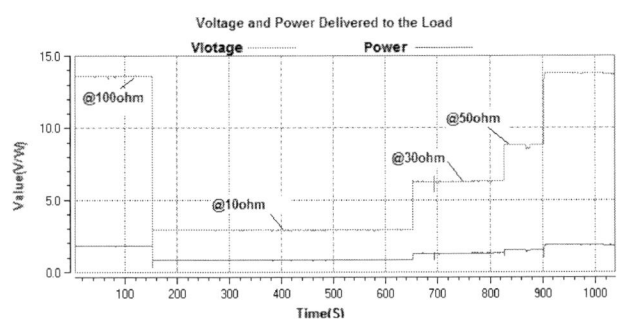

Fig. 9. Output Power for different Load changes

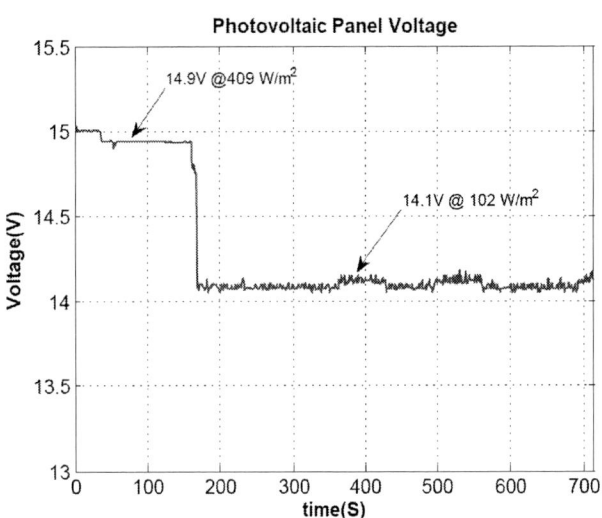

Fig. 10. PVM Voltage for a step variation in the irradiance level

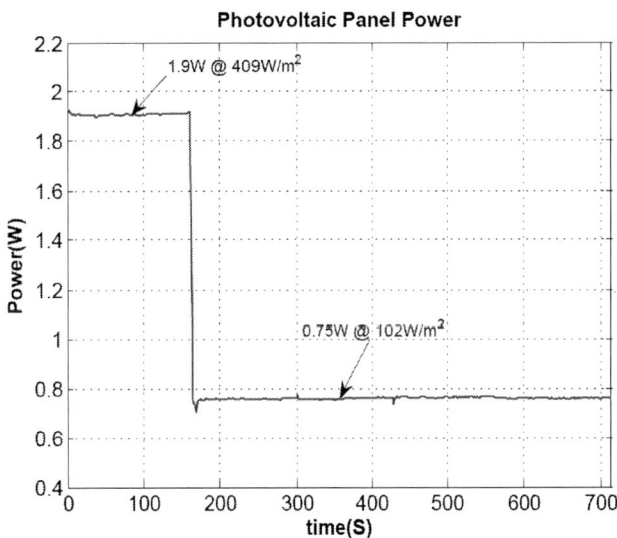

Fig. 11. PVM Power for a step variation in the irradiance level

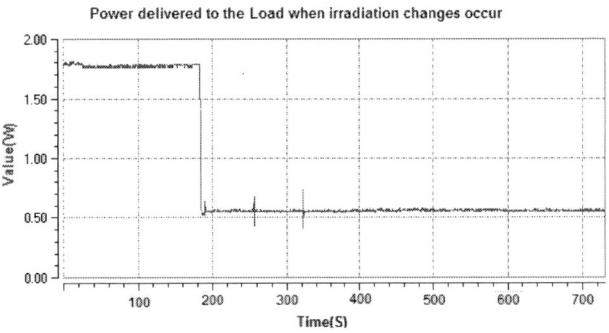

Fig. 12. Load Power for a step variation in the irradiance level

978-1-4244-4782-4/10 $26.00 © 2010 IEEE

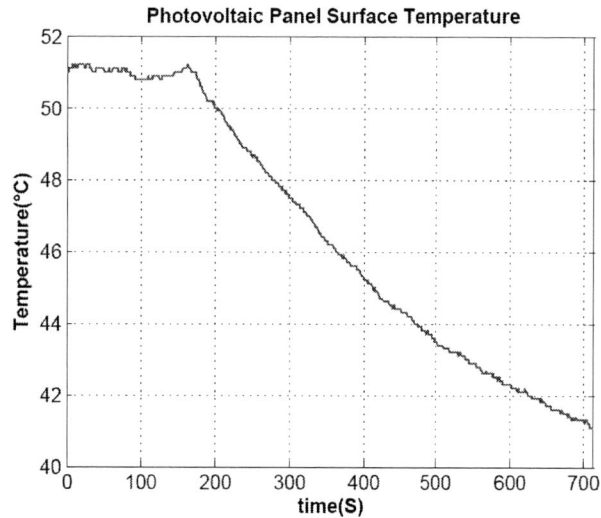

Fig. 13. PVM surface temperature behavior for a step variation in the irradiance level

TABLE I

COMPARISON BETWEEN OPTIMAL VOLTAGE VALUES ESTIMATED (EST), REAL, OBTAINED ON CIRCUIT(OC)

E_i	T	V_{op}(Est)	V_{op}(Real)	V_{op}(OC)
102 W/m^2	38.8 C	14.4202 V	14.85 V	14.2 V
409 W/m^2	49.5 C	14.953 V	15.3 V	15.01 V

TABLE II

COMPARISON BETWEEN MAXIMUM POWER POINTS ESTIMATED REAL AND OBTAINED ON CIRCUIT(OC)

Ei	T	P_{max}(Real)	P_{max}(OC)	Error
102 W/m^2	38.8 C	0.775 W	0.76 W	3.22 %
409 W/m^2	49.5 C	1.89 W	1.86 W	1.58 %

Table I and Table II present the average optimal values for the voltage and power in the PVM at steady state conditions. In both tables, the "real" measures refer to the values for which the module is driven to the maximum power with a direct load matching, while the "on circuit" measures(OC) are the values obtained with the implemented system. $(V_{op}(est))$(Table I) is the theoretical simulated optimal voltage for the given conditions.

VI. CONCLUSION

A novel maximum power point tracking method, capable of estimate the optimal voltage at which the solar cell produces its maximum power, has been presented. The equation for estimating the optimal voltage showed very low and acceptable errors percentage at estimating the optimal voltage and current for different solar cells. The implementation in a low cost microcontroller and its effectiveness for load variations and irradiance variations has been shown. The sliding mode controller was capable of tracking the optimal voltage and

forced the PVM to reach that voltage point in a finite time. It also forced the solar module to keep operating at that voltage for all future time. The sliding mode controller was capable of extracting the maximum power available from the solar module under different temperature, irradiance and load conditions.

ACKNOWLEDGMENT

The authors gratefully acknowledge the contributions of all the members that belong to the Mathematical Modeling and Control of Renewable Energies for Advance Technology and Education ($M_{inds}{}^2$CREATE) Research Team at UPRM.

REFERENCES

[1] J. Lyons and V. Vlatkovic, "Power electronics and alternative energy generation," in *Power Electronics Specialists Conference, 2004. PESC 04. 2004 IEEE 35th Annual*, vol. 1, June 2004, pp. 16–21 Vol.1.
[2] E. Jimenez, E. Ortiz-Rivera, and O. Gil-Arias, "A dynamic maximum power point tracker using sliding mode control," in *Control and Modeling for Power Electronics, 2008. COMPEL 2008. 11th Workshop on*, Aug. 2008, pp. 1–5.
[3] O. Gil-Arias and E. Ortiz-Rivera, "A general purpose tool for simulating the behavior of pv solar cells, modules and arrays," in *Control and Modeling for Power Electronics, 2008. COMPEL 2008. 11th Workshop on*, Aug. 2008, pp. 1–5.
[4] N. Femia, G. Petrone, G. Spagnuolo, and M. Vitelli, "Perturb and observe mppt technique robustness improved," in *Industrial Electronics, 2004 IEEE Inter Symp. on*, vol. 2, May 2004, pp. 845–850 vol. 2.
[5] ——, "Optimization of perturb and observe maximum power point tracking method," *Power Electronics, IEEE Transactions on*, vol. 20, no. 4, pp. 963–973, July 2005.
[6] W. Xiao and W. Dunford, "A modified adaptive hill climbing mppt method for photovoltaic power systems," in *Power Electronics Specialists Conference, 2004. PESC 04. 2004 IEEE 35th Annual*, vol. 3, June 2004, pp. 1957–1963 Vol.3.
[7] H. Al-Atrash, I. Batarseh, and K. Rustom, "Statistical modeling of dsp-based hill-climbing mppt algorithms in noisy environments," in *Applied Power Electronics Conference and Exposition, 2005. APEC 2005. Twentieth Annual IEEE*, vol. 3, Mar 2005, pp. 1773–1777 Vol. 3.
[8] Y. Yusof, S. H. Sayuti, M. Latif, and Z. C. Wanik, "Modeling and simulation of maximum power point tracker for photovoltaic system," in *National Power & Energy Conference (PECon) 2004*, 2004, pp. 88–93.
[9] J. H. Lee, H. Bae, and B. H. Cho, "Advanced incremental conductance mppt algorithm with a variable step size," in *Electronics and Motion Control Conference*, vol. 12 International, Aug 2006, p. 603 607.
[10] W. Libo, Z. Zhengming, and L. Jianzheng, "A single-stage three-phase grid-connected photovoltaic system with modified mppt method and reactive power compensation," *Energy Conversion, IEEE Transactions on*, vol. 22, no. 4, pp. 881–886, Dec. 2007.
[11] I.-S. Kim, M.-B. Kim, and M.-J. Youn, "New maximum power point tracker using sliding-mode observer for estimation of solar array current in the grid-connected photovoltaic system," *Industrial Electronics, IEEE Transactions on*, vol. 53, no. 4, pp. 1027–1035, June 2006.
[12] A. de Medeiros Torres, F. Antunes, and F. dos Reis, "An artificial neural network-based real time maximum power tracking controller for connecting a pv system to the grid," in *Industrial Electronics Society, 1998. IECON '98. Proceedings of the 24th Annual Conference of the IEEE*, vol. 1, Aug-4 Sep 1998, pp. 554–558 vol.1.
[13] E. Ortiz-Rivera and F. Peng, "Analytical model for a photovoltaic module using the electrical characteristics provided by the manufacturer data sheet," in *Power Electronics Specialists Conference, 2005. PESC '05. IEEE 36th*, June 2005, pp. 2087–2091.
[14] J. Gonzalez-Llorente, E. Ortiz-Rivera, A. Salazar-Llinas, and E. Jimenez-Brea, "Analyzing the optimal matching of dc motors to photovoltaic modules via dc-dc converters." in *Applied Power Electronics Conference and Exposition, 2010. APEC 2010. Twenty-Fifth Annual IEEE*, 2010, in press.

Development of the Novel Control Algorithm for the Small Proton Exchange Membrane Fuel Cell Stack without External Humidification

Tae-Hoon Kim, Sang-Hyun Kim, Wook Kim, Jong-Hak Lee and Woojin Choi
Department of Electrical Engineering
Soongsil University
Seoul, Republic of Korea
cwj777@ssu.ac.kr

Abstract— Small PEM (proton exchange membrane) fuel cell systems do not require humidification and have great commercialization possibilities. However, methods for controlling small PEM fuel cell stacks have not been clearly established. In this paper, a control method for small PEM fuel cell systems using a dual closed loop with a static feed-forward structure is defined and realized using a microcontroller. The fundamental elements that need to be controlled in fuel cell systems include the supply of air and hydrogen, water management inside the stack, and heat management of the stack. For small PEM fuel cell stacks operated without a separate humidifier, fans are essential for air supply, heat management, and water management of the stack. A purge valve discharges surplus water from the stack. The proposed method controls the fan using a dual closed loop with a static feed-forward structure, thereby improving system efficiency and operation stability. The validity of the proposed method is confirmed by experiments using a 150-W PEM fuel cell stack. We expect the proposed algorithm to be widely used for controlling small PEM fuel cell stacks.

I. INTRODUCTION

Compared to other types of fuel cells, PEM (proton exchange membrane) fuel cells operate at a relatively low temperature with high efficiency and have a smaller volume. Many studies on PEM fuel cells have been conducted to investigate their applications in portable electronics, residential power generation, and fuel cell vehicles [1]. PEM fuel cell systems are composed of a fuel cell stack that generates electricity by reacting with the fuel gases and the BOP (balance of plant) that assists the stack operation by controlling the fuel supply, pressure, temperature, and humidity.

In small fuel cell systems, the role of the BOP is to consume minimal power while guaranteeing stable operation of the stack. Its main functions include supply of necessary air and hydrogen, heat management in the stack, and water management for the MEA (membrane electrode assembly). Hydrogen is supplied directly to the stack from a compressed

gas cylinder through a pressure regulator. The hydrogen flow rate is decided by instantaneous pressure differences in the stack caused by load changes. Therefore, small PEM fuel systems do not require a separate device to control the hydrogen flow. The use of a pump or motor is not recommended because it increases the overall system size, price, and BOP power consumption. For the same reasons, it is not economically feasible to use an external humidifier for water management. The air needed for reaction and cooling is supplied by a fan without a separate humidifying device. Surplus water in the stack is discharged through a purge valve. Therefore, an economically feasible and efficient method of operation is to manage water and heat by appropriate control of the fan and the purge valve.

Recently, several studies on fuel cell power systems have been published [2–6] showing their potential as power sources for portable electronics. Urbani et al. [2] designed, manufactured, and tested an air-breathing 10-cell PEFC (polymer electrolyte fuel cell) stack; they were able to operate a 12-W DVD player for 100 h without sensitive losses in performance. Dundar et al. [3] built a 70-W fuel cell stack, presented a systematic design procedure for portable applications, and comprehensively analyzed the effects of variables such as reactant gas humidification, cooling, fuel supply, and temperature. However, in the proposed system, a humidifier is used at the input side of the cathode for water management in the membrane, and therefore, the system becomes complicated. Hebling et al. [4] developed several fuel cell systems with power ratings from 10-W to 180-W and implemented digital control algorithms in a digital microprocessor to achieve quick responses to changes in the operating conditions. The hydrogen closed loop achieved current density rates up to 550 mA/cm^2 and a fuel utilization coefficient greater than 0.99 after startup. Vega-Leal et al. [5] focused on the physical implementation of the digital controller for the fuel cell stack. The proposed system includes three control loops assigned to the control of oxygen flow, hydrogen flow, and temperature. Wilhelm et al. [6] developed

978-1-4244-4782-4/10 $26.00 © 2010 IEEE

a PEM-fuel-cell-powered mobile robot. The authors investigated the feasibility of using the robot in a university-level course and succeeded in demonstrating the potential of the technology.

Buchi et al. [7] presented the results of a study on water management for small PEM fuel cell stacks without external humidifiers. They developed a model for PEM fuel cell operation with internal humidification of the gases and investigated the range of operating conditions for a PEM fuel cell using dry H_2/air. It was found that MEA water management could be achieved by maintaining the relative humidity of the exit air between 80% and 100%. Therefore, it should be possible to control small self-humidified stacks by controlling the air flow for fuel and cooling. However, no detailed control algorithm was provided in the article.

Several other studies were conducted on the topic of water management in small self-humidified fuel cell stacks [8–10], and they provide excellent insights into water management. However, practical control algorithm for the stable and reliable operation of small PEM fuel cell systems based on comprehensive analysis of water management was not clearly established in the previously conducted studies.

In this study, we present a novel BOP control method for small self-humidified stacks. The proposed method successfully controls the overall system by calculating and supplying the amount of fuel that allows water and heat in the stack to achieve equilibrium. Using this method, high system efficiency and operating reliability can be achieved. The feasibility of the method was confirmed by experiments using a 150-W PEM fuel cell stack.

II. FUEL SUPPLY AND WATER MANAGEMENT

The air flow rate to be supplied for electrochemical reactions in the fuel cell stack $m_{reac,air}$ is calculated as shown in (1) considering the ratio of oxygen in the atmosphere.

$$\dot{m}_{reac,air} = \frac{P_e}{4 \times F \times V_{cell}} \times \frac{1}{0.21} \tag{1}$$

where P_e is the power of the fuel cell stack, V_{cell} is the voltage per cell of the stack, and F is the Faraday constant.

However, if the value obtained from (1) is used without modification, as oxygen is consumed by the reaction, the voltage loss from the reduced partial oxygen pressure becomes larger, and fuel supply may become difficult because of the water formed in the channel. Therefore, in practical applications, a stoichiometric variable λ_{air} that has a larger value than that defined by (1) should be introduced. According to previous studies, the optimal stoichiometry of air is approximately 2 to 2.4 [11]. In a small PEM fuel cell system with no external humidifier, the reliability of the water management in the MEA can be maintained by appropriate stoichiometric configuration according to the stack conditions.

For the MEA in the PEM fuel cell stack to have sufficient water, the relative humidity of the cathode exit air should be between 80% and 100%, and this can be achieved by selecting an appropriate stoichiometry for the air flow rate according to the operation temperature of the stack [7]. Thus, by considering the water coefficient of entrance air (atmosphere)

Ψ in (3) and substituting it into (2), a stoichiometry that satisfies the water condition in the stack can be calculated.

$$\lambda_{air} = \frac{0.210 \left(P_{sat} - 2P_{exit} \right)}{\Psi P_{exit} - \left(1 + \Psi \right) P_{sat}} \tag{2}$$

$$\Psi = \frac{P_{Went}}{P_{ent} - P_{Went}} \tag{3}$$

where P_{exit} is the total pressure of the exit air, P_{ent} is the total pressure of the entrance air, P_{Went} is the water vapor pressure at the entrance, and P_{sat} is the saturated vapor pressure according to exit-air temperature. In the temperature ($T_{air,exit}$) range from 273.15 K to 373.15 K, the following empirical equation can be used [12] :

$$P_{sat} = f\left(T_{air,exit} \right) = e^{aT^{-1} + b + cT + dT^2 + eT^3 + f \ln(T)} \tag{4}$$

where a = -5880.2206, b = 1.3914993, c = -0.048640239, d = 0.41764768 × 10^{-4}, e = -0.14452093 × 10^{-7}, and f = 6.5459673.

Therefore, taking the stoichiometry into account, the air flow rate necessary for the stack reaction is calculated as shown in (5).

$$\dot{m}_{reac,air} = \frac{P_e}{4 \times F \times V_{cell}} \times \frac{1}{0.21} \times \lambda_{air} \tag{5}$$

III. HEAT MANAGEMENT

Heat management of small PEM fuel cell stacks is mainly accomplished by supplying the fuel air $m_{reac,air}$ and cooling air $m_{cool,air}$ together through the fan. The heating rate that is created by the operation of the stack, Q_{st}, should be equal to the sum of the cooling rate from forced cooling by the fan (Q_{cool}), natural heat dissipation (radiation + convection) (ΣQ_{dis}), and cooling rate from the flow that participates in the electrochemical reaction (Q_{reac}). This relation is shown in (6).

$$Q_{st} = Q_{cool} + \sum Q_{dis} + Q_{reac} \tag{6}$$

Assuming that the water formed through electrochemical reactions in the fuel cell is completely discharged in the vapor state and that the reaction enthalpy is entirely converted to electric energy, the voltage of the fuel cell is 1.25 V (lower heating value basis). The heating rate of the stack can then be expressed as follows using the current of the fuel cell stack, I_{st}, and the number of cells, n_{cell} [13].

$$Q_{st} = \left(1.25 - V_{cell} \right) \times I_{st} \times n_{cell} \tag{7}$$

Natural heat dissipations from the stack are classified into convection Q_{conv} and radiation Q_{rad}, as described in (8) [14].

$$\sum Q_{dis} = Q_{conv} + Q_{rad} \tag{8}$$

The cooling rate from natural convection, Q_{conv}, can be calculated using (9).

$$Q_{conv} = h \times A_{st} \times (T_{st} - T_o) \qquad (9)$$

where h is the convection heat transfer coefficient and is a function of the Nusselt number, Nu_L. The value of h can be calculated using (10).

$$h = \frac{k}{L} \times Nu_L \qquad (10)$$

The Nusselt number changes with the shape of the heating element; it can be calculated by (11) in the case of vertical plates, by (12) in the case of the upper surface of horizontal plates, and by (13) in the case of the lower surface of the fuel cell stack [12, 14].

$$Nu_L = \left\{ 0.825 + \frac{0.387 Ra_L^{1/6}}{\left[1 + \left(\frac{0.5}{Pr} \right)^{9/16} \right]^{8/27}} \right\}^2 \qquad (11)$$

where $Ra_L = g\beta(T_{st}-T_o)L^3/v\alpha$, $Pr = v/\alpha$

$$Nu_L = 0.54 Ra_L^{1/4} \quad (10^4 \leq Ra_L \leq 10^7) \qquad (12)$$

$$Nu_L = 0.27 Ra_L^{1/4} \quad (10^5 \leq Ra_L \leq 10^{10}) \qquad (13)$$

where $L = A_{st}/Pm$

The cooling rate from natural radiation Q_{rad}, can be expressed as (14).

$$Q_{rad} = h_r \times A_{st} \times (T_{st} - T_o) \qquad (14)$$

where h_r is the radiation heat transfer coefficient calculated using (15).

$$h_r = \varepsilon \times \sigma \times (T_{st} + T_o) \times (T_{st}^2 + T_o^2) \qquad (15)$$

Air used as fuel also participates in the cooling of the stack, and the cooling rate is as follows:

$$Q_{reac} = \dot{m}_{reac,air} \times c_p \times \Delta T_{reac} \qquad (16)$$

The forced cooling rate from air supplied by the fan is as follows.

$$Q_{cool} = \dot{m}_{cool,air} \times c_p \times \Delta T_{cool} \qquad (17)$$

where c_p is the specific heat of air at constant pressure, ΔT_{reac} is the temperature difference in the reaction air, and ΔT_{cool} is the temperature difference in the cooling air.

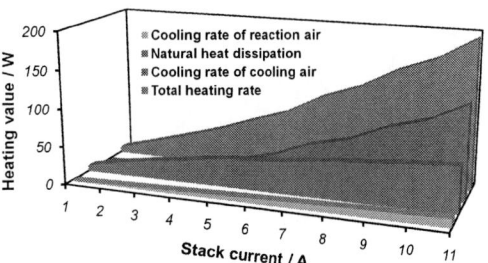

Figure 1. Various heating and cooling rates in 150-W PEMFC stack

Thus, the heating rate of the stack, cooling rate of the reaction air, and heat dissipation by natural convection and radiation can be calculated using (6) to (17) and plotted against the stack current, as shown in Fig. 1. The total air flow rate for the cooling of the 150-W PEM fuel cell stack supplied by the fan can be calculated using (18).

$$\dot{m}_{cool,air} = \frac{Q_{st} - \sum Q_{dis} - Q_{reac}}{c_p \times \Delta T_{cool}} \qquad (18)$$

IV. OPTIMAL SELECTION AND CONTROL FOR BOP

A. Optimal Selction of BOP

Since the fan consumes most of the BOP power in small fuel cell systems, the efficiency and stability of the overall system can be improved by optimal selection and control of the fan. Conditions for selecting the fan include efficiency, ability to be controlled by a microcontroller, ability to be operated without sparks, and reliability. In addition, the fan must have enough capacity to supply the required flow. To select an optimal fan, the air flow rate necessary for operation of the fuel cell stack must first be calculated. This can be calculated as the sum of the air flow rate necessary to carry out the reaction shown in (5) and the air flow rate necessary to achieve cooling according to (18). As shown in Fig. 2, the calculated air flow rate matched the experimental air flow rate during steady-state operation. However, the relationship between the required air flow rate and the stack current is not linear and online calculation of the value is not efficient because of the long calculation time required by the microcontroller. Moreover, it presumes the use of multiple temperature and humidity sensors. Therefore, in this study, we calculated the air flow rate required for each stack current, m_{ref}, in advance and carried out curve-fitting to convert the values into the polynomial expression shown in (19).

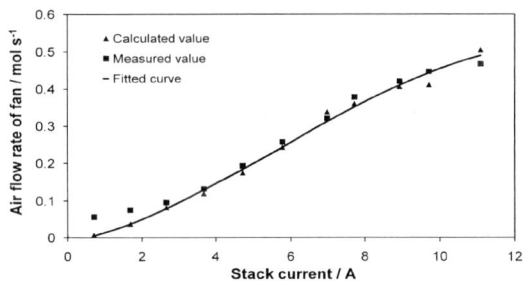

Figure 2. Calculated and measured air flow rate and fitted curve

$$\dot{m}_{ref} = -0.0004I_{st}^3 + 0.0062I_{st}^2 + 0.0207I_{st} - 0.0124 \quad (19)$$

Using the maximum air flow rate calculated above, we selected an optimal fan for the supply of air in the fuel cell system. First, after finding the system resistance curve that gives the static pressure according to air flow rate, we compared the curve with the performance curve of the fan provided by the manufacturer. The intersection of these two curves gives the operating point of the fan in the fuel cell system. Therefore, when the fan is operated at its rated value, the total air flow rate necessary for the fuel cell stack must be satisfied at the operating point. The power consumption must also be small.

Fig. 3 shows the system resistance curve of the 150-W PEM fuel cell system used in the experiment and the performance curves of three fans. From (19), the maximum air flow rate required for operation of the small PEM fuel cell stack in this study is approximately 0.68 m^3/min at a current of 10 A. Because two fans are attached to the left and right of the stack for uniform cooling, the rated flow of each fan should be approximately 0.17 m^3/min. Of the three fans shown in Fig. 3, fan 3 supplies approximately 0.17 m^3/min at the operating point but has a high power consumption of 2.52-W. Fan 2 exhibits the lowest power consumption of 1.68-W but cannot supply the necessary air flow rate. Thus, we selected fan 1, which can supply an air flow rate of 0.17 m^3/min and has a low power consumption of 2.40-W.

The purge valve for the discharge of surplus water should also have a low power requirement. It must withstand the flow of water discharged at the operating pressure. We used a small solenoid valve requiring 2.0-W of power and appropriate for a flow of 13 liter/min at 25 psig.

B. Control of BOP

Speed control of the fan is mandatory to reduce the average power consumption and to quickly supply the standard air flow rate. There are various methods for fan speed control. To perform PI (proportional-integral) control method, we used the internal PWM (pulse width modulation) control method because it has high accuracy, a relatively low noise level, no power loss from external constituents, and high efficiency. The normal transfer function of the angular velocity against the terminal voltage of a BLDC (brushless direct current) motor is expressed in (20) [15]. This transfer function of the fan must be known to design the speed controller of the fan.

$$G(s) = \frac{\omega_m(s)}{V(s)} = \frac{K\omega_n^2}{s^2 + 2\zeta\omega_n s + \omega_n^2} \quad (20)$$

We estimated the transfer function using LabVIEW 8.6 with System Identification Toolkit 4.0 after performing an experiment on fan step response. Fig. 4 shows the response characteristics of the fan against step voltage input, and the response from the estimated model corresponds to the experimental result. As shown in Fig. 4, the fan showed an overdamped characteristic with no overshoot. Table 1 summarizes the extracted fan parameters.

The step response time of the system is approximately 6 s

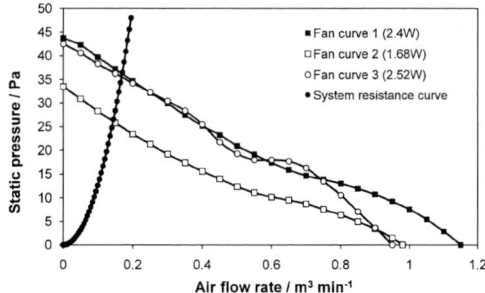

Figure 3. Performance curves of fans and system resistance curve

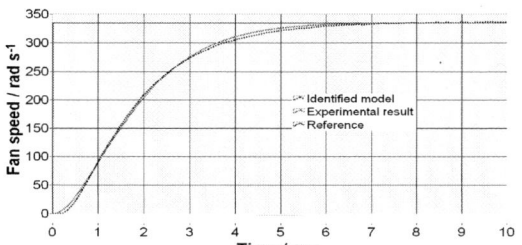

Figure 4. Step response of fan

TABLE I. EXTRACTED FAN PARAMETERS

Fan parameters	Value
Gain, K	67.3
Damping ratio, ζ	1.12
Natural frequency, ω_n	1.12

and such a slow transient response delays the supply of necessary air. To improve the transient response of the BLDC fan, a PI controller defined in (21) [16] was used for the inner control loop, as shown in Fig. 5. The controller was designed to meet a suitable phase margin (45° in this case) at the gain-crossover frequency to guarantee system stability. Fig. 6 shows bode plots of the fan with open-loop and closed-loop control. The proportional and integral gains of the PI controller for the inner control loop are listed in Table 2.

$$C(s) = K_p^{in} + \frac{K_i^{in}}{s} \quad (21)$$

Thus, the closed-loop transfer function of the fan can be obtained by multiplying (20) and (21) in the frequency domain; the transfer function must be converted to a discrete-time linear transfer function for actual implementation in the microcontroller; this is done by using bilinear transformation as shown in (22) [16].

$$H_{CL}(z) = \frac{G(z)C(z)}{1 + G(z)C(z)} = \frac{a_0 z^3 + a_1 z^2 + a_2 z + a_3}{b_0 z^3 + b_1 z^2 + b_2 z + b_3} \quad (22)$$

where a_0 = 0.00035, a_1 = 0.00036, a_2 = -0.00036, a_3 = -0.00035, b_0 = 1, b_1 = -2.97004, b_2 = 2.94064, b_3 = -0.97095.

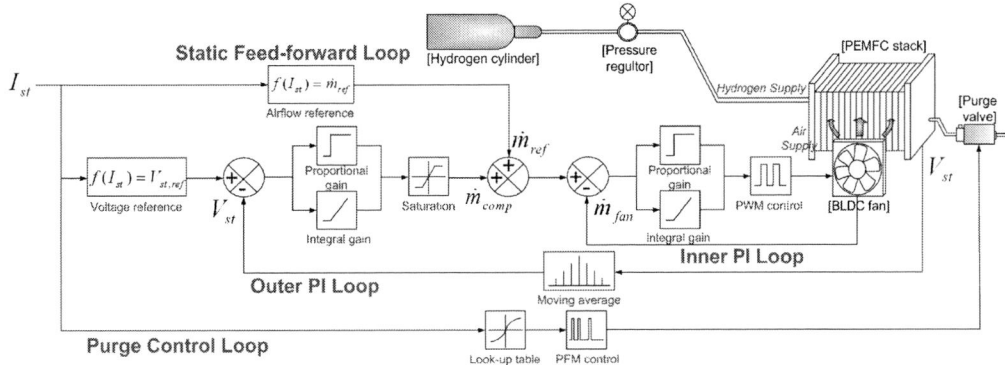

Figure 5. Block diagram of proposed control algorithm for small PEM fuel cell system

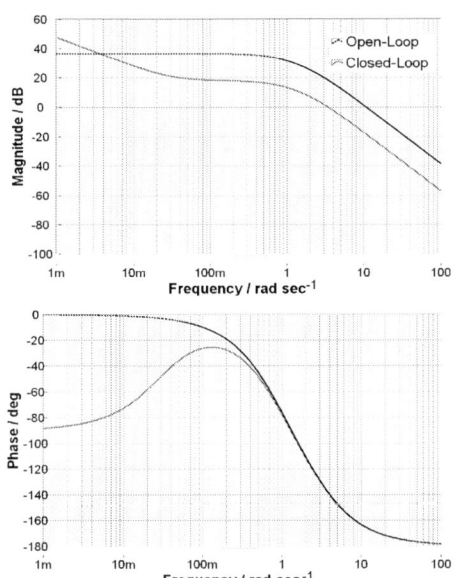

Figure 6. Open-loop and closed-loop bode plots of BLDC fan

TABLE II. GAIN OF INNER PI CONTROLLER

Parameters	Value
Proportional gain, K^{in}_p	0.12
Integral gain, K^{in}_i	0.0032

TABLE III. GAIN OF OUTER PI CONTROLLER

Parameters	Value
Proportional gain, K^{out}_p	0.024
Integral gain, K^{out}_i	2.0e-05

The air flow rate necessary for stack reaction and cooling with consideration of the MEA water management can clearly be calculated using (19) from the stack current. Also, according to the transient response characteristic of the fuel cell stack, the transient response of the voltage has a time delay compared to the response speed of the current when rapid load changes occur. Therefore, the air flow rate calculated using (19) from the stack current must be static feed-forward controlled, as shown in Fig. 5, for a quicker system response. In addition, the outer control loop that controls the voltage feedback of the fuel cell stack is configured to be 10 times slower than the inner control loop that controls the fan speed. A moving-average method is applied to the feedback voltage to form a loop that is insensitive to noise or instantaneous stack voltage changes.

The outer control loop calculates the difference between the feedback voltage and the stack voltage in the steady state to create a correction term. This term is added to the reference air flow rate generated by the static feed-forward loop. Accordingly, the reference air flow rate calculated with the stack current by the static feed-forward control loop is supplied quickly to the stack through the inner control loop to obtain the necessary fuel supply and achieve heat management. Feedback of the stack voltage allows operation of the fuel cell stack at the optimal operating point via the correction supplied by the outer loop. Table 3 lists the PI controller gain for the outer control loop.

However, in small PEM fuel cell systems, because water management in the MEA is also carried out by controlling the air flow rate using the fan, the flow added or reduced by the outer loop must not influence the water management. The condition for sufficient water in the MEA inside the PEM fuel cell stack is that the relative humidity of the cathode exit air must be kept between 80% and 100%, and this should be maintained even if the flow supplied to the stack by the outer loop is changed. Therefore, if there is a correction for the air flow rate from the outer control loop, the supplied air flow must be limited to prevent the relative humidity at the cathode exit from falling below 80%. Similarly, the minimum air flow rate is the value that prevents the relative humidity at the exit from exceeding 100%. In the outer loop control, limitation of the air flow rate is necessary to satisfy these conditions. Here, since changes in the temperature of the cathode exit air are much slower than changes in the voltage, the influence of temperature change does not need to be considered. Fig. 7 shows the relationships between cathode-exit air temperature and relative humidity with various stoichiometries.

On the cathode side of the fuel cell stack, water is formed as a byproduct of the electrochemical reaction, and the amount of water increases linearly as the stack current increases. In the aforementioned control algorithm, the air stoichiometry is selected to keep the relative humidity of the exit air between 80% and 100% to ensure sufficient water circulation in the

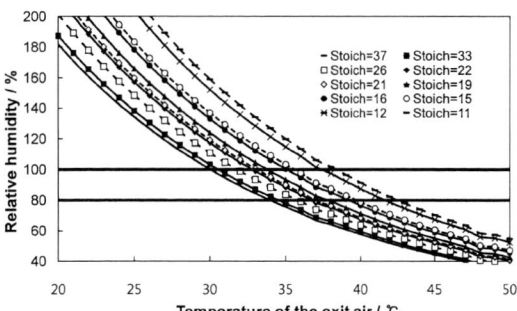

Figure 7. Relationship of relative humidity and temperature with various stoichiometries

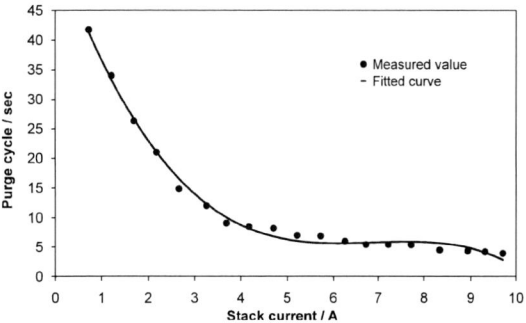

Figure 8. Optimal purge cycle at different load conditions

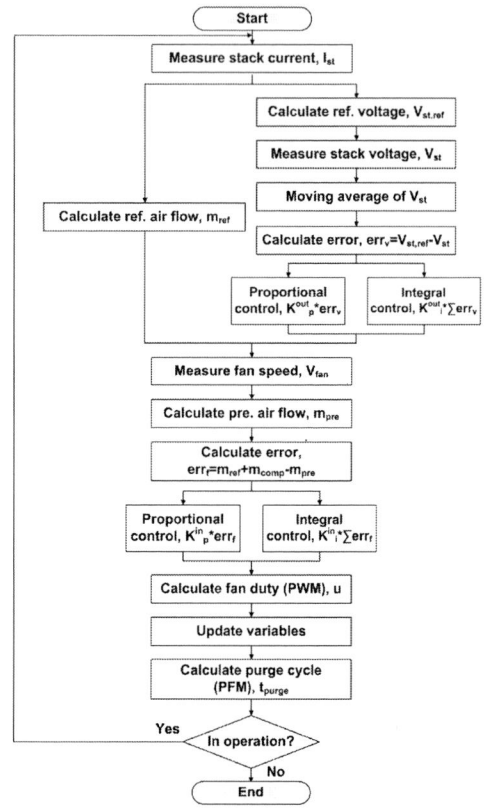

Figure 9. Flowchart for proposed control algorithm for small PEM fuel cell system

MEA by electro-osmotic drag and back diffusion. However, when the stack operates in high-current conditions, water formed by the reaction accumulates in the anode because the amount of water from back diffusion becomes larger than that from electro-osmotic drag. Accumulation of surplus water in the anode causes flooding and to prevent this, discharge of water through the purge valve is required [12]. In this study, the optimal purge cycle of the stack giving the best performance under different load conditions is measured experimentally. The purge valve is controlled by an ATmega128 microcontroller via the PFM (pulse frequency modulation) method based on the polynomial expression obtained by curve-fitting, as shown in Fig. 8 and (23).

$$t_{purge} = -0.160 I_{st}^3 + 3.310 I_{st}^2 - 22.43 I_{st} + 55.84 \qquad (23)$$

V. EXPERIMENTAL

The 150-W PEM fuel cell stack used for the experiment is composed of 24 unit cells and 3 cooling cells. The cell was manufactured by BCS Fuel Cells and has an electrode area of 50 cm^2. Since the use of hydrogen at 99.99% purity is required, high-purity hydrogen stored in a compressed gas cylinder under 100 atm was reduced to 0.2 atm using a pressure regulator before being supplied to the stack. At the end of the anode, a solenoid valve is connected for the purging. Because the system has no separate humidification device, it is appropriate for use as a small power source.

To implement the proposed control algorithm, ATmega128, an 8-bit microcontroller from Atmel, was used to create a control circuit. A sensing circuit, LabVIEW 8.6, and a

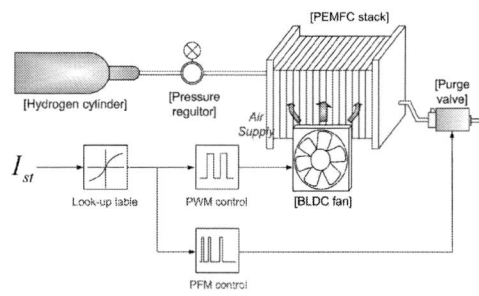

Figure 10. Control of small PEM fuel cell stack by SFFC using look-up table

PCI-6154 were used to record values for the fuel cell stack on a real-time basis. PCI-6154 is a simultaneous sampling multifunction I/O device for PCI bus computers from National Instruments [17]. It is an isolated PCI device featuring four isolated differential 16-bit analog inputs, four isolated 16-bit analog outputs, six DI lines, four DO lines, and two general-purpose 32-bit counter/timers. All A/D converters and D/A converters are capable of handling a max sampling rate of 250 kS/s for each channel. Fig. 9 shows a flowchart of the proposed control algorithm implemented in the microcontroller. A small 150-W PEM fuel cell system and control circuit were developed and used for the experiment.

To test the performance and efficiency of the developed controller, we compare the conventional SFFC (static feed-

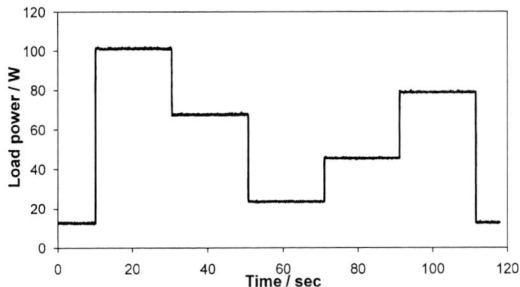

Figure 11. Power profile applied to fuel cell stack

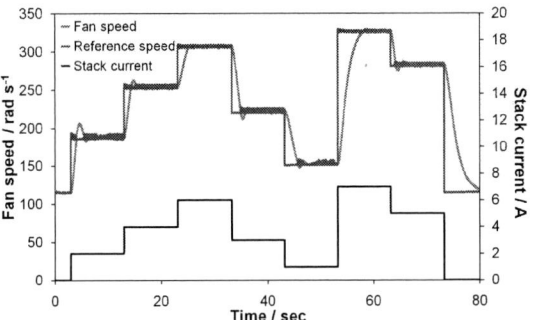

Figure 12. Dynamic response of BLDC fan with PI controller

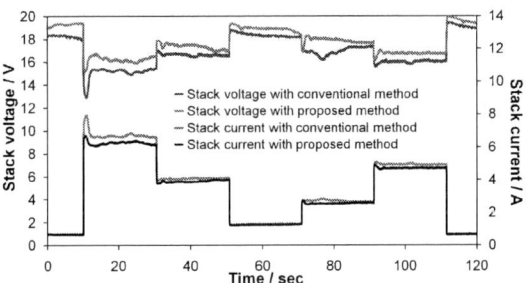

Figure 13. Voltage and current waveforms of stack
for each control method

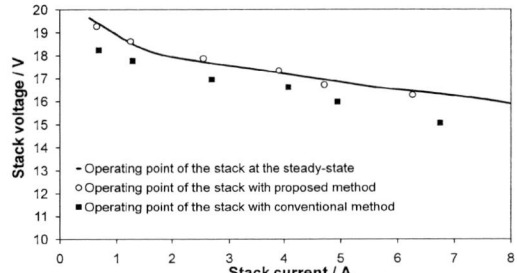

Figure 14. Comparison of operating points for each control method

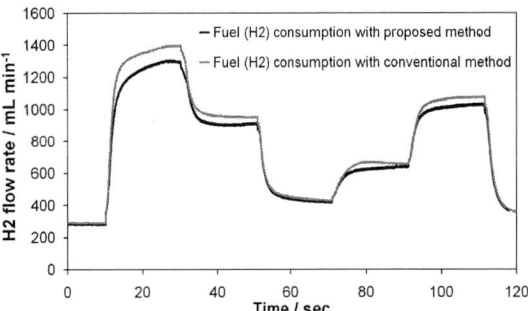

Figure 15. Hydrogen consumption of stack for each control method

forward control) method (Fig. 10) developed using a look-up table [11, 18] and the SFFC + DCC (static feed-forward control + dual closed-loop control) method proposed in this paper by performing identical experiments. In the experiments, the 150-W PEM fuel cell stack is loaded according to the power profile shown in Fig. 11.

VI. RESULTS AND DISCUSSION

Fig. 12 shows the dynamic response waveform of the fan controlled by the PI controller. The fan speed that can supply the required air flow rate calculated from the stack current becomes the command value. As can be seen in Fig. 12, the fan satisfactorily follows the command speed. In addition, the comparison of the step response characteristics shown in Figs. 4 and 12 revealed that using the PI controller shortened the time taken to follow the speed command value, and therefore the design of the fan controller was carried out appropriately.

Fig. 13 shows the voltage and current waveforms of the fuel cell stack using the SFFC and SFFC+DCC control methods when the load varies as shown in Fig. 11. As can be

seen in Fig. 13, both methods show stable tracking performance as load power changes, but the stack voltage of the proposed method is higher than that of the conventional method. In contrast, the stack current of the proposed method is lower than that of the conventional method. This means that the operating points of the two control methods are different although the same power is extracted from the fuel cell stack, as shown in Fig. 14. Moreover, the fuel consumptions are different as shown in Fig. 15. After applying the power profile for 120 s, the electric energy supplied to the load and the number of moles of hydrogen consumed are 6489 J and 0.0635 mol for the conventional method, and 6475 J and 0.0601 mol for the proposed method, respectively. The average stack efficiency was calculated to be 42.24% and 44.53%, respectively. Therefore, the proposed control method allows the stack to run at its optimal operating point and improves the fuel-to-power efficiency.

VII. CONCLUSION

A novel dual closed-loop method with a static feed-forward control was proposed in this paper to control a small PEM fuel cell system, and the algorithm was implemented using a microcontroller. The proposed method improved the slow transient response of a fan via inner PI control and supplied the air flow rate calculated from the stack current via static feed-forward control. Any output voltage deviation from the steady-state value can be corrected by feeding it into the additional outer PI controller to generate a correction term for the air flow rate, thereby improving system efficiency and achieving operation stability. The proposed control algorithm is expected to be useful for controlling small PEM fuel cell stacks.

ACKNOWLEDGMENT

This work (research) is financially supported by the Ministry of Knowledge Economy (MKE) and Korea Institute for Advancement in Technology (KIAT) through the Workforce Development Program in Strategic Technology.

REFERENCES

[1] A. Emadi, and S.S. Williamson, "Status review of power electronic converters for fuel cell applications," J. Power Electronics, vol. 1, pp. 133-144, October 2001.

[2] F. Urbani, G. Squadrito, O. Barbera, G. Giacoppo, E. Passalacqua, and O. Zerbinati, "Polymer electrolyte fuel cell mini power unit for portable application," J. Power Sources , vol. 169, pp. 334-337, March 2007.

[3] F. Dundar, F. Barbir, H. Gorgun, and A. Ata, "Designing PEM fuel cell for portable applications," International Hydrogen Energy Congress, Turkey, July 2005.

[4] C. Hebling, B. Burger, A. Hakenjos, J. Hesselmann, H. Münter, D. Pocza, J.O. Schumacher, U. Wittstatt, M. Zedda, and M. Zobel, "PEM fuel cells for the power supply of electronic appliances," The Fuel Cell World 2003 Proceedings, Lucerne, Switzerland, pp. 143-152, July 2003.

[5] A.P. Vega-Leal, F.R. Palomo, and F. Barragan, "Design of control systems for portable PEM fuel cells," J. Power Sources, vol. 169, pp. 194-197, January 2007.

[6] A.N. Wilhelm, B.W. Surgenor, and J.G. Pharoah, "Design and evaluation of a micro-fuel-cell-based power system for a mobile robot," IEEE/ASME Trans. Mecha., vol. 11, pp. 471-476, August 2006.

[7] F.N. Buchi, and S. Srinivasan, "Operating proton exchange membrane fuel cells without external humidification of the reactant gases," J. Electrochem. Soc., vol. 144, pp. 2767-2772, August 1997.

[8] R. Eckl, W. Zehtner, C. Leu, and U. Wagner, "Experimental analysis of water management in a self-humidifying polymer electrolyte fuel cell stack," J. Power Sources, vol. 138, pp. 137-144, August 2004.

[9] P. Berg, K. Promislow, J. St. Pierre, J. Stumper, and B. Wetton, "Water management in PEM fuel cells," J. Electrochem. Soc., vol. 151, pp. A341-A353, January 2004.

[10] S.U. Jeong, E.A. Cho, H.-J. Kim, T.-H. Lim, I.-H. Oh, and S.H. Kim, "A study on cathode structure and water transport in air-breathing PEM fuel cells," J. Power Sources, vol. 159, pp. 1089-1094, February 2006.

[11] J.T. Pukrushpan, A.G. Stefanopoulou, and H. Peng, Control of Fuel Cell Power Systems: Principles, Modeling, Analysis and Feedback Design, Springer, 2004, pp. 57-90.

[12] F. Barbir, PEM Fuel Cells: Theory and Practice, Elsevier Academic Press, 2005.

[13] J. Larminie, and A. Dicks, Fuel Cell Systems Explained, 2nd ed., John Wiley and Sons, 2003.

[14] F.P. Incropera, and D.P. DeWitt, Fundamentals of Heat and Mass Transfer, 5th ed., John Wiley and Sons, 2002, pp. 534-592.

[15] J.P. Bird, Model of the Air System Transients in a Fuel Cell Vehicle, Masters Thesis, Virginia Polytechnic Institute and State University, 2002.

[16] D. Ibrahim, Microcontroller Based Applied Digital Control, John Wiley and Sons, 2006.

[17] NI 6124/6154 User Manual, National Instruments Corporation, 2008.

[18] T.-H. Kim, W. Choi, "Control of small PEM fuel cell stack by a microprocessor," J. Korean Institute of Power Electronics, vol. 13, pp. 469-475, October 2008.

Stabilization of Constant-Power Loads by Passive Impedance Damping

Mauricio Céspedes,   Troy Beechner,   Lei Xing   and   Jian Sun

Department of Electrical, Computer, and Systems Engineering
Rensselaer Polytechnic Institute, Troy, NY 12180-3590, USA

Abstract– **This paper addresses stability problems of power systems with actively controlled loads that exhibit constant-power behavior. Instability occurs in such systems due to the negative incremental resistance of the constant-power loads (CPL). Existing approaches to stabilizing such systems require modification of the source and/or the load control characteristics, or isolating the CPL from the rest of the system by additional active devices, which are difficult to implement and often conflict with other system requirements such as control bandwidth, size, weight, and cost. In this work, we propose passive damping as a general method to stabilize power systems with CPL. Using a representative system model consisting of a voltage source, an LC filter, and an ideal CPL, we demonstrate that a CPL system can always be stabilized by a simple passive damping circuit added to one of the filter elements, no matter how the original system behaves. Three different damping methods are considered and for each analytical models are developed to define their parameters required for stabilizing the system. The different damping methods are also compared in terms of their stabilization capabilities and impact on other system performances such as filter attenuation. Time- and frequency-domain measurements from an experimental system are presented to validate the proposed methods.**

I. INTRODUCTION

An ideal constant-power load (CPL) consumes a fixed amount of power regardless of what supply voltage is received. It is a mathematical representation of power electronics-enabled loads with point-of-load power conditioning and control. From systems perspective, the most important characteristic of a CPL is its negative incremental resistance - the input current increases when the supply voltage decreases such that the power remains constant. The negative resistance reduces system damping and can lead to instability or unacceptable oscillatory responses when a significant portion of system power is consumed by such CPL.

Constant-power behavior became a concern first for dc distributed power systems such as those found on satellites and other space systems, telecom power stations, and computer power supplies. The use of point-of-load regulators in these systems changed most loads into CPL while the source is also conditioned by switching regulators. In recent years, constant-power behavior also became a major issue for several new applications, including more-electric aircraft [1], electric ships [2], electric vehicles [3] and micro grids [4] where more and more loads are equipped with point-of-load power conditioning to either enable new functions such as fly-by-wire and drive-by-wire, or to improve performance in terms of reducing weight and cost, and increasing efficiency.

Existing methods to stabilize CPL can be categorized into active [6-11] and passive [12-13] techniques. A well established design approach is to start from modeling each component of the system, and use the impedance-based stability criterion [5] to assess system stability. In an active stabilization technique the source [6-8] and/or the loads [9-11] are redesigned to modify their impedance so as to mitigate any instability problems. Most of these approaches however, add complexity to the converter control and require additional sensing circuits. On the other hand, passive stabilization techniques [12-13] use additional resistive damping circuits that can be added at the system integration stage. These techniques can result in unnecessary increase of system weight and size if the damping circuits are not selected properly. In addition, other system performance indexes such as filter attenuation and efficiency can be deteriorated [14].

An active stabilization technique emulates a passive damping circuit by means of active control [7-10]. Active damping was first developed to stabilize voltage sourced converters with input filters in [17] and was realized as a "virtual resistor" principle in [18]. The main advantage is it avoids the power dissipation problem of passive damping techniques. One potential problematic in modifying the source/load regulation to comply with system stability requirements is the deterioration of their dynamic performance [10]. In addition, the stabilization effect is limited by the sensitivity to controller action, and may require additional passive damping circuits for complete stabilization [11].

Passive stabilization techniques provide actual resistive damping to the system oscillations. In [6,8] a purely resistive load is added and contributes to system stability by reducing the effect of the CPL. More efficient stabilization techniques were used in [12-13] by inserting blocking capacitors and inductors in the damping circuit. However, the stability conditions in [12] were not solved analytically and [13] simplified the solution by selecting a relatively large blocking capacitor that is far from optimal.

The need for stabilization techniques could be alleviated if each component of the system could be designed to meet pre-specified impedance requirements that together would guarantee system stability. This was attempted in [15] which established a "forbidden zone" for the source output to load input impedance ratio. The original method was extended in [16] to multiple load scenarios by further limiting the allowable load input impedance region in proportion to its power rating. However, because these forbidden zones are developed from necessary conditions only, they almost always lead to too conservative designs.

978-1-4244-4782-4/10 $26.00 © 2010 IEEE

This paper presents a simple and systematic approach to stabilizing constant-power loads with passive damping techniques. Unlike active damping practices, it is demonstrated a simple passive damper can always stabilize a CPL system. We will develop the general theory based on a simplified system representation consisting of a voltage source, an LC filter, and an ideal CPL. We consider three different damping circuits, and develop the necessary and sufficient conditions required for each to stabilize the system without incurring in unnecessary simplifications. We will also study design optimization of each damper and compare their performance. An experimental system is also presented to validate the proposed approach.

II. SYSTEM MODELING, DAMPING AND STABILITY

A. System Modeling

Practical loads exhibit constant-power behavior only within their control bandwidth, which is limited. To make the results as generally applicable as possible, we consider here an ideal CPL that exhibits negative incremental resistance over an infinite frequency range. The power consumed by the CPL is assumed to be P. To simplify the analysis, we assume that the CPL is powered from a voltage source through a single-stage LC filter as shown in Fig. 1a). The LC filter can represent typical source output impedance, or can be considered part of the input filter of the CPL. Practical sources, particularly those employing active control, may exhibit more complex impedance behavior. However, the stabilization methods developed here are equally applicable and effective, as will be demonstrated by measurements from an experimental system.

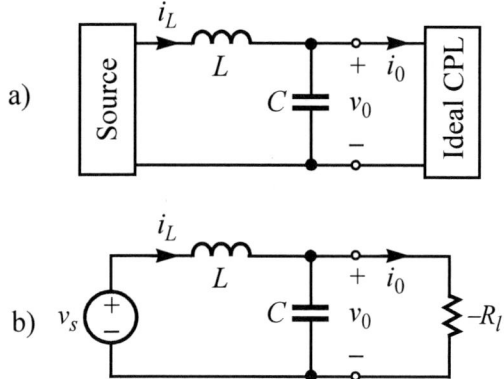

Fig. 1. Circuit diagram of a) The CPL system under consideration and b) its small-signal equivalent circuit.

Small-signal (incremental) impedance of the CPL is modeled as follows

$$Z_l = \frac{\partial v_0}{\partial i_0} = \frac{\partial}{\partial i_0}\left(\frac{P}{i_0}\right) = -\frac{P}{I_0^2} = -\frac{V_0}{I_0} \tag{1}$$

where V_0 and I_0 are the steady-state input voltage and current of the CPL respectively. Defining $R_l = V_0 / I_0$, we obtain the small-

signal equivalent circuit of Fig. 1b). This system is clearly unstable due to the negative damping of the LC filter resonance.

The method proposed here to stabilize the system is to add a passive damping circuit to the LC filter. We will present necessary and sufficient conditions for system stability with the damper circuit added. Optimum damping for minimization of the peak output impedance of the filter is considered [14]. Since we do not incur into any simplification of the stability conditions, the proposed approach yields a criteria for selection of the damping circuit that does not result in unnecessary detriment of system performance.

B. Passive Damping Methods

The three damping methods that will be considered here are depicted in Fig. 2 [14]. For system stability analysis, we will consider the filter and damper to be part of the source. The output impedance of the damped filter with any of the damping circuits can be easily determined and will be denoted as Z_{0d}.

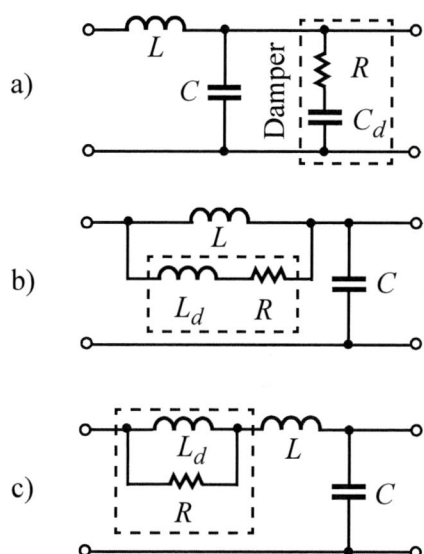

Fig. 2. Circuit diagram of low dissipation damping circuits. a) RC parallel damping; b) RL parallel damping; c) RL series damping.

C. Stability Conditions

Stability of the CPL system requires the impedance ratio Z_{0d}/Z_l meet the Nyquist stability criterion [5], or, alternatively, the characteristic polynomial of $(1 + Z_{0d}/Z_l)^{-1}$ meet the Routh-Hurwitz criterion. The former criterion requires to prepare a complex plot of the impedance ratio and test the number of clockwise encirclements of the $(-1,0)$ point. The latter criterion is more convenient if the purpose is to determine the analytic boundary for stability. For the CPL system with each of the damping circuits under consideration, the characteristic polynomial is of third order, and can be written in a general form as $p(s) = s^3 + a_2 s^2 + a_1 s + a_0$.

Table I summarizes the coefficients of $p(s)$ for each of the three damping circuits. In order to minimize the impact of the input filter on the load converter regulation, the output impedance of the damped filter shall be kept as low as possible. Based on Table I and the Routh-Hurwitz criteria, we can determine the parameters for each damping circuit that are required to stabilize the system. This is accomplished in two steps:

1) We define first a new parameter n as $n = C_d/C$ for the RC parallel damper, and $n = L_d/L$ for the RL parallel as well as the RL series damper. We then replace R by the optimal damping resistance that minimizes the peak output impedance for a given n [14]. This effectively eliminates C_d, L_d, and R from the coefficients given in Table I and replaces them with functions of n. The characteristic impedance of the undamped filter $R_0 = \sqrt{L/C}$ is also used.

2) Next, we apply the Routh-Hurwitz criteria to the coefficients to determine the required n that will stabilize the system by guarantying that the roots of $p(s)$ lie in the left-hand side of the complex plane. Note that the necessary conditions for stability from $p(s)$ are

$$a_2 > 0, \quad a_1 > 0 \qquad (2)$$

while sufficient conditions can be expressed as:

$$a_2 > 0, \quad a_2 a_1 - a_0 > 0. \qquad (3)$$

For the particular case of the RC parallel damper the necessary conditions are:

$$(n+1)\sqrt{\frac{2(n+4)}{(n+2)(3n+4)}} > \frac{R_0}{R_l} \qquad (4)$$

$$\sqrt{\frac{(n+2)(3n+4)}{2(n+4)}} > \frac{R_0}{R_l}, \qquad (5)$$

and sufficient conditions are:

$$(n+1)\sqrt{\frac{2(n+4)}{(n+2)(3n+4)}} > \frac{R_0}{R_l} \qquad (6)$$

$$\left(\frac{R_0}{R_l}\right)^2 - \left[\frac{5n^2 + 20n + 16}{2(n+4)(n+2)(3n+4)}\right]\left(\frac{R_0}{R_l}\right) + n > 0. \qquad (7)$$

The quadratic function on R_0/R_l in (7) has the solutions:

$$s_1 = \sqrt{\frac{(n+4)(3n+4)}{2(n+2)}} \qquad (8)$$

$$s_2 = n\sqrt{\frac{2(n+2)}{(n+4)(3n+4)}} \qquad (9)$$

where it is easy to show $s_1 > s_2$ for all $n > 0$. Therefore the two plausible regions for satisfaction of (7) are respectively:

$$\sqrt{\frac{(n+4)(3n+4)}{2(n+2)}} < \frac{R_0}{R_l} \qquad (10)$$

$$n\sqrt{\frac{2(n+2)}{(n+4)(3n+4)}} > \frac{R_0}{R_l}. \qquad (11)$$

It can be demonstrated the region predicted by (10) violates both of the necessary conditions in (4) and (5) for all $n > 0$. Meanwhile, the region of n for which (11) is valid, satisfies both of the necessary conditions and the remaining sufficient condition in (6). This is verified from the contour plots in Fig. 3.

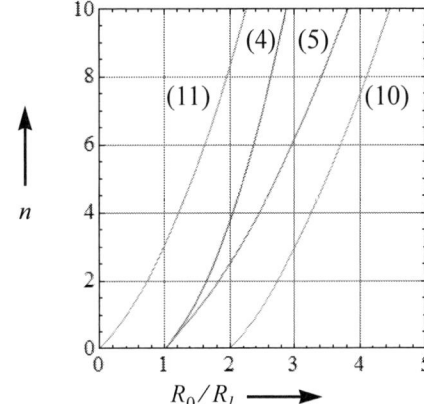

Fig. 3. Example evaluation of stability conditions in RC parallel damper.

TABLE I COEFFICIENTS OF THE SYSTEM IMPEDANCE CHARACTERISTIC POLYNOMIAL

Configuration	a_2	a_1	a_0
RC Parallel Damper	$\dfrac{C + C_d - RC_d \S R_l}{CRC_d}$	$\dfrac{RC_d - L \S R_l}{LCRC_d}$	$\dfrac{1}{LCRC_d}$
RL Parallel Damper	$\dfrac{CR - L_d \S R_l}{L_d C}$	$\dfrac{L + L_d - RL \S R_l}{LCL_d}$	$\dfrac{R}{LCL_d}$
RL Series Damper	$\dfrac{CR(L + L_d) - (LL_d) \S R_l}{LCL_d}$	$\dfrac{L_d - R(L + L_d) \S R_l}{LCL_d}$	$\dfrac{R}{LCL_d}$

III. PERFORMANCE EVALUATION

A. Stabilization Requirement

For a given ratio $R_0 \S R_l$, the following expressions give the required selection of n to guarantee stability:

1) RC Parallel Damper

$$n\sqrt{\frac{2(n+2)}{(n+4)(3n+4)}} > \frac{R_0}{R_l}$$

2) RL Parallel Damper

$$\sqrt{\frac{2(2n+1)}{n(4n+1)(4n+3)}} > \frac{R_0}{R_l} \qquad (12)$$

3) RL Series Damper

$$\frac{1}{2}\left[g(n) + \frac{1}{g(n)} - \sqrt{\left(g(n) + \frac{1}{g(n)}\right)^2 - \frac{4n}{n+1}}\right] > \frac{R_0}{R_l} \qquad (13)$$

with $g(n) = (1+1/n)^2 \sqrt{2(n+1)(n+4)/((n+2)(3n+4))}$.

The stable region for each damper circuit is depicted in Fig. 4 by the shaded area. It can be confirmed (11) indicates that a CPL can always be stabilized by an RC parallel damper. Note a large CPL can also be stabilized with an RL parallel damper by small a small L_d. However, the RL parallel damper reduces the attenuation of the filter which is undesirable. Note the left-hand side of (13) approaches the limit $\sqrt{2/3}$ as n increases, indicating an RL series damper can stabilize the system only if the CPL is such that $R_l > \sqrt{3L/(2C)}$.

B. Power Losses of Damping Resistors

All three damping circuits dissipate power when implemented in ac systems. Only in the particular case of dc systems, the RC parallel damper does not dissipate average power. Following the approach proposed in [19], a proper way to compare the damping resistor losses is to start from a given desired R_0/R_l, calculate the minimum n for stabilization, and the corresponding optimal damping resistor R/R_0 from [14].

Fig. 5 compares the losses of the three dampers at variable fundamental frequency and for $R/R_0 = 0.5$ and $R/R_0 = 0.2$. It can be seen that the RL series damper has the higher losses, followed by the RC parallel damper, and the RL parallel damper is observed to be the most effcient.

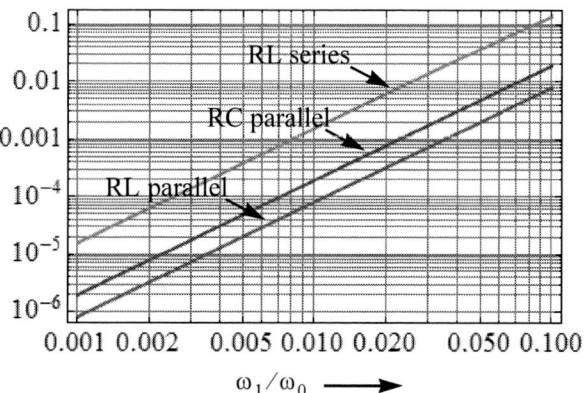

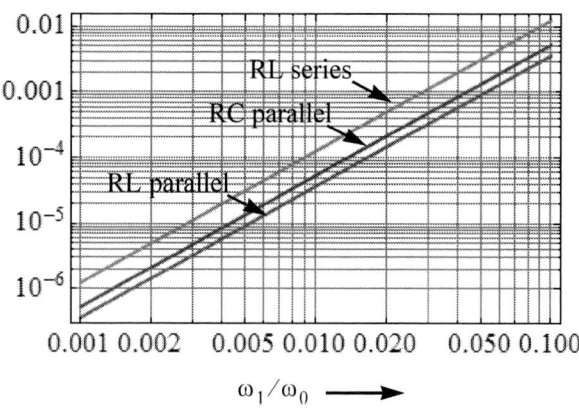

Fig. 5. Comparison of loss scaling factors as defined in [19] for $R_0/R_l = 0.5$ (upper diagram) and $R_0/R_l = 0.2$ (lower diagram)

C. Size of Damping Capacitor and Inductor

We will consider now a comparison of the size of the damping inductor and capacitor. Because the dampers must be designed on a common basis to make their size comparison meaningful, we will require that for a given R_0/R_l the dampers must stabilize the

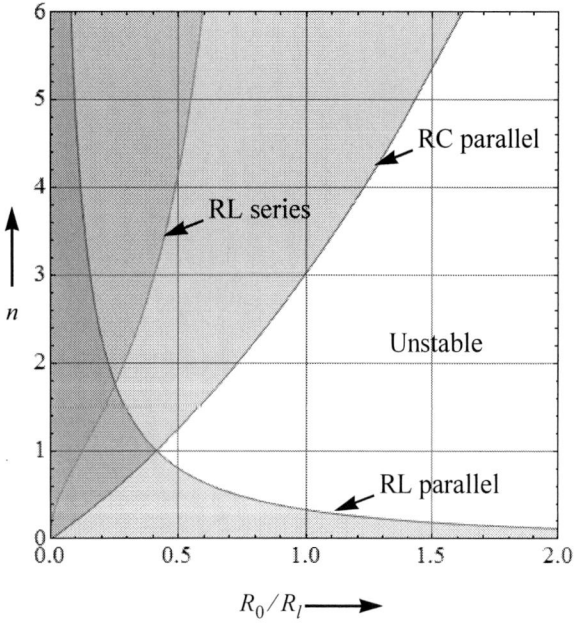

Fig. 4. Stability region for the three damper circuits.

system while retaining the same attenuation as that of the undamped filter. It has been noted the RL parallel damper reduces the effective filtering inductance above the resonant frequency to

$$\frac{n}{n+1}L$$

Therefore both filter inductor and damping inductor have to be increased by a factor $1 + n^{-1}$ in order for the attenuation to be unaffected.

The volume scaling factors are computed in the same manner as in [19]. For the RC parallel damper, it was found the total volume of the filter capacitor and the damping capacitor ($C + C_d$) is $1 + n$ times the volume of the filter capacitor C since both capacitors are rated for approximately the same voltage and their size follows the same ratio as the capacitance. For the RL series damper, since the volume of an inductor is approximately proportional to the product of its inductance and rated current, and both inductors carry approximately the same current at the fundamental frequency, the total volume of the two inductors is also $1 + n$ times the volume of the filter inductor L. The case of the RL parallel damper is slightly more complex because it depends on the ratio of resonance frequency to fundamental frequency. In addition, the factor to compensate for degradation of filter attenuation has to be considered. Fig. 6 shows the variation of the different volume scaling factors, with R_0/R_l changing in the range from 0 to 1. A typical ratio of $\omega_0/\omega_1 = 25$ is used for the RL parallel damper. The RL series damper is observed to have the larger volume scaling factor, followed by the RL and RC parallel dampers.

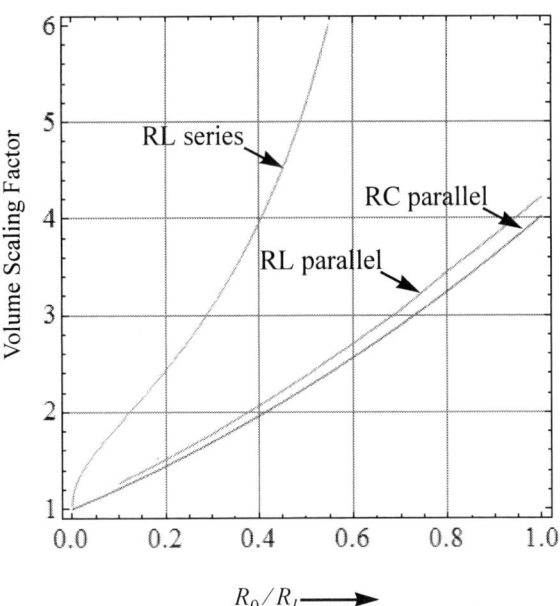

Fig. 6. Total capacitor and inductor volume increase due to the use of different damping methods.

IV. EXPERIMENTAL RESULTS

An experimental system consisting of a 200 V dc source with an LC output filter and a programmable electronic load was set up and measured to validate the stabilization design method. The electronic load can be programmed to operate in different modes, and was initially operated as a 2 kW constant-power load in the experiments. The LC filter has the following parameters: $L = 8$ mH, $C = 150$ μF. Fig. 7 shows the measured unstable start-up of the system without a damper circuit. The signals are as defined in Fig. 1. The source voltage collapsed when the CPL was applied.

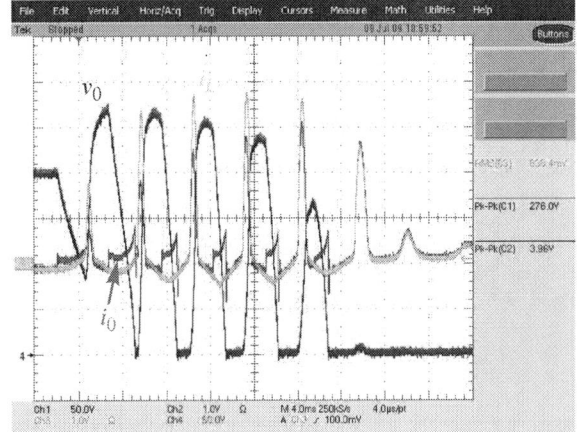

Fig. 7. Start-up transient of the experimental system without a damper.

Measurements of the source output impedance (including the filter) and the load input impedance are shown in Fig. 8. The −180° phase angle of the load input impedance below 500 Hz is an indication of its CPL behavior. The negative input impedance characteristic is lost above 500 Hz due to the limited control bandwidth of the electronic load. The undamped source output impedance magnitude response (dashed black line) shows a high peaking and intersects with the load input impedance, explaining the observed instability problem in Fig. 7.

To stabilize the system, we selected the RC parallel damper and designed it to stabilize the system with 6 dBΩ gain margin. To that end, the value of R_l used in

$$n\sqrt{\frac{2(n+2)}{(n+4)(3n+4)}} > \frac{R_0}{R_l}$$

is set to 10 Ω (20 dBΩ) since the equivalent negative resistor for a 2 kW CPL at 200 V has 20 Ω (26 dBΩ). The minimum n to fulfill the stability requirement is $n = 2$. Hence $C_d = 300$ μF is used, and the corresponding optimal damping resistance is $R = 6.7$ Ω. Fig. 9 shows the measured start-up of the system after the damper was added. The resulting source output impedance was also measured and is represented in Fig. 8 by the solid red lines. Both results indicate stable system operation and validate the damper design.

978-1-4244-4782-4/10 $26.00 © 2010 IEEE 2178

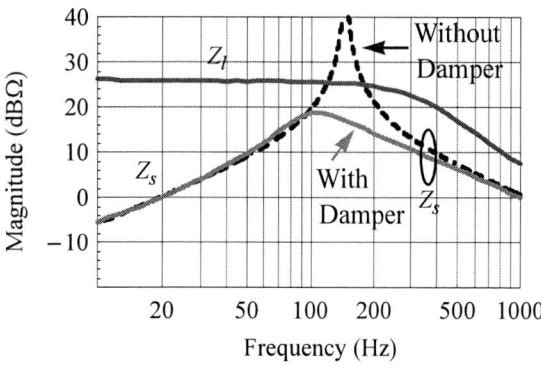

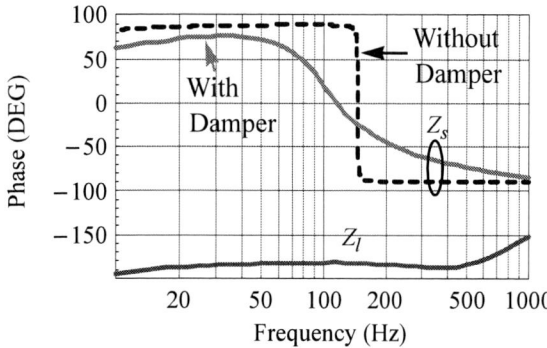

Fig. 8. Frequency response plot of measured load input impedance Z_l and source output impedance Z_s with damper and without damper.

To further verify the boundary for stability, we reduce the system stability margin progressively from 6 dBΩ to the limit where the system becomes unstable under small changes in the load power.

The transient response of the system with 3 dBΩ stability margin in Fig. 10a) shows the experimental system is still stable for a increase in load from 15 A to 16 A. Fig. 10b) shows the voltage collapses in the case where the stability margin is reduced to 2 dBΩ and the load attempts an increase from 16 A to 19 A.

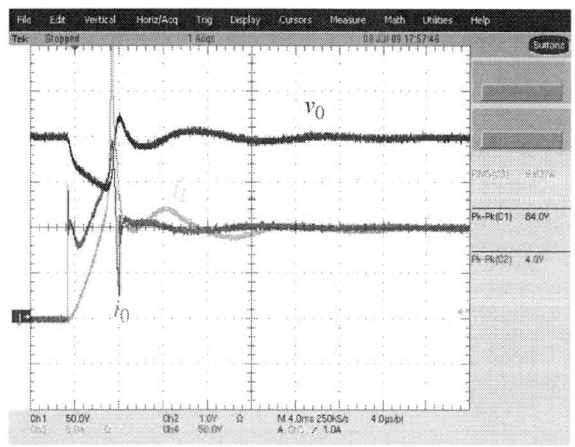

Fig. 9. Start-up transient of the experimental system with an RC parallel damper.

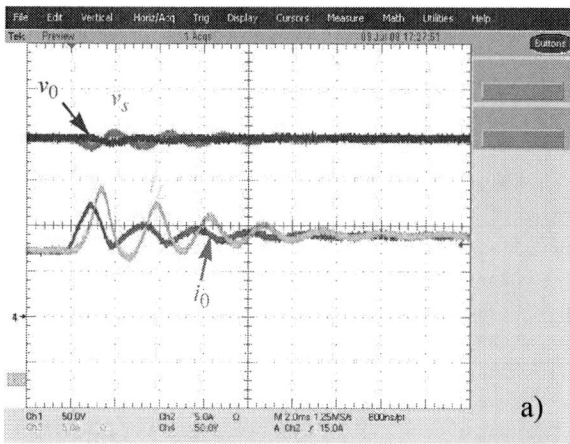

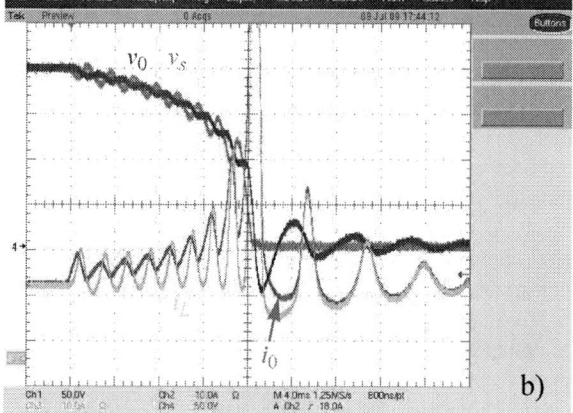

Fig. 10. Transient load variation with reduced system stability margin: a) 3 dBΩ; b) 2 dBΩ

V. SUMMARY

This paper has analyzed unstable system operation resulting from CPL behavior. From an idealized system representation it is demonstrated a simple passive damping approach can always guarantee small-signal stability provided the damping elements are selected to fall within the bounds that are laid down here. In addition, the results are derived for damper topologies that make use of the optimum damping resistor for minimization of the peak output impedance. An experimental set-up demonstrated the validity of the design approach and also served to investigate the practical limits of small-signal stability in the evaluation system.

VI. REFERENCES

[1] M. Gries, O. Wasynczuk, B. Selby and P. T. Lamm, "Designing for large-displacement stability in aircraft power systems," in *Proceedings of SAE 2008 Power Systems Conference*, 2008, pp. 2008-01-2867.

[2] A. L. Julian and R. M. Cuzner, "Design, modeling and stability analysis of an integrated shipboard dc power system," in *Proceedings of IEEE Electric Ship Technologies Symposium*, 2009, pp. 428-432.

[3] A. M. Rahimi and A. Emadi, "An analytical investigation of dc/dc power

electronics converters with constant power loads in vehicular power systems," *IEEE Transactions on Vehicular Technology*, vol. 58, no. 6, pp. 2689-2702, Jul. 2009.

[4] D. P. Ariyasinghe and D. M. Vilathgamuwa, "Stability analysis of micro-grids with constant power loads," in *Proceedings of IEEE International Conference on Sustainable Energy Technologies*, 2008, pp. 279-284.

[5] R. D. Middlebrook, "Input filter consideration in design and application of switching regulators," in *Proceedings of IEEE Industry Applications Society Annual Meeting*, 1976, pp. 366-382.

[6] J. Wang and D. Howe, "A power shaping stabilizing control strategy for dc power systems with constant power loads," *IEEE Transactions on Power Electronics*, vol. 23, no. 6, pp. 2982-2989, Nov. 2008.

[7] X. Wang, D. Vilathgamuwa, and S. Choi, "Decoupling load and power system dynamics to improve system stability," in *Proceedings of International Conference on Power Electronics and Drives Systems*, 2005, pp. 268-273.

[8] A. Rahimi and A. Emadi, "Active damping in dc/dc power electronics converters: a novel method to overcome the problems of constant power loads," *IEEE Transactions on Industrial Electronics*, vol. 56, no. 5, pp. 1428-1439, May 2009.

[9] X. Liu and A. J. Forsyth, "Comparative study of stabilizing controllers for brushless DC motor drive systems," in *Proceedings of 2005 IEEE International Conference on Electric Machines and Drives*, 2005, pp. 1725-1731.

[10] X. Liu, A. J. Forsyth, and A. M. Cross, "Negative input-resistance compensator for a constant power load," *IEEE Transactions on Industrial Electronics*, vol. 54, no. 6, pp. 3188-3196, Dec. 2007.

[11] X. Liu, N. Fournier, and A. J. Forsyth, "Active stabilization of a HVDC distribution system with multiple constant power loads," in *Proceedings of IEEE Vehicle Power and Propulsion Conference*, 2008, pp. 1-6.

[12] D. M. Mitchell, "Damped EMI filters for switching regulators," *IEEE Transactions on Electromagnetic Compatibility*, vol. EMC-20, no. 3, pp. 457-463, 1978.

[13] A. B. Jusoh, "The instability effect of constant power loads," in *Proceedings of 2004 National Power & Energy Conference (PECon)*, 2004, pp. 175-179.

[14] R. W. Erickson, "Optimal single resistor damping of input filters," in *Proceedings of IEEE APEC'99*, 1999, pp. 1073-1079.

[15] C. M. Wildrick, F. C. Lee, B. H. Cho, and B. Choi, "A method of defining the load impedance specification for a stable distributed power system," *IEEE Transactions on Power Electronics*, vol. 10, no. 3, pp. 280-285, May 1995.

[16] X. Feng, J. Liu, and F. C. Lee, "Impedance specification for stable dc distributed power systems," *IEEE Transactions on Power Electronics*, vol. 17, no. 2, pp. 157-162, Mar. 2002.

[17] V. Blasko and V. Kaura, "A novel control to actively damp resonance in input LC filter of a three-phase voltage source converter," *IEEE Transactions on Industry Applications*, vol. 33, no. 2, pp. 542-550, Mar. 1997.

[18] P. A. Dahono, "Damping of transient oscillations on the output LC filter of PWM inverters by using a virtual resistor," in *Proceedings of 4th IEEE International Conference on Power Electronics and Drive Systems*, 2001, pp. 403-407.

[19] L. Xing, F. Feng, and J. Sun, "Optimal damping of EMI filter input impedance," in *Proceedings of IEEE Energy Conversion Congress and Exposition*, 2009, pp. 1685-1692.

An Adaptive External Ramp Control of the Peak Current Controlled Buck Converters for High Control Bandwidth and Wide Operation Range

Liyu Yang, Jinseok Park and Alex Q. Huang

Semiconductor Power Electronics Center (SPEC), Electrical and Computer Engineering Department
North Carolina State University
Raleigh, NC 27606, USA
lyang2@ncsu.edu

Abstract—**Buck converters for wide operation range are frequently seen in various applications. In many cases it is preferable to use one fixed compensator to fit the wide operation range. In this paper, an adaptive ramp control scheme is applied to peak current mode (PCM) controlled Buck converters for wide operation range. The control bandwidth of the PCM Buck converter is improved by adaptively adjusting the external ramp. This paper revisits the small signal model of the PCM control and analyzes the relationship between the external ramp and the system dynamics. Based on the analysis, the proposed adaptive external ramp for PCM is developed to achieve high control bandwidth for wide operation range Buck converters. Assuming one single fixed compensator design, the proposed control scheme can increase the control bandwidth significantly compared to fixed external ramp design. Faster transient response is verified with experimental results.**

I. INTRODUCTION

Voltage mode and peak current mode [1-9] are two classic control methods for Buck converters. Some Buck converters have fixed input voltage and fixed output voltage, also the load current does not change too much. For this kind of application, the feedback control design guidelines for voltage mode control and current mode control are well established [6], and it is relatively easy to design a good compensator and obtain good performance in terms of regulation and transient response. When the operation range is wide or the same controller is used for various applications, normally it is not easy to design one fixed compensator to fit all over the operation range with high performance. Therefore, people normally design different compensators for different operating conditions. These kinds of Buck converters with wide operation range are frequently seen. For example, TI's product TPS54362 has input range from 3.6V to 48V, output range from 0.9V to 18V, and output current up to 3A. Maxim's product MAX8655 has input range from 7V to 28V, output range from 0.7V to 12V, and output current from up to 25A.

However, sometimes it is preferable to use one fixed compensator to fit a wide operation range. If a DC-DC controller is designed with fixed compensator, it is easier for the end users since they don't need to go through the process of calculation, simulation and bench tuning. Another advantage is that the resistors and capacitors in the compensator can possibly be monolithically integrated into the controller IC. By doing so, the controller IC can use less pins, less external resistors and capacitors, and therefore saves the PCB area. A more compact layout can be realized. One example of such controller is LT3493 from Linear Tech, which is a peak current mode controlled Buck converter with internal compensator.

Some control methods have been proposed to design one fixed compensator for wide operation range DC-DC converters. For voltage mode Buck converter with wide V_{in} range, if the PWM ramp amplitude is self-adjusted to be proportional to V_{in}, theoretically the bandwidth is the same for all V_{in} range. Therefore one single compensator is good for all V_{in} range. This method is called V_{in} feed-forward. One example is the LM22674 from National Semiconductor, which is also a monolithic Buck converter with internal fixed type III compensator and V_{in} feed-forward.

For peak current mode (PCM) control, external ramp is widely used to avoid sub-harmonic oscillation when duty cycle D>0.5. If the external ramp can be self-adjusted similarly, can it bring any benefit? This paper proposes such an adaptive external ramp control scheme for PCM Buck converters with wide operation range. The external ramp is adjusted automatically by itself. Assuming one single fixed compensator design, the proposed method can increase the bandwidth significantly compared to fixed external ramp design. Section II provides a review of previous work in this topic. Section III provides the theoretical derivation for the proposed adaptive external ramp method and uses the SIMPLIS software to verify the analysis. A PCM Buck converter is designed with this adaptive external ramp method

---

This work is supported by the Power Management Consortium of SPEC: Fairchild Semiconductor, Intel and International Rectifier.

978-1-4244-4782-4/10 $26.00 © 2010 IEEE

2181

and tested on bench. With control bandwidth improvements brought by the proposed method, the transient response is improved significantly as shown in the experimental results in Section IV. Section V is the conclusion.

II. PREVIOUS WORK IN THE LITERATURE

One single fixed compensator design is preferred in many applications. It is easier for the end customer, and it also provides the possibility to integrate the resistors and capacitors for the compensator into the controller IC to save PCB area and IC pins. Yet, the V_{in}, V_o and I_o range can be wide for the same controller, which is a limiting factor for a single fixed compensator to achieve high bandwidth. For voltage mode control, one solution is to design the ramp amplitude to be proportional to V_{in} value, which is called V_{in} feed-forward. Fig. 1 shows the control diagram of this technique.

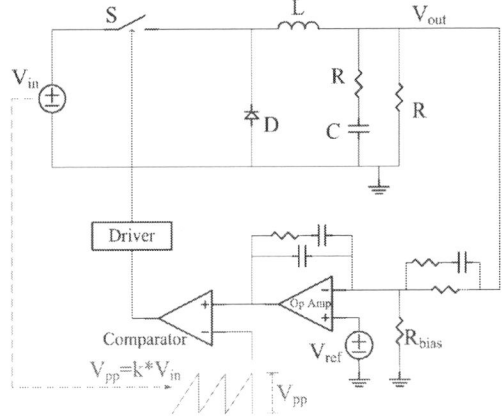

Figure 1. Control diagram of voltage mode controlled Buck converter with V_{in} feed-forward

The loop gain can be expressed in (1).

$$
\begin{aligned}
T &= F_m \cdot G_{vd}(s) \cdot G_{comp}(s) \\
&= \frac{1}{V_{pp}} \cdot V_{in} \cdot \frac{1 + sR_cC}{\Delta(s)/\omega_0^2} \cdot \frac{\omega_I}{s} \cdot \frac{(1+s/\omega_{Z1})(1+s/\omega_{Z2})}{(1+s/\omega_{P1})(1+s/\omega_{P2})}
\end{aligned}
\tag{1}
$$

where $\quad \Delta(s) = s^2 + \dfrac{\omega_0}{Q}s + \omega_0^2, \quad \omega_0 = \dfrac{1}{\sqrt{LC}} \quad$ and

$$
Q = \frac{1}{\omega_0} \cdot \frac{1}{L/R + R_cC}.
$$

As can be seen from (1), if $V_{pp}=k*V_{in}$, the expression of loop gain T is almost the same for all V_{in} values.

The control diagram of peak current mode control and its small signal model are shown in Fig. 2 [1]. The ramp S_n is the switch current, while the ramp S_e is an intentionally added external ramp to eliminate the sub-harmonic oscillation when the duty cycle is greater than 50%.

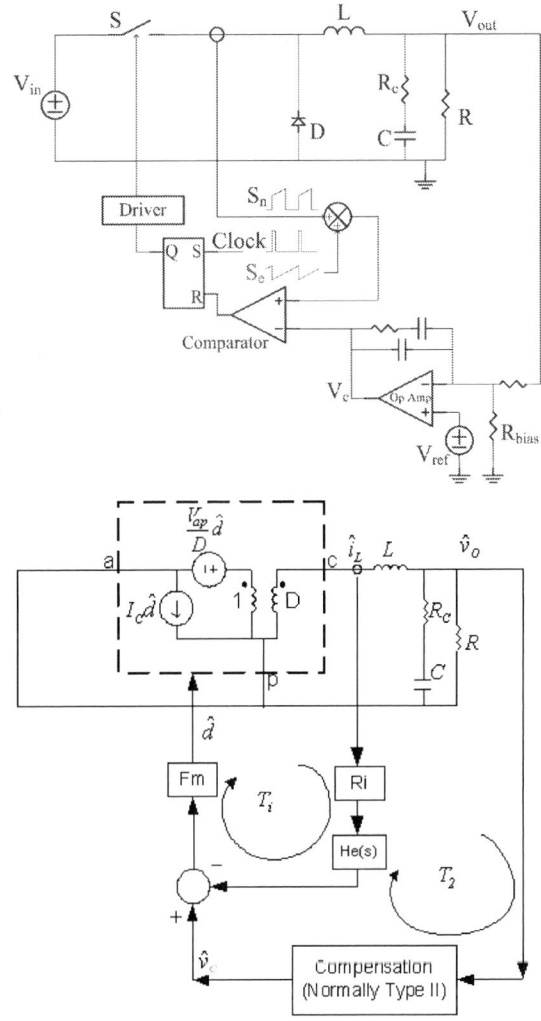

Figure 2. Control diagram of PCM controlled Buck converter and its small signal model

Some papers discussed adaptive external ramp for current mode control before. In [2], a monolithic current mode converter with adjustable-slope compensating ramp is implemented. The external ramp m_a (i. e., S_e) is design to be the same as inductor down slope m_2, i.e., $m_a = m_2 = V_{out}/L$. From stability point of view, this ensures no sub-harmonic oscillation for all V_{out}. However, it didn't do analysis in the frequency domain using small signal model, so it is not clear whether it is a good design in terms of bandwidth. The paper [3] proposes a load current feed-forward mechanism for current mode control. Yet, the analysis is for an application with fixed output voltage. It is not clear whether this mechanism is optimum for the application where one single fixed compensator works for wide range of output voltage. The paper [4] also discusses a method to improve the bandwidth of peak-current controlled voltage regulators by selecting a suitable external ramp. But the analysis is also for the application with well defined input voltage and output voltage.

978-1-4244-4782-4/10 $26.00 © 2010 IEEE   2182

In next section, analysis based on small signal model and transfer functions will be carried out for a wide operation range PCM Buck converter. Based on the analysis, for a given phase margin (60 degrees in this paper), the design guideline is derived to allow maximum bandwidth for wide operation range PCM Buck converters with a single fixed compensator.

III. THEORETICAL DERIVATION

To design the single fixed compensator for wide operation range, one needs to look at the control to output transfer functions $G_{vc}(s) = \dfrac{\hat{v}_o}{\hat{v}_c}$ for all working conditions. Based on the small signal mode in Fig. 2, the $G_{vc}(s)$ function for Buck converter can be derived as in [1].

$$
\begin{aligned}
G_{vc}(s) &= \frac{\hat{v}_o}{\hat{v}_c} \\
&= \frac{T_v}{1+T_i} \\
&= \frac{G_{vd}(s) \cdot F_m}{1 + G_{id}(s) \cdot F_m \cdot R_i \cdot H_e(s)} \\
&\approx \frac{R}{R_i} \cdot \frac{1}{1 + \dfrac{R \cdot T_s}{L}(m_c D' - 0.5)} \cdot F_p(s) \cdot F_h(s)
\end{aligned}
\tag{2}
$$

where $F_p(s) = \dfrac{1 + sCR_c}{1 + \dfrac{s}{\omega_p}}$, $\omega_p = \dfrac{1}{CR} + \dfrac{T_s}{LC}(m_c D' - 0.5)$,

$m_c = 1 + \dfrac{S_e}{S_n}$, $F_h(s) = \dfrac{1}{1 + \dfrac{s}{\omega_n Q} + \dfrac{s^2}{\omega_n^2}}$, $\omega_n = \dfrac{\pi}{T_s}$ and

$Q = \dfrac{1}{\pi \cdot (m_c D' - 0.5)}$.

A Buck converter with the following working conditions is used as an example: V_{in}=6V-24V, V_o=1.2V-5V, I_o=0.5A-5A, f_{sw}=200kHz, C=100uF, R_c=10mΩ, L=10uH, and current sense gain R_i=0.1V/A. The wide operation range includes 4x V_{in} change, ~4x V_o change and 10x I_o change.

A fixed external ramp of S_e=R_i*($V_{o,max}$/L)=50mV/us is used to ensure no sub-harmonic oscillation for all working conditions. The $G_{vc}(s)$ Bode plots in Fig. 3 covers the eight corners of the wide operation range as listed in Table I.

TABLE I.   WORKING CONDITIONS OF THE EIGHT OPERATION CORNERS

	V_{in} (V)	V_o (V)	I_o (A)	Trace style
Corner 1 (trace 1)	24	5	5	Solid line, thin and red
Corner 2 (trace 2)	24	5	0.5	Dotted line, thin and red
Corner 3 (trace 3)	24	1.2	5	Dashed line, thin and red
Corner 4 (trace 4)	24	1.2	0.5	Dotted-dashed line, thin and red
Corner 5 (trace 5)	6	5	5	Solid line, thick and blue
Corner 6 (trace 6)	6	5	0.5	Dotted line, thick and blue
Corner 7 (trace 7)	6	1.2	5	Dashed line, thick and blue
Corner 8 (trace 8)	6	1.2	0.5	Dotted-dashed line, thick and blue

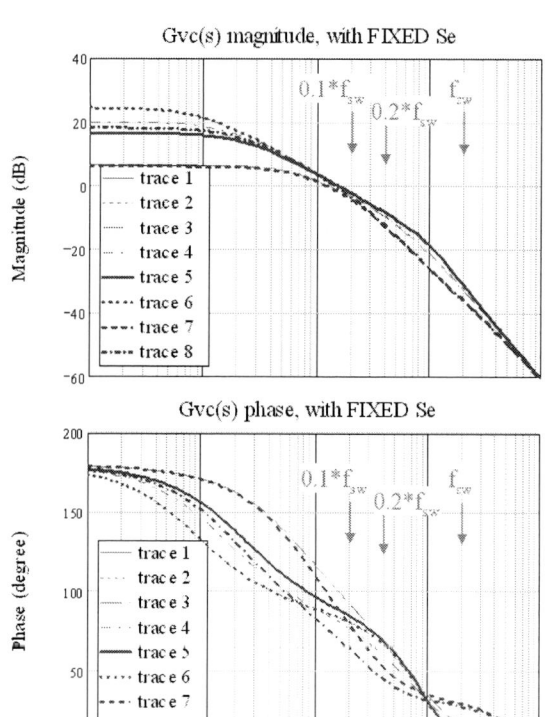

Figure 3. $G_{vc}(s)$ curves with fixed external ramp.

If the $G_{vc}(s)$ for each trace is identical or similar to each other, it will be very easy to design a single fixed compensator and obtain good control performance over the wide operation range. However, as can be seen from Fig. 3, there is quite a lot difference for these curves. The conventional design guidelines recommend to design the bandwidth to be $0.1*f_{sw}$ to $0.2*f_{sw}$. If we look at this frequency range (20kHz to 40kHz) in Fig. 3, the difference in the magnitude is 2 to 5dB, while the difference in the phase is 20 to 30 degrees. Therefore, this variation is a limiting factor to obtain high control bandwidth using a single fixed compensator.

As pointed out in [9], if the ESR zero is not sufficiently lower than the switching frequency, the PCM Buck needs to have at least 60 degrees phase margin to obtain optimum load transient response. To ensure that the loop gain for all working conditions have at least 60 degree phase margin, a fixed compensator $G_{comp}(s) = \dfrac{2650}{s} \cdot \dfrac{1+s/2000}{1+s/1000000}$ is used. The compensation zero is placed close to the dominant pole ω_p, while the compensation pole is placed at the ESR zero [6].

The Bode plots in Fig. 4 are the loop gain T curves, where $T = G_{vc}(s) * G_{comp}(s)$, while Fig. 5 is the same sets of curves but they are zoomed in to 10kHz-100kHz to illustrate at least 60 degree phase margin for all loop gains. The bandwidth cannot be pushed higher, otherwise the phase margin will be less than 60 degree for the working condition of V_{in}=6v, V_o=1.2v and I_o=0.5A (trace 8, high-lighted with thicker line in Fig. 5).

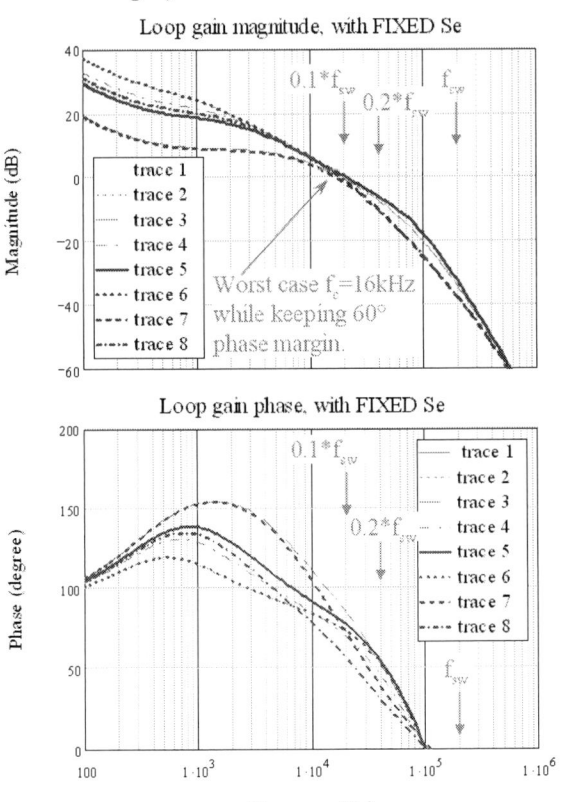

Figure 4.   Loop gain curves with fixed external ramp.

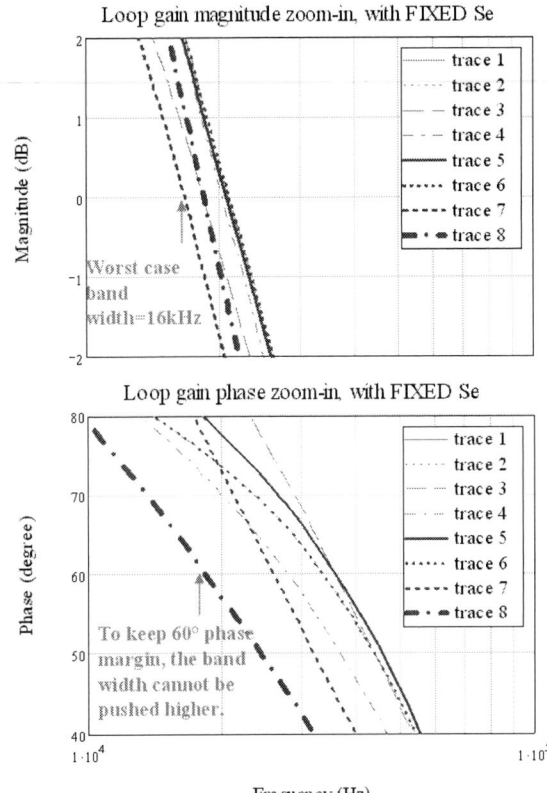

Figure 5.   Loop gain curves with fixed external ramp, zoomed-in.

As can be seen from Fig. 5, the worst case bandwidth is 16kHz (trace 7, V_{in}=6V, V_o=1.2V, I_o=5A), which is only about 1/12 of the switching frequency. If the external ramp can be adjusted automatically instead of fixed, can it improve the control bandwidth?

If the external ramp S_e can be adjusted, it means $m_c = 1 + S_e / S_n$ can be changed. Based on Equation (2), changing m_c will affect the $G_{vc}(s)$ function in three aspects: the DC gain $\dfrac{R}{R_i} \cdot \dfrac{1}{1 + \dfrac{R \cdot T_s}{L}(m_c D' - 0.5)}$, the pole frequency $\omega_p = \dfrac{1}{CR} + \dfrac{T_s}{LC}(m_c D' - 0.5)$ for $F_p(s)$, and the quality factor $Q = \dfrac{1}{\pi \cdot (m_c D' - 0.5)}$ for $F_h(s)$. To see these three effects, the approximated $G_{vc}(s)$ function in (2) can be separated into two terms as listed in (3) and (4).

$$G_{vc\_first\_term}(s) = \frac{R}{R_i} \cdot \frac{1}{1 + \dfrac{R \cdot T_s}{L}(m_c D' - 0.5)} \cdot F_p(s) \qquad (3)$$

$$G_{vc\_second\_term}(s) = F_h(s) = \frac{1}{1 + \dfrac{s}{\omega_n Q} + \dfrac{s^2}{\omega_n^2}} \qquad (4)$$

The first term is a one-pole one-zero transfer function. The pole is generated by the paralleling output capacitor and load resistor while the inductor is like a current source. The zero is from the output capacitor and its ESR. The second term is from the sample-and-hold effect. These two terms can be plotted separately to show their effects to the big variation in $G_{vc}(s)$. The first term is plotted in Fig. 6. In the intended crossover frequency range ($0.1*f_{sw}$ to $0.2*f_{sw}$), we can see that all the $G_{vc\_first\_term}(s)$ curves have only negligible variation in terms of magnitude, while the variation in phase is about 10 to18 degrees.

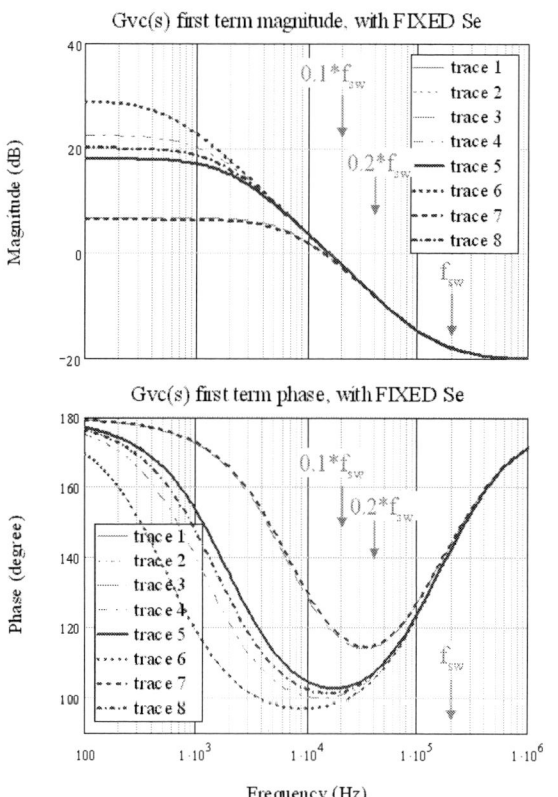

Figure 6. Bode plot for the first term of $G_{vc}(s)$ with fixed external ramp.

But, as can be seen from Fig. 7, in the intended crossover frequency range ($0.1*f_{sw}$ to $0.2*f_{sw}$), the $G_{vc\_second\_term}(s)$ in Equation (4) has about 2 to 4dB variation in magnitude and 19 to 22 degrees variation in phase using fixed external ramp

S_e. The variation of the second term is more significant than that of the first term.

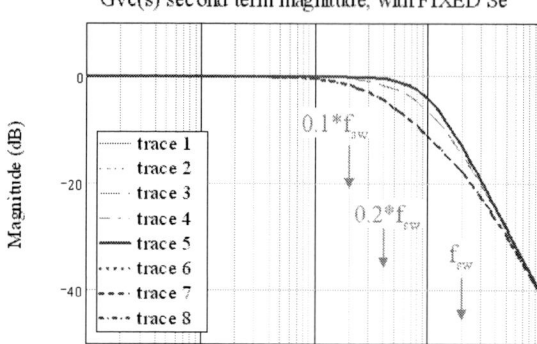

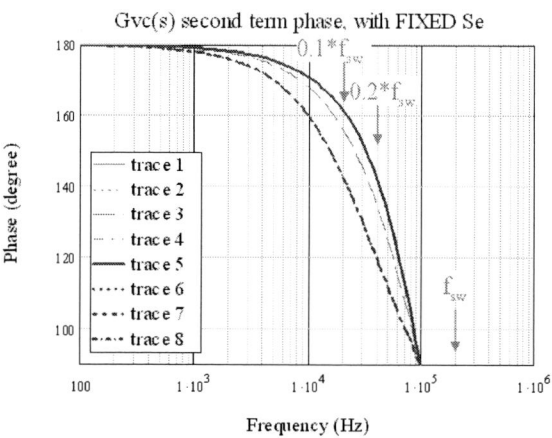

Figure 7. Bode plot for the second term of $G_{vc}(s)$ with fixed external ramp

As mentioned before, the large variation in magnitude and phase of the $G_{vc}(s)$ shows difficulty to design one fixed compensator with high bandwidth. The question is how to eliminate the variation. The variation in the first term $G_{vc\_first\_term}(s)$ shows up almost only in the phase, which is mainly determined by the frequency of the pole

$$\omega_p = \frac{1}{CR} + \frac{T_s}{LC}(m_c D' - 0.5)$$ of $F_p(s)$. Once the

output capacitor is calculated and selected, this pole is mainly determined by the load resistance R, which is normally a design specification instead of a design freedom. So it is not something that can be controlled by the designer.

However, the designer can have more freedom for the second term $G_{vc\_second\_term}(s)$. By observing Equation (4), noticing that $\omega_n = \dfrac{\pi}{T_s}$ is the same for all cases, if the quality factor Q can be designed in such a way that it is the same for all working conditions, the variation for the second term can be eliminated. Therefore, the external ramp S_e should be adaptive such that $Q = \dfrac{1}{\pi \cdot (m_c D' - 0.5)}$ is a constant, which means $m_c D'$ should be a constant. According to [5], Q should be less than 1 to ensure the current loop won't oscillate. On the other hand, if Q is too small, the external ramp S_e is much larger than the sensed current ramp S_n. The control loop will behave more like voltage mode control instead of current mode control, and the advantages of current mode control is lost. A good choice is $Q = 2/\pi$, which means $m_c D' = 1$. An interesting result is that S_e is proportional to V_o with this choice of Q, which is easy to implement the adaptive S_e by using voltage-controlled-current-source to charge a capacitor and to reset its voltage at the beginning of each cycle. This seems to reach the same implementation as that in [2]. But, different from [2], to make S_e proportional to V_o is not the only choice to minimize $G_{vc}(s)$ variation. As long as Q can be kept the same for all operating conditions, the variation of $G_{vc\_second\_term}(s)$ can be eliminated. Q can be other values instead of $2/\pi$, but it is more difficult to implement.

With this adaptive S_e control scheme, $G_{vc\_second\_term}(s)$ will be the same for all working conditions as shown in Fig. 8.

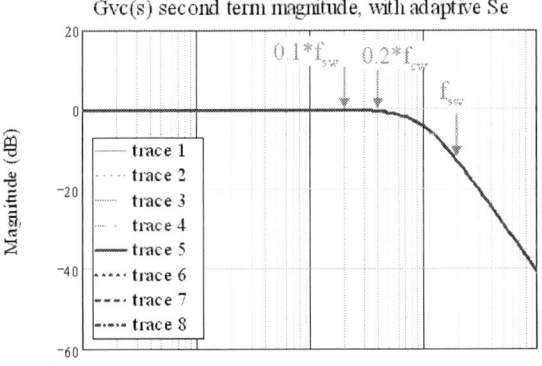

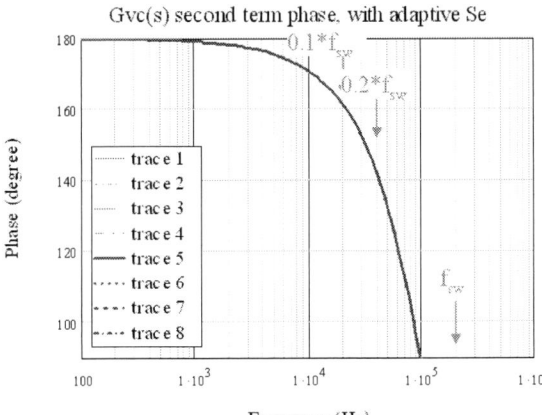

Figure 8.　Bode plot for the second term of $G_{vc}(s)$ with adaptive external ramp

And the variation for the first term is about the same as that of fixed external ramp in the frequency range of $0.1*f_{sw}$ to $0.2*f_{sw}$. So the big variation in $G_{vc}(s)$ is reduced in the frequency range of $0.1*f_{sw}$ to $0.2*f_{sw}$ as shown in Fig. 9. Looking at this frequency range, the difference in the magnitude is negligible, while the difference in the phase is about 8-20 degrees. Both are less than those shown in Fig. 3.

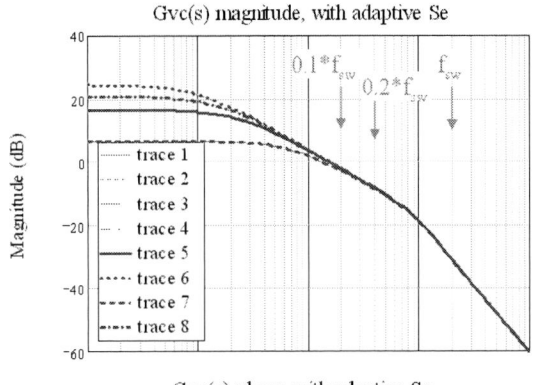

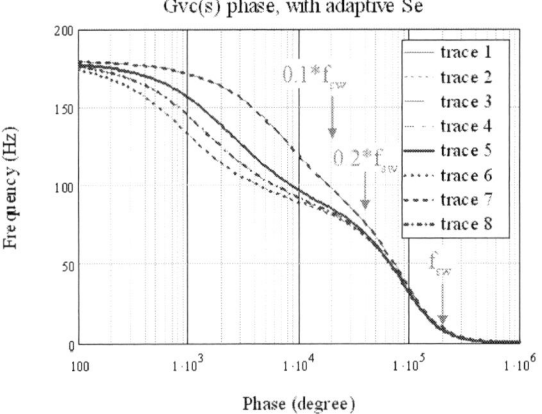

Figure 9.　$G_{vc}(s)$ curves with adaptive external ramp

Because the variation of $G_{vc}(s)$ is reduced, it allows a new compensator $G_{comp\_new}(s) = \dfrac{4290}{s} \cdot \dfrac{1 + s/2000}{1 + s/1000000}$ to be used while keeping at least 60 degree phase margin for all working conditions. The new set of loop gain curves and the zoom-in view are shown in Fig. 10 and Fig. 11, respectively. Now the worst case bandwidth is improved to 31kHz for trace 3 (V_{in}=24V, V_o=1.2V, I_o=5A) and trace 7 (V_{in}=6V, V_o=1.2V, I_o=5A), which is about twice as high as before. And that will improve load transient response.

The above analysis is verified using the SIMPLIS software, which has been considered as a powerful tool in recent years to verify transfer functions of switch mode power supply. The reason is that it can simulate transfer function directly from time domain simulation without using the average models. All the loop gain curves in Fig. 4 and Fig. 10 have been verified with SIMPLIS simulation, and one of the plots is shown in Fig. 12 for the case of V_{in}=6V, V_o=1.2V and I_o=0.5A. Up to 100kHz ($0.5*f_{sw}$), the small signal model match SIMPLIS result very well.

978-1-4244-4782-4/10 $26.00 © 2010 IEEE

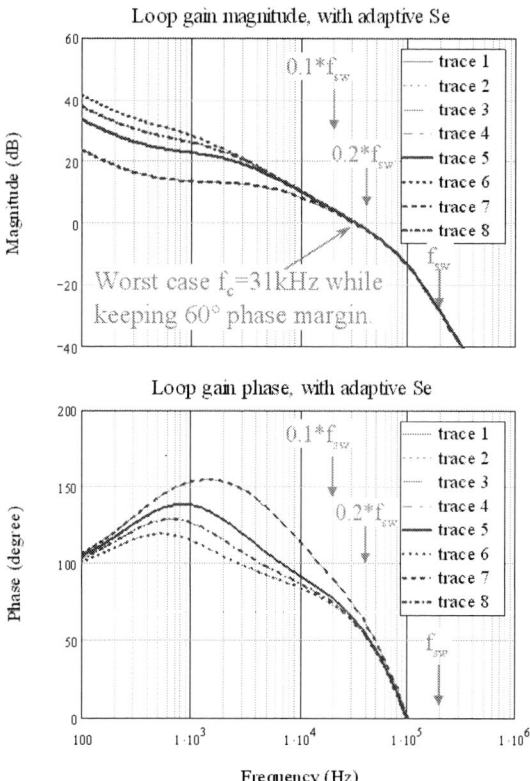

Figure 10. Loop gain curves with adaptive external ramp

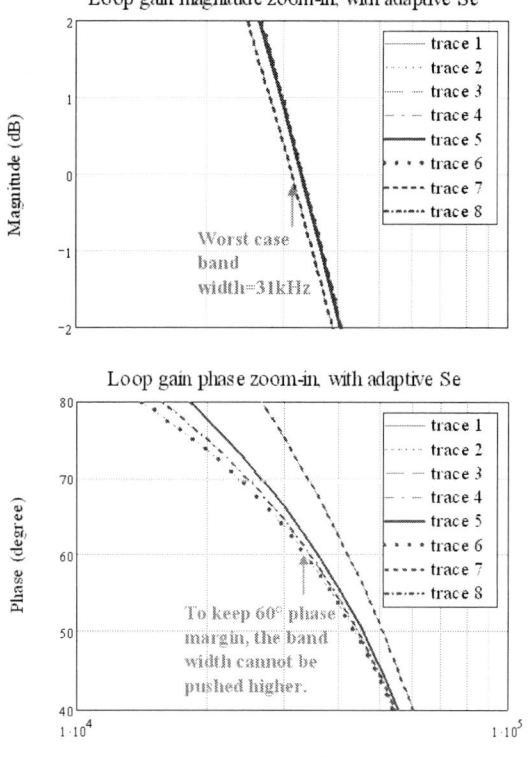

Figure 11. Loop gain curves with adaptive external ramp, zoomed-in.

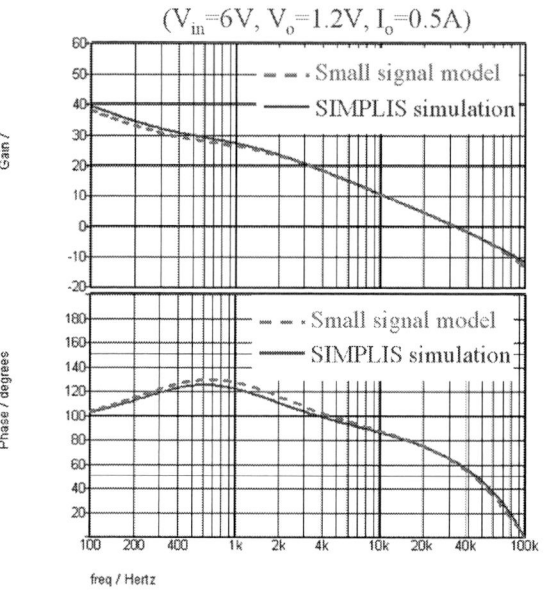

Figure 12. Loop gain comparison between small signal mode (red dashed line) and SIMPLIS simulation (blue solid line).

IV. EXPERIMENTAL RESULTS

To demonstrate the advantage of the proposed adaptive external ramp control, a demonstration board has been build and tested on the bench with the same wide operation range mentioned in Section III: V_{in}=6V-24V, V_o=1.2V-5V, I_o=0.5A-5A, f_{sw}=200kHz, C=100uF, R_c=10mΩ, L=10uH, and current sense gain R_i=0.1V/A. Fig. 13 shows a photo for this PCB board. The control loop is fixed for all the operation range except the value of a resistor R_{bias} connecting between the inverting input of the error amplifier and ground for setting different V_o. The amplitude of the external ramp is automatically controlled by the circuit itself to extend the control bandwidth.

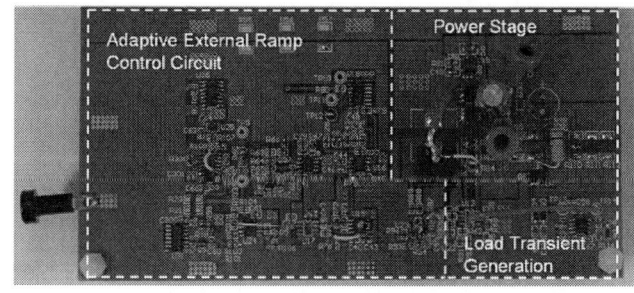

Figure 13. Photo of the demonstration board with adaptive external ramp control.

The steady state operation waveform for the case of V_{in}=6V, V_o=1.2V, I_o=5A is shown in Fig. 14. While the steady state operation waveform for the case of V_{in}=24V, V_o=5V, I_o=0.5A is shown in Fig. 15. They demonstrate that the PCM controller with adaptive external ramp can work stably for wide operating range with the proposed automatically self-adjusted external ramp method.

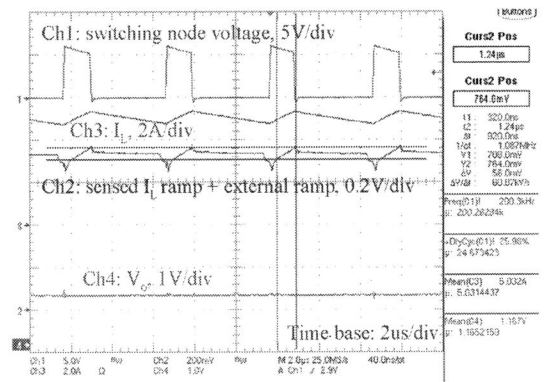

Figure 14. Waveform of V_{in}=6V, V_o=1.2V, I_o=5A.

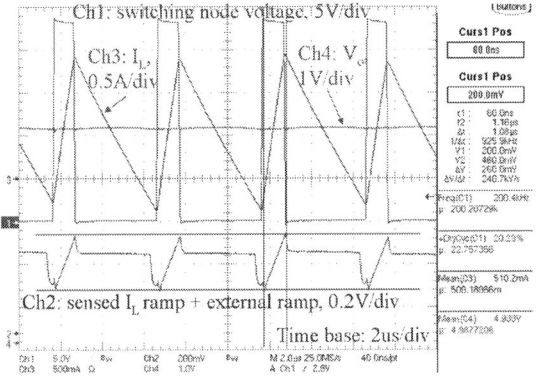

Figure 15. Waveform of V_{in}=24V, V_o=5V, I_o=0.5A.

The load transient waveform of I_o=0.6A➜1.8A➜0.6A is shown in Fig. 16 for the proposed adaptive external ramp control scheme. For comparison, the load transient waveform using fixed external ramp under the same condition is shown in Fig. 17. As illustrated in Fig. 16 and Fig. 17, the transient response of the adaptive external ramp control scheme is much better than that of fixed external ramp because of the higher crossover frequency f_c.

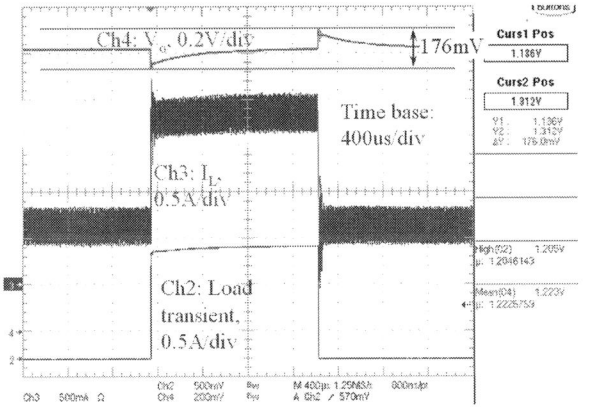

Figure 16. Load transient of adaptive S_e control.

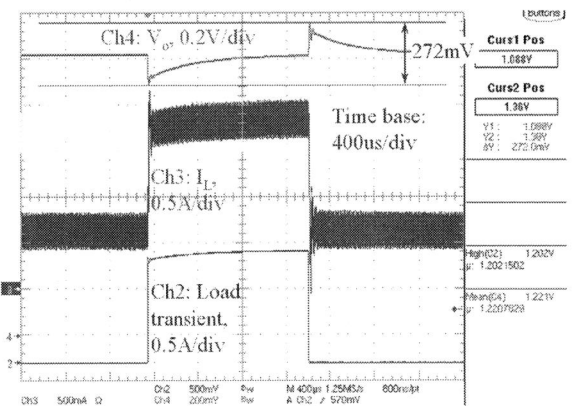

Figure 17. Load transient of fixed S_e control

V. CONCLUSIONS

An adaptive external ramp control scheme is proposed to extend the loop gain bandwidth for wide operation range Buck converters using peak current mode control. Theoretical analysis shows that the Q of the second term of $G_{vc}(s)$ needs to be identical in order to minimize $G_{vc}(s)$ variation for wide operation range PCM Buck converters. By doing so, using a fixed compensator, the control bandwidth of the PCM Buck converter can be extended significantly compared to that of using fixed external ramp. SIMPLIS simulation verifies the theoretical analysis, and experimental results show the benefit of the proposed adaptive external ramp control method.

REFERENCES

[1] R. B. Ridley, " A new, continuous-time model for current mode control," IEEE Transactions on Power Electronics, vol. 6, No. 2, pp. 271-280, April 1991.

[2] C. Yang, C. Wang and T. Kuo, " Current-mode converters with adjustable-slope compensating ramp," APCCAS 2006, pp. 654-657

[3] J. Y. Guo, " Investigating feedforward mechanism in current mode control," APEC 2006, pp. 1008-1013.

[4] Y. Qiu, J. Sun, M. Xu, K. Lee and F. C. Lee, " Bandwidth Improvements for Peak-Current Controlled Voltage Regulators," IEEE Transactions on Power Electronics, vol. 22, No. 4, pp. 1253-1260, July 2007.

[5] R. B. Ridley, " A More Accurate Current-Mode Control Model," available at http://www.ridleyengineering.com/downloads/curr.pdf.

[6] D. M. Sable, R. B. Ridley and B.H. Cho, " Comparison of Performance of Single-Loop and Current-Injection Control for PWM converters that Operate in Both Continuous and Discontinuous Modes of Operation," IEEE Transactions on Power Electronics, vol. 7, No. 1, pp. 136-142, January 1992.

[7] F. D. Tan and R. D. Middlebrook " A Unified Model for Current-Programmed Converters," IEEE Transactions on Power Electronics, vol. 10, No. 4, pp. 397-408, July 1995.

[8] F. J. Azcondo, Ch. Bracas, R. Casanueva, D. Maksimovic " Approaches to modeling converters with Current Programmed Control," IEEE Workshop on Power Electronics Education 2005, pp. 98-104

[9] B. Choi, " Step load response of a current-mode-controlled DC-to-DC converter," IEEE Transactions on Aerospace and Electronic Systems, vol. 33, No. 4, pp. 1115-1121, October 1997

Masterless Multirate Control of Parallel DC-DC Converters

Anthony Kelly, Karl Rinne, Eamon O'Malley
Powervation
Limerick, Ireland

Abstract— **This paper introduces a new masterless multirate control scheme for current share in DC-DC converters. Comprising of a multirate digital controller, decoupling interactions between control loops, a single wire digital communication bus, facilitating simple and reliable communication between devices, and a modular current share architecture. Results show excellent static and dynamic current matching, and have been commercially proven.**

I. Introduction

Many power applications, such as networking and microprocessor power, apply parallel, interleaved DC-DC converters. There are numerous examples in the literature [[1]-[8]] covering a wide range of approaches. Droop methods are the simplest, sharing current passively [[10]]. They lend themselves well to a modular approach, but are limited in performance. Active control of the supplied current allows more accurate matching [[9]] at the expense of complexity.

Active schemes often require a master to set the output voltage or current which the slave devices must follow. However, the master represents a single point of failure in a complex system. In contrast, masterless schemes are more robust against failure and have a distinct advantage in mission critical applications.

Interaction between the control loops of parallel DC-DC converters is problematic, and is compounded by the interaction of current and voltage control loops involved in active current share [[8],[11]].

This paper introduces a new masterless control scheme for current share in DC-DC converters, comprising of a multirate digital controller, decoupling interactions between control loops, a single wire digital communication bus, facilitating simple and reliable communication between devices, and a modular current share architecture.

Aspects of this work have Patents Pending.

II. Modular Current Share Architecture

Parallel DC-DC converters are illustrated in Figure 1. Due to output voltage variations and power stage mismatches, the paralleled devices do not share currents equally. Some devices will be more stressed than the others, leading to reduced reliability, and possibly operating devices beyond their operating range.

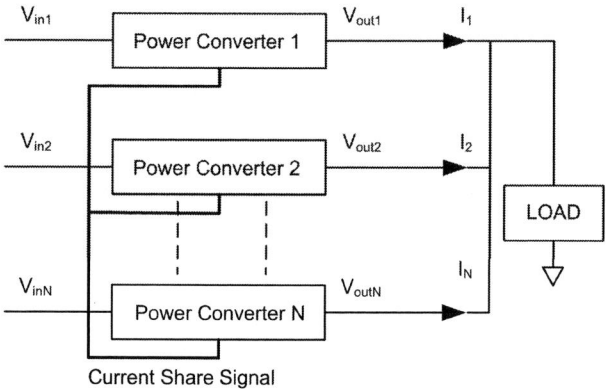

Figure 1 Parallel DC-DC converters

It is desirable for reliability that each of the parallel DC-DC converters employs a modular architecture so there is not a specific reliance any converter. This addresses the single point of failure in master based schemes in which a master DC-DC converter typically controls the output voltage regulation whilst the others are typically current controlled to achieve current sharing.

In order to achieve current balance each power converter connects to a current share signal and actively modifies its current to match the reference current indicated by the current share signal. The current share signal typically represents the maximum, minimum or average current being delivered by a power converter in the system [[1],[9]]. Devices which are

978-1-4244-4782-4/10 $26.00 © 2010 IEEE

not at the reference current move towards the reference current under the control of a current balance control loop which exists in each device. As the power converter currents change under the influence of the current share control loop, the reference current communicated over the current share signal changes, and the currents supplied by the multiple parallel power converters equalize over time. The speed of communication and control is a priority, as this equalization is limited by the bandwidth of the current share loop [[12]].

Being a masterless, modular current share system, all power converters connect to the current share bus. The output voltage of each power converter is determined by its setpoint which is modified by the current share system in order to share current equally. This outer current loop, inner voltage loop method is highly modular [[9]] and reduces interaction between current and voltage loops which is often problematic [[11],[12]].

Figure 2 shows the DC-DC converter in detail, illustrating the current share scheme. The current share communications block is a single wire digital bus. Each device on the bus outputs its maximum and minimum current values, and by a process of arbitration at the bus interface level, the maximum and minimum currents being delivered to the load are made available to the average value estimator (avg).

The average value estimator estimates the average current in the system, providing a reference setpoint for the current share loop. The current share controller (ki) calculates a voltage setpoint multiplier which modifies the desired output voltage setpoint so as to balance the currents between power converters. Each device sharing via the current share bus operates in the same way.

III. COMMUNICATION SCHEME

A digital single wire communication scheme is required in order to robustly achieve accurate communication of each converter's average current. Employing a standardized serial interface is impractical due to bandwidth limitations, and undesirable because of the increased complexity involved in clocking.

The primary challenge to be addressed in a single-wire digital communication scheme is timing latency. The latency of transmission and average value estimation depends on the number of devices connected to the bus. It must also be considered that the total number of connected devices may not be known to each one, therefore requiring cumbersome configuration. The problem is reduced by the following insight: instead of letting all slaves in the power system submit their individual current contribution, only obtain the information from the marginal slaves i.e. those slaves which contribute the highest and the lowest output current. An estimate of the average system current can then be calculated as :

$$\hat{I}_{o,ave} = \frac{I_{o,\max} + I_{o,\min}}{2} \qquad (1)$$

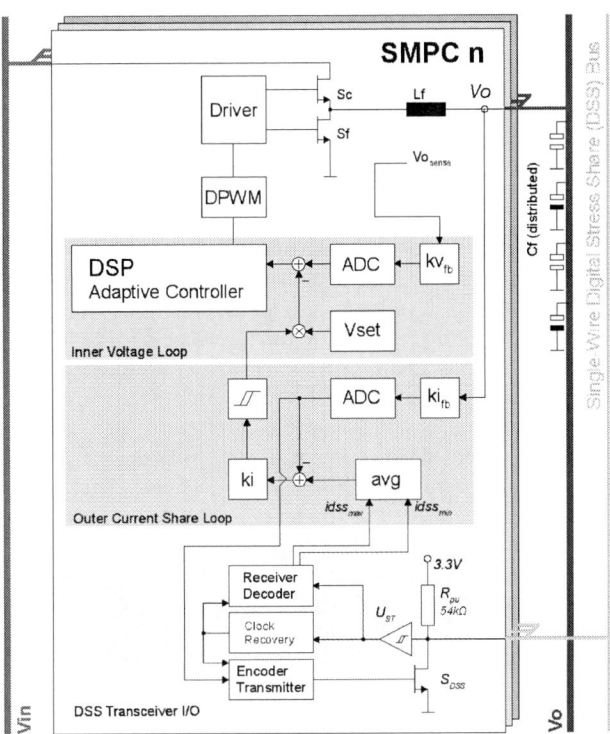

Figure 2 DC-DC Converter Detail

Where $I_{o,\max}$ and $I_{o,\min}$ are the maximum and minimum currents being supplied by devices respectively; as received over the current share bus. $\hat{I}_{o,ave}$ represents an estimate of the average current supplied by the devices to the load.

The communication problem is reduced considerably by employing a protocol in which $I_{o,\max}$ and $I_{o,\min}$ are automatically determined on the bus. In this way the latency of transmission and averaging is fixed and independent of the number of devices.

IV. CONTROL METHOD

A. Average Value Estimator

The purpose of the average value estimator is to provide a reference current which each device must supply. The maximum and minimum system currents, available from the current share comm's block are summed to produce a point average estimate. The average current estimation derived from equation (1) is a spot estimate and is therefore noisy. It may be improved upon by applying a filter to yield a time averaged value. An N-tap FIR average is employed whose transfer function H(z) is:

$$H(z) = \frac{1 - z^{-N}}{1 - z^{-1}} \qquad (2)$$

Although each device contains its own estimator, independently deriving its average share current, the current share information received by each device via its DSS block is the same. This ensures that each device has the same target current from the average value estimator.

B. Multirate Filter

The current share communications block operates at a 20kHz rate and its point estimate of the average system current is filtered in order to provide a reference current for all power converters. Averaging the reference current in this way reduces the continuous interaction between power converters and reduces noise on the current reference signal. The average value is produced through a process of decimation, i.e. down-sampling and filtering, that is conveniently carried out in a Hogenauer decimator, yielding a reference current at a 1.25kHz rate, which all devices track.

Interaction between control loops is problematic in current share schemes [[8],[11],[12]]. It is common to minimize the interaction between control loops by ensuring the control bandwidth of the current share loop is much less than the voltage loop, and therefore they can be considered to be independent of each other.

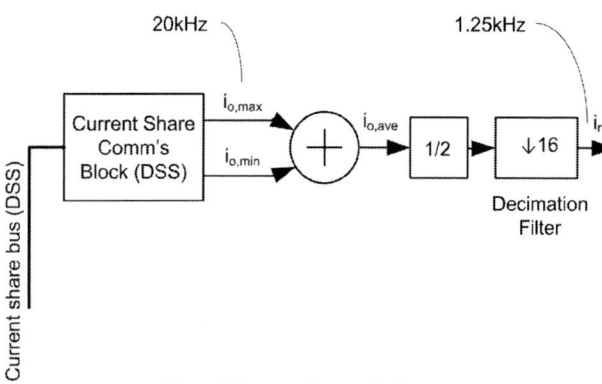

Figure 3 Average Current Estimation

Figure 3 shows the average current estimator used in the present design. It can be seen that the filtered average current estimate from the decimation filter is at a lower sample rate than the signals received by the DSS block. Employing multirate signal processing in this way ensures that the target current setpoint reference which comes out of the averager (i_{ref}) changes slowly compared to the sample frequency of the current loop controller, thus ensuring a degree of decoupling between device interaction.

The frequency domain transfer function of the average current estimator is given by equation (2) and illustrated as a frequency plot in Figure 4. The normalized frequency of 1 corresponds to the input sample rate over two, i.e. 10kHz; the notches occur at the input sample rate over the decimation factor, i.e. 20kHz/16 = 1.25kHz .

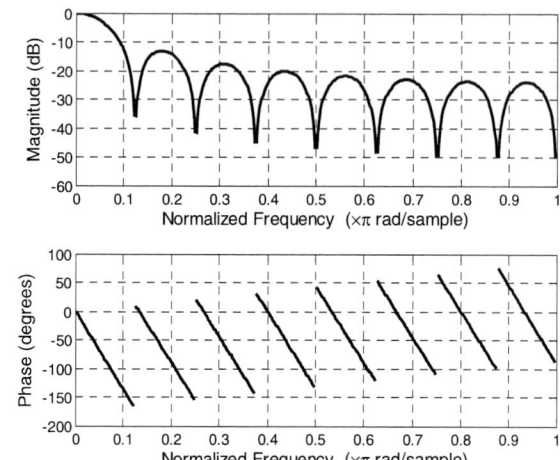

Figure 4 Frequency/Phase Response of the Average Current Estimator

C. Current Controller

The current share controller shown in Figure 5 consists of a compensator $C_c(z)$ in a negative feedback loop with the feedback gain k_f, and an outer current control loop with the devices' measured current (i_{device}). The current share controller operates at the higher sampling rate of 20kHz which allows fast control of each device's output current so as to share current well during load transients.

A drawback of an outer current and inner voltage loop is that of voltage setpoint accuracy. Because the outer current loop controls the voltage setpoint it may deviate from the required voltage setpoint more than is required for current share alone. The present design deals with this traditional drawback by incorporating feedback of the setpoint deviation so that movements away from the desired voltage setpoint are penalized, resulting in improved voltage setpoint accuracy in this type of control scheme.

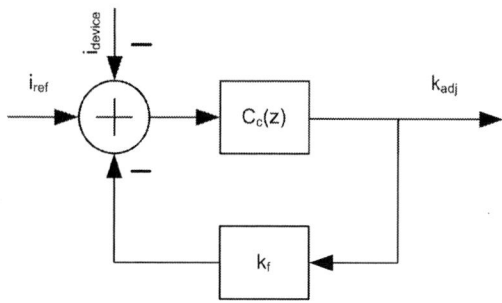

Figure 5 Current control scheme

In the z-domain the current share controller consisting of $C_c(z)$ and k_f can be rearranged into a closed expression $C`_c(z)$ as follows:

978-1-4244-4782-4/10 $26.00 © 2010 IEEE

$$C'_c(z) = \frac{C_c(z)}{1 + k_f C_c(z)} \tag{3}$$

with:

$$C_c(z) = \frac{b_0 z + b_1}{z - 1} \tag{4}$$

we have:

$$C'_c(z) = \frac{b_0}{1 + k_f b_0} \frac{z + \dfrac{b_1}{b_0}}{z + \left(k_f b_1 - 1\right)} \tag{5}$$

Therefore, utilizing an integral controller with real zero as $C_c(z)$, the negative feedback gain k_f shifts the controller pole, resulting in a lower DC gain, thus ensuring a balance between current share accuracy and voltage setpoint accuracy.

V. RESULTS

Four Point of load devices were configured to share current into a single load as illustrated in Figure 1. Static measurements of current sharing between four power converters at 60A total load, Figure 6, show 1.3A worst case matching of the load current.

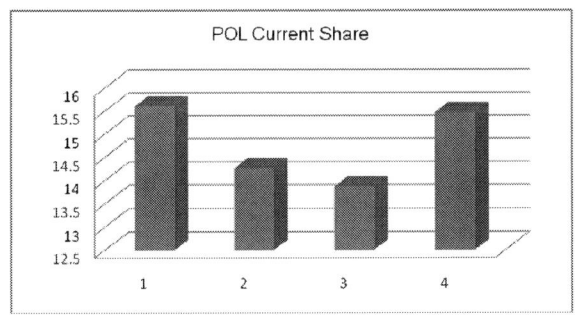

Figure 6 Static current sharing of four POLs sharing at 60A load

Dynamic operation illustrated in, Figure 6, shows no excessive overshoot or noise from the current loop and illustrates the dynamic setpoint adjustment.

VI. CONCLUSION

A new masterless control scheme for current share in DC-DC converters has been introduced. Practically important considerations such as fault tolerance, interaction between control loops and robust communication have been dealt with. System measurements show excellent static and dynamic current matching between parallel converters, illustrating that multirate digital techniques may be practically and usefully applied to the problem of current sharing in parallel DC-DC converters.

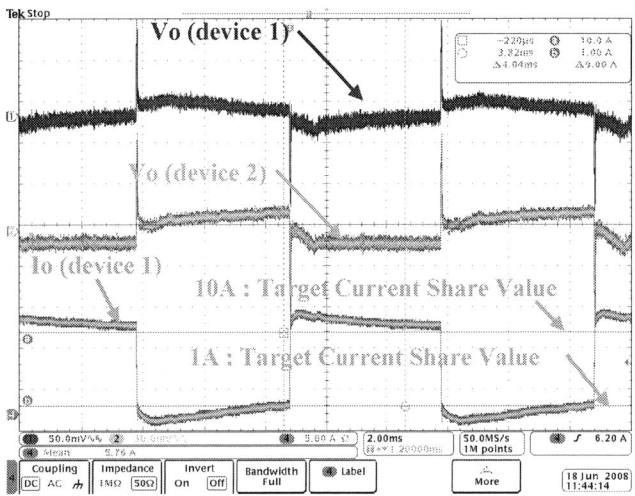

Figure 7 Two POLs sharing current during load current steps

ACKNOWLEDGMENT

The authors would like to acknowledge the valuable assistance and advice of their colleagues Basil AlMukhtar and Adrian Ward in the preparation of this material.

REFERENCES

[1] Balogh, L. Paralleling Power - Choosing and Applying the Best Technique for Load Sharing. in Unitrode Design Seminar. 2003.

[2] Ridley, R.B., B.H. Cho, and F.C.Y. Lee, Analysis and interpretation of loop gains of multiloop-controlled switching regulators [power supply circuits]. Power Electronics, IEEE Transactions on, 1988. 3(4): p. 489-498.

[3] Choi, B., et al. Control strategy for multi-module parallel converter system. in Power Electronics Specialists Conference, 1990. PESC '90 Record., 21st Annual IEEE. 1990.

[4] Siri, K., C.Q. Lee, and T.E. Wu, *Current distribution control for parallel connected converters. I.* Aerospace and Electronic Systems, IEEE Transactions on, 1992. 28(3): p. 829-840.

[5] Zhang, Y., R. Zane, and D. Maksimovic. Current Sharing in Digitally Controlled Masterless Multi-phase DC-DC Converters. in Power Electronics Specialists Conference. 2005.

[6] Kelly, A., Current Share in Multiphase DC-DC Converters Using Digital Filtering Techniques. Power Electronics, IEEE Transactions on, 2009. 24(1): p. 212-220.

[7] Hyunsu, B., et al., *Digital Resistive Current (DRC) Control for the Parallel Interleaved DC-DC Converters.* Power Electronics, IEEE Transactions on, 2008. **23**(5): p. 2465-2476.

[8] Xiaogao, X., et al. Analysis and Design of N paralleled DC/DC Modules with Current-Sharing Control. in Power Electronics Specialists Conference, 2006. PESC '06. 37th IEEE. 2006.

[9] Huang, Y. and C.K. Tse, *Circuit Theoretic Classification of Parallel Connected DC-DC Converters.* Circuits and Systems I: Regular Papers, IEEE Transactions on [Circuits and Systems I: Fundamental Theory and Applications, IEEE Transactions on], 2007. **54**(5): p. 1099-1108.

[10] Irving, B.T. and M.M. Jovanovic. Analysis, design, and performance evaluation of droop current-sharing method. in Applied Power Electronics Conference and Exposition, 2000. APEC 2000. Fifteenth Annual IEEE. 2000.

[11] Yang, Z., R. Zane, and D. Maksimovi. Dynamic Loop Analysis for Modular Masterless Multi-Phase DC-DC Converters. in Computers in Power Electronics. 2006.

[12] Juanjuan, S., et al., High-Frequency Dynamic Current Sharing Analyses for Multiphase Buck VRs. Power Electronics, IEEE Transactions on, 2007. 22(6): p. 2424-2431.

FPGA-Based Spectral Envelope Preprocessor for Power Monitoring and Control

Zachary Remscrim, James Paris, Dr. Steven B. Leeb, Dr. Steven R. Shaw,
Sabrina Neuman, Christopher Schantz, Sean Muller, Sarah Page

Abstract—Smart Grid and Smart Meter initiatives seek to enable energy providers and consumers to intelligently manage their energy needs through real-time monitoring, analysis, and control. We have developed an inexpensive FPGA implementation of a spectral envelope preprocessor. This FPGA permits cost-effective and richly detailed power consumption monitoring for individual loads or collections of loads. It permits a flexible trade-off between data transmission, storage, and computation requirements in a power monitoring or control system. The information from the FPGA can be used to coordinate the operation of power electronic controls.

I. BACKGROUND

Power electronics and power electronic controls are pro-liferating in consumer electronics. There is an increasing expectation that advanced power conditioning electronics will play a role in managing and coordinating power consumption not simply for a particular load, e.g., a variable speed drive in an air conditioning plant, but also in response to the dynamic needs and capability of the utility system. Loads that can respond not only to their own tasking but also to the needs of the utility are implicit in many visions of a "smart grid."

There is a need for flexible, inexpensive metering tech-nologies that can be deployed in many different monitoring scenarios. Individual loads may be expected to compute in-formation about their power consumption. They may also be expected to communicate information about their power consumption through wired or wireless means. Switch gear like circuit breaker panels may eventually be expected to provided detailed submetering information for different loads on different breakers or clusters of breakers and controls. New utility meters will need to communicate bidirectionally, and may need to compute parameters of power flow not commonly assessed by most current meters.

Both vendors and consumers will likely find innumerable ways to mine information if made available in a useful form. However, appropriate sensing and information delivery systems remain a chief bottleneck for many applications, and metering hardware and access to metered information will likely limit the implementation of new electric energy conservation strategies in the near future. The U.S. Department of Energy has identified "sensing and measurement" as one of the "five fundamental technologies" essential for driving the creation of a "Smart Grid" [7]. Consumers will need "simple, accessible..., rich, useful information" to help manage their electrical consumption without interference in their lives [7].

Digital technology has been in use for over 20 years for measuring and metering power flow. A few examples from an enormous array of metering and measurement approaches for monitoring power can be found in [8]–[11], and the references in these documents. Digital power monitoring has also made its way to the "plug" and "power strip" level, e.g., see [13]. Many different schemes for storing or communicating infor-mation are still under exploration—see [12] and its references for example. Most of these solutions deploy computation hard-ware that is either substantially complicated in both hardware and firmware, e.g., [10] where a DSP and a micro-controller work together to coordinate computation of real, reactive, and apparent power, or where fully integrated custom chips are specifically developed for a particular application.

The "spectral envelope" representation of observed current and voltage signals used in the non-intrusive load monitor or NILM [2], [14], [15] can be a very flexible basis for computing and tracking all sorts of useful metrics about power consumption. Spectral envelopes estimate real and reactive power consumption and harmonic contents. Even for waveforms with substantial high frequency content, the time-varying spectral envelopes can be relatively band-limited. Thus, spectral envelopes are often a natural way to "compress" useful data about load current and power consumption, easing communication requirements.

This paper describes an integer-arithmetic implementation of a spectral envelope preprocessor for an inexpensive FPGA. The spectral envelope FPGA coordinates data acquisition and computes spectral envelopes without the need for floating-point computation. Hence, the FPGA can be used in a two-IC suite (with an analog-to-digital converter) to inexpensively acquire load consumption data. This data minimizes the need for "downstream" computation later in the signal processing workflow. Of course, further computation can be used to track, trend, price, or control energy consumption. The FPGA can directly control communication as well, providing wired or wireless access, or storage on flash memory or other media. The spectral envelope FPGA is a cost-effective building block that can be used to enable a huge array of power monitoring and control applications, ranging from the individual load up through the breaker panel or utility service entry level and beyond. It can be used to provide necessary power consumption information for coordinating power electronic controls.

This paper describes the design of this key building block and shows results from a prototype. The next section describes the approach for using spectral envelopes for load identifi-cation and how this data is computed by the preprocessor. The following section presents the FPGA-based spectral en-velope preprocessor and the techniques used to implement the preprocessor cost-effectively. Finally, the paper examines the performance of prototype hardware and describes further enhancements for expanded monitoring applications.

978-1-4244-4782-4/10 $26.00 © 2010 IEEE

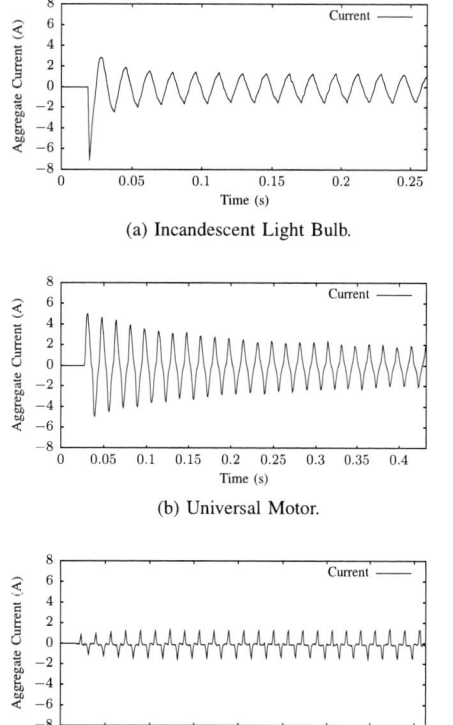

(a) Incandescent Light Bulb.

(b) Universal Motor.

(c) Computer Power Supply.

Fig. 1. Turn-on transients for an incandescent light bulb, a universal motor, and a computer power supply.

II. UTILITY OF SPECTRAL ENVELOPES

Typical turn-on current transients are shown in Fig. 1 for an incandescent lamp, a universal motor (as in a vacuum cleaner or hand tool), and a personal computer. Dynamic changes in the power and harmonic consumption of a load, e.g., during turn-on or turn-off transients, can serve as a fingerprint for identifying load operation [1]. For example, an observed turn-on transient or exemplar from a training observation produced by one of a collection of loads can be used to identify the load in an aggregate current measurement. An analogous procedure can be performed using turn-off transients. All that is needed, in principle, to determine the operating schedule of a collection of loads is to record the aggregate current drawn by those loads and then match each observed transient to the turn-on or turn-off fingerprint of a particular load in the collection.

However, direct examination of current waveforms may not be practical for many stages of some applications, including many components in energy scorekeeping, monitoring, or conservation systems. Direct operations on the current waveform require sample rates adequate to capture the highest harmonic content of interest [5]. In some metering, monitoring, and control applications it is more practical to either store data for a period of time and examine it later, or to transmit data to another location for interpretation and control. In either of these cases, it is convenient to have a useful representation of the data that avoids excessive storage or communication bandwidth requirements.

Spectral envelopes provide a useful separation between data collection and analysis. They permit a small, inexpensive system with low processing power to collect data continuously. A system with larger available processing power, potentially physically remote from the data collection front-end, can either review a storage device at a later time or continuously process a relatively low bandwidth information stream over a convenient communication channel, wired or wireless.

A. Spectral Envelope Definition

The spectral envelopes of current represent the harmonic content of the input waveform for each line-locked period of the service voltage. Given N samples $i[n]$ of a waveform $i(t)$ over one period, the samples can be expressed in terms of their spectral content by

$$i[n] = \frac{1}{N} \sum_{k=0}^{N-1} \left(a_k \sin\left(\frac{2\pi kn}{N}\right) + b_k \cos\left(\frac{2\pi kn}{N}\right) \right) \quad (1)$$

where the spectral envelopes a_k and b_k for that period are defined by

$$a_k = \sum_{n=0}^{N-1} i[n] \sin\left(\frac{2\pi kn}{N}\right) \quad (2)$$

and

$$b_k = \sum_{n=0}^{N-1} i[n] \cos\left(\frac{2\pi kn}{N}\right). \quad (3)$$

Here, k denotes the multiple of the line frequency to which a particular spectral envelope corresponds; for example, $k = 1$ corresponds to the 60 Hz component and $k = 3$ to the 180 Hz component. The values of these spectral envelopes are calculated for each period of the line voltage; the values at period m will be denoted $a_k[m]$ and $b_k[m]$. With this definition, spectral envelopes can naturally be calculated from the real and imaginary parts of the Discrete Fourier Transform (DFT) [16] of $i[n]$ over each period of the line voltage.

B. Spectral Envelope Compression

Given the DFT coefficients over one period of the service voltage, it is possible to exactly reconstruct or preserve all of the information in the raw current samples over that period. Of course, simply recording all of the DFT coefficients will not reduce the data handling requirements—if there are N samples of current, the DFT will produce N DFT coefficients. While it may appear that storing the complex DFT coefficients would take twice as much space as storing the raw current samples, because they are complex numbers, note that the DFT will be symmetric (because the raw current samples are real), so only $\frac{N}{2}$ of the N complex numbers need to be stored. The data requirements for storing all meaningful DFT coefficients is the same as storing all raw current samples at the same level of precision.

In situations where the significant or relevant current drawn by an electrical load consists predominantly of the fundamental frequency (the frequency of the service voltage, for example, 60 Hz) and a small collection of the line frequency harmonics (such as the 3rd, 5th, 7th), it is reasonable to record, for each period of the service voltage, only a few DFT coefficients [2]. These relatively few DFT coefficients can be used to reconstruct the original current samples with a relatively small error. Furthermore, the time varying values of the DFT coefficients themselves can be used directly as fingerprint signatures for the loads, or to track important quantities associated with load operation, with reasonable accuracy.

Because only a few DFT coefficients may be needed to accurately represent the current waveforms, this "spectral" approach to the representation of the waveforms serves as a form of compression. Consider current and voltage samples that are collected at a 7.68 KHz sample rate. The sampling rate is chosen to be sufficiently high to provide adequate anti-aliasing without filtering effects and to provide accurate detection of voltage zero crossings to enable line-locked data collection. This corresponds to 128 samples per 60 Hz line-cycle ($N = 128$), and so 64 meaningful complex DFT coefficients need to be stored to perfectly reconstruct arbitrary current samples. However, for many applications, including load monitoring for diagnostics, only a limited number of DFT coefficients need be stored. In the prototype system discussed in this paper, just 4 coefficients, or 6.25% of the full set of already compact harmonically-related DFT coefficients, were needed.

Of course, other reductions of the data could be applied, e.g., simply recording average aggregate real power once per second (where the average is taken over each second interval), leads to further compression. Such data would not reflect the detailed short term variations that would occur in real power, nor would it reflect any of the behavior of the higher harmonics. Time-varying DFT coefficients or spectral envelopes strike a balance between the need to store or transmit as little data as possible and the need to maintain sufficiently detailed data to perform load monitoring.

C. Physical Interpretation of Spectral Envelopes

Spectral envelopes can be directly interpreted as other meaningful physical quantities under some conditions. In situations where the utility voltage is relatively harmonic free and "stiff" (with constant peak amplitude), the real part of the first DFT coefficient of the current waveform, or fundamental frequency spectral envelope a_1, scales to "real power" in steady-state. Similarly, the imaginary part b_1 scales to "reactive power."

If the voltage is not stiff or harmonically pure, then it is still possible to accurately estimate real and reactive power by also storing a corresponding set of DFT coefficients of the voltage. That is, if the 1st, 3rd, 5th, and 7th DFT coefficients of current are stored, then these same DFT coefficients of voltage should also be stored. For a discrete-time system, real power is given by

$$P = \frac{1}{N} \sum_{n=0}^{N-1} i[n]v[n],$$

where $i[n]$ and $v[n]$ are the samples of current and voltage, respectively, over one period of length N. Let I_k and V_k denote

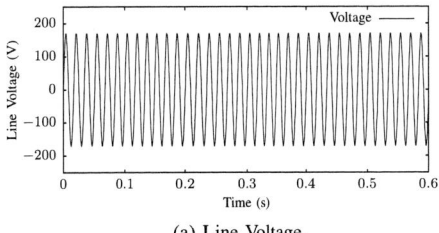

(a) Line Voltage.

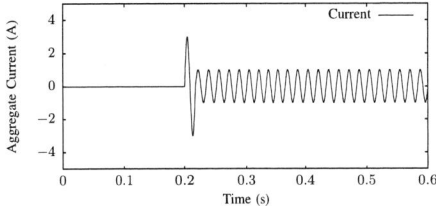

(b) Aggregate Current.

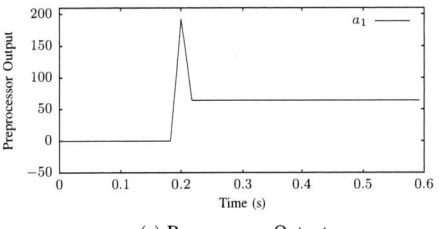

(c) Preprocessor Output.

Fig. 2. Sample preprocessor output. The first two plots depict the line voltage and aggregate current while a simulated device is being turned on. The third plot shows the corresponding preprocessor output.

the (complex) amplitudes of the kth harmonics of current and voltage, respectively. Using the Plancherel Theorem, real power can then be expressed in terms of the DFT of current and voltage,

$$P = \frac{1}{N^2} \sum_{k=0}^{N-1} I_k V_k^*.$$

Reactive power can be calculated in an analogous fashion. Thus, if only a small number of DFT coefficients of current are not approximately zero, then an accurate approximation of real and reactive power could be obtained by storing only the few significant DFT coefficients of current, and the same set of DFT coefficients of voltage.

Figure 2 shows the line voltage, aggregate current, and preprocessor output during the turn-on of a device that draws exclusively real power. The only non-zero preprocessor envelope is the envelope a_1 that corresponds to real power. It is important to note that, because the only non-zero DFT coefficient of current is the 1st coefficient (fundamental), this envelope is a scaled version of real power, regardless of the harmonic content of the voltage waveform. The only harmonics of the voltage waveform that affect real and reactive power are those harmonics that are also present in current.

III. FPGA-BASED SPECTRAL ENVELOPE PREPROCESSOR

To calculate, store, and communicate a relevant subset of DFT coefficients for power monitoring and energy scorekeep-

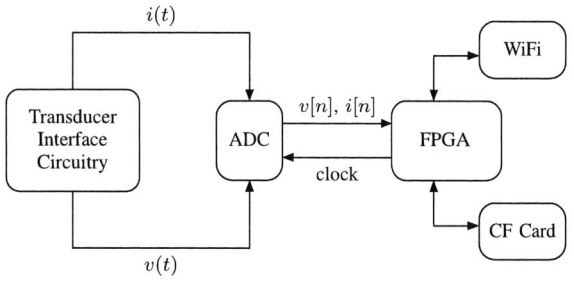

Fig. 3. Spectral envelope preprocessor block diagram. An ADC samples the line voltage and the aggregate current, which are then used by the FPGA to produce spectral envelope coefficients. These coefficients can be stored in a Compact Flash card and also transmitted via WiFi on demand.

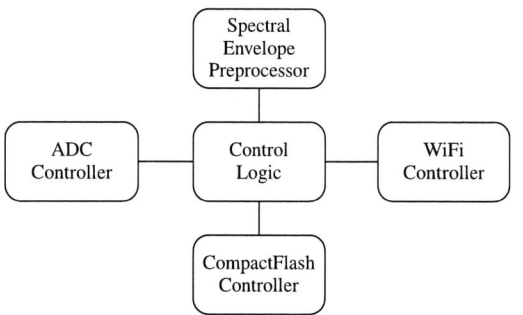

Fig. 4. FPGA block diagram. The preprocessor calculates spectral envelope coefficients. The ADC controller controls the sampling scheme of the ADC. The CF controller interfaces with a Compact Flash card to enable spectral envelopes to be stored and later recalled. The WiFi controller interfaces with an IEEE 802.11 WiFi transceiver to transmit spectral envelope data.

ing, a prototype FPGA (Field Programmable Gate Array) was constructed to implement a spectral envelope preprocessor. This system makes use of a low-cost FPGA (Altera Cyclone I, EP1C3T100C8). The spectral preprocessor consists of four subsystems: the first, to obtain current and voltage samples; second, to compute spectral envelope coefficients; third, to store computed spectral envelope coefficients; and fourth, to transmit the spectral envelope coefficients to another computation or display platform for further analysis. Each of these subsystems will be considered in detail. Figure 3 shows the overall block diagram of the system.

Data flows through the system as follows. The transducer interface circuitry first measures the line voltage and aggregate current, producing the signals $v(t)$ and $i(t)$. These signals are sampled and quantized by an analog-to-digital converter (ADC) that produces the samples $v[n]$ and $i[n]$. The FPGA processes these samples to compute spectral envelopes. The spectral envelope coefficients can be stored in a Compact Flash card for later use. The system also includes an IEEE 802.11b/g WiFi transceiver that allows any collection of the spectral envelope coefficients to be transmitted to another computation device for analysis. The FPGA provides control logic for each of the subsystems. Figure 4 shows a block diagram of the system implemented in the FPGA.

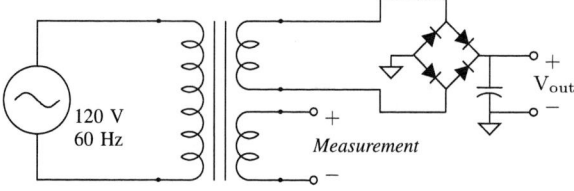

Fig. 5. Power supply and measurement schematic. The FPGA system receives both line voltage measurement and low-voltage supply through a transformer with dual secondary coils.

A. Current and Voltage Measurement

Current and voltage measurements from at least one voltage channel and at least one current channel are used to compute spectral envelopes. The system is easily expanded to measure more channels, supporting three-phase electrical services, for example. The prototype system uses an LEM LA 55 current transducer [18] to measure aggregate current and a transformer to measure the line voltage. A transformer with dual secondary coils is used in the prototype. This provides one coil for measurement, and a second coil for powering the preprocessor. The two coil arrangement provides a voltage sense with very little phase distortion, ensuring accurate calculation of in-phase and quadrature spectral components. Figure 5 illustrates this utility connection.

B. ADC Controller

In many signal processing applications, a computationally efficient algorithm like the Fast Fourier Transform (FFT) computes the complete spectral analysis of a sampled waveform. However, in situations like power monitoring, where a relatively small number of spectral coefficients may contain all or most needed information, needed spectral coefficients can be computed more efficiently by a traditional DFT implementation, i.e., by mixing observed waveform samples directly with the stored samples of basis sinusoids. In this approach, basis sinusoids are stored in a memory and multiplied by observed samples of a waveform. If there are N samples stored in memory for each basis sinusoid, then it is necessary to acquire N samples of the current and voltage waveforms for each line voltage period.

The FPGA coordinates the operation of the ADC rto obtain the samples $i[n]$ and $v[n]$ of the current and voltage waveforms $i(t)$ and $v(t)$. To provide a known number of waveform samples per line period, the FPGA "locks" to the line voltage waveform. That is, the FPGA varies the sample rate to track with variations in the line voltage frequency. The sampling clock is derived from the output of a digital phase-locked loop (PLL) on the FPGA that tracks the line voltage frequency.

The phase of the sampling is set such that the first voltage and first current samples are taken at the negative to positive zero crossing of the line voltage. The goal is to multiply each entry in the basis sinusoid by the value of the waveform to be analyzed at the corresponding point in time. It is essential that the entire process is line locked to the line frequency in order for the estimated spectral envelope coefficients of current to correspond to "in-phase" and "quadrature" components

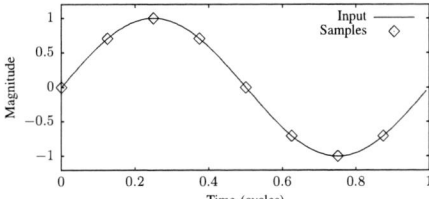

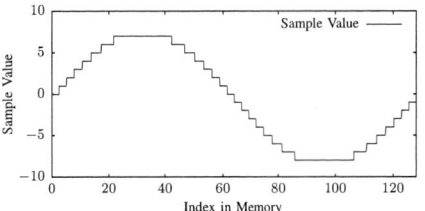

(a) Basis sinusoid $\widehat{s_1}$ used to compute a_1

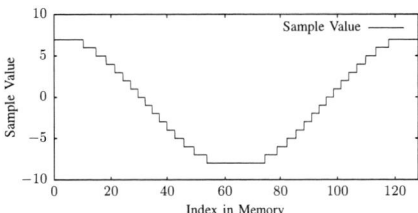

(b) Basis sinusoid $\widehat{c_1}$ used to compute b_1

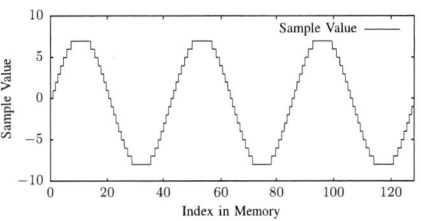

(c) Basis sinusoid $\widehat{s_3}$ used to compute a_3

Fig. 6. Sampling scheme for the line voltage and current. Samples are taken at points in time that correspond to the samples stored in the basis sinusoid memory. For clarity, this figure only depicts 8 sample points per cycle, while the prototype system actually uses 128 sample points per cycle.

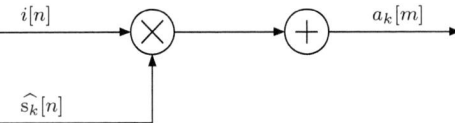

Fig. 7. Signal flow graph for the calculation of a spectral envelope. Raw current values $i[n]$ are multiplied by the appropriate elements of each basis sinusoid $\widehat{s_k}$, and the results are accumulated over one period to produce each spectral envelope value.

of current with respect to the fundamental of the voltage waveform. This sampling scheme is illustrated in Figure 6. The sample times and values are indicated by diamonds.

C. Envelope Preprocessor

References [8]–[12] and their associated references describe various metering schemes that compute real, reactive, and apparent power, and also harmonic distortion in one form or another. References [1]–[4] describe specific implementations. In [1], multiple phase-locked loops, analog multipliers and integrators were used to estimate spectral envelope coefficients. In [2], a design using multiplying digital-to-analog converters, low-pass filters, and a single phase-locked loop was presented. In [3], a expensive digital signal processing board was used to perform the calculations. In [4], the processing power of a personal computer was used for spectral envelope coefficient estimation.

All of these systems can provide accurate estimates of spectral envelope coefficients or related quantities. They serve as essential building blocks of various types of metering systems. They are often expensive and dedicated. The FPGA-based system discussed in this section is an inexpensive single-chip solution that can estimate spectral envelope coefficients for stand-alone use or as part of a turn-key building block in more complex systems. The FPGA computes spectral envelopes using integer arithmetic on stored basis waveforms and observed waveform samples.

The FPGA-based spectral envelope preprocessor calculates the spectral envelopes of current, $a_1[m], b_1[m], a_3[m], \ldots$, where m indexes the periods of the line voltage. Figure 7 shows the computation performed to produce estimates of a single spectral envelope coefficient, $a_k[m]$. The system multiplies $i[n]$, the samples of current, with samples of a basis sinusoid, and sums the result over a single period of the line voltage. Specifically, if N denotes the number of current

Fig. 8. Basis sinusoids. This figure depicts three examples of basis sinusoids, used to calculate real spectral envelopes of in-phase fundamental frequency content (a_1), quadrature fundamental frequency content (b_1), and in-phase third harmonic content (a_3). The basis sinusoids shown here are sampled at 4 bits for illustration—the actual prototype spectral envelope preprocessor uses 10 bit samples.

samples $i[n]$ over one line period, then

$$a_k[m] = \sum_{j=0}^{N-1} i[mN+j] \cdot \widehat{s_k}[mN+j] \qquad (4)$$

$$b_k[m] = \sum_{j=0}^{N-1} i[mN+j] \cdot \widehat{c_k}[mN+j]. \qquad (5)$$

Here $\widehat{s_k}$ and $\widehat{c_k}$ represent the stored integer basis sinusoids, which are pre-calculated to b bits of precision as

$$\widehat{s_k}[n] = \text{floor}\left(\sin\left(\frac{2\pi kn}{N}\right) \cdot 2^{b-1}\right) \qquad (6)$$

$$\widehat{c_k}[n] = \text{floor}\left(\cos\left(\frac{2\pi kn}{N}\right) \cdot 2^{b-1}\right) \qquad (7)$$

Each spectral envelope coefficient has a different basis sinusoid associated with it; for example, calculation of $a_3[m]$ involves multiplying $i[n]$ by $\widehat{s_3}$, which consists of discrete-time samples of a sinusoid at three times the line frequency, with its phase locked to the line voltage.

Figure 8 depicts examples of basis sinusoids. For illustration purposes, these sinusoids are sampled at only 4 bits, while the prototype system makes use of 10 bit samples.

Figure 9 shows a block diagram of the FPGA-based spectral envelope preprocessor. The preprocessor takes as input the

978-1-4244-4782-4/10 $26.00 © 2010 IEEE   2198

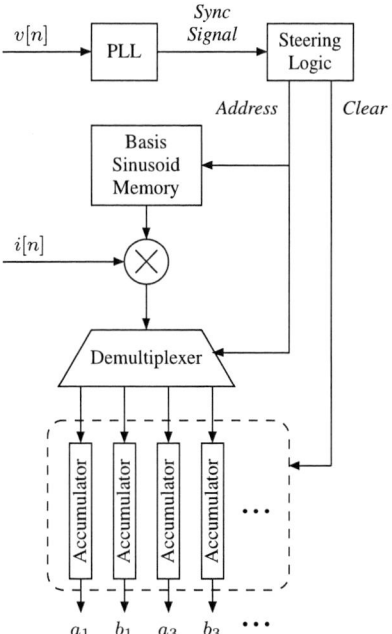

Fig. 9. Preprocessor block diagram.

discrete-time samples of $i[n]$ and $v[n]$ and produces estimates of the spectral envelope coefficients a_k and b_k of the current $i[n]$.

The voltage samples $v[n]$ are used as input to a phase-locked loop (PLL), which synchronizes the entire computation to the line voltage. As noted earlier, the computation process is synchronized to the line voltage so that the calculated spectral envelope coefficients correspond to meaningful physical quantities in steady-state operation (real power, reactive power, etc.). The output of the PLL is sent to a block of steering logic on the FPGA that produces the address for the basis sinusoid memory, as well as a clear signal for the accumulators. The basis sinusoid memory consists of the samples of the various basis sinusoids. The address produced by the steering logic specifies a single sample time of a single basis sinusoid. The sample of the basis sinusoid that is retrieved from the basis sinusoid memory is then multiplied by the current sample $i[n]$. The result of this multiplication is then passed through a demultiplexer which sends the result to the appropriate accumulator by using the address produced by steering logic to determine which spectral envelope coefficient is currently being calculated.

There is one accumulator for each estimated spectral envelope coefficient. The accumulators are all cleared at the end of each period of the line voltage through the use of the clear signal produced by the steering logic. For every sample $i[n]$, the address produced by the steering logic will select each of the basis sinusoids in turn, so that every sample of current is multiplied by the appropriate sample of each of the basis sinusoids.

This FPGA-based implementation provides a great deal of flexibility. For example, the subset of spectral envelope coefficients that are being estimated can be changed by altering the entries in the memory to correspond to a different set of basis sinusoids. This implementation is also efficient in terms of FPGA resource utilization. It uses only a single PLL and a single multiplier, as opposed to previous hardware implementations that often used multiple PLLs and/or multipliers [1], [2]. This system can function with only a single multiplier because the FPGA is capable of multiplying each sample $i[n]$ by the corresponding sample of each of the basis sinusoids and forwarding each result to the appropriate accumulator, before the next sample $i[n + 1]$ arrives. The multiplier consumes substantial logic elements on the FPGA. It utilizes 24% of all resources used by the envelope preprocessor and 13% of the resources used by the complete system. By using only one multiplier, the design is capable of fitting in a small, low-cost FPGA.

There are several ways to configure and deploy the spectral envelope preprocessor for any given application. For situations where the voltage waveform is relatively sinusoidal and "stiff," the spectral envelopes of current can be interpreted as scaled physical quantities in steady state. As discussed in §II-C, under these assumptions, the a_1 envelope of current in steady state corresponds to a scaled estimate of real power or "P". The b_1 envelope of current corresponds to reactive power or "Q". In situations where the voltage is not stiff and/or not sinusoidal, the FPGA could be tasked to also compute the spectral envelopes of voltage as well as current. This more complete set of spectral envelopes could be stored or transmitted to a computation platform or metering instrument that can quickly compute estimates of real or reactive power or other quantities of interest. Alternatively, the FPGA can be reconfigured to compute quantities like real and reactive power itself. In practice, we have found the basic computation of the spectral envelopes of current, assuming a stiff voltage source, to yield information that is directly useful for load monitoring and demand-side load control and diagnostics [17].

D. CF Controller

This FPGA subsystem stores spectral envelope data on an erasable memory like a Compact Flash (CF) storage card. The subsystem is capable of storing spectral envelope data as it is produced, as well as retrieving the spectral envelope data from any point in time, on demand. To interface with the CF card, the "True IDE" interface mode is used [19]. This interface mode is universally supported by Compact Flash storage cards and it allows the system to be easily adapted to interface with other mass storage devices that use the IDE interface standard, such as an IDE hard drive. While it would be possible to impose a filesystem on the CF card (i.e. FAT32), the current design treats the CF card as a single large, raw block of storage, for simplicity. Due to the low data rate of the spectral envelope coefficients (for the prototype preprocessor with 8 spectral envelopes, each stored at 24 bits of resolution, the data rate is 2.8 KB/s), even a moderately sized CF card could store the spectral envelope data for a substantial length of time. For example, in the prototype system, a 1 GB CF card would suffice to record data continuously for approximately 4.3 days.

E. WiFi Controller

This subsystem facilitates the transmission of spectral envelope coefficients to a PC or other computation platform for further analysis or display. It makes use of an IEEE 802.11b/g WiFi transceiver and TCP/IP. The current design is capable of supporting both ad-hoc and access point (infrastructure) networks.

The transmitted data is retrieved from the CF card as needed. The WiFi subsystem can operate in two different modes. In the first mode, the system streams spectral envelope coefficients as they are generated. In the prototype, this corresponds to a data rate of 2.8 KB/s. In the event of a momentary interruption in the connection to the PC, the system will automatically buffer data from the last successfully transmitted packet and resume transmission from that point when a connection is reestablished. The system will then send data at the highest available transmit speed (54 Mb/s for 802.11g), until the system catches up to the freshly produced spectral envelope data. In the second mode, an application on the PC requests data by specifying a range of time; the system then transmits all data from the desired range of time, at the maximum possible transmission speed.

IV. FLEXIBILITY

The design presented above is just one example of an FPGA-based load monitoring interface. The modularity of the design, and the versatility of FPGAs, makes it simple to change the transmission system, for example, to wired Ethernet (IEEE 802.3) or Bluetooth (IEEE 802.15.1) or Zig-Bee, or to change the storage system to, for example, a microSD card. An FPGA permits the interconnection of a wide variety of different subsystems to form a complex utility monitoring system. Even a small, low-cost FPGA is capable of implementing both the spectral envelope preprocessor as well as the required interface logic to control the various subsystems. Thus, an FPGA can serve as the backbone of an inexpensive, complete utility or load monitoring system.

V. PROTOTYPE RESULTS

The FPGA-based system discussed above calculates, stores, and transmits spectral envelope data. Figure 10 shows a picture of the prototype hardware.

To make use of this data, a monitoring or control system typically includes a subsystem to receive and use the spectral envelope data. For example, a homeowner could use a personal computer to collect spectral envelope data from the FPGA preprocessor installed in or near a circuit breaker panel. The prototype includes a PC-based software application that can interface with the FPGA-based preprocessor via wired or wireless communication channels, retrieve spectral envelopes, and display spectral envelope data. Using this spectral envelope data, the PC-side application can disaggregate the operating schedule of individual loads from measurements made on an aggregate power feed serving multiple loads. The application is self-training and identifies loads in essentially real-time. Screenshots of this program are shown in Figure 11 and Figure 12.

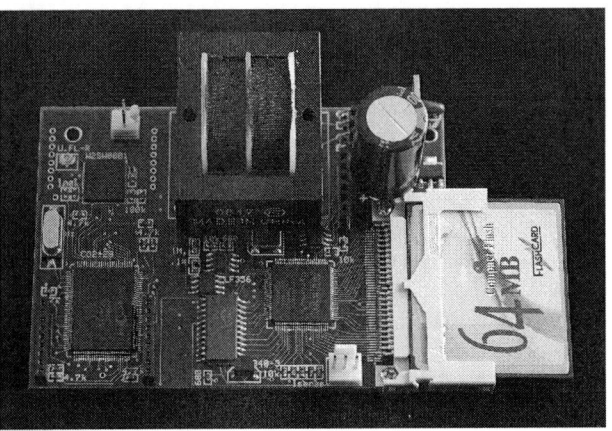

Fig. 10. Picture of the prototype FPGA-based system.

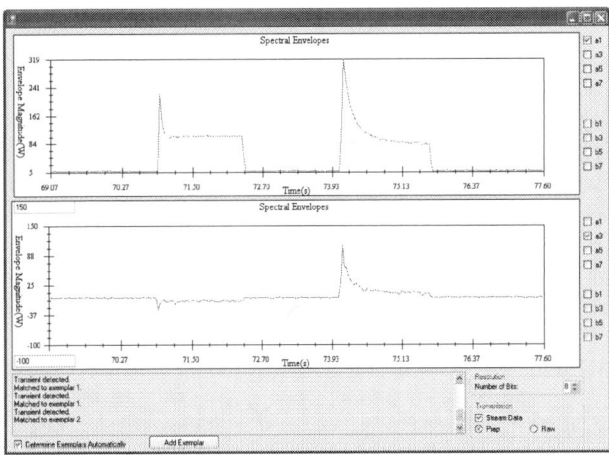

Fig. 11. Screenshot of the prototype software. The upper plot displays real power and the lower plot displays third harmonic content. This spectral envelope data corresponds to the light bulb and motor whose raw current values were shown in Figure 1. The lower left section of the screenshot contains the output of the classifier, which has correctly identified both of the loads.

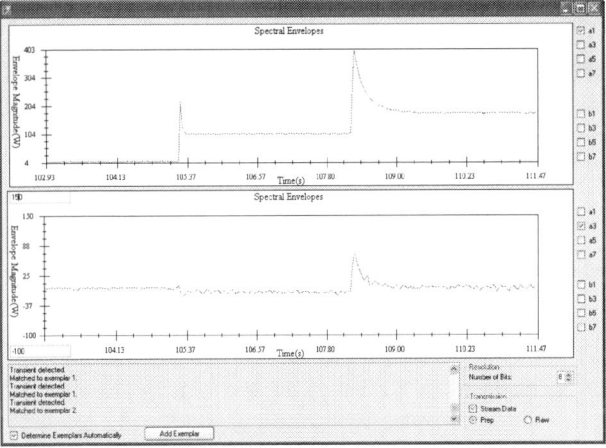

Fig. 12. Screenshot of the prototype software. The data shows the same light bulb and motor as Figure 11, with the difference that the motor starts while the light bulb is still turned on. As shown in the lower left corner of the screen, the system still classifies both devices correctly.

This software communicates via TCP/IP with the FPGA-based preprocessor. The software can retrieve any subset of recorded data, as well as issue commands, such as changing the sampling resolution of the ADC. Once spectral envelope data is retrieved, this software makes use of the Expectation-Maximization (EM) algorithm [6] to classify and recognize the operation of individual loads.

This software is only one example of the many possible ways to use spectral envelope data. For example, a system that uses this data to control a set of generators in a micro-grid is currently in development. Data could be retrieved by a web application that displays a live stream of data on a webpage. Other embedded systems could communicate with the FPGA-based system and use the retrieved spectral envelope data to control electrical loads.

VI. APPLICATIONS

The FPGA preprocessor can provide a turn-key component for creating utility monitoring, energy score-keeping, and diagnostic applications for a variety of systems. The preprocessor is relatively simple compared to microprocessor or DSP-based data acquisition systems. The concepts and hardware illustrated in this paper could be incorporated into individual loads, circuit breakers, or circuit breaker panels to provide energy consumption information for both monitoring and control.

The simplification in data storage and transmission bandwidth requirements afforded by the FPGA can also be extended to other domains. For example, it is possible to extend the non-intrusive monitoring concept beyond the realm of electrical distribution. We have used a single acoustic sensor to monitor the flow of water to a collection of hydraulic loads. This acoustic data is not "line locked" to any particular "utility frequency." However, fingerprint acoustic signatures can still be developed by examining spectral content at frequencies dictated by resonant properties of the pipe. These fingerprints permit recognition of hydraulic loads or events in the water distribution system, as shown in Figure 13.

We are exploring extending the nonintrusive monitoring concept to gas mains, water mains, and to various occupancy detection schemes using air flow measurements and acoustic signals. We are also exploring nonintrusive electrical load detection by looking at both quasi-static and radiated signals associated with the operation of different loads.

Acknowledgements

This research was funded by the Grainger Foundation, the BP-MIT Alliance, the Office of Naval Research under the ESRDC program, the MIT Sea Grant College Program, and by the MIT Center for Materials Science and Engineering. The authors also gratefully acknowledge the advice and support of Dr. Manny Landsman.

REFERENCES

[1] S. B. Leeb, "A conjoint pattern recognition approach to nonintrusive load monitoring," Ph. D. dissertation, Dept. Elect. Eng. Comput. Sci., Mass. Inst. Technol., Cambridge, MA, Feb. 1993.

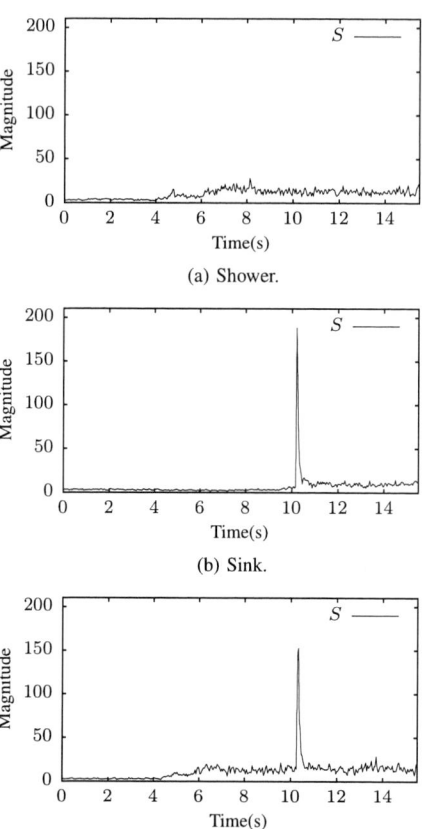

(a) Shower.

(b) Sink.

(c) Shower and sink.

Fig. 13. Sample data showing the application of the FPGA-based preprocessor to acoustic signals from hydraulic loads. The plots show spectral envelope content S at 192 Hz for a shower, sink, and a combination of both.

[2] S. B. Leeb, S. R. Shaw, and J. L. Kirtley, "Transient event detection in spectral envelope estimates for nonintrusive load monitoring," IEEE Trans. Power Del., vol. 10, no. 3, pp. 1200-1210, Jul. 1995.

[3] S. R. Shaw, C. B. Abler, R. F. Lepard, D. Luo, S. B. Leeb, and L. K. Norford, "Instrumentation for high performance nonintrusive electrical load monitoring," Trans. ASME J. Sol. Energy Eng., vol. 120, no. 3, pp. 224-229, Aug. 1998.

[4] S. R. Shaw, "System identification techniques and modeling for non-intrusive load diagnostics," Ph. D. dissertation, Mass. Inst., Technol., Cambridge, MA, Feb. 2000.

[5] J. Paris, et al. "Scalability of Non-Intrusive Load Monitoring for Shipboard Applications," ASNE Day 2009, National Harbor, Maryland, April, 2009.

[6] A. P. Dempster, N. M. Laird, D. B. Rubin, "Maximum Likelihood from Incomplete Data via the EM Algorithm," Journal of the Royal Statistical Society, Series B 39:1-38, 1977.

[7] "The Smart Grid: An Introduction," U. S. Department of Energy, http://www.oe.energy.gov/1165.htm.

[8] U. S. Patent 5,212,441, "Harmonic-Adjusted Power Factor Meter," May 18, 1993.

[9] U. S. Patent 5,229,713, "Method for Determining Electrical Energy Consumption," July 20, 1993.

[10] U. S. Patent 5,548,527, "Programmable Electrical Energy Meter Utilizing a Non-Volatile Memory," August 20, 1996.

[11] U. S. Patent 6,615,147, "Revenue Meter with Power Quality Features," September 2, 2003.

[12] U. S. Patent 7,525,423, "Automated Meter Reading Communication System and Method," April 28, 2009.

[13] http://www.p3international.com/products/special/P4400/P4400-CE.html.

[14] Shaw, S. R., S. B. Leeb, L. K. Norford, R. W. Cox, "Nonintrusive Load Monitoring and Diagnostics in Power Systems," IEEE Transactions on Instrumentation and Measurement, Volume 57, No. 7, July 2008, pp. 1445-1454.

[15] Khan, U. A., S. B. Leeb, M. C. Lee, "A Multiprocessor for Transient Event Detection," IEEE Transactions on Power Delivery, Volume 12, Number 1, pp. 51-60, January 1997.

[16] W. M. Siebert Circuits, Signals and Systems, The MIT Press, 1986.

[17] Laughman, C. R., "Fault Detection Methods for Vapor-Compression Air Conditioners Using Electrical Measurements," MIT BT Ph. D. thesis, September 2008.

[18] http://www.lem.com/docs/products/la%2055-p%20e.pdf

[19] "CF+ & CF Specification Rev. 4.1", http://compactflash.org

Sigma-Delta Modulation of Multi-Phase High Frequency Converters

Jonathan W. Kimball, *Senior Member*, Kyle Roger Eckler, and Luke Watson, *Student Member*
Department of Electrical and Computer Engineering
Missouri University of Science & Technology
Rolla, MO, USA
kimballjw@mst.edu

Abstract— **A power conversion control architecture is proposed that merges advanced digital modulation techniques, high frequency resonant converters, and multi-phase converters. If a converter's switching frequency is high enough, the converter may be turned on and off quickly to modulate power flow. The proposed system is composed of several high frequency converters connected in parallel, only a few of which are active at a given time to regulate the output voltage. The number of active converters is determined through a sigma-delta modulation process. The specific converter enable signals are generated by a balancing algorithm that ensures approximately equal effective duty ratio and switching frequency for all phases. An output voltage regulator, here implemented as a PI loop, completes the system. The method has been simulated for an eight-phase closed-loop system. Experimental results are shown for a four-phase open-loop system with an external command that corresponds to the effective duty ratio of the power converters.**

I. INTRODUCTION

Trends in the power converter industry include a push to higher switching frequencies for reduced size, a migration to digital control techniques, and the use of multi-phase converters. In the present work, these trends converge in a new control scheme for a multi-phase high frequency power converter. Specifically, sigma-delta modulation is used to switch converters on and off at a high rate to regulate output voltage.

If the switching frequency of a power converter is high enough, e.g. 10 MHz, the converter may be switched on and off just as fast as a conventional converter switches. This approach was first proposed in [1] for class E converters, and has also been applied to class Φ_2 converters [2, 3]. In all previous studies, voltage hysteresis control was used to enable and disable converters. In [1], a Vernier scale with hysteresis was proposed for a multi-phase converter. Hysteresis control is a type of asynchronous delta modulation [4].

Hysteresis control is relatively easy to implement in an analog controller, but far more difficult to achieve in a digital controller. Hysteresis is an inherently asynchronous process, whereas digital controllers that act on analog signals must use synchronous processes. Digital controllers are preferable, though, for many reasons, ranging from user interface to precision to field upgradeability. Digital controllers are also able to implement advanced, nonlinear control methods that are impractical in an analog circuit. The synchronous nature of digital processes corresponds to sampled-data techniques.

A conventional digital controller for a power converter samples voltages (and possibly currents) at a fixed rate and uses that information in a discrete-time feedback controller. Instead of direct action on the voltages (as in hysteresis and other geometric techniques), linear or nonlinear feedback controllers in the z domain are used to create a duty cycle command. The duty cycle is the input to a uniformly sampled pulsewidth modulation (PWM) process. A typical uniform PWM process with n bits of resolution is constructed with an n-bit counter that increments at 2^n times the output frequency and a comparator.

Sigma-delta ($\Sigma-\Delta$) modulation, a type of synchronous delta modulation [4], was used in [5] to enhance the precision of a digital PWM process. In a digital controller with high switching frequencies, the clock rate for the counter can become unmanageable. For example, a 1 MHz PWM signal with 8 bits of resolution would require a base clock frequency of 256 MHz—achievable, but highly undesirable. Twelve bits of resolution would require a base clock frequency of 4 GHz, which is achievable only at a prohibitively high cost. Now suppose that the duty cycle command is encoded with m bits of resolution, but the PWM process is only capable of n bits of resolution (where $m > n$). For example, a 4-bit PWM process with a 16 MHz clock could be used to generate 1 MHz switching signals, even though the desired PWM resolution is 12 bits. With $\Sigma-\Delta$ modulation, the extra resolution in the input (m-n bits) is an error term that is driven to zero over time by varying the duty ratio of the basic PWM process, so a low value for n and a low clock frequency may be used. A first-order $\Sigma-\Delta$ modulator is shown in Fig. 1. The feedback on the error ensures that the process has sufficient resolution over a

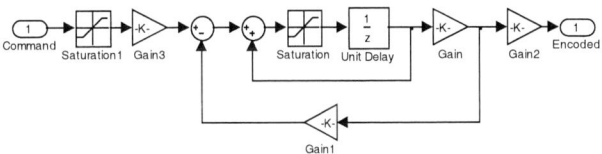

Fig. 1. First-order $\Sigma-\Delta$ modulation process. Gains implement left and right shifts.

This work was supported in part by the U.S. National Science Foundation under grant #0812121. Mr. Watson is supported by a Fellowship from the U.S. Department of Education.

long time period.

The present work combines the basic principles of $\Sigma-\Delta$ modulation, as applied in [5] to power converters, with high-frequency power converters that are alternately enabled and disabled. The combination has the potential for high efficiency and small size, due to the high frequency resonant power converter, with all of the positive attributes of digital control, such as user interface, reconfigurability, and the potential for sophisticated nonlinear controllers. A similar concept was applied to a single-phase class D resonant converter in [6]. The present work extends this concept substantially to address multi-phase power converters. Section II describes the new modulation process. Section III shows simulation results of the modulator combined with Φ_2 power converters. Section IV shows experimental results of the combined system. Important VHDL code is included in an appendix.

II. PROPOSED MODULATION PROCESS

A multi-phase VHF power converter is not directly amenable to PWM, so a new modulation process has been developed for the different requirements. The assumption in [5] is that the actuator is a conventional hard-switched converter, such as a buck converter, whose duty ratio must be controlled. For highly resonant topologies, such as class E and class Φ_2 converters, duty ratio variation should be avoided. Rather, the converter should be turned on and off as

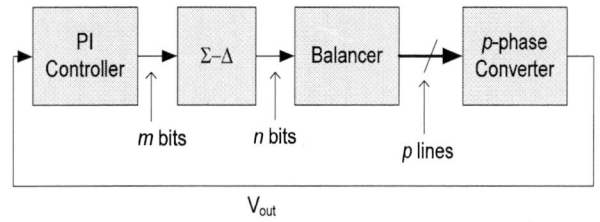

Fig. 2. Proposed control and modulation process.

in [1]. If several converters are connected in parallel, a fraction of the converters may be active at any time. If there are p converters total and q are active, the effective duty ratio of the converter system is q/p. This effective duty ratio can be determined through a $\Sigma-\Delta$ process.

Fig. 2 shows the proposed modulation process. A conventional first-order $\Sigma-\Delta$ process has been used. Higher-order $\Sigma-\Delta$ processes may be used instead with minimal changes. The output of the $\Sigma-\Delta$ process, y, is encoded in n bits. The power converter system contains $p = 2^n$ identical converter phases.

The block in Fig. 2 labeled "Balancer" converts the integer-valued command y into p enable (binary) signals. If a simple thermometer code were used here, some phases would be enabled almost all of the time, some would rarely be enabled, and some would switch frequently. Instead, the phases are treated as equals in a ring with a beginning point that shifts over time. The process is shown schematically in Fig. 3. Corresponding VHDL code, synthesizable with Quartus II for an Altera EP2C50F484I8 FPGA, is given in the appendix for the case $p = 4$.

Fig. 4 shows computed switching frequency for a simulated system similar to that shown in Fig. 2, operating open-loop with a sinusoidally varying effective duty ratio. In this case, $m = 12$, $n = 3$, and $p = 2^3 = 8$. The beginning point in the ring shifts at a frequency that is a multiple of the sampling frequency, which is fixed at 1 MHz. As this multiplier increases, the effective switching frequency of each phase decreases monotonically. The effective switching frequency varies less than 2% among the phases, as indicated by the standard deviation. All eight phases have the same long-term average duty ratio.

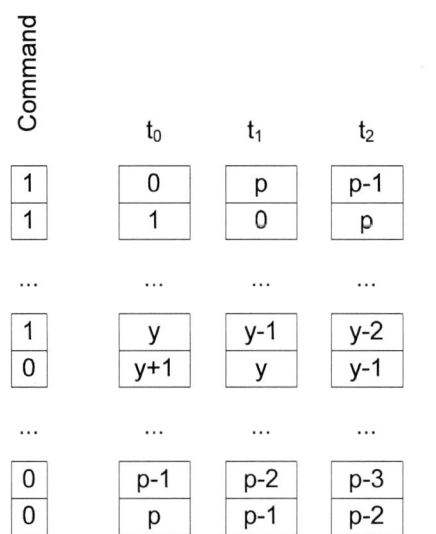

Fig. 3. Balancing method, indicating which phases are enabled as time progresses for a commanded value y.

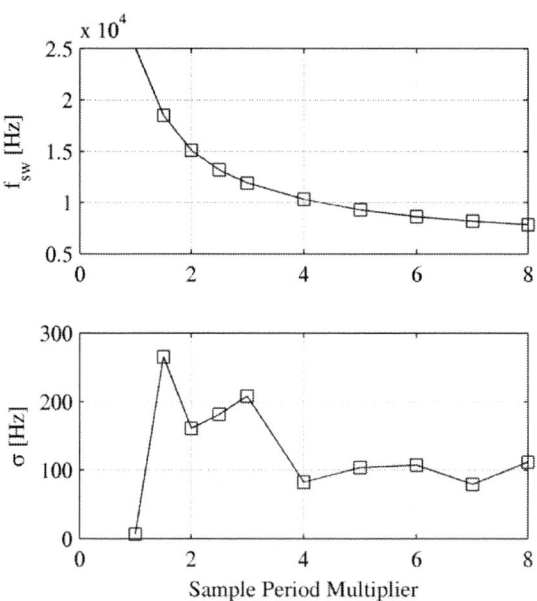

Fig. 4. Effective switching frequency f_{sw} and its standard deviation σ for an 8-phase balanced $\Sigma-\Delta$ process with a sampling frequency of 100 kHz.

III. INTEGRATION WITH HIGH FREQUENCY CONVERTERS

The output of the $\Sigma-\Delta$ modulator described above is a set of enable signals for high frequency converter phases that are connected in parallel. For the present study, class Φ_2 boost converters were used; other converter types with fast start-up dynamics may be used as well. A class Φ_2 converter can be approximated as a current source with a magnitude that depends on switching frequency, component values, and terminal voltages. Fig. 5 shows simulation results for a system with eight class Φ_2 converters, a proportional-integral (PI) regulator for the output voltage, a $\Sigma-\Delta$ modulator to convert the 12-bit PI output into a 3-bit command, and a balancer to convert the 3-bit command into eight balanced enable commands. The straight-line ripple results from a varying number of active phases, and the small-signal ripple (visible only as a thick line in this zoomed-out view) results from the high-frequency switching of the individual phases. The dip at the beginning is a start-up transient in the PI loop. This method incorporates conventional digital control (a PI loop), advanced modulation techniques ($\Sigma-\Delta$), and high frequency power conversion (class Φ_2 converters) in a multi-phase system. More sophisticated voltage regulation methods, such as energy-based methods [7], may be easily incorporated.

IV. EXPERIMENTAL RESULTS

An experimental system was used to validate the simulation results discussed above. First, the digital circuit was validated. The system was composed of an analog command on a 3.3 V scale, a 12-bit analog-to-digital converter (Analog Devices AD7276), a 2-bit first-order $\Sigma-\Delta$ process, and a four-output balancer. The system was clocked at 1.85 MHz, using a 24 MHz clock source and a divide-by-13 block. The logic was implemented in an Altera EP2C50F484I8N FPGA on a Microtronix Firefly II module.

Fig. 6 shows a typical set of waveforms. Here, the analog input is 1.49 V. The mean duty ratio of the four enable signals is 0.4266. The system oscillates between enabling one converter and enabling two converters. Which one or two converters are enabled, though, shifts over time. Therefore, all

four enable signals have duty ratios near the mean, with a total standard deviation of 0.0027. Although the update rate is 1.85 MHz, the effective switching frequency of the enable signals is approximately 154 kHz. The balancer is updated every other sample of the $\Sigma-\Delta$ loop.

Fig. 7 shows the variation in duty ratio as the input command varies. As expected, the effective duty ratio D_{eff} is linear with the command x, such that

$$D_{eff} = 0.302x - 0.0233 \tag{1}$$

The gain, 0.302, comes from the fact that the full-scale input is 3.3 V. The offset is thought to result from voltage drop in the connections.

Next, this controller was applied to a four-phase power converter. Each phase was a class Φ_2 converter, built according to the schematic in Fig. 8. Parameters are shown in Table 1. Each converter switches at 10.18 MHz and has an "enable" input. Fig. 9 shows gate and drain voltages for the main MOSFET (Q1) and the rectifier input voltage (i.e. voltage on C1). There are multiple resonances in the system to achieve proper voltage waveshapes [2]. One of the challenges in building a multi-phase converter with this

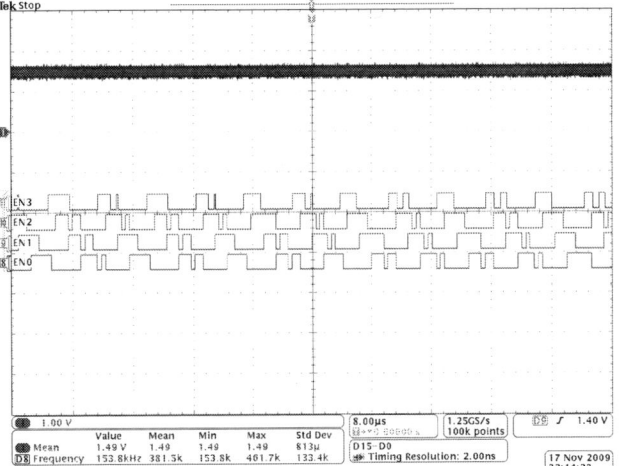

Fig. 6. Sample waveforms of sigma-delta modulator with balancer. From top: analog command, enable signals for phases 3, 2, 1, and 0.

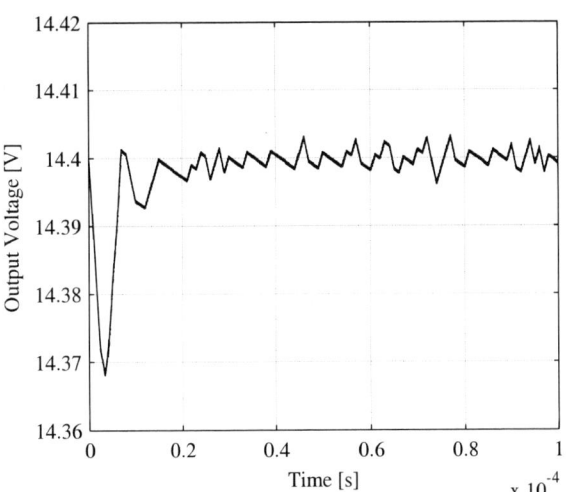

Fig. 5. Simulated output voltage for eight-phase converter with $\Sigma-\Delta$ modulation.

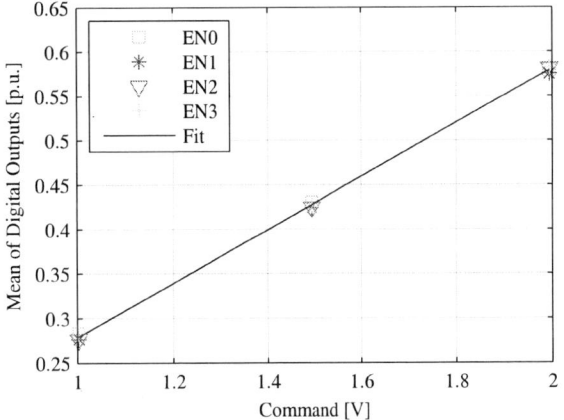

Fig. 7. Relationship between the analog command and the duty ratios of the enable signals (digital outputs).

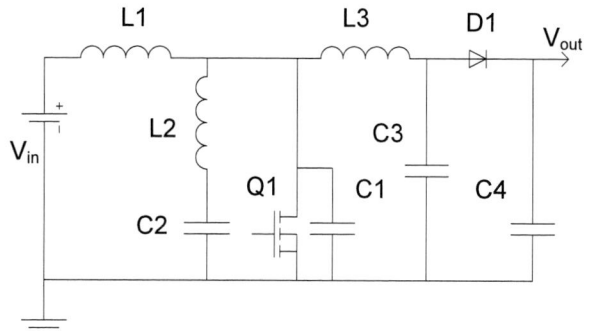

Fig. 8. Schematic of a single Φ_2 boost converter.

TABLE 1. PARAMETERS FOR Φ_2 BOOST CONVERTERS.

Input Inductor L1	35.5 nH
Resonant Inductor L2	18.5 nH
Output Inductor L3	35.5 nH
MOSFET Extra Capacitor C1	6.8 nF
Resonant Capacitor C2	3.3 nF
Output Resonant Capacitor C3	1.5 nF
Output DC Capacitor C4	5 µF
Main MOSFET Q1	Si7726DN
Output Rectifier D1	MBRA340T3

topology is tuning all of the tanks. The boards used in the final test were all built with components whose nominal values are given in Table 1. A compromise switching frequency and a compromise duty cycle were selected to give the most uniform performance.

Results from the four-phase testing are shown in Fig. 10. In this test, the output was connected to an electronic load configured as a constant voltage (9.5 V). As the input command varied, the output current varied. The relationship is expected to be linear, but is nonlinear in the experimental system. A quadratic curve fit is indicated on Fig. 10. Possible explanations include the variation between the phases and an observed delay of approximately 80 ns between the enable signals and the gate waveforms. That is, the gate driver did not start switching until approximately 80 ns after being enabled. This time delay is manageable but not negligible on this time scale, so it may partially explain the nonlinearity in the system response.

V. CONCLUSIONS

A new control method for multi-phase high-frequency converters was derived and demonstrated. The method combines conventional digital control, Σ–Δ modulation, and VHF conversion techniques. Simulations indicate that the converter phases can be modeled as current sources and the output voltage varies linearly with commanded duty ratio. Experimental results show a nonlinear but monotonic input-to-output characteristic. Future work will improve the individual phases and the start-up characteristics to improve linearity.

VI. ACKNOWLEDGMENTS

This work was supported in part by the U.S. National Science Foundation through the FREEDM Systems Center

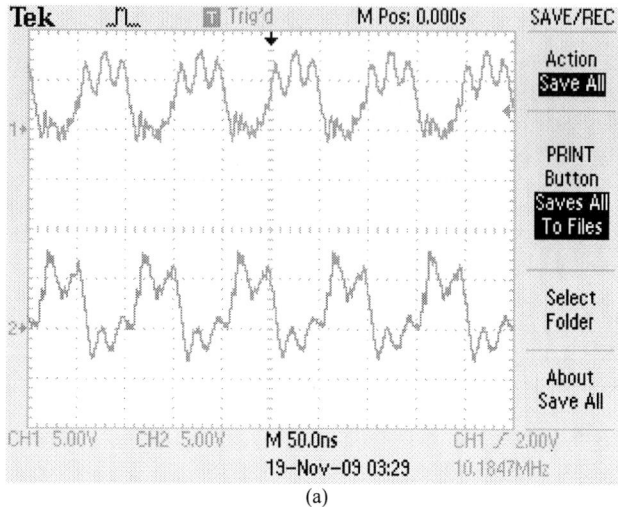

(a)

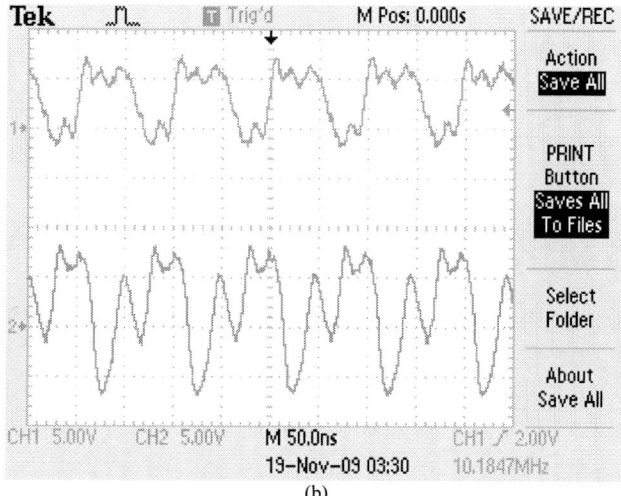

(b)

Fig. 9. Key waveforms in Φ_2 converter. (a) Gate-source (channel 1, top) and drain-source (channel 2) voltages of Q1. (b) Gate-source (channel 1, top) voltage of Q1 and rectifier input voltage (channel 2, voltage on C3).

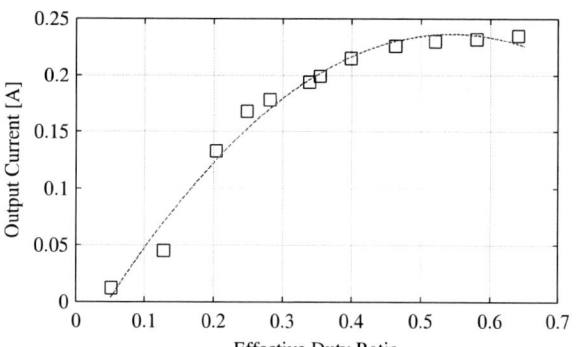

Fig. 10. Output current of the four-phase system as a function of effective duty ratio (given by (1)).

(award #0812121). Mr. Watson is being supported on a Fellowship from the U.S. Dept. of Education.

REFERENCES

[1] J. M. Rivas, R. S. Wahby, J. S. Shafran, and D. J. Perreault, "New architectures for radio-frequency dc-dc power conversion," *IEEE Trans. Power Electronics*, vol. 21, pp. 380-393, March 2006.

[2] R. C. N. Pilawa-Podgurski, A. D. Sagneri, J. M. Rivas, D. I. Anderson, and D. J. Perreault, "Very-high-frequency resonant boost converters," *IEEE Trans. Power Electronics*, vol. 24, pp. 1654-1665, June 2009.

[3] J. M. Rivas, Y. Han, O. Leitermann, A. D. Sagneri, and D. J. Perreault, "A high-frequency resonant inverter topology with low-voltage stress," *IEEE Trans. Power Electronics*, vol. 23, pp. 1759-1771, July 2008.

[4] R. E. Steele, *Delta Modulation*. New York: Wiley, 1975.

[5] Z. Lukić, N. Rahman, and A. Prodić, "Multibit $\Sigma-\Delta$ PWM digital controller IC for dc-dc converters operating at switching frequencies beyond 10 MHz," *IEEE Trans. Power Electronics*, vol. 22, pp. 1693-1707, Sept. 2007.

[6] H. Koizumi, "Delta-sigma modulated class D series resonant converter," in *Proc. Power Electronics Specialists Conference*, 2008, pp. 257-262.

[7] S. R. Sanders and G. C. Verghese, "Lyapunov-based control for switched power converters," *IEEE Trans. Power Electronics*, vol. 7, pp. 17-24, Jan. 1992.

APPENDIX: VHDL CODE FOR BALANCER

```
LIBRARY ieee;
USE ieee.std_logic_1164.all;
USE ieee.std_logic_unsigned.all;

ENTITY balancer_2by4 IS
    PORT
    (
        clk    : IN STD_LOGIC;
        reset  : IN STD_LOGIC;
        valin  : IN STD_LOGIC_VECTOR(1 DOWNTO 0);     -- 2-bit value to encode
        divider  : IN STD_LOGIC_VECTOR(3 DOWNTO 0);   -- clock divider for rotation
        valout : OUT STD_LOGIC_VECTOR(3 DOWNTO 0)     -- 4-bit pseudo-thermometer code
    );
END balancer_2by4;

ARCHITECTURE behavioral OF balancer_2by4 IS
    SIGNAL counter   : STD_LOGIC_VECTOR(3 DOWNTO 0);
    SIGNAL rotation  : STD_LOGIC_VECTOR(1 DOWNTO 0);
    SIGNAL y         : STD_LOGIC_VECTOR(3 DOWNTO 0);
    SIGNAL rot0, rot1, rot2, rot3 : STD_LOGIC_VECTOR(3 DOWNTO 0);
BEGIN
    rot0 <= "00" & rotation;
    rot1 <= ("00" & rotation) + 1;
    rot2 <= ("00" & rotation) + 2;
    rot3 <= ("00" & rotation) + 3;

    y(0) <= '1' when (rot0(1 DOWNTO 0) < valin) else '0';
    y(1) <= '1' when (rot1(1 DOWNTO 0) < valin) else '0';
    y(2) <= '1' when (rot2(1 DOWNTO 0) < valin) else '0';
    y(3) <= '1' when (rot3(1 DOWNTO 0) < valin) else '0';

    valout <= y;

PROCESS(clk, reset) IS
BEGIN
    if (reset = '0') then
        counter <= "0000";
        rotation <= "00";
    elsif (rising_edge(clk)) then
        counter <= counter + 1;
        if (counter = divider) then
            rotation <= rotation + 1;
            counter <= "0000";
        end if;
    end if;
END PROCESS;

END behavioral;
```

978-1-4244-4782-4/10 $26.00 © 2010 IEEE

Specialized Digital Signal Processor for Control of Multi-rail/Multi-phase High Switching Frequency Power Converters

James Mooney, Mark Halton and Abdulhussain E. Mahdi
Department of Electronic & Computer Engineering
University of Limerick
Limerick, Ireland
E-mail: james.mooney@ul.ie

Abstract—**This paper describes a Digital Signal Processor (DSP) whose architecture has been purposely developed for executing power converter control algorithms. It is intended for use in the control of multi-phase or multi-rail switching power converters. The DSP's novel dual multiplier-accumulator datapath allows multiple operations to be executed in a single clock cycle, thus shortening required execution times of power control algorithms compared with standard low-end DSPs. The processor has been designed using the Verilog Hardware Description Language (HDL), implemented on an FPGA development board and tested in a closed-loop system controlling three independent 500 kHz synchronous buck converters.**

I. INTRODUCTION

Digital controllers are used in systems where flexibility, programmability and reliability are required. Through the implementation of auto-tuning algorithms, digital controllers can improve power conversion efficiency and maintain accurate voltage regulation by adapting to changes in the power converter's components over time [1-7]. Three main hardware blocks form the basis for digital power control: an Analog-to-Digital Converter (ADC), a compensator and a Digital Pulse Width Modulator (DPWM). This paper focuses primarily on the digital hardware implementation of the compensator.

Although digital compensators with fixed algorithms have been developed for power converter applications, i.e. Control Law Accelerators (CLAs)) [8-12], in many cases these are unsuitable in practice due to their lack of flexibility. A Digital Signal Processor (DSP)-based compensator is inherently flexible in that it can be programmed to implement a wide range of algorithms, as demonstrated in [13]. Indeed, multiple independent power converters can be simultaneously controlled by a single DSP, executing different control algorithms for each converter [14]. These requirements are common in power supplies for computers and telecommunications systems [15]. However the rate at which the control algorithms must be executed is increasing with the

trend towards using higher switching frequencies, higher sampling frequencies and multi-sampling techniques [16-18]. Existing power control DSPs do not have sufficient hardware resources to meet the increasing performance requirements of multi-rail/multi-phase converters with switching frequencies up to and beyond 2 MHz. Using a high-end DSP or multiple CLAs would fulfill these requirements but these are not cost-effective solutions for Point-of-Load (POL) converter applications.

This paper presents an optimal power controller solution with increased computational resources compared with low-end DSPs but without the superfluous resources of a high-end one. The proposed DSP's architecture has been specifically developed based on analysis of control algorithms and typical multi-rail/multi-phase converter application requirements.

II. DIGITAL SIGNAL PROCESSOR DESIGN

The proposed DSP has three primary components: the program controller, the register file memory and the datapath, as illustrated in Fig. 1. This section describes the unique features of each of these components, which enable the DSP to perform as an effective multi-rail/multi-phase digital power controller. The processor is characterized by its dual datapath architecture, which permits multiple operations to be executed in parallel.

A. Datapath

The datapath component effectively consists of two interconnected datapaths, which receive data from a common register file and share a number of functional elements. Each datapath has a multiplier-accumulator (MAC) unit to execute multiplication, addition/subtraction or combined multiply-accumulate operations in a single DSP clock cycle. These operations are the main operations found in power control algorithms. The execution time of the control algorithms is therefore reduced by carrying out two such operations simultaneously. Data movement operations may also be

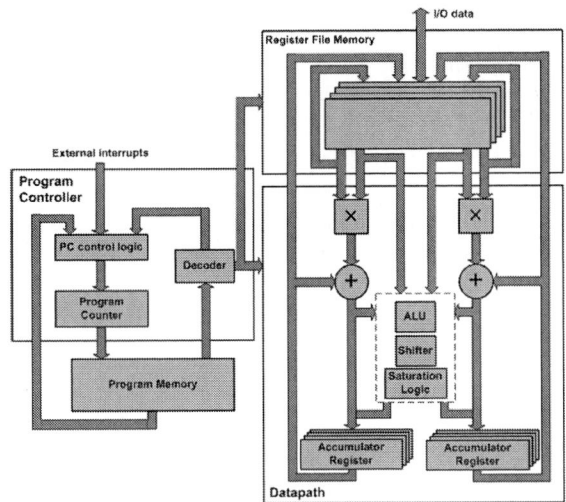

Figure 1. Simplified digital signal processor architecture overview

executed in parallel with computational operations for updating filter delay lines. Other less frequently required functional elements are shared between the two datapaths. These include barrel shifters that shift data by multiple bits per clock cycle, an Arithmetic Logic Unit (ALU) and saturation logic. The internal saturation logic limits the output of the accumulator registers to the maximum representable value to prevent errors due to overflow. The datapath also includes separate dedicated saturator logic that can limit data values to any given threshold in a single clock cycle operation.

The datapath has a 16-bit fixed-point number representation. This provides sufficient precision for power converter control applications because of the limited range and resolution of the sampled voltage and current signals from the ADC. The resolution of the duty cycle value presented to the DPWM is also typically less than 16 bits [19-21]. The datapath's accumulator registers have 32-bit resolution in order to preserve accuracy between successive multiply-accumulate operations.

B. Register File Memory

The register file memory stores the coefficients, delay-line values and other data required during the execution of the control algorithms. It is divided into four segments to achieve a concise addressing scheme, which is desirable to minimize the size of the program memory and the instruction decoding hardware. This restricts access to only the selected segment in any single instruction. The segmentation does not have a negative impact on algorithm execution speed in a multiple power converter application because usually only the data associated with one converter needs to be addressed while the algorithm for that converter is executing. A different segment can be selected before executing the algorithm for the next converter by means of a context switch instruction. This register file memory architecture eliminates the need to frequently access data from an external memory source, which would reduce the time available to execute computational operations.

C. Program Controller

The Program Counter (PC) control logic, together with data from a number of configuration registers manage standard instruction sequencing and pipelining hazards such as branching or interrupts. The output of the program counter is the address in the program memory of the next instruction to be executed.

The interrupt control hardware governs how the computational resources of the DSP are time-multiplexed to execute control algorithms for multiple independent power converters. Each control algorithm can be triggered by an external interrupt signal, which may come from the ADC interface, the DPWM or some other timing synchronization hardware block. A control algorithm cannot be interrupted by another interrupt signal and simultaneously occurring interrupts are serviced in terms of their fixed priority setting. This allows converters with different switching frequencies to be controlled by the DSP. The interrupt signal also automatically executes a context switch operation to select the appropriate segment of the register file before any other operation of the algorithm is executed.

When no control algorithm is being executed by the DSP, it operates in background mode, which provides the option of executing monitoring and communications instructions. Control algorithms for power converters with low switching/sampling frequencies can also be executed in this mode. Fig. 2 shows an example of where the DSP executes control algorithms to regulate two independent voltage rails. Background code is executed between each algorithm.

Similar to commercial DSP implementations, an assembly language-based instruction set was created in order to facilitate human programming of control algorithms on the DSP. An example excerpt of code from a control algorithm is illustrated in Fig. 3. Each instruction may contain two independent sub-instructions, which are executed concurrently on separate datapaths. Operations that involve the use of shared datapath elements may only be specified in one of the sub-instructions. An assembler program was also developed to convert the instruction set mnemonics to the binary code that is loaded into the program memory. The program controller's decoder interprets the binary instructions in the program memory to generate the signals that allow the specified control algorithm to be executed on the processor's datapath.

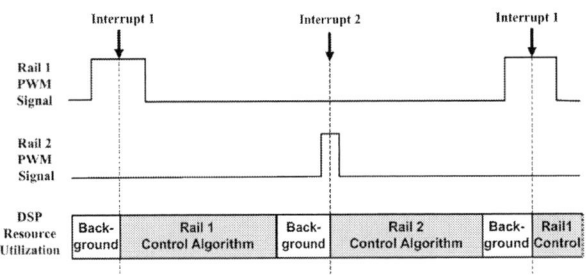

Figure 2. Control of multiple power converter rails using DSP interrupts

```
.
.
.
; b2 * e(n-2) + ACC1,   a2 * d(n-2) + ACC2
MACU  R9  R2          MACU  R12 R5

; b1 * e(n-1) + ACC1,   a1 * d(n-1) + ACC2
MACU  R8  R1          MACU  R11 R4

; b0 * e(n) + ACC1
MACU  R7  R0

; Sum b's and a's
ADD  ACC1 ACC2
.
.
.
```

Figure 3. Excerpt from assembly language-based control algorithm with
concurrent MAC operations

III. EXPERIMENTAL VERIFICATION

A. Functional Verification

Functional verification of the Verilog-specified DSP design was performed using the ModelSim simulator in conjunction with MATLAB/Simulink/PLECS.

A three-pole, three-zero (3P3Z) control algorithm (1) was used to regulate the output voltage of two independent single-phase buck converter models:

$$d(n) = b_0 e(n) + b_1 e(n-1) + b_2 e(n-2) + b_3 e(n-3) \quad (1)$$
$$+ a_1 d(n-1) + a_2 d(n-2) + a_3 d(n-3),$$

where $d(n)$ is the required duty cycle value, $e(n)$ is the latest sampled output voltage error, $e(n-1)$, $e(n-2)$ and $e(n-3)$ are the output voltage errors for one, two and three previous iterations respectively. $d(n-1)$, $d(n-2)$ and $d(n-3)$ are the duty cycle values for one, two and three previous iterations respectively and the compensator's coefficients are b_0, b_1, b_2, b_3, a_1, a_2 and a_3.

Fig. 4 is a screenshot of a selection of the control and data signals involved in executing a single iteration of the 3P3Z algorithm on the DSP. The figure shows that by performing multiple operations in parallel, the processor can execute the algorithm in nine DSP clock cycles. Between successive iterations of the algorithm the DSP is in background mode and can execute monitoring, communications or miscellaneous user defined operations. In this case the processor loops in an idle state in background mode.

B. Digital Hardware Implementation

The Verilog code was synthesized using the integrated synthesizer in Altera's Quartus II design software, targeting implementation on a Cyclone II FPGA device. The area optimization synthesis technique was selected to minimize logic usage.

Table I compares the utilization of the FPGA's resources for the DSP's main constituent elements. In addition to the logic elements and registers listed in the table, the DSP also makes use of some of the FPGA's embedded hardware components. The DSP's two 16 x 16 bit multipliers in the datapath are implemented using the FPGA's embedded multipliers, while the program memory is implemented with the embedded M4K memory blocks.

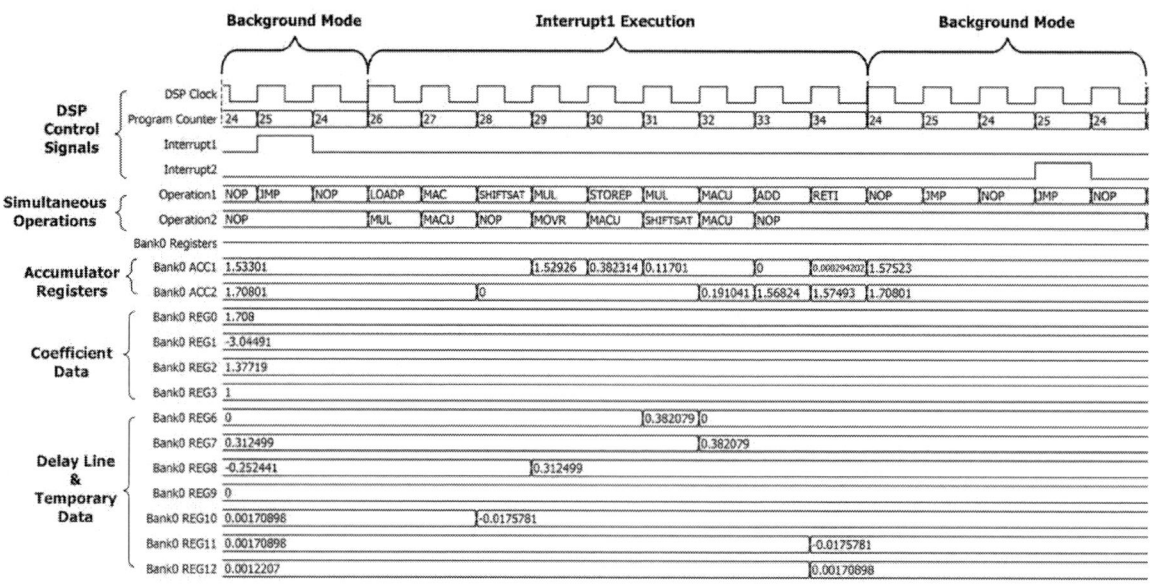

Figure 4. DSP control and data signals during simulation of three-pole, three-zero control algorithm

978-1-4244-4782-4/10 $26.00 © 2010 IEEE   2209

TABLE I.   FPGA RESOURCE UTILIZATION FOR SYNTHESIZED DSP CORE

Hardware Modules	Logic Elements	% of Total Logic Elements	Registers	% of Total Registers
Register File Memory				
Register File	3549	65 %	1296	86 %
Datapath				
Computational Elements	609	11 %	0	0 %
Routing Multiplexers	815	15 %	0	0 %
Program Controller				
Decoder	148	3 %	0	0 %
PC & Interrupt Control	347	6 %	213	14 %
Total	5468	100 %	1509	100 %

The DSP utilizes approximately 16% of the FPGA's total 33,216 logic elements. The FPGA is intended only as a prototyping platform for the DSP, as it is envisaged that the DSP would ultimately be integrated into a System on Chip (SoC) for multi-rail/multi-phase power converter control. For this reason it necessary to minimize the logic resources of the DSP to minimize the cost of the final SoC solution.

Table I indicates that a major proportion of the logic utilized is allocated to the register file memory block. The register file memory consists of four segments, each with sixteen 16-bit registers. Read access to four registers and write access to two registers is required in each clock cycle. The input de-multiplexers and output multiplexers therefore contribute to a significant proportion of the overall logic requirements. In order to reduce the logic requirements, the number of registers per bank can be reduced. However this confines the DSP to executing basic algorithms, which require fewer coefficients and data variables, resulting in a lower performance compensator solution. An alternative method to reducing the logic requirements is to restrict data access between the two datapath sections. This also leads to reduced flexibility in terms of the possible algorithms that can be executed and can render the DSP unsuitable for power converter control. Investigations into an optimal register file memory architecture are on-going.

The synthesis process yielded a maximum DSP clock frequency of 45 MHz due to the critical path from the register file memory, through the datapath to the accumulator register. Thus a 3P3Z algorithm, which takes nine DSP clock cycles to execute, can be run at a frequency of 5 MHz. This means that a buck converter with a switching frequency of 5 MHz could be controlled by the DSP compensator, where the output voltage is sampled once per cycle. Alternatively, a 1 MHz switching converter could be controlled using over-sampling techniques with an over-sampling factor of five. The DSP could also be used to control three separate converters, each with a switching frequency of 1.66 MHz.

C. Application to DC-DC Converter

The synthesized DSP design was combined with the necessary digital interface hardware to allow data acquisition from the voltage-sampling ADCs. A DPWM was also implemented on the FPGA and interfaced to the processor. The digital system illustrated in Fig. 5 was applied to control a prototype power supply system consisting of three single-phase 12 V - to - 1.5 V buck converters, each with a 500 kHz switching frequency. The DSP was programmed to execute a 3P3Z algorithm to regulate the output voltage of each converter.

Fig. 6 illustrates how the processor is time-multiplexed over the 2μs switching period to execute the three control algorithms and some background code for testing purposes. In this case the DSP clock frequency has been reduced to 15 MHz to simplify synchronization with the ADC interface. The figure also verifies the ability of the processor to regulate the output voltage of one of the converters in the presence of a load current step. It should be noted that the compensator used here was not designed specifically for the prototype power supply, hence the long settling time that is evident in Fig. 6.

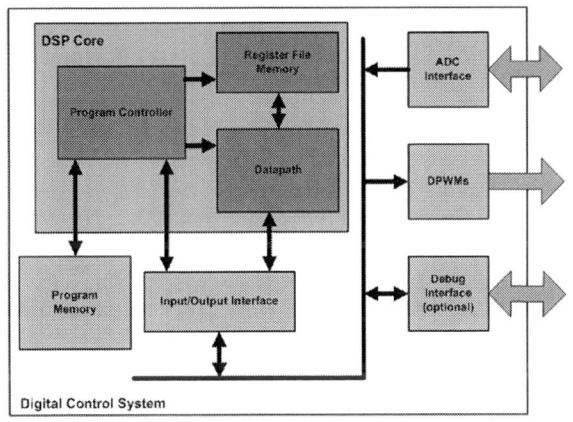

Figure 5.   Digital control system implemented on FPGA for application to three-rail dc-dc converter prototype

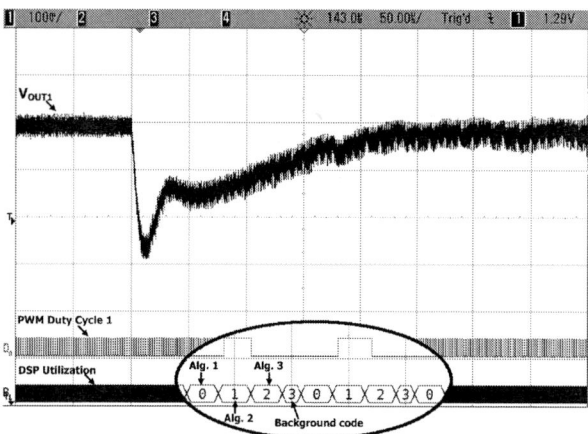

Figure 6. DSP utilization while regulating three independent 500 kHz buck converters

IV. CONCLUSION

Digital Signal Processors are used in power control applications where controller flexibility is required. This paper has proposed a specialized dual-datapath processor architecture, whose novel design enables it to control multiple power converters in a cost-effective manner compared with existing compensator solutions. It is expected that its performance specifications will meet the demands of future controllers for power supply systems which will have a larger quantity of rails/phases with higher switching frequencies. The operation of the DSP has been verified experimentally using an FPGA platform and a three-rail power converter prototype.

ACKNOWLEDGMENT

Special thanks to Simon Effler, (Circuits and Systems Research Centre, University of Limerick) for his assistance with the experimental verification, by providing the DPWM and prototype power supply system.

REFERENCES

[1] D. Maksimovic, R. Zane, and R. Erickson, "Impact of digital control in power electronics," *International Symposium on Power Semiconductor Devices and ICs*, 2004, pp. 13-22..

[2] A. V. Peterchev and S. R. Sanders, "Digital Multimode Buck Converter Control With Loss-Minimizing Synchronous Rectifier Adaptation," *IEEE Trans. on Power Electronics*, vol. 21, pp. 1588-1599, 2006.

[3] J. Morroni, R. Zane, and D. Maksimovic, "Design and Implementation of an Adaptive Tuning System Based on Desired Phase Margin for Digitally Controlled DC-DC Converters," *IEEE Trans. on Power Electronics*, vol. 24, pp. 559-564, 2009.

[4] J. A. Abu Qahouq, L. Huang, and D. Huard, "Efficiency-Based Auto-Tuning of Current Sensing and Sharing Loops in Multiphase

Converters," *IEEE Trans. on Power Electronics*, vol. 23, pp. 1009-1013, 2008.

[5] A. Parayandeh and A. Prodic, "Digitally controlled low-power DC-DC converter with segmented output stage and gate charge based instantaneous efficiency optimization," *IEEE Energy Conversion Congress and Exposition*, 2009, pp. 3870-3875.

[6] Z. Lukic, Z. Zhao, S. Ahsanuzzaman, and A. Prodic, "Self-Tuning Sensorless Digital Current-Mode Controller with Accurate Current Sharing for Multi-Phase DC-DC Converters," *IEEE Applied Power Electronics Conference and Exposition*, 2009, pp. 264-268.

[7] A. Kelly and K. Rinne, "A self-compensating adaptive digital regulator for switching converters based on linear prediction," *IEEE Applied Power Electronics Conference and Exposition*, 2006, pp. 712-718.

[8] L. Ka and D. Alfano, "Design and implementation of a practical digital PWM controller," *IEEE Applied Power Electronics Conference and Exposition*, 2006, p. 6.

[9] V. Yousefzadeh and S. Choudhury, "Nonlinear digital PID controller for DC-DC converters," *IEEE Applied Power Electronics Conference and Exposition*, 2008, pp. 1704-1709.

[10] J. Quintero, A. Barrado, M. Sanz, and A. Lazaro, "Digital control with asynchronous Linear-non-Linear compensator," *IEEE Applied Power Electronics Conference and Exposition*, 2008, pp. 491-497.

[11] J. Zhang and S. R. Sanders, "A Digital Multi-Mode Multi-Phase IC Controller for Voltage Regulator Application," *IEEE Applied Power Electronics Conference*, 2007, pp. 719-726.

[12] A. Prodic and D. Maksimovic, "Design of a digital PID regulator based on look-up tables for control of high-frequency DC-DC converters," *IEEE Workshop on Computers in Power Electronics*, 2002, pp. 18-22.

[13] E. O'Malley and K. Rinne, "A 16-bit fixed-point digital signal processor for digital power converter control," *IEEE Applied Power Electronics Conference*, 2005, pp. 50-56 Vol. 1.

[14] J. Mooney, A. Mahdi, A. Kelly, and K. Rinne, "DSP-based controller for multi-output/multi-phase high switching frequency DC-DC converters," *IEEE Workshop on Control and Modeling for Power Electronics*, 2008, pp. 1-6.

[15] R. Miftakhutdinov, L. Sheng, J. Liang, J. Wiggenhorn, and H. Huang, "Advanced bus control circuit for intermediate bus converter," *IEEE Applied Power Electronics Conference*, 2008, pp. 1515-1521.

[16] L. Corradini, S. Saggini, and P. Mattavelli, "Analysis of a high-bandwidth event-based digital controller for DC-DC converters," *IEEE Power Electronics Specialists Conference*, 2008, pp. 4578-4584.

[17] L. Corradini and P. Mattavelli, "Modeling of Multisampled Pulse Width Modulators for Digitally Controlled DC–DC Converters," *IEEE Trans. on Power Electronics*, vol. 23, pp. 1839-1847, 2008.

[18] S. Effler, Z. Lukic, and A. Prodic, "Oversampled digital power controller with bumpless transition between sampling frequencies," *IEEE Energy Conversion Congress and Exposition*, 2009, pp. 3306-3311.

[19] T. Carosa, R. Zane, and D. Maksimovic, "Scalable Digital Multiphase Modulator," *IEEE Trans. on Power Electronics*, vol. 23, pp. 2201-2205, 2008.

[20] M. Scharrer, M. Halton, and T. Scanlan, "FPGA-Based Digital Pulse Width Modulator With Optimized Linearity," *IEEE Applied Power Electronics Conference and Exposition*, 2009, pp. 1220-1225.

[21] W. Xiaopeng, Z. Xin, P. Jinseok, and A. Q. Huang, "Design and implementation of a 9-bit 8MHz DPWM with AMI06 process," *IEEE Applied Power Electronics Conference and Exposition*, 2009, pp. 540-545.

Computer-Aided Design for Class-E Switching Circuits Taking into Account Optimized Inductor Designs

Natsumi Sagawa†, Hiroo Sekiya†‡, and Marian K. Kazimierczuk‡

†Graduate School of Advanced Integration Science, Chiba University, Chiba, 263–8522 Japan
‡Department of Electrical Engineering, Wright State University, Dayton, OH, 45435–0001 USA
Email: †sagawa@graduate.chiba-u.jp   ‡sekiya@faculty.chiba-u.jp

Abstract—This paper presents a computer-aided design for the class-E switching circuits taking into account optimized inductor designs. The proposed design algorithm provides the element values for satisfying the multiple constrain conditions. Simultaneously, the optimal inductors, which have a minimum volume for achieving the permissible power conversion efficiency, are obtained. A design example of the multi-resonant dc/dc converter with the class-E ZVS/ZDS conditions is given. By carrying out the circuit experiment, it is shown that the designed converter satisfied class-E ZVS/ZDS conditions, specified output voltage, and permissible power conversion efficiency simultaneously, which verify the validity and effectiveness of the proposed algorithm. The measurement power conversion efficiency achieved 94.0 % for 4.92 W (7.0 V/0.70 A) output power and 1 MHz operation.

Index Terms—computer aided design, Class-E ZVS/ZDS switching conditions, multiple constrain conditions, inductor design, multi-resonant dc/dc converter

I. INTRODUCTION

High power conversion efficiency is strongly required to the switching dc/dc converters, dc/ac inverters, ac/dc rectifiers, and power amplifiers. There are two dominant factors of power losses, which are: (i) conduction losses at each element and (ii) switching losses caused by output capacitances of switching elements.

The conduction losses are caused on two major factors. One is the on-resistances of the switching elements, and the other is the equivalent series resistances (ESRs) of the magnetic components. The ESR of the magnetic component depends on inductance value, core, length and thickness of the wires, gap length, number of the winding layer, and so on. Generally, a large core volume provides a low ESR and a small core volume gives a high ESR for a fixed inductance value. Since the designers usually require both a high power conversion efficiency and a small circuit scale, the magnetic element design is regarded as an optimization problem. Therefore, there were many researches of the magnetic component designs [8]–[11]. It is possible to calculate the ESRs of the magnetic components by following these researches. In most researches, however, inductors are designed from the specified inductance value and the information of the inductor current. Therefore, if inductance value is fixed, it is easy to estimate its accurate ESR value. However, if the inductance value cannot be fixed, it is impossible to obtain the accurate ESR value.

The switching techniques such as ZVS (zero-voltage switching), ZCS (zero-current switching), and class-E ZVS/ZDS (zero derivative switching) [1]-[7], [10], [11] are good strategies for reducing the switching losses. However, it is a problem to determine the element values for satisfying these switching conditions. Recently, the computer-aided design techniques for the class-E switching circuits were proposed [6], [7]. These design algorithms give the accurate element values for satisfying multiple constrain conditions with low computational cost. However, it is difficult to obtain the accurate design parameters, when the power conversion efficiency is one of the multiple constrain conditions, in particular. This is because that the algorithms do not have a inductor design function. Therefore, ESRs of magnetic components should be set with unsure values, which lead to the difficulty in obtaining the accurate power conversion efficiency.

The ESRs of the magnetic components affect the waveforms and power conversion efficiency. Conversely, the waveforms affect the element values including the magnetic components. Because of this nesting structure, it is difficult to estimate the accurate power conversion efficiency in the design procedure. Therefore, the derivations of the element values and the inductor designs should be carried out simultaneously. When it is realized, we can obtain accurate design parameters even if the power conversion efficiency is one of the design conditions.

This paper presents a computer-aided design for the class-E switching circuits taking into account optimized inductor designs. The proposed design algorithm provides the element values for satisfying the multiple constrain conditions. Simultaneously, the optimal inductors, which have a minimum volume for achieving the permissible power conversion efficiency, are obtained. By including the inductor designs in the element-value-derivation process, it is possible to estimate the accurate ESR values in the design process. Therefore, the element values can be obtained even if the power conversion efficiency is one of the multiple constrain conditions. A design example of the multi-resonant dc/dc converter with the class-E ZVS/ZDS conditions are given. By carrying out the circuit experiment, it is shown that the designed converter satisfied class-E ZVS/ZDS conditions, specified output voltage, and permissible power conversion efficiency simultaneously, which verify the validity and effectiveness of the proposed algorithm.

978-1-4244-4782-4/10 $26.00 © 2010 IEEE

II. PREVIOUS DESIGN ALGORITHM

A. Numerical Design algorithm for Satisfying Multiple Constrain Conditions

The algorithms presented in [6] and [7] give element values for achieving the multiple constrain conditions numerically. Figure 1 shows a flow chart of this algorithm. Usually, converters, inverters, and power amplifiers are driven by periodic signals with operating frequency f. Let us consider a dynamic circuit described by a set of differential equations :

$$\frac{d\boldsymbol{x}}{d\theta} = f(\theta, \boldsymbol{x}, \lambda), \tag{1}$$

where $\theta \in \boldsymbol{R}$, $\boldsymbol{x} \in \boldsymbol{R}^n$, and $\lambda \in \boldsymbol{R}^m$ are angular time $\theta = \omega t = 2\pi f t$, state variations, and system-parameters, respectively. For simplicity,

$$
\begin{aligned}
f : \boldsymbol{R} \times \boldsymbol{R}^n \times \boldsymbol{R}^m &\rightarrow \boldsymbol{R}^n \\
(\theta, \boldsymbol{x}, \lambda) &\mapsto f(\theta, \boldsymbol{x}, \lambda)
\end{aligned}
\tag{2}
$$

is periodic in θ with period θ_T:

$$f(\theta + \theta_T, \boldsymbol{x}, \lambda) = f(\theta, \boldsymbol{x}, \lambda). \tag{3}$$

We also assume that (1) has a solution $\boldsymbol{x}(\theta) = \varphi(\theta, \boldsymbol{x_0}, \lambda)$, where $\boldsymbol{x_0}$ is an initial condition: $\boldsymbol{x}(0) = \varphi(0, \boldsymbol{x_0}, \lambda) = \boldsymbol{x_0}$. By the periodic hypothesis (3), we can naturally define the mapping T as follows:

$$
\begin{aligned}
T : \boldsymbol{R}^n &\rightarrow \boldsymbol{R}^n \\
\boldsymbol{x_0} &\mapsto T(\boldsymbol{x_0}, \lambda) = \varphi(\theta_T, \boldsymbol{x_0}, \lambda).
\end{aligned}
\tag{4}
$$

If a solution $\boldsymbol{x}(\theta) = \varphi(\theta, \boldsymbol{x_0}, \lambda)$ is periodic with period θ_T, the point $\boldsymbol{x_0}$ is a fixed point of T:

$$T(\boldsymbol{x_0}, \lambda) = \boldsymbol{x_0}. \tag{5}$$

For circuit designs, some design conditions, e.g. switching conditions, specified output power, specified ripple ratio, and so on, are specified. If the number of conditions is $M(\leq m)$, the multiple constrain conditions, which consist of conditions g_p, are expressed as

$$G(\boldsymbol{x_0}, \lambda) = \begin{bmatrix} g_1(\boldsymbol{x_0}, \lambda) \\ g_2(\boldsymbol{x_0}, \lambda) \\ \vdots \\ g_M(\boldsymbol{x_0}, \lambda) \end{bmatrix} = \boldsymbol{0}, \qquad \in \boldsymbol{R}^N. \tag{6}$$

In this case, we can find M design parameters. Therefore, the other $(m - M)$ parameters must be given as the design specifications. We recognize that the design problem boils down to solving the algebraic equations (5) and (6). These equations are rewritten as follows:

$$F(\boldsymbol{x_0}, \lambda) = \begin{bmatrix} T(\boldsymbol{x_0}, \lambda) - \boldsymbol{x_0} \\ G(\boldsymbol{x_0}, \lambda) \end{bmatrix} = \boldsymbol{0}, \qquad \in \boldsymbol{R}^{n+M} \tag{7}$$

where $T(\boldsymbol{x_0}, \lambda)$, $\boldsymbol{x_0}$, and λ are expressed as $T(\boldsymbol{x_0}, \lambda) = [T_1(\boldsymbol{x_0}, \lambda), T_2(\boldsymbol{x_0}, \lambda), \cdots, T_n(\boldsymbol{x_0}, \lambda)]^T$, $\boldsymbol{x_0} = \boldsymbol{x}(0) = [x_1(0),$

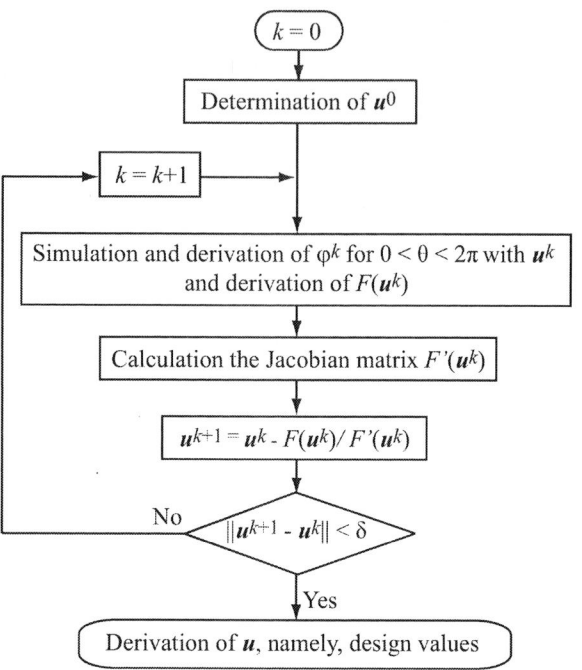

Fig. 1. Flow chart of the design algorithm for achieving multiple constrain conditions.

$x_2(0), \cdots, x_n(0)]^T$, and $\lambda = [\lambda_1, \lambda_2, \cdots, \lambda_m]^T$. Moreover, we define $\lambda_u \in R^M$ as

$$\lambda_u = \{\lambda_{u1}, \lambda_{u2}, \cdots, \lambda_{uM} \mid \lambda_{uk}(k = 1, 2, \cdots, M)$$
$$\text{are unknown design parameters in } \lambda.\} \tag{8}$$

We solve the equations (7) by Newton's method as shown in Fig. 1, which is the general algorithm to solve the algebraic equations. The unknown values of (7) are expressed as $\boldsymbol{u} \in R^{n+M}$: $\boldsymbol{u} = [\boldsymbol{x}_0^T, \lambda_u^T]^T$. By solving $F(\boldsymbol{u}) = 0$, the unknown parameters $\boldsymbol{u}$ can be found, and the design values, that is, λ_u are determined. The Newton's method is the iterative computations

$$\boldsymbol{u}^{k+1} = \boldsymbol{u}^k - \frac{F(\boldsymbol{u}^k)}{F'(\boldsymbol{u}^k)} \tag{9}$$

for $\|\boldsymbol{u}^{k+1} - \boldsymbol{u}^k\| < \delta \ll 1$ in order to find the unknown values, where $F' \in R^{(n+M)\times(n+M)}$ means the Jacobian matrix of F and k is the iteration number. The Jacobian matrix is calculated as shown in [7]. Then $\boldsymbol{u}^{k+1}$ is a solution of (7).

If λ_u includes magnetic-component parameters, the accurate ESRs of the magnetic components cannot be obtained. This is because λ_u cannot be fixed during the iterative calculations. Therefore, it is difficult to obtain the accurate design parameters, when the power conversion efficiency is one of the multiple constrain conditions, in particular.

B. Inductor Designs with Optimum Core and Wire Selection

The algorithm presented in [11] gives a selection procedure of the optimal core and wire for the resonant inductor. The optimal inductor was defined as the inductor, which satisfies the following conditions.

1) The total power loss P_t of the inductor is lower than P_{tmax}.
2) The maximum magnetic flux density B_m is lower than the saturation magnetic flux density of the core material B_{sat}.
3) The peak current density J_m through the wire is lower than the maximum current density J_{max}.
4) The number of the wire layer N_l is lower than the maximum number of the layer N_{lmax}.
5) The core volume is the smallest achieving the above conditions.

It is assumed that the current through the inductor is the sum of the sinusoidal waveform current with the operating frequency f and the dc-bias current

$$i_L = I_m \sin\theta + I_{av}, \tag{10}$$

where I_m and I_{av} are the amplitude of the ac component and the average value of the inductor current, respectively. From inductance value L, current information in (10), and permissible maximum power loss P_{tmax} are needed for the inductor design.

In resonant converters, inverters, and amplifiers, the magnitude of the ac-component of the magnetic flux density B_{acm} on the resonant inductors becomes large because the large-swing currents flow through the resonant inductors, which is expressed as

$$B_{acm} = \frac{L I_m}{A_c N}. \tag{11}$$

A large B_{acm} and high frequency generate high core loss. For reducing the core loss, selecting the large core size is a good strategy. However, we suffer from the increase in the circuit scale. Therefore, we need to select the minimum core volume whose core loss is smaller than P_{tmax}.

Figure 2 shows the core and wire selection algorithm. First of all, we pick the smallest core in the core candidates. The number of turns at the core is calculated by using the inductance per one turn A_L

$$N = \sqrt{\frac{L}{A_L}}. \tag{12}$$

By using N, the maximum magnetic flux density B_m is obtained from

$$B_m = \frac{L(I_m + I_{av})}{A_c N}, \tag{13}$$

where A_c is the core cross-sectional area. When B_m is greater than B_{sat}, the operation of the inductor includes the core-saturation region. In this case, the smallest core in the core candidates except the selected cores is picked and the calculations in (12) and (13) are repeated. The current density of the winding wire is expressed

$$J_m = \frac{I_m}{A_w}, \tag{14}$$

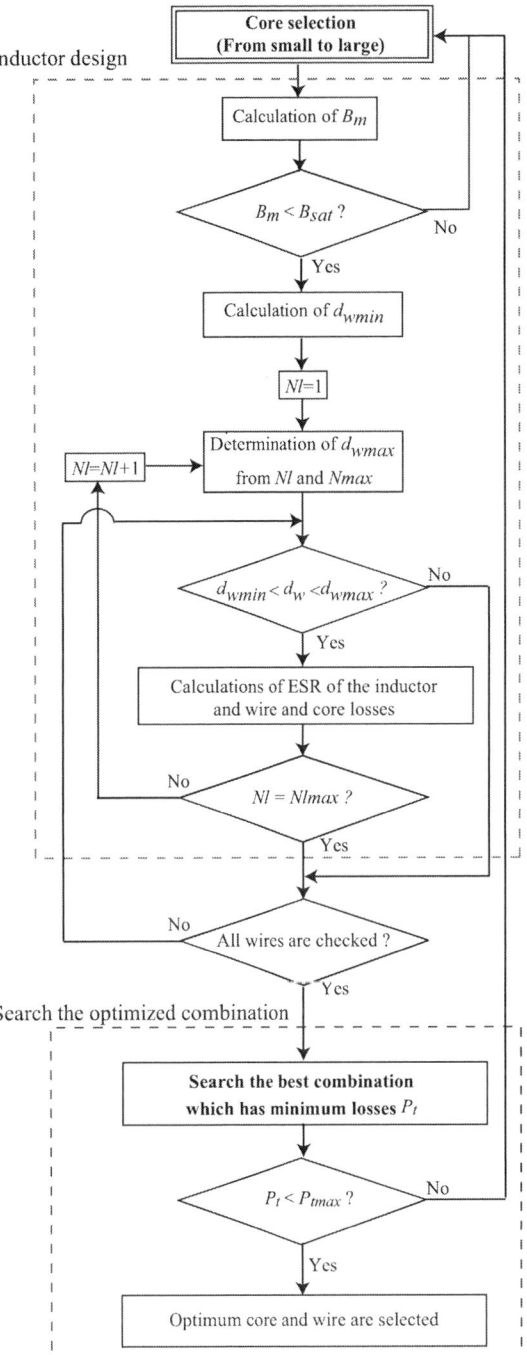

Fig. 2. Optimum core- and wire-selection algorithm.

where A_w is the cross-sectional area of the winding wire. From J_{max}, the minimum wire diameter d_{wmin} is expressed as

$$d_{wmin} = 2 \sqrt{\frac{I_m + I_{av}}{\pi J_{max}}}. \tag{15}$$

We cannot use the winding wire whose diameter is equal or thinner than d_{wmin}. The number of layer N_l is selected from 1 to N_{lmax}. From N_l and the core mechanical parameters, the maximum wire diameter d_{wmax} can be obtained.

Subsequently, the wires whose diameters are in the range of $d_{wmin} < d_w < d_{wmax}$ can be used for the inductor design. It is calculated the core and wire losses for each combination of core, wire, and number of the layer. By calculating d_{wmin} and d_{wmax}, the wire candidates are narrowed down, which reduces the computational cost. It can be narrowed down the wire candidate. It is good strategy to reduce the computational cost. When all wires candidates are checked in N_l, the inductor with $(N_l + 1)$ layer is considered. These computations are repeated up to $N_l = N_{lmax}$ as shown in Fig. 2.

In order to obtain the power losses, core loss P_c, ac winding loss P_{wac}, and dc winding loss P_{wdc} are calculated. The core loss P_c is obtained from

$$P_c = af^c B_{acm}^d V_c, \qquad (16)$$

where V_c is the core volume and a, c, and d are the coefficients given by manufacturers. The dc winding loss is

$$P_{wdc} = r_{wdc} I_{rms}^2, \qquad (17)$$

where I_{rms} means root-mean-square of the current expressed as

$$I_{rms} = \sqrt{\frac{I_m^2}{2} + I_{av}^2}. \qquad (18)$$

In addition, $r_{wdc} = \rho_w l_w / \pi d_w^2$ is the dc winding resistance, where $\rho_w = 6.88 \times 10^{-8}$ Ωm is the resistivity of the copper and l_w is the length of the wire. The ac winding resistance is expressed as

$$r_{wac} = F_R r_{wdc}, \qquad (19)$$

where F_R is the ratio of the ac-to-dc winding resistance, which is derived in [8], [11]. From the ac winding resistance in (19), the ac winding losses expressed as

$$P_{wac} = r_{wac} \frac{I_m^2}{2} = F_R r_{wdc} \frac{I_m^2}{2}, \qquad (20)$$

Finally, the total power losses of the inductor P_t is

$$P_t = P_{wac} + P_{wdc} + P_c. \qquad (21)$$

From P_t, the ESR of the inductor r_L can be obtained as

$$r_L = \frac{P_t}{I_{rms}^2} = \frac{P_{wac} + P_{wdc} + P_c}{\frac{I_m^2}{2} + I_{av}^2}. \qquad (22)$$

The total power losses of the inductor for the selecting core are calculated for all the wires candidate. After that, we select the wire, which generates the minimum inductor loss P_{tmin}. If $P_{tmin} > P_{tmax}$, the smallest core in the core candidates except selected cores should be picked. In case of the $P_{tmin} < P_{tmax}$, it is stated that the optimal inductors, which have a the smallest volume for achieving the permissible power losses, can be obtained.

In this design, the inductors are designed from the specified inductance value and the information of the current through them. However, the results of inductor designs are not reflected on the converter designs. Because of the interdependent relation between the element values and the inductor designs, the derivations of the element values and the inductor designs should be carried out simultaneously.

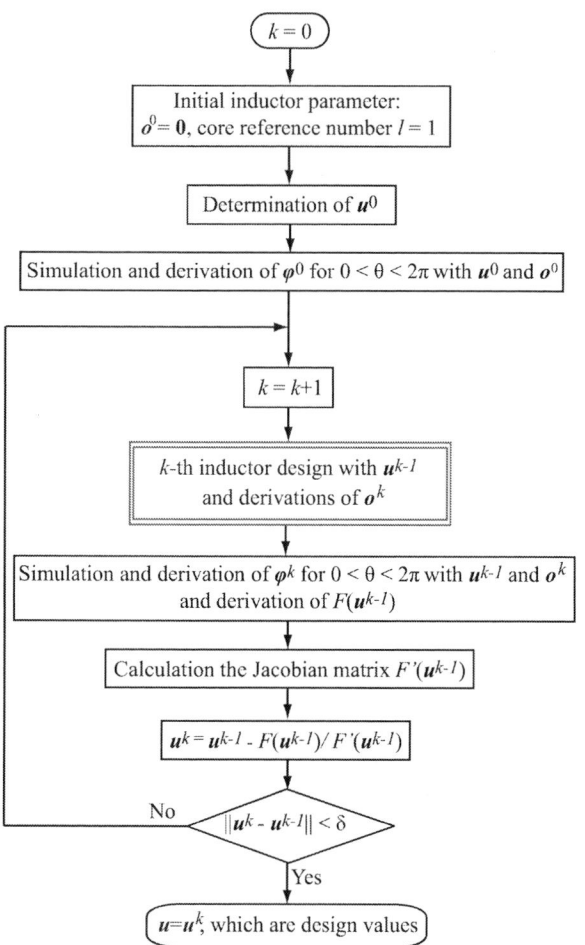

Fig. 3. Flowchart of the proposed design algorithm.

III. PROPOSED DESIGN PROCEDURE

In this paper, we present a computer-aided design for the class-E switching circuits taking into account the optimized inductor designs. The proposed algorithm provides the design values for achieving multiple constrain conditions with the optimized inductors, which have a minimum volume for achieving the permissible power-conversion efficiency.

Figure 3 shows a flow chart of the proposed algorithm. The proposed algorithm is based on the algorithm presented in [7]. Our idea is that the inductor designs are built into the Newton's method for obtaining the design parameters. In Fig. 3, the variation l is the reference number of the core candidates, which is labeled on the increasing order of the core volume. The variation l ranges in $1 \leq l \leq M_c$, where M_c is a number of the core candidates. The core l means the l th smallest core in the core candidates. Since the minimum volume core for achieving the permissible power conversion efficiency should be selected, the initial value of l is 1. Additionally, we express the ESR values of the inductors are o, which is renewed in the iterative calculations of the Newton's method.

Using the design parameters u^{k-1} and the ESR values

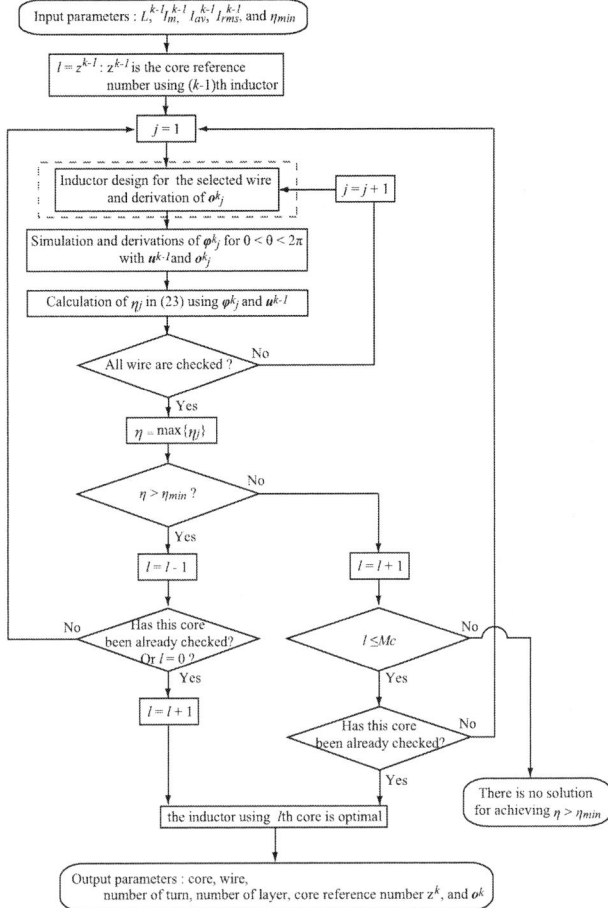

Fig. 4. Block diagram of the optimized inductor design.

o^{k-1}, the waveform vector φ^{k-1} can be obtained. From u^{k-1} and φ^{k-1}, the inductance value L^{k-1} and current informations $I_m^{k-1}, I_{av}^{k-1}, I_{rms}^{k-1}$ are obtained, which are the input parameters for kth inductor design. By using these parameters, the kth inductor design is performed, which provides the renewed ESR values o^k. From u^{k-1} and o^k, F(u^{k-1}) are calculated. Following the algorithm of the Newton's method, the renewed unknown-parameter vector u^k can be obtained. By carrying out the iterative calculations for k, the design parameters $u = u^k$ is obtained by using the stop condition $\|u^k - u^{k-1}\| < \delta << 1$. This stop condition means that the same states are maintained twice in the iterations. Therefore, it is guaranteed that the inductor designs are also converged when the Newton's method stops. As a result, the optimal inductor is obtained simultaneously when the Newton's method is converged. In addition, because this algorithm is based on that in [7], the high accurate element values can be obtained with low computational cost.

Figure 4 shows the detailed algorithm of "k th inductor design and derivation of o^k" in Fig. 3. This algorithm provides the ESR value o^k from the $L^{k-1}, I_m^{k-1}, I_{av}^{k-1}, I_{rms}^{k-1}$ and permissible power conversion efficiency $\eta_m in$. First, we set $l = z^{k-1}$, where z^{k-1} is the core reference number of the core used in $(k-1)$th

inductor design. In Fig. 4, j is iterative number of the inductor design at a certain core, which is added by changing the wire. For each j, the power conversion efficiency of the converter η_j is calculated by using o^k and u^{k-1} expressed as

$$\eta_j = \frac{P_{o_j}}{P_{in_j}} = \frac{V_{o_j}^2/R}{V_I I_{in_j}}. \tag{23}$$

Where, V_I is the input voltage, I_{in_j} is the j th input current, V_{o_j} is the j th output voltage, and R is the road resistance. After the calculations of ESRs for all the wire candidates, we pick the maximum value as $\eta = \max\{\eta_j\}$. By comparing η with η_{min}, we obtain the smallest inductor for achieving the permissible efficiency η_{min} for u^{k-1} as shown in Fig. 4. The ESRs of the designed inductors becomes o^k and the core reference number $l = z^k$ is saved for the $(k+1)$ th inductor design. In this inductor design algorithm, the first inductor design needs the significant computational cost since a proper core should be chosen from many core candidates. The core candidates are, however, narrowed down dramatically after second inductor design because the variations of u^k is small in the Newton's method. This is an important technique for the computational-cost reduction.

IV. Design Example

In this section, we show a design example of the multi-resonant dc/dc converter with the class-E ZVS/ZDS conditions.

A. Multi-Resonant dc/dc Converter

The circuit topology of the multi-resonant dc/dc converter is shown in Fig. 5(a). The multi-resonant dc/dc converter consists of dc-voltage source V_I, back current blocking capacitance C_d, MOSFETs S_1 and S_2 as switching devices with their anti-parallel diodes D_1 and D_2, shunt capacitances C_r and C_{S_2}, resonant inductor L_r, dc-feed inductor L_C, output capacitor C_f, and load resistance R.

Figure 5(b) shows the equivalent circuit of the multi-resonant dc/dc converter. In this figure, R_{S1} and R_{S2} are the equivalent resistances of the MOSFETs S_1 and S_2. The switch resistances R_{S1} and R_{S2} are given as

$$R_{Sk} = \begin{cases} r_{S_k} & \text{for MOSFET is in the on state,} \\ \infty & \text{for MOSFET is in the off state,} \end{cases} \tag{24}$$
$$k = 1 \text{ and } 2.$$

Moreover, r_{L_r} and r_{L_C} express the ESRs of the inductors L_r and L_C, respectively. In this design example, the ESRs of the capacitors are neglected since they are much smaller than that of inductors.

The example waveforms of the multi-resonant dc/dc converter with the class-E ZVS/ZDS conditions are shown in Fig. 6(a). The switch S_1 is driven by driving signal D_{S1} at the operating frequency f and the fixed duty ratio D_{S1}. When the switch S_1 is in the off state, C_r produces the switch voltage v_{S1}. At the turn-on instant, v_{S1} achieves ZVS and ZDS, namely, both the voltage and the slope of the voltage are zero. These conditions are called the class-E ZVS/ZDS conditions. The

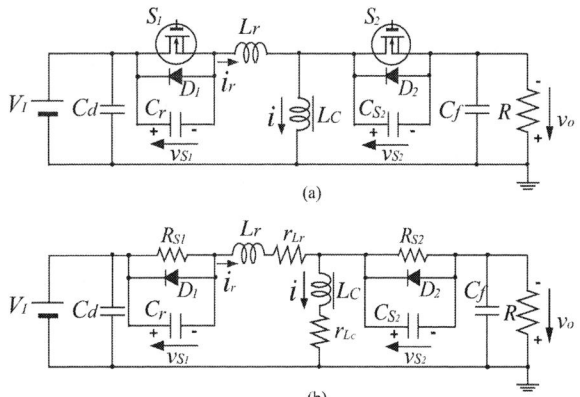

(a)

(b)

Fig. 5. Multi-resonant dc/dc converter. (a) Circuit topology. (b) Equivalent circuit.

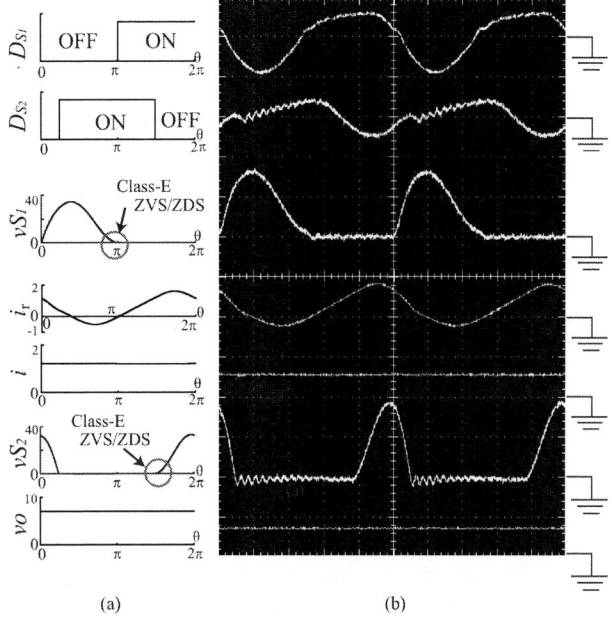

(a)　　　　　　　　　(b)

Fig. 6. Waveforms of the proposed converter. (a)Numerical waveforms. (b)Experimental waveforms. Vertical : D_{S_1}, D_{S_2}, v_{S_1} : 20 V/div, i_r, i : 2 A/div, v_{S_2} : 20 V/div, v_o : 10 V/div. Horizontal : 200 ns/div.

switch voltage v_{S1} tunes to the ac current i_r by L_r-C_r series-resonant circuit. The dc component of i_r flows through L_C, which is i. On the other hand, ac component of i_r flows through the switch S_2 and the shunt capacitance C_{S2}, which shapes the switch voltage v_{S2}. If the switch S_2 turns off when the current which flows through the switch S_2 is zero, the slope of v_{S2} is also zero. Therefore, these conditions are also the class-E ZVS/ZDS conditions. D_{r2} is given to achieve these conditions. Because of C_f, the output voltage v_o becomes almost constant, which is V_o.

As far as we know, there is no example that the class-E ZVS/ZDS conditions are applied in this circuit topology.

B. Design Specifications

The design specifications are given as operating frequency f = 1 MHz, input voltage V_I = 10 V, output voltage V_o = 7.0 V, load resistance R = 10 Ω, permissible minimum efficiency η_{min} = 94 %, loaded quality factor of the resonant inductor $Q_L = \omega L_r/R$ = 2.0, dc-feed inductor L_C = 300 μH, and ripple ratio of the output voltage γ = 1 %, where output voltage and the ripple ratio are expressed as

$$V_o = \frac{1}{2\pi}\sqrt{\int_0^{2\pi} v_o(\theta)^2 d\theta}, \qquad (25)$$

$$\gamma = \frac{v_{o_{max}} - V_o}{V_o}, \qquad (26)$$

where $v_{o_{max}}$ is the maximum value of the output voltage. To calculate the power conversion efficiency η from (23), we define the input current I_{in} as

$$I_{in} = \frac{1}{2\pi}\int_0^{2\pi} i_r(\theta)d\theta. \qquad (27)$$

In addition, the switch on-resistance r_S = 70 mΩ is given since IRFZ24N MOSFETs are used as the switching devices S_1 and S_2. Since the inductance of L_C is given as the design specification, we can make the inductor L_C and measure its ESR before the design calculations. The inductance value and ESR of L_C are measured by LCR meter of HP-4284A as L_C = 295 μH and r_{L_C} = 24 mΩ, respectively. These values are used for the design calculations.

In this design example, we apply the ungapped toroidal core as shown in Fig. 7 for the inductor design. The ungapped core avoids the magnetic flux fringing, which realize the inductors with lower ESR than the gapped core. Since we design the ungapped toroidal inductor, the ratio of ac-to-dc resistance F_R in (19) is obtained using the expression in [8]. It is supposed that core loss of the resonant inductor L_r becomes large. For reducing the core loss, we select the Micrometals iron-powder core material 2 because it has the low relative magnetic permeability, μ_r = 10. The core candidates are given in Table I. For the optimal inductor design, the following specifications are given: J_{max} < 4 A/mm², B_{sat} = 0.2 T, and maximum winding layer N_{lmax} = 2. The wire of the inductors are selected from American Wire Gauge (AWG) standardization as given in Table II.

C. Design and Experimental Results

Following the proposed algorithm in Section III, the T50-2 core and AWG18 wire are selected for the optimal inductor L_r. In this case, the multi-resonant dc/dc converter achieves 94.2 % power-conversion efficiency. The 26 turns and the 2-layers are needed to make the inductor. In this case, core loss and wire losses are P_c = 79.3 mW, P_{wdc} = 6.92 mW, and P_{wac} = 56.7 mW, respectively. Since the I_{rms} = 0.885 A, the ESR of the inductor is r_{L_r} = $(P_{wac} + P_{wdc} + P_c)/I_{rms}$ = 183 mΩ. Table III gives the summary of the designed inductor L_r. In case of the T37-2 core, all wires cannot be accepted because of $\eta < \eta_{min}$. Simultaneously, the design algorithm provides the element values of L_C, C_r, C_{S_2}, and C_f as given in Table IV.

978-1-4244-4782-4/10 $26.00 © 2010 IEEE

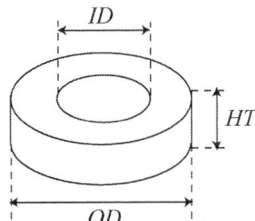

Fig. 7. Parameters of toroidal-core.

TABLE I
MICROMETALS RF TOROIDAL CORE PARAMETERS [12]

Part No.	A_L (nH/N²)	OD (mm)	ID (mm)	HT (mm)	V_c (cm³)
T37-2	4	9.53	5.21	3.25	0.147
T50-2	4.9	12.7	7.7	4.83	0.358
T68-2	5.7	17.5	9.4	4.83	0.759
T94-2	8.4	23.9	14.2	7.92	2.16
T106-2	13.5	26.9	14.5	11.1	4.28
T130-2	11	33	19.8	11.1	5.78

TABLE II
AWG COPPER WIRE PARAMETER

Gauge	Diameter(mm)	Gauge	Diameter(mm)
18	1.02	23	0.573
19	0.912	24	0.511
20	0.812	25	0.455
21	0.723	26	0.405
22	0.644	27	0.361

TABLE III
PARAMETERS FOR THE DESIGN OF RESONANT INDUCTOR L_r IN CASE OF
MULTIPLE-STRAND WINDING.

Inductance L_r	3.31 μH
Core No.	T50-2
Wire diameter d_w	AWG 18
Number of turns	26
Number of layer	2
Core loss P_c	79.3 mW
AC loss P_{wac}	56.7 mW
DC loss P_{wdc}	6.93 mW
ESR r_{L_r}	183 mΩ

Figure 6 shows the numerical and the experimental waveforms. In the laboratory measurement, the power conversion efficiency of the multi-resonant dc/dc converter achieved 94.0 % for 4.92 W output power at 1 MHz operation, which agree with the numerical predictions well. It is seen from Fig. 6 and Table IV that both the numerical and experimental waveforms satisfied class-E ZVS/ZDS conditions, specified output voltage, and permissible power conversion efficiency. The dc-supply voltage V_I and the dc-supply current I_r were obtained from the digital multimeter of Iwatsu VOAC7532. The output voltage v_o was measured by the the digital multimeter of Agilent 3458A.

In this paper, we state the optimization result of the resonant inductor L_r only. In this design example, the ESR of L_C is much lower than that of L_r as shown in Table IV. Additionally,

TABLE IV
NUMERICAL VALUES AND MEASURED ONES OF THE MULTI-RESONANT DC/DC
CONVERTER AT THE NOMINAL STATE.

	Calculated	Measured	Difference
L_r	3.31 μH	3.37 μH	1.75 %
C_r	2.74 nF	2.74 nF	0.00 %
C_{S_2}	1.60 nF	1.58 nF	−1.25 %
L_C	300 μH	295 μH	−1.67 %
C_f	2.22 μF	2.24 μF	0.448 %
R	10.0 Ω	9.92 Ω	−0.80 %
r_{L_r}	183 mΩ	183 mΩ	0.00 %
r_{L_C}	24.0 mΩ	24.0 mΩ	0.00 %
r_S	70.0 mΩ	70.0 mΩ	0.00 %
D_{S_1}	0.50	0.50	0.00 %
D_{S_2}	0.62	0.62	0.00 %
f	1.00 MHz	1.00 MHz	0.00 %
V_I	10.0 V	10.0 V	0.00 %
V_o	7.00 V	6.99 V	−0.143 %
P_o	4.90 W	4.92 W	0.408 %
η	94.2 %	94.0 %	−0.213 %

the volume of the inductor L_C is much smaller than that of L_r. Therefore, we can focus on only the optimization of the resonant inductor L_r. The multiple optimizations for more than two inductors is one of the important problems we should be addressed.

V. CONCLUSION

This paper has presented a computer-aided design for the class-E switching circuits taking into account optimized inductor designs. The proposed design algorithm provides the element values for satisfying the multiple constrain conditions. Simultaneously, the optimal inductors, which have a minimum volume for achieving the permissible power conversion efficiency, are obtained. A design example of the multi-resonant dc/dc converter with the class-E ZVS/ZDS conditions is given. By carrying out the circuit experiment, it is shown that the designed converter satisfied class-E ZVS/ZDS conditions, specified output voltage, and permissible power conversion efficiency simultaneously, which verify the validity and effectiveness of the proposed algorithm. The measurement power conversion efficiency achieved 94.0 % for 4.92 W (7.0 V/0.70 A) output power and 1 MHz operation.

REFERENCES

[1] N. O. Sokal, A. D. Sokal, "Class E - A new class of high-efficiency tuned single-ended switching power amplifiers," *IEEE Jounal of Solid State Circuits*, vol. SC-10, no. 3, pp. 168-176, Jun. 1975.

[2] A. Grebennikov and N. O. Sokal, "Switchmode RF power amplifiers," Elsevier, Burlington, MA, USA, 2007.

[3] T. Onodera, Y. Masuda, and A. Nakajima, "High-efficiency switching regulator using sub class E switching mode," *Third International Telecommunications Energy Conference (INTELEC' 81)*, London, England, May 1981, pp. 132-137.

[4] J. Zhang, X. Xie, X. Wu, G. Wu, and Z. Quian, "A novel zero-current-transition full bridge DC/DC converter," *IEEE Trans. Power Electron.* vol. 21, no. 2, pp. 354-360, Mar. 2006.

[5] R. C. N. Pilawa-Podgurski, A. D. Sagneri, J. M. Rivas, D. I. Anderson, D. Perreault, "Very-High-Frequency Resonant Boost Converters," *IEEE Trans. Power Electron.* vol. 24, No. 6, pp. 1654-1665, June. 2009.

[6] H. Sekiya, I. Sasase and S. Mori, "Computation of design values for class E amplifiers without using waveform equations, " *IEEE Trans. on Circuits and Systems*, vol.CAS-49, no.7, pp.966-978, Jul. 2002.

[7] H. Sekiya, T. Ezawa, Y. Tanji, "Design procedure for class E switching circuits allowing implicit circuit equations, " *IEEE Trans. on Circuits and Systems*, vol. 55, no. 11, pp. 3688-3696, Dec, 2008.

[8] K. W. E. Cheng and P. D. Evans, "Calculation of winding losses in high-frequency toroidal inductors using single strand conductors," *IEE Proceedings-Electric Power Applications*, vol. 141, no. 2, pp. 52-62, 1994.

[9] H. Njiende, N. Frohleke, and J. Bocker, "Optimized size design of integrated magnetic components using area product approach," *2005 European Conference on Power Electronics and Applications (EPE2005)*, Dresden, Germany, Sept. 2005, pp. 1-10.

[10] H. Sekiya, and M. K. Kazimierczuk, "Optimal core and wire of resonant inductor for class-E amplifier," International Workshop on Vision, Communications and Circuits (IWVCC 2008), Xi'an, China, pp. 3–6, Nov. 2008.

[11] M. K. Kazimirkzuk and H. Sekiya, "Design of ac resonant inductors using area product method," *IEEE Energy Conversion Congress and Exposition. (ECCE)*, Sept. 2009.

[12] Micrometals Inc., *Iron Powder Cores for Power Conversion and Line Filter Applications*, issue L, Feb. 2007.

Characterization of IGBT Modules for System EMI Simulation

Tao Qi, Jeff Graham[*], and Jian Sun

Department of Electrical, Computer and Systems Engineering
Rensselaer Polytechnic Institute, Troy, NY 12180-3590, USA
[*]Fairchild Controls Corporation, Frederick, MD 21701-5721, USA

Abstract– EMI filtering is a critical driver for volume and weight for many applications, particularly in airborne and other mobile platforms. Because of the lack of ability to accurately predict system EMI behavior, EMI filter design usually cannot start until prototype EMI measurement results become available. This often leads to costly schedule delay and disruption, and the resulting design is suboptimal at the best. To solve this problem, systematic EMI modeling method is needed to enable the development of optimal system EMI solutions concurrent with the design of the rest of the system.

This paper proposes a piece-wise linear model for IGBT (Insulated Gate Bipolar Transistor) modules for EMI simulation with sufficient accuracy and simplicity. Instead of physics-based models the proposed model is composed of a simple piece-wise linear circuit with parameters that can be extracted from the datasheet or simple device measurement. While the model is simple and easy to use, it can achieve sufficient accuracy required for EMI modeling. The proposed model is applied to a three-phase PFC and validated by comparison of model prediction with experimental measurements.

I. INTRODUCTION

Electromagnetic interference (EMI) is a common problem for systems involving switching power converters. EMI filtering is often a major driver for volume and weight. The impact of EMI filters on overall system volume and weight is especially a concern for mobile applications, such as more-electric aircraft (MEA), electric ships, and electric and hybrid-electric vehicles (HEV). Optimal system solutions to EMI that minimize the size and weight of EMI filtering elements are of critical importance.

Due to the lack of abilities to accurately model system EMI behavior, existing EMI filter design methods are mostly empirical, requiring measurements of unfiltered system EMI emission as the starting point [1]. The resulting design is suboptimal at the best from EMI perspective, for two reasons: 1) The rest of the system (such as circuit topology, control, and packaging schemes) is not optimized for EMI performance; and 2) optimizing the EMI filter design would require expensive and time-consuming design iterations. Designers are also often caught by last-minutes surprises in such an "EMI-last" approach, leading to the use of even less optimal EMI solutions and costly schedule delay and disruption. These problems can be solved by concurrent design of EMI filtering solutions in parallel with the design of the rest of the system, which requires analytical system EMI models that can be developed while the system is being designed.

Various efforts have been reported in the literature on the development of EMI models for power electronic systems. The major components of a typical motor drive system that affect EMI include: semiconductor devices and their driving circuitry, pulse-width modulation and control, filter inductors and capacitors, bus bars, interconnections, heat sink, cable, and motor. Of these, PWM and control are relatively straightforward to model; cables can be modeled as transmission lines and parameterized either analytically or based on impedance measurements; inductors and capacitors can be adequately modeled by a low-order impedance network in most cases; and similar techniques can be applied to bus bars. Induction motor can be modeled by multi-stage LC circuits with mutual couplings [2-6]. The main challenge has been in the modeling of semiconductor power devices and the associated driving circuitry, as well as board-level interconnections. It is commonly believed that accurate EMI prediction requires the use of physics-based device models [7].

The focus of this paper is the modeling of IGBT and IGBT modules for system-level EMI prediction. An IGBT as other semiconductor devices can be modeled by a physics-based or a behavioral model. Reference [8] provides a review of IGBT models most of which are physics-based models. Some of the physics-based models [9-11] use analytical equations to explain the effects which are not obvious in the simple BJT-MOSFET representation. This type of model focuses on carrier behavior and involves partial differential equations such as Poisson's equation Although accurate, these models are not appropriate for EMI simulation due to their complexity. Other type of physics-based models [12-13] are represented by equivalent circuits, which are more attractive for circuit simulation.

In [14] and [15], an IGBT is described as the combination of a MOSFET and a bipolar transistor with each part modeled by a subcircuit. Reference [16] improves the model in [15] by including temperature effect.

The parameters in these models are all process related and difficult to find for the users. In [17], a behavioral model is built using a nonlinear static block and a linear dynamic block. Model parameters are extracted from measurement by using least-squares methods. Compared with the physics-based models, these models are much simpler, but still require many parameters that are difficult to obtain.

The simplest behavioral model of an IGBT is an ideal switch which is however too simple for EMI simulation. Reference [18] presented a first-order piece-wise linear IGBT behavior model for time-domain simulation of motor drive EMI. This paper uses a similar model and extends it also to IGBT modules consisting of multiple IGBT devices. Compared to the model in [18], the model proposed in this paper

1) uses constant capacitance which is compatible with the circuit simulator and more efficient for EMI simulation;

2) doesn't include saturation voltage because this value is negligible, which reduce the number of the electrical devices in the model;

3) uses linear off-state resistance to on-state resistance transition process which is an inside characteristic of ideal switch in Saber.

The rest of the paper is organized as follows. Section II and Section III discuss the structure and parameterization methods of the proposed model of single IGBT and the IGBT module. The application of this model in a three-phase PFC system is presented in Section IV followed by a summary in Section V.

II. PIECE-WISE LINEAR IGBT MODEL

A. Structure of IGBT Model

The single IGBT is modeled in this work as a switch S in series with an on-state resistor R_{on} and coupled with an output capacitor C_0 as shown in Fig. 1. The gate control circuit is modeled by a pulse signal with rise and fall transient time t_r and t_f respectively. Additionally, terminal-to-ground capacitance C_{cg} and C_{eg} are included to model CM characterization.

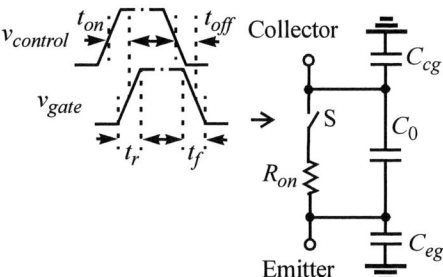

Fig. 1. IGBT model

The output capacitor C_0 models the capacitance between the collector and emitter including that of the anti-parallel diode. It affects mainly the switching-off transient behavior.

Instead of using the gate resistance and capacitance, the rising and falling delay of the IGBT switching transient is modeled by setting the t_r and t_f time of an ideal piece-wise linear switch. The CM capacitance at the gate control terminal is not included in the model because it does not contribute to the CM current.

The rest of this section will discuss the parameterization methods of this model. The on-state resistance can be easily determined from the datasheet or simple measurement, hence will not be discussed here.

B. Output Capacitance

The output capacitance can also be determined from device datasheet. The circuits in Fig. 2 can be used to experimentally

determine C_0.

With a resistive load, the collector-emitter voltage exhibits a first-order transient during switching off from which the output capacitance can be determined. With an inductive load, the emitter-collector voltage and the collector current will both exhibit an oscillatory response during switching off of the IGBT. The output capacitance can be calculated from:

$$C_0 = Q/\Delta v \qquad (1)$$

where Q is the charge accumulated in the IGBT which can be calculated by integrating the current and Δv is the increase of the emitter-collector voltage.

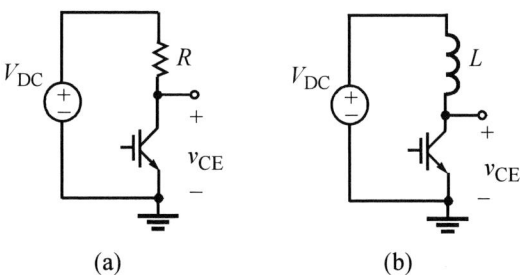

(a)   (b)

Fig. 2. Switching characteristic measurement setup with a) resistive load and b) inductive load.

Fig. 3 shows the turn-off transient of an IGBT module CMFCGF50X60CTAPG from Microsemi Corporation which contains six NPT IGBTs with rated C-E voltage of 600 V and collector current of 50 A, with a resistive and an inductive load, respectively. The DC supply voltage in the measurement is 70 V, and the resistance and inductance for R and L are 350 Ω and 150 μH. The parasitic parameters can make the load function like other types. For example, the ESL of the resistor will cause the oscillatory response of emitter-collector voltage if the ESL value is large. Therefore, the parasitic parameters of the load should be included in measurement. IR2110 from International Rectifier is used as the gate driver.

For the resistive load, the emitter-collector voltage rises from 7.8 V to 63.4 V in 1.76 μS. The RC constant is calculated from

$$\tau = \Delta t / \ln\left[\frac{1 - v_1/V_{DC}}{1 - v_2/V_{DC}}\right] \qquad (2)$$

which is 0.785 μS according to the measured parameters, and the output capacitance is calculated to be 2.24 nF.

For the inductive load, the emitter-collector voltage rises from 11.2 V to 68 V in 279 nS and the current falls from 496 mA to 388 mA in this period. The current waveform is quite linear because it is at the beginning of the oscillation and thus the integration of current is calculated by $(i_1+i_2)\Delta t/2$. The output capacitance is calculated to be 2.17 nF from equation (1).

The simulation is made with the output capacitance of 2.2 nF and the comparison with the measurements is shown in Fig. 3.

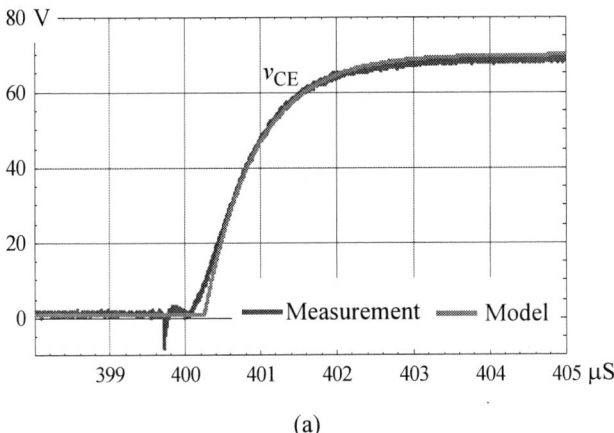

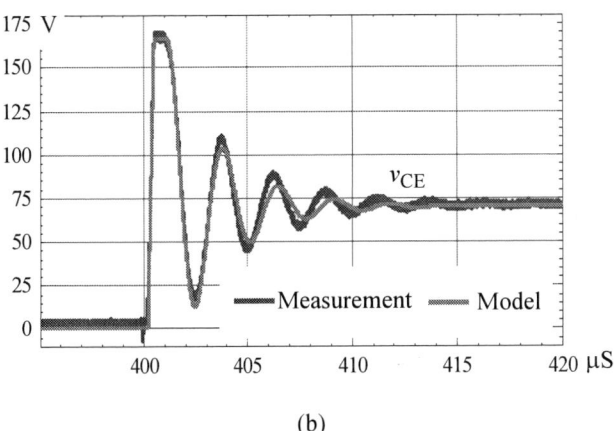

Fig. 3. Collector-emitter voltage measurement and model simulation for output capacitance calculation with a) resistive load and b) inductive load.

C. Switching Time and Delay

The switch S in the proposed equivalent circuit model shown in Fig. 1 is characterized by four parameters. The propagation delay times are defined as t_{on} and t_{off}, describing the time from the control signal switching transient to the gate voltage switching transient of the rising edge and falling edge respectively. The delay is caused by the path of control and driver circuit and the input characteristic of the IGBT. The switching time t_r and t_f are the time when the gate voltage v_{gate} changes from low voltage to high voltage and vice versa respectively. In this process, because the gate-emitter voltage has the switching transient, the on-state impedance of collector-emitter terminals also changes from high to low and low to high respectively, according to the transconductance characteristic of IGBT. In the model of ideal switch in Saber, the changing of the impedance is linear over the switching time which is also used in this work.

These parameters can be determined from IGBT datasheet. Consider, for example, IRAMY20UP60B from International Rectifier, in which the drive circuit is contained inside. The datasheet gives the following parameters: $t_{on} = 590$ nS, $t_{off} = 700$ nS.

These parameters can also be determined from actual measurements of an IGBT switching operation. Based on the definition, the switching time and propagation delay can be determined by measuring the gate control signal and the waveform of the gate voltage. However, in some cases the drive circuit is contained inside of the IGBT module and the gate terminal of the IGBTs is not available as the foot of the module (IRAMY20UP60B, for example) and the waveform of the gate voltage can not be measured.

To make the testing procedure generally applicable to IGBT modules in which two IGBTs are usually connected in series, consider the circuit shown in Fig. 4a) where the upper IGBT is turned on while the lower one is kept off. In this switching transient, the rising transient of v_a is caused by the charging of DC voltage through the on-state resistance of the upper switch and thus the waveform of the rising edge is determined by the rising time t_r of the upper switch and the parasitic capacitance of the middle of the bridge. With the capacitance calculated from the methods introduced in the previous section, the rising time can be obtained from the best curve fitting method. In this example, t_r is set to be 600 nS and t_f is set to be 450 nS. The comparison of model simulation and measurement is shown in Fig. 4b).

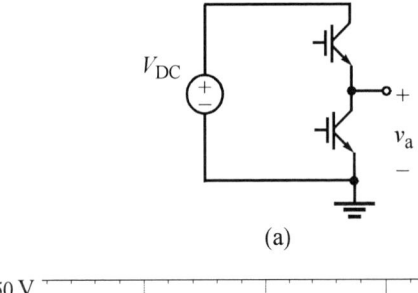

(a)

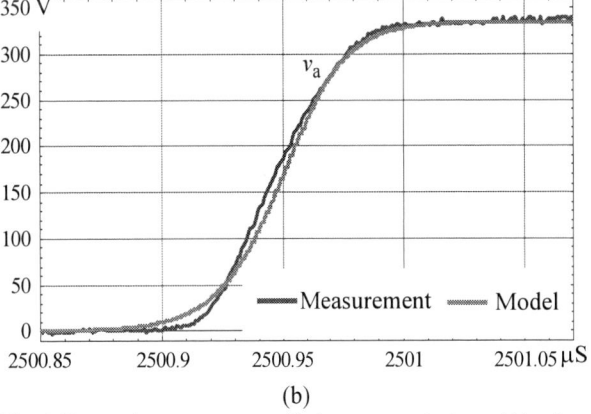

(b)

Fig. 4. Turn-on time measurement with a) measurement setup and b) voltage waveform measurement of middle of bridge.

III. MODELING OF IGBT MODULES

Fig. 5 shows the circuit of a three-phase IGBT module composed of six IGBTs. The parasitic capacitance coupled in each terminal is shown in the figure. In addition to the parasitic capacitance between the collector-emitter terminals, the module also

978-1-4244-4782-4/10 $26.00 © 2010 IEEE   2222

includes parasitic capacitance between each terminal to the base (or heat sink). Such parasitic capacitance is part of the CM capacitance which generates CM current. C_0 is the output capacitance in collector-emitter terminals of each IGBT and the value is considered to be the same for the same manufacture process. C_{pb}, C_{mb}, and C_{nb} are separately the capacitance from top, middle, and bottom of the bridge to the base and the values are considered to be different.

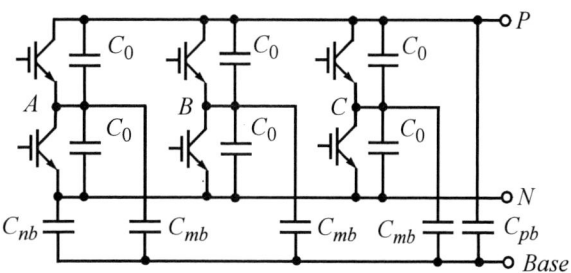

Fig. 5. Three-phase IGBT module

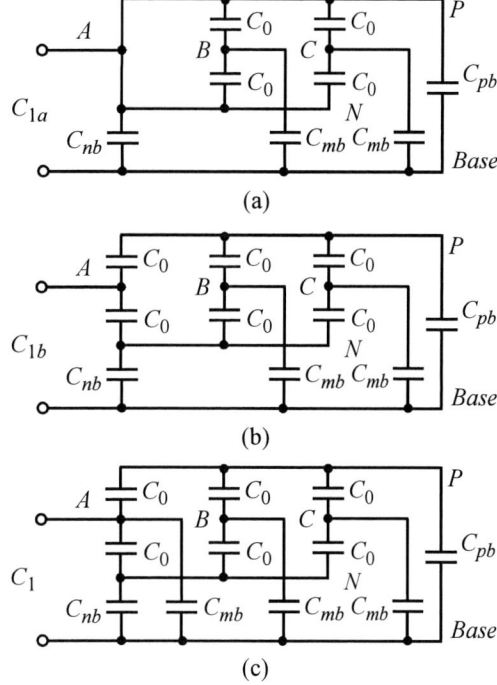

Fig. 6. a) step 1 of calculation for measurement 1; b) step 2 of calculation; and c) equivalent circuit for measurement 1.

The parameterization method should be able to distinguish and determine each of the capacitance in the model. There are four values of capacitance in the model and thus four measurements are required to obtain all the capacitance. The first measurement is to measure the capacitance from the middle of the bridge to the base with all the IGBTs off. Fig. 6 shows the equivalent circuit and the calculation steps of the first measurement.

The capacitance for each step can be calculated to be

$$C_{1a} = C_{pb} + C_{nb} + \frac{4C_{mb}C_0}{C_{mb} + 2C_0} \tag{3}$$

$$C_{1b} = \frac{2C_{1a}C_0}{C_{1a} + 2C_0} \tag{4}$$

$$C_1 = C_{1b} + C_{mb} \tag{5}$$

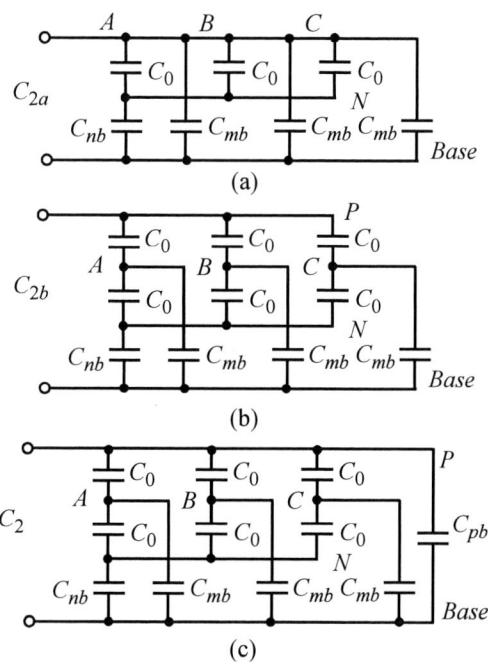

Fig. 7. a) step 1 of calculation for measurement 2; b) step 2 of calculation; and c) equivalent circuit for measurement 2.

The second measurement is to measure the capacitance from the top of the bridge to the base. The equivalent circuit and the calculation steps are shown in Fig. 7. The capacitance for each step can be calculated to be

$$C_{2a} = 3C_{mb} + \frac{3C_{nb}C_0}{C_{nb} + 3C_0} \tag{6}$$

$$C_{2b} = \frac{3C_{2a}C_0}{C_{2a} + 3C_0} \tag{7}$$

$$C_2 = C_{2b} + C_{pb} \tag{8}$$

The third measurement is to measure the capacitance from the bottom of the bridge to the base. The equivalent circuit is similar to the second measurement which is shown in Fig. 8. The calculation steps are not shown in the figure due to the similarity to the second measurement.

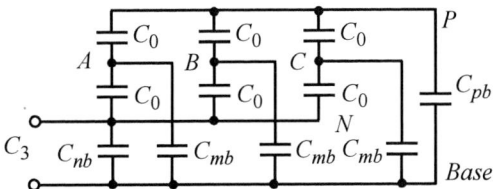

Fig. 8. Equivalent circuit for measurement 3

Similarly, the calculation can be expressed as

$$C_{3a} = 3C_{mb} + \frac{3C_{pb}C_0}{C_{pb} + 3C_0} \qquad (9)$$

$$C_{3b} = \frac{3C_{3a}C_0}{C_{3a} + 3C_0} \qquad (10)$$

$$C_3 = C_{3b} + C_{nb} \qquad (11)$$

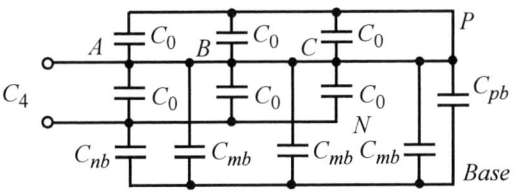

Fig. 9. Equivalent circuit for measurement 4

The last measurement is firstly to connect the top and the middle of the bridges and short the upper IGBTs and measure the capacitance between collector-emitter terminals of the lower IGBTs. The equivalent circuit is shown in Fig. 9, and the capacitance obtained from this measurement can be calculated as

$$C_4 = 3C_0 + \frac{C_{nb}(C_{pb} + 3C_{mb})}{C_{nb} + C_{pb} + 3C_{mb}} \qquad (12)$$

For IGBT module IRAMY20UP60B the measured capacitances are $C_1 = C_2 = C_3 = 80$ pF, and $C_4 = 1250$ pF. Using equations (3)-(12), capacitances are calculated to be $C_0 = 270$ pF, $C_{pb} = C_{nb} = 32$ pF, and $C_{mb} = 10$ pF.

IV. SYSTEM EMI MODEL

A. System Description

The proposed model for IGBT module is applied to a 600 W three-phase PFC boost converter shown in Fig. 10. All the components and the values are listed in the figure. The 5.4 μF capacitors at the input side are used as DM filter for normal operation of the PFC converter. The 0.68 μF capacitors are used to stabilize the voltage on the input voltage sensing transformer. Although functions as sensing circuit, these capacitors are in the EMI loop and thus need to be modeled. The load in this example is a resister for demonstration and the output DC voltage is 350 V. The nonlinear control method [19] is applied in this PFC converter with switching frequency of 36 kHz.

LISNs are placed at the input side so as to provide standard reference for EMI measurement. A 7 nF capacitor is placed between the bus minus terminal to the LISN ground to provide lower impedance than the other nonideal impedance from bus terminal to LISN ground and thus the CM loop characteristic can be easily described. The heat sink of the IGBT module is floating in this system.

B. EMI Modeling

All the passive components are modeled according to the impedance measurement from 150 kHz to 30 MHz with impedance analyzer, including the input inductor, input filters (5.4 μF and 0.68 μF capacitors), output capacitor, resistive load and input wires. When modeling these devices the CM impedance from the terminals to the LISN ground are also measured and included in the system EMI model. The inductance of the long wires on the PCB board and the capacitance between large area of copper plates are also included. The capacitance from the heat sink to the LISN ground is also measured and included in the model as part of the CM loop.

The complex control circuit including all the control details are used for EMI modeling so that the current waveform can be the same as the measurement. It is based on the consideration that the input current waveform will affect the charging time and switching transients of the voltage at the middle of the bridge and thus directly affect the voltage source spectrum in the equivalent EMI model. Of course, a simplified control circuit can be more

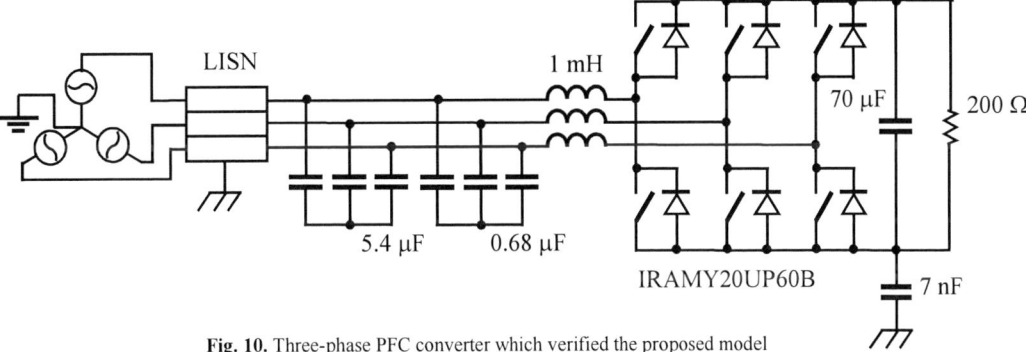

Fig. 10. Three-phase PFC converter which verified the proposed model

efficient for simulation if it can be equivalent to the original control circuit. The detailed model for the IGBT module in this example is shown in Fig. 11. All the parameters are obtained from the previous sections.

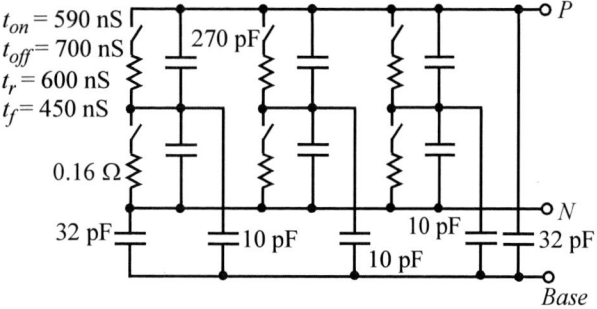

Fig. 11. Detailed model for IGBT module

C. EMI Prediction and Measurement

The input line current and CM current EMI spectrum simulated in model are compared with the measurements shown in Fig. 12. The result shows that the EMI model can be accurate up to 15 MHz with the error less than 10 dB.

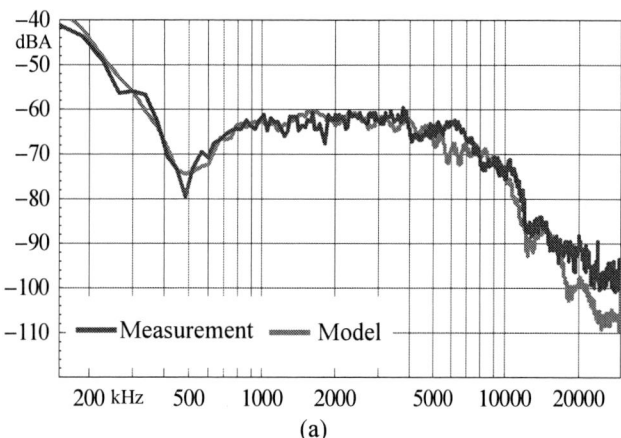

(a)

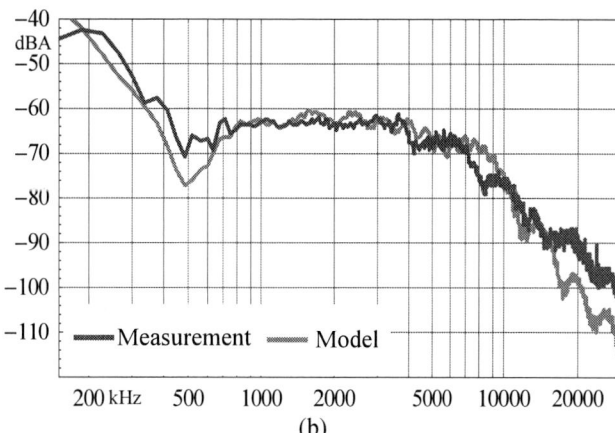

(b)

Fig. 12. a) Input line current spectrum comparison; b) Input CM current spectrum comparison of simulation and measurement

V. SUMMARY

A piece-wise linear model for IGBT module is proposed for system EMI simulation. The structure of the model is simple and can be easily used by the simulator such as Saber. All the parameters can be obtained from the datasheet or simple measurement. With this model applied in a three-phase PFC system, the system model can predict the EMI spectrum with the error less than 10 dB up to 15 MHz.

REFERENCES

[1] R. L. Ozenbaugh, *EMI Filter Design,* 2nd ed. New York: Marcel Dekker, 2001.
[2] E. Gubia, P. Sanchis, A. Ursua, J. Lopez and L. Marroyo, "Frequency domain model of conducted EMI in electrical drives," *IEEE Power Electronics Letters*, vol. 3, no. 2, pp. 45-49, 2005.
[3] A. Boglietti and E. Carpaneto, "Induction motor high frequency model," in *Proceedings of Industry Applications Conference*, vol. 3, pp. 1551-1558, 1999.
[4] A. F. Moreira, T. A. Lipo, G. Venkataramanan and S. Bernet, "High-frequency modeling for cable and induction motor overvoltage studies in long cable drives," *IEEE Transactions on Industry Applications*, vol. 38, no. 5, pp. 1297-1306, 2002.
[5] G. Grandi, D. Casadei and U. Reggiani, "Analysis of common- and differential-mode HF current components in PWM inverter-fed AC motors," in *Proceedings of Power Electronics Specialist Conference*, vol. 2, pp. 1146-1151, 1998.
[6] B. Mirafzal, G. L. Sikbinski and R. M. Tallam, "Determination of parameters in the universal induction motor model," in *Proceedings of Industry Applications Conference*, pp. 1207-1216, 2007.
[7] A. R. Hefner, "Modeling buffer layer IGBTs for circuit simulation," *IEEE Transactions on Power Electronics*, vol. 10, no. 2, pp. 111-123, 1995.
[8] K. Sheng, B. W. Williams, and S. J. Finney, "A review of IGBT models," *IEEE Transactions on Power Electronics*, vol. 15, no. 6, pp. 1250-1266, Nov 2000.
[9] B. J. Baliag, "Analysis of insulated gate transistor turn-off characteristics," *IEEE Electron Device Letters*, vol. EDL-6, no. 2, pp. 74-77, Feb 1985.
[10] A. R. Herfner, "Analytical modeling of device-circuit interactions for the power insulated gate bipolar transistor (IGBT)," *IEEE Transactions on Industry Applications*, vol. 26, no. 6, pp. 995-1005, Nov/Dec 1990.
[11] A. R. Herfner, "Improved understanding for the transient operation of the power insulated gate bipolar transistor (IGBT)," *Power Electronics Specialists Conference, PESC '89 Record*, vol. 1, pp. 303-313, 1989.
[12] Z. Shen, T. P. Chow, "An analytical IGBT model for power circuit simulation," *Power Semiconductor Devices and ICs*, pp. 79-82, Apr 1991.
[13] X. Kang, E. Santi, et al., "Parameter extraction for a physics-based circuit simulator IGBT model," in *Procedings of IEEE APEC '03*, pp. 946-952, 2003.
[14] F. Mihalic, K. Jezernik, et al., "IGBT SPICE Model," in *IEEE Trans. on Industrial Electronics*, vol. 42, no. 1, February 1995.
[15] R. Kraus, P. Turkes, and J. Sigg, "Physics based models of power semiconductor devices for the circuit simulator spice," in *Proc. IEEE PESC' 98*, vol. 2, pp. 1726-1731, 1998.
[16] R. A. Florin Udrea, M. D. Silva, et al., "Advanced SPICE modeling of large power IGBT modules," in *IEEE Trans. on Industry Applications*, vol. 40, no. 3, pp. 710-716, May/June 2004.
[17] J. H. Hsu and K. D. T. Ngo, "A behavioral model of the IGBT for circuit simulation," *Power Electronics Specialists Conference, PESC '95 Record*, pp. 865-871, 1995.
[18] J. L. Tichenor, S. D. Sudhoff, and J. L. Drewniak, "Behavioral IGBT modeling for predicting high frequency effects in motor drives," in *IEEE Trans. on Power Electronics*, vol. 15, no. 2, March 2000.
[19] Z. Bing, X. Du, and J. Sun, "Three-phase PFC current control using DC-rail current as feedback," *Energy Conversion Congress and Exposition*, pp. 1212-1219, September 2009.

A Mathematical Model for Online Electrical Characterization of Thermoelectric Generators using the P-I Curves at Different Temperatures

Eduardo I. Ortiz-Rivera, *Member IEEE*
Dept. of Electrical and Computer Eng.
Univ. of Puerto Rico-Mayaguez
Mayaguez, Puerto Rico 00681
Email: eduardo.ortiz@ece.uprm.edu

Andres Salazar-Llinas
Dept. of Electrical and Computer Eng.
Univ. of Puerto Rico-Mayaguez
Mayaguez, Puerto Rico 00681
Email: andres.salazar@ece.uprm.edu

Jesus Gonzalez-Llorente
Dept. of Electrical and Computer Eng.
Univ. of Puerto Rico-Mayaguez
Mayaguez, Puerto Rico 00681
Email: jesus.gonzalez@ece.uprm.edu

Abstract—Abstract-This paper presents a proposed mathematical model to describe the electrical characteristics for a thermoelectric generator. The TEG model considers the use of second order polynomial equations, the boundary conditions, and characteristic shape for the operation of a TEG. The proposed TEG model has the following advantages: 1) suitable for sustainable energy and power applications, 2) emulate the typical shape for the electrical characteristics for a TEG, 3) useful for circuit analysis at the academic level, 4) able to replicate the realistic constraints in a TEG. Finally, the paper shows simulations to validate the proposed TEG mathematical model.

I. INTRODUCTION

Since the discovery of the thermoelectric effect by Thomas Johann Seebeck in 1821, the thermoelectric generators have received great interest for the production of energy. Works produced by Jean Charles Athnase Peltier, William Thonsom, USA Atomic Energy Commision have been key in the development and trends of thermoelectric generators. But, what is a thermoelectric generator (TEG)? TEG's are devices that convert the heat, or difference on temperature, in electric energy. They use the configuration of various solid state thermopars based on the Seebeck effect principle. TEG's are not too efficient ranging between 5-10%. They usual capacity is from 20W up to 2.2kW. As a brief history, Thomas Seebeck was the first scientist to establish the basic principle for a TEG. He discovered that a conductor generates a voltage when subjected to a temperature gradient. But it is not until 1940, when the boom of semiconductors technology that the first applications appear. Some examples of a TEG applications are the Isotope fueled SNAP generator (1975-to present), and the radioisotope thermoelectric generator used by the NAVY at Fair Rock Island from 1966-1995. At the present, most of the TEG applications are for telecommunications and navigation. Until now, most of the TEG mathematical models are good to describe their thermal reactions but not necessary suitable to characterize the TEG electrical characteristics suitable for circuit analysis and power system simulations. To solve the last problem, this paper proposes a mathematical model to describe the electrical characteristics of a TEG based on the power, current, voltage, and temperature relationships.

II. PROPOSED THERMOELECTRIC GENERATOR (TEG) MATHEMATICAL MODEL

The proposed electrical TEG mathematical model considers the boundary conditions and shape of the TEG P-V curve. The power, P, and the voltage, V, can be obtained by (1) and (2) respectively. The static TEG internal resistance, R, and static TEG internal conductance, G, can be calculated from the proposed TEG model as given by (3) and (4). The variables V_X, I_X, and P_{max} are the open-circuit voltage, short-circuit current, and maximum power for the TEG at a given cooling water temperature, T. The proposed TEG model is continuous and differentiable with respect to I as given by (5). This information will be useful for the design of optimal dynamic algorithms for maximum power tracking applications. As an example, the TEG maximum power can be calculated using (5) or (7) equal to zero then solve for the optimal current, Iop, (6) or the optimal voltage Vop, (8); next substitute Iop and Vop in (1) to solve for the maximum power (9). Interestingly a very simple online algorithms can be implemented to estimate P_{max} by measuring the open-circuit voltage and short-circuit current as given by (9). Also, the optimal internal impedance, Rop, and optimal internal conductance, Gop, can be calculated using the last procedure as given by (10) and (11). As a remark, (12)-(13) prove that all the calculated values are unique for a given temperature including a unique value for P_{max}!

Also, the proposed mathematical model recognizes that a TEG varies with temperature changes affecting the TEG dynamics and performance. Equations (14), (15), (16) describe the temperature effects in V_X, I_X, and P_{max} in an accurate way using the open-circuit voltage and maximum power at a given temperature T_1 where $V_X(T_1) = V_1$ and $P_{max}(T_1) = P_1$ and the open-circuit voltage and maximum power at a given temperature T_2 where $V_X(T_2) = V_2$ and $P_{max}(T_2) = P_2$. This information will be useful for the development of online monitoring software to measure in real time the performance for a TEG.

This model is empirical formulate based on the measurements of the operation of several TEG's and the observed data given by the literature. The main advantage of (1) is that for any TEG, it can be described in terms of the values obtained by the TEG V-I and P-I characteristic curves useful for electrical engineers that not necessary are experts in topics related to thermodynamics, and thermoelectric materials. Also, the proposed TEG static model can be extended to a dynamical model for circuit analysis using (14) and (15).

III. TEG MODEL VERIFICATION AND SIMULATIONS

The data to verify the proposed fuel cell model was obtained from the paper wrote by Chu and Katodani [1]. Figure 1 shows the V-I and P-I characteristics of a TEG at different temperatures (293 Kelvin and 318 Kelvin) of the cooling water [1]. The insulating layer thickness between the cooling jacket and the cold side electrode was 40 μm [1]. It could be seen that the maximum power reaches 52W at around 6A current [1].

Figure 2 shows the V-I and P-I characteristics using the proposed TEG Model using the given data of the figure 1. The results provided by [1] where examined and simulated using the proposed TEG model giving outstanding results and very similar to the results provided by [1]. An additional simulation was done only for a TEG with a cooling water temperature of 318K as given in the figures 3 and 4.

Additionally, the proposed TEG model can provide additional information related to R-I and G-I curves useful to design algorithms for power control. Figures 5 and 6 show the R-I and G-I curves for a TEG at 318K. For this case, Rop and Gop are 1.4167 Ω and 0.7059 S respectively given a TEG at 318K with P_{max} at 52W the It is essential to understand that the optimal internal impedance, Rop, and optimal internal conductance, Gop, has a direct relationship with the maximum power and both are unique. In other words, if a resistive load with the same value as the optimal internal impedance is connected to the TEG then the maximum power is transferred. This information could be used to maximize the efficiency of a TEG power system when load matching is required. Another interesting approach of the proposed TEG model is that just given P_{max} and Ix, the P-I curves can be calculated by substituting 9 in (1) as given by (16). Figure 7 is used to prove the validity of the statement and the comparison is shown in figure 8.

Finally, the TEG could be useful for circuit analysis. As an example consider a dynamic TEG connected to a resistive load, R. Figure 9 shows the diagram for a dynamic TEG. Lx and Cx are the internal inductance and capacitance. For our case, the resistive load is connected in parallel to the capacitor. Now using our model, we can calculated a 2nd order differential equation to describe the dynamics of the voltage across the resistive load, V_R as described by (20). The last example maybe is very simple but very powerful given that one of the interests of several auto manufacturers and researchers is to replace the alternator which would reduce load on the engine thus increasing gas mileage. It is expected that within 5 to 10 years alternators won't be found in new cars with internal combustion engines. The use of the proposed TEG mathematical model could be a catalytic agent in the advance of new technologies for transportation and aerospace applications. Figure 10 show a practical aerospace application for TEG's.

$$P = V \cdot I = V_X \cdot I - V_X \cdot \frac{I^2}{I_X} \quad \forall \ P \in [0 \ \ P_{max}] \quad (1)$$

$$V = V_X - V_X \cdot \frac{I}{I_X} \quad \forall \ V \in [0 \ \ V_X] \quad (2)$$

$$R = \frac{V}{I} = \frac{V_X}{I} - \frac{V_X}{I_X} \quad \forall \ I \in [0 \ \ I_X] \quad (3)$$

$$G = \frac{I}{V} = \frac{I_X}{V} - \frac{I_X}{V_X} \quad (4)$$

$$\frac{\partial P}{\partial I} = V + I \cdot \frac{\partial V}{\partial I} = V_X - V_X \cdot \frac{I}{I_X} = 0 \quad (5)$$

$$Iop = \frac{I_X}{2} > 0 \quad (6)$$

$$\frac{\partial P}{\partial V} = I + V \cdot \frac{\partial I}{\partial V} = I_X - 2 \cdot I_X \cdot \frac{V}{V_X} = 0 \quad (7)$$

$$Vop = \frac{V_X}{2} > 0 \quad (8)$$

$$P_{max} = \frac{I_X \cdot V_X}{4} > 0 \quad (9)$$

$$Rop = \frac{Vop}{Iop} = \frac{V_X}{I_X} \quad (10)$$

$$Gop = \frac{Iop}{Vop} = \frac{I_X}{V_X} \quad (11)$$

$$\frac{\partial P^2}{\partial^2 I} = -2 \cdot \frac{V_X}{I_X} < 0 \quad (12)$$

$$\frac{\partial P^2}{\partial^2 V} = -2 \cdot \frac{I_X}{V_X} < 0 \quad (13)$$

$$V_X = V_2 + (T - T_2) \cdot \left(\frac{V_2 - V_1}{T_2 - T_1} \right) \quad (14)$$

$$P_{max} = P_2 + (T - T_2) \cdot \left(\frac{P_2 - P_1}{T_2 - T_1} \right) \quad (15)$$

$$Ix = 4 \cdot \frac{P_{max}}{Vx} = 4 \cdot \frac{P_1 \cdot T_2 - P_2 \cdot T_1 + T \cdot (P_2 - P1)}{V_1 \cdot T_2 - V_2 \cdot T_1 + T \cdot (V_2 - V1)} \quad (16)$$

$$\frac{dI}{dt} = \frac{V_X}{L_X} - \frac{Vi}{L_X} - \frac{V_X \cdot I}{L_X \cdot I_X} \quad (17)$$

$$\frac{dVi}{dt} = \frac{I}{C_X} - \frac{Ii}{C_X} \quad (18)$$

$$P = I \cdot V = 4 \cdot \frac{P_{max}}{I_X} \cdot \left(I - \frac{I^2}{I_X} \right) \quad (19)$$

$$\ddot{V}_R = \frac{V_X}{L \cdot C} - \left(\frac{R \cdot I_X + V_X}{R \cdot L \cdot C \cdot I_x} \right) \cdot V_R - \left(\frac{1}{R \cdot C} + \frac{V_X}{L \cdot I_X} \right) \cdot \dot{V}_R \quad (20)$$

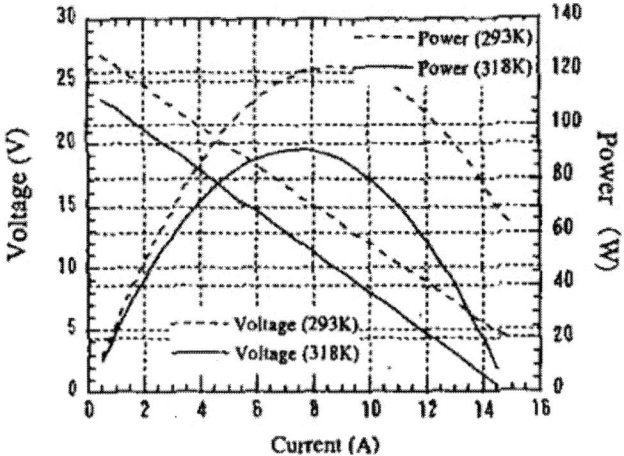

Fig. 1. P-I & V-I Curves [1].

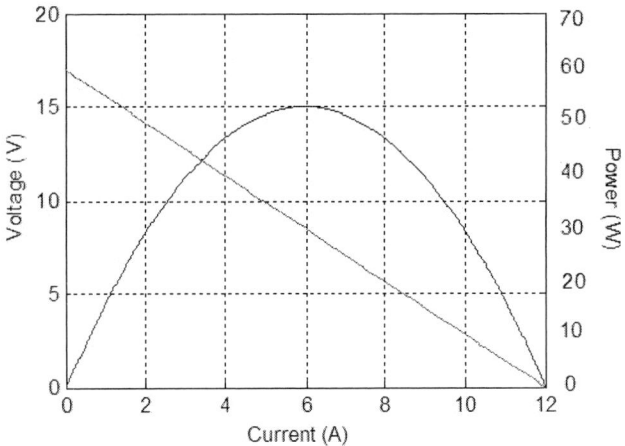

Fig. 4. Proposed TEG Model at 318K.

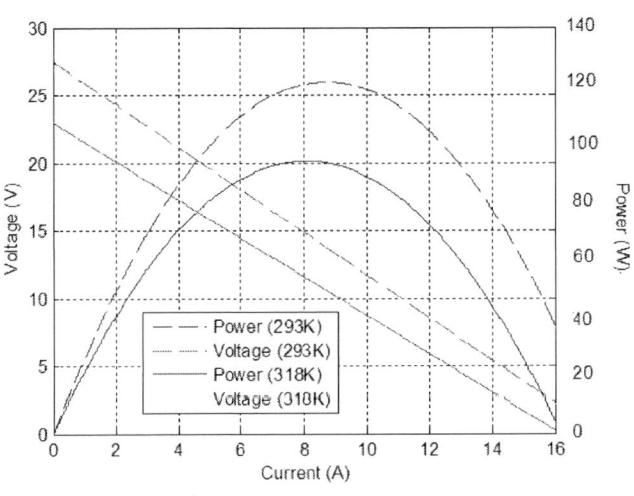

Fig. 2. TEG Model P-I & V-I Curves.

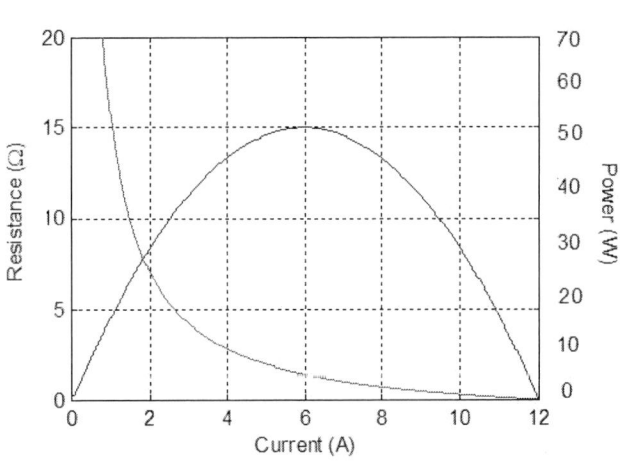

Fig. 5. TEG Model P-I & R-I Curves.

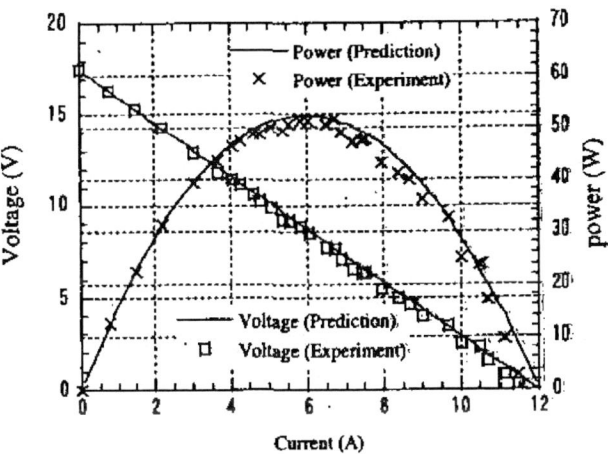

Fig. 3. More TEG Curves at 318K [1].

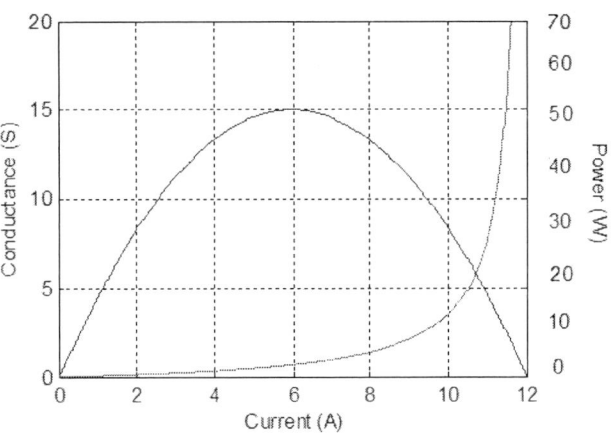

Fig. 6. TEG Model P-I & G-I Curves.

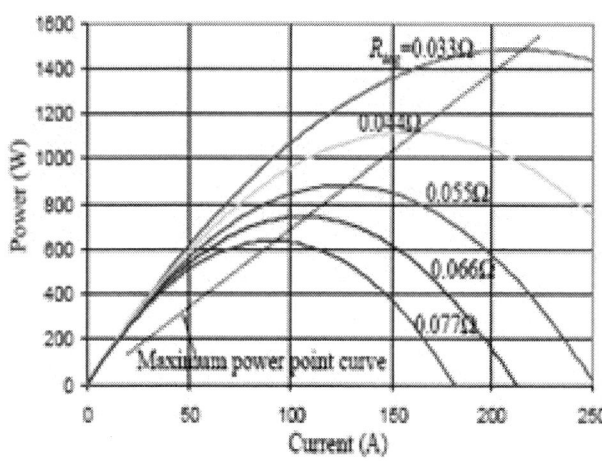

Fig. 7. P-I & V-I Curves [3].

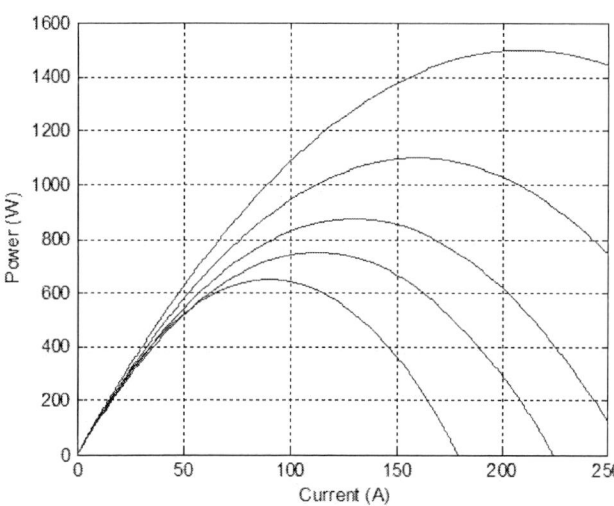

Fig. 8. TEG Model P-I & V-I Curves.

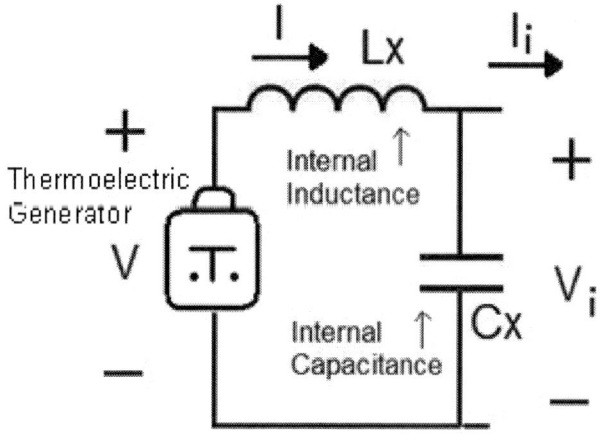

Fig. 9. Diagram for a dynamic TEG.

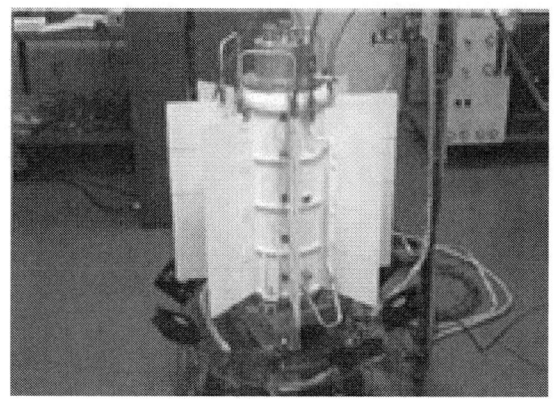

Fig. 10. Multi-Mission Radioisotope TEG [2].

IV. CONCLUSION

This paper proposed a thermoelectric generator model based on the electrical characteristics, P-I curve, and V-I curve for a TEG. The proposed TEG model can be use for steady-state analysis or transient analysis. The proposed TEG model has the following advantages: 1) suitable for sustainable energy and power applications, 2) emulate the typical shape for the electrical characteristics for a TEG, 3) useful for circuit analysis at the academic level, 4) able to replicate the realistic constraints in a TEG. The proposed TEG model can replicate the typical performance of a TEG but also additional information like the internal resistance useful for load matching design. Also, the proposed model will help to provide a more realistic representation of TEG dynamic behavior in time. Also, a major benefit of this task will be to obtain accurate models that describe the electrical characteristics for a TEG suitable for power system analysis.

ACKNOWLEDGMENT

The authors gratefully acknowledge the contributions of all the members that belong to the Mathematical Modeling and Control of Renewable Energies for Advance Technology and Education ($M_{inds}{}^2$CREATE) Research Team at UPRM. Also, we thanks all the support of Dr. Ramon Vazquez, Dean of the UPRM's College of Engineering.

REFERENCES

[1] Chu, R. C.; Kadotani, K.; Shintani, T.; Tanimura, T.; Hatanaka, T.; Nishio, S.; *Thermoelectric generator utilizing boiling-condensation (experiment and modeling*, Thermoelectrics, 2003 Twenty-Second International Conference on - ICT, 17-21 Aug. 2003 Page(s):546 - 549.

[2] Department of Energy and NASA, "Multi-Mission Rdioisotope Thermoelectric Generator", Space Radioisotope Power Systems, www.ne.doe.gov/pdfFiles/MMRTG.pdf As cited on November 24, 2009

[3] Rae-Young Kim; Jih-Sheng Lai; "A Seamless Mode Transfer Maximum Power Point Tracking Controller for Thermoelectric Generator Applications" Industry Applications Conference, 2007. 42nd IAS Annual Meeting. Conference Record of the 2007 IEEE 23-27 Sept. 2007 Page(s):977 - 984

[4] Rae-Young Kim; Jih-Sheng Lai; "Aggregated modeling and control of a boost-buck cascade converter for maximum power point tracking of a thermoelectric generator" Applied Power Electronics Conference and Exposition, 2008. APEC 2008. Twenty-Third Annual IEEE, 24-28 Feb. 2008 Page(s):1754 - 1760

[5] Tsuyoshi, A.; Kagawa, S.; Sakamoto, M.; Matsuura, K.; "A study of commercial thermoelectric generation in a processing plant of combustible solid waste" Thermoelectrics, 1997. Proceedings ICT '97. XVI International Conference on 26-29 Aug. 1997 Page(s):555 - 558

[6] Chung-Yen Hsu; Chun-Kai Liu; Heng-Chieh Chien; Sheng-Liang Kuo; Chih-Kuang Yu; "Stacked Thermoelectric Generator Module Integrated with Partial Electric Conducted Interposer Structure" Electronics Packaging Technology Conference, 2008. EPTC 2008. 10th 9-12 Dec. 2008 Page(s):355 - 360

[7] Anatychuk, L.I.; Razinkov, V.V.; Rozver, Yu.Yu.; Mikhailovsky, V.Ya.; "Thermoelectric generator modules and blocks" Thermoelectrics, 1997. Proceedings ICT '97. XVI International Conference on 26-29 Aug. 1997 Page(s):592 - 594

[8] Heng Xiao; Xiaolong Gou; Chen Yang; "Simulation analysis on thermoelectric generator system performance" System Simulation and Scientific Computing, 2008. ICSC 2008. Asia Simulation Conference - 7th International Conference on 10-12 Oct. 2008 Page(s):1183 - 1187

[9] Rae-Young Kim; Jih-Sheng Lai; "A Seamless Mode Transfer Maximum Power Point Tracking Controller For Thermoelectric Generator Applications" Power Electronics, IEEE Transactions on Volume 23, Issue 5, Sept. 2008 Page(s):2310 - 2318

[10] Hasebe, M.; Kamikawa, Y.; Meiarashi, S.; "Thermoelectric Generators using Solar Thermal Energy in Heated Road Pavement" Thermoelectrics, 2006. ICT '06. 25th International Conference on 6-10 Aug. 2006 Page(s):697 - 700

A Novel method for Permanent Magnet Demagnetization Fault Detection and Treatment in Permanent Magnet Synchronous Machines

Amir Khoobroo, *Student member IEEE*
Renewable Energy and Vehicular Technology Lab
University of Texas at Arlington
#130, Nedderman Hall, 416, S. Yates St.,
Arlington, TX, USA, 76019
Email: amir.khoobroo@mavs.uta.edu

Babak Fahimi, *Senior Member IEEE*
Renewable Energy and Vehicular Technology Lab
University of Texas at Arlington
#130, Nedderman Hall, 416, S. Yates St.,
Arlington, TX, USA, 76019
Email: fahimi@uta.edu

Abstract—Due to relatively high power density, negligible rotor losses, high efficiency, ease of control, and being almost maintenance free Permanent magnet synchronous machines (PMSM) are widely used in industrial applications. In this paper a new method of fault detection and treatment for 5-phase PMSM under partial demagnetization has been introduced. In the proposed technique various rotor demagnetization scenarios have been considered. As the optimum performance is of great importance in industrial applications the excitation of stator phases has been modified to attain the maximum output torque per input phase current in the event of demagnetization. Field reconstruction method has been used in conjunction with optimization methods to detect the fault and find the proper excitation.

I. INTRODUCTION

Fault tolerance has become a design criterion for adjustable speed motor drives (ASMD) which are used in high impact applications. A fault tolerant ASMD is expected to continue its intended function in the event of a failure compliment to its remaining components. There are different types of fault that may happen in the ASMD. In case of a PMSM motor drive, besides the regular monitoring of the current and voltage levels, maximum operating temperature is also limited due to the thermal limitations of the permanent magnets and stator windings. This thermal limit can be exceeded due to poor ventilation or excessive currents caused by short circuits. As a result, the short circuit fault, if not cleared properly, can lead to demagnetization of the rotor magnets. Based on this, the faults in the PMSM should be detected and cleared immediately to avoid further damage to the system. As the continuity of the service is of great importance, the next step would be to calculate the best excitation possible for the stator phases of the machine to harvest maximum torque. A vast amount of the research has been done on the techniques of fault detection [1- 9]. Most of these researches solely focus on detection of the fault rather than the aftermath. Also, most of the methods just apply to the specific faults on the stator windings and are not applicable to the other types of the fault.

Field reconstruction method (FRM) is a recent method that improves the computational time necessary to determine the distribution of magnetic field components. FRM is an alternative to Finite Element Analysis (FEA) with comparable accuracy, but with significantly shorter computational time. The literature and the applications of the method have been addressed [10-16]. This method uses the field created by a single slot on the stator, along with the field generated by the permanent magnets (windings) on the rotor of the AC machine, to find the field distribution and electromagnetic force components.

In this paper the field reconstruction method (FRM) has been applied to a PMSM drive. The FRM has been used to estimate the flux linking each stator tooth so the stator phase flux linkages. These phase fluxes are calculated for fault detection purpose. In addition, to assure post fault optimal operation, FRM has been used in conjunction with optimization methods to find the optimal stator currents.

II. FIELD RECONSTRUCTION

In order to verify the field reconstruction method a model has been developed for the PMSM using MAGNET from Infolytica©. A 10 hp, 5-phase, 6-pole, 30 slot surface mounted permanent magnet synchronous machine is used in this study. The 3D model of the machine is shown in Fig. 1. In this modeling the following assumptions are made:

- No deformations in the permanent magnets or stator teeth due to internal forces.
- Concentrated stator windings.
- No end coil effect.

978-1-4244-4782-4/10 $26.00 © 2010 IEEE

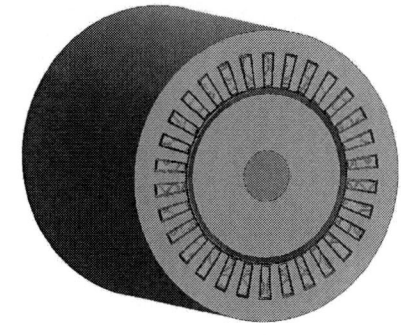

Figure 1: 3D view of the PMSM model

It can be shown that the magnetic field components need to be known in order to calculate the torque. The FEA methods are conventional tools to calculate the magnetic field components. However, they are time consuming and, hence, inadequate for iterative procedures such as those used in optimization processes. FRM has been used to analyze the magnetic field components. It is shown in [12] that for an unsaturated PMSM the magnetization curve can be considered to be linear and, as such, the superposition rule is applied to the field components as follows:

$$B_t = B_{tpm} + B_{ts} \qquad (1)$$

$$B_n = B_{npm} + B_{ns} \qquad (2)$$

Where, B_{npm}, B_{tpm} B_{ns} and B_{ts} denote the normal and tangential field components due to the permanent magnets and stator currents respectively. The resultant magnetic field created by the stator windings is the sum of the field created by each individual stator slot current. The normal and tangential field components due to the stator currents can be written as:

$$B_{ns} = \sum_{k=1}^{L} B_{nsk} \qquad (3)$$

$$B_{ts} = \sum_{k=1}^{L} B_{tsk} \qquad (4)$$

To evaluate (3) and (4) the local flux densities created by the current in the kth slot is expressed as follows:

$$B_{tsk}(\phi_s) = I \cdot f_1(\phi_s) \qquad (5)$$

$$B_{nsk}(\phi_s) = I \cdot f_2(\phi_s) \qquad (6)$$

Where f_1 and f_2 associated with the geometry. A single magnetostatic FEA is needed to find these basis functions. Having these basis functions for a typical slot say 1st carrying current I_0 (5) and (6) can be rewritten as:

$$B_{tsk} = (I/I_0)B_{ts0}(\phi - k\gamma) \qquad (7)$$

$$B_{nsk} = (I/I_0)B_{ns0}(\phi - k\gamma) \qquad (8)$$

Therefore by performing a single off-line FEA for a single slot contribution of the stator to the rotating field components can be calculated for any normal working condition. In the second step, permanent magnet contribution to the field, over

one pole pitch, is computed using an FEA analysis for the unexcited stator condition. Having these two components the magnetic field components can be obtained in the middle of the airgap. The field reconstruction flowchart is shown in Fig. 2. In order to verify the accuracy of this method the tangential and normal components of the magnetic field in the middle of the airgap obtained from FRM have been compared to those from FEA. Figures 3 and 4 depict the accuracy of field reconstruction method. The accuracy of the method can be improved further by increasing the resolution of the basis functions.

There are a variety of ways to calculate the electro-mechanical force in electrical machines [17]. Among the existing options Maxwell Stress Tensor (MST) method is utilized here. According to MST the force components densities in the airgap can be calculated using the following formulae:

$$f_t = B_n B_t / \mu_0 \qquad (9)$$

$$f_n = (B_n^2 - B_t^2)/2\mu_0 \qquad (10)$$

In which B_n and B_t are normal and tangential components of the magnetic flux density. Thus, the force components would be as follows:

$$F_t = \oint_\Gamma \vec{f_t} \cdot dl \qquad (11)$$

$$F_n = \int_0^{2\pi} f_n r d\phi \qquad (12)$$

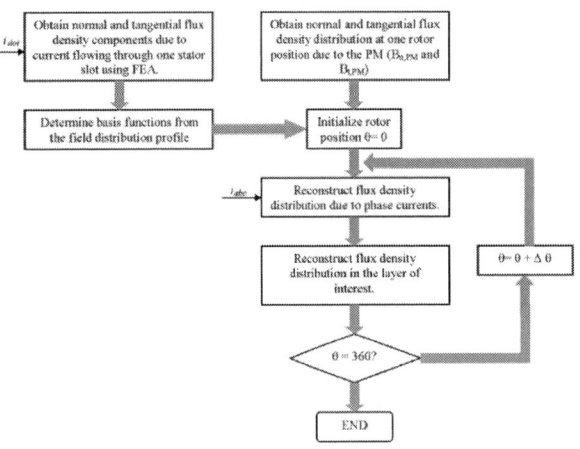

Figure 2: Field reconstruction method flowchart

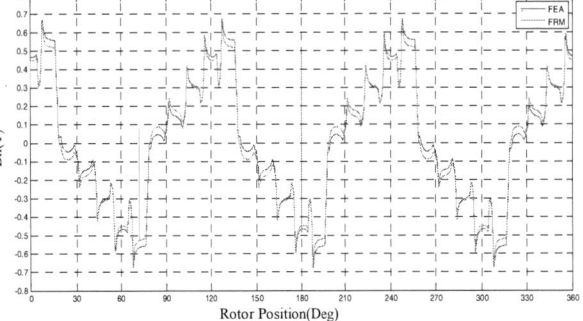

Figure 3: Comparison of FEA and FRM, Normal field component

978-1-4244-4782-4/10 $26.00 © 2010 IEEE

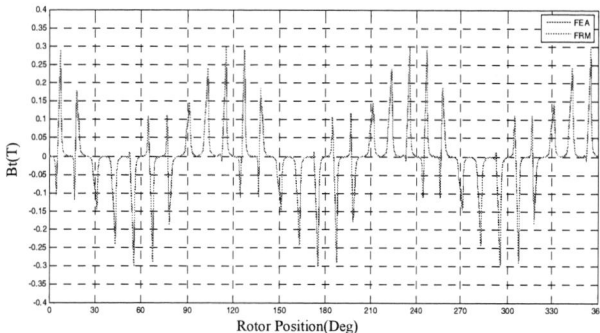

Figure 4: Comparison of FEA and FRM, Tangential field component

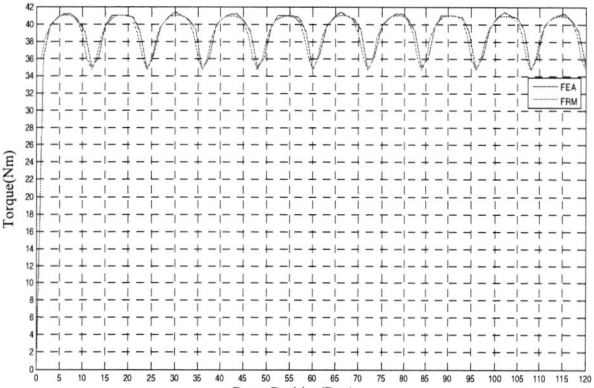

Figure 5: Comparison of torque obtained from FEA and FRM

Where, r is the integration contour. It is obvious that for torque calculations magnetic field components should be known. The MST method is quite effective provided that the FEA solutions are precise. The torque comparison for FEA and FRM is shown in Fig. 5. In the following sections the FRM method is used for detection purposes to estimate the flux passing through each stator tooth. Also, this method has been combined with the MATLAB optimization tool box to achieve the optimal waveforms in case one or more of the stator phases are disengaged.

III. MAGNETIC FLUX CALCULATION

In this section, first the method used for calculating the flux flowing through each stator tooth will be derived and then used to calculate the flux linking each stator phase.

A. Stator teeth fluxes

In the first step, using the field components in the middle of the airgap, the flux passing through the stator tooth would be calculated. The magnetic field distribution in the first quadrant of the PMSM model is shown in Fig. 6. According to this figure a dominant majority of the flux lines that exist in the airgap would enter the stator tooth from the top surface. So, the flux in each stator tooth can be calculated using the magnetic field components in the airgap. There would be a slight error in this calculation because of the leakage flux (i.e. some flux lines would enter the stator tooth from the tooth side surfaces instead of top surface). These flux lines are not accounted for, in the calculation and therefore result in error.

The flux density components are projected on the axis passing the middle of each tooth. This can be done using the following equation:

$$B_{proj}(j) = \sum_{i=1}^{K} \{B_{n,i} \cos(\phi_i - \theta_j) - B_{t,i} \sin(\phi_i - \theta_j)\} \qquad (13)$$

Where ϕ and θ are the position of the field components in the airgap and the position of the projection axis in the model respectively. The indices $i = 1...K$ and $j = 1...L$ refer to the number of field component solutions in the airgap covering one stator tooth and the respective stator teeth, respectively. Based upon normal field components, the flux in the airgap, which is almost equal to the flux in the stator tooth, can be calculated as:

$$\Phi = \iint_{S} \vec{B}_{proj} . d\vec{S} \qquad (14)$$

The above integration is performed on the surface which is concentric to the rotor surface and passes through the stator teeth as shown in Fig.7.

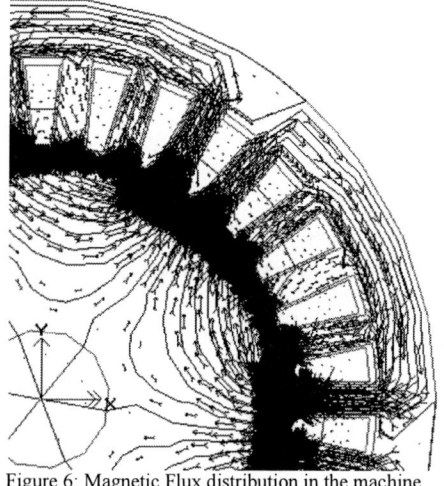

Figure 6: Magnetic Flux distribution in the machine

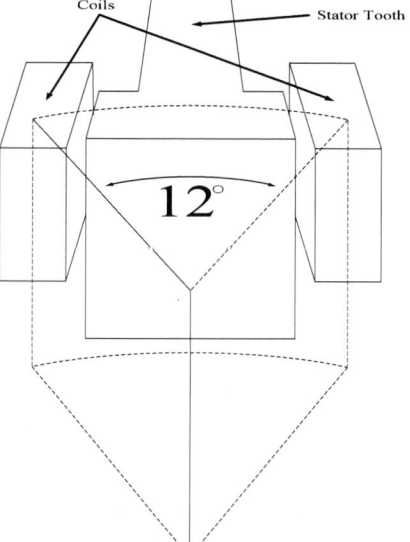

Figure 7: Integration surface

B. 5 phase flux linkages

In order to calculate the stator phase flux linkage, the flux is calculated for one pole and then multiplied by the number of pole pairs. Fig. 8 depicts the flux corresponding to the first pole which passes through the A1-A2 frame (linking A1-A2). The flux linkage of this winding is as follows:

$$\lambda_{A1-A2} = N(\Phi_2 + \Phi_3 + \Phi_4 + \Phi_5 + \Phi_6) \quad (15)$$

Phase A flux linkage is as follows:

$$\lambda_A = 3N(\Phi_2 + \Phi_3 + \Phi_4 + \Phi_5 + \Phi_6) \quad (16)$$

Where, N represents the number of conductors in each coil. This equation could be generalized into the following form for a machine with q stator tooth per pole and 2P magnetic poles (P represents the number of magnetic pole pairs):

$$\lambda_A = PN * \sum_{k=1}^{q} \Phi_k \quad (17)$$

The same analysis can be carried out for phases B, C, D and E. Fig. 9 depicts flux passing through stator teeth 1 to 5 calculated using the observer compared to the FEA. Fig. 10 depicts phase "A" flux linkage calculated using the proposed method, compared to that of the FEA.

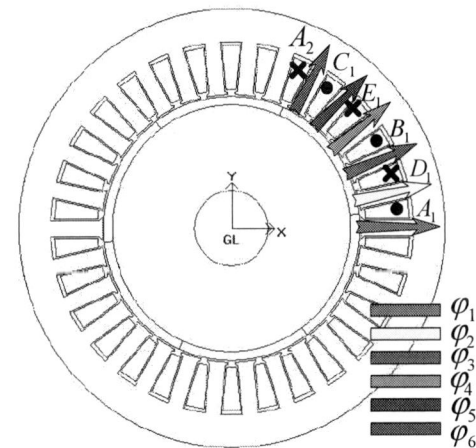

Figure 8: Flux assignment to stator teeth

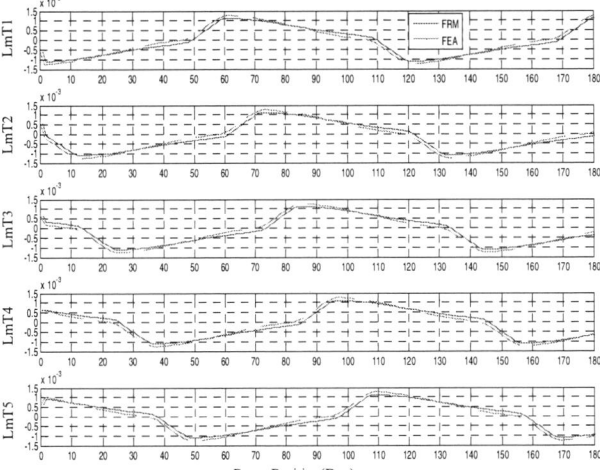

Figure 9: Comparison of stator teeth flux obtained from FEA and FRM

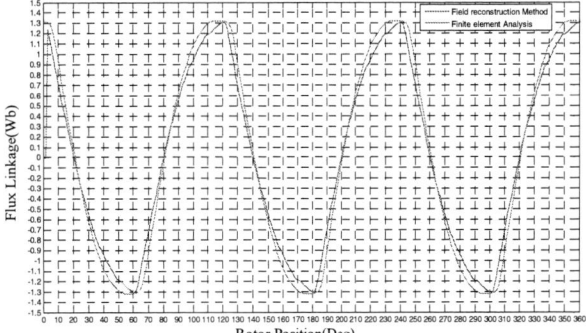

Figure 10: Comparison of Phase A flux linkage obtained from FEA and FRM

It can be seen that, while much faster, the flux observer is quite accurate compared to FEA.

IV. DEMAGNETIZATION DETECTION

The most important task in fault detection is to find unique signatures that can be detected in case of the fault occurrence. For this purpose, generally the quantities such as current, voltage, etc are normally monitored. In case of a healthy machine, while the phase voltages are balanced the sum the voltages and therefore the flux linkages would be zero. In this study, $S = \sum_{i=1}^{5} \lambda_i$ has been monitored to detect the fault. In case of a healthy machine this sum would be zero while in case of a demagnetized rotor it will no longer be zero. Table 1 depicts possible demagnetization faults in the PMSM under study. Figs. 11 and 12 depict the frequency spectrum of S for single and two pair magnet partial demagnetization. It is shown that in case of the rotor magnet demagnetization a set of frequencies would be present in the FFT spectrum of S as summarized in Table 1.

Table 1: Demagnetization scenarios

Fault Scenarios	Place	Frequencies to detect
Single magnet demagnetization	N2	$f_1 = 4.89$
	S3	$f_2 = 9.78$
Double magnet demagnetization	N1N2	$f_3 = 14.67$
	S1S2	$f_4 = 19.56$
Triple magnet demagnetization	N1N2N3	$f_5 = 24.46,$
	S1S2S3	$f_6 = 29.35$
2 Pair demagnetization	N1N2S1S2	$f_7 = 34.24$ $f_8 = 39.13,....$

These frequencies could be used for detection purposes. According to the figures and data from table and considering that the frequency of the stator sinusoidal current is known the frequency spectrum could be determined using the following formula:

$$f_{dem} = \frac{k}{2P} \cdot f_e \qquad k = 1,2,3,..... \quad (18)$$

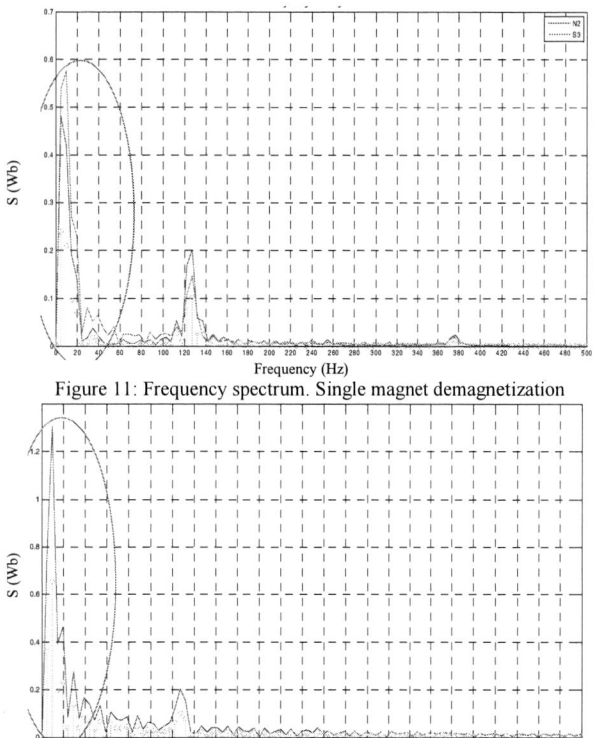

Figure 11: Frequency spectrum. Single magnet demagnetization

Figure 12: Frequency spectrum. Double magnet pair demagnetization

Where, P and f_e are the number of magnetic pole pairs and stator current frequency, respectively. The magnetic flux density components will be calculated using field reconstruction method for 1 electrical cycle. Then these components would be used to determine flux passing each stator tooth which will finally be used to calculate the flux linkages of the stator phases. Next step is to compare the expected flux linkage with the actual quantities. By applying the FFT and low pass filtering the signature frequencies can be extracted. This procedure is demonstrated in Figure 13.

V. FAULT TREATMENT

Depending on the service continuity strategy, various scenarios can be deployed after the fault is detected. In case, the service can be provided by another module the machine could be stopped and the magnets being replaced. In case of an emergency application in which service discontinuity is not possible the stator applied currents can be modified in a way the maximum possible average torque could be squeezed out of the machine shaft. Of course the presence of the harmonics in the current would result in extra torque pulsations. For this purpose the field reconstruction method would be used in conjunction with the optimization methods to attain the optimal current waveforms. Fig. 14 depicts the output mechanical torque of the machine for healthy and demagnetization fault cases in case sinusoidal currents are applied to stator phases. It is shown that the average torque has decreased almost 25% as a result of the magnet demagnetization. Also torque ripple has been increased almost 30%. Using the optimization methods the optimal

waveforms are determined in case, demagnetization occurs. The Matlab optimization toolbox is linked to the FRM code.

For each rotor position, the optimization code calculates a set of currents based on the optimization criteria. These currents are used to calculate the magnetic field components in the machine. Then, using the magnetic field components the torque is calculated using (11). In case the calculated torque complies with the target values, the currents would be stored and a new rotor position would be considered. Fig. 15 depicts the optimal stator phase currents and the output torque of the machine.

The optimization criteria can be chosen to achieve the following cases regarding the target application:

- Maximum average torque
- Maximum average and Minimum torque ripple
- Minimum torque ripple

Here, the optimization process is targeted towards the maximum average torque. It can be seen that the average torque is about 3% less than that of the healthy with sinusoidal stator currents. The torque ripple is increased as expected. Different optimization scenarios can be considered and the optimal currents for each case can be achieved and stored in look up tables in the control unit. Based on the application, the appropriate currents can be applied to stator phases in case the fault is detected.

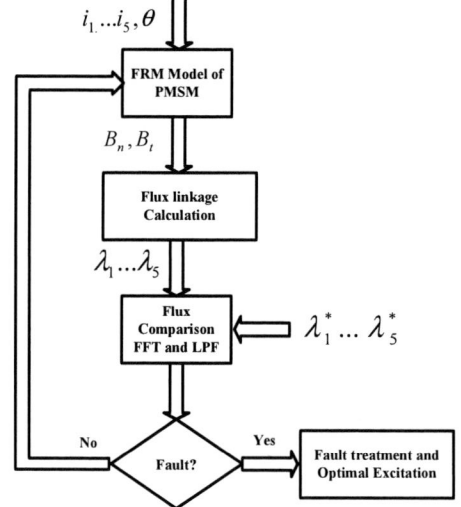

Figure 13: Frequency spectrum. Double magnet pair demagnetization

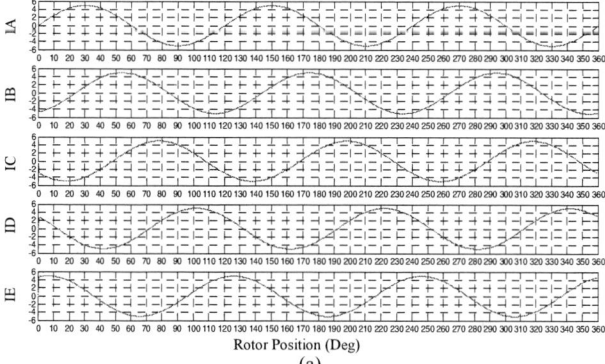

(a)

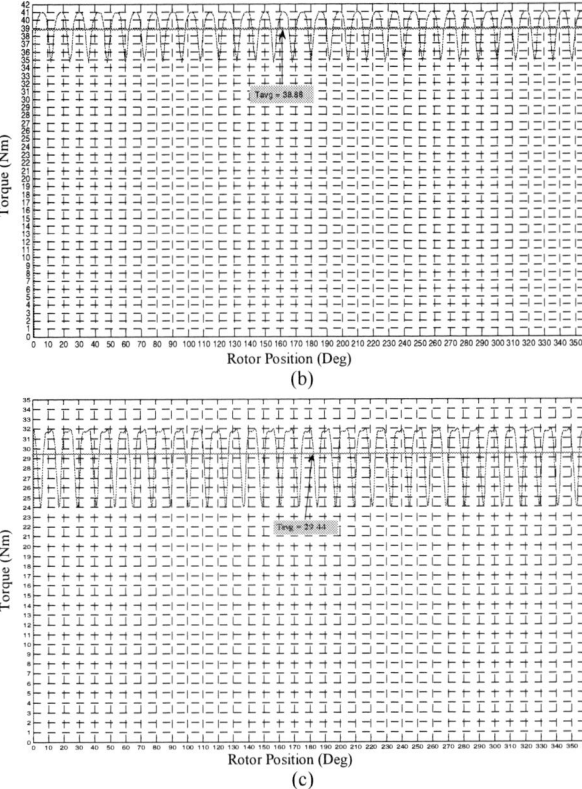

(b)

(c)

Figure 14: Torque Analysis. (a) Sinusoidal stator currents, (b) Healthy machine, (c) Partially demagnetized magnets

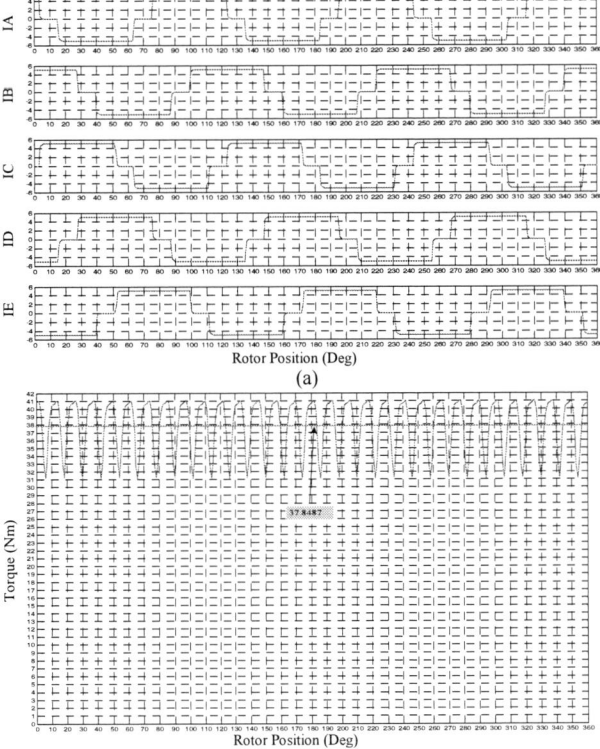

(a)

(b)

Figure 15: Torque Analysis. (a) Optimal stator currents, (b) Output torque, partially demagnetized magnets

VI. CONCLUSION

Field reconstruction method as a powerful magnetic field analysis tool can be used to detect the faults in the PMSM. Demagnetization of the magnets in PMSM is one of the major issues with this type of the machine especially at higher power ratings. It can occur due to high temperatures resulting from poor ventilation or excessive short circuit currents and of course the aging of the magnets. In this paper a new way of demagnetization fault detection has been presented. The flux linkages of the stator phases are calculated using the FRM and used to monitor the faults. FRM can be linked to the optimization tools to achieve the optimal currents which yield maximum average torque and/ or minimum torque ripple. After detecting the fault the optimal currents could be applied to improve output torque characteristic of the machine.

REFERENCES

[1] J. W. Bennett, B. C. Mecrow, A. G. Jack, D. J. Atkinson, S. Sheldon, B. Cooper, G. Mason, C. Sewell, D. Cudley, "A prototype electrical actuator for aircraft flaps and slats", *IEEE Int. Conf. on Electrical machines and Drives*, pp. 41 - 47 , 2005.

[2] N. Ertugrul, W. Soong, G. Dostal, D. Saxon, "Fault tolerant motor drive system with redundancy for critical applications", *IEEE Power Electronics Specialists Conf.*, pp. 1457 - 1462, 2002.

[3] G. J. Atkinson, B. C. Mecrow, A. G. Jack, D. J. Atkinson, P. Sangha, M. Benarous, "The design of fault tolerant machines for aerospace applications", *IEEE Int. Conf. on Electric machines and Drives*, pp. 1863 - 1869, 2005.

[4] Z. Jingwei, N. Ertugrul, L. S. Wen, "Detection and Remediation of Switch Faults on a Fault Tolerant Permanent Magnet Motor Drive with Redundancy", *IEEE ICIEA Conf.* pp. 96 - 101,2007.

[5] B. C. Mecrow, A. G. Jack, J. A. Haylock, J. Coles, "Fault tolerant permanent magnet machine drives", *IEEE Int. Conf. on Electrical machines and Drives*, pp. 433 - 437, 1995.

[6] C. Opera, C. Martis, "Fault Tolerant Permanent Magnet Synchronous Machine for electric power steering systems", *Int. Sym. on Power Electronics, Electrical Drives, Automation and Motion (SPEEDAM)*, PP. 256 – 261, 2008

[7] J. R. Riba Ruiz, J. A. Rosero, A. Garcia Espinosa, L. Romeral, "Detection of Demagnetization Faults in Permanent-Magnet Synchronous Motors Under Nonstationary Conditions", *IEEE Trans. on Magnetics*, vol. 45, No. 7, pp. 2961 – 2969, 2009

[8] J. Rosero, L. Romeral, J. A. Ortega, J.C. Urresty, "Demagnetization fault detection by means of Hilbert Huang transform of the stator current decomposition in PMSM", *Int. Sym. on Industrial Electronics (ISIE)*, pp. 172 – 177, 2008

[9] J. A. Rosero, J. Cusido, A. Garcia, J. A. Ortega, L. Romeral, "Study on the Permanent Magnet Demagnetization Fault in Permanent Magnet Synchronous Machines", *IEEE conf. on Industrial Electronics*, pp. 879 – 884, 2006

[10] A. Khoobroo, B. Fahimi, S. Pekarek, "A New field reconstruction method for permanent magnet synchronous machines", *Int. Conf. on Ind. Electron.*, pp. 2009 – 2013, 2008.

[11] A. Khoobroo, B. Fahimi, "A new method of fault detection and treatment in five phase permanent magnet synchronous machine using field reconstruction method", *IEEE Int. Conf. on Electric machines and Drives*, pp. 682 – 688, 2009

[12] W. Zhu, B. Fahimi, S. Pekarek, "A field reconstruction method for optimal excitation of permanent magnet synchronous machines", *IEEE Trans. on Energy Conversion*, vol. 21, no.2, pp. 305 – 313, June 2006.

[13] B. Fahimi, "Qualitative approach to electromechanical energy conversion: Reinventing the art of design in adjustable speed drives", *Int. Conf. on Electrical machines and Systems*, pp. 432 – 439, Oct 2007.

[14] W. Zhu, B. Fahimi, S. Pekarek, "Optimal excitation of permanent magnet synchronous machines via direct computation of electromagnetic force components", *IEEE Int. Conf. on Electrical machines and Drives*, pp. 918 - 925 , May 2005

[15] A. Kioumarsi, M. Moallem, B. Fahimi, "Mitigation of Torque Ripple in Interior Permanent Magnet Motors by Optimal Shape Design", *IEEE Trans. on Magnetics,* vol. 42, no.11, pp. 3706 – 3711, Nov 2006.

[16] W. Jiang, M. Moallem, B. Fahimi, S. Pekarek, "Qualitative Investigation of Force Density Components in Electromechanical Energy Conversion Process", *IEEE Conf. on Ind. Electron.,* pp.1113-1118, Nov. 2006.

[17] A. Belahcen, "Overview of the calculation methods for forces in magnetized iron cores of electrical machines," *presented at the Seminar on Modeling and Simulation of Multi-Technological Machine Systems,* vol. 29, pp. 41–47, Nov. 1999.

Series connection of IGBT

The-Van NGUYEN, Pierre-Olivier JEANNIN, Eric VAGNON, David FREY, Jean-Christophe CREBIER

Grenoble Electrical Engineering Laboratory, CNRS UMR 5529 INPG/UJF

BP 46, F – 38402 SMH Cedex,

Grenoble, France

The-Van.NGUYEN@g2elab.grenoble-inp.fr

Pierre-Olivier.JEANNIN@g2elab.grenoble-inp.fr

Abstract—This article analyzes the effects of parasitic capacitances in the series connection of IGBT, which exist naturally due to gate driver and power circuit geometry. Two solutions, that can be combined, are proposed to minimize these effects in order to achieve a better voltage balancing. The first one is based on gate driver self-powering technique. The second one is based on a vertical structure assembly of IGBT connected in series. The performance offered by these two complementary solutions is investigated and validated on a series connection of three IGBT in a chopper converter. Both simulation and experimental results show the effectiveness of our approaches.

I. INTRODUCTION

The demand of high voltage switches is steadily growing, especially for applications in electric distribution system (FACTS), and railway traction (high speed train). However, the silicon IGBTs are limited to 6,5kV, with poor switching performances. The series connection of power switches allows to improve the switching performances (using switches with voltage ratings en the range of 4.5kV or lower) and to operate at higher voltage. Nevertheless, the series association of power IGBTs is very difficult; the main problem is to ensure an equal voltage sharing among the components during static and dynamic transient states.

Various voltage balancing methods have been suggested for IGBT connected in series. Several are based on active voltage control employed to limit the voltage during turning off time and to control the dynamic voltage sharing during switching transitions. In [1, 2, 3, 4, 5], numerous control strategies are proposed, including active voltage control and delay balancing. The active clamping circuit [6] is another technique which insures both protection and voltage balancing. In fact, series association of IGBT presents a structural voltage unbalance. All the previous solutions tried to equilibrate voltages among switches by performing complex controls on gate circuits. They lead to unequal gate driver signals to adjust voltage balance.

In this paper, we study the causes of this structural unbalance. In the next section, we explain the influence of parasitic capacitances on the voltage imbalance. Secondly, we show how to design both power and gate drive circuits to

guarantee a natural equal voltage sharing among all the IGBT connected in series. We proposed two solutions. The first one is based on the gate driver self-powering principle; it permits reducing the value of parasitic capacitances in each IGBT driver circuit. The second solution proposed allows reducing gradually the values and the effects of parasitic capacitances between each IGBT and the ground in power circuit, by placing series connected IGBT in a 3D vertical structure. Finally, we demonstrate the benefits of both solutions proposed in an experimental chopper with three IGBT connected in series. Experimental results and simulation results are compared and analyzed.

II. ANALYSIS OF VOLTAGE UNBALANCE CAUSES: EFFECT OF PARASITIC CAPACITANCES

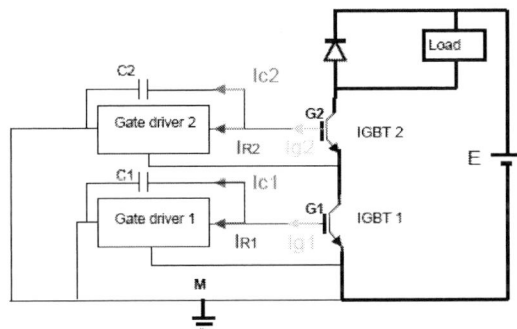

Figure 1. Parasitic capacitances in control circuits

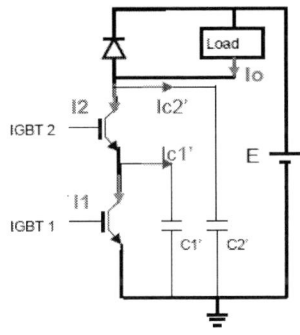

Figure 2. Parasitic capacitances in power circuit

Parasitic capacitances appear between different elements of the electric circuit and the ground. These capacitances have various origins: they can be inherent in the geometry of the circuit (e.g. capacitance between components and their heat sink connected to the ground, or capacitance between printed circuit board and ground...), or they are due to discrete components used to provide electrical isolation (optocouplers or transformers, floating power supplies...).

A. Parasitic capacitances in control circuit

Classical techniques are based on external supplies that are connected to gate drivers. This structure needs electrical isolators such as HF transformers and optical coupling, and thus it contains a parasitic capacitance. This capacitance has a certain influence on switching operation of the IGBT connected in series. In the case of a chopper with 2 IGBT connected in series, controlled by two gate drivers using external supplies (Fig. 1) C1 and C2 represent the two parasitic capacitances in both control circuits. They normally have the same values. Ic1, Ic2, Ir1, Ir2, Ig1 and Ig2 are successively the currents passing through these two parasitic capacitances, the gate drivers and the gates of the IGBTs. At the switching instant, we can demonstrate that:

$$\frac{dV_{G1M}}{dt} < \frac{dV_{G2M}}{dt}$$

The currents flowing through capacitances C1 and C2 are given by the following formula:

$$Ici = C_i \frac{dV_{GiM}}{dt} \tag{1}$$

Thus, we deduce that: $Ic1 < Ic2$

Besides, as a result of using identical components in both gate drivers, we normally have:

$$I_{r1} = I_{r2} \tag{2}$$

And:

$$I_{gi} = I_{ri} + I_{ci} \tag{3}$$

Therefore, the absolute value of the current in the gate of IGBT 2 is higher than the current in the gate of IGBT1:

$$Ig1 < Ig2$$

Consequently, during switching transition, the voltage variation speed dV_{CE}/dt of IGBT 2 is higher than the one of IGBT 1. As a result, the V_{CE} voltage in steady state is higher across IGBT 2 than across IGBT1 [8].

We have shown that the parasitic capacitances between gate drivers and the ground modify the switching speed of IGBTs connected in series. If IGBTs are considered identical,

this is the main reason for the voltage unbalance in both dynamic and static state of these components.

B. Parasitic capacitances in power circuit

Since the collector of each IGBT must be isolated from the ground, there are several critical and additional parasitic capacitances in the power circuit. Fig. 2 illustrates two parasitic capacitances C1', C2' existing between the two IGBT and the ground in a chopper. During switching transient, the voltage V_{CE} of each IGBT changes rapidly, thus there are currents passing through these capacitances. They can be given by:

$$I_{C'i} = C'_i * \sum_{i=1}^{n} \frac{dV_{CEi}}{dt} \tag{4}$$

Here I0, I2 and I1 are successively the currents going across the load, IGBT 2 and IGBT 1. We have:

$$I_0 = I_2 + I_{c2'} \tag{5}$$

$$I_2 = I_1 + I_{c1'} \tag{6}$$

Thus: $\qquad I_2 > I_1$

Moreover, research has shown an analytical formula which calculates switching speed of MOSFET power device according to its current I1 [8]:

$$\frac{dV_{DS}}{dt} = \frac{(I1 + gm.(V_{th} - U_t))}{(C_{GD}.(1 + R_G.gm) + C_{DS} + C_{DS\_ext})} \tag{7}$$

According to the formula (7), the higher is the current passing through the MOSFET; the higher is the switching speed of MOSFET. Since the IGBT has a similar dynamic behaviour with the MOSFET, except for the tail current, thus by using the formula (7), we confirm that the switching speed of IGBT 2 is higher than that of IGBT 1. Therefore, the voltage across IGBT 2 is higher than that of IGBT 1 in static phase.

From the above analysis on the effects of parasitic capacitances, we can conclude that the higher is the rank of the IGBT in the series connection (IGBT 2 on Fig. 2), the higher is its switching speed, and higher is its voltage during the static phase. Therefore, in this paper we focus on solutions to minimize these effects of parasitic capacitances.

III. SOLUTIONS

A. Self-powering principle

We have explained the negative effect of parasitic capacitances connected between ground and control circuitry on the voltage balance among the power switches. These

978-1-4244-4782-4/10 $26.00 © 2010 IEEE

parasitic elements are due to the external power supplies and the necessary insulation components. The self-powering technique has the advantage over the classical supply to eliminate these parasitic capacitances. This technique, presented in Fig.3, is based on five components [9]: an auxiliary high voltage MOSFET, a blocking diode, a bias diode, an avalanche diode and a storage capacitor. In fact, it takes advantage of converter's dv/dt, at every main power switch's turn OFF; it uses part of the energy flowing in the main power devices to recharge periodically the storage capacitor. The energy stored is then used to supply the gate driver while the main switch is turned ON and this until it turns OFF again. The capacitor size must be set accordingly to the gate driver consumption and the switching frequency of the converter. Since this gate driver powering technique has no connection or coupling to the ground, it does not add parasitic capacitance to the driving circuitry. Indeed, all elements are only connected to the power terminals of the main switches

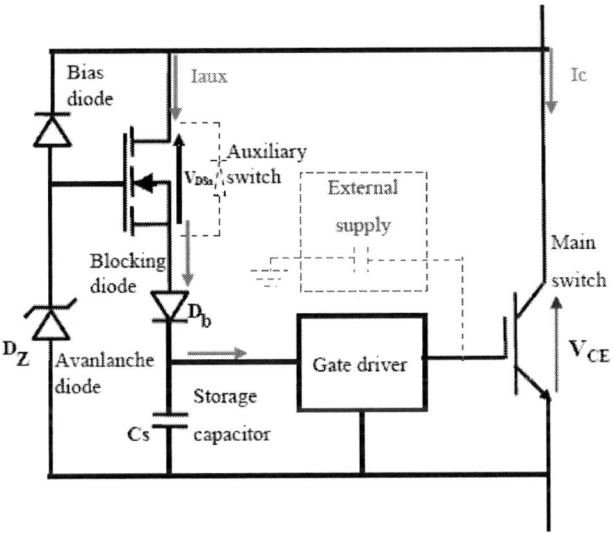

Figure 3.   Self-powering topology around mains switch

B. Realization of a converter using a 3D structure

Equations 4, 5, 6 and 7 shows that the higher are the values of parasitic capacitances in the power circuit, the higher are the values of currents passing through these capacitances. This leads to greater differences between collector's currents of the IGBTs that are connected in series. Furthermore, the unequal IGBT's current affects the device's switching speed (formula 7), and consequently causes voltage unbalances across the series connected power devices. These parasitic capacitances are due to the power structure design and the safety requirements which impose to connect the heat sink to the ground; however, it would be interesting to minimize their value and influence. Hence, our approach aims to put IGBT connected in series in a vertical structure.

Fig. 4 presents a chopper with 3 IGBT connected in series, two realizations of this converter are then proposed: classical horizontal structure and 3D vertical structure.

In the first structure, the 3 IGBT are located in a same board so we deduce that $C'1=C'2=C'3$. In the second one, we can assume that $C'1>C'2>C'3$. This is due to the different distances between IGBT dies and the ground. Moreover, in the classical horizontal structure, we can see that the higher is the rank of the IGBT in the series connection, the faster its collector's potential changes. Therefore, T2 and T3 will be faster than T1. The solution of 3D vertical structure allows extracting less currents passing through parasitic capacitances of T2 and T3 (since $C'2$ and $C'3$ values are reduced), and it permits to obtain the better equality of switching speeds among the IGBTs.

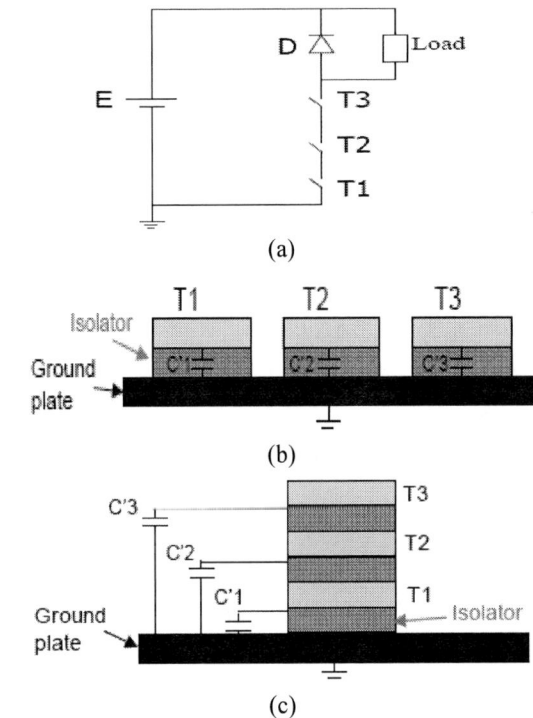

Figure 4.   (a) Chopper   (b) Horizontal structure   (c) 3D Vertical structure.

IV.   EXPERIMENTAL RESULTS

In order to validate our solutions for obtaining a good voltage balancing in IGBT series association, a chopper is considered for analysis and demonstration; Fig. 5 (a) shows the circuit diagram. The series-connected IGBT used in the test are HGTG30N60A (600V, 75A), the dc-bus voltage is set to 900V; the switching frequency is 20 kHz. On the gate driver, we have tested two supply techniques: the self-powering and an external supply (Fig.5 (b)), both of them will be implemented to prove advantage of the self-powering over the external supply in ensuring voltage balancing. Two power structures of this chopper (Fig. 5 (c, d)) have also been realized in order to validate the benefits of the 3D structure over the flat structure.

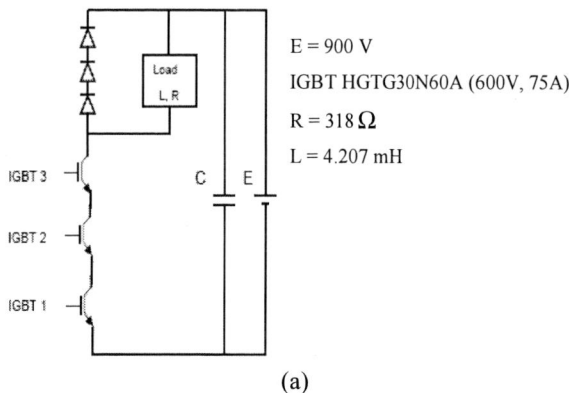

E = 900 V

IGBT HGTG30N60A (600V, 75A)

R = 318 Ω

L = 4.207 mH

(a)

(b)

(c)

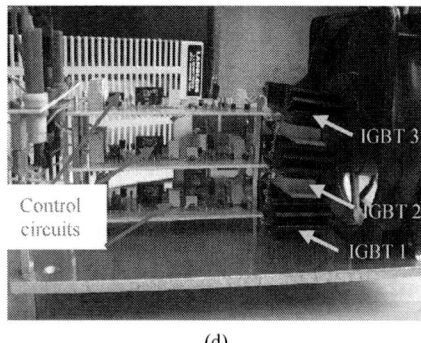

(d)

Figure 5. (a) Chopper (b) Control circuit (c) Horizontal structure (d) Vertical structure

A. Self-powering technique

In this part, we use the self-powering and the external supplies for gate driver powering of the 3 IGBT in the horizontal structure of chopper. Fig. 6 shows the turn-off switching waveforms of 3 IGBT in both cases.

We can see that the delay times between the three waveforms are very short in both cases, approximately less than 10 ns. When the external supply is used to supply the control circuit, the switching speeds of 3 IGBT are very different; however, by using self-powering, this difference of switching speed is significantly reduced (Table 1). Therefore it confirms the effects induced by the parasitic capacitances existing in driver circuit in the case of external supply. As we explained in the previous section, the parasitic capacitances in gate driver produce an effect on accelerating switching speed of IGBT located at higher voltage levels in the series connection (IGBT1 on Fig. 5a). We can observe on Fig. 6a that the V_{CE} voltage of IGBT 3 raises faster than that of IGBT 2 and IGBT 1. When we compare the self-powering solution with, the classical external supply technique, a great improvement between the balances of the raising speeds of the Collector to Emitter voltage's IGBT can be observed.

We can notice that there is still some unbalance. It is due to the effect of other parasitic capacitances located in the optocoupler or between the PCB tracks and ground. .

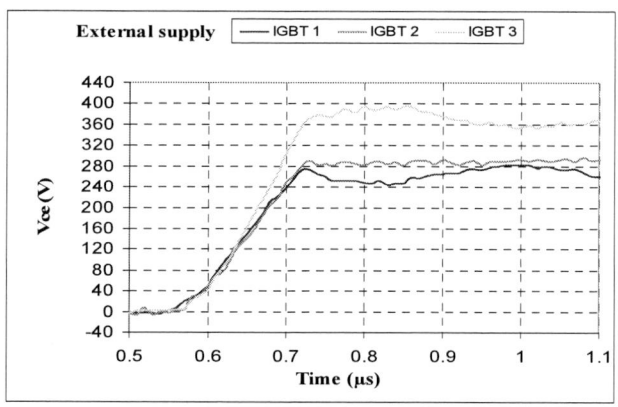

(a) External Supply

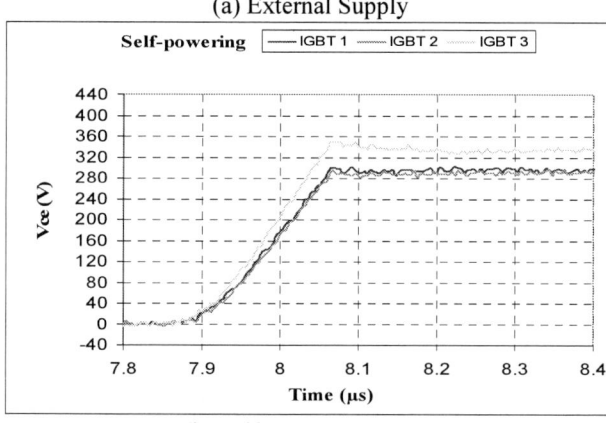

(b) Self-powering supply

Figure 6. Turn-off switching waveforms of 3 IGBTs connected in series (experimental waveforms).

TABLE I.   SWITCHING SPEED IN TWO CASES

IGBT — Switching speed (kV/µs)	IGBT 1	IGBT 2	IGBT 3
External supply	1.88	2	2.64
Self-powering	1.68	1.73	1.86

In order to show that parasitic capacitances are involved in voltage unbalance, we carried out some time domain simulations in Simplorer® software with both cases of external supply and self-powering. Fig. 7 shows the simulation schematic, in which C1', C2', C3' represent the parasitic capacitances between each IGBT collector terminal and the ground; C1, C2, C3 represent the parasitic capacitances in gate drivers while using external supply. They are considered negligible while using self-powering. The value of these capacitances is approximately calculated by using the formula for parallel-plate capacitor.

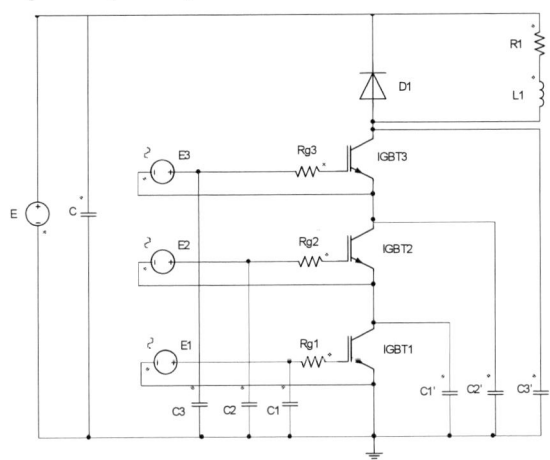

Figure 7.   Simulation schematic in Simplorer

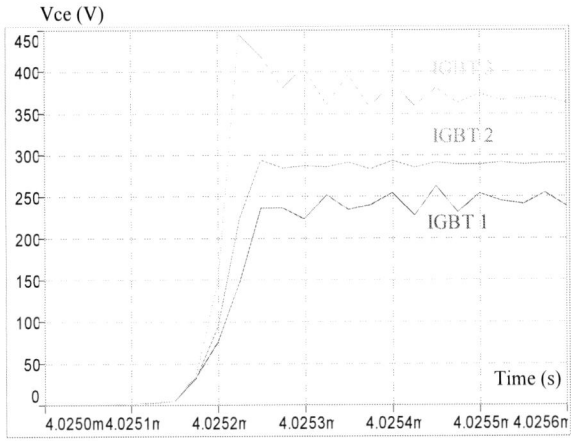

(a) External supply

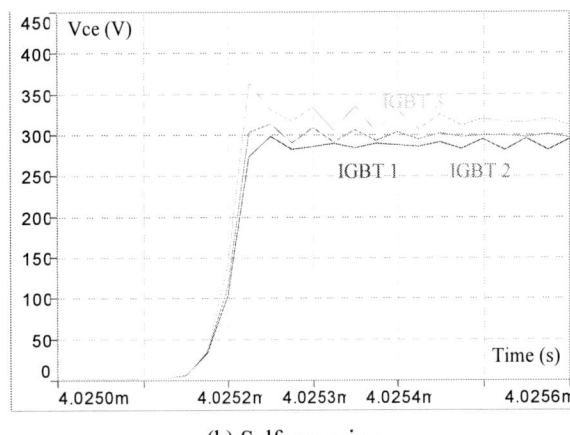

(b) Self-powering

Figure 8.   Simulation results of switching operation of 3 IGBT connected in series (a) external supply and (b) self-powering.

The simulation results (Fig.8) show that the capacitances impact greatly on the voltage unbalance. On Fig. 8a the value of C1, C2 and C3 is 60 pF, and on fig 8-b theses capacitances does no longer exist. The value of C1', C2' and C3' are 8.4 pF in both cases.

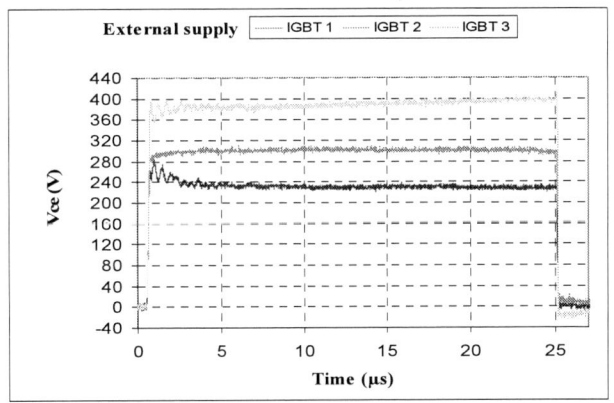

(a) External Supply

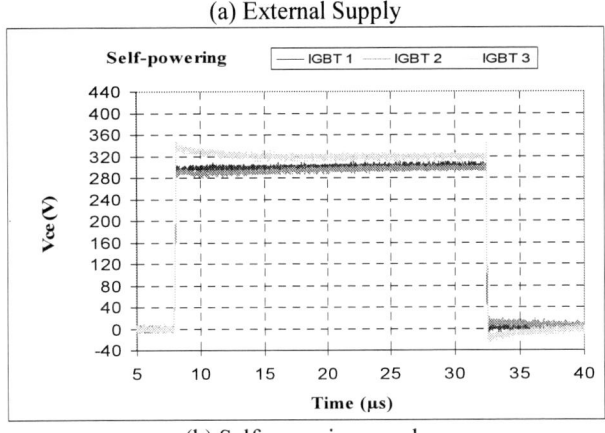

(b) Self-powering supply

Figure 9.   Voltage sharing of IGBT connected in series (experimental waveforms).

978-1-4244-4782-4/10 $26.00 © 2010 IEEE

Fig. 9 presents the voltage sharing among these 3 IGBT in both cases. It can be seen that the self-powering technique improves significantly the balance voltage of the IGBT series connection in static phase. In fact, the dynamic operation decides the voltage sharing in static phase, more the IGBT's switching speed is fast, and more it supports a high voltage in steady state. The self-powering offers a better balance of switching speeds of IGBT connected in series, and it permit to obtain a better voltage balance in static phase: the V_{CE} of the IGBT 3 is approximately 20V more than that the one of IGBT 1 and 2. However, with external supply, there are about 90V and 60 V difference between IGBT 3 and IGBT 2, IGBT 2 and IGBT 1 respectively

Now we consider the turn-on transient of the 3 IGBT, the experimental results in Fig. 10 show in both cases that an over voltage happened on IGBT 3. This phenomenon is due to control signal's delay. If the 3 IGBT do not receive the gate signals at the same time an over voltage will appear. On figure 10, IGBT 1 and IGBT3 start the turn-on transient before the IGBT3, the consequence is an over voltage on IGBT3. To avoid this drawback, the synchronisation of the control signals must be improved.

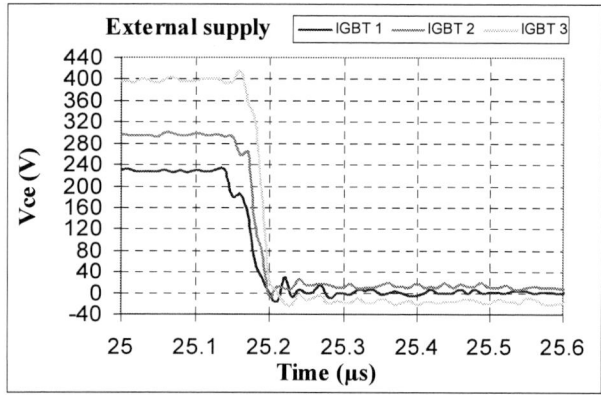

(a) External Supply

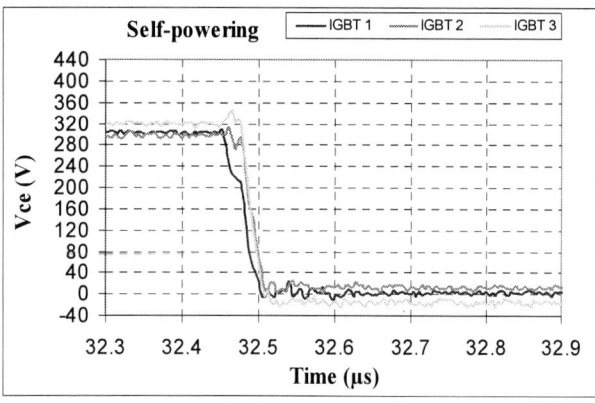

(b) Self-powering supply

Figure 10. Turning-on operation of IGBTs connected in series

All previous results showed that the self-powering solution provides significant improvements versus the classical external supply in balancing the voltages of IGBTs connected in series. However, it would be interesting to take into account

how the parameters impact the efficiency in series association using the self-powering technique. We have observed the effects of the storage capacitor by changing its value; we have used successively three values 330 nF, 100 nF and 68 nF and measured the efficiency of the chopper according to these capacitors.

We measured the efficiency at switching frequency of 40 kHz; Fig. 11 shows that when the capacitance of storage capacitor (Cs on Fig. 3) decreases, the converter's efficiency increases slightly. Regarding to voltage balancing, we find that for the capacitance value of 100 nF and 330 nF, we obtain a better balance compared to that of case C = 68 nF. So we can conclude that the choice of storage capacity is always based on a compromise between voltage balancing and the overall efficiency of the converter.

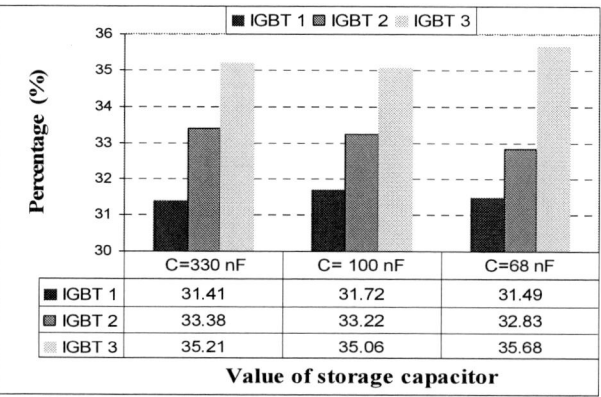

	C=330 nF	C= 100 nF	C=68 nF
IGBT 1	31.41	31.72	31.49
IGBT 2	33.38	33.22	32.83
IGBT 3	35.21	35.06	35.68

Value of storage capacitor

(a) Voltage sharing

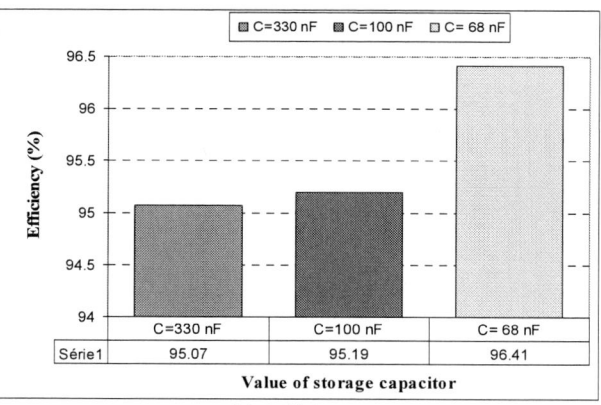

	C=330 nF	C=100 nF	C= 68 nF
Série1	95.07	95.19	96.41

Value of storage capacitor

(b) Efficiency

Figure 11. Volatage balance and efficiency during turning on operation of IGBT connected in series

B. Vertical structure 3D

In order to validate the performances of the 3D vertical structure solution, we used the self-powering technique for gate drivers; the voltage sharing in vertical structure will be compared with that of horizontal structure. Fig. 12 shows the turn-off waveforms in both structures. We can observe that with the vertical structure, the peak transient voltage experienced by IGBT 3 is less than that in the horizontal structure (338 V against 348 V); therefore, the voltage sharing after switching operation is better. According to the formula

978-1-4244-4782-4/10 $26.00 © 2010 IEEE

(4), the parasitic capacitances C'2 and C'3 generate more capacitive currant than C'1 due to the high value of the $\sum_{i=2,3}^{3}\dfrac{dV_{CEi}}{dt}$. Therefore these capacitances C'2 and C'3 will greatly impact the voltage unbalances among the 3 IGBT. The proposed 3D structure reduces these capacitances. The values are C'1=4.58 pF; C'2=0.79 pF; C'3=0.43 pF in the 3D structure against C'1=C'2=C'3=8.4 pF in horizontal structure.

The 3D structure permits an improvement in turn-off switching of series association of IGBT.

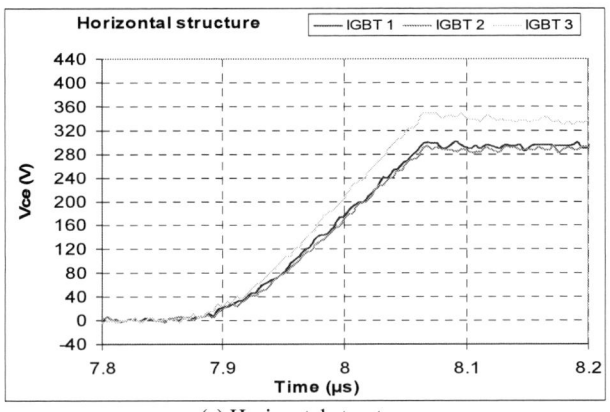

(a) Horizontal structure

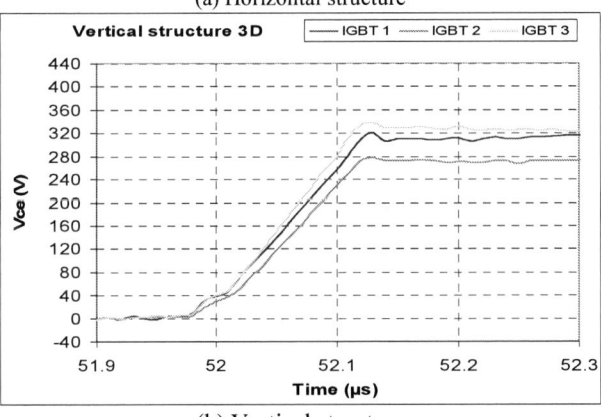

(b) Vertical structure

Figure 12. Turn-off waveforms of 3 IGBT connected in series (experimental waveforms).

V. CONCLUSION

This paper has presented, analyzed and validated two solutions proposed to improve the voltage sharing among IGBT connected in series. The main purpose of this article is to minimize the effects of the parasitic capacitances on voltage balancing by acting on design of the gate driver and the converter. The first solution, based on self-powering technology, offers a very good balancing voltage in steady state phase. The second solution is to minimize the value of parasitic capacitances in power circuit, by putting IGBT series association in a vertical structure; we demonstrated that this structure has advantages over the horizontal structure during the IGBT turn-off.

VI. REFERENCES

[1] Bruckmann. M, Sommer. R, Fasching. M, Sigg. J., "Series connection of high voltage IGBT modules", Industry Applications Conference, 1998. Thirty-Third IAS Annual Meeting. The 1998 IEEE, vol. 2, pp. 1067-1072.

[2] P.R. Palmer, H.S. Rajamani and N. Dutton, "Experimental comparison of methods of employing IGBTs connected in series", IEE Proc.-Electr. Power Appl., vol. 151, No. 5, September 2004.

[3] Soonwook Hong, Venkatesh Chitta, David A. Torrey, "Series connection of IGBT's with active voltage balancing", IEEE Transactions on Industry Application, vol. 35, No. 4, July/August 1999.

[4] Alfio Consoli, Salvatore Musumeci, Giovanna Oriti, Antonio Testa "Active voltage balancement of series connected IGBTs'", Industry Applications Conference, 1995. Thirtieth IAS Annual Meeting, IAS '95., Conference Record of the 1995 IEEE, vol. 3, pp. 2752-2758.

[5] Christian Gerster, "Fast high-power, high voltage switch using series connected IGBTs with active gate-controlled voltage balancing", Applied Power Electronics Conference and Exposition, 1994. APEC '94. Conference Proceedings 1994., Ninth Annual, vol. 1, pp. 469-472.

[6] J.Saiz, M. Mermet D. Frey, P.O. Jeannin, JL. Schanen, P. Muszicki, "Optimisation and integration of an active clamping circuit for IGBT series association", Industry Applications Conference, 2001. Thirty-Sixth IAS Annual Meeting. Conference Record of the 2001 IEEE, vol 2, pp. 1046-1051.

[7] R. Guidini, D. Chatroux, Y. Gwon, D. Lawxe, "Semiconductor Power Mosfets Devices In Series", EPE1993, Brighton, Great Britain.

[8] PO. Jeannin, D. Frey, JL. Schanen, "Sizing Method of External Capacitors for series association of Insulated Gate Components", EPE2001, Graz, Austria.

[9] Nicolas Rouger, Jean-Christophe Crébier, and Stéphane Catellani, "High-efficiency and fully integrated self-powering technique for intelligent switch –based FLYBACK converters", IEEE Transactions on Industry applications, vol. 44, No. 3, May/June 2008.

Three Phase Linear Permanent Magnet Energy Scavenger Based on Foot Horizontal Motion

Igor Stamenkovic, Nikola Milivojevic, Cong Zheng and Alireza Khaligh

Energy Harvesting and Renewable Energies Laboratory (EHREL), Electrical and Computer Engineering Department (ECE); Illinois Institute of Technology, Chicago, IL, USA – EML: khaligh@ece.iit.edu; URL: www.ece.iit.edu/~khaligh

Abstract – **Energy scavenging from human body motion is gaining more attention. Most of research in energy harvesting from foot motion has focused on heel strike and foot vertical motion. This paper presents a linear electric machine topology and power electronic converter for energy harvesting from horizontal foot movement. Micro level aspect of the whole system makes generator topology unconventional. Therefore, design methodology comprises numerical and analytical modelling. Numerical analysis is based on comprehensive three dimensional finite element analysis (FEA), which provides possibilities for inclusive optimizations presented in this paper. Final results demonstrate that the proposed generator topology is a promising candidate for energy scavenging applications from human foot motion.**

Index terms — **Renewable energy, energy scavenging, linear electrical machines, machine design, 3D finite element analysis.**

I. INTRODUCTION

Harvesting kinematic energy of the natural movement of wind, tidal and hydro are renewable way to meet ever growing demand for electrical energy [1-3]. One of the ever growing examples of renewable energy application is the concept of energy scavenging from kinematic energy of the human body motion. The main objective is to power large number of portable electronic devices, such as cellular phones, PDAs, digital cameras, and so forth that are used on the daily basis.

Energy scavenging utilizes human biomechanical walking pattern as the sustainable energy resource. In this way, the energy that is already available through regular metabolic cycles can be harvested through electromechanical conversion. Horizontal foot movement, up-down centre of gravity movement and arm swing are recognized as potential energy resource [4]. It should be noted that the walking pattern is different among people in different ages. The difference comes from different step length and an increased double support stance period [5]. However, the cadence remains very similar. Consequently, available kinematic energy can be harvested in a quite similar way. During walking, the swing phase makes the kinematic change in both horizontal and vertical direction. Studies show that the average toe clearance results to maximal horizontal velocity of 3.6 to 4.5 m/s. On the other hand, vertical displacement occurs twice during one step, and reaches the speeds of up to 1 m/s. This is with the assumption that the mean cadence for 15 healthy and fit elderly individuals is 105 steps per minute. Therefore, the kinematic change for vertical direction occurs 210 times per minute. Since the walking pattern makes either vertical or horizontal linear displacements, it is obvious that linear electric machine is the best solution for energy converter [16-19,23, 24]. Electromagnetic conversion system consists of electric generator and power electronic system that either store energy or directly recharge the portable electronic device. Energy storage devices are usually either batteries or ultra capacitors. Overall system must be on the micro level due to apparent volume constraint. For example, the available volume should remain in the range between 10 cm^3 to 15 cm^3 for the electromagnetic system places in the shoe in order to remain the same comfort level during the walk. Majority of research community was looking at various scavenging methods to harvest vertical movements of the body motion [6-9, 11-15, 20-22]. On the other hand, few proposed systems that harvest kinematic energy released at horizontal foot movement only [4, 10] suggest that potential electric output is not enough to power portable electronic devices. This paper focuses on substantial improvement of power density of the energy converter suitable to harvest energy from horizontal foot movement. In addition, proposed design methodology takes into consideration unconventionality of the generator magnetic circuit by applying comprehensive three dimensional finite element method to analyse and optimize the final design of the electric generator.

In this paper, various linear generator topologies suitable for the energy scavenging application are presented in Section II. The main criterion is power density based on the predefined application requirements. Section III outlines hybrid design methodology comprising three dimensional finite element analysis (3D FEA) and analytical modelling. Basic 3D parameterization and design results are shown in Section IV. Finally, section V presents power electronic converter topology suitable for this application and demonstrates achievable output power and power density.

II. LINEAR GENERATOR TOPOLOGIES

The first reported application includes permanent magnet (PM) tubular linear generator for small-scale application in 1995 [6-8]. The design methodology comprises analytical analysis and two dimensional finite element method. Permanent magnets made of SmCo are attached to the underside of a polyimide membrane. Experimental results showed that 25 mm^3 device could generate 0.3 μW at excitation frequency of 4.4 kHz [7]. Similar topology consisted of a printed circuit board structure with Al coils and movable NdFeB permanent magnets mounted on a Kapton membrane [22]. The initial prototype device generated 0.2 μW power for 6.8 μm vibrations at 360 Hz.

A variation of tubular topology with fully micro-fabricated tubular geometry device uses electroplated permanent magnets and a spiral silicon spring structure [20]. The vibratory magnet was fabricated using sputtering process. The 0.45cm^3 device demonstrated a peak voltage of 40 mV and an output power of 100 µW at 60 Hz. Scherrer et al explored the possibility of using low temperature co-fired ceramics (LTCC) to fabricate a compact, multilayer screen printed coil to power a slotless linear energy harvesting device [21]. The coil is composed of 96 tape layers and had a total of 576 turns. It was held in position between two copper beryllium springs and was designed to move vertically in response to an input excitation, thereby cutting the flux lines of four externally mounted magnets. The springs also act as a pathway for extracting the electrical energy out of the coil and through the housing. The theoretical maximum output power was predicted to be 7 mW, when operated at the resonant frequency of 35 Hz. Lastly, silicon-based tubular linear generator comprising two resonant structures with the possibility to convert low frequency mechanical vibrations is prototyped [15]. Experiment results were reported from a millimetre-scale system, which used a single 50 mm × 15 mm × 0.4 mm styrene cantilever with a 3-turn Cu coil and an oscillating PM. For an input vibration at 1 Hz, frequency up-conversion to the 25 Hz cantilever resonance was demonstrated, and the device generated 4 nW of power.

Further investigation in tubular design comprises air-cored topology and hybrid system with piezoelectric beam. Air-cored prototype having 0.5 cm^3 was built [23]. The prototype has total volume of 30 cm^3 in combination with spring system and whole housing and demonstrates mean output power of 35 µW. The topology comprising tubular electromagnetic converter and piezoelectric beam was designed to harvest up-down movement of the centre of gravity [24]. Proposed topology has moving mass with permanent magnets and four serpentine piezoelectric beams attached on four sides of the frame. The copper windings are fixed in the middle of the moving mass. Proposed analytical model shows that with vibration amplitude of 3 cm and the vibration frequency of 2 Hz, 37 mW and 6 mW could be generated from the electromagnetic and piezoelectric parts, respectively.

On the other hand, a toothless tubular linear generator using off-the-shelf spring, wire, and permanent magnet was developed within volume equal to 23.5 cm^3 [9]. Results show that with the vibration amplitude of 2 cm and the vibration frequency of 2 Hz, 400 µW can be generated. Further investigation included a 500 cm^3 toothed single sided PM synchronous linear generator [10]. By applying dynamic switch of the electrical load impedance to maintain optimum output power, 95 mW is achieved at 6 Hz. Similar silicon-based topology comprising of micro-machined paddle and four NdFeB permanent magnets is also investigated [16-19]. The coil is located on a silicon cantilevered paddle designed to vibrate laterally in the plane of the wafer. Discrete magnets were located within etched recesses in two Pyrex wafers that are bonded to each face of the middle wafer. Two different scenarios applied to a 100 mm^3 topology generated 0.1 µW and 0.12 µW of power at 1.6 and 9.5 kHz, respectively [16]. On the other hand, cantilever-style topology with traditional fabrication achieved 2.85 µW at 357 Hz in a 60 mm^3 structure [19]. Further investigation included micro-fabricated electroplated Cu coils and electroplated $Co_{50}Pt_{50}$ micro-magnets that replaced conventional winding and permanent magnets [17]. A $25 \times 10 \times 1$ mm^3 prototype with cantilever beams comprising coils patterned on their surfaces and a fixed divergent magnetic field is designed, which provides 0.4 nW at 0.64 µm vibration [14].

Another investigation included slotless double sided PM synchronous linear cantilever beam device. One design comprised pair of NdFeB permanent magnets connected via U-shaped back iron, which provides a steady field across an air gap. This magnetic assembly was put on a cantilever beam and vibrated with respect to a stable coil winding. With vibration amplitude of 25 µm at 322 Hz, the 240 mm^3 device demonstrated 0.53 mW power [11]. Another 840 mm^3 design, composed of four permanent magnets, demonstrated 157 µW of power generation when attached to the engine block of a car [12].

Other research group investigated toothed double sided PM synchronous linear generator enclosed in an aluminium outer case [4]. Permanent magnets are mounted on both sides of the stator in this topology. During a stride, the moving coil is forced to slide back and forth in the air gap and collides with outer case. To reduce the energy loss from inelastic collisions, two compression springs are placed at the two ends of the outer case. Experimental results show that output power ranges from 70 to 90 mW.

In conclusion, various research activities include several linear generator topologies with some variations – tubular, single and double sided toothless and toothed linear generators [4-24]. Linear generators with toothed topology show highest power density potentials, mainly because of smaller air gap comparing to toothless topologies [25]. In addition, double sided topology improves power density of the single sided topology and it better utilizes rotor electromagnets. For these reasons, this paper further investigates toothed double sided PM linear generator for the application of energy scavenging of horizontal foot movement.

III. PROPOSED LINEAR ELECTRIC DRIVE

Design requirement for the linear generator that harvest horizontal displacement during human walking is the volume constraint equal to 10 cm^3. It is assumed that the bottom surface is 36 x 30 mm^2, and the height is 9.26 mm. Linear displacement is equal to 12 mm along the longest dimension. The proposed topology is double sided PM generator with two stators and one moving part. Stator comprises stator back iron and permanent magnets. Successive permanent magnets on the same side have opposite magnetization. Magnetization axis of each permanent magnet is perpendicular to the stator back iron. Permanent magnets are NdFeB. Moving part or slider comprises toothed rotor back iron and electromagnets. Successive electromagnets have opposite current direction. Fig. 1 presents the proposed generator topology.

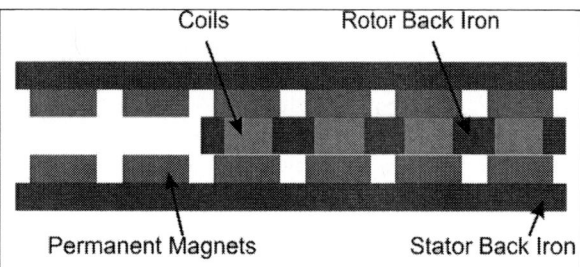

Fig. 1. Toothed double sided permanent magnet generator.

According to Fig. 1, this particular topology has twelve permanent magnets and four coils. Equivalent air gap of the proposed topology is larger than actual air gap because of the permanent magnets, which have similar magnetic permeability as air. Unlike conventional iron-cored electric machines, equivalent air gap with this topology is not order of magnitude smaller in respect to the rest of the machine. As a consequence, some of the features of the proposed magnetic circuit are going to be similar to air-cored machine. Therefore, the comprehensive magnetic analysis of the machine should be conducted with three-dimensional magnetic topology [26].

Proposed topology can be single phase or three phases. Since three phase arrangement is more complex, proposed power electronics topology presents three phase system. In order to generate maximum output power, phase current and back EMF waveform should be in phase. Although vector control would be ideal for maximizing power output of the generator drive, this is rather expensive solution because it requires feedback position sensor. In addition, implementation brings certain complexity. On the other hand, brushless direct current (BLDC) control offers good results with much simpler feedback demands [27]. Fig. 2 shows low cost power controller comprising 6-pack MOSFET active rectifier topology.

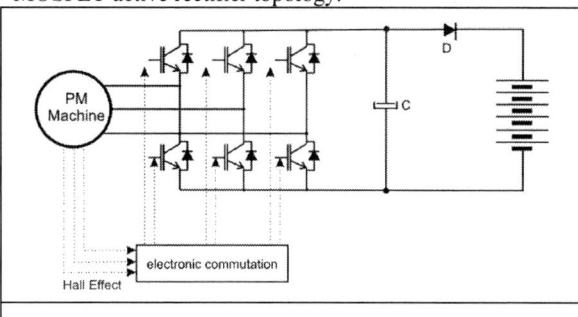

Fig. 2. Low cost power controller based on the BLDC control strategy.

IV. MACHINE DESIGN METHODOLOGY

Design methodology of an unconventional generator topology shown in Fig. 1 comprises numerical and analytical modeling. Hence, such methodology is usually referred as hybrid design methodology [28]. Numerical modeling consists of three dimensional finite element analyses (3D FEA). This method solves electromagnetic circuit on the basis of Maxwell equations. Input data includes definition of the geometry of magnetic circuit, material properties and boundary conditions. Initial

solution generates mesh, which comprises large number of mesh and field elements. Final solution has various magnetic values in those elements. Based on this solution, FEA calculates additional electromagnetic values of interest, such as flux linkage or torque [29]. The purpose of analytical analysis is to estimate power and torque output and densities. Based on the flux linkage, derived from 3D FEA model, back electromotive force (EMF) can be calculated from analytical model. Output power is then derived from back EMF and appropriate current. This is further explained in section V.

Accuracy of the numerical solution is directly linked with the number of mesh elements. On the other hand, higher number of mesh elements increases the need for computational resources and simulation time. Hence, design of the optimal mesh is a tradeoff between solution accuracy and computational resources. For this reason, different components of the machine, or even different slices of the same component have different predefined maximal mesh element size (MES), which directly refines the mesh of those components and slices. In the case of this particular topology, the most sensitive area is equivalent air gap. Therefore, slices of permanent magnets and coils that face air gap have the smallest MES equal to 0.5 mm. Remaining of the permanent magnets and coils have MES equal to 0.8 mm. Fig. 3 shows 3D FEA model with final mesh.

Fig. 3. 3D FEA model of proposed toothed double sided PM generator and its mesh.

Analytical analysis uses discrete value of flux linkage in respect to the relative position of the moving part. Discrete values are first extrapolated with the sum of sinusoidal functions of various harmonics, as it is shown in Eq. 1.

$$\lambda(x) = \sum_{i=1}^{8} a_i \cdot \sin(b_i \cdot x + c_i) \quad (1)$$

Continual function shown in Eq. 1 is a function of linear displacement x. On the other hand, induced electromotive force (EMF) is equal to the first derivative of the flux linkage multiplied by the linear velocity. As said in the introduction, it is expected that linear movement of the grown up person produces an average velocity equal to 5 m/s, which will displace moving part to the other end of the generator. Spring balance will enable returning the slider back to the starting position at the same or similar velocity. These two velocities are taken to be different, and also approximated to be constant. In that case, velocity is the function of the direction of the movement, and not the function of linear displacement. Hence, back EMF is given

978-1-4244-4782-4/10 $26.00 © 2010 IEEE

in Eq. 2.

$$\varepsilon(t) = -\frac{\delta\lambda}{\delta t} = -\frac{\delta\lambda}{\delta x} \cdot \frac{\delta x}{\delta t} = -\frac{\delta\lambda}{\delta x} \cdot v(t) = -\frac{\delta(\lambda \cdot v(t))}{\delta x} \quad (2)$$

In Eq. 2, two different velocities are considered – 5 m/s when the moving part is going from starting to end position, and 4 m/s on the way back. Achievable current conservatively takes into consideration thermal management in respect to wire diameter d with the rule-of-a-thumb equation given in Eq. 3.

$$I_{max} = 4870 \cdot d^2 \quad (3)$$

Equations 4 and 5 presents achievable output power in the case of single phase and three phase winding arrangement, respectively. Parameter E represents RMS value of back EMF. Coefficient n represents the number of coils per phase. Equation 6 presents output power density.

$$P = n \cdot E \cdot I \quad (4)$$

$$P = 3 \cdot n \cdot E \cdot I \quad (5)$$

$$p = \frac{P}{V} \quad (6)$$

V. Basic Design Optimizations And Results

Numerical 3D FEA model shown in Fig. 3 provides possibilities to perform parameterization of various properties of the magnetic circuit. These properties can be of geometrical, magnetic, or electric nature. This paper presents two of such parameterizations: (i) the number of permanent magnets, which ultimately defines the number of independent magnetic circuits; (ii) the number of coil phases. The comparison criterion is magnetic flux density derived from FEA.

Fig. 4 shows shaded plot and direction of the magnetic flux density of the stationary part of the generator with ten permanent magnets.

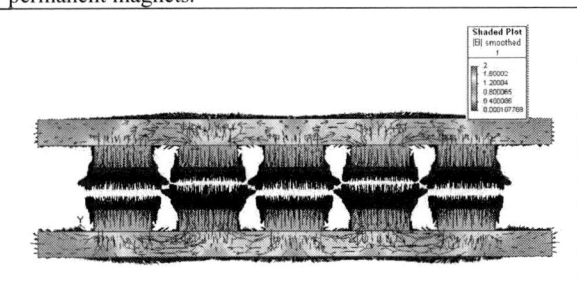

Fig. 4. Shaded plot and direction of the magnetic flux density of the stationary part of magnetic circuits with ten permanent magnets.

According to Fig. 4, there are two independent magnetic circuits, where halves of the permanent magnets in the middle belong to different circuits. Largest air gap flux density is under second and fourth permanent magnets, and is equal to 0.95T, while the average value is 0.8T. One of the drawbacks of such arrangement is non-sinusoidal back EMF generated in the electromagnets.

On the other hand, Fig. 5 shows shaded plot and direction of the magnetic flux density of the stationary part of the generator with twelve permanent magnets.

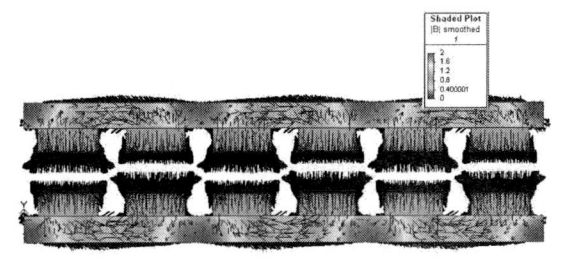

Fig. 5. Shaded plot and direction of the magnetic flux density of the stationary part of magnetic circuits with twelve permanent magnets.

According to Fig. 5, proposed topology has three different magnetic circuits, where all air gaps have similar magnetic flux density equal to 1.1T. Such arrangement provides quasi sinusoidal back EMF.

Fig. 6 shows shaded plot and direction of the magnetic flux density of one magnetic circuit with single phase coil.

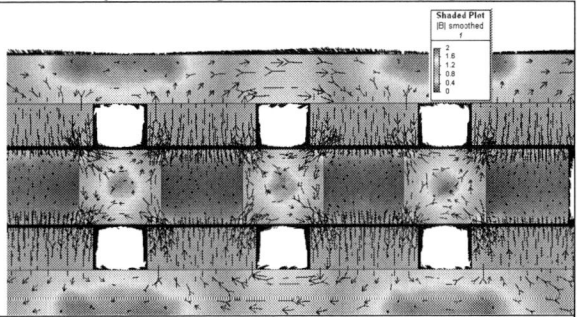

Fig. 6. Shaded plot and direction of the magnetic flux density of one magnetic circuit with single phase coil.

According to Fig. 6, magnetic flux density in the rotor back iron is on average equal to 0.95T. However, since the width of the coil is equal to the width of the back iron, not all of the copper is equally coupled with the magnetic flux. Fig. 7 shows shaded plot of the magnetic flux density of permanent magnets and coils.

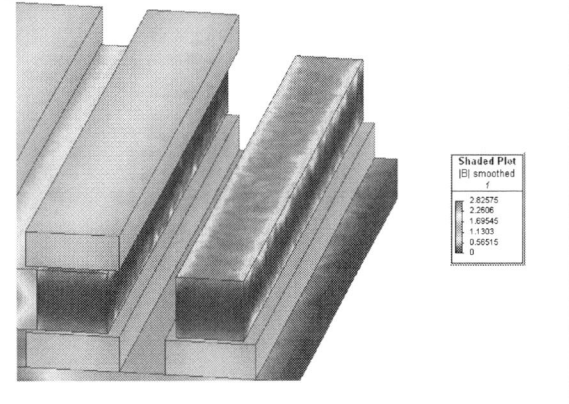

Fig. 7. Shaded plot of the magnetic flux density of single phase generator.

According to Fig. 7, middle portions of the coil are linked with substantially smaller flux density, e.g. in the range 0.6-0.8T. For this reason, three phase topology would enable more even distribution of magnetic flux density. In both cases the amount of copper is the same. Fig. 8 shows shaded plot and direction of the magnetic flux density of the three phase generator with twelve permanent magnets.

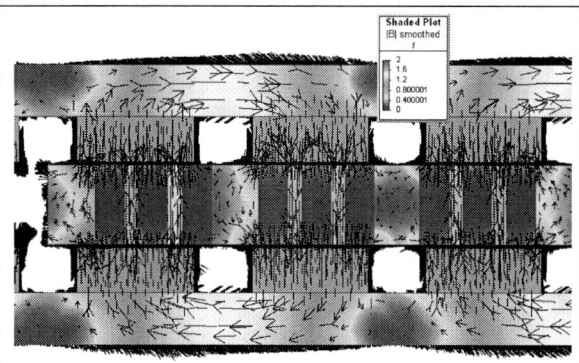

Fig. 8. Shaded plot and direction of the magnetic flux density of three phase generator.

According to Fig. 8, distribution of magnetic flux along all slices of the coil is even. This enables more effective use of active material, i.e. copper, permanent magnet, and iron. Average air gap flux density is around 1.15T.

Final solutions included single phase and three phase winding arrangements. In the case of single phase topology, each coil has 398 turns. Figs. 9 and 10 show flux linkage and back EMF, respectively. Fig. 10 shows back EMF for four periods.

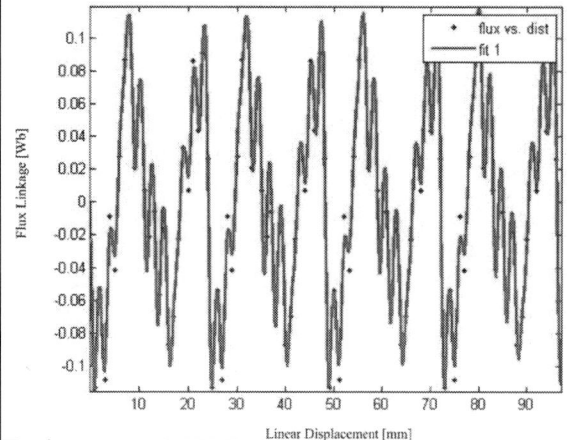

Fig. 9. Discrete (dotted) and extrapolated (solid) function of flux linkage in respect to the relative position of the slider in the case of single phase generator.

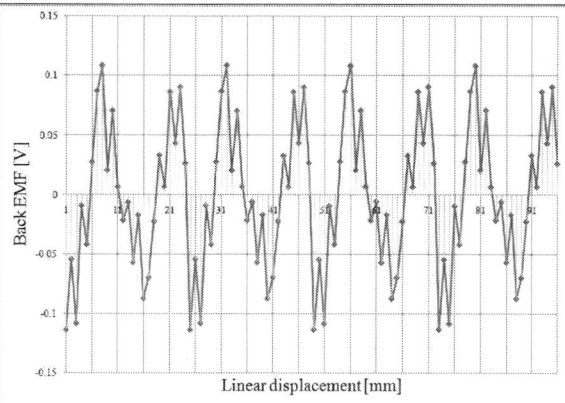

Fig. 10. Waveform of back electromotive force per phase in the case of single phase generator.

In the case of three phase generator topology, each coil per phase has 133 turns. Figs. 11 and 12 show flux linkage and back EMF, respectively. Fig. 12 shows back EMF for four periods.

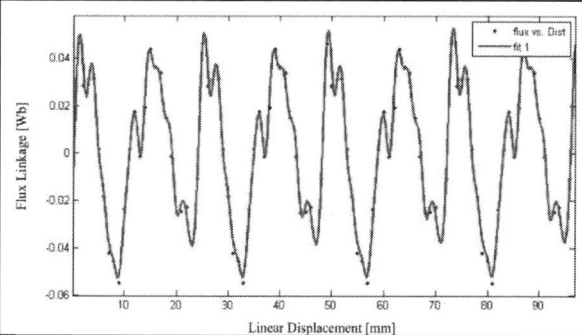

Fig. 11. Discrete (dotted) and extrapolated (solid) function of flux linkage in respect to the relative position of the slider in the case of three phase generator.

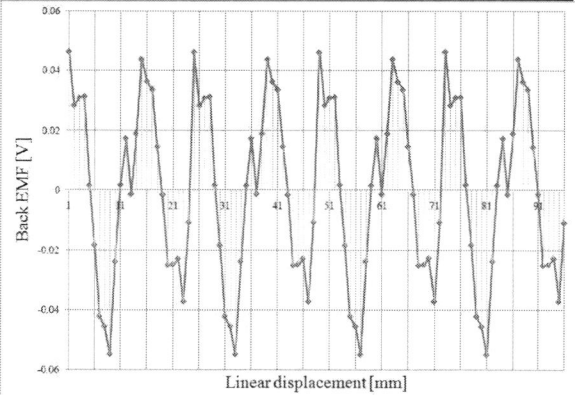

Fig. 12. Waveform of back electromotive force per phase in the case of three phase generator.

Back EMF RMS value in the case of single phase generator is equal to 85mV. Maximum achievable current is equal to 0.158A. Achievable output power is equal to 53.79mW, which means that output power density is 5.379mW/cm^3. In the case of three phase generator, back EMF RMS value is 31.1mV, achievable output power is 59.05mW, and output power density is 5.905mW/cm^3.

Therefore, employing a three-phase structure approximately 59mW output power can be generated from a 10cm^3 three-phase linear permanent magnet generator, based on human foot horizontal motion. This power is enough to energize many portable devices, such as PDAs and many sensor nodes. Since it is possible to charge a cellular phone with 70mW output power, a 12 cm^3 structure can be designed to energize a cellular phone. This is a promising structure to energize next generation of portable electronic devices.

VII. CONCLUSIONS

This paper proposes a linear permanent magnet electric generator suitable for harvesting energy released during human foot motion. Two different topologies, single-phase and three-phase are introduced. Both topologies have been modelled and analyzed through comprehensive three dimensional finite element analysis. Three-phase structure demonstrates higher output power, which is more suitable structure for energy harvesting applications.

ACKNOWLEDGMENTS

The authors wish to acknowledge and thank Infolytica Corporation for technical support of their FEA software.

REFERENCES

[1] Q. Z. Yang, B. Song, "Eco-design for product lifecycle sustainability," Proc. Industrial Informatics, Aug. 2006, pp. 548-553.

[2] A. Farrel, "Making decisions about sustainability: joining social values with technical expertise," Proc. Technology and Society Technical Expertise and Public Decision, June 1996, pp. 188-197.

[3] A. Khaligh and O. C. Onar, Energy Harvesting: Solar, Wind, and Ocean Energy Conversion Systems, CRC Press, 2009.

[4] P. Niu, P. Chapman, "Design and Performance of Linear Biomechanical Energy Conversion Devices," IEEE PESC 2006.

[5] D. A. Winter, A. E. Patla, J. S. Frank, S. E. Walt, "Biomechanical Walking Pattern Changes in the Fit and Healthy Elderly," Research Report.

[6] C. B.Williams, R. B. Yates, "Analysis of a micro-electric generator for microsystems," Proc. Transducers'95/Eurosensors IX, vol. 1, pp. 369–372, Jun. 1995.

[7] C. Shearwood, R. B. Yates, "Development of an electromagnetic micro generator," Electron. Letter, vol. 33, no. 22, pp. 1883–1884, Oct. 1997.

[8] C. B. Williams, C. Shearwood, M. A. Harradine, P. H. Mellor, T. S. Birch, R. B. Yates, "Development of an electromagnetic micro-generator," IEE Proc. Circuits, Devices Syst., vol. 148, no. 6, pp. 337–342, Dec. 2001.

[9] R. Amirtharajah, A. P. Chandrakasan, "Self-powered signal processing using vibration-based power generation," IEEE J. Solid-State Circuits, vol. 33, no. 5, pp. 687–695, May 1998.

[10] K. Sasaki, Y. Osaki, J. Okazaki, H. Hosaka, K. Itao, "Vibration-based automatic power-generation system," Microsyst. Technol., vol. 11, no. 8–10, Aug. 2005.

[11] M. El-hami, P. Glynne-Jones, N. M. White, M. Hill, S. Beeby, E. James, A. D. Brown, J. N. Ross, "Design and fabrication of a new vibration-based electromechanical power generator," Sens. Actuators A, vol. 92, no. 1–3, pp. 335–342, Aug. 2001.

[12] P. Glynne-Jones, M. J. Tudor, S. P. Beeby, N. M. White, "An electromagnetic, vibration-powered generator for intelligent sensor systems," Sens. Actuators A, vol. 110, no. 1–3, pp. 344–349, Feb. 2004.

[13] E. P. James, M. J. Tudor, S. P. Beeby, N. R. Harris, P. Glynne-Jones, J. N. Ross, N. M. White, "An investigation of self-powered systems for condition monitoring applications," Sens. Actuators A, vol. 110, no. 1–3, pp. 171–176, Feb. 2004.

[14] M. Mizuno, D. G. Chetwynd, "Investigation of a resonance micro generator," J. Micromech. Microeng, vol. 13, no. 2, pp. 209–216, Mar. 2003.

[15] H. Kulah, K. Najafi, "An electromagnetic micro power generator for low-frequency environmental vibrations," in Proc. 17th Int. Conf. MEMS (MEMS 2004), Maastricht, The Netherlands, Jan. 2004, pp. 237–240.

[16] E. Koukharenko, S. P. Beeby,M. J. Tudor, N.M.White, T. O'Donnell, C. Saha, S. Kulkarni, S. Roy, "Micro electromechanical systems vibration powered electromagnetic generator for wireless sensor applications," Microsyst. Technol., vol. 12, no. 10–11, Sep. 2006.

[17] S. Kulkarni, S. Roy, T. O'Donnell, S. P. Beeby, M. J. Tudor, "Vibration based electromagnetic micropower generator on silicon," J. Appl. Phys., vol. 99, no. 8, p. 08P511, 2006.

[18] S. P. Beeby, M. J. Tudor, E. Koukharenko, N. M. White, T. O'Donnell, C. Saha, S. Kulkarni, S. Roy, "Micromachined silicon Generator for Harvesting Power from Vibrations," Proc. Transducers 2005, pp. 780–783.

[19] S. P. Beeby, M. J. Tudor, R. N. Torah, S. Roberts, T. O'Donnell, S. Roy, Experimental comparison of macro and micro scale electromagnetic vibration powered generators, Microsyst. Technol., vol. 13, no. 11–12, pp. 1647–1653, Jul. 2007.

[20] C. T. Pan, Y. M. Hwang, H. L. Hu, H. C. Liu, "Fabrication and analysis of a magnetic self-power micro generator," J. Magn. Mater, vol. 304, no. 1, pp. e394–e396, Sep. 2006.

[21] S. Scherrer, D. G. Plumlee, A. J. Moll, "Energy scavenging device in LTCC materials," 2005 IEEE Workshop on Microelectronics and Electron Devices (WMED '05), pp. 77–78.

[22] C. Serre, A. Pérez-Rodríguez, N. Fondevilla, J. R. Morante, J. Montserrat, J. Esteve, "Vibrational energy scavenging with Si technology electromagnetic inertial microgenerators," Microsyst. Technol., vol. 13, no. 11–12, pp. 1655–1661, Jul. 2007.

[23] T. von Büren, G. Tröster, "Design and optimization of a linear vibration-driven electromagnetic micro-power generator," Sens. Actuators A, vol. 135, no. 2, pp. 765–775, Apr. 2007.

[24] A. Khaligh, P. Zeng, X. Wu and Y. Xu, "Hybrid energy harvesting topology for human-powered mobile electronics," in Proc. 34th Conf. of the IEEE Industrial Electronics Society, Orlando, FL, Nov. 2008, pp. 448-453.

[25] S. Chevailler, "Comparative study and selection criteria of linear motors," Ph.D. Thesis no. 3569, Ecole Polytechnique Federale de Lausanne, 2006.

[26] C. Ferreira, J. Vaidya, "Torque analysis of permanent magnet coupling using 2D and 3D finite element methods," IEEE Trans. on Magn., vol. 25, no. 4, July 1989.

[27] N. Milivojevic, I. Stamenkovic, N. Schofield, A. Emadi, "Electrical Machines and Power Electronic Drives for Wind Turbine Applications," Proc. Industry Electronics Conference (IECON), Nov. 2008, pp. 2326-2331.

[28] J. R. Bumby, R. Martin, M. A. Mueller, E. Spooner, N. L. Brown, B. J. Chalmers, "Electromagnetic design of axial-flux permanent magnet machines," IEE Proceedings Electr. Power Appl., Vol 151, No. 2, March 2004.

[29] Infolytica Corporation, Online Documentation Centre, 2008.

Bidirectional Communication Techniques for Wireless Battery Charging Systems & Portable Consumer Electronics

W.P. Choi [1,2], W.C.Ho [1], X. Liu [1] and S.Y.R. Hui [2], *Fellow IEEE*

[1] R&D Group
Convenientpower HK Ltd.
Hong Kong
Email: wcho@convenientpower.com

[2] Center for Power Electronics
City University of Hong Kong
Hong Kong
Email: eeronhui@cityu.edu.hk

Abstract— **Simultaneous power and signal transfer in a wireless/contactless system is an emerging technology for portable electronics. This paper aims at exploring simple and cost effective ways for such applications for a wide range of the portable consumer electronic devices such as mobile phones. Frequency-shift keying (FSK) and amplitude-shift keying (ASK) techniques for an inductive battery charging platform have been successfully implemented for simultaneous power and signal transfer in an 10W wireless charging system. The proposed techniques provide effective bidirectional communication between the charging platform and the loads and form the basis for load identification and battery monitoring in wireless charging system. Considerations of these methods for compatibility among a wide range of portable products are discussed.**

I. INTRODUCTION

Wireless or contactless power transfer systems have been reported in many applications. Examples include electrical transmission system [1], industrial automation and control systems such as robotics and factory automation [2], and the electric vehicle battery system [3]. Recently, more developments of wireless/contacless battery charging systems have been reported [4-8] for portable devices such as mobile phones, games and toys, etc. Without direct electrical contact, indirect detection of system status such as voltage, current and temperature of the secondary device without interrupting power transfer becomes a challenge. To have a robust and closed loop control of the system, it is necessary to have a bidirectional communication mechanism so that load identification and load monitoring can become possible. Some previous methods involve radio frequency communication in the wireless power system [2], using inductive and capacitive signal coupling for communication [9] and overlapping high-frequency signal on power line [10]. Others use separate power and information coils [11] or used separate power and communication circuitries for two-way communication [12]. However, for low-cost objective, power and signal transfer over only one coil with the simple communication techniques

such as frequency-shift keying or amplitude-shift keying remain the most effective ways to develop a simple bidirectional communication in the contactless battery charging platform. This is the focus of this paper, which presents a solution that meets a range of regulatory requirements being set by the international Wireless Power Consortium (WPC) [13] for a wireless charging system as shown in Fig.1.

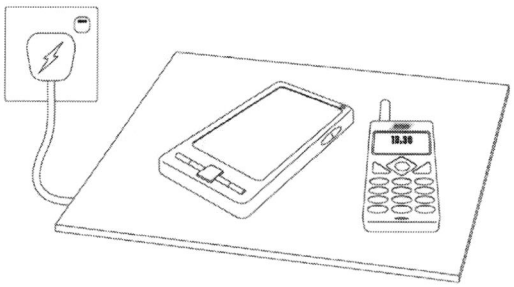

Fig.1 Wireless charging system for portable electronic products.

II. BIDIRECTIONAL COMMUNICATION TECHNIQUES

Simple bi-directional communication techniques for wireless charging systems include at least (1) FSK, (2) ASK and (3) a combination of FSK and ASK. In this section, the basic principles of FSK and ASK are first summarized in the context of a wireless bi-directional communication for the wireless charging system shown in Fig.1. In the Section III, practical implementations of method (1) and (3) are used to demonstrate the feasibility of these methods.

A. Frequency-Shift Keying Modulation

The FSK technique involves the modulation of a carrier and a signal. Fig.2 shows a schematic of a wireless charging system using this principle. The handshaking communication of this wireless charging platform acquires the frequency shift key (FSK) technique by using one demodulator at transmitter, and one modulator/demodulator at receiver. No modulator is needed on the transmitter communication board as the modulation will be performed by modulating inverter's PWM frequencies for achieving signal transmission. The experimental setup has been built up based on the block diagram shown in Figure 2. The modulating technique modulates the inverter switching frequency (*fs*) with a modulating frequency (*fm*) as shown in the following equation:

$$f_S = f_C \pm f_M \qquad (1)$$

When *fs=fc+fm*, the signal represents "0", whilst *fs=fc-fm* represents "1". These "1" and "0" signals can be demodulated from the switching frequency *fs* as shown in Fig.3. This handshaking application is capable of bidirectional data communication by means of frequency-shift keying (FSK) modulation and demodulation. In this application the primary modulation is directly performed by signals ("0" or "1") outputting from a control unit or MCU. The time sequence can be set to be 10ms for primary signal and power transmission and 10ms for secondary signal transmission having ceased power transmission. While primary is transmitting signal, secondary board has to disable signal transmission at the same time, vice and versa. All these functions can be done by the MCU programming.

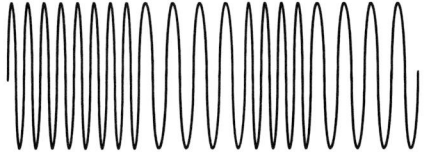

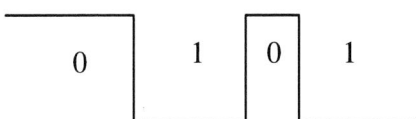

Fig.3 Typical waveforms of the FSK modulation

B. Amplitude Shift Key Modulation

The handshaking communication of this wireless charging platform can use the amplitude shift key (ASK) technique by using one sensing resistor and two low speed comparators in the transmitter circuit, and one loading resistor and one comparator in the MCU in the receiver circuit. The block diagram of the experimental setup is shown in Fig. 4.

(i). Communication from Transmitter to Receiver:

    1.) Data are transferred by varying the dc input voltage V_{TI} (Fig.4 and Fig.5) to the high frequency inverter (controlled by MCU in transmitter)

    2.) Close the switch S1 in the receiver which is controlled by MCU in the receiver.

    3.) Compare V_{S1} with the internal voltage reference in receiver's MCU.

    4.) After comparison, data (D1) as shown in Fig.5 can be read by the receiver's MCU.

(ii). Communication from Receiver to Transmitter

    1.) Set the DC-DC converter in Fig.4 to fully turn-on, so that dc input voltage to the high frequency invert (V_{TI} in Fig. 4) is constant.

    2.) Data are transferred by the switching (on or off) actions of switch S1 in the receiver (Fig.4 and Fig.6) as shown.

    3.) Signal from the sensing resistor R_S is amplified.

    4.) The amplified signal from amplifier C1 is filtered by one low pass filter F1, and a high pass filter F2, so that two signals V_{FL} and V_{FH} as in Fig.6 can be obtained.

    5.) Signals V_{FL} and V_{FH} are further compared by comparator C2 so that the data D2 can be obtained.

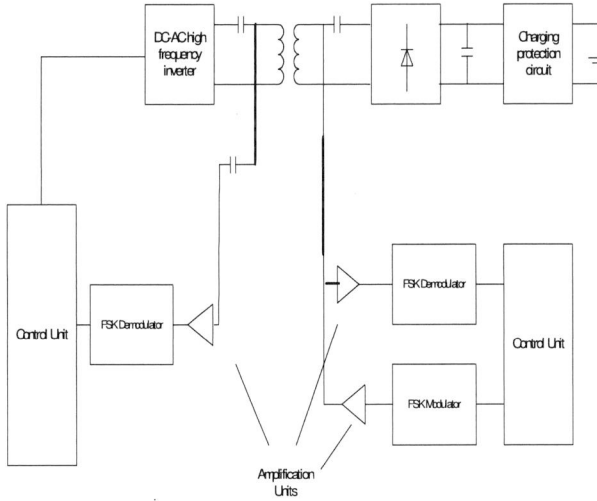

Fig.2 Schematic of a wireless charging system using FSK for bi-directional communication between primary (charging) and secondary (load) circuits.

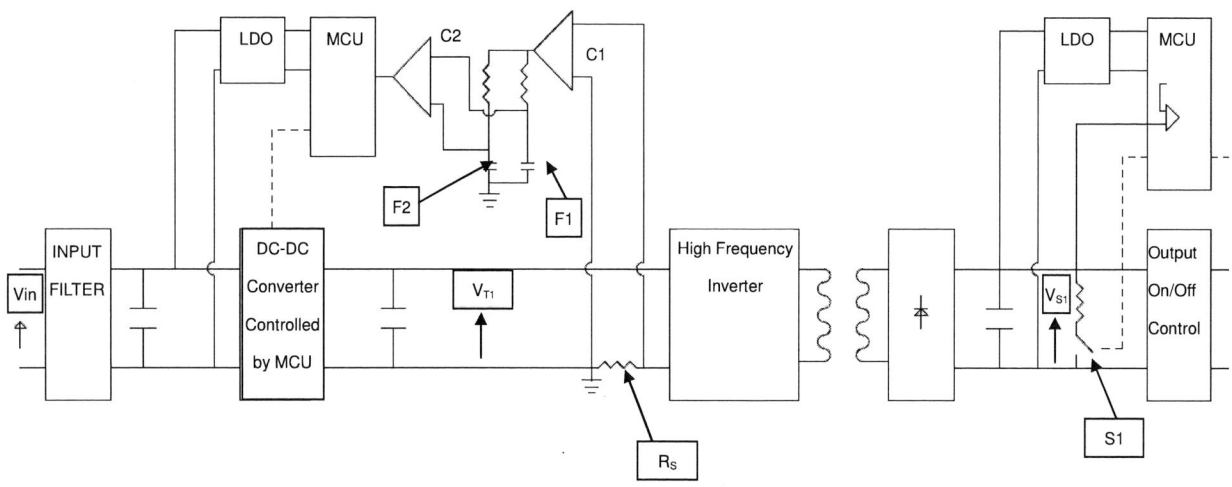

Fig.4 shows a block diagram of the ASK data communication concept

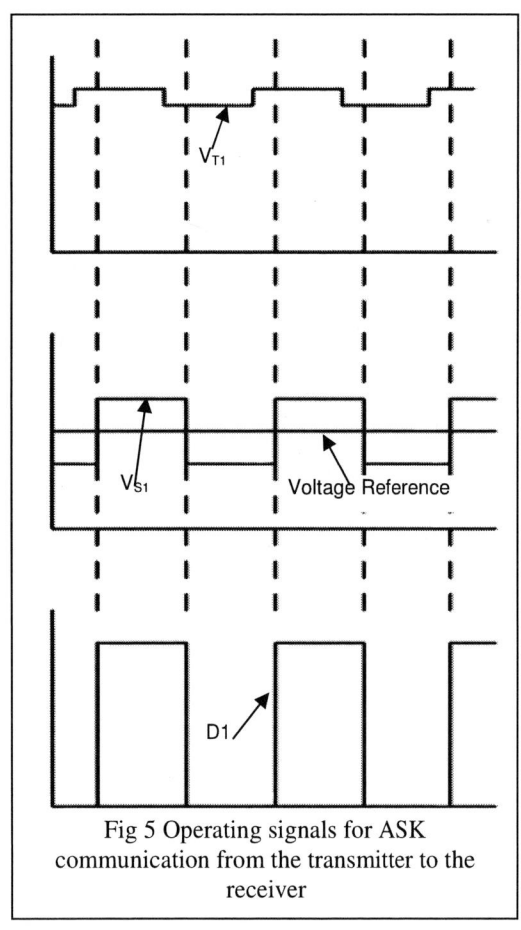

Fig 5 Operating signals for ASK communication from the transmitter to the receiver

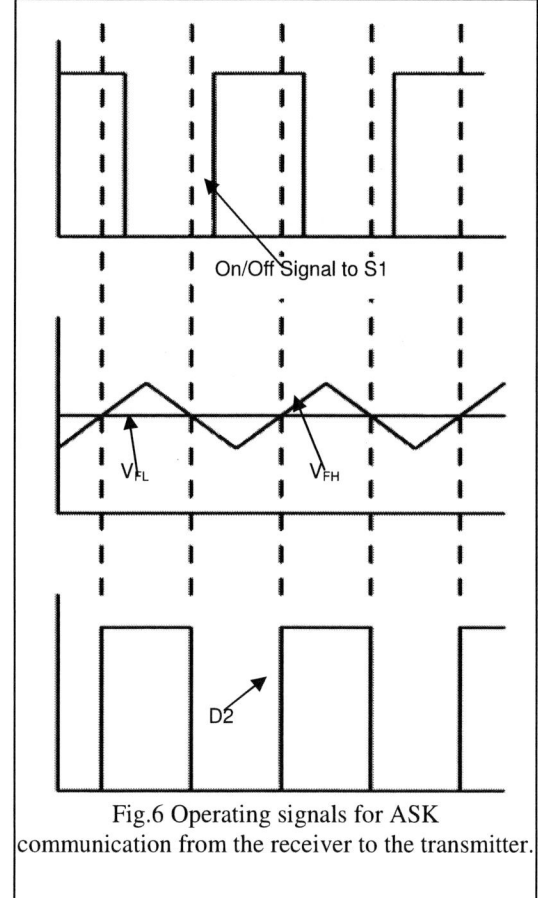

Fig.6 Operating signals for ASK communication from the receiver to the transmitter.

III. PRACRTICAL IMPLEMENTATION

Bidirectional communication in the wireless charging system based on (1) the FSK method and (2) a combination of FSK and ASK methods have been implemented and tested.

A. FSK method

(i) Signal from the Receiver (Rx) to the Transmitter (Tx)

Fig.7 shows the implementation of the receiver-to-transmitter communication using the FSK technique. A receiver (Rx) signal of about 4 kHz generated from the receiver MCU modulates the 460 kHz carrier after the modulator and is injected into the tertiary winding as shown in Figs. 8(a) and 8(b). With the transmitter stops transferring power for several tens of millisecond, the modulated signal can then be detected at the primary coil as shown in Fig. 8(c). The demodulator in the transmitter demodulates the detected signal as displayed in Fig.8(d). The transmitter MCU completely receives the receiver signal and performs the predefined instructions accordingly.

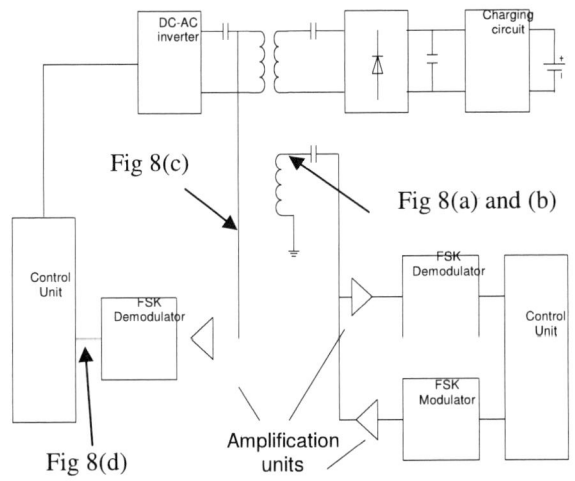

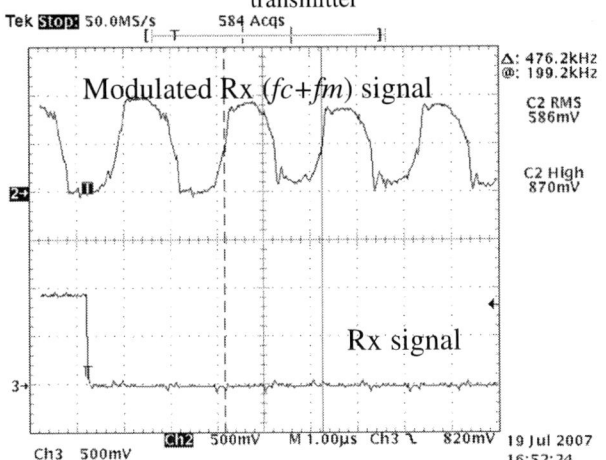

Fig.7 FSK signal transmission from the receiver to the transmitter

Fig.8(a) Receiver and modulated receiver Rx (fc+fm) signals

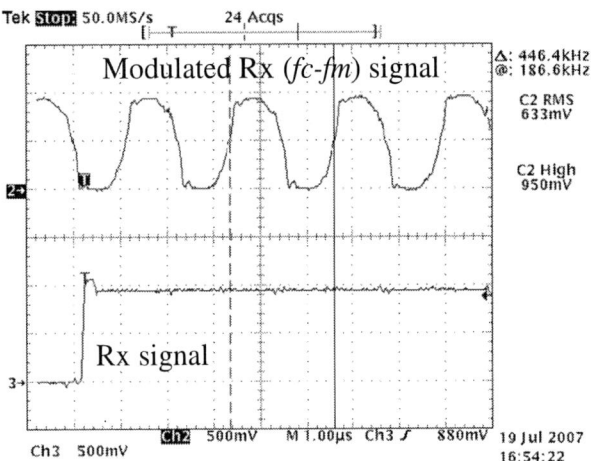

Fig.8(b) Receiver and modulated Rx (fc-fm) signals

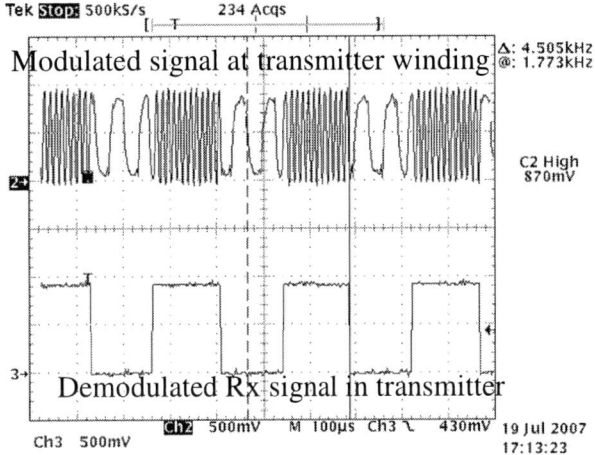

Fig.8(c) Modulated and demodulated Rx signals in the transmitter

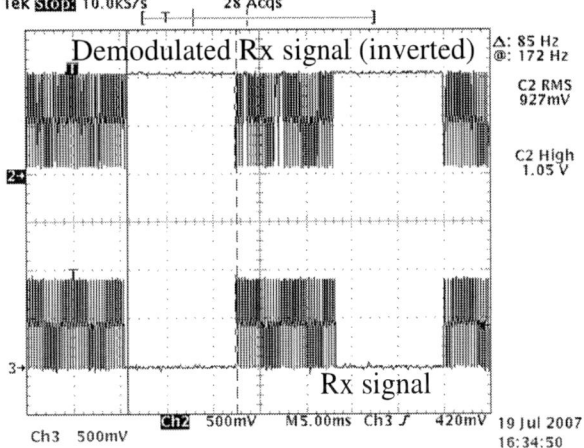

Fig.8(d) Receiver Rx and Demodulated Rx signals

(ii) Signal from the Transmitter (Tx) to the Receiver (Rx)

Fig.9 shows the implementation of the transmitter-to-receiver communication using the same approach as in Fig.7. However, there is no need to have a modulator in the transmitter circuit. The transmitter MCU itself modulates the operating frequency around 440 kHz +/- 10 kHz as in Figs. 104(a) to 10(c) and directly drives the inverter to transfer the power to the secondary winding, and also the modulated signal to the tertiary winding. The demodulator in the receiver then demodulates the transmitter signal as in Fig. 10(d) and inputs the signal to the receiver MCU.

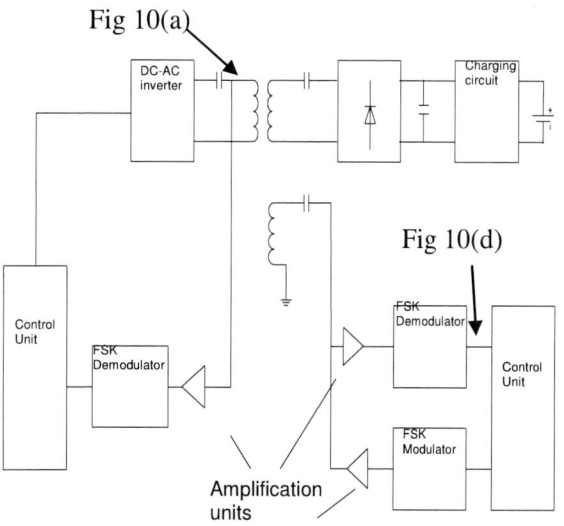

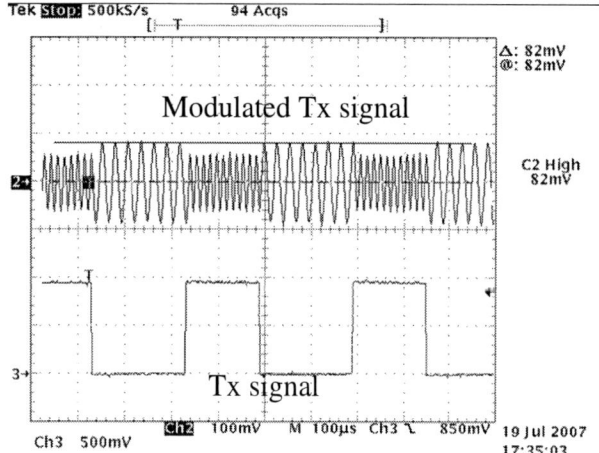

Fig..9 FSK signal transmission from the transmitter to the receiver

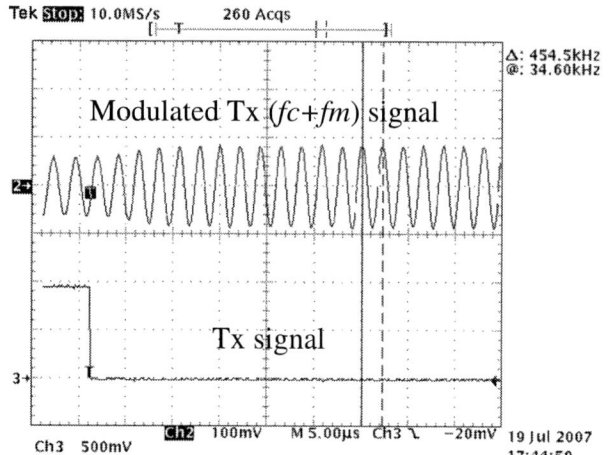

Fig. 10(b) Transmitter Tx and Modulated Tx (fc+fm) signals

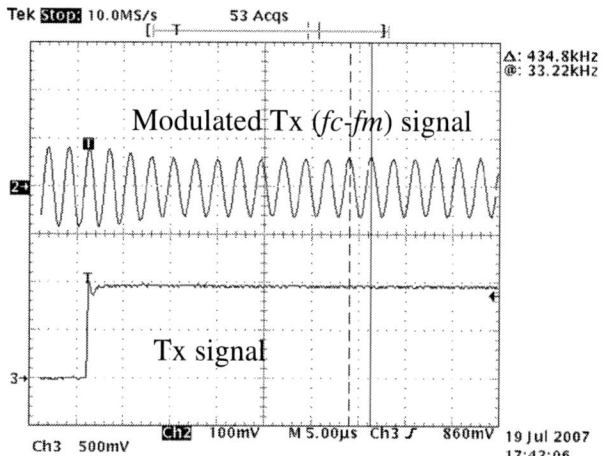

Fig. 10(c) Transmitter Tx and Modulated Tx (fc-fm) signals

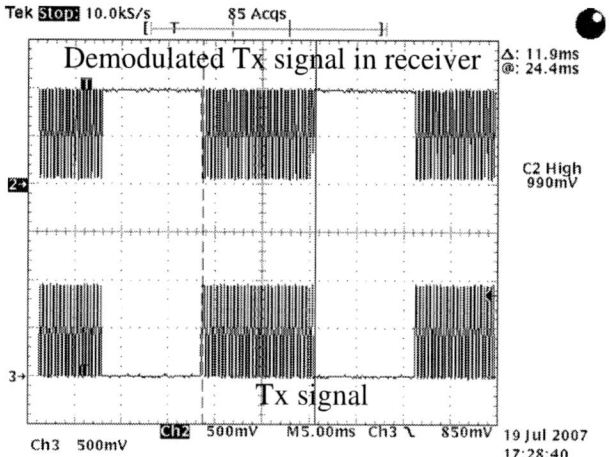

Fig. 10(d) Transmitter Tx signal and the demodulated Tx signal in the receiver

Fig. 10(a) Transmitter Tx and modulated Tx signals

978-1-4244-4782-4/10 $26.00 © 2010 IEEE

B. Combined use of FSK and ASK

In this test, communication from the transmitter to the receiver was done by using frequency-shift keying (FSK) and communication from the receiver to the transmitter was done by using amplitude-shift keying (ASK).

Fig.11 shows the block diagram of the system. In this system, it should be noted that by using FSK from transmitter to the receiver and ASK from receiver to transmitter, the receiver only requires a simple FSK demodulation circuit and a simple switch and a resistor for ASK communication. The advantage is that no extra coils in the receiver are needed. This makes the receiver design very simple and minimizes the cost involved. For mobile phone applications, extra cost and space requirements are very critical and not favorable.

For communication from the transmitter to the receiver, a FSK modulating signal is generated from the transmitter MCU. A FSK demodulation circuit is used to demodulate the signal from the same receiver coil which is used to transfer power as

well.. FSK results of from the transmitter to the receiver have been recorded previously in Figs.10 (a)-(d), and will not be repeated here.

For communication from the receiver to the transmitter, an ASK modulating signal is provided by the receiver MCU together with a simple transistor switch S1 and the resistor. The same demodulating circuit described in section II-B is used to demodulate the signal in the transmitter.

ASK results of from the receiver to the transmitter are shown in Fig.12. An Rx signal (green signal in Fig.12) from the receiver MCU controls the switching of S1 in Fig.11. The signal is sensed by R_S in Transmitter and amplified by amplifier C1. The amplified signal is then filtered by a low pass filter F1 and a high pass filter F2. The measured waveforms (V_{FL}, V_{FH} Pink and Purple signal) are shown in Fig.12. V_{FL} and V_{FH} are then compared by comparator C2 and the final demodulated signal D2 can be obtained. The details of these waveforms are expanded in Fig.12(b).

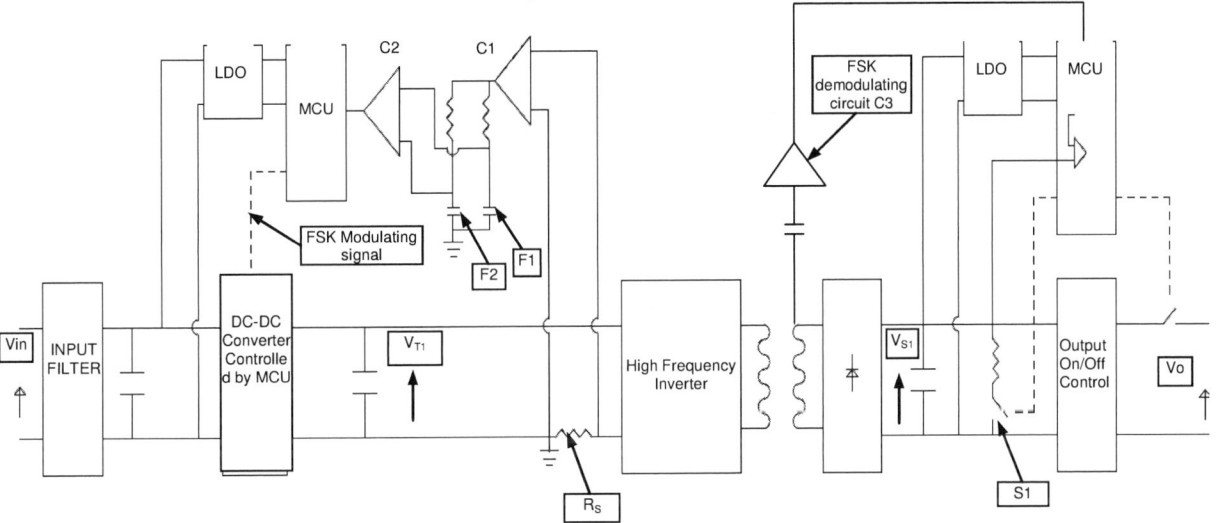

Fig.11 Block diagram of the test – FSK from transmitter to receiver and ASK from receiver to transmitter.

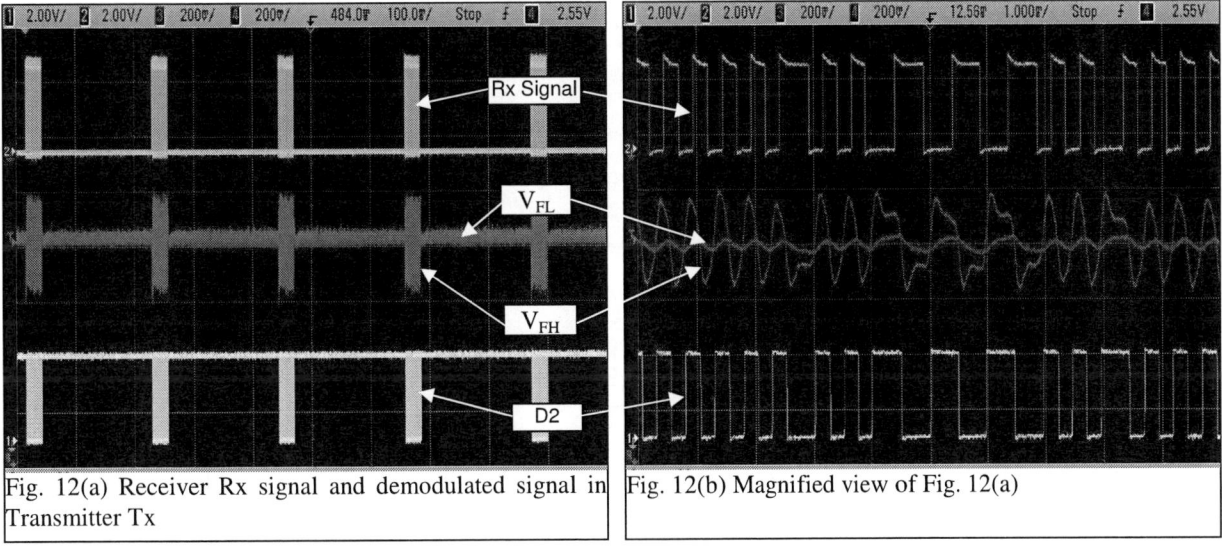

Fig. 12(a) Receiver Rx signal and demodulated signal in Transmitter Tx

Fig. 12(b) Magnified view of Fig. 12(a)

IV. CONCLUSIONS

This paper presents simple and yet effective bidirectional communication techniques suitable for wireless battery charging pads and portable consumer products. Based on the FSK and ASK techniques, bidirectional signal transfer between the primary charging system and the secondary load has been achieved successfully. The use of FSK technique and also a combination of FSK and ASK techniques for such application have been demonstrated. These techniques provide useful tools for load identification, compatibility check and load condition monitoring for wireless charging systems.

ACKNOWLEDGMENT

The authors are grateful to the Hong Kong Research Grant Council for its support for project CityU 114708. The support from Convenientpower HK Ltd. is also acknowledged.

REFERENCES

[1] H. Ayano, K. Yamamoto, N. Hio, and I. Yamato, "Highly Efficient Contactless Electrical Energy Transmission System", IEEE IECON02.

[2] G. Scheible, J. Schutz, and C. Apneseth, "Novel Wireless Power Supply System for Wireless Communication Devices in Industrial Automation Systems", IEEE IECON02.

[3] 3.C.S. Wang, O.H. Stielau, and G.A. Covic, "Design and Considerations for a Contactless Electric Vehicle Battery Charger", IEEE Industrial Electronics 2005.

[4] Y. Jang, Milan .M. Jovanovic "A Contactless Electrical Energy Transmission System for Portable-Telephone Battery Chargers", IEEE Industrial Electronics 2003

[5] S.C. Tang, S.Y. Ron Hui, and Henry Shu-Hung Chung, "Coreless Plannar PCB Transformers – A Fundamental Concept for Signal and Energy Transfer", IEEE Power Electronics 2000.

[6] X. Liu, and S.R. Ron Hui, "Equivalent Circuit Modeling of a Multilayer Planar Winding Array Structure for Use in an Universal Contactless Battery Charging Platform", IEEE Power Electronics 2007.

[7] X. Liu, and S.R. Ron Hui, "Simulation Study and Experimental Verification of a Universal Contactless Battery Charging Platform with Localized Charging Features", IEEE Power Electronics 2007.

[8] X. Liu, and S.Y. Ron Hui, "Optimal Design of a Hybrid Winding Structure for Planar Contactless Battery Charging Platform", IEEE Power Electronics 2008.

[9] A. Esser, "Contactless Charging and Communication System for Electric Vehicles", IEEE Industry Applications 1993.

[10] A. Kawamura, K. Ishioka, and J. Hirai, "Wireless Transmission of Power and Information through one High-Frequency Resonant AC Link Inverter for Robot Manipulator Applications", IEEE Industry Applications 1996.

[11] T. Bieler, M. Perrottet, V. Nguyen, and Y. Perriard, "Contactless power and Information Transmission", IEEE Industry Applications 2002.

[12] U. K. madawala, J. Stichbury, and S. Walker, "Contactless Power Transfer with Two-way Communication", IECON04.

[13] International Wireless Power Consortium Website: http://wirelesspowerconsortium.com/

PROPOSAL OF A DC-DC CONVERTER WITH WIDE CONVERSION RANGE USED IN PHOTOVOLTAIC SYSTEMS AND UTILITY POWER GRID FOR THE UNIVERSAL VOLTAGE RANGE

Jonas Reginaldo de Britto, Fábio Vincenzi Romualdo da Silva, Ernane Antônio Alves Coelho, Luis Carlos de Freitas, Valdeir José Farias and João Batista Vieira Júnior

Universidade Federal de Uberlândia – UFU/Faculdade de Engenharia Elétrica – FEELT
Av. João Naves de Ávila, 2121 - Campus Santa Mônica
CEP 38400-902, Uberlândia, MG - Brasil
e-mail: jonasdebritto@gmail.com, fabiovince@yahoo.com.br, ernane@ufu.br, freitas@ufu.br, valdeir@ufu.br, batista@ufu.br

Abstract – This paper describes the design and implementation of a Boost connected in cascade with a Quadratic Buck converter, with just one active switch, used to drive a LED string for public and private lighting purposes. The proposed converter supports a wide input voltage range that makes it possible to be supplied from 12 V_{DC} to 340 V_{DC} (240 V_{AC}), without any additional adjustment. It can be used in hybrid electrical systems supplied by photovoltaic systems or 110/220 V_{AC} from the utility power grid. Experimental results from an 11 W prototype are also presented.

Keywords – DC-DC converter, LED string, photovoltaic systems and wide voltage conversion range.

I. INTRODUCTION

The interest and importance in renewable energy has been aroused due to the Kyoto agreement on the global reduction of greenhouse emissions. Photovoltaic (PV) is one of the important renewable energy sources [1-3], the cost of the photovoltaic is on falling trend and it is expected to fall further as demand and production increases.

Photovoltaic systems (PVS) are ideally suited for distributed resource applications. Otherwise, it can be connected to the utility power grid and the totality or just the excess of the generated energy can be donated or sold to the electrical power grid industry [4-7].

PVS combined with lighting can be used to avoid buying electrical energy or it can be a solution for places that has not a utility power grid. Although, PV systems can only produce electricity during the day, but they can be matched with energy storage devices, such as batteries, to make a system that can supply the load continuously or just during the time of high peak demand of the grid. This procedure allows a reduction in the size of the PV panel and of the battery bank and, consequently, a decrease of the system cost, which is an advantage for urban areas use [8].

Incandescent lamp and PVS are not a good combination because this type of lamp has low efficiency. Incandescent lamps are being replaced by fluorescent lamps, compact lamps, low and high pressure sodium lamps or metal halide

lamps due to the higher efficiency they provide. Nowadays, with the advance of power LEDs, used in private and public lighting, there have been many efforts on the part of researchers to develop electronic circuits to drive matrix or LED strings. So, the use of PVS to supply power LEDs are a promising combination for near future.

According to [9], white power LEDs present the following advantages:

• extremely long life of 100,000 hours;

• no need for an external reflector because it is enclosed in the lamp casing to a predetermined beam width;

• extreme robustness because there are no glass components or filaments. So, they are virtually insensitive to vibration and movement;

• they are not filled with toxics gases, do not have any explosion, broken or contamination possibility.

For efficiency reasons, LED strings cannot be supplied via series resistors, so they are driven by switched mode converters with current control. For this reason, standard non-isolated dc-dc converters topologies (Buck, Boost, Buck-Boost) are been adapted and used to feed a constant current into a LED load. Previous works had presented proposals of circuits for drive string of LEDs with current regulation and pulse modulation based dimming techniques [10-13].

However, the referred dc-dc converters are not suited for the proposed application because it must step-up when supplied by the PVS and produce a large step-down when supplied by the utility. When a dc-to-dc conversion application requires a large conversion range, conventional PWM converter topologies must operate at extremely low duty ratios, which limit the operation to lower switching frequencies because of the minimum ON-Time of the switch. A scheme that provides wider conversion ratios is the cascade connection of converters. This scheme is a multistage approach that consists of two or more converters connected in cascade [14-18].

Therefore, this work proposes a Boost connected in cascade with two Buck converters (Boost-Buck2), with just one active switch, used to drive a LED lamp. The proposed converter presents a wide voltage conversion range which make it possible to be supplied from a 12 V photovoltaic system or from a 110/220 V utility power grid. Figure 1

Identify applicable sponsor/s here. *(sponsors)*

978-1-4244-4782-4/10 $26.00 © 2010 IEEE

depicts the LED Lamp used in a hybrid system and Figure 2 shows the schematics of the proposed LED Lamp.

Finally, equations, theoretical wave forms of the operation stages, additional experimental results and more discussions are also presented.

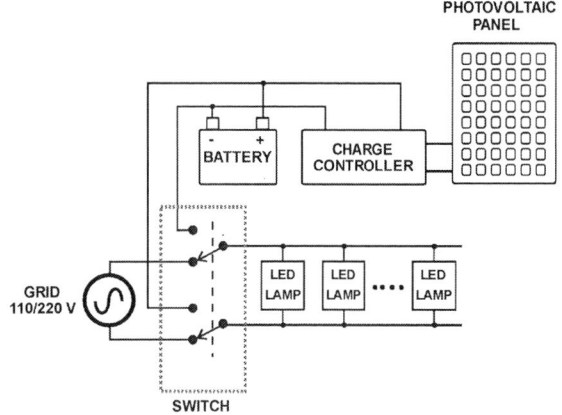

Figure 1. LED lamp used in a hybrid electrical energy system.

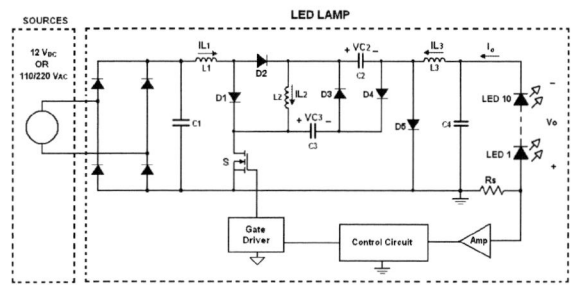

Figure 2. Schematic of the LED Lamp composed by a Boost-Buck2 converter, control circuit and LED string.

II. PRINCIPLE OF OPERATION

Figure 3 depicts the schematic diagram of the proposed Boost-Buck2 that can be used in applications that require voltage step-up or large voltage step-down. In addition, it presents a current source input and output characteristic.

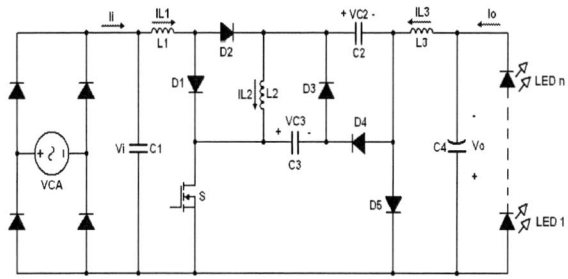

Figure 3. Schematic of the Boost-Buck2 converter.

In the following discussion, for simplicity, we assume that voltage ripples in the capacitor and current ripples in the inductors are entirely negligible.

The boost-buck2 converter has two stages of operation on CCM as described below:

First Stage: According to Figure 4, this occurs when the switch S is turned-on, thus the diodes D2, D3 and D5 are turned-off while diodes D1 and D4 are turned-on. The inductor L1 stores energy supplied by input voltage Vi while inductors L2 and L3 store energy supplied by capacitors C2 and C3, respectively, and the LED lamp is supplied by capacitor C4.

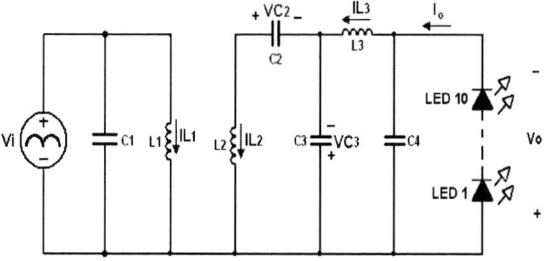

Figure 4. First operation stage.

Second Stage: According to Figure 5, this occurs when the switch S is turned-off, thus the diodes D1 and D4 are, also, turned-off while diodes D2, D3 and D5 are turned-on. The capacitors C2 and C3 store energy supplied by inductors L1 and L2, respectively, while the LED lamp is supplied by inductor L3 and capacitor C4.

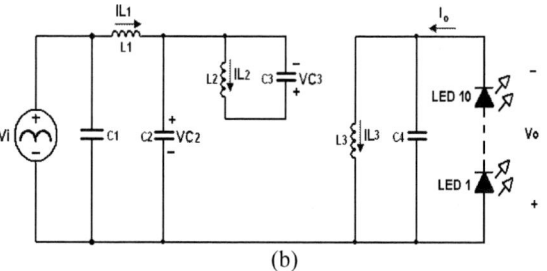

(b)

Figure 5. Second operation stage.

The current through inductors L1 and L3 are nonpulsating which provides, to this topology, an input and output current source nature. The output current source attribute is essential for LEDs driving because the luminous flux of an LED is proportional to the forward current (ILED) and it is controlled by regulating the dc current through it. Besides, there is a direct relationship between the ILED, the LED junction temperature, the permanent loss of light and the reduction of the lifetime. Therefore it is very important to keep the ILED within the nominal value.

Figure 6 presents the theoretical waveforms of the voltage and current through switch S and Figure 7 depicts the theoretical waveforms of the voltage and current through inductors L1, L2 and L3.

978-1-4244-4782-4/10 $26.00 © 2010 IEEE

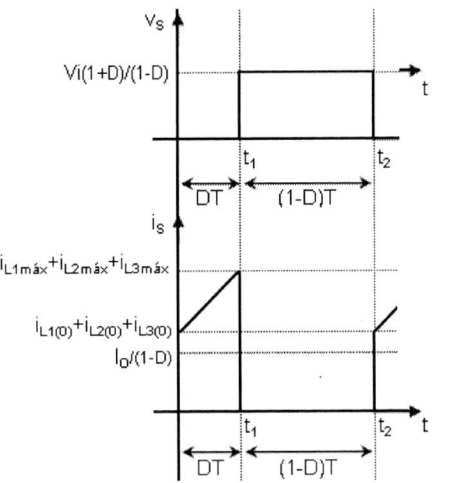

Figure 6. Theoretical waveforms of the voltage and current through switch S.

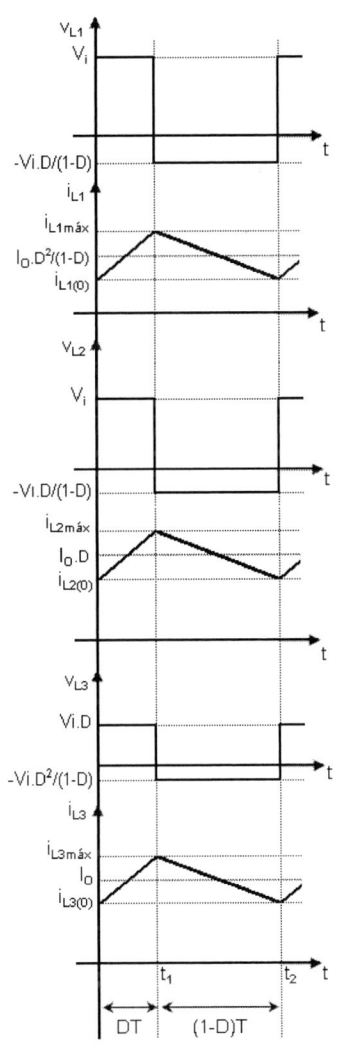

Figure 7. Theoretical waveforms of the voltage and current through inductors L1, L2 and L3.

III. EQUATIONS

Static gain (G) is derived assuming that the converter operate in the continuous conduction mode (CCM). All capacitor voltages and inductor currents are dc quantities with a relatively small superimposed ac ripple. Turn-on and turn-off transitions of all diodes are synchronous with switching transitions of the switch S.

Assuming average voltage on L1 inductor, in steady state, equal to zero, we have

$$V_i DT = \left(V_{C2} - V_i\right)\left(1-D\right)T \qquad (1)$$

Where D is the duty ratio and T is switching period.

Solving (1) for V_{C2}, (2) is obtained.

$$V_{C2} = \frac{V_i}{\left(1-D\right)} \qquad (2)$$

Applying the analysis prior to the inductors L2 and L3.

$$V_{C3} = DV_{C2} \qquad (3)$$

$$V_O = V_{C4} = DV_{C3} \qquad (4)$$

From (2) and (3), the output voltage is

$$V_O = V_{C4} = \frac{D^2 V_i}{\left(1-D\right)} \qquad (5)$$

Thus, the static gain of the Boost-Buck^2 converter is given by (6).

$$G = \frac{V_O}{V_i} = \frac{D^2}{\left(1-D\right)} \qquad (6)$$

Using the principle that the average current through capacitor C4 is zero, for steady state operation, we have

$$I_{C4}DT = I_{C4}\left(1-D\right)T \qquad (7)$$

From the analysis of the operation stages, (7) can be written as

$$\left(I_o - I_{L3}\right)DT = \left(I_{L3} - I_o\right)\left(1-D\right)T \qquad (8)$$

Solving (8) for I_{L3}, we have

$$I_{L3} = I_o \qquad (9)$$

978-1-4244-4782-4/10 $26.00 © 2010 IEEE

Applying the analysis prior to the capacitor C3, we have

$$\left(I_{L3} - I_{L2}\right)DT = I_{L2}\left(1-D\right)T \qquad (10)$$

Solving (10) for I_{L2}, we have

$$I_{L2} = DI_{L3} \qquad (11)$$

From (9), I_{L2} is given by

$$I_{L2} = DI_o \qquad (12)$$

Applying the analysis prior to the capacitor C2, we have

$$I_{L2}DT = I_{L1}\left(1-D\right)T \qquad (13)$$

Solving (13) for I_{L1}, we have

$$I_{L1} = \frac{I_{L2}D}{\left(1-D\right)} \qquad (14)$$

From (12), I_{L1} is given by

$$I_{L1} = \frac{I_o D^2}{\left(1-D\right)} \qquad (15)$$

Finally, the static gain against duty cycle plot is depicted in Figure 8. It shows that if the duty cycle is bigger than 0.6 the converter step-up the voltage, which means that the Boost characteristics overwhelms. By the other hand, if the duty cycle is less than 0.6, the proposed topology step-down the voltage, in a quadratic way, which means it is operating as a quadratic buck converter.

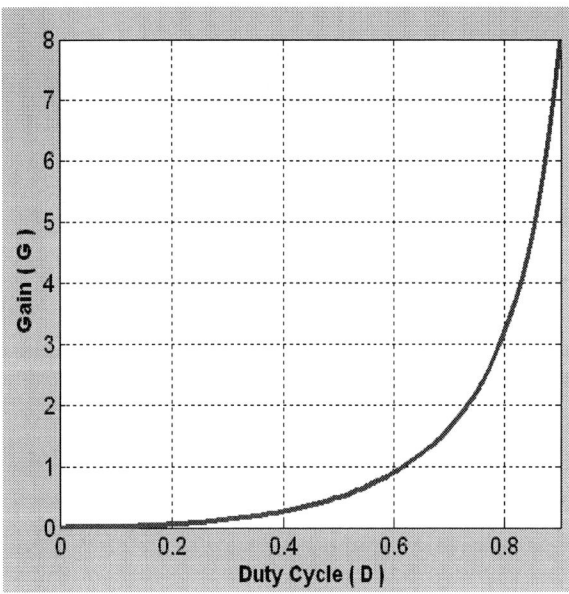

Figure 8. Static gain against duty cycle plot.

IV. EXPERIMENTAL RESULTS

In order to prove the performance of the Boost-Buck2 converter, a prototype was built with the parameters described in the Tables I and II.

TABLE I
PARAMETERS OF THE LED LAMP

Parameters	Values
LEDs	10 XLamp® XR-E
DC Forward Current	350mA
Forward Voltage (@350mA)	31V
Viewing Angle	90º
Luminous Flux	1000 lm

TABLE II
PARAMETERS OF THE BOOST-BUCK2 CONVERTER

Parameters	Values
Switch	SPP17N80C3
Switching frequency	100 kHz
Inductor L1	28 mH
Inductor L2	23 mH
Inductor L3	7 mH
Capacitor C1	94 uF
Capacitor C2	100 nF
Capacitor C3	330 nF
Capacitor C4	100 nF
Diodes D1, D2 , D3, D4 and D5	HFA08TB60

Figure 9 shows the waveform of the voltage between drain and source (VDS) on switch S and the wave form of the current in inductors L1 (IL1) and L3 (I$_O$), when the converter is supplied by a 240 V$_{AC}$ (utility power grid).

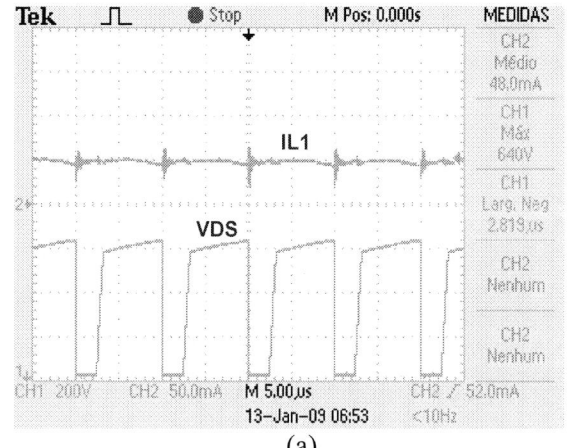

(a)

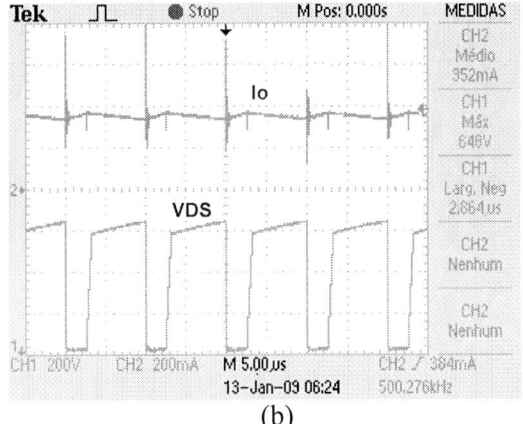

(b)

Figure 9. Waveforms for Vi=240 V_{AC} : (a) Voltage VDS on Switch S - 200 V/div and Current IL1 in the inductor L1 - 50 mA/div (b) Voltage VDS on Switch S - 200 V/div and   Current I_O in the LED string - 200 mA/div.

Inductors oppose instantaneous changes in current through them, so the spikes in the inductor L3 (I_O) current, depicted in Figure 9 (b), are not real. They are caused by interference between the two oscilloscope's channels when Vi=240 $V_{AC.}$

Figure 10 depicts the waveform VDS on switch S and the current in inductors L1 (IL1) and L3 (I_O), when the converter is supplied by a 12 V_{DC} (photovoltaic system).

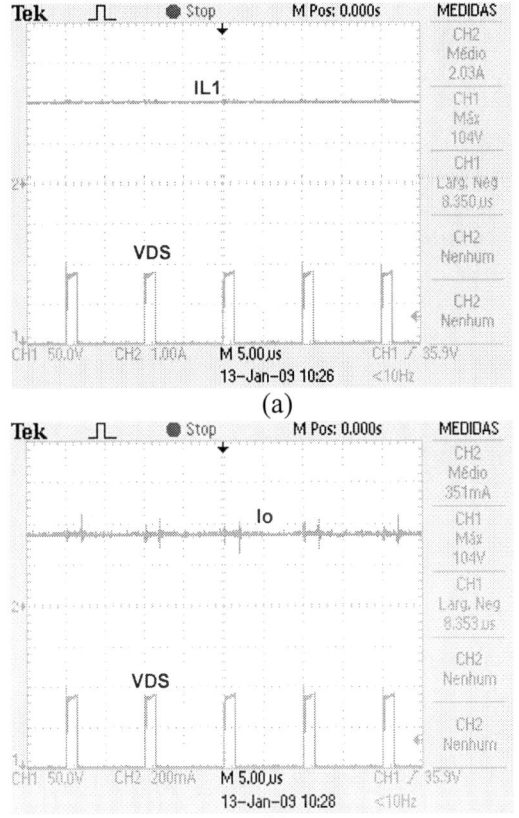

(a)

(b)

Figure 10. Waveforms for Vi = 12 V_{DC} : (a) Voltage VDS on Switch S - 50 V/div and Current IL1 in inductor L1 – 1 A/div (b) Voltage VDS on Switch S - 50 V/div and Current I_O in the LED string - 200 mA/div.

Finally, Figure 11 shows the efficiency against AC input voltage plot. The presented result suggests the use of 24 or 48V PVS in order to get a higher efficiency form the proposed converter.

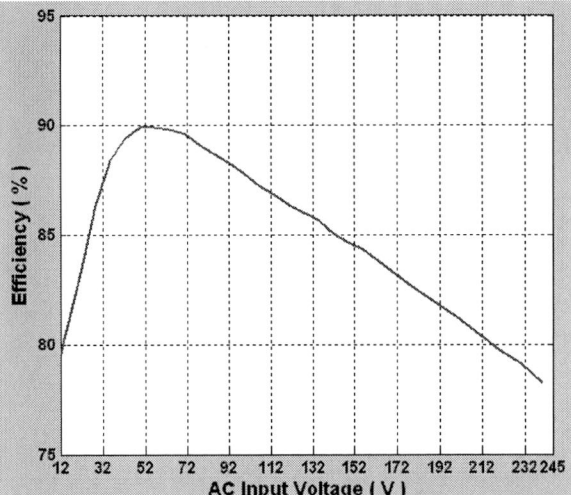

Figure 11. Efficiency against AC input voltage plot.

V. CONCLUSION

This paper presented the implementation of a Boost-Buck Quadratic converter with large input voltage range that can be supplied by a 12 V photovoltaic system or by the 110/220 V electrical grid, ideal to drive a LED lamp used for private or public lighting purposes. In addition, the converter presents just one switch and only two operation stages.

When the converter is supplied by a photovoltaic system, the utility power grid user saves money. The energy is bought only when the photovoltaic system is under maintenance or during clouded days. Besides, the use of renewable energy sources avoids the greenhouse effect and contributes to the carbon credit generation.

Experimental results from an 11 W prototype demonstrate that the converter imposes the desired current in the LED lamp for the maximum and minimum input voltage operation which validates the proposal.

ACKNOWLEDGEMENTS

The authors would like to tank CNPq, CAPES and FAPEMIG for the financial support on this project and many others.

REFERENCES

[1] B. Kroposki and R. DeBlasio, "Technologies for the new millennium: photovoltaics as a distributed resource", *in Proc. IEEE Power Engineering Society Summer Meeting*, vol. 3, pp. 1798-1801, July 2000.

[2] N. L. Castaner and S. Silvestre, "Modeling Photovoltaic System", *John Wiley & Sons Ltd*, 2002.

[3] M. P. Choi and A. Tan, "Photovoltaics Demonstration Projects", *Proc. of EMPD 98*, Vol. 2, 1998, pp.637-643.

[4] D. C. Martins, R. Demonti, "Aproveitamento Viável de Módulos Fotovoltáicos Através do Envio da Energia à Rede Comercial Utilizando Conversores Estáticos de Energia", *Revista Eletrônica de Potência*, vol. 1, no. 10, pp. 59-66, Junho 2005.

[5] M. C. Cavalcanti, G. M. S. Azevedo, B. A. Amaral, F. A. S. Neves, D. C. Moreira, K. C. de Oliveira, "A Grid Connected Photovoltaic Generation System with Compensation of Current Harmonics and Voltage Sags", *Revista Eletrônica de Potência*, vol. 11, no. 2, pp. 93-101, Julho 2006.

[6] K. C. A. de Souza, R. F. Coelho, D. C. Martins, "Proposta de um Sistema Fotovoltaico de Dois Estágios Conectado à Rede Elétrica Comercial", *Revista Eletrônica de Potência*, vol. 12, no. 2, pp. 129-136, Julho 2007.

[7] M. G. Molina, D. H. Pontoriero, P. E. Mercado, "An Efficient Maximum-Power-Point-Tracking Controller for Grid-Connected Photovoltaic Energy Conversion System", *Revista Eletrônica de Potência*, vol. 12, no. 2, pp. 147-154, Julho 2007.

[8] E. M Sá Jr., S. Daher, F. L. M. Antunes, C. M. T. Cruz, K. M. Silva, A. R. Figueira, "Photovoltaic System for Supplying Public Lighting as Peak Demand Shaving", *Revista Eletrônica de Potência*, vol. 12, pp. 113-120, Julho 2007.

[9] B. Cook, "New Developments and Future Trends in High-Efficiency Lighting", *Engineering Science and Education Journal*, pp. 207-217, October 2000.

[10] F. Bernitz, O. Schallmoser, Wolfram Sowa, "Advanced Electronic Driver for Power LEDs with Integrated Colour Management", *Industry Applications Conference 41st IAS Annual Meeting. IEEE*, vol. 5, pp. 2604 – 2607, October 2006.

[11] M. Rico-Secades, A.J. Calleja, J. Cardesín, J. Ribas, E.L. Corominas, J.M. Alonso, J. García, "Driver for High Efficiency LED Based on Flyback Stage With Current Mode Control for Emergency Lighting System", *Industry Applications Conference 39th IAS Annual Meeting*, vol. 3, pp.1655-1659, October 2004.

[12] A. J. Calleja, M. Rico-Secades, J. Cardesín, J. Ribas, E.L. Corominas, J.M. Alonso, J. García, "Evaluation of a High Efficiency Boost Stage to Supply a permanent LED Emergency Lighting system", *Industry Applications Conference, 39th IAS Annual Meeting*, vol. 2, pp.1390-1395, Oct 2004.

[13] G. Sauerlander, D. Hente, H. Radermacher, E. Waffenschmidt, J. Jacobs, "Driver Electronics for LEDs", *Industry Applications Conference 41st IAS Annual Meeting. Conference Record of the 2006 IEEE*, vol. 5, 8-12, PP. 2621 – 2626, October 2006.

[14] J. A. Morales-Saldaña., E. E. Carbajal-Gutierrez, J. Leyva-Ramos: "Modeling of switch-mode DC-DC cascade converters", *IEEE Trans. Aerosp. Electron. Syst.*, 38(1), pp. 295–299, 2002.

[15] H. Matsuo, K. Harada: "The cascade connection of switching regulators", IEEE Trans. Ind. Appl., 12(2), pp. 192–198, 1976.

[16] D. Maksimovic, S. Cuk: "Switching converters with wide dc conversion range", *IEEE Transaction on Power Electronics*, 6(1), pp. 151–157, 1991.

[17] V. M. Pacheco, A. J. Do Nascimento, V. J. Farias, J. B. Vieira, L. C. Freitas: "A quadratic buck converter with lossless commutation", *IEEE Trans. Ind. Electron.*, 47(2), pp. 264–272, 2000.

[18] E. E. Carbajal-Gutierrez., J. A. Morales-Saldaña, J. Leyva-Ramos: "Modeling of a single-switch quadratic buck converter", *IEEE Trans. Aerosp. Electron. Syst.*, 41(4), pp. 1451–1457, 2005.

[19] S. Bowling, "A Digital Constant Power LED Driver", *Application Notes AN1138*, Microchip Technology Inc, 2007.

Characterization of a 5 kW Solid Oxide Fuel Cell Stack Using Power Electronic Excitation

John J. Cooley *Student Member, IEEE*, Eric Seger, Steven Leeb *Fellow, IEEE*,
and Steven R. Shaw *Senior Member, IEEE*

Abstract—**Fuel cells have attracted great interest as a means of clean, efficient conversion of chemical to electrical energy. This paper demonstrates the identification of both non-parametric and lumped circuit models of our stack in response to a test signal introduced by control of a power electronic circuit. This technique could be implemented on-line for continuous condition assessment of the stack, as it delivers power. The results show typical data from the stack, comparison of model and measured data, and whole-stack impedance spectroscopy results using a power electronic system to provide excitation. Run-time excitation currents for the spectroscopy measurement are generated by a hybrid power system controlling the flow of power from the fuel cell and a secondary power source to a fixed resistive load. The hybrid power system generates small-signal currents at the fuel cell terminals while the load current itself is largely unaffected by the impedance spectroscopy measurement.**

Index Terms—**Fuel Cells, Impedance Spectroscopy, Prognostics, Power Electronics**

I. INTRODUCTION

There is an increasing realization that the commercial viability of fuel cells depends on work to enhance reliability and durability [1], [2]. Much of the effort to enhance fuel cell robustness is appropriately focused on materials development using traditional materials science methodologies, e.g. single cell or even single component testing in controlled environments thought to be similar to the conditions inside a stack. However, there is also interest in understanding degradation phenomena that can occur as fuel cells are integrated into real systems. As an example, in [3], Ramschak et al provide a method to detect the failure of a single cell within a stack by analyzing the harmonic distortion on the stack voltage. Similarly, in [4] Gemmen et al study the impact of inverter load dynamics on a fuel cell, with the conclusion that stack / inverter interaction is significant in the operating conditions and long term behavior of the stack.

In our SOFC stack, and in many similar fuel cell applications, it is neither feasible nor desirable to remove the stack from service for the purpose of connecting impedance spectroscopy instrumentation. However, in principle, it is not necessary to remove the load provided that a sufficiently rich test signal can be introduced in addition to the load, as in [5]. This paper demonstrates the use of power electronics to impose a test signal while delivering power to a load. This characterization consists of calculations of whole stack impedance spectroscopy and time-domain model parameters, using both the switching waveform, or "ripple", of the power electronics connected to the stack and an exogenous excitation. This method requires only instrumentation at the stack electrical terminals, and could be integrated with the controls of existing power electronics to

provide non-invasive, low cost stack prognostics. The underlying motivation of this work, not directly addressed in this paper, is that we may ultimately be able to improve reliability and mitigate materials challenges through controls at the electrical terminals that are richly informed of the state of the stack.

The paper begins with an overview of electrochemical impedance spectroscopy (EIS) and associated system identification considerations in section II-B. In section II-C we suggest a lumped parameter, time-domain model and identification procedure for the small signal response of the stack. In section III, we discuss the design considerations and circuit modeling of the hybrid power system used to generate the signals for impedance spectroscopy. The experimental setup is described in section IV, and results are provided in section V.

II. FUELCELL OPERATION AND MODELING

A. Fuel Cell Overview

Figure 1 is a conceptual illustration of the energy conversion mechanism in a solid oxide fuel cell. The cell comprises three layers. The cathode (right) is a porous, electrically conductive material. Molecular oxygen is reduced to oxygen ions in the cathode, with electrons supplied by the external circuit. These oxygen ions move readily from the cathode through a dense electrolyte, which is ion-conducting but is an electronic insulator. At appropriate temperatures, typically in the vicinity of 750 C, the electrolyte becomes conductive to oxygen by means of oxygen vacancies in the lattice structure of the material. The anode layer is another porous, electrically conductive cermet material. Oxygen ions arriving from the electrolyte serve to oxidize fuel and release their electrons to the external circuit. Typical materials for the cathode/electrolyte/anode structure include lanthanum strontium maganate (LSM), ytria stabilized zirconia (YSZ), and nickel/YSZ cermet, respectively. While the overall reaction in Fig. 1 shows hydrogen as a fuel and water as a product, a basic advantage of SOFC technology is that the electrolyte is an oxygen ion conductor. This allows the use of fuels containing carbon, as opposed hydrogen-conducting fuel cell technologies.

Fig. 2 shows a photograph of the actual stack used for testing in this paper. The stack is a 5kW nominal, Fuel Cell Technologies / Siemens Alpha-8 tubular solid oxide fuel cell using city natural gas as a fuel. The vents at the top are for intake and exhaust, and this particular unit was also configured with a recuperator that could be used to heat water for a combined heat and power application. This unit is designed for three-phase grid-tie operation. However, for purposes of this study we were able to access and connect power electronics to the terminals of

978-1-4244-4782-4/10 $26.00 © 2010 IEEE

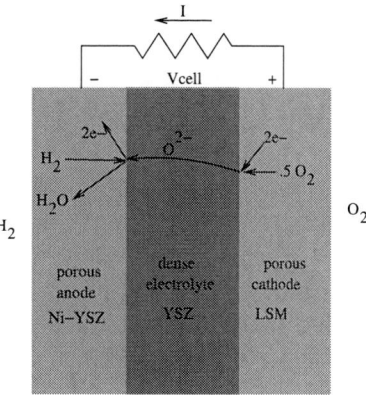

Fig. 1: Conceptual diagram of SOFC energy conversion.

the stack and monitor the response of the stack to test signals imposed by those power electronics.

Fig. 2: A 5 kW Siemens / Fuel Cell Technology stack used for testing.

B. Fuel Cell Impedance Spectroscopy

Electrochemical impedance spectroscopy models the AC electrical terminal response of a fuel cell (or other electrochemical system) in the vicinity of an operating point as a linear impedance $Z(j\omega)$. In particular, for cell voltage and current

$$v_c(t) = V_o + v(t) \tag{1}$$
$$i_c(t) = I_0 + i(t), \tag{2}$$

at a DC operating point V_0, I_0, the impedance captures the frequency domain relationship between the small signal quantities $v(t)$, $i(t)$. Use of this model presumes that the cell responds linearly over the range of excitation in the vicinity of the bias point, i.e. that excitation at a single frequency produces a response at that frequency.

Impedance spectroscopy results are generally presented using a Nyquist plot showing real and complex parts of the impedance

with frequency as an implicit argument. An electrochemist can recognize the shapes characteristic of processes in the Nyquist diagram [6]. Practitioners often extend this non-parametric analysis by fitting lumped-parameter circuit models, in the frequency domain, and in some cases associate physical processes with individual circuit elements. In [7], a parameterized impedance spectroscopy model is used to synthesize an equivalent circuit of an SOFC. Other examples include the analysis of a PEM cell in [8] and the application to an SOFC cell in [7]. Frequencies of 0.01Hz to 1MHz are generally used for studying SOFC systems [2]. For a survey of impedance spectroscopy in fuel cells, see [9].

Under sufficiently rich excitation, an estimate $\hat{Z}(j\omega)$ of the impedance response can be extracted from the terminal voltage and current of a cell. In particular, an impedance estimate is

$$\hat{Z}(j\omega) = \frac{\hat{V}_c(j\omega)}{\hat{I}_c(j\omega)}, \quad |\omega| > 0, \tag{3}$$

where $\hat{V}_c(j\omega)$ and $\hat{I}_c(j\omega)$ are estimates of the spectral content of the electrical terminal responses $v_c(t)$ and $i_c(t)$. The process of estimating spectral content of signals using sampled data and discrete-time Fourier transform techniques, including windowing and other considerations, is reviewed in [10] among others. The excitation $i_c(t)$ imposed at the electrical terminals must be broadly exciting, in the sense of having significant power at frequencies where it is desired to have a good estimate of $Z(j\omega)$. If $\hat{I}_c(j\omega)$ at some frequency is small or dominated by noise, the variance in $\hat{Z}(j\omega)$ can be large. In practice, we avoid this by not evaluating $Z(j\omega)$ for frequencies where the signal content in the $I_c(j\omega)$ is small in comparison to a threshold.

C. Parametric Modeling and Identification

In addition to impedance spectroscopy, it is sometimes useful to model fuel cell responses using a parameterized model, often in the form of a differential equation that represents specific physical processes. For example, Hall [11] develops a transient model of a tubular SOFC including electrochemical, thermal, and mass flow elements. Wang et al. [12] develop a dynamic model for a proton exchange membrane fuel cell using electrical circuit elements, and Pasricha et al. [13] provide a dynamic electrical terminal model of a proton exchange membrane fuel cell. A challenge in developing parametric, physically-baed models of fuel cells is to restrict the phenomena in the model to those which are well supported by the observations.

With preliminary, non-parametric observations in mind, we propose a very simple three-parameter model of the stack, i.e.

$$v(t) = V_{oc} - Ri(t) - Ls\, i(t), \tag{4}$$

where $v(t)$ is the stack voltage, $i(t)$ is the stack current, V_{oc} is the open circuit stack voltage, R is a resistance, L is a inductance, and s is the $\frac{d}{dt}$ operator.

The parameters of (4) are conveniantly estimated using the operator substitution technique in [10]. The low-pass filter operator

$$\lambda = \frac{1}{1 + s\tau}. \tag{5}$$

can be manipulated to isolate s, i.e.

$$s = \frac{1 - \lambda}{\lambda \tau}. \tag{6}$$

Substituting s into (4) and rearranging so λ appears in the numerator provides

$$\lambda \tau V_{oc} - \tau \lambda i(t) R + (\lambda - 1) i(t) L = \tau \lambda v(t). \tag{7}$$

This is appealing because $\lambda \tau$, $\lambda i(t)$, and $\lambda v(t)$ can be evaluated using a discrete-time implementation of λ applied to the data. These quantities can be arranged in a least-squares tableau to obtain estimates for the parameters V_{oc}, R, and L. Setting $\lambda \tau V_{oc}$ to the final value, we form the following equations

$$\begin{pmatrix} \tau & -\tau \lambda i[1] & (\lambda - 1) i[1] \\ \tau & -\tau \lambda i[2] & (\lambda - 1) i[2] \\ \vdots & \vdots & \vdots \\ \tau & -\tau \lambda i[n] & (\lambda - 1) i[n] \end{pmatrix} \begin{pmatrix} V_{oc} \\ R \\ L \end{pmatrix} = \begin{pmatrix} \tau \lambda v[1] \\ \tau \lambda v[2] \\ \vdots \\ \tau \lambda v[n] \end{pmatrix} \tag{8}$$

to estimate the parameters of (4).

III. POWER ELECTRONICS

We can demonstrate the concept of run-time electrochemical impedance spectroscopy (EIS) in a hybrid power system with off-the-shelf power converters. A simplified connection diagram for our EIS-capable hybrid power system is shown in Figure 3a. In our system, the control signal drives the trim pin of the Buck converter module in the fuel cell leg (the upper leg in Figure 3a).

A. Small-signal Behavior

Conceptually, the hybrid system enables run-time fuel cell diagnostics by providing a means for exciting the fuel cell with a small-signal current originating at the secondary source (the battery in this case), while the load current itself is largely unaffected by the EIS measurement. The small-signal current paths corresponding to this behavior are depicted in Figure 3b.

We can analyze the small-signal behavior of such a system starting from Middlebrook's linearized canonical models of CCM-operated power converters [14]. A parallel development could be carried out if the converters operate in DCM by using the corresponding models for DCM-operated converters [15].

1) Middlebrook's Linearized Models of Power Converters:
In reference [14], Middlebrook develops linearized circuit models that can be used to represent the input, output and control properties of many switching power converters.

To that end, Middlebrook demonstrates how CCM-operated converters can be manipulated into one fixed topology and DCM-operated converters into another fixed topology in references [14], [15], [18]. For example, the basic elements of a typical power converter are shown in Figure 4a. In Figure 4b, the buck converter has been replaced with the linearized canonical circuit model developed by Middlebrook in [14]. [1]

The canonical circuit model consists of three pieces (in boxes): an ideal transformer that represents the converter's ideal voltage and current transformation[2], an effective low-pass filter

[1]According typical conventions, the hats (^) denote small-signal quantities.
[2]the straight line and the wavy line drawn on the transformer element in Figure 4b are intended to indicate DC and AC respectively.

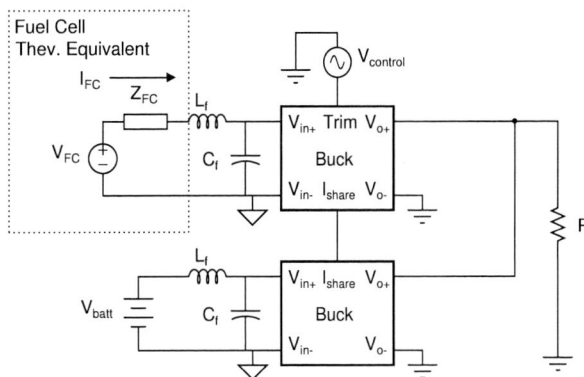

(a) A simplified connection diagram of the hybrid system built from off-the-shelf components.

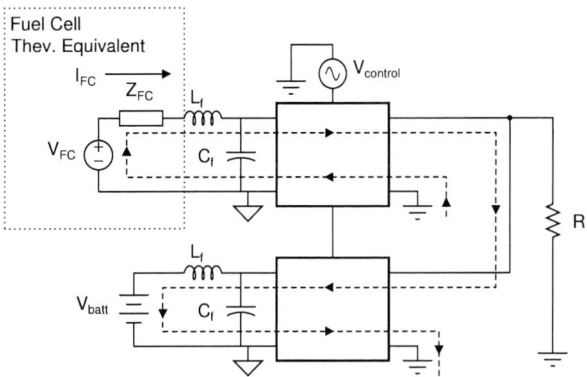

(b) Small-signal current paths for EIS excitation signals.

Fig. 3: A hybrid power system with EIS functionality built from off-the-shelf components.

at the output that includes the effects of the energy storage elements involved in the switching action of the converter, and dependent current and voltage sources that capture the effect of the control signal, $\hat{d}$. Reducing a converter to this "fixed topology" means that a linearized input-output and control description of any converter reduces to looking up the, perhaps frequency-dependent, values for each of the model parameters as in Table I [14], [17]. In [19], the author shows how the values in Table I for a *generalized load* can be taken from similar canonical model parameters that were previously calculated for a converter driving a *fixed load R*.

TABLE I: Canonical Model Parameters for the Buck, Boost and Buck-Boost with a general load [17], [19]

Converter	$M(D)$	L_e	$e(s)$	$j(s)$
Buck	D	L	$\frac{V}{D^2}$	I
Boost	$\frac{1}{D'}$	$\frac{L}{D'^2}$	$V\left(1 - \frac{sLI}{D'^2V}\right)$	$\frac{I}{D'^2}$
Buck-Boost	$-\frac{D}{D'}$	$\frac{L}{D'^2}$	$-\frac{V}{D^2}\left(1 - \frac{sDLI}{D'^2V}\right)$	$-\frac{I}{D'^2}$

978-1-4244-4782-4/10 $26.00 © 2010 IEEE

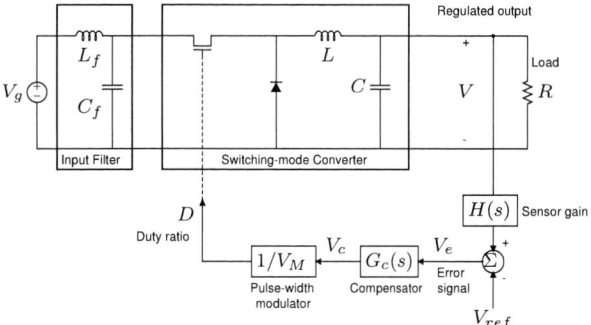

(a) Basic Elements of a switching-mode regulator. The LC input filter and buck converter are shown as typical realizations [16].

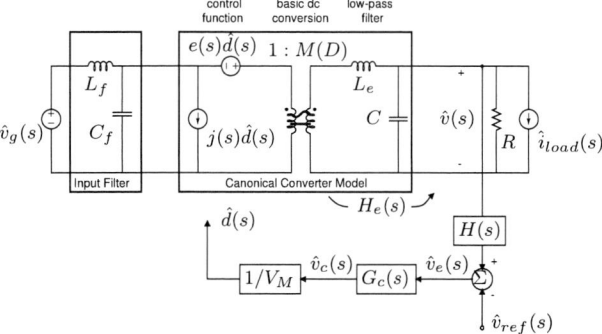

(b) A small-signal regulator model using Middlebrook's linearized canonical circuit model of the power converter [16]. Hats (ˆ) denote small-signal quantities.

Fig. 4: Canonical circuit modeling developed in references [14], [16] and [17].

2) A Linearized Model of the Hybrid Power System:

Having configured the system in Figure 3 so that its small-signal behavior is similar to that of two parallel converters under voltage-mode feedback control, we can build the corresponding linearized model of the hybrid power system shown in Figure 5.

In reference [19], the author uses a linear superposition and replacement of dependent sources approach to derive the closed-loop transfer functions describing the small-signal behavior of a hybrid power system like that in Figure 5. For instance, taking the input $\hat{v}_{ref1}$ as the control signal, assuming two identical converters, and neglecting the effects of the input filters on the system dynamics, the author shows that the transfer functions of interest here are:

$$\frac{\hat{v}}{\hat{v}_{ref1}} = \frac{1}{H} \frac{\lambda T}{1 + 2T\lambda} \tag{9}$$

$$\frac{\hat{i}_{o2}}{\hat{i}_{o1}} = -\frac{\lambda T}{HZ_e} \left(\frac{T(2\lambda - 1)}{1 + 2T\lambda} - 1 \right) \tag{10}$$

$$\frac{\hat{i}_{o1}}{\hat{v}_{ref1}} = \frac{T}{HZ_e} \left(\frac{\lambda T(2\lambda - 1)}{1 + 2T\lambda + 1 - \lambda} \right) \tag{11}$$

$$\frac{\hat{i}_{in1}}{\hat{v}_{ref1}} = j(s)F_m G_c(s) \left(1 - H \frac{\hat{v}}{\hat{v}_{ref1}} \right) + M(D) \frac{\hat{i}_{o1}}{\hat{v}_{ref1}}, \tag{12}$$

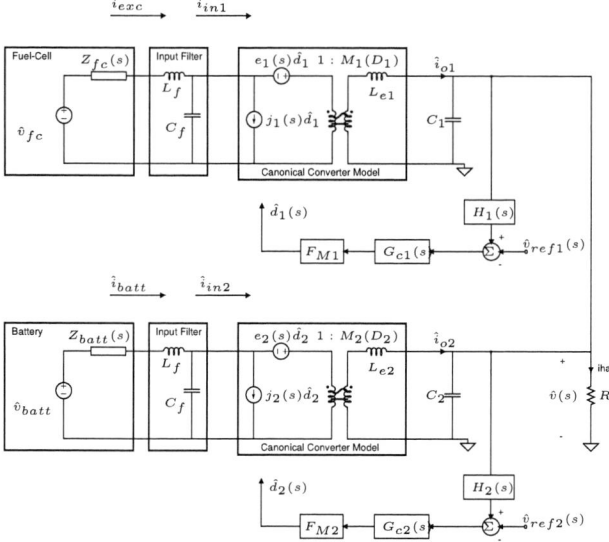

Fig. 5: A small-signal hybrid system modeled using the canonical circuit model of CCM-operated power converters.

where

$$Z_e = sL_e \tag{13}$$

$$\lambda' = Z_e + Z_e \| R \tag{14}$$

$$\lambda = \frac{Z_e \| R}{\lambda'} \tag{15}$$

and the loop transfer function is defined as

$$T = HG_c F_M e(s) M(D). \tag{16}$$

Such a model can be validated by comparing the calculated expressions in (9)-(12) to simulations (LTSPICE) of the system in Figure 5 as shown in Figures 7, 8, and 9. The magnitude and phase plots of $\hat{i}_{o2}/\hat{i}_{o1}$ in Figure 7, confirm our intuition that, at low frequency, the currents out of the two converters are equal and opposite (small-signal currents flow out of one and into the other). This behavior corresponds to the time-domain data shown in the scope shot of Figure 6, taken from the experimental system of Figure 3.

Figures 8 and 9 show that the transconductance from the control voltage, $\hat{v}_{ref1}$, to input current, $\hat{i}_{in1}$, is large and the corresponding load voltage perturbation, $\hat{v}$, is small. This amounts to the desired characteristic of an EIS-capable hybrid power system that the load voltage will be largely unaffected by the run-time EIS behavior.

B. Input Filters and System Stability

The design of the input filters in Figure 5 presents some interesting power electronics design challenges. The classic results concerning the effect of a "post-facto" input filter on converter performance are derived by Middlebrook from an application of the extra element theorem in [16], [17], [20]. The resulting design constraints are typically quoted as a set of impedance inequalities that, if met, ensure negligible degradation of converter performance. However, in our hybrid power system, designed to enable EIS of the fuel cell, we have a

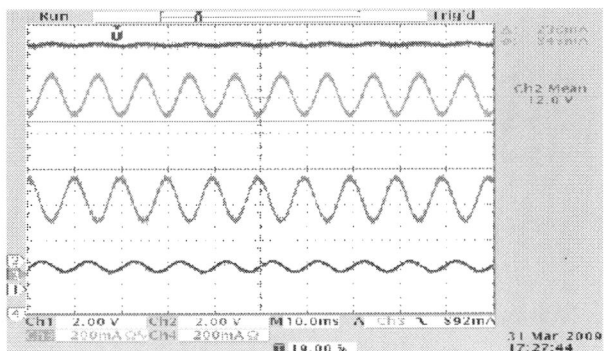

Fig. 6: An oscilloscope screen shot showing the battery and fuel cell currents during run-time EIS ($\approx$100 Hz). Top to bottom: load voltage (ch2), fuel cell current (ch3), battery current (ch4), control signal (ch1). The excitation current flows out of the battery terminals and into the fuel cell terminals while the load voltage is largely unaffected by the run-time EIS measurement.

more complicated set of design constraints that must be met. Specifically, the input filters must not only be designed for system stability but must also pass excitation currents from the converter inputs to the fuel cell terminals. Meanwhile, the analytical results presented by Middlebrook in [16], [20], must be extended so that they may be applied to the hybrid power system (paralleled converter) case.

1) Middlebrook's Application of the EET for Input Filter Design: The treatment of an input filter as a "post-facto" element in a power converter design is a likely outcome of natural design processes. However, this treatment is also *analytically* advantageous. The converter can be designed without the input filter and then the extra element theorem (EET) applied to determine the perturbation on the converter dynamics without ever analyzing the full system. The extra element theorem, best summarized by Middlebrook in [21], allows us to replace one cumbersome and uninsightful calculation, with a few simple and elegant calculations.

The extra element theorem follows from an application of the principle of "null double injection" to a linear circuit [21]. Upon addition of an extra element to the circuit, the transfer function of interest, completely defined by an input and output variable in the circuit, can be modified by using the calculated impedance seen at the "extra element port" under *two* special cases. The first special case corresponds to null-double injection and is the impedance seen at the extra element port when the transfer function input variable is directed in such a way that the transfer function output variable is nulled (equal to zero). The result is the "null-condition" impedance, $Z_{n-c}(s)$. The second special case corresponds to the open-loop behavior and is the impedance at the extra element port when the transfer function input signal is deactivated (set to zero), leading to $Z_{o-l}(s)$. Fundamentally, the extra element theorem uses the unique information obtained about the circuit by calculating those two special-case impedances to derive the circuit's interaction with the extra element itself. The primary result of the ensuing mathematical manipulations is a statement of the correction

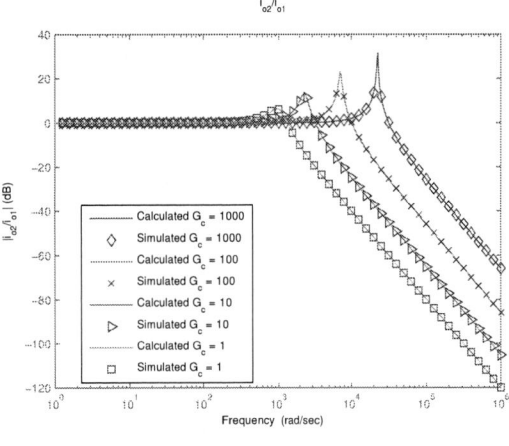

(a) Magnitude

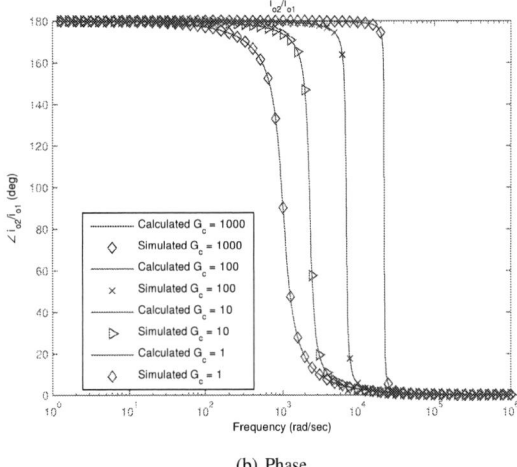

(b) Phase

Fig. 7: i_{o2}/i_{o1}.

factor that multiplies the original transfer function. For a series extra element (one that replaces a short-circuit in the original circuit), the correction factor is

$$CF = \left(\frac{1 + \frac{Z_o(s)}{Z_{n-c}(s)}}{1 + \frac{Z_o(s)}{Z_{o-l}(s)}} \right), \qquad (17)$$

in which $Z_{n-c}(s)$ is the special-case impedance calculated for the null condition, $Z_{o-l}(s)$ is the special-case impedance calculated for the open-loop condition and $Z_o(s)$ is the impedance of the extra element itself.

While the converter transfer function can be defined by any arbitrarily defined input variable and any corresponding output variable, some notable converter transfer functions are represented within the dashed box in Figure 10 adapted from [17]. The converter transfer function, $G_{vd}(s)$, is usually of particular interest because it is "in the loop," i.e. the dynamics of $G_{vd}(s)$ directly impact the stability of the regulator. Upon addition of an input filter, $G_{vd}(s)$ is modified by multiplication with the correction factor in (17), in which $Z_{n-c}(s)$ and $Z_{o-l}(s)$

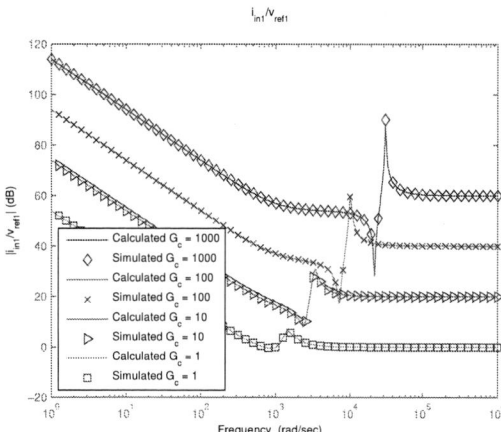

Fig. 8: $|i_{in1}/v_{ref1}|$.

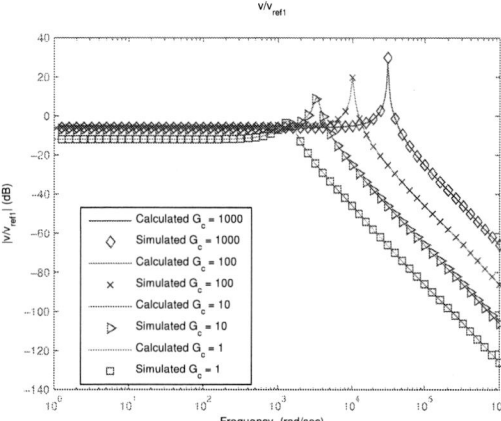

Fig. 9: $|v/i_{ref1}|$.

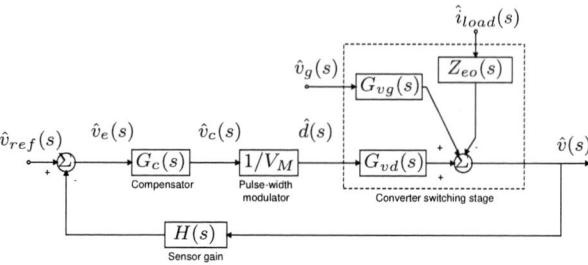

Fig. 10: A dynamical block diagram of a voltage-mode feedback regulated converter. [17]

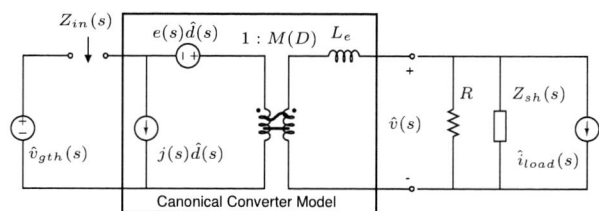

Fig. 11: The linearized converter model with three independent inputs, $\hat{v}_g$, $\hat{d}$, and $\hat{i}_{load}$. for calculating the special-case impedances in the extra element correction factors.

through L_e is zero, the voltage across it is also zero and the zero-valued (nulled) output voltage appears at the secondary winding of the ideal transformer. Therefore, the input voltage and current are simply $-e(s)\hat{d}(s)$ and $j(s)\hat{d}(s)$, respectively and the input impedance in this case is $Z_{n-c}(s) = -e(s)/j(s)$.

In [19], the author shows that the three converter transfer functions represented in the dashed box of Figure 10, can be corrected using the generalized results in Table II. Those results hold for CCM-operated converters, and the special-case impedances can be found by looking up the canonical model parameters in a table such as Table I.

can be calculated from the circuit in Figure 11.

The null-condition does not generally allow us to simplify the circuit topologically, or even to easily write down a closed-form expression of the control signal that leads to the nulled output signal.[3] But, the null-condition often allows us to make observations about the circuit that simplify the calculation, not of the control signal itself, but of the impedance at the extra element port as a result of the conditions that the control signal must impose on that circuit to null the output.

For example, to calculate $Z_{n-c}(s)$ for correcting $G_{vd}(s)$, the transfer function from $\hat{d}$ to $\hat{v}$, in the circuit of Figure 11, we deactivate the other independent inputs, $\hat{v}_g$ and $\hat{i}_{load}$, and null the output $\hat{v} \rightarrow 0$. The analysis is simplified by realizing that for a nulled output, the small-signal voltage across the load impedance is zero so no small-signal current flows through the load. Therefore, no current flows through L_e or through the secondary winding of the ideal transformer. The primary winding current is therefore also zero. Because the current

[3]Note that it would (generally) be a misinterpretation of the null-condition to simply short-circuit the output of the converter in Figure 11 and, in most cases, would lead to different and incorrect results.

TABLE II: Generalized Input Filter Design Constraints adapted from [19]

Special-case	Impedance	Generalized Value	Transfer Function
open-loop	$Z_D(s)$	$\dfrac{sL_e + R\|Z_{sh}(s)}{M(D)^2}$	All
null-condition	$Z_N(s)$	$\dfrac{-e(s)}{j(s)}$	$G_{vd}(s)$
	$Z_e(s)$	$\dfrac{sL_e}{M(D)^2}$	$Z_{eo}(s)$
	$Z_g(s)$	∞	$G_{vg}(s)$

The correction factor can be used to directly evaluate the degradation of converter transfer functions. However, it is immediately obvious from the expression of the correction factor in (17), that if the following inequalities are met, the input filter will have a negligible impact on the converter dynamics [16], [20]:

$$|Z_o| \ll |Z_{n-c}| \quad (18)$$

$$|Z_o| \ll |Z_{o-l}|. \quad (19)$$

978-1-4244-4782-4/10 $26.00 © 2010 IEEE

Meeting the first inequality will ensure that the filter output impedance is always less than the negative incremental resistance presented by the inputs of a regulated converter. For instance, from Tables I and II, $Z_{n-c}(s)$ for the Buck converter is $-V/ID^2$. The same result can be derived for a lossless ($P_{out} = P_{in}$), perfectly-regulated converter ($V_{out} = V = \text{const.}$) with a fixed load ($I_{out} = I = \text{const.}$) as follows:

$$
\begin{aligned}
Z_{n-c}(s) &= \frac{\partial V_{in}}{\partial I_{in}} = \frac{\partial}{\partial I_{in}}\left(\frac{P_{out}}{I_{in}}\right) \\
&= -\frac{P_{out}}{I_{in}^2} = -\frac{V}{ID^2}.
\end{aligned} \tag{20}
$$

A typical plot of the three impedances of interest in Figure 12 illustrates the design choices required to meet the inequalities in (18)-(19). In practice, meeting the inequality in (18) is often achieved for LC filter designs by using a damping leg (a series RC) shunting the input terminals to decrease the magnitude peaking in the LC filter output impedance. Meeting the second inequality (19) is usually achieved by setting the frequency of the 2nd-order peak in the input filter output impedance below that of the 2nd-order dip in the output filter input impedance (represented by $Z_{o-l}(s)$).

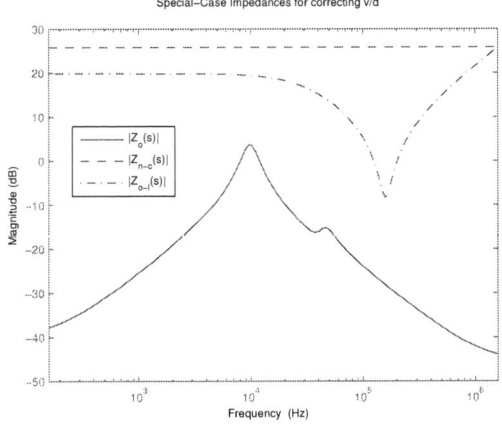

Fig. 12: A typical frequency plot of the special case impedances, $Z_{n-c}(s)$, $Z_{o-l}(s)$, and the input filter output impedance, $Z_o(s)$, for a single converter system.

While the concepts above were reviewed and developed for a single-converter system, the same concepts will be extended, in Section III-B2, to the two-converter case corresponding to the hybrid power system in Figure 5.

2) The 2EET Applied to the Hybrid Power System: In the hybrid power system of Figure 5, each converter is furnished by an input filter. Ignoring, for now, the particular feedback loops in that system, we can consider the circuit as a whole rather than as two separate converters. That system, like the single-converter system in Section III-B1 can be characterized by its open-loop transfer functions from any independent input to any output variable. Of particular interest, are the converter transfer functions that will be "in the loop" upon addition of feedback control. In Figure 5, those are the transfer functions from each

duty ratio, $\hat{d}_1$ and $\hat{d}_2$, to the output voltage, $\hat{v}$, because of the voltage-mode feedback control depicted there.

Now, we must consider the effect on the converter transfer functions upon the simultaneous addition of *two* input filters to the overall system. To that end, the author in [19] applies the two extra element theorem to the system in Figure 5. In [22], Middlebrook presents the two extra element theorem (2EET), the principle result of which is the correction factor:

$$
CF^{(i)} = \frac{1 + \frac{Z_1}{Z_{N1}|_{Z_2=0}^{(i)}} + \frac{Z_2}{Z_{N2}|_{Z_1=0}^{(i)}} + K_N^{(i)} \frac{Z_1 Z_2}{Z_{N1}|_{Z_2=0}^{(i)} Z_{N2}|_{Z_1=0}^{(i)}}}{1 + \frac{Z_1}{Z_{D1}|_{Z_2=0}^{(i)}} + \frac{Z_2}{Z_{D2}|_{Z_1=0}^{(i)}} + K_D^{(i)} \frac{Z_1 Z_2}{Z_{D1}|_{Z_2=0}^{(i)} Z_{D2}|_{Z_1=0}^{(i)}}},
\tag{21}
$$

where Z_1 and Z_2 are the output impedances of the first and second input filters respectively.[4] The interaction parameters can be written (they each have two possible forms) [22]:

$$
K_N^{(i)} = \frac{Z_{N1}|_{Z_2=0}^{(i)}}{Z_{N1}|_{Z_2=\infty}^{(i)}} = \frac{Z_{N2}|_{Z_1=0}^{(i)}}{Z_{N2}|_{Z_1=\infty}^{(i)}}
\tag{22}
$$

$$
K_D^{(i)} = \frac{Z_{D1}|_{Z_2=0}^{(i)}}{Z_{D1}|_{Z_2=\infty}^{(i)}} = \frac{Z_{D2}|_{Z_1=0}^{(i)}}{Z_{D2}|_{Z_1=\infty}^{(i)}},
\tag{23}
$$

In [19], the author shows that analysis of the circuit in Figure 5 leads to the following special-case impedances for calculating the correction factor of the open-loop transfer function $\hat{v}/\hat{d}_1$:

$$
Z_{N1}|_{Z_2=0}^{(1)} = \frac{-e_1(s)}{j_1(s)}
\tag{24}
$$

$$
Z_{N2}|_{Z_1=0}^{(1)} = \frac{sL_{e2}}{M_2^2(D_2)}
\tag{25}
$$

$$
Z_{D1}|_{Z_2=0}^{(1)} = \frac{sL_{e1} + \frac{sL_{e2}Z_L}{Z_L + sL_{e2}}}{M_1^2(D_1)}
\tag{26}
$$

$$
Z_{D2}|_{Z_1=0}^{(1)} = \frac{sL_{e2} + \frac{sL_{e1}Z_L}{Z_L + sL_{e1}}}{M_2^2(D_2)},
\tag{27}
$$

where Z_L is the total impedance shunting the converter outputs, i.e. $Z_L = R||1/(s(C_1 + C_2))$, in Figure 5. The additional special-case impedances required to calculate the interaction parameters, $K_N^{(1)}$ and $K_D^{(1)}$, are

$$
Z_{N1}|_{Z_2=\infty}^{(1)} = \frac{-e_1(s)}{j_1(s)}
\tag{28}
$$

$$
Z_{D1}|_{Z_2=\infty}^{(1)} = \frac{Z_L + sL_{e1}}{M_1^2(D_1)}.
\tag{29}
$$

The correction factor of the second open-loop transfer function of interest, $\hat{v}/\hat{d}_2$, can be similarly derived or inferred from the correction factor for the first by symmetry arguments. This leads

---

[4]N and D historically represent "numerator" and "denominator" [17]

to:

$$Z_{N1}\big|_{Z_2=0}^{(2)} = \frac{sL_{e1}}{M_1^2(D_1)} \tag{30}$$

$$Z_{N2}\big|_{Z_1=0}^{(2)} = \frac{-e_2(s)}{j_2(s)} \tag{31}$$

$$Z_{D1}\big|_{Z_2=0}^{(2)} = \frac{sL_{e1} + \frac{sL_{e2}Z_L}{Z_L+sL_{e2}}}{M_1^2(D_1)} \tag{32}$$

$$Z_{D2}\big|_{Z_1=0}^{(2)} = \frac{sL_{e2} + \frac{sL_{e1}Z_L}{Z_L+sL_{e1}}}{M_2^2(D_2)} \tag{33}$$

and the additional special-case impedances required to calculate the interaction parameters, $K_N^{(2)}$ and $K_D^{(2)}$, are

$$Z_{N2}\big|_{Z_1=\infty}^{(2)} = \frac{-e_2(s)}{j_2(s)} \tag{34}$$

$$Z_{D2}\big|_{Z_1=\infty}^{(2)} = \frac{Z_L + sL_{e2}}{M_2^2(D_2)}. \tag{35}$$

Note that from the results above, the "numerator interaction parameter" equals one ($K_N^{(i)} = 1$) for each of the two transfer functions. This fact, which is characteristic of the hybrid power system in Figure 5, simplifies the numerical computation of the correction factors, $CF^{(i)}$, because, in that case, the numerator is exactly factorable as follows:

$$CF^{(i)} = \frac{\left(1 + \frac{Z_1}{Z_{N1}\big|_{Z_2=0}^{(i)}}\right)\left(1 + \frac{Z_2}{Z_{N2}\big|_{Z_1=0}^{(i)}}\right)}{1 + \frac{Z_1}{Z_{D1}\big|_{Z_2=0}^{(i)}} + \frac{Z_2}{Z_{D2}\big|_{Z_1=0}^{(i)}} + K_D^{(i)}\frac{Z_1 Z_2}{Z_{D1}\big|_{Z_2=0}^{(i)} Z_{D2}\big|_{Z_1=0}^{(i)}}}. \tag{36}$$

In analogy to the impedance inequalities from (18) and (19), the expression for the correction factor in (21) or (36) suggests that the ith open-loop converter transfer function will not be impacted significantly if the following impedance inequalities are met. Recall that meeting these impedance qualities may be sufficient but not necessary to preserve stability of an otherwise stable regulated power system.

$$|Z_1| \ll |Z_{N1}\big|_{Z_2=0}^{(i)}| \tag{37}$$

$$|Z_2| \ll |Z_{N2}\big|_{Z_1=0}^{(i)}| \tag{38}$$

$$|Z_1| \ll |Z_{D1}\big|_{Z_2=0}^{(i)}| \tag{39}$$

$$|Z_2| \ll |Z_{D2}\big|_{Z_1=0}^{(i)}| \tag{40}$$

3) Input Filter Design Approach: The input filters in an EIS-capable hybrid power system may be designed to achieve several goals simultaneously:

1) Attenuate converter switching ripple
2) Avoid converter instability
3) Pass or even amplify excitation signals

Goals 1) and 2) are typical of design goals when adding an input filter onto a regulator. Goal 3) is unique to the EIS-capable hybrid system, because the filter must be designed to allow excitation currents to flow from the converter input to the terminals of the fuel cell up to a specified frequency.

For this example, we consider the input filter shown in Figure 13, which includes both the internal input filter components provided on the off-the-shelf Buck converter from Figure 3a as

well as the external input filter components that we added, L_{f1} and C_{f1}. The internal input filter components are:

$$C_{f3} = 8.8 \ \mu F \tag{41}$$

$$L_{f3} = 2.2 \ \mu H \tag{42}$$

$$C_{f4} = 26.4 \ \mu F. \tag{43}$$

Having set the pass band and rollover frequencies, largely by

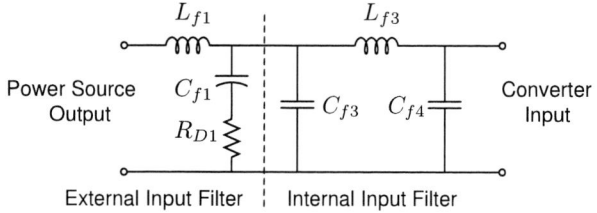

Fig. 13: The input filter for the fuel cell leg.

choosing L_{f1}, the filter transfer function is shown in Figure 14.

The damping leg formed by C_{f1} and R_{D1} in Figure 13 is intended to limit the magnitude peaking in the output impedance of the filter. However, as the impedance of the damping leg decreases it provides a shunt path that diminishes the transmission of excitation currents to the fuel cell terminals. Moreover, due to natural bandlimiting in the system, the designer may actually want to exploit the resonance at the edge of the pass band in Figure 14 to achieve some current amplification at that frequency. Both of these considerations qualitatively lower-bound the damping resistor, R_D, a constraint which directly contends with the impedance inequalities in (37)-(40).

Figure 15 shows a magnitude plot of the special-case impedances for correcting $\hat{v}/\hat{d}_1$ as well as the output impedances from the filters used in our system. Note that the two resonances in Z_o (solid line) correspond to the two resonances

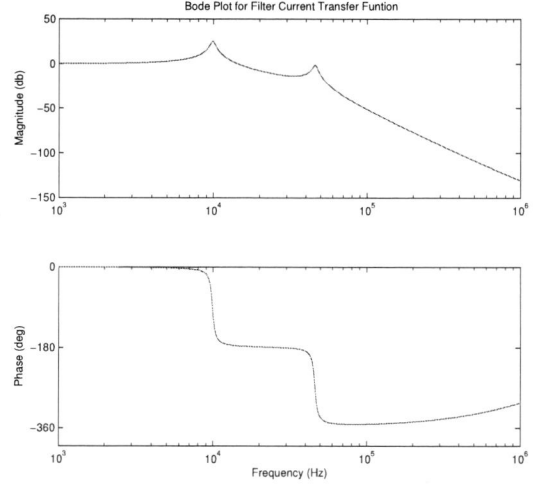

Fig. 14: The current transfer function for the fuel-cell leg input filter

978-1-4244-4782-4/10 $26.00 © 2010 IEEE

in the filter transfer function of Figure 14. Also note that the plot of special-case impedances suggests that the hybrid system of Figure 5 actually lower-bounds the bandwidth of the input filter to ensure negligible impact on converter dynamics. Because the impedance inequalities in (37)-(40) are not strictly met, as is evidenced by the plot in Figure 15, we need to examine the quantitative impact of the input filters on the converter open-loop transfer functions. In this Section, we assume that the feedback regulated system in Figure 5 is stable without the input filters connected, and that we simply need to verify that adding those input filters does not lead to instability.

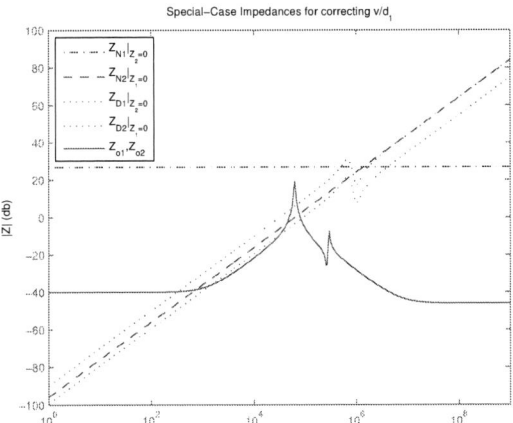

Fig. 15: A frequency plot of the special case impedances for correcting $\hat{v}/\hat{d}_1$ and the input filter output impedances for checking the impedance inequalities in (37)-(40). System parameters: $V_{FC} = 28$V, $V_{batt} = 48$V, V_{out}=12V, $R = 2\Omega$, $L_e = 1\mu$H, $C_e = 1\mu$F

A plot of the correction factors, $CF^{(i)}$, for the ith converter open-loop transfer function from (21) is the most direct way of analyzing the effect of the input filter on system stability. We are generally interested in the additional phase lag in the loop transfer function upon addition of the input filter. More specifically, we are interested in the phase margin, or the phase relative to -180° at the unity gain (0db-crossover) frequency of the entire loop transfer function (including the feedback network) upon addition of the input filters. However, since we assume that we are checking that the input filters do not cause an already stable system to become unstable, we simply need to check the *additional* phase lag which is explicitly shown in the multiplicative correction factor.

For instance, the correction factors, $CF^{(1)}$ and $CF^{(2)}$, for the converter open-loop transfer functions, $\hat{v}/\hat{d}_1$ and $\hat{v}/\hat{d}_2$ respectively, are bode plotted in Figure 16. [5] From the plots, we see that $CF^{(2)}$ introduces a significant additional phase lag near 10^5 rps. However, the phase lag will not degrade the phase margin unless that phase lag occurs at the cross-over frequency of the entire regulator loop transfer function. In some cases, i.e. when the impedance inequalities in (37)-(40) are grossly

---

[5]The simulated data overlayed in the plots of Figure 16 was extracted from LTSPICE by comparing simulations of the open-loop transfer functions with and without the input filters in place.

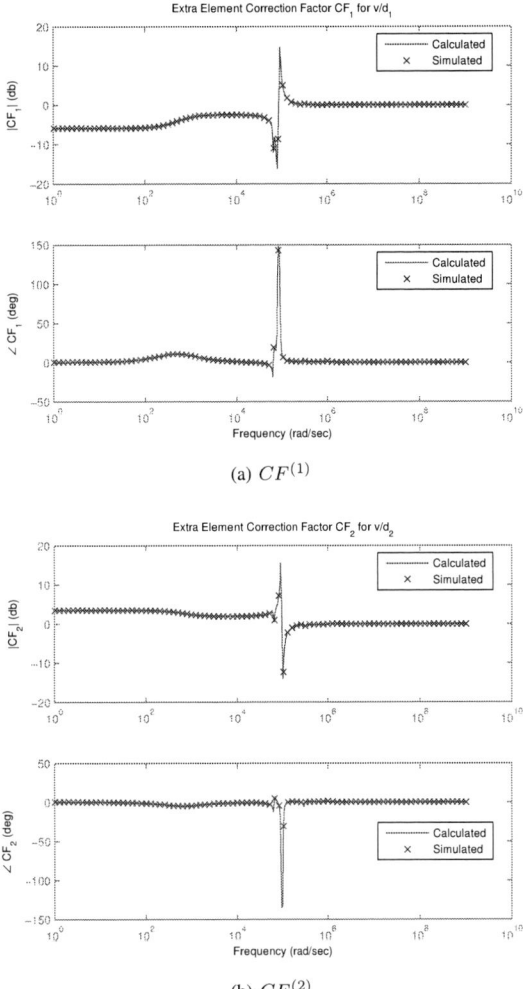

(a) $CF^{(1)}$

(b) $CF^{(2)}$

Fig. 16: Bode plots of correction factors $CF^{(1)}$ and $CF^{(2)}$ for open-loop transfer functions $\hat{v}/\hat{d}_1$ and $\hat{v}/\hat{d}_2$, respectively. System parameters: $V_{FC} = 28$V, $V_{batt} = 48$V, V_{out}=12V, $R = 2\Omega$, $L_e = 1\mu$H, $C_e = 1\mu$F

violated, the correction factor will contribute phase lag for a wide band of frequencies likely causing instability. Because the phase lag in this example is contributed for only a narrow range of frequencies we would not expect the voltage-mode feedback loop to become unstable.

The values for the external input filter components were:

$$C_{f1} = 100 \ \mu F \tag{44}$$
$$R_{D1} = 10 \ \Omega \tag{45}$$
$$L_{f1} = 6 \ \mu H. \tag{46}$$

These were also the values for the filters used in the system of Figure 3 represented by L_f and C_f. Stability of the real system was verified experimentally.

IV. EXPERIMENTAL SETUP

Figure 17 shows an overall schematic of the Siemens 5kW stack, connections to the built-in power electronics and storage,

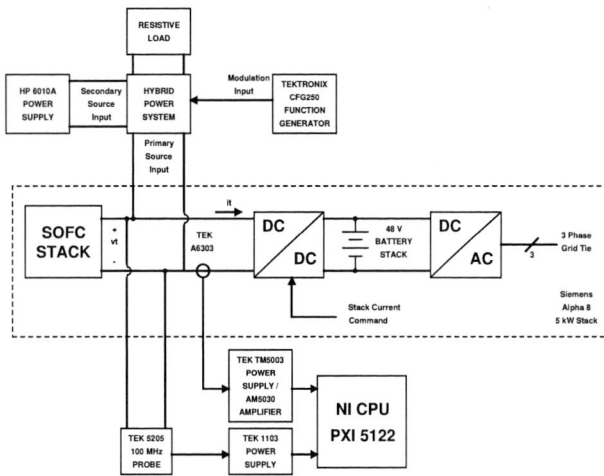

Fig. 17: Schematic illustration of stack, power electronics, and measurements. Components within the dashed line are within the physical envelope of the Siemens Alpha 8 unit.

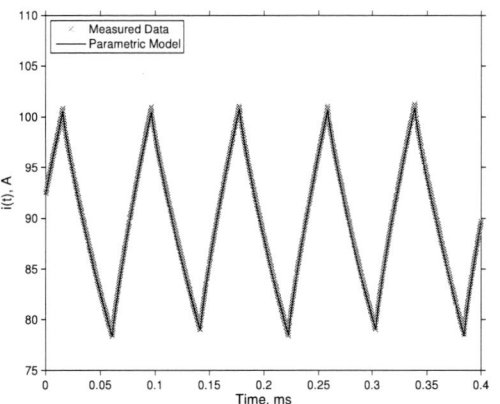

Fig. 18: Measured and predicted stack current as a function of time.

and the locations of our measurements. Under steady-state operation, the unit is remotely configured to regulate current from the stack. This power is then put on the grid through a three-phase inverter. The stack current is measured using a Tektronix A6303 current probe, while the voltage is measured using an isolated, differential Tektronix 5205 probe. Signals from both probes are recorded using a National Instruments data acquisition system with a PXI-5122 14-bit analog to digital converter. Sampling was conducted at a minimum of 2MS/s to avoid under-sampling issues. Figures 19a and 19b show typical data collected from this test setup under steady state operating conditions. The current and voltage levels in Figures 19a and 19b, nominally 100A and 28V, were typical of the stack load during testing.

V. RESULTS

Figure 19 shows typical data collected from the test setup in Figure 17 with a 1kHz exogenous excitation imposed by control of the test power electronics. The triangular ripple current in

Fig. 19 at roughly 12 kHz is due to the operation of the front-end boost converter in the Siemens power management system. The current and voltage levels in Fig. 19, nominally 90A and 28V, were typical of the stack load during testing.

Figure 20 shows Nyquist plots of the impedance $\hat{Z}(j\omega)$ obtained from the response of the stack to the built-in power electronics ripple and the power electronic test signal. The plots were prepared according to the convention for electrochemical impedance spectroscopy results. Fig. 20a shows a overall plot representing impedances for all frequencies with significant content. The discete clusters correspond to harmonics of the triangular boost-converter switching waveform, while the more continuous low-frequency data shows the response to the test signal. As the frequency of the harmonics increases, the amplitude decreases, and the variance in the impedance estimate increases. Fig. 20b is an expanded view of the low frequency portion corresponding to the exogenous excitation. The arc shape of the curve in Fig. 20b is consistent with the series connection of parallel RC elements often used in equivalent circuit models of fuel cells.

Data corresponding to a 1kHz power electronic excitation were used to identify the parametric model in II-C. The parameter estimates were $V_{oc} = 34.1V$, $R = 0.0690\Omega$, and $L = 0.43\mu H$. These results compare favorably to those in [5], where the values for these parameters based on data taken months earlier were found to be $V_{oc} = 34.7V$, $R = 0.0677\Omega$, and $L = 0.471\mu H$. The decrease in voltage and increase in resistance are likely due to the gradual degradation of stack performance observed over this time period. The latest parameters were used for an output-error prediction of the time-domain current waveform in response a 5.4 kHz excitation. This cross-validation result is shown in Fig. 18.

Acknowledgments

The authors gratefully acknowledge the support of NSF award #0547616 and MSU-HiTEC. MSU-HiTEC is funded by the United States Department of Energy (USDOE) under Award No. DE-AC06-76RL01830. However, any opinions, findings, conclusions, or recommendations expressed herein are those of the author(s) and do not necessarily reflect the views of the DOE.

The authors would like to thank The Grainger Foundation, the U.S. Department of Energy (DOE) and ARPA-E for their generous support.

REFERENCES

[1] S. K. Mazumder, K. Acharya, C. L. Haynes, R. Williams, M. R. v. S. Jr., D. J. Nelson, D. F. Rancruel, J. Hartvigsen, and R. S. Gemmen, "Solide-oxide-fuel-cell performance and durability: Resolution of the effects of power-conditioning systems and application loads," *IEEE Transactions on Power Electronics*, vol. 19, no. 5, pp. 1263–1278, September 2004.
[2] B. W. Q. Huang, R. Hui and J. Zhang, "A review of ac impedance modeling and validation in sofc diagnosis," *Electrochimica Acta*, vol. 52, no. 28, pp. 8144–8164, November 2007.
[3] P. P. T. S. E. Ramschak, V. Peinecke and V. Hacker, "Detection of fuel cell critical status by stack voltage analysis," *Journal of Power Sources*, vol. 2006, pp. 837–840, October 2006.
[4] R. S. Gemmen, "Analysis for the effect of inverter ripple current on fuel cell operating condition," *Transactions of the ASME*, vol. 125, pp. 576–585, May 2003.
[5] E. Seger and S. R. Shaw, "In-situ electrical terminal characterization of a 5 kw solid oxide fuel cell stack," *Submitted for publication in the IEEE Transactions on Energy Conversion*, 2010.

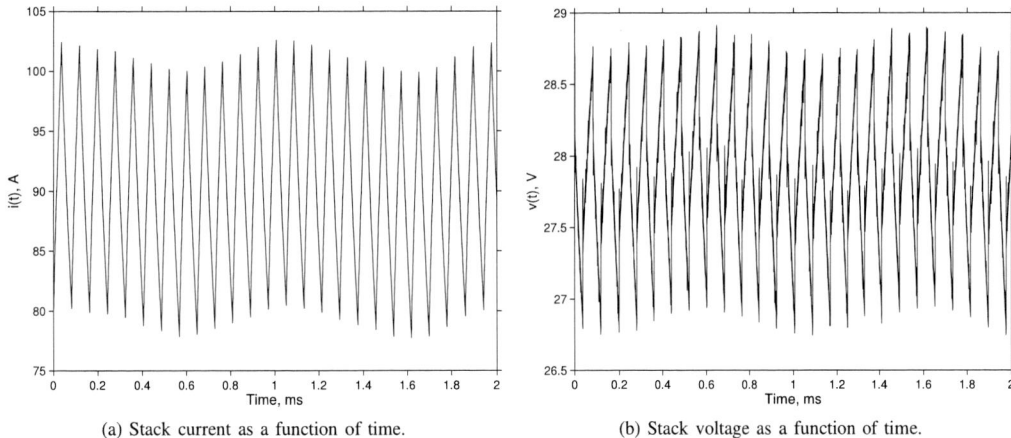

(a) Stack current as a function of time.

(b) Stack voltage as a function of time.

Fig. 19: Stack current and voltage, measured as indicated in Fig. 17. The triangle current waveform in 19a is due to the operation of the DC/DC converter in the system. The corresponding voltage of the stack appears in 19b.

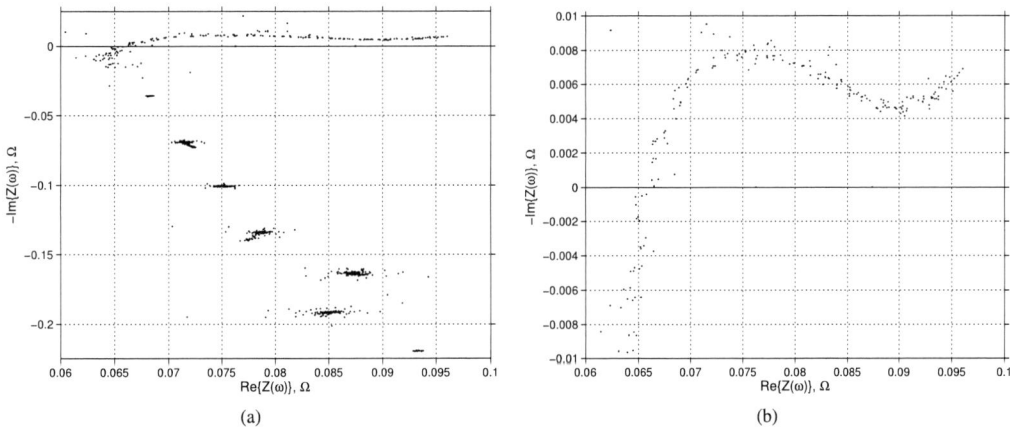

(a)

(b)

Fig. 20: Whole stack impedance spectroscopy results. (a) Stack response to ripple current and power electronic test signal. (b) Low-frequency portion of stack response showing response to power electronic test signal.

[6] C. Gabrielli, "Use and application of electrochemical impedance techniques," vol. Technical Report 43.

[7] N. Wagner, W. Schnurnberger, B. Müller, and M. Lang, "Electrochemical impedance spectra of solid-oxide fuel cells and polymer membrane fuel cells," *Electrochimica Acta*, vol. 43, no. 24, pp. 3785–3793, 1998.

[8] L. G. A. T. Romero-Castañón and U. Cano-Castillo, "Impedance spectroscopy as a tool in the evaluation of mea's," *Journal of Power Sources*, vol. 118, pp. 179–182, May 2003.

[9] E. Barsoukov and J. R. MacDonald, *Impedance Spectroscopy Theory, Experiment, and Applications.* John Wiley and Sons, Inc., 2005.

[10] R. Johansson, *System Modeling and Identification.* Prentice-Hall, Inc., 1993.

[11] D. J. Hall and R. G. Colclaser, "Transient modeling and simulation of a tubular solid oxide fuel cell," *IEEE Transactions on Energy Conversion*, vol. 14, no. 3, pp. 749–753, September 1999.

[12] M. H. N. C. Wang and S. R. Shaw, "Dynamic models and model validation for pem fuel cells using electrical circuits," *IEEE Transactions on Energy Conversion*, vol. 20, no. 2, pp. 442–451, June 2005.

[13] S. Pasricha and S. R. Shaw, "A dynamic pem fuel cell model," *IEEE Transactions on Energy Conversion*, vol. 21, no. 2, pp. 484–490, June 2006.

[14] R. Middlebrook and S. Cuk, "A general unified approach to modeling switching-converter power stages," *International Journal of Electronics*, vol. 42, pp. 521–550, June 1977.

[15] R. Middlebrook, "A general unified approach to modelling switching

dctodc converters in discontinuous conduction mode," *IEEE Power Electronics Specialists Conference*, vol. 32, no. 3, pp. 36–57, 1977.

[16] ——, "Input filter considerations in design and application of switching regulators," *IEEE Industry Applications Society Annual Meeting*, no. 76CH1122-1-IA, p. 366382, 1976.

[17] R. W. Erickson and D. Maksimovic, *Fundamentals of Power Electronics, Second Edition.* Springer Science+Business Media, LLC., 2001.

[18] R. Middlebrook, "Modelling and analysis methods for dc-to-dc switching converters," *IEEE International Semiconductor Power Converter Conference*, pp. 90–111, 1977.

[19] J. J. Cooley, Ph.D. dissertation, Massachusetts Institute of Technology, To be published 2010.

[20] R. Middlebrook, "Design techniques for preventing input-filter oscillations in switched-mode regulators," *Proc. Fifth National Solid-State Power Conversion Conference (Powercon 5)*, p. A3.1A3.16, May 1978.

[21] ——, "Null double injection and the extra element theorem," *IEEE Trans. on Education*, vol. 32, no. 3, pp. 90–111, August 1989.

[22] ——, "The two extra element theorem," *Proc. IEEE Frontiers in Education*, pp. 702–708, Sept. 1992.

Photovoltaic Parallel Resonant DC-link Soft Switching Inverter using Hysteresis Current Control

Young-Ho Kim

School of Information and Communication Engineering Sungkyunkwan University Suwon, South Korea.

Email: tomeito@skku.edu

Jun-Gu Kim

School of Information and Communication Engineering Sungkyunkwan University Suwon, South Korea.

Email: hopuler@skku.edu

Young-Hyok Ji

School of Information and Communication Engineering Sungkyunkwan University Suwon, South Korea.

Email: huma81@skku.edu

Chung-Yuen Won[†]

School of Information and Communication Engineering Sungkyunkwan University Suwon, South Korea.

Email: won@yurim.skku.ac.kr

Yong-Chae Jung*

Department of Electronic Engineering Namseoul University Cheonan, South Korea

Email: ychjung@nsu.ac.kr

Abstract—**This paper proposes a photovoltaic parallel resonant dc-link soft switching inverter using hysteresis current control. The proposed system includes a dc-link switch for soft switching. When the dc-link switch is turned off, the inverter switches are turned on and off with zero voltage switching using hysteresis current controller. So, all of the switches in the proposed inverter are turned on and off with soft switching. As a result, the proposed circuit can reduce the switching loss. The proposed inverter is verified through the theoretical analysis, simulation and experiment.**

I. INTRODUCTION

Today, with the continuing increase of fossil fuel price, the photovoltaic (PV) that directly uses the energy from the sun to generate electricity is becoming more and more interesting from the environmental [1]. The PV system consists of the PV array and the PV power conditioner, and the PV power conditioner usually can be subdivided into both a dc-dc converter to control the dc-voltage and a voltage-source inverter to connect to the ac utility grid line. The inverter in PV system is in need of the development of smaller, lighter more efficient and more reliable systems. The increase of the switching frequency reduces the size of the PV inverter but increases the switching losses [2-3]. To solve these problems, many zero-voltage switching (ZVS) or zero-current switching (ZCS) circuits have presented [4]. Thus, we propose the photovoltaic parallel resonant dc-link soft switching inverter using hysteresis current control for reducing the switching loss. In first stage, a soft switching boost converter performs ZVS and ZCS to improve the efficiency of a boost. Under the condition of zero voltage and zero current by inductor and capacitor resonance, the soft switching can reduce the voltage stress and switch loss produced at the switch. In second stage, the inverter switches are turned on and off with ZVS when the dc-link switch is turned off. In order to control the inverter switches, a proposed hysteresis current controller is used. A conventional hysteresis current controller exhibits several undesirable features, such as variable switching frequency of the inverter stage and drastically increasing switching frequency when the band width is narrow [5]. For solving these problems, novel hysteresis control method is proposed. When the proposed hysteresis current controller is used with dc-link switch, the switching frequency of the inverter is limited because the inverter switches are synchronized with dc-link switch. So, the proposed inverter can decrease the switching stress and loss because all of the switches are turned on and off with ZVS and ZCS. In this paper, the theoretical explanation and the detailed operational principles are explained, along with informative simulation and experimental results.

II. PROPOSED SOFT SWITCHING INVERTER

A. Proposed inverter circuit

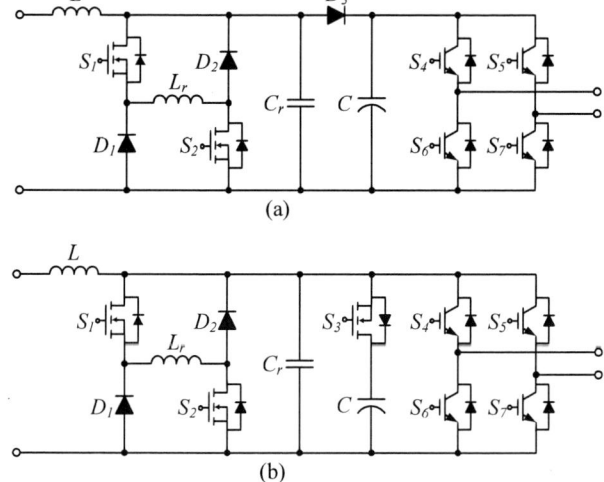

Figure 1. (a) The conventional H-bridge inverter using soft switching boost converter (b) The proposed photovoltaic parallel resonant dc-link.

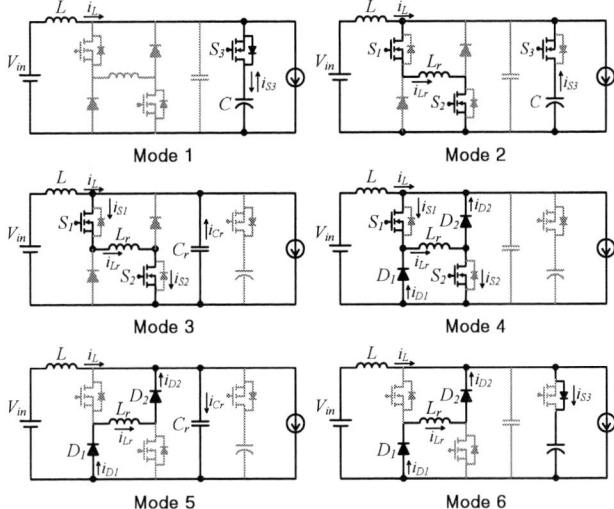

Figure 2. Mode transitions and equivalent circuits of proposed soft switching inverter.

Figure 1 (a) shows a conventional H-bridge inverter using soft switching boost converter [6]. Additionally, the soft-switching boost converter has a switch, inductor, capacitor and two diodes circuit against the conventional boost converter. So, the converter stage switches are turned on and off with ZVS and ZCS. But the inverter stage switches of the conventional inverter are turned on and off with hard switching. Therefore the conventional inverter switches generate the switching loss and stress.

Figure 1 (b) shows the proposed photovoltaic parallel resonant dc-link soft switching inverter using hysteresis current control. The converter stage equals the conventional soft switching boost converter, but the second stage differs from the conventional H-bridge inverter stage because the dc-link switch is included. The additional switch in dc-link is turned on and off with ZVS because of the anti-parallel diode and resonant capacitor C_r. Using the switch in dc-link, the switch S_4, S_5, S_6 and S_7 of inverter stage are turned on and off with ZVS [7].

B. Operation and Mode Transition of proposed Soft Switching inverter

The mode transitions and equivalent circuits of proposed soft switching inverter during one switching period are depicted in Figure 2. The gate pulse timing sequences of the switch S_1, S_2 and S_3 are shown in Figure 3, and the operating voltage and current waveforms of each component of the proposed soft switching inverter are also shown in Figure 3. The operating principle in each switching mode transition of the proposed soft switching inverter is explained as follows.

- Mode 1: The switch S_3 is turned on with ZVS. The dc-link capacitor begins to discharge the energy. The main inductor L current decreases linearly.

- Mode 2: Switch S_1 and S_2 are turned on with ZCS. The dc-link capacitor begins to discharge the energy.

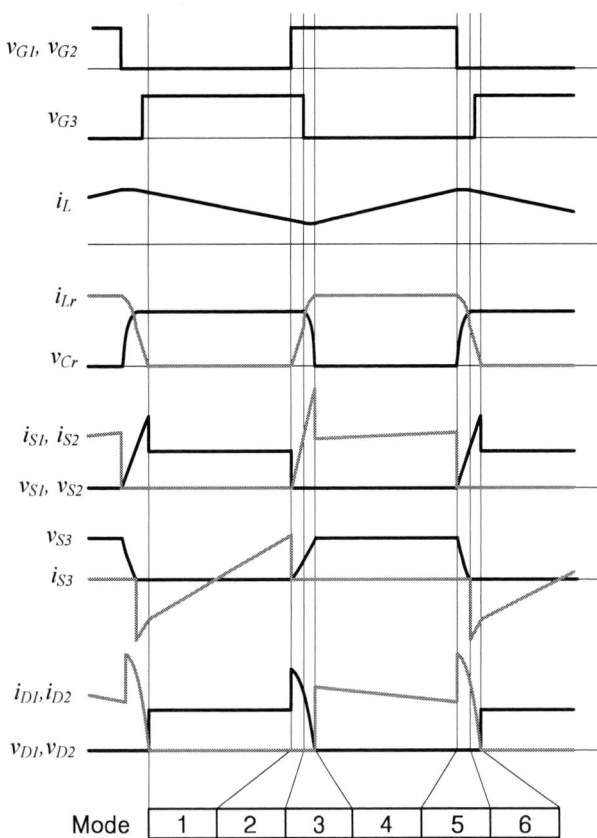

Figure 3. Current and voltage waveforms of the proposed soft switching inverter

- Mode 3: The resonance between resonant inductor L_r and resonant capacitor C_r is started. The main inductor current is minimized.

- Mode 4: The resonant inductor current flows through the two freewheeling paths. When the resonance between resonant inductor L_r and resonant capacitor C_r is finished, the current of resonant inductor L_r flows through diode D_1 and D_2. The current of resonant inductor L_r is increased to accumulate energy in resonant inductor L.

- Mode 5: The switch S_1 and S_2 are turned off with ZVS. The resonance between resonant inductor L_r and resonant capacitor C_r is started. The resonant capacitor C_r is charged by the current flowing through inductor L and resonant inductor L_r. When the voltage of resonant capacitor C_r equals the output voltage, this mode is finished.

- Mode 6: The current of main inductor L decreases linearly. The dc-link capacitor is charged by main inductor L and resonant inductor L_r. The energies of the main inductor L and resonant inductor L_r are transferred to dc-link capacitor C through switch S_3 parallel diode. When the diode is turned off with the zero-voltage, this mode is finished.

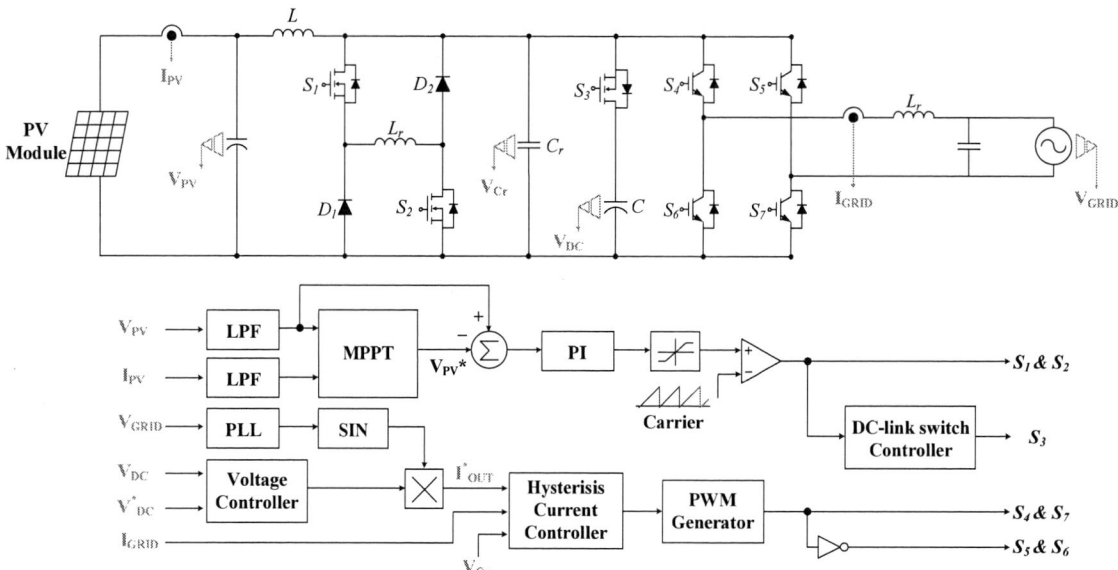

Figure 4. Configuration of the proposed system and control circuit.

III. CONTROL SYSTEM ALGORITHM AND IMPLEMENTATION

A. Converter Swtich Control

The problem for nonlinear characteristic in the PV generation systems is the amount of the electric power generated by solar arrays changing with weather conditions, temperature and the intensity of the solar radiation. A maximum power point tracking (MPPT) method or algorithm, which has quick-response characteristics and is able to make good use of the electric power generated in any weather, temperature and solar radiation, is needed to solve the aforementioned problem [8].

Various MPPT control methods have been discussed in detail in [9]. The reference voltage calculated through current and voltage injected from PV panel is derived from the MPPT algorithm. The perturb-and-observe (P&O) algorithm is used to extract maximum power from PV arrays and deliver it to the inverter [10], [11].

The PV voltage and PV current are required to calculate the PV output power for P&O MPPT. The instantaneous voltage error of MPPT resulting signal is fed to a PI controller. The integral term in the PI controller improves the tracking by reducing the instantaneous error between the reference and the actual current. The resulting error signal is fed to a limiter. A maximum signal and minimum signal of the PI controller resulting value are limited to the constant signal values. The resulting limiter signal is compared with a triangular carrier signal, and intersections are sought to produce PWM signals for the converter switch S_1, S_2.

The dc-link switch controller generates the PWM signal synchronized with the PWM signal of the switch S_1, S_2. After the PWM signal of the switch S_1, S_2 is fallen to zero, the dc-link switch controller calculates the delay time for

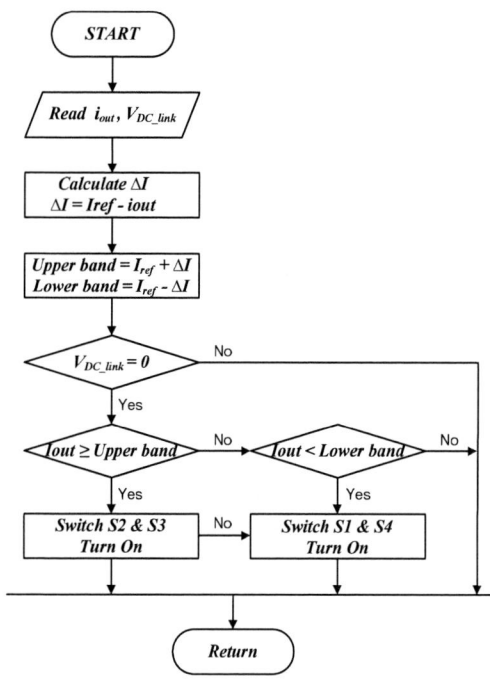

Figure 5. Flowchart of the proposed hysteresis current control.

the resonance between resonant inductor L_r and resonant capacitor C_r. The dc-link switch S_3 is turned on after delay time and is turned off after the PWM signal of switch S_1, S_2.

B. Inverter Switch Control using Hysteresis Current Controller

The conventional hysteresis current controller exhibits several undesirable features, such as variable switching frequency of the inverter stage [11]. The proposed hysteresis current controller has a limited switching frequency because the operation of inverter switches is synchronized with dc-link switch. The proposed hysteresis current controller requires the output reference current, grid current and resonant capacitor voltage to calculate PWM signal of inverter switches. To generate the output reference current, phase locked loop (PLL) for generating an in-phase sinusoid and voltage controller are used [12].

In the grid-connected mode of operation, a fast and accurate PLL method is essential because the current reference is based on it. However, for a single-phase system, because of the lack of a quadrature signal, acquiring the phase-angle information is more difficult. The commonly used zero-crossing detection method does not provide instantaneous phase-angle information of the grid and is sensitive to multiple zero crossings caused by noise. Therefore, the main issue of designing a single-phase PLL is how to generate the virtual quadrature signal. The virtual to phase method PLL is used the proposed hysteresis current control [13].

Figure 5 shows the flowchart of the proposed hysteresis current control. After sensing the grid current and resonant capacitor voltage, the proposed hysteresis current controller calculates the delta current using grid current and output reference current, and determines the upper band and lower band. When the dc-link voltage is zero, the delta current is compared by upper band and lower band. The switch S_4, S_7 is turned on when the output current is bigger than the upper band. The switch S_5, S_6 is turned on when the output current is smaller than the lower band. When the output current is not bigger and smaller than lower band or upper band, the switch condition is maintained.

IV. EXPERIMENTAL RESULTS AND EVALUATIONS

A. Simulation Results

In order to verify that the proposed inverter can be practically implemented in a PV system, simulations were performed by using PSIM. It also helps to confirm the PWM switching strategy which then can be implemented in a DSP.

Figure 6 (a) shows the voltage and current waveforms of the switch S_1, S_2 and S_3. The switch S_1, S_2 are turned on with ZCS and turned off with ZVS. The switch S_3 is turned on and off with ZVS. Figure 6 (b) shows the voltage and current waveforms of the switch S_4, S_5, S_6 and S_7. When the dc-link switch S_3 is turned off, the switches are turned on and off with ZVS. Therefore, all the switching devices are operated in the soft switching state when the switches are turned on and off.

Figure 7 (a) shows a schematic diagram of the output current waveforms using hysteresis current control. The inverter switching frequency is limited converter switching frequency using the proposed hysteresis current controller. Figure 7 (b) shows a schematic diagram of the waveforms of the 60Hz output current and voltage waveforms using the proposed hysteresis current control. It is confirmed that output current waveform is synchronized with output voltage waveform.

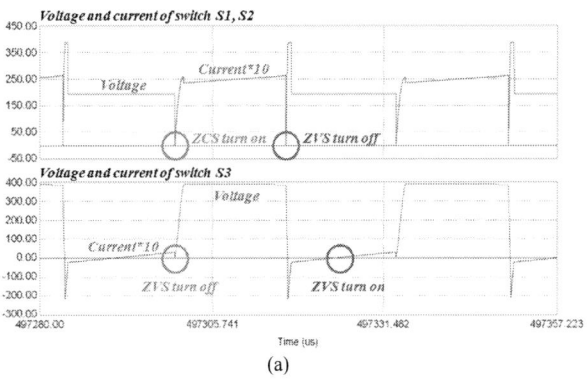

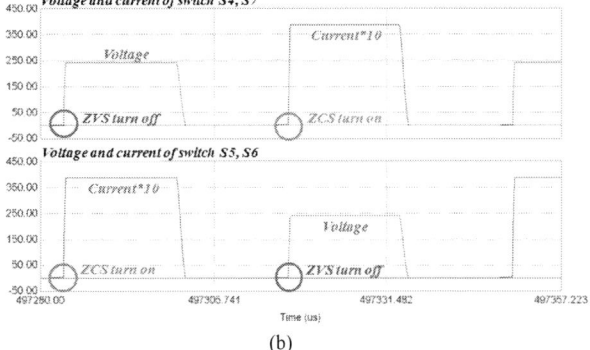

(a)

(b)

Figure 6. (a) Voltage and current waveforms of converter stage switches (b) Voltage and current waveforms of inverter stage switches.

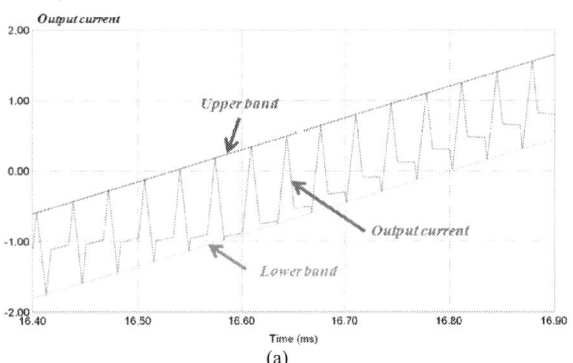

(a)

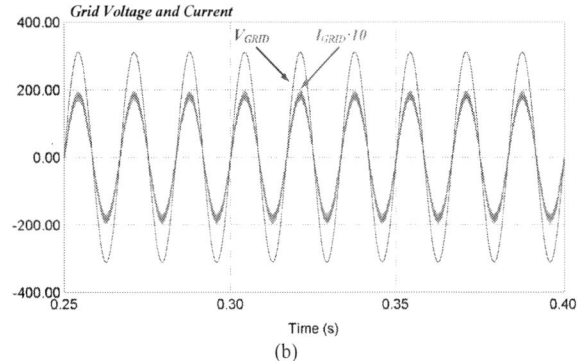

(b)

Figure 7. (a) Output current waveform using hysteresis currnet control (b) 60Hz output current and voltage waveform using hysteresis currnet control.

978-1-4244-4782-4/10 $26.00 © 2010 IEEE   2278

TABLE I.   CIRCUIT PARAMETERS

Input Voltage	100[V]
DC-link Voltage	130[V]
Main Inductor	1000[μH]
Resonant Inductor	10[μH]
Resonant Capacitor	47[nF]
DC-link Capacitor	2200[μF]
Converter Frequency	30[kHz]
Inverter Output Frequency	60[Hz]

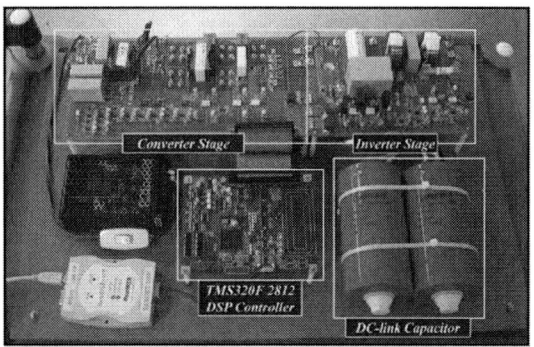

Figure 8.   Prototype of the proposed soft switching inverter.

B.   Experimental Results and Discussions

The simulation results are verified experimentally by using a DSP TMS320F2812.

Table I shows the proposed photovoltaic parallel resonant dc-link soft switching inverter specifications and its controller parameters.

Figure 8 shows prototype experimental set up for proposed system. The experimental set up consists of the converter stage, inverter stage, DSP controller and dc-link capacitor. The DSP, type TMS320F 2812, is used to control the P&O MPPT, PLL, dc-link voltage, dc-link switch and hysteresis current controller. In converter stage, the switch S_1, S_2 and S_3 are driven by the switching frequency of 30 kHz. In the inverter stage, the switch S_4, S_5, S_6 and S_7 are driven by variable switching frequency, but the inverter switching frequency is limited by converter stage switching frequency because of the proposed hysteresis current control synchronized with dc-link switch S_3.

Figure 9 (a) shows the experimental results for voltage and current of the switch S_1, S_2. The switches operate with soft switching when the switches are turned on and off. Figure 9 (b) shows the experimental results for voltage and current of the switch S_3. When the switch turned on and off, the switch operates with ZVS. Figure 9 (c) shows the experimental results for voltage and current of the switch S_4, S_7. When the dc-link voltage $V_{dc\text{-}link}$ is zero, the switches are turned on and off with ZVS. Figure 9 (d) shows the experimental results for output voltage and current. The output current is controlled by the proposed hysteresis current control.

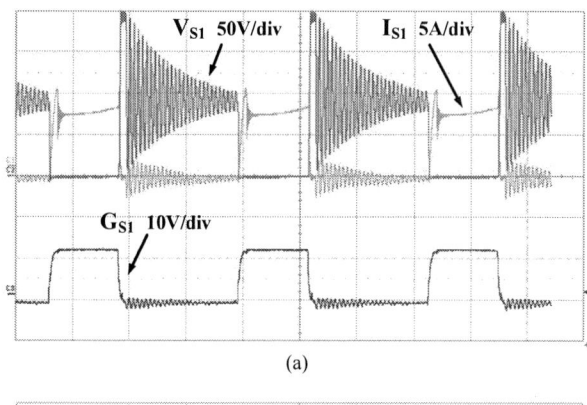

(a)

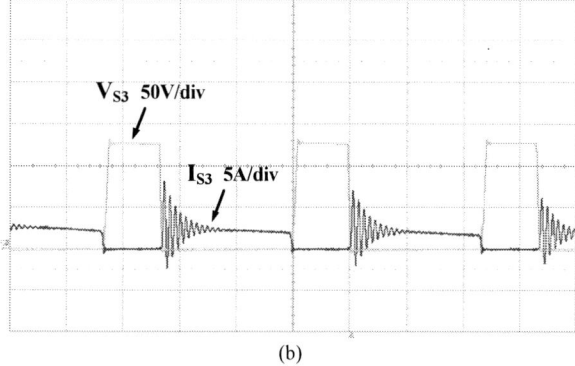

(b)

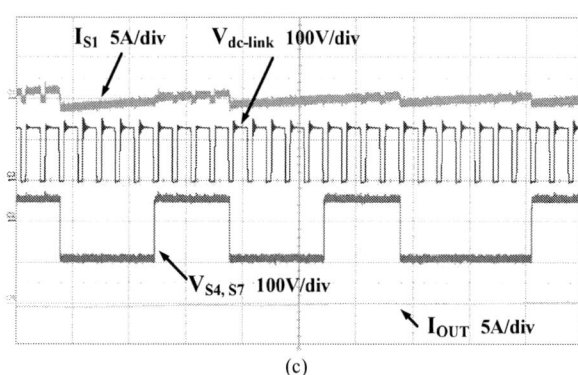

(c)

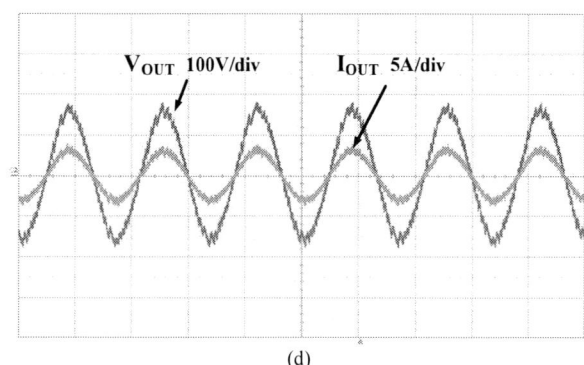

(d)

Figure 9.   (a) Voltage and current experimental waveforms of converter stage switch S_1, S_2. (b) Voltage and current experimental waveforms of converter stage switch S_3. (c) Voltage and current experimental waveforms of inverter stage switches. (d) 60Hz output current and voltage experimental waveforms using hysteresis current control

V. CONCLUSION

In this paper, a photovoltaic parallel resonant dc-link soft switching inverter using hysteresis current control is proposed and analyzed. The circuit topology, modulation law, and operational principle of the proposed inverter were analyzed in detail. A hysteresis current algorithm is implemented in DSP TMS320F 2812 to optimize the performance of the inverter. Simulation and experimental results indicate that all the switching devices are operated in the soft switching state when the switches are turned on and off.

ACKNOWLEDGMENT

This work is outcome of the fostering project of the Specialized Graduate School supported financially by the Ministry of Knowledge Economy (MKE).

REFERENCES

[1] T. Shimiizu, O. Hashimoto, G. Kimura, "A novel high-performance utility-interactive photovoltaic inverter system," *IEEE Trans. Power. Electron.*, vol. 18, no. 3, pp. 704-711, March 2003.

[2] T. Ninorniya, M. Nakahara, T. Higashi, K. Harada, "A unified analysis of resonant converters," *IEEE Trans. Power. Electron.*, vol. 2, no. 2, pp. 735-738, June 1996.

[3] R. L. Steigerwald, "A comparison of half-bridge resonant converter topologies," *IEEE Trans. Power Electron.*, vol. 3, no. 2, pp. 174–182, April 1988.

[4] Y. C. Jung, J. G. Cho, G. H. Cho,"A new zero voltage switching resonant DC-link inverter with low voltage," Industrial Electronics, Control and Instrumentation, 1991. Proceedings. IECON'91., 1991 International Conference on 28 Oct.-1 Nov. 1991, vol. 1, pp. 308-313.

[5] Lajoie-Mazenc. Michel, Villanueva. Carlos, Hector. Jean, "Study and Implementation of Hysteresis Controlled Inverter on a Permanent Magnet Synchronous Machine," *IEEE Trans. Ind. Applications.*, vol IA-21, no. 2, pp. 408-413, March 1985.

[6] Gil-Ro Cha, Sang-Hoon Park, Chung-Yuen Won, Yong-Chae Jung, Sang-Hoon Song, "High Efficiency Soft Switching Boost Converter for Photovoltaic System," Proceeding of IEEE EPE-PEMC 13th, Sept, 2008, pp. 391-395.

[7] Young-Ho Kim, Gil-Ro Cha, Young-Hyok Ji, Jae-Hyung Kim, Yong-Chae Jung, Chung-Yuen Won, "ZVS resonant DC-link inverter using soft switching boost converter," Proceeding of ICIT 2009, IEEE International Conference on 10-13 Feb. 2008, pp. 1-5.

[8] N. Mutoh, T. Inoue, "A control method to charge series-connected ultraelectric double-layer capacitors suitable for photovoltaic generation systems combining MPPT control method," *IEEE Trans. Ind. Electron.*, vol. 54, no. 1, pp. 374–383, Feb. 2007.

[9] S. Liu, R. A. Dougal, "Dynamic multiphysics model for solar array," *IEEE Trans. Energy Conv.*, vol. 17, no. 2, pp. 285-294, Jun. 2002.

[10] N. Femia, G. Petrone, G. Spagnuolo, M. Vitelli, "Optimization of perturb and observe maximum power point tracking method," *IEEE Trans. Power Electron.*, vol. 20, no. 4, pp. 963-973, July 2005.

[11] F. Zare, G. Ledwich, "A hysteresis current control for single-phase multilevel voltage source inverters: PLD implementation," *IEEE Trans. Power Electron.*, vol. 731-738, no. 5, pp. 731-738, Sept. 2002.

[12] J. Selvaraj, N. A. Rahim, "Multilevel Inverter For Grid-Connected PV System Employing Digital PI Controller," *IEEE Trans. Ind. Electron.*, vol 56, no. 1, pp. 149-158, Jan. 2009.

[13] J. Eloy-Garcia, S. Arnaltes, J. L. Rodriguez-Amenedo, "Direct power control of voltage source inverters with unbalanced grid voltages," *IET Trans. Power. Electron.* vol 1, no. 3, pp. 395-407, Sept. 2008.

Supercapacitor-based Hybrid Storage Systems for Energy Harvesting in Wireless Sensor Networks

S. Saggini[*], F. Ongaro[*], C. Galperti[**], P. Mattavelli[***]

[*] DIEGM, University of Udine, Udine, Italy. E-mail: stefano.saggini@uniud.it, fabio.ongaro@uniud.it
[**] DEI, Politecnico of Milano, Milano, Italy. E-mail: galperti@elet.polimi.it
[***] DTG– University of Padova, Vicenza, Italy. E-mail: paolo.mattavelli@unipd.it

Abstract – **This paper investigates a power management architecture based on a hybrid accumulator system that utilizes the supercapacitor cell as storage element for energy harvesting applications. The supercapacitor guarantees a longer lifetime in terms of charge cycles, it presents itself as a "green" technology compared to batteries and it has a wide range of operating temperature. The drawbacks of this type of solution are the low energy density and the leakage current that reduce the performance for very low power applications. In this paper a photovoltaic scavenger based on the supercapacitors is investigated and it can work only with supercapacitors, or together with the lithium battery cell in order to obtain a good compromise in terms of energy density and lifetime. A dedicated power management strategy is also proposed. Experimental results with a 5W photovoltaic energy source are reported.**

Keywords – **DC-DC converters, power management, wireless sensor network, super-capacitors, battery charger**

I. INTRODUCTION

The pervasive computing and network of wireless sensors are applications based on energy-autonomous system [1-4]. For example, a network of wireless sensors consists of a large number of micro-sensors distributed in an area of interest. Each node monitors the local environment and it shares this information with the other neighboring nodes by using a wireless link. The nodes should not require any maintenance and thus, they have to be energetically autonomous without the need of batteries replacement. In many application scenarios the targeted node lifetime ranges typically between 2 to 5 years and the need of energy harvesting is a primary issue in order to grant effectiveness of the wide-spread diffusion of this technology.

In principle, all energy sources should be exploited to extract the available energy; among the others, the solar one [1-5] is generally the most effective in outdoor applications. Solar cells exhibit a strong non-linear electrical characteristic and the extraction of energy is even more difficult in non-stationary environments. Variable operating conditions can be associated with weather changes (e.g., cloudy and non optimally radiating solar power environments) and aging effects or efficiency degradation in the solar panel (e.g., dust in the cell surface). Moreover, the energy transfer mechanism is strongly influenced by the illumination conditions, such as the incidence angle of the sunlight, which varies along the day especially if the sensor node is in a mobile system.

Most of the solar energy harvesting solutions for wireless sensor nodes present a simple on/off-threshold charge mechanism relying on a diode connecting the cell with the rechargeable battery [6]. Unfortunately, a diode-based solution is extremely low cost, but the working point of the cell is set by the battery voltage and it cannot be adjusted to maximize the energy transfer in changing environment. This problem is addressed by substituting the diode-based solution with a Maximum Power Point Tracker (MPPT) system. This approach requires the development of adaptive systems to transfer the energy generated by the solar cell into storage elements, such as batteries or supercapacitors, while maintaining the working point of the PV cell around the optimal one.

In general terms, traditional MPPT circuits differentiate themselves in the design of the power converting electronics and/or in the control strategy. In most applications [6-9], the charging of high-density batteries in presence of fluctuating power sources remains an open issue. For example, Fig.1 reports the power obtainable by a photovoltaic panel, having 2W as the nominal power, in a mobile system constituted by a buoy on the sea. As can be observed, even if the weather condition is good, the irradiation is in the tropical condition and the temperature is maintained stable by the water, the available power has a high fluctuation due to the wave motion that changes the instantaneous orientation. This situation is usually not compatible with the charging of high density accumulators, like the lithium cell, because a precise charging profile, in terms of charging current and final voltage, must be implemented in order to prolong the lifetime [10] and the conditions reported in Fig. 1 determine a large variation of the charge and discharge cycles.

The solution proposed in this paper utilizes an hybrid accumulator architecture that combines the advantages of the supercapacitors in terms of the charge speed and instantaneous output power and the lithium cells for the stored available energy. In this architecture, the instantaneous power demand is supplied by the supercapacitor if the power generated by the converter is less than the power required for the load and the battery charge. If the technologies of the supercapacitors improve their performance in terms of leakage current and energy density, the structure can be simplified and only supercapacitors can be utilized.

978-1-4244-4782-4/10 $26.00 © 2010 IEEE

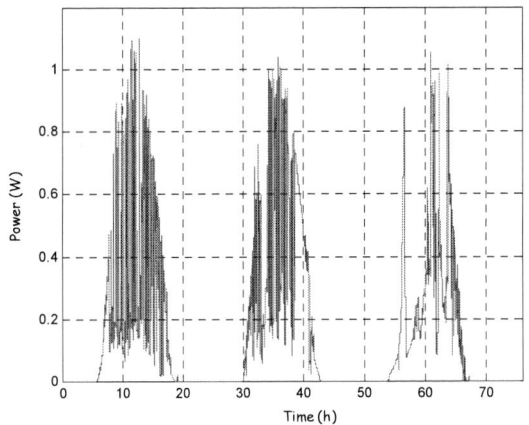

Fig. 1 – Power obtainable by a photovoltaic cell mounted on a buoy on the sea

II. PROPOSED POWER MANAGEMENT ARCHITECTURE

The proposed power management architecture is reported in Fig. 2, where three dc-dc converters are connected in parallel to the dc power bus. The first converter, denoted "*dc-dc PV converter*" interfaces the PV panel to the dc power bus, the second, denoted "*dc-dc battery converter*" connects the dc bus to the battery and the third, denoted "*dc-dc supercap converter*" connects the dc bus to the supercapacitor. The internal power bus is the main power supply of the electronic system utilized by sensor node. The power range of the proposed photovoltaic scavenger is in the range of 5W. The average power obtained by this source during a day is about 150mW, that is sufficient for our application where the average load power is 100mW. Moreover, the energy stored in the battery is designed in order to supply the load for the whole day.

The architecture reported in Fig. 2 is aimed to the maximization of the power conversion efficiency from the PV source to the load. In fact, when the power source is available, the energy flows directly from the source to the load through

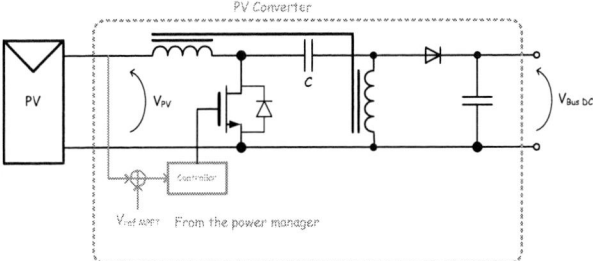

Fig. 3 – Coupled inductor Sepic converter utilized for Photovoltaic conversion with input voltage controller

the dc-dc PV converter. Moreover, with the power architecture of Fig. 2, it is possible to parallel different power sources. In fact, other energy sources can be connected by a converter to the same bus without changing the system architecture or the power management control. Thus, by including a hybrid accumulator system, the battery can be protected from overcharging and discharging stresses in order to prolong its lifetime. In fact, the extra power required by the load or generated by the fluctuating source is smoothed by the supercapacitors elements. In our case of study the dc bus is at 3.3V (V_{BusDC}), the solar panel has a voltage of about 6V (V_{PV}), the maximum voltage utilizable by the supercapacitor is of 2.5V and the lithium battery has a nominal voltage of 3.7V ($V_{Battery}$).

The *dc-dc PV converter* is based on the Sepic topology as reported in Fig. 3. This solution guarantees small input current ripple and it is compatible with the voltage of the photovoltaic cell because, in CCM operation, the conversion ratio can be stepped down to the bus voltage:

$$\frac{V_{BusDC}}{V_{PV}} = \frac{D}{1-D} \tag{1}$$

where D is the converter duty-cycle. In the implementation, a coupled inductor was used, thus reducing the magnetic element to one, as shown in Fig. 3. The input voltage of the converter is controlled by a feedback loop and the reference is determined either by the source MPPT or by the control on the supercapacitor voltage, depending on the state of the power management algorithm.

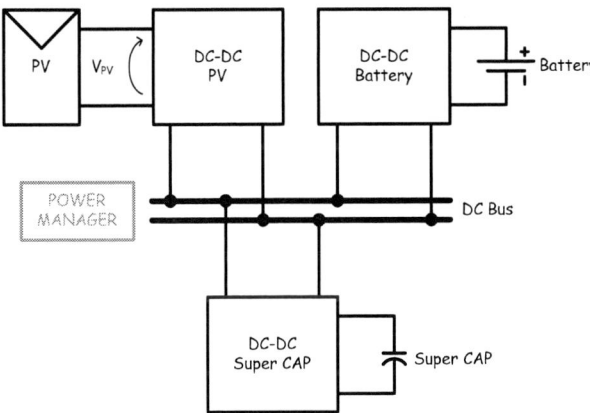

Fig. 2 – Power conversion system of the scavenger with super-capacitors

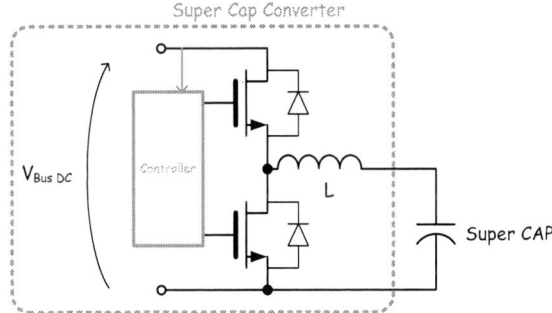

Fig. 4 – Bidirectional Buck converter connected to the super-capacitor

The *dc-dc supercap converter* is a bidirectional converter based on the synchronous buck topology, as reported in Fig.4; it operates in step-down mode when the power flows from the dc bus to the supercapacitor, and in step-up mode when the power flows in the opposite direction. The converter controls the dc bus voltage and it behaves as a sink or a source depending on the instantaneous power conditions. By utilizing the dc-dc converter, the supercapacitors are connected in parallel without requiring an overvoltage protection system on each element that is needed in the serial connection [11].

If the consumption of the wireless node is high during the period of time without irradiation, a lithium battery is used decreasing the size of the energy storage unit. During the charge of the battery the system works as in the previous description, and the supercapacitor system allows a stable charge process. When irradiation is not present and the supercapacitor system is almost discharged, the operation of the battery converter must be inverted by the power manager. In this case the energy required for the super-capacitor is the energy utilized to compensate the input energy variations with respect to the load requirement and the battery charge profile.

The *dc-dc battery converter* is a bidirectional Buck-Boost converter, as shown in Fig. 5, to charge and to utilize the battery at the same time depending on the working conditions. This choice is dictated by the required voltage level of the battery with respect to the dc power bus; in fact, the lithium battery has a nominal voltage of 3.7V and the dc bus voltage is 3.3V. This converter operates like a Buck-Boost by turning on the switch M2 and M3 during the "on phase", and by turning on the switch M1 and M4 during the "off phase". In CCM operation the conversion ratio is:

$$\frac{V_{BusDC}}{V_{Battery}} = \frac{D}{1-D} \qquad (2)$$

This type of converter has a current controlled loop and the power management algorithm decides the current reference value, which is either positive or negative depending on the working conditions of the battery (charge or discharge).

III. POWER MANAGEMENT ALGORITHM

The proposed power management algorithm for the complete hybrid storage system is based on different control states defined by the charge state of the supercapacitor. The supercapacitor energy is used to switch from a state to another one because it inherently monitors the power balance of the conversion system. In Fig. 6 the state machine of the proposed power management is reported.

The first state is the state of "*Turn-off*": from this state, the system wakes up only if some sources can deliver the energy required to start up the control electronic and, consequently, the system can proceed to the *soft-start* state. During this state the energy comes from the source and it is directed to the supercapacitor and to the controllers in order to maintain the

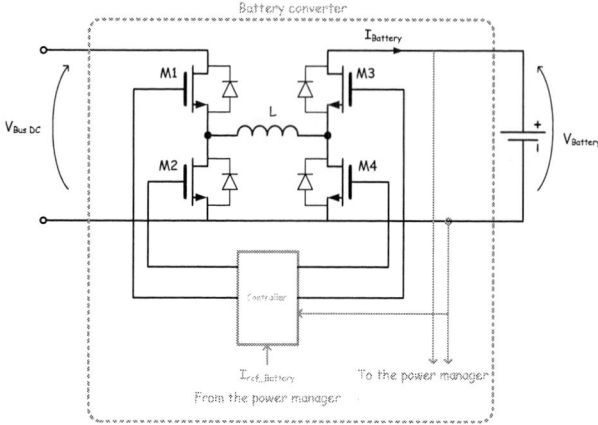

Fig. 5 – Bidirectional Buck-Boost converter connected to the Battery

bus voltage soft-start. In this state, the control strategies are based on the MPPT of the sources and on the dc bus voltage regulation. The state machine switches from the soft-start state to *battery charge* state, when the dc bus voltage has reached its nominal values and the energy of the supercapacitor exceeds the reference energy E_{refBC}.

In these conditions, the dc bus voltage is controlled by voltage loop using the dc-dc supercap converter and the voltage, thus the energy, in the capacitor is regulated at E_{refBC} by changing the charging current of the battery if it is less than the maximum input current of the charge profile of the battery. When the battery charging current reaches the maximum value, the supercapacitor voltage becomes the degree of freedom during this state. When the supercapacitor voltage increases, the algorithm moves towards the *Over-Power* state. In this case, the source MPPT is switched to a control loop on the supercapacitor voltage that regulates the maximum energy to E_{refOP}, thus reducing the input power. Instead, when the input power reduces and it is less than the load power, the supercapacitor voltage decreases and the system moves to the *Battery Help* state. In this state, the control maintains the voltage stable for the dc bus, the MPTT for the source and it uses the battery as a source in order to maintain the voltage of the supercapacitor at E_{refBH} by modulating the battery current level. At the same time, the maximum current output level of the battery is limited by the control in order to prolong the battery lifetime. Moreover, in order to avoid extra discharge of the battery, the output current limit falls to zero if the voltage of the battery is lower than a threshold value.

Finally, if the input power is lower than the output power required by the load and the battery is discharged, the voltage across the supercapacitor decreases until it reaches the threshold voltage associated with the energy E_{th1} and the system goes to the *Turn-off* state. Figure 6 summarizes the operation of the state-machine.

978-1-4244-4782-4/10 $26.00 © 2010 IEEE   2283

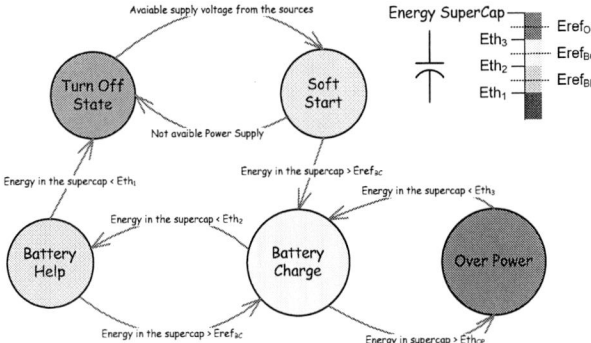

Fig. 6 – Block diagram of the state machine for the power management

The value of the supercapacitor is based on the study of the instantaneous power variation of the source or the load in two main states of the algorithm: *battery help* and *battery charge*. If the system is correctly designed these two states switch during the day and the night.

By neglecting the converter losses, the instantaneous power on the supercapacitor in the battery charge state can be expressed as:

$$p_{SC}(t) = p_{Source}(t) - p_{Load}(t) - p_{Charge}(t) \quad (3)$$

and in the battery help state

$$p_{SC}(t) = p_{Source}(t) + p_{Battery}(t) - p_{Load}(t) \quad (4)$$

If there is a load power increase or an input power reduction during the battery charge state, the battery charging current is reduced in order to regulate the voltage across the supercapacitor and under the worst case, it is reduced to zero. Under these conditions, (3) becomes:

$$p_{SC}(t) = p_{Source}(t) - p_{Load}(t) \quad (5)$$

In the battery help state, when the battery generates the maximum power available for a given current limit I_{max}, the power can be expressed as follow:

$$p_{SC}(t) = p_{Source}(t) + I_{max}V_{battery} - p_{Load}(t) \quad (6)$$

The following analysis is based on the hypothesis that the instantaneous power generated or accumulated on the super-capacitor can be considered an uncorrelated random variable for event with a distance of T_{sample}. This interval depends on the fluctuation of the load and source power. Based on a statistical approach, the power generated or accumulated by the supercapacitor is described by a probability density function f_{Psc} that is different depending on the state of the power manager (Battery Help or Battery Charge). Using f_{Psc}, it is possible to express the probability of extracting power from the supercapacitor at a given time t as:

$$P(p_{SC}(t) < 0) = \int_{-\infty}^{0} f_{p_{SC}}(p)\,dp \quad (7)$$

Let's denote $p_{SCgen}(t)$ the power generated by super-capacitor:

$$p_{SCgen}(t) = -p_{SC}(t). \quad (8)$$

Thus, the probability that $p_{SCgen}(t)$ is greater than power P at a given time t is

$$P(p_{SCgen}(t) > P \mid p_{SC}(t) < 0) = \frac{\int_{-\infty}^{P} f_{p_{SC}}(p)\,dp}{P(p_{SC}(t) < 0)} \quad (9)$$

The extracted energy E_{SCgen} is defined from the extracted power p_{SCgen}:

$$E_{SCgen} = \sum_{n=1}^{N} p_{SCgen}(t + n\,T_{sample})T_{sample} \quad (10)$$

where N is the number of subsequent events where $p_{SC}(t) < 0$. From (10), the probability density function of E_{SCgen} is:

$$P(E_{SCgen} > E \mid p_{SC}(t) < 0) =$$

$$= \frac{\sum_{n=1}^{\infty} P(p_{SC}(t) < 0)^n \int_{-\infty}^{E/T_{sample}} f_{\sum_{k=1}^{n} p_{SC}}(p)\,dp}{P(p_{SC}(t) < 0)}(1 - P(p_{SC}(t) < 0)) \quad (11)$$

where $f_{\sum_{k=1}^{n} p_{SC}}$ is the probability density function of the sum of $p_{SC}(t)$ when $p_{SC}(t) < 0$.

Then, the probability density function of the extracted energy E_{SCgen} is derived differentiating (11):

$$f_{Esc}(e) = [1 - P(p_{SC}(t) < 0)] \sum_{n=1}^{\infty} P(p_{SC}(t) < 0)^{n-1} f_{\sum_{k=1}^{n} p_{SC}}\left(\frac{e}{T_{sample}}\right) \quad (12)$$

For example, let's consider the case where the input power is constant ($p_{Source}(t) = 2.5\,W$), the system is working in battery charge mode and, when the load is active, the power load $p_{Load}(t)$ is a random variable with an uniform distribution from 0 to 4 W with a sample time of T_{sample}=10 s. Using (12), the probability density function f_{Esc} is reported in Fig. 7: it can be noted that in 90% of the cases, the energy supplied by the supercapacitor is less than 25J. Thus, in order to avoid to use the battery during the day, i.e. to avoid the battery help state, we impose:

$$E_{Ref_{BC}} - E_{Th2} > 25J \quad (13)$$

Assuming E_{refBC}=2V and E_{Th2}=1V, the supercapacitor value C should be greater than 16.67 F.

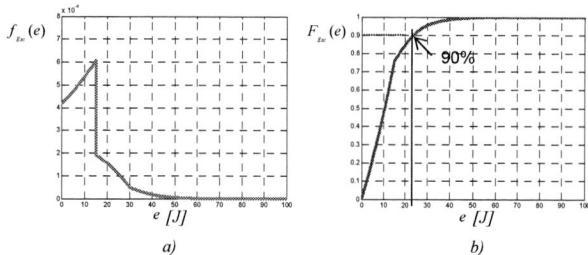

Fig. 7 – a) Density probability function of the energy required by the super-capacitor during the battery charge mode b) Cumulative distribution function of the energy required by the super-capacitor during the battery charge mode

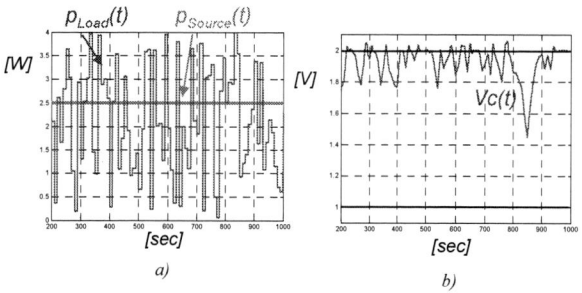

Fig. 8 – a) Simulation of the power load requirement during the battery charge mode b) Super-capacitor voltage in the same condition with the two voltage levels referred to Battery charge mode (2V) and Battery help mode (1V)

The same reasoning should be applied during the night taking into account that the battery is able to supply a maximum current I_{max}=600mA. In these conditions, the input source is zero and the maximum power supplied by the battery is 2.2W. Moreover, when the load is active, the power load $p_{Load}(t)$ is assumed to be a random variable with an uniform distribution from 0 to 2.5 W with a sample time of T_{sample}=10 s. Using (11), we found that in 90% of the cases the energy supplied by the supercapacitor is less than 3J. Thus, in order to avoid the turn-off state, we impose:

$$E_{\mathrm{Ref_{BH}}} - E_{\mathrm{Th1}} > 3J \qquad (14)$$

Assuming E_{refBH}=1V and E_{Th1}=0.2V, the supercapacitor value C should be greater than 6.25 F.

IV. SIMULATION RESULTS

The converter connected to the input power is designed on the maximum input power coming from the source, which is 5W, the nominal power of the photovoltaic panel. The supercapacitor converter is designed to work in a range of power that depends on the maximum instantaneous power required by the load or generated by source. In our application the power fluctuation of the load and the source is approximately the same (5W). The converter operates in bidirectional mode with a minimum capacitor voltage of 200mV. In order to select the appropriate values of inductance and frequency of the sepic and supercapacitor converters, Figures 9a and 9b are used.

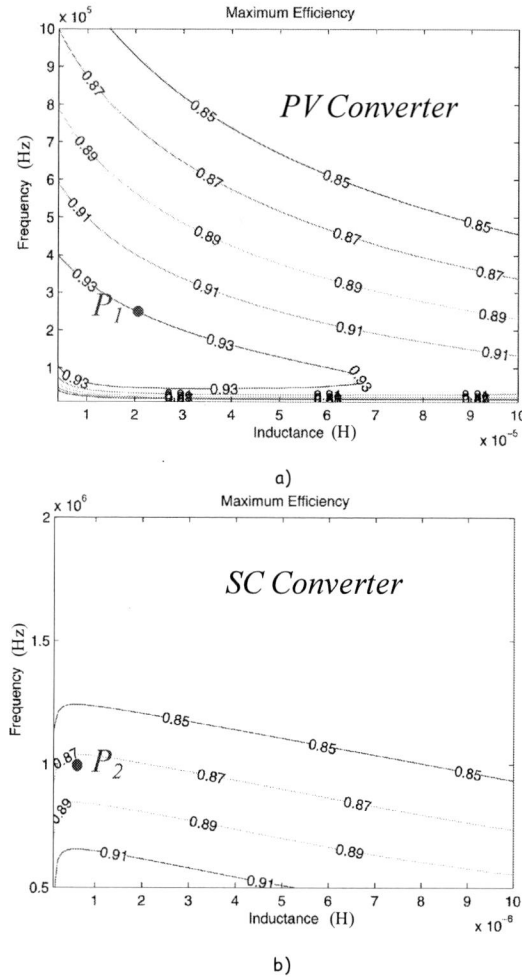

Fig. 9 – a) Efficiency of the Sepic converter as a function of inductance and frequency for the nominal voltage and current and P_1 the value of the utilized parameters (L=22μH, F_{sw}=250kHz) b) The same for the bidirectional Buck converter and P_2 the value of the utilized parameters (L=0.5μH, F_{sw}=1MHz)

Using MATLAB scripts, one for each converter, the losses for every operating conditions are calculated: the inputs are the inductance, the frequency and the input current, while fixed parameters are the input and the output voltages. The program calculates the converter operating conditions (continuous or discontinuous conduction mode); detailed mosfets conduction and switching, DCR, dead times, ground resistance (valued in 50mΩ) and drivers losses are evaluated. As the conclusion of this calculation, the maximum value of the converter efficiency within the input power range is obtained and reported in Figs. 9a and 9/b; Figure 9a is used as the design criteria for the selection of L and f_{sw} for the sepic converter and point P_1 of Fig. 9a has been chosen (L=22 μH, F_{sw}=250 kHz); for the super-capacitor converter the inductor is chosen to optimize the efficiency, but the frequency of operation is chosen higher in order to reduce the ripple of the DC Bus voltage. Thus point P2 of Fig. 9b has been chosen (L=0.5μH, F_{sw}=1 MHz).

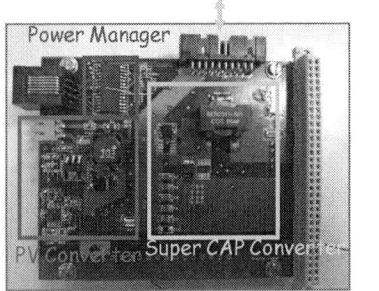

Fig. 10 - Bidirectional Buck-Boost and Sepic converter connected

The control system of both converters is based on the stabilization of the input voltage; in the Sepic case, the regulation is realized with a Proportional-Integral (PI) controller, which allows a dominant pole compensation, being the bandwidth requirement not very restrictive; the bidirectional Buck regulator instead, is a Proportional-Integral-Derivative (PID), which allows a faster stabilization of the input voltage level.

V. EXPERIMENTAL RESULTS

The power management system was experimentally verified on a prototype built using discrete Component Off The Shelf (COTS) reported in Fig. 10. The power manager is implemented by a microcontroller of the Microchip family (PIC18LF2620). The controller utilized for the sepic converter is a commercial IC (TPS43000). The converter works at 250kHz and it utilizes a coupled inductor of L=22μH with a coupling factor of 0.95 (Coilcraft MSD1260 series). The supercapacitor converter is a Buck converter that utilizes a 500nH inductor (Coilcraft SER2000 series) and it works at 1MHz, while the controller IC is ISL8118. In Fig. 11 the main converter waveforms are reported: channel 1 (CH1) is the gate of the low side transistor of the input converter, CH2 the signal of the bus DC voltage, CH3 the current to the supercapacitor and CH4 the gate of the high side transistor of the supercap converter.

The efficiency of each stage has been measured: Fig. 12 reports the efficiency of the input stage with a fixed and

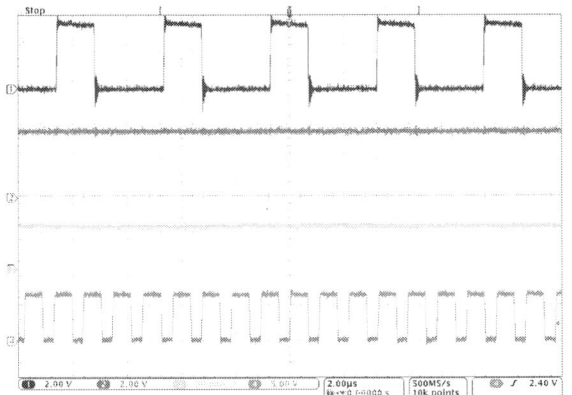

Fig. 11 – CH1: gate of the low side transistor of the input converter, CH2: signal of the bus DC voltage, CH3: sinking current to the supercapacitor, CH4: gate of the high side transistor of super cap converter.

controlled input voltage. Figure 13a shows the efficiency of the supercapacitor converter in Buck operating mode, varying the input power at different supercapacitor voltages; Figure 13b shows the efficiency in Boost mode, where the power is extracted from the supercapacitor; the results reported in

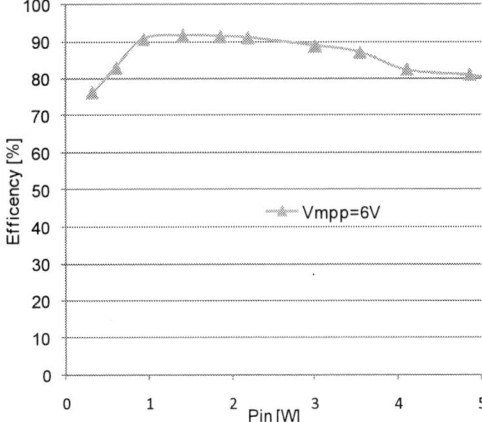

Fig. 12 – Efficiency of the Sepic input power stage, with a controlled input voltage of 6V

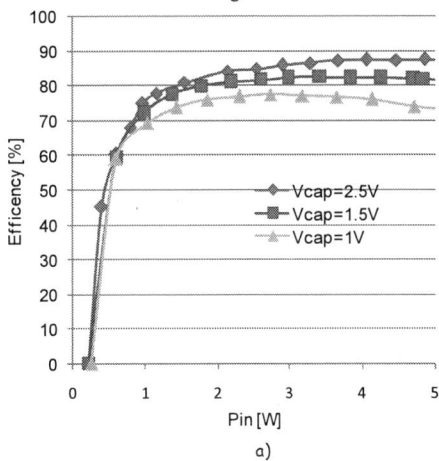

a)

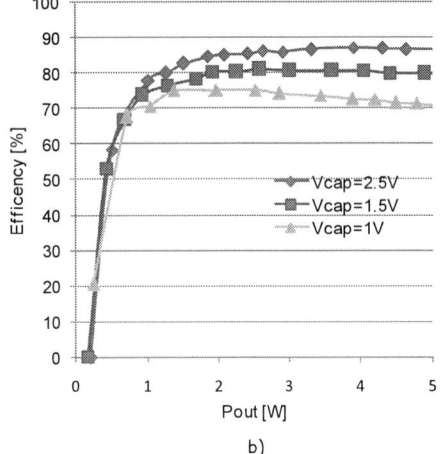

b)

Fig. 13 – a) Efficiency of the supercap power stage a) in Buck mode (sinking mode), b) in Boost mode (sourcing mode), over input power

Figs 12-13 are in agreement with the estimated efficiency reported in Fig. 9.

As a verification of the transient behaviour, the source is turned off and Fig. 14 reports the main waveforms of the proposed system without batteries, with a 50F supercapacitor as primary energy storage and with the load power constant and equal to 3W; in channel 1 (CH1) the input voltage of the sepic stage is reported, during its normal MPPT operation, in CH2 the dc bus voltage, in CH3 the he supercapacitor current and in CH4 its voltage. As can be seen, when the input voltage decreases and no energy flows in the dc bus, the supercap converter provides the holding up og the dc voltage with the supercapacitor energy.

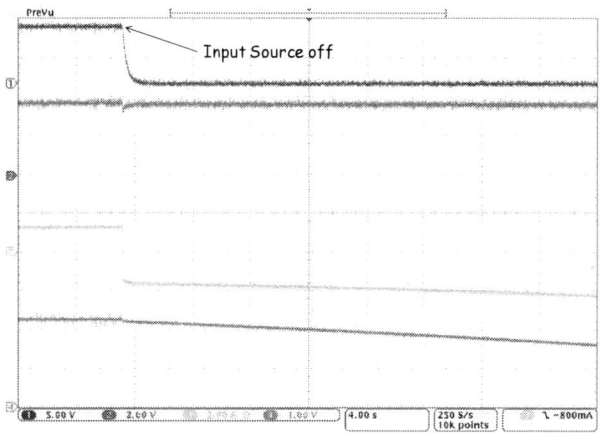

Fig. 14 – Transient response of the circuit when no energy flows from the input source in the dc bus. CH1: input voltage, CH2: dc bus, CH3 super cap current, CH4 super cap voltage.

VI. CONCLUSIONS

In this paper a power management system based on hybrid storage architecture composed by supercapacitors and lithium-ion batteries is proposed. This architecture, combined with the power management algorithm, ensures a controlled battery charge and the maximum power point tracking from the energy sources even in presence of large fluctuations of the sources and of the load, thus prolonging the battery lifetime. Moreover, the experimental results show high efficiency in the proposed conversion systems.

VII. REFERENCE

[1] G. Ottman, H. Hofmann, A. Bhatt, G. Leisutre "Adaptive Piezoelectric Energy Harvesting Circuit for wireless Remote Power Supply", IEEE Transaction on Power Electronics, Volume 17, Sept. 2002, pp. 669-676.

[2] "Energy Harvesting Proejects", Collections of articles on IEEE Pervasive Computing,January-March 2005 pp. 69-71.

[3] J.AParadiso, T.Staner,"Energy scavenging for mobile and wireless electronics", IEEE Pervasive Computing Vol 4, January-March 2005, pp. 18-27.

[4] S.Roundy, E,S Lesland, J.Baker, E.Carleton, E.Reilly, E.Lai, B.Otis, J.MRabaey, P.KWright, V.SunDararajan,"Improving power output for vibration based energy scavengers", IEEE Pervasive computing vol 4. January-March 2005, pp. 28-36.

[5] C.B.Williams, R.B. Yates, " Analysis Of a Micro-electric generator for Microsystems" in the 8th international conference on the solid state sensor and actuators, vol.1 25-29 June 1995 pp.369-372.

[6] V. Raghunathan, A. Kasal, J. Hsu, J. Friedman, M. Srivastava, "Design Consideration for Solar Energy Harvesting Wireless Embedded System", IEEE Internetional Conference on Information Processing in sensor networks (IPSN)2005, 15 April 2005 pp. 457-452.

[7] N. Lujara, J.D van Wyk, P.N. Materu, "Power electronics loss models of DC-DC converters in photovoltaic applications" in IEEE Internetional Symposium and on industrial Electronics, 1998. Volume 1, 7-10 July 1998, pp.35-39.

[8] M. Veerachary, T. SenJyu, K. Uezato, "Neural-Network-Based Maximum Power Point Traking of a Coupled-Inductor Interleaved-Boost Converter Supplied PV System Using Fuzzy Controller", IEEE Transaction on Industrila Electronics Vol.50, No.4, August 2003, pp. 749-757.

[9] J.H.R Enslin, D.BSnyman, "Combinated Low Cost , High Efficient Inverte, Peak Power Tracker and Regulator for PV Applications", IEEE Transaction on Power Electronics Vol.6 No.I January 1991, pp. 73-82.

[10] I. Kim; P.S. Ji, U.D. Han, C. Lhee, H.G. Kim,"State estimator design for solar battery charger", IEEE International Conference on Industrial Technology, 2009. ICIT 2009, 10-13 Feb. 2009.

[11] C. Alippi, C. Galperti, "Energy sotrage mechanisms in low power embedded systems: Twin batteries and supercapacitors", VITAE international conference on Wireless Comunication, Vehicular Tecnology, Information Theory and Aereospace & Electronic Systems Technology, 17-20 May 2009, pp 31-35.

[12] S. Saggini, P. Mattavelli, "Power management in multi-source multi-load energy harvesting systems", European Conference on Power Electronics and Applications, EPE 2009, 8-10 Sept. 2009, pp 1–10.

The Faulty Module Bypass for Thermoelectric Generation

Wei Qian, Fang Z. Peng and Sangmin Han
Dept. of Electrical and Computer Engineering
Michigan State University
East Lansing, MI, 48824, USA
qianwei@egr.msu.edu

Abstract—**This paper proposes the solutions of bypassing the faulty modules for the thermoelectric generation (TEG). To obtain sufficient voltage, several TEG modules usually have to connect in series. For a single string of TEG modules, when one module fails, the output power of the whole string is lost. After comparing several low-cost bypass approaches, one bypass circuit based on the buck-boost converter is analyzed. By transferring energy among the serial modules, the terminal voltage across the serial modules can be automatically balanced. Consequently, the maximum output power can still be achieved in the remaining modules, compared to the other approaches of only bypassing the faulty module. The output power stress can be shared among the serial modules according to their output power capacity, so that the utilization of diverse cells is more effective and the lifetime of the modules is prolonged. The experimental results of a laboratory prototype demonstrate its bypass and balance features.**

I. INTRODUCTION

TEG has drawn more and more attention [1]-[9], as there are more concerns over the energy consumption and the environmental impact from the IC Engine powered vehicles. As the name implies, the TEG converts the exhaust heat in a vehicle to electricity to charge the battery and to power the auxiliary load. The electric energy originates from the TEG modules. One TEG module can be modeled as a voltage source in series with a resistor, whose open-circuit voltage is proportional to the temperature gradient across it. Each module can only produce limited voltage and power, so several modules have to be connected in series and in parallel to supply enough voltage and power to the load. Plus, since the produced voltage as well as the load varies, a dc-dc boost converter usually interfaces the thermoelectric generator with the electric load.

The problem arises when one or more TEG modules fail. For instance, *m* strings of modules are connected in parallel, and each separate string consists of *n* TEG modules in series. Due to the heat gradient and the electrical property, one may somehow present normally an open circuit. This leads to the open circuit in the whole string. Thereby, the system loses the

power from this string. Further, if these TEG modules are connected in an $m \times n$ matrix as shown in Fig. 1, the reliability seems enhanced thanks to the redundancy. However, after one module is gone, the load current distributes among the modules in parallel with the faulty one. Provided that the internal voltage and internal resistance of each module were the same, the m-1 modules had to carry more current stress than the others. Eventually, as more modules fail, the modules paralleled with them will age much faster. In reality, the internal voltage and resistance of a TEG module are affected by the temperature gradient between its hot plate and cold plate. Under the worst case condition, some modules may turn to absorb electric power from the other modules. So the simple serial and parallel connection also has limits in achieving the full power utilization from the rest modules.

Hence, this paper proposes and compares several approaches to bypassing the faulty TEG modules. By adding the faulty module bypass circuit, the new current path for the remaining modules is created. One of these solutions based on the buck-boost converter is implemented. Besides the bypass function, it can also balance the terminal voltage and distribute the output power stress of the individual TEG modules.

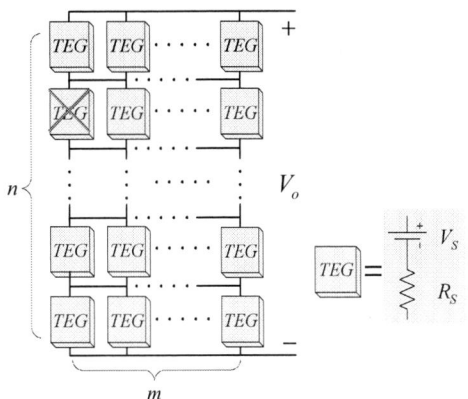

Fig. 1. The matrix connection of TEG modules

II. THE SOLUTIONS FOR FAULTY MODULE BYPASS

Firstly, the bypass circuit (abbreviated as BPS in Fig. 2) can be realized by a diode (or by a P-channel BJT) paralleled to each TEG module. In the normal operation, the diode (or P-channel BJT) exhibits open circuit to the TEG module. When one module becomes open-circuit, the open-circuit voltages of the $(n-1)$ modules cause the diode (or P-channel BJT) forward conducting. These two low-cost methods need no extra gate drive circuit. Their shortcoming is the rest modules can not reach their maximum output power capacity, because the equivalent serial voltage $(n-1) \times V_s$ in the faulty string is lower than that in the intact strings, assuming ideally the same V_s.

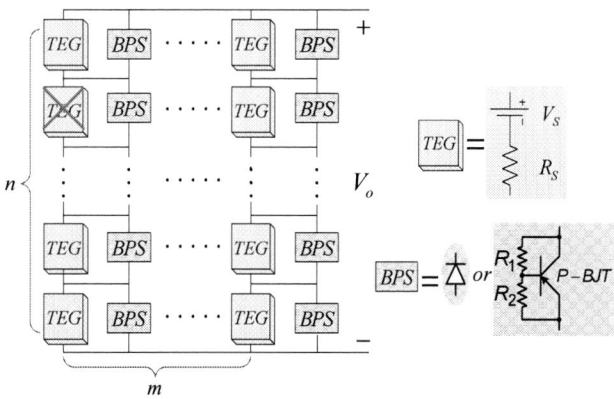

Fig. 2. The basic concept of the faulty module bypass circuit

Secondly, it is preferable to actively balance the terminal voltages of the TEG modules besides the bypass function. Fig. 3 shows two topologies that were developed for the battery equalization: the switched-capacitor structure [10] and the buck-boost based structure [11]. In the switched-capacitor structure as shown in Fig. 3(a), the synchronized switches connect to the upper and lower contact point complementarily. When one TEG module fails, the load current goes through the capacitor across the faulty module in half of the switching cycle; in the other half cycle, this capacitor gets charged by the energy transferred from the intact modules one by one through the capacitors. In the buck-boost based structure in Fig. 3(b), the energy is transferred by the inductors to charge the capacitor across the faulty module. Both of them have the merits of achieving the terminal voltage automatic balance without the voltage sensors and the closed-loop control that otherwise add the cost of the TEG system. The following sections will focus on the buck-boost based bypass circuit.

III. THE BUCK-BOOST BASED BYPASS CIRCUIT

The buck-boost based bypass circuit for n serial TEG modules can be decomposed as $(n-1)$ modular bypass units as illustrated in Fig. 4. The pair of switches S_{jp} and S_{jn} are turned on complementally by 1/2 duty ratio, so the terminal voltages of two TEG modules are the same in the bidirectional buck-boost converter if the voltage loss is negligible. Combining the $(n-1)$ modular bypass units together, the bypass circuit operates in two alternate switching states. Take a 5-cell bypass circuit as shown in Fig. 3(b) for instance. In the switching

state I as shown in Fig. 5 (a), the switches $S_{(2k-1)p}$ and $S_{(2k)n}$ (k=1 and 2) are turned on for 1/2 duty cycle; in the switching state II in Fig. 5 (b), the complementary switches are on. When the terminal voltage across one cell turns to be lower than that across the other cells, its load current is partially shifted to the other cells even in the normal operation with no faulty modules. When one cell fails as open circuit, the load current is temporally supplied by the capacitor paralleled with the faulty cell in half the switching cycle. In the other half cycle, this capacitor gets charged and the load current is supplied cell by cell from the good cells. It is notable that the switch pattern is not unique for this circuit, but as can be seen in the analysis on the capacitor current later, the capacitor voltage ripples can be minimized using this switching pattern, since the charge and discharge of the capacitors are interleaved.

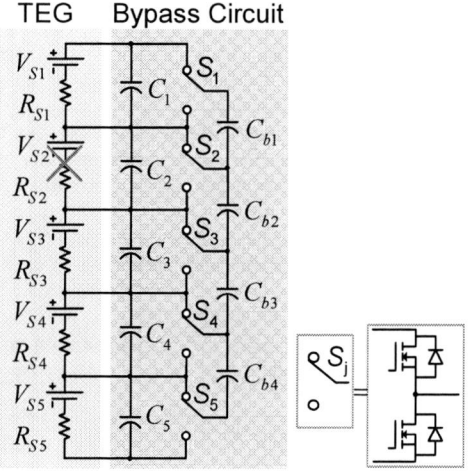

(a) The switched-capacitor structure.

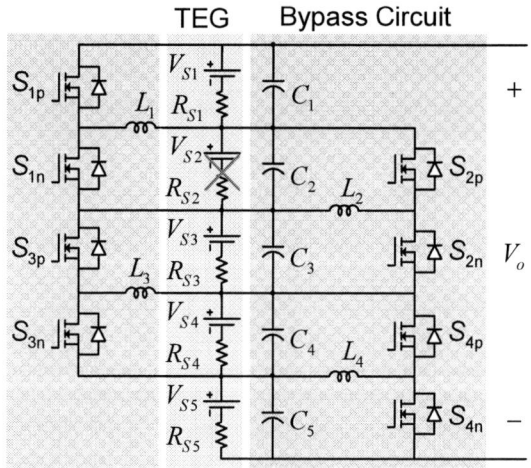

(b) The buck-boost structure

Fig. 3. Some topologies of the bypass and balance circuit.

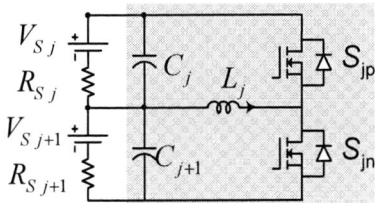

Fig. 4. The modular buck-boost bypass and balance circuit

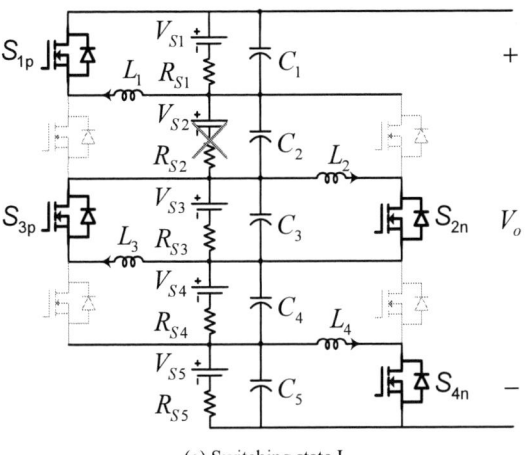

(a) Switching state I

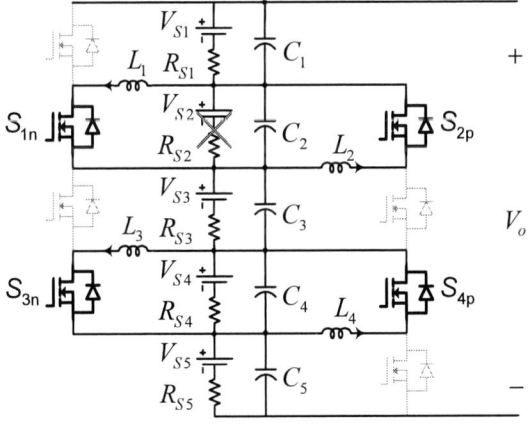

(b) Switching state II

Fig. 5. The switching states of the 5-cell bypass circuit

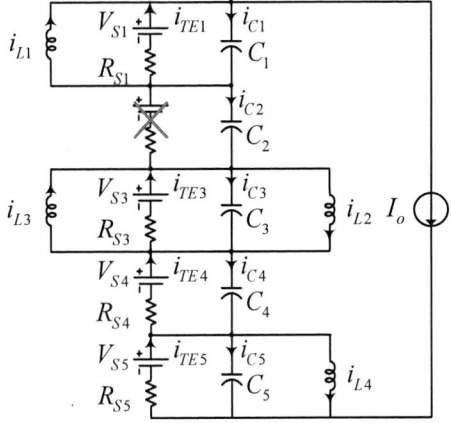

(a) The equivalent circuit of switching state I

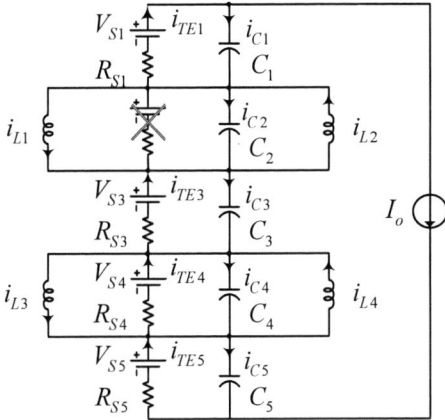

(b) The equivalent circuit of switching state II

Fig. 6. The equivalent circuit of the 5-cell bypass circuit

The following analysis will give the first image of how the load current is shared by the remaining modules. If the second TEG module in Fig. 5, for example, is gone, the equivalent circuits for the two switching states are shown in Fig. 6. To simplify the analysis, the TEG internal source voltage and resistance are assumed equal. Provided the ideal switching devices and negligible dead time, the currents through each TEG module are ideally equal, which can be expressed as the sum of the load current I_o and the extra current I_x resulted from the faulty module:

$$I_{TE1} = I_{TE3} = I_{TE4} = I_{TE5} = I_o + I_x \tag{1}$$

Let us consider the average values of the capacitor currents in every half switching cycle. The average capacitor currents and inductor currents can be calculated by jointly solving the two switching states. According to the defined direction in Fig. 6(a), in the switching state I, the average capacitor currents are:

$$\bar{i}_{C1} = \bar{i}_{C3} = \bar{i}_{C5} = -I_x \tag{2}$$

$$\bar{i}_{C2} = -I_o \tag{3}$$

$$\bar{i}_{C4} = I_x \tag{4}$$

In the switching state II in Fig. 6(b), the average capacitor currents are:

$$\bar{i}_{C1} = \bar{i}_{C3} = \bar{i}_{C5} = I_x \tag{5}$$

$$\bar{i}_{C2} = I_o = \bar{i}_{L2} - \bar{i}_{L1} - I_o \tag{6}$$

$$\bar{i}_{C4} = -I_x \tag{7}$$

The average currents through the inductors are:

$$\bar{i}_{L1} = -2I_x \tag{8}$$

$$\bar{i}_{L2} = 6I_x \tag{9}$$

$$\bar{i}_{L3} = 4I_x \tag{10}$$

$$\bar{i}_{L4} = 2I_x \tag{11}$$

Using the same approach, it can be found that the overall inductor current stress is less when the faulty TEG module is closer to the center. In other words, the small inductors should be designed based on the worst case in which the faulty module is either in the top or in the bottom.

From (6), (8) and (9) one can get:

$$I_x = I_o / 4 \tag{12}$$

So the current stress on the intact TEG modules is:

$$I_{TEj} = 5I_o / 4 \quad (j = 1, 3, 4, 5) \tag{13}$$

IV. THE OUTPUT POWER AND CURRENT STRESS

This section first analyzes the maximum output power if k (k=0, 1, …, $n-1$) arbitrary faulty TEG modules gradually turn to open circuit in an n-cell string. Later, the general analysis on the current stress discusses the merits of the bypass and balance circuit. Based on the assumption in the above section, ignoring the power loss in the circuit, the output power and input power should satisfy:

$$
\begin{aligned}
P_o &= V_o \cdot I_o \\
&= P_{in} \\
&= [(n-k)(V_s - I_{TE} \cdot R_s)] \cdot I_{TE}
\end{aligned} \tag{14}
$$

So the maximum output power can be got as:

$$P_{o\_max} = (n-k)V_s^2 / (4R_s) \tag{15}$$

Hence, theoretically the remaining $(n-k)$ modules can still achieve their individual maximum output power of $V_s^2 / (4R_s)$, as if no faulty modules exist in the string.

Since the output voltage is: $V_o = n(V_s - I_{TE} \cdot R_s)$, the current stress on the remaining modules can be calculated from (14):

$$I_{TE} = \frac{n}{n-k} I_o \tag{16}$$

This explain (13) from another point of view.

Generally, considering different internal voltage and resistance in individual modules, the input power can be represented as:

$$P_{in} = \sum_{j=1}^{n-k} (V_{sj} - I_{TEj} \cdot R_{sj}) I_{TEj} \tag{17}$$

Since the terminal voltage of each cell is equalized to V_o / n if the voltage drop in the bypass circuit is negligible, the current through each TEG modules can be written as:

$$I_{TEj} = (V_{sj} - V_o / n) / R_{sj} \quad (j = 1, ..., (n-k)) \tag{18}$$

From (18), when k=0, some merits of the bypass and balance circuit can be observed in the normal operation of no faulty modules as well. The individual module currents are automatically distributed according to their output capacity. That is, the modules with higher internal voltage and lower

internal resistance undertake more current and thereby more output power. The current drawn from a TEG module is reversely proportional to its internal resistance. So less power loss will be dissipated on the module with large internal resistance, compared the simple serial connection of TEG modules. The life time of the TEG modules can be prolonged. The chance is reduced that such module turns to consume electric power from other modules.

V. EXPERIMENTAL RESULTS

A laboratory prototype was fabricated to verify the bypass and balance functionality. The test configuration is shown in Fig. 7. The capacitance of $C_1 \sim C_5$ across each cell is 2483.5 μF. The inductance of $L_1 \sim L_4$ is around 4.7 μH. The switching frequency is 30 kHz. Two cases of faulty condition were conducted. In the first test, only the last four cells No. 2 ~ No. 5 are powered by four TEG simulators, which simulate the V-I characteristics of the TEG modules. As mentioned before, this is one of the worst case in which the top module is open. When these four modules are directly connected in serials to power a 1.2 Ω resistive load, their terminal voltages, as listed in Table 1, are different. When the buck-boost based bypass and balance circuit is employed, the test results are recorded in Table 2. The terminal voltages of each cell are much more balanced than those in the simple serial connection, with a little voltage difference resulted from the device voltage drop. The serial string can still power the load thanks to the bypass and balance circuit, even if one module is gone. Some corresponding experimental waveforms are shown in Fig. 8. Fig. 8(a) shows the terminal voltage of each cell. Fig. 8(b) shows the output voltage V_o, the output current I_o and the current drawn from one of the cells I_{TE3}. Fig. 8(c) shows the typical waveforms of the inductor currents and the voltage across the switching devices. In the second test, both the No. 1 and No. 5 modules are gone, as shown in Fig. 7. The test results are listed in Table 3. The output power in this case is 3/4 of the power in the first case. The corresponding experimental waveforms are shown in Fig. 9. The two cases demonstrate the aforementioned bypass and balance features.

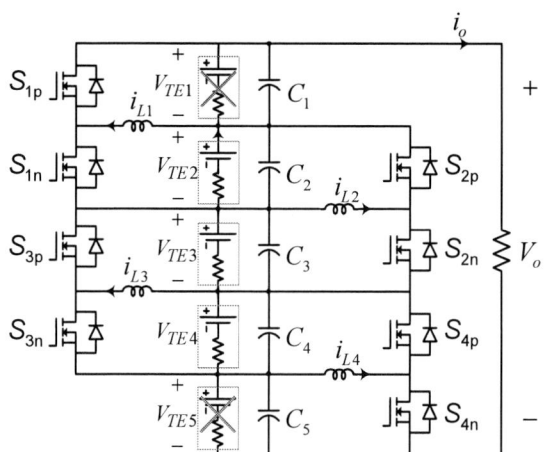

Fig. 7.   The test configuration of the buck-boost bypass circuit

Table 1 The terminal voltages of the four TEG simulators without connecting to the bypass circuit

V_{TE2}	4.370 V
V_{TE3}	3.536 V
V_{TE4}	2.092 V
V_{TE5}	2.130 V

Table 2 Test data with one faulty module

Terminal voltage (V)		Module current (A)	
V_{TE1}	2.187	I_{TE1}	0
V_{TE2}	2.480	I_{TE2}	16.736
V_{TE3}	2.539	I_{TE3}	13.801
V_{TE4}	2.450	I_{TE4}	6.334
V_{TE5}	2.712	I_{TE5}	9.672
Output voltage (V)		Output current (A)	
V_o	12.368	I_o	9.405
P_o/P_{in}		98.3%	

Table 3 Test data with two faulty modules

Terminal voltage (V)		Module current (A)	
V_{TE1}	2.232	I_{TE1}	0
V_{TE2}	2.667	I_{TE2}	11.557
V_{TE3}	2.801	I_{TE3}	10.689
V_{TE4}	2.778	I_{TE4}	13.375
V_{TE5}	2.407	I_{TE5}	0
Output voltage (V)		Output current (A)	
V_o	12.529	I_o	7.156
P_o/P_{in}		91.6%	

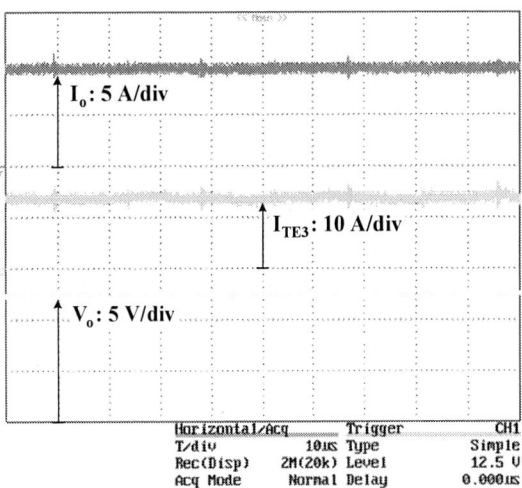

(b) Output voltage, current and the input current of one module

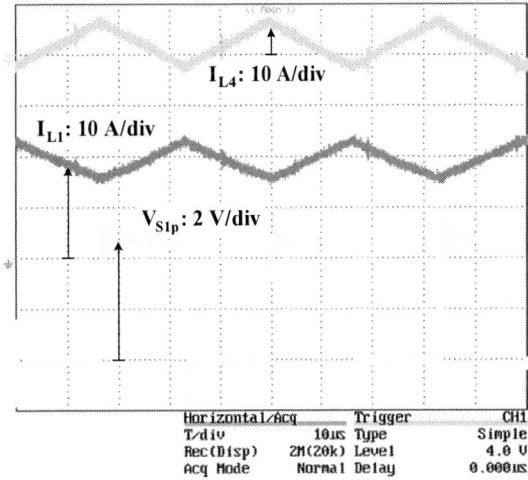

(c) Typical waveforms of the inductors and the switches

Fig. 8. Experimental results for the case of one faulty module

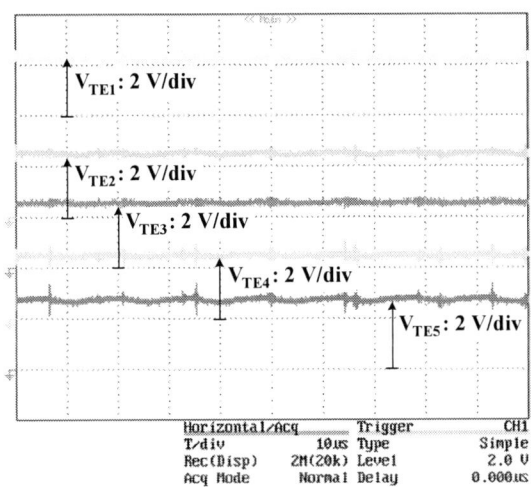

(a) Terminal voltages of the five cells

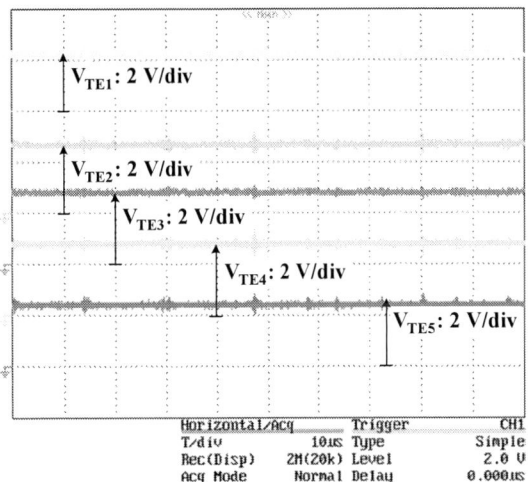

(a) Terminal voltages of the five cells

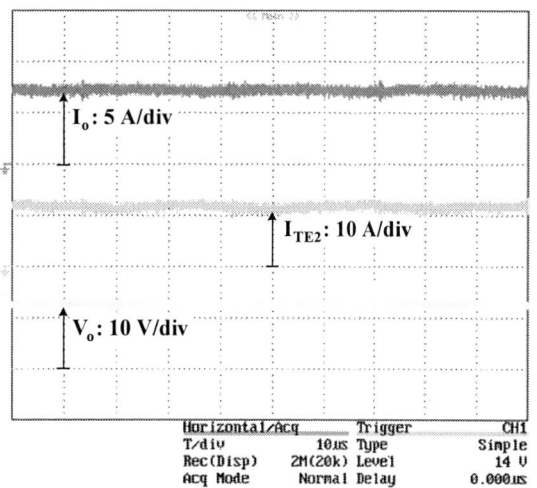

(b) Output voltage, current and the input current of one module

Fig. 9.   Experimental results for the case of two faulty modules

VI.   CONCLUSION

This paper introduces several approaches to bypassing the faulty TEG modules in a series connected string. Among them, the modular buck-boost based bypass circuit was analyzed in detail. The experimental results have verified its functionality and features. Besides bypassing the open-circuit modules, this sensorless solution is able to balance the terminal voltages automatically. So ideally the maximum output power from each of the remaining modules can still be achieved. Furthermore, the current and output power undertaken by individual modules can be distributed according to their output power capacity, which is associated with their temperature gradient and their electric property. Compared to the simple serial connection, the utilization of the series connected TEG modules is more effective, because less power is consumed by the module with large internal resistance. The life time of the serial modules can be prolonged.

REFERENCES

[1]   S. B. Riffat and X. Ma, "Thermoelectrics: a review of present and potential applications," Applied Thermal Engineering, 2003, vol.23, pp. 913-935.

[2]   J. Yang, "Potential applications of thermoelectric waste heat recovery in the automotive industry," in Proc. IEEE Int. Conf. Thermoelectron., 2005, pp. 19–23.

[3]   M. Rahman and R. Shuttleworth, "Thermoelectric power generation for battery charging," in Energy Management and Power Delivery, 1995. Proceedings of EMPD '95., International Conference on, 1995, vol.1, pp. 186-191.

[4]   S. Lineykin and S. Ben-Yaakov, "Modeling and analysis of thermoelectric modules," Industry Applications, IEEE Transactions on, , 2007, vol. 43, pp. 505-512.

[5]   I. Doms, P. Merken and C. Van Hoof, "Comparison of dc-dc converter architectures of power management circuits for thermoelectric generators," in Power Electronics and Applications, 2007 European Conference on, 2007, pp. 1-5.

[6]   J. M. Damaschke, "Design of a low-input-voltage converter for thermoelectric generator", Industry Applications, IEEE Transactions on Volume 33, Issue 5, Sept.-Oct. 1997, pp.1203-1207.

[7]   L. Chen, D. Cao, Y. Huang and F. Z. Peng, "Modeling and power conditioning for thermoelectric generation", Power Electronics specialists Conf., 2008, 15-19 June 2008, pp. 1098-1103.

[8]   H. Nagayoshi, T. Kajikawa and T. Sugiyama, "Comparison of maximum power point control methods for thermoelectric power generator," in Thermoelectrics, 2002. Proceedings ICT '02. Twenty-First International Conference on, 2002, pp. 450-453.

[9]   R. Y. Kim and J. Lai, "A seamless mode transfer maximum power point tracking controller for thermoelectric generator applications", Power Electronics, IEEE Transactions on, Volume 23, Issue 5, Sept. 2008, pp.2310- 2318.

[10]   C. Pascual and P. T. Krein, "Switched capacitor system for automatic series battery equalization", Applied Power Electronics Conference and Exposition, 1997. APEC '97 Conference Proceedings 1997., Twelfth Annual, Volume 2, 23-27 Feb. 1997, vol.2, pp. 848 - 854.

[11]   K. Nishijima, H. Sakamoto and K. Harada, "A PWM controlled simple and high performance battery balancing system," in Proc. IEEE Power Electronics Specialists Conf., June 2000, pp. 517-520.

Maximum Power Point Tracking Feasibility in Photovoltaic Energy-Conversion Systems

Sairaj V. Dhople, Ali Davoudi, Gerald Nilles, and Patrick L. Chapman
Grainger Center for Electric Machinery and Electromechanics
Department of Electrical and Computer Engineering
University of Illinois at Urbana-Champaign
Urbana, Illinois 61801, USA
sdhople2@illinois.edu

Abstract— **Maximum power point tracking (MPPT) algorithms control power-electronic circuits employed in photovoltaic (PV) - energy systems to maximize the energy harvested over all ambient conditions. Irrespective of algorithm utilized, there are likely to be operating conditions under which MPPT can not be realized. This phenomenon is explored in the context of a boost converter and a generalized dc load. Experimental and simulation results demonstrate the infeasibility of MPPT when loads violate limits that are derived from the operational constraints of the boost converter.**

I. INTRODUCTION

A wide range of stand-alone and grid-tied PV applications employ switch-mode dc-dc converters. A typical set-up of a PV energy-conversion system is shown in Figure 1. The energy sourced by the PV module is processed by an MPPT converter and delivered to a load (utility grid, battery, etc.) through a power-conditioning converter. Neglecting switching artifacts, in an averaged sense, the power-conditioning circuit and load can be represented with an equivalent dc load, *Equivalent Load*, as shown in Figure 2.

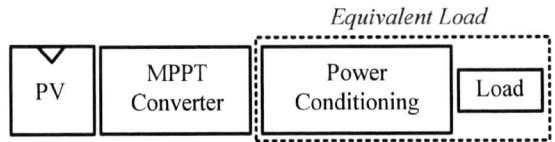

Figure 1. PV energy-conversion and conditioning system

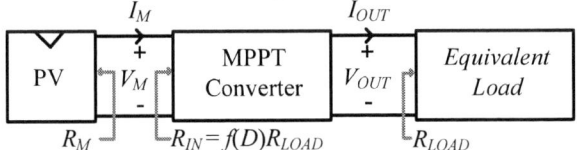

Figure 2. Energy-conversion chain derived by replacing power-conditioning converter and load with an equivalent load

Figure 3 depicts the *I-V* characteristic of a typical PV source, for different values of insolation, *S*, and ambient temperature, *T*. For a given set of ambient conditions, operating the PV module at terminal voltage V_M, ensures the extraction of maximum power. However, as Figure 3 suggests, the optimal operating points (denoted as pairs (V_M, I_M)) are a function of ambient conditions. This necessitates continual tracking by the MPPT converter.

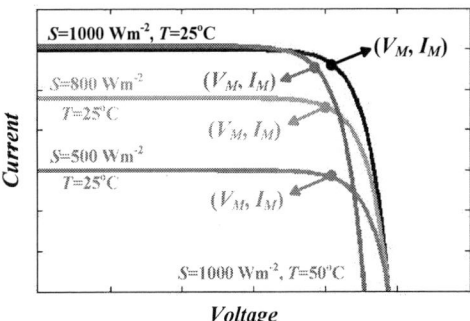

Figure 3. I-V curves of a typical PV module for different ambient conditions

The optimal PV impedance that yields maximum power is denoted as R_M and defined as the ratio of the maximum power point (MPP) voltage, V_M, to MPP current, I_M:

$$R_M = \frac{V_M}{I_M} \qquad (1)$$

Maximum power point tracking techniques control the MPPT converter to match the converter input impedance, R_{IN}, to the optimal source impedance, R_M. The impedance transfer function, $f(D)$, of the MPPT converter (refer to Figure 2) is a function of the duty cycle, D, and relates the input impedance and output load (R_{IN}, R_{LOAD}, respectively) by:

$$R_{IN} = f(D)R_{LOAD} \qquad (2)$$

Since R_M changes with ambient conditions, MPPT controllers seek the optimal duty cycle, D_M, so that:

$$R_{IN} = R_M = f(D_M)R_{LOAD} \qquad (3)$$

Innumerable algorithms have been proposed to track the MPP of a PV module (see [1]-[4] and references therein for recent work in this area). Interested readers can find a comprehensive comparison of the most widely used techniques for MPPT in [5]-[6]. For any MPPT algorithm, there might be ambient conditions or loads that make it impossible to achieve MPPT as dictated by (3). This phenomenon is defined as *tracking ineffectiveness*. Tracking ineffectiveness arises due to intrinsic operational constraints of dc-dc converters, and is independent of the MPPT algorithm. A theoretical assessment of tracking ineffectiveness as a function of converter topology can be found in [7], in which MPPT is implemented with buck, boost and buck-boost converters for a constant-resistance load. There are noted to be ambient conditions and loads for which MPPT can not be realised with the boost and buck topologies, irrespective of MPPT algorithm. This is because, buck converters are only able to step down input voltages and boost converters are only able to step up input voltages. In contrast, the buck-boost converter is not constrained as the other topologies for the case studies investigated.

In this work, tracking ineffectiveness is addressed for a boost converter that realizes MPPT in the context of the general setup depicted in Figure 2. The analysis extends the studies in [7] by considering constant voltage and current loads (in addition to resistive loads). MPPT infeasibility due to discontinuous current conduction mode (DCM) is also assessed. In Section II, the impedance transfer function and critical inductance are derived for the boost converter. Section III describes minimum and maximum limits on generalized dc loads based on operational traits of the boost converter such that MPPT is feasible and the converter operates in continuous current conduction mode (CCM). Theoretical results are verified with simulations and experimental studies in Sections IV and V, respectively.

II. OPERATIONAL TRAITS OF THE BOOST CONVERTER

The impedance transfer function and the critical inductance of the boost converter are derived. Subsequently, these constructs are used to establish load bounds to ensure feasible MPPT.

A. Impedance Transfer Function

Consider implementing MPPT with a boost converter and an arbitrary load, modeled as voltage source, V_{LOAD}, or current source, I_{LOAD}, or resistance, R_{LOAD}, as shown in Figures 4 (a), (b), and (c), respectively. The average output voltage and current are denoted as V_{OUT} and I_{OUT}, respectively. The average input voltage and current are enforced to the optimal values, V_M and I_M, respectively by the MPPT controller.

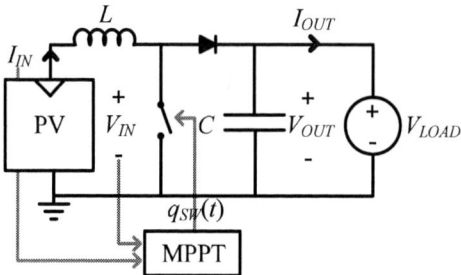

(a) Constant-voltage load

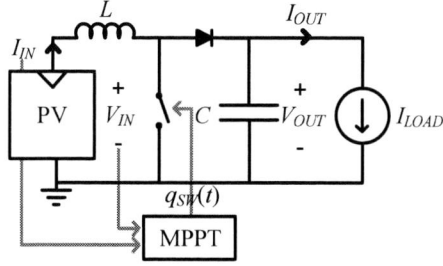

(b) Constant-current load

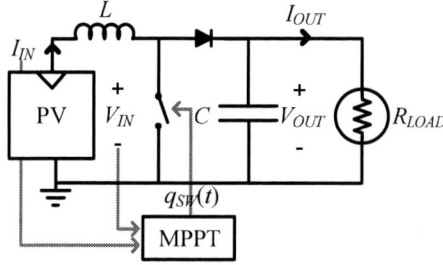

(c) Constant-resistance load

Figure 4.   PV source powering voltage (a), current (b) or resistive (c) loads through a boost converter.

Assuming that the converter is lossless and operates in CCM, the following impedance transformation applies:

$$f(D_M) = \frac{R_M}{R_{OUT}} = (1 - D_M)^2 \qquad (4)$$

The optimal duty cycle that yields maximum power for a given set of ambient conditions and load, D_M, can be solved from (4):

$$D_M = 1 - \sqrt{\frac{R_M}{R_{OUT}}} \qquad (5)$$

Tracking ineffectiveness arises as the boost-converter duty cycle is constrained to the closed set [0,1]. Equation (5) suggests that MPPT is feasible only when

$$R_{OUT} \geq R_M \qquad (6)$$

B. Critical Inductance

Transfer efficiency is defined as the ratio of the average power drawn from the PV source to the theoretical maximum available under a given set of ambient conditions. Switching action induces PV-current ripple which degrades transfer

978-1-4244-4782-4/10 $26.00 © 2010 IEEE   2295

efficiency in boost converters with low input-filter inductance or switching frequency [8]. Hence, it is beneficial to avoid DCM under all ambient conditions to limit the PV-current ripple.

For a given converter topology, the inductance that ensures operation at the boundary of DCM and CCM for a given switching frequency and load is defined as the critical inductance. For the boost converter, the peak-to-peak input current ripple can be expressed as the following function, based on a straight-line approximation for the inductor current during the period when the active switch conducts.

$$\Delta I_{IN} = \frac{V_{IN} D}{L f} \qquad (7)$$

In (7), f denotes the switching frequency, L, the inductance, D, the duty cycle and V_{IN}, the input voltage. At the boundary of DCM and CCM, the peak-to-peak current ripple is twice the average input current. Substituting the current ripple and transforming to output parameters provides the following expression for the boost-converter critical inductance, L_{CRIT}:

$$L_{CRIT} = \frac{D(1-D)^2 R_{OUT}}{2f} \qquad (8)$$

To avoid DCM with MPPT, the boost inductance, L, should be larger than the critical inductance,

$$L \geq \frac{D_M (1-D_M)^2 R_{OUT}}{2f} \qquad (9)$$

Thus, the output load is bounded by above to prevent operation in DCM:

$$R_{OUT} \leq \frac{2fL}{D_M (1-D_M)^2} \qquad (10)$$

With ripple correlation control [9] chosen for MPPT, in critical conduction mode, the peak current is (approximately) the MPP current, I_M, while the PV voltage swings between the open-circuit voltage, V_{OC}, and (approximately) MPP voltage, V_M. For simplicity and ease of analysis, the average input voltage is assumed as the average of V_M and V_{OC}. The average input current is one half of I_M. Thus, the PV source is not loaded at the MPP. The duty cycle in this case can be approximated as:

$$\begin{cases} D_M = 1 - \sqrt{\dfrac{R_{IN}}{R_{OUT}}} \\ R_{IN} = \dfrac{V_{IN}}{I_{IN}} = \dfrac{(V_M + V_{OC})/2}{I_M / 2} \end{cases} \qquad (11)$$

The discussion so far suggests that ensuring MPPT feasibility over all ambient conditions of interest requires careful design of the boost inductance and choice of switching frequency. Other concerns such as cost and dynamic response also deserve attention.

III. LOAD CONSTRAINTS

The loading condition which ensures feasible MPPT (6) and the constraint imposed by the desire to operate in CCM (10) are assessed in this section for constant voltage, current and resistive loads.

A. Constant-Voltage Load

Many power-electronics loads are represented as constant voltages. For an ideal, lossless converter, applying conservation of energy dictates that the average output current, I_{OUT}, satisfies (12) in a boost converter for a constant-voltage load, V_{LOAD},

$$I_{OUT} = \frac{V_{IN} I_{IN}}{V_{LOAD}} \qquad (12)$$

For the constant-voltage load and the ones that are discussed subsequently, the average input voltage and current are V_M and I_M in CCM. In critical conduction mode, they are determined based on (11). In either case, the output load is derived as:

$$R_{OUT} = \frac{V_{OUT}}{I_{OUT}} = \frac{V_{LOAD}^2}{V_{IN} I_{IN}} \qquad (13)$$

Substituting (13) in (6) and (10) yields,

$$\begin{cases} V_{LOAD} \geq V_{LOAD-MIN} = V_M \\ V_{LOAD} \leq V_{LOAD-MAX} = \dfrac{V_{IN}(R_{IN})}{(R_{IN} - 2fL)}, R_{IN} > 2fL \end{cases} \qquad (14)$$

In (14), V_{IN} and R_{IN} are the average input voltage and PV impedance in critical conduction mode, determined through (11). Constant-voltage loads that allow feasible MPPT and avoid DCM are bounded below by $V_{LOAD-MIN}$ that arises due to the voltage-transfer ratio of the boost converter, and bounded above by $V_{LOAD-MAX}$ that is based on CCM operation.

B. Constant-Current Load

Consider delivering power to a constant-current load, I_{LOAD}. The average output voltage, V_{OUT}, satisfies:

$$V_{OUT} = \frac{V_{IN} I_{IN}}{I_{LOAD}} \qquad (15)$$

The converter loading can be expressed as:

$$R_{OUT} = \frac{V_{OUT}}{I_{OUT}} = \frac{V_{IN} I_{IN}}{I_{LOAD}^2} \qquad (16)$$

Substituting (16) in (6) and (10) yields,

$$\begin{cases} I_{LOAD} \geq I_{LOAD-MIN} = \dfrac{I_{IN}(R_{IN} - 2fL)}{(R_{IN})}, R_{IN} > 2fL \\ I_{LOAD} \leq I_{LOAD-MAX} = I_M \end{cases} \qquad (17)$$

With a constant-current load, the lower bound, $I_{LOAD-MIN}$ is imposed to avoid the possibility of DCM, while the upper bound, $I_{LOAD-MAX}$, is imposed due to the current-transfer ratio of the boost converter.

C. Constant-Resistance Load

Substituting the load resistance, R_{LOAD}, in (6) and (10) yields,

$$\begin{cases} R_{LOAD} \geq R_{LOAD-MIN} = R_M \\ R_{LOAD} \leq R_{LOAD-MAX} = \dfrac{\left(R_{IN}\right)^3}{\left(R_{IN} - 2fL\right)^2}, R_{IN} > 2fL \end{cases} \quad (18)$$

Equation (18) suggests that the load should be larger than the optimal MPP impedance, R_M to ensure feasible MPPT. Analogous to the case of constant voltage and current loads, the avoidance of DCM (10) further bounds the load to a maximum of $R_{LOAD-MAX}$.

For the boost converter, we have established limits on general dc loads such that the goal of maximum power point tracking is achieved. One limit arises because the average output voltage is greater than the input voltage. As a consequence, the average output current is lower than the input current. The other limit is imposed by the desire to operate the converter in CCM under all possible ambient conditions. Effectively, these limits constrain realizable loads based on the desired application of the converter. The next section provides simulation results that enable visualization of the bounds listed for a representative set of boost-converter specifications.

IV. SIMULATION STUDIES

The constraints imposed by tracking ineffectiveness on feasible loads are explored through simulations. Table I lists the PV-module parameters. The boost converter has inductance, 500 μH, filter capacitance 100 μF and switching frequency, 10 kHz. For the given specifications, the quantities, $V_{LOAD-MIN}$, $V_{LOAD-MAX}$ are plotted in Figure 5, $I_{LOAD-MIN}$, $I_{LOAD-MAX}$ are plotted in Figure 6, and $R_{LOAD-MIN}$, $R_{LOAD-MAX}$ are plotted in Figure 7. For each load, a higher filter inductance relaxes the DCM bound as expected by (9).

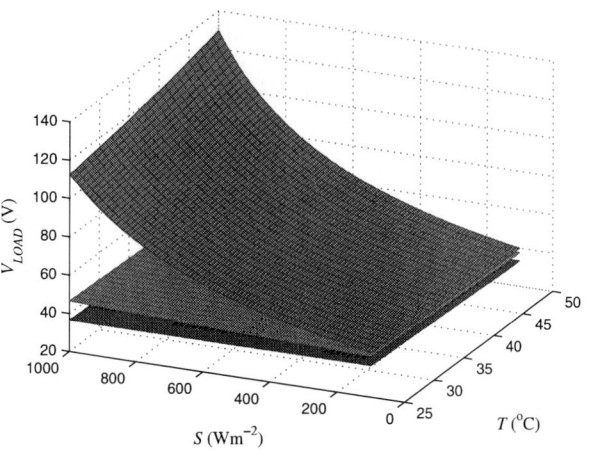

Figure 5. $V_{LOAD-MIN}$ (■), $V_{LOAD-MAX}$ (▩) $L = 100\mu H$, $V_{LOAD-MAX}$ (▩) $L = 500\mu H$.

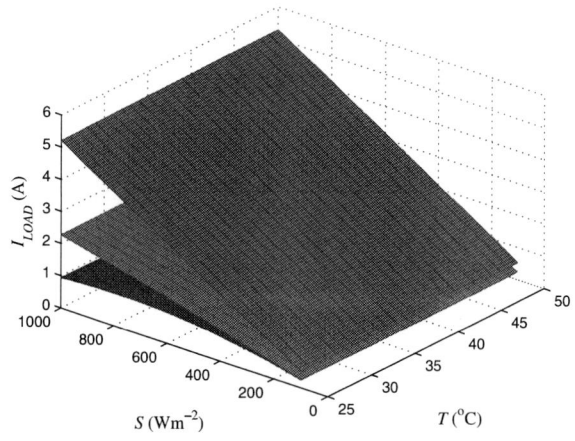

Figure 6. $I_{LOAD-MIN}$ (■) $L = 500\mu H$, $I_{LOAD-MIN}$ (▩) $L = 100\mu H$, $I_{LOAD-MAX}$ (▩)

TABLE I
PARAMETERS OF PV MODULE

Symbol	Quantity	Value
V_{OC}	Rated open-circuit voltage	44.8 V
I_{SC}	Rated short-circuit current	5.5 A
I_M	Rated current	5.2 A
V_M	Rated voltage	36.6 V
α	Temperature coefficient for current	0.065 % / °C
β	Temperature coefficient for voltage	160 mV / °C

Substituting the relevant PV-module and converter specifications in (14) indicates that a voltage load of 112 V forces the converter to operate at the boundary of CCM and DCM in standard test conditions. This is verified in Figure 8 (a) which depicts the PV current in steady-state. The high current ripple renders MPPT infeasible and as a result, the average power harvested from the module is approximately 118 W, much lesser than the rated power of the module (190 W).

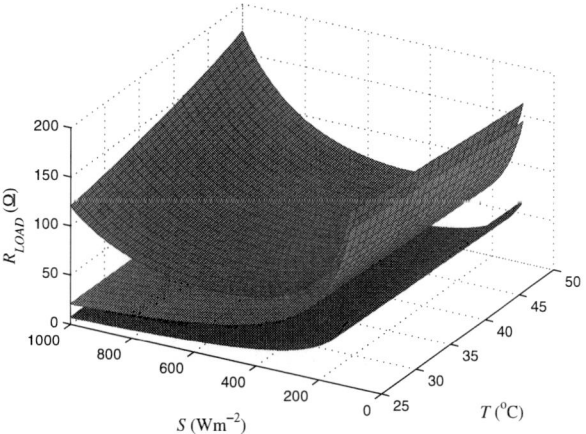

Figure 7. $R_{LOAD-MIN}$ (■), $R_{LOAD-MAX}$ (▩) $L = 100\mu H$, $R_{LOAD-MAX}$ (▩) $L = 500\mu H$.

Equations (17)-(18) indicate that a current load of 0.93 A and a resistive load of 120 Ω cause operation at the boundary of CCM and DCM. This is verified by the steady-state PV currents shown in and Figures 9 (a) and (b), respectively.

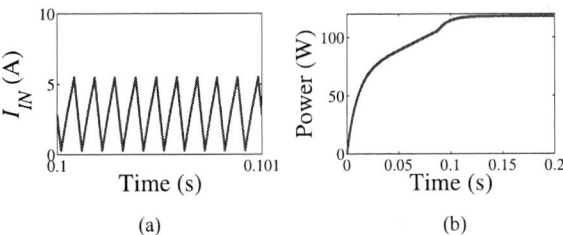

(a)   (b)

Figure 8.   PV current (a) and average PV power (b), V_{LOAD} = 112 V

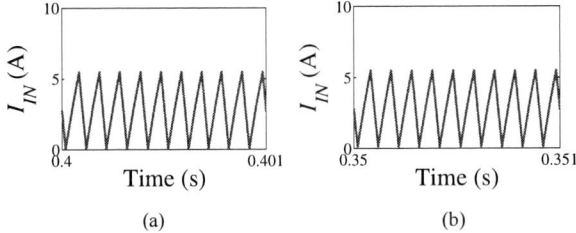

(a)   (b)

Figure 9.   PV current, I_{LOAD} = 0.93 A (a), R_{LOAD} = 120 Ω (b)

V.   EXPERIMENTAL STUDIES

The infeasibility in MPPT with a boost converter is examined for a resistive load. The boost converter specifications are attached in Table II. A maximum limit on the duty ratio is enforced so that the output voltage is no more than the voltage rating of the filter capacitor.

TABLE II
CONVERTER SPECIFICATIONS - EXPERIMENT

Symbol	Quantity	Value
L	Boost inductance	4.45 mH
C	Output capacitance	100 µF
V_{CAP}	Voltage rating of capacitor	450 V
D_{MAX}	Maximum allowed duty ratio	90%
f	Switching Frequency	33 kHz

Figures 10-12 depict the voltage across the active switch of the boost converter for 100, 250 and 500 Ω loads, under similar ambient conditions. The duty ratio ratio required for MPPT increases with load as suggested by (5).

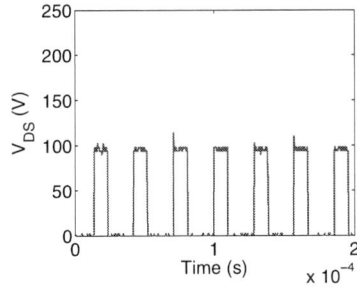

Figure 10: MOSFET drain-source voltage,
R=100Ω, S=626Wm^{-2}, T=22°C, D_M=66.6%

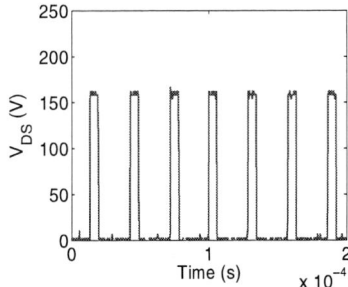

Figure 11: MOSFET drain-source voltage,
R=250 Ω, S=693 Wm^{-2}, T=21°C, D_M=80.5%

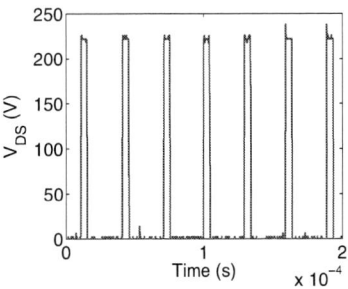

Figure 12: MOSFET drain-source voltage,
R=500Ω, S=663 Wm^{-2}, T=21°C, D_M=88.0%

For each load, the PV current and voltage are depicted in pairs in Figures 13-15. The expected (plotted as dashed lines) and experimental values match up, demonstrating feasible MPPT. Given the maximum allowed duty ratio of 90 %, MPPT can not be guaranteed for loads greater than 500 Ω under similar ambient conditions. This serves as an interesting case study of tracking ineffectiveness imposed by device ratings.

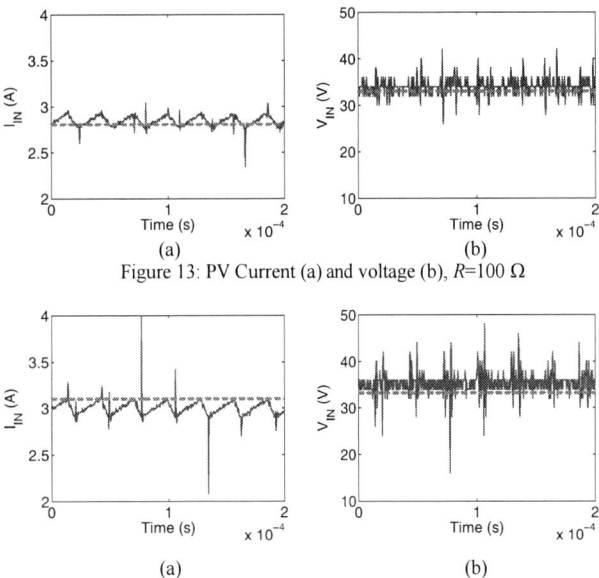

(a)   (b)
Figure 13: PV Current (a) and voltage (b), R=100 Ω

(a)   (b)
Figure 14: PV Current (a) and voltage (b), R=250 Ω

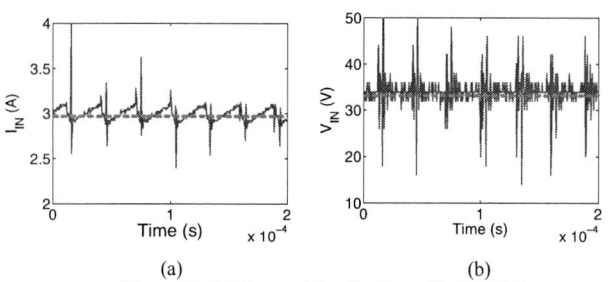

(a)   (b)

Figure 15: PV Current (a) and voltage (b), R=500 Ω

VI. CONCLUSIONS

The limitations on MPPT due to the operational constraints of dc-dc converters are examined in the context of a boost converter. The analysis acknowledges constant resistance, voltage and current loads. Allowed load values are bounded due to the fundamental operational traits of the boost converter and by the desire to operate in CCM. Simulations help visualize the load bounds over the expected range of ambient conditions. Similar tools could be applied to analyze more complicated circuit topologies in stand-alone and grid-tied power electronic converters.

REFERENCES

[1] D. Sera, R. Teodorescu, J. Hantschel, and M. Knoll, "Optimized maximum power point tracker for fast-changing environmental conditions," *IEEE Trans. Industrial Electronics*, vol. 55, pp. 2629-2637, July 2008.

[2] R. Gules, J. Pacheco, H. L. Hey, and J. Imhoff, "A maximum power point tracking system with parallel connection for PV stand-alone applications," *IEEE Trans. Industrial Electronics*, vol. 55, pp. 2674-2683, July 2008.

[3] M. Fortunato, A. Giustiniani, G. Petrone, G. Spagnuolo, and M. Vitelli, "Maximum power point tracking in a one-cycle-controlled single-stage photovoltaic inverter," *IEEE Trans. Industrial Electronics*, vol. 55, pp. 2684-2693, July 2008.

[4] V. V. R. Scarpa, S. Buso, and G. Spiazzi, "Low-complexity MPPT technique exploiting the PV module MPP locus characterization," *IEEE Trans. Industrial Electronics*, vol. 56, pp. 1531-1538, May 2009.

[5] T. Esram and P. L. Chapman, "Comparison of photovoltaic array maximum power point tracking techniques," *IEEE Trans. Energy Conversion*, vol. 22, pp. 439-449, June 2007.

[6] D. P. Hohm and M. E. Ropp, "Comparative study of maximum power point tracking algorithms," *Progress in Photovoltaics: Research and Applications*, vol. 11, pp. 47-62, Jan. 2003.

[7] J. M. Enrique, E. Durán, M. Sidrach-de-Cardona, and J. M. Andújar, "Theoretical assessment of maximum power point tracking efficiency of photovoltaic facilities with different converter topologies" *Solar Energy*, vol. 81, pp. 31-38, Jan. 2007.

[8] N. D. Benavides and P. L. Chapman, "Modeling the effect of voltage ripple on the power output of photovoltaic modules," *IEEE Trans. Industrial Electronics*, vol. 55, pp. 2638-2643, July 2008.

[9] T. Esram, J. W. Kimball, P. T. Krein, P. L. Chapman, and P. Midya, "Dynamic maximum power point tracking of photovoltaic arrays using ripple correlation control," *IEEE Trans. Power Electronics*, vol. 21, pp. 1282-1291, Sep. 2006.

Realization of a General LED Lighting System Based on a Novel Power Line Communication Technology

Chushan Li, Jiande Wu, Xiangning He
College of Electrical Engineering
Zhejiang University
Hangzhou, China
eewjd@zju.edu.cn

Abstract—**LED is regarded as one of the best potential light sources for next-generation lighting. In LED applications, Power Line Communication (PLC) is one of the popular communication techniques, which is used to control the complex lighting system. In this paper, a general LED lighting system based on a novel PLC technology, which is named as P-BUS, is presented. It is made up of four main parts: the master controller, the slave controller which controls the LED module, the DC-DC module and ancillary devices. The communication mode, protocol architecture, circuit structure and frame definition of P-BUS technique are covered and discussed. The control method of the system is introduced. Compared with other communication methods, the proposed P-BUS has many advantages such as fewer communication wires and components, simpler circuit topology and strong interference rejection. A prototype is given to show the advantages of the proposed solution and global system performance is also analyzed.**

I. INTRODUCTION

LED has been regarded as one of the best potential light sources for next-generation lighting as it has many distinctive advantages such as high efficiency, good reliability, long life, variable color and power consumption [1]. As the lumen efficiency of LED increases rapidly in recent years, applications for LED in lighting products migrate from niche markets to general lighting [1-2]. Many general lighting applications based on LED technology have emerged in recent years [1, 3-5].

In LED applications, PLC is one of the chief communication techniques used to control the complex lighting system. Compared with conventional communication techniques such as RS485, CAN, Ethernet, there is no need to install additional transmission line in PLC [5]. Also, in comparison to wireless techniques such as Zigbee, PLC system is simpler to design because of less interference. However, PLC techniques which are widely used in industrial control fields such as Lonworks [6] still have some drawbacks; too many components are needed to realize the communication and the price of the whole PLC system is high.

This work is sponsored by the Power Electronics S&E Development Program of Delta Environmental & Education Foundation

A LED general lighting system based on a novel power line communication technique, which is named as P-BUS, is proposed in this paper. The proposed P-BUS is a simple and flexible PLC technique in a DC system. It has some advantageous features. Firstly, only a single transistor with its drive circuit and a few passive components are needed to realize the communication for each device connected to the bus. Secondly, the reliability of the communication is high and the performance of interference rejection is good due to fewer transmission wires and components. The P-BUS technique is implemented to the general LED lighting fields in this paper. An LED lighting control system prototype based on P-BUS is developed. The LED dimming and the remote monitor functions are realized. System performance is also analyzed to show that the lighting system based on P-BUS technique has fairly high efficiency.

II. SYSTEM CONFIGURATION

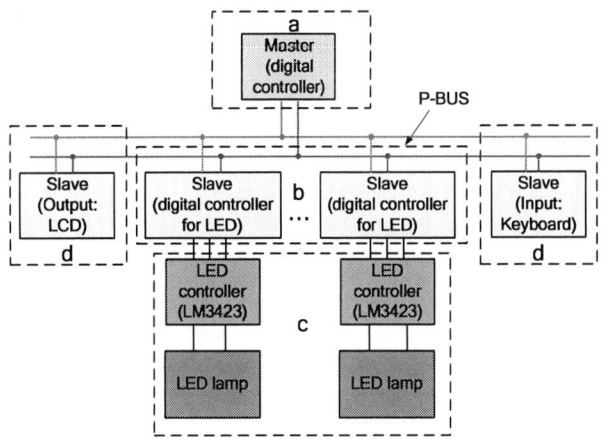

Figure 1. LED lighting system based on P-BUS

Figure 1. shows the block diagram of the proposed LED lighting system. The whole system can be divided into several parts:

a. *Master controller:* The communication mode of the LED system based on P-BUS is a master-slave configuration. Master controller takes charge of the entire system's operation. It supplies the power for all the devices on P-BUS, initially starts the communication process and sends various commands to all slave controllers. The master controller is called as the master node in this paper. The details about the P-BUS communication and LED lighting system operation will be given in the section III and section IV.

b. *Slave controller which is used to control the LED module:* The slave controller has the P-BUS interface. It receives the control commands which are sent by the master controller from the P-BUS. It then translates the commands to a particular duty cycle which is given to the LED controller in the DC-DC module, and then sends the LED status back to the master controller through P-BUS. The slave controller is called as the slave node in this paper.

c. *DC-DC module:* LM3423, a LED driver controller from National Semiconductor is used to regulate the LED module, which is operated in constant-current mode [7]. In the presented prototype, the input voltage varies from 9V to 36VDC. The output is a string of fifteen 1W LEDs in series. Boost topology is used in this circuit.

d. *Other devices on P-BUS:* these devices all have the P-BUS interface and are also slave nodes. The keyboard is used to input orders and transfer them to the master node. The LCD displays the status of the LED and offers a man-machine interface when keyboard being used.

III. P-BUS COMMUNICATION INTRODUCTION

P-BUS is a PLC (power line communication) technique used in DC system. As shown in Figure 1. a pair of wires (red and blue) is needed to transfer energy and signals. It is called bus in this paper. The red one is the power wire and the other is the ground wire. Master-slave mode is employed in the proposed P-BUS system. Master digital controller is the master node, as shown in Figure 1. . The system has only one master node. All the other devices which directly connect to the bus are slave nodes.

There are two layer architectures in the P-BUS protocol, which are the physics layer and the data link layer. The physics layer implements the transmission of "bit" data and exchanges the data in form of "frame" with data link layer. The data link layer receives frames from the physics layer. After dealing with these frames, it creates response frames and hands them to the physics layer where the frames will be sent back.

Figure 2. shows the basic structure of P-BUS interface on the physics layer. A digital controller, a P-channel MOSFET M and its drive circuit, a resister R is necessary for the signal transmission in the master node. A digital controller, a PNP transistor Q and its drive circuit, a diode D, a capacitor C is necessary for the signal transmission in the

slave node. Resister R_L represents the DC-DC part and the LED lamp.

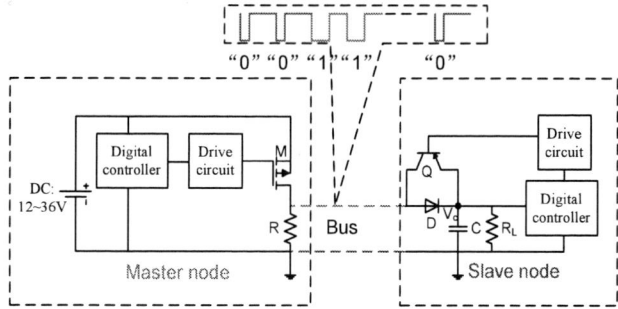

Figure 2. Basis structure of P-BUS interface

As shown in Figure 2. , the red dashed line is the power wire and the blue dashed line is the ground wire. The bit transferred on P-BUS is defined as the following: a 40μs low voltage level followed by an 80μs high voltage level means "1", a 20μs low voltage level followed by a 100μs high voltage level means "0". When there is no communication in the bus, the power wire is always high. The bus only transfers energy to the slave nodes during the periods of no communication. The master node uses switch mechanism to send "0" and "1". When Mosfet M turns off, the power wire is pulled down by resistor R. When Mosfet M turns on, the power wire is pulled up to the input voltage V_{DC}. The master node can set the bit value on the bus by controlling the length of the Mosfet M's off time.

The data sending mechanism of the slave node is much more complicated than the master node. As shown in Figure 3. (a), when the master node needs the response data from slave nodes, it keeps on sending bit "1" to the bus. If the slave node is going to send "1", it takes no action and the master node will read "1" from the bus, which will be considered to be "sent" by the slave node. If the slave node is going to send "0", as shown in Figure 3. (b), it turns on transistor Q after a 20μs delay from the detection of a negative edge on the bus. At this time, as shown in Figure 3. (c), current flows from capacitor C in the slave node to the ground through the resistor R in the master node, the voltage level of the bus will be changed to high. The master node will read "0" from the bus at the moment. So if the master node sends "1" but receives "0", it considers that it is the slave node sending bit "0". The slave node uses this modification mechanism to send "0" and "1".

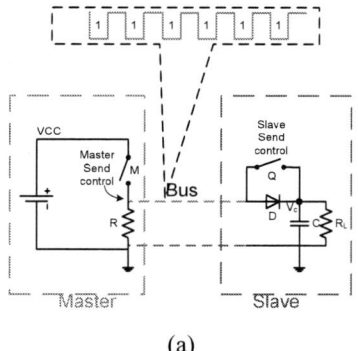

(a)

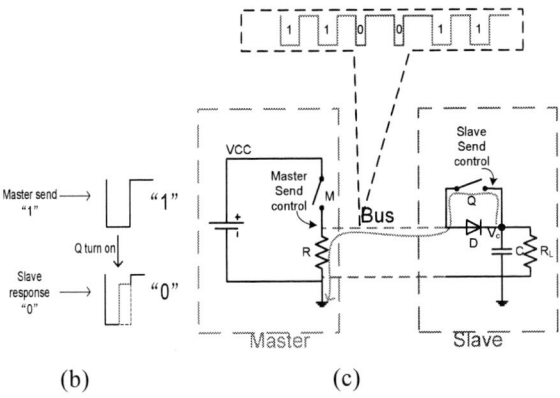

Figure 3.   (a) Master continues sending "1" (b) Slave turns on transistor Q
(c) Slave "modify" the Bit on Bus to "0"

Both the master and the slave nodes receive data by checking the length of the low level on bus. 20µs is "0" and 40µs is "1". The length of the high level is not critical during the communication. The communication waveforms are shown in Figure 4. . The channel 4(upper) is the waveform in the power wire. The channel 2(lower) is the waveform in the transistor Q's base. Figure 4. (a) shows when master node sending. Figure 4. (b) shows when master node keeping on sending "1" and one bit being modified to "0" by a slave node.

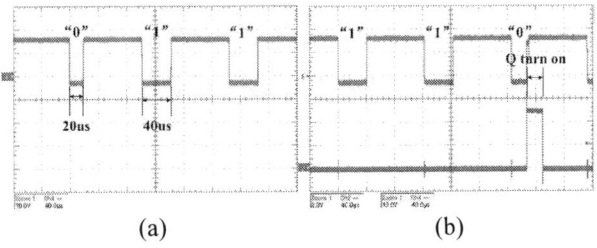

Figure 4.   (a) communication waveform: Master node sending (b)
communication waveform: Slave node response

On the data link layer, the structure of a data frame is defined. A single frame has two parts: one is for the master sending data to the slave(s) and the other part is for the slave response. The master node can communicate with one or several slave nodes in one frame, this can speed up the control of the system. Check codes are introduced into the frame to enhance the communication reliability. TABLE I. is an example of one frame. The master node can check 32 slaves' condition in one frame.

TABLE I.   EXAMPLE OF ONE FRAME: CHECK SLAVE'S CONDITION

Master sending	Command code		length	data	Check code
	0x00 (8 bit)		0x00 (8 bit)		0xFF (8 bit)
Slave response	Node1	Node2	Node3		Node32
	0-normal 1-offline				

IV.   SYSTEM CONTROL INTRODUCTION

The operation of the entire system is controlled by the master controller which is called master node in P-BUS. As shown in Figure 5. , a polling mechanism is used to manage each slave node including the LED node. Before the communication with each existing devices, the master will check firstly if there is a new node (LED module or other device) which has just connected to the bus and if there is a node which has emergency to report such as open circuit of the LED string. In order to distinguish devices connected to the bus, master controller allocate a unique address to each device. New node logging in the system has two steps: First, master node sends the command to slave node asking for its ID number which is unique for every product and stored in slave controller's EEprom. Second, after ID number received, master sends the command including the ID number and the address allocated to the slave node; then the process of logging is complete. If a node has emergency to report, the master controller will communicate with it first to deal with the condition.

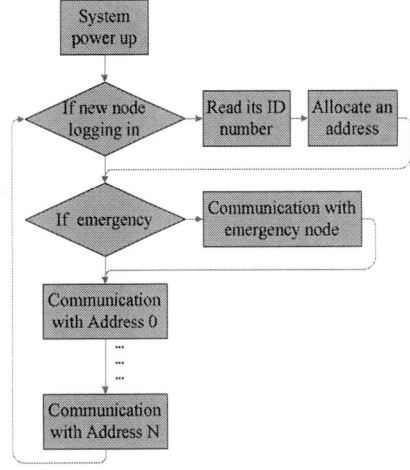

Figure 5.   flow chart of the master controller

As shown in Figure 1. , one LED module is made of three parts: a slave controller, a boost converter controlled by LM3423 and a series of LED lamps. LM3423 has a digital interface. It has a FAULT pin that will be pulled high if the output is abnormal such that there is an open circuit in the output LED string. It also has a DIM pin that can control the luminance of the output LED by giving a certain duty cycle PWM wave to this pin.

Figure 6.   shows that how the LED module is being controlled. The dimming command is sent by master controller to the slave controller through P-BUS. The slave controller receives the command, translates it into a particular duty cycle and gives it to LM3423 through its DIM pin. The slave controller also watches the LED status by checking the FAULT pin's level and feed it back to the master controller. All these are done in one frame in P-BUS protocol. If a fault is happened, another command will be sent by master to let the slave node pull the EN pin low and shut down the LED module.

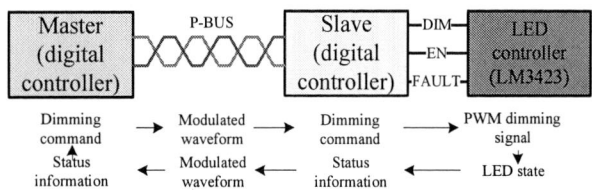

Figure 6. LED control loop

In the prototype presented in this paper, a linkage control among several LED modules is designed to facilitate the consumer to control several LED lamps. A short circuit protection is also included to ensure the safety of the hardware.

V. SYSTEM PERFORMANCE ANALYSIS

An LED lighting system prototype based on P-BUS technique is developed in this paper. Some basic specifications are listed below: The input voltage is 12~36V; the maximum number of slave nodes is 64; the LED lamp is 1W/350mA; the numbers of LED in one LED module is 15; the DC-DC stage efficiency is above 92%.

Figure 7. shows the LED module. As few components are needed for P-BUS communication, the size of the module can be minimized. Figure 8. shows the operation of the whole system. There are two LED modules here.

Figure 7. the LED module (include: P-BUS interface and DC-DC module)

Figure 8. the system overview (include: master node, LED module, LCD, keyboard, LED lamp)

Efficiency is important for an LED lighting system. The proposed LED lighting system using P-BUS technique has fairly high system efficiency. In the system, there are four main parts which consume power.

a. *Mosfet M in master node:* as shown in Figure 2. , mosfet M is turned on and off in order to create voltage change in the bus. Its switching frequency is the same as the communication rate. In the prototype proposed in the paper, the communication rate is about 8.3kb/s. Moreover, P-channel mosfet are used in the prototype, which has larger R_{on} than N-channel mosfet. The R_{on} is about 0.3Ω in this prototype.

b. *Communication bus:* Power loss is mainly on the wire resistance. The resistance of a 100m twisted wire is about 2Ω.

c. *Diode D in slave node:* as shown in Figure 2. , the diode D is necessary for P-BUS interface. The power loss on it can not be ignored because current supplied to LED flow through it. The efficiency of a LED module is listed in TABLE II. . As shown in TABLE II. , efficiency increases as the input voltage becomes higher.

TABLE II.   THE EFFICIENCY OF AN LED MODULE

Vin(V)	Iin(A)	Pin(W)	Vout(V)	Iout(A)	Pout(W)	Efficiency
24.400	0.855	20.5	50.9	0.373	18.986	0.910
20.100	0.917	18.4	49.7	0.335	16.650	0.903
16.300	1.155	18.7	49.7	0.340	16.898	0.898
12.400	1.624	19.8	49.9	0.357	17.814	0.885
9.600	2.211	21.2	50.2	0.365	18.323	0.863

d. *DC-DC circuit (boost):* The efficiency of this part can be calculated precisely. With properly designed controller for the parameter of the converter, the efficiency in the worst case is still above 92%.

TABLE III. is the experimental data for the lighting system prototype efficiency. There is one LED module on the P-BUS; the LED lamp is a series of 15 LEDs.

TABLE III.   TESTING DATA OF THE LIGHTING SYSTEM PROTOTYPE EFFICIENCY

Input conditions	No long wire used; System input voltage: 26.0V Vin=24.0V (same as Table 2)	140m long wire used System input voltage: 25.6V Vin=20.0V (same as Table 2)
Without LED module	Static current: 0.16A Static power loss P_S: 4.2W	Static current: 0.16A Static power loss P_S: 4.2W
LED module lighting	Operation current: 0.96A Total loss P_T: 26.0W LED operation loss P_A: $P_A= P_T - P_S$=21.8W	Operation current: 1.04A Total loss P_T: 26.6W Wire loss P_W: 2.8W LED operation loss P_A: $P_A= P_T - P_S$=22.4W
Output Power P_O	19.0W	16.7W
Efficiency P_O/ P_A	87.2%	74.6%

As seen from the testing result, the system static power loss is about 4.2W. In the results below, the static power loss are not included.

The efficiency of the LED module with diode D is around 90% to 91%. If the loss on the mosfet M is included, the efficiency of the LED module operation is 87.2%, if a long wire (140m, R=2.8Ω) is used, the efficiency of the LED operation is 74.6%

In practical applications, a long wire will not typically be used in the LED lighting system so the efficiency should be higher than 75%. Furthermore, as the number of LED modules on the P-BUS increases, the static power loss will be a less dominant part in the total loss of the system. It is believed that the final efficiency of the system will always exceed 80%.

VI. CONCLUSION

In this paper, a general LED lighting system based on P-BUS is presented. It is made up of four main parts: the master controller, the slave controller which control the LED module, the DC-DC module and others slave devices. The communication mode, protocol architecture, circuit structure in physics layer and frame definition of P-BUS technique has been covered. The control method of the system is introduced. System performance of the prototype is also analyzed to show that the lighting system based on P-BUS technique has fairly high efficiency. Compared with other communication method, P-BUS has many advantages such as fewer communication wires and components, simple circuit topology and good performance of interference rejection. P-BUS is expected to enter into the market in the near future and become an important technique in general LED lighting field

Acknowledgment

The author would like to thank the support of National Semiconductor Company.

REFERENCE

[1] Po-Yen Chen, Yi-Hua Liu, Yeu-Torng Yau, and Hung-Chun Lee, "Development of an energy efficient street light driving system," Sustainable Energy Technologies, 2008. ICSET 2008. IEEE International Conference on, 2008, pp. 761-764.

[2] M. Wendt and J. Andriesse, "LEDs in Real Lighting Applications: from Niche Markets to General Lighting," Industry Applications Conference, 2006. 41st IAS Annual Meeting. Conference Record of the 2006 IEEE, 2006, pp. 2601-2603.

[3] Feng Tian, Xiang Wu, and Jingao Liu, "Research and realization of LED display system based on combining CAN bus with DMX512 standard," Audio, Language and Image Processing, 2008. ICALIP 2008. International Conference on, 2008, pp. 813-818.

[4] O. Mukhtar, S. Golam, E. Liu , and T. Korhonen, "Development of high-rate control system for LED based general lighting applications," Power Line Communications and Its Applications, 2005 International Symposium on, 2005, pp. 417-421.

[5] J. Lee, K. Nam, S. Jeong, S. Choi, H. Ryoo, and D. Kim, "Development of Zigbee based Street Light Control System," Power

Systems Conference and Exposition, 2006. PSCE '06. 2006 IEEE PES, 2006, pp. 2236-2240.

[6] Yushui Huang, Cunying Wan, and Zhiqiang Zhou, "Intelligent Community System Based on LonWorks Technology," Computational Intelligence and Industrial Application, 2008. PACIIA '08. Pacific-Asia Workshop on, 2008, pp. 237-240.

[7] National Semiconductor. Datasheet for LM3421, LM3423. USA: National Semiconductor Corporation, USA, 2008

Solid-State Lamp with Integral Occupancy Sensor

John J. Cooley *Student Member, IEEE*, Dan Vickery, Al-Thaddeus Avestruz *Member, IEEE*,
Amy Englehart, James Paris, and Steven B. Leeb *Fellow, IEEE*

Abstract—**Previous work demonstrated a retrofit proximity detector for fluorescent lamps using the lamp's own stray electric fields. This paper extends the retrofit sensor system to a solid-state (LED) lamp. The design and implementation of a suitable driver ("ballast") for the LED lamp is presented. Design considerations for the ballast include those relevant to lighting (e.g. color cast and dimming levels) as well as those relevant to sensing of human occupants. Two electro-quasistatic modeling approaches for the lamp sensor are discussed. Experimental data from the LED lamp sensor are presented and compared to one of the proposed quasistatic models.**

I. INTRODUCTION: ENERGY EFFICIENT LIGHTING TECHNOLOGY

New illumination sources and new power electronic controls for lighting have the potential to produce energy efficiency gains of 240 percent in the residential sector and 150 percent in the commercial sector [1]. In 2007, lighting accounted for 15.6 percent and 23.3 percent of all electricity consumed in the residential and commercial sectors, respectively, in the USA, [1]. Efficiency gains from lighting sources and active control can substantially reduce overall energy consumption. In particular, solid-state lighting promises improved energy efficiency and long lifetime [2].

A fluorescent lamp with an integral occupancy sensor was demonstrated in [3], [4] and [5]. By exploiting the lamp's own stray electric field, the sensor system is able to detect changes in the electric field below the lamp. Unlike standard proximity sensors which require building planners to design for a proximity detection system separate from the lighting system, this proximity sensor is essentially a drop-in replacement for a standard commercial fluorescent lamp ballast.

This paper describes the extension of the lamp sensor to a solid-state (LED) lamp of similar construction to a fluorescent lamp. The design and implementation of a suitable driver ("ballast") for the LED lamp is presented here. Design considerations for the ballast include those relevant to lighting. For example, the ballast may be designed to achieve consistent color cast across dimming levels. The ballast must be designed to support the lamp sensor electronics by driving the lamp with an alternating voltage. An overview of the lamp sensor from [3]–[5], as well as two electro-quasistatic modeling approaches, are discussed in Section II. The design and implementation of the ballast for the LED lamp is discussed in Section III. Finally, experimental data from the lamp sensor built around the LED lamp are presented and compared to one of the proposed quasistatic models in Section IV.

II. LAMP SENSOR OPERATION AND MODELING

A. System description

The block diagram of the lamp sensor is shown in Figure 1. A suitable lamp and electrode arrangement is depicted in Figure 2. A typical fluorescent or LED lamp may include an electronic ballast that drives the bulbs with an alternating voltage at frequencies in the range of 20 to 50 kHz. As long as the lamp is driven with an alternating voltage, we have found that an electrostatic proximity sensor (lamp sensor) can be developed to measure the resulting stray electric fields below the lamp. In this section, we describe the operating principles of the lamp sensor that may be applied to either a fluorescent or LED lamp. Section III describes a driver specifically suited to operating LED lamps with the proximity detector.

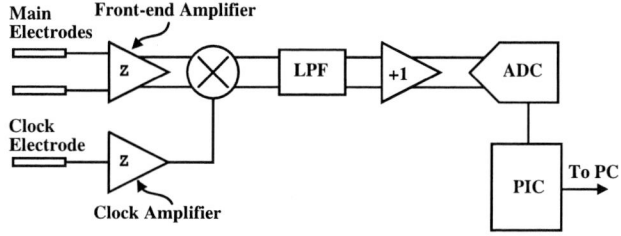

Fig. 1. The block diagram of the lamp sensor. Transimpedance amplifiers are marked with a 'Z'.

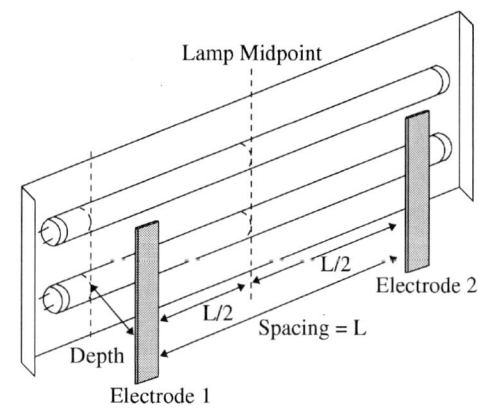

Fig. 2. A diagram of a typical two-bulb fluorescent lamp and electrode configuration. The electrodes are spaced symmetrically about the center of the lamp.

The lamp sensor measures low frequency ($\sim$0.1-5Hz) changes caused by human targets below the lamp in the high-frequency ($\sim$50kHz) alternating stray electric fields coupling from the lamp. The lamp sensor measures the electric fields in front of the lamp with a fully-differential transimpedance front-end amplifier driven by the two electrodes depicted in Figure 2. The front-end amplifier is fully-differential because both its inputs and outputs are differential. When the electric fields measured by the two electrodes are equal, the differential measurement is balanced and the front-end output voltage is "nulled" to zero volts. Therefore, the front-end can have very high gain without saturating its output in the absence of a detection. This very high gain is necessary to amplify the effects of small imbalances in the capacitive system below the lamp. Furthermore, the fully-differential structure of the front-end amplifier relaxes the need for a specific ground reference in the lamp sensor system.

As indicated in Figure 1, the lamp sensor uses synchronous detection to reject stray signals that differ in phase or frequency from the lamp's own excitation signal. Synchronous detection is achieved with a separate measurement of the signal source. The lamp sensor system multiplies this reference signal with the main signal using a fully-differential multiplier. Whereas low-frequency occupant motion below the lamp naturally modulates the high-frequency signal measured by the front-end, the output of the multiplier (demodulator) contains the low-frequency modulations centered on zero frequency (dc). A low-pass filter (LPF) attenuates the high-frequency residue from the multiplier output to yield only the low-frequency signal. The synchronous detector and modulation-demodulation scheme also has the advantage of bypassing potentially overwhelming low-frequency 1/f noise in the front-end amplifier. Bypassing 1/f noise is critical because the signals of interest (occupancy detections) are very low in frequency. The signal of interest is amplified having been naturally up-modulated by the lamp sensor system to high-frequency (~50kHz). The amplified high frequency signal is then downmodulated by the multiplier shown in Figure 1 after amplification. Meanwhile, the low-frequency 1/f noise generated within the front-end amplifier is up-modulated by the multiplier to high-frequency and attenuated by the ensuing LPF. This signal detection scheme is not unlike chopper stabilization of op-amps for elimination of low-frequency noise, detection in an AM radio, or the so-called "lock-in" amplification technique for frequency specificity. Note that this system comprises a suppressed-carrier AM modulation-demodulation signal chain. That is, when there is no detection, there is also no carrier signal measured by the sensor. This is fundamentally why a "clock electrode" (depicted in Figure 1) is necessary to obtain a phase and frequency reference for the desired signal.

Measurement of the electric field can be taken as a measurement of the two electrode potentials (at electrode 1 and electrode 2 in Figure 2) referenced to some arbitrary potential. This corresponds to a voltage-mode measurement and requires a high input impedance measurement device. Alternatively, the electric field measurement can be taken as a measurement of the current shunted between the electrodes when they are connected by a short-circuit. This corresponds to a current-mode measurement and requires a low input impedance measurement device. In the lamp sensor, a fully-differential current-mode amplifier is used as the front end amplifier depicted in Figure 1. Reference [6] details the behavior and operating principles of fully-differential amplifiers like the one used here. In [5], we reported 11ft. of detection range between the lamp and the closest edge of a human target. For a detailed description of the lamp sensor operation, see references [4], [5].

B. Capacitive Modeling

The electric field measurement may be modeled as a measurement within a capacitive network. Such a capacitive network may be comprised of the capacitances connecting all of the conducting objects in the system. This abstraction is a starting point for qualitatively understanding the behavior of the lamp sensor system and its key features. Implicit in this abstraction is that the objects of interest can be represented by conducting surfaces. While some objects in the system are obviously well-represented as conducting surfaces (e.g. the electrodes), the fluorescent lamp bulbs as well as the human target are also represented this way. References [7]–[9] describe the implications of approximating a human as a conducting shell. Reference [5] motivates the treatment of the bulb surfaces as voltage or low-impedance sources rather than current or high-impedance sources in the capacitive system. Reference [10] discusses the approximation of lumping each of the bulbs into "positive" and "negative" source ends. Furthermore, the floor below the lamp is taken as a conducting plane and could be arbitrarily chosen as the potential reference (ground) for the system.

Having these assumptions, we can build a capacitive model like the one proposed in Figure 3 by accounting for capacitances between the conducting objects in the lamp sensor system. Accurate modeling of the lamp sensor system using this modeling approach typically requires the use of capacitance extraction software. Note that the capacitive model pictured in Figure 3 differs from the electrostatic model described and used in Section IV in at least one key way. Most notably, the low-potential source ends of the lumped signal sources in Figure 3 are not grounded as they are in the model described in Section IV.

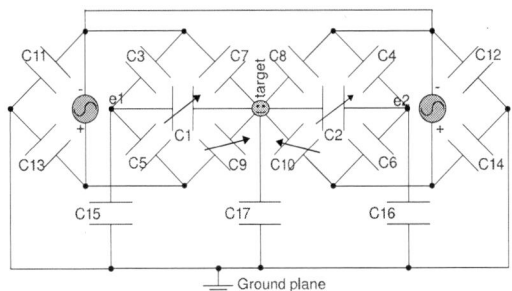

Fig. 3. Capacitive Model of fluorescent lamp sensor.

C. Quasi-Analytical Modeling

Another approach to modeling the lamp sensor system is to solve Laplace's equation, $\nabla^2 \Phi = 0$, directly for the potential everywhere in the system in order to get an equation for the electrode voltage. To that end, a reduced model of the lamp sensor may be developed by modeling the important conducting surfaces in the system as charged conducting spheres. An example of such a reduced model is shown in Figure 4. Modeling the conducting surfaces as conducting spheres is advantageous because the analytical expressions for the potential outside (and inside) a charged conducting sphere are readily available. Moreover, through linear superposition of potential (scalar) fields, the problem of determining the potential everywhere in a system with one sphere can be extended to the problem of determining the potential everywhere in a system with K spheres.

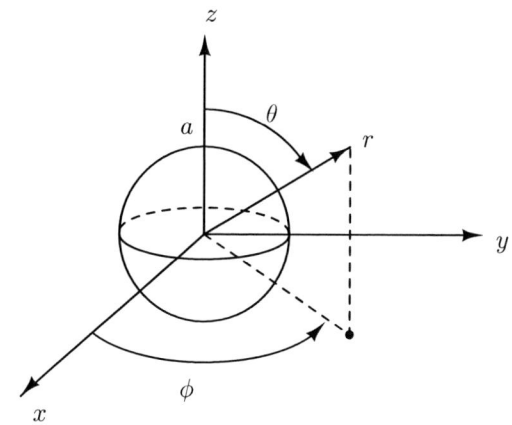

Fig. 5. Coordinate system of one sphere with radius a. r, θ and φ are standard spherical coordinates; that is, they are respectively, radius, pitch angle and azimuthal angle with the origin being the center of the sphere.

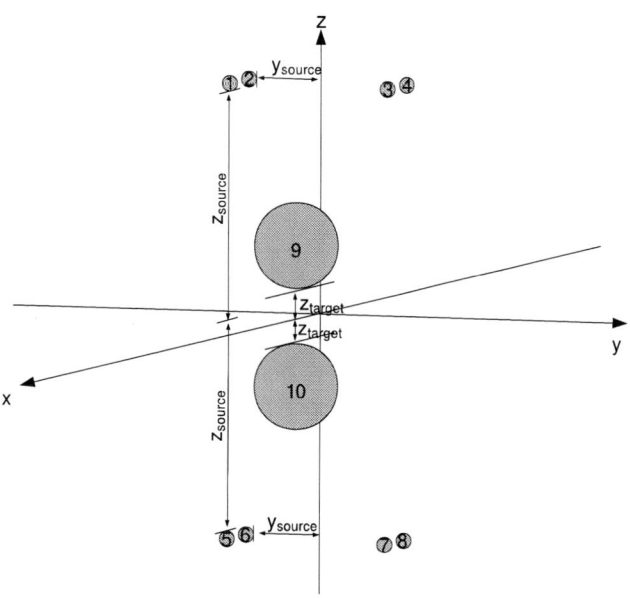

Fig. 4. Sphere model used to represent lamp sensor and target. Spheres 1, 2, 3 and 4 represent the signal sources created by the lamp. Sphere 9 represents the target. Modeling the floor as a conducting plane is achieved using the method of images, and hence spheres 5, 6, 7, 8 and 10.

D. Modeling one conducting sphere

For a charged sphere with radial boundary a as shown in Figure 5, the solution of the potential field everywhere outside the boundary a is well-known as the summation in (1). The summation contains Legendre polynomials, $P_n^m(x)$'s, where x is replaced by the pitch-angle function $\cos\theta$ as follows [11]–[14]:

$$\Phi(r,\theta,\varphi) = \sum_{n=0}^{R} \sum_{m=-n}^{n} \left(\frac{A_{nm}}{r^{n+1}} P_n^m(\cos\theta) e^{jm\varphi} \right), \quad (1)$$

For a single charged conducting sphere, equation (1) can be rewritten in vector form as

$$\Phi(r,\theta,\phi) = p^T \alpha, \quad (2)$$

where α is a vector containing the A_{nm} coefficients:

$$\alpha^T = (A_{00}, A_{1,-1}, A_{1,0}, A_{1,1}, A_{2,-2}, A_{2,-1}, ..., A_{R,R}) \quad (3)$$

The vector p contains the rest of each term in the summation of (1), so that the first few entries of p are

$$p_0 = \frac{1}{r} P_0^0(\cos\theta)$$

$$p_1 = \frac{1}{r^2} P_1^{-1}(\cos\theta) e^{-j\phi}$$

$$p_2 = \frac{1}{r^2} P_1^0(\cos\theta)$$

$$p_3 = \frac{1}{r^2} P_1^1(\cos\theta) e^{j\phi}$$

$$p_4 = \frac{1}{r^3} P_2^{-2}(\cos\theta) e^{-2j\phi}$$

$$\vdots$$

Taking the inner product of the vectors p and α above (as described by (2)) results in $\Phi(r,\theta,\phi)$ from (1). Having fixed the coordinate reference for (r,θ,ϕ), the Legendre polynomials, $P_n^m(\cos\theta)$, can be "looked up" to form the vector p. Then, the A_{nm} coefficients are unknowns that must be found before we can find the potential everywhere in the system, $\Phi(r,\theta,\phi)$. For a given order of the multipole expansion described by equation (1), we take N as the number of terms in the summation leaving us with as many unknown A_{nm} coefficients to find. To that end, $M > N$ independent equations can be formed by "sampling" the potentials at M points in the system leading, each one yielding an equation like the one in (1). The sampled potentials are known potentials, otherwise known as boundary conditions on the conducting surfaces in the system. The resulting M equations can be written in matrix form as follows:

$$b = \mathbf{P}\alpha, \quad (4)$$

978-1-4244-4782-4/10 $26.00 © 2010 IEEE   2307

where P is an $M \times N$ matrix, b is the vector containing the M potentials from the left side of (2):

$$b = \begin{bmatrix} \Phi(r_1, \theta_1, \varphi_1) \\ \Phi(r_2, \theta_2, \varphi_2) \\ \cdot \\ \cdot \\ \cdot \\ \Phi(r_N, \theta_N, \varphi_N) \end{bmatrix} \quad (5)$$

and $\mathbf{P}$ is a matrix whose rows are the N transposed p vectors from the right side of (2):

$$\mathbf{P} = \begin{bmatrix} p_0^T \\ p_1^T \\ \cdot \\ \cdot \\ \cdot \\ p_N^T \end{bmatrix} \quad (6)$$

Then, a least-squares formulation for the solution to α is:

$$\alpha = (\mathbf{P}^T \mathbf{P})^{-1} \mathbf{P}^T b \quad (7)$$

so that, upon substituting this result for α into (2), the value of the potential at (r, θ, ϕ) can be found directly as follows:

$$\boxed{\Phi(r, \theta, \phi) = p^T \left((\mathbf{P}^T \mathbf{P})^{-1} \mathbf{P}^T b \right)}. \quad (8)$$

Having fixed the geometry of the system and the boundary values on the conducting surfaces by defining the p vectors in equation (6), and the values comprising the vector b in equation (5), as well as the order of the multipole expansion, one can directly compute the potential at any point (r, θ, ϕ) using equation (8). While the vector of A_{nm} coefficients given by equation (7) is valid for computing the potential from any equation of the form in (1), choosing the particular vector p effectively establishes the coordinate reference for (r, θ, ϕ) within the system by choosing one particular equation of the form in (2).

E. Modeling K Charged Conducting Spheres

The development above, for calculating the potential in a system with one charged sphere, can be extended to calculate the potential in a system with K spheres. The potential due to each sphere is summed as follows:

$$\Phi = \Phi_1 + \Phi_2 + \ldots + \Phi_k + \ldots + \Phi_K. \quad (9)$$

Each potential, Φ_k, is a summation like the one in (1). For each summation leading to Φ_k, the variables r_k, θ_k and φ_k are the standard spherical coordinates with their origin at the center of the kth sphere.

The total potential from (9) for the K sphere case can also be re-written in vector form in analogy to equation (2) as follows:

$$\Phi = p_1 \alpha_1 + p_2 \alpha_2 + \ldots + p_K \alpha_K \quad (10)$$

Now, the charge distribution on each sphere depends on the charge distribution on the other spheres. Therefore, in the calculation of each Φ_k for eqn. (10), we take into account

the boundary conditions on all K spheres by extending b to include the boundary conditions on all K spheres and $\mathbf{P}$ to include known position terms for all K spheres. Lastly α is extended to include unknown coefficients for all K spheres. The augmented matrices described above can be written in matrix equation form corresponding to the equation in (4) as follows:

$$\begin{bmatrix} b_1 \\ b_2 \\ \cdot \\ \cdot \\ \cdot \\ b_K \end{bmatrix} = \begin{bmatrix} \mathbf{P_{11}} & \mathbf{P_{12}} & \cdots & \mathbf{P_{1K}} \\ \mathbf{P_{21}} & \mathbf{P_{22}} & \cdots & \mathbf{P_{2K}} \\ \cdot & \cdot & \cdots & \cdot \\ \cdot & \cdot & \cdots & \cdot \\ \cdot & \cdot & \cdots & \cdot \\ \mathbf{P_{K1}} & \mathbf{P_{K2}} & \cdots & \mathbf{P_{KK}} \end{bmatrix} \begin{bmatrix} \alpha_1 \\ \alpha_2 \\ \cdot \\ \cdot \\ \cdot \\ \alpha_K \end{bmatrix}, \quad (11)$$

where

$$b_m \in \mathbb{R}^{M \times 1} \quad (12)$$

$$\alpha_m \in \mathbb{R}^{N \times 1} \quad (13)$$

$$P_{mn} \in \mathbb{R}^{M \times N}. \quad (14)$$

Finally, in the K-sphere case, $\Phi(r, \theta, \phi)$ may be directly computed as in (8) by using the augmented vector b and matrix $\mathbf{P}$ shown in (11).

The analytical model above can be validated by comparison between analytical results and simulations from Maxwell 2D as in Figure 6. Figure 6 is a plot of the percent error between the two methods for the potential around a charged conducting sphere. The data for this case show that the magnitude of the error between the simulation and the analytical model is no larger than 0.5%. In Section IV, an analytical model of the lamp sensor system will be compared to measured results.

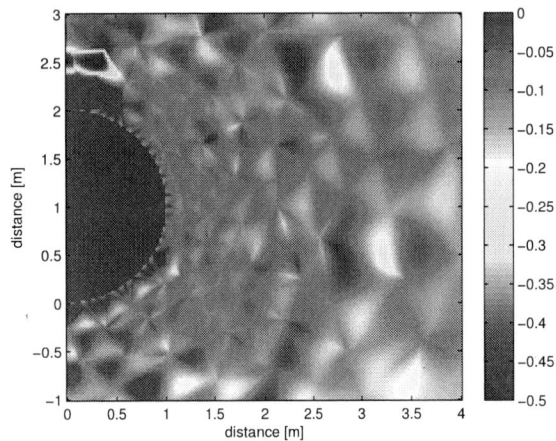

Fig. 6. The percent error between Maxwell 2D® results, Φ_{Mxwl}, and least-squares method results, Φ_{LstSq}. The conducting sphere has radius 1 m and a surface potential of 100 V referenced to the zero-potential at $r = \infty$

III. POWER ELECTRONIC LED DRIVE AND DIMMING

Power electronic drives represent a significant efficiency improvement over linear power sources. In LEDs, light output is best controlled by regulating current. Some difficulties in using the terminal voltage as the control include a negative temperature coefficient that leads to thermal runaway and a poorly behaved exponential relation to power output.

The salient characteristic of a linear regulator is that a voltage is dropped across an element that continuously carries current, resulting in an inescapable power dissipation. Linear current regulators can use active devices in either open-loop (e.g. current mirror) or closed loop (e.g. op-amp with a pass transistor). A voltage source with a series resistor is also used in driving LEDs-the underlying approximation is that the voltage drop across the resistor is large, hence approximating a current source, which means that it is guaranteed that power is dissipated as heat by the resistor. This approximation weakens at low current levels when dimming, resulting in a drifting current level and hence light output.

Pulse width modulation of the current is an effective approach to driving an LED. Not only is there an advantage in efficiency, but a number of studies have suggested that there may be an advantage to the quality of lighting [15], [16]. There are a number of choices for the design of a switching current source. Our design was based on several requirements, which include a bipolar source for proximity sensing, square current pulses for color quality, and a wide dimming range.

A. Power Circuit Implementation

We chose a buck converter with a hysteretic current controller and a post-inverter. The circuit is designed to operate from a 170 V dc bus (nominal rectified line voltage) driving two parallel strings of 40 mA ac LED modules, which are illustrated in Figure 7. Currently, we use commercially available modules with 14 LEDs in a diode bridge. These modules are meant to be used in low-voltage 60 Hz sockets meant for halogen lighting and hence a diode bridge with slow recovery. The problems with the reverse recovery when using a high frequency inverter were avoided by using a fast recovery bridge. The partioning of these LEDs into bridged segments in these commercial modules are by no means optimal from the perspective of cost, power efficiency, and lamps sensor operation, but represent a good proof-of-principle. A future implementation will optimize a taper in the partioning of LEDs into bridged segments with an electric field distribution that is suitable for the lamp sensor, while minimizing the number of necessary diode bridges.

A schematic of the power system is shown in Figure 8. The buck converter was designed to switch at hundreds of kilohertz, while the inverter was designed to operate in the 20–30 kHz range. Because the full-bridge inverter is driven by a current source with free-wheel diode, D_2, no dead-time circuitry for the switches is necessary, which simplifies the design. Transformer T_1 is a 1 : 1 isolation transformer that allows an electrical configuration, where the opposing corner terminals are nominally at the same potential as the floor and fixture, which may result in an improved sensitivity in the lamp sensor. Transformer T_2 is a small current balancing transformer, similar to those which are used in fluorescent lamp ballasts. The volt-second demands on this transformer

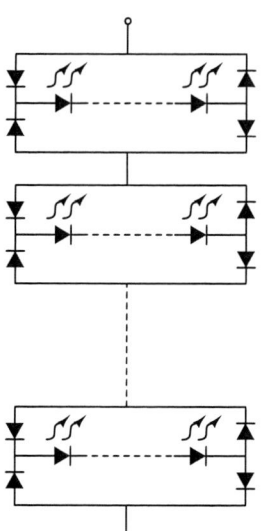

Fig. 7.   Bidirectional LED modules.

are small because they corresponds to the average ensemble mismatch among LEDs in each string.

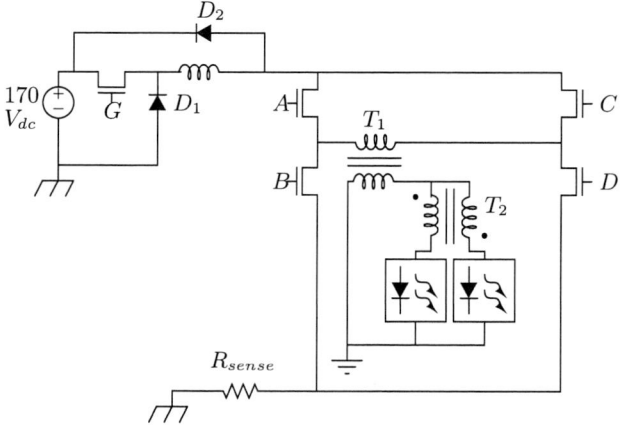

Fig. 8.   Schematic of the bipolar LED driver. Transformer T_1 isolates the inverter output. Transformer T_2 matches the currents between the two LED bulbs.

The buck converter controller shown in Figure 9 was set for a nominal current ripple of 10%. The inverter switches (A,B,C,D) and the gate drive enable (G_{EN}) are controlled by a DSPIC33FJ2560GP710 microcontroller with a fast PLD to create the inverter switching patterns.

Dimming is performed by symmetric tri-state PWM at the inverter. At high duty cycles ($> 60\%$), the buck converter is run at continuous current ($G_{EN} = 1$ always). This means that during the tri-state, we let the buck converter regulate the current into a short, i.e. either A and B ON, or C and D ON, and the complementary switches off. Table I shows inverter switching pattern for $D > 60\%$. CLK corresponds to an internal clock that determines the polarity of the inverter current and PWM is an internal state which corresponds to either the tri-state (LOGIC 0) or on-state (LOGIC 1). Figure

978-1-4244-4782-4/10 $26.00 © 2010 IEEE

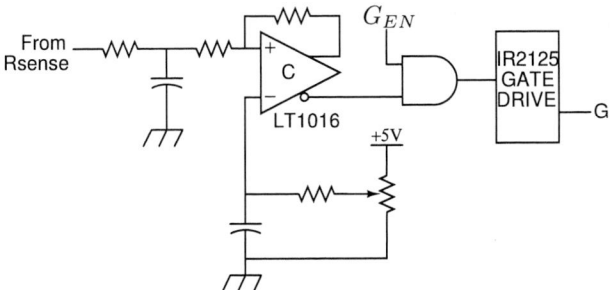

Fig. 9.   Hysteretic Current Controller.

CLK	PWM	A	B	C	D	Status
1	0	1	1	0	0	LED OFF
1	1	1	0	0	1	LED POS Current
1	0	0	0	1	1	LED OFF
0	0	0	0	1	1	LED OFF
0	1	0	1	1	0	LED NEG Current
0	0	1	1	0	0	LED OFF

TABLE I
SWITCHING LOGIC PATTERN WITH NO PRE-CHARGE. BUCK CONVERTER
IS ALWAYS ENABLED.

CLK	PWM	A	B	C	D	G_{EN}	Status
1	0	0	0	0	0	0	LED OFF
1	0	1	1	0	0	1	Pre-Charge
1	1	1	0	0	1	1	LED POS Current
1	0	0	0	0	0	0	LED OFF/Discharge
0	0	0	0	0	0	0	LED OFF
0	0	0	0	1	1	1	Pre-Charge
0	1	0	1	1	0	1	LED NEG Current
0	0	0	0	0	0	0	LED OFF/Discharge

TABLE II
SWITCHING LOGIC PATTERN FOR LOW DUTY CYCLES WITH
PRE-CHARGE.

10 illustrates the results at 96% inverter duty-cycle. *End-to-End Lamp Voltage* is the differential voltage across a single LED string. *LED Current* corresponds to the primary current of T_1 and closely represents the sum of the currents into the two parallel LED strings. A filtered version of *Clock*, with the spikes from switching transient pickup eliminated, is used for synchronization by the lamp sensor.

converter regulate and turning either A and B ON, or C and D ON, with the complementary switches off. Shorting the output of the buck converter in this way results in the fastest precharge for a given input voltage. Table II and the timing diagram in Figure 11 shows the switching pattern when inductor current pre-charging is used. By turning off the buck converter and recovering the inductor current, conduction losses are reduced for low duty ratios ($D < 60\%$). Pre-charging has the advantage of allowing a square LED current without needing very large inductor di/dt, obviating the need for small inductors at a very high switching frequency and the associated switching losses. The pre-charge time can be pre-determined by the rise time, $T_{rise} = LI_{pk}/V_{in}$, where L is a conservative estimate of the inductor value for all operating points and V_{in} is the input dc voltage to the buck converter, which could either be measured by the microcontroller, or set for worst-case low line voltage. Figure 12 shows the low-duty cycle waveforms with pre-charging.

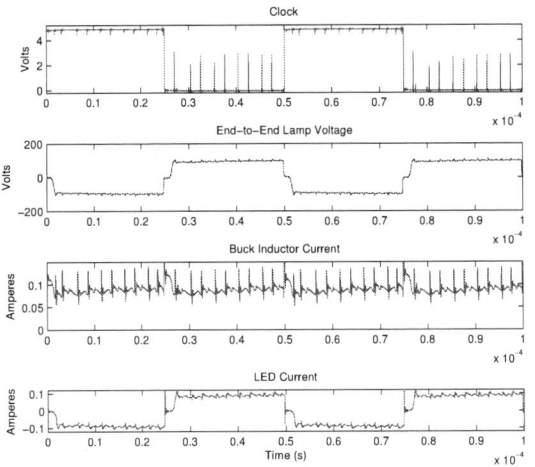

Fig. 10.   Waveforms at 96% Duty-Cycle with Continuous Inductor Current.

At low duty cycles ($< 60\%$), during the tri-state, we turn the buck converter off by gating switch G_{EN} off along with the inverter switches (A–D). This turns diodes D1 and D2 ON, hence recovering the inductor current. Prior to inverter current turn-on, we pre-charge the inductor by letting the buck

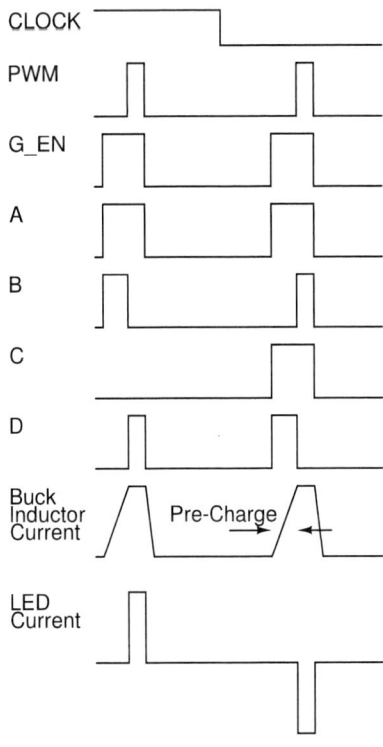

Fig. 11.   Timing Diagram Using Pre-Charging.

978-1-4244-4782-4/10 $26.00 © 2010 IEEE

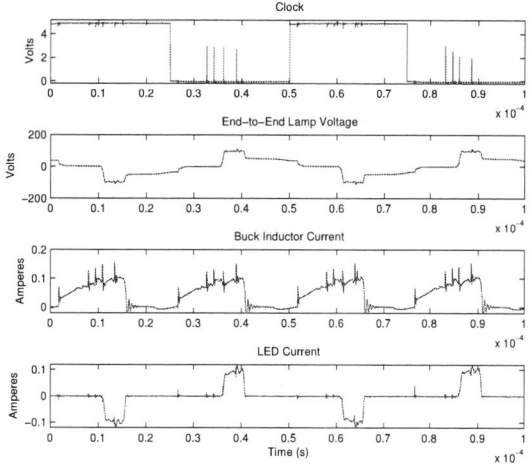

Fig. 12. Waveforms at 10% Duty-Cycle with Inductor Current Pre-Charge.

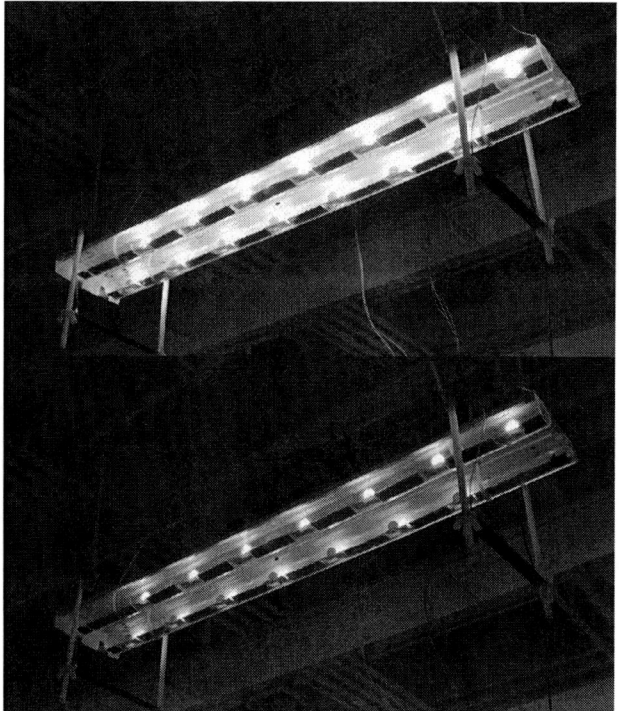

Fig. 13. A photograph of the LED lamp. Top: bright, Bottom: dim.

IV. EXPERIMENTAL SETUP AND RESULTS

The experimental setup of the LED lamp and lamp sensor electronics is shown in Figure 14. The lamp sensor output was measured as a conducting sphere (the target) was passed under the lamp. Data was taken with the target fixed at 20 cm intervals in the y-dimension as depicted in Figure 14. For each interval, the lamp sensor output data was averaged for 20 seconds. The experiment was iterated for three lamp power settings "bright" (Duty ratio of 96%), "medium" (Duty

ratio of 60%)", and "dim" (Duty ratio of 20%). A photograph of the lamp under the "bright" and "dim" settings is shown in Figure 13. The experiment was also repeated for three x-displacements: 0 cm, 22 cm, and 45 cm.

The analytical approach described in Section II-C was also used to model the lamp sensor system. The electrostatic model of the lamp sensor consisted of conducting spheres representing the source nodes, electrodes and the target. The calculated difference between the electrode potentials was taken to be proportional to the output of the lamp sensor. Also, the floor was taken to be a conductor, so the model also consisted of image spheres below the plane of the floor. Finally, the potential of the "Lo" source nodes and of the plane of the floor was assumed to be earth ground.

To compare the analytical approach to measured data, the signal source parameters used in the electrostatic model were first calibrated. A "training run" consisted of taking measured data from the lamp sensor for known x and y displacements. Then, an iterative least-squares optimization method was used to infer the effective signal source parameters based on the measured data.

In the least-squares optimization method, the signal source parameters were first guessed. Then, the results of predicting the lamp sensor output using the electrostatic modeling approach from Section II-C and the guessed signal source parameters were compared to the measured data. Depending on the squared error between the predicted and measured data, the signal source parameters were perturbed and the process was repeated. This iterative process continued until the squared error between the predicted and measured lamp sensor output were less than a certain threshold. The signal source parameters from the last iteration were then taken as the actual effective signal source parameters in the lamp sensor system.

The measured and fitted data in Figure 15 correspond to training runs for each of the three lamp power settings. For each lamp power setting, we have different effective signal source parameters. Those effective signal source parameters were used to predict the lamp sensor response for different x-displacements. Figures 16(a), 16(b), and 16(c) compare those predictions to measured lamp sensor data at x-displacements of 22 cm and 45 cm for each of the three lamp power settings (dim, medium, and bright). The results in Figure 16 show that the modeling approach described in Section II-C yields predictive power for design-oriented estimation of the lamp sensor output voltage in response to a target.

V. CONCLUSION

Reference [17] is one of many references that discusses curtailed demand and its value in the energy market. In curtailed demand, the supply company reduces the effective energy demand by reclaiming unused or wasted energy. Reference [17] cites the sophisticated planning and knowledge of building characteristics necessary to implement curtailed demand and suggests that it can only be effective if "sensing and switching can be done cheaply" and with "a high level of automation."

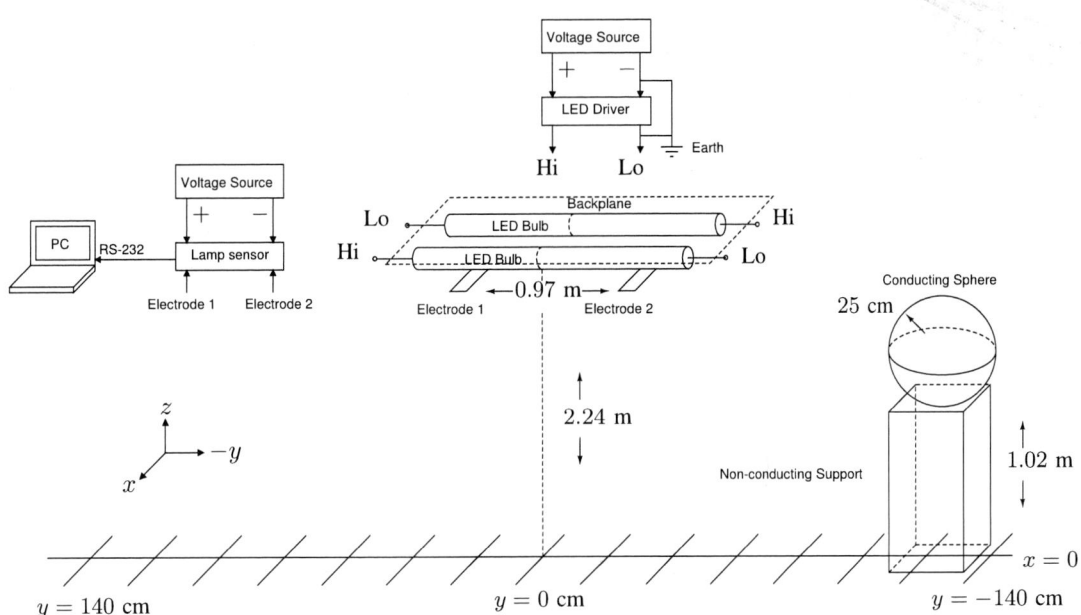

Fig. 14. The experimental setup.

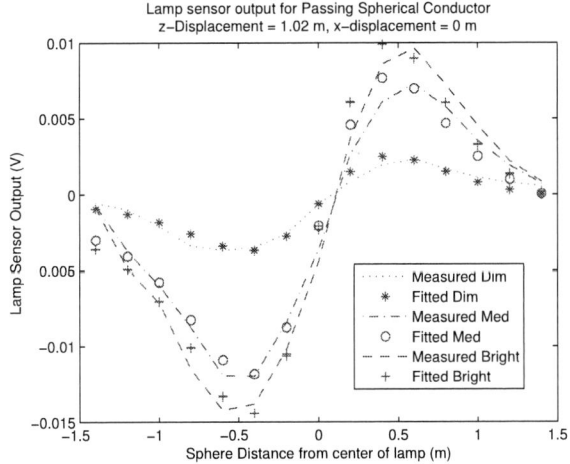

Fig. 15. Training runs (source calibration) $x = 0$ cm

In particular, there is a great interest in controlling lighting to optimize energy consumption. Lighting in commercial and residential spaces consumes a significant portion of the end use demand for delivered energy in the United States. In 2005, lighting consumed 0.73 Quadrillion Btu (QBtu) in the residential sector and 1.18 QBtu in the commercial sector [18].

This paper has presented a ballast or solid-state lamp driver that is suitable for dimmable operation of solid-state or LED lighting. The ballast described in this paper also creates the correct lamp current waveforms to permit a solid-state lamp to function effectively as a proximity sensor for occupants. This opens the door to distributed, autonomous control for lighting. That is, light fixtures can automatically alter their illumination based on the presence or absence of occupants, and any other important environmental variables such as time of day, through the actions of an embedded controller. The proximity sensor does not require motion or other intrusive occupant behavior to function. It is sensitive directly to the dielectric presence of an occupant. Interfacing the lamp sensor with a dimming ballast creates a smart auto-dimming lamp that can use the lamp sensor's occupancy detections to appropriately dim or brighten.

This paper has also demonstrated an electro-quasistatic model that accurately predicts the behavior of the proximity sensor for both fluorescent and solid state lamps. The model can be used by building designers to predict detection range given a particular configuration of luminaire. It can also be used to select a luminaire design to achieve needed detection range.

The incorporation of automatic proximity detection in solid state lighting could be a "game changing" addition to solid-state lamps that accelerates their acceptance.

Acknowledgments

The authors would like to thank The Grainger Foundation, the MIT Energy Initiative (MITei), and the U.S. Department of Energy (DOE) for their generous and essential support and funding. This work was partially supported by the Center for Materials Science and Engineering at MIT as part of the MRSEC Program of the National Science Foundation under grant number DMR-08-19762.

References

[1] "Annual energy outlook 2009," Energy Information Administration: United States Department of Energy, March 2009.
[2] D. Steigerwald, J. Bhat, D. Collins, R. Fletcher, M. Holcomb, M. Ludowise, P. Martin, and S. Rudaz, "Illumination with solid state lighting technology," *Selected Topics in Quantum Electronics, IEEE Journal of,* vol. 8, no. 2, pp. 310–320, Mar/Apr 2002.

[3] J. J. Cooley, A.-T. Avestruz, S. B. Leeb, and L. K. Norford, "A fluorescent lamp with integral proximity sensor for building energy management," in *Power Electronics Specialists Conference, 2007. PESC 2007. IEEE*, June 2007, pp. 1157–1163.

[4] J. J. Cooley, A.-T. Avestruz, and S. B. Leeb, "An autonomous distributed demand-side energy management network using fluorescent lamp sensors," in *Power Electronics Specialists Conference, 2008. PESC 2008. IEEE*, June 2008.

[5] J. J. Cooley, "Capacitive sensing with a fluorescent lamp," Master's thesis, Massachusetts Institute of Technology, Cambridge, Massachusetts, 2007.

[6] J. J. Cooley, A.-T. Avestruz, and S. B. Leeb, "A design-oriented analytical approach for fully-differential closed-loop op-amp circuits."

[7] W. Buller and B. Wilson, "Measuring the capacitance of electrical wiring and humans for proximity sensing with existing electrical infrastructure," *Electro/information Technology, 2006 International Conference on*, pp. 93–96, 2006.

[8] P. Shahidi, A.V.; Savard, "A volume conductor model of the human thorax for field calculations," in *Proceedings of the Twelfth Annual International Conference of the IEEE*. Engineering in Medicine and Biology Society, November 1990, pp. 615–616.

[9] J. R. Smith, "Electric field imaging," Ph. D. Thesis, Massachusetts Institute of Technology, 1999.

[10] J. J. Cooley, "Capacitive sensing with a fluorescent lamp and applications."

[11] J. A. Stratton, *Electromagnetic theory*. McGraw-Hill book company, 1941.

[12] R. M. Fano, L. J. Chu, and R. B. Adler, *Electromagnetic fields, Energy and forces*. John Wiley and Sons, Inc., 1963.

[13] M. Zahn, *Electromagnetic Field Theory: a problem solving approach*. Krieger Publishing Company, 2003.

[14] J. D. Jackson, *Classical electrodynamics*. Wiley, 1999.

[15] M. Dyble, N. Narendran, A. Bierman, and T. Klein, "Impact of dimming white leds: Chromacity shifts due to different dimming methods," in *Fifth International Conference on Solid State Lighting, Proceedings of SPIE 5941*. Bellingham, WA: International Society of Optical Engineers, 2005, pp. 291–299.

[16] Y. Gu, N. Narendran, T. Dong, and H. Wu, "Spectral and luminous efficacy change of high-power leds under different dimming methods," in *Sixth International Conference on Solid State Lighting, Proceedings of SPIE 6337, 63370J*, 2006.

[17] K. Taylor, J. Ward, V. Gerasimov, and G. James, "Sensor/actuator networks supporting agents for distributed energy management," in *29th Annual IEEE International Conference on Local Computer Networks*, Bellingham, WA, November 2004, pp. 463 – 470.

[18] "Annual energy outlook 2007," Energy Information Administration: United States Department of Energy, February 2007.

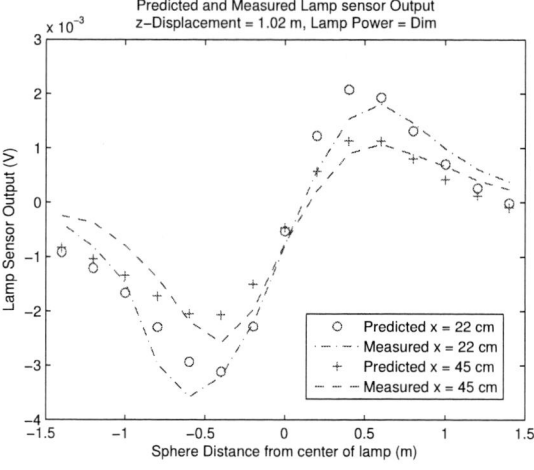

(a) Dim

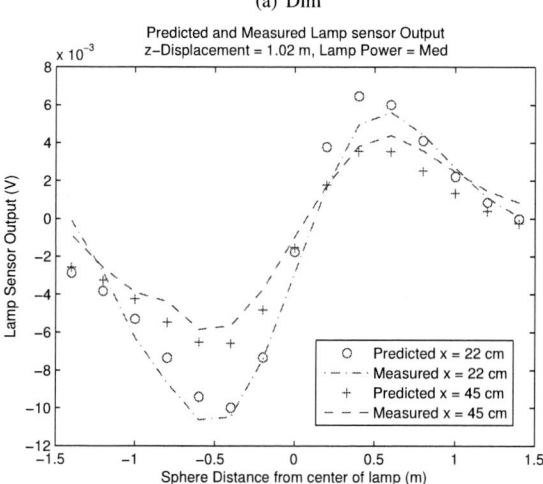

(b) Medium

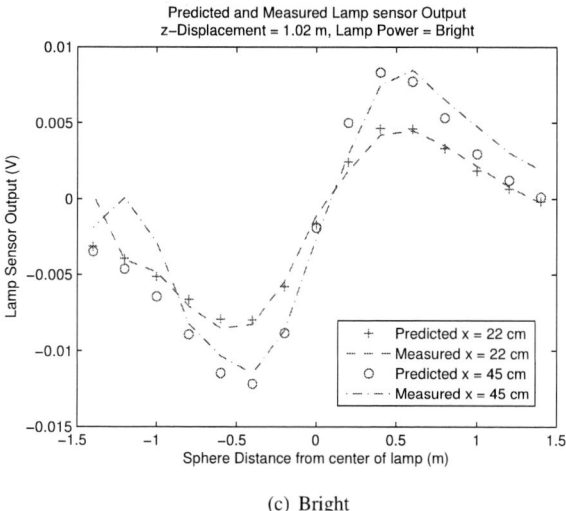

(c) Bright

Fig. 16. LED lamp sensor measured and predicted responses.

A 0.9 PF LED Driver with Small LED Current Ripple Based on Series-input Digitally-controlled Converter Modules

Qingcong Hu, *Student Member, IEEE*, and Regan Zane, *Senior Member, IEEE*
Department of Electrical, Computer and Energy Engineering
University of Colorado at Boulder
Boulder, CO, USA
{qingcong.hu, zane}@colorado.edu

Abstract—This paper introduces a digitally-controlled off-line LED driver based on low-voltage series-input converter modules. The series-input connected modular architecture is adopted to utilize low-voltage high-frequency circuits and low-profile components in off-line applications. Input current is regulated to be constant using a simple controller to achieve 0.9 power factor (PF) without a traditional power factor correction (PFC) stage. Additionally, a bidirectional buck second stage is used to reduce 120 Hz current ripple in the LEDs with minimal energy storage requirements and high efficiency. Experimental results are presented for a series-input system with three modules that drive 6, 8 and 7 LEDs with less than 15% line-frequency current ripple using small filter capacitors.

I. INTRODUCTION

Solid-state light-emitting diodes (LEDs) are expected to achieve significant growth in the lighting industry. Although some trends in high-brightness LEDs are towards high-power, high-current devices, thermal management and efficiency considerations in most commercial, industrial and street-lighting applications [1] still require a large number of LEDs to be used in a single fixture. Anticipated numbers include 50 to 200 LEDs in a fixture driven in the 200 mA to 400 mA range. Although most solutions put many LEDs in a long series string with a high-voltage power supply for each string, the series-input modular structure has been investigated to utilize low-voltage high-frequency electronics and low-profile components in high-voltage applications [2, 3]. The common duty cycle approach in [3] simplifies the feedback loop design and achieves automatic input voltage sharing and LED current copying, providing a solution for applications with a DC source.

This paper introduces an off-line LED driver based on low-voltage series-input converter modules, with a block diagram shown in Fig. 1. The common duty cycle control approach is utilized to distribute the line voltage and realize LED current copying. The input current is regulated to be constant to achieve 0.9 power factor (PF) without a traditional

power factor correction (PFC) stage. An additional bidirectional second stage can be used to reduce the 120 Hz ripple of LED current with a relatively small energy storage capacitor.

II. CONSTANT INPUT CURRENT REGULATION TO ACHIEVE 0.9 PF

The power factor of a circuit can be calculated by

$$ PF = \frac{P_{in,avg}}{V_{in,rms} \times I_{in,rms}}, \qquad (1) $$

where $V_{in,rms}$ and $I_{in,rms}$ are the rms value of input voltage and input current of the circuit, and $P_{in,avg}$ is the average input power. In order to achieve close-to-unity PF, normally a PFC stage with a large bulk capacitor is placed after the ac source (line) and before the application circuit. The PFC stage often requires input voltage and current sensing and can be realized using a wide range of control approaches [4]. For the solid-state lighting industry, recent energy star requirements place

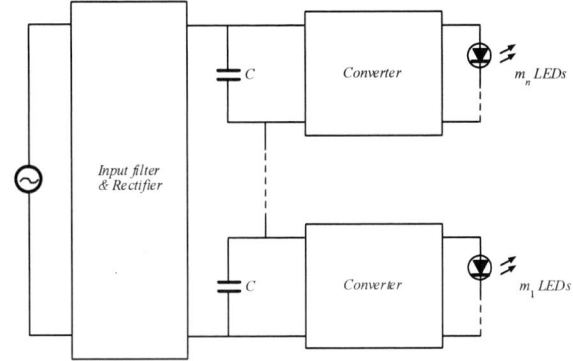

Figure 1. Off-line LED driver based on low-voltage series-input converter modules.

This work was co-sponsored by the National Science Foundation (under Grant No. 0348772) and industry sponsors of the Colorado Power Electronics Center.

limits on the power factor at PF $\geq$ 0.7 for residential lighting and PF $\geq$ 0.9 for commercial lighting [5]. The drawbacks to achieving high-quality PF with very low total harmonic distortion (THD) include increased circuit complexity, reduced efficiency and increased requirement for either bulk energy storage or large line-frequency current ripple in the LED. These drawbacks provide motivation to minimize the PFC circuit while just meeting the requirements in the specification.

According to (1), when the input current is regulated to be a constant value, the resulting PF is equal to 0.9. This leads to a relatively simple controller that meets the PF requirements for lighting products and minimizes the line-frequency output current ripple. Fig. 2 shows a buck-boost converter able to sense and regulate its input current, where the input voltage v_{in} is the rectified ac line voltage scaled down by the number of converter modules in Fig. 1. A boost converter can also be used if the target LED string voltage is larger than the peak input voltage.

A simple integration circuit is sufficient to realize the function of input current regulation. Since in a series-input system the duty cycle needs to be transmitted between modules, a digital control approach is preferred. One solution for measuring the input current in a digital system is to sample the inductor current at the center point of the transistor on-time, which together with duty cycle can be used to calculate input current.

As analyzed in [3], when series-input converters with the same structure operate in continuous conduction mode (CCM) with the same duty cycle, in steady state the line voltage automatically distributes among the modules according to the ratio of loads, and the output currents of slave modules are the same as that of the master. These benefits hold with the proposed 0.9 PF converters given that they operate in CCM during the majority of the 120 Hz ac line cycle.

III. SECOND STAGE TO REDUCE 120-HZ LED CURRENT RIPPLE

Although the method of constant input current regulation achieves 0.9 PF with reduced line-frequency output current ripple when compared to a unity PF circuit, the output current still has a significant line-frequency ripple component, especially for LED load, whose current variation is much larger than voltage variation. There are many motivations to limit low-frequency current ripple in LEDs, including reduced efficiency, life degradation due to thermal cycling and human factors.

One simple method to reduce LED current ripple is to use a large bulk capacitor at the output of the first stage and in parallel with the LEDs. However, the size of the capacitor makes this approach impractical when trying to reduce the ripple to low values (e.g. less than 50%). A traditional solution is to use a bulk capacitor at the output of the first stage and to place a buck converter second stage between the bulk capacitor and the LED output. This approach requires all LED power to be processed by the two series stages, and may result in significant efficiency reduction.

Another method is to keep the LED string at the output of the first stage and utilize a bidirectional second stage with a smaller bulk energy storage capacitor, as shown in Fig. 3. A simulation is conducted to compare the capacitance required by this method and the method of placing a bulk capacitor parallel with the LED string. The simulation uses a string of 8 LEDs, whose target operation voltage and current are 3.2 V and 200 mA, respectively. A rectified ac current source of $|0.1\,\pi\sin(2\pi\,60\,t)|$ A is driving the LED string. The capacitances necessary to meet several LED current ripple requirements are shown in Table 1, which indicates that the capacitance required by the bidirectional second stage is much less than the parallel bulk capacitor method.

The topology of the second stage can be determined by the application requirement. With a buck second stage, the energy storage capacitor C_{esc} may have lower voltage rating but larger capacitance value, while a boost second stage allows smaller capacitance with higher voltage rating. With both topologies, the energy storage capacitor voltage v_{esc} should be as close to LED string voltage v_{LEDs} as possible to improve the efficiency of the second stage. A bidirectional buck second stage will be used as an example. This second stage operates as a buck converter to accumulate the excess energy in its storage capacitor C_{esc} when instantaneous input power is larger than the target power of the LED string. When the input power is low, this stage operates as a boost converter in the reverse direction to restore the energy back to the LED string. As a result, the energy storage capacitor should be large enough to handle the excess energy during a half-line ac cycle and the second stage only processes the line-frequency ripple energy from the first stage. The control of the second

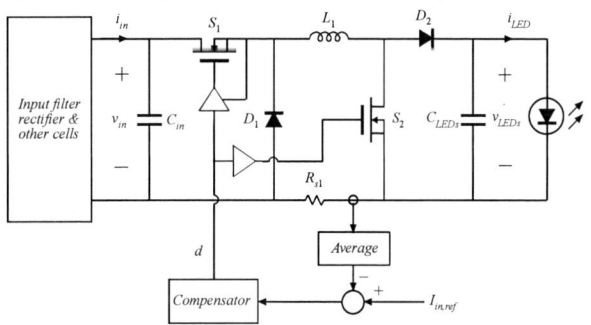

Figure 2. Buck-boost converter with input current regulation.

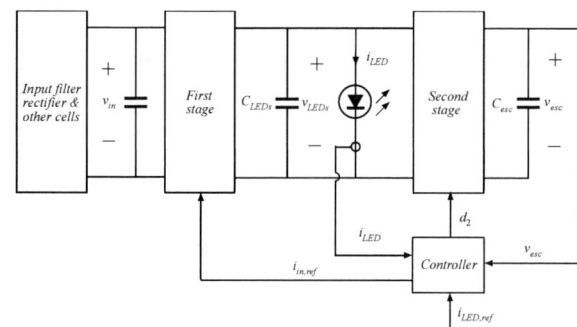

Figure 3. Block diagram of second stage control loops.

TABLE I. CAPACITANCE NECESSARY TO MEET SEVERAL LED CURRENT RIPPLE REQUIREMENTS BY THE METHOD OF BULK CAPACITOR PARALLEL WITH LED STRING AND THE METHOD OF BIDIRECTIONAL SECOND STAGE

LED Current Ripple Requirement (%)	Capacitance Necessary (μF)	
	Bulk capacitor parallel with LED string	Bidirectional second stage
10	1000	30
25	400	18
50	100	2

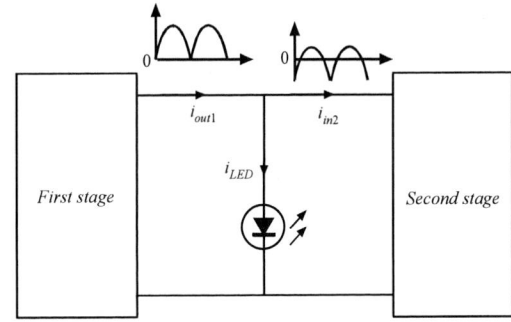

Figure 4. Ideal relation of first stage output current, LED current and second stage input current.

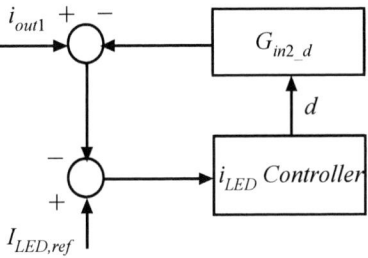

Figure 5. Block diagram of the LED current control loop: control-to-input current transfer function of second stage is used to build the loop gain.

stage can be adjusted to process only the amount of power required to just meet the LED current ripple requirements for a given application, which can result in improved efficiency when compared to a traditional series two-stage approach when some line-frequency ripple is allowed in the LED current, e.g. 10% to 40%.

Two control loops are necessary in order to utilize the second stage. One is to regulate the LED current, and can be called an LED current loop. The other is to balance the energy accumulated and restored by the second stage in one ac line cycle in order to control the capacitor voltage v_{esc}. A block diagram of the second stage control loops is shown in Fig. 3.

A. LED Current Loop

The goal of the LED current loop is to regulate the average LED current. As the input current of the first stage is constant, the module input power follows a rectified sinusoidal wave, whose average value minus converter losses should equal to the average power of the LED string in steady state. As a result, the second stage has to store energy when the input power is larger than the required LED power and to supply energy to the LED string when the input power is low. The model of second stage functionality is illustrated in Fig. 4, which illustrates the relation of the first stage output current i_{out1}, LED current i_{LED} and second stage input current i_{in2}. As the sum of i_{in2} and i_{LED} is constant in a short period, the control-to-input current transfer function of the second stage can be used to build the transfer function of LED current loop. The block diagram of this loop is shown in Fig. 5, which shows the case when the second stage operates as a buck converter.

If the second stage is realized as a synchronous converter operating in CCM, the controller design of the LED current loop can be relatively simple. The reason is that the transition between sinking and sourcing current can be automatically finished by the circuit itself, so that no additional judgment of mode transition is required. However, as the input current of the second stage is quite small for some period in an ac line cycle, operation in CCM leads to bidirectional inductor current of second stage in one switching period, which results in higher conduction loss. Larger inductor value may help reduce bidirectional current, but the larger series resistance leads to higher loss as well.

Operation in DCM helps reduce the power loss due to reverse inductor current in one single switching period. The drawback is the requirement of judgment to switch between sinking and sourcing current. This function for mode transition can be realized by analyzing the duty cycle of the second stage, or other signals which can indicate the variation of first stage output current.

Analyses also indicate that the control-to-input current transfer function is very simple when the second stage operates in DCM. For example, the dc gain of the control-to-input current transfer function of a buck second stage in DCM can be expressed as

$$G_{in2\_d0} = \frac{(V_{LEDs} - V_{esc})DT_s}{L}, \qquad (2)$$

while the dominant pole is determined by the capacitor parallel with LED string and the equivalent resistance of the LED string. Similar result can be achieved when the second stage operates as a boost converter in reverse direction. A simple PI compensator is sufficient to stabilize the loop.

B. Voltage Loop

An additional loop is necessary to balance the energy stored and supplied by the second stage in one ac line cycle.

This goal can be achieved by regulating the voltage of energy storage capacitor v_{esc}, which should be as close to the LED string voltage v_{LEDs} as possible in order to improve the efficiency of the second stage. Such a control loop is shown in Fig. 6(a), which utilize self-balance method to control v_{esc} by adjusting the average LED current $i_{LED,avg}$. As the external input control signal of the whole system is reference LED current $I_{LED,ref}$, the average LED current $i_{LED,avg}$ has to be fed back to first stage to control the input current, which is shown in Fig. 6(b).

A simplified control is shown in Fig. 7, which utilizes the external input LED current reference $I_{LED,ref}$ for LED current regulation in the second stage, and generates the reference signal for input current $I_{in,ref}$ directly by regulating the energy storage capacitor voltage v_{esc}. This method reduces the number of control loops and simplifies the design.

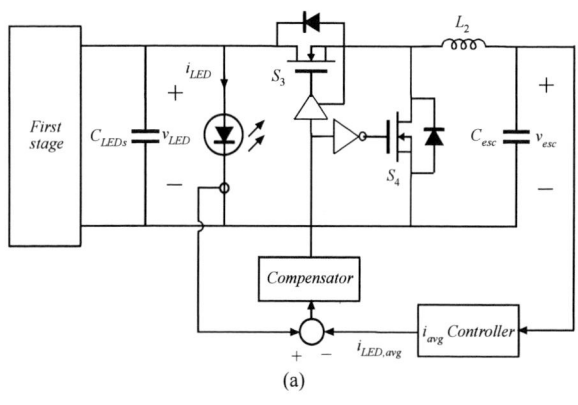

(a)

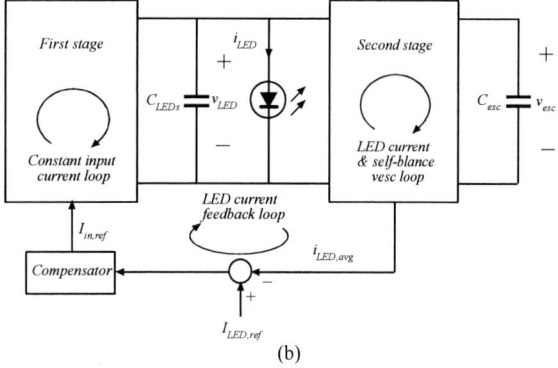

(b)

Figure 6. (a) Block diagram of second stage self-balance loop to regulate v_{esc} by adjusting average LED current. (b) Block diagram of the whole module adopting self-blance loop in (a).

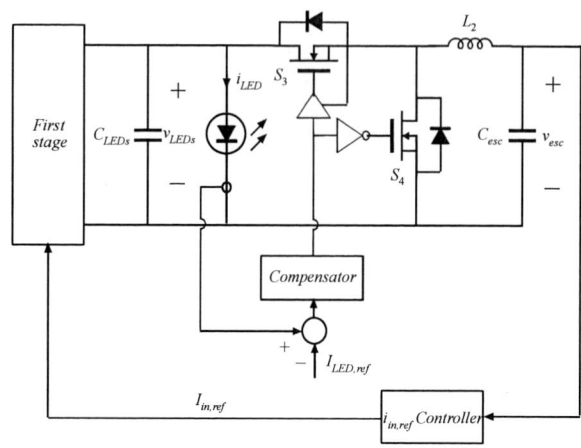

Figure 7. Simplified closed-loop control design regulating the LED current directly with external input reference signal

IV. EXPERIMENTAL RESULTS

A prototype was developed to experimentally verify the performance of the proposed modular architecture and control approach. The circuit has separable converters and load stages to employ the modularity of the architecture. Each cell is controlled via an onboard floating FPGA and corresponding sensing circuitry and ADCs. A two-stage structure, with four-switch buck-boost first stage and bidirectional buck second stage, as shown in Fig. 8, was implemented for the experiments. The values of input capacitor, first stage inductor and output capacitor of first stage, second stage inductor and energy storage capacitor are 0.47 µF, 22 µH, 10 µF, 10 µH and 47 µF, respectively. The switching frequency is about 780 kHz. The devices used are shown as Table 2. Three cells were used, with six, eight and seven LEDs as load, respectively.

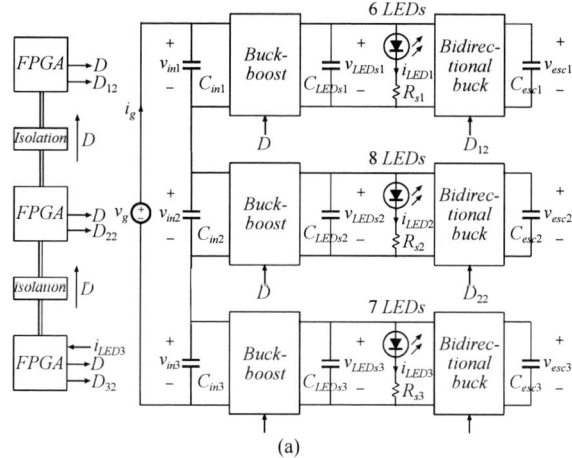

(a)

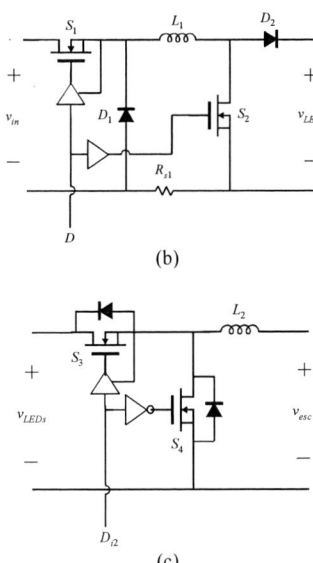

(b)

(c)

Figure 8. (a) Prototype for experiment. Line voltage v_g is rectified ac input signal. FPGAs are used to control the modules and transmit duty cycle. Three modules are used, driving six, eight and seven LEDs, respectively. (b) Non-inverting buck-boost first stage used in experiment. (c) Bidirectional buck second stage used in experiment.

TABLE I.   DEVICES USED IN EXPERIMENT PROTOTYPE

Device	Part Number	Description
FPGA	XC3S500E	Xilinx Spartan 3E
Diode	PDS540	Schottky, 40 V, 5 A
MOSFET	STN3NF06L	N_channel, 60 V, 4 A
ADC	AD7825	8 bit 4 channel Multiplexed
HB Driver	LM5101	Half Bridge Driver
Isolator	ISO7221	Dual Digital Isolator 2MSPS
LED	Luxeon K2	700 mA

A. Constant Input Current Regulation

A buck-boost converter with constant input current regulation is implemented and test in experiment. The waveforms of the rectified input voltage v_{in}, input current i_{in}, output voltage v_{LEDs} and LED current i_{LED} are shown in Fig. 9. The input current is regulated to 200 mA over the majority of the ac line cycle, and reaches zero when the input voltage goes to zero and the duty cycle saturates. The measured power factor of this circuit is 0.93 to 0.95 when the input voltage is 20 Vac to 40 Vac, and the average current of an 8-LED string is 100 mA to 300 mA.

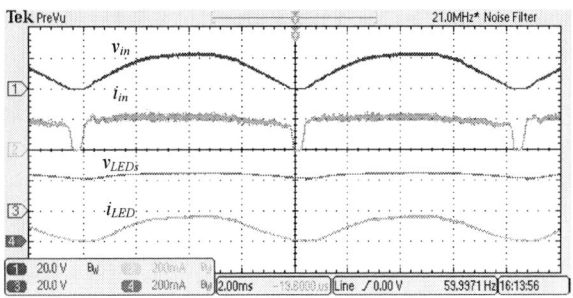

Figure 9.   Experimental input voltage, input current, output voltage and LED current of a 0.9 PF buck-boost converter

B. Line Voltage Sharing and LED Current Copying of series-input 0.9 PF converters

Series-input system with two buck-boost converters are built in experiment with constant input current regulation. The common duty cycle approach is utilized to control the system. Those two converters drive six and eight LEDs respectively, and the input voltages and LED currents are shown in Fig. 10. It can be seen that the input voltage ratio is about 3:4, just the same as the load ratio, and the LED currents are approximately the same for the two converters, indicating line voltage sharing and output current copying.

C. Closed-loop Module with Second Stage to Regulate LED Current

A module with second stage and a voltage loop (Fig. 7) to regulate LED current was implemented in experiment. A selection of experimental results from the realized system is given here. The experimental results use a module that drives 8 LEDs at 200 mA current, with first stage output capacitor $C_{LEDs} = 11 \ \mu F$ and energy storage capacitor $C_{esc} = 47 \ \mu F$. Waveforms of rectified input voltage v_{in}, energy storage capacitor voltage v_{esc} and the LED current i_{LED} are shown in Fig. 11, demonstrating minimal 120-Hz LED current ripple. As a comparison, the second stage was removed and the first stage output capacitance was increased to $C_{LEDs} = 57 \ \mu F$, with the results shown in Fig. 12, where the large 120-Hz LED current ripple can be seen.

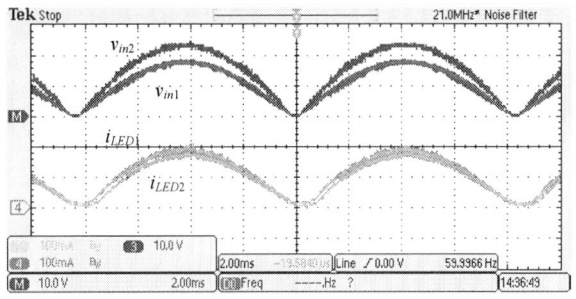

Figure 10. Experimental input voltages and LED currents of two series-input converters driving 6 and 8 LEDs

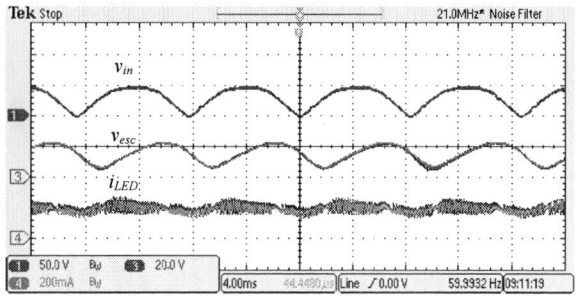

Figure 11. Experimental input voltage, energy storage capacitor voltage and LED current of a module with second stage.

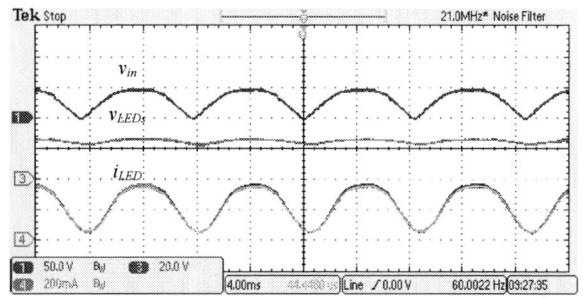

Figure 12. Experimental input voltage, output capacitor voltage and LED current of a 0.9 PF converter without second stage.

D. Series-input Connected Modules with Second Stages

A series-input system with three modules was implemented in experiment. The master module utilizes the simplified feedback loop in Fig. 7 to regulate LED current and generate first stage duty cycle. The slave modules use the duty cycle from master for the first stage, and adopt the self-balance method in Fig. 6(a) to regulate their own energy storage capacitor voltage and LED current. The input voltages and LED currents of two of the modules are shown in Fig. 13. These two modules drive 6 and 8 LEDs as load respectively. The ratio of their input voltages is about the same as that of loads, while their LED currents are about the same without ac line-frequency ripple.

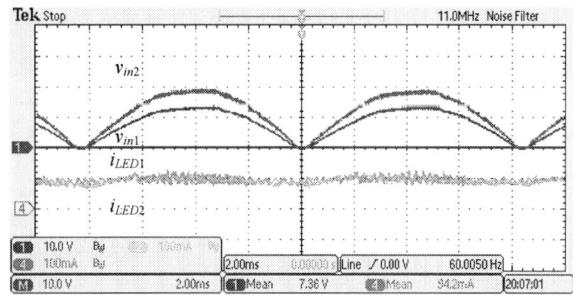

Figure 13. Experimental input voltages and LED currents of two modules in a series-input system. The two modules drive 6 and 8 LEDs as load respectively.

The waveforms of input voltage, input current, energy storage capacitor voltage and LED current of one module in the system are shown in Fig. 14. As illustrated in the waveforms, the input current is almost constant during the ac line cycle, expect when the input voltage across zero point, which provides a PF higher than 0.9. The line frequency ripple on LED current is removed by storing the excess energy in the second stage output capacitor C_{esc} when input power is high and suppling the energy back to LED string when input power is low, which process is indicated by the waveform of v_{esc}.

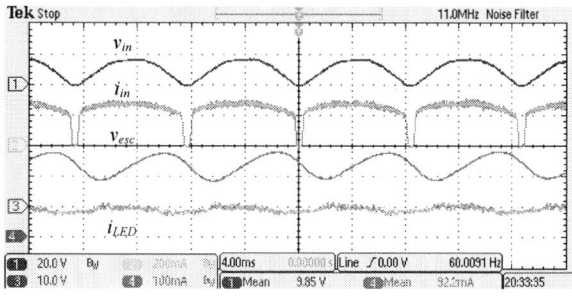

Figure 14. Experimental input voltage, input current, energy storage capacitor voltage and LED current of one module in the series-input system.

V. CONCLUSION

This paper introduces a digitally-controlled off-line LED driver based on low-voltage series-input converter modules. The common duty cycle control approach is utilized to distribute the input rectified voltage and realize LED current copying. The input current is regulated to be constant to achieve 0.9 power factor. A bidirectional buck second stage is able to reduce the 120 Hz LED current ripple with minimal bulk energy storage. The series-input system with 0.9 PF, as well as second stage has been implemented in experiment and results are demonstrated with modules driving 6, 8 and 7 LEDs.

REFERENCES

[1] S. Hui, and Y. Qin, "A general photo-electro-thermal theory for light emitting diode (LED) systems," in Proc. IEEE 2009 Appl. Power Electron. Conf., pp. 554-562, 2009.

[2] J. Patterson, and R. Zane, "Series input modular architecture for driving multiple LEDs," in Proc. IEEE 2008 Power Electron. Spec. Conf., pp. 2650-2656, 2008.

[3] Q. Hu, and R. Zane, "LED drive circuit with series input connected converter cells operating in continuous conduction mode," in Proc. IEEE 2009 Appl. Power Electron. Conf. pp. 1511-1517, 2009.

[4] R. Erickson, and D. Maksimovic, Fundamentals of power electronics second edition, springer, New York, 2001.

[5] "ENERGY STAR program requirements for solid state lighting luminaires," on line: http://www.energystar.gov/ia/partners/product_specs/program_reqs/SSL_prog_req_V1.1.pdf

[6] R. Giri, V. Choudhary, R. Ayyanar, and N. Mohan, "Common-duty_ratio control of input-series connected modular DC-DC converters with active input voltage and load-current sharing," IEEE Trans. Ind. Appl., vol. 42, pp. 1101-1111, Jul./Aug. 2006.

[7] C. Hsieh, C Yang, and K. chen, "A charge-recycling buck-store and boost-restore (BSBR) technique with dual outputs for RGB LED

bucklight and flashling Module," IEEE Trans. Power Electron., vol. 24, pp. 1914-1924, Aug. 2009.

[8] J. W. Kimball, J. T. Mossoba, and P. T. Krein, "a stabiling, high-performance controller for input series-outpu parallel converters," IEEE Trans. Power Electron., vol. 23, pp. 1416-1427, May 2008.

[9] D. Gacio, J. M. Alonso, A. J. Calleja, J. Garcia, and M. Rico-Secades, "A universal-input singal-stage high-power-factor power supply for HB-LEDs based on integrated buck-flyback converters," in IEEE 2009 Appl. Power Electron. Conf., pp. 570-576, 2009.

A Novel Dimmable Electronic Ballast for Compact Fluorescent Lamps Using Phase-Cut Incandescent Lamp Dimmers with Wide Dimming Range and Low Dimming Level Lamp Ignition Capability

John Lam , *IEEE Member*
Queen's Center for Energy and Power Electronics Research
esearch (ePOWER)
Queen's University
Kingston, Ontario, Canada
e-mail address: john.lam@ece.queensu.ca

Praveen K. Jain , *IEEE Fellow*
Queen's Center for Energy and Power Electronics Research
(ePOWER)
Queen's University
Kingston, Ontario, Canada
e-mail address: praveen.jain@queensu.ca

Abstract—A new dimmable electronic ballast that is compatible with standard incandescent lamp dimmer with wide dimming range is proposed in this paper for compact fluorescent lamps (CFL). The proposed electronic ballast is able to improve the poor dimming performance in the conventional CFL and at the same time, the dimmed light output can be maintained at its particular level even when the dimmer is switched off and on again. The proposed circuit consists of a compact, high power factor single switch resonant inverter as the power stage and a novel controller that controls the duty ratio of the switch to provide dimming function and the ability for the lamp to ignite at very low dimming level without any lamp flickering. Experimental results are given on a 13W CFL to verify with the proposed circuit.

I. INTRODUCTION

With increased concern regarding the world's depleting energy resources, power electronics is now playing a key role in providing improved energy-saving technology. In the lighting industry, it has been proposed in many countries to gradually replace conventional incandescent lamps with compact fluorescent lamps (CFL) to save energy [1]-[2]. In North America, for example, incandescent lamps will be phased out by 2012 [1]. However, one of the major differences between incandescent lamp and CFL is that incandescent lamp behaves as a resistive load at the AC side whereas the CFL behaves like a capacitive load. This means that the power factor drawn by the incandescent lamp is much higher than the CFL. Line current with poor power factor drawn by typical consumer CFL means that reactive power is consumed in the power conversion process. Consequential to the poor power factor drawn from conventional CFLs, they do not provide much advantage, in terms of energy-saving, compared to the incandescent lamp.

With the replacement of incandescent lamps by CFLs in the near future, it is important for CFLs to be compatible with the existing dimmers designed for incandescent lamps. This allows existing wiring connections to remain the same, and therefore eliminates extra hardware or labor costs when the replacement occurs. Figure 1 shows the operating waveforms of a standard phase cut dimmer. It reduces the load average power by decreasing the conduction time of the line voltage. Assuming that the load is resistive, the line current is then essentially a pure sinusoidal current that is in phase with the line voltage. A conventional CFL electronic ballast, however, consists of a large DC link capacitor placed in front of a half-bridge resonant inverter [3]-[4]. As a result, the line current drawn by these CFLs has very short conduction time. Consequentially, these CFLs do not work very well with standard phase-cut dimmers. In addition, standard CFLs will not ignite when the phase cut dimmer is switched on if it is set at low dimming level. In order to design a dimmable CFL that can work well with phase-cut incandescent lamp dimmers, the power factor correction (PFC) function must be included in the electronic ballast of the CFL. This allows the load to appear more resistive from the AC input side point of view. Several works [5]-[8] have been proposed to improve the dimming performance of CFL with standard phase cut dimmers.

The above discussions conclude that in order to enhance the dimming performance of a CFL with phase cut dimmers, PFC is an essential stage in the ballast circuit design. This paper proposes a new dimmable high power factor electronic ballast that is highly compatible with incandescent lamp phase-cut dimmers. It also has the ability to ignite the lamp and maintain the dimmed light output at low dimming level. At the same time, high power factor is achieved at the input. The goal of the proposed solution is to allow the CFL to

978-1-4244-4782-4/10 $26.00 © 2010 IEEE

perform in a very similar fashion like the incandescent light bulb. In this paper, detailed descriptions of the proposed electronic ballast are provided. The stability of the overall system including the negative lamp impedance appears when dimming is also discussed. The proposed circuit was tested on a 13W CFL to verify its feasibility.

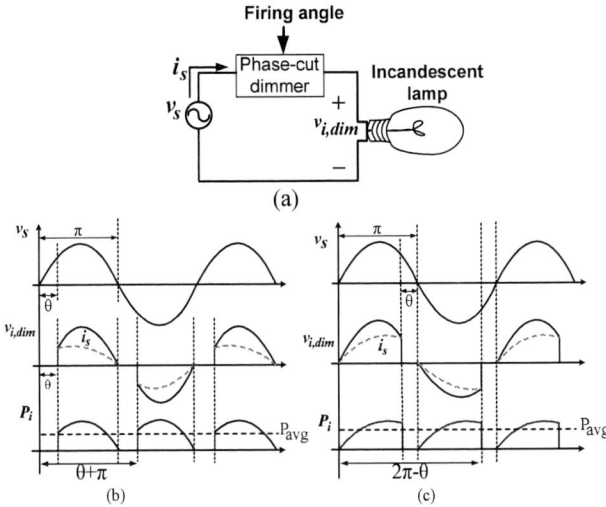

(a)

(b)   (c)

Figure 1 Key waveforms in standard phase cut dimmer

II. ANALYSIS OF PROPOSED SYSTEM

The proposed dimmable electronic ballast system is shown in Figure 2. The power circuit consists of a single-stage single switch resonant inverter with an integrated SEPIC PFC. The detailed descriptions of this circuit have been analyzed in [9]. The proposed control circuit in the proposed ballast system is to control the duty ratio of the MOSFET during dimming. The duty ratio should also be adjusted to provide the required output voltage to ignite the lamp even when the lamp is not at its full power condition.

The operating principles of the control circuit is illustrated in Figure 3. In the proposed controller, the rectified voltage (V_{rect}), which is given by (1) as a function of α with V_p represents the peak line voltage and f_L represents the line frequency, is scaled down and fed to the controller and compared with a DC signal. The purpose is to obtain a pulse V_{u1} at the output of the comparator to show the status of the firing angle applied at the incandescent lamp dimmer. During dimming, part of the sinusoidal line voltage is chopped, and the discontinuous time of V_{rect} increases. The pulse width of V_{u1} then increases. The RC circuit placed after the comparator is to provide the relative magnitude of the pulse width (V_{RC}). Consequently, the magnitude of the pulse width, $|V_{RC}|$, is directly proportional to the magnitude of the firing angle of the dimmer. V_{RC} is then amplified and subtracted from the reference signal. The purpose of utilizing duty ratio control to provide dimming with the phase cut dimmer at the same time is to control the lamp output power without using the full dimming range on the dimmer, which allows relatively high power factor to be achieved during dimming.

$$V_{ret}(\alpha) = \int_{\alpha}^{\pi} \frac{1}{\pi} V_p \sin(2\pi f_L t) d(2\pi f_L t) = \frac{V_p}{\pi}(1 + \cos(\alpha)) \quad (1)$$

In order to ignite the lamp even when the phase cut dimmer is adjusted to a lower power level, the duty ratio of the control circuit has to be adjusted again to provide the lamp ignition function. This function is performed by the block diagram shown in the dotted red line in Figure 2. The operating principles are given as follows: first, it compares the average values of V_{rect} and V_{dc} through two RC networks (RC_1 and RC_2); these RC networks are critical in determining the initial rise of both V_{rect} and V_{dc}. In the event when the dimmer switch is turned on, it is natural that V_{rect1} will increase prior to the increase in V_{dc1} providing that $RC_1 > RC_2$. Transistor (T_1) is then switched on and block B2 is connected to the control circuit, which forces the reference signal (V_{ref1}) to increase as given by (2), where G is the scale down factor, $V_{u1,max}$ is the peak voltage at the output of comparator u1. Hence, the duty ratio will be increased during this instant to ignite the lamp. Until $|V_{dc1}| > |V_{rect1}|$, block B2 will be disconnected and the reference signal is decreased and is given by V_{ref2} as shown in (3).

$$V_{ref1} = V_{dc,\max}G + K\frac{\alpha}{\pi}V_{u1,\max}\left(K\frac{\alpha}{\pi}V_{u1,\max} - 1\right) \quad (2)$$

$$V_{ref2} = V_{dc,\max}G - K\frac{\alpha}{\pi}V_{u1,\max} \quad (3)$$

According to the proposed control concept, there should exist a particular duty ratio that corresponds to a particular firing angle, namely, $d(\alpha)$. $d(\alpha)$ can be derived by assuming that the power losses in the SEPIC converter is negligible, then $p_{avg}(\alpha) \approx V_{dc}(\alpha)I_{dc}(\alpha)$. According to Figure 2, the DC-link capacitor (C_2) voltage (V_{dc}) should be equal to the reference signal (V_{ref2}). The average output current ($I_{dc}(\alpha)$) is equal to the average current flowing diode D_b. Hence, $I_{dc}(\alpha)$ can be obtained by averaging i_{Db} over one line period as shown in (4), where T_s is the switching period; Δ_1 is the discharge period of the PFC inductor current. The average output power can then be obtained as shown in (5). The average input power of the SEPIC converter can be expressed in the form of ((6), where $R_i(\alpha)$ represents the input mean resistance of the SEPIC converter and is given by ((7). By setting equals ((6) to (5), the duty ratio can be expressed as a function of α as given in (8), where L_e is the equivalent inductance of the SEPIC converter and is given in (9).

$$I_{dc}(\alpha) = \frac{1}{\pi}\int_{\alpha}^{\pi}\frac{d\Delta_1 V_p T_s}{2L_{eq}}\sin(\omega_L t)d(\omega_L t) = \frac{d\Delta_1 V_p T_s}{2L_{eq}\pi}(1+\cos(\alpha)) \quad (4)$$

$$p_{o,avg}(\alpha) = V_{dc}(\alpha)I_{dc}(\alpha) = \frac{d(\alpha)\Delta_1 V_p T_s}{2L_{eq}\pi}(1+\cos(\alpha))\left(\frac{GV_{dc,\max} - \frac{\alpha K}{\pi}}{G}\right) \quad (5)$$

$$p_{i,avg}(\alpha) = \frac{V_i^2(\alpha)}{R_i(\alpha)} = \frac{d^2(\alpha)V_p^2 T_s}{4L_{eq}\pi}\left(\pi - \alpha + \frac{\sin(2\alpha)}{2}\right) \quad (6)$$

$$R_i(\alpha) = \frac{d(\alpha)T_s}{2L_{eq}} \tag{7}$$

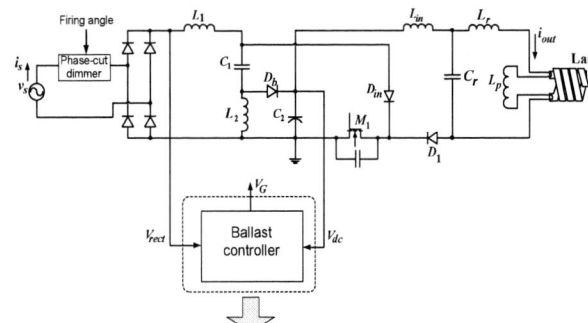

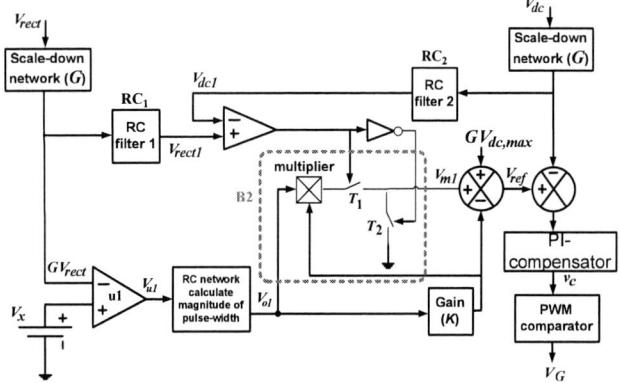

Figure 2 Proposed dimmable electronic ballast

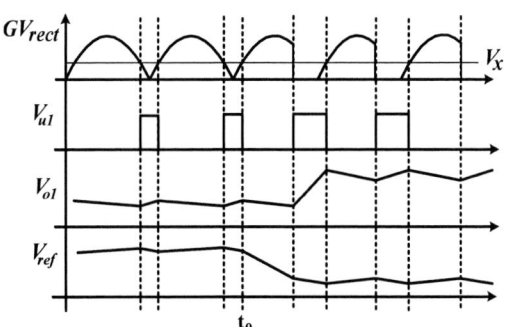

Figure 3 Some key waveforms at the control circuit

$$d(\alpha) = \frac{2(1+\cos(\alpha))\left(\dfrac{GV_{dc,\max} - \dfrac{\alpha K}{\pi}}{G}\right)\Delta_1}{V_p\left(\pi - \alpha + \dfrac{\sin(2\alpha)}{2}\right)} \tag{8}$$

$$L_e = \frac{L_1 L_2}{L_1 + L_2} \tag{9}$$

III. STABILITY ANALYSIS

Although fluorescent lamp behaves as a resistive load when it operates at high frequency, the lamp impedance is observed to be negative from the low frequency envelope point of view [9]-[10]. Figure 4 illustrates the typical lamp V-I characteristic during dimming. As the lamp power decreases, the lamp current decreases at a much faster rate than the increase in the lamp voltage. Due to the presence of the negative lamp impedance at low frequency, the ballast output impedance should offset the effect of the lamp negative impedance so that the overall system is stable when the lamp output power changes during dimming.

Figure 5 shows the block diagram of the proposed resonant inverter, where z_O represents the small signal output impedance of the resonant circuit and z_L represents the lamp impedance. The lamp current is then calculated as shown in (10). According to the Nyquist stability criterion discussed in [12], in order to have a stable lamp current, the stability requirement in here is not to have z_L/z_O encircle the point (-1,0) on the complex plane. z_O, which is the small signal output impedance of the proposed ballast is obtained according to the phasor transformation technique given in [9].

$$\hat{i}_o(s) = \frac{\hat{i}_{in}(s)}{1 + \dfrac{z_L(s)}{z_o(s)}} \tag{10}$$

From [9], any inductor and capacitor's small signal characteristics in the resonant circuit can be described by (12) and (13) respectively, where ω_s represents the steady-state switching frequency. Figure 6 shows the small signal circuit of the proposed resonant inverter, where the input inverter current is modeled as a small signal input current $\hat{i}_{in}(s)$. $z_O(s)$ is then obtained from Figure 6 and is given in (11) at the bottom of this page, where R_{lamp} is the steady-state lamp resistance at the switching frequency. To further simply the analysis [10], it can be deduced that the overall ballast stability is affected mainly at the low frequency range. This simplification can be performed because (1): the lamp small signal impedance appears to be negative only at low frequency range, at high operating frequency, the lamp is a resistive load; (2) the DC-link capacitor places a low frequency pole in the ballast system that governs the overall system crossover frequency. The resulting simplified circuit is then shown in Figure 7. To complete the circuit analysis, the load is replaced by the lamp small signal impedance expression as given in (14) [12], which consists of a low frequency RHP zero that explains the negative impedance phenomena of the lamp at low frequency range. In (14), K_l represents the slope of the V-I curve, which is negative. The detailed explanations of this RHP zero in the lamp impedance have been reported in [12]-[13].

$$\hat{z}_o(s) = \frac{s^3(L_r L_p C_r) + s^2(3j\omega_s L_r L_p C_r) + s(L_p - 3\omega_s^2 L_r L_p C_r) + (L_p j\omega_s - j\omega^3 L_r L_p C_r)}{s^3\left(\dfrac{L_r L_p C_r}{R_{lamp}}\right) + s^2\left(L_t C_r + \dfrac{3L_r L_p C_r}{R_{lamp}}j\omega_s\right) + s\left(\dfrac{L_p}{R} + 2j\omega_s L_t C_r - \dfrac{3L_r L_p C_r}{R_{lamp}}\omega_s^2\right) + \left(1 - L_t C_r \omega_s^2 - j\dfrac{L_r L_p C_r}{R_{lamp}}\omega_s^3\right)} \tag{11}$$

978-1-4244-4782-4/10 $26.00 © 2010 IEEE

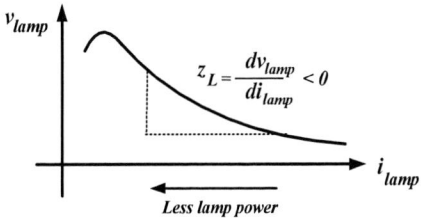

Figure 4 Typical V-I plot of fluorescent lamp

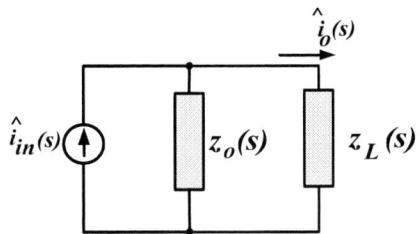

Figure 5 Block diagram of the inverter with lamp impedance

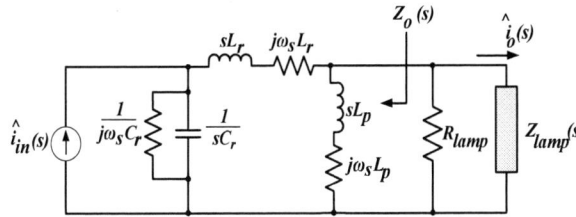

Figure 6 Small signal circuit of proposed resonant inverter

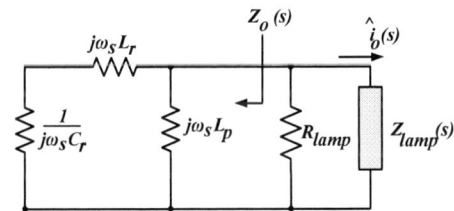

Figure 7 Simplified circuit for very low frequency modeling

$$v_L = L\frac{dI_L}{dt} + j\omega_s L I_L \tag{12}$$

$$i_c = C\frac{dV_c}{dt} + j\omega_s C V_c \tag{13}$$

$$z_{lamp}(s) = K_l \frac{\dfrac{s}{z_L} + 1}{\dfrac{s}{p_L} + 1} \tag{14}$$

The compensator design in the control circuit should be designed to meet the minimum stability requirement. The control-to-output transfer function $v_o/d(s)$ of the power circuit can be expressed as given in (15), where r_L represents the AC load impedance of the SEPIC stage and j_2 is given in (16). Here, r_L represents the total impedance of z_O and z_L. From (16), it is observed that as α increases, the DC gain, which is mainly governed by j_2, decreases during dimming. The

compensator should then be designed according to the maximum firing angle condition (α_{max}). The overall system stability is then verified through calculations and simulation results that will be demonstrated in the next section.

$$\frac{\hat{v}_{dc}(s)}{\hat{d}(s)} = \frac{j_2 r_L}{sC_2 r_L + 1} \tag{15}$$

$$j_2 = \frac{V_{rect}(\alpha)d(\alpha)T_s}{2V_{dc}(\alpha)L_{eq}} \tag{16}$$

IV. EXPERIMENTAL RESULTS

An experimental prototype of the proposed system has been built on a 13W 4-pins Dulux CFL to verify with the theoretical concept. The diameter of the prototype circuit is 53mm with the proposed controller implemented using analog circuitry. TABLE I. lists the parameters of the power circuit components. The subtractor, op-amp and PI-compensator are implemented using LM258 whereas the multiplier used in the prototype is AD633. The two transistors used in the prototype are npn transistors, Q2N2222. Proper biasing resistors are required to ensure the transistors work as switches. The comparator using in the prototype is ADCMP370 to minimize overall circuit size. The PWM saw-tooth is generated by the 555 timer IC with the switching frequency chosen to be 66kHz and the line voltage range is 90-120V_{rms} 60Hz. The maximum firing angle of the dimmer is 125°.

TABLE I. List of Power Circuit Components

	Circuit value/part numbers
L_1	8.2mH
L_2	1.2mH
C_1	33nF
C_2	22µF
D_{in}, D_b, D_1	MUR160
M_1	STP4NK60Z
L_{in}	0.39mH
C_r	3.3nF
L_r	1mH
L_p	3.3mH

Figure 8 shows the lamp voltage and current during ignition process. Figure 9, Figure 10 and Figure 11 show the line current, lamp voltage and lamp current at different dimming levels. The power factor achieved is 0.985, 0.79 and 0.49 at full power, 50% dimming and 10% dimming respectively. Figure 12 shows the lamp voltage and lamp current during dimming transition. The lamp is first dimmed down from full power to about 10% of the dimming level. The dimmer is then switched off for 3 seconds and then switched on again. It is observed that the ignition voltage is enough to strike the lamp and that the lamp current is maintained at the particular dimming level in a very stable manner. This supports that the proposed controller is able to adjust the duty ratio to provide lamp ignition and dim the lamp at low dimming level. To ensure that the MOSFET ratings are working within reasonable region, Figure 13 show the MOSFET voltage at full power and during dimming respectively. It is observed that the peak value of this voltage is less than 400V.

Figure 14 shows the nyquist plot of z_l/z_o, which illustrates that it does not encircle the point (-1,0) and is internally stable. However, as R_{lamp} increases during dimming, the curve moves closer to the (-1,0) point as shown by the green curve. The stability margin can be improved by introducing a compensator to regulate V_{dc}. Figure 15 presents the closed loop bode plot of the designed system, with the PI compensator designed according to the open loop transfer function given by (15). It is observed from Figure 15 that a phase margin of about 53° is achieved at the lowest dimming condition. The crossover frequency is only 20Hz, which means that the transient response is quite slow in this design. This is acceptable as fast transient response is not a mandatory requirement in dimmable electronic ballast designs.

Figure 16 compares the power factor performance among the proposed circuit, a commercial dimmable 15W CFL and a 40W incandescent lamp. The power factor achieved from the proposed circuit is very close to the incandescent lamp case and is much better than the current dimmable CFL. Figure 17 shows the input power with respect to the firing angle of the dimmer.

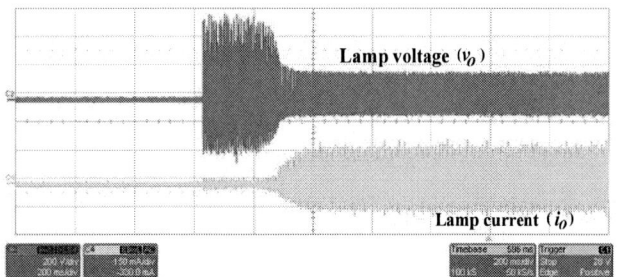

Figure 8 Measured waveforms at lamp ignition (v_o: 200V/div; i_o: 150mA/div; time: 200ms/div)

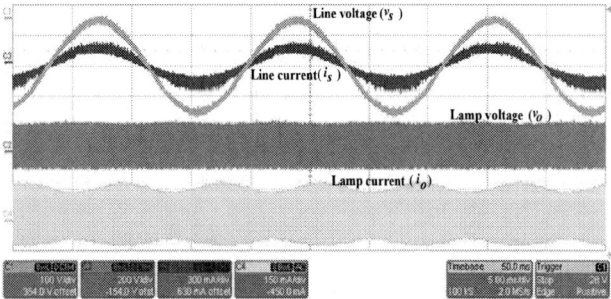

Figure 9 Measured waveforms at full power (v_s: 100V/div; i_s: 0.3A/div; v_o: 200V/div; i_o: 150mA/div; time: 5ms/div)

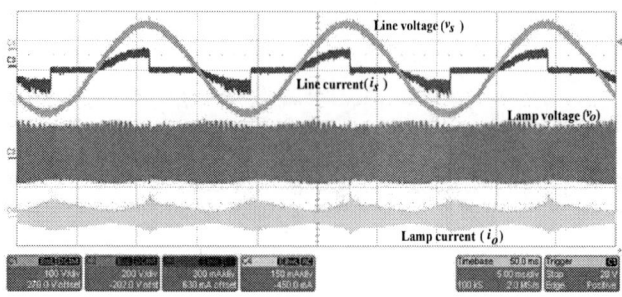

Figure 10 Measured waveforms at 50% dimming (v_s: 100V/div; i_s: 0.3A/div; v_o: 200V/div; i_o: 150mA/div; time: 5ms/div)

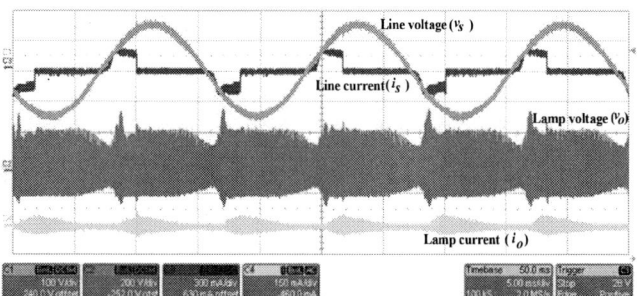

Figure 11 Measured waveforms at 15% dimming (v_s: 100V/div; i_s: 0.3A/div; v_o: 200V/div; i_o: 150mA/div; time: 5ms/div)

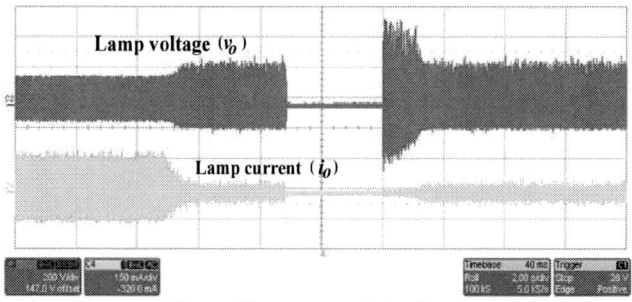

Figure 12 Lamp voltage and lamp current during dimming transition with dimmer switched off and on (v_o: 200V/div; i_o: 0.15A/div; time: 2s/div)

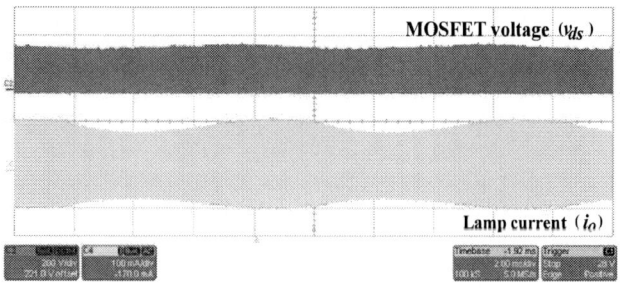

(a)

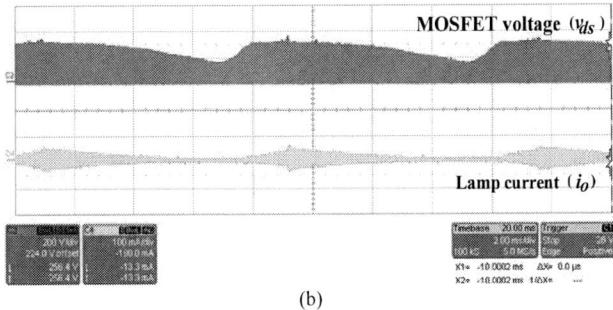

(b)

Figure 13 MOSFET voltage and lamp current: (a) full power; (b) low dimming level (v_{ds}: 200V/div; i_o: 0.1A/div; time: 2ms/div)

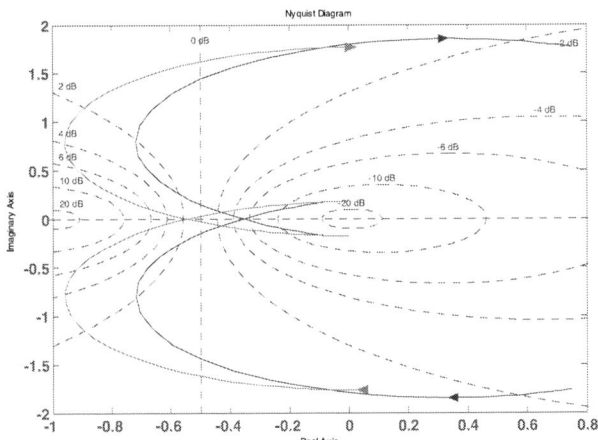

Figure 14 Nyquist plot of z_L/z_O

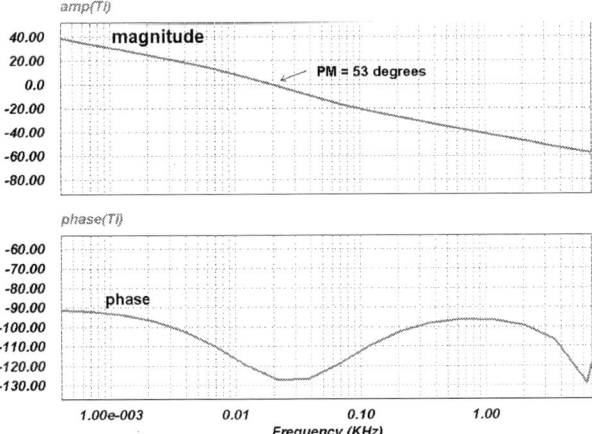

Figure 15 Closed-loop Bode plot

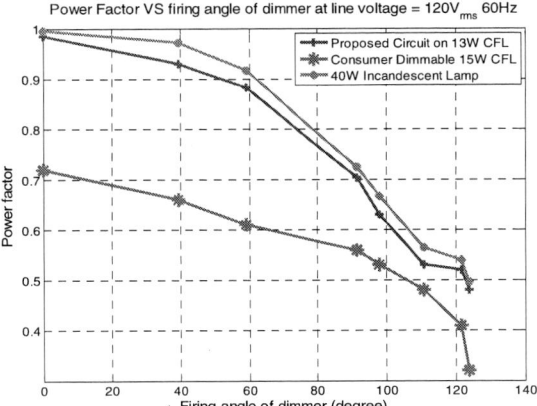

Figure 16 Power factor performance comparison

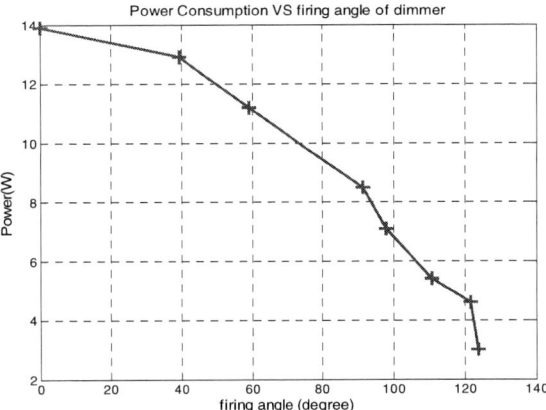

Figure 17 Input power VS dimmer firing angle

V. CONCLUSIONS

To enhance the dimming performance of CFL with incandescent phase-cut dimmer, it has been justified that the PFC stage shall not be omitted in the CFL electronic ballasts. A new dimmable electronic ballast with wide dimming range and the capability to ignite the lamp at low dimming level has been proposed in this paper for CFL applications. It has been demonstrated in this paper that the dimming performance of the proposed dimmable electronic ballast is comparable to that of the incandescent lamp. The detailed explanations of the system operating principles have been provided in this paper. Stability analysis has been given to justify the design of the ballast system. Experimental and simulation results have been provided on a 13W CFL to support the feasibility of the proposed work.

REFERENCES

[1] Olivier. G, Benhaddadi, R., "How green are compact fluorescent lamps?" *IEEE Canadian Review*, Dec, 2007, pp. 21-22.

[2] "Compact fluorescent ballasts" Universal Lighting Technologies, pp. 1-22.

[3] Cosby, M.C., Jr. Nelms, R.M., "A resonant inverter for electronic ballast applications" *IEEE Trans on. Industrial Electronics*, vol. 41, no. 4, Aug 1994, pp. 418-425

[4] Kazimierczuk, M.K.; Szaraniec, W.; "Electronic ballast for fluorescent lamps" *IEEE Trans. on Power Electronics,* vol. 8, no. 4, Oct. 1993, pp. 386 – 395.

[5] Lee, S.T.S.; Henry Shu-Hung Chung; Hui, S.Y.; "TRIAC dimmable ballast with power equalization," *IEEE Trans. on Power Electronics,* vol. 20, no. 6, Nov. 2005 pp. 1441 – 1449.

[6] Tam, P.W.; Hui, S.Y.R.; Chung, S.H.; "An Analysis and Practical Implementation of a Dimmable Compact Fluorescent Lamp Ballast Circuit Without Integrated Circuit Control," *in Proceedings of the IEEE Power Electronics Specialists Conference, 2006,* pp. 1 – 8.

[7] J. Janczak, "TRIAC dimmable integrated compact fluorescent lamp," *J. Illum. Eng. Soc.,* pp. 144–151, 1998.

[8] W. Ki, J. Shi, E. Yau, P. Mok, and J. Sin, "Phase-controlled dimmable electronic ballast for fluorescent lamps," in *Proceedings of the IEEE 1999 Power Electronic Specialty Conf.,* pp. 1121–1124.

[9] Lam, J.; Jain, P.K.; Agarwal, V.; "A novel SEPIC type single-stage single switch electronic ballast with very high power factor and high efficiency," *in Proceedings of the IEEE Power Electronics Specialists Conference, 2008.* pp. 2861 – 2866.

[10] Azcondo, F.J.; Diaz , F.J.; Branas, C.; Casanueva, R, "Microcontroller Power Mode Stabilized Power Factor Correction Stage for High Intensity Discharge Lamp Electronic B allast" *IEEE Trans. on Power Electronics,*2007,pp 845-853

[11] Rim, C.T. Cho, G.H. "Phasor transformation and its application to the DC/AC analyses offrequency phase-controlled series resonant converters (SRC)" Power Electronics, IEEE Trans on Apr 1990 Vol. 5, no. 2, pp. 201-211.

[12] Deng, E. Cuk, S. "Negative incremental impedance and stability of fluorescent lamps" *in Proceedings of the 1997 Applied Power Electronics Conference and Exposition, APEC '97,* pp. 1050-1056.

[13] Yin, Y. Zane, R. Glaser, J. Erickson, R.W. "Small-signal analysis of frequency-controlled electronic ballasts" *IEEE Trans. on Circuits and Systems I: Fundamental Theory and Applications,* Aug. 2003 Vol: 50, no. 8, pp. 1103- 1110.

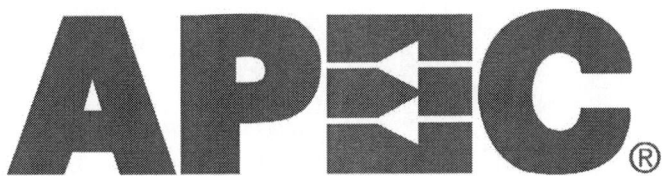

26th Annual IEEE Applied Power Electronics Conference and Exposition
March 6-10, 2011 at the Fort Worth Convention Center, Fort Worth, TX USA

Announcement and Call for Papers

APEC 2011 continues the long-standing tradition of addressing issues of immediate and long-term interest to the practicing power electronic engineer. Outstanding technical content is provided at one of the lowest registration costs of any IEEE conference. APEC 2011 will provide a) the best power electronics exposition, b) professional development courses taught by world-class experts, c) presentations of peer-reviewed technical papers covering a wide range of topics, and d) time to network and enjoy the company of fellow power electronics professionals in a beautiful setting. Activities for guests, spouses, and families are abundant in the Fort Worth area.

Papers of value to the practicing engineer are solicited in the following topic areas:

AC-DC and DC-DC Converters
Single- and Multi-Phase AC-DC Power Supplies, DC-DC Converters (Hard- and Soft-Switched)

Devices and Components
Semiconductor Devices, Magnetic Components, Capacitors, Batteries, Sensors, Interconnects, Device Integration

Manufacturing and Business Issues
Production Processes, Quality, Design for Manufacturability, Material Procurement, Supplier Qualification

Power Electronics for Utility Interface
Power Factor Correction, Power Quality, Electronics and Controls for Distributed Energy Systems

System Integration
Packaging, Thermal Management, EMI and EMC

Power Electronics Applications
Automotive and Transportation, Aerospace, renewable energy harvesting, Lighting (incl. LED), UPS, Power Generation and Transmission, Telecommunications, Military, Portable Power

Motor Drives and Inverters
AC and DC Motor Drives, Single- and Multi-Phase Inverters, PWM Techniques, sensor integration, Fault tolerant operation

Modeling, Simulation, and Control
Device, Component, Parasitics, Circuit and System, CAD /CAE Tools, Sensor and Sensor-less Control, Digital Control

Please note the following time frames (exact dates TBD and posted at www.apec-conf.org):

July, 2010	**Deadline for submission of digests**
October, 2010	**Notification that a paper was accepted or declined**
November, 2010	**Final papers and author registrations are due**

Digest Preparation: Prospective authors are asked to submit a digest explaining the problem that will be addressed by the paper, the major results, and how this is different from the closest existing literature. Papers presented at APEC must be original material and not have been previously presented or published. The principal criteria in selecting digests will be the usefulness of the work to the practicing power electronic professional. Reviewers value evidence of completed experimental work. Authors should obtain any necessary company and governmental clearance prior to submission of digests. Please visit www.apec-conf.org for all details on digest and final manuscript format.

If a digest is accepted, authors must submit a final manuscript before the deadline or the manuscript cannot be published in the Proceedings or presented at the conference. Final manuscripts may be subject to charges if their papers are over the page or file-size limit. At least one of the authors listed on a paper must be registered for either a Full Registration or for the Technical Sessions Only registration. A person registered at the Student rate may claim registration credit for only one paper. Authors registering at any non-Student rate, including the IEEE Life Member Rate, may claim registration credit for as many papers they wish.

Reviews: APEC relies upon a peer review process to ensure the quality of the technical content. To help maintain the high quality of the program, please contribute a few hours to review digests in your area of expertise by registering at www.apec-conf.org (under "Participating in APEC").

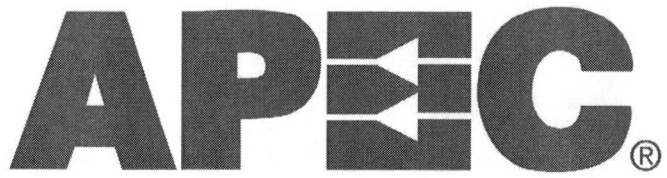

Calls for Special Presentations, Professional Education Seminars, and Exhibitor Seminars will be posted at www.apec-conf.org.

Website: www.apec-conf.org	**APEC**	**APEC Sponsors**
Email: apec@courtesyassoc.com	**2025 M Street**	Power Sources Manufacturers Association
Phone: +1-202-973-8664	**Suite 800**	IEEE Industry Applications Society
Facsimile: +1-202-331-0111	**Washington, DC 20036**	IEEE Power Electronics Society

AUTHOR INDEX

A

Abdel-Rahman, Osama	2073
Abe, S.	30
Abu Qahouq, Jaber A.	19, 120, 1723, 1778, 1800
Acero, J.	92, 439, 1328
Agarwal, Pankaj	512
Agelidis, Vassilios G.	295
Aggeler, Daniel	1584
Agostinelli, Matteo	170
Agostini Junior, Eloi	1911
Aguilar, D.	605
Ahmad, Hani	1871
Ahmadi, Damoun	1038
Ahmed, S.	63, 881
Ahmed, Tarek	1825
Ahsanuzzaman, S.M.	980
Akhavan Fomani, Armin	132
Akin, Bilal	1990
Al Mamun, Mostafa	1261
Alahmad, Mahmoud	672
Alcalá, Janeth	1651
Alderman, Arnold	525
Alesi, Larry	849
Al-Hoor, Wisam	627, 1723
Alico, Jurgen	1113
Aller, J.M.	343, 1139
Almukhtar, Basil	1922
Alonso, J.M.	743
Alonso, Rafael	92, 439
Alou, P.	271, 723, 729, 781
Amin, Mahmoud M.N.	1640
Ang, Simon S.	750
Arias, Manuel	196
Arnet, Beat	474
Aroca, J.	1792
Arthur, Stephen	401, 1598
Asadi, Peyman	1578
Avestruz, Al-Thaddeus	444, 580, 2305
Ayana, Elias	1804
Ayyanar, Raja	2149
Azcona, R.	1300
Azongha, S.	216

B

Badstuebner, U.	773
Bae, Chae-Bong	112
Baek, Seunghun	1666
Baggu, Murali M.	2121
Bai, Sanzhong	1145
Baiju, M.R.	1963
Baker, Jonathan	143
Bakhshai, A.	149, 155, 768, 1209
Bakkaloglu, Bertan	1871
Balda, Juan Carlos	321
Ball, Arthur	533
Bao, Jianyu	1097
Bao, Weibing	1097
Barbaroux, Jean	248
Barbi, Ivo	550, 1911
Barlow, Fred	1108
Barrado, A.	1026, 1131, 1279
Barragán, L.A.	309, 439
Barreto, L.H.S.C.	837
Basu, Supratim	244
Batarseh, Issa	143, 627, 1723, 2073
Bates, John	474
Batschauer, Alessandro L.	909
Bazzi, Ali M.	256
Beaupre, Richard A.	1591
Beaupre, Richard A.	1603
Beechner, Troy	2174
Benavente, C.	729
Bendl, Jiri	895
Benfatto, I.	1622, 1810
Ben-Yaakov, Sam	928
Berzoy, Alberto	343
Bhattacharya, Subhashish	761, 1010, 1243, 1666
Bianco, A.	1287
Biela, J.	773, 1397, 1584, 1865
Bing, Zhonghui	336
Birdane, E.	1197
Boel, R.K.	1711
Boroyevich, D.	355, 408, 881, 1272, 1378, 1487, 1521
Brockerhoff, Philip	1970
Bucher, A.	557, 1763
Bueno, A.	1139
Bueno, Alexander	343
Bull, Chris	398
Burdío, J.M.	92, 309, 439, 1328
Burgos, R.	355, 408, 881

C

Cai, Jun	1018
Campa, L.	743

Canales, Francisco 550
Cao, Dong 1365
Cao, Lingling 920
Cao, Yue 968
Cárdenas, Víctor 1651
Carp, C. 1177
Carretero, Claudio 92
Casady, Jeff 1838
Celanovic, Ivan L. 961
Céspedes, Mauricio 2174
Ceyhan, Adil 1197
Cha, Hanju 1659
Chai, Jianyun 2104
Chan, Walker R. 961
Chan, Yick Po 571
Chang, Soon-Jyh 1043
Chang, Wei-Hsu 1727
Chang-Chien, Le-Ren 1043
Chapman, Patrick L. 2138, 2294
Cheah, Sze Kwan 1804
Chen, Baifeng 887, 1704
Chen, Chien Liang 619
Chen, Chingchi 676
Chen, Ching-Jan 1727
Chen, Dan 1727
Chen, Dong 818
Chen, Emil 1081
Chen, Fu-Zen 188
Chen, Guozhu 361, 1514
Chen, Henglin 642, 691
Chen, Jhih-Han 420
Chen, Jifeng 2091
Chen, Lihua 676, 1119, 1124
Chen, Min 935, 1204, 1340, 1674
Chen, Qianhong 920
Chen, Qing Su 948
Chen, W. 594, 1238, 1358
Chen, Yan 1534
Chen, Yang 1578
Chen, Yu 1435, 1441, 1464
Chen, Zheng 1572
Chen, Zhong 1448, 1471, 1616, 1627
Cheung, Chun 1081
Cheung, Victor Sui-pung 491
Chiang, T.-Y. 1413
Chinthavali, Madhu S. 1108
Chiu, Chen-Hua 1727
Chiu, Huang-Jen 948
Cho, B.-H. 1373
Cho, Un-Kwan 1561
Choi, Hangseok 36
Choi, Mun-Gi 1833

Choi, Seung-deog 1990
Choi, Sewan 1934
Choi, Sungjin 512
Choi, W.P. 2251
Choi, Woojin 466, 2166
Choi, Woo-Young 42, 1494
Chomat, Miroslav 895
Choo, Fook Hoong 2143
Chou, Welly 2057
Chowdhury, Badrul H. 2121
Chuang, S.-A. 1413
Chung, Bong-Gun 1698, 1833
Chung, Henry Shu-hung 491, 1214, 1904
Ci, Song 672
Clifford, Zachary 444
Cobos, J.A. 271, 723, 729, 781
Cochran, Travis 2154
Coelho, Enane Antônio Alves 2258
Cooke, Philip 183
Cooley, John J. 444, 2264, 2305
Corradini, L. 277
Corrêa, M.B.R. 239
Corzine, Keith 58, 452
Costabeber, A. 1287
Cox, Robert 1547
Crebier, Jean-Christophe 248, 2238
Crespo, M. 743
Cruz, C.M.T. 837
Cui, Xizhi 1610, 2042

D

da Câmara, Raphael A. 837
da Silva, Fábio Vincenzi Romualdo 2258
Das, Pritam 564, 1222
Davoudi, Ali 2138, 2294
Day, Jon 1243
de Britto, Jonas Reginaldo 2258
De Doncker, Rik 696
de Freitas, Luiz Carlos 2258
de M. Fernandes, Eisenhawer 1984
de Nie, Robert 1
De Novaes, Yales R. 550
Deboy, G. 1397
DeCarlo, Ray 480
Delgado, Eladio C. 1603
Delicado, Bernardo 1300
Deng, Yan 575, 1069, 1266
Deng, Zhe 2021
Deng, Zhiquan 1018
Dhople, Sairaj V. 2138, 2294
Dias, José A.A. 755

Diaz, D.	723, 729
Dick, Christian P.	696
Dickson, Andrew	702
Ding, Xiaodong	519
Ding, Yi	1555
Djabbari, Ali	1056
Domahidi, Alexander	1995
Dominguez-Garcia, Alejandro	256
Dong, Yan	79
Donlon, John F.	392
dos Santos Girio, J.A.	1306
dos Santos, Jr., E.C.	1191
dos Santos, Euzeli	755, 1183
Du, Chengrui	861
Du, Weijing	823, 1392
Du, Xiaoli	1732
Du, Yu	1145, 1666
Duerbaum, T.	557, 1763
Duryea, Timothy P.	686
Dutta, Sumit	761, 1666

E

Eckler, Kyle Roger	2202
Edrington, C.S.	216
Effler, Simon	315, 1087
Egan, Michael G.	787
Egelkraut, S.	231
Elasser, Ahmed	1598
El-Barbari, Said	1487
Elmes, John	143
Emadi, Ali	1957
Emanuel, Alexander E.	2096
Endredy, John	1172
Englehart, Amy	2305
Enjeti, Prasad	63
Erb, Dylan C.	2066
Eren, Suzan	149, 768
Ertl, H.	986

F

Fahimi, Babak	68, 1498, 2231
Falcones, Sixifo	2149
Fan, Haifeng	210
Fan, S.-Y.	1842
Fang, Xiong	1745
Farias, Valdeir José	2258
Fei, Wanmin	1034, 1732
Feng, Yupeng	915, 1093
Feng, Zhuomin	935
Ferdowsi, Mehdi	58, 452, 2111

Fernández, A. 196, 1313, 1792
Fernández, C. 1131, 1279
Fernández, Carlos 1300
Fernández, Cristina 1026
Ferrieux, J.-P. 1817
Filho, Faete 968
Firmansyah, E. 30
Fleming, F. 216
Foley, Raymond 525
Forsyth, A.J. 1306
Frey, D. 1817, 2238
Frey, L. 231
Friedli, Thomas 1527
Fu, Dianbo 940
Fu, Jizhen 702, 1482
Fu, P. 1622, 1810
Fujita, Atsushi 1825
Fukuda, Kenji 2030
Fukushima, Kentaro 289

G

Gacio, D. 743
Galperti, C. 2281
Gamboa, Gustavo 143
Gao, Feng 1555
Gao, G. 1622
Gao, G. 1810
Gao, Mingzhi 1204
Gao, Mingzhi 1674
Garces, Luis 1295
Garcia, J. 743
García, O. 271, 723, 729, 781
Gargoom, A. 162, 2132
Garrett, Jerome 1598
Gazel, Nicolas 1272
Ge, Baoming 1124
Ge, Qiongxuan 1419
Geng, Hua 2126
Gillmor, Colin 1384
Gimeno, A. 729
Glaser, John S. 401, 654
Gong, Jinwu 887, 1704
Gonzalez, M.C. 271, 781
Gonzalez-Llorente, Jesus 666, 1062, 2161, 2226
Gowda, Arun V. 1591, 1603
Graham, Jeff 2220
Green, R. 1568
Grogan, S.A.S. 873
Guépratte, K. 1817
Guerrero, Josep M. 380
Guo, Rong 1172

Guo, Suxuan 887, 1704
Guo, Zhiqiang 662
Gupta, Ranjan K. 901

H

Ha, Dong Sam 2154
Habetler, T. 343, 1139
Halton, Mark 1075, 1087, 2207
Hamilton, Christopher 143
Han, Baikhee 512
Han, Byung M. 303
Han, Jung Hee 1108
Han, Sangmin 2288
Han, Yunlong 1534
Hang, Lijun 2021
Haque, M.E. 162, 2132
Harada, Shinsuke 2030
Harada, Yosuke 289
Harbaugh, Mark 425
Harris, John H. 1048
Hartmann, L.V. 239
Hartmann, M. 986
Hartnett, Kevin J. 787
Haruni, A.M.O. 162, 2132
Hasan, Jaber 750
Hayes, John G. 787
He, Chao 361
He, Xiangning 575, 681, 801, 1069, 1266, 1454, 1534, 2080, 2300
He, Yanhui 915, 1093
He, Yingjie 1692
Hegarty, Tim 1056
Heldwein, Marcelo L. 909
Henze, C.P. 605
Herbert, Edward 1048
Herbsommer, Juan A. 398
Hewson, Christopher R. 2050
Hirose, Fumitoshi 1879
Ho, W.C. 994, 2251
Holmes, D.G. 873
Hong, Xiaoyuan 1238, 1358, 2021
Hosoda, Hiromi 1261
Hsu, G.-W. 1849
Hu, Haibing 519, 627, 1785, 2073
Hu, Qingcong 2314
Hu, Yuequan 203
Huang, Alex 761, 849, 1010, 1145, 1172, 1477, 1666, 1875, 2181
Huang, Hong 1770
Huang, Jing-Yi 1043
Huang, Xiucheng 433, 823, 1392
Huber, Laszlo 203
Huemer, Mario 170

Huh, Dong-Young 1885, 1949
Hui, Joanne 155
Hui, S.Y.R. 86, 594, 994, 1346, 2251
Hung, Chung-Wen 420
Husain, Iqbal 2007
Hutchens, Chris 1056
Hwu, K.I. 507, 710, 737, 1942
Hyeon, B.-C. 1373

I

Ide, Kozo 103
Inman, Daniel J. 2154
Ioinovici, Adrian 1214, 1904
Ishii, Kenichiro 74
Ishizuka, Yoichi 1879
Itoh, Jun-Ichi 1684
Izquierdo, D. 1300

J

Jacobina, C.B. 755, 1183, 1191, 1984
Jain, P.K. 14, 149, 155, 499, 544, 768, 1209, 1248, 1334, 2321
Jakobsen, Uffe 98
Jang, Minsoo 295
Jang, Sang-Ho 1833
Jang, Yungtaek 23
Jayakanthan, Gnanavel 794
Jeannin, P.-O. 248, 1817, 2238
Jeon, Yong-Seog 1949
Jeong, In Wha 1166
Jeong, Yu-Seok 303
Ji, Biao 1448, 1471
Ji, Feng 1448, 1471
Ji, Young-Hyok 2275
Jia, Liang 124
Jiang, Dong 408
Jiang, Wei 68
Jiménez, O. 309
Jimenez-Brea, Emil 666, 1062, 2161
Jing, Wei 1010
Jovanović, Milan M. 23, 203
Jung, Ha-Jin 1678
Jung, Sungyoon 2002
Jung, Yong-Chae 1885, 2275

K

Kanai, Takeo 1101
Kang, Sung-In 1885
Kang, Yong 1272, 1435, 1441, 1464
Kato, Koji 1684

Kazimierczuk, Marian K. 2212
Kelleher, Paul 1922
Kelley, Robin 1838
Kelly, Anthony 328, 1922, 2189
Keogh, Bernard 1384
Kerkman, Russ J. 634
Kesler, Metin 374
Khaligh, Alireza 1755, 2066, 2245
Khazarei, Mostafa 452, 58
Khoobroo, Amir 2231
Kim, Deuk-Soo 1678
Kim, Eun-Soo 1698, 1833, 1885, 1949
Kim, Gi-Taek 1678
Kim, Gyeong-Hun 2085
Kim, Hyun-Cheol 112
Kim, Jang-Mok 112
Kim, Jin-Tae 540
Kim, Joo-Hoon 1698, 1885, 1949
Kim, Jun-Gu 2275
Kim, Rae-Young 1890
Kim, Sang-Hyun 466, 2166
Kim, Sungmin 103
Kim, Tae-Hoon 466, 2166
Kim, Teahoon 512
Kim, Wook 466, 2166
Kim, Young-Gook 112
Kim, Young-Ho 2275
Kim, Young-Ju 2085
Kimball, Jonathan W. 1508, 2121, 2202
Kirlin, R. Lynn 1749
Kirtley, Jr., James L. 1547
Kisacikoglu, Mithat C. 458
Knight, Andy 2013
Koellner, Walter 1158
Kolar, J.W. 773, 986, 1397, 1527, 1584, 1865
Kong, Na 2154
Kong, Pengju 1424
Koo, Gwan-Bon 540
Krein, Philip T. 256
Krishnamurthy, Mahesh 1957
Kutkut, Nasser 627
Kwon, Bong-Hwan 1494
Kwon, Soon Kurl 1230
Kye, Moon-Ho 1698, 1833

L

Lai, Jih-Sheng 42, 387, 474, 619, 1056, 1494, 1890
Lai, Pengjie 1927
Lai, Rixin 355, 408
Lam, John 2321
Lamar, Diego G. 196

Lascu, Cristian	1749
Laughman, Christopher	1547
Lázaro, A.	1026, 1131, 1279
Leão, J.F. Araujo	239
Lee, Beomseok	2002
Lee, C.K.	86
Lee, Fred C.	79, 176, 533, 940, 1424, 1927
Lee, Hoi	686
Lee, Jae-Sam	1885, 1949
Lee, Jong-Hak	466, 2166
Lee, Jun-Young	303
Lee, Kwang-Ho	1698, 1949
Lee, Sangwon	1934
Lee, Ting-Peng	948
Lee, Tzung-Lin	380
Lee, Yuang-Shung	619
Leeb, Steven B.	444, 580, 1547, 2194, 2264, 2305
Lei, Qin	844, 854, 1002
Leslie, Scott	474
Li, Chushan	681, 2300
Li, Duo	935
Li, Hong	1740
Li, Hui	210, 223, 807
Li, Jian	176
Li, Jin	1521
Li, Jiping	2036
Li, Jun	1010
Li, Ming	1745
Li, Qiang	79, 533
Li, S.N.	594
Li, Weichen	1454, 2080
Li, Wuhua	801, 1069, 1454, 2021, 2080
Li, Xiao	861
Li, Yaohua	1419
Li, Yong	349
Li, Yongdong	1736, 2104
Liang, Xiaoguo	794
Liang, Zhigang	849, 1477
Liao, XiaoZhong	662
Lim, Michele	533
Lim, Sungkeun	1875
Lima, A.M.N.	239, 1984
Lin, Cheng-Tsung	420
Lin, D.Y.	1346
Lin, Fei	1740
Lin, Hai	818
Lin, Hung-Chih	2154
Lisi, Gianpaolo	1056
Litvinov, Alexander	474
Liu, Chih-Wen	420
Liu, Chui Pong	571
Liu, Congwei	1419

Liu, Jingbo 425
Liu, Jinjun 915, 1093, 1521, 1633, 1692, 2116
Liu, Kun 1674
Liu, Liming 223
Liu, Ting 2116
Liu, X. 86, 994, 2251
Liu, Yan-Fei 124, 702, 1482
Liu, Zeyuan 1018
Lo, Yu-Kang 948
Loh, Poh Chiang 1555, 2143
López del Cerro, F.J. 1300
Lopez, Osvaldo 398
Lorduy, Abad 1026
Losee, Peter 401, 1598
Lu, Kaiyuan 98
Lu, Ying 1785
Lu, Zhengyu 519, 1238, 1358, 1542, 2021, 2091
Lucena, Carlos 1026
Lucía, O. 92, 309, 439, 1328
Lukic, Srdjan 1145
Lukić, Zdravko 1, 315
Luo, Fang 1272
Luo, Yingpeng 1616, 1627

M

Ma, Chongguang 1340
Ma, Dongsheng 284, 813
Maciel, A.M. 1191
Maddaleno, Franco 1166
Mahdi, Abdulhussain E. 2207
Maksimović, Dragan 188, 277, 980
Mankani, A.D. 1622, 1810
Mantooth, H. Alan 321
Mao, Jingxin 1740
Mao, Xiaojing 1405
Mariéthoz, Sébastien 1995
Marsili, Stefano 170
Martin, Daniel 619
Marxgut, Christoph 1865
März, M. 231
Massoud, Ahmed 63
Matocha, Kevin S. 401
Matsuo, Hirofumi 1879
Mattavelli, P. 953, 1287, 2281
Mazumdar, Joy 1158
McGrath, B.P. 873
Melkebeek, J.A.A. 1711
Meng, Peipei 433, 642, 691
Metwally, M.K. 414
Meyer, Eric 124
Miaja, P.F. 715

Miftakhutdinov, Rais 1897
Milivojevic, Nikola 1957, 2245
Millan, Ignacio 92, 439, 1328
Miller, Greg 1081
Millner, Alan 1572
Min, Chen 1460, 1503
Ming, Zhengfeng 1919
Ming, Zheng-Feng 109
Mishima, Tomokazu 1230
Mo, Qiong 1674
Moghe, Rohit 1158
Mohammed, O.A. 1640
Mohan, Ned 901, 1804
Mohapatra, Krushna K. 901
Molina Cardozo, Diogenes D. 321
Moon, Sang-Cheol 540
Mooney, James 2207
Morari, Manfred 1995
Moschopoulos, Gerry 564, 829, 1222, 1320
Motto, Eric R. 392
Mourra, O. 1313, 1792
Mousavi, Ahmad 564, 1222
Muller, Sean 2194
Murai, H. 648
Muralidhar, Gautam 19
Mussa, Samir A. 909

N

Nakano, Shinya 74
Nakaoka, Mutsuo 1230, 1825
Nam, Kwanghee 2002
Narveson, Brian 525
Nasadoski, Jeffrey J. 401
Navarro, D. 309
Neely, Jason 480
Negnevitsky, M. 162, 2132
Neuman, Sabrina 2194
Neumeyer, C. 1810
Ng, W.M. 594, 1346
Ng, Wai Tung 132
Ngo, K.D.T. 533, 2036
Ngo, Phong 813
Nguyen, Do Hung 750
Nguyen, The-Van 2238
Ni, Guang-Zheng 109
Nilles, Gerald 2294
Ning, Puqi 1378
Ninomiya, T. 30, 289
Nishi, Mariko 1879
Nondahl, Thomas 425
Noquil, Jonathan 398

Norford, Les K. 1547
Norigoe, Isami 289
Núñez, Ciro 1651

O

Oh, J.S. 1810
Ohashi, Hiromichi 648, 1101, 2030
Oliveira, Jr., D.S. 837
Oliveira, Alexandre C. 1984
Oliver, J.A. 271, 723, 729, 781
O'Malley, Eamon 328, 1922, 2189
Ó'Mathúna, Cian 525
Omori, Hideki 1825
Onar, Omer C. 1755, 2066
Ongaro, F. 2281
Orabi, Mohamed 1778, 1800
Orietti, E. 953
Orji, Uzoma A. 1547
Ortega, F.J. 729
Ortiz-Rivera, Eduardo I. 666, 1062, 2161, 2226
Otsuki, Etsuo 74
Ozdemir, Engin 367, 374
Ozdemir, Sule 367
Ozpineci, Burak 458

P

Page, Sarah 2194
Pahlevaninezhad, Majid 149, 768
Pallo, Nathan A. 961
Pan, S. 499, 1248
Pang, H.M. 973, 1857
Parayandeh, Amir 980
Pardo, J.M. 729
Paris, James 580, 1547, 2194, 2305
Park, Jinseok 2181
Park, Jun-Ho 1885
Park, Minwon 2085
Park, Sung-Yeul 387
Parkhideh, Babak 1666
Parto, Parviz 1578
Pasquesoone, Gregory 2007
Pautsch, Adam G. 1591
Pawellek, A. 557
Pei, Xuejun 1435, 1441, 1464
Pei, Yunqing 1610, 2042, 2060
Pekarek, Steve 480
Peng, Fang Z. 818, 844, 854, 1002, 1119, 1124, 1365, 2288
Peng, Li 1435, 1441, 1464
Pepper, Michael 143
Perin, Arnaldo J. 909

Perreault, David J. 961
Pilawa-Podgurski, Robert C.N. 961
Pokryvailo, A. 1177
Pong, M.H. Bryan 973, 1857
Pong, Man Hay 571
Poon, Ngai Kit 571
Poucand, M. 1306
Praça, P.P. 837
Priewasser, Robert 170
Prodić, Aleksandar 1, 315, 980, 1113, 1256

Q

Qi, Tao 2220
Qian, Hao 474, 1056
Qian, Wei 2288
Qian, Zhaomin 1674
Qian, Zhaoming 642, 691, 818, 823, 935, 1002, 1204, 1340, 1392
Qian, Zhijun 2073
Qiu, Weihong 1081
Quesada, Isabel 1026

R

Radić, Aleksandar 1, 315
Rahimian, Mina M. 1990
Ramamurthy, Anand 1243
Rauch, M. 231
Ray, William F. 2050
Remscrim, Zachary 444, 1547, 2194
Ren, Xiaoyong 920
Ren, Zheng 1204, 1674
Restrepo, J. 343, 1139
Reutzel, Evan 1430
Rico-Secades, M. 743
Rinne, Karl 328, 1075, 1087, 1922, 2189
Ritenour, Andrew 1838
Rivas, Juan M. 654
Rocha, Nady 1183
Rodríguez, A. 715
Rodríguez, M. 715
Rodríguez-Valdez, Carlos D. 634
Rosas, Emanuel 1651
Royak, Semyon 425
Ruan, Xinbo 920, 1214, 1405
Rylko, Marek S. 787

S

Sabate, Juan 1487
Sadakata, Hideki 1825
Sagawa, Natsumi 2212

Saggini, S.	953, 1287, 2281
Saha, Bishwajit	1825
Salah Morsy, Ahmed	63
Salazar-Llinas, Andres	666, 1062, 2161, 2226
Salem, T.E.	1568
Salmon, John	2013
Samsi, Rohan	183
Sanders, Seth	1430
Sanz, M.	1279
Satoh, K.	392
Sayed, Khairy Fathy	1230
Scanlan, Tony	1075
Scapellati, C.	1177
Schantz, Christopher	1547, 2194
Scharrer, Martin	1075
Schletz, A.	231
Schmidt, Peter	425
Schreier, Ludek	895
Schulz, Martin	1970
Schutten, Michael	1598
Schweizer, Mario	1527
Sebastián, J.	196, 715
Seger, Eric	2264
Seidlitz, Steve	1804
Sekiya, Hiroo	2212
Sen, P.C.	702, 1482, 1719
Sepahvand, Hossein	58, 452
Sha, Deshang	662
Shao, Jianwen	601
Sharif, Hamid	672
Shaw, Steven R.	2194, 2264
Shen, Guoqiao	861
Shen, John	627
Shen, Weixiang	2143
Sheng, Honggang	1572
Sheng, Z.	1622, 1810
Sheridan, David	1838
Shi, Lei	1448, 1471
Shi, Wei	1785
Shih, Frank	948
Shim, Won-Sul	1678
Shinohe, Takashi	1101, 2030
Shiny, G.	1963
Shoyama, M.	30, 289
Shuai, Peng	550
Silva, C.E.A.	837
Simanjorang, Rejeki	648
Singh, Bhim	1976
Singh, Sanjeev	1976
Slepchenkov, Mikhail	1166
Slowey, John	525
Smedley, Keyue	47, 264, 1166

Smith, Chris 474
Smith, Greg 1056
Solovitz, Stephen A. 1591
Somani, Apurva 901
Somayajula, Deepak 2111
Song, Bo 2060
Song, Byeong-Mun 2085
Song, Z.Q. 1810
Spiazzi, G. 953
Stamenkovic, Igor 1957, 2245
Steimer, Peter K. 1865
Steiner, Reto 1865
Stephan, H. 1817
Steurer, M. 216
Stevanovic, Ljubisa D. 401, 1591, 1603
Straeussnigg, Dietmar 170
Stum, Zachary 1598
Su, Gui-Jia 1152
Su, Jen-Ta 420
Su, Y.-H. 1842
Su, Y.P. 86
Sugimura, Hisayuki 1230
Sul, Seung-Ki 103, 611, 1561
Sullivan, Charles R. 1048
Sumiyoshi, Shinichiro 1825
Sun, Jian 336, 2174, 2220
Sun, Jianjun 887, 1704
Sun, Julu 533, 1927
Sun, Pengwei 387, 474
Sun, Yi 176
Sung, Kyungmin 1101
Suzuki, K. 392
Szczesny, Paul 1295

T

Takahashi, T. 392
Takao, Kazuto 1101, 2030
Takeda, Takashi 648
Tamyurek, B. 1197
Tan, Kai 1419
Tan, Kuan Khoon 1555, 2143
Tanaka, Yasunori 1101
Tang, Lixin 1152
Tang, Yu 867
Tao, J. 1622, 1810
Tao, X.H. 594
Thomas, Brinda A. 588
Thompson, Chris 1243
Titiz, Furkan Kaan 696
Tjokrorahardjo, Andre 1352
Todd, R. 1306

Todeschini, Grazia 2096
Tolbert, Leon M. 458, 968, 1108
Toliyat, Hamid A. 1990
Tomioka, S. 30
Tomita, Koji 103
Tonicello, F. 1313, 1792
Torrico-Bascopé, R.P. 837
Tran, Manh Hung 248
Trowler, Derik 321
Trzynadlowski, Andrzej M. 1749
Tschirhart, Darryl J. 14, 544, 1334
Tseng, S.-Y. 1413, 1842, 1849
Tsukakoshi, Kenta 289

U
Ucar, Mehmet 367
Undeland, Tore M. 244
Urciuoli, D.P. 1568
Urriza, I. 309

V
Vafakhah, Behzad 2013
Vagnon, Eric 2238
Vaks, Nir 480
Valdivia, V. 1131, 1279
Vasić, M. 271, 723, 729
Veillette, Robert J. 2007
Vickery, Dan 444, 2305
Vieira, Jr., João Batista 2258
Viola, Julio C. 343
Visairo, H. 271, 781
Vodyakho, O. 216
Volfson, Oleg 138
Vu, Trung-Kien 1659
Vyncke, T.J. 1711

W
Wada, Keiji 1101
Waldron, Finbarr 525
Wang, Dong 124
Wang, Fred 355, 408, 881, 1272, 1378, 1487, 1572
Wang, Gangyao 761, 1666
Wang, Huai 491, 1904
Wang, Jin 676, 1038
Wang, Jun 1266
Wang, K.-C. 1413, 1849
Wang, Ke 1745
Wang, Kunrong 7
Wang, Laili 1610, 2042, 2060

Wang, Meng 794
Wang, Mingliang 2036
Wang, Peng 1555, 2143
Wang, Ruxi 1378
Wang, Shunqing 1627
Wang, Shuo 940, 1272
Wang, Siran 1097, 1238, 1358, 1542, 2021, 2091
Wang, Yen-Ching 380
Wang, Yousheng 818
Wang, Yue 915, 1093, 1745
Wang, Zhan 807
Wang, Zhaoan 1610, 1692, 2042, 2060
Wang, Zhengshi 575
Wang, Zhiqiang 361, 1514
Watson, Luke 2121, 2202
Wegner, Hagen 1384
Wei, Jukui 867
Wen, Jun 47
Wichakool, Warit 580, 1547
Wijeratne, Dunisha 829
Wilson, Jr., Thomas G. 183
Wolbank, T.M. 414
Won, Chung-Yuen 2275
Wood, R.A. 1568
Wu, Bin 1034, 1732, 2126
Wu, Chun-Hsun 1043
Wu, D. 1306
Wu, Guan-Hong 948
Wu, Haimeng 575
Wu, Jiande 575, 681, 2300
Wu, Jinlong 915, 1093
Wu, W.-C. 1842
Wu, Xinke 642, 691
Wu, Zhichao 223

X

Xiao, Xi 1736
Xie, Chuan 361, 1514
Xie, Huikai 2036
Xie, Shaojun 867
Xing, Lei 2174
Xing, Yan 1785
Xu, Biwen 1340
Xu, Chunchun 1295
Xu, Dehong 861
Xu, Dewei 2126
Xu, L. 1622
Xu, L.W. 1810
Xu, Ligang 920
Xue, Jianren 1785
Xue, Tao 519

Y

Yamada, Yusuke	1879
Yamaguchi, Hiroshi	648
Yamazaki, Mikio	648
Yan, W.	1346
Yang, Bing-Zhong	109
Yang, Binjian	1266
Yang, Bo	801
Yang, C.-M.	1849
Yang, Geng	2126
Yang, Jianyou	642
Yang, Joonhyun	512
Yang, Liyu	1172, 2181
Yang, Shuitao	844, 854, 1002
Yang, Xinyi	1340
Yang, Xu	1610, 2042, 2060
Yao, Kai	1405
Yao, Wei	1204, 1674
Yao, Wenxi	519, 2091
Yau, Y.T.	507, 710, 737, 1942
Yazdani, D.	1209
Ye, Shaoshi	1238, 1358
Ye, Zhihong	1405
Yim, Jung-Sik	1561
Yin, Zhenggang	1419
Ying, Yucheng	533
Yisheng, Yuan	1460, 1503
Yoo, Hyunjae	611
Yoon, Young-Doo	103
York, Ben	1890
Yoshihiura, Y.	392
Yoshino, Teruo	1261
You, Xiaojie	1740
Young, George	1384
Youssef, Mohamed	1778, 1800
Yu, Haidong	1498, 2025
Yu, In-Keun	2085
Yu, Wen Long	948
Yu, Wensong	387, 474, 1056
Yu, Wen-Song	42
Yuan, Wei	433, 823, 1392
Yuan, Xibo	2104

Z

Zane, Regan	2314
Zawodniok, Maciej	1508
Zeltser, Ilya	928
Zeng, Jie	761
Zha, Xiaoming	887, 1704
Zhang, Di	1487
Zhang, Guoxing	433

Zhang, Hui 1108
Zhang, Jianhui 1056
Zhang, Jing 1514
Zhang, Jiucai 672
Zhang, Jun 861
Zhang, Junming 433, 823, 1392
Zhang, Leqiang 1745
Zhang, Xin 1214
Zhang, Xuan 2116
Zhang, Yanli 1034
Zhang, Yi 284
Zhang, Yingqi 1719
Zhang, Zhe 935
Zhang, Zhiliang 702, 1482
Zhang, Zhongchao 1097
Zhao, April 132
Zhao, Guopeng 1610, 1633, 1745, 2042, 2060
Zhao, Jing 1266, 1534
Zhao, Rongxiang 1534
Zhao, Tiefu 761, 1666
Zhao, Yi 801, 1069, 1454, 2080
Zhao, Zhenyu 1256
Zhaoming, Qian 1460, 1503
Zheng, Cong 2245
Zheng, Feng 1919
Zheng, Jerry 349
Zheng, Sheng 691, 818
Zheng, Trillion Q. 1740
Zheng, Yuzhen 2080
Zhixin, Xu 1266
Zhong, W.X. 994
Zhou, Liang 47, 264
Zhou, Linyuan 2116
Zhou, Xia 2091
Zhou, Xiaohu 849, 1145, 1666
Zhou, Xin 1477
Zhu, Guangyong 7
Zhu, Haipeng 2143
Zhu, Hao 1736
Zhu, Yinyu 1616, 1627
Zou, Ke 676, 1038
Zou, Yunping 1692
Zumel, P. 1131, 1279

CURRAN ASSOCIATES INC.
proceedings
.com

9781424447824